Human Physiology

Third Edition

Rodney Rhoades, Ph.D.

PROFESSOR AND CHAIRMAN
Department of Physiology and Biophysics
Indiana University School of Medicine

Richard Pflanzer, Ph.D.

ASSOCIATE PROFESSOR OF BIOLOGY AND
PHYSIOLOGY/BIOPHYSICS
Indiana University School of Medicine
Purdue University School of Science

®

Saunders College Publishing

Harcourt Brace College Publishers

Fort Worth   Philadelphia   San Diego   New York   Orlando   Austin
San Antonio   Toronto   Montreal   London   Sydney   Tokyo

Text Typeface: Palatino
Compositor: York Graphic Services, Inc.
Acquisitions Editor: Edith Beard Brady
Executive Editor: Julie Levin Alexander
Developmental Editor: Cathleen E. Petree
Art Developmental Editor: Ray Tschoepe
Managing Editor: Carol Field
Project Editor: Laura Shur
Copy Editor: John Beasley
Manager of Art and Design: Carol Bleistine
Associate Art Director: Sue Kinney
Cover Designer: Lawrence R. Didona
Text Artwork: Rolin Graphics
Photo Researcher: Amy Ellis Dunleavy
Director of EDP: Tim Frelick
Production Manager: Joanne Cassetti
Marketing Manager: Sue Westmoreland

Cover Artwork: Greg Purdon

Printed in the United States of America

Human Physiology

ISBN: 0-03-005159-2

Library of Congress Catalog Card Number: 95-074829

678   032   98765432

Comparison of Metric and English Units of Measurement

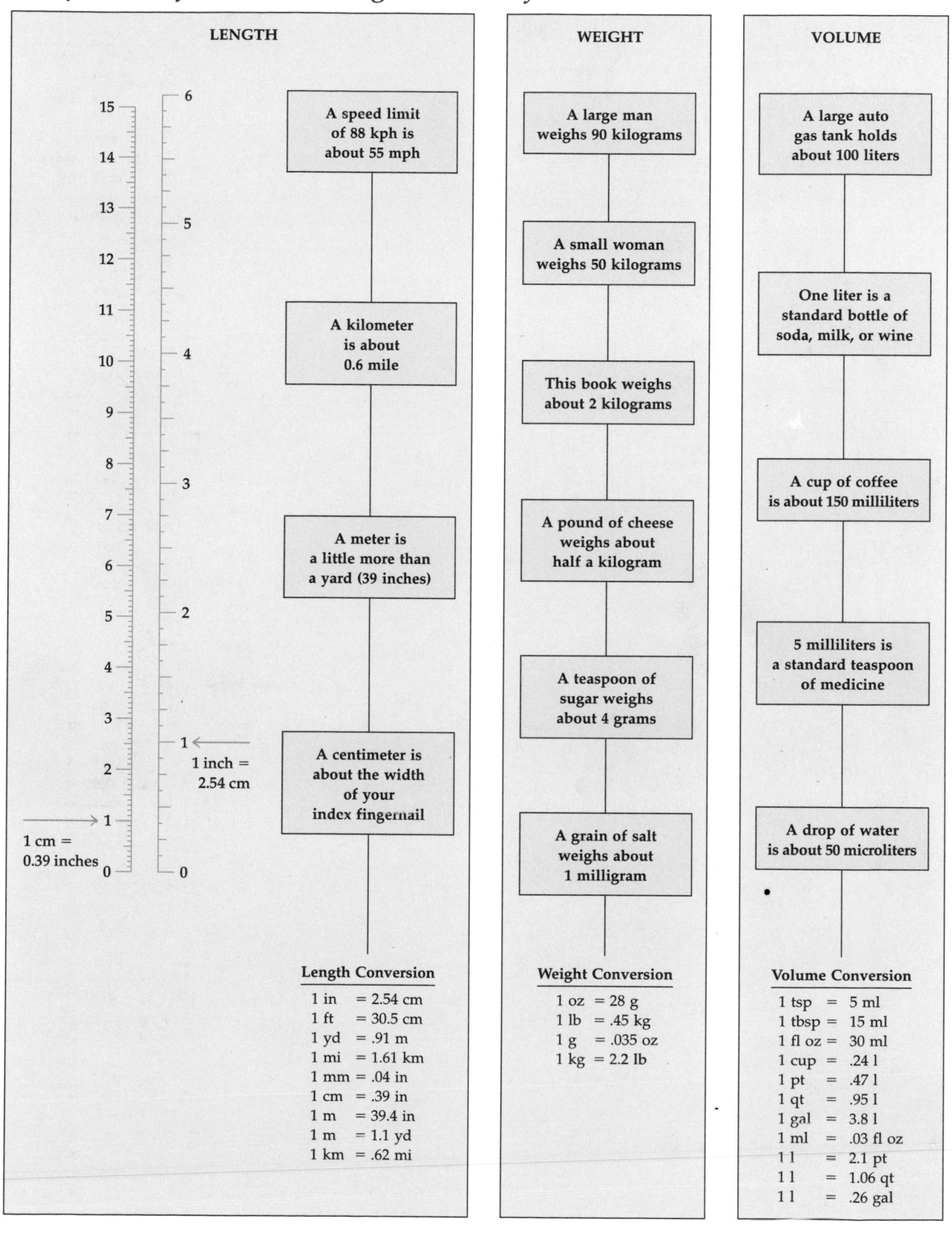

LENGTH

A speed limit of 88 kph is about 55 mph

A kilometer is about 0.6 mile

A meter is a little more than a yard (39 inches)

A centimeter is about the width of your index fingernail

1 inch = 2.54 cm

1 cm = 0.39 inches

Length Conversion

1 in	=	2.54 cm
1 ft	=	30.5 cm
1 yd	=	.91 m
1 mi	=	1.61 km
1 mm	=	.04 in
1 cm	=	.39 in
1 m	=	39.4 in
1 m	=	1.1 yd
1 km	=	.62 mi

WEIGHT

A large man weighs 90 kilograms

A small woman weighs 50 kilograms

This book weighs about 2 kilograms

A pound of cheese weighs about half a kilogram

A teaspoon of sugar weighs about 4 grams

A grain of salt weighs about 1 milligram

Weight Conversion

1 oz	=	28 g
1 lb	=	.45 kg
1 g	=	.035 oz
1 kg	=	2.2 lb

VOLUME

A large auto gas tank holds about 100 liters

One liter is a standard bottle of soda, milk, or wine

A cup of coffee is about 150 milliliters

5 milliliters is a standard teaspoon of medicine

A drop of water is about 50 microliters

Volume Conversion

1 tsp	=	5 ml
1 tbsp	=	15 ml
1 fl oz	=	30 ml
1 cup	=	.24 l
1 pt	=	.47 l
1 qt	=	.95 l
1 gal	=	3.8 l
1 ml	=	.03 fl oz
1 l	=	2.1 pt
1 l	=	1.06 qt
1 l	=	.26 gal

Human Physiology

Third Edition

Dedication

The third edition of this book is dedicated to the memory of Dr. Leon K. Knoebel, who died December 10, 1994. Professor Knoebel taught gastrointestinal physiology to our undergraduate, dental, medical, and graduate students. Ken was an outstanding teacher serving Indiana University for 39 years. He was also a good friend and colleague and will be sorely missed.

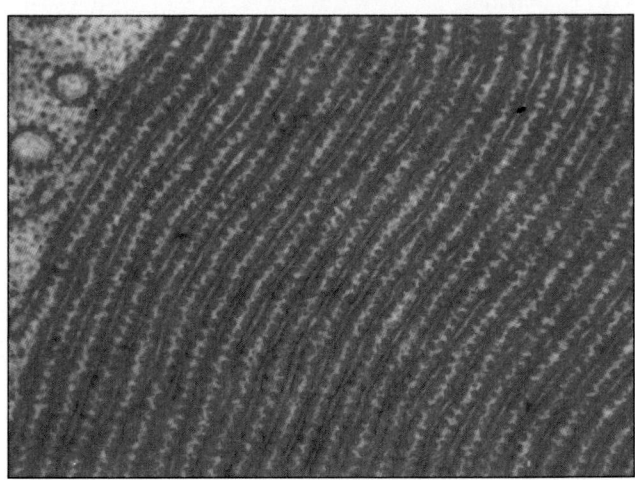

Preface

An integrated circuit is an electronic device in which many different kinds of linked systems function together for a specific purpose. Like the illustrations of the integrated circuit with icons representing component systems used in this textbook, the body systems function in an integrated manner to maintain a relatively stable internal environment. This stability is a requisite for good health. The maintenance of a stably functioning internal environment despite external influences—such as temperature, pathogens, or trauma—defines the concept of homeostatic balance. Homeostasis (*homeo* = same; *stasis* = standing) and physiological control are recurring themes of *Human Physiology*, Third Edition.

Physiology (from the Greek: *physio* = nature; *logy* = the study of) originally meant an inquiry into nature and included the systematized study of both living and nonliving matter. Today, the focus of the study of physiology is the integration and coordination of the systems of living organisms for the purpose of explaining and predicting homeostatic mechanisms that operate in the survival of any organism.

The human body is a complex organism comprising several major systems—nervous, musculoskeletal, endocrine, circulatory, respiratory, gastrointestinal, and reproductive, to mention a few. The study of physiology, therefore, is most logically presented by the "systems" approach. However, we believe it is important, especially for the beginning student of physiology, to realize that the body does not function as a set of independent systems, and that no single system is more important than another in maintaining life. Rather, it is the interactions and integration of the multi-systems to maintain homeostasis that are essential to life.

Human physiology is a fascinating discipline. Most of us, professor and student alike, have a natural curiosity about how our bodies work. The recent explosion of scientific information has afforded the life and health sciences tremendous opportunities and tools for better understanding the world within us. As educators and authors, we face an important challenge in presenting to students a vast amount of scientific information as a cohesive body of unifying principles and concepts, instead of as a catalogue of disconnected facts and mysteries. *Human Physiology*, Third Edition, develops and explains these underlying principles and concepts while acknowledging the excitement of unsolved problems.

The Third Edition, more than previous ones, presents important current questions in physiological and biomedical research and provides the background tools for approaching these problems with possible solutions. One of the biomedical research successes between editions of *Human Physiology* was the approval of a genetically engineered drug, which is inhaled through an atomizer, that cuts down the mucus development in cystic fibrosis patients. Another is the discovery that mutations in a particular segment of DNA are associated with colon cancer as well as Huntington's disease, a finding that will improve detection and treatment of people with these diseases. Recently, thrombopoetin, which induces the body's production of platelets, was genetically engineered, an accomplishment especially important to patients such as those with AIDS with low platelet counts who are at a risk for uncontrolled bleeding. Also recently discovered was the gene for PKD, polycystic kidney disease, which will lead to more informed decisions and an improved quality of life for inheritors of this disease.

The four years between editions have witnessed incredible changes and upheavals in the health care and biomedical research professions in movements promoting both streamlined budget consciousness and accelerated but improved quality. Never before has there been a greater need for well-trained, motivated, inquiring scientific thinkers.

Student Audience

We have written this text for students who are planning careers in the life or health sciences. It should be very useful to students of diverse backgrounds and levels of preparation. We have not presumed, for instance, that all students have had extensive experience with math, chemistry, physics, and biology. The background information on basic chemistry, physics, and molecular and cell biology in the first six chapters will help those students who have had little or no college-level science. Students who have stronger backgrounds in the sciences will find the basic science coverage a useful review.

In addition, underlying principles of human anatomy have been introduced where a knowledge of basic structure is necessary for understanding function. The emphasis of this text, however, is on function rather than on structure.

Our objectives in writing this text are the following:

- to develop concepts and principles of physiology in a logical, clear, and concise manner;
- to improve student ability to reason scientifically;
- to promote understanding and appreciation of normal homeostatic body functions;
- to engender excitement about physiological research and other related career interests.

The organization and approach of *Human Physiology*, Third Edition, have evolved through many years of teaching the subject to students of various backgrounds and abilities. Our research, along with that of many colleagues and other scientists, helped us a great deal in shaping the content of this text.

The field of physiology is burgeoning with new discovery and change. No single author, or even two authors, can possibly be the authority on the base amount of information in a textbook on human physiology. An accurate and current presentation of the most important principles and research issues and details requires the collaborative enterprise of a team of research specialists. We are fortunate to have the skills and knowledge of ten other contributing authors. Please get to know them in the "About the Authors" section that follows the Preface.

We sincerely hope that this presentation of human physiology will capture the interest of student and professor alike and stimulate them to pursue the discipline beyond the limits of this book.

Organization

The book is organized so that topics progress from the molecular-cellular level to integrated organ function. Chapter 1, designed to develop an appreciation of science, explains the scientific method, introduces basic concepts such as homeostasis, and provides a historical perspective. This chapter lays the foundation for Part I, which is basic cellular functions and includes the essential chemistry, biochemistry, physics, and mathematical principles required to enhance understanding of fundamental physiological concepts.

Part II focuses on the body's internal environment and analyzes the nature of biological control systems including the properties of specialized cells from the nervous and endocrine systems that regulate body function.

Part III discusses the coordinated functions of other body systems that all stem from the integration of specialized tissue function and includes chapters that present concepts dealing with integrated systems function, such as temperature regulation, exercise physiology, environmental physiology, and sexual physiology.

Content Features

- Chapter 1 provides an historical perspective of the science of physiology. This helps to define the discipline and offers the student an exciting sense of direction for mastering the text.
- Chapter 1 clearly explains the scientific method and offers a review of basic computations, measurements, graphic analysis, and anatomical terminology that serve as tools for learning physiology.
- Early coverage of molecular biology and cell physiology (Chapters 2–6) provides a strong basis for understanding systemic functions. The concepts that are derived from cellular and molecular biology are applied throughout the text.
- Physiological control systems (Part II) are presented before organ functions (Part III) to establish the concept of homeostasis early.
- Metabolism and energy transformation are covered in Chapter 6 in order to present intermediary metabolism as a fundamental concept rather than as a process specific to nutrition.
- There are separate chapters on the Autonomic Nervous System, Acid-Base Balance, Exercise Physiology, Temperature Regulation, Calcium and Phosphate Metabolism, Environmental Physiology, Body Defenses and the Immune Response, and Sexual Physiology. These chapters provide unique, in-depth coverage of these topics, which are often not found in competitive books.

New to the Third Edition

- Review questions, called Problem Solving in Physiology, which promote critical thinking skill development, have been added to the end of the chapter problem sets. Answers to all problems are provided in Appendix A.
- The outline at the beginning of each chapter has been reduced in length and contains action statements that help focus on important chapter concepts.
- Opening vignettes have been added to each chapter as a means of providing an interesting, relevant, and thought-provoking introduction to chapter content.
- Each chapter contains at least one new focus box adding important current information about clinical and research areas of physiology relevant to chapter content. Each box concludes with a question that requires some critical thinking initiative. Answers to the questions are in the Instructor's Manual/Test Bank.
- Two new chapters have been added: Chapter 28, Body Defenses and the Immune Response, and Chapter 29, Environmental Physiology. These chapters expand and update areas covered in other chapters of the Second Edition as well as address new areas of applied physiology.
- New, comprehensive coverage of feedback control systems now appears in Chapter 1 (including introductory coverage of hormonal mechanisms), accompanied by new line art comparing positive and negative feedback.
- The information concerning diffusion and osmosis has been expanded in Chapter 4.
- In Chapter 13, more focus has been given to the functions of the thyroid hormones and less on their synthesis.
- The role of calcium is expanded to make the discussion of the regulation of skeletal muscle clearer in Chapter 16.
- The control of breathing section has been moved to Chapter 21, and now follows gas exchange and transport, so the student can better understand the overall regulation of the respiratory system.
- The main coverage of buffers has been moved from Chapter 2 to Chapter 25 where the concept can be readily applied.

Learning Aids

Many pedagogical aids incorporated in the Second Edition have been retained in the Third Edition to help the student learn concepts in physiology. They are the following:

- Icons represent each system. The individual icons open each chapter. In addition, each part-opening design utilizes the icons to highlight the systems that are to be discussed in the upcoming chapters. This unity of design visually emphasizes the integrated nature of physiology.
- Each chapter begins with an abbreviated outline of the chapter's content and ends with a bulleted summary.
- Each chapter concludes with a set of review questions (including, in most chapters, quantitative problems and critical thinking skill problems), a list of suggested readings, and key terms that are referenced to specific pages of first-time usage within the chapter.
- Throughout the text, boldface type is used to emphasize important new terms.
- Most chapters contain focus boxes to encourage interest in clinical applications and physiological research. The focus boxes demonstrate relevance of the chapter content, help the student understand applied and/or abnormal physiology, and suggest directions future research may take.
- The artwork and photographs emphasize and reinforce important physiological concepts. Numerous flowcharts clarify complex physiological processes; for examples, see Figures 8–26 and 10–19.
- Three appendices offer the following: detailed answers to support the chapter review questions; tables of normal physiological values (e.g., blood pressure, blood gases, body water content); and medical terminology, which includes Latin and Greek prefixes, suffixes, and terms for body parts for the purpose of building a medical vocabulary.
- The index has been greatly expanded and provides extensive cross-referencing.
- A large glossary with phonetic pronunciations for many of the most difficult terms appears at the back of the text.
- Front endpapers contain tables of Decimal Factors and Prefixes, Conversion Factors and Equivalents for the Metric and English systems of measurement.
- The rear endpapers contain an alphabetized table of commonly used clinical abbreviations.

Ancillaries

A number of excellent supporting aids accompany the text. These include: an *Instructor's Manual with Test Bank,* written by David Cotter of Agnes Scott College, 150 carefully selected overhead transparencies from the text, a *Laboratory Manual,* written by Thomas H. Dietz and Joseph Woodring, both of Louisiana State University; and a *Study Guide,* written by Florence C. Ricciuti. In addition, the *Test Bank* is computerized for the IBM and MacIntosh computers. For the first time, selected pieces of line art from this edition of HUMAN PHYSIOLOGY will be available on the *Saunders General Biology Videodisc IV.* A *Directory* with descriptions, barcodes, and reference numbers for each image and *LectureActive Software Package Version IV,* which enables quick and easy creation of custom lectures, accompany the videodisc. This same selection of line art will also be available in *MediaActive,* a CD-ROM format.

Acknowledgements

We want to express our deepest thanks and appreciation to all the contributing authors without whose expertise, talent, and effort this book would not have been possible. We would also like to give thanks for a job well-done to the editorial staff and management of Saunders College Publishing for their help and guidance in significantly improving the first and second editions of this book. Special thanks go to our Developmental Editor, Cathleen Petree, whose patience and editorial talents were absolutely essential to the successful completion of the second and third editions of this book. We are indebted, as well, to Laura Shur, Project Editor, Carol Bleistine, Art Director, Ray Tschoepe, Art Developmental Editor, and Amy Ellis Dunleavy, Photo Researcher, for their advice and guidance during the production phase. We also thank Martha Colgan, the Developmental Editor of our First Edition, for her hard work in expanding and greatly improving the index for the Second Edition. Finally, we would like to thank Julie Levin Alexander, our Executive Editor, and Edith Beard Brady, Consulting Acquisitions Editor, for their support and commitment to this book, and for sharing with us their administrative talent in managing the human and material resources of this very complex project.

A very special thanks goes to Nancy Christian for typing portions of the manuscript and for her administrative assistance without which we surely would have suffered. Nancy's assistance made our work as authors and editors much easier.

Last, we thank our families and friends for their tolerance of our preoccupation with the textbook for extended periods of time. Our very warm gratitude goes to our wives, Diane Pflanzer and Judy Rhoades, without whose love, understanding, and support this project could not have been completed.

Reviewers

We express our sincere appreciation to the following faculty members and scientists for their conscientious reviews and assessments of the manuscript. Their suggestions, comments, and insights have been invaluable.

Reviewers of the First Edition:

Jerome Yochim, *University of Kansas*
Arnold J. Sillman, *University of California, Davis*
Rick Turnquist, *Augustana College*
Andy Anderson, *Utah State University*
Byron A. Schottelius, *University of Iowa*
John T. Fales, *University of Missouri, Kansas City*
John P. Harley, *Eastern Kentucky University*
L. Stephen Whitley, *Eastern Illinois University*
S. J. Coward, *University of Georgia*
David Noyes, *The Ohio State University*
Pegge Alciatore, *University of Southwest Louisiana*
Stephen Williams, *Glendale Community College*
Kenneth H. Bynum, *University of North Carolina, Chapel Hill*

Reviewers of the Second Edition:

Steve Simasko, *SUNY at Buffalo*
Jack Wood, *Western Michigan University*
Ralph Ferges, *Palomar College*
John Scheide, *Central Michigan University*
Mark Beidebach, *California State University, Long Beach*
Daniel Gibson, *Worcester Polytechnic Institute*
Lin Aanonsen, *MacAlester College*
Mary Lynne Stephanou, *Santa Monica College*
Sunny Boyd, *University of Notre Dame*
Pegge Alciatore, *University of Southwest Louisiana*
Audrey Mackey, *Austin Community College*
Carl Hammen, *University of Rhode Island*
John G. Moner, *University of Massachusetts, Amherst*
Katherine Mechlin, *Wright State University*
Fred D. Hinson, *Western Carolina University*
Sheldon Lustick, *Ohio State University*
Harry Bernheim, *Tufts University*
Lawrence Wit, *Auburn University*
Leon Goldstein, *Brown University*

Robert Slechta, *Boston University*
Barbara Howell, *SUNY at Buffalo*
Stuart Coward, *University of Georgia*
Pinapka Murthy, *Fayetteville State University*
Edmund Tong, *Wheaton College*
Leonard Lundquist, *Lander College*

Reviewers of the Third Edition:

Robert Hazelwood, *University of Houston*
John Moner, *University of Massachusetts, Amherst*
Sunny Boyd, *University of Notre Dame*
Margaret Gould, *Georgia State University*
Harold Falls, *Southwest Missouri State University*
Katherine Mechlin, *Wright State University*
Carl Hoegler, *Marymount College*
Jim Goldinger, *SUNY at Buffalo*
Frederick Hagerman, *Ohio University*
Steve Overmann, *Southeast Missouri State University*
George Veomett, *University of Nebraska at Lincoln*
Ron Wiley, *Miami University*
Kim Cooper, *Arizona State University*
Don Whitmore, *The University of Texas at Arlington*
John Hranitz, *University of Central Oklahoma*
Lewis Lutton, *Mercyhurst College*
James Junker, *Campbell University School of Pharmacy*
J.H.U. Brown, *University of Houston*
Thomas Adams, *Michigan State University*
Alan P. Brockway, *University of Colorado at Denver*
Jerry Stinner, *University of Akron*
Evelyn H. Schlenker, *University of South Dakota School of Medicine*
Andrew Pellett, *Louisiana State University Medical Center School of Allied Health*
Henry Kermott, *St. Olaf College*

Theresa Bacon-Bagley, *Grand Valley State University*
Stephen T. Bishoff, *University of South Carolina at Sumter*
Leonard Beuving, *Western Michigan University*
Lois Jane Heller, *University of Minnesota at Duluth School of Medicine*
Richard L. Hébert, *University of Ottawa*
Vicki Motz, *University of Lowell*
Kathleen M. Susman, *Vassar College*
Judy Van Liew, *SUNY at Buffalo*
Charles Costa, *Eastern Illinois University*
Hugo Lane, *Wake Forest University*
Bruce Grayson, *University of Miami*
James L. Blank, *Kent State University*
Daniel Lemons, *City College of CUNY*
Roger Theis, *University of Oklahoma Health Sciences Center*
Jay Farber, *University of Oklahoma Health Sciences Center*
S.S. Hull, *University of Oklahoma Health Sciences Center*
David Bruce, *Wheaton College*
Michael Lynes, *University of Connecticut at Storr*
T. Earle Bowen, *University of Kentucky, A.B. Chandler Medical Center*
Ian Tizard, *The Texas Veterinary Medical Center of Texas A&M University*
Brenda Alston-Mills, *North Carolina State University*
Sandra D. Michael, *SUNY at Binghamton*
Michael Philips, *University of Florida College of Medicine*

RODNEY A. RHOADES
RICHARD G. PFLANZER
October 1995

About the Authors

Rodney A. Rhoades is Professor and Chairman of Physiology and Biophysics at Indiana University School of Medicine, Indianapolis. He received his Ph.D. in Physiology at The Ohio State University. Respiratory physiology is the focus of Dr. Rhoades' research. Current work concerns lung vascular biology, and includes signal transduction steps in smooth muscle growth and proliferation, oxidative stress and lung vascular injury, hypoxia and endothelial/smooth muscle interaction, and hypoxia and gene expression.

Richard G. Pflanzer is Associate Professor of Biology at the Purdue University School of Science and Associate Professor of Physiology and Biophysics at Indiana University School of Medicine. He received his Ph.D. in Physiology from Indiana University in 1969, with research in comparative cardiovascular physiology. Dr. Pflanzer has authored many articles and books related to teaching physiology at the undergraduate, graduate, and professional levels.

Contributors

Rey S. Elizondo is a Professor of Life Sciences and Dean of the College of Sciences and Engineering at The University of Texas at San Antonio. He received his Ph.D. in Physiology from Tulane School of Medicine in 1967. Temperature reg- ulation, control of metabolism, and pathophysiology of fever are the subjects of Dr. Elizondo's research.

Janice C. Froehlich is an Associate Professor of Medicine and Physiology/Biophysics at Indiana University School of Medicine. She received her Ph.D. in Physiological Psychology from Syracuse University. The main focus of Dr. Froehlich's research is the elucidation of neurochemical, hormonal, molecular, and genetic factors that contribute to the development of alcoholism and drug abuse.

Joe R. Haeberle is an Associate Professor of Molecular Physiology and Biophysics at the University of Vermont. He earned his Ph.D. in Physiology at Indiana University School of Medicine in 1981. Dr. Haeberle is currently investigating the molecu- lar mechanism by which actin- and myosin-associated proteins regulate the contraction of vascular smooth muscle.

Stephen A. Kempson is a Professor of Physiology and Biophysics at Indiana University School of Medicine. He received his Ph.D. in Biochemistry from the University of London, U.K. His research work is focused on cellular regulation of membrane transport systems in the kidney.

Leon K. Knoebel was a Professor of Physiology and Biophysics at Indiana University School of Medicine until his death on December 10, 1994. He earned his Ph.D. in Physiology from the University of Rochester in 1955. Dr. Knoebel's research interests included gastrointestinal physiology and the absorption of fat.

Walter C. Low is a Professor of Physiology, Neuroscience, and Neurosurgery at the University of Minnesota Medical School. He received his Ph.D. from the University of Michigan in 1979. Dr. Low's current research involves the development of gene and cell transplantation therapies for the treatment of neurological disorders.

Richard A. Meiss is Professor of Physiology and Biophysics and of OB/GYN at Indiana University School of Medicine. His Ph.D. degree was awarded by the University of Illinois at Urbana-Champaign in 1969. Dr. Meiss' current research focuses on the mechanical properties of smooth muscle and the interactions between muscle cells and tissues.

Daniel E. Peavy is an Associate Professor of Physiology at Indiana University School of Medicine, Indianapolis. He received his Ph.D. in Physiology from the University of California at Davis in 1976. Dr. Peavy's research interests focus on insulin signal transduction and the biochemical defects in the pathway that lead to insulin resistance and diabetes.

George A. Tanner is a Professor of Physiology and Biophysics at Indiana University. He earned his Ph.D. in Physiology from Harvard University and his current research interest is the pathophysiology of polycystic kidney disease.

Wiltz W. Wagner, Jr. is the V.K. Stoelting Professor of Anesthesiology and Professor of Physiology/Biophysics and Pediatrics at Indiana University. He received his Ph.D. in physiology and biophysics from Colorado State University in 1974. Dr. Wagner's research is focused on pulmonary microcirculation, investigating how red and white blood cells cross the capillary bed and how the capillary bed responds to increased demand for oxygen during exercise and during exposure to high altitude.

Contents Overview

Contents

33

Sexual Physiology   952

Focus Boxes

These focus boxes are intended to apply basic concepts and to illustrate how diseases are characterized by altered functional changes. For most chapters, there are two types of focus boxes, clinical and research.

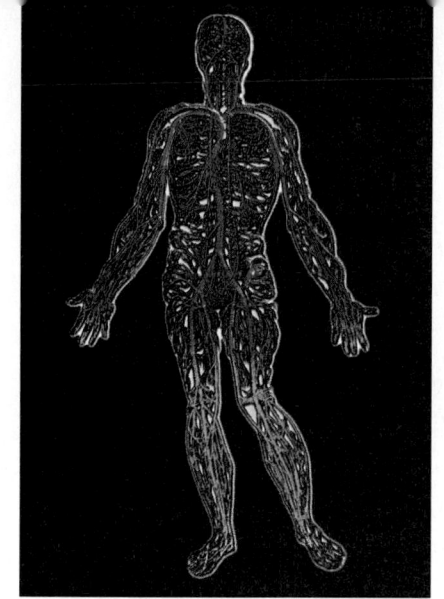

Enhanced images of the circulatory system after Vesalius original. *(Photo Researchers, Inc./© Scott Camazine)*

Chapter 1

The Science of Physiology

Key Concepts

Physiology is a branch of biology that deals with function and coordinated activities of cells, tissues, and organs. Discoveries made during the European Renaissance laid the foundation for modern advances in physiology.

The scientific method is the use of principles and procedures for the systematic pursuit of new knowledge. The scientific method involves formulation of a hypothesis, collection of data through observations or experiments, and testing of hypotheses. Quantitative methods involve measurements, computation, graphic analysis, and other types of data analysis.

The body is organized into cells, tissues, organs, and coordinated systems.

Homeostasis is an integrative process of maintaining a constant internal environment for cellular and organ function.

Chapter Outline

Each year Dave plans a backpacking trip that includes 10 other men. This is an annual event the group has enjoyed and looked forward to for the last five years. A combination of prior training, careful planning, good weather, and common sense has enabled Dave's group to enjoy spectacular views, and great comradeship. This year Dave has planned a 10-hour climb of Mt. Hood (elevation of 11,235 ft). This is their first winter climb, and Dave realizes that, as a result of sudden winter storms, inexperience in winter climbing, or plain bad luck, this could be their most dangerous trip. Anyone in the climbing party could sustain serious injury.

The climbing party arrived at the lodge (6,000 ft) the day before the climb. Their climbing guide informed them the climb would start the next day at 12 midnight so that most of it could be made at night when the snow is still firm and there is less chance of an avalanche. The guide checked all of the climbing gear (including helmets, ice axes, and climbing ropes) and the group's clothing which should whisk away perspiration but retain body heat. The guide also checked each climber's compass, map, and altimeter. He stressed the importance of making precise and accurate readings of their compass and altimeters to pinpoint exact locations on their maps. He pointed out that in the event of a storm this could make the difference between becoming fatally lost or returning safely to the lodge. He also stressed the importance of the altimeter, which not only uses barometric pressure to measure the height above sea level, but also can be used to predict changing weather conditions. Each climber had written down altimeter readings both the night before and the morning of the climb. The morning reading showed a slight increase, indicating a change in the weather pattern.

The evening of the climb, Dave's group had their ritual steak dinner. However, the guide warned everyone that alcohol, nicotine, and caffeine could accentuate altitude sickness—a serious condition that could become life threatening.

The climbing party left the lodge, as planned, at 12 midnight. Dave shivered from both the cold and excitement as he checked his backpack one last time. The first part of the climb followed a ski trail from the lodge. At the top of the trail (7,800 ft), the group stopped to drink some water and eat a snack. Each checked the compass heading to the lodge, and everyone marked on his map this reading of 7,800 ft and a compass reading of 165°. If a whiteout occurred, each climber could return to the lodge by following his compass heading. The next part of the climb was much steeper, and Dave sensed the pounding of his heart rate, labored breathing, and increased sweating. Upon reaching a ridge at 8,900 ft, the group took a much-needed rest. After melting some snow and eating some food, each again took careful compass and altimeter readings. At this point, the climbing guide was watching the climbers carefully to see if anyone was walking unsteadily, or with an uneven gait. He asked how everyone was feeling and asked if anyone had a headache or felt sick—two early signs of altitude sickness.

As the group ascended to 9,800 ft the wind began to pick up. Clouds started forming around the mountain, and Dave noticed that his altimeter was rising. What started as a beautiful sunny morning was now becoming dark, windy, and cold. Suddenly the clouds dropped over them and it began to snow. Dave was shocked to realize that he could not see anything beyond his hand. The wind and snow became worse, and the group stopped and roped each other together. As the temperature dropped and the snow started to accumulate, the guide yelled to the group that they had to dig in and build a snow cave. Everyone suddenly realized how terrifying and life threatening this severe storm was. The storm continued and the guide reassured them that they were carrying everything they needed to wait it out. The storm lifted at 2:00 pm the next day, and as the group dug themselves out they experienced a glorious sight: a clear sky and brilliant sun shining on 3 ft of new snow. Although Dave's group had a spectacular view and a great feeling of triumph, the climbers would not reach the summit because of the high snow that closed their

pass and created the possibility of an avalanche up ahead. Although disappointed about not reaching the summit, Dave's group was grateful for the preparation that had gone into the climb, and knew that the careful and accurate measurements made by the climbing team would make possible their safe return.

Just as careful and accurate measurements are critical to the success of a mountain climb, they are likewise the key to science and medicine. The hypothesis of a scientist must be supported by data derived from careful and accurate experimental measurements, and the diagnosis of a physician depends on accurate data received from the clinical laboratory as well as from basic research. The ultimate goal of physiological research is to come full circle for a deeper understanding of human physiology in the natural environment.

The ancient Greeks considered the universe to be composed of four elemental substances: air, water, earth, and fire. The Greeks further believed each human being was a miniature universe in whom these substances appeared in the form of four "humors," or fluids: *blood* (corresponding to air); *phlegm* (representing water), *black bile* (corresponding to earth), and *choler* or *yellow bile* (representing fire) (Fig. 1–1).

According to the Greek view, the fluids were not present in each person in equal amounts; one fluid usually predominated and characterized the individual's temperament. For example, one temperament was the "sanguine" type. We still use this word today to describe someone with a cheerful, confident personality. In ancient times, such a personality would have been attributed to a predominance of blood over the other three humors within that person's body, as evidenced perhaps by a ruddy complexion (*sanguine* comes from the Latin *sanguineus*, meaning "bloody"). A "melancholy" person was thought to have more black bile than any of the other humors, whereas a "choleric" person was considered to have a predominance of yellow bile (the humor associated with fire), which produced an excitable, easily angered temperament. A "phlegmatic" person generally was calm and unemotional, even sluggish, supposedly because of a predominance of phlegm, the cold, waterlike humor. The Greeks thought that the particular quantities of each humor and the resulting proportion, or balance, determined a person's physical and psychological makeup.

According to this classical scheme, while a normal predominance of one fluid within each person's unique humoral balance produced a characteristic personality, an abnormal lack or excess of one or more

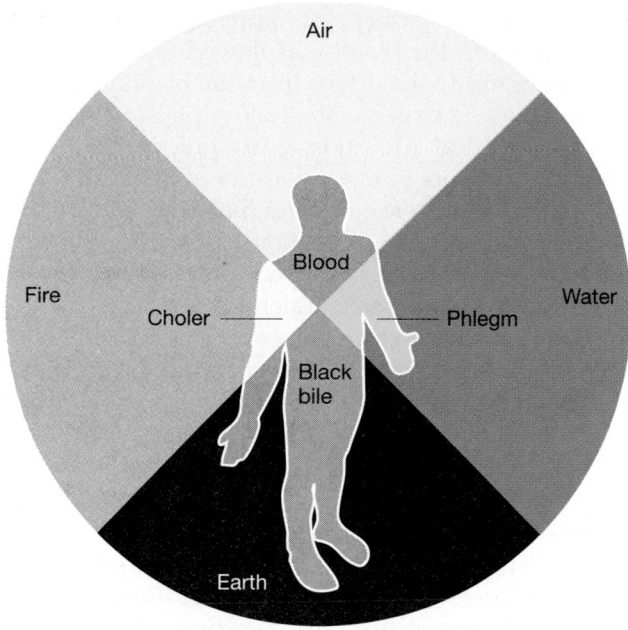

Figure 1–1

The four bodily "humors" of classical and medieval philosophy and physiology. The Greeks thought that the elements of the universe were represented in each human body by blood, phlegm, black bile, and choler. According to this scheme, a predominance of one humor produced each person's temperament.

humors resulted in disease. Doctors from ancient times until well into the Renaissance shaped their therapeutic methods around attempts to restore the proper balance of humors within the sick person's body. Purging, bloodletting, and similar treatments were all intended to adjust humors.

In contrast to the ancient Greek philosophers and physiologists, today we organize the universe of matter into 105, rather than 4, elemental substances, and we attribute each person's innate uniqueness to genetic makeup rather than to relative quantities of bodily fluids. Yet despite the revolutions in our conceptions of human physiology, some classical principles influence modern science. One enduring principle is the idea that the human body, if not a world in miniature, is composed of elements from the universe that obey the laws of nature. Another "modern" idea cherished by the ancient philosophers is that of a harmonious balance within the body. While today we speak of homeostasis and recognize it as a self-regulatory process involving not only fluids but dynamic processes, we share the classical view that health is the evidence of balance among body functions and substances. Finally, we have inherited the Renaissance attitude—which echoed the ancient Greek belief—that the human body is worthy of the most rigorous study and eloquent description.

The History of Physiology

The story of the development of physiological understanding in modern culture is fascinating. The following brief survey of the last 2000 years of physiological theory and research provides a context for the study of twentieth-century physiology in all its complexity and sophistication. Further, the discussion communicates a sense of wonder at the vast advances physiologists have achieved since human beings first began to record observations and theories about the workings of the human body.

Physiological Knowledge in the Classical and Medieval Periods was Steeped in Dogma

Since so many Western attitudes about science, philosophy, and the place of human beings in the universe have been influenced by the traditions of the classical Greeks, that period seems a logical starting point to examine the development of physiology as we understand it today. The studies of the Greek philosophers and the investigations of the Greek physiologists at the great museum in Alexandria influenced European medical and physiological practice until well into the Renaissance because of the mediating influence of Galen. Galen's writings preserved the ancient studies and, in turn, were preserved through the medieval period for examination by Renaissance scholars.

The Observations of Aristotle

Although he is most often thought of as a philosopher, Aristotle (384–322 B.C.) did important work in biology and was one of the first to describe the blood vessels as a system with the heart at the center. Unlike other investigators of his time, he used a bloodless method of killing animals for dissection, finding that the vessels were easier to see and trace when the blood remained.

Some of Aristotle's concepts were corrected through the discoveries of later investigators. Aristotle thought of the heart as both the seat of the intellect and a furnace that heated the blood to provide needed warmth. (Since warmth disappeared from the body so soon after death, it was thought to be the cause or source of life.) Just as the blood needed a source of heat, so too the furnace needed ventilation. According to Aristotle, this was the purpose of breathing; the lungs were a ventilation system and the air a cooling agent.

Aristotle correctly observed that blood flowed between arteries and veins, but he could not see the microscopic capillaries that we now know provide the connection between them. He explained the flow by saying, in essence, that the vessels were themselves made of blood and that they disappeared when blood flow stopped.

> . . . the channels of the blood vessels may be compared to the mud which a running stream deposits, they are as it were deposits left by the current of blood in the blood vessels. . . . Thus just as in the irrigation system the biggest channels persist whereas the smallest ones quickly get obliterated by the mud, though when the mud abates they reappear; so in the body the largest blood vessels persist while the smaller ones become flesh in actuality, though potentially they are blood vessels as ever before.

The Theories of Herophilus and Erasistratus

At the Museum at Alexandria, a famous center of Greek culture and learning in what is now Egypt, two men pursued a number of biological investigations in a continuation of Aristotle's search for an understanding of blood flow. Physiological research flourished at Alexandria partly because it was the only place that permitted dissection of human cadavers for the study of anatomy and physiology.

Herophilus (c. 335–280 B.C.), who is considered by some to be the first physiologist in the Western tradition, identified the brain as the seat of the intellect and demonstrated that the walls of the arteries are thicker than those of the veins. Using the newly invented water clock, he measured the pulse and showed that it varied when disease was present.

Erasistratus (310–240 B.C.) began his physiological studies as an assistant to Herophilus. Erasistratus believed that blood was made in the liver from food and was delivered to the organs of the body by an "ebb and flow" in the veins. He thought of the arteries as air vessels, not blood vessels. The air, called *pneuma*, was thought of as a living force taken in through the lungs to the left side of the heart, where it was transformed into a "vital spirit" and then transported, in "airy" form, through the arteries to the rest of the body. (To explain why blood, not air, flowed from a cut artery, in apparent contradiction with his theory, Erasistratus postulated the existence of tiny connections between veins and arteries that sometimes opened to let blood flow into the arteries.)

The "Pneumatology" of Galen

In a simple experiment with a goose feather, Galen (c. A.D. 130–201) demonstrated that blood, not air, flowed through the arteries. Galen created a hollow tube by cutting off the ends of the feather. Then he inserted the tube into a tied-off artery, from which blood immediately flowed into the tube. Since, by being tied off, the artery was supposedly separated from the veins, the blood that flowed from it had to have been inside the artery from the beginning.

Galen was a Greek who studied physiology both in Greece and at Alexandria. His practice as a surgeon to gladiators in his native city of Pergamum allowed him to observe the internal structure of the body at a time when dissections for the purpose of study were forbidden. Later he was appointed physician to a Roman emperor. Galen is known for his voluminous writings on philosophy, medicine, and physiology, in which he commented and expanded on the work of the investigators who preceeded him. Even though many of his ideas have been disproven, we still consider Galen an important figure because of the intellectual sophistication of his physiological schemes, the breadth and scope of his explanations, and the length of time and extent to which his teachings were accepted. His was the prevailing view of human physiology from his own time until the Renaissance.

Like Erasistratus, Galen believed that the blood was produced from food, in the liver (Fig. 1–2). He thought that the blood took on "natural spirits" and carried them through the veins to the bodily organs, which needed the spirits to perform different functions. After its supply of spirits was depleted, the

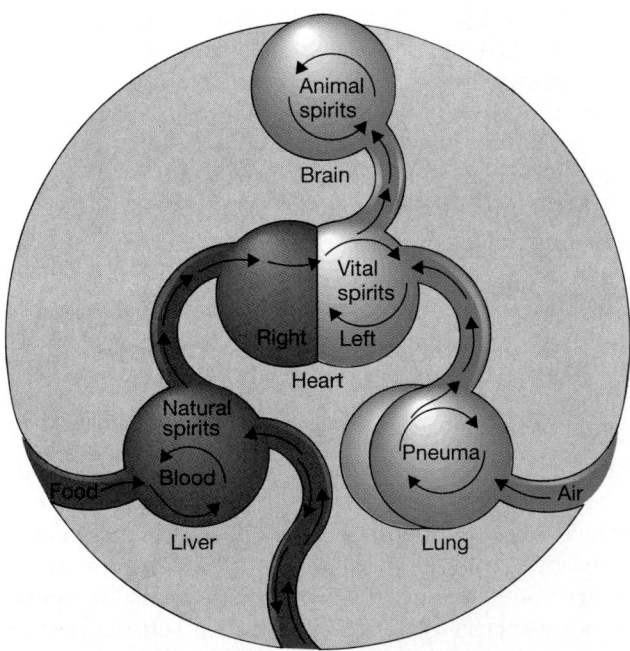

Figure 1–2

In the "pneumatology" of Galen, blood was made in the liver from food and carried "natural spirits" to the bodily organs, traveling both ways in blood vessels. *Pneuma* made from air in the lungs met blood in the heart and was transformed into "vital spirits" and then "animal spirits."

blood returned along the same venous pathways to the liver to be resupplied. (This idea that the blood flowed both ways in the vessels became one of the most revered of Galen's teachings. When Renaissance anatomists correctly challenged it because of their knowledge of valves, they found it very difficult to overthrow.)

According to Galen, some of the blood containing the "natural spirits" went first to the right side of the heart and then to the left side, where it contacted *pneuma*. *Pneuma* was a substance produced from air in the lungs and carried into the left side of the heart. When the "natural spirits" contacted *pneuma*, they were transformed into "vital spirits," a higher form of *pneuma*. These "vital spirits" were carried on up to the brain, where they were further transformed into yet a higher form of *pneuma* called "animal spirits. " These various forms of *pneuma* drove the various processes of the body, according to Galen.

Despite his insistence on the importance of observation in the study of physiology, Galen often made incorrect assumptions about human anatomy based on his dissections of animals. As mentioned previously, however, the logic and elegance of his theories about bodily functions, which unified both the traditions he inherited and the knowledge available to his contemporaries, were convincing. His copious writings further established his authority, and throughout the late classical and medieval periods Galen's physiology was accepted. It prevailed largely unquestioned until the technological advances and humanistic spirit of the Renaissance permitted a fresh look at the body.

The Discoveries of the European Renaissance Period Revolutionized Physiology

In the revival of classical learning that characterized the European Renaissance, Galen's writings about anatomy, physiology, and medicine first began to be widely translated from the Greek and Arabic into Latin. Although the writings were known to exist, and the general outlines of Galen's physiology had dominated medieval and early Renaissance medicine, only in the sixteenth century did the most important writings become available for close scrutiny by European physicians. By this time, the Galenic physiology had already taken hold of the imagination and come to be regarded almost as religious doctrine. Thus, Renaissance professors of medicine taught the anatomy of Galen, and anyone who departed from Galen's concepts was considered a secular heretic. To explain the discrepancies between the anatomy described by Galen and that visible in a dissected corpse, a professor might claim that the body simply had changed since Galen studied it.

The task of the Renaissance physiologists was thus to study Galen's physiology, discover its errors, and replace it with a modern scheme. Many independent-minded investigators contributed to the revolution in physiological knowledge that occurred in the late Renaissance, although they too defaulted to many of Galen's tenets whenever they could not observe or explain for themselves certain structures and functions. In his teachings, writings, and drawings, Andreas Vesalius corrected many of the inaccuracies of Galen's anatomy while perpetuating others and retaining much of Galen's erroneous physiology. William Harvey, in his discovery of the circulation, finally revised Galen's ancient notion that the air and the blood met in the heart, though he still viewed the lungs, as had Galen, merely as cooling organs.

The Drawings of Vesalius

Andreas Vesalius (1514–1564) was educated as a doctor and, at the age of 23, became a professor of surgery and anatomy at the great medical school in Padua, in what is now Italy. He had been taught the physiology of Galen and in turn began teaching it to his students. Unlike other professors of his time, however, who taught almost exclusively from Galen's writings, Vesalius centered his lectures not on the ancient texts but on human cadavers that he himself dissected as he lectured. (Previous instructors might not have used a human cadaver or might have lectured from afar while another person conducted the dissection.)

Another pedagogical device that Vesalius introduced into the study of physiology was the anatomical drawing. Galen's texts had been published with few or no illustrations; Vesalius created six anatomical "tables" that showed labelled parts of the body and allowed students to follow his lectures.

Despite his new approach to the teaching of anatomy, Vesalius considered himself part of the tradition of Galen and initially intended to simply explicate Galen's teachings. Hence, in the beginning he unwittingly reproduced some of Galen's erroneous assumptions about human anatomy. For example, Galen had thought that an organ called a "rete mirabile," similar to one he saw in hoofed animals, existed at the base of the human brain. This organ was supposedly where "vital spirits" were changed into "animal spirits." Vesalius showed this organ in his anatomical drawings, although, like Galen, he had never seen one in a human body.

Gradually, however, Vesalius began to realize that Galen's physiological dogma had been based not on an earlier version of the human body but on the bodies of apes, pigs, and other animals. He decided to write a new anatomy text that would reflect his own personal observations of a human body. The result was *De Humanis Corporis Fabrica* ("The Structure of the Human Body"), often simply called the *Fabrica* and now considered the first modern anatomy textbook. Published in 1543, the text contained plentiful illustrations and corrected many of Galen's errors.

Despite Vesalius's insistence on verification of anatomical details and his intention to correct Galen's anatomy, he perpetuated some of Galen's mistaken ideas. One ancient notion that reappeared in Vesalius's text was that the blood passed directly from the right to the left ventricle of the heart. Vesalius did not recognize what we now know as the pulmonary circulation (the movement of blood between the lungs and the heart) despite evidence of this movement that was accumulating in his time. He claimed, as had Galen, that air traveled from the lungs to the left ventricle of the heart, where it cooled the heart furnace and endowed the blood in the heart with "vital spirits."

The Theories of William Harvey

The picture of the blood flow handed down to the early Renaissance physiologists showed the heart as both a furnace and a meeting place for blood and air. We know now that the heart is a pump, not a furnace, and that the lung, not the heart, is where blood and air meet. Renaissance anatomists thought that blood moved directly from the right ventricle of the heart to the left to be imbued with "vital spirits" brought from the lungs by the *pneuma*. Modern theory explains that before blood can get from the right ventricle to the left it must first leave the heart and pass through the lungs. Before Harvey, Renaissance physiologists accepted Galen's view that blood returned along the same pathways it had come in its travels through the body. We now know that valves prevent this from happening and that blood flows only one way through each vessel. When it has gone as far as it can in one direction through the arteries, it crosses by way of the capillaries over into the veins to return to the heart.

Various pieces of this new picture were available to the Renaissance physiologists even as they taught the traditional scheme of Galen's *pneuma* and spirits. They knew, for example, that valves existed in the veins, and that the large vessels that connect the lungs to the heart contained flowing blood, not air. Only William Harvey, however, was able to make sense of these details and deduce from them the notion of the heart as a pump that maintains a *circular* movement of blood.

Born in England, William Harvey (1578–1657) was educated at Cambridge University and then at

Figure 1–3

William Harvey, the discoverer of the circulation.   *(The Bettman Archive).*

Padua, where he attended the lectures of the famous anatomy professor Fabricius (Fig. 1–3). Fabricius was renowned for his construction of a special amphitheater that allowed large groups of students to observe dissections. He is noted for his observation of valves as well, which later played a critical role in his famous student's conception of the circulation (Fig. 1–4). For his part, Fabricius dismissed the valves as merely slowing, not preventing, the backflow of blood through the veins. He held firmly to the physiology of Galen while patronizing the student whose physiological theories would soon turn Galen's views upside down.

After completing his course at Padua, Harvey returned to England and began a career as a medical doctor (serving as royal physician) and lecturer. He also wrote a brief study of the circulation that was to become a classic in the history of science, revolutionizing the Western view of the movement of blood. *De Motu Cordis et Sanguinis in Animalibus* ("On the Motion of the Heart and of Blood in Animals"), published in 1628, described his search for the true role of the heart and the blood pathways.

The key features of Harvey's theories of blood flow that marked clear breaks from the physiology of

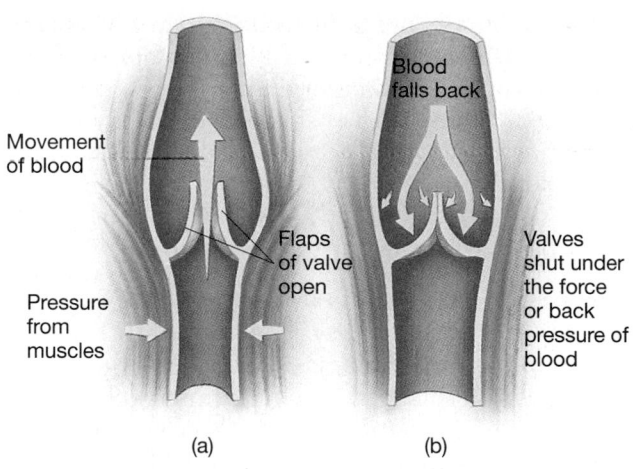

Figure 1–4

Valves in veins. The diagram shows how valves open to allow blood to flow in one direction but then close to prevent blood flow in the opposite direction.

Galen, Vesalius, and Fabricius were (1) the heart was a pump that worked by contracting to expel blood, rather than by expanding to fill with blood; (2) the blood went from the right ventricle of the heart first to the lungs, not directly to the left ventricle; (3) the air stayed in the lungs, where it met the blood, rather than coursing to the heart to meet the blood; and (4) the blood moved in a circular path through the body, taking different routes coming and going rather than traveling both directions in the same vessels.

Fifteen centuries earlier, Galen had used a goose feather to refute Erasistratus's claim that arteries were air vessels. Now Harvey used a similar experiment to refute Galen's claim that blood flowed both ways in a vessel (Fig. 1–5). He tied off the upper part of his arm to allow blood to flow into the arm through the arteries but to prevent blood already in the arm from flowing back out through the veins. As a result, blood coming into the arm through the arteries poured into the veins below the block but could not pass through them, causing them to swell. If, as Galen had claimed, blood could have flowed back through the arteries through which it had entered the arm, the veins would not have swelled.

Thus, Harvey's discovery of the circulation came about through a combination of creative speculation and simple experiments. Not only did he acknowledge the evidence he saw; he went on to conceive an overall scheme that made sense of that evidence (Fig. 1–6). While others of his time knew of valves and the pulmonary circulation, usually they tried to fit these details into Galen's traditional physiology. Here Harvey describes how he combined direct observation with creative reasoning to break with the traditional view:

Figure 1–5

William Harvey's experiment showing direction of blood flow in veins and the connection between arteries and veins. A tourniquet allowed blood to flow through the arteries into the lower arm but prevented blood from flowing out of the veins. The veins below the tourniquet swelled up, indicating that blood was flowing into them from the arteries. When the swellings were removed, they reappeared in the end of the vein farthest from the heart, showing that blood was flowing into the veins from the arteries instead of from the heart.

If I started from the root of these vessels and tried with all the skill that I could muster to pass a probe in the direction of the small vessels, I was unable to do so over any great distance because of the obstacles provided by the valves; on the other hand, it was very easy to pass a probe from without inwards, that is, from the small branches towards the root of the veins. . . .

Since calculations and visual demonstrations have confirmed all my suppositions . . . to wit, that the blood is passed through the lungs and the heart by the pulsation of the ventricles, is forcibly ejected to all parts of the body,

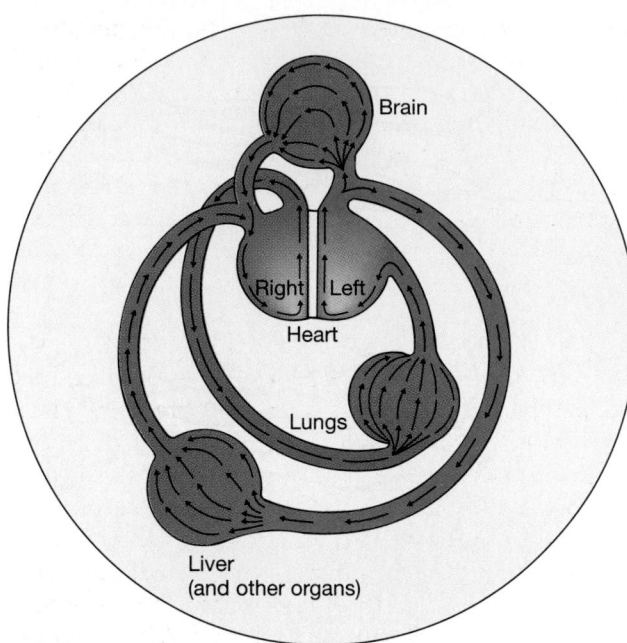

Figure 1–6

A simplified diagram of the circular movement of the blood as envisioned by Harvey.

therein steals into the veins and the porosities of the flesh, flows back everywhere through those very veins from the circumference to the centre . . .

I am obliged to conclude that in animals the blood is driven round a circuit with an unceasing, circular sort of movement, that this is an activity of function of the heart which it carries out by virtue of its pulsation, and that in sum it constitutes the sole reason for that heart's pulsatile movement.

Another unique aspect of Harvey's studies was his use of *quantitative* methods, which included the use of the metric system for his measurements. He calculated that the total amount of blood pumped over the course of an hour, for example, was much greater than what the body could hold or produce from food. He reasoned that, instead of each beat pumping a new quantity of blood, the same blood must be passing through the heart repeatedly. This idea reconciled the ceaseless beating of the heart with the small amount of blood that Harvey determined to be present in the body.

Harvey's discovery of the circulation was critical for our modern conception of the workings of the body, but alone he could not supply all of the details of the picture. While recognizing the pulmonary circulation, Harvey did not know its true purpose, gas exchange. He still thought of the lungs as a cooling system and continued to view blood as an elemental, irreducible substance and an innately vital force that provided the impulse for the heartbeat. We now

know that the autonomic nervous system governs the beating of the heart and that blood comprises various types of cells and other matter. The existence of capillaries, which Harvey suspected but could not confirm, was discovered by the Italian physiologist Malpighi.

The Microscopic Studies of Malpighi

Harvey's explanation of the flow of blood required the existence of some physical connection between the arteries and veins so that, when blood reached the end of its course at the "circumference" of the body, it could immediately enter the veins for its return to the "centre." Harvey was unable to see the capillaries that we now know provide this connection. The invention of the microscope and the brilliant experiments of Marcello Malpighi (1628–1694) provided our first glimpse of these structures, in the lungs of a frog.

Modern Advances in Physiology Opened New Frontiers

By the end of the seventeenth century, the picture of blood flow and heart action that we accept today had been set forth by Harvey and completed by Malpighi. Several other major discoveries awaited the work of eighteenth- and nineteenth-century chemists and physiologists, who would explore the role of the lungs and the substance of air, the existence and processes of cells, and the phenomenon of homeostasis.

In the twentieth century, explorations of bodily control systems, genetic processes, and bioenergetics remain the frontiers of physiological research. The pace of science has quickened. One example is the discovery of atrial natriuretic factor. In 1981 scientists showed that if part of the heart (the atrium) was ground up, made into an extract, and injected into rats, a large amount of sodium was excreted by the kidneys. This was an important breakthrough because some mechanism of this type had long been hypothesized to explain certain aspects of water and salt balance in the body. Within three years, atrial natriuretic factor had been purified, its amino acid sequence defined, assay methods developed, and synthesis by recombinant DNA technology accomplished. Scientists learned where it was made, how it was made, how it was metabolized, and what it did. The astonishing rapidity of this work was possible because of the rapid communication facilitated by international meetings and publications, the availability of sophisticated research methods, and the large number of scientists working in this promising area.

The Study of What We Breathe

The Greeks thought that the air was an elemental, irreducible, and living substance drawn into the lungs to cool the heart and to become *pneuma*. Galen taught that this *pneuma* was transformed into a succession of "spirits" that drove the life processes.

We now recognize that air is made up of several gases. It is not transformed in the heart but contributes oxygen to the blood in the lungs. We can trace this modern understanding to the work of seventeenth- and eighteenth-century physicists and chemists. A series of experiments by European physiologists and physicists (including Torricelli, Borelli, Boyle, Hooke, and Lower) had shown, by the mid-seventeenth century, that: (1) despite its intangibility, air had weight and therefore substance; (2) part, not

(a)

(b)

(c)

Figure 1–7

Joseph Priestley's experiment with a candle and a mouse, showing that something was removed from air during combustion and breathing. These observations later led to the discoveries of oxygen and photosynthesis.

all, of air was necessary for both breathing and combustion; (3) the purpose of breathing was not to cool the body's "furnace," wherever it might have been, but simply to fill the lungs with air; (4) air did not move to the heart but stayed in the lungs, where it met blood that had traveled through the pulmonary circulation from the heart; and (5) in the lungs, air transformed blood in some way, as evidenced by the color change of blood from dark to bright red when air contacted it.

An English chemist, Joseph Priestley (1733–1804), reaffirmed the earlier findings of John Mayow that the actions of both combustion and breathing used up or removed some part of air. Using an experiment similar to Mayow's, Priestley placed a burning candle in a sealed jar (Fig. 1–7). The candle soon went out and could not be re-lit while still in the jar. Priestley then placed a green plant in the jar with the candle. After several days, he was able to relight the candle. He repeated the test with a mouse. After a period in the sealed jar, the mouse died. After a green plant was left in the jar for a few days, however, another mouse was able to live in the jar for a longer period. Priestley deduced that some substance present in air was depleted by the processes of burning and breathing and that the substance could be resupplied by a plant. (This observation led to the discovery of photosynthesis.)

At the time, Priestley did not recognize the meaning of his findings. Although he succeeded in isolating the pure form of oxygen from a mercury compound, he called it "dephlogisticated air," according to a prevailing theory that an element called "phlogiston" was present in air and other combustible matter, and was destroyed by combustion. Priestley shared his findings, however, with the French scientist Antoine Lavoisier (1743–1794), who then identified the essential substance as oxygen.

Lavoisier thought of breathing as a physiological version of combustion because both processes required the same element. He claimed that the purpose of oxygen was to sustain the body's innate heat by allowing blood, the "combustible element," to burn. Lavoisier considered body heat the source of life and the lungs both a ventilation system and the place where combustion occurred. Later chemical advances would show that oxygen was involved in a complex trail of chemical reactions distributed throughout all the cells of the body.

The Advancement of Cell Theory

The Renaissance English scientist Robert Hooke (1635–1703), working with an early microscope, first detected the presence of small compartments he called "cells" in a piece of cork. Although he coined the term we now use to describe the basic unit of all life, it was not until two centuries after Hooke that the theory of the cell as we know it today arose in the writings of the German biologists Matthias Schleiden and Theodor Schwann. The cell theory is the broad basis of all modern biology and physiology and includes the following key concepts: (1) all organisms are made up of cells and their products; (2) new cells arise only from pre-existing cells; (3) all cells have the same fundamental chemical makeup and metabolic processes; and (4) the activities and processes of the organism as a whole result from the interdependent and cooperative workings of groups of cells. The concept that cells originate only from other cells, and not from nonliving matter as had been thought previously, was stated first by Rudolf Virchow (1821–1902) and then confirmed by Louis Pasteur's (1822–1895) experiments with bacteria.

By the late nineteenth century, chemists knew that the breaking of chemical bonds that occurred when substances such as air and food were taken into the body produced chemical energy. Louis Pasteur's research into the process of fermentation helped increase chemists' understanding of the energy-releasing processes that go on in each cell. Researchers since have identified high-energy phosphate bonds (ATP) as the repository for this energy in the cell. Twentieth-century scientists now recognize that oxygen is the substance that is linked to the utilization of ATP. The release of energy from ATP allows cell function to continue. Thus, ATP, not blood (as Lavoisier thought) is the substance whose "ignition" fuels the life of cells. Table 1.1 lists major advances in science, with emphasis on those advances that have profoundly influenced physiology and the practice of medicine.

The Idea of Homeostasis

Before the invention of the thermometer, physicians and physiologists thought that body temperature varied from one individual to the next, and that, within each individual, the temperature could vary with such factors as age, location, climate, and time of day. Early scientists thought that the temperature of the body was subject to external manipulation and thus was unstable. The advent of the thermometer allowed physiologists to measure each person's tem-

Table 1–1
Major Advances in Medicine

Name	Discovery	Date†
Priestley	Discovered oxygen	1774
Laennec	Invented the stethoscope	1816
Hutchins	Invented the spirometer	1840
Morton, Wells, Jackson	Introduced general anesthesia	1846
Koch	Discovered the tuberculosis bacillus	1882
Roentgen*	Discovered X-rays	1895
Abel, Takamine	Discovered epinephrine	1897
Banting/Best and MacLeod	Discovered insulin	1923
Einthoven*	Discovered electrocardiogram	1924
Fleming*	Discovered penicillin	1928
Hench, Kendall, Reichstein*	Introduced corticosteroids to medicine	1949
Waksman*	Discovered streptomycin antituberculus	1947
Clark	Invented platinum PO_2 electrode	1953
Enders, Weller, Robinson;*	Isolated polio virus	1955
Salk, Rabin*	Introduced polio vaccine	
Crick, Watson*	Discovered the double helix of DNA	1962
Ikeda	Introduced flexible fiberoptic bronchoscope	1966
Swan, Ganz	Introduced right heart catheter (Swan-Ganz catheter)	1970
Axerod, Katz, Euler*	Discovered neurotransmitters	1971
Cohen, Levi, Montalcini*	Discovered epidermal growth factor	1986
Bishop, Varmas*	Discovered growth-regulating genes	1989

*A Nobel prize was awarded for their work. The Nobel prize was first awarded in 1901.

†Dates denote when inventions or discoveries were first announced, or introduced in medicine.

perature against the same scale and observe that, apart from minute differences and variations due to abnormal (disease) states, the internal temperature of all human beings fell within a narrow range regardless of what climate they lived in.

This realization led the French physiologist Claude Bernard (1813–1878) to develop the idea of the *"milieu intérieur,"* a constant internal state in which stable conditions of temperature and chemical conditions prevail inside the cell despite changes in the external environment. The American physiologist Walter Cannon (1871–1945) termed this process *homeostasis.* Other twentieth-century scientists continue to reveal the mechanisms by which the body maintains the optimal conditions for its own functioning.

The Study of Physiology

In a broad sense, science is a systematic method for discovering new knowledge. Regardless of the scientific discipline (e.g., chemistry, physics, biology, or physiology), all scientists use a similar approach in their research. This approach employs a rigorous thought process in a rational, systematic, and error-minimizing way to discover new knowledge. Science is systematic because scientists attempt to organize their ideas so that anyone who wishes to build upon their work can obtain the necessary information. By using a set of commonly accepted rules and procedures, scientists strive to provide precise and accurate knowledge about the universe.

The Scientific Method is a Systematic Pursuit of Knowledge

All of the basic ideas in this textbook about how the body works has been gained by the use of the **scientific method**—a systematic approach of explaining how the natural world, including ourselves, works. Although many techniques are used, all are based on the idea of objective observations and reproducible data. Use of the scientific method requires a reasoning process to devise experiments.

Reasoning Processes

The thought processes that scientists use can be grouped into two general categories: **deductive reasoning** and **inductive reasoning.** The two types of reasoning can be thought of as opposites. In deductive reasoning, one proceeds from a general rule to a specific conclusion, whereas in inductive reasoning one goes from a specific observation to a general conclusion (Fig. 1–8).

In deductive reasoning, a scientist begins with supplied, or "given" information, called a **premise,**

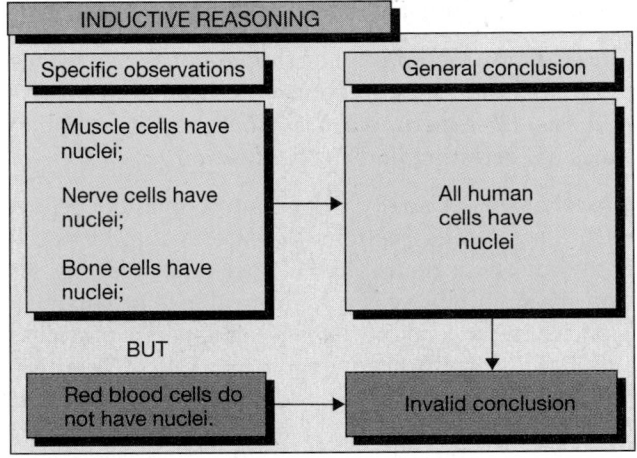

Figure 1–8

A schematic view of the deductive and inductive reasoning processes. In deductive reasoning, one proceeds from a premise or general law to a specific conclusion. In inductive reasoning, one goes from specific observations to a general conclusion, which can be disproved by a single exception.

often stated in the form of an absolute rule using a word such as "all" or "always." From that premise or general rule, the scientist draws a conclusion about a specific example that matches the premise. Frequently, the premise is actually a set of related premises, one general and one more specific. Following is an example of deductive reasoning:

Premise(s): (1) All tissues are composed of cells.
(2) Muscle is a type of tissue.

Conclusion: Muscle is composed of cells.

In inductive reasoning, a scientist proceeds from one or more specific observations to a general conclusion. A conclusion drawn by inductive reasoning is true only until a specific observation is found that disproves it; one specific exception is enough to invalidate the conclusion. Following is an example of inductive reasoning:

Observation(s): (1) Muscle cells have nuclei.
(2) Epithelial cells have nuclei.

(3) Nerve cells have nuclei.

(4) Muscle, epithelial, and nerve cells are all human cells.

Conclusion:　All human cells have nuclei.

However, there is one exception that invalidates this conclusion. Mature human red blood cells lack nuclei; therefore, not all human cells have nuclei. This one example is enough to require that the conclusion be modified or discarded.

Steps in the Scientific Method

The scientific method can be thought of as a sequence of steps, broadly classified as (1) observation, (2) hypothesis, (3) experiment, (4) conclusion, and, sometimes, (5) retesting and (6) publication.

Observation:　Usually the scientific method begins with a set of observations that are not understood. For example, a neurophysiologist who is doing vision research observes a high incidence of cataracts (opacity in the lens of the eye that causes cloudiness and light scattering) in people who have been exposed to sunlight for long periods. Next, the scientist most likely begins to mull over the observation and wonder *why* cataracts are more prevalent in these individuals. Usually at this stage, the scientist also talks to colleagues about the finding and searches the scientific literature for reports of the same observation. Next, the neurophysiologist may speculate about various possible causes of the cataracts, wondering, for example, whether ultraviolet rays (part of sunlight) cause the higher incidence of cataracts.

Hypothesis:　Following this speculation, the neurophysiologist will form a **hypothesis,** which is a statement that tentatively explains the observation. In formulating the hypothesis, the scientist attempts to make a statement that is *testable*, realizing that it may not be true or that it can lead to new hypotheses. Following is an example of a hypothesis:

Hypothesis:　Exposure to ultraviolet light causes cataracts in the lens of the eye. Such a statement will lead to the next logical step: testing the hypothesis.

Experiment:　A hypothesis is tested in an **experiment,** in which carefully specified conditions are created deliberately to produce the observations in question. The purpose of an experiment is to determine whether a cause-and-effect relationship exists between an observation and a variable (the hypothesized cause of the observation).

Ideally the experiment is designed in such a way that only one explanation exists for the observation. To accomplish this goal, most experimenters set up two sets of conditions. One set is the experimental

set, in which the variable is present. The other set of conditions, called the **control,** is identical to the experimental set of conditions except that the variable is omitted. In essence, the control provides a standard to which the experimental group can be compared so that the effect of the variable can be observed. The control must contain conditions identical to the experiment; any variations or differences that creep into the design are termed **uncontrolled variables.**

Figure 1–9 illustrates an example of a controlled experiment used to test the cataract hypothesis. Since the hypothesis states that ultraviolet rays cause cataracts, ultraviolet rays are considered the variable. One group of animals, the experimental group, is exposed to light containing the variable, ultraviolet rays. Another group, the control group, is exposed to light from which the ultraviolet rays have been filtered, or removed. Apart from this difference, all other conditions are the same for the two groups: time of day, temperature of the environment, length of exposure, and so on. If one group were to be exposed inadvertently to the light for a longer period, the length of exposure would become an uncontrolled variable.

After the experiment is completed, results are collected. Results in this experiment might show, for example, that cataracts appeared in 20% of the animals in the experimental group, but in none of the control group. Statistical methods are used to determine significant differences of the data.

Ultraviolet light　　　　Ultraviolet light

UV filter

Experimental group　　　Control group

Figure 1–9

A controlled experiment to test the hypothesis that ultraviolet light causes cataracts. The experimental group of animals is exposed to ultraviolet light for a certain period of time (+uv light). The control group is exposed to light from which ultraviolet rays have been filtered out (−uv light). All other conditions are the same for both the control and the experimental groups.

Conclusion: From the results, a **conclusion** can be drawn in which the original hypothesis is either accepted or rejected. The conclusion can be firm or tentative, depending on the results. For example, the results of the experiment with ultraviolet light point to a conclusion that mice exposed to ultraviolet light have a greater incidence of cataracts than mice that are not exposed to ultraviolet light. This conclusion in turn allows the neurophysiologist to accept the original hypothesis that ultraviolet rays cause cataracts.

Retesting: To be able to accept the hypothesis more confidently, the physiologist can repeat the experiment, introducing another variable to determine whether any other conditions in addition to ultraviolet rays cause cataracts. For example, the researcher might dilate the pupils of the experimental mice, thereby increasing the area of the lens that is exposed to the ultraviolet rays. The rationale for this procedure would be that such an increase in exposure should cause a greater incidence of cataracts if ultraviolet light does in fact cause cataracts.

In such an experiment, four groups might be tested: Group I, the control group, not exposed to ultraviolet rays; Group II, exposed to ultraviolet light but without dilation of the pupils; Group III, exposed to ultraviolet light with pupils dilated; and Group IV, with pupils dilated but exposed to light that does not contain ultraviolet rays.

The results of such an experiment might show, for example, that Groups I and IV had no cataracts, Group II had a 20% incidence of cataracts, and Group III had a 30% incidence of cataracts. From these results, the physiologist could conclude that greater exposure to ultraviolet light causes a higher incidence of cataracts. This conclusion strengthens the original hypothesis that ultraviolet rays cause cataracts.

Publication: Finally, the neurophysiologist is likely to compile the data from these experiments and submit a paper describing them to a scientific journal. The publication process is an important part of the scientific method. The sharing of new information is consistent with the goals of modern science, which strive to maintain an ongoing dialogue among scientists in order to foster greater advancement of knowledge. Many experiments have important ramifications for the health of human beings. For example, the finding that ultraviolet light causes cataracts has prompted government agencies to place warnings on sunglasses that do not filter out ultraviolet rays because sunglasses cause the pupils to dilate, thereby increasing the wearer's exposure to the harmful rays.

Not all investigations produce such clearcut and dramatic results. Many experiments fail because of uncontrolled variables, equipment malfunction, or poor design. In other cases, experimental results do not support the hypotheses, and new hypotheses and experiments must be formulated. When experiments fail, scientists often talk to colleagues and return to the literature, reviewing the work of other investigators for helpful clues. Often, a finding in a related field provides insights not available from a study of the scientist's own specialty. The history of science contains many examples of experiments that seemed at first to have failed, only to lead to a major breakthrough as the researcher pondered the reasons for the failure.

Measurement, Computation, and Graphic Analysis Provide a Quantitative Approach

In the study of physiology, **quantitative methods** are invaluable for understanding bodily functions and processes. A quantitative approach aided William Harvey in his discovery of the circulation of the blood, revealing that the total amount of blood pumped by the heart was too great for the body to hold unless the same blood was pumped over and over again. Often in physiology it is the comparison of amounts, rates, times, and other measurable evidence of the body processes that reveals relationships between function and structure.

Units of Measurement

The metric system is almost exclusively the system used for physiological measurements. Length is expressed in **meters,** mass in **grams,** and volume in **liters.** A length of 1 meter is equivalent to 39.37 inches. A mass of 1 gram is equivalent to 0.035 ounces, and a volume of 1 liter is the same as 0.26 gallons. These metric terms usually are abbreviated for convenience to **m** for meters, **g** for grams, and **L** for liters. A length of 1000 meters would be written as 1000 m.

Prefixes are used to indicate multiples or fractions of these basic units, and the prefixes are abbreviated also. The prefix **kilo,** abbreviated as **k,** indicates a multiple of 1000 of the basic unit. A length of 2000 meters (m), for example, is equivalent to 2 × 1000 m, or 2 kilometers (km). The combinations of prefixes and units that are most commonly used in physiological measurements are summarized and defined in Table 1–2.

These prefixes avoid cumbersome notations across the enormous range of sizes encountered in living organisms. A human being with a height of 1.83 m (equivalent to 6.0 feet) is composed of cells that for the most part have a diameter of about 10 μm. The cells can be discerned only with the aid

Table 1–2
Commonly Used Units of Measurement

The meter (m) is the unit of length.
  1 kilometer (km) = 1×10^3 = 1000 m
  1 decimeter (dm) = 1×10^{-1} = 0.1 m
  1 centimeter (cm) = 1×10^{-2} = 0.01 m
  1 millimeter (mm) = 1×10^{-3} = 0.001 m
  1 micrometer (μm) = 1×10^{-6} = 0.000001 m
  1 nanometer (nm) = 1×10^{-9} = 1 Angstrom
The gram (g) is the unit of mass.
  1 kilogram (kg) = 1×10^3 = 1000 g
  1 milligram (mg) = 1×10^{-3} = 0.001 g
  1 microgram (μg) = 1×10^{-6} = 0.000001 g
  1 nanogram (ng) = 1×10^{-9} = 0.000000001 g
  1 picogram (pg) = 1×10^{-12} = 0.000000000001 g
  1 femtogram (fg) = 1×10^{-15} = 0.000000000000001 g
The liter (L) is the unit of volume.
  1 deciliter (dL) = 1×10^{-1} = 0.1 L
  1 milliliter (mL) = 1×10^{-3} = 0.001 L
  1 microliter (μL) = 1×10^{-6} = 0.000001 L
  1 nanoliter (nL) = 1×10^{-9} = 0.000000001 L

of a microscope because the human eye cannot see objects smaller than about 100 μm. The cells contain amino acid molecules that are only 1 nm in diameter, and the individual atoms of the amino acids may have diameters that are less than 0.1 nm.

Scientific Notation

In physiology, as in many of the sciences, it is often awkward or inconvenient to express measurements using standard digits (e.g., the diameter of a blood vessel as 10,000 μm, a blood glucose concentration of 0.005 moles per liter). Instead of writing out all the "zeroes," scientists use a system of **scientific notation,** in which quantities are expressed as products of a number and a power of ten. Powers of ten are expressed with **exponents,** which are small numbers written above and to the right of the base number, ten. The exponent indicates the number of times the base should be used as a factor in the expression. For example, $10^2 = 10 \times 10 = 100$. Thus,

    10,000 μm = 1×10^4 μm

    0.005 moles/liter = 5×10^{-3} moles/liter

    0.001 grams = 1×10^{-3} grams

(Note that negative exponents indicate the number of times 10 is used as a *divisor* for quantities less than one. For example, $10^{-3} = 1/10 \times 1/10 \times 1/10$.) Numbers without zeroes can also be expressed in scientific notation (e.g., $46,782 = 4.6782 \times 10^4$).

Following are some simple rules for converting a number expressed in scientific notation back to the original number:

1. For positive exponents: Shift the decimal point to the right by the number of places indicated by the exponent.

2. For negative exponents: Move the decimal point to the left by the number of places indicated by the exponent.

3. When multiplying exponential quantities that have the same base, add the exponents ($5^3 \times 5^2 = 5^5$).

Ratios and Proportions

A **ratio** is an expression that compares two numbers or quantities by division. For example, to compare the numbers 100 and 10, we divide 100 by 10 to get 10/1. If we are describing the number of experimental subjects that were exposed to high concentrations of ozone, we might say that of 100 rats exposed to ozone, 10 died. The ratio of animals that died to the total number tested was 10/100 or 1/10 ("one to ten"). In other words, 1 out of every 10 animals died from the ozone exposure. Ratios can be expressed in several different ways with the same meaning:

1:250 means 1 part to 250 parts

1/5 means 1 part out of 5 parts

2 cats to 6 dogs equals a ratio of 1 cat to every 3 dogs

A **proportion** is a mathematical statement of the equality of two ratios. By arbitrarily using the letters A, B, C, and D to express quantities, we can state a proportion in the following way:

$$A \text{ is to } B \text{ as } C \text{ is to } D \text{ or } \frac{A}{B} = \frac{C}{D}$$

This is equivalent to saying:

$$A \text{ times } D \text{ equals } B \text{ times } C$$

For example, if $A = 15, B = 25, C = 3$, and $D = 5$, then

$$\frac{A}{B} = \frac{C}{D} = \frac{15}{25} = \frac{3}{5}$$

$$A \times D = B \times C; 15 \times 5 = 25 \times 3$$

Therefore, it follows that if three of the quantities are known, the value of the fourth can be determined.

For example, assume that an electrocardiogram is being recorded on a moving strip of paper (Fig. 1–10). The speed of the moving paper is 25 mm/sec. If each repeating cycle of the electrocardiogram represents one heartbeat, how many heartbeats are occurring each minute? The problem can be solved as follows:

1. The distance between cycles is 20 mm (as measured from the record).

2. The time interval between cycles (beat to beat) is unknown; it equals x sec.

3. The ratios relating distance to time can be expressed as a proportion:

$$\frac{25 \text{ mm}}{1 \text{ sec}} = \frac{20 \text{ mm}}{x \text{ sec}}$$

4. $25x = 20$; $x = 0.8$ The interval between cycles is 0.8 seconds.

5. The number of cycles (beats) occurring each minute = y cycles (beats):

$$\frac{1 \text{ beat}}{0.8 \text{ sec}} = \frac{y \text{ beats}}{60 \text{ sec}} \qquad 0.8y = 60$$

$$y = 75 \text{ beats/minute}$$

Interpretation of Graphic Data

Much of the study of physiology involves learning about relationships between variables. For instance, there is a definite relationship between the heart rate and the rate of exercise. There is a similar relationship between the body's oxygen consumption and the rate of exercise. Often such relationships can be expressed and understood through graphs.

A **graph** is a diagram that expresses a relationship between two or more quantities. In some cases there is a definite cause-and-effect relationship, whereas in others the association is not as direct, but may be due to a third factor. Graphic presentation of data may not explain the reason for the relationship, but the shape of it can provide clues. A graph puts into visual form abstract ideas or experimental data, so that their relationships become apparent.

Variables. The related quantities displayed on a graph are called **variables.** The simplest sort of graph uses a system of coordinates or axes to represent the values of the variables. Usually the relative size of the

variable is represented by its position along the axis, and numbers along the axis allow the reader to estimate the values. If the relationship being plotted is one of cause and effect, the variable that expresses the cause is called the **independent variable.** Usually this is represented by the horizontal axis (also sometimes called the *x-axis* or the **abscissa**). The variable that changes as a result of changes in the independent variable is the **dependent variable.** It is usually represented on the vertical axis (also called *y-axis* or **ordinate**). The two axes are arranged at right angles to each other and cross at a point called the **origin** (Fig. 1–11).

To show the relationship between two variables that are directly related (at some specific value, such as point A in the figure), the value on the x-axis (X_1) is extended vertically, and the corresponding value on the y-axis (Y_1) is extended horizontally. The point A at which these lines cross is determined by their relationship. If another pair of points (X_2 and Y_2) is chosen, their point of intersection on the graph also can be plotted; this is point B. A line drawn between points A and B can then give information about how all other x and y values on this graph should relate to each other.

Types of Relationships. As you may have guessed, this explanation represents a very simple case in which some important assumptions were made. We first assumed that for every x-value there was only one y-value, and we further assumed that all of the y-variables were directly related to all of the x-vari-

Paper speed: 25 mm/sec

Figure 1–10

A proportion based on the speed of the paper and the distance between points on this electrocardiogram can be used to determine the number of heartbeats per minute.

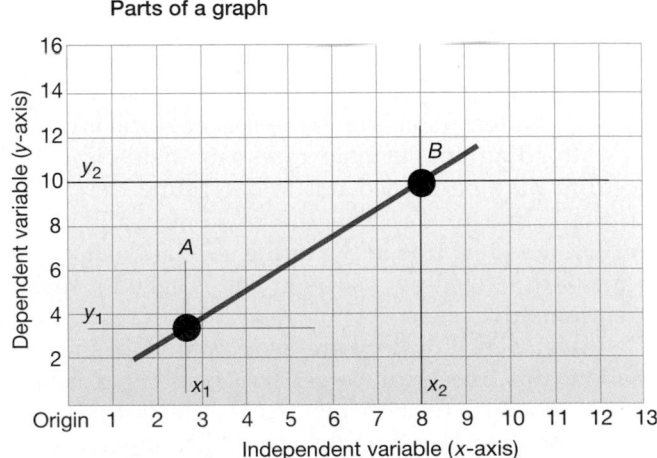

Parts of a graph

Figure 1–11

In a typical line graph, the vertical axis (the *y*-axis or ordinate) shows the values for the dependent variable, whereas the horizontal axis (the *x*-axis or abscissa) shows values for the independent variable. The two axes meet at a point called the origin.

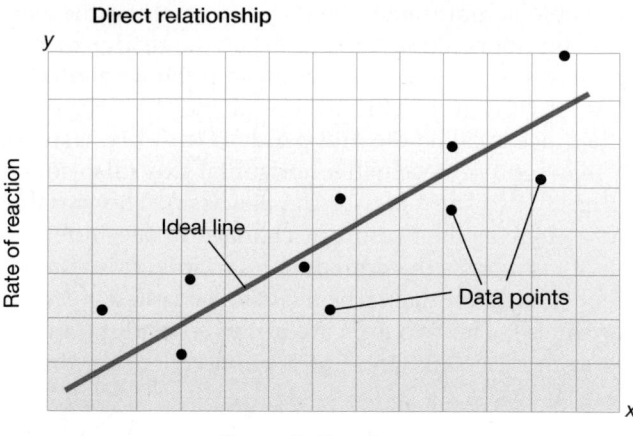

Figure 1–12

In a direct relationship, *y*-values increase as *x*-values increase. Data points frequently do not fall exactly on a straight line but are "scattered" about an ideal line, which is drawn to show the general shape of the relationship. A linear relationship can be shown by a straight line. The equation for a straight line is $y = mx + b$ where mx is the slope and b is the *y*-intercept.

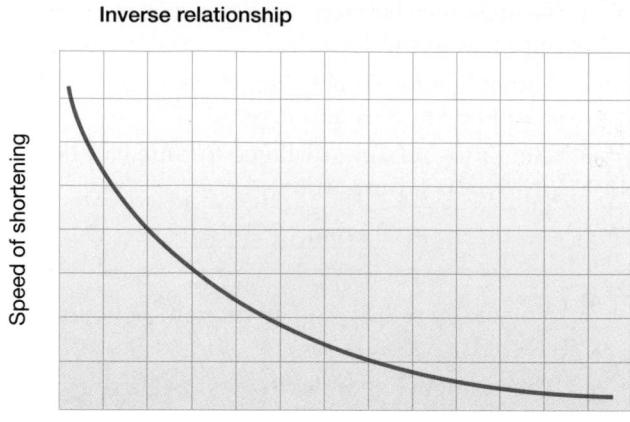

Figure 1–13

In an inverse relationship, *y*-values decrease as *x*-values increase, thus producing a downward-sloping line or curve.

ables. This is the simplest kind of relationship that a graph can represent. It is called a **direct relationship:** the *y*-values get larger as the *x*-values get larger. An example of this type of graph is shown in Figure 1–12. The data plotted here could have come from an experiment in which various concentrations of an enzyme were used to study how fast a particular chemical reaction happened at each concentration.

Some of the imperfections of the "real world" are evident here. The data points show that there is not a perfect relationship between each *x*-value and its corresponding *y*-value. We would say that the data show "scatter." (Sometimes this sort of graph is called a **scatter diagram.**) In many cases a mathematical procedure may be carried out to determine the "best-fitting" line to describe the relationship. This is called the **ideal line** in the figure. A relationship that can be described by a straight line is called a **linear relationship.**

Inverse relationships are also common. In such relationships, the *y*-values get *smaller* as the *x*-values get larger. Such relationships may still be linear if they are described by straight lines. Other inverse relationships may be curvilinear, as in Figure 1–13. The graph shown here summarizes the experimental finding that a muscle exerting a large force must contract more slowly than one exerting only a small force. A graph of this sort is adequate for the process of interpolation within the range of data, but it is poor for extrapolation outside that range.

Interpolation and Extrapolation. If an experimenter had enough confidence in the reliability of the data, two kinds of predictions could be made from such a graph. Predicting data values that fall "between the points" is called **interpolation.** This process is useful if the curve is to be used as a guide for interpreting or testing the reliability of newly obtained data. A more risky procedure is **extrapolation,** which involves extending the ideal line into ranges where experimental data are not present. If there is good reason to believe that the same relationship should hold outside this range, then this prediction could be valid and might permit useful information to be gained.

However, most relationships found in nature are not simple. Over some ranges, a relationship may be direct (and linear, as in the example in Fig. 1–12), and then it may change to another form as a wider range of variables is considered. In this case, extrapolation from a limited amount of data could lead to a wrong conclusion.

Take a few moments to flip through the pages of this book. You will see many graphs. Some express simple relationships over their entire range of data, whereas others are more complicated, expressing several relationships at once. Some, especially those that show the way in which an important variable such as blood pressure varies with time during the course of a heartbeat, present a very complex appearance. In some cases, there are several lines (and possible extra axes) on the graph, each describing some aspect of the idea being presented. Some graphs illustrating other vital processes appear much simpler. Some use bar graphs or pie charts instead of lines to illustrate relationships. Whatever their form, all these graphs are designed to present important relationships in the clearest possible way. When you learn to interpret information presented graphically, you are well on your way to understanding the language of physiology.

Anatomical Planes and Positions Provide Spatial Orientation of the Body

In order to study functions of the human body, one must have a basic knowledge of anatomical terminology since function is often described in terms of bodily structures. Conventional lay terms such as *top*, *bottom*, *above*, *below*, *behind*, *under*, and *on the side of* can be ambiguous when used in complex descriptions of anatomy. To ensure precise understanding, anatomists have devised a special vocabulary that is part of the language of anatomy and physiology. The following survey of anatomical terms is by no means exhaustive, but will introduce the correct meanings and uses of common anatomical terms that appear in subsequent chapters. When encountering new terms, you will find a medical dictionary an invaluable aid for mastering meaning and usage of additional terms.

Structures of the body are always described with reference to a standard static position called the **anatomical position** (Fig. 1–14). In this position, the body is erect, the face and feet are directed forward, and the arms are straight and directed toward the ground, with the hands rotated so that the palms face forward.

Planes

Imaginary planes are used to divide the body into parts (see Fig. 1–14). A **sagittal** plane divides the body into right and left portions. If such a plane passes directly along the midline of the body, it divides the body into right and left halves and is called a **median sagittal (midsagittal) plane**. A **transverse** or **horizontal plane** divides the body into upper and lower portions. A **frontal** or **coronal plane** divides the body into front and back portions.

Positions

Terms of position are used to locate a structure relative to other structures. **Anterior** means "nearer to the front of the body," whereas **posterior** means "nearer to the back of the body." (Sometimes the comparative anatomical term **ventral** and **dorsal** are used in place of **anterior** and **posterior**, respectively.)

Medial means "nearer to the midsagittal plane," whereas **lateral** means "further from the midsagittal plane." **Superior** means "nearer to the head," whereas **inferior** means "nearer to the lower end of the body." (Sometimes the terms **cranial** and **caudal**

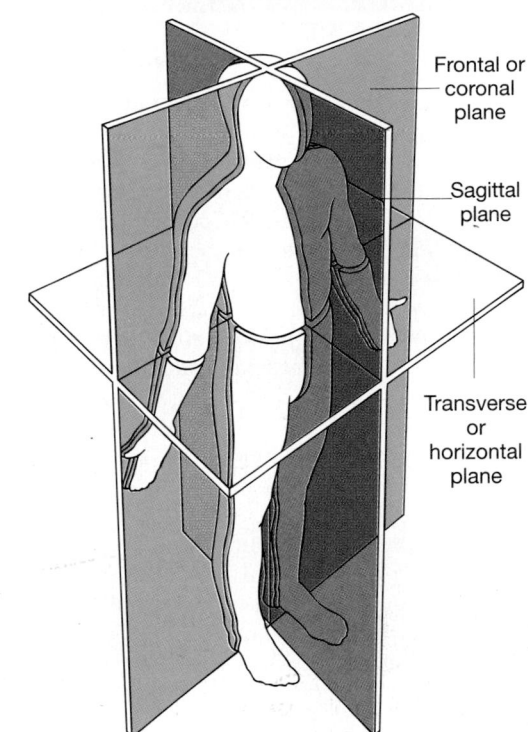

Figure 1–14

A diagram showing basic anatomical planes.

are used instead of **superior** and **inferior**, respectively.)

Internal means "nearer the center of an organ, cavity, or part of the body"; **external** means "farther from the center." **Superficial** means "nearer to the surface" of the body; **deep** means "farther from the surface."

Two special terms are used in describing extremities or structures related to their long axes. **Proximal** means "nearer to the origin or point of attachment," whereas **distal** means "farther from the origin or point of attachment." For example, the humerus (the bone in the upper arm) is attached at its proximal end to the shoulder and at its distal end to the elbow.

General Organization of the Body

Living matter is made up of **protoplasm**, a complex mixture of chemicals that displays the attributes of life. These attributes include (1) organization into specific kinds of structural units; (2) the ability to enter into chemical activities that include the transformation of energy and the maintenance or synthesis of protoplasm; (3) the ability to respond to changes in the environment; and (4) the ability to grow and reproduce.

Radiographic Imaging Techniques

The spectrum of electromagnetic radiation contains wavelengths that are suitable for use in imaging techniques. Visible light, wavelength 10^{-7} to 10^{-6} m, is the most familiar form of electromagnetic radiation. It bounces off the surface of solid structures, allowing us to see them. In contrast, X-rays (wavelength 10^{-11} to 10^{-8} m), gamma rays (10^{-18} to 10^{-10} m), and radio waves (10^{-1} to 10^4 m) pass through most solid objects and can be used to study the internal structure of the human body. Some of the specific techniques are discussed briefly here.

Radiography is a technique for studying internal structures using X-rays. The rays are passed through the body onto a photographic plate to produce a two-dimensional photograph. Some parts of the body, such as bone, absorb more of the X-rays than other parts and can be distinguished in the photograph. The photograph, usually called an X-ray, used to be known as a **roentgenogram** because X-rays were discovered by Wilhelm Roentgen, a German physicist, at the end of the 19th century. Roentgen used X, the symbol for an unknown quantity, because he did not understand completely the phenomenon he had found. In modern times, radiography has become a very useful diagnostic tool. **Contrast radiography** is useful for viewing certain soft tissues that are not seen easily in normal X-rays. For example, the upper gastrointestinal tract can be observed if the patient swallows a drink containing barium prior to an X-ray. Since X-rays cannot pass through barium, the position of the gastrointestinal tract, now lined with barium, can be located in the X-ray photograph. The **computerized axial tomograph (CAT scan)** is a system in which X-rays from many different angles are passed through a selected plane of the body in a fraction of a second. Detectors measure the absorption of X-rays by the various body structures, and all the information is fed into a computer. The computer displays a reconstructed image of the cross-section of the body on a television screen. Hundreds of different cross-sections, like the slices in a loaf of bread, can be obtained and combined to form a three-dimensional image of the contents of large cavities in the body, such as the abdomen or thorax. The **positron emission tomogram (PET scan)** produces very clear images and can identify chemical and physiological changes. It uses gamma rays released by chemicals made especially for this purpose. These chemicals, which are similar to those normally present in the body, are injected into the patient. The PET scan detects the gamma rays released inside the patient and uses them to form an image. This image provides information about where the chemicals are used in the body. One use of the technique has been in studies of the brain to gain information about blood flow and utilization of glucose, an important source of energy.

- **The photograph of an MRI in the focus box shows a section through a human head and neck. What plane or section is represented?**

Protoplasm is organized into cells, which are the basic structural units of the human body and of all life. A **cell** can be defined as a microscopic bit of organized protoplasm surrounded by a membrane called the **plasma membrane.** The adult human body comprises about 100 trillion cells, all of which function collectively to maintain an individual's life.

Each human life begins as a single cell, a fertilized ovum, which then divides to form two cells, four cells, eight cells, and so on. In addition to undergoing numerous cell divisions during development, cells also begin to exhibit specialized functions. There are about 200 different cell types in the body, as determined by differences in both structure and function.

Human Cells Can Differentiate and Exhibit over 200 Modes of Specialization

The process by which a single cell type (the fertilized ovum) develops into many different cell types is known as **cell differentiation.** As you will see, these different cell types are arranged first into **tissues,** or groups of related cells. Tissues in turn are arranged in groups to form organs, and organs in turn form systems. An analogy can be drawn between the cells of the body and a society of individuals. Just as mail carriers, police, doctors, and teachers have specific roles for the common good of the entire society, so do specialized cells such as muscle, nerve, connective tissue, and epithelial cells serve specific functions to promote the survival of the organism. The body can be thought of as a society or social order of cells.

In a multicellular organism such as a human being, labor is divided among groups of specialized cells, each group performing one principal function such as movement, digestion of food, or reproduction. Specialization of cells has been an essential factor in the development of the large size of multicellular animals. The different groups of cells work in concert to maintain the life of the organism, and considerable interdependence exists between the different cell types. For example, most of the cells of ver-

MRI of a herniated brain. *(© Howard Sochurek/ Medichrome/The Stock Shop.)*

Magnetic resonance imaging (MRI) is also known as nuclear magnetic resonance (NMR). It uses low energy radio waves and a strong magnetic field, and can produce excellent cross-sectional views without exposing the patient to high-energy gamma rays or X-rays, or to a contrast medium such as barium. MRI also can provide information about structure/function changes as well as the location of specific chemicals in vivo.

Ultrasonography is based on a different principle. It uses high frequency sound waves (**ultrasound**) to view internal anatomy and is better than X-rays for imaging soft tissues. The equipment consists of a transducer that produces and receives the silent high-frequency sound waves. The transducer is placed against the body and slowly passed over the area to be examined. Sound waves pass through the skin into the body. As they strike various organs, they send echoes back to the transducer. Different kinds of tissue, bone, blood, and other fluids produce different echoes, which are separated and identified by the transducer. The transducer forms a visual representation of the varying echoes as it changes sound waves into electrical energy. This electrical signal is converted into an image on a television screen. Ultrasound is particularly useful for visualizing the pregnant uterus and for detecting abnormalities of fetal organs. It can be repeated as often as required without harm. **Echocardiography** is a form of ultrasound used to examine the pumping action of the heart, especially the position and movement of valves and other structures within the heart.

All of these techniques are painless and are described as **noninvasive** because they allow the internal structures of the body to be studied without physical intervention, such as surgery.

---

tebrates depend on the red blood cells to provide them with oxygen, while at the same time the red blood cells depend on the pumping action produced by the muscle cells of the heart to propel them through the body. Such a division and sharing of labor have allowed multicellular organisms to overcome the limitations of diffusion and surface area to grow to great size, whereas a unicellular organism must carry out for itself all the life processes, and can exist independently only by remaining microscopically small.

Tissues Are Organized Colonies of Specialized Cells

Because multicellular organisms require a division of labor and specialization between cells, the cells in an individual's body differ in size, shape, internal architecture, and function. Some cells are spherical; others are shaped like cubes, discs, stars, or cylinders. Each cell manifests the form and structure best suited to its function (Fig. 1–15). The neuron (a nerve cell), for example, bears numerous cellular extensions that

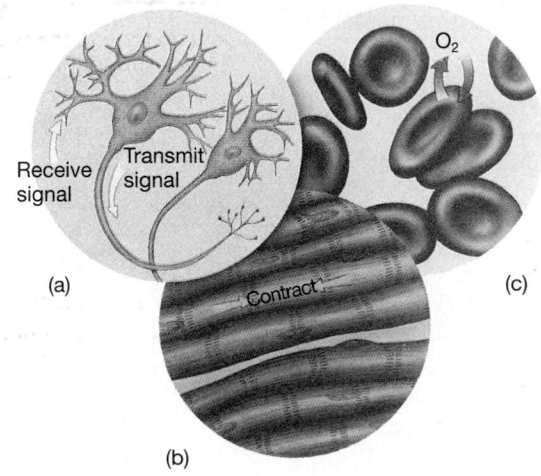

Figure 1–15

Cell shapes are designed to allow optimal functioning of specialized cells: (*a*) The long extensions of the neuron enable it to receive and transmit nervous signals. (*b*) The fibers of a muscle cell allow it to change shape during contraction and relaxation. (*c*) The bi-concave disc shape of a red blood cell assists it in carrying oxygen to the tissues and cells of the body.

receive and transmit the chemical and electrical signals that form the basis for nervous function. A skeletal muscle cell has a long cylindrical shape, which helps the cell move other parts of the body when it contracts. The "bi-concave disc" shape of a red blood cell helps it in its work of transporting oxygen through the body.

When one or more types of specialized cells become closely associated in function, they form a tissue. There are four classes of tissues in animals: epithelial tissue, connective tissue, muscle tissue, and nervous tissue (Fig. 1–16).

Epithelial Tissue

Epithelial tissue consists of closely packed cells arranged in flat sheets ranging from one to several layers in thickness. This type of tissue covers exposed body surfaces and lines body cavities, tubes, and organs. Epithelial tissue protects the body parts and absorbs, filters, and secretes substances. For example, the superficial layers of the skin are comprised of epithelial tissue.

Connective Tissue

The cells that make up connective tissue are somewhat loosely arranged and separated by an intercellular matrix (ground substance). The matrix material, made by the connective tissue cells themselves, frequently contains cellular products such as fibers, soluble proteins, and crystalline complexes. Connective tissue is the most widely distributed tissue in the body and the arrangement is largely what determines growth. It supports, protects, binds, and partitions nearly all bodily components. Spe-

cific examples are cartilage, bone, ligaments, blood, and adipose (fat) tissue.

Muscle Tissue

The cells of muscle tissue are elongated and contain many contractile cells called myofibrils. There are three types of muscle tissue: skeletal muscle, cardiac muscle, and smooth muscle. Skeletal muscle is a specialized tissue for contraction and functions to accomplish movement. It is attached to the bones and is used for conscious, voluntary movements of the body, such as walking. Cardiac muscle is present only in the walls of the heart and produces the automatic pumping contractions of the organ. Smooth muscle is found in hollow internal organs, such as the blood vessels, and allows them to undergo automatic, "involuntary" contraction and expansion.

Nervous Tissue

The chief functional component of nervous tissue are neurons, cells that are specialized for receiving and conducting electrical impulses. The characteristic shape of a neuron is an enlarged main cell body with hairlike extensions termed axons that receive and transmit the nerve impulses. Glia, sometimes called glial cells or neuroglia, provide neurons with structural and metabolic support and protection and also form part of nervous tissue. In addition, bundles of axons called simply nerves are found in all parts of the body.

Organs Are Integrated into Systems

Combinations of the primary tissues make up the organs of the body. An organ is defined as a group of

Figure 1–16

Organizational Level: Specialized cells → tissues → organ → system.

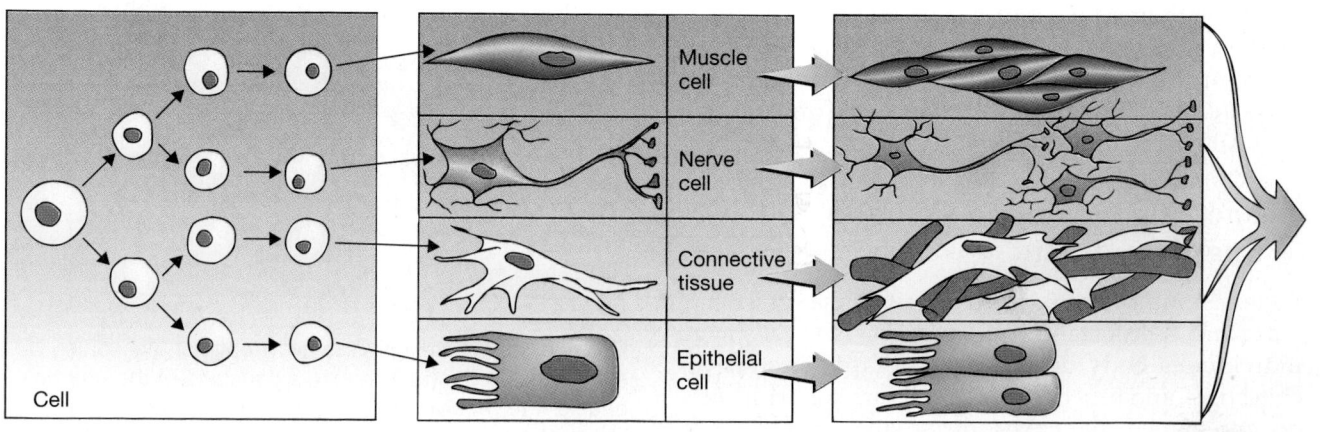

(a) Cell division and growth   (b) Cell differentiation-
                                          specialized cells

(c) Tissues

tissues that have been combined for the performance of a specific function or series of related functions. The stomach, for example, contains all four classes of tissues, organized for the purpose of receiving, storing, and digesting food and for moving partially digested food into the small intestine for further processing. No single tissue can perform as well as several tissues working together.

In a similar manner, organs and other structures that share in the performance of related tasks are grouped into **systems** (see Fig. 1–16). The cardiovascular system, for example, consists of the heart, arteries, veins and a capillary bed. These organs work toward the common purpose of providing nutrients and O_2 to cells, and removing CO_2 and other waste products.

The grouping of organs and related structures into systems provides the approach for this textbook. Keep in mind as you proceed through the various chapters, however, that although many levels of organization and function exist in the human body, all of the body's components function collectively to maintain an individual's integrity and life.

Bodily Fluids Are Divided into Two Compartments

The body can be thought of as mostly a watery solution. Approximately 60% of its weight is water. The body's fluids contain various mineral ions (e.g., sodium, potassium, chloride) and organic substances (e.g., proteins, glucose) dissolved in the body water. Bodily fluids usually are grouped into two major divisions or compartments: **intracellular fluid** and **extracellular fluid.**

The two compartments are separated by the plasma membranes of each of the body's cells, and the chemical compositions of the two compartments differ strikingly. The differences in chemical composition are maintained by the cells themselves and have important effects on the cell membrane potential, cell excitability, cellular metabolism, and the ability of the cell membrane to transport substances in and out of the cell. Much of cell physiology is concerned with the mechanisms for maintaining differences in the ionic composition of intra- and extracellular fluids and the consequences of these differences for cell function.

Intracellular Fluid

The intracellular fluid is the fluid within cells. It is not, therefore, one single continuous compartment, but rather a conglomeration of trillions of cellular subcompartments. Intracellular fluid has a relatively low concentration of sodium ions and a high concentration of potassium ions.

Extracellular Fluid

The extracellular fluid is the fluid outside of the cells. Extracellular fluid, in contrast to intracellular fluid, has a high concentration of sodium ions and a low concentration of potassium ions. The extracellular fluid that bathes our cells is a medium of exchange between one cell and another and between cells and the external environment.

Extracellular fluid consists of four types: interstitial fluid, plasma, lymph, and cerebrospinal fluid. **Interstitial fluid** is the fluid found between body cells; **plasma** is the fluid portion of the blood; **lymph** is the fluid in lymphatic vessels; and **cerebrospinal fluid** is an ultrafiltrate of plasma but lacks many of the plasma proteins. Together these extracellular fluids make up the **internal environment** of the body, whose constant regulation is the purpose of the complex physiological processes that we are about to discuss.

(d) Functional unit (myocardium)   (e) Organ (e.g. heart)   (f) Organ system (e.g. cardiovascular)

Homeostasis: Control of the Internal Environment

The mountain-climbing scenario presented in the chapter opening vignette, provides a look at body functions under extreme environmental conditions, and indicates that the key for survival depends on an important concept in physiology—homeostasis—which will be a major theme of this book. Homeostasis is an intricate, integrative process in which organ function (e.g., blood pressure, ventilation, acid-base, body temperature) is regulated so that the fluid environment (internal environment) that surrounds the billions of cells of our body stays constant despite changes in the external environment (e.g., climbing, altitude, and temperature extremes).

This chapter demonstrates that the concept of homeostasis has served as a major paradigm for the science of physiology for many centuries, that the original development of homeostasis focused on extracellular fluid as the target of regulation but now includes study of the extremes of the external environment, and that regulation and homeostasis integrate interactions and exchanges at all levels. This integrative concept is the key to physiological research, which typically begins with a descriptive observation at the organismal level and leads to a search for underlying cellular and molecular mechanisms.

In a complex, multicellular organism such as a person, most of the living cells of the body are not directly exposed to the external environment (atmosphere, temperature, etc.) but exist in the liquid internal environment of the extracellular fluid. Conditions in this extracellular fluid must be maintained within narrow limits in order to permit our cells to live and function. Conditions such as oxygen tension, pH, osmotic pressure, temperature, and the concentrations of various metabolic substrates, hormones, and waste products are all closely regulated by the cooperative workings of tissues, organs, and systems. At its normal temperature, the human body will function well; at higher or lower temperatures, its processes will fail. When certain concentrations of ions exist in its internal environment, the body will thrive; outside those ranges, it will be unable to function well or to live.

In many diseases, the composition and/or volume of the internal environment becomes abnormal. Consider the person with pulmonary disease, for example, in whom arterial blood is not oxygenated adequately. The oxygen tension in such a person's internal environment will be reduced. Consider the person with renal failure, who cannot get rid of metabolic waste products at an adequate rate. In such a person, toxic waste materials and hydrogen ions will accumulate in the extracellular fluid; essentially, the body will be poisoned by its own internal environment. The functioning of the body's cells will be disturbed under such conditions as these. The organism will be forced to lead a more restricted existence.

The famous nineteenth-century French physiologist Claude Bernard in 1865 was the first to point out that **constancy**, or stability, in the internal environment—the extracellular fluid—of an organism is the requisite condition for a free and independent existence. Coining the phrase *milieu intérieur*, "internal environment," Bernard paved the way for another physiologist at Harvard University, Walter B. Cannon, in 1929 to coin the word **homeostasis.** Cannon is credited for our current concept of homeostasis: the stable state of the internal environment is maintained by integrative processes. Despite the many changes that may occur in the external environment, the temperature and composition of the extracellular fluid remain constant. Factors such as ion concentrations, oxygen content, nutrient and waste product concentrations, and temperature normally do not change appreciably, either minute by minute or day by day.

Indeed, much of the study of physiology is taken up with the analysis of the regulatory processes by which the body maintains the constancy of the internal environment. Some of these regulatory processes are simple, whereas others are highly complex. As you read about each of the various mechanisms in later chapters of this book, bear in mind that each process serves to accomplish, at least in part, one common goal, which is to maintain homeostasis of the internal environment of the body. Thus, for example, although the exchange of gases in the lungs and the secretion of digestive enzymes into the intestine seem to have little in common, each of these processes serves in the end to maintain homeostasis.

A Stable Internal Environment Is Essential for Normal Cell Function

In a single-celled organism (e.g., an amoeba living in a pond), the internal environment of the cell is difficult to control and is subjected to the vagaries of the external world. Changes in temperature, light, and concentrations of various chemicals in the external environment dictate its existence. It has little freedom since the internal environment of the cell is directly exposed to the external environment and is unable to manipulate or use a buffer or mediator between itself and the external world. If the water (its external environment) freezes, the amoeba freezes also. If it does not die, it is at least immobilized.

A more highly developed organism, such as a frog, has a greater degree of independence from external conditions. By virtue of its multicellularity, it has both an external *and* an internal environment. Its internal environment is the extracellular fluid that surrounds its cells; this fluid is *outside* the cells but still *inside* (and therefore part of) the frog. The frog, however, is unable to regulate its own body temperature. When the weather turns cold, it may not freeze, but it will be forced into inactivity. Reptiles such as snakes and alligators must spend a great deal of time simply sitting in the sun after periods of exposure to cold air or water, in order to raise the temperature of their internal environment to a level that permits life processes to continue at full speed.

In contrast to reptiles, mammals have developed a great deal of freedom from external conditions by evolving sophisticated mechanisms for maintaining a stable internal environment. Most noticeably, their internal body temperatures remain virtually constant under a wide range of external temperature conditions, by virtue of physiological, behavioral, and adaptive mechanisms that counteract the effects of cold or heat in the external environment. Mammals are able to maintain a constant body temperature whether the air around them is at 0°C or 45°C. Furthermore, unlike frogs, many mammals are fully active under all climatic conditions; they do not require periods of physiological inactivity while waiting for their bodies to warm. For example, the artic fox in the cold north and fennec fox in the hot North African desert both have distinct morphological features and habits that adapt them to their habitat. The fennec, which avoids the midday heat in its burrow, possesses oversized ears to radiate heat. The artic fox, which faces the need to conserve heat, has reduced ear surface and remarkable insulation that can make shivering unnecessary until ambient temperatures fall to −40°C.

Homeostasis Maintains Protein Conformation

Intuitively, one can appreciate that cells, as living things, can flourish and prosper in certain sets of conditions and not in others. By providing a constant environment for the cells of the body, the various homeostatic mechanisms cooperate to create optimal conditions for cell function.

Beyond this, however, one might ask, "What is the molecular basis of this need for a constant internal environment?" The answer to this question lies in the fact that the primary components of cells, macromolecules called **proteins,** change their shape at different environments. Proteins serve a variety of functions in cells. They form enzymes, which are responsible for the various metabolic processes by which cells obtain energy and synthesize other macromolecules in the cell. They also serve as structural components in the cell interior and the cell membrane. Every protein in the cell has its own distinctive **conformation,** or shape, that allows it to best carry out its own particular function. The shapes, and

therefore the functions, of proteins are affected by such factors as ion concentrations, temperature, and pH. Changes in these factors can change protein conformation. With the right conditions of ion composition, temperature, and pH, proteins assume their proper conformations and carry out their tasks perfectly. Under abnormal chemical, temperature, and pH conditions, however, their shape is altered and their ability to function is diminished or lost. Protein malfunction in turn prevents the cells as a whole from carrying out their own specialized tasks, and the malfunctions accumulate until they become manifest as a disease condition in the organism.

Proteins function outside of cells as well as inside. The internal environment is *extra*cellular fluid. What is the connection? As we shall see in Chapter 4, the connection is everywhere, anywhere cells and their membranes are found. The plasma membranes of cells serve as the boundary between the two body fluid compartments—the extracellular fluid and the trillions of subcompartments of the intracellular fluid. At the same time, they serve as a transport system between the two compartments, constantly permitting or encouraging movement of substances between cells and the extracellular fluid. The composition of that fluid determines whether or not the cells can obtain from it what they need to make proteins and other molecules essential for their existence.

Thus, the molecular basis of the homeostatic process is, in essence, the maintenance of protein conformation. By constantly ensuring that proper conditions exist in the extracellular fluid, homeostatic mechanisms allow proper conditions inside cells, thereby permitting cell proteins to assume their proper shapes and carry out their special functions. The optimal functioning of specialized cells, in turn, ensures the coherent functioning of tissues, organs, and organ systems and the health of the body.

Coordinated Body Activity Requires a Control System

The mechanisms by which the body maintains homeostasis are known collectively as **control systems.** Control systems share several features. These systems use mechanisms called sensors to *detect* conditions in the body and effectors to *change* conditions in the body.

Sensors and Effectors

In order to monitor and control a variable of the extracellular fluid such as temperature, the body must first detect the variable. This is accomplished most often by **sensors,** which determine the level of the variable with respect to a **reference point.** For example, in monitoring a physiological parameter such as

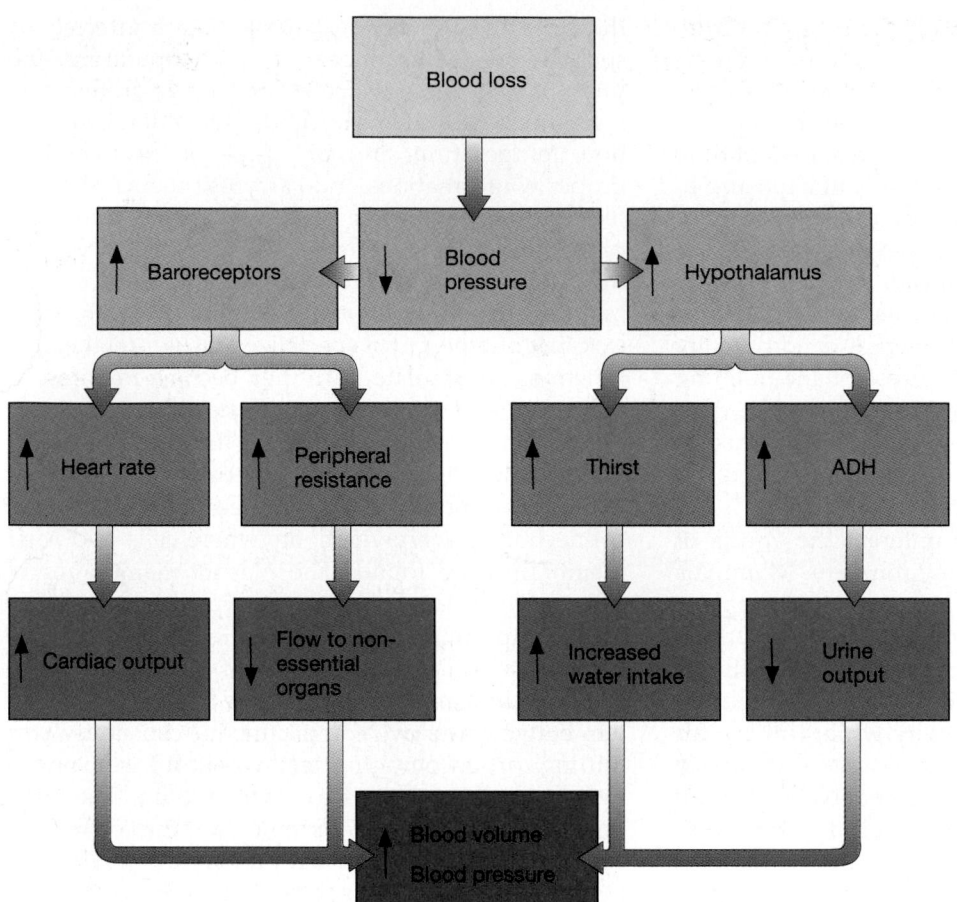

Figure 1–17

Homeostatic control illustrating the concept of negative feedback. Blood loss leads to increased heart rate, increased peripheral resistance (i.e., peripheral blood vessels constrict), thirst, and increased water retention (decreased urine output). Both neuro and endocrine mechanisms are brought into play to restore blood volume.

concentration of calcium in the extracellular fluid, the sensors obtain information about the concentration of calcium present at a particular time and compare it to the level of calcium that *should* be present. If too much or too little calcium is present, this information is sent from the sensors to the effectors, which then attempt to change the parameter to the desired level. Thus, control systems detect discrepancies between *desired* levels or quantities and *actual* levels or quantities.

Two general classes of control systems are active in the body: negative feedback systems and positive feedback systems. **Negative feedback systems** work to restore normal values of a variable and thus exert a stabilizing influence. **Positive feedback systems** destabilize and thus have limited value in maintaining homeostasis. An example of negative feedback is a decrease in blood volume which leads to an increase in heart rate, peripheral vasoconstriction, thirst, and fluid retention (Fig. 1–17). In positive feedback the initial disturbance sets off a chain of events that actually increases the disturbance even more. Examples of positive feedback include fever, which causes increased shivering and an increase in body temperature.

The Nervous System in Homeostatic Control

The nervous system is the major homeostatic control system in the body. For example, it monitors the temperature of the body and attempts to maintain it at 37°C. It has sensors that monitor the mean arterial blood pressure and tries to keep it at approximately 90 mm Hg. It monitors the concentration of hydrogen ions in the extracellular fluid and attempts to maintain a pH of 7.4.

The sensors that are part of the homeostatic function of the nervous system are associated with the autonomic nervous system (see Chapter 10). Sensors called **baroreceptors** are nerve endings that wrap themselves around blood vessels to monitor blood pressure. The information they transmit to the brain causes the blood pressure either to rise or to fall. This adjustment is accomplished by the activation or inactivation of effector systems such as the heart, causing it to change its rate and force of contraction, and the smooth muscles, causing them to dilate or constrict blood vessels.

Chemical sensors, also called **chemoreceptors,** of the autonomic nervous system monitor the concen-

CLINICAL FOCUS

Positron Emission Tomography

Positron emission tomography (PET) is an imaging method based upon the measurement of positron emission from radionuclides such as oxygen-15, fluorine-18, nitrogen-13, and carbon-11. The imaging method is similar to that used for computerized axial tomography scanning (CAT scanning). With this method it is possible to generate cross-sectional images of the body that show the distribution of a particular radiolabeled substance. Depending on the radiolabeled compound used, it is possible to create a map of a variety of physical and biochemical processes that occur in the body. For example, the distribution of $^{13}NH_3$ is related to the distribution of blood flow in organs such as the brain or the heart. Changes in regional metabolism can be monitored by measuring the distribution of $^{18}F-2$-deoxyglucose. Since these images can be generated in real time, it is possible to image changes in regional metabolism and blood flow in the brain, for example, in response to changes in mental activity. Studies of this nature dramatically demonstrate the process of blood flow autoregulation in response to altered cellular activity and metabolism.

PET has been used to monitor the destruction of tumor cells following radiation treatment. Changes in blood flow and metabolic rate can be measured to distinguish actively growing cells from those that have been destroyed by radiation. Patients with Alzheimer's disease (dementia) have a fairly typical pattern on a PET scan, which can be used for diagnostic purposes. Recent experimental studies have used PET as a tool to investigate the physiology of psychiatric disorders such as schizophrenia and epilepsy. PET has also been used to detect coronary artery disease and to distinguish between normal and nonfunctional myocardium following a heart attack.

Because of the very short half-life of most positron emitters (20 to 100 min), it is necessary to generate these isotopes on site. This requires an on-site cyclotron as well as a laboratory for rapid preparation of isotopically labeled compounds under sterile conditions. In spite of the substantial cost of building and operating such facilities, PET may provide a relatively safe and economical alternative to many currently available diagnostic procedures. Experimental studies using PET are providing a wealth of new information concerning the biochemical basis of numerous physiological processes. PET is proving to be a powerful and unique tool for the investigation of normal and pathological brain physiology by providing investigators a means of directly imaging changes in regional brain metabolism associated with specific types of mental activity.

■ **What is the equivalent in nm for an X-ray with a wavelength of 5×10^{-10} m?**

---

trations of hydrogen ions and carbon dioxide in the extracellular fluid. These chemicals affect the pH of the body, which in turn affects the ability of cells to carry out essential enzymatic reactions. Alterations in the levels of hydrogen ions and carbon dioxide are detected by chemoreceptors in the brain and peripheral autonomic nervous system. The information these chemoreceptors transmit to the brain eventually affects the ways in which the lungs and kidneys work to maintain a constant pH in the body.

Sensors within the brain are also capable of monitoring body temperature. These **thermoreceptors,** located in the hypothalamus, detect the temperature of the blood flowing in the brain. When the temperature falls below 37°C, effector mechanisms are activated. These mechanisms reduce the amount of blood that flows to the skin, thus minimizing heat loss. Heat is retained in the body, thereby raising the temperature. In addition, hormones are released to increase the rate of cellular metabolism, which releases energy in the form of heat. Moreover, neural mechanisms go into action to produce shivering, which generates more heat. Together these mechanisms generate and conserve heat in the body until the temperature is raised to 37°C. If too much heat is retained, different effector mechanisms take action to dissipate heat. Blood flow to the skin increases, and sweat glands are activated. Heat is lost from the body, and the temperature falls to 37°C.

The Endocrine System in Homeostatic Control

In addition to the nervous system, the other major homeostatic control system in the body is the endocrine system. In the endocrine system, hormones serve as an important communication link in the effector component of several control systems. Many of the sensors involved are chemical sensors that serve to monitor changes in components in the extracellular fluid. Specialized cells in the pancreas and in the hypothalamus of the brain monitor the glucose concentration. Two hormones produced by the pancreas, insulin and glucagon, produce opposing effects in the body that lower and raise the blood glucose concentration, respectively. Similarly, much of the regulation of the concentration of calcium in the extracellular fluid involves hormones serving as the information link between sensors and effectors.

While there are many examples of homeostatic control mechanisms involving just the nervous system, or just the endocrine system, in a large number of cases the two systems interact and work in concert. For example, when there is a loss of blood volume, a drop in blood pressure is detected by the baroreceptors. This information is relayed by the central nervous system, which in turn directly activates the heart and thirst mechanism. Specific hormones are also released as a result of nervous stimulation. These include epinephrine from the adrenal gland and antidiuretic hormone (ADH). Epinephrine acts on the heart and on blood vessels to produce effects that lead to the restoration of blood pressure to normal. ADH acts on the kidneys to retain body fluids.

CHAPTER REVIEW

Summary

- Early Greeks, like Aristotle, searched for explanations of blood flow, temperature, and other aspects of how the body worked.
- Galen's "pneumatology," based on these early studies, dominated western views of body processes until well into the Renaissance.
- Renaissance scientists corrected many errors of Galen's teachings. William Harvey's discovery of closed circulation and the use of quantitative methods marked an end of Galen's dominance and the beginning of modern physiology.
- The scientific method, which emphasizes experiment, quantitative results, retesting, and information exchange, forms the basis of all modern scientific research. In conducting investigations, scientists use both inductive and deductive reasoning.
- Through measurement, computation, and the use of graphs, physiologists apply quantitative methods to study functional relationships. Physiologists use standard terms of reference to describe planes and positions of the body.

- Epithelial, connective, muscle, and nerve tissues are the four major tissue types in the body.
- Groups of tissues that work together form organs; organs in turn combine their functions to maintain systems.
- The body comprises two fluid compartments: the fluid within each plasma membrane, the intracellular fluid; and the extracellular fluid (plasma, lymph, and interstitial fluid) that surrounds the cells.
- Homeostasis refers to control of the internal environment, and the extracellular fluid forms the "internal environment" of the body, allowing it to maintain constant conditions such as temperature.
- Through homeostasis, or self-regulation, the body protects the shapes of proteins, allowing them to function optimally in cells.
- Many control systems work to maintain homeostasis. They employ sensors to detect changes in bodily conditions and effectors to adjust those conditions.

Review Questions

Identify Each with a Term

1. The units that Harvey used to measure blood pressure.
2. The notation used by scientists to express numbers by a power of ten.
3. The term used for the extracellular fluid that surrounds the cell but is inside the organism.
4. The term used to describe fluid outside of the cell.
5. The term used for receptors that sense chemical changes in the body.
6. The term for a group of specialized cells that become closely associated to perform a specific function.
7. The term used in a line graph to represent the horizontal axis.

Define Each Term

8. homeostasis	11. receptor
9. hypothesis	12. hormone
10. steady state	13. sensors

Choose the Correct Answer

14. The imaginary plane that divides the body into right and left portions is the:
    a. transverse plane   c. frontal plane
    b. sagittal plane   d. horizontal plane
15. The individual who discovered the circulation of the blood was:
    a. Galen   c. Malpighi
    b. Harvey   d. Vesalius

Calculate

16. Express the following in scientific notation.
    a. 1,000,000 cells
    b. 0.0005 mm
17. Convert the following to ordinary numbers.
    a. 5×10^{-3} moles/liter
    b. 1×10^{-4} μm
18. Convert 12 mg to kilograms.

19. Of 100 rabbits exposed to ultraviolet light, 10 developed cataracts. What is the ratio of rabbits that developed cataracts to those that did not develop cataracts?

Answer Each Question in One or Two Sentences

20. What are the major concepts that the cell theory introduced?
21. What advantages does the application of the scientific method have over earlier approaches to the study of physiology?
22. Why is homeostasis so important in cell function?
23. Give an example of a neuroendocrine cell.
24. Give two physiological examples of positive feedback.

Problem Solving in Physiology

25. Why did ancient ideas persist so long?
26. How should we approach new ideas?
27. Plasma calcium is homeostatically regulated. A 45-year-old female has a normal dietary calcium intake of 1,000 mg/day and a normal plasma concentration of 2.5 mM/L. She decides to double her dietary intake and continues this dietary regime of calcium intake indefinitely. With her new steady state what will her plasma calcium concentration most likely be—2.5, 3.0, 5.0, or 6 mM/L?

Suggested Readings

Boas, M. *The Scientific Renaissance 1450–1630*. New York: Harper & Row, 1962.

Broad, W., and Wade, N. *Betrayers of the Truth*. New York: Simon & Schuster, 1982.

Comroe, J.H., Jr. *Retrospectroscope–Insights into Medical Discovery*. Menlo Park, CA: Von Gehr Press, 1977.

Davis, W. *The Serpent and the Rainbow*. New York: Simon & Schuster, 1985.

Dyson, F. *Disturbing the Universe*. New York: Harper & Row, 1981.

Feynman, R.P. *Surely You're Joking, Mr. Feynman!* New York: Bantam Books, 1986.

Fishman, A.P., and Richards, D.W., eds. *Circulation of the Blood–Men and Ideas*. New York: Oxford University Press, 1964.

Knight, B., M.D. *Discovering the Human Body*. New York: Lippincott and Crowell, 1980.

Miller, J. *The Body in Question*. New York: Random House, 1978.

Poncins, G. *Kabloona*. New York: Reynal and Hitchcock, Inc., 1941.

Rothschuh, K.E., M.D. *History of Physiology* (Guenter, B.R., M.D., ed.). New York: Robert E. Krieger Publishing, 1973.

Sagan, C. *Cosmos*. New York: Random House, 1980.

Saunders, J.B., Cordis, D.M., and O'Malley, C.D. *The Anatomical Drawings of Andreas Vesalius*. New York: Bonanza Books, 1982.

Scott, L.M., and Waterhouse, J.M. *Physiology and the Scientific Method*. New Hampshire: Manchester Press, 1987.

Simon, T. *The Heart Explorers*. New York: Basic Books, Inc., 1966.

Thomas, L. *The Youngest Science*. New York: Bantam Books, 1984.

Key Terms

anatomical planes (p. 17)
cell differentiation (p. 18)
cell theory (p. 19)
connective tissue (p. 20)
control systems (p. 23)
data interpretation (p. 15)
epithelial tissue (p. 20)
experimental design (p. 12)

extracellular (p. 21)
graphic interpretation (p. 15)
homeostasis (p. 22)
hypothesis (p. 12)
intracellular (p. 27)
muscle tissue (p. 20)
negative feedback (p. 24)
nervous tissue (p. 21)

organ (p. 21)
physiology (p. 1)
plasma membrane (p. 18)
positive feedback (p. 24)
protein conformation (p. 23)
protoplasm (p. 17)
ratios (p. 14)
receptors (p. 24)

scientific measurements (p. 13)
scientific method (p. 11)
scientific notation (p. 14)
sensors (p. 23)
variables (p. 12)

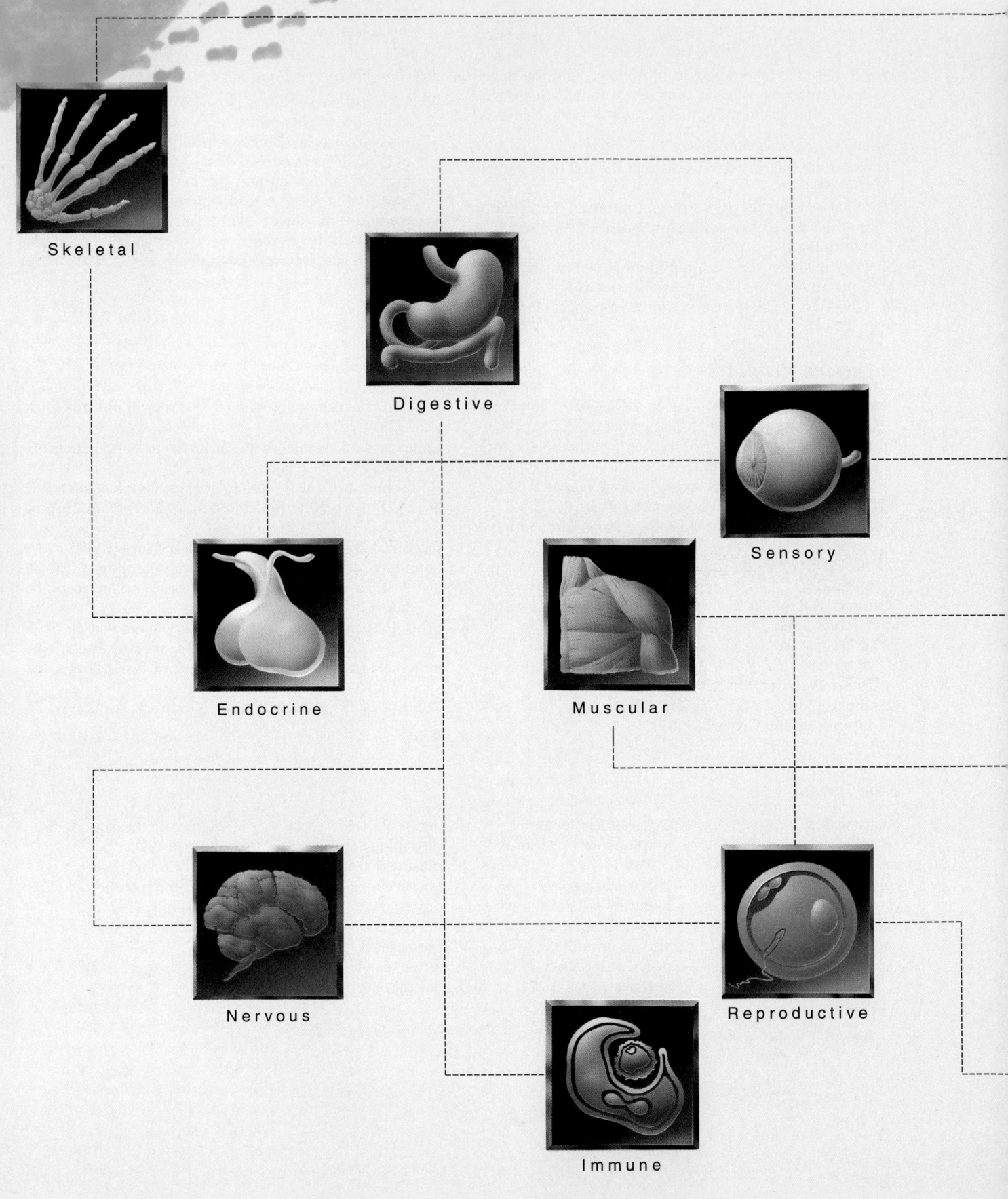

Skeletal

Digestive

Sensory

Endocrine

Muscular

Nervous

Reproductive

Immune

Part I
Cellular Functions

Respiratory

Renal System

Cardiovascular

Thermoregulatory

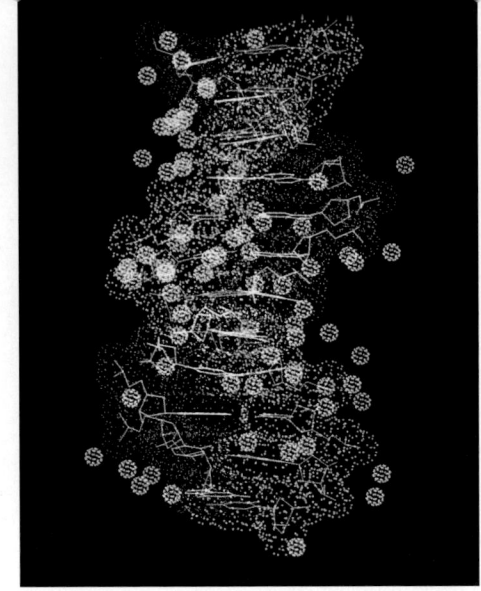

Computer graphics of human DNA.

Chapter 2

Chemical and Physical Principles

Key Concepts

An element cannot be broken down into simpler substances by chemical reactions. An atom is the smallest unit of an element that retains all the chemical properties of the element. All elements are composed of atoms, and all atoms are composed of neutrons, protons, and electrons.

A molecule is formed by the union of two or more atoms derived from the same or different elements. Chemical bonds between atoms are formed by sharing, gaining, or losing electrons. A compound is formed when two or more different elements are combined in a fixed ratio.

A chemical reaction occurs when the bonds between the atoms of a compound are broken and the atoms recombine in different ways to form new compounds. Cells contain enzymes that increase reaction rates by binding the reactant molecules so that specific reactions can occur.

Water is essential for living organisms and is the major chemical compound in most cells. Water is an excellent solvent, and many biologically important reactions require an aqueous environment.

Air is a mixture of several gases, such as oxygen, nitrogen, carbon dioxide, and water vapor. The partial pressure of each gas is directly proportional to the amount of gas present.

Macromolecules play vital roles in cells and organisms. The most important are carbohydrates, lipids, proteins, and nucleic acids.

Living organisms are highly organized systems, and the organization is maintained by constant work. The chemical energy required to perform this work is held in the chemical bonds of molecules involved in metabolic reactions.

Energy released from food molecules is trapped temporarily in the form of adenosine triphosphate (ATP) and can be used to drive synthetic reactions.

Chapter Outline

Overview: Matter and Energy in the Living Cell

Atoms and Elements

Molecules, Groups, and Compounds

Chemical Reactions: Basic Principles

Water and Aqueous Solutions

The Behavior of Gases

Macromolecules

Thermodynamic Principles

The lecture was boring, and Diana began to lose interest. Her thoughts turned to the car she planned to buy after graduation. She hoped to become a high school science teacher, and it suddenly occurred to her that she could use a car analogy to illustrate some fundamental differences between nonliving and living "machines."

For example, combustion of a mixture of gasoline and air in the cylinders of the car's engine converts the gasoline into carbon dioxide and other gases and provides the power that rotates the wheels of the car. And like the engine of the car, the human body requires a regular intake of fuel and air in order to function properly. However, unlike the engine, the body does not use its energy mainly for movement. Some of it is used to make new products that are required for reproduction, for growth and development, for response to stimuli, and for repair of injuries, for example. These are also some of the characteristics that set living things apart from nonliving things. Additionally, the body depends on a continuous supply of water: A human will drink about 265 gallons per year. In fact, the bodies of humans (and most animals) are about 65% water. The fundamental biochemical processes that maintain the living state occur in water. Nonliving things, in contrast, usually contain little water.

This chapter describes the structures of the major atoms and molecules that are present, but unseen, in the body. It also will consider the physical and chemical laws that govern the behavior of the atoms and molecules as they are converted from one form to another during the biochemical processes that sustain life. The physical structure of atoms and molecules determines their chemical properties and the roles they can play in the body. Therefore, understanding the behavior of these atoms and molecules allows an understanding of the principles underlying the behavior of the human body. All atoms contain the same three types of subatomic particles, but the specific arrangement of these particles is different in atoms from different elements. Similarly, the same six elements constitute most of the mass of both a human and a tree. The distinctive properties of a human are different from those of a tree because the six elements are arranged into different molecules in the human compared to the tree. Thus the collective properties of the whole body arise both from the properties of the individual parts and from the precise way those parts are arranged and are able to interact.

L iving organisms are composed of **matter.** Matter is anything that occupies space, has mass, and can assume the form of either a solid (e.g., stone), a liquid (e.g., water), or a gas (e.g., oxygen). We currently organize the matter of the universe into elements. An **element** is a substance that cannot

be broken down into simpler substances by chemical reactions. There are 92 elements that occur naturally, although scientists have made more than a dozen new elements. In nature as a whole, the four most abundant elements are oxygen, silicon, aluminum, and iron. Together, they account for 87% of the atoms in the earth's crust. Elements are commonly referred to by their abbreviations, called **chemical symbols** (e.g., "O" for oxygen and "Fe" for iron).

Elements are composed of atoms. An **atom** is the smallest unit of an element that retains the properties of the element. Atoms of the same element combine in specified ways to form different varieties of the same element. Atoms of different elements combine in specified ways to form compounds. A **compound** is a substance formed by the union of the atoms of two or more elements. Many substances, whether made of one element or more than one element, have molecules (not atoms) as their smallest stable units. A **molecule** is formed by the union of two or more atoms, whether they are the same elements or not. Oxygen atoms, for example, in samples of pure oxygen, form pairs called oxygen (O_2) molecules rather than existing singly. Water, in contrast, is a compound substance; one water molecule consists of two hydrogen atoms combined with one oxygen atom. A molecule or other regular group of atoms is designated by a **molecular formula,** such as H_2O, which shows the number of each type of atom in the group. The subscript "2" after the H, for example, means that two hydrogen atoms are present in a molecule of water. If there is no subscript following a chemical symbol, the symbol represents a single atom.

In living cells, molecules combine further to form large molecules known as **macromolecules.** These macromolecules all contain carbon, whose unique structure allows macromolecules to assume huge proportions and a great variety of shapes. Proteins, sugars, lipids, and nucleic acids are the chief macromolecules found in most living cells. The specific properties of these macromolecules and the organization of their many interactions within the cell form the subcellular structures and facilitate the chemical reactions that are the basis for all physiological processes and for life itself.

Overview: Matter and Energy in the Living Cell

As stated, the predominant elements in the matter of the universe are oxygen, silicon, aluminum, and iron. In contrast, the matter of the human body is largely hydrogen (H), oxygen (O), carbon (C), and nitrogen

RESEARCH FOCUS

Scanning Tunneling Microscope: A New Research Tool for Mapping the Molecular Landscape of the Cell

The scanning tunneling microscope (STM) was developed in 1982 by Drs. Gerd Binnig and Heinrich Rohrer of IBM Zurich. The development of the STM attracted much attention because it allowed scientists to image individual molecules and atoms. The potential of the STM was recognized early on in its development, and the Nobel Prize was awarded four years later to Drs. Binnig and Rohrer.

In biology, the STM offered another advantage: it can image surface details down to the atomic level without using a destructive probe and eventually will redefine the concept of microscopy.*† In some cases the STM will replace techniques such as scanning electron microscopy (SEM) and transmission electron microscopy (TEM).

The STM works by tracing an ultrafine conductive tip over a sample, and detecting topography through tiny variations in a quantum-mechanical "tunneling current" that seeps between the atoms at the surface and the tip (Figure 1). The business end of an STM consists of a piezoelectric transducer that scans the tip over the surface while a computer-controlled feedback loop monitors the tunneling current and applied scan voltage. The feedback loop adjusts the vertical position of the tip to keep the tunneling current constant. The adjustment signal versus the lateral position of the tip provides a way of posi-

tioning the tip in three dimensions that allows topographical images to be generated in which individual atoms can be resolved (Figure 2). An STM image is made up of a series of line scans and displays the path of the tip followed over the surface. A major disadvantage of the STM is that most biological materials are not sufficiently conductive to allow STM images to be made. Recently, this problem has been resolved by the invention of the atomic force microscope (AFM), which can image biological material

Block diagram of the STM. An extremely sharp conductive tip can be moved in three dimensions with an *x,y,z* piezoelectric translator to trace the contours of a surface.

*Hansma, P.K., Elings, V.B., Marti, O., and Bracker, C.E. "Scanning tunneling microscopy and atomic force microscopy: Application to biology and technology." *Science,* 24:209–216, 1988.

†Thompson, M., and Elings, V. "Scanning tunneling and atomic force microscopy: Applications in life sciences." *Am. Laboratory,* 23:(No. 6):36–42, 1991.

(N). Together they account for 99% of the atoms in the body. The marked difference between the chemical compositions of the human body and its external environment indicates that hydrogen, oxygen, carbon, and nitrogen must be uniquely suited for the processes that maintain the living state. In addition to these 4 most abundant elements, 23 others have been found to be essential components of living organisms. The 27 elements essential for life are listed in Table 2–1.

The matter of a living organism is highly organized. It also undergoes frequent transformation. The maintenance of this organized state and the processes of transformation require a constant input of energy. **Energy** is the capacity to produce change. It is mea-

sured by the amount of work performed to achieve a given change. In the universe at large, the energy used for work comes from various sources, such as heat, light, electricity, and chemical reactions. In the cell, the energy used for work is mostly **chemical energy** released by reactions that take place within cells. In the cell, the work to be done consists of moving matter from one place to another and changing it from one form to another—that is, breaking down or building up molecules from other molecules.

The different forms of energy also exist in two different states: **Potential energy** is stored energy; **kinetic energy** is energy that produces motion. Both states of energy enter into the metabolic processes of cells.

(*a*) Schematic view of the STM tip. The tip is mounted on a piezoelectric *x,y,z* scanner. As the tip scans the surface, atomic contours are revealed. (*b*) An image generated by STM shows carbon atoms at the surface.

and other substances that do not readily conduct electrons. The STM and AFM are very similar, with their scanning, feedback, and display systems. However, instead of a conductive tip, the AFM utilizes a tip that detects interatomic forces between the tip and the sample.

■ **Will the atomic force microscope allow scientists to make images of important biological molecules such as proteins?**

The chemical basis of the matter of widely diverse forms of life is strikingly similar. All living cells have fundamentally similar constituents and activities. Thus, both a complex multicellular organism such as a human being and a simple unicellular bacterium such as *Escherichia coil (E. coli)* use the same simple molecules to synthesize, or make, larger molecules. Both human cells and bacterial cells store energy in the form of adenosine triphosphate (ATP). Both cells store genetic information in deoxyribonucleic acid (DNA).

The chemical content of an *E. coli* cell has been analyzed in detail. Its composition is 70% water, 15% proteins, 7% nucleic acids, 2% carbohydrates, 2% lipids, 1% inorganic ions, and 3% other small molecules. The cell is mostly water, and the major macromolecules present are proteins and nucleic acids. The chemical composition of the typical mammalian cell is markedly similar to that of this bacterial cell.

To determine why certain chemical structures dominate all forms of life so regularly, this chapter provides a survey of general chemical and physical principles. Then it focuses more closely on the specific principles that influence the life processes. The chapter begins with the simplest components of matter—atoms and small molecules—and the chemical and physical laws that govern their behavior. It then proceeds to the major classes of giant molecules such as sugars, proteins, lipids, and nucleic acids. A look at their components and properties provides a sound

Table 2–1
*Essential Elements of Living Organisms**

Element	Chemical Symbol
Hydrogen	H
Oxygen	O
Carbon	C
Nitrogen	N
Phosphorus	P
Sulfur	S
Sodium	Na
Potassium	K
Magnesium	Mg
Calcium	Ca
Chlorine	Cl

Trace elements (present in very small amounts)
Manganese (Mn), Iron (Fe), Cobalt (Co), Copper (Cu), Zinc (Zn), Boron (B), Aluminum (Al), Vanadium (V), Molybdenum (Mo), Iodine (I), Silicon (Si), Tin (Sn), Nickel (Ni), Chromium (Cr), Fluorine (F), and Selenium (Se).

*Not all of these elements are essential for every species.

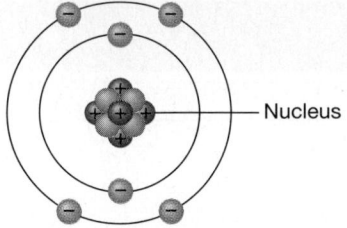

Figure 2–1

The basic structure of an atom. Protons (positively charged particles) and neutrons (neutral particles) are found in the nucleus. The number of protons is the atomic number. Electrons (negatively charged particles) orbit the nucleus. A neutral atom contains equal numbers of protons and electrons.

basis for the description, in later chapters, of their unique roles in the life of cells.

Atoms and Elements

An **atom** is the smallest unit of an element that retains all the chemical properties of the element. In solid matter, atoms are closely packed. In liquid and gaseous matter, atoms are more loosely arranged. The same element can appear in different forms depending on the temperature and pressure of its surroundings. For example, at room temperature and pressure, the element carbon is a solid because its atoms are packed closely together. At higher temperatures, carbon becomes liquid; at still higher temperatures, it takes the form of a gas as its atoms become separated from each other.

Atoms Are Composed of Subatomic Particles

Each atom, regardless of its element, is composed of certain subatomic particles. Three types of particles are important to an understanding of the behavior of atoms in life processes: neutrons, protons, and electrons. **Neutrons** are electrically neutral, whereas **protons** are positively charged and **electrons** negatively

charged. Protons and neutrons stay in the **nucleus,** or innermost area, of the atom, whereas electrons constantly orbit the nucleus (Fig. 2–1). The electrical attraction between the positively charged protons in the nucleus and the negatively charged electrons in orbit around the nucleus stabilizes the atom's structure.

Because atoms are so small, they most often occur together in large groups. In order to count large numbers of atoms conveniently, scientists have come up with a counting method called the **mole.** Just as in everyday speech we talk of a dozen eggs, a dozen pencils, or a dozen pages—all the time meaning 12 eggs, 12 pencils, or 12 pages—scientists speak of a mole of atoms to mean 6.022045×10^{23} atoms. The number 6.022045×10^{23} is called **Avogadro's number.** A mole is a convenient and accurate way to express the same number of atoms of any substance. Thus a mole of sodium atoms would be 6.022045×10^{23} sodium atoms; a mole of chlorine atoms would be 6.022045×10^{23} chlorine atoms.

Uniqueness of Elements: Protons and Atomic Number

Atoms of different elements are distinct from one another because of the different numbers of each subatomic particle they contain. In fact, the essential identity of an atom is found in the number of protons in its nucleus, called the **atomic number.** In the uncharged or electrically neutral atom, the number of electrons is the same as the number of protons. An atom of the element hydrogen has one proton in its nucleus and one electron in orbit around the nucleus. An oxygen atom typically has 8 protons and 8 electrons. A carbon atom has 6 protons and 6 electrons, a nitrogen atom has 7 of each particle, and a uranium atom has 92 of each particle.

Variation Within Elements: Neutrons, Mass Number, and Isotopes

The actual mass of an atom is too small to be measured or expressed conveniently. Instead, scientists refer to the **mass number** of an element. The mass number is the total number of protons and neutrons in the nucleus. Since the number of protons is different for every element, the atomic mass is also different for every elements. For example, the atomic mass of the most common form of hydrogen is 1 because the hydrogen atom has one proton (p) and no neutrons (n) in its nucleus (Fig. 2–2). Oxygen has 8 protons and 8 neutrons, so its mass number is 16. Nitrogen has 7 protons and 7 neutrons, so its mass number is 14.

Moles and grams (the metric unit of mass) can be related in the following way: one mole of an element, or 6.022045×10^{23} atoms, contains x grams of the element, x being equal to the mass number of the element. For example, one mole of sodium atoms (6.022045×10^{23}) weighs 23 grams, 23 being the mass number of sodium. One mole of chlorine atoms weighs 35 grams, since the mass number of chlorine is 35.

While all atoms of the same element have the same atomic number (number of protons), atoms of the same element may have differing mass numbers because they may have differing numbers of neutrons. For example, hydrogen exists in three forms: one form contains no neutrons in its nucleus, another contains one neutron, and another two neutrons. These three forms of hydrogen are called **isotopes** of hydrogen. Since the mass number of an atom takes into account the number of neutrons, different isotopes of an element have different mass numbers.

An isotope is designated with both the atomic number (number of protons) and the mass number (number of protons plus neutrons) written to the left of the chemical symbol. For example, the isotope of hydrogen that has no neutrons has an atomic number of 1 (because all hydrogen atoms have 1 proton) and a mass number of 1, written as $_1^1H$. The isotope of hydrogen that has one neutron (deuterium) would be written as $_1^2H$; the isotope with two neutrons (tritium) as $_1^3H$. Isotopes are important in biomedical research, as will be discussed in Chapter 3.

Ions Are Atoms with Electrical Charges

A neutral, or uncharged, atom has the same number of protons (positive charges) and electrons (negative charges). When a neutral atom gains or loses an electron, it acquires an electric charge and becomes an *ion*. An atom that loses one or more electrons acquires a positive charge because it now has more protons (positively charged particles) than electrons (negatively charged particles). Such an atom is called a **cation** (CAT-eye-on). An atom that gains one or more electrons acquires a negative charge and is called an **anion**.

For example, a potassium cation is formed when neutral potassium loses one electron. The single charge is denoted by a plus sign written to the upper right of the chemical symbol for potassium: K^+. The chloride anion, formed when chlorine gains an electron, is written as Cl^-. A calcium ion, formed by the loss of two electrons, would be denoted by a number 2 and a plus sign: Ca^{2+}. An oxide ion, formed by the gain of two electrons, is written as O^{2-}.

Ions have a strong impact on the ability of certain substances to dissolve in water. The potassium ion is a very important ion in the study of physiology because it is the major cation found inside most cells, while the sodium ion ($Na+$) is the major cation present outside cells in fluids such as plasma. Chloride ions are the most abundant anions in plasma.

Figure 2–2

The electron configurations of four important elements. Hydrogen needs one more electron to complete its only shell. Carbon needs four electrons, oxygen two, and nitrogen three to complete their outer shells. The electron orbits are depicted as concentric circles for the sake of simplicity, but their true movements are more complex.

Atoms Prefer a Full Outermost Shell

The electrons, in their orbit of the nucleus, move in a specified **electron configuration,** consisting of a series of **shells** that can be thought of as concentric circles (although this is a very simplified view of the way electrons actually move). The shells represent different energy levels and can hold different numbers of electrons. The shell with the lowest energy level, located closest to the nucleus, can accommodate two electrons. The shell with the highest energy level can hold eight electrons. The electron configurations of hydrogen, oxygen, carbon, and nitrogen atoms are illustrated in Figure 2–2. Larger atoms have more electrons occupying several different energy levels. For example, the 92 electrons in the uranium atom are arranged in seven different shells, each with a different energy level.

The number of electrons in the outermost shell of an atom determines many of the chemical properties of the element. An atom may be stable and neutral even if its outermost shell is not filled to capacity, but it can fill this outermost shell by interacting in certain ways with other atoms both of the same element and of different elements. An atom of carbon, for example, has four electrons in its outermost shell. Since this is the second shell from the nucleus, it can hold up to eight electrons. The carbon atom will try to fill that shell with four more electrons.

An atom can fill its outermost shell in one of three ways. It can share electrons with one or more atoms that do not have complete outer shells. It can obtain electrons outright from other atoms. Or it can donate all of the electrons in its currently incomplete outer shell to other atoms so that its next-lower shell, which is filled to capacity, becomes its outermost shell.

Molecules, Groups, and Compounds

The process of sharing, gaining, or losing electrons leads to the creation of a **chemical bond** between atoms. A chemical bond represents potential energy, energy being stored in the molecule. The release of this potential energy in chemical reactions for use by the cell is an important aspect of cellular physiology.

Atoms combine in definite proportions to form variations of elemental substances or new substances called compounds. Two atoms from the same element may join to form a "diatomic" molecule (see Fig. 2–3). Oxygen and hydrogen, for example, rarely exist as single atoms; instead, they are found as O_2 and H_2, pairs of atoms bonded together.

Two or more atoms from different elements may come together to form a new compound. Depending on the type of bond they form, the result may be a molecule (a discrete group of atoms) or a larger structure known as a **crystal lattice,** which is made up of atoms arranged in precise patterns. Whether or not separate molecules are formed, atoms bond in defi-

nite proportions, which are indicated by a molecular formula for the compound. For example, even though sodium chloride exists as a crystal lattice made up of many alternating sodium and chloride ions, rather than discrete molecules, the compound has the formula NaCl, indicating that sodium and chloride are always present in a ratio of 1:1.

The molecular formula of a compound shows the relative quantities of atoms; the **structural formula** shows their arrangement in space, for example:

	water	*methane*
Molecular formula	H_2O	CH_4
Structural formula	H—O—H	$\begin{array}{c} H \\ | \\ H-C-H \\ | \\ H \end{array}$

A **chemical group** is a small cluster of bonded atoms that functions almost as a single atom. Chemical groups behave in distinctive ways and have special names. Examples of chemical groups are —OH (the "hydroxide group"), —COOH (the "carboxyl group"), and $—PO_4$ (the "phosphate group"). The dash written before the chemical symbols indicates that the group is not a free molecule but is attached to a larger molecule.

Just as atoms are counted in groups called moles, so are molecules. A mole of molecules, regardless of the substance, is 6.022045×10^{23} molecules. A mole of oxygen is 6.022045×10^{23} molecules. A mole of water is 6.022045×10^{23} H_2O molecules. The **molecular mass** of a compound is the sum of the atomic masses, or mass numbers, of its elements. Molecular mass is expressed in units known as **daltons** (symbol Da), after John Dalton, who proposed the atomic theory of matter. The dalton is defined as $1/12$ of the mass of carbon (mass number = 12). Thus, methane (CH_4) has a molecular mass of $12 + 4 = 16$ Da.

Regardless of whether a molecule is made of atoms from the same element or different elements, the type of bond formed depends on the manner in which the atoms interact. If they complete their outermost shells by sharing electrons, the bond formed is a covalent bond. If the atoms gain or lose electrons outright, the bond formed is an ionic bond.

A Covalent Bond Is Produced by Electron Sharing

Chemical bonds formed when two or more atoms share electrons are known as **covalent bonds.** These bonds are very stable and strong and are the type of

(text continues on page 38)

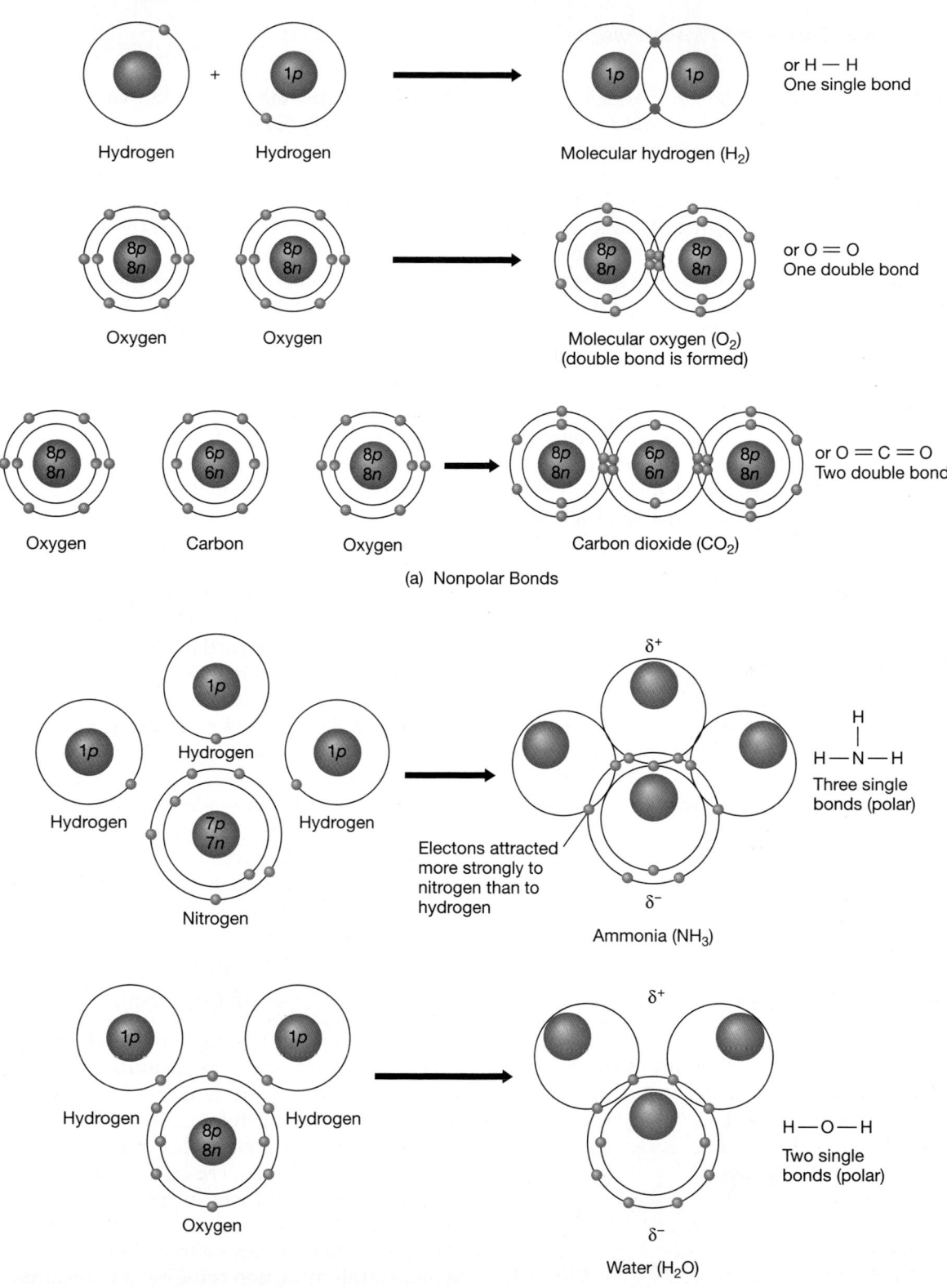

or H — H
One single bond

or O = O
One double bond

or O = C = O
Two double bonds

(a) Nonpolar Bonds

$$H - N - H$$
with H above N

Three single
bonds (polar)

H — O — H
Two single
bonds (polar)

(b) Polar Bonds

Figure 2–3

Covalent bonds are formed when two or more atoms share electrons. (*a*) Nonpolar covalent bonds are formed when the shared electrons are attracted equally to the nuclei of the original atoms. Hydrogen (H_2) and oxygen (O_2) both exist as diatomic molecules formed by single nonpolar bonds. Carbon dioxide (CO_2) has double nonpolar covalent bonds. (*b*) Polar bonds are formed when the shared electrons are attracted more strongly to one nucleus than to another. In ammonia (NH_3) the three shared pairs are more strongly attracted to the nitrogen nucleus, creating a partial negative charge there (symbolized by δ^-) and a slight positive charge (symbolized by δ^+) in the region of the molecule near the three hydrogen nuclei. A similar polarity, formed by two polar bonds, exists in the water molecule (H_2O).

bond most frequently found in biological compounds. Hydrogen, oxygen, carbon, and nitrogen, the four elements most prevalent in living cells, readily form covalent bonds with each other (Fig. 2–3) (see p. 37). Atoms of these four elements all have an incomplete outer shell: hydrogen lacks one electron, oxygen two, carbon four, and nitrogen three. Not only can they form single bonds by sharing one pair of electrons; they can form double and (in the case of carbon) triple bonds. This ability increases their versatility in chemical reactions.

In addition, carbon atoms can form covalent bonds with each other. If a carbon obtains each of the four electrons it needs by sharing a pair of electrons with four other carbon atoms, it will form four distinct covalent bonds. The ability to bond with as many as four other carbon atoms allows carbon

TYPE OF BACKBONE		EXAMPLE
Straight chain		Fatty acids (saturated)
Branched chain		Side chains of amino acids
Chain with double bond		Fatty acids (unsaturated)
6-carbon ring		Steroids
6-carbon ring with double bonds		Side chains of amino acids

Figure 2–4

The ability of carbon to form bonds with four other carbon atoms at once allows it to form a large variety of different linear, branched, and ring structures that serve as the backbones for organic macromolecules.

atoms to form the enormous variety of linear, branched, and cyclic structures that serve as the backbones of large molecules (Fig. 2–4). More than a million compounds are known to contain carbon, and many thousands of them are vital for the maintenance of life. A very large proportion of the molecules found in cells contain carbon.

Covalent bonds and the molecules they form may be either polar or nonpolar (Fig. 2–3). In a **nonpolar covalent bond,** the electrons in the pair are shared equally by each atom in a pair, so the molecule overall is completely neutral. In general, compounds formed with carbon and hydrogen are nonpolar covalent compounds. Their overall shapes are square or rectangular. In a **polar covalent bond,** a pair of electrons is shared unequally; that is, the electrons in the pair are attracted more strongly to the nucleus of one bonded atom than they are to the nucleus of the other bonded atom. The electrons can be thought of as spending more time near the nucleus of one atom, creating a partial negative charge, or pole, at one end of the molecule and a partial positive charge, or pole, at another end. The small charges are symbolized as δ^- and δ^+. Water is an example of a **polar covalent compound.** The two hydrogen atoms form an asymmetrical, triangular shape with the one oxygen atom, with the shared electrons attracted more strongly to the nucleus of the oxygen atom than to the nuclei of the hydrogen atoms. This polarity of certain molecules, and the lack of it in others, has important consequences for chemical reactions, as the section on solutions will show.

An Ionic Bond Is Formed by Gain or Loss of Electrons

When atoms complete their outer shells by either gaining or losing electrons instead of sharing them, they form **ionic bonds** to create **ionic compounds.** When two atoms form an ionic bond, one atom acquires one or more electrons and therefore one or more negative charges. The other atom, by losing the electrons to the first atom, acquires an equal number of positive charges. The equal and opposite electrical charges of the two atoms attract them and hold them together; this attraction is the basis of the ionic bond. A specific example (Fig. 2–5) is the formation of sodium chloride. The neutral sodium (Na) atom has 11 protons and 11 electrons. After donating the lone electron in its outermost shell, it has only 10 electrons left—one less than the number of protons—and a net electrical charge of +1. The neutral chlorine (Cl) atom, with 17 protons and 17 electrons, is able to fill its outermost electron shell by accepting the single electron the sodium atom has given up. The chlorine atom then has 18 electrons—one more electron than

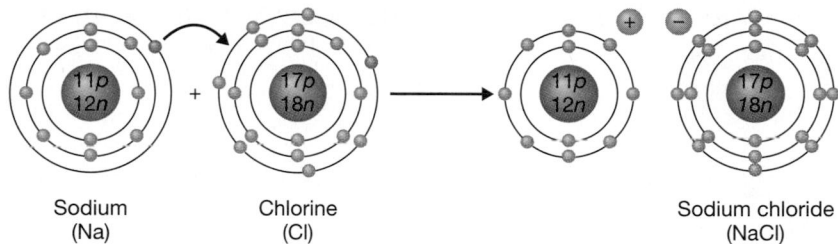

Figure 2–5

Sodium (Na) forms an ionic bond with chlorine (Cl) by donating the single electron in its outer shell to the chlorine atom. The sodium atom then has a complete inner shell, but it has become a cation because it has one more proton than electron. The chlorine atom also has a complete outer shell after gaining the electron from sodium, but it has become an anion because it has one more electron than proton.

protons—and a net electrical charge of −1. An ionic bond has been formed between the oppositely charged sodium and chlorine atoms, and the chemical compound sodium chloride (NaCl) results.

Ionic bonds between atoms do not produce molecules per se; rather, they make crystal lattices, or orderly patterns of cations and anions. As long as the cations and anions are bound together in the lattice, the compound overall is neutral. If the ions become dissociated, as in a solution (see Fig. 2–9a), they exist as charged particles.

Van der Waals Bonds Can Develop Between Two Atoms That Are Close Together

Another type of attractive force develops when any two atoms are close together, about 0.3 to 0.4 nm apart. The **van der Waals bond** forms when the negatively charged electron "cloud" formed by an asymmetrical orbit of electrons in one atom is attracted to the positively charged nucleus of a neighboring atom

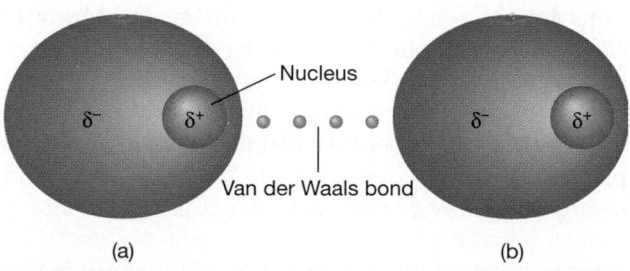

Figure 2–6

Van der Waals bonds form between two atoms because, if the two atoms are close enough, asymmetry in the distribution of electrons in atom (*a*) allows the positively charged nucleus to attract negatively charged electrons in atom (*b*).

(Fig. 2–6). The electrons do not leave one atom for another, however. Van der Waals bonds are a much weaker attractive force than covalent and ionic bonds. The amount of energy needed to break a van der Waals bond is less than 1% of the energy necessary to disrupt a covalent bond between hydrogen and oxygen.

Although individual van der Waals bonds are relatively weak, they become important when large numbers of such bonds form at the same time. This situation often occurs in large molecules when the specific structure of the molecule permits the close proximity of many of the atoms. The van der Waals bonds help to stabilize the three-dimensional shape, which is often important for the function of the molecule.

Hydrogen Bonds Occur Between Polar Molecules

A **hydrogen bond** is similar to a van der Waals bond but slightly stronger. Hydrogen bonds occur among polar covalent molecules, such as water, made up of hydrogen and certain elements such as nitrogen, oxygen, and fluorine. In water, for example, the partial positive charge of the hydrogen atom of one water molecule is attracted to the partial negative charge of the oxygen atom of another water molecule (Fig. 2–7). Hydrogen bonds can occur between molecules or within the same molecule.

Hydrogen bonds are still relatively weak compared to covalent bonds. The input of energy needed to disrupt the hydrogen bonds between water molecules is only 4% of that needed to break the covalent bond between the hydrogen and oxygen in the same molecule. Hydrogen bonds form from the hydrogen atoms in many biological compounds. They are important for maintaining the shapes of large molecules such as proteins and nucleic acids.

Figure 2–7

The polarity of the water molecule allows it to form hydrogen bonds with other water molecules.

Chemical Reactions: Basic Principles

All atoms and molecules are moving continuously and therefore frequently collide with one another. These frequent collisions sometimes disrupt the electron configurations of the atoms involved, thereby breaking the covalent or ionic bonds the atoms have formed with other atoms. When bonds are broken, compounds break down into their elements; when these elements recombine in new ways, new compounds are formed. This process is called a **chemical reaction.** According to the Law of Conservation of Mass, the total number of atoms is the same before and after a chemical reaction; the atoms are simply rearranged into different groups, which represent new compounds with different chemical properties than the original compounds.

A chemical reaction is represented by a **chemical equation,** which shows the relative quantities of each compound in terms of its molecular formula. A simple example is the reaction of perchloric acid ($HClO_3$) and potassium carbonate (K_2CO_3) to form potassium perchlorate ($KClO_3$), carbon dioxide (CO_2), and water (H_2O). The two original compounds are called **reactants;** the resulting compounds are called **products.** Note that the same number of atoms of each element are present both before and after the reaction. This is ensured by providing two molecules (indicated by the number 2 to the left of the molecular formula) of the reactant perchloric acid for each molecule of potassium carbonate.

total atoms	total atoms
2 H	2 H
2 Cl	2 Cl
9 O	9 O
2 K	2 K
1 C	1 C

Sometimes, the reaction proceeds in one direction, giving stable products. Some reactions, however, form unstable products. Either they break down and recombine into the original reactants or break down into new products altogether. An example of this is the reaction of carbon dioxide (CO_2) and water to form carbonic acid (H_2CO_3). The product, H_2CO_3, is unstable, breaking down into the original reactants. Hence, the reaction is called *reversible* and is indicated by a set of opposite arrows. Many metabolic reactions that occur in the cell are reversible.

<table>
<tr><td>water</td><td>carbon
dioxide</td><td>carbonic acid</td></tr>
</table>

original reactants   unstable product

When a reaction reaches a point at which the formation of products is exactly balanced by the reformation of the original reactants, the reaction is said to be in **equilibrium.**

The rate or speed at which a reaction occurs depends on a number of factors, some of which are temperature, concentration, and physical orientation of molecules.

Temperature. An increase in temperature generally accelerates the rate of a chemical reaction by increasing the velocity at which the reactant molecules and atoms are moving. The particles collide more frequently, increasing the possibility that electron configurations will be disrupted. The rate of many reactions doubles when the temperature increases by only 10°C. The metabolic reactions that occur in cells occur at 37°C, a temperature higher than most environments. A further increase in temperature would be harmful to cells because reaction products such as proteins would become unstable, breaking down and losing the ability to perform their functions.

Concentration. At least initially, an increase in the concentration of reactants usually increases the rate of reaction. The presence of more reactant molecules in the same space will increase the frequency with which moving particles collide, thereby disrupting more bonds. Eventually, however, the rate will decrease because the reactants are quickly used up in

Reactants			Products		
2HClO₃	+	K₂CO₃ →	2KClO₃	+ CO₂ +	H₂O
perchloric acid		potassium carbonate	perchlorate	carbon dioxide	water

the formation of products. In the cell, the concentrations of molecules are carefully controlled so that metabolic reactions proceed at optimal rates.

Orientation. The way in which molecules are oriented, or arranged in space, often affects the rate at which their atoms will break off and form new bonds. The atoms also must be near enough for the reaction to occur. Certain important bonds occur only between specific chemical groups (e.g., peptide bonds that form between amino acids in protein molecules).

Some chemical reactions fall into a special category known as **oxidation-reduction reactions,** or **redox reactions.** An example is the reaction in which iron changes to ferric oxide (rust):

$$4\,Fe\ +\ 3\,O_2\ \rightarrow\ 4\,Fe^{3+}\ +\ 6\,O^{2-}\ \rightarrow\ 2\,Fe_2O_3$$

iron   oxygen   ferric oxide (rust)

In this reaction, the iron loses electrons, a process known as **oxidation.** Each iron (Fe) atom loses three electrons, leaving an excess of three positively charged protons and creating iron cations (Fe^{3+}). Each oxygen atom accepts two electrons from the iron atoms; the extra electrons confer two negative charges on each oxygen atom, creating oxide anions (O^{2-}). The process of gaining electrons is known as **reduction.** The charged iron and oxygen atoms then form ionic bonds and produce two molecules of the compound ferric oxide. Three iron atoms lose a total of 12 electrons, which are gained by a total of six oxygen atoms. The iron is being **oxidized** at the same time that the oxygen is being **reduced.** Oxidation and reduction always occur simultaneously, one reactant accepting the electrons removed from another.

Water and Aqueous Solutions

Water is essential for living organisms and is the major chemical compound in most cells, accounting for up to 70% of cell weight. Many biologically important compounds are soluble in water; in fact, the chemical reactions they undergo require an **aqueous** (watery) medium or environment. Here we will consider some special properties of water that make it an essential substance for living cells.

Water Is a Polar Molecule

The water molecule is formed by the covalent bonding of two hydrogen atoms to one oxygen atom. To complete its outer shell, oxygen needs two electrons. It obtains these electrons by sharing an electron with each of two hydrogen atoms. These bond to the oxy-

gen atom to form a triangular structure. The covalent bonds formed between the atoms are polar bonds; that is, the shared electrons are attracted more strongly to the oxygen nucleus than to the hydrogen nuclei. This creates a small negative charge in the area of the water molecule near the oxygen nucleus and a small positive charge near the two hydrogen nuclei. Because it has these oppositely charged areas at opposite ends, the water molecule is said to be **polar** (see Fig. 2–3).

Hydrogen Bonds Occur Between Water Molecules

The polarity of water molecules leads them to form hydrogen bonds with each other when the partial positive charge at one end of one molecule attracts the partial negative charge at the opposite end of another molecule (Fig. 2–7). Hydrogen bonds are responsible for the higher melting and boiling points of water compared to other substances with similar structures that do not form hydrogen bonds (e.g., ammonia [NH_3]). In solid water (ice), for example, the molecules are arranged in a regular three-dimensional pattern so that each molecule is hydrogen-bonded to four other water molecules. When ice melts by heating, and liquid water forms, only about 15% of the hydrogen bonds are broken. Thus, liquid water contains clusters of molecules that have retained a highly organized structure. Energy in the form of heat is required to change ice to liquid water and liquid water to a gas because these changes require the disruption of the hydrogen bonds between water molecules. Similar changes in the physical state of ammonia require less energy because there are no hydrogen bonds to be broken.

Water Is an Excellent Solvent

In chemistry, a **solution** is a type of mixture in which one substance, called the **solute,** is uniformly distributed throughout another substance, usually a liquid, called the **solvent.** A solvent dissolves, or breaks up, the solute without reacting chemically with it: that is, electrons are not shared or exchanged between the two substances. The polarity and hydrogen-bonding tendencies of water make it an excellent solvent of both ionic compounds and polar covalent compounds.

Any substance that dissolves in water is said to be **hydrophilic** ("water-loving"). Ionic compounds and polar covalent compounds are generally hydrophilic. Nonpolar covalent compounds and groups are not easily soluble in water; they are said to be **hydrophobic** ("water-hating").

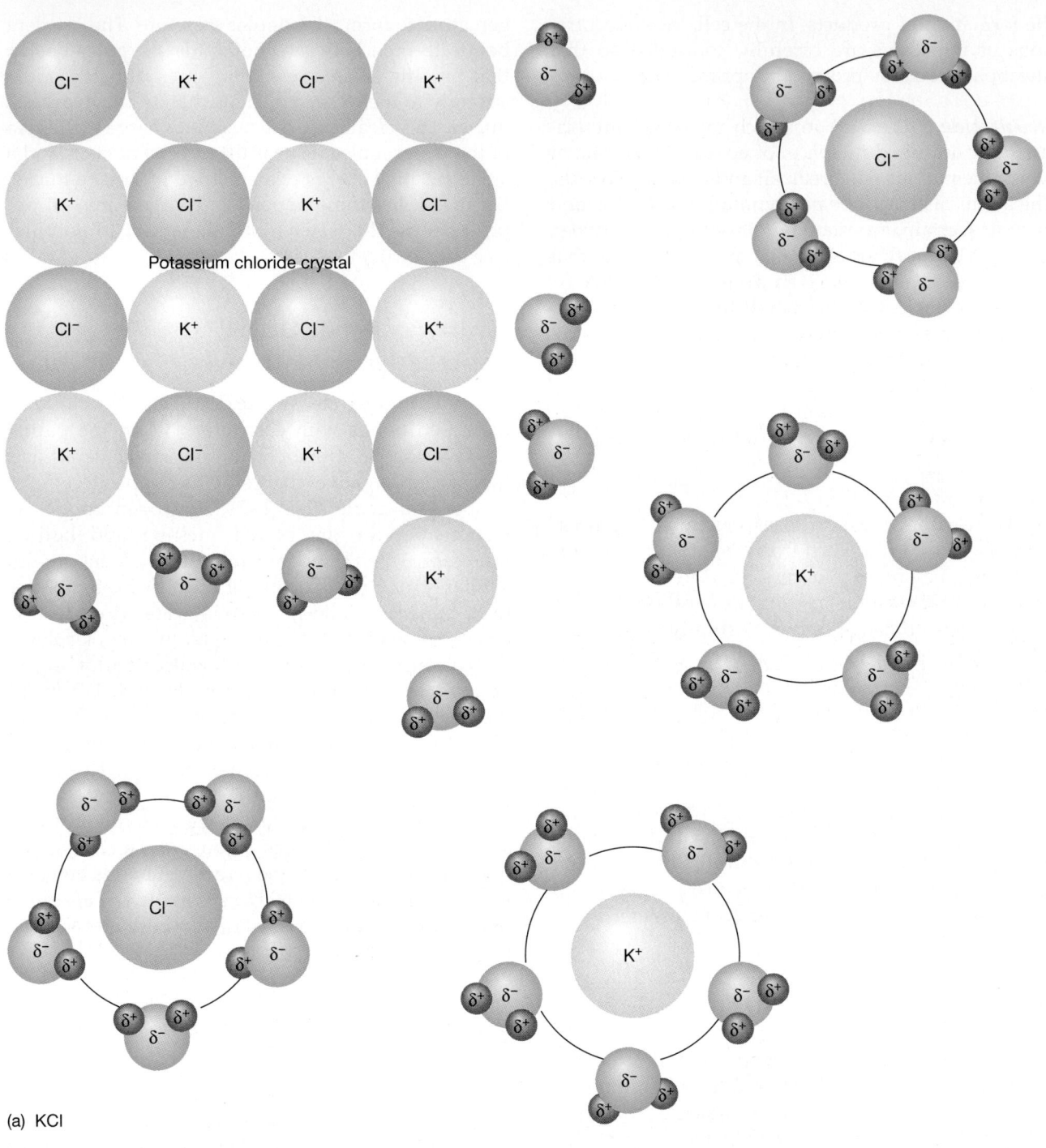

Potassium chloride crystal

(a) KCl

(b) Ethanol

Hydrogen bond

Figure 2–8

(a) Compounds such as potassium chloride (KCl) that form crystal lattices stabilized by ionic bonds dissolve readily in water because the small polar water molecules can slip in between and surround the K⁺ and Cl⁻ ions. Some water molecules point their positively charged poles toward the Cl⁻ anions, forming shells around them. Other water molecules point their negatively charged poles toward the K⁺ cations, forming shells around them. (b) The ethanol molecule consists mostly of nonpolar covalent bonds, but the —OH (hydroxide) group is polar, so it forms hydrogen bonds with water. These bonds allow ethanol and other polar covalent compounds to dissolve in water.

Solutions of Ionic Compounds

Water weakens the ionic bonds that occur in ionic compounds. The ionic bonds in potassium chloride, for example, are readily broken when solid potassium chloride is added to water because the polar water molecules are able to form shells around the potassium and chlorine ions (Fig. 2–8*a*). Some water molecules orient their positively charged hydrogen atoms toward the negatively charged chlorine atoms; others point their negatively charged oxygen atom toward the positively charged potassium atoms. The net force of all the water molecules attracting each ion overcomes the bonds between the ions. As a result, the potassium and chloride ions break away, or **dissociate,** from one another. However, in the process they do not take or give back the electrons they lost or donated in the first place, so they still have unequal numbers of protons and electrons. Therefore, they remain ions in solution as they were in the crystal lattice.

Solutions of Polar Covalent Compounds

Uncharged polar molecules also dissolve readily in water because they can form hydrogen bonds with water molecules. Ethanol is an example (Fig. 2–8*b*). The hydrogen atom in the —OH group of ethanol has a small positive charge because the shared electrons in the covalent bond are more strongly attracted to the oxygen atom. The polar —OH group provides the hydrogen atom for the hydrogen bond, while the water provides the oxygen atom.

Solutions of Nonpolar Compounds: Hydrophobic Interactions

Uncharged molecules that are not polar are poorly soluble in water because they have no positive or negative poles to attract the polar water molecules. Thus, water molecules have no effect on the covalent bonds that hold the solute molecules together. The polar water molecules can form neither shells around the atoms nor hydrogen bonds with them (Fig. 2–9*a*).

Hydrophobic substances are common in biological processes. Important examples are the chains of carbon atoms with only hydrogen atoms attached (called *hydrocarbon chains*). Many important macromolecules are partially composed of these hydrocarbon chains. When the macromolecules are placed in water, the hydrocarbon chains—the hydrophobic portions—form clusters or "pockets" to minimize their contact with the water molecules (Fig. 2–9*b*). This arrangement causes the least disruption of hydrogen bonds in the water. By creating such collections of molecules, hydrophobic interactions play an important role in maintaining the structures of cer-

(a)

(b)

Figure 2–9

(*a*) Nonpolar covalent molecules and groups, such as those formed from only carbon and hydrogen (hydrocarbons), do not dissolve in water because they lack positive and negative poles. The polar water molecules are unable to attract their atoms away from each other since they cannot form hydrogen bonds with them or shells around them. (*b*) Instead, nonpolar molecules cluster together in hydrophobic "pockets"; these reactions are called hydrophobic interactions. Such pockets allow water to retain its hydrogen-bonded structure.

tain cell organelles and of the plasma membrane that surrounds the cell. The strong tendency to cluster means that these hydrophobic groups can create a sturdy mass.

Concentrations of Solutions Are Described in Several Ways

We have said that some chemical reactions require an aqueous medium in order to occur and that the concentration of reactants can be an important factor in determining the rate of reaction. A solution may contain several different solutes, which may then react with one another; indeed, the purpose of their being in solution is to take part in important metabolic reactions. The concentration of reactants in these metabolic processes is determined by their concentration in solutions. Scientists have devised convenient ways of expressing these concentrations in order to understand the rates at which certain reactions occur both in the laboratory and in the living cell.

The concentration of a solution can be expressed in several ways. One way is to compare the number of moles of solute to the kilograms of solvent: this is known as **molality.** In aqueous solutions, this is the same as comparing moles of solute to liters of solution, since one kilogram of water is one liter. For example, if one mole (58.5 g) of sodium chloride (the solute) is dissolved in one liter of water, concentration is expressed as 1 mole per kilogram (abbreviated as 1 mol/kg). This is a "1 molal" solution, abbreviated as "1 m" solution.

An alternative way to express concentration is **molarity.** Molarity is a unit of concentration, based on the number of moles of solute per liter of solution. A solution with a concentration of 1 mol/L is called a "1 molar" solution, abbreviated as "1 M" solution. The solution is said to have a molarity of 1. A solution with a concentration of 2 mol/L is a 2 M solution.

Very low concentrations can be expressed with the same prefixes used in metric measures. A 1 mM solution of sodium chloride contains 1 millimole (58.5 mg) of sodium chloride per liter of solution. The solutions that occur in living organisms have a variety of concentrations, but most are in the millimolar range. Human blood plasma, for example, contains sodium and potassium at concentrations of approximately 140 mM and 4 mM, respectively.

Osmolarity is another expression of concentration that considers the total number of solute **particles** rather than the relative weights of specific solutes. Osmolarity is discussed in the section on osmosis in Chapter 4.

The pH Scale Describes the Concentration of H^+ Ions

Water (H_2O or HOH) has a slight tendency to dissociate into hydrogen ions (H^+) and **hydroxide (OH^-) ions.** This dissociation process can be represented as:

$$\boxed{HOH} \rightleftharpoons \boxed{H^+} + \boxed{OH^-}$$

At the same time, the ions tend to reassociate, forming new water molecules. When the rate of dissociation equals the rate of reassociation, the reaction is said to be in **equilibrium.** At this point, the net concentrations of HOH molecules and of H^+ and OH^- ions remain constant, even while individual molecules and ions may be breaking down and reforming. The dissociation of water reaches equilibrium very rapidly, so that only very few of the water molecules are dissociated at any instant.

Ideally, the concentrations of H^+ and OH^- ions in water are exactly equal and very small, making the water a neutral compound. In reality, however, the concentrations of these ions are not equal because of the presence of solutes in the water that, by dissociating in it, increase or decrease the amount of H^+ ions relative to the number of OH^- ions. A solution in which the H^+ concentration is greater than the OH^- concentration is said to be **acidic.** A solution in which the number of OH^- ions is greater than the number of H^+ ions is a **basic** solution.

We can express the acidity or basicity of a solution by specifying the concentration of H^+ ions. Because this concentration is very small, scientists have devised a logarithmic method of expressing it, called **pH.** This expression is the negative of the base-10 logarithm (log) of the H^+ concentration:

$$pH = -\log [H^+]$$

where [ ] indicates the H^+ concentration in moles per liter.

In one liter of neutral water, there are 1×10^{-7} moles of H^+ ions (and 10^{-7} moles of OH^- ions). The pH of neutral water then is:

$$
\begin{aligned}
pH &= -\log (1 \times 10^{-7}) \\
&= -[\log (1.0) + \log (10^{-7})] \\
&= -[(0.00) + (-7)] \\
&= 7
\end{aligned}
$$

Suppose the concentration of H^+ ions were to increase 10-fold, to 1×10^{-6} mol/L. The pH of the solution would then be:

$$
\begin{aligned}
pH &= -\log (1 \times 10^{-6}) \\
&= -[\log (1.0) + \log (10^{-6})] \\
&= -[(0.00) + (-6)] \\
&= 6
\end{aligned}
$$

The solution is more acidic, and the pH has decreased. Conversely, if the H^+ concentration were to decrease, the solution would become more basic, and the pH would increase.

The pH scale goes from 0 to 14. A solution with a pH of 0 is highly acidic; one with a pH of 14 is highly basic. In terms of pure numbers, the scale is small, so that a 1000-fold increase in H^+ concentration produces a pH decrease of only 3 units. It is important to remember, therefore, that a *small decrease* in pH represents a *large increase* in the H^+ concentration, and vice versa.

Acids and Bases Change the pH of Solutions

An **acid** is any substance whose dissociation in water releases H^+ ions. The hydrogen ion does not contain any neutrons and can be thought of as essentially equivalent to a proton. Therefore, an acid is often referred to as a **proton donor** because it gives off protons in the form of H^+ ions. A **base** is a **proton acceptor,** a substance that accepts H^+ ions and thereby decreases the H^+ concentration in a solution.

Acids: Proton Donors

The addition of an acid to a solution greatly increases the concentration of free H^+ in the solution, producing a more acidic solution and a decrease in the pH. The dissociation of an acid, HA, can be represented symbolically by the following equation:

$$HA \rightarrow H^+ + A^-$$

where A^- represents the anion that remains when the hydrogen cation is formed. The strength of an acid is a function of its tendency to give up H^+ ions. In the case of a strong acid, such as HCl, all the molecules will dissociate in water and the dissociation will not be reversible:

$$HCl \rightarrow H^+ + Cl^-$$
hydrochloric acid
(strong acid)

Weak acids, in contrast, give up H^+ ions less rapidly and do not dissociate completely in water. Thus, a solution of a weak acid contains both dissociated ions and complete molecules. Most of the acids produced in living organisms are weak acids.

Bases: Proton Acceptors

Usually a base gives off hydroxide ions (OH^-) in the process of dissolving in water and these decrease the H^+ concentration by accepting free H^+ ions. The pH will increase. Sodium hydroxide (NaOH) is a strong base that dissociates completely in water to yield free OH^- ions. The hydroxide anions then accept H^+ ions, and the result is water:

$$NaOH \longrightarrow Na^+ + OH^-$$
sodium hydroxide

$$OH^- + H^+ \rightleftharpoons H_2O$$

This reaction occurs when NaOH is added to water. The OH^- ions combine with the free H^+ ions already present in the water, decreasing the concentration of H^+ in the solution. In the case of a weak base, such as the bicarbonate ion (HCO_3^-), H^+ also is used up as follows:

$$H^+ + HCO_3^- \rightleftharpoons H_2CO_3 \rightleftharpoons \boxed{H_2O} + CO_2$$

Bases use up H^+ ions and decrease the H^+ concentration, thereby producing basic solutions.

If a weak base such as $NaHCO_3$ is present in solution with a strong acid such as HCl, the base will absorb the H^+ ions released by the acid. The following four processes will be occurring:

Dissociation of the strong acid

$$HCl \longrightarrow H^+ + Cl^-$$

Dissociation of the weak base

$$NaHCO_3 \rightleftharpoons Na^+ + HCO_3^-$$

Formation of a weak acid

$$H^+ + HCO_3^- \rightleftharpoons H_2CO_3$$

Formation of a neutral salt

$$Na^+ + Cl^- \longrightarrow NaCl$$

These processes can be summarized in the following equation:

$$HCl + NaHCO_3 \rightarrow H_2CO_3 + NaCl$$

| hydrochloric acid (strong acid) | sodium bicarbonate (weak base) | carbonic acid (weak acid) | sodium chloride (neutral salt) |

Buffers: pH Control

The purpose of a **buffer system** is to minimize the change in pH that occurs when an acid or base is added to any solution. The pH of blood is controlled very closely and does not vary outside a small range, usually 7.38 to 7.42. This is a remarkable achievement because the chemical reactions in the cells of the body produce more than 10,000 nmol of H^+ ions each day. An average blood pH of 7.4 represents an H^+ concentration of only 40 nmol/liter, indicating that highly efficient mechanisms must exist to absorb the H^+ ions added to the blood. These mechanisms will be addressed in detail in Chapter 26.

The Behavior of Gases

When a liquid is heated sufficiently, it begins to evaporate; its molecules break apart and begin moving about rapidly in space. The substance has passed into the gas phase. Scientists describe four physical properties of any gas: (1) the **pressure** (P) or the force the gas molecules exert on their surroundings; (2) the **volume** (V) or the amount of space the gas molecules occupy, (3) the **amount,** or number (n) of moles of gas; and (4) the **temperature** (T) expressed in Kelvins (273 plus the Celsius temperature).

The Ideal Gas Law Relates the Physical Properties of Any Gas

Three factors can affect the volume of a gas: pressure, temperature, and amount. At a constant temperature, the same amount of gas will decrease in volume with an increase in pressure. At constant pressure, the same amount of gas will increase in volume with an increase in temperature. At a constant temperature and pressure, the same amount of gas will remain at constant volume; the only factor that would change the volume would be an increase in the number of moles of gas present.

These relationships can be summarized as follows: (1) volume is inversely proportional to pressure; (2) volume is directly proportional to temperature; and (3) volume is directly proportional to amount. They are further summarized in an equation known as the **Ideal Gas Law:**

$$PV = nRT$$

in which R is a constant, known as the **ideal gas constant,** equal to 0.082 liter · atm/deg · mol for any gas.

Partial Pressure Is Directly Proportional to the Amount of Gas Present

Unlike solids and liquids, which do not all readily dissolve in one another, all gases are completely soluble in all other gases. Air is a solution of several gases, including oxygen, nitrogen, carbon dioxide, and water vapor. According to the **law of partial pressures,** the total pressure in a container holding a solution of gases is the sum of the partial pressures of the component gases. **Atmospheric (or barometric) pressure,** which is the pressure exerted by air on the surface of the earth, is the sum of the partial pressures of all the gases that are present in air.

The partial pressure of a gas is proportional to the amount, in moles, of the gas solution it makes up. Oxygen, for example, makes up 21% of air. Atmospheric pressure is 760 mm Hg (so determined be-

cause, at sea level, air causes liquid mercury, Hg, to rise 760 mm in a closed tube). The partial pressure of oxygen, or P_{O_2}, is 21% of 760 mm Hg, or 160 mm Hg. If all other gases were removed from a container of air, with only oxygen remaining, it would still exert a pressure of 160 mm Hg. (Atmospheric pressure at sea level is used as a standard unit of pressure called the "atmosphere." One atmosphere—1 atm—is equal to 760 mm Hg.)

Partial pressures are designated as P_{O_2}, P_{CO_2}, P_{N_2}, P_{H_2O}, and so on. The idea of partial pressure is very useful for dealing with concentration gradients and other aspects of physiology. Because partial pressures are directly proportional to the amount of gas present, they can be used to determine the number of moles of any constituent gas in a solution of gases.

Gas Molecules Are in Continuous Random Motion

According to the *kinetic theory of gases,* gas molecules are in continuous, random motion, colliding with one another and with their surroundings. The pressure of the gas arises from the forces of the collisions with the surroundings. The speed of the molecules is directly related to the temperature; if the temperature increases, the molecules move faster.

Macromolecules

The unique properties of the carbon atom allow it to form four bonds at once and lead to a huge variety of molecules of different shapes and enormous sizes. These molecules are called **macromolecules,** and they play important roles in the life processes of all cells and organisms. The most important macromolecules in cell physiology are carbohydrates, lipids, proteins, and nucleic acids. The following brief survey provides a glimpse of these compounds and a review of their chief functions in cells.

Carbohydrates Form Branched Chains

Carbohydrates are often referred to as **sugars.** All **carbohydrates** are composed of carbon, hydrogen, and oxygen atoms. Carbohydrates are classed in three broad groups: monosaccharides, disaccharides, and polysaccharides. In addition, they form compounds with proteins and lipids to form glycoproteins and glycolipids.

Monosaccharides

The simplest type of carbohydrates are called **monosaccharides.** The general chemical formula for a monosaccharide is $(CH_2O)_n$, where n is a whole num-

Figure 2–10

The structures of the monosaccharides glyceraldehyde, erythrose, ribose, and fructose.

ber indicating the number of carbons. A monosaccharide with three carbons, such as glyceraldehyde, is a **triose** (Fig. 2–10). A monosaccharide with four carbons is a **tetrose;** one with five carbons a **pentose,** and one with six carbons a **hexose.**

In structural formulas of sugars and other "organic" molecules, the carbon atoms often are numbered for identification. Specific carbons often have specific roles based on their position in the molecule and on the other atoms attached to them. A carbon atom, such as the second carbon atom ("carbon 2") in the glyceraldehyde molecule, is described as **asymmetrical** when it is linked covalently to four differ-

ent chemical groups. Carbon 2 in glyceraldehyde (Fig. 2–10) is linked to a hydrogen (H), an aldehyde (—CHO) group, a hydroxide (—OH) group, and —CH₂OH.

Ring Structures. Monosaccharides can have different forms, either linear chains or rings. In solutions, monosaccharides most often form rings. One common hexose is **glucose,** $C_6H_{12}O_6$, or $(CH_2O)_6$. In glucose, ring formation occurs when the aldehyde (—CHO) group on carbon 1 reacts with the hydroxyl (—OH) group on carbon 5 to form a stable covalent bond, giving rise to a six-member ring structure (Fig. 2–11).

Although drawn as a flat structure, the plane of the ring is perpendicular to the plane of the page, and the hydrogen atoms and hydroxyl groups lie above and below the plane of the ring. The edge of the ring nearest the reader is indicated by a thicker line between the carbon atoms. In simplified drawings of ring structures, the letter *C* indicating a carbon atom is omitted, but the carbons are understood to be present wherever two "sides" of the ring meet. Furthermore, the letter *H* indicating hydrogen atoms is also omitted but understood to be present wherever needed to supply each carbon with a total of four bonds. In the most simplified drawings, only the —OH groups are identified by letters (Fig. 2–11).

Pentoses form five-member ring structures, as do some hexoses such as fructose (Fig. 2–12). Two important pentoses are ribose and deoxyribose, which are present in nucleic acids.

When a hexose forms a five-member ring, carbon 1 is left outside the ring because the ring forms

Figure 2–11

The linear and ring structures of the sugar glucose, a common hexose. When the —OH group on carbon 5 comes near the =O group on carbon 1, the two carbons are joined by the —O— atom, and a ring is formed.

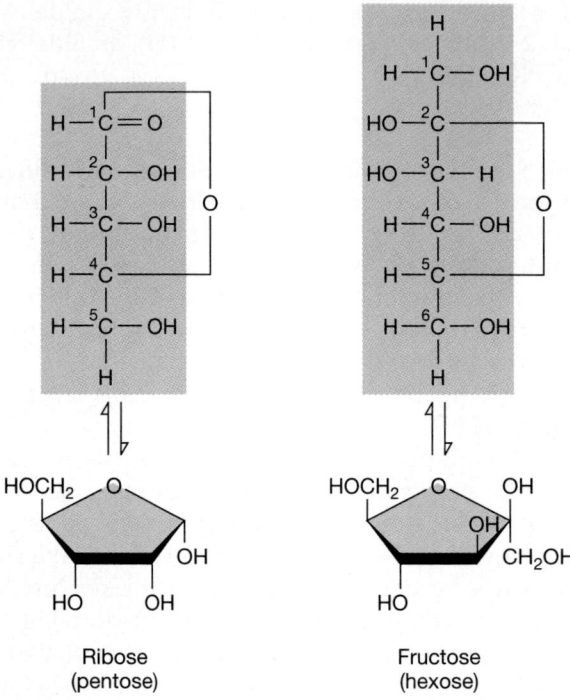

Figure 2–13 area:

Figure 2–13

Maltose and isomaltose are isomers. Both are formed from two molecules of glucose; both have the same molecular formula. They have different structural formulas, however, and therefore constitute two different disaccharides.

Figure 2–12

The cyclic forms of ribose (a pentose) and fructose (a hexose) both contain five-member rings.

when the carbonyl (C=O) group on carbon 2 reacts with the hydroxyl (—OH) group on carbon 5 (Fig. 2–12).

Monosaccharides have a unique ability to form chemical bonds at more than one carbon atom. One monosaccharide can combine with several others to create branched chains of monosaccharide units. In contrast, other macromolecules such as proteins and nucleic acids can assemble only in linear, unbranched chains.

Disaccharides

Two monosaccharides can combine to form a **disaccharide.** The most common disaccharides are **maltose** (two glucose molecules), **lactose** (one glucose molecule plus one galactose molecule), and **sucrose** (glucose plus fructose).

Isomaltose, like maltose, is formed from two glucose molecules, and both isomaltose and maltose have the same chemical formula. However, they have different structures because their monosaccharide units join at different carbon atoms. In maltose, the glucose molecules join at carbons 1 and 4; in isomaltose, they join at carbons 1 and 6 (Fig. 2–13). Maltose and isomaltose are called **isomers,** molecules with the same chemical formula but different structural arrangements of the atoms.

Polysaccharides

Most of the carbohydrates found in nature occur as **polysaccharides,** molecules composed of many monosaccharide units arranged in branched chains. Polysaccharides are distinguished from one another mainly by their monosaccharide units and the length and amount of branching of their chains. Glucose is the most common monosaccharide unit of polysaccharides. Glucose is the fuel that most animal cells require, and it can be stored in certain tissues in the form of the polysaccharide **glycogen.** This important polysaccharide is especially abundant in the liver, where it can account for up to 10% of the wet weight of the tissue. It consists of a chain of glucose units that are linked by chemical bonds between carbons 1 and 4 of adjacent glucose units. Branches occur about every 10 glucose units, arising from a bond between carbons 1 and 6 (Fig. 2–14). The molecular mass of glycogen has been estimated to be in the range of 3 $\times 10^5$ to 3×10^6. This means that the number of glu-

Figure 2–14

A branch point in the polysaccharide glycogen.

cose units present in a glycogen molecule could be as high as 16,667.

Sugar Combinations

Many of the polysaccharides present in living organisms occur in covalent combination with proteins or lipids. **Glycoproteins** and **glycolipids** are present in cell membranes. The polysaccharide chains in these compounds tend to be short but highly branched, containing several different monosaccharide units arranged in complex sequences. They are thought to be important in recognition processes such as those involved in the action of hormones. A hormone released into the blood will be distributed to all the cells in the body, but will affect only those cells that have an external site that the hormone "recognizes" and binds. The carbohydrate portions of membrane glycoproteins and glycolipids are thought to be important parts of these binding sites for hormones.

Lipids Have Low Solubility in Water

Lipids are sometimes called **fats** and are distinguished from the other cellular macromolecules because they are poorly soluble in water and highly soluble in organic lipids such as ether and chloroform. They are composed of carbon, hydrogen, and oxygen atoms, the same elements that make up carbohydrates; however, the proportions of these elements differ in lipids. The lipids contain only small amounts of oxygen relative to carbon and hydrogen.

Three major classes of lipids will be considered here: fatty acids, lipids containing glycerol, and steroids, or lipids that do not contain glycerol.

Fatty Acids

A **fatty acid** is a long chain of covalently linked carbon atoms to which only hydrogen atoms are attached, except for a carboxylic acid (—COOH) group at one end. The long hydrocarbon chain is hydrophobic, or poorly soluble in water, whereas the —COOH group is highly water soluble and reacts readily with other chemical groups, forming covalent bonds. Most of the fatty acids in animal cells are covalently linked to other molecules via the carboxylic acid group. Two of the most abundant fatty acid molecules in animal cells are **palmitic acid** and **stearic acid** (Fig. 2–15).

Specific fatty acids are distinguished from one another by the number of carbon atoms in their hydrocarbon chains and the number of carbon-carbon double bonds. When only single bonds exist between all the carbons in a fatty acid chain, the fatty acid is said to be **saturated,** i.e., saturated with hydrogen, because as many hydrogen atoms as possible are

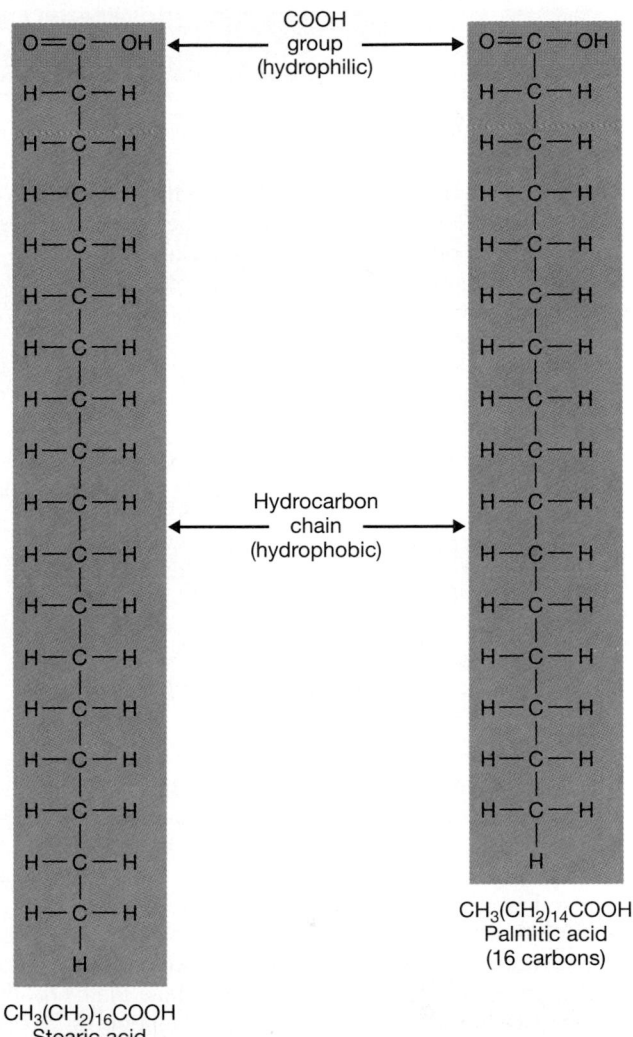

$CH_3(CH_2)_{16}COOH$
Stearic acid
(18 carbons)

$CH_3(CH_2)_{14}COOH$
Palmitic acid
(16 carbons)

Figure 2–15

The structures of stearic acid and palmitic acid, two common saturated fatty acids. The molecules consist mostly of hydrophobic hydrocarbon chains. The carboxylic acid (—COOH) group at one end, however, is water-soluble or hydrophilic. Fatty acids are sometimes given the general symbol RCOOH, R representing the hydrocarbon chain.

bonded to each carbon. When one or more double bonds are present between carbon atoms, the fatty acid is **unsaturated.** The most abundant unsaturated fatty acids in animal cells are oleic acid, linoleic acid, and arachidonic acid, which contain one, two, and four double bonds, respectively:

oleic acid $CH_3(CH_2)_7CH{=}CH(CH_2)_7COOH$

linoleic acid $CH_3(CH_2)_4CH{=}CHCH_2CH{=}CH(CH_2)_7COOH$

arachidonic acid $CH_3(CH_2)_4(CH{=}CHCH_2)_4(CH_2)_2COOH$

R E S E A R C H   F O C U S

An Ether Phospholipid

Platelet-activating factor (PAF) is an ether phospholipid that has striking and diverse physiological effects. It is an analog of phosphatidylcholine (Fig. 2–17c). Unlike phosphatidylcholine, PAF has a long hydrocarbon chain only on carbon 1. At carbon 2 , instead of a second hydrocarbon chain, PAF has an acetyl group (see diagram below). This increases the water solubility of the molecule and allows it to function in an aqueous environment. It is termed an ether phospholipid because the hydrocarbon chain at carbon 1 is attached by an ether linkage (—C—O—C—).

$$H_3C-\overset{\overset{\displaystyle O}{\parallel}}{C}-O-\underset{\underset{\displaystyle C^3H_2-O-\overset{\overset{\displaystyle O}{\parallel}}{\underset{\underset{\displaystyle O^-}{\mid}}{P}}-O-CH_2-CH_2-N^+(CH_3)_3}{\mid}}{\overset{\overset{\displaystyle H_2C^1-O-CH_2-(CH_2)_{14}-CH_3}{\mid}}{C^2}}-H$$

platelet-activating factor

PAF is of interest because it is the first known example of a biologically active phospholipid. Its identity was discovered in 1979. Prior to this period it was assumed that the principal role of phospholipids was to provide a physical barrier as a component of plasma membranes and intracellular membranes.

PAF was found originally in the blood of animals undergoing a severe allergic reaction and was shown to be a factor that activated blood **platelets.** Platelets are blood cells that participate in the formation of blood clots (Chapter 17). However, the term *platelet-activating factor* is now considered a misnomer because PAF affects the cells of many different tissues. In some situations the action of PAF is achieved at very low concentrations such as 1.0 pM (10^{-12} M). PAF has received most attention for its role in **asthma** and **endotoxin shock.** PAF is one of a group of substances produced during an asthmatic reaction and its presence leads to inflammation and narrowing of air passages. Release of PAF is stimulated also during a bacterial invasion. **Toxic-shock syndrome,** for example, occurs because fragments of destroyed bacteria act as endotoxins and induce production of PAF. The latter lowers blood pressure and decreases the amount of blood pumped by the heart, leading to shock. Both asthma and endotoxin shock, when severe, can lead to death.

In addition to these adverse effects in abnormal situations, PAF also may have important beneficial roles in normal physiological processes such as reproduction. For example, PAF secreted by the fertilized egg is helpful for implanting the egg in the wall of the uterus. In the late stages of pregnancy, PAF is produced in increasing amounts by the fetal lung. This may help to stimulate production of fetal lung **surfactant,** a substance that prevents lung collapse in newborns (see Chapter 20). Release of PAF into the fluid surrounding the fetus may contribute to the initiation of labor because PAF is a potent stimulator of uterine contractions.

A major development in the field occurred in 1991 when the structure of the PAF **receptor** was identified. The receptor is a small protein of 342 amino acids which is present in the plasma membrane of many types of cells. Attachment of PAF to the receptor is a necessary first step in the sequence of events by which PAF changes the behavior of a cell. Researchers are hopeful that the design and development of new drugs to block binding of PAF to its receptor could lead to new treatments for asthma and endotoxin shock, and to new ways for preventing conception and premature birth.

■ **How does PAF differ structurally from phosphatidylcholine?**
Give examples of lipids that are not phospholipids.

Fatty acids are a valuable source of fuel for cells. Considerably more energy can be obtained from the breakdown of 1 g of a fatty acid than from 1 g of carbohydrate. Arachidonic acid is an intermediate structure produced during the synthesis of an important group of molecules called **prostaglandins,** which affect a wide variety of physiological processes.

Glycerol Compounds and Phospholipids

Glycerol is a molecule with three hydroxyl groups (Fig. 2–16). These hydroxyl groups can form bonds with the carboxylic acid groups of fatty acids. The attachment of one, two, or three fatty acids to glycerol produces a **monoacylglycerol,** a **diacylglycerol,** or a **triacylglycerol,** respectively. Both mono- and diacylglycerols are important natural compounds, but the triacylglycerols (Fig. 2–16) are the most abundant and represent an important means by which cells are able to store fatty acids until they are needed. Cells can convert excess carbohydrate and protein to fatty acids and then store the fatty acids in triacylglycerols. The continued accumulation of triacylglycerol deposits in the body, however, can lead to obesity. Triacylglycerols often are referred to as **triglycerides.**

Another group of glycerol-containing lipids are the **phospholipids.** The parent compound of this group is **glycerol-3-phosphate,** a glycerol molecule

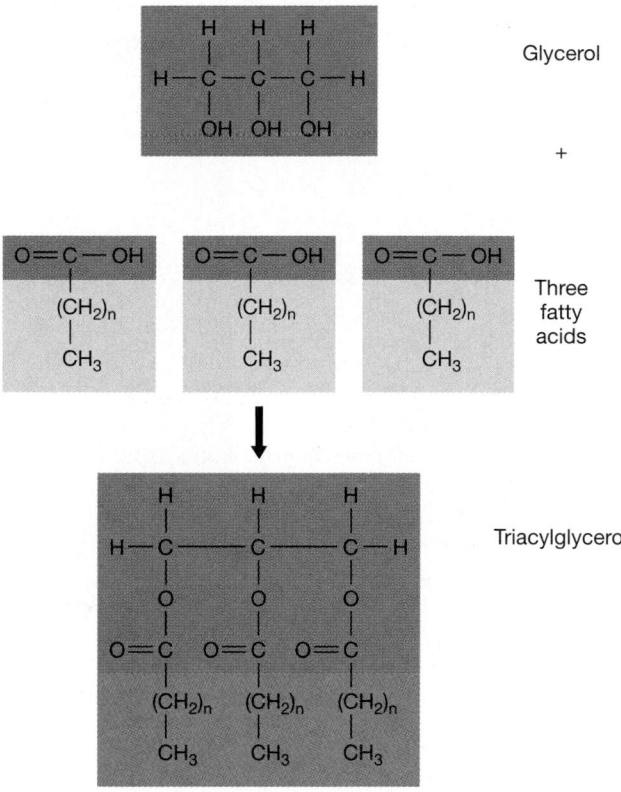

Glycerol

+

Three
fatty
acids

Triacylglycerol

Figure 2–16

Steps in the formation of a triacylglycerol, also termed a triglyceride, from glycerol and three fatty acids. The fatty acids may be identical or different. *n* stands for a whole number, the number of carbon atoms in the parentheses representing all but the last carbon in the fatty acid hydrocarbon chain. For stearic acid, *n* is 16 (see Fig. 2–15).

to which a phosphate ($-PO_4$) group has been attached (Fig. 2–17*a*). A phospholipid is formed from this basic structure by (1) the addition of fatty acids to the free hydroxyl groups on carbons 1 and 2 to give **phosphatidic acid** and (2) the addition of another chemical group to the phosphate on carbon 3. In Step 1, the fatty acid added to carbon 1 of the glycerol unit is usually saturated and contains 16 to 18 carbon atoms, whereas the fatty acid added to carbon 2 is usually unsaturated and contains 16 to 20 carbon atoms. The most common groups added in Step 2 are choline, ethanolamine, serine, and inositol. The phospholipids that result are called phosphatidylcholine, phosphatidylethanolamine, phosphatidylserine, and phosphatidylinositol, respectively (Fig. 2–17*b*).

At physiological pH (7.40), the phospholipids are in the charged forms shown, and one end of the molecule is a polar (hydrophilic) region. In contrast, the long hydrocarbon chains of the two fatty acid groups constitute a highly hydrophobic region of the molecule (Fig. 2–17*c*). The dual nature of these phospholipids—hydrophilic at one end and hydrophobic at the opposite end—plays an important structural role in the plasma membrane surrounding all cells (see Chapter 4). The membrane is made up of a "lipid bilayer": two layers of phospholipid molecules oriented so that their hydrophobic hydrocarbon chains point toward each other, away from the aqueous surroundings both inside and outside the cell (Fig. 2–17*d*). The hydrophilic parts of the phospholipids in each layer are oriented outward. This arrangement represents a very stable physical structure with unique properties, as will be discussed in Chapter 4.

Sphingomyelin is the only phospholipid not derived from glycerol. The basic unit of sphingomyelin is **sphingosine**:

$$CH_3(CH_2)_{12}-CH=CH-CHOH-CHNH_3^+-CH_2OH$$

Sphingosine contains an unsaturated hydrocarbon chain and is converted to sphingomyelin by the addition of a long-chain fatty acid, a phosphate group, and a choline group. Overall, the structure of sphingomyelin is very similar to that of phosphatidylcholine (Fig. 2–17*c*), since it has two hydrophobic hydrocarbon chains and a hydrophilic phosphorylcholine group.

Steroids

Some of the most important lipids do not contain glycerol; these are the **steroids.** The basic structure of a steroid is four interlocking rings of carbon atoms (Fig. 2–18) and is quite different from that of other lipids. The male and female sex hormones and some hormones secreted by the adrenal glands, such as aldosterone, are steroids. The term **sterol** is often used when a hydroxyl group is present. **Cholesterol** is an example of a sterol that plays an important role in the structure of plasma membranes. Like the phospholipids, cholesterol has a polar hydrophilic region (the hydroxyl group) and a nonpolar hydrophobic region (the rest of the molecule).

Glycolipids

Glycolipids are lipids, also derived from sphingosine, that contain monosaccharide units. They have two hydrocarbon chains, as sphingomyelin does, but between one and seven monosaccharide units usually are present instead of the phosphorylcholine group. The monosaccharides make up a hydrophilic region of the molecule. Glycolipids occur in cell membranes.

Glycerol-3-phosphate

Phosphatidic acid

A phospholipid

(a)

Figure 2–17

Steps in the formation of a phospholipid. (*a*) Two different fatty acids, symbolized by RCOOH and R′COOH, join with glycerol-3-phosphate to form phosphatidic acid. R and R′ represent the long, hydrophobic hydrocarbon chains of the fatty acids. (*b*) Another specific group, either choline, ethanolamine, serine, or inositol, is added to phosphatidic acid to form a specific phospholipid. (*c*) An example of a specific phospholipid: phosphatidylcholine, formed by the addition of choline to phosphatidic acid. (*d*) In water, phospholipid molecules become oriented with their hydrophobic tails as far from the water as possible, forming either a sphere or a double layer.

(b)

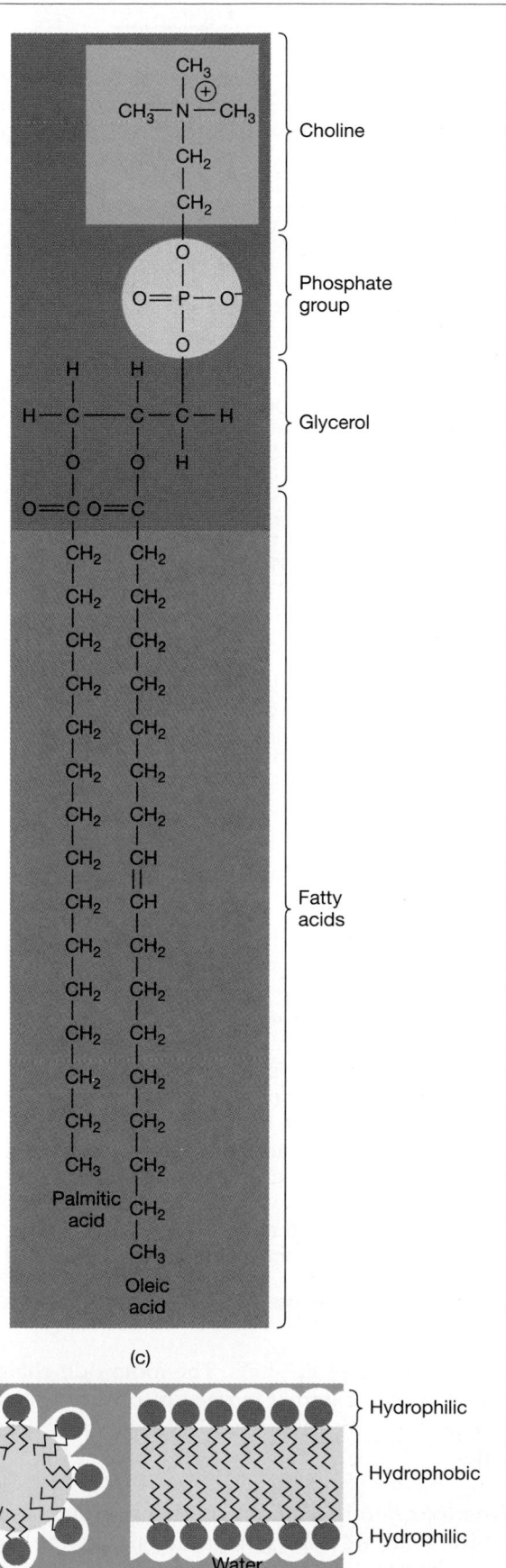

Choline

Phosphate group

Glycerol

Fatty acids

Palmitic acid

Oleic acid

(c)

Hydrophilic

Hydrophobic

Hydrophilic

Water   Water

(d)

Proteins Are Unbranched Chains of Amino Acids

Proteins are some of the largest macromolecules in the cell. The molecular mass of a protein is usually between 5000 and several million. Proteins carry out remarkably diverse functions in the cell. For example, enzymes, the molecules that catalyze specific biochemical reactions, are proteins. Carrier molecules that transport small molecules and ions are proteins. Many hormones are proteins, as are antibodies that combine with foreign bacteria. In addition, proteins are a major component of the muscles, allowing coordinated movement.

Two key aspects of protein molecules are important for an understanding of cell function: (1) the nature of the amino acids themselves, based on their side chain groups, and (2) the shape of the protein macromolecule, called its *conformation.*

Amino Acids

The basic units of protein molecules are **amino acids.** The general structure of an amino acid is illustrated in Figure 2–19. The "alpha" (α) carbon atom is asymmetrical, bonded to four different chemical groups. It is called the **alpha-carbon** to distinguish it from carbons that may be present in the **side chain (R) group.** Twenty different amino acids are available to make proteins, and the same 20 amino acids occur in all proteins whether they are made by bacterial, plant, or animal cells.

Amino acids are distinguished from one another by their side chain groups. These groups have important properties that determine the properties of the protein molecules they make up.

Proteins are formed from linear unbranched chains of amino acids. The amino acids are covalently linked to one another via **peptide bonds,** formed between the alpha-carboxyl (—COOH) group of one amino acid and the alpha-amino (—NH$_2$) group of an adjacent amino acid (Fig. 2–19). A **dipeptide** is a chain of two amino acids, a **tripeptide** a chain of three, and a **polypeptide** a chain of eight or more amino acids. Protein molecules typically are polypeptide chains of between 50 and 2500 amino acids. The exact number of amino acids and the sequence in which they combine are characteristic features of a specific protein that determine its properties and functions.

The amino acids are divided into three broad groups, based on the properties of their side chains at physiological pH (Fig. 2–20): amino acids with ionic polar side chains, those with nonionic polar side chains, and those with nonpolar side chains.

Figure 2–18

Basic structure of steroids, and examples: cholesterol and aldosterone are both formed from a basic structure of four interlocking rings.

Steroid structure

Simplified form

Cholesterol

Aldosterone

Figure 2–19

The general structure of an amino acid. A dipeptide is formed from the attachment of two amino acids, a tripeptide from three. Proteins are chains called polypeptides, made up of many amino acids. R, R', R", and R"' symbolize the specific side chain groups that distinguish the amino acids from each other.

Ionic Polar Side Chains.   Ionic side chains found in amino acids contain terminal —COOH or —NH₂ groups, which donate or accept, respectively, H⁺ ions in solution at pH 7.4 to become COO⁻ or —NH₃⁺. Side chains that donate H⁺ ions are, of course, acids (e.g., aspartic acid). Side chains that accept H⁺ ions are bases (e.g., lysine and histidine). Note that the —NH₂ and —COOH groups on the alpha carbon are involved in formation of peptide bonds and cannot

donate or accept H⁺ ions. The ionic side chains are found on the external surface of protein molecules, where they can interact with the polar water molecules.

Nonionic Polar Side Chains.   Some groups found in amino acids are polar but have no net charge at physiological pH. Examples are the —NH₂ group of glutamine, the —OH group of serine, and the —SH

Figure 2–20

Structures of some amino acid side chain groups. Ionic side chains such as those that produce the amino acids aspartic acid, lysine, and histidine are hydrophilic (water-soluble), as are nonionic polar side chains such as those that form cysteine, serine, and glutamine. Nonpolar groups such as those that form leucine, phenylalanine, and methionine are not water-soluble. (The —NH₂ and —COOH groups of the alpha-carbon are shown as if they were involved in the formation of peptide bonds.)

group of cysteine. These groups may be exposed on the protein surface, as are ionic side chains, or they may be buried within the protein molecule, where they are able to form hydrogen bonds. The side chain —SH groups of adjacent cysteines also interact to form a covalent bond known as a **disulfide bridge:**

Nonpolar Side Chains.   Some side chain groups are nonpolar and therefore hydrophobic. They are usually "hidden" from water by sequestration inside the protein structure where they cluster together via hydrophobic interactions. The side chains of leucine, methionine, and phenylalanine are good examples.

Protein Structure

There are different levels of organization in the structure of a protein molecule. These levels are important to understand because protein shape affects protein

functioning in cells. The sequence in which the amino acids are linked together determines the primary structure of a protein; the way in which the resulting chain bends or folds is the secondary structure. Further bending and folding create the tertiary structure, and, in some complex proteins, the physical relation of separate chains makes up the quaternary structure.

Primary Structure.   The chemical bonds that are important for maintaining the **primary structure,** or the sequence of amino acids, in a protein molecule are the covalent peptide bonds between the adjacent amino acids. By convention, the amino- or *N*-terminus is considered to be the beginning of the chain, and the carboxyl- or *C*-terminus is the end (see Fig. 2–19).

The hormone insulin was the first protein for which the primary structure was discovered. It consists of two separate polypeptide chains, one containing 21 amino acids and the other containing 30 amino acids. Two disulfide bridges hold the chains together, and a third disulfide bond joins two cysteines in the shorter chain (Fig. 2–21). Eighteen different amino acids are present in the molecule.

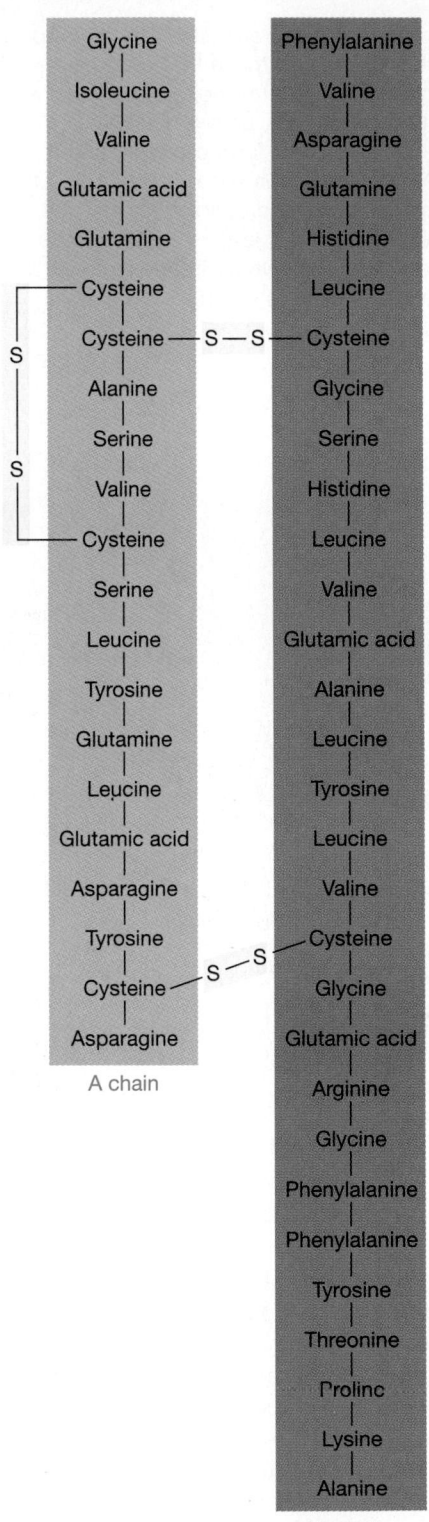

Figure 2–21

The primary structure of the hormone insulin, a protein (polypeptide), consisting of two chains of amino acids in specific sequence joined by two disulfide bridges.

Secondary Structure. A polypeptide chain is usually flexible and, under biological conditions, becomes folded into a specific shape or **conformation.** This conformation is called the **secondary structure** of the protein.

The folding occurs spontaneously because the side chain groups of the different amino acids react in a variety of ways with each other and with water. For example, hydrophobic interactions develop because nonpolar side chains tend to be pushed together in the interior of the molecule, where they can avoid contact with the aqueous environment. Hydrogen bonds develop when two peptide bonds come close together as the polypeptide folds (Fig. 2–22). Any polar side chains buried in the interior also become involved in hydrogen bond formation (Fig. 2–22). The ionic side chains and many of the polar side chains tend to be arranged near the surface of the molecule, where they can interact with the polar water molecules. Finally, covalent disulfide bridges between cysteine-SH groups frequently stabilize the folded conformation. These various interactions depend on the specific amino acids that are present and the sequence in which they are arranged in different proteins. They produce a unique secondary structure for each protein.

All of the specific conformations, however, can be broadly classed as either alpha-helixes or beta-sheets (Fig. 2–22c and d). An **alpha-helix** is a type of conformation that results when a single polypeptide chain coils about itself to form a uniform cylindrical structure. Each complete turn of the helix is occupied by 3.6 amino acids. This structure is stabilized mainly by hydrogen bonds between the atoms of the peptide bonds. A **beta-sheet** results when the polypeptide chain runs back and forth upon itself. It is stabilized, like the alpha-helix, by hydrogen bonds between atoms of peptide bonds in neighboring segments of the chain.

Alpha-helixes and beta-sheets are stable structures that are more rigid than extended polypeptide chains. Both of these types of conformation may occur within the same protein molecule, and they may be joined by unfolded segments that adopt neither conformation.

Tertiary Structure. The next level of protein structure, the **tertiary structure** (Fig. 2–22e), is formed by the folding of the alpha-helix upon itself or by stable associations of beta sheets and alpha-helixes so that a beta-sheet is covered on both sides by alpha-helixes. Ionic and polar side chains of amino acids tend to be on the outside of the structure, whereas nonpolar side chains tend to be buried on the inside. The tertiary structure is stabilized primarily by the hydrophobic

(text continues on page 58)

KEY:

- ● Carbon atom
- ◐ Nitrogen atom
- ◯ Oxygen atom
- ○ Hydrogen atom
- ◯ R group

Figure 2–22

(*a* and *b*) The hydrogen bonds formed within a folded polypeptide chain are important for stabilizing the secondary structure. Hydrogen bonds can form between —NH and —CO groups attached to the alpha carbon atom between specific side chain groups. (*c* and *d*) The secondary structure of a protein can be either an alpha-helix or a beta-sheet. (*e*) Tertiary structure of a hypothetical protein, maintained by several different types of bonds. (*f*) The quaternary structure of hemoglobin.

interactions between the nonpolar groups; hydrogen bonds (formed as in the secondary structure) and ionic bonds between charged side chain groups are also important. Proteins with tertiary structures are typically globular.

Quaternary Structure. Many large globular proteins consist of two or more separate polypeptide chains. The specific way in which the individual folded chains or subunits fit together to form the biologically active protein is known as the **quaternary structure. Hemoglobin,** the protein in red blood cells that binds and carries oxygen, is comprised of 574 amino acids arranged in four separate subunits (Fig. 2–22*f*). The subunits are held together by noncovalent interactions. One molecule of oxygen can bind reversibly to each of the four subunits.

The final conformation, or shape, of the intact protein depends on the interactions among the amino acid side chain groups. These in turn depend on the primary structure, or amino acid sequences, of the polypeptide chains. A change in just one amino acid in the polypeptide chain can alter the physiological function of the protein. The disease **sickle-cell anemia** results when a single amino acid is changed in two of the subunits of hemoglobin. The function of the altered hemoglobin is abnormal and seriously threatens the survival of the individual.

Nucleic Acids Are Long, Unbranched Chains of Nucleotides

A **nucleotide** is a chemical structure consisting of a pentose sugar, one of a set of special groups called bases, and one to three phosphate groups. Nu-

Figure 2–23

The structures of ribose (the pentose sugar that forms part of ribonucleic acid or RNA) and deoxyribose (the pentose sugar that forms part of deoxyribonucleic acid or DNA).

cleotides exist both as free molecules and as subunits of **nucleic acid** molecules. As free molecules, they are important in cell metabolism. As subunits of **ribonucleic acid (RNA)** and **deoxyribonucleic acid (DNA),** they are essential elements of genetic processes and heredity.

Components of Nucleotides

A nucleotide is composed of three smaller subunits: a sugar (either ribose or deoxyribose), a phosphate group or groups, and a specific chemical group called a **base.** The subunits are linked by covalent bonds.

Sugars. The nucleotides of ribonucleic acids contain the pentose **ribose;** the nucleotides of deoxyribonucleic acids contain the pentose **deoxyribose** (Fig. 2–23).

Bases. The bases that form a part of nucleotides are nitrogen-containing ring compounds derived from either **purine** or **pyrimidine** (Fig. 2–24). They are

PYRIMIDINE BASES

Pyrimidine   Thymine (T)   Cytosine (C)   Uracil (U)

PURINE BASES

Figure 2–24

Structures of the pyrimidine- and purine-derived bases that are present in nucleotides. The single-letter abbreviations for the bases are shown in parentheses.

Purine   Adenine (A)   Guanine (G)

Figure 2–25

Assembly of the units of a nucleotide. First, the pentose sugar ribose (or deoxyribose) attaches to a base such as adenine to form a nucleoside (in this case, adenosine). Next, a phosphate group (—PO₄) attaches to the nucleoside to form a nucleotide, adenosine monophosphate (AMP). The attachment of two more phosphate groups produces a different nucleotide, adenosine triphosphate (ATP).

called bases because they are proton acceptors. In DNA, the principal pyrimidine-derived bases are **thymine** and **cytosine;** in RNA, **uracil** and **cytosine.** The principal purine-derived bases in both DNA and RNA are **adenine** and **guanine.**

Phosphate Groups. A free nucleotide molecule can have either one, two, or three phosphate groups.

Formation of Nucleotides and Their Roles in Cell Metabolism

The basic structure of a nucleotide can be symbolized as follows:

Nucleotide = Base + Sugar + Phosphate

The base-plus-sugar group that makes up part of a

nucleotide is called a **nucleoside.** Thus the basic formula for a nucleotide can also be represented as:

$$\boxed{\text{Nucleotide}} = \boxed{\text{Nucleoside}} + \text{Phosphate}$$

The base adenine becomes a nucleoside—**adenosine**—through the formation of a covalent bond between one of its ring nitrogen atoms and carbon 1 of ribose (Fig. 2–25) (see p. 59). The nucleoside adenosine is converted to the nucleotide **adenosine monophosphate (AMP)** when a phosphate group is attached covalently to adenosine at carbon 5 of the ribose unit.

The addition of phosphate groups to adenosine monophosphate produces two new nucleotides: **adenosine diphosphate** or **ADP** (when one phosphate group is added to AMP) and **adenosine triphosphate** or **ATP** (when two phosphate groups are added to AMP) (Fig. 2–25).

Other bases besides adenine can form nucleosides and nucleotides. Some examples are shown in Figure 2–26.

Nucleotides are important subunits for nucleic acids, the genetic storage molecules, but free nucleotides have important roles in the cell energy processes as well. Adenosine triphosphate (ATP) is critical as the storage form of the chemical energy required for many of the biochemical reactions in the cell. GTP is required for protein synthesis, UTP for glycogen synthesis, and CTP for phospholipid synthesis.

In addition, some compounds derived from nucleotides fulfill important functions in the cell. One example is **cyclic AMP** (Fig. 2–26), which mediates the action of many hormones in cells. Adenine nucleotides also are components of many **coenzymes,** compounds required for many biochemical reactions. An important coenzyme is **nicotinamide adenine dinucleotide,** or **NAD** (Fig. 2–26).

Nucleic Acids

Nucleic acids are composed of monophosphate nucleotide units covalently linked in long unbranched chains. The different nucleotides are linked by covalent bonds between the sugar of one nucleotide and the phosphate of the next (Fig. 2–27a). The phosphate group of the first nucleotide is attached to the hydroxyl group on carbon 3 of the pentose sugar of the next nucleotide. Thus, each sugar in a nucleic acid has a base attached to carbon 1 and phosphate groups

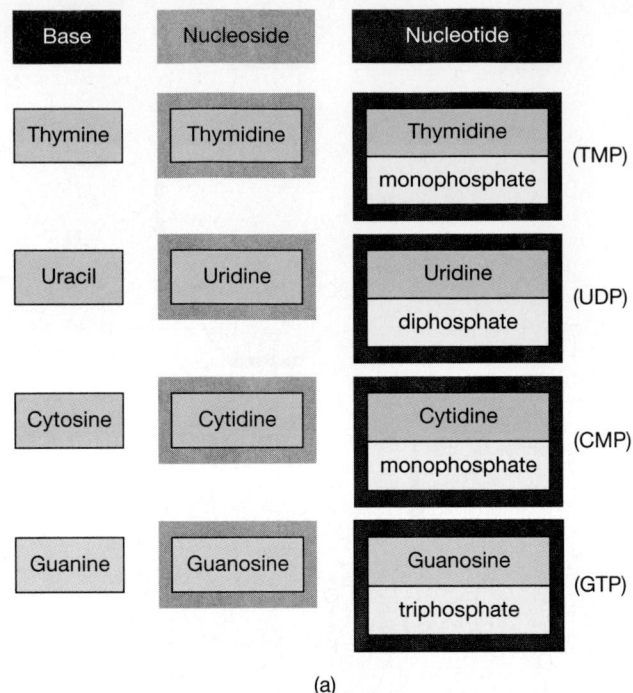

(a)

(b) Cyclic adenosine monophosphate (cyclic AMP)

(c) Nicotinamide adenine dinucleotide (NAD)

Figure 2–26

(*a*) Examples of some nucleosides and nucleotides and the structures of (*b*) cyclic AMP and (*c*) NAD, two important derivatives of adenine nucleotides.

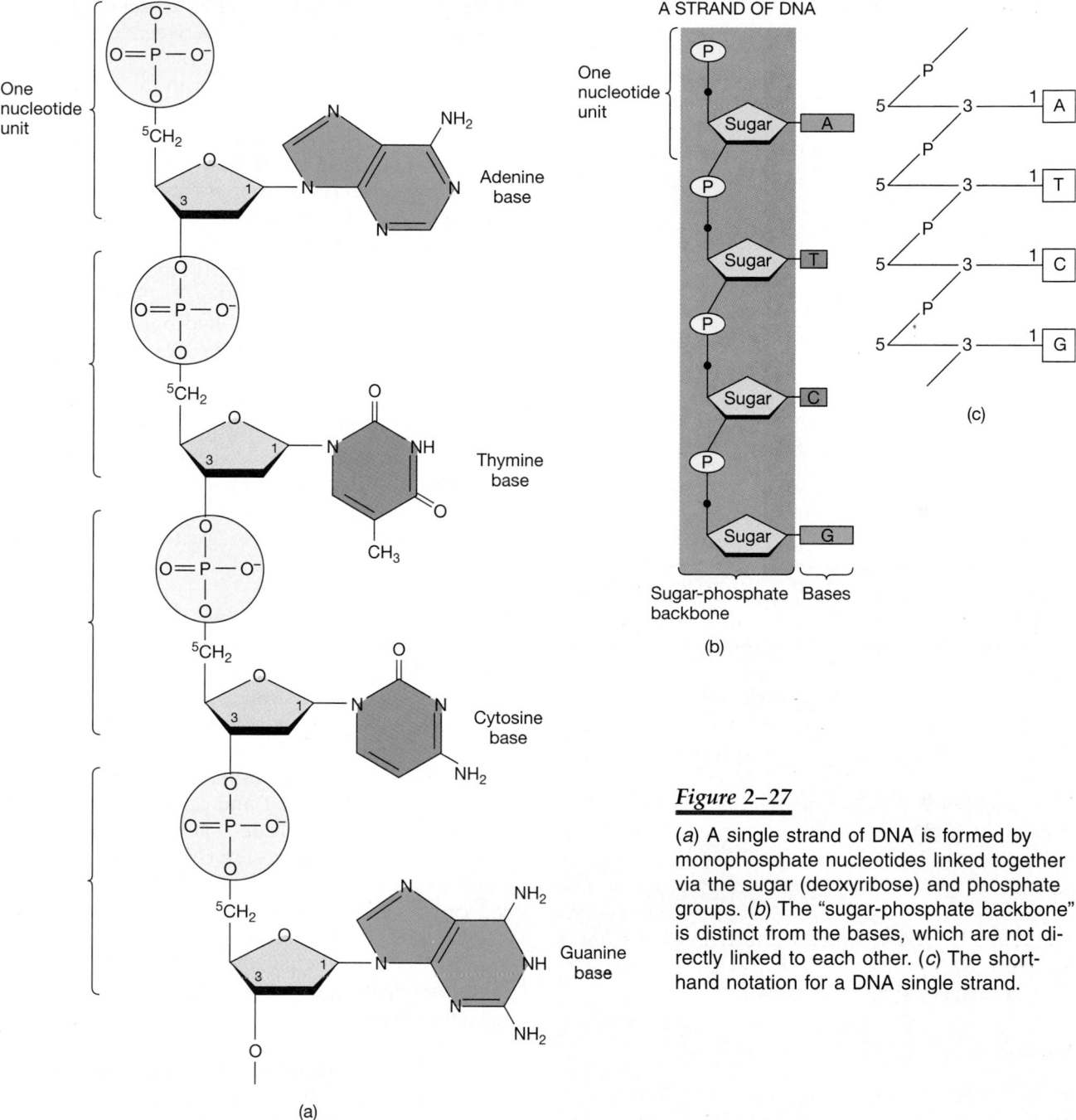

A STRAND OF DNA

One nucleotide unit

Sugar-phosphate backbone Bases

(b)

(c)

One nucleotide unit

Adenine base

Thymine base

Cytosine base

Guanine base

(a)

Figure 2–27

(*a*) A single strand of DNA is formed by monophosphate nucleotides linked together via the sugar (deoxyribose) and phosphate groups. (*b*) The "sugar-phosphate backbone" is distinct from the bases, which are not directly linked to each other. (*c*) The shorthand notation for a DNA single strand.

attached to carbons 3 and 5. The sugar-phosphate chain, analogous to the peptide bond in proteins, is a constant feature throughout any nucleic acid molecule and is known as the **sugar-phosphate backbone** of the nucleic acid (Fig. 2–27*b*). The variable part of the nucleic acid, analogous to the variable amino acid side chain groups in a protein, is the sequence of bases.

The precise sequences of the different bases in both DNA and RNA molecules represent the genetic information of the cell and of the organism as a whole. DNA and RNA interact in complex processes that transmit this genetic information to cellular systems responsible for the synthesis of all proteins involved in cell activities. The specific base sequences of DNA and RNA determine the specific amino acid sequences in the protein molecules, thereby dictating protein shape and function. These factors in turn affect the functioning of cells and the shape and characteristics of the organism as a whole.

The Structure of DNA

Certain aspects of the DNA molecule distinguish it from RNA and will become important in later chapters that discuss the replication of genetic material and cell reproduction. The DNA molecule differs from the RNA molecule in that it contains the sugar deoxyribose rather than ribose; it also contains thymine instead of uracil. In addition, DNA differs from RNA in that it consists of two separate nucleotide chains, or "strands," instead of one.

The Double Helix.   The two nucleotide strands in DNA are coiled around each other in a configuration called a **double helix** (Fig. 2–28). Each turn of the double helix contains 10 bases.

The two nucleotide strands that form the double helix are oriented so that their bases are inside the column of the helix, with the bases from one strand situated close to the bases from the other strand. As a result, hydrogen bonds form between a base in one strand and a base in the other. These hydrogen bonds help to hold the strands together in the helix.

Base-pairing.   A hydrogen bond will form only between a purine-derived base and a pyrimidine-derived base. Hence, these interactions, called **base-pairings,** are highly specific and are the most important aspect of the DNA double helix. Adenine (a purine) always forms two hydrogen bonds only with thymine (a pyrimidine) (Fig. 2–29*a*). Guanine (a purine) always forms three hydrogen bonds only with cytosine (a pyrimidine) (Fig. 2–29*b*).

Complementary Strands.   As a result of the base-pairing, the two DNA Strands in the double helix are not identical; instead, they are **complementary** (Fig. 2–29*c* and *d*). Adenine in one strand always lies opposite thymine in the other strand. Cytosine in one strand always lies opposite guanine in the other strand.

Antiparallel Strands.   The two nucleotide strands are **antiparallel** as well as complementary. This term refers to the arrangement of the sugar-phosphate backbone within each strand of DNA. An antiparallel arrangement is depicted in Figure 2–29*d*. Nucleoside *T* at the top of the left strand is linked to nucleoside *C* below it by a phosphate group attached to carbon 3 on the sugar of *T* and carbon 5 on the sugar of *C*. In contrast, nucleoside *A* at the top of the right strand is linked to nucleoside *G* below it by a phosphate-group attached to carbon 5 on the sugar of *A* and carbon 3 on the sugar of *G*. Both the complementary and the antiparallel properties of the DNA strands play an essential role in the mechanism that passes on identical copies of DNA to offspring during cell division.

Length of DNA Molecules.   A striking characteristic of DNA molecules is the length of the nucleotide chains. The bacterium *Escherichia coli*, for example, contains a single molecule of DNA that is 1.2 mm long. It contains 3.4 million nucleotides and has a molecular mass of 23×10^8. In a typical animal cell, the DNA double helixes are themselves further coiled into tight bundles. A single animal cell might contain 1 meter of DNA, with 3×10^9 nucleotides.

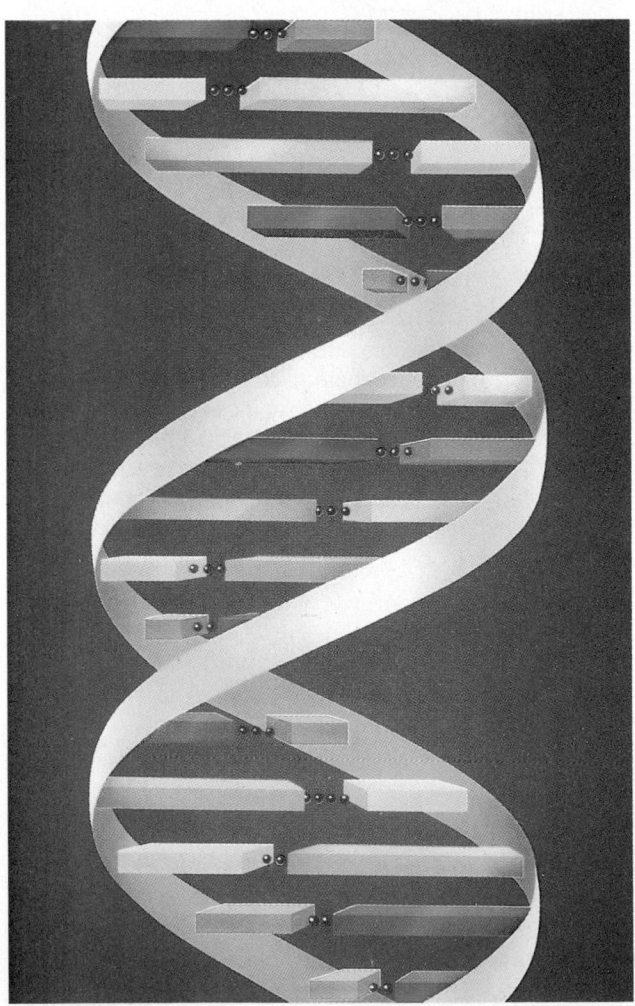

Figure 2–28

The double helix formed when two complementary strands of DNA coil around each other.

Thermodynamic Principles

Living organisms are highly organized, or **ordered,** systems. Maintaining this ordered state requires constant work. The capacity to do work is provided by

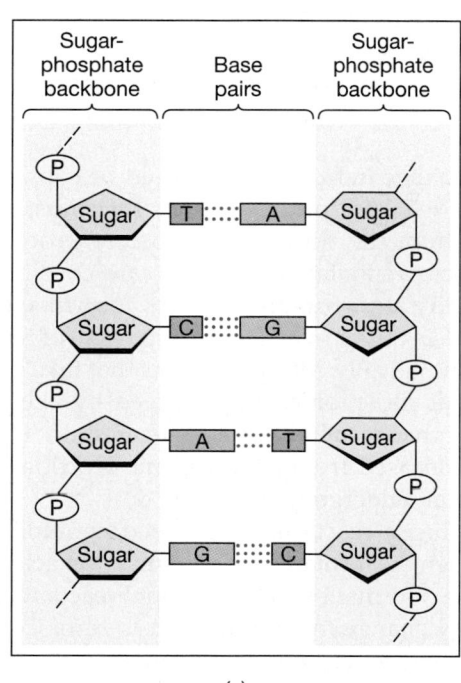

Thymine-adenine (TA) base pair

(a)

Cytosine-guanine (CG) base pair

(b)

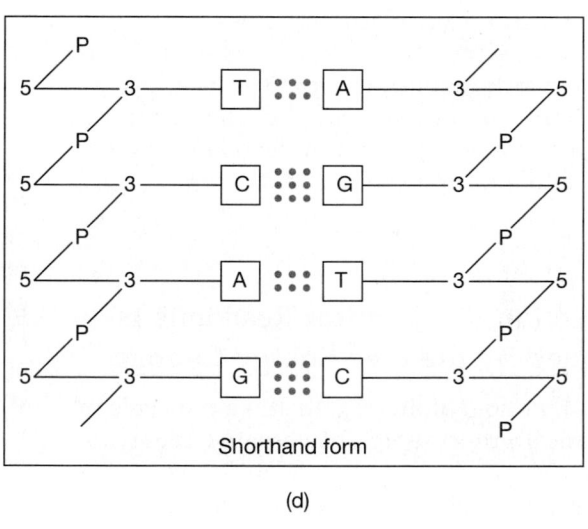

(c)

Shorthand form

(d)

Figure 2–29

Base-pairing between bases of separated DNA strands. (*a*) Thymine and adenine form one base pair. These two bases always form two hydrogen bonds between them. (*b*) Cytosine and guanine form the other base pair. These two bases always form three hydrogen bonds between them. (*c*) The two strands are complementary because specific bases on one strand form hydrogen bonds only with specific bases on the other strand. (*d*) The shorthand notation indicating complementary and antiparallel strands.

energy, which can exist in several different forms, such as radiant energy (e.g., light and heat), mechanical energy, chemical energy, and electrical energy. Various physiological processes use different energy forms. For example, some of the energy released during metabolism is heat that helps to maintain normal body temperature. Mechanical energy is involved in moving a part of the body, such as an arm, by the contraction of skeletal muscles. Chemi-

cal energy is released or absorbed during breakage or formation of chemical bonds in the molecules involved in metabolic reactions. Electrical energy results from the flow of charged particles such as ions and is essential for the transmission of nerve impulses. As elsewhere in nature, energy in the body can be converted from one form to another.

The unit used to measure energy is the **kilocalorie (kcal).** One kilocalorie is defined as the amount of heat that will raise the temperature of 1 kg of water by 1°C. Most forms of energy are readily converted to heat and measured in kilocalories.

The study of energy is known as **thermodynamics** (literally, "heat changes"). **The First Law of Thermodynamics** states that *energy can neither be created nor destroyed;* there is a constant supply of energy in the universe. **The Second Law of Thermodynamics** states that the *disorder* (also called *randomness* or *entropy*) *in the universe is always increasing.*

It is not difficult to appreciate that even simple tasks such as assembling a new bicycle require work to create the necessary order. The parts of a bicycle, when shaken from the package, do not fall to the ground in just the right position and sequence needed for spontaneous assembly of the bicycle. Assembly always requires some work.

Similarly, a living organism, which must impose order on its internal functions as well as on its surroundings, must do work to resist the universal tendency toward disorder. Much of this work takes place in the individual cells of the organism, in the form of reactions that build up macromolecules to support the structures and functions of the organism. This "biological" work requires energy; however, energy does not simply exist in the cell. The cell must transform the potential energy held in the bonds of other molecules into a form it can use to do work.

The Direction of Chemical Reactions Is Determined by the Free Energy Change

The bonds that hold atoms together in a molecule or other chemical group represent potential energy, or stored energy that can be released to do work. According to the laws of thermodynamics, however, not all of the stored energy can be used; some will always be lost because of the tendency toward disorder everywhere in the universe. The term **free energy** is used to describe the energy that can be put to work.

Free energy is given the symbol G and, as with all energy, is measured in units called kilocalories (kcal). While it is difficult to determine the free energy content of a particular molecule, it is relatively easy to calculate the *change* in free energy that occurs in the course of a chemical reaction. The change in free energy is symbolized as **ΔG** (the Δ symbol is the Greek letter "delta").

The concept of free energy is useful for predicting the direction in which a reaction will tend to proceed. Consider the following hypothetical reaction, in which the G values of the reactants and products are used only to illustrate the concept of **free energy change:**

$$A + B \longrightarrow C + D$$

Total G	Total G
100 kcal/mol	60 kcal/mol

$$\Delta G = -40 \text{ kcal/mol}$$

The total free energy of molecules C and D (expressed as kcal per mole) is less than the total free energy of molecules A and B. If the reaction were to proceed from **left** to **right,** then 40 kcal/mol of free energy would be released. By convention, when free energy is released in a reaction, the ΔG is said to be **negative.** The ΔG for the reaction shown here is -40 kcal/mol. This reaction is likely to occur spontaneously because the free energy that is released will increase the disorder of the surroundings, a natural tendency in the universe. Because the reaction releases free energy, it is said to be **exergonic.**

The reverse reaction would proceed by absorbing energy from the surroundings, because the reactants would have less energy than the products:

$$C + D \longrightarrow A + B$$

Total G	Total G
60 kcal/mol	100 kcal/mol

$$\Delta G = +40 \text{ kcal/mol}$$

The change in free energy would be considered **positive,** $+40$ kcal/mol. Reactions that must absorb energy from the surroundings before proceeding will not occur spontaneously; they are considered **energetically unfavorable** and are termed **endergonic.** Figure 2–30 compares these two types of reactions diagrammatically. Many reactions that take place in the cell are endergonic; they proceed by being coupled with other reactions that are exergonic. The free energy released from an exergonic reaction is used to drive an endergonic reaction.

When a reaction has reached equilibrium—that is, when the formation of products is exactly balanced by the reformation of the original reactants—the free energy change for the reaction is zero. Consider the following reaction at equilibrium:

$$AB \rightleftharpoons A + B$$

$$\Delta G = 0$$

The rate at which AB breaks down is balanced by the rate at which A and B recombine to form AB. The reaction can be "driven," or caused to proceed, to the "right" or toward the formation of products A and B, by the removal of product B as soon as it becomes available. This removal prevents the new product B from combining with product A. The removal of B upsets the equilibrium of the reaction, which will be reestablished by the further breakdown of AB. This continual attempt to maintain equilibrium by pro-

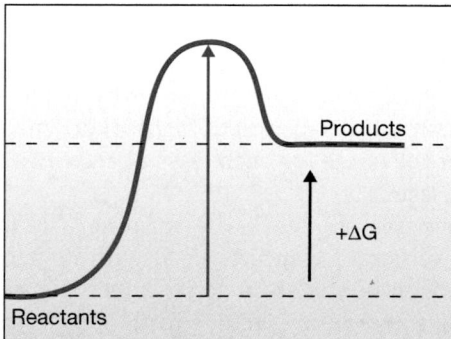

Figure 2–30

Diagrams comparing the free energy changes in endergonic and exergonic reactions. (*a*) In exergonic reactions the products have less free energy than the reactants. Energy has been released in the reaction, and the free energy change (ΔG) is said to be negative. (*b*) In endergonic reactions, the products have more free energy than the reactants, and the free energy change is said to be positive.

viding new product *B* causes the reaction to proceed until all of the *AB* is broken down. In effect, the constant removal of *B* keeps the free energy of the products (mostly *A*) low. Since the reaction tends to proceed spontaneously in the direction from high to low free energy, it proceeds in the direction of the products *A* and *B*. Many metabolic reactions take place with a very small free energy change. They can be driven in either direction easily when one of the reactants or one of the products is removed by involvement in a separate reaction. This provides a cell with the capacity to reverse a metabolic chain of reactions and to respond quickly to specific demands for either reactants or products of the metabolic pathway.

Enzymes Decrease the Activation Energy for a Reaction

Regardless of whether it is endergonic or exergonic, a reaction requires a certain amount of energy to be-

gin. This energy, which is the initial energy required to break the existing bonds and form new ones, is called the **activation energy.** This can be thought of as an energy barrier that reactant molecules must overcome before they can participate in the reaction. The disruption of existing bonds is illustrated in Figure 2–30 by the initial increase in free energy that occurs in both exergonic and endergonic reactions. In a chemical reaction, only those reactant molecules with sufficient internal energy will actually react. The lower the activation energy barrier of a reaction, the more likely it is to proceed because more reactant molecules will have enough energy.

One way to increase the rate of a reaction, as we have said, is to increase the temperature. In the human body, increasing the temperature sufficiently to enable every reaction to occur at a high enough rate would not be feasible. The body must maintain a constant (and relatively low) temperature, too low for many of the reactions necessary for life.

Another way to increase the rate of reaction is to ensure that the reactant molecules approach each other in the proper orientation for specific bonds to occur. But in the body, as elsewhere, molecules are in random, disorderly motion. How then does the body ensure that these reactions occur?

The solution lies in molecules called **enzymes,** which act as natural catalysts to increase the rate of chemical reactions in the body by lowering the activation energy levels for those reactions. They physically bind reactant molecules and hold them in place so that specific reactions can occur.

The Catalytic Role of Enzymes

A **catalyst** is a substance that increases the rate of a chemical reaction without actually participating in it and that is recovered unchanged when the reaction is completed. Catalysts do not influence the direction of a reaction, the final concentrations of the reactants and products, or the free energy change. They merely speed up a reaction that is already thermodynamically feasible.

The degree by which a catalyst can speed up a reaction is very impressive. It is not unusual for a reaction to occur one million times faster than normal when a catalyst is present. The increase is made possible in part because, in many reactions, a single enzyme molecule can catalyze the reaction of 10,000 reactant molecules every second. Such a number is called the **turnover number** for the catalyst. Typically, only a few catalyst molecules are needed to convert a relatively large number of reactant molecules into product.

Catalysts are widely used in industry to increase the rates of reactions and thereby the quantities of

Figure 2–31

A computerized model of an enzyme (protein) that is un-attached to its substrate (left) and one that is attached (right).   *(© Visuals Unlimited.)*

products that can be manufactured in a given amount of time. In the cell, natural catalysts called enzymes control the rates of many important reactions, speeding them up when necessary and slowing them down when necessary.

The Mechanics of Enzyme Action

Enzymes are proteins, and each enzyme has a characteristic three-dimensional shape and surface structure resulting from the folding of the polypeptide chain (Fig. 2–31). Enzymes facilitate the reactions of metabolism in cells by providing binding sites on their surfaces for the reactant molecules. These sites, which may be indentations on the surface of the enzyme, are known as **active sites.** The reactant molecules that bind to the active sites are known as the **substrates** of the enzyme.

We know of more than 1000 different enzymes. About 90% of the proteins present in a cell are enzymes. Given these large numbers, it is remarkable that a key aspect of enzyme action is **specificity.** A particular enzyme will catalyze only a very few

reactions, and sometimes only one reaction, because its active sites will accept and bind only certain reactants.

Although individual enzymes are "tailored" for specific substrate molecules, most enzymes can be classified according to the general types of reactions they catalyze. A **dehydrogenase,** for example, catalyzes the removal of hydrogen from a molecule (oxidation), whereas an **oxidase** catalyzes the addition of oxygen. A **transferase** mediates the transfer of groups of atoms, and an **isomerase** rearranges the atoms in a molecule. A **hydrolase** splits chemical bonds by adding the constituents of water (H and OH) to the separated atoms, and a **ligase** joins ("ligates" or "ties") molecules together by forming new bonds in reactions coupled to breakdown of ATP. Note that the name for each class of enzymes always ends in "-ase."

An enzyme (E) works by forming a temporary **enzyme-substrate complex** (E-S) with its substrates via hydrogen bonds at its active sites. This complex holds the substrate molecules in the proper orientation for the reaction to occur. When the reaction is complete, the E-S complex breaks apart to yield the reaction products (P) and the free, *unchanged* enzyme:

$$E + S \longrightarrow E - S \longrightarrow E + P$$

This equation is represented in diagram form in Figure 2–32, which shows the "lock-and-key model" of enzyme action. Evidence suggests, however, that not all active sites have the rigid conformation shown in this figure. In some cases, the active site may assume a shape complementary to the desired reaction only after the substrates are bound, a process known as **induced fit** (Fig. 2–33).

Formation of an E-S complex increases the frequency with which the reactant molecules collide in the proper orientation for the reaction to occur. This is one of the reasons that an enzyme accelerates the rate of a reaction. In the absence of enzymes, reactant

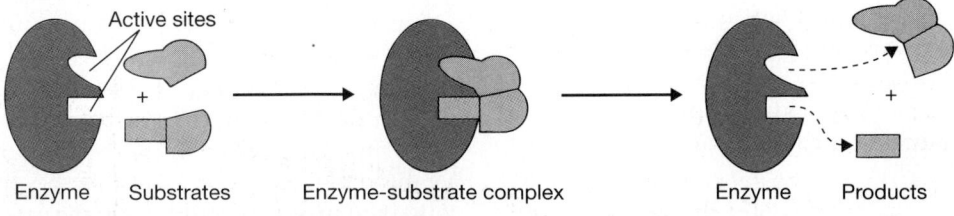

Enzyme   Substrates   Enzyme-substrate complex   Enzyme   Products

Figure 2–32

The "lock-and-key" model of enzyme action. The substrates fit tightly into the active sites on the enzyme molecule. When they have formed products, the enzyme releases them and is unchanged, free to catalyze another reaction.

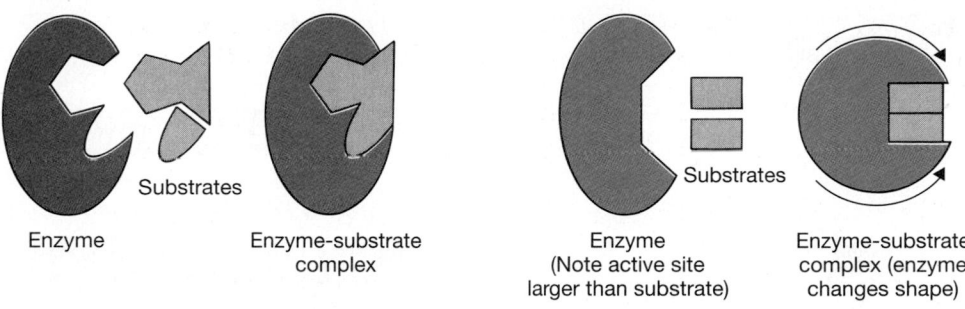

Substrates

Enzyme | Enzyme-substrate complex

Enzyme (Note active site larger than substrate) | Enzyme-substrate complex (enzyme changes shape)

(a) Lock-and-key model | (b) Induced fit model

Figure 2–33

The "lock-and-key" (*a*) model has been revised for some enzymes, in which "induced fit" (*b*) is a more likely scheme. According to this model, the active sites are somewhat larger than the substrates. When the substrates contact them, the active sites change shape to bind the substrates.

molecules do not bind specifically to each other; they collide randomly. Each reactant molecule can potentially react with any number of other molecules that are present. The particular reactant molecules that must come together for a specific reaction are not likely to meet spontaneously at a high rate. By binding only the reactant molecules specific to a desired reaction, an enzyme "favors" that reaction over the other possible reactions in which the molecules could participate.

To further encourage reactions, some enzymes bind a substrate molecule in a manner that distorts the structure of the substrate. This may place a strain on the existing chemical bond—helping to break it—or it may facilitate a chemical reaction between groups that are not normally close enough together to react. These effects contribute to a higher reaction rate.

From a thermodynamic standpoint, enzymes in effect decrease the activation energy necessary for the reaction to occur—that is, the energy needed to break old bonds and form new ones. More reactant molecules have the required energy to overcome this activation energy barrier and can therefore react to form products. Again, enzymes act specifically, decreasing activation energy only for the desired reaction. In the cell, products of reactions are often intermediate products in long chains of reaction. At each step, a product must be directed down one of several alternative pathways. Because an enzyme decreases the activation energy only for the specific reaction that it catalyzes (Fig. 2–34) (see p. 68), the reactant molecules are more likely to attain this level of activation energy and undergo the reaction catalyzed by the enzyme. They are less likely to undergo other possible reactions that require a higher activation energy.

Enzyme Kinetics

Researchers can study reaction rates and other characteristics of enzyme-catalyzed reactions using enzymes that have been isolated from the cell and purified. These in vitro (outside the body) studies of enzyme activity allow comparisons of different enzymes under the same conditions.

The relation between reaction rate and substrate concentration, at a constant enzyme concentration, is a particularly useful measure and is expressed graphically in Figure 2–35 (see p. 68). Initially, the reaction rate increases rapidly as the concentration of substrate increases because at low substrate levels the enzyme is not working at full capacity. Any substrate molecules present are rapidly converted to reaction products. Eventually a maximum reaction rate (V_{max}) is reached when the substrate concentration is high enough. Further increases in substrate concentration produce no change in the reaction rate. A V_{max} is achieved because the enzyme molecule has a limited number of active sites. When substrate molecules occupy all of the sites on the enzyme, it is said to be "saturated" and is working at maximum capacity. The addition of more substrate cannot increase the rate of reaction. The reaction rate in the absence of the enzyme is shown for comparison. It is much slower and increases linearly without exhibiting a V_{max} over the same range of substrate concentration (Fig. 2–35).

Knowledge of the V_{max}, the maximum amount of substrate that can be utilized and turned into reaction product in a given time, allows calculation of the turnover number of the enzyme from the relationship:

$$V_{max} = k[E]$$

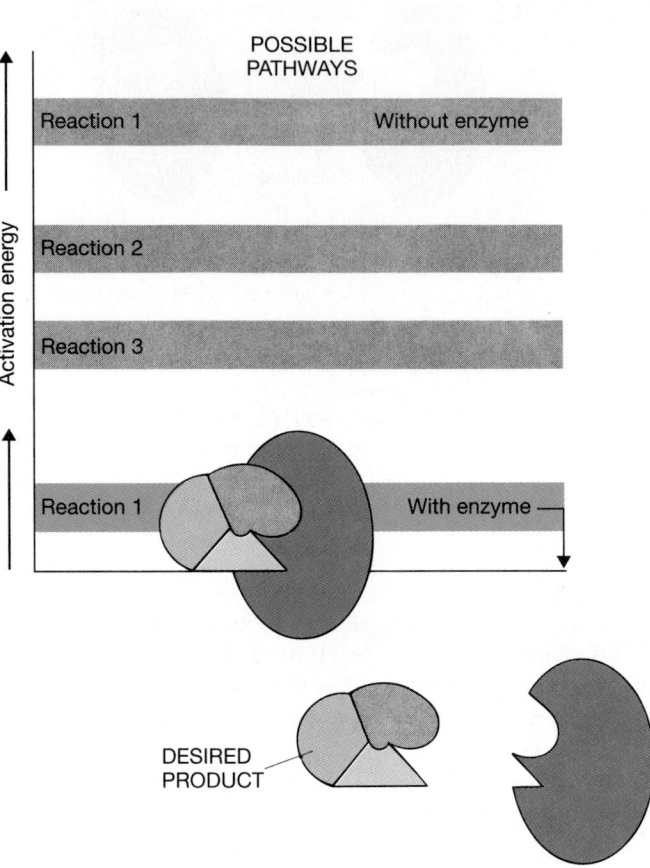

Figure 2–34

Most products of chemical reactions in the body are intermediate products in complex metabolic pathways. Enzymes direct such intermediates down a specific reaction pathway by lowering the activation energy for only the desired reaction. The intermediate product is more likely to participate in a reaction with a lower activation energy than a higher one.

in which k is the turnover number and $[E]$ is the enzyme concentration. For example, for a reaction catalyzed by the enzyme carbonic anhydrase:

$$\boxed{CO_2} + \boxed{H_2O} \xrightarrow{\text{Carbonic anhydrase}} \boxed{H_2CO_3}$$

k can be calculated as follows: When carbonic anhydrase is fully saturated with substrate, just 10^{-6} mol $[E]$ of this enzyme will catalyze the formation of H_2CO_3 at a rate of 0.6 mol/sec (V_{max}). Hence, the turnover number k is 0.6 divided by 10^{-6}, which is 600,000/sec. This is one of the largest known turnover numbers for any enzyme. In contrast, the turnover number for DNA polymerase is only 15/sec, which is very small. Note that the V_{max} and the turnover number are determined by laboratory experiments and provide a measure of the maximum activity of an enzyme. However, most enzymes in the cell are not saturated by substrate and do not operate at their V_{max}. The advantage this provides to the cell is the capacity to deal with fluctuations in the concentrations of its molecules. Thus, when the demand for glucose catabolism is increased rapidly by physical exercise, the enzymes involved have the capacity to

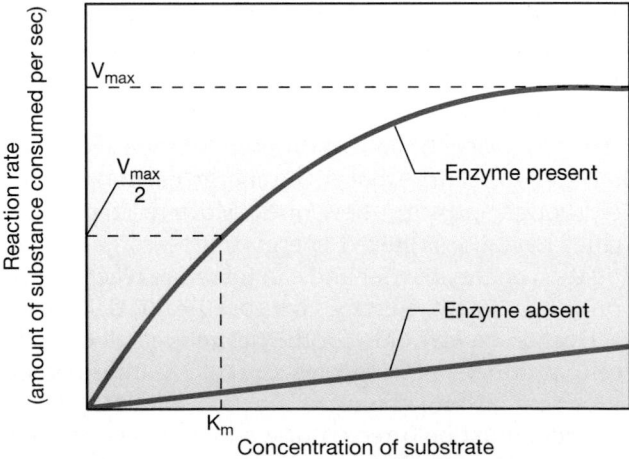

Figure 2–35

When an enzyme is present, a reaction proceeds initially at very high rate, reaching a maximum rate (V_{max}) when the enzyme is saturated with substrate. The Michaelis constant (K_m) is the concentration of substrate that will produce one half of the maximum reaction rate.

respond by processing more substrate molecules in the same amount of time.

Another useful term that can be obtained from the plot in Figure 2–35 is the **Michaelis constant,** K_m, which is defined as the substrate concentration that produces one half of the maximum reaction rate. In the appropriate circumstances, the K_m provides a measure of the affinity of the enzyme for a particular substrate. A high K_m indicates a low affinity, and a low K_m indicates a high affinity. Thus, measurement of V_{max} and K_m values allows quantitative comparisons of the activities and specificities of different enzymes for various substrates. This kind of analysis has greatly aided scientists' understanding of the reactions of cellular metabolism.

The Effects of pH and Temperature

In addition to concentration of substrate, enzyme activity depends on both the temperature and pH of the medium (surroundings). An increase in temperature will increase the rate of most enzyme-catalyzed reactions until temperatures in the range of 50°C to 60°C are reached. At this point, the temperature is high enough to **denature,** or break down the tertiary structure of, the enzyme protein. When this occurs, the enzyme rapidly loses its catalytic activity.

The pH range over which most enzymes can function is relatively narrow but varies from one enzyme to another. **Pepsin,** for example, which breaks down proteins in the stomach, works best at a pH of 2, the **pH optimum** for this enzyme. In contrast, **trypsin,** secreted by the pancreas, is most effective in breaking down proteins at a pH of 8, and **alkaline phosphatase** removes phosphate groups from substrates most effectively when the pH is close to 10. Changes in pH result in addition or removal of hydrogen ions (H^+) from the enzyme, thereby changing the number of positive and negative charges present at the active site and altering the catalytic efficiency of the enzyme.

Enzyme Cofactors: Coenzymes

A **cofactor** is a nonprotein component that some enzyme molecules need to function as catalysts. For example, several enzymes involved in glucose catabolism require a magnesium ion for their activity. Other enzymes use iron, copper, or zinc ions. In some instances, the metal ion actually carries out the catalytic reaction, although the reaction is facilitated by the presence of the enzyme protein. In other cases, the metal ion serves to maintain the enzyme molecule in the structure needed for biological activity. A change in pH can alter the binding of a cofactor and result in a change in the catalytic activity of the enzyme. It

is difficult to remove metal cofactors that are tightly bound to an enzyme without disturbing the protein structure and denaturing the enzyme.

An enzyme cofactor that is an organic molecule is referred to as a **coenzyme** (Fig. 2–36). These types of cofactors can serve as donors or acceptors of the atoms or functional groups added to or removed from the substrate.

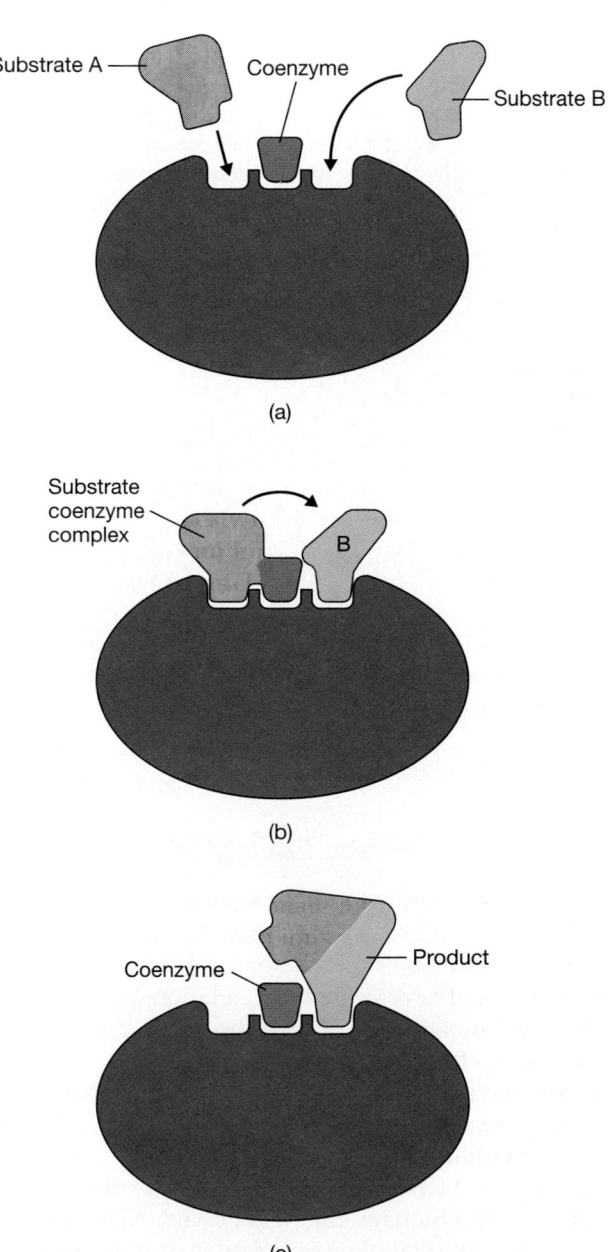

(a)

(b)

(c)

Figure 2–36

Coenzymes are nonprotein molecules that aid enzymes in the attachment of substrates.

CLINICAL FOCUS

Black Death

One quarter of the population of England was killed by **bubonic plague** in 1348. A characteristic feature of this disease is the **bubo,** a painful swelling due to inflammation of lymph glands. The affected glands are located principally in the groin, armpit, head, and neck. The disease came to be known as the **Black Death,** probably because bleeding from small blood vessels in the skin produced discoloration. Inadequate oxygenation of the blood due to associated pneumonia also contributed to the bluish-black appearance of the skin. Bubonic plague is a disease of rodents, and rats are the main source of human infection. The disease, transmitted from rats to humans by fleas, is caused by the bacterium *Yersinia pestis.* The recent discovery that this bacterium contains an enzyme very similar to one normally found in mammalian cells may help to explain why the infection progressed so rapidly and had such lethal consequences for humans in the Middle Ages.

Addition of phosphate groups **(phosphorylation)** to tyrosines on intracellular proteins may be a critical step in regulating the growth and development of cells. For example, the action of several hormones, including insulin, is accompanied by tyrosine phosphorylation on certain proteins. Specific enzymes that add phosphate groups to the tyrosine amino acids of proteins are known as **protein tyrosine kinases.** The phosphate groups can be removed by another class of enzymes known as **protein tyrosine phosphatases.** Thus, the balance between these two groups of enzymes with their opposing ac-

tions may play a critical role in the control of important events in cell physiology.

The bacterium *Yersinia* was found to contain mammalian protein tyrosine phosphatases. This finding was unexpected because the bacterium contains no tyrosine phosphate groups, the normal substrates for these enzymes. It has been speculated that abnormally high levels of protein tyrosine phosphatases in mammalian cells could disrupt normal cell functions. When *Yersinia* invades a human cell and multiplies, it leads to a rapid increase in protein tyrosine phosphatase activity inside the cell. An uncontrolled increase in this enzyme activity results in the sudden removal of phosphate groups from a wide variety of intracellular proteins. This disruption of normal intracellular controls may contribute to the rapid deterioration and death of the infected person.

Fortunately, the development of modern antibiotics (see CLINICAL FOCUS in Chapter 5) has provided an effective means of treating *Yersinia* infections. Bubonic plague has been virtually eliminated from the western world, but outbreaks still occur in parts of South Amrica and Asia. A variant, known as pneumonic plague, appeared in India as recently as 1994.

■ **How does the action of a phosphatase enzyme differ from a dehydrogenase?**
Why is the catalytic activity of an enzyme dependent on pH?

ATP Acts As a Free Energy Store

The intricate processes by which cells transform chemical energy to do work will be examined in detail in Chapter 6. A general overview at this point, however, will reveal the basic relationship—and an important molecule—linking the endergonic and exergonic reactions in the cell.

We have learned that the macromolecule ATP (adenosine triphosphate) is a nucleotide consisting of three phosphate (P_i) groups, a ribose ring, and adenine. When a phosphate group is removed, ATP becomes ADP; when another is removed, ADP becomes AMP. The removal of one phosphate group from ATP releases a large amount of free energy; the attachment of one phosphate group absorbs a large amount of free energy. For these reasons, ATP acts as a store of large amounts of free energy. The bonds among its

phosphate groups are often represented by wavy lines in the ATP structural formula (see Fig. 2–25) and are known as "high-energy bonds."

Exergonic reactions that occur in the cell produce free energy that is used to **phosphorylate** (add a phosphate group to) AMP and ADP:

$$AMP + P_i + Energy \rightarrow ADP$$

$$ADP + Pi + Energy \rightarrow ATP$$

These reactions represent a storage process in which the cell temporarily deposits free energy released in exergonic reactions, such as the breakdown of nutrient molecules, in molecules of ATP (Fig. 2–37) through phosphorylation.

This energy is not stored for long, however, for other reactions that need large amounts of free energy to begin are also occurring: these endergonic reactions are mainly those of biosynthesis. They obtain the free energy they need from the ATP produced by the exergonic reactions:

$$ATP \rightarrow Energy + ADP + P_i$$

leaving ADP and a phosphate group to be used in the next round of energy release and storage.

Figure 2–37

Both endergonic and exergonic reactions occur in the cell. Endergonic reactions obtain the free energy they need to proceed from the free energy spontaneously released from exergonic reactions. This free energy is exchanged in the form of ATP.

CHAPTER REVIEW

Summary

- Matter is composed of elements, the smallest units of which are atoms. Atoms combine to form molecules, and elements combine to form groups and compounds.
- Atoms of the same element have the same number of protons. Uncharged atoms have equal numbers of protons and electrons. Charged atoms (ions) have unequal numbers of protons and electrons.
- Atoms share, gain, or lose electrons to form bonds with other atoms. Bond formation between atoms forms molecules, groups, and compounds.
- Chemical reactions occur when compounds react with other compounds to form new products. The rate of a chemical reaction depends on the temperature, and on the concentration and orientation of reactant molecules.
- The polarity of the water molecule makes water an ex-

cellent solvent for ionic and polar covalent compounds. Nonpolar compounds, in contrast, are insoluble in water.
- In a solution of gases (such as air), the total pressure of the solution is equal to the sum of the partial pressures of the constituent gases.
- The unique covalent bonding capabilities of the carbon atom allow it to form molecules of huge proportions and intricately varied structures. These macromolecules include carbohydrates, lipids, proteins, and nucleic acids. They are essential for many cell processes.
- Enzymes increase the rates of specific reactions by bringing reactant molecules together in the proper orientation. Endergonic reactions in the cell are driven by the free energy released by exergonic reactions.

Review Questions

Identify Each with a Term

1. A reaction that releases free energy.
2. A macromolecule that contains adenine, guanine, thymine, and cytosine.
3. The maximum amount of substrate that can be turned into product in a given time during an enzyme-catalyzed reaction.
4. A macromolecule that contains both protein and carbohydrate.

Define Each Term

5. Molecular formula
6. Atomic number
7. Covalent bond
8. Solution
9. Hexose

Choose the Correct Answer

10. Adenine forms a base-pair with:
    a. thymine
    b. guanine
    c. cytosine
    d. adenine
11. Attachment of serine to phosphatidic acid forms a:
    a. dipeptide
    b. glycolipid
    c. phospholipid
    d. nucleoside
12. A dehydrogenase enzyme catalyzes:
    a. joining of molecules by forming new bonds
    b. addition of oxygen to a molecule
    c. rearrangement of atoms in a molecule
    d. removal of hydrogen from a molecule
13. Nucleic acids are long chains of:
    a. amino acids
    b. nucleosides
    c. fatty acids
    d. nucleotide monophosphates
14. Adenosine triphosphate (ATP):
    a. contains guanine
    b. is a nucleoside
    c. acts as an intracellular store of free energy
    d. is an enzyme that adds phosphate groups to substrates

Calculate

15. What is the molecular mass of water?
16. What is the concentration of hydrogen ions in a solution at pH 5.0?

Answer Each Question in One or Two Sentences

17. How is a peptide bond formed between two amino acids?
18. What are the bonds that help to stabilize the secondary structure of a protein?
19. What does it mean when the nucleotide strands in DNA are described as complementary?
20. Why do solid crystals of potassium chloride dissolve easily in water?
21. How does an enzyme increase the rate of a reaction?

Problem Solving in Physiology

22. Explain the relationships between atoms, elements, molecules, and compounds.
23. Why are hydrogen bonds vital for the maintenance of life?
24. Why can an enzyme increase the rate of one chemical reaction without affecting other chemical reactions in the cell?

Suggested Readings

Branden, C., and Tooze, J. *Introduction to Protein Structure.* New York: Garland, 1991.

Knowles, J.R. "Enzyme catalysis: Not different, just better." *Nature,* 350:121–124, 1991.

Loewy, A.G., Siekevitz, P., Menninger, J.R., and Gallant, J.A.N. *Cell Structure & Function,* 3rd ed. Philadelphia: Saunders, 1991.

Radmacher, M., Tillmann, R.W., Fritz, M., and Gaub, H.E. "From molecules to cells: Imaging soft samples with the atomic force microscope." *Science,* 257:1900–1905, 1992.

Wolfe, S.L. *Molecular and Cellular Biology.* Belmont, CA: Wadsworth, 1993.

Key Terms

acid (p. 45)
activation energy (p. 65)
amino acid (p. 54)
atom (p. 31)
base (p. 45)
carbohydrate (p. 46)

chemical reaction (p. 40)
compound (p. 36)
covalent bond (p. 36)
electron (p. 34)
element (p. 31)
enzyme (p. 65)

equilibrium (p. 40)
fatty acid (p. 49)
ionic bond (p. 38)
kinetic energy (p. 32)
lipid (p. 49)
macromolecule (p. 46)

nucleic acid (p. 58)
oxidation-reduction (redox) reaction (p. 41)
protein (p. 53)
solution (p. 41)

False-color transmission electron micrograph of a section through a single dictyosome forming part of the Golgi apparatus in a mammalian cell. The dictysome is the stack of paired membranes (green with red outlines) curving across the center of the image (original ×31,800). *(© Dr. Gopal Murti/ SPL/Photo Researchers, Inc.)*

Chapter 3

The Structure and Functions of Cells

Key Concepts

DNA is the genetic material of the eukaryotic cell. It contains the instructions that the cell needs in order to perform its function in the body and to reproduce itself. Eukaryotic cells contain organelles that support a chain of chemical reactions by which the cell can obtain energy from food and use it to support processes such as growth and reproduction.

The plasma membrane separates the cell interior from the extracellular fluid so that the distinctive composition of the cell cytosol can be maintained. Intracellular membranes divide the interior of the eukaryotic cell into specific compartments to allow specialized (and separated) activities, such as the nucleus where DNA and RNA are made and where DNA is stored; the rough endoplasmic reticulum (ER) which performs lipid synthesis and protein synthesis; the Golgi complex which modifies the polysaccharide chains added to proteins in the rough ER; and the mitochondria where most of the supply of ATP for the cell is produced.

The cell cycle is the time from the start of one cell division to the start of the next. There are two main phases, interphase and cell division. About 90% of the cycle is spent in interphase. Cell division consists of mitosis and cytokinesis. Entry into different phases of the cell cycle, such as mitosis, is under careful control. Cancer cells do not respond to normal controls of cell growth and division.

A large number of different techniques have been used to study cell structure and function. These include microscopy, utilization of specific antibodies, cell fractionation, and genetic engineering.

Chapter Outline

Overview: Cell Processes and Products
Cell Compartments and Their Functions
The Life Cycle of the Cell
Research Techniques in Cell Physiology

Charles was a busy and successful architect. Some of his associates satisfied their creative urges by the continuous addition of rooms and decks to their houses and by molding their gardens into ornate landscapes. In contrast, Charles devoted much of his spare time to breeding the giant silkmoth **Antheraea polyphemus.** *He had been fascinated by entomology since his high school days, when he had collected and identified more than 100 insects for a science project. Now Charles had the time and resources to breed the silkmoth, which permitted his close observation of the marked changes that occurred during the various stages of the life cycle from egg, to caterpillar, to pupa and, finally, to the exotic adult moth. He knew that the transition from one growth stage to another involved, paradoxically, massive cell death. For instance, most of the muscles die soon after the caterpillar becomes a pupa; only a few are retained to allow the moth to shed the pupal shell. The nerve cells that control the dead muscles are not needed and they also die. What Charles did not know was that the same cells die in every caterpillar, suggesting that signals that drive cells toward death are present in, or are transmitted to, specific cells.*

This organized process of cell death, known as **apoptosis,** *is widespread. It occurs in humans, and other animals, in a variety of physiological states. In the development of the human embryo, apoptotic cell death is important for elimination of unwanted or excess cells. Cell death in the visual system eliminates nerve cells that have formed improper connections. Another important role for apoptosis is to eliminate cells that may be harmful. Certain immature cells in the immune system carry the potential for attacking the tissues. Fortunately, these immature cells are induced to undergo cell death before they can develop further. Apoptosis also functions as a homeostatic mechanism for maintaining the correct number of cells within each tissue. Cells that have already fulfilled their function will die as new cells are generated to replace them. Like the spy who carries a vial of poison in case of capture, cells contain a program for suicide that is set in motion when the cell receives the signal to self-destruct.*

The human body contains about 10^{13} cells, ranging in size from red blood cells (7.5 μm diameter) to huge cells of the nervous system, which may be up to 1 m in length. The combined activity of these cells makes possible the process of life and death as we experience it. Any thorough knowledge of the human life cycle and human physiology requires an understanding of cell structure and organization. This forms the basis for understanding the organization and control systems of tissues, organs, and, ultimately, the intact human.

According to the **cell theory,** which forms the basis of all modern biology and physiology, (1) all organisms are made up of cells and their products; (2) new cells arise only from preexisting cells; (3) all cells have the same fundamental chemical makeup and metabolic processes; and (4) the activities and processes of the organism as a whole result from the interdependent and cooperative workings of groups of cells.

In all of nature there are two broad classes of cells: eukaryotic cells and prokaryotic cells. The primary distinction between these cell types is the presence or absence of a membrane-bounded compartment surrounding the genetic material, the DNA. In **prokaryotic cells,** the DNA is not in a separate compartment from the rest of the cell, whereas all **eukaryotic cells** have a well-defined **nucleus,** or central area, contained within its own nuclear membrane.

Organisms in turn are broadly classified according to whether they have prokaryotic or eukaryotic cells. Bacteria are prokaryotic organisms, for example, while all mammals are eukaryotic organisms. Eukaryotic cells are generally larger (10 to 100 μm in diameter) than prokaryotic cells (1 to 10 μm).

In addition to a membrane-enclosed nucleus, all eukaryotic cells contain many small subcellular structures, most of which are also enclosed in a membrane. These structures are grouped under the general term **organelles.** They are distributed throughout the **cytosol,** the fluid surrounding the nucleus. Together, the organelles and the cytosol make up the **cytoplasm.** The cytoplasm and the nucleus make up the two major compartments of the cell (Fig. 3–1). They are enclosed by a cell membrane called the **plasma membrane.**

Overview: Cell Processes and Products

A **cell** is the smallest distinct unit of life capable of an independent existence. Three features of cells permit independent life: (1) a set of genes that contain the information the cell needs both to direct its own activities (primarily the synthesis of macromolecules) and to pass on to new cells; (2) the machinery for obtaining energy from foodstuffs to sustain growth, reproduction, and movement; and (3) a cell membrane that provides a physical boundary between the cell and its environment.

Cell processes can be thought of in two broad categories: (1) the **synthesis,** or creation from precursors, of macromolecules both for use by the cell itself and for export to other cells; and (2) **degradation,** or breakdown, of other macromolecules for the purpose of obtaining energy. These processes together make up the cell's **metabolism.**

Figure 3–1

The cell has two major compartments: the nucleus and the cytoplasm. The cytoplasm contains the major cell organelles and a fluid called cytosol.

Just as each organ of the body has a unique structure that permits it to perform a special function in a system of the body, each organelle of the cell has a unique structure and function that enables it to participate in the life processes of the cell and allows the cell, in turn, to function in the body. The most important aspect of each cell's role in the body is the type of proteins it synthesizes.

Every cell in the body can potentially make any protein in the body. However, certain cell types make only certain proteins. For example, some cells in the pancreas have as their chief function the production of insulin for use by other cells. Insulin is a type of protein called a hormone. In order to produce insulin, the cell needs instructions. To ensure that future generations of pancreatic cells know how to produce insulin, the cell must preserve those instructions by copying them.

Even if they do not secrete proteins for use by other cells, all cells need certain proteins for their own internal functions. Many of these proteins are **enzymes.** As described in Chapter 2, they assist many important chemical reactions in the cell, controlling both synthetic and degradative processes.

The instructions that tell each cell how to make the specific proteins it needs are contained in the DNA and RNA in the nucleus. The machinery for assembling the proteins is found outside the nucleus, however. When the cell needs to make a protein, the instructions are brought via RNA from the nucleus to the organelles that make the proteins. When the cell needs to divide, the instructions are duplicated in the nucleus so that each new nucleus will have a copy.

Following is a general summary of the major cell processes and products, and the organelles that carry them out. A typical, generalized animal cell with all its organelles is depicted in Figure 3–2*a*.

Instructions for Controlling Cell Functions Are in the Nucleus

Genetic material, as mentioned previously, is necessary for the independent existence of most cells. Contained in the macromolecules of DNA, genetic material is essentially a set of instructions that the cell uses (1) to make the specific proteins and other macromolecules it needs to carry out its own designated function in the body; and (2) to reproduce itself. A set of related organelles in the cell work together to "express" the information contained in the DNA by synthesizing macromolecules according to its instructions.

The **nucleus** is the site where **DNA** is made, stored, and duplicated. This DNA is used not only in cell division but throughout the life of the cell. Recall from Chapter 2 that the DNA molecule is a double strand of base-paired nucleotides. A **gene** is a portion of a DNA molecule containing a specific sequence of nucleotides that serve as a code dictating the sequence of amino acids to be followed when a particular polypeptide is assembled in the cytoplasm.

Directing Synthesis

The nucleus is also the site of **RNA** manufacture. RNA carries genetic information (in the form of instructions about how to make proteins) from the

Plasma
membrane

Cytoplasm

Glycogen
granules

Vacuole

Endocytotic
vacuole
forming

Endocytotic
vacuole

Rough endoplasmic
reticulum with
ribosomes

Nuclear
envelope

Nucleus — Chromatin

Nucleolus

Mitochondrion

Ribosomes

Cilia

Lipid droplets

Transport
vesicles

Golgi
complex

Lysosome

Centrioles

Microtubules

Smooth endoplasmic
reticulum

(a)

Figure 3–2

(*a*) A generalized view of a mammalian cell showing organelles common to all cells (such as the Golgi complex) as well as specialized structures (e.g., cilia) found only in some cells. (*b*) An electron micrograph of a human cell. (×15,000) *(© David M. Phillips/Visuals Unlimited)*

(b)

DNA in the nucleus to the synthesizing organelles in the cytoplasm. RNA travels through **nuclear pores,** or openings, in the nuclear membrane to get from the nucleus to the cytoplasm.

Some RNA forms **ribosomes,** organelles that make proteins according to the instructions carried by other RNA molecules. The ribosomes attach to the **endoplasmic reticulum** when they begin synthesis of proteins. Once made, many of the proteins are transported to the **Golgi complex.** There they are joined with other molecules and transported through the cytosol. Some of the proteins travel to the cell membrane for transport to the outside of the cell. Some proteins stay in the cell to participate in the internal cell processes. For example, some may be enzymes used to digest food for the cell, while others become working parts of membranes in the cell.

Preserving Information

Many cells in the body are continually dividing. Unlike a human being, which can produce offspring and still continue to exist, a "parent" cell is completely replaced by its daughter cells. Therefore, any information to be passed on to the new cells must be prepared ahead of time, before cell division takes place.

The information about synthesis of macromolecules that must be passed on is contained within DNA. This molecule replicates, or copies, itself. Two copies of the same DNA then exist, each to become the nuclear DNA of one of the new cells. DNA replication is discussed in detail in Chapter 5.

Energy Obtained From Food Is Used to Support Life Processes

The second requirement for the independent existence of a cell is the ability to take energy from food and convert it into a useful form to support life processes, including growth and reproduction. The cell has a set of organelles that perform this function by supporting a complex chain of chemical reactions. A system of **vacuoles** conveys food matter in and waste matter out of the cell. **Lysosomes** are small bags of enzymes that surround food molecules that come into the cells in vacuoles. The enzymes in the lysosomes break down the molecules so they can be used by mitochondria to produce ATP, the energy source for the cell. **Mitochondria,** organelles enclosed in a double membrane, are the sites where ATP is produced through the process of aerobic respiration.

Membranes Protect the Cell Interior and Divide It into Different Compartments

The extracellular fluid surrounding the cell is primarily an aqueous solution, as is the cytoplasm inside the cell. Recall from Chapter 1 that each of these two fluid compartments has a distinctive ionic composition that must be maintained. Without the plasma membrane that separates them, the two solutions would merge and their contents mix, becoming indistinguishable.

Just as it is necessary for the cytoplasm to be segregated from the extracellular fluid, the contents of individual organelles within the cell must be protected from each other and from the cytosol. Some cell products, such as digestive enzymes, would destroy other cell products, such as secretory proteins,

if allowed to contact them. Segregation of cytoplasm from the extracellular fluid, and of various cell products within the cell, is the function of the cell's membrane system.

Protecting the Interior: The Plasma Membrane

The outer **cell membrane,** called the **plasma membrane,** is composed of two layers of phospholipid molecules. Recall from Chapter 2 that phospholipids consist of a long hydrophobic, or water-insoluble, tail, and a hydrophilic, or water-soluble, portion. Because of hydrophobic interactions, the phospholipid molecules making up the cell membrane act the way any such molecules would act when placed in an aqueous solution such as that found in cells: their hydrophobic portions cluster together, leaving their hydrophilic portions exposed on the surface to the aqueous environment. The clustering of lipid molecules creates a sturdy barrier to the passage of water-soluble particles, keeping the watery cytosol from flowing out of the cell and the watery extracellular fluid from flowing in.

At the same time, the plasma membrane must permit the passage of certain substances both in and out of the cell. A variety of mechanisms exist for this purpose. Endocytosis involves portions of the membrane breaking off and enclosing material to be taken into the cell, with a reverse process (exocytosis) used to let material out of the cell. These processes are discussed later in this chapter.

Other mechanisms involve the passage of individual molecules through the membrane in very controlled ways via protein molecules called **channels** and **carriers.** Still other mechanisms involve simple diffusion such as that occurring in any aqueous solution. These methods of transmembrane movement are covered in detail in Chapter 4.

Separating Cell Processes: Membranes of Organelles

Since each distinct organelle is bounded by a membrane, it follows that the outer cell membrane of a eukaryotic cell accounts for only a minor proportion (2% to 5%) of the total membrane present in the cell. The most extensive membrane-bound organelle present in most cells is the endoplasmic reticulum (see Fig. 3–2). The membrane of the endoplasmic reticulum of a liver cell makes up more than 50% of the total membrane of the cell. There are seven different subcompartments, or organelles, formed by the intracellular membranes of most eukaryotic cells. Two of these organelles, the nucleus and the mitochondrion, are surrounded by a double membrane. These membranes are composed, as is the plasma membrane, primarily of two layers of phospholipid molecules.

Continuity exists to some extent between the subcompartments in that some of the materials present in one compartment may be passed on to another, but direct physical links between the compartments are probably very limited. Each of the compartments within the cell has a unique role, carried out at the molecular level by the enzymes that are either packed inside the organelle or components of the membrane of the organelle. Since each organelle has a different function, the set of enzymes within a specific organelle will not be found in any of the other organelles. **Catalase,** for example, is found only in organelles called peroxisomes. The enzyme that breaks down glycogen into glucose molecules occurs only in the cytosol. The enzymes that catalyze the sequence of reactions in the citric acid cycle are present only within mitochondria.

Enzymes found in lysosomes, called **lysosomal enzymes,** are essential to normal cell function. The absence of just one of these enzymes, as occurs in some genetic diseases, has severe consequences for the individual. However, this group of **degradative enzymes** can break down all of the macromolecules responsible for the structure and function of the cell. Clearly, these enzymes cannot be allowed direct contact with other organelles or with the components of the cytosol because they would cause irreversible, lethal damage. The cell solves the problem very neatly by keeping the enzymes within lysosomes so that they are always available within the cell. They are allowed access to structures and molecules targeted for degradation (breakdown) only when the lysosomes fuse with endocytotic vesicles containing material to be broken down. This process ensures that the degradative process is confined locally within a specific compartment, the secondary lysosome (see Fig. 3–10).

Similarly, the presence of catalase within peroxisomes ensures that the toxic hydrogen peroxide produced within these organelles is destroyed on the spot before it has a chance to leak into the cytosol, where it would cause serious damage.

The presence of intracellular compartments also permits separation of the processes of synthesis and degradation of molecules. The advantage derived from physically separating these processes is that they can occur simultaneously within the cell and be controlled independently. The metabolism of fatty acids provides a good example. These molecules are continually synthesized by the cell because they are components of the phospholipid molecules needed for the structure of cell membranes. At the same time, the cell is also degrading other fatty acids because they are an excellent source of chemical energy; in fact, they yield more energy than equal weights of carbohydrates such as glucose. The goal of synthesis is the opposite of the goal of degradation, and the two processes would not be compatible if both occurred openly in the cytosol. Eukaryotic cells solve the problem by locating the reactions that degrade fatty acids in the mitochondria and those that synthesize fatty acids in the cytosol.

Finally, compartmentalization of cellular functions allows for intracellular processing and "sorting" of some of the synthesized molecules. We will discuss shortly some of the sorting and processing that occurs in the endoplasmic reticulum (ER) and the Golgi complex. For example, proteins destined to become components of membranes are trapped within the ER membrane during synthesis, whereas proteins destined for export from the cell are discharged into the lumen of the ER, where they are attached to carbohydrates to become glycoproteins. From there, they are transported to the Golgi complex, where the carbohydrate portion of the molecules is chemically modified. The Golgi complex also carries out another sorting step based on the specific structures present in the glycoprotein molecules. It distinguishes between the molecules that must be packaged inside secretory vesicles for export from the cell, and the molecules that are destined to become lysosomal enzymes and must be directed into primary lysosomes for use inside the cell.

Cell Compartments and Their Functions

The Nucleus Is the Library of the Cell

The nucleus (Fig. 3–3) is the site where DNA and RNA are made. The information stored in DNA and RNA molecules directs the synthesis of proteins that determine the shape and function of a cell. (The red blood cell is the only common mammalian cell that lacks a nucleus, and it survives for only a few months.) The nucleus also contains proteins. The approximate composition of the nucleus in a liver cell, expressed as a percentage of the dry weight of the nucleus, is 80% protein, 15% DNA, and 5% RNA.

DNA: The Permanent Files

Recall from Chapter 2 that DNA (deoxyribonucleic acid) is an enormously long, unbranched, double strand of nucleotides arranged in an alpha-helix. A typical animal cell contains more than 1 meter of DNA, all of which must fit inside a nucleus with a diameter of only 0.000006 meter (6 μm). Each section of the chromosomal DNA that carries the informa-

(a)

(b)

(c)

Figure 3–3

(*a*) The cell nucleus is enclosed in a double membrane called the nuclear envelope. Pores in the envelope permit the passage of molecules in and out of the nucleus. The outer layer of the nuclear envelope is continuous with the endoplasmic reticulum, so that the lumen of the ER is continuous with the perinuclear space. In the nondividing nucleus, DNA is visible as chromatin. The nucleolus plays a role in the synthesis of ribosomes from RNA. (*b*) A nuclear pore is formed from the fusion of the two layers of the nuclear envelope. Proteins are thought to be located in the pores. (*c*) An electron micrograph of the nucleus and nucleolus of a glandular cell.   (*© Don W. Fawcett/Photo Researchers.*)

tion necessary for synthesis of a single polypeptide is known as a **gene.** The sum of all the chromosomal genes is termed the **genome** of the cell. All cells in the body, except the reproductive cells, normally contain the same DNA. The DNA molecules remain unchanged throughout the life of the organism unless they are damaged in some way.

The tight packing of DNA (Fig. 3–4) is achieved by specialized proteins called **histones,** which bind to the DNA and organize its structure in the nucleus. Histones are by far the most common protein found in the nucleus. They are rich in amino acids with basic (positively charged) side chains, such as lysine and arginine (see Fig. 2–20). These positively charged

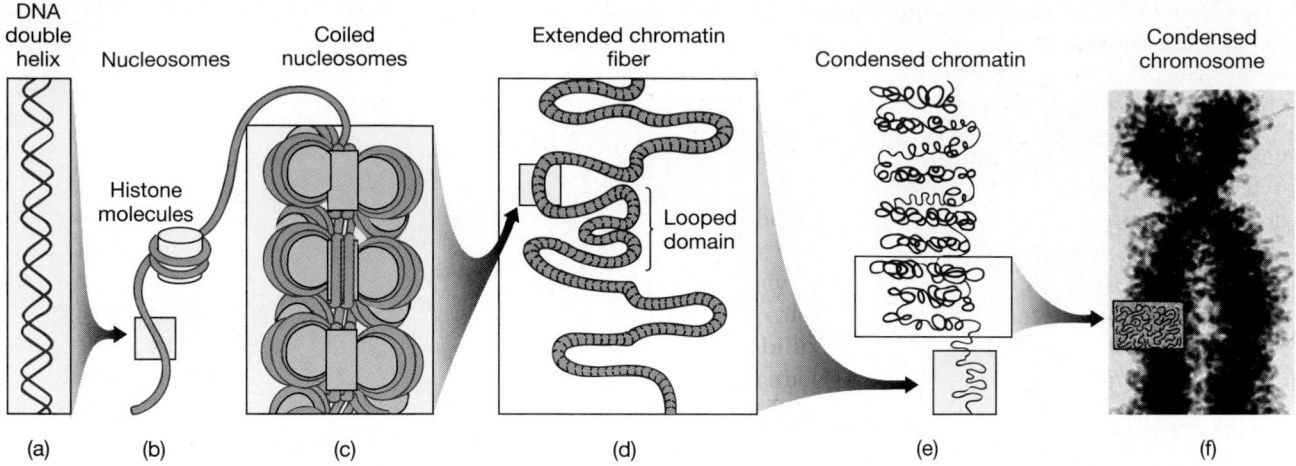

Figure 3–4

The DNA is packed in a series of stages beginning with nucleosomes. In a nondividing cell, DNA appears as chromatin; in a dividing cell, it appears in the more tightly coiled form of chromosomes.

side chain groups may be able to interact with and bind to the negatively charged phosphate groups in DNA (see Fig. 2–27), regardless of the specific nucleotide sequence.

Chromatin.   Most of the DNA is associated with histones and other nuclear proteins, giving rise to a complex known as **chromatin.** The basic packing unit of chromatin is the **nucleosome** (Fig. 3–4*b*), in which parts of the DNA double helix are wrapped around histone molecules and linked by stretches of "free," or **linker,** DNA. The nucleosome is still a relatively extended form of chromatin, and it is unlikely that much of the chromatin in the nucleus is present in this form.

A higher level of organization is produced by the packing together of nucleosomes in a regular array known as a **chromatin fiber.** Loops in these fibers give rise to **looped domains,** which reduce the initial length of the DNA double helix about 500-fold (Fig. 3–4*d*). These loops are supported by a "scaffold" made of proteins.

Chromosomes.   Further levels of folding must occur, perhaps involving close packing of the loops, in order to condense the DNA into a single **chromosome** (Fig. 3–4*e, f*). Thus, each chromosome of a eukaryotic cell consists of a single very large molecule of DNA arranged so that the double helix is folded in a compact and highly organized way. Each species has a characteristic number of chromosomes; human cells, for example, have 23 pairs. When cells are not dividing, the chromosomes are in a relatively uncoiled state and are not readily identifiable as indi-

vidual units. Only when cell division is about to occur do the chromosomes become apparent.

RNA: Selective Retrieval of Information

The RNA (ribonucleic acid) present in the nucleus is destined for export into the cytoplasm, where the machinery for protein synthesis is located. In cells such as rat liver cells, less than 10% of the total RNA in the cell is located at any one time in the nucleus.

An RNA molecule consists of a single strand of nucleotides that represent a copy of a limited region of the nucleotide sequence of one of the strands of DNA. Three major functional types of RNA interact in the cytoplasm to synthesize proteins according to the information stored in DNA: **messenger RNA (mRNA); ribosomal RNA (rRNA);** and **transfer RNA (tRNA).** For now, the related roles of these three types of RNA can be summarized as follows (more detailed discussions will follow in Chapter 5): Ribosomal RNA joins with certain proteins to form ribosomes. These are the protein-making structures that occur in the cytosol or are attached to the endoplasmic reticulum. Messenger RNA contains the information concerning which amino acids should be joined together, and in which specific sequence, on the ribosomes. Transfer RNA seeks out the required amino acids in the cytosol and brings them to the ribosome to be assembled into polypeptide chains.

One of the most obvious structures within the nucleus is the **nucleolus,** which is particularly prominent in cells that synthesize a lot of protein. The nucleolus is not limited by a membrane; it contains proteins and large loops of DNA from which rRNA is rapidly synthesized. Cells begin assembling rRNA and proteins into ribosomes in the nucleolus. However, ribosomes are not mature, or completely formed, until they reach the cytoplasm. This aspect of ribosome formation will be discussed in Chapter 5.

The Nuclear Envelope: A Porous Double Membrane

The interior of the nucleus is separated from the cytoplasm by the **nuclear envelope,** a set of two separate membranes. These membranes are separated by a space about 10 to 50 nm wide termed the **perinuclear space.** The outer membrane of the nuclear envelope often has attached ribosomes that are engaged in protein synthesis. This outer membrane appears to be continuous with the endoplasmic reticulum (Fig. 3–3*a*).

Much exchange occurs between the interior of the nucleus and the cytoplasm. Since DNA and RNA are synthesized only in the nucleus, the individual nucleotides and other parts required for their synthesis must be obtained from the cytoplasm. Conversely, RNA molecules made in the nucleus must move out to the cytoplasm in order to carry out their roles in protein synthesis. The rapid exchange of materials between the nuclear and cytoplasmic compartments probably occurs through the **nuclear pores.** These are open channels in the nuclear envelope that form in areas where the inner and outer membranes are pinched together (Fig. 3–3*b*). A mammalian cell has about 3000 to 4000 pores per nucleus. The diameter of each pore is 9 nm. This size permits the passage of small water-soluble molecules but excludes large macromolecular structures. A number of proteins, including enzymes, are made in the cytosol but their site of action is inside the nucleus. These proteins contain **nuclear import signals** which cause the nuclear pores to open up just enough to allow the proteins to enter the nucleus.

The Cytosol and Organelles Are the Sites of Synthesis, Degradation, and Energy Release

The normal structure, growth, and function of a cell require a constant supply of macromolecules, which are made from simpler molecules, or **precursors.** Synthesis of precursors such as amino acids or nucleotides, for example, is achieved by a set of chemical reactions that are part of the cell metabolism. Also important to intermediary metabolism are the reactions that break down small molecules and trap their chemical energy as ATP, as discussed in the previous section. All the reactions of metabolism take place outside the nucleus (Fig. 3–1).

The Cytosol: A Wealth of Enzymes and Precursors

The cytosol of a mammalian cell, such as a liver cell, is a large compartment. It represents about 50% of the total volume of the cell and contains the small molecules that are precursors for macromolecules. It also contains thousands of enzymes, which account for its high protein content (about 20% by weight). Many of the enzymes are catalysts for the reactions of cell metabolism. Thus, the cytosol resembles a highly viscous fluid rather than a watery solution.

The cytosol contains some structures that are not organelles but are worthy of mention, including glycogen granules and lipid droplets. **Glycogen,** a polysaccharide, is the storage form of carbohydrate. The lipid droplets contain **triglycerides,** which are the storage form of fatty acids (see Chapter 2).

Ribosomes and the Endoplasmic Reticulum: Synthesis of Proteins and Lipids

The endoplasmic reticulum (ER) is like a flattened membrane sheet that is folded repeatedly to form a large interconnected system (see Fig. 3–2). Two regions of the ER are structurally and functionally different, however: the **rough,** or **granular, ER** and the **smooth,** or **agranular, ER** (Table 3–1). The rough ER is so called because ribosomes are bound to its outer surface; smooth ER has no ribosomes attached.

The ER makes new proteins and lipids, including molecules destined to be released outside the cell by a process known as **secretion.** Rough ER is particularly extensive in cells that synthesize and secrete proteins, such as certain hormones. Smooth ER is well developed in cells specializing in lipid synthesis or the synthesis and secretion of steroid hormones. Muscle cells have a highly specialized smooth ER known as the **sarcoplasmic reticulum.** An important function of this structure is to store calcium ions.

Table 3–1
Structural and Functional Differences Between Rough and Smooth Endoplasmic Reticulum (ER)

	Rough ER	Smooth ER
Ribosomes present	Yes	No
Protein synthesis	Yes	No
Lipid synthesis	Yes	Yes
Steroid synthesis	No	Yes

(a)   (b)

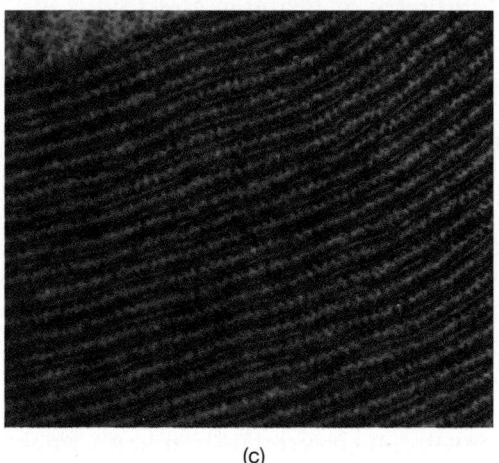

(c)

Figure 3–5

In the rough endoplasmic reticulum, a ribosome synthesizes a polypeptide from amino acids found in the cytosol. (*a*) A protein destined to be a secretory product or an enzyme enters the ER lumen as it is assembled, folding into its various levels of structure. It is enclosed in a transition vesicle for transport to the Golgi complex. (*b*) A protein destined to be membrane protein is partially trapped in the ER membrane as it is synthesized. (*c*) The rough endoplasmic reticulum. (*© Don W. Fawcett/Visual Unlimited.*)

Protein Synthesis on the Rough ER.   Individual amino acids are assembled into polypeptides on ribosomes. The ribosomes occur freely in the cytosol, but many become attached to the ER when they become involved in protein synthesis. The sequence of the amino acids in each polypeptide is dictated by mRNA. When an ER signal sequence appears in the emerging polypeptide, it is recognized and bound by a **signal recognition particle.** This particle directs the whole complex, including the ribosome, to a specific receptor on the ER. Polypeptide chains formed in the ribosomes are directed into the lumen of the ER and separated from the cytosol by the ER membrane (Fig. 3–5). Most of the proteins synthesized in the rough

ER become modified by the addition of monosaccharide units to form glycoproteins. In contrast, the proteins synthesized on free ribosomes in the cytosol are excluded from the ER lumen and remain in the cytosol. They cannot be modified by addition of carbohydrate.

The proteins that are destined to become structural components of membranes (see Chapter 4) are trapped within the rough ER membrane during synthesis. The mechanism for trapping the polypeptide chain may be a specific sequence of amino acids that interacts strongly with the interior of the ER membrane and prevents further movement of the polypeptide into the ER lumen (see Fig. 3–5). The

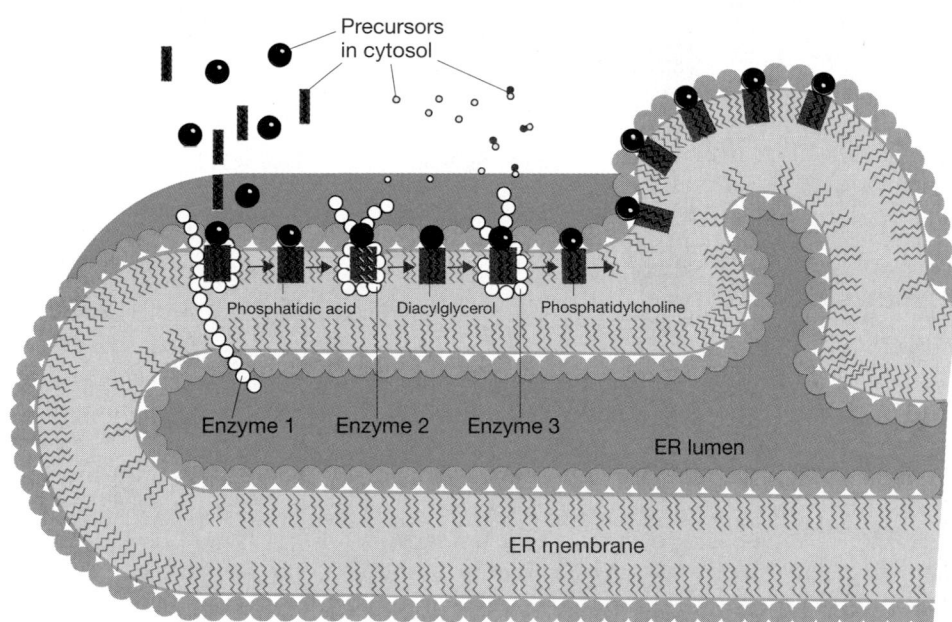

Precursors
in cytosol

Phosphatidic acid   Diacylglycerol   Phosphatidylcholine

Enzyme 1   Enzyme 2   Enzyme 3

ER lumen

ER membrane

Figure 3–6

In the smooth ER, a phospholipid is assembled from precursors in the cytosol through a series of reactions catalyzed by enzymes located in the ER membrane. The developing phospholipid molecule moves through the layer of existing phospholipid molecules that make up one half of the ER membrane.

part of the polypeptide that is synthesized after the trapping sequence is left on the cytosolic side of the ER membrane. Movement of ER membrane to other membrane-bounded organelles and to the plasma membrane will be discussed shortly.

Lipid Synthesis in the Smooth ER. The endoplasmic reticulum in many cells is the major site of the synthesis of lipids, including triacylglycerols, phospholipids, and steroids. Lipids from each of these three general classes form complexes with proteins in the ER lumen. The resulting structures are the compounds known as **lipoproteins.** The lipoproteins that are secreted from the cell and that enter the blood have a critical role in the transport and distribution of lipids, such as cholesterol, that are absorbed by the intestine.

Lipids such as phospholipids and cholesterol are also important structural components of membranes. The successive steps leading to the formation of a phospholipid occur on the ER membrane because the enzyme proteins that catalyze the synthesis are part of the membrane. Thus, synthesis of phosphatidylcholine (see Fig. 2–17c) occurs on the cytosolic surface of the ER membrane; the molecules of phosphatidylcholine that are produced will remain within the ER membrane (Fig. 3–6). Clearly, synthesis of the proteins and lipids that are required for production of new cellular membrane structures is another important function of the ER.

A macromolecule that is to be secreted is first formed in the ER and then moves to the Golgi complex. This occurs via a transition vesicle formed from a portion of the ER membrane. In the Golgi complex, the macromolecule is processed and packaged for export out of the cell or for storage or use within the cell.

The Golgi Complex: Processing and Packaging

The function of the Golgi complex is to modify the macromolecules synthesized in the ER. This is especially important in secretory cells. The Golgi complex processes and packages the macromolecules inside small membrane bags, or **vesicles,** ready for transport to the cell surface or the cytosol.

The Golgi complex is a collection of smooth membranes that look like flattened sacs. Each sac is called a **cisterna.** Commonly, about six cisternae will be stacked together to form a structure known as a **Golgi stack** (Fig. 3–7) (see p. 84). Small vesicles surround and radiate away from the Golgi stack.

The **cis,** or **forming, face** of the stack is oriented toward and closely associated with rough ER. The two structures are thought to be connected by the small membrane-bounded **transition vesicles** that shuttle newly synthesized macromolecules from the rough ER to the Golgi complex (see Fig. 3–7).

The **trans,** or **mature, face** of the Golgi complex, on the other hand, is oriented toward the plasma membrane and away from the nucleus, especially in secretory cells. Membrane-bounded **secretory vesicles** appear to arise from the trans face (see Fig. 3–7) to transport newly processed and packaged macromolecules from the Golgi complex to the plasma membrane for export.

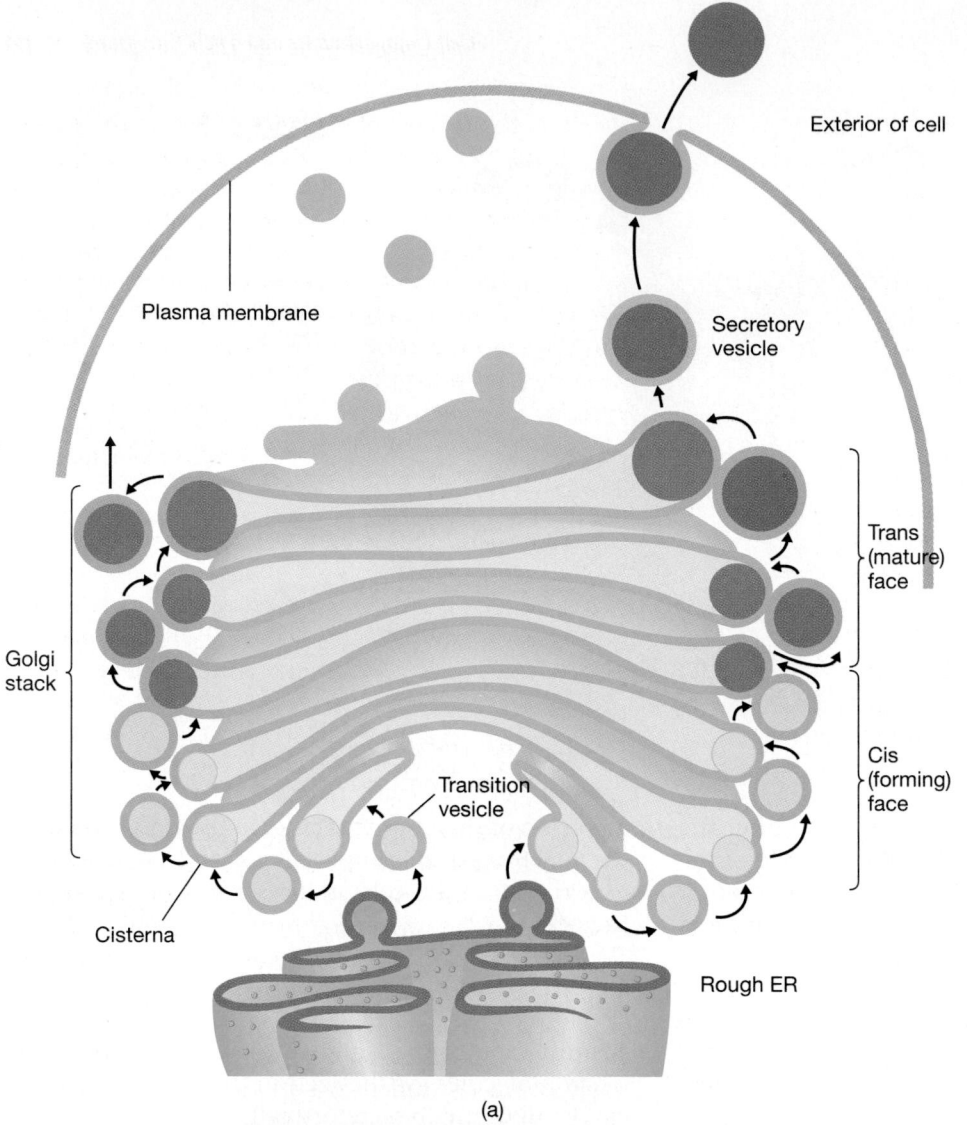

Exterior of cell

Plasma membrane

Secretory
vesicle

Trans
(mature)
face

Golgi
stack

Cis
(forming)
face

Transition
vesicle

Cisterna

Rough ER

(a)

(b)

Figure 3-7

(*a*) One unit of the Golgi complex, called a Golgi stack, is composed typically of about six folded sacs called cisternae. Macromolecules formed in the ER enter the cis face of the Golgi complex and pass through to the trans face via a series of vesicles. In the process, the lipids or proteins are modified. They leave the Golgi complex in secretory vesicles. (*b*) A false-color transmission electron micrograph of a section through one stack of the Golgi apparatus. The Golgi stack is the paired membranes (green outlined in red) in the middle of the image. The large red sphere near the top is a mitochondrion. (×31,800.) *(© Dr. Gopal Murti/SPL/Photo Researchers, Inc.)*

There is good evidence that the Golgi complex modifies the macromolecules destined for secretion. One of its important activities involves modification of the polysaccharide chains that were added covalently to the proteins in the rough ER. The processing of these glycoproteins within the Golgi complex involves a considerable amount of "trimming" as well as the addition of new monosaccharide units.

The form of the Golgi complex can vary considerably. In some cell types it may be compact and limited, whereas in others it may be more spread out, almost like a net. The position of the Golgi complex within the cytosol may also vary. In nerve cells it often surrounds the nucleus, whereas in secretory cells it is usually found between the nucleus and the side of the cell that is engaged in the secretory process. The number of Golgi stacks in a cell varies considerably with the type of cell. The Golgi complex in a liver cell accounts for about 7% of the total cell membrane, and in this cell type is a much less extensive structure than the ER.

Secretory Vesicles: Exocytosis and Membrane Recycling

A secretory vesicle containing a packaged macromolecule moves from the inner face of the Golgi complex to the plasma membrane. The vesicle membrane fuses with the plasma membrane in a process known as **exocytosis,** in which the secretory vesicle is allowed to open and expel its contents into the extracellular fluid. Some secretory vesicles are designed so that the exocytotic step occurs very rapidly whenever the plasma membrane is stimulated from the outside by a signal such as a specific hormone.

At the same time that the products of ER synthesis and Golgi complex packaging are conveyed to the exterior of the cell by exocytosis, the membrane of the secretory vesicle becomes part of the plasma membrane. Recall that some proteins have been synthesized in the ER for use as membrane proteins. They have also been packaged in the Golgi complex and form the membranes of the same secretory vesicles that carry the proteins destined for export. In this way, the new membrane proteins and lipids synthesized and trapped within the ER membrane can be transported to the cell surface and inserted into the plasma membrane, by the fusion step, to replace damaged or "worn out" proteins and lipids.

Not all of the membrane lost from the trans face of the Golgi as secretory vesicles is replaced by new membrane arriving at the cis face in the form of transition vesicles from the ER. Some of the replacement membrane is derived from pieces of the plasma membrane that are internalized and delivered to the trans face of the Golgi by processes such as **endocytosis,**

which is the reverse of exocytosis (Fig. 3–8). The internalized membrane may be returned to the plasma membrane by a subsequent exocytotic step. Exchange of membrane in this manner between the Golgi and the plasma membrane conserves the membranes of the transport vesicles, which may be highly specialized. It is economical for the cell to use them for several series of transport events rather than to synthesize new vesicle membranes each time. This **membrane recycling** mechanism can provide the cell with reusable raw materials because some of the internalized membrane is directed to **lysosomes.** There the component proteins and lipids are taken apart and deposited in the cytosolic pool of precursor molecules to await a new round of synthesis. Membrane recycling also helps to avoid an increase in the surface area of secretory cells when large numbers of secretory vesicles fuse with the plasma membrane.

Lysosomes and Peroxisomes: Digestion of Food and Waste

Lysosomes and peroxisomes are small membrane-bounded bags, about 0.5 μm in diameter. Together they account for less than 1% of the total membrane in a cell, but they are found in almost all cells in large numbers. One mammalian liver cell contains about 300 to 400 of each type of organelle.

Lysosomes. The **lysosome** is an organelle that contains about 50 different digestive enzymes. As a group, these enzymes allow cells to completely break down any of the macromolecules in the body (Fig. 3–9a). Since these enzymes would destroy the cells themselves, they are kept within a specific membrane-bounded organelle. Two general classes of lysosomes have been distinguished: newly formed **primary lysosomes** contain only the digestive enzymes; **secondary lysosomes** contain both enzymes and molecules (substrates) that the cell must break down. (Recall that "substrate" is a general term for any material on which an enzyme is designed to work.)

Primary lysosomes are thought to arise by budding from the trans face of the Golgi complex (Fig. 3–10). Most of the lysosomal enzymes they contain are glycoproteins that originate in the rough ER. When they enter the Golgi complex, the polysaccharide parts of the enzyme molecules present a unique signal to the Golgi complex, which responds by packaging these enzymes into the vesicles that become primary lysosomes.

There are various types of secondary lysosomes. Among the largest are the **digestive vacuoles** that arise after a primary lysosome fuses with an endocytotic vesicle containing particulate material such as

(text continues on page 87)

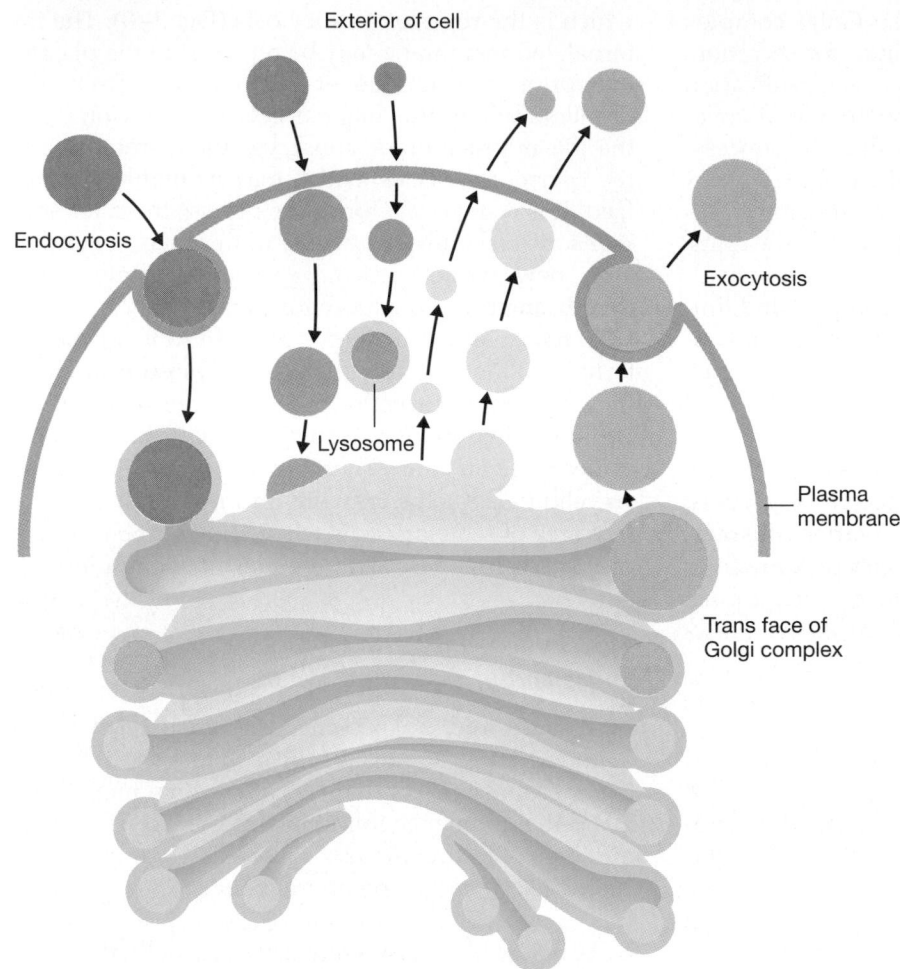

Exterior of cell

Endocytosis

Lysosome

Exocytosis

Plasma
membrane

Trans face of
Golgi complex

Figure 3–8

As portions of the Golgi complex
fuse with the plasma membrane
in the secretion process, portions
of the plasma membrane are returned to the Golgi complex
through various endocytotic
events. Thus, the plasma membrane maintains the same size
despite the continual addition of
membrane from secretory vesicles.

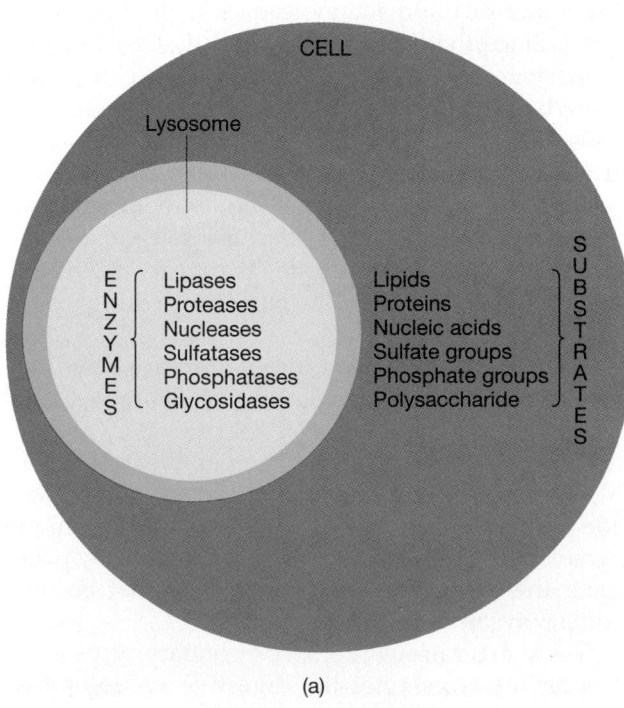

CELL

Lysosome

| E N Z Y M E S | Lipases
Proteases
Nucleases
Sulfatases
Phosphatases
Glycosidases | Lipids
Proteins
Nucleic acids
Sulfate groups
Phosphate groups
Polysaccharide | S U B S T R A T E S |

(a)

Lysosomes

Digestive
enzymes

(b)

Figure 3–9

(*a*) Lysosomes protect the cell contents from degradation
by keeping within their membranes enzymes that can
break down all of the macromolecules in the cell. The
only substrates allowed to contact the lysosomal enzymes are those within food vacuoles or other vacuoles
containing material meant to be digested. (*b*) A transmission electron micrograph of lysosomes.   (© *Don W.
Fawcett/Visuals Unlimited.*)

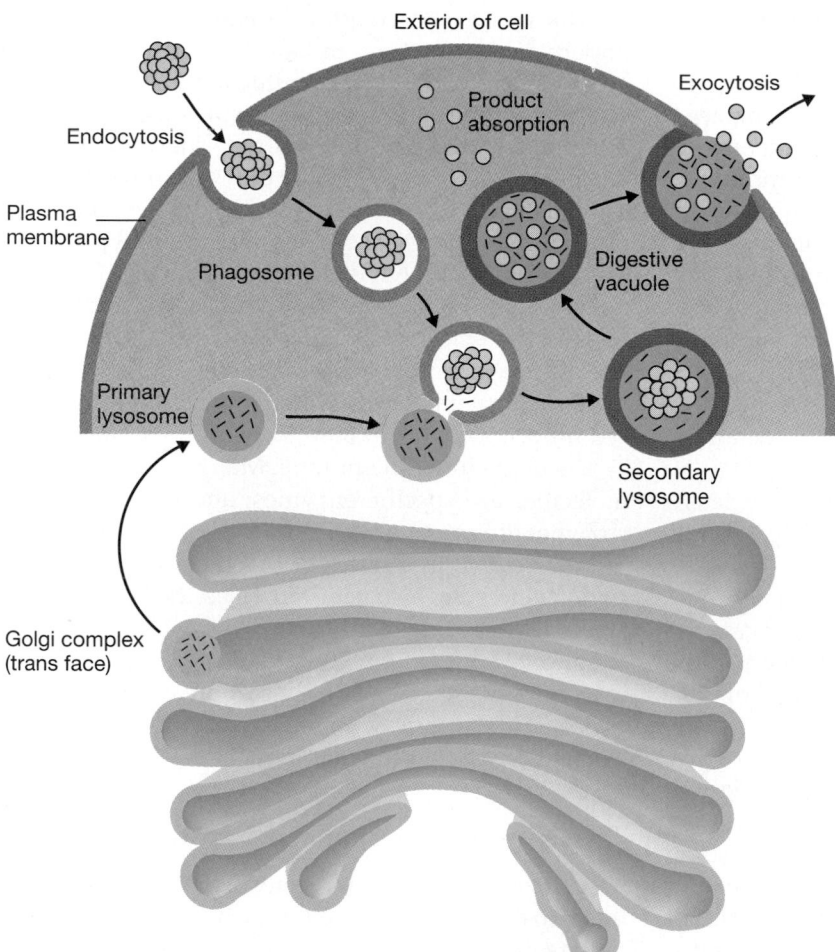

Figure 3–10

Primary lysosomes contain only lysosomal enzymes (proteins that were synthesized in the ER). A primary lysosome fuses with a phagosome carrying material (e.g., nutrient molecules) to be digested, creating a secondary lysosome that contains both enzymes and the nutrient substrates. The vesicle containing digested material is termed a digestive vacuole; from there the products of digestion are either absorbed by the cytosol or expelled to the exterior of the cell by exocytosis.

a bacterium. Internalization of particulate material is a special type of endocytosis known as **phagocytosis.** The resultant endocytotic vesicle is described sometimes as a **phagosome.** The process of fusion allows the digestive enzymes of the primary lysosome to gain access to the material to be broken down, whether it is bacteria or food.

After the lysosomal enzymes have performed their digestive task, the products of digestion can either be released outside the cell by exocytosis, for reuse by other cells, or absorbed from the digestive vacuole and used again within the same cell. Lysosomes are used not only to digest material originating from outside the cell but also to break down the cell's own organelles and other matter in a process known as **autophagy.** This process can be a response to cell injury, or it can occur during "remodeling" of the cell or during adverse nutritional conditions. The secondary lysosomes formed during autophagy are called **autophagic vacuoles.**

Peroxisomes. The **peroxisome** also contains specialized enzymes. These enzymes are synthesized in the cytosol and transported into peroxisomes as they

form, probably as buds from the smooth ER. The enzyme **catalase** is contained almost exclusively within the peroxisomes. Other enzymes within peroxisomes use oxygen (O_2) to carry out oxidative reactions. One of the products of these reactions is hydrogen peroxide (H_2O_2), which is potentially harmful to the cell. Catalase helps to degrade the hydrogen peroxide to harmless water by the following reaction:

$$2 \; H_2O_2 \xrightarrow{\text{Catalase}} 2 \; H_2O + O_2$$

The peroxisomes in many body cells are important in detoxifying a number of molecules.

Mitochondria: Energy Transformation

The mitochondrion is thought of as a "powerhouse" for the cell because its metabolic activities provide the energy needed to sustain the life of mammals and other **aerobic** (oxygen-breathing) organisms.

The principal fuels for most life forms are carbohydrates. After digestion, carbohydrates are stored in cells in the form of glycogen. In the cytoplasm, glycogen is broken down into glucose. The primary mechanism for processing these molecules and extracting the energy stored in them is **aerobic respiration.** This metabolic pathway is completed within the mitochondrion by a process that uses most of the oxygen that mammals take in with each breath. During aerobic respiration, about 36 molecules of adenosine triphosphate (ATP) are produced for each molecule of glucose that is broken down. Without the participation of the mitochondrion, only two molecules of ATP are produced. Clearly, the mitochondrion is the major site of ATP production within the cell. ATP is the form in which the cell is able to trap chemical energy and transfer it to sites where it is needed to drive many synthetic reactions and other energy-requiring processes. These processes are described in detail in Chapter 6.

Mitochondria are much smaller than the nucleus. They are usually cylindrical and about 0.5 to 1.0 μm in diameter. The number of mitochondria per cell varies with the cell type but animal cells usually contain large numbers. A liver cell may have as many as 1700 mitochondria, which would account for one fifth of the cell volume and 40% of the total cell membrane.

The mitochondrion, like the nucleus, is bounded by two separate membranes. The **outer mitochondrial membrane** is smooth and unfolded, but the **inner mitochondrial membrane** is distinguished by numerous well-developed infoldings called **cristae,** which project deep into the interior of the organelle and increase its surface area considerably (Fig. 3–11). The two membranes divide the mitochondrion into two compartments, the narrow **intermembrane space** and the much larger **matrix.**

The inner mitochondrial membrane is a highly impermeable structure, unlike the outer membrane. Most of the "work" of the mitochondrion is carried out by enzymes and other molecules located in the inner membrane and the matrix. The matrix compartment also contains ribosomes and 5 to 10 small double-helical DNA molecules that the mitochondria use to synthesize some of the mitochondrial proteins. Mitochondrial DNA has a circular structure. In mammalian cells, it makes up less than 1% of the total cellular DNA.

The major component of mitochondria, besides water, is protein. Most (67%) of the protein of liver cell mitochondria is located in the matrix, and about 21% is in the inner membrane. Many of these protein molecules are specific enzymes; about 120 different enzymes are present. Some of the enzymes in the matrix participate in a series of reactions known as the **citric acid cycle,** also referred to as the **Krebs cycle** or the **tricarboxylic acid cycle.** This cycle is the final common pathway for the breakdown of all the fuel molecules in the cell, not only carbohydrates but also amino acids and fatty acids. These processes will be discussed in detail in Chapter 6.

The Cytoskeleton: Shape and Structure

Specialized proteins in the cytosol are organized into the **cytoskeleton** of the cell, which is responsible for maintaining the cell shape and the positions of the internal organelles, and for mediating movement of the cell itself or of the organelles within the cell. Spe-

(a)

(b)

Figure 3–11

(*a*) The mitochondrion, the cell's "powerhouse," contains a double membrane. The numerous infoldings of the inner membrane, called cristae, create a large surface area on which the reactions of cellular respiration can take place. (*b*) An electron micrograph of a mitochondrion from an adrenal cortex cell.   *(© Don W. Fawcett/Visuals Unlimited.)*

Table 3–2
Cytoskeletal Fibers

	Diameter (nm)	Structure	Protein Subunit
Microtubules	24	Hollow	Tubulin
Intermediate filaments	10	Hollow	Several types
Microfilaments	6	Solid	Actin
Myosin filaments	15	Solid	Myosin

cific examples of organelle movement that have been discussed already in this chapter are (1) the movement of secretory vesicles from the Golgi complex to the plasma membrane, and (2) the movement of endocytotic vesicles or phagosomes from the plasma membrane to the cell interior.

The cytoskeleton is made up of a dense network of different types of protein fibers, the **microtubules, intermediate filaments,** and **microfilaments.** Their structures and sizes are compared in Table 3–2. Microtubules and intermediate filaments are found in deeper regions of the cell, whereas microfilaments generally are located just beneath the plasma membrane.

Microfilaments also extend into cell processes. Good examples are the fingerlike projections called **microvilli,** which cover the inner surfaces of the intestine and the kidney proximal tubule. Figure 3–12 (drawn to scale) gives an indication of the huge size of the cytoskeletal protein fibers relative to the protein molecules present in the membrane of the microvillus.

Microtubules and microfilaments can be described as dynamic structures because their formation is readily (and often) reversed, and because the distribution of these structures within a cell is subject to change as the physiological conditions change. In contrast, the intermediate filaments are much more

Figure 3–12

Microvilli are found in cells of the digestive system and kidney. The internal structure of a microvillus, drawn to scale, shows that the microtubules and microfilaments that create its fingerlike shape are relatively large structures in the cell. They are composed of and connected by protein molecules (e.g., actin, myosin, tubulin) specialized for contraction and support. For example, a microtubule is formed by a circular collection of tubulin molecules. Tubulin is a protein that can be detected through the use of fluorescent antibodies (see Fig. 3–17).

(a)

(b)

Figure 3–13

(*a*) Microtubules and filaments are found throughout the cell, not only in microvilli. They may be linked by the microtrabecular lattice. (*b*) Fluorescent antibodies show the position of microtubules in a human cell. (See also Figure 3–17.)

stable structures, and their formation may be irreversible. Most of the cytoskeletal fibers are made up of large numbers of a single type of protein molecule or subunit (see Table 3–2).

Myosin filaments (also known as **thick filaments**) are a fourth type of fiber found in almost all cell types. They are most abundant in muscle cells where, together with the thin actin filaments, they form part of the contractile machinery of these cells. The presence of **actin** and **myosin** molecules in non-

muscle cells suggests that nonmuscle cells may possess a contractile mechanism such as that of muscle cells and use it for some internal movements.

The organization of the cytoskeleton is complex. It has been suggested that the major fibers are linked by another set of thin filaments, the **microtrabecular lattice.** These filaments appear to hold the fibers and the internal organelles in their places (Fig. 3–13). They even hold ribosomes in suspension. However the microtubules may have the major role in the overall design of the cytoskeleton because they appear to influence the distribution of microfilaments and intermediate filaments. They may provide the framework on which the rest of the cytoskeleton is erected. The molecular mechanisms by which the cytoskeleton can mediate the movement of cells and the movement of organelles within cells have not been determined.

The Life Cycle of the Cell

The many different cell types present in the human body have different patterns and rates of division. Cells such as red blood cells and nerve cells stop dividing when they reach maturity, whereas certain epithelial cells in the intestine and skin divide rapidly and continuously.

The **cell cycle** is the period of time from the beginning of one cell division to the beginning of the next division (Fig. 3–14). Cell cycle times vary from just eight hours for rapidly dividing cells to more than 100 days for rarely dividing cells.

The cell cycle consists of two main phases: interphase and cell division. **Interphase** is the period between the end of one cell division and the beginning of the next, during which the cell carries out all the normal cell processes of growth and metabolism, including the replication of DNA. **Cell division** is the period during which the nucleus and then the cytoplasm divide to form two new cells, each containing one of the copies of the original DNA produced during the preceding interphase.

The process of cell division, sometimes called the **"M" phase,** involves two steps: mitosis and cytokinesis. In **mitosis,** the nucleus divides into two new nuclei. In **cytokinesis,** the cytoplasm divides, each half taking with it one of the two new nuclei to form a new cell.

Cell Growth and DNA Replication Occur During Interphase

Much of the preparation for cell division occurs during **interphase** (see Fig. 3–14). It occupies 90% or more of the cell cycle. Interphase itself is divided into three phases: the synthesis, or S, phase and two gap, or G, phases.

The period of interphase that directly follows cell division is the **first gap,** or G_1, phase. During this period the new cell begins its growth as the organelles it inherited from the parent cell resume their biosynthetic activities (which slowed during the M phase).

The **S,** or **synthesis, phase** follows the first gap phase. This is the part of interphase during which the cell synthesizes DNA in the nucleus in order to duplicate the DNA in preparation for cell division.

Directly following the S phase is the **second gap phase (G_2),** during which the cell increases protein synthesis in final preparations for division.

The time taken to progress from the beginning of the S phase to the end of the M phase is markedly similar in all cell types. The principal difference between cells with short cell cycles and those with long cell cycles is the amount of time spent in the G_1 phase; slowly dividing cells stay in G_1 for days or years.

Cell Division Occurs During Mitosis and Cytokinesis

The process of mitosis occurs in four stages: prophase, metaphase, anaphase, and telophase. (Fig. 3–15) (see p. 92). Each of the stages merges smoothly with those coming before and after. The stages are distinguished mainly by the different movements of the chromosomes.

Prophase

In the first stage of mitosis, called **prophase,** the DNA duplicated during the preceding S phase condenses into chromosomes. The nuclear membrane breaks

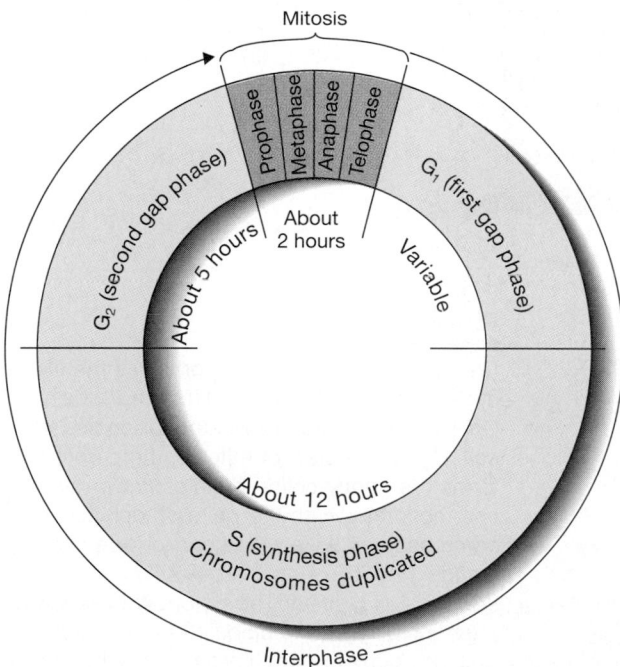

Figure 3–14

The life cycle of the cell contains two phases: Interphase and mitosis. During interphase, the cell carries out normal metabolic activities and duplicates DNA. During mitosis, the cell divides into two new cells in four steps called prophase, metaphase, anaphase, and telophase.

(text continues on page 93)

Nuclear envelope
Chromatin
Nucleolus
Cytoplasm

Plasma membrane

INTERPHASE
Chromatin spread out in indistinct mass. Nucleus and nucleolus distinct.

Centrioles

(a)

MITOSIS

Sets of chromatids

Spindle microtubules

Spindle pole

Polar spindle fibers

Spindle pole

Kinetochore fibers

(b)

(c)

(d)

(e)

PROPHASE
Chromosomes condense and become visible as sets of sister chromatids. Nucleolus and nuclear envelope disappear. Spindle microtubules appear.

METAPHASE
Mitotic spindle complete. Chromatid sets move to spindle equator.

ANAPHASE
Kinetochores divide, freeing sister chromatids as individual chromosomes which then move to opposite poles of the spindle.

TELOPHASE
Two new nuclei form. Division of the cytoplasm often begins now.

Cleavage furrow

Microfilaments

OFFSPRING

CYTOKINESIS
Microfilaments constrict cell at spindle equator, pinching it in two.

(f)

Figure 3–15

(*a*) During the interphase portion of the life cycle of a cell, the chromatin form of the DNA in the nucleus replicates, or copies itself. (*b*) At the start of mitosis, the chromatin forms the tightly coiled bundles known as chromosomes, with the pairs of identical chromosomes linked together to form chromatids. (×450.) *(© M. Abbey/Photo Researchers, Inc.)* (*c–e*) The chromatids separate, after lining up at the spindle equator, such that one copy of each chromosome becomes part of one of the nuclei of the two new cells. Thus, each new cell has the same genetic information contained in the parent cell. (×315.) *(© John D. Cunningham/Visual Unlimited.)* (*f*) Cytokinesis, the physical separation into two cells, begins during telophase.

down, and the contents of the nucleus become mixed with the cytoplasm.

Metaphase

The duplicated chromosomes, not yet separated, line up at the center of the cell and attach to the **mitotic spindle,** which is formed from microtubules.

Anaphase

The third stage begins when the duplicated chromosomes separate. The two sets, each representing a complete copy of the original DNA, are pulled to opposite ends of the cell.

Telophase

During the final stage of mitosis, a new nuclear membrane forms around each set of chromosomes, and they begin to uncoil. **Cytokinesis,** or division of the cytoplasm into two daughter cells, also occurs during this stage. Following cytokinesis, the new cells begin interphase (see Fig. 3–15).

The Cell Cycle Is Driven by an "Engine"

An important factor that induces a cell to enter mitosis from G_2 is known as **maturation promoting factor (MPF).** When MPF is activated, the nucleus responds with chromosome condensation, breakdown of the nuclear membrane, and formation of the mitotic spindle. Upon inactivation of MPF the cell exits from mitosis (via anaphase, telophase, and cytokinesis), the nucleus reassembles, and the new cells enter the next interphase. The cell cycle is driven through the mitotic phase by the periodic activation and inactivation of MPF. MPF and the factors that regulate its activity have been termed the "engine" of the cell cycle. One engine component is **cyclin B,** a protein that increases in abundance during each interphase but is rapidly destroyed at the end of mitosis. In a simple model of the cell cycle engine it is thought that MPF, which is inactive during interphase, is activated by the accumulation of cyclin B. Active MPF drives the cell into mitosis and also induces degradation of cyclin B. Loss of cyclin B leads to inactivation of MPF and the cell proceeds to interphase.

Cyclins are found in all eukaryotes, and there are many different ones with different functions in the cell cycle. For example, cyclin A is important in the control of DNA replication during S phase, and cyclins C, D, E, and F are important in moving the cell cycle from G_1 into S phase.

Cancer Cells Have Abnormal Cell Cycles

The rate of cell division in normal tissues is controlled so that the cells will divide only when new cells are needed. Removal of part of the liver, for example, will stimulate rapid division and growth of the remaining cells. This new growth ceases when the normal liver mass has been restored. The normal functions of tissues and organ systems would be destroyed rapidly if cell division were uncontrolled.

Cancer cells are the cells of any malignant **neoplasm.** "Neoplasm" means "new growth." The term *malignant neoplasm* is used because the cells divide and multiply rapidly and behave abnormally. In contrast to the highly organized growth of normal cells, cancer cells grow upon one another, invading normal tissues and threatening their functions. Cancer cells do not respond to the normal controls of cell division and growth, and the body is simply a very favorable environment to support their growth.

A cancer cell probably develops when the DNA of a cell experiences **mutation,** or alteration, after exposure to radiation, chemicals, or viruses. When the cell divides, all the cells derived from it contain copies of the altered DNA and inherit the defect. An accumulation of several DNA mutations arising independently from various insults may complete the transformation of a normal cell into a cancer cell. See Focus Box on "Cancer."

Research Techniques in Cell Physiology

The information about cell structure and function described in the preceding paragraphs has been obtained through a variety of research methods, some as simple as observation through a microscope, others involving the sophisticated analysis of complex chemical reactions. Following is a brief survey of the research methods still being used to elucidate the details of cell physiology.

Human Cells Grow in the Laboratory

Cells taken out of mammalian tissues will grow and multiply in the laboratory if provided with appropriate nutrients and other factors. Human cells have been used almost routinely for **cell culture** since 1952. Many types of cells can be grown as a single layer attached to glass or plastic surfaces and covered with a liquid **growth medium,** or mixture of nutrients. Under these conditions, the cells are ideally suited for direct study by various techniques of microscopy.

CLINICAL FOCUS

Cancer

In the discussion of cell cycles in this chapter, it is pointed out that different cell types have different rates of cell growth and division. Mature nerve cells, for example, grow no further and cell division ceases. In contrast, epithelial cells such as those in the intestine and skin have a very rapid cell cycle. This produces new cells that are needed to replace the continuous loss of older cells.

Cancer cells also grow and divide rapidly, but their growth is out of control. A mass of cells growing out of control is known as a **neoplasm,** more commonly referred to as a **tumor.** The study of tumors **(oncology)** has revealed that division and growth of the cells are not controlled effectively by the normal mechanisms. Also, tumor cells range from well to less **differentiated** compared to normal cells. A tumor cell that is less differentiated will be one that has lost some or all of the specialized properties of the normal tissue cells from which it is derived. **Benign** tumors are relatively slow growing and remain in a localized area of the body. They are usually removed easily by surgery.

The term *cancer* is used most often to refer to **malignant,** life-threatening tumors. Malignant neoplasms grow more rapidly than benign growths. They also are capable of spreading rapidly to other parts of the body, a process known as **metastasis.** One way this can occur is by some cells breaking away from the original tumor and traveling via the bloodstream or lymphatic system (Chapter 19) to a distant site where they grow to produce a secondary tumor. Cancer of the breast, for example, frequently spreads to bone, lung, or brain tissues. Normal cells and organs are destroyed by the pressure exerted by an expanding malignant tumor. Direct pressure on adjacent nerves produces pain. Life for the normal cells is made even more difficult because the tumor cells compete for the supply of nutrients and other important molecules.

At the present time, cancer is the second leading cause of death in the United States. It is second only to heart disease. The current "war" on cancer began with the National Cancer Act in 1971. At that time, hopes were high for a quick cure. This view was overoptimistic but was very helpful in obtaining the necessary research funds. More than twenty years later, the overall death rate from cancer has not changed, although this statistic obscures the fall in death rates in specific cases such as stomach and uterine cancers in patients under age 65. Major advances have been made, however, by researchers who are trying to understand the origins of cancer, or what turns a normal cell into a malignant one. We now know about **oncogenes,** which transform normal cells into malignant ones and trigger tumor formation, and **tumor-suppressor genes,** which retard growth and division of transformed cells. It is thought now that some cancers arise because mutations activate oncogenes, thus stimulating cell growth and division. These processes continue unchecked because additional mutations inactivate tumor-suppressor genes in the same cell.

One of the best known examples of a suppressor is the **p53** gene, so-called because it codes for a protein that is 53 kDa in size. This protein can suppress tumors by slowing growth and division of the cells, by forcing the cells to self-destruct, and by preventing the damaged DNA from being copied. Each year about 6.5 million people in the world are diagnosed with cancer. About half of these people have mutations in the *p53* gene in their tumor cells. Presumably the mutations interfere with the tumor-suppressor function of the gene.

One of the current goals of cancer research is to learn how to control oncogenes and tumor-suppressor genes. The challenge for the future will be to convert these advances in basic medical research into treatment and prevention of cancer in the clinical setting.

■ **What would happen to the cell cycle if the protein product of the normal *p53* gene activated another gene that inhibits the action of cyclins?**

Alternatively, cells can be grown in suspension in liquid media for the purpose of producing large numbers. Special conditions are used to avoid bacterial growth in the liquid medium, and antibiotics are frequently added as an extra precaution. The growth medium contains glucose, various salts, amino acids, and vitamins and is often supplemented with serum derived from the blood of a horse or calf. (**Serum** is the fluid that remains in the blood after the blood cells have clotted together.) The precise composition of serum is always uncertain, but it is an adequate supplement to the growth medium for routine cell culture.

Cells that are transferred directly from the tissues of an animal to a plastic dish containing artificial growth medium are known as *primary* cultures. As they grow and multiply, by dividing to produce daughter cells, they can be **subcultured**—that is, the cells can be removed and dispersed among many new culture dishes to produce a large number of cells.

One purpose of growing cells in culture is to obtain cells that still exhibit the specific functions of the tissue from which they were removed. Ideally, a culture contains only one specific cell type. Thus, cell cultures have an advantage over studies of intact tissues, in which several different cell types, products, and processes are usually at work, complicating the picture. When they contain only the pertinent cells, cultures are extremely useful for studies of unique cellular functions because the cells are growing under carefully controlled conditions and are far more accessible than when part of an intact tissue inside a living animal.

Most normal cells will divide only a limited number of times in culture before dying. Some researchers believe that studies of this built-in "senescence" will provide important clues about the aging process in humans and other animals. Occasionally, however, a few apparently normal cells will grow indefinitely in culture: they are known as a **cell line.** A cell line, and also a primary culture, can be stored frozen in liquid nitrogen ($-200°C$) for long periods of time, even years. When warmed to $37°C$, some of the cells will resume growth and division. This ability is very useful once a suitable group of cultured cells has been obtained. Some of them can be stored frozen for studies in the future, creating an almost inexhaustible supply of cells.

Among their many research purposes, cell cultures are used in investigations of the interactions that allow similar cell types to "recognize" one another and to form the close associations that lead to the development of a specific tissue. When cells are cultured on a plastic surface, they grow and multiply only to the point at which most of the surface is covered by a monolayer of cells. Further growth is inhibited by close contact with other cells (so-called **contact inhibition**), and the monolayer is said to be **confluent.** Interestingly, cancer cells that grow in an uncontrolled way in the body exhibit the same uncontrolled growth pattern in culture. They are much less sensitive to contact inhibition.

Finally, the study in culture of **mutant** cells, abnormal cells that cannot synthesize a specific protein, often reveals a great deal of information about the role of that protein in normal cells. The specific functions that protein normally direct are absent from the cultured mutant cell. The altered behavior of the mutant cell is a strong clue to the function of the protein in normal cells of the same type.

A word of caution: Cells in culture—especially cell lines that have been cultured for a number of years—are not growing in their normal physiological environment. Therefore, despite the usefulness of the technique, the behavior of cells in culture may be different from those in an intact tissue. The observations made about cultured cells must be checked eventually against cells in their normal situation.

Microscopes Allow Visualization of Cells and Organelles

First used to study cells and tissues in the seventeenth century, the microscope in its modern form is essential in physiological research. A significant advantage of microscopic studies is that the researcher can see the complete cell and its organelles in a relatively undisrupted state.

The Light Microscope

The **light microscope** (Fig. 3–16) is familiar to all students of science. It is extremely useful for examining basic sizes and shapes of cells, and it can be used to detect the presence of nuclei and mitochondria when the cells are stained with dyes that change the color of only these organelles. The tissue of interest is usually **fixed** with a solution that preserves the cell structure and **embedded** in a hard wax to provide support so that thin slices can be prepared for viewing.

Students and researchers are often concerned that these treatments may distort the appearance of the cells. This problem can be avoided by using a different optical system, such as that used in a **phase contrast microscope.** This instrument enhances the contrast between the internal organelles so that mitochondria, lysosomes, chromosomes, nucleoli, and other structures can be seen quite clearly in living, unstained, and nonfixed cells. Light microscopes are rarely used for magnifications greater than 1000 to 1500 times normal size because beyond this point the **resolution,** or clarity of the image, is poor.

The Electron Microscope

The development of the **transmission electron microscope** (see Fig. 3–16) in the 1940s and 1950s made greater magnifications (more than 100,000 times normal size) possible. The use of this type of microscope has provided a wealth of information about the organization of cells and the structure of their organelles.

In an electron microscope, magnetic lenses focus a beam of electrons in much the same way that glass lenses in ordinary microscopes focus a beam of light (see Fig. 3–16). To be seen by an electron microscope, specimens must be fixed and then stained with solutions that introduce atoms such as lead or uranium into the various structures of the cell. The presence of these large atoms permits fewer electrons to pass through the specimen and gives rise to a dark image of the specimen on a bright fluorescent screen like a television screen. This image can then be recorded on a photographic plate to provide a permanent record.

The **scanning electron microscope** (see Fig. 3–16) is especially useful for examining the surface of a

Figure 3–16

Whereas a light microscope focuses a beam of light through curved glass to produce an image, a transmission electron microscope focuses a beam of electrons using an electromagnet. The electrons pass through the specimen to produce an image that can be made visible to the human eye on a screen or photographic film. A scanning electron microscope images the deflection of electrons by the specimen. The photographs show views of the unicellular organism *Euglena.* The light microscope shows the whole cell; the transmission electron microscope shows a thin cross section; the scanning electron microscope shows the cell's outer surface with its pattern of fine ridges.

specimen. These microscopes image the deflection of electrons by the specimen. The **high-voltage electron microscope** uses an energy source of one million volts and stands about 32 feet high. It allows much thicker tissue sections to be viewed with high resolution and in considerable depth, producing an almost three-dimensional view of the cell interior. The use of this technique led to the concept of the microtrabecular lattice in the cytosol.

Cytochemical Studies

Cytochemistry is used in combination with microscopy to produce color or enhanced contrast at the sites of specific molecules. Specific stains for DNA and RNA have been used to demonstrate the location of these molecules within the cell. If the molecule being studied is an enzyme, chemicals can be added to sections of a tissue to cause the enzyme to generate an insoluble reaction product. This material will be deposited wherever the enzyme is located. Researchers can identify these sites precisely by looking at the section through a light microscope or electron microscope. They can see the reaction product readily because it is too dense to allow the passage of light or electrons.

Autoradiography Detects the Location of Specific Molecules

Some isotopes are unstable and tend to break down to a more stable form. Tritium, an isotope of hydrogen, is a specific example. This process is termed **radioactive decay** because it releases high-energy electrons or other particles known as **radiation.** Unstable isotopes that undergo radioactive decay are **radioactive isotopes,** or **radioisotopes.** Although radioisotopes are rare in nature, radioisotopes of phosphorus, iodine, sulfur, carbon, calcium, and hydrogen are readily manufactured for use in cell research.

Since radiation can be detected with extreme sensitivity, the amounts required in the laboratory are very small and well within the accepted safety limits. The use of radioactive molecules (molecules to which a radioisotope has been attached) is a powerful technique that allows researchers to follow the course of almost any process in a cell. A radioactive molecule performs identically to the naturally occurring molecule. Its biochemical properties are unaltered, and the cell reacts as if it were the natural nonradioactive form. However, because it is "labeled" with a radioisotope, any changes in its chemical form or its location can be followed as the molecule is processed within the cell. Researchers worked out the complex metabolic pathways of cells largely with the aid of this technique.

The location of radioactive molecules within cells or tissues is best determined by the technique of **autoradiography.** After incubating the cells with an appropriate radioactive compound to allow its incorporation into the macromolecules of the cells, the researcher washes the cells or slices of tissue to remove unincorporated radioactivity, fixes and embeds them, and cuts them into thin sections for either light or electron microscopy. Each preparation is then overlaid with a film of photographic emulsion and left in the dark. The radioactivity in the preparations reacts with the emulsion in much the same way that photographic film reacts to light. After developing the film, the researcher can determine the position of the radioactivity in the cells from the position of the dark spots on the emulsion when viewed in the microscope. Autoradiography has been very useful in discovering the secretory pathways in cells and in determining the role of the Golgi complex in this process.

Antibodies Are Used to Detect, Isolate, or Measure Specific Molecules

The mammalian **immune system** is an important defense mechanism that recognizes and destroys invading microorganisms and cells not normally present in blood. An important phase of this reaction is the synthesis and release of special proteins, known as **antibodies,** into the bloodstream. A molecule on the surface of the foreign cell triggers their production. A foreign molecule, such as a protein, polysaccharide, or nucleic acid, can elicit a similar response. Any molecule that stimulates the production of antibodies is termed an **antigen.** The antibodies that are released are highly specific molecules that recognize and bind only to the antigen. This reaction may inactivate the cell bearing the antigen, or, alternatively, the antibody-antigen complex on the surface of the cell may stimulate removal of the cell from the bloodstream by phagocytic cells specifically designed for this purpose.

Immunocytochemistry

Mammals can produce an enormous variety of antibodies. The specificity of these molecules makes them powerful research tools because they can be used to pick out specific molecules within a cell. For example, antibodies to tubulin can be used to study the organization of microtubules in a cell. A researcher labels the antitubulin antibodies by attaching a fluorescent molecule. The antibodies then are allowed to react with thin sections of cells or with intact cells that have been treated to allow them to enter. The antibodies will bind only to microtubules, since only the microtubules contain tubulin. Because the antibodies are fluorescent, they can be traced in a microscope fitted with an ultraviolet light source. Only the fluorescent label will be seen, but it suggests the presence of the antitubulin antibodies, which in turn indicate the microtubules to which they are bound (Fig. 3–17*a* and *b*) (see p. 98). Thus, the fluorescent sites within the cell indicate the location of microtubules. If the antibodies to an antigenic molecule are labeled with **ferritin,** an iron-protein complex that does not permit the passage of electrons, the location of the antigen within cells can be seen in great detail in the electron microscope.

Radiolabeled Antibodies

In a method similar to using fluorescent markers, a radioactive isotope can be attached to an antibody to give a **radiolabeled** antibody. The radiolabeled anti-

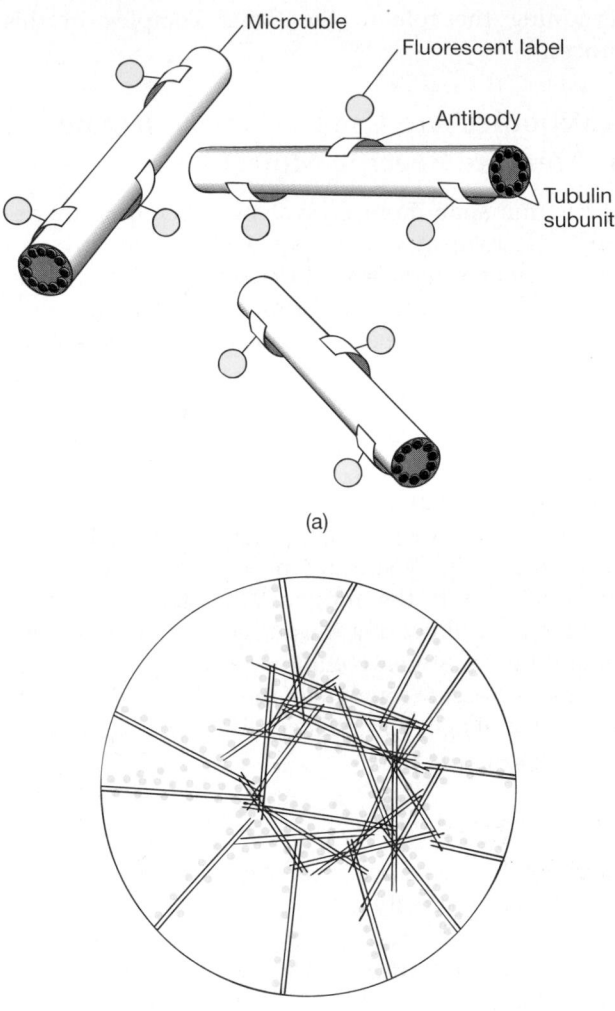

(a)

(b)

Figure 3–17

In the fluorescent-labeled antibody technique, an antibody to tubulin, which is the protein that makes up microtubules, is injected into a cell. The antibody has been labeled with a fluorescent marker. (*a*) The antibody attaches to the tubulin molecules as they form microtubules throughout the cell. A special microscope allows a view of the fluorescent markers, which indicate the presence of the antibodies, in turn showing the presence of microtubules (*b*).

body can be used in combination with autoradiography to pinpoint the location of specific molecules in the cell. Researchers have used this approach to follow the movement and fate of the plasma membrane that is internalized by cells during endocytosis. In this case, the antibody used is one that binds to molecules that are components of the plasma membrane.

Radioimmunoassays

Another use of radioisotopes and immunochemistry provides the means to measure very small amounts $(1 \times 10^{-12}$ mole) of certain molecules with great ac-

curacy. The measuring procedure is known as a **radioimmunoassay.** A specific example of its use is the measurement of cyclic AMP in various tissues, in blood, and in urine.

The basic principle of the method is to extract the cyclic AMP and allow it to bind to an anticyclic AMP antibody. A trace amount of radiolabeled cyclic AMP is added to the mixture, which is then set aside for several hours. The unlabeled and labeled cyclic AMP are identical molecules that compete for binding to the antibody. The amount of radiolabeled cyclic AMP present is always the same. The only variable is the amount of unlabeled cyclic AMP in the sample; this determines how much labeled cyclic AMP can be bound to the antibody. When there is very little cyclic AMP in the sample, a lot of **labeled** cyclic AMP can bind to the antibody because only a few unlabeled cyclic AMP molecules will compete for binding. If there is a lot of cyclic AMP in the sample, however, much more will bind to the antibody and only a small amount of labeled cyclic AMP will be able to bind (Fig. 3–18). A researcher can determine the amount of radiolabeled cyclic AMP bound by the antibody by isolating the antibody-antigen complex and measuring the radioactivity present. The amount of radiolabeled cyclic AMP bound is inversely related to the amount of cyclic AMP present in the sample. The latter is easily calculated.

Affinity Chromatography

Another technique that uses the specificity of antigen-antibody reactions is **affinity chromatography.** Following is an example of how this technique might be used to study a particular antibody. A researcher produces the antibody to be studied by injecting a purified sample of an antigen several times into an animal (such as a goat or a rabbit) in which the antigen is a foreign molecule. When the researcher draws blood from the animal, it now contains the antibodies the researcher wants to use as well as proteins and other antibodies normally present in the blood. The antibody in question must be separated out. The researcher allows the blood sample to clot in order to remove red blood cells. The remaining serum is passed through a column of an insoluble carbohydrate matrix to which an antigen to the antibody is attached. The antibody will bind to the antigen and will be retained on the column while the other serum proteins pass straight through (Fig. 3–19). The antibody can then be washed off the column and recovered separately in a relatively pure form.

A disadvantage of this technique is that antibodies to a single injected antigen are produced by a variety of antibody-producing cells in the animal. Each of these cells produces its own antibody that recognizes and binds to a part of the structure of the anti-

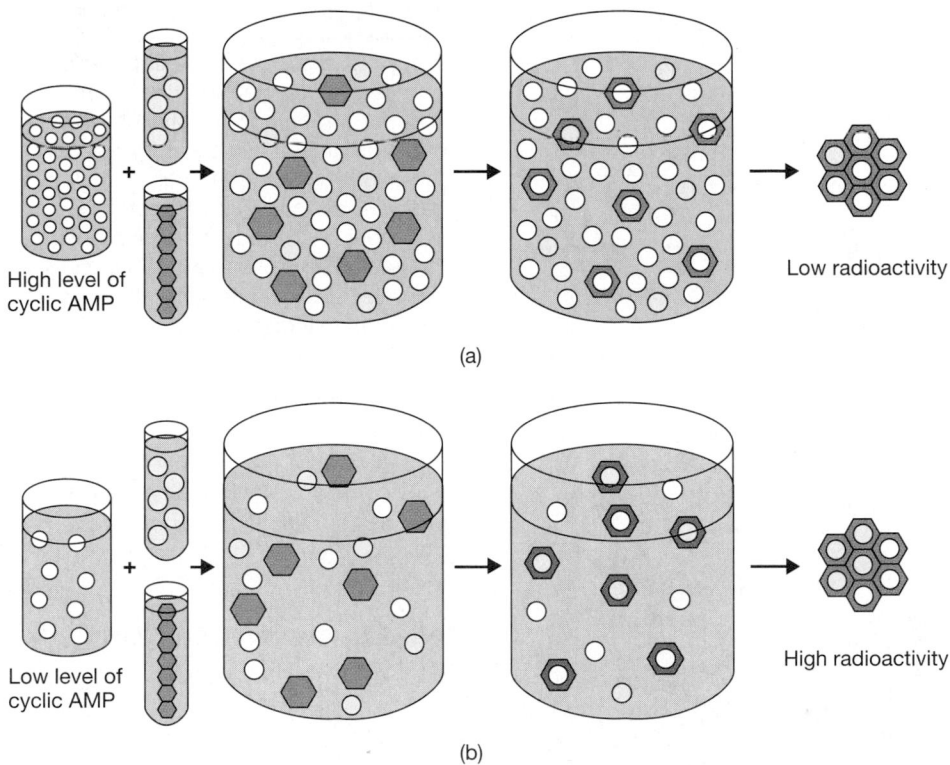

Figure 3–18

A radioimmunoassay to determine the relative quantity of cyclic AMP present in a solution. (*a*) Radioactive (orange) cyclic AMP molecules and antibodies (blue) specific to cyclic AMP are added to a solution containing an unknown amount of (white) cyclic AMP. When the antibody-antigen complex is isolated, the presence of radioactive cyclic AMP is relatively low, revealing that a large amount of cyclic AMP was present in the original solution. (*b*) The same process is followed in another solution, in which the amount of cyclic AMP is also unknown. The same numbers of radioactive molecules and antibodies are used, but the isolated antigen-antibody complex has a high proportion of radioactive cyclic AMP. This indicates that the original solution had a low level of cyclic AMP.

gen, but it is usually not the same part recognized by the antibodies produced by the other cells. Thus, the antibodies specific for a given antigen are not a singular molecular species, and the different antibody molecules vary widely in the efficiency with which they bind to the antigen. The antibody preparation is described as **heterogeneous,** and it has limited use.

Monoclonal Antibodies

The problem of heterogeneity was solved in 1976 by the use of **hybridoma cell lines** to produce large quantities of identical antibody molecules known as **monoclonal antibodies.** The cells that produce antibodies have a limited lifetime outside the body. But researchers can make them grow and multiply indefinitely in a culture dish by joining them to cancer cells, which have an unlimited ability to grow under these conditions. The two joined cells become a single cell that still produces antibodies but now has an unlimited lifetime; the cell line is known as a **hybridoma.** The hybrid cells that produce the specific needed antibody can be selected, and a single hybrid cell is used to start the growth of large numbers of new cells, a procedure known as **cloning.** Since all the new cells were originally derived from the same hybrid cell, the antibody molecules that the new cells produce will be identical.

An additional advantage is that the technique allows a selection step to identify the cell producing the required antibody. This means that researchers do not need to purify the antigen used at the start of the process because cells producing antibodies against the impurities can be eliminated at the selection step. Initial purification of the antigen is a critical step in the conventional procedure for antibody production, as described previously. The high specificity of monoclonal antibodies makes them tools that are as sensitive as radioisotopes for detecting specific molecules in a complex mixture or in a section of tissue.

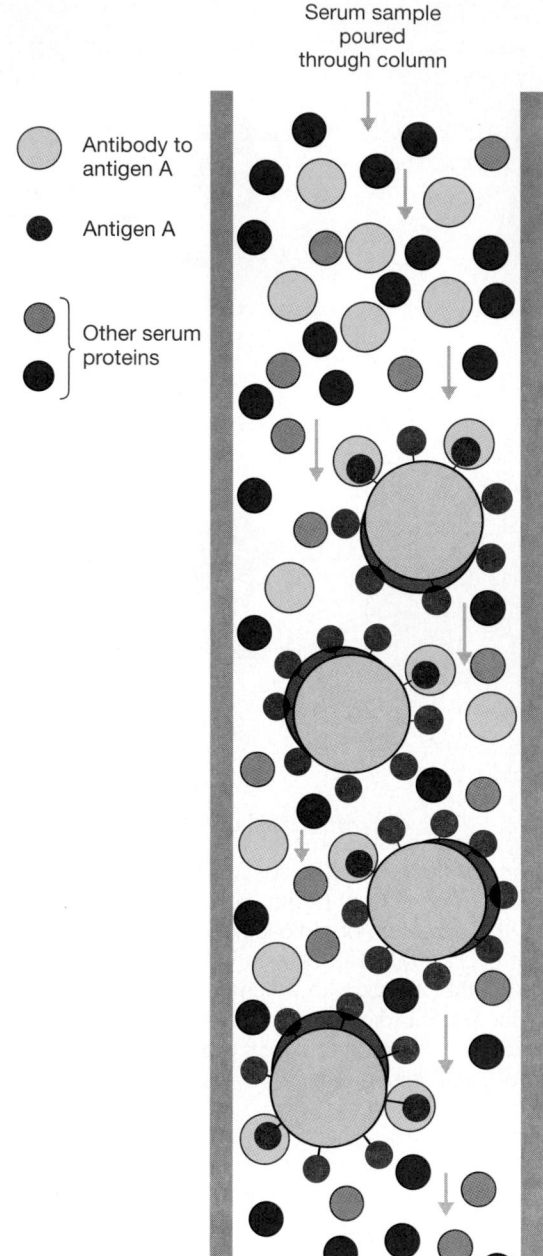

Serum sample
poured
through column

Antibody to
antigen A

Antigen A

Other serum
proteins

◄ *Figure 3–19*

Affinity chromatography. Antibodies are produced when an antigen is injected into the blood of an animal such as a goat or a rabbit. To isolate these antibodies, researchers draw a blood sample, collect the serum, and then pour the serum through a column that contains an antigen specific to the antibody being studied. The antibodies being studied bind to the antigen in the column, whereas other proteins in the serum pass through. The antibodies being studied can then be washed from the column and studied.

factors. In contrast with other column separation techniques, however, HPLC involves the use of great pressures to force the solution through the column in a very short time. HPLC also provides a greater degree of resolution than do other chromatographic techniques.

Cell Organelles Are Isolated by Cell Fractionation

The purpose of **cell fractionation** is to separate the different organelles of the cell in order to study each organelle on its own in a strictly controlled, "cell-free" system. This isolation permits direct, detailed analysis of the structure and the function of the organelle.

Fractionation is begun by a physical disruption of the tissue and its cells, a process that can be controlled to minimize damage to the internal organelles. Before being broken up, the tissue usually is placed in a solution of sucrose, which helps to protect organelle structure. A common method of tissue disruption is **homogenization,** in which the tissue and its cells are broken by being forced between a rotating pestle and the walls of a glass vessel (Fig. 3–20). The clearance between the two surfaces is small enough to break cells but too large to harm most of the internal organelles. Mitochondria, lysosomes, and nuclei remain intact, while large structures such as the endoplasmic reticulum are broken into small pieces, as is the plasma membrane. The resultant thick suspension of broken cells in sucrose is called a **homogenate.**

Once the tissue and its cells have been broken up, centrifugal force is used to separate the organelles. The most widely used technique for separating the cell organelles present in a mixture such as a homogenate is **differential centrifugation.** The homogenate is placed in a special tube that can be spun at high speed in a machine called a **centrifuge.** The centrifugal force generated by these machines can exceed 100,000 times the force of gravity (g). Under the influence of a centrifugal force, the structures present in the homogenate begin to move toward the bottom of the centrifuge tube (see Fig. 3–20). The large and dense structures (e.g., the nuclei) move at the fastest rate and collect as a "pellet" at the bottom of the tube.

HPLC Is Important for Rapid Amino Acid Sequencing

When a protein has been isolated, scientists often wish to determine the exact sequence of amino acids that make it up. Various methods exist for **amino acid sequencing.** Researchers usually begin by breaking the protein into shorter peptide chains, using enzymes that act to split peptide bonds only between specific amino acids. The peptide fragments can then be analyzed by various chromatographic means. One recent advance in this area is the development of **high-performance liquid chromatography (HPLC).** Solutions of amino acids are poured into columns containing densely packed separation substances, which separate amino acids based on size or other

Homogenization

Differential
centrifugation

Homogenate

Step 1   Step 2   Step 3

Nuclei,
plasma
membrane

Mitochondria,
lysosomes
peroxisomes

ER,
small membrane
pieces

Free ribosomes,
soluble molecules

Density gradient centrifugation

Figure 3–20

In cell fractionation, tissue is ground up and then spun by a centrifuge at a set speed. The spinning causes certain organelles to settle out in a pellet so that those organelles can be studied in quantity. The liquids that remain can be spun at a different speed, which isolates a different sample of organelles, and so on. A pellet can be further analyzed through density gradient centrifugation.

The centrifugal force can be adjusted so that only the nuclei collect during one centrifugation, whereas the other organelles remain suspended in the solution or "supernatant."

The supernatant is then poured into a new tube, which is spun at a higher centrifugal force. A different group of smaller organelles is collected as a separate pellet. In this way, the homogenate can be divided by centrifugal force into several "fractions," each containing a specific organelle. The total centrifugal force applied to the homogenate depends not only on the force used but also on the time for which it is applied. The sequence of different forces and times that is commonly used to fractionate tissue homogenates is as follows:

Step	Centrifugation	Pellet Contents
1. Homogenate	$1000 \times g$ for 10 min	Nuclei, large pieces of plasma membrane
2. Supernatant from 1	$10,000 \times g$ for 20 min	Mitochondria, lysosomes, peroxisomes
3. Supernatant from 2	$100,000 \times g$ for 1 hr	Endoplasmic reticulum, small pieces of plasma membrane
4. Supernatant from 3 contains free ribosomes and soluble molecules.		

R E S E A R C H   F O C U S

Gene Therapy

About 1% of all babies are born with an inherited genetic defect. We know of more than 4000 different types of inherited disorders, many of which lead to early death because there is no fully effective treatment. Due to the recent advances in recombinant DNA technology, discussed briefly in this chapter, scientists are on the verge of repairing certain genetic defects by introducing the normal undamaged genes into patients. The first two human gene therapy trials in the USA were approved in September 1990 by the National Institutes of Health (NIH) Recombinant DNA Advisory Committee. This marked the start of a new era for biomedical science, in which genetic engineering will be used to alleviate the suffering caused by genetic diseases and perhaps even serious illnesses such as cancer.

The best candidates for gene therapy are disorders caused by a single damaged gene rather than by multiple genes (recall that every chromosome contains thousands of genes). Ideally the new gene will exactly replace the damaged one and the disease will be cured for life with a single treatment and no side effects. (Random insertion of a new gene anywhere in one of the chromosomes is less ideal because of the risk of activating a cancer-inducing gene, or **oncogene.**) In reality, however, it is extremely difficult to control the fate of DNA introduced into a cell. For this reason, one of the approaches currently being studied is the introduction of a healthy gene into a cell without removal of the defective gene. In this way the healthy gene supplies the missing product of the defective gene and the risk of side effects is reduced. The disadvantage of this approach is that several treatments would be required throughout life, and so it is not a cure. The procedure, known as **gene augmentation,** involves removing cells from the patient, introducing the healthy gene into the cells, and returning the altered cells to the patient. Genetic flaws do not have to be corrected in all of the body's cells, provided that the altered cells can supply enough of the missing product. Bone marrow and skin and liver cells are the best candidates for gene therapy. They withstand removal from the body, can be returned without great difficulty, and survive for a long time after replacement within the body.

An important strategy for introducing a healthy or "therapeutic" gene into the isolated cells uses the native ability of a **virus** to enter cells. The most promising system uses a **retrovirus,** a type of virus that uses RNA as genetic material. Once inside a human cell, the retrovirus converts its RNA to DNA and inserts the DNA into a chromosome, causing the cell to begin making viral proteins. Many retroviruses have been engineered to serve primarily as gene delivery vehicles. Investigators substitute a therapeutic gene for viral genes and delete the instructions for making new viral proteins. The modified retrovirus still enters cells and inserts the therapeutic gene into cellular DNA, but now it has lost the ability to reproduce.

One of the NIH trials is focused on a rare condition known as **severe combined immuno-deficiency,** which often arises because the function of white blood cells, part of the body's defense against disease, is impaired by a damaged gene. The patients in this study are infused with their own white blood cells, which have been previously removed and modified (using the retrovirus technique) to contain the therapeutic gene. The other NIH trial recognizes that gene therapy can also be used to enhance the ability of cells to fight serious diseases such as cancer. A class of white blood cells that penetrate tumors is being modified to contain a new gene that will increase their anti-cancer activity. Their therapeutic potential will be evaluated in patients with advanced cancer.

Gene therapy is an exciting development but is not without problems, both technical and ethical. There is dramatic success in some patients while others show no improvement. The technique will require extensive research in a high technology environment for several more years. Intentional modification of human genes also is disconcerting for many people. However, the consensus among scientists is that human gene therapy can be beneficial if carefully scrutinized and is worth pursuing if it proves successful in combating human disease.

A schematic of the AIDS retrovirus.   *(© Peter Arnold/ Centers for Disease Control.)*

■   **What enzymes are used to construct recombinant DNA?**

Although the pellet fractions are enriched in certain organelles, they are not pure because more than one type of organelle is usually present. This is one of the limitations of differential centrifugation. Different organelles in each pellet can be separated by resuspending the pellet material and placing it as a thin layer on top of a sucrose solution that increases in density from the top to the bottom of the tube (see Fig. 3–20). This technique, **density gradient centrifugation,** allows the separation of organelles that have the same size but different densities. Upon application of a centrifugal force, the organelles with different densities will separate from one another because they will move at different rates through the gradient. At the end of the centrifugation the gradient usually contains two to three separate "zones" or bands of material, each one representing a relatively pure fraction of a specific organelle (see Fig. 3–20). The purity and integrity of a mitochondrial fraction, for example, can be determined by electron microscopy and by checking to see whether enzymes characteristic of lysosomes, peroxisomes, endoplasmic reticulum, and other structures are present.

Isolated fractions of cell organelles are used widely to study the processes that occur within cells, because the process of interest can be isolated and studied without interference from the complex side reactions that occur in the intact cell. The use of a cell fraction that could synthesize polypeptides when provided with mRNA was a crucial step in understanding the mechanism of protein synthesis.

New DNA Molecules Are Formed by Genetic Engineering

Recall that a **gene** is a section of a DNA molecule that contains all the information needed for the synthesis of a single polypeptide chain. The techniques of genetic engineering allow scientists to form new DNA molecules by combining pieces of DNA from markedly different organisms. The new DNA is known as **recombinant DNA.** The cell that contains this new DNA will use it to synthesize a new protein, one that the cell does not normally produce.

One of the early commercial successes of genetic engineering was the insertion of the gene for human

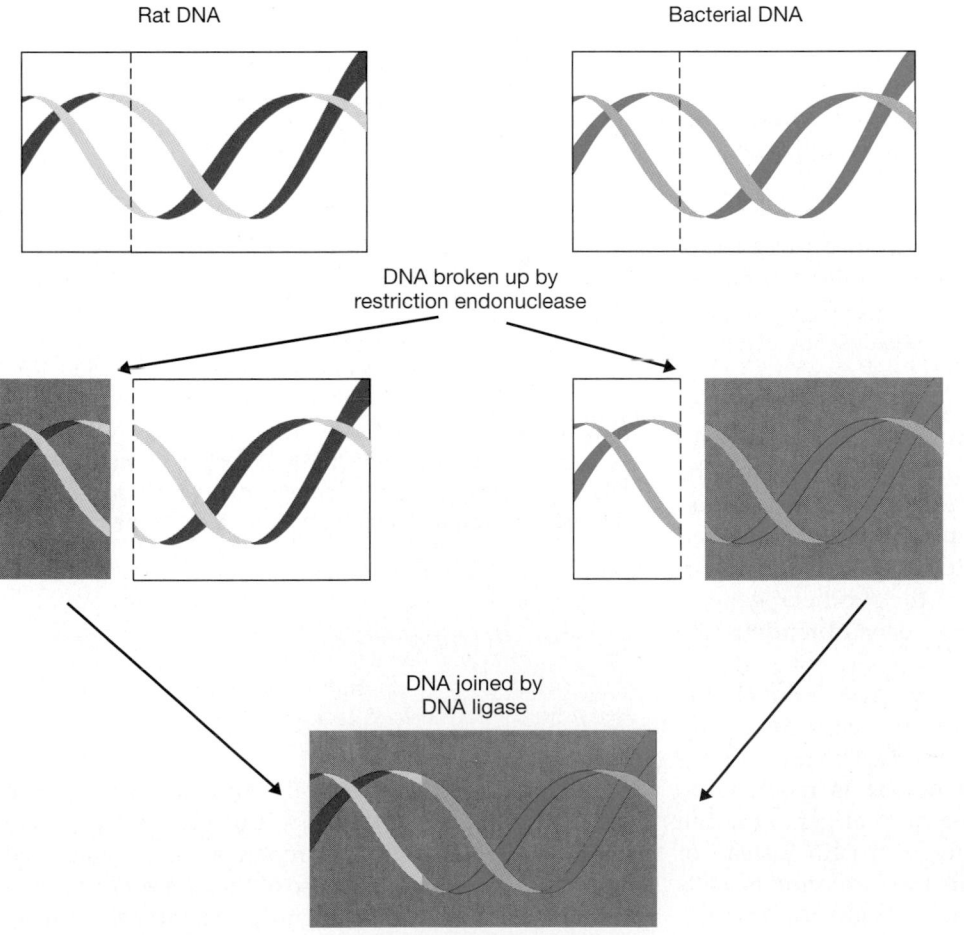

Rat DNA

Bacterial DNA

DNA broken up by restriction endonuclease

DNA joined by DNA ligase

Recombinant DNA

Figure 3–21

Researchers prepare recombinant DNA by splicing together DNA strands from two different organisms.

insulin, a polypeptide, into the DNA of the bacterium *Escherichia coli (E. coli)* so that the bacterium synthesized human insulin in addition to its own proteins. Large amounts of insulin now can be produced at relatively low cost when the *E. coli* are grown on a large scale. The method is one of the most powerful tools for understanding how the information stored in chromosomal genes is used and how it is regulated. Here we briefly review some of the basic techniques, terminology, and potential applications of genetic engineering.

The first step in the preparation of recombinant DNA is to break both strands of the native DNA at specific sites, using enzymes known as **restriction endonucleases.** Next, the broken strands of two different DNA molecules are joined together by a different enzyme called **DNA ligase** (Fig. 3–21) (see p. 103). The combined use of these enzymes allows any DNA fragment to be grafted into the DNA of a bacterium.

If the inserted DNA fragment represents a single gene, these recombination experiments can produce large amounts of the gene because it is copied and reproduced as the bacteria rapidly divide and multiply. This process is known as **DNA cloning.** Researchers recover the gene from the bacterial DNA using the same restriction endonuclease used to break open the original DNA. The amount of recombinant DNA that is now available is sufficient for detailed studies of the precise sequence of nucleotides.

The bacterial DNA used for DNA cloning is not the chromosomal DNA; rather, it is a small accessory DNA molecule known as a **plasmid.** Researchers use these molecules because they are easily separated from the chromosomal DNA and, once the gene of interest has been attached to them, they can be put back inside living bacterial cells (Fig. 3–22). Plasmids that are used in DNA cloning studies are often referred to as **cloning vectors.**

Once the nucleotide sequence of a specific gene has been identified, it is possible to check the accuracy of this analysis by chemically synthesizing the gene from its constituent nucleotides. The gene can be incorporated into a cloning vector and inserted into bacteria for cloning and to see if the bacteria produce a functional protein from the synthetic gene.

The ability to use synthetic genes broadens the scope of these techniques enormously. For instance, after sequencing a gene that produces a functional protein, researchers can synthesize one that is slightly different, a **mutant** gene, or one that makes a nonfunctional protein. If this mutant gene is inserted into a mammalian cell to replace a normal gene, the cell then synthesizes a nonfunctional protein instead of a functional one. Changes in the behavior of cells that contain the mutant gene provide information about the biological role of the normal protein within the cell.

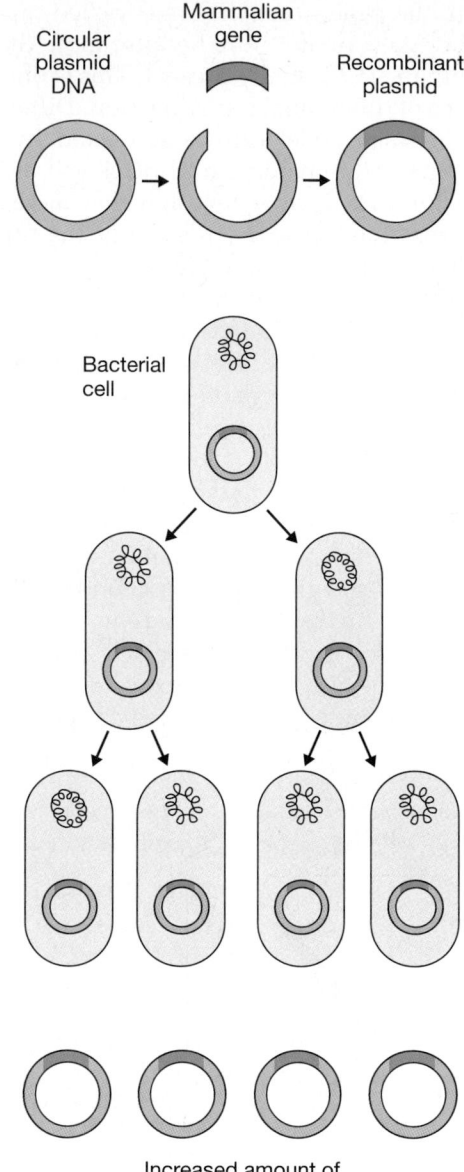

Figure 3–22

Researchers can produce large quantities of recombinant DNA by placing them in bacterial cells, which duplicate the recombinant molecule each time they divide.

At present it is extremely difficult to introduce genes into mammalian DNA, but ultimately it may be possible to genetically repair a mammalian cell that is nonfunctional because of the presence of a mutant gene. In this case, a chemically synthesized fragment of DNA that contains the normal gene would be inserted into the DNA of the cell.

CHAPTER REVIEW

Summary

- Cell functions are controlled from the nucleus where DNA stores the information the cell uses to synthesize proteins. Many of these proteins are enzymes that catalyze specific reactions within the cell. Others are secreted from the cell to perform a specific function elsewhere in the body.
- A plasma membrane surrounds the cell to protect it from fluctuations in the external environment. Intracellular membranes surround specific organelles so that their activities can occur independently without endangering other organelles.
- The nucleus is the site where nucleic acids are synthesized. DNA remains in the nucleus. The information stored in DNA is copied in the form of RNA, which leaves the nucleus and is used by the protein-making organelles in the cytoplasm.

- The cytoplasm consists of various organelles suspended in cytosol. The cytosol is a viscous fluid that contains many enzymes and precursors for macromolecules.
- The life cycle of a cell is divided into interphase (the growth phase) and mitosis (cell division). During interphase, which occupies most of the cell cycle, the cell conducts normal activities and replicates its DNA. During mitosis the nucleus and cytoplasm divide into two, and each daughter cell receives an exact copy of the original DNA.
- Some of the basic research techniques used to study cell physiology include cell culture, light and electron microscopy, use of radioactive isotopes and antibodies, isolation of specific cell organelles, and genetic engineering.

Review Questions

Identify Each with a Term

1. The fluid surrounding the nucleus of a cell.
2. The complex formed between DNA and nuclear proteins such as histones.
3. The process by which secretory vesicles fuse with the plasma membrane and expel their contents outside the cell.
4. The structure that maintains cell shape and mediates cell movements.
5. Division of the cell cytoplasm to form two daughter cells after completion of mitosis.

Define Each Term

6. Eukaryotic cell
7. Genome
8. Autoradiography
9. Density gradient centrifugation
10. Recombinant DNA

Choose the Correct Answer

11. The site of protein synthesis is the:
    a. Nucleus
    b. Lysosome
    c. Nucleolus
    d. Rough endoplasmic reticulum
12. Which of the following organelles is surrounded by a double membrane?
    a. Golgi complex
    b. Nucleus
    c. Peroxisome
    d. Smooth endoplasmic reticulum
13. Intermediate filaments are structural components of the:
    a. Cytoskeleton
    b. Mitochondrion
    c. Ribosome
    d. Chromosome

14. Monoclonal antibodies are produced by:
    a. Cytochemistry
    b. Hybridoma cell lines
    c. Radioimmunoassay
    d. Cloning vectors
15. The growth phase of the cell cycle is known as:
    a. Mitosis
    b. Metaphase
    c. Interphase
    d. M phase
16. The endoplasmic reticulum of a liver cell accounts for what % of the total membrane of the cell?
    a. 2–5
    b. 10–20
    c. more than 50
    d. 100

Answer Each Question in One or Two Sentences

17. List three advantages conferred by the presence of intracellular compartments.
18. Explain how 1 meter of DNA can fit inside the nucleus of a typical animal cell.
19. What are the functions of the three different types of RNA?
20. Why are mitochondria important for aerobic respiration?
21. How do cells from a primary culture differ from cells from a cell line?

Problem Solving in Physiology

22. How can genetic engineering aid our understanding of human physiology?
23. Explain what is meant by the statement "Lysosomes are the suicide system of the cell."
24. Briefly describe the origin and intracellular route followed by a protein destined for export from the cell.

Suggested Readings

Alberts, B., Bray, D., Lewis, J., Raff, M., Roberts, K., and Watson, J.D. *Molecular Biology of the Cell*, 3rd ed. New York and London: Garland, 1994.

Chien, S., ed. *Molecular Biology in Physiology.* New York: Raven Press, 1989.

Darnell, J., Lodish, H., and Baltimore, D. *Molecular Cell Biology*, 2nd ed. New York: Scientific American Books (W.H. Freeman), 1990.

Ellis, R.E., Yuan, J., and Horvitz, H.R. "Mechanisms and functions of cell death." *Ann. Rev. Cell Biol.*, 7:663–698, 1991.

Murray, A., and Hunt, T. *The Cell Cycle.* New York: Freeman, 1993.

Rosenberg, S.A. "Adoptive immunotherapy for cancer." *Scientific American*, 262:62–69 (May 1990).

Ueda, N., and Shah, S.V. "Apoptosis." *J. Lab. Clin. Med.*, 124:169–177, 1994.

Verma, I.M. "Gene therapy." *Scientific American*, 263:68–84 (November 1990).

Key Terms

antibody (p. 97)
antigen (p. 97)
apoptosis (p. 72)
cell cycle (p. 91)
chromatin (p. 80)
chromosome (p. 80)
cyclin (p. 93)
cytokinesis (p. 93)
cytoplasm (p. 74)

cytoskeleton (p. 88)
cytosol (p. 74)
deoxyribonucleic acid (DNA) (p. 75)
endoplasmic reticulum (p. 81)
exocytosis (p. 88)
gene (p. 79)
genome (p. 79)

Golgi complex (p. 83)
hormone (p. 75)
lysosome (p. 77)
metabolism (p. 74)
mitochondrion (p. 88)
mitosis (p. 91)
nucleolus (p. 80)
nucleus (p. 74)
organelle (p. 74)

phagocytosis (p. 87)
plasma membrane (p. 74)
prokaryotic cell (p. 74)
recombinant DNA (p. 103)
ribonucleic acid (RNA) (p. 75)

An electron micrograph of a kidney proximal tubule cell showing microvilli of the apical plasma membrane, extensive infoldings of basal and lateral membranes, and mitochondria between the basolateral infolding. *(© C. Craig Tischer, M.D./University of Florida)*

Chapter 4

Transport Through the Cell Membrane

Key Concepts

The plasma membrane is a selective barrier that separates the cell cytosol from the external environment. It is also important for communication between cells because it contains binding sites for chemical signals.

All cell membranes are composed mainly of phospholipids and proteins. The proteins and lipids can move rapidly within the plane of the bilayer, but movement from one half of the bilayer to the other half is very slow.

Gradients are responsible for the movement of various substances throughout the body, such as the flow of blood and the diffusion of ions and molecules in solution.

Many substances cross the plasma membrane by moving spontaneously down a gradient (passive transport). Other substances that cross the plasma membrane must move against a gradient (active transport). The passage of ions and large molecules is assisted by membrane transport proteins.

Epithelial cells are organized structurally and functionally for transcellular transport of solutes and water.

Chapter Outline

Overview of Plasma Membrane Functions

    The Components and Structure of the Membrane

    Gradients and Movement

    Movement of Substances Through the Membrane

    Epithelial Transport

It was a cold January afternoon and Rod decided to take a break from studying. He pulled his armchair closer to the fire and opened the new novel he had just received as a gift from his sister. It was one of the popular series of fantasies set in medieval times in which a group of friends sets out on an epic quest and succeeds against all odds—ready to begin another hair-raising adventure in the next volume. For Rod, and many others with a vivid imagination, it was gripping and enjoyable escapism: just the thing to take his mind off the impending physiology test. In an early chapter, the heroes take shelter in a castle and wake up the next day to discover that it is under siege. Something would have to be done about this, and Rod became deeply absorbed in the detailed discussion of the possible courses of action.

Medieval castles were usually surrounded by a moat and a high wall to prevent unauthorized entry. Entry and exit were restricted to the main gate, which was guarded and protected by a drawbridge and portcullis. This allowed the owner of the castle to admit friends and exclude enemies. Suppliers of food and other materials could enter when needed. When the gate was closed and the drawbridge was raised, the castle became virtually impregnable. Attack was a major undertaking because a well-designed castle could absorb considerable aerial bombardment, and defenders had protected positions from which to repel attempts to scale the walls. For these reasons, the attackers would sometimes retreat a safe distance and choose the less hazardous approach of besieging the castle until the owner surrendered. If adequate supplies of water, food, and other necessities had been accumulated in the castle, the occupants could continue their lives in safety behind the fortified walls. They could withstand the siege for many months, sometimes a year or two, in the hope that help would arrive or the attackers would eventually go away.

As Rod's thoughts began to focus again on physiology, he realized that, just as the walls of a castle allowed the occupants to survive in a sometimes harsh environment, so the plasma membrane of a human cell acts as a bastion against external influences. No matter what occurs outside, the cell is able to bring in important molecules and ions, release or exclude others, and conduct its internal activities with great precision behind the safety of the plasma membrane. This membrane might seem to be a frail layer of lipids and proteins, but it is an effective barrier to water-soluble substances. Membrane proteins provide passageways by which specific ions or large polar molecules can cross the plasma membrane. In this way, metabolic fuels can enter and waste products can leave the cell. Also, many of these pathways can be opened or closed in response to changing conditions. The plasma membrane, by separating the cytoplasm from the extracellular fluid, allows the cell to maintain its internal environment at the optimal composition and pH for metabolic activity and other intracellular functions. The plasma membrane also ensures that the composition of the cytosol remains markedly different from the extracellular fluid. The difference is important because it is exploited by the cell to drive many membrane transport processes.

In this chapter we will consider first the structure and organization of the main components of plasma membranes. Later sections will discuss the mechanisms by which water, small ions, and solutes can move across membranes and epithelial cells. Although this chapter focuses primarily on the membrane that surrounds each cell, keep in mind that the membranes surrounding the various cell organelles

are similar to the plasma membrane in composition and behavior.

Overview of Plasma Membrane Functions

Every cell is surrounded by a plasma membrane that separates the contents of the cell from the environment. The plasma membrane is a highly selective filter that permits nutrients to enter and waste products to leave the cell. This selective nature allows the composition of the cell cytosol to be very different from its surrounding environment. For example, the concentration of potassium ions inside most cells is about 25 to 30 times higher than the concentration of potassium outside, in the extracellular fluid. Other important differences must be maintained between the two fluid compartments (Fig. 4–1). In addition to this function as a selective barrier, the plasma membrane also has many sites on its external surface that bind specific chemical signals produced by other cells. Thus, the plasma membrane has an important role in cell-to-cell communications.

The Components and Structure of the Membrane

The following observations on the behavior of the major membrane components have been integrated into a generalized model of the structure of cell

Figure 4–1

A simplified animal cell and its plasma membrane, which separates the intracellular fluid (the cytosol) from the extracellular fluid.

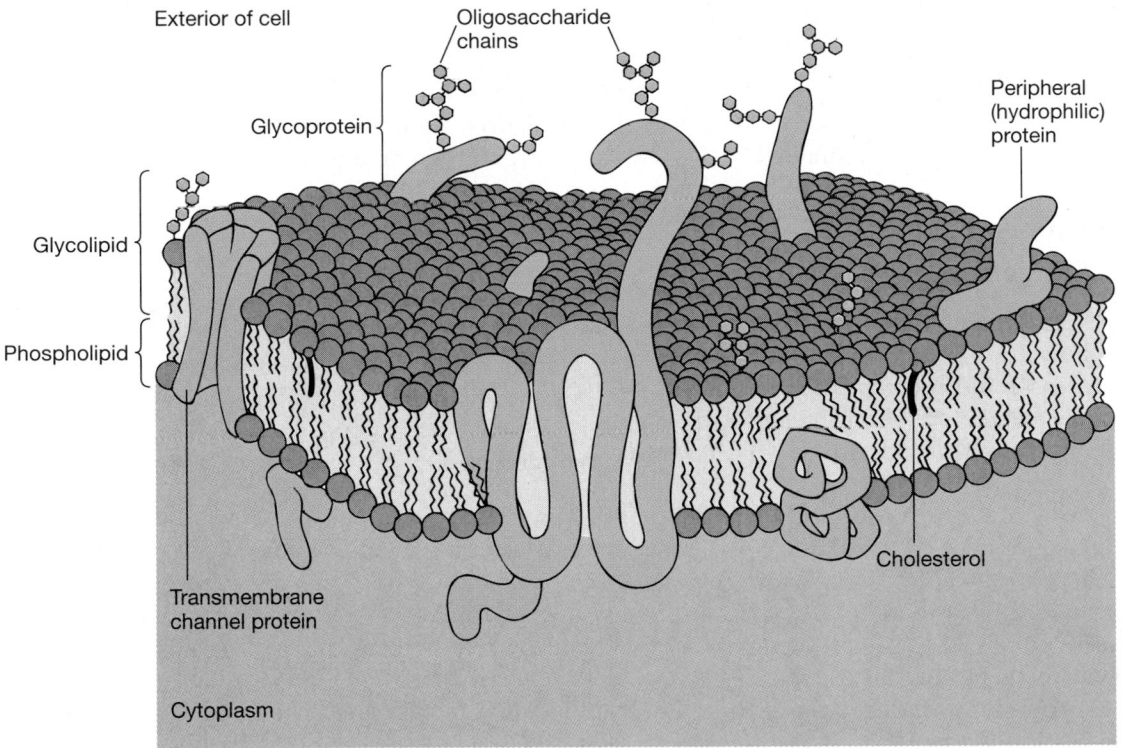

Figure 4–2

The current fluid-mosaic model of the structure of plasma membranes.

plasma membranes (Fig. 4–2). It is often referred to as the **fluid mosaic model** or the **Singer-Nicolson model,** the latter being the names of the scientists who first put forward the concept. Although Figure 4–2 is designed to represent the structure of plasma membranes, the structure of intracellular membranes is very similar.

The plasma membrane consists primarily of lipids and proteins. The core structure of the plasma membrane is a double layer of phospholipid molecules known as the **lipid bilayer** (see Fig. 4–1). Lipids can account for as much as 50% of the mass of a plasma membrane.

Many of the functional properties of the membrane, however, are due to the presence of specific proteins that are closely associated with, and often embedded in, the lipid bilayer. Typically, proteins also account for about 50% of the mass of cell membranes, although this figure can vary from 18% in the membrane of the Schwann cell (a supporting cell in the nervous system) to 76% in the inner membrane of a mitochondrion (Fig. 4–3).

The relative quantities of lipid and protein in a plasma membrane usually reflect some aspect of the function of the cell or organelle that the membrane protects. Thus, the high lipid content of the Schwann cell membrane suits the role of this membrane as an insulator around certain nerve fibers. Conversely, the high protein content of the inner membrane of a mi-

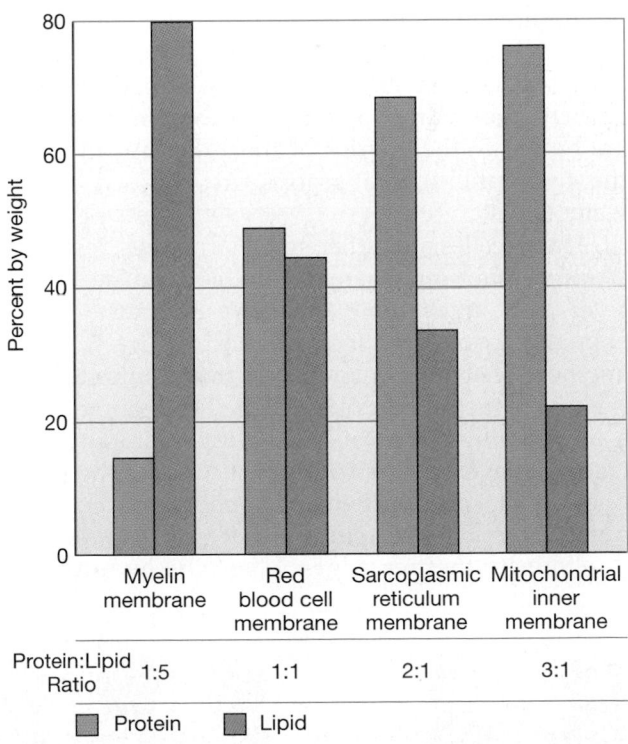

| Protein:Lipid Ratio | 1:5 | 1:1 | 2:1 | 3:1 |

Figure 4–3

The proportions of membrane protein and lipid vary as the function of the membrane changes from primarily an insulator (myelin) to a complex variety of enzymatic and transport functions (mitochondrial inner membrane).

tochondrion reflects the rich complement of enzymes and other proteins vital for the processes of energy transformation that occur in these organelles.

Membrane Lipids Are Amphipathic Molecules

The specific types of lipids found in membranes are **phospholipids, cholesterol,** and **glycolipids.** All of these molecules are insoluble in water. However, they all have hydrophilic, or water-soluble, regions and hence are called **amphipathic** molecules (*amphi* = "both").

Look at the structure of **phosphatidylcholine** (Fig. 2–17c) for an example of a typical amphipathic phospholipid. Phosphatidylcholine has a hydrophilic head group and two long hydrophobic fatty acid tails, each of which may contain 14 to 24 carbon atoms in an unbranched chain. (The 16- to 18-carbon fatty acids are the most common fatty acid chains found in phospholipids; typical examples are **palmitic acid** and **oleic acid.**)

Recall from Chapter 2 that nonpolar molecules and groups, when placed in aqueous solution, cluster together in hydrophobic pockets. Since they consist mainly of hydrophobic fatty acid chains, phospholipids form clusters in order to "hide" their hydrophobic tails from water molecules. They form either spheres, called **spherical micelles,** or **bimolecular sheets (bilayers)** (Fig. 2–17d). In both types of clusters, the hydrophobic tails of the phospholipids are buried in the interior of the structure, and only the hydrophilic head groups are exposed to the water.

The micelles are rather small structures, less than 20 nm in diameter, whereas a bimolecular sheet or bilayer is a much more extensive structure. When enough phospholipid molecules occur together, they are more likely to form a bilayer than a micelle. This is because the interior of a micelle has room for only a limited number of bulky hydrophobic fatty acid tails. A bilayer is a more efficient way for the phospholipids to protect their hydrophobic portions because it tends to close in on itself (forming a circle) to eliminate the free edges where the hydrophobic

tails would be exposed to water. For the same reason, it reseals itself if punctured. Thus, a bilayer is the favored structure in an aqueous solution.

Movement of Phospholipids Within the Bilayer

Studies on the movement of individual phospholipid molecules within a lipid bilayer have revealed that the bilayer is a dynamic structure. Phospholipid molecules move laterally along one layer of the membrane, but crossing from one layer to the other is much more difficult.

Movement Within Molecules: Flexing and Rotation. Within each phospholipid molecule there is a certain amount of movement. The phospholipid molecules are able to rotate about their longitudinal axes (Fig. 4–4a). In addition, at physiological temperatures (37°C), the hydrocarbon tails are mobile and undergo rapid flexing movements. The rate of this movement is higher if the hydrocarbon tail is short and contains double bonds.

Stabilization by Cholesterol. In eukaryotic cell membranes, as much as 20% of the total lipid may be cholesterol. Cholesterol stabilizes the lipid bilayer. Like the phospholipids, cholesterol is an amphipathic molecule; however, its steroid ring structure is more rigid than the long fatty acid tails of the phospholipids. When present in the lipid bilayer, the ring structure of cholesterol can interact with and reduce the motion of the neighboring phospholipid hydrocarbon chains (Fig. 4–4b).

Movement Within Layers. Phospholipid molecules exchange positions with their neighbors in the same half of the bilayer by very rapid lateral diffusion (Fig. 4–5a). By this process a lipid molecule takes only 1 sec to move about 2 μm, the entire length of a large bacterial cell.

Movement Between Layers. Lipids also move from one half of the lipid bilayer to the other in what is called **"flip-flop."** This motion requires that the polar head groups of the phospholipids in one half of the bilayer pass through the hydrophobic interior of

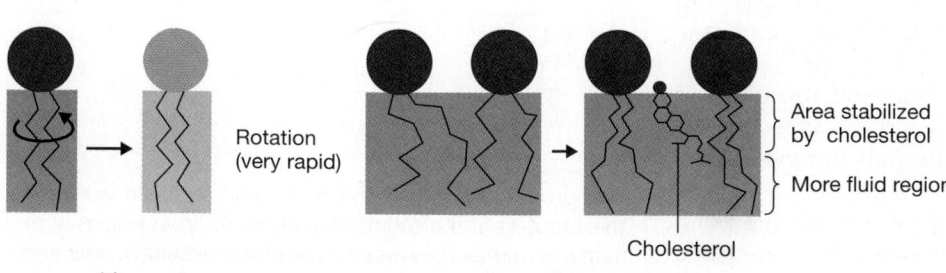

Rotation (very rapid)

Area stabilized by cholesterol

More fluid region

Cholesterol

(a)   (b)

Figure 4–4

Phospholipid molecules (*a*) rotate on their axes and (*b*) undergo flexing motions. The rigid steroid ring structures of cholesterol stabilize certain portions of the bilayer.

the bilayer to reach the other half of the bilayer (Fig. 4–5*b*). This is a difficult task; hence, flip-flop movement is extremely slow.

Lipid Asymmetry in the Bilayer

Phosphatidylcholine is not the only phospholipid in biological membranes. Other important membrane phospholipids are **phosphatidylethanolamine, phosphatidylserine,** and **phosphatidylinositol** (all derived from glycerol), and **sphingomyelin** (derived from sphingosine). These different phospholipids are not distributed equally between the two halves of the lipid bilayer. A good example is the plasma membrane of the red blood cell, in which most of the phosphatidylcholine and sphingomyelin is present in the outer half of the bilayer, whereas most of the phosphatidylethanolamine and all of the phosphatidylserine are present in the inner half (Fig. 4–6).

The persistence of this **lipid asymmetry** in membranes is good evidence that the rate of phospholipid flip-flop is very slow. If flip-flop were rapid, the various phospholipids would probably mix randomly and thus be distributed equally between the bilayers. The phospholipid asymmetry must originate during the initial assembly of the bilayer, but scientists do not understand the mechanism and possible advantages of this arrangement.

Texture of the Plasma Membrane

The net result of the movements of the phospholipids and their interactions with cholesterol is that, at 37°C, the interior of the lipid bilayer of biological membranes is more like a viscous fluid such as olive oil than a rigid crystalline matrix. This fluidity of the membrane structure is probably very important for the normal function of many membrane proteins.

Most Membrane Proteins Penetrate the Lipid Bilayer

Proteins are an important component of most biological membranes. Some proteins, called extrinsic proteins, are found only on the surface of the membranes, whereas others, called intrinsic proteins, partially or completely penetrate the lipid bilayer.

Intrinsic Proteins: Amphipathic Molecules

The proteins that penetrate the bilayer, called **intrinsic (or integral) proteins,** account for about 70% of total membrane protein. Most of them probably span

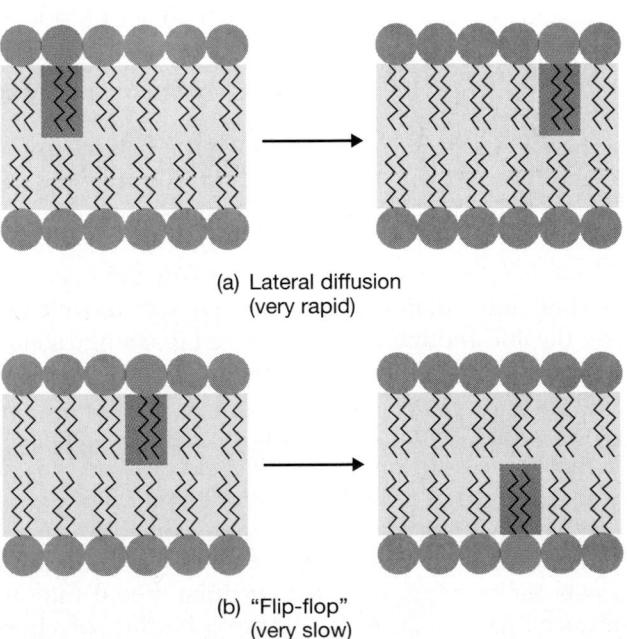

(a) Lateral diffusion
(very rapid)

(b) "Flip-flop"
(very slow)

Figure 4–5

(*a*) Phospholipid molecules diffuse laterally very rapidly within one layer of the plasma membrane. (*b*) Phospholipids move very slowly from one layer to the other.

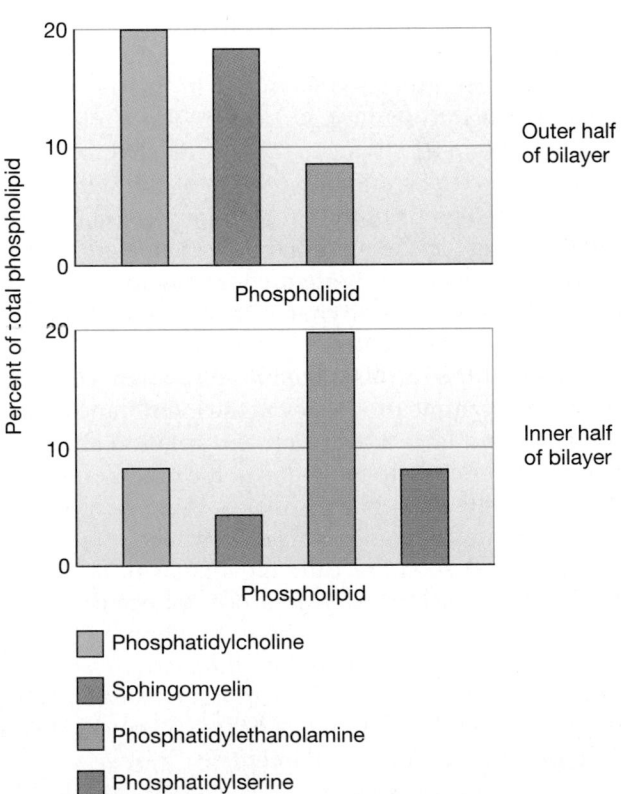

Figure 4–6

Asymmetrical distribution of lipids in the erythrocyte membrane. Note that each half of the bilayer contains 50% of the total lipids present in the membrane.

Figure 4-7

The arrangement of integral and peripheral proteins in a biological membrane.

the bilayer completely, so that parts of the protein are exposed on both sides of the bilayer.

Intrinsic proteins are usually amphipathic molecules. Their hydrophobic regions cluster in the interior of the bilayer with the hydrophobic portions of the lipids, whereas their hydrophilic regions remain exposed to the aqueous medium outside the bilayer (Fig. 4-7).

The three-dimensional structure or conformation (see Chapter 2) of much of the hydrophobic part of the protein is usually an alpha-helix. Any polar (hydrophilic) amino acid side chains in this portion of the molecule are sequestered within the alpha-helix. Only the nonpolar (hydrophobic) side chains are exposed to the hydrophobic interior of the bilayer. In contrast, in regions of the protein that project outside the lipid bilayer, the nonpolar side chains are buried in the interior of the protein conformation, whereas the polar and ionic side chains are exposed to the hydrophilic environment.

Many of the intrinsic membrane proteins display the same dynamic properties as the membrane lipids, such as rapid lateral diffusion and rotation about an axis perpendicular to the plane of the membrane. The fluidity of the lipid bilayer makes these movements possible. However, flip-flop of intrinsic membrane proteins is even more difficult than it is for phospholipids; in fact, it probably does not occur.

The lateral mobility of these membrane proteins can sometimes be prevented by interactions with peripheral components of the membrane (such as other proteins) or with components of the cell ctyoskeleton (microtubules and microfilaments).

Extrinsic Proteins: Hydrophilic Molecules

Membrane proteins that do not penetrate the lipid bilayer are termed **extrinsic** (or **peripheral**) **proteins** (see Fig. 4-7). They are usually not amphipathic molecules. They occur on either the external or cytoplas-

mic surface of the membrane and are attached primarily through interactions (electrostatic and hydrogen-bonding) with the polar portions of intrinsic membrane proteins and with the polar head groups of the phospholipids. These are relatively weak interactions. Consequently, extrinsic proteins can be easily extracted from membranes by relatively simple procedures.

Protein Asymmetry in the Membrane

Both extrinsic and intrinsic membrane proteins, like the phospholipids, are often distributed unequally between the two halves of the bilayer. That is, a specific protein may be associated with only one side of the bilayer. Specific advantages of protein asymmetry are better understood than those of phospholipid asymmetry.

This asymmetry is related to cell function. In the membrane of the red blood cell, for example, the extrinsic protein called **spectrin** is found only on the cytoplasmic (inner) side of the membrane. It is part of a filamentous network inside the cell that may be important in maintaining the biconcave shape while being flexible enough to allow the cell to change shape as necessary during its passage through narrow capillaries.

In contrast, many receptors for specific chemical signals such as hormones are located only on the external surface of the cell membrane and often represent part of an intrinsic membrane protein. If such a protein spans the lipid bilayer, a potential mechanism exists for converting an extracellular signal into an intracellular response. For example, binding of a hormone to the receptor could trigger a conformational change in the protein, especially if the protein is composed of several separate subunits. This conformational change could open a hydrophilic channel between the protein subunits and allow a specific ion to cross the plasma membrane.

Carbohydrate Is on the Surface of the Membrane

Carbohydrate is present in the plasma membrane of all eukaryotic cells and accounts for 2% to 10% of the mass of the membrane. In terms of quantity, carbohydrate is a minor component of the plasma membrane, yet it has important functions in membrane physiology.

Carbohydrates, like the membrane phospholipids and proteins, are distributed unequally between the two layers of the membrane. The carbohydrate molecule is a hydrophilic structure that does not penetrate the lipid bilayer but is found only on the surface of the membrane (see Fig. 4–2). Most of the membrane carbohydrate is bound either to intrinsic membrane proteins, forming glycoproteins, or to lipids in the bilayer, forming glycolipids.

Glycolipids are present only in the external half of the bilayer. The glycoproteins are more abundant than the glycolipids. Most of the proteins that are exposed on the external surface of the membrane have carbohydrate attached, but fewer than 10% of the lipid molecules in the external half of the bilayer have carbohydrate attached.

The amount of carbohydrate present in an individual glycoprotein may be large relative to the amount of protein. An extreme example is **glycophorin,** an intrinsic glycoprotein in the red blood cell membrane. This protein is composed of 131 amino acids and about 100 monosaccharide units. The monosaccharides together make up approximately 60% of the mass of the glycophorin molecule.

The location of membrane carbohydrates on the surface of the plasma membrane is probably an important factor in cell-to-cell recognition processes. These carbohydrates also probably serve as specific membrane receptors for extracellular messengers such as hormones. Unlike amino acids, which form only linear chains, monosaccharide units can be assembled in branching chain structures. This property provides almost limitless possibilities for construction of membrane receptor sites to which only one specific extracellular molecule can bind.

Gradients and Movement

A **gradient** is a difference between two quantities divided by the physical distance (measured in units of length) between them. Gradients are responsible for the movement of various substances throughout the body, including the movement of blood through the circulatory system, air into lungs, and ions through membranes of cells. These gradients take a variety of forms and manifest themselves as pressure gradients, electrical gradients, or concentration gradients. Concentration gradients are part of a larger topic, diffusion and osmosis.

A Pressure Gradient Drives Blood Flow

The movement of blood through arteries and veins is a result of a **pressure gradient.** Just as water will flow through a pipe only if the pressure at one end is greater than the pressure at another end, pressure in one region of the circulatory system must be greater than that at another region in order for blood to flow between them.

The larger the difference in pressure between two points, the greater the flow. In other words, blood flow is directly proportional to the difference in pressure between the two points. The difference in pressure between any two points can be symbolized as $(P_1 - P_2)$, where P_1 is the pressure at the starting point and P_2 the pressure at the final point. (This difference, or pressure change, can also be expressed as ΔP, where the Greek letter Δ or "delta" symbolizes a change.) The relationship between flow, F, and pressure can be represented as:

$$F \propto (P_1 - P_2)$$

where $\propto$ means "is proportional to."

Flow is met with **resistance** in the form of veins and arteries, or, more precisely, their diameters. Vessels with small diameters impede or resist the flow of blood to a greater degree than vessels with large diameters. Other sources of resistance are the length of the vessel and the viscosity of the fluid. Flow is inversely related to resistance, that is, the greater the resistance, the smaller the flow.

$$F \propto 1/R$$

We can express the combined effects of pressure and resistance on flow as:

$$F = (P_1 - P_2)/R$$

The movement of air into the lungs is subject to the same influences. Gradients established by pressure differences between the external environment and the interior of the lung cause air to rush into or out of the lungs. The rate of air flow is greatly affected by the lung airway resistance. In people with asthma, for example, the constriction of the airways, or decrease in their diameters, creates resistance to the flow of air, and breathing becomes difficult.

A Concentration Gradient Drives Diffusion

Diffusion is a process that involves the movement of particles such as ions and molecules in solution. An understanding of these processes in general is essential for an understanding of the movement of particles into and out of cells. In diffusion, a substance such as solute moves through and gradually becomes evenly mixed with another substance such as a solvent.

Diffusion of Solutes in Liquid Solvents

In solutions as well as elsewhere in nature, molecules, ions, and other particles move from one location to another by random thermal motion sometimes called **Brownian motion.** The concentration gradient is one of the main driving forces that determines the magnitude and direction of the net movement of particles. A concentration gradient exists when two adjacent regions have different concentrations of particles. The larger the difference in concentrations, the greater the net movement between them. Particles will tend to move from regions of higher concentration to regions of lower concentration, that is, "down" the concentration gradient. If there is no concentration gradient (a situation that will occur when particles have dispersed throughout the solution), then the net movement of particles will be zero, even though individual particles may move back and forth between the two regions. Such a solution is said to be in **dynamic equilibrium.**

The movement of sugar molecules in water illustrates the effects of a concentration gradient (Fig. 4–8). In Figure 4–8a, the sugar concentration in the lower region of the beaker is higher than elsewhere in the beaker. Although the sugar molecules are moving constantly, and some are moving toward the bottom of the beaker, most will move up and away from the region of concentration until they are evenly dispersed throughout the water (Fig. 4–8d). Even at this point, individual particles will still be moving in all directions, but there will be no net movement and no net change in concentration. The concentration gradient that existed when the particles were more concentrated in one region has been abolished by diffusion.

The driving force for solute diffusion is simply the difference in solute concentration, that is, the concentration gradient, between two regions or compartments. Diffusion of solutes abolishes the concentration differences. At this point, there is no net movement because there is no driving force.

Flux is a general term used to describe the rate of movement of solute molecules. It is expressed in terms of amount of solute moving per unit time, for example, mmol/minute. **Fick's Law** of diffusion states that the flux (J) of an uncharged solute is directly proportional to the driving force (dc/dx) and the area (A) through which diffusion can occur. Thus:

$$J = -DA \cdot dc/dx$$

D is the **diffusion coefficient,** which considers the mobility of the solute molecules in solution. This mobility will vary with both the specific solute and the specific solvent. The driving force (dc/dx) is the concentration gradient—literally, the change in concentration relative to the change in distance. The minus sign is a standard way of showing that the direction of net solute movement is always from the region of high concentration to the region of low concentration, that is, "down" the concentration gradient.

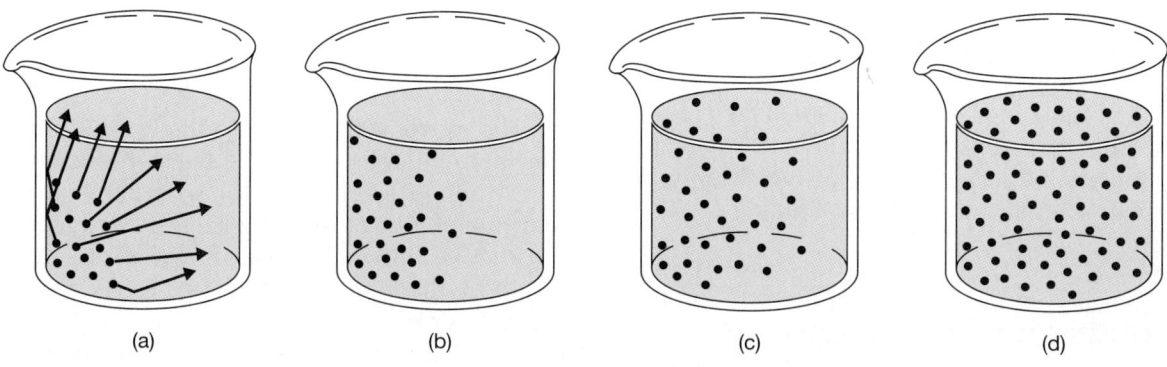

| (a) | (b) | (c) | (d) |

Figure 4–8

Diffusion occurs when one substance moves randomly throughout another and becomes evenly mixed with it. Diffusion occurs only when a concentration gradient exists, such that the substance to be mixed is more concentrated in one region than another. Here sugar molecules move from a region of high concentration to regions of low concentration in a beaker of water. When they are fully dispersed, a gradient no longer exists and diffusion ceases. Although the individual sugar particles are still moving in all directions, there is no net movement in any one direction because there is no gradient.

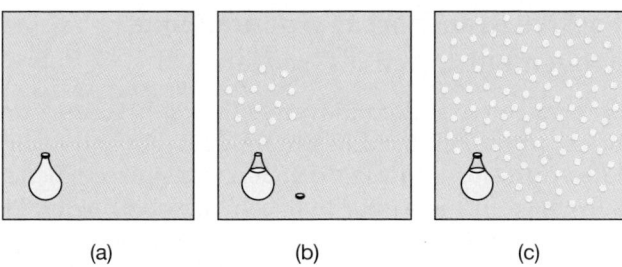

Figure 4–9

A model for the diffusion of gas molecules down a concentration gradient. When a perfume bottle is opened in a closed space, some of the liquid evaporates, and the gaseous perfume molecules pass into the surrounding space. A concentration gradient then exists because the molecules are more concentrated near the bottle than in the rest of the room. When the bottle is recapped, the perfume molecules slowly diffuse throughout the room until they are evenly distributed, and the concentration gradient no longer exists.

Diffusion of Gases in Gaseous Solutions

Gas molecules diffuse in the same way as do solid or liquid solute particles. Imagine a room in which there is no movement of air molecules (Fig. 4–9). If a bottle of perfume is opened, some of the liquid perfume evaporates. The gas molecules come out of the bottle into the room. Slowly, the perfume molecules spread and eventually reach an equilibrium in which the distribution of the scent is even throughout the room. The perfume spreads by diffusion, owing to the concentration gradient that exists because initially the molecules are more concentrated near the bottle than in the rest of the room. The molecules move from this region of high concentration to regions of low concentration.

The movement of gas along a concentration gradient can be pictured as the movement of molecules in two boxes with a common wall (Fig. 4–10a). Suppose one box is filled with oxygen molecules. The molecules move randomly about in the box, bouncing off the walls and ricocheting off each other. Now, suppose a hole is opened in the common wall between the boxes (Fig. 4–10b). A few molecules will travel immediately through the hole into the second box and begin bouncing around. More and more of the molecules will accumulate in the second box. As the concentration builds in the second box, some of the molecules will ricochet back through the hole into the first box. Eventually, the same number of molecules will be traveling in both directions through the hole, and the number of molecules will be the same in both boxes. At that time, the gas in the two boxes will be in dynamic equilibrium (Fig. 4–10c).

This diffusion process can take place even if the overall pressure is the same in each box. It can also occur if there are other kinds of molecules in each box. For example, nitrogen, carbon dioxide, and water vapor might be present in each box in equilibrium. If oxygen is introduced, it will move along its own concentration gradient as if it were the only type of molecule present (Fig. 4–11a, b, and c).

Likewise, different gases can be present in different concentrations in each box. For example, there can be a lot of carbon dioxide in one box and a lot of oxygen in the other. Even though the total pressure is the

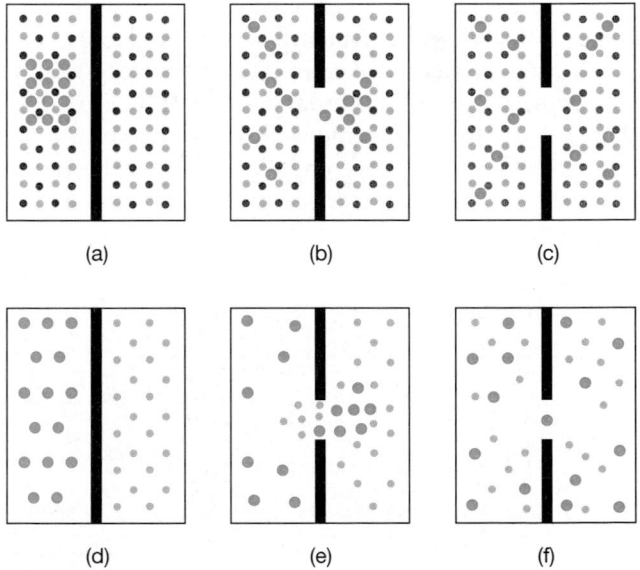

Figure 4–11

(*a, b,* and *c*) In a solution of gases, one constituent gas will diffuse independently along its own concentration gradient to become evenly distributed throughout the solution. (*d, e,* and *f*) Two concentration gradients can exist in two adjoining spaces. Molecules of each gas will move along their own concentration gradients at the same time until both gases are evenly spread throughout the entire space.

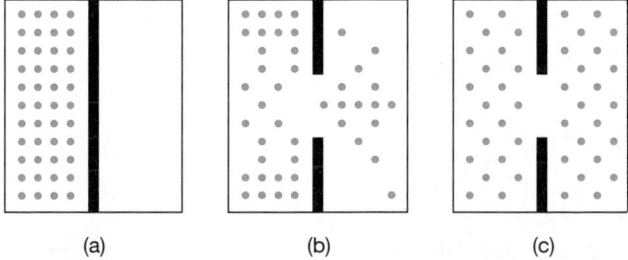

Figure 4–10

In response to a concentration gradient, gas molecules move through a hole in a wall between two boxes until they are evenly distributed.

same in each box, the gases will diffuse simultaneously along their respective concentration gradients and move in opposite directions between boxes as soon as the hole is opened (Fig. 4–11*d, e,* and *f*).

One kind of molecule can move through the spaces between other molecules. There is mostly empty space between the molecules present in air. The movement of perfume molecules between air molecules is analogous to the movement of oxygen molecules through the hole in the wall. In air, there is a very large number of spaces through which molecules can move, and this speeds up any diffusion processes that occur. The wall in the box is analogous also to various membranes in the body through which gas molecules diffuse in response to concentration gradients.

Movement of Substances Through the Membrane

Two types of movement occur as substances pass through the plasma membrane of a cell: (1) **passive,** or **spontaneous,** processes that occur in response to gradients and produce a *less* concentrated state overall; and (2) **active,** or **energy-requiring,** processes that occur in opposition to gradients and produce a *more* concentrated state where a high degree of concentration already existed. **Passive transport** processes include both **simple diffusion,** such as **osmosis,** and **facilitated diffusion. Active transport** processes include **primary** and **secondary active transport** (Fig. 4–12). Note that the same substance—sodium ions, for example—can move one way through the membrane by a passive process and the other way by an active process.

Passive Transport Is Spontaneous Movement down a Gradient

Spontaneous, random movements of substances into or out of the cell include simple diffusion and facilitated diffusion. **Simple diffusion** takes place when a substance, usually one that can dissolve in lipids, passes directly through the lipid interior of the plasma membrane. **Facilitated diffusion** occurs in the presence of a protein that enables the passage of the substance. However, both processes are passive and spontaneous movements that occur in response to a gradient.

Simple Diffusion of Lipid-Soluble Substances

Diffusion in free solution or any other type of mixture occurs because of the random movement that all particles exhibit. This motion is an expression of the thermal energy of the molecules. As a result of diffusion, a substance such as a solute will tend to become distributed in an equal concentration throughout the entire space available.

A Selectively Permeable Membrane. The previous section discussed the diffusion of solute particles in free solution, from a region of high solute concentration to an adjacent region of low concentration. There was no physical division between these two regions; they were more or less continuous.

Diffusion can also occur in the presence of a **selectively permeable membrane.** If a solute is highly concentrated on one side of a membrane but able to pass through the membrane, it will do so, diffusing into the adjoining region of low concentration by the same process that would occur if the membrane were not there (Fig. 4–13). If another solute were present

Figure 4–12

An overview of membrane transport systems.

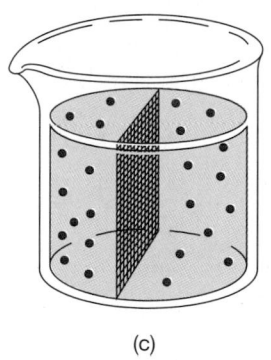

Selectively permeable membrane

(a)   (b)   (c)

Figure 4-13

Diffusion can occur in the presence of a selectively permeable membrane. As long as solute particles can pass through the membrane, they will diffuse from a region of high concentration to one of low concentration.

to which the membrane were not permeable, that solute would remain highly concentrated on one side of the membrane.

The plasma membrane, as a mixture of lipids and proteins, represents a barrier to the diffusion of some solutes into or out of the cell. An inherent, important property of any lipid bilayer is that it is nearly impermeable to ions and most polar molecules (water is a notable exception to this general rule because it readily crosses lipid bilayers even though it is a polar molecule). In contrast, a lipid bilayer is permeable to nonpolar molecules—that is, molecules that can dissolve in its nonpolar interior. The lipid bilayer of the plasma membrane is selectively permeable; it readily permits some substances to cross it but prevents or retards the free passage of others.

Consider the diffusion of glycerol, a polar molecule, from the plasma to the interior of a red blood cell. The diffusion coefficient D is a measure of the mobility of a specific solute molecule within a specific solute. D for glycerol in water is 1.0×10^{-5} cm^2/sec. This is several orders of magnitude larger than the value of D for glycerol inside the lipid phase of the plasma membrane: 1.7×10^{-10} cm^2/sec. This large difference reflects the fact that the mobility of glycerol is greatly diminished once it enters the membrane.

The Permeability Coefficient.

The rate of diffusion of solute through a cell membrane is governed by the thickness of the membrane and by the lipid solubility of the solute. The combined effect of these two factors and the diffusion coefficient can be expressed as the **permeability coefficient *P*.** The value of P varies with the specific solute and the specific cell membrane.

In spite of the barrier represented by the cell membrane, an uncharged solute will still tend to move across it if the solute concentration inside the cell is different from the solute concentration outside—that is, if there is a concentration gradient. In simple diffusion, the solute will always move down the concentration gradient, from the region of high solute

concentration to the region of low concentration, just as in free solution. Fick's Law can be adapted to represent the rate of solute diffusion across a membrane as follows:

$$J = -PA[C_1 - C_2]$$

C_1 and C_2 are the solute concentrations on either side of the membrane. Thus, when C_1 is greater than C_2, there is net movement of solute from Side 1 to Side 2 (Fig. 4-14). As the solute concentration builds up on Side 2, the rate of movement back to Side 1 increases. Eventually the solute concentration gradient will be abolished, and these opposing fluxes will become equal. At this point, the system will be in a dynamic equilibrium—that is, individual solute particles will continue to move from one side of the membrane to the other, but the net movement of solute will cease. This can also be seen from the equation: When $C_1 = C_2$, the value of J is zero.

The Electrochemical Gradient.

When the solute is electrically charged, as are Na^+ and Cl^- ions, for example, the flux of solute is influenced not only by a concentration gradient but also by an electrical gradient. A typical animal cell has a small negative potential (about -70 millivolts) on the inside relative to the outside. This difference tends to cause positively charged solutes to move across the plasma membrane and into the cell while it opposes the entry of negatively charged solutes. This electrical gradient represents an additional factor in diffusion that must be considered when the solute bears a charge. The total gradient is now the sum of the chemical (concentration) and electrical components and is termed the **electrochemical gradient.**

Considered separately, the chemical and electrical gradients often favor solute movement in the same direction. For example, the passive entry of Na^+ into cells is favored by both the concentration gradient (low Na^+ concentration inside the cell) and the electrical gradient (negative potential inside the cell).

On the other hand, electrical and chemical gradients can oppose each other. For example, the passive

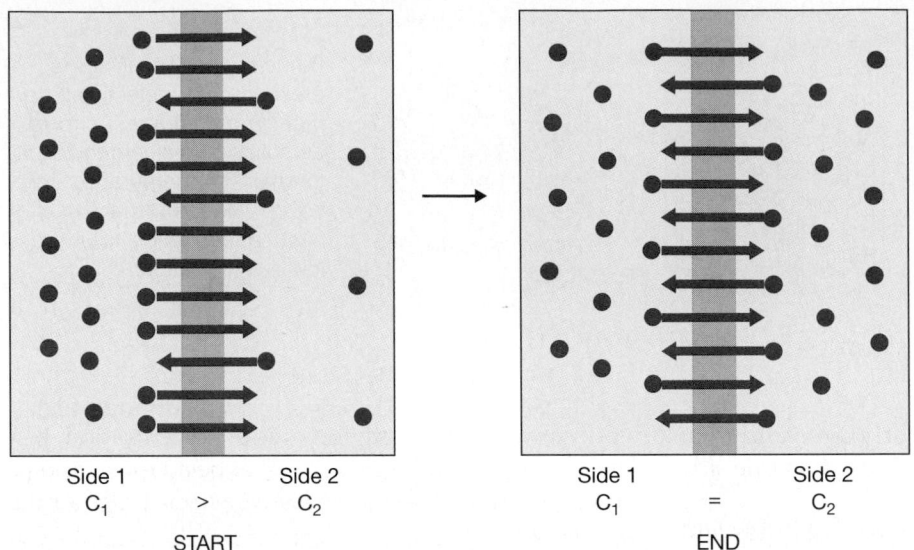

Figure 4–14

Diffusion of solute molecules across a biological membrane. At the start there are 25 molecules on Side 1 and only 7 on Side 2, so there will be net movement of solute from Side 1 to Side 2. The end point is reached when each side has 16 molecules. At this point, there is no net movement of solute across the membrane because equal numbers of molecules are moving in each direction.

leak of K^+ ions out of cells goes against, or "up," the electrical gradient (there is a positive potential outside the cell). However, the concentration gradient is so large—because of the high K^+ concentration inside the cell that the electrochemical gradient force favors the exit of K^+ from the cell.

Overton's Rules.   The principal barrier to solutes that cross cell membranes by simple diffusion is the lipid bilayer. The rate at which a solute diffuses through the membrane depends in large part on two factors: its ability to dissolve in lipids and its molecular size. In general, a nonpolar, and therefore lipid-soluble, solute will diffuse across a cell membrane much more rapidly than a polar (water-soluble) solute of the same size. The effect of the size of the solute molecule is particularly important for polar solutes. Membrane permeability for these molecules, already limited by their polarity, decreases further as their molecular size increases.

These relationships are summarized in **Overton's Rules,** as follows: (1) the permeability of cell membranes to small nonpolar solutes is directly proportional to the lipid solubility of the solute; and (2) the permeability of cell membranes to polar solutes is inversely proportional to the molecular size of the solute. Thus, a small, highly lipid-soluble molecule has the best chance of rapidly penetrating cell membranes by simple diffusion.

An exception to these general statements is water itself. Despite its small size, a water molecule is highly polar (lipid-insoluble) and yet it crosses cell membranes very rapidly. The precise mechanism is not fully understood, but it has been suggested that membranes contain small nonspecific pores that allow water and small water-soluble solutes to cross the bilayer without entering the lipid phase. The dif-

fusion of water will be considered in the section on osmosis.

An electrical charge on the solute is another factor that decreases its lipid solubility. Thus, membranes are almost impermeable to ions, even though an ion represents a solute of small size. Any ionic diffusion that occurs probably uses specific channels in proteins that span the lipid bilayer (see the section on facilitated diffusion later in this chapter).

Weak Acids and Bases.   The very low permeability of cell membranes to solutes bearing an electrical charge has important consequences for the diffusion of **weak acids** and **weak bases.** These compounds, unlike strong acids and bases, do not dissociate completely when dissolved in water (see Chapter 2). Recall that an acid is a proton donor while a base is a proton acceptor. We can write the following general equations to represent the behavior in solution of a weak acid (HA) and a weak base (B):

$$HA \rightleftharpoons H^+ + A^-   \text{(weak acid)}$$
$$B + H^+ \rightleftharpoons BH^+   \text{(weak base)}$$

A solution of a weak acid or weak base will contain a mixture of the undissociated form and the dissociated form, the proportions of these two forms varying with the pH of the solution. In the case of the weak base, for example, a low pH (high H^+ concentration) will shift the equilibrium to the right, resulting in more BH^+. Conversely, a high pH (low H^+ concentration) will shift the equilibrium to the left, resulting in more B and more H^+. The pH at which the concentration of the associated form is exactly equal to that of the dissociated form is defined as the *pK* and is a characteristic property of the weak base or the weak acid.

pH 7.0 | pH 6.0

Interior of kidney tubule cell | Plasma membrane | Fluid in lumen of kidney tubule

Figure 4–15

Diffusion trapping of a weak base. NH_3 can cross the plasma membrane because it is a neutral molecule. When it gains a proton in acidic solution, it becomes the charged ammonium ion NH_4^+, which cannot cross the membrane.

Diffusion Trapping.

The neutral form of a weak base diffuses across a cell membrane whereas the charged form does not, which gives rise to the phenomenon of **diffusion trapping.** This is illustrated in Figure 4–15 using ammonia as an example of a weak base. Ammonia is produced in the cells of the kidney tubules as an end product of the metabolism of amino acids, especially glutamine. Ammonia is neutral and lipid-soluble and easily diffuses out of the cell, moving down its concentration gradient and into the fluid present in the lumen of the tubule. Typically, this fluid has a low pH; it is slightly acidic relative to the cell cytosol. When the ammonia contacts the excess H^+ in this acidic solution, it gains a proton to form the ammonium ion. The ammonium ion is positively charged and lipid-insoluble and cannot diffuse back into the tubule cell. It is effectively "trapped" in the lumen and eventually excreted in the urine.

Protonation of available ammonia in the tubular fluid maintains a low concentration of ammonia outside the tubule cells, and continued production of ammonia by cellular metabolism maintains a relatively high concentration of ammonia inside the tubule cells. The combination of these processes preserves the concentration gradient that favors the passive diffusion of ammonia out of the cells.

Simple Diffusion of Water: Osmosis

When the molecules that are moving in diffusion are water molecules, the movement is known as **osmosis.** Osmosis is the flow of water across a barrier in response to a difference in solute concentration on either side of the barrier. Water flows from the side of the barrier with the lower concentration of solutes (the more dilute region) to the side of the barrier with the higher concentration (the less dilute region). The water molecules simply move down their own concentration gradient, like any other type of molecule. More water molecules are present in the dilute solution than in the concentrated solution, which is why there is net movement of water toward the region of high solute concentration.

Osmosis is involved in the preservation of meat by salting. When bacteria contact salt on the surface of meat, a concentration gradient exists because the salt is more highly concentrated outside the bacterial cell than inside. The plasma membrane of the cell is impermeable to the salt; salt cannot diffuse into the cell to abolish the gradient. The membrane is permeable to water, however, and water flows out of the cell toward the region in which the salt is more concentrated. The bacterial cell dies before it can infect the meat.

Osmotic Concentration.

When discussing osmosis, we use a special measure of concentration in order to better understand the relationship of water and solute particles. The number of solute particles a solute releases in solution is not related to the mass of the solute. Chemical concentration, e.g., molality or molarity, is based essentially on weight and does not always reveal the extent to which solute particles are present in the solution. For example, in a solution of glucose in water, the glucose molecules ($C_6H_{12}O_6$) stay intact rather than dissociating into atoms. In a solution of sodium chloride in water, by contrast, the Na^+ and Cl^- ions dissociate completely and this doubles the total number of particles that are present in the solution.

Osmotic concentration is expressed in units called **osmoles (Osm).** For physiological solutions, we use **milliosmoles (mOsm)** since the concentrations are generally low.

Just as molarity expresses moles of solute per liter of solution, osmolarity expresses osmoles of solute per liter of solution. The relationship between molarity and osmolarity can be summarized as:

$$\text{osmoles} = n \times \text{moles}$$
$$(\text{milliosmoles} = n \times \text{millimoles})$$

where n is the number of particles derived from each molecule of solute when in solution. Note that this relationship applies only to dilute solutions that are assumed to behave ideally. Deviation from ideal behavior occurs as the concentration of the solution increases.

For example, the osmotic concentration of a glucose solution with a chemical concentration of 60 mmol/liter is 60 mOsm/liter because n equals 1 for glucose. Glucose exists as single, intact molecules in water, producing only one solute particle per molecule.

In contrast, the osmotic concentration of a solution of NaCl whose chemical concentration is 60 mmol/liter is 120 mOsm/liter because $n = 2$. Each molecule of NaCl produces two solute particles (Na^+ and Cl^- ions) and both become osmotically active solute particles. A $CaCl_2$ solution with a chemical concentration of 20 mmol/liter will have an osmotic concentration of 60 mOsm/liter because each $CaCl_2$ group dissociates completely in solution into three ions: one Ca^{2+} ion and two Cl^- ions. A 20 mM solution of $CaCl_2$ is said to have an osmolarity of 60 mOsm.

Osmotic Pressure.

Osmotic pressure is a property of a solution that is proportional to the solute concentration. A dilute solution will have a lower osmotic pressure than a concentrated solution. Thus, water will move by diffusion from a region of low osmotic pressure to a region of high osmotic pressure.

The osmotic pressure difference between two solutions of different solute concentrations is illustrated in Figure 4–16. A solution of water containing 1.0 M sucrose is placed inside an inverted thistle tube. The thistle tube is placed in a bath of pure water. The water and the sucrose solution are separated by a membrane placed across the mouth of the funnel. The membrane allows only pure water, not sugar molecules, to pass through. The apparatus is arranged so that, initially, the level of the sucrose solutions is the same as the level of the water in the bath.

The sucrose solution has a higher osmotic pressure than the water because it has a higher solute concentration. Consequently, water will begin to move from the pure water bath across the membrane into the thistle funnel. The movement of water into the sucrose solution gradually dilutes the solution and raises the level of the solution inside the thistle tube.

Eventually, the increased fluid level in the funnel generates a new pressure, hydrostatic pressure, which now opposes the movement of more water into the tube. Soon there is enough hydrostatic pressure to completely stop the osmosis of water into the thistle tube. At this point, the hydrostatic pressure reflects the difference in osmotic pressure between water and the original sucrose solution.

The increase in the fluid level in the funnel can be prevented by applying pressure down the funnel at the start of the experiment. This pressure blocks the osmotic movement of water into the sucrose solution. The pressure applied is the osmotic pressure of 1.0 M sucrose.

If we know the solute concentration of a solution, it is relatively simple to calculate the osmotic pressure. Osmotic pressure is related to the solute concentration by the **van't Hoff law,** which is described by the equation:

$$\pi = RTC$$

in which π is the osmotic pressure, R is the ideal gas constant (0.082 liter · atm/degree · mole), T is the absolute temperature in Kelvins (°C + 273), and C is the total solute concentration expressed as osmoles of solute per liter of solution. For example, the osmotic pressure of a solution of 1.0 Osm/liter at 0°C is calculated as follows:

$$\pi = RTC$$
$$= 0.082 \, \frac{\text{liter} \cdot \text{atm}}{\text{degree} \cdot \text{mole}} \times 273° \times 1.0 \, \text{Osm/liter}$$
$$= 22.4 \, \text{atm}$$

Thus, a 1.0 Osm solution exerts an osmotic pressure of 22.4 atm at 0°C. A similar calculation reveals that human blood plasma, which is normally a 0.31 Osm solution, will have an osmotic pressure of 7.9 atm at 37°C. The osmotic pressure of a solution depends on the total osmolarity of the solution and not on the nature of the specific solute or solutes. The value of C for pure water containing no dissolved solutes is zero; hence, the osmotic pressure of pure water is zero.

Since R and T are constants if there is no temperature change, the osmotic pressure will be directly related to osmolarity. In this case, there is no need to calculate osmotic pressure in order to determine

Thistle tube

1 M sucrose

Diluted sucrose

100% water

(a)   Semipermeable membrane   (b)

Figure 4–16

When a thistle funnel containing a sucrose solution is placed in a bath of pure water, the water moves across a membrane up into the tube, diluting the sucrose solution and raising the level of the solution in the tube. No sucrose flows from the tube into the bath because the membrane is impermeable to sucrose.

which of several solutions has the highest osmotic pressure. A comparison of the osmolarity of the solutions will suffice; the solution with the highest osmolarity will have the highest osmotic pressure.

The osmolarity of a solution can be directly determined by use of a freezing point **osmometer,** even if the identity of the specific solute is unknown. The freezing point of pure water is depressed when solutes are present, and the extent of the depression is proportional to the total amount of all the solutes present. Most osmometers are calibrated to give a readout directly in osmoles or milliosmoles, and they provide a rapid and accurate method for determining the osmolarity of solutions.

Two solutions that have the same osmolarity are said to be **iso-osmotic.** Thus, 300 mOsm NaCl is iso-osmotic with 300 mOsm sucrose. A solution with a higher osmolarity than another is said to be **hyperosmotic;** a solution with a lower osmolarity is said to be **hypo-osmotic** relative to one with a higher osmolarity.

Osmotic pressure is only one property of a solution that is affected by the presence of solute particles. Dissolved solutes lower the vapor pressure and increase the boiling point of solutions. They also lower the freezing point. When salt is spread on a wet road in winter, it dissolves in the water, preventing the water from freezing at 0°C, the point at which pure water would normally freeze. $CaCl_2$ is used instead of NaCl because it releases more solute particles (three rather than two) per molecule and has a greater effect on the freezing point.

Changes in Cell Volume: The Erythrocyte. Nearly all cell membranes are permeable to water. Scientists have recently found specific membrane proteins that act as "water channels" and allow rapid movement of the polar water molecule across the hydrophobic lipid bilayer. Cells must regulate the amount of water that enters. Plant cells and many bacteria have a rigid cell wall that surrounds the plasma membrane and prevents an increase in the volume of the cell due to an influx of water. Animal cells lack an exterior cell wall and have developed other mechanisms to deal with osmotic stress.

We will use the red blood cell, or **erythrocyte,** to illustrate the changes that occur in animal cells when the osmolarity of the external medium changes. The erythrocyte is a favored experimental tool of cell physiologists because of its simple structure. It has no intracellular membranes and is available in large numbers.

Previously in this chapter, we stated that many biological membranes are permeable to solutes as well as to water. This is certainly true of the erythrocyte plasma membrane, but the net movement of the major solutes, such as Na^+, K^+, and Cl^- ions, will be relatively small during osmotic experiments of short duration.

The osmolarity of the erythrocyte cytosol is close to 300 mOsm. Suppose an erythrocyte is placed in an isosmotic solution of 300 mOsm NaCl. The total osmolarity (and hence the osmotic pressure) of the solution inside the cell will be the same as that of the solution outside. There will no concentration gradient favoring osmotic water flow and no net movement of water into or out of the cell; the intracellular volume will not change.

Next, suppose the erythrocyte is placed in hyperosmotic 400 mOsm NaCl. The erythrocyte will no longer be in osmotic equilibrium because the external solution has a higher osmolarity than the cell cytosol. In response to the higher solute concentration outside the cell, water flows out of the cell (Fig. 4–17). However, solutes remain in the cell, and therefore the osmolarity of the cell cytosol increases. Eventually the osmolarity of the cytosol will increase to the point where it is equivalent to that of the external solution, and osmotic flow of water will cease. The cell will return to an osmotic equilibrium with the external medium, but this will occur at the expense of part of the intracellular volume. The decrease in volume is not accompanied by a decrease in surface area, and the cell assumes a spiky, or **crenated,** shape (Fig. 4–17).

When a normal erythrocyte is placed in a hyposmotic solution of 200 mOsm NaCl, the osmolarity of the cytosol is higher than that of the external solution. Water will enter the cell, leading to an increase in intracellular volume (Fig. 4–17). Osmotic flow of water into the cell will continue until the osmolarity of the cytosol has been diluted to 200 mOsm. At this point net movement of water into the cell will cease, and osmotic equilibrium will be reestablished, but with an increase in cell volume (Fig. 4–17).

The erythrocyte has a unique characteristic that enables it to accommodate a considerable increase in cell volume. The normal cell is biconcave, not spherical. When water enters from a hyposmotic solution, the cell simply fills out to become more like a sphere (Fig. 4–17). When the cell has become a perfect sphere, the intracellular volume will have increased by 67% over the volume of the original biconcave cell. In this way, the erythrocyte can undergo a large increase in volume without experiencing a change in the surface area of the cell. Most other types of animal cells are not biconcave and cannot respond in this way to osmotic entry of water. The plasma membrane is not elastic—it cannot stretch—and so the cells can accommodate only a very small increase in intracellular volume. If osmotic entry of water continues, the cells undergo osmotic **lysis**—that is, holes develop

Figure 4–17

Response of a human erythrocyte to changes in the osmolarity of the extracellular medium. The normally biconcave disc becomes crenated when water flows out of the cell because of a greater osmolarity in the surrounding fluid. The cell becomes a sphere when water flows into it from solutions with a lower osmolarity.

in the plasma membrane and the intracellular contents are lost. The cells literally burst open.

Osmotic Fragility.   An erythrocyte will burst open, or **lyse,** if it is placed in a medium that is very dilute (e.g., 100 mOsm NaCl). The difference in osmolarity between the inside and the outside of the erythrocyte is now so large that the cell cannot achieve osmotic equilibrium. Water will continue to enter even when the cell has become a sphere, and so the cell will lyse.

Osmotic fragility is a term used to characterize how a sample of erythrocytes responds to osmotic stress. It represents the osmolarity of NaCl that causes 50% of the erythrocytes to lyse. In the case of normal erythrocytes, 50% of the cells lyse at an osmolarity of 150 mOsm (Fig. 4–18). Some cells, presumably the older ones, are more fragile and lyse at a slightly higher osmolarity than 150 mOsm. Other cells, presumably the younger ones, are less fragile and lyse at a slightly lower osmolarity. However, almost all of the population of cells lyse over a fairly narrow range of osmolarities, as shown by the steep slope of the curve in Figure 4–18.

Osmotic fragility can be an important test for diseases that involve a change in erythrocyte behavior. When the erythrocytes become more resistant to osmotic stress, they are said to be less fragile, and the curve (Fig. 4–18) shifts to the right. When the population is less resistant to osmotic lysis, the cells are said to be more fragile, and the curve shifts to the left.

A specific example of a fragile population are the erythrocytes from mice with **hereditary spherocytosis.** These cells tend to be spherical rather than biconcave. As a result, the cells can no longer accommodate the volume increase caused by osmotic flow of water into the cells. As seen in Figure 4–18, some of the cells lyse when the osmolarity of the NaCl

Figure 4–18

Abnormal erythrocytes are more fragile and lyse in solutions of higher osmolarity compared to normal erythrocytes.

medium decreases only slightly. Fifty percent of the cells lyse at an osmolarity of 225 mOsm, which represents a considerable change from the behavior of the normal cell population. The erythrocytes in outdated blood from a blood bank provide another example of increased osmotic fragility. These cells are less resistant to osmotic stress than erythrocytes in fresh drawn human blood.

After a discussion of some specialized transport mechanisms, we will return to the concept of osmosis and the way in which the cell transports materials to maintain appropriate cell volume.

Facilitated Diffusion of Lipid-Insoluble Substances

A lipid bilayer is essentially impermeable to ions and large water-soluble molecules such as D-glucose, amino acids, and other metabolites. Still, these substances diffuse across the membrane in response to electrochemical gradients. They do so by a process of **facilitated diffusion,** so called because their passage is facilitated by intrinsic membrane proteins that act as **carriers** or **channels.** Ions and other lipid-insoluble substances normally would not be able to pass through the nonpolar interior of the membrane. The proteins enable them to get from one side of the lipid bilayer to the other, in part by shielding them from the lipid molecules. The molecular mechanisms of action of membrane carriers and channels are the subjects of intense investigation.

Simple diffusion and facilitated diffusion both occur spontaneously and passively in response to an electrochemical gradient. Similarly, both processes cease once the gradient has been abolished. Since facilitated diffusion is mediated by a carrier or channel protein, however, a number of characteristics distinguish it from simple diffusion: (1) facilitated diffusion allows a very high rate of solute transport, (2) it is a saturable process; there is a limit to the transport rate that cannot be exceeded, (3) it is a highly specific process, and (4) solute transport can be blocked by competitive inhibitors.

Carrier Proteins.   A hypothesis to explain the operation of a carrier is depicted in Figure 4–19. This model proposes (1) that the carrier protein is composed of at least two subunits, and (2) that binding of solute to a site exposed to the cell exterior triggers an intramolecular rearrangement in the structure of the protein such that the solute binding site is now exposed to the cell interior. The solute is always surrounded by hydrophilic regions of the protein subunits and is never exposed to the hydrophobic lipid phase of the membrane. Once the solute is exposed to the interior of the cell, where the solute concentration is low, it dissociates from the binding site. This step allows the carrier protein to revert back to its original conformation, with the empty binding site again exposed to the cell exterior. The cycle can then be repeated to move another solute molecule into the cell.

Channel Proteins and Gating.   A channel mechanism is particularly important for the rapid transmembrane movement of ions such as Na^+ and K^+.

Figure 4–19

A possible mechanism for carrier-dependent transport. (*a*) Solute concentration is high outside and low inside the cell. A solute molecule binds to the carrier protein at a specific site that is exposed on the exterior surface of the cell membrane. (*b*) Binding of the solute to the carrier triggers a change in conformation, and the solute binding site becomes exposed to the cell interior. (*c*) The solute easily dissociates from the carrier because of the low concentration of solute inside the cell. This dissociation allows the carrier protein to revert to its original conformation. The solute binding site is again exposed to the exterior of the cell and the cycle can be repeated. Probably the changes in protein conformation are much more subtle than depicted here.

Cystic Fibrosis

A disorder that is specifically linked to cell transport and secretion is **cystic fibrosis (CF).** CF is an inherited genetic disorder occurring in children in which the secretions of the body glands are abnormally thick and sticky. They clog up the glands and prevent certain organs from working properly. The pancreas cannot function, and without pancreatic enzymes it is difficult to absorb certain types of food. The liver and lungs also become clogged with sticky mucous, which in turn leads to cystic and fibrotic changes in these organs. In addition, children with cystic fibrosis have abnormally high concentrations of electrolytes (salt) in their sweat and saliva.

The most severe effect of CF is upon the lung. Sticky mucous not only blocks the air passages but also serves as a culture (growth) medium for harmful bacteria. Consequently, the small airways become obstructed, and the bacteria or their toxins cause respiratory infections that often lead to pneumonia. The repeated infection causes scarring and promotes fibrosis of the lung.

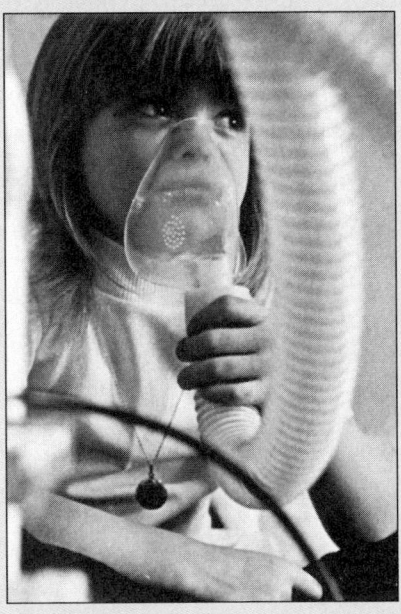

A child with cystic fibrosis using a nebulizer, which disperses medications into a fine mist that can be inhaled. *(Courtesy of the Cystic Fibrosis Foundation.)*

Another complication in CF patients, as a result of the clogged airways, is the low concentration of oxygen in the lungs. This decreases the amount of oxygen that gets into the blood and also triggers a response that causes the pulmonary arteries to constrict. The constriction causes pulmonary hypertension (an abnormal high blood pressure in the lungs) and puts an increased strain on the heart. In fact, 98% of all CF patients die from cardiopulmonary complications.

Prior to the 1940s, babies with CF died early from respiratory failure. In the 1950s, with advances in antibiotic therapy, half of all CF infants lived to be 3 to 5 years old. Today, with strong antibiotics and proper therapy, half of all children with CF live into their twenties, which for everyone else is the prime of life. The disease is an autosomal recessive disorder in Caucasian children and occurs in about one in 1,500 live births. CF is rare in black children and essentially absent among Asiatics.

Scientists are rapidly closing in on the disease process of cystic fibrosis. They have discovered that an abnormal gene leads to a defective transport of chloride ion in the airway epithelial cells. The gene, identified in 1989, has the code for a protein of 1,480 amino acids. The protein is known as the **cystic fibrosis transmembrane conductance regulator** (CFTR). The CFTR protein contains both a chloride channel and a channel regulator. The protein is anchored in the plasma membrane by two hydrophobic membrane-spanning segments which also form a channel. Two different regions of the protein control the function of the channel through interactions with nucleotides, such as ATP, that are present in the cell cytosol. A mutation involving the loss of phenylalanine from one of the nucleotide binding sites is present in 70% of cystic fibrosis patients.

Potential new therapies for cystic fibrosis are now undergoing testing. One approach is to insert a normal gene for CFTR into affected cells in the airways; another approach is to ignore the defective CFTR and instead stimulate the other membrane chloride channels that are present in the cells.

■ **Why would reduced secretion of chloride ions by epithelial cells in the lung airways lead to mucus that is abnormally thick and sticky?**

One hypothesis is that a hydrophilic channel across the membrane is formed between the subunits of a protein. The open channel permits a higher rate of transport than that provided by a carrier. For example, some channels enable ions to pass through them at the rate of 10^8 ions/sec, while the fastest carriers move solutes across membranes at no greater than 10^5 molecules/sec.

It is unlikely that an ion moves through a channel by simple diffusion down the electrochemical gradient. Most likely the ion undergoes specific interactions with charged groups along the sides of the channel. These reactions help to move the ion through the channel. The types of charged groups present in the channel may also be important in determining the specificity of the channel for one ion type.

Channel closed

Channel open

Figure 4–20

A highly simplified view of the gating mechanism for regulating ion transport through a membrane channel protein.

Another difference between a channel and a carrier, in addition to the overall rate of transport, is that a channel can be kept closed by a **"gate"** (Fig. 4–20). A closed channel will not permit movement of ions even though the electrochemical gradient may favor it. The mechanisms that regulate the opening and closing (**"gating"**) of ion channels will be discussed later in Chapter 7.

The High Rate of Facilitated Diffusion. A solute that crosses a cell membrane by facilitated diffusion does so at a rate that is much more rapid than that of a solute crossing by simple diffusion, even though both solutes are of the same molecular size and have the same lipid solubility. Consider a hypothetical example in which there are 17 solute molecules outside a cell and only 1 inside; assume that the solute is electrically neutral. Net movement of solute inside the cell will cease when 8 more molecules have entered —that is, when there are 9 molecules on each side of the membrane. It will take 8 sec to achieve this balance if the solute moves by simple diffusion at a rate of 1 molecule entering the cell per second. In contrast, it will take only 1 sec if the solute moves by facilitated diffusion, which allows solute entry at a rate of 8 molecules/sec.

This example illustrates that the end point of both processes is the same but is achieved faster by facilitated diffusion. In this way the carrier protein serves a function analogous to that of an enzyme in a biochemical reaction. The rates of transport used in this example are arbitrary and are not meant to represent experimentally determined values.

Saturation in Facilitated Diffusion. A facilitated diffusion system can become saturated because the number of binding sites on the carrier for solute molecules is limited. The rate of transport will be at a maximum when all the sites are occupied by solute (Fig. 4–21). A graph drawn to relate transport rate to

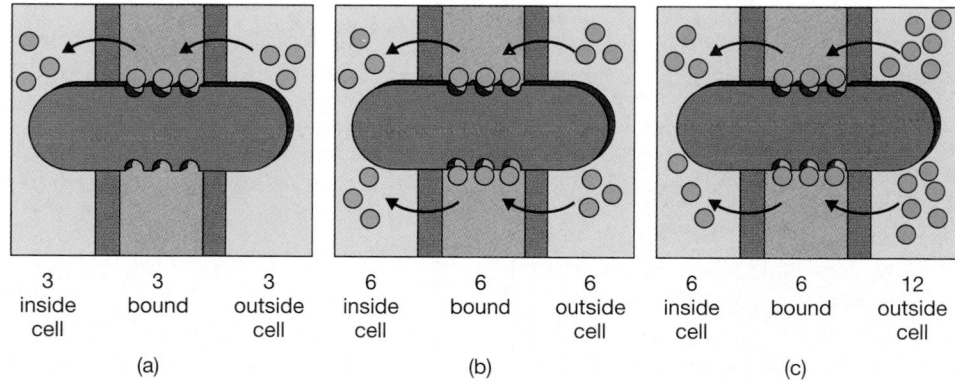

| 3 inside cell | 3 bound | 3 outside cell | 6 inside cell | 6 bound | 6 outside cell | 6 inside cell | 6 bound | 12 outside cell |

(a) (b) (c)

Figure 4–21

The saturation of a membrane carrier with solute. (*a*) Assume that a carrier protein in a membrane has specific binding sites for 6 solute molecules. If 3 solute molecules outside a cell arrive at the carrier, all 3 molecules will be transported inside the cell in one cycle of the carrier. (*b*) If 6 molecules outside the cell arrive at the carrier, all 6 molecules will be moved inside the cell in one cycle. That is, the rate of net transmembrane transport of solute doubles when the amount of solute doubles. (*c*) When 12 solute molecules arrive at the carrier, the rate of solute movement will not increase because the carrier can bind and transport only 6 molecules per cycle. Once the carrier is saturated with solute, any further increase in solute concentration outside the cell will have no effect on the overall rate of solute transport. At this point the rate of solute transport has reached a maximum.

solute concentration (see Fig. 4–22) shows that initially the rate of solute transport increases in proportion to the solute concentration. Once the solute concentration has reached a level at which the carrier binding sites are always full (saturation), however, any further increases in solute concentration will produce no change in the solute transport rate. The rate of transport is now at its maximum; this is termed the V_{max}. The solute concentration that gives rise to one half of the V_{max} is known as the K_m. These kinetic parameters are a very useful way of comparing the characteristics of different transport systems. It is worth adding that transport by simple diffusion is not saturable. The rate of transport increases in proportion to the solute concentration (see Fig. 4–22). The limitation posed by carrier saturation is rarely a problem because the overall rate of transport by this mechanism is so much faster than that by simple diffusion. The V_{max} for transport by facilitated diffusion is analogous to an enzyme-catalyzed reaction, which proceeds at a maximum rate once the enzyme is fully saturated with substrate (see Chapter 2). Any further increase in substrate concentration will have no effect on the reaction rate. The analogy with enzymes is limited, however, because the solute is not permanently altered as a consequence of binding to the carrier protein.

The Specificity of Facilitated Diffusion.   Unlike simple diffusion, facilitated diffusion is a highly specific process. Since transport of the solute involves bind-

ing to a carrier protein, it is likely that there are specific carriers in the cell membrane for each of the solutes that enter or leave the cell by facilitated diffusion. The specificity of the carrier is not absolute. For example, structurally related amino acids are often able to share the same carrier. The transport of sugars across the erythrocyte plasma membrane provides another example. The transport system works best when glucose is the solute being transported, but related sugars such as galactose can utilize the same system to cross the membrane.

Competitive Inhibition in Facilitated Diffusion. Structurally related compounds will compete for the same binding site on a membrane carrier, giving rise to the phenomenon of **competitive inhibition** of solute transport, as characterized by an increase in the K_m without a change in the V_{max} of the transport system. Thus, amino acid A, which shares the same transport system as amino acid B, will act as a competitive inhibitor of the transport of B if both are present simultaneously (Fig. 4–23). The inhibitory effect of A can be overcome by an increase in the concentration of B so that more molecules of B are available to compete with A for the carrier binding sites. In this way the rate of transport of B can be increased to the same V_{max} that was achieved in the complete absence

Figure 4–22

Graphical representation of the saturability of facilitated diffusion. The maximum rate of transport possible is termed the V_{max}, and the K_m is the concentration of solute necessary to reach one half of the V_{max}. Although simple diffusion is not saturable, the rate of transport is much slower at the physiological range of solute concentration.

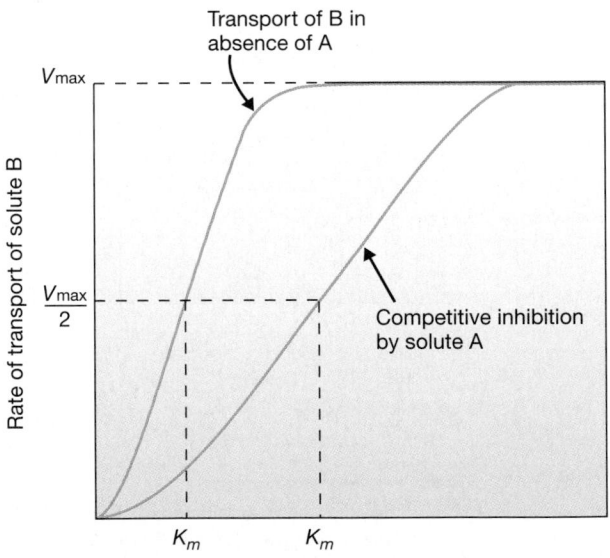

Figure 4–23

Competitive inhibition of facilitated diffusion of solute B by the structurally similar solute A. In the presence of A, the concentration of B must increase to reach the same V_{max} as that attained in the absence of A.

of A, but a higher concentration of B (i.e., an increased K_m) will be required (see Fig. 4–23).

Active Transport Is Energized Movement Against a Gradient

Thus far we have discussed passive transport processes, in which molecules and ions move in or out of the cell from regions of high to regions of low concentration. These movements, while sometimes requiring the presence of a special protein molecule, do not require energy. They occur in response to electrochemical gradients. The result is always a *decrease* in concentration of a once highly concentrated region.

Frequently, however, the cell requires an increase in concentration of an already highly concentrated solute. The solute must move from a less concentrated region outside the cell to a more concentrated region inside—that is, *against* the prevailing gradient. This process requires input of energy. The transport system will continue to operate for as long as the energy supply and the solute are available. Active transport, like facilitated diffusion, requires carrier proteins. The energy supply for active transport is derived from cellular metabolism; hence, active transport systems are susceptible to inhibition by compounds that act primarily as metabolic inhibitors and have no direct effect on the carrier proteins.

The mechanisms of active transport are of two general types depending on whether the movement of solute is linked directly or indirectly to energy-yielding reactions (see Fig. 4–12). When the movement of solute is coupled directly to an energy-yielding reaction, the transport process is termed **primary active transport.** When active transport of a solute is not coupled directly to the energy-yielding reactions, the transport mechanism is described as **secondary active transport.**

Primary Active Transport

The best known example of a primary active transport system is probably the **sodium pump.** This pump is present in the plasma membrane of most animal cells and is also known as the **Na^+/K^+-ATPase pump** because Na^+ is pumped out of the cell at the same time as K^+ is pumped in, and because the ion movements are coupled directly to breakdown of ATP. Removal of one of the phosphate groups from ATP releases free energy that is used to drive the movement of both Na^+ and K^+ against their respective electrochemical gradients. It is chiefly through this active transport system that the cell is able to maintain a high K^+ concentration and a low Na^+ concentration in the cytosol.

The breakdown of ATP is tightly coupled to the transport of Na^+ and K^+; the one cannot occur without the other. For every molecule of ATP that is split into ADP and phosphate, 3 Na^+ are pumped out of the cell and 2 K^+ are pumped in. The molecular events that link ATP breakdown to ion transport are not fully understood, but it seems clear that the enzyme protein is temporarily phosphorylated by attachment of the phosphate group released from the ATP. The phosphorylation step is Na^+-dependent—that is, it occurs only when Na^+ is present, and presumably leads to a change in protein conformation that moves bound Na^+ out of the cell. The subsequent dephosphorylation (removal of phosphate) is K^+-dependent and probably allows the protein to return to its original conformation, moving bound K^+ into the cell. These steps are depicted in the scheme in Figure 4–24. The Na^+/K^+-ATPase is represented as a protein with four subunits. This active transport system is so important to the cell that about one third of the cell's energy supply is used to maintain its activity.

Secondary Active Transport

When transport of a solute is not coupled directly to energy-yielding reactions, it is described as **secondary active transport.** A common example of this kind of mechanism is a transport system that is driven by the energy stored in the electrochemical gradient for another solute. In animal cells, this solute is usually Na^+, which enters cells passively by moving down a very favorable electrochemical gradient.

Symport and Antiport. Recall that there is a low sodium concentration and a negative potential inside the cell relative to the outside. Many different solutes are transported across cell membranes against their electrochemical gradients by coupling to the movement of Na^+. Movement of the solutes in the same direction as Na^+ is known as **symport** or **cotransport;** movement in the opposite direction is known as **antiport** or **exchange** (Fig. 4–25). Solute transport systems that are coupled to Na^+ movement have a very specific requirement for Na^+; other ions of similar size such as K^+ and Li^+ cannot be used in place of Na^+.

The Na^+ gradient across the cell membrane is maintained by the Na^+/K^+-ATPase, which actively pumps Na^+ out of the cell against its electrochemical gradient. As we have just discussed, this process is driven directly by a metabolic energy supply in the form of ATP. Thus, secondary active solute transport that is driven by the Na^+ electrochemical gradient depends ultimately on a supply of ATP but is not linked directly to the reactions that break down ATP.

Cytosol

Extracellular
fluid

(a)

(b)

(c)

(d)

(e)

Figure 4–24

Hypothetical scheme of action of Na$^+$/K$^+$-ATPase. (*a,b*) In the presence of Na$^+$ inside
the cell, ATP binds to and phosphorylates one of the protein subunits. Na$^+$ then binds to
specific sites on the phosphorylated protein. (*c,d*) These events trigger a conformational
change that exposes the bound Na$^+$ to the cell exterior, and Na$^+$ dissociates from the
protein. The binding site for K$^+$ is now exposed, and external K$^+$ binds. (*e*) Binding of
K$^+$ promotes dephosphorylation (removal of the phosphate) of the protein, which allows
it to revert to the original conformation. The conformational change moves the bound K$^+$
inside the cell, where it dissociates from the protein. The cycle now can be repeated.
The conformational changes are greatly simplified in the drawing.

Cytosol

Extracellular
fluid

Symport

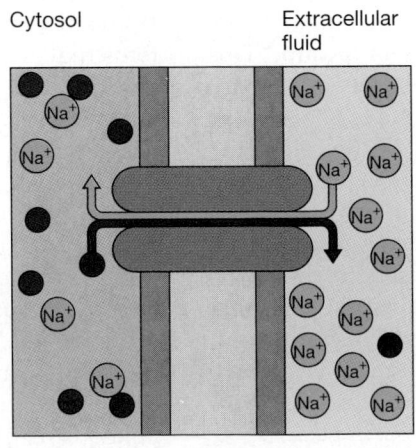

Cytosol

Extracellular
fluid

Antiport

Figure 4–25

Secondary active transport occurs
when solute transport is driven by
the electrochemical gradient for
Na$^+$. Both solute and Na$^+$ bind to
the membrane carrier. The Na$^+$
gradient is maintained by primary
active transport of Na$^+$.

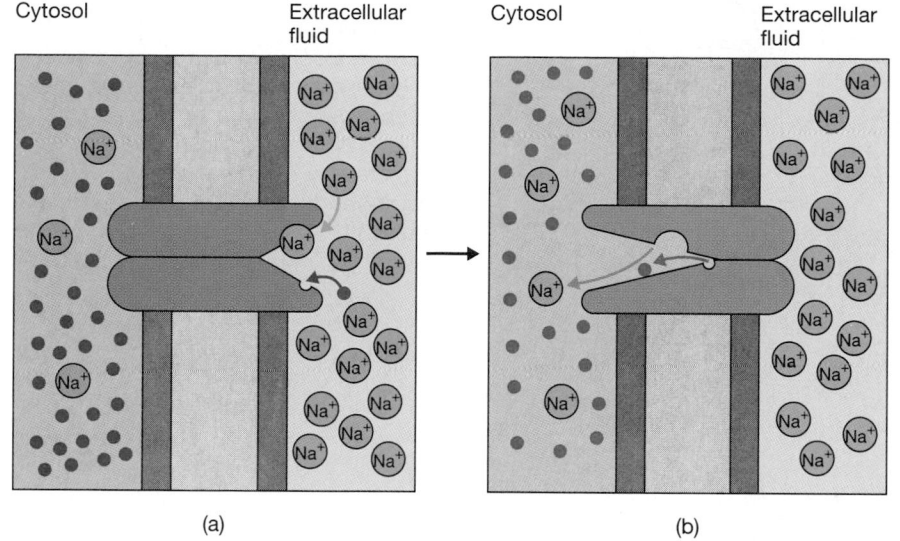

Cytosol | Extracellular fluid | Cytosol | Extracellular fluid

(a)   (b)

Figure 4–26

Hypothetical scheme of action of a two-site carrier for Na^+-coupled solute symport. (*a*) Binding of Na^+ outside the cell to the carrier increases the affinity for the solute, which also binds even though it is present at a low concentration. (*b*) A conformational change in the carrier protein exposes the binding sites to the cell interior. Na^+ dissociates from the carrier because the Na^+ concentration is low inside the cell. This dissociation step leads to a decrease in carrier affinity for solute, so the latter also dissociates from the carrier even though the solute concentration is high inside the cell.

The mechanism of Na^+-coupled transport systems is believed to involve a two-site carrier that binds both Na^+ and the solute. A possible mechanism is illustrated in Figure 4–26. Na^+ outside the cell binds readily to the carrier and encourages binding of solute to the carrier even though the solute concentration is low outside the cell. A conformational change in the carrier protein now exposes the bound Na^+ and solute to the cell interior. Na^+ readily dissociates from its site because the intracellular Na^+ concentration is very low. Loss of Na^+ from the carrier may alter the binding of solute to the carrier and cause the solute to dissociate from the carrier even though there is already a high intracellular concentration of the same solute.

Transport Processes and Cell Volume Are Interdependent

The sodium pump, which uses ATP to pump Na^+ ions out of the cell in exchange for K^+, is a key factor in the mechanism for controlling the osmolarity of the cytosol, and hence the intracellular volume, of animal cells. By effectively adding Na^+ to the extracellular fluid and simultaneously keeping the intracellular Na^+ concentration low, the cell is able to prevent osmotic entry of water and maintain the intracellular volume. The other principal solutes that contribute to the osmotic pressure of cell cytosol include proteins, organic phosphates such as ATP, metabolic intermediates, and K^+ ions.

The importance of the sodium pump in the control of cell volume is evident in the fact that animal cells swell and sometimes burst when the pump is inhibited. Decreased activity of the Na^+ pump probably contributes to the increased osmotic fragility of erythrocytes in stored blood. During storage, the ATP supply tends to fall. Since the Na^+ pump cannot operate without ATP, any Na^+ ions that "leak" (diffuse) back into the cell will not be pumped out. This abnormal buildup in intracellular Na^+ leads to a gradual rise in the osmolarity of the cell cytosol, with osmotic entry of water causing the cell to swell. In this swollen state, the cells are less resistant to osmotic stress than normal biconcave erythrocytes because they cannot accommodate the same increase in intracellular volume.

Typically, animal cells are exposed to a high concentration of external Na^+. There is no net increase in intracellular Na^+ because the rate at which Na^+ leaks into the cell is closely matched by the rate at which Na^+ leaves the cell via the Na^+ pump. This mechanism prevents the external and internal Na^+ concentrations from reaching equilibrium even though Na^+ constantly enters and leaves the cells.

Thus, Na^+ outside the cell can be considered to behave as if it were a nonpermeant solute. Similarly, the rate at which K^+ is pumped into the cell matches the rate at which K^+ leaks out, so the high internal K^+ concentration and the low external K^+ concentration are maintained. Thus, K^+ inside the cell behaves effectively as a nonpermeant solute because, as with Na^+, there is no net movement of K^+ into or out of the cell even though the internal and external concentrations are markedly different.

Some solutes, such as urea and glycerol, penetrate the plasma membranes of an animal cell very rapidly. The cell does not have a mechanism for pumping these solutes back out again. Such **permeant solutes** exert only a transient effect on cell volume, however. Unlike Na^+ and K^+, their external and internal concentrations rapidly equilibrate.

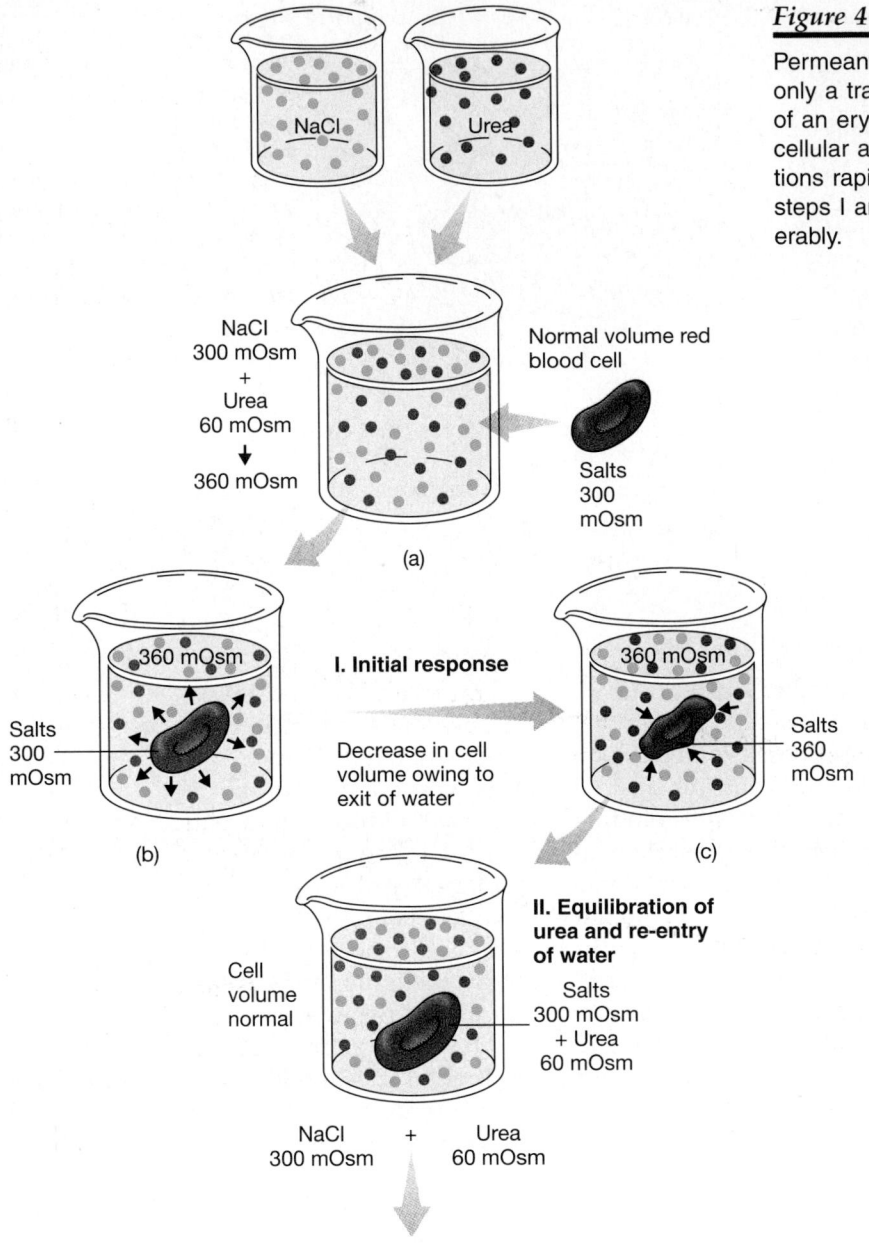

Figure 4–27

Permeant solutes such as urea exert only a transient effect on the volume of an erythrocyte because their intracellular and extracellular concentrations rapidly equilibrate. In reality, steps I and II probably overlap considerably.

Consider an erythrocyte placed in a large volume of a solution containing 300 mOsm NaCl and 60 mOsm urea (Fig. 4–27). Initially the cell will experience a hyperosmotic environment. Water will flow out, and the cell volume will decrease. However, as the urea rapidly enters the cell and the intracellular urea concentration increases toward 60 mOsm, water also will enter and the cell will begin to swell. The final volume of the cell at equilibrium is determined only by the osmolarity of the impermeant solutes in the extracellular medium. In this case, the impermeant solutes (NaCl) have a total osmolarity of 300 mOsm, the same as the cell cytosol, and the cell will assume its original volume.

Na^+ is the most abundant solute present in blood plasma. As an effectively impermeant solute, it is the major determinant of intracellular volume in the animal, even though plasma contains many membrane permeable solutes. A drop in plasma Na^+ concentration will reduce plasma osmolarity, and osmotic flow of water into cells will cause them to swell. Conversely, a rise in plasma Na^+ concentration will lead to cell shrinkage, owing to loss of intracellular water. This is one of the reasons that plasma Na^+ concentration is closely regulated. A change in Na^+ concentration induces a series of rapid responses designed to return it to the normal range.

Close control of the plasma Na^+ level means that most normal human cells are not subjected to large shifts in plasma osmolarity. However, challenges to the internal osmolarity can occur in cells designed to move water and specific solutes from one side of the cell to the other. This is true particularly for epithelial cells (see next section). Changes in internal osmolarity cause water to flow into or out of the cell and lead to changes in intracellular volume. Many cells have mechanisms for rapidly correcting the volume changes, and these mechanisms do not utilize the Na^+/K^+-ATPase pump. Most cells respond to an increase in cell volume by activating pathways that increase the efflux of K^+ and Cl^- from the cell. The loss of solute decreases intracellular osmolarity, which in turn stimulates efflux of water from the cell and leads to a return to normal cell volume. This response is termed a **regulatory volume decrease.** Some, but not all, cells can respond to a decrease in normal cell volume by a complementary process known as a **regulatory volume increase.** In this case, shrinkage of the cell activates pathways which increase influx of solutes such as Na^+ and Cl^-. Gain of solutes increases intracellular osmolarity, causing an influx of water and a return to normal cell volume. These rapid volume regulatory responses may help epithelial cells maintain a near-constant volume even with large changes in the amount of water and solutes passing through the cells.

Epithelial Transport

The transport systems in the plasma membrane of the cell are important for supplying nutrients and removing waste products. In addition to these roles, the specific arrangement of these transport systems in epithelial cells permits the net movement of solutes and water from one side of the epithelium to the other.

Epithelial Cells Are Organized Asymmetrically

This organization is crucial for the function of epithelial cells such as those lining the lumen of the small intestine and the renal proximal tubule, tissues that are specialized for reabsorption. These cells allow **transcellular transport** of water and solutes because the entry and exit pathways are on opposite sides of the cell (Fig. 4–28) (see p. 132).

The entry step is at the apical membrane. This is the membrane where the Na^+-coupled symport systems for amino acids, glucose, and phosphate are located. The exit step is across the basolateral membrane where the Na^+/K^+-ATPase is located. The Na^+ gradient across the apical membrane is maintained because the ATPase in the basolateral membrane constantly pumps Na^+ out of the other side of the cell to keep intracellular Na^+ low.

The asymmetrical distribution of transport systems between the apical and basolateral plasma membranes persists because of the **junctional complexes** between the cells. The junctional complexes occur at the point where the plasma membranes of neighboring cells come into close contact, and they prevent lateral diffusion of the proteins in the cell membrane. Thus, the carrier proteins in the apical membrane are physically prevented from leaving the apical membrane and mixing with the carrier proteins in the basolateral membrane.

In some types of epithelia, these junctions are relatively permeable and allow water and ions to move between the epithelial cells, providing an additional or **paracellular** pathway across the epithelium (see Fig. 4–28). This type of epithelium is termed a **"leaky"** epithelium because water and some solutes can leak between the cells in addition to moving through the transcellular pathway. A good example of this type of epithelium is the proximal tubule of the kidney. In other epithelia, such as the small intestine and the collecting tubule of the kidney, the cell junctions are impermeable; the only route by which water and solutes can cross the epithelia is the transcellular pathway. These types of epithelia are termed **"tight"** epithelia because they do not permit leakage between the cells.

Water movement across epithelia is always passive and is driven by the osmotic gradient. This gradient is set up by the active transport of solutes from the lumen and their addition to the fluid in the intercellular space.

A problem that leaky epithelia face is **backflux** of solute through the paracellular pathway. Thus, in the proximal tubule, the Na^+ that is reabsorbed from the lumen by the transcellular route is added to the fluid in the intercellular space (see Fig. 4–28), tending to increase the Na^+ concentration of the fluid. If the Na^+ concentration increases to the point at which it exceeds the concentration in the fluid in the tubular lumen, a Na^+ concentration gradient will exist. In addition to favoring water reabsorption from the lumen, this Na^+ gradient will drive movement of some of the Na^+ back into the lumen through the paracellular pathway. This backflux will limit the net reabsorption of Na^+ from the lumen. Movement of ions

(a)

(b)

Figure 4–28

(*a*) Transport pathways across a layer of epithelial cells.
(*b*) An electron micrograph of two intestinal epithelial
cells and typical junctional complexes (arrows).
(×70,000.)   (© John Hansen/Photo Researchers, Inc.)

diuretic hormone, which increases the water permeability of the apical membrane of the tubular cells.

In addition to functional specialization of reabsorptive epithelia, there is also marked structural specialization. The surface area of the apical membrane is often increased considerably by numerous **microvilli.** Similarly, the surface area of the basolateral membrane is increased by extensive **infoldings.** Large numbers of mitochondria are present. They are frequently found between the basolateral infoldings (Fig. 4–29), where presumably they provide a plentiful supply of ATP to maintain the activity of the Na^+/K^+-ATPase in the basolateral membrane. In other words, the cells are adapted for the rapid transport of large volumes of fluid containing a mixture of solutes and water.

The Functions of Different Transporters Are Coordinated

We can understand how the different transport systems interact to achieve net movement of a solute across an epithelial cell by considering how inorganic phosphate crosses the proximal tubule in the kidney (Fig. 4–30). The initial step in this process is entry of phosphate into the cell across the apical membrane. This step requires active transport because the negatively charged phosphate ion must move against both an electrical gradient (a negative potential inside the cell relative to the lumen) and a chemical gradient (a higher phosphate concentration inside the cell rela-

through the paracellular pathway is entirely passive and not under any direct control. Ions move in the direction favored by their electrochemical gradient.

The transcellular route achieves net movement from the lumen at one side of the cell to the blood supply at the other side. It is the only route across tight epithelia, and it can be regulated by hormones. This regulation allows almost complete control of the reabsorptive process by specific hormones that are able to regulate either the entry step at the apical membrane or the exit step at the basolateral membrane. For example, the reabsorption of water in the collecting tubule of the kidney is regulated by anti-

Figure 4–29

An electron micrograph of a kidney proximal tubule cell showing microvilli (MV) of the apical plasma membrane, extensive infoldings (IF) of basal and lateral membranes, and mitochondria (M) between the basolateral infoldings. (×10,000.)   *(© C. Craig Tischer, M.D./University of Florida.)*

Figure 4–30

Transepithelial movement of inorganic phosphate (Pi) in the renal proximal tubule.

tive to the lumen). Phosphate enters the cell via a symport mechanism in the apical membrane that couples the movement of phosphate to the movement of Na^+. There is a very favorable electrochemical gradient driving Na^+ entry. Thus, phosphate entry is an example of secondary active transport. Once inside the cell, the phosphate is able to leave by a passive mechanism (possibly facilitated diffusion) located in the basolateral membrane. The electrochemical gradient that opposed phosphate entry at the apical membrane is used now to drive this exit step at the basolateral membrane. After diffusing across the intercellular space, the reabsorbed phosphate enters the blood in the capillary system and is returned to the general circulation. The Na^+ that enters the cell across the apical membrane is pumped out across the basolateral membrane by the Na^+/K^+-ATPase. The Na^+ leaving the cell must be moved against the gradient that enabled it to enter the cell at the apical membrane (see Fig. 4–30). The exit of Na^+ is an example of primary active transport because it is coupled directly to ATP hydrolysis. Na^+ must be removed from the cell in order to maintain a low intracellular Na^+ concentration and a favorable gradient for continued Na^+ entry across the apical membrane.

An analogous arrangement of the appropriate transport systems is used to achieve the reabsorption of glucose and amino acids across the cells of the renal proximal tubule. The solutes enter the cell from the tubule lumen via secondary active Na^+-coupled symport across the apical membrane. Then they exit from the cell via passive facilitated diffusion across the basolateral membrane.

Single Channel Recording

The flow of ions such as Na^+ and K^+ across nerve cell membranes plays a central role in transmission of nervous impulses (see Chapter 7). The ion flow occurs through integral membrane proteins known as **channels.** Each nerve cell has several different types of ion channels, and as many as one million channel proteins of each type may be present in the plasma membrane. Ion flow through large populations of channels, and the resulting electrical activity, has been studied in nerve cells for many years. In 1976, a new method known as the **patch-clamp** technique was introduced. It was developed in Germany by E. Neher and B. Sakmann and is now used in a large number of laboratories. The great advantage of this technique is that it allows the study of a **single** ion channel.

The technique is used in the following way: A clean glass pipette is pressed against an intact cell to form a tight seal. The tip of the pipette has a diameter of about 1 μm and is polished to avoid damage to the cell membrane. Slight suction allows formation of a very stable glass-membrane seal. The ion channels underneath the tip of the pipette can now be studied to determine their regulation by intracellular molecules. The patch of membrane underneath the pipette tip can also be pulled away from the cell by withdrawing the pipette from the cell surface. The membrane patch remains undamaged and functional in the aperture of the pipette. The membrane patch can be detached from the cell because the glass-membrane interactions are stronger than the phospholipid interactions that maintain the integrity of the plasma membrane. The great advantage of using a detached patch is that the plasma membrane surface that faced the inside of the intact cell now becomes accessible to experimental manipulations.

In essence, whenever a channel protein opens within the membrane spanning the pipette tip, there will be a flow of ions from one side of the membrane to the other along their concentration and electrical gradients. This flow of ions generates a tiny electric current that can be measured, in picoamperes (pA), by very sensitive electronic measuring equipment. These currents can be detected because the patch-clamp instrumentation eliminates electrical noise that, if present, would obscure the small single channel currents. Patch clamping isolates a small area of membrane so that only a few channels are present. This allows the contribution from each channel to be separated and analyzed.

Recording single channel activity has allowed new understanding of the action of channel proteins in nerve and other cells. Several unique properties of ion channels have been discovered: (1) Ion channels behave in a simple open-shut manner, i.e. they are either fully open or fully closed. They open for a period of time that is often less than 1 millisecond. (2) Channels open in a random way and so current flows for varying lengths of time. The average duration of current flow is a quantity used to characterize the specific type of channel. (3) The ease with which current flows (known as **conductance**) when a channel opens is also characteristic of the type of channel. (4) Most channels exhibit selectivity with regard to the ions that can pass through them. Some channels allow passage of Na^+ but not K^+, for example. In contrast, others allow flow of K^+ but not Na^+.

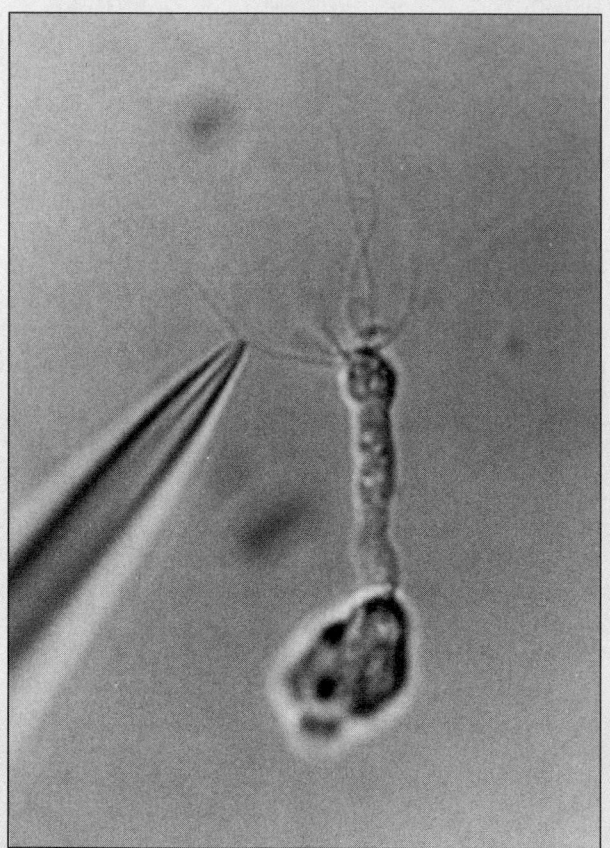

A photograph showing the view in a light microscope of a patch-clamp pipette pressed against the plasma membrane of a cilium which extends from a cell derived from the olfactory epithelium of a toad. Most of the pipette is not in focus and can be seen to the left of the cell. ($\times$2,300.)

■ **What is the mechanism by which ion channels are thought to open and close? What factors cause channels to open?**

CHAPTER REVIEW

Summary

- The core of a plasma membrane is the lipid bilayer, a double layer of lipid molecules. The bilayer is highly impermeable to charged or polar solutes but allows the passage of small lipid-soluble molecules.

- Most membrane proteins are embedded in the lipid bilayer, but a few are not. The latter are loosely associated with one side of the bilayer. The proteins carry out many membrane functions by acting as specific receptors, enzymes, or transporters.

- Carbohydrates are a relatively minor component of the plasma membrane. They are attached to both proteins and lipids and are located exclusively on the extracellular surface of the membrane.

- Gradients of pressure, concentration, or voltage drive the movement of many substances in the body.

- Many solutes can cross the plasma membrane by diffusion down the electrochemical gradient, a process known as passive transport. Lipid-soluble nonpolar molecules pass through the lipid bilayer, but the passage of charged or polar solutes is "facilitated" by specific membrane-spanning proteins.

- Osmosis is the diffusion of water across a plasma membrane in response to a difference in solute concentration on either side of the membrane. The rapid movement of polar water molecules occurs through specific water channel proteins.

- Active transport produces movement of a solute across the plasma membrane even though the solute is moved against its electrochemical gradient. This process requires a specific membrane carrier protein and an input of energy (usually in the form of ATP) derived from cellular metabolism. The energy-yielding reaction can be linked directly or indirectly to the transport step.

- The different types of transport systems act in concert in epithelial cells to achieve directional movement of solutes and water from one side of the cell to the other. This organization is aided by the structural and functional specialization of the cells.

Review Questions

Identify Each with a Term

1. General term for a membrane protein that penetrates the phospholipid bilayer.
2. The process of simple diffusion when the molecules in motion are water molecules.
3. Carrier-dependent transport of solute in response to the electrochemical gradient.
4. Coupled movement of two solutes in opposite directions across a membrane.
5. The process by which a swollen cell returns rapidly to normal size.

Define Each Term

6. Amphipathic molecule
7. Overton's rules
8. Diffusion trapping
9. Secondary active transport
10. Tight epithelium

Choose the Correct Answer

11. Which of the following processes requires a supply of metabolic energy?
    a. Simple diffusion
    b. Facilitated diffusion
    c. Primary active transport
    d. Osmosis

12. Simple diffusion differs from facilitated diffusion because it:
    a. is much slower
    b. can be blocked by competitive inhibitors
    c. reaches a maximum rate
    d. requires expenditure of energy

13. The plasma membrane of a cell separates:
    a. organelles from cytosol
    b. cytosol from chromosomes
    c. cytosol from mitochondrial matrix
    d. extracellular fluid from cytosol

14. In the phospholipid bilayer of a plasma membrane:
    a. phospholipids are distributed asymmetrically
    b. proteins are distributed asymmetrically
    c. both phospholipids and proteins move laterally
    d. all the above are true

15. The plasma membrane Na^+ pump
    a. moves K^+ out of the cell
    b. exchanges 3 Na^+ for 2 K^+
    c. uses energy released by ADP breakdown
    d. moves Na^+ into the cell

16. A membrane that is permeable only to water separates two solutions of glucose dissolved in water. On one side of the membrane (side A) the glucose concentration is 0.1 g/ml. On the other side (side B) the glucose

concentration is 0.5 g/ml. Initially, the rate of water flow will be:
a. most rapid from side A to side B
b. most rapid from side B to side A
c. the same in both directions
d. zero (no flow in either direction)

Answer Each Question in One or Two Sentences

17. List four of the principal features of the fluid mosaic (Singer-Nicolson) model of the plasma membrane.

18. Carbohydrate on the external surface of the plasma membrane is probably important for membrane receptors and cell-cell recognition. Why are carbohydrates particularly suited for these functions?

19. Explain why the erythrocyte can increase its volume by 67% without rupturing.

20. Explain the term regulatory volume decrease.

21. What distinguishes a leaky epithelium from a tight epithelium?

Problem Solving in Physiology

22. Explain the principal difference between osmosis and dialysis and describe a practical application of dialysis.

23. What will happen to the Na^+-coupled transport of glucose, amino acids, and phosphate into an epithelial cell if ATP synthesis in the cell is blocked?

Suggested Readings

Andreoli, T.E. "Molecular aspects of epithelial transporters." *Hospital Practice*, 24:11–12, 1989.

Armstrong, W. McD. "The cell membrane and biological transport." *Physiology*, 5th ed. Selkurt, E.E., ed. Boston: Little, Brown, 1984, pp. 1–26.

Handler, J.S. "Overview of epithelial polarity." *Annual Review of Physiology*, 51:729–740, 1989.

Hille, B. *Ionic Channels of Excitable Membranes*, 2nd ed. Sunderland, MA: Sinauer, 1992.

Lewis, S.A., and Donaldson, P. "Ion channels and cell volume regulation: chaos in an organized system." *News in Physiological Sciences*, 5:112–119, 1990.

Pederson, P.L., and Carafoli, E. "Ion motive ATPases: I. Ubiquity, properties, and significance to cell function." *Trends in Biochemical Sciences*, 12:146–150, 1987.

Pumplin, D.W., and Bloch, R.J. "The membrane skeleton." *Trends Cell Biol*, 3:113–117, 1993.

Quinton, P.M. "Cystic fibrosis: a disease in electrolyte transport." *FASEB Journal*, 4:2709–2717, 1990.

Sharon, N., and Lis, H. "Carbohydrates in cell recognition." *Scientific American*, 268:82–89, 1993.

Singer, S.J., and Nicolson, G.L. "The fluid mosaic model of the structure of cell membranes." *Science*, 175:720–731, 1972.

Stevens, C.F. "Studying just one molecule: single channel recording." *Trends in Pharmacological Sciences*, 5:131–134, 1984.

Walmsley, A.R. "The dynamics of the glucose transporter." *Trends in Biochemical Sciences*, 13:226–231, 1988.

Welsh, M.J., Anderson, M.P., Rich, D.P. et al. "Cystic fibrosis transmembrane conductance regulator: a chloride channel with novel regulation." *Neuron*, 8:821–829, 1992.

Wright, E.M., Hager, K.M., and Turk, E. "Sodium cotransport proteins." *Current Opinion Cell Biol.*, 4:696–702, 1992.

Key Terms

active transport (p. 116)
amphipathic molecule (p. 110)
antiport (p. 127)
carrier protein (p. 123)
channel gating (p. 123)
cholesterol (p. 110)
diffusion coefficient (p. 114)
diffusion trapping (p. 119)

electrochemical gradient (p. 117)
extrinsic protein (p. 112)
facilitated diffusion (p. 116)
fatty acid (p. 110)
fluid mosaic model (p. 109)
intrinsic protein (p. 111)
iso-osmotic (p. 121)
junctional complex (p. 131)
lipid (p. 109)

lysis (p. 121)
osmosis (p. 119)
osmotic concentration (p. 119)
osmotic fragility (p. 122)
passive transport (p. 116)
permeability coefficient (p. 117)
phospholipid (p. 110)
simple diffusion (p. 116)

sodium pump (p. 127)
symport (p. 127)
tight epithelium (p. 131)
transcellular transport (p. 131)
volume regulation (p. 131)

Micrograph of an unwound human DNA molecule. *(© Dr. John A. Olschowka, Ph.D./Dr. Richard Insel, M.D./Dr. Louise Chow/Phototake)*

Chapter 5
Cellular Control Mechanisms

Key Concepts

Different genes are expressed in different cells. The end-products of the expression of most genes are proteins, and the specific proteins made by a cell will determine its structure and function.

Communication between cells allows coordination of the functions of the different organs and tissues in the body. Adjacent cells can communicate via gap junctions, and cells that are far apart release chemical signals.

Chapter Outline

An Intracellular Control System: DNA, RNA and Protein Synthesis

Cell–Cell Signaling: Hormones, Neurotransmitters, and Other Molecules

Cathy, a freshman at a liberal arts college, was earning some much-needed cash by house-sitting for one of her professors, who had decided to travel for a few weeks at the end of the spring semester. The family had left earlier in the day and had given Cathy a list of the items that needed regular attention.

The list looked straightforward except for the mysterious request to avoid bringing any diet soda into the house because these drinks were dangerous for Tommy, an obvious reference to the youngest child.

Cathy knew that some people disliked the taste of diet drinks, but the suggestion that they could be dangerous roused her curiosity. In the student union that evening, she purchased a diet soda and scrutinized the small print on one side of the can. The ingredients were not unusual and included the low-calorie sweetener called aspartame. After further reading, the item Cathy did not understand was the statement in bold letters "Phenylketonurics: Contains Phenylalanine." Since she was heading for the library she decided to stop in the reference section to consult a medical dictionary. She discovered that **phenylketonuria** (PKU) is a disease that allows accumulation of the amino acid phenylalanine and its metabolic products (phenylketones) in the body fluids. Phenylketones interfere with brain development, and the result is severe mental retardation. Fortunately, the disease can be detected during pregnancy and can be controlled by feeding from birth with a diet that contains very low levels of phenylalanine. Diet drinks pose a problem for PKU patients because aspartame contains phenylalanine. The genetic disease PKU is known as an inborn error of metabolism because the genetic defect shows up as a metabolic problem. PKU patients lack the enzyme phenylalanine hydroxylase, which converts phenylalanine to tyrosine. In the absence of this enzyme, all the phenylalanine is metabolized by pathways that are not normally used, and unusually high amounts of phenylketones are produced.

The production of specific proteins for use inside the cell, or elsewhere in the body, is based on instructions stored in genes—specific sequences of DNA. The process requires an intricate series of reactions involving coordinated activities of several intracellular organelles. The genes in the 23 pairs of chromosomes control, via the proteins that are produced, the growth, development, and function of the human body. Because the body is a highly tuned machine, the removal or malfunction of a single genetic "screw" usually has disastrous consequences.

Much of the success of multicellular organisms is enabled by the organization of groups of differentiated cells into specialized tissues and organs. The tissues coordinate to carry out all the functions necessary for the organism's survival.

The specialization and coordination of cell functions require very effective communication, both within each cell and among the different tissues. A good example of the importance of efficient communication is the process involved in avoiding a predator. When the tissues responsible for sight, sound, or smell detect the presence of a predator, the survival of the organism often depends on the almost instantaneous response of other tissues that can carry out rapid evasive movement. Mammals have systems that provide communication and coordination among the different organs. These are the **nervous system** and the **endocrine system.** A large section of the book is devoted to descriptions of these very sophisticated systems.

This chapter emphasizes the ways in which some of the intracellular structures and macromolecules described in the preceding chapters participate in specific control systems that reside and operate within a cell, as well as ways in which these controls are influenced by signals from other cells. We begin by discussing the use of the genetic information stored in DNA, perhaps the fundamental **intracellular control system.** We will learn how this system is subject to regulation primarily by intracellular influences. The latter part of the chapter introduces the concept of **extracellular control systems,** or **cell-to-cell signaling,** the ways in which one cell communicates with and influences the function of another.

An Intracellular Control System: DNA, RNA, and Protein Synthesis

The cells of a complex eukaryotic organism such as a human being have developed a variety of structural and functional specializations, giving rise to different types of tissues, such as epithelial, connective, muscle, and nervous tissue. As previously mentioned in the brief discussion of the functions of the cell nucleus in Chapter 3, the molecular basis for the differentiation of functions and structures of cells is the structure of their proteins. In Chapter 4 we learned, for example, that different cells have different types and amounts of proteins in their membranes. These proteins are distributed in specific ways in the membrane in order to effectively perform their roles as carriers, channels, receptors, or shaping agents. An islet cell in the pancreas produces insulin, but a nerve cell makes none of this protein. All cells in the body make and use Na^+/K^+-ATPase, but different cells make different amounts of it. Most cells in the body produce myosin (a contractile protein), but only muscle cells make it in the abundance necessary to produce the type of motion we associate with visible body movements.

Proteins Are the End Result of Expression of Most Genes

With few exceptions, the markedly different types of cells contain the same **genome** in their nuclei—that is, each cell, regardless of tissue type, has the same DNA molecules, the same "permanent files" or master set of instructions for protein synthesis. These DNA molecules are in groups called **chromosomes.** Every cell in the human body, except the reproductive cells, has the same number (46) of chromosomes containing the same DNA molecules (Fig. 5–1*a*). These DNA molecules are made up of **genes,** and most genes carry the code for a specific protein. (Fig. 5–1*b*). Each gene consists of a certain sequence of nucleotides that dictates the sequence of amino acids in a specific protein produced by the cell. Every cell in the body, except the sex cells, can potentially make every protein found in the body.

It would seem that if all cells in the body contained the same genetic information, they would all make the same proteins in the same amounts. How then do cells become **differentiated,** producing some proteins and not others? How does one cell type, such as a pancreatic cell, know to produce an abundance of insulin, whereas another cell, such as a red blood cell, knows *not* to produce insulin (even though it possesses the gene for insulin), but to produce hemoglobin instead? How does a cell in the kidney know to produce myosin in the controlled amount required for its own shape and internal movements, but not in the abundant amounts that would produce movement of the entire organ? How does a skeletal muscle cell know to produce myosin in abundance, enough to produce movement of other tissues?

(a)

Figure 5–1

(*a*) Human chromosomes. Each nonsex cell in a human body has 46 chromosomes identical to those in all other cells of the same body. Chromosomes are visible during cell division; at other times, the DNA in the nucleus is in a loosely coiled form called chromatin. (*b*) Each chromosome consists of a molecule of DNA wound around proteins called histones. Certain sequences of the nucleotides on each DNA molecule, called genes, contain the code for the order of amino acids in a protein molecule to be made by the cell. The DNA nucleotide sequences are transcribed into RNA nucleotide sequences, which are then translated into amino acid sequences to form a polypeptide.

GENE FOR PROTEIN A

GENE FOR PROTEIN B

GENE FOR PROTEIN C

GENE FOR PROTEIN D

DNA nucleotides

RNA nucleotides

Amino acids

(b)

Genetic Expression: Selective Use of the Permanent Files

The answer to the previous questions lies in the phenomenon of **genetic expression** or, more precisely, the **control of genetic expression.** An individual cell has much more genetic information than it ever needs. Most of the genetic information in the cell is unused, or **suppressed.** At any given time in a particular cell, only a small part of the DNA is being used as a source for information about protein synthesis. In other words, only certain genes are being tapped for information about how to make the specific protein they describe; only certain genes in each cell are **expressed** in the form of **protein products.**

Thus, gene expression is the principal factor that determines the structural, functional, and behavioral characteristics of a cell. In complex eukaryotes, different genes are being expressed as the characteristic protein products of each different cell type. Furthermore, in a given cell, certain genes may be expressed only at certain times (e.g., when the cell receives the appropriate stimulus, possibly a hormone). Some of the mechanisms that control the expression of genes in a cell are discussed in detail in a later section.

The Genetic Code: From Nucleic Acids to Amino Acids

The term **genetic code** refers to the way in which information about protein synthesis is stored in DNA, "translated" into RNA, and then used to make proteins. The information in a DNA molecule specifies the sequence in which amino acids must be assembled to form a functional polypeptide molecule. A relationship exists between the nucleotides of RNA and DNA and the amino acids of a polypeptide chain, so that the order of nucleotides dictates the order of amino acids.

DNA and RNA are each composed of only four different nucleotides, whereas polypeptides can contain up to 20 different amino acids. Since there are only four nucleotides in each type of nucleic acid to specify the 20 different amino acids, we assume that **combinations** of nucleotides, rather than single nucleotides, are used to describe the amino acids. The genetic code must specify, in the sequence in which the four nucleotides are arranged, the sequence in which the 20 common amino acids should be arranged. Think of DNA, RNA, and proteins as three different languages, each with a different alphabet and vocabulary. Protein synthesis then is a process of translation, from the language of nucleic acids to the language of proteins.

The Four Letters of the DNA and RNA Alphabets. DNA molecules are formed from **deoxyribonucleotides,** four different molecules distinguished by four different **bases:** adenine, cytosine, guanine, and thymine (Fig. 5–2*a*). The DNA "alphabet" consists of four letters—A, C, G, and T—representing the four DNA bases (Table 5–1). These four letters can be combined in a total of 64 different "words," or **triplets,** such as CGA, CGT, CGG, and so on.

RNA molecules are formed from **ribonucleotides** —again, four different molecules distinguished by four different bases: adenine, cytosine, guanine, and uracil (see Fig. 5–2*a*). Thus, the RNA alphabet also contains four letters—A, C, G, and U (Table 5–1). These can also be combined in groups of three to make a total of 64 different "words," called **codons,** found on a special type of RNA called mRNA. As we will learn, the codons on an RNA molecule are **complementary** to the triplets on the DNA molecule from which they were transcribed. For example, the DNA triplet CGA has the complementary mRNA codon GCU (uracil is the RNA base complementary to adenine). Each codon represents a "word" in the language of mRNA, translated from the triplet "word" in the language of DNA.

The Twenty Letters of the Protein Alphabet. Each mRNA codon in turn refers to one of the 20 "letters" in the protein alphabet, one of the 20 common amino acids (Table 5–1). Scientists have deduced the individual codons that represent the links between the

---

Table 5–1
DNA, RNA, and Protein "Alphabets"

DNA Bases	*RNA Bases*	*Amino Acids*				
A	A	Alanine	Cysteine	Histidine	Methionine	Threonine
C	C	Arginine	Glutamic acid	Isoleucine	Phenylalanine	Tryptophan
G	G	Asparagine	Glutamine	Leucine	Proline	Tyrosine
T	U	Aspartic acid	Glycine	Lysine	Serine	Valine

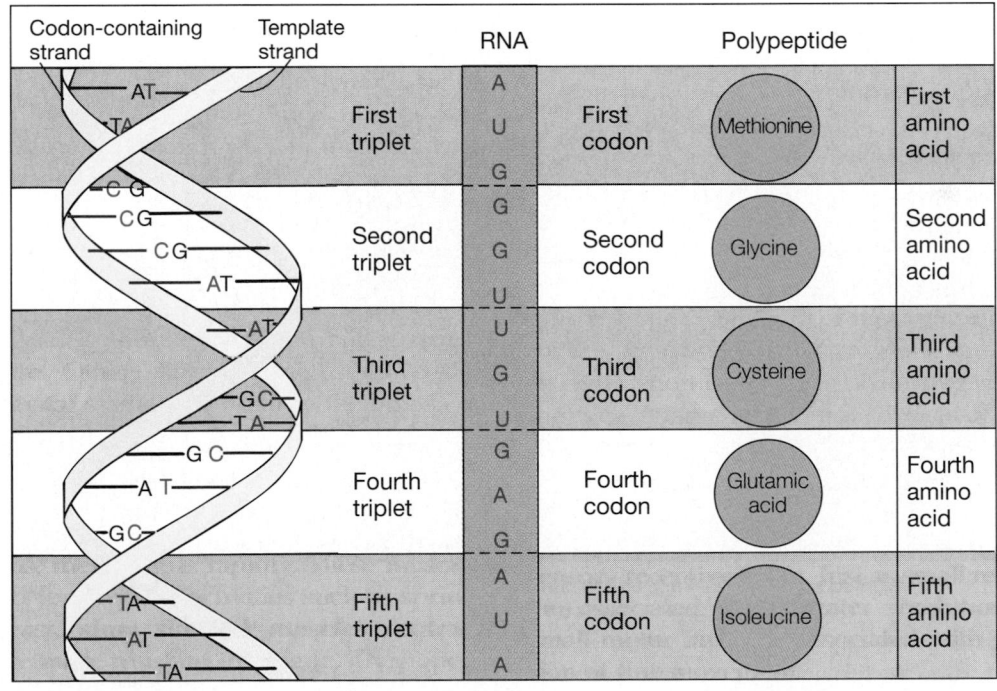

(a)

(b)

Figure 5–2

(a) The DNA "alphabet" consists of four nucleotides, derived from triphosphate nucleotides and distinguished by their bases: adenine, cytosine, guanine, and thymine. The RNA "alphabet" contains nucleotides formed from a different sugar molecule than that in DNA and distinguished by a slightly different set of bases: adenine, cytosine, guanine, and uracil. (b) The genetic code on DNA is copied from the template strand as a complementary strand of RNA, and then a matching chain of amino acids is formed. Each set of three bases on the DNA template strand, called a triplet, is transcribed into a complementary set of RNA bases, called a codon. Each codon refers to a specific amino acid.

Table 5–2
The Genetic Code

DNA Triplets	mRNA Codons		Amino Acids	DNA Triplets	mRNA Codons		Amino Acids
AAA	UUU	1	Phenylalanine	GTA	CAU	34	Histidine
AAG	UUC	2		GTG	CAC	35	
AAT	UUA	3	Leucine	GTT	CAA	36	Glutamine
AAC	UUG	4		GTC	CAG	37	
GAA	CUU	5		TTA	AAU	38	Asparagine
GAG	CUC	6		TTG	AAC	39	
GAT	CUA	7		TTT	AAA	40	Lysine
GAC	CUG	8		TTC	AAG	41	
TAA	AUU	9	Isoleucine	CTA	GAU	42	Aspartic
TAG	AUC	10		CTG	GAC	43	acid
TAT	AUA	11		CTT	GAA	44	Glutamic
CAA	GUU	12	Valine	CTC	GAG	45	acid
CAG	GUC	13		ACA	UGU	46	Cysteine
CAT	GUA	14		ACG	UGC	47	
CAC	GUG	15		ACC	UGG	48	Tryptophan
AGA	UCU	16	Serine	GCA	CGU	49	Arginine
AGG	UCC	17		GCG	CGC	50	
AGT	UCA	18		GCT	CGA	51	
AGC	UCG	19		GCC	CGG	52	
GGA	CCU	20	Proline	TCT	AGA	53	
GGG	CCC	21		TCC	AGG	54	
GGT	CCA	22		TCA	AGU	55	Serine
GGC	CCG	23		TCG	AGC	56	
TGA	ACU	24	Threonine	CCA	GGU	57	Glycine
TGG	ACC	25		CCG	GGC	58	
TGT	ACA	26		CCT	GGA	59	
TGC	ACG	27		CCC	GGG	60	
CGA	GCU	28	Alanine	TAC	AUG	61	Methionine START
CGG	GCC	29					
CGT	GCA	30					
CGC	GCG	31		ATT	UAA	62	STOP
ATA	UAU	32	Tyrosine	ATC	UAG	63	
ATG	UAC	33		ACT	UGA	64	

nucleic acid strands and the polypeptide chains. The DNA triplets and complementary mRNA codons for each of the 20 amino acids are listed in Table 5–2. Figure 5–2b gives examples of the way in which the relationship among the three languages results in an RNA molecule and a polypeptide.

Of the 64 different codons, 61 specify a particular amino acid (Table 5–2). The codon for **methionine (AUG)** has a dual role because it also serves as a **start codon,** which is involved in initiating **translation** of the nucleotide triplets in mRNA. The remaining 3 codons do not specify amino acids; instead, they are used as **stop codons** to terminate translation.

The Vast Language of Proteins. Consider the enormous number of words that can be generated from our alphabet of 26 letters. Then consider the huge number of different polypeptide chains that could be constructed from nearly that many amino acids, without the constraints on "word" length found in verbal

Figure 5–3

A simplified overview of the flow of information in the expression of the genetic code. The information on a DNA molecule is transcribed, or reproduced in coded form, on an RNA molecule. After the RNA molecule is processed, it leaves the nucleus and travels to a ribosome, where the coded information is translated into a sequence of amino acids that forms a polypeptide chain.

language. This will give you some sense of the variety of proteins that can be synthesized in the cell and the great possibilities for specialization of function that this variety allows. Approximately 2000 different proteins are thought to exist in the body.

Think of the difference in meaning one letter change makes in a word—for example, the change from *deride* to *decide.* This is similar to the difference that one amino acid can make in the nature of a protein. No other type of macromolecule allows this great variety and specificity of structure and function. Compared with lipids, carbohydrates, and nucleic acids (which have only a few subunits to arrange), proteins (with 20 different subunits) have an enormous evolutionary advantage because of the variety of different functional macromolecules that can be assembled. Indeed, it is likely that the molecular basis of the evolution of species lies in the ability of cells to make new proteins in response to changes in their environment.

A Consistent and Universal Code. The genetic code is said to be **consistent** because each codon specifies only a single amino acid. It is essentially **universal** because it is the same in diverse types of organisms including plants and humans. Because there are 64 different codons for only 20 amino acids, some amino acids are specified by more than one codon. For example, there are 2 mRNA codons for phenylalanine, 6 for leucine, 3 for isoleucine, etc. (see Table 5–2). For this reason, the code is said to be **degenerate.**

A gene and its polypeptide product are usually regarded as **colinear molecules.** This term means that the codon representing the beginning of the gene determines the amino acid present at the beginning of the polypeptide, while the codon representing the second triplet in the gene determines the second amino acid in the polypeptide, and so on. Many eu-

karyotic genes are interrupted with nucleotide sequences known as **introns,** the functions of which have yet to be determined. Introns will be discussed again later in this chapter.

Let us now look at how the nucleic acids themselves are assembled and translated into proteins. Figure 5–3 gives a simplified overview of the flow of information during these processes. We will discuss **DNA synthesis** (also known as **DNA replication**) and **RNA synthesis** (also known as **DNA transcription**). We will also look at methods of altering newly synthesized RNA, known as **RNA processing,** before it leaves the nucleus, methods that result in the production of functional RNA molecules. Finally, we will examine the process of **protein synthesis** (also known as **RNA translation**), in which the information in RNA is translated into a polypeptide chain.

New Nucleic Acids Are Synthesized by Copying a Template

As discussed in Chapter 2, the macromolecules deoxyribonucleic acid (DNA) and ribonucleic acid (RNA) are each composed of molecules called nucleotides. Nucleotides contain a ribose sugar and one to three phosphate groups, in addition to one of five bases. DNA and RNA are distinguished by the type of ribose sugar and bases their nucleotides contain. All nucleotides forming part of a nucleic acid strand have one phosphate group.

Synthesis of DNA: Replication of the Permanent Files

When a cell divides into two, the contents of both the cytoplasm and the nucleus divide. Division of the nucleus, a process known as **mitosis** (see Chapter 3), ensures that both of the new cells have an identical copy

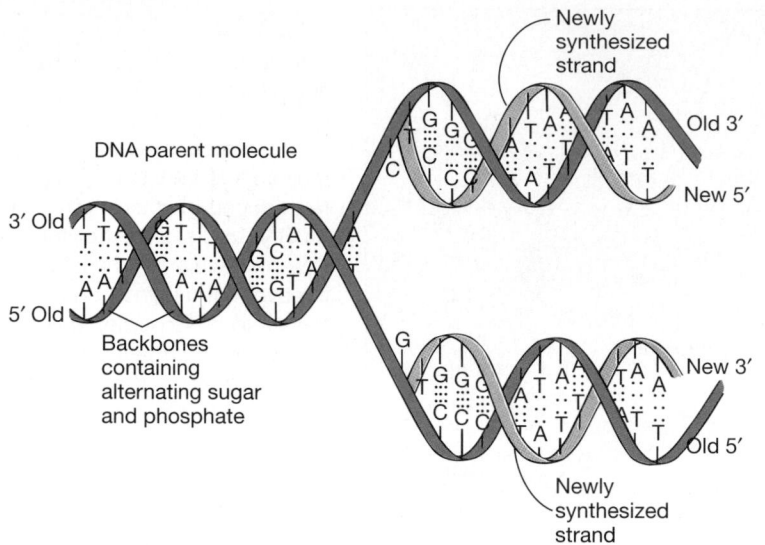

DNA parent molecule

3' Old

5' Old

Backbones
containing
alternating sugar
and phosphate

Newly
synthesized
strand

Old 3'

New 5'

New 3'

Old 5'

Newly
synthesized
strand

Figure 5–4

DNA replication is known as semiconservative replication because each new DNA molecule contains one new strand and one original strand. Thus, single strands of DNA are preserved and passed on for many generations.

of the DNA of the parent cell. Thus, the new cells will be identical in form and function to the parent.

Before the nucleus can divide, the DNA molecules within the nucleus (in the form of **chromatin**) must be copied, or **replicated.** The replication process must be accurate so that the genetic information is not altered, and it must be fast so that it can occur in time for a cell to divide. DNA replication, as mentioned in Chapter 3, occurs during the interphase preceding mitosis. Each chromosome, a long DNA molecule, is copied to produce two identical DNA molecules called **chromatids.**

A DNA molecule consists of the two long strands of nucleotides arranged in a double helix (see Chapter 2). The two strands are held together by hydrogen bonds between the bases of the nucleotides. Adenine always forms bonds with thymine, and guanine always forms bonds with cytosine. The specific base pairing produces two strands that complement each other.

Semiconservative Replication.   During the process of DNA replication, the two strands of the DNA molecule are first separated. Each one then acts as a mold, or **template,** upon which a new complementary strand of nucleotides is assembled. In this way, two new molecules of DNA are produced. Each molecule contains one strand that was the template derived from the parent molecule and one newly synthesized strand (Fig. 5–4). The process is described as **semiconservative** because the structure of the parent DNA molecule is disrupted but the structure of the subunits — that is, the nucleotide sequence of each strand — is conserved for many generations. Maintenance of the nucleotide sequence of DNA is a critical step that ensures that the new cells receive an exact copy of the genetic information in the parent cell.

Separating the Strands of the Double Helix.   Before replication of DNA can take place, the double helix must be unwound to separate the two strands. Unwinding is accomplished by an enzyme known as **DNA helicase. DNA topoisomerases** act downstream of the helicase to relieve the torsion that develops as the double helix unwinds. The separated strands are stabilized by proteins that bind to the single DNA strands and therefore assist the opening of the helix. They are known as **helix-destabilizing proteins.** Each strand then acts as a template for the assembly of a complementary chain. Separation of the two original DNA strands produces a structure called a **replication bubble** at the site of active synthesis of the new strands (Fig. 5–5). Many replication bubbles can exist on a DNA molecule at once, and the replication forks within each replication bubble move in opposite directions until they meet the fork coming from an adjacent bubble.

Replicating the Strands: DNA Polymerase.   DNA replication is catalyzed by an enzyme called **DNA polymerase,** which takes instructions from the DNA strand that is acting as the template for the new DNA strand. The specific reaction catalyzed by DNA polymerase is the joining together of the various nucleotides that are components of DNA. The individual nucleotides must contain deoxyribose and three phosphate groups; otherwise the enzyme cannot use them. Thus, DNA polymerase takes the nucleotides

deoxyadenosine triphosphate (dATP)

deoxycytidine triphosphate (dCTP)

deoxyguanosine triphosphate (dGTP)

deoxythymidine triphosphate (dTTP),

removes two of their phosphate groups to form the monophosphate nucleotides

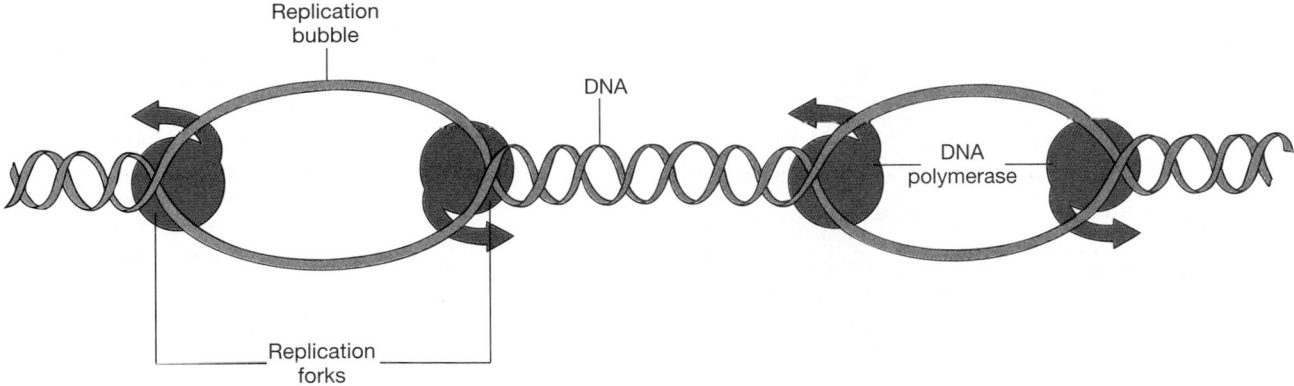

Replication bubble

DNA

DNA polymerase

Replication forks

Figure 5–5

A replication bubble, containing two forks, is formed when double-stranded DNA is separated. DNA replication is bidirectional in mammalian cells because synthesis occurs at both forks in the bubble. Many bubbles are present because replication starts simultaneously at multiple sites. (Only the parent DNA strands are shown here.)

dAMP

dCMP

dGMP

dTMP,

and joins them in the order specified by the complementary bases on the DNA template to form a new strand of DNA. Recall that the base C always forms a pair with the base G, and A with T. Hence, a DNA strand of nucleotides with bases in the order ACAGTG will serve as a template for a set of nucleotides with bases in the order TGTCAC. Base-pairing ensures that the order of bases in the new DNA strand is identical to the order in the original strand, from which the template has recently separated.

Figure 5–6*a* depicts the end of a DNA strand as it is synthesized on a DNA template. The incoming nucleotide dCTP has been selected for attachment to the end of the strand because its base cytosine (C) is the only base that can form hydrogen bonds with guanine (G), the base of the next nucleotide on the DNA template. The preliminary hydrogen bonding step serves to line up the individual nucleotide molecules on the DNA template. DNA polymerase then catalyzes a reaction that covalently attaches dCTP to the end of the new strand. During the reaction, the two phosphate groups on the end of the dCTP molecule are removed, and the rest of the molecule is attached via the remaining phosphate to the hydroxyl group on carbon 3 of the deoxyribose ring of the nucleotide at the end of the new strand (see Fig. 5–6*a*). DNA polymerase can add nucleotides to the DNA chain at a rate of more than 20 per second, and each one is always attached to carbon 3 of the sugar of the last nucleotide. Thus, the new DNA chain grows from the end where there is a free hydroxyl group on carbon 3.

Synthesis in the 5′ to 3′ Direction. A special numbering system is commonly used for nucleotides in order to distinguish between the two ring structures that are present in the molecule—that is, the sugar and the base. By convention, the numbers of the carbon atoms of the sugar are given a superscript (e.g., 3′ or 5′), whereas the carbon atoms of the base are numbered in the usual way. A DNA strand is formed by linking nucleotides together via phosphate groups between carbon 3′ of one nucleotide and 5′ of another. The nucleotide at the beginning of the DNA strand has a free phosphate group on the 5′ carbon. This end of the strand is usually called the **5′ ("five prime") end.** The nucleotide at the other end of the strand has a free hydroxyl on the 3′ carbon. This is termed the **3′ ("three prime") end** of the strand (see Fig. 5–6).

In theory, a new nucleotide could be attached to either the 5′ end or the 3′ end of the chain. In practice, however, DNA polymerase attaches nucleotides only to the 3′ end of a strand, so that the new strand always grows from the **5′ to 3′ direction.**

Antiparallel Strands: Leading and Lagging Strands. In the DNA molecule, the 3′ ends of the two complementary nucleic acid strands are at opposite ends of the double helix (see Fig. 5–4); they are said to be **antiparallel.** A DNA template and its complementary growing strand are also antiparallel. Hence, a template with the nucleotide sequence 3′—TGGCGTATAC—5′ will direct the synthesis of the comple-

(text continues on page 147)

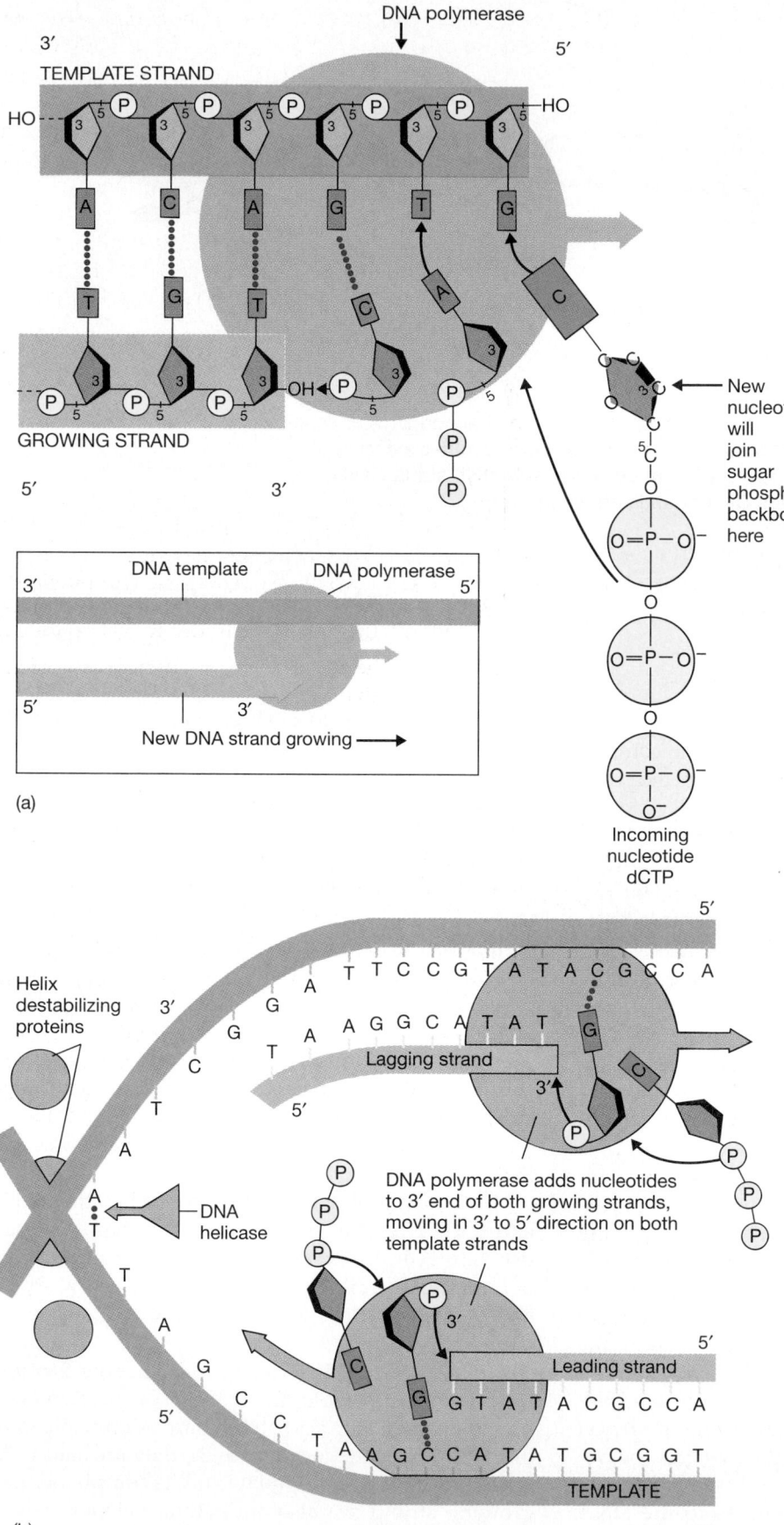

(a)

(b)

Figure 5–6

(*a*) DNA polymerase moves from the 3′ end to the 5′ end of a DNA template, joining each new nucleotide first to the complementary base on the template via hydrogen bonds and then to the previous nucleotide on the growing strand at the 3′ carbon. DNA polymerase adds nucleotides only to the 3′ end of the growing strand (so called because it has a free hydroxyl group attached to the 3′ carbon of the sugar). (*b*) DNA polymerase works on both strands of the parent DNA molecule simultaneously. However, because the parent strands are antiparallel, and because DNA polymerase travels only in the 3′ to 5′ direction on a DNA template, in effect it moves in opposite directions on the two parent strands. When DNA polymerase is moving in the same direction as DNA helicase, the growing strand is termed the *leading strand*. When DNA polymerase is moving away from the replication fork, the growing strand is called the *lagging strand*.

mentary and antiparallel strand 5'—ACCGCATATG —3' (see Fig. 5–6*b*). This complicates the process of DNA replication because DNA polymerase moves along the *template* strands in the 3' to 5' direction only, and synthesizes new DNA by adding nucleotides only to the 3' end of the growing strand.

In the simultaneous replication of both strands of the DNA double helix, one of the new DNA strands, the so-called **leading strand,** is synthesized in a straightforward manner because it can grow in the 5' to 3' direction on the template strand (see Fig. 5–6*b*). It is synthesized continuously and "follows" the DNA polymerase as the polymerase moves from the 3' end of the template strand.

The other new strand of DNA, the **lagging strand,** is formed by a process of **discontinuous synthesis,** in a series of small sections (**Okazaki fragments**) (see Fig 5–6*b*). Each section of the lagging strand is assembled in exactly the same way as a section of the leading strand, in the 5' to 3' direction by a DNA polymerase moving from 3' to 5' on the template. By the time one section is finished, the replication fork has opened up to expose more of the template, allowing another DNA polymerase to synthesize a second fragment of new DNA behind the first

(Fig. 5–7*a*). However, DNA polymerase does not join the segments of the lagging strand. This task is carried out by another enzyme, called **DNA ligase** (Fig. 5–7*b*).

In summary, the replication fork depicted in Figure 5–7*a* is always moving to the left, and the overall growth of both of the new DNA strands is from right to left following the movement of the fork. However, the synthesis of both of the new strands is exclusively in the 5' to 3' direction, which means that the lagging strand is actually synthesized in the direction opposite to the movement of the replication fork. As already pointed out (see Fig. 5–4), this mechanism of DNA replication produces two double-stranded DNA molecules, each consisting of one nucleic acid chain from the "parent" and one newly synthesized chain.

Proofreading in DNA Replication. Replication of DNA is a remarkably accurate process. There is probably only one error per 10^9 base pairs. This efficiency is well within the limits required to maintain the genome of a mammalian cell (about 3×10^9 base pairs). One of the reasons for the high degree of accuracy is that the polymerase is able to check or

(a)

(b)

Figure 5–7

Throughout a DNA molecule, several DNA polymerases may be working at many sites simultaneously. (*a*) The leading strand is synthesized continuously; the lagging strand is synthesized in small segments. DNA polymerase cannot join these segments. (*b*) Another enzyme, DNA ligase, joins lagging strand segments, known as Okazaki fragments.

"proofread" the new chain as it is synthesized, meaning that the polymerase responds to the incorporation of an incorrect nucleotide by immediately removing it and inserting the correct one before moving on to the next. DNA replication is also remarkably rapid. The average human DNA chain contains more than 10^8 nucleotides linked together. Such a chain can be replicated completely within a few hours. This rapidity is possible because many replication forks are present on each DNA molecule that is undergoing replication, so that many different sections of the huge molecule can be replicated simultaneously.

Synthesis of RNA: Selective Transcription of the Permanent Files

DNA does not provide protein-synthesizing information directly to the ribosomes; the information in DNA is first transcribed into RNA. In fact, RNA is synthesized in a process known as **DNA transcription.**

Three kinds of RNA are synthesized: **messenger RNA (mRNA), ribosomal RNA (rRNA),** and **transfer RNA (tRNA).** All three kinds of RNA molecules interact closely in the mechanism whereby a nucleotide sequence in DNA is translated into a specific protein. All three types of RNA are transcribed from the DNA template strand; all three types of RNA consist of ribonucleotides complementary to DNA nucleotides. The nucleotides on mRNA molecules are translated directly into amino acids, while the nucleotides on tRNA and rRNA molecules provide other information crucial to protein synthesis. The interactions of the three RNA molecules will be discussed in more detail soon. Strictly speaking, genes code for functional RNA molecules. Only when the functional RNA is mRNA can it be said that the gene codes for a protein.

The base sequence in the RNA is complementary to the sequence in the DNA template strand, and the base pairs formed ensure that the correct base sequence is copied (see Fig. 5–2). Synthesis of RNA differs from synthesis of DNA in four principal ways that are summarized in Table 5–3.

Transcribing the DNA Template Strand: RNA Polymerase. During transcription of DNA, the new chain of RNA is synthesized by the enzyme **RNA polymerase.** There are three distinct types of RNA polymerase enzymes in eukaryotic cells, and all three are located in the nucleus. **RNA polymerase I** synthesizes the major rRNA molecules in the nucleolus. The other polymerases are active in the portion of the nucleus that is outside the nucleolus. **RNA polymerase II** synthesizes mRNA, and **RNA polymerase III** synthesizes tRNA and small rRNA molecules.

Promoter and Terminator Sites. Transcription of a given gene begins when the RNA polymerase binds to a **promoter** site on the DNA template strand, a discrete sequence of nucleotides immediately preceding the 3' end of the gene (Fig. 5–8). Polymerase action stops when the enzyme reaches the **terminator** site, which is a sequence of nucleotides immediately following the 5' end of the gene (see Fig. 5–8). At this point, the polymerase releases both the finished RNA strand and the DNA template. The promoter and terminator act as recognition sites only; they are not transcribed. The promoter sequence also determines which strand of the double-stranded DNA will be transcribed by the RNA polymerase.

Synthesis in the 5' to 3' Direction: Antiparallel Strands. During synthesis of RNA, the RNA polymerase moves along the DNA template strand in the 3' to 5' direction, adding ribonucleotides to the 3' end

Table 5–3
Principal Differences Between DNA Synthesis and RNA Synthesis

	DNA Synthesis	RNA Synthesis
Sugar required	Deoxyribose	Ribose
Bases required	Adenine	Adenine
	Cytosine	Cytosine
	Guanine	Guanine
	Thymine	Uracil
Fate of DNA template	Incorporated into reaction product	Not incorporated into reaction product
Type of product	Double strand of nucleotides	Single strand of nucleotides

DNA TEMPLATE STRAND

Gene

Promoter — Terminator

3′ — 5′

C G T A A T G C C G A T

TRANSCRIPTION

G C A U U A C G G C U A

5′ — 3′

RNA TRANSCRIPT

Figure 5–8

The synthesis of a molecule of RNA begins when the enzyme RNA polymerase encounters a sequence of DNA nucleotides called a promoter. Ribonucleotides are assembled in an order complementary to the order of nucleotides on the DNA template strand in a process known as transcription, producing an "RNA transcript." Transcription stops when the RNA polymerase encounters a terminator sequence on the DNA template strand.

poration of the sequence 5′—GGUCAUUG—3′ into RNA (see Fig. 5–9). On a single DNA template strand, several RNA molecules may be synthesized at the same time (Fig. 5–10).

RNA Processing: Post-transcriptional Modification

Before they leave the nucleus to direct protein synthesis, newly synthesized RNA molecules are covalently modified in a series of reactions known as **RNA**

of the growing RNA strand. Thus, the new RNA strand grows in the antiparallel 5′ to 3′ direction (Fig. 5–9). A portion of the DNA template with the base sequence 3′—CCAGTAAC—5′ will direct the incor-

Figure 5–9

RNA polymerase works in the 3′ to 5′ direction on a DNA template strand, adding ribonucleotides to the 3′ end of a growing RNA strand. The RNA is made in the 5′ to 3′ direction. Ribonucleotides initially form hydrogen bonds with complementary bases on the DNA template, but when transcription is completed, the RNA transcript is released from the template strand.

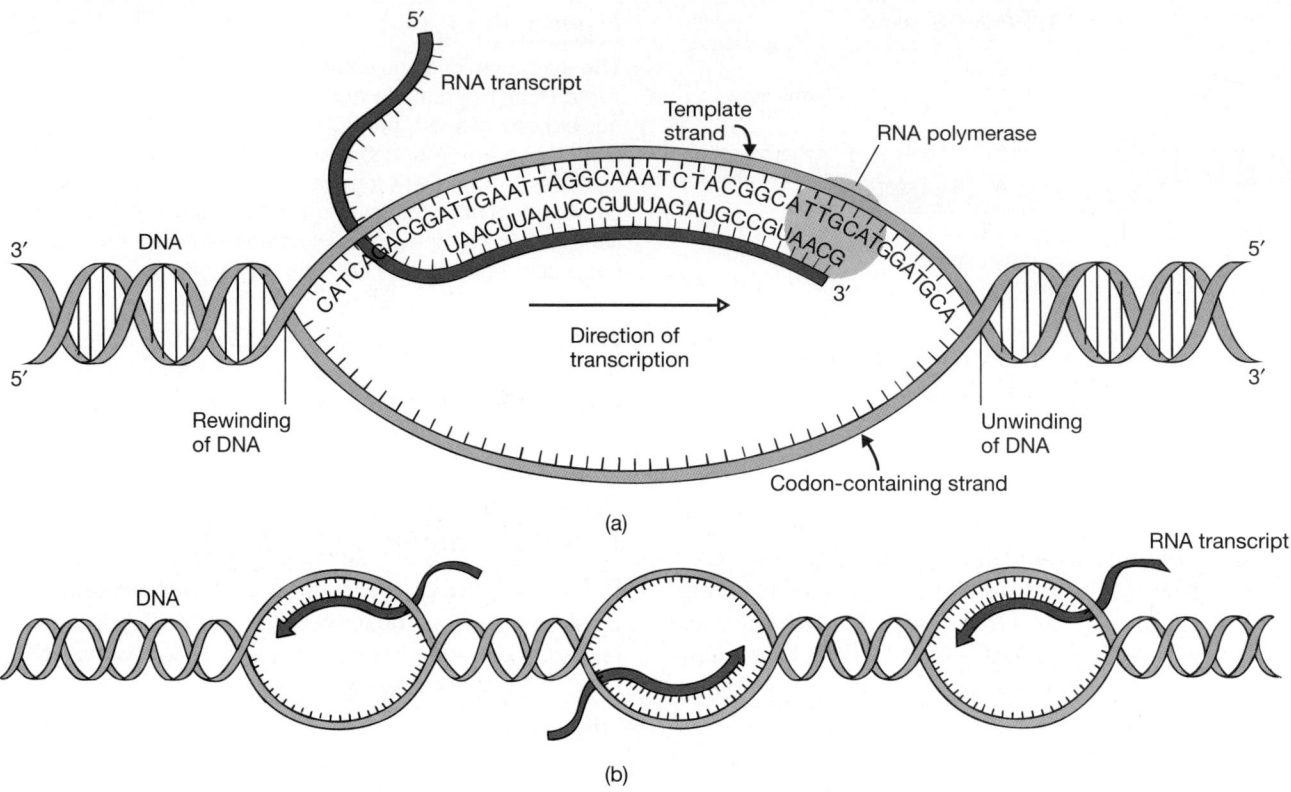

Figure 5–10

Several different RNA transcripts may be synthesized simultaneously on a single DNA template strand.

processing or **post-transcriptional modification.** Several examples are discussed below.

The RNA that is synthesized by the action of RNA polymerase II is known as **premessenger RNA (pre-mRNA).** The synthesis of pre-mRNA accounts for more than 50% of the total RNA synthesis in a typical mammalian cell. Many of these pre-mRNA molecules are destined to leave the nucleus in the form of much smaller mRNA molecules.

The 5′ Cap and the Poly(A) Tail.   Processing of pre-mRNA helps to provide the resultant mRNA with characteristics that distinguish it from rRNA and tRNA, the products of the action of the other RNA polymerases. These characteristics are a **5′ cap** and a **poly(A) tail.** The 5′ cap is formed by addition of **7-methylguanosine** and three phosphate groups, represented in shorthand notation as **Gppp** (Fig. 5–11), to the nucleotide at the 5′ end of the RNA transcript. This is the end of RNA that is synthesized first, and it is capped immediately—before the rest of the molecule is finished. The cap is required later to initiate binding of mRNA to ribosomes in the cytoplasm. The poly(A) tail consists of 100 to 200 adenine residues that are added to the 3′ end of the RNA after transcription is completed (see Fig. 5–11). The function of the poly(A) tail is unknown, but it may serve to sta-

bilize the mRNA structure and/or assist its export from the nucleus.

RNA Splicing: Introns Out.   Newly synthesized RNA molecules in eukaryotic cells contain intervening nucleotide sequences known as **introns,** nucleotides transcribed from DNA nucleotide sequences that contain no useful information (noncoding segments). These sequences are reproduced in the pre-mRNA during transcription and must be removed before the RNA leaves the nucleus. After addition of the 5′ cap and the poly(A) tail, the introns are cut out in order to convert the RNA to mature mRNA that can be used to direct the synthesis of a complete protein. This is another example of RNA processing, and it is known specifically as **RNA splicing** (see Fig. 5–11) because the remaining segments, called **exons,** are joined together to form functional mRNA.

RNA processing removes most of the mass of the hnRNA. The mature mRNA that is produced by these reactions accounts for only 3% of the total RNA present in a cell.

Formation of Functional tRNA.   A molecule of tRNA is relatively small, usually 70 to 80 nucleotides long, compared to the other types of RNA. The sec-

Figure 5–11

As part of the processing of newly synthesized RNA molecules, a "5′ cap" and a "poly(A) tail" are added at either end. In addition, introns (transcripts of meaningless segments of the DNA template strand) are excised, with the remaining RNA segments, known as exons, spliced together.

ondary structure adopted by a tRNA molecule is represented in Figure 5–12. Transfer RNA molecules have two important features. One, called the **amino acid accepting end** or the **aminoacyl attachment site,** is where an amino acid destined to join a growing polypeptide chain attaches for transport to a ribosome. The other, called the **anticodon,** is a set of three nucleotides complementary to a codon on an mRNA molecule. Each amino acid has a specific tRNA molecule designed to convey it to a ribosome and link it in the correct order to another amino acid, in a process to be described.

In a recent major advance in the understanding of the genetic code, molecular biologists have determined that the identity of a tRNA molecule, by which it is assured of linking up with the correct amino acid, is found in a few specific bases in each tRNA molecule. In some molecules, these "identity sites" include the bases that form the anticodon as well as other bases in the molecule. In others, the identity sites do not include the anticodon. Experiments have shown that if the identity elements are removed from a tRNA molecule, the molecule no longer recognizes its amino acid. It may link up with another amino acid or simply fail to function altogether.

Formation of Ribosomal Subunits from rRNA. Ribosomal RNA is formed in the nucleolus, an area of the nucleus. There, rRNA combines with special proteins to form the large and small subunits that will later combine to produce ribosomes. The subunits do not combine in the nucleolus; they are exported into the cytoplasm and remain separated except when they are participating in protein synthesis.

New Proteins Are Synthesized by Translating mRNA

After export from the nucleus, all three kinds of RNA cooperate in protein synthesis in the cytoplasm. Ribosomal RNA makes up part of the ribosome, the organelle responsible for protein synthesis. Messenger RNA carries to the ribosome the codons specifying the amino acids to be used in the polypeptide chain. Transfer RNA obtains the amino acids in the cytosol

Figure 5–12

A molecule of tRNA contains two important sites: the aminoacyl attachment site, where an amino acid attaches to the molecule, and the anticodon, a triplet of bases complementary to a specific codon on a molecule of mRNA. In addition, equally important "identity sites," sometimes including the anticodon, ensure that the molecule attaches to the correct amino acid.

and brings them to the ribosome, where the polypeptide is assembled.

RNA in the Cytoplasm: A Three-Way System

Although all three types of RNA are complementary transcripts of a DNA template, only mRNA nucleotides are directly translated into amino acids. The nucleotides on rRNA and tRNA serve other purposes in the protein synthesizing process, interacting with mRNA to produce a polypeptide. Figure 5–13 gives an overview of the role each type of RNA plays in protein synthesis.

Ribosomal RNA: The Protein-Synthesizing Organelle. The several steps involved in translation of mRNA and synthesis of a polypeptide chain take place on and are coordinated by **ribosomes**. Eukaryotic ribosomes are composed of approximately equal weights of rRNA and protein. The molecular weight of these ribosomes is about 4.5×10^6. Each ribosome is composed of two subunits. Some ribosomes are found free in the cytosol or in clusters called **polysomes** (see Fig. 5–17); other ribosomes are attached to the endoplasmic reticulum.

Transfer RNA: Procuring Amino Acids. A preliminary step in protein synthesis involves coupling amino acids to the tRNA molecules, a process driven by the energy derived from ATP and requiring the catalytic action of an enzyme called **aminoacyl-tRNA synthetase (AAS):**

amino acid + tRNA + ATP $\rightarrow$

aminoacyl-tRNA + AMP

The amino acid is attached to the ribose ring of the nucleotide at the 3' end (the amino acid-accepting end) of the tRNA.

Each of the 20 amino acids has one specific synthetase enzyme and one specific tRNA. The synthetase must be able to "recognize" both the amino acid and the tRNA that is specific for that amino acid (Fig. 5–14). Free amino acids cannot participate in protein synthesis because they are unable to form peptide bonds (—CONH—) spontaneously. Participation is mediated by the tRNA molecules to which the amino acids are covalently linked.

The tRNA molecules help to maintain the accuracy of translation by making sure that the correct amino acid is inserted at the correct location in the

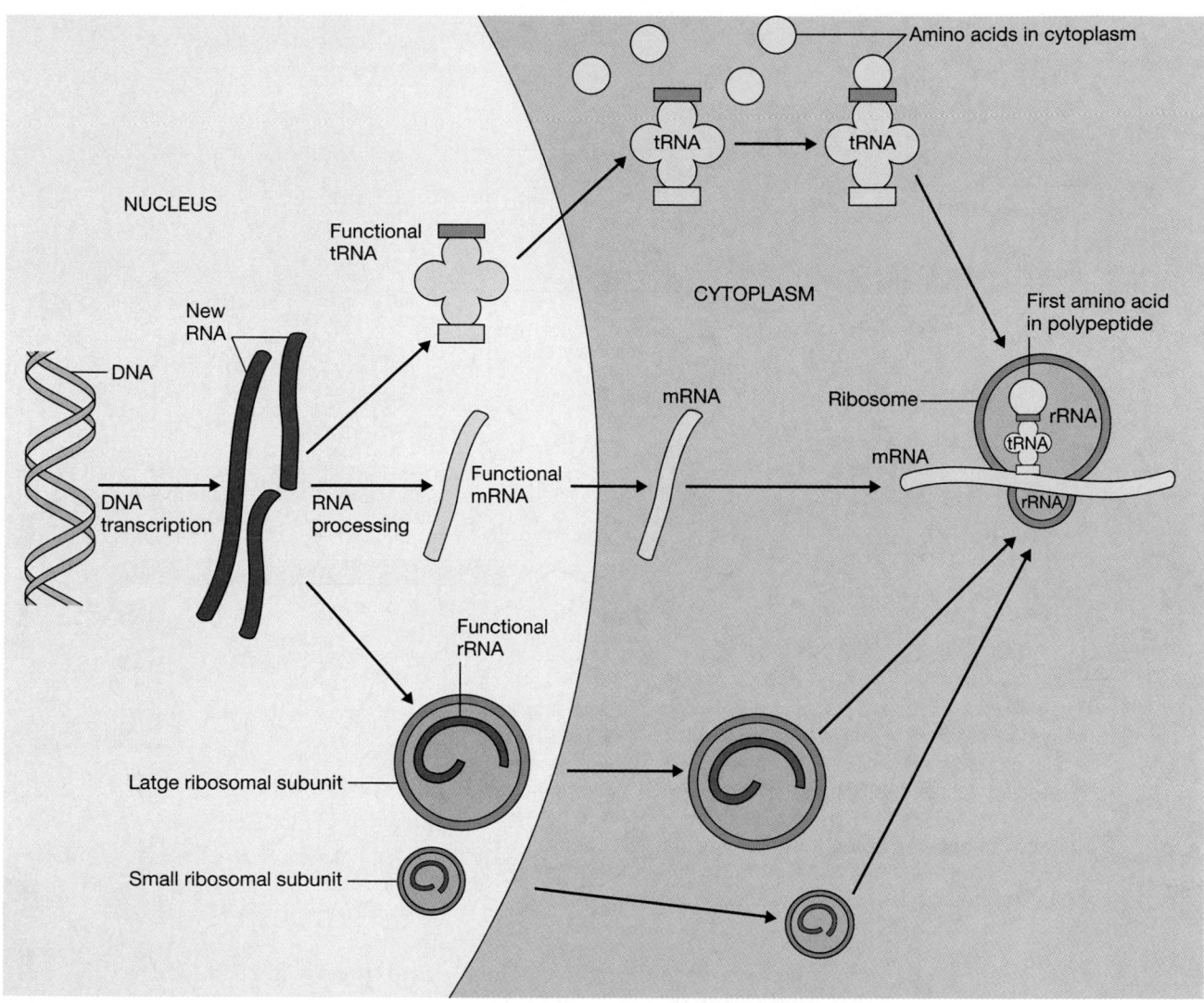

Figure 5–13

The three types of RNA cooperate in the synthesis of a polypeptide. Messenger RNA molecules carry the codons that specify the order of amino acids to be assembled. Transfer RNA molecules obtain individual amino acids in the cytoplasm and ferry them to the mRNA. Ribosomal RNA forms the subunits of ribosomes, the structures on which mRNA and tRNA interact to join amino acids into polypeptide chains.

growing polypeptide chain. The basis of this mechanism is complementary base-pairing between the anticodons in the tRNA molecules (see Fig. 5–14) and the codons in mRNA. Thus, it is the tRNA molecule, not the attached amino acid, that recognizes the codon in mRNA. Only one of the many kinds of tRNA molecules in the cell can form base-pairs with a specific codon in mRNA; this is how the amino acid specified by the codon can be selected accurately and added to the polypeptide chain.

Messenger RNA: Conveying the Amino Acid Sequence. The role of mRNA in the three-way RNA system is to bear the codons that dictate the order of amino acids in the new polypeptide product. Messenger RNA is the focal point of protein synthesis, a meeting place for rRNA and tRNA and the mediator between the information stored in DNA and the final polypeptide product.

Initiation of Protein Synthesis: Formation of the Ribosome

In eukaryotic cells, the synthesis of a polypeptide is initiated at the AUG start codon that is nearest to the 5' cap of the mRNA. The other AUG codons in the mRNA do not act as start codons. They specify incorporation of methionine into the polypeptide. The

Figure 5–14

Each amino acid has a specific tRNA molecule and a specific enzyme designed to join it to the tRNA. For example, the amino acid serine has a specific tRNA molecule designated as tRNA SER. A specific aminoacyl tRNA synthetase (AAS) for serine couples the amino acid to the tRNA molecule via a bond at the aminoacyl attachment site. Later, the amino-acid-tRNA complex will join the mRNA molecule via base pairs formed between the tRNA SER anticodon (AGA) and the mRNA codon for serine (UCU).

AUG start codon allows the ribosome to bind to the mRNA (Fig. 5–15). The ribosome itself is not able to "recognize" the AUG start codon. This process is mediated by a molecule of **initiator tRNA** bound to the ribosome. Specialized proteins called **initiation factors** are also involved in the complex initiation process.

A ribosome has two binding sites for tRNA molecules, denoted as the **P-site** and the **A-site** (see Fig. 5–15. When a tRNA molecule binds to either of these sites, it is held close to the mRNA strand that is also bound. The bound tRNA is oriented so that the nucleotides in its anticodon form complementary base pairs with the nucleotides of the appropriate codon in the mRNA. Thus, the aminoacyl-tRNA molecule cannot fit into the ribosomal binding site unless the appropriate mRNA codon also is present.

During elongation of the polypeptide chain (see the following section), when both tRNA binding sites are occupied by tRNA molecules, the two tRNA molecules form base pairs with adjacent codons in the mRNA (Fig. 5–16). The tRNA in the P-site is attached to the amino acid at the growing end of the polypeptide chain, while the tRNA in the A-site carries the

next amino acid that will be added to the polypeptide.

Elongation of the Polypeptide Chain: Reading the mRNA Strand

Once a ribosomal complex is formed by the conjunction of a tRNA-amino acid complex and an mRNA molecule with ribosomal subunits, translation of successive codons on the mRNA begins. The amino acids specified by the mRNA codons that follow the AUG start codon are joined together in sequence to form the polypeptide. The nucleotides on the mRNA strand are "read" in groups of three, the codons, beginning at the 5′ end. Each codon specifies the amino acid that will be added next to the growing polypeptide chain.

Elongation of the polypeptide begins when a second tRNA molecule brings the amino acid specified by the second mRNA codon to the ribosome. This new tRNA molecule occupies the A-site on the ribosome, right beside the first tRNA molecule, which occupies the P-site (see Fig. 5–16*b*).

(text continues on page 157)

Figure 5–15

Protein synthesis is initiated when a small ribosomal subunit attaches to a start codon (AUG) on an mRNA molecule (*a*). An aminoacyl complex bearing an amino acid (e.g., valine) attaches to the mRNA strand via hydrogen bonds between the tRNA anticodon and the mRNA codon (*b*). A large ribosomal subunit then attaches, forming a complete ribosome (*c*). The ribosome contains two sites for tRNA molecules to occupy, the A site and the P site (*d*).

(a)

(b)

(c)

(d)

Figure 5–16

Elongation of a polypeptide begins when a tRNA molecule bearing another amino acid enters the A site of the ribosome. The amino acid is specific to the second codon on the mRNA strand (*a*). Aided by the enzyme peptidyl transferase and energy from GTP, a peptide bond forms between the two amino acids that now sit side by side on the ribosome. At the same time, the first amino acid dissociates from its tRNA molecule (*b*). The free tRNA molecule leaves the P site, and the ribosome moves to the right along the mRNA strand, moving the second codon and attached tRNA-amino acid complex from the A site to the P site (*c*). A third amino acid-tRNA complex, corresponding to the third codon, now enters the vacant A site (*d*).

In a reaction catalyzed by an enzyme called **peptidyl transferase,** the carboxyl end of the first amino acid, bound to the tRNA on the P-site, is joined via a peptide bond to the second amino acid, bound to the tRNA at the A-site. After this reaction, the now-free tRNA at the P-site is ejected. The ribosome moves to the right along the mRNA by a distance of one codon, and the tRNA at the new end of the polypeptide chain is moved from the A-site to the P-site. The A-site is now vacant and ready to accept the tRNA bearing the next amino acid to be incorporated into the polypeptide. The various steps of this cycle are illustrated in Figure 5–16. It occurs 20 times each second on bacterial ribosomes, and is driven by the chemical energy stored in GTP (see Chapter 2).

Several ribosomes may be attached to the same mRNA strand, each synthesizing a copy of the same polypeptide. As one ribosome moves toward the 3′ end of the mRNA, a second ribosome attaches to the AUG start codon at the 5′ cap and begins synthesizing a second copy of the polypeptide. As more ribo-

somes become attached in this way to a single strand of mRNA, a structure known as a **polysome** (or **polyribosome**) is formed (Fig. 5–17).

Translation of the mRNA ceases when a **stop codon** (see Table 5–2) is encountered. At this point, the completed polypeptide is released, and the ribosome dissociates from the mRNA.

Gene Expression Is Regulated at Several Different Steps

The expression of prokaryotic genes is regulated primarily at the level of transcription, rather than translation (see Fig. 5–13). A specific **repressor** protein can bind to the DNA at a site next to a gene and can block transcription of the gene by RNA polymerase. Repression of the gene can be reversed by an **inducer** molecule that forms a complex with the repressor and prevents binding of the repressor to the DNA, thus allowing normal transcription of the gene by the polymerase. This kind of a mechanism readily explains how the expression of different genes can be turned on or off at different times.

Control at the level of transcription also may be the most important way of regulating gene expression in eukaryotes, but the system is complex and many of the mechanisms still unclear. The action of steroid hormones on the expression of specific genes is one control system in which many of the steps have been worked out through extensive investigations. Steroid hormones include estrogens (the female sex hormones produced by the ovary and the placenta), testosterone (the male sex hormone from the testis), and mineralocorticoids such as aldosterone from the adrenal glands.

The general action of these hormones is as follows. On arriving at the cell, the hormone crosses the plasma membrane and binds to a specific receptor molecule inside the cell. The steroid-receptor complex is able to enter the nucleus and bind to the DNA at a specific site. The binding of this complex either increases or decreases the transcription of specific genes. This changes the amount of mRNA and leads, in turn, to an increase or decrease in the specific protein products of the genes. It is these few specific proteins that produce the changes in cellular function that represent the characteristic response of the cell to the hormone.

While not a general mechanism for regulating gene expression, **amplification** of genes has been well documented in certain cells such as the developing eggs of a species of African clawed toad. In the nuclei of the egg cells, the genes that code for rRNA are repeated about 450 times on the chromosomes. This repetition, or amplification, allows the cell to make a huge amount of one gene product very

First ribosome to start at 5′ end

Figure 5–17

Several ribosomes may be at work at the same time on the same mRNA strand, creating a structure known as a polysome.

RESEARCH FOCUS

Antibiotics

Antibiotics are molecules produced by various species of microorganisms such as bacteria and fungi. An antibiotic released by one species blocks the growth of other species and may eventually destroy them. Hundreds of antibiotics differing in physical and chemical properties and mechanism of action have been found. More than 50 of these are used clinically to control and cure human diseases caused by invading microorganisms. The specific antibiotic selected for use is usually determined by the identity of the microorganism.

Penicillin and closely related molecules are widely used to treat many infectious diseases. Alexander Fleming discovered penicillin in 1928 when he noticed that growth of bacteria was inhibited by a contaminating fungus of the Penicillium family. Penicillin blocks the synthesis of the tough bacterial cell wall that normally keeps bacterial cell membranes intact despite their high internal osmolarity. Other antibiotics, such as *chloramphenicol* and the *tetracyclines,* act on bacterial ribosomes to inhibit protein synthesis. *Aminoglycosides* also act at the ribosomal level, where they interfere with the initiation of protein synthesis. An additional effect of aminoglycosides is to induce misreading of the mRNA code so that abnormal polypeptides are produced. *Rifampin,* a drug used to treat tuberculosis, inhibits RNA polymerase, which is the enzyme responsible for synthesis of RNA from DNA. Other key steps in bacterial metabolism are inhibited by the *sulfonamides.*

A serious medical problem arises when invading bacteria acquire the genes that allow them to resist the antibiotics normally used to destroy them. The resistant cells in the bacterial population multiply unimpaired and eventually predominate because the growth of the drug-sensitive cells is suppressed. The time taken for this to occur varies greatly among various microorganisms and drugs. Antibiotic resistance requires a stable genetic change that can be inherited by each new generation. *Mutation* is often the cause, and the alterations that occur are remarkable. Bacterial mutants have de-

veloped resistance to penicillin by producing an enzyme that destroys the penicillin. Mutants that are highly resistant to rifampin were found to contain a slightly altered RNA polymerase, which fulfills its normal function but is no longer sensitive to inhibition by the drug. Antibiotic resistance can also be acquired by the transfer of genetic material from one bacterial cell to another.

Bacteriophages are viruses that enter bacterial cells; some of these "phages" can carry fragments of bacterial DNA. If the DNA contains a gene for drug resistance that the bacterial cell can transcribe, then the bacterium may become drug resistant and capable of passing on this property to its offspring. The same results can occur when genes for drug resistance are passed from one bacterium to another via a direct bridge between the cells, a process known as *conjugation.* The emergence of antibiotic-resistant bacteria has created a constant need for new drugs. The problem of antibiotic resistance was reviewed recently in *Science,* 264, pp. 359–393, 1994.

The hydrophobic segment of nystatin is able to form a channel in the plasma membrane that allows the loss of important ions and small molecules from the cell. This action is dependent on the interaction of nystatin with **ergosterol,** the principal sterol present in the plasma membranes of yeasts and fungi.

■ *Nystatin* **is an antibiotic used to treat yeast and fungal infections of mucous membranes, intestinal tract, and skin. It has no effect on protein synthesis. The nystatin molecule contains a hydrocarbon chain that is strongly hydophobic. Suggest a mechanism by which nystatin might kill yeast and fungal cells.**

quickly. The accumulation of rRNA in the toad eggs allows the formation of large numbers of ribosomes. These structures facilitate the very rapid protein synthesis that will be needed when the egg is fertilized and starts to divide and grow rapidly. About 10^{12} ribosomes are needed per cell. Without gene amplification, the cell would need several hundred years to make this number of ribosomes.

In some eukaryotic cells, gene expression can be regulated at steps following transcription–for exam-

ple, by modification of the RNA molecules themselves (see Fig. 5–11), or by a control system at the level of translation. The latter provides a means for synthesizing large amounts of a single protein without amplifying the gene that codes for that protein. This occurs when transcription of the gene produces a very stable mRNA molecule that remains in the cytoplasm for days and can be used repeatedly for translation and synthesis of the specific protein product. In other words, the mRNA molecules that are in-

RESEARCH FOCUS

Molecular Chaperones

Many proteins that have been isolated and purified in a test tube will spontaneously refold and reform their native conformation after being unfolded. The ability to fold rapidly into this conformation depends in part on the amino acid sequence in the polypeptide chain. Refolding in a test tube is very inefficient, however, compared to the process in a cell. Furthermore, it often requires protein concentrations and physical and chemical conditions that are very different from those present inside a cell. Studies on protein folding in cells have revealed the existence of two types of proteins that assist this process. The first type includes enzymes that catalyze specific steps. The second type are the **molecular chaperones,** which are proteins that bind to and stabilize an unstable conformation of another protein. They do not modify the substrate protein covalently and they are not part of the finished product. However, controlled binding and release of the substrate by chaperones greatly facilitates correct folding. Chaperones probably recognize structural features that become exposed when polypeptides are not folded properly. Some of these proteins are known as **heat-shock proteins (hsp)** because they appear only during the response of cells to elevated temperature (42°C). A more appropriate general term may be **stress proteins** because many chaperones, including those that do not respond to heat-shock, are induced under other stress conditions such as depletion of ATP. The majority of chaperones are known now to be abundant in cells in the absence of any stress and are essential for survival of the cell even under normal conditions. This reflects the role of chaperones in many steps of cellular metabolism including synthesis and maturation of proteins, protection from environmental stress, rearrangement of macromolecules during normal cycles of assembly and disassembly, and targeting proteins for degradation.

Molecular chaperones are distributed widely, occurring in bacterial, plant, and animal cells. Eukaryotic cells contain families of heat-shock proteins known as hsp60 and hsp70 (the numbers indicate the molecular size [in kDa] of the proteins). Distinctive forms are present in different organelles. For example, the mitochondria contain hsp60 and hsp70 that differ slightly from those in the cytosol, and proteins that fold in the lumen of the endoplasmic reticulum are assisted by a special hsp70 protein called **BIP.** These families of chaperones bind to the exposed hydrophobic regions of newly synthesized, unfolded or misfolded polypeptides. This intimate contact is thought to change the structure of the substrate enough to allow another chance to fold. The hsp70 acts on the protein before it detaches from the ribosome. The hsp60 acts at a later stage and provides a favorable local environment where proteins that remain unfolded can have another chance to refold. Proteins that do not fold correctly are ultimately targeted for destruction. The heat-shock response may be induced when the elevated temperature causes an increase in the amount of misfolded proteins. The cell responds to this situation by producing more chaperones to assist refolding of the proteins.

■ **How could a heat shock selectively induce the synthesis of one or two specific proteins?**

volved in the production of the major proteins of a cell tend to be long-lived. A specific example is the synthesis of the protein hemoglobin in the immature red blood cell. As red blood cells mature, their nuclei are lost and new RNA cannot be synthesized, yet hemoglobin synthesis continues for weeks. This is possible because the mRNA molecules that code for this protein have a much longer lifetime than the other mRNA molecules in the cell. These observations suggest that controlling the lifetime of mRNA molecules in the cytoplasm may be a mechanism for regulating the expression of certain genes.

The final step at which gene expression may be regulated is that following translation. Many newly synthesized polypeptides undergo one or more types of covalent alterations known collectively as **post-translational modification** (not to be confused with the post-transcriptional modification of RNA discussed earlier (see Fig. 5–11). Some of the covalent modifications of polypeptides were discussed earlier in Chapters 2 and 3—for example, the attachment of carbohydrate to produce a glycoprotein and the regulation of the biological activity of a protein by alternate cycles of attachment and removal of a phosphate group. Other examples include the permanent covalent attachment of a coenzyme molecule essential for the activity of an enzyme protein, and the production of a biologically active polypeptide from a much larger inactive polypeptide molecule (known as a **precursor**) through the activity of enzymes that selectively remove small sections. The latter mechanism is involved in the production of the hormone **insulin,** a polypeptide of 51 amino acids, from an inactive precursor called **proinsulin,** a polypeptide of 84 amino acids.

Finally, although the function is not understood, the side chain groups of certain amino acids in the polypeptide may be altered. A hydroxyl (—OH) group is commonly attached to the side chain of the amino acid proline to produce hydroxy, proline, and a methyl (—CH₃) group may be attached to the side chain of amino acids such as lysine and glutamic acid.

Damaged DNA Can Be Repaired

The genetic information stored in DNA must remain essentially unchanged as it is passed on from cell to cell. The replication process is designed to faithfully reproduce molecules that are identical to the parent molecule, but it cannot correct alterations to DNA that occur in the intervals between the replication steps. DNA is very sensitive to damage by factors commonly present in the environment. In fact, it has been estimated that several thousand bases in the genome of a mammalian cell are damaged every day. Changes on this scale in the structure of DNA could have drastic consequences, leading to mistakes in genetic information, malignant cells, and a threat to the long-term continuance of the species. Fortunately, cells have developed repair systems to correct almost any kind of damage, and very few stable changes are allowed to accumulate in the genome of each cell.

Environmental factors that cause spontaneous damage to DNA are examples of extracellular signals that are potential modifiers of an internal cellular control system. Ultraviolet light is an environmental factor that damages DNA. Other factors include a number of common chemicals and the radiation produced by the decay of radioactive isotopes.

The repair mechanisms that correct the damage from ultraviolet light rays (wavelength 20 to 200 nm) are the best understood. When exposed to ultraviolet light, pyrimidine bases, such as thymine, that are adjacent to one another on the same DNA strand tend to become linked together by covalent bonds. The resultant structure, a **thymine dimer,** causes distortion of the DNA chain and blocks replication (Fig. 5–18). A number of enzymes recognize this structure and can repair it.

Figure 5–18

Ultraviolet radiation may cause adjacent thymine bases on a DNA strand to join together, forming a thymine dimer and distorting the shape of the sugar-phosphate backbone.

Photoreactivation

The simplest method of repair is by **photoreactivation,** a process whereby a specific enzyme binds to the DNA section containing the thymine dimer and uses the energy of visible light (wavelength 300 to 600 nm) to break the covalent bonds that hold the dimer together. The DNA reverts to its original structure containing adjacent thymine bases, and the bound enzyme is released.

Excision Repair

Other enzymes act by selectively cutting out or **excising** the section of DNA that contains the thymine dimer. The remaining gap can be filled relatively easily because the complementary strand has all the information for the correct sequence of nucleotides needed to fill the gap. The process is known as **excision repair** and consists of essentially three steps (Fig. 5–19). Repair is initiated by a specific **endonuclease** enzyme that breaks the DNA chain containing the dimer. Next, DNA polymerase simultaneously cuts out the thymine dimer and replaces it with a new seg-

Figure 5–19

The damage caused by a thymine dimer is repaired when DNA endonuclease excises the distorted segment and DNA polymerase joins new nucleotides to replace those removed. DNA ligase joins the new segment to the undamaged portion.

ment that is complementary to the undamaged strand. Finally, the 3' end of the new segment is joined to the rest of the chain by **DNA ligase.**

Base Excision

In a different type of repair mechanism, any nucleotide containing an altered base can be removed. In **base excision, DNA glycosylase** enzymes "recognize" the altered bases and remove them. The remaining sugar-phosphate portions of the nucleotides are also removed and replaced with the correct intact nucleotides by the stepwise actions of an endonuclease, DNA polymerase, and DNA ligase, in a sequence similar to that depicted in Figure 5–19. The correct nucleotide can be inserted because the nucleotide with which it will form a base-pair is already in position in the complementary DNA strand.

Repair by base excision is particularly useful for correcting changes due to spontaneous **deamination,** a term used to describe the loss of NH_3. Deamination of cytosine produces uracil (Fig. 5–20*b*), which is a base not normally present in DNA. The introduction of uracil causes a problem because it resembles the structure of thymine so closely that it can form hydrogen bonds with adenine, just as thymine does. Thus, an AU base pair will form during replication of the strand that contains uracil. After another replication step, a thymine will be placed in the new strand opposite the adenine so that in the new DNA molecule an AT pair base will be present at the position where there was once a GC base pair (see Fig. 5–20*c*). This change in the nucleotide sequence of DNA is known as a **mutation.** It would have disastrous results for the organism because the new sequence would be carefully copied during further DNA replication and passed on to each of the new cells. The mutation must be corrected by prompt removal of the uracil by base excision.

These are a few examples of the variety of DNA-repairing enzymes that recognize and replace abnormal nucleotides, ensuring that the genetic information stored in the nucleotide sequence of DNA remains unaltered during the life of a cell. A built-in proofreading mechanism ensures that the nucleotide sequence is copied very accurately during replication of DNA, so that an identical store of information is passed on to the new cells produced by cell division.

(a)

(b)

(c)

Figure 5–20

In normal DNA replication, the same base pair is copied indefinitely (*a*). A mutation can occur, however, when the base cytosine undergoes deamination to become uracil (*b*). When the strand containing the erroneous uracil is replicated, adenine forms a base pair with uracil. When the strand containing the erroneous adenine is copied, thymine forms a base pair with the adenine (*c*). Thus, a GC base pair has been changed to an AT base pair, changing the basic genetic code for that cell.

Cell-to-Cell Signaling: Hormones, Neurotransmitters, and Other Molecules

The coordination of the different functions of the many tissues in a multicellular animal is achieved because cells have developed communication systems. Cells that are in direct physical contact with one another have structures known as **gap junctions** in their plasma membranes (Fig. 5–21*a*). These junctions allow small molecules to pass directly from the cytosol of one cell to the cytosol of the adjacent cell. Cells some distance apart are able to communicate by sending **chemical signals,** molecules secreted by one cell that bind to the plasma membrane or enter the cytoplasm of the distant cell (see Fig. 5–21*b*). Chemical signaling systems usually involve three separate components: extracellular signals, called first messengers; receptors; and intracellular signals, called second messengers.

Adjacent Cells Communicate Directly Via Gap Junctions

Gap junctions are common structures in tissues of all animals. They consist of a group of small pores or channels, each about 1.5 nm in diameter, that allow small water-soluble molecules with molecular weights under 1000 to move directly from the cytosol of one cell to the cytosol of an adjacent cell (Fig. 5–22). This includes ions, sugars, amino acids, and nucleotides, but excludes the macromolecules.

The major advantages of this direct communication system are its speed and reliability. The "message" must travel only a short distance, with little chance of intervention by outside forces because it does not leave the cytosol. In principle, this system could also allow **metabolic coupling** between different cells present in a tissue. This means that molecules made by only one group of cells within the tissue could be shared with the other cell types present by being passed through the gap junctions.

Gap junctions also may allow electrical communication between cells because they permit the rapid flow of ions. Electrical communication can occur in this way between certain nerve cells. Normally, nerve cells communicate chemically with each other through substances called **neurotransmitters,** which cross from one nerve cell to the other through junctions called **synapses.** The nerve impulse from one nerve cell is passed to the next cell by the release of the chemical neurotransmitter, which crosses the synapse and initiates the nerve impulse in the next nerve cell (this is discussed in more detail later in the chapter).

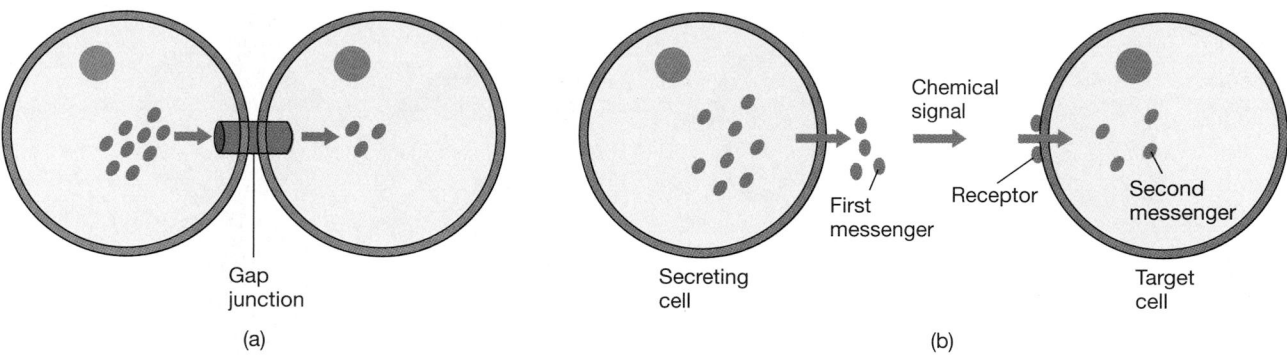

Figure 5–21

The two broad types of communication between cells: (*a*) gap junctions (between adjacent cells) and (*b*) chemical signals (between distant cells). Chemical signals usually involve an extracellular signal produced by one cell, a receptor in or on the target cell, and an intracellular signal in the target cell.

Occasionally, however, nerve cells must communicate with more speed than chemical neurotransmitters can provide. In these situations, a gap junction can act as a special kind of synapse that transmits the nerve impulse electrically rather than chemically. Electrical transmission is much faster than chemical transmission, and sometimes this difference is very important. An electric eel can generate a powerful electric shock only because all the cells in its electrical organ are coupled via electrical synapses and can be synchronized to "fire" at the same time. Chemical transmission would be too slow to achieve the necessary degree of precision. **Electrical coupling** is also important in mammalian tissues such as the heart, for example, in which all the muscle cells must contract simultaneously in order for the heart to pump blood efficiently.

Distant Cells Use Chemical Signals for Communication

Compared with the study of communication between adjacent cells, which is difficult to observe or measure, it is relatively easy to study communication systems that use chemical signals. These signals take the form of specific molecules released by one cell and sent to produce a change in a **target cell.** Consequently, much more is known about such systems, and the remainder of this chapter is concerned with how they operate.

Chemical signaling systems generally involve three components: (1) the signal molecules, also called **extracellular signals** or **first messengers;** (2) **receptors,** molecules in either the membrane or cytoplasm of the target cell; and (3) **intracellular signals** or **second messengers.** Extracellular signals travel from the cell that produced them to the target cell, where they react with receptor molecules. The receptor molecules in turn produce intracellular signals. The cumulative effect of these interactions is to elicit the desired physiological response in the target cell.

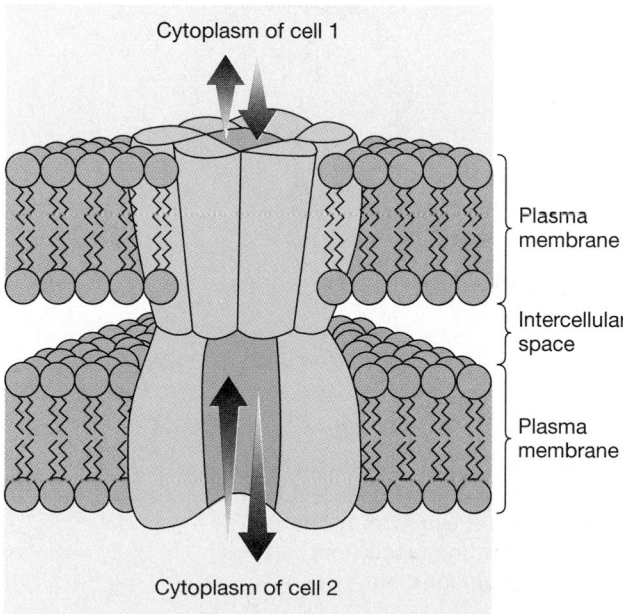

Figure 5–22

A gap junction forms direct connections between the cytoplasm of adjacent cells. Proteins create a tunnel-like structure in the plasma membrane of each cell, and two tunnels of adjacent cells connect to allow materials to pass between them.

Extracellular Signals: First Messengers

Cells use various types of chemical signals to communicate with other cells. **Local chemical mediators** affect only nearby cells and are rapidly destroyed if they are not used. **Neurotransmitters** are produced by nerve cells and also affect cells that are nearby. Finally, **hormones** are sent into the blood by secretory

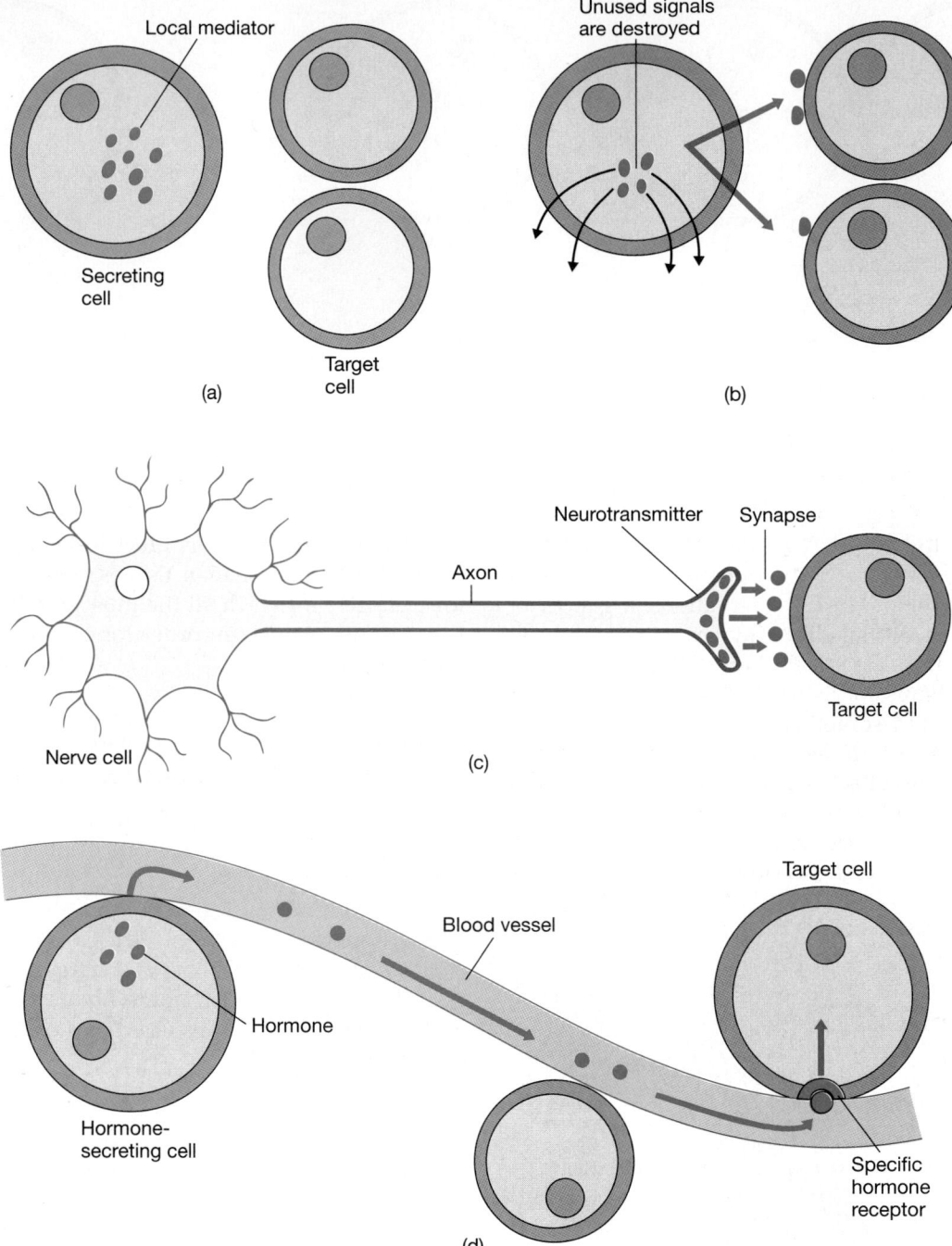

Figure 5–23

The three basic types of chemical signaling systems. (*a* and *b*) Local chemical mediators act only on cells near them and are quickly destroyed if they are not used. (*c*) Neurotransmitters are produced by nerve cells and are delivered via a synapse at a target
cell, which may be a long distance away. (*d*) Hormones travel via the blood from
hormone-secreting cells to target cells that have specific receptors for them.

cells to affect target cells, which may be nearby or far
away from the hormone-producing cells. Figure 5–23
summarizes these three types of signaling systems.

With the exception of the prostaglandins and related compounds, most of these chemical signals are
synthesized by specialized cells where they are stored
until the appropriate stimulus triggers their release.
Specific examples of chemical signals are listed in
Table 5–4, which summarizes the wide variety of
molecules used.

Table 5–4
Examples of Molecules That Serve as Chemical Signals

Type of Molecule	Local Mediator	Neurotransmitter	Hormone
Peptides	—	Neuropeptides	Vasopressin
Polypeptides	—	—	Insulin
Amino acids and derivatives	Histamine	Glycine	Epinephrine
Steroids	—	—	Testosterone
Fatty acid derivatives	Prostaglandins	—	—
Other small molecules	—	Acetylcholine	—

These groups of chemical signals can be classed into two large groups: **water-soluble signals** and **lipid-soluble signals.** Many of the hormones, such as parathyroid hormone and vasopressin, and all of the neurotransmitters, are water-soluble. These molecules are removed or destroyed very rapidly (within seconds, or a few milliseconds for neurotransmitters) after being released. Steroid hormones and thyroid hormones (such as thyroxine), in contrast, are lipid-soluble molecules that can persist in the bloodstream for several hours, or even days in the case of thyroid hormones. Thus, water-soluble signal molecules are used to produce responses that are needed rapidly but only for a short time, whereas lipid-soluble signal molecules are used to elicit responses that may develop more slowly but will last for a long time.

A common feature of the chemical communication systems is the presence of specific **receptors** that allow the signaling molecule to recognize and bind to the target cell. Receptor function is discussed in more detail in a later section.

Local Chemical Mediators. Some signal molecules released by a cell affect only the cells in the immediate vicinity, either because the molecules are destroyed rapidly once released or because they readily bind to specific **receptors** that are present on the first cells that they encounter nearby (see Fig. 5–23*a* and *b*). Destruction and cell binding of these molecules occur so rapidly that only a minor fraction of the amount released will reach the bloodstream. The molecules are called **local chemical mediators;** specific examples include **histamine** and the **prostaglandins.** Histamine is derived from the amino acid histidine, and is stored in the specialized **mast cells** present in connective tissues.

Histamine is released rapidly in response to injury, infection, and allergic reactions. One of its prin-

cipal actions is to cause expansion and leakiness of local blood vessels. This effect allows antibodies and white blood cells access to the site of injury, where they perform their tasks of destroying foreign substances.

Prostaglandins were named after the prostate gland, which was once thought to be their only source. It is now recognized that prostaglandins are formed by most tissues, rather than by specialized cells, and that they are released continuously instead of being stored. They are synthesized from essential fatty acids containing 20 carbon atoms, such as arachidonic acid (see Chapter 2).

Prostaglandins have diverse effects on different tissues. Smooth muscles are major target tissues that either contract or relax, depending on the specific type of prostaglandin. Prostaglandins also mediate inflammatory responses. The effectiveness of aspirin in reducing fever and inflammation is due largely to its ability to inhibit prostaglandin synthesis. Many of the actions attributed originally to the prostaglandins may be due to their highly unstable but potent relatives known as **prostacyclins** and **thromboxanes.** Another metabolic pathway that begins with arachidonic acid leads to formation of the **leukotrienes,** another group of locally acting chemicals that appear to mediate allergic and inflammatory reactions.

Neurotransmitters. A second group of chemical signals, known as **neurotransmitters,** is used by nerve cells to influence their target cells. Although the target cell may be many centimeters from the nerve cell body, the neurotransmitter is delivered directly to the target cell by a long slender extension of the nerve cell called an **axon.** At the **synapse** (the gap between the end of the axon and the target cell), the neurotransmitter is released by the axon and diffuses across to the target cell (see Fig. 5–23*c*). The gap it must cross is very small (only a few nanometers), and it reaches the target cell in less than 1 msec. Receptor molecules that are present on the surface of the target cell in the region of the synapse provide binding sites for the neurotransmitter.

Delivery of the neurotransmitter directly to the target cell, and the presence of specific receptors there, ensure that the effects of the chemical signal are restricted to a very small area, even though the target cell may be many centimeters from the main body of the nerve cell.

A variety of molecules serve as neurotransmitters at chemical synapses. The amino acid **glycine** and small peptides known as **neuropeptides** are important in the central nervous system. Synapses where the target is a muscle cell use **acetylcholine** as the neurotransmitter. The nervous system is a unique communication system because, in addition to transmitting information, it can store a vast amount of information in the cells of the brain.

Hormones.   The cells of a few specialized tissues synthesize and secrete chemicals known as **hormones** into the blood, which then carries them to target cells that are far away from the hormone-secreting cells (see Fig. 5–23*d*). This type of communication system relies on the flow of blood and requires the chemical signal (the hormone) to travel distances that are much greater than the distance a neurotransmitter must travel in crossing a synapse. Consequently, hormone-secreting cells transmit information much more slowly than nerve cells. Unlike the systems discussed previously, most hormones are not delivered directly to the target by the secreting cell. Furthermore, the effects of the hormone are much more diffuse.

Once released, many hormones affect more than one type of cell or tissue in the body. **Parathyroid hormone,** for example, raises the level of calcium ions in the blood by acting on bone to release Ca^{2+} and by acting on the kidney to decrease excretion of Ca^{2+} from the body. The secreted hormone must be able to "recognize" the target cells and must not affect the other cells that it encounters. Accurate recognition is achieved because the target cells have receptor molecules that the hormone can bind to. Nontarget cells lack the specific receptor and are not recognized by the hormone.

The different types of hormone-secreting cells make up the **endocrine system** of the body. This system is influenced in part by the nervous system. One result of this influence is that the secretion of several specific hormones is subject to overall regulation by the nervous system.

Hormones are as varied a group of molecules as the other types of chemical signals. Insulin and parathyroid hormone are polypeptides. Vasopressin (also called antidiuretic hormone) is a small peptide of nine amino acids. Epinephrine is derived from the amino acid tyrosine. The female and male sex hormones (estrogens and testosterone, respectively) are steroids.

Receptors: Mediating Between First and Second Messengers

When a chemical signal arrives at a target cell, it ultimately changes processes such as enzyme-catalyzed reactions, transport of ions, and transcription of genes. The initial event, however, is the interaction of the signal molecules with specific receptors located either on the surface of the target cell or on its cytoplasm.

The Nature of Receptors.   A **receptor** is a macromolecule, usually a protein, to which the signal molecule can bind with high specificity. This means that other molecules do not bind easily to the receptor. This specificity may arise because the specific shape of the signal molecule may facilitate its binding to the receptor. Other molecules with slightly different

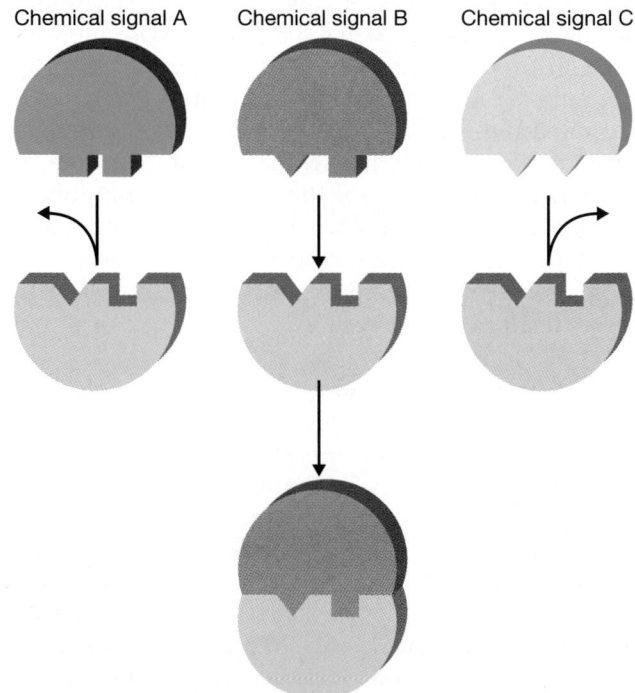

Figure 5–24

Receptor molecules have specific shapes designed to discriminate among various chemical signals. Although chemical signals A, B, and C are very similar in shape, only B fits the shape of the receptor exactly, so only B binds with the receptor and affects the target cell.

shapes may be unable to fit into the binding sites on the receptor. This point is illustrated in Figure 5–24. In addition to specificity, the signal molecule binds to the receptor with **high affinity,** which means that the binding is strong.

The combined purpose of these receptor properties is to ensure that only the signal molecule will bind and remain attached to the receptor. These properties are particularly important for hormone receptors because a hormone that is delivered to a cell via the bloodstream is usually present in a low concentration (10 nM or less) and is only one of the many different types of molecules in the blood that will be "seen" by the receptor.

Binding of a hormone to its receptor in a target cell triggers a sequence of events that leads ultimately to the changes that represent the cellular response to the hormone. The cellular location and action of hormone receptors vary with the type of hormone.

Receptors in the Cytoplasm. The receptors for all steroid hormones and for **vitamin D** (a sterol that promotes calcium absorption in the intestine) are soluble proteins located in the cytosol of the cell (Fig. 5–25). The hormones must penetrate the cell membrane in order to gain access to the receptors. This process is facilitated by the lipid-soluble nature of the hormones. Binding of the hormone to the receptor is followed by movement of the hormone-receptor complex into the nucleus, where it binds to specific acceptor sites on the DNA and regulates the transcription of specific genes.

Thyroid hormones also exert their effects via gene transcription and through a similar mechanism, except that the receptors for these hormones are concentrated in the nucleus even in the absence of hormone. (Regulation of gene expression at the level of transcription was discussed earlier.)

Receptors in the Plasma Membrane. Polypeptide hormones and **catecholamines** such as epinephrine initiate a completely different series of events upon binding to their receptors. The receptors for these hormones are located on the plasma membrane of the target cell. The hormones bind to the receptors, producing a response within the cell without gaining entry to it (see Fig. 5–25). The effects of the hormones are mediated by second messengers that are generated within the cell when the hormones bind to their receptors. (The actions of second messengers are discussed in detail later.)

In summary, a plasma membrane receptor is more than just a binding site. It acts more like a **transducer.** It can convert an extracellular event—the binding of a signal molecule—into an intracellular response that modifies the behavior of the target cell.

Plasma membrane receptors are intrinsic membrane proteins that usually account for less than 1% of the total mass of protein present in the membrane. For these reasons, they are very difficult to identify, isolate, and study. The study of receptors has been aided considerably by the development of techniques for attaching radioisotopes (see Chapter 3) to signal molecules without interfering with the physiological

Figure 5–25

Receptors can be located in either the plasma membrane or the cytoplasm of a target cell. Receptors for water-soluble signals are often intrinsic membrane proteins, whereas receptors for lipid-soluble signals, which can diffuse through the plasma membrane, are often found inside the cell.

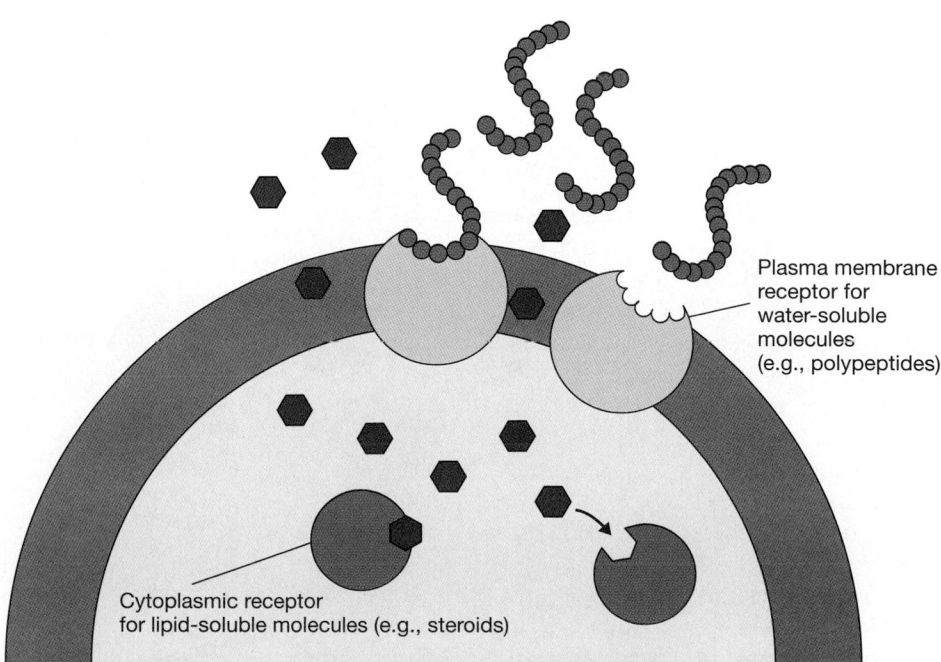

Plasma membrane receptor for water-soluble molecules (e.g., polypeptides)

Cytoplasmic receptor for lipid-soluble molecules (e.g., steroids)

activity of the molecule. Receptors bind the labeled molecules, making it possible to follow some aspects of receptor function, such as numbers, distribution, and the fate of the bound molecules.

Ligands: Agonists and Antagonists. Any compound, such as a signal molecule, that binds with high specificity to a receptor is termed a **ligand.** Any ligand capable of both binding to a plasma membrane receptor and evoking a physiological response is known as an **agonist.** A ligand that binds with high affinity to a receptor but evokes no response is called an **antagonist** because, by occupying receptors, it interferes with (or "antagonizes") the action of an agonist.

Antagonists of hormone receptors can be of great therapeutic importance in medicine. Certain hormones called **catecholamines** act on the heart to increase both the heart rate and the amount of blood it pumps, with the net effect of increasing blood pressure. Too much catecholamine action can cause high blood pressure and angina (chest pain). A catecholamine antagonist called **propranolol** is used to block these effects of catecholamines.

Another example is **spironolactone,** which is an antagonist of the steroid hormone **aldosterone.** Aldosterone increases the retention of Na$^+$ by the kidneys; hence, water is also retained (see Chapter 25). Treatment with spironolactone blocks the action of aldosterone and produces an increase in the excretion of water, which is important in cases of severe fluid retention and edema. Spironolactone is known as a **diuretic,** a substance that increases water excretion.

Variations in Receptor Quantities. One of the original hypotheses regarding receptors was that they were present in cells at fixed concentrations. Quantitative studies of receptor numbers using radiolabeled ligands have revealed, however, that the concentration of many receptors in cells can be regulated. For example, a sustained elevation of the concentration of an agonist circulating in the blood often leads to a decrease in the number of receptors present in target cells. This phenomenon is known as **down regulation** of receptor number and is accompanied by a decrease in the physiological response **(desensitization)** of the cells to the circulating agonist. When the agonist concentration falls, the receptor number rises again. This phenomenon makes many cells particularly sensitive to a **change** in the normal concentration of a chemical signal, as opposed to its absolute concentration.

The mechanism by which agonists can regulate the number of their receptors in target cells remains unclear, but it may involve any one of a number of procedures, including inactivation of receptors, and changes in the rate of synthesis or rate of degradation of the receptor proteins. In the specific case of plasma membrane receptors, binding of agonist molecules to receptors can stimulate internalization of the receptor-agonist complex by an endocytotic process.

Since few receptors have been isolated and studied in great detail, little is known about the interactions that stabilize the binding of signal molecules. The binding is noncovalent and reversible and probably involves contributions by hydrophobic interactions, hydrogen bonds, and ionic bonds (see Chapter

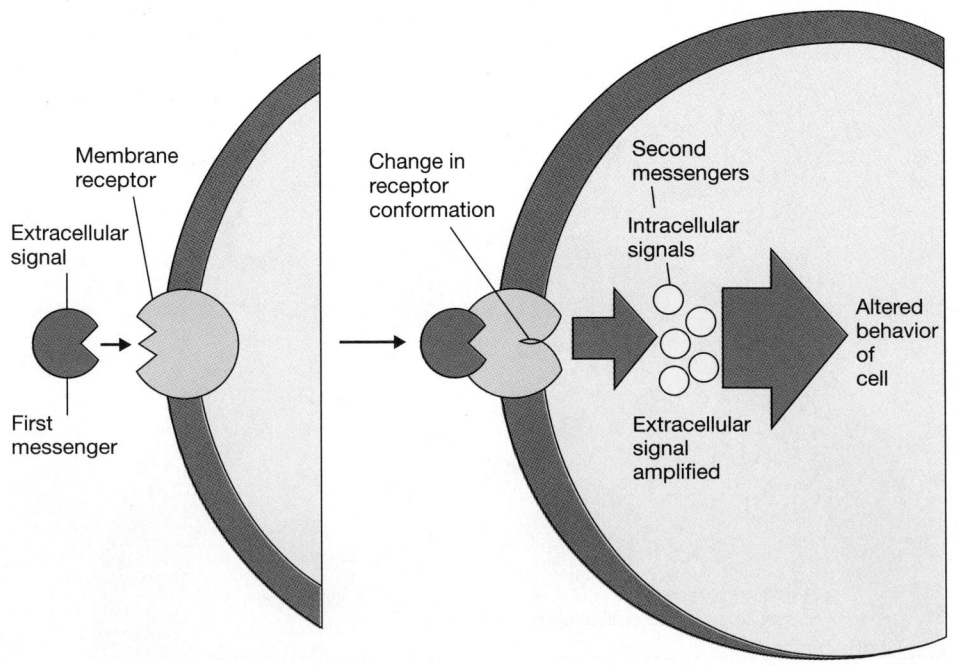

Figure 5–26

When an extracellular signal binds to a receptor molecule, the receptor changes its conformation. This change triggers the activation of intracellular signal molecules, which amplify the extracellular signal and produce the physiological response of the target cell.

2). The remainder of this chapter examines some of the ways by which the extracellular signal—binding of an agonist molecule to a plasma membrane receptor—is converted (transduced) to an intracellular response.

G Proteins

The largest family of plasma membrane receptors are the receptors that are linked to GTP-binding proteins **(G proteins).** These are discussed in more detail in Chapter 12. G-proteins provide a link between the receptor and its target. When an agonist binds to a G-protein-linked receptor it initiates a chain of events that changes the concentration of one or more intracellular "signal" molecules.

Intracellular Signals: Second Messengers

It is generally accepted that the conformation (three-dimensional shape) of a receptor molecule is altered when an extracellular signal (first messenger) molecule binds to it. The change in receptor conformation generates an intracellular signal molecule known as a **second messenger.** This process also increases or **amplifies** the signal because several second messenger molecules are generated for each extracellular signal molecule that is bound to a receptor (Fig. 5–26). The second messengers initiate the intracellular reactions that modify the behavior of the target cell.

Three principal mechanisms are thought to produce second messenger molecules inside cells when an external signal molecule binds to a plasma membrane receptor. These are summarized in Figure 5–27.

(a)

(b)

Figure 5–27

Three examples of intracellular signals, or second messengers. (*a*) An extracellular signal, when it binds to a receptor, may activate adenylate cyclase, generating cyclic AMP from ATP. (*b*) An extracellular signal may cause a channel to open in the plasma membrane, causing an influx of Ca^{2+} ions. (*c*) An extracellular signal may activate phospholipase C, which converts phosphatidylinositol bisphosphate (a membrane phospholipid) into diacylglycerol and inositol trisphosphate.

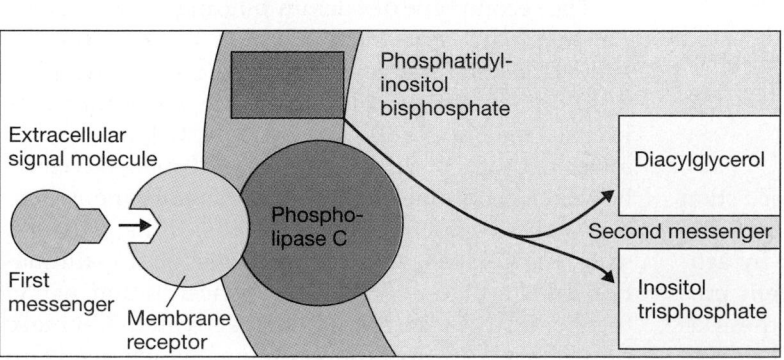

(c)

Cyclic AMP as a Second Messenger.   One of the best understood mechanisms is the production of a second messenger, known as **cyclic AMP,** from ATP by a reaction catalyzed by **adenylate cyclase** (see Fig. 5–27a). The latter is a membrane-bound enzyme that is stimulated by binding of the signal molecule to the receptor. This mechanism will be discussed again in more detail in Chapter 12.

Calcium Ions as Second Messengers.   The conformational change in the receptor caused by an extracellular signal could open a **membrane channel,** which in turn could lead to two possible events. First, a small transient influx of ions into an electrically excitable nerve or muscle cell will briefly change the electrical potential difference between the inside and outside of the cell. This change can trigger an **action potential** (see Chapter 7) that spreads rapidly throughout the membrane of the target cell. Many neurotransmitters work in this fashion. Second, a major influx of a specific ion could raise the intracellular concentration of that ion to the point at which it acts as a second messenger to stimulate an intracellular response. **Calcium ions** are likely candidates for this role. The extracellular concentration of Ca^{2+} is not large (1000 μM), but since the intracellular concentration of free Ca^{2+} is kept so low (0.1 μM)—in part because it is potentially toxic to the cell—a highly favorable concentration gradient exists for passive entry of Ca^{2+}. The electrical gradient across the membrane also favors entry of the positively charged Ca^{2+} (see Fig. 5–27b).

Inositol Trisphosphate and Diacylglycerol as Second Messengers.   More recently, a third system for producing second messengers has been identified. It involves the breakdown of the phosphorylated form of a membrane phospholipid called **phosphatidylinositol bisphosphate.** This compound is produced by the covalent attachment of two phosphate groups to phosphatidylinositol (see Chapter 2). Binding of the signal molecule to the membrane receptor leads to activation of a specific **phospholipase C** enzyme that catalyzes the breakdown of the phosphorylated lipid into **inositol trisphosphate** and **diacylglycerol,** a pair of intracellular second messenger molecules (see Fig. 5–27c).

Action of Second Messengers.   The mode of action of second messengers is summarized in Figure 5–28. Cyclic AMP modifies the behavior of a cell by activating enzymes called **cyclic AMP-dependent protein kinases.** When activated by cyclic AMP, these kinases catalyze reactions that transfer a phosphate group from ATP to the side chain of an amino acid

such as serine in specific protein molecules. Covalent attachment of a phosphate group (phosphorylation) to a protein usually modifies its biological activity (see Fig. 5–28). If the protein is an enzyme, for example, the result of phosphorylation could be either activation or inactivation of its enzymatic capacity. Such a change could lead to the physiological response of the target cell.

Inositol trisphosphate and diacylglycerol are important parts of receptor systems that operate primarily by increasing the level of Ca^{2+} in the cell cytosol. Inositol trisphosphate is involved in the release of Ca^{2+} from intracellular stores such as the endoplasmic reticulum. This effect, coupled with the influx of extracellular Ca^{2+}, can raise the cytosolic concentration of free Ca^{2+} from the resting level of 0.1 μM to somewhere in the range of 1 to 10 μM (see Fig. 5–28). The increased Ca^{2+} becomes bound with high affinity to **calcium-binding proteins** and produces a change in the conformation of these proteins and, hence, a change in their function. There are two types of these proteins.

The first type is the true receptor proteins that have no enzymatic activity but, upon binding Ca^{2+}, alter their association with other proteins. This alteration leads to a change in the function of the other proteins. Specific examples of this type of receptor protein are **troponin C,** found in skeletal and cardiac muscle cells, and **calmodulin,** which is found in almost all cells. Binding of a Ca^{2+}-calmodulin complex by a calmodulin-dependent protein kinase will activate the kinase and lead to protein phosphorylation, a mechanism similar to that resulting from activation of cyclic AMP-dependent protein kinases. The Ca^{2+}-calmodulin system is more versatile than the cyclic AMP system because, in addition to producing physiological changes via protein phosphorylation, the Ca^{2+}-calmodulin complex can directly activate enzymes such as adenylate cyclase and cyclic AMP phosphodiesterase, the enzymes that make and break down cyclic AMP. Clearly, this allows interaction between the cyclic AMP and the Ca^{2+} intracellular signaling pathways.

The second type of calcium-binding proteins is the Ca^{2+}-regulated enzymes that bind Ca^{2+} ions directly. A specific example is the Ca^{2+}-sensitive, phospholipid-dependent protein kinase, also known as **C-kinase** (see Fig. 5–28). The cytosolic C-kinase is not affected by Ca^{2+}. In the presence of diacylglycerol, however, it becomes bound to the membrane, where it is activated by the phospholipids. In its activated state, the C-kinase is extremely sensitive to stimulation by Ca^{2+} ions. When the concentration of cytosolic Ca^{2+} is increased, the activated C-kinase phosphorylates specific proteins, which leads to generation of a physiological response.

Figure 5–28

Second messengers may activate certain enzymes that catalyze the phosphorylation of certain proteins, which in turn produce the physiological response of the cell to the extracellular signal.

In certain cell types, the inositol trisphosphate system, via Ca^{2+}-calmodulin, and the diacylglycerol system, via C-kinase, may act in concert to produce the physiological response. For example, the Ca^{2+}-regulated contraction of smooth muscle occurs in two phases, an initial rapid component followed by a slower but more sustained component. It has been proposed—but not yet proven—that the rapid phase is due to release of Ca^{2+} from intracellular stores, leading to muscle contraction via protein phosphorylation by a Ca^{2+}-calmodulin–dependent protein kinase. In contrast, the slower sustained contraction could be mediated by influx of extracellular Ca^{2+} and activation of a C-kinase, which also leads to muscle contraction via protein phosphorylation.

Removal of Second Messengers. The concentrations of second messengers in the cytosol of an unstimulated cell are usually very low because these molecules are continuously and rapidly destroyed or removed from the cytosol (Fig. 5–29). The cytosolic concentration of a second messenger is altered principally by changes in its rate of synthesis or influx into a cell. This ensures that an increased rate of synthesis or influx will rapidly increase the cytosolic concentration of the messenger, allowing the target cell to respond quickly to the extracellular signal (see Fig. 5–29). The concentration of cyclic AMP is usually only 1 μM, but this can be increased fivefold within seconds after a hormone binds to the plasma membrane receptor and stimulates adenylate cyclase. Synthesis or influx of the second messenger will fall to zero when the extracellular signal is "turned off." Since the rapid rate of destruction or removal of the messenger will continue, the cytosolic concentration of the messenger molecules will decline rapidly to the point where the cell will cease to respond. In other words, the response of the cell to the second mes-

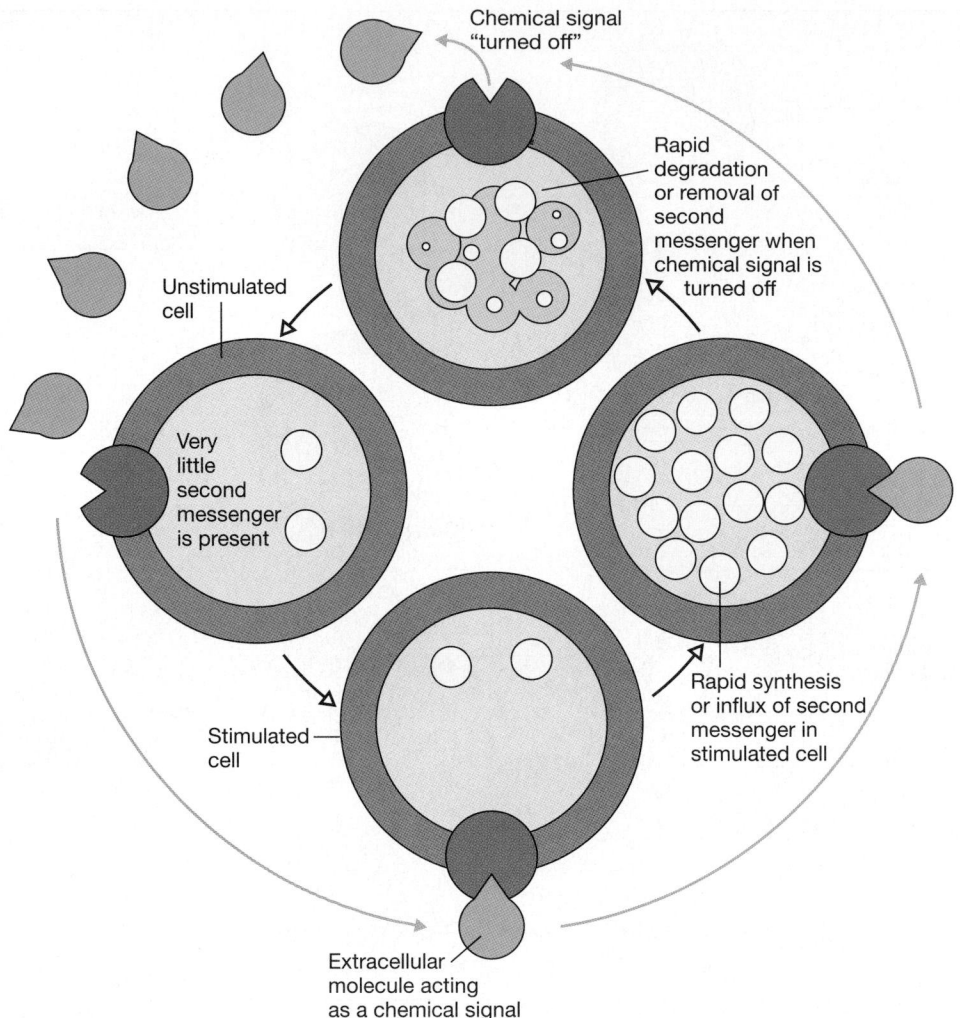

Figure 5–29

A second messenger is rapidly synthesized when a first messenger molecule (an extracellular signal) binds to a receptor. When the extracellular signal is removed, however,
the second messenger is rapidly destroyed or removed from the cell. Thus, the next
time a signal binds to the receptor, the second messenger must be synthesized anew.

senger is reversed rapidly once the second messenger is no longer present (see Fig. 5–29).

Cyclic AMP, for example, is degraded very
rapidly to adenosine 5'-monophosphate (AMP) by a
reaction catalyzed by an enzyme known as **cyclic
AMP phosphodiesterase** (see Chapter 12). Likewise,
inositol trisphosphate and diacylglycerol are rapidly
degraded by specific enzymes.

To give another example, Ca^{2+} is removed from
the cytosol by several mechanisms acting together
(Fig. 5–30). One mechanism actively pumps Ca^{2+} out
of the cell, moving the ion against its electrochemical
gradient. This is achieved by a specific protein,
known as Ca^{2+}-ATPase, which utilizes chemical energy (as ATP) to pump Ca^{2+} out of the cell. Another
mechanism, involving a different protein, is Ca^{2+}/Na^+
antiport. Ca^{2+} leaves the cell in exchange for the Na^+

that enters the cell by diffusing down a favorable electrochemical gradient.

Free Ca^{2+} is also removed by binding to cytosolic calcium-binding proteins and other molecules,
and by sequestration inside intracellular organelles
such as mitochondria and the endoplasmic reticulum.
The large number of different mechanisms that can
regulate cytosolic Ca^{2+} levels may be, at least in part,
a defense system to guard against possible toxic effects of uncontrolled concentrations of this ion.

*Case Study: Glycogen Metabolism and Protein Phos-
phorylation.* Protein phosphorylation is one example of a way in which the production of second messengers can mediate a change in cell function that
represents the cellular response to a chemical signal.
The metabolism of **glycogen** in skeletal muscle cells

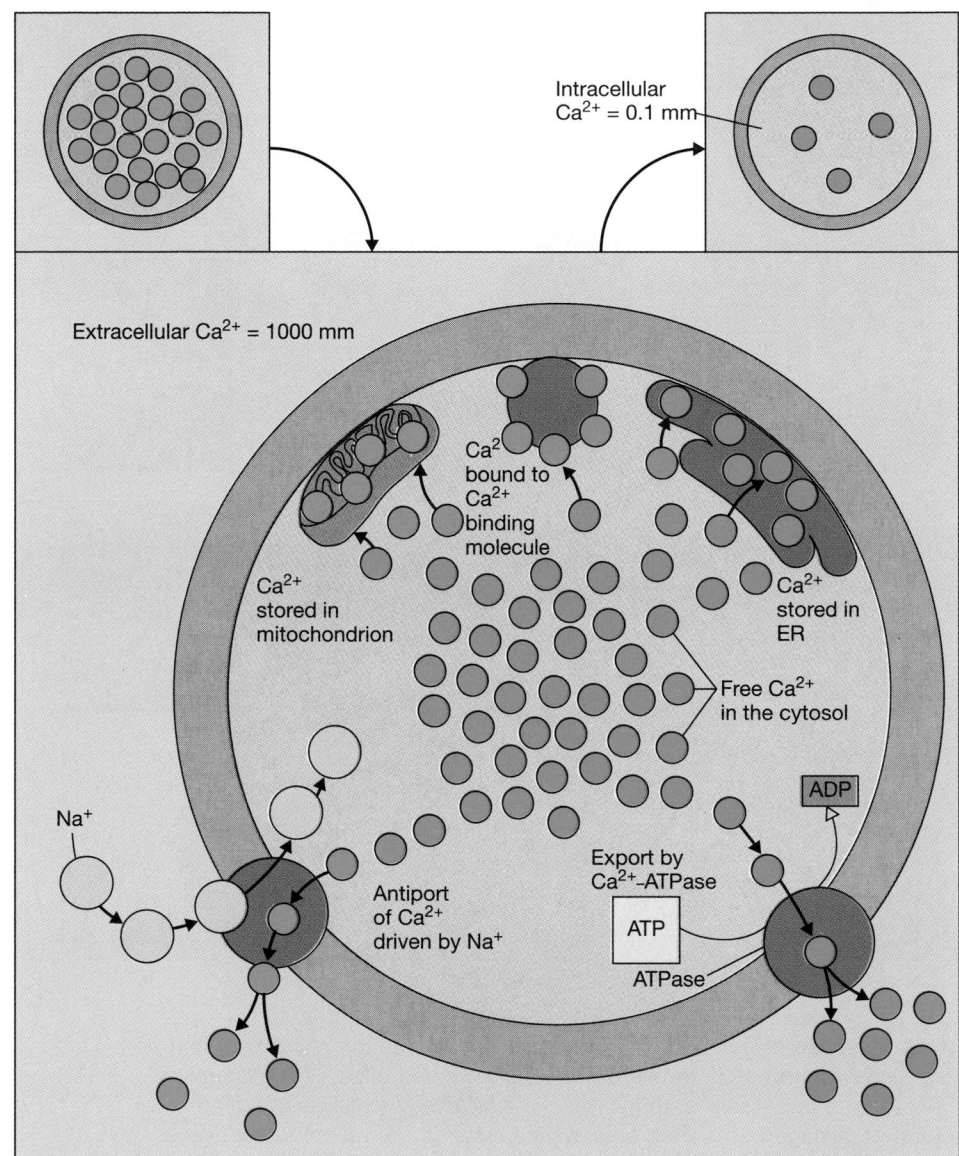

Figure 5–30

Examples of the ways in which a second messenger is removed from a cell. The second messenger calcium ions in the cytosol may be removed by active transport, by antiport, by sequestration inside certain cytoplasmic organelles, or by binding to special molecules.

provides clues to a possible sequence of events. Recall from Chapter 2 that glycogen is a polysaccharide, the form in which glucose is stored until its chemical energy is needed by the muscle cell. The hormone **epinephrine** stimulates glycogen breakdown and shuts off glycogen synthesis—the net effect being to provide the cell with an adequate supply of glucose for strenuous activity.

Epinephrine works by stimulating adenylate cyclase in the plasma membrane and elevating intracellular cyclic AMP. The cyclic AMP activates a protein kinase, initiating a sequence of phosphorylation reactions (Fig. 5–31*a*). Phosphorylation by the protein kinase of Enzyme 1 activates this enzyme; once activated, Enzyme 1 phosphorylates Enzyme 2, which begins to remove individual glucose molecules from glycogen. The same protein kinase phosphorylates a third enzyme (Enzyme 3) that is involved in synthesis of glycogen from glucose. In contrast to Enzymes 1 and 2, phosphorylation of Enzyme 3 causes inactivation.

The overall effect of these protein phosphorylations is to make available a maximum supply of glucose by simultaneously stimulating breakdown of

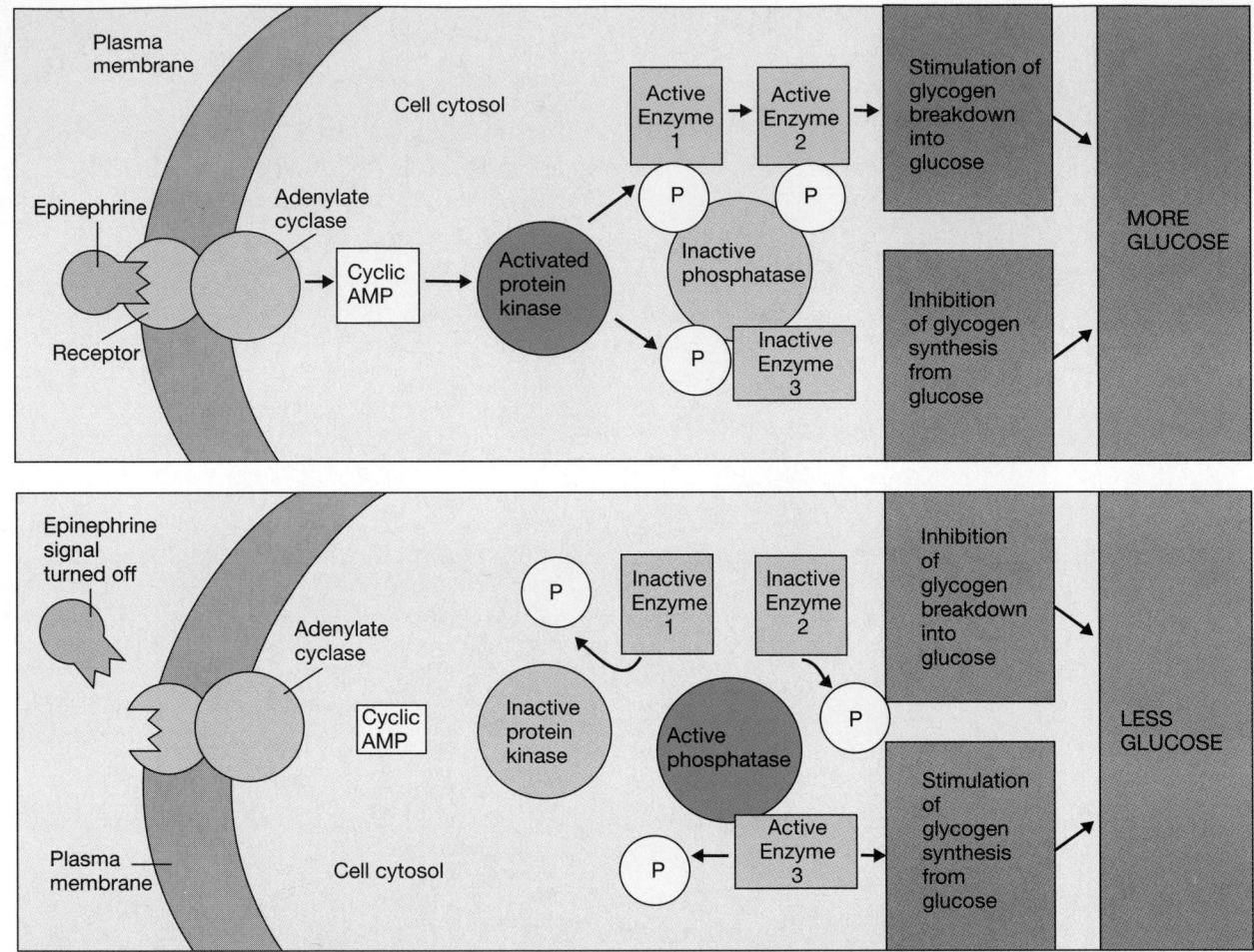

Figure 5–31

An example of the response of a cell to an extracellular signal and the reversal of that response once the signal is turned off. In this example, the desired effect of the chemical signal epinephrine on the target cell is an increase in available glucose as opposed to its storage in the form of glycogen. (*a*) When epinephrine binds to the receptor, adenylate cyclase produces the second messenger cyclic AMP. This, in turn, activates protein kinase, which causes the phosphorylation of three enzymes. Phosphorylated Enzymes 1 and 2 stimulate glycogen breakdown, while Enzyme 3, which normally stimulates glycogen synthesis from glucose, is inactivated by phosphorylation. The net result is the desired response of the target cell, an increase in the availability of glucose. (*b*) When the epinephrine signal is removed, all three enzymes are dephosphorylated by phosphoprotein phosphatase. As a result, glycogen synthesis is resumed and less glucose is available.

glycogen into glucose molecules while inhibiting the use of glucose for synthesis of glycogen.

When the epinephrine signal is turned off, the phosphorylation steps just described are rapidly reversed by reactions catalyzed by the enzyme **phosphoprotein phosphatase** (see Fig. 5–31*b*). In other words, the phosphate groups on each of the three key enzymes are removed by the phosphatase, with the net effect of inhibiting glycogen breakdown (Enzymes 1 and 2 are inactivated) and stimulating glycogen synthesis (Enzyme 3 is activated). The opposing action of the phosphoprotein phosphatase does not interfere with the action of cyclic AMP because the

phosphatase is inhibited while the level of cyclic AMP is elevated.

Regulation of both breakdown and synthesis of glycogen is an important feature of this cyclic AMP control system. It increases the response to an elevated level of cyclic AMP in the cell, the specific response being a readily available supply of glucose molecules. We now know of more than 25 different enzymes regulated by a reversible phosphorylation process. These and other types of protein modifications are so abundant that most of the proteins that mediate physiological processes are probably regulated in this way.

CHAPTER REVIEW

Summary

- Genetic information, the instructions a cell uses to synthesize RNA and proteins, is stored in DNA by a code formed from the specific linear sequence in which the four different nucleotides are arranged. Different cell types arise because the genes expressed in these cells are different.

- DNA is copied before cell division by a replication procedure catalyzed by DNA polymerase. RNA is transcribed from DNA in a process catalyzed by RNA polymerase. Newly synthesized RNA is converted to mature functional RNA by processing in the nucleus.

- Translation of the information in mRNA is coordinated by the ribosomes. Amino acids are linked to specific tRNA molecules that are bound to ribosomes, along with mRNA, during polypeptide synthesis. The amino acid attached to the tRNA is added to the growing polypeptide chain by formation of a peptide bond.

- Expression of genes can be regulated at several levels including transcription, RNA processing, RNA stability, translation, and post-translational modifications of the protein product.

- DNA is very sensitive to damage by factors normally present in the environment. Highly efficient repair mechanisms ensure that alterations in genetic information do not persist.

- Gap junctions between adjacent cells allow communication by direct movement of certain molecules from the cytoplasm of one cell to the other.

- Cells that are far apart release chemical signals that bind to specific receptors in the target cell. When the receptor is at the cell surface, binding of the chemical signal generates an intracellular signal known as a second messenger. The second messenger mediates the cellular response to the extracellular signal.

Review Questions

Identify Each with a Term

1. The term applied to the genetic code to indicate that many amino acids are specified by more than one codon.
2. The term used to describe DNA replication when each new DNA molecule contains one nucleotide strand derived from the parent DNA.
3. A nucleotide sequence in newly synthesized RNA that contains no useful information.
4. A structure that forms a direct connection between the cytoplasm of two adjacent cells.
5. General term for any compound that binds with high specificity to a receptor.

Define Each Term

6. Start codon
7. RNA polymerases
8. Post-translational modifications
9. Antagonist
10. Cyclic AMP phosphodiesterase

Choose the Correct Answer

11. Which of the following describes the function of a nucleolus?
    a. Produces ATP
    b. Synthesizes DNA polymerase
    c. Produces ribosomal subunits
    d. Stores genetic information for protein synthesis
12. Amino acids are carried to the ribosomes in the cytoplasm by:
    a. messenger RNA
    b. mitochondrial RNA
    c. ribosomal RNA
    d. transfer RNA
13. Unwinding the DNA double helix prior to replication is aided by:

a. DNA polymerase
b. DNA topoisomerase
c. DNA endonuclease
d. DNA ligase

14. Chemical signaling systems generally involve:
    a. an extracellular signal molecule
    b. a receptor molecule
    c. an intracellular signal molecule
    d. all of the above
15. The ion that acts as a second messenger is:
    a. calcium
    b. magnesium
    c. potassium
    d. sodium
16. Inositol trisphosphate is an example of a:
    a. second messenger
    b. membrane receptor
    c. first messenger
    d. calcium channel
17. The enzyme responsible for making cyclic AMP is:
    a. cyclic AMP phosphodiesterase
    b. cyclic AMP-dependent protein kinase
    c. adenylate cyclase
    d. DNA glycosylase

Answer Each Question in One or Two Sentences

18. How can only four different nucleotides be used to construct a genetic code that will specify 20 different amino acids?
19. List five principal differences between DNA synthesis and RNA synthesis.
20. If all the cells in an animal contain the same genetic information, how do cells with such marked structural and functional differences arise?
21. What is the difference between an agonist and an antagonist?

22. List the three principal types of extracellular chemical signals.
23. How do second messengers produce a change in the function of a cell?

Problem Solving in Physiology

24. At any given time, only a small part of the DNA of a human cell is being transcribed and expressed in the form of specific proteins. Most of the genetic informa- tion is suppressed. Suggest two ways by which the ex- pression of a gene could be influenced by chromoso- mal structure.

25. Polypeptide hormones are degraded rapidly compared to steroid hormones. Explain why this is consistent with a primary role for polypeptide hormones in con- trolling rapid responses.

Suggested Readings

Alberts, B., Bray, D., Lewis, J., Raff, M., Roberts, K., and Watson, J.D. *Molecular Biology of the Cell*, 3rd ed. New York: Garland, 1994.

Berridge, M.J. "Inositol triphosphate and diacylglycerol: two interacting second messengers." *Annual Review of Biochemistry*, 56:159–193, 1987.

Dreyfuss, G., Matunis, M.J., Pinol-Roma, S., and Burd, C.G. "hnRNP proteins and biogenesis of mRNA." *Ann. Rev. Biochem.*, 62:289–321, 1993.

Hendrick, J.P., and Hartl, F.U. "Molecular chaperone func- tions of heat-shock proteins." *Ann. Rev. Biochem.*, 62:349–384, 1993.

Kabakov, A.E., and Gabai, V.L. "Heat-shock proteins main- tain viability of ATP-depleted cells: what is the mecha- nism?" *Trends Cell Biol.*, 4:193–196, 1994.

Lake, J.A. "The ribosome." *Scientific American*, 245:84–97, August 1981.

Linder, M.E., and Gilman, A.G. "G Proteins." *Scientific American*, 267:56–65, July 1992.

Ptashne, M. "How gene activators work." *Scientific Ameri- can*, 260:41–47, January 1989.

Radman, M., and Wagner, R. "The high fidelity of DNA du- plication." *Scientific American*, 259:40–46, August 1988.

Rasmussen, H. "The cycling of calcium as an intracellular messenger." *Scientific American*, 261:66–73, October 1989.

Rodbell, M., Birnbaumer, L., Pohl, S.L., and Krans, H.M.J. "The glucagon-sensitive adenyl cyclase system in plasma membranes of rat liver. An obligatory role of guanyl nucleotides in glucagon action." *Journal of NIH Research*, 2:63–70, 1990.

Snyder, S.H. "The molecular basis of communication be- tween cells." *Scientific American*, 253:132–140, October 1985.

Key Terms

amplification (p. 157)
antibiotic (p. 158)
anticodon (p. 151)
calcium ions (p. 169)
chromosome (p. 138)
cyclic AMP (p. 169)
DNA polymerase (p. 144)
DNA repair (p. 160)
DNA replication (p. 143)

DNA transcription (p. 143)
exon (p. 150)
gap junction (p. 162)
genetic code (p. 140)
genetic expression (p. 140)
hormone (p. 163)
inositol trisphosphate (p. 169)
intron (p. 143)

ligand (p. 168)
messenger RNA (p. 148)
neurotransmitter (p. 162)
phosphodiesterase (p. 172)
phospholipase C (p. 170)
phosphoprotein phosphatase (p. 174)
post-translational modification (p. 159)

protein kinase (p. 170)
protein synthesis (p. 143)
ribosomal RNA (p. 148)
ribosome (p. 152)
RNA polymerase (p. 148)
RNA processing (p. 143)
second messenger (p. 163)
template (p. 144)
transfer RNA (p. 148)

Mitochondria. *(© Don W. Fawcett/Visuals Unlimited)*

Chapter 6

Energy and Cellular Metabolism

Key Concepts

Living cells transform energy from one form to another. Plants use radiant energy from the sun to photosynthesize high-energy organic molecules. Animals, by eating plants, break down these molecules and convert their energy into ATP, which they use to drive biological processes.

Digestion of the proteins, carbohydrates, and fats present in food produces amino acids, glucose, and fatty acids. Cells degrade these macromolecules by pathways that culminate in the citric acid cycle in the mitochondrial matrix. The high-energy electrons released by this cycle are trapped by coenzymes and fed into the electron transport chain to generate ATP.

Glucose molecules are converted by glycolysis into pyruvate, and the acetyl group of each pyruvate is fed into the citric acid cycle.

Long-chain fatty acids are a rich source of energy. The chain of carbon atoms is cut into 2-carbon fragments, and these fragments (acetyl groups) are fed into the citric acid cycle. The glycerol enters the glycolytic pathway.

Many amino acids are degraded by initial removal of the amino groups followed by entry of the carbon atoms at various points in the pathway of glucose catabolism.

The 2-carbon fragments produced by catabolism are transferred as acetyl groups to coenzyme A. These acetyl groups are the precursors required for synthesis of fatty acids and cholesterol. Depending on metabolic demands, they are either shunted into the citric acid cycle or used in biosynthetic pathways.

Chapter Outline

Basic Principles of Energy Transformation in the Cell

An Overview of Catabolism: Dietary Proteins, Carbohydrates, and Fats

Metabolism of Carbohydrates: The Oxidation of Glucose

Metabolism of Fats

Metabolism of Proteins

Acetyl Coenzyme A: Mediating Between Catabolism and Biosynthesis

The combined social cost of loss of productivity, crime, accidents, damaged health, and death due to alcohol abuse has been estimated to be many billions of dollars annually in the United States. However, Rick Quick neither was aware of nor cared about this. He had spent much of the warm sunny day traveling at 200 mph around the steeply banked track as the racing team prepared the car for the first race of the season. Now he wanted to slake his thirst with some cold beer. He also needed to unwind and, with an off-day tomorrow, one drink with the team would do no harm.

Rick's driving last year had been disappointing, and he knew that the sponsors would lose interest if he did not win some races soon. Depressed by this thought, he had indulged in regular heavy drinking during the winter months. In order to break the habit, he had quietly enrolled as an outpatient in a drug and alcohol treatment program. As part of the treatment, Rick was given **disulfiram** *regularly, and he had been warned that he would be very sick if he drank alcohol while taking this drug. This reaction would help him to stop drinking and pursue other outlets to relieve stress. Since the program began, Rick had avoided alcohol completely, a fact he attributed to a growing realization that he would need peak fitness for any success in the new racing season. The disulfiram was having no apparent effect on its own, and Rick could not imagine that a tasty dark beer now would cause any problems. He would soon realize that he was mistaken. The beer's flavor was as good as he had anticipated and he drank quickly, but he had barely finished when he began to feel hot and a pulsating headache developed. He also experienced weakness and waves of nausea. Rick vomited repeatedly over the next couple of hours and, when this stopped, he fell exhausted into bed and slept.*

The unpleasant reaction that Rick experienced is due to an increased concentration of **acetaldehyde** *in the body. The ethanol present in an alcoholic drink is converted to acetaldehyde by the enzyme* **alcohol dehydrogenase** *in the liver. Normally, this is not a problem because the acetaldehyde is immediately converted to acetic acid by a second enzyme,* **aldehyde dehydrogenase.** *In the presence of disulfiram, however, the second step is blocked and there is a build-up of acetaldehyde in the blood and tissues. In other words, the blockade is due to inhibition of aldehyde dehydrogenase by disulfiram.*

The enzyme-catalyzed conversion of alcohol first to acetaldehyde and then to acetic acid is a reaction sequence known as a metabolic pathway. There are many such pathways in a living cell, and they form an integrated network of chemical reactions by which the cell extracts energy from food molecules and synthesizes the macromolecules that it needs to perform its function in the body. These metabolic processes must be regulated
carefully and must be responsive to possible changes in the external environment of the cell. This chapter is focused on the major metabolic pathways for carbohydrates, fats, and proteins.

Metabolism is the term used to describe the highly organized and integrated set of chemical reactions and energy transformation that sustain a living organism. These reactions are myriad; at least 1000 different reactions can take place within the unicellular bacterium *Escherichia coli.*

Two distinct aspects of the metabolism of a cell are **energy extraction** and **energy use.** In a series of steps known as **catabolism,** the cell extracts energy from its environment, the extracellular fluid. Catabolism is usually achieved in animals by the breakdown of the molecules present in food. This process has the additional purpose of converting complex macromolecules into simple molecules that can be used as "building blocks" to have new macromolecules. In a series known as **anabolism** or, more descriptively, **biosynthesis,** the energy extracted in catabolism is used to assemble the building block molecules into needed proteins, nucleic acids, lipids, and other new macromolecules for the cell.

Every cell is always engaged in both catabolism and biosynthesis, and normally these processes are carefully balanced. The supply of a specific macromolecule is matched closely to the demand for it; ideally, there is no shortage or excess. The fine control that this balance requires is remarkable, both at the cellular level and, on a larger scale, at the level of the whole animal. Consider, for example, that the average adult human can eat about six tons of food over a span of 40 years without losing or gaining weight and without changing the composition of the body.

Although cellular metabolism is a highly complex process, the central pathways are not difficult to understand. Moreover, they are essentially similar in most animal cells. This chapter examines some of the basic principles that drive all metabolic processes and their specific involvement in the metabolism of carbohydrates, fats, and proteins.

Basic Principles of Energy Transformation in the Cell

Cells obtain the energy they need to carry out work from their environment and, ultimately, release some of that energy back to the environment in the form of heat. Some of the ways by which living cells obtain and transform energy are depicted in Figure 6–1.

Figure 6–1

The flow of energy. Plants use radiant energy directly from the sun to photosynthesize high-energy organic molecules. Animals obtain these molecules by eating plants, and then transform their energy into ATP, which they use to do biological work.

Plants Use Radiant Energy to Synthesize Organic Molecules

Plants use light from the sun, a form of radiant energy, for the **photosynthesis** of ATP, the most versatile form of chemical energy (see Chapter 2). The ATP drives biosynthesis within plant cells, including the series of reactions known as **CO_2 fixation,** in which CO_2 from the atmosphere is used in the creation of carbohydrates and other high-energy organic com-

pounds. They are considered high-energy compounds because their many covalent bonds represent solar energy stored as chemical energy. Photosynthesis requires CO_2 and produces O_2 as a byproduct. The process can be summarized as:

$$CO_2 + H_2O + \text{radiant (solar) energy} \longrightarrow$$
low-energy
compounds

$$(CH_2O)_n + O_2$$
organic
molecule
(high-energy
compound)

Plants convert some newly synthesized carbohydrates, by a series of metabolic reactions, to the other types of high-energy organic molecules that they need.

Animals Obtain Chemical Energy from Plant Molecules

The organic molecules made by the plants are a source of chemical energy and building block molecules for the animals that eat the plants. Unlike plants, animals cannot use the energy from sunlight directly. They can obtain this energy only after its transformation by plants.

The organic "fuel" molecules that animals ingest when they consume plants are catabolized by reactions that require atmospheric O_2, and are termed **cellular respiration.** They produce CO_2 and water as byproducts and can be thought of as the reverse of photosynthesis:

$$(CH_2O)_n + O_2 \longrightarrow CO_2 + H_2O + \text{chemical energy}$$
high-energy   low-energy   (ATP)
organic   compounds
compounds

These complex reactions are directed by **enzymes** and designed to release the chemical energy of the organic molecules in a controlled fashion so that it can be "trapped" and stored as ATP. This is a form of chemical energy that cells can use for biosynthesis, active transport, and the generation of force and movement required for muscle contraction.

There Is a Constant Demand for ATP in Living Cells

The transformation of energy by animal cells is not a perfect process. Not all of the energy released from a fuel molecule can be used to form ATP. The untrapped energy is lost to the environment in the form of heat. Because they constantly need ATP and are continually losing heat energy, living cells require constant input of energy in order to survive. ATP is not a long-term storage form of chemical energy. It is used almost as soon as it is formed; the "turnover" of an ATP molecule in a cell is very rapid. At any moment, less than 1 g of ATP is present in the human body. However, in a person who is resting with minimal activity, about 45,000 g of ATP will be formed and used each day. Much more ATP will be required if the person undergoes strenuous physical activity during the day.

Many Factors Influence the Rate and Direction of Chemical Reactions in Cells

Free Energy Change in Cell Reactions

Recall from Chapter 2 that some reactions, called **exergonic reactions,** are highly likely to proceed spontaneously. The driving force for such reactions is the difference in total free energy between reactants and products. If a large difference exists—one in which the reactants contain much more free energy than the products—a large amount of useful energy will be released when the reaction occurs.

Other reactions, called **endergonic reactions,** cannot proceed without input of free energy. Both types of reactions occur in the cell. Endergonic reactions (in which ΔG is said to be positive) obtain the free energy they need by being **coupled** to exergonic reactions (in which ΔG is negative).

One of the first steps in glucose catabolism, for example, is to add a phosphate group (Pi) to carbon atom 6 in the glucose molecule, an endergonic reaction called a **phosphorylation:**

$$\text{Glucose} + \text{Pi} \rightarrow \text{Glucose-6-Pi}$$
$$\Delta G = +3 \text{ kcal/mol}$$

This reaction takes place because the Pi that is added comes from a molecule of ATP. The Pi group has been removed from ATP by a strongly exergonic reaction:

$$\text{ATP} \rightarrow \text{ADP} + \text{Pi} \qquad \Delta G = -7 \text{ kcal/mol}$$

In the coupling of these two reactions, the 7 kcal/mol of free energy that ATP gives up by losing a Pi is used to drive the endergonic phosphorylation of glucose. A summary equation shows the overall reaction and the net change in free energy:

$$\text{Glucose} + \text{ATP} \rightarrow \text{Glucose-6-Pi} + \text{ADP}$$
$$\Delta G = -4 \text{ kcal/mol}$$

The ΔG for the overall reaction is negative, indicating that the phosphorylation of glucose by this mechanism is thermodynamically feasible.

Reversible Reactions in the Cell

Reactions that are highly exergonic will proceed in one direction, toward the formation of products. Recall from Chapter 2, however, that many reactions are **reversible;** that is, the products may immediately recombine by the reverse reaction. The equation below describes a reversible reaction:

$$AB \rightleftharpoons A + B \qquad \Delta G = 0$$

Such a reaction is said to be in **equilibrium** when the rate of formation of the products ($A + B$) is exactly balanced by the rate at which the products recombine to yield the reactant (AB). In this situation there will be no net change in free energy. The smaller the difference in free energy between reactants and products, the less likely the reaction is to proceed to completion; that is, the more likely it is to be in equilibrium.

In an equilibrium reaction, it might be difficult to obtain enough of the Product B, since it tends to recombine with A as soon as it is formed. One way to ensure that such a reaction proceeds "to the right," toward "completion" or the formation of A and B, is to remove any B as soon as it is formed. Then AB will continue to break down, since there will be no B to recombine with A.

Similarly, a metabolic reaction can be driven to completion even though there is no difference in the free energy of the reactants and the products; that is, even though the formation of products is almost exactly balanced by the reformation of reactants. In the cell, metabolic reactions are driven to completion because the product of one reaction (such as B) is constantly being "removed" by participating in another reaction:

$$(1)\ AB \rightarrow A + B$$
$$\downarrow$$
$$(2)\ B + C \rightarrow D$$

Many reactions of metabolism, while not at equilibrium, take place with a very small change in free energy and can be driven in either direction relatively easily. This fact allows a cell to reverse a pathway of metabolism according to the specific demands of any given moment. For example, the synthesis of glucose from precursors takes place by reversal of many of the reactions that are used to catabolize it.

These last points emphasize one of the characteristic features of cellular metabolism: that no single reaction is taking place independently or in isolation. Hundreds of reactions are occurring in the cell at the same time, and most of them are interrelated; the products of one reaction are being used for the next reaction in a metabolic pathway. Even the so-called final products of one pathway are likely to be used in another pathway.

Oxidation–Reduction Reactions in the Cell

Recall from Chapter 2 that **redox reactions** are combinations of **oxidation reactions** (involving loss of electrons) with **reduction** reactions (involving gain of electrons). A compound becomes "reduced" by gaining electrons.

$$Ae^- \quad + \quad B \quad \rightarrow \quad A \quad + \quad Be^-$$

electron donor	electron acceptor	has been oxidized	has been reduced

Hydrogen: The "Electron" in Biological Redox Reactions. Organic compounds do not give up electrons easily. Removal of an electron during oxidation is achieved only by loss of an entire atom. The specific atom given up is almost always hydrogen, and so the oxidation process is actually a **dehydrogenation** reaction and the reduction process a **hydrogenation** process. In general, compounds that are highly reduced—contain many hydrogen atoms—are high-energy compounds. Oxidized compounds contain few hydrogen atoms and are low-energy compounds.

$$AH \quad + \quad B \quad \rightarrow \quad A \quad + \quad BH^-$$

electron donor	electron acceptor	has been oxidized	has been reduced
high energy	low energy	low energy	high energy

Photosynthesis: A Reduction Process. The formation of an organic molecule through photosynthesis can be thought of as a reduction process, in which CO_2 becomes reduced by gaining electrons in the form of hydrogen atoms:

$$H_2O \quad + \quad CO_2 \quad + \text{radiant energy} \rightarrow$$

electron donor	electron acceptor

$$O_2 \quad + \quad (CH_2O)n$$

has been oxidized	has been reduced
	high-energy organic molecule

At the same time, the H_2O becomes oxidized, losing electrons. The resulting O_2, an oxidized molecule, is a low-energy molecule.

Cellular Respiration: An Oxidation Process. Conversely, the breakdown of an organic molecule in cellular respiration can be thought of as an oxidation process (the removal of electrons in the form of hydrogen atoms) that produces the highly oxidized, low-energy compound CO_2:

$$(CH_2O)_n \quad + \quad O_2 \quad \rightarrow \quad CO_2 \quad + \quad H_2O \quad + \text{energy}$$

electron donor (high-energy organic compound)	electron acceptor	has been oxidized	has been reduced

Note that the electrons (H atoms) removed from the organic molecule are accepted by the O_2 molecule, forming water. Oxygen is therefore called the electron acceptor. (Since oxygen is the most common electron acceptor, the loss of electrons is called oxidation.) As we have said, organic fuel molecules such as glucose ($C_6H_{12}O_6$) are highly *reduced* compounds because they contain many hydrogen atoms. Thus, the catabolism, or breakdown, of these molecules is essentially an *oxidative* process, involving the removal of electrons in the form of those many hydrogen atoms. The hydrogen atoms often combine with oxygen to form water. For example, the complete catabolism of one glucose molecule uses up six molecules of oxygen and can be summarized as:

$$C_6H_{12}O_6 + 6\,O_2 \rightarrow 6\,CO_2 + 6\,H_2O$$

Before combining with water, however, the hydrogen atoms that glucose gives up during oxidation are transferred temporarily to molecules called coenzymes, chiefly NAD, which thereby becomes reduced. The reduced forms of NAD and other coenzymes in turn donate electrons to the **electron transport chain,** known also as the **respiratory chain,** which is embedded in the inner mitochondrial membrane (see Chapter 3). The flow of electrons along the respiratory chain occurs with the release of free energy that is used to drive endergonic synthesis of ATP (Fig. 6–2). This process will be discussed later in the chapter.

Figure 6–2

Energy transformation occurs in the mitochondria of animal cells through a series of oxidation-reduction reactions, in which electrons in the form of hydrogen atoms are transferred from highly reduced compounds such as NADH to highly oxidized compounds such as O_2. The energy released in these reactions is used to phosphorylate ADP to give ATP, the energy storage molecule.

Electron Acceptors: Key Molecules in Cellular Respiration. Electrons cannot simply be released but must be accepted by another molecule. Oxygen and coenzymes play critical roles in cellular respiration by accepting the electrons given off as organic molecules are oxidized. Coenzymes accept electrons in intermediate steps in the process. In the final steps, oxygen accepts electrons from the coenzymes.

Coenzymes: Intermediate Electron Acceptors. An example of a biological redox reaction that involves a coenzyme is the conversion of lactic acid into pyruvic acid:

$$
\begin{array}{ccc}
\text{COOH} & & \text{COOH} \\
| & & | \\
\text{CHOH} & \xrightarrow{\text{oxidation}} & \text{C=O} \quad + \quad 2\,\text{H} \\
| & & | \\
\text{CH}_3 & & \text{CH}_3 \\
\text{lactic acid} & & \text{pyruvic acid} \\
\text{electron} & & \\
\text{donor} & & \text{electrons}
\end{array}
$$

The freed hydrogen atoms are usually transferred to a coenzyme (so called because it works closely with an enzyme). The coenzyme most frequently employed as a hydrogen acceptor in metabolic reactions is **nicotinamide adenine dinucleotide,** or NAD (see Chapter 2; Fig. 6–8).

The formation of pyruvic acid by this reaction is a reversible process in the cell. Pyruvic acid can be converted back to lactic acid by the addition of two hydrogen atoms; by accepting the hydrogen atoms, the pyruvic acid becomes reduced:

$$
\underset{\substack{\text{electron} \\ \text{acceptor}}}{\text{Pyruvic acid}} \;+\; \underset{\text{electrons}}{2\text{H}} \;\xrightarrow{\text{reduction}}\; \text{Lactic acid}
$$

The hydrogen atoms for this reaction are provided by the same coenzyme (NAD) involved in the oxidation of lactic acid. Note that during the interconversion of lactate and pyruvate, the NAD is donating or accepting hydrogen atoms from the reactants, thereby becoming either oxidized or reduced. The process can be summarized as:

$$
\underset{\text{acid} \quad \text{NAD}}{\text{Lactic} + \text{Oxidized}} \;\rightleftharpoons\; \underset{\text{acid} \quad \text{NAD}}{\text{Pyruvic} + \text{Reduced}}
$$

This illustrates the point the oxidation and reduction of the organic molecules involved in cellular metabolism are true redox reactions. The oxidation of one molecule, such as lactic acid, is coupled to the reduction of another, such as NAD.

In some oxidative reactions involving removal of hydrogen atoms from an organic compound, the protons and electrons of the hydrogen atoms depart separately or become separated after removal of the atoms. Recall from Chapter 2 that the hydrogen atom contains 1 proton bearing a single positive charge and

1 electron with a single negative charge. Thus, a hydrogen atom that has given up its electron will have a single positive charge, due to the remaining proton, and can be represented as H^+. A single electron can be represented as e^-. By way of example, the oxidation of hydroquinone can be represented in two stages as follows:

$$
\underset{\text{reduced}}{\text{OH} \cdots \text{OH}} \;\xrightarrow{2H^+}\; \underset{\substack{\text{two protons} \\ \text{removed}}}{O^- \cdots O^-} \;\xrightarrow{2e^-}\; \underset{\substack{\text{two electrons} \\ \text{removed}}}{O \cdots O} \;\; \text{oxidized}
$$

Oxygen: The Ultimate Electron Acceptor. Organic molecules vary in their tendencies to give up or accept electrons. Their affinity for electrons can be measured in terms of their oxidation-reduction potential, better known as the **redox potential.** The redox potential of molecules is expressed on a scale in which the redox potential of hydrogen gas (H_2) is 0 millivolts (mV). A molecule that has a negative redox potential on this scale has a low affinity for electrons. A specific example is NADH, the reduced form of NAD. This molecule, with a redox potential of -320 mV, has a low affinity for electrons and will donate them to another electron carrier, thereby reverting to its oxidized form, NAD^+.

At the other end of the scale of redox potentials is molecular oxygen, which, with a redox potential of $+820$ mV, has a high affinity for electrons and will readily accept them, becoming reduced to water in the process. Molecular oxygen is the ultimate acceptor of the electrons donated by NADH to the mitochondrial respiratory chain (see Fig. 6–2).

High-Energy Phosphate Compounds

Earlier in this chapter, we discussed the use of the chemical energy stored in a molecule of ATP to drive an endergonic reaction such as the phosphorylation of a glucose molecule. ATP is an example of a special group of compounds known as **high-energy phosphate compounds** that have a strong tendency to transfer their phosphate to an acceptor molecule. This transfer is accompanied by the release of a large amount of free energy.

The ΔG for a given reaction is influenced by the concentrations of the reactants and products; it is not a very useful term for comparing the energetics of different reactions. For comparisons of reactions, the free energy difference between the reactants and the products is considered under a set of "standard conditions": temperature of 25°C, pressure of 1 atmosphere, and a concentration of 1.0 M for all reactants and products present. Under these conditions, the free energy change for a reaction is called the **standard free energy change** and is given the symbol $\Delta G°$. This term is a constant for a specific reaction. For example:

$$ATP \rightarrow ADP + Pi \qquad \Delta G° = -7 \text{ kcal/mol}$$

In the cell, however, the "standard conditions" are not present. The temperature is 37°C, and the concentration of ATP is very high relative to ADP and Pi. Under these conditions, the free energy change (ΔG) for ATP breakdown by the above reaction is about -12 kcal/mol. The large difference between $\Delta G°$ and ΔG in this example indicates that it is the ΔG that must be used to predict the direction of reactions within the cell.

The standard free energy change ($\Delta G°$) can be used to compare the tendency of different high-energy phosphate compounds to transfer their phosphate group to an acceptor molecule. The $\Delta G°$ values for the phosphorylated compounds found in cells range from about -2 to -13 kcal/mol. ATP is unique in that it occupies a central position in this scale (Fig. 6–3). All of the high-energy phosphate compounds for which $\Delta G°$ is more negative than ATP have a lower affinity for the phosphate group and can form ATP by donating their phosphate group to ADP:

$$ADP + Pi \rightarrow ATP$$

Compounds for which $\Delta G°$ is less negative than ATP have a greater affinity for the phosphate group and serve as acceptors for phosphate groups donated to them by ATP. Specific examples are given in Figure 6–3. Phosphoenolpyruvate and 1,3-diphosphoglycerate are intermediates formed during glucose ca-

tabolism. Phosphate groups removed from these compounds are transferred to ADP to form ATP. By a similar mechanism, phosphocreatine is used to replenish the ATP supply in skeletal muscle during periods of intense activity when the metabolic production of ATP cannot meet the demand. In fact, phosphocreatine is the major phosphorylated compound present in the resting muscle cell. The amount of phosphocreatine is three to eight times greater than the amount of ATP.

The direct formation of a high-energy phosphate compound such as ATP, by a chemical reaction that incorporates a phosphate group, is a process known as **substrate-level phosphorylation.** We shall see that this reaction does not require oxygen, in contrast to the mechanism for ATP formation by the electron transport chain in the mitochondrion (see Fig. 6–2). Phosphorylation of ADP in the mitochondrion is coupled to loss of electrons (oxidation) by the components of the electron transport chain and is termed **oxidative phosphorylation.** Most of the ATP supply for the cell is generated by oxidative phosphorylation. While ADP is the only molecule that can be phosphorylated by mitochondrial oxidative phosphorylation, substrate-level phosphorylation can generate other high-energy phosphate compounds such as GTP (by phosphorylation of GDP) and 1,3-diphosphoglycerate (by phosphorylation of glyceraldehyde-3-phosphate).

The large release of free energy that occurs upon removal of a phosphate group from a high-energy phosphate compound is a property of the whole molecule; the energy does not reside solely in the covalent bond by which the phosphate group is attached. There is nothing special about this bond, even though it may be referred to as a "high-energy bond." Several factors account for the free energy that is released. In the specific case of ATP, one important fac-

Figure 6–3

The standard free energy of hydrolysis of some phosphate compounds important in metabolism. Arrows indicate that phosphate groups can be transferred from high-energy donors to low-energy acceptors.

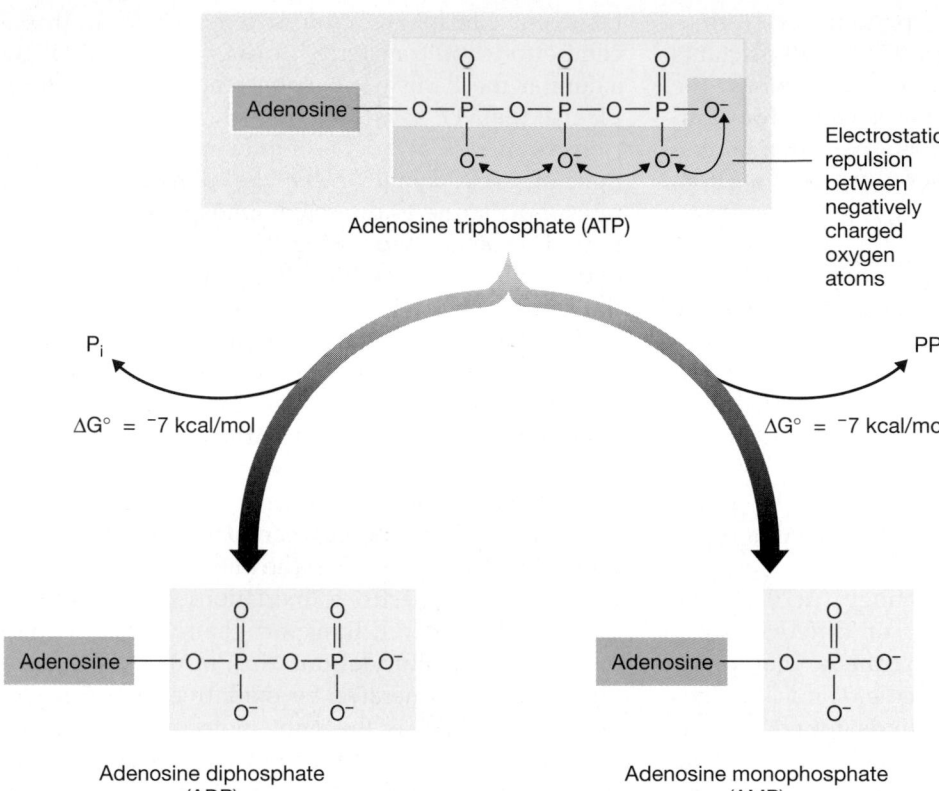

Figure 6–4

Release of phosphate (Pi) or pyrophosphate (PPi) groups reduces the electrostatic repulsion between negatively charged phosphate groups in adenosine triphosphate (ATP).

tor is the reduction of **electrostatic repulsion** that is achieved by loss of a phosphate group (Fig. 6–4). Removal of the terminal phosphate group yields ADP, while the removal of two phosphate groups (which occurs in some reactions) yields AMP. In either case the free energy released is the same.

Enzymes in Cellular Metabolism

In Chapter 2 we learned that an enzyme facilitates reactions by recognizing specific reactants and bringing them close together by binding them to its active sites. For many reactions, the physical orientation of the reactant molecules determines the rate of reaction; certain bonds such as peptide bonds cannot form unless the reacting molecules (e.g., amino acids) are oriented in a specific way (Fig. 6–5). By binding reactants in specific ways, enzymes ensure that they meet in the most favorable orientation for reactions to occur. For example, ATP and glucose are both found by the enzyme **hexokinase,** which facilitates the phosphorylation reaction in which a phosphate group is added to glucose. The enzyme in effect couples an exergonic reaction (removal of a phosphate

No reaction

Figure 6–5

Certain reactions require reactants to have a specific physical orientation. Peptide bonds, for example, will form only when the amino group of one amino acid is near the carboxyl group of another amino acid.

group from the ATP) to an endergonic reaction (addition of a phosphate group to the glucose). Glucose-6-phosphate and ADP are the reaction products.

Enzymes are also important in directing various products of cell reactions down the correct pathways. In Chapter 2 we discussed the ways in which enzymes lower the activation energy for specific reactions. Enzymes lower the activation energy only for specific desired reactions, so that at each step reactants are more likely to participate in those reactions than in others that might be possible.

Coenzymes.   **Coenzymes,** which have already been mentioned as intermediate electron acceptors in cell metabolism, are important features of enzyme action in cell processes. Coenzymes are nonprotein organic molecules that some enzymes require before they can catalyze reactions. **Coenzyme A,** for example, feeds 2-carbon acetyl groups (CH_3CO—) from glucose into the **citric acid cycle,** and NAD serves as a carrier of hydrogen atoms. Many coenzymes, such as NAD, are separated easily from the enzyme. NAD and its phosphorylated version NADP are derivatives of the vitamin **niacin,** while the coenzyme flavin adenine dinucleotide (FAD) is derived from another vitamin called **riboflavin.**

The cell must maintain all of the various metabolic substrates in the required concentrations, ultimately by controlling the rates of the chemical reactions that produce and consume them. All of these reactions are believed to require catalysis by enzymes, and one of the great advantages of an enzyme catalyst is that its activity in the cell can be controlled. One control mechanism, discussed already in Chapter 5, is covalent modification of the enzyme protein by attachment of a phosphate group, a reaction that is itself catalyzed by a specific **kinase** enzyme. Phosphorylation of an enzyme will either activate or inactivate it, and this change can be reversed by removal of the phosphate group with a **phosphoprotein phosphatase** enzyme (Fig. 6–6).

Allosteric Inhibition.   A specific enzyme usually catalyzes each of the distinct reactions of a pathway of metabolism. A control mechanism that regulates the output of an entire pathway is **end-product inhibition,** also known as **allosteric inhibition.** Typically an allosteric inhibitor is the final product of a metabolic pathway that "feeds back" to inhibit the enzyme catalyzing the *first step* of that pathway (Fig. 6–7a). When the first enzyme (e.g., Enzyme *A*) in the pathway is inhibited, the first reaction will not occur. Hence, no substrate will be available for the next reaction, catalyzed by Enzyme *B*, and so on. In this way, the rate of formation of the final product *X* will rapidly decrease when the first step is inhibited.

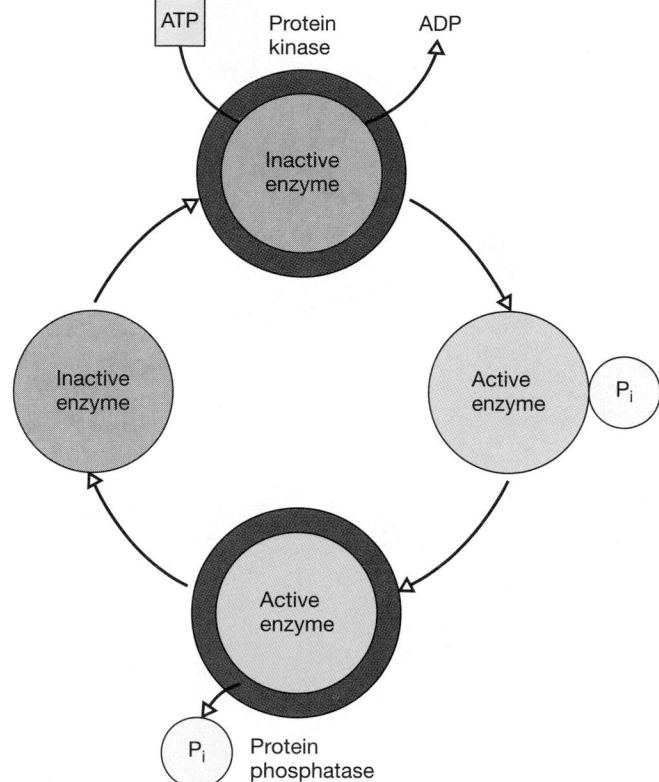

Figure 6–6

The phosphorylation of certain enzymes can activate them, whereas the removal of a phosphate group inactivates them.

An allosteric inhibitor inhibits an enzyme by binding to it at a site that is separate from the active site, causing a conformational change in the protein that impairs the function of the active site (see Fig. 6–7b).

Allosteric inhibition by the end product of a metabolic pathway is a reversible process. As formation of Product *X* declines, owing to inhibition of Enzyme *A* (see Fig. 6–7), the concentration of *X* in the cell will also decline because now it is used and consumed faster than it is produced. The concentration of *X* will fall until there is no longer enough of *X* to inhibit Enzyme *A*. When the inhibitor of Enzyme *A* is removed, the pathway leading to production of *X* will be "turned on" again.

Cells can also control enzyme-catalyzed reactions by increasing or decreasing synthesis of the enzyme proteins, a process that changes the amount of enzyme present in the cell. If a cell requires more of Product *X* produced by the pathway depicted in Figure 6–7, for example, it can synthesize more of Enzyme *A*, Enzyme *B*, and the other enzymes involved in this pathway. This type of control is relatively slow to take effect; it may take several hours or days for the enzyme levels to change significantly. It does not provide the rapid and fine control possible with covalent modification or allosteric inhibition.

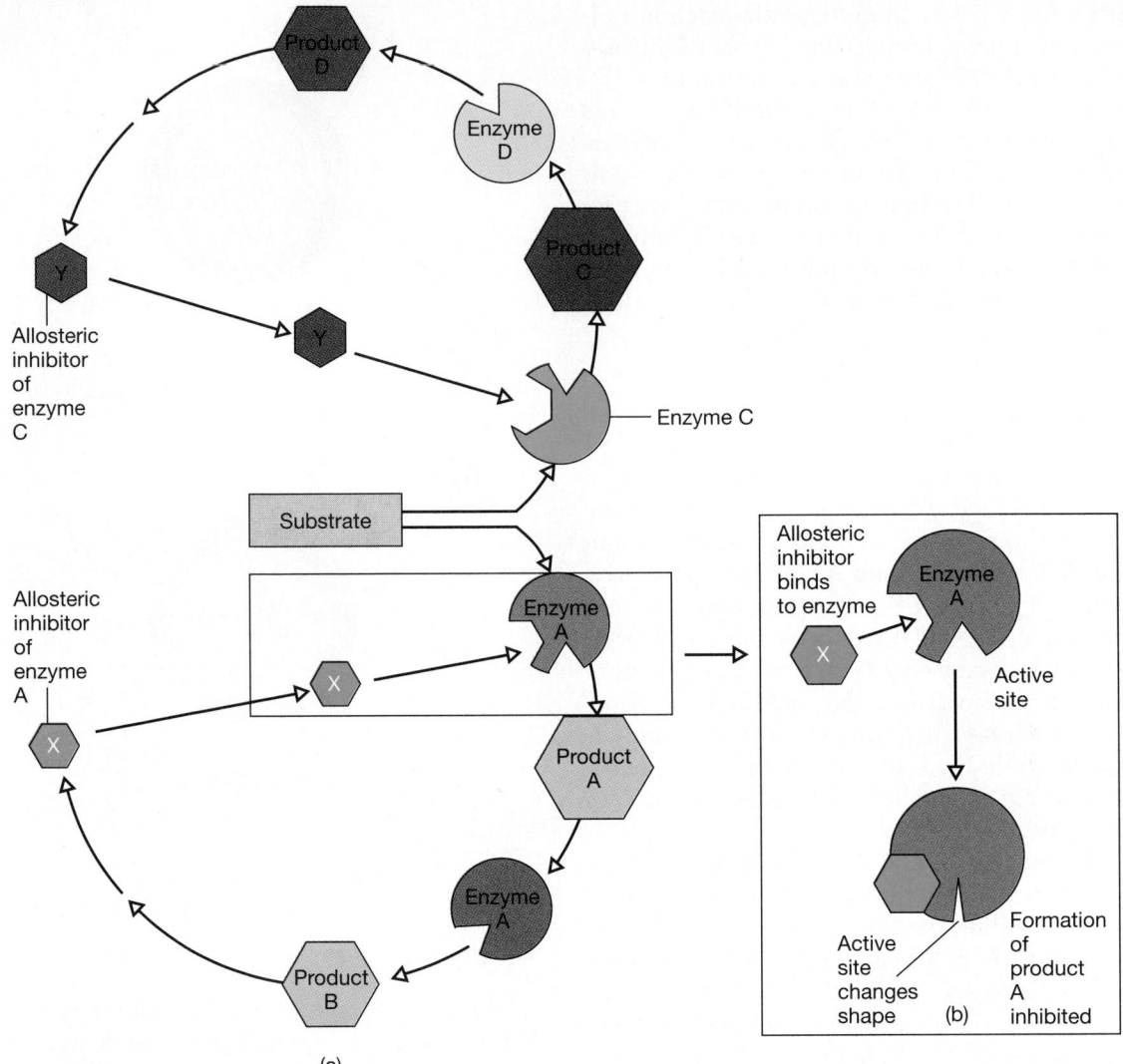

(a)

Figure 6–7

End-product (allosteric) inhibition of the initial steps in two metabolic pathways (*a*). Allosteric inhibitors typically bind to a site on the enzyme molecule that is different from the active site. The binding causes the active site to change shape, preventing the enzyme from catalyzing the first reaction in the pathway (*b*).

Most enzymes in the cell are not saturated by substrate and do not operate at their V_{max}. This provides the cell with the capacity to deal with fluctuations in the concentrations of its metabolites. Thus, when physical exercise rapidly increases the demand for glucose catabolism, for example, the enzymes involved have the capacity to respond by processing more substrate molecules in the same time.

Redox Reactions and Coenzyme Action. The oxidized forms of the coenzymes NAD and NADP have a single positive charge (Fig. 6–8) and, when reduced, they accept electrons and protons from two hydrogen atoms. Since a hydrogen atom consists of one

electron and one proton, two hydrogen atoms are equivalent to two electrons and two protons. NAD^+ and $NADP^+$ accept both electrons but only one of the protons:

$$\underset{\text{oxidized}}{NAD^+} + 2H^+ + 2e^- \rightarrow \underset{\text{reduced}}{NADH} + H^+$$

FAD, like NAD^+, can accept two electrons, but it also accepts two protons. In other words, FAD can accept two hydrogen atoms when it is being reduced (Fig. 6–8):

$$\underset{\text{oxidized}}{FAD} + 2H^+ + 2e^- \rightarrow \underset{\text{reduced}}{FADH_2}$$

(text continues on page 188)

Figure 6-8

The structures of common coenzymes. Arrows indicate the points that accept hydrogen atoms (NAD$^+$, NADP$^+$, and FAD) or acetyl groups (coenzyme A).

Recall that the reduction of these coenzymes is coupled to oxidation of a molecule such as glucose. NAD^+ and FAD are important coenzymes used in catabolic pathways such as the oxidation of glucose and other fuel molecules. $NADP^+$ is important in biosynthetic pathways of metabolism such as the synthesis of fatty acids.

An Overview of Catabolism: Dietary Proteins, Carbohydrates, and Fats

The simple act of consuming food is only a preliminary to the complex process by which the body obtains the energy and matter it needs to live. Even the complex process of digestion is not the ultimate destiny of food substances. Only the individual cells can enact the specialized reactions required to transform the energy and matter contained in food into a form the body can use.

Before examining in detail the complex processes by which cells transform matter and energy, let us look at all of them in general and discuss the ways in which they relate. We will first review the way in which cells obtain food molecules and what they do with those molecules before delivering them to the mitochondria, the organelles finally responsible for the release of energy.

The cells of the human body obtain energy from food that has been ingested in the form of dietary proteins, carbohydrates, and fats. These food substances are broken down in the digestive tract by the action of certain enzymes. Proteins consumed in the form of meat, for example, are broken down into amino acids. Carbohydrates eaten in the form of fruits, vegetables, and grains are broken down into

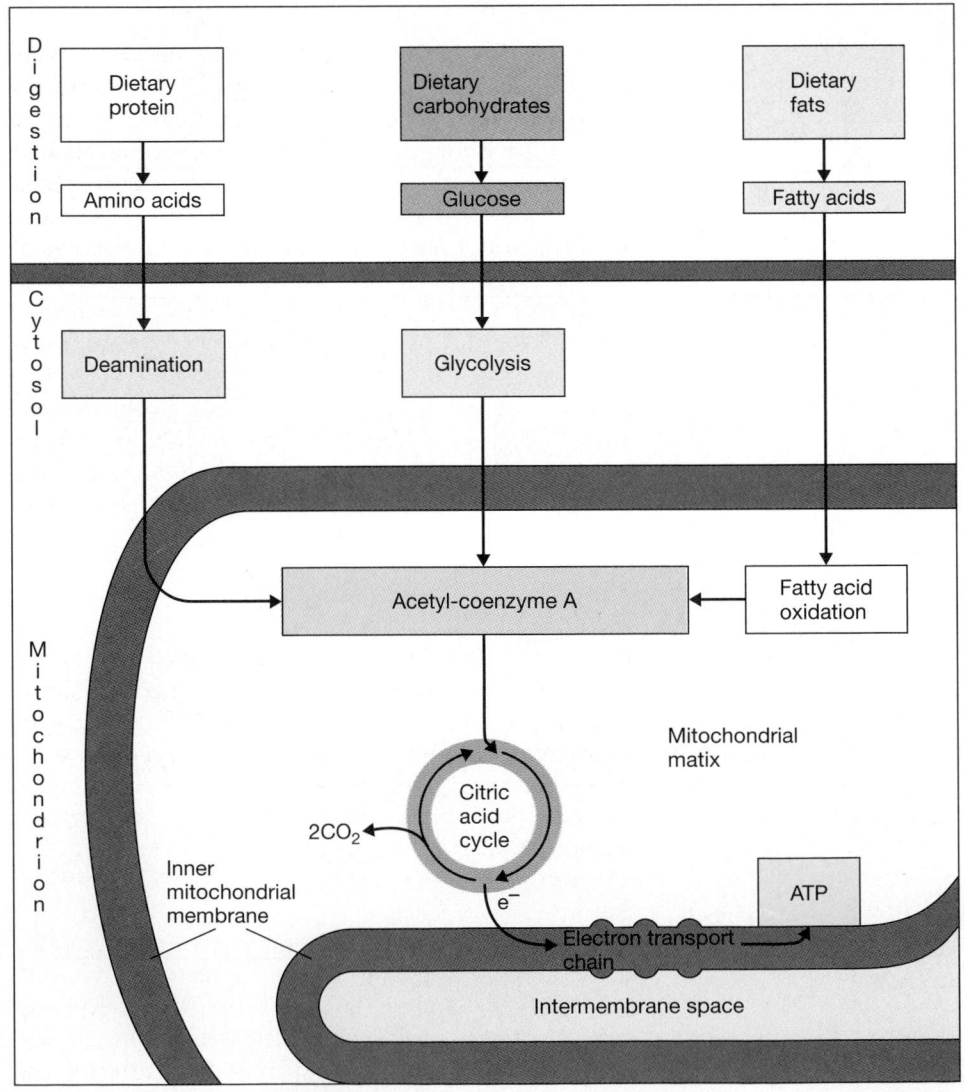

Figure 6–9

An overview of catabolism showing the locations of important processes. (The outer mitochondrial membrane has been omitted.)

simple sugars (monosaccharides) such as glucose. Dietary fats such as margarine and oils are degraded into fatty acids and glycerol. After digestion, these macromolecules, the products of digestion, circulate in the blood to all cells in the body (Fig. 6–9).

The chemical bonds that hold together these various macromolecules represent a store of energy the cells need to tap. They do so by obtaining the macromolecules from the blood and breaking them down. Each type of food molecule is broken down by a specific process. Amino acids obtained by the cell are broken down by **deamination,** which separates nitrogen atoms and carbon atoms. Monosaccharides (most commonly glucose) are broken down by **glycolysis.** These reactions occur in the cytosol of the cell. Fatty acids are broken down by a process known as **fatty acid oxidation** in the mitochondrial matrix (see Fig. 6–9). All of these processes are regulated and facilitated by enzymes.

Most of the molecules that result from the breakdown of amino acids, glucose, and fatty acids are converted to **acetyl (CH$_3$CO—) groups** in the matrix of the **mitochondrion,** the "powerhouse" of the cell. The acetyl groups are fed into a series of reactions known as the **citric acid cycle.** This cycle, occurring in the mitochondrion, is the final common pathway for the breakdown of all the fuel molecules in the cell (see Fig. 6–9).

In the mitochondrial matrix, all of the acetyl groups produced from food molecules are coupled to the terminal—SH group of coenzyme A (see Fig. 6–8) to produce **acetyl coenzyme A.** In this form, the acetyl groups are able to enter the citric acid cycle, which converts each acetyl group to two molecules of CO$_2$ (see Fig. 6–9). This conversion releases high-energy electrons that are temporarily trapped by the coenzymes NAD and FAD. The coenzymes later feed the electrons into the electron transport chain, which is a series of electron carriers, many of them proteins, embedded in the inner mitochondrial membrane. The redox reactions of the electron carriers release the energy used to synthesize ATP.

The final step in this process requires oxygen, making cellular respiration an **aerobic (oxygen-requiring) process.** The electron transport chain uses oxygen as the final acceptor of the electrons, a step that converts the oxygen to water. As the electrons pass from one electron carrier to the next, their energy is released in several small bursts that are used to drive the energy-dependent synthesis of ATP from ADP and inorganic phosphate (Pi). Thus, the ADP becomes phosphorylated; this step is tightly coupled to the flow of electrons. The electron flow will not occur unless the phosphorylation step can proceed. The loss of electrons, as we have said, is defined as oxi-

dation. Since electron transport consists of successive oxidations and is coupled to ADP phosphorylation, the entire process is known as **oxidative phosphorylation.**

Metabolism of Carbohydrates: The Oxidation of Glucose

The metabolism of carbohydrate involves three main processes: (1) **glycolysis,** in which glucose obtained from food is converted to pyruvic acid; (2) **the citric acid cycle,** in which pyruvic acid is oxidized, releasing electrons that are trapped by coenzymes; and (3) **oxidative phosphorylation,** also called **chemiosmotic synthesis** of ATP, in which electrons held by coenzymes are transferred to oxygen, releasing free energy that is stored in ATP.

Glycolysis Is the Conversion of Glucose to Pyruvate

The overall reaction that occurs in this metabolic pathway is the breakdown of a **glucose** molecule, a 6-carbon compound, into two molecules of **pyruvic acid,** a 3-carbon compound. This is accomplished with the net release of enough free energy to phosphorylate two molecules of ADP to ATP by substrate-level phosphorylation. In addition, two molecules of the coenzyme NAD$^+$ are reduced, providing the potential for more ATP synthesis when the reduced NAD molecules later give up electrons to the mitochondrial electron transport chain (see Fig. 6–2). The pathway of glycolysis is represented in condensed form in Fig. 6–10. This pathway consists of 10 reaction steps, each one catalyzed by a specific enzyme, and each one taking place in the cell cytosol. There are two stages to the pathway.

The First Stage of Glycolysis

The first stage (Fig. 6–11) consumes two ATP molecules for phosphorylating and converting the glucose molecule to a form that can be used to drive net synthesis of ATP. There are two phosphorylation steps (Steps 1 and 3), and neither can be reversed under the conditions that are present in the cell. The second phosphorylation (Step 3) is catalyzed by **phosphofructokinase** and is the primary point for controlling the rate of glycolysis.

Phosphofructokinase is regulated allosterically by both ADP and ATP, which have opposite effects: ADP stimulates the enzyme and ATP inhibits it. The significance of these opposing effects is as follows. When ATP is converted to ADP during biosynthesis, the

Figure 6–10

A summary of the pathway of glycolysis.

concentration of ADP will be high relative to ATP. The ADP will stimulate phosphofructokinase and increase the rate of glucose breakdown to generate more ATP. When the ATP supply is replenished, the relatively high ATP level compared to ADP will act to decrease the breakdown of more glucose by inhibiting phosphofructokinase.

The Second Stage of Glycolysis

The second stage is represented in Figure 6–12. Many of the intermediate products are shown in their ionized form, which is probably the form that the molecules assume in aqueous solution. Thus, pyruvic acid is shown as the **pyruvate** ion. The 6-carbon molecule from the first stage (fructose-1, 6-bisphosphate) is split into two 3-carbon molecules called **dihydroxyacetone phosphate** and **glyceraldehyde-3-phosphate** (Step 4). The molecule of dihydroxyacetone phosphate undergoes a rearrangement **(isomerization)** of its atoms and becomes glyceraldehyde-3-phosphate (Step 5). Thus, two molecules of glyceraldehyde-3-phosphate are ultimately derived from fructose-1,6-bisphosphate. Each molecule of glyceraldehyde-3-phosphate is converted first to *1,3-diphosphoglycerate* (Step 6) and then to *phosphoenolpyruvate* (Steps 7 to 9). These are high-energy phosphate compounds that can donate a phosphate group to ADP to generate ATP (see Fig. 6–3). Formation of 1,3-

Figure 6–11

The first stage of glycolysis.

Figure 6–12

The second stage of glycolysis.

diphosphoglycerate in Step 6 is not only a substrate-level phosphorylation; it is also an oxidation reaction. The protons (H^+) and electrons that are removed from glyceraldehyde-3-phosphate during oxidation are accepted by NAD^+, which is reduced to NADH.

The phosphorylation of ADP by phosphoenolpyruvate in Step 10 is essentially irreversible under the conditions present in the cell.

The second stage of glycolysis produces a total of two NADH molecules and four ATP molecules from each glucose molecule that is metabolized. Since two ATP molecules are consumed in the first stage, there is a net gain by the cell of two ATP molecules as a result of glycolysis.

The yield of free energy from glucose as it is catabolized by the glycolytic pathway to pyruvic acid is low. The standard free energy change ($\Delta G°$) is about -20 kcal/mol, and only two molecules of ATP are gained from the process. Compare this with the $\Delta G°$ of -686 kcal/mol when glucose is fully oxidized to CO_2 and water:

$$C_6H_{12}O_6 + 6\,O_2 \rightarrow 6\,CO_2 + 6\,H_2O$$

Glycolysis yields less than 3% of the maximum amount of free energy that can be obtained from glucose. However, it is important to realize that this pathway does not require molecular oxygen—that is, glycolysis can proceed under **anaerobic conditions.** This capability is important in tissues such as skeletal muscle when the demand for ATP rapidly increases at a time when oxygen is in low supply, such as during strenuous exercise. The skeletal muscle cells are able to meet their increased requirement for ATP by increasing the rate of glycolysis, producing ATP despite the fact that little or no oxygen is available.

However, under these anaerobic conditions, the pyruvic acid produced by glycolysis is converted to lactic acid (Fig. 6–13), which accumulates because it cannot be metabolized further. When exercise stops,

the lactic acid in the muscle cells is transferred to the liver, where it is converted back to pyruvic acid and broken down by aerobic metabolism to CO_2 and water. This requires additional oxygen, which is obtained by continued excessive breathing after muscular activity has ceased, a phenomenon know as **oxygen debt.**

Remember that two molecules of pyruvic acid are generated from every molecule of glucose that enters the glycolytic pathway (see Fig. 6–10). If the oxygen supply is plentiful—as is usually the case in most cells—the pyruvic acid produced by glycolysis is fed rapidly into the **citric acid cycle,** a pathway that takes place in the **mitocondrial matrix.**

The Citric Acid Cycle Provides Reducing Power as NADH and FADH₂

Pyruvic acid produced in glycolysis next enters the mitochondrial matrix, where it is coupled to **coenzyme A** in preparation for processing via the **citric acid cycle** (also called the **Krebs cycle** or the **tricarboxylic acid cycle**). The first step in the cycle is the formation of **acetyl coenzyme A.**

The Formation of Acetyl Coenzyme A

In a reaction catalyzed by **pyruvate dehydrogenase** (see Fig. 6–13), an acetyl group (CH_3CO—) is transferred from pyruvic acid to the —SH group of coen-

Figure 6–13

The metabolic fate of pyruvic acid. When oxygen is present, pyruvic acid reaches the citric acid cycle as acetyl coenzyme A. When oxygen is absent, pyruvic acid is converted to lactic acid. (CoA = coenzyme A.)

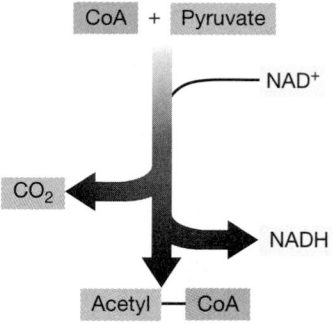

Figure 6–14

The formation of acetyl coenzyme A from pyruvic acid and coenzyme A (CoA).

zyme A to form **acetyl coenzyme A.** The acetyl group contains two of the three carbons present in pyruvic acid. (The third carbon of pyruvic acid is lost as CO_2, which is ultimately exhaled via the lungs.) The reaction also removes protons and electrons from pyruvic acid and coenzyme A; these are accepted by NAD^+, which is reduced to NADH. The net result of the pathway up to this point is that glucose, a compound containing 6 carbon atoms, is fed into the citric acid cycle as two 2-carbon units in the form of acetyl groups (Fig. 6–14).

The 2-carbon acetyl groups enter the citric acid cycle in a reaction catalyzed by **citrate synthetase,** which transfers the acetyl group from acetyl coenzyme A to a 4-carbon compound called **oxaloacetic acid.** One product of this reaction is the 6-carbon compound **citric acid** (hence the name for the cycle). The other reaction product is free coenzyme A, which becomes available to accept another acetyl group from another pyruvic acid:

$$\text{oxaloacetic acid} + \text{acetyl coenzyme A} \rightarrow$$
$$\underset{\text{4 C atoms}}{} \qquad \underset{\text{2 C atoms}}{}$$
$$\text{citric acid} + \text{coenzyme A}$$
$$\underset{\text{6 C atoms}}{}$$

The pathway of the citric acid cycle is a true cyclic process because the last reaction regenerates the 4-carbon oxaloacetic acid, which is free to accept an incoming acetyl group from acetyl coenzyme A. These steps are summarized in a simplified view of the citric acid cycle in Figure 6–15.

The Reduction of NAD^+ and FAD

The reactions of the citric acid cycle regenerate free coenzyme A and produce two molecules of CO_2 for every 2-carbon acetyl group that enters the cycle. The process can be summarized as:

$$CH_3CO\text{—coenzyme A} \rightarrow 2\,CO_2 + \text{coenzyme A}$$

The carbon atoms lost in one round of the cycle, in

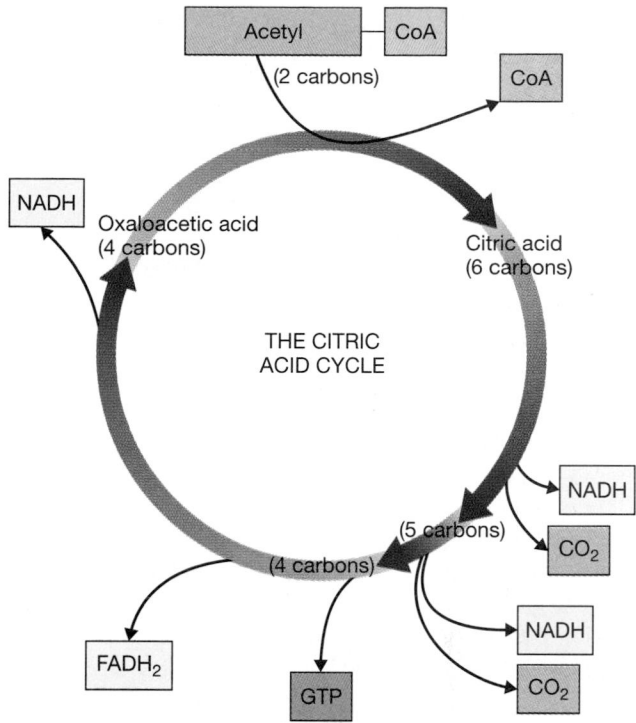

Figure 6–15

A summary of the citric acid cycle.

the form of CO_2 exhaled by the lungs, are those that entered as an acetyl group in a previous round. Since CO_2 is much more oxidized than the CH_3CO— (acetyl) group, the latter must become oxidized during the cycle. These oxidation steps are coupled to the reduction of the electron-carrying coenzymes NAD^+ and FAD. There are four such reactions that generate, from each acetyl group entering the cycle, a total of three molecules of NADH and one molecule of $FADH_2$. In addition, one high-energy phosphate molecule (GTP) is formed by substrate-level phosphorylation (see Fig. 6–15).

Each reaction of the citric acid cycle is shown in detail in Figure 6–16. There are nine consecutive steps beginning with the entry of an acetyl group (Step 1). In Steps 2 and 3, H and OH are interchanged to form **isocitrate.** Steps 4 and 5 are oxidation-reduction reactions. The net result is that the 6-carbon isocitrate is oxidized to the 4-carbon **succinyl coenzyme A** with the release of two CO_2 molecules. At the same time, two molecules of NAD^+ are reduced to NADH. Step 4 is catalyzed by **isocitrate dehydrogenase.** Step 6 is a substrate-level phosphorylation reaction that utilizes the free energy released by breakdown of succinyl coenzyme A to drive phosphorylation of GDP to GTP:

$$\text{succinyl coenzyme A} + \text{GDP} + \text{Pi} \rightarrow$$
$$\text{succinate} + \text{GTP} + \text{coenzyme A}$$

Figure 6–16

A detailed look at the citric acid cycle.

This is the only reaction in the cycle that produces a high-energy phosphate compound directly. The GTP formed in this reaction is used to phosphorylate ADP to produce ATP:

$$ADP + GTP \rightarrow ATP + GDP$$

The next three reactions involve the 4-carbon compounds of the cycle and culminate in the regeneration of oxaloacetic acid. There are two oxidation reactions (Steps 7 and 9) coupled to the reduction of one molecule of FAD to $FADH_2$ and one molecule of NAD^+ to NADH.

Molecular oxygen is not required for the reactions of the citric acid cycle, but it is essential for oxidation of the reduced coenzymes by the mitochondrial electron transport chain (see Fig. 6–2). A supply of oxidized coenzymes is needed for continued operation of the cycle. Hence the cycle cannot operate under anaerobic conditions because under such conditions all the coenzyme molecules remain in a reduced form. In contrast, a supply of NAD^+ is always available for anaerobic glycolysis in the cytosol because the reduction of pyruvate to lactate is coupled to oxidation of NADH to NAD^+ (see Fig. 6–13).

Allosteric Inhibition in the Citric Acid Cycle

The citric acid cycle contains several control points. The cycle is slowed when the ATP supply is adequate and accelerated when the ATP supply is low. When the supply of ATP is high, ATP itself acts as an allosteric inhibitor of citrate synthetase to reduce the rate of formation of citrate (see Step 1, Fig. 6–16) and slow down the cycle. When the supply of ATP is low

The Role of Oxygen in Tissue Injury

In most mammalian cells, oxygen is used largely by the mitochondria, the site where most of the ATP is synthesized. Oxygen is essential for efficient production of ATP, and during this production process (cellular respiration) most of the oxygen is converted to water. A small but significant amount (1% to 2%) of the oxygen consumed during cellular respiration is converted to highly reactive oxygen intermediates. These intermediates can injure and even kill cells, and most cells have developed protective mechanisms. The margin of safety is a narrow one, however, and the cellular defenses can be overwhelmed if the generation of oxygen intermediates is increased. It has been proposed that reactive oxygen intermediates may contribute to many illnesses, including inflammatory diseases, toxic reactions, cancer, and aging disorders.

Molecular oxygen is the terminal electron acceptor in the electron transport chain of the inner mitochondrial membrane (Fig. 6–17). The transfer of 4 electrons (and 4 protons) to molecular oxygen produces complete reduction to water, a safe product:

$$O_2 + 4e^- + 4H^+ \longrightarrow 2H_2O$$

In contrast, partial reduction generates hazardous compounds. For example, transfer of a single electron to oxygen produces the superoxide anion:

$$O_2 + e^- \longrightarrow O_2^-$$

Superoxide is highly reactive because addition of the electron to one of the oxygen atoms gives the oxygen an unpaired electron in its outer electron shell (see Chapter 2).

One of the mechanisms for protecting the cell from the action of superoxide anions involves the enzyme superoxide dismutase, which is present in all aerobic cells. This enzyme is refered to as a scavenger of oxygen intermediates because it coverts the superoxide anion to hydrogen peroxide and oxygen, as follows:

$$2O_2^- + 2H^+ \longrightarrow H_2O_2 + O_2$$

The hydrogen peroxide produced by this reaction is scavenged by the enzyme catalase, present in peroxisomes (Chapter 3). The products of catalase action are water and oxygen:

$$2H_2O_2 \longrightarrow 2H_2O + O_2$$

Peroxidase, an enzyme less common than catalase, scavenges hydrogen peroxide by using a reducing agent (AH_2):

$$H_2O_2 + AH_2 \longrightarrow 2H_2O + A$$

Hydrogen peroxide also may create the extraordinarily reactive hydroxyl radical, usually written as $\cdot OH$. This nomenclature denotes that there is an unpaired electron in the outer shell of the oxygen atom but that, unlike superoxide, there is no excess of electrons compared to protons, and hence the structure has no overall negative charge. The hydroxyl radical differs from the hydroxide ion because although hydroxide has an excess of electrons (hence an overall negative charge) there are no unpaired electrons. Production of hydroxyl radicals from hydrogen peroxide can arise from interaction with compounds containing iron, as follows:

$$H_2O_2 + Fe^{2+} \longrightarrow \cdot OH + OH^- + Fe^{3+}$$

In the normal liver cell, the average concentrations of these toxic molecules have been estimated to be 10^{-12} to 10^{-11} M for O_2^-, 10^{-9} to 10^{-7} M for H_2O_2, and 10^{-15} to 10^{-14} M for $\cdot OH$. Both hydrogen peroxide and superoxide can diffuse from their intracellular production sites and affect the entire cell. Hydrogen peroxide can also cross biological membranes. In contrast, the highly reactive $\cdot OH$ probably only interacts near its production site.

Unless destroyed by scavengers, these oxygen-derived molecules can injure and kill cells. They damage cell membranes by reacting with phospholipids, destroy enzymes and other proteins by oxidizing sulfhydryl groups, and induce breaks in the strands of DNA molecules.

Reactive oxygen intermediates have been shown to play a role in the development of injury to several different tissues, including the kidney. Studies on the cellular mechanisms of acute renal failure, defined as an abrupt decrease in kidney function, have revealed the possibility of increased production of reactive oxygen intermediates. A small proportion of patients who receive gentamicin, an antibiotic used to treat infection, develop acute renal failure. Studies in animals have revealed that gentamicin increases the kidney's production of superoxide and hydrogen peroxide. Furthermore, the presence of scavengers blocks the development of renal failure during gentamicin treatment. Thus, reactive oxygen intermediates appear to play an important role in acute renal failure induced by gentamicin.

Oxygen-derived radicals are not always bad news. Some cells in the body make use of these intermediates to carry out their normal functions. For example, phagocytic cells, an important part of the body's defense against invading organisms, make use of the toxic action of these intermediates to kill the invaders. The generation of reactive oxygen intermediates is markedly enhanced in phagocytes that have surrounded and engulfed a foreign organism. Furthermore, defective production of reactive oxygen intermediates by phagocytic cells results in the body's increased susceptibility to infections.

■ **If an oxygen intermediate were to oxidize the sulfhydryl groups of an enzyme important in the citric acid cycle, explain how this might kill the cell.**

Figure 6–17

The oxidation of NADH and FADH$_2$ by the electron transport chain. Each of three electron transport complexes has one phosphorylation site where enough free energy is released to synthesize one molecule of ATP. The ATP is not synthesized directly. The free energy released at each site is used to pump protons (H$^+$) across the inner mitochondrial membrane.

and the concentration of ADP relatively high, ADP acts as an allosteric activator of isocitrate dehydrogenase to increase the rate of production of alpha-ketoglutarate (Step 4) and speed up the cycle. These and other mechanisms cooperate to match precisely the rate of the reactions of the cycle to the requirement of the cell for ATP.

The reactions of the citric acid cycle generate two molecules of ATP (from GTP) when the two acetyl groups from one glucose molecule are processed. There is also a net gain of two molecules of ATP during glycolysis (see Fig. 6–10), which gives a total of four ATP molecules synthesized directly during the complete oxidation of glucose ($C_6H_{12}O_6$) to six molecules of CO_2. Most of the chemical energy of the glycose molecule is conserved, by transfer of electrons, in the reduced coenzymes. Two NADH are produced by aerobic glycolysis of glucose to two pyruvates (see Fig. 6–10). Two more NADH are produced during formation of two acetyl coenzyme A molecules (see Fig. 6–14), and six NADH are produced when two acetyl groups are fed into the citric acid cycle (see Fig.

6–15). Thus, the complete oxidation of a single molecule of glucose generates 10 molecules of NADH; in addition, two molecules of FADH$_2$ are formed. The subsequent transfer of electrons from these reduced coenzyme molecules to O_2 drives the synthesis of 32 more molecules of ATP.

Electron Transfer Is Coupled to Synthesis of ATP

The transfer of electrons from reduced coenzymes to O_2 takes place on the electron transport chain, which is an integral part of the inner membrane of the mitochondrion. These oxidation steps (loss of electrons) are coupled to phosphorylation of ADP to form ATP. The process produces most of the ATP supply for the cell and, since molecule oxygen is an essential reactant, it is an aerobic pathway called **oxidative phosphorylation**. The overall process by which oxidation phosphorylation generates ATP is called the **chemiosmotic synthesis of ATP**.

Cytochromes: Electron Carriers

The driving forces for oxidative phosphorylation are the negative redox potentials of NADH and $FADH_2$, which allow these molecules to readily donate electrons to proteins serving as **electron carriers** in the electron transport chain. Many of these electron carriers are proteins called **cytochromes,** and their redox potentials increase sequentially from the beginning to the end of the chain.

The importance of this increase in redox potentials is that each cytochrome has a greater affinity for electrons than the previous one in the chain. This increase facilitates the passage of electrons from one cytochrome to the next, and ultimately, to oxygen. Each cytochrome becomes reduced when it accepts electrons and oxidized when it donates the electrons to the next cytochrome. The components of the electron transport chain are organized into three major complexes (Fig. 6–17).

The flow of electrons from NADH to molecular oxygen represents an oxidation-reduction reaction that can be summarized as:

$$1/2 \, O_2 + NADH + H^+ \rightarrow H_2O + NAD^+$$

The standard free energy change $\Delta G°$ for this reaction is -53 kcal/mol. It is an exergonic reaction; the free energy released is used to drive the endergonic phosphorylation of ADP to produce ATP (see Fig. 6–4). Under standard conditions, about 7 kcal/mol of free energy can be stored in one ATP molecule, so the free energy released by the above reaction could be stored in seven molecules of ATP.

Phosphorylation Sites

The electron transport chain ensures that the free energy is not released all at once, but even so, not all of this free energy can be used to synthesize ATP. This is because under physiological conditions there are only three sites in the chain where the transfer of electrons releases enough free energy to synthesize a molecule of ATP. These sites are termed **phosphorylation sites,** and one is present in each of the three electron transport complexes (see Fig. 6–17). Thus, in the cell, three molecules of ATP are synthesized for each molecule of NADH that is oxidized in the electron transport chain.

The redox potential of $FADH_2$ is less negative than that of NADH, and it cannot donate electrons to the beginning of the electron transport chain in Complex I. $FADH_2$ donates its electrons to the chain at a point between Complexes I and II; they do not pass through the first phosphorylation site (see Fig. 6–17). Just two ATP molecules are synthesized when each molecule of $FADH_2$ is oxidized.

The Flow of Protons: The Chemiosmotic Mechanism

The free energy released at the phosphorylation sites is harnessed by the electron transport complexes in order to pump protons (H^+) from the mitochondrial matrix to the intermembrane space (Fig. 6–18). The movement of positively charged protons, unaccompanied by negatively charged anions, sets up an electrical gradient across the inner mitochondrial membrane. In addition, the accumulation of protons in the intermembrane space builds a chemical gradient. The result of the net electrochemical gradient is a passive flow of protons from the intermembrane space back to the matrix. This flow can occur only through a specific protein in the inner membrane that bears the enzyme **ATP synthetase.** Just as the flow of sodium ions down an electrochemical gradient can be used to do "work" such as moving another solute against its

Figure 6–18

The chemiosmotic mechanism of ATP synthesis. Protons (H^+) are pumped from the matrix to the intermembrane space, then returned to the matrix by passive flow down the electrochemical gradient. ATP synthetase uses the passive proton flow to drive the synthesis of ATP.

Table 6–1

Production of 36 ATP Molecules by Complete Oxidation of One Glucose Molecule (2 molecules of pyruvic acid are produced from each glucose molecule and both enter the citric acid cycle)

Glycolysis (cytosol):
    2 **ATP** and 2 NADH
Shuttle (inner mitochondrial membrane):
    2 mitochondrial $FADH_2$ produced from 2
    cytosolic NADH
Conversion of pyruvate to acetyl CoA
(mitochondrial matrix):
    2 NADH
Citric acid cycle (mitochondrial matrix):
    2 **ATP** (from 2 GTP), 6 NADH and 2 $FADH_2$
Oxidative phosphorylation (mitochondria):
    24 **ATP** from 8 mitochondrial NADH
    8 **ATP** from 4 mitochondrial $FADH_2$
Total: 36 **ATP**

electrochemical gradient (secondary active transport, Chapter 4), the flow of protons is coupled by ATP synthetase to drive the endergonic synthesis of ATP from ADP and a phosphate group (see Fig. 6–18). In other words, the ATP synthetase transforms the energy of the proton electrochemical gradient into chemical energy in the form of ATP.

The proton pumps use the free energy released by electron transport to maintain the proton gradient, which in turn maintains the flow of protons through ATP synthetase. This is known as the **chemiosmotic mechanism** of ATP synthesis because it links chemical reactions to transport processes. The newly formed ATP is transferred out of the mitochondrion and used to drive metabolic reactions in the rest of the cell.

A "balance sheet" summarizing the pathways involved and the yield of 36 ATP molecules from complete oxidation of one glucose molecule is presented in Table 6–1. Only one additional comment is needed: the inner mitochondrial membrane is impermeable to NADH (and NAD^+), so the cytosolic "pool" of this coenzyme never mixes with the mitochondrial pool. Cytosolic NADH produced by glycolysis is oxidized by a shuttle mechanism that, in many cell types, transfers electrons and protons to FAD inside the mitochondrion. As a result of this mechanism, each cytosolic NADH molecule ultimately produces only two ATP molecules.

The Pentose Phosphate Pathway Is an Alternative for Glucose Oxidation

The purpose of glucose oxidation, via glycolysis and the citric acid cycle, is to produce ATP as an energy source for biosynthesis. Another pathway for the oxidation of glucose exists in the cytosol, called the **pentose phosphate pathway** or the **hexose monophosphate shunt.** The main purpose of the pentose phosphate pathway is to generate NADPH, the reduced form of the coenzyme $NADP^+$ (see Fig. 6–8), to drive the oxidation-reduction reactions in various biosynthetic pathways such as the synthesis of fatty acids and steroids. It also provides the pentose sugars that are essential components of nucleotides and nucleic acids (see Chapter 2). This process uses different enzymes from those used in glycolysis.

The pathway starts with glucose-6-phosphate, a hexose. After several reactions, the final result is the reduction of two molecules of $NADP^+$ to NADPH and the formation of one molecule of ribose-5-phosphate, a pentose (Fig. 6–19). The carbon atom removed from glucose-6-phosphate is lost as CO_2. Excess ribose-5-phosphate not needed for biosynthetic reactions is recycled by conversion to fructose-6-phosphate and glyceraldehyde-3-phosphate, two of the intermediates of the glycolytic pathway. These intermediates can be used to synthesize glucose-6-phosphate, and the cycle begins again. Six rounds of the cycle result in the complete oxidation of one molecule of glucose-6-phosphate to six molecules of CO_2 with the generation of NADPH:

$$\text{glucose-6-phosphate} + 12\,NADP^+ \rightarrow$$
$$6\,CO_2 + 12\,NADPH$$

In contrast to aerobic oxidation of glucose, the pentose-phosphate pathway does not use ATP, oxygen, or the citric acid cycle. It does not produce ATP, and the only electron acceptor involved is $NADP^+$. Which of these pathways is used for glucose catabolism seems to depend on whether or not the cell is heavily engaged in biosynthesis. Skeletal muscle, for example, catabolizes glucose solely via glycolysis and the citric acid cycle. Adipose and liver tissue, on the other hand, catabolize a large fraction of glucose via the pentose-phosphate pathway. These tissues are important sites of fatty acid synthesis, a reaction pathway that consumes a large amount of NADPH and is located in the cytosol where the pentose phosphate pathway also occurs.

Note the markedly different roles of NADH and NADPH. NADH is oxidized by the electron transport chain to generate ATP, whereas NADPH is oxidized directly in biosynthetic reactions.

Figure 6–19

Figure 6–19

A summary of the pentose phosphate pathway.

Glucose

Glucose-6-phosphate
(A hexose)

Fructose-6-phosphate

Glyceraldehyde-3-
phosphate

$2NADP^+$

2 NADPH

CO_2

Ribose-5-
phosphate
(a pentose)

Recycled
via
gluconeogenesis

Metabolism of Fats

Fats stored in adipose tissue provide a rich energy source for the body because they can be oxidized to produce ATP. The body can store more fat than carbohydrate, and, as we shall see, much more energy can be obtained from fats than from carbohydrates. The amount of ATP produced by a given weight of fat exceeds that produced by the same weight of carbohydrates. Since a mobile animal, unlike a stationary plant, must carry its energy store with it, the amount of ATP available per unit weight of the energy store is an important factor. This is a major reason that animals use fat as an energy store.

Catabolism of a neutral fat, or **triacylglycerol,** molecule begins with the release of the glycerol backbone and the three long-chain fatty acids. Glycerol is readily converted to glyceraldehyde-3-phosphate,

which can be used for glycolysis. The fate of the fatty acids is illustrated in Figure 6–20, using palmitic acid as an example. Coenzyme A is attached to the carboxyl (—COOH) group and, in a series of complex reactions within the matrix of a mitochondrion, a fragment of two carbon atoms is removed. This fragment, which also contains coenzyme A, is actually a molecule of acetyl coenzyme A (see Fig. 6–20). The 16-carbon chain of palmitic acid is reduced to 14 carbons by the removal of acetyl coenzyme A. The 14-carbon fatty acid is recycled through the same series of reactions and another two carbon atoms are removed as acetyl coenzyme A. In this way, the original long-chain fatty acid is converted into several molecules of acetyl coenzyme A. While NADH and $FADH_2$ are produced by this series of reactions, no ATP is generated. All the ATP derived from fatty acid catabolism is produced when the acetyl coenzyme A

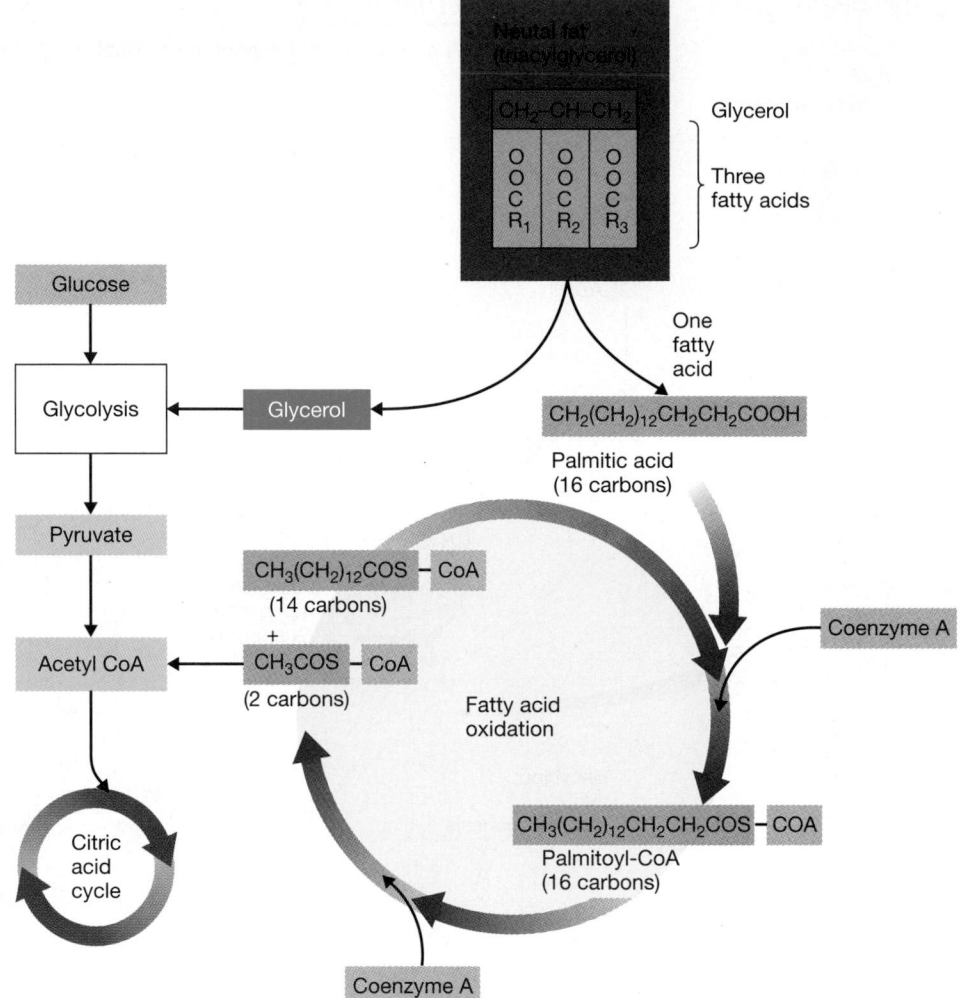

Figure 6–20

The catabolism of fatty acids to acetyl coenzyme A (acetyl CoA). The long hydrocarbon chains of fatty acids are represented by the symbols R_1, R_2, and R_3.

groups are fed into the citric acid cycle (see Fig. 6–20). The catabolism of palmitoylcoenzyme A, the activated form of palmitic acid, can be summarized as follows:

$$\text{palmitoyl-coenzyme A} + 7\ \text{FAD} + 7\ \text{NAD}^+$$
$$+ 7\ \text{coenzyme A} \rightarrow$$
$$8\ \text{acetyl coenzyme A} + 7\ \text{FADH}_2 + 7\ \text{NADH}$$

When this yield of acetyl coenzyme A and reduced coenzymes is processed via the citric acid cycle and electron transport, it can be calculated that 131 ATP molecules will be generated. Since two high-energy phosphate bonds are used to activate the fatty acid (attachment of coenzyme A), the net gain from oxidation of 1 molecule of palmitic acid is 129 molecules of ATP. This exceeds by far the 36 ATP molecules generated by oxidation of one molecule of glucose, rein-

forcing the point that fats are a much richer energy source than carbohydrates.

Fatty acids are synthesized in a different subcellular compartment (the cytosol) and by a different pathway from that used for this catabolism. The pathway begins with acetyl coenzyme A, and the carbon chain is elongated by a succession of reactions that add two carbon atoms at a time. The energy source for these reactions is ATP, with NADPH from the pentose-phosphate pathway providing reducing "power" for oxidation-reduction reactions. Synthesis of palmitic acid can be summarized by these equations:

$$8\ \text{acetyl coenzyme A} + 7\ \text{ATP} + 14\ \text{NADPH} \rightarrow$$
$$\text{palmitic acid} + 7\ \text{ADP} + 14\ \text{NADP}^+ + 8\ \text{coenzyme A}$$

Metabolism of Proteins

The building blocks of proteins are amino acids. In contrast to glucose and fatty acids, which can be stored until needed, excess amino acids cannot be stored and instead are used as a metabolic fuel. Many of the amino acids are catabolized by removal of the amino group ($-NH_2$) in the form of ammonia, a process known as **deamination.** Ammonia is potentially toxic and so is converted to urea and excreted from the body in the urine (Fig. 6–21).

The carbon atoms of the amino acids can enter the pathway of glucose oxidation, where they are converted to CO_2 and water with the generation of ATP. Some amino acids enter the pathway as pyruvate, some as acetyl coenzyme A, and others as intermediates in the citric acid cycle (see Fig. 6–21).

About half of the basic set of 20 amino acids required by humans are described as **essential amino acids** because they cannot be synthesized; they must be ingested as food. The other needed amino acids are termed **nonessential amino acids** because they can be synthesized by relatively simple pathways.

Some of the steps by which amino acids are incorporated into a polypeptide chain during protein synthesis require a source of energy. ATP provides the energy needed for attachment of each amino acid to its tRNA molecule, and GTP is the energy source required to move a ribosome from one mRNA codon to the next (see Chapter 5).

In addition to their use in protein synthesis, certain amino acids are the precursors for a number of important molecules. These molecules include the **purines** and **pyrimidines** used for synthesis of DNA and RNA. **Histamine** is a powerful locally acting vasodilator synthesized from the amino acid histidine. The hormones **epinephrine** and **thyroxine** are formed from the amino acid tyrosine, and the nicotinamide ring of NAD^+ is derived from the amino acid tryptophan.

Acetyl Coenzyme A: Mediating Between Catabolism and Biosynthesis

It should be possible to appreciate from this discussion of cellular metabolism that the citric acid cycle, together with mitochondrial electron transport, is the principal pathway for supplying the cell with ATP. Very little of the chemical energy present in the macromolecules of the cell is made available until small units of these macromolecules are fed into the citric acid cycle (Fig. 6–22).

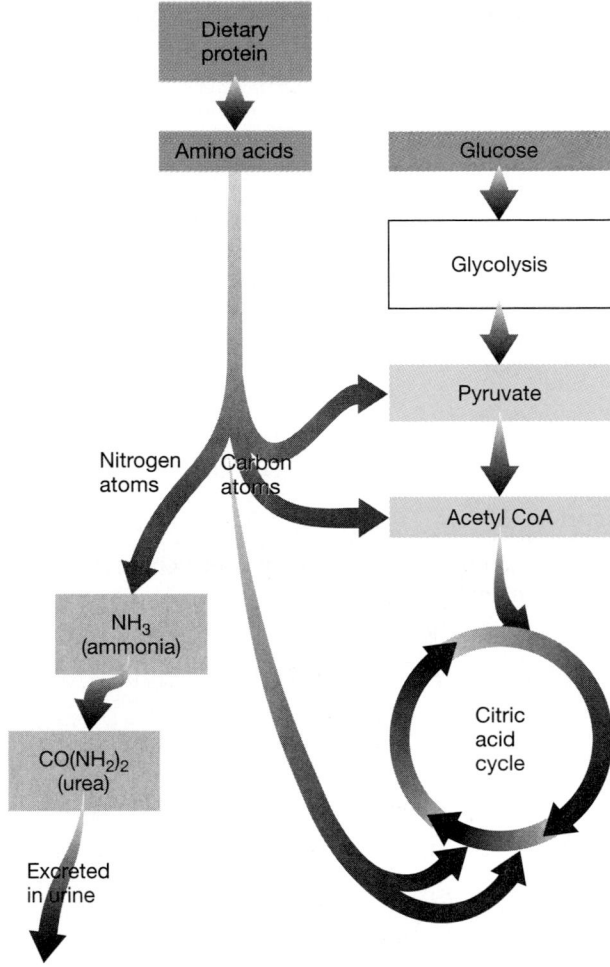

Figure 6–21

During catabolism, carbon atoms from amino acids are used to form pyruvate, acetyl groups, and various products of the citric acid cycle. Nitrogen atoms from amino acids are eventually excreted from the body in the urine.

Although the carbon fragments of several amino acids can enter the citric acid cycle directly, almost all the other intermediate compounds produced by catabolism are fed into the cycle as acetyl groups coupled to coenzyme A. The acetyl groups are also the 2-carbon precursors used in the synthesis of fatty acids and cholesterol. Steroid hormones are derived from cholesterol. Thus, coenzyme A is required not only to derive energy from fuel molecules, but also to provide the many carbon atoms required for biosynthesis of a large variety of organic molecules. Carbon atoms derived from catabolism of fats and proteins can be shunted through acetyl coenzyme A and utilized in biosynthesis (see Fig. 6–22), depending on the metabolic demands at any given time. In other words, acetyl coenzyme A provides a place for the different metabolic pathways of the major macro-molecules of the cell to intersect. Note that in mammals, acetyl coenzyme A cannot be converted to pyruvate; hence fats cannot be converted to carbohydrates.

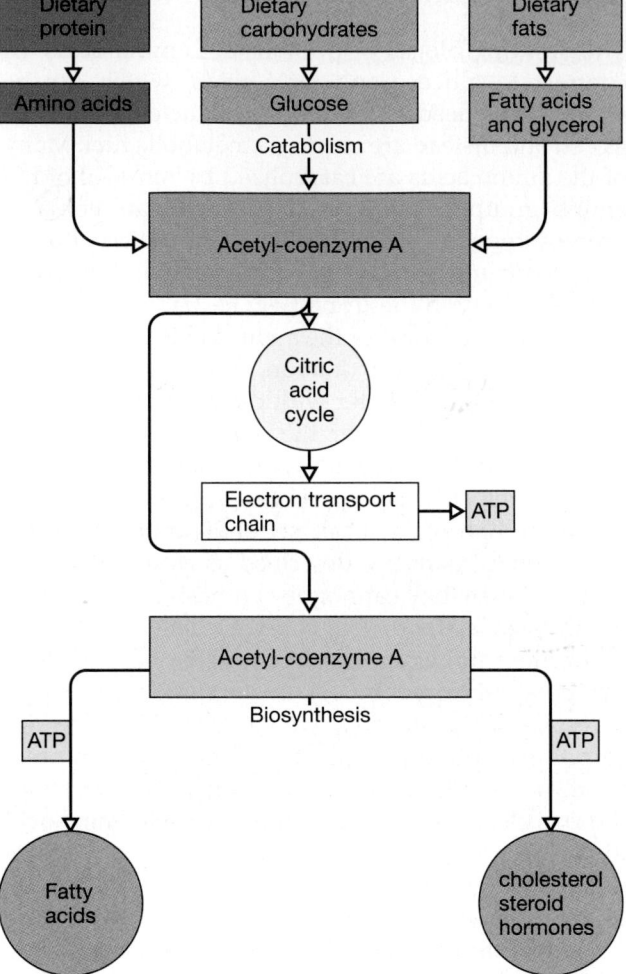

Figure 6–22

An illustration of the central role of acetyl coenzyme A in cellular metabolism of mammals.

CHAPTER REVIEW

Summary

- Plants transform radiant energy from the sun into carbohydrates, a form of chemical energy that animals can use.
- Animals degrade the carbohydrates and transform the chemical energy into ATP. The ATP is used for processes such as biosynthesis and movement.
- ATP turnover in an animal cell is very rapid. Some of the chemical energy of a fuel molecule cannot be trapped as ATP and is released as heat.
- Endergonic reactions in the cell are driven by free energy released from exergonic reactions. All chemical reactions

in the cell are catalyzed by enzymes, and many reactions are reversible.
- The citric acid cycle and electron transport chain in the mitochondrion are the final common pathways in the catabolism of all fuel molecules, whether amino acids, glucose, or fatty acids.
- The reactions of glycolysis split one glucose molecule into two molecules of pyruvic acid and generate two molecules of ATP.
- When one pyruvate is fed into the citric acid cycle as

CLINICAL FOCUS

Heat Generation by Brown Fat

Some of the energy released by oxidation of food molecules cannot be used to drive synthesis of ATP in the mitochondria. Although it is released as heat, it is not wasted. It is a critical source of heat for humans and other warm-blooded animals that maintain their internal temperature higher than that of their environment. In some animals, including humans, the ability of newborns to generate heat **(thermogenesis)** is vital for survival. Thermogenesis is also important for mammals during arousal from hibernation and for adapting to cold climates. Mammals are equipped for thermogenesis by the presence of brown fat **(adipose)** tissue in which fatty acids are the primary source of fuel for oxidative metabolism. Fat cells are termed **adipocytes,** and brown adipocytes are characterized by the presence of high numbers of mitochondria. Furthermore, the inner membrane of these mitochondria contains very high amounts of a specific protein known as the **uncoupling protein** (UCP). The UCP facilitates the return of H^+ to the mitochondrial matrix by facilitated diffusion across the inner membrane, thus bypassing the ATP synthetase. The diffusion is rapid enough to reduce the H^+ gradient without dissipating it completely. The leak of H^+ removes the controls on oxidative phosphorylation that are dependent on the relative amounts of ADP and ATP, and electron transport runs at a high rate whether ATP is needed or not. Consequently, the free energy produced by oxidation of fatty acids in brown adipocytes is "uncoupled" from synthesis of ATP and is released as heat. This is carried from the brown adipose tissue by the circulatory system and distributed to other parts of the body. Brown adipose tissue may represent only a small proportion of the body weight, but the thermogenesis is so efficient that body temperature can be maintained without shivering.

The level of UCP *in vivo* can be modulated by various physiological stimuli. A prominent example is the exposure of rats to cold, which triggers a marked increase in the UCP level and converts the mitochondria of brown adipocytes to heat-producing machines. It was found that exposure of rats to 4°C for just 15 min produced a 10-fold increase in the rate of transcription of the UCP gene. The mechanisms that control expression of the UCP gene in large mammals, including humans, change soon after birth. The UCP level and the amount of brown fat decrease during the first few days of life. The gene remains responsive, however, because the amount of brown adipose tissue and its activity in adult humans can be changed by environmental conditions. Overeating, for example, stimulates the thermogenic response, which may be a device to rapidly oxidize excess fuel molecules so that they are not stored as fat. In contrast, exposure to a hot climate switches off the thermogenic response and promotes degeneration of brown adipose tissue.

■ **Explain the effect an uncoupling agent will have on the rate of oxygen consumption by a fat cell.**

acetyl coenzyme A, the cycle generates three NADH, one FADH$_2$, and one GTP.
- The oxidation of reduced coenzymes by the electron transport chain drives phosphorylation of ADP to form ATP, a process known as oxidative phosphorylation. Complete oxidation of one glucose molecule generates 36 ATP molecules.
- Glucose oxidation by the pentose phosphate pathway generates NADPH, which is needed to drive redox reactions in biosynthetic pathways.
- Fats are a much richer energy source than carbohydrates.

Oxidation of one molecule of palmitic acid generates 129 molecules of ATP.
- The carbon atoms of amino acids are fed into various points of the glucose oxidation pathway. The nitrogen atoms are eventually excreted in the urine in the form of urea.
- The mitochondria, via the citric acid cycle and electron transport chain, are the principle source of ATP in the cell. Acetyl coenzyme A in the mitochondrial matrix is the central molecule that connects the different metabolic pathways.

Review Questions

Identify Each with a Term

1. The general term for the energy-requiring process by which simple molecules are assembled into new macromolecules needed by the cell.
2. A chemical reaction that cannot proceed without input of free energy.
3. The coenzyme used most frequently as a hydrogen acceptor in metabolic reactions.
4. The metabolic pathway by which glucose can be synthesized from molecules that are not carbohydrates.
5. The compound excreted in urine that represents the means for disposing of nitrogen produced by deamination of amino acids.

Define Each Term

6. Chemiosmotic mechanism
7. Pentose phosphate pathway
8. Oxidation phosphorylation
9. Allosteric inhibitor
10. High-energy phosphate compound

Choose the Correct Answer

11. When normal metabolic production cannot meet the demand for ATP in a skeletal muscle cell (e.g., during intense activity), new ATP is provided by another mechanism using:
    a. phosphocreatine
    b. GTP
    c. glucose-6-phosphate
    d. lactate dehydrogenase
12. Complete oxidation of carbohydrates such as glucose requires:
    a. oxidative phosphorylation
    b. glycolysis
    c. citric acid cycle
    d. all of the above
13. An organic molecule that functions in metabolism as an acceptor of acetyl groups is:
    a. NADP
    b. FAD
    c. coenzyme A
    d. pyruvate

14. How many molecules of ATP are synthesized for each molecule of NADH oxidized in the electron transport chain?
    a. 1
    b. 2
    c. 3
    d. 7

Calculate

15. How many molecules of ATP are gained from the complete metabolism of one molecule of pyruvic acid (A) under anaerobic conditions and (B) under aerobic conditions?
16. How many molecules of (A) NADH and (B) $FADH_2$ are gained from the complete oxidation of one molecule of palmitic acid, a 16-carbon fatty acid?

Answer Each Question in One or Two Sentences

17. What is the fundamental difference between the roles of the reduced coenzymes NADH and NADPH?
18. In what metabolic processes does acetyl coenzyme A participate?
19. How does ATP slow down the citric acid cycle?
20. What are the major steps in the process by which animals are able to use energy from the sun to synthesize the macromolecules they need?
21. Why do animals use fat as an energy store?

Problem Solving in Physiology

22. A specific cell in a multicellular organism synthesizes a molecule of glucose that is destined for use by another cell at a distant location in the organism. Explain how movement of this glucose molecule from the synthesizing cell to the utilizing cell could represent a flow of mass, or energy, or information.
23. Explain briefly (a) why reactions between organic chemicals are slow under mild conditions at room temperature; (b) how these reactions can be accelerated in an organic chemistry laboratory; (c) why the same methods would not work for organic reactions inside living cells; (d) how cells achieve a rapid rate for intracellular organic reactions.

Suggested Readings

Babcock, G.T., and Wikstrom, M. "Oxygen activation and the conservation of energy in cell respiration." *Nature*, 356:301–309, 1992.

Capaldi, R.A. "Mitochondrial myopathies and respiratory chain proteins." *Trends in Biochemical Sciences*, 13:144–148, 1988.

Cross, C.E., Halliwell, B., Borish, E.T., Pryor, W.A., Ames, B., Saul, R., McCord, J., and Harman, D. "Oxygen radicals in human disease." *Annals of Internal Medicine*, 107:526–545, 1987.

Fothergill-Gilmore, L.A. "The evolution of the glycolytic pathway." *Trends in Biochemical Sciences*, 11:47–51, 1986.

Halliwell, B. "Oxidants in human disease: some new concepts." *FASEB Journal*, 1:358–364, 1987.

Klingenberg, M. "Mechanism and evolution of the uncoupling protein of brown adipose tissue." *Trends in Biochemical Sciences*, 15:108–112, 1990.

Lehninger, A.L., Nelson, D.L., and Cox, M.M. *Principles of Biochemistry*, 2nd ed. New York: Worth, 1993.

Ricquier, D., Casteilla, L., and Bouilland, F. "Molecular

studies of the uncoupling protein." *FASEB Journal,* 5:2237–2242, 1991.

Schlenk, F. "The ancestry, birth and adolescence of adenosine triphosphate." *Trends in Biochemical Sciences,* 12:367–368, 1987.

Shah, S.V. "Role of reactive oxygen metabolites in experimental glomerular disease." *Kidney International,* 35:1093–1106, 1989.

Wallace, D.C. "Diseases of the mitochondrial DNA." *Ann. Rev. Biochem.,* 61:1175–1212, 1992.

Key Terms

acetyl coenzyme A (p. 192)

adenosine triphosphate (ATP) (p. 179)

aerobic (p. 189)

anabolism (biosynthesis) (p. 178)

anaerobic (p. 192)

catabolism (p. 178)

chemiosmotic mechanism (p. 198)

citric acid cycle (Krebs cycle, tricarboxylic acid cycle) (p. 192)

coenzyme (p. 185)

cytochrome (p. 197)

deamination (p. 189)

electron carrier (p. 197)

electron transport chain (p. 197)

energy transformation (p. 178)

essential amino acids (p. 201)

fat (p. 188)

glycolysis (p. 189)

high-energy phosphate compound (p. 182)

metabolic pathway (p. 180)

nonessential amino acids (p. 201)

oxidative phosphorylation (p. 185)

pentose phosphate pathway (p. 198)

photosynthesis (p. 179)

redox reactions (p. 181)

standard free energy change (p. 183)

substrate-level phosphorylation (p. 183)

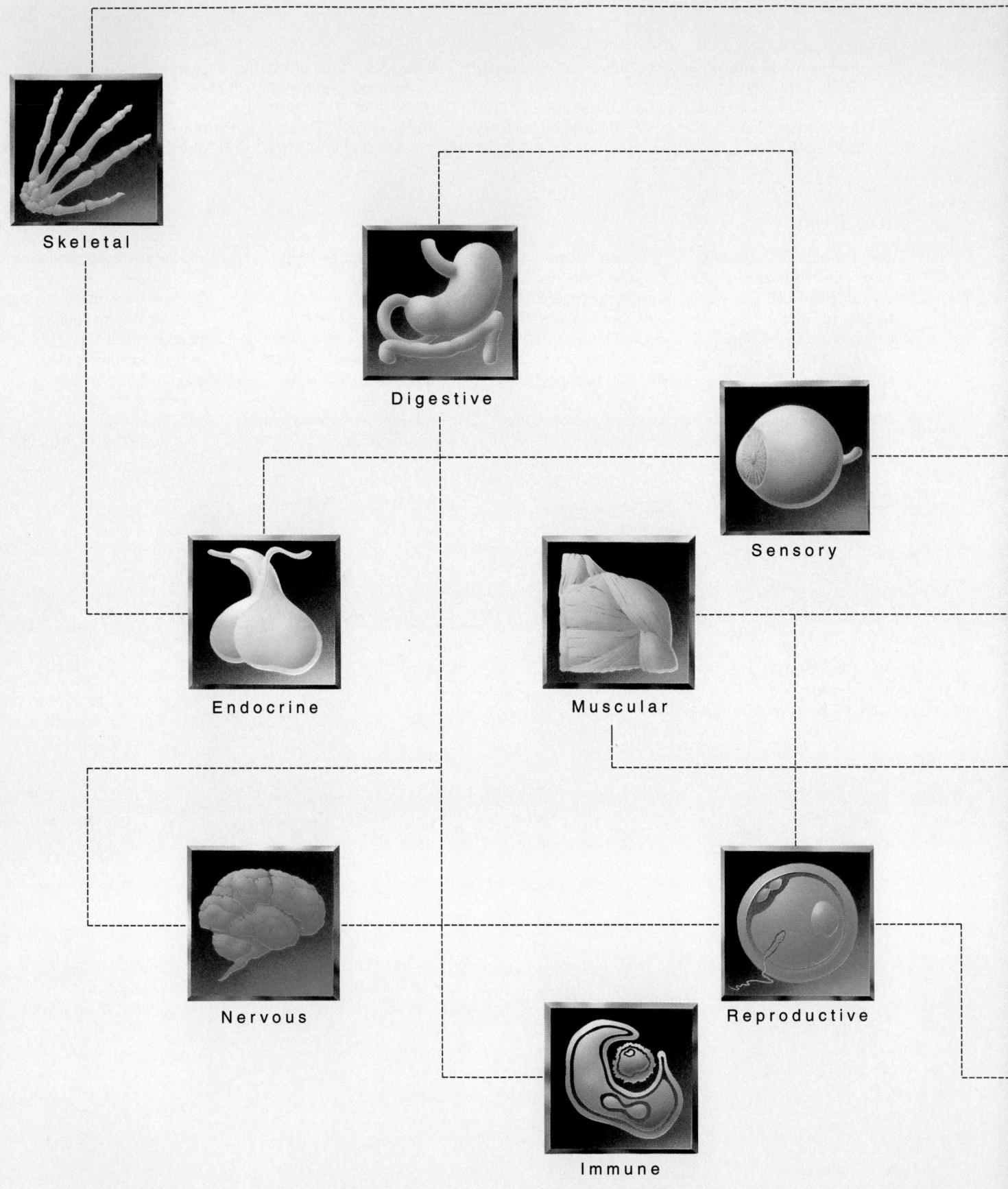

Skeletal

Digestive

Sensory

Endocrine

Muscular

Nervous

Reproductive

Immune

Part II
Physiological Control Systems

Respiratory

Renal System

Cardiovascular

Thermoregulatory

Scanning electron micrograph of
a human neuron in culture.
(© CNRI/SPL/Science Source)

Chapter 7

Functional Organization of the Nervous System

Key Concepts

The nervous system is functionally organized and designed to receive, store, and transmit information. It receives and detects information from the external environment and transduces this information into electrical impulses.

Neurons and glia are the principal types of cells found in the nervous system. Glial cells play a variety of roles, ranging from the transport of molecules in maintaining a constant extracellular environment to phagocytosis. Neurons transmit information to other neurons by an electrochemical process that involves the propagation of electrical impulses and the release of neurotransmitters.

The propagation of electrical impulses along nerve fibers is mediated by the action potential, a transient depolarization of the membrane. Changes in voltage-gated ion conductances across the nerve membrane are responsible for the initiation and propagation of the action potential.

Neurons communicate with one another by the process of synaptic transmission. This process involves the depolarization of the nerve terminal, which causes the activation of voltage-gated calcium channels, the fusion of transmitter-containing vesicles with the terminal membrane, and the release of transmitters into the synaptic cleft.

Neurotransmitters can be grouped according to their molecular weight. The synthesis of low molecular weight transmitters typically takes place in the nerve fiber terminal where they can be rapidly recycled for release. High molecular weight transmitters are synthesized in the region of the cell soma and are transported to the terminal region for release.

Neurons process incoming information by summating postsynaptic membrane potential changes, which are spatially distributed across the soma and dentrites of the neuron. This summation process also occurs temporally in relation to the frequency at which synapses are activated.

Chapter Outline

Ever since Jim was a teenager he had enjoyed sports, especially tennis. In high school and college he made the varsity tennis teams and even lettered as a freshman. As recently as a year ago, Jim and his wife would play at least once a week at the local racquet club. But now, Jim's life has changed drastically. He first began to notice subtle changes in his eyesight, with everything being just a little out of focus. Now Jim is having great difficulty seeing, and he has lost much of his motor coordination skills. His doctors tell him that he has a disease called multiple sclerosis, in which his immune and nervous systems are at odds with one another. For some reason, cells in the immune system are attacking cells in his brain as if they were foreign to his body. The cells that are being destroyed produce myelin, which is needed to help nerve cells conduct electrical impulses; without myelin, Jim's nerve cells can't function properly.

For the moment, Jim is not totally disabled; he can still move around by himself, and there are days when he feels pretty good. Someday researchers may be able to replace the myelin-producing cells that have been destroyed, but the first step is to get the immune system to work cooperatively with the brain.

The nervous system is one of the major control systems in the body. It coordinates the activity of other organ systems so that they operate in concert with one another. During exercise, for example, the nervous system activates selective muscle groups for movement; it causes the heart to pump faster to supply more blood to various parts of the body via the circulatory system, and it causes the respiratory system to increase the rate and depth of breathing to provide the body with more oxygen. These aspects of nervous system function represent the motor output to skeletal, cardiac, and smooth muscles of target organs.

The nervous system also detects and transmits sensory information from the external environment to the brain. We respond to our environment based on what we see, hear, touch, smell, and taste. In general, these sensations are processed in separate pathways within the nervous system. At higher levels, however, there is a convergence of various sensory inputs. Such is the case with information from the visual and auditory systems, which converges in areas of the brain involved in the interpretation of language.

The picture emerging from the study of the nervous system is that specific areas of the brain are responsible for discrete functions. In this chapter we will focus on the functional organization of the brain and how the cellular components of the nervous system transduce and transform neural information as it is transmitted from one area of the brain to another.

Overview: The Central and Peripheral Nervous Systems

The nervous system is composed of the **central and peripheral nervous system** (Fig. 7–1). The central nervous system consists of the **brain** and **spinal cord,** and the peripheral nervous system consists of **peripheral nerves** and **ganglia.** The brain is encased in the skull and, when exposed, it appears as a convoluted mass of tissue that makes up the **neocortex.** The neocortex is proportionally larger in humans than in most other species. It is functionally divided

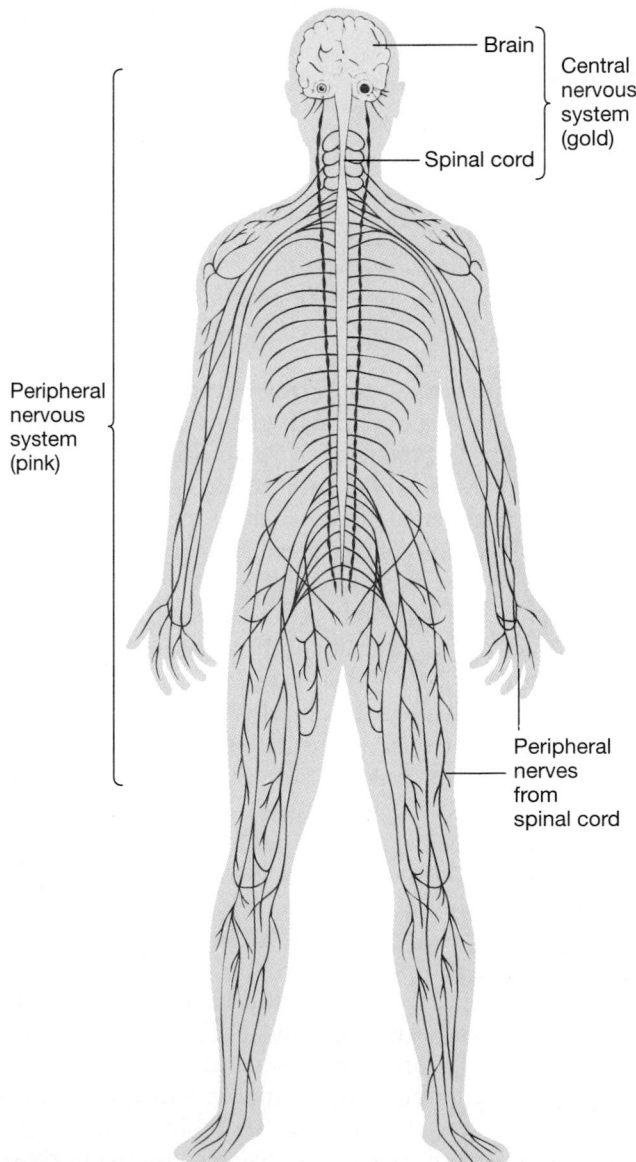

Figure 7–1

The human nervous system, consisting of the brain, spinal cord, and nerves that extend from each.

Figure 7–2

The functional areas of the neocortex.

(Fig. 7–2) into areas that control movement **(motor cortex)**, areas that respond to sensations felt on the surface of the skin **(somatic sensory cortex)**, visual images seen by the eye **(visual cortex)**, and sounds detected by the ear **(auditory cortex)**. Other areas of the neocortex are responsible for language **(Wernicke's area** and **Broca's area)**, and personality **(frontal cortex)**.

Beneath the neocortex are subcortical areas of the brain involved in a variety of different functions. The **thalamus** (Fig. 7–3*a* and 7–3*b*), for example, relays information from the sense organs to specific sensory cortical areas. Collectively, the sense organs, thalamus, and sensory cortices form the **sensory systems.** These systems will be discussed in Chapter 8. Other subcortical areas such as the **hypothalamus** (Fig. 7–3*a*) control the secretion of hormones for growth, reproduction, and metabolism; and others such as the **striatum** (Fig. 7–3*b*) are involved in the initiation and coordination of movement.

The neocortical and subcortical components of the brain rest upon the **brain stem** (see Fig. 7–1)—an area that plays important roles in several functions. By receiving information from the motor cortex and transmitting it to the spinal cord, the brain stem aids in maintaining body posture. It also transmits information to and from the **cerebellum,** another area involved in locomotor coordination. Collectively, these areas form the **motor systems,** which will be discussed in Chapter 9.

Areas within the brain stem also control the breathing and heart rates, in which changes occur automatically and are not typically under conscious control. These functions of the brain stem are thus part of the **autonomic nervous system,** which will be discussed in Chapter 10. Emerging from the brain stem are cranial nerve fibers (see Fig. 7–1) connected to the body's internal organs. These cranial nerves are sensory and motor fibers that innervate the face and head. Sensory nerves transmit information *from* the internal organs conveying information about their condition or state of function. Motor nerves transmit information *to* the internal organs to control how they operate. They also receive and transmit information to the muscles and skin of the head and neck. The **spinal cord** is located caudal to the brain stem (see Fig. 7–1), and it also transmits information to and receives information from the body's internal organs, as well as from skeletal muscles and skin.

The peripheral nervous system is comprised of a **somatic** portion and an **autonomic** portion. Within the **somatic nervous system,** sensory nerves transmit information from sense receptors to the central nervous system, and motor nerves transmit instructions about appropriate responses to skeletal muscles. Within the **autonomic nervous system,** sensory nerves transmit information about the condition of internal organs to the central nervous system, and motor nerves instruct glands and involuntary mus-

Neocortex

Parietal lobe

Corpus callosum

Thalamus

Hypothalamus

Occipital lobe

Cerebellum

Temporal lobe

Fourth ventricle

Midbrain

Medulla

Pons

(a)

Basal ganglia:

STRIATUM — Lentiform nucleus — (1) Caudate nucleus / (2) Putamen / (3) Globus pallidus

Thalamus

Lateral ventricle (anterior horn)

Corpus callosum

Caudate nucleus (head)

Insula

Internal capsule

(b)

Tail of caudate

Caudate nucleus

Posterior

Thalamus

Striatum

Lateral

Anterior

Putamen and globus pallidus

(c)

Figure 7–3

(*a*) A medial view of the left half of the brain, showing some of the functional areas of the neocortex and the subcortical structures. (*b*) A horizontal section through the cerebrum at the level of the basal ganglia and thalamus, showing the striatum. (*c*) The striatum comprises the caudate nucleus, putamen, and globus pallidus.

cles about appropriate responses. Motor nerves in the autonomic nervous system are either sympathetic or parasympathetic in function. **Sympathetic nerves** generally prepare the body for stressful situations, while **parasympathetic nerves** prepare the body for such processes as food consumption and digestion.

In general, the nervous system is organized to receive information from sensory organs (such as the eyes and ears), interpret this information, and respond to these sensory inputs by transmitting information to organ systems (such as our skeletal muscles) to initiate the appropriate responses (Fig. 7–4).

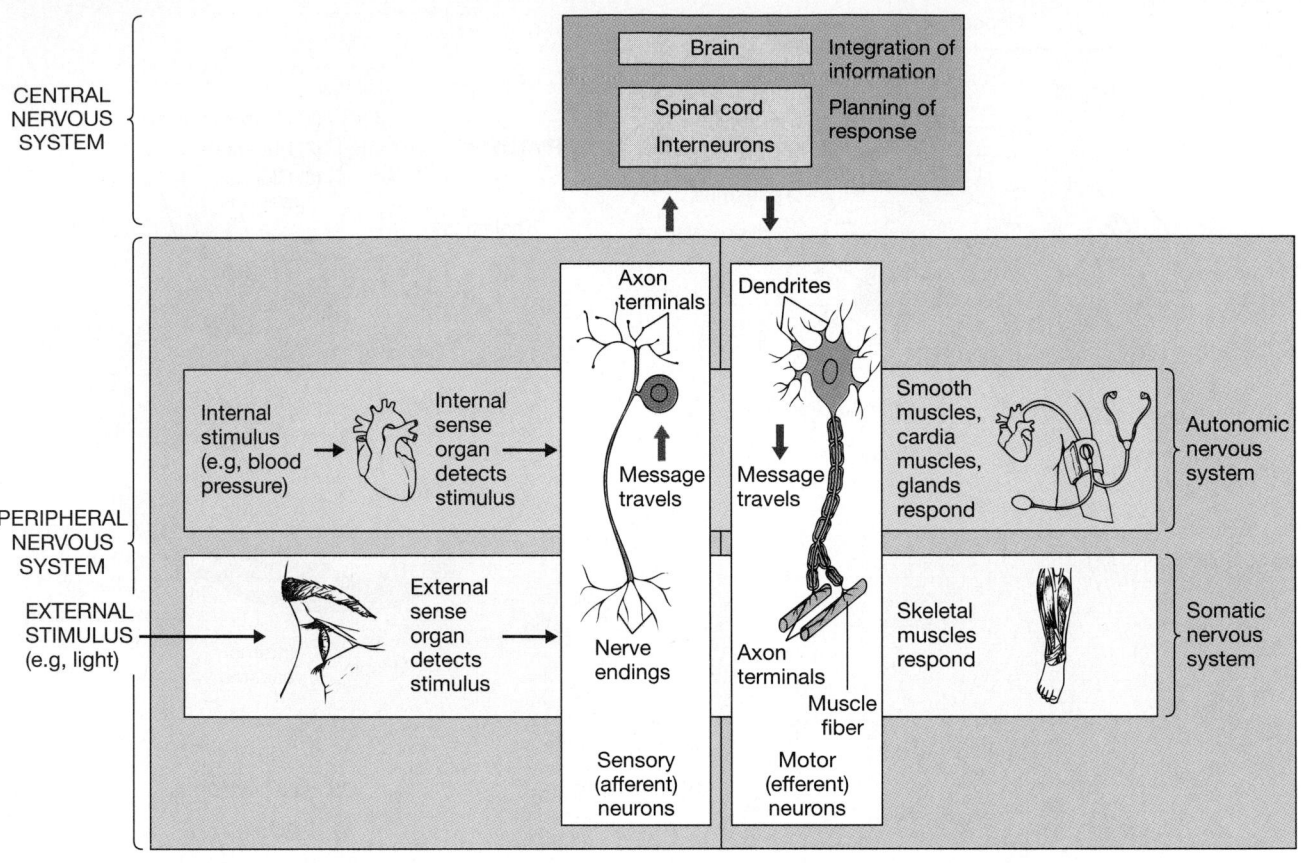

Figure 7–4

The functions of, and flow of information through, the various components of the nervous system.

Cells of the Nervous System

The nervous system contains a complex organization of over a trillion cells. Many cells rapidly communicate with each other by producing electrical impulses that are sent from one cell to another. We will direct our attention in the remainder of this chapter to the basic mechanisms by which cells in the nervous system operate and communicate with each other.

The nervous system is composed of two types of cells: neurons and glia (or glial cells). **Neurons** are designed to transmit information rapidly from one cell to another. They are responsible for the functions normally associated with the nervous system, such as sensory perception and movement. **Glia,** on the other hand, help maintain the environment surrounding neurons and aid in their ability to transmit information rapidly.

Glia Are Multifunctional Cells

There are three basic types of glial cells in the CNS: oligodendroglia, astroglia, and microglia (Fig. 7–5a). **Oligodendroglia** wrap themselves around axons to form **myelin,** layers of lipid membrane, which insulate the axon to prevent the passage of ions through the axonal membrane (Fig. 7–5b and 7–5c). In between the myelinated regions of the axon are **nodes of Ranvier.** The exposed nerve membranes at these nodes contain a high concentration of voltage-gated sodium channels. It is in these nodes that electrical impulses called action potentials are generated and conducted along the length of myelinated axons. Typically, one oligodendroglia will myelinate many axons in the central nervous system. In the peripheral nervous system, **Schwann cells** play a similar role to oligodendroglia in forming myelin around nerve fibers. In contrast to oligodendroglia, however, one Schwann cell will myelinate only one axon in the PNS. This difference in geometry is not trivial; the one-to-one Schwann cell sheath has an intact basement membrane that allows the sheath to act as a regeneration tube for damaged nerve processes. Oligodendroglia sheaths cannot guide regeneration, and damage in the CNS is therefore usually irreparable.

Astroglia are star-shaped cells that exhibit the greatest diversity of function among glial cells in the CNS. One type of astroglia, the **fibrous astrocyte,** is

(text continues on page 214)

(a)

(b)

(c)

(d)

Figure 7–5

Types of glial cells and their functions in protecting neurons. (*a*) Astroglia make contact with blood vessels and neurons. (*b*, *c*) Oligodendroglia form myelin sheaths around axons. AX = axon; MS = myelin sheath. (*d*) Astroglia remove cellular debris from the central nervous system.

R E S E A R C H   F O C U S

Gene Therapy for Brain Tumors

Oncogenes and tumor suppressor genes are involved in the process of cell division during development. Under normal conditions these genes are highly regulated to control cell proliferation. Recently, researchers have shown that the oncogene *c-myb* encodes a nuclear phosphoprotein involved in the regulation of cell proliferation and differentiation. The expression of the *c-myb* gene is greatest in cells as they prepare to enter and traverse the G1/S transition phase of the cell cycle. Its expression subsequently decreases as cells terminally differentiate. C-MYB

protein binds to specific DNA sequences and transactivates transcription of *DNA polymerase-alpha* and *cdc2*, two genes that are critical for DNA synthesis. *c-myb* mRNA levels are exceptionally high during the development of the nervous system, and in cancerous glial cells called glioblastomas. One approach researchers are now using in attempts to suppress the proliferation of glioblastoma cells is gene therapy with antisense oligonucleotides.

Antisense oligonucleotides are specific nucleic acid sequences that can interfere with the transcription of DNA into mRNA, and the translation of mRNA into protein. It has been estimated that antisense oligonucleotides 15 to 18 bases in length are sufficient to bind to specific genes and disrupt their function. Researchers have now developed an 18-base antisense oligonucleotide to the initiation region of the *c-myb* genes for codons 2–7. The administration of this antisense to human glioblastoma cells has suppressed their ability to proliferate in tissue culture and in mice that received injections of the tumor cells. Ultimately, this type of gene therapy may be attempted in treating patients with brain tumors.

Human glioblastoma cells in tissue culture.   *(From Atlas of Brain Tumors, by K. Tabuchi and A. Nishimot., Springer Verlag, 1988)*

■ **Before using antisense oligonucleotides in brain tumor therapy, what additional tests should be conducted to evaluate the safety and efficacy of this approach?**

---

found in areas containing predominantly nerve fibers. It is called fibrous because of the large number of intermediate filaments found in glial cells. **Protoplasmic astrocytes** are similar to fibrous astrocytes but have fewer filaments. They are found in areas containing predominantly nerve cell bodies, dendrites, and synapses. Both types form "glial end-feet" on blood vessels (see Fig. 7–5*a*). The anatomical relationship between glial end-feet and blood vessels was once thought to represent the blood-brain-barrier that prevented unwanted substances in the circulatory system from penetrating the brain. However, more recent studies have found that the tight junctions between the endothelial cells that make up the blood vessels are the key cellular components of the blood-brain-barrier.

During injury to the brain, another type of astrocyte called a **reactive astrocyte** appears. It removes degenerating debris by the process of phagocytosis. Reactive astrocytes proliferate after the degenerated neurons have been removed and, in conjunction with

fibroblasts, form **glial scars** that are typically seen after brain injury.

Astrocytes regulate the concentration of K^+ ions in the extracellular space by absorbing and redistributing them to other astrocytes. This is accomplished by the transport of K^+ ions through the gap junctions that link networks of glial cells. This type of spatial buffering is needed to maintain the constant extracellular environment of K^+ that is critical to the function of neuron-generated electrical impulses.

Astrocytes also regulate the concentration of chemicals called **neurotransmitters** that are released by nerve cells during the process of **synaptic transmission.** This uptake by glial cells is needed to terminate synaptic transmission and prevent the flooding of neurotransmitters between nerve cells.

Astrocytes are also thought to be involved in the maintenance of interstitial pH, which is required for the optimal function of many cellular processes. As in other organ systems, the enzyme **carbonic anhydrase** plays a critical part in acid-base balance. In the central nervous system, this enzyme is found mainly in astrocytes and oligodendroglia.

Microglia are found in the nervous system near blood vessels. Typically their occurrence is somewhat rare in healthy brain tissue. After injury, however, they are found to migrate to the site of damage, where their main function is the removal of cellular debris by phagocytosis. Unlike other glial cells, recent studies have demonstrated that microglia originate from outside the brain, most likely from bone marrow.

Neurons Are Conductors of Nervous Impulses

Neurons are also called **nerve cells.** They have four functionally distinct regions: the soma, dendrites, axon, and axon terminals (Fig. 7–6). The **soma** is the cell body of a neuron and its metabolic center. It contains the components necessary to fabricate and package proteins used in other parts of the cell to maintain a variety of cellular functions. **Dendrites** are fiberlike structures that branch out from the cell body. Both dendrites and soma form the surface of a neuron, which receives information from other nerve cells.

The **axon** is a long fiberlike structure that extends from the soma to make contact with other nerve cells. The **initial segment** of the axon is a sensitive region that triggers the discharge of electrical impulses. In many nerve cells, sections of the axon are surrounded by myelin. The areas in between the segments of insulation are called **nodes of Ranvier.** The axon ends in many **axon terminals,** which make contact with other nerve cells at junctions called **synapses** in order to transmit information to other cells.

Types of Neurons

Neurons can be classified in two ways: by their structure, as multipolar neurons, bipolar neurons, or unipolar neurons; and by function, as sensory neurons, motor neurons, or interneurons.

Classification by Structure. A **multipolar neuron** has several short dendrites and one long axon (see Fig. 7–6). A **bipolar neuron** has one axon and one dendrite, on opposite sides of the soma. A **unipolar neuron** has one short branch that gives off an axon and a dendrite. (A **nerve** is a bundle of axons from different neurons.)

Classification by Function. A **sensory neuron,** also called an **afferent** ("going toward") **neuron,** carries information about external or internal stimuli from sensory receptors to the central nervous system. A **motor neuron,** also called an **efferent** ("leading away") **neuron,** carries movement instructions to muscle from the central nervous system, in response to sensory stimuli. An **interneuron,** found in the intermediate zone of the spinal cord, for example, connects sensory and motor neurons and integrates their functions. Motor neurons are typically multipolar neurons, whereas sensory neurons are often unipolar.

Specialized Neuronal Structures

As with other types of cells, the nucleus of nerve cells contains the genetic information needed for the synthesis of specific proteins. Some proteins, for example, are involved in the formation of new functional connections with other cells as a result of signals from afferent inputs or during the process of regeneration.

In neurons the nucleus is relatively quite large. A substantial portion of the genetic information contained within the genome is continually transcribed by neurons. Based on hybridization studies, it is estimated that one third of the genome is actively transcribed, producing more mRNA than any other organ in the body. Because of the high level of transcriptional activity, the nuclear chromatin is in the form of **dispersed chromatin** (Fig. 7–7). In contrast, the chromatin in non-neuronal cells in the brain such as glia is found in clusters on the internal face of the nuclear membrane. The neuronal nucleus also contains one or two nucleoli, which often have attached electron-dense bodies. One of the dense bodies associated with the nucleolus in females is the Barr body, which is the condensed chromatin of the inactive X-chromosome.

Most of the proteins formed by free ribosomes and polyribosomes remain within the cell, while proteins formed by rough endoplasmic reticulum are exported. Polyribosomes and rough endoplasmic reticulum (RER) are found predominantly in the soma of neurons. Axons contain no RER and are unable to synthesize proteins. Dendrites, on the other hand, contain ribosomes and RNA, and are thought to play an important part in the modulation of dendritic function.

The smooth endoplasmic reticulum is involved in the intracellular storage of calcium. Smooth endoplasmic reticulum within neurons binds calcium and maintains the intracellular cytoplasmic concentration at a low level of 0.1 μmoles. Prolonged elevation of intracellular calcium has been shown to lead to neuronal death and degeneration.

In neurons the Golgi apparatus is found in the soma. As in other types of cells, this structure is engaged in the terminal glycosylation of proteins synthesized in the RER. The Golgi apparatus takes proteins produced for exportation in the RER and forms

(a)

(b)

(c)

Figure 7–6

(*a*) The general structure of a neuron. (*b*) Types of neurons. Multipolar neurons have several short dendrites and one long axon, while bipolar neurons have one axon and one dendrite. Unipolar neurons have one short branch that gives off an axon and a dendrite. (*c*) SEM of a human neuron in tissue culture.   *(Petit Format/Science Source.)*

protein-containing vesicles. These vesicles are released into the cytoplasm, where some are carried by axoplasmic transport to the axon terminals.

The Neuronal Cytoskeleton. The transport of proteins from the Golgi apparatus and the highly spe-

cialized form of the neuron are dependent on the internal framework of the cytoskeleton. The neuronal cytoskeleton is made of microfilaments, neurofilaments, and microtubules (Fig. 7–8). **Microfilaments** are composed of actin, a contractile protein commonly found in muscle. They are 4-5 nm in diameter, and in neurons they are found in dendritic spines and growth cones. **Neurofilaments,** on the other hand, are found in axons and dendrites and are thought to provide structural rigidity. They are not found in the growing tips of axons and dendritic spines, which are more dynamic structures. Neurofilaments are about the size (10 nm diameter) of the

Figure 7–7

Types of axonal transport. Materials move by anterograde fast transport from the soma to the axon terminal and by retrograde transport from the axon terminal to the soma.

Figure 7–8

The cytoskeleton structure of an axon, consisting of microtubules and neurofilaments that aid in the transport of organelles and secretory products to and from the axon terminal.

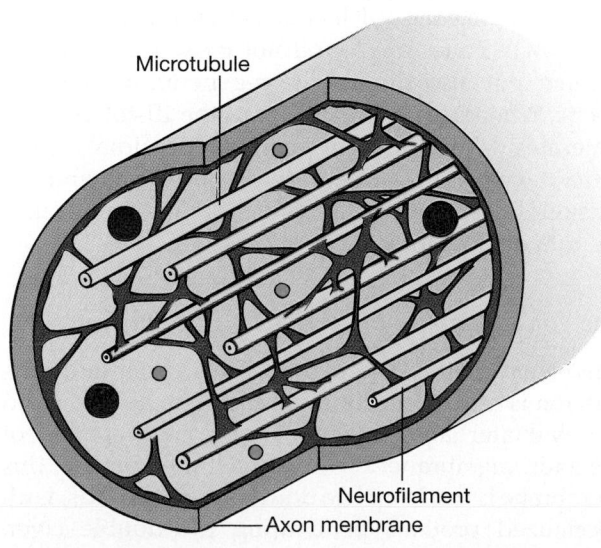

intermediate filaments found in other types of cells. However, intermediate filaments in other cell types consist of one protein, while neurofilaments are com-

posed of three proteins (70 kd, 140 kd, and 220 kd in size). The core of the filament consists of the 70 kd protein, similar to intermediate filaments in other cells. The two other neurofilament proteins are thought to be side arms that interact with microtubules.

Microtubules are responsible for the rapid movement of material in axons and dendrites. They are 23 nm in diameter and are composed of tubulin with molecular weights of 52 and 56 kd. In neurons, microtubules have accessory proteins called **microtubule-associated proteins (MAPS).** Dendrites have high molecular weight MAPS, while axons have low molecular weight MAPS; these two types of MAPS are thought to determine whether material is distributed to dendrites or to axons.

In neurons, mitochondria are highly concentrated in the region of axon terminals. They produce adenosine triphosphate (ATP), which is required as a source of energy for many cellular processes. In the axon terminal the mitochondria not only provide a source of energy for the process of synaptic transmission but also provide substrates for the synthesis of certain neurotransmitter chemicals such as the amino acid glutamate. In addition, enzymes involved in the degradation of other types of neurotransmitter chemicals are embedded in the outer membrane of the mitochondria. Thus the role of mitochondria in the neuron is multifunctional and varied.

Axonal and Dendritic Transport Mechanisms. The shapes of most cells in the body are relatively simple in comparison to the complexity of neurons, with their elaborate axonal and dendritic processes. Because of the length of nerve cell processes, however, neurons have developed specialized mechanisms to transport proteins, organelles, and other cellular material along the length of axons and dendrites needed for the maintenance of the cell. These transport mechanisms are capable of moving cellular components along fiber processes in an **anterograde** direction away from the soma, or in a **retrograde** direction toward the soma (see Figure 7–7). **Kinesin,** a microtubule associated protein, is involved in anterograde transport of organelles and vesicles via the hydrolysis of ATP. Retrograde transport, on the other hand, is mediated by another microtubule associated protein called **dynein.**

In the axon, anterograde transport occurs at both a slow and a fast rate. The rate of **slow axonal transport** is 1 to 2 mm/day. Structural proteins such as actin, as well as neurofilaments and microtubules, are transported at this speed. The rate of **fast axonal transport** is 400 mm/day. Fast transport mechanisms are utilized by organelles, vesicles, and membrane glycoproteins needed at the synaptic terminal. Another feature that distinguishes fast from slow transport mechanisms is that fast transport requires Ca^{2+}, glucose, and ATP (i.e., it is dependent upon oxidative metabolism).

In dendrites, anterograde transport occurs at a rate of 0.4 mm/day, and like fast transport, it also requires ATP. Dendritic transport appears to be involved in the movement of ribosomes and RNA, suggesting that protein synthesis occurs within dendrites.

In retrograde axonal transport, material is moved from terminal endings to the cell body. This provides a mechanism for the cell body to sample the environment around its synaptic terminals. In some neurons, maintenance of synaptic connections depends on the **transneuronal** transport of trophic substances, such as nerve growth factor (NGF), across the synapse. After transport to the soma, NGF activates mechanisms for protein synthesis.

Nerve Fiber Development and Regeneration. One of the major features distinguishing nerve cell differentiation and growth from that of other cell types is the outgrowth of the axon from the nerve cell body, in a specific direction and along a specific pathway, to form synaptic connections with specific targets. The growth of axons is determined largely by interactions between the growing axon and the tissue environ-

ment. At the leading edge of a growing axon is the growth cone. **Growth cones** are structures that give rise to protrusions called filopodia. Growth cones contain actin, a contractile protein, and are quite motile, with filopodia extending and retracting at a rate of 6 to 10 μm/min. Newly synthesized membranes in the form of vesicles are also found in the growth cone and fuse with it as it extends. As the growth cone elongates, microtubules and neurofilaments are added to the distal end of the fiber and partially extend into the growth cone. They are transported to the growth cone via the process of slow axonal transport.

The direction of axonal growth is directed in part by **cell adhesion molecules (CAMs).** CAMs are glycoproteins that are expressed on cell surfaces to promote cell adhesion. Neuron-glia-CAM (Ng-CAM) is expressed in postmitotic neurons and is particularly prominent in growing neurites (axons and dendrites); these neurites migrate along certain types of glial cells that provide a guiding path to target sites. The secretion of tropic factors by target cells also influences the direction of axon growth. Once the proper target site is reached, and synaptic connections are formed, the processes of growth cone elongation and migration are terminated.

During development, axon terminals are found on target cells in excess of the numbers seen after the nervous system has matured. Cells in the lateral geniculate nucleus of the visual system, for example, receive both ipsi- and contralateral inputs during development. In the adult, however, these cells receive inputs no longer from both eyes but only from either the ipsilateral or contralateral input. This loss of synaptic contacts is a result of a selection process whereby the most active inputs predominate and survive at the expense of less active synaptic contacts.

Growth cones are found not only during development but also during the regeneration of nerve fibers. When a nerve fiber is cut, the distal end degenerates while the fiber segment proximal to the soma develops growth cones for elongation and extension. Regeneration occurs at a rate of 1 mm/day, the rate of slow axonal transport.

The Ion Channels of Nerve Cell Membranes

The electrical impulse that initiates synaptic transmission is called an **action potential** (to be discussed in detail later) and is dependent on the properties of the axon membrane. As are all cell membranes, this membrane is made up of a double layer of lipids, with specialized proteins penetrating the double layer. These proteins regulate the movement of ions across the membrane, which in turn creates the action potential. Certain ions (Na^+, K^+, Cl^-, and Ca^{2+}) can

(a)

(b)

(c)

Figure 7–9

Types of ion channels. (*a*) Passive channels are specific to individual ions. (*b*) Chemically activated channels, also known as receptors, depend on activation by a chemical transmitter. (*c*) Voltage-activated channels open or close when they detect a change in the membrane potential.

cross the membrane only through protein pores in the membrane that form **ion channels.** These channels typically allow only specific ions to pass, while blocking others because of their size, charge, or state of hydration. There are three basic types of ion channels: passive, chemically activated, and voltage-activated channels (Fig. 7–9).

Passive ion channels are found in membranes throughout all areas of the nerve cell. Each passive channel is identified according to the specific ion it allows through (e.g., Na$^+$ channel, K$^+$ channel, Cl$^-$ channel, and Ca^{2+} channels).

Chemically activated ion channels are located predominantly on dendrites and the soma. These channels are generally closed by "gates" to prevent the flow of ions through the membrane. Chemical

transmitters bind to sites on these protein channels and open the gate across the channel to permit the flow of ions through the channel. These chemically activated channels are also known as **receptors.**

Voltage-activated ion channels, found in the membranes of axons and soma, are opened when they detect a certain voltage. They are responsible for generating and propagating the action potential.

The Membrane Potential and the Action Potential

The movement of ions across their channels is caused by a difference in the electrical charge or differences in the concentration of ions between one side of the membrane and the other. The imbalance in ionic concentration is called the **concentration gradient.** The resulting imbalance in electrical charge is called the **membrane potential.**

Resting cells, cells not in the process of conducting nervous impulses, have a stable membrane potential. The onset of an **action potential,** in which the membrane potential temporarily changes, indicates a cell in the process of nervous conduction.

Electrical Potentials Are Produced Across the Membrane of Resting Cells

The membrane potential exists because of a difference in the number of positive and negative charges across the membrane. These charges are attributed to the **cations** (positively charged ions) and **anions** (negatively charged ions) that occur on each side of the membrane. When there is a difference in the net charge between the inside and outside of a cell, an **electrical potential (E)** is established. This electrical potential is measured in units called **volts (V)** and is determined by the charge difference between two points. The charge at two points in the extracellular fluid is the same, for example, so the charge difference and electrical potential are therefore zero. In contrast, since there is a charge difference between a point in the extracellular space and a point within the cell, an electrical potential is established. In many nerve cells, this electrical potential difference across the membrane, E_m, is approximately -0.060 V or -60 mV (Fig. 7–10). The convention that has been adopted for measuring the voltage differences across the membrane is the use of the extracellular voltage as a reference. A minus sign thus indicates that the inside of the cell is more negative than the extracellular fluid.

The electrical potential can be viewed as the electrical equivalent to pressure. Differences in pressure between two points cause currents of water and wind to flow. In a similar fashion, an electrical potential results in the movement of ions or electrons. In physical systems, the movement of electrons in a copper wire is the basis of **current flow.** In biological systems, current flow is due to the movement of ions. This movement occurs when anions are attracted to positively charged regions and cations to negatively charged regions. The flow of ions is called **ionic current (I)** and is measured in units called **amperes (A).** The ionic sodium current, I_{Na^+}, that moves through a single channel, for example, is approximately 12×10^{-12} A.

The ease with which ions can flow through an ion channel is called **conductance (G)** and is measured in dimensional units called **siemens (S).** Several factors affect the conductance of an ion through a channel. The first is the size of the ion relative to that of the channel. If the channel is large in comparison to the size of the ion, the ion will pass with relative ease and the channel therefore has a high conductance. A smaller channel increases the probability of the ion colliding with the walls of the channel, making it more difficult for the ion to pass. Smaller channels therefore have a lower conductance. Other factors, such as the charge of an ion and its state of **hydration** (the addition of water molecules), can also affect its conductance through the channel. Each type of channel has a specific conductance for its associated ion. Sodium ion channels, for example, have a conductance of 15×10^{-12} S, while calcium ion channels

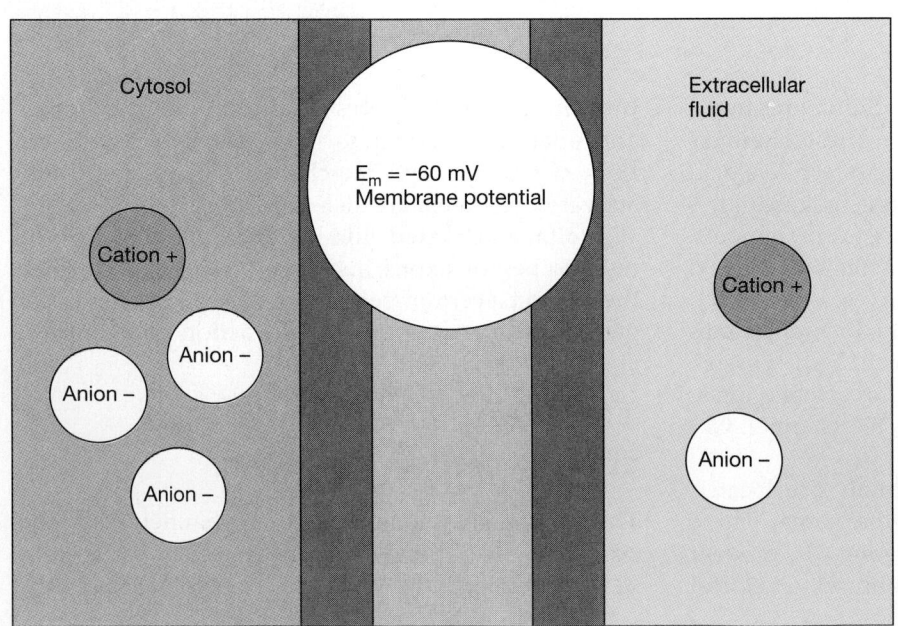

Figure 7–10

The membrane potential (E_m) of -60 mV indicates that the inside of the cell is more negative than the extracellular fluid.

Cytosol

Extracellular fluid

$E_m = -60$ mV
Membrane potential

Cation +

Anion –

Anion –

Anion –

Cation +

Anion –

Low conductance

(a)

High conductance

(b)

Figure 7–11
Membrane conductance for a particular ion depends on the number of ion channels that occur in the membrane.

have a conductance of 1×10^{-12} S. Consequently, a sodium ion is able to move through its channel more easily than calcium ions can move through a calcium channel.

The relationship between electrical potential, ionic current, and conductance is given by **Ohm's Law.** This law states that the rate of ion flow through a channel is directly proportional to the conductance of the channel and the magnitude of the electrical potential, $I = G \times E$. Larger electrical potentials and greater channel conductances produce a faster flow of ions across the membrane. This is similar to the flow of water through a tube. Larger pressures and tubes with bigger diameters will allow more water to flow.

The overall effect of all of the channels for a particular ion is called the **membrane conductance** for that ion. Membrane conductance of an ion, therefore, depends on the number or density of a particular type of channel within a region of the membrane (Fig. 7–11). Membrane conductance can vary from one region of a nerve cell to another. The membrane conductance for sodium ions, for instance, is greater in the region of the initial segment of the axon and nodes of Ranvier than in other areas of the cell. As we will learn, this regional difference in the number of Na^+ channels aids in the initiation and propagation of the action potential, the basis of nervous impulses.

Concentration Gradients and Diffusion

The distribution of ions in the extra- and intracellular spaces of nerve cells is similar to that of most cells in the body. The inside of nerve cells has a relatively high concentration of K^+ ions and a low concentra-

tion of Na^+. In contrast, the extracellular fluid contains a relatively low concentration of K^+ ions and a high concentration of Na^+. The high intracellular concentration of K^+ ions in comparison to the extracellular compartment produces a concentration gradient that causes K^+ to diffuse out of the nerve cell through the passive K^+ channels (Fig. 7–12). Simi-

Figure 7–12
Concentration gradients for sodium and potassium. Sodium (Na^+) ions are more highly concentrated outside than inside the cell, so they tend to passively diffuse into the cytosol, whereas K^+ ions (more concentrated inside the cell) tend to diffuse into the extracellular fluid.

larly, the high extracellular concentration of Na$^+$ causes Na$^+$ to diffuse into the cell through passive Na$^+$ channels. If the diffusion of these ions continued, the concentration gradients would eventually equilibrate, and the intracellular concentrations of K$^+$ and Na$^+$ would be equal to their respective extracellular concentrations. The action of the Na/K-ATPase pumping mechanisms, however, works to transport Na$^+$ ions back out to the extracellular compartment and K$^+$ back into the intracellular compartment (see Chapter 4). Consequently, the Na/K pumping mechanisms maintains the Na$^+$ and K$^+$ gradients despite the continued diffusion of the ions across the membrane.

The Membrane Potential of a Cell Is a Consequence of Ionic Permeability and the Equilibrium Potentials of Certain Ions

To understand how the membrane potential is established, let us first consider a hypothetical cell that contains only K$^+$ ions and large negatively charged organic anions, represented as A$^-$ (Fig. 7–13). These anions are charged amino acids and proteins that are too large to pass through the ion channels of the membrane. Assume that inside a nerve cell are relatively high concentrations of K$^+$ and A$^-$ ions, and that initially these ions do not exist in the extracellular fluid. As we discussed earlier, the establishment of concentration gradients would cause K$^+$ and A$^-$ to diffuse out of the cell through passive channels. In our hypothetical cell, however, only K$^+$ ions can diffuse across the membrane. If K$^+$ and A$^-$ are initially contained within the cell in equal concentrations, then before diffusion starts there will be no net charge (Fig. 7–13a). As K$^+$ begins to diffuse outward, fewer K$^+$ ions are available inside the cell to balance the negatively charged A$^-$ ions. Consequently, the interior of the cell becomes more negative than the outside of the cell (Fig. 7–13b). As we learned earlier, when the cytosol and the extracellular fluid have different numbers of charged ions, an electrical potential develops and the charged ions are attracted to the region that is oppositely charged. The electrical force established by the negatively charged interior of the cell attracts positive ions, causing them to flow from the extracellular fluid across the membrane to the cytosol. Therefore, even though the concentration gradient for K$^+$ alone would continue to force K$^+$ ions outward, an electrical force gradually develops, attracting K$^+$ ions back into the cell (Fig. 7–13c). As more and more K$^+$ ions leave the cell, a greater electrical force develops. Eventually, because of the opposing concentration gradient (Fig. 7–13d), the elec-

trical force becomes sufficient to draw K$^+$ ions into the cell at the same rate that they leave the cell. The voltage at this particular electrical force is known as the equilibrium potential. The **equilibrium potential** for any ion is the potential at which the efflux and influx of an ion are the same so that the net flux across the membrane for that ion is zero. The equilibrium potential can be expressed as the product of several constants and the logarithm of the ratio of the extra- and intracellular concentration of a particular ion, and can be described by the **Nernst equation.**

$$E_{ion} = \frac{RT}{zF} \times \log_e \frac{[ion]_o}{[ion]_i}$$

where R is the gas constant, T is the absolute temperature, z is the ion valance, and F is the Faraday constant. At 20°C this expression for K$^+$ is reduced to

$$E_K = 58 \times \log_{10} \frac{[K^+]_o}{[K^+]_i}$$

Typically the ratio of $[K^+]_o/[K^+]_i$ results in a value of E_K of about −75 mV.

Thus, for this simple cell that contains only K$^+$ and A$^-$ and allows only K$^+$ to cross the membrane, the membrane potential is equal to the equilibrium potential of K$^+$. Nerve cells in general, however, contain other ions that affect the membrane potential of the cell. The two other major ions that contribute to the membrane potential are Na$^+$ and Cl$^-$. Both Na$^+$ and Cl$^-$ have very high extracellular concentrations in comparison to their intracellular concentrations and therefore passively diffuse into the cell. Na$^+$ and Cl$^-$ do not diffuse across the cell membrane as easily as K$^+$ ions. When the nerve cell is at "rest"—that is, when it does not generate an action potential—its membrane is most permeable to K$^+$. This is the reason that the resting membrane potential is near the equilibrium potential for K$^+$. (Typically, the K$^+$ equilibrium potential, E_{K^+}, is approximately −75 mV; the resting membrane potential, E_m, is −60 mV; and the Na$^+$ equilibrium potential, E_{Na^+}, is +55 mV.) The factors that affect the resting membrane potential, E_m can be expressed in terms of the intra- and extracellular concentrations of Na$^+$, K$^+$, and Cl$^-$ ions, as well as their membrane **permeabilities,** P_{Na}, P_K, and P_{Cl}. This relation is given by the **Goldman equation.**

$$E_m = \frac{RT}{zF} \times \ln \frac{P_k [K^+]_o}{P_k [K^+]_i} + \frac{P_{Na} [Na^+]_o}{P_{Na} [Na^+]_i} + \frac{P_{Cl} [Cl^-]_i}{P_{Cl} [Cl^-]_o}$$

We will learn later that, during the action potential, there are changes in the relative permeabilities of K$^+$ and Na$^+$ that account for changes in the membrane potential.

(a)

(b)

(c)

(d)

Figure 7–13

A membrane potential develops when a difference in electrical charge between the cytosol and the extracellular fluid arises, owing to the movement of ions down a concentration gradient. (*a*) Initially, there is no net charge within the cell because the number of anions and cations is equal. (*b*) K$^+$ ions flow out of the cell down their concentration gradient, while anions must remain inside because the membrane is impermeable to them. The cytosol now has a net negative charge. (*c*) K$^+$ ions next flow into the cell because they are attracted to the negative charge of the cytosol, although other K$^+$ ions are still flowing out, owing to the concentration gradient. (*d*) Equal numbers of K$^+$ ions flow in and out of the cell when the intracellular charge is very negative. The membrane potential when this occurs is the equilibrium potential (E$_K$) for K$^+$.

The Action Potential Is a Temporary Change in the Membrane Potential

Action potentials are electrical impulses transmitted by nerve cells. They can be thought of as a change in the voltage of the membrane potential that causes it to go from its negative resting state to a positive value for a very brief time. This type of change in the membrane potential is typically found in nerve axons. Figure 7–14 shows the changes in the membrane potential that characterize the action potential. The membrane potential is initially at its resting level of -60 mV. As the membrane potential becomes more positive (a process called **depolarization**), it reaches a **threshold** value at approximately -45 mV. After reaching threshold, the membrane potential rapidly

changes to more positive values during a **rising phase**. It reaches a **peak** at $+25$ mV and begins to **repolarize**. In repolarization, the membrane becomes **hyperpolarized,** or more negative than in the original resting state. It remains hyperpolarized for a time before returning to the resting level of -60 mV.

What causes such a drastic change in the membrane potential? The voltage changes seen in the action potential are a result of the opening and closing of voltage-sensitive ion channels that control the influx of Na$^+$ and efflux of K$^+$. These channels are sensitive to the voltage across the membrane. When they detect that the voltage has reached a threshold level, they open their gates to allow the passage of these ions. The sodium channel appears to have three states of operation, a **resting state,** an **activation state,** and

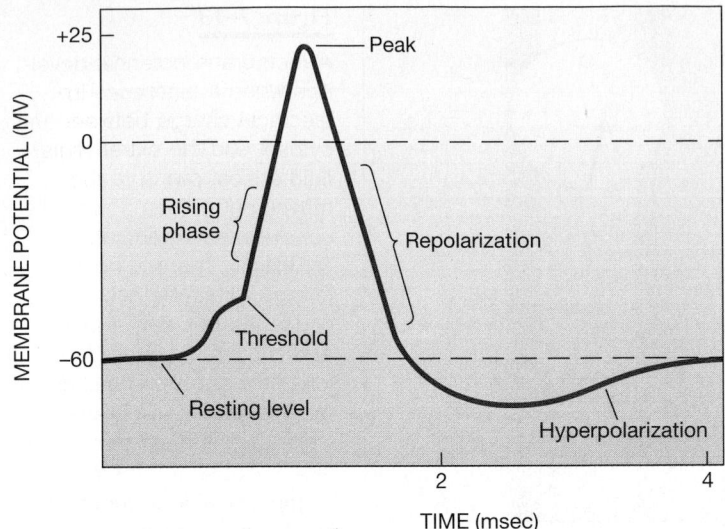

Figure 7–14

A graph showing the various components of the action potential.

an **inactivation state** (Fig. 7–15). When the nerve cell is at rest, the sodium channel is in its resting state, and its gates are closed to prevent the influx of Na^+ ions. If the nerve cell is stimulated and the membrane potential is depolarized to the threshold value of approximately −45 mV, the sodium channel switches from its resting state to the activation or open state. Sodium ions flow into the cell until the inactivation gate closes the channel. The closing of the inactivation gate is a time-dependent phenomenon and is

independent of the opening of the activation gate. Thus, Na^+ channel activation and inactivation can be thought of as two separate gates that open and close independently. The inactivation gate remains closed for a period of time, after which both gates return to their resting state. The properties of the sodium channel, therefore, are dependent on voltage and time. Once the threshold value of voltage is reached, it triggers this series of events that is carried out sequentially.

Figure 7–15

Mechanism of the voltage-sensitive Na^+ channel. (*a*) In the resting state, the voltage-activated gate is closed. (*b*) In the activation stage, triggered by a change in the membrane potential, the voltage-activated gate is open, allowing Na^+ ions to flow into the cell. (*c*) In the inactivation state, the time-dependent gate is closed.

RESTING STATE

(a)

SLOW ACTIVATION STATE

(b)

Figure 7–16

Mechanism of the voltage-sensitive K^+ channel. The opening of the K^+ channel is entirely voltage-dependent and is slower than the Na^+ channel.

A threshold value of voltage also opens voltage-sensitive K^+ channels, but the process is entirely different from the opening of Na^+ channels (Fig. 7–16). Unlike the Na^+ channel, there is no inactivation gate in the K^+ channel. Also, the K^+ channels open more slowly than the Na^+ channels and remain open until they sense a particular voltage. The closing of these channels is not a time-dependent process; it depends only on the voltage, the membrane potential. When the membrane is depolarized, the K^+ channels remain in the open state; when the membrane potential repolarizes, the K^+ channels close.

Ion Channels and the Action Potential

The influx of Na^+ ions and efflux of K^+ ions produce the action potential. In the resting state, voltage-sensitive Na^+ and K^+ ion channels are closed (Fig. 7–17). The simultaneous activation of many sodium channels in the membrane of an axon causes an influx of Na^+ ions. This influx of positive charges causes the membrane potential to become more positive, producing a gradual depolarization of the membrane. Once the threshold value of the membrane potential has been reached (typically $-45\,mV$ at the initial segment), a series of events is triggered, leading to the **initiation** or **generation** of the action potential. At the threshold level of the membrane potential, more voltage-sensitive sodium channels are activated, resulting in a greater influx of Na^+ ions. This influx of positive charges depolarizes the membrane still further, leading to a further influx of Na^+ and consequently a further depolarization.

This **regenerative process** (Fig. 7–18) is the basis for the rising phase of the action potential once threshold has been reached. The regenerative process can occur at different speeds depending on how much the membrane is depolarized and how quickly the membrane potential achieves the threshold value. Greater depolarizations of the membrane potential above the threshold can lead to the opening of more sodium channels, in turn depolarizing the membrane even faster.

Note that not all depolarizations will result in an action potential. If the depolarization is too small, too few voltage-sensitive sodium channels will be activated to perpetuate the regenerative process and no action potential will be produced. Thus, the action potential is an *"all-or-none"* phenomenon.

At the peak of the action potential (see Fig. 7–17) the membrane is much more permeable to Na^+ than to K^+; consequently, the value of the membrane potential is closer to the Na^+ equilibrium potential than to the K^+ equilibrium potential. After the peak has been reached the inactivation channels close, Na^+ influx is blocked, and the membrane potential begins to repolarize. The membrane potential repolarizes because at this point the membrane conductance for K^+ is greater than that for Na^+. As the sodium channels begin to become inactivated, the potassium channels begin to become activated. This increase in potassium conductance causes the membrane potential to become more negative and contributes to the repolarization phase of the action potential. Finally, the prolonged opening of K^+ channels causes a continued efflux of K^+ ions. This removal of positive charges

MEMBRANE POTENTIAL (mV)

① ② ③ ④ ⑤ ①

TIME

D | D | R | R

① RESTING STATE | SLOW ② RISING PHASE | RAPID ③ RISING PHASE | EARLY ④ REPOLARIZATION | ⑤ HYPERPOLARIZATION | ① RESTING STATE

K⁺ | K⁺ | K⁺ | (K⁺)→ | K⁺ | K⁺

K⁺ | K⁺ | (K⁺)→ | (K⁺)→ | (K⁺) | K⁺

Na⁺ | ←(Na⁺) | ←(Na⁺) | ←(Na⁺) | Na⁺ | Na⁺

Na⁺ | ←(Na⁺) | ←(Na⁺) | Na⁺ | Na⁺ | Na⁺

Figure 7–17

The phases of the action potential and corresponding movements of ions and ion channels.

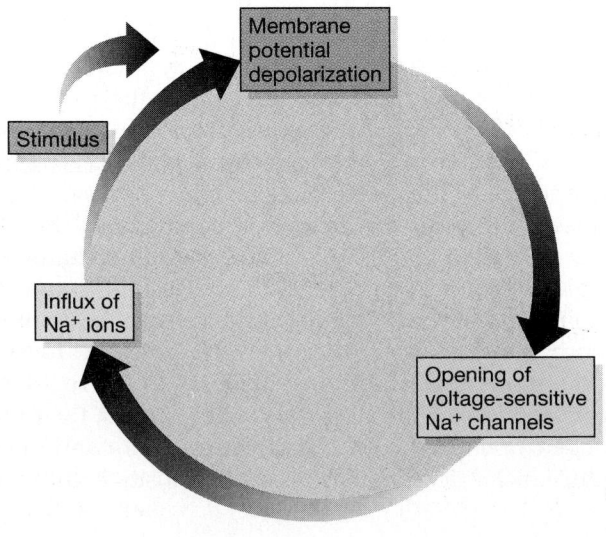

Stimulus

Membrane potential depolarization

Opening of voltage-sensitive Na⁺ channels

Influx of Na⁺ ions

from the cell in turn causes the membrane potential to remain hyperpolarized briefly before returning to the resting level.

Refractory Periods

For a period of time just after the initiation of an action potential, the axon is unable to generate a second action potential regardless of how much the membrane is depolarized. This phase is known as the **absolute refractory period** and typically lasts about

Figure 7–18

The regenerative nature of the process that initiates the action potential.

Figure 7–19

A graph showing the absolute and relative refractory periods that follow the action potential.

millisecond after the onset of the action potential (Fig. 7–19). After this period, an axon is capable of initiating a second action potential, but only if the membrane is depolarized to a large degree. During this **relative refractory period,** the threshold voltage needed to initiate an action potential becomes much greater.

What causes this increase in threshold? In the absolute refractory period, the Na^+ inactivation gates are still in their closed state and have not been reset. Consequently, no matter what the voltage difference across a sodium channel, the channel will not open. In the relative refractory period, many (but not all) of the sodium channels have now been reset. Thus a greater depolarization is needed to trigger the regenerative process required to initiate the second action potential. The significance of the absolute refractory period is that its time interval determines the fastest frequency at which an axon can generate action potentials.

Propagation of the Action Potential

Thus far we have considered how an action potential is initiated. After initiation it **propagates,** or moves, along the axon from the region of the initial segment down to the terminal endings. How does this propagation occur?

Let us first consider the axon as a tube with many membrane regions (Fig. 7–20). In its resting (nonconducting) state, concentrations of various ions are maintained by various active transport systems (Fig. 7–20a). Assume that the first membrane region generates an action potential so that Na^+ ions flow into

the axon at this point (Fig. 7–20b). The sodium ions come from the extracellular fluid surrounding adjacent membrane regions. The removal of Na^+ ions from adjacent regions causes these regions to depolarize. Since current flow follows the path of a loop, the influx of Na^+ ions in the region of the action potential is followed by the movement of positive charges into the adjacent areas of the membrane. The depolarization of adjacent membrane regions is therefore aided by the movement of positively charged ions within the nerve fiber along its length. If this depolarization is sufficient to reach threshold, it will initiate an action potential in the second region. In this manner, each successive region is depolarized to threshold and generates an action potential. As a consequence of these mechanisms, the action potential propagates in a direction from the soma to the axon terminal.

The Velocity of the Action Potential

The propagation of an action potential along the axon occurs at a constant speed. In unmyelinated nerve fibers, the conduction velocity is proportional to the diameter of the axon. The larger the diameter of the axon, the greater the speed of propagation. This is because axons with large diameters do not offer as much resistance to the flow of ions along the length of the axon. In myelinated axons, the velocity of propagation is determined not only by the diameter of the axon but also by the distance between the nodes of Ranvier.

The formation of myelin around an axon prevents the penetration of ions needed for the conduction of

(a)

(b)

Figure 7-20

The propagation of the action potential. (*a*) In a resting axon, ion concentrations are maintained by active transport. (*b*) The action potential travels as adjacent regions of the axon are successively depolarized by the movement of Na⁺ ions.

the action potential. Between the segments of myelin, however, are nodes of Ranvier, where the membrane of the nerve axon is exposed and contains large numbers of Na⁺ channels. It is in the nodes of Ranvier that the membrane is depolarized and action potentials are generated (Fig. 7–21). The generation of an

action potential in one node of Ranvier causes the membrane potential in the adjacent node of Ranvier to depolarize and also to generate an action potential. In this way, the propagation of an action potential along a myelinated nerve axon appears to jump (*saltus* in Latin) from one node of Ranvier to the next

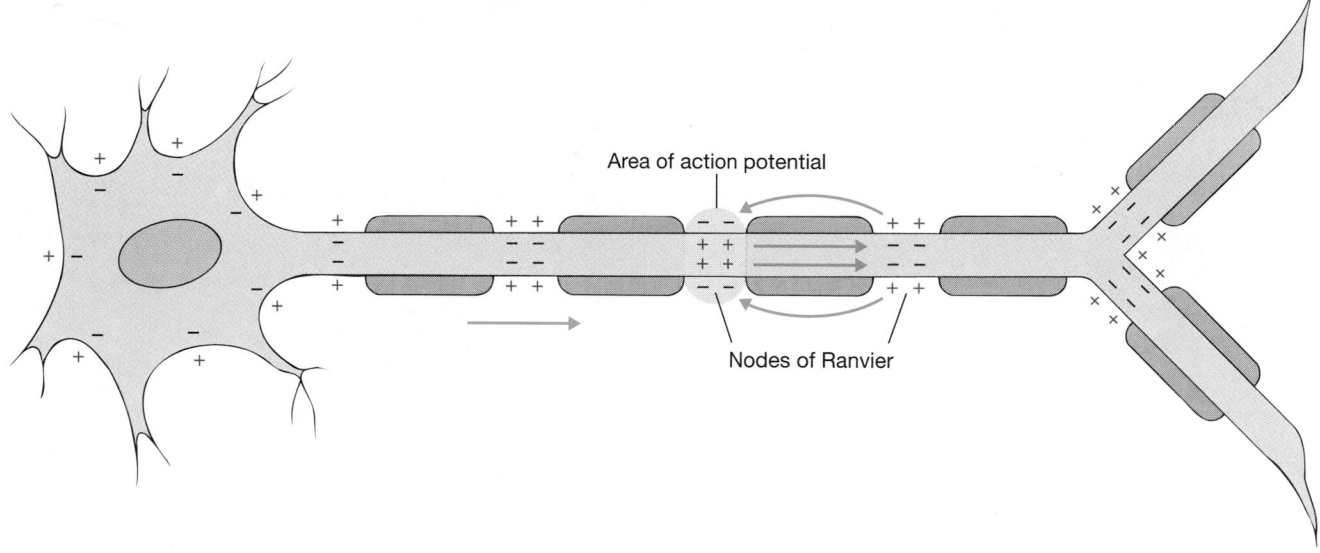

Figure 7–21

Saltatory conduction. In a myelinated axon, the impulse jumps from one node of Ranvier to the next.

in the process of **saltatory conduction.** Thus the greater the distance between nodes of Ranvier, the greater the velocity of action potential propagation.

When the action potential reaches the end of the axon, it invades the **synaptic terminal.** This structure is one of the key elements in the communication between nerve cells. At the synaptic terminal, the action potentials transmitted along the axon cause the release of chemicals used in the communication between neurons in a process called **synaptic transmission.**

Communication Between Nerve Cells: Synaptic Transmission

The communication between nerve cells occurs at junctions called **synapses.** In the nervous system, there are two types of synapses, **electrical synapses** and **chemical synapses.** Electrical synapses consist of **gap junctions** that permit the direct passage of ions and other small molecules from one cell to another. Electrical synapses transmit information in both directions. In chemical synapses, on the other hand, there is a space, called a **synaptic cleft,** between one cell and the other that prevents the direct passage of ions. In order for ions to flow into the second cell, chemical transmitters must be released from the first cell into the synaptic cleft, where they bind with a receptor located on the second cell.

Electrical Synapses Mediate Communication Between Adjacent Cells by the Passage of Molecules Between Cells

The structure of the electrical synapse is similar to the bridge junctions or gap junctions seen in other cells (see Chapter 3). Under the electron microscope, the membranes of two adjoining cells appear to be fused together. Studies have shown that the pre- and postsynaptic cells are connected by a protein channel called a **connexon** that spans the gap between them (Fig. 7–22). The connexon is made of six protein subunits called **connexin** that are arranged into a hexagonal assembly.

Electrical synapses are found between axons and soma, axons and dendrites, dendrites and dendrites, and soma and soma. These synapses are pathways for the direct conduction of ions from one cell to another. In addition, the channels are large enough to allow the passage of molecules such as cAMP, sucrose, and small peptides. Thus, these "electrical" synapses serve as channels for both electrical and metabolic communication.

In terms of electrical communication, these synapses synchronize the activity of many adjoining cells as well as provide a pathway for a rapid communication between cells. As we will see in a later chapter, electrical synapses in the form of gap junctions are also found in cardiac cells of the heart, many

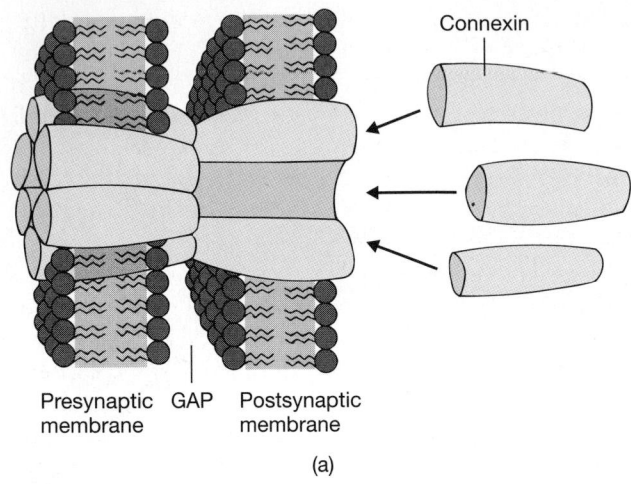

Connexin

Presynaptic
membrane

GAP

Postsynaptic
membrane

(a)

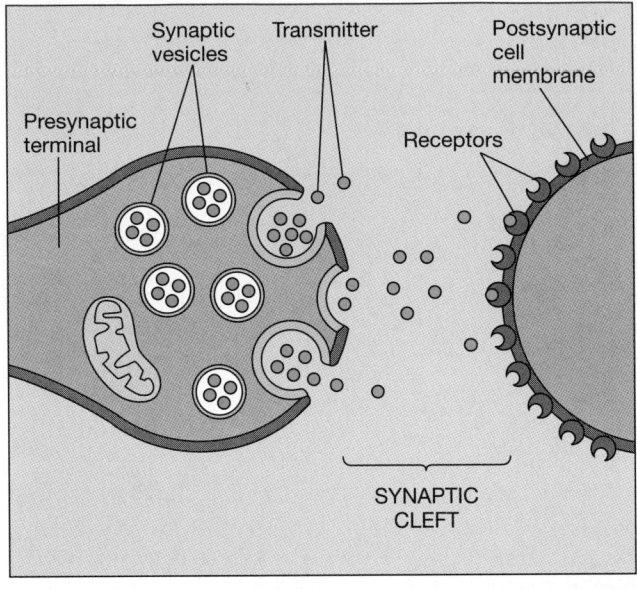

Synaptic
vesicles

Transmitter

Postsynaptic
cell
membrane

Presynaptic
terminal

Receptors

SYNAPTIC
CLEFT

(d)

(b)

SV

SC

(c)

(e)

smooth muscle cells, and other cells that display a synchronization of activity.

Chemical Synapses Mediate Communication Between Distant Cells by Transmitter-Receptor Interactions

Although electrical synapses are found in many areas of the nervous system, the predominant type of synapse is the chemical synapse. In chemical synapses, at least two cells participate—the cell producing the chemical signal, called the **presynaptic cell,** and the target cell receiving the signal, called the **postsynaptic cell.** The presynaptic component of the synapse consists of the terminal ending, which contains vesicles called **synaptic vesicles** (Fig. 7–22d). These vesicles, filled with chemicals referred to as **neurotransmitters** (or **transmitters**), fuse with the presynaptic membrane. The fusion of the vesicle with the presynaptic cell membrane causes the release of the chemical neurotransmitters into the **synaptic cleft.** The transmitters in turn act upon receptors located in postsynaptic cell membranes. Synaptic contacts can be formed at different sites on the postsynaptic cell. Those between axons and dendrites are called **axodendritic synapses.** Contacts between axons and somata are called **axosomatic synapses,** and contacts between two axons are called **axoaxonic synapses.** In the peripheral nervous system, the postsynaptic cell can also be a muscle or a glandular cell.

Presynaptic Mechanisms of Chemical Transmission

The process of synaptic transmission involves four phases: the synthesis of the transmitter, the storage and release of the transmitter, the interaction of the transmitter with the postsynaptic receptor, and the termination of synaptic transmission (Fig. 7–23). The biochemical mechanisms for synthesis differ for each neurotransmitter and will be considered in detail later in the chapter. In general, however, each type of transmitter is packaged and stored in vesicles, which are then positioned for the release of their chemical transmitters into the cleft.

The release of the neurotransmitter is produced by the action potential when it invades the terminal ending. The resulting change in the membrane potential activates voltage-sensitive Ca^{2+} channels, causing an influx of Ca^{2+} ions into the terminal. By mechanisms not yet clearly understood, the Ca^{2+} ions cause the synaptic vesicles to fuse with the presynaptic membrane and release transmitter molecules into the synaptic cleft by the process of **exocytosis.**

Each synaptic vesicle releases a fixed, or **quantum,** number of neurotransmitter molecules. Acetylcholine-containing vesicles, for example, have approximately 10,000 molecules per vesicle. The number of synaptic vesicles that fuse with the synaptic membrane to release their chemical transmitters depends on the concentration of Ca^{2+} ions within the synaptic terminal. Greater concentrations of Ca^{2+} within the terminal cause more vesicles to release their contents into the synaptic cleft.

The amount of Ca^{2+} within the terminal is regulated by the activity of the cell (that is, the number of action potentials it has generated). As action potentials invade the synaptic terminal with greater frequency, there is a residual increase in Ca^{2+} ions within the terminal. This increase in Ca^{2+} leads to a greater release of the neurotransmitter from the synaptic terminal. In this manner, the frequency of action potentials produced by the nerve cell can govern the amount of chemical transmitter it releases.

The amount of transmitter released can also be affected by extrinsic processes such as synaptic inputs

◀ *Figure 7–22*

Types of synapses. An electrical synapse (a) consists of a gap junction formed by proteins in adjacent cell membranes. The individual subunits that make up each half of the junction are known as connexins. (b) Electron micrograph of a gap junction between smooth muscle cells in the wall of the uterus. (c) Several gap junctions shown at a lower magnification (approximately ×36,000). (Electron micrographs courtesy of Puri and Garfield, from *Biology of Reproduction,* 27:967–975, 1982.) (d) Chemical synapses consist of a presynaptic axon terminal, a synaptic cleft, and a postsynaptic cell membrane. Synaptic vesicles release transmitter molecules that diffuse across the synaptic cleft and bind to receptors on the target (postsynaptic) cell membrane. (e) Electron micrograph of a chemical synapse, showing fusion of a vesicle with a presynaptic membrane, approximately ×125,000. SV = synaptic vesicles; SC = synaptic cleft. *(Courtesy of Dr. John Heuser.)*

(a)

(b)

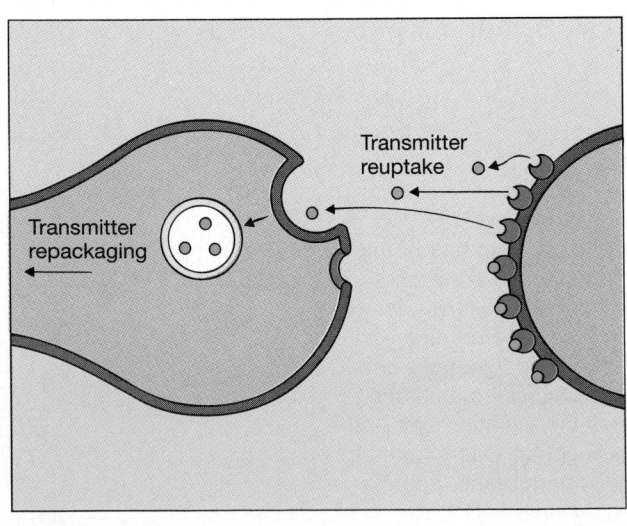

(c)

from other cells that alter the calcium levels within the terminals. The mechanisms of these axoaxonic inputs will be discussed later in this chapter.

Postsynaptic Mechanisms of Chemical Transmission

After a chemical transmitter is released into the synaptic cleft, it can bind to a receptor located on the postsynaptic cell. The coupling of the transmitter with the receptor causes the receptor to open, permitting the passage of specific ions through the membrane. The receptor can directly activate the ion channel or indirectly activate it through a second messenger such as cAMP, cGMP, or IP_3.

The movement of charged ions into or out of a postsynaptic cell can affect its membrane potential. For instance, the activation of a neurotransmitter-sensitive ion channel that permits the influx of Na^+ into the cell results in more positive charges inside of the cell. This in turn depolarizes the cell, making the membrane potential more positive (Fig. 7–24). Likewise, the efflux of positive charges from the interior of the cell or the influx of negative charges into the cell hyperpolarizes it, causing the membrane potential to become more negative.

The properties of these chemically activated ion channels are quite different from those of the voltage-activated ion channels responsible for the action potential. First of all, the chemically activated channels remain open as long as the transmitter is bound to the receptor. Secondly, the chemically activated channels are generally not sensitive to changes in the membrane potential.

The activation of a single receptor results in the influx or efflux of one or more types of ion. The number of ions that flow through the channel per unit time is referred to as the **single channel current.** The **synaptic current** is made up of all the single channel currents through the membrane at the synapse.

The flow of synaptic current resulting from the release of transmitters from one vesicle produces a

Figure 7–23

In the presynaptic terminal, chemical transmitter molecules are synthesized by various enzymes and packaged into secretory vesicles (*a*). When calcium ions flow into the cell, the secretory vesicles fuse with the cell membrane, and the transmitters are released into the synaptic cleft (*b*). Some transmitter molecules bind to receptor proteins on the postsynaptic membrane, whereas others are pumped back into the presynaptic terminal by a reuptake mechanism (*c*).

Figure 7–24

Ion flow and postsynaptic potentials. (*a*) Resting membrane potential. (*b*) When a transmitter binds to a receptor, the receptor depolarizes the membrane by allowing cations to flow into the cell. Chemical transmitters can also produce hyperpolarization by causing cations to flow out (*c*) or anions to flow in (*d*) to the cell.

change in the membrane potential referred to as a **unitary postsynaptic potential** (Fig. 7–25). These postsynaptic potentials can summate, or act together, to depolarize or hyperpolarize the membrane. Postsynaptic potentials that depolarize the membrane tend to excite the nerve cells to discharge action po-

tentials and are therefore called **excitatory postsynaptic potentials (EPSPs).** In contrast, postsynaptic potentials that hyperpolarize the membrane potential tend to prevent or inhibit the nerve cell from generating an action potential and are therefore called **inhibitory postsynaptic potentials (IPSPs).**

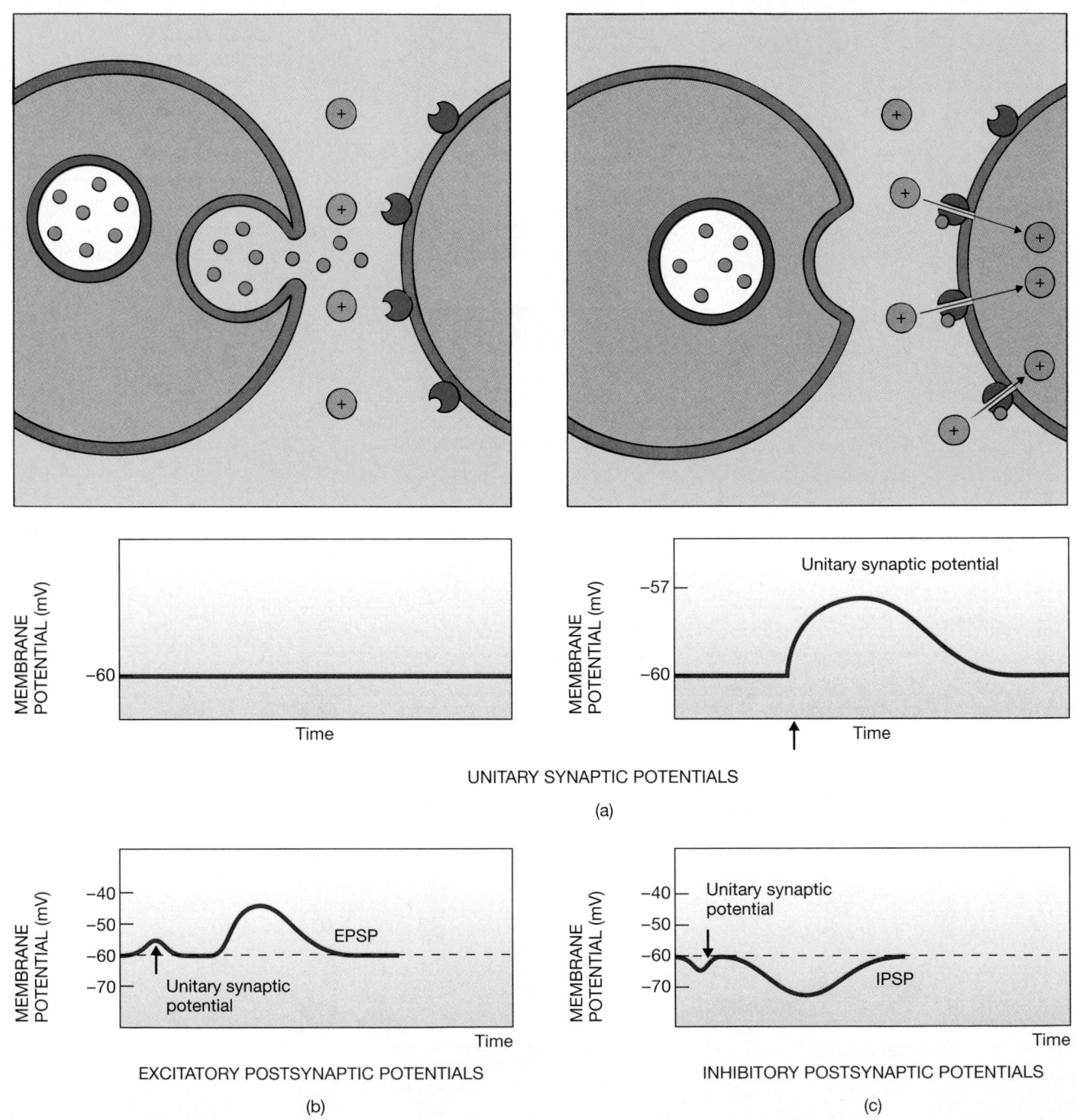

Figure 7–25

Postsynaptic potentials. (*a*) Unitary synaptic potentials. (*b*) Excitatory postsynaptic potentials (EPSPs). (*c*) Inhibitory postsynaptic potentials (IPSPs).

Excitatory Postsynaptic Potentials. A number of ions are capable of producing EPSPs. We will first focus on the mechanisms involved in producing EPSPs by Na^+ ions alone. The synaptic activation of a receptor that selectively allows Na^+ to flow results in the influx of Na^+ into the interior of the cell. The addition of more positive charges into the cell therefore depolarizes the membrane. The influx of Na^+ ions is due to two factors: a concentration gradient and an electrical gradient. As we learned earlier, the concentration of Na^+ ions is greater outside of the cell than inside; this concentration gradient moves Na^+ ions into the cell. In addition, a membrane potential of -60 mV establishes a strong electrical gradient for the influx of Na^+ (Fig. 7–26). The equilibrium potential for Na^+ is typically $+55$ mV. Recall that the equilibrium potential is the membrane potential at which the movement of ions resulting from electrical forces is equal to the movement of ions—in the opposite direction—resulting from the concentration gradient. Since the value of the resting membrane potential is much less than the equilibrium potential for Na^+, the electrical force will not be sufficient to counteract the inward flux of Na^+ ions. In fact, the electrical driving force for Na^+ aids in the inward influx of Na^+ ions.

The electrical driving force acting on a particular ion is the difference between the membrane potential and the equilibrium potential for that ion ($E_m - E_{ion}$). In the case of Na^+, this driving force is $[-60$ mV $- (+55$ mV$)]$ or -115 mV. The negative value of the electrical driving force for Na^+ aids in attracting Na^+ ions into the cell. Thus, the concentration gradient and the electrical gradient both act to produce an influx of Na^+.

Excitatory postsynaptic potentials can also be produced by the opening of channels that are somewhat nonselective and result in the simultaneous flow of both Na^+ ions into the cell and K^+ ions out of the cell. This is characteristic of a type of acetylcholine receptor found at the junction between nerve and muscle cells. How can an ion channel that allows positively charged ions such as K^+ to flow out of a cell cause the membrane to depolarize? This occurs because more Na^+ ions flow into the cell than do K^+ ions out. At a resting membrane potential of -60 mV, the driving forces that influence the influx of Na^+ ions are greater than those that affect the efflux of K^+ ions. Like the Na^+ ion, the forces acting on K^+ pro-

Figure 7–26

The net gradient for ion flow comprises the electrical gradient (*a*) and the chemical (concentration) gradient (*b*). The net gradient for Na^+ flow into the cell is greater than the net gradient K^+ flow out of the cell (*c*).

(a)

(b)

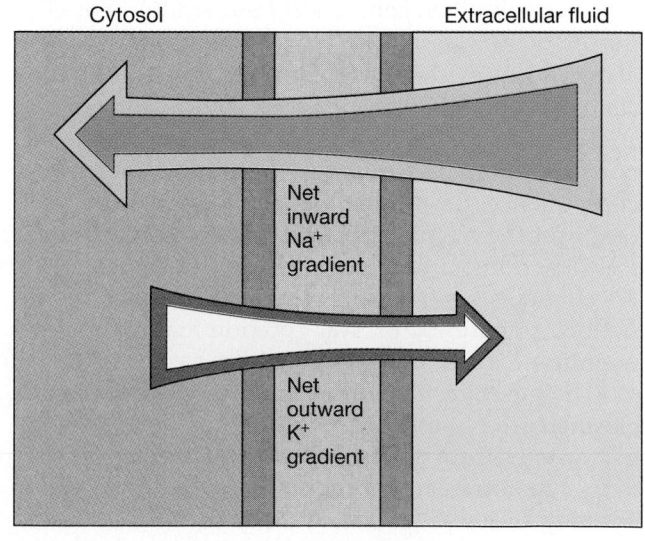

(c)

duce K^+ fluxes that act in the same direction, but in the case of K^+ we have an outward flux. Moreover, the outward K^+ flux is not as great as the inward Na^+ flux.

The electrical force resulting from the difference between the resting potential and the equilibrium potential for K^+ is [-60 mV $- (-75$ mV)] or $+15$ mV. The positive value of the driving force results in an efflux of K^+ ions. Together, the concentration and electrical forces act to move K^+ ions from the interior of the cell outward. From the calculation of electrical driving forces, however, we can see that the resulting magnitude of the K^+ efflux is much smaller than the magnitude of the Na^+ influx (see Fig. 7–26). Thus, when this type of acetylcholine receptor is activated, the total influx of Na^+ ions will be substantially greater than the efflux of K^+ and will therefore result in the depolarization of the membrane.

Inhibitory Postsynaptic Potentials. As with the EPSPs, there are a number of ways that an IPSP can be produced. One mechanism is the synaptic activation of a receptor that opens its channel to K^+ ions. Because the concentration and electrical gradients for K^+ cause it to move outward, the interior of the cell loses positive charges and becomes more negative. Consequently, the membrane potential hyperpolarizes, and an inhibitory postsynaptic potential develops. Other receptors that produce IPSPs are those that permit Cl^- to flow across the membrane. The increase in the flow of negative charges into the cell also acts to hyperpolarize the membrane potential.

Closing of Ion Channels by Receptors. Our discussion of transmitter-receptor interactions has thus far focused only on the opening of ion channels. There is also a class of receptors, however, that close their ion channels when coupled to a transmitter. The closing of these chemically activated ion channels can also produce depolarization or hyperpolarizing changes of the membrane potential.

For example, the closing of Na^+ channels that are normally open prevents the influx of Na^+ ions. As a consequence, the efflux of K^+ is the predominant effect, and the membrane hyperpolarizes. In a similar fashion, transmitter-receptor interactions that close K^+ channels prevent the efflux of K^+ ions. In this situation, the influx of Na^+ predominates, and the membrane depolarizes.

The mechanisms of transmitter-activated channel closings are typically carried out by second messenger systems from within the postsynaptic cell and last longer than transmitter mechanisms that involve the opening of ion channels. The details of the ways in which channels close vary with different types of re-

ceptors and will be discussed in relation to their associated neurotransmitters.

Termination of Synaptic Transmission

The termination of synaptic transmission occurs when the transmitter is removed from the synaptic cleft. This process is accomplished in most neurotransmitter systems by the transport of the transmitter back into the presynaptic terminal. A **reuptake** mechanism pumps most transmitters back into the presynaptic terminal. Other transmitters are removed from the synaptic cleft by degrading enzymes, and the metabolic products are then transported back into the presynaptic terminal. The pumping mechanism is specific for each type of transmitter substance or metabolite and can be affected by selective drugs.

Recycling of Neurotransmitters

After a transmitter has been transported back into the presynaptic terminal, it is again packaged into vesicles for storage in preparation for release. In cases in which the metabolite is transported into the presynaptic terminal, the transmitter is resynthesized from the metabolic precursor and then packaged and stored into vesicles. The repacking of the neurotransmitter is an energy-dependent process requiring ATP. The details of this process will be discussed later in relation to the neurotransmitter, norepinephrine.

Diffuse and Discrete Chemical Synapses

Chemical synapses have been shown to have either discrete or diffuse actions. In **discrete synapses,** the chemical neurotransmitter is released from restricted areas of the presynaptic terminal, called **active zones,** into a small synaptic cleft; only 30 nm separate the pre- and postsynaptic membranes (Fig. 7–27). An example of a discrete synapse is the junction between nerve and striated muscle cells, which tend to activate a small region on the muscle fiber.

In **diffuse synapses,** on the other hand, transmitter release is not limited to specific active zones, and the distance between the presynaptic and postsynaptic membrane can be as much as 150 nm. These synapses have the form of beads on a chain, called **varicosities.** The varicosities are an extension of the axon and form an overlapping network of synapses **en passant,** or synapses in passage. As the action potential invades each varicosity, vesicles fuse with the presynaptic membrane, and the action potential continues on to the next varicosity. The effect of this type of synapse is to activate a large surface area of one cell or a large number of cells in a diffuse manner. The diffuse synapse is typical of nerve terminals of

(a)

(b)

Figure 7–27

(*a*) In discrete synapses, transmitters are released only through active zones of the axon terminal and travel across very small clefts to target cells. (*b*) In diffuse synapses, axon terminals have varicosities that release transmitters over a wider range.

Table 7–1
Low-Molecular-Weight Transmitters

Acetylcholine	$CH_3 - \overset{\overset{\displaystyle O}{\|}}{C}OH_2CH_2\overset{\oplus}{N}(CH_3)_3$
Dopamine	
Norepinephrine	
Serotonin	
Glutamate	$\underset{\|}{NH_2}$
	$HOOCCH_2CH_2CHCOOH$
GABA	$HOOCCH_2CH_2CH_2NH_2$

the sympathetic autonomic nervous system and of nerve cells containing catecholamines within the central nervous system.

Neurotransmitters and Associated Receptors

Chemical neurotransmitters can be broadly classified into two groups: **low-molecular-weight transmitters** and **neuropeptide transmitters** (see Tables 7–1 and 7–2). The low-molecular-weight transmitters are synthesized within the presynaptic terminal. The necessary synthesizing enzymes are produced within the soma and transported to the terminal region. Neuropeptides, on the other hand, are fabricated in the soma and carried via axonal transport mechanisms to the synaptic terminal.

Although the processes of transmitter synthesis, release, degradation, uptake, and physiological effect

are not completely understood for the peptides, they are fairly well characterized for many of the low-molecular-weight transmitters.

Low-Molecular-Weight Transmitters Are Produced at Axon Terminals

Acetylcholine

One of the first low-molecular-weight transmitters to be studied was **acetylcholine (ACh).** It is found not only in the central nervous system but also in the peripheral nervous system, where it acts as the chemical transmitter between nerve and muscle. ACh is synthesized from acetyl CoA and choline with the catalytic enzyme, choline acetyltransferase (Fig. 7–28). ACh is packaged and stored in vesicles within the presynaptic terminal, where it is positioned for release near the active zones. The fusion of the synaptic vesicle with the presynaptic membrane causes the release of ACh into the synaptic cleft. It then diffuses across the cleft and binds with receptors on the postsynaptic membrane.

Table 7–2
Neuropeptides

Neuroactive Peptides: Mammalian Brain Peptides Categorized According to Principal Tissue Localization

Hypothalamic-releasing hormones
  Thyrotropin-releasing hormone
  Gonadotropin-releasing hormone
  Somatostatin
  Corticotropin-releasing hormone
  Growth hormone-releasing hormone

Neurohypophyseal hormones
  Vasopressin
  Oxytocin
  Neurophysin(s)

Pituitary peptides
  Adrenocorticotropin
  β-Endorphin
  α-Melanocyte-stimulating hormone
  Prolactin
  Luteinizing hormone
  Growth hormone
  Thyrotropin

Gastrointestinal peptides
  Vasoactive intestinal polypeptides
  Cholecystokinin
  Gastrin
  Substance P
  Neurotensin
  Methionine-enkephalin
  Leucine-enkephalin
  Insulin
  Glucagon
  Bombesin
  Secretin
  Somatostatin
  Motilin

Others
  Angiotensin II
  Bradykinin
  Sleep peptide(s)
  Calcitonin
  CGRP (calcitonin gene-related peptide)
  Neuropeptide Y

Table 7–3
Types of Neurotransmitters and Receptors

Neurotransmitter	Receptor	Membrane Conductance	Membrane Potential	2nd Msgr
glutamate	kainate	increase g_{Na}, g_K	EPSP	
	quisqualate	increase g_{Na}, g_K	EPSP	
	NMDA	increase g_{Ca}	EPSP	
acetylcholine	nicotinic	increase g_{Na}, g_K	EPSP	
	muscarinic M1	decrease g_K	EPSP	IP$_3$ and DAG
	muscarinic M2	increase g_K	IPSP	cAMP
serotonin	5HT-1A	increase g_K	IPSP	cAMP
	5HT-1B			
	5HT-1C	increase g_{Cl}	IPSP	IP$_3$
	5HT-1D			
	5HT-2	decrease g_K	EPSP	IP$_3$
	5HT-3	increase g_{Na}, g_K	EPSP	
GABA	GABA-A	increase g_{Cl}	IPSP	
	GABA-B	increase g_K	IPSP	cAMP?
dopamine	D-1		EPSP?	cAMP
	D-2		IPSP	cAMP
norepinephrine	alpha-1	increase g_K	IPSP (CNS) [contraction (PNS)]	IP$_3$, DAG
	alpha-2	decrease g_{Ca}		cAMP
	beta-1		[heart acceleration]	cAMP
	beta-2		[dilation (PNS)]	cAMP

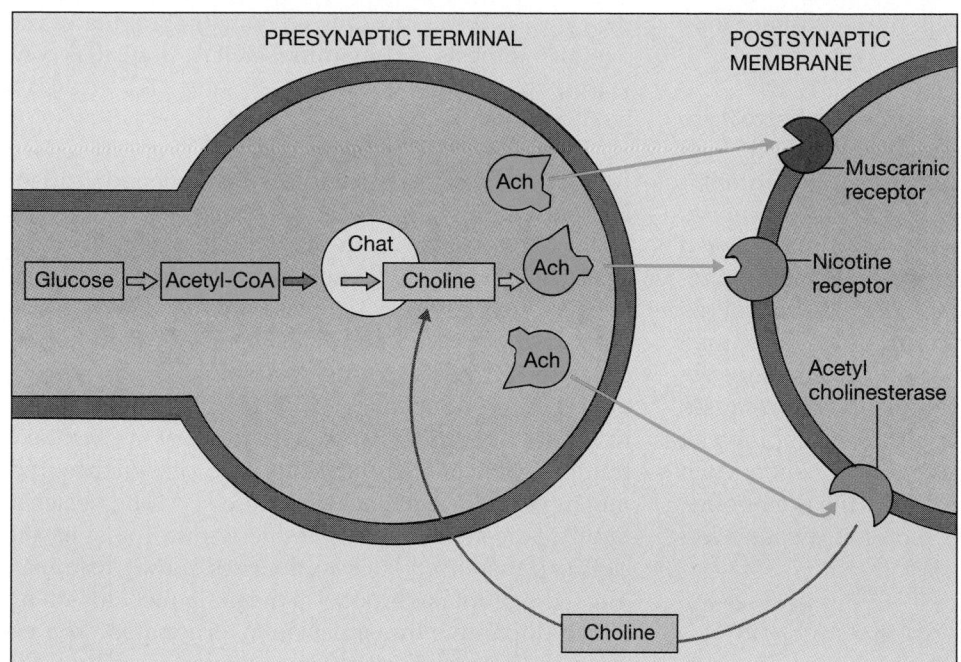

PRESYNAPTIC TERMINAL

POSTSYNAPTIC MEMBRANE

(a)

(b)

Figure 7–28

(a) Acetylcholine is synthesized from acetyl CoA in the axon terminal and binds to two types of receptors: muscarinic and nicotinic receptors. After binding to acetylcholinesterase, acetylcholine is degraded to choline and taken up into the presynaptic terminal for resynthesis of ACh. *(b)* The nicotinic acetylcholine receptor consists of five subunits. When acetylcholine binds to two of the subunits, the protein opens to form an ion channel into the postsynaptic cell.

In the central and peripheral nervous systems, there are two types of receptors for acetylcholine: nicotinic receptors and muscarinic receptors. Nicotinic acetylcholine receptors are sensitive to nicotine, while muscarinic receptors respond to the drug muscarine.

Nicotinic Acetylcholine Receptors. Nicotinic acetylcholine receptors have been well characterized and found to consist of five subunits: beta, gamma, delta, and two alphas. The five components are thought to be arranged so that an ion channel is formed at the central core of the receptor (Fig. 7–28*b*). The active binding sites for the nicotinic receptor are on the two alpha subunits. Both binding sites must be occupied by an acetylcholine molecule in order for the receptor to become activated. Once activated, the receptor opens its gate to permit the simultaneous influx of Na^+ ions and the efflux of K^+ ions (see Table 7–3).

As we have seen, the driving forces for Na^+ influx are much greater than for K^+ efflux. Consequently, the excess positive charge movement into the cell causes the membrane to depolarize. The ion channel of the nicotinic receptor remains in the open state until ACh uncouples from its receptor. After ACh dissociates from its receptor, the receptor channel closes and Na^+ and K^+ are no longer able to pass through the channel. ACh is then free to diffuse within the synaptic cleft, where it binds with the membrane-bound enzyme acetylcholinesterase, **AChE.** The AChE enzyme degrades ACh by hydrolysis to produce choline and acetate. Choline is then taken up into the presynaptic terminal by a high-affinity uptake mechanism and recycled to be used again to synthesize ACh. This process of cholinergic nicotinic transmission is found at the neuromuscular junction in addition to various locations in the central nervous system.

Muscarinic Acetylcholine Receptors.

The structure and function of the second type of acetylcholine receptor, the muscarinic receptor, are quite different from those of the nicotinic acetylcholine receptor. Two types of muscarinic receptors, M1 and M2, have been identified. Both are composed of 7 membrane-spanning domains and both exert their actions through a G protein. The activation of M1 receptors, however, results in a decrease of K^+ conductance via phospholipase C, while M2 receptor activation causes an increase in K^+ conductance via the inhibition of adenylate cyclase. As a consequence, when ACh binds to an M1 receptor it depolarizes the membrane, and when it binds to an M2 receptor it causes a hyperpolarization.

Biogenic Amines

Biogenic amines are a class of low-molecular-weight transmitters characterized by the presence of an amine group. A subgroup of the biogenic amines, the catecholamines, have a catechol ring. This subgroup includes the transmitters **dopamine, norepinephrine,** and **epinephrine.** The synthesis of each catecholamine follows a similar biochemical pathway that starts with the synthesis of dopamine.

Dopamine.

Dopamine is synthesized from the amino acid tyrosine, which is converted to DOPA by the enzyme tyrosine hydroxylase (TH). Tyrosine hydroxylase is the controlling enzyme that regulates the overall synthesis of dopamine. DOPA, in turn, is converted to dopamine by the enzyme DOPA decarboxylase (Fig. 7–29).

Based on their pharmacological properties, two subtypes of dopamine receptors have been identified: D1 and D2 (Fig. 7–30). **D1 receptors** are coupled to G_s proteins, leading to the activation of adenylate cyclase (AC), while **D2 receptors** are coupled to G_i proteins, which cause a decrease in AC activity. The activation of D2 receptors has been shown to hyperpolarize the postsynaptic membrane by increasing potassium conductance. After uncoupling with the receptor, dopamine is transported by a reuptake pumping mechanism back into the presynaptic terminal, where it is repackaged into synaptic vesicles. Eighty percent of the dopamine within the synaptic cleft is transported back to the presynaptic terminal. Thus, the reuptake process is the main mechanism by which dopamine transmission is terminated. The remaining 20% of dopamine is degraded within the cleft by the enzyme catechol-O-methyl transferase (COMT).

The synaptic transmission of dopamine is greatly affected by commonly used illegal drugs. Cocaine, for example, inhibits the reuptake of dopamine into the presynaptic terminal, and amphetamine increases the release of dopamine into the synaptic cleft. Both drugs effectively increase the levels of dopamine within the cleft to activate dopamine receptors.

Norepinephrine.

Norepinephrine (NE) is another member of the catecholamine family that is found

PRESYNAPTIC TERMINAL

POSTSYNAPTIC MEMBRANE

Tyrosine hydroxylase

DOPA decarboxylase

Re-uptake

Tyrosine

DOPA

DA

DA

DA

DA

DA

DA

Dopamine receptor

Figure 7–29

Dopamine is synthesized in the axon terminal and released to the synaptic cleft, to be pumped back into the presynaptic terminal.

(a)

(b)

(c)

Figure 7–30

(*a*) Norepinephrine (NE) binds to two types of receptors on the postsynaptic membrane: $\alpha 1$ and β_1. On the presynaptic membrane, binding of NE to α_2 receptors causes norepi-nephrine release to be inhibited (*b*), while binding of NE to β_2 receptors causes more NE to be released (*c*).

CLINICAL FOCUS

Alzheimer's Disease: The Loss of Acetylcholine Neurons and Memory

Alzheimer's disease is characterized by the loss of recent memories. Initially, a person's more distant memories are preserved, but as the disease progresses even the earliest of memories are lost. Eventually, the victim is unable to recognize even close family members. As the disease progresses further, other deficits begin to appear. Along with loss of memory, speech is often impaired; gradually, difficulties in reading, writing, and performing complex movements occur. In later stages, Alzheimer victims are unable to perform even simple tasks such as eating, dressing, and caring for themselves. Most patients die within 5 to 15 years after the onset of the disease.

ChAT immunocytochemistry of cholinergic neurons.
(Courtesy of Dr. Walter Low.)

Postmortem studies of brain tissue from Alzheimer's patients have revealed a selective loss of neurons that synthesize acetylcholine. These acetylcholine neurons are located at the base of the brain (the nucleus basalis of Meynert, the diagonal band of Broca, and the medial septal nucleus) and send axons to all areas of the neocortex and hippocampal formation. These areas of brain are responsible for sensory perception, movement, speech, and language, as well as learning and memory (see Chapters 8, 9, and 11). The loss of these neurons is thought to cause the gradual deterioration of memory, language, and motor functions.

Of all individuals with Alzheimer's, 15% to 20% have inherited the disorder. Research indicates that the genetic form of this disease follows an autosomal dominant pattern of inheritance and that offspring have a 50% chance of inheriting the disease from a parent.

Environmental factors also may play a role in Alzheimer's disease. Studies have shown abnormally high concentrations of aluminum in degenerating neurons of Alzheimer's patients; however, the role of aluminum still needs to be clarified. In addition, infectious agents such as viruses may be involved. Slow-acting viruses have been found to cause other forms of dementia in humans. At this time, there is not treatment to prevent or slow the progression of Alzheimer's disease. Drugs that increase the amount of acetylcholine have been considered in the treatment of Alzheimer's patients to improve memory function.

■ **How might you alter the process of acetylcholine synthesis and metabolism to achieve this objective?**

---

throughout the central nervous system and also at the junction between nerves and smooth muscles in the autonomic nervous system. Norepinephrine is synthesized from dopamine by the enzyme dopamine beta-hydoxylase (DBH) (Fig. 7–30a).

The formation of NE is quite labile and is regulated by mechanisms that can change the enzymatic activity associated with NE synthesis over either a short or long period of time. The short-term changes involve the modulation of tyrosine hydroxylase. Typically the activity of TH is regulated by both dopamine and norepinephrine by the process of end-product inhibition (see Chapter 5). When the enzyme is phosphorylated, however, the process of end-product inhibition is less effective. In addition, the phosphorylation of tyrosine hydroxylase increases the affinity of the enzyme to the tyrosine substrate. Thus the phosphorylation of the tyrosine hydroxylase can increase its enzymatic activity. The phosphorylation of tyrosine hydroxylase has been shown to occur by cAMP-dependent protein kinase and Ca^{2+}/calmodulin mechanisms (see Chapter 5). The short-term increases in the synthesis of NE occur within minutes and are easily reversible.

Long-term increases in NE can be caused by factors such as stress, since stress can lead to increases in tyrosine hydroxylase and dopamine beta-hydroxylase enzymes. Environmental factors can therefore play a significant role in modifying nerve cell function on a long-term basis.

The binding of NE to alpha receptor in the central nervous system opens K^+ channels and causes a hyperpolarization of the postsynaptic cell. In the peripheral nervous system, two types of norepinephrine receptors have been identified: alpha receptors and beta receptors. Alpha-1 receptors are found on the smooth muscles of blood vessels. When they are activated, an increase in Ca^{2+} ion influx occurs, which in turn causes the contraction of smooth muscles. Alpha receptors also stimulate the hydrolysis of phosphatidylinositol, a lipid in the postsynaptic membrane that activates the intracellular messenger diacylglycerol. Diacylglycerol in turn activates protein kinase C, which initiates a variety of cellular functions (see Chapter 5).

Beta-2 receptors are also found on smooth muscles. Their activation, however, results in a relaxation of smooth muscles by mechanisms that remain to be determined. In addition to activating ionic channels, the binding of NE to beta-2 receptors activates metabolic pathways by converting ATP to cAMP. As previously discussed, this second messenger in turn activates a cAMP-dependent protein kinase, which regulates a number of important cellular functions (see Chapter 5). Beta-1 receptors are found in the heart, kidney, and adipose tissue, where they are responsible for the acceleration of heart rate, renin secretion, and lypolysis, respectively.

Alpha-2 and beta-2 receptors are located on the membrane of the presynaptic terminal and appear to regulate the amount of NE that is released (Fig. 7–30b). These receptors are often referred to as **autoreceptors.** As the amount of NE within the synaptic cleft is increased, more alpha-2 autoreceptors become activated. These autoreceptors then inhibit the release of NE from the presynaptic terminal in a process of **feedback inhibition.** The activation of beta-2 receptors, on the other hand, increases the release of NE in a process called **feedback excitation.**

After uncoupling from its receptor, NE is taken back up into the presynaptic terminal, where it is repackaged into vesicles in preparation for release again into the synaptic cleft. As with dopamine, this reuptake process removes approximately 80% of the NE from the synaptic cleft and is the main mechanism that terminates NE synaptic transmission. The remaining NE is degraded within the cleft by the enzyme COMT.

Serotonin. Another common biogenic amine is **5-hydroxytryptamine (5-HT),** or **serotonin.** It is found throughout the brain, but is primarily synthesized in the region of the brain stem. The synthesis of 5-HT begins with the amino acid tryptophan, which is converted to 5-hydroxytryptophan (5-HTP) by the enzyme tryptophan hydroxylase (Fig. 7–31). In turn, 5-HTP is converted to serotonin by 5-HTP decarboxylase.

After its release into the synaptic cleft, serotonin can interact with several types of serotonin receptors. The best characterized serotonergic receptors are the 5HT-1A, 5HT-1C, 5HT-2, and 5HT-3 receptors. Activation of either the 5HT-1A or 5HT-1C receptors results in IPSPs. The inhibitory potential of 5HT-1A receptor activation, however, is mediated by an increase in K^+ conductance via cAMP as the second messenger, while that produced by 5HT-1C activation is carried out by an increase in Cl^- conductance via IP_3. Activation of either the 5HT-2 or 5HT-3 receptors results in the generation of EPSPs. The exci-

Figure 7–31

Serotonin neurotransmission.

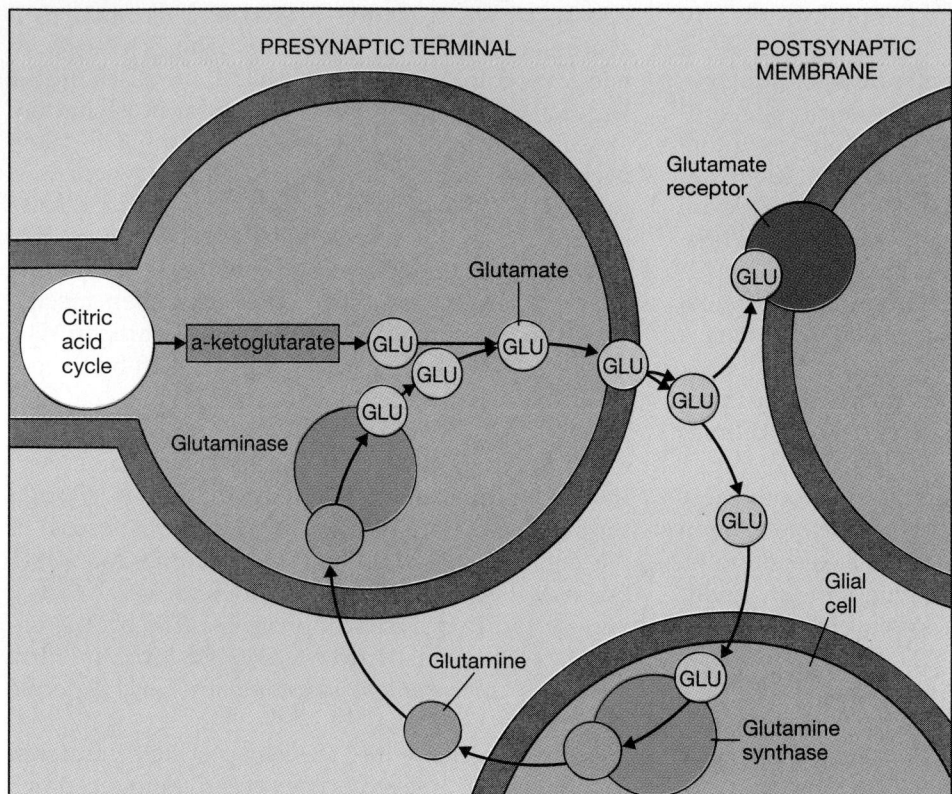

PRESYNAPTIC TERMINAL

POSTSYNAPTIC MEMBRANE

Glutamate receptor

Citric acid cycle

a-ketoglutarate

Glutamate

GLU

Glutaminase

Glutamine

Glutamine synthase

Glial cell

Figure 7–32

Glutamate is synthesized from a product of the citric acid cycle and is converted to glutamine in a glial cell before reuptake.

tatory potential produced by the activation of the 5HT-2 receptor is due to a decrease in K^+ conductance via IP_3, while the 5HT-3 receptor is directly coupled to an ion channel that allows the influx of Na^+ ions and the efflux of K^+ ions.

After uncoupling from its receptor, serotonin is transported back into the presynaptic terminal. As with NE and DA, approximately 80% of the serotonin within the synaptic cleft is removed by this reuptake process. The remainder of the serotonin is degraded by an enzyme called monoamine oxidase (MAO).

Amino Acids.   A variety of amino acids also satisfy the conditions necessary to be classified as neurotransmitters. We will consider only glutamate and gamma amino butyric acid, since they are found in great abundance throughout the nervous system.

Glutamate.   **Glutamate** is synthesized from alpha-ketoglutarate by way of the citric acid cycle (Fig. 7–32). It is one of the most potent excitatory neurotransmitters in the nervous system. There are three subtypes of glutamate receptors: kainate, quisqualate, and N-methyl-D-aspartate (NMDA). Activation of the *kainate* and *quisqualate receptors* produces excitatory postsynaptic potentials by opening ion channels that increase Na^+ and K^+ conductance. *NMDA receptor* activation results in an increase in Ca^{2+} conductance. This receptor, however, is blocked by Mg^{2+} when the membrane is in the resting state and becomes unblocked when the membrane is depolarized. Thus the NMDA receptor can be thought of as both a ligand and a voltage-gated channel.

Synaptic transmission by glutamate is terminated in part by its reuptake into the presynaptic terminal. Much of the glutamate within the synaptic cleft, however, is transported into glial cells. There, it is converted to glutamine by the enzyme glutamine synthase. The glutamine in turn is transported back into the presynaptic terminal, where it is then reconverted to free glutamate and repackaged into vesicles. Glial cells, therefore, play a critical role in regulating glutamate neurotransmission.

Gamma Amino Butyric Acid.   Another amino acid neurotransmitter found throughout the central nervous system is **gamma amino butyric acid (GABA).** GABA is a potent inhibitory neurotransmitter synthesized from glutamate by the enzyme glutamic acid decarboxylase (GAD) (Fig. 7–33).

There are two types of GABA receptors: $GABA_A$ and $GABA_B$. The **$GABA_A$** receptor is a ligand-gated Cl^- channel, and its activation produces inhibitory postsynaptic potentials by increasing the influx of Cl^- ions. This receptor is composed of 2 alpha and 2 beta subunits, each of which spans the membrane 4 times. Its activation requires the binding of GABA to the beta subunits. The increase in Cl^- conductance is

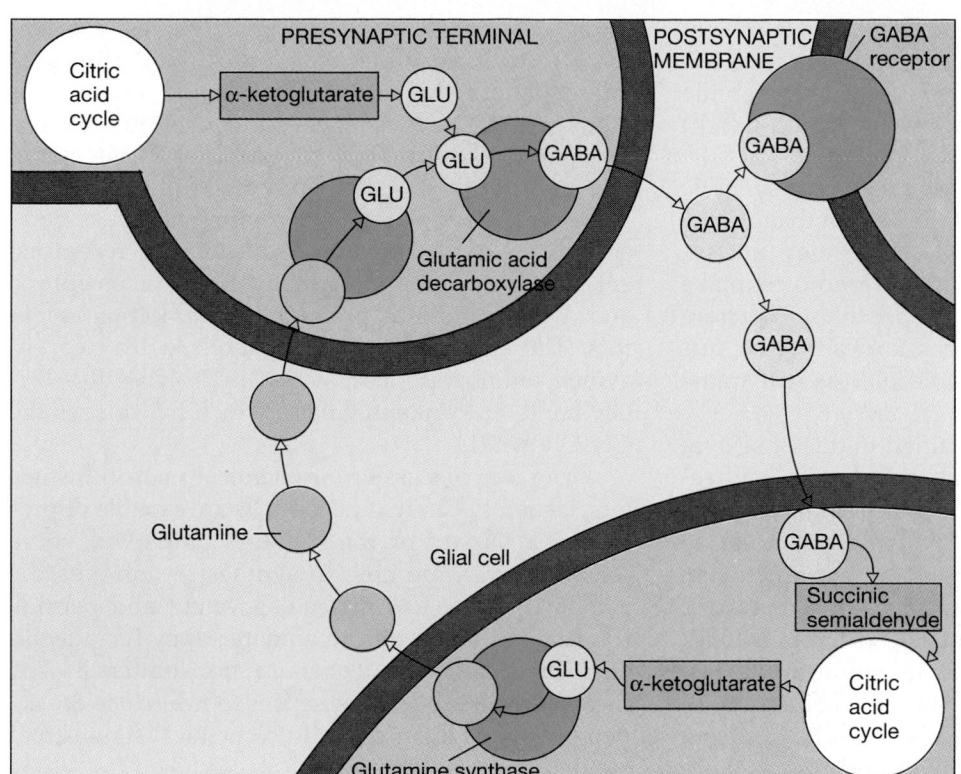

PRESYNAPTIC TERMINAL

POSTSYNAPTIC MEMBRANE

GABA receptor

Figure 7–33
The GABA synapse.

facilitated by drugs called **benzodiazepines,** which bind to the alpha subunits. Benzodiazepines such as librium are used as anticonvulsants and sedatives.

The activation of the **GABA$_B$** receptor also produces inhibitory postsynaptic potentials. With this receptor, however, the IPSP results from an increase in K$^+$ conductance via the activation of a G protein. The synaptic transmission of GABA is terminated by its reuptake into the presynaptic terminal and by its transport into glial cells. The mitochondria within glial cells convert GABA into succinic semialdehyde by the enzyme GABA-T. At the same time, this enzyme is coupled to the conversion of alpha-ketoglutarate to glutamate. Glutamate in turn is converted to glutamine by glutamine synthase and is then transported to the presynaptic terminal. Within the presynaptic terminal, glutamine is converted into glutamate and subsequently into GABA to be packaged into synaptic vesicles.

High-Molecular-Weight Neuropeptides Are Produced in the Soma

Neuropeptides are chemical transmitters that consist of chains of amino acids. The processing of neuropeptides differs considerably from that of low-molecular-weight transmitters. Neuropeptides are synthesized in the soma of neurons rather than in the synaptic terminal. In addition, neuropeptides are created when large proteins, or **polyproteins,** are broken down. The various peptides are packaged within secretory vesicles and carried to the terminal area by mechanisms of fast axonal transport. Within the synaptic terminal, vesicles containing peptides are found to coexist with vesicles containing low-molecular-weight transmitters. Table 7–4 lists the

Table 7–4

Colocalization of Low-Molecular-Weight Transmitters and Neuropeptides

Low-Molecular-Weight Transmitter	*Neuropeptide*
Acetylcholine	Vasoactive intestinal peptide
Norepinephrine	Somatostatin Enkephalin Neurotensin
Dopamine	Cholecystokinin Enkephalin
Adrenalin	Enkephalin
Serotonin	Substance P Thyrotropin-releasing hormone

low-molecular-weight and neuropeptide transmitters that have been found within the same synaptic terminals. It is thought that the released neuropeptide modulates the actions of the low-molecular-weight neurotransmitter.

At the terminal ending, the process of synaptic transmission of peptides is different from that of low-molecular-weight transmitters. Once they are released into the synaptic cleft, there are no reuptake mechanisms to recycle the neuropeptides. Therefore the process of peptide transmission cannot be sustained as it is for the low-molecular-weight transmitters.

Neuropeptides can be classified in one of several families of peptides based on their amino acid sequence and function (Table 7–5). The family of **neurohypophyseal peptides** is structurally similar to those found in the posterior pituitary and function to regulate plasma osmolarity and lactation. **Secretins** and **gastrins** are peptides that are structurally similar to the peptides and hormones found in the gastrointestinal system (see Chapter 26). The family of **insulins** is similar in structure to the insulin hormone and is responsible for the growth and maintenance of nerve cells, while the **somatostatins** are structurally similar to growth hormone.

Opiates are peptides that bind to opioid receptors. They appear to be involved in the regulation of pain information. Opioid peptides include met-enkephalin, leu-enkephalin, dynorphin, and beta-endorphin (see Table 7–5). Structurally, they share homologous regions consisting of the amino acid sequence Tyr-Gly-Gly-Phe. The opiates are derived from three propeptides: **proenkephalin, proopiomelanocortin,** and **prodynorphin.** Proenkephalin gives rise to met- and leu-enkephalin; proopiomelanocortin gives rise to beta-endorphin; and prodynorphin is the precursor of dynorphin. There are several opioid receptor subtypes. Beta-endorphin binds preferentially to **mu receptors;** enkephalins bind preferentially to **delta receptors,** and dynorphin binds preferentially to **kappa receptors.** The enkephalins are metabolized by two enzymes: aminopeptidase, which hydrolyzes the Tyr-Gly bond, and enkephalinase, which hydrolyzes the Gly-Gly bond.

From our discussion of chemical neurotransmitters, it can be seen that nerve cells are capable of producing a variety of transmitters. Individual nerve cells, however, are able to synthesize only specific combinations of low-molecular-weight and peptide transmitters based on the enzymes they have available. This defined set of chemical transmitters is used by a neuron at all of its synapses to transduce the action potentials it generates into chemical signals that are detected by target cells.

Neuronal Integration: Temporal and Spatial Summation

The dendrites and somata of nerve cells have many different types of receptors embedded within their membranes. The synaptic activation of these receptors can simultaneously produce excitatory and inhibitory responses. If a nerve cell receives both excitatory and inhibitory information from many synaptic inputs, how does it decide, so to speak, whether or not to initiate an action potential? The decision is made by adding all of the postsynaptic potentials, using two different processes: temporal summation and spatial summation. If the addition of all of the synaptic potentials results in a depolarization of the membrane potential to approximately $-45\,mV$ near the initial segment or trigger zone of the axon, then an action potential will be generated.

Temporal summation involves the addition of postsynaptic potentials that arise from the activation at one synaptic input (Fig. 7–34). The first time that the synapse is activated, it may produce a depolarization that is not sufficient to cause the membrane potential to reach threshold. This depolarization by itself would decay back to the resting level. If, however, a second action potential invades the terminal, the release of neurotransmitters causes the postsynaptic membrane to again depolarize. Moreover, if

---

Table 7–5
Families of Peptides

Opioid: opiocortins, enkephalins, dynorphin, FMRF amide

Neurohypophyseal: vasopressin, oxytocin, neurophysins

Tachykinins: substance P, physalaemin, kassinin, uperolein, eledoisin

Secretins: secretin, glucagon, vasoactive intestinal peptide, gastric inhibitory peptide, growth hormone–releasing factor, peptide histidine isoleucineamide

Insulins: insulin, somatomedins, relaxin, nerve growth factor

Somatostatins: somatostatins, pancreatic polypeptide

Gastrins: gastrin, cholecystokinins

(text continues on page 249)

RESEARCH FOCUS

Brain Slices and Long-Term Potentiation

Physiological studies of the central nervous system have been greatly advanced by the development of new methods in which tissue from the brain is removed, cut into thin slices, and placed into incubation chambers for experimentation. In this **brain slice** technique, nerve cells are maintained in an extracellular environment similar to that found in a normal brain. The slices can also be perfused with various chemical compounds to determine their effects on neuronal activity and cell-to-cell communication. Another advantage of the slice technique is that it maintains the synaptic organization between nerve cells within the slices of tissue (see figure).

The development of the brain slice technique has enabled scientists to study various cellular mechanisms of neuronal communication. One cellular phenomenon

called **long-term potentiation (LTP)** has been extensively studied using brain slices. LTP is the long-lasting enhancement of synaptic strength and was first observed in the **hippocampal formation,** an area of the brain long known to be involved in processes of learning and memory.

LTP in the hippocampus can be induced by a brief high-frequency stimulation of nerve fibers that synapse upon target pyramidal neurons. The resulting enhancement is specific only to those synapses that have been activated.

In the hippocampal formation, LTP occurs at synapses that use glutamate as the neurotransmitter. The underlying basis of LTP appears to be a persistent alteration in the **quisqualate (Q)** receptor function, most likely either an increase in the number of Q receptors within the synapse, or a change in single channel properties. Whatever the changes in Q receptor function, they occur as a result of a sequence of events. The first step in the induction of LTP is the depolarization of the postsynaptic cell. This is a result of the initial activation of Q receptors by glutamate. In contrast, the **N-methyl-D-aspartate (NMDA)** receptor is not initially activated by glutamate since the channel is still blocked by Mg^{2+} ions when the membrane potential is at the resting level (see text). With greater depolarizations of the postsynaptic membrane, however, Mg^{2+} ions are removed from the channel, permitting the influx of Ca^{2+} ions through the NMDA receptor channel as a result of glutamate binding to this receptor.

The importance of the NDMA receptor is seen in studies where it is blocked by the antagonist **D-2-amino-5-phosphonovalerate (APV).** Perfusions of APV into the incubation chamber can prevent the induction of LTP in hippocampal slices. In addition, the removal of Ca^{2+} from the incubation medium, and hence the removal of extracellular Ca^{2+}, also prevents LTP.

Influx of Ca^{2+} activates a number of possible intermediate steps that lead to LTP. The most important biochemical pathway is the activation of **Ca^{2+}/Calmodulin** CAM-dependent protein kinase II (CaM kinase II). Perfusion of drugs such as trifluoperpazine, which block the action of CaM, also block LTP. Moreover, the intracellular injection of synthetic peptides that inhibit Ca^{2+}/CaM-dependent substrate phosphorylation by CaM kinase II prevents the induction of LTP within the injected cell but not in adjacent hippocampal neurons.

(a)

(b)

(*a*) Hippocampal formation and brain slice. (*b*) Glutamate receptors quisqualate (Q) and N-methyl-D-aspartate (NMDA) involved in long-term potentiation.

■ **Since the influx of Ca^{2+} ions is critical to the induction of LTP in brain slices, how might you determine whether it has similar effect in laboratory animals?**

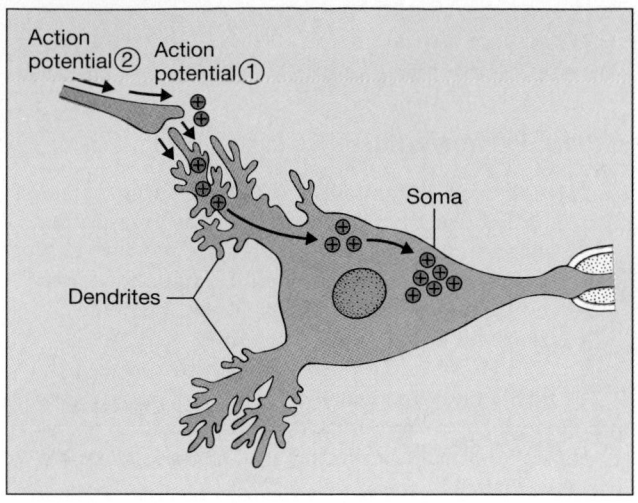

Figure 7–34

Temporal summation of EPSPs.

(a)

(b)

Figure 7–35

Spatial summation of EPSPs.

the time between the first and second action potentials is short enough, the two postsynaptic membrane potentials together will produce a depolarization that may be sufficient to reach the threshold voltage and discharge an action potential.

Another summation process that involves the activation of more than one synaptic input is the process of **spatial summation.** In this case, two synaptic inputs, which alone are insufficient to cause the membrane potential to reach threshold (Fig. 7–35), are activated simultaneously, thus resulting in a depolarization sufficient to achieve a threshold value and discharge an action potential. The action potential discharge by a nerve cell, therefore, is based on the summation of excitatory and inhibitory postsynaptic potentials that occur in time over the space of the neuron's receptive surface.

CHAPTER REVIEW

Summary

- The nervous system is composed of the brain, brain stem, spinal cord, and peripheral nerves. The first three make up the *central nervous system (CNS)*, whereas the latter constitute the *peripheral nervous system (PNS)*. The brain is subdivided into functional regions involved in sensory input from the skin, eye, ears, or other sensory organs. Other areas are involved with locomotor activity as well as other functions. This functional partitioning of brain areas is found throughout the nervous system. A general organizational theme of the nervous system is that the brain receives sensory input and provides the appropriate output for such functions as speech, movement, and the control of internal organs.

- Glia, or glial cells, aid in maintaining the environment of the brain. They also act as scavengers in the brain to remove cellular debris. Another important function of glial cells is the formation of myelin around the axon of nerve cells to increase the conduction and transmission of electrical impulses. *Neurons* are the basic cellular elements in the nervous system involved in the transmission of information. The membranes of neurons have structural and functional properties specialized for the integration and rapid transmission of information from one nerve cell to another.

- Neurons have extensive processes called dendrites and axons that are involved in the reception and transmission of information. The cytoskeletal structure of these processes is made of microtubules and neurofilaments. In addition to providing the structural framework for the neuron, these components are involved in the transport of cellular material from the somal to the outer reaches of axonal and dendritic processes. The fast axonal transport process will carry organelles and vesicles at a rate of 400 mm/day, while slow axonal transport will carry material at a rate of 1.0 to 2.0 mm/day. These processes play important roles in the maintenance and regeneration of nerve fibers. The membranes of nerve cells are composed of passive, voltage-activated, and chemically activated ion channels. These channels are made of proteins embedded within the membrane and permit specific ions to move through the membrane. Chemically activated ion channels, also known as receptors, are opened (or closed) when specific chemicals bind to receptor sites on the channel. Voltage-activated ion channels open (or close) when the membrane is depolarized. Passive ion channels remain in the open state and play a role in establishing the membrane potential.

- A difference in the number of positive and negative ions between the inside and outside of a cell produces the membrane potential. In general, the inside of a cell is more negatively charged than the extracellular space, and thus establishes an electrical driving force for the movement of ions across the membrane. Differences in the concentration of ions between the inside and outside of cells establish concentration gradients, which act as chemical driving forces for the movement of ions across the membrane. The summated effects of electrical and chemical driving forces determine whether an ion will move through the membrane in an inward or outward direction.

- The axon of nerve cells transmits information by generating action potentials. Action potentials are initiated and propagated along the axon as a result of the influx and efflux of Na^+ and K^+ ions, respectively. These ionic currents that flow across the axonal membrane are a result of the opening and closing of voltage-sensitive ion channels for Na^+ and K^+, which are embedded within the membrane.

- Action potentials that reach the axon terminal initiate the process of synaptic transmission. At electrical synapses, the cytoplasm between two cells is in direct contact. Thus ions will move directly from one cell to another. At chemical synapses, synaptic transmission occurs when an action potential depolarizes the membrane within the synaptic terminal to activate voltage-sensitive Ca^{2+} channels. The opening of these channels causes the influx of Ca^{2+} into the terminal, which in turn leads to the fusion of synaptic vesicles with the preterminal membrane. As a consequence, chemical transmitters contained within the vesicles are released into the synaptic cleft, where they eventually bind to receptors located on the membrane of postsynaptic cells.

- The binding of chemicals called neurotransmitters with their receptors affects the ionic conductances of the target cells. These receptors are chemically activated ion channels that permit the passage of specific ions. Well-characterized neurotransmitters include acetylcholine, dopamine, norepinephrine, serotonin, glutamate, and gamma amino butyric acid. Less well understood transmitters known as neuropeptides are synthesized in the soma.

- The influx and efflux of ions through the chemically activated channels produce excitatory or inhibitory postsynaptic potentials. These responses are integrated by the postsynaptic cell to produce a summated response. If the summation of these alterations in the postsynap-

tic membrane causes a depolarization that reaches threshold levels, an action potential will be generated in the postsynaptic nerve cell. The transmission of information between nerve cells, therefore, is an electrochemical process that features the conduction of an electrical impulse coupled to the secretion of chemical transmitters.

Review Questions

Identify Each with a Term

1. The term used for a positively charged ion.
2. The term used for a negatively charged ion.
3. The term used when the interior of the nerve membrane becomes more positively charged than when in its resting state.
4. The term used when the interior of the nerve membrane becomes more negatively charged than when in its resting state.

Define Each Term

5. membrane potential
6. equilibrium potential
7. action potential
8. saltatory conduction

Choose the Correct Answer

9. What is the direction of the driving forces for the movement of Na^+ ions when a nerve cell is at rest?
   a. inward electrical gradient
   b. outward electrical gradient
   c. inward chemical gradient
   d. both (a) and (c)
10. The repolarization of the action potential is a result of:
    a. an increase in sodium ion conductance
    b. a decrease in sodium ion conductance
    c. an increase in potassium ion conductance
    d. both (b) and (c)
11. A peripheral nerve fiber will regenerate at a rate of:
    a. 0.1 mm/day
    b. 0.4 mm/day
    c. 1.0 mm/day
    d. 400 mm/day
12. Which of the following statements is *not* true about glial cells?

a. Astroglia remove degenerating debris.
b. Each oligodendroglia cell forms myelin around one axon.
c. Microglia remove degenerating debris.
d. Glial cells transport glutamate from the synaptic cleft into their cytoplasmic compartment.

Answer Each Question in One or Two Sentences

13. Describe the membrane conductance changes that occur during the absolute refractory period of the action potential.
14. What membrane conductance changes can produce EPSPs?
15. What membrane conductance changes can produce IPSPs?
16. What effects do the concentration and electrical gradients have on the movement of ions across the nerve membrane?

Problem Solving in Physiology

17. If the extra- and intracellular concentrations for K^+ are 4 mM and 120 mM, respectively, what is the equilibrium potential for K^+?
18. The forearm of a patient is electrically stimulated by two electrodes to activate the motor nerves that cause the thumb to contract. The first electrode is placed 15 cm from the tip of the thumb while the second electrode is placed 30 cm away. The first electrode causes the thumb to contract 2.5 msec after the electrical stimulus. The second electrode causes the thumb to contract 4.5 msec after the stimulus. What is the conduction velocity of the motor nerves that innervate the thumb?

Suggested Readings

Cajal, S.R. "A new concept of the histology of the central nervous system," 1892. D.A. Rottenberg (trans.). In D.A. Rottenberg and F.H. Hochberg (eds.) *Neurological Classics in Modern Translation*, New York: Hafner, 1977.

Dale, H.H., Feldberg, W., and Vougt, M. "Release of acetylcholine at voluntary motor nerve endings." *Journal of Physiology*, 86:353–380, 1936.

Eccles, J.C., Fatt, P., and Koketsu, K. "Cholinergic and inhibitory synapses in a pathway from motor-axon col-

laterals to motoneurons." *Journal of Physiology*, 126:524–562, 1954.

Hodgkin, A.L., and Huxley, A.F. "A quantitative description of membrane current and its application to conduction and excitation in the nerve." *Journal of Physiology*, 117:500–544, 1952.

Kandel, E.R., Schwartz, J., and Jessel, T.M. *Principles of Neural Science*. Amsterdam: Elsevier Press, 1991.

Katz, B., and Miledi, R., "The timing of calcium action during neuromuscular transmission." *Journal of Physiology,* 189:535–544, 1967.

Neher, E., and Sakmann, B. "Single-channel currents recorded from membrane of denervated frog muscle fibres." *Nature,* 260:799–802, 1976.

Nitta, A., Itoh, A., Hasegawa, T., and Nabeshima, T. "Beta-amyloid protein-induced Alzheimer's disease animal model. *Neuroscience Letters,* 170:63–66, 1994.

O'Dell, T.J., Huang, P.L., Dawson, T.M., Dinerman, J.L., Snyder, S.H., Kandel, E.R., and Fishman, M.C. "Endothelial NOS and the blockade of LTP by NOS inhibitors in mice lacking neuronal NOS." *Science,* 265:542–546, 1994.

Resnicoff, M., Sell, C., Rubini, M. Coppola, D., Ambrose, D., Baserga, R., and Rubin R. "Rat glioblastoma cells expressing an antisense RNA to the insulin-like growth factor-1 (IGF-1) receptor are nontumorigenic and induce regression of wild-type tumors." *Cancer Research,* 54:2218–2222, 1994.

Sherrington, C.S. *The Integrative Action of the Nervous System,* 1906. Reprint, New Haven: Yale University Press, 1947.

Zalutsky, R.A. and Nicoll, R.A. "Comparison of two forms of long-term potentiation in single hippocampal neurons." *Science,* 248:1619–1624, 1990.

Key Terms

acetylcholine (p. 237)
action potential (p. 218)
autonomic nervous system (p. 210)
axon (p. 215)
axonal transport (p. 218)
brain stem (p. 210)
central nervous system (p. 209)
cerebellum (p. 210)

dendrite (p. 215)
depolarization (p. 223)
electrical potential (p. 220)
glia (p. 212)
ion channel (p. 219)
membrane potential (p. 222)
microfilament (p. 216)
microtubule (p. 217)
myelin (p. 212)

neocortex (p. 209)
nerve (p. 215)
neuron (p. 215)
neurotransmitter (p. 214)
norepinephrine (p. 240)
peripheral nervous system (p. 209)
Schwann cell (p. 212)

spinal cord (p. 209)
synapse (p. 229)
thalamus (p. 210)
visual cortex (p. 210)

Scanning electron micrograph of
inner hair cells of the human ear.
(© Prof. P. Motta/Dept. of Anatomy/
University of La Sapienza, Rome/
Science Photo Library/Photo
Researchers, Inc.)

Chapter 8

Sensory Systems

Key Concepts

Sensory organs detect specific features of a stimulus and decompose it into its elemental components. This information is then transmitted to higher centers in the brain, where the elemental components are used to reconstruct neural representations of the stimulus which are typically in the form of maps.

The transduction of sensory information involves the influx or efflux of ions across cell membranes that lead to alterations in the membrane potential and the discharge of action potentials.

Chapter Outline

General Principles of Sensory Transduction
Sensory Systems and Their Processes

*From the moment she awakens in the morning until she falls asleep at night, Marianne is bombarded with information from the outside world. She awakens to the sound of an alarm clock that is detected by her **auditory system.** Her eyes focus on the clock to send information to her **visual system** and confirm that it is time to get up. Marianne then climbs out of bed into the warmth of a hot shower that is sensed by nerve endings of the **somatic sensory system.** At the breakfast table, the smell of toast and bacon comes wafting to her nose and **olfactory system,** and the flavor of a rich cup of coffee stimulates the taste buds of her **gustatory system.** Throughout the day Marianne is continually exposed to information from the external environment that must reach the brain so that she can respond accordingly. The detection and transmission of this information are the functions of humans' **sensory systems.***

*Information about the external and internal environments is detected by **sensory receptor cells** located within various **sense organs** (Table 8–1). Information about the external environment is transduced by **extero** receptors, while information about the internal environment is transduced by **intero** receptors. (These latter sensory receptors are discussed in Chapter 10.) Sensory information from exteroreceptors is* transmitted to the neocortex by different pathways, depending on the type of sensation (Fig. 8–1). Somatic sensation from the body and limbs, for example, is transmitted from receptor cells to the spinal cord, where it then ascends to the higher levels of the nervous system. Auditory information detected by the ears, on the other hand, is transmitted from sensory receptor cells to the brain stem, where it ascends to eventually reach the neocortex. In general, a sensory receptor will respond only to a specific type of stimulus. After detecting the stimulus, it transduces this information into electrical impulses that encode the stimulus intensity. In this chapter we will consider the mechanisms by which sensory receptors detect and transduce information for touch, vision, hearing, taste, and smell.

General Principles of Sensory Transduction

Sensory receptors are activated when they detect a specific stimulus. This specific stimulus is called an **adequate stimulus** (Fig. 8–2) and is unique to each sensory receptor. In the visual system, for example, the photoreceptors detect colors of light but are insensitive to frequencies of sound. Likewise, the auditory receptors in the ear respond only to sound and are insensitive to light.

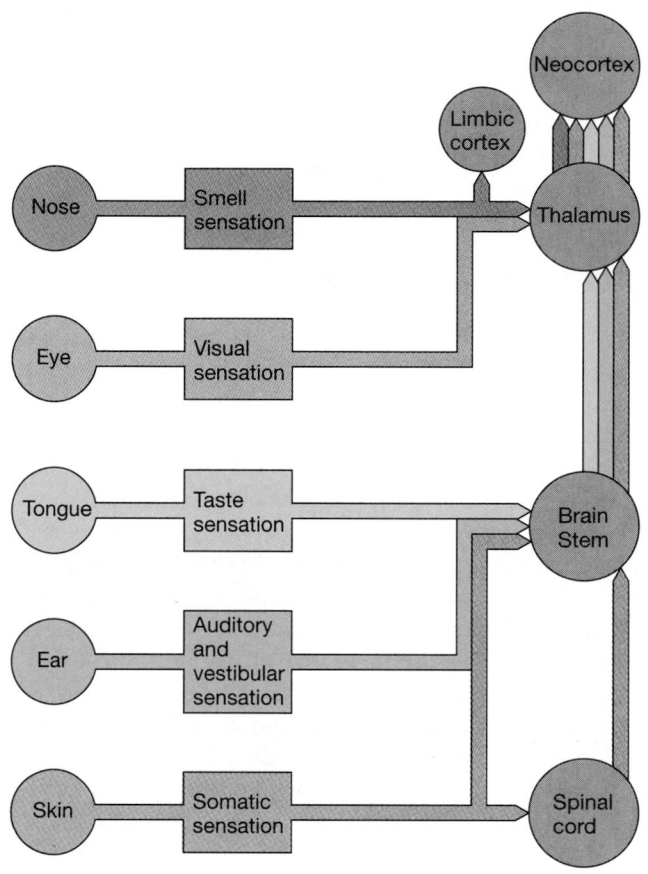

Figure 8–1

Sensory information travels to the brain via specialized pathways for each type of sensation.

Table 8–1

Sensory Modalities and Receptor Cells

Sensory Mode	Receptor	Sense Organ
Vision	Rods and cones	Eye
Hearing	Hair cells	Ear (organ of Corti)
Rotational acceleration	Hair cells	Ear (semicircular canals)
Linear acceleration	Hair cells	Ear (utricle and saccule)
Smell	Olfactory neurons	Olfactory mucous membrane
Taste	Taste receptor cells	Taste buds
Touch—pressure	Nerve endings	Skin
Warmth	Nerve endings	Skin
Cold	Nerve endings	Skin
Pain	Naked nerve endings	Skin
Joint movement and position	Nerve endings	Various
Muscle length	Nerve endings	Muscle spindle
Muscle tension	Nerve endings	Golgi tendon organ

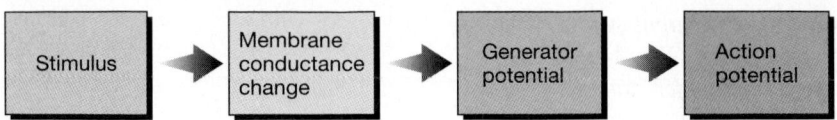

Figure 8–2

All sensory transductions involve an adequate stimulus, a generator potential resulting from a membrane conductance change, and one or more action potentials.

An adequate stimulus will produce a change in the membrane potential of the sensory receptor cell. This change in the membrane potential is called a **generator** or **receptor potential.** In some sensory receptors, such as somatic sensory receptors, the generator potential is a **depolarization** of the membrane. In others, such as the photoreceptors of the eye, it is a **hyperpolarization** of the membrane. The generator potential, in turn, produces an **action potential** or a series of action potentials. These action potentials can be generated by the receptor cell itself or by a neuron connected to the receptor cell. They transmit information about the nature of the stimulus to the central nervous system.

The transduction mechanisms of all sensory systems, therefore, involve (1) an adequate stimulus, (2) a generator potential, and (3) the initiation of action potentials. In order to examine these mechanisms in greater detail, let us consider a somatic sensory receptor called a **Pacinian corpuscle,** which responds to touch, or **tactile,** stimuli. This type of receptor is located beneath the skin and consists of a free nerve ending encapsulated by layers of connective tissue (Fig. 8–3). The nerve ending is wrapped with myelin along the length of the fiber. The fiber itself extends into the spinal cord, where it forms synapses with other nerve cells.

The adequate stimulus for the Pacinian corpuscle is **pressure** applied to the skin. This pressure causes the layers of connective tissue and free nerve endings to compress. The compression of the free nerve ending causes an opening of Na^+ channels embedded within the nerve membrane (see Fig. 8–3b). The resulting influx of Na^+ ions, in turn, depolarizes the membrane and produces a generator potential. If the generator potential is of sufficient amplitude, it will then depolarize the membrane in the region of the first node of Ranvier to threshold and initiate an action potential (Fig. 8–3c).

The Intensity of a Stimulus Is Encoded by the Frequency of Action Potentials

The number of action potentials generated per unit time is a function of the intensity of the pressure. Greater pressures produce a greater number of action potentials in a single fiber. This is known as the **frequency code of stimulus intensity** (Fig. 8–4a). The increase in the number of action potentials generated with greater pressure is due to the continued opening of the Na^+ channels and continued depolarization of the generator potential above the threshold value. As a consequence, after the repolarization of one action potential, another is generated.

A second method of informing the brain about the intensity of the stimulus is a **population code.** With this method of coding, more sensory receptors of the Pacinian corpuscle type are activated as the pressure becomes greater. This occurs because the increased pressure affects a greater area beneath the skin (Fig. 8–4b).

The Quality of a Stimulus Is Encoded by the Pathway of Transmission

The type of information sent to the brain is also coded in the way that nerve fiber pathways leading to the brain are arranged physically. The skin, for example, contains sensory receptors for temperature in addition to pressure. The information about the temperature of the skin reaches the brain by a different nerve fiber pathway than does information about pressure. In this way, information about the quality, or type, of stimulus is maintained within each pathway without becoming mixed with other types of stimuli. This aspect of the nervous system, the mechanisms of coding for the **type** of stimulus detected by a sensory receptor, is called the **labeled-line code of stimulus quality** (Fig. 8–5).

Nerves Adjust to Repetitive Stimuli by Mechanisms of Adaptation

Often, when a stimulus is continuously applied, the brain at some point no longer consciously perceives it. This adjustment happens, for example, with background noise such as the ticking of a clock. After a period of time it is unnoticed. This phenomenon is called **sensory adaptation.** The adaptation to a sensory stimulus can be caused by mechanisms either within the brain or at the receptor site. Adaptation mechanisms that work at receptor sites are the most

(a)

(b)

(c)

Figure 8–3

(*a*) A Pacinian corpuscle, an example of a sensory receptor for touch, consists of a neuron wrapped in layers of connective tissue, with a myelinated nerve ending (axon). (*b*) A tactile (pressure) stimulus causes the neuron within the connective tissue to change shape, allowing Na$^+$ to enter the cell. (*c*) Pressure stimulus produces a depolarizing generator potential and an action potential.

clearly understood and best illustrated by the Pacinian corpuscle.

When pressure applied to the Pacinian corpuscle is continuously maintained, the free nerve ending eventually reverts back to its original shape (Fig. 8–6) even though the layers of connective tissue remained deformed. Since the Na$^+$ channels embedded within the nerve membrane are opened only when the nerve fiber is deformed, they close when the form of the nerve is again circular. Consequently, even though

(a)

LESS PRESSURE

MORE PRESSURE

(b)

Figure 8-4

(*a*) According to the frequency code of stimulus intensity, more pressure produces more action potentials from the same receptor. (*b*) According to the population code of stimulus intensity, more pressure produces more action potentials because more receptors are affected.

pressure on the Pacinian corpuscle is still maintained, it ceases to produce a generator potential when it resumes its normal shape and thus no longer transmits information about the pressure of the stimulus.

Mechanisms of adaptation are also found at the molecular level. During sunny days, our eyes quickly adapt to the brightness of the outdoors by the

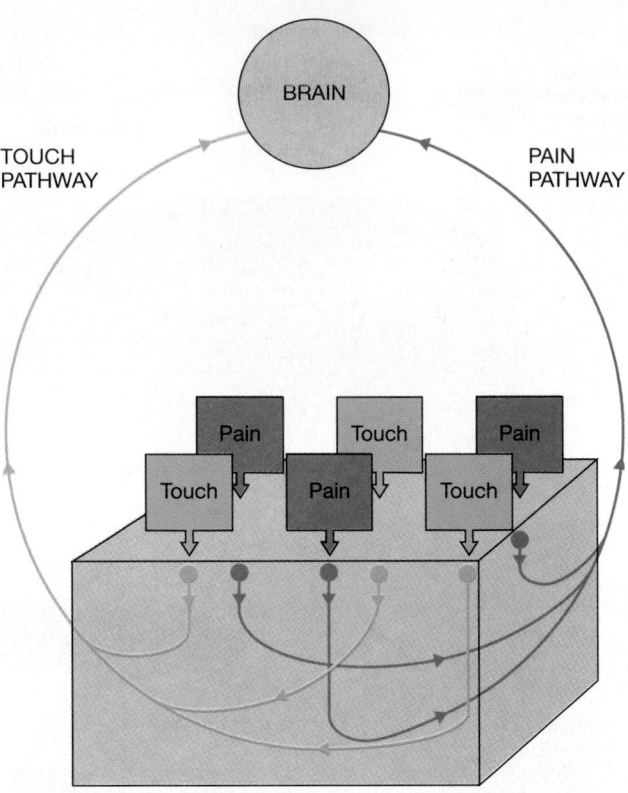

Figure 8-5

The labeled-line code of stimulus quality. Although different stimuli affect the skin, each type of stimulus has its own pathway to the brain, ensuring clarity of sensory messages.

"bleaching" of the photoreceptors. The underlying basis of this bleaching process is the removal of molecular receptors that capture light.

Sensory Systems and Their Processes

Now that we have an understanding of the general features of sensory systems, let us consider each sensory system in more detail.

The Somatic Sensory System Transduces Information About Pain, Temperature, and Pressure

The **somatic sensory system** is sensitive to external stimuli that affect the skin or surface of the body. This sensory system also conveys information about the position of the body's limbs in relation to the rest of the body.

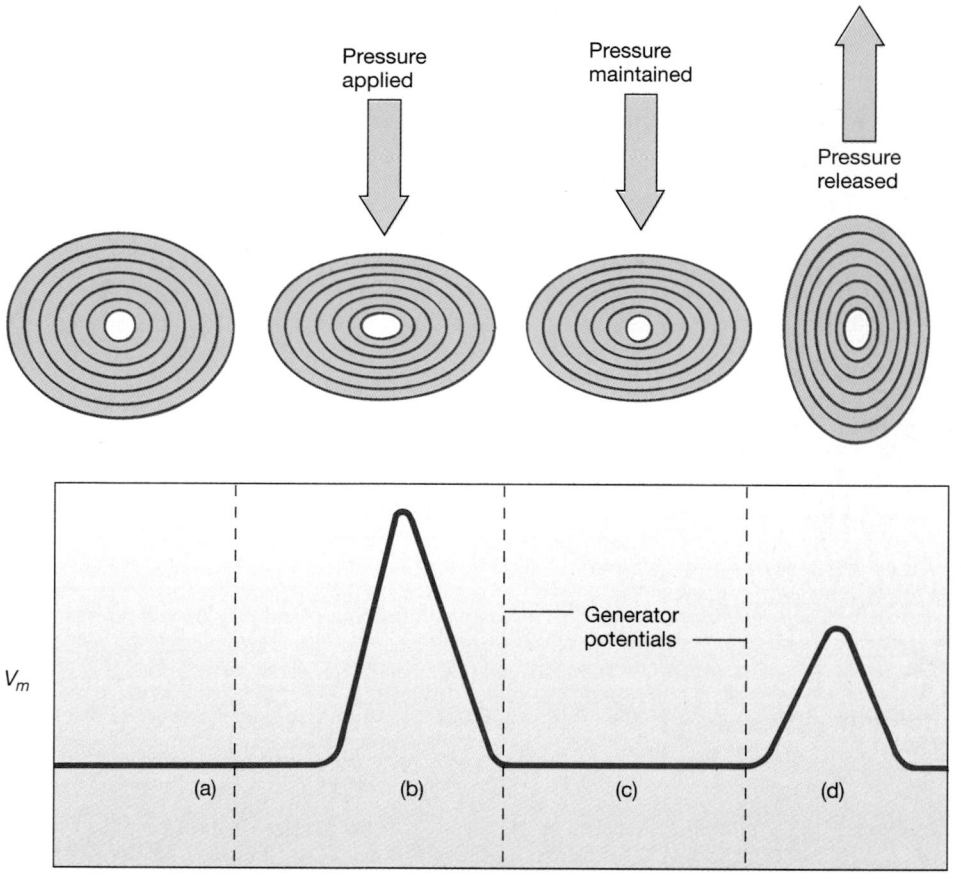

Figure 8–6

The process of sensory adaptation in a Pacinian corpuscle. (*a*) When not stimulated, the nerve ending inside the connective tissue layers has a rounded shape. (*b*) When pressure is applied, the entire corpuscle is pressed down and flattened, including the nerve ending inside. The elongation opens Na$^+$ channels into the nerve ending as shown in Figure 8–3. (*c*) As pressure is maintained, the corpuscle remains flattened, but the nerve ending inside resumes its resting shape, and Na$^+$ channels close; hence, no action potential is generated. (*d*) When pressure is released, the entire corpuscle, including the nerve ending, "springs back," becoming compressed in a different direction.

Types of Somatic Sensory Receptors

The major types of somatic sensory receptors are (1) **tactile receptors** activated by the mechanical stimulation of the body's surface, (2) **thermal receptors** activated by changes in temperature on the surface of the body, (3) **nociceptors,** or *pain receptors* activated by noxious (harmful) stimuli, and (4) **proprioceptive** receptors activated by the movement of the limbs. Each type of sensation is detected by a variety of sensory receptors.

Tactile Receptors: Touch, Pressure, and Vibration.

Tactile receptors are called **mechanoreceptors.** These receptors are responsible for detecting touch, pressure, and vibrations applied to the skin (Fig. 8–7). We have already examined the functional properties of one type of pressure-sensing mechanoreceptor, the **Pacinian corpuscle.** In addition to pressure, however, the Pacinian corpuscle is capable of detecting vibrations. The mechanisms that allow it to detect vibrations are extensions of its ability to rapidly adapt to applied pressures.

Recall that when pressure is applied to the Pacinian corpuscle for a length of time, it quickly stops generating action potentials. As we discussed earlier, this occurs when the Pacinian nerve ending reverts back to its original shape (see Fig. 8–6). As the applied pressure is released, the layers of connective tissue surrounding the nerve ending spring back to their original ellipsoidal form and then become compressed again in the opposite direction because of the elastic nature of the connective tissue (see Fig. 8–6*d*). This recompression deforms the nerve ending and again causes Na$^+$ channels to open and produce a generator potential and action potentials. In this way, both the application and release of pressure result in the discharge of action potentials. This oscillation of pressure is the basis of a vibratory stimulus. It should be clear from this example that the Pacinian corpuscles must adapt very quickly. This ability to quickly adapt to a stimulus classifies these corpuscles as **rapidly adapting receptors.**

Another rapidly adapting tactile mechanoreceptor is the **hair receptor.** The hair receptor is a hair follicle on the surface of the skin that has a nerve fiber wrapped around its base (see Fig. 8–7). The bending of the hair cell such as that caused by a slight breeze produces a mechanical displacement of the nerve fiber at the base of the follicle. This displacement, in

Figure 8–7

Types of sensory receptors in the skin.

Ruffini ending

Free nerve ending

Epidermis

Dermis

Pacinian corpuscle

Hair receptor

turn, opens Na^+ channels, resulting in a generator potential and the initiation of action potentials. This type of receptor is very sensitive to a gentle touch across the surface of the skin.

Another group of tactile receptors, called **slowly adapting tactile mechanoreceptors,** continue to generate action potentials as long as the stimulus is applied. An example of a slowly adapting mechanoreceptor is the **Ruffini ending** (see Fig. 8–7). This receptor is sensitive to stretching of the skin such as that produced during a massage.

Thermal Receptors: Heat and Cold.   Sensory receptors that respond to temperature are called **thermal receptors.** There are two types of thermal receptors: one type responds to temperatures below 30°C, while the other type responds to temperatures above 30°C. These thermoreceptors are capable of adapting to the external environment. When exposed to cold weather, for example, the cold thermoreceptors do not become activated until temperatures lower than 30°C are reached, while warm thermoreceptors become activated well before 30°C.

Nociceptors: Painful Touch and Temperatures.   *Nociceptors* are free nerve endings that detect painful stimuli (Fig. 8–7). In general, there are two types of nociceptors: **mechanical nociceptors** activated by intense mechanical stimulation such as a knife cut on the arm or a slap to the head, and **heat nociceptors** that respond to temperatures above 45°C. The transduction of pain is activated by a chemical process that also involves the immune system.

Proprioceptors: Position and Movement.   The position of the body's limbs is detected by **proprioceptors.** One type of proprioceptor detects the stationary position of the limbs in space with respect to the other parts of the body. Other proprioceptors transmit dynamic information about limb movement to convey the sense of movement, or **kinesthesia.** The brain needs this information to determine where the arms and legs are located in order to calculate how much further they need to go to complete a certain movement.

The operation of these receptors can be tested by trying to bring the tips of your left and right index fingers together with your eyes closed. If your proprioceptors are working properly, your fingers should touch without your having to watch and visually guide them together.

The sense of stationary or static position is transmitted to the brain by mechanoreceptors located in joint capsules, cutaneous mechanoreceptors, and mechanoreceptors in muscles that are specialized to transduce the stretching of the muscle. Extremes of joint angles are sensed by **joint capsules,** while intermediate angles are transduced by the **muscle spin-**

ACTION POTENTIALS FROM
STATIC PROPRIOCEPTORS

(a)

ACTION POTENTIALS FROM
DYNAMIC PROPRIOCEPTORS

(b)

Figure 8–8

Action potential patterns produced by static (*a*) and dynamic (*b*) proprioceptors.

dle receptor. The functional properties of these receptors and muscle tension receptors will be discussed in detail in Chapter 9, in conjunction with the motor system.

The static proprioceptors produce a continuous frequency of action potentials in response to different joint positions. If the joint is left in one particular position, the receptor will generate action potentials at one specific frequency (Fig. 8–8*a*). This type of action potential response is called a **tonic discharge.** The dynamic proprioceptor generates action potentials only with a change in direction of movement. The burst of action potentials produced is very brief (Fig. 8–8*b*). This type of action potential response is called a **phasic discharge.**

The information from somatic sensory receptors is transmitted to the central nervous system by several different pathways organized according to three general principles: (1) Each type of somatic sensation has its own pathway; (2) most pathways cross over from one side of the brain to the other; (3) the nerve cells within each nucleus are topographically organized according to the location of their sensory receptors on the surface of the body. As we shall see, these three principles also apply to the organization of other sensory systems.

The Dorsal Root Ganglion: A Cluster of Cell Bodies

Each of the somatic sensory receptors previously described is associated with a peripheral nerve axon of a particular diameter. The cell bodies of these sensory receptors are located within a cluster of cells immediately outside the spinal cord called the **dorsal root ganglion** (Fig. 8–9). The fibers with the largest diameters are those associated with touch, pressure,

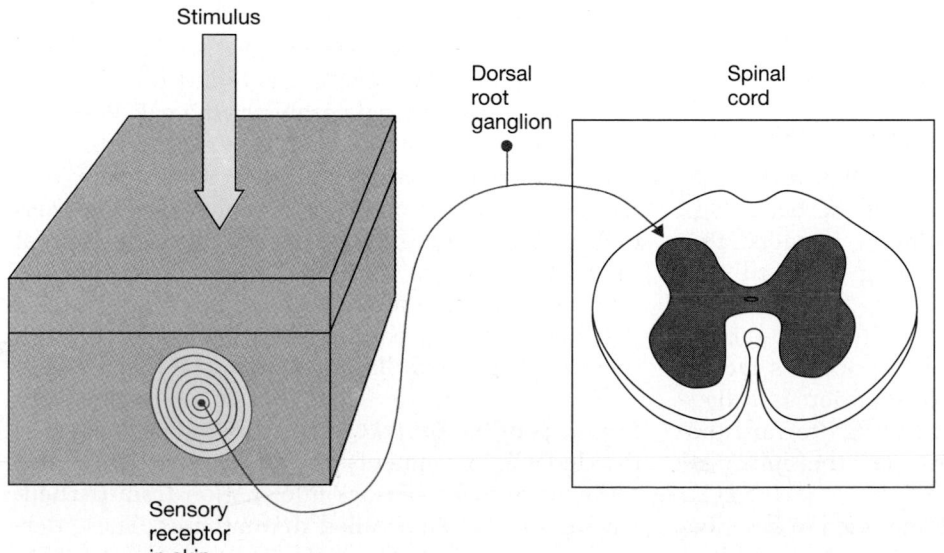

Figure 8–9

The dorsal root ganglion is a cluster of cell bodies located just outside the spinal cord on the dorsal side.

Table 8–2
Sensory Fibers

Type	Diameter (μm)	Conduction Velocity (m/sec)	Sensory Information
I$_a$ (A$_\alpha$)	22	120	Primary muscle spindle
I$_b$ (A$_\alpha$)	22	120	Primary touch and pressure
II (A$_\beta$)	13	70	Secondary muscle spindle, touch, and pressure
III (A$_\delta$)	5	15	Lightly myelinated fibers for touch, pressure, and pain
IV (C)	1	2	Unmyelinated fibers for pain and temperature

and vibration sense. These are heavily myelinated fibers that range from 13 to 22 μm in diameter and have action potential conduction velocities of 70 to 120 meters per second (Table 8–2). These fibers release the neurotransmitter glutamate at their terminal endings. The fibers with the smallest diameters are the lightly myelinated and unmyelinated fibers that convey pain and temperature information. These fibers range from 1 to 5 μm in diameter and have action potential conduction velocities of 2 to 15 meters per second. These fibers release the neuropeptide substance P at their terminal endings.

Two major pathways transmit somatic sensory information from the surface of the body to the neocortex: the **dorsal column pathway** and the **anterolateral pathway** (Fig 8–10c). These pathways sequentially transmit sensory information to nerve cell nuclei in the **spinal cord, brain stem, thalamus,** and **neocortex** (see Fig. 8–1). In order to understand how these pathways operate, we first need to consider the functional organization of the areas of the nervous system that are involved in the processing of somatic sensory information.

The Spinal Cord in Somatic Sensory Function

The spinal cord is shaped like a tube that runs from the base of the head to the small of the back (Fig. 8–10a). This longitudinal orientation is called the **rostro-caudal axis** (*rostral* meaning "toward the nose" and *caudal* meaning "toward the tail" in Latin). The spinal cord is also oriented so that one part of the tube is facing the back and the other is facing the stomach. This transverse orientation is called the **dorso-ventral axis** (*dorsal* meaning "toward the back" and *ventral* meaning "toward the belly" in Latin).

Four other terms also used quite often to describe the location of nuclei and fiber tracts in the brain are:

medial meaning "close to the midline" of the body; *lateral* meaning "to the side" of the body; *anterior* meaning "toward the front" of the body; and *posterior* meaning "toward the back" of the body.

A transverse section of the spinal cord shows that it has a central gray area and a peripheral region that is white (see Fig. 8–10b). The gray matter is composed of nerve cells of the spinal cord, whereas the white matter is composed of axons. The myelination of axons is what makes the peripheral area appear white. The gray region has four protrusions shaped like butterfly wings that are symmetrical about the midline. The dorsal protrusions, called the **dorsal horns,** contain nerve cells involved in the processing of sensory information. The ventral protrusions are called the **ventral horns** and contain nerve cells that innervate skeletal muscles and are thus involved in motor function. This latter group will be discussed in Chapter 9.

Nerve fibers that carry sensory information in the spinal cord follow one of the two major pathways already mentioned: the dorsal column pathway or the anterolateral pathway (see Fig. 8–10c). Nerve fibers carrying information about touch and pressure traverse the dorsal column pathway, whereas those carrying information about pain and temperature traverse the anterolateral pathway. When pain and temperature fibers enter the spinal cord, they cross over to the midline; consequently, the anterolateral pathway transmits information from the opposite (contralateral) side of the body. In contrast, fibers of the dorsal column pathway do not cross over until they reach the medulla.

Segments of the Spinal Cord.   The spinal cord is subdivided into 31 segments (Fig. 8–11). Each spinal cord segment receives sensory information from particular areas of the skin called **dermatomes.** These dermatomes contain receptors for all of the somatic sen-

Figure 8–10

(*a*) The spinal cord is described frequently in terms of its transverse orientation (i.e., dorsal and ventral). (*b*) A cross-section of the spinal cord reveals gray and white areas as well as dorsal and ventral horns. (*c*) Two major pathways through the spinal cord that transmit sensory information to the brain are the dorsal column pathway and the anterolateral pathway.

sations. The loss of sensory information by certain dermatomes can thus be used as a clue to the location of damage to the spinal cord. In the uppermost region of the cord are 8 cervical segments (C1 to C8). These segments receive sensory information from the back of the head, neck, shoulders, part of the arms, and hands. Next are 12 thoracic segments (T1 to T12), which receive sensory inputs from parts of the arms and hands and trunk of the body. Following the tho-

racic segments are 5 lumbar segments (L1 to L5), which receive sensory inputs from the waist, thighs, upper and lower legs, and parts of the feet. The remaining segments of the spinal cord are five sacral segments (S1 to S5) and one sacro-coccygeal segment. These receive sensory input from the back of the legs, buttocks, and anus.

Sensory fibers of the dorsal column system enter the spinal cord and ascend toward the brain in the

Spinal
cord
segments

Dermatomes
viewed
from the front

Dermatomes
viewed
from the side

Figure 8–11

The 31 segments of the spinal cord are linked to areas of the skin known as dermatomes.

dorsal portion of the spinal cord (Fig. 8–12). These fibers continue to the medulla oblongata, where they form synapses with neurons in the **dorsal column nuclei.** Fibers from the dorsal column nuclei, in turn, cross the midline and become part of the **medial lemniscal** pathway. These fibers reach the **thalamus** on the opposite side of the brain. Neurons in the thalamus, in turn, extend their fibers to the cortex.

Lesions of the Spinal Cord. In accidents that damage the spinal cord, the loss of cutaneous sensation appears in patterns that are characteristic of the level of the lesion. For example, an accident that severs half of the spinal cord on the right side of the body at the waist causes a loss of pain and temperature sensation on the left leg and a loss of touch and vibration sense on the right leg (Fig. 8–13).

This pattern occurs because tactile information is transmitted in the spinal cord along the same side of the body that receives information about touch and

Figure 8–12

The dorsal column, medial lemniscal pathway.

pressure. Pain and temperature information, on the other hand, crosses the midline once it enters the spinal cord (see Fig. 8–13). Obviously, accidents that transect all or half of the spinal cord will also result in some form of paralysis. This loss of motor control after spinal cord lesions will be considered in Chapter 9.

The Thalamus in Somatic Sensory Function

The **thalamus** is one of the brain areas that receives most types of sensory information. It receives not only somatic information, but also visual, auditory, and taste information. The region of the thalamus that receives somatic sensory information is the **ventral posterior lateral (VPL)** nucleus. Neurons in the VPL of the thalamus then transmit sensory information to the neocortex.

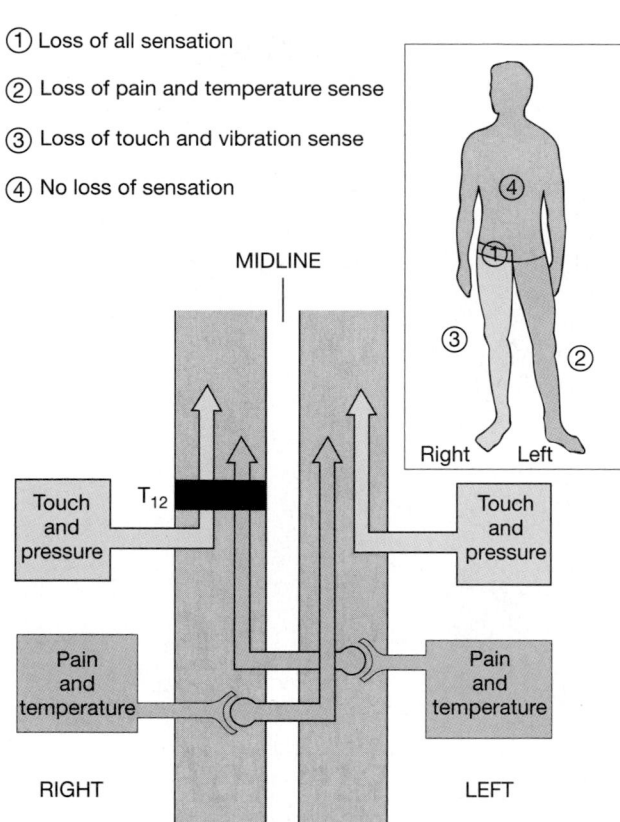

① Loss of all sensation

② Loss of pain and temperature sense

③ Loss of touch and vibration sense

④ No loss of sensation

MIDLINE

Touch and pressure — T₁₂

RIGHT

Touch and pressure

Pain and temperature

Pain and temperature

LEFT

Right · Left

Figure 8–13

Some effects of spinal cord lesions. Damage to the spinal cord on the right side of the body near the waist causes the left leg to lose pain and temperature sensation, while the right leg loses the sense of touch and vibration.

The Somatic Sensory Cortex

The **neocortex** is the part of the brain that evolved most recently (*neo* meaning new). It is divided into discrete areas that receive somatic, visual, auditory, and gustatory sensations as well as areas for the control of movement and other functions (Fig. 8–14). The area of the neocortex that receives somatic sensations is called the **somatic sensory cortex.**

A cross-section of the somatic sensory cortex shows that it is organized like a map of the body (Fig. 8–15). Sensory information from the legs is sent to the medial portion of the cortex. Adjacent but more lateral regions of the cortex receive sensory information from the torso. The most lateral regions of the cortex receive sensory information from the arms, hands, and face.

PRIMARY MOTOR CORTEX
(precentral gyrus)

PRIMARY SOMATIC
SENSORY CORTEX
(postcentral gyrus)

PREMOTOR AREA

Central sulcus

SECONDARY SOMATIC
SENSORY CORTEX

VISUAL
ASSOCIATION
AREA

PRIMARY
VISUAL
AREA

BROCA'S
SPEECH
AREA

PRIMARY
AUDITORY
AREA

PRIMARY
GUSTATORY
AREA

AUDITORY
ASSOCIATION
AREA

Figure 8–14

The primary somatic sensory cortex and other areas of the neocortex.

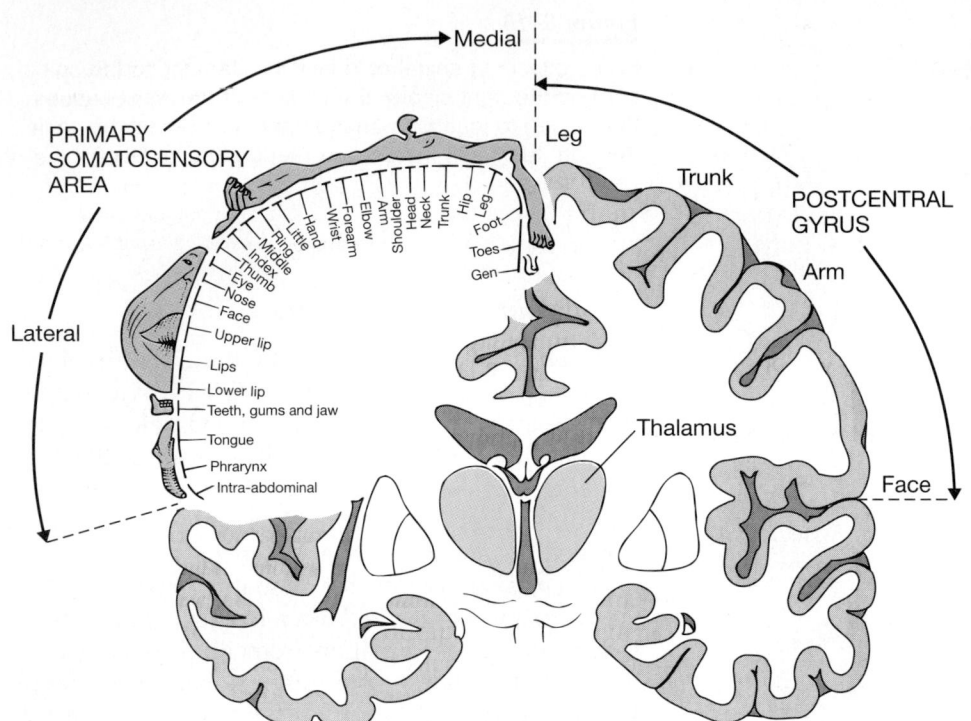

Figure 8–15
The somatotopic organization
of the primary somatic sensory cortex.

This type of topographical organization of sensory information from various regions of the body on to the sensory cortex is called a **somatotopic organization.** The map that is created in the somatosensory cortex is a distortion of the true proportions of the body. The regions of the body that have the greatest density of sensory receptors are represented by a greater proportion of the cortex. The areas in the cortex representing the thumb and fingers of the hand, for instance, occupy substantially more space than that represented by the torso.

Receptive Fields of Neurons in the Somatic Sensory System

Stimuli detected by a sensory receptor are transmitted to various groups of nerve cells. The most effective location of a stimulus, however, differs for each cell within a cluster. For instance, pressure applied to an area of the skin directly above a Pacinian corpuscle will activate the associated dorsal root ganglion cell (Fig. 8–16).

If pressure is applied to another location on the surface of the skin, another dorsal root ganglion cell is activated; however, the first dorsal root ganglion cell is not affected by pressure applied to this second location on the skin. Thus, only a small area on the receptive surface of the skin can activate a nerve cell in the dorsal root ganglion. This small area of skin represents the **receptive field** for that cell. The receptive field can be described as a circular area on the skin where pressure is applied to excite the cell. We

Figure 8–16

A cell in the dorsal root ganglion has a specific receptive field on the skin.

Figure 8–17

Annular receptive fields of neurons in the dorsal column nuclei are formed by inhibitory interneurons. (*a*) Activation of Pacinian corpuscle #1 excites dorsal column cell #1, while activation of Pacinian corpuscle #2 inhibits dorsal column cell #1 via an inhibitory interneuron (red). (*b*) Mapping of receptive field on the surface of the skin for dorsal column cell #1. Pressure stimulus #1 activates this cell, while pressure stimuli #2 and #3 inhibit dorsal column cell #1. *(c)* One-dimensional view of receptive field for dorsal column cell #1. (*d*) Two-dimensional view of receptive field for dorsal column cell #1.

(a)

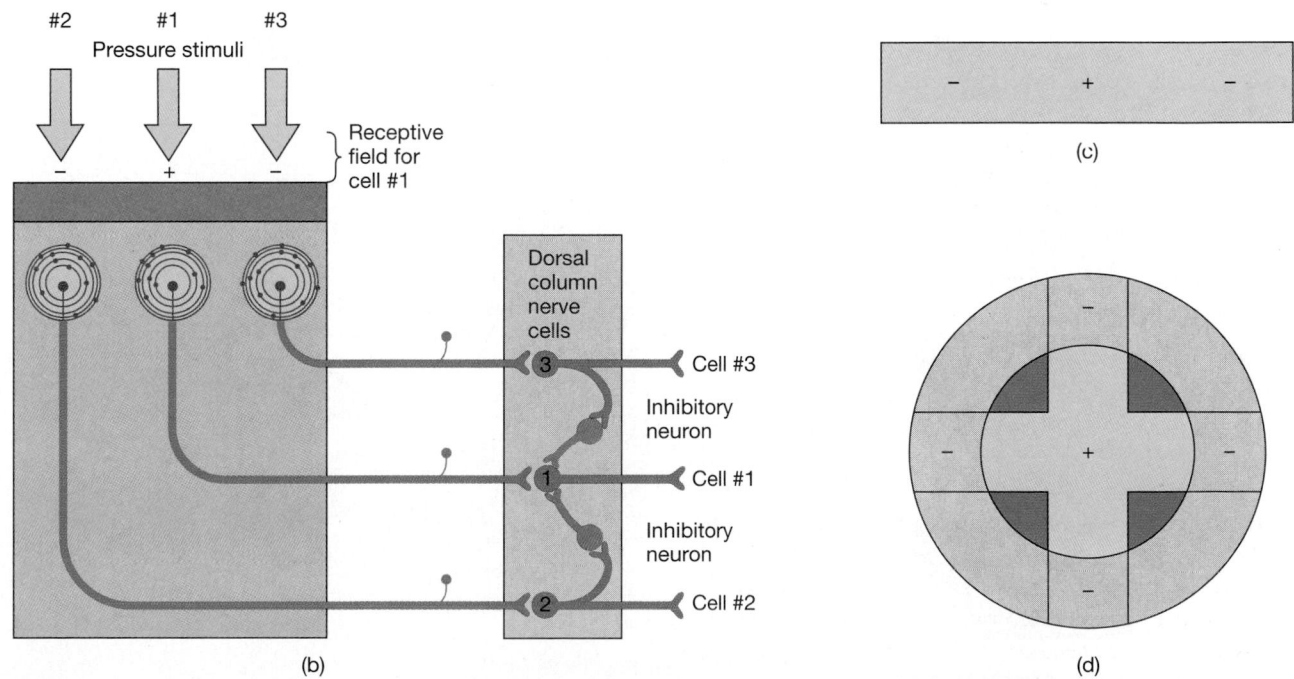

(b)   (c)   (d)

will denote the ability to excite the cell with a "+," and the receptive field area of the skin by a circle.

Nerve cells in the dorsal column nuclei, the next cluster of cells along the dorsal column pathway, have different receptive fields than those in the dor-

sal root ganglia. Dorsal column nerve cells generate a low rate of spontaneous action potentials. This is their tonic baseline activity. When pressure is applied to an area of the skin, a dorsal column nerve cell (cell #1 in Fig. 8–17*a*) is excited to produce a higher frequency of action potential discharge than its baseline rate of activity. However, when the pressure is applied to a different location, the frequency of the action potentials is decreased to less than the baseline rate.

This inhibition, "−," of the dorsal column nerve cell is a result of activating a second Pacinian corpuscle, which in turn activates a different nerve cell (cell #2) within the dorsal column. The activation of this second dorsal column nerve cell leads to the activation of an **inhibitory interneuron.** This inhibitory neuron, in turn, prevents the excitation of the first dorsal column neuron. There are also other areas of the skin that, when pressed, will inhibit the response of dorsal column nerve cell #1 (see Fig. 8–17b). When the excitatory and inhibitory areas on the skin are mapped for nerve cell #1, an annulus or doughnut-shaped receptive field is revealed with an excitatory center and an inhibitory ring (Fig. 8–17d).

This annular receptive field is also found for nerve cells in the thalamus and certain areas of the somatosensory cortex. The cells in the somatosensory cortex that exhibit the annular receptive field are all located in area 3 of the **primary somatosensory cortex** (see Fig. 8–15). This is a narrow strip of cortical tissue that is also called **Brodmann area 3** (named for the neuroanatomical studies by K. Brodmann in 1909). In Brodmann areas 1 and 2, the receptive fields are no longer annular. The stimulus that most effectively activates nerve cells in areas 1 and 2 is pressure applied to the skin in the form of a rectangle (Fig. 8–18) rather than a ring. The effectiveness of a rectangular stimulus is the result of nerve cells in area 3 that have overlapping receptive fields. The output fibers of these cells converge upon a nerve cell on area 1 (Fig. 8–19). Consequently, the simultaneous activation of all three input cells is needed in order to excite the one cell in area 1. This excitation occurs by

Figure 8–18

The different types of receptive fields for cells of different nuclei.

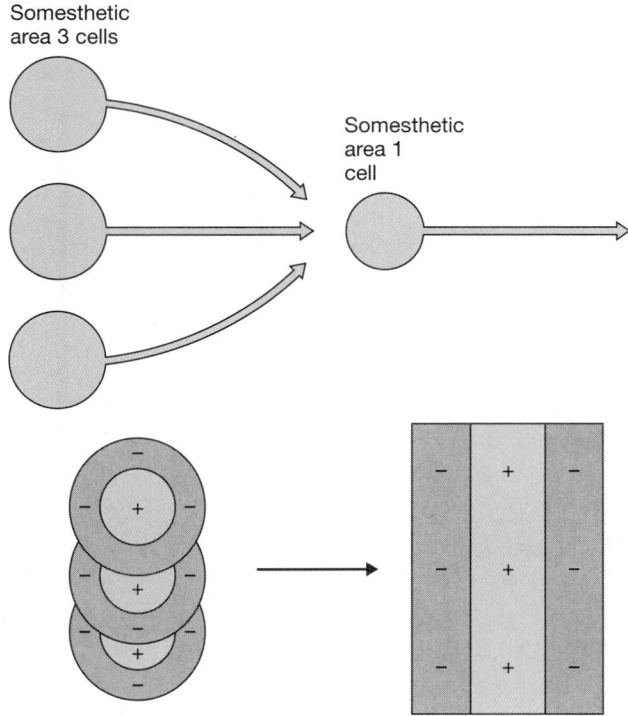

Figure 8–19

The receptive fields in the secondary somatosensory areas are rectangular.

Figure 8–20

The somatic sensory cortex is organized so that all of the cells specialized for one sensation are grouped in a vertical column.

the process of spatial summation. This process is described in Chapter 7.

Although sensory information of a particular type of stimulus is transmitted from area 3 to 1, and area 1 to 2, in any one Brodmann area of the somatosensory cortex, information from one type of somatic sensation tends to predominate. The nerve cells in Brodmann area 1, for example, respond to rapidly adapting cutaneous receptors; those in area 2 respond to deep pressure; those in area 3a respond to stretch receptors in skeletal muscle; and those in area 3b respond to rapidly and slowly adapting cutaneous receptors. In this way the somatic sensory cortex has several somatotopic maps of the body.

The most effective shapes of stimuli needed to activate nerve cells in area 2 can become quite complex. In essence, the physical features of an object in the environment are first broken down into its fundamental components by the peripheral sensory receptors. This information is then assembled to reconstruct the holistic features of the object that are capable of activating nerve cells in higher centers of the brain. In our discussion of the visual system, we will see that a similar synthesis and reconstruction of the external environment takes place for visual information.

The somatic sensory cortex is also organized so that all of the cells that respond to one type of sensation are located together within vertical columns. Each nerve cell within a vertical column of the somatic sensory cortex, for example, is activated by Pacinian corpuscles in one fingertip (Fig. 8–20). The location of the receptive fields for cells within a vertical column is also similar. Thus, the vertical columns in neocortex represent basic units of sensation-specific and location-specific function.

The Anterolateral Pain and Temperature Pathway

As do the nerve fibers that transmit tactile sensation, the fibers that send pain and temperature information enter the dorsal portion of the spinal cord (Fig.

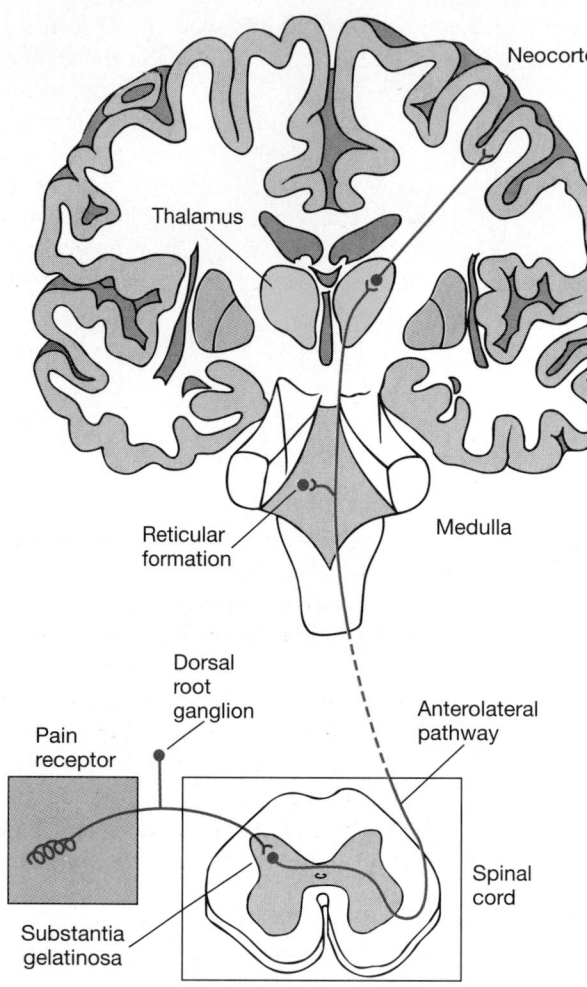

Figure 8–21

Pain and temperature information travels to the brain via the anterolateral pathway.

Figure 8–22

The perception of pain can be inhibited by the action of an interneuron.

8–21). After entering the spinal cord, however, the majority of them form synaptic connections with nerve cells located in the dorsal horn. This group of nerve cells is collectively called the **substantia gelatinosa.** These cells, in turn, cross the midline and ascend toward the brain as a bundle of fibers along the anterior and lateral portion of the spinal cord, forming the **anterolateral pathway.** These fibers have inputs to the reticular formation of the brain stem and the thalamus. The reticular formation serves as an alerting system and activates other areas of the brain.

Inputs to the thalamus terminate in the **ventrobasal (VB)** nuclei. This information in turn is sent to the primary somatic sensory cortex in a somatotopic fashion, and to the frontal cortex. The pain information that reaches the primary sensory cortex is associated with acute pain, whereas the information that reaches the frontal cortex is associated with chronic pain.

The perception of pain can be altered by other sensory inputs from the periphery or controlled by descending inputs from the brain stem and other higher brain centers. This gating of pain information is carried out by an inhibitory interneuron located in the dorsal horn of the spinal cord (Fig. 8–22). As previously discussed, the pain fibers from the periphery release the peptide called substance P as the neurotransmitter. In the absence of substance P release, pain cannot be detected by the brain. The interneuron acts to inhibit the release of substance P by a mechanism called **pre-synaptic inhibition.** The axon terminal of the interneuron forms a synapse upon the terminal of pain fibers. The terminals of the interneurons release the peptide called **enkephalin,** which in turn inhibits the release of substance P from the terminals of the pain fibers.

Thus, the interneuron is the key element in altering the sensation of pain. It can be activated by sensory fibers associated with touch, pressure, and vibration, or by fibers from higher brain centers. The influence of other tactile inputs is seen in the reduction of pain caused by rubbing the skin surrounding a region of injury; this causes the pain to subside more quickly. The influence of higher brain centers on pain perception is seen in cases in which individuals involved in emergency situations do not realize that they have been injured until much later. The mechanisms that prevent the perception of pain under these highly stressful conditions are part of an autonomic "fight or flight" response that will be discussed in Chapter 10.

Another aspect of pain sensation is the perception that pain actually caused by the distress of internal organs is coming from the surface of the body. This is often seen in cases of heart attacks in which the area of the chest and left arm become painful. This type of **referred pain** will also be considered in Chapter 10.

The Visual System Detects Information About Color, Shape and Movement of an Object

The visual system allows us to determine the shape and color of objects such as books, and the form of the letters printed within their pages. Parts of the visual system also allow us to recognize the grouping of letters in the form of words, and the grouping of words to represent ideas. These latter functions will be discussed in Chapter 11 when we consider the localization of language centers in the brain. In this section, we will consider how the visual system detects the shape and color of objects and the movement of objects in the external environment. The features of shape, color, and movement are processed by different cell groups within the visual system, beginning with those found in the eye.

The eye is the sensory organ of the visual system that transduces light into electrical impulses, which the nervous system then uses to transmit information to the brain. The light perceived is part of the continuum of **electromagnetic radiation** that ranges from radio waves to X-rays (Fig. 8–23). Each type of electromagnetic radiation is characterized by a **wavelength.** Radio waves have long wavelengths, whereas X-rays have very short wavelengths. The colors of light that we see have wavelengths in between radio waves and X-rays, and range from 400 to 700 nm. These wavelengths correspond to the range of colors seen in a rainbow.

In addition to its wavelike nature, light can also be thought of as a composite of particles called **photons.** These photons have wavelengths and energies associated with different colors. All objects have the capability of selectively absorbing certain photons of light, and those photons not absorbed are reflected back into the environment. The reflected photons are those that give an object its color and that interact with the eye.

Anatomical Organization of the Eye

Figure 8–23 is an illustration of a cross-section of the eye. It shows the cornea, iris, lens, vitreous humor, and retina. Together the **cornea** and **lens** act like the lens of a camera. They operate to bend the light rays and focus them on to the retina. Focusing a camera changes the distance between the lens and the film. Our eyes accomplish this feat by changing the shape of the lens; **ciliary muscles** tighten and fatten the lens

to focus on objects that are near by, or relax to focus on objects that are far away. The **retina** is the part of the eye containing receptor cells called **photoreceptors** that absorb photons. As the illustration shows, photons must first pass through the cornea and lens before reaching the retina. The most sensitive part of the retina is a region called the **fovea;** consequently, the lens attempts to focus the image of an object onto the fovea. The image projected onto the retina, however, is upside down. The **inversion** of the image is similar to that produced by a camera lens onto a roll of film. Even though the retina detects the form of objects in an upside-down fashion, the brain reorganizes this information so that we interpret the outside world as right-side-up.

The Nerve Cells of the Retina.   We can best understand how the eye transduces the form of an object by first examining the functional organization of the nerve cells within the retina. In addition to photoreceptors, the retina contains other types of nerve cells called **ganglion cells, bipolar cells,** and **horizontal cells** (see Fig. 8–23). Photoreceptors, either **rods** or **cones,** are responsible for absorbing the photons and are situated at the back portion of the retina, behind these other types of cells. Photons must therefore pass through several layers of cells before reaching the layer of photoreceptors (Fig. 8–23*b*).

Rods are cylindrically shaped photoreceptors (Fig. 8–23*c*) that contain one type of pigment capable of absorbing photons representing a broad range of wavelengths. Rods display a high degree of convergence onto other cells and thus as a group are very sensitive to low levels of illumination. They therefore operate well during the night.

Cones are conically shaped and absorb photons associated with a more narrow range of wavelengths for color. Cones display a low degree of convergence onto other cells and therefore as a group are not as sensitive as rods but have better spatial resolution. There are three types of cones, each responsive to photons of a different range of wavelengths. These cones form the basis of color perception and are used for day vision.

Photoreceptors and Phototransduction.   The photoreceptor pigment that absorbs photons is called **visual pigment.** This pigment is embedded within the membranes of **discs,** lamellar structures in the **outer segment** of the photoreceptor. Also present in the disc membrane is a protein called **transducin** and an enzyme called **phosphodiesterase** (Fig. 8–24*a*). The pigment in a rod cell disc is **rhodopsin.** When the rhodopsin molecule absorbs a photon, its three-dimensional form changes. In its new form, rhodopsin is now capable of activating transducin. Transducin,

(text continues on page 272)

Figure 8–23

(*a*) General structure of the eye and the visible portion of the electromagnetic spectrum. (*b*) Organization of photoreceptors (P, rods and cones), bipolar (B), horizontal (H), amicrine (A), and ganglion (G) cells in the retina. (*c*) Photomicrograph of the different cell types of the retina. (*d*) The general structure of a rod cell and a cone cell. Note the membranous discs within the outer segments of both cells. The membranes of these discs contain the pigment that, when activated by photons, produces visual sensation. (© Ed Reshke/Peter Arnold, Inc.)

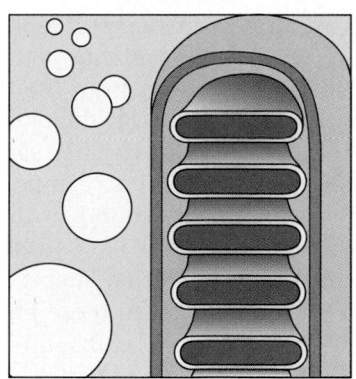

Figure 8-24

(*a*) In the absence of light, sodium channels in the plasma membrane of red cells are kept open by the presence of cGMP, and sodium ions enter the cell. (*b*) When a photon of light enters a rod cell, it causes a rhodopsin (pigment) molecule to activate transducin, which in turn activates phosphodiesterase. The enzyme hydrolyzes cytosolic cGMP of the rod to 5'-GMP. The reduction in the amount of cGMP causes Na^+ channels in the rod plasma membrane to close, reducing the amount of Na^+ entering the cell and producing a hyperpolarization.

(a)

(b)

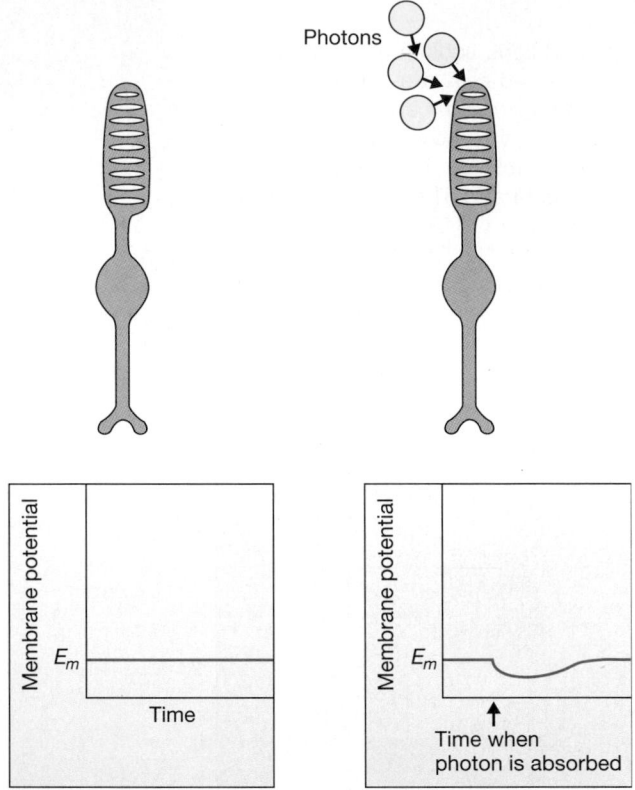

Photons

Figure 8–25

The hyperpolarization of a rod cell membrane caused by a photon is eventually transduced into actin potentials that travel to the optic nerve.

in turn, is now able to activate phosphodiesterase, which hydrolyzes cyclic GMP (cGMP) to 5′ -GMP. In this process, the net effect is the reduction in the amount of intracellular cGMP.

What is the significance of decreasing the cGMP concentration? Normally, cGMP opens Na^+ channels in the photoreceptor plasma membrane. When cGMP is decreased, Na^+ channels begin to close, thus reducing the influx of positive ions into the cell. This results in a hyperpolarization of the membrane. Thus, the photoreceptor transduces the photon of light into a more negative membrane potential (Fig. 8–25). In order to understand how this hyperpolarization is transformed into a series of action potentials, we must first consider the organization and response of nerve cells in the retina.

Interaction of Cells in the Retina.   We can best understand the way in which visual information is processed in the retina by considering the interactions between photoreceptors, bipolar cells, horizontal cells, and ganglion cells. Photoreceptors form synaptic connections with **bipolar cells** (Fig. 8–26). Some

bipolar cells, in turn, form synaptic connections with **ganglion cells.** These cells generate action potentials even in the absence of input from the bipolar cells. The bipolar cells, however, can depolarize the ganglion cells, resulting in a greater frequency of action potentials. Ganglion cell activity can be inhibited by a fourth type of nerve cell found in the retina, the **horizontal cell.** This cell receives its inputs from photoreceptors located in the retina lateral to the region of excitation (Fig. 8–26). When these photoreceptors detect light, they inhibit the horizontal cell.

The inhibition of the horizontal cell, in turn, causes a depolarization of adjacent photoreceptors (most likely by preventing the release of an inhibitory transmitter) and the subsequent inhibition of ganglion activity. Depending on where light impinges on the retina, ganglion cells can thus be activated or inhibited.

The areas of the retina that excite or inhibit a ganglion cell make up the receptive field for that ganglion cell. Therefore—as in the case of the dorsal column nuclei cells of the somatosensory pathway—the retinal ganglion cells have receptive fields that are shaped like an annulus. For some ganglion cells, the center of the annulus is excitatory, while the periphery is inhibitory (Fig. 8–27). These neurons are called **"on"** center and **"off"** surround cells. Other ganglion cells have **"off"** centers and **"on"** surrounds.

The spatial pattern of activated ganglion cells in the two-dimensional plane of the retina represents the form of an object whose image is projected onto the surface of the retina. The image of a pencil, for example, would activate a selective array of ganglion cells (Fig. 8–28). The image of the pencil is thus encoded by the spatial location of activated ganglion cells within the retina.

Pathways for Visual Information

Other groups of neurons in the visual pathway have receptive fields that differ from those of retinal ganglion cells. In order to fully appreciate the visual receptive fields that are exhibited by neurons in other areas of the brain, we first need to consider the anatomical pathways that connect these regions together.

The ganglion cells of the retina are the only cells that send axons outside of the eye (Fig. 8–29a). These fibers make synaptic connections with neurons in the lateral geniculate nucleus of the thalamus. Fibers from ganglion cells situated in the retina near the nose (nasal region) cross the midline and form connections with the lateral geniculate nucleus on the opposite side of the brain. In contrast, fibers from gan-

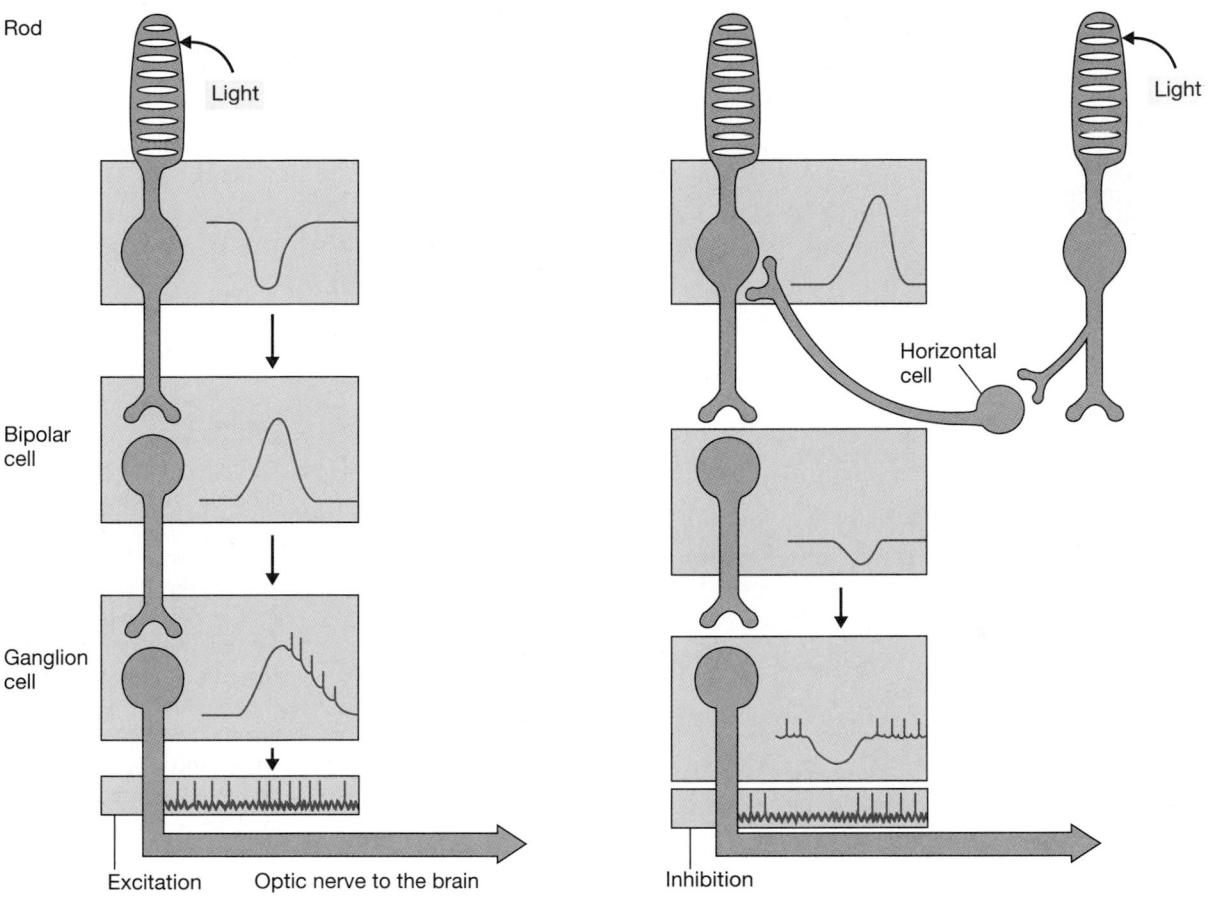

Figure 8–26

Visual information is processed in the retina by interactions between photoreceptors, bipolar cells, horizontal cells, and ganglion cells. Photoreceptors synapse with bipolar cells, causing bipolar cells to depolarize when the photoreceptors detect light. The bipolar cells can depolarize ganglion cells. Ganglion cell depolarization can be inhibited by a horizontal cell.

glion cells in the retina located farthest away from the nose (temporal region) do not cross the midline and thus form connections with the lateral geniculate nucleus on the same side of the brain.

The lateral geniculate nucleus contains six distinct layers of neurons (Fig. 8–29b). The first, fourth, and sixth layers contain cells that respond to light impinging upon the eye on the opposite side of the brain. The second, third, and fifth layers contain cells that respond to light impinging upon the eye on the same side of the brain. Nerve cells in these layers, in turn, have axons that synapse upon neurons in the primary visual cortex (Brodmann area 17).

The cells of the visual cortex are organized in the form of a map of the external visual field (see Fig. 8–29c). An area in the center of the visual field projects to the fovea of the retina. Since the fovea is the region in the retina of greatest acuity and contains the greatest density of cones, the associated areas in the visual cortex also occupy the largest amount of cortical space. Regions of the visual field that are adjacent to the center are represented in the visual cortex in areas next to the representation of the fovea. Likewise, the cortical representation of the peripheral visual field is found adjacent to the para-foveal cortical representation. In this manner, the visual field is precisely mapped onto the visual cortex and is referred to as a **visuotopic organization.** Since the visual field is also precisely mapped onto the retina, the visual cortex can also be thought of as a **retinotopic map.**

Within each region of the visual cortex, the cells are organized into functional columns. Each column of cells responds to light impinging upon either the right or left eye. These columns are therefore referred to as **ocular dominance columns.**

(text continues on page 276)

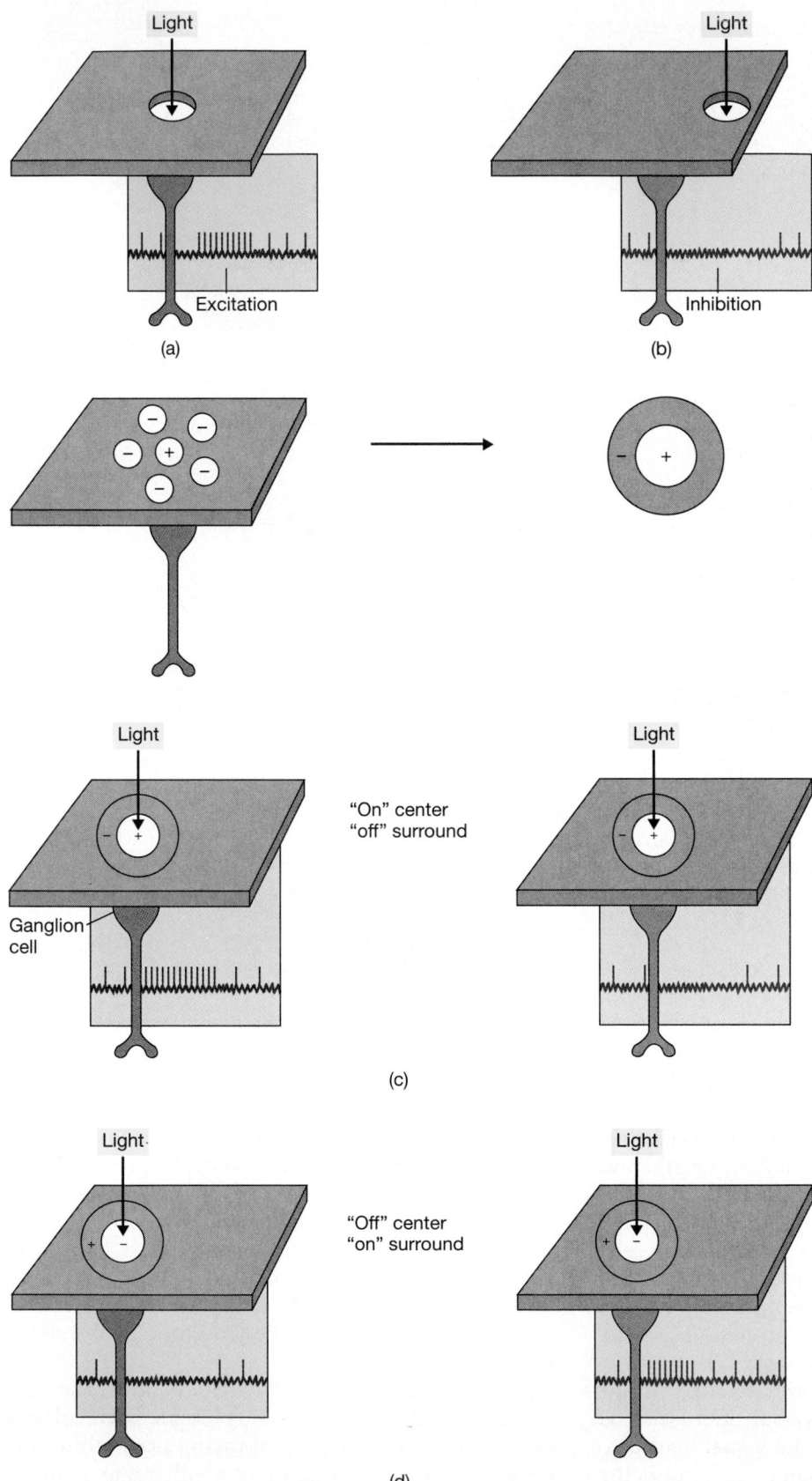

Figure 8–27

The effects of light on ganglion cell activity. (*a*) Light near the region of a ganglion cell produces depolarization and action potentials, whereas light in regions farther from the cell may inhibit the formation of action potentials (*b*). (*c*) A ganglion cell with a receptive field whose center is excitatory and whose surrounding ring is inhibitory is an "on-center and off-surround" cell. (*d*) Other ganglion cells have "off" centers and "on" surrounds.

Figure 8–28

The location of activated ganglion cells encodes the image of an object.

Left eye

Right eye

Nasal retina

Temporal retina

Optic nerve (ganglion cell axons)

Optic tract

Optic chiasm

Midline

Lateral geniculate nucleus

Area 17

(a)

Lateral geniculate nucleus

Right visual cortex

LEFT

RIGHT

Occipital dominance columns

(b)

Fovea

Visual field

Left

Right

Projection of the fovea

Projection of the fovea

Primary visual cortex

(c)

Figure 8–29

(*a*) Ganglion cells extend their axons outside of the eye through the optic nerve. Ganglion cell fibers in the temporal retina cross the midline at the optic chiasma and form connections with cells in the lateral geniculate nucleus on the opposite side of the brain. (*b*) In the lateral geniculate nucleus, visual information is sorted according to which eye it originated in and then sent to the visual cortex, where cells are arranged in ocular dominance columns, as well as in the form of a map of the external visual field (*c*).

RESEARCH FOCUS

Environments That Enhance Brain Development

Research with laboratory animals reveals that exposure to sensory-enriched environments can change the structure and chemistry of the brain. Rats raised in environments containing toys, stimulating objects, and other rats exhibit an increased thickness of their neocortex. In addition, the cell bodies and nuclei of neurons are larger, dendrites are longer, and the area of synaptic contacts is greater. These rats also have more protein, more glial cells, larger capillaries, and an increased ratio of RNA to DNA.

The most prominent change in these stimulated animals is found in the visual association cortex (area 18); however, most areas of the cortex, as well as the cerebellum and hippocampus (Chapters 9 and 11), are also affected. Except for the frontal cortex, where the right hemisphere is affected more than the left, both hemispheres appear to be equally influenced by the enriched environment.

The more varied the enriched environment and the longer the rat remains there, the longer it retains its increased cortical dimensions after being moved to a less stimulating environment. These effects have been found in both young and very old rats. Research also indicates that rats exposed to enriched environments perform maze tasks significantly better than those raised in less stimulating environments.

■ **What are the possible implications of these laboratory studies on the rehabilitation of patients with brain injury?**

Rat in a sensory-enriched environment, Columbia University.   *(© Hank Morgan/Science Source/Photo Researchers, Inc.)*

Receptive Fields of Neurons in the Lateral Geniculate Nucleus and Visual Cortex

The receptive fields exhibited by cells in the lateral geniculate nucleus are similar to those of the ganglion cells — that is, annular with either excitatory centers and inhibitory rings (or with inhibitory centers and excitatory rings). In the primary visual cortex, however, the visual stimulus that most effectively activates the neurons in this region of the brain is light that impinges upon the retina in the form of a rectangular bar (Fig. 8–30).

Why have the receptive fields changed from annular to rectangular in going from cells in the LGN to those in the primary visual cortex? The cells in the primary visual cortex receive inputs from cells in the

LGN that have overlapping annular receptive fields. Figure 8–30a shows three neurons in the LGN that synapse onto a nerve cell in the visual cortex. The activation of one cell in the LGN alone is not sufficient to activate the neurons in the primary visual cortex; the simultaneous activation of all three is required to turn on the visual cortex cell. Since the three cells in the LGN have annular receptive fields that overlap in a linear array, the combination of these receptive fields has the form of a rectangular bar.

The nerve cells in the primary visual cortex are also sensitive to the orientation of the rectangular bar of light in addition to its form. Some columns of cells are most sensitive to a rectangular bar of light oriented vertically (see Fig. 8–30b). With this orientation, they generate the greatest frequency of action

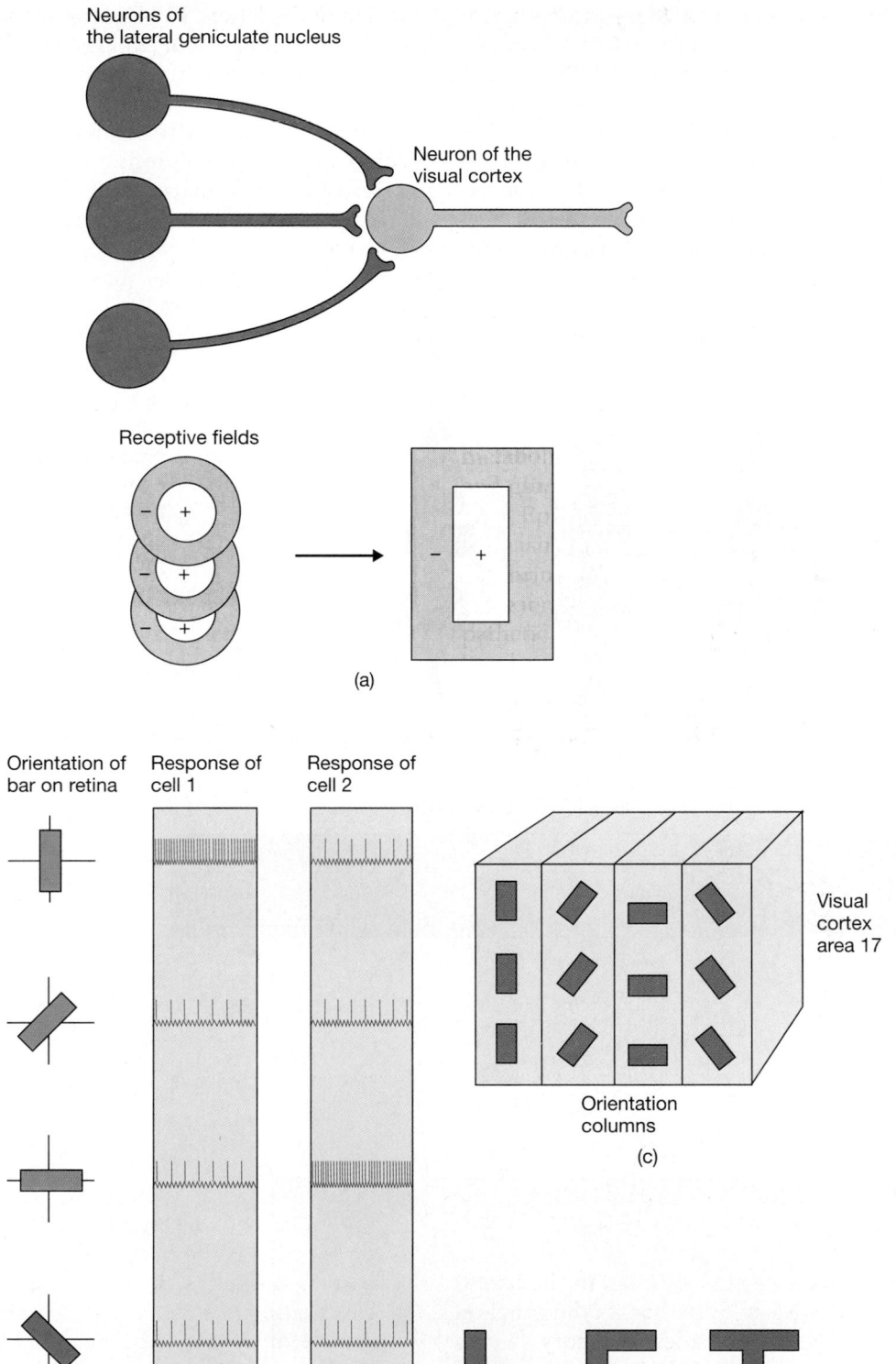

Figure 8–30

(*a*) The cells in the primary visual cortex receive information from LGN cells with over-lapping receptive fields; thus, the visual cortex cells have rectangular receptive fields. (*b*) Some visual cortex cells respond only to stimuli with the correct orientation. (*c*) Orientation-sensitive cells are arranged in columns in the visual cortex. (*d*) Some examples of complex stimuli needed to activate neurons in the secondary visual cortices.

potentials. Other columns of cells are most responsive to a rectangular bar of light oriented horizontally, and still other columns of cells are sensitive to bars of light oriented at different angles. These columns of cells are referred to as **visual orientation columns** (see Fig. 8–30c). Cells in vertical orientation columns, for example, would be activated by the letter "I," whereas cells in horizontal orientation columns would be activated by the horizontal component of the letter "H." Studies with laboratory animals have shown that young animals raised in visual environments with vertical bars but without horizontal bars do not perceive horizontal bars when tested later in life. These results indicate that the horizontal orientation columns do not exist in the visual cortex of these animals and demonstrate the importance of the visual environment in the development of visual perception.

(a)

(b)

VISIBLE LIGHT PORTION
OF THE ELECTROMAGNETIC SPECTRUM

Figure 8–31

(a) The three types of cones—red, blue, and green—allow us to see and discriminate among colors. Each type of cone absorbs photons over a particular range of the visible spectrum, though the three ranges overlap (b).

Stimuli more complex than rectangular bars are required to activate neurons in the secondary visual cortices (areas 18 and 19). The most effective forms of visual stimuli capable of activating neurons in visual areas 18 and 19 have complex shapes that are combinations of simple rectangular bars (see Fig. 8–30*d*). Cells in these areas of the visual cortex combine the features of the receptive fields from the cells in the primary visual cortex. In this way, information about the form of the object being viewed is synthesized. Thus, like the somatosensory system, the photoreceptors of the visual system detect the basic features of a visual object. As this information is transmitted to the higher visual areas of the brain, the image is pieced together by the brain to reconstruct the basic features of the image.

Color Vision: A Three-Way Process

Our ability to discriminate between different colors is due to the three types of photoreceptor cones in the retina that form the basis of the **trichromatic theory** of color vision. Each cone has a different type of visual pigment capable of absorbing different wavelengths of light. The blue cone, for instance, absorbs photons with a wavelength of 443 nanometers. It also absorbs photons with wavelengths slightly greater than 443 nanometers and slightly less than 443 nanometers, but it does so less efficiently. Thus, the blue cones are most sensitive to blue light (Fig. 8–31). The other two types of cones are the **green cones** and **red cones.** These cones are most sensitive to photons with wavelengths of 535 and 570 nanometers, respectively; they absorb photons with other wavelengths less efficiently. The three photoreceptor cones, therefore, have overlapping probabilities of absorbing photons of various wavelengths.

If the retina contains only three cones, how can we perceive more than three colors? Each cone absorbs photons of a particular wavelength to a different degree. Consequently, each photoreceptor will be hyperpolarized to a different level in response to photons of a particular wavelength. The differential hyperpolarization of only one type of photoreceptor would give the ability to perceive colors, but in a somewhat limited way. Assume, for example, that the retina contained only red cones. If we were to view a picture painted with colors of 540 and 605 nanometers, we would not be able to tell that two different colors were used. This is because both colors emit photons that are absorbed by the red cone with equal probability (Fig. 8–32*a*). The addition of a green cone, however, results in a greater absorption of the 540 nanometer photons, and thus allows us to distinguish

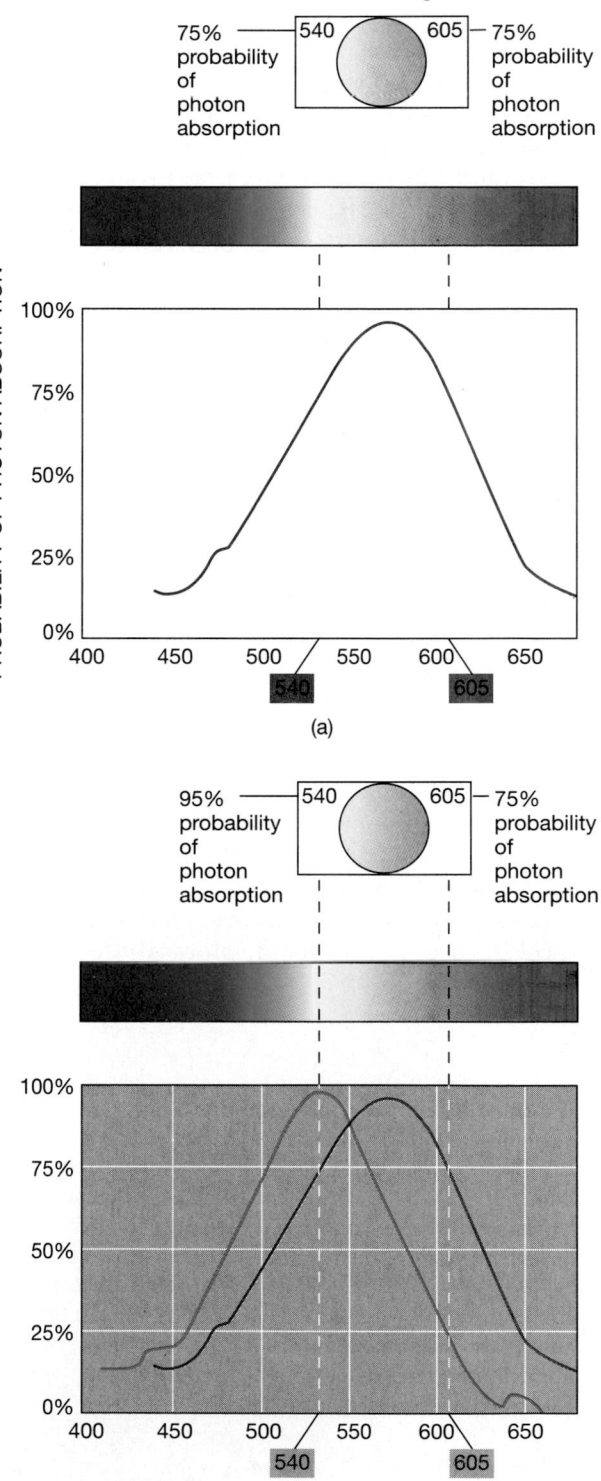

Figure 8–32

(*a*) A person with only red cones may not be able to distinguish two colors for which the rod has the same probability of absorbing photons. (*b*) The presence of a second type of cone allows the two colors to be distinguished by increasing the probability that the rod will absorb photons at one wavelength.

between the two colors. With three photoreceptor cones, our ability to distinguish between different colors becomes even better.

Color vision that relies on one type of cone is called **monochromatic color vision;** color vision that relies on two types of cones is called **dichromatic color vision.** Most people exhibit **trichromatic color vision.** Some people, however, are **color blind** and are not able to tell the difference between certain colors. Color blindness is a sex-linked trait that affects males predominantly. These males are usually **dichromats,** possessing only two types of cones, and represent approximately 2% of the population. Half of these dichromats do not have green cones, and the other half do not have red cones. These individuals cannot see the colored numbers embedded within the design of Figure 8–33.

The trichromatic theory of color vision alone is not sufficient to explain completely how we perceive color. After the selective absorption of photons by cones, cellular responses that are also responsible for the perception of color come into play. Ganglion cells, for example, differ in the way they process information from cones. Small ganglion cells are able to distinguish information about different colors. They in turn transmit this color information to small cells in the lateral geniculate nucleus (Fig. 8–34). These so-called **parvocellular neurons** in the geniculate transmit information to neurons in the visual cortex that respond selectively to color information. Large ganglion cells, on the other hand, process information from cones in such a way that they lose their ability to respond selectively to specific colors. They in turn transmit information to large cells in the lateral geniculate nucleus. These so-called **magnocellular neurons** transmit information to the primary and secondary visual cortices, where it is analyzed to extract features about an object's movement.

Neurons that selectively respond to color are divided into three different groups: (1) **broad-band cells,** (2) **single-opponent cells,** and (3) **double-opponent cells** (Fig. 8–35). Broad-band cells are excited by a single color presented to the center of the receptive field and inhibited when that color is in the periphery. Single-opponent cells are also excited when one color of light is in the center of the receptive field; however, they are inhibited when another color is in the periphery. For example, one type of cell excited by red in the center of the receptive field is inhibited by green in the periphery. The opposing colors are usually paired in combinations of red and green, blue and yellow, or black and white, and oppose one another across the compartmental boundaries of the receptive field.

Double-opponent cells differ from single-opponent cells in that the color pairs oppose each other within the boundaries of the receptive field. In this type of cell, for example, red in the center of the field might excite the cell, while green within the center would inhibit the cell. In addition, green in the peripheral boundary of the field might excite the cell, while red in this region would be inhibitory.

Figure 8–33

A rough guide to red-green color-vision defects (Ishihara, 1968). More severe than green weakness, the phenotype green blindness is due to the absence of the green color-vision pigment in cone cells. The two green defects are controlled by alleles. A parallel situation relates the two defects of the red visual pigment. What you see in the patterns of dots is the diagnostic criterion:

Perception	*Phenotype*
42	Normal
4 (and less clearly a 2)	green-weak
4 only	green-blind
2 (and less clearly a 4)	red-weak
2 only	red-blind

This color plate is not intended for use in the diagnosis of color blindness. A diagnosis of color blindness must be based on the original plates, which appear in Ishihara, S.: *Tests of Colour-Blindness*, 38 plate edition, Kanehara & Co., LTD, Tokyo, Japan (1968).

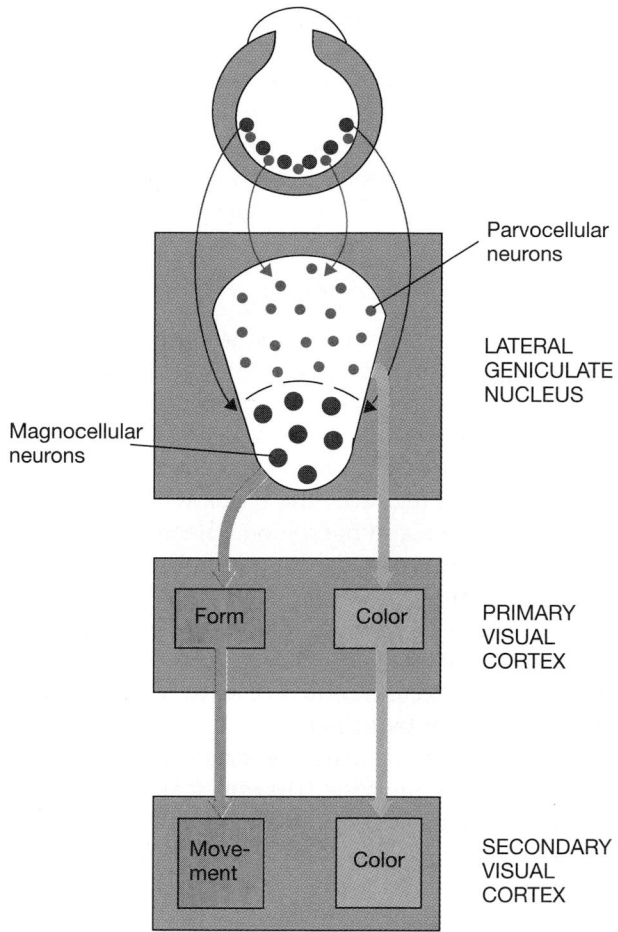

Parvocellular neurons

Magnocellular neurons

LATERAL GENICULATE NUCLEUS

Form | Color

PRIMARY VISUAL CORTEX

Movement | Color

SECONDARY VISUAL CORTEX

Figure 8–34

Small ganglion cells distinguish among different colors and transmit the information to small cells, or parvocellular neurons, in the lateral geniculate nucleus. Large ganglion cells transmit information about an object's movement to large, or magnocellular, neurons in the lateral geniculate.

The double-opponent cell can also be thought of as transmitting information about the contrast between two colors. In the previous example, the cell would become maximally excited when red is present in the center of the field and green is in the periphery. The activation of this type of cell would therefore communicate to the brain that two contrasting colors are adjacent to one another in the receptive field.

Some cells in the visual cortex display the characteristics of double-opponent cells with annular receptive fields. Other cells have rectangular receptive fields. The response properties of these cells are a summation of the annular properties displayed by converging neurons (Fig. 8–36). The annular patterns of these converging neurons overlap and thus give rise to a bar-shaped receptive field. These cells, however, do not appear to be associated with orientation columns and respond to rectangular forms of color in any orientation. The color-sensitive cells are clus-

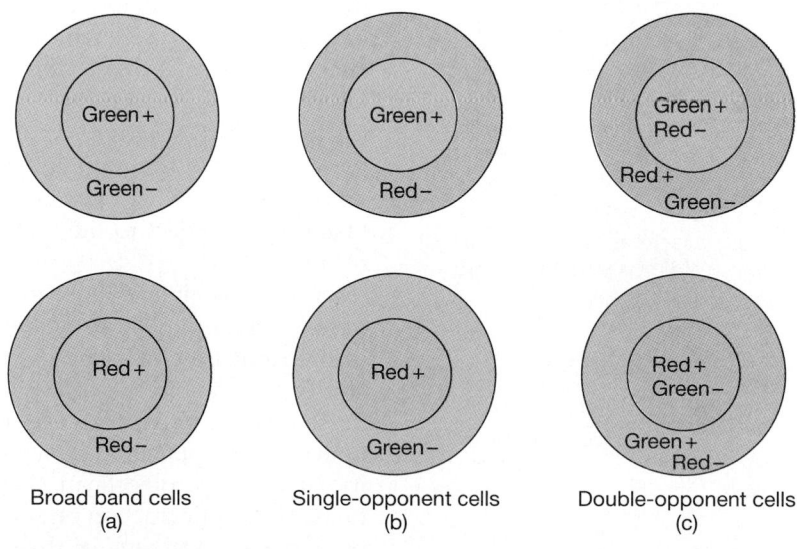

Broad band cells
(a)

Single-opponent cells
(b)

Double-opponent cells
(c)

Figure 8–35

Neurons that respond selectively to color are classified as broad-band, single-opponent, or double-opponent cells, dependent on the effects of different colors.

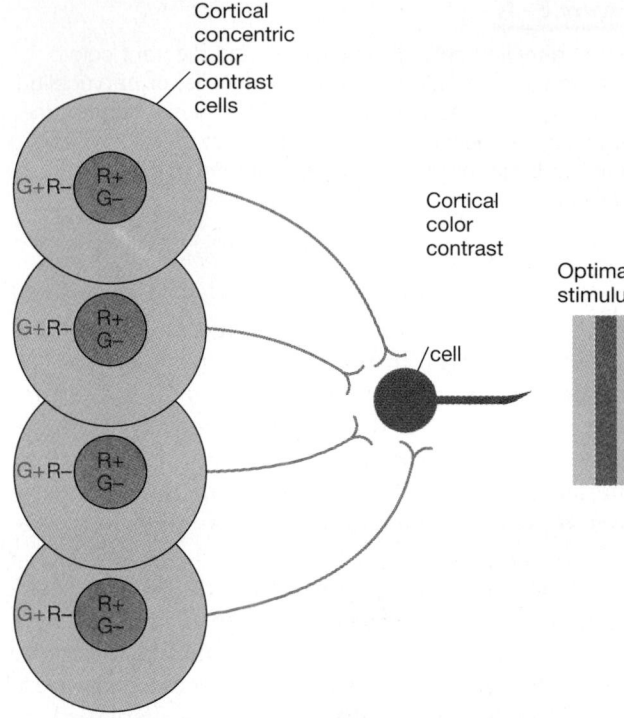

Figure 8–36

The annual receptive fields of some cells in the visual cortex overlap to give bar-shaped receptive fields for color.

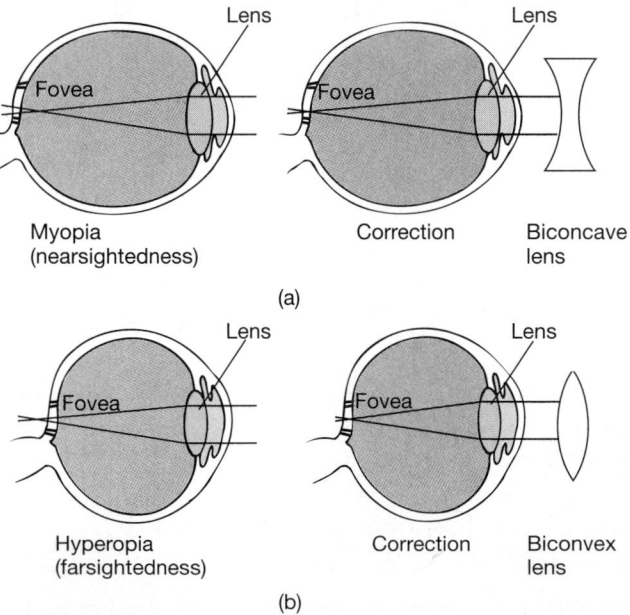

Figure 8–37

In a nearsighted (myopic) person, the lens focuses an image in front of the retina, requiring correction with biconcave lenses. In a farsighted (hyperopic) person, the lens focuses the image behind the retina, requiring a biconvex lens for correction.

tered in functional columns that are parallel to orientation columns (Fig. 8–34). This arrangement suggests that the color and form of an object are analyzed independently. In support of this theory are observations of stroke patients who are unable to determine the color of an object even though they recognize its form.

Disorders of Visual Sensation

As mentioned earlier, the most sensitive part of the retina is the **fovea** (Fig. 8–37). In this part of the retina, the ganglion cell fibers are sparse so that photons can easily reach the photoreceptors. In addition, all of the photoreceptors in the fovea are cones. The fovea, therefore, is the region of the retina with the greatest spatial resolution and the region where the lens of the eye attempts to focus the incoming images of light. Some individuals, however, have lenses that cannot properly focus the light onto the fovea. Lenses without enough focusing power cause the image to be focused at some point behind the retina. Individuals with this type of lens disorder are farsighted and require biconvex eye glasses or contacts to adjust the focal point. In some cases, the lens focuses the image in front of the fovea. Individuals with this type of lens disorder are nearsighted and require biconcave lenses to extend the focal length.

Visual disturbances can also occur when accidents or disease damage parts of the visual pathway. These disturbances are characterized by a loss of visual perception in some part of the **visual field** (Fig. 8–38). The left and right eyes have overlapping visual fields; this overlapping region is called the nasal portion of the visual field. Images in the nasal portion of the visual field are projected to the temporal portion of the retina in the left and right eye. Images in the temporal portion of the right visual field project to the nasal portion of the retina of the right eye only. Likewise, images in the temporal portion of the left visual field project to the nasal portion of the retina of the left eye.

Trauma such as a blow to the eye can sometimes damage the optic nerve. Damage to the optic nerve of the right eye, for example, results in a loss of the ability to see objects in the right visual field. Disorders such as pituitary tumors can damage nerve fibers in the optic chiasm, which results in an inability to perceive objects in the left and right visual fields. The optic tract on one side of the brain can also be damaged. Damage of the optic tract on the right side of the brain results in defects in the left temporal and right nasal visual field.

Lesions can also occur in the secondary visual cortices (areas 18 and 19), resulting in **visual agnosia.** Patients with visual agnosia have difficulty recog-

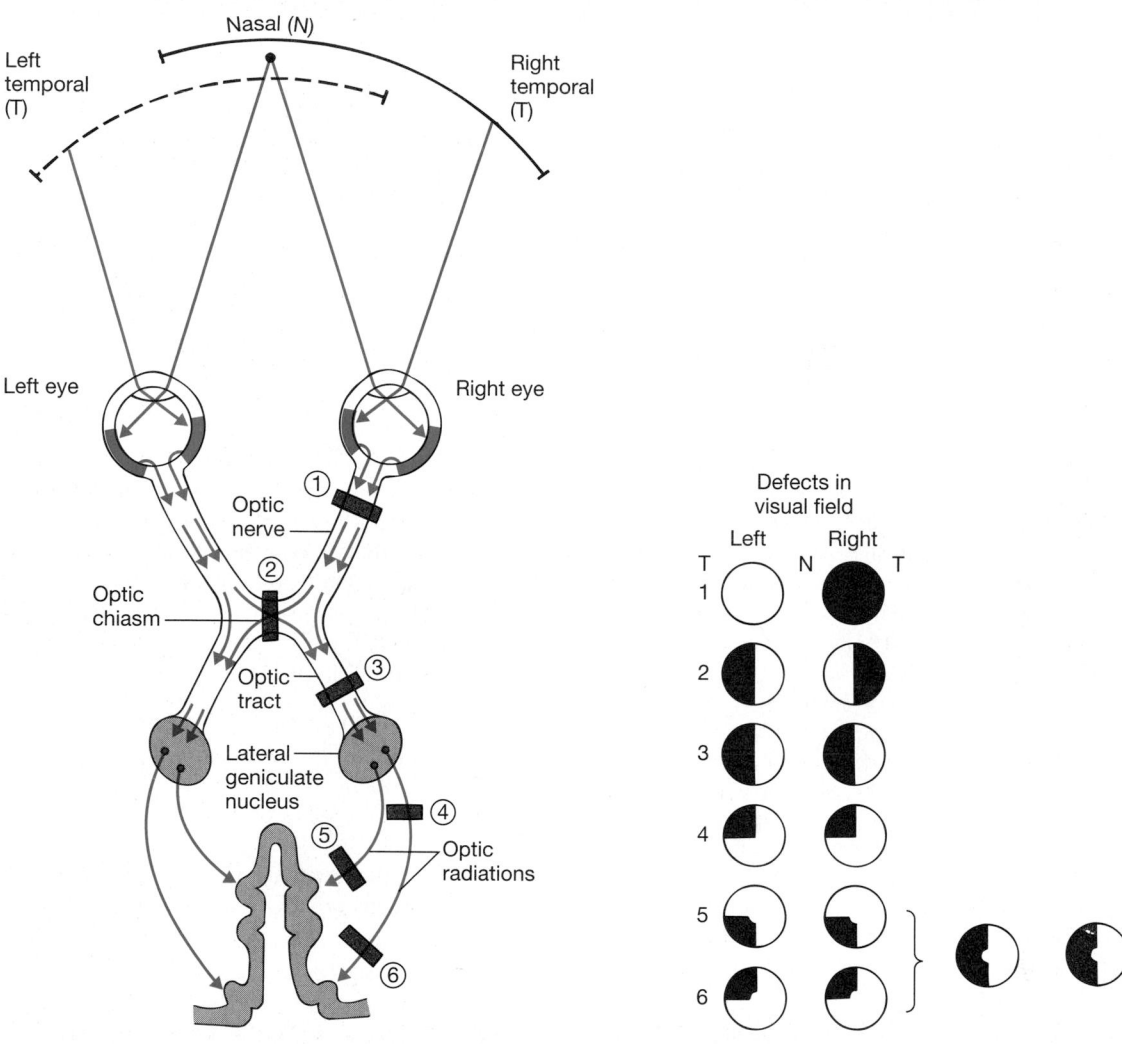

VISUAL FIELD

Figure 8–38

Damage to some parts of the visual pathway may result in disturbances in the visual fields of the eyes.

nizing the faces of friends, familiar objects, or colors (see Table 8–3).

The Auditory System Analyzes the Frequency Components of Sound and Its Location

The auditory system detects complex sounds such as spoken words and breaks them into their basic sound frequencies. The fundamental frequency components are transduced into action potentials by receptor cells in the ear and transmitted to the brain. Auditory areas of the brain subsequently recombine the basic

sound features of the words, which are then interpreted in areas of the brain responsible for language (see Chapter 11).

Sound Transduction of the Functional Organization of the Ear

Sound is the adequate stimulus of the auditory system and is transmitted through the air by the compression and expansion of air molecules in the form of air pressure waves. The sense organ for the transduction of sound is the ear. The ear consists of the **outer ear, middle ear,** and **inner ear** (Fig. 8–39). The outer ear channels the sound waves to the **ear drum,**

(text continues on page 285)

CLINICAL FOCUS

Ocular Dominance and Critical Periods of Development

In the lateral geniculate nucleus (LGN) and visual cortex are clusters of nerve cells that respond to the stimuli of light impinging upon either the left or right eye. The cells in the LGN are arranged in several layers, with each layer specific for inputs from one eye. The number of layers in the LGN is species dependent. Rodents, for example, have two layers, while monkeys have six. The segregation of left and right eye inputs emerges gradually during development from an initial pattern where inputs from both eyes are intermingled. In cats in which this phenomenon has been studied in detail, ganglion cell fibers from the contralateral eye enter the LGN at embryonic day 32 (about half of a cat's gestational period). The ipsilateral inputs arrive a few days later. During the next four and a half weeks the inputs from the two eyes gradually segregate; by birth the eye-specific layers are present as a result of a retraction of fibers from the nondominant input.

Laboratory studies have shown that competition between the ipsilateral and contralateral fibers is involved in the segregated pattern of eye inputs. The loss of one eye, for example, results in the expansion of fibers from the preserved eye into those layers typically occupied by inputs from the lost eye. Other studies have demonstrated that the segregated patterns are a result of activity-dependent processes. This has been shown by blocking, with the drug tetrodotoxin, the initiation and propagation of action potentials by ganglion axons. The administration of this drug in prenatal cats results in a loss of eye-specific layers within the LGN.

In the visual cortex the pattern of left-eye and right-eye inputs takes the forms of ocular dominance columns. These columns exhibit alternative left-eye and right-eye inputs. The width of each column is about 0.5 mm in the monkey.

The closure of one eye at various times after birth alters the pattern of ocular dominance columns. Closure of one eye at birth in monkeys greatly reduces the width of the associated ocular column in visual cortex; closure at 4 weeks after birth or later causes no alterations in the ocular dominance columns. These results reveal the modifiable nature of ocular dominance columns and the fact that their formation depends on a postnatal critical period of development.

Visual stimuli of various types have also been shown to alter the functional properties of cells in the visual cortex. Kittens raised in an environment of vertical black and white stripes develop cells in the visual cortex that respond preferentially to vertical bars of light. Animals raised in visual environments that minimize movement perception have fewer movement selective cells.

■ **What are the consequences of critical periods of development for other sensory systems?**

Eye-specific layers of the lateral geniculate nucleus. (*a*) Normal pattern of eye-specific layers in the adult cat. (*b*) Overlap of eye-specific layers in the prenatal cat. (*c*) Adult pattern of eye-specific layer in the cat after prenatal removal of one eye. (*d*) Adult pattern of eye-specific layers after prenatal administration of TTX to both eyes.

Table 8–3
Visual Agnosia

Type of Agnosia	Visual Defects
Agnosia for color	
Color agnosia	Associating colors with objects
Color anomia	Naming colors
Agnosia for form and pattern	
Object agnosia	Naming objects
Agnosia for drawings	Recognizing drawn objects
Prosopagnosia	Recognizing faces
Agnosia for depth and movement	
Visual spatial agnosia	Stereoscopic vision
Movement agnosia	Discerning movement of objects

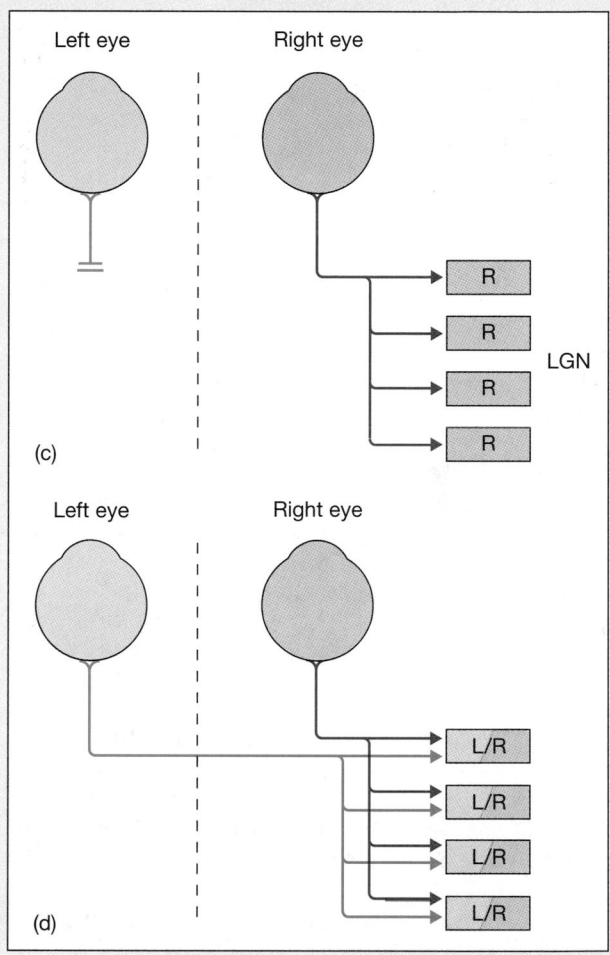

located at the interface between the outer and middle ear. The sound waves cause the ear drum to vibrate much like the vibration of the speakers of a stereo system. The vibration of the ear drum, in turn, causes the small bones or **ossicles** within the middle ear to move. These bones, called the **malleus, incus,** and **stapes** (also known as the **hammer, anvil,** and **stirrup**), are connected in series and transfer the vibrations of the ear drum to the inner ear.

The stapes, the last in the series of ossicles, is connected to the inner ear at the **oval window,** a membranous structure that vibrates with the movement of the ossicles. The inner ear consists of a fluid-filled chamber called the **cochlea.** The cochlea is a cone-shaped structure wrapped in a coil much like the shape of a snail's shell. Along the length of the cochlea is a membrane called the **basilar membrane** (Fig. 8–39*d*). This membrane is a flexible structure that vibrates when sound impinges upon the ear drum. The vibrations of the basilar membrane are in the form of traveling waves (Fig. 8–40*b*) that deform the shape of the membrane. As the traveling wave progresses down the length of the membrane, the amplitude of the deformation becomes greater and greater until it reaches a peak. As the traveling wave progresses further, its amplitude diminishes. The location of the peak amplitude of the traveling wave along the basilar membrane depends on the fre-

(a)

(b)

(c)

(d)

(e)

◀ *Figure 8–39*

(*a*) The general anatomy of the ear. (*b*, *c*) The cochlea is a coiled structure containing three fluid-filled chambers formed by a system of membranes. (*d*) A cross-section of the cochlea showing the fluid chambers and membranes. (*e*) The cochlea as if "unwound," showing the path of a sound pressure wave from the oval window through the scala vestibuli, around the apex of the cochlea, into the scala tympani, and back to the round window at the base of the cochlea. (Some pressure waves do not go all the way to the apex.)

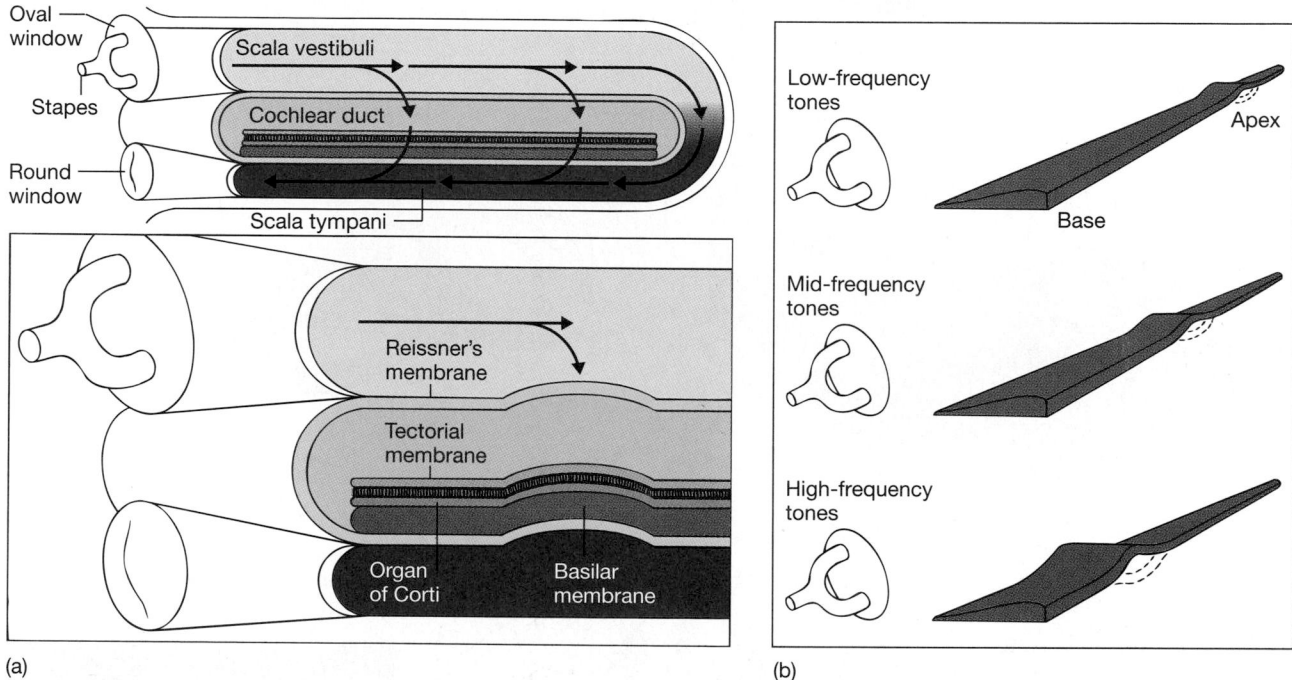

(a)

(b)

Figure 8–40

(*a*) The cochlear membranes shown in a different cross-sectional view. Sound causes the fluid in the scala vestibuli to exert pressure on Reissner's membrane, which in turn causes the basilar membrane to bulge. This movement sets up vibrations in the form of waves in the basilar membrane (*b*). The waves peak at different locations on the membrane depending on the frequency of the sound.

quency of the sound. The membrane is designed such that it is wider at the base in comparison to the apex. This design causes high-frequency sounds to produce traveling waves that reach their peak amplitudes near the base of the cochlea. Low-frequency sounds cause the traveling waves to reach their peak near the apex of the cochlea. Intermediate frequencies of sound cause peak amplitudes to occur at locations along the membrane between the base and apex of the cochlea. Thus, the basilar membrane is designed mechanically to respond to different frequencies of sound by changing its degree of depolarization at different points along its length.

The deformations of the basilar membrane are transduced into action potentials by **hair cells.** The hair cells sit upon the basilar membrane and have fiberlike elements called **stereocilia** that protrude

from the cell (Fig. 8–41*a*). These cilia are embedded in another membranous structure called the **tectorial membrane.** When the basilar membrane vibrates and reaches a peak amplitude, the stereocilia are bent in relation to the hair cell (Fig. 8–41*b*). This bending of the stereocilia causes ion channels of the hair cells to open, leading to a depolarization of the membrane potential. This depolarization is analogous to the generator potential described for the Pacinian corpuscle of the somatosensory system. Unlike the Pacinian corpuscle, however, the hair cell is not capable of producing its own action potential. When the hair cell is depolarized, Ca^{2+} ions flow into the cell and release transmitters that in turn depolarize auditory nerve fibers (see Fig. 8–41*c*). The auditory nerve fiber then generates action potentials that are transmitted to the brain to convey information relating to frequencies of

Vestibular membrane

Tectorial membrane

Stereocilia

Scala vestibuli

Organ of Corti

Scala tympani

Hair cells

Area of enlargement

Tectorial membrane

Stereocilia

Resting state

AREA OF ENLARGEMENT

ACTIVE STATE

Scala vestibuli

Hair cells pushed against tectorial membrane

Reissner's membrane

Basilar membrane vibrating

Stereocilia bent

Hair cells pushed up against tectorial membrane

Basilar membrane vibrating

(a)

(b)

Stereocilia

Tectorial membrane

Membrane receptor

Auditory nerve ending

Auditory nerve fiber

Basilar membrane

Depolarization of hair cells causes release of neurotransmitters

Bending of stereocilia

Increased action potential generation by auditory nerve fiber

Basilar membrane vibrating

(c)

(d)

sound. Hair cells are also responsible for transducing information regarding rotational and linear acceleration needed for balance and equilibrium (see Table 8–1). These hair cells are part of the vestibular system and will be discussed in Chapter 9.

Auditory Pathways and Receptive Properties of Auditory Nerve Cells

The peripheral fibers of the auditory nerve make contact with the cochlear nucleus in the brain stem (Fig. 8–42). Information from the cochlear nucleus is transmitted to both sides of the brain and eventually reaches the medial geniculate nucleus of the thalamus. From the medial geniculate nucleus, information is transmitted to the primary auditory cortex in the temporal cortex.

The receptive field of neurons along the auditory pathways is characterized by the use of tones with different frequencies and intensities. A tone is a sound wave of one frequency. The sound produced by playing a single note on the piano, for example, produces a pure tone. This tone, in turn, produces a sound pressure wave that causes the compression and expansion of air molecules at one frequency (Fig. 8–43). If many notes are played simultaneously, the sound produced is a combination of all of the individual frequencies. Tone can be represented graphically by an oscillating waveform; the peak of the waveform represents the compression of the air molecules, and the valley of the waveform represents the expansion of the molecules. The difference between the peak and the valley is the **amplitude (A_O)**, or intensity of the sound wave. The time between the peaks is the **time period** of the tone **(T)** and is measured in units of seconds. Its inverse is the **frequency (f)** of the tone and is measured in units of **Hertz (Hz)**. The amplitude and frequency of a tone can be used to characterize the receptive properties of nerve cells in the auditory pathway. If the frequency of the tones is varied, one frequency will be found that most effectively activates a particular auditory nerve fiber. This frequency is known as the **characteristic frequency (f_O)** for the nerve fiber. Different auditory nerves have different

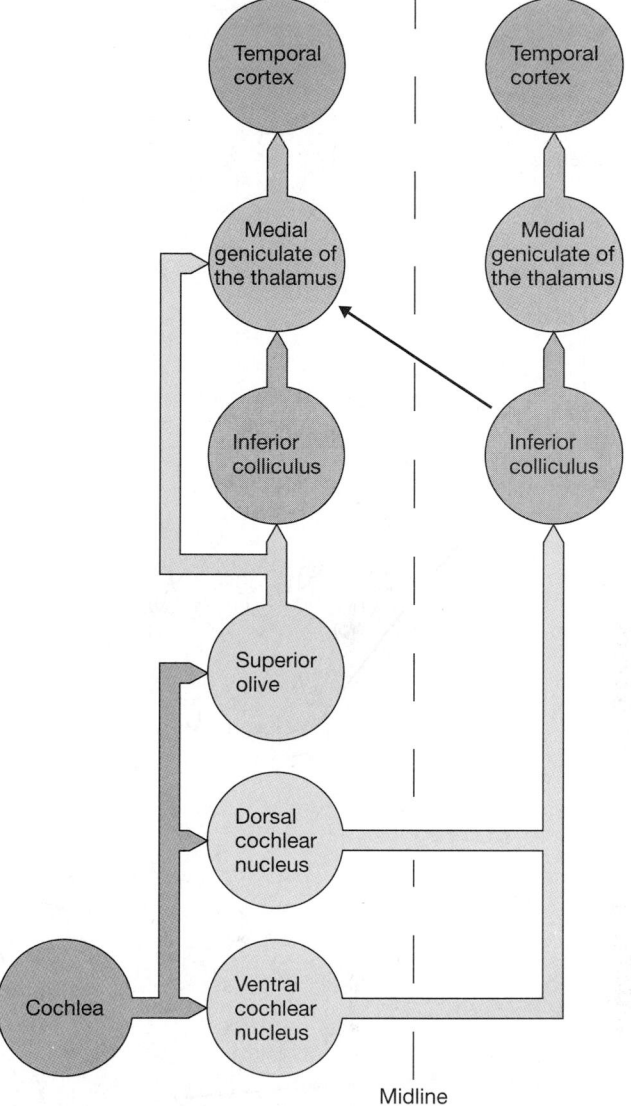

Figure 8–42

Auditory pathways, originating in the cochlea on one side of the brain and terminating bilaterally in the temporal cortices.

characteristic frequencies. These auditory nerve fibers have a **tonotopic organization** such that those with a high characteristic frequency innervate hair cells near the base of the cochlea, while those with a

◄ *Figure 8–41*

(*a*) When the basilar membrane changes shape as waves travel along it, hair cells above the membrane, which form the organ of Corti, are pushed closer to the tectorial membrane so that (*b*) stereocilia that protrude from the hair cells bend (*c*). The bending causes hair cell ion channels to open, leading to membrane depolarization, inflow of Ca^{2+}, and release of a neurotransmitter that causes action potentials to arise in auditory nerve fibers. (*d*) SEM of inner hair cells of the human ear. (×1705). *(© Dr. G. Bredberg/ SPL/Photo Researchers, Inc.)*

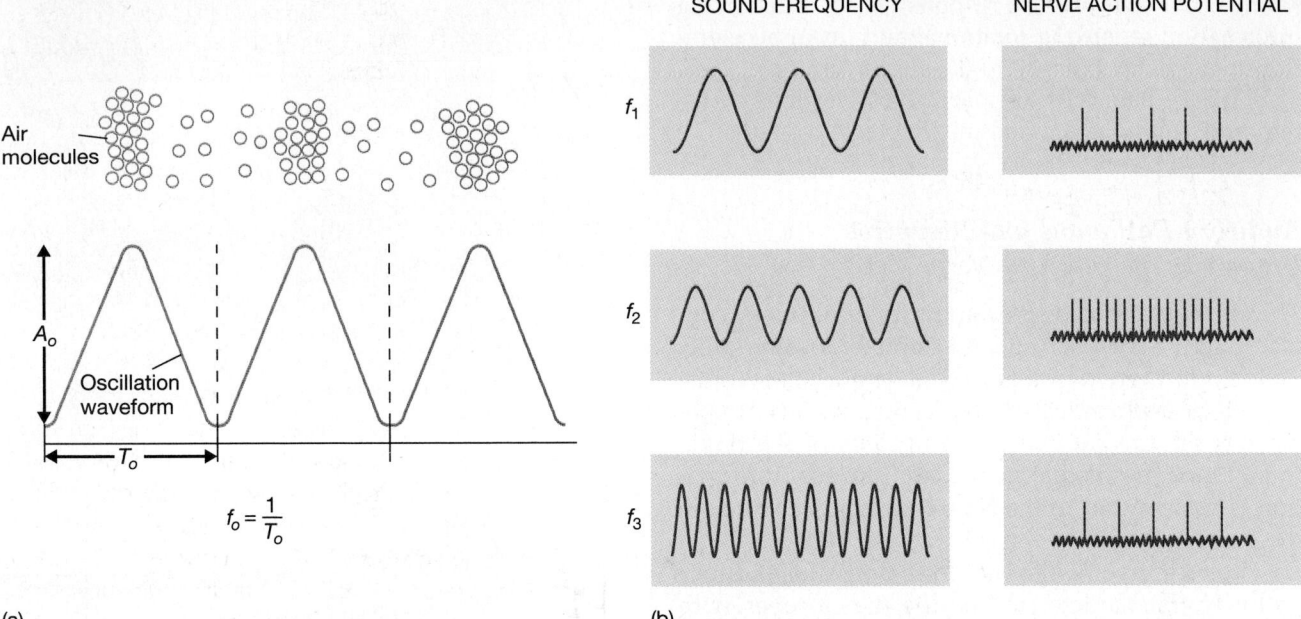

(a)

(b)

SOUND FREQUENCY   NERVE ACTION POTENTIAL

f_1

f_2

f_3

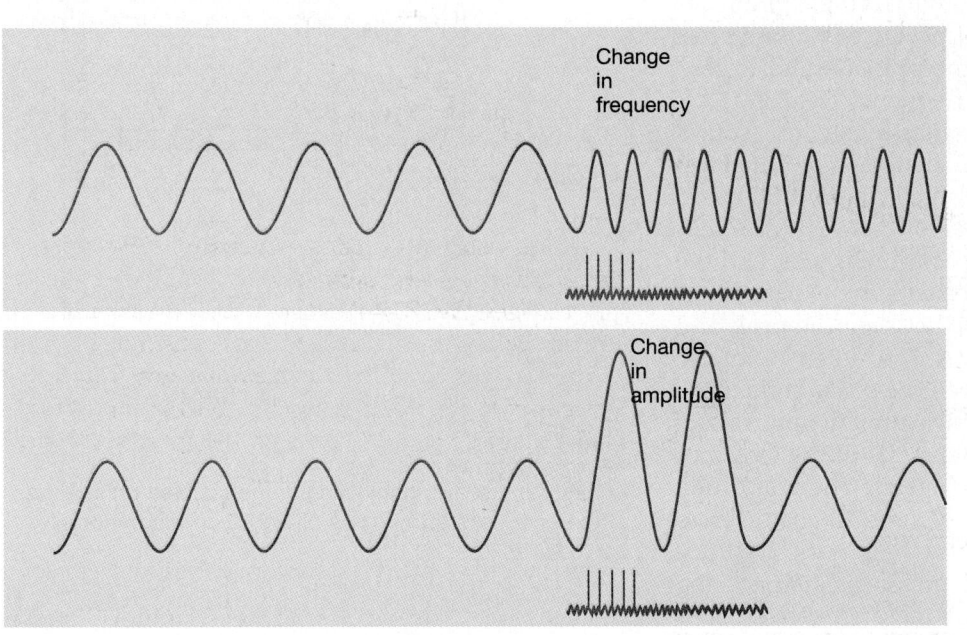

(c)

Change in frequency

Change in amplitude

Figure 8–43

(a) A single tone produces a sound pressure wave that alternately compresses and diffuses air molecules at a particular frequency and amplitude. The characteristic frequency of a nerve fiber is the frequency that most effectively activates the fiber (b). In this example, f_2 would be the characteristic frequency. (c) Some nerve fibers generate action potentials only in response to a change in the frequency or amplitude of a tone.

low characteristic frequency innervate hair cells near the apex of the cochlea.

Neurons of the auditory system also respond in a variety of different ways to a particular tone. Some neurons generate action potentials only when the tone is on; other neurons are prevented from generating action potentials with the onset of the tone. Some neurons generate action potentials when the frequency of the tone is suddenly changed, whereas others generate action potentials only when the amplitude changes. The variety of responses suggests that nerve cells of the auditory system extract the ba-

sic features of a sound from its complex waveform. The fundamental features of a sound are reassembled by the convergence of neurons. These neurons converge onto other neurons that respond only to the reassembled complex auditory stimuli. For example, certain neurons in the auditory cortex of monkeys respond only to specific sounds used by that species of animal. Animal studies of auditory responses suggest that there are neurons in the human auditory cortex that respond only to the word *go*, and others that respond to the word *stop*.

Sound Localization

In addition to decoding the frequency and intensity of sound information the auditory system is also designed to determine where sounds originate. This function of the auditory system allows us to orient toward a sound stimulus and respond accordingly. In crossing a street at a busy intersection, for example, the noise of a speeding car coming around the corner would cause us to stop, look in the direction of the oncoming car, and avoid a collision. The localization of sounds in space is accomplished by comparing differences in the timing and intensity of sounds as they reach each ear. Acoustic waves that originate on the right side of the head would first reach the right ear before traveling on to the left ear. The time delay between the two ears is call the **interaural time difference.** Sounds that originate along the midline would produce no interaural time difference. Sounds that originate off of the midline have been shown to produce interaural differences of 42 μsec for every 20 degree shift in space. These interaural time differences form the basis of an auditory field for sound localization as shown in Figure 8–44*a*. Each line represents an interaural time associated with a specific location in space.

In studies that have examined the mechanisms of sound localization in barn owls, researchers have found neurons that discharge in response to interaural timing differences. These neurons are found in regions of the medulla comparable to the human medial superior olive and anterior lemniscal nucleus (Fig. 8–44*b*). These neurons receive inputs from the cochlear nucleus on both sides of the brain and respond to specific frequencies and timing differences in the arrival of those frequencies to each ear. It is thought, therefore, that the mechanism underlying these responses is the coincident arrival of information from each ear to a specific neuron. The summation of incoming activity at these so-called "coincidence cells" (Fig. 8–44*c*) in turn causes them to discharge. The concurrent arrival of information on to specific cells in spite of interaural differences is thought to be due to propagation delays from the cochlear nucleus. For sound sources originating in the midline, for example, the propagation delays from the left and right ear are the same. In contrast, sound sources located closer to the right ear would result in the transmission of sound information through pathways with longer propagation delays than information coming from the left ear in order to reach a "coincidence cell" at the same time (Fig. 8–44*c*). Likewise sound sources located closer to the left ear would result in the transmission of information with longer propagation delays than pathways from the right ear.

The concept of topographic maps of neurons in sensory systems can now be extended to include a mapping system in the auditory system for interaural time differences. Since these neurons overlap in their responses to specific frequency of sound, the maps for tonotopic organization and interaural time differences can thus be thought of as a mapping of frequence and time information along orthogonal axes in the auditory system (see Fig. 8–44*a*).

Along with timing differences, sound information reaching the left and right ears from the same source will be perceived as differences in intensity. In part this is due to the attenuation of the acoustic wave as it propagates through air, and as it is absorbed by the side of the head. In animals such as the barn owl, the ears are slightly offset to aid in the detection of intensity differences. A source of sound equidistant to both ears would produce a sound intensity difference. However, for every 20 degree shift in space, there is a 4 decibel (db) change in intensity difference (see Fig. 8–44*a*). A decibel is a unit of measure for sound (based on the logarithm of the ratio of a sound's intensity with respect to a reference value). A sound with an intensity of 10 db is about equivalent to a soft whisper. Together, interaural intensity and time differences form the basis of an "auditory field" for the localization of sound. Each location is represented by a unique combination of interaural intensity and time differences. This auditory field is similar to the concept of the visual field for the location of objects within visual space.

Cells that respond to intensity differences are found in the lateral superior olive and posterior lemniscus (see Fig. 8–44*b*). Information about interaural intensity and time differences is transmitted through different pathways in the medulla and finally converges in the inferior colliculus to form a neural map of the auditory field. Cells in the inferior colliculus thus respond to unique combinations of sound intensity and time differences to allow us to locate their place of origin. This information, in turn, is ultimately transmitted to areas of brain involved in moving our head and torso toward the direction in which the sound originates.

Disorders of Auditory Sensation

A loss of hearing can occur in a variety of ways. Hair cells can be destroyed when the eardrum is suddenly exposed to a loud noise such as an explosion, which is thought to crush hair cells between the basilar and tectorial membranes. The loss of hair cells is associated with frequencies of sound that make up the noise. The resulting hearing disorder is called **sensory neural hearing loss.** Another disturbance of hearing of the sensory neural form is called **tinnitus,** a condition in which a constant ringing in the ear is perceived even in the absence of sound waves. This is due to the inappropriate discharge of action po-

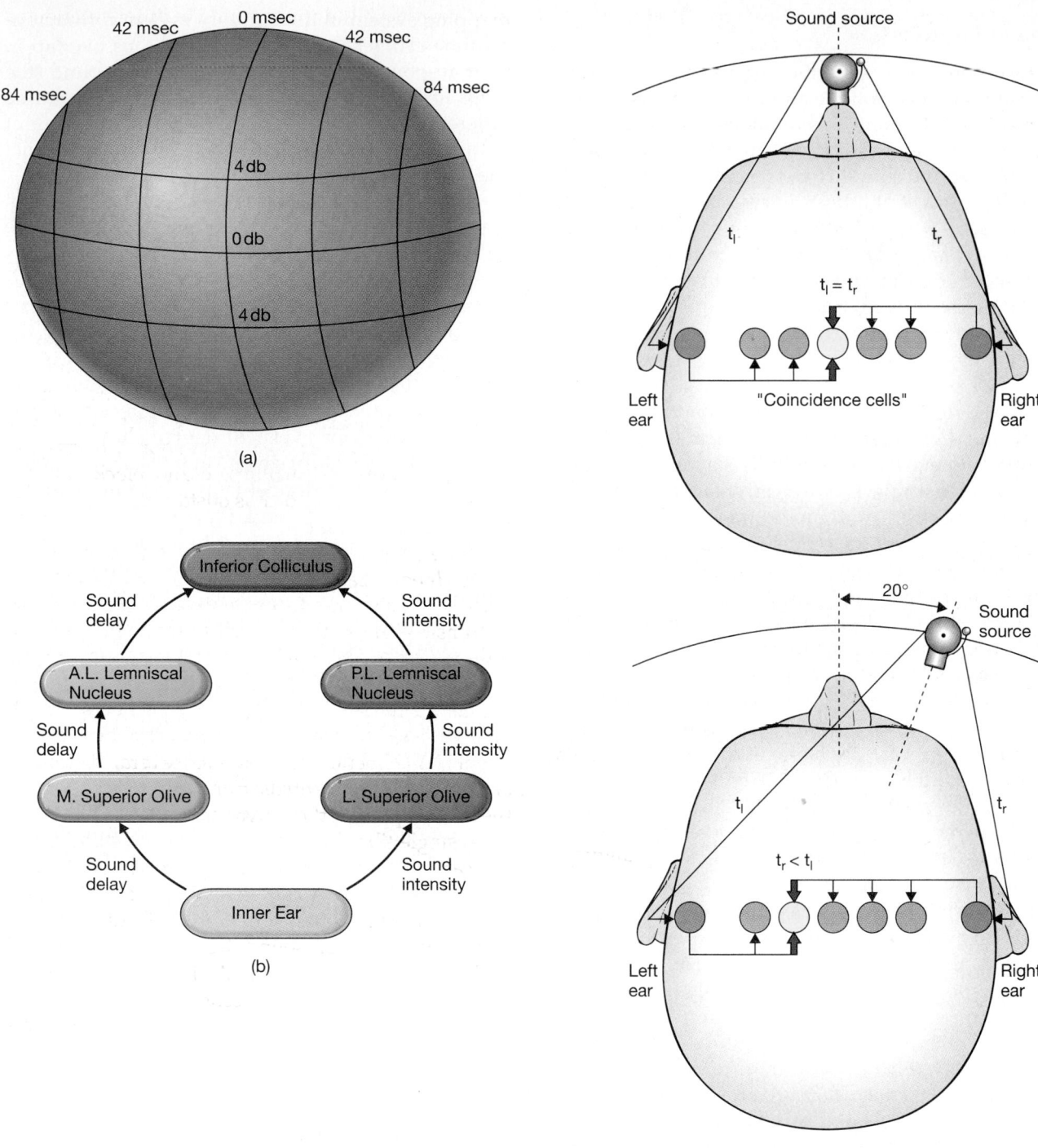

Figure 8–44

(a) The auditory field for sound localization is characterized by time and intensity differences for sound to reach the right and left ears. Vertical isocontours represent time differences. Sound sources originating in the midline reach both ears at the same time and thus have a 0 μsec time difference. Moving the sound source by 20° to the right or left produces a time delay of 42 μsec. Differences in sound intensity reaching the left and right ear are represented by the horizontal isocontours. Moving the sound source by 20° in either direction produces a 4 db difference in sound intensity. (b) Auditory pathway for the transmission of sound delays and intensity differences. (c) Response of coincidence cells to sound sources originating at different locations in the auditory field.

tentials by particular nerve fibers associated with specific frequencies. The exact cause of this disorder is unknown.

Another type of hearing disorder is **conductive hearing loss,** which results from damage to the middle ear. This can result from an inflammation of the small bones in the middle ear, or from abnormal bone growth caused by a process called **otosclerosis.** Conductive hearing loss produces a general reduction in hearing ability. Hearing by conduction through the skull, however, is still functional and can be tested by placing a tuning fork over the bony structure behind the ear to elicit the sensation of sound.

Central hearing losses are typically due to brain tumors and lesions of the auditory pathway within the central nervous system. Lesions in the primary auditory cortex (area 41) on one side of the brain have little effect because of the bilateral projection of sound information. However, lesions in the auditory association cortex (area 22) or Wernicke's area produce **auditory sensory aphasia,** a condition in which voices can be heard but the words appear to have no meaning.

Hearing disorders in infants can lead to lifelong language disorders. Ear infections, for example, that chronically impair the ability of infants to hear spoken words hinder the development of auditory nerve cell connections needed for hearing and speech. If these auditory connections are not formed within the first few years after birth, they will never be made. This developmental disorder illustrates the importance of the auditory environment and the formation of critical nerve cell connections underlying language skills.

The Gustatory System Gives Us An Appreciation for Different Taste Sensations

Our ability to taste food is the result of the activation of taste receptor cells in the tongue. These receptors detect chemicals associated with four basic taste qualities: sweet, salty, sour, and bitter. The tip of the tongue contains receptors that are sensitive to sweet substances (Fig. 8–45a). The lateral region near the tip of the tongue is sensitive to salty substances; the lateral region near the back of the tongue is sensitive to sour substances. The middle region at the back of the tongue is sensitive to bitter substances. Taste receptors are also located in the palate and pharynx.

Taste receptors are called **gustatory receptors** and are a type of epithelial cell clustered with other cells called **supporting** and **basal** cells (see Fig. 8–45b). Together these cells compose the **taste bud.** Taste receptors degenerate every 10 days and are replaced by new receptor cells derived from the differentiation of basal and supporting cells.

Transduction of Taste Sensations

Various mechanisms appear to underlie the transduction of different taste stimuli. Sweet-tasting compounds exert their effect by binding to receptors that utilize cAMP as a second messenger (Figure 8–45c). Elevations in cAMP, in turn, cause the closure of K^+ channels, thus depolarizing the membrane. The binding of bitter compounds to receptors in the base of the tongue results in an increase in intracellular Ca^{2+} via the second messenger IP_3. The transduction of saltiness is mediated by the direct passage of sodium ions through passive cation channels, thus altering the membrane potential of the taste receptor cell (see Fig. 8–45c). Acids responsible for sour tastes appear to penetrate the cell membrane and block voltage-gated K^+ channels, and thus sustain the depolarization of the membrane.

Increases in intracellular Ca^{2+} by IP_3 or by membrane depolarization induced activation of voltage-gated Ca^{2+} channels ultimately result in the release of transmitters from taste bud cells. The binding of transmitters to receptors located on postsynaptic fibers would then cause the initiation of action potentials in the afferent fiber. Afferent fibers branch many times before they innervate a taste bud; consequently, one fiber often innervates several taste buds, and the action potentials from a single fiber may therefore represent the activation of many taste buds.

A single fiber can respond to stimuli from different categories of chemicals, but it responds best to one type of chemical group. This property of taste fibers is similar to the responses of fibers in the auditory pathway for sound frequencies and visual pathways for colors of light. As in the other sensory pathways, the higher brain centers of the taste pathway use this type of information to make comparisons and contrast the taste qualities of consumed substances.

Gustatory Pathways

Taste receptor cells are innervated at their base by nerve fibers that form part of the VII, IX, or X cranial nerve (see Fig. 8–45b). Taste buds in the front of the tongue are innervated by the VII cranial nerve; those in the back of the tongue are innervated by the IX cranial nerve; and those in the pharynx are innervated by the X cranial nerve. Thus, information about bitter tastes is transmitted by way of a different pathway than information about sweet, sour, and salty tastes. Fibers innervating the taste buds in the palate are part of the X cranial nerve.

Figure 8–45

(*a*) Taste buds sensitive to the various tastes are grouped in specific regions of the tongue. (*b*) Taste buds consist of taste cells surrounded by supporting and epithelial cells. (*c*) Receptors on taste cells respond to sweet, bitter, salty, and sour compounds. Binding of sweet compounds to receptor (R_s) elevates cAMP levels, which leads to closure of K^+ channel. Binding of sour compounds to receptor (R_B) elevates IP_3 levels and increases intracellular Ca^{2+}. Salty taste is transduced by entry of Na^+ through passive ion channels. Sour compounds produce H^+ which penetrates the cell membrane and causes the closure of K^+ channels. (*d*) Pathways for the transmission of gustatory sensation.

All afferent fibers synapse on cells in the brain stem in the **gustatory nucleus.** Neurons in the gustatory nucleus, in turn, send their axons to the thalamus on the same side of the brain and terminate in the **ventral posterior medial (VPM) nucleus** of the thalamus. From the VPM, taste information is transmitted to the gustatory area of the neocortex just in front of the somatic sensory region of the tongue.

Alternate pathways for taste sensation transmit information to the limbic system and hypothalamus (see Chapter 11). These pathways are responsible for the affective and emotional aspects of taste. For instance, some individuals become nauseated by the taste of certain foods. This can happen even if the food was at one time a favorite dish. The consumption of spoiled food, for example, can lead to food poisoning. In some cases, a single exposure to the spoiled food can condition an individual to avoid it in the future.

The Olfactory System Allows Us to Detect Odor Molecules in the Air

Our ability to smell odors is a result of the activation of **olfactory receptor cells** in the nose. Olfactory receptor cells are contained within the **olfactory mucosal membrane** of the nasal cavity of the nose and surrounded by supporting and basal cells (Fig. 8–46*a*). As do the receptor cells in taste buds, the olfactory receptor cells degenerate and are replaced by basal cells that become new receptor cells. The regenerative cycle for olfactory cells, however, is 60 days.

Human beings can detect seven general types of odors: camphoraceous, musk, floral, peppermint, ethereal, pungent, and putrid. These odor molecules bind to olfactory receptor cells along ciliated fiber processes that extend from the cell body and into the mucosa of the nose (see Fig. 8–46*a*). The coupling of odor molecules to receptors within the cilia activates Gs proteins to regulate levels of intracellular cAMP (Fig. 8–46*b*). Increases in cAMP levels, in turn, result in the activation of Na^+ channels leading to the influx of Na^+ ions and the depolarization of the membrane. When threshold is reached, an action potential is initiated and is propagated along axons that pass through the **cribiform plate** and terminate upon **mitral cells** located in the **olfactory bulb**. The bundle of fibers from mitral cells forms the **olfactory tract**. These fibers directly innervate areas of cortex known as the **limbic system,** and the orbitofrontal cortex indirectly via the thalamus (see Fig. 8–46*a*). The olfactory innervation of the limbic system is exceptional in that it represents the only sensory system that directly innervates areas of cortex.

The olfactory system plays an obvious role in stimulating appetite and eating behavior. The activation of eating behavior is carried out by a motivational area of the brain called the **hypothalamus** and will be considered in Chapter 11, on central integrative systems.

The olfactory system also plays a role in mating behavior. Perfumes and colognes are examples of the use of artificial odors to attract members of the opposite sex. In animals such as mice, the odor of a female stimulates the release of testosterone in males and enhances the growth rate of the male sexual organ in adolescents. The odor of a female also causes the release of hormones in the adult male that in turn activate the sebaceous glands to give off a male odor to signal its readiness for mating. Male odors also in-

(a)

(b)

Figure 8–46

(*a*) Olfactory receptor cells in the nasal cavity contact mitral cells in the olfactory bulb. Mitral cell fibers form the olfactory tract, which innervates the limbic system and the orbitofrontal cortex. (*b*) Receptor in olfactory cilia. Binding of odor molecule to receptor (R) leads to activation of stimulatory G-protein (G$_s$) and adenylate cyclase (A$_c$). Elevation of cAMP causes opening of cation channel and influx of Na^+ ions to depolarize the olfactory receptor cell.

crease the rate of sexual maturation in adolescent female mice and accelerate the estrous cycle of adult females.

In humans, the olfactory system is poorly developed; thus, visual, tactile, and auditory stimuli play more important roles in human mating behavior.

CHAPTER REVIEW

Summary

- We detect information from the external environment with receptors that transduce sensory stimuli into action potentials, which are then conducted to the central nervous system. Sensory systems detect stimuli from the outside world in the form of light, sound, heat, pressure, chemical substances, and odors. Each type of stimulus is capable of activating a specific type of sensory receptor.

- The intensity of a stimulus is coded by the number of action potentials generated in a frequency code and by the number of receptors activated in a population code. The quality of a sensory stimulus is coded in the labeled-line code of stimulus quality. In sensory adaptation, receptors stop responding to a stimulus after a time.

- Somatic receptors on the surface of the skin detect pressures applied to the skin as well as temperature changes and painful stimuli. Pressure sensation is transmitted ipsilaterally in the spinal cord along the dorsal column pathway.

- Pain and temperature sensation are transmitted contralaterally along the anterolateral pathway of the spinal cord. Relay stations that process pressure sensation are located in the dorsal column nuclei of the medulla and the thalamus. In contrast, pain and temperature information are transmitted directly to the thalamus.

- From the thalamus, somatic information is sent to the somatosensory cortex. This cortex is organized in the form of a somatotopic map of the body with the lower limbs located in the medial region of the cortex and the upper extremities and face in the lateral region.

- Nerve cells that process somatic information from the periphery to the somatosensory cortex have receptive fields. These receptive fields are areas of the skin that, when stimulated, can either excite or inhibit a particular nerve cell. The pattern of excitation or inhibition forms unique shapes that are characteristic of neurons at different levels along the somatic sensory pathway.

- Sensory receptor cells respond to the basic features of a complex stimulus. In this way, they decompose the stimulus into its fundamental components. Contrasts and comparisons are made of the basic features by groups of nerve cells in each sensory pathway. These features are eventually reassembled by nerve cells in the sensory areas of the neocortex.

- Sensory receptors for vision are called photoreceptors and are located in the retina of the eye. Rod photoreceptors are used for night vision while cones are involved in color vision.

- There are three types of cones used in the process of color. These are red, green, and blue cones because of their sensitivity to specific wavelengths or colors of light. Light in the form of photons is absorbed by photoreceptors, causing a hyperpolarization of the photoreceptors.

- Information is then transmitted to ganglion cells via bipolar cells. Visual information is then transmitted from ganglion cells in the retina to the lateral geniculate nucleus of the thalamus. Information from the lateral geniculate is then transmitted to the visual cortex. The receptive fields of ganglion cells are annular with "on" centers and "off" surrounds or with "off" centers and "on" surrounds. This type of receptive field is also exhibited by neurons in the lateral geniculate. In the primary visual cortex, neurons are more sensitive to bar-shaped stimuli. Ganglion cells detect the basic features of a visual image. These features are assembled to reconstruct the total image at the higher levels of the visual system.

- The auditory system transforms sounds into action potentials. The external ear channels sound to the ear drum, causing the ear drum to vibrate. The ear drum is connected to a series of small bones called the malleus, incus, and stapes, which also oscillate in response to sound pressures. The mechanical displacement of the stapes is transferred to a fluid-filled structure called the cochlea. The cochlea contains sensory receptor cells called hair cells. The movement of fluid within the cochlea also displaces the basilar membrane. The displacement of the basilar membrane activates the hair cell, which in turn excites fibers of the auditory nerve.

- Each auditory nerve responds to a narrow range of sound frequencies. In addition, the nerves are tonotopically organized so that those innervating hair cells near the base of the cochlea respond best to high frequencies while those innervating the apex respond best to low frequencies. At the higher levels of the auditory system, neurons respond best to more complex sound frequencies.

- Taste information is detected by the gustatory system using taste buds located in the tongue. Taste buds near the back of the tongue respond best to bitter tastes, those in the middle respond best to sour and salty tastes, and those on the tip respond best to sweet tastes. The chemical activation of a taste receptor in turn excites afferent fibers that innervate the receptor. This information is eventually transmitted to the gustatory area of the neocortex near the somatic sensory region associated with the tongue.

- The sense of smell is conveyed by olfactory receptors in the nose. These receptors are innervated by mitral cells whose axons constitute the olfactory tract. This pathway provides inputs to the limbic system and thalamus. Information reaching the thalamus is then transmitted to the frontal cortex.

- The sense of smell is evoked when odor-producing molecules interact with the cilia of the olfactory receptors. Humans can detect camphoraceous, musk, floral, peppermint, ethereal, pungent, and putrid odors. In humans, the sense of smell is poorly developed in comparison to other animals.

Review Questions

Identify Each with a Term

1. The term used to describe the functional column of cells in the visual cortex that respond to visual input to one eye.
2. The term used to describe the sensory receptor that detects limb position.
3. The term used to describe a sensory receptor that discharges tonically in response to a sensory stimulus.
4. The term used to describe the functional organization of sensory inputs from the surface of the body to the neocortex.

Define Each Term

5. generator potential
6. conductive hearing loss
7. labeled-line code of stimulus quality

Choose the Correct Answer

8. The constant exposure to loud, high-frequency noise near 20,000 Hz would result in the loss of hair cells on the basilar membrane:
   a. near the base of the cochlea
   b. near the apex of the cochlea
   c. at a location midway between the base and the apex
9. The sensory information carried by fibers in the optic tract on the right side of the brain is derived from:
   a. the entire right visual field
   b. the nasal component of the left and right visual fields
   c. the temporal component of the left and right visual fields
   d. the nasal component of the right visual field and the temporal component of the left visual field
10. The absorption of a photon of light by a photoreceptor results in:
    a. the depolarization of the photoreceptor
    b. the inhibition of the phosphodiesterase enzyme
    c. the conversion of cGMP to 5′-GMP
    d. all of the above

Answer Each Question in One or Two Sentences

11. What are the general mechanisms of sensory transduction?
12. How is stimulus intensity encoded by the nervous system?
13. What is the concept of a receptive field?
14. What is the trichromatic theory of color vision?

Problem Solving in Physiology

15. What design changes would you make in the inner ear to improve our ability to hear at frequencies greater than 20,000 Hz?
16. What are the differences in sensory deficits seen in patients with spinal cord lesions? Consider both complete lesions and hemisections.

Suggested Readings

Hubel, D.H., and Wiesel, T.N. "Receptive fields, binocular interaction and functional architecture in the cat's visual cortex." *Journal of Physiology,* 160:106–154, 1962.

Hubel, D.H., Wiesel, T.N., and LeVay, S. "Plasticity of ocular dominance columns in monkey striate cortex." *Philosophical Transactions of the Royal Society of London,* 278:377–409, 1977.

Hudspeth, A.J. "Transduction and tuning by vertebrate hair cells." *Trends in Neuroscience,* 6:366–369, 1983.

Mogdans, J., and Knudsen, E.I. "Representation of interaural level difference in the VLVp, the first site of binaural comparison in the barn owl's auditory system." *Hearing Research,* 74:148–164, 1994.

Moulton, D.G. "Spatial patterning of response to odors in the peripheral olfactory system. *Physiological Reviews,* 56:578–593, 1976.

Mountcastle, V.B. "Modality and topographic properties of single neurons of cat's somatic sensory cortex." *Journal of Neurophysiology,* 20:408–434, 1957.

Penfield, W., and Rasmussen, T. *The Cerebral Cortex of Man. A Clinical Study of Localization of Function.* New York: Macmillan, 1950.

Shatz, C.J., "Impulse activity and the patterning of connections during CNS development." *Neuron,* 5:745–756, 1990.

Stewart, W.B., Knauer, J.S., and Shepherd, G.M. "Functional organization of rat olfactory bulb analyzed by the 2-deoxyglucose method." *Journal of Comparative Neurology,* 185:715–734, 1979.

Takahashi, T.T., and Keller, C.H. "Representation of multiple sound sources in the owl's auditory space map." *Journal of Neuroscience,* 14:4780–4793, 1994.

von Bekesy, G. *Experiments in Hearing.* New York: McGraw-Hill, 1960.

Key Terms

auditory system (p. 283)
bipolar cell (p. 272)
cochlea (p. 285)
dermatome (p. 252)
dorsal root ganglion (p. 259)
ear drum (p. 283)

gustatory system (p. 293)
hair receptor (p. 257)
inner ear (p. 283)
mechanoreceptor (p. 257)
middle ear (p. 283)
nociceptor (p. 257)
olfactory system (p. 295)

outer ear (p. 283)
pain receptor (p. 258)
photoreceptor (p. 258)
proprioception (p. 258)
retina (p. 269)
rhodopsin (p. 269)
sensory system (p. 256)

somatic sensory system (p. 256)
thermal receptor (p. 258)
visual field (p. 282)
visual system (p. 269)

Neurons of the spinal cord.
(© Cabisco/ Visuals Unlimited)

Chapter 9
Motor Systems

Key Concepts

The motor system conveys information to muscles that allows us to interact with the external environment by the activation of skeletal muscles. The functions are carried out by the activation of selective muscle groups.

The spinal cord contains the neural circuitry needed for reflex activity and locomotion. The interaction of neurons within the spinal cord coordinates the activation of synergistic and antagonistic muscle groups to carry out intended movements, as well as reflexive movements initiated by sensory receptors.

Conscious initiation of movement results from the flow of information from other parts of the brain to the spinal cord. Inputs from the brain stem to the spinal cord help us maintain our balance and posture. Inputs from the motor cortex are responsible for the control of more complex movements.

Specific sensory systems provide input to the motor system. The vestibular system monitors the movement of the body with respect to angular and linear acceleration. Sensory receptors in muscles monitor and regulate muscle force, length, and velocity of shortening. These sensory-motor systems are needed for making appropriate movements and necessary adjustments in movement.

Damage to the motor system produces movement disorders. These deficits can be manifested in the form of paralysis, uncontrolled movements, and a lack of coordination.

Chapter Outline

Muscles: Controlled by Nerves

    The Spinal Cord in the Motor Response System

    The Brain in the Motor Response System

    Vestibular Control of Equilibrium and Balance

    Disorders of the Motor System

The first thing Dave remembered after the accident was his coach asking him if he could move his toes and fingers. He was still dazed and couldn't quite remember where he was, other than the fact that he was lying on the ground wearing his football helmet. He tried to move his fingers, but couldn't tell if he had succeeded because all sensation in his hands and arms was gone. He wasn't even sure if he was wearing the rest of his uniform. When the paramedics arrived, they immobilized his head and lifted him onto the stretcher. He had vague memories of an initial silence in the crowd as he was carried off the field, followed by applause and encouragement when he was taken into the ambulance.

   It has been two weeks since the accident, and he is still quadriplegic. Now he gets around with the aid of a wheelchair, but someone else has to push it. He lost the use of his arms and legs, but he can still talk, still breathe, still think. This morning, for the first time, he was able to move one of his fingers and big toe. The doctors tell him that, when he was tackled and hit the ground, part of his spinal cord was crushed. Other players with similar spinal cord injuries have recovered so that they can now walk again; Dave may be one of them.

*The act of moving requires the coordinated activation and inhibition of various muscle groups. This is seen in activities such as reaching, walking, running, jumping, and typing. Each activity uses different sets of muscle groups to produce a movement in a particular direction, with specific velocity and force. The planning, initiation, execution, and termination of these movements are carried out by the **motor system.***

The repertoire of motor responses ranges from **reflex** (automatic, involuntary) **actions** such as the knee jerk to complex, **voluntary** activities such as playing the piano. Reflex responses are controlled in the spinal cord, while more complex motor behaviors are organized in other areas of the CNS. The range of complexity of motor behaviors reflects the **hierarchical** (or **serial**) **organization** of the motor system (Fig. 9–1). The spinal cord is the lowest level of the hierarchy; it is followed in order by the **brain stem, motor cortex (Brodmann area 4),** and the **premotor and supplemental areas** of the cortex (Brodmann area 6). Other noncortical areas that play im-

portant roles in motor function are the **cerebellum** and **basal ganglia** (Fig. 9–2).

In the hierarchical organization, the highest motor levels influence the levels below them. For instance, the premotor area has inputs to the motor cortex; the motor cortex, in turn, provides inputs to the brain stem; and the brain stem provides inputs to the spinal cord. This hierarchical arrangement connects different areas of brain together serially in a descending fashion.

In addition to this serial organization, the motor system also has a **parallel organization.** In this arrangement, information from the motor cortex and brain stem, for example, can simultaneously and independently affect the spinal cord (see Fig. 9–1). With both the parallel and serial organization of the motor system, the targets of the efferent (outgoing) information are the **motor neurons** of the spinal cord. These motor neurons are the nerve cells that directly activate the skeletal muscles responsible for movement.

The body contains three major types of muscle: skeletal muscle, cardiac muscle, and smooth muscle. **Cardiac muscles** are the muscles of the heart; they pump blood through the circulatory system (see Chapter 18). **Smooth muscles** surround certain blood vessels to control blood pressure and are also associated with a variety of other organ systems. In general, they maintain the homeostatic state of the body. These are primarily controlled by the autonomic ner-

(a)

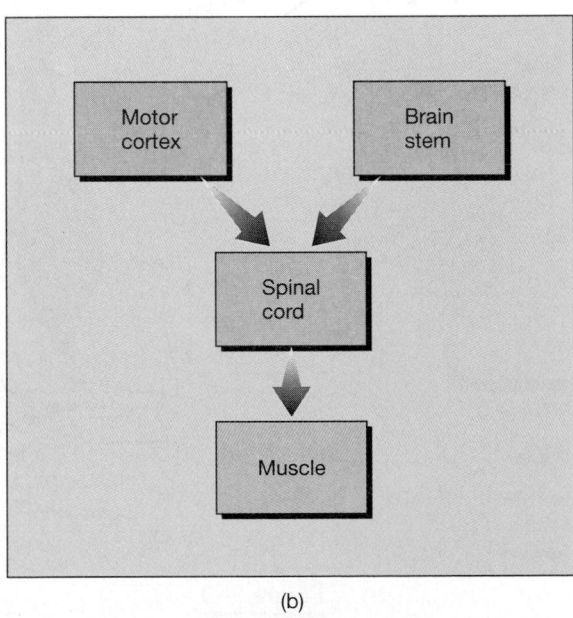

(b)

Figure 9–1

The motor system is organized in both hierarchical and parallel fashions.

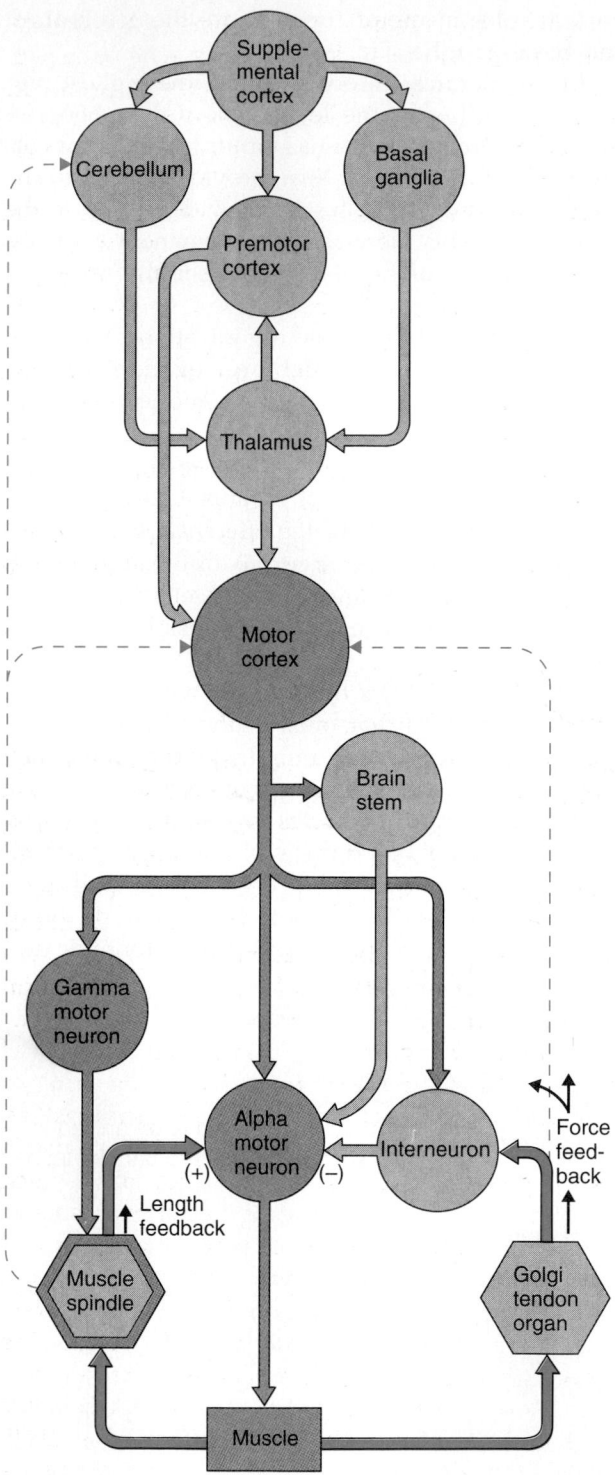

Figure 9–2

An overview of the motor system and the flow of information among the various components.

vous system as discussed in Chapter 10. In this chapter, we will focus on the areas of brain that control the **skeletal muscles,** also called **striate** or **striated muscles,** which are responsible for locomotion, fine movement, and the maintenance of balance and posture.

Muscles: Controlled by Nerves

Motor Areas of the Brain Transmit Information About Movement

Although the skeletal muscle system is discussed in detail in Chapter 16, we will present some general aspects of muscle structure and function here in order to better understand the interactions of muscles and neurons to be discussed shortly.

General Structure of Skeletal Muscle

A skeletal muscle is made up of **muscle cells,** which, because of their elongated shape, are called **muscle fibers.** Muscle fibers are **multinucleate;** one muscle fiber contains many nuclei. A **muscle** comprises many muscle fibers arranged in a parallel way. Two types of muscle fibers make up a skeletal muscle: **extrafusal fibers** and **intrafusal fibers** (Fig. 9–3). Extrafusal fibers are the strongest and most common

Figure 9–3

A skeletal muscle consists of extrafusal and intrafusal muscle fibers attached to tendons, which connect the muscle to a bone. Skeletal muscles work in antagonistic pairs of flexors and extensors.

fibers and form the exterior of the muscle. Inside the muscle are the intrafusal fibers, arranged parallel to the extrafusal fibers. Both extra- and intrafusal fibers are attached to the tendon at either end of the muscle. By a process to be discussed in more detail in Chapter 16, muscle fibers shorten when activated, resulting in contraction of the entire muscle and producing movement of a bone.

Types and Actions of Skeletal Muscle

Most skeletal muscles are attached to bones at two points: an **origin** and an **insertion** (Fig. 9–3), by strong connective tissue structures called **tendons.** As skeletal muscles **contract**—typically meaning that they shorten, although sometimes they remain the same length—they move bones, frequently causing joints to open or close. Skeletal muscles are usually arranged in **antagonistic** pairs or groups such that contraction of one muscle or muscle group will cause lengthening of another. In fact, all movements are controlled by antagonistic muscles working in opposite ways.

Antagonistic muscles are arranged in parallel fashion across a joint where two bones meet. The muscle that, by contracting, causes the joint to stretch out is called an **extensor,** while the muscle that, by contracting, causes the joint to close up is called the **flexor.** A flexor for a joint is said to be **synergistic** to another flexor for the same joint, while a flexor and an extensor for the same joint are said to be **antagonistic.** An example of an antagonistic pair producing flexion and extension are the bicep and tricep muscles, respectively.

Skeletal muscles produce movements of the bones of the **axial skeleton,** consisting of the skull, backbone, and ribs. These muscles are important in maintaining posture. Other skeletal muscles move the bones of the **distal skeleton,** primarily the limbs.

Skeletal muscles can be categorized as either fast muscles or slow muscles. **Fast skeletal muscles** contract rapidly, are poorly vascularized (poorly supplied with blood vessels), and operating under anaerobic conditions, they fatigue rapidly. These muscles are best suited for physical activities such as sprinting. In contrast, **slow skeletal muscles** contract slowly and are more resistant to fatigue. They operate under aerobic metabolism. These muscles are best suited for activities such as maintaining posture.

Motor Neurons Cause Skeletal Muscles to Contract

The contraction of skeletal muscles necessary for movement is controlled by nerve cells in the spinal cord called **alpha motor neurons** (see Fig. 9–2). These neurons form synaptic connections with skeletal muscles at **neuromuscular junctions,** where the neurotransmitter **acetylcholine** is released from the nerve fiber terminal. Within the neuromuscular junction, acetylcholine binds to cholinergic **nicotinic receptors** that are located on the surface of the skeletal muscle membrane. As in the CNS, activation of the nicotinic receptor causes a depolarization of the membrane. In the case of skeletal muscles this depolarization leads to a sequence of events resulting in the contraction of the muscle (see Chapter 16).

The motor neurons that innervate fast and slow types of muscles have different functional properties. Motor neurons that innervate fast skeletal muscles discharge action potentials with high frequencies and have rapid action potential conduction velocities. In contrast, motor neurons that innervate slow skeletal muscles have low rates of action potential discharge and exhibit slow conduction velocities.

The Motor Unit: The Myoneural Junctions of a Single Motor Neuron

The motor neurons that innervate skeletal muscles and cause them to contract have cell bodies located in the ventral horn of the spinal cord. Their axons leave the spinal cord through the ventral root and find their way to the appropriate skeletal muscle.

One motor neuron can form synaptic connections with many muscle fibers. An individual motor neuron and all the muscle fibers that it innervates form a **motor unit** (Fig. 9–4). The motor unit is the functional unit of the **neuromuscular system.**

Motor units come in various sizes. A small motor unit consists of a motor neuron that innervates few muscle fibers. This type of motor unit is responsible for the precise control of very fine movements such as contraction of a finger or darting movements of the eyes. A large motor unit consists of a motor neuron that innervates many muscle fibers. Such a unit is responsible for gross movements such as contraction of the legs or the maintenance of posture. The concept of motor unit size parallels that of the size of sensory receptive fields. Just as small receptive fields are associated with greater resolution of stimuli, small motor units are associated with greater precision of fine movement.

The Coding of Contractile Force

The amount of force that the muscle generates during contraction is controlled by the nervous system in two ways. First, a motor neuron can control the tension developed by a single muscle fiber with the frequency of action potentials it generates (Fig. 9–5). This code is the motor system's **frequency code** of

Figure 9–4

(*a*) The motor unit consists of a motor neuron and all the muscle fibers it innervates through myoneural junctions.
(*b*) Photomicrograph of a nerve fiber innervating a skeletal muscle.
(*© J. & L. Weber/Peter Arnold, Inc.*)

Figure 9–5

A motor neuron generates action potentials that cause a muscle fiber to contact and exert force.

contractile force. Secondly, more motor neurons and therefore more motor units can be activated to increase a muscle's force of contraction. This code is the motor system's **population code** of contractile force.

The Spinal Cord in the Motor Response System

The spinal cord plays a crucial role in the neuromuscular system. Some automatic body movements, called reflex movements, are controlled solely by the spinal cord and have no control from higher structures. Moreover, locomotion, or walking, can be produced by spinal cord neurons alone.

Motor Neurons and Muscle Fibers Represent Functional Units

Motor neurons are organized in the spinal cord in two ways: (1) the motor neurons in the dorsal area of the ventral horn are responsible for flexor movements, while cells in the ventral region are responsible for extensor movements (Fig. 9–6); and (2) those motor neurons in the dorsolateral region of the ventral horn innervate muscles in the extremities while those in the ventromedial region of the ventral horn innervate the axial muscles of the body to maintain posture.

Motor neurons that innervate a single muscle are functionally grouped in the spinal cord in **motor neuron pools.** Motor neurons within a single pool are located in several adjacent segments of the spinal cord (Fig. 9–7). The activation of a motor neuron pool thus coordinates the contractile action of the muscle. Mo-

Figure 9–6

Motor neurons that activate flexors originate in the dorsal area of the ventral horn of the spinal cord, while those that activate extensors originate in the ventral region of the ventral horn.

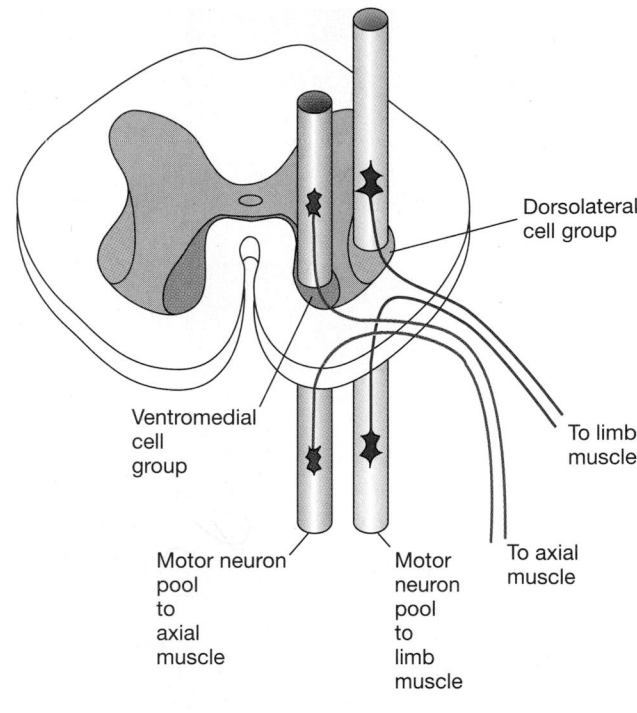

Figure 9–7

A motor neuron pool is a group of motor neurons that innervate a single muscle.

tor neurons within a pool are those recruited to increase contractile force by the population code.

Interneurons of the Spinal Cord Coordinate the Response of Antagonistic Muscles

The interneurons of the spinal cord also play an important role in movement. These cells are located in the intermediate zone of the spinal cord. Those located in the lateral part of this zone have axons that synapse **ipsilaterally** (on the same side of the body) with motor neurons that innervate distal limb muscles. Interneurons lying closer to the midline have axons that synapse with motor neurons on both sides of the spinal cord that control muscles for posture. These neurons form synapses with alpha motor neurons and mediate excitatory or inhibitory actions. The excitatory effects are associated with the activation of synergistic muscles, while the inhibitory effects are associated with suppressing the activation of antagonistic muscles.

Interneurons also send their axons up and down the spinal cord to form synaptic connections in the regions several cord segments away from their cell bodies. They connect with motor neurons that control the contraction of several muscle groups. In this way, the activities of muscle groups are coordinated in maintaining posture and for reaching movements.

Locomotor Generators in the Spinal Cord Control Walking Movements

The act of walking requires the coordinated contraction and relaxation of flexor and extensor muscles of the legs. This rhythmic alteration between flexion and extension is produced entirely by the interconnections of neurons in the spinal cord. These neurons and their interconnections compose the **locomotion generators** for walking. Animals whose spinal cords have been severed can walk without inputs from higher centers. This phenomenon is seen in chickens, for example, which remain capable of running for a short period after decapitation. Experimental animals with spinal cord transections are also capable of walking on treadmills. The speeds at which they walk are dictated by the speed of the treadmill. In these experimental animals, if one limb is prevented from walking, the other limb will continue to walk. The locomotor generator for each limb, therefore, does not

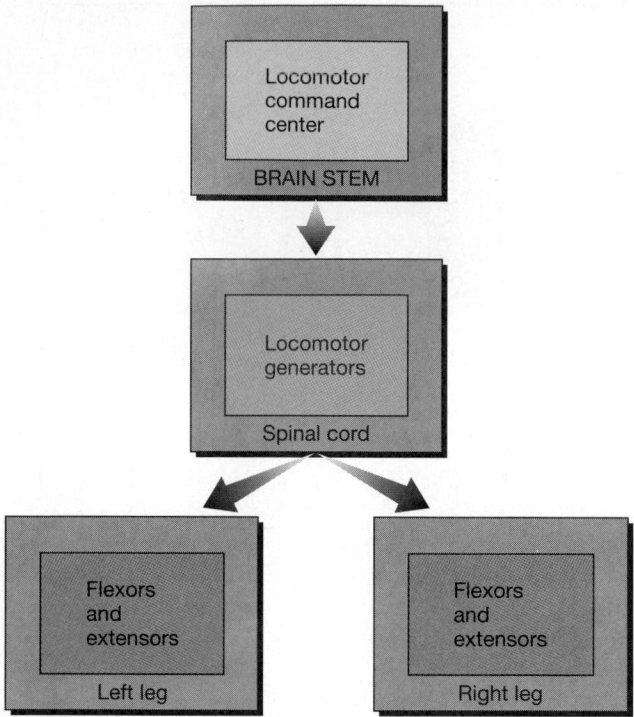

Figure 9–8

Locomotor generators and the locomotor command center.

require activity from the other limb. However, when all limbs are active, the generators for limb movement are coupled to produce a coordinated response (Fig. 9–8).

The locomotor generators in the spinal cord are under the control of a "locomotor command center" in the brain stem. Electrical stimulation of this command center causes animals with spinal cord lesions to walk on a treadmill. Weak stimulation causes walking while stronger stimulation can cause the animal to run.

During the walking process in normal animals, the locomotor generators coordinate the flexion and extension of leg muscles to carry out the **stance phase** and the **swing phase** of walking. In the stance phase, the foot is on the ground and supporting the weight of the body. In the swing phase, the foot is off the ground and swinging forward. If sensory feedback from muscle spindles and Golgi tendon organs is prevented from reaching the spinal cord during the stance phase, the step cycle can be stopped during the extension of the leg. In other phases of the walk cycle, however, the absence of sensory information does not inhibit the process of locomotion.

Sensory information can also change the activation of one motor response to another. The activation of tactile receptors on the top of the foot during the swing phase of walking, for example, results in a flexion response that is appropriate for stepping over an object. Sensory information to the spinal cord is therefore capable of changing ongoing motor programs and reflexes.

Sensory Fibers in Skeletal Muscle Convey Information About Muscle Length, Force, and Velocity of Shortening

Sensory neurons also have fiber processes within skeletal muscles. They transmit information about the state of contraction and tension in muscle fibers to the CNS so that it can revise its instructions as necessary to the motor neurons producing the contraction or tension. Sensory neurons that innervate intrafusal muscle fibers form what is known as a muscle spindle, while sensory neurons that innervate tendons form Golgi tendon organs.

The Muscle Spindle: A Length Detector

The sensory receptors that detect the length of the muscle and its velocity of contraction are called **muscle spindles** (Fig. 9–9). These structures are formed by sensory neurons that entwine intrafusal muscle fibers.

The muscle spindle is composed of two types of intrafusal fibers: **nuclear bag fibers** and **nuclear chain fibers** (see Fig. 9–9a). The nuclear bag fiber is innervated by **Type Ia nerve fibers,** which transmit information about muscle length and velocity of contraction to the CNS. Nuclear chain fibers are innervated by **Type II nerve fibers,** which transmit information about muscle length.

The muscle spindle acts as a **stretch receptor**— that is, it increases its discharge of action potentials when the intrafusal fibers stretch. Conversely, it decreases its discharge when the fibers shorten.

Gamma motor neurons innervate contractile elements at the poles of the muscle spindles (see Fig. 9–9a). The activation of these contractile elements causes the central region of the muscle spindle to stretch as a rubber band would. In this manner, the gamma motor neurons can regulate the sensitivity of the Type Ia sensory nerves of the muscle spindles. Figure 9–9c shows the action potential activity of Type Ia muscle spindle sensory nerves as a function of the length of the extrafusal muscle fiber. In the absence of gamma motor neuron activation, a change in the length of the extrafusal muscle fibers results in a slight change in the frequency of action potentials generated by the sensory neurons. With gamma motor neuron activation, however, a slight change in the length of the extrafusal muscle fibers causes a great

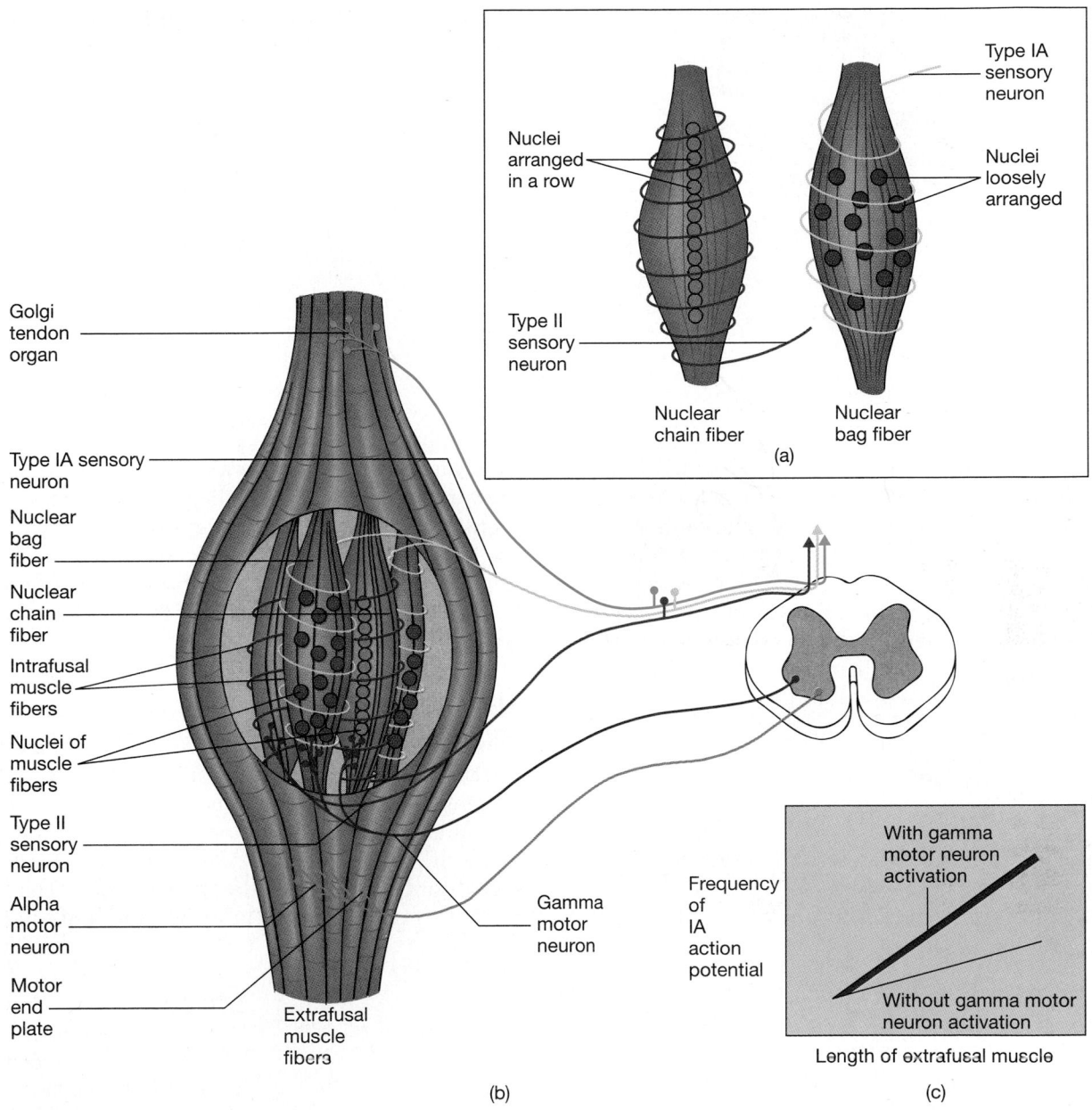

Nuclei arranged in a row

Type II sensory neuron

Type IA sensory neuron

Nuclei loosely arranged

Nuclear chain fiber

Nuclear bag fiber

(a)

Golgi tendon organ

Type IA sensory neuron

Nuclear bag fiber

Nuclear chain fiber

Intrafusal muscle fibers

Nuclei of muscle fibers

Type II sensory neuron

Alpha motor neuron

Motor end plate

Extrafusal muscle fibers

Gamma motor neuron

(b)

Frequency of IA action potential

With gamma motor neuron activation

Without gamma motor neuron activation

Length of extrafusal muscle

(c)

Figure 9–9

(*a*) Intrafusal muscle fibers consist of nuclear bag fibers and nuclear chain fibers. (*b*) These fibers and the sensory neurons that innervate them make up a muscle spindle, a receptor for muscle length and state of contraction. Golgi tendon receptors innervate tendons and transmit information about the force exerted by the muscle. Gamma motor neurons activate intrafusal fibers, while alpha motor neurons activate extrafusal muscle fibers. (*c*) When activated by gamma motor neurons, Type IA sensory neurons generate action potentials in response to changes in extrafusal muscle length more frequently than in the absence of gamma motor neurons.

change in the frequency of generated action potentials. In this way, the activation of gamma motor neurons maintains the sensitivity of the muscle spindles even as the extrafusal muscle fibers shorten during contraction.

Golgi Tendon Organs: Tension Receptors

A second type of muscle receptor is the **Golgi tendon organ.** This type of receptor is located in series with the extrafusal muscle fiber (see Fig. 9–9*b*) and

Figure 9–10

The stretch reflex occurs when a muscle spindle detects a tap on the knee and informs the spinal cord, which activates the leg extensor and inhibits the flexor, causing the knee joint to open up.

transmits information about the force or tension produced by the contraction of the muscle. As greater tension develops when the muscle contracts, the Golgi tendon organ generates more action potentials. Since the Golgi tendon organ and muscle spindle are somatic sensory receptors, they transmit their information about muscle force, velocity, and length to the spinal cord, where it is then forwarded to the somatic sensory cortex or used within the spinal cord for reflex action.

Spinal Cord Reflexes Are Activated by Sensory Receptors

Spinal cord reflexes represent the most basic of motor responses. These reflexes are carried out entirely within the spinal cord and are modified by inputs from high centers to generate complex movements. They are also used to help diagnose disorders of the motor system. Additionally, sensory motor reflexes are found in the brain stem. Many of these reflexes involve the constriction of the pupils in response to visual inputs or the movement of the eyes in response to vestibular sensations.

The Stretch Reflex

The **stretch reflex** is commonly called the **knee jerk reflex.** A stretch reflex, for example, is activated by tapping the **patellar tendon** below the knee to stretch the muscle spindles. Action potentials conducted along the muscle spindle sensory neurons enter the spinal cord via the dorsal roots (Fig. 9–10). The fiber from the muscle spindle branches after entering the

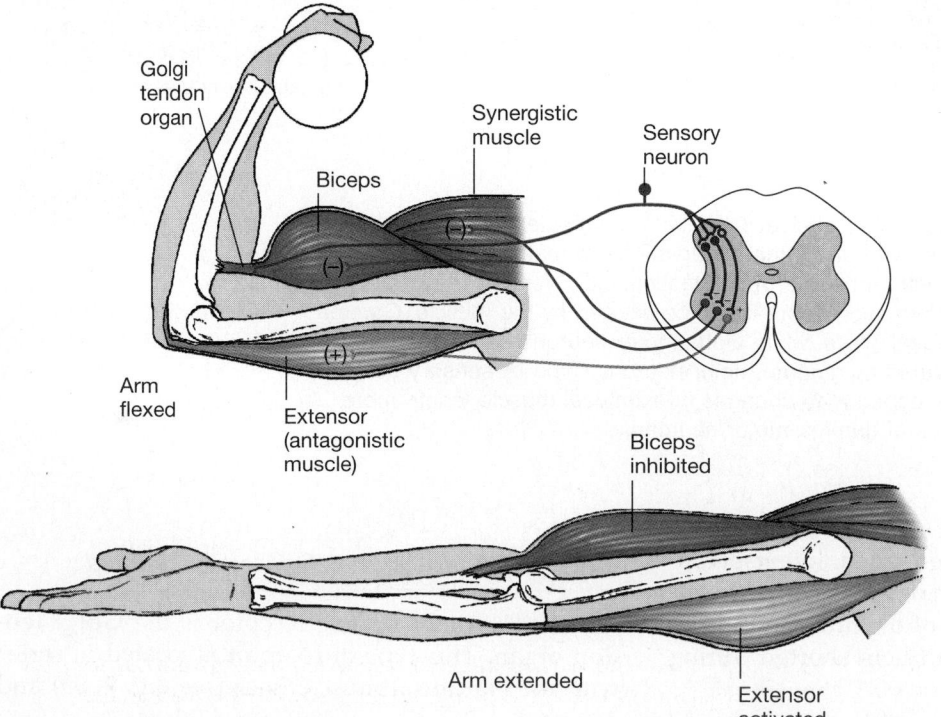

Figure 9–11

Detection of excessive weight by the Golgi tendon organ initiates the inverse myotatic reflex causing a flexed bicep muscle to relax and an extensor in the arm to contract so that the arm extends and drops a weight that might be harmful.

spinal cord, with the ascending branch joining the dorsal column pathway. The other branch forms synaptic connections with motor neurons in the ventral horn.

These motor neurons activate muscle fibers, causing the leg to extend quickly or kick out. In order for the leg to extend, however, the antagonistic muscles that cause the leg to flex must be inhibited. This inhibition comes about by the activation of inhibitory interneurons within the ventral horn. These interneurons receive information from muscle spindle sensory neurons and in turn form inhibitory synaptic connections with motor neurons of flexor muscles. Consequently, the activation of these interneurons prevents the contraction of the flexors. By inhibiting flexors and activating extensors, the interconnections of nerve cells within the spinal cord produce well-coordinated motor responses.

The Inverse Myotatic Reflex

A reflex that involves the Golgi tendon organ is the **inverse myotatic reflex.** This reflex is seen when a person attempts to lift more weight than he or she can actually carry. Weightlifters, for example, often add increasing amounts of weight to barbells when lifting the weights in a maneuver called the "curl." In this maneuver, a person lifts the barbells by contracting the bicep muscles. If the weight is too great for the biceps to lift, however, they suddenly relax, and the barbells are dropped.

The inhibition of the biceps occurs because the Golgi tendon organ detects excessive force, which might damage the muscle. With increasing force, the Golgi tendon organ begins to discharge action potentials (Fig. 9–11). As with the muscle spindles, these action potentials are transmitted to the dorsal column nuclei and subsequently to the somatic sensory cortex. In the spinal cord, however, information about excessive muscle tension is transmitted to inhibitory interneurons that turn off the motor neuron innervating the biceps. At the same time, interneurons turn off any muscles (flexors) synergistic to the biceps and turn on any muscles (extensors) that are antagonistic to the biceps, so that the arm extends and drops the barbells. This reflex thus protects the muscle from damage that might occur when a person attempts to lift too heavy a weight.

In addition to regulating muscle tension, this reflex is also thought to involve the smooth onset and termination of muscular contraction in walking. In the motion of stepping, the flexors of one leg begin to contract while its extensors relax. The flexion of one leg needs to be coordinated with the extension of the opposite leg. To achieve this, sensory information from the Golgi tendon organ from the leg being flexed is transmitted to the other side of the spinal

cord to activate the extensors and inhibit the flexors of the opposite leg. Although sensory inputs play an important role in the coordination of walking, the motor "program" for walking is contained within the spinal cord.

The Flexor Withdrawal Reflex

A reflex that involves cutaneous receptors is the **flexor withdrawal reflex.** An example of this reflex is activated by pain receptors on the bottom of the foot (Fig. 9–12). When you step on a tack, for instance,

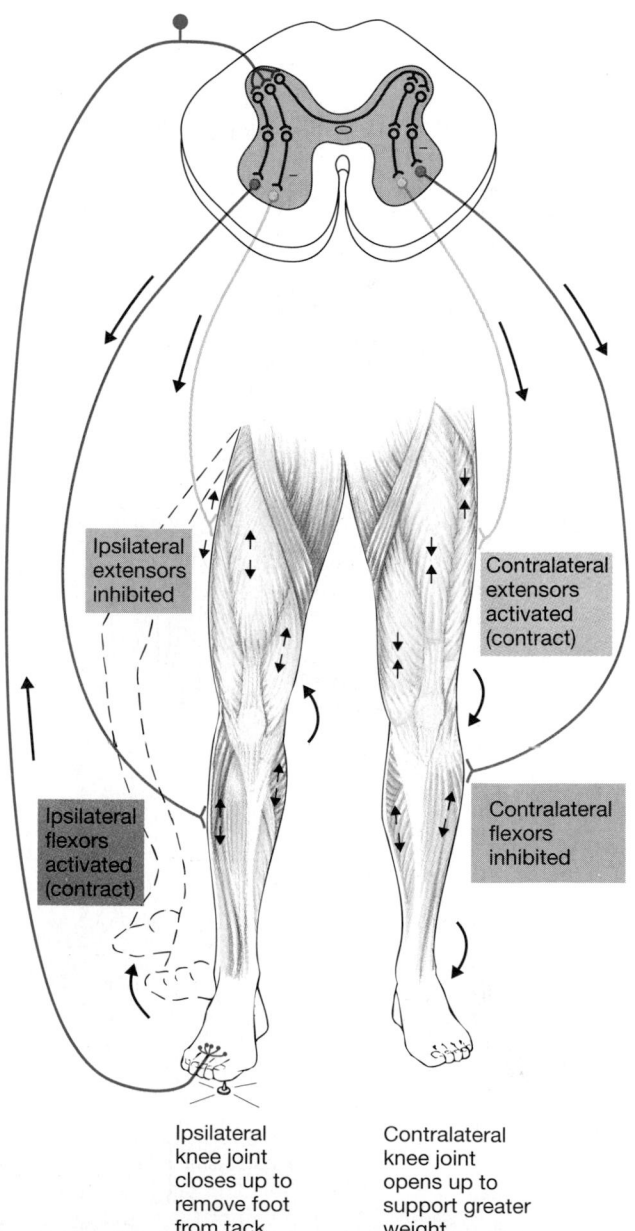

Ipsilateral extensors inhibited

Contralateral extensors activated (contract)

Ipsilateral flexors activated (contract)

Contralateral flexors inhibited

Ipsilateral knee joint closes up to remove foot from tack

Contralateral knee joint opens up to support greater weight

Figure 9–12

The withdrawal reflex in one leg requires the cooperation of the opposite leg, arranged by interneurons in the spinal cord.

pain information is transmitted to the spinal cord and causes a contraction of the flexors to remove the foot from the tack. At the same time, extensors of the leg that would normally keep the foot on the tack are inhibited. In order to support the rest of the body, however, the extensor muscles in the other leg are activated while its flexors are inhibited.

The Brain in the Motor Response System

Several important structures in the brain represent the highest control centers for the motor system (see Fig. 9–2). The brain stem controls spinal cord output in maintaining posture and balance of the body. Various areas of the neocortex organize and interpret complex sensory input and program the responding movements. The cerebellum and the basal ganglia also plan and program specific types of motor responses.

These components of the motor system interact with each other to carry out specific functions. The desire to move activates the supplemental motor cortex, basal ganglia, and cerebellar components of the motor system. Information from these areas is transmitted to the premotor cortex, where movements are

planned and programmed. From the premotor cortex, information about the planned movement is transmitted to the motor cortex. The motor program is then carried out when neurons in the motor cortex instruct those in the spinal cord to activate the skeletal muscles. The force and velocity of contraction, and the length of muscle shortening, are monitored by Golgi tendon organs and muscle spindles. This sensory information is fed back to the motor neurons at the spinal level and also transmitted back to the motor cortex and cerebellum.

Comparisons are made between the intended movement, as designed by the motor program, and the actual movement. Deviations from the intended movement are then corrected by transmitting another set of instructions to the motor neurons in the spinal cord.

The Brain Stem Controls Posture

One of the primary roles of the brain stem is maintaining the body's posture and balance. Nerve cells within different clusters in the brain stem send axons that terminate in the spinal cord. Three pathways from the brain stem provide input to the motor neurons of the spinal cord: (1) the ventromedial pathway, (2) the lateral reticulospinal pathway, and (3) the rubrospinal pathway.

BRAIN STEM

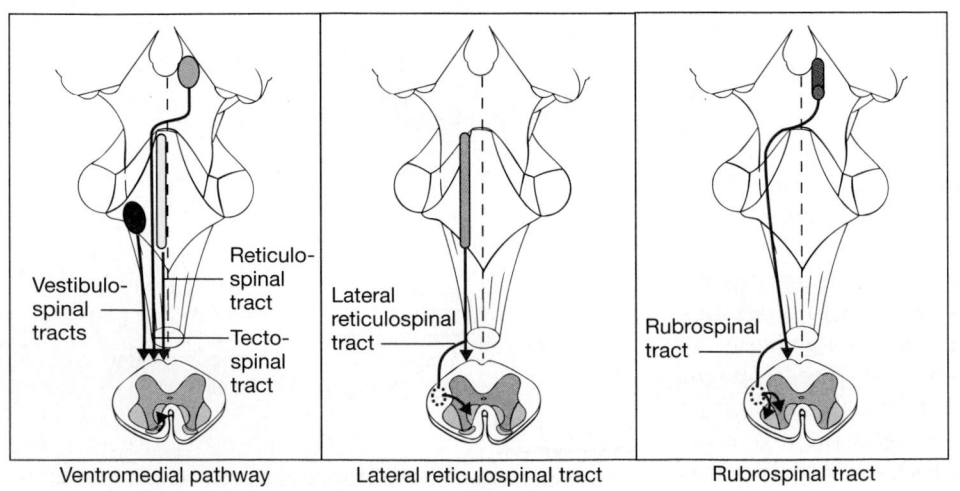

| Ventromedial pathway | Lateral reticulospinal tract | Rubrospinal tract |
| (a) | (b) | (c) |

Figure 9–13

Pathways from the brain stem to the spinal cord: (*a*) the ventromedial pathway, (*b*) the lateral reticulospinal tract, and (*c*) the rubrospinal tract.

Figure 9–14

A summary of the effects of the brain stem on spinal motor neurons.

The Ventromedial Pathway

The **ventromedial pathway** impinges on motor neurons that control axial muscles. This pathway has three major components, all of which descend along the ventral and medial region of the spinal cord (Fig. 9–13*a*). The first component is the **vestibulospinal tract,** which originates in the vestibular nucleus and carries information for the reflex control of equilibrium (discussed in a later section). The second component of the ventromedial pathway is the **tectospinal tract,** which originates in the **tectum,** a structure involved in the coordinated control of head and eye movements. The third component is the **medial reticulospinal tract,** which originates in the **reticular formation,** a structure involved in maintaining posture by the activation of extensor muscles.

The Lateral Reticulospinal Tract

The second major descending pathway from the brain stem to the spinal cord is the **lateral reticulospinal tract.** Nerve fibers in this pathway are derived from the lateral reticular nucleus and descend the spinal cord in the lateral region of the cord (Fig. 9–13*b*). These fibers innervate flexors in the control of posture.

The Rubrospinal Tract

The rubrospinal tract has fibers that originate in the red nucleus of the brain stem. These fibers descend along the dorsal and lateral border of the cord to innervate motor neurons that control distal flexor muscles (Fig. 9–13*c*). The effects of the major descending brain stem pathways on flexor and extensor muscles in the maintenance of posture and equilibrium are summarized in Figure 9–14.

The Motor Cortex Controls Reaching and Fine Voluntary Movement

The control of reaching and fine voluntary movement is a result of instructions transmitted via descending pathways to the spinal cord from the **motor cortex.** The motor cortex occupies a cortical region rostral to the somatic sensory cortex (Fig. 9–15). The cells in the motor cortex are organized in a somatotopic manner similar to that of the somatic sensory cortex (Fig. 9–16). Cells in the medial region of the motor cortex cause the contraction of muscles in the leg. Nerve cells located in more lateral regions of the motor cortex activate the muscles of the torso, arm, hand, and face.

The parts of the body involved with fine movement, such as the fingers, occupy more space in the motor cortex than parts of the body involved in gross

Figure 9–15

The functional areas of the cerebral cortex involved in motor control.

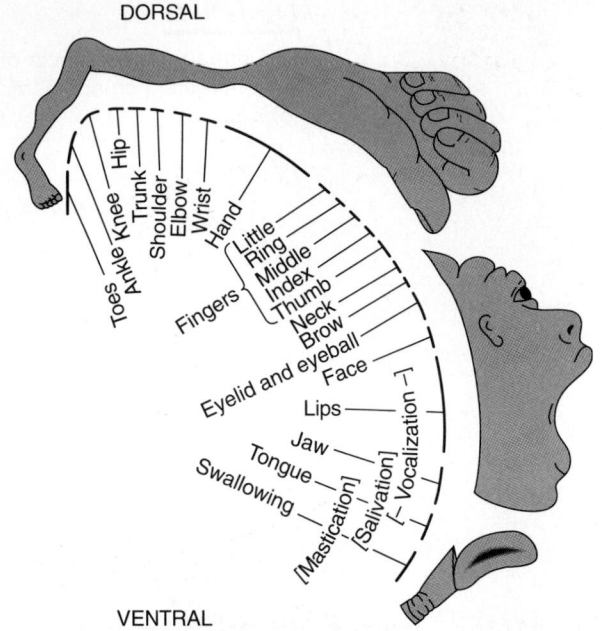

DORSAL

VENTRAL

Figure 9–16

The somatotopic organization of the motor cortex.

Figure 9–16

The somatotopic organization of the motor cortex.

movement, such as the torso. This relationship is in keeping with the general principle that the amount of cortical space devoted to different parts of the body is in proportion to each part's sensory sensitivity or degree of fine motor control.

Organization of the Motor Cortex

The cells of the motor cortex are organized in functional columns called **cortical efferent zones (CEZ).** All of the cells within one efferent zone are involved in the contraction of a given muscle. Neurons in the motor cortex are also organized horizontally into six different layers. Those located in layer V provide the

(a)   Lateral corticospinal tract

(b)   Ventral corticospinal tract

Figure 9–17

Descending pathways form the motor cortex: the lateral and ventral corticospinal tracts.

output from the motor cortex. Pyramid-shaped nerve cells in this layer send axons to the spinal cord to synapse with spinal motor neurons and interneurons. These cortical neurons are involved primarily in controlling distal muscles. They excite both alpha and gamma motor neurons and thus allow the muscle spindles to remain sensitive to changes in muscle length.

Descending Pathways from the Motor Cortex

Information from the motor cortex is transmitted to the spinal cord and brain stem by the **corticospinal pathway** and **corticobulbar pathway,** respectively. Inputs to the brain stem ultimately influence axial muscles near the midline for the maintenance of posture, while inputs to the spinal cord control the distal limb muscles. Approximately 30% of the corticospinal and corticobulbar fibers are from neurons in area 4; another 30% originate from the premotor cortex (area 6). The remaining fibers are derived from the somatic sensory cortex.

As these axons descend from the motor cortex, they form the **pyramidal tract.** When the pyramidal tract reaches the level of the brain stem, the majority of the fibers that continue to the spinal cord cross the midline of the body and continue their descent along the **lateral corticospinal tract** (Fig. 9–17). Nerve fibers exit from the lateral corticospinal tract at various levels of the spinal cord to innervate motor neuron pools that control distal limb muscles. The direct connections from motor cortex to spinal cord thus permit the independent control of individual muscles.

In addition to the lateral corticospinal tract, a minority of the fibers from the motor cortex descend to the spinal cord without crossing the midline of the body. These fibers form the **ventral corticospinal tract** and primarily innervate motor neurons in the medial region of the ventral horn associated with axial muscles of the body (see Fig. 9–17).

Sensory Feedback to the Motor Cortex

Neurons in the motor cortex are informed of the consequences of movement through sensory feedback pathways. They receive input from either the muscles they project to or from areas of skin surrounding the muscle. This long loop of sensory feedback results in the alteration of information from the motor cortex to the spinal cord to correct any deviations from the intended movement. Sensory feedback to the motor cortex occurs by way of the somatic sensory cortex (Fig. 9–18).

The nerve cells in the sensory cortex are connected to those in the motor cortex in a topographic manner. Cells in the sensory cortex receiving proprioceptive input from muscles in the thumb, for example,

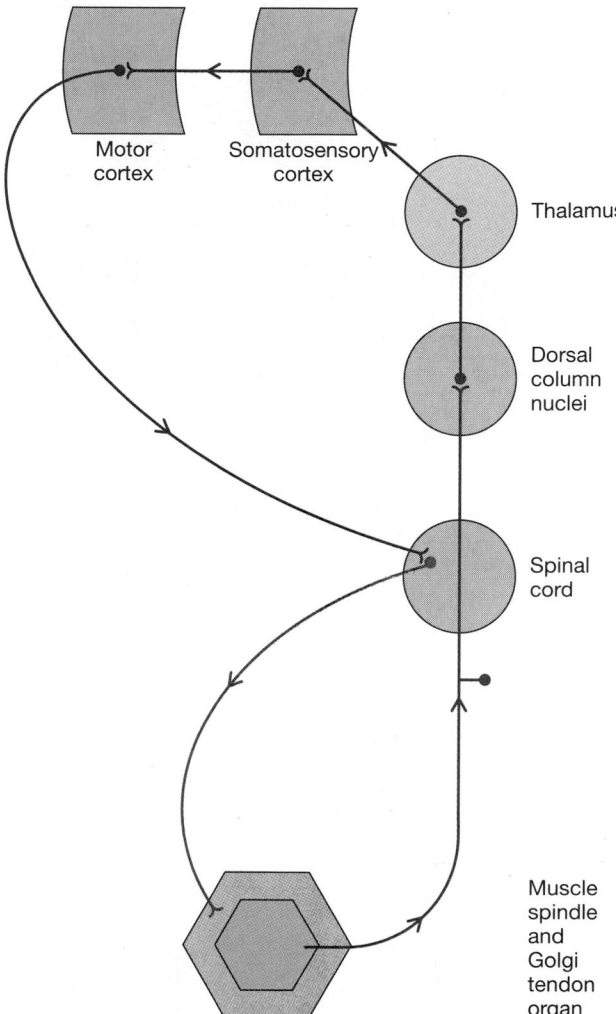

Figure 9–18

Sensory feedback pathways to the motor cortex.

are connected with cells in the cortical efferent zone responsible for the contraction of these same muscles. In this way, feedback is provided by the sensory system to inform the cells in the motor cortex whether the instructions they have transmitted for the muscular contraction have been faithfully executed. If not, then the sensory information that is sent back to the motor cortex causes the cells in the CEZ to modify their activity and thus modify the activity of the motor neurons.

Cortical Coding of Reaching Movements

Studies correlating nerve cell activity in the motor cortex with reaching movements have been carried out predominantly in monkeys. These monkeys were trained to move a lever to different locations in two-dimensional space while recordings were made of the action potential activity of individual nerve cells in

(a)

(b)

(c)

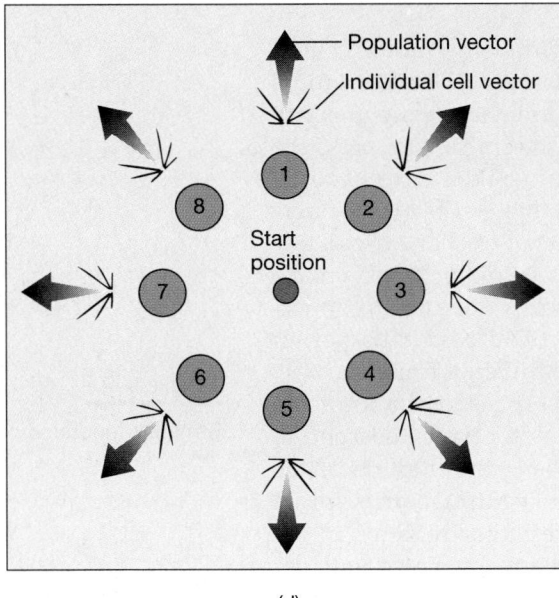

(d)

Figure 9–19

(*a*) Eight directions of possible arm movements in a 2-dimensional plane, starting from a center position. (*b*) Activity of single neuron in motor cortex correlated with arm reaching direction. (*c*) Vector for single cell indicating preferential direction and intensity of activity. (*d*) Population vector representing the summation of individual cell vectors that encodes the direction of arm movement.

motor cortex (Fig. 9–19*a* and 9–19*b*). During a reaching motion, the activity of a cell in the motor cortex is best correlated with a specific direction of reaching in space. This preferred direction can be represented by a **directional cell vector,** which indicates not only direction but also the frequency of action potential discharge (Fig. 9–19*c*). From the direction vector of individual cells, a **population vector** can be generated, representing the preferred direction of a cluster of cells in motor cortex. This population vector is highly correlated with direction of movement (Fig. 9–19*d*) and represents a *distributed code* such that each neuron carries part of the information for movement

and that the activity of the cells as a group uniquely determines the direction of movement in space.

The activity of nerve cells in the motor cortex encodes directional information and information about the force required to make a particular reaching movement. With an increase in force there is a corresponding increase in the frequency of action potentials produced by individual nerve cells in the motor cortex. This results in an increase in the length of the cell vector and also the population vector. Thus these vectors are most likely representations of the direction of movement and the force of muscular contraction needed to carry out a particular movement.

Coding of contractile force is carried out by two types of neurons. First, dynamic neurons code for the rate of force development. These neurons change their activity when there is an increase in the level of applied force. A second type of neuron codes for steady-state force. These neurons discharge action potentials throughout the entire period during which force is maintained.

The pyramidal cells in the motor cortex begin generating action potentials approximately 50 milliseconds before a movement. Nerve cells in other areas of the brain involved in motor control become active much earlier than the motor cortex. These areas of the brain, called the **supplemental and premotor cortex, posterior parietal cortex, cerebellum,** and **basal ganglia,** are involved in the planning and programming of movement.

The Supplemental and Premotor Areas Program Movement

The motivational factors that cause specific movements most likely originate in subcortical areas of the brain such as the **hypothalamus.** This area of the brain is involved with processes such as hunger, thirst, and sex, and will be discussed in Chapter 11. The motivational areas of the brain that respond to the need to accomplish a particular task transmit this information to the **supplemental and premotor cortex** (Brodmann area 6), which designs the necessary movements to accomplish the desired task.

In the act of reaching for a glass of water to bring to the lips, the nervous system must determine which motor program will activate certain muscles, and when and how much they need to be contracted. The components of the motor program are thought to be developed by supplemental and premotor cortices (area 6).

Stimulation of the supplemental and premotor areas produces movements more complex than those produced by stimulation of the motor cortex. Activation of the premotor area causes movements of the torso or the opening and closing of the hand. In contrast, stimulating the motor cortex produces small, twitching movements. The supplemental area has direct inputs to the motor cortex for the control of distal limb muscles. However, it also has inputs to motor neurons in the spinal cord that control axial muscles involved with the maintenance of posture.

An example of the role of the supplemental area in the programming of a motor behavior is seen in studies of simple and complex movements of the hand. In these studies, blood flow to certain brain regions is monitored. The amount of blood flow is a measure of the activity of nerve cells in each area of

(a) Simple finger flexion (performance)

(b) Finger movement sequence (performance)

(c) Finger movement sequence (mental rehearsal)

Figure 9–20

The supplemental cortex is involved in the planning of complex movements. (*a*) During simple finger flexion, blood flow (indicated by red stippling) increases in the motor cortex, while (*b*) during a complex sequence of finger movements, blood flow is also increased in the supplemental cortex. (*c*) When the same sequence is mentally rehearsed, blood flow increases in the supplementary cortex but not in the primary motor cortex.

the brain. With simple movements, blood flow increases in the motor and somatic sensory cortices, but no increase is found in the supplemental motor area (Fig. 9–20). With more complex movements, blood flow also increases in the supplemental motor area.

Interestingly, when patients are told to rehearse a movement in their minds, blood flow increases only in the supplemental motor area and not in the motor and somatic sensory cortices. These studies suggest that the supplemental motor area is involved in the programming of complex movements.

The Posterior Parietal Cortex Integrates Sensory Stimuli for Purposeful Movement

The **posterior parietal cortex** is necessary for the processing of sensory stimuli leading to purposeful movement. It is located immediately behind the somatic sensory cortex (see Fig. 9–15). This cortex receives both somatic and visual sensory information and transmits it to the supplemental and premotor areas.

Three types of neurons have been identified in the posterior parietal cortex. **Arm projection neurons** generate action potentials when the arm reaches for an object. **Hand-eye coordination neurons** are most active when the eye fixates on an object that is being touched; **hand manipulation neurons** increase their firing rate when the hand explores an object. These types of neurons thus become active only with very specific behavioral motor responses.

Patients with lesions of this cortex caused by strokes or trauma are unable to perform previously learned movements in the appropriate sequence. They appear to synthesize the spatial coordinates of objects in abnormal ways and behave as if their movements are not in accord with the coordinates of the objects in space. For example, when drawing a clock, a patient with a posterior parietal lesion places all of the numbers on one half of its face.

The Cerebellum and the Basal Ganglia Coordinate Movements

The basal ganglia and cerebellum (Fig. 9–21) are the major subcortical components of the motor system. They both receive inputs from the neocortex and transmit information back to the cortex by way of the thalamus. The inputs to the basal ganglia, however, are from the entire cortex, while those of the cerebellum are primarily from sensory and motor areas. In addition, the basal ganglia do not receive direct sensory information from somatic receptors, nor do they transmit descending information directly back to the spinal cord as does the cerebellum. These differences suggest that the basal ganglia are involved in more complex motor functions, while the cerebellum is

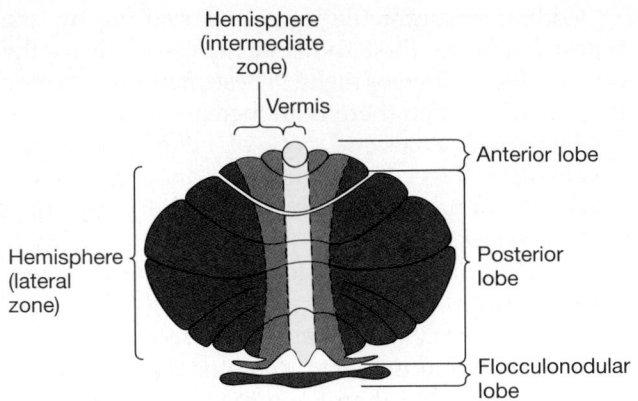

Figure 9–21

A dorsal view of the cerebellum.

more involved with the control of movement that requires constant monitoring by sensory feedback.

The Cerebellum: Planning, Coordination, and Posture

The cerebellum ("little brain" in Latin) has three basic functions: (1) the planning of a movement, (2) the control of posture and equilibrium, and (3) the control of smooth limb movement. The cerebellum accomplishes the latter two functions by comparing information concerning an intended movement with sensory feedback about the actual movement, and adjusting its output to compensate for differences between the two. It participates in the planning of a movement by receiving information from motor (supplemental and premotor) and parietal cortices and then uses these inputs to initiate a planned movement.

These cerebellar functions are associated with specific anatomical locations. The cerebellum has three major lobes: the **anterior lobe,** the **posterior lobe,** and the **flocculonodular lobe** (see Fig. 9–21). The flocculonodular lobe is involved in the maintenance of equilibrium and posture, while the midline and intermediate regions of the anterior and posterior lobes are involved in limb movement. The planning and initiation of motor programs are the responsibility of the lateral region of the anterior and posterior lobes.

The most prominent nerve cell in the cerebellum is the **Purkinje cell,** found throughout the cerebellum and packed tightly into a single cellular layer (Fig. 9–22). The axons of Purkinje cells form synaptic connections with neurons in one of three deep cerebellar nuclei, the **dentate nucleus,** the **interpositus nucleus,** or the **fastigial nucleus.** At the synapses

Figure 9–22

Cell types and circuits in the cerebellum.

between Purkinje cells and the neurons of the deep cerebellar nuclei, the transmitter GABA is released, causing a hyperpolarization of the postsynaptic neurons. The effect of Purkinje cells in the deep cerebellar nuclei is therefore inhibitory.

The deep nuclei receive information from Purkinje cells located in different regions of the cerebellum. The fastigial nucleus receives information from Purkinje cells in the flocculonodular lobe and is therefore involved in the maintenance of balance and posture. The fastigial nucleus transmits its information to the vestibular nucleus, which in turn controls motor neurons that innervate axial muscles (see Fig. 9–22).

The interpositus nucleus receives information from Purkinje cells in the medial and intermediate regions of the anterior and posterior lobes of the cerebellum and is therefore involved in the control of limb muscles. The interpositus sends its information to the thalamus and red nucleus of the brain stem. The red

nucleus in turn controls motor neurons innervating distal limb muscles.

The dentate receives information from Purkinje cells in the lateral regions of the anterior and posterior lobes of the cerebellum and is therefore involved in the planning and initiation of movement. The dentate transmits its information to the thalamus, which in turn forwards information to the motor and premotor cortex. The dentate also provides inputs to the red nucleus.

Since the outflow of information from the cerebellum must eventually pass through the deep cerebellar nuclei, these nuclei constitute the output of the cerebellum. Research indicates that the input of information to the cerebellum is simultaneously transmitted to the deep cerebellar nuclei (Fig. 9–23). After the incoming information is processed by the cerebellum, it is transmitted by way of the Purkinje cells to the deep nuclei. The outputs from the deep nuclei are thus modulated by the inhibitory actions of the

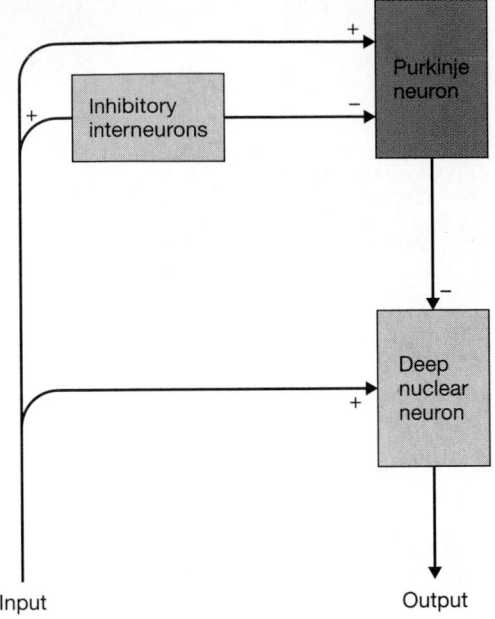

Figure 9–23

Information input to the cerebellum is simultaneously transmitted to the deep cerebellar nuclei.

Purkinje cells in the comparison of intended and actual movements. Neurons of the deep nuclei then transmit information to correct for any deviation from the intended motion.

The inputs to the Purkinje cells of the cerebellum traverse two routes: the climbing fiber and parallel fiber pathways. Climbing fibers originate in the inferior olivary nucleus (Fig. 9–24*a*). Axons from neurons in this nucleus form synaptic connections around the soma and dendrites of Purkinje cells. Each Purkinje cell receives synaptic connections from only one climbing fiber. The activation of this input causes the Purkinje cell to generate complex patterns of action potentials (Fig. 9–25).

The parallel fiber input to the Purkinje cell originates from **granule cells.** These fibers synapse with the dendrites of Purkinje cells. Each Purkinje cell receives inputs from many granule cells, and the activation of the parallel fibers causes the Purkinje cells to generate a simple pattern of action potentials.

Inputs to the inferior olivary nucleus come from pathways originating in supplemental and premotor areas of the cortex as well as from the spinal cord, while inputs to granule cells are derived from so-

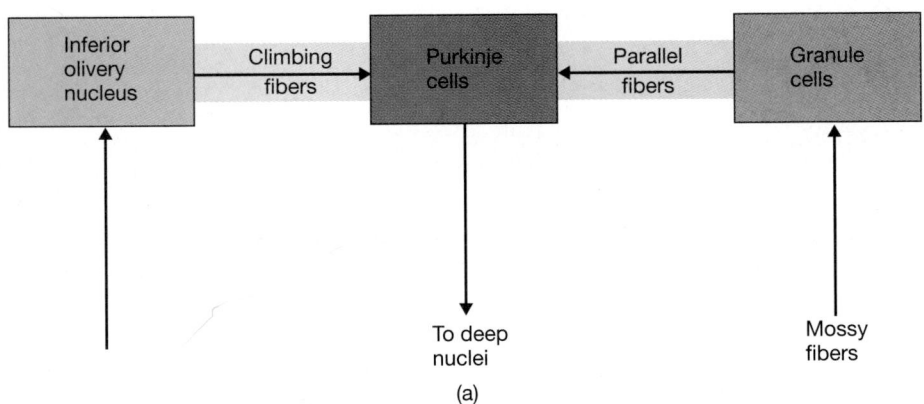

(a)

Figure 9–24

Climbing fibers to the cerebellum originate in the inferior olivary nucleus, while parallel fibers originate in the granule cells (*a*). Activation of climbing fibers generates complex pattern of action potentials (*b*), while activation of parallel fibers generates simple action potential patterns (*c*).

(b)   (c)

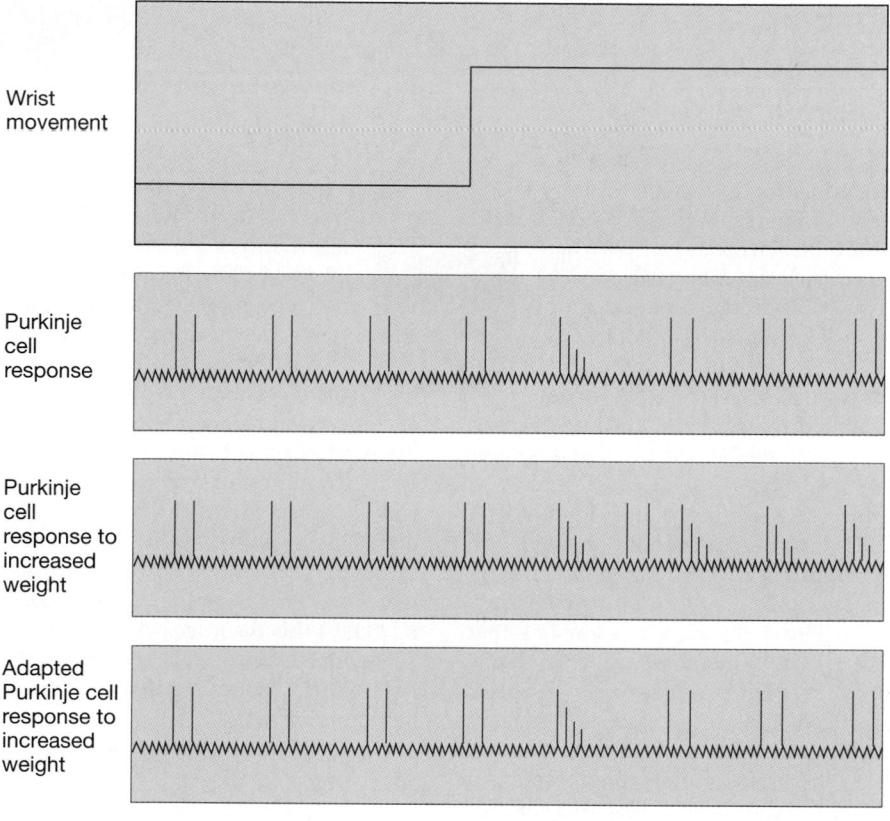

Figure 9–25

Purkinje cells can change their pattern of activity during adaptation to new motor responses. Complex action potential patterns show the involvement of climbing fibers in the learning of a new motor skill. As an animal adapts to the skill, simpler action potential patterns emerge.

called **mossy fibers.** These mossy fibers also originate in the neocortex as well as in the spinal cord. The mossy fibers and olivary climbing fibers, however, respond quite differently during movement. Sensory stimuli and voluntary movements enhance the activity of mossy fibers but have little effect on climbing fiber activity. Climbing fibers are therefore thought to modulate the responsiveness of Purkinje cells to mossy fiber and hence to parallel fiber inputs. This is seen most clearly during the learning of motor behaviors.

When monkeys are trained to maintain a lever in one position with their hands, Purkinje cells generate a simple pattern of action potentials (Fig. 9–25). The pattern is characteristic of that produced by Purkinje cells activated by parallel fibers. When different weights are placed on the lever to change its position, the animal must compensate for this added force in order to maintain its original position. When more weight is added, Purkinje cells begin to produce

complex (rather than simple) patterns of action potentials. These complex patterns of Purkinje cell action potentials are due to their activation by climbing fibers. As the animal learns to adapt to the new weight, the complex pattern of action potentials diminishes and the simple pattern reemerges. Thus, the Purkinje cells are capable of changing their pattern of activity during the adaptation of new motor responses.

In addition to information concerning the force of muscle contraction, the activity of nerve cells in the cerebellum can encode the direction of reaching. Similar to that described for cells in the motor cortex, individual cells in the cerebellum become activated when a particular direction of movement is made. This preferential direction of movement gives rise to directional vectors for cells and population vectors for cell clusters within the cerebellum. These cerebellar vectors therefore represent the neural code for the direction of limb movement.

CLINICAL FOCUS

"Brain Attack" and Stroke

Stroke can be defined as the loss of blood flow to the brain. This condition can result when blood vessels weaken and rupture within the brain, or when they become clogged by a thrombus (blood clot) or an embolus of air. In addition, strokes can occur during heart attacks, when there is a transient loss of blood flow to all areas of the brain. Approximately 500,000 new cases of stroke occur each year in the United States. It is the third leading cause of death in this country behind heart disease and cancer.

Patients who survive a stroke are often left with severe disabilities. The type of disability depends on which areas of the brain are affected. Often patients are paralyzed, and unable to speak and communicate. Currently, the American Heart Association estimates that there are approximately 2.9 million survivors of stroke in this country, many of whom are severely disabled.

Research has focused on developing ways to prevent the death of neurons within the brain following a stroke. When blood flow to the brain is interrupted, there is a loss of oxygen and glucose, which are essential for normal brain function. Among the functions that are disrupted is the removal of excitatory amino acid neurotransmitters such as glutamate. Abnormally high levels of extracellular glutamate, in turn, cause a prolonged excitation of postsynaptic neurons and an increase in the influx of calcium ions. The excessive accumulation of intracellular calcium has now been shown to be a critical step in the process of stroke-induced neuronal death.

In understanding the mechanisms and pathophysiology of stroke, researchers are developing new strategies to prevent stroke-induced cell death. These strategies include such approaches as the use of calcium channel blockers to inhibit calcium influx, and hypothermia to reduce the metabolic demands of the brain. In laboratory animals these approaches have been shown to be quite effective in preventing the loss of neurons and their corresponding neurological deficits.

The time in which these strategies are effective in preventing cell death, however, is soon after the onset of the stroke. Unfortunately for the vast majority of stroke patients, the warning signs of stroke are not recognized by family and friends, and as a consequence, there is a considerable delay between when the stroke occurs and when treatment is initiated. Public education and awareness of these warning signs is therefore critical in reducing the morbidity and mortality of strokes. In developing public awareness of strokes, the term *"brain attack"* has been adopted to stress the urgency for medical attention. This is similar to the programs involved in educating the public about the signs of "heart attacks," which have significantly reduced the number of deaths related to this disorder.

A brain attack (stroke) is a medical emergency. If you have the following signs and symptoms, get medical help immediately:

- *Sudden weakness or numbness of the face, arm, or leg on one side of the body*
- *Sudden dimness or loss of vision, particularly in only one eye*
- *Loss of speech, or trouble talking or understanding speech*
- *Sudden, severe headaches with no apparent cause*
- *Unexplained dizziness, unsteadiness or sudden falls, especially along with any of the previous symptoms*

■ **If the influx of calcium ions is important for the induction of long-term potentiation as discussed in Chapter 7, why is it detrimental in the case of stroke?**

The Basal Ganglia: Planning of Movements

Other major areas of the brain involved in motor control are the basal ganglia. The basal ganglia include the **caudate, putamen,** and **globus pallidus** (Fig. 9–26). The caudate and putamen receive inputs to the basal ganglia, while the globus pallidus provides the output. Inputs to the basal ganglia are from the entire neocortex, thalamus, and substantia nigra of the brain stem. The primary input, however, is from the neocortex. The extensiveness of this input suggests that the basal ganglia are involved in other functions besides motor activities. In fact, diseases of the basal ganglia often produce cognitive abnormalities.

The major output of the basal ganglia is to the prefrontal and premotor cortices by way of the thalamus (Fig. 9–26c.) Through this pathway, the basal ganglia can modulate the descending components of the motor system. The basal ganglia also have outputs to the substantia nigra. The interconnections between basal ganglia and substantia nigra play a prominent role in diseases of the motor system. Nerve fibers

Figure 9–26

The caudate and putamen of the basal ganglia receive inputs, while the globus pallidus provides the output. The basal ganglia receive inputs from the neocortex, thalamus, and substantia nigra.

(a)

(b)

(c)

from the substantia nigra that terminate in the basal ganglia release dopamine as the neurotransmitter. The degeneration of these dopamine fibers is responsible for the motor disorder called Parkinson's disease. Patients with parkinsonism exhibit rigidity, a rhythmic tremor at rest, and difficulty in initiating movement. These symptoms can be alleviated, in part, with drugs such as L-dopa, which act as precursors to increase the synthesis of dopamine.

Vestibular Control of Equilibrium and Balance

The vestibular system aids in maintaining the body's balance by detecting the position and motion of the head in space. The sensory organs of the vestibular system are the **semicircular canals** (Fig. 9–27). Three semicircular canals make up the vestibular apparatus: the **superior, inferior,** and **horizontal semicircular canals.** Each canal is situated perpendicular to the other two. With this arrangement, each semicircular

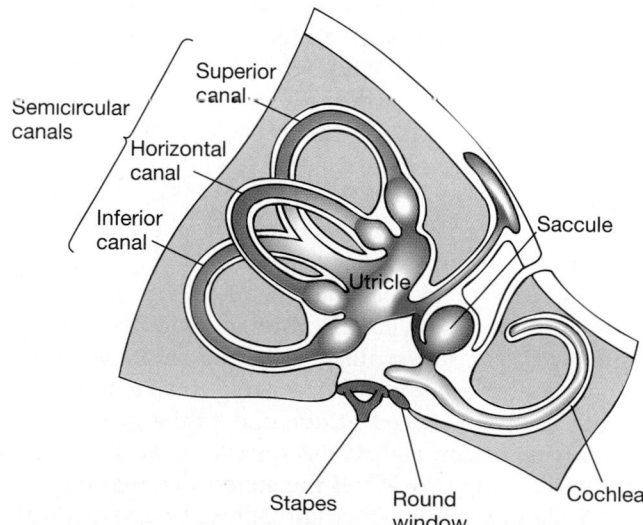

Figure 9–27

The vestibular apparatus consists of the semicircular canals, including the inferior, superior, and horizontal canals.

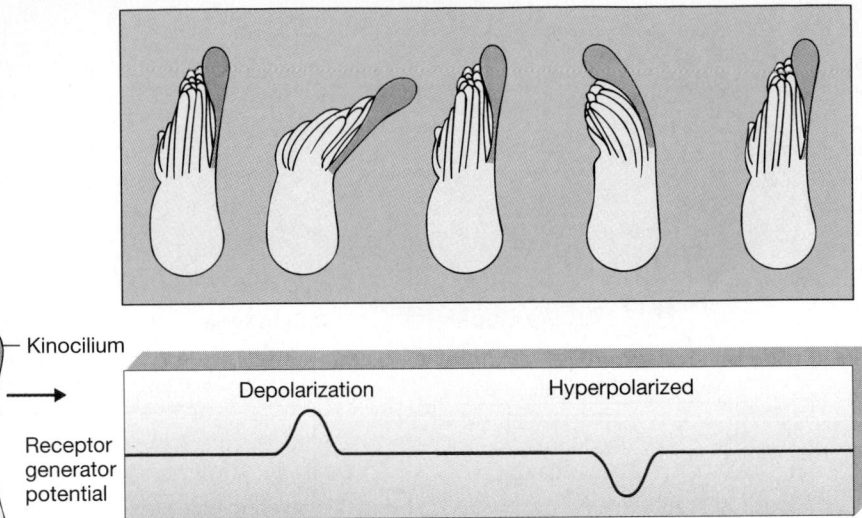

- Kinocilium

Receptor generator potential

Sensory nerve activity

Figure 9–28

The cilia in the hair cells of the vestibular apparatus are arranged in order of increasing length, with the largest cilium, the kinocilium, at one end. When the cilia are bent in the direction of the kinocilium, the hair cell membrane becomes depolarized; when the kinocilium bends the other cilia toward the smallest cilium, the membrane becomes hyperpolarized.

canal detects the angular acceleration in one of three planes in space. The vestibular apparatus also consists of the **utricle** and **saccule** (see Fig. 9–27). These so-called **otolithic organs** detect linear acceleration in the horizontal plane by the utricle, and in the vertical plane by the saccule.

The semicircular canals and otolithic organs are filled with the same endolymph fluid that surrounds the hair cells of the cochlea. As does the cochlea, the semicircular canals and otolithic organs have hair cells responsible for the transduction of sensory stimuli. The appropriate stimulus for these vestibular hair cells is the acceleration of the head. As the head begins to move, the inertia of the endolymph fluid causes the stereocilia to bend (Fig. 9–28). This bending of the stereocilia causes the hair cell either to depolarize or to hyperpolarize.

The type of membrane potential change in the hair cell depends on the direction in which the stereocilia bend. The stereocilia of hair cells in the vestibular apparatus are aligned according to size, with the smallest cilium placed at one end of the group and the largest cilium, called the *kinocilium,* at the opposite end (see Fig. 9–28). If the stereocilia are bent toward the kinocilium, the hair cell will depolarize. If the stereocilia are bent toward the smallest cilium, the hair cell will hyperpolarize.

These hair cells, like those in the cochlear nucleus, are not capable of generating action potentials. They do, however, excite or inhibit the **vestibular nerve fibers** that innervate them. As the hair cell depolar-

izes, it releases a neurotransmitter that excites the innervating vestibular nerve (see Fig. 9–28). The hyperpolarization of a hair cell causes the vestibular nerve to produce fewer action potentials. This may be due to a decrease in the amount of an excitatory transmitter that is tonically being released from the hair cell.

How do these characteristics of vestibular hair cells allow them to detect the accelerated movement of the head? The hair cells are arranged in a polarized manner on both sides of the brain. In the horizontal semicircular canals, for example, the stereocilia on each hair cell are positioned so that the kinocilium is nearest to the nose or front of the head, while the smallest cilium is nearest the back of the head (Fig. 9–29). As the head rotates from right to left (counterclockwise), the inertia of the endolymph makes it appear to be moving from left to right (clockwise) with respect to the stereocilia. The fluid's relative motion causes the stereocilia on the left side of the brain to depolarize the hair cells, while the stereocilia on the right side of the brain hyperpolarize the hair cells. Correspondingly, the vestibular nerves on the left side of the brain increase their rate of action potential generation, while the nerves on the right side of the brain decrease their rates of action potential generation. This information is thus transmitted to inform the brain that the head is rotating clockwise.

The information from the vestibular nerves is transmitted to the vestibular nucleus located within

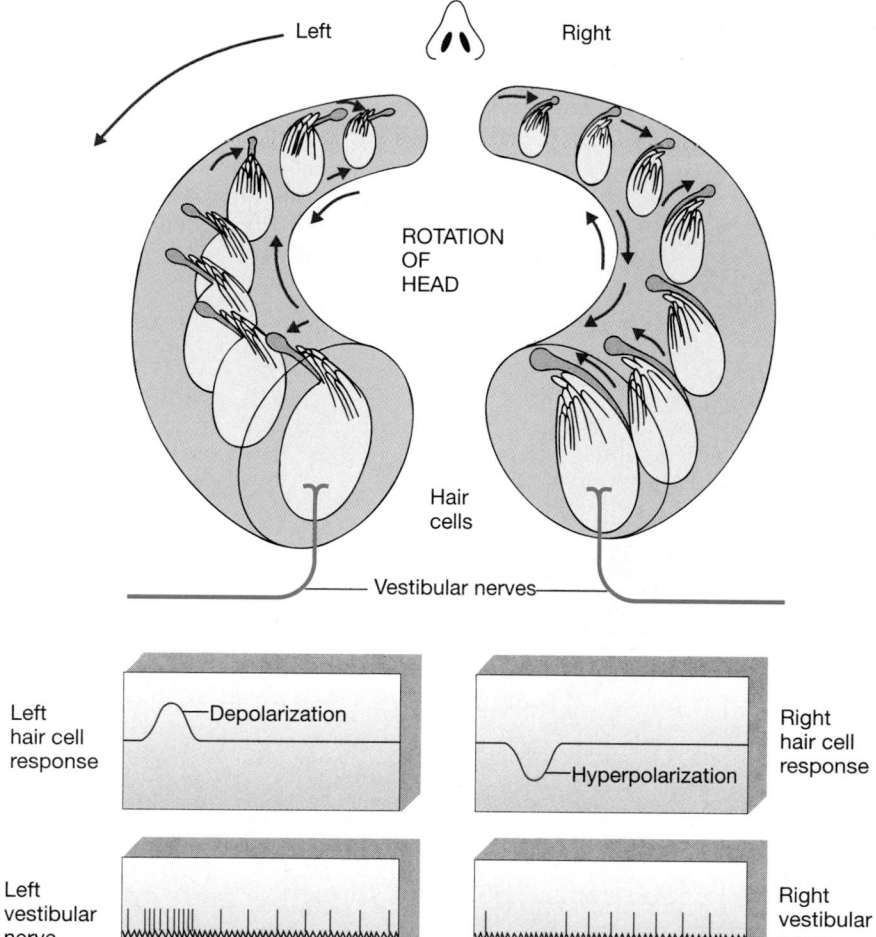

Left   Right

ROTATION
OF
HEAD

Hair
cells

Vestibular nerves

Left
hair cell
response

Depolarization

Right
hair cell
response

Hyperpolarization

Left
vestibular
nerve
response

Rotation

Right
vestibular
nerve
response

Rotation

Figure 9–29

In the horizontal semicircular canal, the kinocilia of the hair cells are oriented toward the nose. When the head rotates in a counterclockwise direction, the endolymph in effect flows in the opposite direction relative to the cilia. On the left side of the brain, the hair cells depolarize because the fluid motion causes the cilia to bend toward the kinocilium, while the opposite effect takes place on the right side. The vestibular nerves detect the depolarization and hyperpolarization and thereby inform the brain about the direction of rotation.

the brain stem. From there, it is sent down to the spinal cord to act on nerve cells that regulate movement, and up to the cerebellum and somatosensory cortex. The cerebellum and somatosensory cortex use this information to coordinate muscular activity in order to maintain balance.

The vestibular system also helps the eyes to fixate on a spot in the visual field to provide a stable reference point during acceleration. This process is part of the vestibulo-ocular reflex, and is achieved by the transmission of vestibular information for the control of eye movements. Outputs from the vestibular nucleus project to the motor nucleus of the III cranial nerve ipsilaterally and the VI cranial nerve contralaterally. These brain stem nuclei control muscles that cause the eyes to turn inward or outward, respectively. The movement of the head in a clockwise direction causes a slow counterclockwise movement of both eyes to allow them to fixate on a point of reference. As the head continues to turn, there is a fast movement of the eyes (*saccade*) in the direction of ro-

tation. This allows the eyes to fixate on a new point of reference during rotation. The slow and fast movements of the eyes during accelerated angular rotation are called *nystagmus*. The fast saccade in the direction of rotation is conventionally chosen to describe the direction of the nystagmus.

In addition to the transduction of information about the movement of the head, the vestibular apparatus also transduces information about the stationary position of the head in space with respect to gravity. The stereocilia of the hair cells in the utricle are embedded in an **otolithic membrane** that contains calcium carbonate crystals called **statoconia** (Fig. 9–30). The statoconia give the otolithic membrane mass, so that when the head is tilted to the left or right, the gravitational force causes the weighted otolithic membrane to bend the stereocilia. The stereocilia are arranged on the hair cells so that when the head is tilted to the left, the hair cells depolarize, and when the head is tilted to the right, the hair cells hyperpolarize.

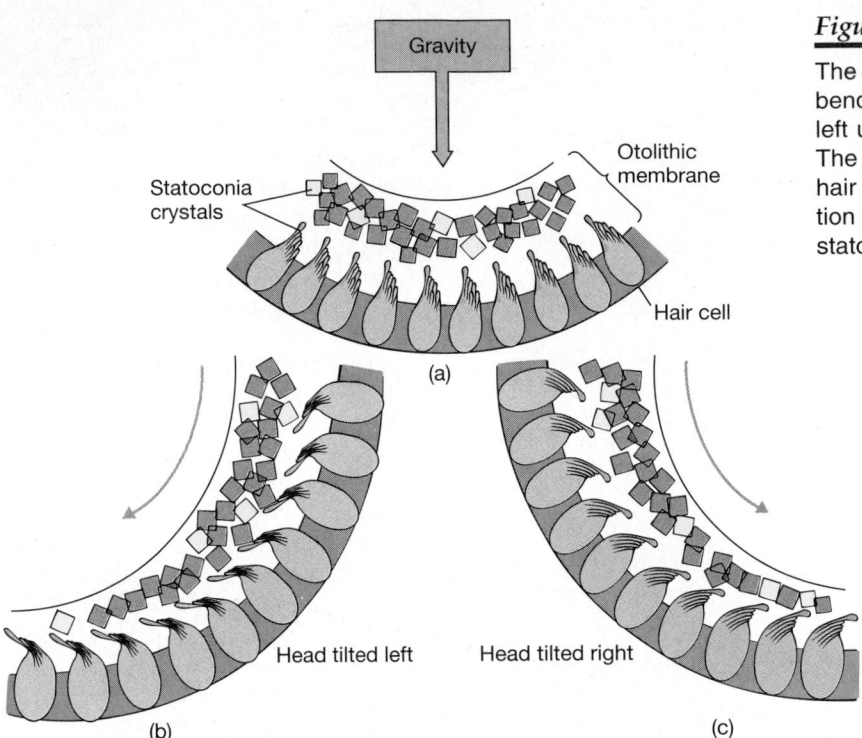

(a)

Gravity

Statoconia crystals

Otolithic membrane

Hair cell

Head tilted left (b)

Head tilted right (c)

Figure 9–30

The stereocilia of the hair cells of the utricle bend when the head bends to the right or left under the weight of statoconia crystals. The depolarization or hyperpolarization of hair cells informs the brain about the direction of bending. Arrows indicate direction of statoconia movement.

Disorders of the Motor System

Because the different parts of the motor system have such distinctive roles, lesions in these areas result in characteristic movement disorders. These lesions often occur because of trauma, stroke, or tumors.

Lesions of the Corticospinal Tract Produce Paralysis

Upper-motor-neuron lesions result from damage to descending pyramidal cell fibers (Fig. 9–31). These lesions are characterized by paralysis on the side of the body opposite the site of the lesion, increase in muscle tone, hyperactive reflexes, and the extension of the big toe and fanning of the other toes in response to stroking the bottom of the foot. This latter sign is called the **Babinski sign.** Muscle atrophy associated with this type of lesion is not prominent.

   Corticospinal tract lesions occur within the spinal cord and disrupt the inputs to motor neurons. These lesions are characterized by loss of strength and movement in muscle groups, loss of strength in voluntary muscle contraction, and the Babinski sign. These effects are ipsilateral to (occurring on the same side of the body as) the side of the lesion.

   Lower-motor-neuron lesions can occur with spinal cord injuries that directly affect the motor neurons rather than their inputs. These lesions are characterized by ipsilateral hypoactive reflexes, paralysis limited to specific groups of muscles, and flaccid

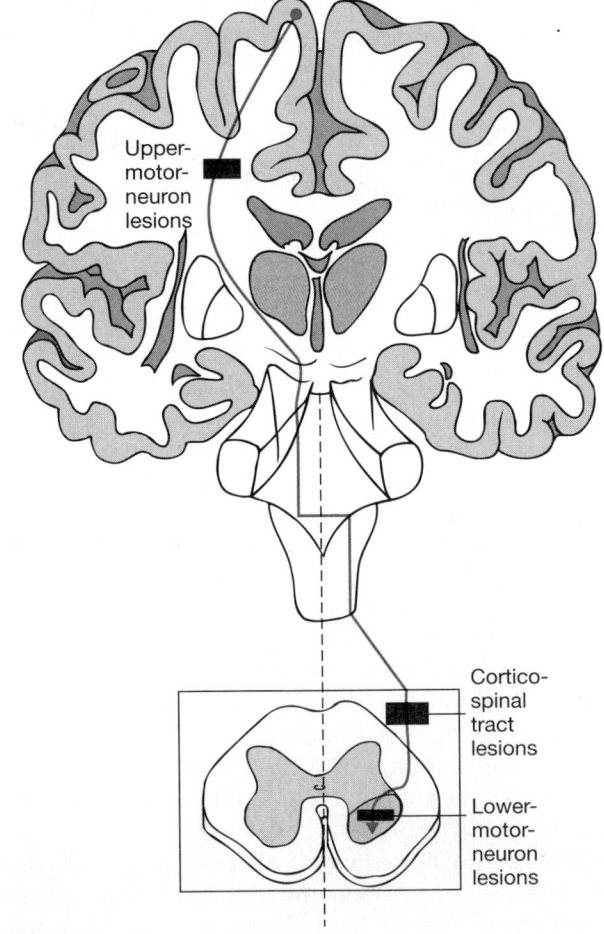

Upper-motor-neuron lesions

Cortico-spinal tract lesions

Lower-motor-neuron lesions

Figure 9–31

Lesions of the motor system result in characteristic motor disorders.

(text continues on page 325)

RESEARCH FOCUS

Huntington's Disease

Huntington's disease (HD) is a neurological disorder characterized by depression, dementia, psychiatric problems, and a progressive loss of motor control. In 1872, George Huntington described the motor components of this disorder as an "irregular and spasmodic action of certain muscles, as of the face, arms, etc. These movements gradually increase, when muscles hitherto unaffected take on the spasmotic action, until every muscle of the body becomes affected." Ultimately, the "patient presents a spectacle which is anything but pleasing to witness."

In addition to noting the motor abnormalities, Huntington made key observations regarding its hereditary nature. HD is passed from generation to generation by an autosomal dominant mode of genetic transmission. Both males and females are equally susceptible and each offspring of an affected parent has a 50% chance of inheriting the gene. Symptoms of HD can begin at any time; the average age of onset, however, is about 40 years. At present there is no treatment that can cure the disease or slow its course of action. Symptoms gradually worsen, and death occurs an average of 16 years after onset.

The cause of the abnormalities seen in HD is the progressive loss of neurons in the caudate and putamen of the neostriatum. These neurons receive nerve fiber inputs from the neocortex. They, in turn, innervate the globus pallidus and use GABA as their neurotransmitter. In HD, therefore, a major inhibitory input to the globus pallidus is missing.

The reason for the degeneration of neurons in the caudate and putamen is still unknown. One theory speculates that fiber inputs from the neocortex, which release the transmitter glutamate onto neurons in the caudate and putamen, may be the source of the problem. Glutamate is an excitatory transmitter that when released in excess can damage postsynaptic neurons. Exactly how the HD gene may be involved in this process remains to be determined.

Although the precise role of the HD gene in the destruction of caudate and putamen neurons is not known, the recent application of molecular techniques has determined the chromosomal location of the gene. Using genetic linkage analysis and DNA markers, James Gusella, Michael Conneally, and their collaborators have localized the HD gene to the short arm of chromosome 4. In genetic linkage analysis, two traits are said to be linked when they are inherited together by several members of a family. This linkage typically occurs when the two genes lie close together on the same chromosome. The linkage of two genes is considered proven when the odds in favor of linkage are greater than 1000:1. These odds are generally expressed as the log of the odds (LOD), so that odds of 1000:1 give an LOD score of 3.0.

Human DNA has many areas in which nucleotide sequences may vary from person to person. These polymorphic areas can be used as a source of genetic markers and are detected by restriction enzymes that cut specific nucleotide sequences. The resulting restriction fragment length polymorphisms **(RFLPs)** can be used as markers not only for the HD gene but for other genetic disorders as well.

Using an RFLP marker for chromosome 4, it was found that the marker was tightly linked to the HD gene with an LOD score of 6, indicating a million-to-one odds in favor of linkage. Recently, by using other markers for chromosome 4, the exact site of the HD gene has been located and its nucleotide sequence identified. A striking feature of the HD gene is a repeating sequence consisting of the nucleotides CAG. Each Huntington patient has been found to exhibit a varying number of this sequence. Moreover, the greater the number of sequence repeats is correlated with an earlier onset of the disease process.

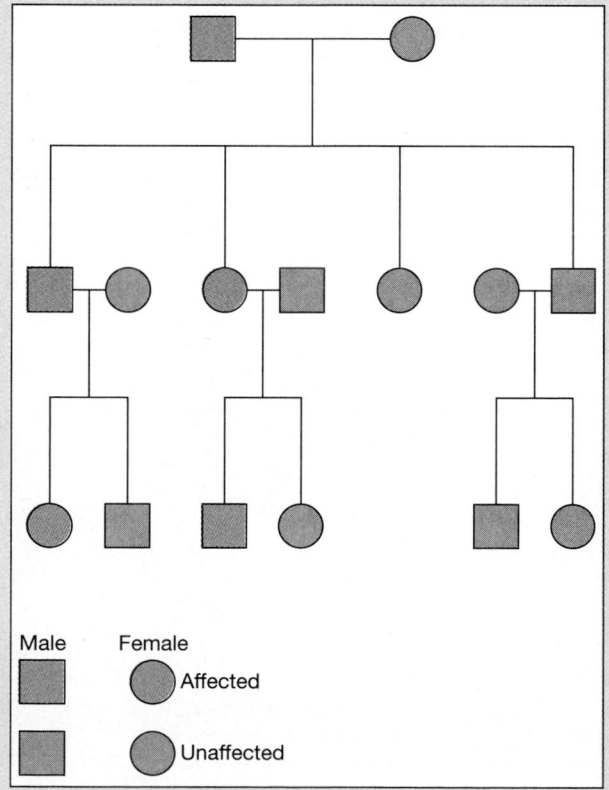

Male   Female

■ Affected

■ Unaffected

Pedigree of an HD family.

■ **What strategies might be used to counteract the genetic deficits of the HD?**

Parkinson's Disease and Brain Cell Transplants

Parkinson's disease is a disorder of the motor system characterized by slowness of movement, tremor of the fingers and hands (most prominent at rest), absence of facial expression, slowness of speech, stooped body posture, and a shuffling gait. In extreme cases, patients are bedridden and incapable of feeding and caring for themselves.

This disorder frequently appears in persons between the ages of 50 and 65 years. Currently in the United States, approximately 1 million patients are afflicted with parkinsonism, with 50,000 new cases reported each year.

Examination of brain tissue from Parkinson's patients has revealed a loss of dopamine-producing neurons in an area called the substantia nigra, located in the brain stem nucleus. These neurons send axons to form connections with other neurons in the neostriatum, an area of brain involved in locomotor activity. The resulting depletion of dopamine in the neostriatum appears to be the major cause of this disorder.

The reason for the degeneration of dopamine cells in the substantia nigra is unknown. A form of juvenile parkinsonism is thought to be genetically inherited. Evidence for a genetic component for adult parkinsonism is more controversial. Epidemiological studies reveal a strong correlation between parkinsonism and agricultural communities. It is thought that pesticides may be a contributing factor in the development of Parkinson's disease.

One form of treatment for Parkinson's disease—the administration of drugs such as L-dopa that become precursors for the synthesis of dopamine—has been found to be an effective form of therapy for many patients. Some patients, however, do not respond to drug therapies, or experience debilitating side effects such as hallucinations.

In animal models of Parkinson's disease, scientists have taken a new approach in which nerve cells that produce dopamine are transplanted directly into the neostriatum. These transplanted neurons form synaptic connections with those of the host. In addition, they are capable of correcting the locomotor abnormalities associated with Parkinson's disease. The sources of these dopamine neurons in laboratory animal experiments are fetuses of the same species. Because of social issues concerning the use of human fetal tissue, however, medical researchers are also exploring the possible use of alternative sources of dopamine cells for transplantation.

Recent studies suggest that the lack of dopamine results in an abnormal outflow of neural activity from the striatum, which is responsible for the motor deficits associated with Parkinson's disease.

■ **What approaches might you try to alter or negate this abnormal activity?**

Photomicrograph of nigral dopamine neuronal loss in a genetic animal model of Parkinsonism. (*a*) Substantia nigra neurons in a normal mouse. (*b*) Substantia nigra in a mouse model of Parkinson's disease. Note degeneration of dopamine neurons. (*Courtesy of L.C. Triarhou, B. Ghetti, and W.C. Low.*)

(limp) muscles with prominent atrophy (decrease in size).

Lesions of the Cerebellum Cause Disturbances in Motor Coordination

Lesions of the cerebellum cause ipsilateral disturbances. Lesions occurring in the lateral cerebellum result in a lack of limb coordination. Lesions occurring in the vermis, the central region of the cerebellum (Fig. 9–21), produce ataxia (loss of coordination); those localized in the flocculonodular lobe produce a disturbance in equilibrium as well as ataxia. Diseases of the cerebellum are also characterized by inaccurate range and direction of movement, diminished resistance to passive movement, inability to rapidly move a limb and to stop sharply, and tremor with movements. Interestingly, chronic alcoholism also affects the cerebellum. Excessive, long-term consumption of alcohol leads to the degeneration of Purkinje cells. As a consequence, chronic alcoholics often exhibit a wide base ataxic gait.

CHAPTER REVIEW

Summary

- The motor system controls locomotion and fine movement, and maintains body posture and equilibrium, by acting on motor neurons in the spinal cord that innervate skeletal muscles. The components of the motor system (the spinal cord, brain stem, and motor cortex) interact in both hierarchical (serial) and parallel fashion.

- A motor neuron and the muscle fibers that it innervates (by forming myoneural junctions) constitute a motor unit. A motor neuron causes a muscle to contract or relax.

- Motor neurons have cell bodies located in the gray matter of the spinal cord, in the ventral horn. The spinal cord contains interneurons, which play a role in coordinating the responses of antagonistic and synergistic muscles. Walking movements are created by locomotor generators in the spinal cord. These locomotor programs are driven by the locomotor command center in the brain stem and can be modified by sensory feedback.

- Sensory fibers within skeletal muscles detect muscle length, force, and velocity of muscle shortening. Reflex actions are initiated when sensory information is sent from muscle, tendon, or skin receptors to the spinal cord, which causes a reflexive movement in the appropriate muscles. Reflexes include the stretch, myotatic, and withdrawal reflexes.

- Complex motor behaviors require inputs from the motor cortex, supplemental and premotor cortices, cerebellum, and basal ganglia. The maintenance of posture and balance is carried out by brain stem inputs to the spinal cord. Three pathways provide motor input to the spinal cord: the ventromedial, lateral reticulospinal, and rubrospinal pathways.

- The motor cortex is responsible for fine movements. It is a central motor command center that instructs the motor neurons to activate their respective skeletal muscles. The motor cortex is organized in a somatotopic fashion similar to the somatic sensory cortex. Regions of the body containing small motor units represent the largest portion of the motor cortex. The cortex contains neuronal populations with summated activities (population vectors) responsible for the direction of limb movement.

- The supplemental and premotor cortices are involved in the planning of movements; they transmit this information to the motor cortex. The posterior cortex is involved in the processing of sensory information leading to purposeful movement. This cortex receives somatic and visual sensory information and transmits this information to the supplemental and premotor areas.

- The cerebellum has three motor functions: the initiation and planning of movement, the coordination of limb movement, and the maintenance of posture and equilibrium. The cerebellum carries out these functions in concert with the basal ganglia. The basal ganglia are also involved in the planning of movements. They have inputs from all neocortical areas in addition to the motor cortex and are involved in functions such as cognition and movement.

- The vestibular system provides sensory information to maintain body posture and balance. The semicircular canals detect changes in angular movement, while the otolithic organs detect changes in linear movement. Sensory information from the semicircular canals and otoliths is transmitted by the vestibular nerve to the vestibular nucleus of the brain stem and then to motor neurons in the spinal cord, which activate the appropriate muscles to maintain balance.

- Lesions in different areas of the motor system produce unique movement disorders. Upper-motor-neuron lesions of the corticospinal tract result in bodily paralysis and increased muscle tone, with hyperactive reflexes and the extension of the big toe on the side opposite to the lesion. Corticospinal lesions in the spinal cord produce loss of strength and movement in specific muscle groups, loss of strength in voluntary contraction, and extension of the big toe on the same side as the lesion. Lower-motor-neuron lesions of the spinal cord result in paralysis restricted to specific groups of muscles, hypoactive reflexes, and flaccid muscles. Cerebellar lesions are characterized by a lack of coordination and equilibrium.

Review Questions

Identify Each with a Term

1. The term used to describe the type of muscle that decreases the angle of a joint.
2. The term used to describe a functional group of nerve cells in the motor cortex that activate the same skeletal muscle.
3. The type of sensory receptor that detects changes in muscle length and velocity of muscle shortening.
4. The sensory organ that helps maintain balance and equilibrium by sensing when the body undergoes angular acceleration.

Define Each Term

5. extensor
6. alpha motor neuron
7. Babinski sign
8. utricle

Choose the Correct Answer

9. If fibers of gamma motor neurons are cut, the contraction of extrafusal muscle fibers will cause:
   a. a decrease in type Ia nerve fiber activity
   b. a decrease in type Ib nerve fiber activity
   c. an increase in type II nerve fiber activity
   d. both (a) and (b)
10. Which of the following statements about the cerebellum is *not* true?
    a. Purkinje cells of the cerebellum release GABA as their neurotransmitter.
    b. Lesions on one side of the cerebellum cause a contralateral disturbance.
    c. Nerve fibers from the cerebellar dentate nucleus project to thalamus on the contralateral side of the body.
    d. Lesions of Purkinje cells near the midline of the cerebellum result in a disturbance in balance and equilibrium.
11. The alpha motor neurons in the spinal cord responsible for maintaining posture are located in the:
    a. medial aspect of the dorsal horn
    b. lateral aspect of the dorsal horn
    c. medial aspect of the ventral horn
    d. lateral aspect of the ventral horn
12. An example of a large motor unit is the innervation of:
    a. muscles of the thumb
    b. muscles responsible for the movement of the eyes
    c. axial muscles controlling posture
    d. muscles responsible for speech

Answer Each Question in One or Two Sentences

13. How are alpha motor neurons involved in increasing the force of muscle contraction?
14. What are the functional properties of motor neurons that innervate fast and slow muscles?
15. What is the role of gamma motor neurons in regulating the sensitivity of muscle spindle activity?
16. Compare and contrast motor units versus motor neuron pools.

Problem Solving in Physiology

17. Why do unilateral lesions of the cerebellum result in motor deficits on the ipsilateral side of the body while unilateral motor cortex lesions produce deficits on the contralateral side?
18. How can some lesions of the spinal cord result in the paralysis of both legs while functions such as urination and defecation are maintained?

Suggested Readings

Asanuma, H. "Cerebral cortical control of movement." *Physiologist,* 16:143–166, 1973.

Camarata, P.J., Latchaw, R.E., and Heros, R.C. "Brain attack": The rationale for treating stroke as a medical emergency. *Neurosurgery,* 34:144–158, 1994

Evarts, E.V. "Pyramidal tract activity associated with a conditioned hand movement in the monkey." *Journal of Neurophysiology,* 29:1011–1027, 1966.

Fortier, P.A., Kalaska, J.F., and Smith, A.M. "Cerebellar neuronal activity related to whole-arm reaching movements in the monkey." *Journal of Neurophysiology* 62:198–211, 1989.

Georgopoulos, A.P., Kettner, R.E., and Schwartz, A.B. "Primate motor cortex and free arm movements to visual targets in three-dimensional space. II. Coding of the direction of movement by a neuronal population." *Journal of Neuroscience,* 8:2928–2937, 1988.

Graybiel, A.M., Aosake, T., Flaherty, A.W., and Kimura, M. "The basal ganglia and adaptive motor control." *Science,* 265:1826–1831, 1994.

Grillner, S., and Shik, M.L. "On the descending control of the lumbosacral spinal cord from the 'mesencephalic locomotor region.'" *Acta Physiological Scandinavia,* 87:320–333, 1973.

Grillner, S., and Zangger, P. "How detailed is the central pattern generation for locomotion?" *Brain Research,* 88:367–371, 1975.

Liddell, E.G.T., and Sherrington, C. "Recruitment and some other features of reflex inhibition. " *Proceedings of the Royal Society of London,* 97:488–518, 1925.

Lundberg, A. "Integration in a propriospinal motor centre controlling the forelimb in the cat." In *Integration in the Nervous System,* edited by H. Asanuma and V.J. Wellson, 47–64. Tokyo: Igaku-Shoin, 1979.

Matthews, B.H.C. "Nerve endings in mammalian muscle." *Journal of Physiology,* 78:1–53, 1933.

Mitchell, S.J., Richardson, R.T., Baker, F.H., and DeLong, M.R. "The primate globus pallidus: Neuronal activity related to direction of movement." *Experimental Brain Research,* 68:491–505, 1987.

Mountcastle, V.B., Lynch, J.C., Georgopoulos, A., Sakata, H., and Acuna, C. "Posterior parietal association cortex of the monkey: Command functions for operations within extrapersonal space." *Journal of Neurophysiology,* 38:871–908, 1975.

Sherrington, C. *The Integrative Action of the Nervous System,* 2nd ed. New Haven: Yale University Press, 1947.

Key Terms

Flourescent labeled neurons
from superior cervical ganglion.
(© Nancy Kedersha/UCLA/Science
Photo Library/Photo Researchers, Inc.)

Chapter 10

The Autonomic Nervous System

Key Concepts

The autonomic nervous system is organized into central and peripheral components. The central autonomic nervous system regulates the outflow of information to the peripheral components in order to coordinate the activity of the body's organs.

The peripheral component of the autonomic nervous system is composed of sympathetic and parasympathetic subdivisions. The sympathetic subdivision coordinates the body's response to stress, while the parasympathetic subdivision is responsible for vegetative responses.

The status of internal organs is monitored by sensory receptors that transmit information to initiate or coordinate autonomic actions such as respiration, urination, and defecation.

One of the primary functions of the autonomic nervous system is to regulate the body's internal environment by controlling parameters such as water volume, pH, O_2, CO_2, blood pressure, and body temperature—and thus maintain conditions that are optimal for cells to function.

The hypothalamus controls the activity of different organ systems within the body so that they act in a coordinated fashion.

The adrenal gland is controlled, in part, by hormones released from the hypothalamus and secreted into the circulatory system which, in turn, cause the adrenal gland to secrete norepinephrine and epinephrine into the circulatory system.

Chapter Outline

Organization of the Autonomic Nervous System

Control of Bodily Organs by the Sympathetic and
    Parasympathetic Systems

Reflexes Governed by the Autonomic Nervous System

Regulatory Systems of the Autonomic Nervous System

Autonomic Functions of the Hypothalamus

Hypothalamic Control of the Adrenal Gland

328

It's finals week and Mario is exhausted. After two all-nighters and too much coffee, he finally collapses into bed and quickly falls into a deep sleep. As he awakens in the morning he is half conscious of the birds outside the window, and the warmth of his blanket. Refreshed and relaxed, he turns toward the clock and instantly realizes that he has overslept and now has just five minutes to get to his anatomy final. Mario's heart is beating wildly as he struggles to get out of bed and into his clothes. Maniacally he ravages the desk for a pencil. His breathing becomes more rapid as he sprints out of the dorm. He finally reaches the lecture hall, dripping with perspiration, just as the proctor begins to hand out the exam.

This scenario illustrates the dramatic changes in bodily functions that occur in the "fight-or-flight" response. The realization that he is late for his final exam activates the autonomic centers that coordinate Mario's stress responses. The adrenal glands release adrenalin into the circulatory system. Heart rate and respiration increase in order to supply energy to parts of the body that have to react to the stressful situation. In this case, more blood and oxygen are pumped to skeletal muscles to allow Mario to run. The increased heat generated by his muscles while he is running is then dissipated by the activation of sweat glands beneath the surface of the skin. In this way, the autonomic nervous system can coordinate the body's organs to respond to stressful conditions. These coordinated responses are especially important in life-threatening situations; they therefore have tremendous survival value not only for the individual, but also for the species.

The **autonomic nervous system (ANS)** coordinates bodily functions needed for survival. It senses the need for food and activates the digestive processes; it senses the need for water and activates processes for the retention and consumption of fluids. The ANS aids in removing waste products from the body and it prepares the body for the stress of life-threatening conditions.

The ANS is also a regulatory system that maintains the environment necessary for cells to function properly. It maintains the body's temperature at 37°C, mean blood pressure at 90 mm Hg, pH at 7.4, and blood plasma osmolarity at 300 mOsm. These are the optimal conditions of the **internal environment** for the functioning of the organs and their cells.

The actions of the ANS are not primarily under direct voluntary control. Unlike the **somatic nervous system,** with which we can consciously command skeletal muscles to move (e.g., the muscles in the fin-

gers of the hand to pick up a pencil), the actions of the ANS are all subconscious. It is difficult to consciously will the heart to beat faster or command the blood pressure to increase. Instead, autonomic processes transmit sensory information from **"visceral" organs** to the central ANS, which in turn sends instructions to the smooth muscles and other cells of those organs, producing an appropriate automatic response.

This chapter examines the functional organization of the ANS, and the ways in which the ANS controls other organ systems to supply the necessary nutrients to the body, to remove waste products of metabolism, to coordinate the response to stress, and to regulate the internal environment.

Organization of the Autonomic Nervous System

The ANS has both central and peripheral components. The central components of the ANS are the **hypothalamus,** the **brain stem,** and the **spinal cord** (Fig. 10–1). The peripheral components consist of the nerves that innervate the organs of the body, classified as either parasympathetic or sympathetic nerves. Autonomic nerve fibers include both afferent (sensory) and efferent (motor) neurons.

The Peripheral Autonomic Nervous System Regulates the Activity of Other Organs in the Body

The peripheral ANS is functionally composed of the **sympathetic nervous system** and the **parasympathetic nervous system.** The sympathetic division of the ANS coordinates the body's response to stress, while the parasympathetic division coordinates the body's more "vegetative" activities such as digestion. The sympathetic and parasympathetic nervous systems not only have distinct anatomical differences, but also release different neurotransmitters at their target sites. The sympathetic nervous system simultaneously activates more organ systems than its parasympathetic counterpart. It plays an important role during stress responses when there is a need to coordinate increases in heart rate, blood pressure, and respiration rate. By contrast, in parasympathetic responses, different organ systems are activated more independently.

Nerve fibers of the sympathetic nervous system emerge from the spinal cord at the thoracic and lumbar levels Fig. 10–1b). The cell bodies of these fibers are located in the **interomedial lateral nuclei (IML)**

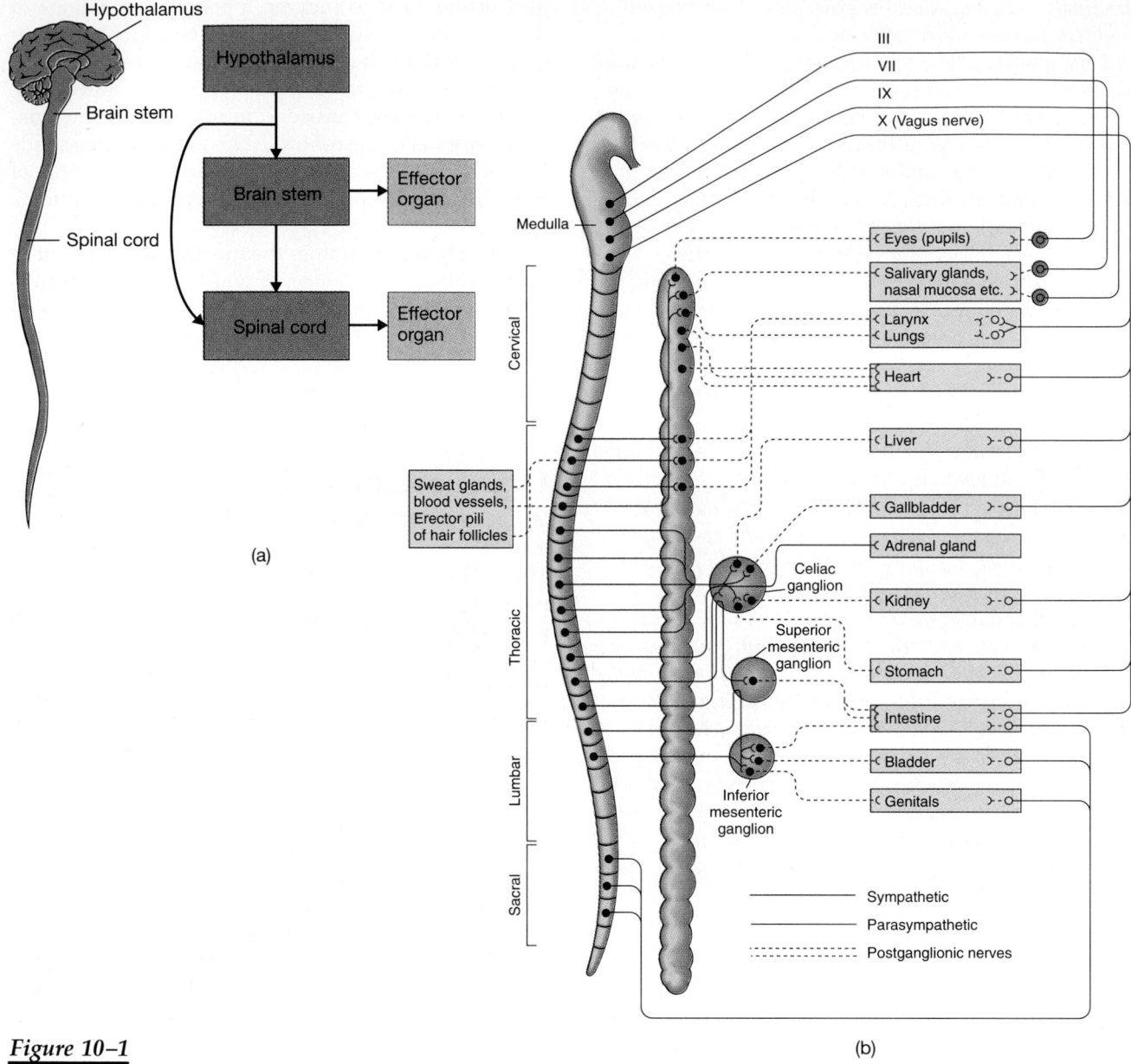

Figure 10–1

(*a*) The principal components of the autonomic nervous system (ANS) are the hypothalamus, brain stem, and spinal cord. (*b*) The ANS comprises the sympathetic and parasympathetic nervous systems. Sympathetic pathways are shown in pink, parasympathetic pathways in blue. Dotted lines indicate postganglionic nerves.

of the spinal cord (Fig. 10–2). These cells are called **preganglionic nerve cells** and have short axons that innervate cells in **ganglia** located near the spinal cord. The cells within the ganglia are **postganglionic nerve cells.** The postganglionic cells have long axons and synapse with cells in other organ systems.

Fibers from the parasympathetic nervous system exit from the brain stem and also leave the spinal cord at the level of the sacrum (see Fig. 10–1*b*). These preganglionic fibers have long myelinated and unmyelinated axons and innervate postganglionic

nerve cells clustered near or in the target organ. As a consequence, the postganglionic parasympathetic nerve cells have very short axons.

As Figure 10–1*b* illustrates, most organs are innervated by both sympathetic and parasympathetic nerve fibers. The effects of each system on the target organ, however, are quite different. One system increases the activity of the target organ, while the opposing system decreases its activity. For example, the sympathetic innervation of the heart increases the heart rate, while parasympathetic innervation causes

Figure 10–2

Nerve fibers of the sympathetic nervous system originate in the interomedial lateral nuclei (IML) of the spinal cord. These preganglionic nerve cells innervate postganglionic nerve cells located within ganglia near the spinal cord.

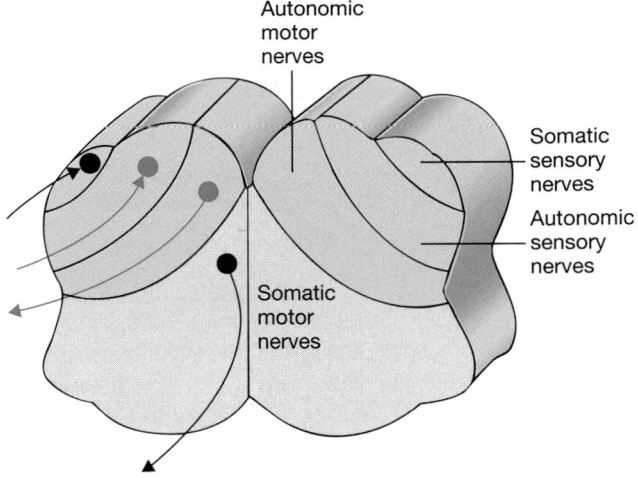

Figure 10–3

Autonomic sensory and motor nerves enter and leave the brain stem from different regions than somatic nerves do.

the heart rate to decrease. In contrast, sympathetic innervation of the small intestines decreases intestinal motility, while parasympathetic innervation causes it to increase. From this description, the actions of the sympathetic and parasympathetic nervous systems seem to oppose each other. However, the sympathetic and parasympathetic inputs to a target organ are not active at the same time; as one system is activated, the activity of the other is diminished. Later in this chapter, we will examine the mechanisms that coordinate this type of activity.

The Central Autonomic Nervous System Orchestrates the Coordinated Activity of Organs to Function in Concert

The organization of the ANS in the spinal cord is somewhat similar to that of the somatic nervous system. The preganglionic nerve cells are located in the interomedial lateral cell column and have fibers that leave the spinal cord from the ventral roots (see Fig. 10–2), while sensory fibers that carry information from the target organ enter the spinal cord through the dorsal roots. The cell bodies of these sensory fibers are located in the dorsal root ganglia. This spinal organization is identical for the sympathetic and the parasympathetic components of the autonomic nervous system. Sympathetic fibers, however, emerge from the thoracic and lumbar segments of the spinal cord, while parasympathetic fibers emerge from sacral segments of the cord.

The organization of the autonomic nervous system in the brain stem is different from that in the

spinal cord. The autonomic sensory fibers coming into the brain stem are segregated from the somatic sensory fibers, which enter more laterally than autonomic sensory fibers (Fig. 10–3). On the other hand, autonomic efferent nerves leave the brain stem in more lateral regions than somatic efferents.

Neurotransmitters in the Autonomic Nervous System Exhibit a Unique Pattern of Distribution

The two major neurotransmitters in the autonomic nervous system are **acetylcholine** and **norepinephrine**. Acetylcholine is released at the synaptic terminals of all preganglionic neurons (Fig. 10–4). In postganglionic nerve cells, norepinephrine is found in the terminals of sympathetic fibers, while acetylcholine is found in the terminals of parasympathetic fibers. Exceptions to this rule are postganglionic sympathetic fibers that innervate sweat glands, and blood vessels of skeletal muscles that release acetylcholine. Nerve fibers that transmit acetylcholine are sometimes called **cholinergic** fibers; those that transmit norepinephrine are sometimes referred to as **adrenergic** fibers.

The acetylcholine receptor on the cell body of postganglionic neurons is different from the acetylcholine receptor on target cells of the parasympathetic nervous system. Cholinergic receptors are classified as either **nicotinic** or **muscarinic** acetylcholine receptors, based on whether they are activated by the drug nicotine or the drug muscarine. The acetylcholine receptor on the postganglionic neuron can be activated by nicotine and is thus a nicotinic receptor,

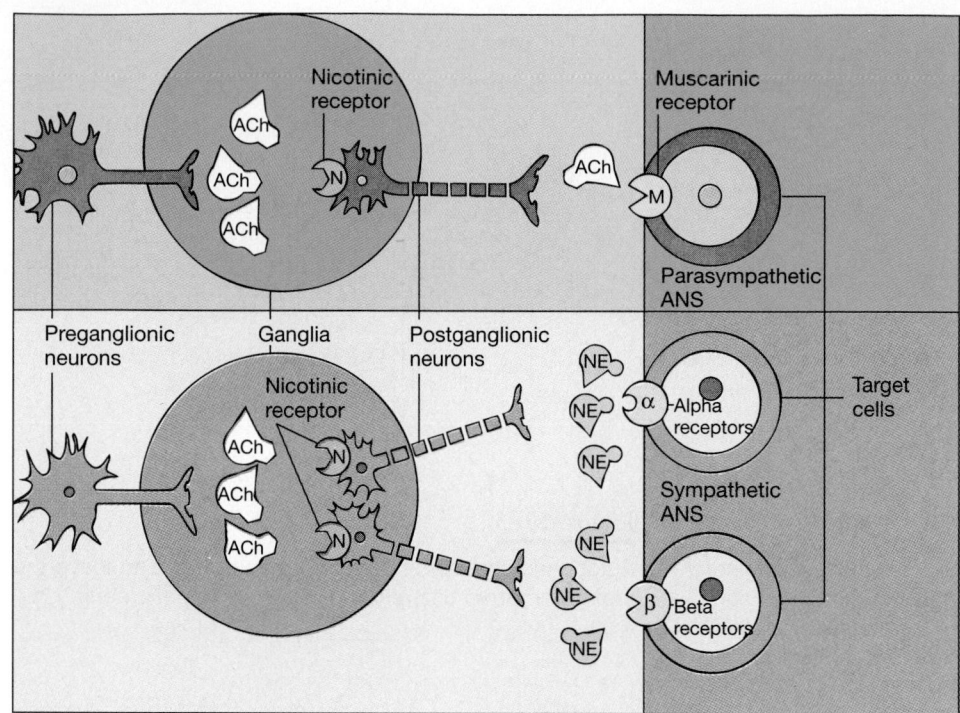

Figure 10–4

All preganglionic neurons release acetylcholine (ACh). Postganglionic parasympathetic nerve cells also release ACh, while postganglionic sympathetic cells release norepinephrine (NE). Exceptions to this, however, are fibers from sympathetic nerve cells that innervate sweat glands, and blood vessels of skeletal muscles that release ACh. Receptors for ACh are either muscarinic (M) or nicotinic (N); receptors for NE are either alpha or beta receptors.

while the acetylcholine receptor on target cells such as smooth muscle is muscarinic. Norepinephrine activates two types of receptors, **alpha receptors** and **beta receptors.** Beta receptors are distinguished from alpha receptors by their greater sensitivity to isoproterenol, a drug similar to norepinephrine.

As discussed in Chapter 7, the alpha adrenergic receptors are subdivided into alpha-1 and alpha-2 subtypes based on their anatomical location. The alpha-1 receptor subtype is located on postsynaptic target cells of sympathetically innervated organs, and alpha-2 receptors are located on the presynaptic terminal of cholinergic nerve fibers.

The beta receptor can also be subdivided into subtypes beta-1 and beta-2. At the beta-1 receptor, epinephrine and NE are more or less equipotent. In contrast, beta-2 receptors respond better to epinephrine than to NE, suggesting that they are normally activated by circulating hormones rather than by sympathetic nerve stimulation. Table 10–1 lists different organs and the effects of acetylcholine and norepinephrine on them.

Control of Bodily Organs by the Sympathetic and Parasympathetic Systems

The actions of the sympathetic and parasympathetic inputs to an organ have opposite effects. These actions operate to control various parameters in the body. For instance, when the blood pressure becomes too high, the parasympathetic inputs are activated to decrease the heart rate. They accomplish this by releasing acetylcholine. When acetylcholine binds to muscarinic receptors located on **pacemaker cells,** which are cardiac cells that regulate heart rate, the cells decrease their rhythmic rate of action potential discharge. This occurs because the muscarinic receptors, once acetylcholine binds to them, allow the K^+ ions to escape from the cells, causing them to become hyperpolarized. This slows their pacemaker activity, decreases the heart rate, and ultimately decreases the blood pressure (see Ch. 18).

When the blood pressure is too low, the sympathetic nervous system steps in. The sympathetic inputs to the heart release norepinephrine. The binding of norepinephrine to beta receptors on cardiac cells causes the cells to depolarize faster, resulting in an increase in heart rate. In addition, norepinephrine also causes the heart to contract with greater force. These mechanisms result in a compensatory increase in blood pressure (see Chapter 19).

Blood pressure is also controlled by the contraction and relaxation of smooth muscles that surround blood vessels (Fig. 10–5). These smooth muscles respond to both norepinephrine and acetylcholine that are released by nerve fibers. In addition, circulating norepinephrine and epinephrine released by the adrenal medulla can also affect the contraction of smooth muscle. Alpha-1 norepinephrine receptors cause contraction of smooth muscles, which increases blood pressure (see Chapter 19). Norepinephrine

(text continues on page 334)

Table 10–1

Summary of Cholinergic and Adrenergic Influences on Effector Organs[*]

Effector Organs	Cholinergic Impulses	Adrenergic Impulses Receptor	
	Response	*Type†*	*Response*
Eye			
Radial muscle of iris	—	α	Contraction (mydriasis)
Sphincter muscle of iris	Contraction (miosis)		—
Ciliary muscle	Contraction for near vision	β	Relaxation for far vision
Heart			
S–A node	Decrease in heart rate	β_1	Increase in heart rate
Atria	Decrease in contractility, and (usually) increase in conduction velocity	β_1	Increase in contractility and conduction velocity
A–V node and conduction system	Decrease in conduction velocity	β_1	Increase in conduction velocity
Ventricles	—	β_1	Increase in contractility, conduction velocity, automaticity, and rate of idiopathic pacemakers
Blood vessels			
Coronary	Dilation	α	Constriction
		β_2	Dilation
Skin and mucosa	—	α	Constriction
Skeletal muscle	Dilation	α	Constriction
		β_2	Dilation
Cerebral	—	α	Constriction (slight)
Pulmonary	—	α	Constriction
		β_2	Dilation
Abdominal organs	—	α	Constriction
	—	β_2	Dilation
Kidney	—	α	Constriction
Salivary glands	Dilation	α	Constriction
Lung			
Bronchial muscle	Contraction	β_2	Relaxation
Bronchial glands	Stimulation		Inhibition (?)
Stomach			
Motility and tone	Increase	α_2, β_2	Decrease (usually)
Sphincters	Relaxation (usually)	α	Contraction (usually)
Secretion	Stimulation		Inhibition (?)
Intestine			
Motility and tone	Increase	α_2, β_2	Decrease
Sphincters	Relaxation (usually)	α	Contraction (usually)
Secretion	Stimulation		Inhibition (?)
Gallbladder and ducts	Contraction		Relaxation
Urinary bladder			
Detrusor	Contraction	β	Relaxation (usually)
Trigone and sphincter	Relaxation	α	Contraction
Ureter			
Motility and tone	Increase (?)	α	Increase (usually)

[*]Modified from Goodman & Gilman, *The Pharmacological Basis of Therapeutics.* 4th ed. New York: Macmillan, 1970.

†Where adrenergic receptor type has been established, α receptors are all α_1 unless indicated otherwise. *continued*

Table 10–1
Summary of Cholinergic and Adrenergic Influences on Effector Organs (continued)

Effector Organs	Cholinergic Impulses	Adrenergic Impulses Receptor	
	Response	Type†	Response
Uterus	Variable‡	α, β_2	Variable‡
Male sex organs	Erection		Ejaculation
Skin			
Pilomotor muscles	—	α	Contraction
Sweat glands	Generalized secretion	α	Slight, localized secretion§
Spleen capsule	—	α	Contraction
Adrenal medulla	Secretion of epinephrine and norepinephrine		—
Liver	Glycogen synthesis	α, β_2	Glycogenolysis
Pancreas			
Acini	Secretion		—
Islets	Insulin secretion	α	Inhibition of insulin secretion
		β_2	Insulin secretion
Salivary glands	Profuse, watery secretion	α	Potassium and water secretion
Lacrimal glands	Secretion	β	Amylase secretion
Nasopharyngeal glands	Secretion		—
Adipose tissue	—	α, β_1	Lipolysis
Kidney	—	β_2	Renin secretion

‡Depends on stage of menstrual cycle, amount of circulating estrogen and progesterone, and other factors. Responses of pregnant uterus different from those of nonpregnant.

§On palms of hands and in some other locations ("adrenergic sweating").

beta-1 receptors, on the other hand, cause relaxation of the smooth muscle and decreased blood pressure. The relaxation of some blood vessels in smooth muscle can also occur by the activation of muscarinic acetylcholine receptors of the parasympathetic nervous system. Thus, smooth muscles are relaxed by the activation of either muscarinic or beta-1 receptors. Reduction of alpha-1 receptor activation can also produce relaxation. This is the mechanism typically found in the increase in blood flow caused by a decrease in the vasoconstriction of blood vessels.

Receptors for norepinephrine and acetylcholine also exist in the membranes of the presynaptic terminals. The alpha-2 receptors are found on synaptic terminals that release acetylcholine. These receptors inhibit the release of acetylcholine. Muscarinic receptors are found on synaptic terminals that release norepinephrine. Activation of these receptors inhibits the release of norepinephrine (Fig. 10–6). These receptors therefore allow the ANS to enhance the effect of either the sympathetic or parasympathetic actions while inhibiting the opposite action by local mechanisms at target organs. For example, when sympathetic inputs to the heart release norepinephrine, causing the heart to beat faster, the norepinephrine also binds to presynaptic alpha-2 receptors located on acetylcholine nerve terminals. The activation of the alpha-2 receptors inhibits the release of acetylcholine, which would normally decrease the heart rate. Likewise, when parasympathetic inputs to the heart release acetylcholine and reduce the heart rate, acetylcholine also binds to muscarinic receptors located on norepinephrine terminals. The activation of these muscarinic receptors decreases the amount of norepinephrine released, which would typically increase the heart rate. Thus, the antagonistic actions of the sympathetic and parasympathetic nervous systems are controlled at the site of the target organ.

A more complex coordination of sympathetic and parasympathetic responses is found at the central lev-

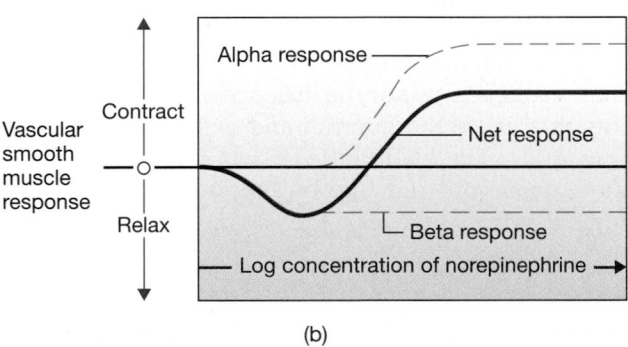

Figure 10–5

Alpha-1 norepinephrine receptors cause smooth muscle cells lining blood vessels to contract, reducing the inner space in the vessel and thereby increasing blood pressure. Beta-1 norepinephrine receptors cause smooth muscle cells to relax, thus widening vessels and decreasing blood pressure (*a*). The net response, however, is a contraction of smooth muscle cells (*b*).

els of the ANS. At the level of the spinal cord and brain stem, these responses are viewed as **autonomic reflexes,** while at the hypothalamic level the responses are viewed as **autonomic control systems.**

Reflexes Governed by the Autonomic Nervous System

Spinal Reflex Arcs Can Activate Autonomic Functions

Earlier in this chapter, we discussed the sensory connections from the target organ to the spinal cord. This pathway transmits information about the status of

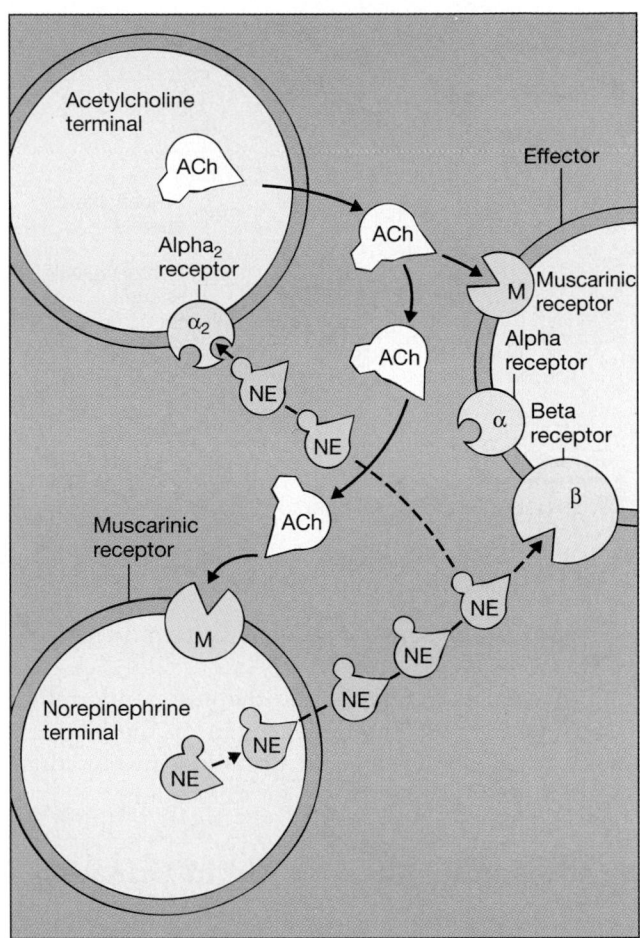

Figure 10–6

Receptors for norepinephrine and acetylcholine located in the presynaptic membranes can inhibit the release of these neurotransmitters, allowing the ANS to enhance the effect of a sympathetic or parasympathetic action.

the target organ using transduction mechanisms similar to those described in Chapter 8. Within the spinal cord, this information is then transmitted up to the brain and to neurons within the IML of the spinal cord. Since the cells in the IML provide the output from the ANS to the target organs, a loop of nerve cell activity is therefore established; this forms the **spinal autonomic reflex arc** (Fig. 10–7). The formation of this neuronal circuit starts with the detection of visceral information by autonomic sensory receptors in the organ. Sensory fibers in turn activate interneurons within the spinal cord. These neurons synapse with the cells of the IML. The IML cells integrate incoming information from many sensory neurons that enter the spinal cord within one spinal cord segment and discharge to activate postganglionic neurons. Finally, the postganglionic cells convey information to the target organ, causing it to make the appropriate responses. This reflex arc is quite similar to other reflex circuits such as the knee-

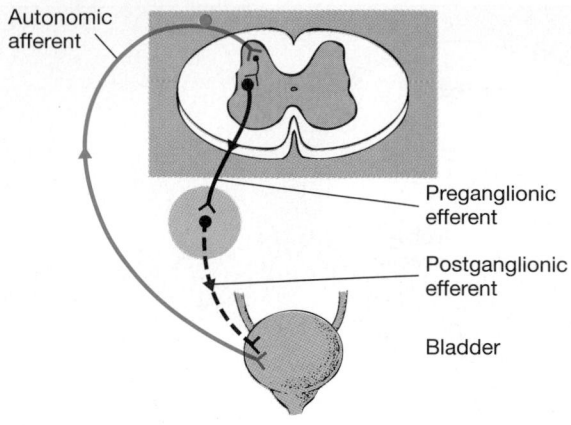

Figure 10–7

A spinal autonomic reflex arc.

jerk reflex of the somatic nervous system (see Chapter 9).

The particular response in the autonomic reflex system varies with the target organ. In the bladder, this reflex is responsible for urination; in the colon and rectum, it is responsible for defecation.

Reflexes in the Bladder Aid in Urination

The bladder stores urine produced by the kidneys (see Chapter 25). It fills at a rate of approximately 50 ml per hour and can store about 150 to 200 ml before the urge to urinate occurs. The ability to keep urine in the bladder is called **continence,** and the emptying of the bladder is called **micturition.**

The bladder (Fig. 10–8*a*) contains a wall of smooth (involuntary) muscle (detrusor) that is innervated by sensory nerve endings. These nerves function as stretch receptors, sensing the enlargement of the bladder. Two groups of muscles close off the exit from the bladder and prevent urine from leaving. The first group is made of smooth muscle and is called the **internal sphincter.** These muscles are innervated by parasympathetic nerve fibers in the pelvic nerves and by sympathetic nerve fibers in the hypogastric nerves. The second group of muscles are striated (voluntary) muscles and are called the **external sphincter.** They are innervated by motor neurons that form the pudendal nerve. These motor neurons are under voluntary control of higher brain centers.

As urine begins to fill and distend the bladder, the stretch receptors become activated (Fig. 10–8*b*) and discharge action potentials. The action potentials from sensory fibers activate preganglionic parasympathetic nerve cells that result in the contraction of the smooth muscles surrounding the wall of the bladder. These actions constitute the urinary bladder reflex in the spinal cord.

In order for the bladder to empty, however, the internal and external sphincter muscles must be placed in a relaxed state. This involves the activation of a "micturition center" in the brain stem by inputs from the sensory stretch receptors that sense the enlargement of the bladder. The output from the micturition center inhibits the preganglionic sympathetic neurons and alpha motor neurons that normally cause the internal and external sphincters to contract. In addition, the micturition center further excites the preganglionic parasympathetic cells that cause the smooth muscles surrounding the bladder to contract and the internal sphincter to relax. Thus, the spinal bladder reflex is aided by the micturition center to coordinate the emptying of the bladder.

Individuals who have been in accidents that crushed their spinal cords can still control the bladder reflex. If the damage is above the level of the sacrum, the spinal reflex is intact. The reflex is initially lost, however, as a result of "spinal shock," but recovers over the next one to five weeks. Urination after spinal cord damage occurs with great frequency and is initiated when the bladder is slightly filled. This occurs because of the absence of descending inputs that cause the internal and external sphincters to contract. The bladder reflex can also be initiated after spinal cord damage by tapping the lower abdomen, which activates the segmental cutaneous-visceral reflex.

Reflexes in the Rectum Aid in Defecation

The colon and rectum store unabsorbed food products or fecal (waste) material from the small intestine (see Chapter 23). The ability to retain fecal material within the colon and rectum is called **continence,** and the removal of fecal material is called **defecation.** Continence is controlled by spinal reflexes and descending inputs from higher brain centers, and is maintained until approximately 2 liters of feces have accumulated.

As with the bladder, the walls of the colon and rectum are surrounded by smooth muscles that contract to expel fecal material. These muscles are innervated by stretch receptors that detect the filling of the rectum (Fig. 10–9). A set of internal and external muscles surrounding the anal opening prevents fecal material from leaving the rectum. The internal sphincters are smooth muscles not under voluntary control, while the external sphincters are striated muscles, under voluntary control and innervated by motor neurons from spinal segments S2 to S4. Typically, both internal and external sphincters are closed. The tonic contraction of the external sphincter is maintained by the activation of the motor neurons from sensory nerves from the skin and from the external sphincter muscle itself.

(a)

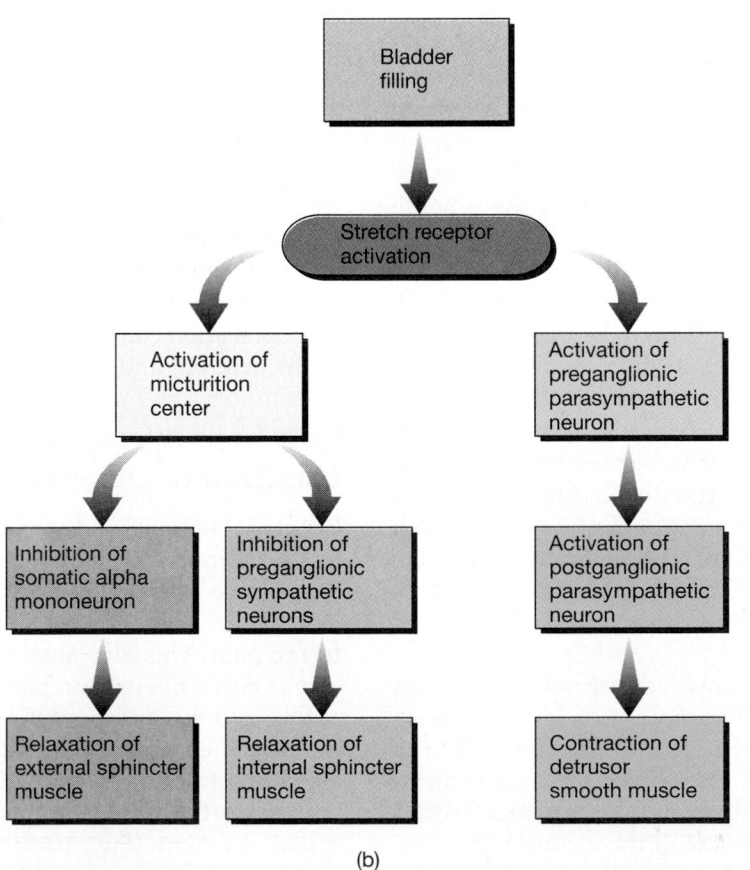

(b)

Figure 10-8

(a) Neural pathways for urination. (b) Mechanisms of urination.

Figure 10–9

The defecation reflex.

As the rectum fills with fecal material, stretch receptors become activated and transmit this information to the spinal cord via the pelvic nerve. Within the spinal cord, these sensory inputs excite preganglionic parasympathetic neurons that in turn excite postganglionic nerve cells innervating the smooth muscles of the colon and rectum. This causes the smooth muscles to contract, forcing fecal material to move toward the anal opening. The sensory stretch receptors also transmit information to higher brain centers to initiate the urge to defecate. These higher centers activate descending inputs to the parasympathetic preganglionic nerve cells to facilitate the defecation reflex. In order to further coordinate the emptying of the rectum, the descending inputs activate preganglionic parasympathetic neurons that lead to the relaxation of the internal sphincters and inhibit the motor neurons of the pudendal nerve that innervates the striated muscles of the external sphincters.

These actions alone, however, are not sufficient for defecation to proceed. The force necessary for the expulsion of feces also comes from the contraction of the abdominal muscles. This contraction causes an increase in the intra-abdominal pressure, which squeezes the contents inside of the rectum and colon. This increase in pressure works in concert with the contraction of the smooth muscles surrounding the colon and rectum, and together they are sufficient to generate the forces necessary for fecal expulsion. Thus, descending inputs from sensory-motor areas of cortex orchestrate defecation by coordinating both autonomic and somatic actions.

The defecation reflex is lost in individuals who have been in accidents that destroy the sacral segments of their spinal cord. However, transections of the spinal cord above the sacrum preserve the reflex, and defecation can occur with the manual expansion of the external and internal sphincters.

Pain from Internal Organs Is Localized to Other Parts of the Body

People who experience stomach pains and heart attacks commonly perceive that their pain comes from the surface of their bodies. Since the pain is attributed to a location other than its source, it is called **referred pain.** This displaced perception of pain is a result of nerve fibers from pain receptors from internal organs and the surface of the body that converge onto the same neurons of the pain pathway. This convergence of information occurs with visceral and somatic pain fibers that enter the spinal cord within the same spinal segment (Fig. 10–10). Pain fibers from the heart, for example, enter the spinal cord at thoracic segments T1 through T3 and synapse with neurons

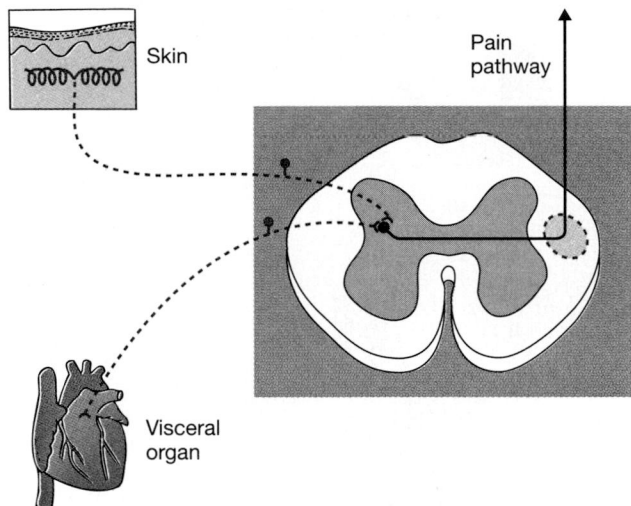

Figure 10–10

An example of a referred pain pathway.

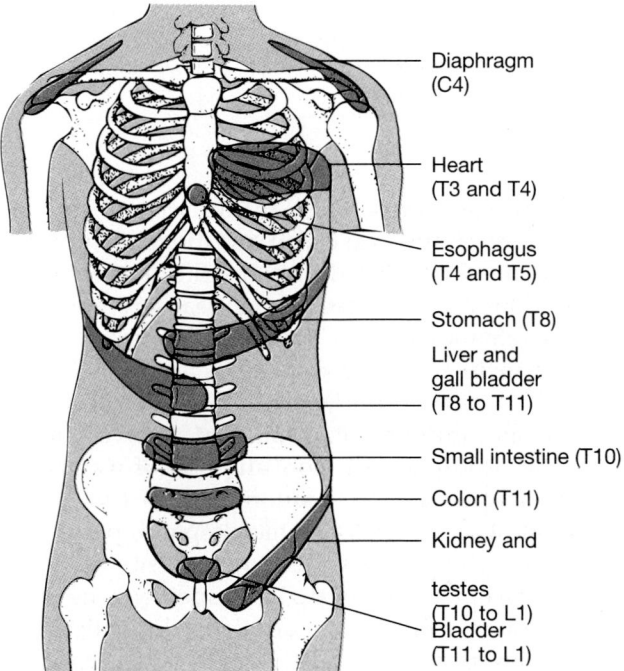

Figure 10–11

Hyperalgesic zones for pain originating in the major internal organs.

Diaphragm (C4)

Heart (T3 and T4)

Esophagus (T4 and T5)

Stomach (T8)

Liver and gall bladder (T8 to T11)

Small intestine (T10)

Colon (T11)

Kidney and

testes (T10 to L1)
Bladder (T11 to L1)

Figure 10–12

An example of a segmental spinal autonomic reflex, in which reddening of the skin signals internal inflammation.

in the **substantia gelatinosa** of the pain pathway (see Chapter 8). Pain information from the left chest and the upper portion of the left arm also synapse with these same neurons. Consequently, since the brain has come to associate pain information from spinal segments T1 through T3 with the left chest and arm, it continues to do so, and thus heart attacks are often perceived as pain in the left arm and chest.

The painful areas on the surface of the body that are associated with inflammations of internal organs are called **hyperalgesic zones.** Figure 10–11 shows the hyperalgesic zones for pain originating in the heart, stomach, kidney, liver, gallbladder, and other internal organs. The hyperalgesic zones are useful in diagnosing disorders of visceral organs.

In addition to the increased pain associated with the hyperalgesic zone, the inflammation of internal organs often causes the skeletal muscles in this zone to tighten and the skin to redden (Fig. 10–12). These signs are often used to help determine which organ is in distress. The reddening of the skin and the tightening of the muscles are results of reflex connections. These connections occur between sensory fibers from the internal organ that synapse with neurons activating skeletal muscles to cause contractions. The inputs from sensory fibers also inhibit preganglionic sympathetic neurons and thus decrease vasoconstriction and increase blood flow to the skin. The segmental spinal reflex connections, therefore, occur within the appropriate spinal cord segments where visceral sensation is received and transmitted out through somatic and autonomic pathways.

Regulatory Systems of the Autonomic Nervous System

Actions of the autonomic nervous system involving the brain stem and hypothalamus are part of **autonomic regulatory systems** that are more complex than spinal cord reflexes. These regulatory systems maintain a constant internal environment for the optimal functioning of the various organs and their cells. Regulatory systems of the body have characteristics similar to those found in a modern home. For example, home thermostats have sensors to determine if the ambient temperature is warmer or colder than the selected temperature. If the ambient temperature is warmer than the desired temperature, the **sensor system** activates the air conditioner to introduce cooler air into the house. On the other hand, if the ambient temperature is colder than the desired temperature, the sensor system activates the heater to provide warmer air. The air conditioner and heater in this example are **effector systems** that work to change the air temperature. The change in air temperature provided by the effector systems always attempts to minimize the difference between the actual and desired temperatures. This type of regulatory system is referred to as a **negative feedback system** and is the basis of control systems in the body (see Chapter 1).

Thirst Is Regulated by Osmoreceptors and Receptors for Water Volume

A large portion of the human body consists of water. Water is lost in the form of sweat and urine. It also leaves the body as water vapor in air exhaled from the lungs. In order to maintain a constant amount of water within the body, thirst mechanisms motivate us to drink water; in addition, the ANS and kidneys act in concert to retain water that would normally be excreted as urine. The amount of water in the blood affects its **osmolarity** (see Chapter 2). Consequently the regulation of thirst, water volume, and osmolarity is interrelated.

The osmolarity of the blood is a critical factor in maintaining the internal environment of the cells in the body. As discussed in Chapter 4, hyper- and hypoosmotic fluids cause cells to shrink or swell and alter their function. The desired osmolarity for blood plasma is 300 mOsm. Increases in plasma osmolarity above this value can occur from excessive dehydration and sweating. This increase in osmolarity is detected by osmoreceptors in the **supraoptic nucleus** and the **paraventricular nucleus** of the hypothalamus (Fig. 10–13; see also Fig. 11–1). In response to in-

Figure 10–13

The hypothalamus detects increases in the osmolarity of the blood and corrects them by activating mechanisms that increase the water in the body, such as water consumption and water retention by the kidney, controlled by the secretion of ADH.

creases in osmolarity, these nuclei secrete the hormone **vasopressin,** also called **antidiuretic hormone (ADH),** into the blood. Eventually this hormone acts on the kidneys and causes them to reabsorb more water. The retention of water reduces the plasma osmolarity to maintain it at 300 mOsm.

The activation of hypothalamic osmoreceptors also initiates thirst mechanisms that involve somatic responses leading to the consumption of water. The intake of water in turn reduces plasma osmolarity.

In addition to central osmoreceptors, peripheral stretch receptors act as **water-volume sensors.** These sensory receptors are located in large veins near the heart. Water volume can decrease dramatically if blood is lost through hemorrhages or other accidents. As the water volume decreases, the volume receptors activate centers in the brain stem (Fig. 10–14). These centers activate sympathetic neurons associated with the renal nerve and cause the kidneys to release the enzyme **renin** into the blood stream. In the circulatory system, renin converts the peptide angiotensinogen to angiotensin I (AI). AI in turn is converted to **angiotensin II** (AII), which circulates to the brain and causes the hypothalamus to release ADH into the circulatory system. Eventually, ADH reaches the kidneys to increase water reabsorption. Release of hypothalamic ADH by either osmoreceptors or AII

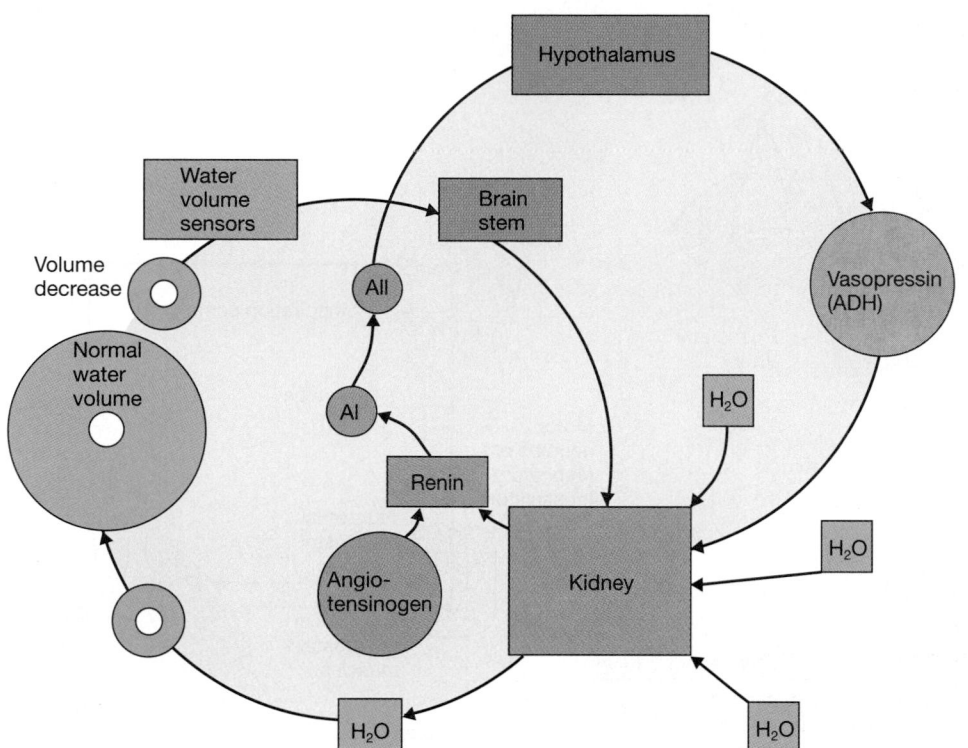

Figure 10–14

Water volume deficiencies are detected by sensors near the heart, setting off a series of reactions and mechanisms that increase the water content of the body.

is therefore the key pathway leading to the retention of water by the kidney.

As the feeling of thirst is satisfied, water consumption subsides. This happens, however, long before the water is absorbed into the circulatory system and long before plasma osmolarity is restored to normal. The feeling of being quenched is thought to be triggered by stretch receptors in the stomach. How this information is transmitted to the brain has not yet been determined.

Breathing Is Regulated by Chemoreceptors and Stretch Receptors

Cellular production of energy from glucose requires oxygen. As cells become more active, their need for oxygen and energy increases. Thus, the amount of oxygen available to each cell must be closely regulated to match metabolic demands. In addition, for cells to function properly, the waste products of metabolism, CO_2 and H^+, must be removed from the internal environment. The supply and regulation of oxygen to the body and the removal of CO_2 (and H^+ indirectly) are primary functions of the lungs (see Chapter 20). These functions are carried out in concert with the ANS.

Receptors called **chemoreceptors** in the ANS monitor the amount of O_2 available to the cells and the amount of CO_2 they produce. In the peripheral nervous system there are two types of chemoreceptors: **carotid bodies** and **aortic bodies** (Fig. 10–15*a*). These chemoreceptors are sensitive to levels of O_2 and CO_2. The respiratory muscles respond to the activation of the carotid and aortic bodies by increasing the frequency and depth of breathing. In this way, more oxygen is supplied to the body and more CO_2 is removed (see Chapter 20).

The peripheral chemoreceptors indirectly activate motor neurons that control the muscles for breathing. Peripheral chemoreceptors have inputs to neurons of the respiratory center in the brain stem (see Fig. 10–15*b*) that generate the rhythm of **inspiration** (breathing in) to bring oxygen to the body and **expiration** (breathing out) to remove carbon dioxide. The rhythm-generating neurons in turn have inputs to the motor neurons that control the muscles for respiration (see Fig. 10–15*c*). The muscles used in inspiration are the diaphragm and external intercostal muscles (see Chapter 20).

The carotid and aortic bodies are also sensitive to blood flow. Decreases in blood flow result in a decrease in the amount of oxygen delivered to the chemoreceptors. This decrease activates the chemoreceptors, increases the rate and depth of breathing,

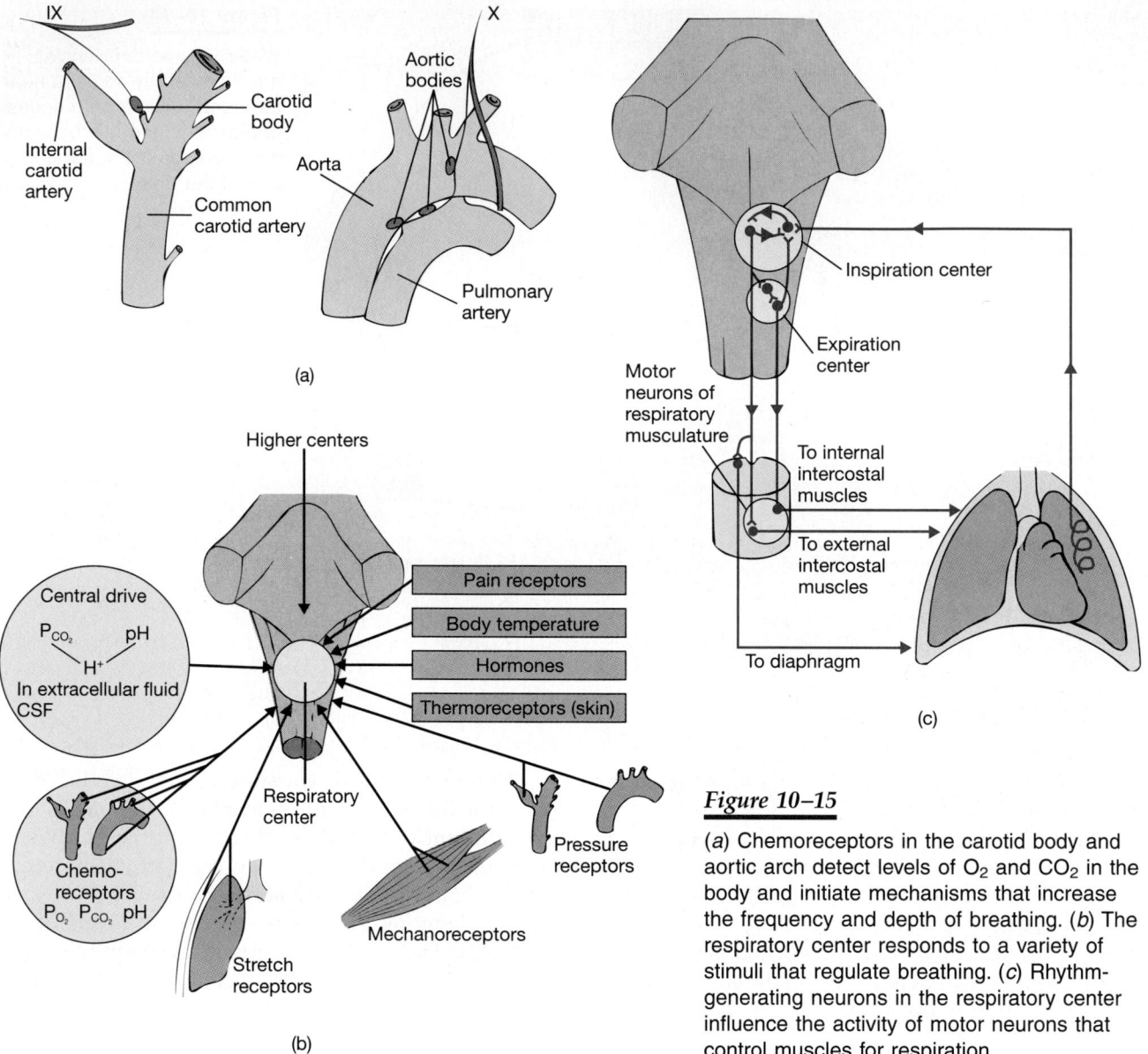

(a)

(b)

(c)

Figure 10–15

(*a*) Chemoreceptors in the carotid body and aortic arch detect levels of O_2 and CO_2 in the body and initiate mechanisms that increase the frequency and depth of breathing. (*b*) The respiratory center responds to a variety of stimuli that regulate breathing. (*c*) Rhythm-generating neurons in the respiratory center influence the activity of motor neurons that control muscles for respiration.

and brings more oxygen to the body. The peripheral chemoreceptors are also sensitive to the temperature of the blood. A rise in temperature stimulates the chemoreceptors and increases the rate of respiration. In some animals, this is an important mechanism for dissipating (releasing) heat and reducing the body's temperature.

In addition to the peripheral chemoreceptors, chemoreceptors are also located within the CNS. As are the carotid and aortic bodies, these central chemoreceptors are sensitive to increases in carbon dioxide and H^+; however, they are not sensitive to decreases in oxygen. The central chemoreceptors are located in the **medulla** of the brain stem and are in contact with the **cerebrospinal fluid (CSF).** As discussed in Chapter 7, the CSF bathes the brain and

provides the extracellular environment for the brain cells. Changes in the chemical composition of the CSF alter the function of nerve cells within the brain.

Most molecules in the blood are isolated from the CSF by the blood-brain barrier (see Chapter 7). However, gases such as oxygen, carbon dioxide, and other lipid-soluble substances are able to penetrate the barrier, going from blood to brain or from brain to blood. The body can therefore regulate carbon dioxide levels in the CSF by removing it in respiration.

The maintenance of sufficient oxygen in the body and the removal of carbon dioxide are thus regulated by chemoreceptors of the autonomic nervous system. Peripheral chemoreceptors detect decreases in oxygen; both peripheral and central chemoreceptors detect increases in carbon dioxide and H^+. Peripheral

CLINICAL FOCUS

The Relaxation Response

The relaxation response is characterized by a decrease in the activity of the sympathetic nervous system, a decrease that results from conditioning and training. During this response, there is a decrease in oxygen consumption, heart rate, blood pressure, and respiration rate, and an increase in the alpha waves of the electroencephalogram (Chapter 11). The relaxation response is not just simple relaxation. In simple relaxation, changes in the rate of respiration, oxygen consumption, and alpha wave activity do not occur.

Learning to induce the relaxation response is a common practice associated with various Eastern religions such as Zen Buddhism. One can induce the relaxation response by sitting in a comfortable position, focusing on an object, and blocking out all other distractions while repeating a word or phrase.

The relaxation response is thought to modify the way in which stressful stimuli affect the sympathetic nervous system. A temporary activation of the sympathetic system is important in preparing the body to respond appropriately to stressful situations by the release of substances such as norepinephrine. The prolonged release of these substances, however, can have deleterious effects that include abnormally high blood pressure and a decrease in immune system function. Research indicates that the relaxation response reduces the body's response to norepinephrine secreted into the circulatory system.

The relaxation response produces changes in the sympathetic nervous system that extend outside of the training period. Clinical studies have shown that the relaxation response can alleviate heart beat disorders, reduce pain, and lower blood pressure in hypertensive patients. If practiced daily, it may alleviate physiological conditions arising from undesirable increases in sympathetic nervous system activity, and thus improve and maintain overall health.

Buddhist monks during morning meditation, Eiheiji Monastery, Japan.   *(© Dieter Blum/Peter Arnold, Inc.)*

■ **What are the potential advantages/disadvantages of these forms of "alternative medicine" in comparison to more conventional medical therapy?**

and central chemoreceptors act together to coordinate the intake of oxygen and removal of carbon dioxide based on the metabolic demands of the body.

Blood Pressure Is Regulated by Baroreceptors

The maintenance of blood pressure is necessary for blood to flow and provide oxygen and nutrients to the organs of the body. The desired average arterial pressure is 90 mm Hg. Blood pressure is sensed by stretch receptors called **baroreceptors** located on structures in the heart called the **aortic arch** and **carotid sinus.** Pressures above 90 mm Hg cause the stretch receptors to expand and discharge more action potentials, similar to stretch receptors described in Chapter 8. Pressures below 90 mm Hg cause the receptors to contract and discharge fewer action potentials.

The response to changes in blood pressure is complex, involving several areas of the brain. When the pressure is greater than 90 mm Hg, the increased action potential activity in the fibers from baroreceptors activates a cardiovascular regulatory center in the brain stem (Fig. 10–16*a*). This center is called a **depressor center** because it causes a reduction in blood pressure. Neurons in the depressor center synapse with preganglionic sympathetic nerve cells to inhibit their activation. This inhibition of sympathetic nerve cells reduces the heart's force of contraction and rate of contraction, and reduces the vasoconstriction of blood vessels. The depressor center

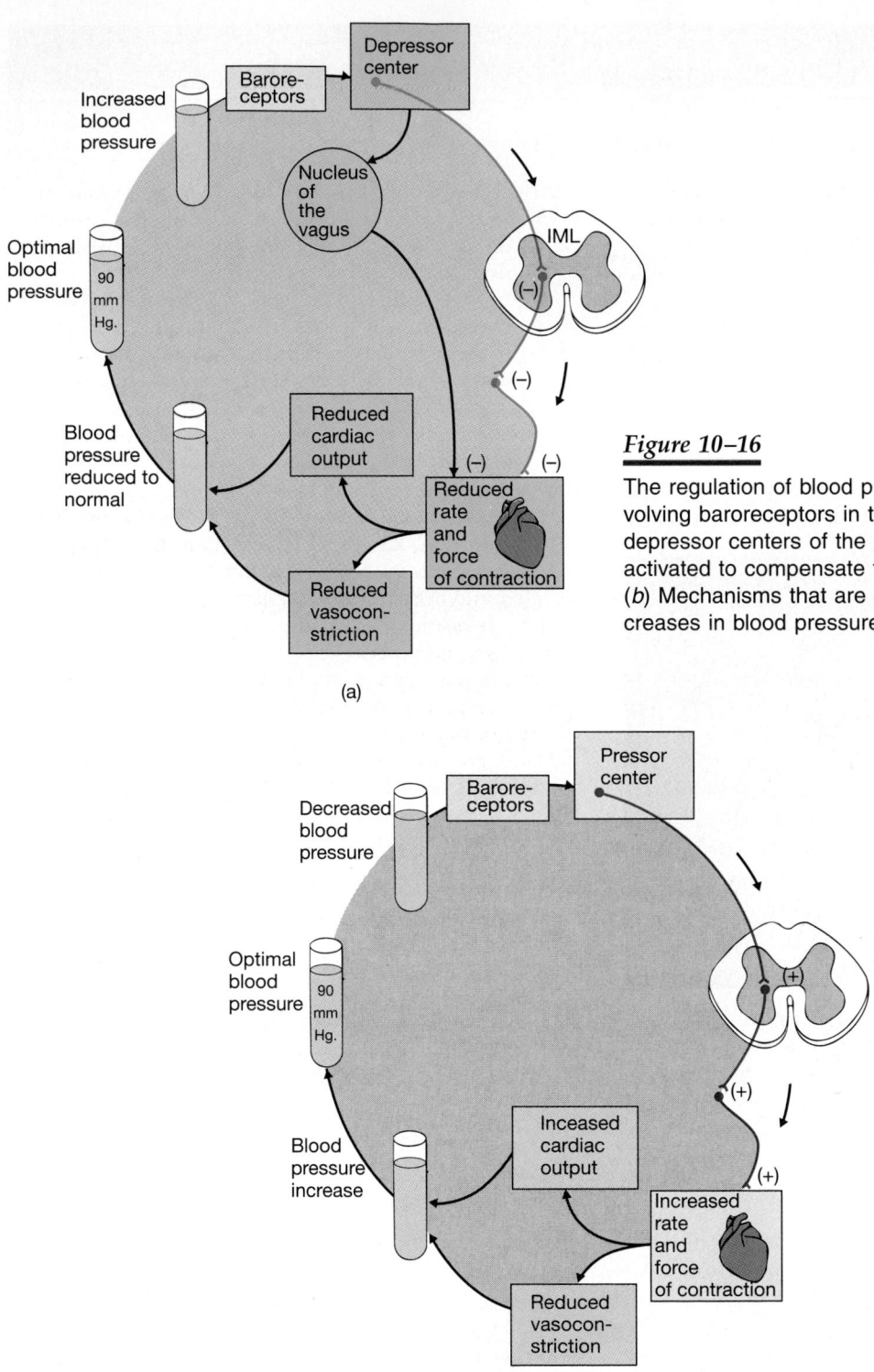

Figure 10-16

The regulation of blood pressure is a complex process involving baroreceptors in the heart and the pressor and depressor centers of the brain. (*a*) Mechanisms that are activated to compensate for increases in blood pressure. (*b*) Mechanisms that are activated to compensate for decreases in blood pressure.

also activates another region of the brain stem called the **nucleus of the vagus,** which directly reduces the heart rate. The net result of these actions is a reduction in blood pressure.

Decreases in blood pressure below 90 mm Hg (Fig. 10–16*b*) will cause the baroreceptors to activate

the cardiovascular **pressor center,** located in the brain stem. The pressor center attempts to increase blood pressure in response to sudden decreases such as those due to hemorrhages. The pressor center in turn stimulates the preganglionic sympathetic nerve cells to increase heart rate and the force of the heart's contraction. In addition, it causes blood vessels to constrict. The net effect of these actions is to increase blood pressure in response to sudden decreases detected by the baroreceptors.

The hypothalamus affects the regulation of blood pressure by its excitatory inputs to the pressor center and preganglionic sympathetic neurons to increase blood pressure. This action of the hypothalamus, however, is only one of many functions in its regulation of autonomic activity.

Autonomic Functions of the Hypothalamus

The activation of different regions of the hypothalamus produces a variety of coordinated autonomic responses. The activation of the **dorsal hypothalamus,** for example, increases blood pressure, intestinal motility, and intestinal blood supply but decreases the blood supply to the skeletal muscles (Fig. 10–17). These responses are associated with feeding behaviors. Activation of the **ventral hypothalamus** increases blood pressure and the blood supply to the skeletal muscles but decreases intestinal motility and blood flow to the intestines. These responses are associated with defense behaviors or the so-called **"fight or flight responses"** (see Table 10–2). This response prepares the body for the stress of threatening situations. For example, at the onset of intense pain or sudden fright, blood vessels to the skeletal

Table 10–2

Physiological Changes in the Fight or Flight Response

Fight or Flight Response

1. Increased blood pressure
2. Increased heart rate
3. Increased force of heart contraction
4. Increased heart conduction velocity
5. Increased depth and rate of respiration
6. Shift of blood flow distribution away from the skin and splanchnic regions and more to skeletal muscles and heart (and, of course, brain)
7. Mobilization of liver glycogen to glucose (glycogenolysis)
8. Mobilization of free fatty acids from adipose tissue (lipolysis)
9. Contraction of spleen capsule (increased hematocrit)
10. Mydriasis (widening of pupil)
11. Accommodation for far vision (relaxation of ciliary muscle)
12. Widening of palpebral fissure (eyelids open wide)
13. Piloerection
14. Inhibition of gastrointestinal motility and secretion, contraction of sphincters
15. Sweating ("cold" sweats as skin blood vessels are constricted)

Figure 10–17

The hypothalamus coordinates the functions of different organs. Activation of the dorsal regions of the hypothalamus results in "feeding" behaviors, while activation of ventral regions produces the "fight or flight" response to a threat.

muscles dilate, and heart rate increases along with an increase in the force of heart contraction. These actions increase blood flow to the skeletal muscles in preparation for running or fighting. In addition, energy sources are mobilized to provide more fuel for the skeletal muscles and heart. The liver produces more glucose, for example; adipose tissue is converted to free fatty acids; and the rate and depth of respiration are increased to provide more oxygen to the cells in the body.

Hypothalamic Control of the Adrenal Gland

The coordinated actions of the hypothalamus during the stress response are mediated in part by the activation of the adrenal gland. As shown in Figure 10–18, the adrenal gland secretes catecholamines (epinephrine and norepinephrine) and glucocorticoids in response to stress. The mechanisms by which the hypothalamus causes the release of these adrenal hormones, however, are quite different. The adrenal gland is composed of an outer cortical area called the **adrenal cortex** and a central area called the **adrenal medulla** (see Chapter 14). The cells in the adrenal cortex secrete **glucocorticoids** (as well as other steroids), while the medullary cells secrete catecholamines. The secretion of glucocorticoids is mediated by a cascade of hormones, beginning with the release of **corticotropin-releasing hormone (CRH)** from the hypothalamus. CRH, in turn, causes the release of **adreno-**

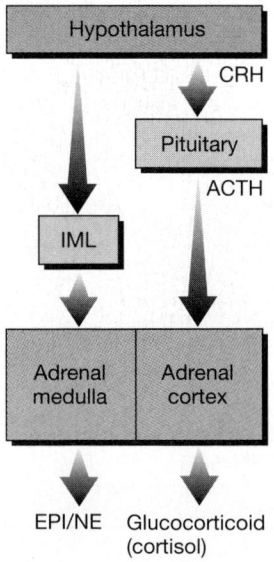

Figure 10–18

Hypothalamic control of the adrenal gland during the stress response.

corticotropic hormone (ACTH), which then causes the secretion of glucocorticoids from the adrenal cortex. In contrast, the release of catecholamines from the adrenal medulla is carried out by the direct connection of nerve fibers from the hypothalamus to the IML, and by nerve fiber from the IML to cells in the medulla. The catecholamine-secreting cells of the adrenal medulla therefore function as postganglionic sympathetic cells. The cells of the adrenal medulla are homologous to postganglionic neurons since they are directly innervated by preganglionic sympathetic nerve fibers. As previously mentioned, the cells of the adrenal gland, when activated, release the catecholamines **epinephrine** and **norepinephrine** into the circulatory system. (Epinephrine-synthesizing cells represent 80% of the cells in the adrenal medulla while the remaining 20% synthesize norepinephrine.)

The catecholamines that circulate in the blood affect the same receptors and target organs as postganglionic sympathetic nerve cells (Fig. 10–19). Because they enter the blood, however, their actions are not as discrete as those generated by sympathetic nerve cells. Circulating catecholamines cause the air passages in the lungs to dilate and increase the supply of oxygen to the skeletal muscles, heart, and brain. In the cardiovascular system, circulating catecholamines cause the heart to beat faster and with greater force; they also cause the constriction of arterioles in the skin, mesentery, and kidney. In contrast, the arterioles of the skeletal and coronary (heart) muscles dilate. The net effect of these actions on the cardiovascular system is to provide more blood to skeletal muscles, heart, and brain and to reduce blood flow to the vegetative organs in the body. Since one of the primary functions of blood is to carry oxygen from the lungs to the other organs of the body, the increase in blood flow also provides more oxygen.

Circulating catecholamines also increase the supply of fuel for cells in the form of glucose, free fatty acids, and lactate. The action of catecholamines on the liver causes an increase in **glycogenolysis,** a process that increases the amount of glucose in the blood (see Chapter 14). Circulating catecholamines cause adipose or fat tissue to release free fatty acids. They also cause skeletal muscles to twitch and produce lactate. The net effect of these metabolic actions is to increase the substrates for cells in the skeletal muscles, heart, and brain to maintain their increased activity.

Several factors influence the function of the adrenal medulla. One is an individual's emotional state. As you will see in Chapter 11, the brain center for emotion is the hypothalamus. Stress-induced increases in nerve cell activity in the hypothalamus activate preganglionic sympathetic neurons. These pre-

R E S E A R C H   F O C U S

Chronic Stress and the Loss of Brain Cells

Glucocorticoids are steroids released by the adrenal cortex during stress. The work of Robert Sapolsky and his colleagues at Stanford University has shown that prolonged elevation of glucocorticoids can lead to the degeneration of nerve cells in rodents and monkeys. The area of the brain most sensitive to glucocorticoid damage is the hippocampal formation, a region long known to be involved in learning and memory. Earlier studies by Bruce McEwen and coworkers at the Rockefeller University revealed that the hippocampus is a major target site in the brain for circulating glucocorticoids.

In studies of vervet monkeys housed in primate colonies in Kenya, Sapolsky and his coworkers found that socially subordinate monkeys exhibited gastric ulcers, high incidences of bite wounds, and hyperplastic adrenal cortices suggestive of sustained glucocorticoid release. Moreover, these characteristic signs of social stress were associated with a severe degeneration of nerve cells within the hippocampal formation.

In experiments to determine whether glucocorticoid secretion alone could lead to neuronal loss, glucocorticoid-containing pellets were implanted directly into the hippocampi of non–socially stressed monkeys. These animals ultimately displayed the same type of nerve cell loss as socially stressed monkeys.

Other studies have shown that chronic stress causes premature aging of hippocampal brain activity. Conversely, reducing the exposure of glucocorticoids can protect against the age-related loss of hippocampal neurons. Furthermore, glucocorticoids in concentrations that are not in themselves toxic can impair the ability of hippocampal nerve cells to survive neurological insults such as reductions in blood flow and oxygen to the brain, and seizures. Together these studies suggest that prolonged exposure to glucocorticoids or stress can accelerate the loss of hippocampal neurons as well as increase their susceptibility to neurological insults.

■ **What changes in your lifestyle would you implement to reduce chronic stress?**

 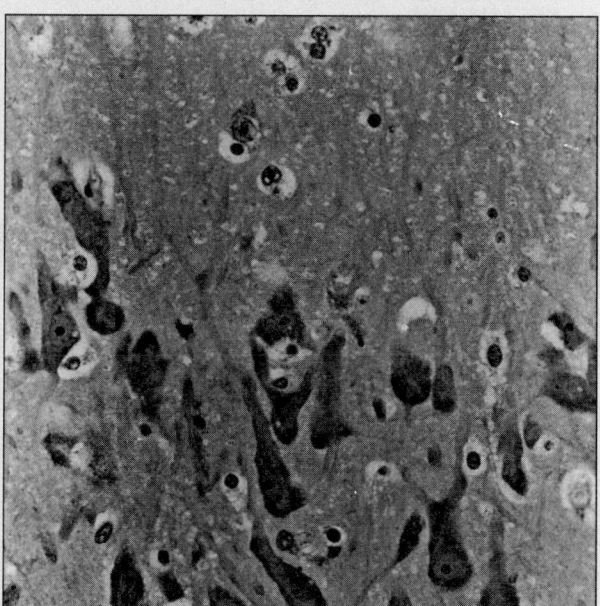

Effects of chronic stress on brain cells. Left panel shows the normal nerve cells of the hippocampal formation. The right panel shows the degenerating hippocampal nerve cells from the brain of a chronically stressed monkey. *(Hideo Uno, M.D., Ph.D.)*

ganglionic neurons in turn innervate and activate the adrenal medulla, causing the secretion of epinephrine and norepinephrine into the blood stream. Other factors such as the extreme loss of blood, hypothermia, hypoglycemia, and hypoxia will also activate the adrenal medulla in attempts to correct functional disorders and maintain a constant internal environment (see Chapter 14).

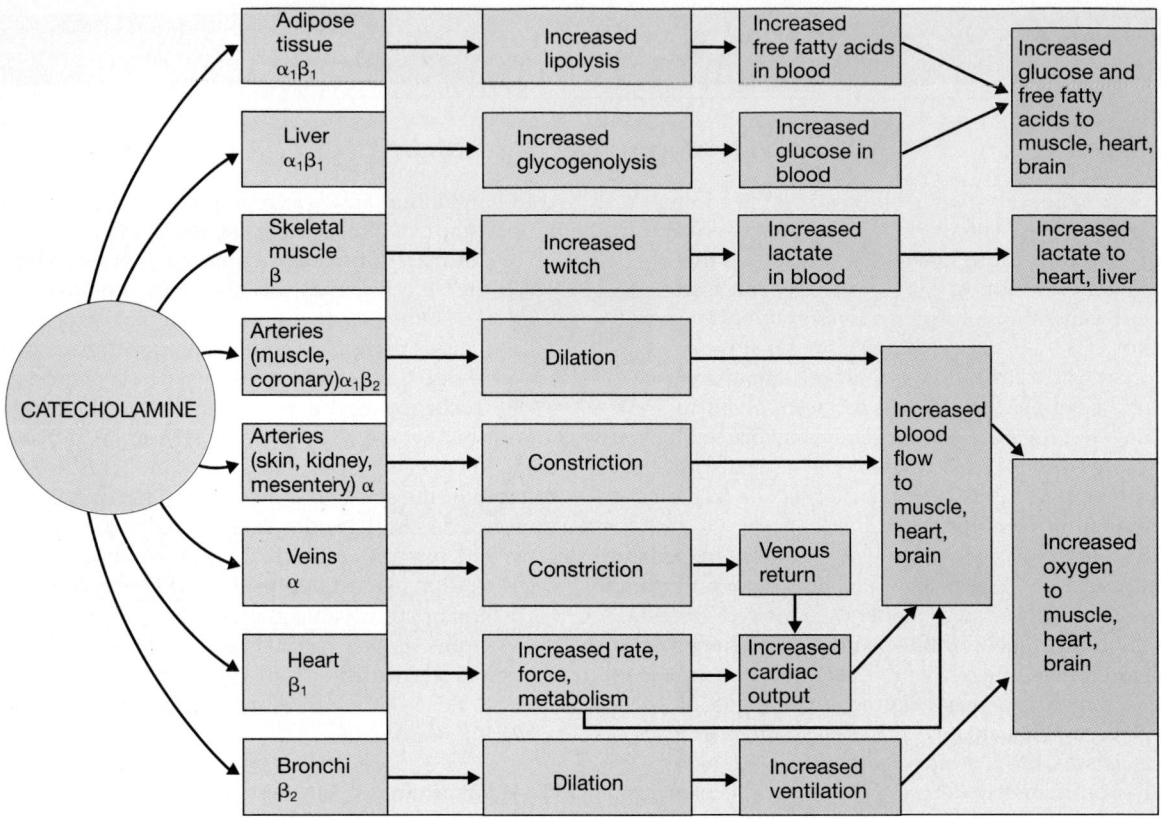

Figure 10–19

Circulating catecholamines produced by the adrenal medulla affect the same receptors and target organs that postganglionic sympathetic neurons do.

CHAPTER REVIEW

Summary

- The autonomic nervous system (ANS), a regulatory system that attempts to maintain a constant internal environment, has both central and peripheral components.
- The central component of the ANS consists of the hypothalamus, brain stem, and spinal cord. The peripheral ANS can be further subdivided into the *sympathetic* and *parasympathetic* nervous systems. The sympathetic nervous system prepares the body to respond to stressful conditions, while the parasympathetic nervous system prepares the body for vegetative functions such as digestion. The sympathetic and parasympathetic nervous systems are anatomically quite different.
- Both preganglionic sympathetic nerve fibers and preganglionic parasympathetic fibers release acetylcholine from their nerve terminals. Postganglionic sympathetic nerve fibers typically release norepinephrine from their nerve terminals, while postganglionic parasympathetic nerve fibers release acetylcholine.

- Organs with vegetative functions are activated by parasympathetic inputs and inactivated by sympathetic inputs. Organs involved in stress responses are activated by sympathetic inputs and inactivated by parasympathetic inputs.
- Autonomic reflexes are carried out primarily in the spinal cord. Reflexes such as the urinary bladder reflex and the defecation reflex are activated by stretch receptors and play important roles in coordinating the actions of smooth muscles and sphincters.
- The convergence of pain information from internal organs and regions on the surface of the body is the basis of referred pain.
- Osmoreceptors that detect plasma osmolarity are located in different parts of the hypothalamus and regulate thirst, osmolarity, and water volume. Brain stem responses involved in respiration are activated by chemoreceptors sensitive to oxygen, carbon dioxide, and

H$^+$. Brain stem responses that control blood pressure are activated by stretch receptors that sense pressure changes.

- The hypothalamus is the center for the "fight or flight" response. It coordinates the variety of bodily changes needed for survival during life-threatening situations. It increases blood flow and oxygen to the brain, heart, and skeletal muscles while reducing blood flow to the organs involved with the digestion of food.

- The adrenal gland coordinates the body's response to stress by releasing epinephrine, norepinephrine, and glucocorticoids into the circulatory system. Epinephrine and norepinephrine in the blood stream have effects similar to the activation of sympathetic nerve fibers; the onset of their actions, however, is delayed, since they must course through the circulatory system.

Review Questions

Identify Each with a Term

1. The phrase used to describe the coordinated physiological changes seen in response to stress.
2. The hormones released from the adrenal medulla during stress.
3. The neurotransmitter released by postganglionic parasympathetic nerve terminals.

Define Each Term

5. sympathetic nervous system
6. parasympathetic nervous system
7. referred pain
8. viscera

Choose the Correct Answer

9. Which type of receptor is found on the target cells of postganglionic parasympathetic nerve fibers?
   a. nicotinic receptor
   b. muscarinic receptor
   c. alpha receptor
   d. beta receptor
10. A drug that enhances the release of norepinephrine from nerve terminals in the heart would:
    a. increase heart rate.
    b. decrease heart rate.
    c. increase the force of contraction.
    d. both (a) and (c)
11. An increase in plasma osmolarity above 300 mOsm
    a. is detected by osmoreceptors in the aortic arch.
    b. causes the release of norepinephrine by the hypothalamus.

   c. increases the reabsorption of water by the kidney.
   d. both (a) and (c)
12. A drug that activates alpha receptors would:
    a. constrict blood vessels in skeletal muscles.
    b. dilate coronary blood vessels.
    c. cause renin secretion from the kidney.
    d. contract bronchial muscles in the lung.

Answer Each Question in One or Two Sentences

13. Describe the conditions of the extracellular environment needed for optimal cellular function, and how these conditions are maintained by the autonomic nervous system.
14. Describe the role of the autonomic nervous system in the process of urination.
15. How does the autonomic nervous system regulate thirst?
16. What are the physiological changes that occur during the "fight or flight" response?

Problem Solving in Physiology

17. What are the differential effects of drugs that inhibit muscarinic vs. nicotinic cholinergic receptors in the autonomic nervous system?
18. Strokes are a leading cause of death in this country. Research has shown that hypertension is a major risk factor for developing a stroke. What drug therapy would you suggest to reduce blood pressure?

Suggested Readings

Appenzeller, O. *The Autonomic Nervous System*, 3rd ed. New York: Elsevier, 1982.

Ciriello, J., Calaresu, F.R., Renaud, L.P., and Polosa, C. *Organization of the Autonomic Nervous System—Central and Peripheral Mechanisms*. New York: Alan R. Liss, Inc., 1987.

Gabella, G. *Structure of the Autonomic Nervous System*. London: Chapman and Hall, 1976.

Gebber, G.L., and Barman, S.M. "Lateral tegmental field neurons of cat medulla: A potential source of basal sympathetic nerve discharge." *Journal of Neurophysiology*, 54:1498–1512, 1985.

Hilton, S.M., and Smith, P.R. "Ventral medullary neurones excited from the hypothalamic and mid-brain defence areas." *Journal of the Autonomic Nervous System*, 11:35–42, 1984.

Irwin, M. "Stress-induced immuno suppression: Role of brain corticotropin releasing hormone and autonomic nervous system mechanisms." *Advances in Neuroimmunology*, 4:29–47, 1994.

Janig, W. "The Autonomic Nervous System." In *Human Physiology*, 2nd ed., edited by R.F. Schmidt and G. Thews, 333–370. Berlin: Springer-Verlag, 1989.

Kregel, K.C., Stauss, H., and Unger, T. "Modulation of autonomic nervous system adjustments to heat stress by central ANG II receptor antagonism." *American Journal of Physiology*, 266:R1985–R1991, 1994.

Reis, D.J., Ruggiero, D.A., and Morrison, S.F. "The C1 area of rostral ventrolateral medulla: A critical brainstem region for control of resting and reflex integration of arterial pressure." *American Journal of Hypertension*, 2:363S–374S, 1989.

Rust, G., Burgunder, J.M., Lauterburg, T.E., and Cachelin, A.B. "Expression of neuronal nicotinic acetylcholine receptor subunit genes in the rat autonomic nervous system." *European Journal of Neuroscience*, 6:478–485, 1994.

Sapolsky, R.M., Uno, II., Rebert, C.S., and Finch, C.E. "Hippocampal damage associated with prolonged glucocorticoid exposure in primates." *Journal of Neuroscience*, 10:2897–2902, 1990.

Schramm, L.P. "Spinal factors in sympathetic regulation." In *The Molecular Basis for the Central and Peripheral Regulation of Vascular Resistance*, edited by A. Magro, 303–352. New York: Plenum Press, 1986.

Uno, H., Tarara, R., Else, J.G., Suleman, M.A., and Sapolsky, R.M. "Hippocampal damage associated with prolonged and fatal stress in primates." *Journal of Neuroscience*, 9:1705–1711, 1989.

Key Terms

adrenocorticotropic hormone (ACTH) (p. 346)

alpha receptor (p. 332)

baroreceptor (p. 343)

beta receptor (p. 332)

carotid sinus (p. 343)

chemoreceptor (p. 341)

corticotropic-releasing hormone (CRH) (p. 346)

epinephrine (p. 332)

"fight or flight" response (p. 345)

glucocorticoids (p. 346)

hypothalamus (p. 329)

interomedial lateral nuclei (IML) (p. 329)

medulla (p. 342)

parasympathetic nervous system (p. 329)

referred pain (p. 338)

sympathetic nervous system (p. 329)

vagus (p. 344)

vasopressin (p. 340)

Human brainwaves during REM sleep (dreaming state).
(© SIU/Visuals Unlimited)

Chapter 11

Central Integrative Systems

Key Concepts

Many bodily functions are driven by central integrative systems within the brain that exhibit circadian rhythms. These integrative systems have their own periodicities but can be entrained by exposure to light-dark cycles.

Motivational systems are activated in response to the body's need to maintain fluid levels and the supply of energy to fuel cellular processes, as well as, for procreation and the activation of reward pathways in the brain.

The development of mechanisms for learning and memory is critical to the adaptation of new situations and environments. Various areas of the brain are responsible for storing and retrieving different types of information.

Language is localized in the dominant hemisphere, typically the left hemisphere for most people.

Many neurological functions in humans are located in either one side of the brain or the other. Spatial and musical abilities are predominantly located on the right side of the brain.

Chapter Outline

Bodily Rhythms, Cyclic Systems, and the Hypothalamus
Motivational Systems: Maintaining an Internal Balance and Ensuring Survival

Learning and Memory Systems
Language Systems
    Laterality of Brain Function

In the movies, a beautiful young woman suffers a bump on her head and suddenly can't remember her name, where she lives, or how she got the fur coat and diamonds. Fortunately, the amnesia is temporary. Just before she is arrested for theft, her fiancé rescues her and, at the instant of his kiss, her memory floods back. A hackneyed plot, but can it really happen?

Researchers are still trying to unlock the mysteries of memory and learning. Why is some memory loss selective—for example, a head trauma victim might not be able to recognize loved ones yet remembers how to drive a car or ride a bicycle? Why do some stroke victims lose the ability to read aloud yet are still able to speak? Why are some patients with "split brains" unable to identify an object they see yet are capable of stating its name if they touch it? Why are patients with damage to language centers of the brain unable to read English, yet Chinese patients are still capable of reading the Chinese language? How do drugs such as cocaine and heroin modify brain chemistry and behavior and cause addiction?

Higher functions such as learning and memory, reading and writing, and reward-motivated behaviors are carried out by regions of the brain that act as central integrative systems. These integrative systems receive, synthesize, store, and transmit neural information that allows us to carry out functions relevant to daily activities.

In the previous chapters, we saw how the nervous system detects and processes sensory information from the external environment and how the motor system puts out information to skeletal muscles for locomotion and fine movement. These sensory and motor functions were described one by one, suggesting that they occur in isolation. In reality, however, human behavior is highly coordinated in order to carry out purposeful actions. The brain areas involved in these coordinated behaviors are part of the **central integrative systems.** Central integrative systems can be subdivided into several broad categories. Motivational systems, for example, are those responsible for our drive to satisfy such basic needs as hunger and thirst. Other systems carry out functions such as learning and memory, which endow us with the ability to acquire and store new information. Still others provide us with the ability to communicate in the form of written and spoken language. In this chapter, we will consider the brain areas and the mechanisms responsible for these integrative functions. We will also examine how the brain brings together and integrates information from its two hemispheres, and how functions are localized on one side of the brain. We begin with how the brain controls bodily rhythms by acting as a master clock.

Bodily Rhythms, Cyclic Systems, and the Hypothalamus

The previous chapter described the role of the hypothalamus in maintaining the body's internal environment. (Its role in neuroendocrine functions will be discussed in Chapter 13.) In this chapter, we will consider the role of the hypothalamus in controlling biological rhythms and motivational systems. The hypothalamus is composed of many different cell clusters (Fig. 11–1), several of which are involved in regulation of body rhythms.

Rhythms of the body vary on daily or monthly cycles. **Monthly cycles** are those seen in the process of ovulation, in which the female body prepares itself for reproduction by the release of gonadotropic hormones from the hypothalamus (see Chapter 30).

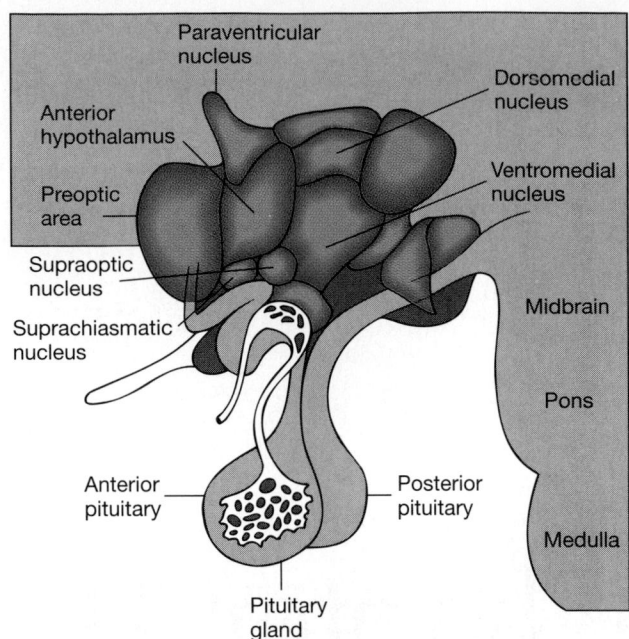

Figure 11–1

The hypothalamus comprises several different cell clusters or nuclei, some of which are involved in central integrative systems of the body.

Diurnal (daily) or **circadian** (*circa* = "approximately" and *dies* = "day") **rhythms** are those that vary on a daily basis.

Biological Clocks in the Brain Regulate Circadian Rhythms

Well over 100 bodily functions and biochemical processes have been found to vary in accordance with a 24-hour schedule. Body temperature, for example, decreases during the night and increases during the day (Fig. 11–2). This rhythmic variation is controlled by the hypothalamus, which influences the secretion of the hormone **thyroxin** by the thyroid gland. This hormone in turn speeds up the rate at which cells consume oxygen, metabolize glucose, and produce heat. The control by the hypothalamus of other hormones such as **ACTH** and **cortisone** has a similar circadian rhythm (see Chapter 14).

Many of the body's rhythms have periods that are longer or shorter than 24 hours. The sleep-wake cycle, for instance, has a 25-hour period. The body's rhythms, however, are entrained and synchronized to external factors such as the 24-hour light-dark cycle, or to our work schedules.

The "master clock" thought to be responsible for coordinating various biological rhythms is the **suprachiasmatic nucleus** of the hypothalamus (see Fig. 11–1). This nucleus receives information about light and darkness directly from the retina of the eyes.

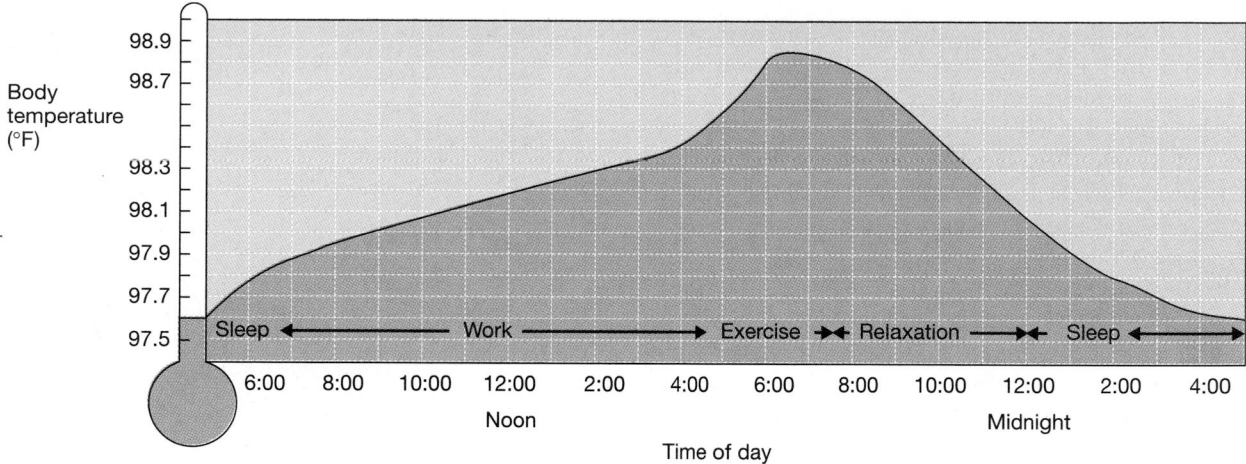

Figure 11–2

Body temperature shows a rhythmic variation throughout the day and night.

By this pathway, the light-dark cycle entrains the body's biological clock. The destruction of this nucleus in experimental animals disrupts their sleep-wake cycle, along with many other bodily functions.

The uncoupling of the body's rhythms can result in fatigue and poor mental and physical performance. On long airplane flights, most of us have experienced the fatigue of "jet lag." When we cross different time zones, our bodily functions are no longer synchronized with the day-night cycle; hormonal fluctuations, body temperature, and the sleep-wake cycle are desynchronized. Eventually, the synchrony of these cycles is reestablished and the body's rhythms again function in harmony.

Sleep Has Different Stages

Despite research efforts of many scientists, the need for a sleep-wake cycle remains a mystery. Sleep has various stages in which the nerve cells are very ac-

tive. According to some theories, the increased activity of certain nerve cells during sleep creates the internal stimuli necessary for the maturation of the brain during development. Other theories suggest that sleep and dreams strengthen the nerve cell connections associated with the memories that were formed during the day.

Stages of Sleep

During one night of sleep, the brain displays several different stages of electrical activity (Fig. 11–3). This activity can be monitored by electrodes placed on the scalp that produce a record, called an **electroencephalogram,** or **EEG,** of brain activity. The EEG of a person who is awake exhibits electrical activity that is low in amplitude and high in frequency. During the onset and progression of sleep, the frequency of the electrical activity decreases while the amplitude increases.

(a)

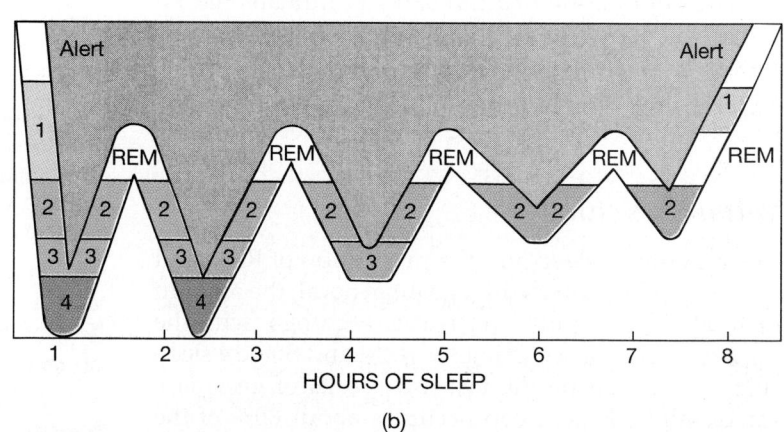

(b)

Figure 11–3

(a) The electrical activity of the brain during the various stages of sleep can be recorded by means of electroencephalograms. (b) During one night's sleep, a person goes through three to five cycles, each cycle including several sequential sleep stages.

Stages 1 to 4 Based on the changes of frequency and amplitude of the brain's electrical activity, four distinct stages of sleep have been identified. Stage 1 of sleep has the lowest amplitude and highest frequency of electrical activity, while Stage 4 has the highest amplitude and lowest frequency. An individual passes through each stage in sequence an average of 3 to 5 times during the night. In the first 45 minutes of sleep, the brain progresses from Stage 1 to Stage 4. During this progression, the skeletal muscles are relaxed, but a sleeper frequently tosses and turns to readjust the sleeping position. In addition, heart rate and blood pressure decrease during Stage 4, while gastrointestinal motility increases. (These latter functions are associated with the parasympathetic activation of the autonomic nervous system, discussed in Chapter 10).

REM Sleep. During the next 45 minutes of sleep, the electrical activity of the brain reverts back from Stage 4 to Stage 1. In Stage 1 of this sleep cycle, the skeletal muscles, except those that control the movement of the eyes, are completely inhibited from moving. In fact, eye movement becomes very rapid during this phase of sleep and it is thus referred to as **rapid eye movement (REM)** sleep. During REM sleep, heart rate and blood pressure also increase, while gastrointestinal motility decreases. These changes are associated with the activation of the sympathetic division of the autonomic nervous system (ANS).

In general, REM sleep is associated with visual dreaming. One night's sleep contains about 3 to 5 REM episodes. Each successive episode is longer and is characterized by increases in the electrical activity of the brain and the number of visual images. Dreams are primarily visual; however, people who are born blind have auditory dreams. Individuals who lose their sight eventually lose their ability to dream visually.

The deprivation of REM sleep in humans has no apparent long-term effects on behavior. When permitted to sleep undisturbed after REM deprivation, subjects initially exhibit more REM sleep. Normal sleeping patterns, however, are soon reestablished.

Variations with Age

The amount of sleep and the proportion of REM and non-REM sleep vary with age. In general, the amount of total sleep required decreases with age, with the greatest changes occurring for REM and Stage 4 sleep (Fig. 11–4). During the last few weeks of gestation, for example, REM sleep occupies about 80% of the total sleep time. This amount decreases to 50% for full-term newborns and levels off at 25% by 10 years of age. The amount of total sleep time spent in Stage 4 also decreases rapidly with age. Unlike REM sleep,

(a)

(b)

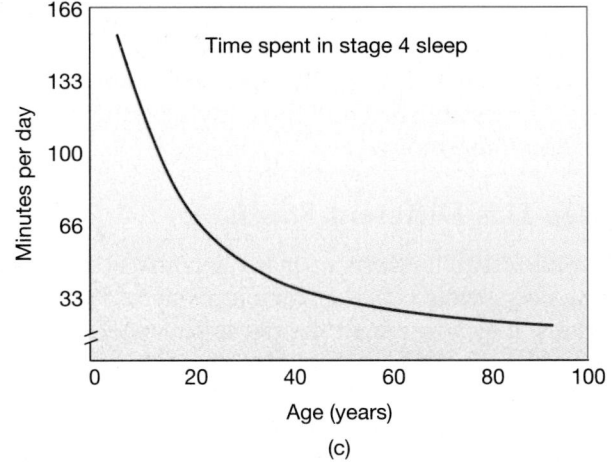

(c)

Figure 11–4

Total sleep time decreases with age, as does the percentage of total sleep time spent in REM sleep and in Stage 4.

however, Stage 4 sleep often disappears after 60 years of age.

Neural Mechanisms of Sleep

The areas of brain responsible for the sleep-wake cycle are the hypothalamus and brain stem. Electrical stimulation of the **preoptic area** (Fig. 11–5) of the hy-

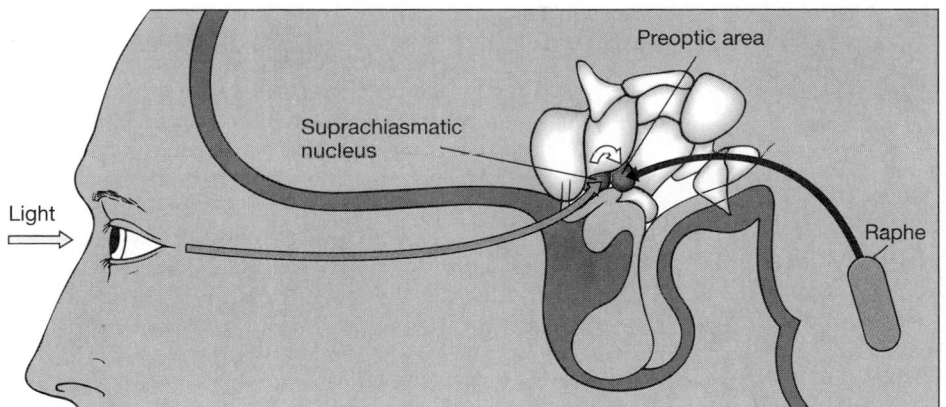

Figure 11–5

The suprachiasmatic nucleus, the body's "biological clock," receives information about light directly from the retina and is thought to regulate the sleep-wake cycle with the co-operation of the preoptic nucleus. The raphe nucleus of the brain stem is thought to be involved in the generation of REM sleep.

pothalamus induces non-REM sleep, and the destruction of this area of brain causes insomnia in laboratory rats. The preoptic area is located near the body's biological clock, the **suprachiasmatic nucleus** (Fig. 11–5), which receives direct input from the retina to provide information about the light-dark cycle. Thus, the suprachiasmatic nucleus is thought to provide inputs to the preoptic area to entrain the sleep-wake cycle.

Sleep also can be induced by injections of serotonin into the preoptic nucleus. A possible internal source of this serotoninergic innervation is the **raphe nucleus** (Fig. 11–5) of the brain stem. The destruction of the raphe causes insomnia in laboratory animals. With small lesions of the raphe, non-REM sleep gradually returns, but REM sleep does not. Thus, it is thought that the raphe is responsible for the generation of REM sleep. The descending fibers from the raphe nucleus innervate the spinal cord and may be responsible for inhibiting the activity of motor neurons of skeletal muscles. In this way, the motor programs that are planned and programmed during dreaming and REM sleep are normally prevented from being carried out.

Disorders of Sleep

Insomnia is a general term for the inability to sleep. Normal causes of insomnia include disruptions of the body's circadian rhythms—such as those due to the travel across time zones. **Narcolepsy** is a severe disorder in which an individual can suddenly lapse into sleep at any time during the middle of the day. Narcolepsy is thought to result from the activation of the REM-sleep generator. Narcoleptic attacks that occur during the waking state have all of the characteristics of REM sleep. These attacks often last from 90 to 100 minutes, the period of time normal for REM sleep. In addition, they are accompanied by vivid hallucinations similar to REM dreams. Finally, nar-coleptic attacks are associated with a lack of muscle tone, except for eye movement.

Narcolepsy is an inherited disorder; it is thought that the defect is the inability to inhibit the REM generator. The primary inhibitory inputs to the REM generator are noradrenergic nerve fibers. Consequently, drugs such as amphetamines that enhance norepinephrine release are used in treating this disorder.

Motivational Systems: Maintaining an Internal Balance and Ensuring Survival

Motivational systems initiate behaviors that satisfy bodily needs. These systems typically operate to ensure the survival of an individual or species. In some cases, motivational systems are driven by signals that represent the imbalance of certain chemicals within the body. Hunger and thirst, for example, are activated by internal signals that communicate the need for food and water in order to provide energy for the body and to maintain a constant level of osmolarity within the body's internal environment. Other motivational systems, such as those responsible for sexual drive, are activated by external signals from members of the opposite sex as well as internal signals from our own bodies. In this section, we will consider the motivational systems of hunger, thirst, and sexual drive, and the brain's so-called "reward system."

Hunger Is Regulated by Levels of Glucose and Triglycerides

The consumption of food is required for the growth, maintenance, and repair of all organ systems, and for the production of energy required by all cells in the body. The hypothalamus plays a central role in regulating food intake by acting as a sensor of glucose

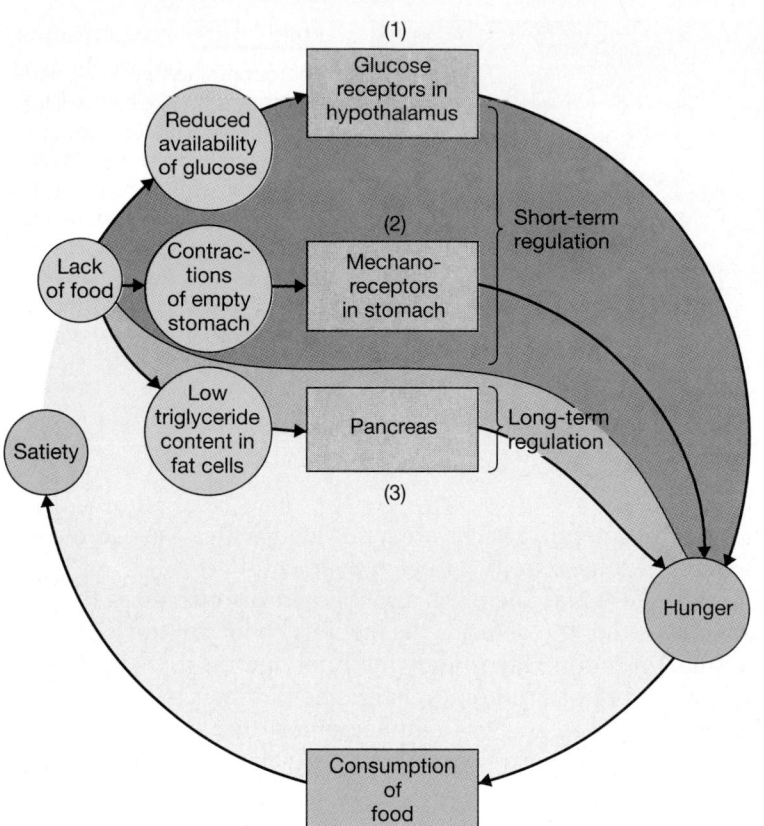

(1)

Figure 11–6

Three monitoring devices cooperate to produce the feeling of hunger and resulting consumption of food: (1) glucose receptors in the hypothalamus detect the lack of available glucose; (2) mechanoreceptors detect the contractions of an empty stomach; and (3) the pancreas detects low levels of triglycerides in fat cells.

availability (Fig. 11–6). This process is part of the motivational system involved in the short-term regulation of food intake.

The long-term regulation of food intake appears to involve the maintenance of body weight. Research has shown that laboratory animals forced to gain weight by eating an excessive amount of food will reduce their food consumption and revert back to their original body weight when left to eat at their own pace. Animals that are forced to lose weight by starvation will also eventually gain weight back to their original level when allowed free access to food. These studies suggest that the body may have a "set-point" for weight, which it attempts to maintain.

The maintenance of body weight is now thought to operate in an attempt to maintain a constant amount of **adipose tissue,** or **fat cells.** Fat cells store **triglycerides,** a major source of energy for the body (see Chapter 2). In adults, the amount of adipose tissue remains relatively constant, and there is a good correlation between body weight and the amount of adipose tissue. The variable that is thought to be regulated in the maintenance of adipose tissue is the volume of individual fat cells. These cells are thought to release a chemical signal in amounts proportional to their triglyceride content. This adipose signal affects the level of **insulin,** a hormone produced by the pan-

creas, in the circulatory system. Research has shown that basal insulin levels are proportional to body adipose tissue. Circulating insulin, in turn, penetrates the blood-brain barrier and binds to receptors in the ventral hypothalamus. The activation of these hypothalamic insulin receptors lowers food intake and, subsequently, body weight. Thus, insulin is involved not only in the control of circulating glucose after a meal, but also in the long-term regulation of body fat and hence of body weight (see Fig. 11–6). What remains to be determined about the long-term control of adipose tissue is how adipose cells regulate the levels of insulin in plasma.

The other aspect of the control of hunger is **satiety,** or the feeling of fullness after a meal. The mechanisms of satiety involve the ventromedial hypothalamus and were discussed in Chapter 10.

Thirst Is Regulated by the Osmolarity of the Blood

The human body consists of approximately 60% water. Water escapes from the body in the form of sweat and urine. It also leaves the body as water vapor in air exhaled from the lungs. In order to maintain a constant amount of water within the body, thirst mech-

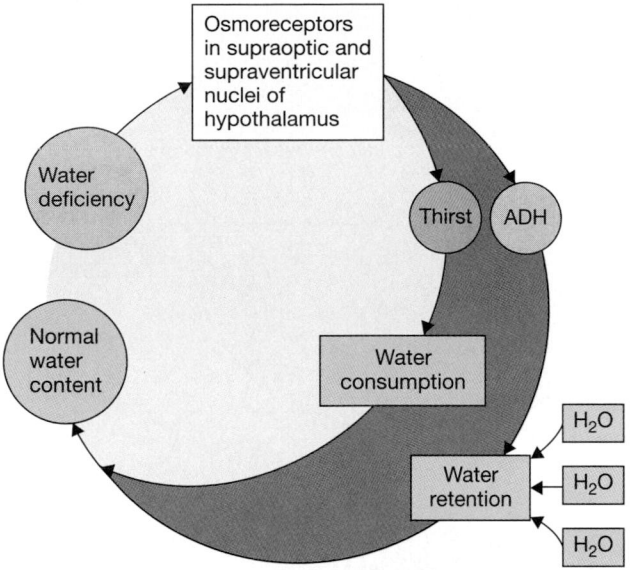

Figure 11–7

Osmoreceptors in the hypothalamus detect water deficiency and spark the release of ADH (which leads the kidneys to retain water) as well as causing the feeling of thirst (leading to the consumption of water). These two mechanisms decrease the osmolarity of the blood to normal levels.

anisms motivate us to drink water; in addition, the ANS and kidneys act in concert to retain water that would normally be excreted as urine. The amount of water in the body affects the amount of water in the blood, its **osmolarity** (see Chapter 2), which must be maintained at a constant level in order for blood cells to function. Consequently, the regulation of thirst, water volume, and osmolarity are interrelated.

The osmolarity of the plasma is a critical factor in the internal environment of the cells in the body. As explained in Chapter 4, hyper- and hypoosmotic fluids cause cells to shrink or swell, altering their function. The desired osmolarity for blood plasma is 300 mOsm. Increases above this value in plasma osmolarity can occur from excessive dehydration and sweating. This increase is detected by osmoreceptors in the **supraoptic nucleus** and **paraventricular nucleus** of the hypothalamus (Fig. 11–7). In response to increases in osmolarity, these nuclei secrete the hormone **vasopressin,** also called **antidiuretic hormone (ADH),** into the blood. Eventually, this hormone acts on the kidneys, causing them to reabsorb more water. This reabsorption of additional water reduces the plasma osmolarity to 300 mOsm.

The activation of hypothalamic **osmoreceptors** also initiates thirst mechanisms that involve somatic responses leading to the consumption of water. The intake of water in turn reduces plasma osmolarity. As

the feeling of thirst is satisfied, water consumption subsides. This happens, however, long before the water is absorbed into the circulatory system and long before plasma osmolarity is restored to normal. It is thought that the cessation of water consumption is triggered by stretch receptors in the stomach. How this information is transmitted to the brain has not yet been determined.

Sexual Drive and Sexual Behavior Can Be Altered by Exposure to Hormones During Development

Mating is a basic drive common to all mammals. It is essential for reproduction and the continuation of the species. Although the sexual behaviors of the males and females of a species are quite different, among different species there is a great deal of similarity in the mating posture of the animals. During mating, males display a **mounting** posture, and the female adopts a receptive posture called **lordosis.** In studies of laboratory animals, the frequency of these two postures is used as a measure of sexual drive. In both males and females, the **anterior hypothalamus** (see Fig. 11–1) appears to be one of the brain centers for sexual drive. Lesions of this area of brain in experimental animals reduce their sexual activity.

The stimulus for activating sexual behavior in many species is the release of chemicals called **pheromones.** Ovulating female monkeys (those who are prepared to conceive), for example, emit pheromones along with other vaginal secretions. These chemical odors are detected by the olfactory system of male monkeys and enhance their sexual drive. Sensory information from the male olfactory system is transmitted to the pyriform cortex and amygdala and indirectly to the hypothalamus (Chapter 8). In lower animals, pheromones play a prominent role in sexual behavior. In humans, other social cues come into play. The use of perfume or cologne as a sexual attractant, however, is still a common practice.

Mounting and lordosis are also used as indicators of differences in sexual behavior or sexual phenotype. Under certain circumstances, both the male and female animals can adopt the sexual behavior of the opposite sex.

The development of a particular pattern of sexual behavior is a result of the presence or absence of particular steroid hormones called **androgens** during **critical periods** of development. The presence of androgens during the fetal and early postnatal periods causes the brain to develop male sexual-behavioral characteristics, while the absence of these steroids re-

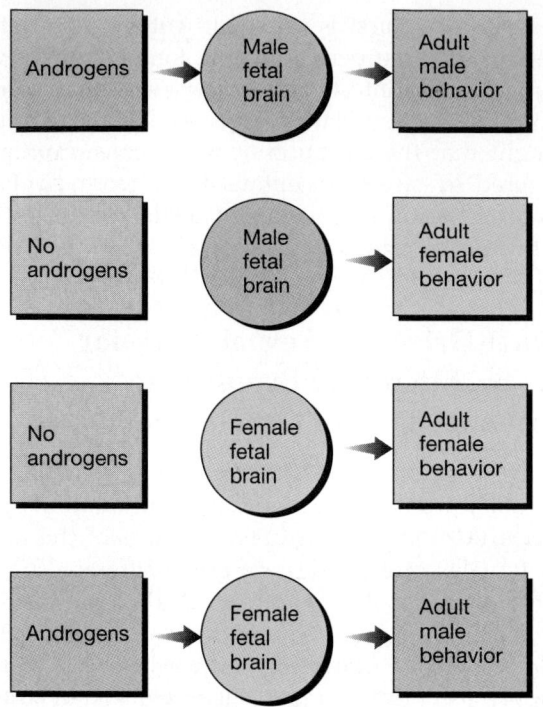

Figure 11–8

In laboratory experiments involving mice, exposure (or lack of it) to androgens during fetal development has been shown to affect adult sexual behavior. If a male fetal brain is exposed to androgens during critical periods of development, the adult will exhibit male behavior. If androgen exposure does not occur, the adult male will exhibit female behavior. If a female fetal brain is not exposed to androgens during critical periods, the adult will exhibit female behavior, but if the female fetal brain is exposed to androgens, the adult will exhibit male behavior.

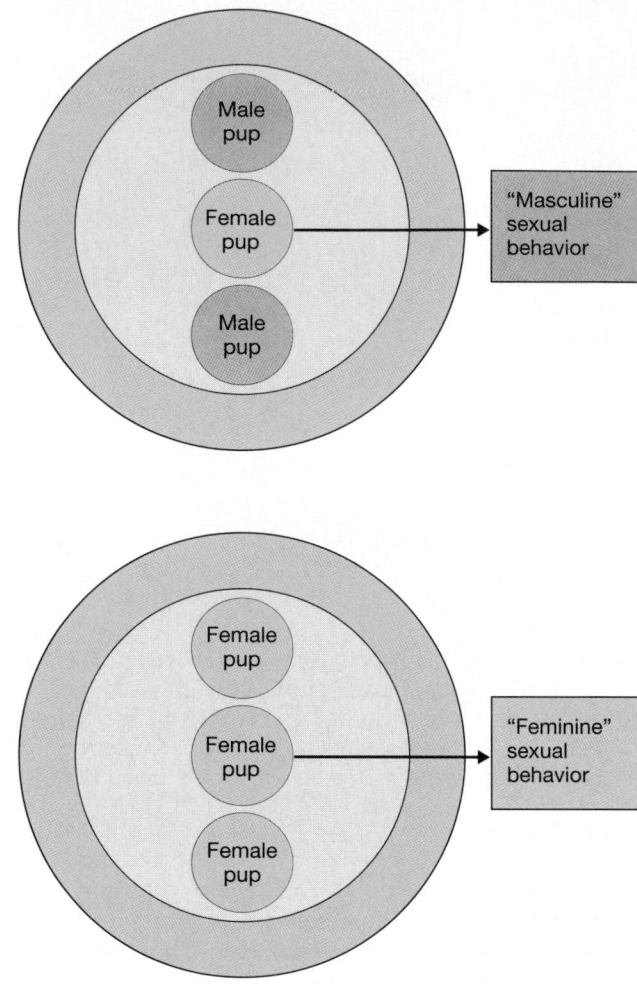

Figure 11–9

Fetal environment can affect the adult sexual behavior of laboratory animals. A female fetus that develops between two males will show male behavior as an adult, while a female surrounded by other females in the fetal stage will develop as a female.

sults in the development of female characteristics (Fig. 11–8). In laboratory mice, for example, the removal of the testis (the primary source of androgens) from males soon after birth causes these animals to develop lordotic behavioral responses as adults. If given ovaries, these males will ovulate as would a normal female. Likewise, the sexual behavior of a female can also be altered if the hormonal environment is altered during critical periods of development. Injections of androgens into newborn female mice cause them to develop male mounting behaviors as adults. Research has shown that the critical time period for the exposure (or lack of exposure) to androgens in order to affect brain-sex alterations in laboratory animals is the first few days after birth. In humans, androgen activity is elevated during the twelfth and twenty-second weeks of gestation and during the first six weeks after birth.

The fetal environment of these laboratory animals also influences the sexual behavior that the mice will display as adults. The positioning of a fetal mouse in

the uterus is a random process. A female fetus, for instance, could be placed in between two male fetuses, two females, or in between a male and a female. Research has shown that female fetuses placed between two male fetuses are exposed to higher androgen levels than if placed between two female fetuses (Fig. 11–9). As adults, these "androgenized" females are more aggressive; they exhibit male mounting behaviors, display irregular estrus (ovulation) cycles, mate later, and cease ovulating earlier.

A male fetus situated between two females in the uterus will have smaller testicles than a male fetus surrounded by other male fetuses. In addition, these animals, when grown, require higher doses of testosterone in order to induce aggressive behavior.

How do androgens affect the genetic programming of nerve cells, leading to differences in "male"

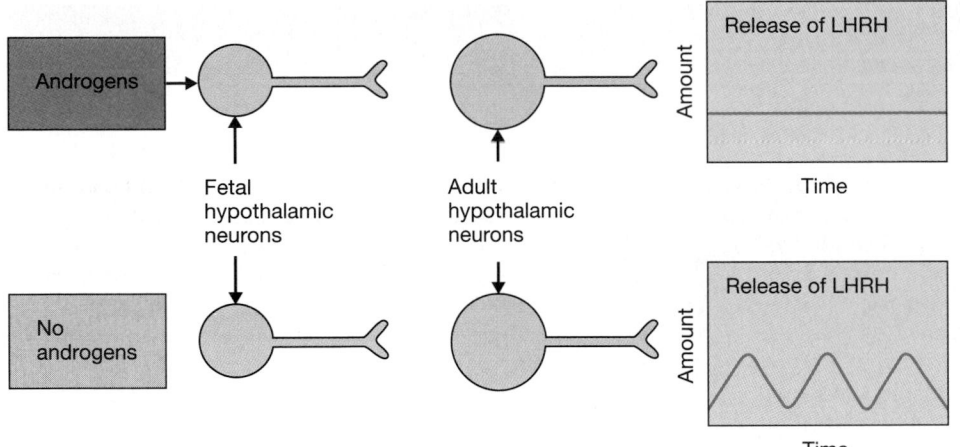

Figure 11–10

When nerve cells in the preoptic area of the hypothalamus are exposed to androgens, they release luteinizing hormone releasing hormone (LHRH) at relatively constant levels as adults, while preoptic nerve cells of animals not exposed to androgen show a cyclic pattern of LHRH release.

and "female" behavior? It is thought that when nerve cells in the preoptic area of the hypothalamus are exposed to androgens, they release **luteinizing hormone (LH)** at relatively constant levels as adults (Fig. 11–10). Nonandrogenized brains, on the other hand, develop a cyclical pattern of LH release. In addition, nerve cells in the preoptic area of nonandrogenized female brains are sensitive to estrogen secreted by ovarian follicles. Androgenized brains are not affected by estrogen. Thus, the function of nonandrogenized preoptic neurons has been altered not only in their release of LH, but also in their sensitivity to estrogen.

Alterations in sexual behavior are also observed in humans. In a condition known as **androgen insensitivity,** males are incapable of responding to androgens. This disorder is a genetic defect transmitted by the mother. Males with this defect develop as females in their external appearance as well as in their psychosexual preference.

Drugs can also affect the androgenic environment of the developing brain. Barbiturates, pesticides such as DDT, and hormones such as **diethylstilbestrol (DES)** (at one time given as a "morning after" birth control pill and also to prevent premature births) have been shown to masculinize the brain.

Learning and Memory Systems

Prior to our birth, and from the moment we are born, we acquire information and knowledge about our environment. In the womb, babies learn to distinguish the sounds of their mothers' voices. Soon after birth, they are conditioned to orient their heads to their mothers' breasts in order to receive milk. Eventually, babies are trained to use a toilet in the excretion of body waste. The acquisition of knowledge is called **learning,** and the retention of knowledge is called **memory.**

Learning Is Associative and Nonassociative

Research has shown that there are two types of learning: **associative learning** and **nonassociative learning.** In nonassociative learning, we learn about a single type of stimulus and adapt to it according to its relevance to our survival. The sound of a ticking clock, for example, may initially draw our attention, but after a period of time we ignore it. This type of behavior, in which we learn to decrease our response to a repeated stimulus, is called **habituation.** In contrast, we can become more sensitized to certain other sounds. In school, for example, we learn to recognize the sound of a fire alarm and respond accordingly. **Sensitization** is thus an increase in response to a stimulus.

In one type of associative learning, an animal acquires an understanding about the relationship or association between one stimulus and another. This learning process is called **classical conditioning.** An example of classical conditioning is the salivation of a dog in response to a stimulus of light. Normally, the dog salivates only in response to a piece of food. After a light stimulus (the **conditioned stimulus** or **CS**) has been paired with the presence of a piece of meat (the **unconditioned stimulus** or **UCS**) in several conditioning trials, however, the light alone is sufficient to cause the dog to salivate (salivation is the **conditioned response** or **CR**). The dog has been **conditioned** to salivate in response to light.

The other type of associative learning is called **operant conditioning.** In this type of learning, animals acquire an understanding of the relationship between their own behavior and subsequent rewards (or punishment). With this type of learning, animals tend to repeat behaviors that are rewarded and avoid behaviors that are punished.

In both classical and operant conditioning, a critical factor is the time interval between the appear-

R E S E A R C H   F O C U S

Drugs of Abuse and the Reward System of the Brain

The reward system of the brain was first discovered when stimulating electrodes were placed into the brains of laboratory animals, which were then allowed to control the amount of electrical current passing through the electrodes. By pressing a bar, animals were able to deliver current to the electrodes. When the electrodes were placed in specific areas of the brain, these animals would press the bar well over 100 times per minute (Figure (a)). When given a choice of receiving food instead of the electrical current, animals would choose the electrical stimulus, starving as a result. The reward areas of the brain thus constitute a powerful motivational system that overrides drives such as hunger, sex, and, in some cases, the avoidance of pain and punishment.

The areas of the brain that are associated with the reward system are shown here (Figure (b)). They include the **substantia nigra, hypothalamus, nucleus accumbens, caudate nucleus,** and the **frontal cortex.** The common thread linking these areas together is the **dopamine nerve fiber pathway** that originates in the substantia nigra. This pathway contains axons that form synaptic connections with each of the reward areas of the brain.

The release of dopamine by these synaptic connections is critical to the reward process. Drugs such as amphetamine and cocaine that enhance the release of dopamine also augment the rate of self-stimulation. Drugs that block dopamine release or its ability to bind to its receptor decrease the rate of self-stimulation. The reward system may therefore play a role in the process of addiction to certain abused drugs.

■ **What strategies might be used to help chronic cocaine users break free from their drug addiction?**

(a)

Rat with a brain-stimulating electrode. *(© Yoav/Phototake.)* The reward system of the brain involves the substantia nigra, the hypothalamus, the nucleus accumbens, the caudate nucleus, and the frontal cortex.

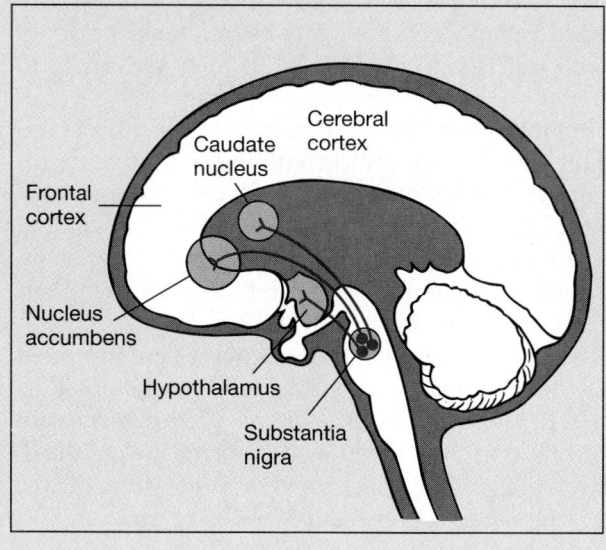

(b)

ances of the two stimuli, or of a behavior and its reward. With classical conditioning, if the time between the appearances of a conditioned and unconditioned stimulus is long, the conditioned response is poor. Likewise, with operant conditioning, if the reward is delayed, the conditioned behavior is weak.

Only certain types of stimuli are capable of producing a conditioned response. The most effective stimuli are those relevant to the survival of the animal. Food poisoning, for example, is a case of conditioning in which nausea acts as a form of punishment. Subsequent exposure to the taste of the food causes an animal to avoid it. The conditioning of food aversion is most effective when taste is used as the stimulus. If other stimuli such as sounds or lights instead of taste are paired with nausea, the animal does not

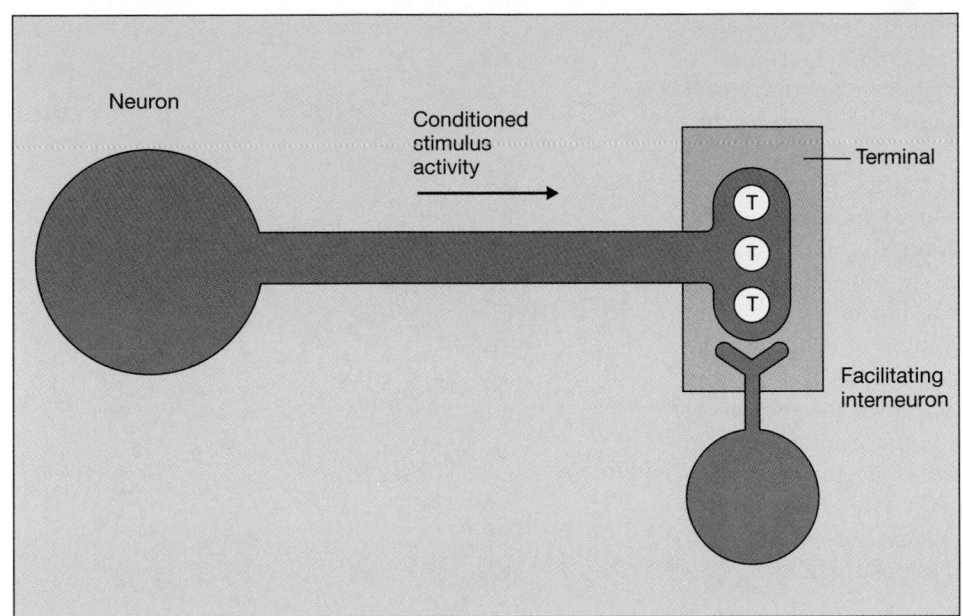

(a)

Figure 11–15

A mechanism for classical conditioning. (*a*) Interaction between a neuron along the conditioned stimulus-response pathway and a facilitating interneuron. (*b*) Enlargement of boxed area shown in figure above, illustrating the effect of no activity along the conditioned stimulus-response pathway. (*c*) Activation of conditioned stimulus-response pathway temporarily raises intracellular levels of Ca^{2+}, activating a calmodulin-activated increase in adenylate cyclase production of cAMP. This activity closes K^+ channels and prolongs the action potential.

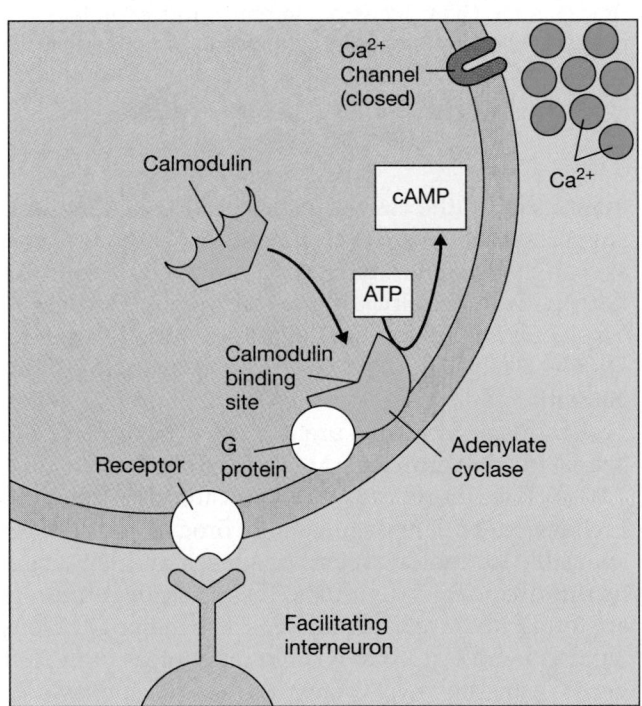

Conditioned stimulus-response pathway (no preceding activity)

(b)

Conditioned stimulus-response pathway (with preceding activity)

(c)

Use and Nonuse of Neural Pathways Result in Maintenance or Atrophy of Connections

During muscle exercise, there is an eventual change in muscle tone and muscle mass. In a similar way, the use and nonuse of neural pathways can strengthen or weaken the connections between nerve cells. Animals raised in "enriched" environments have nerve cells that form more synaptic connections with other nerve cells than would normally appear. In contrast, animals raised in "impoverished" environments develop fewer synapses than normal.

The use and nonuse of pathways can also alter topographical maps in the cortex. The map of the somatic sensory cortex and the motor cortex, for example, can be modified when an animal is trained to use one finger more than the others. The amount of cortex representing the function of the exercised finger expands beyond that which is normally found. In contrast, if the nerve fibers innervating the finger are severed, the cortical space devoted to that finger is gradually eliminated and eventually incorporated into the maps of other fingers.

The results of these studies indicate that the brain is a dynamic organ system, capable of changing the strength of existing synaptic connections and in some cases forming new synapses as part of the process of learning and memory. Many of these alterations in synaptic function can take place only during certain critical periods of development, while other modifications can occur at all ages. These observations argue for continued learning and exposure to new challenges, regardless of age.

Language Systems

Language is a tool that allows the communication of thoughts, knowledge, feelings, and emotions in both written and spoken form. The deciphering of words from complex sounds or from complex visual symbols and the extraction of concepts from these words represent activities at the highest level of sensory integration. Written language allows us to transfer knowledge from generation to generation so that each generation can build upon the achievements of the past.

The development of language has thus played a critical role in the progress of humankind. It is difficult to determine when language was first used; however, fossil evidence suggests that the areas of the brain responsible for language existed over 500,000 years ago. Spoken language obviously developed before written language. Consequently, another leap in human evolution may have been the formation of connections between the visual and language areas of the brain, allowing people to develop symbolic writing as a method of communication.

Comprehension and the Motor Aspects of Language Are Located in the Dominant Hemisphere

The hemisphere in which the language centers are located is called the **dominant hemisphere.** Typically, the speech area in the dominant hemisphere is larger than the same region in the nondominant hemisphere. This difference in size is apparent in the hu-

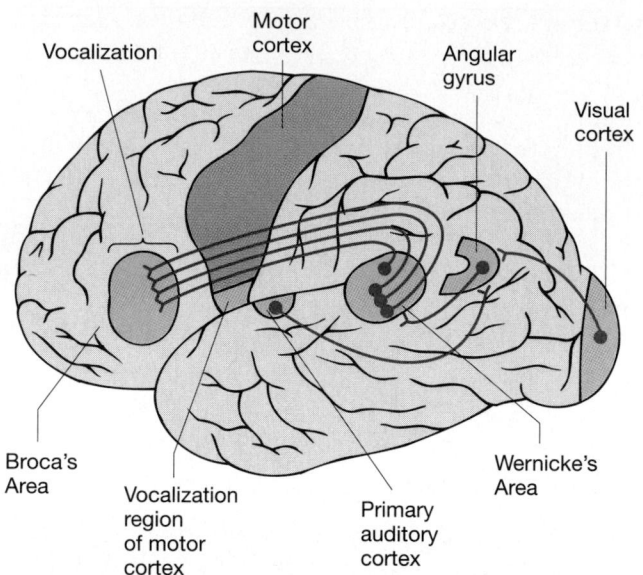

Figure 11–16

Language functions are localized in Wernicke's area and Broca's area. Other areas of the brain involved in language function are the primary auditory cortex, the visual cortex, the vocalization region of the motor cortex, and the angular gyrus.

man fetus by the thirty-first week of gestation and suggests that the development of language and speech may be an innate process. Indeed, infants are born with the ability to learn all human languages. This ability to learn other languages fluently after acquiring a first language, however, appears to diminish with age.

Language functions are located in regions of the brain called **Wernicke's area** and **Broca's area** (Fig. 11–16). Wernicke's area is responsible for the comprehension of language, while Broca's area is responsible for motor aspects of speech and language. In right-handed individuals, these language centers are found in the left hemisphere. The majority of left-handed people also have language centers in the left hemisphere; however, many left-handed individuals have language centers in both hemispheres or only in the right hemisphere.

For our understanding of spoken and written language, both auditory and visual information must reach the language centers. The reading of a word on a page, for example, requires that sensory information about the shape and form of the letters be transmitted from the retina to the primary and secondary visual cortices (see Chapter 8). This information is then transmitted to an association cortex called the **angular gyrus.** This area of brain is thought to be involved with the association of visual, auditory, and tactile sensations (see Fig. 11–16). From the angular gyrus, information about the word is transmitted to Wernicke's area, where it is thought to be recognized or comprehended. From Wernicke's area, information about the word is transmitted to Broca's area, where the motor programs for speech reside. Broca's area in turn contains nerve fibers that transmit information to the regions of the motor cortex responsible for control of movements of the mouth and vocal cords for speaking and of the hand for writing.

The Nondominant Hemisphere Is Responsible for Intonation and the Emotional Aspects of Language

The regions of the nondominant hemisphere of the brain that reside in the location corresponding to the location of Wernicke's and Broca's areas in the dominant hemisphere are responsible for the **affective** aspects of language (aspects related to mood or emotion). The manner in which a person speaks depends largely on his or her mood, and a listener can interpret simple statements in different ways depending on the speaker's tone of voice. The region in the nondominant hemisphere comparable to Wernicke's area is responsible for the comprehension of the tone of voice in speech, while the nondominant region comparable to Broca's area is responsible for the expression of intonation. Damage to these areas of brain causes **aprosodias,** disorders producing the inability to understand or express intonation.

Disorders of Speech and Language Have Anatomical Origins

Disorders of speech and language can also be explained in terms of the connections between the language centers in the dominant hemisphere. Disorders of language resulting from damage to the brain are called **aphasias.** A congenital disorder affecting reading is **dyslexia.**

Aphasia: Damage to the Language Centers

Several types of aphasia occur. In **Broca's aphasia,** damage to Broca's area disrupts the motor programs that control speech and writing. Patients with this disorder lose their ability to speak fluently and grammatically and are unable to express their ideas in writing.

Damage to Wernicke's area produces an aphasia characterized by loss of language comprehension. Patients with this disorder can still speak and write but have difficulty understanding the spoken and written language.

Dyslexia: A Congenital Reading Disability

Dyslexia is more common among boys than among girls, and also more common among left-handers than among right-handers. Dyslexics have difficulty associating letters with their corresponding sounds, and distinguishing letters that are similar in form, such as "p" and "b." Research has revealed that the **planum temporale,** a region that includes Wernicke's area, is reduced in size in dyslexics. In addition, the layers of nerve cells in people with dyslexia have not properly separated in this region, suggesting that the migration of neurons has been altered during development.

Laterality of Brain Function

The localization of language to one hemisphere is only one example of the laterality of brain function. Other functions, such as musical and spatial abilities, also exhibit a lateralization to one hemisphere. Tests show that, in right-handed subjects, the left ear is better at recognizing music, while the right ear is better at recognizing verbal material.

Spatial Abilities Are Localized to One Side of the Brain

Studies of so-called split-brain patients who have undergone surgery to disconnect the two hemispheres have shown that spatial abilities appear to reside in the right hemisphere. This surgical procedure cuts through the corpus callosum and is used as a last resort to control the spread of epileptic seizures. When lesions of the corpus callosum are made, the two hemispheres are no longer able to send information from one side to the other.

Patients with lesions of the corpus callosum who are asked to reassemble a jigsaw puzzle are unable to do so with their right hands, but are able to piece the puzzle together with their left hands. This is because the left hand is controlled by the right hemisphere, which houses the spatial abilities.

The spatial abilities of the right hemisphere are also seen in the processing of the written language. The Japanese, for example, use two writing systems: one consists of 71 letters that form the basis of phonetic (sound-related) symbols, and the second consists of over 40,000 **characters** or ideographic symbols, each one representing a separate idea or object. The first system uses the letters to aid in the pro-

CLINICAL FOCUS

Dyslexia: A Reading Disorder with Neuroanatomical Abnormalities

Dyslexia is a reading disorder of developmental origin. Children with this disorder cannot learn to read with normal proficiency, despite specialized instruction. Although these children appear to have normal intellectual ability and motivation, they have difficulty associating letters with their corresponding sounds, and distinguishing letters that are similar in form, such as "p" and "b."

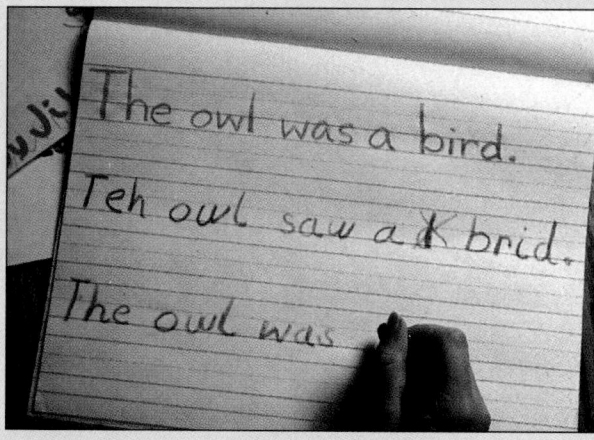

A person with dyslexia learning to write. *(© Will & Deni McIntyre/Science Source/Photo Researchers, Inc.)*

Dyslexics represent about 2% to 16% of school-age children. Male dyslexics outnumber female dyslexics by 3:1, and the disorder is more common among left-handed children. Research indicates that inheritance plays an important role in the transmission of dyslexia.

Neurological signs often related to dyslexia include stuttering, abnormal electroencephalograms (measures of brain electrical activity), and problems involving manual dexterity of one or both hands.

Anatomical abnormalities are also apparent. The brain region required for the interpretation of language, Wernicke's area, is notably smaller in dyslexics, and the cell layers are disorganized. In addition, blood vessels are found in regions where they are not normally present. An abnormally high percentage of dyslexics have dominant right hemispheres.

Current theories argue that testosterone activity in dyslexics during the prenatal period is abnormally high. This alteration in testosterone preferentially affects the development and function of the left hemisphere. As a consequence, handedness and speech (stuttering) are affected, along with the manifestation of dyslexia.

■ **What neurophysiological mechanisms might account for the inability to distinguish between the letters "p" and "b"?**

---

nunciation of a word, while the second uses "pictures" to represent the word. Although both the phonetic and ideographic systems of writing use the language centers of the left hemisphere, the ideographic system also uses the right side of the brain.

Lesions of the angular gyrus near the language centers in the left hemisphere impair reading in the phonetic system of writing but generally not in the ideographic system. Even in cases when ideographic reading is impaired, patients are able to explain the meaning of the word. For the same word written in the phonetic system, they lose this ability. Thus ideographic writing, which uses the language centers in the dominant hemisphere, also relies on the spatial abilities of the nondominant hemisphere.

Male and Female Brains Exhibit Differences in Laterality

Differences in the development of the lateralization of brain function exist between males and females. In tests that evaluate the development of spatial func-

tion, males exhibit a lateralization of spatial tasks to the right hemisphere by 6 years of age. With females, spatial function is equally developed in both hemispheres until the age of 13.

The absence of early hemispherical specialization imparts certain advantages to females. Damage to the left hemisphere in childhood, for example, impairs language development in males more than in females. It appears that the delay in hemispherical specialization allows the transfer of language function from the left to right hemisphere to occur more readily in females.

To some extent, differences in the laterality in brain function between males and females are also seen in adults. Damage to the left hemisphere is more strongly associated with language disorders in adult males than in adult females. Women appear to be less lateralized for language functions. Likewise, damage to the right hemisphere is more strongly associated with nonlanguage disorders in males than in females. These observations suggest that the female brain is more symmetrical in function than that of the male.

RESEARCH FOCUS

Magnetic Resonance Spectroscopy and the Brain

Magnetic resonance imaging (MRI) is a method used to visualize the internal organs of the body in a noninvasive manner (see Focus Box on Imaging of Internal Structures Chapter 1). This approach uses a powerful magnet to align the molecules within the tissue of an organ. The most abundant element in the body is hydrogen. The nucleus of a hydrogen atom possesses a magnetic moment that causes it to behave like a small bar magnet. When an external magnetic field is applied to the hydrogen atom it aligns itself with the magnetic field. When the magnet field is turned off the hydrogen molecules then return to their normal orientation. The rate at which hydrogen atoms align with the magnetic field and return to their original orientation gives rise to so-called "relaxation times." These relaxation times are different for hydrogen atoms embedded within a water environment like CSF, and in lipid and protein environments. These differences in relaxation times result in differences in magnetic signal intensity, which in turn can be detected to produce the types of brain images we see with MRI.

The advancement of MRI technology has provided us with previously unattainable images of the living brain with tremendous detail. Recent developments in MRI now allow us to combine images of the brain with its chemistry. By using different pulsing sequences, the applied magnetic field can be manipulated to produce a spectrum of biochemical constituents within the brain. This form of "magnetic resonance spectroscopy" can be used to detect compounds such as lactic acid in the brain, which forms during ischemic brain injury. The combination of magnetic resonance imaging with spectroscopy will therefore allow us to map the distribution of a variety of neurochemical compounds in the brain, and their alterations during pathophysiological conditions. This approach might also be used to monitor the effectiveness of therapies to prevent nerve cell death and to destroy brain tumors.

■ **What other metabolic markers might be useful in monitoring the health status of the brain by magnetic resonance spectroscopy?**

(a)

(b)

(c)

Magnetic resonance image and spectroscopy of neonatal rat brain after ischemic brain injury. This approach permits the noninvasive monitoring of neurochemical alterations in the brain under pathophysiological conditions. (*a*) MRI of rat brain after ischemic injury. (*b*) Normal chemical spectra before injury. (*c*) MR spectra after injury. PM = phosphomonoesters, PI = inorganic phosphate, PCr = pophocreatine, ATP = adenosine triphosphate. *(Photo courtesy of Elizabeth M. Jansen and Michael Garwood)*

CHAPTER REVIEW

Summary

- Many bodily functions fluctuate on a monthly and/or daily basis. Those that vary daily are called circadian rhythms. Many of these rhythms are entrained to the 24-hour light-dark cycle by the suprachiasmatic nucleus of the hypothalamus.
- The cycle of sleep and wakefulness varies on a 25-hour schedule but is entrained to a 24-hour period. Sleep involves repeated cycles of brain activity, each having similar stages.
- Hunger is the desire for food, a motivational system that operates to satisfy the energy needs of the body and the need for growth, development, and maintenance of the body's organs. Hunger is regulated by both short- and long-term mechanisms that lead to food consumption.
- Thirst is a drive to maintain a constant level of osmolarity within the body's internal environment, through fluid retention and consumption.
- Sexual drives are important for reproduction and the perpetuation of the species. They are activated by the interaction of social environment, hormones, and genetically programmed sexual behaviors. The development of a male or female sexual behavior is influenced by androgens during development.
- The reward system of the brain is a motivational system that can override other motivational drives such as hunger and thirst.
- In nonassociative learning, an understanding of the properties concerning a single type of stimulus leads to either a habituation or sensitization to the stimulus. In associative learning, we acquire an understanding of the relationship either between two stimuli (called classical conditioning) or between a stimulus and a behavior (called operant learning).

- The storage of information progresses from short-term memory to long-term memory. If information is disrupted while in short-term memory, it will not become consolidated into long-term storage. Memory is stored in a variety of memory circuits found in different areas of the brain. Memory circuits have been identified for the learning of motor responses, cardiovascular responses, and visual perceptions.
- Habituation is a decrease in response to a repeated stimulus. Sensitization is an increase in response to a repeated stimulus and is due to an increase in the release of neurotransmitters.
- The connections between nerve cells can be altered by the amount of activity between the cells. The use of neural pathways can strengthen the connections between groups of nerve cells, while the nonuse of pathways weakens their connections.
- Language centers are located in one or the other side of the brain depending on the individual. The hemisphere in which language is located in an individual is called the *dominant hemisphere*. Within the dominant hemisphere, Wernicke's area is responsible for the interpretation of language, and Broca's area for the motor aspects of speech.
- Spatial, artistic, and musical abilities are located predominantly in the right hemisphere. Laterality of brain function develops earlier in males than in females. In the event of damage to one side of the brain, the opposite side is capable of taking over the functions of the damaged hemisphere.

Review Questions

Identify Each with a Term

1. The term used to describe the hemisphere where language is localized.
2. The area of brain that exhibits an abnormal development in people with dyslexia.
3. The term for the sleep disorder in which individuals cannot prevent themselves from suddenly falling asleep during the day.
4. The area of brain responsible for REM sleep.

Define Each Term

5. synaptic facilitation
6. synaptic habituation
7. associative learning
8. REM sleep

Choose the Correct Answer

9. Which of the following statements about speech and language is *not* correct?
   a. Damage to Wernicke's area (22) produces a loss of comprehension for both the written and spoken language.
   b. Patients with damage to Wernicke's area (22) lose the ability to write.
   c. Damage to area 22 in the nondominant hemisphere produces a loss in comprehending the intonation of speech.
   d. Patients with dyslexia have an abnormal development of the planum temporale.
10. Which of the following peptides is secreted by the supraoptic and paraventricular nucleus of the hypo-

thalamus in response to an increase in blood osmolarity?
a. prolactin
b. oxytocin
c. vasopressin
d. beta-endorphin

11. Which of the following is *not* a cellular mechanism that leads to an enhanced response to a conditioned stimulus?
a. activation of voltage-gated potassium channels
b. increase in intracellular levels of calcium
c. amplification of cAMP levels
d. a calmodulin-mediated increase in adenylate cyclase activity

Answer Each Question in One or Two Sentences

12. What are the various stages of sleep? Which stage is typically associated with dreaming?
13. What are the brain areas involved in the control of sleep?
14. How is sexual behavior affected by androgens?

15. What are the differences between short- and long-term memory?

Problem Solving in Physiology

16. In order to prevent seizure activity from spreading from one side of the brain to the other, a patient with a dominant left hemisphere undergoes surgical transection of the corpus callosum to disconnect the cerebral hemispheres. When tested after surgery, the patient could not correctly identify the image of an object projected to the temporal aspect of his left visual field by verbally naming the object, or select the object from among other objects using his right hand. Yet if the image of the object is projected to the nasal aspect of the patient's left visual field he is now able to identify the object verbally and select it with his right hand. Explain these differences in response.
17. What evidence would support the hypothesis of a neurobiological basis for homosexuality?

Suggested Readings

Dement, W., and Kleitmann, N. "Cyclic variations in EEG during sleep and their relation to eye movements, body motility, and dreaming." *Electroencephalography and Clinical Neurophysiology,* 9:673–690, 1957.

Domasio, A.R., and Geschwind, N. "The neural basis of language." *Annual Review of Neuroscience,* 7:127–147, 1984.

Gazzaniga, M.S., and LeDoux, J.E. *The Integrated Mind.* New York: Plenum Press, 1978.

Goy, R.W., and McEwen, B.S. *Sexual Differentiation of the Brain.* Cambridge, MA: MIT Press, 1980.

Kandel, E.R., and Schwartz, J.H. "Molecular biology of learning: Modulation of transmitter release." *Science,* 218:433–443, 1982.

Olds, J., and Milner, P. "Positive reinforcement produced by electrical stimulation of septal area and other regions of rat brain." *Journal of Comparative Physiology and Psychology,* 47:419–427, 1954.

Raichle, M.E. "Visualizing the mind." *Scientific American,* 270:58–64, 1994

Sperry, R.W. "The great cerebral commissure." *Scientific American,* 210:42–52, 1964.

Steriade, M., and Hobson, J.A. "Neuronal activity during the sleep-wake cycle." *Progress in Neurobiology,* 6:155–376, 1976.

Thompson, R.F., Berger, T.W., and Madden, J. IV. "Cellular processes of learning and memory in the mammalian CNS." *Annual Review of Neuroscience,* 6:447–491, 1983.

Weisskopf, M.G., Castillo, P.E., Zalutsky, R.A., and Nicoll, R.A. "Mediation of hippocampal mossy fiber long-term potentiation by cyclic AMP." *Science,* 265:1878–1882, 1994.

Zola-Morgan, S.M., and Squire, L.R. "The primate hippocampal formation: Evidence for a time-limited role in memory storage." *Science,* 250:288–290, 1990.

Key Terms

accumbens nucleus (p. 360)
adipose tissue (p. 356)
androgens (p. 357)
angular gyrus (p. 366)
aphasia (p. 367)
associative learning (p. 359)
Broca's area (p. 366)

circadian rhythm (p. 352)
classical conditioning (p. 359)
dominant hemisphere (p. 366)
dyslexia (p. 367)

electroencephalogram (EEG) (p. 353
habituation (p. 359)
long-term memory (p. 368)
nonassociative learning (p. 359)

operant conditioning (p. 359)
REM sleep (p. 354)
sensitization (p. 359)
short-term memory (p. 361)
Wernicke's area (p. 366)

Scanning electron micrograph of the progisterone crystal (original ×3000). *(© David M. Phillips/Visuals Unlimited)*

Chapter 12

Endocrine Control Mechanisms

Key Concepts

Hormones are chemical signals that are used for communication between cells.

Chemically, hormones are a rather diverse group of substances. Some are derived from metabolism of amino acids, some are peptide or protein in nature, and some are derived from cholesterol.

Binding of a hormone to its receptor and activation of that receptor is the initial step in producing a hormone effect. Some hormones act via second messengers to alter activity of preexisting proteins in target cells. Other hormones act by altering gene expression in target cells to change the amount of a few key proteins.

Chapter Outline

Each spring the engineering students at Purdue University in West Lafayette, Indiana, sponsor a Rube Goldberg contest. The contest is quite popular among engineering students and attracts entries from across the nation. Rube Goldberg was a cartoonist in the early part of this century who specialized in drawing humorous, complicated machines with complex mechanisms that theoretically would perform simple tasks, but with great fanfare. Each year at Purdue the contest task varies, but in recent years the machines that the students design and build have been required to do such things as make a cup of coffee, sharpen a pencil, or moisten a stamp and place it on an envelope.

Often, the initiating event in a Rube Goldberg apparatus is something very simple like releasing a ball to roll down a ramp. The ball then knocks down some dominoes, which in turn starts a wheel spinning, and eventually the final desired result is obtained.

Hormones are much like that ball that rolls down the ramp; they initiate a series or cascade of events in the cell that ultimately leads to the desired outcome of a change in cell activity. The descriptions and figures of mechanisms of hormone action that are to be described later in this chapter may initially look like Rube Goldberg machines on a molecular level, but in fact they are very highly tuned, well-regulated mechanisms that have stood the test of biological time and are used by organisms with varying degrees of complexity, from yeasts to humans.

Much as the Rube Goldberg mechanisms buzz, whir, chug, and then produce their final effect, every cell in our bodies does something similar every minute of every day. While our cells don't necessarily buzz, whir, or chug, they do respond in a well-defined, crisp manner to the signals that hormones provide. Hormones are just one demonstration of how the body keeps everything in order; they signal the row of dominoes to fall, which leads to what we ultimately call homeostasis. Thus, hormones provide signals that contribute to the maintenance of a constant internal environment.

Endocrinology is the branch of biology concerned with the actions of hormones and the organs or glands in which the hormones are produced. Endocrinology often overlaps with areas such as anatomy, biochemistry, and neurophysiology. Included within the field of endocrinology are such topics as the anatomy and physiological function of the endocrine organs, the chemical nature of the hormones these glands produce, the cellular mechanisms by which the hormones produce their biological effect, and clinical manifestations of abnormal endocrine function. This brief list is by no means complete, as there are numerous other exciting areas of study within the field of endocrinology. However, the majority of current research efforts in endocrinology can probably be placed into one of these four categories.

General Concepts of Endocrine Control

The word hormone is derived from the Greek *hormaein*, which means "to excite" or "to arouse." But what exactly is a hormone, and conversely, what is not a hormone? Is the glucose that circulates in the blood a hormone? What about the lactic acid produced by a vigorously exercising muscle, or the carbon dioxide produced by virtually all cells in the body? Why do we consider insulin, produced by the pancreas, and epinephrine (also known as adrenaline), secreted by the adrenal medulla, as hormones?

Over the years, scientists have made many attempts to define a hormone, and much discussion has ensued over the propriety of any particular definition. As new hormones were discovered, definitions of hormones were rewritten to include the newest members of the "hormone family." Today, it is virtually impossible to write an accurate definition that includes the large numbers of known hormones or hormone-like substances, while still excluding other substances that do not function in a manner similar to that of a typical hormone.

Hormones Are Chemical Signals Used In Cell-To-Cell Communication

Despite the complexity involved in describing hormones, at this point in the discussion a broad, working definition is probably useful to help gain a general appreciation of what hormones are and how they function in the body. Hormones are generally considered to be substances secreted into the blood in very small amounts by specialized cells or glands and carried by the bloodstream to other parts of the body, where they interact with specific **receptors** in **target tissue cells** to produce a particular biological response (Fig. 12–1*a*). This is a generalized definition of a hormone, not an absolute one. Several bona fide hormones would probably not fit this definition in its entirety, and likewise, several nonhormonal substances could be viewed as filling most of the criteria of this definition. (Some of the terms used in this definition, such as *receptor* and *target tissue*, will be described in greater detail later in this chapter.)

In recent years, knowledge of the degree of the complexities involved in cell-to-cell communication mechanisms has increased dramatically. Virtually every cell in the body receives information in the form of chemical signals, from either nearby cells or distant cells or organs. We have only begun to characterize and understand the details of these very powerful control systems in many instances. A fuller understanding of these mechanisms may lead to important breakthroughs in the treatment of a wide variety of diseases.

Not all hormonelike substances are carried to their action sites by the blood. Some reach their target cells by **diffusion** and exert their effects by what are known as **autocrine** or **paracrine** mechanisms. In these cases, the hormone product is not secreted directly into the blood, but instead is released into the interstitial fluid surrounding the cells in which it is produced. Autocrine regulation (see Fig. 12–1*b*) involves the secretion of a hormone product that exerts its effects on the cell in which it was produced, while

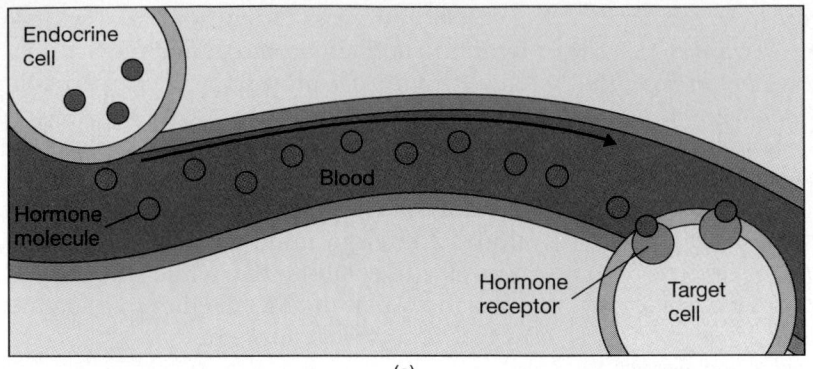

(a)

Figure 12–1

Comparison of endocrine (*a*), autocrine (*b*), and paracrine (*c*) mechanisms of cellular communication.

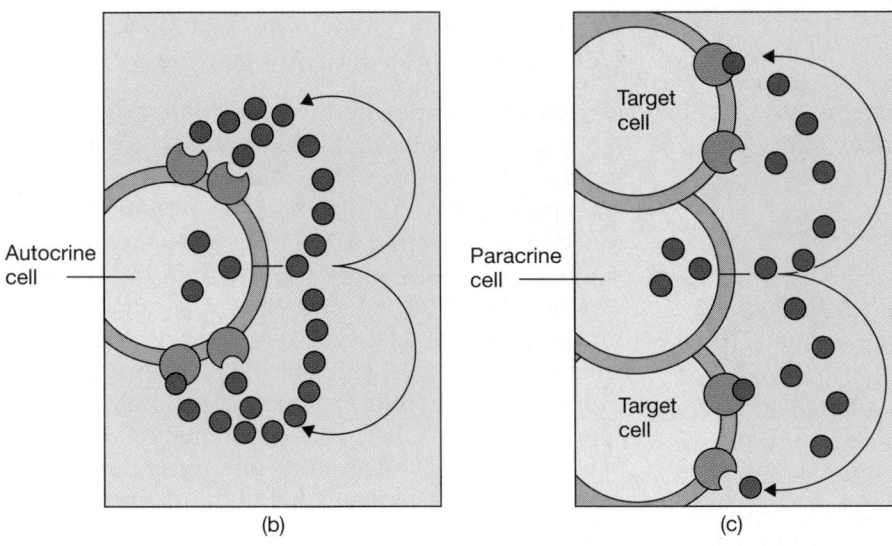

(b)　　　　　　(c)

paracrine regulation (see Fig. 12–1*c*) involves the secretion of a hormone that diffuses and acts on adjacent cells. While our knowledge of classical endocrine control systems is extensive, information about autocrine and paracrine function is rather limited and remains to be more completely characterized. However, it is likely that these play a major role in the establishment and maintenance of a normal, healthy state.

The Nervous System and the Endocrine System Function Together to Promote Homeostasis

As ancient single-cell organisms evolved into today's complex animals, it became necessary and advantageous for cells to specialize in their functioning. Specialization allowed the various cell types to perform unique functions in a cooperative manner and thus contribute to the overall survival of the organism. Along with this specialization came an increasing need for communication between cells or groups of cells as they became more interdependent. Such communication ensures that the activities and functions of various tissues and organs are carried out in a coordinated, purposeful manner.

In the human body, cells communicate through two major systems: the **nervous system,** discussed in previous chapters, and the **endocrine system.** There is a considerable degree of interaction between the two systems. This overlap between the nervous and endocrine systems constitutes an area of study known as **neuroendocrinology.**

Figure 12–2 illustrates the different mechanisms whereby neural, endocrine, and neuroendocrine cells receive and transfer information. Neurons, for example, receive information from other nerve cells in the form of chemical signals (neurotransmitters) that reach the cell by diffusion across synaptic clefts. The neuron in turn releases neurotransmitters at its synapses, which then activate subsequent cells in the neuronal chain. Thus, in the example of neurotransmission shown in Figure 12–2*a*, information in the form of electrochemical signals is transferred from neuron to neuron.

A typical endocrine cell, on the other hand, receives information in the form of chemical messages

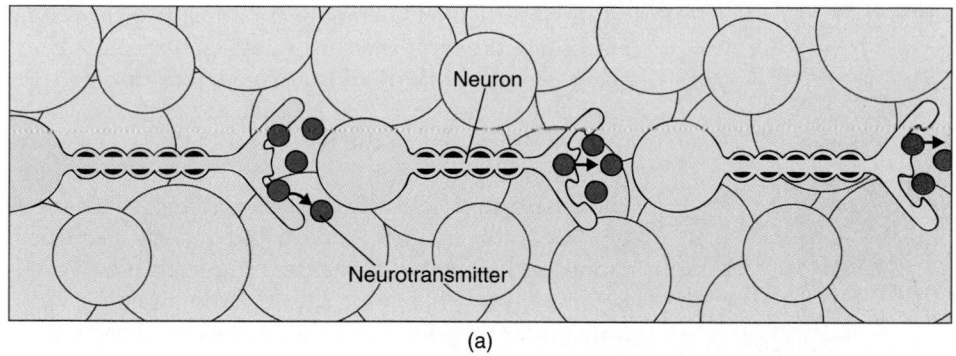

Figure 12–2

Basic mechanisms of neural (*a*), endocrine (*b*), and neuro-endocrine (*c*) communication methods.

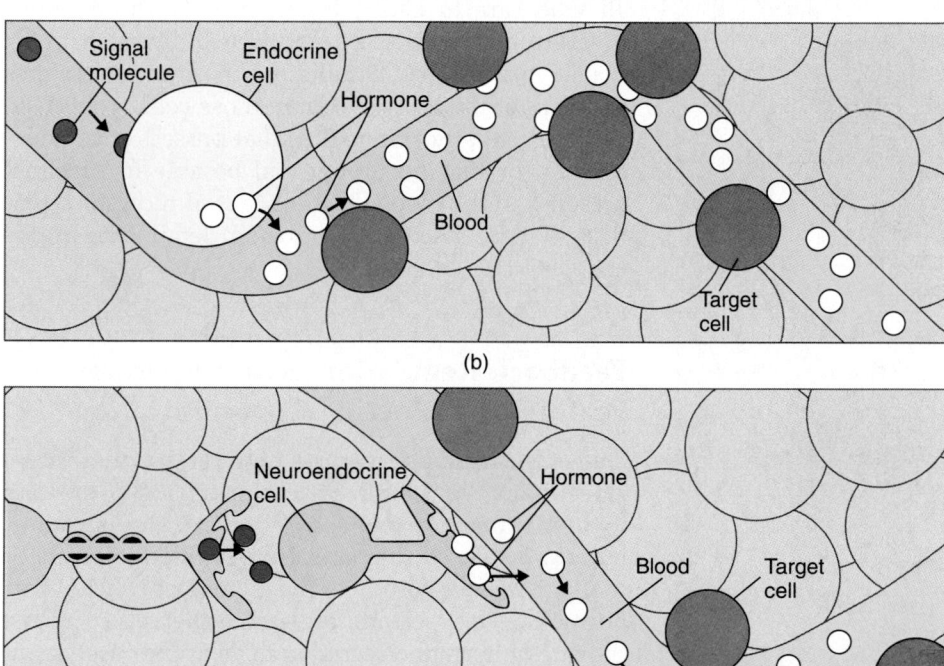

("signal" molecules) from the blood, as illustrated in Figure 12–2*b*. In response to the appropriate input signal, the endocrine cell then secretes its own specific hormonal signal into the blood. The endocrine cell may sense the concentration of another hormone present in the blood, or it may perceive the concentration of some particular ion or nutrient, such as sodium or glucose. When the endocrine cell detects a significant change in the concentration of that particular factor in the blood, it appropriately alters the rate of release of its own secreted hormone product.

Neuroendocrine cells function as an interface between the nervous and endocrine systems, providing a crucial link between these two communication systems. As shown in Figure 12–2*c*, a neuroendocrine cell is directly innervated by nerve cells but, unlike other neurons, in response to stimulation it secretes its product, a hormone, directly into the blood. Thus, neuroendocrine cells act as one-way translators. They convert **electrochemical signals** from the nervous system into **hormonal signals** in the blood.

Figure 12–3 illustrates a basic difference in the functional organization of the nervous and endocrine systems. The ability of one particular neuron to communicate with its target cells—whether they are other nerve cells, muscle cells, or other excitable cells—is generally determined by the anatomical connections within the system. A single neuron can transmit information—in this case, an electrochemical signal—only to those cells with which it makes synaptic contact. For example, in Figure 12–3*a*, nerve cell A can communicate only with cells d and e, nerve cell B only with cells e and f, and nerve cell C only with cell g. Thus, the flow of information within the nervous system is determined by the point-to-point anatomical connections made between different cells. We can make an analogy between the nervous system and an intercom system in an office building.

(a)

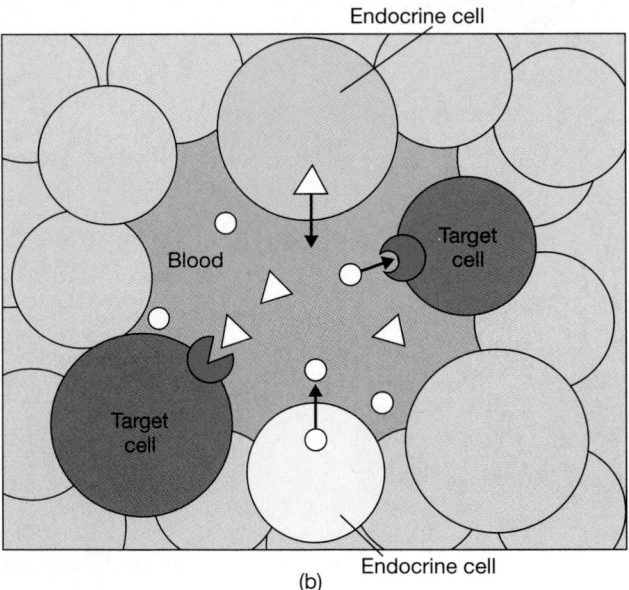

(b)

Figure 12–3

Mechanisms of cellular communication in the nervous (*a*) and endocrine (*b*) systems. Note that anatomical connections determine specificity in the nervous system (*a*), whereas receptor distribution determines specificity in the endocrine system (*b*).

Only those workers who are connected by the wiring of the intercom system can directly communicate with one another.

By contrast, once a hormone is secreted into the blood, it is carried to nearly every cell in the body. This difference is illustrated in Figure 12–3*b*. Whether or not a certain cell will respond to a particular hormone molecule is determined by the presence or absence in the cell of a specific **receptor** for that hormone. A receptor is a molecule in the membrane of or inside the cell that specifically recognizes and

binds a particular hormone. As a result of hormone binding to the receptor, the biological responses characteristic of that particular hormone are initiated. As depicted in Figure 12–3*b* all cells are exposed to the hormones present in the blood. Only those cells having a specific receptor for a hormone can respond to it. For example, only those cells having a "triangular" receptor can respond to the "triangular" hormone, and so on. In this case, we can draw an analogy between a hormone and a radio signal: a radio transmitter broadcasts its signal in every direction for all who care to listen; however, only those radios tuned to the appropriate station will detect that particular radio signal. Similarly, a hormone circulates throughout the body, but only those cells "tuned" to that particular hormone (i.e., that possess specific receptors for that hormone) will be able to sense its presence in the blood. The concept of receptor specificity will be discussed in greater detail later in this chapter.

Feedback Regulation Is an Important Feature of Endocrine Communication

One of the hallmark features of the endocrine system is **feedback regulation.** Endocrine cells have the ability to manufacture and secrete a particular hormone; in most instances, they are also equipped to detect or monitor the magnitude of the biological effect of that hormone. This permits the endocrine cell to adjust the rate of hormone secretion in an appropriate manner to achieve the desired effect, which is the maintenance of homeostasis.

In most instances, feedback regulation involves **negative feedback,** although some cases of **positive feedback** are also known. Figure 12–4*a* illustrates a very simple negative feedback system. In this case, the endocrine cell (A) secretes a hormone, which has an effect on one of its target cells (B). The hormone triggers a specific biological response in the target cell, and the magnitude of the response is in turn sensed, or monitored, by the endocrine cell that first secreted the hormone. If the response is too small, the endocrine cell will produce and secrete more hormone in an attempt to elicit a larger, more normal response. Likewise, if the response of the target cell is too great, the endocrine cell will reduce the amount of hormone it secretes until the response decreases into the normal range. In this case, the particular response being produced can be said to have a negative feedback effect on the amount of hormone secreted by the endocrine cell.

An example of this simple type of negative feedback regulation is the control of blood glucose concentration by the hormone **insulin,** which is pro-

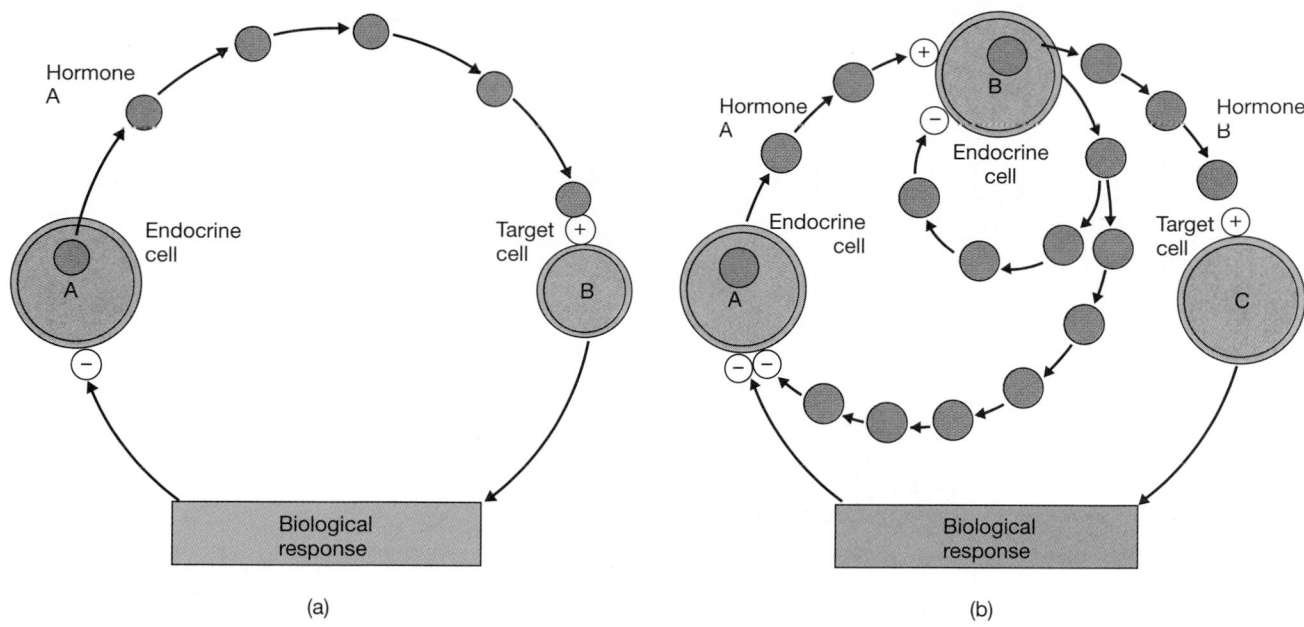

Figure 12–4

(*a*) A simple negative feedback loop. (*b*) A more complex negative feedback system involving two different endocrine cells and two different hormones.

duced by specialized endocrine cells within the pancreas. These cells are triggered to secrete insulin when the amount of glucose in the blood exceeds the normal concentration of about 100 mg of glucose per 100 ml of blood. Therefore, when blood glucose levels are increased above normal, such as might occur soon after a meal, the insulin-secreting cells respond by increasing the secretion of insulin. Insulin in turn stimulates the uptake of glucose from the blood by such tissues as muscle and adipose (fat). The overall response to insulin therefore is a reduction in the amount of glucose circulating in the blood. The stimulus for insulin secretion is therefore lessened, and the secretion rate decreases. (These effects of insulin will be discussed in greater detail in Chapter 15.)

More complex types of negative feedback systems are also prevalent within the endocrine system. An example of one such system is illustrated in Figure 12–4*b*. In this case, endocrine cell A secretes a hormone (A), which has as its target endocrine cell B. In response to hormone A, endocrine cell B increases its production of hormone B. This hormone in turn acts on target cell C to produce some particular biological response. In similar fashion to the example shown in Figure 12–4*a*, the response produced by the final target cell—in this case, cell C—inhibits the production of hormone A. In addition, hormone B itself has negative feedback effects on cell A, and it also has negative feedback effects to limit its own production by cell B. Several examples of this type of

multilevel feedback can be found within the endocrine system, especially in the regulation of secretion of pituitary hormones.

The more complex type of feedback control shown in Figure 12–4*b* offers certain advantages over the simple type shown in Figure 12–4*a*. It permits fine-tuning of the endocrine cells with regard to their hormone-secreting activity. This in turn leads to biological responses that are closely matched to the original stimulus. A nearly constant internal environment results. Additionally, these multilevel control systems provide some degree of back-up in the event that one segment of the overall system functions abnormally. In such a situation, a multilevel control system is better able to achieve the desired biological response than a simple system such as the one shown in Figure 12–4*a*.

Hormones Regulate or Influence a Variety of Processes in the Body

Table 12–1 lists primary human hormones, their tissues of origin, their target cells, and a brief description of the processes they regulate. Note that tissues or organs with endocrine functions can be found in many parts of the body. The specific effect of individual hormones can generally be categorized as relating to the regulation of one of four physiological

(text continues on page 380)

Table 12–1
Summary of the Primary Human Hormones

Hormone	Gland or Source	Target Cells	Primary Biological Effect
Release or release-inhibiting hormones	Hypothalamus	Anterior pituitary	Regulate hormone secretion
Oxytocin	Synthesized in hypothalamus, secreted from posterior pituitary	Breast Uterus	Triggers milk "let down" Stimulates uterine contraction
Vasopressin (antidiuretic hormone) (ADH)	Synthesized in hypothalamus, secreted from posterior pituitary	Kidney Blood vessels	Increases water reabsorption Causes constriction
Growth hormone (GH)	Anterior pituitary	Many; especially bone, fat, and liver	Stimulates growth of skeleton and muscle
Prolactin (PRL)	Anterior pituitary	Breast	Stimulates milk production
Adrenocorticotropic hormone (ACTH)	Anterior pituitary	Adrenal cortex	Promotes adrenal steroid production
Thyroid-stimulating hormone (thyrotropin) (TSH)	Anterior pituitary	Thyroid gland	Promotes thyroid hormone production
Follicle-stimulating hormone (FSH)	Anterior pituitary	Gonads (in females, ovarian follicle cells; in males, Sertoli cells of testes)	Stimulates growth and development
Luteinizing hormone (LH)	Anterior pituitary	Gonads; ovarian follicle cells Leydig cells of testes	Triggers ovulation Stimulates testosterone production
Insulin	Pancreas—islets of Langerhans	Primarily liver, muscle, and fat	Regulates metabolism and blood glucose
Glucagon	Pancreas—islets of Langerhans	Primarily liver	Regulates metabolism and blood glucose
Somatostatin	Hypothalamus Pancreas—islets of Langerhans	Anterior pituitary Other cells of islets	Inhibits growth hormone secretion Regulates insulin and glucagon secretion
Glucocorticoids	Adrenal cortex	Primarily liver, muscle, and fat	Regulate metabolism
Aldosterone	Adrenal cortex	Kidney	Regulates sodium excretion
Epinephrine	Adrenal medulla	Cardiovascular system Muscle, liver, and fat	Stimulates cardiovascular function Regulates energy metabolism
Angiotensin II	Formed by conversion steps involving kidney, blood, and lung	Adrenal cortex	Stimulates aldosterone production
Atrial natriuretic factor (ANF)	Atrial wall of heart	Primarily kidney	Regulates sodium excretion

Table 12–1 continued
Summary of the Primary Human Hormones

Hormone	Gland or Source	Target Cells	Primary Biological Effect
Thyroid hormones (T$_3$, T$_4$)	Thyroid gland	Many cell types	Regulate energy metabolism
Somatomedins			
Insulin-like growth factor I (IGF-I)	Primarily liver	Primarily bone	Promotes bone growth
Insulin-like growth factor II (IGF-II)	Primarily liver	Many tissues	Promotes tissue growth and repair
Parathyroid hormone (PTH)	Parathyroid glands	Bone, kidney	Regulates plasma calcium and phosphate
1,25 Dihydroxyvitamin D$_3$	Conversion in kidney from precursor formed in liver	Gastrointestinal tract; bone	Regulates plasma calcium and phosphate
Calcitonin (CT)	Parafollicular cells of thyroid gland	Bone	Regulates plasma calcium and phosphate
Gastrointestinal (GI) hormones	Various cells of the GI tract	GI tract; gallbladder and pancreas	Regulates digestive processes, including secretion and motility
Gastrin, secretin, cholecystokinin, gastric inhibitory peptide, and somatostatin			
Androgens	Primarily testes; also adrenal cortex	Reproductive tract	Stimulates growth and development
Estrogens	Ovaries and placenta	Reproductive tract; breasts	Stimulates growth and development
Progesterone	Ovaries and placenta	Uterus; breasts	Promotes proper development and function
		CNS	Inhibits ovulation
Placental hormones	Placenta	Various reproductive tissues	Maintains pregnancy
Chorionic gonadotropin, chorionic somato-mammotropin, estrogen, and progesterone			
Mullerian inhibitory factor (MIF)	Testes	Mullerian ducts	Causes regression of ducts in fetus
Inhibin	Testes	Hypothalamus and pituitary	Inhibits FSH secretion
Melatonin	Pineal gland	Reproductive system	Influences onset of sexual maturity
Erythropoietin	Kidney	Bone marrow	Stimulates red cell formation
Bradykinin	Kidney and other tissues	Blood vessels	Causes vasodilation

processes: (1) the production of metabolic energy, its use, and storage; (2) salt and water balance; (3) growth and development; and (4) reproductive function. Later sections of this chapter describe details of the biochemical mechanisms involved in hormonal regulation of a variety of cellular processes.

The descriptions of target tissues and hormonal effects presented in Table 12–1 are considerably abbreviated and simplified. Most hormones have *several* different target tissues. In addition, many hormones produce multiple individual effects within their specific target cells. These multiple effects, when added together, produce the overall response characteristic of that particular hormone. For example, the male sex hormone, **testosterone,** has several different effects in different tissues. It is required for normal sperm formation to occur in the testes and also promotes the growth and development of most parts of the male reproductive tract, such as the prostate gland and seminal vesicles. Testosterone is also responsible for the development of secondary male sexual characteristics at puberty, including beard growth and a deepening of the voice. Thus, this one single hormone has many different effects in a variety of different tissues.

Just as one hormone may influence many different processes, a single process may in turn be regulated by several different hormones. For example, the process of converting blood sugar into glycogen in the liver is regulated by insulin, glucagon, epinephrine, adrenal glucocorticoids, and thyroid hormone. Several other hormones have minor influences as well. The overall effect is an integrated response reflecting the circulating levels of all these hormones.

Thus, endocrine regulation is characterized by single hormones with multiple effects and by processes regulated by several hormones.

The Chemistry of Hormones

Up to this point, the discussion of hormones has been rather general in nature. Let us now begin to examine the hormones in more detail, beginning first with the chemistry of hormones.

Most Hormones Can Be Characterized as Belonging to One of Three Chemical Classifications

Hormones generally fall into one of three different chemical categories: (1) amino acid derivatives, (2) peptides or proteins, and (3) steroids. Figure 12–5 shows an example of each type of hormone.

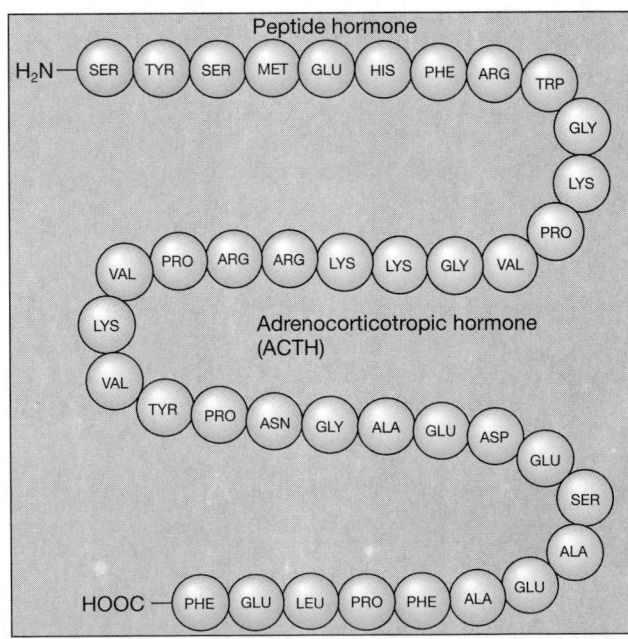

Figure 12–5

Examples of the three types of hormones.

Hormones Derived from Amino Acids

Some hormones are simple derivatives of amino acids either obtained from the diet or synthesized in the body. With just a few relatively simple chemical changes in the molecule, an ordinary amino acid can be converted into the powerful regulatory substance that is a hormone. Examples of hormones that are amino acid derivatives are **thyroxine, epinephrine,** and **melatonin;** their structures are shown in Figures 12–5 and 12–6. Thyroxine consists of four iodide atoms joined to a pair of tyrosines that have been coupled end to end. Epinephrine is derived from tyrosine, and melatonin is formed from the amino acid tryptophan.

Thyroxine: An Iodide Compound.

Although it is an amino acid derivative, thyroxine is synthesized in a somewhat different manner from amines. Details of thyroxine formation will be presented in Chapter 13. Briefly, thyroxine is first formed as part of a much larger precursor protein, known as **thyroglobulin,** which is stored in relatively large amounts within the **thyroid gland.** Because iodide is not always available in the diet, thyroxine is the one hormone in the body that is not solely made from readily available dietary constituents. A system has therefore evolved that allows for the storage of large amounts of thyroxine in its precursor form, thus providing a reservoir of hormone in the event that iodine is not available in the diet for an extended time period (weeks or months). Stimuli for thyroxine secretion result in the rapid breakdown of the thyroglobulin molecule to liberate thyroxine and stimulate the synthesis of new thyroglobulin protein.

Epinephrine and Melatonin: Amines.

Epinephrine and melatonin also represent a subset of hormones known as **amines,** which have a free amino group. Most amines are synthesized, stored, and secreted in a similar way. Within the endocrine cell, an amino acid is converted to an amine by a sequence of specific enzyme-catalyzed steps. The amine is packaged into secretory granules in the cell, where it is held until it is discharged into the blood in response to the appropriate stimulus to the endocrine cell. Stimuli that trigger amine secretion not only cause discharge of preformed granules from the cell but also tend to increase the activity of key regulatory enzymes involved in formation of the hormone. Thus, most stimuli not only increase the rate of secretion of previously formed hormone but also accelerate the rate of formation of new hormone molecules.

Hormones That Are Peptides or Proteins

Many hormones are either peptides or proteins. They can be very small peptides such as **thyrotropin releasing hormone (TRH),** which is a tripeptide produced by neuroendocrine cells in the hypothalamus. It plays a role in the regulation of thyroid hormone production. Peptide hormones can be intermediate in size, such as **adrenocorticotropic hormone (ACTH),** which contains 39 amino acids, or **parathyroid hormone (PTH),** which contains 84 amino acids. There can even be much larger peptides such as **growth**

Figure 12–6

Examples of two hormones (epinephrine and melatonin) that are derivatives of amino acids.

hormone (GH), which in humans is a chain of 191 amino acids. Some are simple, single-chain peptides such as the ACTH and PTH molecules, while others such as growth hormone contain disulfide (S—S) bonds that connect cysteine residues within different portions of the peptide chain. Still others, such as **follicle-stimulating hormone (FSH)**, are composed of different peptide subunits joined together to form the active hormone molecule.

As is true of other proteins in the body, some peptide hormones also contain a significant amount of carbohydrate. One family of **glycoprotein hormones**, which includes **follicle-stimulating hormone (FSH)**, **thyroid-stimulating hormone (TSH)**, **luteinizing hormone (LH)**, and **human chorionic gonadotropin (hCG)**, in addition to having similar basic structures, average about 20% carbohydrate by weight.

Thus, the peptide hormones are quite diverse in terms of their chemical structure, ranging from small, single-chain peptides to very large, multiunit glycoproteins.

Precursors of Peptide Hormones. As are other proteins destined for secretion, peptide hormones are synthesized on ribosomes of the rough endoplasmic reticulum (see Chapter 3). From the rough ER, they are transported to the Golgi complex, where they are packaged into vesicles prior to secretion from the cell. During these transport and packaging steps in the cell, some very important events take place. Many, if not most, peptide hormones are first synthesized in the form of much larger **precursor** molecules. In some cases, the precursor may even contain the amino acid sequence for several hormones or hormone-like substances. The precursor for ACTH, for example, contains the ACTH sequence as well as the sequences for at least four other bioactive peptides. The precursor for TRH contains several repetitions of the TRH sequence, so that a number of TRH molecules may be formed from a single precursor molecule. During the intracellular transport phase of hormone secretion, the precursor molecule is "clipped" by specific proteolytic ("protein-breaking") enzymes to produce the mature, active hormone molecules that will be secreted into the blood.

When an endocrine cell that produces a peptide hormone receives an appropriate stimulus, vesicles containing the mature hormone rapidly move toward the cell surface, and the contents are released by the process of exocytosis (see Chapter 3). Much as they do in other cell types, calcium ions play a key role in coupling a stimulus with secretion from the endocrine cell. In most cases, the amount of hormone available in preformed vesicles is somewhat limited and usually only enough to support high rates of se-

cretion for a matter of minutes. To compensate for this, stimuli that promote peptide hormone secretion usually also promote increased rates of synthesis of the hormone or its precursor.

Figure 12–7

(a) Chemical structure of cholesterol. The letters A through D designate each of the four rings in the molecule. Numbers designate each of the 27 carbon atoms in cholesterol. (b) Examples of steroid hormone structure. Each steroid shown represents one of the five classes of steroid hormones.

Hormones Derived from Cholesterol: Steroids

The third major category of hormones are the steroid hormones. Figure 12–7a shows the structure of cholesterol, which is the precursor for all of the steroid hormones. Differences in biological activity between the various steroid hormones are the result of sometimes very subtle alterations or substitutions in the cholesterol backbone. Figure 12–7a also shows the designation of each of the four rings (A to D) and the numbering convention used to designate each of the 27 carbon atoms in the cholesterol molecule. These are used when the various steroids are described according to strict chemical nomenclature. For simplicity, however, in most instances the steroids are referred to by their common names.

In the body, the major sites of steroid hormone production are the adrenal cortex, ovaries, testes, and the placenta. Hormonally active metabolites of vitamin D, which are also derived from cholesterol and have structures similar to the steroid hormones, are produced by sequential modifications occurring first in the liver and then in the kidney.

Types of Steroid Hormones.

The steroid hormones can be grouped into five categories according to the type of biological process they regulate: (1) glucocorticoids, (2) mineralocorticoids, (3) androgens, (4) estrogens, and (5) progestins. Glucocorticoids, produced by cells in the cortex of the adrenal gland, are involved in regulating glucose and energy metabolism (see Table 12–1). Mineralocorticoids are the principal steroids produced by the outer portion of the adrenal cortex and are involved in regulating sodium balance by the kidneys. Androgens, the male sex hormones, are produced primarily in the testes of the male, although the adrenal cortex produces and secretes physiologically significant amounts of androgens as well. Estrogens, the female sex hormones, are secreted by the ovaries and the placenta. The progestins are involved in the maintenance of pregnancy and are also produced by the ovaries and the placenta.

Figure 12–7b shows examples from each category of steroid hormone. This figure shows that differences in structure of these steroids are relatively small. In the body, however, these small changes in structure are translated into markedly different biological activities. This phenomenon is a result of the high degree of receptor specificity, causing them to recognize and respond to one particular hormone and not to another.

Pathways in Steroid Hormone Synthesis.

Conversion of the cholesterol precursor into each of the steroid hormones occurs by a number of well-defined steps, each catalyzed by a specific enzyme within the steroid-producing cell. Figure 12–8 is a simplified diagram of the primary synthetic pathways involved in the formation of each of the major steroids. It should be evident from this figure that several common steps are involved in the formation of each of the five classes of steroids. For example, both cortisol and aldosterone are formed from progesterone, and thus the conversion of pregnenolone to progesterone is common to the synthesis of both steroids. Similarly,

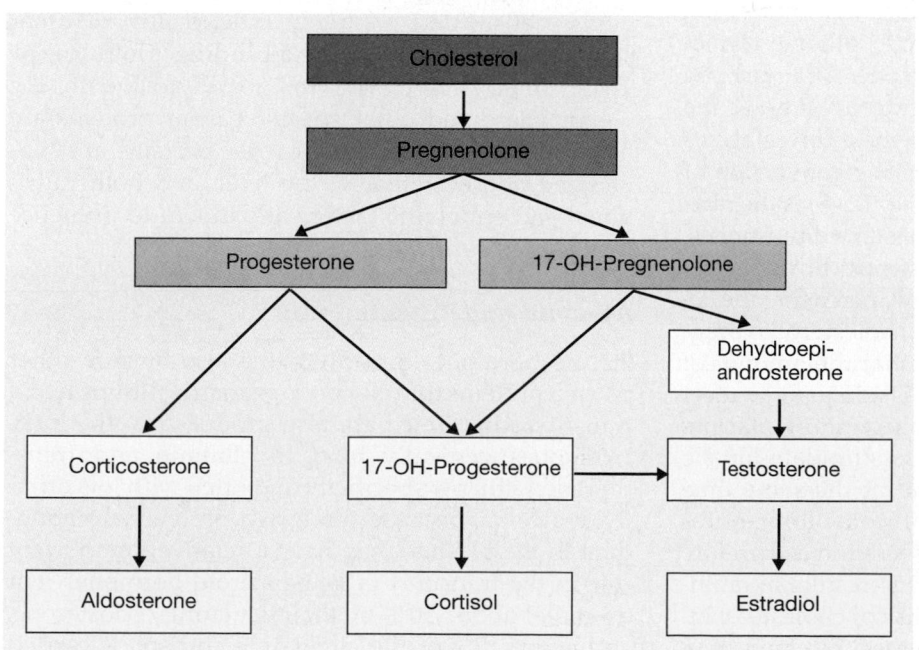

Figure 12–8

Simplified diagram of the pathways of steroid hormone synthesis. Note that the first step in synthesis of all steroids is the conversion of cholesterol to pregnenolone.

testosterone, an androgen, also serves as an intermediate in the formation of **estradiol,** an estrogen.

Given these common steps, what factors determine which hormone a particular steroid-producing endocrine gland will produce? As indicated previously, each step in steroid hormone biosynthesis is catalyzed by a specific enzyme. The relative amount of each of these enzymes in the endocrine cell determines the particular steroids produced. For example, cells in the outer portion of the adrenal cortex are the site of most aldosterone production in the body. These cells have relatively high amounts of the enzymes needed to convert pregnenolone to aldosterone, but have little or none of the enzymes needed to convert pregnenolone or progesterone to cortisol, testosterone, or estradiol. Similarly, the steroid-producing cells of the testes possess the enzymes needed to form testosterone, but not those needed to form significant amounts of cortisol or aldosterone. Likewise, they have relatively low amounts of the enzyme needed to convert testosterone into estradiol. Estrogen-producing cells of the ovary, however, have a set of steroid-metabolizing enzymes similar to that of the testes, but they also have relatively high amounts of the enzyme needed to transform testosterone into estradiol. Thus, each of these tissues produces and secretes different steroid hormones because of the different enzymes they contain.

Factors Affecting the Rate of Steroid Synthesis. Unlike peptide or protein hormones, which are stored in secretory vesicles, steroid hormones are synthesized and secreted only on demand. They are not stored in the endocrine cell prior to secretion, but exit the cell immediately after formation. The rate of steroid hormone synthesis therefore determines the rate of hormone secretion. As with other enzyme-catalyzed reactions, steroid hormone synthesis can be influenced by alterations in the activity of the "rate-limiting enzyme" (the enzyme that catalyzes the slowest step) in the pathway. The conversion of cholesterol to pregnenolone (see Fig. 12–8) is the rate-limiting step in the synthesis of all steroid hormones. Factors that increase the rate of steroid hormone secretion therefore do so primarily by increasing the activity of the enzyme that converts cholesterol to pregnenolone. In addition, to ensure that the amount of cholesterol inside endocrine cells is adequate to meet the needs for steroid hormone biosynthesis, factors that increase steroid synthesis also stimulate uptake of cholesterol from the blood and at the same time cause a release of free cholesterol from storage sites within the cell. Steroid hormone secretion is therefore increased by two mechanisms. One results in an increase in the uptake and availability of cholesterol inside the cell, and the other stimulates the conversion of cholesterol to pregnenolone.

Some Hormones Are Carried in the Blood Via Transport Proteins

Once a hormone is released from the endocrine cell, it travels via the blood, in most cases, to reach its site of action in the body. This can present two problems. The first and most important is that several hormones are only slightly soluble in the blood. For example, because they are lipid in nature, steroid hormones have limited solubility in aqueous solutions. Thyroxine, the principal hormone secreted by the thyroid gland, is also relatively insoluble in the plasma. Therefore, some mechanism is needed to increase the solubility of these hormones in the aqueous environment of the blood. The second problem is that, because of the small size of several of the hormones, they would be readily lost from the kidneys into the urine if allowed to exist free in solution in the plasma. Large peptide or protein hormones, because they generally are freely soluble in water and large enough to not pass through the kidneys into the urine, do not present either of these problems.

Specific Transport Proteins

To solve the problems of solubility and excessive urinary loss, thyroxine and the steroids are bound to large "carrier" proteins for transport in the blood. Generally these hormones bind to two types of carrier proteins: (1) specific transport proteins that bind one particular hormone with a relatively high degree of specificity and a relatively high affinity; and (2) albumin and transthyretin, which bind hormones without high specificity or affinity.

An example of the first type of carrier protein is **cortisol-binding globulin,** or **CBG,** which primarily binds cortisol and relatively little, if any, of other steroids. Similarly, **thyroxine-binding globulin,** or **TBG,** binds and carries mainly thyroxine in the blood. These and other specific carrier proteins are listed in Table 12–2. These carrier proteins are synthesized and secreted by the liver, and both nutritional and endocrine factors are known to influence their rate of synthesis.

Albumin and Transthyretin

Steroid hormones and thyroxine also bind to other plasma proteins for transport, primarily **albumin** and **transthyretin,** which are also produced in the liver. Hormones generally bind to albumin and transthyretin with low specificity and often with low affinity. However, because these two proteins are abundant in the plasma, they have a relatively important role in the transport of some steroid hormones. For example, about 50% of the circulating aldosterone, and about 10% of the circulating cortisol, is carried by binding to albumin.

Table 12–2
Circulating Carrier Proteins

Name	Principal Hormone Carried
Specific carrier proteins	
Cortisol-binding globulin (CBG)	Cortisol, some aldosterone
Thyroxine-binding globulin (TBG)	Thyroxine (T$_4$)
Sex steroid-binding globulin (TeBG)	Testosterone and estradiol
General carrier proteins	
Albumin	Many steroids, thyroxine
Transthyretin	Thyroxine, some steroids

Free and Bound Hormone: A Dynamic Equilibrium

The binding of steroid or other hormones to these plasma proteins occurs according to the principles of a simple chemical equilibrium. Thus, the free and the bound forms of the hormone exist in the plasma in a dynamic equilibrium with one another. In most instances, the equilibrium favors the bound form, and the majority of the hormone therefore exists in the bound state. However, a finite amount will also be present in a free, or unbound, state. Only the hormone present in the free state is able to move from the blood across the walls of capillaries (the small blood vessels that bring blood into contact with most cells) to reach cells and interact with hormone receptors. The free hormone is therefore the physiologically active form of the hormone. As free hormone is lost from plasma, it is replaced by hormone newly released from carrier proteins. Similarly, if a sudden large burst of hormone secretion occurs, most of the newly secreted hormone molecules will become bound to the carrier proteins, and the free hormone concentration will increase only slightly. Thus, in addition to facilitating hormone solubility and transport, carrier proteins also serve as a reservoir and buffer system to help maintain a relatively constant concentration of the free hormone in the blood.

Peripheral Transformation, Degradation, and Excretion of Hormones Are Important Considerations Regarding Hormone Action

In most cases, the form of the hormone secreted by the endocrine cell is the form directly responsible for producing the biological effects of the hormone; hormones need not be transformed to be fully active.

Two notable exceptions to this general rule are thyroxine and testosterone. Both hormones are converted "peripherally" (within some of their target tissues) to forms displaying biological potency many times greater than the original hormone. Details of the conversion of thyroxine and testosterone to their more active forms will be presented in Chapters 13 and 30, respectively. Also possibly included on this list of peripherally transformed hormones are **angiotensin II** and **1,25 dihydroxyvitamin D$_3$**, both of which are formed by sequences of conversion steps involving several different tissues.

The concentration of any one hormone circulating in the blood is determined by both its rate of secretion and the rate at which it is removed or metabolized to an inactive form. For most hormones, the major pathways for removal or degradation occur in the liver or kidneys.

Steroid hormones are primarily degraded by the liver and the metabolites secreted by the liver into the bile, eventually being eliminated in the feces. The liver first converts the steroid into a relatively less active form and then, to make the molecule more water soluble, couples the steroid to a polar sulfate or glucuronide group in a process known as conjugation. The **conjugated** steroids are then primarily excreted in the urine with smaller amounts being secreted by the liver into the bile. Other hormones, such as epinephrine, are inactivated by specific degrading enzymes circulating in the blood and are then rapidly excreted in the urine. Determination of the rates of urinary excretion of a variety of steroid hormones or epinephrine metabolites has proven to be a useful indirect index of the rate of secretion of the active hormone and has provided physicians with an important diagnostic tool.

Many of the larger peptide hormones, such as insulin and prolactin, are taken into their target tissue cells by a process known as **receptor-mediated endocytosis.** After the hormone binds to its receptor on

CLINICAL FOCUS

The Enzyme-Linked Immunosorbent Assay

As described in the text, the radioimmunoassay (RIA) allowed for significant advancements in clinical medicine and in research. Before the RIA, assays of hormone concentrations relied on the use of whole animals or tissues/cells derived from animals. In general, these assays lacked sensitivity and specificity, and were somewhat costly to perform. The development of the RIA overcame a number of these problems, but not all.

The classic radioimmunoassay, as the name implies, utilizes radioisotopes. In recent years the disposal of even low-level radioisotopes such as used in an RIA has become an environmental concern. As the number of disposal sites for radioactive waste disposal has dwindled to near zero, the economic cost to the producers of these wastes has skyrocketed. Thus the ultimate cost of these assays to the patient and the researcher has increased as well. In addition, a typical RIA is somewhat labor-intensive in that it may take one technician several days to process a hundred or so samples, which contributes to the final cost per sample to be evaluated. Moreover, by nature of the assay, the time cannot be reduced. Thus, the time to obtain results may also be an issue in some instances.

The **Enzyme-Linked Immunosorbent Assay** (*ELISA*) provides an answer to many of the problems currently posed by use of the RIA. Results from an ELISA ultimately depend on development of a colored or fluorescent product that can be detected using optical devices, and thus radioisotopes are not involved. An ELISA is typically automated, which allows for analysis of a large number of samples in any one period of time. Furthermore, the time between starting an ELISA and obtaining a final result is typically reduced as compared to a RIA.

ELISA assays are typically performed in formed polystyrene (plastic) plates approximately the size of a postcard. Each plate contains a grid of 96 small wells (8 × 12). Each well is approximately the diameter of a standard pencil and twice the depth of a pencil eraser. The polystyrene plastic has a positive charge and attracts most proteins and a variety of other negatively charged substances. As a result, the hormone to be assayed (as well as numerous other proteins) binds to the charged plastic surface. Unbound materials are washed out of the well, leaving a variety of proteins bound to the plastic surfaces of the well. An antibody that recognizes the hormone in question is added, and after various washing and reagent addition steps a colored, fluorescent, or luminescent product is formed. The amount of product formed is determined by optical means and is in direct proportion to the amount of hormone present in the initial sample. These results are compared with a set of samples that are analyzed concurrently in which varying but known amounts of hormone are added to a series of wells. A *standard curve* in which the amount of hormone added (x-axis) versus the amount of product formed (y-axis) is then used for determination of the amount of hormone in the unknown samples.

A typical ELISA is specific, sensitive, and economical. Robotic instruments can perform the various pipetting, addition, and withdrawal steps involved, and the wastes that are generated can be disposed of in the regular trash. Thus, wherever possible and practical the ELISA is replacing the RIA.

There are several instances in clinical medicine in which the measured hormone concentration as determined using either RIA or ELISA is very high, and yet the patient does not display the expected signs and symptoms of an elevated hormone level.

■ **What might be the basis for an RIA or ELISA test to yield high results, and why aren't these apparently high hormone levels expressed in the body in terms of a clinical problem?**

---

the surface of target cells, the hormone-receptor complex is internalized or taken into the cell. Once internalized, the hormone is separated from its receptor. In most cases, the hormone molecule is then proteolytically degraded, and the majority of the internalized receptors are recycled back to the cell surface. Many peptide hormones are taken up and degraded by the kidneys, although the exact contribution made by the kidneys can vary considerably depending on the hormone.

Measurement of Hormone Concentrations in the Blood Provides Important Information to the Clinician and the Researcher

As indicated earlier, hormones are present in extremely low concentrations in the blood, usually in the range of 10^{-9} to 10^{-12} M. To put these low concentrations into perspective, consider that a hormone concentration of 10^{-9} M corresponds roughly to one hormone molecule in the blood per 50 billion water molecules. Thus, exceedingly sensitive techniques are needed to allow the measurement of circulating hormone concentrations. The measurement of circulating hormone concentrations, either in the basal (resting) state or after an appropriate stimulus has been given, is an important step in the diagnosis of many different endocrine disorders. Because of the similarity in structure and activity of many of the hormones, hormone assays (measurements) must be highly specific if meaningful results are to be obtained.

In the late 1950s, two American scientists, Solomon Berson and Rosalyn Yalow, developed an assay technique that is both extremely sensitive and highly specific. The technique can be performed quite rapidly and inexpensively. The assay that Berson and Yalow developed is known as the **radioimmunoassay (RIA).** The basic technique, described in Chapter 3, has been modified and adapted in a number of different ways, but these newer assays are still based on the original principal of **competition binding,** in which the amount of radioactive hormone that binds to an antibody reveals how much nonradioactive hormone is present. This assay technique can be used to measure the concentrations of hormones in the plasma and in various other bodily fluids, such as amniotic fluid or cerebrospinal fluid. In addition, the concentration of a variety of drugs and vitamins can also be measured by this technique. As a result, the radioimmunoassay has revolutionized the field of endocrinology, both clinically and in the area of basic research. For their efforts in the development of the technique, Berson and Yalow were awarded the Nobel Prize in Physiology and Medicine in 1977.

Mechanisms of Hormone Action

For convenience, hormones can be considered to use one of two different general mechanisms of action to produce their effects. One mechanism, used by amine and peptide hormones, involves membrane receptors and the generation of an intracellular signal or "second messenger" (see Chapter 5). These hormones take effect by altering the activity of proteins that already exist in the cell. The other mechanism, used by steroid hormones and thyroxine, uses intracellular receptors and involves alterations in the expression of particular genes within the nucleus of the cell. These latter hormones work by initiating production of new proteins in the cell. These two models of hormone action will be considered in more detail shortly. First, let us look at the nature of hormone receptors.

Receptors Are a Key Element in the Initiation of Hormone Action

As was described earlier, a **receptor** is a molecule inside the cell or in the plasma membrane that specifically recognizes and binds one particular hormone. Binding of the hormone to the receptor initiates a sequence of events that ultimately results in the biological effects typical of that hormone. Therefore, specificity within the endocrine system is determined at the level of the hormone receptor.

Receptors are themselves proteins, or in some cases glycoproteins, and thus have definite three-dimensional structures. As illustrated in Figure 12–3b, only those hormones with a structure complementary to the structure of a specific receptor can bind to and activate the receptor. A very simple key and lock analogy works here: the key functions like a hormone, and the lock like a receptor.

The ability of receptors to discriminate between hormones is not absolute, however. Similarities in structure between hormones can result in an overlap of hormonal activities. For example, cortisol primarily produces effects typical of a glucocorticoid, but it also interacts weakly with aldosterone receptors and therefore is considered a weak mineralocorticoid. Because of the similarities in structure among the steroid hormones, this type of overlapping activity is fairly common among these hormones. Similar, though fewer, examples of overlapping activities can be found among the peptide hormones. Growth hormone and prolactin have similar structures, and each displays a weak degree of the other's activity. Thus, while receptors are responsible for the specificity that exists within the endocrine system, the inability of the receptor to discriminate absolutely between hormones having similar structures leads to some overlap of activities. Normally, these weak secondary activities are of little or no physiological consequence. However, in an abnormal situation in which the circulating concentration of a particular hormone is markedly elevated, these secondary activities of the hormone may be strong enough to produce noticeable biological effects.

The binding of a hormone to its receptor involves weak chemical forces and is usually readily reversible. When a hormone binds to its receptor, the three-dimensional conformation of the receptor is altered. In this regard, a hormone functions in a manner similar to that of an allosteric modifier of an enzyme. The alteration in structure of the receptor is the first step in the sequence of events leading to the biological effects of the hormone. The magnitude of the biological response is therefore proportional to the number of receptors occupied by hormone molecules. However, receptors for any particular hormone are present only in a finite or limited number in their target tissue cells. It follows, therefore, that once a certain hormone concentration is reached at which all receptors are occupied, no greater response can be produced by further increases in the hormone concentration. Thus, for each biological effect of a hormone, there is a maximal response that can be produced by that hormone.

Current evidence indicates that all the biological effects of hormones are the result of interactions with their respective receptors. The hormones themselves do not directly affect other cellular constituents, such as enzymes, but instead use their receptors as intermediaries to produce the desired effects. The receptor therefore serves as a **signal transducer,** converting a specific extracellular hormonal signal into an intracellular signal that will redirect or alter cell function.

Given the role of the receptor in signal transduction, what is the nature of the intracellular signal generated by the receptor, and how does a single message or signal within the cell lead to a variety of different effects of the hormone?

The "Second Messenger" Model of Hormone Action Describes How Hydrophilic Hormones Produce Their Effect

Peptide hormones are generally hydrophilic ("water-loving") and therefore not soluble in the lipid layer of the plasma membrane. As a result, they are unable to readily penetrate the cell membrane and gain access to the interior of the cell. These hormones must therefore use mechanisms involving intracellular sig-

nals, or second messengers, to produce their effects. For this class of hormone, the receptor is an intrinsic membrane protein with its hormone-binding component exposed to the extracellular fluid. As illustrated in Figure 12–9, binding of the hormone to that portion of its receptor exposed to the outside of the cell results in generation of a second messenger. Many of the membrane receptors are glycoproteins oriented such that the carbohydrate portion of the molecule is exposed to the outside of the cell. As discussed in Chapter 5, several second messengers are known. Details of the mechanisms involved in the formation of each and the role they play in hormone action will be discussed shortly.

These surface receptors are not rigidly fixed in place in the membrane, as was once thought, but instead show considerable lateral mobility. In addition, as discussed in the earlier section on hormone degradation, a number of peptide hormones have been shown to undergo receptor-mediated endocytosis, in which both the receptor and the hormone are taken into the cell in an endocytotic vesicle. Binding of several peptide hormones to their receptors has also been shown to induce a clustering of the occupied receptors into patches in the plasma membrane. This clustering is believed to facilitate the endocytotic uptake of the hormone-receptor complexes.

As was indicated in Chapter 5, three intracellular second messenger systems have been identified and characterized: (1) the adenylate cyclase/cyclic AMP system, (2) the phosphatidylinositol and diacylglycerol system, and (3) receptor-linked ion channels. For some peptide hormones, such as insulin, the mechanism of action has yet to be fully determined, so there may be additional, as yet unrecognized, second messengers.

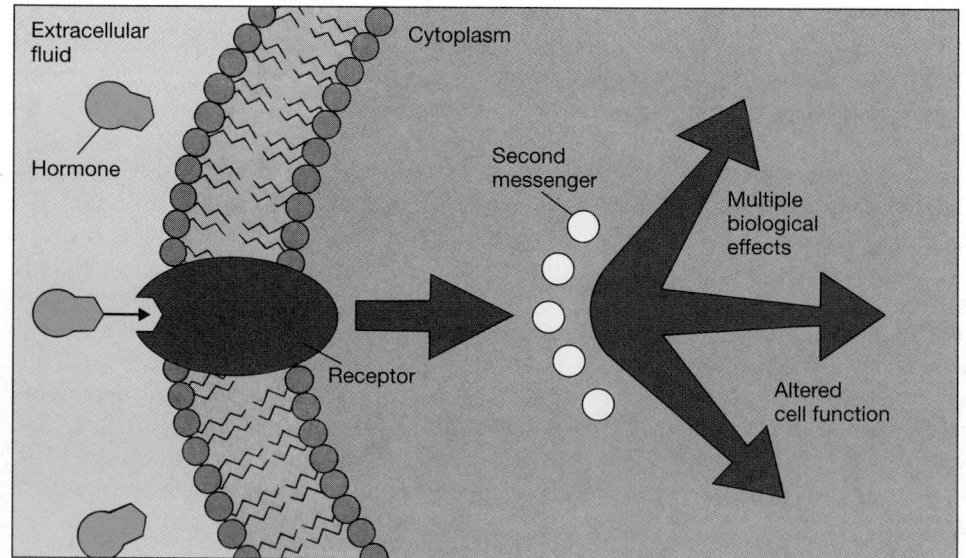

Figure 12–9

The general action mechanism of hormones with membrane receptors.

Figure 12–10

Conversion of ATP to cAMP by the enzyme adenylate cyclase, and the subsequent conversion of cAMP to AMP by phosphodiesterase.

Cyclic AMP as a Second Messenger

Cyclic 3′,5′-adenosine monophosphate, or more simply **cyclic AMP (cAMP),** was one of the first second messengers to be recognized and has been extensively studied and characterized. As shown in Figure 12–10, this cyclic nucleotide is generated from cellular ATP by the action of the enzyme **adenylate cyclase,** which is located on the inner surface of the cell membrane. Cyclic AMP is rapidly inactivated by conversion to AMP, a process catalyzed by one or more enzymes known as **phosphodiesterases.** As a result of its rapid inactivation, any increase in cytoplasmic cAMP will be short-lived, allowing for the rapid termination of any effects it mediates.

The activity of adenylate cyclase is coupled to membrane receptors by another group of integral membrane proteins known as **G-proteins,** so named because they bind and require **guanosine triphosphate (GTP)** to function. One type of G-protein stimulates adenylate cyclase, while a second type of G-protein inhibits it. These are referred to as the **G_s** and **G_i proteins,** respectively (Fig. 12–11). Binding of hormones to membrane receptors (designated as **R_s** in

Figure 12–11

Illustration of the coupling of membrane receptors to adenylate cyclase by the G_s and G_i proteins.

Table 12–3
*Some Hormones That Use cAMP as Second Messenger**

Hormone	Target Tissue
Adrenocorticotropic hormone (ACTH)	Adrenal cortex
Epinephrine	Heart skeletal muscle, and adipose
Glucagon	Liver
Luteinizing hormone (LH)	Testis and ovary
Parathyroid hormone (PTH)	Kidney and bone
Thyroid-stimulating hormone (TSH)	Thyroid gland
Follicle stimulating hormone (FSH)	Testis and ovary

*This is only a partial list of hormones known to act via cAMP. Many other hormones and local mediators too numerous to list also stimulate adenylate cyclase and raise intracellular cAMP concentrations.

the figure) coupled to adenylate cyclase through G_s proteins results in stimulation of adenylate cyclase and an increase in intracellular cAMP concentrations. Similarly, binding of a hormone to receptors (R_i receptors) linked to the G_i proteins will result in inhibition of adenylate cyclase and a reduction in cAMP. Table 12–3 shows an abbreviated list of hormones that produce their effects at least in part as a result of changes in the intracellular concentration of cAMP. Most hormones act through a stimulatory receptor and therefore produce their effects by increasing intracellular cAMP. These changes in cellular cAMP occur within a matter of just a few seconds after the hormone binds to its receptor. Likewise, the intracellular effects of cAMP occur very rapidly; hence, the

overall system is well suited to producing very rapid, acute effects within the cell.

A number of agents, including some toxins, are known to bypass the receptor and act directly on various components of the adenylate cyclase system. A toxin released by the bacteria that causes cholera directly activates the G_s protein (see Fig. 12–11). This in turn activates adenylate cyclase, resulting in the continual formation of cAMP, even in the absence of the external hormonal signal normally required for adenylate cyclase activation. In the intestine, cAMP is known to be a potent activator of fluid secretion by the mucosal cells that line the intestinal walls. The effect of cholera toxin on the G_s protein therefore accounts at least in part for the severe diarrhea charac-

Figure 12–12

Example of how an increase in intracellular cAMP produced in response to a hormone can result in many different biological responses in the cell. The particular set of kinases shown in this figure is purely hypothetical. Some cells may have fewer kinases, some more, and some may have a different mixture altogether.

teristic of this disease. This is just one example of how the molecular basis of a number of disease states may lie in a disruption of normal events involved in cell-to-cell signaling.

Given that certain hormones act by altering cellular cAMP concentrations, how can a change in the concentration of this single compound then lead to the many different effects that some hormones are known to produce within one cell type? As was described in Chapter 5, cAMP is known to produce its effects by activating enzymes known as **cyclic AMP-dependent protein kinases.** Each particular protein kinase catalyzes the **phosphorylation** of (addition of a phosphate group to) a limited number of specific cellular proteins. Phosphorylation of these proteins results in alterations in their activities, which in turn leads to specific changes in cell function. As indicated in Figure 12–12, the activation of several different protein kinases within a cell by cAMP can result in a variety of different intracellular effects. In this way, binding of a hormone to its receptor on the surface of cells and generation of a single intracellular signal can regulate a variety of different processes within the cell. In addition, the particular set of kinases present in cells may vary from one tissue to the next, such that responses elicited as a result of increases in cAMP vary from tissue to tissue.

Although not as widespread as the cAMP second messenger system, **cyclic GMP (cGMP)** serves as a second messenger in some tissues. Mechanisms for cGMP formation, second messenger action, and removal from the cell are essentially analogous to that for cAMP.

Inositol Trisphosphate and Diacyglycerol as Second Messengers

Although turnover of membrane lipids was first suggested as a possible mechanism for intracellular signaling in the 1950s, it is only within the past few years that the importance of this mechanism has become widely appreciated. **Phosphatidylinositol** is a minor phospholipid constituent of the plasma membrane and is primarily located in the inner half of the bilayer. Upon the donation of two phosphate groups from ATP, phosphatidylinositol is converted to **phosphatidylinositol 4,5-bisphosphate (PIP₂)**, which serves as the immediate precursor for generation of two intracellular second messengers. As shown in Figure 12–13, PIP₂ is split to form **diacylglycerol (DG)** and **inositol trisphosphate (IP₃)**. This cleavage is catalyzed by a membrane-bound enzyme known as **phospholipase C.** Several other enzymes function to replenish the supply of membrane PIP₂, therefore preventing its depletion.

As with adenylate cyclase, activity of phospholipase C is functionally coupled to membrane hormone

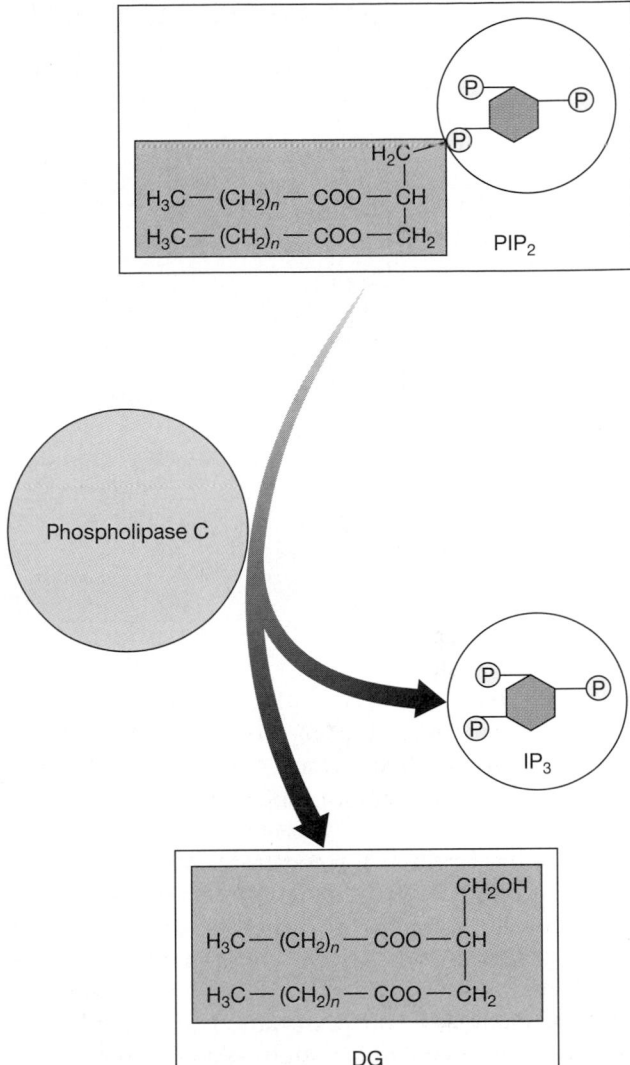

Figure 12–13

Pathway of conversion of phosphatidylinositol 4,5-bisphosphate (PIP₂) to inositol trisphosphate (IP₃) and diacylglycerol (DG) by phospholipase C.

receptors by a GTP-binding protein (G-protein). There appears to be only a stimulatory G-protein, since as yet no inhibitory G-protein has been identified for this system. Figure 12–14 illustrates the functional relationship between a receptor, G-protein, and phospholipase C. As indicated, activation of phospholipase C results in hydrolysis of PIP₂ to form IP₃ and DG. These two substances both play a role as intracellular second messengers, although they produce their effects by different mechanisms.

The IP₃ released from intracellular membranes acts inside cells by causing release of calcium ions from storage sites in the endoplasmic reticulum, thereby raising the concentration of free cytosolic calcium. As described in Chapter 5, calcium serves an important role as an intracellular signal or mediator

Figure 12–14

Diagram of the formation and the mechanisms of action of inositol trisphosphate (IP_3) and diacylglycerol (DG) as intracellular second messengers.

in a variety of different processes. Thus, the cellular effects of one branch of this mechanism ultimately result from increased cytoplasmic calcium concentrations.

The other product formed from PIP_2 hydrolysis, diacylglycerol (DG), remains associated with the membrane. It activates a membrane-bound protein kinase known as **C-kinase.** In a manner analogous to that described for cAMP-dependent kinases, C-kinase catalyzes phosphorylation of a number of specific cellular proteins, thereby altering their activity and producing specific changes in cell function. Although many details of this two-branch IP_3/DG pathway have yet to be fully characterized and understood, it appears that the two mechanisms seem to act together to regulate a number of different processes.

The overall biochemistry of inositol phosphate metabolism is in fact much more complicated than described in the preceding paragraphs. A number of multiply-phosphorylated inositol phosphates, such as IP_4, IP_5, and IP_6, have now been identified in mammalian cells. However, details of the metabolic pathways leading to their formation and their postulated functions are not completely understood.

Ions as Second Messengers

A third signal transduction mechanism used by membrane receptors is that of receptor-linked ion channels. In certain hormone/receptor systems, the change in configuration of a receptor caused by hormone binding results in opening of specific, gated, ion channels. Other membrane proteins such as G-proteins may also be involved in coupling the

changes in receptor configuration to the functional changes in ion channels.

Opening of ion channels in the membrane generates a signal in one of two ways. The first, which operates mainly in electrically active cells such as nerve

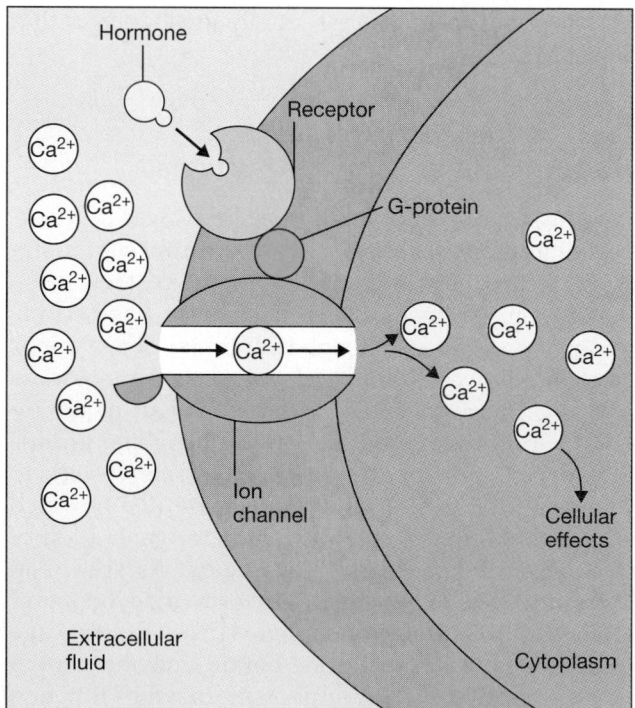

Figure 12–15

Coupling of membrane receptors to membrane calcium channels.

Extracellular fluid

Cytoplasm

ATP

Enzyme

Kinase region

Hormone binding region

ADP

Phosphorylated enzyme

P

Biological effects

Figure 12–16

Possible mechanism of action of hormones that have receptors with intrinsic protein kinase activity.

cells and muscle, involves a small and transient flux of ions, which alters the membrane potential without any appreciable change in cytoplasmic concentrations of the ions. The second mechanism involves a major influx of ions into the cytoplasm, which in turn initiates a cellular response. This mechanism operates primarily in cells that are not electrically active. Because of the important role of calcium ion as an intracellular signal, it is not surprising that a number of membrane receptors are linked to calcium channels, as illustrated in Figure 12–15. Hormone binding activates the receptors, causing calcium channels to open and allowing calcium ion to flow into the cell, down their electrochemical gradient. Once inside the cell, calcium functions as a second messenger to produce a variety of effects by activating a number of calcium-dependent proteins (see Chapter 5).

As indicated earlier, the mechanism of action of some peptide hormones is currently unknown. However, insulin and certain peptide growth factors have unique receptor structures and activities that may be related to their mechanism of action. As illustrated in Figure 12–16, receptors for these agents are transmembrane proteins that exhibit an intrinsic protein kinase activity. Thus, a single receptor molecule has both an external hormone binding site and an internal enzymatically active region. As Figure 12–16 depicts, interaction of hormone with the external binding site results in activation of the protein kinase activity of the receptor. In this regard, it is easy to see how these hormones act as allosteric modifiers of receptor function, converting the relatively inactive receptor kinase into a more active form. Presumably,

the kinase catalyzes phosphorylation of cytoplasmic proteins, altering their activity and producing the desired biological effects. At present, the degree to which these events are responsible for the biological effects of insulin and the growth factors is not completely understood.

The "Gene Expression" Model of Hormone Action Describes How Hydrophobic Hormones Produce Their Effect

Unlike peptide hormones, which are generally hydrophilic, steroid hormones and thyroxine are hydrophobic, or lipid soluble, and therefore can readily pass through plasma membranes and other membranes within cells. As a result, these hormones do not require a second messenger system to produce their effects, but instead can enter cells and bind to intracellular receptors. In addition to this basic difference from peptide hormones, the manner in which these hormones regulate cellular function and the time course of changes they produce differ considerably from those of hormones that utilize membrane receptors and second messengers.

Figure 12–17 depicts the general mechanism of action for thyroid hormones and the steroids. After dissociating from carrier proteins in the plasma, these hormones freely move across the cell membrane into the cytoplasm. For many years, scientists believed that receptors for steroid hormones existed free in the

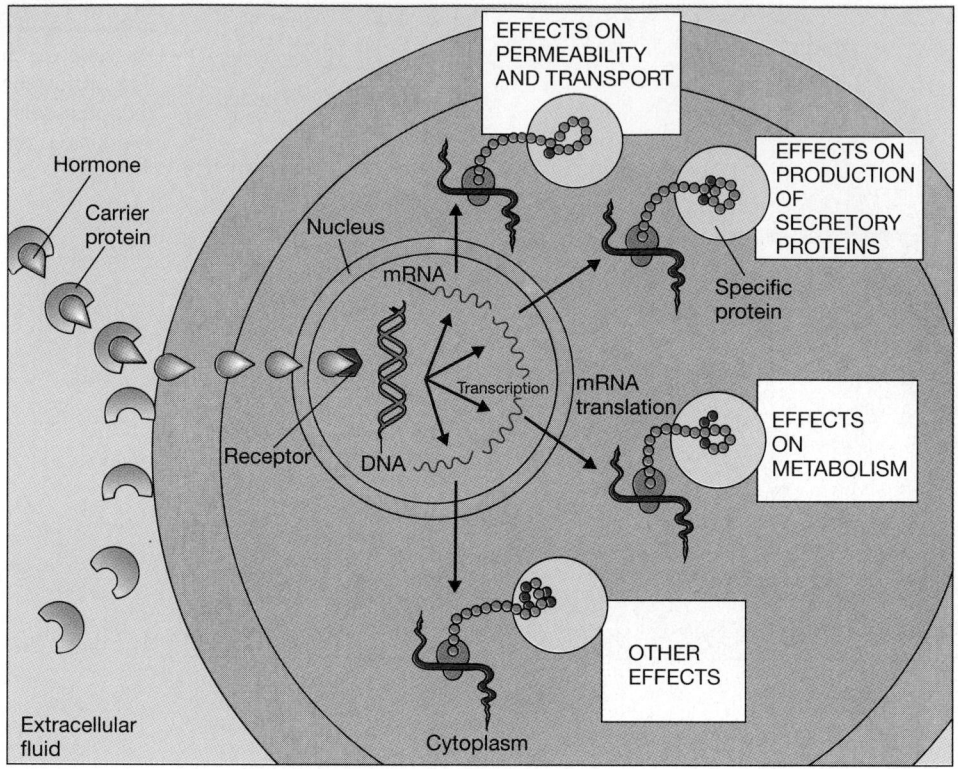

Figure 12–17

Mechanism of action of hormones with intracellular receptors, indicating the effects on gene expression and the possible ways in which cell function might be altered.

cytoplasm of target cells. However, recent evidence suggests that these receptors exist either on the nuclear envelope or within the nucleus of the cell. Binding of a hormone molecule to the receptor causes a conformational change in the receptor so that the hormone/receptor complex then interacts with specific "acceptor sites" on the nuclear DNA. This interaction initiates transcription of mRNA molecules that code for specific proteins. After transcription, the mRNA precursors are processed within the nucleus before moving into the cytoplasm of the cell (see Chapter 5). Because of the increased abundance of these particular mRNA molecules, synthesis of a few specific proteins is increased. Increased synthesis of these particular proteins leads to the altered cellular function characteristically produced by the particular hormone. As indicated in Figure 12–17, these newly synthesized proteins may be involved in a variety of different cellular functions.

The cellular response to a particular hormone acting via these mechanisms usually results from increased synthesis of several proteins. Although each of these proteins may serve a different role in the cell, each contributes in some way to producing the final coordinated biological effect. For example, aldos-

terone, a mineralocorticoid produced by the adrenal cortex, has the effect of increasing sodium transport across tubule cells of the kidney. As shown in Figure 12–18, sodium first enters the cell from the lumen of the tubule (right side of Figure 12–18), by diffusion through specific sodium channels and is then pumped out on the opposite (left) side of the cell by an energy-requiring sodium pump. Aldosterone both increases the synthesis of several mitochondrial citric-acid-cycle enzymes involved in ATP production, making more energy available to power the sodium pump, and stimulates synthesis of sodium channel proteins, making more sodium available to the sodium pump. There is some evidence that aldosterone also increases the synthesis of sodium pump proteins themselves. The overall effect is to increase the transfer of sodium across kidney tubule cells. Thus, as is typical of other steroid or thyroid hormones that take effect by regulating events within the nucleus, aldosterone alters the function of its target cells by changing the rate of synthesis and therefore the amount of several key proteins in the cell. Recall that, by contrast, hormones that act through membrane receptors using second messengers generally do so by modifying the activity of preexisting

Figure 12–18

Mechanism of action of aldosterone to increase sodium transport across kidney tubule cells.

proteins, using such mechanisms as phosphorylation or dephosphorylation.

The Time of Action for Hormones Using the "Second Messenger" Model and Those Using the "Gene Expression" Model Differs Considerably

Another basic difference between the cell-surface and intracellular receptor systems involves the time course over which responses occur. In general, responses to hormones that use the second messenger system can be observed in a matter of seconds or minutes. In addition, once the hormonal stimulus is withdrawn, cell function rapidly returns to its basal (resting) state. In contrast, responses to steroid or thyroid hormones are usually not evident for many minutes or hours, and in some cases, the response may even take days to develop. Likewise, after removal of the hormonal signal, it may take an equal period of time for the effect to diminish and the cells to return to their original state. With these two mechanisms, the endocrine system is able to closely regulate a number of cellular processes on both a short-term (acute) and a long-term (chronic) basis.

Amplification of Hormone Signals Is an Important Aspect of Hormone Action

One characteristic feature within the endocrine system is amplification, or increase in signal strength. As a result of this amplification, binding of just a few hormone molecules to their receptors can result in a dramatic change in cell function. Signal amplification

(text continues on page 397)

Growth Factors, Cellular Communication, and Oncogenes

In recent years, there has been a nearly exponential increase in the number of growth factors identified and at least partially characterized. As the name implies, exposure of certain cells or tissues to specific growth factors stimulates growth in those cells. In addition, growth factors also stimulate a variety of metabolic processes in their target tissues. A typical growth factor is peptide in nature and generally exerts its hormone-like actions via autocrine or paracrine mechanisms. In most instances, the signal transduction pathways utilized by growth factors are similar if not identical to the mechanisms used by peptide and protein hormones described in the accompanying sections of this chapter.

The area of cellular communication mechanisms and signal transduction has been the subject of active research for many years. Much attention has been focused on these processes with the hope that understanding the pathways responsible for regulating normal cell growth and function would lead to greater insight into conditions of uncontrolled growth, such as in cancer. In recent years, this assumption has been supported by the finding that altered versions of protein-encoding genes that function in signal transduction pathways related to cell growth are present in certain forms of animal cancer. These altered genes are known as *oncogenes* (from the Greek *onkos*, meaning mass or tumor). The oncogenes arise as a result of a virus picking up either a fragment or a corrupted copy of a normal gene from its host cell. As a result of normal viral activity, the gene fragment is inserted into subsequent host cells, where the oncogene product, or *oncoprotein,* is produced.

Normally, when a growth factor reaches its target cells, the signal transduction pathway leading to growth stimulation is only transiently activated. Presumably, growth-inhibitory signals are generated at some point to prevent cellular proliferation from proceeding indefinitely. In cells containing oncogenes, however, pathways are continuously activated, causing cells to endlessly grow and replicate. In many instances, oncogenes encode proteins (oncoproteins) that are very similar to the gene products found in normal cells, except that these proteins have lost important regulatory effects on their activity, and therefore do not require activation by an external signal. Thus, this situation is clearly one in which the molecular basis of disease results from a disruption of the normal events involved in cell-to-cell signaling and communication.

In fact, oncogenes (and their oncoproteins) have been identified that are associated with various components of hormone-signaling pathways. For example, epidermal growth factor, or EGF, normally binds to and activates a cell-surface receptor that is a transmembrane tyrosine kinase similar to that illustrated in Figure 12–16. The *erbB* oncogene (the name is an abbreviation of the virus in which it was discovered) has been shown to code for a protein analogous to the EGF receptor, but lacking the extracellular binding region of the normal EGF receptor. In the absence of the extracellular EGF binding domain, the intracellular tyrosine kinase activity of this altered receptor remains continuously activated, and therefore those processes that the receptor normally regulates are under continuous stimulation. Numerous other examples are also readily evident: the **ras** oncogene encodes a protein analogous to a normal G-protein except that an amino acid substitution results in continuous activation of the counterfeit G-protein; the **erb-A** oncoprotein is an unregulated version of the thyroid hormone receptor; the **src** family of oncogenes code for a group of unregulated tyrosine kinases; and the **raf/mil** oncoproteins are cytoplasmic serine/threonine kinases that lack normal regulated function.

The study of oncogenes is likely to remain an area of intense research for many years. As a result of studying oncogene biology, we have learned much about structure/function relationships of various components of hormone action pathways. For example, studies of the erbB oncoprotein suggest that in a normal EGF receptor, the extracellular domain of an unoccupied receptor continuously inhibits the tyrosine kinase activity of the cytoplasmic portion of the receptor. Normally, when EGF binds to the receptor, this inhibitory effect is released, leading to activation of the cytoplasmic kinase portion of the receptor. In addition, the insight about normal signaling processes that we have gained from the study of oncogenes should permit the design of improved pharmaceutical agents that will restore normal growth.

■ **Suppose you were working for a pharmaceutical company trying to develop new drugs to counteract these oncoproteins and oncogenes. As a starting point, what might be the general types of cell proteins you hope to target?**

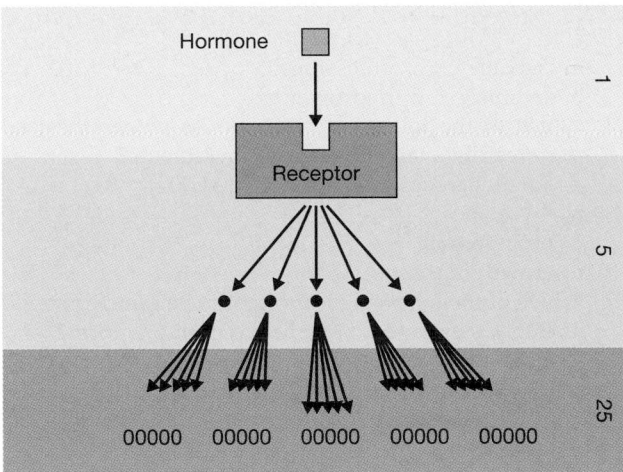

Figure 12–19

Illustration of how signal amplification might occur with regard to hormone action.

original signal consisting of one hormone molecule has now been multiplied many times over. Amplification occurs because each step in the sequence involves enzymes that, when active, are responsible for forming not just one but many "signal" molecules. For example, when glucagon binds to its receptor, it activates adenylate cyclase. Adenylate cyclase catalyzes the formation of many cAMP molecules, each of which in turn activates a particular protein kinase. Each cAMP-activated protein kinase can now catalyze the phosphorylation of many copies of a particular enzyme, leading to an alteration in the activity of many enzyme molecules. Thus, just one glucagon molecule can result in the activation of many individual cell proteins. Amplification allows hormones to be effective regulators of cell function even though they are present in the blood in extremely small amounts.

occurs at nearly every step in hormone action, from binding of a hormone molecule to its receptor to the final biological effect produced by the hormone. An example of signal amplification is illustrated in Figure 12–19. It easy to see that after just a few steps, the

CHAPTER REVIEW

Summary

- Hormones provide a chemical means for communication between cells or groups of cells. The endocrine system and the nervous system are the two primary means for communication in the body, and these two systems function in a coordinated manner to promote homeostasis. Neuroendocrine cells provide a means by which information in the nervous system can be transferred into endocrine signals.

- In general, hormones regulate processes related to either energy metabolism, salt and water metabolism, growth and development, or reproduction. Most hormones act on several tissues and produce several different specific effects.

- Chemically, hormones generally fit into one of three categories: amino acid derivatives, peptide or protein hormones, and steroid hormones, which are derived from cholesterol. Steroids can be further subdivided into glucocorticoids, mineralocorticoids, androgens, estrogens, and progestins.

- Steroid hormones and thyroid hormones travel in the bloodstream by binding to specific carrier proteins and to albumin and other plasma proteins. Peptide and protein hormones generally do not require carrier proteins for transport.

- The radioimmunoassay allows for the measurement of hormone concentrations in the blood. Development of

the assay was an extremely important advancement in the field of endocrinology.

- The receptor is the molecular entity of the cell that recognizes the hormone and initiates its biological effects. Receptors are largely responsible for the specificity that exists within the endocrine system, although the specificity is not absolute in every case.

- Peptide and protein hormones generally act via plasma membrane receptors present on the cell surface. Binding of a hormone to its receptor on the cell surface results in the generation of intracellular second messengers, which alter or regulate cell function. Currently known second messengers include cyclic AMP, inositol trisphosphate, diacylglycerol, and certain ions, most notably calcium.

- Steroid hormones exert their effects via intracellular receptors located on or in the nucleus. The hormone/receptor complex interacts with the DNA to alter transcription of messenger RNA molecules coding for specific proteins. Alterations in the rate of synthesis of these specific proteins lead to the particular cellular response characteristically produced by the hormone.

- Signal amplification occurs at nearly every step in hormone action, allowing hormones to be effective regulators of cell function even though they are present in the blood in extremely small amounts.

Review Questions

Identify Each with a Term

1. The terms for the mechanisms by which hormones reach their target cells solely by diffusion, rather than via the bloodstream.
2. The term for the cell type that links the nervous system and the endocrine system.
3. The precursor for synthesis of all steroid hormones.
4. The assay system commonly utilized to measure hormone concentrations in the blood.
5. The enzyme responsible for the formation of cAMP.
6. The names of the two second messengers formed as a result of the hydrolysis of phosphatidylinositol 4,5-bisphosphate (PIP$_2$) by phospholipase C.

Define Each Term

7. hormone receptor
8. neuroendocrinology
9. mineralocorticoids
10. conjugation of steroids
11. G$_s$ protein
12. receptor-mediated endocytosis
13. phosphodiesterase

Choose the Correct Answer

14. Which of the following is *not* a steroid hormone?
    a. aldosterone   c. cortisol
    b. epinephrine   d. progesterone
15. Which of the following is *not* one of the five categories of steroid hormones?
    a. progestins
    b. estrogens
    c. prostaglandins
    d. glucocorticoids
16. Which of the following is a glycoprotein?
    a. parathyroid hormone (PTH)
    b. growth hormone (GH)
    c. thyroxine (T$_4$)
    d. follicle-stimulating hormone (FSH)
17. Which of the following does *not* appear to use cAMP as a second messenger?
    a. glucagon
    b. epinephrine
    c. thyroid-stimulating hormone (TSH)
    d. insulin
18. Which of the following exerts its effects primarily via the gene expression model of hormone action?
    a. growth hormone
    b. aldosterone
    c. parathyroid hormone
    d. luteinizing hormone (LH)
19. Hormones are involved in regulating processes related to:
    a. production, use, and storage of metabolic energy
    b. reproductive functions
    c. growth and development
    d. salt and water balance
    e. all of the above
20. The ion thought to be involved in stimulus-secretion coupling of hormone secretion by endocrine cells is:

a. calcium   c. potassium
b. sodium   d. hydrogen ion

21. Which of the following hormones is *not* synthesized from readily available dietary constituents?
    a. adrenocorticotropic hormone (ACTH)
    b. thyroxine
    c. progesterone
    d. growth hormone
22. Which of the following is thought to be transformed in its target tissues to a more active hormone form?
    a. insulin
    b. follicle-stimulating hormone (FSH)
    c. estradiol
    d. testosterone
23. Receptors for steroid hormones are located:
    a. within the nucleus
    b. in the cytoplasm of nontarget cells
    c. always near the site of synthesis of the hormone
    d. in the nuclear membrane
    e. either A or D, not B or C
24. Which of the following is *not* a general characteristic of peptide hormones?
    a. are transported in the blood largely free in solution
    b. are synthesized and secreted on demand
    c. utilize cell membrane receptors and intracellular second messengers
    d. are usually first synthesized as a larger precursor protein

Answer in One or Two Sentences

25. What is the role of hormone receptors in hormone action, and how do receptors convey specificity within the endocrine system?
26. What is the general mechanism of action of steroid and thyroid hormones?
27. Describe the adenylate cyclase/cAMP second messenger system and explain how an increase in intracellular cAMP can lead to different effects in different cell types.
28. Describe the phosphatidylinositol second messenger system.

Problem Solving in Physiology

29. Hormones circulate in the bloodstream in vanishingly small amounts, yet they regulate the intracellular function of cells where the concentration of vital components such as metabolic intermediates like ATP may exist in the millimolar or micromolar range (10^{-3} M to 10^{-6} M). How can just a few hormone molecules regulate or influence other molecules that are present at a million or a billion times higher concentration?
30. There are numerous diseases that result from a lack of hormone action. Starting from the initial synthesis and secretion to the final action of the hormone, list the steps that potentially could be responsible for a deficiency in hormone action.
31. Compare the time of action for peptide versus steroid hormones.

Suggested Readings

Berridge, Michael J. "The molecular basis of communication within the cell." *Scientific American*, 253:142–152, 1986.

Berridge, Michael J., and Irvine, Robin F. "Inositol phosphates and cell signaling." *Nature*, 341:197–205, 1989.

Druker, Brian J., Mamon, Harvey J., and Roberts, Thomas M. "Oncogenes, growth factors, and signal transduction." *New England Journal of Medicine*, 321:1383–1391, 1989.

Goodman, H. Maurice. *Basic Medical Endocrinology*, 2nd ed. New York: Raven Press, 1994.

Griffin, James E., and Ojeda, Sergio R. *Textbook of Endocrine Physiology*. New York: Oxford University Press, 1992.

Hanly, Rochelle M., and Steiner, Alton L. "The second-messenger system for peptide hormones." *Hospital Practice*, 24(8):59–70, 1989.

Lewis, Deborah L., Lechleiter, James D., Kim, Donghee, Nanavati, Chaya, and Clapham, David E. "Intracellular regulation of ion channels in cell membranes." *Mayo Clinic Proceedings*, 65:1127–1143, 1990.

Linder, Maurine E., and Gilman, Alfred G. "G proteins." *Scientific American*, 267(1):56–65, 1992.

Rasmussen, Howard (ed.). "Cell Communication in Health and Disease." *Readings from Scientific American Magazine*. New York: W.H. Freeman and Company, 1991.

Snyder, Solomon H. "The molecular basis of communication between cells." *Scientific American*, 253:132–141, 1986.

Key Terms

adenylate cyclase (p. 389)
aldosterone (p. 378)
autocrine (p. 373)
cholesterol (p. 302)
cortisol (p. 382)
cortisol binding globulin (p. 384)
cyclic AMP (cAMP) (p. 389)
cyclic AMP-dependent protein kinases (p. 391)
cyclic GMP (cGMP) (p. 391)

diacylglycerol (DG) (p. 391)
endocrine system (p. 374)
estrogen (p. 383)
G-proteins (p. 389)
glucocorticoid (p. 383)
glycoprotein hormones (p. 382)
hormone receptors (p. 373)
human chorionic gonadotropin (hCG) (p. 382)

inositol triphosphate (IP_3) (p. 391)
neuroendocrine (p. 375)
paracrine (p. 373)
phospholipase C (p. 391)
progesterone (p. 379)
radioimmunoassay (RIA) (p. 387)
receptor-mediated endocytosis (p. 385)

receptors (p. 373)
second messengers (p. 395)
steroid hormones (p. 383)
target tissue (p. 373)
testosterone (p. 380)
thyroxine binding globulin (p. 384)
transthyretin (p. 384)

Photo micrograph of a follicle
cell from the thyroid gland.
*(© Biophoto Associates/Photo
Researchers, Inc.)*

Chapter 13
The Pituitary Hormones

Key Concepts

The pituitary consists of two main lobes (anterior and posterior) that have different embryonic origins and different regulatory mechanisms. Ultimately, the hypothalamus of the brain serves to regulate hormone secretion from both lobes.

The anterior pituitary is regulated in a typical endocrine fashion and secretes six different hormones into the general circulation. The posterior pituitary is a typical neuroendocrine gland and contains the terminals of neuroendocrine cells whose cell bodies are in the hypothalamus.

Growth hormone is one of six hormones produced by the anterior pituitary. It affects skeletal growth as well as protein and carbohydrate metabolism. Thyroid-stimulating hormone (TSH) is also produced by the anterior pituitary and regulates thyroid hormone production by the thyroid glands. Thyroid hormones in turn regulate metabolism in most tissues in the body.

Chapter Outline

Anatomy of the Pituitary Gland and Its Relation to the Hypothalamus

Regulation of Pituitary Hormone Secretion by the Hypothalamus

Actions and Effects of the Pituitary Hormones

On a Saturday night during a recent break from college, Julia's parents told her that they were going out to hear the local symphony. They had an extra ticket and asked if she would be interested in going with them. Although her musical tastes are far from classical, she decided to go along; she really didn't want to stay home and study.

Julia and her parents arrived shortly before the symphony was to start. The musicians were all in place and going through their warm-up routines, producing a jumble of notes and chords played at different volumes and tempos. Julia thought to herself, "My parents call rock music just a bunch of noise, so how can they consider this any better?" Just then, the musicians concluded their warm-up, the hall fell silent, and the conductor walked onto the stage. He acknowledged the applause of the audience, turned to the orchestra, and, as he began to move his baton, the hall became filled with a harmony of sounds unlike anything Julia had ever experienced before.

As the symphony progressed, there was an obvious control that the conductor had over each section of the orchestra and the individual musicians involved. At various times he would motion for one group of instruments to play forcefully, while directing another group to soften their play. Julia recognized that, while the conductor does not produce the individual notes that create the music, he or she does play a significant role in ensuring that the amplitude and timing of individual instruments is correct.

In a similar way, the pituitary gland plays a central role in regulating several other endocrine glands. Depending on the body's physiologic status at any particular time, the pituitary will signal to other endocrine glands to increase or decrease their level of production of hormone, much like the conductor signals to the musicians in his orchestra.

As will be described in this chapter, the pituitary is itself regulated by the hypothalamus of the brain; thus it is really the hypothalamic/pituitary unit that serves as the endocrine "conductor." Together they ultimately regulate a variety of functions related to metabolism, growth, reproduction, and water and mineral balance in the body.

The pituitary gland plays a central role in hormonal regulation of a wide variety of processes and has been one of the most extensively studied of all endocrine glands. As we will describe, pituitary function is largely controlled by the hypothalamus. The interaction of the pituitary and hypothalamus is therefore an excellent example of interaction between the nervous and endocrine systems.

We begin with a discussion of pituitary hormones and regulation of pituitary function by the hypothalamus. Later sections of this chapter describe actions of two pituitary hormones: growth hormone and thyroid-stimulating hormone.

Anatomy of the Pituitary Gland and Its Relation to the Hypothalamus

During Development the Pituitary Is Formed from Two Different Tissue Sources

In humans, the pituitary measures about 1 cm (1/2") in diameter and is located just below the hypothalamus at the base of the brain (Fig. 13–1). As described in Chapter 10, the hypothalamus serves as an integrative center for a variety of different bodily processes. The hypothalamus also controls pituitary function and therefore serves as an important regulator for the endocrine system. The pituitary is not a single gland, but consists of three separate glands, or

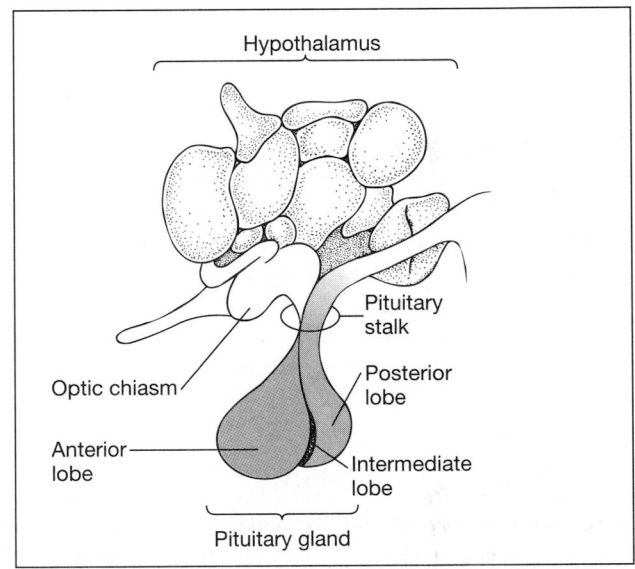

Figure 13–1

The location of the pituitary gland relative to the brain and hypothalamus.

lobes: the **anterior lobe,** the **intermediate lobe,** and the **posterior lobe.** The pituitary is connected to the hypothalamus by a thin segment of tissue known as the **pituitary stalk.** The hormones secreted by each of the three different lobes of the pituitary are listed in Table 13–1. As this table indicates, the pituitary gland secretes at least nine different hormones. However, as will be described later, the pituitary does not itself synthesize all nine.

Regulation of hormone secretion by each of the lobes of the pituitary involves different mechanisms. Cells of the anterior pituitary behave as true endocrine cells (see Fig. 12–1*a*) in that they receive signals or information via the blood and in turn release their own hormones directly into the blood. The posterior pituitary, in contrast, is an extension of the ner-

Table 13-1
Hormone Secreted by the Pituitary Gland

Hormones secreted from the anterior lobe
    Luteinizing hormone (LH)
    Follicle-stimulating hormone (FSH)
    Prolactin (PRL)
    Adrenocorticotropic hormone (ACTH)
    Growth hormone (GH)
    Thyroid-stimulating hormone (TSH)
Hormones secreted from the intermediate lobe
    Melanocyte-stimulating hormone (MSH)
Hormones secreted from the posterior lobe
    Oxytocin
    Antidiuretic hormone (ADH, also known as
        vasopressin)

(a)

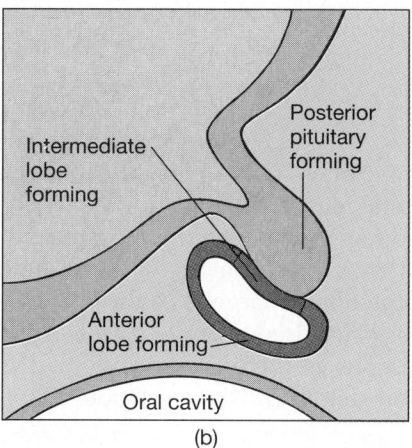

(b)

vous system and functions like a group of neuroendocrine cells. In humans, the intermediate lobe is rudimentary and does not exist as a distinct lobe per se, but consists of only a few cells interspersed along the borders of the anterior and posterior lobes. Although the intermediate lobe serves an important role in a number of lower vertebrates, it has little or no endocrine function in humans.

Both Vascular and Neural Components Are Important for the Regulation of Pituitary Function

An examination of the embryologic development of the pituitary illustrates how both the endocrine and neuroendocrine portions of the gland are formed. As Figure 13-2 indicates, during early development a pocket of cells known as **Rathke's pouch** pinches off from the roof of the primitive mouth cavity. At about the same time, a finger-like projection of neural tissue extends downward from the base of the hypothalamus. Rathke's pouch meets this neural protrusion of the hypothalamus, and the two structures fuse to form the anterior and posterior lobes of the pituitary, respectively.

Neural connections between the hypothalamus and the posterior lobe of the pituitary are maintained during development. As Figure 13-3 shows, the posterior pituitary contains the axons and terminals of neurons that have their cell bodies within the hypothalamus. These neurons terminate close to numerous capillaries located throughout the posterior pituitary. As indicated in Figure 13-3, the cell bodies of these neurons are found within two distinct areas

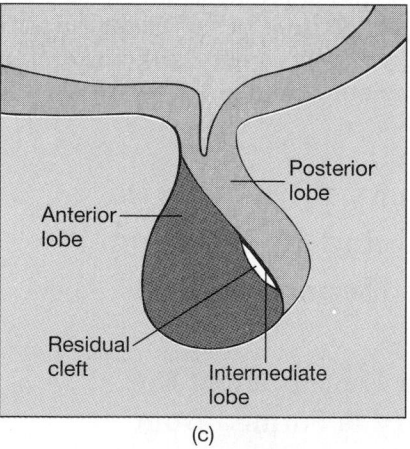

(c)

Figure 13-2

Formation of the anterior, posterior, and intermediate lobes of the pituitary during embryological development. The normal sequence of development is from (*a*) through (*c*). Note that the posterior lobe arises from neural tissue, while the anterior lobe is formed from Rathke's pouch. The intermediate lobe forms from the portion of Rathke's pouch that makes contact with the neural tissue. The residual cleft (*c*) is the remnant of what was once the interior of Rathke's pouch.

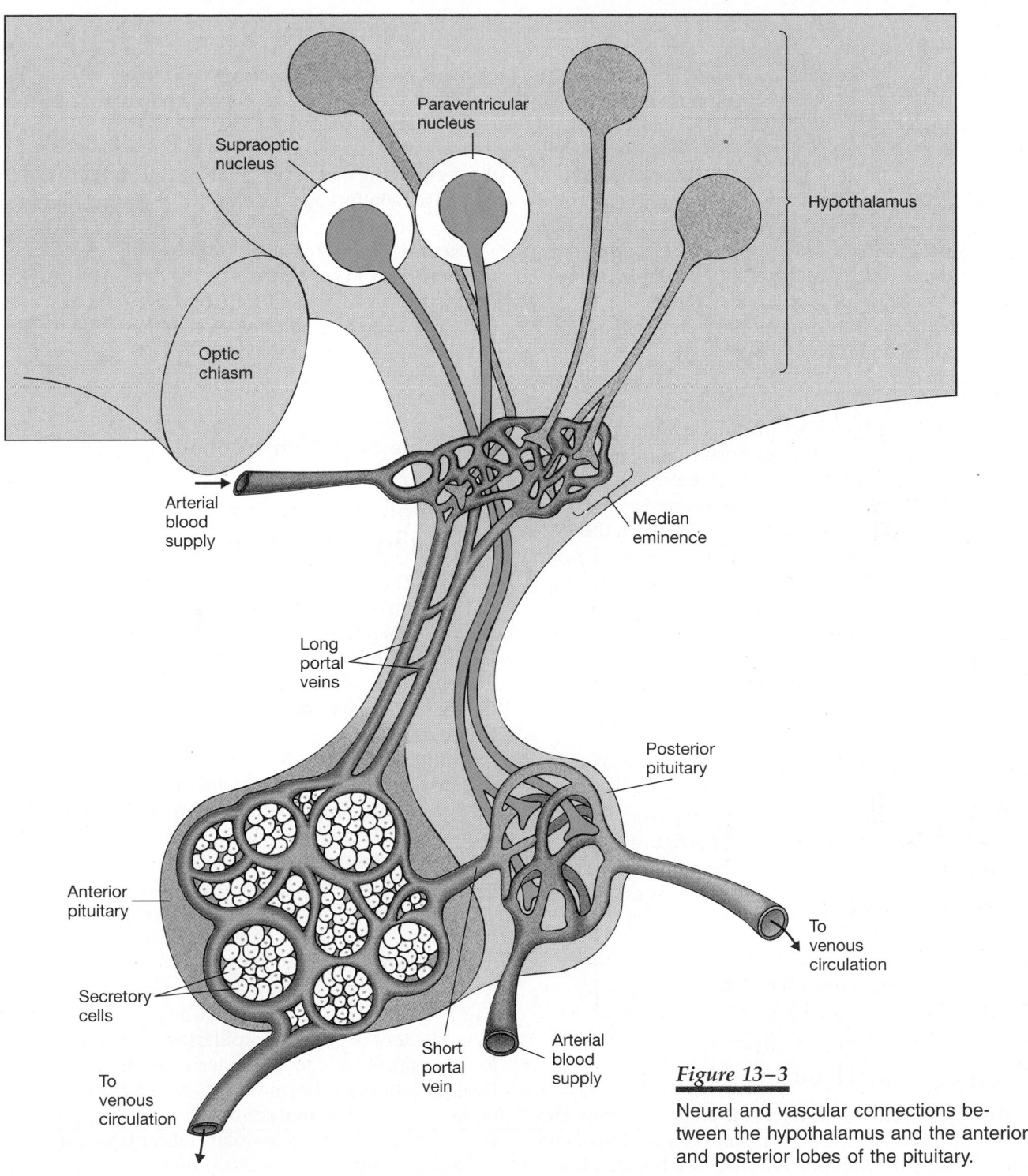

Figure 13–3

Neural and vascular connections between the hypothalamus and the anterior and posterior lobes of the pituitary.

of the hypothalamus, the **supraoptic nucleus** and the **paraventricular nucleus.** The significance of these anatomical features will become evident when the regulation of posterior pituitary hormone secretion is discussed in the next section.

In contrast to the posterior pituitary, function of the anterior pituitary is regulated not by direct innervation, but instead by hormones that are pro-duced in the hypothalamus and are then carried to the anterior pituitary via blood vessels. As shown in Figure 13–3, arterial blood reaching the hypothalamus enters a specialized region known as the **median eminence,** where vessels branch into a network of primary capillaries. From these primary capillaries, the blood enters the **long portal veins** and is carried down through the pituitary stalk to the anterior lobe of the pituitary, where a second capillary network is located. From this second capillary network in the anterior pituitary, blood leaves the gland through the veins and then mixes with the systemic venous blood.

This complex of blood vessels, which consists of the primary capillary network, the secondary capillary network, and the vessels that connect them, is known as the **hypothalamic-pituitary portal system.** As will be described later, the hypothalamic-pituitary portal system carries factors from the hypothalamus to the anterior pituitary that serve to regulate the secretion of anterior pituitary hormones.

In addition to the long portal veins that connect the hypothalamus and anterior pituitary, a less prominent portal system connects the posterior and anterior lobes of the pituitary. This vascular system is referred to as the **short portal system.** Some evidence suggests that the short portal veins carry substances secreted by the posterior lobe to the anterior lobe, where they regulate hormone secretion from the anterior pituitary. The exact degree to which this system is involved in regulating anterior pituitary function is not well understood at this time, however.

Regulation of Pituitary Hormone Secretion by the Hypothalamus

The mechanisms by which the hypothalamus regulates secretion of hormones from both the posterior and anterior lobes of the pituitary are shown in diagrammatic form in Figure 13–4. This figure illustrates the functional significance of neural and vascular connections between the hypothalamus and the pituitary shown in Figure 13–3.

Hormone Secretion by the Posterior Pituitary Occurs by a Typical Neuroendocrine Regulatory Mechanism

Neural connections between the hypothalamus and posterior pituitary are shown on the right side of Figure 13–4. As indicated previously, cell bodies of neurons that send projections to the posterior pituitary are located primarily in the supraoptic and paraventricular nuclei of the hypothalamus. As indicated in Table 13–1, the posterior pituitary secretes two hormones: **oxytocin** and **antidiuretic hormone (ADH,** or vasopressin). These two hormones are actually synthesized within cell bodies of neurons in the supraoptic and paraventricular nuclei of the hypothalamus. Oxytocin and ADH are transported to the posterior pituitary through axons of the same neurons that pro-

duce them. Generation of action potentials in these nerves results in release of oxytocin or vasopressin from nerve terminals adjacent to capillary beds in the posterior pituitary, much the same way that neurotransmitters are released into a synapse. The hormones enter the blood vessels and are then carried throughout the body. Cells that produce oxytocin and ADH are therefore good examples of neuroendocrine cells.

Neural inputs to the supraoptic and paraventricular nuclei from other areas of the hypothalamus or from other areas of the brain modulate electrical activity of the neuroendocrine cells in these two nuclei. In this manner, a variety of stimuli are capable of regulating oxytocin and ADH secretion.

Hormone Secretion by the Anterior Pituitary Occurs by a Typical Endocrine Regulatory Mechanism

As Table 13–1 indicates, cells of the anterior pituitary synthesize six different hormones. Secretion of these hormones is under the control of several different **releasing/inhibiting hormones** produced in the hypothalamus. All of the releasing/inhibiting hormones that have been chemically characterized thus far are either small peptides or amino acid derivatives. As illustrated in Figure 13–4a, these hormones are synthesized within the cell bodies of neuroendocrine cells located in a number of different sites in the hypothalamus. Axons of these neuroendocrine cells converge in the median eminence of the hypothalamus, where they terminate on or near the primary capillary network of the hypothalamic-pituitary portal system. Generation of action potentials in these neuroendocrine cells causes secretion of particular releasing/inhibiting hormones from their nerve terminals. Upon release, the releasing/inhibiting hormones enter the primary capillaries and travel via the long portal vessels to the anterior pituitary. The releasing/inhibiting hormones then exit from the secondary capillaries and act on the endocrine cells of the anterior pituitary to regulate (stimulate or inhibit) secretion of anterior lobe hormones.

Table 13–2 lists the hypothalamic releasing/inhibiting hormones involved in regulating anterior pituitary secretion. Note that some stimulate secretion of a particular hormone, while others inhibit hormone secretion. As a group, they are therefore referred to as "releasing/inhibiting hormones." The system would be relatively easy to understand if there were a simple one-to-one relationship between each anterior pituitary hormone and a single releasing hormone or a pair of releasing and inhibiting hor-

Cell bodies of neurons that produce releasing/inhibiting hormones

Hypothalamus

Arterial inflow

Primary capillaries in median eminence

Releasing hormones
Anterior pituitary hormone

Long portal veins

Releasing/ inhibiting hormones

ANTERIOR PITUITARY

Secretory cells that produce anterior pituitary homones

Anterior pituitary hormones

Venous outflow

Paraventricular nucleus
Supraoptic nucleus

Cell bodies of neurons that produce posterior pituitary hormones

POSTERIOR PITUITARY

Primary capillaries

Posterior pituitary hormones

Venous outflow

Arterial inflow

Prolactin	Gonadotropic hormones (FSH and LH)	Thyroid-stimulating hormone	ACTH	Growth hormone

(a)

Oxytocin
Antidiuretic hormone (ADH)

(b)

Figure 13-4

An illustration of the mechanisms involved in regulating of anterior and posterior pituitary function. (*a*) Neuroendocrine cells of the hypothalamus transmit releasing/inhibiting hormones to the capillaries in the median eminence. These hormones travel via the long portal veins to the secretory cells of the anterior lobe, where they stimulate or inhibit the release of anterior pituitary hormones into the blood. (*b*) Neurons that originate in the supraoptic and paraventricular nuclei of the hypothalamus synthesize oxytocin and ADH and transmit them through axons that release the hormones into capillaries in the posterior pituitary. The hormones are then distributed to the general circulation.

Table 13–2
Releasing Hormones Involved in Regulating Anterior Pituitary Secretion

Hypothalamic-releasing Hormone	*Primary Effect on Anterior Pituitary**
Corticotropin-releasing hormone (CRH)	Stimulates ACTH secretion
Thyrotropin-releasing hormone (TRH)	Stimulates TSH secretion
Gonadotropin-releasing hormone (GnRH)	Stimulates LH and FSH secretion
Somatostatin	Inhibits GH secretion
Growth-hormone-releasing hormone (GHRH)	Stimulates GH secretion
Prolactin-releasing factor (PRF)**	Stimulates prolactin secretion
Prolactin-inhibitory hormone (PIH)	Inhibits prolactin secretion

*As described in the text, several of the releasing hormones influence the secretion of more than one anterior pituitary hormone. For simplicity, only the primary effects of the releasing hormones are presented here.

**The chemical structure of the substance which stimulates prolactin secretion has not been established. By convention, it is referred to as a "factor" until its structure is known.

mones. However, a single hypothalamic hormone may influence the secretion of several anterior pituitary hormones. For example, **somatostatin** regulates the secretion not only of growth hormone but also of **prolactin** and **TSH.** Similarly, GnRH regulates the secretion of both LH and FSH. The degree to which multiple activities of these hypothalamic releasing/inhibiting hormones regulate hormone secretion from the anterior pituitary has not yet been fully established. The general belief, however, is that secretion of any particular hormone from the anterior pituitary depends on the relative amounts of the various hypothalamic releasing/inhibiting hormones that reach the anterior pituitary.

Cell bodies of neurons that synthesize releasing/inhibiting hormones are not randomly distributed throughout the hypothalamus, but instead each releasing factor is synthesized by neurons located in different, discrete nuclei of the hypothalamus. For example, GnRH is produced primarily in one nucleus of the hypothalamus, while CRH is produced in another. Although they are synthesized in different regions of the hypothalamus, both hormones are ultimately released into the capillary network in the median eminence. Secretion of releasing/inhibiting hormones from these hypothalamic neurons depends on the input they receive from neurons located in other areas of the hypothalamus or in higher brain centers.

Actions and Effects of the Pituitary Hormones

Now that we have examined the anatomical and physiological bases for regulating hormone secretion from the anterior and posterior lobes of the pituitary,

a logical question might be, "What are the actions of the pituitary hormones?" The following sections describe specific biological effects of the pituitary hormones. Note the wide variety of physiological processes that the pituitary hormones regulate.

Two Hormones Are Secreted from the Posterior Pituitary

Antidiuretic Hormone (ADH): An Increase in Water Retention

Antidiuretic hormone, or **ADH,** is one of two small peptide hormones secreted by the posterior pituitary. It is nine amino acids in length and is synthesized primarily in the supraoptic nucleus of the hypothalamus, although smaller amounts are produced in the paraventricular nucleus as well.

The actions of ADH will be discussed in detail in Chapters 23 and 24, when the regulation of fluid balance by the kidney is described. Briefly, however, the primary effect of ADH is to increase water retention by the kidney. The net result of this effect is a decrease in urine volume and an increase in extracellular fluid volume. ADH can also act to constrict blood vessels (vasoconstriction). The other name for ADH, **"vasopressin,"** stems from this vasoconstrictor effect, but this response occurs only when the hormone is secreted in relatively large amounts.

The two primary physiological stimulators of ADH release are (1) an increase in osmolality of the blood or extracellular fluid, and (2) a large decrease in blood volume, as in a hemorrhage (heavy bleeding). How do changes in osmolality or blood volume trigger ADH release? Specialized cells within the hypothalamus known as **osmoreceptors** are very sensitive to increases in osmolality of the extracellular fluid. Even a slight increase in osmolality will stimulate osmoreceptor cells. Osmoreceptors in turn activate neuroendocrine cells of the supraoptic nucleus and trigger ADH release from the posterior pituitary. As a result of promoting water retention by the kidney, ADH causes a dilution of the extracellular fluid, which returns plasma osmolality to normal.

A decrease in blood volume also stimulates ADH release. However, stimulation of ADH release requires a 10% or greater blood volume decrease, which is a severe loss of blood. As a result of increasing fluid retention by the kidneys, ADH acts to restore blood volume to normal. In addition, a large loss of blood can result in sufficient ADH secretion to cause constriction of blood vessels. By constricting blood vessels, this effect serves to lessen the decrease in blood pressure that would occur with severe hemorrhage.

In addition to changes in osmolality and blood volume, several other factors are known to influence ADH release. For example, alcohol is a strong inhibitor of ADH release. When ADH secretion is inhibited, urine volume increases, and excess fluid is lost from the body. This accounts, in part, for the increase in urine volume and the thirst experienced following the consumption of alcoholic beverages. Nicotine, barbiturates, and certain anesthetics are known to stimulate ADH release. Overproduction of ADH caused by these agents results in excess fluid retention.

In the absence of ADH secretion, which can occur as a result of either a genetic abnormality or damage to the posterior pituitary, the kidneys are unable to conserve water, and large quantities of fluid are lost in the urine (a process called *diuresis*). This condition is known as **diabetes insipidus.** Diabetes insipidus is distinctly different from what is commonly termed "diabetes," which is actually diabetes mellitus and will be discussed in Chapter 15.

Oxytocin: Regulation of Breast Milk Release

Oxytocin, the other hormone secreted by the posterior pituitary, is also a small peptide hormone (nine amino acids in length) and is very similar in structure to ADH. Like ADH, oxytocin is synthesized in the cell bodies of neurons in both the supraoptic and paraventricular nuclei of the hypothalamus, but its primary site of synthesis is in the paraventricular nucleus.

The principal action site of oxytocin is the female breast. In the breast, oxytocin stimulates contraction of specialized smooth muscle cells, which results in the transfer of milk from its site of synthesis in structures known as **alveoli** into the larger ducts of the breast. The net result of this effect is that milk is made available for a nursing infant. Actions of oxytocin will be described in Chapter 32, when the hormonal control of lactation is discussed. In addition to its effects on the breast, oxytocin can also stimulate contraction of smooth muscle in the uterus.

The primary stimulus for oxytocin secretion in a female is suckling by a nursing infant. When an infant nurses, touch sensors in the mother's nipple are activated and neural signals are transmitted to her brain. This information is processed in the brain, and the oxytocin-producing cells in the paraventricular nucleus are stimulated, resulting in oxytocin release from the posterior pituitary.

A variety of psychological stimuli can also influence oxytocin release. The mere sound of a baby crying can trigger oxytocin release in nursing mothers, while fear or apprehension can strongly inhibit its secretion. Thus, in addition to the neural mechanism that is triggered by suckling, input from other areas of the brain is also important in regulating oxytocin secretion.

Five Cell Types Are Responsible for the Secretion of Six Hormones from the Anterior Pituitary

Five different hormone-secreting cell types are found in the anterior pituitary. These are listed in Table 13–3 along with the hormone(s) produced by each cell type. **Somatotropes,** which produce growth hormone, are the most numerous cells of the anterior pituitary. Normally, about 50% of the cells present in the anterior pituitary are somatotropes. **Corticotropes,** the ACTH-secreting cells, compose about 20% of the cells, and the TSH-producing cells, the **thyrotropes,** about 5%. **Lactotropes** and **gonadotropes** are the other two cell types in the anterior pituitary.

The size and activity of the latter two cell types vary under different physiological conditions. In lactating (nursing) females, the lactotropes, which secrete prolactin, are large and numerous, reflecting the high rate of prolactin secretion in these individuals. In males

Table 13–3

Hormone-Producing Cells of the Anterior Pituitary

Name of Cell Type	Principal Hormone Produced
Somatotropes	Growth hormone (GH)
Corticotropes	ACTH
Thyrotropes	TSH
Lactotropes	Prolactin
Gonadotropes	LH and FSH

or nonlactating females, fewer lactotropes are present, and they are smaller as well. The gonadotropes, which secrete both LH and FSH, account for only about 5% of the cell population of the anterior pituitary of males. In females, their size varies in accordance with cyclical changes in LH and FSH secretion that occur during the monthly menstrual cycle.

As Table 13–3 indicates, these five cell types are responsible for production of six hormones in the anterior pituitary. Gonadotropes are responsible for the synthesis and secretion of *both* LH and FSH. As Chapter 31 will describe, concentrations of LH and FSH in the blood do not always change in parallel during the monthly ovarian cycle. Thus, although they are produced within the same cell, their secretion is regulated independently. At present, the mechanisms by which a single releasing hormone (GnRH) is able to differentially regulate the secretion of two hormones (LH and FSH) are not well understood.

The following sections describe the secretion and actions of each anterior pituitary hormone. The physiology of LH, FSH, and prolactin will be presented in detail in Chapters 31 and 32. Similarly, the physiology of ACTH will be presented primarily in Chapter 14. Therefore, these hormones will only be discussed briefly here. Instead, we will emphasize the remaining two hormones, growth hormone and TSH.

Luteinizing Hormone (LH) and Follicle-Stimulating Hormone (FSH): Regulation of Reproductive Function

LH and FSH are collectively referred to as the **gonadotropins.** This name reflects the fact that the target tissue of these hormones is the **gonads,** or the ovaries and testes in females and males, respectively. In general, the gonadotropins have two primary effects: (1) to promote the development and maturation of sperm and egg, and (2) to stimulate production of sex steroid hormones by the gonads. The principal sex steroids in males and females are **testosterone** and **estradiol,** respectively. As Table 13–2 indicates, secretion of gonadotropins is under the control of GnRH from the hypothalamus.

Prolactin: Regulation of Milk Synthesis

Prolactin has only minor function in males. In females, the primary effect of prolactin is to stimulate milk production by the breast. The target cells for prolactin are milk-producing alveolar cells of the breast. (Specific mechanisms by which prolactin stimulates milk formation will be presented in Chapter 32.) Prolactin secretion is under the dual control of a **prolactin-releasing factor (PRF)** and a **prolactin release-inhibiting hormone PIH,** which has been shown to

be dopamine. Similar to the situation for oxytocin described previously, prolactin release increases in response to stimulation of the mother's nipple by a suckling infant.

Adrenocorticotropic Hormone (ACTH): Regulation of Cortisol

The primary effect of **adrenocorticotropic hormone (ACTH)** is to regulate the synthesis and secretion of the steroid hormone **cortisol** from the adrenal cortex (a portion of the adrenal gland, an endocrine gland located near the kidney). Cortisol produced by the adrenal gland in turn has numerous effects on metabolism in a variety of different tissues. As the name indicates, ACTH also has **trophic** or "growth-promoting" effects on cells of the adrenal cortex. As Table 13–2 indicates, the secretion of ACTH is regulated by **corticotropin-releasing hormone (CRH),** which is produced in the hypothalamus. The specific effects of ACTH on the adrenal cortex and the effects of adrenal steroids on metabolism will be covered in detail in the following chapter.

ACTH is actually synthesized as part of a much larger precursor protein that contains not only ACTH but also several other biologically active peptides. This precursor has been named **proopiomelanocortin peptide,** or **POMC peptide,** to reflect the fact that it contains the sequences for endogenous (internally synthesized) opioid compounds (see Chapter 7), the sequence for **melanocyte-stimulating hormone (MSH),** and the sequence for ACTH.

Growth Hormone (GH): Body Growth and Metabolism

As was indicated in Chapter 12, **human growth hormone (hGH or GH)** is a peptide hormone consisting of 191 amino acids. The structure of GH is remarkably similar to that of a peptide hormone secreted by the placenta during pregnancy, **human chorionic somatomammotropin (hCS).** Growth hormone is also somewhat similar to prolactin, although to a much lesser extent. GH, hCS, and prolactin compose a family of related peptide hormones collectively referred to as the **somatomammotropic hormones,** named to indicate their effects on body growth and breast development. The sections that follow will focus on the effects of GH on body growth and metabolism. The physiology of hCS and prolactin is presented in later chapters.

Regulation of Growth Hormone Secretion. As Tables 13–2 and 13–3 indicate, GH secretion from the somatotropes of the anterior pituitary is under the dual control of growth-hormone-releasing hormone

Figure 13–5

Pattern of GH secretion that might be seen in a normal individual.

Table 13–4

Partial Listing of Factors or Conditions Known to Influence Growth Hormone (GH) Secretion

Stimulators
  Deep sleep*
  Low blood glucose concentration
  Stress
    Physical trauma
    Infection
    Psychological stress
  Amino acids, especially arginine
Inhibitors
  REM sleep*
  High blood glucose concentration
*See Chapter 11 of this text.

(GHRH) and somatostatin, both of which are produced by the hypothalamus. As indicated previously in Table 12–1, somatostatin is also produced in the pancreas, where it is thought to play a local role in regulating pancreatic hormone secretion. This particular action of somatostatin will be described further in Chapter 15. At this point we should emphasize, however, that the somatostatin produced by the hypothalamus and that produced by the pancreas serve different physiological functions, and also that the secretion of somatostatin from each site is independently regulated.

As Figure 13–5 shows, growth hormone is not secreted in a steady continuous fashion throughout the day, but instead is secreted in a pulsatile manner. The most consistent period of GH secretion occurs about one hour after the onset of deep sleep (Fig. 13–5). This can occur during the normal sleep period at night or during a daytime nap. The phase of sleep associated with dreams, known as REM sleep (see Chapter 11), initiates the return of GH secretion to basal pre-sleep levels. The exact significance of this sleep-related surge in GH secretion is not known, but it has been suggested that it may be important in stimulating processes of tissue growth and repair during sleep.

In addition to the sleep-induced increase in GH secretion, a series of pulses often occurs anywhere from two to four hours after a meal (see Fig. 13–5). The relative frequency and the size of these pulses tend to increase about the time of puberty. Some of the other agents or conditions known to influence GH secretion are listed in Table 13–4. Hypoglycemia (low blood glucose) stimulates GH secretion, while high blood glucose inhibits GH secretion. A useful test that physicians can perform to evaluate whether a patient's pituitary is capable of secreting GH involves the administration of a dose of insulin, followed by the monitoring of GH concentrations in the blood. As Chapter 15 will describe, insulin decreases the blood

glucose concentration. In a normal individual in this test situation, the hypoglycemia caused by insulin triggers a pronounced increase in GH secretion. In contrast, the absence of increased GH secretion in response to insulin indicates to the physician that pituitary GH secretion is impaired. As Table 13–4 also shows, various types of stress, both physical and emotional, stimulate GH secretion. Several amino acids, especially arginine, also stimulate GH secretion.

How is such a wide variety of stimuli, such as those listed in Table 13–4, able to regulate GH secretion? As Figure 13–6 illustrates, GHRH and somatostatin serve as the final common pathway for regulating GH secretion. Inputs from various areas of the brain and hypothalamus regulate GHRH and somatostatin release in the median eminence, thus regulating GH secretion from the anterior pituitary.

The Effects of Growth Hormone. The overall effect of GH is to promote tissue growth. In this regard, GH is considered to be an **anabolic hormone.** (Those hormones that tend to have the opposite effect, promoting breakdown of tissues, are known as **catabolic hormones.**)

Another important anabolic hormone is insulin, a hormone produced by the pancreas. While many of the effects of GH are similar to those of insulin, some are exactly opposite to those of insulin and actually antagonize or impair the actions of insulin. GH is therefore said to have both **insulinlike** effects and **anti-insulin** or **"diabetogenic"** effects.

The most striking and obvious effect of GH is that it stimulates linear growth of the skeleton. In addition, it also stimulates growth of a number of tissues

Figure 13-6

A variety of factors can influence growth hormone secretion by regulating the production of GHRH and somatostatin by the hypothalamus.

cle growth, GH stimulates liver protein and RNA synthesis as well. These effects on muscle and liver are similar to those of insulin and are known as the **insulin-like effects** of GH.

However, as indicated previously, some effects of GH antagonize those of insulin. In adipose tissue, GH decreases glucose uptake and stimulates breakdown of fat stores (Fig. 13–7), effects that are opposite those of insulin. Other effects of growth hormone, which decrease glucose uptake into muscle and increase gluconeogenesis in the liver, tend to raise the blood glucose concentration. These effects also antagonize those of insulin, which normally causes a decrease in the blood glucose concentration.

As indicated in Figure 13–7, the net effect of GH on metabolism in adipose tissue and in muscle is to decrease fat storage and promote accumulation of muscle protein. The overall effect of the hormone is therefore to promote accumulation of lean body mass.

Growth Hormone Stimulation of Somatomedins. Early studies of GH indicated that not all effects of the hormone observed in the whole animal were the result of direct actions of GH on individual tissues. For example, although GH was known to promote bone growth when injected into young animals, it was without effect when directly added to pieces of bone incubated in a test tube. Later, scientists determined that GH stimulated the production of a group of peptide substances known as **somatomedins,** and that the somatomedins were directly responsible for some of the effects of the hormone.

The liver is the predominant and probably most important production site of somatomedins, although many other tissues produce them as well. The two most abundant somatomedins produced by the liver are known as **insulin-like growth factor I (IGF-I)** and **insulin-like growth factor II (IGF-II).** This particular nomenclature for the somatomedins grew from the observation that, in addition to insulin itself, the plasma of most animals contains peptide substances exhibiting many of the growth-promoting activities of insulin. IGF-I and IGF-II are very similar to each other in structure and also bear a remarkable resemblance to **proinsulin,** which is the precursor for insulin found in the pancreas (see Chapter 15). This similarity in structure between the two IGFs and proinsulin accounts for much of the insulin-like activity of the two peptides.

IGF-I production by the liver appears to be primarily determined by the concentration of GH in the blood. As a result, IGF-I is markedly decreased in the blood of individuals deficient in GH and markedly increased in many patients who have higher than normal concentrations of GH in their blood. In addi-

in the body. Some effects are due to direct actions of GH on the tissue, while others result from the stimulation by GH of **somatomedin** production by the liver (Fig. 13–7). A later section of this chapter discusses the physiology and actions of the somatomedins.

As Figure 13–7 shows, GH directly stimulates the uptake of amino acids from the blood into muscle cells and also stimulates protein synthesis in muscle. In addition to these effects, which stimulate mus-

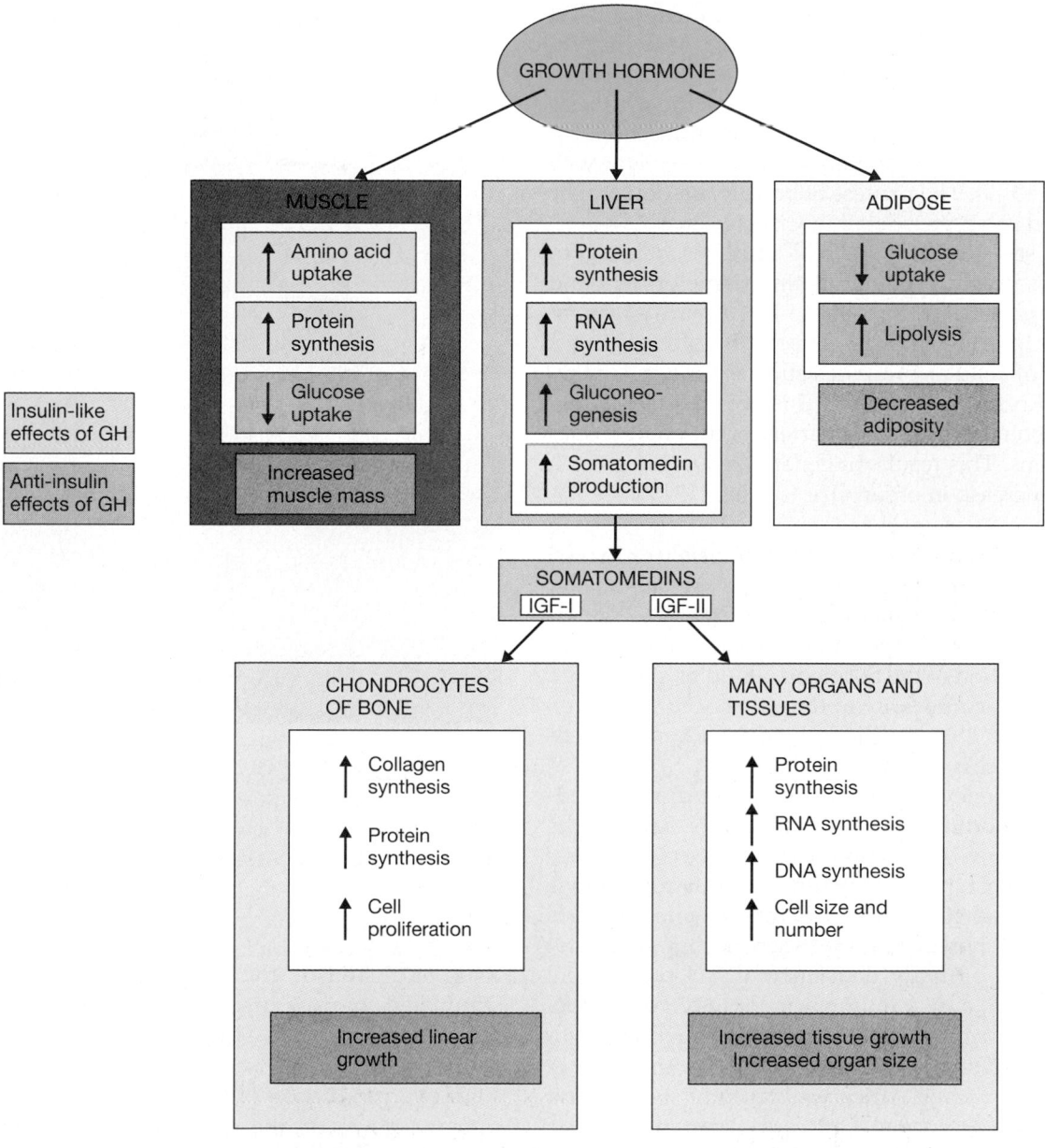

Figure 13–7

Growth hormone has direct effects on muscle, liver, and adipose tissue, and indirect effects that occur as a result of somatomedin production. The overall effect of growth hormone is to promote skeletal growth and the accumulation of lean body mass.

tion to GH, insulin and thyroid hormones also tend to promote IGF-I production. In contrast to IGF-I, the production of IGF-II by the liver is much less dependent on the concentration of circulating GH. A deficiency in circulating GH usually results in a decrease in IGF-II to only about half of the normal value.

The full spectrum of activities of the IGFs has not yet been fully characterized and investigated. In general, however, the IGFs have two types of effects, as indicated in Figure 13–7. The first type of effect involves a stimulation of events leading to an increase in linear growth of the skeleton. Of the two growth factors, IGF-I is primarily responsible for producing this particular effect. IGF-I stimulates skeletal growth by increasing the formation of cartilage in specialized regions near the ends of growing bone known as **epiphyseal plates.** The cartilage deposited in epiphyseal plates will eventually be replaced by bone mineral, and the bone will therefore increase in length. IGF-I promotes bone growth by stimulating those cells involved in cartilage formation, cells known as **chondrocytes.** Specifically, IGF-I stimulates the synthesis of collagen as well as total cell protein in chondrocytes, and promotes proliferation of these cells as well (see Fig. 13–7). The process of bone formation will be described in greater detail in Chapter 26.

The second type of IGF effect is related to processes involved in stimulating tissue growth and tissue repair. IGF-II appears to be primarily responsible for producing these effects. By stimulating a number of cellular processes such as protein synthesis and RNA synthesis, IGF-II is able to elicit a general anabolic response in a variety of different tissues and organs. This results in generalized tissue growth and an increase in organ size (see Fig. 13–7).

Conditions of Growth Hormone Deficiency or Excess. Given the role of GH in stimulating skeletal and soft tissue growth, one might imagine that the consequences of abnormalities that involve either under- or oversecretion of the hormone would be significant. The particular clinical symptoms that occur, however, depend on the age of the individual when the deficiency or excess occurs.

If a deficiency in GH secretion exists during childhood, the condition of **dwarfism** results. Because the deficiency in GH secretion usually occurs as the result of a defect in pituitary function, this condition is often referred to as **pituitary dwarfism.** Scientists now realize, however, that dwarfism may result not only from a pituitary deficiency of GH production, but also from either an impairment in IGF-I production by the liver or by an inability of target tissues to respond to somatomedins produced by the liver. A tribe of Pygmies in Africa was recently studied and found to have normal GH concentrations in their blood, but low circulating IGF-I concentrations due to a genetic impairment in growth factor production by the liver. Other forms of dwarfism appear to result from an impairment in the responsiveness of target tissues. Thus, dwarfism can result from abnormal function in either the anterior pituitary, the liver, or the target tissues. In contrast to the situation in children, however, if the deficiency in GH production occurs during adulthood after normal bone growth has occurred, generally no overt clinical symptoms will be evident.

Overproduction of GH can result in one of two clinically significant conditions: (1) **gigantism** or **acromegaly.** Gigantism results when increased GH secretion occurs during childhood. In these individuals, the skeleton is stimulated to grow excessively. A photograph of a person displaying an extreme case of gigantism is shown in Figure 13–8. Generally, in

Figure 13–8

The world's tallest woman, Sandy Allen (7' 7 1/4", 480 lbs.), pictured at home with her 12-year-old brother. *(© Bettina Cirone/Photo Researchers.)*

cases of gigantism, the long bones of the body are stimulated to grow disproportionately, and as a result these persons have relatively long arms and long legs.

If overproduction of GH begins after the time of puberty, however, the condition of acromegaly results. Around the time of puberty, the growth centers, or epiphyseal plates, of the long bones of the body "close," meaning that they become unresponsive to hormonal stimulation. Thus, bones of the arms and legs cease to grow at this time. However, the bones of the hands, feet, skull, and lower jaw do not "close" and can still be stimulated to grow. As a result of the oversecretion of GH, these bones continue to grow and enlarge, leading to the distinctive physical appearance of acromegaly shown in Figure 13–9. Typically, individuals with acromegaly have an enlarged lower jaw, large hands and feet, and also generally large facial features due to the stimulation of soft tissue growth resulting from increased somatomedin production. In extreme situations, the diabetogenic effects of GH can also lead to overt cases of diabetes mellitus in individuals suffering from acromegaly.

C L I N I C A L   F O C U S

G-Proteins and Acromegaly

As described in the previous chapter, cell-to-cell communication mechanisms are vital for survival of higher organisms. While these regulatory mechanisms open important avenues that allow greater flexibility and survival potential, they also present the opportunity for error in the communication process. Perhaps it is not surprising that a variety of human diseases result from an abnormality in normal cell communication mechanisms.

Growth hormone (GH) from the pituitary is an important regulator of growth of the skeleton during adolescence. Overproduction of GH during adolescence leads to gigantism, and during adulthood overproduction leads to acromegaly. Most commonly, gigantism or acromegaly results from pituitary tumors involving GH-producing cells.

Regulation of GH secretion as well as growth of GH-producing cells involves changes in intracellular cAMP concentrations. These changes in intracellular cAMP are coupled to extracellular signals by G-proteins. In Chapter 12, the role of G-proteins in generating hormone second messengers was described. Mutations of G-proteins that result in either continuous activation or loss of activity have been identified in several human diseases.

Recently, it has been demonstrated that in approximately 40% of GH-secreting pituitary tumors there is elevated basal adenylate cyclase activity and elevated cAMP levels. These changes result from a tissue-specific mutation in the G-protein gene that leads to a single amino acid substitution in the final protein product. This amino acid substitution occurs at a position in the molecule critical for normal G-protein function. The mutation results in continuous activation of the G-protein and increased cAMP production. Thus, in this situation, the disease of acromegaly results from a simple, single amino acid substitution in a pituitary G-protein that leads to an abnormally activated G-protein. This is yet another example of how an alteration in the normal process of cellular communication can lead to disease.

Similarly, a defect in G-protein function in thyroid glands appears to be responsible for certain types of thyroid dysfunction.

■ **At what step(s) are G-proteins involved in the regulation of thyroid hormone secretion?**

Figure 13–9

Hand of someone with acromegaly (left) placed next to a normal hand (right).   *(© Custom Medical Stock Photo.)*

Thyroid-Stimulating Hormone (TSH): Regulation of Thyroid Hormones

As its name implies, the primary target tissue of thyroid-stimulating hormone (TSH) is the thyroid gland. In the thyroid, TSH stimulates cell growth and secretion of thyroid hormones. Thyroid hormones in turn affect a variety of metabolic processes in the body. Therefore, this section provides a description not only of the physiology of TSH, but also of the thyroid gland and actions of thyroid hormones.

Anatomy of the Thyroid Gland.   The gross anatomy of the thyroid gland is shown in Figure 13–10. The thyroid is one of the largest endocrine glands, weighing approximately 20 grams in a normal adult. As Figure 13–10 indicates, the thyroid consists of two lobes that lie on either side of the trachea, just below the larynx. A thin band of tissue known as the **isthmus** connects the two lobes. The gland has an abundant blood supply and actually exhibits one of the highest blood flow rates of any tissue or organ in the body. The thyroid gland also has a tremendous capacity for growth. With an appropriate stimulus, the gland can become greatly enlarged, a condition known as **goiter.** An example of an extreme case of goiter is shown in Figure 13–11. In this particular case, the goiter was caused by a dietary deficiency in iodide. A later section of this chapter describes the mechanism by which iodide deficiency causes goiter.

The thyroid gland consists of cells that are arranged in closely packed follicles or spheres of cells

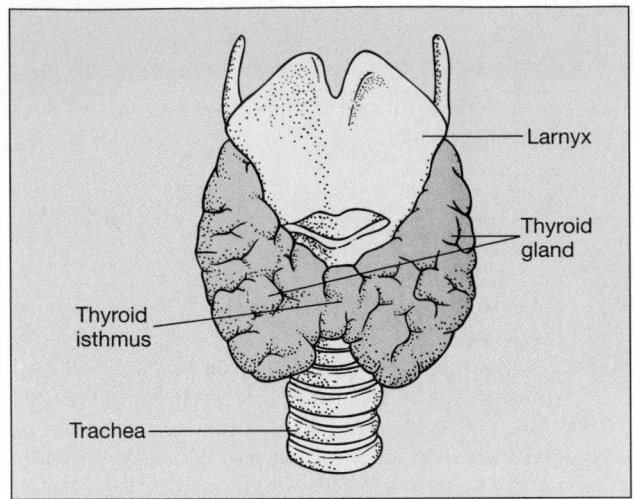

Figure 13–10

Gross anatomy of the thyroid gland showing its location relative to the trachea and the larynx.

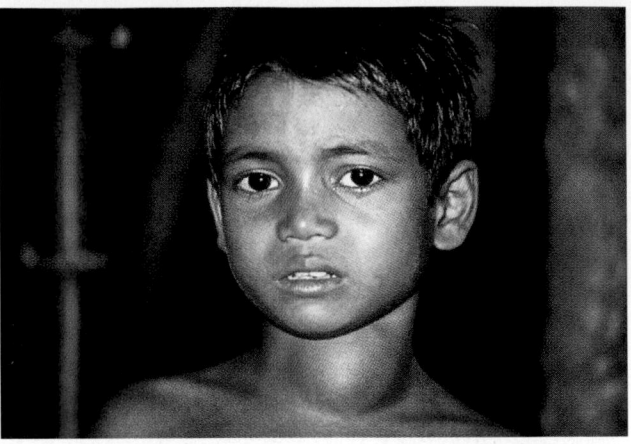

Figure 13–11

An individual with a goiter.   *(© John Paul Kay/Peter Arnold, Inc.)*

(Fig. 13–12), and in a normal healthy adult there are about 3 million of these thyroid follicles. Cells that form a follicle are usually cuboidal in shape, and the interior of the follicle is filled with a protein-containing material termed **colloid.** The major component of colloid is a protein known as **thyroglobulin,** which serves as the precursor of thyroid hormones. A considerable amount of colloid is usually present within each follicle, so that colloid is normally the major constituent of the total mass of the thyroid gland. An extensive capillary network surrounds each follicle.

Synthesis and Secretion of Thyroid Hormones. The major compounds relating to thyroid hormone production and metabolism are shown in Figure 13–13. **Thyroxine (tetraiodothyronine, or T_4)** is the primary hormone product of the thyroid gland. **Triiodothyronine (T_3)** is also produced by the thyroid gland but in lesser amounts. (In target tissues, T_4 is converted into T_3, as will be discussed later.) The other two iodinated compounds shown in Figure 13–13, **monoiodotyrosine (MIT)** and **diiodotyrosine (DIT),** are found primarily within the thyroid follicle cell, although small amounts are also secreted into the blood.

Each of the compounds shown in Figure 13–13 contains iodine as an integral part of the molecule. In comparison to other hormones, thyroid hormones are unique in that their synthesis requires a constituent (iodine) not always present in the diet. As a result, mechanisms have evolved that permit concentration of iodide in the gland and storage of large amounts of thyroglobulin, the thyroid hormone precursor. Normally, enough thyroglobulin may be stored in the gland to sustain thyroid hormone secretion for 2 months or more in the complete absence of any further thyroglobulin synthesis.

Figure 13–12

(*a*) Representation of a typical cross-section through a portion of the thyroid gland. Note the numerous blood vessels and the relative abundance of colloid in the gland. (*b*) Photomicrograph of a section through the thyroid gland. (×425). *(© Biophoto Associates/Science Source)*

Figure 13–13

Structures of the primary iodinated compounds found in the thyroid gland and in the blood. Note the difference in number and location of the iodides between T_4 and T_3.

Thyroglobulin itself is a large glycoprotein synthesized by thyroid follicle cells (Fig. 13–14). As with other glycoproteins that are secreted, the protein component of thyroglobulin is synthesized in the rough endoplasmic reticulum of the cell, and the carbohydrate portion of the molecule is added in the Golgi complex. From the Golgi complex, thyroglobulin is then secreted from the apical side of the cell into the lumen of the follicle, where it is iodinated and stored as part of the colloid.

The first step in thyroid hormone formation involves the uptake of iodide (I^-, the ionic form of the element iodine) into the thyroid follicle cell across its basal membrane from the plasma (Fig. 13–14). Uptake involves an active transport process that pumps iodide into the cells and can produce an intracellular iodide concentration 30 times higher than that in the extracellular fluid. This transport system is so effective at pumping iodide into the follicle cell that the process is often referred to as the **"iodide trap."**

Once inside the cell, free iodide diffuses down a concentration gradient toward the apical surface of the cell and the interior of the follicle. Soon after leaving the follicle cell, iodide undergoes **oxidation** and **organification** reactions. Enzymes that rapidly oxidize iodide and convert it to iodine (I) are present on the surface of the follicle cell membrane that faces the colloid. Iodine formed by these enzymes then serves as a substrate for other enzymes that catalyze addition of iodine to tyrosine residues within the thyroglobulin protein. Tyrosine residues containing either one or two iodine atoms can be formed, and thus at this point thyroglobulin contains both MIT and DIT as part of its protein structure.

The next major step in thyroid hormone formation involves coupling two iodinated tyrosines within thyroglobulin to form either T_3 or T_4. If two DIT residues are coupled, then T_4 will be formed. If, however, DIT and MIT residues of thyroglobulin are coupled, then T_3 is produced. It should be emphasized that at this point T_3 and T_4 are *not* free, but are still present as part of the thyroglobulin peptide. In later steps to be described, thyroglobulin is broken down to release the free hormones.

When follicle cells are stimulated to produce thyroid hormones, numerous microvilli are formed on the apical surface. These microvilli extend out from the cell and engulf a portion of colloid by pinocytosis (see Fig. 13–14). This small droplet of colloid is taken into the follicle cell and moved toward the basal surface of the cell. During this movement, the colloid droplet meets with lysosomes, and the structures fuse to form a **phagolysosome** (see Fig. 13–14). Phagolysosomes move toward the basal end of the cell and, as they do, thyroglobulin molecules are broken down by proteolytic enzymes to produce free T_4 and T_3. T_4 and T_3 are released into interstitial fluid, where they are picked up by nearby capillaries and carried by the circulation throughout the body.

Proteolytic breakdown of thyroglobulin also results in release of small amounts of free MIT and DIT, which are present due to incomplete coupling reactions. MIT and DIT are subject to metabolism within the follicle cell by very active **deiodinases** present in the cytoplasm. These deiodinases remove iodide from MIT and DIT to produce iodide and tyrosine. The iodide can be recycled for use in further hormone synthesis, thereby conserving the iodide. Very little free MIT or DIT is actually released from the thyroid gland.

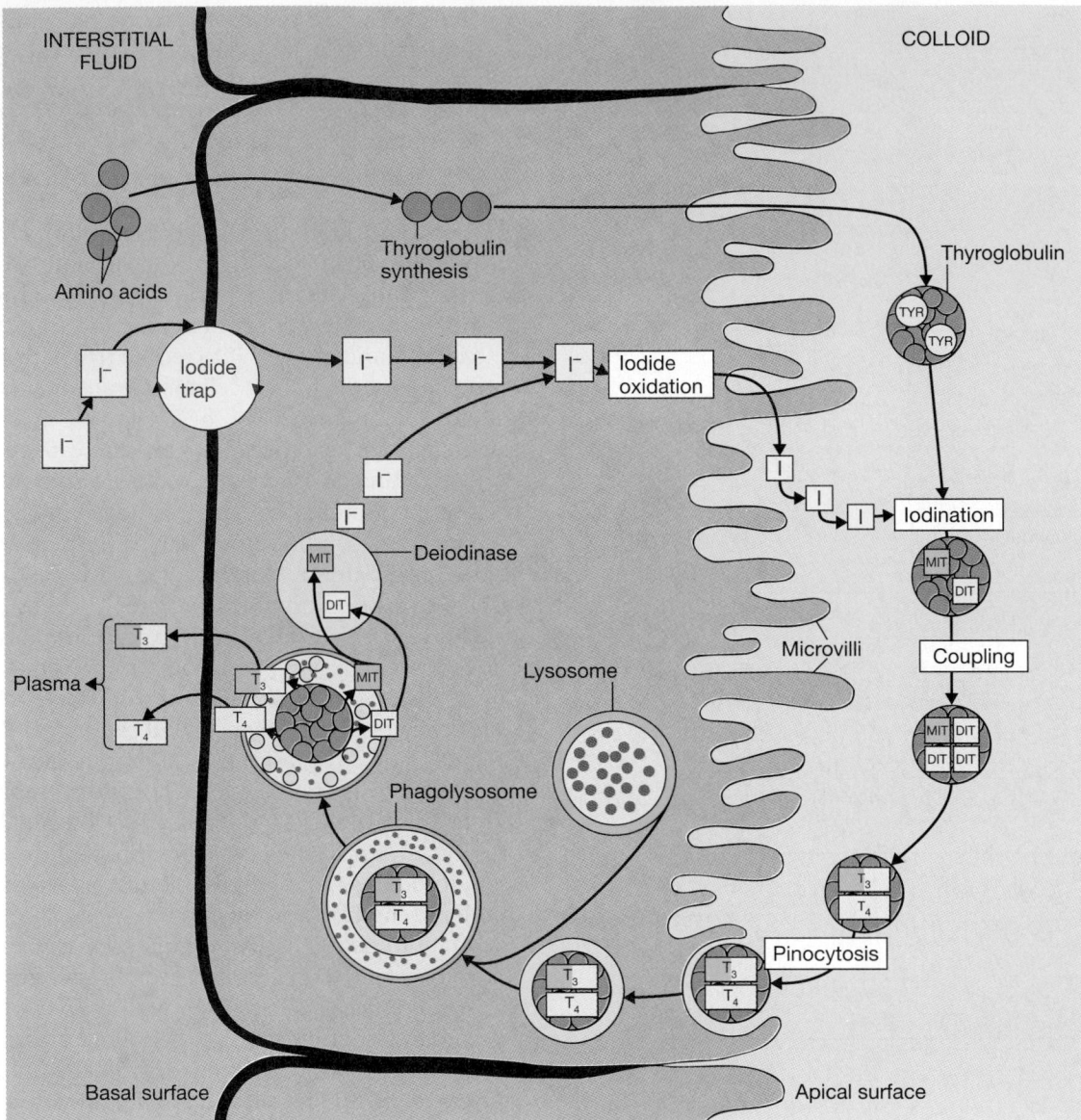

Figure 13–14

The major steps involved in thyroid hormone formation and release.

Regulation of Thyroid Hormone Synthesis and Secretion. The primary regulator of thyroid follicle cell activity and thyroid hormone secretion is **thyroid-stimulating hormone,** or **TSH,** which is produced by thyrotropes of the anterior pituitary (see Table 13–3). Specific receptors for TSH are present on basal surfaces of follicle cells. These receptors are coupled to adenylate cyclase, and hence binding of TSH to its receptors leads to an increase in intracellular cAMP. The specific effects of TSH on thyroid hormone production are listed in Table 13–5. Note that TSH stimulates virtually every aspect of thyroid hormone formation. In addition to these specific effects on thyroid hormone formation, TSH also stimulates growth of the thyroid follicle cell. Thus, when TSH

secretion is increased, follicle cells are stimulated to grow and become elongated and more columnar in shape. Likewise, when TSH secretion is low, the cells regress and become somewhat flattened.

A diagram outlining the regulation of thyroid hormone secretion is presented in Figure 13–15. The relationship between the hypothalamus, anterior pituitary, and thyroid gland indicated in this figure is known as the **hypothalamic-pituitary-thyroid axis.** As described previously, thyroid hormone secretion is under the control of TSH from the pituitary. Secretion of TSH in turn is regulated primarily by TRH produced by the hypothalamus. Secretion of TRH can be influenced by a variety of factors, including inputs from higher centers within the CNS and from tem-

Table 13–5

Effects of TSH on Thyroid Hormone Production

In the thyroid follicle cell, TSH stimulates:
1. Iodide uptake by active transport mechanisms
2. Thyroglobulin synthesis
3. Reactions resulting in the oxidation and organification of iodide
4. Microvilli formation and engulfment of colloid at the apical cell surface
5. Movement of lysosomes from the basal toward the apical surface of the follicle cell
6. Movement of phagolysosomes from apical to basal surfaces of cells
7. Activity of the deiodinase enzymes

perature regulatory centers in the hypothalamus. Temperature regulatory centers in turn receive information concerning changes in body temperature and changes in environmental temperature. Also indicated in Figure 13–15 is the important negative feedback effect that thyroid hormones exert upon the anterior pituitary to limit TSH secretion. As a result, thyroid hormones exert negative feedback effects that limit their own production. In addition, some of the biological effects of thyroid hormones result in increased heat production in the body (see the following), which also has the negative feedback effect of limiting further thyroid hormone formation.

Prior to the introduction of iodized salt into the diet, the occurrence of goiter, or enlargement of the thyroid gland, was fairly common. Figure 13–15 provides the framework to understand how a deficiency of dietary iodide can lead to an enlargement of the thyroid gland. As described previously and indicated in Figure 13–15, iodide is a necessary component for thyroid hormone formation. In the absence of iodide, thyroid hormones cannot be produced, and in the absence of thyroid hormone secretion, negative feedback effects of T_4 and T_3 on TSH release are also absent (see Fig. 13–15). TRH and TSH secretion therefore increases, and the growth-promoting effects of TSH on thyroid follicle cells cause the gland to increase in size, despite the inability to actually synthesize active hormone. As a result, goiter is produced with a prolonged absence of dietary iodine. A number of agents known as **goitrogens** inhibit iodide uptake or organification reactions, resulting in a blockage of thyroid hormone synthesis and the formation of a thyroid goiter. Vegetables of the turnip family are known to contain goitrogens and can produce goiter if eaten raw in excessively large quanti-

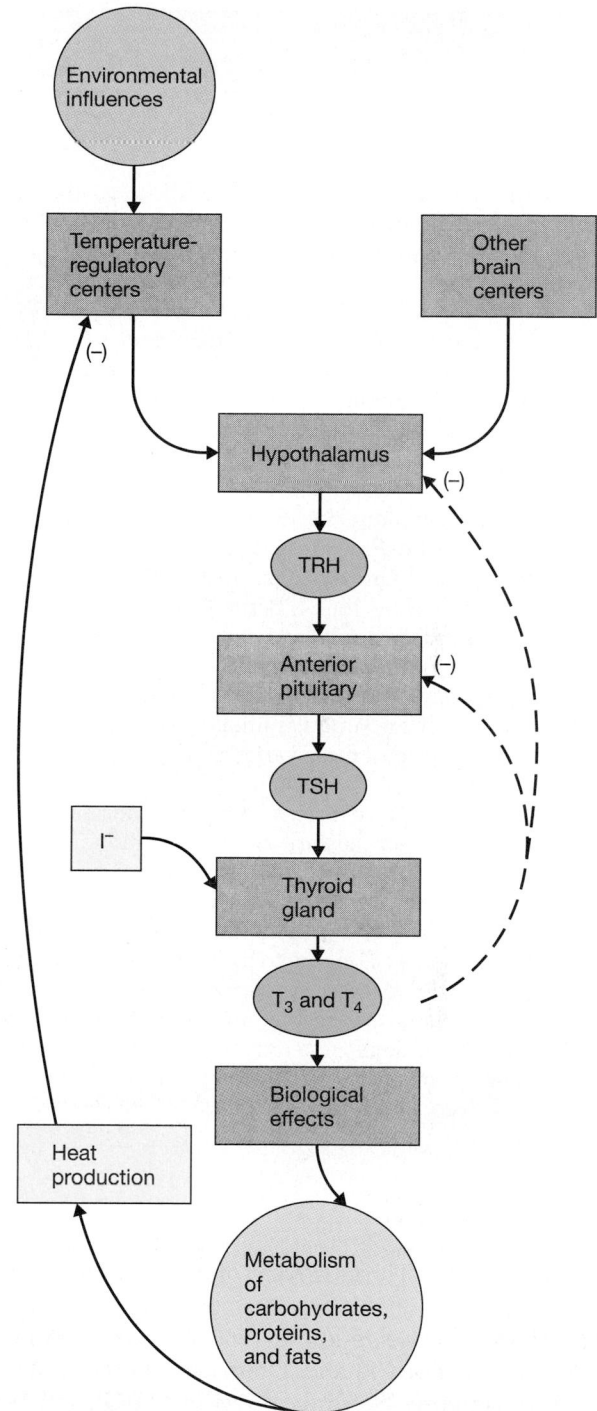

Figure 13–15

Primary steps involved in regulating thyroid hormone production.

ties; however, cooking destroys the goiter producing activity of goitrogens.

Thyroid Hormone Transport and Metabolism. After secretion from the thyroid gland, T_3 and T_4 are carried in the blood reversibly bound to a number of

RESEARCH FOCUS

Dwarfism in Animals

As indicated in the accompanying sections, the endocrine system can have a profound influence on growth and body size. Normal growth requires a sequence of events involving growth hormone (GH) secretion, GH action on the liver, IGF-I secretion by the liver, and IGF-I action on tissues to promote growth. In addition, the appropriate secretion of and response to various other hormones must also take place. An impairment at any one of these steps can lead to reduced growth and short stature.

Among mammals, the domestic dog exhibits perhaps the highest degree of variation in body size of any species. Selective breeding has resulted in both very large breeds, such as the St. Bernard, which may reach 200 lbs or more in adulthood, and very small breeds, such as the Chihuahua, which may weigh less than 2 lbs. Even within one particular breed, such as the poodle, there may be a large variation in size. The poodle breed is composed of three varieties—Standard, Miniature, and Toy.

Recently, a group of investigators addressed the question of whether the variation in size among the different varieties of poodle could be explained on the basis of their particular endocrine status. They measured both GH secretion and plasma IGF-I concentrations in groups of poodles belonging to the Standard, Miniature, and Toy varieties. All three groups of dogs showed similar GH secretion, suggesting that the differences in stature among groups were not due to differences in the capacity of their pituitary glands to produce and secrete GH. However, significant differences in plasma IGF-I concentrations were observed among groups. Standard poodles had significantly greater amounts of circulating IGF-I than did their Miniature or Toy counterparts. When the body weight of each animal used in this study was graphed against the concentration of IGF-I in their blood, a highly significant correlation was obtained. That is, the larger the animal, the higher the circulating IGF-I concentration.

Although these results are consistent with the notion that the variation in stature among these dogs was the direct result of differences in circulating IGF-I, this has not yet been proven. There may well have been some other primary change that directly led to both smaller stature *and* lower IGF-I secretion. However, it is tempting to speculate that in breeding the different varieties of poodle, the breeders have in fact been selecting for differences in the ability of these animals to produce IGF-I. (*Source:* Eigenmann, J.E., Patterson, D.F., and Froesch, E.R. *Acta Endocrinologica* 106:448–453, 1984.)

As the accompanying text describes, a defect in GH action may result from a problem in the overall GH pathway at any number of steps. Some of the more common defects are described, but, in theory, an impairment in GH action could occur at many other steps.

■ **Starting with GHRH production in the hypothalamus and ending with IGF-I action in peripheral tissues, can you make a list of all possible points at which a problem in the GH pathway might lead to reduced growth?**

---

plasma proteins. These carrier proteins serve to increase solubility of thyroid hormones and to buffer acute changes in secretion, as was discussed earlier in Chapter 12. Both T_4 and T_3 are associated primarily with **thyroxine-binding globulin (TBG),** but are also bound to a lesser extent to **albumin** (see Table 12–2). The relative affinity or strength with which the two hormones bind to these proteins differs greatly, however. T_4 has a much higher affinity and therefore binds more tightly to the plasma proteins than does T_3. As a result, T_3 is more rapidly degraded and removed from the plasma. Thus, not only is T_4 secreted from the thyroid gland in much greater amounts than T_3, but it is also removed from the blood much more slowly than T_3.

Target tissues for thyroid hormones have the capacity to convert T_4 into T_3. Although T_3 and T_4 are capable of producing exactly the same biological effects, their relative strengths or potencies differ considerably. Within a target cell, T_3 is much more active than T_4. It is currently thought that most, if not all, of the biological effects of T_4 are due to its conversion to T_3 within target tissue cells.

Effects of Thyroid Hormones on Metabolic Processes.
Thyroid hormones exert numerous effects on metabolic processes in a wide variety of different tissues and cells; these are summarized in Table 13–6. The effects of thyroid hormones are so wide-spread that virtually no tissue or organ system escapes the adverse results of thyroid hormone deficiency or excess, which are especially evident during fetal development. In general, the effects of thyroid hormones on metabolic processes are slow in onset and long-last-

Table 13–6
Summary of Effects of Thyroid Hormones

Stimulate calorigenesis in most cells
Increase cardiac output
    Increase rate of cardiac contractions
    Increase strength of cardiac contractions
Increase oxygenation of blood
    Increase rate of breathing
    Increase number of red blood cells in the
    circulation
Effects on carbohydrate metabolism
    Promote glycogen formation in liver
    Increase glucose uptake into adipose and
    muscle
Effects on lipid turnover
    Increase lipid synthesis
    Increase lipid mobilization
    Increase lipid oxidation
Effects on protein metabolism
    Stimulate protein synthesis
Promote normal growth
    Stimulate growth hormone (GH) secretion
    Promote bone growth
    Promote IGF-I production by liver
Promote development and maturation of
    nervous system
    Promote neural branching
    Promote myelinization of nerves

ing, as compared to the effects of other hormones. Thus, thyroid hormones exert relatively long-term regulatory influences over metabolism.

One of the primary effects of thyroid hormones is the stimulation of **calorigenesis,** or heat production, in the body (see Table 13–6). This response occurs after a delay of several hours or days and is reflected by an increase in oxygen consumption. This particular effect is evident in most tissues, with brain, spleen, and testes being the most notable exceptions. To provide for an increase in oxygen consumption by the tissues, thyroid hormones also have effects on the cardiopulmonary system. In the heart, thyroid hormones increase cardiac output (the amount of blood being pumped) by increasing both the rate and strength of cardiac contractions. The amount of oxygen carried by the blood is also enhanced by thyroid hormones as a result of increases in both the resting rate of breathing and the number of red blood cells present in the blood.

Thyroid hormones influence several aspects of carbohydrate metabolism, although many of the effects that are produced depend on or are modified by other hormones. For example, normal concentrations of thyroid hormones increase glycogen formation in the liver (see Table 13–6), but this response occurs only when insulin is also present. Thyroid hormones also increase uptake of glucose from blood into adipose tissue and muscle and thereby act to potentiate the effect of insulin in this regard. In addition, many of the effects of epinephrine are enhanced by thyroid hormones as a result of their enhancement of the responsiveness of the adenylate cyclase/cAMP system.

Thyroid hormones have a number of effects on lipid metabolism in both liver and adipose tissue (see Table 13–6). Thyroid hormones stimulate virtually all aspects of lipid metabolism, including the synthesis, mobilization, and oxidation of lipids. In general, however, oxidation of lipids is affected more than is their synthesis, such that in conditions of hormone excess the net effect is a decrease in size of most fat stores in the body.

Thyroid hormones exert important regulatory effects on protein metabolism in the body as well. When present in normal physiological concentrations, thyroid hormones increase protein synthesis and promote an overall accumulation of protein (see Table 13–6). However, when thyroid hormones are present in excess, they tend to cause a decrease in protein synthesis and an increase in protein breakdown. The result of these two effects is an overall loss of protein from the body. Effects of thyroid hormones on protein metabolism are therefore described as **biphasic.** Thyroid hormones have an additional effect on the pituitary, one that influences protein metabolism indirectly. In the anterior pituitary, thyroid hormones are known to stimulate somatotropes and increase GH secretion. As a result, owing to their influence on pituitary GH secretion, thyroid hormones also affect protein metabolism and growth.

In addition to specific effects on protein metabolism, thyroid hormones play a role in regulating the overall growth and development of several tissues. In humans, thyroid hormone is required for normal skeletal growth. This is due in part to the stimulatory effects of thyroid hormones on GH production and in part to thyroid hormone stimulation of maturation of the epiphyseal growth centers in bone. In addition, as was noted in the earlier section describing somatomedins, thyroid hormones influence IGF-I production by the liver, which further contributes to promoting normal growth. Thyroid hormones are also required for normal development and maturation of teeth, hair follicles, and skin.

Development and maturation of the CNS are markedly affected by thyroid hormones. This effect of thyroid hormones is particularly evident in their absence. Branching of axons and dendrites that normally occurs during fetal and early neonatal development occurs to only a limited degree in the absence of thyroid hormones. Normal myelinization of nerves is also impaired in the absence of thyroid hormones. As a result, severe mental retardation can occur if thyroid hormone is deficient during the period of fetal and early neonatal development. In adults, thyroid hormones influence mental alertness and responsiveness to external stimuli. The velocity of conduction of action potentials in peripheral nerves has also been shown to vary in response to an excess or deficiency of thyroid hormone secretion.

Conditions of Abnormal Thyroid Hormone Secretion.

Many of the specific effects of thyroid hormones that were discussed previously are readily evident in conditions of either thyroid hormone deficiency or thyroid hormone excess.

Hypothyroidism occurs when there is a deficiency in thyroid hormone production. In most instances, this occurs as a result of a defect in the thyroid gland itself. However, in some cases the defect occurs in either the hypothalamus or pituitary, resulting in a deficiency in TSH production. Because of a slower metabolic rate and a reduced rate of heat production, hypothyroid individuals cannot tolerate cold temperatures. Water tends to accumulate in the skin of these individuals, leading to a condition termed **myxedema,** in which the skin has a thickened, puffy appearance. This change is most evident in the facial features. The heart rate and the strength of cardiac contractions are reduced, contributing to an overall reduction in the cardiac output. There is also a general slowing of all intellectual functions, leading to a feeling of lethargy and some degree of speech impairment.

Excess secretion of thyroid hormones results in **hyperthyroidism.** When tissues are presented with excessive quantities of thyroid hormones, a complex of biochemical and physiological events occurs. Hyperthyroid individuals exhibit such symptoms as a markedly increased heart rate, an insensitivity to heat, and considerable weight loss owing to the lipid-mobilizing and protein-catabolic effects of excess thyroid hormones. In contrast to the lethargy that occurs in hypothyroidism, hyperthyroid individuals tend to be highly responsive to external stimuli. The most common form of hyperthyroidism is a condition known as **Grave's disease,** which occurs as a result of an interesting set of circumstances. In affected individuals, an abnormality in the immune system leads to the production of antibodies that recognize and bind to sites on the surface of thyroid follicle cells. When these antibodies bind to the follicle cell, the TSH receptor is activated, as though TSH itself were present. This results in a marked increase in thyroid hormone production, and a goiter forms as a result of the TSH-like effects of the antibodies on thyroid cell growth. The cause of the immune system abnormality and the mechanism by which the antibody activates the TSH receptor are not completely understood, however.

As indicated previously, a deficiency in thyroid hormone secretion during fetal development can have a dramatic impact, owing to the hormone's role in promoting normal development of the nervous system. A form of mental handicap known as **cretinism** can occur as a result of an untreated thyroid hormone deficiency. In addition to mental impairment, linear growth is also impaired in these individuals, so that dwarfism is also evident. If therapy with thyroid hormone is begun very soon after birth, severity of the impairment can be dramatically reduced. For this reason, most infants are checked very soon after delivery for the presence of adequate thyroid hormones in their blood.

C H A P T E R   R E V I E W

Summary

- The pituitary is a small gland located just below the hypothalamus at the base of the brain and consists of two main lobes: the anterior lobe and the posterior lobe.
- The posterior pituitary is derived from neural tissue and contains the terminals of neuroendocrine cells that originate in the hypothalamus. Cells of the anterior pituitary are derived from nonneural tissue and are regulated by hypothalamic hormones carried from the hypothalamus to the anterior pituitary via the hypothalamic-pituitary portal blood vessels.

- The posterior pituitary secretes two hormones, ADH and oxytocin. These two hormones are synthesized in cell bodies of neuroendocrine cells in the hypothalamus and are then transported down their respective axons to the posterior pituitary where they are released into capillaries when the hypothalamus receives an appropriate stimulus.
- Cells of the anterior pituitary synthesize and secrete at least six different hormones: LH, FSH, prolactin, ACTH, GH, and TSH. Secretion of each hormone is under the

control of releasing/inhibiting hormones produced by the neuroendocrine cells in the hypothalamus and released into blood vessels of the median eminence. The releasing/inhibiting hormones travel to the anterior lobe via the hypothalamic-pituitary portal system, where they act to regulate secretion of the six anterior pituitary hormones.

- The primary action of ADH is to increase water retention in the kidneys and, secondarily, to stimulate vasoconstriction. Oxytocin acts primarily on the breast and causes release of milk from the alveoli into the collecting ducts.
- LH and FSH, collectively referred to as the gonadotropins, regulate development of sperm and egg in males and females, respectively, and also regulate production of sex steroids by the gonads. Gonadotropin secretion is regulated by GnRH produced in the hypothalamus.

- Prolactin stimulates milk production by the alveolar cells of the female breast. Prolactin secretion increases in response to suckling stimuli.
- ACTH regulates the synthesis and secretion of cortisol from the adrenal cortex and also stimulates growth of the adrenal cortex. ACTH secretion is regulated by CRH.
- Growth hormone (GH) secretion from the anterior pituitary is regulated by GHRH and somatostatin. GH effects are generally anabolic, although specific effects can be either insulinlike or diabetogenic. GH affects metabolism in liver, muscle, and adipose tissue and stimulates production of somatomedins by the liver. Somatomedins stimulate both skeletal growth and tissue growth.
- Thyroid hormones promote calorigenesis in most cells of the body, with specific effects on carbohydrate, lipid, and protein metabolism. In addition, thyroid hormones have important influences on the maturation of the nervous system during fetal development.

Review Questions

Identify Each with a Term
1. The fetal structure that during development gives rise to the anterior pituitary.
2. The specialized region of the hypothalamus that contains the primary capillaries of the hypothalamic-pituitary portal system.
3. The hypothalamic hormone that inhibits growth hormone secretion by the anterior pituitary.
4. The name of the condition resulting from a deficiency in ADH secretion or action.
5. The name of the condition resulting from excess GH secretion during adulthood.
6. The component of colloid that serves as the precursor of thyroid hormones.

Define Each Term
7. hypothalamic-pituitary portal system
8. corticotropes
9. gonadotropins
10. somatomedins
11. "iodide trap"

Choose the Correct Term
12. Which of the following is *not* secreted by the anterior pituitary?
    a. oxytocin
    b. growth hormone
    c. ACTH
    d. FSH
13. Which of the following is *not* a stimulus for GH secretion?
    a. deep sleep
    b. high blood glucose
    c. infections
    d. amino acids
14. The insulinlike effects of GH include
    a. increased protein synthesis in muscle

b. increased lipolysis in adipose tissue
    c. increased gluconeogenesis in liver
    d. all of the above
15. Which of the following are the most numerous hormone-producing cells of the anterior pituitary?
    a. gonadotropes
    b. corticotropes
    c. lactotropes
    d. somatotropes
16. Which "hormone:site of synthesis" pair listed below is incorrect?
    a. oxytocin:supraoptic nucleus
    b. corticotropin-releasing hormone:hypothalamus
    c. somatostatin:somatotropes
    d. FSH:gonadotropes
17. Overproduction of GH can result in:
    a. cretinism
    b. acromegaly
    c. gigantism
    d. either b or c
18. An abnormal enlargement of the thyroid gland is termed:
    a. myxedema
    b. thyroid isthmus
    c. goiter
    d. colloid
19. In the thyroid gland, the "iodide trap":
    a. is responsible for the conversion of thyroglobulin into T_3 and T_4
    b. removes iodide from MIT and DIT
    c. is located on the apical surface of the cell and is directly responsible for transporting iodide from the intracellular space into the colloid space
    d. is located on the basal surface of the cell and is responsible for transporting iodide from the extracellular space into the thyroid follicle cell
20. Intolerance to cold, myxedema, and reduced cardiac output are typically seen in:

a. hyperthyroidism
b. hypothyroidism
c. Grave's disease
d. acromegaly

Answer in One or Two Sentences

21. Describe the processes involved in the synthesis and secretion of posterior pituitary hormones.
22. What are the primary stimuli for secretion and the primary actions of antidiuretic hormone?
23. Describe the hypothalamic-pituitary-thyroid axis.
24. What are the differences in physical appearance among dwarfism, gigantism, and acromegaly? What factors account for these differences?
25. What effects do thyroid hormones have on carbohydrate, lipid, and protein metabolism?

Problem Solving in Physiology

26. Goiter, or enlargement of the thyroid gland, may be associated with either the condition of hypothyroidism or that of hyperthyroidism. Explain how this same physical manifestation in the thyroid gland can occur in these two opposite conditions.
27. The anterior pituitary is an endocrine gland and the posterior pituitary is a neuroendocrine gland, yet both are regulated by the hypothalamus. Explain the mechanisms underlying the regulation of the two lobes of the pituitary.
28. Most hormones are not stored in the body to any great degree, yet the majority of the total volume of the thyroid gland consists of the thyroid hormone precursor, thyroglobulin. What is unique about thyroid hormones that makes precursor storage advantageous?

Suggested Readings

Berne, Robert M., and Levy, Matthew N. *Physiology,* 3rd ed. St. Louis: Mosby, 1993.

Fregley, Melvin J., and Luttge, William G. *Human Endocrinology: An Interactive Text.* New York: Elsevier Biomedical, 1982.

Ganong, William F. *Review of Medical Physiology,* 15th ed. Los Altos: Lange Medical Publication, 1991.

Goodman, H. Maurice. *Basic Medical Endocrinology,* 2nd ed. New York: Raven Press, 1994.

Griffin, James E., and Ojeda, Sergio R. *Textbook of Endocrine*

Physiology, 2nd ed. New York: Oxford University Press, 1992.

Guyton, Arthur C. *Human Physiology and Mechanisms of Disease,* 4th ed. Philadelphia: W.B. Saunders Company, 1987.

Norman, Anthony W., and Litwack, Gerald. *Hormones.* Orlando: Academic Press, 1987.

Tepperman, Jay, and Tepperman, Helen M. *Metabolic and Endocrine Physiology,* 5th ed. Chicago: Year Book Medical Publishers, 1987.

Key Terms

acromegaly (p. 412)
adrenocorticotropic hormone (ACTH) (p. 408)
anabolic hormone (p. 409)
anterior pituitary gland (p. 401)
antidiuretic hormone (ADH) (p. 404)
calorigenesis (p. 419)
catabolic hormone (p. 409)
chondrocytes (p. 412)
cretinism (p. 420)

diabetes insipidus (p. 407)
diiodotyrosine (DIT) (p. 414)
dwarfism (p. 412)
follicle-stimulating hormone (FSH) (p. 408)
gigantism (p. 412)
goiter (p. 413)
growth hormone (p. 408)
hypothalamic-pituitary portal system (p. 404)
IGF-I (p. 410)
IGF-II (p. 410)

luteinizing hormone (LH) (p. 408)
median eminence (p. 403)
monoiodotyrosine (MIT) (p. 414)
myxedema (p. 420)
osmoreceptors (p. 406)
oxytocin (p. 404)
paraventricular nucleus (p. 408)
posterior pituitary gland (p. 401)

prolactin (p. 406)
Rathke's pouch (p. 402)
somatomammotropic hormone (p. 408)
somatomedin (p. 410)
supraoptic nucleus (p. 403)
thyroglobulin (p. 414)
thyroid hormone (p. 413)
thyroid-stimulating hormone (TSH) (p. 416)
triiodothyronine (T_3) (p. 414)

The zona fasiculations of the
adrenal cortex (original ×450).
(© R. Calentine/Visuals Unlimited)

Chapter 14

The Adrenal Glands

Key Concepts

The adrenal glands consist of two distinct endocrine glands, each under separate regulation and each producing different hormone products.

Cells of the adrenal cortex produce steroid hormones, predominantly cortisol and aldosterone. Cortisol affects metabolism and generally results in the release of stored fuels. Aldosterone acts on the kidneys and influences sodium and potassium balance in the body.

The adrenal medulla functions as part of the sympathetic nervous system and secretes primarily epinephrine. Epinephrine has effects on the cardiovascular and pulmonary systems as well as affecting a variety of metabolic processes in the body.

The net effect of the adrenal hormones cortisol and epinephrine is to increase the body's ability to effectively respond to stress.

Chapter Outline

Gross Anatomy of the Adrenal Glands
  The Adrenal Cortex
  The Adrenal Medulla
  Summary of Effects of
    Adrenal Hormones

The voice of the public address announcer comes over the loudspeaker system: "All entrants in the 100-yard dash please report to the starting line." This is the moment for which Yolanda has been training and that she has anticipated for so many years. It is the district track meet and she has qualified for the finals. As she walks toward the starting line, she notices that her heart rate has picked up, and she feels as though she is breathing a bit heavier and more rapidly than normal. Normally Yolanda is not an overly anxious person, and this is the first time she has experienced this type of response. Yolanda enters the starting blocks, sets herself, and BANG, the starter's gun goes off. Memory of the race itself is little more than a blur, since it all ended so quickly. Although she ran a personal best, Yolanda came in third, behind a teammate and a runner from another school.

Later, in talking with Tammy, her teammate, Yolanda related how her heart rate had increased so dramatically just before the race. Tammy indicated that she had been so nervous about the race that she had not been able to eat anything since the previous day. Both wondered about the physiologic mechanisms that would cause Yolanda's heart rate to increase even before she had begun to exercise, and also how Tammy had managed to maintain her energy level sufficiently to win the race, despite not having eaten for over 24 hours.

In this chapter the role of the adrenal glands in physiological regulation will be presented. The adrenal glands produce several hormones. One is cortisol, a steroid that acts on metabolism to help maintain blood glucose levels. Another is epinephrine (adrenaline), which affects many organs and tissues in the body, including the cardiovascular system. The net effect is that these adrenal hormones enable us to respond to and withstand a variety of stressful situations.

This chapter discusses the physiology of the adrenal glands and the adrenal hormones. Humans have two adrenal glands, each one composed of two distinct endocrine tissues that differ in cell type, hormone products, and regulation. The two endocrine components of the adrenal gland are the outer portion, or **cortex,** which constitutes about 80% of the gland, and the inner portion, or **medulla,** which forms the remaining 20%. The cortex produces and secretes several different steroid hormones, while the medulla produces and secretes catecholamines. The first section of this chapter focuses on the adrenal cortex, and a discussion of the adrenal medulla follows in the latter sections.

Gross Anatomy of the Adrenal Glands

The shape of each adrenal gland roughly resembles that of a pyramid, and each of the two adrenals in humans rests like a cap just above the upper end of each kidney (Fig. 14–1). There is no direct physical connection between each adrenal and kidney, however. The arterial blood supply entering each adrenal gland is also separate from that entering the kidneys. In a normal adult, each adrenal gland weighs about 3 to 4 grams and is about 5 cm (2 inches) across at its widest point.

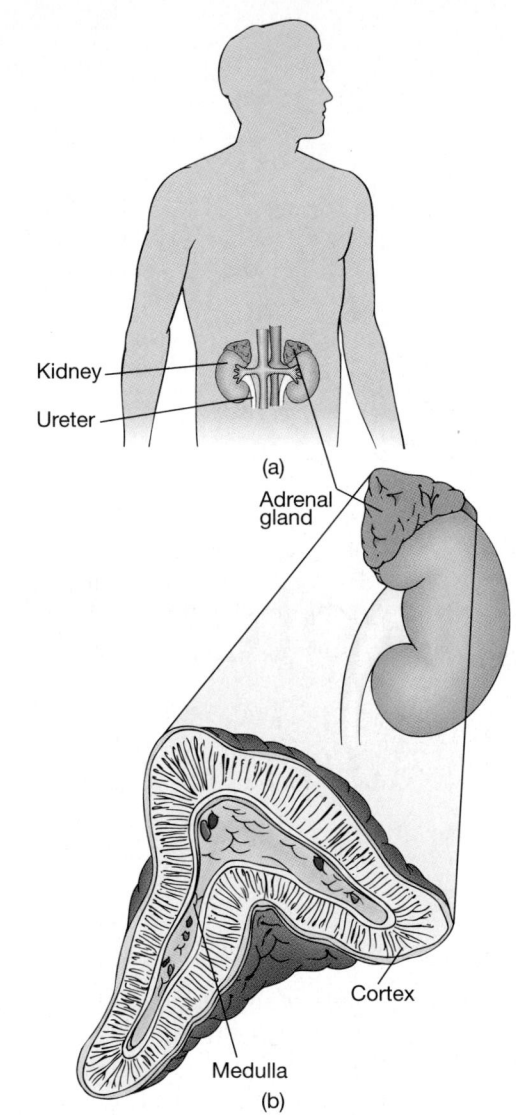

Figure 14–1

(*a*) Location of the adrenal glands relative to the kidneys. (*b*) The adrenal gland shown in cross-section, with both the cortex and medulla visible.

The Adrenal Cortex

Cells of the Adrenal Cortex Show Three Specific Zones, and Each Zone Produces a Specific Class of Steroid Hormone

Based on the arrangement of cells as seen under a microscope, the adrenal cortex can be divided into roughly three different zones. Each zone secretes a different class of steroid hormone. The outermost region of the adrenal cortex is termed the **zona glomerulosa** (Fig. 14–2) and consists of cells

Figure 14-2

The adrenal cortex: three zones. *(© Ed Reschke/ Peter Arnold, Inc.)*

arranged in "whorls" **(glomeruli).** Cells of the zona glomerulosa produce **mineralocorticoids.** The middle region of the cortex is widest and is termed the **zona fasciculata** because of the arrangement of cells in "fascicles" or cords separated by venous sinuses (small channels for venous blood) (see Fig. 14–2). Cells of the zona fasciculata produce **glucocorticoids.** The innermost region, which borders the medulla, is known as the **zona reticularis,** so named because cells are arranged in the form of a network, or reticulum. This zone of the cortex produces and secretes primarily **androgens.**

Numerous lipid droplets are present within cells of the adrenal cortex, especially in cells of the zona fasciculata. These lipid droplets represent stored cholesterol to be used to synthesize adrenal steroid hormones, as will be discussed later.

Activity of the Adrenal Cortex Is Under the Control of ACTH Produced by the Pituitary

Activity of the adrenal cortex is controlled by adrenocorticotropic hormone (ACTH), which is produced by corticotropes of the anterior pituitary. As was described in Chapter 13, ACTH secretion is in turn regulated by corticotropin-releasing hormone (CRH), which is produced by neuroendocrine cells of the hypothalamus and carried to the anterior pituitary by the hypothalamic-pituitary portal vessels (refer to Figs. 13–3 and 13–4). This relationship between CRH, ACTH, and the adrenal cortex is illustrated in Figure 14–3 and is known as the **hypothalamic-pituitary adrenal axis.**

ACTH has both **tropic** effects and steroidogenic effects on cells of the adrenal cortex. The tropic effects of ACTH include promoting the general growth and function of all three zones of the adrenal cortex. Thus, in the absence of ACTH, adrenal cortex cells become smaller, and the production of all three classes of steroid hormone is dramatically reduced. Steroidogenic effects of ACTH involve regulation of glucocorticoid secretion by cells of the zona fasciculata. As will be described later in this chapter, secretion of mineralocorticoids from the zona glomerulosa is primarily regulated by factors other than ACTH.

Three Steroids Are the Primary Products of the Adrenal Cortex

In all, some 30 to 50 different steroids have been isolated from the adrenal cortex. However, only three are produced and secreted in quantitatively significant amounts. These are **cortisol, aldosterone,** and **dehydroepiandrosterone (DHEA),** which are shown

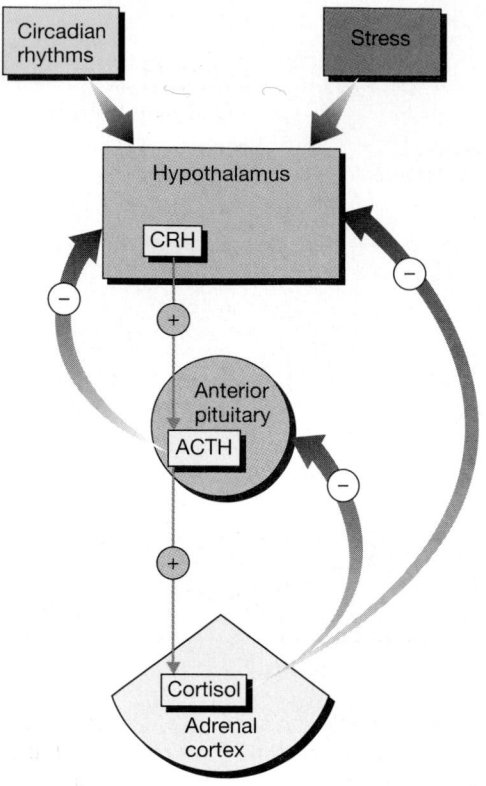

Figure 14-3

The hypothalamic-pituitary adrenal axis. Stimulatory effects are shown by the solid arrows; inhibitory effects by the dashed arrows.

androgen production by the adrenal is of little or no physiological significance, since far greater amounts are produced by the testes. In females, however, androgens produced by the adrenal gland account for most of that found circulating in the blood.

The general pathways involved in steroid hormone biosynthesis were previously described in Chapter 12 (see Fig. 12–8). For reference, a simplified diagram of the pathways is also presented here (Fig. 14–5). Recall that steroid hormones are synthesized and secreted on demand and are not stored. Therefore, regulation of steroid hormone secretion occurs at the level of hormone synthesis. As was described in Chapter 12, the first step in the synthesis of all steroid hormones begins with enzymatic conversion of cholesterol to pregnenolone. In most instances, this step in steroid synthesis is slowest, or rate-limiting. Once pregnenolone is formed, subsequent steps proceed fairly rapidly. Therefore, the enzyme that catalyzes conversion of cholesterol into pregnenolone represents the primary site of regulation of steroid synthesis (see Step A, Fig. 14–6).

Increased supply of cholesterol inside cells also enhances steroid synthesis. Most cholesterol that cells of the adrenal cortex use to make steroids is taken up from the blood, where it is carried in the form of lipoproteins. Cells of the adrenal cortex can remove these lipoproteins from the blood and bring them into the cell, making cholesterol available for steroid synthesis (see Step B, Fig. 14–6). These cells store any cholesterol not immediately used for steroid synthesis as lipid droplets. Later, stored cholesterol can be rapidly mobilized and used for steroid synthesis when needed (see Step C, Fig. 14–6).

Cells within each of the three zones of the adrenal cortex contain enzymes specifically suited for producing the steroid hormone characteristic of that zone. For example, only cells of the zona glomerulosa contain enzymes required for conversion of cor-

in Figure 14–4. Of the three, cortisol, a glucocorticoid, is produced in the greatest amount and aldosterone, a mineralocorticoid, is produced in the least amount. This chapter focuses on the physiology of these two steroids. The third, dehydroepiandrosterone (DHEA), is itself a rather weak androgen, but it can be converted in peripheral tissues to testosterone, a much more powerful androgen (see Chapter 31). In males,

Figure 14-4

Chemical structures of the principal steroids of the adrenal cortex.

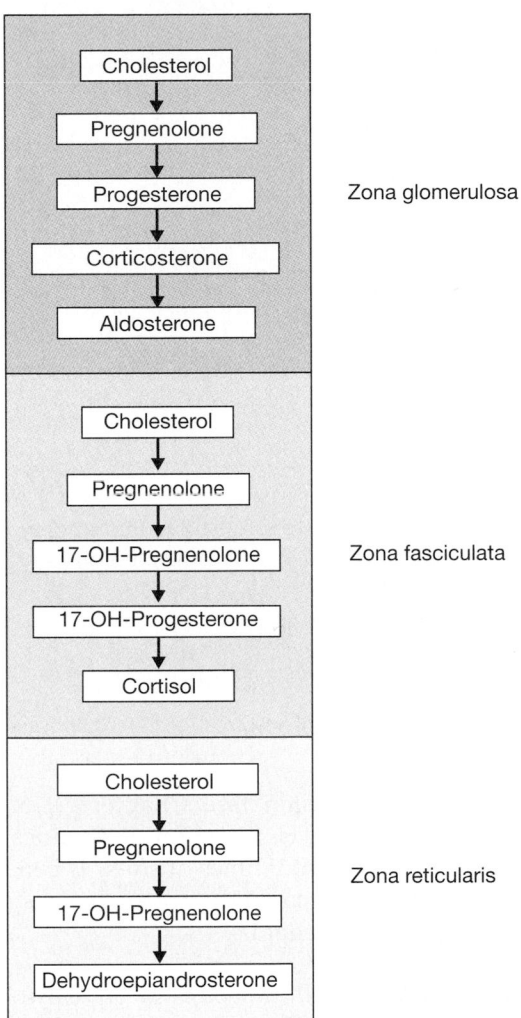

Figure 14–5

Pathways of steroid synthesis in the adrenal cortex.

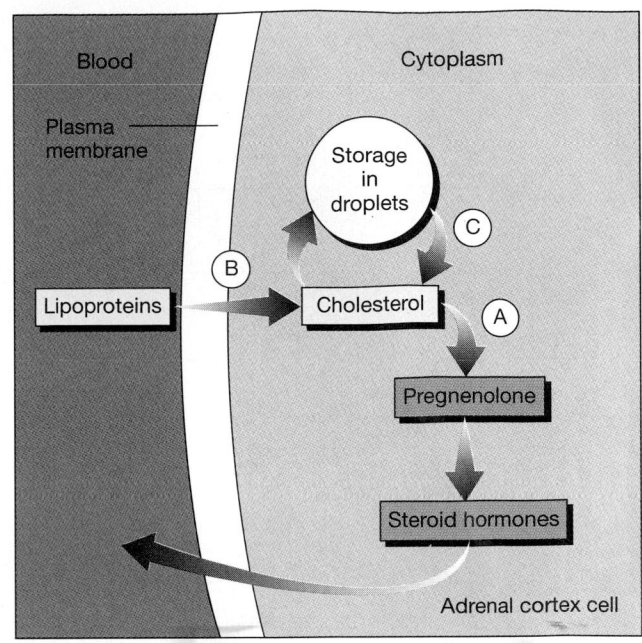

Figure 14–6

The regulatory steps in biosynthesis of adrenal cortex steroids. Step A, the conversion of cholesterol to pregnenolone, is the primary site of regulation of steroid synthesis. Steps B and C, the uptake of cholesterol from the bloodstream and its removal from intracellular storage sites, respectively, are secondary sites of regulation.

ticosterone to aldosterone, and thus only these cells can produce aldosterone (see Fig. 14–5). These cells lack the enzyme required for the conversion of pregnenolone to 17-OH-pregnenolone and so are unable to synthesize either cortisol or DHEA. Similarly, cells of the zona fasciculata contain the enzymes that favor the production of cortisol, and cells of the zona reticularis exhibit high levels of the enzymes favoring DHEA synthesis.

Glucocorticoids Play an Important Role in Metabolic Regulation

Regulation of Glucocorticoid Secretion

As illustrated in Figure 14–3, secretion of glucocorticoids from cells of the adrenal cortex is regulated by a sequence of steps beginning with CRH secretion by

the hypothalamus. CRH release by the hypothalamus is regulated by inputs from a variety of higher brain centers. Therefore, much of the regulation of glucocorticoid secretion is due to these inputs to the hypothalamus. In general, glucocorticoid secretion from the adrenal glands is responsive to two types of stimuli. The first type involves a pattern of secretion that varies over the 24-hour period of a day, and the second category involves increased secretion in response to a variety of specific stimuli.

The Circadian Rhythm of Glucocorticoid Secretion. Figure 14–7 shows typical changes in plasma cortisol concentrations that might occur over a 24-hour period in a normal individual. As indicated, plasma cortisol concentrations are highest around the time of awakening in the morning and lowest around midnight. Because this general pattern of secretion is repeated approximately every 24 hours, it is termed a circadian rhythm. These changes in cortisol secretion are due to similar circadian variations in CRH and ACTH secretion. This rhythm in cortisol secretion has its origin in higher brain centers that in turn influence the hypothalamic-pituitary system. These circadian rhythms are related primarily to the sleep-wake patterns of the individual rather than by environ-

Figure 14–7

Graph of the daily rhythm of the plasma cortisol concentration.

Table 14–1

Types of Stress Known to Affect Cortisol Secretion

Physical stress
  Hypoglycemia
  Trauma
    Broken bones
    Burns
    Surgery
    Cold exposure
    Infection
  Heavy exercise
Psychological stress
  Acute anxiety
    Anticipation of stressful situations; surgery, college exams, or an airplane flight, for example
    Novel situations
Chronic anxiety

mental light-dark cycles. Changes in a person's sleep-wake cycle, such as working nights and sleeping days, will result in a temporal shift in the daily rhythm of cortisol secretion.

Stress-Induced Glucocorticoid Secretion. A variety of different stress stimuli can also significantly increase cortisol synthesis and secretion (see Fig. 14–3). Increased secretion in response to stress occurs in addition to the normal circadian rhythm of secretion described previously. Table 14–1 lists several different types of stress known to cause increased glucocorticoid secretion. Generally these can be either physical or psychological in nature. Hypoglycemia (low blood glucose) is an important stimulator of cortisol secretion and provides for important adaptive changes that occur with fasting. Practically any type of physical trauma or injury, such as a broken bone, severe burn, infection, or surgery, will result in increased cortisol secretion. Strenuous exercise, such as in competitive athletics, also results in increased cortisol secretion.

Several types of psychological or emotional stress can also stimulate cortisol secretion. Acute anxiety is probably the strongest psychological stimulus for cortisol secretion. Examples might include anticipation associated with a surgical operation or final exams in school. Encountering a new social situation often triggers increased cortisol secretion. In contrast, chronic anxiety is considerably less important as a stimulus for cortisol secretion.

As described in Chapter 13, ACTH is initially synthesized in the anterior pituitary as part of the larger POMC peptide. When ACTH secretion is stimulated, several other bioactive peptides, including the mor-

phinelike beta-endorphin (see Chapter 7), are secreted along with ACTH. Secretion of peptides such as beta-endorphin during times of stress is beneficial since they function as an endogenous analgesic and increase the pain threshold.

Negative Feedback Relationships in Glucocorticoid Secretion. As is true of most other hormones, cortisol secretion is regulated by a series of negative feedback loops. The dotted lines in Figure 14–3 show that cortisol exerts dual negative feedback effects to inhibit secretion of ACTH from the pituitary. First, cortisol decreases the activity of CRH-producing neurons in the hypothalamus, thereby decreasing CRH release. Second, cortisol decreases the sensitivity of corticotropes to CRH. As a result, CRH is less effective in stimulating ACTH release when cortisol is present than when it is absent. Thus, the combination of a decrease in CRH secretion as well as a decrease in sensitivity to CRH contributes to reduced ACTH secretion, leading ultimately to decreased cortisol secretion. In addition, as indicated in Figure 14–3, ACTH itself exerts a negative feedback effect on the hypothalamus to reduce CRH secretion.

Transport of Cortisol in the Blood

Normally, about 80% of blood cortisol is bound to a specific carrier protein known as **cortico-steroid-binding globulin.** Another 15% of the cortisol is bound to albumin, and the remaining 5% normally exists free in solution. As with other hormones, only free cortisol molecules are able to interact with receptors and produce physiological effects characteristic of the hormone.

Physiological Effects of Glucocorticoids

Glucocorticoids play an important role in our day-to-day survival. Actions of glucocorticoids enable us to survive the potentially damaging effects of wide fluctuations in environmental conditions, as well as a number of noxious or potentially noxious factors. As a result, glucocorticoids are considered essential for life. Even a modest period of fasting would not be possible without the actions of cortisol. Before we look at specific effects of glucocorticoids, let us briefly review the general mechanism of steroid action.

Glucocorticoids produce their effects via the general mechanism for steroid hormone action described in Chapter 12 (see Fig. 12–17). Recall that steroid hormones are lipid soluble; they readily pass through plasma membranes and bind to receptors located in the nucleus. Hormone-receptor complexes then bind to specific sites on DNA, altering transcription of mRNA molecules that code for specific proteins. As a result, synthesis of a few specific proteins is appropriately increased or decreased, and the desired changes in cell function occur. Given these mechanisms, what are the specific changes in cell function that are produced by cortisol?

Cortisol exerts its primary effects on three different tissues in the body: liver, skeletal muscle, and adipose tissue. Specific physiological effects of the hormone in these three tissues are summarized in Table 14–2. The overall effect of cortisol under most con-

Table 14–2
Summary of Effects of Cortisol on Metabolism

Liver
    Increases gluconeogenesis
    Increases glycogen synthesis
Skeletal muscle
    Decreases protein synthesis
    Increases protein degradation
    Decreases glucose uptake
Adipose tissue
    Decreases glucose uptake
    Increases lipid mobilization

ditions can be summarized as being **catabolic,** owing to its promotion of protein breakdown in tissues. At very high concentrations, such as might be used therapeutically in medical treatment, glucocorticoids also produce effects that are **anti-inflammatory** and **immunosuppressive.**

Cortisol in the Liver: Increased Blood Glucose. In liver, the net response to cortisol is increased **gluconeogenesis** (see Table 14–2), primarily as a result of increased conversion of amino acids into glucose (for a discussion of gluconeogenesis, refer to Chapter 6). Increased glucose formation results from specific effects of the hormone (Fig. 14–8). First, cortisol increases the activity of several enzymes that catalyze

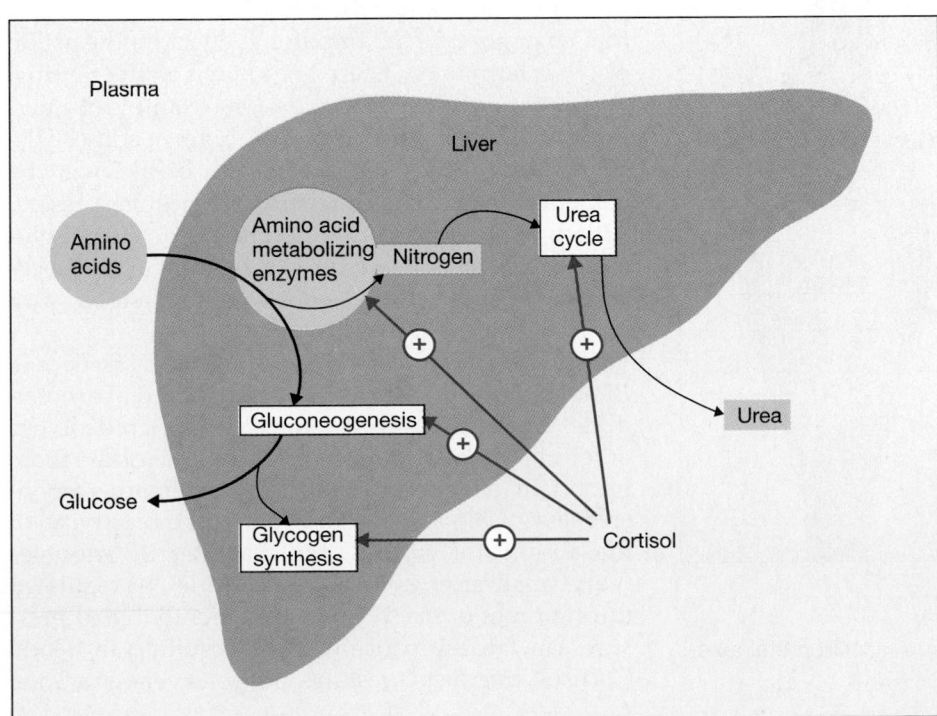

Figure 14–8

Summary of the effects of cortisol on liver metabolism.

key steps in the gluconeogenic pathway. As a result, glucose formation from a variety of substrates is increased. Second, cortisol increases the activity of several enzymes involved in amino acid metabolism, facilitating the use of amino acids as substrates for gluconeogenesis. Third, glucocorticoids promote gluconeogenesis in liver by stimulating the activity of enzymes of the urea cycle. This effect increases the ability of the liver to dispose of nitrogen liberated during metabolism of amino acids. Together, these specific effects of cortisol permit increased conversion of amino acids into glucose (see Fig. 14–8). As a result, glucocorticoids tend to increase blood glucose. Glucocorticoids also tend to promote formation of liver glycogen. This is due in part to increased glucose formation in the liver via gluconeogenesis and in part to a stimulation of enzymes involved in glycogen formation.

Cortisol in Skeletal Muscle: Decreased Protein. In skeletal muscle, cortisol has a catabolic action and tends to promote the net loss of protein. This occurs as a result of two different effects (Table 14–2 and Fig. 14–9). First, cortisol inhibits muscle protein synthesis. As a result, uptake of amino acids from blood and their subsequent incorporation into muscle protein is reduced. Second, cortisol increases degradation of existing muscle protein. Amino acids released during the degradation of muscle protein are then free to enter the blood. As a result of these two effects of cortisol, there is a net transfer of amino acids from muscle protein into the blood. Liver can then use these amino acids for the synthesis of glucose via gluconeogenesis. The combined effects of cortisol on liver and muscle help maintain blood glucose at the expense of muscle protein. These catabolic effects of cortisol on muscle occur in skeletal muscle *only,* not in cardiac muscle. Furthermore, the catabolic actions of cortisol and the anabolic actions of other steroids should not be confused. Anabolic steroids used to increase muscle mass consist primarily of androgens and not of glucocorticoids. Cortisol also decreases glucose uptake by skeletal muscle, an effect that opposes that of insulin.

Cortisol in Adipose Tissue: Decreased Glucose Uptake. In adipose tissue, cortisol decreases glucose uptake (see Table 14–2). As in muscle, this effect is opposite to that of insulin and thus represents an anti-insulin effect of the steroid. Glucocorticoids also promote mobilization of lipid from some adipose tissue stores in the body. Not all adipose tissue is affected equally, however. This differential sensitivity of adipose tissue to glucocorticoids can best be seen when the glucocorticoids are secreted in excess. When this occurs, fat stores in the legs and arms decrease, and fat is redistributed to the trunk and shoulder-blade region. This leads to a very distinct physical appearance, which will be discussed in a later section describing the effects of cortisol hypersecretion.

The Permissive Actions of Glucocorticoids. In a number of instances, although cortisol itself does not directly regulate a particular metabolic process or initiate events inside the cell, it must be present for other hormones to produce their full effect. Cortisol therefore tends to have an amplifying effect on the action of other hormones. These are known as the **permissive actions** of glucocorticoids. One example of a permissive action of glucocorticoid occurs in adipose tissue, where stimulation of lipid breakdown by epinephrine is enhanced when cortisol is also present. Although numerous other permissive actions of glucocorticoids are known, the exact nature of the mechanisms for these effects is not well understood.

Effects of Glucocorticoids on Blood Vessels and Blood Cells. In addition to its effects on carbohydrate, lipid, and protein metabolism, cortisol has other important effects in the body. One of the more important effects of cortisol is to enhance responsiveness of blood vessels, or **vascular reactivity.** In the absence of cortisol, the diameter of arterioles (very small arteries) is not appropriately regulated during times of stress, and consequently blood pressure can fall dramatically, even resulting in death. Cortisol treatment restores responsiveness of the

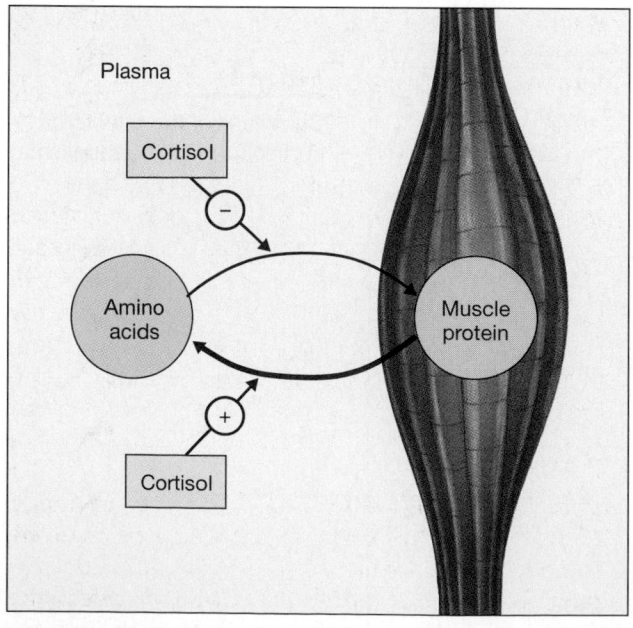

Figure 14–9

Summary of the effects of cortisol on muscle protein metabolism.

blood vessels and therefore is said to restore vascular reactivity.

Glucocorticoids also affect various blood cell types and lymphoid tissues. Cortisol decreases the number of eosinophils and basophils and increases the number of neutrophils, red blood cells, and platelets in the blood (see Chapter 17 for a further discussion of blood cell types). As will be discussed, high doses of glucocorticoids also influence inflammation and the immune response.

Pharmacological Effects of Glucocorticoids

Effects of cortisol described in the foregoing paragraphs are those generally observed in response to physiological amounts of the hormone. At much higher concentrations, these effects are magnified, and other effects are produced as well, some of which have proven very useful for medical purposes. These are known collectively as **pharmacological effects** of glucocorticoids and other closely related steroids.

Perhaps the two most important pharmacological effects of glucocorticoids are their **anti-inflammatory** effects and their **immunosuppressive** effects. In large doses, glucocorticoids profoundly inhibit the inflammatory response to tissue injury or infection by inhibiting virtually every step in a normal inflammatory response, including vasodilation, increased capillary permeability, and increased phagocytosis. High doses of glucocorticoids also suppress the immune system, resulting in decreased antibody production. The number of circulating lymphocytes decreases, owing to increased lymphocyte destruction in lymphoid tissues, and the size of lymph nodes also decreases. As a result of these pharmacological effects, glucocorticoids are particularly useful in medicine for reducing the inflammatory response to injury or infection, or for suppressing the rejection response to transplanted organs. Although the immunosuppression produced by glucocorticoids is of great value following organ transplantation, it does leave the individual more susceptible to infections, necessitating treatment with antibiotics as well as with steroids.

Mineralocorticoids Play an Important Role in Regulating Mineral Metabolism

Regulation of Mineralocorticoid Secretion

Aldosterone is the principal mineralocorticoid produced in the zona glomerulosa of the adrenal cortex. The two primary regulators of aldosterone secretion are **potassium ion** and a peptide hormone known as

angiotensin II. The kidneys play a key role in determining plasma levels of each of these two factors, and are also the major action site of aldosterone in the body.

Regulation by Angiotensin II. **Angiotensin II** is formed by a sequence of steps that begin in the kidney and are collectively known as the **renin-angiotensin system.** These steps are outlined in Figure 14–10 and begin with the secretion of **renin** by specialized cells in the kidney known as **granular cells.** The location and function of the granular cells will be described in detail in Chapter 24. Renin is a proteolytic enzyme that acts upon a large protein, **angiotensinogen,** which is synthesized by the liver and secreted into the blood (see Fig. 14–10). In the blood, renin splits a specific peptide bond in angiotensinogen, resulting in the formation of **angiotensin I,** which is a peptide chain of 10 amino acids. Angiotensin I is carried via the blood to the lungs, where **angiotensin-converting enzyme** removes two amino acids from the end of angiotensin I to form **angiotensin II,** a peptide of eight amino acids (see Fig. 14–10). From the lungs, angiotensin II travels via the blood to the tissues of the body.

In the adrenal glands, cells of the zona glomerulosa contain specific receptors for angiotensin II. Binding of angiotensin II to these receptors triggers increased synthesis and secretion of aldosterone. The exact mechanisms whereby angiotensin II binding leads to increased aldosterone synthesis are not known, although it appears that the phosphatidylinositol pathway provides the second messengers that couple hormone binding with the cellular effects of increased steroid synthesis (refer to Fig. 12–14).

As indicated in Figure 14–10, the amount of aldosterone produced by the adrenal cortex is related to renin release from the kidney. Therefore, those factors that regulate **renin secretion** ultimately serve to regulate **aldosterone secretion** by the adrenal cortex. The primary stimulus for renin secretion is a decrease in blood pressure or blood flow to the kidneys. Conversely, an increase in blood pressure or blood flow to the kidneys inhibits renin release. In addition, sympathetic innervation of granular cells modulates the response in renin secretion to these and other stimuli.

Regulation by Potassium Ion. The second major regulatory factor involved in controlling aldosterone secretion is the concentration of **potassium ion.** Aldosterone secretion is stimulated by an *increase* in the plasma potassium concentration. Potassium ions appear to directly affect cells of the zona glomerulosa. These cells are extremely sensitive to very small changes in the plasma concentration of potassium; an

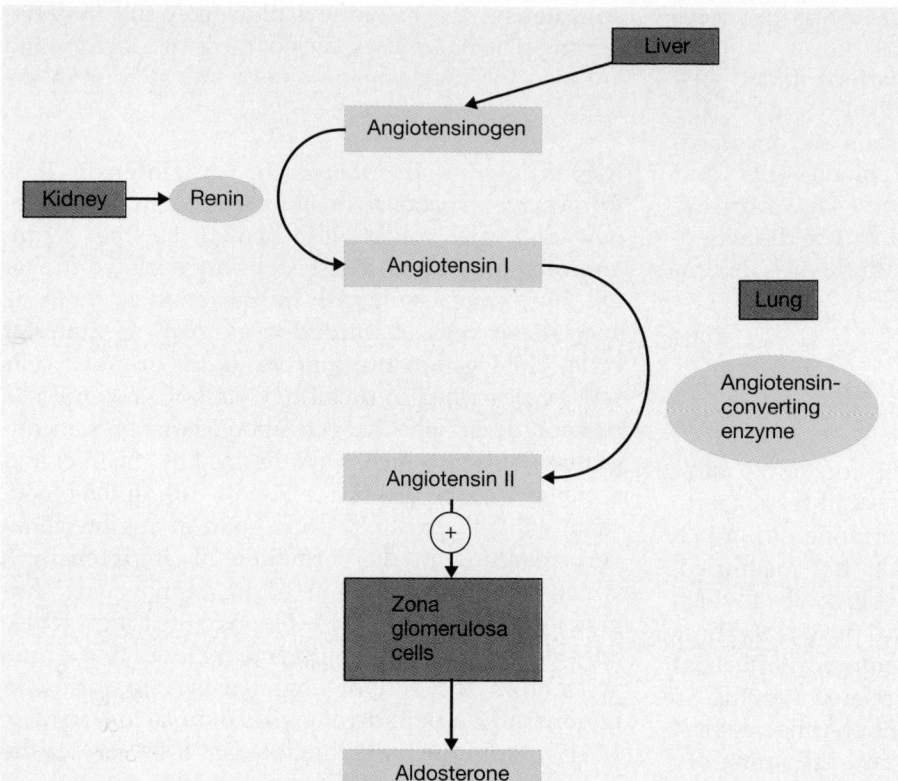

Figure 14–10

Summary of the renin-angiotensin system.

increase in potassium of as little as 5% above normal is enough to increase aldosterone secretion. As the following section will describe, one of the actions of aldosterone is to promote potassium excretion by the kidney and thus to lower plasma potassium concentrations. Thus, aldosterone and the kidneys make up two components of a feedback loop that regulate plasma potassium concentration.

Effects of Mineralocorticoids

The net effect of aldosterone is to increase extracellular fluid volume. This increase in extracellular fluid volume results in increased blood volume, which tends to raise blood pressure and cardiac output. Thus, aldosterone indirectly affects blood pressure and blood flow in the cardiovascular system and is an important component of processes involved in regulating fluid balance (see Chapter 25).

Kidneys are the primary action site for aldosterone. In response to aldosterone, kidneys retain more sodium in the body while excreting more potassium. Thus, in response to aldosterone, urinary sodium decreases, while urinary potassium increases. The specific mechanisms involved in these effects of aldosterone are presented in Chapter 23 and 24.

In addition to its effects on aldosterone secretion, angiotensin II has other important effects related to the maintenance of blood pressure and extracellular fluid volume. These include a stimulation of smooth muscle contraction in blood vessels, activation of thirst centers in the brain, and stimulation of ADH release from the posterior pituitary. Each change that results (decreased blood vessel diameter, increased water intake, and increased fluid retention by the kidneys) contributes to maintenance of a normal blood volume and blood pressure. These and other effects of angiotensin, as well as other characteristics of the renin-angiotensin system, will be described in Chapter 24, which discusses in detail the regulation of fluid and electrolyte balance.

Abnormalities of Adrenal Cortex Function Lead to Altered Physiologic Regulation

Several common diseases result from either deficiencies or oversecretion of adrenal cortex steroids. As with other endocrine disorders, the symptoms expressed result from either absence or magnification of effects of the hormones involved. A discussion of these conditions is therefore a useful way to review

CLINICAL FOCUS

Designer Steroids

Steroid hormones are widely used in medicine today, not only for purposes of replacement therapy when the body's own hormone production is lacking, but also for anti-inflammatory and immunosuppressive effects that certain steroids produce. Because of similarities in structure, most naturally occurring steroids have some activity that overlaps with steroids of other classes. For example, cortisol has primarily glucocorticoid-like activity, but can produce mineralocorticoid-like and androgen-like effects at high doses. Similarly, aldosterone has primarily mineralocorticoid activity, but does possess a weak amount of glucocorticoid activity.

When treating patients with steroids or any other drug, it is beneficial if extraneous or undesirable activities are minimized. Recognizing this, the pharmaceutical industry has produced hundreds of synthetic steroids, each having its own spectrum of hormonal activity. For example, dexamethasone is a synthetic steroid widely used in medicine. On a molecule-per-molecule basis, it has much more glucocorticoid-like activity than does the natural steroid cortisol, but unlike cortisol, dexamethasone has essentially no mineralocorticoid activity. Prednisone is another steroid that is commonly prescribed. It has largely glucocorticoid activity, but also exhibits a significant degree of mineralocorticoid action, making it useful in cases where the adrenal glands have been removed or their function is impaired and the physician wants to provide some degree of replacement of both hormones. Furthermore, some steroids have been modified in such a way as to slow their rate of removal from the bloodstream and thus prolong the time course over which effects are produced. Others have been modified in such a way as to limit their solubility. This is particularly useful for steroids having anti-inflammatory activity, as they can be formulated in a cream or ointment and be used topically to treat problems like skin rashes, poison ivy, or small superficial wounds.

Another form of steroid that has been modified to enhance certain biological effects are anabolic steroids. While glucocorticoids of the adrenal cortex generally tend to be catabolic in nature, anabolic steroids possess predominantly androgen-like activities and thus promote acquisition of muscle mass. Although they are occasionally used clinically, for the most part they are used (illegally) by athletes wishing to build up their muscles and therefore their strength. Although they obviously produce the desired short-term benefits, they also produce damage to other organs—particularly the liver—which accumulates over time. It is not uncommon to hear on the news of a prominent athlete who has died at a relatively early age as a result of the use of anabolic steroids. This is clearly a situation in which the use of designer steroids is not advantageous, and is in fact highly detrimental to the well-being of the individual.

■ **Based on your knowledge of hormone/receptor interactions, binding of steroid hormones to carrier plasma proteins, and degradation of steroid hormones, can you make a list of possible things to consider when trying to design a "new" form of steroid for therapeutic purposes?**

---

the actions of these hormones, as well as to become familiar with endocrine disease states that are fairly common in the general population.

Excess Secretion of Glucocorticoids

A common disorder of adrenal cortex function involves increased secretion of glucocorticoids. This can occur either as a result of a primary abnormality in steroid hormone production by the adrenal cortex or as a result of overproduction of ACTH by the pituitary, causing excessive stimulation of the adrenal cortex. In either case, physiological changes that occur are due to the effects of excess cortisol. Excess cortisol secretion results in the condition known as **Cushing's syndrome.** Figure 14–11 illustrates the general appearance of a patient with Cushing's syn-

drome. Note the thin arms and legs, due in part to the loss of muscle mass as a result of the protein-catabolic effects of excess cortisol. Fat is redistributed from the extremities to the trunk, also contributing to a thinning of the arms and legs. Loss of fat in the extremities is accompanied by increased fat in the face, the trunk, across the shoulder blades, and at the base of the neck. Accumulation of fat in the face gives it a characteristically rounded appearance. As a result of the protein-catabolic effects of cortisol, connective tissue is lost from the skin, causing it to become thinner. As a result, blood vessels are located closer to the surface, which is most often apparent in the red cheeks that characterize Cushing's patients. In addition, as the skin of the abdomen is stretched with increased fat deposition, it thins even further, and reddish-purple stretch marks, or **striae,** form, owing

Many of the effects of cortisol antagonize the actions of insulin. Because a deficiency in insulin action results in diabetes, severe cases of Cushing's syndrome can actually produce diabetes if left untreated for a prolonged period of time.

Insufficient Secretion of Adrenal Cortex Steroids

A deficiency in steroid production by the adrenal cortex leads to the clinical condition termed **adrenal insufficiency.** If not treated, this particular disease state may lead to death. Adrenal insufficiency can be of two general causes. Primary adrenal insufficiency, or **Addison's disease,** results from disease of the adrenal cortex that impairs both glucocorticoid and mineralocorticoid production. Secondary adrenal insufficiency results from inadequate ACTH production. Since ACTH has only minor effects on mineralocorticoid production by the adrenal cortex, there is little or no change in aldosterone secretion in secondary adrenal insufficiency; it is primarily glucocorticoid secretion that is deficient in this situation.

Addison's disease is the more common form of adrenal insufficiency. Typical symptoms (Table 14–3) reflect a loss of both glucocorticoid and mineralocorticoid action. Due to loss of aldosterone, a person suffering from Addison's disease typically exhibits low plasma sodium and high plasma potassium levels. Low blood pressure is a common finding, in part due to electrolyte and water loss leading to reduced blood volume caused by the lack of aldosterone, and also due to the absence of the stimulatory effect of cortisol on vascular reactivity. Disturbances of electrolyte balance may lead to muscle weakness and easy fatigue as well as other symptoms, including vomiting and loss of appetite. Low blood glucose is also common in persons with Addison's disease, owing to an absence of the stimulatory effect of cortisol on gluconeogenesis in the liver.

In addition, hyperpigmentation of the skin and gums may occur in patients with Addison's disease

Figure 14–11

A man with Cushing's syndrome.   *(© VU/Visuals Unlimited.)*

to the increased visibility of underlying blood vessels. Loss of connective tissue in small blood vessels makes them more fragile and easily broken than normal, resulting in an increased susceptibility to bruising.

Hypertension (high blood pressure) is also commonly present in persons with Cushing's syndrome and in many cases is the primary contributing factor leading to death. These patients also suffer from increased susceptibility to infection and poor wound healing due to suppression of the inflammatory and immune systems caused by excess cortisol. Excess cortisol also impairs bone metabolism, and thus patients with Cushing's syndrome may also suffer from bone fractures and osteoporosis.

Table 14–3
Typical Findings in Addison's Disease

Low plasma sodium, high plasma potassium
Low blood pressure
Muscle weakness, fatigue
Vomiting, loss of appetite, dehydration
Low blood sugar
Excess pigmentation of skin in some patients

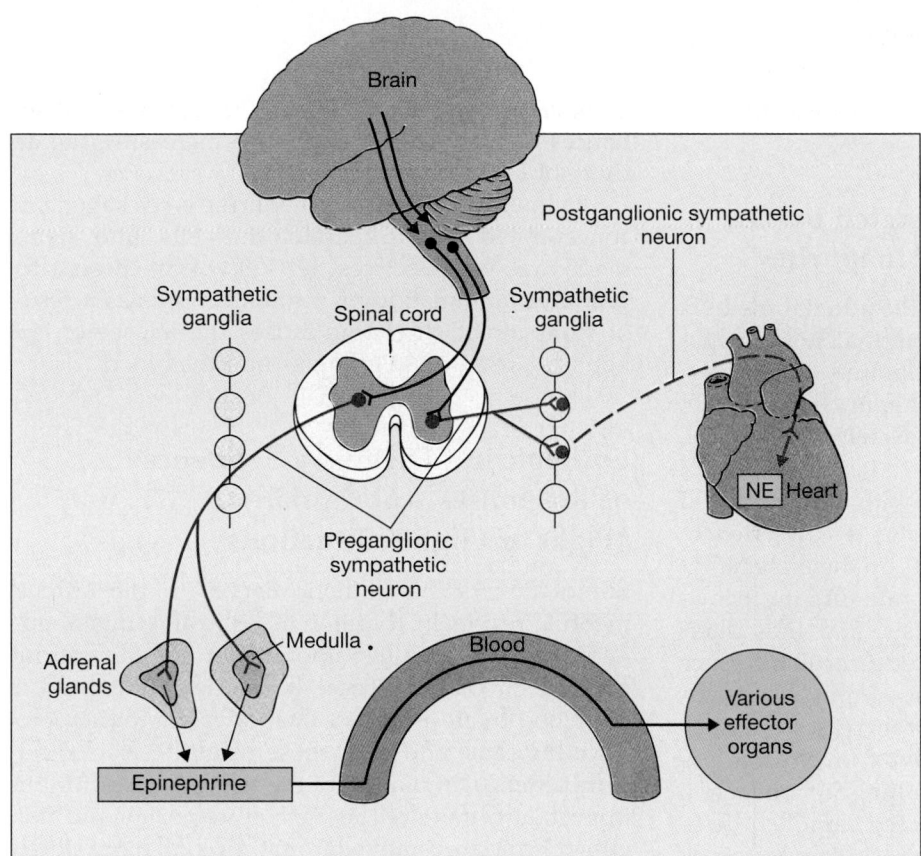

Figure 14–12

Figure illustrating the anatomical analogy between cells of the adrenal medulla and the sympathetic postganglionic neurons. Note that a particular postganglionic fiber (dotted orange line) has effects on one specific effector organ, such as the heart. In contrast, cells of the adrenal medulla, because they secrete epinephrine into the blood, are able to influence the activity of various effector organs throughout the body.

when the abnormality results from a primary loss of adrenal cortical function (see Table 14–3). Recall from Figure 14–3 that in the absence of glucocorticoids, their negative feedback effect to inhibit ACTH secretion is lost, and plasma ACTH therefore increases dramatically. The structure of ACTH is similar to melanocyte-stimulating hormone (MSH), which increases pigmentation in the skin. Although the MSH-like activity of ACTH is weak, greatly increased plasma concentrations of ACTH that occur in some forms of Addison's disease are sufficient to cause increased skin pigmentation.

The Adrenal Medulla

The adrenal medulla, which secretes epinephrine and norepinephrine, is an important component of the sympathetic nervous system. As a result, the sympathetic nervous system and adrenal medulla are often referred to together as the **sympathoadrenal system.**

The Medulla Is Centrally Located in Each Adrenal Gland

As shown in Figure 14–1, the medulla is located in the interior portion of each adrenal gland. Within the adrenal gland, blood flows from the outer cortex inward to the medulla. As a result, blood that reaches the medulla contains high concentrations of glucocorticoids. This vascular orientation is important since it has been shown that glucocorticoids can influence hormone synthesis in the medulla.

The functional unit of the adrenal medulla is the **chromaffin cell,** which functions as a neuroendocrine cell. The name "chromaffin" originated from the fact that these cells are readily stained with solutions containing chromium. Functionally, chromaffin cells of the adrenal medulla are analogous to **postganglionic fibers** of the sympathetic nervous system (Fig. 14–12). Recall that in the sympathetic nervous system, preganglionic fibers form synapses with postganglionic fibers in the sympathetic chain ganglia or in one of the outlying sympathetic ganglia. Postganglionic fibers then project to a particular target organ such as the heart, as illustrated in Figure 14–12. The arrangement is slightly different for the adrenal medulla, however. Preganglionic fibers to the medulla pass through the sympathetic chain without synapsing and directly innervate chromaffin cells of the adrenal medulla. In response to stimulation, chromaffin cells secrete epinephrine directly into the blood and epinephrine is carried to various effector organs in the body. Thus although chromaffin cells of the medulla are analogous to sympathetic postganglionic fibers, they differ in that chromaffin cells influence the activity of a variety of effector organs throughout the body. In contrast, sympathetic postganglionic neurons generally influence the activity of single effector organs.

The Primary Hormone Secreted by the Adrenal Medulla Is Epinephrine

The synthesis of epinephrine in the adrenal medulla is outlined in Figure 14–13. Note that norepinephrine, another important catecholamine, is an intermediate in the synthesis of epinephrine. However, the adrenal medulla normally secretes only small amounts of norepinephrine.

Epinephrine synthesis begins with the amino acid tyrosine (see Fig. 14–13). Tyrosine is converted to **DOPA** (dihydroxyphenylalanine) by the enzyme **tyrosine hydroxylase.** This is the rate-limiting (slowest) step in epinephrine synthesis, and thus alterations in the activity of this enzyme directly influence the rate of epinephrine synthesis. Input of sympathetic stimulation to the medulla increases epinephrine synthesis by increasing activity of tyrosine hydroxylase. Subsequent steps resulting in conversion of DOPA to norepinephrine proceed rapidly. The final step in epinephrine synthesis is conversion of norepinephrine to epinephrine, which is catalyzed by the enzyme **phenylethanolamine-N-methyl transferase (PNMT).** This enzyme is stimulated by cortisol, and thus high concentration of this steroid in

blood flowing from the cortex to the medulla enhances epinephrine synthesis by increasing the activity of PNMT.

Following synthesis, epinephrine is packaged and concentrated within chromaffin cells into dense membrane-bounded vesicles known as **chromaffin granules.** Stimulation of chromaffin cells as a result of sympathetic discharge causes the release of epinephrine from these granules into the blood.

Epinephrine Triggers a Sequence of Responses Appropriate to "Fight or Flight" Situations

Stimulation of sympathetic nerves to the adrenal medulla results in secretion of large amounts of epinephrine and smaller amounts of norepinephrine into the circulation. These hormones are carried in the blood to all tissues of the body, where they produce the same effects as those resulting from direct sympathetic stimulation of the tissue, except for one important difference. Because epinephrine and norepinephrine are removed from the blood more slowly than from the area of the sympathetic nerve terminal, effects of the circulating hormones are more prolonged than effects resulting from direct sympathetic stimulation.

The adrenal medulla is part of a general response system that triggers a sequence of events appropriate to **"fight-or-flight"** situations. The net result of this response is to increase the capability of the body to perform vigorous muscle activity. Thus, if faced with a life-threatening situation, the body can respond more effectively, bettering the chances of survival.

Examples of some actions of epinephrine are listed in Table 14–4. Many of the effects of epinephrine occur within the cardiovascular system and serve to increase and redistribute blood flow between various tissues. Epinephrine increases both the heart rate and strength with which the heart contracts; as a result, it increases the amount of blood pumped per minute (the cardiac output). Epinephrine also produces changes that affect the distribution of blood flow among different tissues. Epinephrine causes dilation of blood vessels in skeletal muscle while simultaneously causing constriction of blood vessels in the skin and in visceral organs such as the gastrointestinal tract. These changes in diameter of blood vessels cause a shunting of blood away from the skin and internal organs to the muscles, which is advantageous to an animal faced with a "fight-or-flight" situation. In addition, epinephrine affects the CNS, increasing the state of mental alertness, which is also advantageous in a life-threatening situation.

Figure 14–13

Pathways of epinephrine synthesis in adrenal chromaffin cells. Conversion of tyrosine to DOPA, catalyzed by tyrosine hydroxylase, is normally the rate-limiting step. The enzyme PNMT catalyzes the conversion of norepinephrine to epinephrine.

Table 14–4
Summary of the Actions of Epinephrine

Effects on the cardiovascular system
   Increased cardiac output
      Increased heart rate
      Increased strength of cardiac contractions
   Vasodilation in skeletal muscle
   Vasoconstriction in internal organs and skin
Effects on other tissues
   Relaxation of smooth muscle in
      gastrointestinal tract, urinary bladder,
      airways of lung
   Increased mental alertness
Effects on metabolism
   Increased glycogenolysis in liver and muscle
   Increased lipolysis in adipose tissue
   Decreased insulin secretion
   Increased glucagon secretion

Furthermore, epinephrine increases the ability of blood to transport oxygen. By causing relaxation of smooth muscle in airways of the lungs, epinephrine increases inflow of oxygen to the lungs. In some species, epinephrine also causes the spleen to contract, which increases the circulating blood volume and the number of red blood cells in the circulation. Both factors contribute to an increased oxygen-carrying capacity of the circulatory system.

Epinephrine affects metabolic processes in several different tissues, most notably the liver, skeletal muscle, and adipose tissue (see Table 14–4). The net result of epinephrine action is an increased breakdown of stored fuels into metabolizable substrates. The substrates that are released may serve as an energy source for local metabolism within the particular cell or tissue where they are formed, or they may be released into the blood for general distribution. In liver and muscle, epinephrine stimulates the breakdown of glycogen, while in adipose tissue the hormone stimulates breakdown of stored triglycerides into fatty acids and glycerol (see Table 14–4).

Epinephrine indirectly affects metabolism as a result of its effects on pancreatic hormone secretion. Insulin and glucagon are secreted by the pancreas and play key roles in regulating metabolism. Insulin has metabolic effects that oppose those of epinephrine, and glucagon has effects similar to those of epinephrine. Epinephrine decreases insulin secretion while increasing glucagon secretion. Thus, this effect of epinephrine on the pancreas serves to indirectly promote its own metabolic actions. Discharge of the sympathetic nervous system also leads to an increased basal metabolic rate as a result of increased cellular metabolism throughout the body. This effect is primarily due to release of norepinephrine from sympathetic nerve endings, rather than release of epinephrine from the adrenal medulla.

A summary of the metabolic effects of epinephrine is presented in Figure 14–14. In liver, epinephrine stimulates glycogenolysis and thus increases glucose release from liver into the blood. Epinephrine also stimulates glycogenolysis in muscle, but muscle cannot release free glucose into the blood because it lacks the enzyme necessary to convert glucose-6-

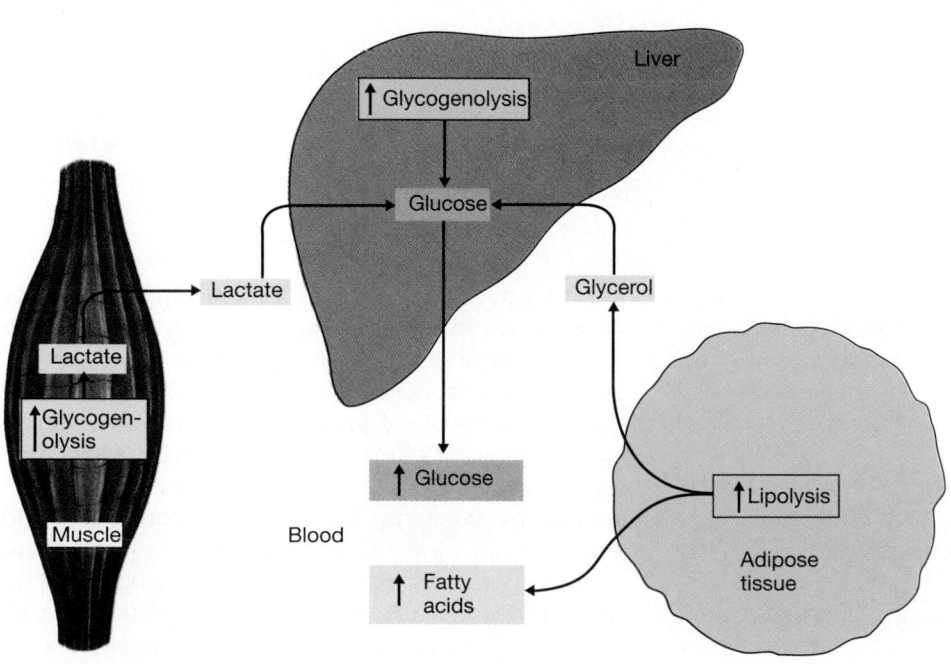

Figure 14–14

Summary of the principal effects of epinephrine on energy metabolism. The primary effects of epinephrine are to stimulate glycogenolysis in liver and muscle, and lipolysis in adipose tissue. The net result is an increase in blood glucose and blood fatty acid concentrations.

R E S E A R C H   F O C U S

The Beta Adrenergic Receptor

When epinephrine interacts with one of its target cells, it does so through a group of receptors known as adrenergic receptors. The two principal adrenergic receptors are α-adrenergic receptors and β-adrenergic receptors. These two receptor types differ in their structure and in the type of hormone that they bind. Epinephrine has a higher affinity (strength of binding) to β-adrenergic receptors and lesser affinity for α-adrenergic receptors. Conversely, norepinephrine has a higher affinity for α-adrenergic receptors and a lesser affinity for β-adrenergic receptors.

The β-adrenergic receptor is one of the most intensively investigated hormone receptors, and information gained from studying this one receptor type has helped us understand many of the events in second messenger cascades. Research has shown that it is a member of a larger family of cell surface receptors that share common structures and common coupling to downstream events in hormone signaling pathways.

A striking feature about β-adrenergic receptor structure that it shares with a number of other receptors (including rhodopsin, the light receptor in the retina of the eye) is that it contains seven transmembrane domains. The peptide chain that forms the receptor crosses back and forth through the plasma membrane seven times. As it does so, loops of peptide chain are exposed on both the outer and inner surfaces of the cell membrane. On the outside of the cell, these exposed parts of the peptide chain form the actual binding site for epinephrine. When epinephrine binds to its receptor, a change in shape of the receptor protein occurs that results in activation of a portion of the receptor exposed to the cell interior. A common feature of β-adrenergic receptors and other receptors with a seven-transmembrane domain structure is that they are coupled to G-proteins (see Chapter 12). This finding is so consistent that when new receptors are discovered to have a seven-transmembrane structure, it is common to assume that their actions are coupled to cellular events via G-proteins.

One important site of action for epinephrine is on the heart, where, via β-adrenergic receptors, it increases both heart rate and strength of contractions. Recently, using techniques of molecular biology, investigators were successful in manipulating the genes of mice to give them extra copies of the β-adrenergic receptor in their heart (1, 2). Animals having increased numbers of receptors showed increased force of cardiac contractions in proportion to the relative number of receptors present. Although considerably more research must be done before it becomes a reality, these results suggest the possibility of using gene therapy in certain forms of heart disease. One such situation might be congestive heart failure, in which the normal strength of cardiac contractions is considerably reduced. By giving the heart extra copies of the β-adrenergic receptor gene, the hope is that the heart would beat more forcefully and alleviate the need for treatment with drugs.

■ **If all goes as planned, the general approach described above should produce some very powerful therapeutic effects. But supposing all doesn't go as planned, what other steps in the β-adrenergic pathway might be possible targets for gene therapy?**

References

Seachrist, L. "Gene therapy to spark a failing heart." *Science*, 264:507–508, 1994.

Milano, C.A., Allen, L.F., Rockman, H.A., Dolber, P.C., McMinn, T.R., Chien, K.R., Johnson, T.D., Bond, R.A., and Lefkowitz, R.J. "Enhanced myocardial function in transgenic mice overexpressing the beta₂-adrenergic receptor." *Science*, 264:582–586, 1994.

phosphate into plasma glucose. Instead, glucose is metabolized to lactate, which is then released. Lactate can be taken up by liver and converted to glucose via gluconeogenesis. As glucose is metabolized to lactate in muscle, ATP, which the muscle can use for contraction, is produced.

In adipose tissue, epinephrine stimulates **lipolysis,** which is the process by which triglycerides are broken down to fatty acids and glycerol. The fatty acids released can be utilized by other tissues to obtain energy. Heart and skeletal muscles in particular are able to fill a considerable amount of their energy

needs by burning fatty acids; as a result, these tissues do not have to use glucose. This ability to use fatty acids as an alternative energy source is important, since glucose is the primary fuel used by the nervous system. Provision of fatty acids is therefore said to have a "sparing effect" on blood glucose use. Glycerol released from adipose tissue as a result of lipolysis can be taken up by liver and converted into glucose.

The net effect of epinephrine on metabolism is increased blood glucose concentration. The increase occurs as a direct result of increased glycogenolysis in liver and increased gluconeogenesis from lactate and glycerol in liver, and as an indirect result of the sparing effects of fatty acids on blood glucose. The effect of epinephrine to inhibit insulin secretion and stimulate glucagon secretion from the pancreas also plays an important part in elevating blood glucose.

Summary of Effects of Adrenal Hormones

When we consider together the effects of hormones produced by the adrenal cortex and adrenal medulla, it is clear that the adrenal glands play an important part in the body's ability to effectively respond to stress. Epinephrine produces effects on metabolism and on the cardiovascular system and lungs that result in an increased supply of oxygen and metabolic fuels for the muscles and brain. Cortisol stimulates gluconeogenesis in the liver and thus helps to maintain blood glucose during prolonged periods of stress. As a result of either direct discharge of sympathetic nerves to the kidney or the release of epinephrine from the adrenal medulla, renin release can be increased during stress. An increase in renin results in increased aldosterone secretion. In response to increased aldosterone, the kidney retains more sodium and water, resulting in a greater blood volume. This is obviously beneficial when the body suffers blood loss. In addition, the constriction of blood vessels in the skin caused by epinephrine serves to limit blood loss in the case of superficial wounds. Thus, the hormones of the adrenal glands produce a variety of effects that enable the body to adapt to potentially harmful situations and therefore serve to promote survival.

CHAPTER REVIEW

Summary

- Humans have two adrenal glands, one located just above each kidney. Each adrenal consists of two distinct endocrine glands; the adrenal cortex and the adrenal medulla.
- The adrenal cortex primarily produces three steroids: cortisol, aldosterone, and dehydroepiandrosterone (DHEA). These steroids are produced in distinct zones of the adrenal cortex known as the zona glomerulosa, zona fasciculata, and zona reticularis, respectively.
- The adrenal cortex is controlled by ACTH from the anterior pituitary, which is in turn regulated by CRH secretion from the hypothalamus. Together this forms the hypothalamic-pituitary adrenal axis. ACTH has both tropic and steroidogenic effects on the adrenal cortex.
- Cortisol is secreted in a circadian rhythm, with plasma levels being highest around 8 A.M. and lowest around midnight. Both physical and psychological stress also stimulate cortisol secretion. Cortisol has negative feedback effects on both the hypothalamus and the anterior pituitary.
- The overall effect of cortisol on metabolism is catabolic. In liver, cortisol produces effects that tend to raise blood glucose concentrations. In skeletal muscle, cortisol promotes protein loss, and in adipose tissue, it decreases glucose uptake and promotes the mobilization and redistribution of fat stores.
- In addition, cortisol enhances vascular reactivity, and in large doses glucocorticoids have anti-inflammatory and immunosuppressive effects that have proven useful in medical practice.

- Aldosterone is the principal mineralocorticoid produced by the adrenal cortex. Aldosterone secretion is regulated by the renin-angiotensin system and by changes in plasma potassium concentrations. Aldosterone acts on the kidneys to increase sodium retention and potassium excretion.
- When excess glucocorticoids are secreted, the result is Cushing's syndrome, in which there is a distinct physical appearance due to magnified effects of cortisol on protein and fat metabolism. A deficiency in steroid production by the adrenal cortex leads to the condition known as Addison's disease, in which there are abnormalities in electrolyte balance and blood pressure regulation.
- The adrenal medulla is coupled to the sympathetic nervous system and consists of neuroendocrine cells that produce and secrete catecholamines. Epinephrine is the primary product of chromaffin cells of the adrenal medulla.
- Epinephrine produces a variety of different effects that together prepare the individual for a fight-or-flight situation. Epinephrine affects the cardiovascular and pulmonary systems to increase delivery of oxygen to the muscles. In addition, epinephrine affects metabolism in liver, muscle, and adipose tissue in a manner to increase supplies of metabolic fuels such as glucose and fatty acids for the rest of the body.

Review Questions

Identify Each with a Term

1. The zone of the adrenal cortex responsible for synthesis of glucocorticoids.
2. The primary mineralocorticoid secreted by the adrenal cortex.
3. Name of the proteolytic enzyme secreted by the kidney that converts angiotensinogen into angiotensin I.
4. The condition that results from excess cortisol secretion.
5. Name of the enzyme that catalyzes the rate-limiting step in epinephrine synthesis.
6. Name of cells in the adrenal gland responsible for epinephrine synthesis.

Define Each Term

7. zona glomerulosa
8. corticotropin-releasing hormone
9. pharmacological effects of glucocorticoids
10. angiotensin II
11. Addison's disease
12. lipolysis
13. renin-angiotensin system

Choose the Correct Answer

14. Dehydroepiandrosterone is primarily secreted from which region of the adrenal gland?
    a. zona fasciculata
    b. zona glomerulosa
    c. zona reticularis
    d. medulla
15. Of the following conditions, when would plasma cortisol levels be highest?
    a. at the onset of deep sleep
    b. just prior to awakening
    c. when ACTH levels are lowest
    d. right after a meal, when blood glucose levels are high
16. Cortisol has a number of effects on carbohydrate, protein, and fat metabolism. Cortisol:
    a. decreases gluconeogenesis in liver
    b. increases glucose uptake into adipose tissue
    c. decreases urea production in liver
    d. decreases protein synthesis in muscle
17. Angiotensin-converting enzyme is responsible for conversion of:
    a. angiotensinogen into angiotensin I

    b. angiotensin I into angiotensin II
    c. renin into angiotensin I
    d. angiotensin II into aldosterone
18. Effects of epinephrine on the cardiovascular system include:
    a. reduced strength of cardiac contractions
    b. increased heart rate
    c. redistribution of blood flow to internal organs
    d. vasoconstriction in skeletal muscle
19. Which of the following is *not* a metabolic effect of epinephrine?
    a. increased glycogenolysis in liver
    b. decreased insulin secretion
    c. increased lipolysis in adipose tissue
    d. decreased urea synthesis
20. Which of the following "hormone class:hormone" pairs is *incorrectly* matched?
    a. glucocorticoid:cortisol
    b. catecholamine:epinephrine
    c. mineralocorticoid:dehydroepiandrosterone
    d. all of the above pairs are correctly matched

Answer in One or Two Sentences

21. What is meant by the terms "tropic" and "steroidogenic" effects of ACTH?
22. How does the hypothalamic-pituitary-adrenal axis function?
23. What is the relationship of the sympathetic nervous system and the adrenal medulla?
24. How do characteristics of Cushing's syndrome relate to the physiological effects of cortisol?

Problem Solving in Physiology

25. Excess skin pigmentation may occur in conditions of adrenal cortical *hyper*secretion and *hypo*secretion. Explain the physiological situations that could result in either high or low adrenal cortex steroid production and excess skin pigmentation.
26. Glucocorticoids decrease glucose uptake by adipose tissue cells, and thus would be expected to reduce fat storage in the body. However, persons with excess glucocorticoids, as in Cushing's disease, show varying degrees of obesity. Explain these two apparently conflicting observations.

Suggested Readings

Evans, Ronald M. "The steroid and thyroid hormone receptor superfamily." *Science*, 240:889–895, 1988.

Ganong, William F. *Review of Medical Physiology*, 12th ed. Los Altos: Lange Medical Publications, 1985.

Goodman, H. Maurice. *Basic Medical Endocrinology*. Chapter 4, Adrenal Glands. New York; Raven Press, 1994.

Griffin, James E., and Ojeda, S.R. *Textbook of Endocrine Physiology*. New York: Oxford University Press, 1992.

Hedge, George A., Colby, Howard D., and Goodman, Robert L. *Clinical Endocrine Physiology*. Chapter 7, Adrenal Cortex. Philadelphia: W.B. Saunders, 1987.

Norman, Anthony W., and Litwack, Gerald. *Hormones*. Orlando: Academic Press, 1987.

Tepperman, Jay, and Tepperman, Helen M. *Metabolic and Endocrine Physiology*, 5th ed. Chicago: Year Book Medical Publishers, 1987.

Walters, Marian R. "Steroid hormone receptors and the nucleus." *Endocrine Reviews*, 6:512–543, 1985.

Wilson, Jean D., and Foster, Daniel W., eds. *Williams Textbook of Endocrinology*, 7th ed. Philadelphia: W.B. Saunders Company, 1985.

Key Terms

ACTH (p. 425)
Addison's disease (p. 434)
androgens (p. 425)
angiotensin I (p. 431)
angiotensin II (p. 431)
angiotensin-converting
enzyme (p. 431)
angiotensinogen (p. 431)
chromaffin cell (p. 435)
circadian rhythm (p. 435)

corticosteroid-binding
globulin (CBG) (p. 428)
CRH (p. 425)
Cushing's syndrome
(p. 433)
DHEA (p. 425)
fight-or-flight (p. 436)
hypothalamic-pituitary-
adrenal axis (p. 425)

lipolysis (p. 438)
permissive actions (p. 430)
postganglionic fibers
(p. 435)
renin-angiotensin system
(p. 431)
sympathoadrenal system
(p. 435)

tyrosine hydroxylase
(p. 436)
urea cycle (p. 429)
vascular reactivity (p. 430)
zona fasciculata (p. 425)
zona glomerulosa (p. 424)
zona reticularis (p. 425)

Photomicrograph of an islet of Langerhans. *(© David M. Phillips/ Visuals Unlimited)*

Chapter 15
The Endocrine Pancreas

Key Concepts

The endocrine pancreas produces the hormones insulin and glucagon, which play major roles in regulating fuel homeostasis in both the fed and fasted states.

Insulin is secreted primarily in response to an increased blood glucose level. Glucagon is secreted in response to a decreased blood glucose level.

Insulin directs the storage of excess nutrients in the form of glycogen, triglycerides, and protein. The major tissue targets of insulin are muscle, liver, and adipose tissue. Glucagon directs the movement of stored nutrients into the bloodstream. Liver is the primary physiological target of glucagon.

In the fed state, the actions of insulin predominate in tissues, and nutrients are stored. In the fasted state, the actions of glucagon predominate and stored nutrients are mobilized.

Diabetes mellitus occurs when there is a deficiency in insulin action, as a result of either an impairment in insulin secretion, or an impairment in insulin action in its target tissues.

Chapter Outline

Islets of Langerhans: Functional Units of the Endocrine
      Pancreas
      Synthesis and Secretion of Pancreatic Hormones
      Metabolic Effects of Pancreatic Hormones
      Overview of Metabolic Regulation by Pancreatic
            Hormones: The Fed State Versus the Fasted State
Diabetes Mellitus

Smokey has been a part of the family for practically as long as James can remember. He is a grey tabby cat that the family got from the loca Humane Society many years ago. A neutered male, ɩ ɩ his younger years he loved to roam the neighborho˷ ' and the adjacent fields. In recent years, however, he ha˷ become more of a house cat, but given his age of about 14, he deserves to be a bit lazy.

About two years ago James noticed that Smokey was beginning to show considerably less energy, and his litter box was almost always wet, although everyone in the house seemed to be doing their job of changing the litter. In addition, it seemed as though his water bowl was always empty, despite everyone's claim that they had been attentive and filled it on schedule.

James read a few books on cat health and at first thought that Smokey might have a kidney problem. He decided it was time to take Smokey to a veterinarian. The vet did a thorough physical plus some blood work, and finally the answer came back: "Smokey has diabetes."

Initially it was a shock to learn that the family pet had diabetes, similar to the situation when it is learned that a family member has the disease. The vet prescribed insulin and instructed James as to how to give Smokey his injections. He gets two shots per day of a form of insulin, which lasts 6 to 12 hours in its time of action. In addition, just like humans, he is being treated by restricting food intake in the hope that weight loss will alleviate his diabetes. He is doing reasonably well; however, it is likely that the diabetes will limit his lifespan, as it does in humans.

This chapter will describe the role of insulin and other hormones of the pancreas in metabolic regulation. Insulin does more than just control blood sugar levels, and the pancreas plays a pivotal role in regulating metabolic functions in the body. Diabetes mellitus is a fairly common disease; the symptoms that Smokey exhibited as well as his treatment regimen are very similar to those seen in humans, as will also be described in this chapter.

Islets of Langerhans: Functional Units of the Endocrine Pancreas

The endocrine pancreas consists of groups of cells known as the **islets of Langerhans,** which are imbedded in the exocrine portion of the gland. Figure 15–2 shows a cross-section of the pancreas and the relationship of the islets to the exocrine portions of the gland. Islets comprise only about 1% to 2% of the total mass of the pancreas, although the average human pancreas has about one million islets. Each islet is richly supplied with blood vessels, and the hormones islet cells secrete enter these blood vessels directly. In contrast, cells of the exocrine pancreas are arranged to form blind-ended sacs and secrete their products into a duct system that eventually empties into the small intestine.

Islets are composed of four major cell types, listed in Table 15–1. Each cell type synthesizes and secretes a different hormone. As later sections will describe, **glucagon** and **insulin,** which are produced in the **alpha cells** and **beta cells,** respectively, are important hormones involved in the regulation of blood glucose concentrations and fuel homeostasis. **Somatostatin** produced by **delta cells** is identical to somatostatin produced in the hypothalamus. Recall that hypothalamic somatostatin functions as a release-inhibit-

Three pancreas is located in the abdominal cavity adjacent to the upper part of the small intestine, as shown in Figure 15–1. The pancreas serves two functions, which are carried out by different groups of cells within the organ. These two groups of cells are designated as the **exocrine** and **endocrine** portions of the pancreas.

The **exocrine pancreas** is responsible for secretion of fluid and various enzymes involved in the process of digestion. These secretions of the exocrine pancreas travel through a duct and empty into the upper part of the small intestine. The regulation and function of the exocrine pancreas will be described in Chapter 22.

The **endocrine pancreas,** the subject of this chapter, represents an anatomically small part of the pancreas. It secretes hormones that are important regulators of energy metabolism and fuel homeostasis in the body. This chapter describes the physiology of those hormones. Although the exocrine and endocrine portions of the pancreas are usually discussed in separate chapters in physiology textbooks, as they will be here, bear in mind that the pancreas as a whole serves functions related to the overall process of digestion, uptake, and use of metabolic fuels.

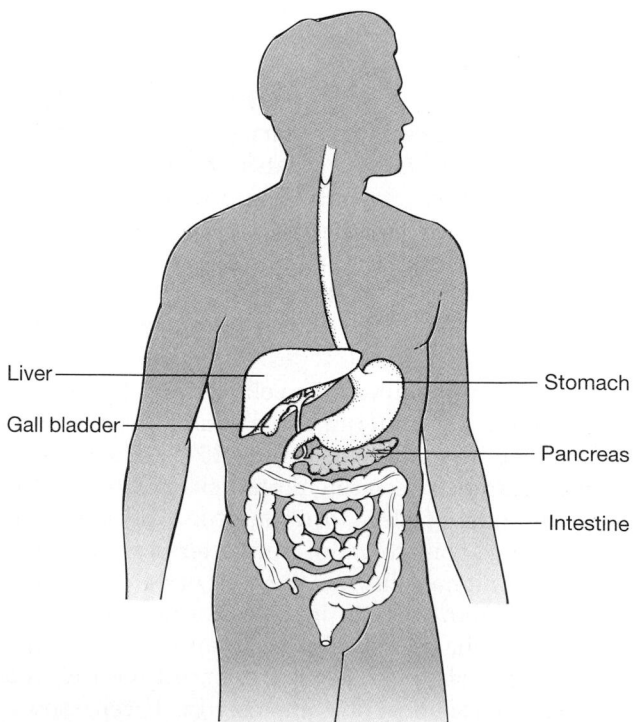

Liver

Gall bladder

Stomach

Pancreas

Intestine

Figure 15–1

Location of the pancreas in the abdominal cavity.

Figure 15–2

(*a*) Relationship of the islets of Langerhans to the (*b*) exocrine cells of the pancreas. (*c*) Photomicrograph of an islet of Langerhans.   *(© Peter Arnold, Inc./A.F. Michler.)*

Connective tissue

(a)

(b)

Islets of Langerhans

(c)

Cells of exocrine pancreas

ing hormone, regulating secretion of growth hormone from the anterior pituitary (Chapter 13). In the pancreas, somatostatin regulates hormone secretion by the alpha and beta cells. The physiological function of **pancreatic polypeptide,** produced by the **F cells,** has not been fully characterized at this time, and will not be discussed further.

Each individual islet shows an orderly arrangement of the different cell types. As Figure 15–3 illustrates, insulin-secreting beta cells tend to be located more toward the center of the islet and are generally the most numerous (see also Table 15–1). The less numerous glucagon-secreting alpha cells are located toward the edges of islets, forming a rim. The delta cells that produce somatostatin are scattered in between this rim of alpha cells and the core of beta cells (see Fig. 15–3). Somatostatin inhibits hormone secretion from both alpha cells and beta cells and, therefore this arrangement of cells may be important with regard to paracrine actions of somatostatin. F cells show roughly the same distribution as the delta cells, although they are considerably fewer in number (see Table 15–1).

Islets also receive innervation from both the **sympathetic** and **parasympathetic** divisions of the autonomic nervous system (ANS). The importance of these neural inputs in regulating hormone secretion will be described in later sections.

Table 15–1

Major Cell Types of the Islets of Langerhans and the Hormones They Produce

Name	Hormone Produced	Percentage of Total Islet*
Alpha cell	Glucagon	25
Beta cell	Insulin	60
Delta cell	Somatostatin	10
F cell	Pancreatic polypeptide	1

*The remaining 4% consists of connective tissue and blood vessels.

(a)

(b)

(c)

(d)

Figure 15–3

(*a*) A typical islet of Langerhans, showing the anatomical relationship between the major hormone-producing cell types. (*b–d*) Consecutive serial sections from an islet of a normal human pancreas. Tissue sections were treated with antibodies to either insulin, glucagon, or somatostatin to determine the location of insulin-secreting beta cells (*b*), glucagon-secreting alpha cells (*© Biophoto Associates/Science Source.*) (*c*), or somatostatin-secreting delta cells (*© Carolina Biological Supply Co./Phototake, NYC ×75.*) (*d*) by the technique of immunofluoresence microscopy. (*© Vu/Cabisco ×70.*)

Synthesis and Secretion of Pancreatic Hormones

As mentioned previously, the islets of Langerhans secrete three hormones whose actions have been well studied: **insulin, glucagon,** and **somatostatin.** Insulin is secreted in response to a number of factors, most importantly increased blood glucose. It promotes the uptake of glucose from the blood into cells, thereby decreasing blood glucose. Conversely, glucagon is secreted when blood glucose is low, it acts to raise the glucose concentration. Somatostatin is secreted in response to several factors as well, including increased blood glucose and glucagon levels.

Insulin Is Secreted in Response to Increased Glucose in the Blood

As indicated in Figure 15–4, insulin is a peptide hormone that consists of an A-chain and a B-chain held together by two disulfide (S—S) bonds. In addition, there is a third disulfide bond within the A chain. These disulfide linkages are essential for the biological activity of the hormone.

Synthesis of Insulin by Beta Cells of the Pancreas

Insulin is derived from a larger precursor molecule known as **proinsulin.** Proinsulin is first synthesized in the rough endoplasmic reticulum of beta cells. Proinsulin consists of a single peptide chain that contains both the A and B chains of insulin linked by a third connecting peptide segment. As the hormone is packaged into secretory vesicles within the beta cell, proinsulin is converted into insulin by proteolytic enzymes that clip the peptide chain in two places. The peptide segment that is removed is known as **C-peptide.** When insulin is secreted into the blood, an equal amount of C-peptide is also secreted. C-pep-

Figure 15–4

Insulin is a two-chain peptide consisting of 51 amino acids. There are 21 amino acids in the A-chain and 30 in the B-chain. The two chains are held together by a pair of disulfide (S—S) bonds, and a third disulfide bond is present in the A-chain.

tide does not appear to have any biological function other than its role in insulin synthesis. However, C-peptide measurements are useful clinically, as they provide an indirect index of insulin production.

Secretion of Insulin in Response to Various Stimuli

Several factors promote insulin secretion (Table 15–2), including increased blood glucose, amino acids, fatty acids, gastrointestinal hormones, neural and pharmacological stimuli, and other hormones secreted by the pancreas. When insulin secretion is stimulated, insulin-containing vesicles move toward the plasma membrane of the beta cell by a microtubule-microfilament system. The membrane of the vesicle fuses with the plasma membrane, and the contents of the vesicle (insulin and C-peptide) are released by exocytosis. In addition to stimulating release of preformed insulin, factors such as glucose that stimulate insulin secretion also increase proinsulin synthesis and the conversion of proinsulin to insulin.

After release from the beta cells into the surrounding capillary network, most of the insulin circulates free in the blood, although a small amount is loosely associated with plasma proteins such as albumin.

Increased Blood Glucose. The most important physiological stimulus for insulin secretion is an increased concentration of glucose in the blood. Figure 15–5 shows average changes in plasma concentrations of glucose and insulin in a group of normal individuals who consumed 75 grams of glucose at time zero following an overnight fast. Note that with a normal overnight fasting blood glucose concentration of 80 to 90 mg/100 ml (prior to ingestion of the glucose),

insulin secretion is low. As the plasma glucose concentration increases above about 100 mg/100 ml, however, insulin secretion is stimulated. Notice in Figure 15–5 that after glucose ingestion, the concentration of glucose in plasma rises rapidly and is closely matched in time by an increase in the concentration of circulating insulin. In this case, the plasma insulin concentration increases to a peak value approximately 10 times that observed before glucose ingestion. With administration of larger amounts of glucose, the blood glucose concentration can be driven even higher, and insulin secretion may reach values as high as 30 times the basal value shown in Figure 15–5. Thus, an increase in blood glucose can elicit a very rapid and large increase in insulin secretion.

As a result of the glucose-induced stimulation of insulin secretion, plasma insulin concentrations are generally highest in the period immediately follow-

Table 15–2

Factors Affecting Insulin Secretion from the Pancreas

Stimulators	Inhibitors
Increased blood glucose	Somatostatin
Amino acids	Epinephrine
Fatty acids	Norepinephrine
Gastrointestinal hormones (gastrin, secretin)	
Acetylcholine	
Sulfonylureas	
Glucagon	

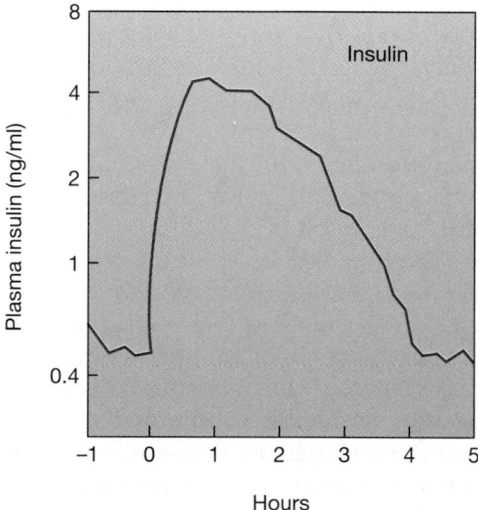

Hours

Figure 15–5

Time course of changes in plasma glucose and insulin after an oral glucose load. (Redrawn from *Williams Textbook of Endocrinology*, Wilson & Foster, eds. [Philadelphia: Saunders, 1985.] Figure 25–8, p. 996). Values shown are the averages from a group of normal, healthy volunteers. After a 12-hour overnight fast, each person consumed 75 grams of glucose at time zero. Blood samples were drawn at various times, and the concentrations of glucose and insulin in the plasma were determined. Values for glucose and insulin are given in terms of milligrams (mg) and nanograms (ng), respectively. Note that the values for insulin are expressed as a logarithmic scale.

Figure 15–6

Role of insulin in the regulation of blood glucose concentration.

ing a meal, especially if the meal is rich in carbohydrate. As later sections will describe, the overall effect of insulin is to increase glucose uptake and use by tissues and thus to lower blood glucose. Therefore, the secretion of insulin in response to glucose and the accompanying action of insulin to decrease blood glucose are part of a feedback loop involved in the maintenance of a constant blood glucose concen-

tration. Figure 15–6 shows the basic characteristics of this loop.

Amino Acids, Fatty Acids, and Gastrointestinal Hormones. Several other factors influence insulin secretion, although none is as physiologically important as glucose. A partial list of some of these other factors is given in Table 15–2. Several **amino acids,** especially arginine, are known to stimulate insulin secretion. Thus, ingestion of a high-protein meal results in increased insulin secretion. Similarly, fatty acids can also stimulate insulin secretion.

The gastrointestinal tract secretes a number of hormones that act to coordinate the various processes involved in digestion of food (see Chapter 22). A number of these gastrointestinal (GI) hormones, including gastrin and secretin, also stimulate insulin secretion by the pancreas (Table 15–2). It is well known that the administration route of glucose influences the amount of insulin secreted by the pancreas. If glucose is given orally, a greater increase in insulin secretion occurs as compared to when the same amount of glucose is given intravenously. This difference is due to a stimulatory effect of GI hormones on insulin secretion. As Chapter 22 will describe, GI hormones are generally secreted in response to the presence of food in the GI tract. Soon after food enters the GI tract, the blood glucose concentration begins to rise. The effect that GI hormones have to stimulate insulin secretion therefore provides an anticipatory signal to beta cells that the blood glucose concentration is likely to increase in the near future.

Autonomic Nervous System.   Neural inputs to islets appear to play a role in regulating insulin secretion. Note in Table 15–2 that acetylcholine stimulates insulin secretion. Recall from Chapter 10 that acetylcholine serves as the neurotransmitter for postganglionic fibers in the parasympathetic division of the autonomic nervous system. Activation of parasympathetic neurons to the islets, as might be expected to occur during digestion, also results in increased insulin secretion. This, too, provides an anticipatory signal to the pancreas that an increase in blood glucose is likely to occur. Table 15–2 also indicates that epinephrine and norepinephrine inhibit insulin secretion. Activation of sympathetic fibers to the islets or release of epinephrine from the adrenal medulla, as might occur in response to stress, results in an inhibition of insulin secretion. The significance of stress-induced inhibition of insulin secretion was discussed in Chapter 14.

Drugs and Other Islet Hormones.   A class of drugs known as **sulfonylureas** also promote insulin secretion (see Table 15–2). These drugs can be taken orally and are widely used in the treatment of some forms of diabetes mellitus, as will be discussed in a later section of this chapter.

Table 15–2 also indicates that two other islet hormones, glucagon and somatostatin, influence insulin secretion. Glucagon is known to augment glucose-stimulated insulin secretion. By contrast, somatostatin inhibits insulin secretion. This suggests that alpha cells and delta cells exert paracrine influences on insulin secretion within each islet.

Glucagon Is Secreted in Response to Decreased Glucose in the Blood

Synthesis of Glucagon by Alpha Cells of the Pancreas

Glucagon is a single-chain polypeptide consisting of 29 amino acids. It is first synthesized as part of a larger precursor molecule that is later converted into the active hormone. In this case, the immediate precursor, **proglucagon,** is about five times the size of glucagon itself. Interestingly, proglucagon is also synthesized in certain cells of the small intestine and in the hypothalamus. However, in these two tissues, proglucagon is not converted into glucagon. Instead, several other **glucagon-like peptides** are formed. The function of these peptides and the regulation of their production are not well understood at this time. It is clear, however, that the regulation of secretion of the glucagonlike peptides differs from that of glucagon secretion by the alpha cells of the pancreas.

Secretion of Glucagon in Response to Various Stimuli

Decreased Blood Glucose.   As with insulin, the major physiological regulator of glucagon secretion is the blood glucose concentration. However, in direct contrast to insulin, a *decrease* in blood glucose concentration stimulates glucagon secretion (Table 15–3). Therefore, circulating glucagon concentrations are highest during periods of fasting, when the blood glucose levels tend to be lowest. Conversely, an increased blood glucose concentration, such as might occur after a meal, tends to reduce glucagon secretion.

Thus, alterations in blood glucose either above or below normal result in opposite changes in insulin and glucagon secretion. Insulin and glucagon have opposite effects on metabolism, lowering and raising blood glucose, respectively. It is not surprising, therefore, that the regulation of secretion of these two hormones by glucose involves a reciprocal relationship.

Amino Acids, Fatty Acids, and the Autonomic Nervous System.   In addition to glucose, a number of substances that regulate insulin secretion also influence glucagon secretion (see Table 15–3). Amino acids stimulate glucagon secretion; as with insulin, arginine is the most potent stimulus among these. As a result of the stimulation of insulin and glucagon se-

Table 15–3

Factors Affecting Glucagon Secretion from the Pancreas

Stimulators	Inhibitors
Low blood glucose	Fatty acids
Amino acids	Somatostatin
Acetylcholine	Insulin
Norepinephrine	
Epinephrine	

cretion by amino acids, the plasma concentration of both hormones increases after a high-protein meal. The parallel increases in both insulin and glucagon secretion in response to amino acids serve as a protective mechanism to ensure that blood glucose levels are maintained following ingestion of meals rich in protein but low in carbohydrate, as will be discussed in later sections describing the actions of these two hormones. An increase in circulating fatty acids inhibits glucagon secretion, whereas a decrease in fatty acid concentrations increases glucagon secretion (see Table 15–3).

In addition, the ANS also has a role in regulating glucagon secretion. Both the sympathetic and parasympathetic divisions promote glucagon secretion.

Insulin.   Insulin also affects glucagon secretion from alpha cells (see Table 15–3). Normally, high blood glucose tends to inhibit glucagon secretion, but this effect depends on the presence of insulin. In the absence of insulin, alpha cells cannot detect the elevated blood glucose, and glucagon secretion continues at a high rate. This fact is important in those forms of diabetes mellitus in which insulin-secreting beta cells are destroyed. Not only is insulin secretion deficient; glucagon secretion is *inappropriately* high due to the absence of the inhibitory effect of insulin on glucagon secretion.

Somatostatin Is Secreted in Response to Increased Glucose and Glucagon in the Blood

As described in Chapter 13, somatostatin is a small peptide hormone. As was the case for insulin and glucagon, somatostatin is also first synthesized as part of a larger precursor molecule. The immediate precursor, **prosomatostatin,** is then converted into the smaller somatostatin peptide. Factors regulating somatostatin secretion from the delta cells are similar to those regulating insulin and glucagon secretion. The primary stimuli of somatostatin secretion are (1) increased blood glucose, (2) increased plasma glucagon, and (3) amino acids.

Metabolic Effects of Pancreatic Hormones

Even while at rest or sleeping, the body is continually using energy to drive such vital processes as ion transport, synthesis of various cellular proteins, and the mechanical activity involved in respiration or cardiac contractions. Physical activity increases energy requirements above those in the basal (resting) state.

However, although energy use is continuous, the intake of energy in the form of food is intermittent. Thus, excess fuels taken in with a meal must be stored for subsequent use between meals. As a result, mechanisms that coordinate and regulate fuel homeostasis have evolved. The following sections discuss the actions of insulin and glucagon, two primary hormones involved in regulating fuel homeostasis in the body.

As described earlier, alterations in the blood glucose concentration provide the primary signal regulating insulin and glucagon secretion. Glucagon and insulin have reciprocal effects that tend to raise and lower the blood glucose concentration, respectively. Thus, these two hormones serve as key components of feedback loops whereby the blood glucose concentration is maintained within fairly narrow limits. As we have learned in previous chapters, other hormones such as cortisol, growth hormone (GH), and epinephrine also have effects that influence the blood glucose concentration. An appropriate question to ask at this point might be, "What is the physiological basis for this tight regulation of blood glucose concentrations?"

The answer to that question lies in the fact that the central nervous system (CNS) relies almost solely on glucose for its metabolic needs. Although other tissues, such as muscle and liver, can use alternate substrates for fuel, the nervous system ordinarily must derive all of its energy from glucose. Thus, if the blood glucose concentration were to fall to low levels, the brain could not obtain energy, and death would soon follow. Conversely, although a marked elevation of blood glucose would not produce such immediate deleterious effects, it can lead to wasteful loss of glucose in the urine as well as concurrent losses of large volumes of water. In a normal individual, the blood glucose concentration following an overnight fast is usually in the range of 75 to 115 mg of glucose per 100 ml of blood. Following ingestion of a large amount of glucose, such as with a high-carbohydrate meal, the concentration may reach as high as 200 mg glucose per 100 ml of blood. However, as shown in Figure 15–5, in a normal individual the blood glucose concentration quickly returns toward the basal level, owing to the secretion of insulin and its effects on glucose utilization.

Insulin Has Effects on Metabolism that Are Anabolic in Nature

Insulin has been termed the **hormone of nutrient abundance.** When influx of nutrients is high, insulin provides the signal that directs storage of excess fuels while at the same time it suppresses mobilization of pre-existing fuel stores. Insulin therefore has **anabolic** effects.

The specific cellular mechanisms by which insulin produces its effects has yet to be determined, despite considerable research efforts in this area in the past several years. The insulin receptor is located in the plasma membrane, as might be expected, since insulin is a peptide hormone. Although the structure of the receptor has been characterized biochemically, it is not currently known how the binding of insulin to its membrane receptor is transduced into the intracellular events that it produces. Research has shown that the insulin receptor itself has protein kinase activity, and thus insulin may produce some of its effects by stimulating its receptor to phosphorylate specific regulatory proteins within its target cells. Thus, insulin may act via a mechanism similar to the one depicted earlier in Figure 12–16.

The major target tissues of insulin are skeletal muscle, adipose tissue (fat), and liver.

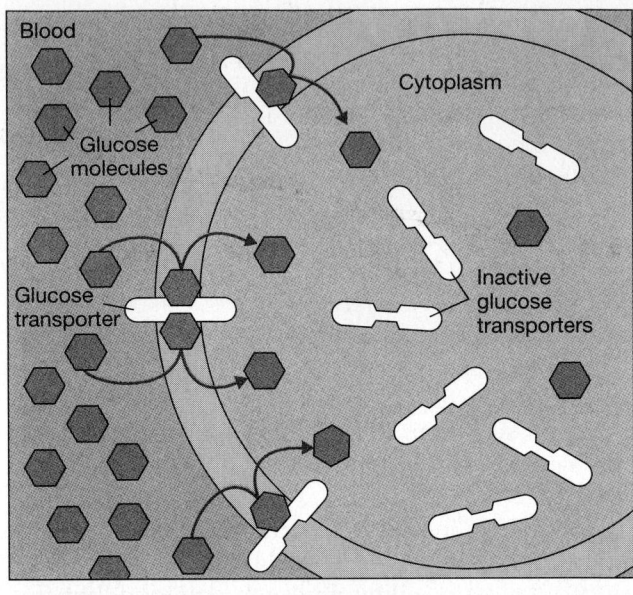

(a)

Effects of Insulin on Carbohydrate Metabolism

The most obvious action of insulin is that it very rapidly and effectively lowers the blood glucose concentration. Most tissues in the body depend on insulin for the uptake of glucose from the blood. In the absence of insulin, glucose does not readily enter cells and therefore cannot be metabolized and used for energy. Two primary exceptions to this rule are the brain and the liver, whose cells are readily permeable to glucose even in the absence of insulin. In tissues such as muscle and adipose, glucose enters cells primarily by facilitated diffusion. (Recall that facilitated diffusion differs from active transport in that it is not energy-dependent and cannot move glucose against a concentration gradient. Instead, facilitated diffusion involves specific carrier proteins that "shuttle" glucose across the membrane at a faster rate than would occur by diffusion alone.) As indicated in Figure 15–7a glucose transporters are located both in the plasma membrane and in intracellular sites. Insulin promotes movement of transporters from their intracellular location to the plasma membrane (see Fig. 15–7b). By increasing the number of transporters present in the plasma membrane, insulin is able to increase the rate of glucose uptake into cells.

In addition to stimulating glucose uptake into cells, insulin also stimulates metabolism and use of glucose. In most cells, particularly those in liver and muscle, insulin stimulates glycogen synthesis while inhibiting glycogen breakdown (Fig. 15–8). The net result is an increase in the amount of glycogen stored. Insulin also stimulates a number of enzymes in the glycolytic pathway. Glucose metabolism via glycolysis therefore increases in most insulin-sensitive cells. In addition to stimulating glucose uptake and use by peripheral tissues such as muscle and adipose tissue,

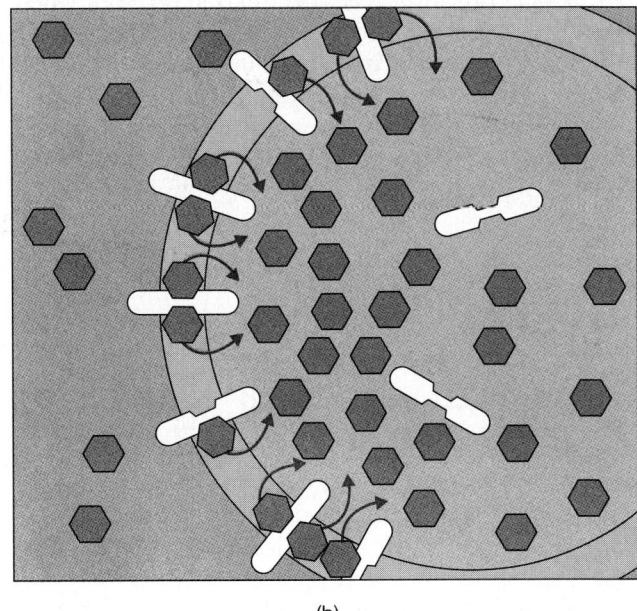

(b)

Figure 15–7

Illustration of the effect of insulin to stimulate glucose uptake into cells. This figure depicts the situations that might exist in the absence (*a*) or presence (*b*) of insulin. Note that insulin increases glucose uptake by increasing the number of glucose transporters in the cell membrane.

insulin decreases glucose output by the liver in two different ways. First, insulin decreases the activity of several key enzymes in the gluconeogenic pathway. (Recall that gluconeogenesis is the process by which nonglucose substrates such as amino acids are converted into glucose.) Second, insulin promotes use of amino acids in peripheral tissues (see the following

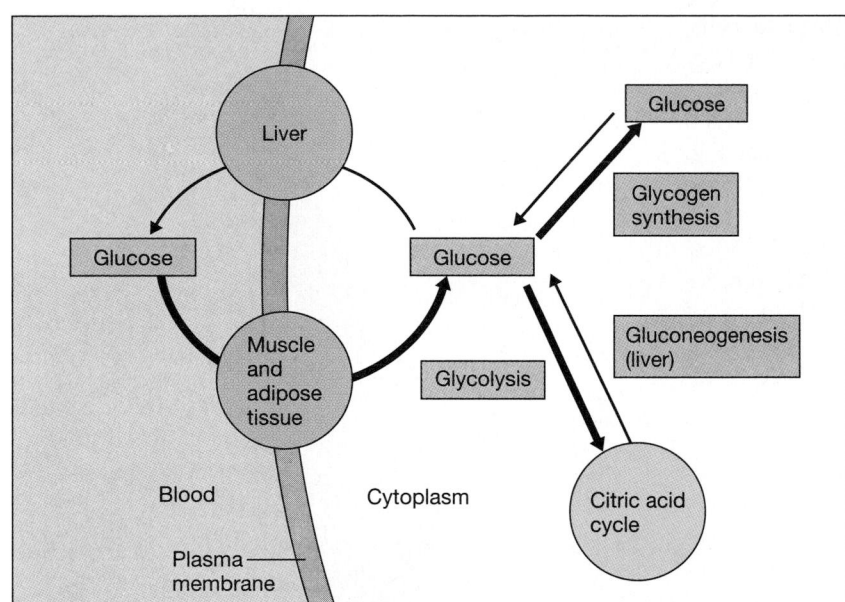

Figure 15–8
Summary of the effects of insulin on carbohydrate metabolism. Pathways or processes in which there is increased flow in the presence of insulin are shown with the heavy arrows; those in which there is reduced flow are shown by the light arrows.

section on protein metabolism) and so decreases the supply of amino acids, which are normally a major substrate for gluconeogenesis in liver.

Effects of Insulin on Lipid Metabolism

As a result of several specific effects of insulin on lipid metabolism, it promotes storage of excess fuel as triglycerides. In both liver and adipose tissue, insulin stimulates synthesis of fatty acids. In adipose tissue, fatty acids combine with glycerol-phosphate and are stored as triglycerides (Fig. 15–9). By contrast, liver stores only a small amount of the fatty acids it produces. Instead, the majority of fatty acids are packaged as lipoproteins and released into the blood. These lipoproteins are taken up by adipose tissue, and the fatty acids are then converted and stored as triglycerides (see Fig. 15–9). In adipose tissue, insulin stimulates the activity of **lipoprotein lipase,** an enzyme which promotes uptake of lipoproteins from the blood (see Fig. 15–9). The effect of insulin to stimulate glucose uptake by adipose tissue also tends to

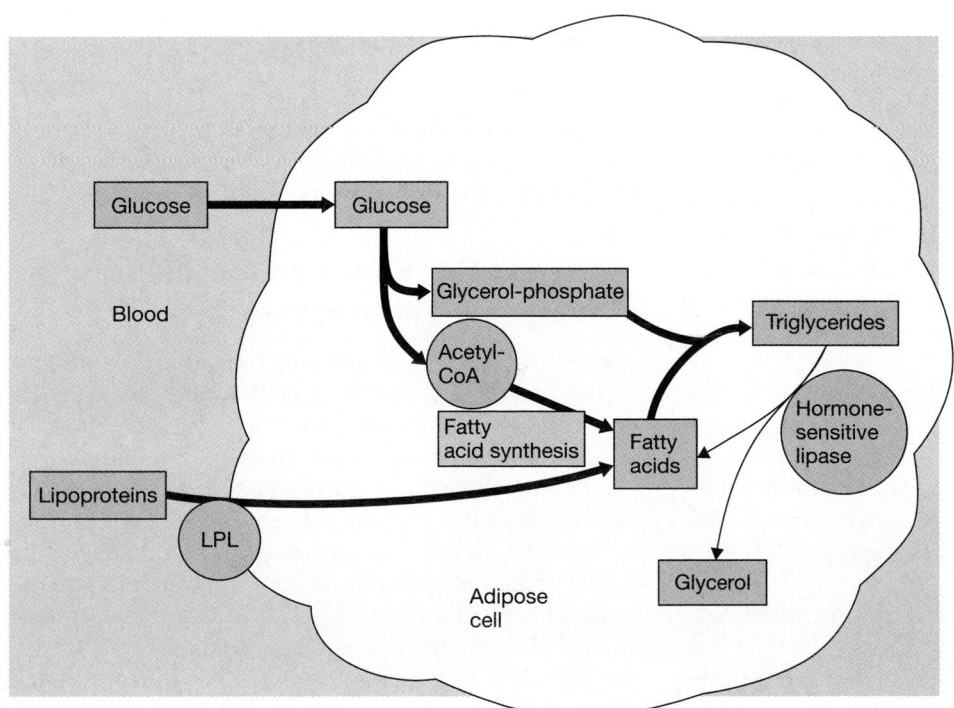

Figure 15–9
Summary of the effects of insulin on adipose cell metabolism. Pathways or processes in which there is increased flow in the presence of insulin are shown with the heavy arrows; those in which there is reduced flow are shown by the light arrows. The net effect of insulin is to promote triglyceride storage. (LPL = Lipoprotein lipase.)

promote triglyceride storage. Inside adipose cells, glucose can be either metabolized and used for fatty acid synthesis or used to form glycerol-phosphate. Glycerol-phosphate serves as the backbone to which fatty acids are attached to form triglycerides (see Fig. 15–9). Thus, by increasing glucose entry into adipose cells, insulin promotes synthesis of both the fatty acid and glycerol-phosphate components of triglycerides.

In addition to its stimulatory effect on fat synthesis, insulin also inhibits lipolysis (breakdown of triglycerides). Breakdown of triglycerides into their component fatty acids and glycerol is catalyzed by an enzyme known as **hormone-sensitive lipase** (see Fig. 15–9). Insulin inhibits the activity of this enzyme both in liver and in adipose tissue, thus decreasing lipolysis in these two tissues. Therefore, the overall effect of insulin on lipid metabolism is to promote triglyceride storage while inhibiting triglyceride breakdown.

Effects of Insulin on Protein Metabolism

Insulin also has profound effects on protein metabolism in the body. These effects are most readily seen in skeletal muscle and liver, although the *anabolic* effects of insulin, promoting the accumulation of protein, occur in most tissues.

The effects that insulin has on muscle protein metabolism are of particular significance since approximately 40% of the total body protein is present in muscle. Therefore, much of the influence that insulin has on overall protein balance in the body results from its effects on muscle protein metabolism. In muscle, insulin stimulates active transport of amino acids from the blood into individual muscle cells (see Fig. 15–10). As a result, more amino acids are available for muscle protein synthesis. Insulin also stimulates the process of protein synthesis (see Fig. 15–10), in part through an increased number of ribosomes in each muscle cell and in part due to increases in the activity of individual ribosomes. At the same time, insulin strongly inhibits protein degradation in muscle (see Fig. 15–10). Therefore, the overall effect of insulin in muscle is to promote accumulation of muscle protein.

Insulin has similar effects on protein metabolism in the liver and adipose tissue; it increases protein synthesis and decreases protein degradation. In addition, as described earlier, insulin also inhibits gluconeogenesis in liver. As a result, fewer amino acids are converted into glucose, and more amino acids are available for protein synthesis.

As a result of its anabolic effects on protein metabolism, insulin produces a "positive nitrogen balance," meaning a net accumulation of protein in the body. By contrast, when insulin is deficient, as in di-

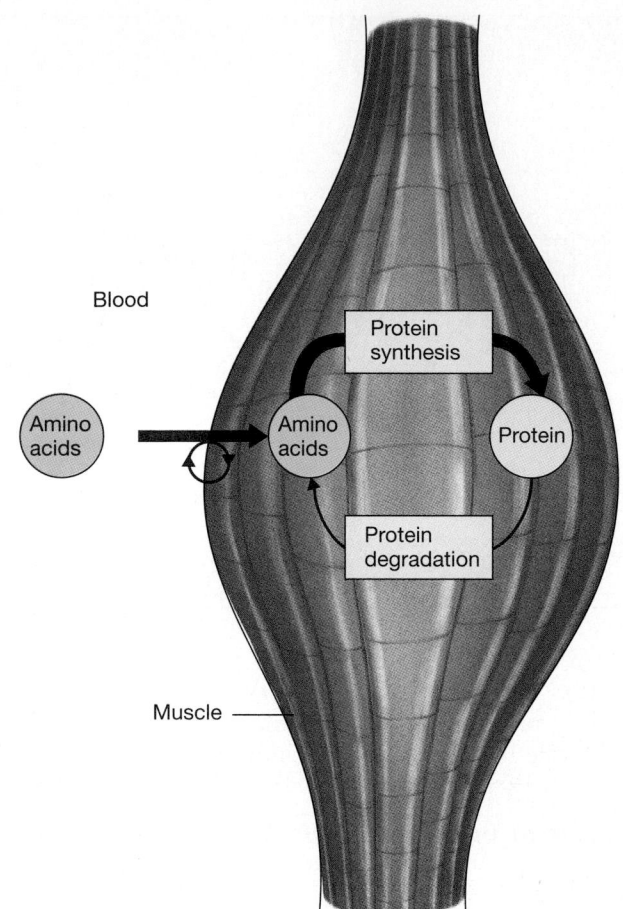

Figure 15–10

Summary of the effects of insulin on muscle protein metabolism. Pathways or processes in which there is increased flow in the presence of insulin are shown with the heavy arrows; those in which there is reduced flow are shown by the light arrows.

abetes mellitus, there is a net loss of protein, or a negative nitrogen balance. Insulin therefore serves as an important regulator of tissue growth.

Glucagon Has Effects on Metabolism that Oppose Those of Insulin

As will be evident in sections that follow, glucagon affects many of the same metabolic pathways as insulin. However, in each instance, the net effect of glucagon is opposite that of insulin.

The mechanism of glucagon action is well established. Since it is a peptide hormone, glucagon interacts with a receptor on the plasma membrane of its target cells. This receptor is coupled to adenylate cyclase and results in an increase in the concentration of cAMP inside the cell. As described earlier in Chapter 5, cAMP activates several protein kinases inside the cell, leading to changes in the phosphorylation

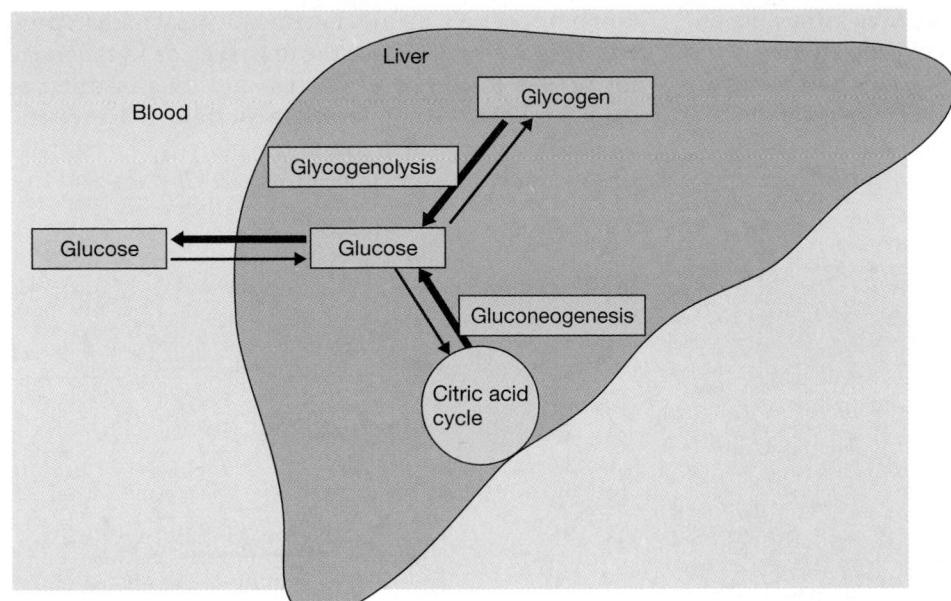

Figure 15–11

Summary of the effects of glucagon on liver carbohydrate metabolism. Pathways or processes in which there is increased flow in the presence of glucagon are shown with the heavy arrows; those in which there is reduced flow are shown by the light arrows.

state and activity of various regulatory proteins and enzymes.

The principal target site of glucagon is liver, where its overall effect on metabolism is catabolic. Some effects of glucagon on adipose tissue have also been reported, but these usually require a relatively high concentration of the hormone and thus their physiological significance is uncertain.

Effects of Glucagon on Carbohydrate Metabolism

The overall effect of glucagon action is an increase in blood glucose. This occurs as the result of three separate effects that glucagon has on the liver (Fig. 15–11). First, glucagon stimulates conversion of glycogen into glucose **(glycogenolysis)** in liver while inhibiting glycogen synthesis (see Fig. 15–11). The net result is an increase in the conversion of liver glycogen into glucose, which enters the blood. The second means by which glucagon raises blood glucose is by stimulating the conversion of nonglucose substrates into glucose **(gluconeogenesis)** in the liver (see Fig. 15–11). Third, as the following section will describe, glucagon affects lipid metabolism, resulting in glucose "sparing," which tends to raise blood glucose. To illustrate the opposing effects of insulin and glucagon on carbohydrate metabolism, compare Figure 15–8 with Figure 15–11.

Effects of Glucagon on Lipid Metabolism

Effects of glucagon on lipid metabolism oppose those of insulin. In liver, glucagon stimulates hormone-sensitive lipase and thereby stimulates lipolysis (Fig. 15–12). In addition to being oxidized and used directly for energy production in muscle, fatty acids can be partially metabolized and converted in liver into **ketones** (also known as **ketone bodies** or **ketoacids**). The ketones, which are four-carbon compounds, are released by the liver and can be taken up and used by peripheral tissues for energy. Glucagon stimulates the formation of ketones **(ketogenesis)** in liver. Skeletal muscle and heart both derive a considerable amount of their energy needs from the oxidation of ketones. In fact, ketones are a major source of energy for heart. By stimulating ketogenesis in liver, glucagon provides increased amounts of ketones for use by muscle and heart. This has the effect of sparing glucose, which, when coupled with other effects of glucagon, also serves to raise blood glucose.

Effects of Glucagon on Protein Metabolism

In liver, glucagon is a potent stimulator of protein degradation (Fig. 15–13). As free amino acids are released when liver proteins are broken down, they can be used for glucose synthesis via gluconeogenesis. Glucagon has an additional effect of increasing the supply of amino acids for gluconeogenesis by stimulating transport of amino acids from blood into liver. As more amino acids are converted into glucose, more ammonia, a by-product of amino acid metabolism, must be excreted. To accomplish this, glucagon also stimulates **urea synthesis** (see Fig. 15–13).

As indicated earlier, amino acids stimulate secretion of *both* insulin and glucagon from the pancreas. In almost every case, actions of insulin are opposite to those of glucagon. Depending on the particular metabolic state, it is desirable to have either insulin or glucagon present in high concentrations in the blood, but usually not both, since their opposing actions would cancel out one another. However, the apparent paradoxical effect of amino acids to stimulate *both* insulin and glucagon secretion serves an important purpose. When one consumes a meal high in protein, it is desirable to use the amino acids in the meal for protein synthesis in tissues, and thus insulin secretion is beneficial under these conditions. However, when the meal is also low in carbohydrate, very little glucose enters the blood from the GI tract, and the

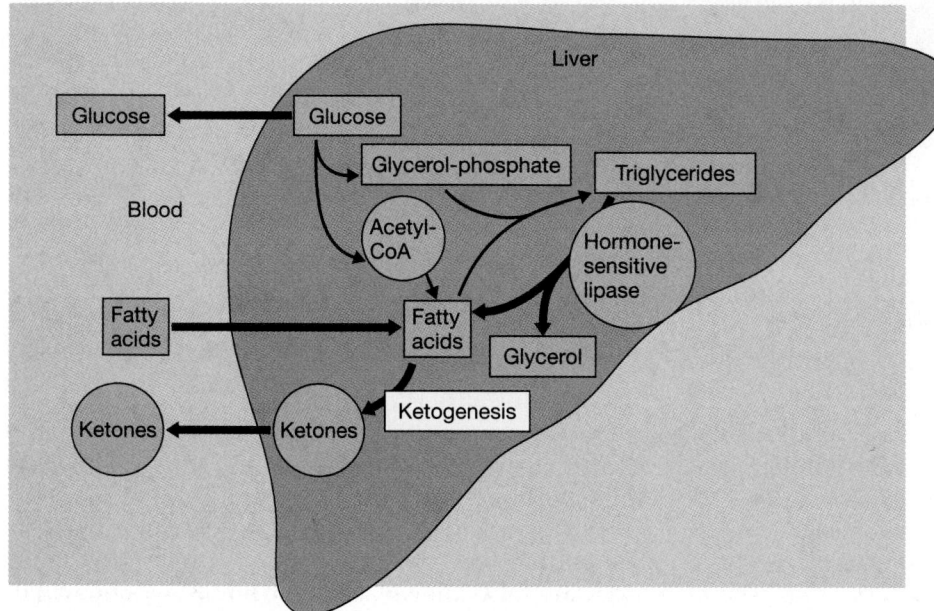

Figure 15–12

Summary of the effects of glucagon on liver fat metabolism. Pathways or processes in which there is increased flow in the presence of glucagon are shown with the heavy arrows; those in which there is reduced flow are shown by the light arrows.

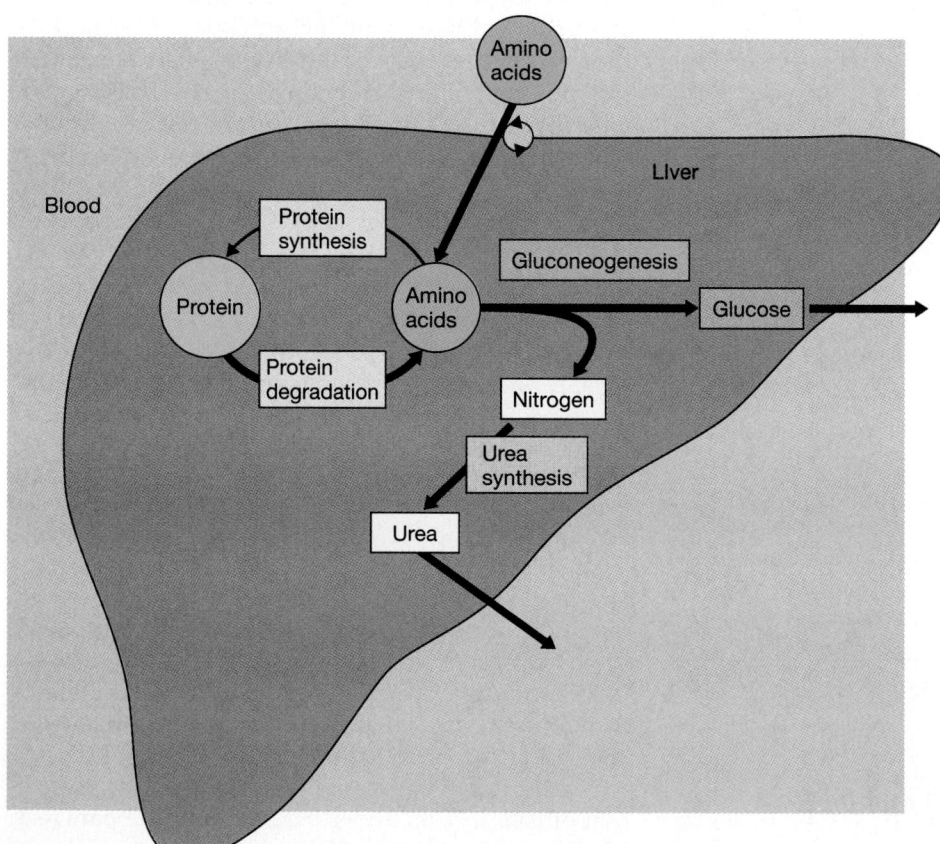

Figure 15–13

Summary of the effects of glucagon on liver protein metabolism. Pathways or processes in which there is increased flow in the presence of glucagon are shown with the heavy arrows; those in which there is reduced flow are shown by the light arrows.

actions of insulin to decrease blood glucose would cause the blood glucose concentration to drop to very low levels. It is therefore beneficial for glucagon to also be secreted in this situation. The effects of glucagon oppose the hypoglycemic (blood-glucose-lowering) actions of insulin, and therefore the blood glucose concentration is maintained near normal, while at the same time insulin is able to promote efficient use of amino acids for protein synthesis in various tissues.

The Insulin/Glucagon Ratio in the Blood Is an Important Determinant of Metabolic Flow

Because of opposing actions of insulin and glucagon, it has been suggested that the **insulin-to-glucagon (I/G) ratio,** rather than absolute concentrations of either hormone is the primary factor determining metabolic status in the body. Thus, when the I/G ratio is high (insulin high, glucagon low), effects of insulin predominate, and an anabolic state exists. Excess fuels are stored as glycogen and triglycerides, and tissue protein synthesis increases. Conversely, when the I/G ratio is low (insulin low, glucagon high), effects of glucagon predominate, and tissues are in a catabolic state.

Shortly after a high-carbohydrate meal, an individual's molar ratio of insulin to glucagon can be as high as 30. When one awakens in the morning after an overnight fast, the I/G ratio is about 2. If one fasts for a day or two, the I/G ratio may decrease to 0.5 or less. Thus, in going from a fully fed state to a fasted state, the I/G ratio may change by a factor of 50 to 60.

As an example of the importance of the I/G ratio in determining the relative metabolic state, consider the situation that can exist in cases of diabetes mellitus in which insulin secretion is deficient. In the absence of insulin, blood glucose increases dramatically. Normally, a high blood glucose concentration inhibits glucagon secretion (refer to Table 15–3), but in the absence of insulin this inhibitory effect is lost, and therefore plasma glucagon levels increase. In such cases, the I/G ratio is extremely low. The catabolic effects of glucagon greatly predominate, leading to increased glycogenolysis, lipolysis, ketogenesis, and gluconeogenesis. Blood glucose concentrations increase to extremely high levels, and a negative nitrogen balance exists. Experimental treatment of diabetic patients with somatostatin, which inhibits glucagon secretion, reduces the severity of the catabolic state (somatostatin also inhibits insulin secretion, but in this case insulin secretion is already extremely low). Thus, the I/G ratio is more important in determining the overall metabolic state than is the absolute concentration of either hormone.

Overview of Metabolic Regulation by Pancreatic Hormones: The Fed State Versus the Fasted State

As described earlier, use of energy by the body is continuous; whereas the intake of energy in the form of food is intermittent. This situation requires that fuels be stored in the body for use during between-meal periods or for periods when food is not readily available.

Excess Fuel Is Stored in the Body in Several Forms

Table 15–4 shows the chemical nature and tissue distribution of stored fuels in an average 155-lb (70 kg) person. Note that most calories in the body are stored in the form of triglycerides in adipose tissue. By comparison, relatively little of the total stored fuel consists of glycogen in either liver or muscle. Since liver glycogen can be broken down directly into free glucose, which is then released into the blood, one might expect that it would be a better form in which to store excess calories. However, compared to glycogen, triglycerides are considered a much more efficient means of storing excess fuel for several reasons. First, one gram of pure fat contains about twice the calories as one gram of carbohydrate. In other words, triglycerides have about twice the **caloric density** of glycogen. Secondly, glycogen is very hydrophilic, whereas fat is hydrophobic. Thus, glycogen attracts a considerable amount of water, which adds a significant amount of weight to the tissues in which glycogen is stored. If we stored all of our excess calories as glycogen and not as fat, the average person

Table 15–4
Distribution of Stored Fuels in the Body

Storage Form	Site of Storage	*Energy Stored (kcal)*
Fat (triglyceride)	Adipose tissue	141,000
Glycogen	Muscle	480
Glycogen	Liver	280
Protein	Muscle	24,000

would weigh about 700 lb. Obviously, this would be a very inefficient way in which to store fuel, since the transport of this excess fuel would represent a considerable burden.

Note also in Table 15–4 that muscle protein represents about 17% of the total stored fuels in the body. During a prolonged fast, the body can and does call upon its muscle protein reserves as a source of stored energy.

Flow of Nutrients after a Meal Is Largely Directed by Insulin

By examining nutrient flow in both the fed and fasted states, we can review the effects of insulin and glucagon on metabolism and at the same time begin to appreciate nutrient flow between organs.

Figure 15–14 summarizes the metabolic state expected in liver, muscle, and adipose tissue of some-

one who has just eaten a meal rich in carbohydrate. Remember that, in this situation, the concentration of insulin in the blood is high, whereas glucagon is low. In liver, insulin promotes conversion of glucose into glycogen, so that liver glycogen stores are increased (see Fig. 15–14). Insulin also stimulates fatty acid synthesis from glucose in liver. Some of the fatty acids are stored as triglycerides, but most are packaged as lipoproteins and exported to adipose tissue (see Fig. 15–14). In adipose tissue, lipoprotein lipase (LPL) is stimulated by insulin, resulting in increased uptake of lipoproteins from the blood. Entry of glucose into adipose tissue is also stimulated under these conditions, as is synthesis of fatty acids and glycerol-phosphate (see Fig. 15–14). Triglyceride formation from fatty acids and glycerol-phosphate is also stimulated, resulting in expansion of fat stores in adipose tissue. In skeletal muscle, glucose uptake is stimulated, as is glycogen synthesis (see Fig. 15–14). Thus, the amount

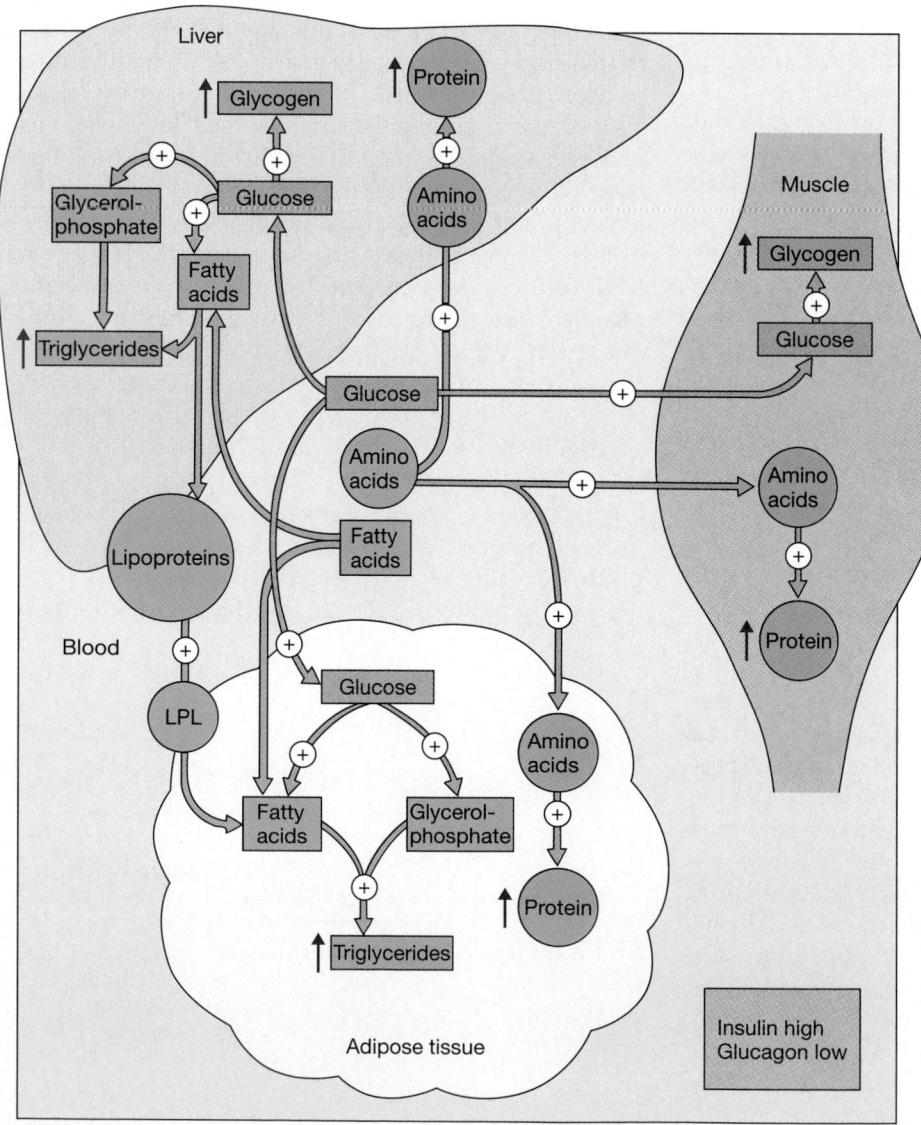

Figure 15–14

Summary of nutrient flow immediately after a meal. Those processes accelerated due to the prevailing high insulin/low glucagon concentrations are indicated by the +.

of glycogen stored in muscle increases under these conditions. Uptake of amino acids into liver, muscle, and adipose tissue also increases in the fed state in response to insulin. In each tissue, insulin also stimulates the conversion of amino acids into protein while at the same time inhibiting protein degradation. Thus, protein stores in each of the three tissues increase (see Fig. 15–14). Any fatty acids taken in with a meal are either taken up directly by adipose tissue and stored as triglycerides or taken up by liver and then packaged as lipoproteins before export to adipose tissue.

Glucagon Influences the Flow of Nutrients During Fasting

Figure 15–15 summarizes the state of nutrient flow expected during a brief period of fasting. In this situation, the expected blood insulin levels are low

while glucagon levels are high. In liver, glycogen breakdown and gluconeogenesis are stimulated in an attempt to maintain blood glucose levels near normal (see Fig. 15–15). In the absence of insulin, breakdown of muscle protein is increased, and amino acids released by muscle are used in liver to synthesize glucose via gluconeogenesis. To facilitate use of amino acids by liver, glucagon also stimulates amino acid uptake and urea synthesis in liver (see Fig. 15–15). In the absence of insulin, glucose entry into muscle and adipose tissue is greatly reduced, which also contributes to maintenance of blood glucose levels. In adipose tissue, lipolysis predominates over lipid synthesis under these conditions. A result is increased release of fatty acids from adipose tissue, which leads to increased fatty acids in the blood (see Fig. 15–15). Fatty acids can be used by muscle to obtain energy and therefore provide muscle with an alternate fuel to use instead of glucose. Glycogen breakdown in

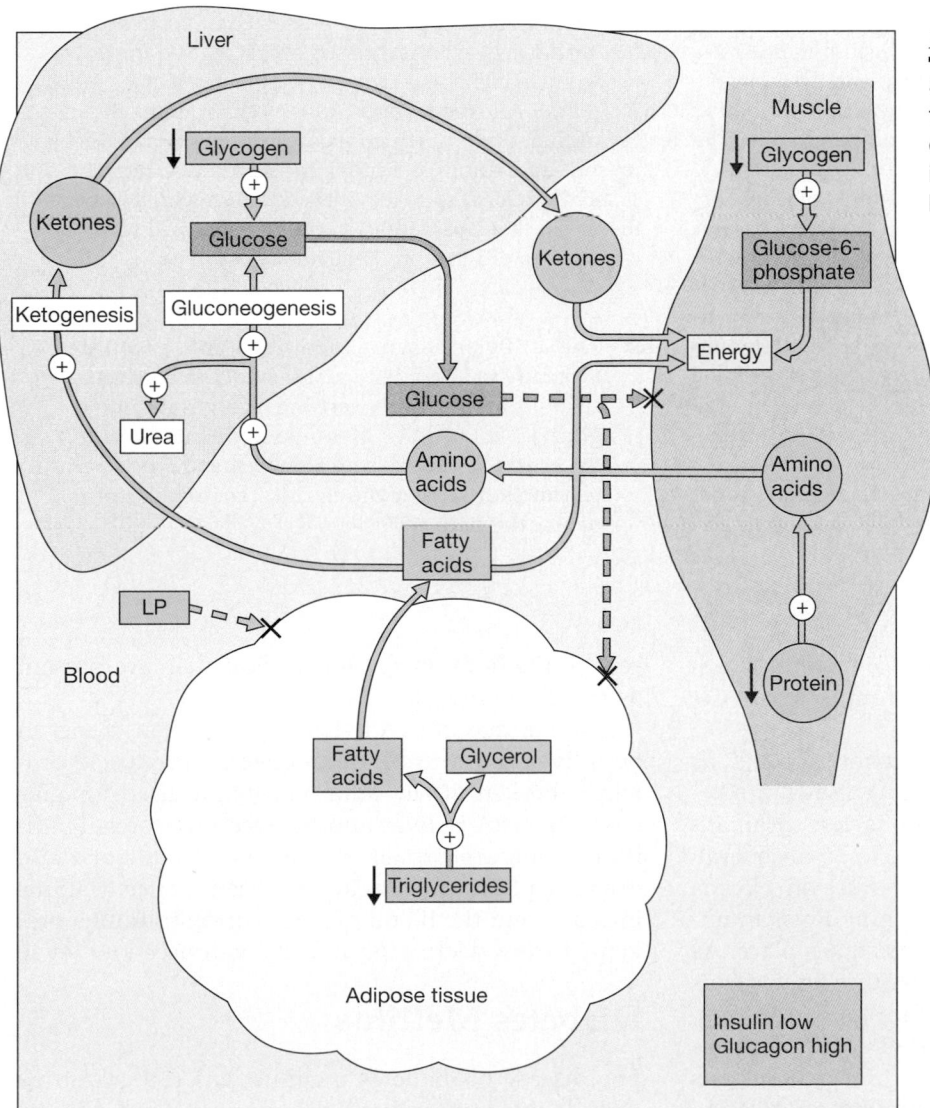

Figure 15–15

Summary of nutrient flow during fasting. Processes accelerated under the conditions of low insulin/high glucagon are indicated by the + . (LP = lipoproteins.)

RESEARCH FOCUS

Advances in Insulin Delivery

Although approximately 90% of all individuals with diabetes are Type II diabetics and many do not require insulin therapy, some 20 million Type I diabetics worldwide do require insulin injections to remain alive. Insulin was first discovered and partially purified in Toronto, Canada, in 1921, and became available commercially a short time later, in 1923. Four researchers, F.C. Banting, C.H. Best, J.B. Collip, and J.J. Macleod, were instrumental in the discovery of insulin, although historically Banting and Best have received most of the credit for the work.

Following the initial discovery of insulin, a second major advancement in the treatment of diabetes was the development in the late 1930s of insulin preparations with extended duration of action. Regular insulin is short-acting, so, initially, patients had to administer the hormone many times during the course of a day and night. The extended-action insulin formulations were designed to reduce the solubility of insulin in body fluids. These alterations allow insulin to enter the circulation slowly. Thus a single injection can supply the body's needs over a longer period of time compared to regular insulin.

Advancements in electronics and biotechnology have been applied to the field of insulin delivery as well. One example from this area is the development of what is known as an insulin pump. A typical insulin pump is approximately the size of a small hand calculator. The primary components of the device are a small, motor-driven unit that pushes on the plunger of a syringe filled with a solution of insulin. A programmable computer chip controls movement of the plunger and thus delivery of insulin. Insulin is delivered to the body via a thin piece of tubing that ends with a needle inserted under the skin and taped in place. Other attempts are under-way to create what in essence is an "artificial pancreas." By coupling the output of an insulin pump to a glucose sensor, the device could accomplish what the beta cells do in terms of regulating insulin secretion in response to changing blood glucose levels. However, it may be several years before such a device is perfected to the extent that it can be widely used in patients.

Another approach to the problem of insulin delivery and the subject of considerable research effort is islet transplantation. Although the method has been used only experimentally in humans, there is considerable hope that the technique will one day have widespread use, since in theory such a procedure could "cure" diabetes. One obstacle to adoption of this technique, as is the case with many transplant procedures, is rejection of transplanted tissue by the host's immune system. The situation in most Type I diabetics is complicated by the fact that even if a perfect tissue match is used for donor tissue (in which case tissue rejection by the recipient would not be expected), conditions that caused the initial autoimmune destruction of beta cells may still be present and result in similar destruction of the transplanted tissue, despite its virtual identity with tissues of the recipient. Considerably more research will likely be required before this problem can be resolved.

■ **The artificial pancreas described above would be a rather simplified version of a true beta cell. In addition to simply sensing changes in blood glucose, what other blood components should the artificial pancreas be able to sense if it were to function in a manner fully analogous to a pancreatic beta cell?**

muscle also increases in the absence of insulin, which provides another source from which muscle can derive energy.

In liver, glucagon stimulates ketogenesis, and, as a result of increased conversion of fatty acids into ketones, plasma ketone concentrations rise dramatically. Ketones can be used by muscle as an energy source, which again has a sparing effect on plasma glucose. If fasting continues for several days, an important adaptation involving ketones takes place. As stated earlier, the CNS normally requires glucose as its energy source. However, with prolonged fasting, the CNS actually adapts and begins to use ketones for energy. This adaptation is important, as it lessens the need to convert muscle protein into glucose. As Table 15–4 shows, considerably more calories are present in the body in the form of fat than are present as protein in muscle.

To summarize: with fasting; glycogen stores in both liver and muscle are depleted and muscle protein is broken down; amino acids are used for glucose synthesis in liver and triglyceride stores in adipose tissue are broken down. As a result of these changes, plasma fatty acid and ketone concentrations increase, and the blood glucose concentration is prevented from decreasing to a dangerously low level.

Diabetes Mellitus

The disease of diabetes mellitus was recognized as early as the first century A.D., when the term *diabetes*, which stems from the Greek word for "siphon," was applied to a disease characterized by a marked increase in urine volume. Note that the term *diabetes* is commonly used (as it will be here) to refer to diabetes mellitus and should not be confused with diabetes insipidus, which is an entirely unrelated endocrine abnormality caused by a deficiency in ADH secretion or action (refer to Chapters 13 and 23 for discussions of diabetes insipidus).

Diabetes mellitus is by far the most common of all endocrine disorders and is a worldwide health problem. In the United States, about 6 million individuals have been diagnosed as having diabetes, and estimates suggest there may be another 4 to 5 million more who are borderline diabetics not yet diagnosed. Diabetes is the third leading medical cause of death and the second leading cause of blindness in the United States. Therefore, research on the causes and cures of diabetes is an area of intense interest.

Diabetes mellitus is not a single disease as was once thought. Instead, it is now well known that diabetes comprises a heterogeneous group of disorders that differ in both cause and severity. The common characteristic of every form of diabetes is high blood glucose concentration, which in fact is the primary means by which the disease is diagnosed.

Elevation in blood glucose in diabetes is the result of a deficiency, either relative *or* absolute, in insulin action. The deficiency in insulin action is most commonly due to either (1) inadequate insulin secretion by beta cells of the pancreas, or (2) a relative lack of response by target tissue cells to insulin. Inadequate insulin secretion results in the form of diabetes referred to as **insulin-dependent diabetes mellitus (IDDM)** or Type I diabetes. The other common form is caused by a relative lack of insulin action in its target tissues and is known as **non-insulin-dependent diabetes mellitus (NIDDM)** or Type II diabetes. Other classifications of diabetes are also well known, but make up only about 2% of all cases. Often, these forms are secondary to or associated with other conditions. One example is **gestational diabetes,** a transient condition that occurs in about 3% of all pregnant women (see Clinical Focus box).

Type I, or Insulin-Dependent Diabetes Mellitus Results from an Inability of the Pancreas to Produce Adequate Amounts of Insulin

As mentioned, Type I diabetes (IDDM) is characterized either by very low levels of insulin or by an absolute lack of insulin secretion by the pancreas. To compensate for this deficiency and control their disease, individuals with IDDM must receive daily insulin injections. About 10% of all diabetics suffer from this form of the disease.

The loss of beta cell function usually occurs during adolescence, although less frequently it occurs later in life. Recent investigations suggest that the loss of beta cell function is due to an autoimmune disorder that causes the immune system to attack and destroy the pancreatic beta cells. The causes of the autoimmune disorder are not well understood, but at least two major components appear to contribute to the disease state. The first is a genetic component by which certain individuals have an increased susceptibility to the disease. However, the genetic component is not by itself sufficient to cause the autoimmunity. The effects of an as yet unidentified environmental component are also required to produce the disease. What that agent might be is not currently known, although viruses have been suggested as likely candidates. The environmental component in some way triggers the autoimmune response, resulting in beta cell destruction.

Type II, or Non-Insulin-Dependent Diabetes Mellitus Results from an Inability of Insulin Target Tissues to Respond to the Hormone

About 90% of all diabetics suffer from the form of the disease known as **Type II diabetes** (NIDDM). Type II diabetes results when target tissues of insulin lose responsiveness to the hormone. Reasons for loss of responsiveness are not currently understood but are an area of considerable research interest. Normal or often higher-than-normal concentrations of insulin may be present in the blood, but cells of target tissues such as adipose and muscle simply do not respond to the hormone as they normally would, a condition referred to as **insulin resistance.** In many cases, persons with Type II diabetes are also obese. The exact relationship between the obesity and insulin resistance is not clear, but the severity of the disease can often be reduced considerably if the patients are placed on a diet and lose weight. In addition to diet, persons with Type II diabetes are often treated with oral hypoglycemic agents known as **sulfonylureas,** which, as indicated in Table 15–2, stimulate insulin secretion from beta cells. In addition to stimulating insulin secretion, sulfonylureas also appear to augment the actions of insulin in tissues and therefore help overcome the insulin resistance in target tissues.

The cause of insulin resistance in Type II diabetes is poorly understood. As indicated previously, there does appear to be some link with obesity, but obesity per se is not sufficient to cause Type II diabetes. There does, however, appear to be a strong genetic component to Type II diabetes. Chances are nearly 100% that if one genetically identical twin develops Type II diabetes, the other will also, even if they are raised sep-

CLINICAL FOCUS

Diabetes and Pregnancy

Even in a normal woman, pregnancy presents challenges to a number of physiological systems in the body. If a pregnancy is complicated by diabetes in the mother, the therapeutic challenge to patients, physicians, and health care professionals becomes even greater. In the past, pregnancy in women with diabetes was often associated with increased rates of death of the developing fetus or newborn, and in some cases even the mother. However, most physicians and health care professionals now recognize the extra challenge that diabetes poses during pregnancy, and with intensive treatment and monitoring, the survival rate in infants of diabetic mothers is almost identical to that in nondiabetic pregnant women.

During a normal pregnancy, an integrated series of hormonal changes occur that help to provide glucose for the developing fetus. Glucose readily moves across the placenta from the mother's blood to that of the fetus. As pregnancy progresses, the maternal blood concentration of several hormones that have anti-insulin effects increases. The hormones that have these effects are largely produced by the placenta. The result of these anti-insulin effects is that glucose entry into maternal tissues is reduced, making more glucose available for the fetus.

In a pregnant woman with preexisting diabetes, these anti-insulin effects will usually exacerbate the diabetic state and require close monitoring and adjustments in ongoing therapy. In previously normal women, these hormonal changes can lead to a transient form of diabetes known as **gestational diabetes.** Estimates of the incidence of gestational diabetes range from 3% to 12% of all pregnancies. Gestational diabetes is similar to Type II diabetes in that it is typified by insulin resistance and can often be treated by modest dietary restriction. In some instances, treatment with insulin may be required.

In most cases, the problem of gestational diabetes is resolved at the time of delivery, with the birth of the child and expulsion of the placenta. (Recall that the placenta is the major source of anti-insulin hormones during pregnancy.) In some women, however, diabetes may persist after delivery and develop into a full-blown case of the disease. It is presumed that these women were borderline or developing diabetics at the onset of their pregnancy and that their condition only worsened during the nine months prior to giving birth.

Because of the complications that diabetes can introduce during pregnancy, women are carefully monitored for evidence of diabetes, such as glucose in the urine. The prospect of diabetes also contributes to a physician's concern regarding excessive weight gain during pregnancy. Although the prognoses of both the pregnant woman with diabetes and her infant are usually quite good, managing such women requires vigilance on the part of both the patient and a well-trained health care team.

Insulin and most other peptide hormones do not readily move from the mother's blood to the fetus, or vice-versa. However, a number of small molecules such as glucose do move across.

■ **What do you suppose might be the consequences to the developing fetus of hyperglycemia in a poorly controlled pregnant diabetic mother? Would hyperglycemia in the mother's blood affect insulin secretion by the pancreas of the developing fetus?**

arately under entirely different environmental, social or economic conditions.

Diabetes Mellitus Can Result in both Short- and Long-Term Complications

Several immediate complications can arise if the glucose concentration in the blood is not checked by insulin; these problems include hyperglycemia, ketoacidosis, and electrolyte imbalance. In addition, long-term complications of the condition include vascular problems, vision problems, and impairment of nerve function.

Acute Complications of Diabetes

As indicated previously, in the absence of either insulin secretion or insulin action, the blood glucose concentration increases markedly, a condition termed **hyperglycemia.** If hyperglycemia becomes severe, the concentration of glucose in the blood can exceed the capacity of the kidneys to recapture glucose by active transport, and glucose will be excreted in the urine (a condition called **glucosuria**). Because of osmotic effects, glucose in the urine will also attract considerable amounts of water, which will be excreted along with the glucose. As a result, urine volume and frequency of urination will increase (a condition known as **polyuria**). As increased amounts of water are lost in the urine, **dehydration** of the individual can also occur.

In Type I (insulin-deficient) diabetics, the unopposed actions of glucagon result in increased ketone formation by the liver. The primary ketones that are formed are acids, and when they are produced in sufficiently large amounts, the normal acid/base balance in the body is considerably disturbed, resulting in the condition termed **ketoacidosis.** As with glucose, if ketones reach a high enough concentration in the blood, they will begin to "spill over" into the urine. As this spillover occurs, ketones also carry cations, such as sodium and potassium, with them into the urine. Thus, accompanying severe ketoacidosis is a loss of these ions, resulting in an **electrolyte imbalance** in the body. If left untreated, severe ketoacidosis accompanied by dehydration can rapidly result in coma and death.

Chronic Secondary Complications of Diabetes

In addition to acute complications that may occur, several secondary complications usually accompany longstanding diabetes. These secondary complications often involve gradual changes that develop over a period of years and may shorten the life expectancy of these individuals. The most common secondary complications of diabetes are seen within the vascular system. Changes much like those seen with atherosclerosis lead to the narrowing of larger blood vessels in the brain, heart, and lower extremities. The resulting reduction in circulation to these areas may result in stroke, heart attack, or loss of limb, respectively. Lesions in the microvasculature (small blood vessels and capillaries) are also common in diabetics and are most detrimental in the kidney and in the retina of the eye. These changes can result in kidney disease and in the condition termed **diabetic retinopathy.** Diabetic retinopathy is due to the deterioration of blood vessels that nourish the retina. As the blood vessels in the retina break down, scar tissue is formed that eventually interferes with the light-sensing function of the retina. Each year, several thousand diabetics become legally blind as a result of retinopathy.

Another common secondary complication of diabetes involves impairment in nerve function **(diabetic neuropathy).** Fibers of the ANS are often involved, frequently resulting in abnormalities in bladder or gastrointestinal tract function, or in impotence. Diabetic neuropathy also frequently involves peripheral sensory nerves, resulting in loss of feeling in the lower limbs in particular.

C H A P T E R   R E V I E W

Summary

- Functional units of the endocrine pancreas are groups of cells known as the islets of Langerhans. The islets are composed of four major cell types: alpha cells, which secrete glucagon; beta cells, which secrete insulin; delta cells, which secrete somatostatin; and F cells, which secrete pancreatic polypeptide.
- Insulin is a peptide hormone consisting of an A-chain and a B-chain held together by disulfide bonds. The most important physiological stimulus for insulin secretion is an increased blood glucose concentration. Glucagon is a simple polypeptide hormone whose secretion is stimulated by a decreased blood glucose concentration. Somatostatin is a small polypeptide whose secretion is stimulated by increased blood glucose and glucagon.
- Insulin has metabolic effects that are anabolic in nature; its most obvious effect is to rapidly lower blood glucose as a result of increasing glucose uptake into muscle and adipose tissue by facilitated diffusion. Insulin also stimulates glycogen formation and glycolysis in its target cells and decreases glucose output by the liver. In liver and adipose tissue, insulin stimulates fatty acid synthesis and promotes the storage of fatty acids as triglycerides. In adipose tissue, insulin promotes the uptake of lipoproteins into cells and inhibits lipolysis. Insulin stimulates the accumulation of protein in most tissues of the body, thus promoting a positive nitrogen balance.
- Glucagon has catabolic effects on metabolism and acts primarily on liver. Glucagon increases blood glucose by stimulating glycogenolysis and gluconeogenesis in liver. Glucagon also stimulates lipolysis as well as ketogenesis in liver. In addition, glucagon provides increased substrate for gluconeogenesis in liver by stimulating protein degradation and amino acid transport.
- Because of the opposing actions of insulin and glucagon, the status of nutrient flow and metabolism in the body is determined by the relative amounts of the two hormones. A high I/G ratio exists in an anabolic state, and a low I/G ratio is found in a catabolic state.
- The principal storage form of calories in the body is triglycerides (fat). Fat has a greater caloric density than glycogen and is hydrophobic. It is therefore a more efficient storage form than glycogen.

- Shortly after a meal, the actions of insulin on metabolism predominate over those of glucagon. As a result, glycogen stores are increased in liver and muscle, triglyceride stores are increased in liver and adipose tissue, and protein synthesis is promoted in all three tissues.
- In the fasting state, glucagon levels are high, insulin levels are low, and nutrient flow is exactly opposite to that in the fed state. Glycogen stores in liver and muscle are depleted; muscle protein is broken down, and the amino acids used for gluconeogenesis in liver and triglyceride stores in adipose tissue are broken down.
- Diabetes is a common endocrine disorder caused by either an absolute or relative deficiency in insulin action. Type I diabetes, or insulin-dependent diabetes mellitus (IDDM), is caused by destruction of insulin-secreting beta cells of the pancreas. Type II diabetes, or non-insulin-dependent diabetes mellitus (NIDDM), is caused by lack of responsiveness of insulin target tissues to the hormone, a condition termed insulin resistance.
- Hyperglycemia, glucosuria, polyuria, and dehydration are common acute complications of diabetes. Type I diabetics are also prone to ketoacidosis. Chronic secondary complications of diabetes include diabetic retinopathy and diabetic neuropathy.

Review Questions

Identify Each with a Term

1. The small clusters of cells that make up the endocrine portion of the pancreas.
2. Cells of the pancreas responsible for insulin secretion.
3. Name of the precursor molecule of insulin.
4. Name of the enzyme in adipose tissue that promotes uptake of lipoproteins from the blood.
5. Name of the process in liver whereby nonglucose substrates are converted into glucose.
6. The type of diabetes characterized by a relative lack of insulin action in its target tissues.
7. The part of the pancreas responsible for secretion of fluid and digestive enzymes into the small intestine.

Define Each Term

8. C-peptide
9. hormone-sensitive lipase
10. glycogenolysis
11. ketogenesis
12. caloric density
13. Type I diabetes mellitus (IDDM)
14. sulfonylureas
15. hyperglycemia

Choose the Correct Answer

16. Which cell type is most numerous in the islets of Langerhans?
    a. alpha cell
    b. beta cell
    c. delta cell
    d. F cell
17. Which of the following is incorrectly paired?
    a. delta-cell:somatostatin
    b. alpha cell:pancreatic polypeptide
    c. beta cell:insulin
    d. all of the above pairs are correct
18. Insulin secretion from the pancreas increases in response to:
    a. ingestion of a high protein meal
    b. decreased blood glucose concentration
    c. activation of the sympathetic nervous system
    d. somatostatin
19. Among the effects of insulin on carbohydrate metabolism are:
    a. inhibition of glycogenolysis
    b. stimulation of glycolysis
    c. stimulation of glycogen synthesis
    d. all of the above
20. In the liver, glucagon:
    a. stimulates ketogenesis
    b. inhibits glycogenolysis
    c. stimulates glucose uptake
    d. all of the above
21. In adipose tissue, insulin:
    a. stimulates hormone-sensitive lipase and lipoprotein lipase
    b. promotes lipolysis
    c. stimulates urea synthesis
    d. stimulates glucose uptake
22. Which of the following is not associated with insulin-dependent diabetes mellitus?
    a. autoimmune disease
    b. high blood glucose concentrations
    c. ketoacidosis
    d. low blood glucagon

Answer in One or Two Sentences

23. What is the anatomical arrangement of the cell types of the islets of Langerhans, and how might the arrangement of cells be involved in the regulation of hormone secretion?
24. Why are triglycerides more efficient than glycogen as a storage form of energy?
25. What are the basic differences in characteristics of Type I and Type II diabetes?
26. What are some of the more common secondary complications of diabetes?
27. Describe the mechanism whereby insulin stimulates glucose uptake into cells.

Problem Solving in Physiology

28. Two kinds of treatments for patients with diabetes are insulin injections and tablets taken orally. What distinguishes between these two types of diabetics and what is the basis for their different methods of treatment?
29. Numerous substances are transported across cell membranes by active transport processes, yet insulin stim-

ulates glucose uptake into cells by a facilitated diffusion mechanism. How do active transport and fa-cilitated diffusion differ, and how does insulin promote facilitated uptake of glucose?

30. Growing children with diabetes often show somewhat reduced muscle growth. What are the effects of insulin on muscle protein and how might these effects lead to reduced muscle size in diabetes?

Suggested Readings

Atkinson, M.A., and Maclaren, N.K. "What causes diabetes?" *Scientific American*, July:62–71, 1990.

Diabetes in America. A report of the National Diabetes Data Group, U.S. Department of Health and Human Services. National Institutes of Health Publication No. 85–1468, Bethesda, 1985.

Ganong, William F. *Review of Medical Physiology*, 15th ed. Los Altos: Lange Medical Publications, 1991.

Hedge, George A., Colby, Howard D., and Goodman, Robert L. *Clinical Endocrine Physiology*. Philadelphia: W.B. Saunders Company, 1987.

Notkins, A.L. "The causes of diabetes." *Scientific American*, 241:62–73, 1979.

Olefsky, Jerrold M. "The insulin receptor. A multifunctional protein." *Diabetes*, 39:1009–1016, 1990.

Rifkin, Harold, and Porte, Daniel, Jr. *Ellenberg and Rifkin's Diabetes Mellitus: Theory and Practice*, 4th ed. New York: Elsevier, 1990.

Wilson, Jean D., and Foster, Daniel W., eds. *Williams Textbook of Endocrinology*, 8th ed. Philadelphia: W.B. Saunders Company, 1992.

Key Terms

alpha cell (p. 443)
beta cell (p. 443)
C-peptide (p. 445)
diabetes mellitus (p. 458)
diabetic neuropathy (p. 461)
diabetic retinopathy (p. 461)
endocrine pancreas (p. 443)
exocrine pancreas (p. 443)

glucagon (p. 443)
glucosuria (p. 460)
hormone-sensitive lipase (p. 452)
hyperglycemia (p. 460)
insulin (p. 443)
insulin/glucagon (I/G) ratio (p. 455)

insulin resistance (p. 459)
islets of Langerhans (p. 443)
ketoacidosis (p. 461)
ketones (p. 453)
lipoprotein lipase (p. 451)
lipoproteins (p. 451)
pancreas (p. 443)
polyuria (p. 461)

proinsulin (p. 451)
somatostatin (p. 443)
sulfonylureas (p. 448)
Type I (IDDM) diabetes (p. 449)
Type II (NIDDM) diabetes (p. 449)

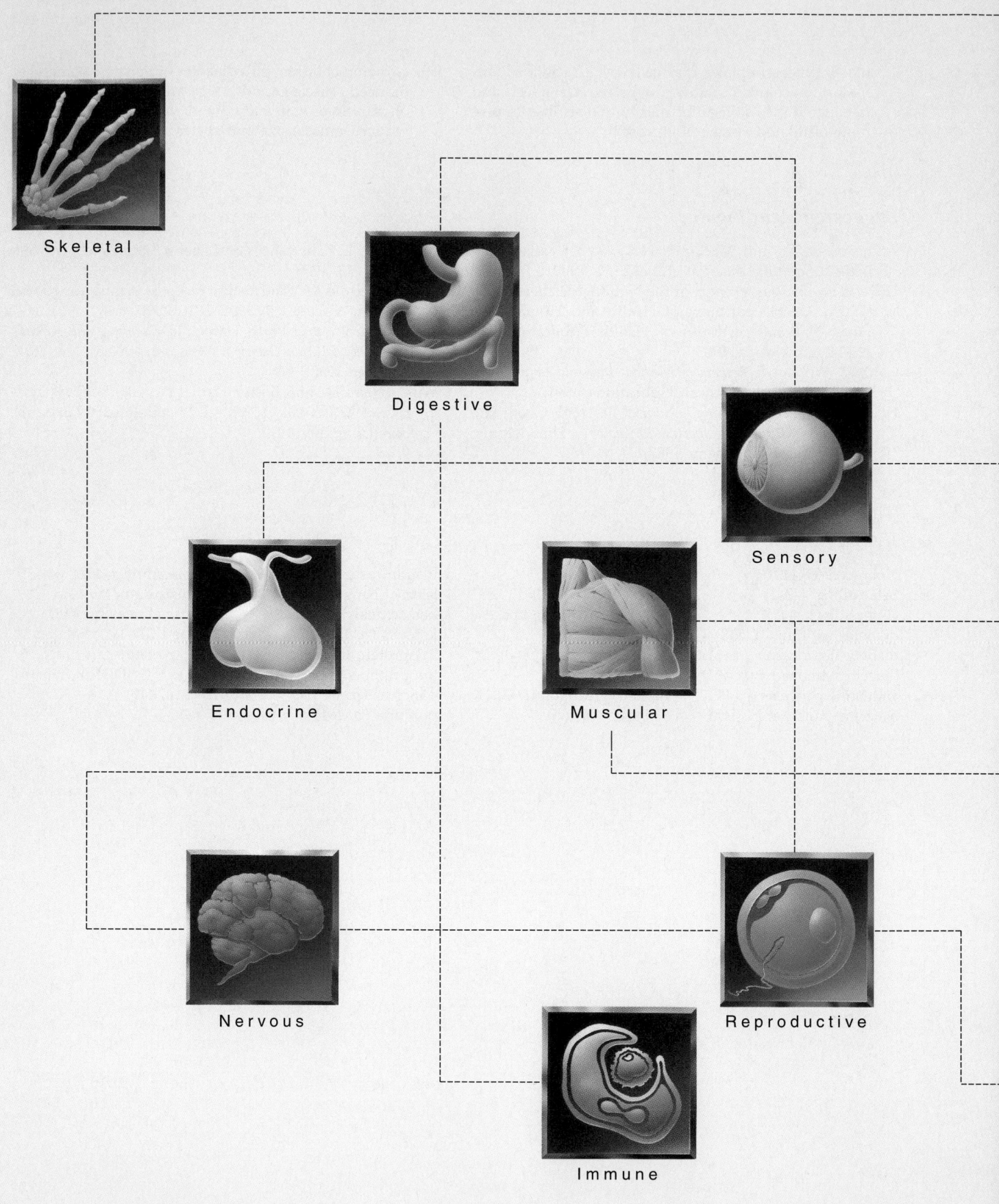

Skeletal

Digestive

Sensory

Endocrine

Muscular

Nervous

Immune

Reproductive

Part III

Integrative Organ Functions

Respiratory

Renal System

Cardiovascular

Thermoregulatory

Transmission electron
micrograph of cardiac muscle
showing the intercattel discs.
(© Don W. Fawcett/Visuals Unlimited)

Chapter 16
Muscle

Key Concepts

Muscles in the human body fall into three major categories: *skeletal*, a voluntary, striated, type used in functions such as locomotion and breathing; *smooth*, an unstriated type associated with blood vessels and visceral organs and involved with involuntary internal processes; and *cardiac*, an involuntary striated type that makes up the heart muscle and powers the circulation of blood.

Striated muscles are composed of two sets of overlapping protein myofilaments, called *actin* and *myosin*, whose relative sliding motion, driven by the action of ATP on the myosin cross-bridges, produces *shortening* and generates *force*. Smooth muscles, although they have a less-regular internal structure, operate with a similar contractile protein system.

Muscles can shorten at a constant force (*isotonic* contraction) or develop force without shortening (*isometric* contraction). The *force-velocity curve* characterizes isotonic contraction, and the *length-tension curve* summarizes the relationships involved in isometric contraction.

The link between cell membrane *action potentials* and muscle *contraction* is provided by *calcium ions*. Skeletal and cardiac muscle have *actin-linked* control systems in which calcium acts on the thin filaments to release the inhibition of actin-myosin interaction. In smooth muscle, which has a *myosin-linked* control system, calcium ions cause the activation of the thick-filament myosin to promote contraction.

The ultimate fuel for muscle contraction is adenosine triphosphate, *ATP*. Its energy is derived from *glucose* and *fatty acids* by either aerobic or anaerobic metabolism. Muscles, according to their particular bodily role, are usually adapted to favor one or the other kind of metabolism. Cardiac muscle has a predominantly aerobic metabolism.

Chapter Outline

Overview: The Roles and Types of Muscle

The Structure of Muscle

The Function of Muscle

The Control of Muscle

The Metabolism of Muscle Contraction

*Pete is riding his bicycle through rush-hour traffic. He is aware, of course, that the **skeletal muscles** of his legs are providing the power that moves the bicycle forward. But many other muscles are at work to keep him moving safely on his way. The muscles of his arms, under the precise control of his central nervous system and his body's organs of balance, are performing hundreds of adjustments each minute to keep his cycle upright and on course. Other skeletal muscles are keeping his body erect and are making air move in and out of his lungs.*

Deep within Pete's body, muscles of other types are also at work. The blood that carries oxygen and *nutrients to his muscles and removes their waste products is pumped by the pulsing action of his **heart muscle**. The blood vessels supplying skeletal muscles and other vital organs are regulated by the **smooth muscle** in their walls, ensuring that an adequate supply of blood goes to the organs that need it most at the moment—the skeletal muscles and the brain.*

Suddenly the horn of a truck sounds behind him. The reactions of all of Pete's muscles are rapid and, thanks to his nervous system, automatic. Tiny skeletal muscles in his middle ears quickly contract, protecting delicate hearing structures from the sudden loud noise. His heart speeds up and the contractions of its muscle become more powerful; his blood pressure rises, while the smooth muscles that control blood distribution allow even more of it to flow to the skeletal muscles. Even though the extra energy that his skeletal muscles must supply to move him out of harm's way outstrips the energy being brought to them by the bloodstream, specialized metabolic pathways and internal energy stores in the muscles allow an emergency surge of power. All of these muscles working in concert allow him quickly to avoid the path of the speeding truck. As the truck roars past, Pete's body's systems gradually settle back to their nonemergency levels.

There are many types of muscle in the body. While we are most familiar with those large muscles attached to bones and responsible for our ability to move, breathe, and manipulate our environment, many other muscles are used for bodily activities of which we are usually unaware. All of our muscles have important features in common. They may be similar in structure, function, and in various other ways. This chapter covers all of these aspects of muscle and elucidates the critical roles of muscle in the body economy. Because skeletal muscle is the type best understood, most of the chapter will be devoted to it.

Overview: The Roles and Types of Muscle

The range of activities that muscles carry out in the body is extremely broad, so it is not surprising that muscles show a wide range of functional adaptations that specifically suit their many tasks in the overall function of the body.

Muscle Serves as the Output for the Central Nervous System

As described in Chapter 10, our central and peripheral nervous systems function by gathering information and instructing muscles to take useful action in response. This coordination between nerve and mus-

cle enables us to walk, talk, digest our food, defend our bodies, propagate our species, and do almost everything else that we do. In addition to effecting changes prescribed by the nervous system, muscles also help to regulate what goes into that system by adjusting the sensitivity of our sense organs. Some muscles are under neural control, while others are more independent. Very few muscles, however, are totally independent. The degree of neural control serves to adapt various muscles to their special roles.

Our other important control system, the endocrine system, also has muscle as one of its important effectors, or means of expression, although this function may not be very obvious when our bodies are functioning well.

Muscle is a Biological Motor

Muscles can do what they do because they are a sort of **biological motor.** They enable our movements and bodily locomotion and help us to maintain posture. Like mechanical or electrical motors, they consume fuel while they do useful work, and waste some of the energy that they consume as heat. Muscles produce extra heat because they, like other motors, are not completely efficient. That is, they cannot turn all of the energy they consume into mechanical work. The heat they produce as a result of this inefficiency may at times be very important in overall bodily function. Under most conditions of activity, most of our body heat is produced as a result of the contraction of muscles. When the body temperature falls, brisk physical exercise (which may be relatively efficient) or shivering (which is very inefficient from a mechanical standpoint) provides the necessary warmth. During heavy exercise or under the stress of high environmental temperatures, our problem may involve getting rid of the excess heat produced by muscular contraction. Many of our conscious "cooling" mechanisms also involve the use of muscle, as may become apparent when the work of fanning to cool off produces more heating than cooling.

Muscle Aids in Regulating Physiological Processes

Muscles also help to regulate many important body functions. Muscles control the movement of substances through the tubular structures (such as blood vessels and intestines) of the body and expel those substances from the body at the proper time. For example, the regulation of blood pressure involves a complex interaction between the heart muscle, which pumps the blood, and other muscles that control the diameter of the blood vessels. Other regulatory mechanisms in which muscles play a central role include

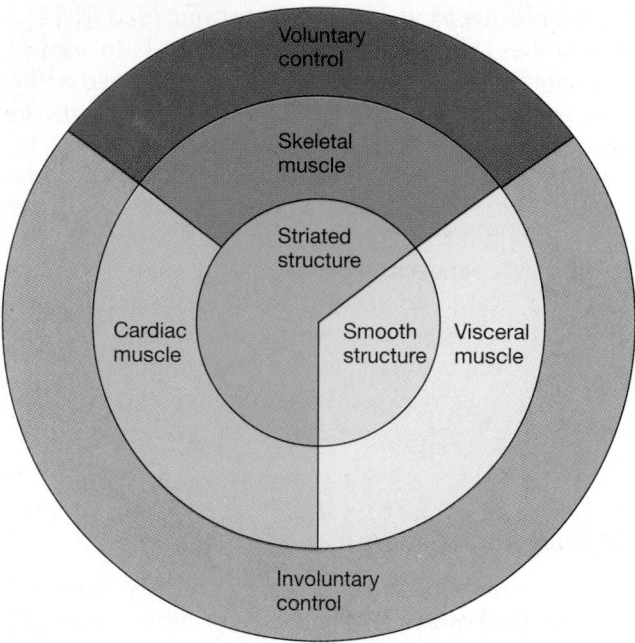

Figure 16–1

A simplified view of the classification of human muscle. Although there may be some difficulties with each type of categorization, muscle may be classified by three different means: by *location* (skeletal, cardiac, or visceral); by *structure* (smooth or striated); and by *mode of control* (voluntary or involuntary). Some of these categories and functions may overlap.

the maintenance of an upright posture and the control of the body temperature.

Muscle Can Be Classified in Several Ways

Several different sets of criteria are used to place muscles in separate categories relating to location and function, microscopic structure, or mode of control and action. These topics serve as convenient ways to describe and discuss particular aspects of muscle, but as with any attempt at classification, some exceptions will have to be noted. The most usual sort of classification is shown in Figure 16–1.

Classification of Muscles by Location and Function

The most obvious muscles in the human body are called **skeletal muscles.** These are the large muscles used, among other things, for locomotion and for maintaining body posture. They usually attach to the skeleton and move the jointed bones with respect to one another. In some instances, the skeletal attachments may be indirect or may exist at only one end of the muscle (as is the case with the tongue, which

is a skeletal muscle), or they may not exist at all (as is the case with the upper end of the esophagus). The bulk of skeletal muscles, however, are attached to bones at both ends.

Muscles that line the walls of the internal organs (the viscera) are called **visceral muscles.** Muscles of this type are also located in nonvisceral organs such as blood vessels and in sensory organs such as the eye, where they aid in focusing and adapting to different levels of lighting.

The location of **cardiac muscle** is obvious from its name; it makes up the pumping muscle mass of the heart. Cardiac muscle is found only in the heart (although it may extend a little way into the large vessels that enter and leave that organ). While confined to a relatively small anatomical region, its critical role in body function in health and disease gives great importance to its understanding and study.

Classification of Muscles by Structure

When viewed at the level of microscopic structure, muscle falls into two general categories: striated muscle and smooth muscle. The term *striated* means "striped" and refers to the regular cross-striped appearance of the muscle cells under a microscope. This appearance is an important clue to the way in which these muscles function on a molecular level.

Skeletal muscle is **striated muscle.** Cardiac muscle is also striated muscle. While its cells are considerably smaller than those of skeletal muscle, its molecular function is similar in many respects. Many of the details of the function of skeletal muscle have been found to apply to cardiac muscle also. In addition, cardiac muscle also shares some functional similarities with visceral muscle.

Visceral muscle cells lack the striations found in the other muscle types and are therefore referred to as **smooth muscle.** Their microscopic structure provides fewer clues to their function than in the case of skeletal muscle, although many aspects of their internal processes are becoming better understood as research in this field progresses.

Classification of Muscles by Mode of Action and Control

Because most movements of skeletal (striated) muscle occur in response to a conscious (willed) effort, these muscles may be called **voluntary muscles.** However, many actions of skeletal muscle, such as the acts of breathing or walking, are almost automatic and are thus in a sense involuntary. The strictest basis for this type of classification is that skeletal muscles, in order to function, must receive a specific signal or signals from the central nervous system (CNS). In the event of an accident or illness that disconnects a skeletal muscle from its nerve supply, paralysis results.

On the other hand, the actions of most smooth (visceral) muscles are **involuntary.** In many cases these involuntary actions are initiated by the **autonomic nervous system (ANS)** in response to some internal reflex adjustment that may take place automatically. Smooth muscle is also capable of acting on its own, sometimes in response to hormonal or environmental factors. Voluntary control of some smooth muscle functions is also possible; for instance, specially trained persons using the techniques of "biofeedback" can cause blood vessels to dilate or constrict. In the majority of situations, however, the function of smooth muscle is completely involuntary.

Cardiac muscle is also an involuntary muscle, and no signals from the nervous system are required to start or maintain its contraction. The heart will continue to beat indefinitely after all nerve connections to it are removed, although the rate and strength of the heartbeat are no longer under external control. Because of the special nature of the cardiac muscle, it is not usually described as being either voluntary or involuntary.

Smooth muscle is itself subject to further classification in the area of its mode of control. Some smooth muscle is called **multiunit smooth muscle;** this muscle is closely controlled by the nervous system. Since the nervous elements involved in its control are part of the ANS, it is not subject to voluntary control. Other smooth muscle, that which makes up the bulk of our visceral organs, is called **unitary smooth muscle.** This sort of muscle is less strictly controlled by the nervous system, and large numbers of its cells function together as a single unit.

A summary of the three classifications is provided in Figure 16–1. When allowance is made for the exceptions noted previously, the classifications are useful for discussing and comparing the various muscle types. Within each type are important variations, and these will be taken up as the specific details of each muscle type are discussed.

The Structure of Muscle

The study of muscle structure can be done at two different levels. At one level, large-scale structural aspects of muscle can be observed and described without microscopic aid. The position, size, shape, and attachment points of a skeletal muscle give important information about its function. At another level, light microscopes and electron microscopes reveal details of the fine structure responsible for the actual mechanism of contraction. Muscle is an excellent example of the way in which the structure and function of a biological tissue are intimately related. For this reason we will discuss in detail the structure of striated muscle before going on to other muscle types. Striated muscle can serve as a basis of comparison for the description of other muscle types.

Skeletal Muscle Structure and Function Are Closely Related

Contraction, the major function of a skeletal muscle, involves its shortening and the resulting movement of parts of the skeleton. The large-scale structure of skeletal muscle reflects this function, and the organization of the cells and tissues reflects the structural and metabolic adaptations necessary for functions in which large forces are generated and large amounts of energy are expended.

Gross Structure of Skeletal Muscle

In mechanical terms, the skeleton works with the muscles as a **lever system.** The term *lever system* refers to the fact that groups of muscles working together with bones permit a larger range of movement than could muscles acting on their own. Muscles attach to bones by means of very strong connective tissue structures called **tendons** (Fig. 16–2). The more stationary (and **proximal**) attachment is called the **origin** of a muscle, and the other (**distal**) end is called the **insertion.** The wide or thick central portion of a muscle is called the **belly.** Some muscles, the **biceps** for example, have a double origin (*biceps* means "two heads") and a single insertion, while others have a narrow origin and a very wide insertion. The variable anatomy of skeletal muscles represents adaptations to the specific functions of the individual muscles. Muscles with many parallel (side-by-side) fibers can exert a large force but cannot shorten rapidly or by a very great amount. On the other hand, long muscles with many fibers in series (end-to-end) are able to shorten rapidly but develop less force.

Because most skeletal muscles are attached to bones at either end, the range of motion of a muscle itself may be limited by the skeletal system. The needed range of motion must then be multiplied by the lever system provided by the skeleton. This arrangement has some important consequences. Large amounts of muscle shortening, as we shall see, can significantly affect the amount of force a muscle can exert, and limiting the range of muscle shortening helps to compensate for these effects. However, this arrangement also means that the muscle must exert a much greater force at its insertion than is actually produced at the end of the limb. For instance, in the human forearm, the lever is hinged at the elbow. the hand is approximately seven times as far from the

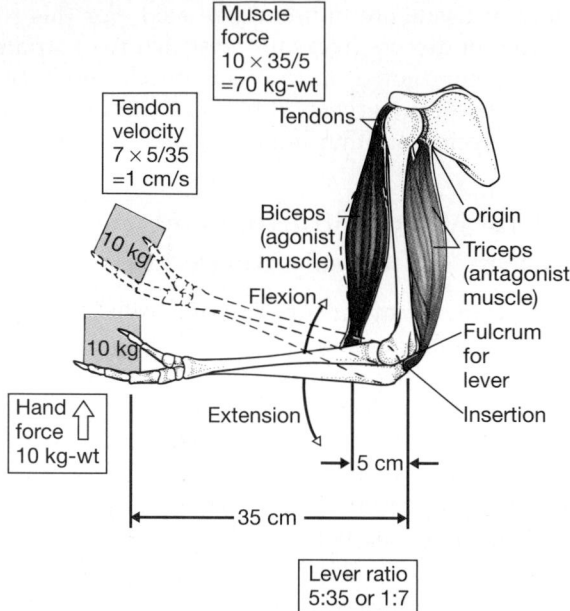

Muscle force
10 × 35/5
=70 kg-wt

Tendons

Tendon velocity
7 × 5/35
=1 cm/s

Origin

Biceps (agonist muscle)

Triceps (antagonist muscle)

Flexion

Fulcrum for lever

10 kg

10 kg

Hand force
10 kg-wt

Extension

Insertion

|← 5 cm →|

|← 35 cm →|

Lever ratio
5:35 or 1:7

Figure 16-2

Antagonistic muscle arrangements and the skeletal lever system. The major muscles that move the upper arm work in opposition to each other. The biceps *flexes* (bends) the arm, while the triceps *extends* it. The tendon of the biceps attaches at a point approximately 5 cm away from the fulcrum of the lever system (the elbow joint), and the hand is 35 cm away. Therefore, the upward force exerted by the muscle must be seven times greater than the downward force supplied by a weight in the hand. However, any shortening of the muscle is multiplied by seven, meaning that the hand moves seven times as fast and far as the lower end of the biceps muscle.

joint as is the insertion of the biceps. This means that if the hand is holding an object with a mass of 10 kilograms, the tendons of the biceps must exert seven times as much force (see Fig. 16–2). This situation can lead to serious muscle and tendon injury in athletes and in people who do heavy manual labor. On the other hand, it means that the hand can move seven times as fast as the biceps can shorten. Such an increase in speed is often valuable. For instance, it allows the leg muscles to propel the body at a high rate of speed, or to jump up in the air much higher than the distance over which the muscles themselves can shorten.

As mentioned in Chapter 10, in most situations, (especially in the limbs) skeletal muscles are arranged in so-called **antagonistic pairs.** When a muscle moves a joint in one direction (e.g., **flexion,** in which a joint closes and a limb assumes a more withdrawn position), another muscle or muscles move the joint in the opposite direction, toward a more extended position

(i.e., **extension**). This arrangement is necessary because muscles can only *pull*; they cannot *push*.

Skeletal **muscle cells** (also called **muscle fibers**) are quite large compared to other cells. Typical diameters are on the order of 100 micrometers, while cells from some large muscles may be many centimeters in length. The way in which muscle cells are assembled into the complete muscle is diagrammed in Figure 16–3. In an intact muscle, shown in cross section in Figure 16–3a, the individual muscle fibers are surrounded by a delicate connective tissue sheet called the **endomysium.** Bundles of muscle fibers are grouped into **fasciculi** (shown in longitudinal section in Figure 16–3b) by a connective tissue called the **perimysium,** and the entire muscle is covered by a connective tissue sheet called the **epimysium.** During physical exercise, muscle requires a plentiful supply of oxygen and nutrients, and must rapidly get rid of waste products. Accordingly, muscle is supplied with a plentiful network of blood vessels arranged among the fibers so that no muscle cell is very far away from a blood vessel.

Both the connective tissue sheets and the muscle fibers themselves are connected at either end into a very tough connective tissue extension of the muscle that makes up the tendon. Some tendons are thick and short, like those that provide the insertion of biceps. Some tendons are very long; the muscles that move the fingers are actually located in the forearm, and their tendons run across the wrist joint and the back and palm of the hand to connect the muscles to the bones of the fingers.

Cellular and Molecular Structure of Skeletal Muscle

At the cellular level of organization, there is a remarkably clear relationship between the structure and function of muscle. The molecules that make up the contractile substance of muscle have a dual role. Not only are they the major structural elements of the muscle cell; they are also the enzymes that convert the biochemical energy (derived ultimately from the food we eat) into mechanical energy.

When viewed through a light microscope, skeletal muscle cells show a repeating pattern of dark and light crossbanding. This striped, or striated, pattern gives skeletal muscle one of its names; more importantly, it provides a clue to how striated muscle actually works. Analysis using electron microscopy and other methods has revealed that the dark and light crossbands, or striations, are composed of two types of protein filaments, with each type being arranged in sets side by side across the cell. These protein filaments are called **myofilaments,** and the two types differ from each other in their size and internal struc-

Connective tissue (epimysium)

Bone

Connective tissue (perimysium)

Bundles of muscle (fasciculi)

Connective tissue (perimysium)

Muscle cells (muscle fibers)

Muscle cell

Sarcolemma (cell membrane)

Fascicles

Myofibrils

Nucleus

Connective tissue (endomysium)

(a)

Muscle (deltoid)

Bundle of muscle fibers (muscle fasciculus)

Nucleus

Muscle fiber

Sarcolemma

I-Band   H-Zone   A-Band   Z-Disc

Myofibril

←Sarcomere→

Thick filament

Z-Disc   Z-Disc

Sarcomere

Thin filament

(b)

Figure 16–3

The organization of skeletal muscle. (*a*) Cross-section of the thigh, progressively magnified to show finer details. (*b*) Longitudinal section of the deltoid muscle, showing the bundles of fibers (fasciculi) that make up a whole muscle. Further magnification shows that the individual muscle fibers, or cells, contain long structures called *myofibrils* that are made up of myofilaments and are organized into *sarcomeres.*

(c)

Figure 16–4

The microscopic structure of skeletal muscle. These drawings were made from electron micrographs of skeletal muscle. (*a*) A longitudinal section showing the arrangement of the sarcomeres. The letters H, Z, A, I, and M refer to the parts of the sarcomere diagrammed in Figure 16–5. (*b*) A cross-section of the same tissue. Several cells are shown in the inset at the upper left, and the area in the rectangle is magnified in the rest of the figure. (Modified from Krstic, R.V., *Ultrastructur der Saugertierzelle*. Berlin: Springer-Verlag, 1976.)

ture. The two sets overlap each other in a geometrically repetitive way so that the striation pattern changes as the muscle is lengthened or shortened.

The Structure of a Sarcomere. The two sets of myofilaments are organized into a structure called a **sarcomere;** this is the basic functional and anatomical unit of the contractile machinery of the skeletal muscle. Figure 16–4 shows both a cross section and longitudinal section view of the microscopic structure of skeletal muscle. The center portion of the sarcomere (the **A-band**) is made up of **thick filaments** (composed of the protein *myosin*), and the lateral regions are made up of **thin filaments** (consisting largely of the protein *actin*). A structure called the **M-line** runs

across the middle of the A-band and holds the thick filaments in side-by-side alignment. The thin filament array interdigitates (overlaps) with the thick filament array, and the depth of its penetration depends on the overall amount of shortening of the muscle. The region where only thin filaments are present is called the **I-band.**

Both sets of filaments are arranged in a hexagonal pattern; this is the most compact way in which the filaments could be packed (Fig 16–5*a*). If a cut is made in a region in which sets of thick and thin myofilaments overlap, then the spaces provided by the hexagonal array of the thick filaments will be occupied by an interlocking hexagonal array of the thin filaments. "Crossbridges" from the thick filaments

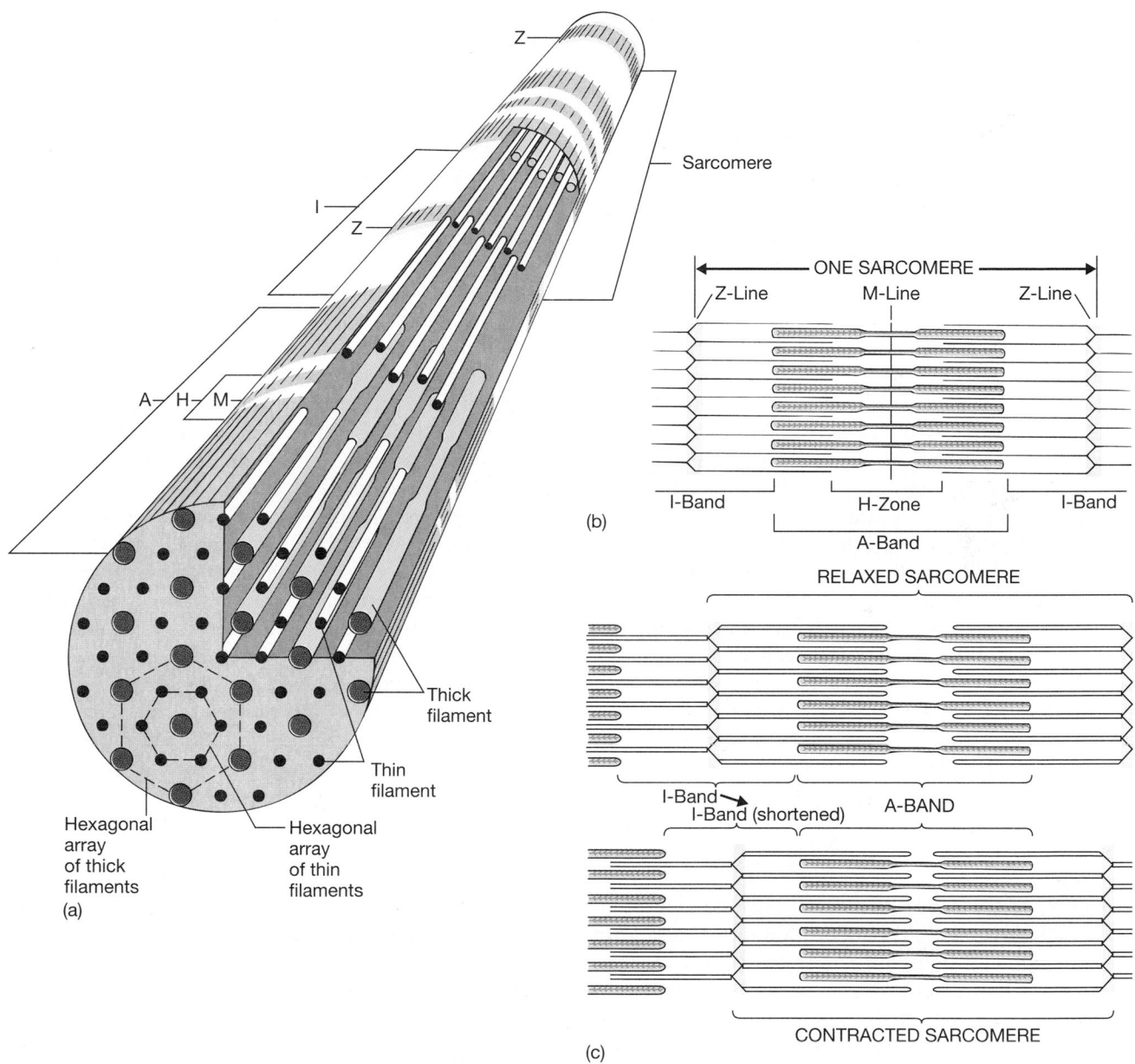

Figure 16–5

The structure of a sarcomere. (*a*) Diagram showing the relationship of the longitudinal and cross-sectional arrangement of the myofilaments. (*b*) The A-band is made up of thick filaments arranged side by side; the thin filaments extend from the Z-line into the A-band. The I-bands are made up of the portions of thin filaments where there is no overlap between thick and thin filaments, and the H-zone is the region of the A-band into which the thin filaments do not extend. The M-line runs down the center of the A-band and appears to help in maintaining the uniform arrangement of the thick filament array. (*c*) During contraction, the width of the A-band stays constant, while the H-zones and the I-bands become narrower and the Z-lines move closer together.

will be seen to reach out directly toward the thin filaments.

The usual names given to the parts of the sarcomere are shown in Figure 16–5*b*. The center of each I-band is marked by a structure called the **Z-line,** which serves to mark the boundary of the sarcomere. (Sometimes this structure is called the **Z-disc** in order to emphasize its three-dimensional nature.) The function of the Z-line is to connect the ends of the thin filaments from one sarcomere to those of the next one in line. The Z-lines also have cross-connections that keep the thin filaments in register with each other in a lateral direction. In the center region of the A-band, where the thin filaments of the I-band have not penetrated, is a somewhat lighter region called the **H-zone** (the M-line runs through the center of the H-zone, since the filament overlap is symmetrical).

During the process of contraction, in which the muscle may shorten considerably, *neither the thick nor the thin filaments change their length.* All of the change in the length of the muscle is accounted for by a greater or lesser degree of overlap between the thick and thin filaments. This means that when the muscle shortens, the length of the A-band stays the same while that of the I-band decreases (see Fig. 16–5c). The thin filaments themselves do not change in length. Shortening of the muscle is brought about by a cyclic "rowing" motion of the crossbridges, in which they successively attach to the closest available thin filament site, change their shape and pull the actin filament along, then let go of the actin and re-attach further along.

Hundreds of sarcomeres connected in series (end-to-end) form a functional unit called a **myofibril,** which runs for the length of the cell. Most muscle cells contain a large number of parallel myofibrils, and the intracellular spaces between them are occupied by **mitochondria,** and by molecules of glycogen, a source of chemical energy. Space between the myofibrils is also occupied by the **internal membrane system.**

Structure of the Myofilaments.   A globular protein called *actin* makes up the **thin filaments.** (See Figure 16–6a.) The slightly oblong actin molecules are each about 5.5 nanometers in diameter and are joined end-to-end to form a long structure that resembles a bead-chain. Two such chains are wound about each other so that the pair makes a half-turn every seven actin monomers. This means that the thin filaments are helical structures, with a half-pitch of approximately 35 to 37 nanometers. Each half-turn of seven actin molecules is also associated with a long fibrous protein molecule called *tropomyosin.* Each of these tropomyosin molecules bears a protein complex called *troponin.* The troponin is in turn composed of three subunits whose names are *troponin-I, troponin-T,* and *troponin-C.* Although it is the actin protein of the thin filaments that interacts with the myosin crossbridges of the thick filaments, the four components of the tro-ponin-tropomyosin system (known collectively as the *regulatory protein complex*) have an important function in regulating the contraction of skeletal muscle.

Thick filaments are made up of very large protein molecules called *myosin* (Fig. 16–6b). A thick filament is composed of about 300 to 400 myosin mol-ecules. Each myosin molecule has a long rod-shaped section (a structural portion sometimes called the tail region) and a double globular head at one end. The head portion of the molecule, which has ATP and actin binding sites, contains the biochemical and enzymatic properties that participate in the actual contractile function. Also attached to each head portion are two protein molecules called **light chains** because of their comparatively low molecular weights. These have some secondary regulatory functions in skeletal muscle.

When myosin molecules are packed together to form a thick filament, the tail regions of the molecules make up the structure of the filament itself, while the head portions project from the sides of the filament. The projections of the myosin heads are what form the structures called crossbridges. Because of the way the molecules are packed, the heads project from the filament in a radial fashion, and the projection makes a full turn for every six mysoin molecules. The angle between the successive myosin heads is thus 60 degrees, and they project out of the filament every 14.3 nanometers. In this way, they form a helical pattern (like a spiral staircase) that winds down the length of the thick filament.

The tail regions of molecules at either end of the filament point toward its center, making the thick filament symmetrical about its center, with myosin heads pointing in opposite directions in the two halves of the filament. This is important in creating the forces involved in contraction, because when both sets of filaments are assembled into a sarcomere (Fig. 16–6c) action of the myosin heads will pull actin filaments into the sarcomere from both directions. The construction of the thick filaments also means that there is a bare zone in the middle of each thick filament; this is an area containing only the tail portions of myosin molecules. The assembled myofilaments are held in place by the proteins of the Z-discs, the M-line, and by accessory proteins (not shown) called **nebulin** and **titin** that are associated with the thick and thin filament arrays.

The Internal Membranes of a Muscle Fiber.   These membrane structures participate in the control of contraction and relaxation. (See Fig. 16–7 for their anatomical arrangements.) The **plasma membrane** and a thin layer of connective tissue fibers form the outer covering of the cell, called the **sarcolemma.** Extending inward from the sarcolemma into the depths of the cell is a set of membrane-lined passageways, the **transverse tubules** (often called **t-tubules** for short). Depending on the type of muscle being considered, the t-tubules may enter the fiber either at the region where the A- and I-bands overlap, or at the level of the Z-discs. The t-tubules are an inward ex-

(a)

(b)

(c)

<u>*Figure 16–6*</u>

The structure and assembly of myofilaments. (*a*) The thin filaments are made up of two chains of *actin* protein monomers that are twisted one-half turn for every seven monomers. The regulatory protein complex, made up of *troponin* (composed of its three subunits) and the long molecule *tropomyosin,* is associated with every seven actin monomers, with the tropomyosin extending along the groove formed by the twisting of the actin chains. (*b*) The individual *myosin* molecules that make up the thick filament are paired structures, with two adjacent head portions and long tails that are wound about each other. The head portions each contain an ATP binding site and an actin binding site, and associated with each of the heads are two protein *light chains* that have regu-

latory functions. The thick filaments are composed of bundles of such individual myosin molecules packed together so that the tail portions form the "backbone" of the filaments and their heads project outwards in a spiral fashion around the filaments. The center part of the filament contains only the tail regions, because the molecules are oriented oppositely in either end of the filament, a *bipolar* arrangement. (*c*) When the filaments form a sarcomere, thin filaments extend into the thick filament array so that action of the crossbridges will draw the thin filaments inwards from both directions; this is because of the bipolar arrangement of the myosin molecules. Notice that the spacing of the crossbridges is different from the spiral period of the thin filaments.

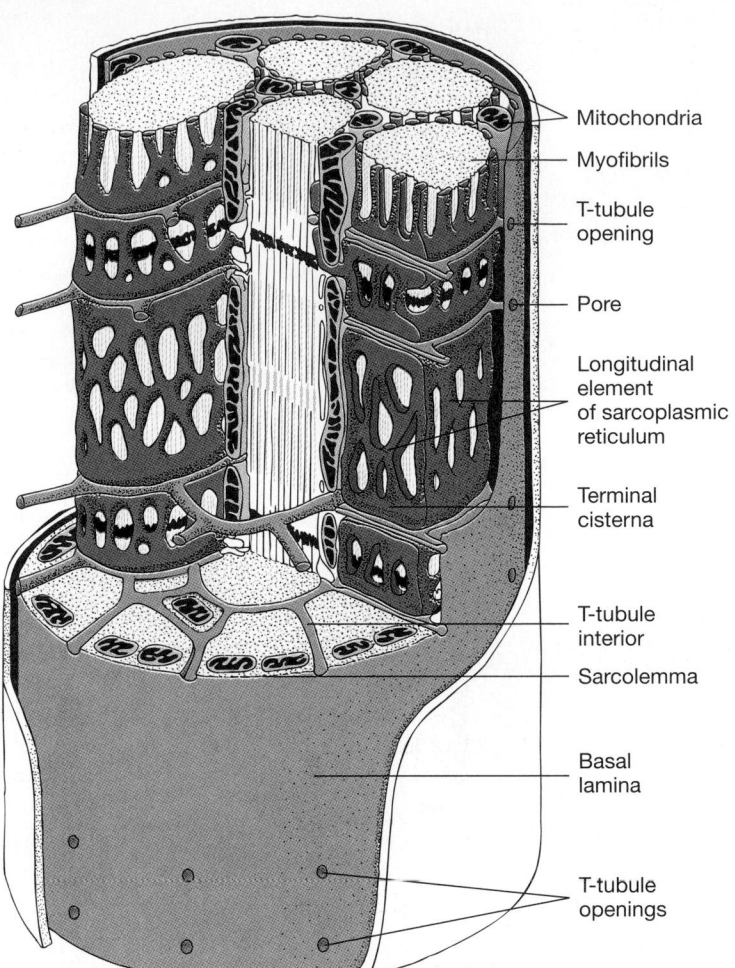

Figure 16–7

The internal membranes of skeletal muscle, showing the structure of the *t-tubule system* and the *sarcoplasmic reticulum.* (Modified from Krstic, R.V., *Ultrastructur der Saugertierzelle.* Berlin: Springer-Verlag, 1976.)

Mitochondria

Myofibrils

T-tubule opening

Pore

Longitudinal element of sarcoplasmic reticulum

Terminal cisterna

T-tubule interior

Sarcolemma

Basal lamina

T-tubule openings

tension of the plasma membrane, and the interior of the t-tubule system is continuous with the extracellular space. Closely associated with the t-tubular system is another set of membranes that run at right angles to the t-tubules and down the length of the sarcomere. This is the **sarcoplasmic reticulum.** Unlike the t-tubules, the sarcoplasmic reticulum is not connected to the extracellular space; however, it does come into close contact with the t-tubules as they course across the fiber. This region of contact, which consists of the t-tubules and the enlarged end portions of the sarcoplasmic reticulum (called **lateral sacs** or **terminal cisternae**), is called a **triad** in skeletal muscle because it is made up of one t-tubule and the terminal cisternae from two adjacent sections of sarcoplasmic reticulum sarcomeres. Periodic structures (sometimes called **"feet"**) appear to join the two membrane systems at the triad. These internal membrane systems function in the control of the contraction and relaxation of the muscle. The function of the "feet" structures is not fully understood, but they are thought to play a role in communicating the signal for contraction from the t-system to the interior of the fiber.

Cardiac Muscle Structure Is Similar to that of Skeletal Muscle

The cellular structure of heart muscle is similar in many ways to that of skeletal muscle, but there are also some very important differences (Fig. 16–8). The cells of heart muscle are small compared with those of skeletal muscle. They are from 5 to 15 micrometers in diameter and may be from 20 to 30 micrometers long. The cells are connected end-to-end and, to a lesser extent, side-by-side. Some of the cells may branch so that one end connects to two other cells. At the regions of connection between the cells is a specialized structure called the **intercalated disc.** This is an area of extremely close contact in which intercellular space is essentially absent; many projections from one cell extend into another, and the other cell has the mirror image of this arrangement. There is a strong mechanical connection between cells in this region, as well as a close electrical connection. The mechanical connection is provided by a cellular structure called a **desmosome** (or tight junction), and the electrical connection is made through a **gap junction** (a structure found also between smooth muscle cells).

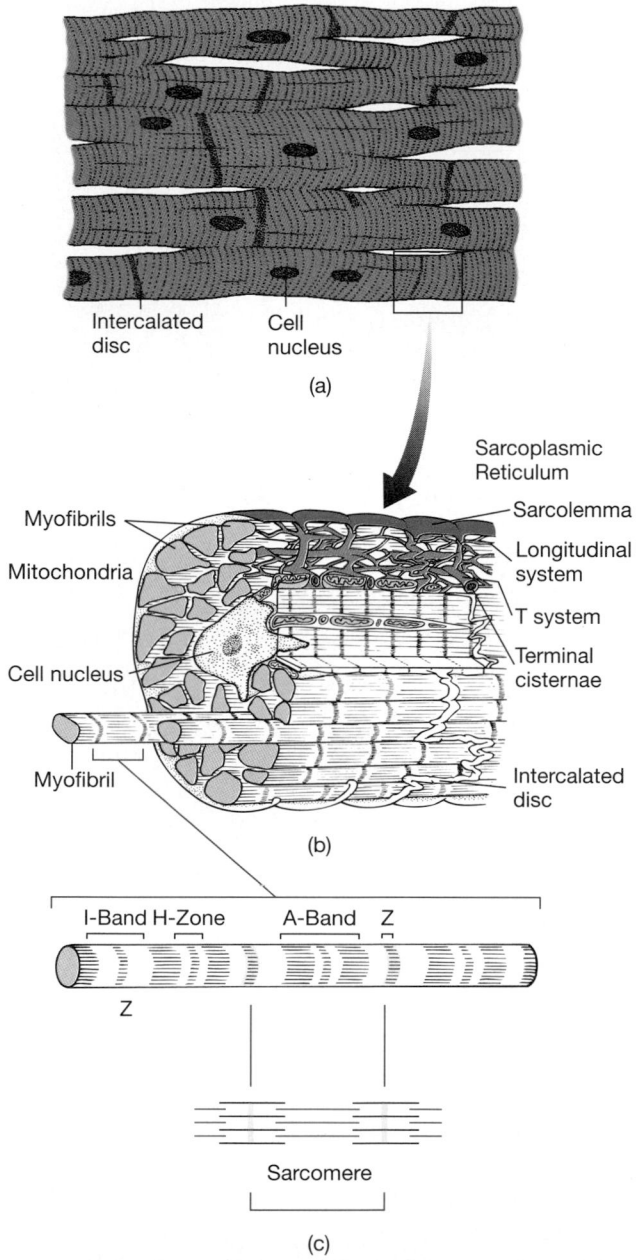

Intercalated disc   Cell nucleus

(a)

Sarcoplasmic Reticulum

Myofibrils

Mitochondria

Cell nucleus

Myofibril

Sarcolemma

Longitudinal system

T system

Terminal cisternae

Intercalated disc

(b)

I-Band H-Zone   A-Band Z

Z

Sarcomere

(c)

(d)

Figure 16–8

The structure of cardiac muscle. (*a*) The cells of the muscle are small, sometimes branched, and they attach end-to-end. (*b*) Seen at higher magnification, bundles of *myofibrils* run the length of the cells, and a *sarcoplasmic reticulum* and *t-tubule* system, which function much as they do in skeletal muscle, are present. (*c*) The sarcomeres are organized in the same way as they are in skeletal muscle. (Modified from Braunwald, E., Ross, J., and Sonnenblick, E.H. *Mechanisms of Contraction of the Normal and Failing Heart.* Boston: Little, Brown, 1968.) (*d*) Electron micrograph of cardiac muscle, showing *sarcomeres, mitochondria* and two *intercalated discs*. (© *Don W. Fawcett/Visuals Unlimited.*)

The specialized contact structure of the intercalated disc allows a tissue made up of small cells to function in many ways as though it were a single large cell, and this forms the cellular basis of the coordinated action of the muscle in the beating heart.

The intracellular organization of cardiac muscle is the aspect in which cardiac muscle is most like skeletal muscle. The structure and arrangement of skeletal and cardiac myofilaments are quite similar, as is their protein composition. The similar contractile structure gives rise to the same sort of striated appearance that skeletal muscle has, and many of the mechanical responses are similar. Because the muscle of the heart has a largely **aerobic** metabolism, it contains numerous mitochondria. A transverse tubular system is present in muscle from many areas of the heart, although the tubules are larger in diameter and usually make close contact with only one portion of longitudinal sarcoplasmic reticulum. This results in a structure called a **dyad,** rather than a **triad** as in skeletal muscle.

Smooth Muscle Is Made of Small Cells without Striations

The Ultrastructure of Smooth Muscle

Although it is very similar to skeletal and cardiac muscle in the molecular structure of its contractile proteins, smooth muscle has a very different cellular and tissue structure. (Drawings made from electron micrographs are shown in Fig. 16–9). Its cells are quite small, 5 to 15 micrometers in diameter and perhaps 200 to 500 micrometers long, depending on which type of muscle is examined. The cells have a single nucleus, centrally located, and the ends taper off to a small diameter. In many smooth muscle tissues, numerous portions of cells are connected very closely together by a structure called the **nexus,** or **gap junction.** This structure has been described in

Thin filament
Nucleus
Thick filament
Plasma membrane
Smooth endoplasmic reticulum
Mitochondria
Glycogen granules
Dense bodies

Smooth muscle cell
Plasma membrane
Nucleus
Parts of Schwann cell
Axon   Epinephrine vesicles

(a)   (b)

Figure 16–9

The structure of smooth muscle. (*a*) Drawing based on an electron micrograph of a longitudinal section of parts of several smooth muscle cells. Most of the cell organelles tend to cluster at the ends of the nucleus. The rectangular inset shows a lower-power view of cells in both longitudinal- and cross-section. (*b*) A nerve muscle contact between a smooth muscle cell and a small bundle of autonomic nerve fibers. There is no specialized region of contact. (Modified from Krstic, R.V., *Ultrastructur der Saugertierzelle.* Berlin: Springer-Verlag, 1976.)

some detail in chapters 5 and 7. These close cell-to-cell contacts in smooth muscle allow for coordination of smooth muscle activity (as in the heart, which also has gap junctions between cells). In the case of the smooth muscle of the uterus, gap junctions increase greatly in number and area at the end of pregnancy, just at the time at which labor is to begin. Their presence is reflected in the greatly increased coordination and effectiveness of the contractions that result in childbirth.

A network of connective tissue fibers (collagen and elastin) surrounds the cells and helps to link them together. It also gives strength to the whole tissue, particularly in those organs that may be subject to a high degree of stretch. The force of contraction

of the individual cells is transmitted to the whole tissue by strong cell-to-cell connections and by the connective tissue network.

Mechanical Arrangements of Smooth Muscle

Smooth muscle makes up much of the walls of organs such as the stomach, intestines, and uterus. It is generally arranged in layers. In the small intestine, for example, one layer (the inner one) is arranged in a circular fashion around the lumen, and the outer layer runs lengthwise (longitudinally). When the circular layer contracts, the intestine is reduced in diameter, and when the longitudinal layer contracts, the intestine becomes shorter. Unlike skeletal mus-

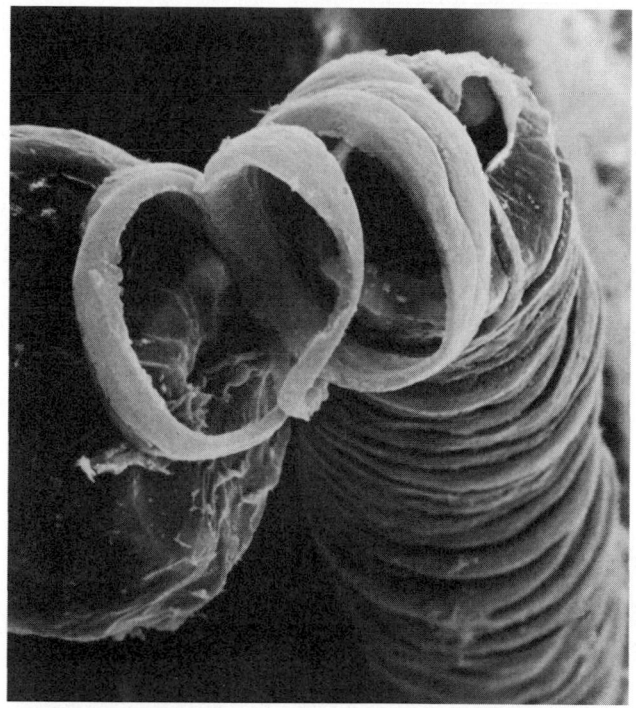

(a)

Blood cells   Smooth muscle

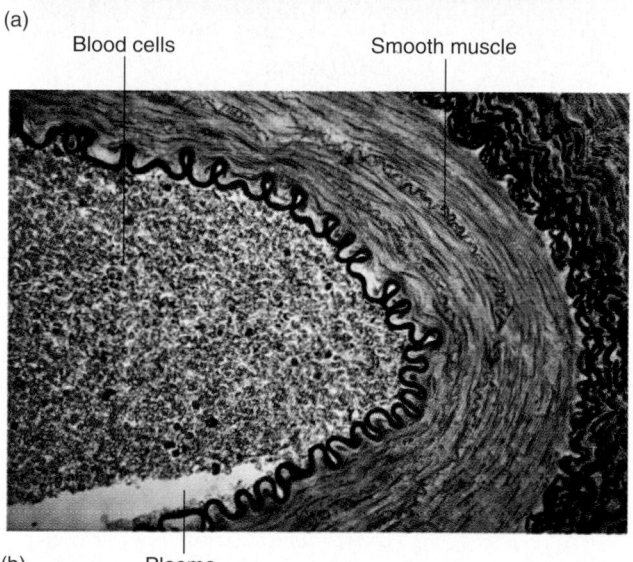

(b)   Plasma

Figure 16–10

Special arrangements of smooth muscle. (*a*) An example of the smooth muscle cells in the wall of a blood vessel. This is a scanning electron micrograph (SEM) of a small blood vessel called an arteriole. At the cut end of this vessel, some of the smooth muscle cells have been spread apart so that they may be seen better, while further down the vessel the cells are in their natural position. Contraction and shortening of the smooth muscle cells, which run in a circular direction, would cause the arteriole to become smaller in diameter. It is by this means that smooth muscle can regulate the flow of blood to the tissues. (SEM courtesy of Dr. A. Evan.) (*b*) Photomicrograph of a cross-section of an artery. Note again the circular arrangement of the smooth muscle. *(© Stan Elems/Visuals Unlimited.)*

cles, which usually have antagonistic muscles to stretch them out after contraction, smooth muscles must return to their pre-contraction length in a different way. An increase in the volume of the contents of the organ re-stretches the muscles in the walls. This usually comes about by the entry of new fluid (e.g., by newly formed urine in the bladder) or by movement of contents from elsewhere along a tubular structure (as in the intestine). Coordinated movements of smooth muscle organs such as the intestine often involve the alternate contraction and relaxation of the two muscle layers. In some saclike organs such as the uterus, muscle fibers are oriented in several directions, reducing the overall volume of the organ when the muscle contracts. The musculature of most blood vessels is simpler, usually consisting only of a circular layer of muscle. In very tiny blood vessels (e.g., an arteriole, which is pictured in Fig. 16–10), a single cell may wrap completely around the circumference of the vessel. Contraction of this muscle cell leads to a narrowing of the blood vessel and a reduction in the amount of blood flow. Such smooth muscle plays a vital role in the regulation of blood flow and pressure.

The Function of Muscle

Most of what muscle does is obvious. We are aware that it shortens and develops force to lift an object, and there is some apparent relationship between how heavy a load is and how fast it can be lifted. These everyday impressions have been refined into a more precise set of concepts that are used to describe, quantify, and compare the way in which muscle functions. Before we approach this more exact analysis of muscle function, some basic definitions are necessary.

Skeletal Muscle Contraction Depends on External Conditions

First of all, the term **contraction** has a special meaning. In everyday usage, it means "to shorten"; in the terminology of muscle physiology, it means any form of muscle activity in response to stimulation, whether it involves shortening or not. For instance, in an **isometric contraction** (Fig. 16–11*a*) (*iso* = same; *metric* = measure), the muscle does not do any external shortening; instead, it develops **force**, or **tension**, while pulling against immovable attachments. In an **isotonic contraction** (*tonic* = tension), the force in the muscle remains constant while the muscle shortens. In an **auxotonic contraction**, the force continually increases as the muscle shortens (e.g., pulling back a bowstring or stretching a rubber band involves this sort of contraction). A **meiotonic** contraction, how-

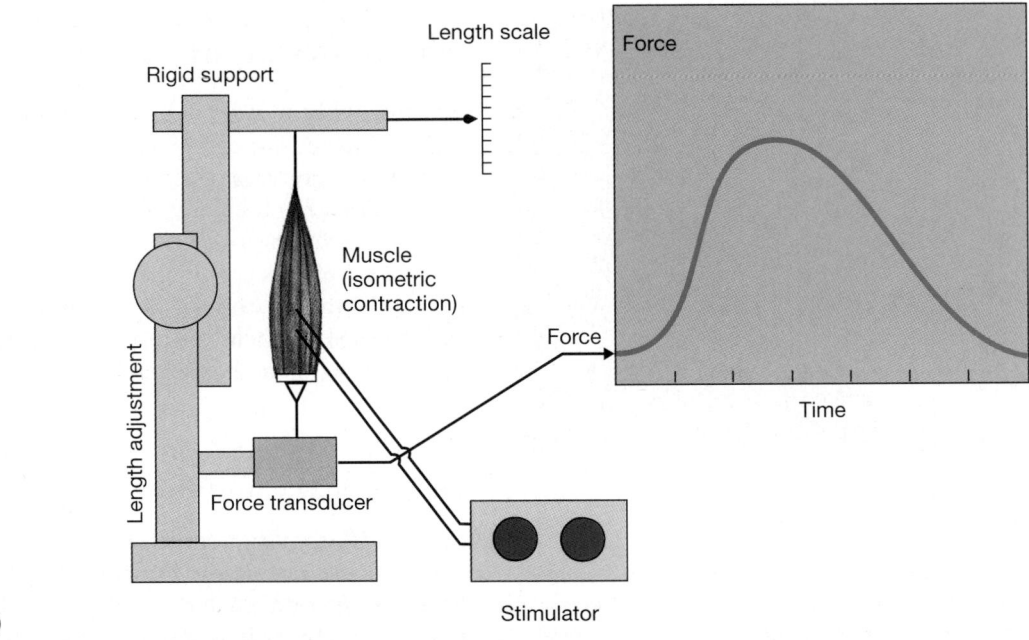

Figure 16–11

(*a*) Diagrams of isotonic and isometric conditions of contraction in the human arm.
(*b*) Typical apparatus for measuring isometric contraction. The length of the muscle may
be varied between contractions, and the record of the force of contraction is viewed on
the face of an oscilloscope. The muscle is stimulated with an electric shock.

ever, is one in which the force *lessens* as the muscle shortens. Many useful mechanical devices, such as gearshift levers and typewriter keyboards, allow this type of contraction in order to provide a clear indication that an action has been performed. These de-

fined conditions represent special circumstances, and realistic contractions usually involve some combination of them. For purposes of analysis, however, contraction conditions can be set up to emphasize the specific properties desired.

Isometric Contractions in the Laboratory: Clues to the Sliding-Filament Hypothesis of Contraction

Under laboratory conditions, an isolated skeletal muscle (and other muscle types as well) can be made to produce contractions conforming to any of these definitions. The relationships between muscle and its external mechanical environment are usually studied by using some "standard" experimental conditions. The simplest type to setup (as diagrammed in Fig. 16–11*b*) can record **isometric contractions.** Here the muscle is securely attached to a device called a **force transducer,** which produces an electrical signal that varies in direct proportion to the force the muscle exerts. The other end of the muscle is attached rigidly to a fixed support (whose position the experimenter may adjust). All connections, and the force transducer itself, are made very stiff so as to prevent any shortening of the muscle. The force signal from the transducer is fed to a device such as a **chart recorder,** which produces an ink-on-paper graph of the force as it changes with time. (An **oscilloscope,** which produces a similar, although temporary, record on a **cathode ray tube,** may also be used.) The muscle is kept alive by being immersed in a physiological saline solution containing all of the necessary ionic and organic substances (along with oxygen and an energy source) that it requires.

The Muscle Twitch.

A single electrical stimulus to a skeletal muscle sets off a series of electrical and chemical events that result in a single, brief contraction called a **twitch.** On the graph drawn by the chart recorder, the force rises rapidly to a peak and then declines a bit more slowly back to the resting value. For a given stimulus sufficient to activate all of the muscle fibers, and under a constant set of temperature and metabolic conditions, all twitches made at the same resting muscle length will produce similar contraction records.

The Isometric Length-Tension Curve.

The mechanical factor with the most profound effect on an isometric contraction is the resting *length* at which the muscle is held. Figures 16–12 and 16–13 illustrate these length-dependent relationships. Making a series of isometric twitches, and changing the muscle length *between* (not during) the contractions, allows an important relationship to be observed. First the muscle is held at a very short length (shorter than it would ever be in the body) and then stimulated. If it is then progressively stretched out between contractions, the peak amount of contraction force will increase each time until some length is reached at which the twitch force is greatest. From this point on, further lengthening will result in two effects. First, force

will begin to be measured in the muscle even before the stimulus is applied. This is called the **resting (or passive) force** and is due mostly to the connective tissue that holds the muscle together, acting much like a rubber band as it is stretched. Secondly, the amount of force (the so-called active force) that the muscle can develop over and above the resting force now diminishes as the muscle is lengthened further. At an extreme degree of stretch (provided that the muscle has not torn), there will be a very high resting force and very little active force in response to stimulation.

This relationship, in which the force capability of the muscle is greatest at some intermediate length, is called the **isometric length-tension curve.** Usually the maximum force capability occurs at a length that is close to the natural length of the muscle in the body. This length is often termed the **optimal length** or L_o, while the peak force at this length is designated P_o or F_o.

While this length-tension behavior is very important to an understanding of how muscle works, most skeletal muscles are limited in how far they can be stretched out by their attachments to the skeleton, and the length-tension curve is of relatively little importance in normal function. Heart muscle, however, does not have its length changes limited by skeletal attachments, and its resting length is set by the amount of blood that returns to fill the heart between beats. The length-tension curve thus allows the heart to make a more forceful contraction if it receives a larger filling of blood between contractions.

Optimal Length and the Overlap of Myofilaments.

Why does the resting length have such an important effect on the contraction of the muscle? The answer to this question has provided some important clues to the basic physiology of skeletal muscle, clues that relate the structure of the muscle to its function. Careful measurements with the electron microscope have shown that the A-bands do not change in width as the muscle is lengthened or shortened, but the width of the I-bands increases as the muscle is stretched (and vice-versa). The spacing between the Z-lines also changes in proportion to the muscle length. Other measurements have shown that the length of the myofilaments themselves also does not change. What does change, however, is the amount of overlap between thick and thin filaments. Stretching the muscle decreases this amount of overlap; in some muscles, the sarcomeres may be pulled completely apart. These observations (when compared with the length-tension curve of the living muscle) provided evidence that the overlap of the myofilament controls the amount of force that the muscle can develop (see Figs. 16–12, 16–13). During delicate experiments with single fibers of skeletal muscle in specially controlled

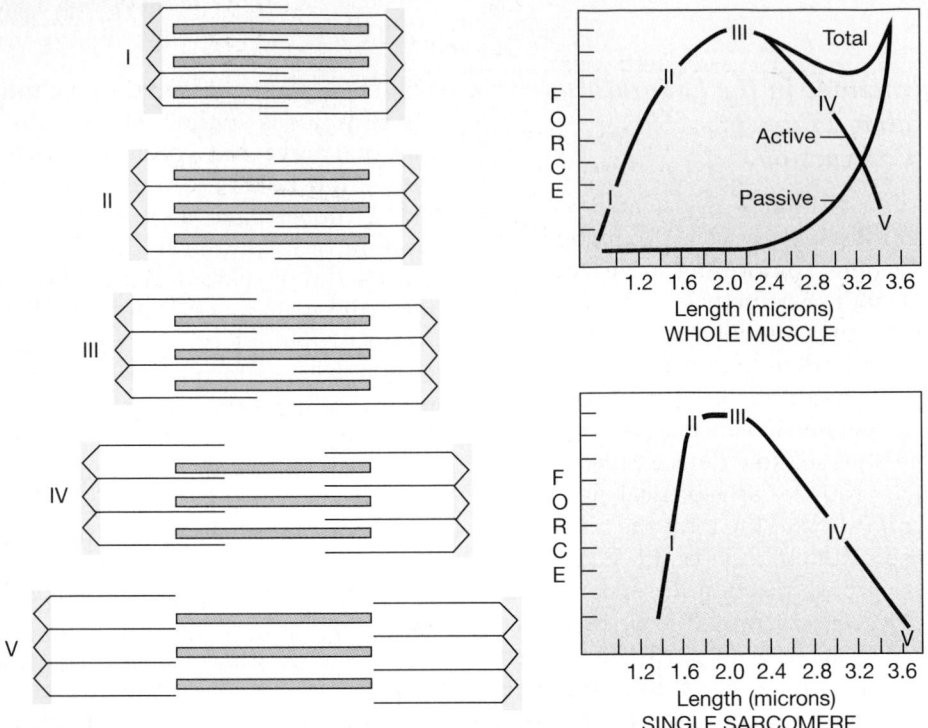

Figure 16–12

The length-tension curve. The upper graph shows how the resting length affects the tension (also called force) produced by a muscle during an isometric contraction. The curve labeled "Total" represents the "Active" force plus the "Passive" force. The key to understanding the basis for the length-tension relationship lies in the sarcomere diagrams to the left. They show the relative amounts of overlap in the sarcomeres of the whole muscle at various resting lengths (as shown by the roman numerals I to V). This may be seen better in the lower graph, which is the length-tension curve of a single sarcomere. This curve is made up of straight lines that correlate very well with the overlap of myofilaments; this correlation is one of the main pieces of evidence for the sliding-filament hypothesis. In the upper curve the collective effect of millions of sarcomeres obscures the simple relationship.

isometric contractions, the sarcomere length of the living muscle was measured as the length of the muscle was varied. These studies showed that the amount of force developed was indeed directly related to the amount of myofilament overlap at lengths greater than L_o. Over a small region of the length-tension curve near L_o, the force was independent of the length, but as the muscle was shortened to lengths less than L_o, the force decreased.

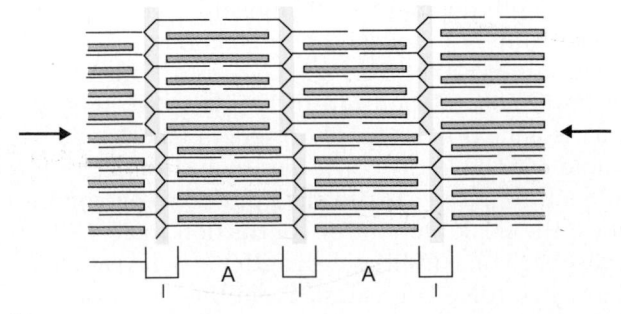

(a)

Figure 16–13

The shortening of sarcomeres in series. The upper portion represents a "snapshot" of two sarcomeres somewhere along a myofibril. There is about 50% overlap of thick and thin filaments at this length. The lower portion shows sarcomeres from a shortened muscle. Here the overlap is complete, and three sarcomeres now fit in the space of two. The shortening of all of the sarcomeres in a muscle is added together, and the tiny amount of shortening of each sarcomere results in a very large shortening of the whole muscle.

(b)

Myofilament Overlap and the Amount of Force.
These findings are interpreted in the light of the structural evidence for the existence of crossbridges and the biochemical evidence that it is the head (crossbridge) portions of the myosin molecules that interact with actin and catalyze the release of energy from ATP. When the muscle is stretched out beyond L_o, there is incomplete overlap between the thick and thin filaments, and not all of the cross-bridges on the thick filaments have the chance to attach to an actin filament. Therefore, when the muscle produces force, the amount of force is proportional to the myofilament overlap. In the region of the curve near L_o, overlap is complete, and force is at its greatest. Because of the **bare zone** at the center of the thick filaments (where there are only myosin "tails" and no crossbridges), shortening the sarcomere does not allow any more crossbridges to attach to the thin filament as it slides further in. This is why the force does not change with length in this region. As the sarcomere shortens further, myofilaments from opposite sides of the sarcomere begin to interfere with each other, and less force can be produced. At the extreme shortening, where force has virtually disappeared, the thick filaments have been pushed against the Z-lines, and there is no further opportunity for shortening or developing force.

It has also been shown that the processes that activate the myofilaments do not function as well at short lengths, and this factor also reduces the force. In experiments with whole muscles, the regions of the length-tension curve are not as clear-cut, because the millions of sarcomeres making up the whole muscle are all at somewhat different places on this curve. For this reason, and because of the effects of the connective tissue, the length-tension curve for a whole muscle is a smooth curve without a flat region at its peak. Nevertheless, it is well established that the effects of length on muscle function do reflect a fundamental property of the molecular mechanisms responsible for muscle contraction.

The Sliding-filament Model of Contraction in Other Muscle Types.   The account of muscle function just given is usually called the **sliding-filament hypothesis.** This is a very well-established body of knowledge that integrates much of what is known about skeletal muscle into a consistent framework. Because cardiac muscle is similar to skeletal muscle in structure and function, the same basic hypothesis appears to be valid there as well. Less is known about the relationship between structure and function in smooth muscle because of the different cell arrangement and lack of obvious structural regularity. However, what is known about the details of smooth muscle function and structure is at least consistent with a sliding-filament hypothesis, and as more information becomes available such a mechanism will probably be experimentally supported.

Isotonic Contractions in the Laboratory: Clues to the Work of Muscle as a Motor

Analysis of isotonic contractions can reveal other aspects of muscle contraction. Because muscles shortening at a constant force are doing work, it is during isotonic contraction that the functions of muscle as a **motor** are best emphasized.

The equipment necessary to produce isotonic contraction from an isolated muscle is somewhat more complex than that used in isometric experiments. A force transducer is still used to measure the tension produced, but now the changes in muscle length must be measured, and a means of keeping the force constant during shortening must be provided. These requirements may be met by a **lever system** (Fig. 16–14). The muscle is attached to one end of a very light (but rigid) lever, and the desired load (such as a brass weight) is hung from the other end. The muscle is anchored at its other end to the force transducer. A support is placed under the weight so that when the muscle is at rest it is not stretched out. The position of this support sets the resting length of the muscle. If the muscle length is appropriately set (see Fig. 16–14), there will be no passive tension (recall this definition from the previous section). The force that the weight provides during contraction is called the **afterload,** because the muscle bears this force only after beginning to contract. (If the muscle is stretched out by the attached weight so that some **resting tension** is present, this resting force is called the **preload.**)

The Mixed Contraction.   When a stimulus is applied to an afterloaded muscle, the resulting contraction is at first **isometric.** It remains isometric until the muscle has developed an amount of force equal to the afterload. Then the muscle begins to shorten and lift the load. Now the force will remain essentially constant and equal to the afterload as long as the load is lifted clear of its support. this is the **isotonic** portion of the contraction. As relaxation begins, the load is lowered (still isotonically) to the support. When the support prevents further lowering of the load, the muscle length no longer changes, and relaxation becomes isometric, just as the first portion of the contraction was. Only if no load at all were applied would the contraction be entirely isotonic. Under most circumstances, then, an isotonic contraction actually consists of an isometric portion followed by the isotonic shortening. Such a contraction is sometimes called a **mixed contraction.** With a large afterload, the isometric portion lasts relatively longer, because the muscle takes longer to build up enough force to lift the load.

Figure 16–14

A typical apparatus for measuring isotonic contraction. The equipment is a bit more complicated than that shown in Figure 16–11*b* because the muscle is now free to shorten if it develops enough force, and the length change must be recorded along with the force. The afterload attached to the lever provides the force during isotonic shortening; early after the stimulus is applied the muscle force is too low to lift the load, and that portion of the contraction is isometric. When force equal to the afterload is built up, the load is lifted, and the length change of the muscle is to be traced out by the pointer mounted on to the lever to which the muscle is attached.

Several important relationships can be expressed during isotonic (mixed) contractions (Fig. 16–15). First, the lighter the load, the sooner it can be lifted. (This information could actually also be obtained from the force record of an isometric contraction.) Secondly, the larger the load, the less the muscle is able to shorten. In fact, if the afterload is greater than what the muscle can lift, the contraction will be completely isometric.

The Force-Velocity Curve. Most importantly, the speed at which a muscle shortens is related to the amount of the afterload (see Fig. 16–15). This speed is measured at the very beginning of shortening, since the muscle continually slows down during most of the shortening. The lighter the afterload, the greater the speed (velocity) of shortening. As the afterload is made larger and larger (in subsequent contractions), the speed with which the muscle begins its shortening becomes less and less. The relationship is described by the **force-velocity curve.** The highest velocity of shortening, called V_{max}, occurs with zero load. While useful contractions do not occur under conditions of no load, a measurement of V_{max} pro-

vides a good indication of the maximal rate at which the energy-converting processes of a muscle can operate. This information is often used for comparisons among different types of muscle. At the other extreme of loading, the lowest velocity of shortening is *zero*, and at this point the contraction is completely isometric.

The Power of a Muscle. The force-velocity curve emphasizes the function of muscle as a biological motor. It provides a basis for studying how our muscles interact with the loads and forces encountered in our environment. We use muscles to do **work**, which is defined as *moving a force through a distance.* A concept related to this is the notion of **power output,** defined as the **rate** of doing work. Analysis of the information expressed in the force-velocity curve shows that the value obtained by the corresponding value of velocity is the power output of the muscle. A graph of the way in which the power output of muscle varies with the afterload force (Fig. 16–16) shows that at the maximum force that the muscle can attain, the power output is zero. Because the isometric muscle does not shorten, it does no work and therefore produces no

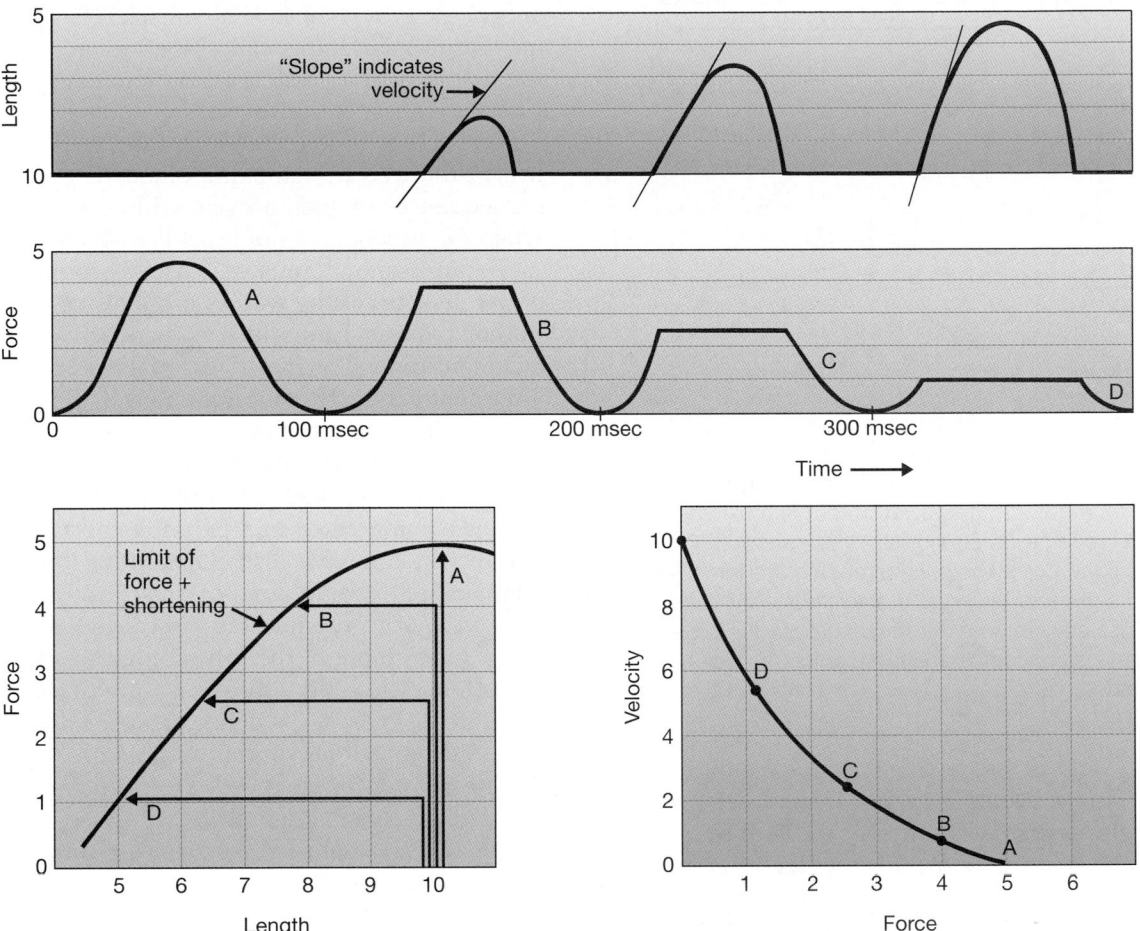

Figure 16–15

Force, velocity, length, and time in isotonic contraction. Isotonic contractions can be recorded in several ways. (The letters A, B, C, and D refer to the same contraction in each of the viewpoints.) At the top of the figure is the usual display of force and length as they change with time. The same four contractions are also shown on the *length-tension curve* axes at the lower left. Here the dimension of time is not represented, but it can be seen that the length-tension curve provides the limit to shortening at each of the different forces. The *force-velocity curve* at the lower right shows how the velocity of shortening (as taken from the sloping tangent lines in the uppermost data display) varies with different forces.

power. A related situation occurs at V_{max}, the velocity at no load. Here shortening is the most rapid, but no force is exerted. Again, this means that the power output is zero. Between these two extremes, the power output passes through a maximum when the force is at about 30% of its maximal value.

The Efficiency of Muscular Contraction. For many skeletal muscles, it is at about the point of maximal power output that the **efficiency** is the greatest. (Efficiency is defined as *the amount of work done for a certain input of energy.*) This relationship has some practical consequences for the way in which we do work (or play). An athlete (a long-distance runner, for in-

stance) who needs to maximize endurance will run for most of the race at a pace that is less than "all out." Only in the final stages of the race will the runner strive for the greatest speed. This high-speed running is done much less efficiently than the running in the earlier part of the race and cannot be sustained as long. In human-powered machines, especially those designed to be operated for long periods of time, the force-velocity and power output properties of exercising muscle need to be considered. Machines such as bicycles (and now airplanes) that get their power from human muscle are designed so that the load that they present to the muscles is near to the optimum afterload on the force axis of the force-

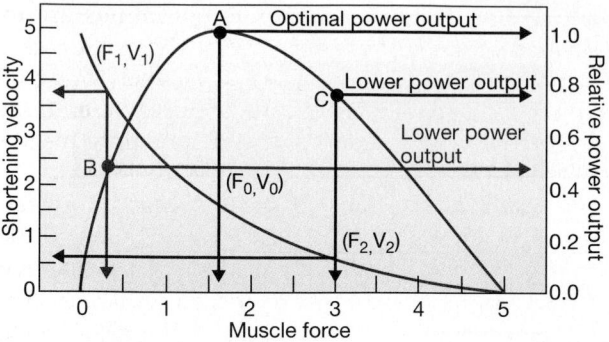

Figure 16–16

How the power output of muscle varies with the afterload. At the highest and lowest values of force, the power output is zero (see text). At an afterload of around one-third the maximum (at A), the power output is maximal. Decreasing the force (at B) increases the shortening velocity but lowers the power output; increasing the force (at C) decreases the shortening and also decreases the power output. The efficiency of the muscle is also reduced at the force-velocity combinations at which the power output is reduced.

velocity curve. Many bicycles, in fact, are provided with a number of "gear ratios" so that the rider can match the muscle force to the leg-muscle force-velocity curves as the vehicle goes up and down hills, for example. While such considerations are not always important in everyday activities, all muscular activity is subject to the limitations described by the length-tension and force-velocity curves.

Cardiac and Smooth Muscle Operate Under Specialized Mechanical Conditions

The contraction of heart muscle also can be described by length-tension and force-velocity curves, but its mechanical situation in the body is much different from that of skeletal muscle. Because it is not attached to the rigid bones of the skeleton, it does not have such a strictly limited range of lengths over which it must operate. In addition, unlike skeletal muscle, which is set by its attachments to operate at lengths near L_o, cardiac muscle generally operates at lengths somewhat less than L_o. This gives it some length "reserve" with which to produce more forceful contractions when they are needed.

Cardiac muscle, in normal body function, also relaxes differently from skeletal muscle that has undergone isometric contraction. During isotonic relaxation, a skeletal muscle is stretched back to its original length by the load it has lifted or by antagonist muscles. The relaxation is thus mostly isotonic. Ventricu-

lar muscle (a type of cardiac muscle), however, at its maximum shortening, has the afterload *removed* by the action of the aortic and pulmonary valves (see Chapter 18), and the relaxation is mostly *isometric*. The muscle is re-lengthened by the re-filling of the heart produced by returning blood. The mechanical consequences of such a cycle will become apparent when the function of the heart is discussed.

Because smooth muscle is usually found as a part of the makeup of the wall of a hollow organ or tube (rather than as a muscle stretched between two attachment points), it rarely can make a "typical" isotonic contraction. Most contractions in which smooth muscle shortens are done against a decreasing load as fluid or soft contents are propelled or contents of an organ are expelled. Smooth muscle of the blood vessels is maintained in a partially shortened condition in which force is also constant for long periods. At the other extreme, smooth muscle sphincters contract isometrically and stay contracted most of the time. Even though the physiological situation of smooth muscle is very different from that of skeletal and cardiac muscle, its actions too are described by force-velocity curves similar to those of skeletal muscle. the differences are mostly in speed (smooth muscle is much slower than skeletal muscle) and in the shape of the length-tension curve (smooth muscle typically can function over a very wide range of lengths).

The Control of Muscle

Muscle is subject to a variety of bodily influences that control its function. Some of these controlling factors provide very precise and specific control of some types of muscle. Other control mechanisms may have a more subtle but still important control over the functions of various muscle types.

Cell Membrane Activity Is Important in Controlling Muscle Function

As with nerve cells, the cells of muscle are surrounded by an electrically excitable **plasma membrane.** The characteristics of this membrane vary widely among the muscle types. In the case of most skeletal muscles, the function of the membrane can be understood in much the same terms used to describe the way in which neuronal membranes behave. We encounter additional complexity in the membrane properties of heart muscle, and a wide variety of membrane functions are found in smooth muscle. However, most of the basic features of membrane control can be illustrated with skeletal muscle as an example.

Action Potentials in the Plasma Membrane

The surface membrane of most skeletal muscle cells can be excited to produce **action potentials.** As in neurons, the action potential is caused by time- and voltage-dependent changes in the membrane permeability to sodium and potassium ions. These changes may be started by a brief electrical shock to the membrane, but stimulation usually occurs by way of the **motor neurons** that normally innervate voluntary muscle. The details of neural activation will be discussed later on. At present we will simply consider that stimulation has taken place and that an action potential is present in a muscle cell plasma membrane. This action potential spreads rapidly along the muscle fiber. As in a nerve fiber, the current from the active region causes depolarization of resting membrane areas nearby. Since muscle cells lack the myelin sheath found in many nerves the propagation of the activity is somewhat slower than that found in myelinated neurons. The relative slowness with which action potentials travel *along* the muscle cell plasma membranes is not especially important from a practical standpoint, since the muscle contraction itself is considerably slower than the speed of the action potential. A much more important problem involves the spread of activation *inward* from the surface plasma membrane to the depths of the fiber where the contractile material is located.

Action Potentials in the Transverse Tubular System

As cells go, those of skeletal muscle are quite large. The deepest parts of a muscle fiber may be as much as 50 micrometers away from the surface membrane. While this does not appear to be very far, it does pose an important problem. Suppose that muscle activation were due to the release of a chemical substance at the plasma membrane. This substance could then diffuse into the depths of the fiber. If this were actually the case, muscle contractions would have to be much slower than they are because diffusion over the distances involved would take a significant amount of time. If local electrical currents from the active plasma membrane flowed to the depths of the cell and caused its activation, they would not have sufficient strength to affect more than a few of the surface myofibrils.

In skeletal muscle these potential problems are overcome by the special conduction that takes place in the **transverse tubular system (t-system).** The membrane-lined tubules of this system, whose structure was outlined previously, penetrate the interior of the fiber at the level of each sarcomere. Recall that the insides of these t-tubules are open to the extra-cellular space, and their membrane linings are extensions of the surface membrane. When they enter each sarcomere of the cell, they effectively bring the extracellular space and the muscle plasma membrane very close to the interior of the fiber. As does the surface membrane, the t-tubule membranes conduct action potentials. When an action potential sweeps along the surface membrane, it is also conducted into the depths of the fiber along the t-system and arrives at the innermost portions of the cell with very little delay. This mechanism overcomes the slowness of diffusion and the weakness of the membrane currents, and this step in the activation of skeletal muscle becomes very rapid.

Calcium Ions in the Sarcoplasmic Reticulum

The t-tubule system does not actually open into the interior cytoplasm of the cell, since the tubules are completely lined with cellular membrane. Instead, they come into close contact with the membranes of the **sarcoplasmic reticulum (SR),** a closed membrane system. This means that it is a structure that maintains a separate region of the cell interior. This enclosed region is used to store and release the actual chemical substance that controls the contractile activity of the muscle proteins. This substance is **ionized calcium** (calcium ions), and in the resting muscle it is highly concentrated within the SR. It is contained within the saclike areas called **terminal** (or **lateral**) **cisternae.** The longitudinal elements of the SR communicate with the terminal cisternae at each end of the sarcomere and form a highly branched network surrounding the myofilament bundles. The interior of the longitudinal elements is continuous with the terminal cisternae.

As this description suggests, each sarcomere has its own separate portion of sarcoplasmic reticulum. The junction between a t-tubule and the terminal cisternae of two adjacent sarcomeres (the **triad**) is the place where the electrical signal from the outside of the fiber actually is communicated with the interior of the fiber. Just how the signal passes from the t-tubule to the SR at this point is not well understood, although it is assumed that the local movement of electrical charges plays a role. It is well known, however, that when this signal has been passed, the SR is caused to release a large portion of its stored calcium ions into the cytoplasm very near the myofilaments. These calcium ions free the myofilaments from their resting state (by a process that will be discussed shortly), and contraction takes place. Relaxation is associated with the movement of calcium ions back into the longitudinal elements of the SR. This ionic movement takes place up a steep concentration gradient via an ATP-dependent calcium ion pump present in

Terminal cisternae   T-Tubule

Plasma membrane

Longitudinal sarcoplasmic reticulum

Terminal cisterna

Myofilaments

(a)   (b)   (c)   (d)

Figure 16–17

How calcium moves during a contraction. (*a*) With the muscle at rest, the calcium ions are stored in the terminal cisternae of the sarcoplasmic reticulum. (*b*) As activation begins, calcium is released from the sarcoplasmic reticulum into the region of the myofilaments. (*c*) Activation ends as calcium is taken back up into the longitudinal elements of the sarcoplasmic reticulum. (*d*) Immediately after the contraction, the calcium moves back into the terminal cisternae.

this region of the SR membrane. While this mechanism (diagrammed in Fig. 16–17) may seem complicated, it quite effectively overcomes the diffusion and conduction obstacles posed by the large size of skeletal muscle fibers. It is capable, in some muscles, of allowing activation to be repeated as many as 60 to 100 times per second. Muscles without such a mechanism would be limited to very slow rates of contraction.

Calcium Ions Are a Vital Link in the Control of Muscle

The regulation of muscle contraction involves controlling the interaction of the actin and myosin myofilaments. In all types of muscle, this control is carried out through the action of calcium ions, though the actual process differs greatly between smooth muscle on one hand and cardiac and skeletal muscle on the other.

The Molecular Mechanisms of Muscle Contraction

When an action potential, through the sequence of events described previously, causes the SR of skeletal muscle to release a quantity of stored calcium ions, these ions diffuse to the region of the myofilaments,

where the resting calcium concentration is normally very low (see Figure 16–18). Here they bind to the troponin-C (TnC) subunit of the regulatory protein complex, which is attached to the tropomyosin molecule on the thin (actin) myofilament. When calcium ions bind to troponin-C, they cause a conformational change in the other parts of the troponin molecule, and this in turn causes the tropomyosin molecule to change its position along the "groove" of the actin filament. At rest, the tropomyosin interferes with the interaction of actin and myosin, but with the tropomyosin position slightly shifted, the actin filament becomes able to interact with the myosin crossbridge. Because the tropomyosin molecule covers seven adjacent actin monomers, it can exert a large degree of control over the actin-myosin interaction.

The contraction resulting from this interaction of actin and myosin in a series of steps is diagrammed in Figure 16–19. In resting skeletal muscle (with calcium ions not yet released), an ATP molecule is bound to each myosin molecule head that can potentially form a crossbridge with the actin filament. The ATP molecule has been "split," but its energy cannot be released until the actin is allowed to interact with the myosin. When the presence of calcium ions permits this interaction, the myosin head attaches to the actin filament at an angle of approximately 90° and the

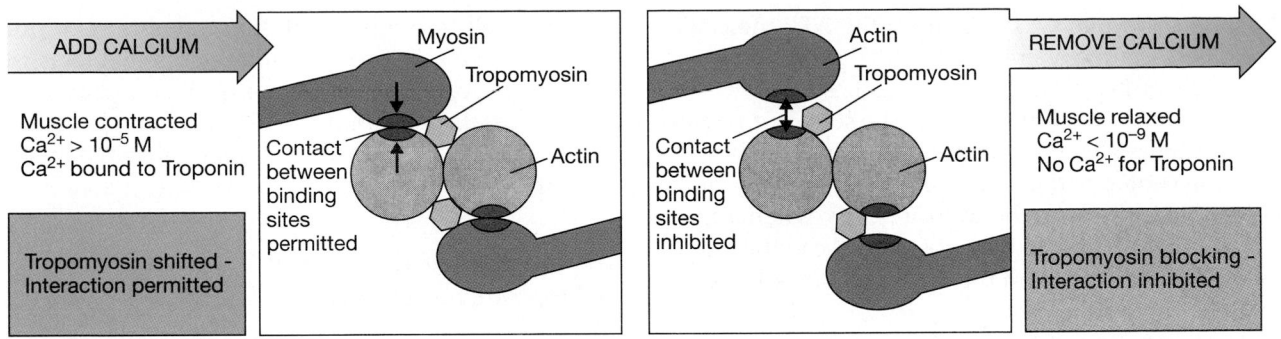

Figure 16–18

How calcium and the regulatory proteins control the interaction between actin and myosin. When internal calcium concentration is very low (the relaxed state, lower left), the position of tropomyosin along the thin filament (seen in cross-section) prevents inter-action between the myosin heads and the active sites on the actin molecules. When cal-cium is released from the sarcoplasmic reticulum, the ions bind to the Tn-C subunit of troponin. This causes a shift in the position of tropomyosin, which allows the interaction between the myosin heads and the actin monomers.

Membrane depolarization by action potential causes release of calcium ions from the sarcoplasmic reticulum.

Contraction begins

If ATP is taken away, crossbridges will remain attached, and rigor mortis will occur.

ATP

① ATP binds and crossbridge attaches

④ Crossbridge detaches

As long as Ca²⁺ and ATP are present, cycle continues. Z line is pulled closer and closer to thick filaments.

Membrane repolarization allows uptake of calcium ions by the sarcoplasmic reticulum.

If Ca²⁺ is taken away, the cycle stops here, and muscle relaxes.

Z-line — ATP

Z-line — ADP + Pᵢ

② Crossbridge attached, not rotated

Pᵢ

Crossbridge attached; rotates Z-line moves closer to thick filament.

③ Z-line — ADP

ADP

Figure 16–19

The crossbridge cycle of skeletal muscle. The series of events goes on continuously as long as calcium and ATP are present.

phosphate ion that was split from the ATP molecule is released. As the energy of the ATP molecule is released, the angle of attachment becomes closer to 45°. This change in attachment angle (the so-called power stroke) causes the actin filaments to be pulled in toward the center of the sarcomere, and the ADP molecule is released from the myosin. Only when a new ATP molecule binds to the myosin head can the crossbridge detach itself from the actin and become ready for a new cycle.

As long as enough ATP and calcium ions are present, the cyclic process of attachment, angle change, ATP binding and hydrolysis, and detachment can take place. Repeated crossbridge cycles result in the actin filaments being pulled further into the myosin filament array, and the whole muscle becomes shorter. The force that the muscle is working against is important in determining how fast the crossbridge cycle will run. If the ends of the muscle are held stationary (as in isometric conditions), the crossbridge cycles will still continue to some extent, since some internal movement is allowed by the elastic nature of the muscle. When stimulation of the muscle is stopped, no additional calcium ions are released from the SR, and calcium ions that diffuse free from the troponin or that are free in the cytoplasm are taken back up by the SR. Under these conditions, tropomyosin resumes its blocking function. Crossbridges will not reattach, and relaxation takes place. If the supply of ATP becomes depleted, attached crossbridges will not be able to detach, and the muscle becomes stiff and unable to relax. An extreme example of this situation is the condition known as *rigor mortis*, which occurs after death as the supplies of ATP in the muscle become exhausted

Actin-Linked Regulation of Muscle Contraction

Muscle regulation that is accomplished through some action on the thin filaments is called *actin-linked regulation*. This type of regulation is also found in heart muscle, where the steps involved are quite similar to those in skeletal muscle. In cardiac muscle, however, the amount of calcium ions released in response to an action potential is usually not sufficient to activate all of the possible actin molecules, and a maximal contraction is not obtained. This provides cardiac muscle with a "reserve capacity" for contraction, because there are several ways in which extra calcium ions can be released to meet additional requirements. In the specialized action potentials of cardiac muscle, one of the ions that flows into the cell is calcium. Because of the small size of cardiac muscle cells, it is possible for a significant amount of calcium ions to enter the cell through the plasma membrane during the course of an action potential. Factors that increase the duration of action potentials, or the number of action potentials per minute, or the amount of calcium ions entering with each action potential, can increase the force of contraction.

Myosin-Linked Regulation of Muscle Contraction

Regulation of contraction that takes place via thick filaments is called *myosin-linked regulation*. Smooth muscle is controlled in this way. Its thin filaments lack the troponin protein complex, although they do contain tropomyosin on the thin filaments. In smooth muscle, one of the two protein light-chains on the head of each myosin molecule has a critical regulatory function. Interaction between myosin and actin can take place only when a phosphate group (from ATP) is bound to this light-chain. Since the thin filaments of smooth muscle lack the troponin regulatory complex, they are always ready for interaction with myosin, and it is the myosin, not the actin, that must be activated.

The activation and regulation processes of smooth muscle can be divided into three groups of reactions that take place more or less simultaneously. These are shown in Figure 16–20. The first group (region 1) is the step where control of contraction begins. Calcium ions are released into the cytoplasm (upper left), either from the sarcoplasmic reticulum or from outside the cell via the cell membrane. These ions bind to a protein called **calmodulin,** which is closely associated with an enzyme called myosin light-chain kinase (MLCK). This binding activates the MLCK and allows it to catalyze one of the reactions shown in region 2. This reaction is the *phosphorylation* of the regulatory myosin light chains; when the terminal phosphate group of an ATP molecule is transferred to a regulatory light-chain molecule, the catalytic function of myosin becomes activated and it is able to interact with actin. A **phosphatase** enzyme simultaneously acts to remove other similar phosphate groups, but during activation the phosphorylation reaction occurs at such a high rate that the dephosphorylation reaction is relatively unimportant.

The interaction between actin and activated myosin is shown in region 3. This is a crossbridge cycle similar to that of skeletal muscle, involving the cyclic attachment, rotation, and detachment of crossbridges to produce force and movement. This cycle will also continue as long as there is a supply of ATP from the cellular energy stores and as long as myosin remains in its activated state. Notice that the ATP molecule that gives up its phosphate group to the light-

Figure 16–20

The major steps in the regulation of smooth muscle contraction. A contraction is initiated by the release of calcium into the cytoplasm, and the myosin molecules are activated by the steps in regions 1 and 2 so that the crossbridge cycle (shown in region 3) begins to operate. Relaxation begins when the calcium is taken back up and the phosphatase enzyme (in region 3) is no longer working against the MLCK and can dephosphorylate (and thus deactivate) the myosin molecules.

chain to activate the myosin molecule is not the same ATP whose energy is consumed in the contraction cycle.

Relaxation takes place when calcium ions are taken up from the cytoplasm (upper right) by the action of calcium pumps in the sarcoplasmic reticulum or plasma membrane. The MLCK becomes inactive and can no longer catalyze the phosphorylation reaction. The phosphatase reaction is then able to dephosphorylate the myosin molecules, and so it can no longer interact with actin molecules. The cross-bridge cycle comes to a halt and the muscle relaxes. In addition to these basic processes, smooth muscle also contains other regulatory processes responsible for its ability to maintain contraction for a long period of time without consuming large amounts of energy. These processes appear to involve modifications in the properties of crossbridges; in some smooth muscles crossbridges become "latch bridges" that detach very slowly and therefore spend most of their time in the attached, force-producing state. Thus, while the details of regulation of smooth muscle are

R E S E A R C H   F O C U S

Measuring Muscle Contraction at a Molecular Level

Although the overall process of muscle contraction is very complex, the fundamental events that change chemical energy into mechanical motion are relatively simple, involving the interaction of the heads of the myosin molecules with actin filaments. In recent years, techniques have been worked out to allow actual mechanical measurements to be made on single filaments.

One of these methods (called an *in vitro* motility assay) uses myosin molecules and actin filaments that have been separately isolated in intact form from muscle tissue. In order to make them visible, the actin filaments are linked with modified antibody molecules (rhodamine-phalloidin) that give off light ("fluoresce") when they are illuminated with ultraviolet light under a powerful microscope. The single myosin molecules, separated out from the thick filaments, are chemically attached by their tail region to a specially coated glass surface so that the head portion is free to project into a bathing solution containing ATP as an energy source. When actin filaments are added to the bathing solution (which also contains necessary nonmuscle chemicals), the filaments begin moving over the myosin-coated surface, propelled by the actin-myosin crossbridges. A highly sensitive video system attached to the microscope observes and records the movement. The sliding speed is about 5 microns per second with myosin from fast skeletal muscle, and it is about 25 times slower if myosin from smooth muscle is used. The speed of sliding depends on the properties of the myosin and is correlated with the speed of the intact muscle that it is taken from. The actin and its source are much less important, although it is the orientation of the monomers of the actin filament that determine the direction of movement. The actin filaments can be taken from nonmuscle sources with little effect on the rate of sliding.

The effects of the regulatory proteins (troponin and tropomyosin) can be studied by including them on the thin filaments, and researchers have found that their presence allows calcium ions to control the interaction between actin and myosin. When smooth muscle myosin is used, it must be phosphorylated (see text) in order for movement to take place. Mixtures of smooth and skeletal muscle myosin cause movement at intermediate speeds. Nonphosphorylated smooth muscle myosin molecules retard the movement caused by phosphorylated myosin molecules in the mixture coating the glass surface.

A further refinement of this technique allows the force of a single crossbridge to be measured. In this case, a single actin filament is chemically attached to a microscopic needle, which is then brought near to the myosin-coated surface. The force that a myosin molecule generates when it interacts with the actin filament is sufficient to cause a microscopic bending of the needle. This bending can be translated into a measurement of the mechanical force, which is about 0.5 picoNewtons (one trillionth of a Newton). When the value for a single crossbridge is scaled upward to the dimensions of an intact muscle, there is rather good agreement.

Use of the *in vitro* assay has answered many questions arising from experiments on intact muscle and has provided unique insight into biochemical processes that would be very difficult to interpret using strictly chemical means. Several laboratories are conducting experiments that are probing individual steps in the crossbridge cycle in an effort to work out the details of actin-myosin interaction.

■ **While these molecular methods can provide important insights into muscle function, they have some important potential drawbacks. Try to make as complete as possible a list of the positive and negative aspects of this approach to the function of muscle.**

much different from the regulatory processes in skeletal muscle, the control over contraction and relaxation is still exercised by calcium ions.

The Pattern of Stimulation Regulates Muscle Function

One of the important results of the way in which muscle is controlled is seen in the types of contractions produced in response to various types of stimuli. Many muscles (especially skeletal and cardiac) can respond to a single stimulus with a single action potential. This usually results in a single brief contraction. Such a response is called a **twitch.** The rapid contraction and relaxation of skeletal muscle during a twitch are the result of two processes that take place at the same time. As we have seen, the result of the action potential is the sudden release of stored calcium ions from the SR. While these calcium ions are diffusing to the myofilaments and activating the contraction, the processes that take the calcium ions back up are also operating. Even when the amount of calcium ions released is potentially enough for a maximal contraction (as is usually the case in skeletal muscle), their rapid uptake limits the amount of force that can develop. In most cases, the muscle does not have a chance to develop its full amount of force before relaxation processes have set in. Crossbridge cycling, necessary for the buildup of force, takes time, and this time is not available.

Summation of Action Potentials: Producing a Sustained Contraction in Skeletal Muscle

Rapid relaxation and low force mean that twitch contractions are not particularly useful for everyday tasks. However, this limitation is readily overcome by a repetitive type of contraction. The action potential of twitch-type skeletal muscles is very brief compared with the duration of the twitch contraction. Even though the action potential has the usual absolute and relative refractory periods typical of excitable membranes in general, these also are brief compared

to the twitch time. This means that the muscle can be restimulated before its relaxation is complete, and the force produced by the second stimulus can add to the force left from the first.

This process is called temporal **summation,** and it results in a contraction, called a **tetanus,** that lasts longer and produces more force than a single stimulus could produce. Inside the muscle, the number of calcium ions available to activate the myofilaments has been increased by the second stimulus. Stimuli can be repeated at regular intervals in order to keep the internal calcium ions continuously available to the myofilaments. This continuous supply in turn keeps the muscle from relaxing completely, and the result is a sustained contraction that produces significantly more force than that developed in a twitch (Fig. 16–21). If the stimuli are spaced relatively far apart, the force will rise and fall somewhat between stimuli, but if the stimuli are close enough together, the developed force will be steady. This is called a **complete,** or **fused, tetanus.** Just how many times a

Figure 16–21

The relationship between twitches and a tetanus. Upper portion: A single motor nerve action potential produces a single twitch contraction (first event). Applying a stimulus during relaxation (second event) produces a contraction whose force adds to that of the first. Two stimuli delivered close together (third event) also produce an addition of force (this is made possible because of the short refractory period of the muscle cell membrane). Lower portion: When action potentials are made to occur closer and closer together, less and less time is available for relaxation, and the force produced by the individual twitches adds up. When the rate of stimulation is high enough, a smooth tetanus can be produced. A gradual build-up of twitches is not necessary for a tetanic contraction; a tetanus will be produced immediately if the stimulus rate is high at the outset.

second a muscle must be stimulated in order to produce a fused tetanus depends on the relative rates of contraction and relaxation of the muscle. The lowest frequency that will produce a fused tetanus is called the **tetanic fusion frequency.** It is around 20 to 60 times per second for most skeletal muscles.

Useful contractions—those that get us through daily life—are usually a mixture of twitches and partly fused tetanic contractions. These are distributed among the many individual fibers that make up a complete muscle, so that the contraction is "shared." The question of which fiber does what to produce a partial contraction of a muscle is a matter for the CNS to decide.

Distancing of Action Potentials: Producing Discrete Twitches in Cardiac Muscle

Cardiac muscle is a twitch-only type of muscle. A sustained and maximal tetanic contraction would be of little use in the task of pumping blood, and heart muscle has some special properties that guard against contractions of this sort. The action potentials of heart muscle are quite long; they typically last almost as long as the contraction itself. This means that the membrane is **refractory** during most of the twitch, so that restimulation is not possible while the muscle has any significant tension. At higher rates of stimulation (high heart rates), the duration of both the action potential and the contraction are reduced. This allows for closely spaced twitches, with time for complete relaxation between them.

Even though cardiac muscle cannot develop a partial tetanus, the properties of its contractions can be adjusted by other means. In contrast to the situation in skeletal muscle, the amount of calcium released in response to a single action potential is generally not sufficient to provide for full activation. A number of factors can change the amount of calcium made available for each contraction. Making the heart contract more times per minute (up to a point, of course) will result in the cellular accumulation of calcium and an increase in the strength of the contraction. Because calcium is one of the ions that enter the cell during each action potential, changes in this calcium current can cause changes in the contraction. This is known as the *force-frequency relationship.* Some drugs, those of the **digitalis** family, for example, increase the overall internal supply of calcium, and more is released from the SR with each beat. Drugs called **calcium blockers** can reduce the strength of contractions, and the chemical **epinephrine** (both a drug and a hormone) can increase the strength of contraction by permitting the accumulation of internal calcium. All of the examples given here are agents that affect the *contractility* of the heart muscle. These influences are called **inotropic agents** (*ino* = sinew; *tropic* = having an effect). They permit changes in the properties of the contraction that do not depend on the possibility of a tetanic activation.

Along with the changes in contractility, changes in muscle length (as we have seen previously) can have an important effect on the strength of the contraction of cardiac muscle. (Recall the relationships expressed in the length-tension curve.) When the heart is functioning properly in the body, these two types of influences work together to produce a very effective degree of control of its pumping function.

Phasic and Tonic Contractions: A Variety of Responses in Smooth Muscle

Smooth muscle can contract in a variety of different ways. Some smooth muscles can give twitchlike contractions in response to single stimuli or brief bursts of stimuli. The contractions are called **phasic** and are often found in smooth muscles that propel or move materials. Other smooth muscles give long and sustained contractions in response to single or continued stimulation by nerves, drugs, or hormones. These contractions are called **tonic** and are likely to be found in smooth muscle organs, such as sphincters or bladders, that must remain contracted for long periods of time. Many smooth muscles are almost never in a state of complete relaxation, and the steady force that they generate is called **tonus.** The level of tonus is often set by the concentration of specific hormone such as norepinephrine.

In many cases, the contraction of smooth muscle is associated with electrical activity of the cell membranes. Most phasic contractions are associated with **action potentials.** In some tissues, such as intestinal muscle, with rhythmic contraction patterns, there is a kind of electrical activity called the **slow wave.** The slow waves, which are many seconds in duration, periodically depolarize the membrane to the threshold for action potential generation; this mechanism acts to produce periodic contractions. Other smooth muscles show action potential activity that may be caused by a pacemaker cell and conducted to adjacent cells by means of the gap junctions previously mentioned.

The Central Nervous System Is Important in Regulating Muscle Function

Almost all of the muscle in the body is under the direct or indirect control of the CNS, although smooth muscle and cardiac muscle do have a degree of autonomous function that is suited to their specialized roles. Because neural control is so highly developed in the case of skeletal muscle, most of the general process can be understood by a study of how the action of a motor nerve regulates the function of a skeletal muscle.

Nervous Control of Skeletal Muscle

Skeletal muscle depends on the somatic division of the CNS for the control of its contraction. As described in Chapter 10, under normal conditions skeletal muscle will not contract unless it is stimulated by a **motor neuron,** which has its origin in the CNS. Connection between the motor nerve fibers, which are large myelinated axons, and the muscle fibers takes place at a structure called the **myoneural junction.** Because in some cases this structure spreads out somewhat on the muscle fiber, it is also called a **motor end-plate.**

The Myoneural Junction and the Motor Unit.
The myoneural junction is a special case of a **synapse.** As in the vast majority of synapses in the CNS, the transmission of the impulse from the nerve to the muscle is by chemical means. The particular transmitter substance at the myoneural junction is **acetylcholine (ACh).** Unlike some CNS synapses, only **excitation** of the postsynaptic membrane is possible at the myoneural junction. Because many of the myoneural junctions in a muscle lie at its surface, they have been extensively studied, and the process of transmission is well understood. In fact, much of what we know about synapses in general has been learned by the study of this highly specialized synapse.

The structure of the myoneural junction provides keys to an understanding of its function. As a motor axon from the motor nerve bundle nears the muscle, it branches to innervate a number of individual muscle fibers. The assemblage of a motor axon, together with all of the muscle fibers that it innervates, is called a **motor unit.** Since all of the muscle fibers served by a single branched axon are activated together when an impulse passes down the axon, the fineness of control of a muscle depends on just how many muscle fibers a single axon controls. Muscles capable of fine and delicate movements, such as the **extraocular** muscles that move the eyeball, have a very few (or even only one) muscle fibers served by a single axon, while muscles used for rapid and relatively coarse movements, such as those of the posterior thigh, have many muscle fibers controlled by a single axon. Each muscle fiber normally receives innervation from only a single axon terminal, although some multiply innervated muscle fibers do exist.

Synaptic Transmission at the Myoneural Junction.
As the motor axon branches, the myelin sheath becomes reduced to the cytoplasm and cell membranes of a single **Schwann cell** (the type of cell that forms the myelin sheaths around motor axons), and this covers the axon as it makes its final approach to the muscle (Fig. 16–22). The terminal region of the axon lies in a "groove" on the surface of the muscle. This arrangement increases the amount of nerve and muscle cell portions that makeup the junctional region. The motor axon terminal contains small vesicles that carry the transmitter substance. Also present in this presynaptic region is a large number of mitochondria. These provide the metabolic energy for the synthesis of the transmitter substance and the ionic pumping mechanisms necessary for the recovery processes that follow the passage of an impulse. The muscle plasma membrane making up the postsynaptic side of the junction is folded into a series of deep grooves called **junctional folds,** which are lined with the receptors for the ACh transmitter substance. The receptors are complex protein molecules with dual function. Each receptor has two binding sites for ACh and also functions as an ion channel. Normally, the receptor is not permeable to ions but when the transmitter substance has attached to the binding sites, sodium and potassium ions can pass through the channel down their electrochemical gradients. When the ACh is no longer bound to the receptor, the channel again becomes impermeable to ions. The postsynaptic membrane also contains an enzyme called **acetylcholinesterase.** This enzyme is essential for breaking down the transmitter to an inactive form once it has done its job of causing the channel to open.

Between the time when a motor nerve is stimulated and when the muscle contracts, a long series of events takes place (Fig. 16–23). The action potential travels down the motor axon by the mechanisms previously described. When the action potential spreads into the terminal regions of the axon branches, membrane calcium channels open in response to the depolarization, allowing the entry of calcium ions into the neuron. This increase in the intracellular calcium ions causes the synaptic vesicles containing the acetylcholine to move to the region of the presynaptic membrane. Here, the vesicle membranes fuse with the plasma membrane. The fusion releases the contents of the vesicle into the synaptic cleft. These acetylcholine molecules diffuse across the cleft to the postsynaptic region, where they attach to the specific binding sites on the receptor molecules. When this binding takes place, the membrane channels associated with the receptors become permeable simultaneously to both sodium and potassium ions. This causes the membrane potential to move away from the resting potential (about −80 mV) and toward a new potential of about −15 mV, which is called the **end-plate potential.** (This is the potential expected from a membrane permeable to both sodium and potassium at the same time.) The post-synaptic mem-

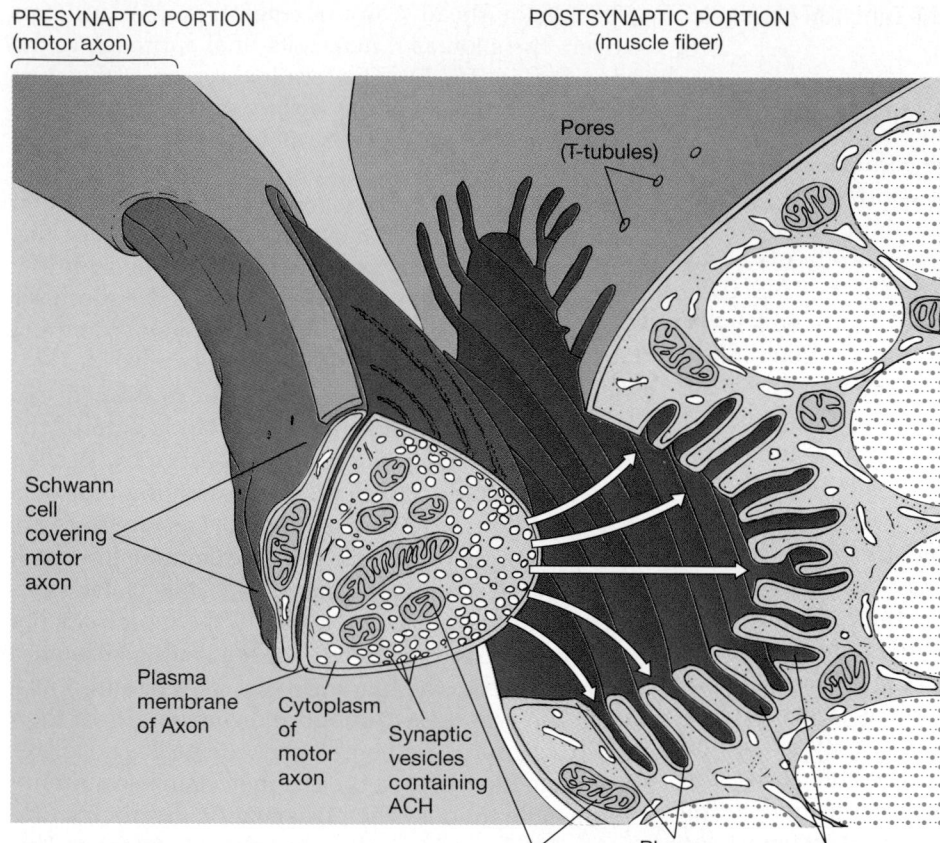

PRESYNAPTIC PORTION
(motor axon)

POSTSYNAPTIC PORTION
(muscle fiber)

Pores
(T-tubules)

Schwann
cell
covering
motor
axon

Plasma
membrane
of Axon

Cytoplasm
of
motor
axon

Synaptic
vesicles
containing
ACH

Mitochondrion

Plasma
membrane
of muscle
fiber

Junctional
folds
containing
receptor
molecules

(a)

(b)

Figure 16–22

(*a*) An expanded view of the myoneural junction. This drawing shows the essential features of the myoneural junction of skeletal muscle. (Modified from Krstic, R.V., *Ultrastructur der Saugertierzelle*. Berlin: Springer-Verlag, 1976.) (*b*) SEM of a region of (hamster) striated muscle fiber showing the branching motor axon and two myoneural junctions. *(© Don W. Fawcett/Science Source.)*

brane is unlike the rest of the plasma membrane in one important aspect. Since its channels are chemically activated and both permeability changes occur together, it cannot produce an action potential. What the end-plate potential does, however, is cause a small local current (the **end-plate current**) to flow across the membrane at this region. This inward current flows into the fiber and begins to spread down its length. The corresponding outward current is forced to flow across adjacent areas of the general plasma membrane, and it depolarizes these regions so that they reach their threshold for action potential generation. The action potentials produced by the end-plate potential then propagate down the length of the fiber and activate the contraction in ways that have already been described. After the passage of the action potential, the muscle membrane returns to its resting condition.

If the action of the transmitter substance were not terminated, the end-plate depolarization would continue. This would prevent the muscle plasma membrane from being stimulated again, and subsequent

(a)

(b)

Figure 16–23

How the end-plate potential produces a muscle action potential. The region of membrane (the postsynaptic membrane) immediately opposite the axon terminal does not produce an action potential. Instead, its depolarization causes local currents to flow through the interior of the muscle cell and across the adjacent muscle cell membrane. This flow of current depolarizes the membrane to a threshold value and produces an action potential. This action potential then travels the length of the muscle cell membrane and enters via the t-tubule system to activate the contraction.

synaptic transmission would fail. Hydrolysis of the ACh by the postsynaptic acetylcholinesterase terminates ACh action and produces its components, acetate and choline. The choline is actively taken back up into the presynaptic terminal, where it is made back into ACh. The acetate fragment, which is a chemical common throughout the body, is not saved by any special mechanism. These events are summarized in Table 16–1.

Factors that Block Myoneural Transmission. All of these events take place in only a few milliseconds of time, and the muscle can be restimulated many times per second. The presence of so many steps critical to the process, however, indicates that many things could go wrong. **Myoneural blockade** is the term given to this transmission failure. Some blockade can take place presynaptically. The presence of too many magnesium ions or too few calcium ions near the axon terminal will prevent release of the transmitter substance. Some drugs, called **hemicholiniums**, interfere with uptake of choline, and the transmitter substance becomes depleted. **Botulinum toxin,** an extremely deadly poison produced by a bacterium (*Clostridum botulinum*) present in spoiled food, also causes a failure of transmitter release, and poisoning with this substance results in paralysis and death.

Postsynaptic blockade is also possible. The substance **curare** (used on blowgun darts by aboriginal hunters in South America) can bind to the acetyl-

choline receptors, in preference to acetylcholine. However, its binding does not produce a permeability change or depolarization. It is also resistant to breakdown by the acetylcholinesterase. When it is presenting sufficient quantity, therefore, myoneural transmission cannot take place.

A close chemical relative of acetylcholine is the drug **succinylcholine,** which is used during surgical procedures to produce muscle relaxation. This drug is broken down only very slowly by acetylcholinesterase. Another possible site for blockade is the acetylcholinesterase enzyme. This enzyme is inhibited by drugs such as **physostigmine** and **eserine,** which are similar to the active (organophosphate) compounds in chemical warfare agents, the so-called nerve gases. When the action of cholinesterase is inhibited, the end-plate membranes remain depolarized, and the muscle action potential mechanism cannot be reset.

In all of the situations mentioned, the myoneural blockade would result in death from respiratory failure. The diaphragm, a skeletal muscle, would fail to respond to nerve impulses, and along with other skeletal muscles, it would be paralyzed. Carefully controlled doses of blockers such as succinylcholine can be used as surgical muscle relaxants as long as respiration is maintained artificially.

Other factors also affect the myoneural junction. As is true of other synapses, it is subject to **fatigue** due to depletion of the transmitter substance. How-

Table 16–1
Sequence of Events During Neuromuscular Transmission

Event	Possible Blocking Agent	Event	Possible Blocking Agent
Action potential arrives at axon terminal.	Nerve-blocking agents such as procaine or tetrodotoxin	End-plate region (postsynaptic membrane) becomes depolarized; end-plate potential is set up.	
Extracellular calcium ions enter axon terminal.	Low extracellular calcium or high magnesium concentration	Local end-plate currents cause adjacent muscle membrane to depolarize.	
Transmitter vesicles migrate to axon membrane and fuse with it.		Muscle action potential is triggered.	
Acetylcholine is released from the transmitter vesicles into synaptic cleft.	Botulinum toxin	Acetylcholine diffuses away from receptors and is hydrolyzed by cholinesterase enzyme.	Eserine or other cholinesterase inhibitors
Acetylcholine molecules diffuse across synaptic cleft.		Choline is taken up by presynaptic terminal (now repolarized).	Hemicholinum drugs
Acetylcholine molecules bind to receptor proteins in postsynaptic membrane on muscle.	Curare, succinylcholine	Acetylcholine resynthesized and vessicles re-filled.	
Ion channels in postsynaptic membrane open and allow the flow of sodium and potassium ions.			

ever, the transmission normally has a large **safety factor.** That is, more transmitter is released than is necessary for transmission, and the end-plate potential is normally larger than necessary to stimulate the muscle membrane. In addition, the process of contraction in the muscle is subject to fatigue far sooner than is the process of myoneural transmission. An abnormal situation, though, is present in the disease known as **myasthenia gravis.** In this condition, it appears that an immune reaction against the acetylcholine receptors reduces their numbers, and the end-plate potentials may not be large enough to stimulate the muscle fibers. Weakness or paralysis results from this condition, but effective therapy, at least in the short term, is possible. Careful administration of cholinesterase inhibitors allows the acetylcholine to

remain active in the synaptic cleft for a longer time, and what receptors there are may be bound to and activated several times in succession. This allows sufficient end-plate current to flow to permit near-normal muscle function.

Nervous Control of Smooth and Cardiac Muscle

The two general types of smooth muscle, multiunit and unitary, differ considerably in the ways they are controlled by the nervous system. Multiunit smooth muscles are generally very densely innervated by autonomic nerve fibers and have rudimentary motor end-plates on each cell in the tissue. Myoneural transmission in these muscles is much like that found in

CLINICAL FOCUS

Myasthenia Gravis

Myasthenia gravis (meaning "muscle weakness associated with pregnancy") is an illness associated with muscular weakness and extreme fatigue following moderate exercise. Despite its name, it is not closely linked to childbearing, although it often makes its appearance in young women in their early childbearing years. In men its onset is typically after middle age (in their fifties or later). The condition stems from a failure of neuromuscular transmission rather than from any defects in the muscle contraction mechanism itself. While no cure has been found for the disease, several kinds of therapy can give patients some symptomatic relief.

Scientists now know that myasthenia gravis is one of the so-called autoimmune diseases. For unknown reasons, some people produce antibodies that react with the acetylcholinesterase receptors in the postsynaptic (motor end-plate) plasma membrane. This immune reaction destroys many of the receptors, and adequate end-plate potentials can no longer be produced. This means that a muscle action potential does not result from a nerve impulse, and so the muscle fails to contract. For a person afflicted with the condition, the result is that extreme effort may be required for even ordinary movements. In extreme cases, even breathing becomes very difficult.

Specific diagnosis of the condition involves study of the neuromuscular junction and immune system. Careful observation of the relationships between the nerve stimulation and muscle response can reveal delays in the transmission process even before it fails completely, and the presence of antibodies specific to the acetylcholine receptors may be detected in the blood of affected persons.

Several forms of treatment are possible. Drugs that inhibit the action of acetylcholinesterase allow the remaining postsynaptic receptors to function for a longer period of time, and adequate end-plate potentials can be produced. Unfortunately, the effect is only temporary, and the drug must be readministered repeatedly. This approach can lead to harmful side effects and, in any case, treats only a symptom of the underlying cause, with no effect on the general course of the disease.

Like other autoimmune diseases (such as rheumatoid arthritis), myasthenia gravis can undergo spontaneous remissions in its early stages, although the symptoms usually return. Removal of the thymus gland (which is known to play an important role in the immune system) can result in substantial improvement of the condition, especially if it is done early in the course of the disease.

Other therapies, also directed at the immune system, may be applied. Administration of gamma globulin may produce temporary improvement. Steroid drugs can be used to reduce the severity of the problem for a time, but they cause harmful side effects in many cases. Immunosuppressant drugs reduce the function of the entire immune system and can produce an improvement in the symptoms, although the patient will then be at risk from other infections and diseases. The technique of plasmapheresis, in which the proteins (including antibodies) are removed from the blood by a special transfusion method, can produce temporary improvement. But the procedure, though relatively safe, is time-consuming and expensive.

While an understanding of the effects of the disease can be obtained through study of the neuromuscular junction, its actual cause lies elsewhere. With immunological methods, myasthenia gravis can be produced in experimental animals, especially in rodents and monkeys. Since the animal condition very much resembles the human disease, studies involving animal immune responses offer hope for development of a more satisfactory treatment.

■ **Newborn infants of mothers with myasthenia gravis often show symptoms of the disease when they are born, but the symptoms disappear in a short time. Can you suggest why this is the case?**

---

skeletal muscles. The structures involving this muscle are quite small. They include some very small blood vessels, the **arterioles,** and the muscles that make our hair stand on end, the **piloerector muscles.** Because of the small size and the plentiful innervation of the tissues, the fact that the cells do not communicate with one another is of little practical consequence.

Most tissues composed of **unitary** smooth muscle are also innervated by postganglionic fibers of the autonomic nervous system. There is nothing approach-

ing a one-to-one connection between the nerve supply and the muscle cells. Instead of a nerve-muscle connection such as a motor end-plate, the innervating fibers course among the smooth muscle cells. Periodically the nerves have swellings (called **varicosities**) that hold vesicles containing various transmitter substances (e.g., acetylcholine or norepinephrine). Action potentials traveling down the nerves cause the release of the transmitter substance, which can then diffuse across the intercellular spaces to the vicinity of the muscle cells. Because there are so many cells in the tissue, only a few of them are directly affected by the transmitter substance. The cells that depolarize and produce actin potentials, however, can affect neighboring cells through gap junctions. The gap junctions form low-resistance connections between adjacent cells; depolarizing one cell can cause an adjacent cell to depolarize by means of **local current flow.** This has the effect of spreading the activity throughout the tissue, and even though most individual cells are not directly innervated, the tissue tends to function as though they were. This is the source of the term *unitary,* and a tissue that behaves in this way is called a **functional syncytium** (*syn* = together; *cytium* = cells). Many unitary smooth muscle tissues are spontaneously active and can contract rhythmically on their own. In these cases, the effect of the autonomic innervation may be to regulate (increase or decrease) the ongoing function.

Heart muscle also receives an extensive autonomic innervation, but the function of the nerves in heart muscle is not to initiate contractions. The muscle of the heart does not contain motor end-plates, and its contractions begin directly in the muscle tissue, not in response to an external stimulus. As in some smooth muscles, the function of nerves in the heart is to regulate the ongoing timing and strength of contraction. Cell-to-cell communication is also of very great importance in the function of heart muscle, because, as is true of smooth muscle, it is composed of cells that are very small in relation to the size of the tissue. The process of the control of heart muscle contraction is sufficiently specialized to be treated separately in Chapter 18.

The Metabolism of Muscle Contraction

Because muscle does work, it must consume fuel in the form of energy-yielding biochemical compounds. The fuel most directly consumed by the actin-myosin contractile system is the universal high-energy compound, ATP. While muscle cells require energy for ion pumping, growth, and cell maintenance, as do other cells, their major function is that of contraction, and this function consumes more energy than does any other type of cell. Not surprisingly, the metabolic adaptations of muscle are specialized to provide an adequate and constant supply of ATP for the contractile process. The metabolic aspects of muscle contraction covered in this chapter are part of the broader subject of energy metabolism in general. Many essential details of these processes are found in Chapter 6, which provides a basis for the somewhat specialized processes in muscle.

Energy for Contraction Comes from Several Sources

As we have seen, ATP is broken down during the crossbridge cycle and provides the energy that is turned into mechanical work. If chemical inhibitors of the general cell energy metabolism are used to prevent replenishment of the ATP supply, a contracting muscle will shortly become exhausted and enter a state similar to **rigor mortis.** However, a muscle maintains activity for a longer period of time than simple chemical measurements of the ATP content would indicate. This is because of a reserve "energy pool" in the cell in the form of a compound called **creatine phosphate.** This high-energy compound is capable of transferring its energy to ATP very readily, so that as soon as the supply of ATP to the myofilaments begins to fall, it is rapidly replenished. The ADP from the contraction process is rephosphorylated to ATP, and the creatine phosphate becomes creatine. This creatin is itself rapidly rephosphorylated by ATP derived from the general metabolism of the cell. The creatine phosphate pool provides a "buffer" for the rapid supply of ATP for the work of contraction, as well as a link to the cellular sources of ATP. Depending on the type of muscle fiber, this cellular ATP is produced by one or both of two common biochemical pathways, as diagrammed in Figure 16–24.

Glycolysis: The Anaerobic Pathway

Glycolysis, the first of these pathways, yields a relatively small amount of ATP from the metabolic fuel (glucose) in a series of reactions that take place in the cytoplasm of the cell. Recall from Chapter 6 that since this reaction pathway takes place in the absence of oxygen, it is called **anaerobic.** For each molecule of glucose metabolized by this pathway, only two molecules of ADP are "recharged" to ATP.

The Citric Acid Cycle: Aerobic Metabolism

When sufficient oxygen is present, however, the end products of glycolysis (and those of fatty acid metabolism) can enter an **aerobic (oxidative)** reaction pathway, the citric acid cycle. These reactions, which take place in the mitochondria of the cell, yield an additional 36 molecules of ATP from a metabolized glucose molecule.

Given the much higher ATP production of aerobic metabolism, it is reasonable to ask why muscles would also have the less efficient anaerobic pathways. The answer lies in the rates at which energy is consumed and at which metabolic fuels can be supplied. Under conditions of normal activity, the blood supply to a muscle can bring the energy-supplying

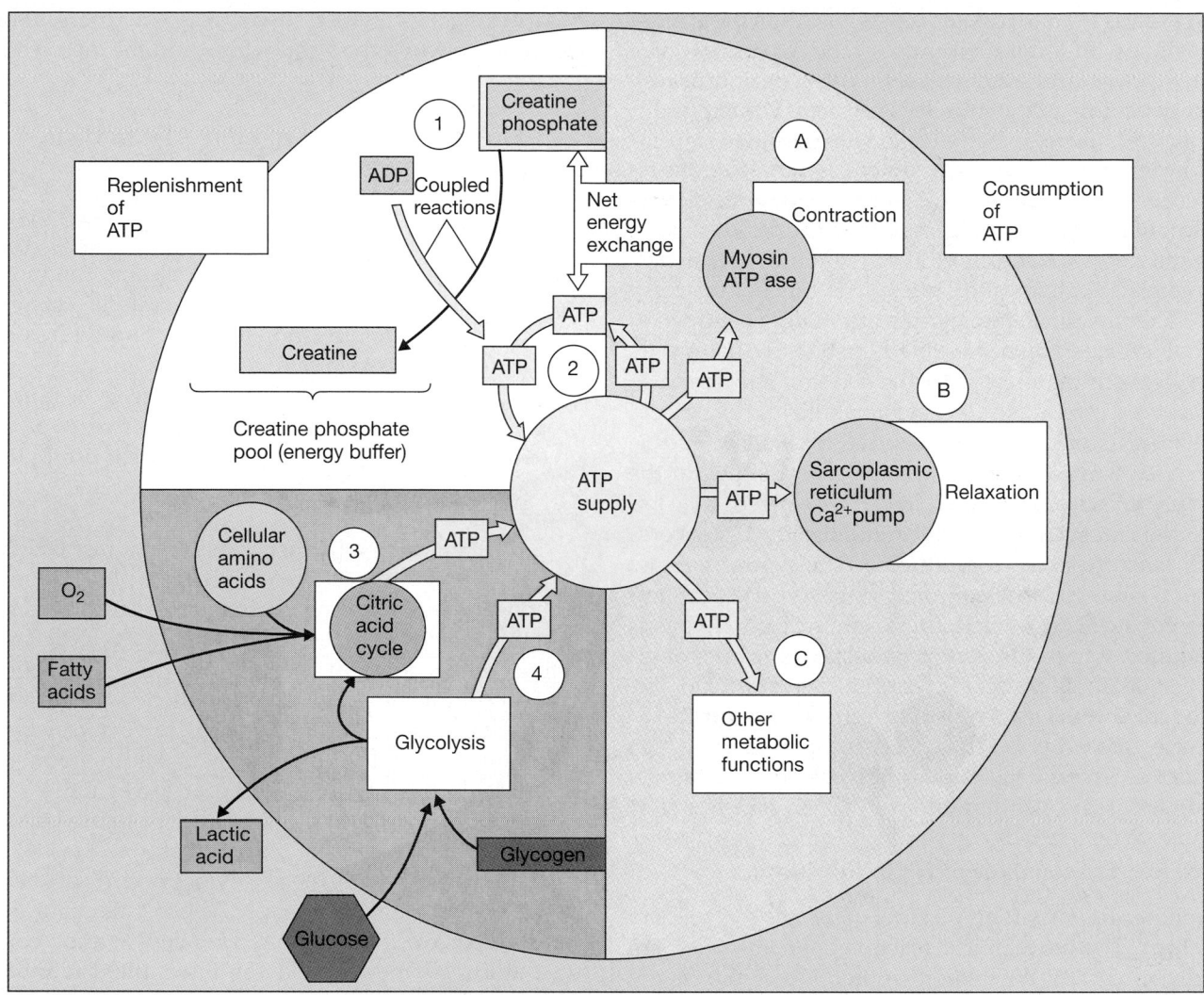

Figure 16–24

Summary of a cell's mechanisms for replenishing its supply of ATP, with the numbers indicating the order of replenishment. The cell uses ATP for a variety of energy-related function. The letters A, B, and C indicate the order in which cellular functions consume ATP.

molecules (fatty acids and glucose) to a muscle at a rate that allows aerobic metabolism to supply almost all of the ATP needed by the contractile system. During moderate exercise as well, the blood supply and the oxygen it delivers can keep up with the needs of the muscle.

The Need for an Anaerobic Pathway

When the level of activity reaches about 70% of the maximum possible, however, aerobic metabolism is no longer able to supply sufficient ATP. At this point, glycolytic (anaerobic) metabolism begins to take over to provide the ATP. The reactions of the glycolytic pathway run more quickly than the aerobic ones, and they can rapidly supply the needed amounts of ATP from glucose in the blood and from the glycogen (a polymer of glucose) stored in the muscle itself.

The Oxygen Debt

While rapid, however, these anaerobic reactions are not very efficient, and they produce large amounts of a temporary byproduct, **lactic acid.** When the bout of heavy exercise is over, the body is left with an overabundance of lactic acid and is deficient in its stores of creatine phosphate and glycogen. **Muscle fatigue** often results from prolonged exercise in which the energy supply fails to keep up with all of the energy demands. In a fatigued muscle, the maximum force that can be exerted decreases; after a period of rest, the capability of the muscle is restored. Paradoxically, chemical measurements made on fatigued muscle show that the total ATP content is not severely reduced; instead, the failure to contract appears to lie somewhere in the processes that couple the membrane depolarization to the activation of the myofilaments. Whatever the defect, it is quickly restored.

The metabolic imbalances that follow contraction are corrected by the aerobic metabolism that occurs in the resting muscle after the activity. The liver also plays an important role in metabolizing lactic acid that is circulating in the blood as a result of muscle activity. Some of this lactic acid serves as fuel for the heart muscle. The lactic acid is oxidized to compounds that can provide ATP, and the glycogen stores are resynthesized from glucose. These processes are responsible for the continued high oxygen consumption of a resting muscle after exercise. This oxygen is required to run the aerobic reactions until the original state of the muscle is restored. The burden of lactic acid and the glycogen deficit that occur during exercise make up the so-called **oxygen debt,** which must be "paid back" quantitatively by postcontraction oxygen consumption.

Muscle Efficiency and Heat Production

Another important by-product of muscle metabolism is **heat.** The biochemical reactions of contraction are only about 20% efficient; the other 80% of the energy is degraded into heat. This is the heat that the body must get rid of during heavy exercise; it is also the heat that warms the body by shivering or other exercise in the cold.

Muscle Types Are Adapted for Special Tasks

In light of the many functions that muscles must perform, it is not surprising that there are specialized types of fibers adapted to particular sorts of tasks. These adaptations involve structural and biochemical specializations that are interrelated and that fall into some fairly distinct categories. Table 16–2 gives the characteristics of these specialized muscle types.

Some muscle tasks, such as maintaining body posture, require that muscles maintain tension for long periods with a low expenditure of energy. Other tasks, such as sprinting in a race or running after a bus, require rapid contractions that may be made at a high energy cost. Between these extremes are muscles that require a mixture of metabolic and structural specializations. In general, muscle fibers fall into two broad categories, **white fibers** and **red fibers.** (For example, consider the light [white] and the dark [red] meat from poultry.)

Red Muscle Fibers and Aerobic Metabolism

The color differences among muscle fibers arise because of the differing content of a protein called **myoglobin,** which imparts a red color to the tissue. The presence of myoglobin also gives a clue to the most important distinction between the muscle classes. Myoglobin is a protein (similar to the protein **hemoglobin** found in red blood cells) that is capable of binding, storing, and releasing oxygen. It is found most abundantly in those muscle fibers that depend on oxidative (aerobic) metabolism, where it promotes the rapid diffusion of oxygen in times of heavy demand.

Red fibers can be further classified as **fast twitch** and **slow twitch** types. They both have plentiful mitochondria and a rich blood supply. Their speeds of contraction are correlated with their rates of ATPase activity—that is, with the rate at which the actin-myosin contractile proteins can break down ATP and release its energy. The slow fibers are very resistant to fatigue when fulfilling their usual function of maintaining posture or working at lower rates. Attempts to use such muscles for rapid, sustained tasks will cause them to fatigue rather quickly as their ATP consumption outstrips the ability of the citric acid cycle to replenish the ATP supply. The fast fibers of the red group have a higher ATPase activity; they contract and relax more quickly and are suited to moderate endurance-type activities.

White Muscle Fibers and Anaerobic Metabolism

White muscle fibers, on the other hand, are adapted for fast and powerful contractions, but they fatigue quickly. They contain the enzymes necessary for glycolytic (anaerobic) metabolism. Glycolysis, a process that operates faster than oxidative metabolism, can use the stored glycogen at a high rate to produce ATP. Since their immediate contractile function does not depend heavily on their oxygen supply, these muscles are not as well supplied with a dense capillary network as are the red types. The white muscles show a high ATPase activity and very rapid contraction and relaxation. The rapid contraction appears to be due to specializations in the myosin molecules that permit them to convert the energy in ATP more rapidly. The rapid relaxation is due in part to a more aggressive calcium-pumping mechanism present in the sarcoplasmic reticulum. As might be inferred from the anaerobic character of their metabolism, the cells contain fewer mitochondria than do the red types.

Table 16–2
Classification of Skeletal Muscle

	Fast		Slow
Feature	White	Red	Red
Mechanical			
Contraction speed	Fast	Fast	Slow
Force capability	High	Medium	Low
S.R. Ca^{2+} pumping	High	High	Moderate
Motor axon velocity	100 M/sec	100 M/sec	85 M/sec
Biochemical			
ATPase activity	High	High	Low
Source of ATP	Anaerobic glycolysis	Oxidative phosphorylation	Oxidative phosphorylation
Glycolytic enzymes	High	Moderate	Low
No. of mitochondria	Low	High	High
Myoglobin content	Low	High	High
Glycogen content	High	Moderate	Low
Rate of fatigue	Fast	Moderate	Slow
Structural			
Diffusion distance	Large	Small	Moderate
Fiber diameter	Large	Moderate	Small
Number of capillaries	Few	Many	Many
Sarcomere structure	Very regular	Less regular	Irregular Z-lines
Functional			
Role in body	Rapid, powerful movements	Medium endurance	Postural endurance
Example	*Latissimus dorsi*	*Vastus lateralis* (a mixed-fiber muscle)	*Soleus*

Smooth Muscle Is Highly Economical in Its Energy Usage

While the basic chemical processes of skeletal muscle are shared by smooth muscle, there are some important adaptations that allow smooth muscle to perform its specialized functions. The crossbridges of smooth muscle cycle much more slowly than those of skeletal muscle, and they spend much more of their cycle time in the attached state. This allows smooth muscle to maintain high levels of force with a much lower rate of ATP hydrolysis, although this property means that smooth muscle can shorten only very slowly compared to skeletal muscle. While some ATP must be used in skeletal muscle for noncontractile processes such as ion pumping, this amounts to only a small portion of the total energy consumption. In smooth muscle, however, the ATP that must be used in the regulatory phosphorylation of myosin light-chains and in control of calcium concentration is a significant portion of the total ATP usage. This is more than offset, however, by the slower crossbridge cycle, and aerobic metabolism is sufficient to keep up with the demand.

Patterns of Innervation and Muscle Growth

These various characteristics are summarized in Table 16–2. A close look at the features compared there will reveal many correlations among function and biochemical makeup. Note that the characteristics listed are for fibers, not whole muscles. Most whole muscles contain a mixture of these fiber types, so they are not strictly confined to a single type of activity. Studies on athletes have shown that, in spite of particular types of training programs, the fiber composition of a given muscle remains relatively constant. For an individual, the specific mixture of fiber types begins to be established before birth and becomes set in childhood. In some newborn mammals, the fibers are predominantly the slow type. Patterns of innervation and usage during early development play an important role in deciding the ultimate distribution of fiber types. Experiments have been done with animals in which the nerve supply to fast and slow muscles was surgically reversed. These studies have shown that the re-innervated muscles take on characteristics related to the type of nerve supplying them. That is, formerly slow muscles begin to take on the contraction characteristics of fast muscles.

The Pattern of Use Affects the Development of Muscle

It is well known that the pattern of use is important in determining the size and strength of a muscle. An increase in the mass of a muscle (without an increase in the actual number of cells) is called **hypertrophy.** Such hypertrophy is most readily brought about by exercises that emphasize isometric or slow isotonic contractions repeated many times. Muscles subjected to such routines will increase in their cross-sectional area and in the amount of contractile protein they contain. The relative distribution of fiber types will remain relatively constant. A negative aspect of this type of muscle adaptation (to a specific set of exercise conditions) is that the vascular supply to the muscle does not increase in proportion to the increase in muscle mass, nor is there a large improvement in the function of the heart. This means that the endurance of the hypertrophied muscle is not greatly increased, and its improved function is largely limited to the sort of exercise that produced the hypertrophy.

Heart muscle is also subject to hypertrophy in response to increased demands. This occurs normally during athletic conditioning but may also be a result of some disease process. For example, if the improper

functioning of a heart valve makes the heart do extra work to maintain the blood pressure, a so-called enlarged heart will result. Such enlargement can at least partially compensate for the condition that caused it, but it also increases the metabolic demands of the heart and may make it more prone to disturbances in the heartbeat rhythm (see Chapter 18).

Exercise that emphasizes isotonic contractions at moderate workloads produces a different type of modification in the muscle. The popular "aerobic" exercise regimens produce less marked increases in strength than the isometric exercises mentioned previously, but do result in an increased vascular supply to the muscle. This enhances the ability to produce increased performance for extended periods of time, such as for the duration of a crosscountry race. Such exercises also provide conditioning for the heart muscle and can increase the efficiency of its pumping.

Adaptation in the opposite direction is also possible. Long-term immobilization of a muscle (such as might be produced by wearing a cast) causes **disuse atrophy,** in which the mass and strength of the muscle decrease. When a muscle is paralyzed by injury to its motor nerve and cannot contract on its own, it is also subject to atrophy. If there is a probability that nerve function may eventually be restored, periodic electrical stimulation of the muscle may prevent some of the atrophy.

Less severe atrophy occurs when a regular form of exercise is discontinued, and the muscle adapts to the lower level of use. This may occur during the "offseason" periods in the life of an athlete and is a factor in the "deconditioning" observed after long periods of weightlessness during space travel. Both of these situations can be at least partially reversed by a proper exercise program.

CHAPTER REVIEW

Summary

- Muscle functions as a motor for locomotion, for the pumping of blood, and for the regulation of internal body processes. It also regulates much of the input to the nervous system.
- Muscle may be classified in three ways: on the basis of its anatomical location (as skeletal, visceral, or cardiac); on the basis of its structure (as striated or smooth); and on the basis of its mode of control (as voluntary or involuntary). There is some overlap in these categories.
- Skeletal muscle fibers are composed of sarcomeres, which are the fundamental units of contraction.
- The sarcomeres are composed of thick (myosin) and thin (actin) myofilaments joined by crossbridges. Contraction

of the individual sarcomeres results in shortening or force development of the whole muscle.
- Cardiac muscle is striated, and the sarcomeres are also similar to those of skeletal muscle. The small cells are closely coupled together both mechanically and electrically.
- Smooth muscle is composed of small unstriated cells. They contain contractile proteins, although no obvious sarcomere arrangement is present, and the internal membrane systems are not very extensive. Cells are coupled together mechanically and electrically.
- Muscles can shorten and exert a constant force (under isotonic conditions), or they can exert a force that does

not result in movement (isometric conditions). Most practical contractions involve a mixture of these two conditions.

- The length of a muscle before contraction has an important effect on the amount of force the muscle can exert when stimulated, and there is a particular length for optimal force. This relationship is described by the length-tension curve.
- The interaction between the actin filaments and the myosin crossbridges produces mechanical motion. The speed of this process and the overall amount of force generated are influenced by muscle length and the force during shortening.
- The crossbridge cycle is controlled by presence or absence of calcium ions and requires ATP.
- The speed at which muscle shortens is related to the amount of force exerted, a relationship described by the force-velocity curve. The power output of muscle also depends on the relationships presented in this curve.
- Cardiac muscle has a contractile mechanism similar to that of skeletal muscle, although the means of external control is different. Smooth muscle also appears to function with a sliding filament mechanism.
- Release of calcium in the region of the myofilaments allows the contractile proteins to interact in the crossbridge cycle. This action is exerted via control proteins on the actin (thin) filaments and is called actin-linked regulation. Cardiac muscle also has a similar actin-linked regulatory process.
- Smooth muscle is under the control of a myosin-linked regulatory process, in which calcium ions activate the myosin filaments by promoting the phosphorylation of the mysoin crossbridges.
- Skeletal muscle contractions are controlled by the plasma membrane. A single brief activation of a skeletal muscle results in a brief contraction called a twitch, and repeated activation results in a sustained contraction called a tetanus. Cardiac muscle is also controlled by its plasma membranes, but only twitch contractions are possible. A long-lasting contraction of smooth muscle is often called a tonic contraction, or tonus.

- The electrical activity of the skeletal muscle plasma membrane is controlled by the central nervous system (CNS). The link between a motor axon and a muscle fiber is called the myoneural junction, or motor end-plate. A nerve action potential causes the release of acetylcholine from the nerve terminals to the muscle end-plate membrane, producing the end-plate potential. This in turn causes the muscle membrane to produce an action potential, which travels along the muscle fiber and activates contraction.
- Some smooth muscles are controlled by the autonomic nervous system (ANS), although they do not have well-developed myoneural junctions. Cardiac muscle functions under the control of specialized cells in the heart, and the nervous system acts to regulate both the rate and strength of the heartbeat.
- The energy for muscle contraction is supplied by glucose and fatty acids and is ultimately used by the muscles in the form of ATP. Glycolysis, an anaerobic process, produces relatively little ATP from each glucose molecule and produces lactic acid as a waste product. Such a pathway produces energy quickly but inefficiently and results in an oxygen debt in the form of lactic acid.
- Some skeletal muscles are specialized for powerful and rapid movements, but they become fatigued rapidly. These muscles have a primarily anaerobic metabolism, are composed of large fibers, and are light in color. They are termed fast, white muscles. Other muscles contract more slowly but can sustain activity for long periods of time. These muscles have a primarily aerobic metabolism, with many mitochondria and a rich blood supply. This group, which contains both slow and moderately fast muscle fiber types, is dark-colored; these are the red muscles.

Review Questions

Identify Each with a Term

1. The basic contractile unit of skeletal muscle.
2. The protein strands that form the basic contractile mechanism of all types of muscle.
3. The type of muscle associated with blood vessels and internal organs.
4. The type of muscle activity in which force is developed without any change in muscle length.
5. The connection between a motor neuron and a skeletal muscle fiber.
6. The chemical fuel most directly consumed by the muscle contraction process.
7. The relationship between muscle force and speed of shortening.

Define Each Term

8. twitch
9. tetanus
10. sarcoplasmic reticulum
11. troponin
12. crossbridges
13. end-plate potential
14. acetylcholine esterase
15. rigor mortis
16. myoglobin
17. disuse atrophy

Choose the Correct Answer

18. In the sliding-filament hypothesis for muscle contraction:
    a. movement occurs due to shortening of the myofilaments.
    b. The cyclic interactions between myosin crossbridges and actin filaments cause the movement to occur.
    c. Movement is caused by the action of calcium ions on the thick filaments.

19. In an isotonic contraction of skeletal muscle:
    a. no movement occurs.
    b. the muscle shortens at a constant force.
    c. force increases as the muscle shortens.

20. Increasing the length of a skeletal muscle prior to an isometric contraction:
    a. may either increase or decrease the force that can be developed, depending on at what length the muscle was originally held.
    b. has no effect on the force that the muscle can develop.
    c. can only decrease the force of contraction.
21. When a weight is held in the hand, the lever action of the bones of the arm causes the flexor muscles of the upper arm:
    a. to exert a force greater than that provided by the weight.
    b. to shorten more rapidly than the speed at which the weight itself is moved.
    c. to move the weight even though they exert less force than the weight provides.
22. The role of the transverse tubules (t-tubules) in the excitation of skeletal muscle contraction is:
    a. to provide an inward path for the spread of muscle action potentials.
    b. to serve as the storage site for calcium ions.
    c. to connect the sarcomeres end-to-end.
23. The primary role of calcium in the activation of skeletal and cardiac muscle is:
    a. to cause depolarization of the muscle cell plasma membrane.
    b. to remove the inhibition of the reaction between the actin filaments and the myosin filaments.
    c. to activate myosin molecules so that they can interact with actin filaments.
24. The primary role of calcium in the activation of smooth muscle is:
    a. to cause depolarization of the muscle cell plasma membrane.
    b. to remove the inhibition of the reaction between the actin filaments and the myosin filaments.
    c. to activate myosin molecules so that they can interact with actin filaments.
25. During the process of transmission at the myoneural junction, the role of acetylcholine is:
    a. to cause the depolarization of the presynaptic (nerve) cell membrane.
    b. to prevent depolarization of the muscle cell membrane.
    c. to cause the depolarization of the postsynaptic (endplate) membrane of the muscle cell.
26. Which statement best describes the control of smooth muscle by the nervous system?
    a. Smooth muscle is completely inactive unless it is stimulated by a motor nerve.
    b. Nerve stimulation may either initiate mechanical (contractile) activity or it may modify ongoing activity.

c. Nerve stimulation is usually inhibitory and serves mostly to prevent contraction.
27. During the aerobic metabolism of exercising skeletal muscle:
    a. the primary source of energy is glucose, and relatively little ATP is formed.
    b. the production of ATP is associated with the consumption of oxygen.
    c. large amounts of lactic acid are produced.
28. An oxygen debt in exercising skeletal muscle:
    a. is due to the build-up of lactic acid produced by anaerobic metabolism.
    b. can usually be "paid back" by anaerobic metabolism.
    c. occurs only when the muscle is at rest.
29. Which set of characteristics is found in those skeletal muscle fibers best adapted to produce fast and powerful movements?
    a. resistance to fatigue, red color, mostly aerobic metabolism
    b. small diameter fibers, red color, irregular sarcomere structure
    c. little resistance to fatigue, large fiber diameter, white color, mainly anaerobic metabolism

Answer in One or Two Sentences
30. What are the three types of muscle in the body?
31. What conditions can lead to hypertrophy and atrophy in skeletal muscle?
32. How does the speed of contraction of cardiac muscle compare with that of smooth and skeletal muscle?
33. How does the drug curare block transmission at the skeletal muscle myoneural junction?
34. Why does skeletal muscle produce more force during a tetanus than during a twitch?
35. What internal cellular event leads to relaxation in all types of muscle?

Problem Solving in Physiology
36. Why would a skeletal muscle that had the contractile and metabolic properties of cardiac muscle serve poorly as a postural muscle?
37. If the calcium uptake mechanism of a skeletal muscle sarcoplasmic reticulum were impaired, what effect would this have on the muscle contraction?
38. If two skeletal muscles were of the same overall length and had the same metabolic properties, but one had sarcomeres twice as long as the other, which one would be likely to shorten faster under a very light load?
39. Why does shivering help to produce body warmth?
40. What type of skeletal muscle would be least affected by a reduction in its blood supply during heavy exercise? Why would this be?

Suggested Readings

Aidley, D.J. *The Physiology of Excitable Cells.* New York: Cambridge University Press, 1978.

Bagshaw, C.R. *Muscle Contraction,* 2nd ed. New York: Chapman and Hall, 1993.

Cole, W.C., and Garfield, R.E. "Ultrastructure of the myometrium." In Wynn, R.M., and Jollie, W.P., eds., *Biology of the Uterus,* 2nd ed. New York: Plenum Medical Book Co., 1989.

Gabella, G. "Structure of intestinal musculature." In Schultz, S.G., and Wood, J.D., eds. *Handbook of Physiology—The Gastrointestinal System I.* Bethesda, MD: The American Physiological Society, 1989.

Hall, Z.W., and Sanes, J.R. "Synaptic structure and development; The neuromuscular junction." In Jessell, T.M., Kandel, E.R., Lewin, B., and Reid, L., eds. "Signaling at the Synapse," *Cell 72* (Neuron 10) suppl 99–122, 1993.

Jankovic, J., and Brin, M.F. "Therapeutic uses of botulinum toxin." *New Engl. J. Med.* 324:1186–1194, 1991.

Junge, D. *Nerve and Muscle Excitation,* 3rd ed. Sunderland, MA: Sinauer Associates, Inc., 1992.

Kao, C.Y. "Electrophysiological properties of uterine smooth muscle." In Wynn, R.M., and Jollie, W.P., eds. *Biology of the Uterus,* 2nd ed. New York: Plenum Medical Book Co., 1989.

Katz, A.M. *Physiology of the Heart,* 2nd ed. New York: Raven Press, 1992.

Keynes, R.D., and Aidley, D.J. *Nerve and muscle,* 2nd ed. New York: Cambridge University Press, 1991.

Matthews, G.G. *Cellular Physiology of Nerve and Muscle,* 2nd ed. Boston: Blackwell Scientific Publications, 1991.

Meiss, R.A. "Mechanical properties of gastrointestinal smooth muscle." In Schultz, S.G., and Wood, J.D., eds. *Handbook of Physiology—The Gastrointestinal System I.* Bethesda, MD: The American Physiological Society, 1989.

Peachey, L.D., and Adrian, R.H. *Handbook of Physiology—Muscle.* Bethesda, MD: The American Physiological Society, 1983.

Rall, J.A. "Energetic aspects of skeletal muscle contraction: implications of fiber types." In Terjung, R.L., ed. *Exercise and Sport Science Reviews,* Vol. 13. New York: Macmillan, 1985.

Ruegg, J.C. *Calcium in Muscle Contraction: Cellular and Molecular Physiology,* 2nd ed. New York: Springer-Verlag, 1992.

Shephard, R.J. *Physiology and Biochemistry of Exercise.* New York: Praeger, 1985.

Squire, J.M., ed. *Molecular Mechanisms in Muscular Contraction.* Boca Raton, FL: CRC Press Inc., 1990.

Stern, M.D., and Lakatta, E.G. "Excitation-contraction coupling in the heart: the state of the question." *FASEB Journal* 6:3092–3100, 1992.

Woledge, R.C., Curtin, N.A., and Homsher, E. *Energetic Aspects of Muscle Contraction.* New York: Academic Press, 1985.

Key Terms

actin (p. 466)
afterload (p. 483)
antagonistic pair (p. 470)
auxotonic contraction (p. 479)
contractility (p. 494)
crossbridge (p. 472)
end-plate potential (p. 495)

force-velocity curve (p. 484)
hypertrophy (p. 504)
involuntary muscle (p. 469)
isometric contraction (p. 479)
isotonic contraction (p. 479)
length-tension curve (p. 481)

motor unit (p. 495)
muscle fiber (p. 470)
myofilament (p. 470)
myoglobin (p. 502)
myosin (p. 466)
meiotonic contraction (p. 479)

sarcomere (p. 472)
sliding-filament hypothesis (p. 483)
tendon (p. 469)
tension: active, passive, resting (p. 481)
voluntary muscle (p. 468)

A scanning electron micrograph
of an angry macrophage
engulfing a yeast cell.
*(© Biology Media/Science Source/
Photo Researchers, Inc.)*

Chapter 17

Functions of the Blood

Key Concepts

Blood helps maintain homeostasis by being a vehicle for transportation and communication among various cells, tissues, and organs of the body.

Blood is a liquid connective tissue composed of approximately 5 million red blood cells per microliter, 7 thousand white blood cells per microliter, and 3 hundred thousand platelets per microliter, suspended in 3 liters of intercellular fluid called plasma.

Plasma is by weight 93% water and 7% solutes, of which most (86%) are proteins.

Red and white blood cells and platelets originate from pluripotential hemopoietic stem cells found mainly in the red bone marrow.

Red blood cells lack internal organelles, contain a red pigment called hemoglobin, and transport oxygen from the lungs to body tissues and carbon dioxide from body tissues to the lungs for elimination.

Each of several types of human blood are characterized by the presence or absence of specific antigens on the surface of the red blood cell.

All white blood cells are nucleated and play various important roles in defending the body against foreign chemicals and microorganisms.

Thrombocytes (platelets) reduce blood loss by helping blood to clot and injured blood vessels to heal.

Mechanisms of hemostasis that prevent or minimize blood loss include constriction of blood vessels, platelet aggregation at

site of blood vessel damage, coagulation (clotting) of blood, and clot retraction.

Chapter Outline

General Functions of Blood

Properties of Whole Blood

Plasma

Hematopoiesis: The Formation of Blood Cells

Erythrocytes: Red Blood Cells

Blood Types

Leukocytes: White Blood Cells

Thrombocytes (Platelets)

Hemostasis: Mechanisms to Prevent Blood Loss

For a period of several weeks, Mary Ann, a 45-year-old computer programmer, noticed that she seemed to tire more easily than she used to. She said she just didn't seem to have the energy to do anything except get up in the morning, go to work, return home in the evening, and go to bed. She breathed comfortably at rest, but experienced difficulty in breathing when she exerted herself and had a hard time catching her breath. Mary Ann felt that she was not as mentally alert as she should have been and began to realize that her job performance was being adversely affected. Fellow workers noticed her pallor and her inattentiveness as well as her tendency to become drowsy at meetings and seminars. Her appetite was depressed, and she believed that perhaps the changes in her mental and physical condition were the results of an inadequate diet. However, her attempt to increase eating resulted in nausea. Eventually, increased heart palpitations and chest pain caused her to seek advice from her physician. A complete physical and gynecological examination, including blood studies, confirmed a preliminary diagnosis of anemia.

Anemia is a condition characterized by an inability of the blood to carry an adequate amount of oxygen to meet the needs of the body's cells. All of the body's cells require oxygen for normal function and thus are dependent on blood to absorb oxygen from inspired air in the lungs and to carry the oxygen to them. Therefore, anemia affects all of the body's systems, although some are more noticeably affected than others.

In Mary Ann's case, oxygen delivery to her brain gradually diminished as her anemia developed and her mental sharpness, alertness, and attentiveness deteriorated.

The mental fatigue that often occurs in the anemic person is frequently accompanied by physical fatigue, as with Mary Ann. Physical weakness and early onset of fatigue develop as a result of inadequate oxygen delivery to skeletal muscles. Mary Ann's pallor (lack of color, especially in the face) indicated the anemia was associated with a lack of oxygenated hemoglobin, as a result of either an inadequate number of circulating red blood cells or an insufficient quantity of normal hemoglobin within the red blood cells.

It has been known for centuries that abnormalities or deficiencies of blood are associated with illness, and can even be the cause of death. The ancient Greeks considered blood to be one of four elemental substances (called fluids or humors) of the human body (see Chapter 1). Because of the life-sustaining qualities of blood, many magical properties have been accorded to it by civilizations of the past, beliefs that continue in primitive societies and cults of the modern world.

In this chapter, we will examine some physical and chemical properties of blood, formation and functions of each type of blood cell, normal and abnormal chemistry of the blood, and why one person's blood differs from another's.

General Functions of Blood

Circulation of blood through the vasculature provides a transportation and communications system between the body's cells, serving to maintain a relatively stable internal environment for optimum cellular activity. The role of the blood in homeostasis largely involves transport of molecules and heat, de-fense against foreign agents, and regulation of extra-cellular fluid pH and osmolarity.

Strictly speaking, blood is not a bodily fluid like tears, saliva, and urine; instead it is a living connective tissue composed by volume of approximately 45% cellular elements and 55% intercellular fluid. Blood is normally confined to the blood vessels and chambers of the heart, and is circulated throughout the body by the heart's pumping action. The lives of all cells depend on circulating blood because it helps maintain the relatively stable internal environment essential for normal cell function. When blood circulation ceases, so does the delivery of oxygen and nutrients, and cells of the brain, heart, liver, and other organs begin to die within a few minutes.

Blood Transports Molecules and Heat

Blood is the primary means by which substances are conveyed from one area of the body to another. Nutrients, such as glucose, amino acids, and fatty acids, electrolytes, and water are absorbed from the gastrointestinal tract and carried by the blood to various body tissues. Oxygen is absorbed by blood as it passes through the lungs and is transported to respiring tissues. Carbon dioxide produced during cellular respiration is absorbed by the blood as it circulates through tissues and is transported to the lungs, where carbon dioxide is eliminated in expired air. In addition, other cellular waste such as urea, uric acid, and excess water are carried by the blood to the kidneys for excretion. Hormones and other chemical messengers produced within the body are transported by blood from sites of production to target cells. In essence, blood serves as a vehicle for moving many different substances to various areas within the body for a variety of purposes.

Body metabolism produces a considerable quantity of heat as a by-product, and the excess heat must be dissipated into the environment (see Chapter 27) in order to prevent overheating. Blood conducts heat from the body's core to the upper respiratory passageways and the skin where dissipation of heat occurs.

Blood Cells Defend Against Foreign Agents

Some of the blood cells are phagocytic, that is, they are capable of ingesting and rendering harmless substances foreign or toxic to the body and microorganisms that are pathogenic or disease-producing. Other blood cells produce and release chemicals (antibodies) which react with and nullify the harmful effects of foreign agents, or chemicals that regulate blood flow and blood clotting during a defensive response to injury.

Blood Proteins Help Maintain Normal Extracellular Fluid pH and Osmolarity

Blood plays a major role in the regulation of extracellular fluid pH. Chemical buffers present in blood convert strong acids and bases into weak acids and bases, thereby minimizing large shifts in pH during the course of daily metabolism. (Mechanisms of buffering are discussed in Chapter 25.) The blood also transports acidic and basic substances to organs of excretion. Normal blood is slightly alkaline, with a pH between 7.35 and 7.45.

Most proteins in the blood do not readily pass through capillary membranes. As a result, plasma proteins create an osmotic pressure which influences the movement of other solutes and water between blood and interstitial fluids, and in turn, the osmolarity of extracellular fluids. This role of the blood proteins is discussed in greater detail later in this chapter and in Chapters 19 and 24.

Properties of Whole Blood

Whole Blood Consists of Cells, Platelets, and Plasma

Whole blood is an opaque red liquid connective tissue consisting of microscopically visible, formed elements—the **red corpuscles** (erythrocytes), the **white corpuscles** (leukocytes), and the **platelets** (thrombocytes), suspended in an intercellular fluid called

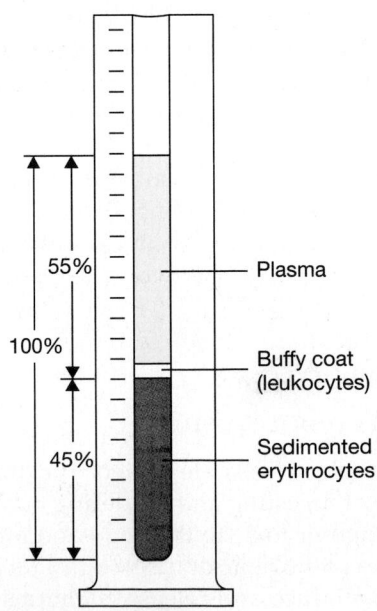

Figure 17–1

Sedimented blood.

plasma. The various components of whole blood can be separated using fractionation techniques such as centrifugation. The components, called **blood fractions,** include red corpuscles (packed RBCs), white corpuscles, platelets, and the protein fractions of the plasma (e.g., gamma globulin: the protein fraction that contains antibodies). A blood fraction, rather than whole blood, is often used clinically to alleviate symptoms of disease. For example, packed RBCs are sometimes used to reduce the effects of severe anemia, or gamma globulin to increase resistance to disease.

Blood accounts for approximately 6 to 8% of adult body weight. Normal adult **blood volume** ranges from 4.5 to 5.5 liters for a female and 5.0 to 6.0 liters for a male. Total blood volume varies, being altered by changes in body water balance (see Chapter 25), pregnancy, hemorrhage, and other factors. **Normovolemia** is a blood volume within normal range. An above-normal blood volume is called **hypervolemia,** and a below-normal volume is **hypovolemia.**

The Specific Gravity and Viscosity of Blood Are Determined by the Cells and Solutes

The **specific gravity (SG)** of a liquid is the ratio between the weight of a given volume of the liquid and the weight of an equal volume of pure water. The specific gravity of water is 1.000; the specific gravity of normal adult blood ranges from 1.050 to 1.060. The specific gravity of blood depends on the number of blood cells and the concentration of chemicals normally dissolved in the plasma. Diseases that alter cell numbers or chemical concentrations of the blood may change the specific gravity.

Viscosity is a measure of a fluid's resistance to flow. The greater the viscosity the greater the fluid's resistance to flow. If we assume the viscosity of pure water to be 1.00, then whole blood viscosity by comparison is approximately 3.50 to 5.50 and that of blood plasma is 1.90 to 2.60; mean values for an adult are 4.5 and 2.2 respectively. Blood viscosity is determined by the number of cells present—cells resist being moved—in whole blood and the mutual attraction of molecules dissolved in the plasma. When the total number of cells increases above normal or when the plasma concentration of large molecules such as protein is elevated above normal, the viscosity of the blood increases, forcing the heart to work more strenuously to maintain a normal rate of flow. A prolonged pathological increase in blood viscosity overloads the heart and may weaken its ability to pump. It also tends to elevate systemic blood pressure, which damages blood vessels and increases the possibility of hemorrhage, or it reduces blood flow to the point where circulation through certain organs (e.g., kidneys) becomes inadequate.

Erythrocyte Sedimentation Rate Is a Convenient Indicator of the Level of Inflammatory Processes

If a small quantity of blood is removed from the body, mixed with a chemical that prevents clotting (an anticoagulant), and placed in a vertical graduated cylinder, the blood cells, having a higher specific gravity than the plasma, will slowly sink to the bottom, leaving behind a transparent amber upper layer of plasma. The leukocytes are lighter than the erythrocytes, and, as a result, the leukocytes form a thin whitish layer (called the buffy coat) between the upper plasma and the lower sedimented erythrocytes (Fig. 17–1). The rate at which the erythrocytes settle is called the **sedimentation rate (SR).** Normal sedimentation rates for adults are: male, 2 to 8 mm per hour, female, 2 to 10 mm per hour.

Sedimentation of erythrocytes proceeds in three phases: formation of rouleaux (clumping of red cells together like a stack of coins), rapid settling, and final packing of the red cell mass. The formation of rouleaux does not occur with cells of abnormal shape such as those in sickle cell anemia, and the sedimentation rate is therefore slower. Increases in the concentration of plasma proteins associated with inflammatory processes tend to increase the sedimentation rate of blood by facilitating rouleaux formation. Often, the erythrocyte sedimentation rate is a convenient and simple indicator of the level of inflammatory processes occurring in an individual. The SR tends to be elevated in acute and chronic infections such as arthritis (and other inflammatory diseases), tuberculosis, rheumatic fever, and toxemia. The SR usually decreases in polycythemia (too many RBCs), allergies, and hyperglycemia (elevated blood sugar).

The Hematocrit Is the Percentage of the Blood Volume Made Up of Erythrocytes

In any sample of whole blood, the ratio between the volume occupied by all of the red cells and the volume of the sample is called the **hematocrit ratio.** The numerical value of the ratio is multiplied by 100 to express the ratio as a percentage (Fig. 17–2). Normal values for adults are: males, 47% plus or minus 5%; females, 42% plus or minus 5%. Normal hematocrit values vary with age, sex, and the part of the body from which the blood sample was taken. The hematocrit value of arterial blood is slightly less than that

$$\text{Wintrobe hematocrit \%} = \frac{\text{red cell volume}}{\text{whole blood volume}} \times 100$$

Normal hematocrit ratios	
Adult female	37-47%
Adult male	42-52%

Figure 17–2

The hematocrit.

of venous blood. The hematocrit varies from tissue to tissue, and the capillary hematocrit is generally lower than either the systemic arterial or systemic venous hematocrit.

To determine hematocrit, venous blood is drawn into a siliconized syringe (to minimize cell damage and prevent platelet activation), rendered incoagulable with oxalate, deposited in a Wintrobe tube (see Fig. 17–2), and centrifuged for 10 minutes at about 1500 g (g = relative acceleration due to gravity = 980 cm/sec²). After centrifugation, the hematocrit is calculated by dividing the measured red cell volume by the measured whole blood volume and multiplying the result by 100 to obtain percentage.

Clinically, it is often more convenient to use the **microhematocrit** method rather than the Wintrobe method to determine hematocrit. A small-bore (1 mm) heparinized (to prevent coagulation) capillary tube is filled with capillary blood taken from a fingertip puncture. The blood is then subjected to centrifugation and the hematocrit determined.

After centrifugation of the blood, about 4% of the plasma always remains trapped among the erythrocytes. To obtain a **corrected hematocrit,** the value of the **apparent hematocrit** is simply multiplied by 0.96.

Abnormal hematocrits occur in diseases that alter the number of circulating erythrocytes, the volume of plasma, or both. In anemias characterized by a reduction in the number of circulating erythrocytes, the hematocrit may fall to 25, whereas in polycythemia (too many erythrocytes), the hematocrit may approach 70. The hematocrit is useful in diagnosing and assessing progress in treatment of diseases that alter the number of circulating normal erythrocytes or conditions that alter the normal ratio between erythrocyte and plasma volumes (e.g., dehydration). The hematocrit is also used to compute red blood cell indices, which are indicators for abnormal erythrocyte size, shape, and color. Red blood cell indices will be discussed later in the chapter.

Plasma

The Chemical Composition of Plasma Is Normally Variable

Plasma is the complex fluid in which blood cells and platelets circulate. It is defined as whole blood minus the formed elements (cells and platelets). Plasma may be separated from whole blood by centrifugation, but, if left to stand, it coagulates within minutes, forming a gel plus a clear fluid called serum. Serum is similar in composition to plasma, except that it lacks coagulation factors.

Normal plasma is transparent and light yellow in color, and is about 93% water and 7% solutes. One liter of human plasma contains about 930 grams of water, 60 grams of protein, 8 grams of inorganic solutes such as Na^+, K^+, Cl, HCO_3, and Ca^{++}, and 2 grams of nonprotein organic substances such as glucose, glycerol, and fatty acids. Plasma contains dissolved gases (O_2, N_2, CO_2, hormones, enzymes, vitamins, pigments, minerals), a variety of cell waste products (urea, uric acid, etc.), and cell nutrients such as amino acids. Major constituents of human plasma are shown in Table 17–1.

The data in the table reveal a range of normal values for each constituent suggesting variability in chemical composition of plasma. In fact, not only does the exact chemical composition of plasma vary from individual to individual, but a person's plasma composition varies slightly from minute to minute as substances are exchanged across capillary walls between plasma and interstitial fluids.

Figure 17–3 is a diagram of the three major fluid compartments in the body: **the blood plasma, the interstitial fluid,** and **the intracellular fluid.** The interstitial fluid constitutes the immediate environment of all living cells of the body and is separated from the intracellular fluid by the plasma membrane. The capillary wall separates blood plasma from interstitial fluid. Normally, the three fluid compartments are in osmotic equilibrium. The exchange of water and small molecules, such as electrolytes, between plasma and interstitial fluid is very rapid: about 70% of the plasma fluid is exchanged with the interstitial fluid in one minute. Because of this rapid exchange, the

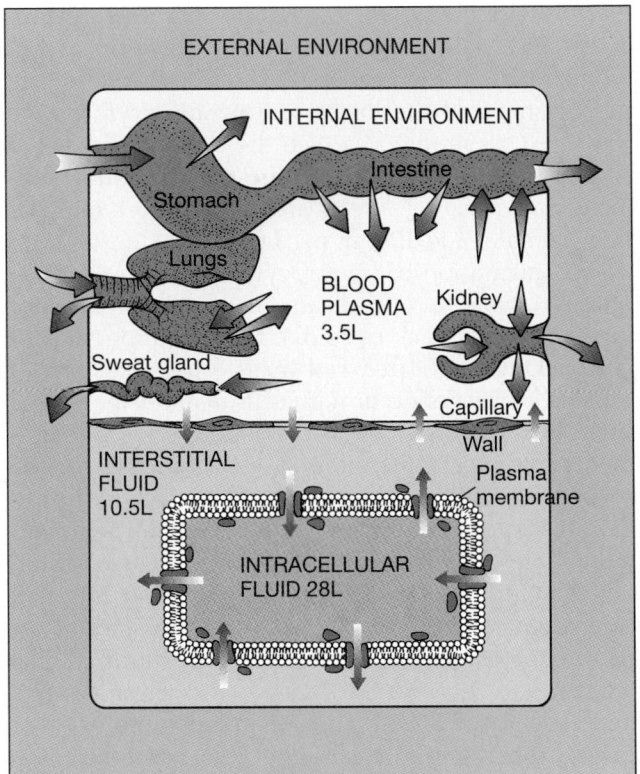

Figure 17–3

The body's internal environment contains three major fluid compartments—blood plasma, interstitial fluid and lymph, and intracellular fluid. The chemical composition of plasma changes as various organs exchange solutes and water between the external and internal environments. The plasma compartment is separated from the interstitial fluid compartment by the capillary wall, which in most areas of the body allows for rapid exchange of water and small solutes between the two compartments. The interstitial and intracellular fluids are separated by the plasma membranes, most of which allow for the rapid passage of water through protein pores. Normally, the three fluid compartments are in osmotic equilibrium because water is able to shift rapidly from one compartment to another.

Table 17–1

Normal Constituents of Blood

Compound[a]	Concentration and Fraction[b]
Acetoacetic acid + acetone (ketone bodies)	0.2–2.0 mg/100 ml (S)
Adenosine triphosphate	1 μm (E)
Aldosterone	3–10 mg/dl (P)
Amino acid, total	300 mg/dl (P), 120 mg/dl (E)
Ascorbic acid	0.4–1.5 mg/dl (fasting B)
Bicarbonate	25 mM
Bilirubin	Direct: 0.1–0.3 mg/dl (S); indirect: 0.2–1.2 mg/dl (S)
Calcium	8.5–10.5 mg/dl; 4.3–5.3 mEq/L (S)
Carbon dioxide	26–28 mEq/L (S)
Chloride	100–106 mEq/L (S); 102 mm (P); 78 mm (E)
Cholesterol	150–240 mg/dl (S)
Cobalt	17 μm (E)
Copper, total	100–200 μg/dl (S); 17 μm (E)
Cortisol (17-hydroxycorticoids)	5–18 μg/dl (P)
Creatine	40 μm (P): 600 μm (E)
Creatinine	0.7–1.5 mg/100 ml (S); 450 μm (P); 160 μm (E)
2,3-Diphosphoglycerate (DPG)	4 mm (E)
Glucose (folin)	80–120 mg/dl (fasting, B)
Glucose (true)	70–100 mg/dl (B)
Iodine, protein-bound	3.5–8.0 μg/dl (S)
Iron	50–150 μg/dl (S)
Lactic acid	0.6–1.8 mEq/L (B); 1 mm (P)
Lipids, total	450–1000 mg/dl (S)
Magnesium	1.5–2.0 mEq/L; 1–2 mg/dl (S)
Nitrogen, nonprotein	15–35 mg/dl (S)
Phosphatase, acid total[c]	Male: 0.13–0.63 sigma units/ml (S); female: 0.01–0.56 sigma units/ml (S)
Phosphatase, alkaline[c]	2.0–4.5 Bodansky units/ml (S)
Phospholipids	145–225 mg/dl (S)
Phosphorus, inorganic	3.0–4.5 mg/dl (adult, S); 0.9–1.5 mm/L (adult, S)
Potassium	4 mm (P); 111 mm (E)
Protein	
Total	6.0–8.0 mg/dl (S)
Albumin	3.5–5.5 mg/dl (S)
Globulin	1.5–3.0 gm/dl (S)
Sodium	136–145 mEq/L; 310–340 mg/dl (S); 140 mm (P); 6 mm (E)
Sulfate	0.5–1.5 mEq/L (S)
Transaminase (serum glutamic oxaloacetic transaminase [SGOT][c])	10–40 units/ml (S)
Urea nitrogen (blood urea nitrogen, BUN)	8–25 mg/dl (B)
Uric acid	3.0–7.0 mg/dl (S)

[a]The substances and the values listed for them in this table give an indication of the biochemical complexity of blood and are not to be construed as absolute. Blood levels of specific components can vary among normal individuals or in the same individual at different times. Furthermore, the normal values for specific blood constituents can differ greatly among different laboratories, depending on the technique used for their determination.

[b]Abbreviations used for the several blood fractions are: B = whole blood; P = plasma; S = serum; E = erythrocytes.

[c]Activities of over 50 specific enzymes have been detected in various blood fractions. *Source:* From D. Jensen, *The Principles of Physiology* (2nd ed.). New York: Appleton-Century-Crofts, 1980. Table 32–2.

water and electrolyte concentrations of plasma and interstitial fluid are nearly identical and vary little despite considerable variation in the uptake and release of substances by cells (see Chapter 24). Protein accounts for the major concentration difference between plasma and interstitial fluid because protein molecules are generally too large to readily pass through the capillary wall. The greater concentration of proteins in the plasma establishes an osmotic gradient that, as we shall see, governs the distribution of water between the blood and the interstitial fluid.

Plasma Proteins Supply Cell Nutrition, Serve as Carriers and Buffers, Help Clotting, and Govern Water Distribution

Separation techniques can be used to divide the plasma proteins into the following fractions: **albumin, globulins,** (alpha-1, alpha-2, beta-1, beta-2, and gamma), and **fibrinogen** (Fig. 17–4). Plasma proteins are large molecules with high molecular weights (44,000 to 1,300,000) and configurations (arrangement of atoms) ranging from spherical (B-lipoprotein) to ellipsoid (albumin and globulins). Due to their large size, plasma proteins are classified as **colloids.** Plasma proteins collectively function in several different ways.

First, plasma proteins serve as an important reserve supply of amino acids for cell nutrition. If necessary, cells of the reticulo-endothelial system—macrophages in liver, gut, spleen, lung, and lymphatic tissue—can ingest plasma proteins, break them down, and release component amino acids into the blood so that other cells may use the amino acids for synthesis of new protein.

Second, plasma proteins serve as carriers for other molecules (see Fig. 17–4). Many types of small molecules bind to specific plasma proteins and are transported from absorptive organs (e.g., intestine) or storage organs (e.g., liver) to other tissues for utilization. Iron, for example, is carried in the blood combined with a transport protein called transferrin. Many of the ions in the blood bind to plasma proteins (e.g., Ca^{++}), as do drugs, pigments, and hormones. Lipids (fats) absorbed into the lymph and blood by the intestine are water-insoluble; yet, when combined with proteins in lymph and blood, they can be made soluble for transport (see Chapter 23).

Third, plasma proteins act as buffers helping to maintain a stable blood pH. The normal range for the pH of blood is 7.35 to 7.45; that is, normal blood is slightly alkaline. Proteins are amphoteric: they are able to bind H^+ or OH^- depending on whether they exist as weak acids or weak bases, which in turn depends upon pH. Generally, plasma proteins function as weak bases, binding excess H^+, and thus help to keep the blood slightly alkaline (see Chapter 26).

Fourth, several plasma proteins, some of which are zymogens and others not, interact in specific ways

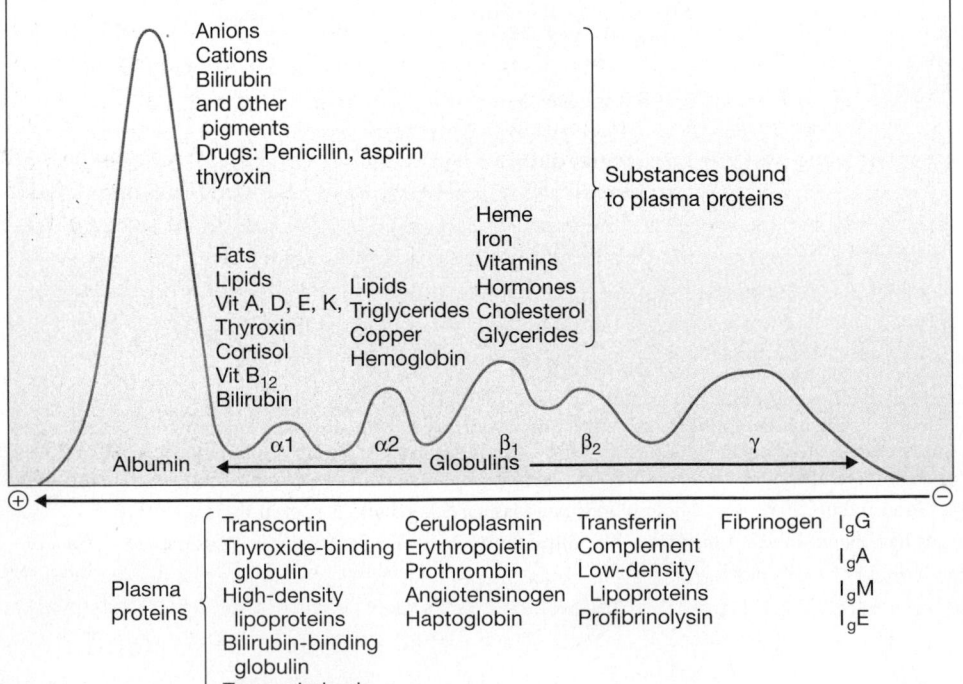

Figure 17–4

Electrophoretic survey of plasma proteins and substances bound to plasma proteins. Electrophoresis is a fractionation technique based on the movement of electrically charged particles, dissolved or suspended in a fluid, along a voltage gradient. Comparative mobilities of charged particles are a function of molecular size, shape, and electric charge, and the magnitude of the voltage gradient. In neutral or alkaline solutions, plasma proteins are electrically charged and can be fractionated by electrophoresis. The charged proteins migrate toward the anode (+ electrode) but at different velocities.

to cause blood to coagulate. Coagulation of blood is part of the body's response to vascular injury and helps protect against the loss of blood and invasion by microorganisms and viruses.

Fifth, plasma proteins produce a **colloid osmotic (oncotic) pressure** that governs the distribution of water between the blood and the interstitial fluid.

The total osmotic pressure of normal plasma is about 7.3 atm (1 atm = 760 mmHg), or 5550 mmHg. The total osmotic pressure is the osmotic pressure plasma would exert if separated from pure water by a membrane permeable only to water. Solutions with the same osmotic pressure as plasma are called **isotonic.** An erythrocyte placed in an isotonic solution, such as 0.85% NaCl (clinically rounded off to 0.9% NaCl and called physiological saline) will neither gain nor lose water (see Chapter 4). Solutions with a higher osmotic pressure than plasma are called **hypertonic,** and those with a lower osmotic pressure are called **hypotonic.** An erythrocyte will lose water and shrivel (crenate) if surrounded by a hypertonic solution, whereas it will gain water, expand, and hemolyze (break open and lose hemoglobin) if surrounded by a hypotonic solution. In either case, red blood cells are distorted or destroyed and fail to function normally, thus it is important when administering fluid intravenously to ensure that the tonicity of the fluid is correct for the circumstances.

Since plasma, interstitial fluid, and intracellular fluid are normally in osmotic equilibrium, homeostasis of all three depends critically on regulation of the osmotic pressure of the plasma. Any deviation from normal extracellular fluid osmotic pressure causes cells to gain or lose volume, and, like red blood cells, those cells may malfunction or die.

About 99.5% of the total osmotic pressure of plasma is due to small molecules such as electrolytes, urea, glucose, and others, which readily pass through capillary membranes along with water. Thus, the osmotic pressure due to these solutes is about the same in interstitial fluid as in plasma. The contribution of plasma proteins to the total osmotic pressure of plasma is about 0.5%: normal colloid osmotic pressure (oncotic pressure) of the plasma is 28 mmHg. Nevertheless, oncotic pressure plays an important role in regulating the distribution of water between plasma and interstitial fluid because plasma proteins do not readily pass through capillary membranes. Small amounts of protein that do pass from blood to interstitial fluid do not collect in the interstitial fluid because of cell uptake and removal by the lymphatic system; therefore, a protein concentration gradient from blood to interstitial fluid exists, is maintained, and osmotically governs the movement of water and small molecules between plasma and interstitial

fluid. Mechanisms of capillary exchange are presented in greater detail in Chapter 19.

In addition to functioning collectively as previously described, the individual types of plasma proteins have specific properties and functions.

Albumin

Approximately 60% of the total plasma protein is albumin, one of the smallest proteins (mol. wt. = 69,000) in the plasma. Being small and numerous, albumin accounts for about 80% of the plasma protein osmotic pressure; thus, reductions in plasma albumin content that may occur in nutritional diseases (e.g., kwashiorkor), liver disease, and kidney disease may result in a loss of fluid from the blood and a gain of fluid in the interstitial space (peripheral edema).

In addition to its osmotic role, albumin helps many substances dissolve in the plasma by binding to them; thus albumin plays an important role in plasma transport. Substances bound by albumin include drugs such as barbiturates and penicillin, pigments such as bilirubin and urobilin, hormones such as thyroxin, and other substances such as fatty acids and bile acid salts.

Globulins

Approximately 40% of the total plasma protein is globulin: 4% is alpha-1 globulin, 8% is alpha-2 globulin, 7% is beta-1 globulin, 4% is beta-2 globulin, and 17% is gamma globulin.

Alpha-1 globulins form **glycoproteins** (protein joined to carbohydrate) and small amounts of lipoproteins (protein joined to lipid). **High density lipoproteins (HDL)** function in lipid transport carrying fats to cells for use in energy metabolism, membrane construction, and hormone formation. HDLs also appear to prevent cholesterol from invading and settling in the walls of arteries (see Focus Box). Other proteins in the alpha-1 group include **thyroxine-binding globulin, cortisol-binding globulin (transcortin),** and **vitamin B$_{12}$-binding globulin (transcobalamin),** all of which transport their respective substrates.

Alpha-2 globulins include **haptoglobin,** a protein that combines with free hemoglobin preventing its excretion by the kidney, and **ceruloplasmin,** a copper-containing oxidase enzyme. Other proteins in the alpha-2 group are: **prothrombin,** a protein involved in blood coagulation; **erythropoietin,** a hormone that stimulates erythrocyte production; and **angiotensinogen,** a hormone precursor involved in regulating blood pressure, body fluid, and electrolyte balance.

CLINICAL FOCUS

Blood Cholesterol

One of the primary risk factors for development of heart and vascular disease is an abnormally high level of cholesterol in blood serum. Excess cholesterol tends to be deposited, in the form of fatty plaques, on the lining of blood vessels (atherosclerosis)—much the way mineral scale builds up on the inside of a water pipe. These deposits reduce blood flow and oxygen delivery to tissues downstream, and increase the chances of internal blood clotting (thrombosis) and heart attack. Excess cholesterol also accumulates inside and outside the cells that form the walls of arteries; this results in a loss of vascular elasticity and a hardening of the arterial wall (arteriosclerosis), thereby increasing the blood pressure and heart work.

Elevated blood triacylglycerols (triglycerides) also contribute indirectly to heart and vascular disease because triacylglycerols influence the way in which the body processes cholesterol. For reasons not yet clear, very high levels of blood triacylglycerols also cause the pancreas to become severely inflamed (pancreatitis). Cholesterol and triacylglycerols are lipids which, by themselves, are insoluble in plasma and thus share a common system of transport in the blood: the lipoproteins.

Transport proteins, minus the triacylglycerols and cholesterol they normally carry, are called apolipoproteins. Apolipoproteins serve as a package for lipids so they can be transported in a polar environment, and also as messengers, delivering the lipid to specific cells (e.g., adipose cells) for storage or use.

Four major categories of lipoprotein participate in the transport of triacylglycerols and cholesterol: (1) chylomicrons, (2) very low density lipoproteins (VLDL), (3) low density lipoproteins (LDL), and (4) high density lipoproteins (HDL).

Chylomicrons transport dietary cholesterol and triacylglycerols from the intestine to adipose tissues and liver. Chylomicrons contain mostly triacylglycerols and a little cholesterol.

VLDL transport triacylglycerols and cholesterol made by the liver to adipose and other tissues. Adipose cells remove the triacylglycerols from the BLDL and leave behind the apolipoprotein and cholesterol that form LDL.

LDL transports cholesterol to peripheral tissues and regulates endogenous cholesterol levels in those tissues. High levels of LDL promote atherosclerosis and arteriosclerosis, and thus the cholesterol contained in LDL is sometimes referred to as "bad" cholesterol.

HDL transports cholesterol from peripheral tissues to the liver, where the cholesterol is removed from the blood. HDL thus help to protect against cholesterol-related heart and vascular disease. The cholesterol in HDL is sometimes referred to as "good" cholesterol. HDL removes cholesterol from arterial walls thus opposing the action of LDL. High levels of HDL have been shown to slow the development of arteriosclerosis and to actually reverse the disease process.

Both the total amount of cholesterol in the blood and the ratio between LDL and HDL cholesterol are important in determining the risk of developing heart and vascular disease. Cholesterol management strategies are geared toward: (1) lowering total blood cholesterol to below 200 mg/dl and maintaining the lowered level, (2) lowering the LDL-cholesterol to below 120 mg/dl and (3) increasing the HDL-cholesterol so as to decrease the LDL/HDL ratio.

One way to lower LDL-cholesterol and raise HDL-cholesterol is to modify dietary intake of fatty acids. A diet rich in saturated fatty acids (SFAs) such as those found in animal fats (e.g., butter, cheese, cream, meats) is associated with high plasma cholesterol levels and the high LDL. Dietary substitution of polyunsaturated fatty acids (PUFA) and monounsaturated fatty acids (MUFA), such as those found in fish and vegetable oils, for saturated fatty acids tends to decrease total cholesterol and LDL-cholesterol levels, and to raise HDL cholesterol levels. The mechanisms involved are not yet clear and are the subject of much research.

■ **In addition to reducing the development of coronary arterial disease, substituting PUFAs and MUFAs for SFAs in the diet reduces the incidence of heart attack by decreasing the chance for coronary thrombosis. How?**

The **beta globulins** (beta-1 and beta-2) include most of the apolipoproteins, which are carrier proteins for lipids. Beta-1 lipoprotein, called **low-density lipoprotein (LDL),** carries cholesterol and fats to tissues for use in manufacturing steroid hormones and building cell membranes, but it also favors the deposition of cholesterol in arterial walls and thus appears to play a role in disease of the blood vessels and heart (see Focus Box). Other substances transported by beta globulins include phospholipids, glycerides, lipid-soluble vitamins (A, D, E, K), and metals such as copper and iron. **Transferrin,** a copper and iron transporting protein, is among the beta globulins.

The **gamma globulin** fraction contains the **immunoglobulins** (antibodies) (Ig). There are five major classes of immunoglobulin: **IgG, IgA, IgM, IgD,** and **IgE.** More than 99% of all imunoglobins in the plasma are either class G, A, or M. Class D and E immunoglobulins as free proteins are rare in plasma. The role of immunoglobulins in defense against foreign substances and disease-causing organisms will be discussed later in Chapter 28.

The Albumin/Globulin Ratio

The total amount of plasma protein normally remains stable and contains about twice as much albumin as globulin. A typical adult normally consumes and replaces about 15 grams of albumin and 5 grams of globulin every 24 hours, but the amounts vary according to body needs for specific proteins. For example, inflammatory diseases cause an increase in the production of immunoglobins, which is accompanied by a decrease of equal magnitude in the production of albumin. The A/G ratio is reduced but the total amount of protein in the plasma remains stable (A + G).

Fibrinogen

Fibrinogen is a soluble plasma protein manufactured in the liver. When blood clots, fibrinogen is converted into an insoluble protein called fibrin, which forms the foundation of the blood clot. Coagulation of blood will be discussed later in the chapter.

Hematopoiesis: The Formation of Blood Cells

The formed elements of the blood, sometimes referred to as corpuscles (little bodies), include the **erythrocytes** (red cells), **leukocytes** (white cells), and **thrombocytes** (platelets). Of the formed elements, only the leukocytes are true cells, possessing a nucleus and other organelles; however, we shall follow convention and refer to the erythrocyte as though it were still a true cell. Leukocytes include **agranular leukocytes** or **agranulocytes** (**monocytes** and **lymphocytes**), and **granular leukocytes** or **granulocytes** (**basophils, eosinophils,** and **neutrophils**).

All of these blood cells and structures originate from primitive cells known as **pluripotential hemopoietic stem cells (PHSC),** found mainly in the bone marrow (Fig. 17–5). (Specifically, they are found in the red marrow of certain bones, although other sites are possible, particularly in the fetus.) These uncommitted stem cells are called "pluripotent" because they are capable of developing into any one of the various blood cells. In the process known as **hematopoiesis,** which is generally ongoing throughout life, these uncommitted stem cells (except those giving rise to lymphocytes) become committed progenitor cells called colony forming units (CFU) because they generate colonies of specific blood cell types. Some CFUs are multipotential, able to give rise to colonies containing more than two types of mature blood cells; some CFUs are bipotential (able to give rise to two types of mature blood cells), and other CFUs are unipotential, that is, committed to the development of a single kind of blood cell (see Fig. 17–5).

Hematopoiesis is a general term for the development of all types of blood cells. Figure 17–5 gives an overview of the process. **Erythropoiesis** is a term that specifically describes the development of red blood cells from uncommitted stem cells; **leukopoiesis** is the term for the development of the white blood cells. The specific steps in each of these processes are described and illustrated in the relevant upcoming discussions of each cell type.

Erythrocytes: Red Blood Cells

The Erythrocyte Shape Is Optimal for Its Function of Gas Transport

The erythrocyte, or **red blood cell (RBC),** is a cell shaped like a biconcave disc with a diameter of about 8 μm and a maximum thickness of about 2 μm (Fig. 17–6). The shape of the cell results in a high ratio of surface-area to volume, favoring more efficient diffusion into and out of the cell. The principal function of the erythrocyte is to transport oxygen and carbon dioxide in the blood; as we will learn, its shape is optimal for this function.

The erythrocyte lacks a nucleus and the other organelles, and therefore is not a true cell. The lack of internal organelles increases the erythrocyte's capacity for gas transport. The interior of the erythrocyte consists of a homogeneous matrix containing a loosely arranged meshwork of fibrous and globular proteins that form a cytoskeleton. The cytoskeleton is anchored to the inner surface of the plasma membrane so as to give the erythrocyte its discoid shape. Erythrocytes are pliable, elastic cells that bend, twist, and flex as they pass through tiny blood vessels; after such changes, they normally regain their discoid shape very quickly. Genetic abnormalities (such as **hereditary spherocytosis** or **hereditary elliptocytosis**) that affect the synthesis of cytoskeleton proteins may result in red blood cells that have abnormal spherical or elliptical shapes. Such cells do not func-

Figure 17–5

An overview of hematopoiesis.

Figure 17–6

(*a*) Typical dimensions of erythrocytes, or red blood cells. (*b*) A scanning electron micrograph of a red blood cell.   (*© Ken Elward/Science Source/Photo Researchers, Inc.*)

tion normally and are destroyed at a faster-than-normal rate, resulting in anemia.

Normal human blood contains approximately 5 million red blood cells per microliter of whole blood. Normal adult ranges are: male, 5.4 million/μl plus or minus 0.8 million/μl; and female, 4.8 million/μl plus or minus 0.6 million/μl.

The Oxygen-Carrying Capacity of Blood Is Dependent on Hemoglobin Concentration

Erythrocytes contain **hemoglobin,** a pigment molecule that gives the cell a pale red color. Each erythrocyte contains 200 to 300 million molecules of hemoglobin, accounting for about one-third of the mass of the red blood cell. Hemoglobin readily associates and dissociates with oxygen and carbon dioxide and is responsible for the erythrocyte's ability to transport these gases. Hemoglobin also plays an important role as a chemical buffer in the body's defense against changes in hydrogen ion concentration (pH).

The hemoglobin molecule (Fig. 17–7) consists of a **globin molecule** and four attached **heme groups.** Each heme group contains an iron atom with which a molecule of oxygen may associate and dissociate; hence, each hemoglobin molecule can transport a maximum of four molecules of oxygen. When blood passes through the lungs, hemoglobin becomes saturated with oxygen, forming **oxyhemoglobin** and becoming bright red in color. Then, when this blood passes through body tissues, some of the oxygen dissociates from hemoglobin. **Reduced hemoglobin** is formed, and the blood turns a darker red.

Normal blood contains approximately 15 grams of hemoglobin per deciliter (100 ml) of whole blood. Normal adult ranges are: male, 16.0 g/dl plus or minus 2.0 g/dl; female, 14.0 g/dl plus or minus 2.0 g/dl.

About 98% of the oxygen carried in the blood is in the form of oxyhemoglobin. The **oxygen-carrying capacity** of blood is defined as the amount of oxygen the hemoglobin in 100 ml of whole blood is capable of transporting. Each gram of hemoglobin can combine with 1.34 ml of oxygen. Since normal blood contains approximately 15 grams of hemoglobin per deciliter, the oxygen-carrying capacity of normal adult blood is approximately 20 ml of O_2 per deciliter (15 g/dl $\times$ 1.34 ml O_2/g = 20.1 ml O_2/dl).

Not All Hemoglobin Molecules Are Alike

Several forms of hemoglobin have been found in human blood. Two types (**type A** in adults and **type F** in the fetus) are normal hemoglobins, whereas others are considered abnormal. Differences in the globin molecules give rise to the various types (the heme groups of all types are alike).

The globin part of hemoglobin consists of two dissimilar pairs of peptide chains, each pair designated by the Greek letter alpha (α), beta (β), gamma (γ), or delta (δ). A heme group is bound to each of the four chains. Ninety-six percent of adult hemoglobin is **type A_1,** consisting of two alpha chains and two beta chains. Two percent of adult hemoglobin is **type A_2,** consisting of two alpha and two delta chains, and two percent is type F, containing two alpha and two gamma chains.

Hemoglobin F is the predominant type in fetal blood. Fetal hemoglobin has a greater affinity for oxygen than does adult hemoglobin. This greater affinity permits fetal red cells to absorb adequate oxygen at the lower PO_2 (partial pressure of oxygen) prevalent in the fetal environment. At birth, when the lungs become the infant's organ of gas exchange, hemoglobin A begins to replace hemoglobin F. By the time the child is six months old, the replacement is nearly complete.

The types of hemoglobin present in the blood are determined genetically. Abnormalities in the types of hemoglobin present (**hemoglobinopathies**) usually result from altered DNA templates that produce minor variations in the amino acid sequence or composition of either the beta chain (e.g., **sickle cell disease**) or the alpha chain (e.g., **hemoglobin H disease**). **Sickle cell hemoglobin (type S),** for example, differs from hemoglobin A because of the substitution of a single amino acid in one of the beta chains. Oxyhemoglobin S, like normal hemoglobins A and F, is sol-

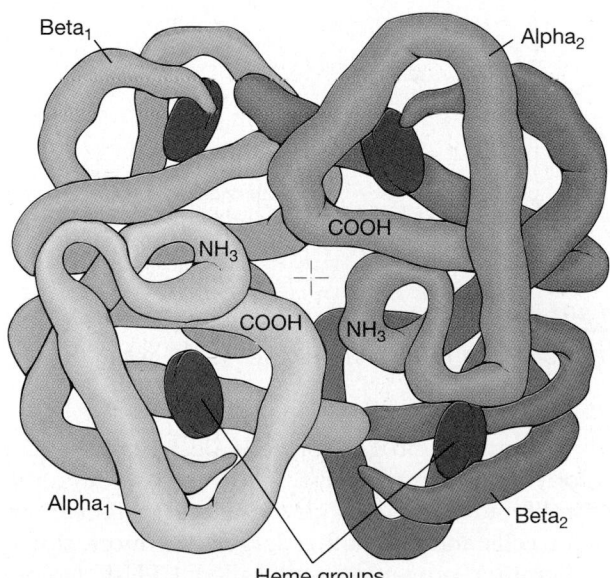

Figure 17–7

A model of the hemoglobin A molecule.

Figure 17–8

Blood from a patient with sickle cell anemia.
(© Dr. Donald L. Rucknagel, Director of the Comprehensive Sickle Cell Center/Children's Hospital Medical Center/Cincinnati, Ohio.)

uble in the intracellular fluid of the erythrocyte; however, deoxygenated hemoglobin S is insoluble, forming fibrous precipitates that change the red blood cell shape from a biconcave disc to a sickle shape (Fig. 17–8). Sickle cells hemolyze (lyse, or burst) readily, producing anemia. Sickle cells often become trapped in capillaries, blocking the flow of blood and producing swelling of tissue, pain, and tissue hypoxia (inadequate oxygen supply). These conditions, if severe, can result in tissue death.

Erythropoiesis Is the Formation of Red Blood Cells

Erythropoiesis is the formation of erythrocytes from pluripotential hemopoietic stem cells. In the fetus, red blood cells are formed from stem cells in the yolk sac, liver, spleen, and red bone marrow. In the adult, red blood cell production is confined to the red marrow of the humerus (upper arm bone), femur (thigh bone), the ribs, sternum (breastbone), scapulae (shoulder blades), skull, vertebrae, and pelvis. In certain diseases during adult life, extramedullary erythropoiesis (red blood cell formation occurring outside the bone marrow) may be seen in the spleen and liver. Each day, the adult marrow produces about 230 billion red blood cells. If necessary, the marrow can increase its production of red blood cells six-fold.

An erythrocyte exists for approximately 4 months, after which it is removed from the blood and destroyed by macrophages in the spleen, liver, and marrow. Normally, the rate of destruction equals the rate of production, so that the number of circulating erythrocytes is fairly constant.

Erythropoiesis begins with the differentiation of **pluripotential hemopoietic stem cells** (Fig. 17–9)

Figure 17–9

Erythropoiesis: The formation of red blood cells.

into committed progenitor cells designated CFU-B (colony forming unit-burst), which have a high rate of proliferation. CFU-Bs give rise to erythrocyte precursor cells and to another larger and more slowly proliferating progenitor cell called CFU-E (colony forming unit-erythroid). Erythrocyte progenitor cells develop into precursor cells that proceed through a sequence of maturation stages during which they lose their nuclei and acquire hemoglobin.

Upon losing its nucleus, the erythrocyte precursor becomes a **reticulocyte.** Both mature erythrocytes and reticulocytes are released by the marrow into the blood. Released reticulocytes mature into erythrocytes about 1.5 days after entering the circulation. Normally, blood contains about 1 reticulocyte for every 100 erythrocytes, their production serving as an index of erythropoietic activity (i.e., the higher the reticulocyte proportion in the blood, the higher the rate of red blood cell production in the marrow).

Iron and Other Substances Are Required for Erythropoiesis

Several substances are required for proper formation of red blood cells and hemoglobin: **amino acids, iron, copper,** and **vitamins B₂ (riboflavin), B₁₂ (cyanocobalamin), pyridoxine,** and **folic acid** are essential. Deficiencies in any one of these substances may lead to anemia.

Iron is required for the production of heme; in fact, approximately two-thirds of the body's iron content is found in hemoglobin. Most of the remaining iron is stored as **ferritin,** an iron-protein complex found in the liver, and in the intestinal epithelium, spleen, and marrow. Stored iron is readily made available for hemoglobin synthesis via a protein called **transferrin** found in the plasma. Transferrin combines with iron that has been absorbed from the intestinal tract or released from storage. It then transfers the iron to erythrocytes that are developing in the bone marrow. Transferrin also transports iron released from worn-out red blood cells to the bone marrow to be reused for hemoglobin synthesis or to the liver to be stored.

Nearly all body cells require iron because iron-containing enzymes are required in oxidative metabolism. The shedding of cells by the lining of the intestinal and reproductive tracts as well as the skin, coupled with smaller amounts of iron in sweat and urine, result in an average daily loss of about 4 mg of iron. Iron loss is normally replaced by dietary intake. The recommended daily allowance is 15 mg, although typically only about 4 mg of iron per day is absorbed, primarily in the small intestine.

Iron is absorbed by active transport, a process that occurs until all the plasma transferrin in the plasma is saturated with iron. When transferrin is saturated, iron absorption practically ceases, and the remaining dietary iron is excreted in the feces. Conversely, when iron stores are low, iron absorption increases until iron stores and plasma iron are replenished. Thus, a feedback mechanism involving absorption, transport, and storage maintains a stable supply of iron for hemoglobin synthesis.

Erythropoietin Regulates Red Blood Cell Production

Hemopoietins are hormones that regulate the production of specific kinds of blood cells. Erythropoietin is a hormone that regulates erythropoiesis. A reduction in oxygen tension in systemic arterial blood stimulates the kidneys to release erythropoietin, which travels via the blood to the marrow. In the marrow, erythropoietin stimulates the differentiation of erythrocyte progenitor cells and shortens the maturation time of precursor cells thereby increasing the rate of red blood cell formation and release (Fig. 17–10). Under conditions of maximal stimulation (such as when the tissues seriously lack oxygen, a condition termed severe **hypoxia**), the erythropoiesis can increase five-fold to correct the deprivation.

The synthesis of erythropoietin in turn is influenced by sex hormones. **Androgens** (male sex hormones) such as testosterone stimulate production of erythropoietin; in part, this stimulation may account for the higher RBC count normally seen in adult males.

Old Erythrocytes Are Destroyed by Macrophages

The erythrocyte has no nucleus, mitochondria, endoplasmic reticulum, or other organelles required for cell maintenance and thus has a very limited lifespan. As erythrocytes age, changes in their plasma membranes make them susceptible to recognition and ingestion by **macrophages** in the spleen, liver, and

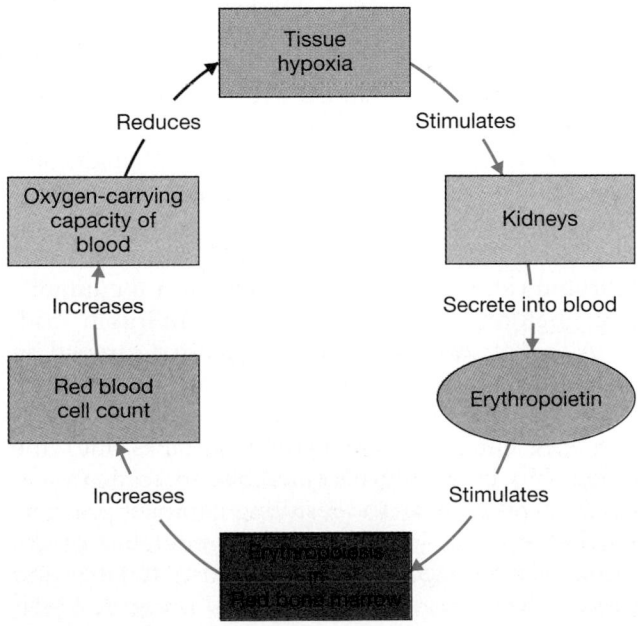

Figure 17–10

The mechanism of erythropoietin action.

Figure 17–11

The metabolism of hemoglobin.

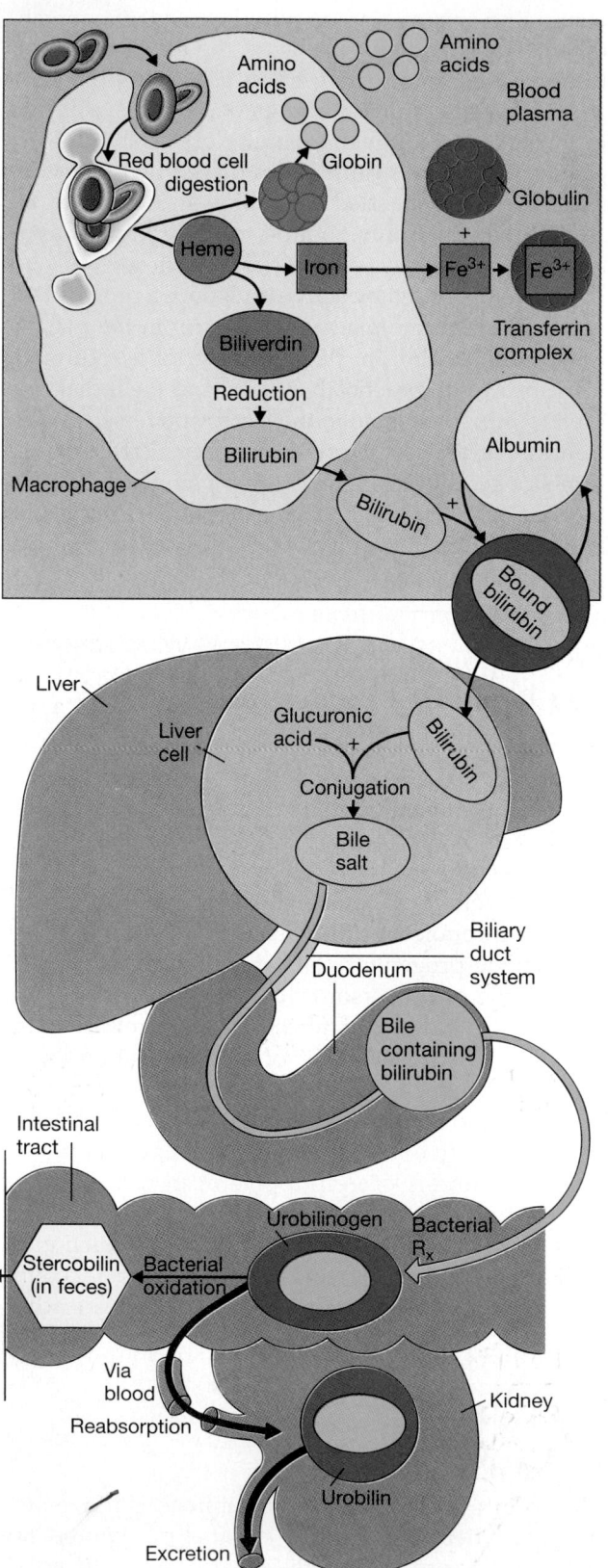

marrow. About 230 billion red cells are destroyed each day.

When the red cell is destroyed by a macrophage, hemoglobin is broken down, and important components of the molecule are recycled within the body (Fig. 17–11). The peptide components are reduced to individual amino acids, which may be reused by other cells for protein synthesis. The heme portions are broken down into iron (Fe^{3+}) and **biliverdin,** a greenish pigment, which is reduced to **bilirubin** and released into the plasma. In the plasma, bilirubin binds to **albumin,** a carrier protein, and is transported to the liver. The liver cells take up the bilirubin, join it to the **glucuronic acid,** and excrete the complex as a bile salt into the biliary duct system which empties into the small intestine. Intestinal bacteria convert bilirubin into **urobilinogen,** most of which is eliminated in the feces in the form of **stercobilin,** a brown pigment that colors stools. Some urobilinogen is absorbed from the intestine and excreted by the kidney into the urine, where it becomes oxidized to **urobilin,** the amber coloring matter in urine. The iron is released into the plasma, where it combines with transferrin and travels to other cells for storage or reuse.

Bilirubin is very toxic to the nervous system and, if allowed to accumulate in the body, can cause neural damage. An elevation of blood bilirubin above normally low levels is called **hyperbilirubinemia.** This condition may result from an excessively high rate of cell destruction, liver disease, or blockage of the liver biliary system. **Icterus,** or **yellow jaundice,** a symptom of hyperbilirubinemia, is caused by an elevation in plasma bilirubin to the point where the eyes, skin, and mucous membranes appear yellow. Hyperbilirubinemia is common in newborn infants, who often become jaundiced after the first day of life. The chief cause is an immature liver unable to excrete bilirubin as fast as it is being formed. Fortunately, bilirubin is a photosensitive pigment that decomposes when exposed to near-ultraviolet light; thus, when the jaundiced infant's skin is exposed to the appropriate wavelength of light in an incubator, blood bilirubin can be kept at low levels until the infant's liver matures enough to handle the bilirubin load.

Although most old erythrocytes are destroyed by macrophages, a few erythrocytes actually disintegrate within the blood. When the red blood cell membrane is damaged, hemoglobin escapes into the blood, but it quickly becomes bound to a protein called **haptoglobin.** The resulting complex prevents the kidney from excreting the hemoglobin in the urine. Macrophages in the liver then remove and process the hemoglobin, as indicated in Figure 17–11.

Red Blood Cell Indices Are Used to Assess Anemia

About 98% of the oxygen transported by blood is carried within the red blood cells where oxygen becomes loosely bound to hemoglobin, a protein pigment that colors the blood red. A deficiency in the amount of normal hemoglobin present in each of the red blood cells or a deficiency in the number of circulating red blood cells reduces oxygen delivery to tissues and impairs systemic function.

Anemia is a condition characterized by an abnormally low oxygen-carrying capacity of the blood, resulting from marked deficiencies in the number of circulating red cells, the amount of hemoglobin in the blood, or both. Anemia is considered to be present if the hemoglobin is less than 12 g/dl or the hematocrit is less than 37%. Because the oxygen-carrying capacity of their blood is reduced, anemic persons may experience physical weakness, shortness of breath, and difficulty in performing mental work. Chronic physical and mental fatigue are common symptoms.

Anemia may result from a failure of red blood cell production or an excessive rate of cell loss, or both. Anemias are classified on the basis of cell morphology (shape and size) and red blood cell indices.

With respect to morphology, red blood cells are described in specific terms. A **normocytic** erythrocyte is normal in size and shape; a **normochromic** erythrocyte is normal in color (and thus in hemoglobin content). A **microcytic** erythrocyte is smaller than normal; a **macrocytic** erythrocyte larger than normal. A **hypochromic** erythrocyte is paler in color than normal (indicating insufficient hemoglobin); a **hyperchromic** erythrocyte is deeper in color than normal. **Anisocytosis** is a variation in size, **poiklocytosis** a variation in shape. The optimum situation is for erythrocytes to be normocytic and normochromic.

Red blood cell indices are used to detect abnormalities in erythrocyte size, shape, and color. Three commonly used indices are the **mean corpuscular volume** index **(MCV)**, the **mean corpuscular hemoglobin (MCH)**, and the **mean corpuscular hemoglobin concentration (MCHC)**.

The MCV index is a quantitative assessment of the red blood cell volume reported in cubic micrometers. It is computed by the following formula:

$$MCV = \frac{\text{hematocrit} \times 10}{\text{RBC (millions per } \mu\text{l)}}$$

The normal adult value of MCV is 87 μm^3 plus or minus 5 μm^3. A low MCV indicates microcytosis, and a high MCV indicates macrocytosis.

The MCH index is used to determine the concentration of hemoglobin within the erythrocytes. It is reported in micromicrograms ($\mu\mu g$) picograms (pg) and is computed as follows:

$$MCH = \frac{\text{g/dl hemoglobin} \times 10}{\text{RBC (millions per } \mu\text{l)}}$$

The normal adult value of MCH is 29 $\mu\mu g$ plus or minus 5 $\mu\mu g$. A low MCH indicates microcytosis, hypochromia, or both. A high MCH suggests hyperchromia.

The MCHC is an index of the proportion of hemoglobin measured (weight/volume) per average red blood cell in a sample of blood. It is calculated by the following formula:

$$MCHC = \frac{\text{g/dl hemoglobin} \times 10}{\text{hematocrit}}$$

The normal adult value for MCHC is 34% plus or minus 2%. A low MCHC means hypochromia, but a high MCHC indicates a loss of RBC volume without a proportionate loss of hemoglobin (microcytosis). Table 17–2 summarizes normal red blood cell values for adult males and females.

In addition to changes in red cell morphology and indices, most forms of anemia are characterized by an elevated reticulocyte count. Increased numbers of circulating reticulocytes indicate increased bone marrow activity with early release of reticulocytes. This is common in hemolytic anemias, for example, when red blood cells are destroyed faster than marrow can produce them.

Anemia is a sign of disease. There are as many types of anemia as there are many underlying causes. Only a few of the more common anemias will be discussed here.

Iron-deficiency anemia is caused by lack of dietary iron, inadequate absorption of iron, loss of iron stores, or abnormalities in iron utilization. Deficiencies in iron metabolism leads to a reduction in he-

Table 17–2
Normal Red Cell Values (Adult)

RBC Measurement	Male	Female
Red cell count (millions/μl)	5.4 ± 0.8	4.8 ± 0.6
Hemoglobin (gm/dl)	16.0 ± 2.0	14.0 ± 2.0
Hematocrit (%)	47.0 ± 5.0	42.0 ± 5.0
MCV (μm^3)	87 ± 5	87 ± 5
MCH ($\mu\mu g$)	29 ± 5	29 ± 5
MCHC (%)	34 ± 2	34 ± 2

moglobin production and the release of microcytic, hypochromic red blood cells from the marrow. Good dietary sources of iron include liver, red meat, egg yolks, carrots, and spinach.

Pernicious anemia occurs because of a lack of intrinsic factor and a resultant deficiency in the absorption of vitamin B_{12}. The lining of the stomach secretes a glycoprotein called the **intrinsic factor,** which is essential for the absorption of vitamin B_{12} in the small intestine. Vitamin B_{12} and the intrinsic factor combine to form the **erythrocyte maturation factor,** which is required for maturation of red blood cells in the marrow. Genetic lack of the intrinsic factor or inadequate dietary intake of vitamin B_{12} leads to the impairment of erythrocyte maturation. The marrow releases large nucleated red blood cell precursors called **megaloblasts** that contain more hemoglobin than normal red blood cells but that do not function normally. Pernicious anemia is characterized by the presence in the blood of macrocytic, hyperchromic, immature red blood cell precursors and a reduction in the numbers of normal erythrocytes.

Aplastic anemia occurs as a result of a decrease in the production of all blood cells in the marrow (**aplasia** denotes faulty or incomplete development). Blood-forming cells do not mature at a normal rate in this disorder. The resultant anemia is characterized by reduced numbers of normal red blood cells. Aplastic anemia can be caused by radiation, cytotoxic drugs used for the treatment of malignant disease, exposure to chemicals (arsenic, DDT, benzene) that poison the marrow, or genetic failure of bone marrow development.

Hemolytic anemia occurs when the rate of red blood cell destruction exceeds the capacity of the marrow to replace lost cells. Hemolytic anemias are usually accompanied by jaundice and elevated serum bilirubin. Causes of hemolytic anemia include damage to the red blood cell or red blood cell defects (such as hereditary spherocytosis and sickle cell disease) that cause early destruction by macrophages and various extracellular agents (toxins, antibodies) that cause hemolysis.

Thalassemias are a group of disorders caused by inheritance of one or more genes that limit the rate of synthesis of alpha or beta chains of hemoglobin. The impairment leads to a hypochromic, microcytic anemia, the severity of which is determined by the number and kind of genes affected. **Thalassemia major** is usually homozygous and most often affects beta chain synthesis (**B-thalassemia major, Cooley's anemia**). Beta-thalassemias occur most frequently in inhabitants of the Mediterranean area (*thalassa* = "sea"), and alpha-thalassemias are more frequent among Asians and Central Africans.

Hemorrhagic anemias occur after acute or chronic blood loss, such as associated with hemorrhoids, bleeding ulcers, excessive menopausal bleeding, etc., and are usually characterized by normocytic, normochromic erythrocytes.

Polycythemia Is an Increase in Red Blood Cell Count

Polycythemia (*poly* = "many"; *cyte* = "cell") is a relative or absolute increase in the number of circulating red blood cells above 6.2 million per cubic millimeter. It may result from an increase in the concentration of erythrocytes without an increase in the total red blood cell mass (**relative polycythemia**) due to dehydration and hemoconcentration, or it may result from an increase in the concentration of erythrocytes accompanied by an increase in total cell mass (**polycythemia vera**) due to the hyperactivity of the bone marrow.

Secondary polycythemia (erythrocytosis) is a physiological polycythemia that occurs in response to arterial hypoxia and its stimulation of erythropoiesis. Exposure to chronic hypoxic (low oxygen) environments, such as normally found at high altitudes, is often accompanied by secondary polycythemia. It may also occur in emphysema, pulmonary fibrosis, carbon monoxide poisoning, and other abnormal states in which oxygenation of the blood is reduced.

Polycythemia vera (*true polycythemia*) is a chronic, slowly progressive disease involving excessive production of all blood cells produced in the marrow (**hyperplasia**). In most cases the cause is unknown (**idiopathic**), while in others the polycythemia may be evidence of tumors in the bone marrow, kidney, or brain. Polycythemia vera is not associated with lack of oxygen in the blood.

Polycythemias can be life-threatening because they increase the viscosity of the blood, placing an increased load on the heart and blood vessels. They also make an individual more susceptible to thrombosis, the intravascular formation of blood clots.

Blood Types

Blood Types Are Determined by Molecules Embedded in the Surface of the Red Blood Cell

All human blood contains erythrocytes, leukocytes, platelets, and plasma; however, not all blood is the same. When some types of blood are mixed, red blood cells aggregate into clumps (a process called **agglutination**). Agglutinated cell masses can become lodged in small blood vessels, blocking the flow of blood and seriously impairing tissue and organ function (e.g., kidneys, heart, brain). If agglutination is widespread (major agglutination), death can occur. Since many circumstances require transfusion of whole blood from one person to another, the donor type must be carefully selected so as to avoid major agglutination in the recipient.

Incompatibilities of blood types are due to the presence of antigens embedded in the erythrocyte plasma membrane, and specific antibodies in the blood plasma. An **antigen** is a molecule or a part of a molecule, recognized by the body as foreign. An antigen (*anti* = "against," *gene* = "to originate") stimulates specific defense mechanisms of the body (see Chapter 28) that include the production and secretion of an antibody, a protein that reacts with the antigen normally to nullify its harmful effects. When antibodies react or bind to erythrocyte antigens, the affected red blood cells clump together or agglutinate; therefore, erythrocyte antigens are called **agglutinogens** and the corresponding antibodies are called **agglutinins**. The presence or absence of erythrocyte antigens is determined genetically; that is, blood type is inherited.

Several hundred types of erythrocyte antigens have been identified. Nearly all of them are weakly antigenic; that is, they do not provoke a strong immune response when given to a person who does not normally possess it. The recipient would not have, nor develop large amounts of the corresponding antibody. However, a few erythrocyte antigens are strongly antigenic, causing the recipient to reject, by agglutination, the donated erythrocytes. Three strong antigens that may be present in erythrocytes are the **A, B,** and **D agglutinogens.** The corresponding antibodies, normally absent if the antigen is present, are designated as **agglutinin a (anti-A), agglutinin b (anti-B),** and **agglutinin d (anti-D).** The presence or absence of agglutinogens and agglutinins allows the blood to be classified into types or groups. We will discuss only the blood types of the ABO and D (Rh) blood groups.

The ABO Blood Group Contains Four Major Blood Types

The presence or absence of the A and B antigens and antibodies determines blood type in the ABO blood group as follows (Fig. 17–12): **Type A** blood contains the A antigen on its red blood cell membranes. The

THE ABO BLOOD GROUP

Blood Types:	A	B	AB	O
RBC types				
Plasma antibody (serum)	Anti-B b	Anti-A a	Neither a nor b	a b
Agglutination reactions	A + a Agglutination	B + b Agglutination	AB + a or b Agglutination	O + a or b No agglutination

Figure 17–12

The ABO blood group.

Table 17–3
The ABO Blood Group

Blood Type	RBC Antigen (Agglutinogen)	Plasma Antibody (Agglutinin)	Population Distribution (Caucasian)
A	A	b	41%
B	B	a	10%
AB	A and B	Neither a nor b	4%
O	Neither A nor B	a and b	45%

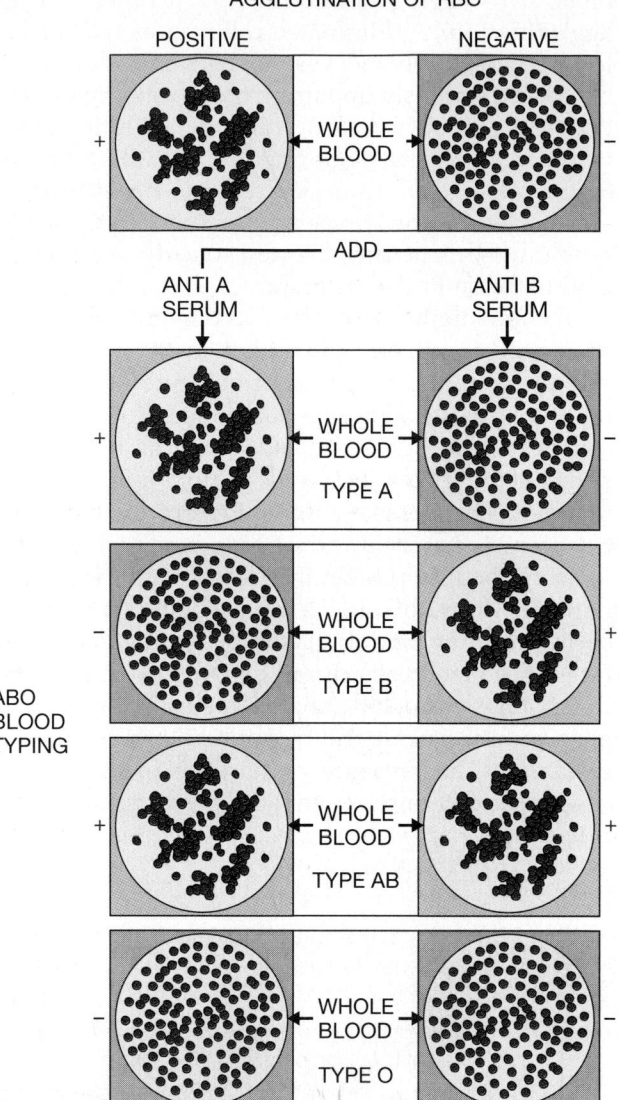

Figure 17–13

ABO blood typing.

plasma contains anti-B. **Type B** blood contains the B antigen on the plasma membranes of its red cells. The plasma contains anti-A. **Type AB** blood contains both the A and B antigens on its red blood cell membranes, but the plasma lacks both anti-A and anti-B. **Type O** blood contains neither the A nor the B antigens, but the plasma contains both anti-A and anti-B. The blood types of the ABO group and their constituents are listed in Table 17–3.

When Type B blood is transfused into a Type B recipient, no reaction occurs because the recipient's body does not recognize the blood as being foreign. When type B is transfused in a Type A recipient, two reactions occur:

1. The anti-A in the donor's plasma reacts with the A antigen on the recipient's red cells (**minor agglutination**).

2. Anti-B in the recipient's plasma reacts with the B antigen on the donor's red cells (**major agglutination**).

The first reaction is of little consequence because the anti-A in the donor's plasma (serum) becomes greatly diluted on entering the recipient's blood and thus becomes inconsequential. Agglutination of recipient cells occurs, but the agglutination is minor. The second reaction, however, is of major consequence because larger quantities of recipient anti-B (compared to much smaller quantities of donor anti-A) are available to react with the B antigens of the donor's cells. When large numbers (about 5 million per microliter of blood) of donor red cells enter the recipient's body, widespread agglutination occurs, and a transfusion reaction develops in the recipient as the donor's cells are agglutinated by the recipient's antibodies. A **transfusion** reaction is characterized by general discomfort, anxiety, difficulty in breathing, flushing of the face, pain in the chest

and neck, and other variable symptoms leading to shock—as evidenced by rapid feeble pulse, cold clammy skin, fall in blood pressure, nausea, and vomiting. These acute symptoms usually occur within the first two hours. Depending on the amount of blood administered, more severe complications involving kidney, heart, lung, liver, and brain damage may develop, leading to death several days later.

Blood may be typed and cross-matched to determine its compatibility with other blood. **Blood typing** for the ABO group involves placing serum containing a known antibody (anti-A or anti-B) on a microscope slide and mixing it with a small quantity of the blood to be typed (Fig. 17–13). The presence or absence of red blood cell agglutination determines the type of unknown blood (Table 17–4). For exam-

Table 17–4
Hemagglutination Test for Blood Type

Whole Blood	Agglutination Observed with	
Blood type	Anti-A serum	Anti-B serum
A	Yes	No
B	No	Yes
AB	Yes	Yes
O	No	No

ple, since its cells contain A antigens, type A blood would not agglutinate with test serum samples containing anti-B.

The blood type may be confirmed through testing of agglutinin in the serum of the blood whose type is to be confirmed. Such a procedure is called **cross-matching.** A small quantity of serum to be tested is placed on four microscope slides and mixed with red blood cells of type A, B, AB, and O respectively. The presence or absence of red blood cell agglutination indicates the type of antibody present in the test serum, thus confirming the presence or absence of A and B antigens on the red blood cells in the blood from which the test serum was taken (Table 17–5). Cross-matching is clinically useful to confirm compatibility of a potential donor's red blood cells because the donor's erythrocytes may display an antigen, other than A or B, with which existing recipient antibodies would react.

Because Type AB blood contains both the A and B antigens and neither anti-A, nor anti-B, persons with Type AB blood may act as recipients for any of the other types (A, B, or O). Historically, such per-

Table 17–5
Confirmation of Blood Type by Cross-Matching

Serum Antibody	Test Cells		
	A	B	O
a	+	–	–
b	–	+	–
a + b	+	+	–
Neither a nor b	–	–	–

(+) = Agglutination.
(–) = No agglutination.

sons have been referred to as "universal recipients." By the same reasoning, persons with Type O blood have been called "universal donors," because their erythrocytes contain neither the A nor the B antigen to oppose the recipients; anti-A and anti-B. In fact, there are no such things as a "universal donor" and a "universal recipient." The presence of other antigens on the erythrocyte membrane or other erythrocyte antibodies in the plasma precludes this possibility. One such antigen, present in about 85% of the population, is the Rh factor. As we shall see, Type O^+ blood cannot be donated to an AB^- recipient.

The Presence or Absence of Erythrocyte Antigen D Determines Rh Blood Group

The Rhesus blood group, first described in 1940, is named after the monkey in which the antigens were discovered. As with the A and B antigen, the presence or absence of the Rh antigen is determined genetically. Most of the Rh antigens are weakly antigenic and thus of little clinical importance; however, one of the Rh antigens, called **antigen D,** is strongly antigenic. The D antigen is commonly called the **Rh factor.**

If the Rh factor is present on the red cell, the blood is said to be **positive;** if it is absent, the blood is said to be **negative.** Antibodies to the Rh factor are not normally present in the plasma of Rh negative blood. However, on exposure to the Rh antigen (through transfusion or mixing of fetal and maternal blood during childbirth), Rh negative persons will produce antibodies against the Rh antigen, so that the next time negative blood is exposed to positive blood, the positive cells will be destroyed. As a general rule, Rh positive blood should never be transfused into an Rh negative recipient; however, assuming compatibility in the ABO and other blood groups, Rh negative blood may be given to an Rh positive recipient. Blood of the type A^+, for example, may not be given to an A^- recipient, but A^- could be given to an A^+ recipient.

An important consequence of the Rh blood group concerns the Rh negative mother who bears an Rh positive fetus for the first time. There are usually no blood complications if the mother has never been exposed to Rh positive blood. The mother would posses few, if any, Rh antibodies (anti-D), and the fetal erythrocytes containing the antigen D cannot cross the placenta and enter the maternal blood to stimulate the mother's immune system. At birth, however, the placental barrier is destroyed, allowing fetal and maternal blood to mix. When fetal erythrocytes enter the maternal circulation, they stimulate the mother's immune system to produce Rh antibodies, which remain in her blood for several years. Upon conceiving and carrying a second Rh positive fetus, the maternal Rh antibodies can cross the placenta, enter fetal blood, and cause the fetal erythrocytes to agglutinate, resulting in a form of hemolytic anemia called **erythroblastosis fetalis.** Unless the agglutinated fetal erythrocytes are replaced, the fetus could die from hypoxia and other complications. Fortunately, the chance of developing erythroblastosis fetalis may be significantly decreased by administration of Rh antibodies (RhoGAM) to the mother within 72 hours following birth of an Rh positive child. The antibodies react with any positive fetal erythrocytes that have entered the mother's immune system. As a result of this procedure, called desensitization, the circulating titer (concentration) of maternal Rh antibodies usually will not increase to the point of interfering with the successful carrying of a second Rh positive fetus.

Leukocytes: White Blood Cells

Leukocytes (*leuko* = "white"; *cyte* = "cells") are called **white blood cells (WBC)** because they lack pigmented (color) molecules such as hemoglobin. White blood cells must be stained in order for the different types to be distinguished and to be more visible when they are counted.

Leukocytes Are Classified into Two Main Categories: Granular and Agranular

There are several types of leukocytes (Fig. 17–14). Each leukocyte is nucleated; some contain large cytoplasmic vesicles, or granules, while others do not. These granules are mostly lysosomes. Leukocytes are classified according to the shapes of their nuclei and the presence or absence of cytoplasmic granules.

Leukocytes containing large granules are called granular leukocytes, or granulocytes. Granulocytes also have a complex lobulated nucleus (a nucleus consisting of several lobes, or segments) and are therefore called **polymorphonuclear leukocytes.** The types of granulocytes, named according to the staining properties of their granules are the neutrophil, the eosinophil, and the basophil.

Leukocytes without a granular cytoplasm are called agranulocytes. They have a simple, unlobulated nucleus and thus are designated as **monomorphonuclear leukocytes.** The two basic types of agranulocytes are the monocyte and the lymphocyte.

On average, whole blood contains 7000 to 10,000 leukocytes/μl; this figure represents the **combined white blood cell (WBC) count.** Because leukocytes have different functions, however, it is diagnostically useful to count the white blood cells according to type; this count is called the **differential white blood cell count.** When a differential count is performed, the number of each type of leukocyte is usually expressed as a percentage of the total or combined white blood cell count. A neutrophil count of 65%, for example, means that neutrophils make up 65% of all leukocytes present in a blood sample. Normal differential counts for the leukocytes are given in Figure 17–14.

Leukocytes function collectively to combat foreign substances that enter the body. They are part of a system of defensive cells that **phagocytize** (engulf and destroy) material, detoxify poisons, produce antibodies, and release chemical messengers, enzymes, and other substances. Each type of leukocyte performs different functions in body defense, but each function is necessary for an integrated and effective defense.

Leukocytes are not confined to the blood or lymph; they may also be found in other tissues, particularly in loose connective tissue. They are motile cells, traveling via **ameboid movement** (movement similar to that of an amoeba, in which the cell pro-

Figure 17–14

Types of leukocytes.

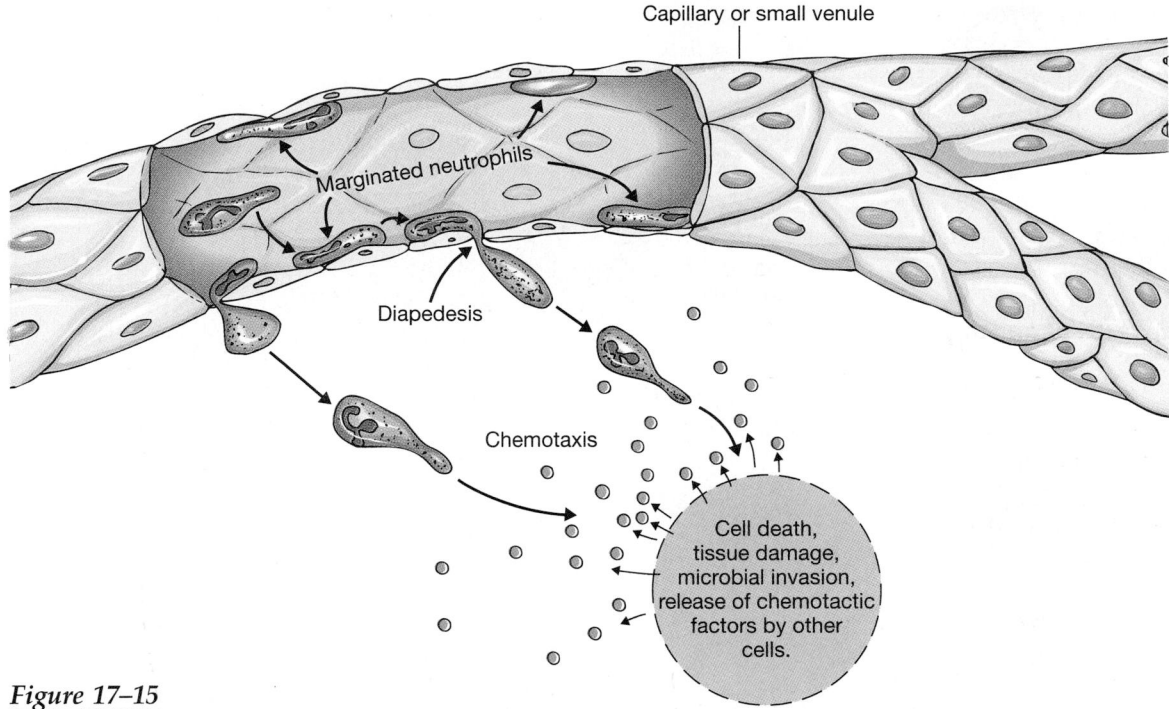

Capillary or small venule

Marginated neutrophils

Diapedesis

Chemotaxis

Cell death, tissue damage, microbial invasion, release of chemotactic factors by other cells.

Figure 17–15

Leukocyte diapedesis and chemotaxis.

jects protoplasmic extensions ahead and then follows them). They are capable of passing through the capillary wall (a process called **diapedesis**) to enter the spaces around the blood vessels whenever necessary. Leukocytes are attracted to sites of injury, inflammation, or bacterial invasion when chemical substances are liberated by damaged cells or bacteria, and when immune complexes form during immune responses. The attraction and subsequent movement of leukocytes toward these chemical substances is known as **chemotaxis** (Fig. 17–15).

Granulocytes: Granular Leukocytes

Neutrophils.   The neutrophil functions as the body's first line of defense against invasion by **pyrogenic** (pus-forming) **bacteria.** Neutrophils are the most numerous white blood cells. They are very motile and actively phagocytic. They can ingest and destroy several kinds of microbes, small particulate matter, and fibrin (from blood clots). Most granules in the cytoplasm of neutrophils are lysosomes containing **hydrolytic enzymes** that digest ingested matter. Other cytoplasmic granules contain **phagocytins,** a group of antibacterial proteins. Neutrophils are usually the first white cells to arrive at the site of a bacterial invasion, and they arrive in large numbers. In the process of defending the body against microbial invasion, many neutrophils die (Fig. 17–16), forming what is called **pus** at the invasion site. Dying neu-

trophils also release digestive enzymes into surrounding connective tissue. To a limited extent, these enzymes damage tissue and contribute to localized pain and swelling at a site of inflammation. After the invading or traumatizing factor has been controlled, many neutrophils remain to assist in healing.

Eosinophils.   The eosinophil is less motile than the neutrophil and, although capable of phagocytosis, it is not as active in this regard. The eosinophil does not, for example, normally phagocytize bacteria. The lysosomes of the eosinophils contain many enzymes, such as **oxidase, peroxidases,** and **phosphatases,** whose presence indicates that the primary function of eosinophils is detoxification of foreign proteins and other substances. As do neutrophils, eosinophils exhibit chemotaxis; however, their attraction to traumatized tissues appears to depend on the presence of antibodies specific to foreign proteins. Antigen-antibody complexes have been shown to cause eosinophils to travel from the blood into connective tissues, where they phagocytize and destroy the complexes.

Normally, eosinophils are more abundant in connective tissues other than the blood—in particular, the lung, mammary glands, omentum, and inner wall of the small intestine. They are observed to increase in number in allergic and autoimmune reactions, during decomposition of body protein, and in certain parasitic infections. It is believed that blood basophils

Figure 17–16

(*a*) The processes of phagocytosis, bacterial lysis, and neutrophil death. (*b*) Scanning electron micrograph of an angry macrophage engulfing a yeast cell. (*© Biology Media/Science Source/Photo Researchers, Inc.*)

and mast cells (connective tissue basophils) release **eosinophilic chemotactic factors** that attract eosinophils to an area of tissue damage.

Basophils.   Basophils are the rarest of the leukocytes and are usually found in greater numbers outside the blood in loose connective tissue. They show little ameboid movement and do not contain lysosomes. Many of their cytoplasmic granules contain either **heparin,** an anticoagulant, or **histamine,** a vasodilator substance that also increases capillary permeability. Basophils release histamine in areas of tissue damage in order to increase blood flow, attract neutrophils, and facilitate the movement of leukocyte emigration from blood to interstitial fluids in the area of damage. The effects of histamine release have been experienced by everyone who has suffered from the common cold: invasion of the nasal tissue by viruses induces basophils to gather at the invasion site and release histamine, which causes swelling and congestion of the tissue. Antihistamines and vasoconstrictors contained in decongestant nasal sprays may provide relief but actually interfere with the body's normal defense and may prolong infection.

Agranulocytes: Agranular Leukocytes

The two types of agranulocytes are the monocyte and the lymphocyte. Monocytes constitute 1 to 6% and lymphocytes 20 to 40% of the circulating white blood cells. Lymphocytes are second in number only to the neutrophil. All lymphocytes appear similar, but they differ in their development, functions, and life span. There are two functionally distinct types of lymphocytes: **T-lymphocytes** and **B-lymphocytes.**

T-lymphocytes.   T-lymphocytes originate from stem cells in the bone marrow (Fig. 17–17). They enter the blood and complete their development as they pass through the **thymus gland,** a lymphoid organ in the thorax. T-lymphocytes are the most numerous type of lymphocyte in the blood. They continually migrate between the spleen, lymph nodes, and connective tissues. As T-lymphocytes pass through the thymus, receptors are inserted in them that allow the T-lymphocyte to recognize antigens (as discussed earlier, chemicals foreign to the body that evoke a protective response, the "immune response").

These lymphocytes are said to be "preconditioned" or "competent." Competent T-lymphocytes respond to antigens either by attacking them directly, or by releasing chemicals called lymphokines that attract granulocytes to the area and stimulate B-lymphocytes and other T-lymphocytes. Activated T-lymphocytes may also release nonspecific toxins that neutralize the antigens. T-lymphocytes become activated when antigens contact them and bind to specific membrane receptors. Activated T-lymphocytes give rise, through cell division, to clones of lymphocytes that are responsive to the activating antigen,

PLURIPOTENTIAL
HEMATOPOIETIC
STEM CELL

(COLONY FORMING UNITS)
MYELOID PROGENITOR CELLS

LYMPHOID PROGENITOR CELL

CFU-Eo

CFU-Ma

CFU-GM

CFU-G

CFU-M

Myeloblast

Myeloblast

Myeloblast

Monoblast

Lymphoblast

Promyelocyte

Promyelocyte

Promyelocyte

Promonocytes

Prolymphocyte

Eosinophilic
myelocyte

Basophilic
myelocyte

Neutrophilic
myelocyte

Lymphocytes

Eosinophilic
metamyelocyte

Basophilic
metamyelocyte

Neutrophilic
metamyelocyte

Monocytes

T-lymphocytes

B-lymphocytes

Bands

Bands

Bands

Plasma   Memory
cells   cells

Antibodies

Eosinophil

Basophil

Neutrophil

CD8+   CD4+

In tissue
macrophages

T₈ Suppressor
cell

T₈ Cytotoxic
cell

T₄ Helper
cell

T₄ Inducer
cell

Figure 17–17

The processes of leukopoiesis.

thus conferring cell-mediated immunity. T-lymphocytes also play important roles in helping B-lymphocytes function normally.

B-lymphocytes.   B-lymphocytes develop in the bone marrow and migrate by way of the blood to lymphoid tissues in the spleen, tonsils, lymph nodes, and walls of the small intestine, where they remain for variable periods of time before recirculating in the blood and lymph. When stimulated by an antigen, some B-lymphocytes become specialized to make and secrete **antibodies,** proteins that react with the antigens and help destroy them. Other B-lymphocytes become specialized to recognize the antigen should it re-enter the body at a later date; these B-lymphocytes are called **memory cells.** B-lymphocytes can live for many years and are the basis of the body's **humoral** or **antibody-mediated immunity.** Mechanisms of humoral and cell-mediated immunity will be discussed in greater detail later in Chapter 28.

Monocytes.   **Monocytes** are the largest of the leukocytes, some of them being nearly three times the size of an erythrocyte. Monocytes originate in the bone marrow, taking about 3 days to develop before being released into the blood. They remain in circulating blood for only a few days, after which they migrate to connective tissues in various organs. There they develop into **macrophages,** large phagocytic cells that form part of the body's defense against microorganisms and harmful chemicals.

Some tissue macrophages, such as those in the liver **(Kupffer cells)** or lung **(alveolar macrophages)** do not move about as much as do macrophages in loose connective tissue, but all macrophages display ameboid movement and are actively phagocytic. They contain a variety of digestive enzymes and are very effective in destroying bacteria, detoxifying harmful chemicals, and cleaning up cellular debris in areas of tissue damage.

Leukopoiesis Is the Generation of White Blood Cells

Leukopoiesis, the formation of white blood cells, begins with the **pluripotential hemopoietic stem cell (PHSC)** in the marrow (see Fig. 17–17). PHSCs develop into **lymphoid progenitor cells,** which give rise to T cell precursors and B cell precursors, and **myeloid progenitor cells** (CFU-GM, CFU-Eo, CFU-Ma) which give rise to the granulocytes, monocytes, and mast cells (tissue basophils). Granulocytes and monocytes develop to maturity in the marrow. Lymphocytes usually begin their lives in the marrow, but most leave before maturing. Potential T-lymphocytes mature in the thymus and the spleen. Potential B-

lymphocytes develop and mature in lymphoid tissue of the intestines, the spleen, and the bone marrow.

Uncommitted stem cells can leave the marrow and become **free stem cells** moving from organ to organ via the blood and lymph. Thus, hematopoietic centers in the marrow may serve to maintain lymphoid tissue via "seeding" with free stem cells. In turn, lymphoid organs may become sites of **extramedullary hematopoiesis** when the marrow fails to function normally.

Granulopoiesis (formation of granulocytes) has been shown to be stimulated by the presence of **leukocyte-inducing factor (LIF),** a specific factor in plasma that can be demonstrated when leukocytes are depleted. Various other chemicals released into the blood when tissue is damaged also stimulate leukopoiesis.

Granulocyte, monocyte, and megakaryocyte formation has been shown to be regulated by chemicals called colony stimulating factors (CSFs), which are glycoproteins secreted by a variety of cell types both within and outside of the bone marrow. Six CSFs, their target cells, and the colonies of mature blood cells they regulate are listed in Table 17–6.

Lymphopoiesis, the formation of lymphocytes, is regulated by an array of diffusable signaling molecules called **lymphokines** or **interleukins,** chemicals secreted by macrophages and white blood cells to communicate with one another (*inter* = "between," *leuko* = "white") as they coordinate a defensive response to an antigen. To date, thirteen interleukins (IL-1, IL-2, etc.) have been characterized as to their source, target cells, and effects. These molecules and other regulators of white blood cell production are discussed in greater detail in Chapter 28.

The life span of a leukocyte depends on the role it plays in defense. Neutrophils and other actively motile and phagocytic leukocytes tend to live from minutes to a few days because they are continually defending the body against invasion by bacteria and dying in the process. On the other hand, B-lymphocytes and some T-lymphocytes may live and continue to function normally for years before being destroyed. Old leukocytes are destroyed by the liver, spleen, marrow, and lymph nodes.

Leukocytosis Is an Increase in White Blood Cell Count

Leukocytosis is an abnormal increase in the total number of circulating leukocytes—sometimes as many as 400,000 cells per microliter. Leukocytosis may have physiological or pathological causes. All combined white blood cell counts above 50,000 cells/μl indicate the latter. **Physiological leukocytosis** (WBC count 10,000 to 20,000/μl) is non-

Table 17–6
Colony Stimulating Factors

Factor	Principal Target Cell	Regulated Colony
Monocyte-colony stimulating factor (M-CSF)	Colony forming unit-monocyte (CFU-M)	Monocyte
Granulocyte-colony stimulating factor (G-CSF)	Colony forming unit-granulocyte (CFU-G)	Neutrophil
Granulocyte-monocyte-colony stimulating factor (GM-CSF)	Colony forming unit-granulocyte-monocyte (CFU-GM)	Neutrophil monocyte
Multi-colony stimulating factor (Multi-CSF) (IL-3)	Colony forming unit-granulocyte, eosinophil, monocyte, megakaryocyte (CFU-GEMMe)	Neutrophil, eosinophil, monocyte, megakarocyte
Megakaryocyte-colony stimulating factor (Meg-CSF)	Colony forming unit-megakaryocyte (CFU-Me)	Megakaryocyte
Erythropoietin	Colony forming unit-erythrocyte (CFU-E)	Erythrocyte

pathological and occurs in newborn infants and in pregnancy, emotional disturbances, menstruation, and severe exercise. **Pathological leukocytosis** occurs in bacterial and viral infections, metabolic and hormonal disturbances, allergies, and malignancies.

Leukemia is a malignant disease characterized by the proliferation of hematopoietic cells that are immature and therefore functionally impaired. Environmental factors (e.g., radiation, benzene), genetic factors, and RNA viruses have been implicated. Leukemias are classified by the course and duration of the illness and the abnormal type of cells and tissue involved. **Acute leukemias** (e.g., **acute lymphocytic leukemia, acute myelogenous leukemia**) are characterized by rapid onset, massive numbers of immature leukocytes, severe anemia, and other signs. **Chronic leukemias** (e.g., **chronic granulocytic leukemia, chronic lymphocytic leukemia**) are characterized by gradual onset and increased numbers of mature leukocytes. Chronic leukemias may lead to acute leukemias. Although symptoms and complications vary with the type of leukemia, they generally include anemia, excessive bleeding due to impairment of thrombocytes, and local and systemic infection due to impaired granulocyte function.

Leukopenia Is a Reduction in White Blood Cell Count

Leukopenia is an abnormal reduction in the total number of circulating leukocytes. It may result from bone marrow defects, arrest of cell development and maturation, or excessive destruction of leukocytes. **Agranulocytosis** is a type of leukopenia characterized by a marked reduction in neutrophils. Often, it is caused by slowed functioning of the bone marrow due to irradiation or cytotoxic drugs. As a result of agranulocytosis, the body's defenses against bacterial invasion are considerably weakened and the chance of dying from bacterial infection is greatly increased.

Thrombocytes (Platelets)

A **platelet** or **thrombocyte** (*thrombus* = "clot") is a cell fragment split from a large cell called a **megakarocyte** (Fig. 17–18). One megakaryocyte gives rise to about 6000 thrombocytes. Since thrombocytes are not actually cells produced by a mitotic division of a parent cell, the term **platelet** is a more commonly used.

Megakarocytes derive from pluripotential hemopoietic stem cells in the bone marrow. PHSCs become committed progenitor cells called **megakaryocyte colony forming units (CFU—Me)** which in turn develop into colonies of mature megakaryocytes (see Fig. 17–18). The development of megakaryocytes **(megakaryocytopoiesis)** is regulated by several interleukins (IL-3, IL-6, IL-11) and thrombopoietin, a hormone that stimulates both megakaryocyte development and the release of platelets from megakaryocytes.

Most megakaryocytes remain in the marrow, releasing platelets into the circulating blood. Some megakaryocytes enter the blood and travel to other organs (particularly the lung) where they remain and produce platelets.

Pluripotential hematopoietic stem cell

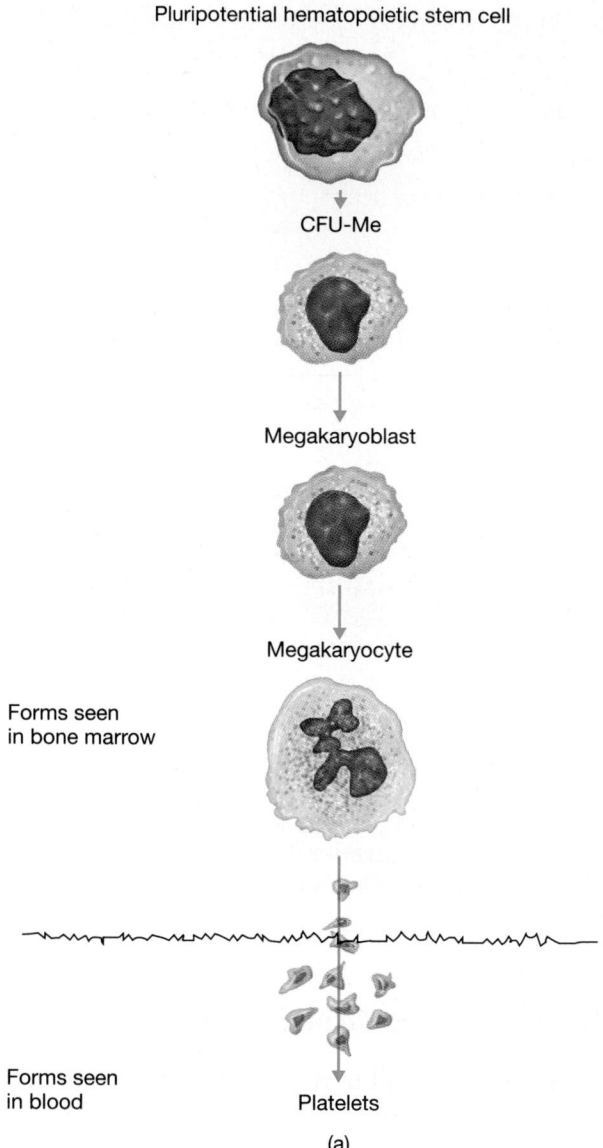

CFU-Me

Megakaryoblast

Megakaryocyte

Forms seen
in bone marrow

Forms seen
in blood

Platelets

(a)

Figure 17–18

(*a*) The development of platelets. (*b*) Photomicrograph of thrombocytes (platelets).   (© *Manfred Kage/Peter Arnold, Inc.*)

Platelets are small, membrane-bounded bodies without nuclei, about 2 to 4 micrometers in diameter. They are packed with granules, vesicles, microfilaments, microtubules, and occasionally mitochondria. They are normally present in numbers ranging from 150,000 to 350,000/μl of blood and play several important roles in hemostasis. Following injury, chemicals contained within or absorbed in platelets are released to the lining of the blood vessels, where they stimulate contraction of the injured vessels to minimize blood loss. Because of their adhesive properties, they clump together at the site of vascular damage, effectively plugging the defective end. In addition, they participate in the formation of factors that initiate coagulation of the blood, and they release **platelet-derived growth factor (PDGF),** a protein that promotes repair after vascular injury. The role of platelets in hemostasis will be discussed later.

Circulating platelets have a lifespan of approximately 1 to 2 weeks. If not consumed in the process of blood coagulation, platelets eventually are destroyed by macrophages in the liver and spleen. The spleen is also an important storage organ for platelets. In certain kinds of stress, such as a hemorrhage or burns, sympathetic stimulation of the spleen results in the release of large numbers of stored platelets into circulating blood.

Hemostasis: Mechanisms to Prevent Blood Loss

The term **hemostasis** (*hemo* = "blood"; *sta* = "remain") refers to mechanisms that minimize or prevent the loss of blood when a blood vessel is opened. Hemostasis is vital because unchecked hemorrhage (blood loss) eventually leads to cardiovascular collapse and death. Four interrelated events constitute hemostasis: (1) local vasoconstriction, (2) formation of a platelet aggregate (clump), (3) formation of a blood clot, and (4) clot retraction and dissolution.

Vasoconstriction Is an Autonomic Nervous System Response

When a blood vessel is injured, its immediate response is to constrict and thereby reduce blood flow. This initial vasoconstriction is due to local spasm of

Aggregated
Platelets   Erythrocyte   Platelet

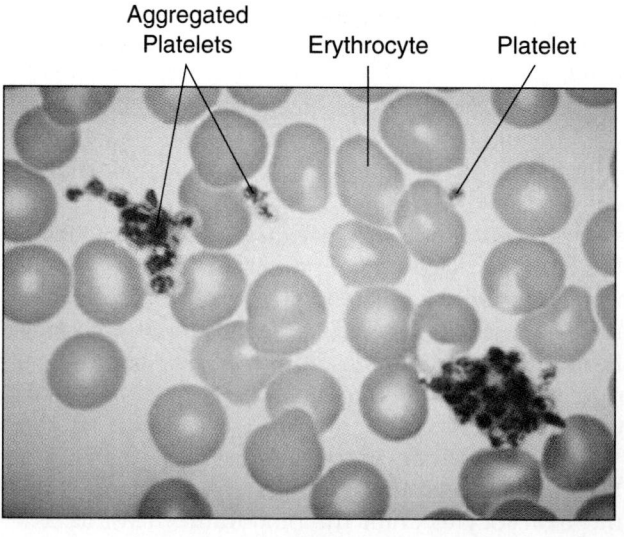

(b)

the smooth muscle in the wall of the blood vessel and to sympathetic reflexes. In smaller vessels, vasoconstriction can be maintained by the release of vasoconstrictor chemicals from platelets that begin to accumulate at the damaged site.

Platelet Aggregation Provides the Framework for Coagulation

Damaged cells of the injured blood vessel release **adenosine diphosphate (ADP),** which attracts platelets and causes them to clump together at the damage site. Platelets coming in contact with exposed collagen of the vascular wall **degranulate** (release stored chemicals), releasing ADP, **serotonin** (a vasoconstrictor), and **platelet factors** necessary for coagulation of blood. The release of ADP attracts more platelets and causes them to swell and become sticky; they adhere in increasing numbers to the damaged site and form a plug (a **platelet aggregate**). Activated platelets also produce **thromboxane A_2,** a powerful vasoconstrictor and platelet aggregator derived from platelet **prostaglandin H_2.** In addition, the spherical platelets extend **pseudopodia** (footlike extensions of their cytoplasm) to nearby exposed collagen, anchoring the platelet plug and establishing a framework upon which coagulation can proceed.

Coagulation Requires the Production of Fibrin

Coagulation is the process by which some of the blood loses its fluid consistency and becomes a clot (a semisolid mass similar in consistency to gelatin). In the formation of a clot, an enzyme called **thrombin** converts **fibrinogen,** a soluble plasma protein, into an insoluble plasma protein, **fibrin.** Fibrin is a threadlike protein. Fibrin aggregates form a meshlike network at the site of vascular damage, trap red blood cells and plasma, and form a clot (Fig. 17–19). The complex sequence of chemical events that produce fibrin are divided into three phases or stages designated as **Stage I, Stage II,** and **Stage III** (Fig. 17–20). Common synonyms for blood clotting factors involved in each stage are given in Table 17–7.

Stage I: Formation of a Prothrombin Converting Factor

Stage I may begin when blood comes in contact with injured tissue (the **extrinsic pathway**) or it may be initiated in the absence of tissue damage (the **intrinsic pathway**).

When blood comes in contact with injured tissue, clotting is initiated via the extrinsic pathway, owing

Figure 17–19

Formation of a clot. Fibrin forms long threads in which blood cells, platelets, and plasma become trapped.

to the release of the **tissue lipoprotein thromboplastin** (Factor III). Tissue thromboplastin interacts with a plasma protein, **proconvertin** (Factor VII), and **calcium ions** to form an agent that activates the **Stuart factor** (Factor X). The activated Stuart factor, in the presence of **calcium ions,** forms complexes with **accelerin** (Factor V) on phospholipid micelles provided by tissue thromboplastin to form the **prothrombin-converting factor.**

Intrinsic coagulation may occur inside or outside of the body. In either case, the first step is the activa-

Figure 17–20

A schema of blood coagulation.

tion of the **Hageman factor** (Factor XII). In the body, activation of the Hageman factor may occur from collagen, fibrin, or platelet membrane during platelet aggregation. In addition, it can apparently be activated under conditions of stress, anxiety, fear, and other states. Outside the body (e.g., in a test tube), activation of the Hageman factor occurs when blood comes in contact with foreign substances whose common property appears to be a negative surface charge.

The Hageman factor activates the plasma enzyme, **plasma thromboplastin antecedent** (**PTA,** Factor XI), which, in the presence of calcium ions, activates a plasma protein, the **Christmas factor** (Factor IX). Activated Christmas factor interacts with another protein, **antihemophilic factor** (Factor VIII), on the surface of phospholipids and in the presence of calcium, to form a complex that activates the Stuart factor. The succeeding steps in the formation of prothrombin converting factor are the same as for the extrinsic mechanism.

Stage II: Conversion of Prothrombin to Thrombin

Prothrombin is a plasma globulin manufactured by the liver and normally present in circulating plasma. It is the inactive precursor of an active enzyme called **thrombin.** Thrombin is not normally present in plasma unless blood is clotting. In the presence of calcium ions and the prothrombin-converting factor, prothrombin is enzymatically split into two fragments—one inert and the other possessing the properties of thrombin. Initially, the conversion of prothrombin proceeds too slowly to produce significant amounts of thrombin needed for coagulation. Thrombin itself, however, increases its own rate of formation by converting a labile plasma protein, **proaccelerin** (Factor V), into **accelerin,** which then accelerates the formation of thrombin. Thrombin also activates the antihemophilic factor and is needed to activate a fibrin-stabilizing factor in Stage III.

Table 17–7
Clotting Factors

International Committee Designation	Synonyms	Location
Factor I	Fibrinogen	Plasma
Factor II	Prothrombin	Plasma
Factor III	Tissue thromboplastin	Tissue cells
Factor IV	Calcium ion	Plasma
Factor V	Proaccelerin	Plasma
	Prothrombin accelerator	
	Accelerator globulin	
	Labile factor	
Factor VI	Obsolete	
Factor VII	Serum prothrombin conversion accelerator (SPCA)	Plasma
	Proconvertin	
	Autoprothrombin I	
	Stable factor	
Factor VIII	Antihemophilic factor (AHF)	Plasma
	Platelet cofactor I	
	Thromboplastinogen	
	Antihemophilic factor A	
Factor IX	Plasma thromboplastin component (PTC)	Plasma
	Christmas factor	
	Platelet cofactor II	
	Antihemophilic factor B	
	Autoprothrombin II	
Factor X	Stuart-Prower factor	Plasma
	Stuart factor	
	Autoprothrombin III	
Factor XI	Plasma thromboplastin antecedent (PTA)	Plasma
Factor XII	Hageman factor	Plasma
	Contact factor	
Factor XIII	Fibrin-stabilizing factor	Plasma
	Plasma transglutaminase	
	Laki-Lorand factor	
Platelet factor	Platelet factor 3	Platelets

Stage III: Conversion of Fibrinogen to Fibrin

Fibrinogen is a soluble plasma protein produced by the liver and normally circulating in the plasma. Thrombin converts fibrinogen to **fibrin** by cleaving two pairs of small polypeptides (fibronopeptides) from each fibrin molecule, leaving a **fibrin monomer** (a single link in a chain of molecules that make up fibrin). Fibrin monomers spontaneously polymerize, forming an insoluble threadlike protein known as fibrin. In addition, thrombin activates another plasma enzyme, **fibrin-stabilizing factor (FSF, Factor XIII)**, which in the presence of calcium ions stabilizes the fibrin polymer through covalent bonding of the fibrin monomeric units.

Clot Retraction Is Caused by Thrombosthenin

After a clot forms, it retracts over a period of several hours and extrudes serum. Retraction serves to draw wound surfaces together and open the vessel if it has been occluded by the clot, thereby increasing blood flow and promoting tissue repair and clot dissolution. Retraction is caused by a platelet factor, **thrombosthenin,** a contractile protein that, to shorten, requires the presence of thrombin and adenosine triphosphate (ATP).

Fibrinolysis Causes the Dissolution of a Clot

Blood clots are not permanent. After hemorrhage has been checked and tissue repair is well underway, the clot is gradually dissolved by breaking down fibrin into soluble fragments by the enzyme, **plasmin.** Plasmin is derived from **plasminogen,** an inactive beta globulin normally present in plasma. The formation of plasmin from plasminogen requires an enzyme called **activator.** Activator is normally absent from plasma but its precursor, **proactivator,** is present and may be converted into activator by **cytofibrokinase** (a tissue enzyme), **staphylokinase** and **streptokinase** (bacterial enzymes), **plasma kinase** (Hageman factor), and other enzymes found in urine **(urokinase),** tears, and saliva. **Tissue plasminogen activator (TPA)** formed by endothelial cells appears to be responsible for most of the physiological fibrinolysis. TPA has been synthesized by recombinant DNA techniques and is being used to treat coronary thrombosis (see Focus Box).

Normally, a balance between deposition of fibrin and fibrinolysis limits coagulation to the area of vascular injury. Additional limits on excessive clotting include plasma enzymes, which inactivate each clotting factor shortly after each is activated, and anticoagulants.

Anticoagulants Inhibit Blood Clotting

An **anticoagulant** is a chemical that blocks or inhibits the coagulation process. Some anticoagulants are normally produced in the body and are referred to as **physiological anticoagulants.** Others, called **therapeutic anticoagulants,** are manufactured. and can be administered to block coagulation.

Unless there is trauma to the blood vessels or to the blood itself, coagulation does not normally occur. Although tissue breakdown and platelet destruction are normal occurrences in the absence of trauma, intravascular clotting does not usually occur because (1) the amounts of tissue and platelet procoagulants released are very small, and (2) natural coagulation inhibitors such as **antithromboplastin, antithrombin III,** and **heparin** are present.

Antithrombin III inhibits activated Stuart factor, and to a lesser extent, inhibits other enzymes required for production of prothrombin converting factor. Antithrombin III also inhibits thrombin, thereby blocking the conversion of fibrinogen to fibrin. Heparin, produced by basophils and mast cells of connective tissues, acts in anticoagulation by combining with antithrombin III and greatly potentiating its effects. Acting alone, heparin is not a significant anticoagulant. Although heparin is produced in the body, it is normally absent from blood.

Therapeutic anticoagulants, on the basis of action, may be classified into three general categories: (1) agents that depress the action of procoagulants; (2) agents that remove calcium from the blood by precipitation or chelation; and (3) agents that inhibit the synthesis of prothrombin in the liver.

In cases of excessive coagulation, heparin is widely used as a therapeutic anticoagulant. It inhibits the conversion of fibrinogen to fibrin and also inhibits the production of thrombin. Its effects can be countered by administering protamine sulphate.

Agents that remove calcium ions from the blood are used only to prevent coagulation in blood outside the body, since removal of adequate amounts of calcium to prevent clotting would interfere with the normal role of calcium in the heart, skeletal muscle, and nerves. **Sodium citrate** chelates (binds and makes unavailable) calcium ions, whereas **sodium oxalate** precipitates calcium ions out of solution.

Liver cells synthesize four clotting factors: prothrombin, (factor II), factor VII, Christmas factor (factor IX), and Stuart factor (factor X). Vitamin K is required and acts as an enzyme cofactor during synthesis of the clotting factors. **Bishydroxycoumarin** (Dicumarol), an agent originally isolated from spoiled sweet clover, inhibits coagulation by interfering with the liver's use of vitamin K, and hence, clotting factor production.

Abnormalities of Hemostasis Occur for Many Reasons

Thrombocytopenia is an abnormally low number of circulating platelets in blood. If the platelet count falls below $50,000/\mu l$, there is an increased risk of internal hemorrhage associated with accidental trauma or surgery. Platelet counts around $20,000/\mu l$ are associated with multiple small bruises **(purpura),** hemorrhagic spots **(petechial)** in the skin, and sometimes spontaneous bleeding from mucosal surfaces. Potentially fatal hemorrhage in the intestines or brain can occur with platelet counts less than $10,000/\mu l$.

Thrombocytopenia can result from decreased production of platelets by the bone marrow due to toxins, radiation, or infection. It can also result from sequestration of platelets such as in congestive **splenomegaly,** or from increased destruction of platelets by autoimmune process as in **idiopathic thrombocytopenic purpura (ITP).** Disorders characterized by increased coagulation of blood and thus

CLINICAL FOCUS

Clot Busters

Heart muscle (myocardium) contracts and relaxes more than 100,000 times per day and requires an abundant blood supply to meet metabolic requirements. Blood is supplied to heart muscle by way of branches of the left and right coronary arteries. Partial or complete blockage of a coronary artery interrupts the normal blood supply to heart muscle producing ischemia (a below-normal reduction in the blood supply of a tissue or organ) which may lead to death (infarction) of the affected muscle cells. Ischemic heart disease, commonly called coronary heart disease (CHD), frequently manifests with chest pain (angina pectoris) that worsens with exertion, difficulty in breathing, sweating, nausea, vomiting, pallor, restlessness, cold extremities, and a slowing of the heart rate with a fall in blood pressure. These manifestations characterize what is commonly referred to as a "heart attack." If the infarct is large, the heart may stop, resulting in sudden death.

Typically the underlying cause of myocardial infarction is a coronary blood vessel narrowed by atherosclerosis and occluded by a thrombus (clot). Damage to the heart muscle distal to the occlusion begins within minutes but the muscle deteriorates gradually instead of dying immediately; therefore, damage can be minimized if blood flow can be restored quickly (ideally within one hour after symptoms of a heart attack first appear).

One way to restore blood flow is by means of angioplasty, a surgical procedure, whereby a balloon is introduced by catheter into the blocked coronary artery and inflated at the site of obstruction, flattening the obstruction against the arterial wall (see Chapter 19). Another way to improve coronary blood flow is to surgically by-pass the blockage by grafting a vein around it.

A nonsurgical approach to the problem of quickly restoring blood flow through a vessel blocked by a thrombus uses a thrombolytic agent, a "clot buster," to dissolve the clot. One of the most recent thrombolytic agents is recombinant tissue plasminogen activator (TPA), synthesized using recombinant DNA technology. Although effective, it is very expensive to use (more than $2000 per dose). Another older, but effective and less costly (about $200 to $300 per dose), thrombolytic agent is streptokinase, manufactured using more conventional technology.

Heart attack patients who are candidates for thrombolytic therapy should be treated with "clot busters" within 60 minutes of the onset of symptoms and within 30 minutes of arriving at emergency rooms in order to derive maximum benefit from the drugs. Studies have shown that "clot busters" administered after 12 hours or more offer the patient little or no benefit. Clearly, when heart attack symptoms appear, time is of the essence.

Thrombolytic therapy, coronary angioplasty, and coronary by-pass surgery, administered separately, or in combination, are used to treat coronary heart disease, after symptoms of ischemia appear,

■ **Why do some persons regularly take aspirin to prevent heart attack?**

---

increased platelet consumption, such as disseminated intravascular coagulation, also produce thrombocytopenia.

Disseminated intravascular coagulation (DIC) involves widespread coagulation that produces thrombosis in small blood vessels, increased fibrinolysis, and depletion of coagulation factors, which in turn, collectively result in generalized bleeding. Causes of DIC include bacterial infections that cause widespread endothelial damage, disseminated cancers that cause increased release of procoagulants, and complications of pregnancy such as the release of amniotic fluid (procoagulant material) into the circulation.

Hemophilia is an inability of the blood to properly coagulate due to a lack of a coagulation factor. **Hemophilia A** involves a deficiency in Factor VIII and **hemophilia B** (Christmas disease) involves a deficiency in Factor IX. Hemophilias A or B are transmitted genetically and affect only males; females carry the gene but do not show symptoms. **Hemophilia C** is not sex-linked and results from a deficiency in Factor XI. About 85% of hemophilia is Type A. Hemophilia is characterized by spontaneous or traumatic subcutaneous hemorrhage, blood in the urine, bleeding in the mouth, lips, and tongue, and within the joints.

Absence or lack of adequate amounts of other factors also interferes with normal coagulation. **Afibrinogenemia,** a lack of circulating fibrinogen, may be hereditary or may result from liver disease. **Hypoprothrombinemia,** a lack of prothrombin, may be congenital, a result of liver disease, or a complication of vitamin K deficiency. Vitamin K is required for liver manufacture of prothrombin. Newborn infants, for example, lack store of vitamin K at birth. Vitamin

K is normally synthesized by bacteria in the gastrointestinal tract, but infants are born with sterile gastrointestinal tracts. Therefore, hypoprothrombinemia may develop, and life-threatening hemorrhage may occur. The disorder is called **hemorrhagic disease of the newborn.**

Hemorrhagic disease may also occur because of an increase in heparin release into the blood (**hyperheparinemia**). Allergies, collagen diseases, and disorders of the liver or bone marrow are sometimes accompanied by hyperheparinemia and a reduction in the ability of blood to coagulate effectively.

CHAPTER REVIEW

Summary

- Blood transports molecules and heat from one area of the body to another, defends against foreign agents, and assists in the maintenance of extracellular fluid pH and osmolarity.
- Blood is a connective tissue containing erythrocytes, leukocytes, thrombocytes, and plasma.
- Average adult blood volume is 5 liters.
- Plasma is the intercellular fluid of the blood. Plasma is about 93% water and 7% solutes (inorganic and organic).
- Plasma proteins are divided into the following fractions: albumin, globulins, and fibrinogen. Plasma proteins serve as amino acid reserves, carriers, and buffers, and also participate in blood coagulation and the regulation of fluid distribution within the body. The immune globulins play important roles in the body's defense.
- Hematopoiesis is the generation of blood cells from pluripotential hemopoietic stem cells (PHSC).
- PHSCs develop into committed progenitor cells called colony forming units (CFU), which generate colonies of specific blood cell types.
- The erythrocyte is a biconcave, anucleate cell that numbers about 5 million/μl of whole blood.
- Erythrocytes contain hemoglobin, which transports oxygen. One gram of hemoglobin can transport 1.34 ml of O_2. The amount of hemoglobin in blood is about 15 g/dl; therefore, the oxygen-carrying capacity of blood is about 20 ml of O_2 per deciliter.
- Erythropoiesis is the formation of erythrocytes. Normally, about 230 billion red blood cells are produced each day. Substances required for erythropoiesis include vitamin B_{12}, folic acid, iron, and protein. Erythropoiesis is controlled by a renal hormone called erythropoietin.
- About 230 billion red blood cells are destroyed each day. When an erythrocyte is destroyed, most parts of the hemoglobin are conserved. The remainder is excreted as bilirubin in bile.
- Anemia is an abnormal reduction in the oxygen-carrying capacity of the blood. Anemias are characterized by changes in one or more of the following red blood cell indices: MCV, MCH, MCHC.
- Polycythemia is a relative or absolute increase in the number of circulating red blood cells above normal. Two main types are polycythemia vera and secondary polycythemia.
- The ABO blood group contains blood types based on the presence or absence of antigens A and B on the red blood cell membrane.
- Major agglutination occurs when donor red blood cells containing a strong antigen possessed by the recipient are infused into the recipient's body.
- Minor agglutination occurs when donor plasma containing antibodies against the recipient's red cell antigens is infused into the recipient's body.
- The Rh blood group contains blood types based on the presence or absence of antigen D. If antigen D is present on the red blood cell membrane, the blood type is Rh^+. If antigen D is absent, the blood type is Rh^-. The plasma of an Rh^- person may or may not contain anti-D.
- Leukocytes are blood cells that do not contain hemoglobin. Two principal categories are granulocytes and agranulocytes. On the average, whole blood contains about 7000 to 10,000 leukocytes/μl. Their principal function is body defense.
- T-lymphocytes confer cell-mediated immunity, and B-lymphocytes confer antibody-mediated immunity.
- Leukopoiesis, the formation of leukocytes, occurs primarily in the bone marrow, although some leukocytes mature outside the marrow in lymphoid organs.
- Thrombocytes, or platelets, are produced from megakaryocytes. One megakarocyte gives rise to about 6000 platelets.
- Platelets play several roles in hemostasis.
- Local vasoconstriction helps minimize blood loss when a blood vessel is opened.
- Platelets clump together, forming a plug at the site of vascular damage. Platelets also release procoagulants.
- Coagulation of blood, or formation of a blood clot, at the site of vascular injury helps seal the opened vessel. Coagulation involves the formation of fibrin thread, which is the foundation of the clot.
- Fibrin formation occurs in three states: Stage I—formation of prothrombin-converting factor, Stage II—conversion of prothrombin to thrombin, and Stage III—conversion of fibrinogen to fibrin.
- After formation of a clot, clot retraction helps draw wound surfaces together.
- Clots are gradually dissolved by enzymes that break down fibrin.
- Abnormalities of hemostasis can result from platelet defects, coagulation defects, and/or excessive production of anticoagulants. Principal abnormalities include thrombocytopenia and hemophilia.

Review Questions

Identify Each with a Term

1. The percentage of volume occupied by red blood cells in a sample of whole blood.
2. The pressure created by the presence of proteins in the plasma.
3. A solution with an osmotic pressure lower than the plasma.
4. The expansion of the red blood cell and its loss of hemoglobin.
5. The formation of blood cells from primitive stem cells.
6. A relative or absolute increase in the number of circulating red blood cells.
7. The process whereby leukocytes pass through a capillary wall.

Define Each Term

8. MCHC   13. hemostasis
9. MCV   14. phagocytosis
10. MCH   15. thrombin
11. anemia   16. heparin
12. LDL

Choose the Correct Answer

17. The largest WBC, this blood cell is part of the body's macrophage defense system:
    a. neutrophil   d. basophil
    b. eosinophil   e. lymphocyte
    c. monocyte
18. The most common WBC, this highly motile and phagocytic cell forms the primary line of cell defense against bacterial invasion:
    a. eosinophil   d. monocyte
    b. neutrophil   e. basophil
    c. lymphocyte
19. Numerically, which of the following is the largest?
    a. monocyte count
    b. eosinophil count
    c. basophil count
    d. lymphocyte count
    e. neutrophil count
20. During Stage I of the blood coagulation process:
    a. thrombin is formed.
    b. fibrin is formed.
    c. prothrombin is converted into thrombin.
    d. prothrombin converting factor is formed.
    e. fibrin is dissolved.
21. Which of the following whole blood types can be safely infused into a B^+ recipient?
    a. AB^- d. AB^+
    b. A^+ e. A^-
    c. B^-
22. On destruction of red blood cells and metabolism of hemoglobin:

a. all components of hemoglobin are conserved except iron, which is excreted.
b. all components of hemoglobin are conserved.
c. all components of hemoglobin are excreted.
d. iron and protein (amino acids) are conserved and the remainder is excreted as waste in the urine or feces.
e. all of the globin is conserved, all of the heme is excreted.

23. Which of the following values is abnormal:
    a. hematocrit—43%
    b. RBC—5 million/μl
    c. WBC—8 thousand/μl
    d. O_2 carrying capacity—20 ml O_2/dl
    e. MCHC—29%

Calculate

24. The total blood volume in a heparinized, centrifuged test tube is 9 ml, and the packed red blood cell mass volume is 4 ml. Calculate the apparent hematocrit.
25. The hemoglobin content of a blood sample is 16 g/dl. Calculate the oxygen-carrying capacity of the blood.
26. Given a hematocrit of 45% and an RBC count of 5 million/μl, calculate the MCV.
27. Assume that 230 billion red blood cells are destroyed and replaced each day, that the RBC count is 5 million/μl, and that the total blood is 5 liters. How many days would the body require to replace all erythrocytes currently present?

Problem Solving in Physiology

28. Starvation that involves an inadequate intake of dietary protein eventually leads to wasting of skeletal muscle and reduction of plasma protein, which leads to fluid shifts within the body. Would you expect tissue spaces (interstitial fluid compartment) to shrink or swell? Why?
29. If an AB recipient receives O blood, minor agglutination will occur, but if an O recipient receives AB blood, major agglutination will occur. Why?
30. A student whose blood type is A is given unknown packed red blood cells and corresponding serum from each of the four ABO blood types. Using only his own red cells and serum, how can he correctly identify each of the unknown bloods?
31. The flow of blood through the heart and blood vessels is determined by the pressure causing the flow and resistance opposing flow ($F = P/R$). How does polycythemia influence blood flow and why can polycythemia be life-threatening?

Suggested Readings

Bloom, W., and Fawcett, D.W. *A Textbook Of Histology*, 12th ed. London: Chapman and Hall, 1994.

deSauvage, F.J. et al. "Stimulation of Megakaryocytopoiesis and Thrombopoiesis by the c-Mpl Ligand." *Nature*, 369 (16 June 1994).

Gleich, G.J., Adolphson, C.R., and Leiferman, K.M. "The Biology of the Eosinophilic Leukocyte," *Annual Review of Medicine*, 44 (April 1993).

Golpe, D.W., and Gasson, J.D. "Hormones That Stimulate the Growth of Blood Cells." *Scientific American* (July 1988).

Metcalf, D. "Thrombopoietin." *Nature*, 369 (16 June 1994).

Spangrude, G.J. "Biologic and Clinical Aspects of Hematopoietic Stem Cells." *Annual Review Of Medicine*, 45 (April 1994).

Spivak, J.L. "Recombinant Erythropoietin." *Annual Review of Medicine*, 44 (April 1993).

Stamler, J. and Neaton, J.D. "Benefits of Lower Cholesterol." *Scientific American Science and Medicine*, 1, No. 2 (May–June 1994).

Key Terms

agglutination (p. 525)
albumin (p. 514)
anemia (p. 504)
antibody (p. 524)
antibody-mediated immunity (p. 532)
anticoagulant (p. 536)
antigen (p. 525)
atherosclerosis (p. 539)
B cell (p. 532)
basophil (p. 530)
cell-mediated immunity (p. 532)
clot (p. 535)

colony forming unit (p. 517)
colony stimulating factors (p. 532)
eosinophil (p. 528)
erythropoiesis (p. 517)
extrinsic clotting pathway (p. 535)
fibrinolysis (p. 538)
globulin (p. 514)
hematopoiesis (p. 517)
hemoglobin (p. 519)
immunoglobulin (p. 517)

interleukin (p. 532)
intrinsic clotting pathway (p. 535)
intrinsic factor (p. 524)
leukocyte (p. 517)
leukocytosis (p. 532)
leukopenia (p. 533)
lymphocyte (p. 528)
macrophage (p. 521)
monocyte (p. 528)
neutrophil (p. 528)
oxygen-carrying capacity (p. 519)

phagocytosis (p. 530)
plasma (p. 510)
pluripotential hemopoietic stem cell (p. 517)
polycythemia (p. 524)
progenitor cell (p. 517)
red cell indices (p. 510)
reticulocyte (p. 521)
Rh factor (p. 527)
T cell (p. 530)
thrombocyte (p. 533)
thrombus (p. 539)

A normal coronary artery.
(© Cabisco/Visuals Unlimited)

Chapter 18

The Heart

Key Concepts

The heart consists of two separate pumps that generate the pressure that drives the unidirectional flow of blood through the pulmonary circulation, where gas exchange occurs, and through the systemic circulation, where exchange of nutrients, metabolites (including heat), and hormones occurs.

Each beat of the heart, or cardiac cycle, involves a number of interrelated and highly coordinated events including electrical activation of the atria and ventricles, contraction and relaxation of the atria and ventricles, closing and opening of the cardiac valves, and filling and emptying of the atria and ventricles.

Chapter Outline

Overview of the Cardiovascular System
The Heart: A Muscular Pump

The last thing Joe remembered was a gripping, crushing pain in his chest as he was shoveling snow from the driveway. Then, there were bright lights everywhere, frantic noises and voices, and strange faces—faces shrouded in tightly fitting green caps and face masks. It all seems very real, and yet, at the same time like a dream. Slowly the sounds and the lights faded away as he lost consciousness. One of the doctors yelled, "clear,"

and Joe's body arched momentarily and then went limp. This was repeated once, twice, and then the doctors stopped. The piercing background squeal suddenly stopped, replaced by a periodic beep, beep, beep. The source of this noise was a monitor with a green line scrolling across the face, tracing out Joe's EKG in synchrony with the beeping sound. Someone exclaimed, "normal rhythm," and some of the tension melted away from the group of physicians and nurses crowded around him. Then, a catheter was threaded into a vein in his neck, and on another monitor there was an image of the end of the catheter moving, probing, retracing its path and then advancing again. Suddenly a new image flowed from the end of the catheter and then branched into several smaller tributaries. A dye had been injected through the catheter into Joe's coronary artery. One of the physicians pointed to the screen at one branch of the spreading image where the image had abruptly stopped. One of Joe's coronary arteries, partially occluded by atherosclerotic plaque, had become completely occluded by a blood clot, totally blocking blood flow to part of his heart. He had just experienced a heart attack. The absence of blood flow to the heart muscle had caused the electrical activity in this region of the muscle to become erratic. This erratic electrical activity eventually deteriorated into ventricular fibrillation, a random chaotic spread of electrical current in the heart. By placing electrodes on Joe's chest and applying a sudden shock to his heart, the physician had halted this chaotic electrical activity, allowing Joe's heart to slip back into its normal mode of electrical activity. Then a nurse injected something into the catheter that had been inserted into his jugular vein. Again, the doctor pointed to the monitor, as another stream of light flowed from the end of the catheter. This time the dye didn't stop where it had before; it continued as a narrow stream of light for maybe a centimeter, and then abruptly widened again. Although still partially occluded, blood had begun to flow through the blocked branch of Joe's coronary artery. He had just received an injection of tissue plasminogen activator, a clot-dissolving human enzyme made by infecting bacterial cells with human DNA. Again, the stream of light flowed out from the catheter, and then the narrow stream had widened even more. As Joe slowly drifted back to consciousness, he again noticed sounds around him that were gradually getting louder, more real, more urgent. The darkness slowly lifted, and once again he looked up into those masked faces. While Joe's thoughts were still confused and dreamlike, he slowly came to the realization that he had had a heart attack. Once again his thoughts blurred and the sounds and the lights faded. And just before Joe slipped into a deep sleep he vowed to himself that, this time, he would take better care of that all-important machine—his heart.

In this chapter we will consider the entire cardiovascular system with a special focus on function of the heart.

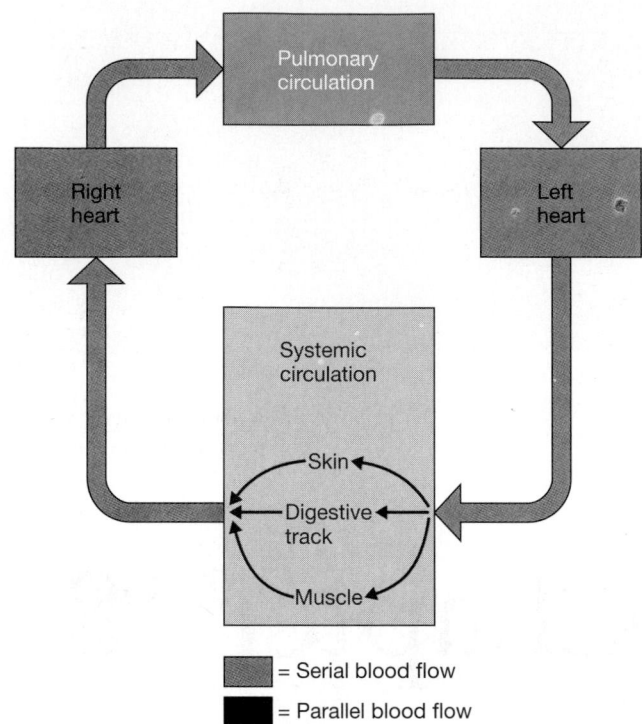

Figure 18–1

Arrangement of the circulatory system.

Overview of the Cardiovascular System

William Harvey demonstrated in the seventeenth century that the cardiovascular system forms a closed loop through which blood is pumped by the heart. As shown in Figure 18–1, this loop consists of two pumps, the **left heart** and the **right heart,** and two vascular systems, the **pulmonary circulation** and the

Before we begin looking in detail at the physiology of the cardiovascular system, it will be helpful to consider its overall design and function. The cardiovascular system consists of a pump (the **heart**), which propels blood throughout the body by way of a series of tubes **(blood vessels).** The **circulating blood** provides a vehicle for the rapid movement of cells and molecules from one part of the body to another.

systemic circulation. From an anatomical perspective, there is only one heart; however, from a physiologic perspective there are two independent pumps, as shown in Figure 18–1. While in practice, you will see that they are both mechanically and electrically interdependent, it is useful to visualize them independently when considering the path of blood flow through the body.

The left and right hearts (brown) and the pulmonary (gray) and systemic circulations (light grey) are arranged in **series** (pink arrows), which means that blood must flow through these four components in sequence, one after the other. Blood flows, via the pulmonary circulation, through the lungs, where oxygen is added and carbon dioxide is removed. After passing through the left heart, this oxygenated blood then flows, via the systemic circulation, to all the cells of the body, which extract oxygen from the blood and add carbon dioxide to it. The blood then flows through the right heart and the cycle repeats.

Notice in Figure 18–1 that the systemic circulation consists of many separate circulatory pathways that supply blood to various organ systems within the body (red arrows). Notice also that these individual **vascular beds,** which make up the systemic circulation, are arranged in **parallel** (red arrows) with one

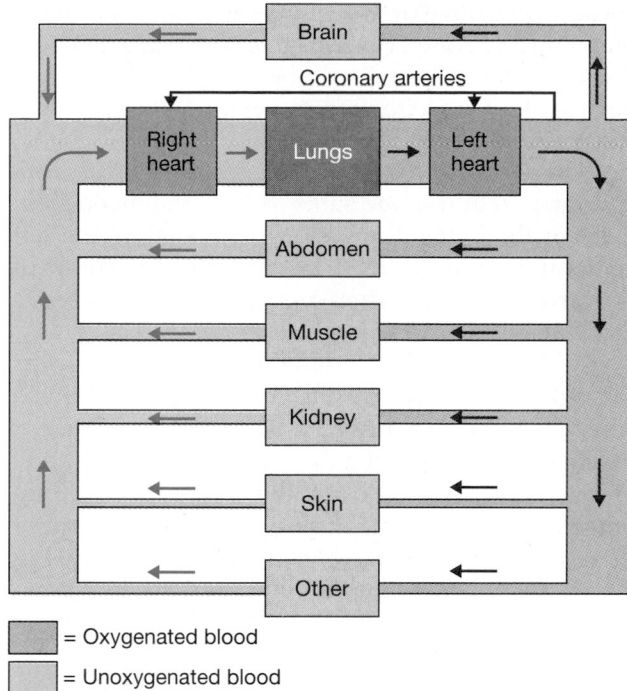

Figure 18–2

Relative distribution of blood flow in the cardiovascular system at rest. Vessel diameter is proportional to blood flow.

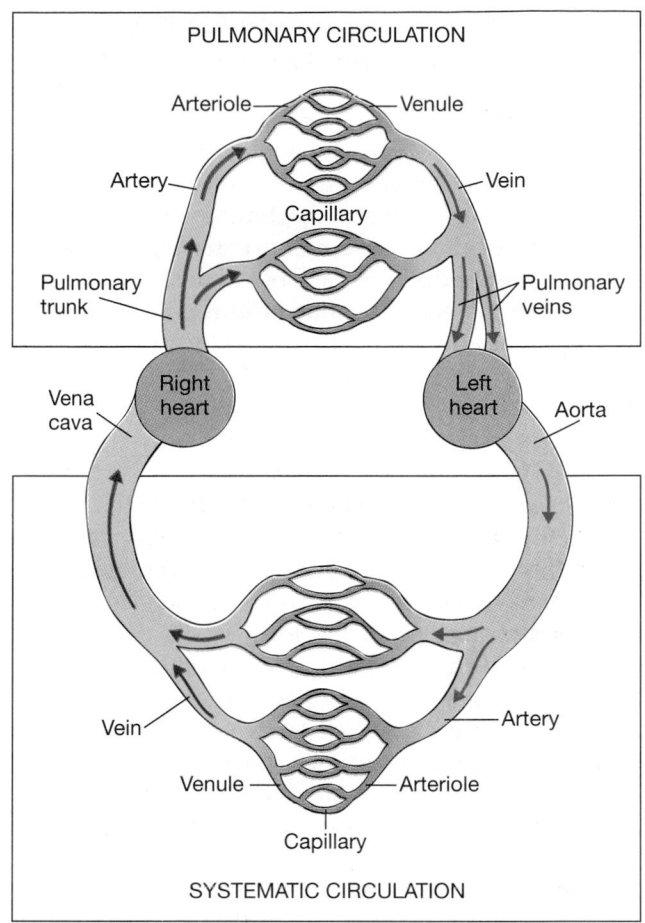

Figure 18–3

Branching pattern of blood vessels in the cardiovascular system.

another. This means that the total volume of blood flowing through the systemic circulation is partitioned, or divided, among the different vascular beds. As we will see later, our bodies are able to dynamically adjust the distribution of blood flow among the various tissue beds. For example, during exercise, the blood flow to skeletal muscle and to the skin can be increased severalfold, while the flow to other organ systems, such as the digestive tract, is simultaneously reduced. We will discuss in the next chapter the mechanisms that allow for this optimization of blood flow in response to the needs of different parts of the body. The relative distribution of the **total blood flow** among the various organs of the body during rest is shown in Figure 18–2.

The Systemic Circulation Carries Oxygenated Blood to the Tissues of the Body

The vessels that carry blood *from* the heart make up the **arterial system;** the vessels that carry blood *back to* the heart make up the **venous system** (Fig. 18–3). The systemic circulation begins with a single artery, the **aorta,** which receives all the blood the heart pumps out. The aorta branches into a number of smaller **arteries,** which partition and direct blood flow to the various organs of the body. Once it enters an organ, an artery soon branches into progressively smaller arteries, which in turn branch into smaller **arterioles.** This arteriolar tree distributes blood flow within individual organs. The arterioles then branch into an extensive network of **capillaries,** the smallest blood vessels in the body. It is here in the capillaries that exchange occurs between the blood and the interstitial fluid. Blood then flows from the capillaries into the venous system, which channels the blood back to the heart.

The venous system begins at the point where the capillaries reunite to form larger vessels called **venules,** which in turn unite to form larger **veins.** All the veins from the upper portion of the body drain into a single large vein called the **superior vena cava,** and the veins from the lower portion of the body drain into the **inferior vena cava.** The two venae cavae unite to return all of the blood to the right heart.

The Pulmonary Circulation Carries Unoxygenated Blood Through the Lungs for Gas Exchange

A similar pattern of branching occurs in the **pulmonary circulation** (see Fig. 18–3). The blood leaves the right heart by way of the **pulmonary trunk,** which soon divides into the **pulmonary arteries.** The pulmonary arteries distribute the blood to the two lungs. Notice that the entire **cardiac output** passes through the lungs before being distributed throughout the systemic circulation.

As in the systemic circulation, the pulmonary arteries progressively branch into smaller and smaller vessels. The **pulmonary capillaries** are the site of gas exchange between the blood and the air in the lungs. As the blood flows through the pulmonary capillaries, its oxygen content increases and its carbon dioxide content decreases as a result of diffusional exchange with the air in the lungs. This oxygenated blood then flows from the pulmonary capillaries through progressively larger venules and veins to the **pulmonary veins** and then into the left heart.

Not all arterial blood has a high oxygen content. Both the **pulmonary venous blood** and the **systemic arterial blood** are oxygenated. The distinguishing factor between arteries and veins is that arteries carry blood *from* the heart *to* capillaries, and veins carry blood back *to* the heart *from* capillaries.

Blood Flow Is Determined by the Blood Pressure and the Resistance to Blood Flow

We now know that the role of the cardiovascular system is to provide for a continuous flow of blood through the tissues. Let's consider for a moment what the term *flow* means and how it might be quantified.

Blood flow can be defined as the volume of blood that moves past a particular point in the cardiovascular system during a given period of time—for example, the amount that flows through the left heart in one minute. A practical example of fluid flow with which we are all familiar is the flow of water through a garden hose. If we were to collect the water coming out of a hose over a period of 1 minute, we could calculate the flow of water through the hose. If, after 1 minute, we had 5 liters of water in our bucket, then the calculated flow would be 5 liters in 1 minute, or 5 L/min. We have measured the volume of water that moved past a particular point in the garden hose (in this case, the nozzle) during a period of time (1 minute).

Does this number tell us anything about the flow of water elsewhere in the hose? The answer is yes,

because the flow of water will be the same at all points in the hose. We can generally state that the flow in a closed system will be the same through all portions that are arranged in **series.** In terms of the cardiovascular system, this means that we could measure the flow of blood from the heart into the aorta and know that it is the same as the total blood flow through the lungs, the systemic circulation, the left heart, and the right heart, because these portions of the cardiovascular system are all connected in series and constitute a closed system.

Could we also say what the blood flow was through the kidneys on the basis of this one measurement? The answer is no, because the individual vascular beds that make up the systemic circulation are arranged in parallel, and therefore only a portion of the total systemic blood flow goes through each of these vascular beds (see Fig. 18–2). As illustrated in Figure 18–4, the combined flow through the vascular beds arranged in parallel with one another must be equal to the flow through portions of the system that are in series with these vascular beds. Furthermore, as mentioned previously, the proportion of the total systemic blood flow channeled through a particular vascular bed can change in response to the changing needs of the body, even though the total blood flow through the systemic circulation remains constant.

The force that causes blood to flow through the vessels can be measured as **blood pressure.** Blood pressure measurements are usually expressed in units of **mm Hg** (millimeters of mercury). For example, average **aortic blood pressure** is 90 mm Hg. This means that the pressure inside the aorta (relative to

Figure 18–4

A comparison of flow through tubes arranged in series and in parallel. Blood flow through all portions of a closed system that are arranged in series (A, B, and C) must be the same (5 L/min). Blood flow through the individual vessels arranged in parallel (small vessels in section B) must add up to the total blood flow through the system (5 × 1 L/min = 5 L/min).

atmospheric pressure outside the aorta) is equal to the downward pressure exerted by a column of mercury 90 mm tall. Notice that the pressure exerted by a column of mercury, or any other fluid, is determined by the height of the column and not the width of the column. Why is this?

A more familiar unit of pressure is pounds per square inch (psi). Automobile tires, for example, have an internal air pressure of about 35 psi. A blood pressure of 100 mm Hg is equivalent to 2 psi. Blood pressure is a measure of the driving force propelling blood through the blood vessels. Just as the high air pressure within a tire will cause air to rush out of the tire through an open valve, the differences in blood pressure within the circulatory system will cause blood to flow from one point to another. Flow does not occur in response to any particular pressure (P), but rather to a pressure difference or **pressure gradient** ($P_1 - P_2$ or ΔP) between two points. This is analogous to net diffusion between two points; the rate of diffusion is determined by the magnitude of the **concentration gradient** and not by the absolute concentration of a diffusing substance. Thus, the blood flow between two points in the circulatory system is proportional to the *difference* in blood pressure between these same two points and not the *absolute* pressures at either point. This concept is illustrated in Figure 18–5.

The reason that a physical force or pressure is required to maintain blood flow through the circulatory system is that frictional forces tend to impede blood flow. Friction exists among the molecules that make up the blood and between the blood and the walls of the vessel. Resistance is a quantity that summarizes the net effect of all the frictional components that oppose blood flow in a particular situation. Thus, the greater the resistance, the lower the blood flow for any given pressure difference. The equation that describes the relationship between blood flow (F), driving pressure ($P_1 - P_2$ or ΔP), and resistance (R) is:

$$F = \frac{P_1 - P_2}{R}$$

As we will see in the next chapter, the resistance to blood flow through a vessel is determined by a number of factors, including the length and diameter of the vessel, as well a the consistency or viscosity of the blood itself.

The Heart: A Muscular Pump

The major role of the heart is to maintain an adequate level of blood flow throughout the cardiovascular system by pumping blood under pressure into the vascular system. The next several sections will consider the anatomy of the heart and the manner in which cardiac muscle contraction is coordinated within the heart to generate pressure and flow.

The Heart Consists of Two Separate Pumps that Maintain the Unidirectional Flow of Blood

As previously mentioned, the heart consists of two functionally distinct pumps: the left heart and the right heart. The left and right hearts, in turn, each contain two chambers: an atrium (plural, *atria*) and a ventricle. The **atria** are thin-walled chambers that receive blood returning to the heart from either the vena cavae (in the case of the right heart) or the pulmonary veins (in the case of the left heart). Blood then is pumped from each atrium into the associated **ventricle**. The ventricles (particularly the left ventricle) have thicker, muscular walls capable of pumping blood out of the heart at higher pressures. Figure 18–6 shows schematically how the chambers of the heart are organized and the direction of blood flow through the heart.

Valves in the Heart

Notice that there is a valve at the entrance and exit of each ventricle (Fig. 18–7). These valves allow blood to flow in only one direction. We will see later how

0 mmHg   0 mmHg

(a)

150 mmHg   150 mmHg

(b)

150 mmHg   0 mmHg

(c)

Figure 18–5

Relationship between pressure and flow. Flow through vessels A and B is zero because there is no pressure difference anywhere within the vessel. Fluid flow in vessel C is driven by the pressure difference (150 mm Hg − 0 mm Hg = 150 mm Hg) between the ends of the vessel. The arrow indicates the direction of blood flow.

Figure 18–6

Overview of the heart. The direction of blood flow through the chambers of the heart and the major vessels leading into and out of the heart is indicated by the arrows.

these one-way valves operate to ensure unidirectional flow of blood through the heart. The **atrioventricular (AV) valves** are located between the atria and the ventricles. The AV valve associated with the right heart is called the **tricuspid valve,** and the AV valve associated with the left heart is called the **bicuspid** or, more commonly, the **mitral valve.** The valves located at the junction of the ventricles with the pulmonary and systemic arteries are called **semilunar valves.** The semilunar valve between the right ventricle and the pulmonary trunk is called the **pulmonary valve,** and the valve between the left ventricle and the aorta is called the **aortic valve.** The opening and closing of the valves are passive processes that occur in response to the **blood pressure gradient** across the valve. For example, when the pressure in either of the two atria is greater than that in the adjoining ventricles, the AV valves open and allow blood to flow into the ventricle (Fig. 18–8). If the pressure gradient is reversed—with ventricular pressure being greater than atrial pressure—the atrioventricular valves will close in response to the momentary reversal of blood flow and then be held closed by the reversed pressure gradient; this prevents the flow of

blood from the ventricle back into the atrium during ventricular systole.

The Walls of the Heart

The walls of the heart are composed of three anatomically distinct tissue layers (Fig. 18–9). The middle, muscular layer, the **myocardium,** is composed primarily of cardiac muscle cells. (The special properties of this muscle type were discussed in Chapter 16.) The inner surface of the myocardium, which forms the inside wall of the atria and ventricles, is lined with a thin layer of endothelial cells, the **endocardium.** The endothelial layer forms a continuous lining that covers not only the inner surfaces of the heart and the heart valves but also the entire vascular network. The outer surface of the myocardium is covered by the **epicardium,** which consists primarily of connective tissue and fat. The main **coronary vessels** that supply blood to the myocardium (see Fig. 18–7) are located within this tissue layer.

It should be emphasized at this point that the terms "left heart" and "right heart" refer to separate parts of the one anatomical heart as shown in Figure

Pulmonary valve

Left coronary artery

Mitral valve

Aortic valve

Right coronary artery

Tricuspid valve

Figure 18–7

Cardiac valves. Top view of the heart with the major vessels cut away to show the location of the four cardiac valves. Both the aortic and pulmonary valves consist of three crescent-shaped (semilunar) cusps. The left atrioventricular valve consists of two valve cusps (bicuspid or mitral valve), while the right atrioventricular valve consists of three valve cusps (tricuspid valve). All the heart valves are connected to a tough connective tissue framework or cardiac skeleton that lies between the atria and the ventricles.

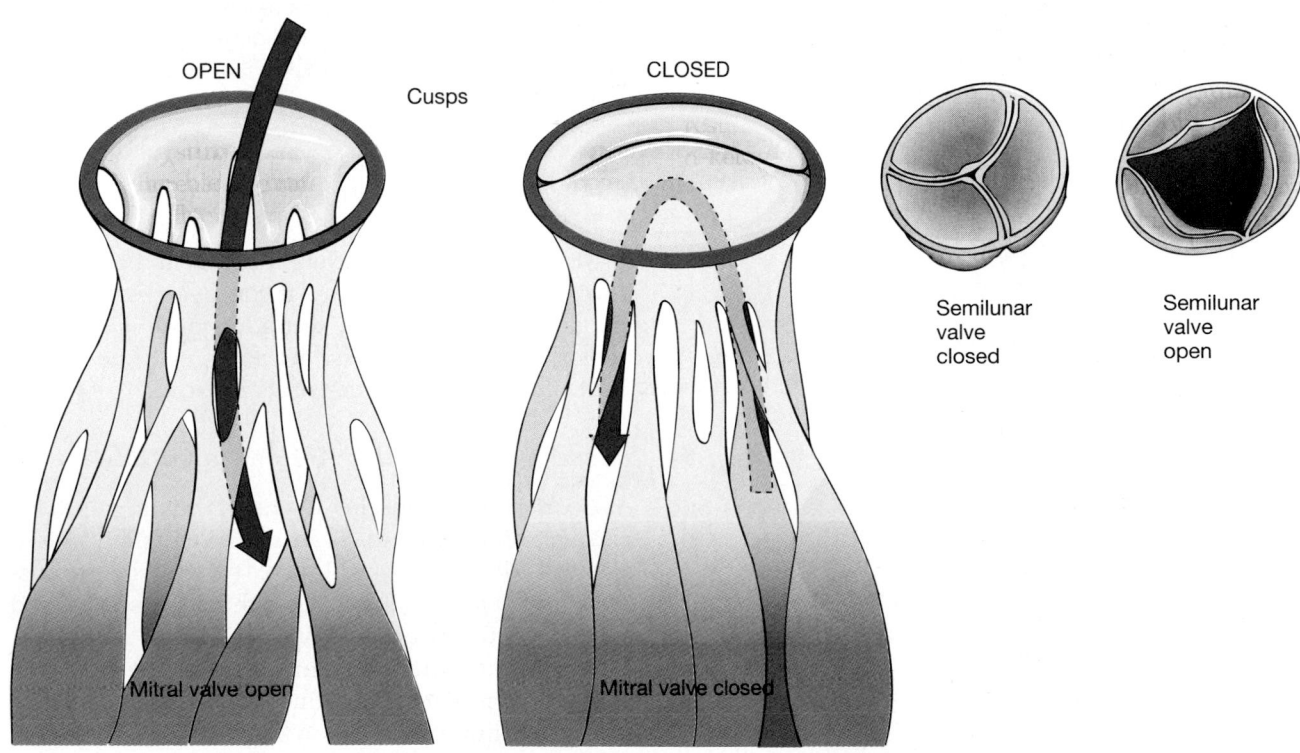

OPEN

Cusps

CLOSED

Mitral valve open

Mitral valve closed

Semilunar valve closed

Semilunar valve open

18–6. While the left and right hearts are distinctly separate pumps, they are hydraulically, mechanically, and electrically coupled to each other, and, in fact, must remain coupled in this fashion if they are to function effectively. The anatomical placement of the left and right hearts within a single organ facilitates both the mechanical and the electrical coupling. This will become apparent when we discuss these aspects of cardiac physiology. While there are unique aspects of both of the pumps, many properties are common to both. The term "heart" will subsequently be used to refer to aspects that are common to both, whereas, "left heart" and "right heart" will be used to refer to specific aspects of these two pumps that are not necessarily common to both.

The Sequential Contraction of the Atria and Ventricles Is Activated by a Wave of Electrical Depolarization that Is Initiated by Pacemaker Cells in the Sinoatrial Node in the Right Atrium

Although the heart is composed primarily of muscle, it must function quite differently from the way in which skeletal muscle works. Different parts of the

Figure 18–8

Cardiac valve operation. Valves in the heart normally open when the pressure gradient across the valve is increasing in the direction that blood normally flows through the heart (forward pressure gradient), as shown in the first drawing. In contrast, a reverse pressure gradient (see middle panel) will force the valve closed. This prevents reverse blood flow in response to reverse pressure gradients that occur as a result of the pumping action of the heart. In the third drawing, a forward pressure gradient forces the semilunar valve to open. Darker colors correspond to higher pressure.

Epicardium   Myocardium   Endocardium

Figure 18–9

Anatomy of the heart wall.

heart must contract at different times, and they must do it in a proper sequence if the heart is to pump blood efficiently. The key to understanding much of the special function of the heart lies in a knowledge of the electrical properties of its muscle.

Phases of the Cardiac Action Potential

As do most other muscles, the heart muscle cells have plasma membranes that can generate action potentials. It is difficult to speak of a "typical" heart muscle action potential, because each of the specialized regions of the heart has special action potentials. While they are similar in many ways to the relatively simple action potentials found in nerves and skeletal muscles, action potentials in the heart involve additional membrane processes. Cells found in the ventricular region have most of the special features that we need to discuss, so a cell of this sort can serve as our typical example.

Figure 18–10 shows such a complex action potential. This is the type of electrical activity found in the **Purkinje fibers** of the ventricles (we will discuss the function of these fibers later). Notice first of all that this action potential lasts hundreds of times longer than one from a nerve. In order to identify the extra voltage changes present, we use numbers to designate the various parts of this action potential. **Phase 0** is the rapid upstroke. This is a very fast depolarization that **overshoots** the zero millivolt level. As in nerve tissue, this rapid depolarization is due to a rapid increase in the membrane permeability to

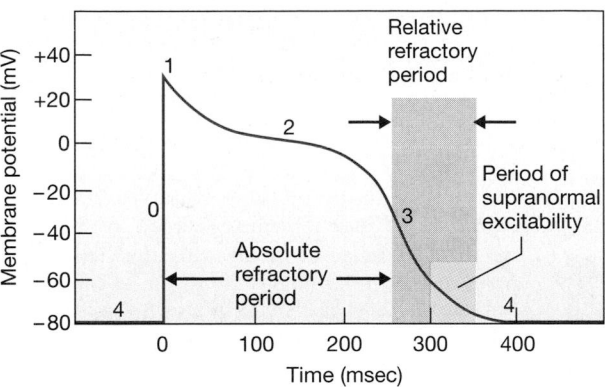

Figure 18–10

A sample action potential from heart muscle, typical of those found in the fast conduction fibers of the ventricle, such as the Purkinje fibers. The numbers above the curve designate the phases described in the text, and the approximate durations of the refractory periods are shown.

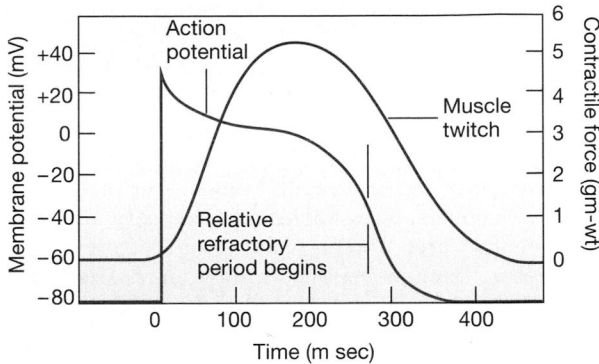

Figure 18–11

The relationship between the action potential and the twitch. The isometric twitch of a sample of heart muscle is shown in relation to the action potential that caused it. To some extent the duration of the twitch is determined by the duration of the action potential. Agents that prolong the plateau of the action potential make the twitch last longer, and the force of contraction may be increased because of the greater opportunity for calcium to enter the cell during the action potential.

sodium ions. Associated with this is a reduction of the potassium permeability to a value lower than when at rest. As the high sodium permeability begins to decline, a small amount of repolarization takes place. This brief event is **Phase 1.** Shortly thereafter, the membrane potential becomes relatively steady at a value near zero millivolts. This is **Phase 2,** also called the **plateau** of the action potential. Here a slow inward membrane current due to the flow of **calcium ions** is balanced against the efflux (outward flow) of **potassium ions,** even though the potassium permeability is reduced at this time. The membrane channels through which calcium passes also allow a small amount of sodium to pass inward. Late in the plateau, the calcium permeability declines, and with the falling membrane potential, the potassium permeability begins to increase. These factors produce a **repolarization,** which is called **Phase 3.** During this time, the membrane potential falls to its original value and remains steady until the next beat. This steady portion is **Phase 4,** which is equivalent to the resting potential of other muscle cells. This sequence of membrane potential changes lasts for about 300 milliseconds, depending on the heart rate.

As in nerve and other types of muscles, these phases of the action potential are associated with refractory periods. These are diagrammed in Figure 18–11. During the **absolute refractory period,** the muscle cannot be stimulated again. The **relative refractory period** then follows, and latest of all is the period of **supernormal excitability.** Here the muscle is actually easier to stimulate because the potassium permeability is still a bit lower than normal. The ac-

tion potential produced will be smaller than normal, because not all of the sodium channels from Phase 0 have had time to "re-set." Normally, the muscle is never stimulated during this time period. If such stimulation were to occur, the result could be very dangerous because the activity could spread to other tissue normally at rest at that time.

The Cardiac Action Potential and Contraction of Cardiac Muscle

Notice in Figure 18–11 that the absolute refractory period lasts almost as long as the isometric twitch. This relationship prevents the muscle from contracting again before it has relaxed, since the muscle will not respond to a second stimulus so soon after the first. This means that a tetanus is not possible in heart muscle; such a long-lasting contraction would prevent the heart from filling with blood for the next beat, making it a poor pump indeed.

Action potentials are necessary to make heart muscle contract. In skeletal muscle, action potentials are produced as a result of stimulation by a motor nerve. Each cell in a skeletal muscle is individually supplied by its own motor nerve terminal, and coordinated activity is due to the timing of impulses from the central nervous system (CNS).

The heart, in contrast, has no motor nerve supply, and its cells are very small. A separate nerve supply to each of the millions of muscle cells in the heart would make it a huge and complex organ. Instead,

CLINICAL FOCUS

Cardiac Imaging

A number of techniques are available for imaging the heart chambers, heart valves, and coronary arteries. A frequently used noninvasive (not requiring surgery) technique is **echocardiography.** High frequency sound waves are used to image the chamber walls and valves in the heart. The basic principle is the same as that used for radar imaging of airplanes. An echocardiogram (see figure) provides a two-dimensional image of a cross-section through the heart and shows the same structural organization you would see if you cut a heart into two pieces. A small, hand-held device both emits and receives reflected sound waves. This device (a **transducer**) is placed against the chest wall; different cross-sectional views are generated depending on the position of the transducer. This imaging method provides information on the thickness and motion of the muscular walls, the size of the heart chambers, and the functioning of the heart valves. Absence of normal motion in a particular area of the chamber wall can be diagnostic of dead cardiac muscle **(infarct).** Excessively thickened walls (cardiac hypertrophy) frequently indicate that the heart is pumping blood against an abnormally high pressure load. Damaged (scarred, torn, or congenitally malformed) valves will frequently show an abnormal pattern of movement on the echocardiogram. Newer echocardiography techniques enable the clinician to measure the velocity of the blood as it moves through the valve orifices. The backward leakage of blood through a damaged valve **(valvular insufficiency)** or the high velocity jet of blood moving through an obstructed valve **(valvular stenosis)** is often detected by echocardiography.

Angiography is an invasive technique that requires the insertion of a catheter into the heart **(cardiac catheterization),** followed by an injection of a radiopaque dye. The movement of the dye-bolus through the heart sequentially delineates the chambers. If the catheter is inserted into one of the coronary arteries **(coronary angiography),** the size of the artery lumen can be visualized (see coronary angiography figure). Obstructions of the coronary artery (e.g., in atherosclerosis and coronary artery spasms) can be easily visualized with this method. Recent developments in computer technology and digital image-enhancement techniques have significantly enhanced the image quality of coronary angiograms.

■ **With aortic valve insufficiency (i.e., a leaky aortic valve) what would be the expected appearance of the left ventricular chamber? How would this differ from aortic valve stenosis and obstruction of the outflow tract?**

the cells of heart muscle communicate with one another directly, and muscle takes over the function of nerve. This means that the heart can be stimulated to contract at a single location, and the message will be passed along to the rest of the heart muscle. Within the individual cells of the heart muscle, the process of activation is similar to that in skeletal muscle. The t-tubule system of the ventricular muscle cells aids in conducting the surface action potential inward, where it causes the release of stored calcium for the activation of the troponin-tropomyosin regulation system. Significant amounts of calcium enter the cell during the plateau of the action potential, and this calcium is added to the internal stores. The entering calcium is also directly responsible for causing some release of internal stored calcium from the sarcoplasmic reticulum. Because of this continual entry of calcium, cardiac muscle cells must (and do) have a way of getting rid of excess internal calcium by exchanging it for incoming sodium.

Initiation of the Action Potential by Pacemaker Cells

The single location from which the stimulus for the heart action arises is called the **pacemaker.** Normally, the pacemaker region is the **sinoatrial node** (also called the **SA node**), a specialized region of muscle cells in the right atrium. A pacemaker action potential is shown in Figure 18–12. These cells do not have a steady resting potential. After each action potential, the cells repolarize to a value called the **diastolic potential.** Very soon after repolarization, the membrane begins a slow depolarization. This depolarization quickly brings the membrane potential to its **threshold,** and another action potential takes place. This action is repeated again and again, as many as 3 billion times in a 70-year lifetime. At resting heart rates, it is repeated approximately 70 times per minute, and it can reach as high as 200 times per minute during heavy exercise. The rate of the heart is regulated by

(a)

(a) Echocardiogram of a normal heart. *(© Phototake/ Phototake, NYC.)* (b) Arteriogram showing a catheter inserted into the right coronary artery. *(© CNRI/Phototake.)* (c) A schematic of the arteriogram.

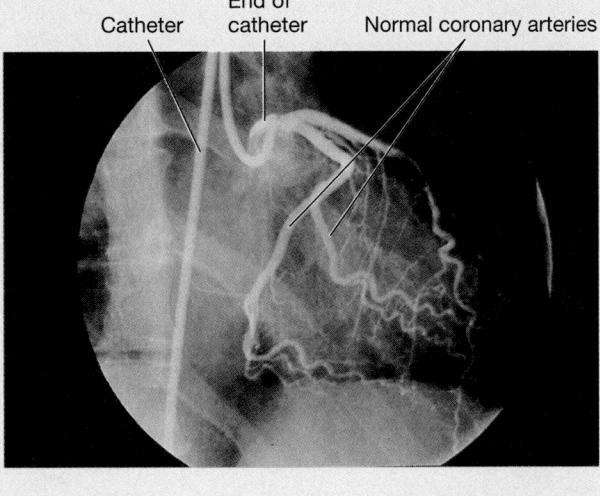

Catheter End of catheter Normal coronary arteries

(b)

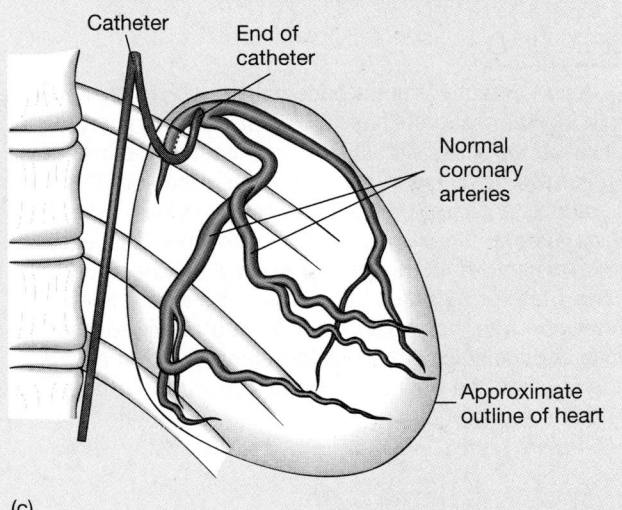

Catheter End of catheter

Normal coronary arteries

Approximate outline of heart

(c)

the rate at which the diastolic potential reaches the threshold.

Normally, only one pacemaker is active at one time in the heart. This is because whichever pacemaker cell "fires" first stimulates other cells that are about to produce an action potential. Therefore, the fastest pacemaker cell predominates. The heart actually contains many cells capable of becoming pacemakers, but their activity is not usually expressed because they are inherently slower. Under some conditions, if these "hidden" pacemakers do not receive a stimulus from other cells, their activity will become apparent. As we will see later, this can be an important safety device.

The rate of the SA node pacemaker is subject to control by the **autonomic nervous system.** The **parasympathetic** nervous system decreases the rate of diastolic depolarization and makes the diastolic potentials more negative (by making the cell membranes more permeable to potassium ions). This

means that it takes longer for the pacemaker cells to reach threshold, and so the heart rate is decreased. The **sympathetic** nervous system speeds up the diastolic depolarization of the pacemaker cells (and the repolarization as well), and allows the cells to reach threshold sooner; this increases the heart rate. Normally, the resting rate of the heartbeat is somewhat slowed by the parasympathetic nervous system, and if parasympathetic activity is blocked, the heart will speed up. The pacemakers of heart transplant patients (whose hearts are not connected to the autonomic nervous system) can still respond to circulating hormones (e.g., epinephrine) to produce an increase in heart rate.

The rate of the heart varies in response to physiological needs, increasing during exercise and decreasing at rest. If the resting rate is very low (fewer than 60 or so beats per minute), the condition is termed **brachycardia;** an elevated resting heart rate (above 100 per minute) is called **tachycardia.** Nor-

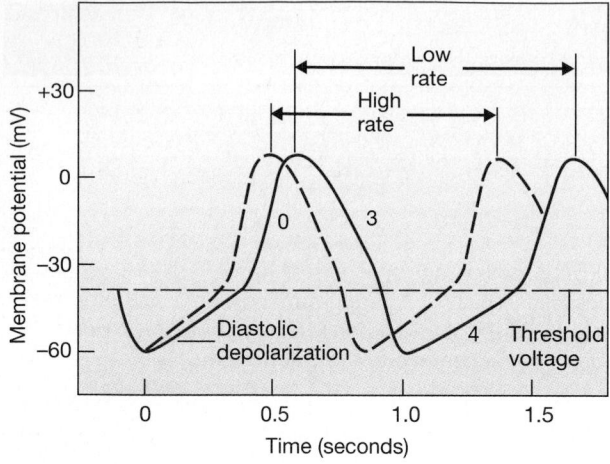

Figure 18-12

An action potential from a pacemaker cell. The slow diastolic depolarization (Phase 4) lets the membrane potential reach the threshold voltage shortly after the cell has repolarized, resulting in another action potential. The solid line shows a pacemaker potential at a low heart rate, which is determined by the spacing between successive action potentials. In this case the low rate corresponds to a heart rate of 65 beats per minute. The dotted line shows the rate that would result from a more rapid diastolic depolarization. The action potentials occur closer together, and the new spacing corresponds to a rate of 72 beats per minute.

mally, these rates are under the control of the SA node pacemaker; in other cases, blockages of conduction elsewhere in the heart may lead to a very serious brachycardia, and abnormal pacemaker activity at locations other than the SA node can lead to tachycardias with rates as high as 300 beats per minute.

Drugs and other influences that affect the rate of the heart are called **chronotropic** influences, while those agents that affect the strength of contractions are called **inotropic.**

The Sequence of Cardiac Activation

Activation of the Atria. The rhythmic activity of the SA node pacemaker is conducted first into the musculature of the right atrium and then to the left atrium (Fig. 18-13). This impulse conduction occurs when the action potential in one cell causes current to flow in an adjacent resting cell. This local current, which flows between the cells in the region of their intercalated disks (see Chapter 16), brings the membrane potential of the resting cell to its threshold. It then produces its own action potential, which in turn stimulates the next adjacent cell. This process occurs (with varying degrees of efficiency) in all regions of the heart. Conduction of action potentials between sepa-

rate cells in heart muscle is similar in many ways to conduction along a nerve axon (a single cell), in which active regions stimulate the resting regions next to them. Cell-to-cell conduction in heart muscle is made possible by the low intercellular resistance associated with the intercalated disks.

Conduction through the atria is quite rapid. Atrial action potentials are conducted quickly because the muscle cells are relatively large and have a very negative resting potential. In addition, some specialized atrial conducting pathways conduct very rapidly. The rapid conduction and fast action potentials result in a very rapid Phase 0 depolarization with a large overshoot in the cells, which produces a large local current flow. These conditions favor the rapid conduction of the impulse throughout the tissue.

Conduction of the Action Potential from the Atria to the Ventricles. At the lower portion of the right atrium is another region of specialized cells called the **atrioventricular (AV) node.** These cells are quite small and have resting potentials that are not as negative as those of the atrial cells. As action potentials pass through this region, they become weaker and travel more slowly. This slowed impulse travel is called **decremental conduction.** If the distance is great enough, or the impulse becomes weak enough, it may die out altogether. Under normal conditions, the impulse is slowed but not stopped at the AV node. The delay this causes ensures that the atria finish their contraction before the ventricles begin theirs. The AV node is the only normal connection between the atria and the ventricles, and its delay serves as a guard against very high atrial heart rates (>200 per minute) being conducted to the ventricles.

Activation of the Ventricles. After the impulse crosses the AV node, it enters the **Bundle of His** (also called the **common bundle**). This conducting structure is composed of muscle cells specialized for very rapid conduction, and here the impulse travels at its greatest speed. The common bundle soon divides into two **bundle branches,** one for each of the ventricles. Conduction is very rapid here as well.

The bundle branches divide further in each ventricle, eventually becoming an array of **Purkinje fibers,** which form a distribution system over the endocardial surface and even penetrate slightly into the main muscle mass of the ventricles. The Purkinje fibers are also specialized muscle fibers with a high speed of conduction.

From the Purkinje fibers, the impulse passes into the muscle of the ventricle wall, stimulating contraction. The ordinary ventricular muscle fibers are somewhat slower conductors than the specialized conducting tissues, but by this time the activity is well

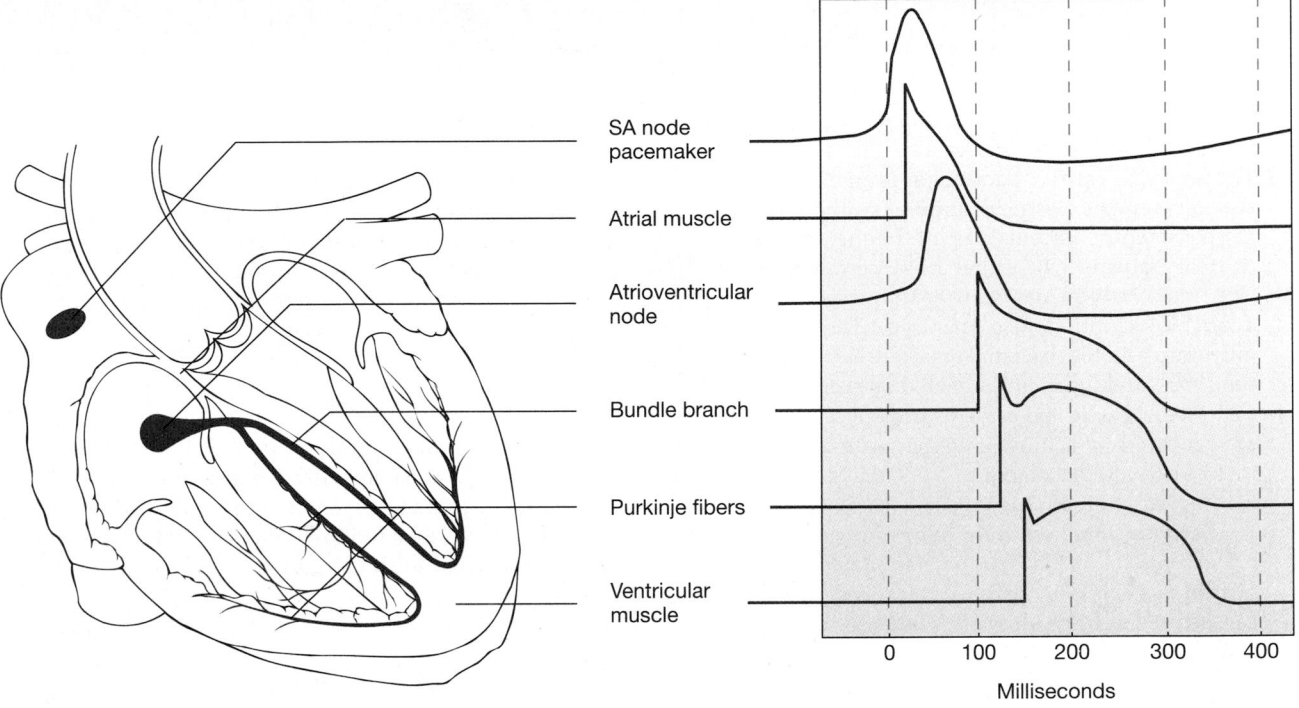

SA node pacemaker

Atrial muscle

Atrioventricular node

Bundle branch

Purkinje fibers

Ventricular muscle

0   100   200   300   400

Milliseconds

Figure 18–13

The succession of action potentials during the course of one heartbeat. The sequence starts with the *pacemaker potential* at the sinoatrial node. The activity spreads rapidly to the *atrial muscle*. The *atrioventricular* node action potential, similar in form to the SA node potential, is conducted slowly, and there is a longer space between this and the first ventricular potential shown, that of the *left bundle branch*. (This is similar to the slightly earlier action potential of the *Bundle of His*, which is not shown.) Action potentials of the *Purkinje fibers* and the *ventricular muscle* follow quickly in succession. The entire activation process takes approximately two-tenths of a second from the time the action potential begins in the SA node until the last ventricular muscle is activated.

distributed throughout the heart, and the distance remaining is relatively short. The arrangement of the specialized conducting system ensures a nearly simultaneous activation of all of the ventricular muscle.

The overall effect of the distribution of the stimulating impulse by the rapid conduction system is a wave of depolarization that sweeps over the heart from its base to its apex and from the endocardial to the epicardial surface. Because of the rapid conduction of the impulse and the fast rise of the action potential, the boundary between active and resting tissue is very sharp and moves very quickly. During repolarization, there is also an apparent movement of the active areas.

As active cells return to rest, a less-pronounced wave of repolarization sweeps over the heart in the opposite direction. However, the directionality of the repolarization is determined not by the conduction system, but by the fact that the cells activated last have the shortest action potentials and begin to re-

polarize first. This produces an apparent movement in the opposite direction from the wave of depolarization. Because this is not a conducted phenomenon, and because the repolarization is not nearly as rapid as the depolarization, the speed of the repolarization wave is lower and the boundary between recovering and still-active cells is not very sharp.

Abnormal Activation of the Heart. The proper function of the heart as a pump depends on the correct sequence of activation, and this in turn depends on the normal function of the conducting system. For example, failure of the AV node to conduct the impulse will cause **heart block,** in which the ventricles are not stimulated by the atrial impulse. At such times, cells other than those of the SA node can adopt a pacemaker function and provide a source of stimulation, although the rate from such auxiliary pacemakers is usually lower than normal. These "emergency" pacemakers usually lie somewhere in the conduction system, and their activity is normally

Cardiac Pacemakers

The **artificial cardiac pacemaker** is an electronic device that delivers an electrical stimulus to the heart to initiate contraction. Pacemakers are frequently used to treat patients who suffer from permanent or recurrent bradycardias (slow heart rates) with undesirable symptoms (e.g., fainting). The first totally implantable pacemakers were reported in the 1950s, and today an estimated one million patients worldwide have them. Implantable pacemakers are powered by lithium batteries, which may last for 6 to 10 years. Replacement of the battery requires surgical implantation of a new pacemaker.

Early pacemakers were unable to sense the electrical activity of the heart and stimulated it (either the atrium or ventricle) at a fixed rate. Modern pacemakers can sense the heart's normal electrical activity and are activated only when they fail to detect appropriate spontaneous activity.

Such **demand pacemakers** are frequently programmable to respond in a specified manner when a particular arrhythmia is detected. For example, to treat com-

plete **heart block,** a pacemaker is used that senses the depolarization of the atria and then delivers a stimulus to the ventricles after a delay equal to the normal atrioventricular conduction time. This method of pacing retains the normal temporal relationship between atrial and ventricular contraction, and therefore preserves the normal contribution of atrial contraction to ventricular filling. In addition, by sensing the normal rate of atrial depolarization, the pacemaker can respond appropriately to increases in heart rate associated with exercise and stress. Programmable pacemakers can be reprogrammed without being removed, which allows their operating parameters to be revised to fit the patient's changing needs.

■ **Why would it be advantageous for a pacemaker not to stimulate the ventricle whenever a normal impulse was conduced through the AV node?**

---

suppressed by the faster pacemaker rhythm of the SA node. A typical pacemaker from the AV node might have an inherent rate of around 50 beats per minute, while a ventricular pacemaker could have a rate of 30 beats per minute.

Regular muscle tissue rarely shows any spontaneous activity. If there is such a pacemaker in the ventricular region—and if for some reason it becomes active from time to time—its impulse will be conducted both in the normal direction and "backwards" toward the atria. When it stimulates the AV node, it produces refractoriness in the nodal tissue. The next atrial impulse arriving may not be able to be conducted through the AV node, and the ventricles will "drop" a beat. This can be very apparent to the person in whom it happens, because the delay gives the ventricles a chance to become filled with extra blood, and the next heartbeat is larger than usual. Such "skipping a beat" is not rare and is usually of no concern.

However, a dangerous condition may arise if the conducting system of the ventricle is damaged so that conduction along parts of it is slowed. Such damage is not uncommon following a heart attack (a **myocardial infarction**), because areas of the heart muscle may have been deprived of oxygen for some time. This slowed conduction can cause cells that are in the process of normal repolarization to be stimulated by a late-arriving action potential, which may arrive

during the period of supranormal excitability. The resulting stimulation can give rise to even more action potentials out of their proper time sequence, which are then conducted to normally resting cells. The final result may be an uncoordinated stimulation of the ventricular muscle called **ventricular fibrillation.** Such uncoordinated contractile activity is useless for pumping blood, and death will follow shortly. These conduction and rhythm disturbances, and many others related to them, are called **arrhythmias** and are often fatal if not treated immediately.

The Electrocardiogram.   The action potentials and their conduction sequence as just described can be observed only by painstaking study of exposed or isolated hearts. While the knowledge gained is important in understanding the operation of the heart, it is obviously not a realistic approach to the study or treatment of human (or veterinary) patients. Fortunately, the electrical activity of the heart may be harmlessly detected at the body surface by a technique called **electrocardiography.**

The heart may be thought of as an electrical generator lying in a conducting medium made up of the body tissue and fluids. During the phases of the cardiac cycle in which depolarization or repolarization is sweeping over the heart muscle mass, some portions are positively charged and others negatively charged. This potential difference, or **voltage gradi-**

ent (as much as 100 millivolts), causes current to flow in the external medium between these regions of the heart. These currents are strongest close to the heart, and they become weaker at greater distances from the heart. The largest voltage measured at the body surface is only around 1 millivolt; however, this is sufficient to be detected by a sensitive instrument called an **electrocardiograph.** This device converts the minute surface currents to movements of a pen on paper or a spot of light on a cathode ray tube (CRT), producing the familiar **electrocardiogram (ECG or EKG).**

Figure 18–14 is a diagram of a typical electrocardiographic recording. For convenience, the principal features of the recording have been designated by the letters P, Q, R, S, and T. The so-called **P-wave** is caused by the depolarization of the atrial muscle, and the **QRS complex** is caused by the depolarization of the ventricles. The **T-wave** is the result of the repolarization of the ventricles. Repolarization of the atria takes place during the ventricular depolarization, and its weak signal is lost. When the entire heart is either depolarized (active) or is resting, there is no potential difference to set up a current flow, and no deflection is seen in the ECG. The line of no deflection is called the **isoelectric line.** Figure 18–14 also shows the relationship between cellular action potentials and the surface ECG.

It is important to realize what the ECG can and cannot show. First, only electrical events are shown, and one can only *infer* that a contraction of the muscle has taken place (sometimes it has not). The ECG is a record of **voltage** and **time** only. Secondly, only the electrical events that involve large amounts of muscle show up in the ECG record. Because of the small tissue mass involved, activity of the conducting system and nodal tissues can only be inferred, but these inferences can be very revealing. For example, if a P-wave is not followed by a QRS complex, then one can assume that conduction was blocked somewhere between the atria and the ventricular muscle. If the QRS complex is very broad or abnormally shaped, one can assume that some areas of the muscle of the ventricle were activated later than they should have been. This implies a defect in the ventricular conducting system. A very long delay between the P-wave and the QRS complex most probably means that the conduction in the AV node was more decremental than usual.

Further information can be gained about the direction of the spread of electrical activity by recording from electrodes located on different areas of the body. The portion of the electrical activity that the recording electrodes intercept depends on their location and orientation. Clinical recording of the ECG assumes that the heart lies at the center of an equilateral triangle and that the recording connections are

Figure 18–14

The time relationship between the action potentials and the electrocardiogram. All three traces share the same time axis, and the ECG recording may be used to infer when the actual electrical changes took place in the heart muscle.

made at the vertices of this triangle (Fig. 18–15). In practice, the electrical connections to the right and left arms and to the left leg are considered to connect to these vertices, and the standard **limb leads** are recordings between any two of the connections. A recording made between the right and left arms is designated as **Lead I** and is most sensitive to activity spreading laterally across the heart. **Leads II and III** are recordings made between the left leg and the right and left arms, respectively. Records from these leads are most sensitive to activity proceeding from the top to the bottom (base to apex) of the heart. By combining the timing and voltage information recorded from a number of leads, someone skilled in the interpretation of the ECG can build up a very complete picture of the functioning of the heart.

Each Beat of the Heart Involves Integrated Electrical, Biochemical, and Mechanical Events Referred to as the Cardiac Cycle

As discussed in the previous section, the electrical activation of the left and right hearts occurs almost simultaneously. As a result, the sequence of mechanical events occurring during a single beat or cycle of the heart is very similar for both hearts. Because of

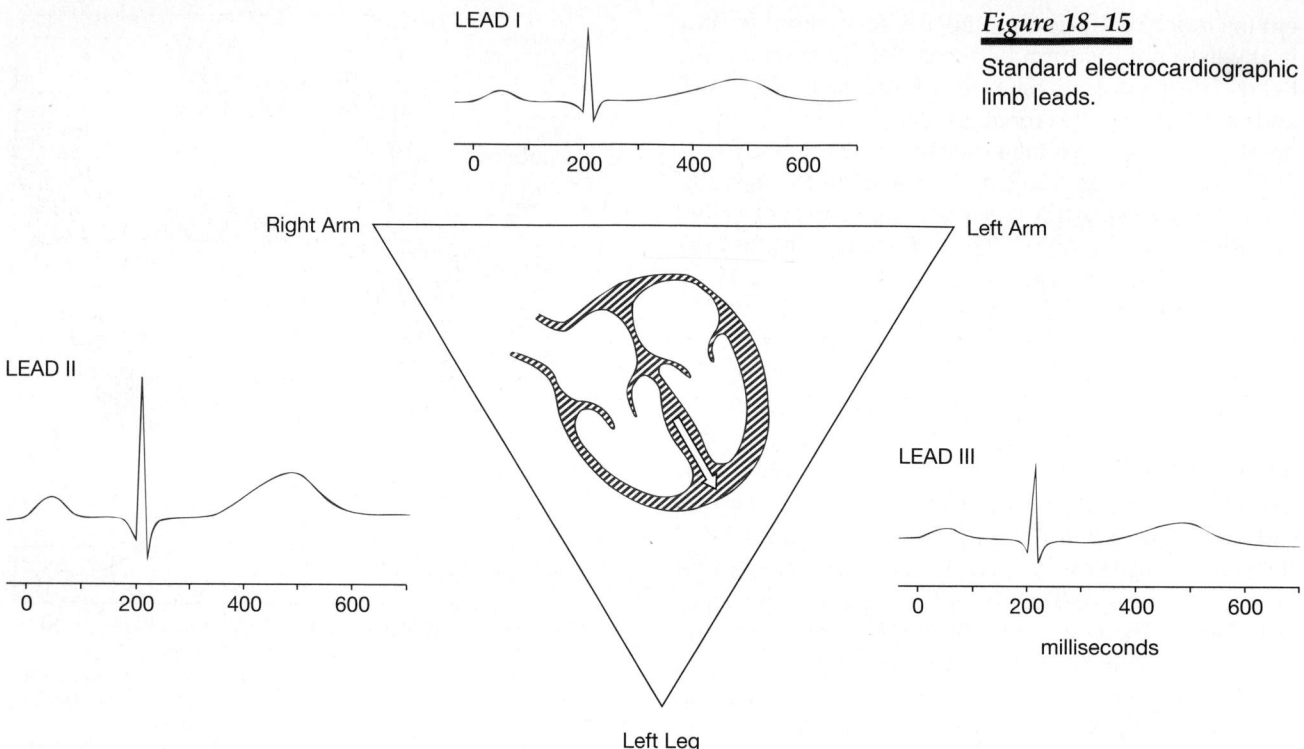

Figure 18–15

Standard electrocardiographic limb leads.

this similarity, portions of the following discussion of the cardiac cycle will focus on either the left or the right heart. Keep in mind that similar events are occurring at the same time in both sides of the heart.

The Ventricular Pump

Let us first focus on the ventricles of the heart, and examine how the ventricles function as pumps and how the heart valves ensure that blood flows in one direction only. Figure 18–16 shows schematically the four basic steps in the pumping process for the right heart; the same basic steps also apply to the left heart. Beginning with Panel (a) **(ventricular filling),** we see that the ventricle is being filled with blood from the atrium. Contraction of the atrial muscle facilitates the movement of blood into the ventricle through the one-way AV valve. Notice that the semilunar valve, between the ventricle and the outflow tract, is closed and prevents the backflow of blood from the outflow tract into the ventricle.

Panel (b) **(ventricular contraction)** shows the beginning of the ventricular contraction phase of the cycle. Notice that as the ventricle contracts, pressure increases inside the ventricle. This initially causes a small backflow of blood into the atrium, which causes the AV valve to close.

As soon as ventricular pressure exceeds the pressure in the outflow vessel tract (Panel [c]), the semilunar valve is forced open, and blood flows out of the ventricle **(ventricular ejection).** In order for the cycle

to repeat, the ventricle must now relax and refill with blood. As the ventricle relaxes **(ventricular relaxation),** pressure inside the ventricle falls, and the semilunar valve is forced closed as blood attempts to flow from the outflow vessel track back into the ventricle (Panel [d]). When ventricular pressure falls below that of the atrium, the AV valve opens and blood flows into the ventricle (Panel [a]). This completes the cycle.

The Cardiac Cycle

You will notice that the ventricles are the main pumping chambers of the heart. The atria serve as reservoirs to receive the continuous flow of blood back to the heart. Although the atria do contract, this adds very little to the filling of the ventricle under normal resting conditions. Now let's look in greater detail at the sequence of events occurring during a single cardiac cycle as they occur in the heart. The sequence of events is depicted in Figure 18–17. Notice that the overall sequence of events occurs almost simultaneously in the left heart and the right heart.

Ventricular Filling. The term **diastole** refers to the period of the cycle during which the ventricles are filling with blood, and **systole** refers to the portion of the cycle during which the ventricles are actively contracting and pumping blood out of the heart. Let's begin our examination of the cardiac cycle with the **ventricular filling** phase of diastole (see Fig. 18–17,

VENTRICULAR FILLING	VENTRICULAR CONTRACTION	VENTRICULAR EJECTION	VENTRICULAR RELAXATION

(a) Ventricle is relaxed and full

(b) Ventricle contracts, but all valves are closed. Thus, pressure increases but volume remains constant

(c) When ventricular pressure exceeds that in outflow tract, valve opens

(d) Ventricle relaxes, but all valves are closed. Thus pressure decreases, but volume remains constant

Figure 18–16

The cardiac cycle. Diagrammatic depiction of a single pumping cycle of the ventricle. This illustrates the relative volume changes that occur in the atrium and ventricle during a single pumping cycle, as well as the opening and closing sequence of the inflow (atrioventricular) and outflow (semilunar) valves. Arrows indicate direction of blood flow. The intensity of blue color indicates relative pressures. Purple indicates contracting myocardium.

Panel 5). At this point in the cycle, the ventricular muscle is relaxing from the previous beat, and the pressure inside the ventricles is falling. When the intraventricular pressure falls below the pressure in the atria (Fig. 18–17, beginning of Panel 6), the forward pressure gradient from the atria to the ventricles forces open the AV valves, and ventricular filling begins (Fig. 18–17, Panel 6). The pressure in the atria exceeds that in the ventricles because the continuous flow of blood into the atria from the veins fills and stretches the elastic walls of the atria. Because the atria are relatively compliant, as compared to the ventricles, the pressure increase in the atria is sufficient to facilitate ventricular filling but remains low enough so that blood can continuously flow into the atria during most of the cardiac cycle.

Ventricular filling occurs rapidly at first (Fig. 18–17, Panel 6), as the accumulated blood in the atria flows into the relaxed ventricles and then more slowly (Fig. 18–17, Panel 1) as blood continues to flow from the veins into the atria. Notice that most of the ventricular filling occurs prior to atrial contraction. Normally, atrial contraction increases the filling of the

ventricle by only about 10% to 30%. This is because adequate time is available during diastole for nearly complete filling of the ventricles prior to the onset of atrial contraction. As heart rate increases, the time available for ventricular filling is reduced, and the atrial contraction becomes an important contributor to complete ventricular filling (Fig. 18–18). Atrial contraction occurs at the end of the ventricular filling phase (Fig. 18–17, Panel 2) and immediately prior to the onset of ventricular contraction. The P-wave of the ECG signals the onset of atrial depolarization. Keep in mind that the electrical depolarization precedes the mechanical contraction of the atria by a few milliseconds because of the latency period required for muscle activation.

The end of ventricular filling corresponds to the end of diastole (Fig. 18–17, Panel 2). The volume of blood in the ventricle at this time is referred to as the **end-diastolic volume.** As we will see in a later section, the end-diastolic volume is an important determinant of cardiac function. Notice also that there is

(text continues on page 561)

◄ *Figure 18–17*

Events of the cardiac cycle. The graph in the upper portion of this figure shows a time-course of the pressure measured in the aorta and left ventricle during a single cardiac cycle. Also shown are the occurrence of heart sounds, left ventricular volume, and an electrocardiogram (ECG). The state of the heart valves during the cardiac cycle is shown at the bottom of the graph. The six panels at the bottom illustrate the direction of blood flow (arrows) within the heart and connecting vessels at each stage of the cardiac cycle. Higher pressures are indicated by darker colors.

a reverse pressure gradient across the semilunar valves during the ventricular filling phase (aortic pressure > left ventricular pressure; pulmonary artery pressure > right ventricular pressure). This reverse pressure gradient forces the semilunar valves to remain closed during this period of the cycle. This prevents blood that was pumped into the aorta during the previous cycle from flowing back into the heart. Congenital or pathological abnormalities of the semilunar valves allow blood to squirt back into the ventricles during diastole. This condition is referred to as a **valvular insufficiency** and most commonly afflicts the mitral valve. The condition where blood

Figure 18–18

Effect of heart rate on diastolic filling. Atrial contraction makes only a small contribution to ventricular filling under normal conditions (upper panel). At high heart rates (lower panel), such as those occurring during heavy exercise, atrial contraction contributes significantly to ventricular filling because of the reduced time available for filling. The ventricular volume added during atrial contraction is indicated by the dark pink area.

flows backwards through the mitral valve during diastole is called **mitral regurgitation.**

Ventricular Contraction. The initiation of ventricular contraction marks the beginning of systole. Recall that the specialized conduction system activates the ventricles approximately 100 msec after atrial activation. This electrical activation of the ventricles is reflected by the QRS complex of the ECG. Shortly after the occurrence of the QRS complex, the ventricular muscle begins to contract (Fig. 18–17, Panel 3). Blood pressure inside the ventricles increases rapidly as the ventricular muscle contracts. Ventricular pressure quickly rises to a value that is greater than atrial pressure. This reversal of the pressure gradient causes a momentary reversal of the flow of blood through the atrioventricular valves. The valves fill like boat sails and are pushed into their closed position and held there by the reverse pressure gradient, preventing any additional blood flow back into the atria. During the remainder of this phase of the cardiac cycle, both the AV and semilunar valves are closed. This phase is referred to as **isovolumetric ventricular contraction** (see Fig. 18–17, Panel 3) because the volume of the ventricles cannot change while both the inflow and outflow valves remain closed. If the ventricular volume cannot change, then the cardiac muscle fibers are contracting isometrically during this period. When pressure inside the left ventricle has increased to a level that exceeds that in the aorta, there is a forward pressure gradient for blood flow from the ventricle into the aorta. The aortic valve opens in response to this forward flow of blood, and the **ventricular ejection** (see Fig. 18–17, Panel 4) phase of the cardiac cycle begins. The ejection of blood from the left ventricle occurs quite rapidly at first and then slows as pressure builds up in the aorta. Just as pressure increases inside a balloon as it is inflated with air, pressure increases inside the aorta as it is inflated with blood. Keep in mind that even as blood is being pumped from the heart into the aorta, blood is flowing out of the aorta into the systemic vasculature. The rise in aortic pressure during ventricular ejection indicates that the flow of blood into the aorta from the heart is greater than the flow of blood out of the aorta into the systemic circulation.

Ventricular Relaxation. The onset of ventricular relaxation immediately follows ventricular repolarization. Recall that the T-wave of the ECG is a measure of the electrical repolarization of the myocardium. Notice in Figure 18–17 that ventricular pressure begins to fall rapidly after the occurrence of the T-wave and is soon less than the blood pressure in the aorta. This reversal of the pressure gradient (aortic > ventricular), reverses the direction of blood flow in the region of the aortic valve. This reverse blood flow closes the aortic valve, preventing any further flow of blood back into the heart from the aorta. This momentary reversal in blood flow results in a decrease in the aortic pressure that appears as a dip or notch on the aortic pressure tracing that is referred to as the **incisura** (Fig. 18–17). The closure of the semilunar valve marks the end of systole and the beginning of diastole.

The first phase of diastole is referred to as **isovolumetric relaxation** (see Fig. 18–17, Panel 5); relaxation of the ventricles occurs at constant volume because both the AV and semilunar valves are closed. As soon as ventricular pressure falls below atrial pressure, the forward pressure gradient from the atria to the ventricles initiates blood flow into the ventricles through the one-way AV valves (see Fig. 18–17, Panel 6).

Heart Sounds

One of the physical manifestations of the intense mechanical activity occurring with each contractile cycle of the heart is a series of distinct heart sounds that correlate with specific events of the cardiac cycle. These sounds can normally be heard through a stethoscope placed against the chest, directly over the heart. Heart sounds can also be recorded electronically, as shown by the **phonocardiogram** in Figure 18–17.

Although four heart sounds are distinguishable in most healthy individuals, the first two are the most distinct and have a sound similar to a low pitch "lub-dub." The **first heart sound** occurs at the time of closure of the AV valves, and, as previously mentioned, marks the beginning of systole. The **second heart sound** occurs at the time of closure of the semilunar valves and defines the end of systole. These sounds are the result of vibrations in the heart and chest initiated by the closure of the heart valves and the sudden cessation of blood flow. A common example of how vibrations of this sort can occur is seen in the plumbing of older buildings. The rumbling or clanging noise that occurs when a faucet is turned off is the result of vibrations set up in the pipes as the moving column of water is suddenly brought to a halt.

A number of additional heart sounds associated with abnormal conditions of the heart can occur. For example, valvular obstructions can result in a high velocity jet of blood flowing through the narrowed opening of an obstructed valve, which produces higher pitch sounds called **murmurs**. Likewise, valve abnormalities that allow blood to leak backwards (e.g., from the aorta to the ventricle) are often associated with audible murmurs that can be heard during the normal periods of silence. The interpretation of abnormal heart sounds is based on the quality and intensity of the sound as well as on the timing of the sound relative to other normal heart sounds.

The Heart Is Regulated by Changing both the Strength and Rate of Contractions to Produce Whatever Cardiac Output Is Required

As we have already stated, the major purpose of the heart is to maintain a constant flow of blood through the pulmonary and systemic circulations. The amount of blood pumped out of the left ventricle and into the aorta during a period of one minute is defined as the **cardiac output.**

Recall that the left and right hearts are in series and must pump exactly the same volume of blood over any extended period of time. Therefore, if the average output of the left heart is 5 L/min, then the average output of the right heart must also be 5 L/min. Even though the right heart generates significantly lower pressures during systole than the left heart, the volume of blood pumped by the two hearts must be the same.

The average cardiac output for a resting male is about 5 L/min; however, this value can increase or decrease significantly in response to the changing needs of the body or in response to disease. In order to understand how the cardiac output can be altered, we must understand the many factors that determine its magnitude.

We know from our previous discussion that the heart is a pulsatile pump that ejects a volume of blood with each beat or cycle. The volume of blood ejected in a single beat is called the **stroke volume** and is simply the difference between the volume of blood in the heart at the end of diastole (end-diastolic volume or EDV) minus the volume of blood remaining in the heart at the end of systole (end-systolic volume or ESV), as shown in Figure 18–19. The cardiac output is simply the sum of all the stroke volumes ejected from the heart over a period of one minute. If we know the stroke volume (SV) and the heart rate (HR) (number of beats in one minute), then we can calculate cardiac output (CO) as follows:

$$CO = HR \times SV$$

An average adult at rest might have a heart rate of 72 beats/min and a stroke volume of 70 ml/min (70 ml = 0.070 L = 2.4 oz.). on this basis, the cardiac output would be:

$$CO = 72 \text{ beats/min} \times 0.070 \text{ L/beat} = 5.0 \text{ L/min}$$

The total volume of blood contained in all the blood vessels and the heart of a 70-kg male is about 5 L, or

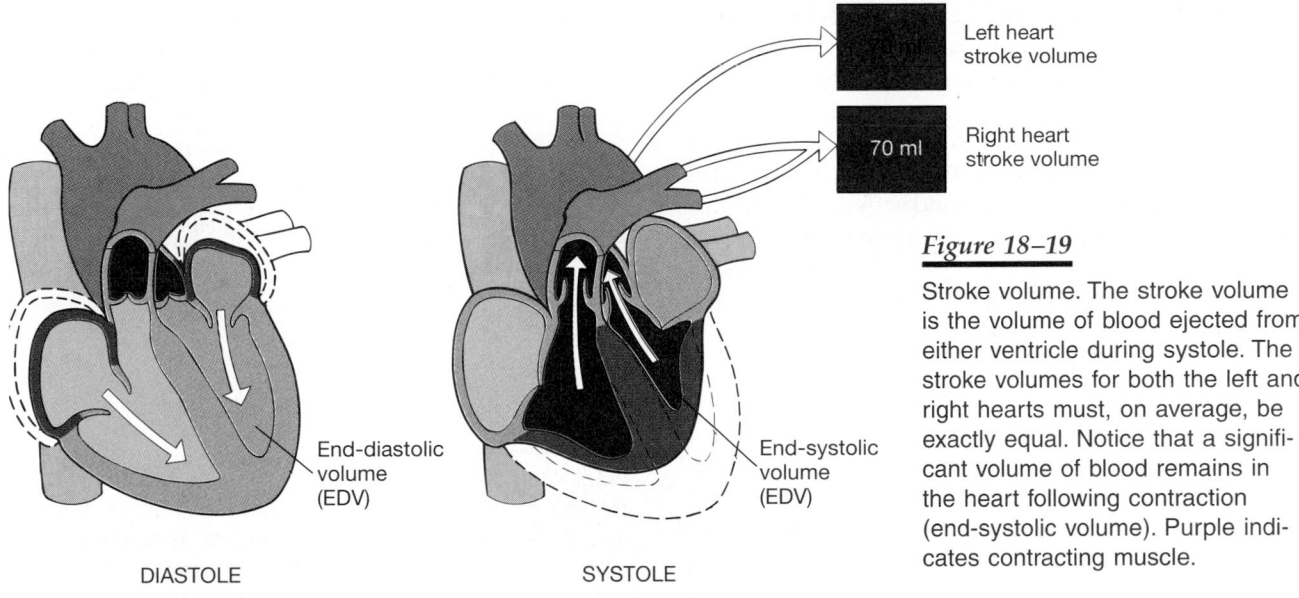

Left heart
stroke volume

70 ml   Right heart
stroke volume

Figure 18–19

Stroke volume. The stroke volume is the volume of blood ejected from either ventricle during systole. The stroke volumes for both the left and right hearts must, on average, be exactly equal. Notice that a significant volume of blood remains in the heart following contraction (end-systolic volume). Purple indicates contracting muscle.

End-diastolic volume (EDV)

End-systolic volume (EDV)

DIASTOLE   SYSTOLE

slightly more than 1 gallon. Therefore, a volume equal to the total blood volume is pumped through the heart each minute. This means that, on average, a blood cell will require 1 minute to travel from the left heart, through the systemic circulation, through the right heart, through the pulmonary circulation, and back to the left heart. It is interesting to consider that over a 70-year life span the heart will pump about 200 million L of blood. This is approximately 53 millions gallons—enough to fill 564 million 12-oz cans, which, if placed end-to-end, would circle the earth twice.

During strenuous exercise, cardiac output may increase four to five times in a physically fit individual and by as much as 7 times in a well-trained marathon runner. From the equation for calculating cardiac output, we can see that changes in either stroke volume or heart rate can alter cardiac output. During strenuous exercise, for example, heart rate may increase to 180 to 200 beats/min. Let's look at each of these factors in turn and the mechanisms by which they are regulated.

Regulation of Heart Rate

You will recall from the previous discussion of the electrical activation of the heart that heart rate is determined by the rate of spontaneous depolarization of the SA node. The rate of SA nodal depolarization can be altered by a number of influences, including stimulation of autonomic nerve fibers that innervate the SA mode, circulating hormones such as epinephrine, plasma electrolyte concentrations, and body temperature.

Let's consider the effects of autonomous nerve activity on heart rate. Figure 18–20 shows how an increase in the firing rate of the sympathetic and parasympathetic cardiac nerve fibers alters the rate of spontaneous depolarization of the SA node. Norepinephrine, the sympathetic neurotransmitter, increases the rate of spontaneous depolarization of the SA nodal cells so that threshold is reached more quickly and the heart rate is increased. Acetylcholine, the parasympathetic neurotransmitter, in contrast, decreases the rate of spontaneous depolarization and, in addition, hyperpolarizes the pacemaker cells. These two effects of parasympathetic stimulation work together to increase the time required for the membrane potential to reach threshold, with a resultant decrease in heart rate. Under resting conditions, both the parasympathetic and sympathetic cardiac fibers are active. If the heart is denervated, either surgically or pharmacologically, the heart rate will increase to about 100 beats/min. The fact that the rest-

Figure 18–20

Effect of autonomic nerve activity on spontaneous depolarization of the SA node.

ing rate for an innervated heart is 70 beats/min indicates that, under resting conditions, the parasympathetic nerves are the dominant regulatory influence on the SA node.

There are definite limits on how much cardiac output can be increased by increasing heart rate alone. First, the upper limit for conduction of impulses through the AV node limits heart rate to about 250 beats/min. Secondly, with rapid heart rates **(tachycardia)** of more than about 150 to 170 beats/min, cardiac output will begin to decrease because of inadequate time for complete filling of the ventricles during diastole (see Fig. 18–24). If the amount of blood returning to the ventricle during diastole is decreased, we might suspect that the amount of blood pumped by the heart (stroke volume) would also be decreased. As we will see in the next section, stroke volume is *very* dependent on ventricular filling. Recall that the normal mechanism for increasing heart rate is to increase sympathetic nerve activity to the heart. In the next section you will see how this increase in sympathetic nerve activity increases the available filling time.

Changes in body temperature also produce changes in heart rate. This is of little use for normal regulation of heart rate, since body temperature remains relatively constant; however, cardiac surgeons will often decrease the body temperature to reduce the heart rate **(bradycardia)** and the motion of the heart during surgery.

Regulation of Stroke Volume

The volume of blood pumped by the heart with each beat is highly regulated both by mechanisms that are **intrinsic** to cardiac muscle itself, and by **extrinsic** factors such as neural stimulation and circulating hormones. Let's begin by looking at the intrinsic mechanisms for regulation of stroke volume.

Intrinsic Regulation of Stroke Volume   The heart has the intrinsic ability to adjust its output (stroke volume) in response to changes in the input (venous return). This property of the heart is called the **Frank-Starling Law of the Heart** in honor of two physiologists, Otto Frank and Ernest Starling. As shown in Figure 18–21, there is a proportional relationship between the diastolic volume of the heart and the stroke volume over a relatively wide range of end-diastolic volumes. If the rate of blood flow to the heart increases, the end-diastolic volume of the heart will also increase, causing an increase in the stroke volume. Simply stated, the Frank-Starling law says that within defined limits the heart will pump whatever volume of blood it receives.

Figure 18–21

Frank-Starling Law of the Heart. This graph illustrates the relationship between stroke volume and changes in ventricular end-diastolic volume. The insets, showing diagrammatic sarcomeres, illustrate the relationship between end-diastolic volume and myofilament overlap. At normal resting ventricular volumes, sarcomere length is less than the optimal length for contraction.

The mechanism underlying the Frank-Starling relationship is the same mechanism that was discussed in Chapter 16 for the length-tension curve. When the end-diastolic volume of the heart is increased, the length of the cardiac muscle fibers in the wall of the ventricle is also increased (see Fig. 18–21). Recall from Chapter 16 that cardiac muscle, like skeletal muscle, has a well-defined length-tension relationship, which indicates that there is an optimal length for muscle contraction (see Fig. 16–13). Although skeletal muscle is generally constrained by the skeletal attachments so that it operates near the optimal length for contraction, this is not true of cardiac muscle. In fact, in a normal resting heart, the muscle fibers are *shorter* than their optimal length. Therefore, an increase in end-diastolic volume increases the muscle fiber length toward the optimal length for contraction. This increases the vigor of ventricular contraction, resulting in the expulsion of a greater stroke volume from the heart. Conversely, a decrease in end-diastolic volume results in an immediate reduction in the vigor of contraction and a decrease in the stroke volume. The Frank-Starling mechanism regulates each ventricle independently and adjusts the vigor of contraction, on a beat-by-beat basis. This ensures that each ventricle pumps out the same volume that it receives. This mechanism regulates each ventricle independently.

Although the full significance of this mechanism for regulation of cardiac output will become more apparent as we discuss the overall regulation of the cardiovascular system, one aspect can be appreciated now, based on our previous discussion of the serial arrangement of the two hearts and the systemic and pulmonary circulations. Because of this arrangement, it is vital that the two hearts each pump exactly the same volume of blood over any extended period of time. If this were not the case, excess blood would soon accumulate either in the lungs or in the systemic circulation. If the output of the right heart exceeded that of the left heart by only 1 ml/beat, within 90 minutes the entire blood volume would have accumulated in the lungs. Although such an extreme situation would never actually occur, this example demonstrates how rapidly even small imbalances in the stroke volumes of the two hearts could lead to the accumulation of blood in one portion of the cardiovascular system. An increase in the amount of blood in the pulmonary vasculature can compromise gas exchange in the lungs, with serious and often fatal results. The ability of the two hearts to independently self-regulate their output to match their input ensures that the average cardiac output of the two hearts remains exactly equal. Later, we will discuss how the Frank-Starling mechanism contributes to the maintenance of normal arterial blood pressure.

The Frank-Starling mechanism also provides intrinsic regulation of cardiac output in response to changes in aortic pressure. Consider the effect on the heart of a sudden increase in mean aortic pressure from 90 mm Hg to 120 mm Hg. In order for the heart to eject blood, the ventricular pressure must now increase to 120 mm Hg before the aortic valve will open, and blood must be ejected against a greater pressure. As a result of this increased pressure load, the muscle fibers shorten more slowly, and less blood will be ejected. If less blood is ejected (reduced stroke volume), more will remain in the heart at the end of systole (increased end-systolic volume). This extra blood in the heart, when added to the normal diastolic filling from the pulmonary vein, results in an increase in the end-diastolic volume of the heart. The Frank-Starling mechanism provides for a more forceful contraction on the next beat of the heart and results in an increase in the stroke volume back toward normal. In this manner, the Frank-Starling mechanism helps to maintain a constant cardiac output in the face of changes in aortic blood pressure.

If aortic blood pressure remains chronically elevated, the heart can self-regulate its capacity to pump blood in a more permanent fashion. Just as skeletal muscle will **hypertrophy** in response to exercise, cardiac muscle will also respond to chronically elevated pressure or demands for increased cardiac output by increasing the size and contractile protein content of the individual muscle cells of the heart.

Cardiac hypertrophy of both ventricles is readily apparent in athletes (marathon runners, for example) who perform intensive aerobic exercise. Cardiac hypertrophy can also be a response to an abnormal load placed on the heart because of disease. In many cases, this results in selective hypertrophy of a single chamber of the heart. For example, blockage of the pulmonic valve **(pulmonic stenosis)** results not only in a systolic murmur but also in hypertrophy of the right ventricle, because the right heart must generate higher-than-normal pressures to force blood through the partially occluded valve. The mechanism by which an increased load on the heart ultimately leads to increased production of contractile proteins by the myocardial cells is not currently understood.

Extrinsic Regulation of Stroke Volume. A number of factors extrinsic to the heart can also influence the vigor of ventricular contraction without changing end-diastolic volume. Any changes in the vigor of cardiac contraction that occur independently of changes in end-diastolic volume are referred to as changes in **contractility**. A change in contractility is mechanistically different from the altered vigor of contraction seen with changes in muscle length. Changes in the contractility of the heart are the direct result of changes in the rate and extent of calcium movement into the cytoplasm, as indicated diagrammatically in Figure 18–22. Recall from Chapter 16 that the concentration of calcium ions in the cytoplasm determines the degree of muscle activation. Increased firing of cardiac sympathetic nerve fibers results in an increase in both the rate and extent of calcium movement into the cytoplasm, from both the sarcoplasmic reticulum and from outside the cell. This results in a more rapid and more forceful contraction of the ventricles and an increase in the stroke volume ejected from the heart with each beat. As shown in Figure 18–23, an increase in stroke volume due to an increase in contractility will result in a reduced end-systolic volume.

Sympathetic stimulation also increases the rate of relaxation by increasing the rate at which the sarcoplasmic reticulum removes calcium from the cytoplasm. Because sympathetic stimulation increases both the rate of contraction and the rate of relaxation, the duration of systole is decreased. This effect can indirectly lead to a further increase in cardiac output. Sympathetic stimulation of the heart generally results in an increase in heart rate. Recall from the previous section that increases in cardiac output due to increases in heart rate were limited at higher rates by inadequate filling time during diastole (see Fig. 18–24, upper and middle panels). The effect of shortening the duration of systole, at any given heart rate,

| ● Contracting crossbridge | Inactive crossbridge | Ca²⁺ ions | Ca²⁺ diffusion |

Plasma membrane (sarcolemma)

Extracellular

T-tubule

Intracellular

Contractile filaments

Sarco- plasmic reticulum

NORMAL CONTRACTILITY
(a)

INCREASED CONTRACTILITY
(b)

Figure 18–22

Relationship between contractility and intracellular calcium. An increase in contractility is the result of an increase in the cytoplasmic calcium concentration. This is the result of both an increased release of calcium from the sarcoplasmic reticulum and an increased influx of calcium into the cell from the extracellular space. The increased concentration of intracellular calcium results in the activation of additional crossbridges, with a resultant increase in the vigor of cardiac muscle contraction. Active crossbridges are indicated in red.

is to increase the duration of diastole, and consequently, increase ventricular filling time (Fig. 18–24, lower panel). Cardiac drugs are available that increase the contractility of the heart, primarily by increasing the intracellular concentration of calcium. One of the oldest and most widely used drugs is **digoxin**, a plant extract.

Increased firing of the cardiac parasympathetic nerve fibers decreases the contractility of the heart, but not by a direct action of the parasympathetic neurotransmitter (acetylcholine) on cytoplasmic calcium concentrations. It appears that acetylcholine blocks or antagonizes the effects of sympathetic stimulation on the heart. In the absence of sympathetic nerve activity, increased firing of ventricular parasympathetic nerve fibers has only a small depressant effect on contractility.

In summary, stroke volume can be altered by two basic mechanisms. Changes in end-diastolic volume result in changes in stroke volume (Frank-Starling law). Changes in contractility, on the other hand, are the result of changes in intracellular calcium and do not require a change in end-diastolic volume. Since these two mechanisms for changing stroke volume are based on independent mechanisms, they can add together to produce even greater changes in stroke volume. The interaction of these two mechanisms is shown graphically in Figures 18–23 and 18–25. Notice that changes in stroke volume due to changes in contractility can occur at any end-diastolic volume. Likewise, changes in stroke volume due to changes in end-diastolic volume can occur at any level of contractility. Figure 18–26 summarizes the different mechanisms that we have talked about by which the

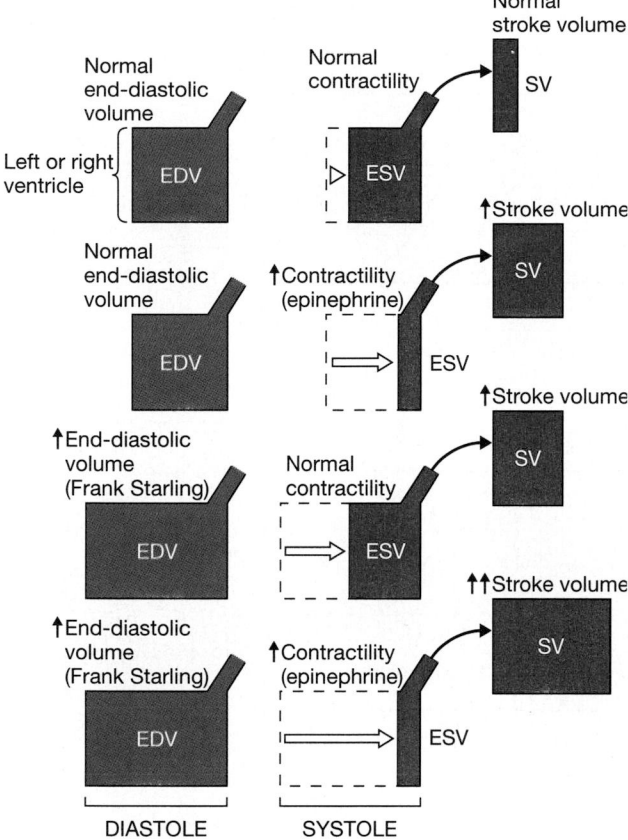

Figure 18–23

Changes in stroke volume due to changes in contractility are mechanistically different from those occurring as a result of increased end-diastolic volume. Therefore, the two mechanisms can operate simultaneously to increase stroke volume, as shown in the lower panel. EDV = end-diastolic volume; ESV = end-systolic volume; SV = stroke volume.

cardiac output can be altered. Cardiac output can be increased by increases in stroke volume and/or heart rate. Stroke volume can be increased by increases in end-diastolic volume (Frank-Starling law) or by increases in contractility. Both contractility and heart rate can be increased by increases in the level of sympathetic nerve activity to the heart or by increases in the release of epinephrine from the adrenal medulla. A reduction in aortic pressure increases stroke volume by reducing the load on the heart. Although an increase in either heart rate or stroke volume will result in a proportional increase in cardiac output, the metabolic costs of increasing cardiac output by these two mechanisms are not equivalent, as we will see in the next section.

Energy Consumption by the Heart

Contraction of all muscle types is associated with the hydrolysis of ATP. Cardiac muscle derives ATP primarily from the metabolism of circulating glucose and fatty acids. An important difference between cardiac muscle and other muscle types is that cardiac muscle must contract continuously, without interruption, throughout the life of an individual. Recall from Chapter 16 that skeletal muscle is able temporarily to sustain high levels of contractile activity by anaerobic metabolism of circulating glucose and stored glycogen. Because skeletal muscle can metabolize glyogen stored within the muscle cells, the supply of oxygen and glucose to the muscle by the circulating blood does not limit the production of ATP. Anaerobic metabolism results in the accumulation of lactate and the depletion of muscle glycogen. Upon cessation of muscular activity, lactate levels are reduced and glycogen stores are replenished.

Cardiac muscle, in contrast, cannot stop contracting to accumulate stored glycogen for support of anaerobic metabolism. The heart must instead rely upon continuous, oxidative metabolism of circulating glucose and fatty acids to produce ATP. This means that the ability of the heart to contract is strictly limited by the availability of oxygen supplied by the coronary circulation. As we will see in the next chapter, a number of diseases can prevent adequate cardiac blood flow. In conditions in which the heart is receiving inadequate oxygen, we might anticipate that cardiac function would be impaired. This is indeed the case, and one goal in the treatment of heart failure is to maintain adequate cardiac output while minimizing the oxygen requirements of the heart. Although at first glance this might appear to be an unresolvable situation, it turns out that the relationship between oxygen supply and cardiac output is not the same under all conditions. For example, at high heart rates and low stroke volumes, the heart uses more nutrients, and therefore more oxygen, to maintain a given cardiac output than at lower heart rates and larger stroke volumes. Blood pressure also affects the energy cost of maintaining cardiac output. The energy cost of producing a given cardiac output increases as blood pressure increases. The most economical method for increasing cardiac output is to increase the end-diastolic volume (Frank-Starling law) without increasing either heart rate or blood pressure. Figure 18–27 illustrates the relationship between increases in cardiac output and increases in oxygen consumption. In the next chapter, we will see how this information can be useful in the design of effective therapies for the clinical treatment of heart failure.

Figure 18–24

Increased cardiac output with sympathetic stimulation (effect of increasing the rate on contraction and relaxation of the heart). (*Top*) Changes in left ventricular volume in the resting heart of a marathon runner and calculated cardiac output (right). (*Middle*) Effect of tachycardia without sympathetic stimulation on ventricular volume and cardiac output. Tachycardia of this nature results from cardiac arrhythmias such as atrial tachycardia. (*Bottom*) Tachycardia induced by sympathetic stimulation and associated cardiac output. The increased rate of contraction and relaxation increases the time available for diastolic filling of the ventricle, leading to increased stroke volume and cardiac output compared to tachycardia (middle).

Figure 18–25

Summary of factors that affect stroke volume.

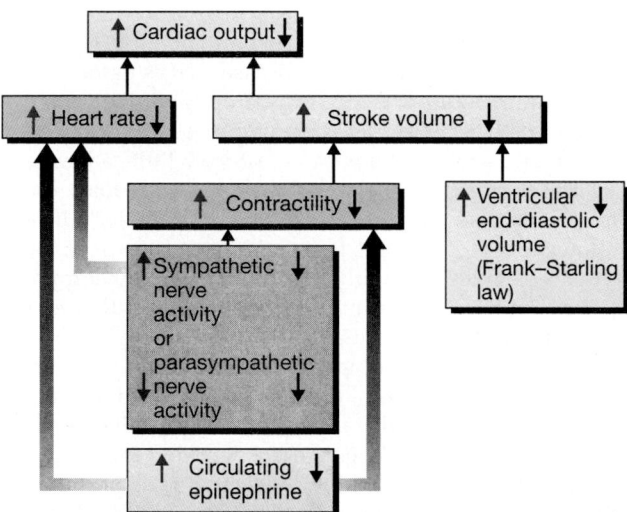

Figure 18–26

Summary of mechanisms that affect cardiac output.

Figure 18–27

Effect of heart rate, stroke volume (SV), and arterial pressure on oxygen consumption by the heart.

New Tools for Investigating Myosin and Other Molecular Motor Proteins

A number of new research tools have become available over the past 5 to 10 years that promise to revolutionize our understanding on how cardiac muscle myosin converts chemical energy to mechanical work, and how this process is altered by disease. One of the most recent, and most exciting advances was the publication of the crystallographic structure of myosin. By analyzing the diffraction pattern produced by directing a beam of X-rays through a crystal of purified myosin, scientists have deduced the atomic structure of the myosin molecule. Using sophisticated modeling programs, three-dimensional representations of the myosin molecule can be generated (see Fig. (b)). It is now possible to begin to visualize exactly how myosin binds to actin, and how this interaction is coupled to the breakdown of ATP to generate molecular motion and force (Fig. (a)).

Another recent advancement has been the development of *in vitro* motility assays that allow scientists to examine in detail the interaction of single actin filaments with normal and abnormal myosins. In one version of this assay, purified myosin molecules are attached to the surface of a glass coverslip as shown in the diagram (Fig. (c)). To this are added chemically modified actin filaments that fluoresce (i.e. give off red light) when exposed to green light. The light emanating from the actin filaments can be visualized with a light microscope, appropriate color filters, and a very sensitive video camera of the sort first developed for night-time recognizance by the military (i.e. night-vision devices). In the presence of ATP, the actin filaments move over the surface with a velocity that is determined primarily by the source of the myosin (Fig. (d)). For example, cardiac myosin moves actin slightly slower than fast skeletal muscle myosin and about three times faster than smooth muscle myosin. Motility assays of this sort have revolutionized the study of "motor proteins."

Analysis of the molecular structure of myosin suggests that certain regions are important for binding to actin, while others are important for ATP hydrolysis. These ideas can be tested using molecular genetics to introduce specific mutations at predefined sites in the myosin molecule, or by examining naturally occurring mutations. The effects of these alterations can then be examined under controlled conditions using a motility assay to evaluate the effect on filament velocity or force production (Fig. (d)), along with more standard biochemical assays to measure the rate of ATP hydrolysis. A number of different methods have been devised within the last five years to measure the force generated on a single actin filament, and, more recently, to measure the force produced by a single myosin molecule

(1–10 pN); this is roughly equal to the gravitational attraction between your body and the textbook you are now reading. One of these techniques utilizes a laser beam to grab onto single actin filaments floating in solution, and to hold them stationary (somewhat like the tractor-beam from Star Trek) while a single myosin molecule pulls on the filament.

These investigations of the molecular structure and function of myosin are beginning to yield clues about how specific mutations lead to heart failure in conditions such as hypertrophic cardiomyopathy. Hypertrophic cardiomyopathy is a disease that is associated with abnormal growth (i.e., hypertrophy) of the ventricles of the heart. This devastating disease appears usually in late adolescence with no, or very little, advance warning. Approximately 3% of these individuals die each year from sudden cardiac death. Roughly half of all hypertrophic patients have an inherited form of the disease known as familial hypertrophic cardiomyopathy. Evidence is accumulating that suggests that mutations leading to the alteration of a single amino acid out of the several thousand amino acids that make up one molecule of myosin, may be responsible for many of the symptoms of this inherited form of the disease.

Basic and clinical scientists are working together to unravel the mystery of how molecular structure leads to whole heart function. Scientists are currently developing transgenic animal models using molecular genetics techniques to introduce stable mutations at predefined locations in the nuclear DNA that are transmitted from generation to generation. As scientists struggle to understand the relationships between molecular structure and cardiac function, it is evident that our current understanding of how the heart works is the culmination of years of basic and clinical research that began with the pioneering studies by William Harvey in 1857. It is also clear that further advances will depend not only on new developments in molecular biology, but also on further refinements of our knowledge concerning the operation of the heart as a mechanical pump.

■ 1. **How is it that a single actin filament that has a diameter of 10 nm can be visualized with a light microscope that has a resolution limit of slightly less than 1 μm?**
2. **What advantages would a motility assay have as an experimental method compared to an intact beating heart or dissected strips of contracting cardiac muscle?**

(*a*) Schematic diagram showing the orientation of the myosin filament and myosin cross-bridge relative to the actin filament. (*b*) Atomic model of the crossbridge region of the myosin molecule. This model is based on the crystal structure of myosin determined by X-ray diffraction techniques to a resolution of 2.8 angstroms. The light chain subunits of myosin are shown in green and yellow. The brown, magenta, and purple regions indicate different functional domains of the cross-bridge.   © *Science, Vol. 261 Cover, July 2, 1993.* *Photo courtesy of Ivan Rayment and Adam Steinberg.* (*c*) Single actin filaments moving over a lawn of myosin molecules can be visualized using fluorescent dyes and ultrasensitive videomicroscopy. (*d*) Actual photograph of the fluorescent image of a single actin filament moving over myosin, as shown schematically in Figure c, at a velocity of 2 μm/sec in the direction indicated by the arrow.   (© *Dr. Joe Haeberle, University of Vermont*)

CHAPTER REVIEW

Summary

- The function of the cardiovascular system is to rapidly transport blood throughout the body through an extensive network of blood vessels.
- The cardiovascular system can be viewed as a closed loop consisting of the left and right hearts and the pulmonary and systemic circulations.
- Exchange between the blood and the cells occurs at the level of the capillaries.
- Blood flow is the volume of blood that moves past a particular point in the cardiovascular system during a period of time. Blood flow is identical in all portions of the cardiovascular system arranged in series.
- Blood pressure is a measure of the driving force that causes blood to flow. Resistance is a quantity that summarizes the net effect of all the frictional forces opposing blood flow. The relationship between pressure, flow, and resistance is: Flow = ΔPressure/Resistance.
- The left and right hearts both consist of a thin-walled atrium and a muscular ventricle. A series of one-way cardiac valves ensures that blood flows in only one direction through the heart. Blood flows from the atria to the ventricles through the atrioventricular valves (right heart = tricuspid valve; left heart = bicuspid or mitral valve). Blood flows out of the ventricle through the semilunar valves (right heart = pulmonic valve; left heart = aortic valve).
- The wall of the heart consists of a middle, muscular layer (myocardium), an inner layer (endocardium), and an outer layer (epicardium). The coronary vessels supply blood flow to the heart.
- The sequential activation of different parts of the heart is governed by the conduction of a cardiac action potential from the atria to the ventricles.
- Because the absolute refractory period is as long as the muscle twitch (300 milliseconds), tetanus is not possible in cardiac muscle.
- The action potential is initiated by pacemaker cells located in a region of the right atrium called the sinoatrial node. The rate of diastolic depolarization determines the rate at which the heart beats.
- The action potential is conducted from the right atrium to the ventricles via the atrioventricular node (AV node).

- The action potential is then conducted through the Bundle of His, the bundle branches, and the Purkinje fibers to the ventricular muscle cells. This specialized conduction system ensures the nearly simultaneous activation of all of the ventricular muscle.
- Electrocardiography is a method for measuring the electrical activity of the heart.
- The cardiac cycle is the sequence of mechanical events that occur during a single contraction of the heart. The sequence of events is the same for both the left and right hearts, which beat simultaneously.
- The ventricles are the primary pumping chambers of the heart. Ventricular contraction and ventricular ejection occur during systole. The systolic period begins with the occurrence of the first heart sound and ends with the occurrence of the second heart sound.
- Ventricular relaxation and ventricular filling occur during diastole. Diastole begins with the second heart sound and ends with the first heart sound.
- The volume of blood ejected from the heart with each beat is called the stroke volume. The sum of all the stroke volumes ejected from the heart over a period of one minute is equal to the cardiac output.
- Heart rate is determined by both sympathetic and parasympathetic innervation of the SA node.
- The intrinsic ability of the heart to respond to an increase in end-diastolic volume with an increase in stroke volume is referred to as the Frank-Starling law of the heart, which ensures that the volume of blood pumped by the left and right hearts is equal.
- A change in the vigor of contraction that is not dependent on a change in end-diastolic volume is referred to as a change in contractility. Contractility changes are the result of changes in the concentration of intracellular calcium ion.
- Cardiac muscle derives ATP from the metabolism of circulating glucose and fatty acids. The ability of the heart to contract is strictly limited by the supply of nutrients carried to the cardiac muscle by the coronary circulation. The energy cost of maintaining a given level of cardiac output is related to a number of factors, including heart rate, stroke volume, and arterial pressure.

Review Questions

Identify Each with a Term

1. A change in the vigor of cardiac contraction that is independent of end-diastolic volume.
2. The volume of blood ejected from the heart during a single cardiac cycle.
3. The phrase that refers to the period of the cardiac cycle during which both the atrioventricular and semilunar valves are closed and the ventricles are contracting.

4. The volume of the ventricles immediately prior to activation of ventricular contraction.

Define Each Term

5. pacemaker
6. blood flow
7. systole
8. Frank-Starling Law of the Heart

9. electrocardiogram
10. arrhythmia

Choose the Correct Answer

11. Which of the following is *not* a determinant of blood flow?
    a. pressure gradient
    b. blood vessel radius
    c. colloid osmotic pressure
    d. resistance

12. Which event occurs after the first heart sound and before the second heart sound?
    a. ventricular ejection
    b. P-wave of the ECG
    c. closure of the mitral valve
    d. atrial contraction

13. Which of the following represent parallel blood flows?
    a. total systemic blood flow and total pulmonary blood flow
    b. blood flow to muscle and blood flow to the skin
    c. right heart output and left heart output
    d. total capillary blood flow and total venous blood flow

14. An increase in the concentration of circulating epinephrine would:
    a. decrease stroke volume
    b. decrease heart rate
    c. increase cardiac output
    d. decrease contractility of the heart

15. In a normal, healthy heart, stroke volume would be increased by:
    a. increased sympathetic stimulation of the heart
    b. increased parasympathetic stimulation of the heart
    c. decreased contractility
    d. decreased end-diastolic volume

16. Contractility of the heart is *not* altered by which of the following?
    a. a change in end-diastolic volume
    b. a change in cytoplasmic calcium concentration
    c. the drug digitalis
    d. a change in the level of sympathetic stimulation to the heart.

17. The exchange of nutrients and gases between the blood and the tissues occurs:
    a. in the aorta
    b. in the arteries
    c. in the arterioles
    d. in the capillaries

18. The normal location of the pacemaker in the human heart is in the:
    a. atrioventricular node
    b. sinoatrial node
    c. left ventricle
    d. Purkinje fibers

19. Cardiac muscle cannot contract in a tetanic fashion because:
    a. the long absolute refractory period prevents the muscle from becoming restimulated while still contracting.
    b. the action potentials travel too slowly along the conducting tissue to restimulate the muscle.
    c. contraction is possible only when the heart is filled with blood.
    d. the autonomic nervous system blocks rapid action potentials.

20. In the electrocardiogram, the T-wave is associated with:
    a. atrial depolarization
    b. atrial repolarization
    c. ventricular depolarization
    d. ventricular repolarization

Calculate

21. End-diastolic volume of the left ventricle is 110 ml and end-systolic volume is 40 ml. What is the stroke volume?

22. Stroke volume is 70 ml and heart rate is 80 beats/min. What is cardiac output?

23. Examination of an electrocardiogram shows that two successive QRS complexes are 0.8 seconds apart. What is the heart rate in beats per minute?

Answer Each Question in One or Two Sentences

24. How does vasoconstriction of resistance arterioles result in redistribution of blood flow to tissues?

25. Why is atrial contraction more important at elevated heart rates?

26. What is the underlying molecular mechanism associated with changes in cardiac contractility?

27. Trace the spread of activation of heart muscle, naming (in proper order) each of the structures through which the impulse passes.

28. Can an electrocardiogram be used to determine whether a change in contractility has resulted in an increased cardiac output?

Problem Solving in Physiology

29. Average resting cardiac output is 5.0 to 5.5 L/min and averaging resting heart rate is around 70 beats/min, giving an average resting stroke volume of 70–80 ml. If the maximum attainable heart rate is 200 beats/min and the maximum increase in stroke volume is slightly less than 2-fold, then the upper limit for cardiac output should be around 25 to 30 L/min (5- to 6-fold increase in cardiac output). How is it, then, that well-trained marathon runners can achieve cardiac outputs of 35 to 40 L/min (6- to 7-fold increase in cardiac output)? *Hint: Many well-trained athletes have resting heart rates of only 40 to 50 beats/min and relatively normal resting cardiac outputs.*

30. a. If there is a potential pacemaker in the heart just below the AV node, under what conditions would its activity become apparent?
    b. Which portion of the electrocardiogram would be most affected by a disruption of impulse conduction in the ventricles? What sort of changes might be expected?

31. A patient is suffering from cardiac failure due to diffuse occlusion of the coronary arteries. On the basis of the material presented in the previous chapter, why might the use of digoxin to increase cardiac contractility not be the best approach to treating this individual

since this should increase the contractility of the heart, increase mean arterial pressure, and consequently increase blood flow through the coronary arteries? Of course the use of vasodilators to dilate the coronary arteries might be effective if the coronary arteries were not already maximally dilated in response to local tissue factors. Why else might vasodilator drugs be effective? (*Hint: consider the information summarized in Fig. 18–27.*)

Suggested Readings

Berne, Robert M., and Levy, Matthew N. *Physiology,* 3rd ed. St. Louis: C.V. Mosby, 1992.

Cantin, M., and Genest, J. "The heart as an endocrine gland." *Scientific American,* 254(2):76–81, 1986.

Goerke, Jon, and Mines, Allan H. *Cardiovascular Physiology.* New York: Raven Press, 1988.

Harken, A.H. "Surgical treatment of cardiac arrhythmias." *Scientific American,* 269(1):68–74, 1993.

Honig, C.R. *Modern Cardiovascular Physiology,* 2nd ed. Boston: Little, Brown and Co., 1988.

Jarvik, R.K. "The total artificial heart." *Scientific American,* 244(1):74–80, 1981.

Little, Robert C., and Little, William, C. *Physiology of the Heart and Circulation,* 4th ed. Chicago: Year Book Medical Publishers, 1989.

Robinson, T.F., Factor, S.M., and Sonnenblick, E.H. "The heart as a suction pump." *Scientific American,* 254(6):84–91, 1986.

Solaro, J.R., "Myosin and why hearts fail" *Cell* 85:1945–1946, 1992.

Teitz, Carol C. *Scientific Foundations of Sports Medicine.* Philadelphia: B.C. Decker Inc., 1989.

Key Terms

aorta (p. 545)
aortic valve (p. 548)
arrhythmia (p. 556)
arterial system (p. 545)
arterioles (p. 545)
atrioventricular node (AV node) (p. 554)
atrioventricular valve (p. 548)
blood pressure (p. 546)
bradycardia (p. 553)
bundle branches (p. 554)
bundle of His (p. 554)
cardiac cycle (p. 557)

cardiac output (p. 562)
diastole, diastolic pressure (p. 559)
ECG, EKG (p. 557)
endocardium (p. 548)
epicardium (p. 548)
first heart sound (p. 562)
Frank-Starling Law of the Heart (p. 564)
heart sounds (p. 562)
limb leads (I, II, III) (p. 557)
mitral valve (p. 548)
myocardium (p. 548)

P-wave (p. 557)
pacemaker (p. 562)
pulmonary circulation (p. 544)
pulmonary valve (p. 548)
Purkinje fiber (p. 550)
QRS-complex (p. 557)
second heart sound (p. 562)
semilunar valve (p. 548)
sinoatrial node (SA node) (p. 552)
stroke volume (p. 562)

systemic circulation (p. 544)
systole, systolic pressure (p. 559)
T-wave (p. 557)
tachycardia (p. 553)
tricuspid valve (p. 548)
vascular resistance (p. 547)
venous system (p. 545)
ventricle (p. 547)
ventricular fibrillation (p. 556)

A single capillary surrounded by cells and interstitial space. The cells in the lumen of the capillary are red blood cells. *(© Don W. Fawcett/Visuals Unlimited)*

Chapter 19

Circulation

Key Concepts

The vascular system is both a conduit for the flowing blood and a dynamic system that controls the distribution of blood in the body.

The elastic arteries dampen the pulsatile outflow of blood from the heart to provide a more continuous flow of blood to the tissues. The arterioles are the major site of resistance to blood flow, and, are a major determinant of both the blood flow within a particular tissue bed as well as the distribution of blood flow among different tissue beds.

Capillaries are the major site of molecular exchange between the blood and the extracellular fluid. The balance of hydrostatic and oncotic forces (i.e., Starling's forces) determines the rate of fluid absorption or secretion by the capillaries.

The veins contain the major portion of the circulating blood volume, and because of the surrounding smooth muscle and venous one-way valves, play a major role in regulating the distribution of the blood volume within the body.

The lymphatic system returns fluid from the extracellular space to the general circulation, but only after it has been filtered through the lymph nodes.

Arterial pressure is regulated by the medullary cardiovascular control center that integrates sensory input from many sources and subsequently modulates the autonomic nerve activity to the heart and blood vessels to maintain a relatively constant arterial blood pressure.

Any number of factors can cause an inability of the heart to maintain cardiac output resulting in a variety of compensatory responses of the body, which can lead to cardiovascular disease.

Chapter Outline

Overview of the Vascular System

The Arterial System: Carrying Blood to the Tissue

The Capillary System: The Exchange of Molecules

The Venous System: Returning Blood to the Heart

The Lymphatic System

Regulation of Systemic Arterial Pressure

Cardiovascular Disease

The alarm goes off. Hannah wakes and jumps out of bed. In a matter of seconds, she has completely altered the hemodynamic state of her cardiovascular system. If appropriate adjustments are not made within seconds, this will be the last time she responds to her alarm. While asleep, her heart was beating at a leisurely rate of 60 to 70 beats per minute and her mean blood pressure was relatively constant at about 90 mm Hg. Upon awakening, the demands on the heart increase immediately. Her conscious brain requires increased

575

blood flow. Standing up places further demands on the heart. Skeletal muscles in her legs and back require increased blood flow. In addition, gravity causes a significant fraction of her blood to pool in the distensible blood vessels in her feet and legs. At the very moment her heart needs to increase the rate at which it pumps blood, there is less blood available to be pumped. A grave situation indeed—a potential cardiovascular crisis! It's no wonder that the majority of heart attacks occur in the morning. This momentary crisis occurs every morning of Hannah's life and is usually resolved within seconds. Without a second thought on her part, a host of homeostatic feedback mechanisms are activated that adjust the activity of her heart, as well as the physical properties of her blood vessels, to accommodate these new conditions. Sympathetic nerves innervating the heart dramatically increase the rate at which it pumps blood. Sympathetic nerve activity also causes the smooth muscles that encircle her blood vessels to contract and prevent peripheral pooling of blood. Epinephrine, released from her adrenal gland, reaches the heart within seconds and further stimulates her heart.

But it doesn't end here; the day has just begun. As Hannah eats her breakfast, blood flow to her stomach and intestines increases significantly. Then when she bolts out of the door and sprints to the bus stop, there will be a dramatic increase in the blood flow to skeletal muscles in her legs, as well as to the skeletal muscles involved in respiration. As you might guess, there is a limit to how much blood the heart can pump. Decisions have to be made—priorities must be set—which activities are essential for life, which activities are important at the moment, and which activities can be put off until later. Within a blink of an eye, these decisions are made by the autonomic nervous system.

So Hannah missed her bus, it's only a mile to work, and it's a beautiful summer day. Actually, it's a little warm after that sprint to the bus stop, and she feels flushed and begins to perspire. Her exercising muscles have generated considerable heat that must be dissipated from the body—another job for the circulating blood. A major cardiovascular readjustment takes place and blood flow to the skin is increased to dissipate the extra metabolic heat. To accommodate this demand on the heart to generate increased blood flow, smooth muscles surrounding veins throughout the body contract to move additional blood volume back to the heart, while blood flow is further restricted to the gastrointestinal system, the kidneys, and other nonessential organs. As Hannah sits at her desk, recovering from her jaunt to work, blood flow to her legs is maintained for a short time as her muscles replenish depleted stores of glucose. Eventually, blood flow is redistributed back to the stomach and intestines so that the digestion of her breakfast can continue. Once again, in response to the cardiovascular

control centers in the brain stem which continuously monitor the status of her cardiovascular system, her autonomic nervous system has made the appropriate adjustments to maintain life.

And so it goes, the heart beats continuously day and night to provide the life-sustaining flow of blood to all the cells in the body. So important is the continuous flow of blood to the brain that cessation of flow for even a few seconds will result in irreversible tissue damage. Likewise, the heart itself is absolutely dependent on the continuous blood flow through the coronary arteries, which in turn is the result of the heart's pumping action. This delicate balance and interplay of physiologic processes may continue for a century or more, as long as it is not interrupted for more than a few seconds at any time during that period.

Overview of the Vascular System

Recall from our previous discussion that the major subdivisions of both the systemic and pulmonary vascular systems are: arteries, arterioles, capillaries, venules, and veins. As we will see, the exchange of substances between the blood and the interstitial fluid occurs as blood passes through the capillaries. In large part, the role of the arteries, arterioles, venules, and veins is to direct the flow of blood to and from the capillaries. Far from being passive conduits, however, these vessels also play an active role in determining what portion of the total cardiac output reaches each of the various tissue beds in the body. In addition, the venules and veins help to ensure the efficient operation of the heart as a pump, by maintaining an optimal distribution of blood volume between the venous and arterial systems.

The Arterial System: Carrying Blood to the Tissue

The aorta and the arteries are conduits that direct the flow of blood from the heart to the various tissue beds within the body. The aorta, a large vessel with an average internal diameter of 2.5 cm (1 inch), must be of sufficient size to accommodate the combined blood flow to all the tissue beds. The arteries that branch off the aorta to distribute blood to the various tissues have internal diameters on the order of 0.4 cm. Relative to the diameter of the arterioles and capillaries, these are large vessels that offer comparatively little resistance to the flow of blood (see Fig. 19–5).

The arterial system plays a second role in the operation of the cardiovascular system. The walls of the arteries are made of smooth muscle and elastic tissue. When the ventricle contracts and suddenly ejects a volume of blood, the aorta and arteries stretch to accommodate the ejected blood. During diastole, when the heart is no longer ejecting blood, the elastic arteries recoil and force blood through the downstream blood vessels.

Blood Pressure in the Arteries Propels the Blood Through the Capillaries and Back to the Heart

Most people are aware that blood is under pressure within the arterial system. Cutting a large artery results in a serious medical emergency because blood is rapidly lost from such a wound. The high pressure inside the artery forces blood out through the cut. This high arterial blood pressure is necessary for the normal operation of the cardiovascular system. As mentioned previously, blood pressure is a measure of the driving force that causes blood to flow through the blood vessels. The magnitude of the driving pressure must be adequate to overcome the **resistance** to blood flow, which is a result of the friction between the flowing blood and the walls of the blood vessels, and the friction between the individual molecules and cells that make up the blood. We all deal with frictional forces every day. It is necessary to apply pressure to one side of a heavy box to slide it across the floor because of the friction between the bottom of the box and the floor. Likewise, pressure is necessary to move blood through the vascular system.

We can estimate how much pressure must be generated by the heart to overcome this resistance by using the equation presented in the previous chapter that describes the relationship between flow (F), pressure (P), and resistance (R).

$$F = \frac{P_1 - P_2}{R}$$

Since we are looking at flow from the aorta to the vena cava, it is the pressure difference between the aorta (P_1) and vena cava (P_2) that determines flow (i.e., aortic pressure − vena cava pressure = pressure difference). We can simplify our analysis by recognizing two points. First, the pressure in the vena cava (P_2) is small compared to that in the aorta and can be considered to be zero for our purposes. Secondly, pressure within the aorta (P_1) is approximately equal to the **arterial pressure** measured in any of the major systemic arteries; this equality indicates that the resistance to flow through the arteries is relatively small. Knowing that blood pressure in the vena cava is approximately zero and that arterial pressure

equals aortic pressure, we can modify the equation for blood flow as follows:

$$F = \frac{P_1 - P_2}{R}$$

$$F = \frac{(\text{aortic pressure} - \text{vena cava pressure})}{R}$$

$$F = \frac{(\text{arterial pressure} - 0)}{R}$$

$$F = \frac{\text{arterial pressure}}{R}$$

Under normal resting conditions, the arterial pressure required to provide adequate tissue flow is approximately 100 mm Hg. Figure 19–1 illustrates the values for blood pressure at various points in the circulatory system. Notice first that the systemic arterial pressure starts out about five times higher than the pulmonary arterial pressure, and yet the pressures in both systems return to very nearly the same value by the time the blood reaches the veins. The decrease in blood pressure reflects the loss of energy because of the resistance to blood flow. The greater the resistance to flow, the greater the pressure drop as blood flows through the vessels. On this basis, we can reason that the resistance to blood flow must be about five times greater in the systemic circulation than in the pulmonary circulation.

Notice that the greatest pressure drop, in the systemic circulation, occurs at the level of the arterioles.

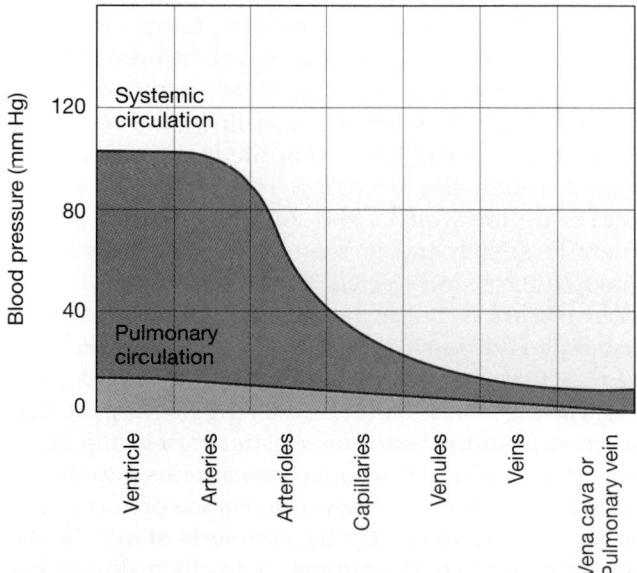

Figure 19–1

Illustration of the range of blood pressures throughout the systemic and pulmonary circulations.

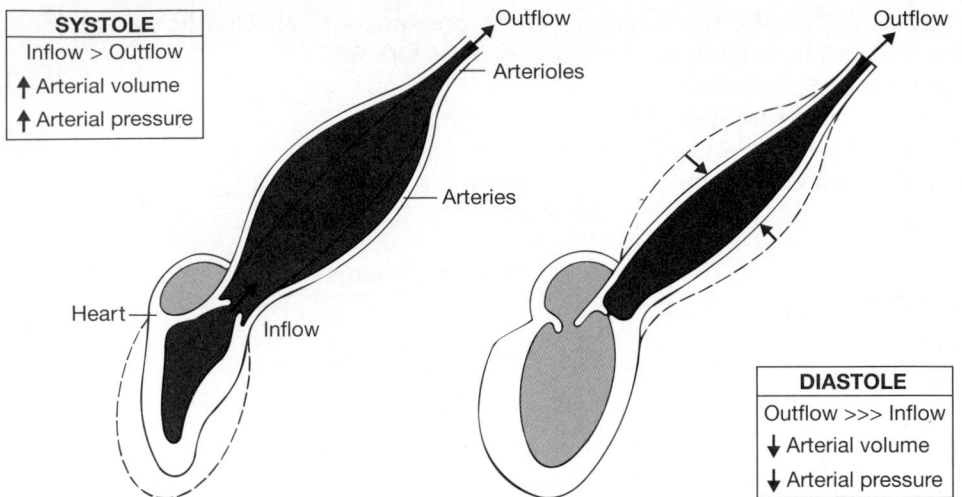

SYSTOLE
Inflow > Outflow
↑ Arterial volume
↑ Arterial pressure

Outflow
Arterioles
Arteries
Heart
Inflow

Outflow

DIASTOLE
Outflow >>> Inflow
↓ Arterial volume
↓ Arterial pressure

Figure 19–2

Role of the arteries as a high-pressure storage reservoir. The flow of blood into the arteries (from the heart) during systole exceeds the flow out of the arteries (through the arterioles), leading to an increase in arterial volume and pressure. During diastole, the elastic recoil of the arterial walls provides the driving force to propel blood out of the arteries.

This tells us that a major portion of the overall resistance to blood flow occurs at this point in the circulation. As we will see, this has important implications for the distribution of blood flow within the body.

Pulse Pressure

If we were to directly measure arterial blood pressure, the first thing we would notice is that it is pulsatile. Recall from the previous chapter that arterial pressure is lowest at the end of ventricular diastole and rapidly increases to a peak during ventricular systole. When the ventricle contracts, a volume of blood (the stroke volume) is ejected rapidly from the heart into the arterial vessels. As shown in Figure 19–2, the flow of blood into the arteries during systole is greater than the flow of blood out of the arteries into the capillaries. Blood flows out of the arteries more slowly because of the high resistance to flow through the arterioles. The elastic tissue that makes up the walls of the aorta and arteries allows them to stretch and increase in diameter to accommodate the blood ejected from the ventricle. Just as the pressure increases inside an elastic balloon as it is inflated with air, the pressure inside the arterial system increases as the vessels inflate with blood. During diastole, when blood is no longer flowing into the arteries from the heart, the elastic recoil of the arteries forces blood to flow out of the arteries into the arterioles. In other words, a portion of the pressure generated by the heart during systole is stored in the stretched walls of the arteries and is then slowly dissipated during diastole as blood flows out of the arterial system. The elastic properties of the arteries help to convert the *pulsatile* flow of blood from the heart into a more *continuous* flow of blood through the rest of the circulation.

As blood flows out of the arteries, the pressure progressively falls until it reaches a minimum called **diastolic pressure** Fig. 19–3. The peak arterial pressure, which occurs during contraction of the ventricle, is called **systolic pressure.** Your nurse or physician may have measured your blood pressure as "120 (mm Hg) over 80 (mm Hg)." These two numbers refer to the systolic and diastolic pressures, respectively, and are normally written as 120/80 mm Hg. The difference between systolic and diastolic pressures (120 − 80 = 40 mm Hg) is referred to as the **pulse pressure.**

As you might guess, the amplitude of the pulse pressure indicates the vigor of contraction by the ventricle; it also tells us something about the elasticity of the arteries. As shown in Figure 19–3, the magnitude of the pulse pressure is determined by two factors: (1) the stroke volume and (2) the distensibility of the arteries. As we discussed in the previous chapter, the stroke volume of the heart is dynamically regulated and can change significantly from beat to beat. In order for the heart to inject an increased stroke volume into the arteries, it must generate greater pressure to further stretch the elastic walls of the arteries. A sudden increase in pulse pressure therefore indicates an increase in the stroke volume ejected by the heart. Although the distensibility of the arteries can also affect the magnitude of the pulse pressure, arteriolar distensibility is relatively constant from day to day. However, progressive arterial disease, such as atherosclerosis, results in a decrease in the distensibility of the arteries and therefore an increase in pulse pressure. Analogously, greater pressure is required to blow up a thick-walled, less distensible balloon than a thin-walled, easily distensible one. A progressive increase in pulse pressure as an individual ages can be diagnostic of atherosclerotic arterial disease com-

Figure 19–3

Illustration of the dependence of pulse pressure on stroke volume and aortic distensibility.

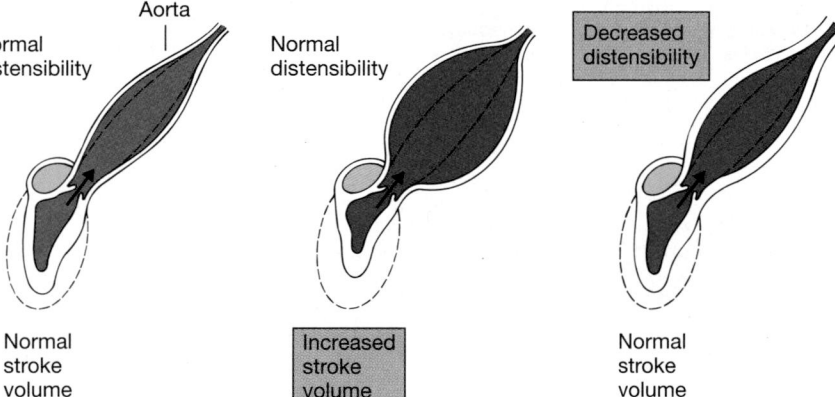

monly referred to as "hardening" of the arteries (see Clinical Focus: Atherosclerosis and Myocardial Infarction).

Measurement of Arterial Pressure

Everyone has had blood pressure measurements taken at one time or another. The physician or nurse places a cuff around the arm, inflates it with air, and then slowly deflates it while listening to the brachial artery with a stethoscope. Figure 19–4 illustrates how this process can measure both systolic and diastolic pressures. When the cuff is fully inflated (see Fig. 19–4a), the artery under the cuff is fully collapsed by the pressure exerted by the cuff, and no blood can flow through the artery. The pressure in the cuff is then gradually lowered. When the cuff pressure falls below the peak systolic pressure (Fig. 19–4b), blood is able to momentarily flow through the partially collapsed artery. This momentary, high-velocity spurt of blood through the artery can be heard through a stethoscope placed over the artery (downstream from the cuff) as a soft tapping sound. When this sound first occurs, the cuff pressure is recorded as systolic pressure. As the cuff pressure is progressively lowered (Fig. 19–4c), the duration of blood flow during each cardiac cycle gets longer and the nature of the sounds heard with the stethoscope changes in a characteristic manner. When the cuff pressure has decreased below the diastolic pressure (Fig. 19–4d), the vessel remains open at all times and flow is again continuous. Just after the cuff pressure falls below di-astolic pressure, the sounds disappear entirely. The cuff pressure at which the sounds disappear is recorded as diastolic pressure.

Mean Arterial Pressure

Although arterial blood pressure is constantly fluctuating, we can mathematically smooth out the peaks and valleys to determine a single value for blood pressure that would produce the same flow as the pulsatile pressure that actually exists. This **mean arterial pressure (MAP)** is calculated with the following equation:

MAP = 1/3 pulse pressure + diastolic pressure

As you can see in Figure 19–4, MAP is not the mathematical average of systolic and diastolic pressure, but rather an approximation of the geometric mean. By calculating the mean blood pressure, we can significantly simplify the calculation of blood flow because it is the MAP that is related to blood flow. Hence, we can further modify the equation for flow as follows:

$$F = \text{arterial pressure}/R = \text{MAP}/R$$

Without calculating the MAP, it would be rather difficult, for example, to evaluate how a change in blood pressure from 120/90 to 160/70 would alter blood flow. If we calculate the MAP in each case:

MAP = 1/3(120 − 90) + 90 = 100
MAP = 1/3(160 − 70) + 70 = 100

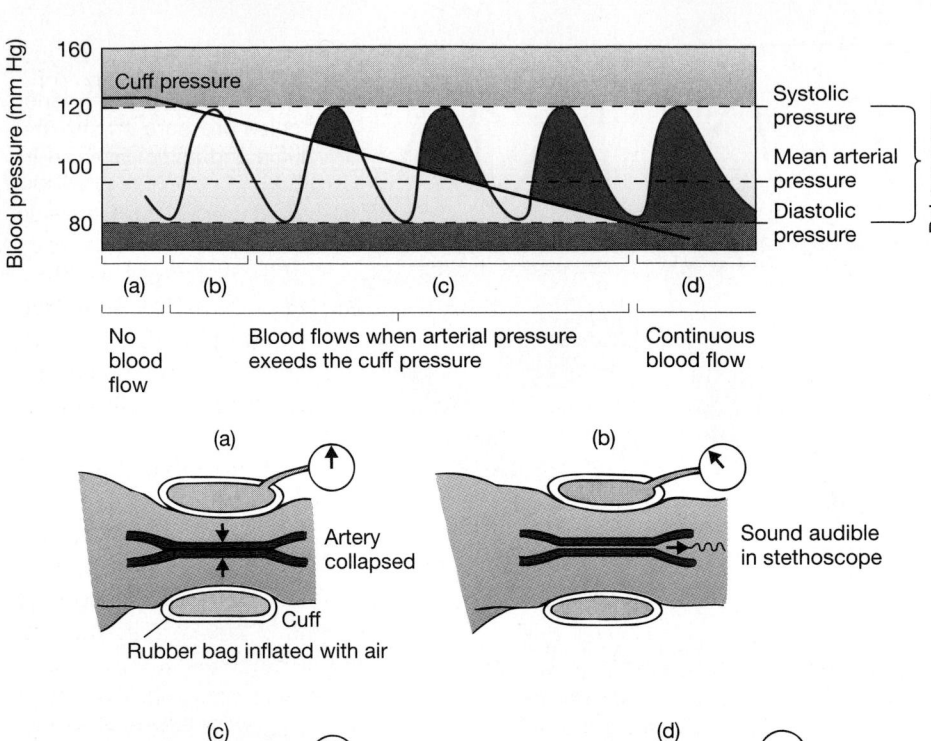

(a)

Artery
collapsed

Cuff
Rubber bag inflated with air

(b)

Sound audible
in stethoscope

(c)

Louder noise

(d)

No sound

Figure 19–4

Measurement of systolic and diastolic blood pressures by sphygmomanometry. The shaded areas indicate when blood is able to flow through the brachial artery as the cuff pressure is lowered.

we quickly realize that flow remains the same. As we will see in a later section, our bodies have homeostatic systems that sense MAP and then change the functional state of the heart or the vasculature, or both, to maintain MAP constant. If we recall that the purpose of the cardiovascular system is to maintain a constant **flow** of blood to the tissues, it should not be too surprising to see that our bodies regulate MAP and not pulsatile pressure.

The Major Site of Resistance to Blood Flow Is the Arteriole

Recall that soon after the arteries reach the various tissue beds within the body, they divide into smaller arterioles. As shown in Figure 19–5, the diameter of the arteriole is considerably smaller than that of the artery. This sudden decrease in diameter results in a relatively large increase in the resistance to blood flow.

The Diameter of Arterioles

Although a number of factors contribute to the resistance of flow through blood vessels, the diameter of a blood vessel is one of the more important con-

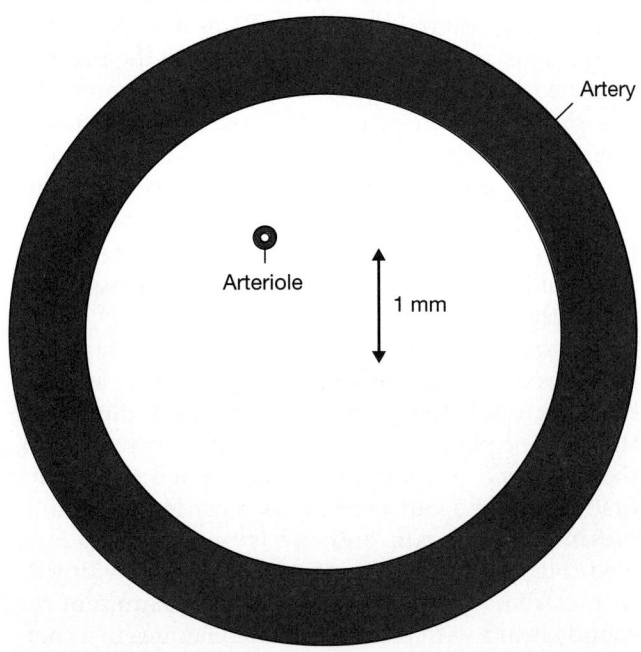

Figure 19–5

Comparison of the relative diameters and wall thicknesses of arteries and arterioles.

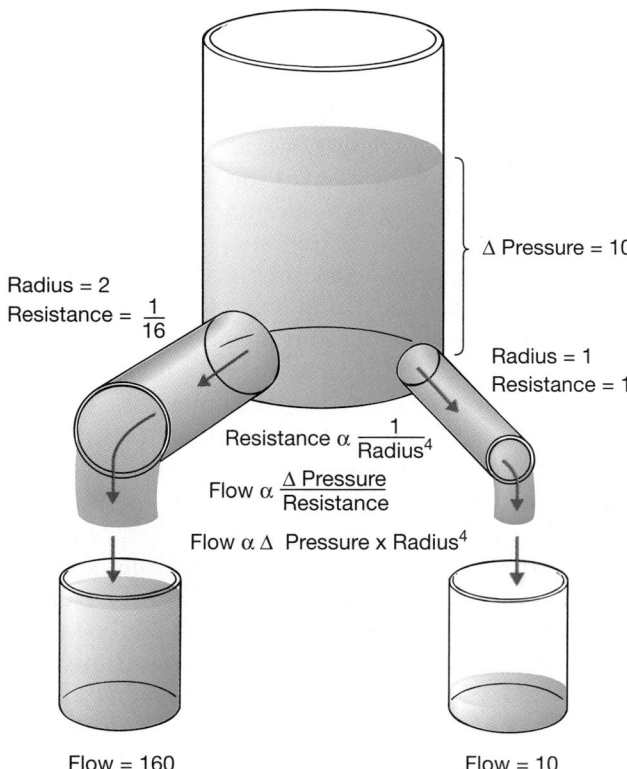

Radius = 2
Resistance = $\frac{1}{16}$

Δ Pressure = 10

Radius = 1
Resistance = 1

Resistance $\alpha \dfrac{1}{Radius^4}$

Flow $\alpha \dfrac{\Delta\ Pressure}{Resistance}$

Flow $\alpha \Delta$ Pressure x Radius^4

Flow = 160

Flow = 10

Figure 19–6

The effect of vessel radius on resistance and flow.

tributing factors, for two reasons. First, small changes in diameter cause relatively large changes in resistance. The resistance to flow through a blood vessel is inversely proportional to the radius raised to the fourth power:

$$\left(R\alpha\frac{1}{radius^4}\right)$$

Therefore, halving the diameter of an arteriole increases the resistance to flow by a factor of 16 (Fig. 19–6). Secondly, arterioles are able to dynamically adjust their diameters. The walls of arterioles contain relatively more smooth muscle cells than other blood vessels in the body. Moreover, these smooth muscle cells are arranged so that when they contract, they cause the diameter of the vessel to decrease, as illustrated in Figure 19–7 (see also Fig. 19–13). These two properties (high resistance and variable diameter) of the arterioles provide a mechanism by which the body can adjust the blood flow through different organs or tissues.

Recall from the previous chapter that the blood vessels supplying different tissues in the body are arranged into parallel circuits (see Fig. 18–2). Because of this arrangement, the total cardiac output is distributed among the various tissues, with the sum of all the individual flows equal to the cardiac output. The amount of flow through the kidney, for example, is determined by the resistance to blood flow through the kidney relative to the resistance to flow through the other systemic beds. Because the resistance to flow through the arterioles is so much greater than that through the other vessels, the resistance to blood flow through a tissue is determined almost entirely by the resistance of the arterioles. Therefore, by adjusting arteriolar diameter, individual tissue beds can regulate the local flow of blood. Figure 19–8 shows how relative changes in arteriolar resistance can produce shifts in blood flow. The overall resistance to the flow of blood from the aorta back to the heart (previously designated as R) is referred to as the **total peripheral resistance** or **TPR**. This terminology reflects the fact that essentially all this resistance is located in the peripheral regions of the circulatory system at the level of the resistance arteriole. Substituting TPR for R, our flow equation becomes: $F = MAP/TPR$.

(a)

Decreased resistance

Smooth muscle cell

Interior of arteriole

Increased resistance

Smooth muscle relaxation

Normal resting tone

Smooth muscle contraction

(b)

Figure 19–7

(*a*) Smooth muscle cells wrap around the wall of an arteriole. (*b*) Changes in resistance of an arteriole due to contraction or relaxation of the smooth muscle cells in the vessel wall.

Figure 19-8

The effects of arteriolar resistance on distribution of blood flow among capillaries arranged in parallel. Note that total flow in arteries, arterioles, capillaries, and veins must all remain equal because they are arranged in series within a closed system. The arteriolar smooth muscle has been represented by a single band of smooth muscle in these operational diagrams (Figs. 19–8– 19–12). See Figures 19–7 and 19–13 for more realistic depictions of the continuous distribution of smooth muscle cells along the entire length of the arteriole.

The Water System Analogy for the Regulation of Blood Flow

A good analogy for this aspect of the vascular system is a public water system. One large pipe (the aorta) carries water (the blood) from the water company's pump (the heart) to a series of distribution lines (arteries) that direct water to individual houses (organs). Within each house are a number of spigots (arterioles) that can be adjusted (variable resistance) to control the flow of water to different appliances (tissue beds). In this system, the distribution of water flow among the various houses and appliances is determined by local demand. The water company (the CNS) has only to ensure that the pressure in the main line (aorta) remains constant, and the flow can then be distributed among the various appliances (tissue beds) according to local demand. If you want

to wash the dishes, your clothes, the dog, and the car all at the same time, you don't have to notify the water company of your intentions; you open the appropriate spigots (decrease the local resistance to flow).

Such a system makes the water company's job of providing adequate flow to all the users rather easy because it doesn't have to monitor the usage by individuals to set the pumping rate (cardiac output). It simply monitors the water pressure (MAP) and adjusts the output of the pump (cardiac output) to keep the pressure constant. As long as the demand for water does not exceed the capacity of the water company's pump (the heart), all the consumers (the organs) will receive whatever flow they demand. If the demand on the system does exceed the capacity of the pump, then pressure will fall, and flow to the consumers will decrease. We have all experienced such a drop in water pressure during a morning shower. Since many other people are also taking showers at the same time, the demand may temporarily exceed the capacity of the pump, and both pressure and flow will decrease. If you wanted to remedy this loss of pressure and flow, you could call your neighbors (other organs) and ask them to turn off their spigots (increase local resistance to flow), while at the same time opening your spigot further (decreasing local resistance) to direct a greater portion of the limited flow to your house. This would be a reasonable thing to do if you had an urgent need (your house was on fire) compared to that of your neighbors (watering the grass). Notice that this cooperative approach to distribution of *limited* flow requires communication (the ANS) among the various users (the organs). However, as long as the capacity of the pump (the heart) exceeds the demand of the users (the organs), local regulation (by the arterioles) is adequate to adjust flow according to local needs.

Let us summarize the points made by this analogy. First, the individual organs and tissue beds can adjust local blood flow to meet demand by changing the resistance to flow through changes in the diameter of the arterioles. Second, the body can correctly adjust cardiac output to match the demand for flow simply by monitoring mean arterial pressure and adjusting the cardiac output to keep pressure constant. Third, as long as the demand for flow does not exceed the pumping capacity of the heart, individual tissue beds can function autonomously. Fourth, if demand does exceed flow, then extrinsic regulatory mechanisms must be used to redistribute flow to prevent a loss of pressure and flow to tissues with critical needs.

In order to fully understand how blood flow is distributed in the body, we must understand (1) how the resistance of the arterioles changes in response to local demand, (2) how the tissue beds in the body coordinate activity through extrinsic regulatory mechanisms, and (3) how the body monitors and adjusts cardiac output to maintain mean arterial pressure. Let's first look at the mechanisms by which local factors change arteriolar resistance.

Local Factors Such as Blood Pressure and Metabolites Autoregulate Blood Flow to Some Vascular Beds

The smooth muscle content of the arteriolar wall provides the means for dynamic regulation of arteriolar diameter and consequently of arteriolar resistance. Therefore, the contractile state of the smooth muscle determines the extent of local blood flow. If we were to administer a drug that relaxes vascular smooth muscle (a **vasodilator**) to a resting individual, we would find that the diameter of most of the arterioles in the body would increase. This shows that the normal resting state of the arteriolar smooth muscle is one of maintained contraction, referred to as a resting vascular "**tone.**"

Resting tone is important since an increase in activity of an organ demands greater blood flow and thus relaxation of the arteriolar smooth muscle to produce dilation of the arteriole. Several factors contribute to this level of resting tone. Like the sinoatrial node of the heart, the resting membrane potential of vascular smooth muscle undergoes spontaneous depolarizations. As discussed in Chapter 16, such depolarizations are characteristic of single-unit smooth muscle and result in the initiation of a burst of action potentials, which in turn lead to the activation of muscle contraction. As we will see in the next few sections, both mechanical and chemical factors can modify this spontaneous activity.

Stretching the wall of an arteriole, and of the smooth muscle cells included in it, causes an increase in the rate of spontaneous membrane depolarizations and leads to an increased level of contraction. The increase in contractile activity in response to stretch is often referred to as **myogenic activity.** Let's examine how this might contribute to the local regulation of blood flow. An increase in arterial blood pressure would initially cause all arteries and arterioles to increase their diameters by stretching the elastic elements (including smooth muscle cells) in the artery walls. If we recall the equation for blood flow ($F = $ MAP/TPR), we can see that an increase in pressure increases flow not only by increasing the driving force (MAP), but also by increasing the vessel radius and decreasing the resistance to flow. Both the physical stretching of the smooth muscle and the increased blood flow stimulate myogenic contraction of the smooth muscle. This myogenic contraction decreases the diameter of the vessel and decreases blood flow back toward the initial level. Therefore, the increased myogenic activity of the smooth muscle in response to increased arterial pressure tends to minimize increases in both vessel diameter and flow in response to increased pressure. Conversely, a decrease in blood pressure initially decreases flow and unstretches the vascular smooth muscle. Again, both effects lead to decreased myogenic activity. The ensuing vasodilation minimizes the overall reduction in flow in response to a reduction in arterial pressure.

The sequence of events associated with an increase in blood pressure is shown in Figure 19–9. Experiments have shown that the blood flow to many organs remains relatively constant despite large changes in blood pressure. This ability of individual vascular beds to maintain constant blood flow is referred to as **autoregulation.** Although the mechanisms underlying autoregulation remain to be fully described, considerable evidence exists to suggest that myogenic activity is a contributing factor. As we will see in the next section, a number of other factors also contribute to autoregulation of blood flow.

Blood flow to most tissues increases in proportion to the metabolic demand of the tissue (**active hyperemia**). This is dramatically illustrated in skeletal muscle, in which blood flow may increase as much as tenfold in response to intense exercise. Although some of this increase is mediated by factors external to the muscle, most of it is mediated by the change in the chemical composition of the interstitial fluid.

Consider the chemical changes that occur as a result of cellular metabolism. The oxygen concentration decreases as a result of cellular use of oxygen for energy production. A wide range of metabolic end-products are generated, including carbon dioxide, adenosine, hydrogen ions, and heat. In addition, the potassium ion concentration of the interstitial fluid increases with increased cellular metabolism because of potassium loss from inside the cells. All of these changes (increased carbon dioxide, decreased oxygen, increased adenosine, increased hydrogen ion, increased potassium ion concentration, and increased temperature) have been shown to promote vasodilation of the arteriolar smooth muscle (Fig. 19–10). Vasodilation results in increased local blood flow, which tends to increase the oxygen concentration and reduce the metabolite concentrations in the interstitial fluid back toward resting values. Realize that local vasodilation in response to increased metabolism can be an effective method for regulating local blood flow only if the blood vessels are partially contracted at rest (resting tone).

Chemical changes such as these also probably contribute to the autoregulatory response to changes in blood pressure that were discussed previously. For

(text continues on page 585)

Figure 19–9

Autoregulation of capillary blood flow by myogenic activity of arteriolar and precapillary sphincter smooth muscle. Flow remains constant as perfusion pressure changes because of changes in vessel resistance due to myogenic activity.

Figure 19–10

Autoregulation of capillary blood flow by local tissue factors. At high metabolic levels, the increased amount of vasodilator substances and decreased amounts of vasoconstrictors shift the balance such that there is vasodilation. At low metabolic levels the balance is shifted so as to cause vasoconstriction. Relative concentrations are indicated by the size of the label.

example, a decrease in arterial pressure tends to decrease tissue blood flow, promoting the accumulation of chemical substances associated with metabolism and the depletion of oxygen. These chemical changes tend to promote vasodilation and increased blood flow.

One extreme form of chemically mediated autoregulation is seen following the complete occlusion (blockage) of blood flow to a tissue. When the occlusion is removed, an extreme reactive vasodilation occurs (reactive hyperemia). This is apparently due to the depletion of oxygen and the accumulation of metabolites that occur during the absence of blood flow. Next time you have your blood pressure measured, watch the changes in the blood flow to the skin in your forearm during and after inflation of the pressure cuff. Upon inflation of the blood pressure cuff, blood flow to the forearm is diminished. If the cuff is left inflated for several seconds and then deflated, the skin of the forearm will flush as a result of the reactive vasodilation of the arteriolar smooth muscle associated with the subcutaneous vasculature. As the increased blood flow restores the normal chemical balance of the interstitial fluid, vascular tone increases and blood flow to the skin returns to normal.

A number of other vasoactive substances produced locally by cells may also modify the level of vascular tone. For example, histamine released from mast cells during an inflammatory response (see Chapter 17) stimulates vasodilation. Prostaglandins may also be released from cells to cause either vasodilation or vasoconstriction, depending on the specific prostaglandin released. Vasoactive substances are also released from the endothelial cells that line the blood vessels. The first of these to be described was endothelium-derived relaxing factor (EDRF) which has been shown in many cases to be nitric oxide (NO). In addition to being a very potent vasodilator, NO also appears to be a neurotransmitter in the central nervous system. An endothelium-deprived vasoconstrictor has been described called endothelin. The specific roles of many such locally released vasoactive agents remain to be clarified.

Nerves and Circulating Hormones Coordinate the Blood Flow Through All the Vascular Beds with the Pumping Activities of the Heart

Earlier in this discussion we noted that some form of communication between isolated organs and tissues is important in the distribution of blood flow, particularly under conditions in which cardiac output is limited. Massive and complete vasodilation of all the vascular beds in the body would result in a drastic fall in blood pressure because the heart would not be able to pump fast enough to maintain such a high level of flow to all the tissues (recall the water system analogy).

This is precisely what happens with heat exhaustion. The vasculature to the skin is so extensively vasodilated (to dissipate excessive heat from the body) that blood flow to other critical organs (e.g., the brain, heart, and kidneys) is seriously compromised. A less extreme example of vasodilation occurs during strenuous exercise, when blood flow to the muscles may increase to greater than 15 L/min. In both of these situations, the heart is not able to provide normal flow to the remaining tissues, and therefore redistribution of flow must occur. As pointed out in the water system analogy, redistribution of flow requires some form of communication between the various users. The communication link in our bodies involves the CNS and ANS as well as circulating hormones.

Most arterioles in the body are richly innervated by postganglionic sympathetic nerve fibers. (Exceptions include the coronary vessels, which supply the heart, and the cerebral vessels in the brain; these two vascular beds are predominantly regulated by local metabolic factors.) The sympathetic neurotransmitter norepinephrine binds to alpha receptors on the vascular smooth muscle cells and stimulates contraction (Fig. 19–11). Under resting conditions, the systemic arterioles are constantly being stimulated by sympathetic nerve impulses originating from vasoactive centers in the medulla. This continuous neural activity contributes to the basal level of resting tone, which we have previously discussed.

Although the parasympathetic neurotransmitter acetylcholine is a vasodilator, parasympathetic nerve fibers are not generally involved in the neural regulation of blood flow. This presents us with a rather one-sided control system that can only stimulate vasoconstriction. However, this does not represent a serious problem for two reasons. First, in tissues such as the skin that rely heavily on sympathetic activity for normal control of blood flow, there is a high level of tonic sympathetic nerve activity and consequently a high level of resting vascular tone. Under these conditions, vasodilation can be accomplished by a reduction of the level of sympathetic nerve activity. Second, if one examines the conditions under which neural regulation of blood flow plays an important role, we see that a major function of neural stimulation is to restrict flow to nonessential organs during physiological stress. Organs whose uninterrupted function, and therefore uninterrupted blood flow, are essential for life (e.g., the brain and the heart) have less sympathetic innervation than organs that can function intermittently (e.g., skin, liver, spleen, and

Figure 19–11

Opposing actions of the neurotransmitter norepinephrine and the hormone epinephrine on the contractile state of vascular smooth muscle. While some tissues, like skeletal muscle, contain both receptor types, most tissues are dominated by one receptor type. For example, the blood vessels of the skin contain predominantly alpha receptors, whereas the blood vessels of the heart (coronary arteries) contain predominantly beta receptors.

gastrointestinal tract). Therefore, a major role of sympathetic innervation of the resistance vessels is to prevent cardiovascular collapse, under conditions of physiological stress, by reducing the blood flow to certain noncritical organs.

A number of endocrine substances have vasomotor activity. Epinephrine, which is released from the adrenal medulla, can bind to both **alpha** and **beta receptors** on smooth muscle cells. The binding of epinephrine to beta receptors of smooth muscle produces relaxation and vasodilation of the arterioles. Although epinephrine can also bind to alpha receptors and produce vasoconstriction, this occurs only at relatively high concentrations of the hormone. The concentrations of epinephrine found in the circulating blood due to medullary stimulation are generally assumed to be too low to allow binding to alpha receptors. While all vascular smooth muscle contains alpha receptors, beta receptors are found predominantly on vascular smooth muscle cells in the coronary vessels of the heart and in the blood vessels in skeletal muscle and liver. Therefore, the effect of circulating epinephrine is to antagonize the vasoconstrictor activity of sympathetic stimulation (see Fig. 19–11). Moreover, the antagonism of norepinephrine-mediated vasoconstriction is most pronounced in the heart, skeletal muscle, and liver, owing to the greater numbers of beta receptors associated with the resistance vessels of these tissues. During periods of elevated sympathetic stimulation, when both norepinephrine and epinephrine are being released, blood flow is shifted away from vascular beds containing predominantly alpha receptors and to vascular beds

containing both alpha and beta receptors. Figure 19–12 summarizes the effects of these vasoactive agents on arteriolar resistance.

One of the most potent vasoconstrictor substances is **angiotensin II** (see Chapter 23). Angiotensin II may play a role in the regulation of the renal circulation and may be a contributing factor in hypertension. The major role of angiotensin II is probably the regulation of aldosterone production (see Chapter 23).

Atrial natriuretic factor is a potent vasodilator secreted from cells in a variety of tissues, including the atria of the heart. This peptide causes both vasodilation and increased excretion of salt and water by the kidneys and may play a role in the maintenance of normal blood volume. The function of this peptide in the overall control of vascular tone, however, remains to be determined.

The Capillary System: The Exchange of Molecules

The Exchange of Molecules Between the Blood and the Extracellular Space Occurs Primarily in the Capillary Beds

We have reached an important point in our discussion of the cardiovascular system. Thus far, we have focused on how the cardiovascular system functions to generate blood flow and to distribute this flow appropriately to the various tissues. We are now ready

Figure 19–12

Summary of local tissue factors that produce either vasodilation or vasoconstriction of vascular smooth muscle. The resting level of vascular tone represents a balance between vasoconstrictor and vasodilator substances. An increase in the amount of vasoconstrictors and/or a decrease in the amount of vasodilators shifts the balance to produce vasoconstriction (upper panel) relative to resting tone (middle panel). A decrease in the amount of vasoconstrictors and/or an increase in the amount of vasodilators shifts the balance to produce vasodilation (lower panel) relative to resting tone (middle panel).

to consider how molecules are exchanged between the blood and the interstitial fluid. The **capillaries** are an extensive network of very thin-walled blood vessels that allow for the rapid diffusional exchange of molecules. Each capillary is only about 1 mm long, and yet together they represent the major point of communication between the interstitial fluid and the blood.

The capillaries are well suited to this task. The capillary network is so extensive that most cells in the body are within 0.02 mm of a capillary. Because the distances are so small, molecules can rapidly move by diffusion from a cell to the nearest capillary. Once in the blood, molecules are then rapidly transported back to the heart.

A second aspect of the capillary that suits them to this task of molecular exchange is that the capillary walls are only one cell thick. This further facilitates rapid diffusional exchange between the blood and the interstitial fluid. The exchange processes that occur at the level of the capillary represent a dynamic balance between diffusional, osmotic, and hydrostatic forces. We will examine the dynamics of capillary exchange in some detail in the next few sections.

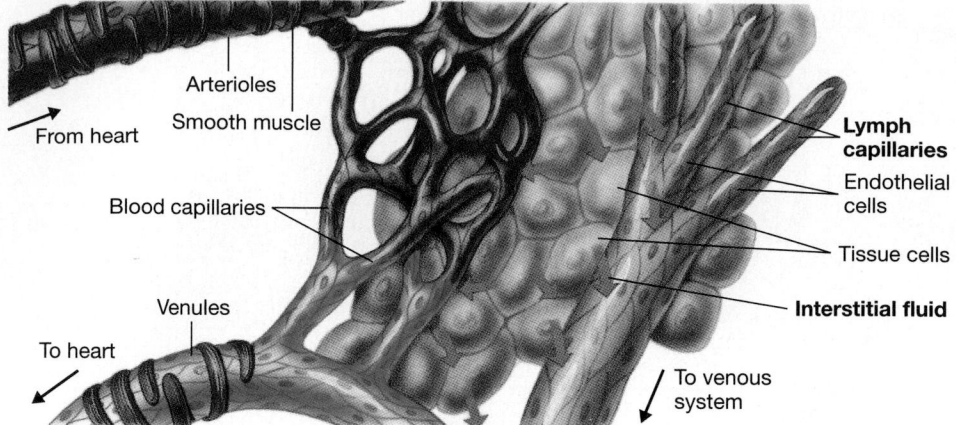

Figure 19–13

Diagrammatic representation of the microcirculation. The diameter of the capillaries in humans ranges from about 4 to 12 μm.

Blood Flow Through Capillaries Is Controlled by the Contractions of Vascular Smooth Muscle at the Level of the Arteriole and the Precapillary Sphincter

The anatomy of the **microcirculation** is shown in Figure 19–13. The capillary beds are at the center of the microcirculatory system, which includes the **arterioles, capillaries,** and **venules.** The capillaries (5 to 10 μm in diameter) branch out from the arterioles and then empty into the venules. Although the capillaries themselves contain no smooth muscle, in many cases there is a small cuff of smooth muscle called a **precapillary sphincter** at the beginning of the capillary. The contractile state of the arteriolar and precapillary sphincter smooth muscle determines blood flow through the capillaries. If we were to look at the flow of blood through the capillaries with a microscope, we would see that the blood flow is often sporadic. This is the result of the contraction and relaxation of the precapillary sphincter **(vasomotion),** which tends to be either completely opened or completely closed. This vasomotion is heavily influenced by the metabolic state of the tissue. For example, during exercise, the number of open capillaries in skeletal muscle will increase several-fold as precapillary sphincters vasodilate to increase blood flow to the muscle.

The skin contains two distinct capillary beds. The more superficial capillary bed is similar to that discussed above. The second consists of a series of muscular, larger diameter, arteriovenous shunts that are found predominantly in the fingers, palms of the hands, toes, and the face. These latter vessels are not involved in exchange of molecules, but rather are specialized for exchange of heat between the blood and the surface of the body. Reduced sympathetic tone to these vessels causes vasodilation, which, in turn, shunts a portion of the circulating blood closer to the surface of the body to facilitate the loss of heat from the body. This represents one of the major systems for maintenance of normal body temperature as discussed in Chapter 27.

The Absence of Smooth Muscle Cells Distinguishes the Capillary from Other Small Vessels

As shown in Figure 19–14, the capillary wall is made up entirely of endothelial cells and contains no smooth muscle. The flat, thin, endothelial cells are joined at their edges like patches in a patchwork quilt. Small pores **(clefts)** between the endothelial cells allow for the passage of molecules. The size and number of these **endothelial pores** vary greatly from tissue to tissue. The capillaries in the brain, for example, may not contain any pores, which account for the impermeability of the "blood-brain barrier" to a variety of substances. The capillaries in the kidney, on the other hand, contain relatively large pores that allow the passage of large molecules. Although the endothelial pores represent an obvious point of exchange, in most capillaries they represent a very small percentage of the capillary surface area. As we will see in a later section, direct diffusion of molecules across the endothelial cell membranes accounts for the major portion of molecular exchange.

Molecules Exchange Across the Capillary Wall by Transcytosis, Free Diffusion, and Bulk Flow

There are three fundamentally different mechanisms by which substances can cross the capillary wall. These are **transcytosis, diffusion,** and **bulk flow** (Fig. 19–15).

(a)

(b)

Figure 19–14

(*a*) Cross-sectional view of a single large capillary. The capillary wall is made up of a single layer of endothelial cells. (*b*) Scanning electron micrograph of a single capillary surrounded by cells and interstitial space. The cells in the lumen of the capillary are red blood cells (erythrocytes).   (*© Don W. Fawcett/Visuals Unlimited.*)

Transcytosis and Endocytosis

Relatively large molecules can cross the capillary wall by an active process called **transcytosis** in which molecules cross the epithelial cells either in discrete shuttling vesicles or through transient vesicle-derived channels. True **endocytosis** (e.g. phagocytosis and pinocytosis) is a less frequent occurrence. The extent of transcytosis and endocytosis varies greatly among different vascular endothelia. Pinocytosis is prominent in the postcapillary venules of lymph nodes, whereas, transcytosis is the dominant event in most other capillaries.

Diffusion

By far the most important mechanism for exchange of water and dissolved substances is diffusion. We can get some idea of this exchange by comparing the rate of blood flow through a capillary with the rate of water exchange by diffusion across the capillary wall. The rate of exchange is as much as 100 times greater than the rate of blood flow down the length of the capillary. Since water is moving across the capillary wall in both directions, there is little net gain or loss of water from the circulation by diffusion; however, the exchange rate is very high.

Of course, many solute molecules dissolved in blood and interstitial fluid are also exchanged by diffusion. Blood entering the systemic capillaries has a relatively high concentration of oxygen and glucose and a relatively low concentration of carbon dioxide. The concentrations of oxygen and glucose in the interstitial fluid are lower because these substances are being continuously taken up by the cells. Metabolism of these substances results in the production of carbon dioxide, which diffuses out of the cells into the interstitial space. As a result of these concentration

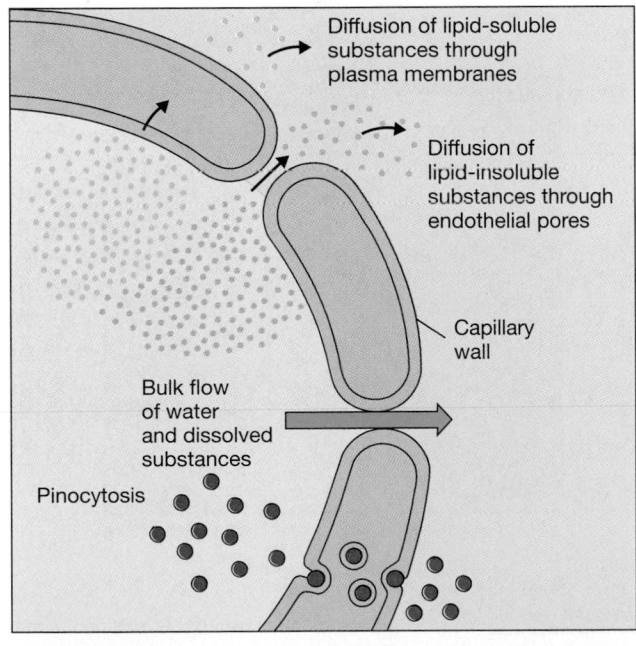

Figure 19–15

Diagrammatic summary of transport pathways across the capillary wall. Movement of substances is in response to either a hydrostatic pressure gradient (bulk flow) or a diffusional gradient.

gradients, there is net movement of oxygen and glucose out of the capillary and net movement of carbon dioxide into the capillary (Fig. 19–16). Concentration gradients for other nutrients, metabolites, and hormones result in net transfer of these substances between the interstitial fluid and the blood. In all cases, the direction of the concentration gradient determines the direction of exchange. Analogously, heat generated by metabolism is transferred by convection down its thermal gradient from the cell to the blood.

Molecules can diffuse across the capillary walls either through the water-filled pores or directly through the endothelial cells. The path a molecule takes is determined by its relative solubility in water and lipids. Molecules that are only slightly soluble in lipid (e.g., Na^+, K^+, Cl^-, and glucose) will more likely diffuse through pores. Molecules such as oxygen, carbon dioxide, and urea, which are more lipid-soluble, can diffuse directly through the plasma membranes of the endothelial cells. Water molecules are able to cross either through pores or by diffusion through the endothelial cells. Recent studies have demonstrated the existence of specific water channels in some tissues.

Because pores occupy less than 1% of the capillary wall area, the surface area available for diffusion of lipid-soluble substances is more than 100 times greater than that for lipid-insoluble substances. We can begin to appreciate how well suited this system is for the very rapid and continuous exchange of oxygen and carbon dioxide, which are both lipid-soluble.

A number of physical factors contribute to the rapid diffusional exchange of substances across the capillary wall through either pores or the endothelial cell. First, the distances that molecules must move are very small. Most metabolically active cells are within 20 μm of a capillary. Secondly, the total surface area of the capillary walls is extensive. Both diffusion distance and surface area influence the rate of net diffusion. The total surface area of all the capillary walls may be as high as 7000 sq. ft. Imagine spreading the entire blood volume of the capillaries (approximately 250 ml or 1 cup) over an area 1.5 times larger than a basketball court; this gives you some idea of the ideal conditions for diffusion that exist in the capillaries.

A third factor is the relatively slow rate of blood flow that exists within the capillaries. The velocity of blood flow through the capillaries is greatly reduced because of the large total cross-sectional area of all the capillary beds (Fig. 19–17). The velocity of blood flow is inversely proportional to the cross-sectional area of the vessel(s) through which it is flowing. We see this principle in action whenever we observe the flow of water down a river. The velocity is greatest where the stream narrows (as in rapids) and slowest where the stream widens; total flow, however, is the same at both points. When a stream empties into a lake, the velocity of water movement in the lake is imperceptible because of the dramatic increase in cross-sectional area. For precisely the same reason,

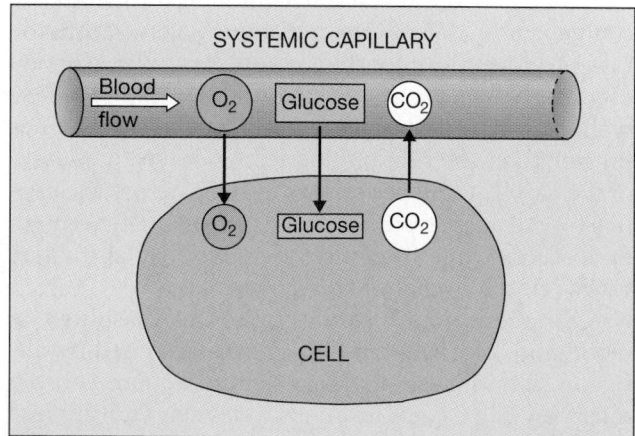

Figure 19–16

Transfer of substances between the cells and the systemic capillary blood in response to diffusional gradients. The size of the label indicates the relative concentrations of a substance.

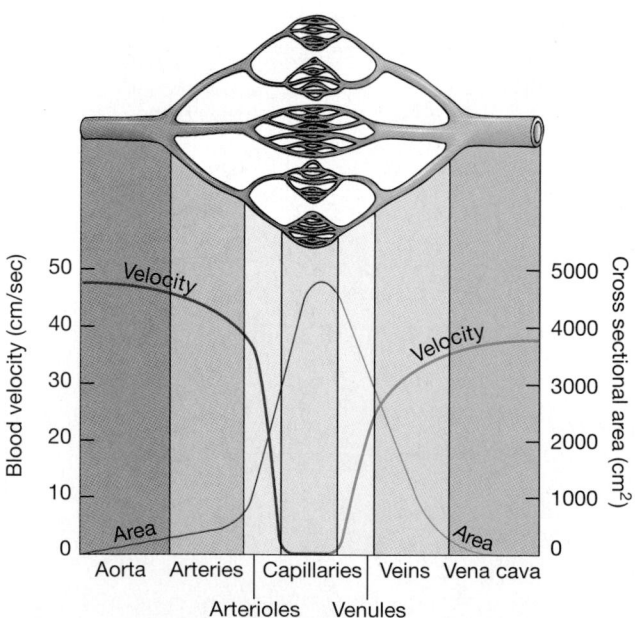

Figure 19–17

Vascular cross-sectional area and velocity of blood flow. Cross-sectional area is the total area of all the vessels added together at each level in the circulatory system. Blood velocity is that which would be measured in a single vessel at each level in the circulatory system.

the average velocity of blood flow in the aorta may be more than 1000 times greater than that in the capillaries. The velocity of blood flow in a capillary ranges from 0.1 to 1 mm/sec, compared to 40 cm/sec in the aorta. The dramatic reduction in flow velocity in the capillaries means that the time for exchange is substantially increased.

Bulk Flow and Oncotic Pressure

Exchange of water (and dissolved solutes) also occurs through endothelial pores by **bulk flow** in response to the pressure gradient between the inside and outside of the capillary (Fig. 19–18). As we will see, the pressure gradient is always from inside the capillary to outside, which causes water to flow out of the capillary. Although the total exchange of fluid by this means is relatively small (except in the kidney), it represents an extremely important mechanism for the maintenance of the circulating blood volume. This exchange is important because it occurs in response to a pressure gradient rather than to a diffusional gradient.

The **hydrostatic pressure** gradient that drives water through the endothelial pores is simply the difference between the hydrostatic pressure inside the capillary (capillary blood pressure) and the hydrostatic pressure outside the capillary (interstitial hydrostatic pressure). The hydrostatic pressure outside the capillary is extremely difficult to measure, and

there is some uncertainty about the actual value. Although it may in fact be slightly subatmospheric (negative), for purposes of our discussion we will assume that it is zero. Given this assumption, the hydrostatic pressure gradient (capillary hydrostatic − interstitial hydrostatic pressure) is equal to the capillary hydrostatic pressure. When a blood vessel (usually a superficial vein) is cut, blood flows out of the vessel in response to the pressure difference between the inside and the outside of the vessel. Since water can move through the capillary pores by bulk flow, we can reason that water and dissolved solutes should be forced out of the vessels. Clearly, if this were the whole story we would have a great deal of trouble maintaining a normal circulating blood volume. There is in fact an opposing force that minimizes the loss of fluid by bulk flow.

The opposing force that prevents this loss of vascular volume is the **osmotic pressure** due to the plasma proteins (proteins in the blood plasma). Although all dissolved solutes contribute to the **total osmotic pressure** of blood, we can ignore the contribution of all solutes that are freely exchanged across the capillary wall because these solutes are at diffusional equilibrium. Only the plasma proteins that do not freely cross the capillary wall contribute to the osmotic pressure gradient between the inside and outside of the capillary. The terms **oncotic pressure** or **colloid osmotic pressure** (π) refer to that portion of the total osmotic pressure of blood that is due to the presence of plasma proteins. Recall from Chapter 4 that net diffusion of water across a selectively permeable membrane will occur in response to a concentration gradient for water. The capillary wall functions as a selectively permeable membrane because of its limited permeability to plasma proteins. Recall that the major plasma protein is albumin. Most plasma proteins cannot leave the capillary and therefore establish an oncotic pressure gradient that favors the net diffusion of water from the interstitial space to the intravascular space. You have probably guessed by now that the oncotic pressure balances the hydrostatic pressure so that there is normally only a very slow leakage of water from the circulating blood. Water is lost from the capillary system at a rate of approximately 3 to 4 ml/min (5 L/day), and is returned to the general circulation via the **lymphatic system,** as discussed in a later section.

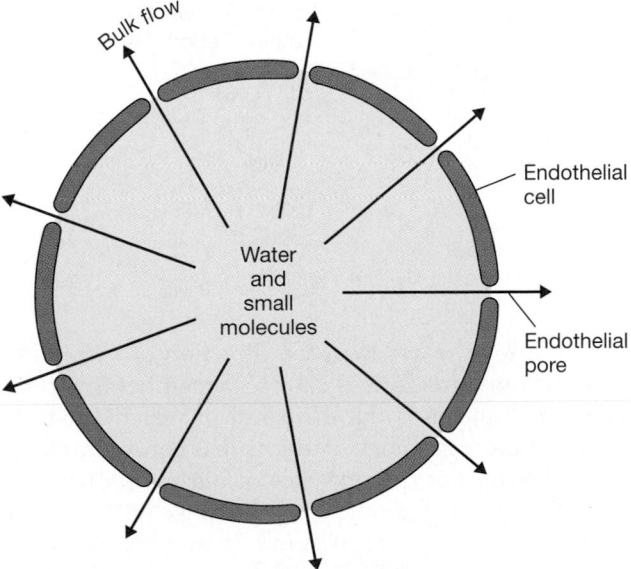

Figure 19–18

Bulk flow of water and dissolved substances out of the capillary through endothelial pores in response to a hydrostatic pressure gradient. Hydrostatic pressure inside the capillary is always greater than the pressure of the interstitial fluid surrounding the capillary.

Filtration and Absorption: Starling's Hypothesis

This balance of hydrostatic and oncotic pressures across the capillary endothelium was originally described by Starling in what is now called **Starling's hypothesis** (Fig. 19–19).

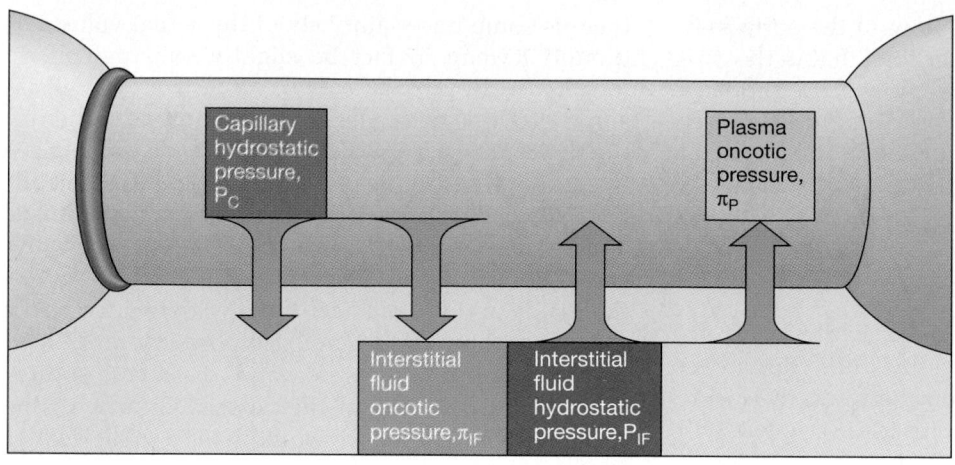

Filtration forces

Absorption forces

$$(\boxed{P_C} + \boxed{\pi_{IF}}) \quad - \quad (\boxed{P_{IF}} + \boxed{\pi_P})$$

(a)

Arterial end of capillary

Venous end of capillary

$$(\boxed{35} + \boxed{3}) - (\boxed{28} + \boxed{0}) = \boxed{+10}$$

$$(\boxed{15} + \boxed{3}) - (\boxed{28} + \boxed{0}) = \boxed{-10}$$

| Net filtration | | Net absorption |

(b)

Figure 19–19

Summary of hydrostatic and osmotic forces causing movement of water across the capillary wall. The ideal capillary shown here represents the average of all capillaries in the body. Pressures within individual capillaries vary considerably. (Modified from *Human Physiology,* 4th ed, 1985, Vander, Sherman, Luciano (eds.). Fig. 11–53, p. 346.)

Let's begin by looking at an idealized average capillary. (The situation in any particular capillary might differ significantly from this model, but the principles remain the same.) The oncotic pressure gradient is determined by the difference between the oncotic pressure of the capillary blood and that of the interstitial fluid. The oncotic pressure of blood is normally about 28 mm Hg. Although the capillaries are relatively impermeable to plasma proteins, some proteins do leak out of the capillaries. This small amount of protein in the interstitial fluid results in a tissue oncotic pressure of about 3 mm Hg. The oncotic pressure gradient (28 − 3 = 25 mm Hg) causes a net flow of water into the capillary. Remember that the os-

motic flow of water is in the direction of increasing osmotic pressure. This is easy to remember if you recall that osmosis is the diffusion of water down its concentration gradient. Water is less concentrated in blood because of the presence of plasma proteins.

If we examine a typical capillary, we will measure a hydrostatic pressure of about 35 mm Hg at the arterial end of the capillary and about 15 mm Hg at the venous end. The pressure drop along the length of the capillary reflects the resistance to flow of blood through the capillary. The hydrostatic pressure causes fluid to move out of the capillary; this is opposed by the oncotic pressure that moves water back into the capillary (see Fig. 19–19). For the sake of this discus-

Arteriole | Capillary | Venule

VASODILATION

P_C +50 $\quad$ π_P −28

π_{IF} +3 $\quad$ P_{IF} 0

+25

Net filtration

P_C +30 $\quad$ π_P −28

π_{IF} +3 $\quad$ P_{IF} 0

+5

Net filtration

VASOCONSTRICTION

P_C +15 $\quad$ π_P −28

π_{IF} +3 $\quad$ P_{IF} 0

−10

Net absorption

P_C +15 $\quad$ π_P −28

π_{IF} +3 $\quad$ P_{IF} 0

−24

Net absorption

Figure 19–20

Effect of changes in arteriolar smooth muscle contraction on the balance of hydrostatic and osmotic forces in the capillary. Vasodilation promotes filtration, and vasoconstriction promotes absorption of fluid.

sion, we can say that a positive pressure is one causing loss of capillary water and a negative pressure is one causing water reabsorption into the capillary. At the arterial end of the capillary, hydrostatic pressure (35 mm Hg) exceeds oncotic pressure (−25 mm Hg), leading to a net **filtration pressure** of 10 mm Hg. On the other hand, at the venous end of the capillary, hydrostatic pressure (15 mm Hg) is less than oncotic pressure (−25 mm Hg), leading to a net **absorption pressure** of −10 mm Hg. At the middle of the capillary, a hydrostatic pressure of 25 mm Hg is exactly balanced by an opposing oncotic pressure of −25 mm Hg. In this example, filtration exactly equals absorption, which means there will be no change in the overall capillary volume.

We stated earlier that capillary flow often changes dramatically, following the contraction of smooth muscle in the terminal arterioles and precapillary sphincters. These changes in flow are associated with similarly large changes in intracapillary pressure. As shown in Figure 19–20, when precapillary smooth muscles are fully dilated, the associated capillaries

show filtration along their entire length. Fully contracted precapillary muscles lower the capillary pressure enough to cause absorption of fluid along the entire length of the capillary. Changes in arteriolar pressure can also lead to changes in capillary pressure to favor either filtration or absorption. We can get some idea of what the "average" state of the capillaries is by looking at the net loss or gain of fluid from the cardiovascular system. Under normal conditions, approximately 2 to 5 L of fluid is returned from the interstitial space to the circulation by the lymphatic system during a single 24-hour period. Thus, the average situation slightly favors filtration over absorption.

Filtration and absorption provide an important mechanism of the dynamic maintenance of the circulating blood volume. Although the cardiovascular system is a closed system at the "macro" level, it is an open system at the level of the microcirculation. The constancy of the circulating blood volume in fact represents a dynamic balance between forces tending to move water into and out of the capillary. Recall that cardiac output is determined in part by filling of the heart (end-diastolic volume). Cardiac filling, in turn, is greatly influenced by the total volume of circulating blood. Any significant reduction in blood volume leads to a reduction in cardiac output. As we will see later, transfer of fluid from the extravascular space into the capillary represents an important and automatic compensatory response to abnormally low blood pressure as occurs, for example, with blood loss **(hemorrhage).** The distribution of fluid can also be affected by disease. Heart failure ultimately leads to an increase in capillary pressure, causing the accumulation of fluid in the interstitial space (edema). Other diseases may also lead to edema, because of a decrease in plasma protein concentration.

The Venous System: Returning Blood to the Heart

The capillaries empty into thin-walled vessels called **venules,** which represent the beginning of the venous system. The venules contain endothelial clefts like those in the capillaries. Although fewer in number, these allow for additional exchange of substances. The capillaries and venules together are often referred to as the **exchange vessels.** The venules then converge to form progressively larger veins.

The Veins Provide a Conduit for the Return Flow of Blood to the Heart, and Function as a Variable Capacity Reservoir for Blood

The physical nature of the veins is considerably different from that of the arteries and arterioles. The veins are to the arteries as a paper bag is to a balloon. The veins, like a paper bag, can accommodate large changes in volume with little change in internal pressure up to a point after which pressure increases rapidly with further increases in volume. In addition, the total volume of the venous system is several times greater than that of the arteriolar system. These characteristics allow the veins to function as a low-pressure reservoir for blood; the veins are often referred to as the **capacitance vessels.** At any one time, approximately 75% of the total blood volume is contained within the veins, compared to 20% in the arterial system (Fig. 19–21).

This large capacity in and of itself is not significant. What is important is that the veins can reduce their capacity through contraction of the smooth muscle in their walls. This occurs in response to increased sympathetic nerve activity. As the venous smooth muscle contracts, blood is transferred from the veins to the heart and the arterial system. As we will see in a later section, this ability of the veins to

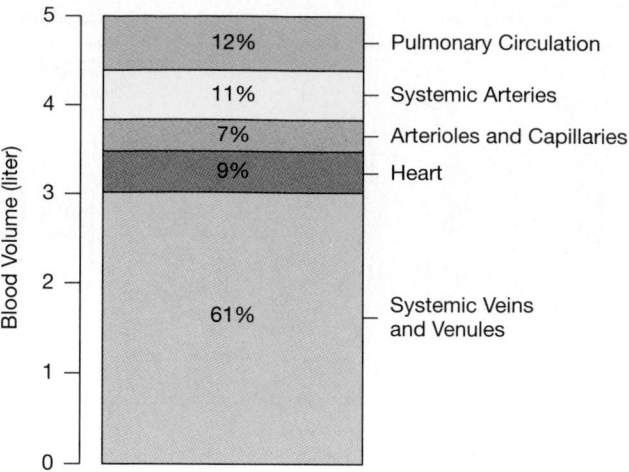

Figure 19–21

Distribution of the total blood volume among the various regions of the circulatory system. Recall from Figure 18–1 that the total blood flow through each region is the same because these regions are arranged in series with one another.

dynamically alter their capacity provides a mechanism by which the body can maintain end-diastolic volume of the heart. Recall that the distribution of blood volume can change dynamically as a result of changes in arteriolar resistance. For example, during exercise, the volume of blood in muscle increases as capillary beds open up in response to dilation of precapillary resistance vessels. Other factors such as postural changes (e.g., standing up from a recumbent position) can result in passive redistribution of blood volume in response to gravity. Contraction of the veins can compensate for these changes. We will say more about this in a later section when we discuss the overall regulation of cardiovascular function.

The Return Flow of Blood to the Heart Is Facilitated by the Large Diameter of the Veins, One-Way Venous Valves, Skeletal Muscle Contraction, and Respiration

The pressure in the venules and small veins is only about 10 mm Hg. Since the pressure at the right atrium is essentially zero, there is a pressure drop of only 10 mm Hg between the venules and the heart, indicating that the resistance to venous blood flow is low. This low resistance is due to the large diameter of the veins. Recall that the resistance to blood flow decreases as the fourth power of the radius.

Several other factors also contribute to the flow of blood back to the heart. Some veins contain one-way valves that function in a way similar to those in the heart (Fig. 19–22). These valves serve two major functions. First, by preventing backflow of blood away

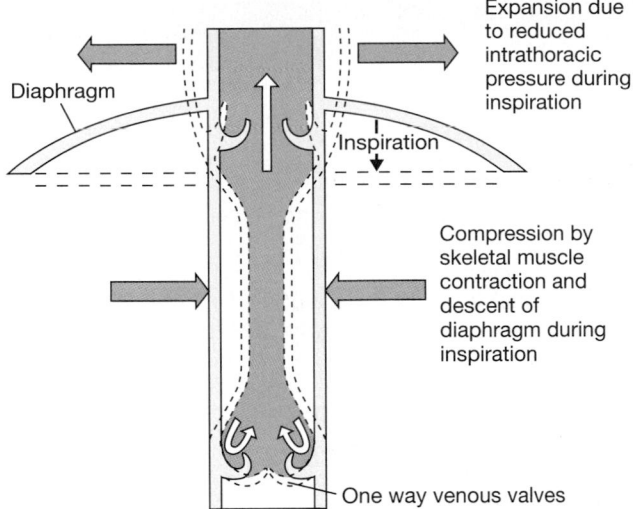

Figure 19–22

Factors promoting venous return to the heart. The combination of one-way venous valves and the compression/decompression of veins by contraction/relaxation of skeletal and respiratory muscles enhances the flow of blood from the veins back to the heart.

(a)

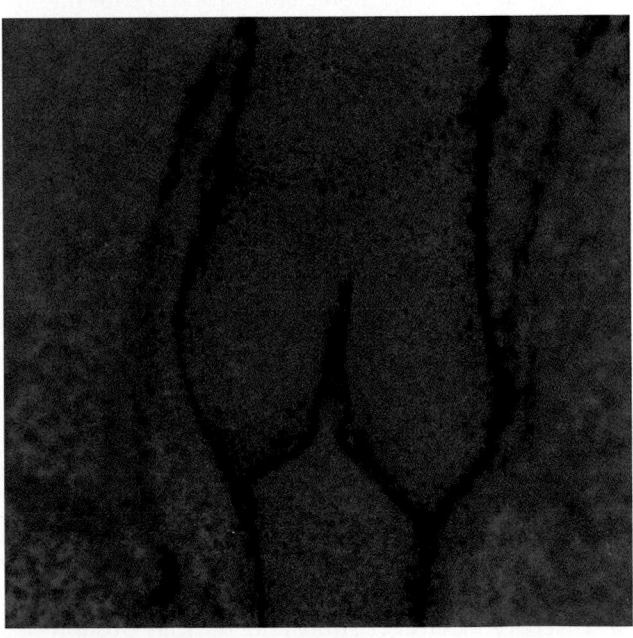

(b)

Figure 19–23

(a) Diagram of lymphatic capillary, showing entry of fluid and proteins and the pumping action of surrounding smooth muscle and lymphatic valves. (b) Photomicrograph of lymphatic valve. (© *John D. Cunningham/Visuals Unlimited.*) See Figure 19–13 for review of the relationship of the lymphatic capillary to that of surrounding blood capillaries and interstitial cells.

from the heart, they tend to minimize the pooling of blood in the feet and legs upon standing. Second, they allow the contractile activity of the muscles surrounding the veins to contribute to the flow of blood back to the heart. Contraction of the surrounding muscle tends to squeeze the veins. Blood is forced out of the locally constricted vein and, because of the valves, is constrained to flow toward the heart. So you can see that the combination of venous valves and skeletal muscle contraction constitutes a pump that assists the heart in maintaining the unidirectional flow of blood through the circulatory system.

An additional pumping activity further facilitates the flow of blood back to the heart. The muscular activity associated with normal respiration acts through the venous system to pump blood back to the heart. When the diaphragm descends during inspiration, the contents of the abdomen are compressed along with the abdominal veins. As before, this compression of the veins forces blood back toward the heart. Furthermore, the intrathoracic pressure decreases during inspiration; this provides a negative pressure that draws air into the lungs. Since the right atrium is located within the thoracic cavity, this negative pressure also increases the flow of blood from the veins into the right heart.

The Lymphatic System

The lymphatic system is a second system of vessels that differ anatomically from the blood vessels (Figs. 19–13, 19–23). The function of the lymphatic vessels

C L I N I C A L   F O C U S

Edema, Lymphatic Pathology and Varicose Veins

Edema

Any condition that causes blockage or interruption of normal lymph flow will lead to peripheral **edema.** It results from an accumulation of protein (which has leaked out of the capillaries) in the interstitial spaces. The osmotic pressure caused by the presence of this extra protein promotes the retention of extra vascular water, causing swelling or edema. Obstruction of lymph flow may result from injury, inflammation, parasitic infection, or surgery. Lymph nodes are surgically removed during a radical mastectomy in an attempt to arrest the spread of cancer cells from the breast to the rest of the body via the lymphatics. This may lead to temporary edema in the patient's arms; new lymph vessel growth usually alleviates the problem. Another cause of impaired lymph flow is parasitic infection, as shown in the figure on filariasis. **Filariasis** is the result of a parasitic infection of the lymphatic system. The parasitic larvae are transmitted by mosquitoes, and the adult worms reside in the lymph nodes. The resultant blockage of lymph flow results in progressive swelling of tissues in the affected region. As shown in the figure, affected limbs may swell to several times their normal size. The appearance of such swollen limbs has led to the more descriptive name for this disease—elephantiasis.

The other major cause of edema is accelerated leakage of fluid from the blood to the interstitial space. If the rate of leakage exceeds the rate of fluid removal by the lymphatics, edema will develop. A frequent cause of excessive lymph formation is congestive heart failure (see Fig. 19–34). When the heart fails, blood backs up in the veins and capillaries due to the reduced rate of blood transfer by the heart to the arterial system. The distension of the veins causes increased venous (and therefore increased intracapillary) hydrostatic pressure, followed by increased filtration of water (see Figs. 19–19 and 19–20). Increased fluid leakage may also occur as a result of reduced concentrations of plasma proteins, secondary to either renal failure or severe protein malnutrition (see Figs. 19–19 and 19–20). With a decrease in plasma protein concentration, the edema is not localized

and fluid accumulates in the most readily expandable areas of the body, such as the peritoneal cavity (see figure on kwashiorkor).

Varicose Veins

Varicose veins are usually superficial veins that have become excessively dilated, creating a distended and tortuous appearance. The primary cause for this abnormal distention is the failure of the venous valves, or **valvular incompetence.** The one-way venous valves normally facilitate the flow of blood back to the heart and prevent pooling of blood in the peripheral veins. Valvular incompetence is precipitated by factors that increase the volume of blood in the veins. Dilation of the veins by the presence of excessive blood volume impairs the proper closing of the valve leaflets; the resulting valvular incompetence compromises the flow of blood from the veins to the heart and leads to further engorgement. Prolonged distention of the veins in this fashion eventually causes permanent morphologic changes in the vessel wall, with a resultant increase in diameter and a loss in elasticity. Varicose veins can be removed surgically, and this may prevent or alleviate problems such as discomfort and ulceration.

Numerous factors contribute to the formation of varicose veins. Prolonged standing promotes venous pooling because the weight of the column of blood between the feet and the heart pushes out on the walls of the peripheral veins, causing a progressive increase in the internal diameter. Imagine filling a balloon with water and then lifting it by one end; the weight of the water and the compliance of the balloon will allow all the water to remain near the bottom. The degree of distension will be greatest at the very bottom of the balloon because the outward pressure at any point in the balloon is directly proportional to the height of the column of water above that point. Recall that skeletal muscle activity associated with walking and running normally compresses the veins and pushes blood back to the heart. This is equivalent to squeezing the bottom of the water-filled balloon with your hands to force the water back up into the elevated end. The low level of muscle activity associated with passive standing reduces the

---

is to return whatever fluid and plasma protein has leaked out of the capillaries into the interstitial space, back to the circulating blood. Despite the selectively permeable nature of the capillary wall and the balance of Starling's forces, there is a net loss of protein and fluid from the blood to the interstitial space. The extent of this loss increases significantly during exercise and with certain diseases. The normal operation of the lymphatics prevents the edema that would otherwise occur.

Lymphatic Vessels Begin as Blind Sacs that Empty into Lymphatic Vessels

The lymphatic vessels, like the capillaries, are made up of a single layer of epithelial cells. As shown in Figure 19–23a, the lymphatic system begins as a series of sacs with low internal hydrostatic pressure. These terminal sacs have relatively large endothelial pores, allowing for the movement of both fluid and protein into the lymphatic system. The **lymph** capil-

Filariasis causes the formation of edema and swelling of the afflicted areas as illustrated here. *(© Science VU/Visuals Unlimited.)*

Children suffering from kwashiorkor, San Salvador, Brazil. *(© Richard Frieman/Photo Researchers, Inc.)*

pumping effects of skeletal muscle and promotes pooling of blood in the peripheral veins. Varicose veins often form during pregnancy. The growing fetus causes compression of all the internal organs, including the inferior vena cava. The compression increases the resistance to venous return and therefore increases the pressure inside the capillaries, and inside the veins between the capillaries and the compression point; the increased internal pressure leads to venous distension. An analogous process is the increase in arterial pressure that follows constriction of the arterioles. **Hemorrhoids** are varicosities of the veins in the anal region. Formation of hemorrhoids is promoted by pregnancy for the same reasons discussed previously, and also by chronic constipation, which leads to excessive straining (attempted expiration against a closed glottis), and hence to elevated venous pressure during defecation.

■ **Do postural changes affect blood flow?**

laries drain into larger vessels, which themselves drain into the subclavian veins.

Lymph Flow Is Facilitated by One-Way Lymphatic Valves and the Pumping Effect of Skeletal Muscle Contraction

Lymph flow is driven by much the same mechanism as the one that propels venous blood back to the heart. The lymphatic vessels have an extensive series of one-way valves like those in the veins (Fig. 19–23*b*). The pumping action comes from phasic contractions of the smooth muscle associated with lymphatic vessels, as well as compression of the vessels by skeletal muscle activity. In 24 hours, the lymphatics may return a volume of fluid back to the heart equal to the total circulating blood volume (5 L). In addition, approximately one quarter to one half of the total plasma protein is returned to the blood during this period. The lymphatic system represents the only means by which plasma proteins can be returned to the blood. The lymphatics also transport substances absorbed from the gastrointestinal tract—primarily fat and fat soluble vitamins—to the circulating blood. Periodic swellings of the lymphatic vessels, called **lymph nodes,** are packed with lymphocytes and phagocytes. As the lymph percolates through this meshwork of reticular cells, microorganisms and other foreign material are removed from the lymph by phagocytosis.

Regulation of Systemic Arterial Pressure

Recall from our previous discussions that the body maintains normal tissue flow by monitoring and maintaining mean arterial blood pressure (MAP). As pointed out in the water system analogy, this approach simplifies the regulatory process and at the same time allows for a great deal of local regulation of blood flow by individual tissues. You will also recall that under conditions in which the demand for flow exceeded the capacity of the pump, it was necessary to selectively decrease the flow to certain users in order to maintain flow to others. Based on our accumulated knowledge of how the heart and vasculature function, we are now in a position to investigate the mechanisms by which the body achieves the necessary coordination of the heart and vasculature to maintain arterial pressure.

As we discuss how the body operates to maintain blood pressure, it will be easy to lose sight of the fact that the ultimate goal is to maintain blood flow, not arterial pressure. If the goal were only to maintain arterial pressure, this could be achieved with minimal effort by reducing blood flow to all the tissues to a trickle; this in fact does happen following extensive hemorrhage and can result in irreversible tissue damage to the kidneys and other organs, followed by cardiovascular collapse and death. Maintaining arterial pressure provides a simple method for adjusting cardiac output to meet the blood flow requirements of the tissues.

As previously noted, the relationship between cardiac output (CO), mean arterial pressure (MAP), and total peripheral resistance (TPR) is described by the equation:

$$CO = MAP/TPR$$

Algebraic rearrangement gives us the equation:

$$MAP = CO \cdot TPR$$

We can see from this equation that two factors determine the magnitude of the mean arterial pressure. One (CO) is determined by the status of the heart and

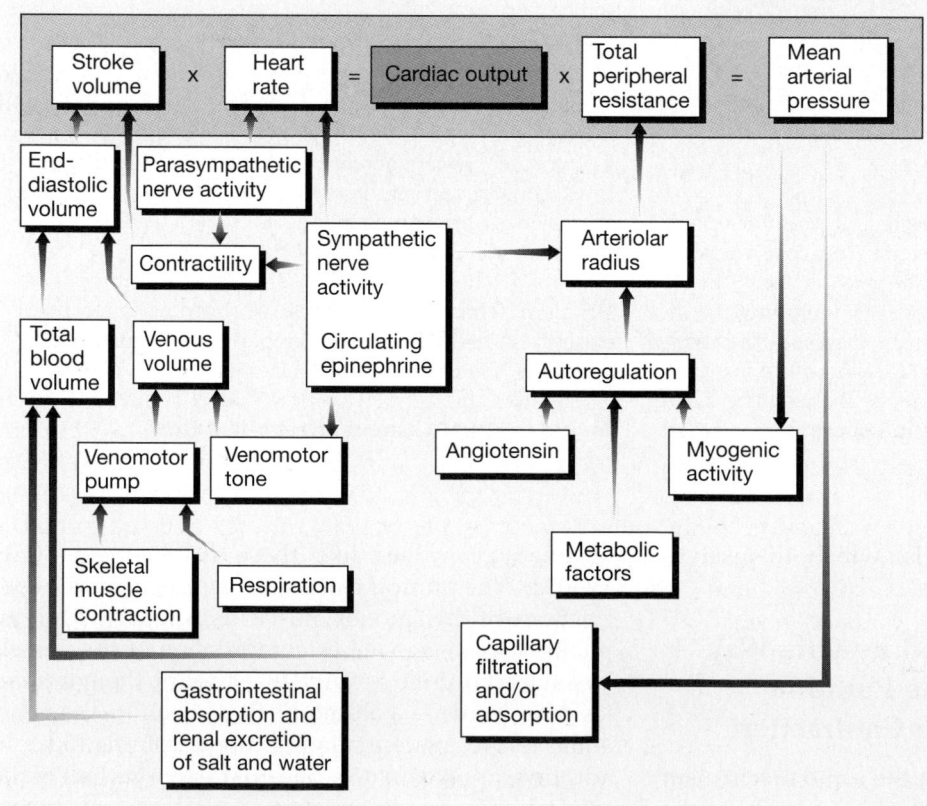

Figure 19–24

Summary of the factors that can alter cardiac output and arterial pressure.

the other (TPR) by the operational status of all the resistance vessels in the body. Figure 19–24 summarizes all the factors that we have discussed that can change either cardiac output or total peripheral resistance. This figure indicates only where relationships exist, without delineating how a change in one component alters related components. The ways in which these various parameters are integrated to achieve normal cardiovascular function are the subject of the next several sections.

Before proceeding, let us look briefly at the meaning of total peripheral resistance. The **total peripheral resistance** is the total resistance to the flow of blood from the aorta back to the right atrium. Keep in mind, however, that the major portion of the resistance to blood flow is posed by the arterioles and precapillary sphincters. More importantly, changes in peripheral resistance are due entirely to changes in the diameters of these vessels. Therefore, when we talk about changes in total peripheral resistance, we will be referring to changes in the overall resistance posed by all the resistance vessels in the body. Although other factors such as hematocrit can alter the viscosity of the blood—and consequently, the resistance to flow—these changes do not constitute a mechanism for the homeostatic maintenance of arterial pressure. We will therefore consider the viscosity of the blood to be constant under most conditions.

Cardiac Output Is Determined by the End-Diastolic Volume of the Heart, Which in Turn Is Determined by the Venous Return

These terms must be defined clearly if we are to understand how the factors interact. The terms **cardiac output** and **venous return** obviously refer to blood flow from the heart to the capillaries and from the capillaries to the heart, respectively. What is not obvious is whether they refer to the average flow over a period of time or to the instantaneous flow at any given point in time.

If we use these terms to describe average flow over a period of many minutes, then cardiac output and venous return must be equal since this is a closed system. Frequently, however, we will consider transient changes in flow that occur in selected portions of the system. Consider for a moment the changes following an increase in venomotor tone. Prior to this event, cardiac output and venous return were equal. As the veins contract and shift blood volume from the peripheral veins to the heart, venous return will, for a short time, be slightly higher than cardiac output because of the extra volume of blood flowing back to the heart from the peripheral veins. As long

as venous return to the heart exceeds cardiac output, the volume of blood in the heart will increase. The increase in end-diastolic volume will then lead to an increase in stroke volume and cardiac output (according to the Frank-Starling law). After the venoconstriction, cardiac output and venous return will again be equal (but greater than before the venoconstriction occurred because of the increased cardiac end-diastolic volume). We might then state that an increase in venous return (temporarily) produced an increase in cardiac output, leading to an increase in venous return (average).

This sounds like a circular agreement if it is not made clear that in the first case the term *venous return* refers to instantaneous flow, while in the second it refers to average flow. A clearer description of the same sequence of events is that the redistribution of blood volume from the peripheral veins to the heart produces an increase in end-diastolic volume, leading to an increase in cardiac output and venous return (average). The important point is that cardiac output is very sensitive to changes in end-diastolic volume, which in turn depends on the total blood volume and the distribution of blood within the vascular system.

This point is illustrated by the system diagrammed in Figure 19–25. When the pump is activated, water will be transferred from the large container on the left to the narrow container on the right. As the water rises in the narrow container, water will begin to flow back through the connecting tube at the bottom. If the rate of the pump is increased, the water level in the narrow chamber will again increase while the level in the larger container goes down. This will continue until the hydrostatic pressure gradient is sufficient to drive water back through the connecting tube at exactly the same rate as the pump rate. Notice, however, that turning up the pump rate even more has little additional effect, because the pumping rate becomes limited by the water level in the large container on the left. Therefore, the capacity of the pump becomes limited by the filling rate of the input tube. Two methods by which the actual pumping rate could be increased are shown in the bottom two panels. Adding more water to the left container would raise the water level and increase the filling rate of the pump (bottom left); this could also be accomplished by raising the container as illustrated on the right. By analogy to the cardiovascular system, adding more water would be analogous to increasing the circulating blood volume (e.g. by ingesting liquid) and raising the container would be analogous to redistribution of available blood volume (e.g. by venous constriction). The bottom line is that it does not matter whether you increase end-diastolic volume by increasing total blood volume by shifting

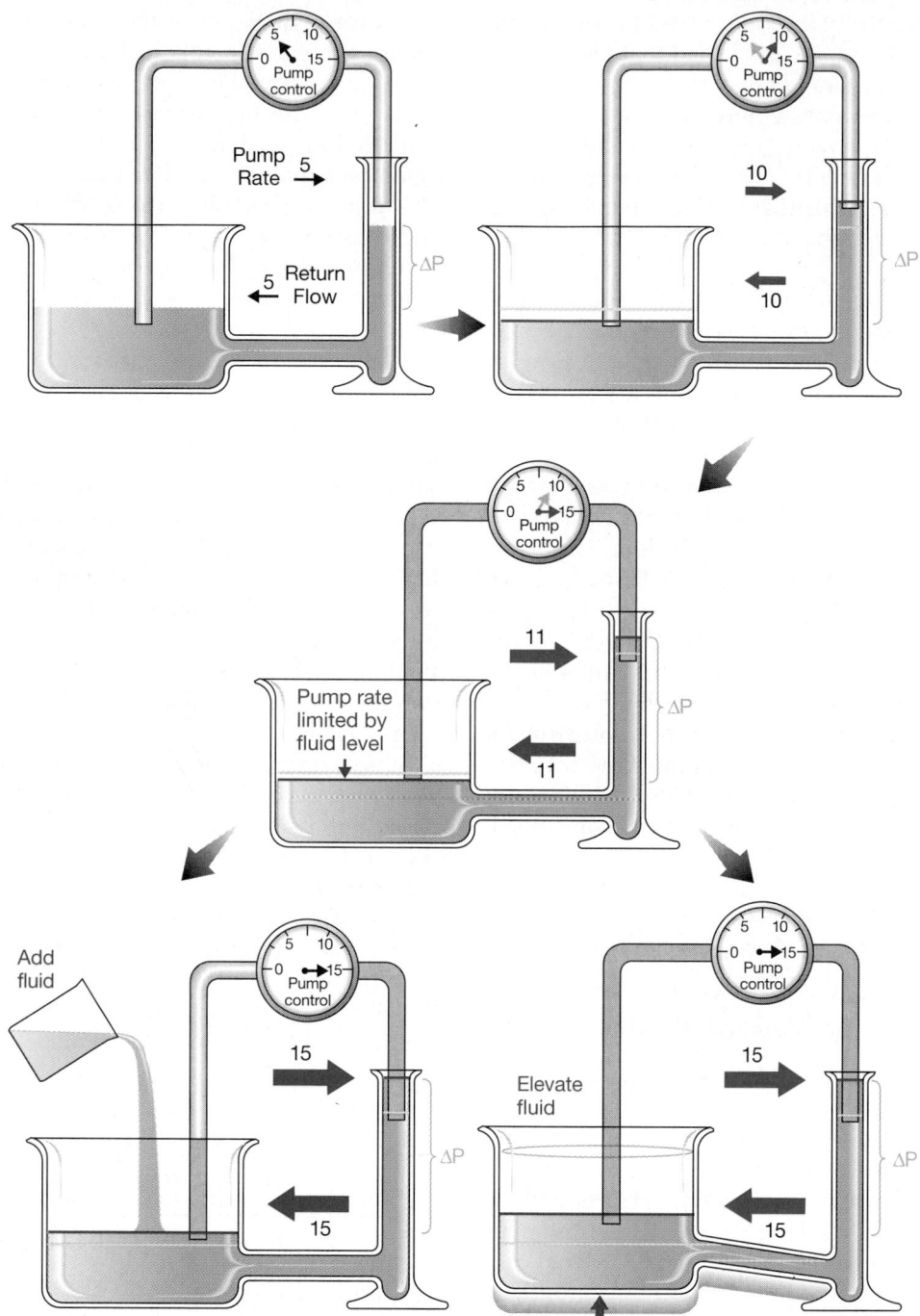

Figure 19–25

Demonstration of how venous return (return flow) can limit cardiac output (pump rate). The upper left two panels demonstrate that an increase in pump rate will shift blood volume from the venous (large reservoir on left) to arterial vessels (small reservoir on right) until a new steady-state is reached where return flow = pump rate. The upper right panel demonstrates that cardiac output (pump rate) is ultimately limited by venous return (return flow) because of the dependence of cardiac output on end-diastolic volume (level in venous reservoir is below pump intake line) even though the potential cardiac output (pump control) might be greater. The lower two panels demonstrate how this can be remedied by either increasing total blood volume (lower left) or by redistribution of existing blood volume by venoconstriction (lower right).

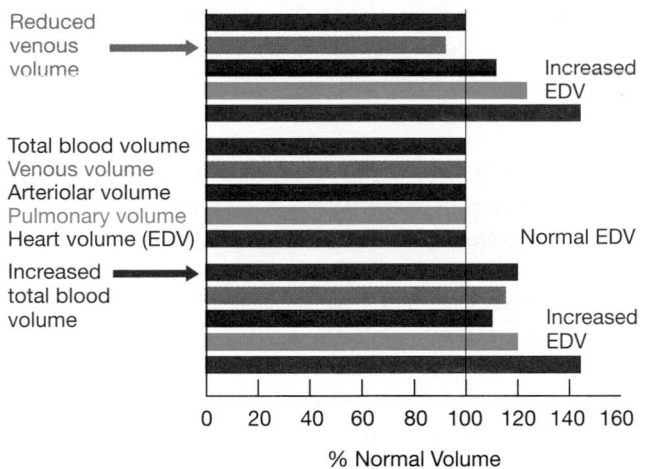

Figure 19–26

Graphic illustration of how end-diastolic volume (EDV) of the heart can be increased either by increasing total blood volume (right) or by shifting blood volume by venoconstriction (left).

blood volume to the heart by venoconstriction (Fig. 19–26); the effect on cardiac output will be the same. So you can see that the pumping capacity of the heart can be influenced equally by changes in *total blood volume* and by changes in the *distribution of blood volume* in the circulatory system.

Arterial Baroreceptors Are the Sensing Element of a Negative Feedback System That Regulates Heart Rate So That Mean Arterial Pressure Is Maintained Relatively Constant

You know from previous discussion of reflexive control mechanisms that an essential part of any such system is the presence of a receptor that can monitor the variable to be held constant—in this case, mean arterial pressure. The receptors that perform this operation in the cardiovascular system are called **baroreceptors.**

Location of Baroreceptors

Arterial baroreceptors are located in the wall of the arch of the aorta and in the carotid sinus (Fig. 19–27). The **carotid sinus** is a thin-walled, highly innervated region of the carotid artery located high in the neck at the point where the carotid artery branches into two smaller arteries.

Activation of Baroreceptors

The baroreceptor nerve endings function as stretch receptors and are distributed within the wall of the

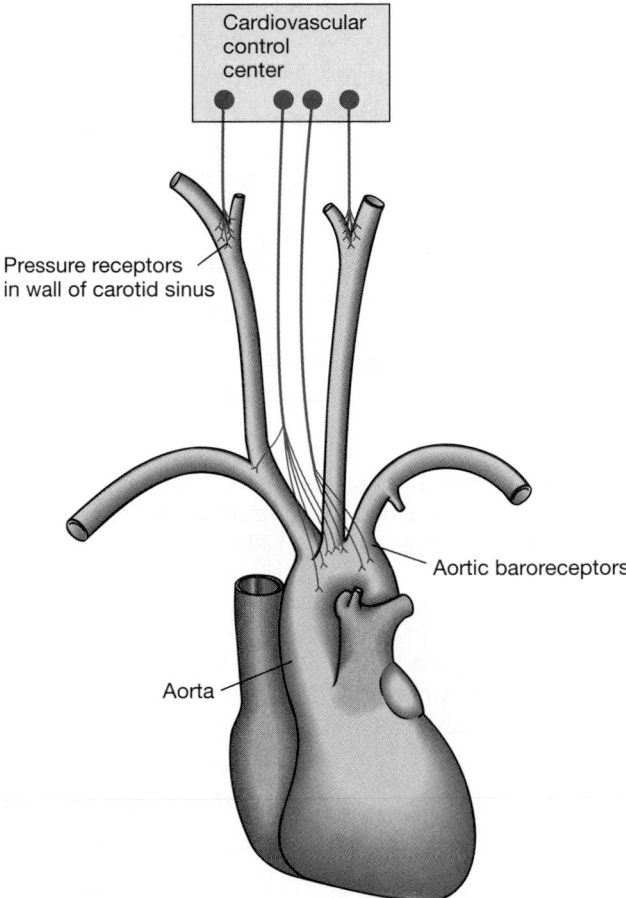

Figure 19–27

Location of the arterial baroreceptor nerve endings in the systemic arterial system. Afferent nerves from the baroreceptors travel to the medullary cardiovascular control centers.

elastic arteries. You have already encountered stretch receptors in other systems in the body, and the characteristics of all stretch receptors are similar. These receptors respond to stretch with an increase in the rate of action potential firing. Because of the elastic nature of the arterial wall, an increase in arterial pressure causes the arterial wall and the associated receptors to stretch. Stretching the nerve endings initiates the firing of action potentials, which are transmitted by the associated afferent (sensory) neurons to the cardiovascular control center.

Patterns of Baroreceptor Activation by Changes in Blood Pressure

Figure 19–28 shows the pattern of action potential firing, with changes in both mean arterial pressure and pulsatile arterial pressure. Notice that action potential firing increases in response to the pulsatile pressure changes during a single heart beat. This tells us that these receptors respond very quickly to changes in pressure. Therefore, the baroreceptors are effective

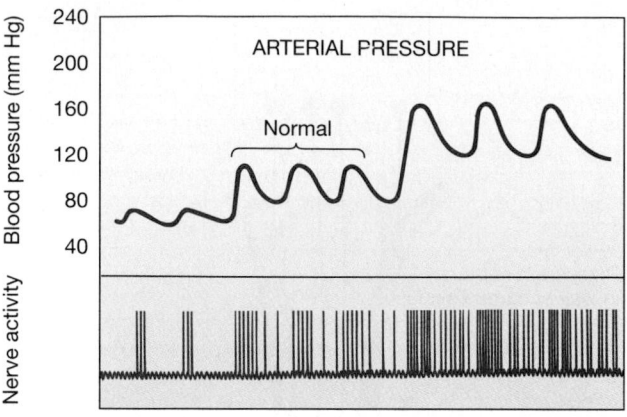

Figure 19–28

Changes in carotid sinus nerve activity as a function of changes in arterial blood pressure (modified from Berne and Levy, *Physiology*, 2nd ed., 1988, Fig. 33–10, p. 519).

in monitoring acute changes in blood pressure. As mean arterial pressure increases, the overall firing rate of the receptor increases.

The Cardiovascular Control Center in the Brain Determines the Level of Autonomic Nerve Activity to Both the Heart and the Blood Vessels

The afferent nerves from the baroreceptors travel to the brain, where they synapse with neurons that make up the **cardiovascular control centers.** The primary control center for the baroreceptor reflex is located in the medulla. The medullary cardiovascular control center is a complex and diffuse network of neurons that integrates neural inputs from the baroreceptors with inputs from other control centers in the brain. For now we will concentrate on how changes in the level of input from the baroreceptors lead to changes in autonomic nerve activity to the heart and vasculature.

Cardiac Output and Total Peripheral Resistance Are Regulated by the Cardiovascular Control Center So That Mean Arterial Pressure Remains Relatively Constant

You will recall from our previous discussions that the heart is innervated by both parasympathetic and sympathetic nerves, and that the blood vessels are innervated primarily by sympathetic nerves. Preganglionic efferent nerve fibers from the cardiovascular centers synapse with the cell bodies of both parasym-

pathetic and sympathetic neurons that innervate the heart and blood vessels (Fig. 19–29). Changes in baroreceptor activity produce reciprocal changes in parasympathetic and sympathetic activity to the heart and blood vessels. An increase in the rate of nerve activity from the baroreceptors leads to a decrease in the rate of stimulation of sympathetic nerves and an increase in the rate of stimulation of parasympathetic nerves. Conversely, a decrease in baroreceptor nerve activity leads to increased sympathetic and decreased parasympathetic activity.

Cardiac Output and Total Peripheral Resistance

Let's now examine the overall response of the system to a change in blood pressure (Fig. 19–30). If arterial pressure increases, the baroreceptor firing rate increases. This leads to a decrease in sympathetic nerve activity to the heart and blood vessels and an increase in parasympathetic nerve activity to the heart. This, in turn, causes heart rate to decrease and contractility to decrease, leading to a reduction in cardiac output. The decreased sympathetic nerve activity to the blood vessels promotes dilation of the heavily innervated vascular beds such as the skin, muscle, and splanchnic circulations, leading to a reduction of total peripheral resistance. The reduction in both cardiac output and total peripheral resistance causes a reduction of arterial pressure (MAP = CO · TPR). Clearly, the response of the system is to minimize changes in arterial pressure by changing both cardiac output and total peripheral resistance in the appropriate directions. If arterial pressure were to decrease, decreased baroreceptor firing would lead to increased cardiac output and increased total peripheral resistance. As before, this compensatory response would tend to restore arterial pressure back to normal.

Venomotor Tone

In our previous discussion of the venous system, we pointed out that contraction of venous smooth muscle (**venomotor tone**) occurs in response to sympathetic stimulation. Changes in venomotor tone also play a role in the baroreceptor reflex to maintain arterial blood pressure. In response to increased arterial pressure and increased baroreceptor firing, there is a decrease in sympathetic activity to the veins, resulting in venodilation. Venodilation results in a shift of blood volume from the heart and arteries to the peripheral veins. This leads to a decrease in end-diastolic volume, which decreases cardiac output according to the Frank-Starling law. The decrease in cardiac output then leads to a decrease in mean blood pressure back toward normal.

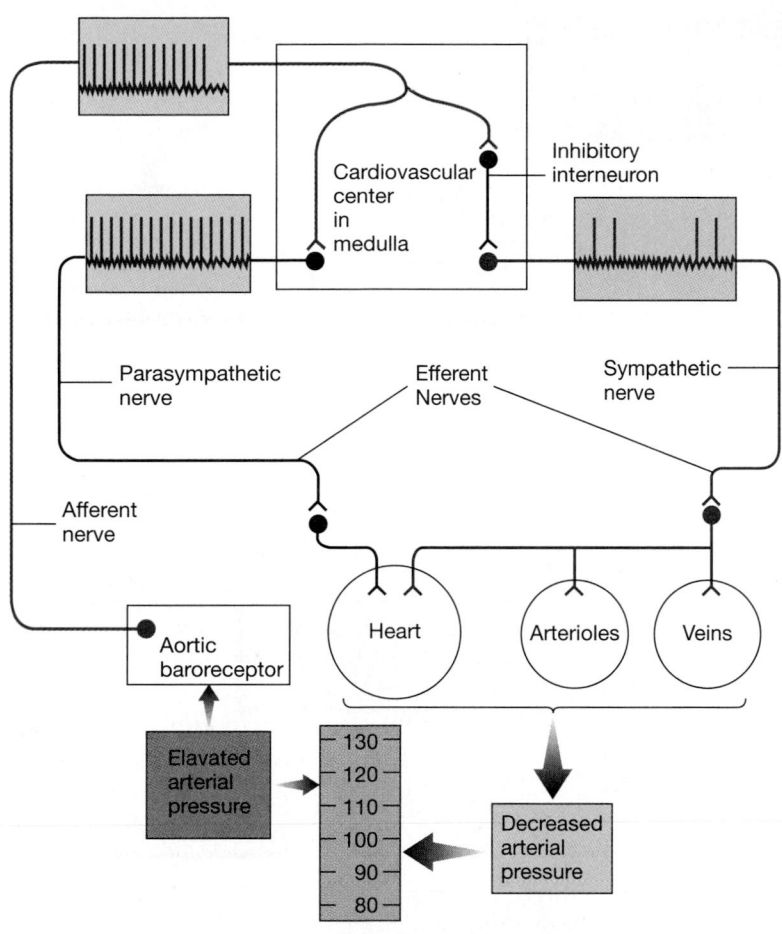

Figure 19-29

Increased aortic blood pressure leads to increased afferent nerve activity from the baroreceptors, which results in reciprocal changes in efferent sympathetic and parasympathetic nerve activity to the heart and blood vessels with a resultant decrease in aortic blood pressure back to normal.

Long-Term Regulation of Blood Pressure

Many physiologists contend that regulation of blood volume is one of the most important aspects of blood pressure regulation, particularly on a long-term basis. Whereas reflexive mechanisms such as those just discussed play a dominant role in short-term pressure regulation, the maintenance of adequate circulating blood volume is primarily under the control of the kidneys (which mediate the excretion of salt and water from the body) and the gastrointestinal system (which mediates absorption of salt and water). You will see more clearly how important the maintenance of blood volume is when we discuss the ways in which hemorrhage affects cardiovascular performance. Also, we will discuss the way in which venomotor tone and alteration of blood volume by changes in filtration and absorption in the capillaries can temporarily compensate for inadequate blood volume. If cardiac output is compromised by inadequate end-diastolic volume, this can be corrected by (1) increasing the total circulating blood volume, (2) shifting of blood volume from peripheral vascular beds to the heart by increasing venomotor tone, or

(3) increasing the reabsorption of interstitial fluid by the capillaries (see Fig. 19-20). These mechanisms are equivalent in terms of their effect on the heart. Remember, however, that they are not equivalent in terms of overall physiological homeostasis. For example, you will see that, during hemorrhage, the body compensates for loss of blood by severely decreasing flow to organs such as the kidneys. While this may temporarily serve to maintain blood pressure and blood flow to the heart and brain, the kidney can be damaged severely as a result of prolonged compensation of this nature.

Hemorrhage, Hypotension, and Exercise All Cause Primary Changes in Either Cardiac Output or Total Peripheral Resistance That Must Be Matched by Compensatory Changes Elsewhere

We have now examined the structural and functional characteristics of the individual components of the cardiovascular system. We have discussed how the

Figure 19–30

Summary of cardiovascular changes in response to increased arterial blood pressure and increased baroreceptor nerve activity. This is an example of negative feedback. An increase in blood pressure leads to compensatory changes that tend to restore blood pressure back toward normal.

baroreceptor reflex functions to maintain arterial blood pressure by initiating changes in cardiac output and total peripheral resistance. As we have stated on several occasions, the purpose of the cardiovascular system is to maintain adequate blood flow to the tissues to support cellular metabolism. What we have begun to see is that not all tissues are equal in terms of their need for blood flow and that, under some conditions, blood flow is directed from some tissue beds to others to sustain temporarily elevated levels of metabolism (e.g., exercising muscle). Moreover, blood flow to certain tissues such as the brain and heart must be maintained, even at the expense of other tissues, because of the body's need for their uninterrupted function. Perhaps the best way to see

how all of these factors operate together is to look at the response of the cardiovascular system to changes in its normal resting state. To illustrate these points we will examine three cardiovascular disturbances: postural hypotension, hemorrhage, and exercise.

Hemorrhage: Loss of Blood

The immediate effect of blood loss (hemorrhage) is a drop in arterial blood pressure. This occurs as a result of an associated decrease in cardiac output. If blood volume is removed from the vascular system, the end-diastolic volume of the heart will decrease and cardiac output will be reduced, as described by the Frank-Starling law. We know from the equation

CLINICAL FOCUS

Systemic Hypertension

The term **systemic hypertension** refers to a complex disease state resulting in chronically elevated arterial blood pressure. Hypertension is a leading cause of death and disability among adults living today in the United States and most other industrialized societies. More than half of all heart attacks and two thirds of all strokes occur in hypertensive individuals.

Although the precise definition of hypertension is somewhat elusive, studies have shown that individuals with resting blood pressures between 140/90 and 160/95 have a significantly greater risk of cardiovascular disease. **Essential hypertension,** the most common form of hypertension, is due to a primary disturbance of the cardiovascular system. Other forms of hypertension are secondary to disease processes such as renovascular disease.

Blood pressure is determined by cardiac output and total peripheral resistance. Hypertension is associated with abnormalities in both of these parameters. In many patients, cardiac output is initially high with normal peripheral resistance. With the progression of the disease, cardiac output falls and peripheral resistance increases. The initial elevation of cardiac output may have multiple causes, including abnormal retention of salt and water by the kidneys. The progressive increase in vascular resistance is most likely a response to the elevated pressure and involves structural changes in the wall of the blood vessel.

While the exact cause of hypertension is not known, a number of factors are known to contribute to the development of the disease. Hereditary predisposition to high blood pressure has been demonstrated. It is also well established that excessive dietary sodium intake is linked to hypertension, particularly in those individuals genetically predisposed. Other factors linked to hypertension include obesity, excessive alcohol consumption, smoking, and diabetes.

Hypertension has been called the "silent killer" because the uncomplicated disease state is almost always asymptomatic. Individuals may be unaware for as long as 10 to 20 years that their elevated blood pressure is causing progressive cardiovascular damage. Frequent blood pressure measurements are necessary to detect the early onset of the disease. If left untreated, hypertension can lead to stroke, peripheral vascular disease, myocardial ischemia, and heart failure. Frequently, hypertension can be treated successfully with a combination of nondrug therapies including restriction of dietary sodium, weight reduction, exercise, and moderation of alcohol and coffee consumption. Antihypertensive drug therapy includes the use of diuretics to reduce plasma volume, adrenergic blocking drugs to reduce sympathetic stimulation of the cardiovascular system, and vasodilator agents to reduce systemic resistance.

■ **Why doesn't systemic hypertension lead to increased loss of fluid from the capillaries and formation of edema?**

$MAP = CO \cdot TPR$ that restoration of arterial pressure requires an increase in either cardiac output or total peripheral resistance. The decrease in arterial pressure following hemorrhage immediately causes a reduction the rate of baroreceptor nerve activity. This leads to increased sympathetic nerve activity to the heart and increases in both heart rate and contractility. The associated increase in cardiac output increases arterial pressure back toward normal. The decreased baroreceptor activity also results in increased sympathetic nerve activity to the resistance arterioles.

As we know from previous discussions, not all vascular beds receive the same degree of sympathetic innervation. The vascular beds in the skin, skeletal muscle, gastrointestinal tract, and kidneys are all innervated by sympathetic nerves. Arteriolar vasoconstriction of these vascular beds minimizes the fall in blood pressure by increasing total peripheral resistance. The vascular resistance to the heart and brain does not increase because of the sparse sympathetic innervation and because of the high degree of local control present in these vascular beds.

We can now begin to appreciate the purpose of the selective innervation of different vasculature beds in the body. Blood flow to those tissues whose continuous function is not essential for life can be temporarily reduced so that a limited blood flow can be distributed to those organs, such as the brain and heart, that must function continuously. Loss of blood flow to the heart would weaken the heart and prevent it from participating in the restoration of blood pressure. Loss of blood flow to the brain would disrupt the function of the cardiovascular centers responsible for the mediation of the compensatory response. The kidneys can sustain a short-term reduction in blood flow, but long-term reductions lead to renal failure and death. As you will see in more detail later, the reduction of blood flow to the kidney has the added benefit of maintaining blood volume by reducing the formation of urine. Blood flow to the skin, muscles, and the gastrointestinal system can be substantially reduced with little ill effect. The net result of these changes in cardiac output and peripheral resistance is to transfer limited blood flow to the heart and brain.

Figure 19–31 provides an overview of the compensatory responses to hemorrhage. You can see that the compensatory response consists of a rapid component that includes increased heart rate, increased cardiac contractility, increased total peripheral resistance, and venoconstriction. This is clearly a temporary response that sacrifices the immediate needs of some tissues to maintain those of other, more critical tissues such as the heart and brain. A second, more prolonged response consists of fluid reabsorption from the interstitial compartment to restore total blood volume. While this tends to alleviate the hemodynamic crisis, complete recovery involves several additional processes. Complete restoration of total body water involves control of fluid ingestion and kidney function. Restoration of the blood requires stimulation of erythropoiesis to increase the hematocrit and synthesis of plasma proteins.

One of the more pronounced changes that occurs with hemorrhage is a reduction in stroke volume; this results in a weak or shallow pulse. Although the increase in heart rate and contractility can partially compensate for this reduction, both of these compensatory mechanisms are limited by the diastolic filling of the heart. Recall from earlier discussions that, even under normal conditions, increasing cardiac output by increasing heart rate is limited by the progressively diminished diastolic filling time. As blood is transferred to the arterial system to maintain blood pressure following hemorrhage, less blood is available for filling of the heart. So you can see that the extent of compensation by increased heart rate, increased contractility, and increased peripheral resistance becomes self-limiting. The only way to overcome this limitation is to increase the end-diastolic volume of the heart. As was previously demonstrated in Figures 19–25 and 19–26, this can be accomplished by increasing the total blood volume or by redistributing blood volume.

Obviously, a blood transfusion is an effective way of returning the system to normal and is the treatment of choice for hemorrhage. There are, however, compensatory mechanisms that allow the body to recruit blood volume from other sources. One response to hemorrhage is generalized venoconstriction. Although this does not increase the total circulating blood volume, it does increase the volume of blood available to the heart. Consider what happens as the veins constrict. Because of the reduction in vessel di-

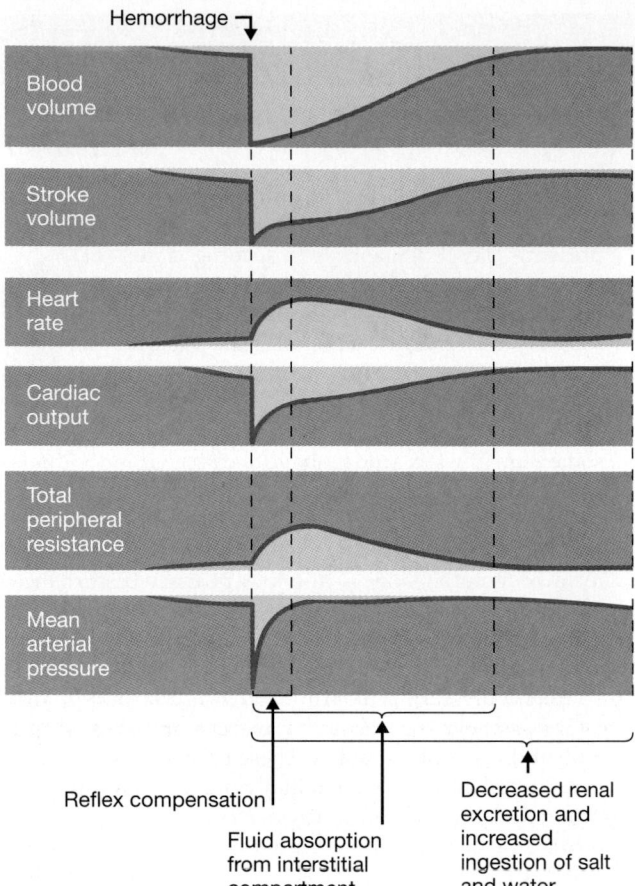

Figure 19–31

Summary of the hemodynamic responses of the cardiovascular system to mild hemorrhage. A rapid, reflexive compensation is followed by a slower response to return blood volume to normal by increasing fluid reabsorption and decreasing fluid excretion.

ameter, blood shifts from the veins to the heart and arterial system. From the standpoint of the heart, this is as effective as an increase in total blood volume.

There is another mechanism by which the body is able to create a real increase in total blood volume. Consider what is happening to capillary pressure, particularly in vascular beds where arteriolar vasoconstriction has taken place (see Fig. 19–20). Mean arterial blood pressure and venous pressure have been reduced because of the hemorrhage, and capillary pressure therefore must also be reduced. The plasma protein concentration, however, has not changed. According to Starling's hypothesis, reabsorption of fluid will be favored, and interstitial fluid volume will be transferred into the capillaries. As much as 1 liter of fluid per hour can be transferred to the vascular compartment by this mechanism following severe hemorrhage.

This reabsorption of fluid does not occur immediately and may take as long as 12 to 24 hours to reach completion, depending on the extent of hemorrhage. For example, if you donate a pint of blood, within 12 hours your circulating blood volume will be restored almost to normal. The fluid transferred from the interstitial compartment is free of red blood cells, and what little plasma protein this fluid contains is filtered during reabsorption. Remember that the capillary wall is relatively impermeable to protein. As you might expect, hematocrit and plasma protein concentration are reduced as a result of fluid reabsorption. The dilutional effect of fluid reabsorption limits the extent of this compensatory mechanism. As interstitial fluid is transferred into the blood, the protein concentration of the interstitial fluid increases while that of the blood decreases. This progressively reduces the osmotic gradient, which is the driving force for fluid reabsorption.

The mechanisms just described can adequately compensate for blood losses of as much as 1 to 1.5 liters (2 to 3 pints) with few ill effects. Greater losses of blood, however, can lead to severe tissue damage, irreversible circulatory collapse, and death. This progressive deterioration of cardiovascular function is referred to as **circulatory shock.** Depending on the extent of blood loss and the elapsed time, a point is reached after which no known therapy, including massive transfusions, can prevent death. Although the details of this process are complex, it is clear that contributing factors include irreversible damage to many organ systems, including the kidneys, heart, and brain, as a result of inadequate blood flow. A phenomenon that has gained general recognition recently is that of "reperfusion injury." This refers to the events that occur following restoration of blood flow to any tissue that has experienced a prolonged episode of ischemia. At least part of this process involves the release of molecular free radicals from immune cells after reperfusion. These free radicals (e.g., oxygen free radicals) are extremely reactive molecules that cause cellular damage by oxidation of proteins, lipids and nucleic acids. These effects can be partially protected against by the use of free radical scavengers and antioxidants. Natural antioxidants present in the body include Vitamin C and Vitamin E.

Hypotension: Low Blood Pressure

From our previous discussion, we know that hemorrhage produces a form of hypotension that results from a loss of blood volume. Recall that one of the compensatory processes for hemorrhage involves the translocation of blood volume from the veins to the heart and arterial system by venoconstriction. From the point of view of the heart, this is entirely equivalent to a transfusion of an equal volume of blood. As you can see, many of the regulatory and compensatory mechanisms that we have been discussing involve blood volume redistribution.

An everyday example of the effects of hypotension are fainting spells that might occur when someone stands abruptly after waking, or stands for a long time in the hot sun while wearing warm, heavy clothing. The problems here are gravity and inadequate peripheral vascular tone, which result in inadequate filling of the heart. From our earlier analogy in Figure 19–25, this would be the equivalent of lowering the large reservoir on the left until the water was below the level of the input pipe going to the pump.

In the first example given above, the sleeper probably has a heart rate under 60 beats per minute, and the veins in the legs are dilated. When this person stands up, blood pools in the veins of the legs and feet, and suddenly no blood is available to fill the heart. Cardiac output drops precipitously, leading to inadequate blood flow to the brain, causing unconsciousness. In the second instance, the person is standing quietly, so that heart rate slows and blood pressure drops. Because the person is also overheated, the subcutaneous blood vessels are dilated to bring blood close to the surface of the body for heat dissipation. This reduction in peripheral resistance lowers arterial pressure. The lack of venomotor pumping activity, results in the transfer of blood volume away from the heart. This, in turn, leads to reduced end-diastolic volume, reduced cardiac output, and a further reduction of arterial blood pressure. The fall in arterial blood pressure ultimately results in insufficient blood flow to the brain. The body involuntarily compensates by fainting, which adjusts the posture to promote greater blood flow to the heart and brain. Such problems are normally avoided by a combination of compensatory mechanisms. Normal levels of activity result in periodic contraction of the skeletal muscles in the legs, which prevents pooling of blood in the veins of the lower extremities. In addition, normal levels of activity, both mental and physical, are associated with slight increases in sympathetic nerve activity and therefore with increases in heart rate and venomotor tone.

Exercise and Cardiovascular Homeostasis

The cardiovascular changes occurring during exercise are in many ways the opposite of those occurring during hypotension and hemorrhage. With hypotension, the major disturbance is reduced arterial blood pressure resulting from reduced cardiac output. The compensatory response (increased cardiac output and increased vascular contraction) is mediated primarily by decreased activation of the arterial baroreceptors in response to reduced arterial blood pressure. During both mild and intense exercise, there is an increase in arterial pressure. Under resting conditions an increase in arterial pressure tends to decrease cardiac output and decrease constriction of systemic vascular beds via the baroreceptor reflex. Yet we know that cardiac output always increases during exercise; well-trained athletes may experience an increase in cardiac output from 5 L/min to as much as 35 L/min during maximal exercise.

It is now well accepted that the cardiovascular changes observed during exercise are not in fact mediated by the baroreceptor reflex. Although a number of local and humoral factors have been cited as contributing to vasodilation of the skeletal muscle vasculature, reflexes originating from within the contracting muscles are now thought to lead to activation of sympathetic nerves to the heart and peripheral vasculature, to further adapt the cardiovascular system to the demands of exercise.

Let's first consider what demands exercise places on the cardiovascular system. Exercise, by definition, is an increase in the rate and extent of contraction of skeletal muscle. We can correctly assume that this is associated with a substantial increase in the metabolic rate of muscle. A good measure of the level of muscle metabolism is oxygen consumption. During maximal exertion, muscle oxygen consumption for the well-trained athlete may increase by as much as 50 to 60 times resting levels. This means that a significant increase in blood flow must occur (up to 20 times resting flow) to deliver oxygen to the muscle and to remove metabolic byproducts. If we consider that resting muscle blood flow is about 1 to 1.5 L/min, then the blood flow to skeletal muscle alone during exercise will be as high as 15 to 20 L/min. Neglecting the needs of the rest of the body, it is clear that cardiac output will have to increase substantially to support this increased demand by exercising muscle (Fig. 19–32).

The necessary cardiovascular changes required for exercise actually begin even before the onset of muscular activity. The mental anticipation of exercise leads directly to increased sympathetic activity and decreased parasympathetic activity to the heart. There is also an increase in sympathetic activity to the peripheral vascular beds such as the skin, kidney, and gastrointestinal tract, which increases vascular resistance and directs blood flow to muscle. In certain animals—and possibly humans—there may be sympathetic fibers to skeletal muscle that release acetylcholine and stimulate vasodilation of precapillary arterioles to further increase muscle blood flow before exercise.

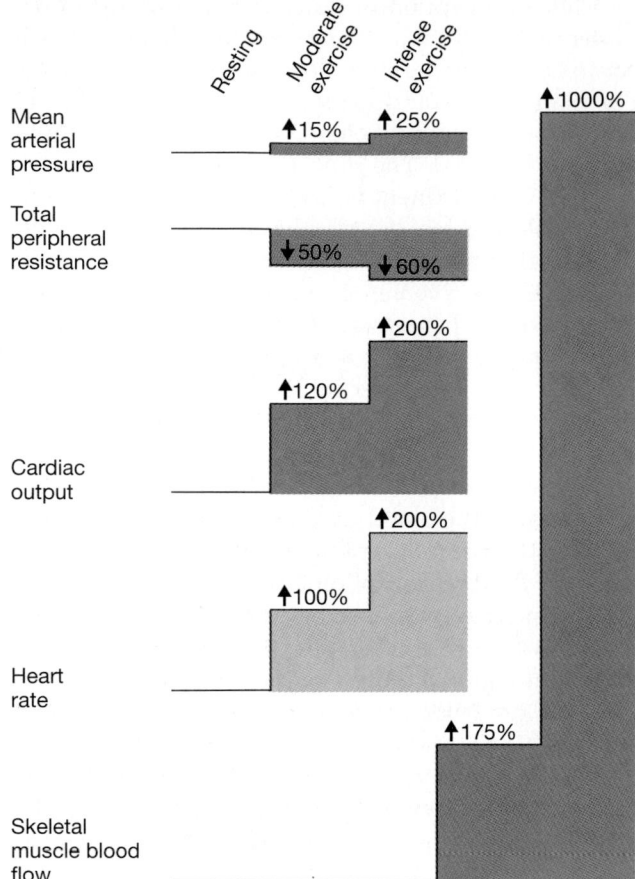

Figure 19–32

Summary of the hemodynamic changes in the cardiovascular system in response to mild and intense exercise expressed as percent change from resting.

Once muscular exercise begins, a number of additional mechanisms come into play. The release of vasodilator substances from the muscle cells such as potassium and adenosine leads to vasodilation of the precapillary resistance vessels. As exercise progresses and body temperature begins to increase, there is a reflexive vasodilation of the subcutaneous vessels to allow for dissipation of excess body heat. Chemoreceptors within the muscle respond to the release of metabolites and initiate a reflexive increase in sympathetic nerve activity to the heart and resistance vessels of the gastrointestinal tract and kidneys. The level of afferent nerve activity is roughly proportional to the level of muscle work being performed and therefore provides a means for matching cardiac performance with the requirements of the active muscle.

Consider for a moment how the situation during exercise differs from that during compensation for hemorrhage. In both cases, there is increased stimulation of heart rate and contractility. In both cases, blood is being shunted from inactive to active tissues.

With hemorrhage, cardiac output is limited by inadequate cardiac filling; this is clearly not the case with exercise, since cardiac output may reach 35 L/min. The difference with exercise is that there has been no reduction of blood volume. Compensation for blood loss due to hemorrhage requires a net transfer of blood from the venous system to the heart and arterial side of the circulatory system in order to maintain adequate perfusion pressures for the heart and brain. This process is self-limiting because of decreased cardiac filling with depletion of venous blood volume. We can deduce that, during exercise, there is little change in arterial blood volume, even as cardiac output increases, because mean blood pressure increases only slightly. No transfer of blood volume from the venous to arterial systems occurs because total peripheral resistance has decreased in proportion to the increase in cardiac output; blood flow from the arterial side to the venous side increases in direct proportion to the increase in cardiac output. If we think back to the analogy in Figure 19–25, the decreased total peripheral resistance with exercise would be analogous to increasing the diameter of the connecting return tube. As long as water returns to the bucket as fast as it is removed, the only limitation to the pumping rate is the maximum rate at which the pump can operate. For the heart, this is determined by the minimum time required for contraction of the ventricle and diastolic filling. The effective upper limit for heart rate is about 180 beats/min.

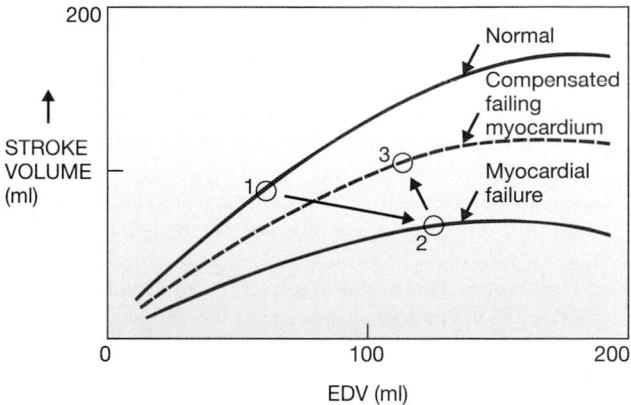

Figure 19–33

Change in the relationship between end-diastolic volume and stroke volume as a normal heart (1) progresses into heart failure (2). The decrease in stroke volume with myocardial failure is lessened by the increase in EDV that occurs as a result of the reduced stroke volume. Increased sympathetic nerve activity to the heart can compensate for mild failure by increasing the contractility (dotted curve), which in turn decreases the end-diastolic volume back toward normal (3). At this point (3), stroke volume has returned to normal because of the combined effects of increased EDV and increased contractility.

cular system and then impose a reduction in contractility on the heart. We know that the decrease in contractility will decrease stroke volume and cardiac output. Furthermore, we can reason that a decrease in cardiac output will lead to a decrease in mean arterial pressure.

Cardiovascular Disease

Heart Failure Refers to a Variety of Primary Cardiac Pathologies in which the Heart Is Unable to Maintain Normal Cardiac Output

As the name implies, heart failure is the result of the inability of the heart to maintain adequate cardiac output. Although many factors may cause failure of the heart, the resulting changes in cardiovascular performance are similar.

Cardiovascular Effects of Heart Failure

As before, let us first specify what cardiovascular parameter undergoes a change. In this case, the problem is a reduction in cardiac contractility. Recall that this means that the ability of the heart to pump blood at any end-diastolic volume is decreased (Fig. 19–33, bottom curve). Let's start with a normal cardiovas-

Compensation for Heart Failure

As the end-diastolic volume of the heart increases, the Frank-Sterling mechanism tends to increase cardiac output. Recall that in hemorrhage, the kidneys respond to the reduction in blood pressure by increasing the retention of water and sodium. Exactly the same process of fluid retention occurs with heart failure. The increase in circulating blood volume further increases the end-diastolic volume of the heart to increase cardiac output (see Fig. 19–33). Unfortunately, this compensatory mechanism is limited. Recall that increases in end-diastolic volume lead to increased vigor of contraction because, in a normal resting heart, the cardiac muscle is shorter than its optimal length for contraction (refer to Fig. 18–12). With progressive heart failure, additional increases in volume stretch the muscle beyond its optimal length. Once this point is reached, further increases in heart volume actually reduce the ability of the heart to pump blood. This in turn leads to further increases in venous and cardiac volume, and a vicious cycle

Atherosclerosis and Myocardial Infarction

Atherosclerosis is a disease process of arteries that results in the progressive occlusion of the lumen of the vessel. This process is seen most frequently in the aorta and in the cerebral and coronary arteries. Although the precise cause of atherosclerosis is still widely debated, the progression of events is well documented. The initial event probably involves damage to the endothelial lining of the vessel wall. This is followed by proliferation of the underlying smooth muscle cells that make up the muscular wall of the vessel. A progressive reduction in the size of the vessel lumen results from the thickening of the vessel wall. As the disease progresses, lipid from the blood accumulates along with fibrous tissue **(atherosclerotic plaque)**, resulting in the progressive occlusion of the vessel lumen.

Occlusion of the coronary arteries in this manner **(coronary artery disease)** is the most frequent cause of cardiac dysfunction. The occlusion of the vessel lumen gradually reduces blood flow to the heart. As the disease progresses, individuals may begin to experience periodic episodes of inadequate coronary blood flow, often associated with chest pain referred to as **angina pectoris.** Episodes of angina are frequently associated with increased physical exertion or emotional stress. A heart attack is the result of an acute coronary occlusion, which may occur because of either clot formation at the site of the plaque **(thrombus)** or intense spasm of the coronary smooth muscle near the plaque. Since the heart receives its blood supply from the coronary vessels, occlusion can lead to injury and death of that portion of the myocardium supplied by the occluded vessel. The area of damaged myocardium is referred to as an **infarct.**

An infarct can reduce the heart's ability to pump blood in a number of ways. Depending on the extent and location of the infarct, the pumping capability of the heart may be compromised because of the loss of functional muscle. The infarct may also initiate cardiac arrhythmias. Either or both of these situations can lead to decreased cardiac output. This, in turn, may lead to reduced arterial blood pressure, depending on the body's ability to compensate by increasing peripheral resistance. Autoregulation of coronary blood flow compensates for small drops in arterial pressure; larger decreases in arterial pressure result in reduced coronary blood flow. This, in turn, further diminishes the pumping ability of the heart, leading to more reductions in cardiac output and arterial pressure. Unless halted, this downward spiral will lead to collapse of the cardiovascular system and death.

The body's compensatory responses are limited, and therefore the consequences of advanced coronary artery disease can be severe. Arterial pressure can be increased following a heart attack only by increasing cardiac contractility, heart rate, or peripheral vascular resistance. Unfortunately, these compensatory responses all require that the heart do more physical work, which in turn increases the need for oxygen delivery to the heart. Since the primary problem is inadequate blood flow and insufficient oxygen delivery to the heart, these compensatory responses frequently only advance the extent of myocardial damage.

As with all cardiac disorders, reversal of the primary disturbance is desirable. Several therapeutic methods are available to increase coronary blood flow. **Nitroglycerin** provides relief from angina by dilating the coronary arteries and increasing coronary blood flow. A class of drugs known as **calcium antagonists,** or **calcium channel blockers,** which block the entry of calcium into the smooth muscle cells, also provide relief by dilating the coronary arteries. Agents that dissolve blood clots in the coronary arteries are being used to unblock arteries occluded by thrombus formation. If progressive occlusion of a coronary vessel is detected prior to the onset of a heart attack, measures are available to alleviate the occlusion. **Coronary bypass surgery** involves the grafting of a vein (removed from elsewhere in the patient's body) around the occluded segment of the coronary artery so that blood flow can bypass the diseased area. Another approach is to insert a catheter with a small balloon at the tip into the occluded coronary artery. When the catheter tip is positioned within the occluded portion of the vessel, the balloon is inflated and the atherosclerotic coronary plaque is fractured, allowing the vessel to dilate. This procedure is referred to as **coronary angioplasty.** These treatments do not stop the progression of the disease but merely relieve the symptoms temporarily.

Prevention is clearly the best treatment. Although the cause of atherosclerosis is unknown, a number of risk factors have been associated with an increased in-

A small catheter is introduced into the blocked coronary artery. A guide wire containing a small balloon at its tip is pushed through the blockage. When the balloon is in place, it is inflated to flatten the blockage against the arterial wall.

cidence of the disease. These include smoking, hypertension, elevated plasma cholesterol, lack of exercise, and diabetes. Current evidence suggests that a reduction of cholesterol in the diet as well as regular exercise may offer some protection against the progression of the disease.

■ Based on previous discussions about the dependence of blood flow on pressure and vessel diameter, you might conclude that angina could be effectively treated either by dilation of the coronary arteries or, when that would not provide adequate flow due to the presence of atherosclerotic narrowing of the coronary arteries, by increasing arterial pressure. In light of our previous consideration of how increased pressure affects oxygen consumption by the heart (Fig. 18–27), why might increasing arterial pressure be a less effective and possibly counterproductive therapy?

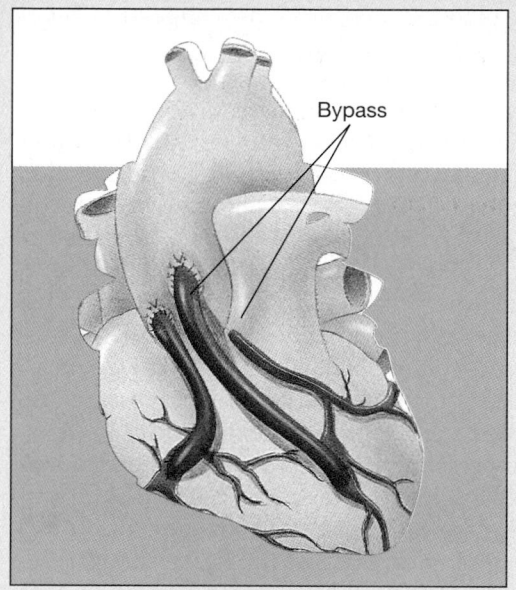

Tissue sections from an atherosclerotic vessel.
(a) Early stage of atherosclerosis with slight thickening of the arterial wall indicated by the arrow.
(b) Progressive accumulation of smooth muscle cells, lipid, and calcium results in plaque formation. (c) The progressive accumulation of plaque results in almost complete occlusion of the vessel lumen and a substantial increase in the resistance to blood flow.

(a)

(b) Lumen narrowed

(c) Thrombus

ensues. This progressive congestion of the veins and heart with blood is described as **congestive heart failure.**

Clearly the reduced blood pressure will stimulate the baroreceptor reflex, which results in increased sympathetic and decreased parasympathetic stimulation of the heart as well as sympathetic stimulation of the vasculature. The result is increased contractility, increased heart rate, increased arteriolar constriction, and increased venomotor tone. As you know, these changes are very similar to those seen with hemorrhage. The major difference concerns the distribution of blood volume. Following hemorrhage, heart rate and vigor of contraction increase to transfer venous volume to the arterial system in order to maintain arterial pressure. Recall that full compensation is limited by inadequate filling of the heart. With cardiac failure, just the opposite occurs. Because of the reduced pumping capacity of the heart, blood accumulates in the heart and veins. The heart fails to maintain the flow of blood to the arterial system. This, of course, results in a fall in arterial pressure and an increase in the volume of blood in the venous system and in the heart.

Left Heart Failure Versus Right Heart Failure

Until now, we have considered the heart as a single unit. In reality, either or both ventricles can undergo failure. Failure of the right ventricle leads to an accumulation of blood in the right heart and the venous system. As blood accumulates in the veins and ultimately in the capillaries, capillary filtration increases because of the increase in capillary pressure. This leads to the accumulation of excessive fluid in the tissues in the form of edema (Fig. 19–34). The formation of edema is especially pronounced in the lower extremities because of the contributing effects of gravity.

If, on the other hand, the left heart fails, blood will accumulate in the pulmonary circulation. The formation of pulmonary edema can be particularly serious because this accumulation of fluid impairs gas exchange. This can lead to a vicious cycle; poor gas exchange will futher exaggerate the loss of ventricular contractility, which in turn will lead to further edema.

Treatment of Heart Failure

How can cardiac failure be treated medically? Obviously, if contractility could be returned to normal, the remaining symptoms would be alleviated. Administration of the drug **digitalis** tends to improve contractility by increasing intracellular calcium concentrations. The excess fluid accumulation can be

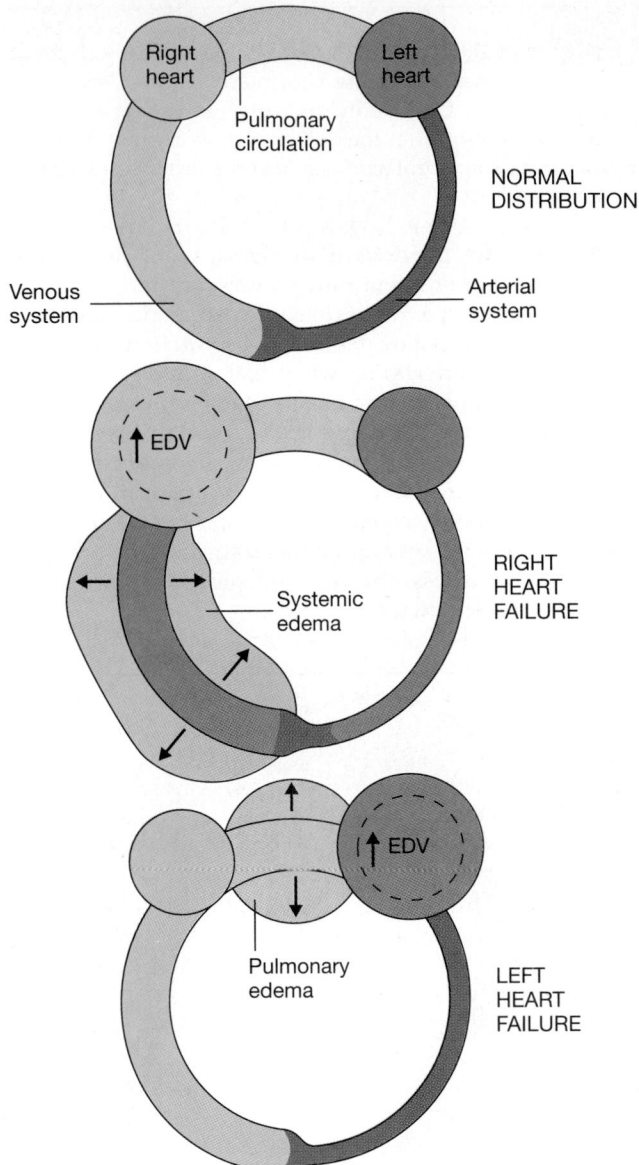

Figure 19–34

Blood volume redistribution in the cardiovascular system due to failure of either the left or right heart. The increased volume in either the pulmonary or systemic veins leads to increased capillary pressure and loss of fluid to the interstitial space (edema).

reversed by the administration of drugs that enhance the loss of sodium and water by the kidneys (i.e. antidiuretics). This latter treatment must be used with caution, since excessive reduction in volume would compromise cardiac output as a result of reduced end-diastolic volume. A more recent therapy involves the use of ACE inhibitors (angiotensin converting enzyme inhibitors) that inhibit the enzyme that converts angiotensin I to angiotensin II (see Chapter 14). At present it is not clear what the mechanism is for the beneficial effects of these drugs.

CHAPTER REVIEW

Summary

- The purpose of the vascular system is to direct blood flow throughout the body and to facilitate the exchange of materials between the blood and the interstitial fluid.
- The arteries distribute blood to the various tissues in the body.
- Blood flow = mean arterial pressure/total peripheral resistance; mean arterial pressure = 1/3 pulse pressure + diastolic pressure.
- Arterioles are the major resistance vessels. By changing their diameter (via smooth muscle contraction), they can redistribute blood flow among different tissues.
- In many tissues, blood flow is autoregulated according to demand.
- In other tissues, blood flow is regulated by the autonomic nervous system.
- Epinephrine binds to beta receptors to produce vasodilation in tissues with large numbers of beta receptors such as the heart, skeletal muscle, and the liver.
- Capillaries are thin-walled vessels that contain no smooth muscle and are optimized for diffusional exchange.
- The precapillary sphincter is a small cusp of smooth muscle at the beginning of some capillaries that frequently modulates capillary flow in an all-or-none fashion.
- Substances move between the capillary blood and the extracellular fluid by transcytosis, diffusional exchange, or bulk flow.
- Starling's hypothesis describes the balance of hydrostatic and oncotic pressures that prevent fluid loss in the capillaries.
- The veins not only serve as a conduit for the return flow of blood from the capillaries to the heart, but also provide a variable capacitance storage reservoir for blood.
- The flow of blood in the veins is facilitated by one-way valves and the pumping action of skeletal muscle contraction and respiration.
- The lymphatic vessels return any fluid or protein that leaks from the capillaries back to the circulating blood and transport absorbed fat digestion products from the gut to the blood.
- Cardiac output is matched with tissue blood flow by maintaining mean arterial blood pressure constant in much the same way that municipal water utilities match water supply with demand by maintaining water line pressure constant. Baroreceptors provide sensory input to the cardiovascular center in the medulla, which in turn alters the autonomic outflow to the heart, resistance arterioles, and capacitance veins to maintain blood pressure constant.
- Average cardiac output and venous return must be equal.
- Total blood volume affects cardiac output by changing end-diastolic volume of the ventricles.
- Hemorrhage leads to reduced blood pressure due to reduced cardiac output as a result of reduced ventricular end-diastolic volume. The normal compensatory response to hemorrhage is vasoconstriction of resistance vessels and capacitance veins and increased cardiac contractility and heart rate.
- Hypotension can also result from sudden postural changes or prolonged quiet standing. The compensatory response is vasoconstriction of resistance vessels and capacitance veins and increased cardiac contractility and heart rate.
- During exercise, the blood flow to active skeletal muscle may increase 10 to 15 times normal and cardiac output may increase to as much as 35 L/min.
- Failure of the left heart to maintain normal cardiac output leads to an accumulation of blood in the lungs which inhibits gas exchange; failure of the right heart leads to accumulation of blood in the systemic veins and edema formation. Compensatory responses include increased baroreceptor activity, increased heart rate, increased total peripheral resistance, vasoconstriction of capacitance veins, and increased retention of salt and water by the kidneys.

Review Questions

Identify Each with a Term

1. The vasodilation that occurs following occlusion of blood flow to a tissue.
2. The maintained contraction of vascular smooth muscle under resting conditions.
3. The stretch-induced contraction of vascular smooth muscle.
4. The small cuff of smooth muscle at the beginning of a capillary.
5. the pressure receptor located in the carotid sinus.

Define Each Term

6. colloidal osmotic pressure
7. mean arterial pressure
8. Starling's hypothesis
9. vasomotion
10. cardiovascular control center

Choose the Correct Answer

11. Diastolic blood pressure is:
    a. equal to the cuff pressure at which flow through the brachial artery begins.
    b. lower than capillary blood pressure.

c. equal to mean blood pressure minus systolic pressure.

d. equal to the cuff pressure at which flow through the brachial artery is continuous.

12. Which of the following causes vasoconstriction of the arterioles in the skin?
    a. decreased arterial pressure
    b. increased tissue metabolism
    c. increased concentration of CO_2 in the blood
    d. increased circulating epinephrine

13. In which of the following organs are the blood vessels *not* regulated by sympathetic innervation?
    a. skin   c. liver
    b. brain   d. gastrointestinal tract

14. A decrease in vessel radius from 10 mm to 5 mm will increase the resistance to blood flow by how much?
    a. 4 times   b. 6 times   c. 8 times   d. 16 times

15. Which of the following blood vessels is *not* surrounded by smooth muscle cells?
    a. artery   b. arteriole   c. capillary   d. venule

16. Which of the following vascular beds presents the greatest overall resistance to blood flow?
    a. aorta   b. arterioles   c. capillaries   d. veins

17. Which of the following cause reabsorption of water into the capillary?
    a. plasma proteins
    b. capillary hydrostatic pressure
    c. interstitial fluid oncotic pressure
    d. soluble electrolytes

18. In a normal individual, the balance of filtration and absorption forces is such that over a 24-hour period:
    a. filtration of water exactly equals absorption of water.
    b. there is *no* net transfer of water across the capillary wall.
    c. there is net absorption of water.
    d. there is net filtration of water.

19. On average:
    a. left heart output is slightly greater than right heart output.
    b. systemic blood flow exceeds pulmonary blood flow.
    c. cardiac output equals venous return.
    d. resistance to blood flow is greater in the pulmonary circulation than in the systemic circulation.

20. An increase in arterial blood pressure causes which of the following?
    a. a decrease in baroreceptor afferent nerve activity
    b. a decrease in efferent parasympathetic nerve activity from the medullary cardiovascular control center
    c. an increase in total peripheral resistance

d. a decrease in sympathetic nerve activity to the resistance arterioles

Calculate

21. An individual's blood pressure is measured as 120/90 mm Hg. What is the mean arterial pressure?

22. Mean arterial blood pressure is 100 mm Hg and total peripheral resistance is 20 mm Hg $\cdot$ min $\cdot$ L^{-1}. What is the total blood flow?

23. Capillary hydrostatic pressure is 35 mm Hg, capillary colloidal osmotic pressure is 28 mm Hg, and interstitial fluid colloidal osmotic pressure is 3 mm Hg. Calculate the net filtration or absorption pressure.

24. Pulse pressure is 50 mm Hg and systolic pressure is 120 mm Hg. What is the diastolic pressure?

Answer Each Question in One or Two Sentences

25. How do the elastic properties of the arterial blood vessels contribute to the continuous flow of blood through the capillaries?

26. What determines the distribution of blood flow among the various tissue beds in the body?

27. How does myogenic activity of vascular smooth muscle minimize the changes in tissue blood flow in response to changes in arterial pressure?

28. Exercise and hemorrhage both stimulate increased heart rate and increased contractility. Why is cardiac output much lower following hemorrhage than during exercise?

29. What cardiovascular changes occur following heart failure to compensate for the loss of contractility?

Problem Solving in Physiology

30. Cardiac transplants are becoming increasingly more common. One apparent dilemma that presents itself in this situation is that the transplanted heart has no nervous innervation. Are these hearts able to respond appropriately to exercise? Would an individual with a transplanted heart have problems with orthostatic hypotension?

31. In a standing individual, the mean blood pressure at the level of the aortic arch might be 90 mm Hg, whereas, the pressure measured in an artery in the foot could be as high as 170 mm Hg. If blood flow through blood vessels results in a pressure drop that is proportional to the resistance to blood flow, how is it that blood pressure increases in the feet? Is more pressure required to move blood through the systemic circulation in a standing position compared to a supine position? Will the cardiac output be the same in both the standing and supine positions?

Suggested Readings

Berne, Robert M., and Levy, Matthew N. *Physiology*, 3rd ed. St. Louis: C.V. Mosby Company, 1992.

Eisenberg, M.S., Bergner, L., Hallstrom, A.P., and Cummins, R.O. "Sudden cardiac death." *Scientific American*, 254(5):37–43, 1986.

Goerke, Jon, and Mines, Allan H. *Cardiovascular Physiology*. New York: Raven Press, 1988.

Goldstein, G.W., and Betz, A.L. "The blood-brain barrier." *Scientific American*, 255(3):74–83, 1986.

Grady, P. "Can heart disease be reversed?" *Discover*, March 1987, p. 54.

Hall, E.D., and Hooper, C. *Free Radicals: Research on Biochemical Bad Boys Comes of Age*, J.NIH Res., 1:101–106, 1989.

Honig, C.R. *Modern Cardiovascular Physiology*, 2nd ed. Boston: Little, Brown and Co., 1988.

Levy, R.I., and Moskowitz, J. "Cardiovascular research: Decades of progress, a decade of promise." *Science*, 217:121–129, 1982.

Little, Robert C., and Little, William, C. *Physiology of the Heart and Circulation*, 4th ed. Chicago: Year Book Medical Publishers, 1988.

Mukerji, B., Alpert M.A., and Mukerji, V. "Cardiovascular changes in athletes." *American Family Physician*, 40:169, 1989.

Ross, R. "The pathogenesis of atherosclerosis: A perspective for the 1990s." *Nature*, 362:801–809, 1993.

Key Terms

angina pectoris (p. 582)
arterial pressure (p. 571)
arterioles (p. 580)
atherosclerotic plaque (p. 582)
atrial natriuretic factor (p. 588)
autoregulation of blood flow (p. 585)
baroreceptor (p. 603)
cardiovascular control center (p. 604)

carotid sinus (p. 603)
circulatory shock (p. 609)
congestive heart failure (p. 612)
diastolic pressure (p. 578)
edema (p. 598)
endothelium-derived relaxing factor (EDRF) (p. 587)
filtration pressure (p. 595)
hemorrhage (p. 606)
hypotension (p. 609)

infarct (p. 582)
mean arterial pressure (MAP) (p. 579)
microcirculation (p. 589)
precapillary sphincter (p. 590)
pulse pressure (p. 578)
Starling's hypothesis (p. 593)
systolic pressure (p. 576)
tone (p. 585)

total peripheral resistance (TPR) (p. 581)
transcytosis (p. 591)
vasoconstrictor (p. 589)
vasodilator (p. 585)
vasomotion (p. 590)
venous capacitance (p. 596)
venous return (p. 601)
venules (p. 596)

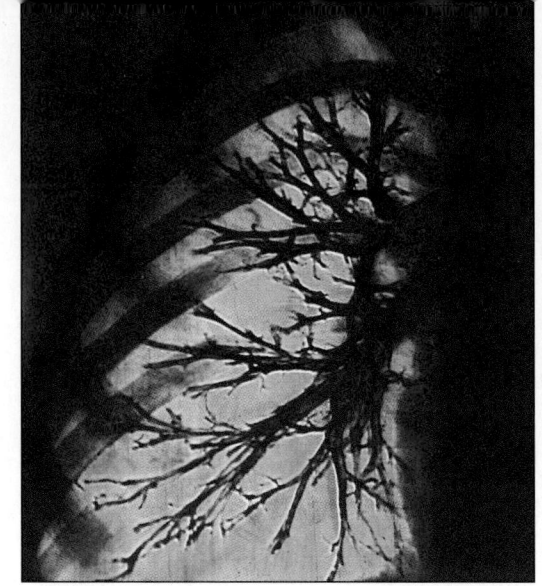

A false-color arteriograph of a normal right lung showing the pattern of bronchial and bronchiolar branching.
(© CNR/SPL/Photo Researchers, Inc.)

Chapter 20
Respiration

Key Concepts

The lungs are organs of gas exchange, and the architectural design is constrained by the principles of gas diffusion. The site of gas exchange are the 300 to 500 million alveoli.

Ventilation, the process of moving air in and out of the lungs, optimizes the gas composition in the alveoli for gas exchange. Not all of the air that enters the lungs reaches the alveoli, and some of the ventilation is wasted.

The lungs are highly elastic, and a pressure change is required across them before they can become inflated. Work is required to expand the lungs and chest wall. A pressure-volume curve is used to assess the elastic properties of the lung. Loss of elastic properties severely hampers the efficiency of ventilation.

Chapter Outline

The Structure of the Lung

Lung Pressures and Ventilation

Mechanical Properties of the Lung

Jonathan is a 12-year-old boy who has increased shortness of breath when playing hockey. This has become such a problem that he tells his coach he would prefer to play goalie instead of center, much to his father's disappointment. Jonathan has had several colds this winter with frequent coughing in class, which resulted in his being sent home with his cold for a couple of days. The coughing started with the beginning of hockey practice. Recently, the coughing occurs at night, awakening his parents; Jonathan feels better after he has a glass of water. In April, during hockey practice, Jonathan had a very serious episode of shortness of breath that prompted his coach to call his parents.

After a series of tests on Jonathan's lungs at the hospital he is diagnosed with asthma—which simply means shortness of breath and comes from the Greek word for panting. Asthma is one of several disorders that can affect our breathing. In this chapter we will examine how the lungs work: the mechanical properties of breathing, how pressures are generated to expand the lung, and how lung function is measured under normal and abnormal conditions such as asthma. Finally, we will examine airflow and the work of breathing, and see how work is dramatically altered with conditions such as asthma.

In our coverage of respiration, two unifying themes emerge. First, the architecture of the lung is constrained by the principles of gas diffusion. Second, the lungs have a tremendous ability to supply O_2 and to remove CO_2 from the blood, and this ability for efficient gas exchange has played an important role in the success and extraordinary adaptability in the mammalian species.

The process of exchanging oxygen and carbon dioxide between an organism and its environment is known as **respiration.** Respiration takes place in two stages. The first, known as **gas exchange,** is the process of bringing air from the external environment to the cells. More specifically, gas exchange involves the steps of bringing oxygen from the external environment and transporting it to the cell while carbon dioxide is carried from the cell to the external environment. Gas exchange involves three steps and includes (1) *ventialion*—moving air in and out of the lungs, (2) gas diffusion and uptake, and (3) gas transport by the blood. The second stage of respiration is known as **cellular respiration,** which consists of a series of complex metabolic reactions that break down molecules of fuel, releasing carbon dioxide and energy. Driving the body's chemical reactions, especially those of muscular contraction, requires large amounts of energy, and little energy is available until the food we eat is digested and **oxidized** (burned with oxygen). Oxidation occurs when

the lungs extract oxygen from the air and the blood transports that oxygen to the cells, where it can take part in cellular respiration. Recall from Chapter 6 that oxygen is required in the final step of cellular respiration to serve as an electron acceptor in the process by which cells obtain energy. The mammalian respiratory system is so efficiently designed to supply oxygen and remove carbon dioxide that gas exchange and transport rarely limit our activity.

A review of the human respiratory system can be divided into three main sections: (1) ventilation, and the mechanics of breathing, (2) gas transfer and transport, and (3) the control of breathing. This chapter discusses ventilation and the "pumping machinery" —that is, the physical design of the lung, air movement in and out of the lungs, mechanical properties, and the work of breathing. Chapter 21 primarily treats gas uptake, transport, and the control of breathing—that is, circulatory design, gas diffusion between the lung and blood, transport of respiratory gases by the blood to the cells, and how the control of breathing is connected to our metabolic as well as our voluntary activities.

The Structure of the Lung

Movement of gases in and out of cells occurs by simple diffusion. In unicellular organisms, the process by which oxygen and carbon dioxide diffuse between the cell and the surrounding environment is very simple (Fig. 20–1*a*). No special respiratory structures are

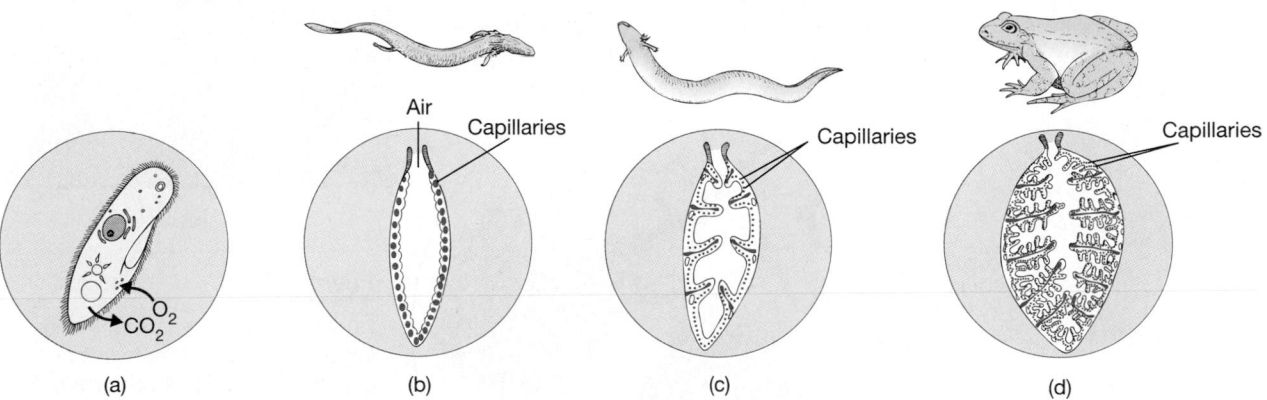

Figure 20–1

Structural design for gas exchange. (*a*) In unicellular organisms, no special organ is required. In higher animals, lungs serve as the gas exchange organ. Note the development of the multialveolar lung, which increases the internal surface area. (*b*) The uni-alveolar lung of the European olm (Proteus). (*c*) The multi-alveolar lung of the mud eel (Siren) with simple septa. (*d*) The multi-alveolar lung of the frog (Rana) with complex septa. Red dots in walls or septa represent pulmonary capillaries.

required. In more complex organisms, however, cells deep within the body cannot exchange enough oxygen and carbon dioxide to the environment because the diffusion distance is too great. As a result, special gas exchange organs have developed. These include tracheal tubes in insects, gills in fish, and lungs in higher animals such as humans. All of the respiratory organs mentioned—from the simplest to the most complex—have three features in common. First, they all have very thin walls so that gas diffusion can readily occur. Second, they are always moist so that oxygen and carbon dioxide can dissolve in fluids. Third, they are richly supplied with blood vessels to ensure efficient gas exchange and transport of the respiratory gases to the cells.

Lung Architecture Is Constrained by the Principles of Diffusion

As the metabolic demands increased, with larger and more complex organisms, lungs began to have a greater internal surface area for the exchange of oxygen and carbon dioxide. This was accomplished by an increase in lung size and by the formation of **septa,** or out-pockets, in the lung, which allow for more blood to flow through the lung and more air to be exposed to the blood (see Fig. 20–1*b, c, d*). Even the most primitive lung, such as that of the salamander (consisting of only a simple gas bag with numerous septa containing blood in the walls) represented a quantum jump in efficiency of gas exchange. The frog lung (Fig. 20–1*d*) has more internal surface area, which permits more blood to come into contact with air for gas exchange.

In the mammalian lungs, these septa evolved into microscopic air sacs called **alveoli** (singular, *alveolus*) (Fig. 20–2*a, b*). Each alveolus is surrounded by a network of capillaries that bring blood into close contact with air inside the alveolus. Oxygen diffuses across the thin walls of the alveolus into the blood, while carbon dioxide diffuses from the blood into the alveolus. The human lung contains 300 to 500 million alveoli with an internal surface area about the size of a tennis court, which is roughly 40 times greater than the external body surface area.

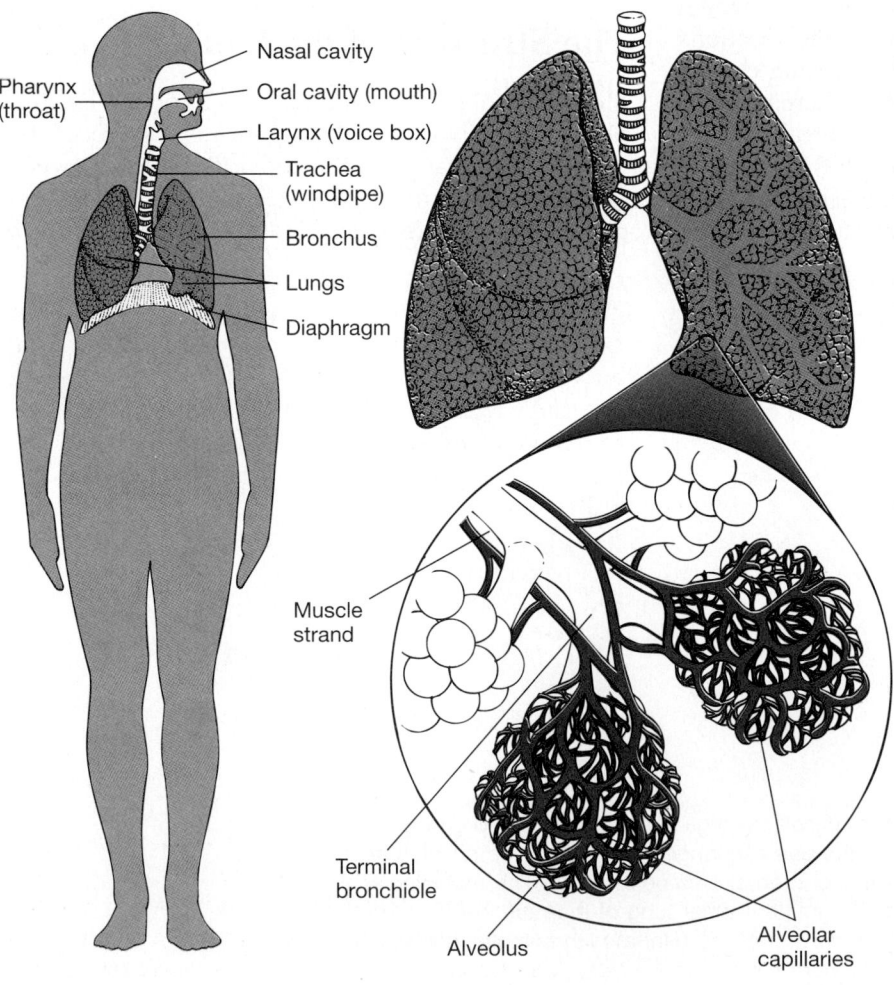

Nasal cavity
Pharynx (throat)
Oral cavity (mouth)
Larynx (voice box)
Trachea (windpipe)
Bronchus
Lungs
Diaphragm

Muscle strand

Terminal bronchiole

Alveolus

Alveolar capillaries

(a)

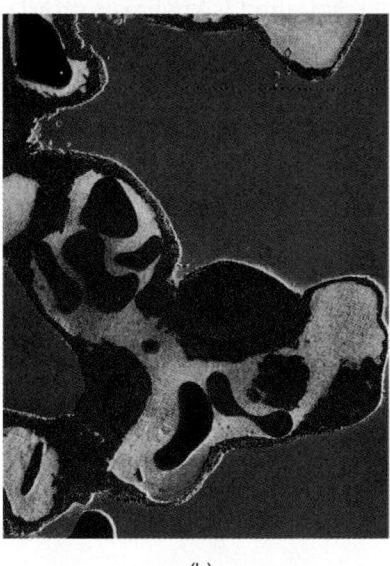

(b)

Figure 20–2

(*a*) Human lung with each alveolus enmeshed in a network of capillaries. (*b*) SEM of normal lung tissue showing an alveolar septum and the partition between the alveoli where gas exchange takes place. The capillaries are colored yellow in this image and are filled with red blood cells. (×2200.) *(© SECCHI-LECAQUE-ROUSSEL-UCLAF/CNRI/SPL/ Photo Researchers, Inc.)*

A marina can provide a simple analogy for the creation of a larger surface area. To dock more boats without enlarging the marina, smaller piers must be built. In nature, this strategy is followed as well. In species with high metabolic rates, the demand for gas exchange is also high. Accordingly, smaller and smaller alveoli are packed together in the lung, thereby increasing the internal surface area for more gas exchange. Figure 20–3, in which metabolic rate is plotted against alveolar size (diameter), shows this graphically. The manatee (sea cow), which has a slow metabolic rate, has large alveoli. The shrew and bat, on the other hand, are like metabolic blast furnaces and have very small alveolar diameters.

The Airway Tree Divides Repeatedly to Enhance Airflow

The human gas exchange consists of two lungs, each divided into several lobes (Fig. 20–4). The lungs comprise two treelike structures, the **vascular tree** and the **airway tree,** which are embedded in elastic tissue. The vascular tree consists of arteries and veins tied together by capillaries. (Details of the vascular tree will be discussed in Chapter 21.) The airway tree consists of a series of highly branched hollow tubes that decrease in diameter and become more numerous at each branching. The main airway (the **trachea**), in turn, branches into two **bronchi** (singular, *bronchus*), one of which enters each lung. Within each lung, these bronchi branch many times into progressively

smaller bronchi, which in turn branch into terminal **bronchioles** analogous to twigs of a tree. The trachea, bronchi, and bronchioles are part of the **conducting** zone that carries gas to and from alveoli (Fig. 20–5). The respiratory bronchioles and alveoli constitute the **respiratory zone** and are analogous to leaves on a tree, having a similar function as exchange units between the organism and the environment.

The conducting units (trachea, bronchi, and bronchioles) have three important functions: (1) to warm and humidify the air, (2) to distribute air to the gas exchange surface of the lung, and (3) to serve as part of the body defense system. Since the inside of the lung is literally exposed to the external environment, the conducting unit contains a **mucociliary transport system** that keeps microorganisms, dust particles, and noxious gases from entering the alveoli (Fig. 20–6). The conducting units are lined with cilia that beat synchronously. These cilia are buried in a carpet of sticky mucus that traps foreign material. The mucoloculary carpet "floats" on top of the beating cilia. For example, particles $>10\mu$m are trapped on the hairs and moist mucus lining of the nasal passage. However, small particles ($<10\mu$m) get inhaled into the lungs. At every bend in the airway these small particles fail to make the turn and stick on the mucus film. The mucus is "streamed" toward the throat by the beating cilia and can be swallowed or coughed up when its presence is noticed as it stimulates receptors in the throat.

The cilia are greatly affected by irritant gases like tobacco smoke. When they come in contact with irritants, they become partially paralyzed, slowing the mucociliary transport system. As a result, microorganisms and excess mucus accumulate in the lower airways (e.g., the bronchioles). Often, the excess mucus plugs the lower airways, and a cough is required to dislodge and remove it. If excess mucus accumulates over a long period of time, bacterial infection will often develop. When this occurs, the mucus is no longer a light yellow but instead is green as a result of the infiltration and rupturing of white blood cells.

Figure 20–3

Alveolar diameter (in micrometers) plotted against resting metabolic rate (oxygen consumption in milliliters per unit body weight per unit time) in mammalian lungs of different species. Note the inverse relationship between alveolar diameter and metabolic rate.

Lung Pressures and Ventilation

The lungs are housed in an airtight chest cavity, the **thoracic cavity,** and are separated from the abdomen by a large, dome-shaped muscle, the **diaphragm** (Fig. 20–7). The thoracic cage is made up of 12 pairs of **ribs,** a **sternum** (breast bone), and a set of internal and external intercostal muscles that lie between the ribs. The rib cage is hinged to the vertebral column

(text continues on page 621)

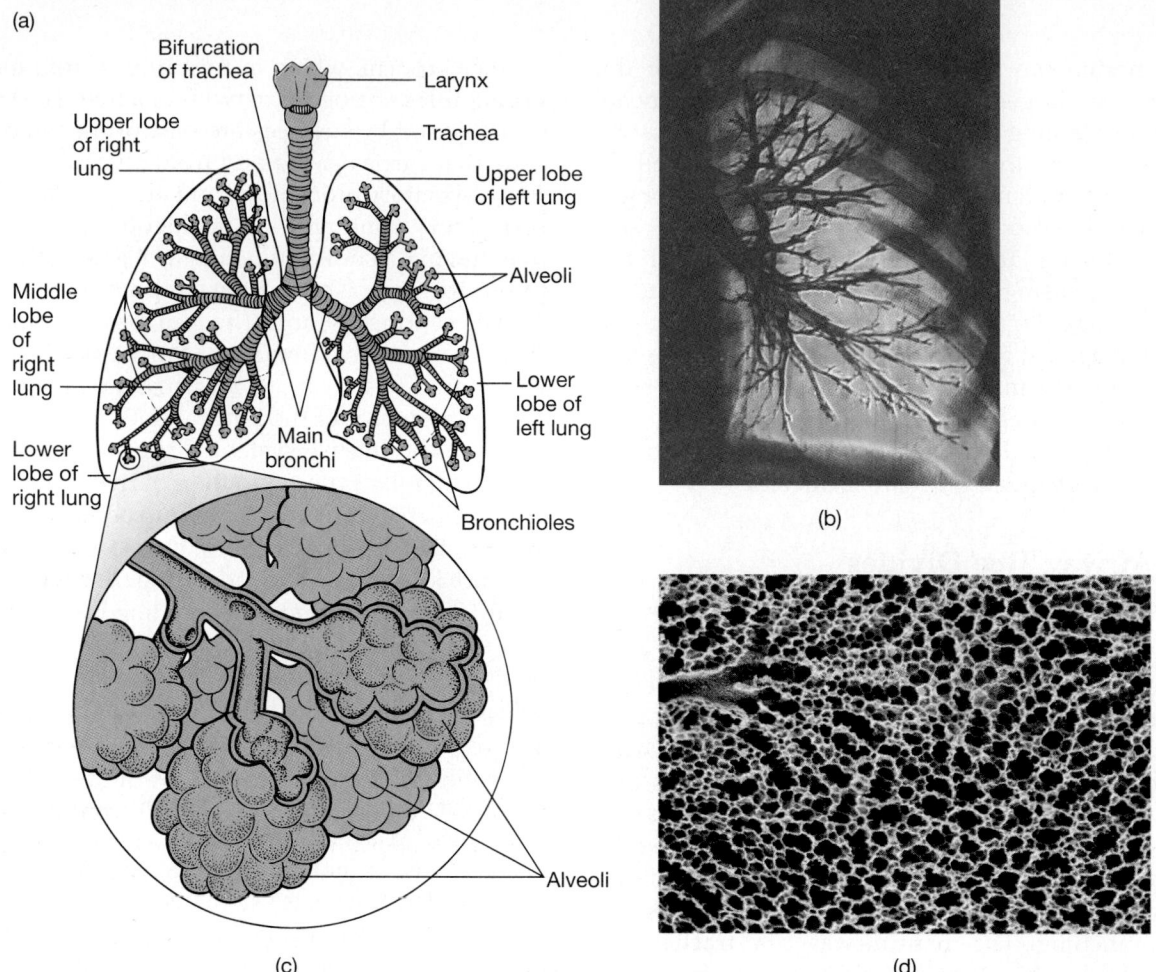

Figure 20–4

(*a*) Organization of the human lung. The right lung consists of three lobes (upper, middle, and lower) and the left lung consists of two lobes (upper and lower). (*b*) False-color arteriograph of a normal right lung showing the pattern of bronchial and bronchiolar branching.   (© CNRI/SPL/Photo Researchers, Inc.) (*c*) The trachea subdivides to form bronchi, which branch repeatedly, leading to bronchioles and then to alveoli. Note that alveoli are not all the same size. (*d*) SEM of a normal human lung with numerous alveoli arising from two alveolar ducts.   (© Fawcett/Gehr/Science Source/Photo Researchers, Inc.)

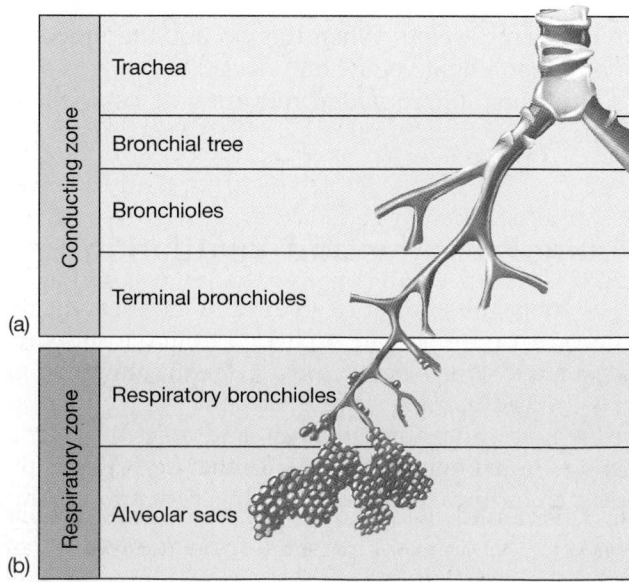

Figure 20–5

Tree-like pattern of airway, showing the conducting zone and the respiratory zone. The conducting zone (*a*) is illustrated from the trachea to the terminal bronchioles, and the respiratory zone (*b*) is illustrated from the respiratory bronchioles to the terminal bronchioles.

Figure 20–6

Mucociliary transport system. Mucus layer traps inhaled particles. The mucus layer is propelled by cilia toward the trachea to remove bacteria and particulate matter.

(backbone), allowing it to rise and lower during breathing. The space between the lungs and chest wall is the **pleural space,** which contains fluid and serves as a lubricant so that the lungs can slide against the chest wall.

Respiratory Muscles Affect Lung Pressures

Inflation of the lungs causes air to rush in **(inspiration)** and is the result of an active muscular process. Consider for a moment the breathing cycle. The main muscle during breathing is the **diaphragm.** The lungs inflate automatically when the diaphragm contracts and enlarges the thoracic cavity. This is accomplished two ways. First, when the diaphragm (which is attached to the lower ribs and sternum) contracts, the abdominal contents are pushed down, thus enlarging the pleural space. As the diaphragm pushes down on the abdominal contents, the rib cage is also pushed outward, further enlarging the pleural space. The increase in pleural space volume causes a decrease in pressure in the pleural space, which in turn causes the lung to expand and fill with air. This key pressure-volume relationship in breathing is based on **Boyle's Law** which states that at a constant temperature the volume (V) of any quantity of gas varies inversely as the pressure (P) of the gas. Boyle's law may be expressed as:

$$V \; \alpha \; \frac{1}{P}.$$

Muscular Movements During Inspiration

During normal inspiration the diaphragm is the only respiratory muscle used. However, during forced inspiration, in which a large volume of air is taken, the **external intercostal** muscles (see Fig. 20–7) are used as a second set of inspiratory muscles. These muscles also pull the rib cage upward and outward. To see how this works, stand sideways in front of a mirror and take a deep breath. The chest wall moves upward while the sternum moves outward, thereby enlarging the thoracic cavity.

The last group of muscles used in inspiration are the **accessory muscles.** These include the neck and chest muscles and are brought into play only during deep and heavy breathing, such as during intense exercise.

Muscular Movements During Expiration

Breathing out **(expiration)** is a much simpler process. At the end of a normal inspiration, the diaphragm relaxes, the thoracic cavity decreases, and the lungs deflate. Again, the lung is passive in breathing. Although expiration is purely passive during normal breathing, with exercise or with forced expiration the expiratory muscles become active. The expiratory muscles are the muscles of the **abdominal walls** and, to a lesser extent, the **internal intercostal muscles** (see Fig. 20–7). Contraction of the abdominal muscles helps push the diaphragm up, and the internal intercostal muscles pull the rib cage down. These expiratory muscles are important and necessary for coughing and for endurance running. This is one of the reasons that competitive long-distance runners do exercises to strengthen their abdominal muscles as part of their training.

Lung Pressures During the Breathing Cycle

In Chapter 18, which discusses the cardiovascular system, the units for pressure were millimeters of mercury (mm Hg). In respiration, both mm Hg and

(a)

(b)

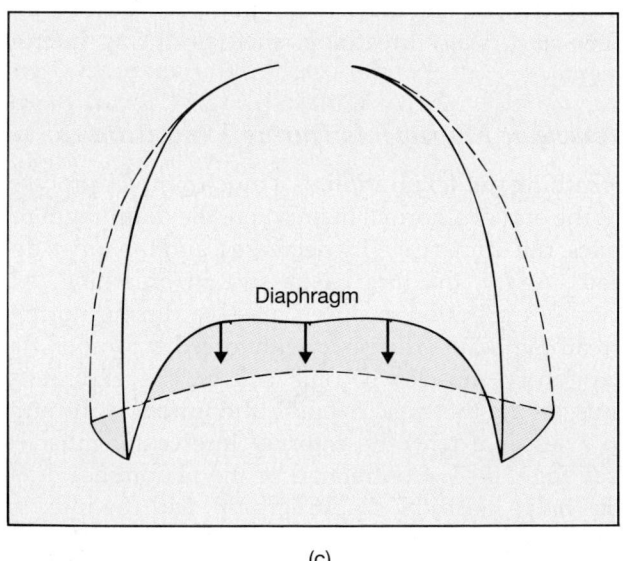

(c)

Figure 20–7

Functions of the rib cage and respiratory muscles during breathing. (*a*) During inspiration, the diaphragm contracts and the rib cage is elevated, enlarging the thoracic cavity by increasing both vertical and the anterior-posterior (A-P) diameter. During forced inspiration the external intercostals are also used. (*b*) During expiration, the diaphragm relaxes and the rib cage drops, thereby decreasing the space in the thoracic cavity. The internal intercostal muscles are also used with forced expiration. (*c*) The lever action of the diaphragm on abdominal contents pushes the rib cage up and outward during inspiration.

centimeters of water (cm H_2O) are used. The reason cm H_2O is convenient is because of the small pressure changes involved. The following conversion equations are used: 1 cm = 0.74 mm Hg; 1 mm Hg = 1.36 cm H_2O. The atmosphere, a sea of air we breathe and live in, exerts a pressure (P) at sea level known as **barometric pressure** (P_B) or atmospheric pressure

and is equal to 760 mm Hg (Fig. 20–8). The total pressure or barometric pressure is the sum of individual **partial pressure** of gas in the atmosphere. Partial pressure is the pressure of each gas in the atmosphere (e.g. O_2, N_2, CO_2). Thus, P_B can be written as:

$$P_B = P_x + P_y + P_z.$$

Figure 20–8

At sea level, atmospheric pressure can push a column of mercury (Hg) to a height of 76 cm. This is also known as 1 atmosphere of pressure (760 mm Hg).

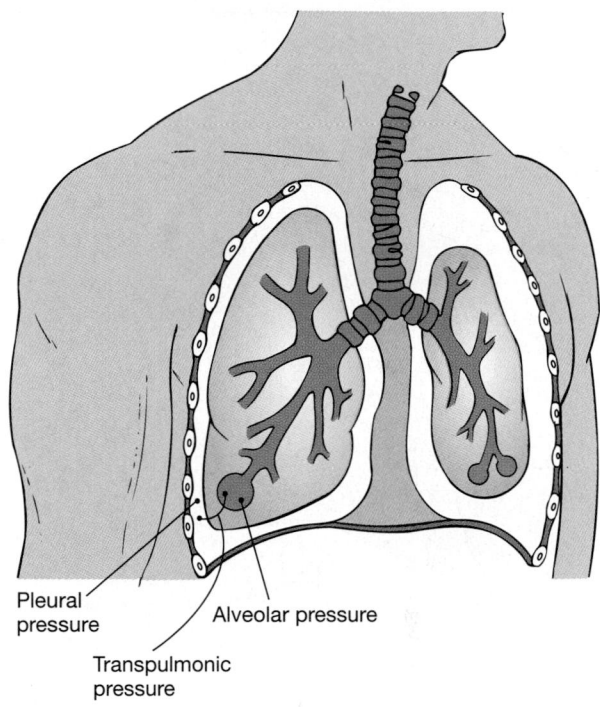

Figure 20–9

The basic pressures involved in breathing are atmospheric pressure, transpulmonic pressure, pleural pressure, and alveolar pressure.

when P_B is barometric pressure or total pressure and P_x, P_y, P_z are partial pressures of the individual gases. Sometimes partial pressure of a gas is referred to as **gas tension** and the partial pressure or gas tension of any gas can be determined by measuring barometric pressure and the fractional concentration (percent of total) of the gas. For example the partial pressure of oxygen (P_{O_2}) is determined as follows:

$$P_{O_2} = P_B \cdot F_{O_2}$$

where P_{O_2} is the partial pressure of O_2 in mm Hg, P_B is barometric pressure in mm Hg, and F_{O_2} is the fractional concentration of O_2. At sea level the partial pressure of O_2 is 160 mm Hg since O_2 is 21% (F_{O_2} = 0.21) of the atmospheric air (760 mm Hg × 0.21).

Respiratory physiologists also commonly measure lung pressures relative to atmospheric pressure, or **relative pressure.** The unit for relative pressure is cm H_2O, and is the pressure relative to the atmospheric pressure. For example, in the middle of expiration, the pressure inside the alveoli (alveolar pressure) is 5 cm H_2O. This means the pressure is 5 cm H_2O *above* atmospheric pressure. Conversely, in the middle of inspiration alveolar pressure is −2 cm H_2O.

The negative sign does not indicate that the pressure is negative but rather that alveolar pressure is 2 cm *below* atmospheric pressure. It is important to remember that when relative pressures are used, atmospheric pressure is set at zero. Therefore, if alveolar pressure is zero, this means that alveolar pressure equals atmospheric pressure. So, unless otherwise specified, the pressures of breathing are relative pressures, and the units are cm H_2O.

Three important relative pressures involved in breathing are associated with lung inflation and air flow (Fig. 20–9). First is **pleural pressure,** which is the pressure in the pleural fluid between the lung and chest wall. The second, **alveolar pressure,** is the pressure inside the alveoli. The third, **transpulmonary pressure,** is the pressure gradient across the lung wall. Transpulmonary pressure (P_{TM}) is measured by subtracting pleural pressure (P_{pl}) from alveolar pressure (P_A). Thus, transpulmonary pressure ($P_{TP} = P_A - P_{pl}$) is what keeps the lungs from collapsing and is also responsible for distending (stretching) the lung during inflation. For that reason, transpulmonary pressure is often called the **distending pressure.** One way of remembering transpulmonary pressure is "In minus Out," where P_A is the pressure inside the lung and P_{pl} is the pressure outside the lung.

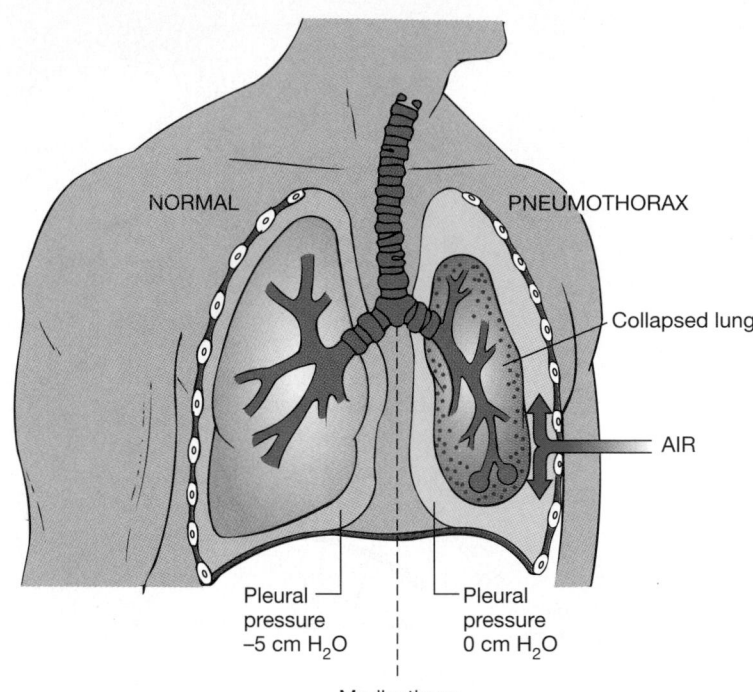

NORMAL

PNEUMOTHORAX

Collapsed lung

AIR

Pleural
pressure
−5 cm H₂O

Pleural
pressure
0 cm H₂O

Mediastinum

Figure 20–10

(*a*) Pneumothorax. When air enters the pleural space through a hole in the chest wall or the lung, the pleural pressure equals atmospheric pressure (zero), and the lung collapses. Note that the mediastinum membrane prevents the other lung from collapsing during a pneumothorax. (*b*) X-ray of male lung shortly after the development of spontaneous pneumothorax. The opaque area in the left lung represents air-filled region and the absence of lung tissue. (From: *Organ Physiology: Structure and Function of the Lung*, Fraser and Paré, W.B. Saunders, 1977).

A key question is: why is pleural pressure negative or subatmospheric? The reason is that the lung and chest wall are both **elastic** (i.e., capable of being stretched). At the end of expiration, the lung and chest wall are stretched in equal but opposite directions. Consequently, the elastic lung has the potential to recoil inwardly, and the elastic chest wall has the potential to recoil outwardly. These opposing forces cause the pleural pressure to be below atmospheric pressure. During quiet breathing, pleural pressure is negative and becomes more negative with deep inspiration. Only during forced expiration does pleural pressure become positive.

Since the pleural pressure is subatmospheric, any time the chest wall or the lung is punctured, air rushes into the pleural cavity. As a result, the pressure difference across the lung is eliminated (transpulmonary pressure equals zero), and the stretched lung collapses. When atmospheric air enters the pleural fluid the lungs collapse. This condition is known as a **pneumothorax** (Fig. 20–10), and can occur when the chest wall is punctured, as with a knife or gunshot wound, or when the lung is ruptured, as in an abscess and severe coughing.

The pressure-volume relationships can be followed during a normal breathing cycle (Fig. 20–11). At the end of tidal volume (end of expiration), the respiratory muscles are relaxed and there is no air flow. At this point, alveolar pressure is zero (equal to atmospheric pressure) and pleural pressure is −5 cm H₂O. Transpulmonary pressure at the end of a normal tidal volume would therefore be 5 cm H₂O [0 − (−5) = 5]. Remember that the transpulmonary pressure is always positive in normal breathing and is the pressure difference across the lung wall (P_{A-Ppl}). The positive transpulmonic pressure keeps the lungs inflated, and the more positive the transpulmonary pressure becomes, the more the lung inflates.

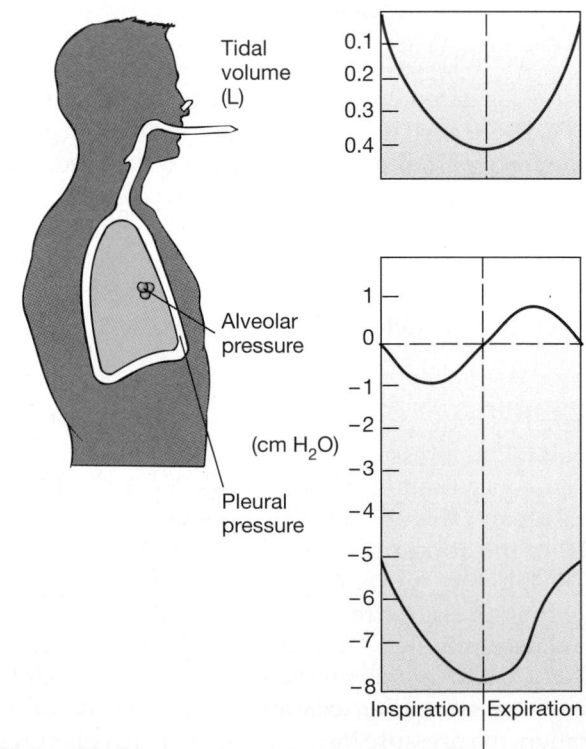

Tidal
volume
(L)

0.1
0.2
0.3
0.4

Alveolar
pressure

1
0
−1
−2
−3
−4
−5
−6
−7
−8

(cm H₂O)

Pleural
pressure

Inspiration | Expiration

Figure 20–11

The relationship between pleural pressure, alveolar pressure, and tidal volume during a breathing cycle.

Changes in Lung Pressures During Inflation

Inspiration is initiated by contraction of the dome-shaped diaphragm. The diaphragm moves down and at the same time pushes the rib cage out, this process enlarges the thoracic cavity, and causes the pleural pressure to become even more negative (subatmospheric). As the lung inflates, the alveoli are enlarged causing alveolar pressure to drop below atmospheric pressure. This produces a pressure difference or gradient between the mouth and alveoli ($P_{atm} - P_A$), which causes air to flow into the lungs down to the alveoli. Air flow stops at the end of inspiration because alveolar pressure again equals atmospheric pressure (left side of Fig. 20–11).

Air flow is similar to blood flow in that air moves from a point of high pressure to one of low pressure, and flow is proportional to the pressure difference between two points (P_{atm-PA}). Thus, the greater the contraction of inspiratory muscles, the more the thoracic cavity increases, the more negative the pleural pressure becomes, the greater the lung inflation and the greater the volume of air brought into the lungs. The sequence of events in inspiration is as follows:

1. The inspiratory muscles contract.
2. The thoracic cavity expands.
3. Pleural pressure becomes more negative.
4. Transpulmonary pressure becomes more positive.
5. The lung inflates.
6. Alveolar pressure becomes subatmospheric.
7. Air flows into the lungs.

Changes in Lung Pressures During Expiration

During expiration, the inspiratory muscles relax, pleural pressure becomes less negative, transpulmonary pressure decreases, and the lungs deflate, which is a reversal of inspiration (right side of Fig. 20–11). As the lung deflates, alveoli decrease in size. Consequently, alveolar pressure increases and exceeds atmospheric pressure, thus causing air to flow out of the alveoli through the airways. Remember, with breathing under resting conditions, expiration is completely passive (muscles relax and the stretched lung recoils back to normal). It is only during forced expiration (e.g., exercise) that the expiratory intercostal and abdominal muscles contract and actively compress the thoracic cavity. The sequence of events in expiration is as follows:

1. Respiratory muscles relax.
2. The thoracic cavity decreases in size.
3. Pleural pressure becomes less negative.
4. Transpulmonary pressure decreases.

5. Lung size decreases.
6. Alveolar pressure becomes greater than atmospheric pressure.
7. Air flows out of the lung.

Typical pleural and alveolar pressures during a normal breathing cycle are shown in Table 20–1. Note from the values presented in Table 20–1 and Figure 20–11 that it takes only a small pressure gradient to move a typical volume (500 ml) in and out of a normal lung.

Lung Volumes and Airflow Are Measured by Spirometry

The volume of air breathed in and out of the lungs is measured by a **spirometer** (Fig. 20–12). This device is a simple volume recorder consisting of a double-walled cylinder in which an inverted bell is immersed in water to form a seal. The bell is attached by a pulley to a pen that writes on a rotating drum. As air enters the spirometer from the lungs, the bell rises. Because of the pulley arrangement, the pen is lowered. Thus, a downward pen deflection represent expiration, and an upward pen deflection represents inspiration. The recording is known as a **spriogram,** which measures two parameters: (1) airflow and (2) the volume of air moved in and out of the lungs.

The volume of air entering or leaving the lungs during a single breath is called **tidal volume** (V_T). Under resting conditions, this volume is approximately 500 ml (see Fig. 20–12). The resting tidal volume rep-

Table 20–1
Pressure Changes During a Normal Breathing Cycle

	Pleural Pressure (cm H_2O)	Alveolar Pressure (cm H_2O)
End expiration	−5	0
Mid inspiration	−7	−1
End inspiration	−8	0
Mid expiration	−6	+1
End expiration	−5	0

Figure 20–12

Lung volumes and capacities measured by a spirometer. With inspiration the pen shows an upward deflection on the spirograph, and with expiration a downward deflection. Note that the functional residual capacity and residual volume cannot be measured directly with a spirometer.

resents only a fraction of the total air that the lungs can bring in.

The volume of air remaining in the lungs at the end of a normal tidal volume (end of expiration) is referred to as the **functional residual capacity (FRC)** and amounts to about 2.4 L. Note the use of "volume" in the first term and "capacity" in the second. When referring to the subdivisions of the lung, the terms *volume* and *capacity* are both used. **Capacity** is used when a volume can be broken down into two or more smaller volumes; for example, **expiratory reserve volume (ERV)** and **residual volume (RV)** = FRC. These lung volumes and capacities are summarized in Table 20–2.

Forced Vital Capacity

One of the most useful tests to assess the overall ability to move air in and out of the lungs (ventilation) is called **Forced Vital Capacity (FVC).** This is the maximum amount of air that can be forcefully and rapidly exhaled after a deep breath (maximal inspiration) and measures approximately 5 liters for an adult male (Fig. 20–13). For example, a well-trained athlete can have a forced vital capacity of more than 7.0 L, whereas an asthmatic may have a forced vital capacity no greater than 3 L. The constricted bronchioles of asthmatics can collapse during forced ex-

piration, trapping air in the lungs and leading to a decrease in forced vital capacity. A similar problem exists with individuals who smoke or have **bronchitis,** a condition in which the lining of the bronchioles becomes inflamed and swollen, thereby reducing airway diameter.

Figure 20–13

Spirogram of forced vital capacity (FVC). The subject inspires maximally to total lung capacity and then exhales forcefully and as completely as possible into a spirometer. From forced vital capacity another measurement can be obtained, i.e. the forced expiratory volume in one second (FEV_1). Both FVC and FEV_1 are measured in liters.

Table 20–2

Definitions of Standard Lung Volumes, Capacities, and Pulmonary Function Tests for a Typical Young Adult

Term	Value	Definition
V_T (Tidal Volume)	0.5 liter	Volume of air inhaled or exhaled with each breath
VC (Vital Capacity)	4.8 liters	Volume of air that can be exhaled after maximal inspiration (VC = IRV + V_T + ERV)
RV (Residual Volume)	1.2 liters	Volume of air remaining in the lungs after maximal expiration
ERV (Expiratory Reserve Volume)	1.2 liters	Maximal volume of air that can be exhaled at the end of a tidal volume
IRV (Inspiratory Reserve Volume)	3.1 liters	Volume of air that can be inhaled at the end of a normal inspiration
FVC (Forced Vital Capacity)	4.8 liters	Volume of air that can be exhaled as forcibly and rapidly as possible after maximal inspiration
IC (Inspiratory Capacity)	3.6 liters	Maximal volume of air that can be inhaled after a normal expiration (IC = IRV + V_T)
FRC (Functional Residual Capacity)	2.4 liters	Volume of air remaining in the lungs at the end of a tidal volume (FRC = ERV + RV)
TLC (Total Lung Capacity)	6.0 liters	Total volume of air in the lungs after maximal inspiration (TLC = VC + RV)
RV/TLC (Residual Volume/Total Lung Capacity)	20%	Percent of total lung capacity made up of residual volumes
FEV_1 (Timed Forced Expiratory Volume in One Second)	3.8 liters	Volume of VC forcibly exhaled in 1 second
FEV_1/FVC (Timed Forced Expiratory Volume-Forced Vital Capacity) × 100	80%	Percent of VC that can be forcibly exhaled in one second

Forced vital capacity is determined primarily by four factors: (1) strength of chest and abdominal muscles, (2) airway resistance, (3) lung size, and (4) elastic properties of the lung. Any condition that decreases the strength of the respiratory muscles (e.g., poliomyelitis), decreases lung volume (e.g., tuberculosis), increases airway resistance (e.g., bronchitis), or makes the lung stiffer will lead to a significant decrease in forced vital capacity.

From FVC, another important determination that can be obtained is the forced volume exhaled in one second. This volume is termed **forced expired volume** (FEV_1) has the least variability of the measurements obtained from a forced expiratory maneuver and is considered one of the most reliable measurements (Fig. 20–13). Another useful way of expressing FEV_1 is as a percentage of FVC (i.e., FEV_1/FVC × 100), which corrects for differences in lung size. Normally, FEV_1 is 80% or more of the FVC (i.e., 80% of an individual's forced vital capacity can be exhaled in the first second). Lung diseases are often classified as either *obstructive* or *restrictive disorders*. With an obstructive disorder, *expiratory flow* is obstructed, and with a restrictive disorder, *lung inflation* is restricted.

In obstructive lung disorders, FVC and FEV_1 are both reduced and the ratio of FEV_1/FVC is reduced drastically below normal (<80%). Examples include asthma and emphysema. In restrictive disorders both FVC and FEV_1 are also reduced. However, the ratio of FEV_1/FVC is normal or slightly higher. The ratio of FEV_1/FVC is often used to differentiate between obstructive and restrictive lung disease. Examples of restrictive disorders include pulmonary edema and pulmonary fibrosis.

To summarize, the measurement of FVC is most commonly and easily obtained from a forced expiratory maneuver. From a FVC determination, we can readily measure FEV_1 and FEV_1/FVC ratio. These measurements, by themselves, provide a considerable amount of information about ventilatory function during normal and abnormal conditions.

Residual Volume and Functional Residual Capacity

Even after an individual expires as forcefully and completely as possible (forced vital capacity), some air, approximately 1.0 L, remains in the lungs and is

RESEARCH FOCUS

Bronchitis and Emphysema

By far the most common chronic lung disease in the United States is **chronic obstructive pulmonary disease (COPD),** which includes both chronic bronchitis and emphysema. The current estimate is that over 10 million Americans suffer from one or both of these diseases. COPD is now the fifth leading cause of death in the Untied States (over 75,000 deaths occurred last year), and the morbidity and mortality for COPD are rising at an alarming rate. Costs related to the care and treatment of COPD patients are staggering. Each year the direct costs exceed $4 billion.

The predominant clinical feature of both emphysema and chronic bronchitis is a chronic obstruction to air flow. Surprisingly, patients frequently have an abnormally high lung volume (i.e., their total lung capacity is increased), but their biggest difficulty is exhausting air from their lungs. The latter is reflected in their pulmonary function tests, which show abnormally low values for FVC, FEV_1, and FEV_1/FVC (see Table 20–1 for normal values).

Both diseases share similar symptoms; cough, wheeze, shortness of breath, and production of large amounts of mucus and sputum. Despite their similarities, the two diseases have marked structural differences and are brought about by different mechanisms. Chronic bronchitis is characterized by excessive mucus production, which plugs small airways. This comes about when the lining of the small airways becomes irritated from exposure to inhaled irritants such as tobacco smoke, which leads to excess mucus. In emphysema, alveoli are destroyed by the release of protolytic enzymes. The enlarged areas result from coalescing when walls are de-

stroyed (see inset). Late in COPD disease, patients frequently experience hypoxemia (low blood oxygen) and hypercapnia (high blood carbon dioxide). These two conditions, especially hypoxemia, lead to pulmonary hypertension (high blood pressure in the pulmonary circulation), which in turn causes right heart failure. Based on changes in blood oxygen levels, COPD patients can be categorized into two types. Those with predominantly emphysema are referred to as "pink puffers" because their oxygen levels are usually satisfactory and their skin remains pink. They develop a pursed-lipped style of breathing with puffing breaths. Those with predominantly chronic bronchitis are called "blue bloaters" because low oxygen levels give their skin a blue cast and because fluid retention from heart failure gives them a bloated look.

Smoking is the single most common cause of chronic bronchitis and emphysema. Actually, COPD is uncommon in nonsmokers. While pollution and occupational hazards are less important than smoking, they can act to worsen the condition. COPD is four times more common in men than in women, but as smoking habits of women change, this difference may diminish.

Recently, scientists have discovered a rare type of emphysema that occurs in families even in the absence of smoking. These individuals develop severe emphysema at a very early age (20 or 30 yrs). Scientists have discovered that this rare type of emphysema is due to the absence of a specific immunoglobin, alpha-1, in the blood. Immunoglobins are proteins that ordinarily act as antibodies, but alpha-1 also carries a substance known as alpha-1 antitrypsin, which inhibits the actions

called residual volume. When air becomes trapped in the lungs during expiration because of airway plugging or airway closure, the residual volume becomes abnormally high.

Another often-measured lung capacity is functional residual capacity. This is the volume of air remaining in the lungs after a normal expiration and amounts to about 2.4 L.

Because the lungs cannot be emptied completely following forced expiration, neither residual volume nor functional residual capacity can be measured directly by simple spirometry. These two volumes are measured indirectly by a dilution technique based on the following relationship:

$$C_1 \times V_1 = C_2 \times V_2$$

in which C is concentration and V is volume.

The technique used to measure these two volumes involves the dilution of helium, an inert and insolu-

ble gas. The subject is connected to a spirometer filled with 10% helium in air (see Fig. 20–14). Consequently, the initial 10% concentration of helium in the spirometer is C_1, and the volume in the spirometer is V_1. Since the lung initially contains no helium, after the subject breathes the helium-air mixture, the helium concentration in the lungs becomes the same as in the spirometer after equilibration. Because helium is insoluble and not taken up by the blood, the concentration of helium in the lung after equilibration is C_2, and the unknown volume in the lung is V_2. By conservation of mass, $C_1 \, V_1 = C_2 V_1 + C_2 V_2$ and by rearranging:

$$V_2 = \frac{V_1(C_1 - C_2)}{C_2}$$

Starting the test at precisely the right time is important. If the test begins at the end of a normal tidal volume (end of expiration), the volume of air re-

Scanning electron photomicrograph of a normal human lung showing the contrast between normal and enlarged alveolar spaces due to emphysema. The enlarged alveolar spaces are caused by overinflation. (Photo by Janice Nowell. From *Pathophysiology*, 2nd ed., by Smith and Thier. W.B. Saunders, 1985.)

of certain enzymes (proteases) involved in the breakdown of proteins. The absence of alpha-1 antitrypsin is an inherited trait that leaves the affected individual open to the breakdown of lung proteins by his or her own proteases, particularly the one known as elastase. Nor-

mally there is a balance between the elastase activity and alpha-1 antitrypsin so that no detrimental breakdown of lung tissue takes place. When alpha-1 antitrypsin is present in low amounts or absent altogether, increased destruction of lung tissue leads to emphysema.

Not only can scientists explain how alpha-1 antitrypsin–deficient individuals develop early emphysema, but they have recently discovered how an alpha-1 antitrypsin–elastase imbalance is involved in emphysema in individuals who do not inherit an alpha-1 antitrypsin deficiency. Smoking can upset the balance in two ways. The first way is by increasing in the amount of elastase. White cells are the main products of elastase. The number of white cells is significantly higher in cigarette smokers, and this leads to excess release of elastase. The second way that cigarette smoking interferes with this balance is by inactivating alpha-1 antitrypsin.

Although great strides have been taken toward conquering COPD, much research remains to be done. Future directions include the production of alpha-1 antitrypsin using recombinant DNA techniques. Another promising approach is the use of synthetic antiproteases to slow down lung damage.

A final comment: The most important therapy for COPD patients is to stop smoking.

■ **Why do emphysema patients breathe very deeply and slowly?**

$$V_2 = \frac{V_1 (C_1 - C_2)}{C_2}$$

Before equilibration

After equilibration

Figure 20–14

Dilution technique to measure an unknown volume. The dilution of helium is used to measure functional residual capacity. Dots represent helium before and after equilibration.

maining in the lung represents functional residual capacity. If the test begins at the end of a forced vital capacity and the subject starts breathing the helium-air mixture, then the test will measure residual volume.

Some Inspired Air Does Not Reach the Alveoli

So far, the lung volumes discussed have been primarily absolute volumes. However, ventilation is a dynamic process. If a person has a V_T of 500 ml and the breathing rate is 14 times a minute, then the total volume of air that enters the lungs each minute is $500 \times 14 = 7000$ ml per minute or 7 liters per minute. This volume of air is known as **minute ventilation.** ($\dot{V} = V_T \cdot f$)

Not all of the air from minute ventilation enters the alveoli for gas exchange. The tidal volume (V_T) is distributed between the conducting airways and alveoli (Fig. 20–15). Since gas exchange occurs only in the alveoli and not in the conducting airway (trachea, bronchi, bronchioles), part of the minute ventilation becomes wasted air. For each 500 ml of air inhaled, approximately 150 ml remains in the conducting airways. This volume of wasted air is known as **dead space volume** (V_D). Picture, then, what occurs during a normal breathing cycle. A normal tidal volume of 500 ml is expired. During the next inspiration, another 500 ml is taken in, but the first 150 ml of air entering the alveoli is dead space (old alveolar gas left behind). Thus, only 350 ml of fresh air reaches the alveoli, and 150 ml is left in the conducting airways.

tidal volume = 500 ml

dead space volume = −150 ml

fresh air entering alveoli = 350 ml

The normal ratio of dead space volume to tidal volume (V_D/V_T) is in the range of 0.20 to 0.35. In the example shown, the ratio (150/500) is 0.30. This means that 30% of the tidal volume or 30% of minute ventilation does not participate in the gas exchange and constitutes wasted air.

To determine how much fresh air reaches the alveoli per minute, we subtract the dead space volume from the tidal volume and multiply the result by breathing frequency. For example, the volume of air entering the alveoli from a normal tidal volume of 500 ml with a breathing rate of 14 breaths per minute is (500 ml − 150 ml) × 14 = 4900 ml. This represents the volume of "fresh air" available for gas exchange and is known as **alveolar ventilation** ($\dot{V}_A = V_T - V_D \cdot F$). Alveolar ventilation is the most important variable in gas exchange. This is because minute ventilation does not necessarily represent the amount of oxygen reaching the alveoli, since the amount of dead space varies. For instance, a swimmer using a snorkel breathes through a tube that increases dead space volume. Similarly, a patient connected to a mechanical ventilator also has increased dead space volume. Indeed, if minute ventilation is held constant, then alveolar ventilation is decreased with snorkel breathing or with mechanical ventilation.

The significance of dead space volume, minute ventilation, and alveolar ventilation can readily be seen in Table 20–3. In this experiment, Subject A's breathing is slow and deep, while B's breathing is normal and C's breathing is rapid and shallow. Note that each subject has the same minute ventilation (i.e., each is moving the same volume of total air in and out per minute), but each has marked differences in alveolar ventilation. Subject C has no alveolar ventilation and would die in a matter of minutes, while Subject A has an alveolar ventilation greater than normal. The important message from the examples presented in Table 20–3 is that increasing the **depth** of breathing is far more effective in elevating alveolar ventilation than increasing the frequency or rate of breathing. This fact is important in exercise. In most exercise situations, increased ventilation is accomplished by increases in the depth of breathing more

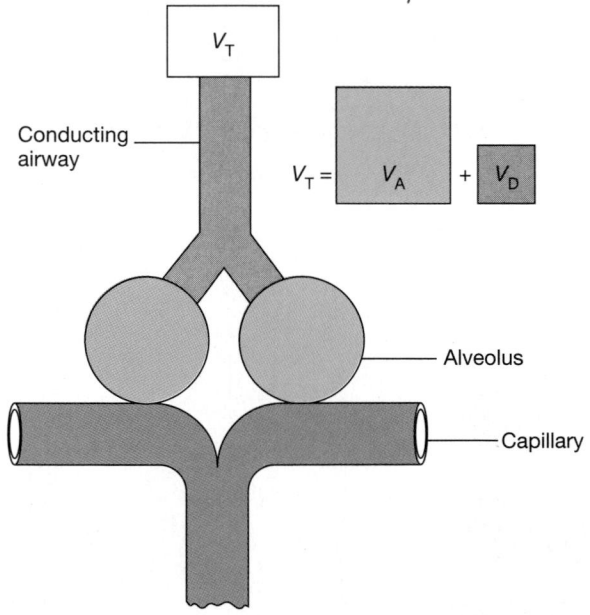

Figure 20–15

Tidal volume (V_T) is a mixture of gas from alveoli and conducting airways. The gas in the airways represents dead space volume (V_D) because this volume of gas does not participate in gas exchange. The volume of gas that participates in gas exchange in the alveoli is represented by V_A.

Table 20–3

Influence of Breathing Patterns on Alveolar Ventilation

Subject	Tidal Volume (ml)	×	Frequency (breaths/min)	=	Minute volume (ml/min)	−	Dead space ventilation (ml/min)	=	Alveolar ventilation (ml/min)
A	1000	×	6	=	6000	−	900 (150 × 6)	=	5100
B	500	×	12	=	6000	−	1800 (150 × 12)	=	4200
C	150	×	40	=	6000	−	6000 (150 × 40)	=	0

than in the rate. For example, a well-trained athlete can often increase alveolar ventilation during moderate exercise with little or no increase in breathing frequency.

Mechanical Properties of the Lung

The lung, airway and vascular trees are embedded in elastic tissue. When the lung is inflated, it must stretch to overcome these elastic components. Thus, the degree of lung expansion at any given time in the breathing cycle is proportional to transpulmonary pressure. How well a lung inflates and deflates with a change in transpulmonary pressure is a measure of its **elastic property.** Obviously, a stiff lung that does not stretch easily will not inflate to the same volume as a highly elastic lung, given the same effort.

Compliance Measures
Lung Distensibility

One important elastic property of a lung is distensibility (i.e., ability to stretch). Distensibility is determined by measuring the changes in lung volume that occur with changes in pressure, plotted on a **pressure-volume curve.** A simple analogy is inflation of a balloon with a syringe (Fig. 20–16). For each change in pressure, the balloon inflates to a new volume. The slope of the line is calculated by measuring the volume change per unit pressure change (Δ volume/Δ pressure). This value is known as **compliance** (C_L = Δ volume/Δ pressure).

A similar pressure-volume curve for the human lung can be generated by measuring a change in lung volume with a spirometer and a change in pleural pressure with a pressure gauge (Fig. 20–17). In practice, a pressure-volume curve is produced by first having the subject inspire maximally to total lung capacity and then expire slowly while lung volume and

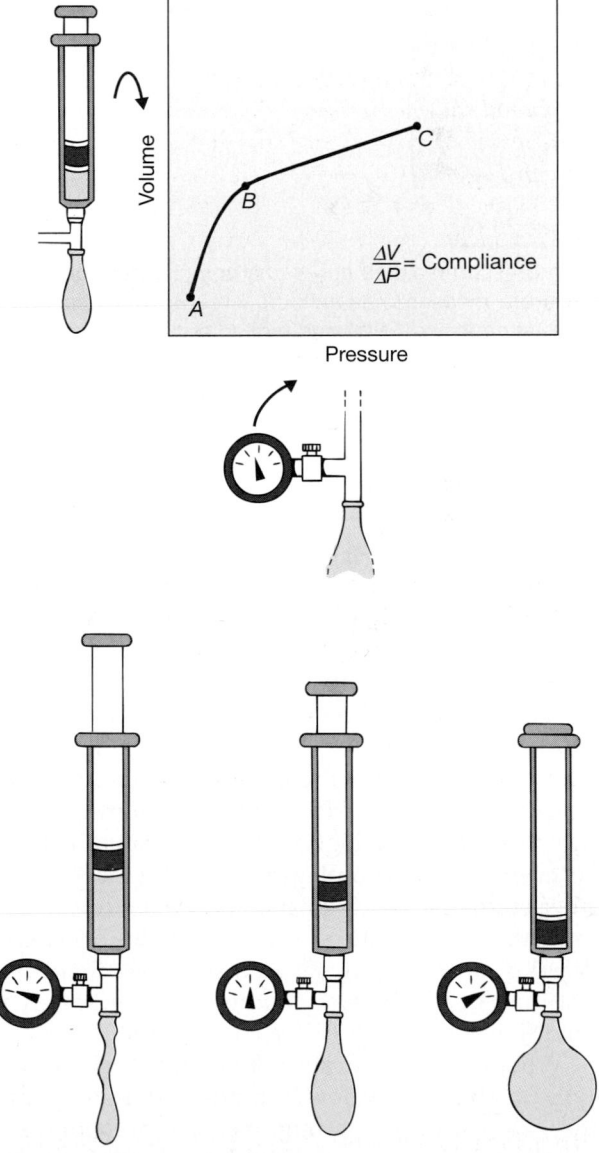

Figure 20–16

Pressure-volume curve. The slope of the line between B and C (change in volume/change in pressure) is compliance.

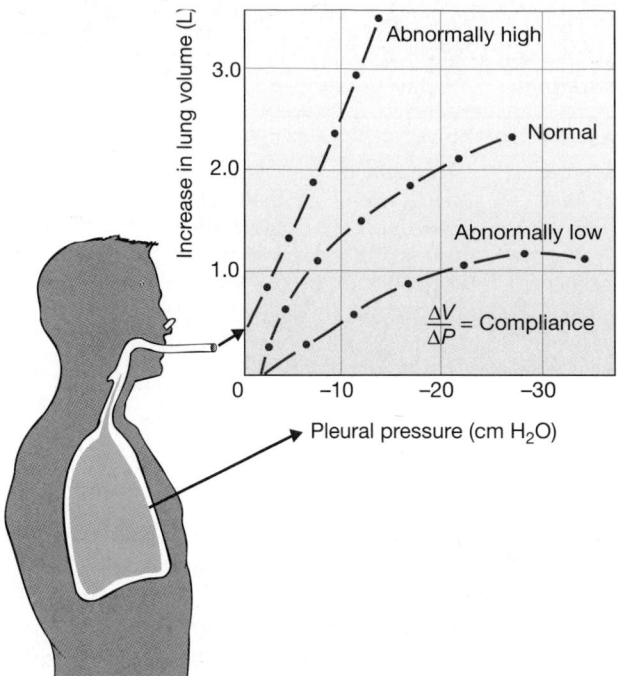

Figure 20–17

Lung pressure-volume curve and compliance. Lung volumes are measured as an increase above functional residual capacity (FRC) and plotted against a change in pleural pressure. A lung with a slope greater than normal would have an abnormally high compliance. A lung with a slope smaller than normal would have an abnormally low compliance.

pleural pressure measurements are being recorded. Compliance (the slope of the pressure-volume curve) provides a measure of lung distensibility. For example, a steep slope (high compliance) indicates that the lung is very distensible and easy to inflate. This means the lung can be inflated to a higher volume with a given amount of effort than can a less distensible lung. In other words, with the same transpulmonary pressure, the lung with the higher compliance will inflate to a higher volume because it is more distensible. Lung compliance indirectly measures another elastic property, lung *stiffness* (i.e., resistance to stretch). A lung with a low compliance (decreased slope) is stiffer and is not as distensible. Remember that lung distinsibility and stiffness are inversely related (i.e., a stiff lung is less compliant).

What is the significance of having an abnormally high or abnormally low compliance? Low compliance, indicating a stiff lung, means more work is required to bring in a normal tidal volume. Any time the lung becomes injured (by infection or by toxic environmental insults), the elastic properties are often altered, and the lung loses its distensibility (becomes stiffer). A very high compliance, however, is just as detrimental as a low compliance. Consider the disease **emphysema,** in which the lung has been overstretched from chronic coughing and congested airways. Lungs of patients with emphysema (which is strongly linked with smoking) have a high compliance (high distensibility) and are extremely easy to inflate. However, getting air out again is another matter! Lungs with abnormally high compliance (see Fig. 20–17) have poor **elastic recoil,** the ability of stretched tissue to return back to its normal position. Thus, a lung affected by emphysema is easily distended but does not recoil back during expiration. Consequently, a lot of effort is required to get air out of the lungs. A simple analogy is a sock or stocking that has become overstretched. It can be easily stretched out but does not recoil to fit the contour of the legs and so sags and looks "baggy." Similarly, smoking-induced emphysema can lead to "baggy lungs" that retain air (abnormally high residual volume).

Surfactant Stabilizes Alveoli

Another property that markedly affects lung compliance is **surface tension** occurring within the inner surface of the alveoli at the gas-liquid interface. Sur-

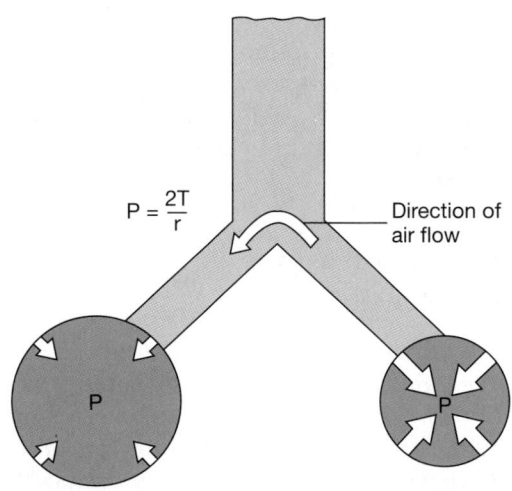

Figure 20–18

Surface tension alters alveoli stability. If surface tension remains constant then alveoli that are interconnected but vary in diameter cannot coexist and the smaller alveoli collapse. The smaller alveoli generate a greater pressure and cause air to flow into larger units. At low lung volumes, these smaller alveoli become unstable and collapse because of the higher surface tension. Surfactant prevents this from occurring. Surfactant promotes alveolar stability by lowering surface tension proportionately more in the small alveoli. The relationship between pressure and surface tension can be expressed as $P = \frac{2T}{r}$ where P = pressure, T = surface tension, and r = radius.

CLINICAL FOCUS

ARDS: The Case of Scot

Scot is a healthy, 20-year-old who is starting his second year of college. He is in very good condition and works out regularly by lifting weights and swimming one mile each day. However, one fall afternoon he began to feel ill. He went to the student health center with complaints of a severe headache and pain in the neck and lower abdomen. He was told that he was coming down with the flu and was advised to go back to the dorm and rest. The next day his symptoms became worse and included fever, cold sweats, skin rash, and nausea. His roommate took him back to the health center, and Scot was placed in one of the student beds for closer observation. When his condition started deteriorating he was rushed to the emergency room at the university medical center. Upon arrival, Scot was in shock with a high fever and a breathing rate of 36 (normal = 12 to 15/min). Because of the danger of respiratory failure, the attending physician immediately had Scot intubated and placed on assisted ventilation. The decision was made to transfer him to the intensive care unit (ICU). Upon arrival at the ICU, Scot's chest X-ray showed patchy infiltrate and pulmonary edema. He was diagnosed as having Adult Respiratory Distress Syndrome (ARDS), with a secondary widespread infection of unknown origin.

Scot's case is similar to the other 200,000 to 250,000 persons who develop ARDS each year. Despite Scot's youth and previous history, anticipated mortality for ARDS is >80%. Scot was lucky and was able to return to his classes after 12 weeks in the hospital and a bill of over $200,000.

ARDS is the name given to diffuse lung injury caused by a wide variety of etiology. Few patients have histories of previous lung disorders. The etiologies include trauma from chest injury (e.g., car accidents), long bone injury, and pelvic injury. During the Vietnam era the trauma that caused ARDS was coined "Da Nang lung." Two other major etiologies include sepsis (presence of pathogen or toxin in the blood) and aspiration of gastric contents. The latter occurs with gastric reflux, and usually occurs at night during sleep. Other etiologies include cardiopulmonary bypass, smoke inhalation, high altitude, and exposure to irritant gases. Exposure to irritant gases was the cause of the widespread ARDS cases that occurred at the disaster in Bhophal, India.

Although the etiology for ARDS is varied, the pathophysiology is nearly identical. ARDS is characterized by decreased lung compliance, pulmonary edema, focal atelectasis, and hypoxemia (low O_2 in the blood), with an inflammatory reaction that leads to an infiltration of neutrophils into the lung. The increase in lung stiffness (i.e., decreased compliance) is due to loss of surfactant, which is unrelated to edema.

Neutrophil aggregation is a key underlying mechanism of ARDS. Aggregation of neutrophils causes capillary endothelial damage by releasing a number of toxic products. These include oxygen free radicals, proteolytic enzymes, arachidonic acid metabolites, (leukotrienes, thromboxane, prostoglandin) and platelet-activating factor (PAF).

New approaches to therapy involve ways to reduce neutrophil chemoattraction and aggregation in the pulmonary capillaries, and ways to reduce the amount of toxic substances released by neutrophils.

■ **The nutritional status of ARDS patients is watched very closely. Why?**

---

face tension is a molecular force created at the gas-liquid interface—specifically, the attractive forces between the liquid and air molecules. These forces create a surface film. Since the surface of the alveolar membrane is moist and is in contact with air, a large gas-liquid interface is produced. Surface tension in alveoli produces a force that pulls inward. This tension tends to cause alveoli to become unstable by increasing alveolar pressure proportionately more in smaller alveoli than in larger ones at low lung volumes (Fig. 20–18).

To counteract the high surface tension, specialized epithelial cells lining the alveoli secrete a detergent-like material that coats the inner surface of the alveoli and drastically reduces surface tension. This material is known as **pulmonary surfactant** and is a mixture of lipids (fat) and protein. The main component of pulmonary surfactant responsible for reducing surface tension is the phospholipid, **phosphatidylcholine**. Thus, pulmonary surfactant lowers surface tension, prevents alveolar collapse, and makes the lung more distensible (compliant).

This is extremely important because the effects of surface tension on lung compliance are just as great as the effects of elastic forces. A striking example of what happens when inadequate amounts of surfactant are present is seen with a disorder known as **respiratory distress syndrome.** This disorder frequently afflicts premature neonates who are born with insufficient amounts of surfactant as well as adults with acute lung injury who have their surfactant destroyed. In the neonate, breathing is extremely labored because high surface tension makes it difficult to inflate the lungs. Because of the high surface tension, these infants also have alveolar collapse, a phenomenon known as **atelectasis.** Forty to fifty thousand babies are afflicted each year with infant respiratory distress syndrome. These infants are at high risk until the lung becomes mature enough to secrete surfactant. Consequently, infant respiratory distress syndrome is still a major problem among newborns in the United States. Adult respiratory distress syndrome (ARDS) presents a similar pattern and also constitutes a major cause of death.

Airflow Is Affected by Gas Density and Airway Diameter

When you take a breath of air into your lungs, you can hear the sound of air rushing in, especially if you breathe rapidly. The sound comes from air flow in the airway tree of which there are two basic types.

Turbulent Air Flow

One type, occurring in the trachea and large bronchi, is **turbulent flow** (Fig. 20–19*a*). Turbulent flow occurs at high flow rates and consists of completely disorganized patterns of air flow in which the molecules literally collide with each other, resulting in a noise that can be heard on inhalation or exhalation. Thus, the faster you inhale or exhale air, the more turbulent the noise.

An important consideration regarding turbulent flow is that turbulence in the large airways is the primary source of resistance to breathing. Turbulent flow is also greatly affected by gas density. If you breathe a helium-oxygen mixture, which is much lighter than air (nitrogen and oxygen), less turbulence is produced, meaning that less resistance and less work are involved in breathing. Gas density also affects the vocal cords in the larynx by changing the pitch of our voices, making us sound like Donald Duck when we talk after breathing a helium-oxygen mixture.

Laminar Air Flow

The second type of air flow in the lungs is **laminar air flow,** which is characterized by streamlined flow that parallels the side of the airways (Fig. 20–19*c*). Laminar flow is silent, because air molecules slide over each other. This type of air flow occurs mainly in the small peripheral airways where air flow is extremely slow.

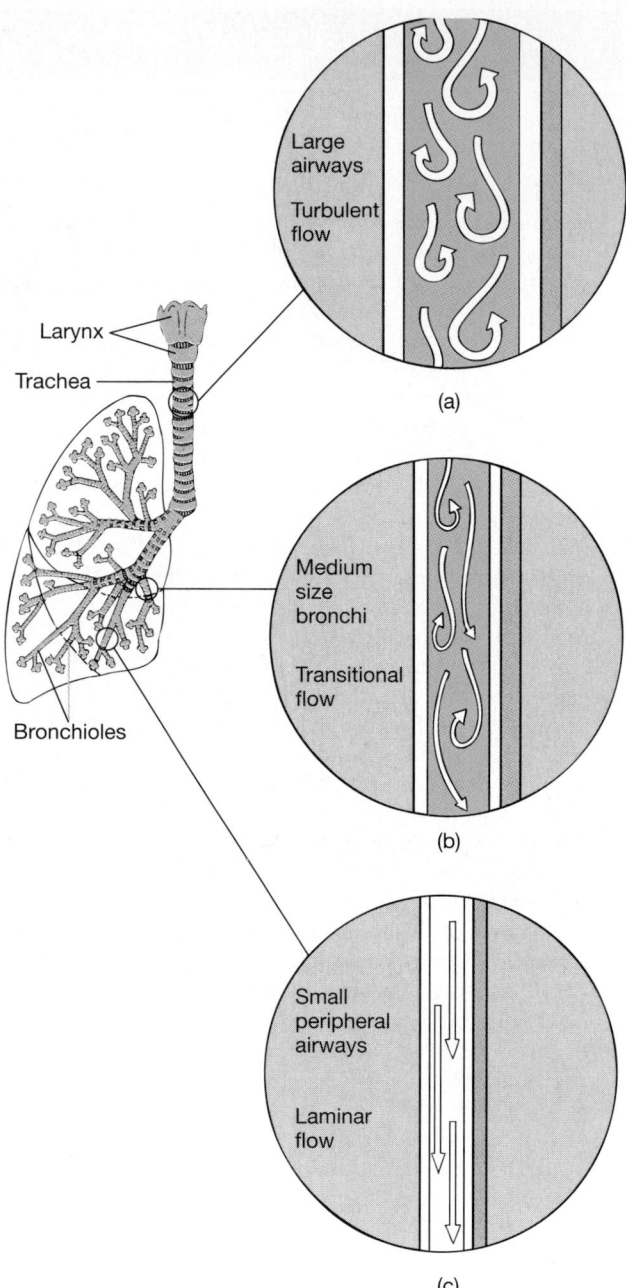

Figure 20–19

Patterns of air flow in the lung. (*a*) Turbulent flow occurs at high flow rates in large airways. (*b*) Transitional flow is a combination of laminar and turbulent flow and occurs in medium-sized bronchi. (*c*) Laminar flow occurs at low flow rates in small peripheral airways.

Laminar flow is greatly affected by airway diameter. Specifically, laminar flow is directly related to the radius raised to the fourth power (Flow $\propto$ radius^4). This means, for example, that if a peripheral airway constricts such that the radius is halved, then airflow in that tube decreases 16-fold.

Asthma

One of the airway diseases that many of us have encountered is asthma—a breathing disorder brought on by narrowing of the small airways, namely the bronchi. The asthmatic person experiences labored breathing, shortness of breath, and a distressing tightness in the chest. Shortness of breath can occur suddenly and without warning. During the asthmatic attack, the person experiences a feeling of suffocation, and must sit up or stand and devote every bit of energy to breathing. The attack can last several minutes to several hours. Afterwards a cough may occur, slight and dry at first but often becoming more marked, with considerable sputum. Following an asthmatic attack, there can be considerable soreness in the chest.

The cellular mechanism of asthma, which causes the shortness of breath, wheezing, and tightness in the chest, is thought to be the release of vasoactive mediators (histamine, prostaglandins, and leukotrienes). The agents cause the bronchial smooth muscles to constrict the airways and also cause an excess secretion of mucus, which plugs the terminal airways. The mucus is usually coughed up at the end of an attack.

One form of asthma, triggered by an allergic reaction, is often called allergic asthma. This type of asthma is most prevalent in children and adolescents. In most cases, these allergic reactions occur in the bronchial tree and are triggered by a foreign protein such as pollen, animal dander, mites, mold, spores, feathers, and, less commonly, food (e.g., milk, nuts, chocolate, and seafood). Certain synthetic drugs, such as aspirin and penicillin, although not proteins, can also trigger asthma. Such agents or allergens react with antibodies and induce the release of the potent vasoactive mediators, most importantly histamine. Underlying allergic asthma is an inherited tendency to develop hypersensitivity to these allergenic agents.

A second form of asthma is idiopathic asthma, in which a specific cause cannot be identified. This form usually occurs in adults over 30 and can be triggered by exercise, emotions, inhalation of cold air, and infections.

About 1% of the entire population (1 in 100) are estimated to have asthma. Asthma is estimated to occur in 5% to 15% of children under the age of 15, and is the leading cause of school absenteeism—accounting for 8 million school days lost. Asthma also has a substantial economical impact. The annual costs for hospital emergency room visits add to more than $2.5 billion. Asthma is more frequent among boys and in men 30 years and older. An important aspect of asthma treatment is the identification, when possible, of the agent that triggers the attack. Medications include oral drugs, which relax airway smooth muscle. Aerosol drugs are also used if the attack is severe.

■ **Why do patients with asthma frequently have their forced vital capacity measured?**

$$F = 1^4 = 1$$

$$F = \left(\frac{1}{2}\right)^4 = \frac{1}{16}$$

This relationship between airflow and radius in the small airways explains why a person who suffers from asthma (a condition caused by spasmodic contraction of the bronchiole smooth muscles) has great difficulty getting air in and out of the alveoli. During asthma attacks, many asthmatics take epinephrine, a hormone that acts on the beta receptor to cause the smooth muscle to relax and in turn dilate the airways. At lower flow rates during expiration—particularly at branches of the airway tree where flow from separate tubes comes together and empties into a single airway—the air flow pattern is a mix. This is referred to as **transitional flow,** a combination of turbulent and laminar flow (Fig. 20–19b). Transitional flow occurs most commonly at bifurcations, and high flow in medium-size airways.

Work Is Required for Breathing

During inspiration, muscular work is involved in expanding the thoracic cavity, inflating the lungs, and overcoming airway resistance. Since work can be measured as force multiplied by distance, the amount of work involved in breathing can be expressed as a change in lung volume (distance) multiplied by the change in transpulmonic pressure (force). Thus, energy is expended during muscular contraction to create a force (transpulmonary pressure) to inflate the lungs. When greater transpulmonary pressure is required to bring the same volume of air into the lungs, more muscular work and greater energy are required.

What factors affect the amount of work involved in ventilating the lungs? First is lung compliance. As discussed earlier, lung inflation requires energy to overcome elastic forces. Thus, the more compliant the lung, the more easily it is stretched and the less energy is required in breathing. When the lung becomes stiff (less compliant), more muscular work is required to increase transpulmonary pressure in order to stretch the lungs to expansion.

A second factor affecting the work of breathing is high surface tension. As pointed out earlier, individuals who lack surfactant have high surface tension and extremely labored breathing. Some of these individuals die from exhaustion and respiratory muscle fatigue.

A third factor affecting the work of breathing is airway resistance, which can be increased in two ways: (1) by an increase in flow rate and (2) by abnormal constriction of the airways. Both of these lead to a greater energy expenditure because greater transpulmonary pressure is required to maintain adequate air flow.

How much energy is involved in breathing? In a normal person, the energy needed for breathing represents approximately 1% of the body's total energy expenditure. Even during heavy exercise, only about 3% of the total energy expenditure is involved. Breathing is most economical when there is a balance of elastic, surface, and resistive forces. This balance requires the least amount of work. It is important to remember that the faster the breathing rate, the greater the amount of work required to overcome the increase in airway resistance. In contrast, the greater the depth of breathing, the greater the amount of work required to overcome elastic and surface forces. Thus, patients with stiff lungs will compensate by breathing small amounts of air (low tidal volume) more rapidly, while patients with severe airway obstruction tend to take deeper breaths more slowly. Both of these breathing patterns help to minimize the amount of work required for breathing under such conditions.

CHAPTER REVIEW

Summary

- Gas exchange is the process of transferring O_2 and CO_2 between the atmosphere and the capillary blood in the lungs and then between the blood in systemic capillaries and the cells.
- Total lung capacity is made of several volumes and overlapping capacities. All can be measured directly with a spirometer with the exception of residual volume. One of the most useful measurements is forced vital capacity.
- Ventilation is the process of moving air in and out of the lungs. Total minute ventilation ($\dot{V}$) is the volume of air moved per minute while alveolar ventilation ($\dot{V}_A$) is volume of fresh air reaching the alveoli per minute.
- Dead space represents volume of air that fills the airways and does not participate in gas exchange. The V_D/V_T ratio represents the fraction of wasted ventilation.
- The lungs are elastic and in order to overcome the elastic forces a pressure gradient across the lungs (transpulmonary pressure, P_L) is required to inflate the lungs.
- A volume-pressure curve is used to assess the elastic properties of lungs. The slope of the volume-pressure curve, compliance, is a measure of lung distensibility (i.e., the ease in which lungs can be inflated or stretched).
- Surface tension at the alveolar gas-liquid interface is a major force affecting lung compliance. Surfactant is lipoprotein that coats the inner surface of the alveoli and acts as a surface-reducing agent that lowers surface tension and stabilizes alveoli.
- Work is required to expand the chest wall, inflate lungs, and overcome airway resistance. Energy costs for the work of breathing represents 1 to 2% of total oxygen consumption.

Review Questions

Identify Each with a Term

1. The volume of air that can be forcibly exhaled after maximal inspiration.
2. The amount of fresh air that reaches alveoli per unit time.
3. A collapsed lung.

Define Each Term

4. pulmonary surfactant
5. hypoventilation

Choose the Correct Answer

6. A change in lung volume per unit transpulmonary pressure change in a volume-pressure curve is a measure of:

a. recruitment   d. elastance
b. compliance   e. hystersis
c. alveolar pressure

7. A patient has V_T of 500 ml, a breathing rate of 16/min, a dead space volume of 150 ml, and a FRC of 3 liters. Minute volume for this patient would be:

a. 8 liters/min   c. 48 liters/min
b. 5.6 liters/min   d. 2.4 liters/min

8. Which of the following is *not* part of the conducting zone of the airway tree?
   a. trachea   c. bronchioles
   b. bronchi   d. alveoli

9. Total lung capacity is the sum of:
   a. residual volume and functional residual capacity
   b. vital capacity and functional residual capacity
   c. vital capacity and tidal volume
   d. residual volume and vital capacity

10. Functional residual capacity is:
    a. the volume of air seen during normal breathing
    b. the volume of air expired after full inspiration
    c. the volume of air remaining in the lungs at the end of normal expiration
    d. the volume of air remaining in the lungs at the end of maximal expiration

11. Minute ventilation represents:
    a. total air movement in the lung
    b. total air movement minus dead space volume
    c. alveolar ventilation minus dead space volume
    d. tidal volume divided by frequency of breathing

12. Which of the following pressures is subatmospheric at rest?
    a. alveolar pressure   c. transpulmonary pressure
    b. pleural pressure   d. tracheal pressure

13. Pleural pressure is the most negative at:
    a. residual volume
    b. functional residual capacity
    c. end of tidal volume
    d. total lung capacity

14. A lung with an abnormally high compliance would:
    a. be easier to inflate
    b. be easier to deflate
    c. require more work to inflate
    d. have greater elastic recoil

Calculate

15. A person has a forced vital capacity of 5 L and a residual volume of 1.2 L. What is the person's total lung capacity?

16. A person has a tidal volume of 500 ml, a frequency of 14 breaths/min, and a dead space volume of 150 ml. What is the person's alveolar ventilation?

17. A person has an alveolar ventilation of 5 L/min, a frequency of 10 breaths/min, and a tidal volume of 700 ml. What is the person's dead space volume?

18. A subject expires into a spirometer for 10 minutes and has an expired volume of 40 L. Her respiration rate is 12/min during the 10-minute period. (a) What is the subject's minute ventilation? (b) What is the subject's tidal volume?

Answer Each Question in One or Two Sentences

19. Why is forced vital capacity so useful as a test in lung spirometry?

20. How does the lung inflate/deflate itself with air during breathing?

21. Why is elastic recoil so important in breathing?

22. How does emphysema change lung compliance?

23. What is the functional importance of pulmonary surfactant?

Problem Solving in Physiology

24. The time span for inspiration and expiration is approximately the same for a normal individual. In conditions of increased airway resistance, such as in patients with asthma, what change would be expected in the ratio of inspiratory to expiratory time?

25. A patient has a total lung capacity of 6 L and a forced vital capacity of 4.0 L. If forced vital capacity is reduced by 30% with no change in total lung capacity (TLC) what would be the ratio of residual volume (RV) to total lung capacity? Would this change in the ratio suggest an obstructive disorder?

26. A well-conditioned student has an oxygen consumption of 250 ml/min and a carbon dioxide production of 200 ml/min. If the student voluntarily hyperventilates to double her alveolar ventilation ($\dot{V}_A$) from 4 to 8 L/min, what will happen to her oxygen uptake and carbon dioxide output? Will they double?

Suggested Readings

Decramer, M. "Respiratory muscle interaction." *News in Physiological Sciences*, 8:121–124, 1993.

Demling, R.H. "Current concepts in the adult respiratory distress syndrome." *Circulatory Shock*, 30:297–309, 1990.

Foster, R.E., Dubois, A.B., Briscoe, W.A., and Fisher, A. *The Lung*, 3rd ed. Chicago: Year Book Medical Publishers, 1986.

Nunn, S.F. *Applied Respiratory Physiology*, 2nd ed. Boston: Butterworth, 1977.

Piiper, J., and Scheid, P. Gas exchange in vertebrates through lungs, gills and skin. *News in Physiological Sciences*, 7:199–203, 1992.

Raffin, T.A. "ARDS: Mechanisms and management." *Hospital Practice*, 65–80, November 1987.

Rhoades, R.A. *Medical Physiology*, Chapters 19, 20, and 21, edited by Rhoades & Tanner. Boston: Little, Brown, 1984.

Sieck, G. Neural control of the inspiratory pump. *News in Physiological Sciences*, 6:260–265, 1991.

Staub, N.C. *Physiology*, 3rd ed., Chapters 33, 34, and 37, edited by R. Berne and M. Levy. St. Louis: C.V. Mosby Co., 1992.

Van Golde, L.M.C., Batenburg, J.J., and Robertson, B. "The pulmonary surfactant system." *News in Physiological Sciences*, 9:13–20, 1994.

West, J.B. *Respiratory Physiology*, 4th ed. Baltimore: Williams and Wilkins, 1990.

Key Terms

airways (p. 619)
alveolar pressure (p. 623)
alveolar surface area
    (p. 630)
alveolar ventilation
    (p. 630)
alveoli (p. 618)
atmospheric pressure
    (p. 623)
atelectasis (p. 634)

barometric pressure
    (p. 622)
bronchi (p. 619)
compliance (p. 631)
dead space (p. 630)
diaphragm (p. 621)
forced vital capacity
    (p. 626)
functional residual capacity
    (p. 626)

gas exchange (p. 617)
intercostal muscle (p. 621)
laminar flow (p. 634)
lung volumes (p. 628)
partial pressure (p. 622)
pleural pressure (p. 623)
pneumothorax (p. 624)
relative pressure (p. 623)
residual volume (p. 627)
respiration (p. 617)

spirogram (p. 625)
spirometry (p. 625)
surfactant (p. 632)
thoracic cavity (p. 619)
trachea (p. 619)
transpulmonary pressure
    (p. 623)
turbulent flow (p. 613)
ventilation (p. 619)

Scanning electron micrograph of a lung alveolus. *(© David M. Phillips/Visuals Unlimited)*

Chapter 21

Pulmonary Circulation, Gas Exchange, and Control of Breathing

Key Concepts

The architecture of the pulmonary circulation is designed for efficient gas exchange.

Pulmonary circulation is a low-pressure, low-resistance system that accommodates all of the cardiac output.

O_2 in the blood is transported by hemoglobin and CO_2 as bicarbonate ions.

Changes in the pulmonary circulation in the neonate provides an independent existence at birth.

Control of breathing is tightly linked to our metabolic and physical activity.

Chapter Outline

Functional Design of the Pulmonary Circulation

Vascular Pressure and Blood Flow

Gas Uptake and Transport

Fetal and Neonatal Circulatory Physiology

Control of Breathing

Jane and Susan are second-semester college students and live in the same dormitory. Susan informs Jane about a mini-marathon to be held on campus in three months, and Susan bets her that she could not finish the race if she entered. Jane decides to take Susan up on the bet and enters her name. Jane runs frequently, but never in a competitive race; she becomes very diligent in her practice and works very hard on her conditioning. A week before the race Jane finishes a 13-mile practice run in her best time and feels very good about her stamina. On the day of the race Jane arrives early to warm up. As she takes her position, she suddenly becomes aware that her heart is pounding, her breathing has increased, and her hands have become sweaty. When the gun fires, Jane and 200 other runners are off! She has an excellent race, and she hears the crowd roar as she staggers across the finish line.

Imagine that Jane could peer inside her body and spy on its activities as she runs. What would happen, for example, to her lungs? One of the first things Jane would notice is that the frequency of electrical impulses that travel down the phrenic nerve to the diaphragm *increases with exercise. As the diaphragm contracts, air rushes into the lungs as part of the first step of gas exchange (ventilation). In the microscopic alveoli, O_2 and CO_2 are busy shuffling between the alveoli and pulmonary capillaries in the next step of gas exchange (gas uptake). With each beat of the heart, about 500 billion red cells race through the lungs, in a fraction of a second, to pick up life-giving O_2. The transport of O_2 (the third step in gas exchange) to the muscle, brain, and every other part of the body is carried by a pigmented protein, hemoglobin. The transport of CO_2 from tissue like exercising muscle, a tissue which has gone from a slow burn state to a metabolic blast furnace, is also carried by hemoglobin as well as by bicarbonate in the blood.*

In this chapter we will see how the design of the pulmonary circulation is tied to gas exchange and that alveolar-capillary units are so efficient that the diffusion of O_2 and CO_2 in the lungs rarely limits our activity. We will examine gas uptake and see how large volumes of O_2 and CO_2 can be exchanged in a very short period of time across an exceedingly thin alveolar-capillary membrane. We will examine how O_2 and CO_2 are transported by the blood and how disorders of pulmonary circulation interfere with this process. Finally, we will examine the control of breathing and see how it is exquisitely tied to our activity.

Functional Design of the Pulmonary Circulation

Pulmonary Vascular Tree Accepts All of the Cardiac Output

The way in which fresh air travels from the atmosphere to the alveoli was presented in the last chapter. The other components of gas exchange to be discussed include the diffusion of gases across the alveolar-capillary membrane, blood flow from the heart to alveoli, and the transport of O_2 and CO_2 by the blood to tissues. Blood is oxygenated in the alveoli and then returned to the left side of the heart to be pumped, along with the oxygen it contains, to the rest of the body. The circulation that carries blood through the lungs, the **pulmonary circulation,** is unique because it is the only circulation in the body that must accommodate all of the cardiac output. The pulmonary vessels are designed to operate at low pressure and to have a large reserve in order to meet the increased blood flow demands of exercise. The blood leaves the right ventricle via the pulmonary artery and follows the same kind of tree-like branching pattern found in the airways (Fig.

(a)

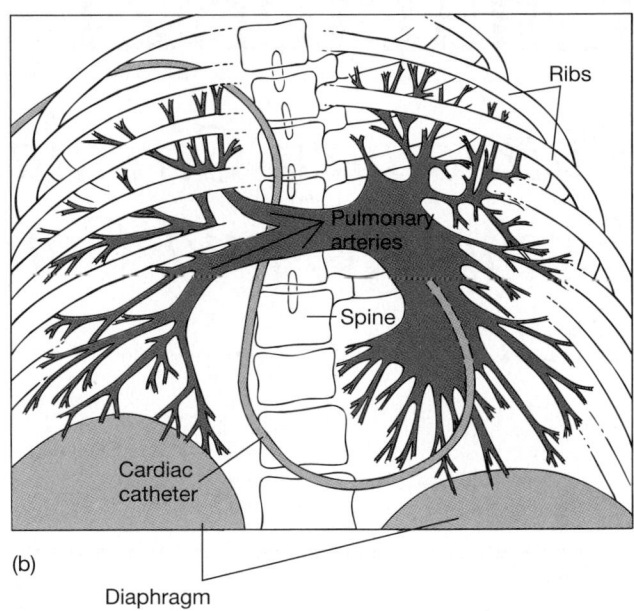

(b)

Diaphragm

Figure 21–1

When a dye that is opaque to X-rays is injected into the pulmonary artery and an X-ray image is made, the dye shows the pathways followed by the blood through the pulmonary arteries. If another X-ray image is taken a few seconds later, the dye-filled veins show the pulmonary vein tree. Artist's rendering of the arteriogram.

21–1). In fact, each time the airway tree branches, the arterial tree branches also, so that the two shadow each other.

Once the arteries reach the alveoli, the vessels break up into a fine network of tiny vessels, the pulmonary capillaries, which form a dense double network embedded in the alveolar wall (Fig. 21–2). This complex system results from the way that the fetal lung develops. As the alveoli grow, each develops its

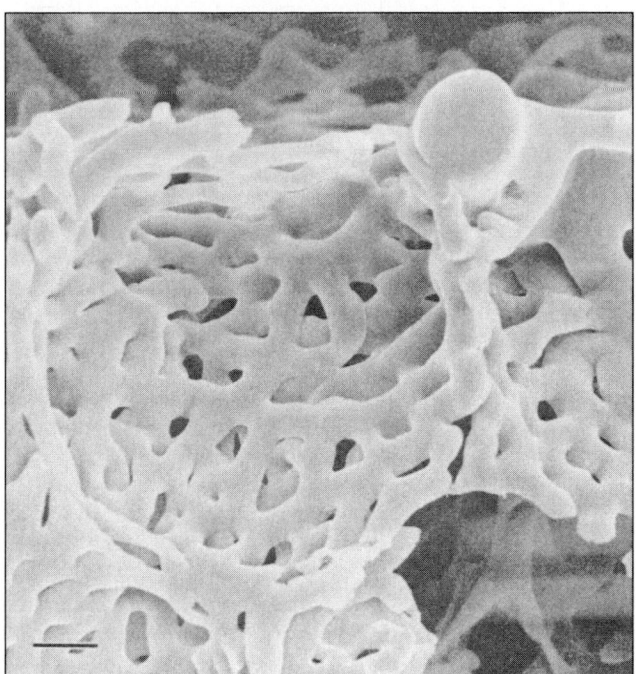

Figure 21–2

Scanning electron micrograph of the alveolar capillary bed shows an incompletely fused double network that offers the red cells a very large array of paths. Bar in lower left is 10 microns long. (With permission from Guentheroth, et al. *Journal of Applied Physiology*, 53:510–515, 1982.)

own set of capillaries. During lung maturation, the capillaries between adjacent alveoli join, but the joining is incomplete (Fig. 21–2). This leaves a complex, partially joined capillary bed with a large number of pathways through which blood can travel in each alveolar wall.

Once the blood leaves the terminal arteriole and enters the capillaries, it does not simply cross a single alveolar wall and then drain into a venule. The arterioles and venules are separated by four to eight alveoli. Thus, the blood can follow many pathways through a given alveolar wall, as well as travel over different alveolar walls; these combinations result in a huge number of possible routes by which the blood can travel across the gas exchange surface to pick up O_2 and unload CO_2.

Each capillary is about equal to the diameter of an erythrocyte (red blood cell). This forces the erythrocytes to pass through the capillaries single file. The erythrocytes are separated from the capillary wall by a thin layer of plasma that acts as a lubricant. The distance across the capillary wall and plasma layer is short, measuring about 1 μm (in more familiar terms, a typical human hair is about 100 μm in diameter, so that distance between gas in the alveolus

and the blood is 1/100th of a hair). The alveolar gas reaches the blood quickly because the distance from the gas to the blood phase is so small.

The passage of blood through the capillaries takes about one second. In that time, the blood unloads CO_2 and picks up all the O_2 it can carry. The blood then enters the smallest venules and returns through the pulmonary venous tree to the left side of the heart, from which the newly oxygenated blood is then pumped to the body.

Another tree-shaped conducting system in the pulmonary system is the lymphatic system. The lymph vessels are auxiliary drainage vessels. Fluid is continually being forced from the capillaries, Starling filtration into the interstitial space. Much of the fluid is reabsorbed by the capillaries, but part of it remains behind and must be drained away through the lymphatics. If the fluid accumulates, the diffusion distance between the alveolar gas and the capillary blood will increase and the uptake of oxygen will decrease. Accumulation of fluid in the lung is called **pulmonary edema.** This is a serious condition that can, in extreme cases, result in death. When edema forms, the lymphatics go into full operation to try to keep the alveoli dry. The lymphatic fluid drains into systemic veins.

Pulmonary Circulation Has Metabolic Functions

The alveoli have a large surface area, as discussed in Chapter 20. Since the capillary bed takes up about 80-90% of the alveolar wall, the surface area of the capillaries is impressively large as well. The inside of the capillary wall is completely covered by metabolically active endothelial cells. Because of the central location of the pulmonary circulation, all of the blood is exposed to the capillary endothelium each time it is pumped through the body. The endothelial cells act upon a number of substances. For example, **angiotensin II,** a potent vasoconstrictor, is an important hormone in the regulation of blood flow, as mentioned in Chapter 19. The hormone reaches the lungs in an inactive form, angiotensin I. During its passage through the pulmonary circulation, angiotensin I is converted to active angiotensin II by an enzyme located on the surface of the capillary endothelium. Angiotensin II is delivered by the blood to the resistance vessels (arterioles) in the systemic circulation to cause constriction and raise blood pressure. Many other vasoactive substances, for example bradykinin, serotonin, and the prostaglandins, are inactivated or removed from the blood by the endothelium. In this way, the pulmonary circulation serves important metabolic as well as gas-exchange functions.

Table 21–1

Pressures in the Pulmonary Artery During Various Conditions

Condition	Systolic/ Diastolic (mm Hg)	Mean (mm Hg)
Rest	25/10	15
Moderate exercise	30/15	20
Maximal exercise	50/25	33
At very high altitude	50/25	33
At very high altitude with maximal exercise	75/45	55
Severe pulmonary hypertension	175/100	125

From Groves, B.M., et al. *J. Appl. Physiol.*, 63:521–530, 1987.

Figure 21–4

Because of the low values of both pulmonary arterial pressure and vascular resistance, most of the blood flows to the lower part of the lung during rest. If a person stands on his or her head, blood will flow to the apexes of the lung.

Vascular Pressure and Blood Flow

Pulmonary Circulation Is a Low-Pressure System

In contrast to systemic circulation, pulmonary circulation is a low-pressure, low-resistance system with highly compliant vessels. A striking contrast between pulmonary circulation and systemic circulations is the pressure profile. Pressure in the pulmonary artery is much lower than in the arteries of the rest of the body. The right ventricle generates a pressure of about 25 mm Hg, which becomes the pulmonary arterial systolic pressure as the ejected blood passes through the pulmonic valve. Once the pulmonic valve closes, the pressure falls to diastolic pressure, just as it does in the systemic circulation. Pulmonary arterial diastolic pressure is about 10 mm Hg. Mean pressure for a normal adult is 15 mm Hg. These numbers are stated the same way as systemic arterial pressure: 25/10, with a mean of 15 mm Hg (Table 21–1). Pulmonary arterial pressure is usually reported as a mean pressure. These pressures shown in Figure 21–3, are from a pressure tracing as a cardiac catheter is passed through the right atrium and right ventricle and into the pulmonary artery.

The low pulmonary arterial pressure is just enough to pump blood to the top of the lung.* Although all of the lung receives some perfusion, even during resting conditions, the majority of the blood flows through the lower lung (Fig. 21–4). This creates a gradient of blood flow, with most of the flow going to the bottom of the lung and little going to the top. Almost all of the capillaries in the lower lung are open and have blood flowing rapidly through them; under these circumstances, the capillary bed is said to be highly "recruited." In the upper lung, only a few capillary pathways are open (recruited), and the blood cells flowing through them travel slowly.

Figure 21–3

Pressure tracing from a cardiac catheter as it is passed from the right atrium into the right ventricle, past the pulmonic valve, and into the pulmonary artery.

*To calculate the height to which the right ventricle can pump the blood, consider the following. Blood has a specific gravity near 1, whereas mercury has a specific gravity of 13.6 — that is, mercury is 13.6 times heavier than water with its specific gravity of 1.0. Multiplying the mean pulmonary arterial pressure of 15 mm Hg by 13.6 gives 204 mm H_2O. Thus, mean pulmonary arterial pressure is sufficient to pump the blood 200 mm high, which is approximately the height of the top of the lung above the heart.

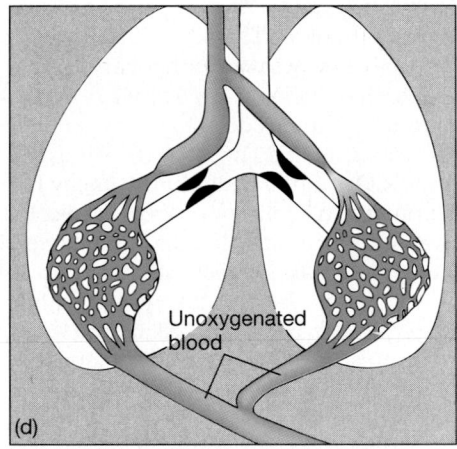

Figure 21–5

The two-alveolus model of the lung can be used to represent two alveoli or two regions of the lung with many alveoli in each region. (*a*) In this figure, all the alveoli are well ventilated so the unoxygenated (blue) blood can flow to any region and pick up oxygen, shown here turning from blue to red. (*b*) In this case, one region of lung is hypoxic. If blood continues to flow past the hypoxic alveoli at normal rates, unoxygenated (blue) blood will enter the pulmonary veins and reduce the proportion of oxygen in the blood. Ventilation (by air) and perfusion (by blood) will be out of balance. (*c*) In some undiscovered way, the pulmonary artery accompanying the hypoxic airway detects the low oxygen and constricts. The narrowed vessel has increased resistance and causes local blood flow to be reduced and directed to other, better oxygenated parts of the lung. In this way, ventilation-perfusion balance is maintained and the amount of oxygen in the arterial blood is kept at high levels. (*d*) If all regions of the lung become hypoxic, as happens at high altitude or in certain diseases of the airways when gas movement is impeded, the arteries detect hypoxia in the airways and constrict. Because all of the arteries constrict, the resistance to blood flow through the lung increases and pulmonary arterial pressure rises.

Hypoxic Vasoconstriction Balances Blood Flow with Ventilation

As long as inspired air is distributed evenly through the lung, venous blood in the pulmonary artery (the blue blood in Fig. 21–5*a*) can flow to any alveolar region and pick up oxygen from the alveolar gas; as the blood flows pass the alveoli, it becomes oxygenated. Inspired air, however, is not always evenly distributed to all alveoli. This causes **regional hypoxia** (low oxygen) to exist from time to time in groups of alveoli. Of course, blood passing through such areas will not be completely oxygenated (see Fig. 21–5*b*). This blood will mix with the rest of the oxygenated blood

CLINICAL FOCUS

Pulmonary Edema

Pulmonary edema is a common and important disorder of the pulmonary circulation. Pulmonary edema is an abnormal accumulation of fluid that leaks from the pulmonary capillaries, and occurs when fluid accumulation exceeds the rate of removal by the lymphatics. The condition is serious, and patients with pulmonary edema have difficulty breathing (tachypnea: rapid and shallow breathing) and suffer from hypoxemia.

Pulmonary edema occurs in two stages. The first is interstitial edema, which causes flooding and engorgement in the interstitial space. The second is alveolar edema, which flood alveoli. In severe pulmonary edema the fluid moves into the small and large airways and is coughed up as frothy sputum. Edematous lungs interfere with the gas exchange (i.e., increase the diffusion distance for O_2 and CO_2) and impede ventilation by interfering with surfactant (i.e., increase surface tension) which makes alveoli shrink in size (decrease surface area for gas diffusion) and makes lungs less compliant (stiffer).

The etiology of pulmonary edema includes: (1) increased hydrostatic pressure, (2) increased capillary permeability, (3) lymphatic insufficiency, (4) decreased interstitial pressure and (5) decreased colloid osmotic pressure. By far the most common cause of pulmonary edema is due to abnormal rise in hydrostatic pressure. Events which lead to a rise in hydrostatic pressure includes heart and/or renal failure. When the left ventricle is unable to empty itself during systole, left atrial pressure rises, which causes hydrostatic pressure to rise. The second most common cause of pulmonary edema is due to increased permeability. Conditions that lead to increased permeability include inhaled toxins (e.g., chlorine, sulfur dioxide, nitrogen oxides), circulatory toxins (e.g., endotoxin), oxygen toxicity, radiation, and adult respiratory distress syndrome.

Lymphatic insufficiency is due to carcinoma of the lymph glands and to silicosis (damage from silicon dust). Events that can cause decreased interstitial pressure include a pneumothorax and hyperinflation of the lung. Starvation (hypoproteinemia) and overtransfusion are the causes of decreased colloid osmotic pressure.

The pattern of edema formation is much different depending on the etiology. For example, in Adult Respiratory Distress Syndrome a patchy pattern of edema occurs around the lung periphery. In heart failure or fluid overload (overhydration or renal failure) a central edema pattern occurs, principally at the base of the lung. These patterns of edema formation help the pulmonary physician to postulate the etiology of pulmonary edema.

■ **Why is salt intake one of the first things restricted in patients with pulmonary edema?**

---

and reduce the total amount of oxygen that can be delivered to the body. The condition of low blood oxygen is called **hypoxemia** (*-emia* refers to blood; e.g., *anemia* refers to decreased red blood cell count).

The lung has a mechanism called **hypoxic vasoconstriction** for balancing the flow or perfusion of blood with the availability of regional ventilation. When a region of lung becomes hypoxic, the small arteries feeding that region sense the hypoxia in the alveolar gas and constrict. Their constriction causes the resistance to local blood flow to rise and forces the blood to flow away from the hypoxic region to other parts of the lung, where resistance is low because oxygen is available (see Fig. 21–5c). In this way, the balance between ventilation and perfusion is maintained, and compensation for disturbances in delivery of inspired air to the alveoli can be made. When an imbalance between ventilation and perfusion occurs due to lung disease, hypoxemia—sometimes of a serious nature—can result.

When someone experiences hypoxia throughout the entire lung—at high altitude, for example—all of the local hypoxic vasoconstrictor responses are triggered (see Fig. 21–5d). This situation causes the resistance to blood flow in the entire pulmonary circulation to increase, leading to a rise in pulmonary arterial pressure. The increase in pressure has an interesting effect. The alveoli in the upper part of the lungs have more oxygen than the alveoli in the lower part of the lungs, because most of the blood flow goes to the lower lung (see Fig. 21–4) and extracts more oxygen from the lower most alveoli. The elevated pulmonary arterial pressure caused by whole lung hypoxia is useful because it enables more blood to be pumped to the upper regions of lung, where capillaries are recruited and thus increase the surface area for gas exchange. Because recruitment occurs in the upper lungs where alveolar oxygen is relatively abundant, the lungs become a more effective gas exchanger. If, however, the high pulmonary arterial pressure is sustained for long periods, there are detrimental effects, since pulmonary hypertension can cause vessel damage and eventually failure of the right ventricle. Therefore, the redistribution of blood

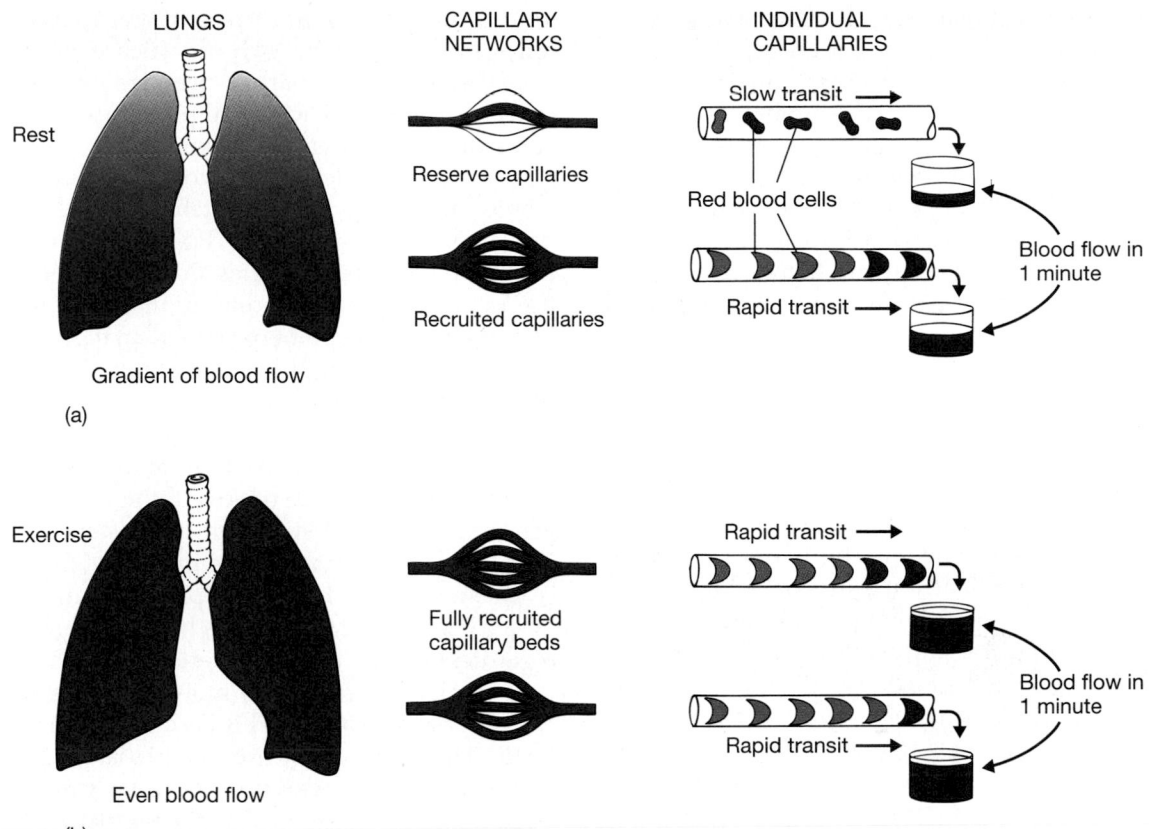

LUNGS

Rest

Gradient of blood flow

(a)

CAPILLARY NETWORKS

Reserve capillaries

Recruited capillaries

INDIVIDUAL CAPILLARIES

Slow transit →

Red blood cells

Rapid transit →

Blood flow in 1 minute

Exercise

Even blood flow

(b)

Fully recruited capillary beds

Rapid transit →

Rapid transit →

Blood flow in 1 minute

Figure 21–6

(a) During resting conditions, most of the capillaries in the lower lung are open (recruited), with blood flowing briskly through them. In the upper lung, few capillaries are recruited and what flow there is moves at a slow pace. (b) With the increased blood flow during exercise, more blood perfuses both the upper and lower lung. In the lower lung, the primary change is that the blood moves more rapidly through the capillaries. In the upper lung, not only does the blood move more rapidly, but there is also capillary recruitment.

from the bottom to the top of the lungs can be useful only if the resulting pulmonary hypertension is either mild or transient (see Table 21–1).

Exercise Recruits Capillaries and Decreases Transit Time

As cardiac output increases during exercise, pulmonary arterial pressure rises due to the increased quantity of blood flowing through the lung. The pressure rise is modest at low levels of exercise such as walking (see Table 21–1). As the level of exercise increases, pulmonary arterial pressure continues to increase, resulting in an adequate driving force to propel the blood to the uppermost parts of the lung (Fig. 21–6). This results in more even perfusion of the lung from top to bottom and improves the ability of the lungs to exchange gas.

As blood flow increases, two important changes take place in the capillaries to improve oxygen uptake from the alveolar gas. The first change involves the number of capillaries with blood flowing through them. In the lower part of the lungs, blood flows through most of the capillary segments (Fig. 21–6a), even during resting conditions when cardiac output is low. In the upper part of the lungs, relatively few capillaries are perfused during rest, which leaves capillaries in reserve that can be recruited as the demand for oxygen increases. As cardiac output rises to higher levels during exercise, pulmonary arterial pressure continues to rise as well, and more blood flow is directed to the upper part of the lungs, where reserve capillaries are recruited (see Fig. 21–6b). In this way, the surface area for gas exchange is increased.

The second change that takes place during exercise to improve oxygen uptake is more rapid movement of the blood through the capillaries. Even though the normal transit time through the capillaries of one second is rapid, the blood is completely oxygenated in only a quarter of a second. Thus, a large part of the capillary **transit time** is wasted. As cardiac output increases with exercise, the transit time decreases until it approaches a quarter-second minimum time. This change causes more blood to be oxygenated every second during exercise. Again, there is reserve in the upper lung, where the transit times are considerably longer than one second at rest. As blood flow is shifted upward during exercise, evidence suggests that not only are there capillaries to be recruited, but also that the slowly flowing blood is speeded up as well.

Gas Uptake and Transport

Once the inspired air approaches the alveoli, the air flow slows nearly to zero. The predominant means of air movement is no longer the bulk flow that moves along a pressure gradient, but diffusion, which causes individual gases to move along a concentration gradient. In the lung, the diffusion distance across an alveolus is small, in the range of 100 to 200 μm. Gas molecules diffuse very quickly over such short distances; hence, diffusion is an effective means of moving oxygen molecules rapidly to and through the alveolar-capillary membrane.

Partial Pressures Move Respiratory Gases Across Alveolar-Capillary Membrane

Recall from Chapters 2 and 20 that partial pressures of gases are used to tell us how much pressure is exerted by one particular gas in a mixture of gases. The relationship between the total pressure exerted by a mixture of gases and the pressure of individual gases is governed by **Dalton's law,** which states that the total barometric pressure (barometric pressure = PB) is equal to the sum of the partial pressures of the individual gases. Air is made up of different gases, mostly nitrogen and oxygen, each of which exerts its own pressure, called its partial pressure or sometimes called gas tension. The partial pressures are the pressures that the individual gases would exert if each gas were present alone in the volume occupied by the whole mixture at the same temperature. Thus, the partial pressure of oxygen (P_{O_2}), according to Dalton's Law, is determined as $P_{O_2} = P_B \times F_{O_2}$, where F_{O_2} is the fractional concentration of oxygen. Since 21% of air is made up of oxygen, then 0.21 is the fractional concentration of 760 mm Hg of air pressure at sea level that is accounted for by oxygen. In this example, the partial pressure (P_{O_2}) exerted by oxygen is 160 mm Hg (760 × 0.21). If all of the other gases in a container of air were removed, the remaining oxygen by itself would still exert a pressure of 160 mm Hg at sea level. Atmospheric pressure (760 mm Hg) is the sum of all the partial pressures in air (Table 21–2). The percentages for each gas remain constant regardless of the barometric pressure.

The air in the lung is moisturized as it passes through the airways until it becomes saturated with water. The water vapor exerts a partial pressure, just as oxygen and nitrogen do. The water vapor in saturated air is directly related to the temperature. In the case of a normal body temperature of 37°C, water vapor is 47 mm Hg. Therefore, to compute the partial pressure of O_2 in moist tracheal air, 47 mm Hg must be subtracted from the barometric pressure: $P_{O_2} = (760 - 47) \times 0.21$. The partial pressures of gases are shown in Table 21–2.

As inspired air reaches the respiratory zone of the lung, the air mixes with alveolar gas (see Chapter 20) that, being relatively rich in carbon dioxide and poor in oxygen, alters the composition of the incoming air. The final composition of alveolar gas has a partial pressure of oxygen, of 104 mm Hg. Even though that

Table 21–2
Partial Pressures of Various Gases in Air*

	Dry Air	Moist Tracheal Air (37°C)	Alveolar Gas	Arterial Blood	Mixed Venous Blood
P_{O_2}	159.1†	149.2†	104†	100	40
P_{CO_2}	0.3	0.3	40	40	46
P_{H_2O}	0.0	47.0	47	47	47
P_{N_2}	600.6	563.5	569	573	573
P total	760.0	760.0	760	760	706

*Usual values in a resting, healthy male at sea level (barometric pressure = 760 mm Hg).

†This is an approximate value and holds only for males breathing air at sea level. Because the total atmospheric pressure at Denver, Albuquerque, or Salt Lake City is about 640 mm Hg, the partial pressure of O_2 in inspired and alveolar gas is significantly below values at sea level.

figure is reduced from the atmospheric partial pressure of oxygen (160 mm Hg), there is still a sufficient partial pressure gradient to drive oxygen from the alveolar gas into the capillary blood. The initial O_2 diffusion gradient is 64 mm Hg. (Alveolar oxygen tension is 104 mm Hg, while mixed venous blood oxygen tension is 40 mm Hg—see Table 21–2.)

Once inspired air has reached the alveolus, and venous blood has reached the capillaries in the alveolar wall, gas exchange can take place. The partial pressure gradients between alveolar gas and capillary blood cause oxygen and carbon dioxide to move in opposite directions (Table 21–2). The rate of movement of molecules by diffusion is governed by Fick's Law (Chapter 2). To accomplish such movement, the gases must cross the alveolar-capillary membrane. The oxygen in the alveolar gas dissolves in the membrane just as it would in water. The oxygen passes through the thin membrane and into the plasma, where it remains in solution. Diffusion continues through the plasma and then through the very thin membrane of the red blood cell. It is inside the red cell that the majority of oxygen is transported to the body.

Oxygen Is Transported by Hemoglobin

Every 100 ml of blood that passes through tissue delivers about 5 ml of oxygen. A normal young adult at rest requires about 250 ml of oxygen per minute. This means that the heart must pump 5 liters per minute [$(250/5) \times 100 = 5000$ ml/min].

A schema of the circulatory delivery is shown in Figure 21–7. During each minute, 7.5 liters of air (minute ventilation) are moved in and out of the lungs. From that, 250 ml/min of oxygen are taken up through the pulmonary capillaries and transported to the tissue capillaries, where 250 ml/min of oxygen are taken up by the tissue. The venous blood is still partially loaded with oxygen after passage through the tissue capillaries (Fig. 21–7). After returning to the right heart, the blood is circulated once again through the lungs.

With exercise, the body's requirement for oxygen increase considerably. During heavy exercise, a normal, physically fit adult might require 12 times the resting amount of oxygen, or 12×250 ml/min = 3000 ml/min. A cardiac output of 20 L/min is required to circulate enough blood to deliver 3000 ml/min of oxygen. In highly trained athletes, the oxygen demands may exceed 5000 ml/min. Cardiac outputs may be more than 30 L/min in these endurance athletes.

Even though these numbers are large, they are much smaller than would be required if, in order to be transported, oxygen had to dissolve in blood. If

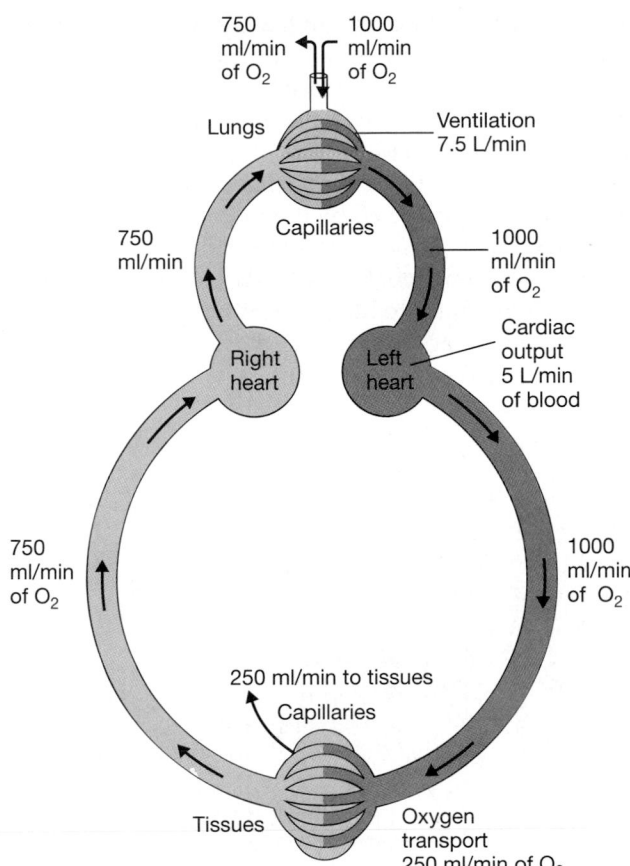

Figure 21–7

Schema of the circulation, showing oxygen deliveries.

such were the case, the cardiac output would have to be more than 1000 L/min. That flow rate would completely fill an automobile gasoline tank in 3 seconds! Such a delivery rate would require a huge heart. Each beat of the heart would cause the blood to exceed the sound barrier.

A remarkable molecule, hemoglobin, solves the problems of transporting oxygen. Hemoglobin is a pigmented protein that permits large amounts of oxygen to be carried to the tissues in an efficient way. The structure of hemoglobin was presented in Chapter 17. Each hemoglobin molecule contains four heme molecules, each of which contains an iron atom. A single iron atom is capable of binding with a molecule of oxygen. When the partial pressure of oxygen is high, oxygen combines loosely with the heme portion of hemoglobin in the lung to form **oxyhemoglobin.** When the oxyhemoglobin reaches the tissue where oxygen pressure is low, the hemoglobin releases the oxygen. Because of the loose way in which O_2 binds with iron atoms, the oxygen is delivered to the tissue fluids as dissolved molecular oxygen.

The capacity of hemoglobin (Hb) for oxygen is much higher than that of plasma. About 98% of the oxygen in the blood is carried by the hemoglobin. Every 100 ml of blood can combine with 20 ml of oxygen. This value is calculated assuming that 1 g Hb carries 1.34 ml O_2 and a normal hematocrit of 15 g/100 ml of blood (1.34 ml O_2/g Hb × 15 g/100 ml blood = 20 ml O_2/100 ml). The maximum amount of O_2 that can be carried by hemoglobin is called **oxygen carrying capacity,** and is frequently written as 20 volumes percent (vol.%). Arterial blood usually has less oxygen bound to hemoglobin than 20 ml O_2/100 ml of blood. The amount actually bound to hemoglobin is called **oxygen content** (where capacity is the amount that can potentially be bound). The ratio of O_2 content/O_2 capacity is **percent O_2 saturation.** The percent O_2 saturation of hemoglobin is calculated as:

$$\text{percent } O_2 \text{ saturation} = \frac{O_2 \text{ content}}{O_2 \text{ capacity}} \times 100.$$

Now we can see how O_2 loads with hemoglobin in the lung. Fig. 21–8a shows an important physiological curve that is "S" shaped and called the **oxyhemoglobin dissociation curve.** If the blood is 100% saturated, then all of the binding sites on the hemoglobin contain an oxygen molecule. If the blood is

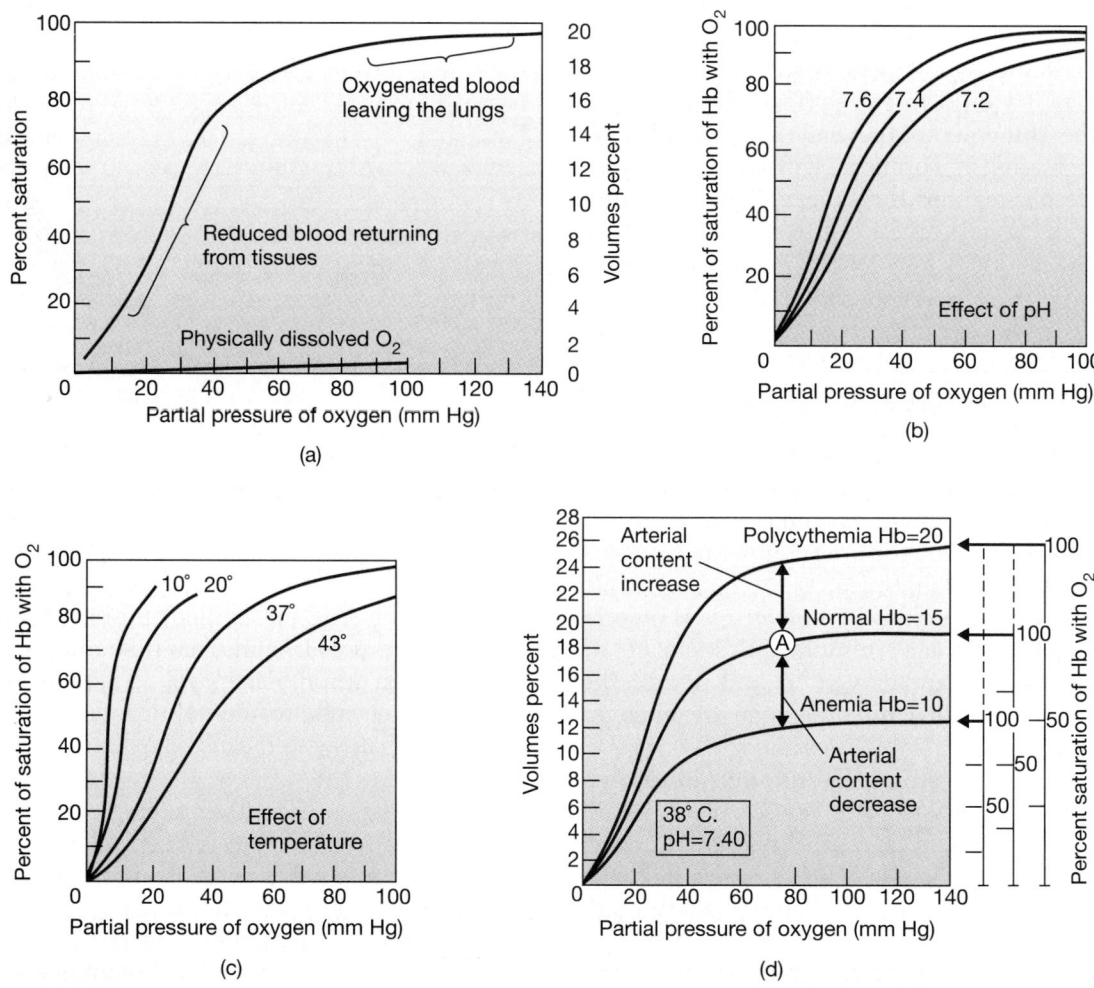

Figure 21–8

(a) The oxyhemoglobin dissociation curve, showing the way in which oxygen binds to hemoglobin in the lung and unloads oxygen in the tissues. Note that as oxygen tension increases in the mid-range, the hemoglobin loads rapidly with oxygen. As the hemoglobin molecules become saturated with oxygen, loading slows as oxygen tension increases. (b and c) Both pH and temperature cause the oxyhemoglobin curve to shift. A shift in the curve causes the hemoglobin to load or unload more easily. In the lungs, where the temperatures are relatively cool and the pH is relatively alkaline, loading is facilitated by the leftward shift of the curve. In an exercising muscle where the environment is relatively warm and acidotic, unloading is facilitated. (d) Oxygen dissociation curve in anemia and polycythemia.

50% saturated, then only half of the hemoglobin binding sites contain oxygen molecules. Saturation gives an indication of how completely the binding sites are filled and how many are left in reserve. Another way to measure the amount of oxygen in the blood is to measure oxygen content—that is, how many milliliters of oxygen are actually carried by each 100 ml of blood. As we have seen before, the unit for content is vol.%. That measure is found on the right-hand axis of Figure 21–8a. Under normal resting conditions, systemic arterial blood is 97% saturated and a content of 19.4 vol.% (20 vol.% × 0.97). Venous blood has a partial pressure of oxygen of 40 mm Hg and a content of 15 vol.% (see Table 21–2). Under certain conditions, the amount of hemoglobin in the blood can vary. The effect is shown in Figure 21–8d. The normal curve in the middle has 15 gm of hemoglobin in every 100 ml of blood and when it is 100% saturated, it carries about 20 ml of oxygen. In cases of anemia, when the number of red blood cells is reduced, there is less hemoglobin in each 100 ml of blood. In the lower curve with only 10 gm of hemoglobin per 100 ml of blood, less oxygen can be carried (1.34 × 10.0 = 13.4 ml of oxygen). However, the blood is still 100% saturated when this amount of oxygen is carried, because all of the hemoglobin binding sites have oxygen attached (see saturation axis in Figure 21–8d). In the upper curve in Figure 21–8d, there are abnormally large numbers of red blood cells present, a condition known as polycythemia. In this case more oxygen can be carried (20.0 × 1.34 = 26.8 ml of oxygen) when the blood is 100% saturated.

As shown in Figure 21–8a dissolved oxygen makes a small contribution to the total carrying capacity of the blood. For example, in a healthy person at rest, with a partial pressure of oxygen in the blood of 100 mm Hg, the resulting value for dissolved oxygen is 0.30 vol.% and the hemoglobin will be 97% saturated. If the hemoglobin is 15 gm per 100 ml of blood, then the oxygen combined with hemoglobin is 0.97 × 15.0 × 1.36 = 19.79 vol.%. The total content is 19.79 + 0.30 = 20.09 vol.%.

Oxygen Delivery During Rest and Exercise

Under normal conditions when alveolar tension is 100 mm Hg, the hemoglobin saturation of systemic arterial blood is 97% and the oxygen content is 20 vol.%. Normal venous blood with its oxygen tension of 40 mm Hg is 75% saturated (15 vol.%). Thus 5 ml of oxygen is delivered for each 100 ml of blood pumped by the heart through the systemic capillaries, which means that the arteriovenous oxygen difference is 5 vol.%. In other words, the tissues extract 5 ml of oxygen for every 100 ml of blood coming through them.

During exercise, the oxygen tension in the muscles drops to lower values than when at rest because oxygen is being used so rapidly by the working muscle. Heavy exercise can reduce muscle oxygen tension to below 15 mm Hg. This increases the gradient that favors the movement of more oxygen from the capillary blood and into the tissue. Thus the delivery of oxygen is increased because of greater extraction. For example, if the venous oxygen tension drops 40 mm Hg to 15 mm Hg, then the content drops to about 4 vol.%. Now much more oxygen is delivered by each 100 ml of blood: 20 − 4 = 16 ml of oxygen, which triples the delivery simply by increasing the extraction.

Some interesting characteristics of hemoglobin facilitate oxygen transport. Blood acidity (see Fig. 21–8b) shifts the position of the oxyhemoglobin dissociation curve to the right. For a given oxygen tension, the amount of oxygen (oxygen content) is lower. This means that for a given oxygen tension in the tissue, the hemoglobin will unload more oxygen. Exercising muscle has greater acidity owing to increased carbon dioxide and lactic acid production. The unloading of oxygen from hemoglobin is therefore facilitated. Similarly, increased temperature also causes a rightward shift of the curve (see Fig. 21–8c). Blood temperature rises when it reaches a hot exercising muscle, and oxygen unloading is enhanced.

A leftward shift in the oxyhemoglobin dissociation curve facilitates oxygen loading. At a given oxygen tension, the content and saturation are higher. As the blood enters the alveolar capillaries, carbon dioxide leaves the blood, the pH becomes more basic, and oxygen loading is enhanced. Similarly, cooler temperatures cause a leftward shift of the curve and, since the lungs are cooler than exercising muscle, a leftward shift is also promoted. Thus, the oxyhemoglobin curve shifts in a remarkable way from right to left and back to the right with each passage around the circulation in order to favor oxygen loading in the lungs and unloading in the tissues.

Carbon Dioxide Is Transported as a Bicarbonate Ion

Carbon dioxide is transported in the circulation more readily than oxygen, because carbon dioxide—being a nonpolar molecule—is highly soluble in lipid and thereby moves easily across plasma membranes. Because of this, blood carbon dioxide tension tends to remain more nearly normal than oxygen in many diseases. Since the maintenance of pH in the body is crucial to so many chemical reactions, it is important that carbon dioxide remain normal. Carbon dioxide in the blood is intimately linked to acid-base balance, as described in Chapter 25.

Figure 21–9

(*a*) The ways in which carbon dioxide is carried by the blood from the tissues to the lungs. (*b*) The carbon dioxide dissociation curve.

In a normal resting adult, 4 vol.% of carbon dioxide are transported to the lungs to be exhaled each minute. Carbon dioxide diffuses from the tissue into systemic capillary and is transported in the blood in three ways (Fig. 21–9).

The largest fraction of carbon dioxide (60% to 70%) is transported as bicarbonate ions (HCO_3^-). These ions are formed by an important series of reactions that take place inside the red blood cell.

First, carbon dioxide and water in the red blood cell combine to form carbonic acid as shown in this equation:

$$CO_2 + H_2O \xrightleftharpoons{\text{carbonic anhydrase}} \underset{\text{carbonic acid}}{H_2CO_3}$$

Ordinarily, this is a slow reaction requiring many seconds for completion, but when catalyzed by the enzyme carbonic anhydrase, the reaction is accelerated by a factor of 5000 and is complete in a fraction of a second.

In the next chemical step, the carbonic acid dissociates into a bicarbonate ion ($H_2CO_3^-$) and a hydrogen ion (H^+):

$$H_2CO_3 \rightleftharpoons \underset{\text{bicarbonate}}{HCO_3^-} + H^+$$
$$\underset{\substack{\text{carbonic}\\\text{acid}}}{}$$

The bicarbonate ion moves from the red cell into the plasma, where it is readily dissolved. This leaves a hydrogen ion free in the red cell, where it rapidly combines with hemoglobin in the following way:

$$H^+ + Hb^- \rightarrow HHb$$

As the bicarbonate ions diffuse into the plasma, chloride ions diffuse into the red blood cells to replace the bicarbonate. This exchange, known as the **chloride shift,** causes red blood cells in venous blood to have a higher chloride content than red blood cells in arterial blood (see Fig. 21–9*a*). These reactions are summarized in Figure 21–9.

Carbaminohemoglobin in CO_2 Transport

Carbon dioxide can combine directly with hemoglobin to form **carbaminohemoglobin.** The carbon dioxide does not bind at the same site on the iron molecules as oxygen but instead binds loosely by a direct reaction with some of the amino groups that form the hemoglobin molecule. Between 15% and 30% of carbon dioxide transport is accomplished in this form (Fig. 21–9*a*). The hemoglobin molecule has a higher affinity for carbon dioxide when it is oxygen-desaturated—exactly the condition that exists in venous blood, in which oxygen is low and carbon dioxide is high. The bond between hemoglobin and carbon dioxide is loose enough to be reversible in the lung. The carbaminohemoglobin reaction is much slower than the enzyme-catalyzed reaction that forms bicarbonate.

Dissolved Carbon Dioxide in CO_2 Transport

Between 7% and 10% of the carbon dioxide is transported in a dissolved state in the plasma and red blood cells. This accounts for only about 0.3 vol.% of the total carbon dioxide transport.

All three of these methods of transport are readily reversible in the pulmonary capillaries, where the

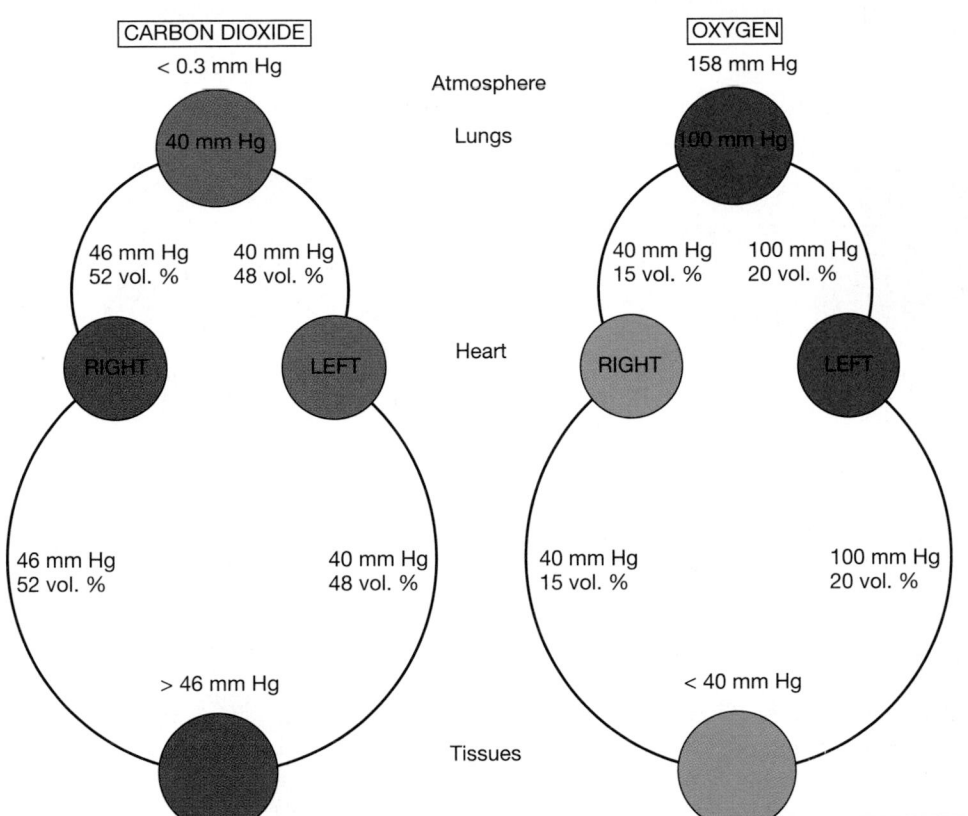

CARBON DIOXIDE

< 0.3 mm Hg

Atmosphere

Lungs

40 mm Hg

46 mm Hg
52 vol. %

40 mm Hg
48 vol. %

RIGHT Heart LEFT

46 mm Hg
52 vol. %

40 mm Hg
48 vol. %

> 46 mm Hg

Tissues

OXYGEN

158 mm Hg

100 mm Hg

40 mm Hg
15 vol. %

100 mm Hg
20 vol. %

RIGHT LEFT

40 mm Hg
15 vol. %

100 mm Hg
20 vol. %

< 40 mm Hg

Figure 21–10

A comparison of the partial pressures of carbon dioxide and oxygen in the circulation during resting conditions. It is important to realize that arterial blood contains carbon dioxide, and that venous blood contains oxygen. That is, the blood is not scrubbed clean of carbon dioxide in the lungs, and not all of the oxygen is removed from the blood by the tissues. Partial pressures are in mm Hg and contents in vol. %.

concentration gradient favors the movement of carbon dioxide from the blood into the alveolar gas. Since the equations read in the reverse as well as the forward direction, the transport cycle is complete when carbon dioxide is given off in the lung.

All of the ways in which carbon dioxide can be transported in the blood are dependent on the partial pressure of carbon dioxide, just as is the case with oxygen. The carbon dioxide dissociation curve, with its narrow normal range of operation of between 40 and 46 mm Hg, is shown in Figure 21–9b. The total carrying capacity of blood for carbon dioxide is considerably higher than for oxygen (compare Figures 21–9b and 21–8a). As blood passes through the systemic capillaries, the carbon dioxide content rises to about 52 vol.%. As the blood passes through the pulmonary capillaries, 4 vol.% of carbon dioxide are unloaded into alveolar gas and the carbon dioxide content falls to 48 vol.%.

An important fact to remember is that the blood does not completely empty itself of carbon dioxide in the lung or of oxygen in the tissues. Even during the most strenuous exercise—when oxygen extraction is high in the tissues and ventilation is high in the lungs—oxygen and carbon dioxide always remain in the blood. The partial pressures in mm Hg and contents in vol.% of these gases in various regions of the circulation are shown in Fig. 21–10.

Fetal and Neonatal Circulatory Physiology

The Placenta Is the Gas-Exchange Organ before Birth

Before birth, the lungs have no respiratory function, because air is unavailable. The fetus must depend on the mother for gas exchange. This is accomplished through the placenta (Fig. 21–11), in which gas exchange vessels from fetus and mother pass close to each other. Such a system is less effective than lungs exposed directly to the atmosphere and results in a low fetal arterial P_{O_2}, on the order of 25 to 30 mm Hg. Blood travels to and from the placenta through the umbilical arteries and veins. The oxygenated blood flows into the right heart, but only 10% to 12% flows through the lungs. About half of the blood flows from the right to the left atrium through a passage called the **foramen ovale**. This oxygenated blood is pumped by the left ventricle out the aorta and perfuses mostly the head and upper body. The remainder of the blood follows the usual pathway through the right heart into the pulmonary artery. The bulk of pulmonary arterial blood, however, flows through the **ductus arteriosus** into the descending aorta to perfuse the lower body. More than half of the cardiac

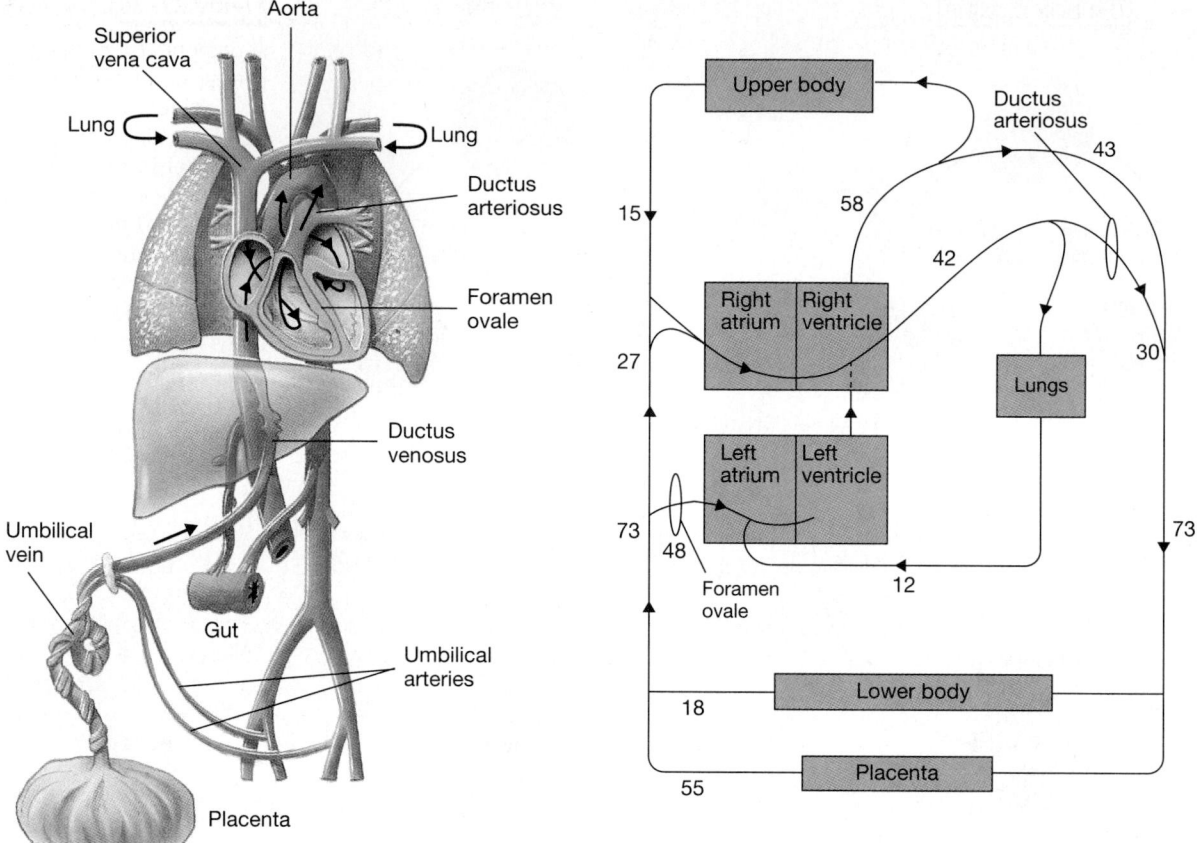

Figure 21–11

In the fetus, the circulation is different from that in the adult in a number of important ways. This figure shows the names of the blood pathways peculiar to the fetus and also shows how the cardiac output is distributed. Note that the partial pressures of oxygen in mm Hg (the numbers in right panel) in the fetus are very low compared to postnatal values.

output passes through the umbilical artery and into the gas-exchanging vessels of the placenta (Fig. 21–11).

The changes that occur in the circulation at birth are almost as important as the act of breathing itself. As soon as the umbilical cord is severed, the increased blood flow in the aorta immediately raises pressure. As the lungs expand, the pulmonary vessels are no longer compressed, and hypoxic vasoconstriction is relieved; these have the combined effect of lowering pulmonary vascular resistance nearly ten-fold. Because left atrial pressure is 2 to 4 mm Hg higher than right atrial pressure, the flap that lies on the left side of the atrial septum closes the foramen ovale. A few hours after birth, the powerful muscular wall of the ductus arteriosus constricts and closes that vessel. Growth of fibrous tissue into the lumen of the ductus over the course of the next few months causes permanent closure.

Congenital Heart Disorders Interfere with Gas Exchange

In some newborns, the ductus does not close after birth. This condition, known as **patent ductus arteriosus** (patent means "open"), can lead to a left-to-right shunt of blood from the aorta into the pulmonary artery. The high aortic pressure causes pulmonary hypertension, which in severe cases leads to irreversible damage to the pulmonary arteries.

Sometimes, the pulmonary hypertension becomes so great that the shunt reverses and flows from right to left. Such a shunt forces venous blood into the aorta and causes hypoxemia. Patients with patent ductus arteriosus can experience right ventricular failure and die. The patent ductus can be readily closed by surgery, and if the diagnosis is made in time, the disorder can be repaired with few untoward effects. The diagnosis is made by listening to the chest with a

CLINICAL FOCUS

Pulmonary Emboli

The design of the pulmonary circulation, with its many branching arteries all feeding into the capillary net, acts as a sieve for blood as it circulates through the body. This design has an important protective effect in that material in the circulation larger than red blood cells will plug small vessels and prevent the flow of blood. In the systemic circulation, such blockages can be life threatening if they occur in the coronary vessels (causing heart attack) or in the brain (causing stroke).

The most common abnormally large objects in the circulation are blood clots. These often form in the veins in the legs, especially when blood flow is slowed because of varicose veins or prolonged bed rest. A blood clot stuck in a vein is called a thrombus. If the clot breaks loose, it is called an embolus, or a thromboembolus. The embolus proceeds through the right side of the heart and out into the pulmonary arteries, moving farther and farther into the lung until it wedges in a branch of the pulmonary arterial tree. The resultant trapping blocks blood flow to part of the lung. Since the gas exchange surface area in the pulmonary circulation is so vast, oxygenation of blood is not adversely affected unless the blockage is massive. Sometimes the thromboembolus is so large that it can block the main pulmonary artery and cause death. Smaller clots can resolve over a period of a month or so, allowing the pulmonary circulation to return to normal.

Any large material in the venous circulation will be filtered by the pulmonary circulation. Bone marrow, from broken bones, for example, sometimes embolize to the lung. If air enters the venous circulation, bubbles will form and cause air embolization of the lung. The air diffuses across the air-blood barrier and is exhaled with alveolar gas, so air emboli do not last long. Much air would have to be injected to cause harm.

In the accompanying figure, the pulmonary circulation has been embolized by liquid mercury (the small white lines in the lung field). The X-ray was taken of a drug user who injected mercury intravenously for a special "kick." It is difficult to tell from an X-ray whether the mercury is in blood vessels or small airways. The differential diagnosis is made by the pool of mercury (arrow) in the bottom of the right ventricle.

■ **Why is heparin administered immediately to a patient with a strong suspicion of pulmonary emboli?**

(X-ray of an embolism created by an injection of liquid mercury. Courtesy of Dr. William Beamish, University Hospital, Edmonton, Alberta.) Fraser and Paré: *Diseases of the Chest*, Saunders.

stethoscope. Blood traveling through the patent ductus makes a sound called a murmur that alerts the physician that an abnormality is present.

Numerous other defects can occur during the development of the fetus. For example, the septum between the atria can be incompletely formed, causing an atrial septal defect. A similar defect in the ventricles is called a ventricular septal defect. Although there are a number of other congenital heart defects, fortunately these disorders as a group are rare and can often be repaired by modern surgical techniques, such as those employing a heart-lung bypass machine.

Control of Breathing

Breathing is an automatic process and occurs without any conscious effort while we are awake and asleep, and even while we are under anesthesia. We can, however, exert some conscious control over our own ventilation by voluntarily changing the rate and depth of breathing (Fig. 21–12). We can also voluntarily stop breathing for a short period of time until carbon dioxide builds up in the blood, which will then stimulate breathing regardless of how hard we try to hold our breath. Clearly, the basic automatic rhythm of breathing can be modified by both neural and chemical stimuli (see Fig. 21–12). Remember that breathing, although automatic, depends entirely on the cyclical excitation of the respiratory muscles. The diaphragm receives a respiratory signal from the **phrenic nerve,** which leaves the spinal cord in the upper half of the neck and passes down through the chest to innervate the diaphragm.

Breathing Has a Basic Rhythm

In a resting individual, the normal rate of breathing is approximately 14 breaths per minute. The control for the basic rhythm is housed in the **medulla,** the lower portion of the brain stem, which also contains the cardiovascular control centers.

Within the medulla are a cluster of neurons, **inspiratory neurons** (Fig. 21–13), responsible for the basic rhythm. These neurons have the property of intrinsic firing (see Chapter 11). The intrinsic rhythm pattern starts with a latent period of several seconds in which there is no activity. Action potentials begin and increase over the next few seconds. During this build-up of action potentials, inspiratory muscles are activated in crescendolike fashion. Finally, the action potentials cease and the inspiratory muscles relax. The cyclic pattern then repeats itself.

The medullary center for controlling the basic rhythm also receives a rich supply of synaptic inputs

Figure 21–12

Recordings of tidal volume and phrenic nerve action potential to the diaphragm. (*a*) Normal breathing pattern with basic rhythm. (*b*) The rate of discharge determines depth of breathing (tidal volume), while the time interval between action potentials determines breathing rate (*c*). (*d*) Both rate and depth of breathing are increased.

from other neurons that come from the cortex and pons. Cortical input transmits the conscious effort of breathing. In automatic breathing, neuronal inputs from the pons modulate the inspiratory center and act as a "cut off" signal to inhibit inspiration.

Another potential "cut off" signal comes from *proprioceptors*. These include **pulmonary stretch receptors** embedded in the smooth muscles of the airways and stretch receptors located in skeletal muscle spindles. When activated, these stretch receptors

Figure 21–13

The respiratory center located in the medulla. The center consists of two clusters of neurons: inspiratory and expiratory neurons.

send impulses (action potentials) to the medulla and either inhibit or stimulate the inspiratory neurons. This feedback mechanism from the airways not only helps stop inspiration but also helps prevent overinflation, which can damage the lung and alter the blood flow through the lung. In the newborn, the pulmonary stretch receptors play a very important role in regulating the basic rhythm and preventing overinflation. However, in the adult, the threshold for these receptors is high and the stretch receptors play a role only under conditions in which there is a very large tidal volume (e.g., heavy exercise). The proprioreceptors located in skeletal muscle play an important role in stimulating ventilation during exercise. The details of how ventilation is stimulated by exercise will be discussed in Chapter 30.

Inhibition of the inspiratory neurons is all that is required for expiration to occur, since expiration is normally a passive process. In more forceful breathing, however, expiration becomes active; **expiratory neurons** (see Fig. 21–13) stimulate the expiratory muscles. These expiratory neurons are also located in

the medulla, in close proximity to the inspiratory neurons, and are quiescent (inactive) during normal breathing. They become active during forced expiration such as exercise.

Failure of the Respiratory Center

The respiratory center located in the medulla is somewhat fragile and can disrupt the respiratory cycle. Two of the more frequent causes of basic rhythm disruption are cerebral concussions and cerebral edema, especially around the medulla. Both of these cause blood vessel collapse, which stops blood flow and causes the respiratory rhythm to cease functioning.

Another frequent cause of respiratory failure is barbiturate overdose. Barbiturates depress the inspiratory neurons and stop the rhythm. General anesthesia and narcotic pain relievers also act in a similar fashion, inhibiting the respiratory center. If too much anesthesia is given, breathing will stop altogether. Once breathing stops, it is extremely difficult for the body to start the cycle up again. Very few drugs can be used to reactivate the inspiratory rhythm. In general, those available to excite the respiratory centers, such as caffeine, are too weak to be of any value. The only effective treatment is to place these individuals on a mechanical ventilator to sustain ventilation until the body itself can rectify the problem. In some cases of severe respiratory depression, it may take weeks and sometimes months before the body can resume breathing on its own.

Changes in Blood Gases Stimulate Ventilation

Although the basic rhythm of ventilation is continuous, the rate and depth of breathing can vary tremendously depending on metabolic demands. First, when muscle cells need more oxygen (i.e., exercise), alveolar ventilation must increase. Second, when cells become more active, they produce more CO_2. The body rids itself of the excess CO_2 by increasing alveolar ventilation. Third, in order for metabolic processes to occur, a constant pH environment must be maintained. The appropriate hydrogen ion concentrations are regulated, in part, by alveolar ventilation. These three chemicals, **oxygen, carbon dioxide,** and **hydrogen ions (H^+),** can profoundly change alveolar ventilation by affecting both the rate and the depth of breathing (see Fig. 22–12).

Central and Peripheral Chemoreceptors

Two sets of **chemoreceptors** are strategically located in the body to detect changes in arterial partial pressures of oxygen and carbon dioxide and changes in

RESEARCH FOCUS

Transit of Leukocytes Through the Pulmonary Circulation

Of all the white blood cells that circulate in the blood, neutrophils make up the largest proportion and are one of the body's chief defenders against invading microorganisms. For unknown reasons, neutrophils travel through the lung far more slowly than red blood cells. This causes neutrophils to concentrate in the small pulmonary blood vessels. From that site, they can either defend the body against inhaled bacteria or move quickly through the circulation to any part of the body.

There are times, however, when neutrophils misuse their powerful bacteria-killing ability and damage the blood vessels, which causes plasma to leak into the tissue. In the lung, edema in alveoli increases the thickness of the blood-gas barrier and impedes the exchange of oxygen and carbon dioxide. In its severe form, this condition is called the Adult Respiratory Distress Syndrome.

Because neutrophils are present in large numbers in the lungs of patients with Adult Respiratory Distress Syndrome, they are implicated in the etiology of this disorder. Therefore, considerable investigative attention has been focused on what causes the neutrophils to stop under normal conditions in the pulmonary circulation. Two hypotheses are currently being investigated. The first, the mechanical impediment hypothesis, suggests that the highly viscous nature of the neutrophils impedes their travel through the pulmonary capillary bed.

When the neutrophils reach capillary segments with diameters smaller than their own, they require considerable time to deform sufficiently to fit through the narrow lumens of the obstructing capillary segments. The second hypothesis, the adhesion hypothesis, is based on the fact that the neutrophil surface membrane has many adhesion glycoproteins that recognize receptors on the cells lining the blood vessel walls. There the binding of the neutrophils to the endothelial cells may slow them down. These adhesion sites are crucial in the cells' response to invading bacteria, because they anchor the neutrophils to the vessel wall, allowing them to then move through it and attack the microorganisms. Their function under normal conditions, however, is unknown.

The mechanical impediment hypothesis is being investigated by treating neutrophils with chemicals that make them either more rigid or more flexible. After treatment, the transit times of the neutrophils through the pulmonary circulation are measured to determine whether the altered viscosity affects transit. To test the adhesion hypothesis, scientists are determining the effect on neutrophil transit times of antibodies that specifically block the functioning of the adhesion molecules.

■ **How does exercise affect neutrophil transit time in the lungs?**

---

hydrogen ion concentration. One set of receptors is located in the medulla. These receptors are referred to as **central chemoreceptors;** they monitor only changes in cerebral spinal fluid hydrogen ions (Fig. 21–14). The other set of receptors, located outside the central nervous system, are called **peripheral chemoreceptors.** These peripheral chemoreceptors are located in the arch of the aorta **(aortic bodies)** and in the neck at the bifurcation of the common carotid artery **(carotid bodies)** (Fig. 21–15). Although they respond mainly to changes in oxygen tension (PaO_2), they are also stimulated by changes in $PaCO_2$ and $[H^+]$ in the arterial blood. These peripheral chemoreceptors are made up of epithelial-like cells that contain neurons in contact with the arterial blood. These receptors sense changes in partial pressure of O_2, CO_2, and $[H^+]$ concentrations in the arterial blood. The afferent (sensory) nerve fibers that arise from these cells extend to the respiratory center in the medulla.

Of the three chemicals that influence aveolar ventilation, carbon dioxide is the most powerful (Fig. 21–16). When arterial carbon dioxide levels increase

above normal, all portions of the respiratory center become excited. An increase of only 2 to 5 mm Hg in arterial carbon dioxide can more than double alveolar ventilation. The second most powerful chemical stimulus to ventilation is arterial pH. Arterial oxygen has the smallest effect on ventilation.

Control of Arterial CO_2

The only effective way the body can regulate short-term carbon dioxide levels in the blood is through alveolar ventilation. If arterial CO_2 levels suddenly become too high, all chemical reactions essentially stop because most enzymatic reactions operate within a precise range of pH. To prevent this, the respiratory center regulates alveolar ventilation so that PCO_2 stays within a narrow range. Increased alveolar ventilation is known as **hyperventilation,** which results in "blowing off" excess CO_2 and thereby lowering arterial CO_2 back to normal. Conversely, if arterial PCO_2 becomes too low, the blood becomes too alkaline (basic), which depresses the respiratory cen-

Figure 21–14

Central chemoreceptor located in the medulla next to the respiratory center measure changes in cerebral spinal fluid pH. Note that the blood-brain barrier is impermeable to hydrogen ions (H^+) and to bicarbonate (HCO_3^-). The central chemoreceptors are much more sensitive than the peripheral chemoreceptors.

ter and causes a decreased alveolar ventilation, or **hypoventilation.** With hypoventilation, CO_2 is retained and arterial P_{CO_2} will be elevated back to normal.

The control of ventilation by carbon dioxide is mediated by both the central and peripheral chemoreceptors. When CO_2 diffuses across the blood-brain barrier (see Fig. 21–14), it combines with water by the enzyme carbonic anhydrase, to form carbonic acid (H_2CO_3), which rapidly dissociates into bicarbonate (HCO_3^-) and hydrogen ion (H^+) in the following manner:

$$H_2O + CO_2 \rightleftharpoons H_2CO_3 \rightleftharpoons HCO_3^- + H^+$$

Since H^+, HCO_3^-, and CO_2 are in equilibrium, any increase in arterial P_{CO_2} will cause a profound increase in the hydrogen ion concentration in the cerebrospinal fluid (CSF), which in turn will increase in alveolar ventilation. The brain is especially sensitive to pH changes because the cerebrospinal fluid lacks

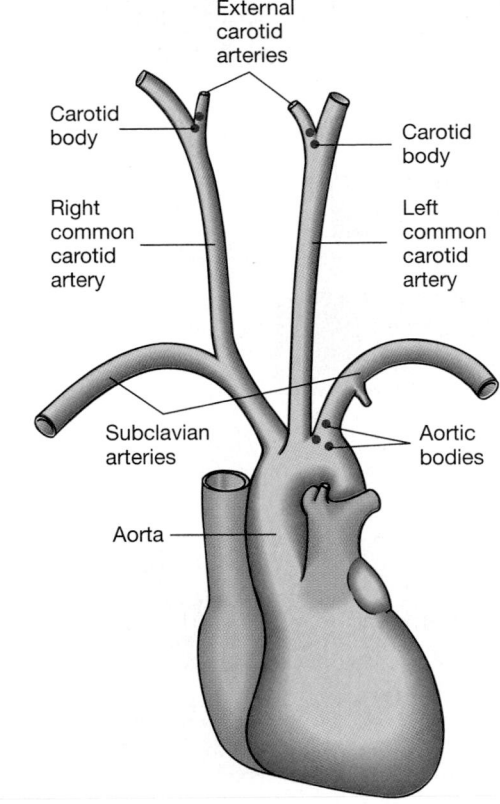

Figure 21–15

Peripheral chemoreceptors located in the aortic arch (aortic bodies) and carotid artery (carotid bodies).

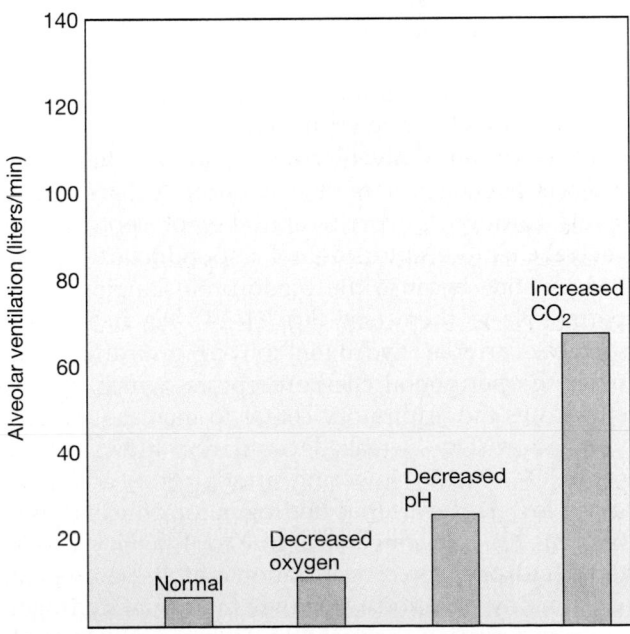

Figure 21–16

Effect of maximal decrease in arterial blood pH, decrease in arterial blood oxygen, and maximal increase in arterial blood CO_2 on alveolar ventilation. Note that elevated arterial CO_2 can increase alveolar ventilation about 15 times above normal.

Figure 21–17

Stimulation of alveolar ventilation by carbon dioxide. The main stimulus is via the central chemoreceptors (solid heavy line). Note that the direct effect of arterial CO_2 on the peripheral chemoreceptors is minor (dashed line).

a protein buffer to neutralize changes in hydrogen ion concentration. Low arterial Pco_2 has an equally strong inhibitory effect on ventilation. Although the central and peripheral chemoreceptors work in concert to regulate CO_2, over two thirds of the stimulation comes from the central chemoreceptor (Fig. 21–17). For further information about pH regulation and buffers, please refer to Chapter 25.

The control of alveolar ventilation by changes in arterial hydrogen ion concentration occurs exclusively through the peripheral chemoreceptors. The central chemoreceptors do not respond to arterial hydrogen ions because the blood-brain barrier is impermeable to them (see Fig. 21–14). An increase in systemic arterial hydrogen ion concentration activates the peripheral chemoreceptors, which in turn stimulates the respiratory center to increase alveolar ventilation. The increased ventilation blows off excess CO_2, thereby lowering arterial Pco_2 and indirectly lowers the arterial hydrogen ion concentration (Fig. 21–18). The converse is true for low H^+; low arterial hydrogen ion concentration inhibits ventilation.

In many situations, a change in arterial hydrogen ion concentration is caused by factors other than altered blood carbon dioxide. For example, a decreased blood pH occurs with severe diabetes, owing to excess fatty acids in the blood, and the acidosis stimulates ventilation through the peripheral chemoreceptors. Another example is during vomiting, in which

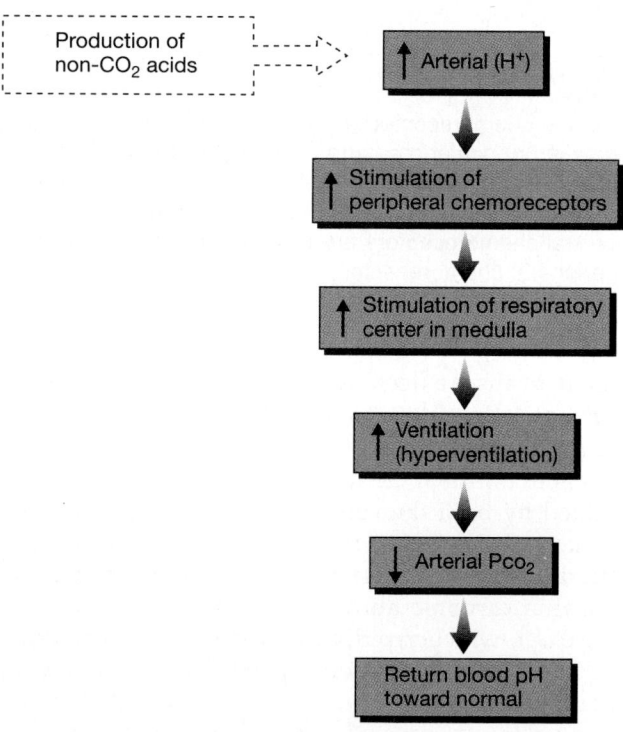

Figure 21–18

Increase in alveolar ventilation by arterial blood pH. Changes in hydrogen ion concentration can occur through the production of non-CO_2 acids such as lactic acid from muscles, or from excess fatty acid in the blood with severe diabetes.

hydrogen ions are lost. With the consequent decrease in arterial hydrogen ion concentration, ventilation is depressed because the alkalosis inhibits the peripheral chemoreceptors.

Hypoxia and Control of Ventilation

As mentioned earlier, of the three chemicals that stimulate ventilation, oxygen has the smallest effect. It may seem surprising that the control of ventilation is so insensitive to reduced oxygen, since most of us think intuitively that breathing is controlled by the body's need for oxygen. The reason that arterial P_{O_2} has so little effect on the ventilation is that gas transfer in the lung is so complete that blood leaving the lung is almost fully saturated with oxygen over a wide range of P_{O_2}; it is not until the arterial P_{O_2} decreases to 60 mm Hg that delivery to the tissues is affected. Thus, sensitive control of ventilation to maintain a constant oxygen concentration in the blood is not required. When the arterial P_{O_2} becomes too low (<85 mm Hg) the blood becomes **hypoxemic.** The receptors that sense the low O_2 tension are the peripheral chemoreceptors. When arterial P_{O_2} is reduced by lung disease, depressed ventilation, or by breathing at high altitude, the peripheral chemoreceptors stimulate ventilation (Fig. 21–19).

During exhaustive exercise, alveolar ventilation can increase 20-fold. The details of how exercise stimulates ventilation will be discussed in Chapter 30.

Breathing Patterns Can Be Abnormal

Sleep Apnea

Ventilation normally decreases during sleep because the body's metabolic demand decreases. Furthermore, skeletal muscles of the tongue, larynx, and pharynx relax, as well as airway smooth muscle in the upper airway, resulting in obstruction. A common consequence of airway narrowing and partial obstruction is snoring. However, in some individuals the degree of obstruction may be sufficient to cause complete occlusion that can lead to **sleep apnea,** a condition in which respiration stops for relatively long periods of time (often 30 to 60 seconds), the decrease in ventilation causes the arterial P_{O_2} to become dangerously low. Obstruction of the upper airway during sleep is often accentuated with flabby fatty tissue that causes additional narrowing. When breathing stops, arterial CO_2 and O_2 changes become exaggerated (high P_{CO_2} and low P_{O_2}). Eventually, it is the rising levels of arterial CO_2 that stimulate the respiratory center. The individual reacts with grunting, gasping breaths and immediately returns to a deeper sleep until the next apneic episode occurs.

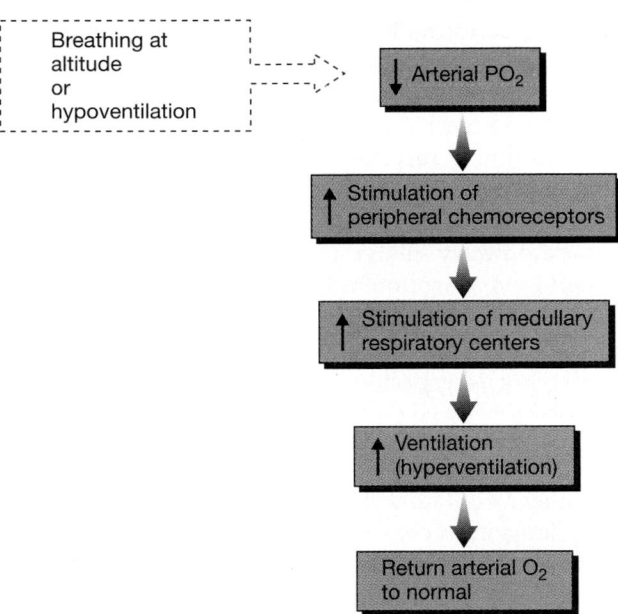

Figure 21–19

Stimulation of ventilation by hypoxemia occurs via the peripheral chemical receptors, mainly the carotid bodies.

Sleep apnea victims may stop breathing anywhere from a dozen to a hundred times a night, although they are usually unaware of the sleep disruptions. This loss of sleep can lead to pronounced daytime drowsiness, a classic symptom of sleep apnea. Some of the victims are involved in automobile accidents because they fall asleep at the wheel. In another form of sleep apnea that is called nonobstructive form, the respiratory muscles stop working, producing a similar cycle of repeated sleep disruptions. This form often causes insomnia rather than excessive daytime sleepiness.

Sleep apnea occurs most frequently in obese people, especially men, who suffer from the obstructive form. The real danger of sleep apnea is that death from cardiac arrhythmia can occur during an episode. No one knows what causes sleep apnea. Some respiratory physiologists think the cause in the nonobstructive form may be reduced sensitivity of the chemoreceptors to carbon dioxide during certain phases of sleep. At present, sleep apnea is being studied extensively because it may be linked to **sudden infant death syndrome (SIDS),** the disease in which a seemingly normal, healthy infant is found dead in its crib. Many of these infants, upon autopsy, have been found to have poorly developed carotid bodies.

Cheyne-Stokes Breathing

Another abnormality in respiration is **Cheyne-Stokes breathing,** named after a Scottish physician (Cheyne) and an Irish physician (Stokes) who first characterized the abnormality. This abnormality is referred to as periodic breathing where breathing becomes uneven, (labored breathing) and is interrupted by periods of no breathing at all (apnea). Rhythmic waxing and waning in respiratory depth also occurs. In most cases, Cheyne-Stokes breathing is related to central asphyxia (a fall in P_{O_2} and a rise in CO_2) in which the respiratory center is depressed. Cheyne-Stokes breathing is encountered in heart failure, in injury involving the brain, and in normal individuals at high altitudes. Cheyne-Stokes breathing is sometimes seen in newborns, especially those born prematurely.

Shallow-Water Blackout

Another ventilatory pattern that effects the control of breathing is hyperventilation. If a swimmer hyperventilates before diving under water, they can hold their breath longer. With hyperventilation, arterial P_{CO_2} is drastically decreased, with no appreciable change in oxygen content. As a result of the low arterial CO_2 tension, the respiratory centers become depressed. At the same time, oxygen is rapidly taken up by the exercising muscles, lowering the arterial oxygen content and P_{O_2}. However, the stimulating effect of the low P_{O_2} is overridden by the strong inhibitory effect of the low P_{O_2}. As a result, the brain becomes **hypoxemic,** causing many swimmers to black out under water and drown, commonly referred to as *shallow-water blackout.*

The Diving Reflex

Another breathing pattern associated with water is the **diving reflex.** The diving reflex is most pronounced in newborns and can be elicited when the face is submerged in water. This reflex causes the individual to gasp; breathing stops (apnea), heart rate decreases, and blood flow to peripheral tissue also decreases (peripheral vasoconstriction). Diverting blood away from the peripheral tissues allows for a heart-brain perfusion. The diving reflex is a protective reflex. A number of cases have been reported in which young children have broken through ice on a pond or fallen into a lake and survived under water

for 20 to 30 minutes. The gasp reflex prevents water from getting into the lungs, and oxygen is conserved for the brain.

The diving reflex by itself is not enough to save victims who remain under water for long periods of time. Lowered body temperature caused by the cold water is equally important. A lower body temperature means a lower metabolic rate, which means less oxygen is required. The reason children survive better than adults is twofold. First, they have a more pronounced diving reflex. Second, their body temperature drops at a faster rate because they have a high surface area to body weight ratio. For further details about the relationship between body temperature and surface area, see Chapter 27.

The Pickwickian Syndrome

A fifth type of altered breathing pattern that affects ventilation is called the **Pickwickian Syndrome.** The disorder occurs in severely obese people who often suffer from hypoventilation. The Pickwickian Syndrome was named after "Joe," the boy in Charles Dickens' novel *The Pickwick Papers.* Joe was excessively sleepy during the day and snored loudly. Pickwickian patients suffer from hypoventilation because of their excess weight. These individuals usually have abnormalities in the control of breathing and often suffer from sleep apnea, which may account for the fact that they are always sleepy during the day.

Yawn

Finally, a respiratory movement that is most common and is associated with slow or shallow breathing is the **yawn.** The yawn is thought to be involved in preventing alveoli from collapsing (atelectases). Recall, from Chapter 20, that at low lung volumes alveoli tend to become very unstable. The yawn is a deep involuntary inspiration with the mouth open, often accompanied with stretching. The yawn is thought to aid the spreading of surfactant to stabilize alveoli.

C H A P T E R   R E V I E W

Summary

- Pulmonary arteries form a fine network of capillaries embedded in the alveolar wall. Gas exchange takes place when the blood is spread over the large alveolar surface area via the capillaries, and the blood loads with oxygen

and unloads carbon dioxide. The newly oxygenated blood returns through pulmonary veins to the left side of the heart to be pumped throughout the body.
- The lymphatic system is an auxiliary network of vessels

that help prevent pulmonary edema by draining excess fluid leaked from the capillaries into the interstitial space.

- Blood pressures in the pulmonary circulation are important measures of how the lungs are functioning. Under normal conditions, the pulmonary arterial pressure is low in comparison to arterial pressure in the rest of the body, usually 25/10 mm Hg with a mean of 15 mm Hg. This low-pressure, low-resistance circuit accommodates all of the cardiac output. The majority of the blood flows to the lower lungs.
- Sometimes, the flow of inspired air is unevenly distributed among the alveoli, causing regional hypoxia (low oxygen). The lung has a mechanism called hypoxic vasoconstriction that causes arteries in a hypoxic region of lung to constrict and direct blood flow to better-ventilated regions.
- The ventilation-perfusion balancing system senses low oxygen and responds by triggering vasoconstriction of the arteries, increasing pulmonary arterial pressure, and routing blood to the upper regions of the lungs, where more oxygen is available. The balancing system is effective even when an individual is exposed to whole lung hypoxia, as happens at high altitudes.
- During exercise, the increased cardiac output causes pulmonary arterial pressure to rise. Much of the increased blood flow perfuses the upper regions, resulting in a more even perfusion of the lungs. The body further compensates for increased oxygen demand by enhancing the oxygen uptake from the alveolar gas in two ways: (1) by increasing the number of capillaries through which the blood flows, thus increasing gas exchange surface area, and (2) by shortening the capillary transit time.
- At rest, the blood delivers 5 ml of oxygen for every 100 ml of blood that circulates through the tissues. Resting cardiac output is 5 liters per minute; therefore, 250 ml/min of oxygen are delivered to the tissues. With exercise, this delivery can be increased to 15 to 20 times.
- Hemoglobin greatly facilitates transport by loosely binding oxygen to its four binding sites in the alveolar capillaries and then releasing the oxygen in the tissues. The oxyhemoglobin dissociation curve describes how hemoglobin is loaded and unloaded with oxygen at a given oxygen tension.
- During rest in a normal adult, 4 vol.% of carbon dioxide are transported to the lungs to be exhaled each minute. There are three ways in which carbon dioxide is transported in the blood: bicarbonate ions, carbaminohemoglobin, and dissolved carbon dioxide.
- Before birth, the lungs of the fetus do not serve any respiratory function. Gas exchange takes place between the mother and the fetus through the placenta. In the fetus, blood is shunted past the lungs by the foramen ovale and the ductus arteriosus.
- After birth, the blood flow through the lungs increases, systemic arterial pressure rises, and the foramen ovale and ductus arteriosus close. Examples of congenital heart disease are patent ductus arteriosus (ductus remains open) and ventricular septal defect (a hole in the ventricular septum).
- Breathing is under automatic and voluntary control. The control center for breathing is located in the brain stem (medulla and pons). Changes in *chemoreceptors* (blood, P_{CO_2}, pH, P_{O_2}) and changes in proprioceptors activity (stretch receptors located within the lung and skeletal muscle spindles) feed information to the medulla to modify the depth and rate of breathing.

Review Questions

Identify Each with a Term
1. The method for measuring the pressures in the chambers of the heart and in the pulmonary artery.
2. Fluid that collects in the lung.
3. The molecule that carries oxygen in the blood.
4. Blood clots that lodge in the lung circulation.
5. The organ that the fetus uses for gas exchange.

Define Each Term
6. ductus arteriosus
7. hypoxic vasoconstriction
8. sleep apnea
9. carbaminohemoglobin
10. hypoxemia

Choose the Correct Answer
11. Which of the following causes the oxyhemoglobin dissociation shift to the right?
   a. low CO_2
   b. low pH
   c. low temperature
   d. low O_2
12. Carbonic anhydrase catalyzes the reaction between:

   a. water and carbon dioxide.
   b. the dissociation of carbonic acid to bicarbonate ions and protons.
   c. the combination of hydrogen ions and hemoglobin.
   d. carbon dioxide and hemoglobin to form carbaminohemoglobin.
13. Blood flow in the lung:
   a. is equal to the total blood flow to the rest of the body.
   b. goes mostly to the upper lung.
   c. is less per minute than in the rest of the body.
   d. is very high in the fetus.
14. Oxygen removal from the alveolar gas can be facilitated by:
   a. pulmonary edema
   b. capillary recruitment
   c. capillary transit times greater than 60 seconds
   d. anemia
15. The pulmonary and systemic circulations have the same:
   a. systolic pressure
   b. volume flow of blood per minute
   c. mean pressure
   d. diastolic pressure during rest

16. Hypoxic vasoconstriction:
    a. helps maintain ventilation-perfusion balance.
    b. directs blood away from hypoxic alveoli.
    c. causes a rise in pulmonary artery pressure at high altitude.
    d. all of the above.
17. Approximately how long does an average red blood cell spend in the pulmonary capillaries during resting conditions?
    a. 60 seconds
    b. 1 second
    c. 0.25 second
    d. 5 seconds
18. The pulmonary circulation:
    a. acts as a sieve for emboli.
    b. converts angiotensin to its active form.
    c. has a large surface area.
    d. all of the above.
19. In the fetus, which of the following is true?
    a. Arterial oxygen tension closely approximates maternal oxygen tension.
    b. All cardiac output flows through the pulmonary circulation.
    c. Blood flows from the right to the left side of the circulation through the patent ductus arteriosus.
    d. All blood flows through the placenta.
20. Carbon dioxide is transported in the blood predominantly in the form:
    a. physically dissolved in plasma
    b. of carbaminohemoglobin
    c. of carbonic acid
    d. of bicarbonate ions
21. The greatest increase in ventilation would occur with a low P_{O_2} in:
    a. CNS
    b. carotid bodies
    c. pulmonary capillaries
    d. alveoli
22. Hypoxia stimulates ventilation primarily through:
    a. central chemoreceptors
    b. carotid bodies
    c. medullary chemoreceptors
    d. direct stimulation of diaphragmatic muscle
23. Control of basic rhythms of breathing is located in the:
    a. medullary center
    b. pneumotaxic center
    c. chemoreceptors located in carotid and aortic arch
    d. respiratory center of the pons

24. The most powerful blood-bone stimulus to ventilation is:
    a. bicarbonate
    b. oxygen
    c. hydrogen ion
    d. carbon dioxide

Calculate

25. What is the partial pressure of oxygen at sea level?
26. At a resting cardiac output of 5 L/min, how much carbon dioxide is delivered to the lungs to be exhaled each minute?
27. If the top of the lung is 15 cm above the heart, what pressure in mm Hg must the right ventricle generate to pump blood to the top of the lung?
28. What is the partial pressure of oxygen in Denver, Colorado?
29. If the arteriovenous oxygen difference is 7 vol.% in an individual at rest, and the cardiac output is 6 L/min, how much oxygen is being delivered to the tissue each minute?

Answer Each Question in One or Two Sentences

30. What is the function of the pulmonary lymphatics?
31. Describe the metabolic functions of the pulmonary circulation.
32. What are the blood pressures in the pulmonary circulation?
33. How does the pulmonary circulation maintain ventilation-perfusion balance?
34. What are the changes that occur in the pulmonary circulation during exercise?

Problem Solving in Physiology

35. The normal transit time of blood in the pulmonary circulation is about 1 second. However, the blood is completely oxygenated in a quarter of a second. How would transit time and the time it takes to completely oxygenate the blood in an individual with anemia be affected?
36. How does hemorrhage affect arterial P_{O_2} and O_2 saturation in a normal individual who has lost 1 liter of blood?
37. Why does pulmonary vascular resistance increase at high lung volumes?

Suggested Readings

Altman, L.K. *Who Goes First?* New York: Random House, 1987.

Forster, R.E., DuBois, A.B., Briscoe, W.A., and Fischer, A.B. *The Lung.* Chicago: Year Book Medical Publishers, 1986.

Guyton, A.C. *Textbook of Medical Physiology,* 8th ed. Philadelphia: W.B. Saunders, 1991.

Mines, A.H. *Respiratory Physiology,* 3rd ed. New York: Raven Press, 1992.

Sieck, G. "Neural control of the inspiratory pump." *News in Physiological Sciences,* 6:260–265, 1991.

Simon, T. *The Heart Explorers.* New York: Basic Books, 1966.

Taylor, A.E., Rehder, K., Hyatt, R.E., and Parker, J.C. *Clinical Respiratory Physiology.* Philadelphia: W.B. Saunders, 1989.

West, J.B. *Respiratory Physiology.* Baltimore: Williams & Wilkins, 1990.

Snapper, J.R. "Pulmonary Edema." *Hospital Practice,* May 87–101, 1986.

Key Terms

alveolar-capillary (p. 641)
angiogram (p. 640)
angiotensin (p. 641)
aortic bodies (p. 655)
apnea (p. 659)
breathing patterns (p. 654)
capillary recruitment
  (p. 642)
capillary transit time
  (p. 646)

carbon dioxide dissociation
  curve (p. 650)
carbonic anhydrase
  (p. 650)
chemorecepters (p. 655)
cartoid bodies (p. 655)
chloride shift (p. 650)
ductus arteriosus (p. 652)
foramen ovale (p. 652)
gas diffusion (p. 640)

gas uptake (p. 646)
hypoxemia (p. 644)
hypoxia (p. 643)
hypoxic vasoconstriction
  (p. 643)
oxygenated blood (p. 647)
oxygen tension (p. 646)
oxyhemoglobin (p. 647)
oxyhemoglobin dissocia-
  tion curve (p. 648)

partial pressure (p. 646)
pulmonary circulation
  (p. 640)
pulmonary edema (p. 641)
pulmonary hypertension
  (p. 644)
septial defect (p. 652)
ventilation-perfusion
  (p. 643)

Scanning electron micrograph of
the surface of an epithelial cell
from the small intestine
illustrating the microvilli.
(© Custom Medical Stock Photo)

Chapter 22

The Gastrointestinal System

Key Concepts

The gastrointestinal system is a group of organs that work collectively to fulfill the process of absorption, which is the major function of this organ system. Absorption is the transfer of water, inorganic ions, and organic nutrients derived from food from the lumen of the digestive tract to the blood and the lymph.

The processing of food for absorption begins in the upper digestive tract, where chewing and salivary secretion prepare food for swallowing. Saliva contains an amylase that initiates the digestion of carbohydrate.

The stomach is a storage organ for food and discharges its contents slowly into the duodenum at a rate that is optimal for absorption by the small intestine. Hydrochloric acid secreted by the stomach aids in the digestion of food and activates pepsin, an enzyme secreted by the gastric mucosa that initiates the digestion of protein.

The exocrine secretion of the pancreas provides bicarbonate for the neutralization of gastric acid and a wide variety of enzymes that almost complete the digestion of food.

Bile secreted by the liver is stored and concentrated in the gallbladder, an organ that empties its contents into the upper small intestine in response to the ingestion of food. The bile salts contained in bile play important roles in (1) the emulsification of fat, and (2) the solubilization of fat digestion products.

The muscular movements of the small intestine facilitate digestion and absorption and move unabsorbed material into the large intestine. Practically all carbohydrate, protein, and lipid and most water and inorganic ions are absorbed during transit through the small intestine.

The absorption of sodium and water is completed in the proximal large intestine. At infrequent intervals, propulsive mass movements of the proximal colon drive contents into the distal colon, where they are stored for varying periods of time. When the rectum is distended, the defecation reflex is activated and feces are moved to the outside of the body.

Pete has just returned home from the annual Thanksgiving Day football game between his high school and its arch-rival located just a few miles up the road. It's been a long, cold afternoon at the stadium and he has expended a lot of energy cheering the home team on to a last-minute 28-27 victory. Pete is very hungry, and as he enters his front door, he is greeted by the aroma of freshly roasted turkey and all of the trimmings. Saliva begins to flow, and rumblings emanate from his abdominal region in anticipation of the feast that lies ahead. Pete takes a seat at the dining room table with the rest of the family and consumes large quantities of food until he is satiated. Then he sits back in the easy chair, and perhaps naps, unaware of the numerous activities that have been initiated and will continue to occur for several hours as the body processes the food that he has eaten.

It is the gastrointestinal system, which consists of the digestive tract and its accessory organs, that processes the food we eat, and in so doing, provides nutrients to the cells of the body. Suppose that you were able to make the journey along with food as it traverses the digestive tract from the mouth to the end of the large intestine. What activities would be encountered as the trip took place?

The fastest part of the voyage would occur after food has been chewed and mixed with saliva and then propelled rapidly from the mouth through the esophagus into the stomach, where it is stored for a while. Transit becomes much slower as the stomach empties its contents into the upper small intestine, and progressively slower as these move first through the small intestine and then the large intestine. At various points along the route, different juices are secreted into the lumen of the tract. Some of these, namely those of the salivary glands, stomach, and pancreas, contain digestive enzymes that reduce large organic molecules of the diet (carbohydrate, protein, fat, etc.) to molecules small enough to be transferred (absorbed) across the wall of the small intestine into the blood and the lymph. Another important secretion that is encountered along the way is

bile, which is secreted by the liver and stored in the gallbladder, and which enters the upper small intestine to play an important role in the absorption of fat. Of course, the journey ends when unabsorbed material is expelled to the exterior of the body.

In this chapter, we will examine how the various processes of the gastrointestinal system function to ensure that the cells of the body are provided continuously with the nutrients they need to survive. You will see that it is the musculature of the digestive tract that not only mixes the contents of the tract with the secretions that enter it, but also propels contents in a downward direction and at a rate that is appropriate for the function to be served in the different regions of the tract. We will examine how the secretions of a variety of exocrine secretory cells of the gastrointestinal system are regulated and the role these secretions play in the digestion and absorption of specific components of the diet. And finally, we will also consider the various mechanisms that are involved in the absorption of water, inorganic ions, and the small organic molecules that result from the digestion of food.

Overview of the Gastrointestinal System

Eating is one of life's pleasures. However, there is more to eating than sensory gratification; food is needed to provide the nutrients required for the maintenance of normal structure and function of the body. Food contains water, inorganic salts, and a variety of structurally complex organic molecules such as carbohydrates, proteins, lipids, and vitamins. Carbohydrates and fats supply most of the fuel (calories) required for the activities of everyday life, whereas proteins are used mainly for the processes of growth and the constant repair of body tissue. Inorganic salts play important roles in many body processes, and vitamins are essential for normal growth and development.

The **gastrointestinal system** is a collection of organs specialized to handle the food we eat (*gastro* = "relating to the stomach"). The primary function of this system is to process the materials contained in food and, in so doing, to supply the cells of the body with nutrients and energy. Four major processes are essential for the nutritive function of the gastrointestinal system: **motility, secretion, digestion,** and **absorption.** The term **motility** refers to the way in which food is moved downward through the various regions of the digestive tract at different rates depending on the functions served by the different organs of the tract. In the **secretion** process, a number of juices produced by exocrine glands enter the tract at various points along the route. The enzymes contained in some of these secretions play a primary role in **digestion,** the conversion of many large organic molecules of the diet to smaller molecules. In **absorption,** digestion products and other nutrients travel across the wall of the small intestine to the blood or the lymph, which in turn distributes them to the cells of the body. Indigestible materials and a number of waste products are moved to the end of the tract and eliminated from the body.

This chapter will consider these processes as they occur in sequence after the eating of food. However, before proceeding to a description of the handling of food by the gastrointestinal system, we will consider some of the general properties of the system as a whole.

Anatomical Features of the Gastrointestinal System Are the Digestive Tract and Its Accessory Structures

The gastrointestinal system (Fig. 22–1) consists of the **digestive tract and accessory structures** that lie outside the tract but are connected to it by ducts.

The digestive tract is a tube about 4.5 m long that traverses the body from lips to anus; it is composed of the **mouth, pharynx, esophagus, stomach,** and **small** and **large intestines.** Both ends of the digestive tract open to the external environment, which means that the lumen of the tract is continuous with, and thus a part of, the external environment. The wall of the digestive tract therefore is an interface between the external and internal environments of the body. This interface permits entry of certain materials into the body and excludes others. Most nutritional materials that enter the body do so by passing through the wall of the tract.

The accessory structures that lie outside the tract are the **salivary glands, pancreas, liver,** and **gallbladder.** These structures either produce or store secretions that aid in the digestion or absorption of food; these secretions are conducted into the lumen of the tract through ducts.

Cellular Arrangements in the Gastrointestinal System

Although the microscopic structure of the digestive tract varies from one region to another, there are common features in the overall organization of that part of the tract that extends from the esophagus to the anus. These structural characteristics are illustrated diagrammatically in Figure 22–2, which shows that the wall of the digestive tract is composed of four layers. From the lumen outward, these are the **mucosa, submucosa, muscularis externa,** and the **serosa.**

From the inside out, the mucosa consists of an **epithelial lining** (invaginations of which form **exocrine**

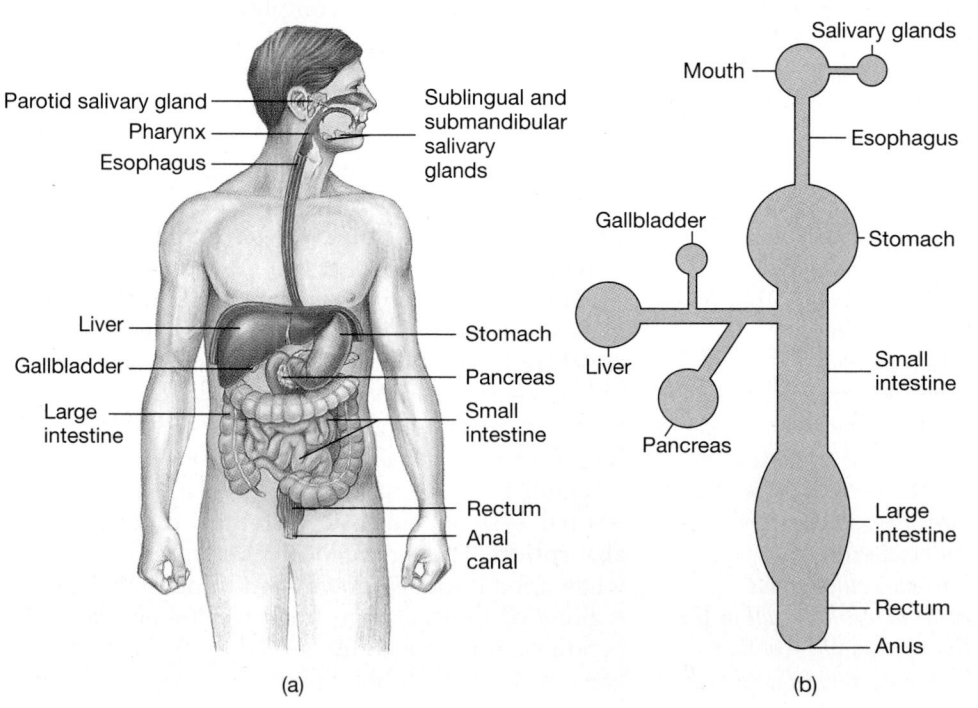

Parotid salivary gland
Pharynx
Esophagus

Sublingual and submandibular salivary glands

Liver
Gallbladder
Large intestine

Stomach
Pancreas
Small intestine
Rectum
Anal canal

Gallbladder
Liver
Pancreas

Salivary glands
Mouth
Esophagus
Stomach
Small intestine
Large intestine
Rectum
Anus

(a)

(b)

Figure 22–1

(*a*) Gross anatomy of the gastrointestinal system. (*b*) Diagrammatic representation of the gastrointestinal system.

Mucosa
Submucosa
Circular muscle
Muscularis externa
Longitudinal muscle
Serosa
Mesentery
Lamina propria
Myenteric nerve plexus
Circular folds
Duct
Lumen
External gland
Mesentery
Acini
Submucosal nerve plexus
Epithelial lining
Submucosal gland
Mucosal gland
Muscularis mucosae

Figure 22–2

Diagrammatic representation of the structure of the gastrointestinal system, showing a cross section of the digestive tract and an exocrine gland external to the tract.

glands), the **lamina propria** (a layer of connective tissue), and the **muscularis mucosa** (a thin layer of smooth muscle).

The submucosa, a layer of connective tissue, contains blood and lymph vessels that branch off into the overlying muscularis externa and the underlying mucosa. Also located in this layer are exocrine glands and a plexus of nerve cells, the **submucosal plexus.** The muscularis externa consists of an inner layer of **circular smooth muscle,** which encircles the tract, and an outer layer of **longitudinal smooth muscle,** which runs lengthwise along the tract. A network of nerve cells, the **myenteric plexus,** lies between these muscle layers. The serosa, a layer of connective tissue, is the outer tunic of the tract. The **mesentery** is an outward fold of the serosa that carries blood vessels, lymphatics, and nerves to and from the tract.

The salivary glands and parts of the pancreas contain clusters of exocrine secretory cells called the **acini** (see Fig. 22–2). The acini lead into a system of small ducts, which converge into larger ducts such that saliva is finally conducted into the mouth and pancreatic juice is conducted into the lumen of the upper small intestine through a main duct. A system of ducts also delivers bile (the exocrine secretion of the liver cells) into the upper small intestine.

The Gastrointestinal Circulation

The circulation of the stomach, small and large intestines, pancreas, and liver is called the **splanchnic circulation** (Fig. 22–3). The heart pumps blood to these organs through arteries that branch from the abdominal aorta. The blood that leaves the stomach, intestines, and pancreas goes next to the liver by way of the portal vein. This segment of the vasculature, which begins and ends in capillaries, is called the **portal circulation.** The significance of this vascular organization is that many absorbed digestion products are subject to metabolic processing by the liver before being distributed to the cells of the body.

The splanchnic circulation receives almost one fourth of the cardiac output. Its resting flow rate is about 1400 ml per minute, which is greater than that of any of the other major peripheral circulations. Flow of blood through the splanchnic circulation increases when a meal is eaten; this increase facilitates the removal of digestion products from the digestive tract.

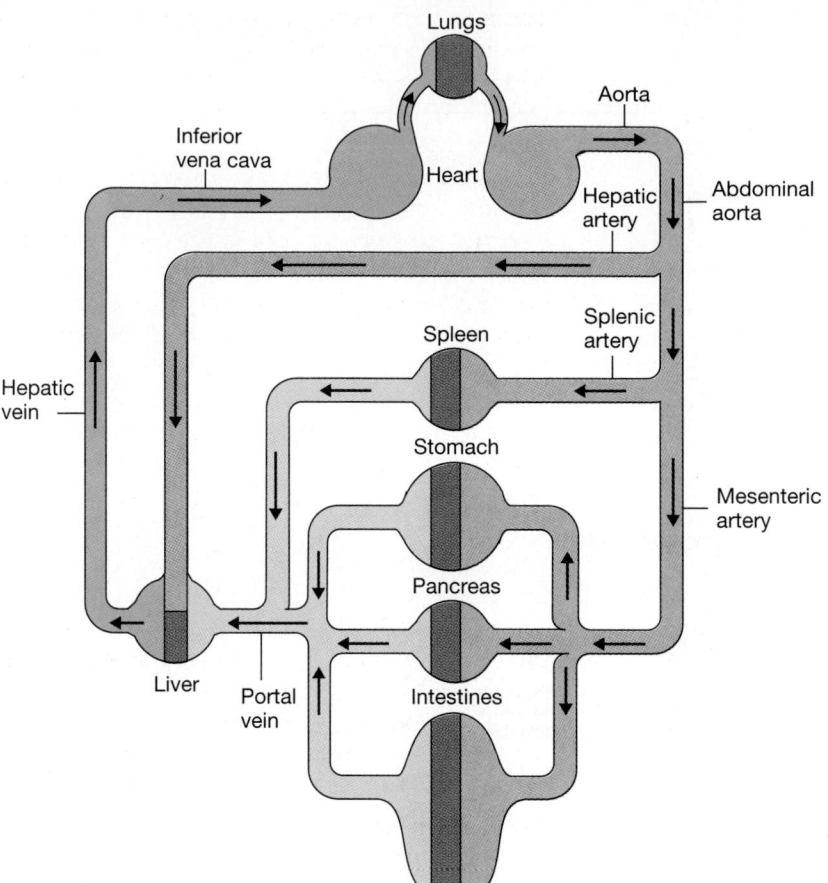

Figure 22–3
The splanchnic circulation.

In addition, extra oxygen and other nutrients are supplied to meet the enhanced demands for energy imposed by the increased muscular, secretory, and absorptive activities that occur after a meal.

The Processes of the Gastrointestinal System Are Carried Out by Various Cell Types

Before proceeding to a detailed discussion of motility, secretion, digestion, and absorption, we will discuss the relationships among the structures just described and the functions they serve.

Motility: Mixing Movements and Propulsive Movements

The motor functions of most of the digestive tract are performed by the smooth muscle layers that extend from the lower two thirds of the esophagus through most of the large intestine. Note, however, that skeletal muscle activity is of primary importance in the mouth and through the upper third of the esophagus at the upper end of the tract, and in the external anal sphincter at the lower end.

The two basic types of motor activity involved in the digestive and absorptive functions of the digestive tract are **mixing movements** and **propulsive movements.** The mixing movements of the stomach and small intestine aid digestion by mixing the digestive secretions with the food that enters from above. These movements also promote absorption from the small intestine and proximal large intestine by bringing their contents into contact with the absorbing surfaces. In the case of propulsion, the rate of movement of contents through the various regions of the digestive tract depends on the function served by each organ of the tract. For example, the transit of food from mouth through pharynx and esophagus is very rapid, because these regions simply provide a conduit for conveying food to the stomach. On the other hand, movement of contents from the stomach and through the small and large intestines is slow, to allow sufficient time for digestion and absorption.

Secretion: Inorganic Ions, Digestive Enzymes, Liver Products, and Mucin

The exocrine glands present throughout the gastrointestinal system contain a variety of secretory cells, each of which produces a characteristic secretion. All

Table 22–1

Exocrine Secretions of the Gastrointestinal System

Site of Secretion	Exocrine Secretory Cell	Secretory Components
Entire gastro-intestinal system	All cells	Water and inorganic salts
Salivary glands	Serous	Amylase
	Mucous	Mucin
Esophagus	Mucous	Mucin
Stomach	Parietal	HCl and intrinsic factor
	Chief	Pepsin
	Mucous	Mucin
Pancreas	Acinar	Amylase, proteases, lipase, and other digestive enzymes
	Duct epithelium	Bicarbonate
Liver	Hepatocyte	Bile salts, bile pigments, and cholesterol
	Duct epithelium	Bicarbonate
Small intestine	Mucous	Mucin
Large intestine	Mucous	Mucin

of these secretions consist of water and inorganic ions, but many also contain organic constituents. Important inorganic ions are hydrogen ions secreted by the stomach and bicarbonate ions secreted by the pancreas. The most notable of the organic components are the **digestive enzymes** of the salivary glands, stomach, and pancreas; the **bile salts, bile pigments,** and **cholesterol** secreted by the liver; and **mucin** produced by all regions of the digestive tract. The secretory cells of the gastrointestinal system, their sites of secretion, and the major components of these secretions are listed in Table 22–1.

Blood is the ultimate source of the gastrointestinal exocrine secretions in the sense that it supplies the constituents necessary for their formation. Some secretory components such as water and certain inorganic ions move directly from blood through and out of the cells at the time of secretion. Other com-

ponents such as digestive enzymes are synthesized by the cells from raw materials supplied by blood, and in many instances the finished products are stored in the cells until an appropriate stimulus causes them to be secreted.

Water and solutes are transferred across secretory cell membranes to the cell exterior by a variety of active and passive transport processes. Stimuli such as nerve impulses or hormones activate active transport systems in the plasma membranes of the secretory cells, and substances are actively transported from the interior to the exterior of the cell, usually into a duct. The active transport of a substance is the initial event that occurs, and water and other dissolved materials follow along passively. The overall result is the formation outside the cell of a secretion that ultimately finds its way into the lumen of the digestive tract, where it performs its function in the digestion and absorption of food.

The secretion that moves across the secretory cell membrane is known as the **primary secretion.** The composition of the primary secretion may be modified during passage through ducts as the result of movement of substances to and from blood across the duct epithelium. This is the case particularly for the salivary glands, pancreas, and liver, each of which has a relatively long duct system connecting the glandular secretory cells with the lumen of the tract.

Digestion: Breakdown of Food into Small Molecules

Carbohydrates, proteins, and fats are too large to be absorbed from the digestive tract in their ingested forms. By the process of enzymatic digestion, these large organic molecules are broken down into smaller molecules that can be absorbed. For example, complex carbohydrates such as starch are broken down to simple sugars. Proteins are degraded to small peptides and amino acids, and triacylglycerol fats are hydrolyzed to the smaller free fatty acid and monoglyceride molecules.

Digestion is carried out in the lumen of the digestive tract by enzymes contained in the secretions of the salivary glands, stomach, and pancreas, and also by enzymes in the epithelial cells lining the absorptive surface of the small intestine. The process of converting the complex molecules of food to simple molecules is a progressive one that starts in the mouth and continues in the stomach and small intestine. The small intestine is the region in which digestion is completed and most absorption occurs. Table 22–2 lists the major digestive enzymes, the substrates upon which they act, and the products of digestion.

Table 22–2
The Digestive Enzymes of the Gastrointestinal System and Their Actions

Source of Enzyme	Site of Action	Enzyme	Substrate	Products of Digestion
Salivary glands	Mouth and gastric lumen	Amylase	Polysaccharides	Oligosaccharides
Stomach	Gastric lumen	Pepsin	Proteins	Peptides
Pancreas	Small intestinal lumen	Amylase	Polysaccharides	Oligosaccharides
		Trypsin, chymotrypsin, and carboxypeptidase	Proteins and peptides	Peptides and amino acids
		Lipase	Triacylglycerols	Monoglycerides and free fatty acids
Small intestinal mucosa	Small intestinal epithelium	Oligosaccharidases	Oligosaccharides	Monosaccharides
		Peptidases	Peptides	Amino acids

Absorption: Transfer of Nutrients to the Blood and Lymph

The absorption of water and dissolved materials is the major function of the gastrointestinal system, and the various processes of this organ system are dedicated for the most part to fulfilling this function. This transfer of water and solutes from the lumen of the digestive tract into the blood or the lymph is accomplished largely by the epithelial cells that line the inner surface of the small intestine. However, significant amounts of water and inorganic salts are absorbed from the proximal large intestine.

The luminal membranes of the intestinal epithelial lining contain a number of transport systems specialized for the absorption of specific components of the diet. Some of these transport systems are capable of moving certain digestion products and inorganic ions from lumen to blood by active transport against electrochemical gradients. Other substances move by diffusion, at varying rates depending on their molecular size and lipid solubility.

Gastrointestinal Function Is Regulated by Neural and Hormonal Mechanisms

Few mechanisms regulate the processes of digestion and absorption. Since this lack of regulation applies to those constituents of the diet that supply calories to the body, the more of these foods that are eaten, the more that will be digested and absorbed. Thus,

most cases of obesity result from eating too much food too often.

In contrast to digestion and absorption, motility and secretion are highly regulated—particularly by neural and hormonal mechanisms. These processes are regulated by stimulation of receptor cells, which in turn triggers a chain of events operating through various pathways to produce specific responses of smooth muscle and exocrine gland cells.

Many of the stimuli that initiate the regulatory events are present in the lumen of the digestive tract. The most important stimuli are digestion products themselves, hydrochloric acid (HCl) secreted by the stomach, and distention by the contents of the tract. However, responses of the gastrointestinal system are not limited to stimuli applied to the tract, because stimulation of many other regions of the body can elicit responses of the gastrointestinal system. For example, both the sight and smell of food stimulate salivary, gastric, and pancreatic secretion. Another example is the profound motor activity (i.e., vomiting) that results from stimulation of the vestibular apparatus (Chapter 9) during motion sickness.

The receptors of the tract include sensory nerve receptors, such as chemoreceptors that respond to chemical agents in the contents of the tract, and mechanoreceptors that respond to distention of the tract. The reflex activity that results from excitation of these nerve receptors plays an important role in the regulation of gastrointestinal secretory and motor activity. In addition, a number of different types of endocrine cells in the digestive tract serve as receptors. When excited by a variety of stimuli, these cells release hormones into the circulation. As do the nerve reflexes, these hormones also participate in the regulation of secretion and motility.

Neural Control of Gastrointestinal Processes

The neural pathways for the control of motor and secretory activity of the gastrointestinal system are summarized in Figure 22–4. Two types of neural pathways are involved in the regulation of gastrointestinal motor and secretory activity: **short reflexes,** mediated through nerve cells that lie entirely in the wall of the digestive tract, and **long reflexes,** mediated over nerves outside of the tract.

Regulation by short reflexes occurs through the myenteric and submucosal nerve plexuses present in the wall of the digestive tract (see Fig. 22–2). These nerve plexuses possess all of the elements required for the local regulation of motor and secretory activity; i.e., they contain (1) sensory neurons that are excited by the intraluminal stimuli mentioned previously, (2) interneurons that conduct impulses up and down the tract, and (3) motor and secretory neurons that innervate the effector smooth muscle and secretory cells. The short reflexes that are mediated over these nerve cells give the digestive tract a degree of autonomy in the sense that muscle contraction and exocrine secretion are to an extent independent of regulatory influences originating outside the wall of the tract. These short reflexes can be either excitatory or inhibitory, depending on the mediator released at the effector cell by motor and secretory neurons. If the mediator is acetylcholine, as is usually the case, the result is excitation of exocrine secretion and contraction of visceral smooth muscle.

Gastrointestinal activity is also regulated by reflexes mediated through nerves outside of the gastrointestinal system. The major function of these long reflexes is to correlate activities, not only between different regions of the gastrointestinal system, but also between this system and other parts of the body. These long reflex arcs are mediated from either the gastrointestinal system or other regions of the body over afferent sensory nerves to the central nervous system (CNS). The central nervous system in turn relays impulses over efferent parasympathetic and sympathetic nerves to the gastrointestinal system. Since the efferent autonomic nerves synapse with cell bodies of the internal nerve plexuses of the digestive tract, impulses over external nerves can modify short reflex activity initiated in the wall of the tract. This modification can be either excitatory or inhibitory, depending on the type of efferent fibers involved. The

Figure 22–4

Neural pathways for the control of motor and secretory activity of the gastrointestinal system. Green lines show short reflex pathways; blue lines show long reflex pathways.

vagus nerve (a parasympathetic preganglionic nerve) is the major motor and secretory external nerve of this organ system, and most vagal efferent fibers are cholinergic and excitatory. The splanchnic nerves (sympathetic postganglionic nerves) are adrenergic, and the norepinephrine they liberate depresses excitability of the nerve plexuses, thereby inhibiting short reflexes mediated through these local nerve networks.

Hormonal Control of Gastrointestinal Processes

Four hormones of gastrointestinal origin—**gastrin, cholecystokinin (CCK), secretin,** and **gastric inhibitory peptide (GIP)**—are important in the control of motility and secretion. All of these hormones are polypeptide hormones consisting of chains of amino acids of varying lengths. The four types of endocrine cells that secrete these hormones occur as single cells throughout the epithelial lining of certain regions of the tract. When stimulated, these endocrine cells release their hormones into the blood. The blood car-

Table 22–3
The Gastrointestinal Hormones

	Gastrin	CCK	Secretin	GIP
Location of endocrine cells	Antrum of stomach	Upper small intestine	Upper small intestine	Upper small intestine
Stimuli for release	Amino acids peptides, acetylcholine	Amino acids, peptides, fatty acids	Acid	Fatty acids glucose
Major physiological actions				
Stomach				
Acid secretion	Stimulates		Inhibits	Inhibits
Gastric emptying	Stimulates		Inhibits	Inhibits
Mucosal growth	Stimulates			
Pancreas				
Enzyme secretion		Stimulates		
Bicarbonate secretion			Stimulates	
Growth		Stimulates		
Biliary system				
Liver bicarbonate secretion			Stimulates	
Gallbladder contraction		Stimulates		
Sphincter of Oddi		Relaxes		
Small intestine				
Ileal motility	Stimulates			
Ileocecal sphincter	Relaxes			
Mucosal growth	Stimulates			
Large intestine				
Motility	Stimulates			

ries the hormones to target organs of the gastrointestinal system, where they produce their motor and secretory effects. All four hormones are released in response to intraluminal stimuli acting directly on the endocrine cells in the luminal lining. Gastrin is also secreted in response to acetylcholine released by nerve impulses—that is, through short reflexes in the intestinal wall activated by intraluminal stimuli and through long reflexes, the efferent limbs of which are in the vagus nerve. The location of the gastrointestinal endocrine cells, the stimuli for the release of their hormones, and the major physiological actions of these hormones are summarized in Table 22–3.

Chewing, Salivary Secretion, and Swallowing

Chewing Reduces the Size of Food Particles and Mixes Them with Saliva

The process of chewing is accomplished by the combined actions of the skeletal muscles of the jaws, lips, cheeks, and tongue. The contraction and relaxation

of these muscles are coordinated by impulses over a number of cranial nerves. Although these muscles can be controlled voluntarily, the act of chewing is usually reflex in nature. The pressure that results when the jaws close following the introduction of a bolus of food into the mouth produces a reflex inhibition of the jaw-closing muscles and a reflex contraction of the jaw-opening muscles. The mouth opens, and the resulting reduction in pressure causes a reflex contraction of closing muscles and relaxation of opening muscles. This sequence of muscular activity continues rhythmically until the bolus of food has been chewed.

A major function of chewing is to reduce mechanically a mouthful of food to particles of a size convenient for swallowing. Although this varies from one person to another, the extent to which food is chewed has little effect on digestion and absorption. These processes are completed even when large chunks of food are swallowed. Chewing, however, does aid in the digestion of food by increasing the surface area on which digestive enzymes can act. Chewing also disrupts the cellulose walls of ingested plant cells, freeing intracellular material for subsequent processing by the gastrointestinal system. This

RESEARCH FOCUS

Peptide Messengers of the Digestive Tract

In recent years, a large number of peptides present in cells of the digestive tract have been identified. Some of these peptides (**messenger peptides**) travel, by various means, from their sources to target organs in different regions of the gastrointestinal system, where they are thought to play important roles in the regulation of function. It is now possible to distinguish not only the sources, but also the targets of these physiologically important molecules, through the use of immunostaining to identify peptide-containing cells, and radioimmunoassay to measure peptides in cells and biologic fluids. It has been established that the gastrointestinal messenger peptides are of endocrine, paracrine, and neural origin.

Endocrine messenger peptides are those released from mucosal endocrine cells of the digestive tract that reach target cells via the systemic circulation. These include the well-established hormones gastrin, cholecystokinin, secretin, and gastric inhibitory peptide, all of which play important roles in the regulation of gastrointestinal function (see Table 22–3). A number of peptides produced in mucosal, endocrinelike cells may also be important in regulating gastrointestinal function, but they have not yet achieved the status of hormone. An example of such a "candidate hormone" is **motilin**, a peptide released from cells in the epithelium of the upper small intestine. It stimulates contractile activity that begins in the stomach and travels caudad down the intestine in two-hour cycles after a meal. This type of muscular activity serves to cleanse the intestine of residual contents.

Paracrine peptide messengers are released from cells and either diffuse through the extracellular space or are transported by the local circulation to their target cells. **Somatostatin,** a peptide present in the wall of the stomach and small intestines, is of paracrine origin. This messenger peptide inhibits gastric acid secretion, gastric and intestinal motility, and the secretion of enzymes and bicarbonate by the pancreas.

Neural peptide messengers (neurocrines) are released from nerve terminals of the internal plexuses of the digestive tract and reach their target cell by diffusion across synaptic junctions. One of these peptides is **vasoactive intestinal polypeptide (VIP),** which is present in cells in the walls of the stomach and intestines.

A number of functions have been ascribed to this neural peptide, including stimulation of salt and water secretion by the small intestine, inhibition of gastric acid secretion, and inhibition of gastric and intestinal motility. **Substance P,** which is distributed from the stomach through the large intestine, excites smooth muscle contraction of the digestive tract, whereas **enkephalin** is inhibitory to motor function. The neural peptide **bombesin** is a potent stimulator of gastric acid secretion.

These are just a few of the peptides that have been identified in cells of the digestive tract. As additional research is done, some will be established as playing roles in the control of gastrointestinal function, whereas others will not. Work is presently being carried out to elucidate the pathways through which these peptides manifest their actions, and as more information becomes available, it becomes increasingly apparent that regulation of gastrointestinal function is not a simple affair. Endocrine, paracrine, and neurocrine peptides are released together during the digestion of food; the evidence indicates that these molecules usually act in concert with one another as well as with the acetylcholine released by parasympathetic pre- and postganglionic nerve fibers, and with the norepinephrine released by postganglionic sympathetic nerve fibers of the digestive tract. Consequently, gastrointestinal function reflects the interplay of numerous endocrine, paracrine, and neural agents.

It is of interest that many of the peptides of the digestive tract are present in other tissues of the body, particularly the brain. Thus, these molecules are not restricted to modifying activity of the gastrointestinal system.

■ **What are the routes traveled by secretin, somatostatin, and vasoactive intestinal polypeptide from their site of origin to their target organs?**

is important, because unlike some farm animals, the human digestive system does not possess **cellulase,** an enzyme that breaks down cellulose, allowing enzymatic disruption of plant cell walls. The maximal force that can be exerted during chewing is much greater than that required for the chewing of ordinary food. More important in the grinding of food is the occlusive contact area between the molars and premolars. Another function of chewing is to move food around the oral cavity and in so doing to stimulate receptors for taste and smell. This is important because much of the satisfaction of eating is derived from the perception of these sensations. In addition, stimulation of these receptors promotes secretion of saliva, which softens and lubricates the bolus of food.

The Long Reflex Secretion of Saliva Lubricates the Mouth and Initiates the Digestion of Starch

The acinar cells of the **parotid, submaxillary,** and **sublingual** glands combine to secrete between 1 and 2 liters of **saliva** every day. The resting rate of secretion is about 0.5 ml per minute. When stimulation is maximal (by acid solutions), the secretory rate increases to about 4 ml per minute. The parotid gland contains mostly **serous cells,** which secrete a watery solution containing inorganic ions and **amylase,** a digestive enzyme that helps break down starch.

In addition to serous cells, the submaxillary and sublingual glands contain **mucous cells,** which secrete **mucin.** When mucin mixes with the watery secretion of the serous cells, a solution of high viscosity called **mucus** is produced, and this gives a thick viscous characteristic to certain types of saliva.

The secretion of saliva is regulated by long reflexes. Afferent impulses originate in chemoreceptors in the mouth and nose when these receptors are stimulated by the taste and smell of food. Another sensory pathway is through stimulation of oral mechanoreceptors by food. Sight of food also stimulates salivary secretion. In all instances, impulses are sent to salivary centers in the medulla, which in turn relay impulses over the two divisions of the autonomic nervous system to the various salivary glands, causing them to secrete their specific juices. Both parasympathetic cholinergic and sympathetic adrenergic stimulation *excite* salivary secretion, in contrast to their usual opposing actions on other effector organs of the body. However, the parasympathetic nerves are the most important regulators of salivary secretion, because the mouth is dry when these nerves are nonfunctional.

Salivary secretion serves a number of functions. Saliva softens and lubricates food, which facilitates swallowing. The moistening properties of saliva facilitate chewing and speech. Saliva contains a high concentration of bicarbonate ion. This alkaline (basic) ion neutralizes acids produced by oral bacteria, thereby preventing these acids from dissolving the enamel of teeth. In addition, the decreased secretion of saliva that occurs during dehydration contributes to the sensation of thirst.

The digestive enzyme salivary amylase is synthesized by the serous cells and stored in these cells as vesicles of enzymes called **zymogen granules.** This enzyme (Fig. 22–5) degrades complex carbohydrates such as starch. Starch is a polysaccharide consisting of many branched and straight chains of the monosaccharide glucose. The major end-products of this enzyme are **oligosaccharides,** molecular fragments consisting of two to nine molecules of glucose. The major oligosaccharide produced is the disaccharide maltose, which is a combination of two glucose molecules. For optimal activity, salivary amylase requires a pH close to neutrality (7), as is found in the mouth. Although food remains in the mouth for only a short time, and although the contents of the stomach are highly acidic, a bolus of food remains intact for a while after entering the stomach, so that digestion can continue inside the bolus for a period of time.

Swallowing Moves Food Rapidly from the Mouth into the Stomach

Once a mouthful of food is chewed and mixed with saliva, it is propelled rapidly from the mouth through the pharynx and esophagus into the stomach. The food bolus moves in response to skeletal and smooth muscle contractions that occur sequentially from above to below. These muscular movements are coordinated mainly by long reflexes mediated through a swallowing center in the brain.

Part of a
starch molecule

Amylase

Maltose   Oligosaccharides

Figure 22–5

Structure of part of a starch molecule and the action of amylase.

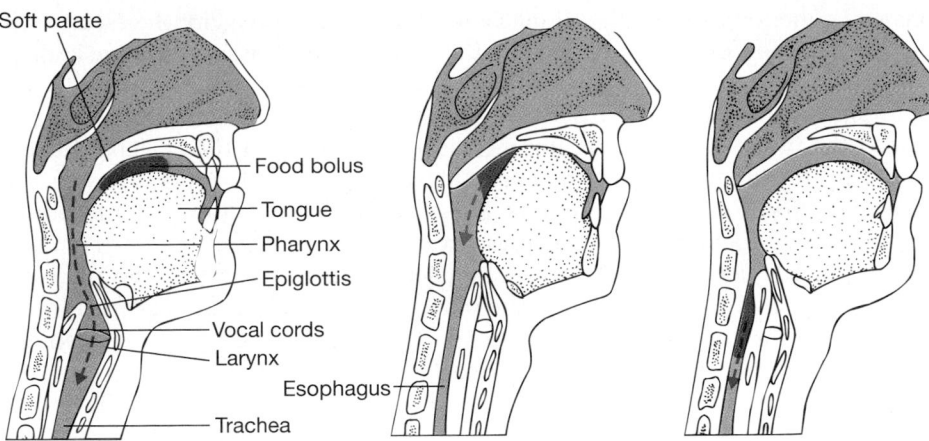

Figure 22–6

Passage of a bolus of food from the mouth through the pharynx into the upper esophagus during a swallow.

Swallowing Movements in the Mouth and Pharynx

A bolus of food moves from the mouth through the pharynx into the esophagus (Fig. 22–6) in about 1 second. The movement is accomplished by the coordinated contractions of a number of skeletal muscles.

The front of the tongue is pushed up against the hard palate, and the food mass, which can be either liquid or solid, is rolled toward the back of the tongue. The bolus is then forced into the pharynx by the contraction of skeletal muscles in the back of the throat. Respiration is inhibited briefly. The soft palate is raised and closes the airways to the nose. Food is prevented from entering the trachea by closure of the glottis (the opening of the vocal cords). The glottis is closed by elevation of the larynx, the region between the base of the tongue and the trachea, and by the approximation of the vocal cords. An auxiliary mecha-

nism that prevents food from entering the respiratory passages is that, as a bolus of food passes over the epiglottis, the bolus presses the epiglottis down over the closed glottis. As these openings close, the constrictor muscles of the pharynx contract and force the bolus into the esophagus.

The Role of the Esophagus in Swallowing

The esophagus, a muscular tube located mostly in the thorax, conducts material from the pharynx to the stomach. The major anatomical features of this organ (Fig. 22–7*a*) are the **upper esophageal sphincter** (a band of skeletal muscle), the **body** of the esophagus (the upper third of which is skeletal muscle and the lower two thirds of which are smooth muscle), and the smooth muscle **lower esophageal sphincter,** the last portion of which lies below the diaphragm in the abdominal cavity.

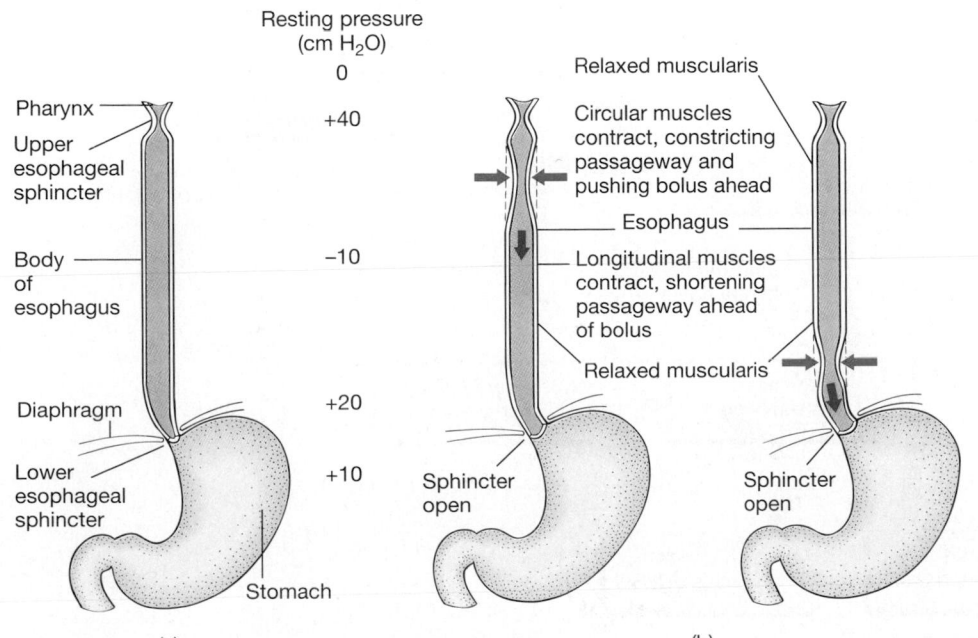

Figure 22–7

(*a*) Representative resting intraluminal pressures in the pharynx, upper esophageal sphincter, body of the esophagus, lower esophageal sphincter, and stomach. (*b*) Movement of a bolus of food through the esophagus by a peristaltic contraction.

In order to understand esophageal function, it is necessary to know what pressures exist in the lumen of the different regions of the esophagus, both at rest and during a swallow. It is also important to know how these pressures relate to those in the pharynx at the upper end of the esophagus and in the stomach below, because it is the existence and direction of pressure gradients that determine not only whether contents will move through the esophagus but in which direction they will move.

When the esophagus is at rest (see Fig. 22–7a), pressure in the lumen of the body of the esophagus is the same as intrathoracic pressure, which is subatmospheric (see Chapter 20). On the other hand, pressures in the upper and lower esophageal sphincters are above atmospheric pressure, indicating that these sphincters are contracted when there is no swallowing activity. Since pressure in the resting pharynx is atmospheric, a pressure gradient would exist between the pharynx and the body of the esophagus if it were not for the zone of high pressure that results from the contraction of the upper esophageal sphincter. If this pressure gradient were allowed to manifest itself, large volumes of air would flow from the pharynx to the esophagus during the normal course of breathing, and much of this air would be swallowed. The above-atmospheric pressure produced by closure of the upper sphincter prevents this from happening.

At the lower end, intragastric pressure (the pressure inside the stomach) is greater than atmospheric pressure, but less than that of the contracted lower sphincter. Consequently, gastric (stomach) contents do not normally reflux into the esophagus. (However, reflux is fairly common despite the barrier imposed by the lower sphincter, and because the contents of the stomach are quite frequently acid, the esophagus is irritated, producing the unpleasant sensation known as **heartburn**.)

Soon after a swallow is initiated, pressure in the pharynx increases as a result of the skeletal muscle contraction that occurs here. Also, pressures in the upper and lower sphincters abruptly drop to atmospheric pressure, showing that these sphincters have relaxed. A bolus of food moves from the pharynx through the relaxed upper sphincter into the esophagus in response to the downward pressure gradient that now exists through these regions.

After the bolus enters the esophagus, the upper sphincter closes, the glottis opens, and respiration resumes. Then pressure rises sequentially from above to below in the body of the esophagus. This sequence of pressure changes is due to the fact that a wave of contraction moves down the length of the esophagus. The contractile wave propels a semisolid food mass ahead of it (Fig. 22–7b) and into the stomach through the previously relaxed lower sphincter. This type of propulsive activity, which is characteristic of most regions of the digestive tract, is called **peristalsis**. The esophageal peristaltic wave travels at a rate of 2 to 4 cm per second and takes about 9 seconds to traverse the esophagus. If a bolus is liquid, it is shot through the esophagus by the initial force of swallowing and travels by force of gravity to the stomach in about 1 second. Once the bolus enters the stomach, the lower sphincter contracts, preventing reflux of gastric contents into the esophagus.

The Swallowing Reflex

The all-or-none swallowing reflex is activated when a bolus of food stimulates mechanoreceptors in the

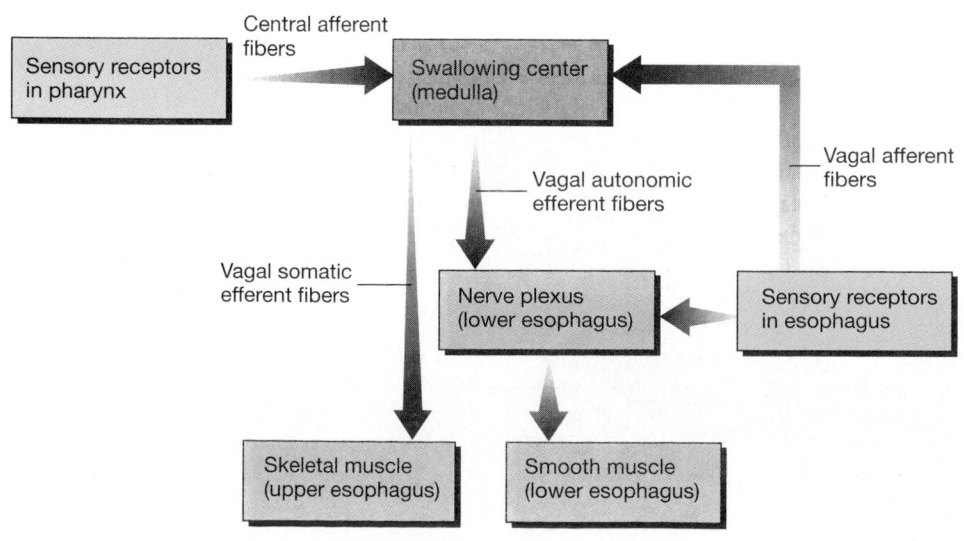

Figure 22–8

Neural pathways for the regulation of esophageal motor activity.

pharynx. Afferent impulses are sent to a swallowing center in the medulla. This center in turn sends out, in the proper sequence, efferent impulses over somatic nerves to the skeletal muscles, and over autonomic nerves to the smooth muscles involved in the swallowing response (Fig. 22–8). Motor impulses that regulate the activity of the skeletal muscles of the pharynx, larynx, upper esophageal sphincter, and upper body of the esophagus travel over various cranial somatic nerves. The esophagus is innervated primarily by the vagus nerves. The vagal efferent nerve fibers that supply the upper esophageal sphincter and upper body of the esophagus terminate at skeletal muscle motor end-plates and thus are somatic (and not autonomic) nerves. The vagal efferent nerve fibers that coordinate both the orderly progress of the peristaltic wave through the smooth muscle portion of the lower esophagus and the activity of the lower esophageal sphincter are autonomic nerves. These vagal preganglionic fibers synapse with cholinergic fibers of the internal nerve plexuses in the smooth muscle wall, and it is through these nerve networks that the vagus nerves manifest their motor effects.

The peristalsis that follows a conscious effort to swallow is **primary peristalsis. Secondary peristalsis** follows stretch of the esophageal wall without being preceded by a swallowing movement. This type of peristalsis is mediated by long reflexes over the vagus nerves from the esophagus to the swallowing center and back to the esophagus, and also by short reflexes in the esophageal wall (see Fig. 22–8). Secondary peristalsis is important because it facilitates the removal of any food that remains in the esophagus following the passage of a primary peristaltic wave.

Any one of a variety of disorders associated with the nerves and muscles involved in swallowing creates difficulty in swallowing, a condition known as **dysphagia.** One example of dysphagia is poliomyelitis, which paralyzes certain cranial nerves that are a part of the swallowing reflex. Another example of a disorder of the swallowing reflex is myasthenia gravis, a condition in which certain skeletal muscles involved in the swallowing response can contract only weakly.

The Processes of Motility and Secretion in the Stomach

The stomach (Fig. 22–9), located between the esophagus and small intestine, functions as a storage organ for food. Without the stomach, it would be necessary to consume many small meals every day, rather than a few large meals, in order to meet the energy re-

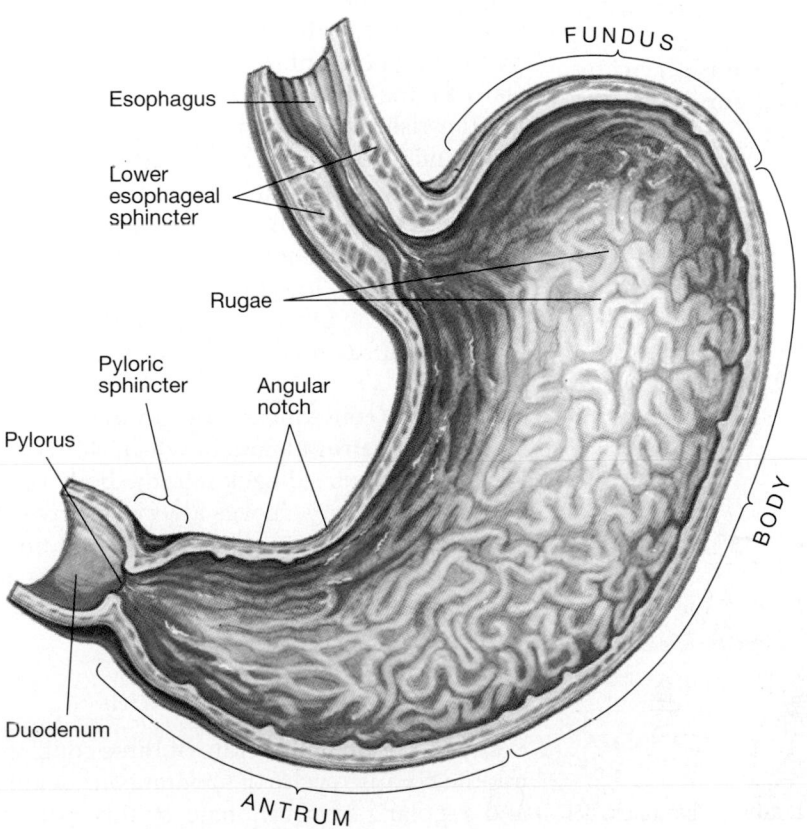

Figure 22–9

Anatomy of the stomach.

FUNDUS

Esophagus

Lower esophageal sphincter

Rugae

Pyloric sphincter

Angular notch

Pylorus

Duodenum

ANTRUM

BODY

quirements of the body. In addition to storing food, the stomach secretes HCl and the enzyme **pepsin,** both of which contribute to the digestion of food. The digestive activity, along with the muscular mixing movements of the stomach, converts large, solid food particles into a liquid suspension of finely divided particles. The small intestine can best perform its digestive and absorptive functions when presented with contents in this form. Intestinal digestion and absorption is also optimized by regulatory mechanisms that ensure that the musculature of the stomach discharges gastric contents into the small intestine slowly; that is, the small intestine is not overwhelmed by a flood of contents from above. The stomach also secretes **intrinsic factor,** a substance required for the absorption of vitamin B_{12} from the small intestine.

The Muscular Activities of the Stomach Associated with Gastric Filling and Emptying Are Regulated by Neural and Hormonal Mechanisms

The motor activity of the stomach serves the following major functions: (1) accommodation of a meal with a relatively small increase in intragastric pressure, (2) mixing of food with gastric (stomach) secretions so that digestion can begin, (3) reduction of the size of food particles, and (4) emptying of gastric contents into the duodenum (the upper region of the small intestine).

The storage function of the stomach is served by the smooth muscle of the **body** and **fundus** of the stomach. These regions of the stomach can adapt to the volume of contents they contain because, as the stomach fills to capacity, intragastric pressure increases only slightly (Fig. 22–10). A property of smooth muscle that allows the stomach to accommodate large volumes with minimal changes in intragastric pressure is its ability to adjust its resting length over a wide range without large changes in resting tension (see Chapter 16). This property of **plasticity** is passive in the sense that it does not depend on neural or hormonal influences. However, a neural mechanism does contribute to the volume adaptation. As a meal is eaten, with each swallow the body and fundus of the stomach relax slightly. This **receptive relaxation** is mediated by a long reflex over the vagus nerves.

Peristalsis in the Stomach

Peristaltic contractions of the stomach mix food with gastric juice, reduce food particle size, and expel gastric contents into the duodenum. When the stomach contains food, peristaltic contractions travel over the stomach from above to below at a frequency of about three per minute. These contractions are much more vigorous in the **antrum** than in the body and fundus because antral muscle is so much thicker than elsewhere (see Fig. 22–9). Consequently, it is primarily the muscular activity of the antrum that generates the forces responsible for emptying the stomach. The driving force behind the emptying process is the pressure differential that develops between the antrum and upper small intestine as a peristaltic wave travels over the antrum. At this time, pressure in the antrum rises momentarily above that in the duodenum, and a few milliliters of stomach contents move into the upper small intestine through the open **pyloric sphincter,** the thick bundle of circular smooth muscle and connective tissue separating the stomach from the small intestine. As the peristaltic wave passes over the pyloric sphincter, this structure contracts and offers enough resistance to prevent further evacuation.

Another consequence of closure of the pyloric sphincter is **retropulsion,** in which stomach contents are forcibly squirted back into the body of the stomach. Retropulsion achieves a very effective mixing of food with gastric juices and also breaks lumps of food into small particles. The process of evacuation and retropulsion continues about three times a minute until the stomach is empty (Fig. 22–11).

Regulation of Stomach Motility

Peristalsis occurs at the rate of three contractions per minute because cycles of spontaneous depolarization and repolarization originate at this rate in muscle

Figure 22–10

Intragastric pressure during filling of a rabbit's stomach.

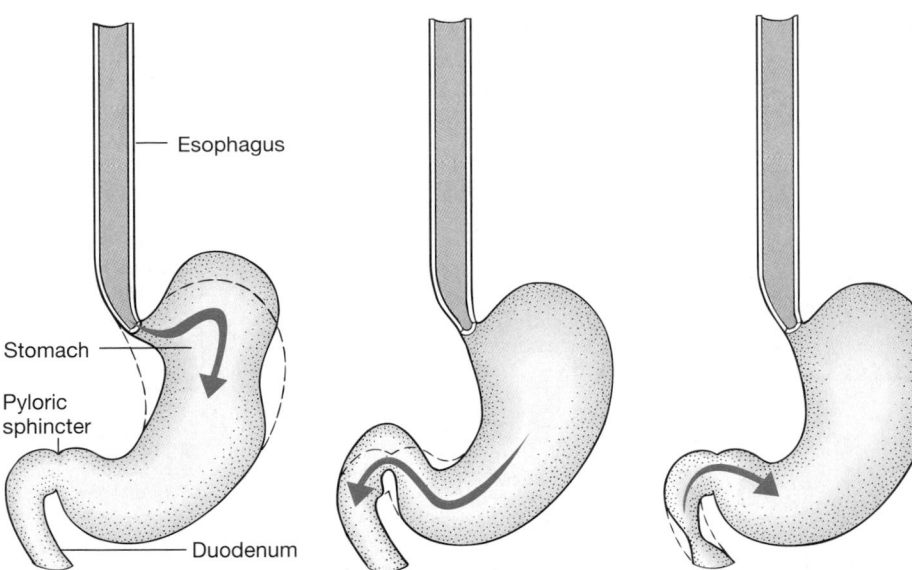

Figure 22–11

Evacuation and retropulsion of gastric contents during passage of a peristaltic wave down the stomach. *(Modified from Vander,* Human Physiology, *McGraw-Hill, 1985, p. 488.)*

pacemaker cells at the top of the stomach. Because this activity is continuous, it has been termed the **basic electrical rhythm** of the stomach. These depolarization-repolarization cycles, which have amplitudes of 10 to 15 millivolts and durations of 1 to 4 seconds (Fig. 22–12), travel as **slow waves** through the gastric muscle from above to below as a ring around the stomach. In the absence of excitatory stimuli, the depolarizations are too small to cause the muscle membrane to reach threshold and the action potentials to be fired to elicit contraction. When excitatory influences (such as release of acetylcholine by nerve fibers or the release of gastrin from the antral mucosa) are sufficient to produce action potentials, these occur at the peak of the slow waves because muscle excitability is closest to threshold at this time. The intensities of excitatory stimuli do not alter the frequency of the basic electrical rhythm and thus do not alter the frequency of peristalsis. On the other hand, intensity of stimulation does affect the magnitude of

contraction; that is, as intensity of stimulation increases, more action potentials are fired and contraction is more vigorous (see Fig. 22–12).

The rate at which the stomach empties is the result of the interplay between mechanisms that either excite or inhibit contraction of gastric smooth muscle. Both neural and hormonal mechanisms are involved in regulating the rate of gastric emptying.

The mechanisms that increase the vigor of gastric contractions originate in the stomach (Fig. 22–13). When the stomach is distended with food, stimulation of mechanoreceptors in the gastric wall activates short reflexes in the wall and long reflexes over the

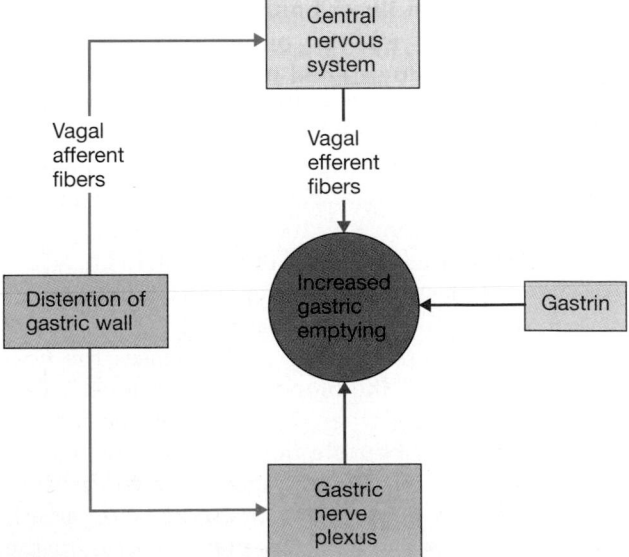

Figure 22–12

Relationship between slow waves, action potentials, and generation of tension in gastric smooth muscle.

Figure 22–13

Excitation of gastric emptying.

(a)

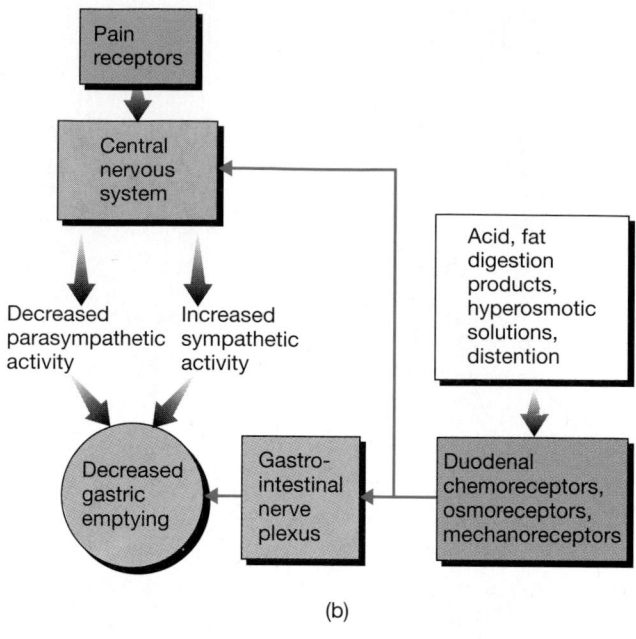

(b)

Figure 22–14

Inhibition of gastric emptying. (*a*) Hormonal pathways.
(*b*) Neural pathways.

vagus nerves. Both types of reflex activity result in increased gastric motor activity. The greater the distention, the more impulses fired and the more vigorous the gastric peristalsis. Peristalsis is also augmented by the hormone **gastrin,** which is released from G cells that are scattered throughout the epithelial lining of the antral mucosa. Gastrin is released from these endocrine cells by intraluminal stimuli that are present during the digestion of a meal.

The excitatory influences that originate in the stomach are kept in check by inhibitory mechanisms that originate in the duodenum (Fig. 22–14*a, b*). These duodenal inhibitory mechanisms prevent the upper small intestine from being overwhelmed by material from the stomach. Gastric emptying is retarded through signals from the duodenum when the contents that enter this region are either high in fat, acidic (having a pH less than 3.5), hyperosmotic, or bulky enough to stretch the duodenal wall. The duodenal mucosa, when contacted by acid, releases the hormone **secretin** into the blood. This hormone is carried to the stomach, where it inhibits motor activity. **Gastric inhibitory peptide,** another hormone that inhibits gastric contractions, is released from the duodenal mucosa by the products of fat digestion. In addition, certain endocrine cells may respond to the intraluminal stimuli just listed by releasing as-yet-unidentified hormones that inhibit gastric motility. A variety of duodenal nerve receptors, such as osmoreceptors, mechanoreceptors, and chemorecep-

tors, respond to these intraluminal stimuli to initiate reflex inhibition of gastric motility. These neural mechanisms, collectively called the **enterogastric inhibitory reflex,** include long reflexes mediated by external nerves and short reflexes through the internal nerve plexuses in the duodenal and gastric walls. In general, long reflex inhibition occurs when parasympathetic (vagal) activity to the stomach is decreased or when sympathetic activity is increased. The reverse is true for excitation of gastric motility.

Although control of gastric movements largely centers on mechanisms that originate in the stomach and duodenum, gastric motility may be influenced by long reflexes initiated from any sensory region of the body. For example, stimulation of visceral and somatic pain receptors inhibits gastric motility (Fig. 22–14*b*). In addition, various emotional states, such as anger, fear, and depression, produce changes in gastric motor activity, although the direction of these changes is not always predictable.

Gastric Secretion of Hydrochloric Acid Is by An Active Transport System and Is Regulated by Neural and Hormonal Mechanisms

The stomach secretes about 2 to 3 liters of gastric juice every day. Most of this juice is produced by exocrine glands located in the body and fundus of the stom-

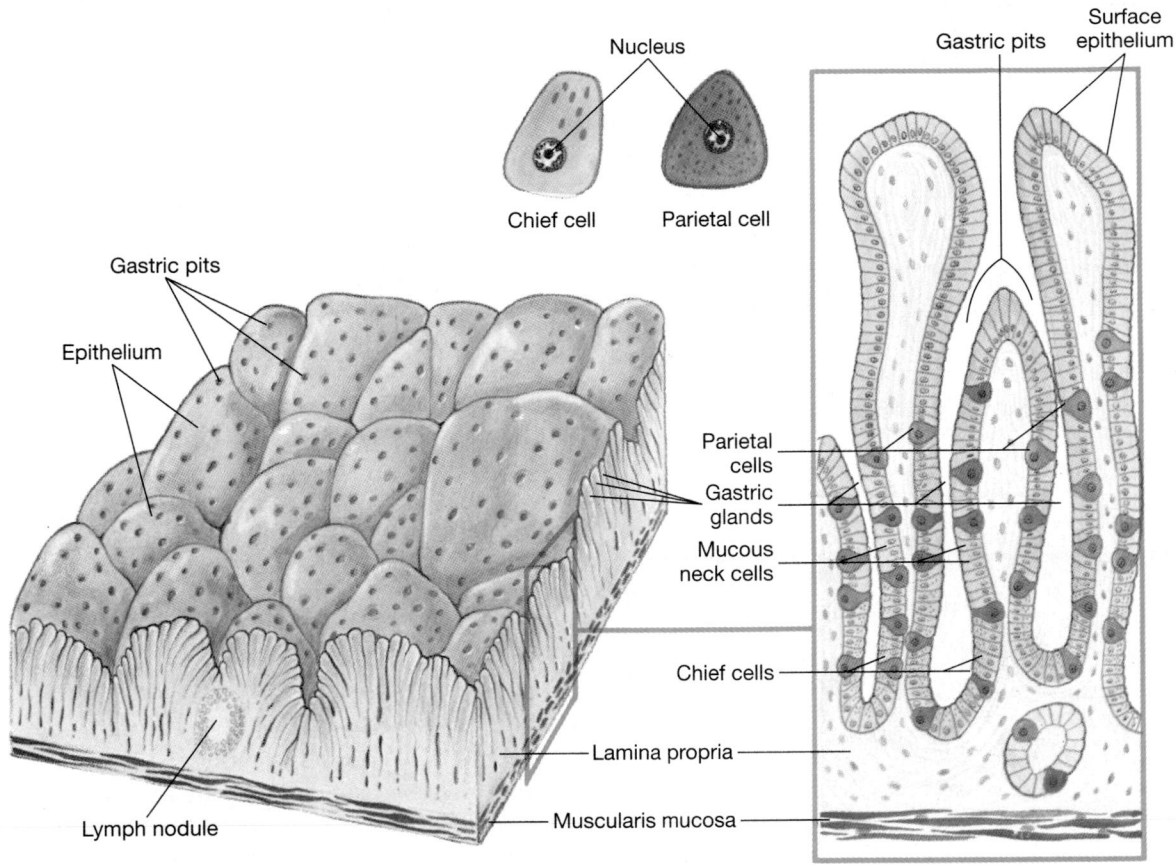

Figure 22–15

Exocrine gland from the body of the stomach.

ach. These exocrine secretory glands contain three types of secretory cells: chief cells, parietal cells, and mucous cells (Fig. 22–15).

The **chief cells,** located in the basal regions of these glands and also in the glands of the antrum, secrete **pepsinogen.** Pepsinogen is the inactive precursor of **pepsin,** an enzyme that initiates the digestion of protein in the digestive tract.

The **parietal cells** secrete the strong (highly dissociated) acid **hydrochloric acid (HCl).** Hydrochloric acid, in addition to converting pepsinogen to pepsin, also provides an optimal pH for the activity of this enzyme. This acid contributes to the breakdown of muscle fibers and connective tissue and kills bacteria that enter the digestive tract through the mouth. Hydrochloric acid is the major factor in the formation of ulcers in the upper digestive tract. The parietal cells also secrete **intrinsic factor.** This substance is necessary for the intestinal absorption of vitamin B_{12}, which is required for formation of normal red blood cells (Chapter 17).

Mucous cells are located at the necks of these glands, in the glands of the antrum, and scattered throughout the gastric epithelial lining. The alkaline, mucus-containing fluid that these cells secrete coats the gastric wall and protects against various forms of damage to the gastric mucosa. Its neutralizing properties provide some protection against chemical injury by HCl; the fluid also protects against mechanical injury by acting as a lubricant.

Secretion of Acid by the Parietal Cells

The parietal cells secrete an isosmotic solution of HCl with a hydrogen ion concentration of about 150 milliequivalents per liter (pH = 0.9). Since the hydrogen ion concentration of parietal cell secretion is about 3 million times that of blood, this ion moves by active transport into the gastric lumen against a large concentration gradient. One idea that has evolved concerning the mechanism of secretion of HCl is that hydrogen ions are derived from the hydrolysis of water in the parietal cells and are actively secreted into the gastric lumen by a pump located in the luminal membrane of the cell (Fig. 22–16). Energy supplied by the hydrolysis of ATP is utilized in the secretory process.

Figure 22–16

Mechanism for the gastric secretion of HCl.

For every hydrogen ion secreted, a hydroxyl ion remains in the cell, and this alkaline ion must be neutralized in order to keep intracellular pH constant. This neutralization is accomplished by the intracellular hydration of carbon dioxide to carbonic acid, a reaction catalyzed by the enzyme carbonic anhydrase, which is present in the parietal cell in high concentrations. Carbonic acid dissociates into a bicarbonate ion and a hydrogen ion, the latter combining with and neutralizing the hydroxyl ion to form water. Bicarbonate moves from the cell interior to blood in exchange for chloride, which is actively transported from the blood to the cell interior by a pump located in the basal membrane. Chloride is then transported into the gastric lumen by a luminal membrane carrier to form HCl in the gastric lumen. Enough water is osmotically carried along with these two ions to produce an isosmotic solution of HCl.

As is the case for gastric emptying, a check-and-balance system operates in which both excitatory and inhibitory mechanisms regulate gastric HCl secretion. Excitation and inhibition of secretion occur in three phases, identified by the region in which a stimulus acts to modify the rate of secretion. These are the **cephalic, gastric,** and **intestinal** phases of gastric acid secretion.

During the cephalic phase (Fig. 22–17*a*), receptors in the head associated with the taste, smell, sight, and chewing of food are stimulated to activate long reflexes that excite the parietal cell. The efferent fibers of these long reflex arcs are in the vagus nerves. Impulses carried over efferent vagal nerve fibers excite neurons of the internal nerve plexus. These internal neurons release acetylcholine in the vicinity of the parietal cells, stimulating these cells to secrete acid. Impulses carried over vagal nerves fibers also act through the internal nerve plexus of the antral wall to cause the release of acetylcholine at the **G-cell (gastrin-secreting cell).** These endocrine cells respond by secreting gastrin into the blood, which carries the hormone to the body and fundus of the stomach, where it excites secretion of acid.

A point that should be made is that *histamine* (Chapter 5), plays a role in the sequence of events that leads to excitation of secretion of the parietal cell. The gastric mucosa is rich in histamine, a molecule that is a potent stimulus of parietal cell secretion. The idea is that both a catylcholine and gastrin act by causing the release of histamine in the vicinity of the parietal cell, and that histamine is the substance that actually causes the parietal cell to secrete.

The gastric phase is triggered by certain chemical agents in gastric contents and by distention of different regions of the stomach when food enters (see Fig. 22–17*b*). For example, the products of protein digestion present in the stomach as the result of the gastric digestion of protein excite the secretion of acid by directly stimulating the parietal cells to secrete acid and also the G cells to secrete gastrin (no nerves are involved). Stimulation of mechanoreceptors in the antral wall by stretch activates short reflexes that result in the release of gastrin. Mechanoreceptors in the body and fundus are excited by stretch, resulting in stimulation of acid secretion through activation of short reflexes in the wall and long reflexes over the vagus nerves. The excitation of acid secretion that oc-

Figure 22–17

Regulation of gastric secretion of HCl. (*a*) Cephalic phase. (*b*) Gastric phase. The solid arrows indicate excitation; the dashed arrow indicates inhibition.

curs as a result of the release of gastrin during the cephalic and gastric phases is checked by an inhibitory mechanism that operates during the gastric phase. Hydrochloric acid secretion is inhibited when acid contacts the antral mucosa, because HCl suppresses the release of gastrin. Thus, HCl acts as an inhibitor of its own secretion, an autoregulatory mechanism that prevents excessive secretion of acid.

Inhibition of secretion also occurs during the intestinal phase. Acid, various digestion products, hyperosmotic solutions, and stretch activate inhibitory mechanisms that originate in the duodenum. Both neural and hormonal pathways operate in the duodenal inhibition of gastric acid secretion, and these are the same pathways involved in the duodenal inhibition of gastric motility (see Fig. 22–14*b*). These duodenal inhibitory mechanisms prevent large quantities of acid from entering the upper small intestine.

Secretion of Pepsin by the Chief Cells

Pepsinogen, the inactive precursor of the protein-digesting enzyme pepsin, is synthesized by and stored in the chief cells as packets called **zymogen granules.** Since pepsinogen cannot hydrolyze proteins, chief cells are safe from being destroyed by digestion of their own protein molecules.

Once pepsinogen enters the gastric lumen, however, both hydrochloric acid and pepsin convert it to

pepsin. Hydrochloric acid initiates the conversion, and the pepsin that is formed carries this process to completion very rapidly. Pepsin hydrolyzes proteins, breaking them into peptide chains of varying lengths. Digestion of protein by pepsin occurs most effectively at a pH of 2 to 3, an acidic environment commonly encountered in the lumen of the stomach during the digestion of a meal.

Since the most effective stimulus for pepsinogen secretion is acetylcholine, the short and long reflexes that occur during the cephalic and gastric phases of digestion and that result in the release of acetylcholine at the chief cells are important in determining pepsin levels in the gastric contents. Note that both acetylcholine and peptides, by virtue of their ability to excite HCl secretion by the parietal cell, play important roles not only in the activation of pepsin but also in the maintenance of the highly acidic environment required for optimal activity of this enzyme. The pathways in the secretion and activation of pepsinogen are summarized in Figure 22–18.

Vomiting Rapidly Expels Gastric Contents Through the Mouth

In vomiting, the expulsion of gastric contents is preceded by a copious secretion of saliva, increased heart rate, sweating, pallor, and nausea, all characteristic

CLINICAL FOCUS

Ulcers

An **ulcer** is an erosion of the mucosa of certain regions of the digestive tract. This erosion can extend deep into the wall of the tract, damaging blood vessels, with resultant bleeding. A **perforated ulcer** extends entirely through the wall of the tract, allowing food, gastrointestinal secretions, and bacteria to escape into the peritoneal cavity. This in turn may result in peritonitis, shock, and even death.

The dominant factors in the development of ulcers are the corrosive action of HCl and autodigestion by pepsin, as evidenced by the fact that ulcers occur mostly in regions of the digestive tract exposed to acid and pepsin. These regions are the upper small intestine (where the most prevalent type of ulcer, **duodenal ulcer,** forms), the stomach (where **gastric ulcers** occur at a frequency about one tenth that of duodenal ulcers), and occasionally the esophagus (when this organ is subjected to reflux of gastric contents for prolonged periods).

A major reason for the development of ulcers is hypersecretion of acid and pepsin, which overwhelms the defenses of the digestive tract. Hypersecretion may result from excessive activity of the mechanisms that excite gastric secretion. Recently, antisecretory drugs such as **cimetidine, rantidine** (blocks the action of histamine),

and **omiprazole** (blocks the H^+ pump) have been used effectively in ulcer therapy. Another reason for ulcer production is lack of protection against acid and pepsin. The most important mechanism that protects against gastric ulcer is the **gastric mucosal barrier.** The permeability characteristics of the cell membranes of the gastric mucosa restrict the entrance of ions, even ions as small as hydrogen ions, to the interior of these cells. Thus, HCl does not usually enter these cells in quantities sufficient to destroy them. Substances such as acetylsalicylic acid (aspirin) and ethanol increase the permeability of these cell membranes, allowing acid to penetrate the cells and destroy them.

Another important protective mechanism is the neutralization of acid by food and by the alkaline secretions of the upper digestive tract. For example, the pancreas secretes an alkaline juice into the duodenum that very efficiently neutralizes acid arriving from the stomach, thereby protecting against the possibility of duodenal ulcer. The principle of neutralization is widely used in the treatment of ulcers. Various commercial alkalies (bases) and buffers are prescribed for this purpose. Another common approach is the eating of five to six well-spaced meals during the day, which, by keeping food in the stomach much of the time, keeps gastric acidity at a relatively low level.

Finally, the rapidity (every 1 to 3 days) with which cells lining this organ divide prevents lasting injury to the mucosa of the stomach.

A recent development in the understanding of ulcer formation concerns **Helicobactor pylori,** a bacterium that lives in the gastric mucosa. It is likely that infection by this bacterium is responsible for either the production or the maintenance of ulcers by weakening of the gastric mucosal defenses. Therapies directed at eradicating the bacterium are particularly effective in reducing the recurrence of ulcers.

Photograph of a duodenal ulcer. *(© Custom Medical Stock Photo.)*

■ **In past years, certain surgical procedures were used in the treatment of ulcer. These included vagotomy of the stomach, removal of the pyloric antrum, and removal of a part of the body and fundus of the stomach. What was the physiological rationale for using these procedures?**

of a general discharge of the sympathetic nervous system.

Vomiting begins with a deep inspiration and closure of the glottis and nasal passages. Contraction of skeletal muscle provides the major force for expulsion; descent of the diaphragm during a forced inspiration and contraction of the abdominal muscles raise abdominal, and thus intragastric, pressure. The increased pressure in the stomach forces gastric contents into the esophagus through the upper esophageal sphincter, which relaxes, and into the mouth. Vomiting is frequently preceded by **retching,** a pro-

Figure 22–18

Pathways in the secretion and activation of pepsinogen.

cess by which contents are forced into the esophagus, but not into the mouth, because the upper esophageal sphincter is not relaxed. A series of retches often leads to vomiting.

The complicated reflex act of vomiting is coordinated by a center located in the medulla. Potent stimuli arise from sensory nerve endings in the throat—for example, when a finger is stuck in the back of the throat. Other receptor areas include all regions of the digestive tract (distention of the stomach and small intestine), other abdominal viscera, and the labyrinths of the ear when rotational movements of the head result in motion sickness. Certain chemical agents, called **emetics,** cause vomiting by stimulating chemoreceptors in the stomach and small intes-

tine or in the brain. When vomiting is prolonged and large quantities of gastric contents are lost, disturbances in water volume (dehydration), electrolyte balance (loss of Na^+, K^+, Cl^- and HCO_3^-), and acid-base balance (metabolic alkalosis) result (Chapter 25).

The Processes of Secretion in the Exocrine Pancreas

The pancreas (Fig. 22–19a) is both an endocrine and an exocrine secretory organ. In addition to containing endocrine cells that secrete the hormones insulin and glucagon (Chapter 15), the pancreas contains ex-

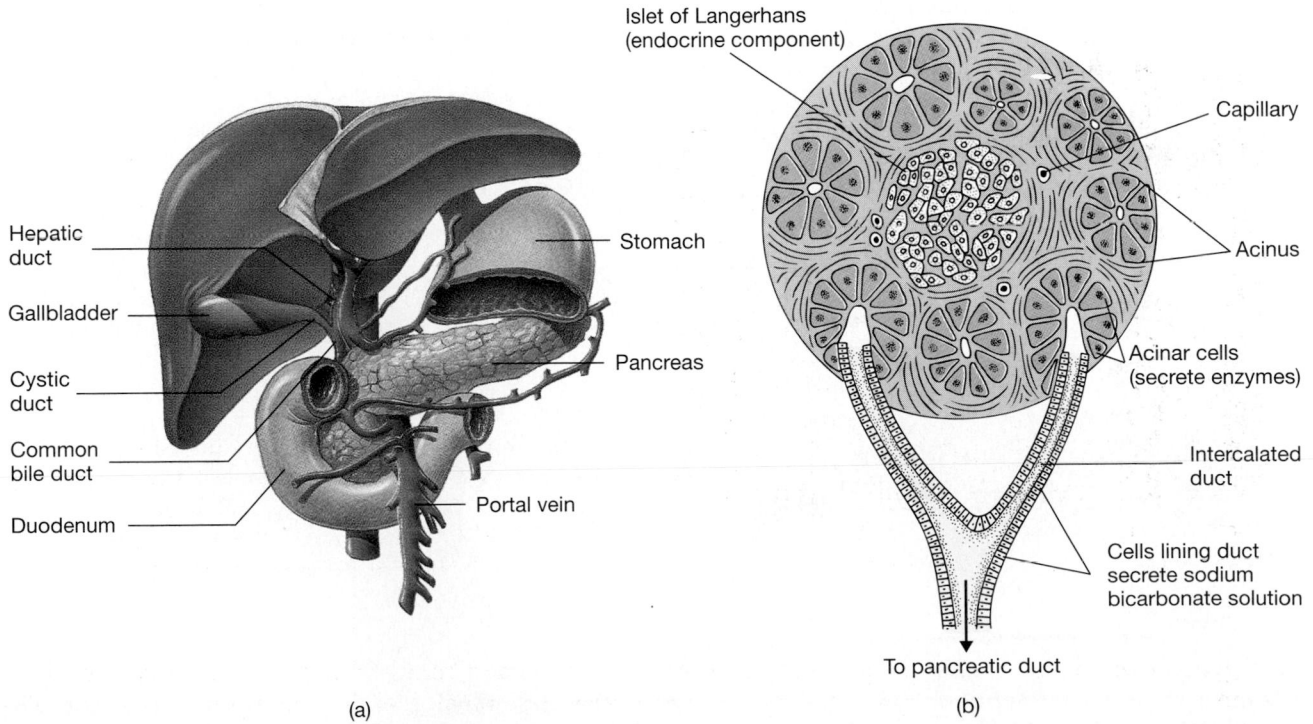

(a)   (b)

Figure 22–19

(a) Gross anatomy of the pancreas, duodenum, and biliary system. (b) Microanatomy of the exocrine pancreas.

ocrine secretory cells called **acinar cells** that are arranged in saclike clusters called **acini** (singular, *acinus*, "rounded lobule") (Fig. 22–19*b*). These cells secrete a small volume of juice containing the digestive enzymes of the pancreas. The acini are connected by small **intercalated ducts** to larger ducts that converge into one duct (sometimes two) that delivers the exocrine secretion of the pancreas into the duodenum. The epithelial cells lining the intercalated ducts make an important contribution to the overall exocrine secretion of the pancreas. These cells secrete a relatively large volume of juice with a high concentration of the alkaline salt sodium bicarbonate. By neutralizing acid that arrives in the duodenum from the stomach, bicarbonate performs two functions: (1) it protects the delicate lining of the upper small intestine from damage by acid, and (2) it provides an environment with the optimal pH for activity of the pancreatic digestive enzymes.

The Secretion of Pancreatic Bicarbonate Is by an Active Transport System

The mechanism of secretion of pancreatic bicarbonate by the intercalated duct cells of the pancreas is shown in Figure 22–20. The bicarbonate is formed when CO_2 diffuses into the cell and is converted to bicarbonate and hydrogen ions in a reaction catalyzed by the enzyme carbonic anhydrase. Bicarbonate ions are actively transported from the duct cells to the duct lumen, whereas hydrogen ions are ac-

tively transported into the interstitial fluid on the blood side of the cell. Hydrogen ions are pumped out of the cell in exchange for sodium ions that passively diffuse into the cell from interstitial fluid. Sodium passively follows the secreted bicarbonate along with enough water to form an isosmotic solution of sodium bicarbonate. Recall from the discussion of secretion of HCl by the gastric parietal cells that bicarbonate enters the blood as a part of this secretory process, and this alkalinizes the blood leaving the stomach. However, since an equivalent amount of hydrogen ion enters the blood during secretion of pancreatic bicarbonate, overall blood pH remains unchanged during the digestion of a meal.

The Pancreas Secretes a Wide Variety of Digestive Enzymes

Pancreatic juice is the most versatile of the digestive secretions because its enzymes are capable of nearly completing the digestion of food in the absence of all other digestive secretions. These enzymes require a pH close to neutrality for optimal activity, and this is provided by the alkaline secretion of the pancreatic intercalated duct cells. The actions of the pancreatic enzymes and the products of their digestive activity are summarized in Table 22–4.

The pancreas secretes three protein-digesting enzymes. **Trypsin** is secreted as inactive **trypsinogen,** which is converted to active trypsin by **enterokinase,** an enzyme present in the luminal membrane of the

Figure 22–20

Mechanism for the pancreatic secretion of bicarbonate. Solid arrows indicate active transport and dashed arrows indicate passive transport.

epithelial lining of the small intestine. Trypsin activates the remaining trypsin and the other proteolytic enzymes; that is, **chymotrypsinogen** is converted to **chymotrypsin** and **procarboxypeptidase** to **carboxypeptidase.** Trypsin and chymotrypsin split certain amino acid linkages in the interior of proteins and peptides, whereas carboxypeptidase liberates amino acids with free carboxyl groups from the end of proteins and peptide chains. The combined actions of gastric pepsin and the pancreatic protein-digesting enzymes degrade proteins to a mixture of small peptides and amino acids.

Under certain conditions, activated proteolytic enzymes are liberated from the acinar cells into the surrounding pancreatic tissue. These enzymes may attack the pancreas itself, producing the condition called **pancreatitis.** This self-digestion process often reduces output of both pancreatic digestive enzymes and bicarbonate.

The pancreas secretes **lipase,** the only enzyme of the gastrointestinal system capable of digesting triacylglycerol fats. This enzyme hydrolyzes triacylglycerols to monoglycerides and free fatty acids. When it is lacking, as in pancreatitis, excessive amounts of undigested fat appear in the feces, a condition known as **steatorrhea.** Other lipid-digesting enzymes are **phospholipase** and **cholesterol esterase.**

The pancreas also secretes an **amylase,** the action of which is similar to that of salivary amylase; that is, it degrades starch to oligosaccharides, mainly maltose. Other digestive enzymes secreted by the pancreas are the nucleic acid–digesting enzymes, **ribonuclease** and **deoxyribonuclease.**

Regulation of Pancreatic Bicarbonate and Enzyme Secretion Is by Hormonal Mechanisms

The hormones **secretin** and **cholecystokinin,** which are released from the duodenal mucosa during the intestinal phase of digestion, play the central role in the control of secretion by the pancreas. Acid is the most potent stimulus for release of **secretin** into the blood. This hormone stimulates the intercalated duct cells of the pancreas to secrete a bicarbonate-rich juice. This secretion neutralizes the acid that causes secretin to be released and so reduces the stimulus for additional secretin release until more acid enters the duodenum from the stomach.

The duodenal mucosa releases **cholecystokinin** when contacted by protein and fat digestion products. Cholecystokinin stimulates the pancreatic acinar cells to secrete digestive enzymes. As a result of

Table 22–4
Digestive Enzymes of Pancreatic Juice

Enzyme	Substrate	Action	Products of Digestion
Proteolytic			
Trypsin, chymotrypsin	Proteins and peptides	Splits amino acid linkages in the interior of proteins and peptides	Small peptides
Carboxypeptidase	Proteins and peptides	Splits terminal amino acids with free carboxyl groups from the ends of proteins and peptides	Peptides and amino acids
Amylolytic			
Amylase	Polysaccharides	Splits internal glucose bonds of polysaccharides	Oligosaccharides
Lipolytic			
Lipase	Triacylglycerols	Splits two fatty acids from triacylglycerols	Free fatty acids and mono-glycerides
Phospholipase	Lecithin	Splits one fatty acid from lecithin	Lysolecithin
Cholesterol esterase	Cholesterol esters	Splits the ester bond in cholesterol esters	Cholesterol and fatty acids
Nucleolytic			
Ribonuclease, deoxyribonuclease	Nucleic acids	Splits nucleic acids	Mononucleotides

Figure 22–21

Pathways in the regulation of exocrine pancreatic secretion.

	Liver Bile	Gallbladder Bile
Table 22–5		
Composition of Human Bile		
Constituent	(%)	(%)
Water	97.48	83.98
Bile pigments	0.53	4.44
Bile salts	0.93	8.70
Fatty acids	0.12	0.85
Cholesterol	0.06	0.87
Lecithin	0.02	0.14
Inorganic salts	0.83	1.02

the actions of these enzymes, more digestion products are produced, which in turn stimulate the pancreas to secrete more enzymes. This process continues as long as food remains in the small intestine to be digested. The pathways involved in the regulation of secretion of pancreatic juice are summarized in Figure 22–21.

The Processes of Secretion in the Biliary System

Bile is secreted by liver cells **(hepatocytes)** into minute channels called **bile canaliculi,** which drain into the **hepatic duct.** When food is not being digested, bile is diverted by way of the **cystic duct** into the **gallbladder,** a small sac that stores bile between meals (see Fig. 22–19). Following a meal, contractions of the gallbladder move bile through the **common bile duct** and the **sphincter of Oddi** into the small intestine, where it plays its role in the digestion and absorption of fat.

The molecules present in bile that are important in these processes are the **bile salts.** Other major components of bile are the **bile pigments** (see Chapter 17). These molecules are breakdown products of hemoglobin and are for the most part excreted in the feces. Two more important organic constituents of bile are the lipid molecules **cholesterol** and **lecithin.** Inorganic ions of bile include sodium, the major cation, and chloride and bicarbonate, the predominant anions.

The Gallbladder Stores and Concentrates Bile Secreted by the Liver

Between meals, bile is prevented from entering the small intestine and is shunted into the gallbladder by the contraction of the sphincter of Oddi, a ring of smooth muscle surrounding the common bile duct in the region where this duct enters the duodenum. When food enters the upper small intestine, fat and protein digestion products stimulate the release of cholecystokinin from the duodenal mucosa into the blood. This hormone stimulates contraction of the gallbladder and relaxation of the sphincter of Oddi. These muscular movements result in the flow of bile into the small intestine.

While bile is stored in the gallbladder, sodium is actively transported from the gallbladder lumen to the blood by the epithelial cells that line the inner aspect of this organ. The osmotic gradient produced by the movement of sodium and accompanying anions (chloride and bicarbonate) causes water to follow across the wall of the gallbladder, with the result that the organic constituents of bile, which are not absorbed, become concentrated in the process (Table 22–5).

Despite the fact that the organic constituents of bile are concentrated by the removal of fluid from the gallbladder, the water-insoluble lipid molecules of bile usually, but not always, remain in solution. The presence in the gallbladder and bile ducts of *gallstones,* one of the most common disorders of the gastrointestinal system, is an example of lipid molecule that does not always remain in solution in bile. Cholesterol, a normal constituent of bile, is the major component of most gallstones. Cholesterol itself is virtually insoluble in water. However, two other constituents of bile, the bile salts and lecithin, together can dissolve cholesterol by forming water-soluble aggregates called **micelles** that contain all three of these molecules.

Whenever cholesterol precipitates out of solution, small crystals of cholesterol are formed, and these can grow gradually into large gallstones. This can happen when the capacity of the bile salts and lecithin to dissolve cholesterol is exceeded—when, for example, the liver secretes higher-than-normal quantities of cholesterol, or when the amount of bile salts secreted is lower than normal. When the bile ducts are obstructed by a gallstone, the flow of bile into the small intestine is retarded or stopped. Since secretion of bile by the liver continues, pressure in the biliary system rises to higher-than-normal levels, and severe pain may result from stimulation of receptors in the wall of the biliary tree. Removal of either the gallbladder **(cholecystectomy)** or of stones from the bile ducts has been the usual treatment for gallstones. More recently, substances with the capacity to dissolve cholesterol gallstones, such as the bile acid **chenodeocycholic acid,** have been given to patients with encouraging results.

Bile Salts Are Recycled Between the Liver and Small Intestine

Bile salts are synthesized by the liver hepatocytes and actively transported by these cells into the bile canaliculi. Once the bile salts enter the small intestine, they play an essential role in the digestion and absorption of fat, a topic that will be discussed later in this chapter. These molecules traverse the small intestine and are absorbed from the lower part of this organ, the **terminal ileum,** by a specialized active transport process. They are then returned to the liver, where the hepatocytes remove them from the blood and re-secrete them into the bile. This recycling of bile salts between liver and small intestine is called the **enterohepatic circulation of bile salts** (Fig. 22–22). The process is interrupted when the bile salts are stored in the gallbladder.

The body's pool of bile salts is circulated a number of times every day. This process is not 100% efficient, because small quantities of bile salts enter the large intestine with each circulation and are excreted in the feces. However, synthesis by hepatocytes of new bile salts from the structurally related molecule cholesterol easily makes up this loss. The amount of bile salts synthesized depends on the portal blood concentration of these molecules; that is, the rate of synthesis is high when portal blood concentration is low, and vice versa. Synthesis is regulated through inhibition by circulating bile salts of one of the steps in the series of reactions leading to the conversion of cholesterol to bile salts (see Fig. 22–22).

Regulation of Secretion of Bile Is by the Portal Blood Concentration of Bile Salts and by Secretin

Bile salts are the major stimuli for the secretion of bile. As these molecules are removed from the portal circulation and actively secreted into the liver canaliculi, water and inorganic ions passively follow along in proportion to the quantity of bile salts transported. Consequently, the greater the concentration of bile salts in the portal blood, the greater the transport of bile salts and the volume of bile secreted. Portal blood concentrations of bile salts are greatest—and thus bile flow is at its highest level—when food is being digested, at which time the bile salts are being circulated between liver and small intestine. Between meals, when they are stored in the gallbladder, the

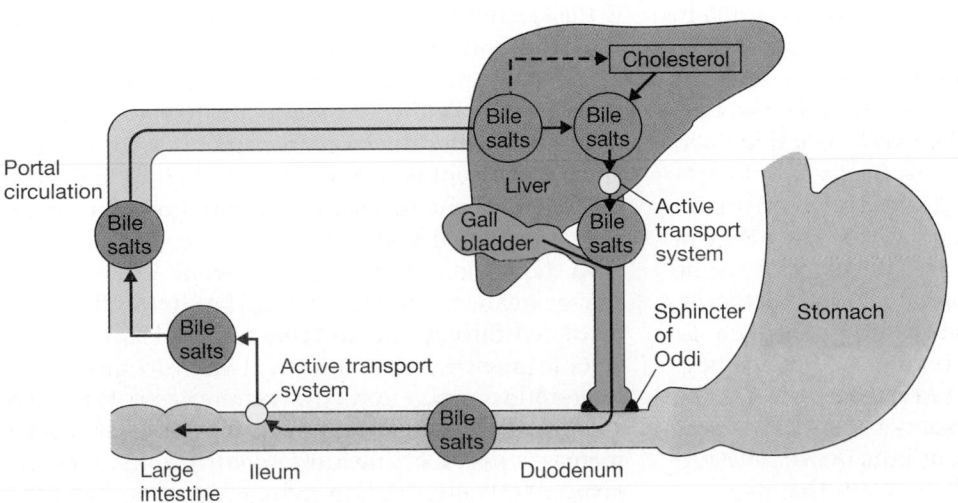

Figure 22–22

The enterohepatic circulation of bile salts as shown by the heavy solid arrows. The dashed arrow represents inhibition of bile salt synthesis.

concentration of bile salts in portal blood is low, and liver bile flow is minimal.

The hormone secretin also stimulates the secretion of bile, but in this instance the cells stimulated are the epithelial cells that line the bile ducts. These cells respond to secretin with an increased output of water and inorganic ions, particularly sodium bicarbonate. This is analogous to secretin stimulation of the pancreas in that secretin also stimulates the duct cells of this organ to secrete a bicarbonate-rich juice practically devoid of organic substances. In both instances, the resulting secretion helps neutralize acid that enters the small intestine from the stomach.

The Processes of Motility, Secretion, and Absorption in the Small Intestine

The small intestine is about 2.8 m long that includes an initial 20- to 25-cm segment, the **duodenum.** The remainder of the tube is divided equally between **jejunum** and **ileum.** The digestion of food is completed by enzymes in the epithelial cells that line the inner surface of this organ, and the products of digestion are absorbed into the blood and the lymph. Smooth muscle contractions of the small intestine facilitate digestion and absorption and move unabsorbed food residues and other excretory products into the large intestine.

The Muscular Movements of the Small Intestine Are Segmenting Contractions (Mixing Movements) and Peristalsis

The most frequently occurring type of muscular movement of the small intestine is the **segmenting contraction,** which is illustrated diagrammatically in Figure 22–23. The segmenting contractions move the contents of different regions of the small intestine back and forth over short distances. In doing so, they mix the contents with the digestive secretions and also bring the products of digestive activity into contact with the absorptive epithelium. The frequency of these movements is determined by the frequency of the slow waves initiated by pacemaker cells in the longitudinal muscle of different regions of the intestinal wall. These basic electrical rhythms progressively decrease from above to below. About 11 segmenting contractions occur every minute in the duodenum during the digestion of food, whereas only about 8 occur per minute for the ileum. This difference in the rate of contraction means that more in-

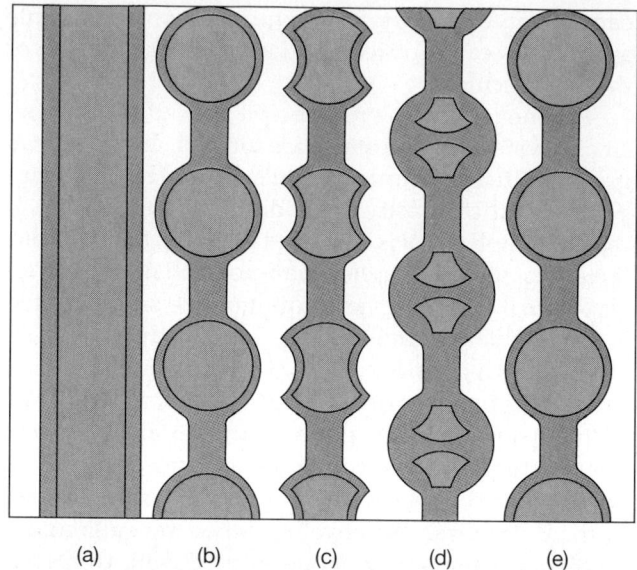

    (a)   (b)   (c)   (d)   (e)

Figure 22–23

Diagram of segmenting contractions of the small intestine. (*a*) Absence of contractile activity. (*b*) The circular muscle contracts at a number of places, dividing a region of the intestine and its contents into a series of segments. (*c,d*) Each segment divides, with adjoining halves and their contents coming together to form new segments. (*e*) The original pattern is re-established.

testinal contents are forced downward than upward, with the net effect of propelling contents toward the large intestine.

Peristalsis is the other major type of motor activity of the small intestine, and it occurs after most of a meal has been absorbed. These waves of contraction move downward slowly and travel only short distances, providing the slow transit required for the completion of the absorptive processes.

The last 2 to 3 cm of the musculature of the ileum is thicker than the rest of the ileum. This region, the **ileocecal sphincter,** is normally closed by contraction. However, during periods of increased motor activity of the ileum, the sphincter relaxes, allowing contents to enter the large intestine. It then contracts again to prevent backflow of contents into the ileum.

Chemical and mechanical stimulation of receptors in the intestinal wall by intestinal contents initiate both short and long reflexes important in the regulation of small intestinal motility. The short reflexes are mediated through the internal nerve plexuses to the smooth muscle in the intestinal wall. In the case of long reflex regulation, the generalized vagal discharge that occurs during the digestion of a meal increases vigor of intestinal contractions, whereas sympathetic stimulation, when it occurs, produces inhibition of motility. Gastrin plays a role in the propulsion of contents from the terminal ileum by increasing motor activity of the ileum and relaxing the ileocecal sphincter.

Water and Solute Move from Blood to Small Intestinal Lumen in Response to Osmotic and Solute Concentration Gradients and by Secretion

About 1 to 2 liters of fluid move from the blood to the intestinal lumen every day. Some of this fluid movement results from the osmotic shift of water from blood to lumen in response to the presence of hyperosmotic contents in the lumen. Intestinal contents can have a high osmotic activity, either because the material that enters the small intestine from the stomach is hyperosmotic or because the rapid breakdown of large molecules to many small molecules by enzymatic digestion produces hyperosmotic contents. Enough water moves into the lumen to produce contents that are isomotic to blood, thereby minimizing osmotic damage to the delicate lining of the small intestine. It is also significant that gastric emptying is retarded when the contents of the duodenum are hyperosmotic. The regulatory mechanisms involved reduce the possibility of large volumes of hyperosmotic contents entering the small intestine, which in turn ensures that there will not be large shifts of water from blood to lumen. If this were to happen, circulatory collapse might occur as the result of depletion of blood volume.

The intestinal mucosa also contributes to the material contained in the intestinal lumen. Mucosal exocrine glands scattered throughout the small intestine secrete water, inorganic ions, and mucus. Local chemical and mechanical stimulation by intestinal contents are effective stimuli for secretion by these glands. In addition, epithelial cells are shed from the intestinal mucosa into the intestinal lumen. This process of sloughing of the epithelium is continuous, resulting in complete renewal of the intestinal epithelium every 1 to 2 days. The disintegration of these cells in the intestinal lumen contributes small quantities of enzymes, some organic material, and inorganic ions to the intestinal contents.

Absorption of Nutrients Occurs Largely from the Small Intestine

Intestinal absorption is the movement of water and dissolved materials from the lumen of the small intestine through the intestinal epithelium into the blood and lymph. Most of the organic constituents of the contents of the small intestine are completely absorbed midway through this organ. Exceptions to this are vitamin B_{12} and the bile salts (both of which are absorbed from the terminal ileum by specific transport processes) and certain excretory products such as the bile pigments. The lower small intestine (the distal jejunum and ileum) can absorb a variety of food digestion products, and thus it acts as a reserve area of absorption in situations in which food escapes absorption upstream. Water and sodium chloride are absorbed throughout the small intestine, but significant quantities of these enter the large intestine, where their absorption is normally completed.

The Absorptive Surface of the Small Intestine

The absorptive surface of the small intestine is highly folded and, in addition, is convoluted by small fingerlike projections called **villi** (singular, *villus*) (Fig. 22–24*a*). Each villus is lined with epithelial cells and contains a rich network of blood vessels and lymph vessels. Another major feature of the absorptive surface is that the luminal aspect of each epithelial cell is covered with much smaller projections, the **microvilli** (see Fig. 22–24*b*). An epithelial cell's microvilli are the luminal membrane of that cell; collectively, the microvilli of all of the epithelial cells lining the small intestine are known as the **brush border membrane.** The small intestine is well suited for the task of absorption, because the surface area available for absorption (Fick's Law—Chapter 2) is greatly increased not only by the intestinal folds but also by the villi and the microvilli (Fig. 22–25).

The brush border membrane contains a variety of enzymes that complete the process of digestion and also a number of transport systems that move digestion products and inorganic ions from the intestinal lumen to the blood. In many instances, this is active transport against an electrochemical gradient. In other cases, molecules diffuse passively from blood to lumen or lumen to blood down concentration gradients at differing rates, depending on the magnitude of the gradient and on the size and lipid solubility of the molecules.

Absorption of Water and Inorganic Ions from the Small Intestine

Approximately 9 liters of water enter the small intestine every day. About 2 liters are ingested, and the remaining 7 liters are derived from the various secretions of the gastrointestinal system. Most of this water is absorbed from the small intestine, as shown by the fact that only about 500 ml of water enters the

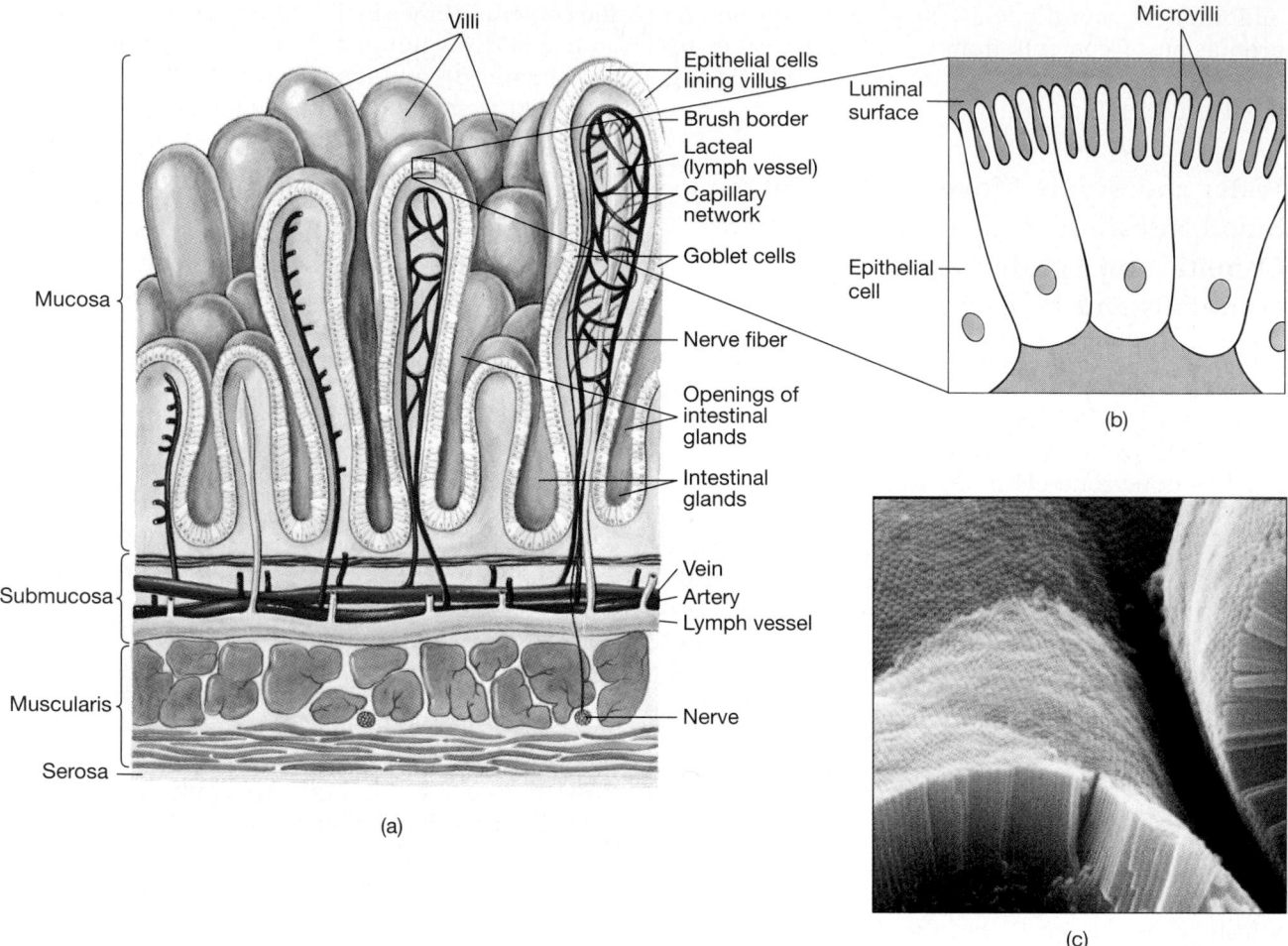

Villi

Epithelial cells lining villus
Brush border
Lacteal (lymph vessel)
Capillary network
Goblet cells

Nerve fiber

Openings of intestinal glands
Intestinal glands

Vein
Artery
Lymph vessel

Nerve

Mucosa

Submucosa

Muscularis

Serosa

(a)

Microvilli
Luminal surface
Epithelial cell

(b)

(c)

Figure 22–24

(a) Structure of the absorptive surface of the small intestine. (b) Microvilli on the luminal surface of intestinal epithelial cells. (c) Scanning electron micrograph of the surface of an epithelial cell from the small intestine, illustrating the microvilli.   *(© Custom Medical Stock Photo.)*

proximal large intestine. Thus, the small intestine is about 95% efficient as a water-absorbing organ. Water is absorbed from the large intestine to such an extent that only about 100 ml is lost in the stool every day. This means that a total of 8900 ml of water is absorbed daily by the combined actions of the small and large intestines; that is, the overall efficiency of water absorption by the intestines is 99%. The handling of water by the gastrointestinal system is summarized in Figure 22–26.

The absorption of water occurs by osmosis throughout the small intestine. As solutes such as sugars, amino acids, other organic molecules and inorganic ions are transported across the intestinal epithelium into the blood, a gradient for the diffusion

of water from lumen to blood is created, and water responds to this gradient by diffusing out of the intestinal lumen into the blood. Since the most abundant solute of intestinal contents is sodium, this ion is of greatest importance in the absorption of water. Sodium transport is an active, energy-requiring process and is polarized in that this ion diffuses down an electrochemical gradient from the lumen across the brush border membrane into the intestinal epithelial cell and is then extruded against an electrochemical gradient into the blood by a sodium pump located in the basolateral membrane of the cell (Fig. 22–27).

Other inorganic ions absorbed into the blood (some actively and others passively) are potassium,

		Increase in surface area (relative to cylinder)	Surface area (sq.cm)
	Simple cylinder	1	3,300
	Intestinal fold	3	10,000
	Villi	30	100,000
	Microvilli	600	2,000,000

Figure 22–25

Three ways by which the surface area of the small intestine is increased from that of a simple cylinder of the same gross dimensions as those of the small intestine (280 cm in length and 4 cm in diameter).

calcium, magnesium, iron, zinc, chloride, phosphate, bicarbonate, and iodide. The intestine possesses special mechanisms for the absorption of iron (see Chapter 17) and calcium (see Chapter 26). The transport of both of these cations is an active process and represents instances of regulated absorption.

Absorption of Carbohydrates from the Small Intestine

Carbohydrates make up about one half of the caloric intake of people in the United States. Plant starch, a polysaccharide, is the major digestible carbohydrate

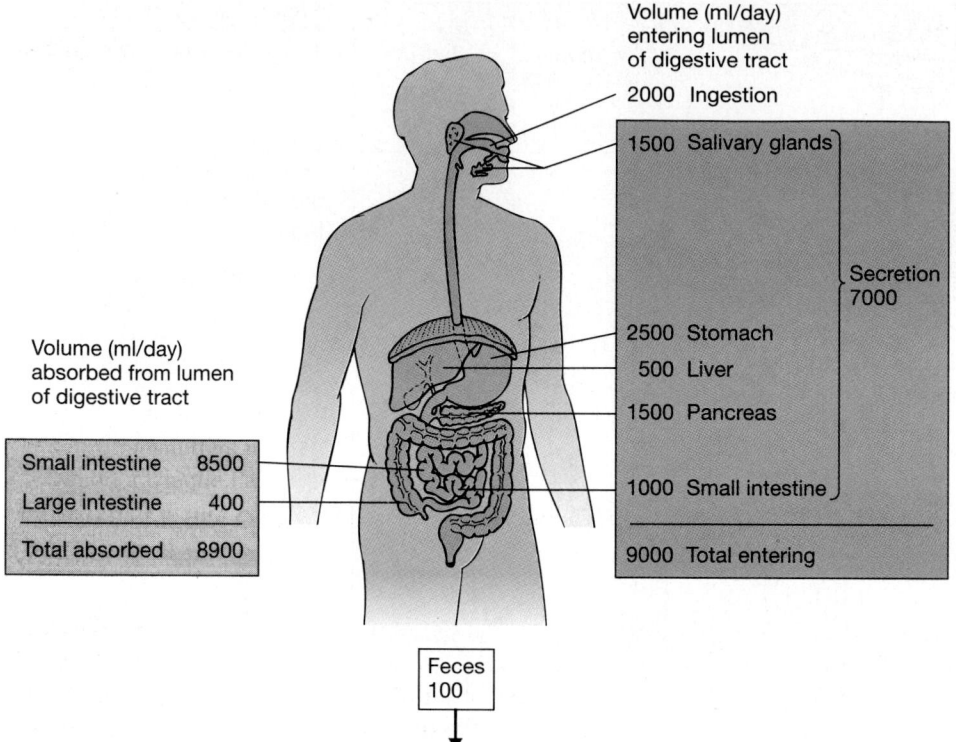

Figure 22–26

Water balance of the gastrointestinal system.

Volume (ml/day) entering lumen of digestive tract

2000 Ingestion
1500 Salivary glands
2500 Stomach
500 Liver
1500 Pancreas
1000 Small intestine
9000 Total entering

Secretion 7000

Volume (ml/day) absorbed from lumen of digestive tract

Small intestine	8500
Large intestine	400
Total absorbed	8900

Feces 100

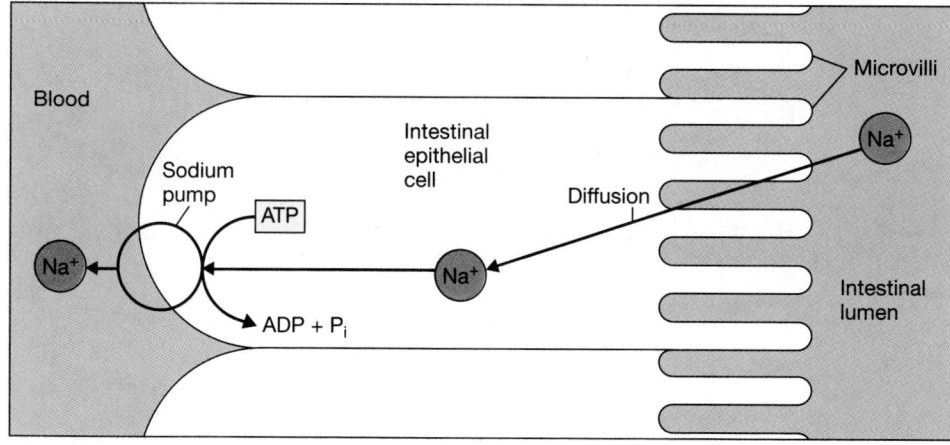

Figure 22–27

The transport of sodium across the intestinal epithelium.

Blood

Sodium pump

ATP

Na^+

ADP + P_i

Intestinal epithelial cell

Na^+

Diffusion

Microvilli

Na^+

Intestinal lumen

---

of the diet. It is composed of many straight and branched chains of the monosaccharide glucose. Starch is hydrolyzed rapidly by salivary and pancreatic amylase to oligosaccharides, molecular fragments consisting of two to nine glucose molecules. The major oligosaccharide produced is maltose, a disaccharide that consists of two glucose molecules. Other oligosaccharides present in the diet in smaller quantities are the disaccharides sucrose, or table sugar (made up of glucose and fructose), and lactose, or milk sugar (glucose and galactose).

Another common carbohydrate in the diet is cellulose, a plant polysaccharide made up of glucose molecules but joined by linkages different from those

of starch. Since humans do not secrete an enzyme that digests cellulose, this carbohydrate is excreted in the feces.

The oligosaccharides present in the lumen of the small intestine are broken down into their constituent monosaccharides by oligosaccharidases located in the microvilli of the intestinal epithelial cell. Maltose is hydrolyzed by the **disaccharidase maltase** to two glucose molecules. Lactose is degraded by **lactase** to one molecule of glucose and one molecule of galactose, and sucrose is broken down by **sucrase** to one molecule of glucose and one molecule of fructose (Fig. 22–28a). The higher-molecular-weight oligosaccharides produced by the action of amylase on starch

CLINICAL FOCUS

Lactase Deficiency

Lactase deficiency is a condition in which there is a lack or absence of the disaccharidase, lactase, in the small intestinal brush border membrane. Since there is little or no lactase available to hydrolyze lactose this disaccharide, it is not absorbed and is excreted **(lactose intolerance).**

Because lactose is osmotically active, this molecule retains water and some dissolved materials in the small intestinal lumen, and these are excreted along with the lactose. A further complication is that small intestinal and colonic bacteria convert lactose to lactic and other organic acids. These molecules not only increase the osmotic load of solute in the intestinal lumen, but also stimulate motility of the intestines. The rapid transit of intestinal contents that results is an additional factor in causing water and dissolved materials that normally would have been absorbed to be excreted. The severe diarrhea that results leads to dehydration and negative electrolyte balance. (See Chapter 24).

■ **Why are milk and milk products eliminated from the diet of persons with lactase deficiency?**

---

are hydrolyzed in the brush border membrane to glucose molecules. Fructose is absorbed passively by facilitated diffusion. On the other hand, glucose and galactose are actively transported into the portal blood by a carrier-mediated process located in the brush border membrane (see Fig. 22–28*b*). The transport of glucose and galactose is coupled (in cotransport) with that of sodium and requires the expenditure of metabolic energy. There is a mobile carrier in the microvilli with binding sites for both sugar and sodium. It is the diffusion gradient for sodium movement into the cell that provides the driving force for moving the sugar- and sodium-loaded carrier from the outer to the inner aspect of the microvillous mem-

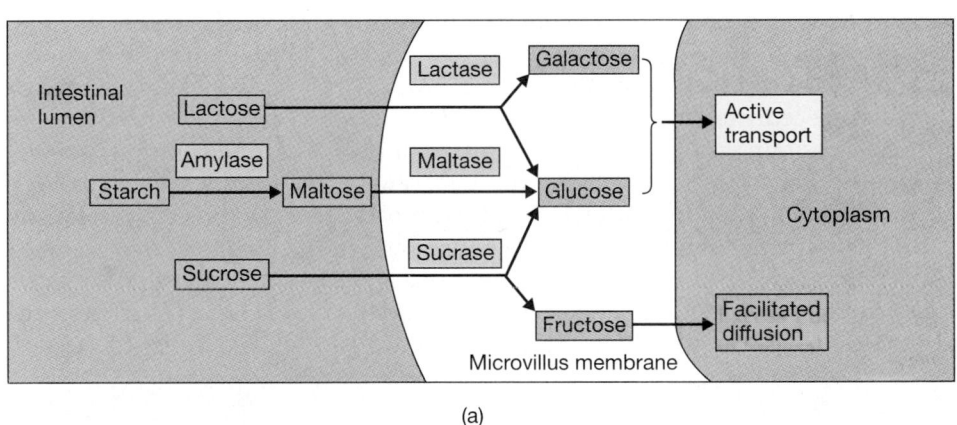

(a)

Figure 22–28

(*a*) Luminal and microvillous membrane digestion of carbohydrates. (*b*) Carrier-mediated transport of glucose and sodium across the intestinal epithelial cell.

(b)

brane, where both sugar and sodium are released to the cell interior. The sugar then moves from the cell interior through the basolateral membrane into the blood by facilitated diffusion. The energy in this system is expended by a sodium pump (see Fig. 22–27), which transports this ion across the basolateral membrane, thereby keeping the intracellular sodium concentration at a level low enough to maintain a diffusion gradient from lumen to cell, and ensuring continued movement of the sugar-sodium carrier.

Absorption of Protein from the Small Intestine

A normal adult requires about 35 to 50 grams of protein daily in order to have essential amino acids and to achieve nitrogen balance. The average American consumes 70 to 90 grams of protein every day. However, the protein presented to the digestive tract for absorption consists not only of dietary protein but also of endogenous protein that approaches the amount of protein derived from food. This **endogenous protein** is in the form of digestive enzymes, epithelial cells, mucoproteins, and plasma proteins that leak into the tract. Most all of the dietary and endogenous protein is digested and absorbed.

The end products of the intraluminal digestion of proteins are amino acids and small peptides. A variety of enzymes that hydrolyze peptides are present in the brush border membrane, and many peptides are hydrolyzed there (by specific **peptidases**) to their constituent amino acids. Amino acids are transported into the portal blood by active processes occurring in the intestinal epithelial cell. Each class of amino acid has its own specific carrier-mediated transport system in the microvilli and, as is the case for certain sugars, amino acid transport is coupled with sodium transport. Thus, clearly there are many similarities between the absorption of carbohydrate and the absorption of protein. However, in contrast to carbohy-

drate absorption, in which only simple sugars are transported across the intestinal epithelium, carrier-mediated mechanisms are available in the intestinal epithelial cell for the active transport of dipeptides and tripeptides. These carrier-mediated processes for the absorption of small peptides may be dependent on sodium.

Absorption of Lipid from the Small Intestine

About 60 to 100 grams of lipid are present in the daily diet of the average adult American. Lipids are a concentrated source of calories in the diet and account for 30% to 40% of the caloric intake. About 90% of the dietary lipid is in the form of water-insoluble triacylglycerols containing mostly long-chain fatty acids (16 to 18 carbon atoms). Other dietary lipids include cholesterol esters, phospholipids, and the fat-soluble vitamins A, D, E, and K.

Both pancreatic lipase and the bile salts are required for the normal digestion and absorption of triacylglycerols. Pancreatic lipase degrades triacylglycerol to one monoglyceride and two free fatty acid molecules (Fig. 22–29). These two digestion products are the form in which fat enters the intestinal epithelial cell.

Since triacylglycerols and their digestion products are insoluble in water, special mechanisms ensure that both digestion and absorption proceed to completion in the aqueous environment of the intestinal contents (Fig. 22–30). For example, when triacylglycerols first enter the small intestine from the stomach, these hydrophobic molecules form large oil droplets by clustering together in such a way that relatively few lipid molecules are exposed to the aqueous environment, with most being inside the droplet. Pancreatic lipase acts only at the oil-water interface. Thus it would seem that only a small fraction of the triacylglycerols in the intestine are subject to the di-

Figure 22–29

Digestion of triacylglycerols by pancreatic lipase.

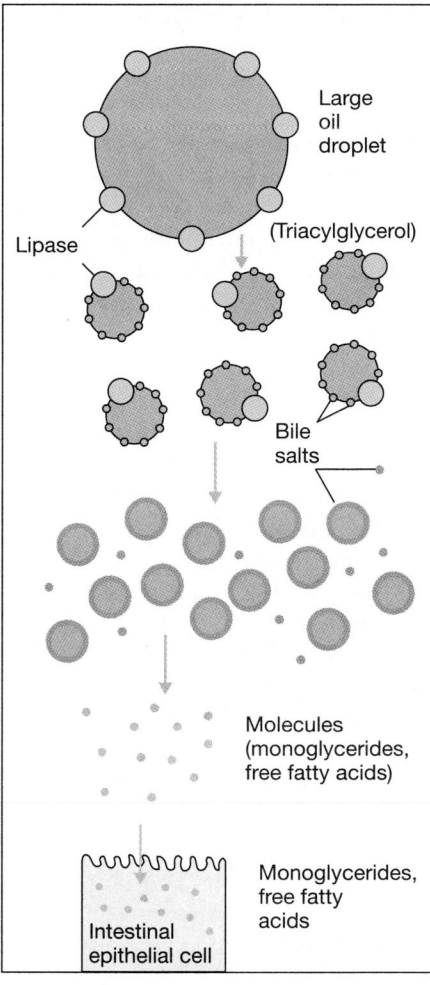

Figure 22-30

Intraluminal and cell penetration phases of triacylglycerol absorption.

gestive activity of this enzyme. If such were actually the case, digestion would not go to completion very rapidly, if at all, and absorption would be very much limited by digestion.

In reality, co-existing in the intestinal contents with the oil droplets is a stable emulsion of triacylglycerols consisting of very small fat particles. The existence of such an emulsion is important because it provides a large surface area on which lipase can act to degrade triacylglycerol and allow digestion to go to completion rapidly. This emulsion is produced by the combined action of three emulsifying agents present in the intestinal contents. These substances are the bile salts supplied by the liver and the monoglycerides and free fatty acids produced by lipase digestion of triacylglycerols.

The bile salts play an additional role in lipid absorption. They render monoglycerides and free fatty acids water-soluble by interacting with these triacyl-

glycerol digestion products to form very small aggregates called **micelles** (see Fig. 22–30). Each micelle contains only about 20 lipid molecules and, unlike the much larger emulsion particles, diffuses rapidly in aqueous solution.

What is the significance of micelle formation with regard to absorption of monoglycerides and free fatty acids? Both digestion products do have a limited solubility in water and are present as single molecules in intestinal contents only in trace amounts. That is, the level of **monomolecular dispersion** for monoglycerides and free fatty acids is very low. The idea is that monoglycerides and free fatty acids are absorbed as individual molecules from the monomolecular dispersion, and that this occurs by diffusion from the intestinal lumen through the lipid portion of the intestinal epithelial cell membrane into the cell (see Fig. 22–30). The function of the micelles is to maintain the monoglycerides and free fatty acids in the monomolecular dispersion phase at saturation levels by serving a reservoir function; that is, as monoglycerides and free fatty acids are removed from the monomolecular phase by absorption, they are immediately replenished by the movement of monoglycerides and free fatty acids from micelles. This allows absorption of individual lipid molecules to occur rapidly. Along this same line, as the micelles are depleted of monoglycerides and free fatty acids by movement into the free phase, the micellar phase is replenished with monoglycerides and free fatty acids from the emulsion phase, where these lipid digestion products are continuously produced.

We should point out that the micelles dissolve cholesterol and the fat-soluble vitamins within their micellar structure, thereby preparing these lipids for absorption in the same way that they prepare monoglycerides and free fatty acids. The bile salts, which are not absorbed until they reach the terminal ileum, remain in the intestinal lumen and perform their absorptive function throughout the length of the intestine until all fat is absorbed. This process is efficient because minimal quantities of bile salts are required to prepare relatively large quantities of lipid for absorption.

Once monoglycerides and free fatty acids enter the epithelial cells, they ar resynthesized to triacylglycerols by intracellular enzyme systems in the smooth endoplasmic reticulum. The newly synthesized triacylglycerols are aggregated into droplets, which become progressively larger during passage through the cells. A layer of phospholipid and protein is added to each droplet, and the final product is known as a **chylomicron.** These particles enter the central lacteal of the villus and are carried by the lymphatic system to the general circulation, where they are made available to the cells of the body. The cel-

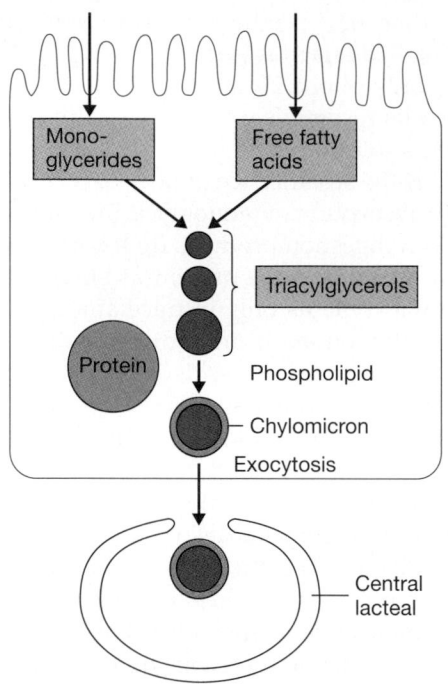

Figure 22–31

Cellular phase of triacylglycerol absorption.

lular phase of lipid absorption is illustrated in Figure 22–31.

Processes of Absorption, Secretion, and Motility in the Large Intestine

The large intestine, which is about 1.2 m long, extends from the ileocecal junction to the anus. This organ is subdivided into the **cecum** and its appendix, the **ascending colon,** the **transverse colon,** the **descending colon,** the **sigmoid colon,** the **rectum,** and the **anus,** with its **internal** and **external anal sphincters** (Fig. 22–32).

The large intestine does not secrete digestive enzymes, and it lacks villi. Water is absorbed from the contents that enter from above; in this way, a firm fecal mass is formed. This relatively dehydrated material is stored in the large intestine for variable periods of time and is ultimately evacuated to the outside of the body by defecation.

Water and Sodium Are Absorbed and Potassium and Bicarbonate Are Secreted by the Large Intestine

About 500 ml of water and a significant amount of sodium chloride enter the ascending colon every day. The primary absorptive process that occurs in the large intestine is the active transport of sodium from the colon. As in the small intestine, this is accompanied by the osmotic absorption of water. This occurs to the extent that normally only about 100 ml of water are lost to the feces every day. Carbohydrate, protein, and fat digestion products are not absorbed from the large intestine.

Bacteria present in the lumen of the large intestine synthesize certain vitamins. Although these nutrients are absorbed from the large intestine, they provide only a small part of the daily vitamin requirement. Other products of bacterial metabolism contribute to the gas **(flatus)** present in the large intestine. Flatus, which is produced at a rate of about 1 to 1.5 liters per day, consists of nitrogen and carbon dioxide, along with small amounts of hydrogen, methane, and hydrogen sulfide. Certain foods, such as beans, contain carbohydrates that cannot be digested in the small intestine. These carbohydrates are metabolized by colonic bacteria with the resultant production of large quantities of gas.

The colonic mucosa contains glands that produce an alkaline, mucus-containing secretion that protects the mucus of the large intestine from chemical and mechanical trauma. Mucus lubricates the fecal mass, thereby facilitating the passage of feces. The bicarbonate of this secretion neutralizes irritating acids produced by bacterial metabolism. The large intestine also secretes potassium. When large volumes of fluid are abnormally excreted in the feces, body potassium can be depleted.

Motility of the Large Intestine Is Characterized by Both Mixing and Propulsive Movements

The musculature of the large intestine is quiescent (inactive) much of the time. Segmenting contractions occur in the proximal colon at a much lower frequency than in the small intestine, and these produce a limited back-and-forth movement of the contents of the colon. Segmenting movements are the most frequent type of muscular activity of the large intestine, and they promote the absorption of water and inorganic ions by exposing the contents to the mucosa. A wave of contraction passes over the proximal colon about 3 to 4 times a day and drives contents into the distal large intestine, where material is stored until defecation occurs. These propulsive **mass movements** usually occur after a meal and are often followed by the desire to defecate. Colonic movements are largely initiated locally by distention of the colonic wall, which triggers short reflexes through the nerve plexuses. However, mass movements sometimes occur when food enters the stomach. Although and by neural and hormonal mechanisms that originate in the duodenum. The gastric chief cell secretes pepsinogen, which, when activated to the enzyme pepsin, initiates the digestion of protein.

- The intercalated duct cells of the pancreas secrete a large volume of a water-electrolyte secretion with a high concentration of bicarbonate, an ion that is largely responsible for the neutralization of acid in gastric contents. The pancreatic acinar cells secrete a wide variety of digestive enzymes, whose combined actions almost complete the digestion of food. Bicarbonate secretion is stimulated by secretin released in response to acid stimulation of the duodenal mucosa, whereas enzyme secretion is by cholecystokinin released in response to fat and protein digestion products.

- Between meals, bile is stored and concentrated in the gallbladder. When food is eaten, this organ contracts in response to the release of cholecystokinin and propels bile into the duodenum. Bile salts are synthesized and actively secreted by the liver. Following transit through the small intestine, these molecules are actively absorbed from the terminal ileum into the portal circulation, which returns them to the liver for resecretion (enterohepatic circulation).

- Segmenting contractions mix the contents of small intestine, thereby facilitating digestion and absorption, and peristalsis propels unabsorbed material into the large intestine.

- The major process involved in the absorption of water from the small and large intestines is the active transport of sodium, which carries water along passively from lumen to blood. Iron and calcium absorption are regulated processes mediated by special transport systems.

- Digestion of carbohydrate to monosaccharides and protein to amino acids is completed by enzymes in the intestinal brush border membrane. Most monosaccharides and amino acids are absorbed into the portal circulation by sodium-dependent active transport systems.

- Water-insoluble monoglycerides and free fatty acids, the end products of pancreatic lipase digestion of triacylglycerol fats, interact with bile salts to form water-soluble aggregates called micelles. Monoglycerides and free fatty acids diffuse passively from monomolecular dispersion into the intestinal epithelial cell, where they are resynthesized to triacylglycerols and passed on to the lymph as chylomicrons.

- Segmenting contractions promote the absorption of water from the proximal large intestine, resulting in the formation of a firm fecal mass. This dehydrated material is moved into the distal large intestine by mass movements and is stored there for variable periods of time before the defecation reflex is elicited by distension of the rectum, when defecation occurs.

Review Questions

Identify Each with a Term

1. The transfer of water and solutes from the lumen of the digestive tract into the blood or lymph.
2. The spontaneous depolarization-repolarization cycles of pacemaker cells in the smooth muscle of the stomach and small intestine.
3. The route traveled in the body by the bile salts.
4. Very small projections of the luminal surface of the small intestinal epithelial cell.
5. The aggregates of bile salts, monoglycerides, and free fatty acids present in the lumen of the small intestine during the digestion and absorption of triacylglycerols.
6. Frequent defecation of highly fluid material.

Define Each Term

7. digestive tract
8. portal circulation
9. short reflex
10. secretin
11. dysphagia
12. G cell
13. cephalic phase
14. emetic
15. trypsin
16. pancreatic lipase
17. peristalsis
18. lower esophageal sphincter
19. steatorrhea
20. flatus

Choose the Correct Answer

21. The various processes of the gastrointestinal system are directly associated with certain types of gastrointestinal cells. Which of the following associations is *not* valid?
    a. motility—esophageal smooth muscle cells
    b. secretion—gallbladder epithelial cells
    c. digestion—pancreatic acinar cells
    d. absorption—small intestinal epithelial cells

22. A gastrointestinal process that is *not* regulated is:
    a. small intestinal absorption of glucose
    b. gastric secretion of HCl
    c. gastric emptying
    d. secretion of pancreatic enzymes

23. The organ of greatest importance in providing the enzymes required for the digestion of food is the:
    a. parotid salivary gland
    b. stomach
    c. pancreas
    d. liver

24. In which of the following regions is intraluminal pressure highest when there is no swallowing activity during rest?

a. pharynx
b. body of the esophagus
c. lower esophageal sphincter
d. stomach

25. A hormone that stimulates secretion by the gastric parietal cell is:
    a. gastrin
    b. pepsin
    c. gastric inhibitory peptide
    d. secretin

26. The cephalic phase of gastric HCl secretion is *eliminated* by:
    a. surgical removal of the pyloric antrum
    b. vagal denervation of the pyloric antrum
    c. vagal denervation of the fundus and body of the stomach
    d. none of the above

27. A component of the secretion of the intercalated duct cells of the pancreas that is of major physiological significance is:
    a. trypsin   c. lipase
    b. NaHCO$_3$ d. bilirubin

28. A close association exists between the absorption of Na$^+$ and the absorption of water from the lumen of the:
    a. small intestine
    b. proximal colon
    c. gallbladder
    d. all of the above

29. The absorption of water from the digestive tract:
    a. is dependent primarily on the active transport of Ca^{2+}.
    b. occurs primarily from the small intestine.
    c. is completed during transit through the small intestine.
    d. occurs primarily from the proximal colon.

30. The intestinal *digestion* of triacylglycerols is facilitated by the incorporation of these molecules into:
    a. large oil droplets
    b. small emulsified fat particles
    c. micelles
    d. chylomicra

Answer Each Question in One or Two Sentences

31. How does the stomach adapt to the wide range of volumes it handles with only small increases in intragastric pressure?

32. What are the mechanisms by which high concentrations of HCl in the (a) stomach and (b) duodenum inhibit gastric secretion of HCl?

33. What is the functional importance of the bicarbonate ion of pancreatic juice?

34. Why are the bile salts essential for normal intestinal digestion and absorption of triacylglycerols?

35. Why is it important that our diet contains indigestible material?

Problem Solving in Physiology

36. A 40-year-old business executive visited his physician with complaints of burning epigastric pain that was most prominent 3 to 4 hours after a meal and that was temporarily relieved by a snack. Radiological studies revealed a duodenal ulcer. The physician prescribed appropriate medication, including antacids and a drug to suppress vagal activity, and also gave dietary advice, such as taking periodic snacks and avoidance of agents known to stimulate gastric secretion. Give reasons for the therapy described.

37. A 55-year-old man tells his physician that he has been losing weight continuously, despite eating large quantities of food. He has also noticed recently that his feces are very bulky and oily. Microscopic examination of the feces revealed the presence of fat droplets and undigested meat fibers. When secretin was infused intravenously, the patient responded by secreting only 10 ml of pancreatic juice in 60 minutes (normal average = 176 ml/hr) and 2 mM HCO$_3^-$ (normal average = 13.5 mM). What do you conclude is the problem of the patient? Give reasons for making your conclusion.

38. Because of ileal obstruction, it was necessary to remove the terminal ileum of a 35-year-old woman. Following the surgery, the fecal fat excretion of the patient was 30 gm/day as compared to an average normal value of 4 gm/day. It was also determined that the concentration of bile salts in her jejunal contents was considerably lower than normal. What is the reason for the marked steatorrhea exhibited by the patient?

Suggested Readings

Davenport, H.W. *Physiology of the Digestive Tract*, 5th ed. Chicago: Year Book Medical Publishers, 1982.

Granger, D.N. Barrowman, J.A., and Kvietys, P.R. *Clinical Gastrointestinal Physiology*. Philadelphia: W.B. Saunders Company, 1985.

Greenberger, N.J. *Gastrointestinal Disorders: A Pathophysiologic Approach*, 3rd ed. Chicago: Mosby, 1991.

Johnson, L.R., ed. *Gastrointestinal Physiology*, 4th ed. St. Louis: C.V. Mosby Co., 1991.

Johnson, L.R., ed. *Physiology of the Gastrointestinal Tract*, 3rd ed. (2 volumes). New York: Raven Press, 1994.

Kamm, M.A., and Lennard-Jones, J.E., eds. *Gastrointestinal Transit*. Petersfield, U.K.: Wrightson Biomedical Publishing, 1991.

Livingston, E.H., and Guth, P.H. "Peptic Ulcer Disease." *American Scientist*, 80:592–598, 1992.

Makhlouf, G.M. "Neural and Hormonal Regulation of Function in the Gut." *Hospital Practice*, 25:79–98, 1990.

Moog, F. "The Lining of the Small Intestine." *Scientific American*, 245:154–176, 1981.

Thompson, J.C., ed. *Gastrointestinal Endocrinology*. New York: McGraw-Hill, 1987.

Van Dyka, R.W. *Mechanisms of Digestion and Absorption of Food*. In Sleisenger, M.H., and Fordtran, J.S., eds. *Gastrointestinal Disease*, 4th ed. Philadelphia: Saunders, 1989.

Weisbrodt, N.W. "Gastrointestinal Motility." *Annual Review of Physiology*, 43:7–51, 1985.

Williams, J.A. "Regulatory mechanisms in pancreas and salivary acini." *Annual Review of Physiology*, 46:361–365, 1984.

Key Terms

absorption (p. 665)
amylase (p. 674)
basic electrical rhythm (p. 679)
bile (p. 688)
bile salts (p. 669)
brush border membrane (p. 691)
cholecystokinin (p. 689)
constipation (p. 699)
defecation reflex (p. 699)
diarrhea (p. 703)
digestion (p. 606)

duodenum (p. 690)
enterohepatic circulation (p. 689)
exocrine gland (p. 666)
gallstones (p. 688)
gastric mucosal barrier (p. 684)
gastrin (p. 671)
gastrointestinal system (p. 665)
hydrochloric acid (HCl) (p. 681)

ileum (p. 689)
lipase (p. 687)
micelle (p. 697)
microvilli (p. 691)
motility (p. 665)
mucosa (p. 666)
pepsin (p. 681)
peristalsis (p. 676)
portal circulation (p. 667)
secretin (p. 687)
secretion (p. 665)

segmenting contractions (p. 690)
short reflexes (p. 671)
splanchnic circulation (p. 667)
trypsin (p. 686)
ulcer (p. 684)
villi (p. 671)
vomiting (p. 683)
zymogen granules (p. 674)

An isolated, perfused proximal convoluted tubule. *(© Biophoto Associates/Photo Researchers, Inc.)*

Chapter 23

The Kidney

Key Concepts

The kidneys regulate the composition and volume of the extracellular fluid.

The kidney glomeruli filter the blood plasma at a high rate, and the kidney tubules reabsorb and secrete substances, so that appropriate amounts of various substances are excreted in the urine.

The kidneys regulate the amount of sodium in the body by excreting just the correct amount to maintain homeostatic balance.

Human kidneys can form a urine which has a higher or lower total solute concentration (osmolality) than plasma.

Micturition (urination) is a complex reflex act controlled by higher brain centers.

Chapter Outline

Functional Anatomy of the Kidney
Processes Involved in Formation of Urine
Tubular Transport of Mineral Electrolytes
Tubular Reabsorption of Water
Functions of Ureters and Urinary Bladder

*Tim is a 10-year old patient. He passes a dark red urine and shows puffiness around his eyes. The patient had a bad sore throat, fever, and upper respiratory tract infection two weeks ago, but these symptoms went away. Physical examination reveals an elevated blood pressure and swelling (edema) of the face and lower extremities. Blood chemistry determinations show elevated plasma levels of waste products (creatinine and urea) and a reduced plasma albumin concentration. The urine is strongly positive for protein and contains many red cells (the cause of the red urine). The facts of this case suggest a post-infection acute **glomerulonephritis**, an immune system—mediated disease in which kidney glomeruli are damaged. In this case, as often happens, the disease was self-limited, and the bloody urine, signs of renal failure, and proteinuria went away over the course of days to weeks. Some people are less fortunate and have persistent glomerular inflammation and scarring, which lead to permanent renal failure.*

The findings in this patient illustrate a few of the many disturbances in the body that result from glomerular disease, the leading cause of renal failure. Our kidneys normally play a major role in regulating the volume and composition of the extracellular fluid. They provide a suitable "internal environment" for the body's cells. This chapter will consider the functional anatomy of the kidneys, renal blood flow, the processes involved in urine formation, and the ways in which the kidneys handle a variety of solutes and water. It will end with a brief discussion of the functions of the lower urinary tract (ureters, bladder, urethra).

Chapter 23 will discuss how renal (kidney-related) and other mechanisms are controlled and integrated to accomplish the important task of regulating the internal environment. Before considering kidney function in detail, we will outline some of the functions of the kidneys.

1. The kidneys regulate the total solute concentration (osmolality) of the body fluids.

2. By controlling the excretion of sodium and water, the kidneys regulate the volume of the extracellular fluid.

3. The kidneys regulate the individual concentrations of numerous electrolytes in the extracellular fluid, including sodium, potassium, calcium, magnesium, chloride, sulfate, and phosphate ions.

4. The kidneys control the plasma bicarbonate concentration and hence the hydrogen ion concentration and so play an important role in acid-base balance. This topic will be discussed in Chapter 25.

5. The kidneys eliminate metabolic waste products such as urea (an end product of protein metabolism), uric acid (an end product of purine metabolism), and creatinine (an end product of muscle metabolism). They also eliminate many foreign compounds from the body, including many drugs and toxins.

6. The kidneys produce a number of special substances. These include: **erythropoietin,** a hormone that stimulates the rate of production, maturation, and release of red blood cells from bone marrow; **renin,** a proteolytic enzyme important in the regulation of extracellular fluid volume and blood pressure; **kallikrein,** a proteolytic enzyme that leads to the formation of kinins, which are vasodilators; and various **prostaglandins** and **thromboxane,** fatty acid derivatives that act as local hormones. Prostaglandins E_2 and I_2 have several actions in the kidneys, including vasodilation, enhancement of renal excretion of salt and water, and stimulation of renin release. Thromboxane is a vasoconstrictor that may be responsible for

reduced renal blood flow in a number of renal diseases.

7. The kidneys have several special metabolic functions. They are responsible for converting the inactive form of vitamin D to its active form, **1,25-dihydroxy-vitamin D_3.** The latter is a hormone that stimulates intestinal calcium absorption. The kidneys synthesize ammonia from amino acids. This is important in acid-base regulation. The kidneys can synthesize glucose from noncarbohydrate sources (e.g., amino acids), a process called **gluconeogenesis.** During a prolonged fast, the glucose added to the blood by the kidneys helps to maintain the blood sugar concentration. The kidneys are an important site of degradation (hence inactivation) of several polypeptide hormones, including angiotensin II glucagon, insulin, and parathyroid hormone.

Functional Anatomy of the Kidney

The functions of the kidneys can only be understood in terms of their unique anatomy. Figure 23–1 presents an overview of the urinary system and the human kidney. The kidneys are bean-shaped organs that lie retroperitoneally in back of the abdominal cavity on either side of the vertebral column. They are drained by the **ureters,** which carry the urine to the **urinary bladder.** The tube draining the bladder is called the **urethra.** If a kidney is cut through (see Fig. 23–1*b*), two parts are easily distinguished: an outer part, called the **cortex,** and an inner part, called the **medulla.** The cortex is reddish-brown and has a granular appearance. It contains all of the **glomeruli** and **convoluted tubules** (proximal and distal). The medulla is a lighter color and is striated (striped). Striations result from the parallel arrangement of **loops of Henle, medullary collecting ducts,** and blood vessels. The medulla is usually divided into an **outer medulla** (closer to the cortex) and an **inner medulla** (farther from the cortex).

The human kidney is divided into about a dozen **lobes.** Each consists of a **pyramid** of medullary tissue plus the cortical tissue overlying its base and covering its sides. The apex of a medullary pyramid forms a renal **papilla,** which drains the urine into a **minor calyx.** Minor calyces unite to form a **major calyx;** major calyces lead to the expanded **renal pelvis,** which is drained by the ureter. The ureter, renal artery, renal vein, nerves, and lymphatic vessels enter or leave the kidney at the **renal hilum,** a depression of its medial surface.

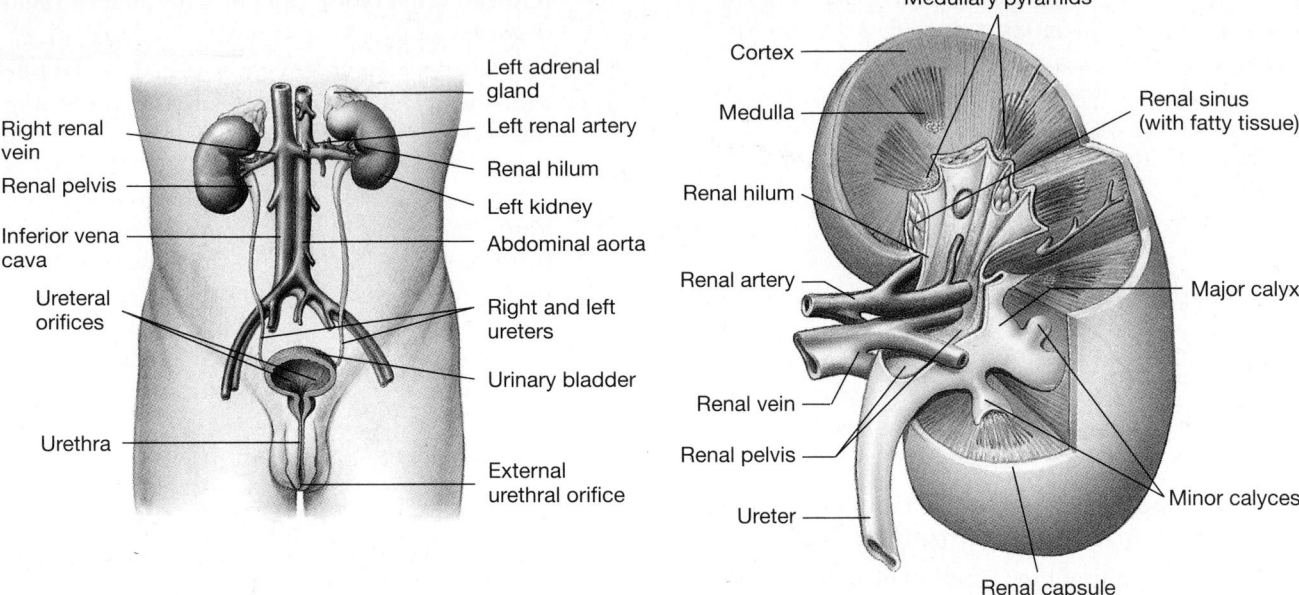

Figure 23–1

(*a*) Diagram of the urinary tract. The urine formed in the kidneys drains into the renal pelvis. It is carried by the ureters to the urinary bladder, which is drained by the urethra. (*b*) Diagram of the human kidney.

The Nephron Is the Basic Unit of Kidney Structure and Function

Each human kidney contains about one million **nephrons.** Each nephron consists of a **renal corpuscle** and its attached **renal tubule** (Fig. 23–2).

The renal corpuscle consists of a tuft of capillaries, the **glomerulus,** surrounded by an expanded, double-walled cup, **Bowman's capsule.** The **urinary space** of Bowman's capsule is continuous with the lumen of the renal tubule. Traditionally, the renal tubule has been divided into several anatomically distinct segments. The first segment is called the **proximal tubule.** This is divided into two parts: the **proximal convoluted tubule,** which coils and twists in the neighborhood of its renal corpuscle, and the **proximal straight tubule,** which plunges toward the renal medulla. The **loop of Henle** is the portion of the nephron between the proximal convoluted and distal convoluted tubules. Its first part is the proximal straight tubule. Next is a **thin limb** (descending and ascending portions), and finally a **thick ascending limb.** This is followed by a short **distal convoluted tubule.** Distal convoluted tubules join **connecting tubules,** which lead to **cortical collecting ducts.** These lead to the **outer medullary collecting ducts,** which pass straight through the outer medulla. In the inner medulla, **inner medullary collecting ducts** unite to form larger and larger ducts, which convey the urine into the minor calyces.

Strictly speaking, the collecting ducts are not part of the nephron. These structures have different embryological origins, and there are many more nephrons than collecting ducts. Functionally, the collecting ducts are not just conduits; they are responsible for determining the final composition and volume of the urine. Since the collecting ducts continue the tasks performed by the nephrons, they are often considered part of the nephron unit.

Not all nephrons are alike (see Fig. 23–2). Nephrons with renal corpuscles in the outer cortex, called **cortical nephrons,** have relatively small glomeruli, short or absent ascending thin limbs, and loops that bend wholly within the cortex or outer

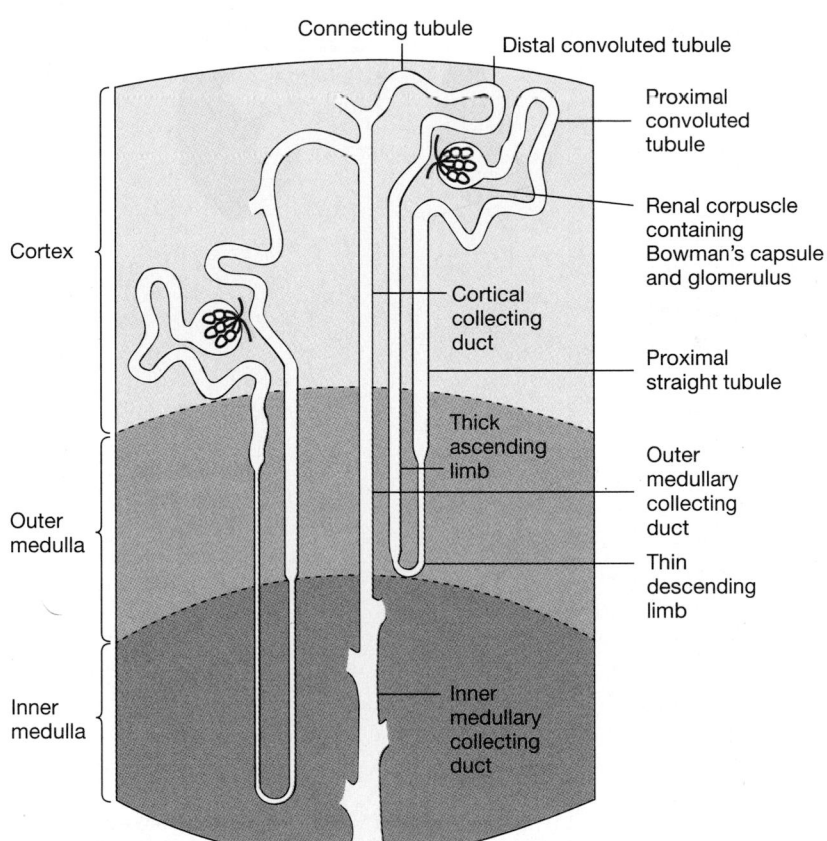

Connecting tubule — **Distal convoluted tubule**

Proximal convoluted tubule

Renal corpuscle containing Bowman's capsule and glomerulus

Cortical collecting duct

Proximal straight tubule

Thick ascending limb

Outer medullary collecting duct

Thin descending limb

Inner medullary collecting duct

Cortex

Outer medulla

Inner medulla

Figure 23–2

Diagram of parts of the nephron and collecting duct system. The nephron on the right is a cortical nephron, the nephron on the left is a juxtamedullary nephron. *(Modified from: Kriz, W., and Bankir, L. "A standard nomenclature for structures of the kidney." American Journal of Physiology, 254:F1–F8, 1988.)*

medulla. Nephrons with renal corpuscles in the inner cortex (i.e., near the medulla), called **juxtamedullary nephrons,** have large glomeruli, long thin limbs, and long loops, some of which may reach the papillary tips. Functional differences between these two nephron populations include differences in filtration rate, tubular transport properties, and renin content. In the human kidney, the majority of nephrons (85%) are of the cortical type, and the rest (15%) are of the juxtamedullary type.

The renal tubule and collecting duct are composed of a single layer of epithelial cells surrounding a tubule lumen. Cell structure and function vary considerably from one segment to another. As an example of the specialized architecture of the renal tubule epithelium, we will consider a cell from the proximal tubule (Fig. 23–3). The cell surface that contacts the fluid in the tubule lumen (the **luminal** or **apical cell membrane**) has numerous **microvilli,** collectively called the **brush border.** The microvilli greatly increase the membrane surface area, thereby facilitating the transfer of materials between cells and tubular fluid. At the apical side of the cell are numerous **endocytic vesicles,** thought to be involved with the reabsorption of proteins. Adjacent cells are linked to each other near the apical end of the cell by a **tight junction (zonula occludens).** In the proximal tubule,

this junction is leaky and permits some movement of small ions and water. The **basolateral cell membrane** surface area is considerably increased by numerous ridges and processes that are formed as adjacent cells interdigitate with each other. The space between adjacent cells, the **lateral intercellular space,** is thought to be an important route for salt and water movement across the epithelial cell layer. Within the cell, numerous mitochondria lie close to the basolateral cell membrane and furnish the energy for an important sodium-potassium pump (Na^+/K^+-ATPase) located in these membranes. The tubule cells rest on a **basement membrane,** which acts as a supportive structure. If the cells are damaged, the basement membrane provides a framework on which regenerating cells can reline the tubule. The overall appearance of the proximal tubule cell is that of a metabolically active cell, and we will see later that it is heavily engaged in transport activities.

Blood Vessels and Nerves Enter or Leave the Kidney at Its Hilum

Each **renal artery** arises from the abdominal aorta and divides into progressively smaller and more numerous branches: segmental, interlobar, arcuate, and cor-

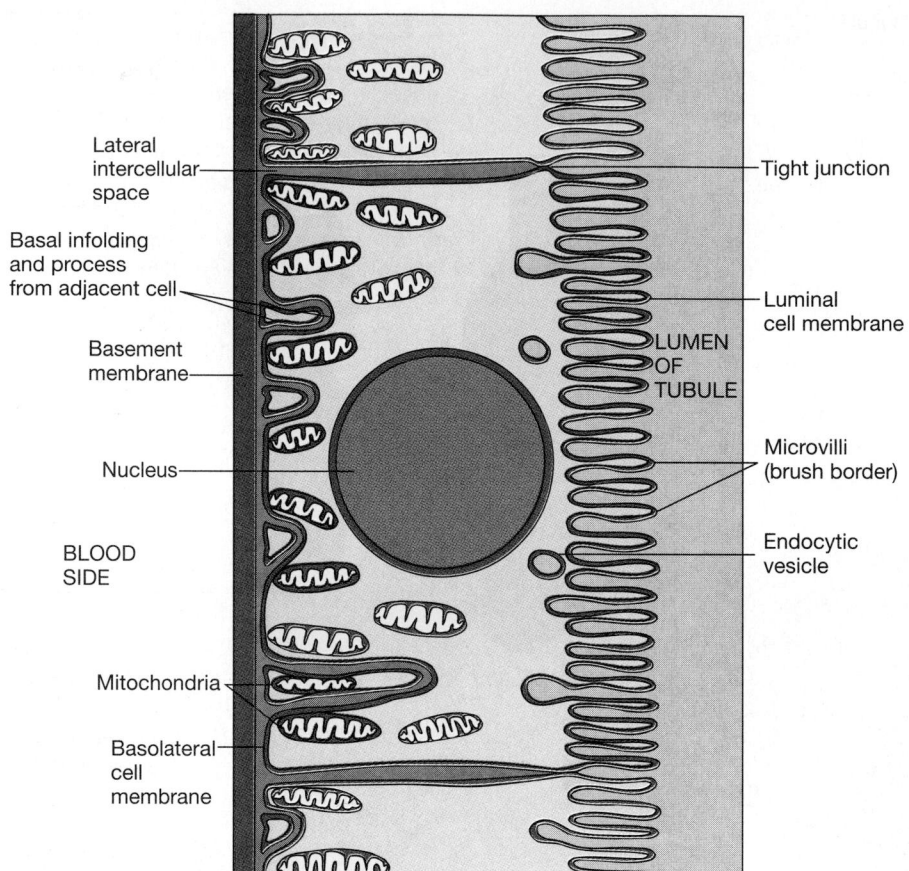

Figure 23–3

Schematic view of a proximal tubular cell.

Lateral intercellular space

Basal infolding and process from adjacent cell

Basement membrane

Nucleus

BLOOD SIDE

Mitochondria

Basolateral cell membrane

Tight junction

Luminal cell membrane

LUMEN OF TUBULE

Microvilli (brush border)

Endocytic vesicle

tical radial arteries (Fig. 23–4). The **cortical radial arteries** course toward the surface of the cortex and supply the short, muscular **afferent arterioles.** Each afferent arteriole leads to a **glomerulus.** The glomerular capillaries re-form the **efferent arteriole,** which usually has a thin muscular coat. This breaks up into the **peritubular capillaries,** which form an extensive network surrounding the tubules in the cortex. The peritubular capillaries lead to veins that generally parallel the course of the arterial vessels and ultimately empty into the **renal vein** (Fig. 23–4).

The blood supply to the kidney medulla is derived from the efferent arterioles associated with juxtamedullary nephrons. These give rise to long, straight capillaries called **vasa recta.** The vasa recta descend into and ascend from the medulla in parallel with the tubular structures of the medulla.

The renal circulation is unusual in that there are two capillary beds in series. The blood pressure is high in the glomeruli and low in the peritubular capillaries. The glomeruli are specialized for filtration, whereas the peritubular capillaries are specialized for uptake of fluid reabsorbed by the tubules.

The kidneys are richly innervated. Sympathetic nerve fibers travel to the kidneys mainly in the Xth, XIth, and XIIth thoracic and Ist lumbar spinal nerves.

The sympathetic efferent fibers predominantly innervate blood vessels and cause vasoconstriction. Sympathetic fibers are also found adjacent to tubular cells and may increase sodium reabsorption. Afferent (sensory) nerve fibers from the renal pelvis and ureter mediate pain of renal origin.

Lymphatic vessels are found in the cortex, but not in the medulla.

The Juxtaglomerular Apparatus Produces Renin and Also Controls the Rate of Glomerular Filtration

Just before it becomes the distal convoluted tubule, the thick ascending limb contacts the afferent and efferent arterioles of its parent glomerulus (see Fig. 23–2). The tubular epithelium at this spot shows a dense crowding of cells and is called the **macula densa** (Fig. 23–5). Adjacent smooth muscle cells in the walls of the arterioles (especially the afferent arteriole) have an epithelial cell-like appearance and may contain granules of renin. These **granular cells** synthesize, store, and release this enzyme into the blood and interstitial fluid. In the region bounded by macula densa and arterioles are closely packed cells

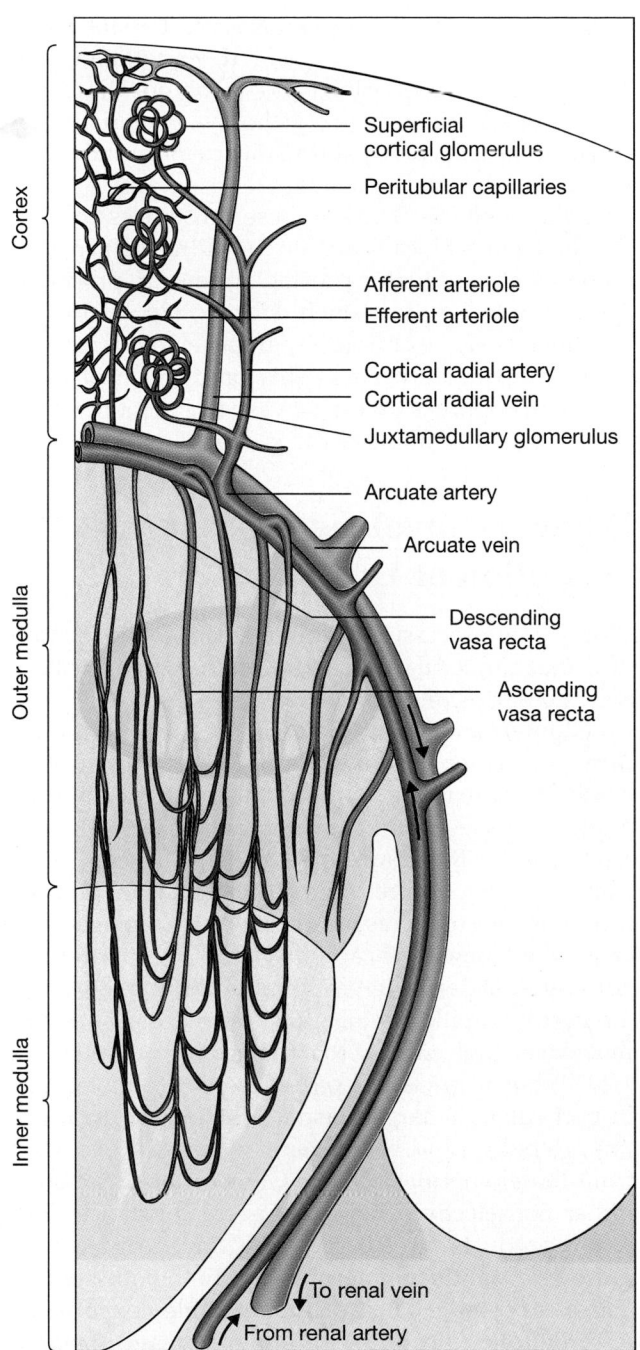

Figure 23–4

Blood vessels within the kidney. *(Modified from: Koushanpour, E., and Kriz, W. Renal Physiology. Principles, Structure, and Function, 2nd ed. New York: Springer, 1986.)*

Labels in Figure 23–4:
- Superficial cortical glomerulus
- Peritubular capillaries
- Afferent arteriole
- Efferent arteriole
- Cortical radial artery
- Cortical radial vein
- Juxtamedullary glomerulus
- Arcuate artery
- Arcuate vein
- Descending vasa recta
- Ascending vasa recta
- To renal vein
- From renal artery
- Cortex
- Outer medulla
- Inner medulla

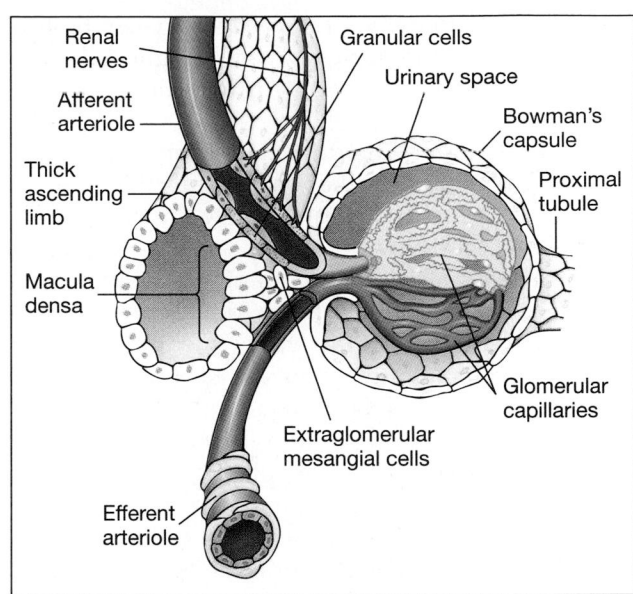

Figure 23–5

The juxtaglomerular apparatus consists of macula densa, extraglomerular mesangial cells, and granular cells. This structure may play a role in controlling glomerular blood flow and filtration rate at the level of a single nephron.

Labels in Figure 23–5:
- Renal nerves
- Afferent arteriole
- Thick ascending limb
- Macula densa
- Efferent arteriole
- Extraglomerular mesangial cells
- Granular cells
- Urinary space
- Bowman's capsule
- Proximal tubule
- Glomerular capillaries

at this site has suggested that changes in tubule fluid composition might be signaled to the blood vessels and lead to changes in glomerular blood flow and filtration rate; this idea is known as the **tubuloglomerular feedback hypothesis.** An increase in flow through the loop of Henle causes constriction of the afferent arteriole, resulting in a decrease of filtration rate.

The Kidneys Have a High Blood Flow

Total renal blood flow in a resting, young adult 70-kg male averages about 1200 ml per minute or 20% to 25% of the cardiac output (5 liters per minute), even though both kidneys weigh less than 0.5% of body weight (300 gm/70 kg). Blood flow per gram of kidney tissue averages 4 ml per minute (1200 ml per minute/300 gm), which is much higher than in other organs considered to be well perfused (e.g., brain, heart, liver).

Why is such a high blood flow to the kidneys necessary? It is needed to sustain a high rate of filtration of plasma in the glomeruli. Essentially all of the blood perfusing the kidneys passes through glomeruli, where it is subject to filtration. Every minute a volume of filtrate is formed that is equal to about 10% of the blood flow rate. The high renal blood flow does not reflect excessive metabolic demands. Although the kidneys use about 8% of the body's oxygen consumption in a resting adult, renal blood flow is more

known as **extraglomerular mesangial cells.** These cells may transfer information from macula densa cells to cells in the blood vessel walls. The entire group of cells made up of macula densa cells, extraglomerular mesangial cells, and granular cells is known as the **juxtaglomerular apparatus.** The intimate relationship between tubule and blood vessels than adequate to deliver the needed oxygen. In fact, renal venous blood typically has a bright red color owing to its high oxygen content, a result of low oxygen extraction.

Blood flow is not distributed uniformly within the kidney. It is highest in the renal cortex, where all of the glomeruli are located and where the blood is filtered. Less blood flows to medullary structures, and flow decreases along the depth of the medulla. The relatively low blood flow helps to preserve the high solute concentration in the kidney medulla.

If arterial blood pressure is varied between 80 and 180 mm Hg, renal blood flow and glomerular filtration rate change relatively little. This is because of **renal autoregulation,** a self-regulating mechanism in the kidney that does not depend on extrinsic nerves or extrinsic hormones. Changes in the caliber of blood vessels upstream to the glomerulus (cortical radial arteries and afferent arterioles) are thought to be primarily involved. If blood pressure is increased, these vessels constrict, thereby increasing renal vascular resistance and minimizing an increase in blood flow. If blood pressure is decreased, these vessels dilate, thereby decreasing renal vascular resistance and minimizing a fall in blood flow. The autoregulatory response has a limited range, and so if blood pressure is decreased below 80 mm Hg, renal blood flow and glomerular filtration rate may fall markedly. The importance of renal autoregulation is that it prevents changes of arterial blood pressure from producing large changes in glomerular filtration rate. Such changes could lead to dramatic changes in sodium excretion, which could lead to abnormalities in extracellular fluid volume (Chapter 24).

Total renal blood flow is decreased in a number of stressful situations, including cold, deep anesthesia, fright, hemorrhage, pain, and strenuous exercise. This decrease is a reflex response due primarily to activation of renal sympathetic nerve fibers, which cause vasoconstriction. The increase in renal vascular resistance causes blood flow to shift to the brain and heart, which are more vital to immediate (short-term) survival.

A number of substances also affect renal blood flow. For example, angiotensin II, endothelin, epinephrine, norepinephrine, and thromboxane cause renal vasoconstriction. Acetylcholine, bacterial pyrogens, kinins (e.g., bradykinin), nitric oxide, and many prostaglandins (e.g., prostaglandins E_2 and I_2) have vasodilator effects. Some of these substances are synthesized and act locally within the kidneys. Increased amounts of vasodilator prostaglandins are produced in circumstances in which kidney blood flow is threatened—for example, by intense stimulation of renal sympathetic nerves. This production of prostaglandins opposes excessive vasoconstriction, which might otherwise damage the kidneys.

Processes Involved in Formation of Urine

Three basic processes are involved in the formation of urine: glomerular filtration, tubular reabsorption, and tubular secretion (Fig. 23–6).

Glomerular filtration is the starting point for urine formation. Because of the high hydrostatic blood pressure in the glomerular capillaries, a filtrate of plasma is pushed out of the capillaries into the urinary space of Bowman's capsule. This process of filtration is often referred to as **ultrafiltration,** and the filtrate is referred to as an **ultrafiltrate** of plasma because the glomerular capillary wall (the **glomerular filtration barrier**) behaves like an ultrafine filter. The glomerular capillary wall allows free passage of small molecules (molecules with a molecular weight below 10,000) but restricts the passage of large molecules. In particular, the large plasma proteins (serum albumin, globulins, and fibrinogen) are virtually excluded from the glomerular filtrate. Glomerular filtration is rather nonselective, and both useful substances and waste products are filtered. The final urine differs radically from the glomerular filtrate in both volume and composition. Most of the valuable constituents of the filtrate (most of the salts, water, and metabolites) are returned to the blood by the tubules.

Figure 23–6

Processes involved in urine formation.

Tubular reabsorption is the process whereby substances are transferred out of the tubular fluid and returned to the capillaries surrounding the kidney tubules. In this way, substances that have been temporarily lost from the plasma in the process of glomerular filtration are returned to the circulation. Reabsorption is a selective process, and the tubules handle different substances differently.

Tubular secretion is movement of substances across the tubule epithelium in a direction opposite to that of reabsorption. Certain substances—for example, the organic anion p-aminohippurate (PAH)—are taken up by the tubule cells from the blood surrounding the tubules and deposited into the tubular fluid. Other substances, such as hydrogen ions and ammonia, are generated within the kidney tubule cells and then secreted into the tubular urine. The terms **reabsorption** and **secretion** simply indicate the direction of transport across the tubular epithelium, either out of or into the tubular urine, respectively.

Excretion refers to what comes out in the urine that leaves the body. In most cases, excretion depends on both glomerular filtration and tubular transport. If a substance is reabsorbed, then less is excreted than is filtered. If a substance is secreted, then more is excreted than is filtered (Fig. 23–7).

Table 23–1 presents data on the daily (24-hour) filtration, reabsorption, and excretion of some normal plasma constituents. Since filtration rate is so high (180 liters per day), enormous quantities of water, salts, and organic compounds are filtered. The bulk (more than 99%) of the filtered water, sodium, chloride, and bicarbonate is usually reabsorbed. The small fraction of these substances that is excreted is of critical importance. This is because in order for a person to stay in balance—that is, to show no net change in the amount of these substances in the body—exactly the right amount should be excreted, in other words, the amount that was added to the extracellular fluid because of dietary intake.

A small change in the percentage of reabsorption of a filtered substance may lead to a large change in the absolute excretion of the substance. For example, if the reabsorbed percentage of filtered sodium decreases from 99.6% (23,900 mmoles per day divided by 24,000 mmoles per day) to 99.2% (23,800 mmoles per day divided by 24,000 mmoles per day), the amount of sodium excreted will double from 100 to 200 mmoles per day. Clearly, tubular reabsorption of electrolytes and water must be carefully regulated to avoid excessive losses or abnormal retention.

Do not memorize the numbers in Table 23–1, but try to appreciate the magnitude of filtration, reabsorption, and excretion of different substances, and recognize that different substances are handled differently by the tubules. Notice that potassium is listed as "reabsorbed" since the excreted amount is less than the filtered amount. Actually, nearly all of the filtered potassium is reabsorbed by proximal portions of the nephron, and most of the excreted potassium is secreted by the collecting ducts. Filtered calcium is mostly reabsorbed. Dietary phosphate is largely excreted in the urine. Filtered glucose is normally completely reabsorbed; it would make little sense to lose

Reabsorbed = Filtered – Excreted

Secreted = Excreted – Filtered

Figure 23–7

The net rate of reabsorption or secretion of a substance can be calculated from the amounts filtered and excreted per unit time.

Table 23–1

*Daily Filtration, Reabsorption, and Excretion of Some Normal Plasma Constituents**

Constituent	Filtered	Reabsorbed	Excreted
Water	167.5 liters	166 liters	1.5 liters
Sodium	24,000 mmoles	23,900 mmoles	100 mmoles
Potassium	720 mmoles	630 mmoles	90 mmoles
Calcium	270 mmoles	266 mmoles	4 mmoles
Magnesium	135 mmoles	120 mmoles	15 mmoles
Chloride	19,500 mmoles	19,400 mmoles	100 mmoles
Bicarbonate	4,500 mmoles	4,498 mmoles	2 mmoles
Phosphate	6 g	5 g	1 g
Glucose	150 g	150 g	0 g
Urea	50 g	25 g	25 g
Uric acid	8 g	7.2 g	0.8 g
Creatinine[†]	1.5 g	0 g	1.8 g

*Data are for a normal, young adult, 70-kg man.

†The amount of creatinine excreted exceeds the amount filtered because of tubular secretion of creatinine.

this valuable metabolite in the urine. Filtered creatinine is excreted without reabsorption. Urea and uric acid are reabsorbed to some extent. For all three of these waste products, production rates in the body and urinary excretion are normally equal.

The rates of excretion in Table 23–1 can vary widely, depending on the diet and other circumstances. Renal handling of each of the substances listed in this table will be discussed in more detail later.

Glomerular Filtration Involves Ultrafiltration of Plasma at a High Rate

The Glomerular Filtration Barrier

The glomerular filtration barrier consists of three layers: capillary endothelium, glomerular basement membrane, and podocyte cell layer (visceral epithelium of Bowman's capsule). Figure 23–8 shows part of this structure as seen with the transmission electron microscope. The endothelial cell layer lining the capillaries has large pores or windows (**fenestrae**) about 50 to 100 nanometers (nm) in diameter. The **basement membrane** consists of a meshwork of fine fibrils embedded in a gel-like matrix. The **podocytes** ("foot cells") have numerous cytoplasmic processes (**foot processes**) that rest on the basement membrane.

The narrow space between adjacent foot processes is about 20 nm wide and is called the **slit pore.** A thin membrane stretches across the slit pore; this **filtration slit diaphragm** may have a fine porous structure. The glomerular filtrate passes through all three layers of the filtration barrier. It probably passes through the endothelial cell fenestrae and the slit pores and not through the thicker cell cytoplasm.

The filtration barrier has a high permeability to water and small molecules but restricts the passage of large molecules (macromolecules). It is convenient to think that this sieving effect is due to the presence of pores in this barrier. Based on studies of the filterability of macromolecules of different molecular size, scientists have determined that the glomerular filtration barrier behaves as though it were penetrated by cylindrical pores 7.5 to 10 nm in diameter. But no one has ever seen pores of these dimensions; the fenestrae and slit pores are too big, and no pores are visible in the basement membrane. It is possible that the pores may be lost during the preparation of tissue for electron microscopy. In any case, most scientists believe that the basement membrane is probably the major size-selective barrier.

Electrical charge also affects the passage of macromolecules through the filtration barrier. The endothelial pores, basement membrane, and surface coat of the podocytes contain negatively charged glycoproteins. These negatively charged structural elements impede the filtration of negatively charged

Figure 23–8

Electron micrograph of the glomerular filtration barrier. Note the pore (fenestra) in the thin capillary endothelium and the slit pores between the foot processes of the podocytes. *(From: Fawcett, D.W. A Textbook of Histology, 11th ed. Philadelphia: Saunders, 1986, p. 765; Courtesy of Dr. Friend.)*

plasma proteins such as serum albumin. The layers of the filtration barrier closest to the capillary lumen, the endothelium and the innermost portion of the basement membrane, are probably the major electrostatic barriers to the filtration of negatively charged plasma proteins.

One of the hallmarks of glomerular disease is the abnormal excretion of plasma proteins. This may be due to (1) an increase in size of glomerular pores or (2) a loss of fixed negative charges from the glomerular filtration barrier. Abnormal quantities of protein are filtered and lost in the urine, as in the child described at the very beginning of this chapter.

Factors That Affect Glomerular Filtration Rate

The factors that affect **glomerular filtration rate (GFR)** are basically the same as those that influence fluid movement across capillary walls anywhere in the body (Chapter 19). The driving force for filtration is the glomerular capillary hydrostatic pressure (P_{GC}). This pressure ultimately depends on the energy im-

parted to the blood by the beating of the heart. Filtration is opposed by the hydrostatic pressure in the urinary space of Bowman's capsule (P_{BS}) and by the intracapillary colloid osmotic pressure (oncotic pressure) due to the plasma proteins (π_{GC}) (Fig. 23–9). Since the glomerular filtrate is essentially protein-free, the colloid osmotic pressure of fluid in the urinary space is zero and can be disregarded. The net filtration pressure gradient is equal to the difference between pressures favoring and opposing filtration. In other words,

$$\text{net filtration pressure} = P_{GC} - P_{BS} - \pi_{GC}$$

The glomerular filtration rate depends not only on the pressure gradient but also on the glomerular filtration coefficient K_f. We can write

$$\text{GFR} = K_f \times \text{net filtration pressure}$$

This equation is analogous to the equation that describes fluid flow rate along a tube, $F = \Delta P / R$ (Chapter 19), except that here we are considering fluid movement across a capillary wall. K_f can be thought of as the conductance (1/resistance) of the glomeru-

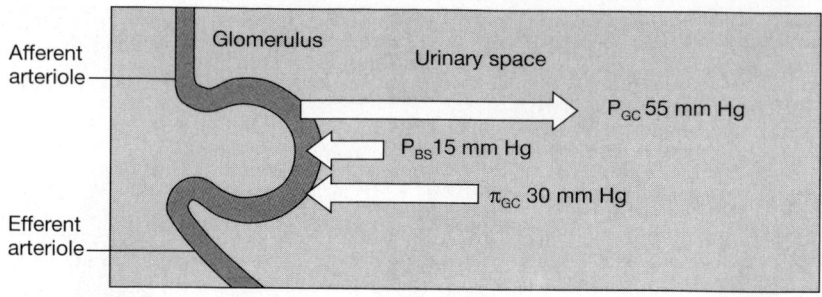

Figure 23-9

Pressures involved in glomerular filtration. The glomerular capillary hydrostatic pressure (P_{GC}) favors filtration, and the hydrostatic pressure in the urinary space of Bowman's capsule (P_{BS}) and the colloid osmotic pressure of proteins in the glomerular capillary plasma (π_{GC}) both oppose filtration. The pressures indicated are for the human kidney and are estimates based on measurements in other mammals.

lar filtration barrier; its magnitude depends on the fluid permeability and surface area of the barrier.

Figure 23–9 summarizes the average pressures thought to exist in the normal human kidney. Glomerular hydrostatic pressure is probably about 55 mm Hg, about twice as high as in most other capillaries in the body. The high pressure in the glomerulus results from the relatively low resistance of wide upstream blood vessels (cortical radial artery, afferent arteriole) and high resistance of narrower downstream blood vessels (efferent arteriole). Glomerular hydrostatic pressure probably changes very little (perhaps 2 mm Hg) along the length of the glomerular capillaries. This is because there are many (30 to 50) parallel capillary loops in a glomerulus, making the resistance to blood flow in the capillaries very low. Hydrostatic pressure in the urinary space of Bowman's capsule is about 15 mm Hg. This is the pressure left after filtration, and it opposes further filtration. It provides the driving force for fluid flow down the entire length of the nephron. The colloid osmotic pressure of glomerular capillary blood plasma averages about 30 mm Hg. This pressure rises along the length of the glomerular capillary, because a protein-free filtrate is forced out of the glomerular plasma, and the plasma proteins are left behind. Consequently, filtration rate declines from the beginning (afferent arteriole) to the end (efferent arteriole) of the glomerular capillaries. In the normal human kidney, outward filtration probably occurs along the entire length of the glomerular capillaries. The net filtration pressure gradient, averaged over the entire glomerulus, is about 10 mm Hg (55 mm Hg − 15 mm Hg − 30 mm Hg = 10 mm Hg).

The rate of filtrate formation in the glomeruli of the kidney greatly exceeds that in all other capillary beds in the human body. The high rate of glomerular filtration in the kidneys, about 180 liters per day, is due to several factors. First, the glomerular filtration coefficient is very high, which reflects, in part, a high fluid permeability of the glomerular filtration barrier. This barrier appears to be very "porous"; it behaves as though it contains many more pores than are present in most other capillaries. In part, the high glomerular filtration coefficient also reflects a large surface area. If all of the glomerular capillaries were spread out as a flat sheet, the total surface area occupied would be about 2 square meters (roughly equal to body surface area). Second, glomerular filtration rate is high because glomerular hydrostatic pressure is unusually high for a capillary bed. Third, the high rate of renal blood flow permits a rapid rate of filtration. This effect on filtration rate is subtle. Consider what would happen, however, if glomerular blood flow were low. In this case, forcing a small quantity of protein-free filtrate out of the capillary would lead to a sharp rise in intracapillary colloid osmotic pressure early along the length of a glomerular capillary. Filtration would soon be brought to a halt; the rest of the capillary would no longer function in filtration. Therefore, a high renal blood flow is necessary for a high glomerular filtration rate.

GFR may be altered in a variety of physiological and pathophysiological conditions. The conditions that affect filtration rate do so by affecting the factors already mentioned—namely, filtration coefficient, glomerular hydrostatic pressure, hydrostatic pressure in Bowman's space, and plasma colloid osmotic pressure.

Filtration coefficient (K_f)—and with it GFR—will be decreased if there is a decrease in either filtration barrier fluid permeability or surface area. K_f

CLINICAL FOCUS

Polycystic Kidney Disease

A *cyst* is basically a fluid-filled sac surrounded by a wall made up of epithelial cells. Cysts can form in practically any epithelial tissue, e.g., liver, pancreas, and ovaries; when numerous cysts form in the kidneys, this produces *polycystic kidney disease (PKD)*. Renal cysts may develop from Bowman's capsules, renal tubules, or collecting ducts.

There are several forms of PKD. *Autosomal dominant polycystic kidney disease (ADPKD)* is transmitted, on average, to half of the progeny of one affected parent. It is the most common lethal inherited illness and affects about 600,000 Americans, as well as millions more worldwide. Symptoms such as pain in the back or side commonly develop when patients are in their 30s or 40s, and about 50% of patients with this disease develop end-stage renal disease by the age of 60. When this happens, patients can be kept alive only by a kidney transplant or dialysis. About 10% of the dialysis population in this country suffers from ADPKD. The costs to Medicare for transplantation and dialysis amount to over a billion dollars annually. The disease is devastating to patients and their families. *Autosomal recessive polycystic kidney disease (ARPKD)* is a rare disorder that affects several hundred babies in the United States every year. Unlike ADPKD, which is a slowly progressive disorder, the recessive disease often leads to rapid renal failure and death in infancy or early childhood. PKD may also be acquired. Certain chemicals (called "cystogens") can produce this disorder. And, in patients who have been on dialysis for a long time, cysts may start to grow in their damaged kidneys.

Many research laboratories around the world are trying to find out how renal cysts form and grow. We know that, when a cyst grows, three concurrent events occur: fluid accumulates in the cyst lumen, the epithelial cells in the cyst wall proliferate, and the extracellular matrix surrounding the cysts is remodelled. Which is the primary problem is not known. Fluid accumulation could occur because of obstruction to outflow from a cystic nephron, addition of fluid because of continued glomerular filtration and cyst wall fluid secretion, or impaired fluid reabsorption. The cells in the wall of cysts may be stimulated to divide because of chemical growth factors or mechanical factors, such as stretch of the cyst wall. Many investigators have demonstrated that the basement membrane surrounding cysts is abnormal. The observation that patients with ADPKD show an increased frequency of cardiac valve abnormalities and cerebral aneurysms supports the view that there is some fundamental systemic defect in basement membrane or extracellular matrix in this disease.

Researchers are studying the properties of cyst epithelial cells obtained from humans with ADPKD and grown in vitro in tissue culture. Investigators have reported that, compared to normal renal tubular cells, these cells show unusual properties such as abnormal growth rates, altered responses to growth factors, and mislocation of the Na/K ATPase from its usual basolateral location on renal epithelial cells to an apical position. Several animal models of polycystic kidney disease are now available, and these allow investigators to probe the pathophysiology of this disorder and test possible drug and dietary therapies. A particularly promising animal model is the Han:SPRD *cy* rat, a mutant strain of Sprague-Dawley rats first discovered in an animal breeding facility in Hanover, Germany. This rat displays a slowly progressive form of ADPKD that closely mimics the human disease.

It has been known since 1985 that a gene responsible for ADPKD in people is located on the short arm of chromosome 16. In 1994, a consortium of European scientists reported (*Cell* 77:881–894, 1994) that they had identified and sequenced this gene. It is an unusually large one and its sequence is unlike that of any other known gene. We do not know whether defects in this gene cause the disease by preventing the production of a normal protein or by leading to a defective protein. This gene is called the PKD1 gene, and mutations in it cause about 85–90% of ADPKD cases. A second gene, the PKD2 gene, is located on chromosome 4, and produces a less severe form of ADPKD. A third gene, which causes ARPKD, has been recently located to chromosome 6. At the moment, identities of the second and third genes are not known. The identification of genes that cause PKD is an important step toward early detection of the disease, unravelling of the basic disorder, and, possibly in the future, gene therapy. Early detection of the disease helps in treating common complications such as hypertension and urinary tract infections.

Another exciting recent development is a report by Woo and colleagues (*Nature* 368:750–753, 1994) on *cpk* mice with autosomal recessive polycystic kidney disease. Mice which have both defective genes and are left untreated develop massively enlarged (cystic) kidneys and ordinarily die at one month of age from renal failure. Mice treated weekly with taxol, an anticancer drug extracted from the bark of the Pacific yew tree, lived for more than 200 days with minimal loss of renal function. This drug stabilizes the cell's microtubule cytoskeleton and interferes with cell division. Whether taxol or similar drugs can be used to slow the progression of polycystic kidney disease in humans remains to be seen.

■ **What factors might contribute to fluid accumulation in a renal cyst?**

may be altered physiologically by a number of hormones (e.g., angiotensin II). In chronic renal failure, glomeruli are destroyed, leading to a reduction in the filtering surface and a diminished GFR.

Glomerular hydrostatic pressure (P$_{GC}$) depends mainly on the arterial blood pressure and the resistances of upstream blood vessels (cortical radial arteries, afferent arterioles) and downstream blood vessels (efferent arterioles). Within the range of operation of renal autoregulation (an arterial pressure of 80 to 180 mm Hg), glomerular hydrostatic pressure changes very little. If, however, blood pressure is reduced below this range, glomerular hydrostatic pressure and GFR may decrease substantially. At a mean arterial blood pressure of 40 to 50 mm Hg (as in shock or hypotension), GFR and, hence, urine output are practically zero, since the glomerular hydrostatic pressure is inadequate to form a filtrate.

Glomerular hydrostatic pressure may be altered by changing the diameter of afferent and efferent arterioles (Fig. 23–10); these blood vessels are affected by sympathetic nerve fibers and by the variety of vasoconstrictor and vasodilator substances mentioned earlier. Constriction of afferent arterioles lowers the downstream glomerular pressure. At the same time, blood flow is decreased. For both of these rea-

sons, GFR decreases. Dilation of afferent arterioles increases glomerular hydrostatic pressure and blood flow, which leads to an increase in GFR. Constriction of efferent arterioles leads to an increase in the upstream glomerular pressure; blood flow decreases. The effect on GFR of efferent arteriolar constriction is complex. An increase in glomerular pressure tends to increase GFR; on the other hand, a fall in blood flow tends to decrease GFR. So, depending on the degree of efferent constriction, GFR may be either increased or decreased. With modest constriction, GFR rises (the increase in glomerular pressure predominates); with severe constriction, GFR falls (the fall in blood flow predominates). Efferent arteriolar dilation leads to a decrease in glomerular pressure and an increase in glomerular blood flow; GFR falls.

If the urinary tract is obstructed (e.g., by stones, prostate enlargement), **hydrostatic pressure in Bowman's space (P$_{BS}$)** will increase, and consequently GFR will decrease. A very high rate of urine output also may be accompanied by a rise in hydrostatic pressure in Bowman's space and a fall in GFR, because an increased pressure head is required to force a large volume flow down the tubules and collecting ducts.

A decrease in plasma protein concentration (e.g., due to intravenous infusion of a large volume of saline) decreases **plasma colloid osmotic pressure (π$_{GC}$)** and leads to an increase in GFR.

The Clearance Concept and the Measurement of Glomerular Filtration Rate

The determination of GFR is one of the most important measurements of kidney function, both clinically and in experimental studies on renal function. This is done by applying the rather ingenious clearance concept. First we will explain what is meant by "clearance."

The kidneys clear substances from the blood plasma. Different substances have different clearance rates. The **renal plasma clearance** of a substance is defined as the rate of excretion of a substance divided by its plasma concentration. By convention, the symbol C_x indicates the clearance of a substance x; U_x the **urine concentration** of the substance x; P_x the **plasma concentration** of the substance x; and $\dot{V}$ the **urine flow rate**. The rate of excretion of a substance is equal to the product of its urine concentration and the urine flow rate, or in symbols, $U_x \times \dot{V}$. From the definition of clearance, we can write the **clearance equation:**

$$C_x = \frac{U_x \times \dot{V}}{P_x}$$

Plasma and urine concentrations of substances are often expressed in terms of milligrams per milliliter,

Figure 23–10

Scanning electron micrograph of a cast of a glomerulus with afferent (af) and efferent (ef) arterioles. Constriction or dilation of the arterioles will change glomerular blood flow and pressure. *(From: Kimura, K., et al. Effects of atrial natriuretic peptide on renal arterioles: morphometric analysis using microvascular casts. American Journal of Physiology 259:F936–F944, 1990.)*

and urine flow is given as milliliters per minute. Substituting these units into the clearance formula, we see that the units of clearance are milliliters of plasma per minute.

$$C_x = \frac{\text{mg } x/\text{ml urine} \times \text{ml urine/minute}}{\text{mg } x/\text{ml plasma}}$$

$$= \text{ml plasma/minute}$$

Renal clearance may be considered as the volume of plasma per unit time that must be completely freed (cleared) of a substance to supply the quantity of a substance excreted in the urine per unit time.

To measure GFR, we need a substance that is cleared from the plasma solely by glomerular filtration. This substance, therefore, should not be reabsorbed or secreted by the kidney tubules. It should not be metabolized, synthesized, or stored by the kidneys. It should pass through the glomerular filtration barrier unhindered; in other words, it should not be a molecule that is too large or that is bound to plasma proteins. It should be nontoxic. Finally, we should be able to measure this substance in plasma and urine using simple analytical methods.

Such an ideal substance is the polysaccharide **inulin,** a fructose polymer, derived from the roots of certain plants; it has an average molecular weight of about 5000. Figure 23–11 shows that the amount of inulin (IN) filtered per unit time, $P_{IN} \times$ GFR, equals the amount of inulin excreted per unit time, $U_{IN} \times \dot{V}$. The inulin clearance, C_{IN}, is defined by the expression $U_{IN}\dot{V}/P_{IN}$ and is therefore equal to the GFR.

The way in which inulin is used to measure GFR can be illustrated by an example. An inulin solution is infused intravenously to achieve a constant plasma inulin concentration and the person is given water to drink to improve the urine output. A timed urine sample is collected, and average urine flow rate is calculated by dividing the urine volume by the duration of collection. In the middle of the urine collection period, a blood sample is taken. Subsequently, plasma and urine samples are analyzed for inulin. Suppose the following values are found: $P_{IN} = 0.30$ mg/ml, $U_{IN} = 30$ mg/ml, and $\dot{V} = 1.25$ ml/min. Substituting in the inulin clearance equation, $C_{IN} = U_{IN} \times \dot{V}/P_{IN} =$ GFR, we get

$$\text{GFR} = \frac{30\text{mg/ml urine} \times 1.25 \text{ ml urine/min}}{0.30 \text{ mg/ml plasma}}$$

$$= 125 \text{ ml plasma/min}$$

You may have noticed that in this example the inulin concentration in the urine is 100 times that in plasma. This is not due to tubular secretion of inulin; rather, it is caused by water reabsorption. As water is reabsorbed by the kidney tubules, filtered inulin is left behind and therefore becomes concentrated in a

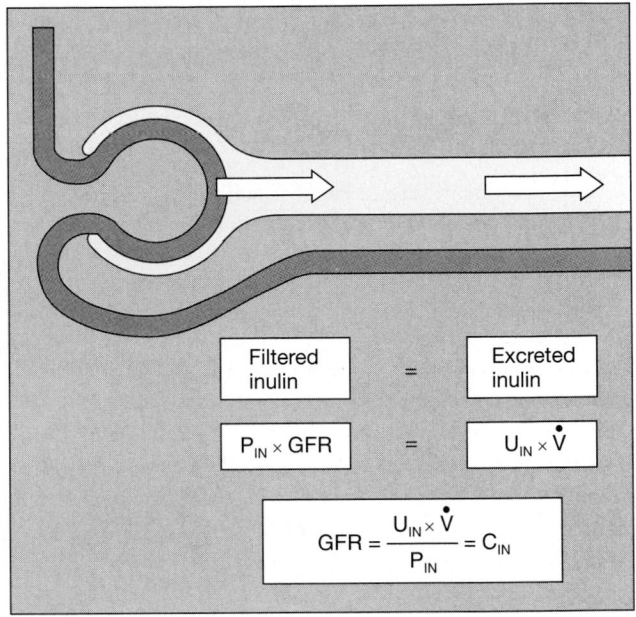

Figure 23–11

Principle behind the measurement of glomerular filtration rate (GFR). Inulin is freely filterable, so the amount filtered per unit time equals its plasma concentration (P_{IN}) multiplied by the GFR. Inulin is neither reabsorbed, secreted, synthesized, metabolized, nor stored by the kidney tubules, so filtered and excreted amounts are equal. Rearranging the equation, we find that GFR is equal to the inulin clearance (C_{IN}).

smaller volume of water. For example, if the inulin contained in 100 ml of filtrate is concentrated in 1 ml of urine, the inulin concentration will be increased 100-fold, and 99 ml of water (or 99% of the filtrate) must have been reabsorbed by the kidney tubules.

GFR in people depends on body surface area; it is usually corrected to a standard body surface area of 1.73 square meters. In normal young adult males, it averages 125 ml/min (180 liters/day), and in normal young adult females, it averages 110 ml/min (165 liters/day). GFR is very low in the newborn, 20 ml/min per 1.73 square meters of body surface area, and attains adult values (when corrected for body surface area) by about one year of age. In men, GFR is maintained at about 125 ml/min to about 45 to 50 years of age, after which it declines about 13 ml/min per decade. A similar decline is seen in older women.

Inulin is the "gold standard" for measuring GFR, but it is not often used clinically for this purpose. Infusing this substance is somewhat inconvenient. Instead, the **endogenous creatinine clearance** is used. Creatinine does not have to be infused since it is normally produced in the body. Its plasma levels are normally quite constant, and it can be easily measured

(text continues on page 719)

RESEARCH FOCUS

Methods Used to Study Kidney Function

Our understanding of renal function has depended on the use of a great variety of experimental techniques. Four approaches have been of particular value: clearance methods, micropuncture, perfusion of isolated kidney tubules, and membrane vesicles.

Renal clearance methods were developed in the 1930s and still provide an excellent way to evaluate overall kidney function in an unanesthetized person or animal. This approach permits the measurement of glomerular filtration rate, renal blood flow, and the rates of tubular reabsorption or secretion of various substances.

Kidney micropuncture was developed in the 1920s in A.N. Richards' laboratory at the University of Pennsylvania. In a classic study (American Journal of Physiology, 71:209–227, 1924), Wearn and Richards punctured Bowman's space in a frog's kidney with a fine glass capillary micropipette. They analyzed the collected fluid and determined that it contained glucose and chloride but no protein. This proved that the glomerulus produced an ultrafiltrate of plasma. Since glucose and chloride were absent from the frog's bladder urine, but were present at the beginning of the nephron, it was also evident that the tubules must have reabsorbed these two substances.

The sites of reabsorption of various substances (chloride, glucose, water, urea, and inorganic phosphate) and secretion of hydrogen ions along the tubules were subsequently identified in Richards' laboratory by micropuncture in frogs and salamanders. The basic approach was to collect minute amounts of fluid from single nephrons, analyze the tubule fluid for the substance of interest, and determine the site of collection by microdissection of the nephron. This technique was later applied to the mammalian kidney in studies reported in 1941, but owing to its great difficulty and to the outbreak of World War II, the method fell into disuse. In the 1950s, Phyllis Bott (Women's Medical College, Philadelphia), Heinrich Wirz (University of Basel, Switzerland), and Carl Gottschalk (University of North Carolina) revived the method. Wirz and Gottschalk used it to prove the existence of the countercurrent mechanism for concentrating the urine. Micropuncture continues to be an extremely valuable method to localize and study tubule functions and to measure pressures within the kidney. A limitation, however, is that not all nephron segments are accessible, because some lie deep within the kidney substance.

A method for isolating and perfusing single tubule segments in vitro was developed by Maurice Burg and his colleagues at the National Institutes of Health in 1966. With this approach any tubule segment can be dissected out of the kidney. One end of the tubule is cannulated with a glass micropipette to perfuse the lumen, and the other end is connected to another micropipette to collect the fluid which has passed through the lumen. A great advantage of this technique is that lumen and bath compositions can be changed at will, allowing the effects on tubule transport processes to be studied. Measurements of cell membrane potentials and of cell ion concentrations can be made with microelectrode or fluorescence techniques.

The membrane vesicle technique can be used to study the properties of transporters in the luminal and basolateral cell membranes of tubular epithelial cells. Basically, kidney tissue is ground up, and then biochemical techniques (precipitation, centrifugation, etc.) are used to separate luminal and basolateral cell membrane preparations. The membranes actually round up to form tiny vesicles, with a space in the middle. Uptake into the vesicle space is conveniently studied by using radioactively labeled substances. This technique has been used, for example, to demonstrate that active uptake of glucose by luminal cell membranes is dependent on sodium cotransport.

This combination of techniques permits the analysis of kidney function at whole organ, individual nephron, cellular, and subcellular levels. Cell culture, enzymatic, and molecular biology techniques are also furnishing new insights into renal cell function.

An isolated, perfused amphibian collecting duct. The duct is held by pipettes at both ends and its lumen is perfused with a physiological salt (Ringer's) solution. A glass microelectrode (upper part of figure) is used to impale a single cell and to measure its basolateral cell membrane potential. *(Photo by G. Tanner and J.-D. Horisberger.)*

■ **Which of the above research methods has been widely used to study renal function in people?**

in plasma and urine. In people, creatinine is actually secreted in addition to being filtered, but the error introduced by tubular secretion is usually small, and the endogenous creatinine clearance does provide a clinically valuable estimate of GFR.

If GFR falls, substances (including creatinine and urea) produced in the body that depend primarily on glomerular filtration for their elimination from the body will accumulate, and their plasma concentrations will rise. The extent to which the plasma concentration of creatinine is elevated may serve as an index of the degree of impairment of filtration rate. Likewise, the degree to which blood urea nitrogen (BUN) is elevated may reflect the degree of impairment of filtration rate; however, urea is less well suited for this assessment than creatinine, because BUN also depends on other factors, such as the rate of protein catabolism in the body and the urine flow rate.

We have devoted considerable attention to the measurement of GFR because it is so important in assessing kidney function. Glomerular filtration is a major process by which the kidneys regulate the volume and composition of the body fluids. A reduction in GFR to far below normal will impair a person's ability to excrete various waste products and the proper amounts of water and mineral electrolytes. In patients, the measurement of GFR provides the most valuable indicator of the severity of renal failure. Also, in studying tubule transport functions, a topic we will turn to next, the measurement of GFR is important. In order to tell how much of a substance is reabsorbed, we must first know how much was filtered. To tell whether a substance is secreted by the kidney tubules, we cannot simply measure the amount excreted; we must know how much was also filtered. Measurements of GFR are thus important to the study of transport by the kidney tubules.

Tubular Reabsorption Returns Filtered Substances Back to the Blood

In the process of glomerular filtration, large quantities of mineral electrolytes, water, and organic compounds are presented to the tubules. Varying amounts of some of these substances are excreted (see Table 23–1). Most of the glomerular filtrate is not excreted, because the tubules selectively reabsorb the constituents of the glomerular filtrate.

Reabsorption may be either active or passive. **Active reabsorption** (active transport) is a process requiring the local expenditure of metabolic energy by the tubular epithelium, and it can effect the net movement of a substance against concentration or electrical gradients, or both. Examples of actively reabsorbed substances are sodium, glucose, and phosphate. **Passive reabsorption** (passive transport) does not depend

directly on the expenditure of metabolic energy, and movement occurs down concentration, electrical, or osmotic gradients. Passively reabsorbed substances include urea, chloride, and water. In this section, we will discuss the mechanisms for reabsorbing various organic compounds. Later we will consider the reabsorption of various mineral electrolytes and water.

Reabsorption of Glucose

Glucose is filtered by the glomeruli and actively reabsorbed by the tubules. The glucose reabsorptive mechanism, which appears to be confined to the proximal tubule, is so effective that normally all of the filtered glucose is reabsorbed, with none excreted in the urine.

The cellular mechanism for glucose reabsorption by the proximal tubule is depicted in Figure 23–12. Glucose is cotransported with sodium across the luminal (brush border) cell membrane. This movement of glucose is energetically "uphill" and leads to accumulation of glucose in the cell. The energy required is derived from the sodium gradient across the luminal cell membrane. The cell sodium concentration is below that in the luminal fluid and is maintained at a low level by the Na^+/K^+-ATPase in the basolateral cell membrane. Also, the cell interior is electrically negative (by about 70 mv) compared to the fluid in the tubule lumen. As sodium moves into the cell down its concentration and electrical gradients, it moves glucose against its concentration gradient. Glucose leaves the cell by moving down a concentration gradient (cell-to-peritubular capillary blood); this movement is via a carrier-mediated, sodium-independent mechanism in the basolateral cell membrane. Overall, glucose reabsorption is an example of "secondary" active transport coupled to "primary" active transport of sodium.

The ability of the tubules to reabsorb filtered glucose is limited (Fig. 23–13). At normal plasma glucose levels (65 to 90 mg/dl, fasting values), the filtered glucose load can be completely reabsorbed by the tubules, and so no glucose is excreted. If the plasma level is increased sufficiently—for example, to 180 to 200 mg/dl—the filtered glucose load can no longer be completely reabsorbed. The plasma level at which glucose first appears in the urine in appreciable amounts is called the **glucose threshold.** Glucose appears in the urine because the tubules have a limited reabsorptive capacity. At high enough filtered glucose loads, all of the glucose carriers become saturated, and any extra filtered glucose is excreted. The maximal rate of glucose reabsorption is called the **tubular transport maximum (Tm) for glucose (G),** usually abbreviated Tm_G. Tm is analogous to the V_{max} (maximum velocity) of enzyme-catalyzed reactions (Chapter 2).

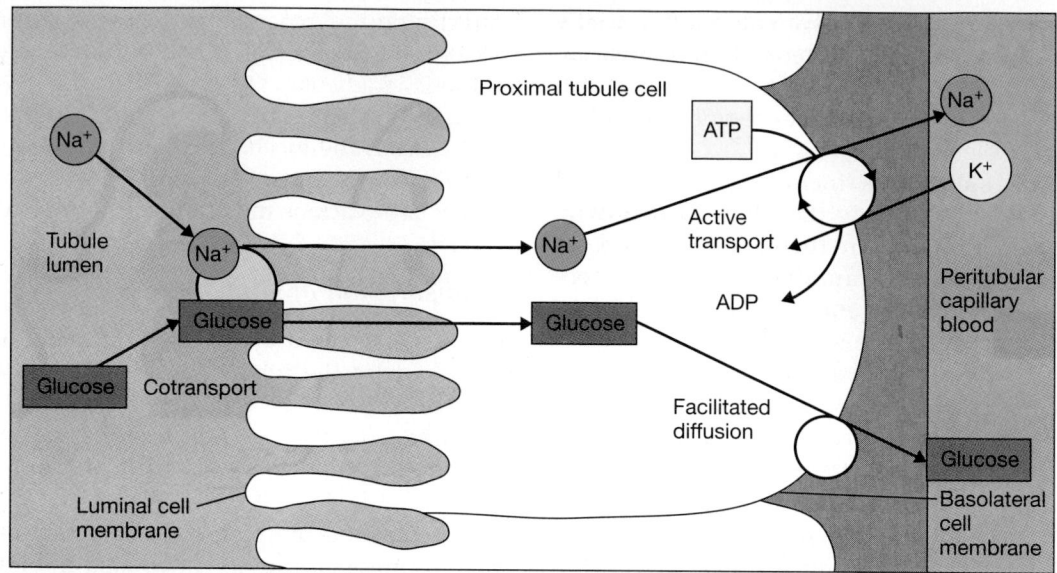

Figure 23–12

A model of glucose reabsorption by the proximal tubule cell. Glucose is co-transported with sodium across the luminal cell membrane, resulting in a high intracellular glucose concentration. Glucose passes through the basolateral cell membrane via facilitated diffusion and then diffuses into the peritubular capillary blood.

Figure 23–13

Effect of increasing the filtered amount of glucose on glucose excretion. In the normal condition (*a*), all of the filtered glucose is reabsorbed, and none is excreted. At threshold (*b*), the amount of glucose filtered is increased to a sufficiently high level (by increasing the plasma glucose concentration) so that not all of the filtered glucose is completely reabsorbed. (*c*) Far above threshold, further increases in the amount of glucose filtered result in increased glucose excretion, because the reabsorptive capacity of the tubules is operating at a maximal rate (glucose Tm). Note that the arrows for reabsorbed glucose in (b) and (c) are identical in size.

Glucose in the urine (**glucosuria**) is often a sign of uncontrolled **diabetes mellitus.** In this condition, a deficiency in the insulin system results in an increased plasma glucose concentration. Filtered glucose exceeds the reabsorptive capabilities of the tubules, and so glucose is excreted in the urine. Many of the early symptoms of diabetes mellitus are due to this phenomenon. Excretion of glucose causes loss of water and salt in the urine. The result is a high urine flow rate (increased frequency of urination), dehydration, and thirst.

Glucose in the urine is not necessarily a sign of uncontrolled diabetes mellitus. Glucosuria may also occur because of a high blood glucose level after eating a heavy meal or in conjunction with emotional stress (which results in increased sympathetic nervous system activity). There are also certain kidney tubule disorders (e.g., familial renal glucosuria, Fanconi syndrome) in which the ability of the tubules to reabsorb glucose is diminished.

The glucose reabsorptive mechanism reabsorbs other simple sugars such as galactose, xylose, and possibly fructose. Since the capacity of the transporting mechanism is limited, two sugars will depress each other's reabsorption if they are simultaneously presented to the tubules. One sugar may displace another from their common carrier, depending on their relative concentrations and affinities for the carrier. This phenomenon is called **competition** for transport. Saturation of transport (existence of a *Tm*) and competition for transport are closely linked. It appears to be a universal law of nature that when there is a limited amount of something, there is bound to be competition.

Reabsorption of Amino Acids

Plasma amino acids are filtered by the glomeruli and extensively reabsorbed by the renal tubules. Normally, only traces appear in the urine. There are several distinct transport mechanisms for amino acids, each reabsorbing a group of structurally similar compounds. The amino acid transport mechanisms, although separate from the glucose reabsorbing mechanism, show several similarities to the latter. Thus, amino acid reabsorption occurs in the proximal tubule. It is active and depends on sodium transport, usually showing a maximal rate of transport and within-group competition.

Reabsorption of Uric Acid

Uric acid, an end product of purine metabolism, is continuously produced in the body and excreted by the kidneys. The amount excreted is normally about 10% of the amount filtered. Hence, uric acid is reabsorbed by the kidney tubules. The kidney tubules also secrete uric acid, but reabsorption normally predominates. Both reabsorption and secretion occur in the proximal tubule by carrier-mediated processes.

Gout is a condition in which plasma uric acid is elevated and uric acid crystals precipitate in the joints, causing inflammation and painful swelling of the affected joints (often in the great toe). Gout may be treated by **uricosuric agents,** drugs that increase uric acid excretion. Probenecid (Benemid), sulfinpyrazone (Anturane), and high doses of aspirin inhibit tubular reabsorption of uric acid. This leads to an increase in uric acid excretion and a fall in plasma uric acid. When uricosuric agents are administered, there is a risk of uric acid precipitation in the urine. This danger can be reduced by increasing urine flow rate (drinking large amounts of water) and by making the urine alkaline (through ingestion of sodium bicarbonate).

Reabsorption of Urea

Urea, formed chiefly in the liver, is the major nitrogen-containing end product of metabolism in human beings. Its rate of production in the body depends on the rate of protein breakdown. Under ordinary conditions, this is determined by the dietary intake of protein. It is important to recognize, however, that protein catabolism, and hence urea production, can also be increased by tissue trauma, internal bleeding, infection, fever, and other conditions. Urea elimination from the body depends on an adequate rate of glomerular filtration.

Figure 23–14 shows that the renal handling of urea along the tubules is fairly complex. The numbers in this figure refer to relative amounts, not concentrations, present at different sites. The amount of urea filtered per unit of time is taken as "100." Fifty percent of the filtered urea is left at the end of the proximal convoluted tubule. The explanation is as follows: as water is reabsorbed, filtered urea becomes more concentrated in the tubular fluid. Since the proximal tubule epithelium is somewhat permeable to urea, urea will diffuse out of the tubular fluid back into the blood surrounding the tubules. Some 50% of the filtered urea is reabsorbed in this portion of the nephron. If we examine fluid from the distal convoluted tubule, we see that more urea is present here than was present at the end of the proximal convoluted tubule (see Fig. 23–14). Hence, urea was added (or secreted) into tubular fluid in the loop of Henle; again, this is thought to be a passive movement. A very high concentration of urea exists in the interstitial fluid of the medulla, and so there is a gradient for urea diffusion into the loop of Henle. Notice that the thick ascending limb, distal convoluted tubule,

Figure 23–14

Movements of urea along the nephron. The numbers indicate relative amounts (100 = filtered urea). The heavy green outline indicates relative impermeability of these nephron segments to urea. Urea is reabsorbed by the inner medullary collecting duct; most of this urea reenters the loop of Henle, and some is removed by the medullary blood vessels (vasa recta).

cortical collecting duct, and outer medullary collecting duct are quite impermeable to urea. Water is reabsorbed by the cortical and outer medullary collecting ducts, and so a concentrated urea solution is delivered to the inner medullary collecting duct. The inner medullary collecting duct is permeable to urea, so urea diffuses out of the duct and is added to the inner medulla. As we will discuss later, this addition of urea is important in terms of the urinary concentrating mechanism. Of the quantity of urea entering the inner medullary collecting duct, 80% is reabsorbed (50% re-enters the loops of Henle to be recycled, and 30% is carried away by the blood vessels of the medulla), and 20% is excreted. In this illustration, there is a high degree of reabsorption of filtered urea (80%), reflecting a high degree of water reabsorption. If, however, water reabsorption is diminished, then urea reabsorption is also lower because a reduction in tubular water reabsorption results in less of a con-

centration gradient for diffusion of urea from tubule fluid back into the blood. Also, less time is available for back-diffusion when the urine flow rate is high. With urine flow rates greater than 2 ml/min, about 40% of filtered urea is reabsorbed and 60% is excreted.

Reabsorption of Proteins and Other Substances

The glomerular filtrate is nearly, but not completely, protein-free. Quantitatively speaking, serum albumin is the major protein in plasma, and so we will consider this substance. Serum albumin concentration in the glomerular filtrate is probably less than 1 mg/dl; compare this to a plasma concentration of 4.5 g/dl. Clearly, the glomerular barrier is very effective in hindering the passage of this macromolecule. If we consider the high rate of filtrate formation per day, the amount of serum albumin filtered per day (approximately 1 mg/dl × 180 liters/day = 1.8 gm/day) is not trivial. Normally, less than 0.03 gm of serum albumin is excreted per day. This comparison suggests that the kidney tubules reabsorb filtered serum albumin. Indeed, there is evidence that the proximal tubules reabsorb a variety of proteins by endocytosis. The urine normally contains less than 0.15 gm of protein per 24-hour urine sample. **Proteinuria** (excess protein in the urine) is usually due to an abnormal leakiness of the glomerular filtration barrier. Heavy proteinuria (more than 4 g/day) is almost always the result of glomerular disease.

Many other organic compounds are reabsorbed by the kidney tubules. Specific tubular reabsorptive mechanisms are known for the Krebs cycle acids (citric, α-ketoglutaric, and malic acids), lactic acid, ascorbic acid (vitamin C), and the ketone body acids (acetoacetic and β-hydroxybutyric acids). Most of these mechanisms have properties similar to the glucose-reabsorbing mechanism.

Proximal Tubules Actively Secrete Many Organic Anions and Organic Cations

Many organic compounds are actively transported from the blood surrounding the kidney tubules into the tubular urine by specific secretory mechanisms. The amount excreted is the sum of filtered and secreted amounts.

A few secretory mechanisms (carriers) transport a variety of organic anions; these compounds are mainly carboxylic and sulfonic acids, which at blood pH (7.4) are mostly ionized and bear a negative charge. Other secretory mechanisms transport a variety of organic cations (organic bases); these are mainly amine and ammonium compounds, which at blood pH bear a positive charge. These active secre-

Table 23–2

Some Organic Compounds Secreted by the Kidney Tubules

Compound	Use
Organic Anions	
Phenol red (phenol-sulfon-phthalein, PSP)	pH indicator dye
p-aminohippuric acid (PAH)	Measurement of renal plasma flow and tubular secretory activity
Penicillin	Antibiotic
Probenecid (Benemid)	Inhibitor of penicillin secretion and uric acid reabsorption
Chlorothiazide (Diuril)	Diuretic drug
Acetazolamide (Diamox)	Carbonic anhydrase inhibitor
Creatinine*	Normal end product of muscle metabolism
Organic Cations	
Histamine	Vasodilator, stimulator of gastric acid secretion
Norepinephrine	Neurotransmitter
Quinine	Antimalarial drug
Quinidine	Antiarrhythmic drug
Tetraethyl-ammonium (TEA)	Ganglionic blocking agent
Creatinine*	Normal end product of muscle metabolism

*Creatinine is an unusual compound, since it is secreted by both the organic anion and cation mechanisms. This is probably because of negatively and positively charged groups in the molecule at physiological pH.

tory mechanisms are present only in the proximal tubule. The secretory mechanisms show a maximal transport rate (*Tm*) at high plasma levels, and show within-group competition for transport. Some of the more lipid-soluble organic acids and bases may also be passively reabsorbed or secreted.

Table 23–2 lists a few of the many secreted organic compounds. Many of these substances are foreign to the body and can be of medical importance. By supplementing filtration, tubular secretion results in rapid excretion. Urinary excretion is an important consideration in adjusting the dosage of some drugs. In patients with renal failure, drug dosage may have to be reduced to avoid excessively high plasma levels.

PAH (*p*-aminohippuric acid) is so avidly secreted by the kidney tubules that it is almost completely cleared (removed) from all of the plasma in one passage of blood through the kidneys. Therefore the clearance of PAH can be used to estimate the rate of plasma flow through the kidneys. The explanation for this follows: the amount of PAH per unit of time delivered to the kidneys in the renal blood flow is equal to the product of the renal plasma flow (RPF) and the plasma PAH concentration (P_{PAH}). (The amount of PAH in red blood cells is negligible.) The amount of PAH per unit of time leaving the kidneys, assuming that all is extracted from the renal plasma and excreted in the urine, is equal to $U_{PAH} \times \dot{V}$ (the PAH excretion rate). From the principle of conservation of matter, we can write:

$$\text{input} = \text{output}$$
$$\text{RPF} \times P_{PAH} = U_{PAH} \times \dot{V}$$

Rearranging,

$$\text{RPF} = U_{PAH} \times \dot{V}/P_{PAH}$$

By definition, the expression on the right is the PAH clearance (C_{PAH}). Therefore, the renal plasma flow equals the PAH clearance. To determine the renal blood flow (RBF), we need to correct the renal plasma flow for the blood hematocrit (hct), using the following equation:

$$\text{RBF} = \text{RPF}/(1 - \text{hct})$$

Renal blood flow can be readily measured in unanesthetized human beings or animals by applying the PAH clearance method. One simply infuses PAH so as to achieve a steady, low plasma level of this compound, collects a timed urine sample and a blood sample, measures the hematocrit, analyzes plasma and urine for PAH, and uses the equations just given. In this example, we assumed that PAH is 100% extracted from the plasma flowing through the kidneys, which, in the normal human kidney, is almost true (90% is closer to the truth). To measure renal blood flow more accurately, it would be best to determine the actual extraction by also measuring the PAH concentration in renal venous blood.

Tubular Transport of Mineral Electrolytes

The inorganic salts of the minerals sodium, potassium, calcium, magnesium, and phosphate are completely ionized in aqueous solution and can conduct electricity, so they are often referred to as mineral electrolytes. Next, we will discuss the renal handling of sodium, potassium, calcium, magnesium, and phosphate ions. Although we will consider the various ions separately, recognize that these ions do not exist in isolation. Positively or negatively charged ions are always accompanied by ions of opposite charge because of the principle of electrical neutrality in solutions.

Sodium Reabsorption Is the Major Activity of the Kidney Tubules

The renal handling of sodium is important for several reasons. First, the kidneys are major regulators of the amount of sodium in the body. Since we take in variable amounts of sodium in our diets, the kidneys must adjust the excretion of sodium to maintain balance. We will consider this topic in detail in the next chapter. Second, sodium reabsorption by the kidney tubules is the major driving force for water reabsorption. Sodium is, for the most part, actively reabsorbed; water follows passively. Third, sodium reabsorption is coupled to the reabsorption of both organic (e.g., glucose, amino acids) and inorganic substances (e.g. chloride, phosphate). Fourth, sodium reabsorption is intimately related to the secretion of hydrogen and potassium ions. Thus, it affects acid-base balance and potassium balance. Finally, in terms of energy demands, reabsorption of sodium is the major operation in the kidneys. Sodium and its accompanying anions (chloride and bicarbonate) constitute the solutes reabsorbed in greatest amounts. Almost all of the filtered sodium is actively reabsorbed by the kidney tubules. Normally, 80% of the oxygen

consumed by the kidneys is devoted to powering sodium reabsorption.

Later, we will consider how sodium reabsorption is affected by various factors and how it is controlled in order to keep the body in sodium balance. For now, we will discuss how sodium is reabsorbed in various parts of the nephron. Figure 23–15 presents a summary of the percentages of filtered sodium (and water) that are reabsorbed under normal conditions by various nephron segments.

Sodium Transport in the Proximal Convoluted Tubule

The logical place to start is the proximal convoluted tubule. This segment reabsorbs 70% of the filtered sodium and water. The cellular mechanisms involved are depicted in Figure 23–16. Sodium enters the cell across the luminal membrane (brush border) and moves down chemical and electrical gradients. The intracellular sodium concentration is low (kept low by the basolateral membrane Na^+/K^+-ATPase), and the inside negativity of the cell also favors sodium entry. The entry of sodium into the cell is coupled to the movement of other solutes. As was discussed earlier, sodium enters the cell together with glucose and amino acids. It is also accompanied by other organic solutes and phosphate. A luminal membrane Na^+/H^+ exchanger results in sodium reabsorption and acidification of the tubular urine. Sodium is actively pumped out the basolateral cell membrane by its Na^+/K^+-ATPase. This is an example of primary ac-

	Proximal convoluted tubule	Loop of Henle	Distal convoluted tubule and collecting duct
Percent of filtered sodium reabsorbed	70	20	9
Percent of filtered water reabsorbed	70	10	19

Figure 23–15

Percentages of filtered sodium and water reabsorbed by different nephron segments under normal conditions. About 1% of filtered sodium and water are excreted.

Figure 23–16

Model of ion and water transport in the proximal convoluted tubule. Solid arrows indicate active transport; dashed arrows indicate passive transport. A luminal membrane Cl^--formate ($HCOO^-$) exchanger participates in chloride reabsorption.

tive transport. The energy needed to move sodium out of the basolateral cell membrane, against combined concentration and electrical gradients, is directly derived from the hydrolysis of ATP.

The reabsorption of sodium and accompanying solutes results in the reabsorption of water. The epithelium of the proximal convoluted tubule has a high water permeability, and so only a very small osmotic gradient is necessary to account for the observed rate of water reabsorption. Removal of solute is thought to lead to an effective osmotic pressure difference of about 2 to 5 mosm/kg H_2O across the tubular wall; this is considered sufficient to drive water reabsorption by osmosis. Since the gradient is so small, proximal tubular fluid is essentially isosmotic to plasma. Sodium and its accompanying anions compose the major osmotically active solutes in the glomerular filtrate and proximal tubule fluid. Since 70% of the filtered sodium is reabsorbed, 70% of the filtered water is also reabsorbed in the proximal convoluted tubule.

The final step in the reabsorption of sodium is the uptake of the reabsorbed fluid by the peritubular capillaries. The forces involved here are usually referred to as **peritubular capillary Starling forces** and involve the hydrostatic and colloid osmotic pressure differences that affect fluid movement across any capillary wall (see Chapter 19). The hydrostatic pressure in these capillaries is low, since the blood has passed through several upstream resistance vessels (cortical radial arteries, afferent arterioles, efferent arterioles). Also, the colloid osmotic pressure of the peritubular capillary plasma is high because the plasma proteins

were concentrated by glomerular filtration. Consequently, there is a net pressure gradient favoring uptake of reabsorbed fluid. Under some circumstances (e.g., loading a person with isotonic saline), peritubular capillary pressures are altered in a direction that diminishes the uptake force. The reabsorbed fluid then accumulates in the interstitium. The lateral intercellular spaces widen, and the tight junctions become expanded and very leaky. This results in backleak of sodium and water into the tubule lumen and an overall reduction in net sodium reabsorption.

Sodium Transport in More Distal Nephron Segments

The loop of Henle reabsorbs about 20% of the filtered sodium and 10% of the filtered water (see Fig. 23–15). Note that more sodium than water is reabsorbed here because the ascending limb is impermeable to water but reabsorbs salt. The transcellular transport of sodium in the thick ascending limb is accomplished by a coupled $Na^+/K^+/2Cl^-$ cotransporter in the luminal cell membrane (which is inhibited by "loop" diuretic drugs such as furosemide and bumetanide) and a Na^+/K^+-ATPase in the basolateral cell membrane. As will be discussed later, the removal of salt along the ascending limb and its deposition in the medulla are essential steps in the mechanism whereby the kidneys produce an osmotically dilute or concentrated urine.

The distal convoluted tubule and collecting duct accomplish the reabsorption of most, but not all, of the remaining salt and water (see Fig. 23–15). Sodium reabsorption here is active, and sodium ions can be reabsorbed against steep concentration and electrical gradients. For example, if a person is depleted of salt, the collecting ducts can lower the final urine sodium concentration to 1 mEq per liter. Aldosterone, a hormone secreted by the adrenal cortex, stimulates sodium reabsorption by the collecting ducts. Most of the reabsorbed sodium is accompanied by chloride; some reabsorbed sodium is replaced by potassium or hydrogen ions. The collecting duct epithelium has an intrinsically low water permeability. Antidiuretic hormone (ADH), from the posterior pituitary, increases collecting duct water permeability in a graded fashion. The collecting ducts are the final site of regulation of sodium and water excretion, so it is not surprising that reabsorption there is under hormonal control.

To summarize and compare reabsorption in the early (proximal convoluted tubule) and late (collecting duct) portions of the nephron, we note that in the proximal tubule, large quantities of salt and water are reabsorbed along small gradients. The proximal tubule has a leaky epithelium (Chapter 4) and cannot sustain large gradients for small ions or water. Reabsorption here can be considered as a bulky, coarse operation. By contrast, the collecting ducts have a smaller capacity for reabsorption and can establish large gradients. The collecting ducts have a tight epithelium, and are the site of fine control; the hormones aldosterone and ADH regulate, respectively, sodium and water reabsorption here.

Potassium Is Filtered, Reabsorbed, and Secreted in the Kidney

The kidneys are normally the major route of potassium excretion from the body and serve as an important site of potassium regulation. Renal handling of potassium involves reabsorption of most of the filtered potassium by the proximal convoluted tubule and loop of Henle. Under conditions of potassium depletion, potassium continues to be reabsorbed along the collecting ducts. Under normal conditions or with potassium excess, potassium is secreted by the collecting ducts, especially by the cortical collecting ducts. The major cell type of the collecting ducts, the **principal cells,** appear to be responsible for this secretion. Most of the excreted potassium usually represents secreted potassium.

A number of factors affect potassium secretion, most of which can be understood by examining the cell model for a cortical collecting duct principal cell in Figure 23–17. According to this model, potassium is taken up by the cell via a Na^+/K^+-ATPase located in the basolateral membrane. This leads to an increase

in intracellular potassium ion concentration. Potassium can then diffuse out of the cell into the lumen and in this way be secreted. The intracellular potassium concentration is a key element in determining the rate of potassium secretion. An increase in extracellular potassium concentration or increased plasma levels of aldosterone enhance cell uptake of potassium and overall secretion. Aldosterone, in addition, increases the luminal membrane permeability to both potassium and sodium, which also favors potassium secretion and sodium reabsorption. There appears to be a **transtubular potential difference** across many nephron segments. If we advance a potential-measuring microelectrode through the tubule wall, we first record a negative deflection of 70 mv when the microelectrode enters the cell. (The reference potential [zero] is always on the blood side of the tubules.) A negative deflection of about 50 mv is recorded when the microelectrode penetrates the lumen of the cortical collecting duct. Therefore, the electrical potential opposing potassium movement out of the cell is much smaller at the luminal, -70 mv $- (-50$ mv$) = -20$ mv, than at the basolateral side of the cell, -70 mv. Consequently, there is less force opposing movement into the lumen, and potassium secretion is favored. Increased amounts of poorly reabsorbed anions in the tubule lumen increase the transtubular potential difference (make the lumen more negative) and so enhance potassium secretion, as does an increased rate of flow through the collecting duct lumen, due to a "sweeping away" effect that maintains a concentration gradient favoring potassium secretion. This effect is probably the main reason that most diuretic drugs increase potassium excretion.

Calcium, Magnesium, and Phosphate Are Reabsorbed by the Kidney Tubules

Renal excretion of **calcium** is important, but represents only about 10% of the total daily output of calcium from our bodies on a normal diet. About 90% of our intake is excreted in the feces. The regulation of plasma calcium involves control of intestinal calcium absorption, renal excretion, and exchanges with bone. The parathyroid hormone and vitamin D play important roles in this regulation. Calcium balance will be discussed in Chapter 26.

Calcium is filtered and reabsorbed by the kidneys. Since about 40% of plasma calcium is bound to plasma proteins, only the nonbound portion (60%) is filtered. About 50% to 60% of the filtered calcium is reabsorbed in the proximal convoluted tubule, and most of the rest in the thick ascending limb, distal convoluted tubule, and collecting ducts. The quantity excreted is about 1% to 2% of the filtered load. Regulation of urinary excretion of calcium occurs pri-

Figure 23–17

A model of potassium and sodium transport by a principal cell in the cortical collecting duct. The two steps in potassium secretion are (1) uptake into the cell via the basolateral membrane Na$^+$/K$^+$-ATPase and (2) passive diffusion across the luminal cell membrane. Sodium reabsorption involves passive diffusion into the cell through a sodium-selective ion channel in the luminal cell membrane, and active pumping of sodium across the basolateral cell membrane.

marily in the distal portions of the nephron. The parathyroid hormone stimulates calcium reabsorption and so acts to conserve filtered calcium for the body.

About two thirds of the **magnesium** in our diet is normally excreted in the feces, and one third in the urine. Approximately 30% of plasma magnesium is bound to plasma proteins and so is not filterable at the glomeruli. The proximal convoluted tubules reabsorb about 20% to 30% of filtered magnesium, and 50% to 60% is reabsorbed in the loop of Henle. Only 3% to 5% of the filtered magnesium is excreted under normal conditions. The kidneys play a major role in regulating the plasma magnesium concentration. Excesses of magnesium are rapidly excreted in the urine; in magnesium-deficient states, magnesium is avidly reabsorbed by the kidney tubules and virtually disappears from the urine.

The kidneys play an important role in regulating the plasma concentration of **inorganic phosphate.** We usually excrete in the urine most of the phosphate ingested in our diet. Filtered phosphate is reabsorbed by sodium-dependent, secondary active transport in the proximal tubule. The reabsorptive mechanism has a limited rate, or *Tm*. The *Tm* is normally exceeded by the quantities of phosphate filtered, and so phosphate is excreted in the urine. The situation here differs from that of glucose, where the *Tm* is reached only when the plasma glucose concentration is raised dramatically. Because the filtered phosphate load and *Tm* are normally so close to one another, the kidneys can participate in regulating the plasma phosphate concentration.

If phosphate intake is increased, plasma phosphate levels will rise, and the filtered phosphate load will be increased. The *Tm* will be exceeded more than usual, and so phosphate excretion will increase,

thereby ridding the body of the extra phosphate. Conversely, if phosphate intake is decreased, plasma phosphate and filtered phosphate amounts will decrease, and all of the filtered phosphate will be reabsorbed, thus conserving phosphate for the body. Phosphate excretion is also regulated by changing the *Tm*. For example, the parathyroid hormone inhibits phosphate reabsorption, thus promoting phosphate loss in the urine.

Tubular Reabsorption of Water

The rate of water excretion in the urine depends on the rates of filtration of water (GFR) and tubular water reabsorption. Since the urine volume is mostly water, we can assume that the renal water excretion rate is simply equal to the urine flow rate. Urine flow rate can vary widely, depending on conditions. If a person drinks a large quantity of water, urine flow rate may be as high as 20 ml/min. The osmolality* of this urine may be as low as 30 to 40 mosm/kg H$_2$O, much more dilute than plasma (approximately 300 mosm/kg H$_2$O). On the other hand, in a dehydrated person, urine flow rate may be as low as 0.3 ml/min. The osmolality of this urine may be as high as 1200 to 1400 mosm/kg H$_2$O, or approximately 4 to 5 times the normal plasma osmolality. A "normal" urine flow rate is about 1 ml/min (1.5 liters/day), and the urine is usually modestly concentrated (600 to 800 mosm/kg H$_2$O).

---

*The terms *osmolality* and *osmolarity* (Chapter 4) are often confused. Both are measures of solute concentration. Osmolality is expressed as osmoles per kg H$_2$O and osmolarity as osmoles per liter solution. Biological fluids are mostly water, so the two terms are practically identical.

The Kidneys Save Water When They Produce a Hyperosmotic Urine

We will consider now the mechanisms for producing an osmotically concentrated (hyperosmotic) or dilute (hypo-osmotic) urine—that is, a urine with a total solute concentration greater than or less than that of plasma. If the body contains excess water (e.g., due to drinking a lot of liquid), it makes sense for the kidneys to get rid of the extra water and excrete the urinary solutes in a large volume of osmotically dilute urine. The importance of excreting an osmotically concentrated urine is more subtle. Recall that the kidneys are always called upon to excrete solutes. These solutes include waste products of metabolism and various mineral electrolytes consumed in the diet. The excretion of these solutes requires the excretion of water, since this is the vehicle in which these substances are dissolved. Suppose someone had to excrete 600 mosm of solutes per day. Suppose also that the person could form only an isosmotic urine—that is, a urine with an osmolality of 300 mosm/kg H_2O. How much water would have to be excreted? It would take 2 kg (or 2 liters) of water to excrete 600 mosm. Suppose now that the same amount of solutes could be concentrated in a urine with an osmolality of 1200 mosm/kg H_2O. How much water would have to be excreted? One-half kg (or 0.5 liter) would contain 600 mosm solute as a 1200 mosm/kg H_2O solution. We can see that by excreting the urinary solutes in an osmotically concentrated (hyperosmotic) urine, the kidneys save water for the body. In the example given, by excreting the same amount of solutes in urine with an osmolality of 1200 mosm/kg H_2O instead of 300 mosm/kg H_2O, the kidneys, in effect, saved 1.5 liters of pure water for the body.

Since water is often scarce, the ability to form an osmotically concentrated urine is an important adaptation to a terrestrial environment. By getting rid of the urinary solutes in a hyperosmotic urine, the need for water intake is diminished. In the animal kingdom, only birds and mammals can produce a urine that is osmotically more concentrated than the blood. These animals are the only classes that have loops of Henle. Furthermore, in general, those mammals with relatively long loops of Henle are able to produce the most concentrated urines. For example, the kangaroo rat, a small rodent that inhabits the southwestern deserts of the United States, has an exceptionally long loop of Henle and can produce urine with an osmolality of 5500 mosm/kg H_2O, some 18 times that of plasma. This mammal's high metabolic rate (due to its small size) results in especially vigorous sodium transport by the thick ascending limb of Henle's loop, a key step in concentrating the urine. These facts from comparative anatomy and physiology suggest an association between loops of Henle and the ability to produce a concentrated urine.

The Countercurrent Hypothesis Explains the Formation of Osmotically Concentrated Urine

The **countercurrent hypothesis** is now universally accepted as the explanation for the mechanism whereby an osmotically concentrated urine is formed. Scientists are still not sure about certain aspects of this hypothesis—in particular, how the osmotic gradient is established in the inner medulla. The term *countercurrent* indicates that fluid flows in opposite directions in adjacent limbs of tubules and blood vessels in the kidney medulla.

There are actually two countercurrent mechanisms. The loops of Henle act as **countercurrent multipliers;** they set up a gradient of osmolality that increases from the junction of the cortex and medulla to the tip of the renal papillae. The blood vessels of the medulla (vasa recta) act as passive **countercurrent exchangers;** they help to preserve the osmotic gradient in the medulla. The collecting ducts act as **osmotic equilibrating devices;** it is here that the urine finally becomes osmotically concentrated. Depending on the plasma level of ADH, the fluid in the collecting ducts tends to equilibrate osmotically with the surrounding interstitium.

If kidney tissue from a water-deprived animal is taken from various levels of the medulla, a progressively higher osmolality is found with increasing depth of the medulla. The highest osmolalities are found at the papillary tip. All of the structures in the medulla participate in this increasing gradient. Also, in a dehydrated animal, with high plasma levels of ADH, the final urine has the same osmolality as tissue fluid at the tip of the kidney papilla.

This gradient is established by countercurrent multiplication in the loops of Henle. Figure 23–18 shows a simplified model for the countercurrent multiplication process. Consider a loop filled with fluid isosmotic to plasma (300 mosm/kg H_2O). Assume that the membrane separating the two loops is impermeable to water. Also assume that the loop is able to establish an osmotic gradient of 200 mosm/kg H_2O between the two limbs of the loop at any level (see Fig. 23–18b). This step is often called the **single effect.** This gradient could be produced by transport of salt out of the ascending limb and its deposition in the descending limb. Water is left behind (since the intervening membrane is water-impermeable), and so the osmolality of the ascending limb fluid decreases and that of the descending limb increases by the same amount. Next, we add new fluid to the loop (see Fig.

Stepwise shift of fluid | Development of "single effect"

A

| 300 | 300 | 300 | 300 | 300 | 300 | 300 | 300 |
| 300 | 300 | 300 | 300 | 300 | 300 | 300 | 300 |

B

| 200 | 200 | 200 | 200 | 200 | 200 | 200 | 200 |
| 400 | 400 | 400 | 400 | 400 | 400 | 400 | 400 |

C

| 200 | 200 | 200 | 200 | 400 | 400 | 400 | 400 |
| 300 | 300 | 300 | 300 | 400 | 400 | 400 | 400 |

D

| 150 | 150 | 150 | 150 | 300 | 300 | 300 | 300 |
| 350 | 350 | 350 | 350 | 500 | 500 | 500 | 500 |

E

| 150 | 150 | 300 | 300 | 300 | 300 | 500 | 500 |
| 300 | 300 | 350 | 350 | 350 | 350 | 500 | 500 |

F

| 125 | 125 | 225 | 225 | 225 | 225 | 400 | 400 |
| 325 | 325 | 425 | 425 | 425 | 425 | 600 | 600 |

G

| 125 | 225 | 225 | 225 | 225 | 400 | 400 | 600 |
| 300 | 325 | 325 | 425 | 425 | 425 | 425 | 600 |

H

| 112 | 175 | 175 | 225 | 225 | 313 | 313 | 500 |
| 312 | 375 | 375 | 425 | 425 | 513 | 513 | 700 |

Figure 23–18

The principle of countercurrent multiplication, based on the assumption that, at any level along the loop, an osmotic gradient of 200 mosm/kg H_2O can be established by salt transport between ascending and descending limbs. *(Modified from: Pitts, R.F. Physiology of the Kidney and Body Fluids, 3rd ed. Chicago: Year Book, 1974.)*

23–18*c*), as fresh fluid flows in from the proximal convoluted tubule. Again allow the 200 mosm/kg H_2O gradient to be established (Fig. 23–18*d*). Repeat the process several times (Fig. 23–18*e* to *h*). What is the final result? Notice that at any level of the loop a gradient of 200 mosm/kg H_2O exists. Along the length of the loop, however, a larger osmotic gradient (about 400 mosm/kg H_2O) has been established. This is what is meant by countercurrent multiplication; namely, a gradient existing at any level of the loop is multiplied by countercurrent flow so as to produce a larger gradient along the length of the loop.

The model just discussed should not be taken too literally. For example, flow through the loop of Henle is not a discontinuous process. Also, ascending and descending limbs do not share a common membrane; rather, there is an intervening interstitial space. The model is realistic, however, in that salt is transported out of the ascending limb across a water-impermeable membrane. The descending limb is water-permeable, and its osmolality is increased primarily by water removal, not by active deposition of salt into its lumen. Notice that fluid leaving the ascending limb is osmotically dilute, even in a kidney that will be forming a concentrated urine. This is so because salt has been removed along the ascending limb.

Another important feature of countercurrent multiplication is that it is an energy-consuming process. In other words, in order to establish a gradient, there must be an energy source. The energy source for operating the countercurrent multiplier is ultimately active sodium transport, which is linked to ATP hydrolysis. Note that the magnitude of the gradient established depends on the size of the single effect and the length of Henle's loop. If the single effect is reduced, or if the loops are short, then a large gradient along the length of the loop cannot be established.

Figure 23–19 shows a model for the operation of the countercurrent mechanism in the kidney. In this model, we are assuming that a maximally concentrated urine is formed. The numbers in blue give the osmotic concentration in mosm/kg H_2O. The boxed numbers in green give the relative amounts of filtered water present at various points along the nephron. The heavy blue shading along the ascending limb indicates a low water permeability. A juxtamedullary nephron, with a long loop of Henle extending to the tip of the papilla, is depicted.

Seventy percent of the filtered sodium and water are reabsorbed along the proximal convoluted tubule. This reabsorption is essentially isosmotic. Fluid entering the descending limb of Henle's loop has an osmolality of about 300 mosm/kg H_2O. The descending limb is water-permeable, and since the interstitium surrounding the tubules is hyperosmotic, water leaves the descending limb and the osmolality increases. Since NaCl is the main solute in the descending limb fluid, its concentration rises.

In the thick ascending limb of Henle's loop, there is a vigorous Na^+/K^+-ATPase whose activity results in sodium chloride deposition in the outer medulla. An osmotically dilute fluid, which contains mainly NaCl and urea, leaves the loop and enters the distal convoluted tubule. As the fluid traverses the cortical collecting ducts in the presence of ADH, water is reabsorbed. This is because water moves along its osmotic gradient from the tubule fluid into the cortical interstitium, where it is carried away by a high blood flow. Before the tubule fluid once again re-enters the outer medulla, it becomes isosmotic, and its volume is substantially reduced. The reabsorption of dilute fluid in the cortex is important because if a large volume of water were presented to the medulla, too

Figure 23–19

Summary of movements of water, ions, and urea in the kidney during elaboration of a maximally concentrated urine (1200 mosm/kg H_2O). Numerals boxed in blue give osmolality in mosm/kg H_2O. Numbers boxed in green give the relative amount of water present at each level of the nephron. Solid arrows indicate active transport; dashed arrows passive transport. The heavy blue outline along the thin ascending and thick ascending limbs of Henle's loop indicates that these segments are relatively water-impermeable.

much water would be added to the medullary interstitium, and the urine could not be maximally concentrated. The cortical collecting ducts are urea-impermeable, so the urea concentration rises.

As the fluid passes through the outer medulla, it becomes hyperosmotic as water is reabsorbed into the medullary interstitium. The urea concentration rises further. Next, when the urine enters the inner medulla, it encounters a urea-permeable segment. Urea diffuses out of the collecting ducts and accumulates in the inner medullary interstitium. The urea in the medulla can be thought of as osmotically balancing urea in the urine, so that other solutes can be osmotically balanced by the salt deposited in the

medulla. As fluid moves down the length of the inner medullary collecting duct, water is reabsorbed. In the presence of high levels of ADH, fluid in the collecting ducts achieves the same osmolality as in the surrounding interstitium. Urine of high osmolality and low volume is finally excreted.

Urea Plays a Special Role in the Concentrating Mechanism in the Inner Medulla

In the inner medulla there are thin, but no thick, ascending limbs. The epithelium of the thin ascending limb is flat, with few mitochondria. It does not look like the type of epithelium usually associated with vigorous salt transport. Urea is added to the inner medulla from the collecting ducts and from the calyceal urine that bathes the papilla. The osmotic pressure due to urea causes the removal of water along the descending limb and thereby concentrates NaCl in this descending limb fluid. When the NaCl-enriched fluid enters the NaCl-permeable (but water-impermeable) thin ascending limb, a concentration gradient for outward, passive movement of NaCl is established. This outward movement drives the countercurrent multiplier in the inner medulla. It results in net addition of solute to this region of the kidney, which is important in establishing the gradient in the medulla that will be used to concentrate urine in the collecting ducts.

The Vasa Recta Are Countercurrent Exchanges

The blood vessels of the medulla, the vasa recta, supply its nutritional needs. They also act as countercurrent exchangers. Countercurrent exchange is a passive process that helps to maintain a gradient established by some other means. For example, countercurrent exchange of heat between blood flowing into and out of the limbs allows the core body temperature to be maintained higher than that of fingers and toes. Blood flowing into and out of the medulla exchanges water and solutes between ascending and descending vasa recta (see Fig. 23–19). This exchange reduces the extent to which blood flowing into the medulla tends to dissipate the osmotic gradient. A relatively low blood flow to the kidney medulla (compared to the cortex) facilitates exchange and reduces washout of the medullary osmotic gradient.

The vasa recta are essential to the operation of the concentrating mechanism because they are responsible for removing water from the medulla. Water enters the medullary interstitium from the descending limbs of the loops of Henle and from the collecting ducts; there must be a pathway for removal of water —otherwise, it would accumulate there. The force for fluid uptake by the ascending vasa recta is their plasma colloid osmotic pressure and elevated concentration of small solutes.

Many Factors Influence the Ability to Form an Osmotically Concentrated Urine

Antidiuretic hormone is necessary for the production of an osmotically concentrated urine (Table 23–3). In the absence of ADH, the urine is not isosmotic to plasma, but is actually markedly hypo-osmotic. The range of urine osmotic concentration, from very di-

Table 23–3
Factors Affecting Urinary Concentrating Ability

Antidiuretic hormone (ADH)
Delivery of NaCl to ascending limb of Henle's loop
Reabsorption of NaCl by ascending limb
Delivery of fluid to medullary collecting ducts
Medullary blood flow
Urea
Length of Henle's loop

lute to very concentrated urine, is normally due to varying plasma levels of ADH. **Diabetes insipidus** is a condition in which there is an ADH deficiency, and as much as 20 liters of dilute urine may be excreted in a day. To survive, a person with a disorder of this magnitude would have to drink 80 glasses of water a day and would spend a large part of the day and night urinating and drinking. In **neurogenic (central, hypothalamic, pituitary) diabetes insipidus,** synthesis or secretion of ADH is inadequate. In **nephrogenic diabetes insipidus,** ADH is present, but the collecting ducts are unresponsive.

An adequate delivery of salt to the ascending limb is necessary for maximal urine concentration. If glomerular filtration rate is abnormally low—for example, following severe hemorrhage—then concentrating ability is impaired.

Not only must delivery of salt be adequate, but the rate of reabsorption of salt along the ascending limb must be maintained. After all, this is the single effect essential for countercurrent multiplication. The "loop" diuretic drugs inhibit salt reabsorption in the thick ascending limb; they decrease concentrating ability and cause a marked increase in salt and water excretion.

The urine can be maximally concentrated only if a trickle of fluid flows through the collecting duct system. If fluid reabsorption is impaired in early portions of the nephron, as occurs during an osmotic diuresis, then a maximal urine osmolality cannot be achieved.

Medullary blood flow can affect the osmotic gradient in the medulla and thus affect concentrating ability as well. An excessive blood flow to this region tends to wash out the gradient and thereby reduces the maximum urine osmolality that can be attained.

Urea plays a critical role in the concentrating mechanism, especially in the establishment of an osmotic gradient in the inner medulla. With a low-protein diet and the resulting decrease in urea production, concentrating ability is impaired.

Finally, the length of Henle's loops affects concentrating ability. Some desert-living mammals have very long loops for their body sizes and a correspondingly enhanced ability to concentrate the urine and save water.

Excretion of an Osmotically Dilute Urine Rids the Body of Excess Water

In principle, the production of an osmotically dilute urine is quite straightforward. All that is needed is the reabsorption of solute across a relatively water-impermeable membrane, with water being left behind. In the thick ascending limb, active reabsorption of sodium (accompanied by chloride) results in considerable dilution of the tubular fluid, as we have already seen. This segment is often called the **diluting segment.** In the absence of antidiuretic hormone, the collecting ducts are very water-impermeable. Even though the medullary interstitium is hyperosmotic (but not as much as in a concentrating kidney), collecting duct water mostly stays in the duct lumen. Continued reabsorption of salt results in further dilution of the urine along the collecting ducts. The end result is the excretion of a large volume of hypoosmotic urine.

Functions of Ureters and Urinary Bladder

The kidneys form urine constantly. Urine is conveyed to the urinary bladder by the ureters and stored there until the bladder is emptied by way of the urethra.

The Ureters Transmit the Urine to the Urinary Bladder

The two ureters are muscular tubes that carry the urine from the pelvis of each kidney to the urinary bladder. Peristaltic movements originating in the pelvis and passing down the ureters force the urine toward the bladder. These muscular contractions are probably myogenic in origin. They help overcome the gradually increasing pressure as urine accumulates in the bladder. The ureters enter the base of the bladder obliquely, thus forming a valvular flap that passively prevents reflux of urine during bladder contraction. The ureters are innervated by sympathetic and parasympathetic nerve fibers. Afferent sensory nerve fibers are present, as evidenced by the excruciating pain felt when a stone is in the ureter.

The Urinary Bladder Stores the Urine and Is Periodically Emptied

The urinary bladder (see Fig. 10–8) is a distensible, hollow organ containing smooth muscle in its wall. The muscle is called the **detrusor,** which means in Latin "that which pushes down." The neck of the bladder **(internal sphincter)** also contains smooth muscle. The body of the bladder and bladder neck are innervated by parasympathetic **pelvic nerves** and sympathetic **hypogastric nerves.** The external sphincter is composed of skeletal muscle and is innervated via somatic nerves, the **pudendal nerves.** Pelvic, hypogastric, and pudendal nerves contain both motor and sensory fibers.

The bladder has two functions: (1) to serve as a distensible reservoir for storage of the urine and (2) to empty its contents at suitable intervals. As the bladder fills, it adjusts its wall tension to its capacity, so that minimal increases in bladder pressure occur. The external sphincter is kept closed via impulses along the pudendal nerves. The first awareness of bladder filling occurs at a volume of about 100 to 150 ml, and the desire to void is usually experienced when the bladder contains 150 to 250 ml of urine. A person becomes uncomfortably aware of a full bladder when the volume is 350 to 400 ml; at this volume, pressure in the bladder is about 10 cm of water. With further increases in volume, bladder pressure rises steeply, in part due to reflex contractions of the detrusor. An increase in volume to 700 ml creates pain and often loss of control. The sensations of bladder filling, of conscious desire to void, and of painful distension are mediated by afferent fibers in the pelvic nerves.

Micturition Is a Complex Act

The periodic complete emptying of the urinary bladder is called **micturition** (or **urination**). It involves coordinated activity over both autonomic and somatic nerve pathways and several reflexes that can either be facilitated or inhibited by higher centers in the brain. The basic reflexes occur at the level of the sacral spinal cord and are modified by centers in the midbrain and cerebral cortex. Distension is sensed by stretch receptors in the bladder wall; these induce reflex contraction of the bladder and relaxation of internal and external sphincters. This reflex is released by removing inhibitory impulses from the cerebral cortex. Fluid flow through the urethra reflexly causes further contraction of the detrusor and relaxation of the external sphincter. Increased parasympathetic nerve activity is responsible for contraction of the detrusor and relaxation of the internal sphincter. The perineal and levator ani muscles relax during micturition; this shortens the urethra and decreases its resistance. Descent of the diaphragm and contraction of abdominal muscles raise intraabdominal pressure, thus aiding in the expulsion of urine from the bladder.

Fortunately, micturition is under voluntary control. In young children, however, it is purely reflex and occurs whenever the bladder is sufficiently distended. At about 2½ years of age, it begins to come under cortical control, and complete control is usually achieved by 3 years of age.

CHAPTER REVIEW

Summary

- The kidneys play a dominant role in regulating the volume and composition of the extracellular fluid, the environment of the living body cells.
- The basic unit of kidney structure and function is the nephron.
- The juxtaglomerular apparatus consists of specialized cells thought to play a role in regulating glomerular blood flow and filtration rate.
- The kidneys have a huge blood flow because of the need to sustain a high rate of plasma filtration.
- Urine formation starts with the filtration of plasma in the kidney glomeruli. Glomerular filtration is favored by the high hydrostatic pressure of the blood in the glomerular capillaries and is opposed by the hydrostatic pressure in the urinary space of Bowman's capsule and by the glomerular capillary colloid osmotic pressure. Glomerular filtration is rather nonselective; proteins are mostly retained in the plasma by the glomerular filtration barrier, but all low-molecular-weight substances are freely filtered. Glomerular filtration rate is most accurately measured by determining the inulin clearance.
- Glucose is reabsorbed by the proximal tubule. Reabsorption is active and sodium-dependent and shows a maximal rate *(Tm)*. Normally, all of the filtered glucose is reabsorbed. Urea is filtered by the kidney glomeruli and is passively reabsorbed by the kidney tubules. Active reabsorption of sodium provides the driving force for tubular reabsorption of water, glucose, amino acids, chloride, and phosphate.
- Some organic compounds are secreted from the blood surrounding the kidney tubules into the tubular urine.
- Most of the filtered sodium is reabsorbed by the kidney tubules; the quantity of sodium excreted plays an important role in body sodium balance. The proximal convoluted tubule reabsorbs the greatest fraction (70%) of the filtered sodium and water; the tubule fluid stays essentially isosmotic to plasma in this nephron segment. The loop of Henle reabsorbs about 20% of the filtered

sodium and 10% of the filtered water. The distal convoluted tubule and collecting ducts reabsorb about 9% of the filtered sodium and 19% of the filtered water. The collecting ducts are the site of final regulation of sodium and water excretion; aldosterone and antidiuretic hormone (ADH) increase sodium and water reabsorption, respectively, by the collecting ducts.

- Potassium is filtered, reabsorbed, and secreted by the kidney tubules. The cortical collecting duct is an important site in the determination of potassium secretion, and hence of excretion.
- The kidneys play a major role in regulating the concentrations of calcium, magnesium, and inorganic phosphate in the plasma.
- The mammalian kidney can form a urine that is osmotically more dilute or concentrated than plasma. The formation of an osmotically concentrated urine depends on the establishment of an osmotic gradient in the medulla by the loops of Henle. These structures act as countercurrent multipliers. The vasa recta act as countercurrent exchangers and remove water from the kidney medulla. The addition of urea to the inner medulla allows for efficient operation of the urinary concentrating mechanism.
- Factors affecting urinary concentrating ability are ADH, salt reabsorption by the ascending limb of Henle's loop, tubule fluid and blood flows, urea, and the length of Henle's loops.
- Production of an osmotically dilute urine involves the reabsorption of solute across a membrane, with water left behind.
- The two ureters are muscular tubes that carry the urine from the kidneys to the bladder. The urinary bladder functions as a reservoir for urine and is periodically emptied (micturition). Micturition is a complex act involving autonomic and somatic nerves, spinal reflexes, and higher brain centers.

Review Questions

Identify Each with a Term

1. The basic unit of kidney structure and function.
2. The liquid collected when plasma is pushed through a very fine filter.
3. The barrier separating glomerular capillary blood from the urinary space of Bowman's capsule.
4. The volume of plasma per unit of time needed to sup-

ply the quantity of a substance excreted in the urine per unit of time.

5. The plasma level at which glucose first appears in the urine in appreciable amounts.
6. The maximal rate of PAH secretion.
7. The process that leads to establishment of an osmotic gradient in the kidney medulla.
8. The disorder in which ADH synthesis or secretion is inadequate.

Define Each Term

9. afferent arteriole
10. juxtaglomerular apparatus
11. proteinuria
12. urethra
13. glomerular filtration rate (GFR)
14. tubular reabsorption
15. tubular secretion
16. filtered load
17. renal autoregulation
18. juxtamedullary nephron

Choose the Correct Answer

19. Which of the following produces renin?
    a. granular cells
    b. intercalated cells
    c. macula densa
    d. podocytes
20. Which of the following produces an increase in renal blood flow?
    a. angiotensin II
    b. norepinephrine
    c. prostaglandin I_2
    d. thromboxane
21. Approximately what percentage of filtered water is usually reabsorbed by the kidney tubules?
    a. 1%
    b. 20%
    c. 70%
    d. 99%
22. In which nephron segment is PAH secreted?
    a. collecting duct
    b. distal convoluted tubule
    c. proximal tubule
    d. thick ascending limb
    e. all of the above
23. Which of the following substances normally has the highest renal plasma clearance?
    a. glucose
    b. inulin
    c. PAH
    d. urea
24. Active reabsorption of glucose across the luminal membrane of proximal tubule cells is accomplished by:
    a. a glucose pump
    b. facilitated diffusion
    c. glucose-sodium cotransport
    d. simple diffusion
25. About 70% of filtered sodium is reabsorbed in the:
    a. collecting duct
    b. distal convoluted tubule

c. loop of Henle
d. proximal convoluted tubule

26. Which of the following is associated with an increased ability to concentrate the urine osmotically?
    a. administration of a "loop" diuretic drug
    b. a low protein diet
    c. an increased medullary blood flow
    d. long loops of Henle

Calculate

27. The following values were obtained in a renal clearance study on a young, 50-kg woman:

Plasma creatinine concentration	0.010 mg/ml
Urine creatinine concentration	0.80 mg/ml
Urine volume	720 ml
Duration of urine collection	12 hr

    What is her glomerular filtration rate in ml plasma/min?

28. If glomerular filtration rate is 125 ml/min, glomerular hydrostatic pressure is 55 mm Hg, hydrostatic pressure in Bowman's space is 15 mm Hg, and mean glomerular capillary colloid osmotic pressure is 30 mm Hg, what is the glomerular filtration coefficient?

29. The following data were collected in a renal clearance study on a young, 70-kg man:

Plasma inulin concentration	0.30 mg/ml
Urine inulin concentration	7.50 mg/ml
Plasma glucose concentration	4.00 mg/ml
Urine glucose concentration	25.0 mg/ml
Plasma PAH concentration	0.02 mg/ml
Urine PAH concentration	2.64 mg/ml
Urine flow rate	5.00 ml/min
Blood hematocrit	0.45

    a. What is the glomerular filtration rate?
    b. What is the glucose clearance?
    c. What is the rate of tubular reabsorption of glucose?
    d. What is the PAH clearance?
    e. Assuming that the PAH clearance is equal to 90% of the renal plasma flow, what is the true renal plasma flow?
    f. What is the renal blood flow?

Draw

30. Draw a nephron and collecting duct system and the associated blood vessels; label afferent arteriole, Bowman's capsule, connecting tubule, cortical collecting duct, distal convoluted tubule, efferent arteriole, glomerulus, inner medullary collecting duct, loop of Henle, macula densa, outer medullary collecting duct, peritubular capillaries, proximal convoluted tubule, proximal straight tubule, thick ascending limb, and thin limb.

Answer Each Question in One or Two Sentences

31. Why does a loss of fixed negative charges from the glomerular filtration barrier lead to proteinuria?
32. What factors account for the very high rate of plasma filtration in the glomeruli of the human kidney?
33. How will the following affect the rate of glomerular fil-

tration: (1) a decrease in K_f, (2) a decrease in glomerular capillary hydrostatic pressure, (3) an increase in hydrostatic pressure in the urinary space of Bowman's capsule, (4) an increase in plasma colloid osmotic pressure, and (5) afferent arteriolar constriction?

34. Why does a fall in glomerular filtration rate lead to a rise in plasma creatinine concentration?

35. Why is uncontrolled diabetes mellitus often accompanied by glucosuria?

36. Why can the clearance of PAH be used to estimate renal plasma flow?

37. What are the pathways for sodium reabsorption across the proximal tubule epithelium?

38. How do sodium and water reabsorption differ in the proximal convoluted tubule and collecting duct?

39. How do the following affect potassium excretion: (1) an increase in plasma aldosterone concentration, (2) an increase in plasma potassium concentration, (3) an increase in transtubular potential difference in the cortical collecting duct (lumen more negative), and (4) an increase in tubule fluid flow rate in the cortical collecting duct?

40. What are the functions of the vasa recta?

41. What special role does urea play in the operation of the countercurrent mechanism in the inner medulla?

42. What are two functions of the urinary bladder?

Problem Solving in Physiology

43. Why is the urea clearance not a suitable measure of glomerular filtration rate?

44. Why are diseases of the kidney medulla often associated with polyuria (excretion of more than 2 L urine/day)?

45. Sodium reabsorption is the main energy consuming process in the kidneys. Explain.

Suggested Readings

Hladky, S.B., and Rink, T.J. *Body Fluid and Kidney Physiology.* London: E. Arnold, 1986.

Koeppen, B.M., and Stanton, B.A. *Renal Physiology.* St. Louis: Mosby-Year Book, 1992.

Kriz, W., and Bankir, L. "A standard nomenclature for structures of the kidney." *Am J Physiol* 254:F1–F8, 1988.

Pitts, R.F., *Physiology of the Kidney and Body Fluids,* 3rd ed. Chicago: Year Book Medical Publishers, 1974.

Rose, B.D. *Clinical Physiology of Acid-Base and Electrolyte Disorders,* 4th ed. New York: McGraw-Hill, 1994.

Seldin, D.W., and Giebisch, G.H. (eds.) *The Kidney: Physiology and Pathophysiology,* 2nd ed. New York: Raven Press, 1992.

Smith, H.W. *From Fish to Philosopher.* Boston: Little, Brown and Company, 1953.

Valtin, H., and Schafer, J.A. *Renal Function,* 3rd ed. Boston: Little, Brown and Company, 1995.

Vander, A.J. *Renal Physiology,* 5th ed. New York: McGraw-Hill Book Company, 1995.

Key Terms

cortical nephron (p. 706)
countercurrent exchange (p. 728)
countercurrent multiplication (p. 728)
diabetes insipidus (p. 731)
endogenous creatinine clearance (p. 717)
excretion (p. 711)
glomerular filtration (p. 710)

glomerular filtration coefficient (p. 714)
glucose threshold (p. 719)
glucosuria (p. 721)
juxtaglomerular apparatus (p. 709)
juxtamedullary nephron (p. 707)
micturition (p. 732)
nephron (p. 706)

peritubular capillary Starling forces (p. 725)
proteinuria (p. 722)
renal autoregulation (p. 710)
renal plasma clearance (p. 716)
tubular reabsorption (p. 711)

tubular secretion (p. 711)
tubular transport maximum (p. 719)
tubuloglomerular feedback hypothesis (p. 709)
ultrafiltration (p. 710)
vasa recta (p. 708)

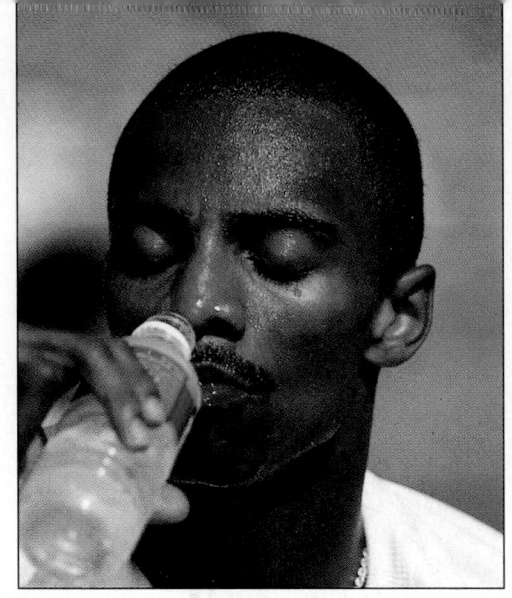

A man drinking to replace water and salt lost by sweating.
(© David Madison)

Chapter 24

Regulation of Fluid and Electrolyte Balance

Key Concepts

Two thirds of the body water is in intracellular fluid and one third is in extracellular fluid; intracellular and extracellular fluids are in osmotic equilibrium.

The kidneys help to keep us in water, sodium ion, and potassium ion balance, by excreting the appropriate amounts of these substances in the urine.

Renal failure results in an abnormal internal environment. Dialysis and renal transplantation are used to treat patients with terminal (end-stage) renal disease.

Chapter Outline

Body Fluids and Fluid Compartments
The Concept of Fluid and Electrolyte Balance
Kidney Failure and Its Treatment

Two young, apparently healthy prisoners in a penal institution surreptitiously obtained a supply of what was described as an antifreeze mixture (ethylene glycol). How much was drunk and for how long is not known, but after eight days, one prisoner was admitted to the hospital with kidney pain on both sides and shortness of breath; red cells and albumin were found in his urine. His blood pressure was 170/100 mm Hg. On the twelfth day, renal clearances of PAH and inulin were essentially zero; during four hours, he excreted less than 5 ml urine. The blood urea nitrogen concentration was 266 mg/dL (normal 8 to 20 mg/dL), and plasma potassium concentration was 8.8 mEq/liter (normal 3.5 to 5.5 mEq/liter). He died suddenly of cardiac arrest. The second prisoner was admitted to the hospital one day after his companion, with essentially the same clinical findings. He was put on an artificial kidney machine, appeared to be doing better, but on the twenty-fourth day became disoriented and died.

These two cases, taken from a classic study by Dr. Jean Oliver and his coworkers (Journal of Clinical Investigation 30:1305–1439, 1951), illustrate acute renal failure produced by a nephrotoxic chemical that damages the kidneys. When the kidneys fail, a host of abnormalities result; the people described above had edema, hypertension, and elevated plasma concentrationsof urea and potassium. Acute renal failure, if untreated by adequate dialysis, may be lethal within a matter of days.

In this chapter we will discuss the volume and composition of the various body fluids. Then we will discuss how we keep water and sodium and potassium ions in balance. We will indicate some of the body disturbances that occur when our kidneys fail and discuss some of the methods used for treating patients with renal failure.

Body Fluids and Fluid Compartments

The fluids of the body can be divided into two categories: **intracellular fluid** (fluid within cells) and **extracellular fluid** (fluid outside of cells). These fluids differ strikingly in composition and in volume. Because most cell membranes are water permeable, however, the total solute concentration inside and outside of our cells is the same.

Body Water Is Distributed in Several Fluid Compartments

The body fluids contain both solutes and water, but in terms of amount or volume occupied, water is the major constituent. Therefore, we will consider the water content of the body first, and later cover the various solutes dissolved in the body fluids.

The percentage of total body weight that is water varies from 45% to 75% in different individuals. This range mostly reflects differences in the amount of body fat. The water content of adipose (fat) tissue is about 10% by weight; most other tissues contain 70% to 75% water by weight. An obese individual has a low percentage of body weight made up of water, and a lean individual has a high percentage. Young adult males average about 60% water by weight; young adult females average about 50% water. The difference is due to females' smaller muscle mass and greater amount of subcutaneous fat. With aging, the percentage of body weight that is water decreases because water-rich muscle tissue tends to be replaced by water-poor adipose tissue.

Total body water is distributed in several divisions or compartments (Fig. 24–1). Approximately two thirds of body water is contained in the **intracellular fluid compartment,** and one third in the **extracellular fluid compartment.**

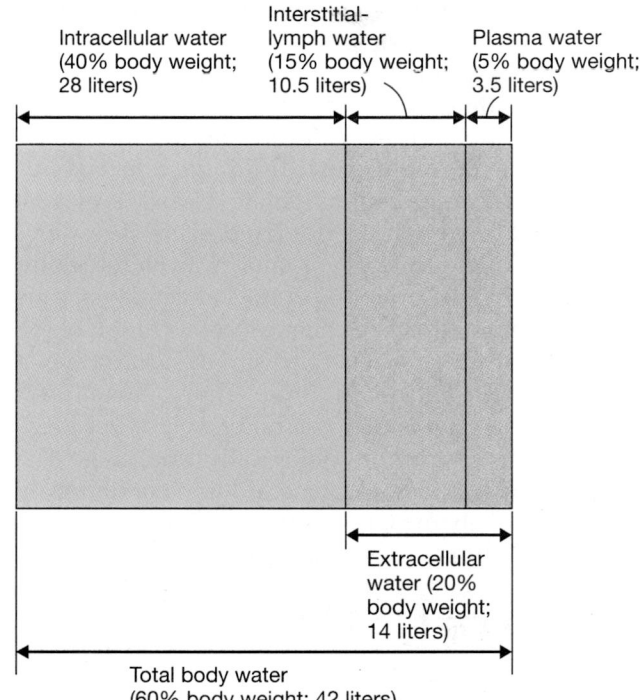

Figure 24–1

Distribution of water in the body of an average, young adult, 70-kg male. In an average, young adult, 55-kg female, total body water is 50% of body weight, intracellular water is 30% of body weight, and extracellular water is 20% of body weight.

The intracellular fluid compartment is not one continuous space, but is made up of trillions of cells. The intracellular space is separated from the extracellular space by cell plasma membranes. If we assume an idealized, young adult, 70-kg male, in whom total body water is 60% of body weight, we can calculate that total body water is 42 kg (or 42 liters), intracellular water is 28 liters, and extracellular water is 14 liters.

The extracellular fluid can be further divided into two major subcompartments, separated from each other by the endothelium of the blood vessels. Blood vessels contain **blood plasma,** the part of the blood exclusive of blood cells and platelets. The plasma is about 93% water (by weight or volume). Plasma water accounts for about one quarter of the extracellular water (3.5 liters in a 70-kg, young adult male). Outside of the blood vessels are the **interstitial fluid** and **lymph.** Interstitial fluid is the fluid directly bathing most body cells, and lymph is the fluid in lymphatic vessels. The interstitial fluid and lymph together contain three quarters of the extracellular water (10.5 liters in a 70-kg male). Blood plasma, interstitial fluid, and lymph are nearly identical in chemical composition, except for a higher protein concentration in plasma.

Figure 24–1 does not show an additional small extracellular fluid compartment, the **transcellular fluid.** This fluid compartment includes specialized fluids such as **cerebrospinal fluid,** the **aqueous humor** of the eye, secretions of the digestive glands, sweat, and renal tubular fluid and urine. These fluids are separated from the plasma by an endothelium as well as by a continuous epithelial cell layer. This epithelium modifies the chemical composition of the transcellular fluid, so it is not a simple ultrafiltrate of plasma, as is interstitial fluid. Transcellular fluid amounts to only 1% to 3% of body weight but is extremely important, physiologically speaking. The transcellular fluids, in general, are continuously formed, and abnormal loss of these fluids or blockage of fluid drainage may have serious consequences for the organism.

Body Fluids Differ in Electrolyte Composition

The body fluids contain many uncharged organic substances (e.g., glucose, urea), but in terms of concentration, electrolytes are quantitatively more important. Electrolytes contribute most to the total solute concentration (osmolality) of the body fluids and consequently are crucial in the distribution of body water.

The concentrations of various electrolytes in plasma, interstitial fluid, and intracellular fluid are summarized in Table 24–1. The intracellular fluid values are based on measurements in skeletal muscle cells. These cells were chosen as a representative cell type because they account for about two thirds of the cell mass in the human body. Concentrations are expressed in terms of milliequivalents per liter solution or per kg H_2O. An **equivalent** is equal to the product of the moles and valence of an ion; one **milliequivalent (mEq)** is equal to 1/1000 Eq. For singly charged **(univalent)** ions, 1 mEq is the same as 1 millimole (mmole). Thus, we can express plasma sodium concentration as 142 mEq per liter or 142 mmole per liter. For doubly charged **(divalent)** ions, 2 mEq is the same as 1 mmole. For example, a plasma Ca^{2+} concentration of 5 mEq per liter is the same as 2.5 mmole per liter. Some electrolytes, such as proteins, are **polyvalent,** so there are several milliequivalents per millimole. The virtue of expressing concentration in terms of milliequivalents per liter is that in any fluid compartment the sum of the positive ions (cations) must equal the sum of the negative ions (anions), because of the requirement of electroneutrality in solutions. This allows us to calculate the concentration of unmeasured anions ("Others") in Table 24–1.

Plasma concentrations are listed in the first column of Table 24–1. Sodium is the major cation in plasma, and chloride and bicarbonate are the major anions. The plasma proteins (mostly serum albumin) bear a net negative charge at physiological pH. The

---

Table 24–1

Electrolyte Composition of the Body Fluids

Electrolytes	(1) Plasma (mEq/liter)	(2) Plasma Water (mEq/kg H_2O)	(3) Interstitial Fluid (mEq/kg H_2O)	(4) Intracellular Fluid (Skeletal Muscle) (mEq/kg H_2O)
Cations				
Na^+	142	153	145	10
K^+	4	4.3	4	159
Ca^{2+}	5	5.4	3	1
Mg^{2+}	2	2.2	2	40
Total	153	165	154	210
Anions				
Cl^-	103	111	117	3
HCO_3^-	25	27	28	7
Protein	17	18	—	45
Others	8	9	9	155
Total	153	165	154	210

electrolytes are actually dissolved in the watery phase of plasma, so concentrations in plasma water are presented in the second column of Table 24–1. These values were calculated by assuming a plasma water content of 93%. For example, for Na^+, 142 mEq per liter plasma divided by 0.93 liter water per liter plasma equals 153 mEq per liter water. Since a liter of water weighs 1 kg, the concentrations could also be expressed per kg of water.

The interstitial fluid (third column of Table 24–1) is an ultrafiltrate of plasma. It contains essentially the same concentrations as plasma of all low-molecular-weight substances, but little protein. The differences in concentrations between the small ions in plasma water and interstitial fluid (compare columns two and three in Table 24–1) arise because of differences in protein concentration between these two fluids. Two factors are involved. The first is an electrostatic effect: because the large plasma proteins are negatively charged and kept within the circulation, there is a lower concentration of small diffusible cations and a higher concentration of small diffusible anions in a plasma ultrafiltrate. Second, Ca^{2+} and Mg^{2+} are also bound to some extent (about 40% and 30%, respectively) to plasma proteins; consequently the total concentrations of these ions in interstitial fluid are lower than in plasma.

If we examine intracellular fluid composition (column four of Table 24–1), major differences between this fluid and the extracellular fluids are apparent. The concentrations of potassium, magnesium, and protein in the cell are much higher than in the surrounding interstitial fluid. The concentrations of sodium, calcium, chloride, and bicarbonate are much lower in the cell. The anions in muscle cells labeled "Others" are mainly proteins and organic phosphate compounds (e.g., creatine phosphate, ATP), and various other anions to which the cell membrane is mostly impermeable.

The high internal potassium concentration and low internal sodium concentration are maintained by activity of the cell membrane Na^+/K^+-ATPase, which extrudes sodium ions and takes up potassium ions (Chapter 4). The lower intracellular concentrations of chloride and bicarbonate can be mostly accounted for by the membrane potential difference, approximately -90 mv, which favors the outward movement of these negatively charged ions.

The distribution of ions within the cell is not uniform; different organelles may have different ion concentrations. For example, the cell nucleus has a higher sodium concentration than the cell as a whole. In skeletal muscle, most of the intracellular calcium is in the sarcoplasmic reticulum and mitochondria, and the cytoplasmic calcium concentration is very low.

Do not be misled by the higher number of milliequivalents per kg H_2O in the cell than in the interstitial fluid (compare totals in columns three and four of Table 24–1). A single protein molecule or phosphate compound may bear several negative charges, so the number of milliequivalents in the cell is greater than the number of milliosmoles. Also, magnesium is divalent (40 mEq per kg H_2O corresponds to 20 mmoles per kg H_2O) and is largely protein-bound within the cell, so it is not effective osmotically. It is generally accepted that the total solute concentrations (osmolalities) in the cell and surrounding extracellular fluid are identical. Water moves freely through most plasma membranes, so there is a condition of osmotic equilibrium between cellular and extracellular fluids.

The Osmolalities of Intracellular and Extracellular Fluids Are Normally Identical

Osmotic pressure is of prime importance in determining the distribution of water between cells and extracellular fluid. Osmotic pressure is directly related to total solute concentration (osmolality). An increased solute concentration lowers the activity (or concentration) of water; water moves from a region of lower osmolality to one of higher osmolality. Such water movement could be prevented by applying a hydrostatic pressure to the more concentrated solution. Animal cells cannot employ such a mechanism, since a hydrostatic pressure difference of a few centimeters of H_2O leads to rupture of the plasma membrane. In multicellular animals, osmotic pressures inside and outside of cells are equal. If solute or water is added to or lost from the extracellular fluid, osmotic equilibrium will be temporarily upset. Water will move into or out of the cells until a new osmotic equilibrium is achieved.

Most of the volume in any body fluid compartment is occupied by water. At a given osmolality, the total volume of (or amount of water in) a compartment is directly related to the amount of solute in the compartment. This important relationship follows from the equation that defines the term *concentration* —namely, concentration = amount/volume. If solute concentration (osmolality) is maintained constant, then the volume must change directly with the amount of solute.

In the extracellular fluid, the amount of sodium present is a major determinant of the volume of water present. This relationship results because sodium and its accompanying anions, chloride and bicarbonate, are the major osmotically active solutes present in extracellular fluid. If we add up the concentrations of these ions in interstitial fluid (see Table 24–1), we get $145 + 117 + 28 = 290$ mmoles per kg H_2O. The corresponding osmolality is about 270 mosm per kg H_2O. (This figure is lower than 290 because of interactions among ions in solution; it is only in infinitely dilute solutions that ions behave as completely independent particles.) Since interstitial fluid (or plasma) osmolality averages 287 mosm per kg H_2O (approximately 300 mosm per kg H_2O), it is apparent that sodium salts account for more than 90% of the osmolality. The osmolality of the extracellular fluid is closely regulated by the thirst sensation and control of renal excretion of water by ADH. Consequently, if osmolality (sodium concentration) is kept constant, it follows that the amount of extracellular sodium will determine the amount of extracellular water. To express it another way, if a person takes in extra salt (NaCl), thirst will result, water will be ingested, and extracellular fluid volume will be increased. If a person loses salt, extracellular fluid volume will be decreased.

Cell volume, also mostly water, is similarly influenced by the amount of contained solute. In the cell compartment, potassium is the major osmotically active solute. Hence, a loss of potassium from the cell results in a decrease in cell volume, and a gain in potassium results in an increase in cell volume. The large amount of impermeable anions (e. g., proteins, organic phosphates) within cells could affect cell volume. In theory, these solutes would cause a redistribution of small, permeable ions so that an excess of solute would be present in the cell, and, therefore, water would tend to enter the cell. This tendency is, however, counteracted by the activity of the Na^+/K^+-ATPase, which effectively excludes sodium and thereby lowers the total amount of solute in the cell. In this way, the Na^+/K^+-ATPase plays a key role in the regulation of cell volume. If Na^+/K^+-ATPase activity is decreased (e.g., by low temperature, oxygen lack, metabolic poisons, or cardiac glycosides), cells gain sodium and water, and they swell.

An interesting phenomenon has been recently described: when certain (e.g., proximal tubule) cells are placed in a hypo-osmotic medium, they initially swell (as expected). Soon, however, they shrink back to their original size. This response is due to loss of potassium, chloride, and water and appears to be caused by activation of membrane ion transport processes. Opposite changes are observed in hyperosmotic media (cells initially shrink and then increase back to normal size). These responses demonstrate that cells have special mechanisms for regulating and preserving their volumes (Chapter 4).

(a) Normal

(b) Add pure water

(c) Add isotonic saline

(d) Add pure NaCl

Figure 24–2

The effects of adding pure water (*b*), isotonic saline (*c*), and pure NaCl (*d*) to the extracellular fluid on osmolalities and volumes of the intracellular fluid (ICF) and extracellular fluid (ECF). (*a*) is the normal condition.

The normal distribution of water between cellular and extracellular compartments changes in a variety of circumstances. Figure 24–2 provides some examples. Note that the height of the boxes corresponds to solute concentration. The width of the boxes corresponds to volume. The area of each box (height × width) corresponds to the total amount of solute in a compartment   (concentration × volume = amount). The dotted lines represent the normal condition and are the same in all four parts of this figure. The solid lines represent conditions after a new osmotic equilibrium has been achieved. Note that the height of the boxes is always identical in intracellular and extracellular fluids, since osmotic equilibrium is attained.

Figure 24–2a shows the normal condition. Osmolalities are identical in cellular and extracellular compartments (300 mosm per kg H_2O). The volume of the intracellular compartment (40% of body weight) is twice the volume of the extracellular compartment (20% of body weight).

In Figure 24–2b, pure water is added to the extracellular fluid; ingestion of water would have the same effect. Since the extracellular fluid osmolality is lowered, water moves into the cells until the osmolality in both compartments is lowered to the same level. The entry of water into the cells increases their volume. Since the cells contain two thirds of the body fluid solutes, two thirds of the added water enters the cells, and one third remains in the extracellular space.

In Figure 24–2c, isotonic saline is added to the extracellular fluid. This can be done by intravenous infusion or ingestion of a 0.9% NaCl solution. Note that the final osmolality is not different from normal. This is not surprising, since isotonic saline is also isosmotic. Note also that cell volume does not change, since, by definition, cells do not shrink or swell when exposed to an isotonic solution. All of the isotonic saline is retained in the extracellular fluid and accordingly increases only this volume.

Although it is not illustrated, a loss of isotonic fluid from the extracellular compartment has predictable consequences: no change in body fluid osmolality or in cell volume and a decrease in extracellular fluid volume. Hemorrhage and diarrhea are two cases in which an isotonic fluid is lost from the body.

Figure 24–2d shows the effect of adding sodium chloride without water to the extracellular compartment. Extra salt intake (e.g., eating salted potato chips) would be a corresponding circumstance. The added salt increases the osmolality of the extracellular fluid, which results in water movement out of the cells. Extracellular fluid volume is increased at the expense of cell volume. At equilibrium, osmolality is uniformly increased. When pure NaCl (or a hypertonic NaCl solution) is added to the extracellular

fluid, the NaCl is diluted by the total body water, even though all of the NaCl stays in the extracellular fluid.

The Distribution of Water Between Blood Plasma and Interstitial Fluid Is Determined by Starling Forces

The relative volumes of plasma and interstitial fluid are primarily determined by the Starling forces that operate across capillary walls. As discussed in Chapter 19, movement of fluid across capillary walls is determined by the balance between capillary and tissue hydrostatic and colloid osmotic pressures. The hydrostatic pressure in the capillaries tends to push fluid out of the capillaries; the colloid osmotic pressure due to the plasma proteins acts to retain fluid within the vascular system. An increase in capillary hydrostatic pressure, a decrease in plasma colloid osmotic pressure, or an increase in capillary membrane permeability to proteins would favor net movement of fluid out of the vascular system and result in a decreased plasma volume and an increased interstitial fluid volume (edema). On the other hand, a decrease in capillary hydrostatic pressure or an increase in plasma colloid osmotic pressure would favor movement of interstitial fluid into the circulation. In most vascular beds, the net force favoring outward movement of the fluid at the arterial end of the capillaries is roughly the same as the net force favoring inward movement of fluid at the venous end of the capillaries. Therefore, little fluid leaves the circulation. Extra fluid that has left the circulation is returned by the lymphatics. Inadequate lymphatic drainage can lead to an appreciable increase in size of the interstitial space.

The Concept of Fluid and Electrolyte Balance

Despite varying daily intake of food and water, the volume and composition of body fluids remain constant. This suggests that we stay in balance with respect to many substances. A person in a **stable balance** maintains the same amount of a specified substance in the body over a period of time. The rates of gain or input (due to intake or production in the body) and loss or output (due to excretion or destruction in the body) of a particular substance are exactly equal, so there is no net accumulation or loss.

If fluids and electrolytes are not in balance, two alternative conditions occur. If input exceeds output, then accumulation of a substance in the body results, and a **positive balance** exists. For example, if the intake of salt exceeds the rate at which it is excreted, then salt and water will accumulate in the body and extracellular fluid volume and body weight will increase. If output exceeds input, then net loss of a substance from the body results, and a **negative balance** exists. For example, if urinary excretion exceeds the daily intake of potassium, a fall in body potassium and plasma potassium concentration results.

The kidneys play a major role in maintaining balance with respect to water, mineral electrolytes, hydrogen ions, and various organic compounds. They accomplish this by adjusting the output of these substances in the urine to match the rate of addition to the body. Renal excretion of water is under the continuous control of the antidiuretic hormone and is normally adjusted to maintain water balance. The kidneys are the primary site of control of electrolyte output. The mineral electrolytes, such as sodium, potassium, and phosphate, are ingested in variable amounts. They are not produced or destroyed in the body, and so, in a balanced system, the amounts taken in are excreted, mostly in the urine. Some substances, such as hydrogen ions or urea, are produced by metabolic reactions in the body. Since these substances should not normally accumulate, they are excreted at a rate matching their production rate.

Water Input and Output Are Normally Equal

Table 24–2 presents a balance chart for water for an average 70-kg adult. We assume that the person is in water balance—that is, total input and output are equal (2500 ml per day). The input of water is normally derived from the diet. In a hospital setting, intravenous infusions may also be a source of water (and electrolytes). The various beverages we drink contain water. This intake of water is largely conditioned by habit but can also be stimulated by the thirst sensation. Water is also contained in solid food. For example, most fruits and vegetables are 80% to

90% water, and a hamburger is about 55% water. Water is also produced by oxidation of foodstuffs in the body. For example, for each mole of glucose oxidized, six moles of water are produced. (Recall from Chapter 6 the following reaction: $C_6H_{12}O_6 + 6\ O_2 \rightarrow 6\ CO_2 + 6\ H_2O$.)

On the output side, we need to consider loss of water by way of the skin, lungs, gastrointestinal tract, and kidneys. Some water is always lost by way of the lungs and skin. This loss is usually not sensed and so is called **insensible water loss.** On a cold day, however, it is possible to perceive this water loss, because exhaled air is saturated with water vapor, and the water condenses to form visible droplets in cold air. Large quantities of water also can be lost from the skin through sweating. In a hot environment or during heavy exercise, as much as 4 liters of water per hour can be lost in perspiration. When we perspire, we lose not only water, but electrolytes, especially sodium chloride.

Gastrointestinal losses of water are normally small. The salivary glands, stomach, liver, pancreas, and intestine add about 7 liters of fluid to the gastrointestinal tract every day, but almost all of this is absorbed, and only 100 ml of water is normally lost in the feces (Chapter 22). With vomiting or diarrhea, large losses of water (and electrolytes) can result.

Finally, the kidneys are a site of water loss. Normally, renal water loss is regulated to maintain water balance. If there is a water deficit, renal excretion of water is decreased. If there is an excess of water in the body, renal excretion of water is increased. Renal excretion of water is controlled by the antidiuretic hormone (ADH).

The data for water balance illustrated in Table 24–2 approximate the magnitude of water inputs and outputs in a normal adult. You should recognize, however, that water intake and output vary widely, even in normal people, depending on conditions such as climate and activity. Children have greater need for water and are more susceptible to dehydration than adults, because children have a higher surface

Table 24–2
Daily Water Balance in an Average 70-kg Adult

Input		Output	
Water in beverages	1000 ml	Skin and lungs	900 ml
Water in food	1200 ml	Gastrointestinal tract (feces)	100 ml
Water of oxidation	300 ml	Kidneys (urine)	1500 ml
Total	2500 ml	Total	2500 ml

area per unit volume and a greater metabolic rate for their size.

The regulation of body water balance involves two basic mechanisms: (1) the thirst sensation and (2) control of renal water excretion by the antidiuretic hormone.

Thirst and the Control of Fluid Intake

Thirst is a conscious sensation associated with a craving for drink and is ordinarily interpreted as a desire for water. The function of the thirst mechanism is to repair a fluid deficit.

People drink largely by habit, and such drinking normally covers water needs. Thirst is an emergency mechanism not ordinarily experienced. In other words, the thirst threshold is usually not reached. Constant thirst in temperate climates and with moderate activity may indicate an underlying disease or disorder.

The thirst sensation is interesting because—unlike many sensations—it reaches the level of consciousness. With thirst, we can sense how body mechanisms generally act to bring about normalization of body conditions. The greater the need for water, the greater the intensity of thirst and the tendency for corrective action.

In human beings, thirst is primarily associated with dryness of the mouth and throat, due to a reflex decrease in secretion by salivary and buccal (cheek) glands. A dry mouth plays a role in the conscious definition of thirst. It is also important in metering water intake and so helps us know when to stop drinking. A dry mouth is not essential in regulating water intake.

Thirst is primarily controlled by centers in the hypothalamus. The hypothalamic centers relay information to the cerebral cortex, where thirst becomes a conscious sensation. Interestingly, the hypothalamus also controls release of ADH and is therefore concerned with regulating both water intake and output. The hypothalamus is concerned with several other functions related to water balance: temperature regulation, food intake, and cardiovascular function.

There are two major stimuli for thirst: cellular dehydration and extracellular dehydration.

Cellular dehydration means a decrease in water within cells, or cell shrinkage. If the effective osmotic pressure of the plasma is increased, water will move out of all body cells. In the anterior hypothalamus there are **osmoreceptor cells;** when these neurons shrink, they signal the cerebral cortex to give rise to the thirst sensation. Not all solutes are effective stimuli for the osmoreceptor cells. For example, urea and ethanol are ineffective because they readily penetrate the osmoreceptor cells and therefore do not cause

them to shrink. Sodium chloride is an effective stimulus. An increase in plasma osmolality of 1% to 2% is needed to reach the thirst threshold.

Extracellular dehydration, which is synonymous with **hypovolemia,** or a decrease in blood volume, stimulates thirst. To be more exact, a decrease in **effective arterial blood volume** or **effective circulating blood volume** is the stimulus here. The effective arterial blood volume is the volume of blood pumped by the heart into the arterial system that perfuses the tissues. People with severe congestive heart failure may experience intense thirst, even though their blood volume is typically greater than normal, because, due to inadequate cardiac function, their effective arterial blood volume is decreased. Thirst is stimulated by loss of isosmotic fluid such as occurs with hemorrhage, diarrhea, or vomiting.

There appear to be several receptors for hypovolemia. The various baroreceptor areas in the cardiovascular system (Chapter 19), such as the arterial baroreceptors in the carotid sinus and aortic arch, and the stretch (volume) receptors in the cardiac atria and great veins in the thorax, are involved. These receptors transmit impulses to the thirst control centers in the hypothalamus. The kidneys may also act as volume receptors. When blood volume is decreased, the kidneys release renin into the circulation. This results in production of angiotensin II, which acts on neurons near the third ventricle of the brain to stimulate thirst. The kidneys not only participate in water balance by adjusting water output, but also affect water intake.

Extracellular dehydration has to be considerable before thirst is stimulated. Blood donors usually do not experience thirst after donating 500 ml of blood (10% of blood volume). If a person loses a lot of blood, however, intense thirst results.

The two stimuli, cellular dehydration and extracellular dehydration, usually work together and reinforce each other. For example, if you have been active in hot weather and have lost a lot of water by sweating, plasma osmolality will be increased and blood volume will be decreased; thirst will be stimulated by both pathways. In some situations, however, the two signals may oppose each other. For example, as stated before, people with congestive heart failure experience thirst because of a decrease in effective arterial blood volume. When they drink water, their kidneys tend to retain it in the body because ADH secretion is also stimulated. The result is that the body fluids become hypo-osmotic. Because the plasma sodium is diluted by excess water, the condition that develops is often called **hyponatremia** (low sodium concentration in the blood). In this situation, plasma osmolality is lower than normal, yet thirst persists. Apparently, the body's attempt to restore an effective arterial blood volume is a more powerful drive, and so plasma osmolality is sacrificed.

Antidiuretic Hormone and the Control of Water Output

Antidiuretic hormone (ADH) plays a key role in controlling water excretion by the kidneys by facilitating the reabsorption of water by the collecting ducts (Chapter 23). This peptide hormone is produced in the hypothalamus and released from the posterior pituitary into the blood (Chapter 13). The two major factors affecting ADH release are the same as those affecting thirst: (1) cellular dehydration and (2) extracellular dehydration.

The main stimulus under ordinary conditions is cellular dehydration, caused by an increase in effective plasma osmolality. When plasma osmolality is increased, osmoreceptor cells in the anterior hypothalamus (distinct from those that cause thirst) shrink and signal the nearby ADH-producing cells to release ADH from their nerve terminals in the posterior pituitary.

The threshold for ADH release is normally exceeded, and so the plasma contains ADH at a measurable level (about 2.5 pg per ml). This plasma level is associated with a modestly concentrated urine (urine osmolality about 600 mosm per kg H_2O). This means that plasma ADH can be induced to fall (by lowering the plasma osmolality) or rise (by raising the plasma osmolality). Thus, changes in plasma osmolality are continuously translated into changes in plasma ADH levels and, in turn, into changes in renal water excretion. The fact that the osmotic stimulus threshold for ADH release is considerably lower than that for thirst is consistent with the view that thirst is mainly an emergency mechanism.

Extracellular dehydration (hypovolemia or a decrease in effective arterial blood volume) stimulates ADH release. The receptors include stretch receptors in the left atrium of the heart and in the pulmonary veins within the pericardium; less stretch results in fewer impulses transmitted to the brain via vagal afferents. An increase in intrathoracic blood volume inhibits ADH release and produces an increase in urine flow. The baroreceptors in the carotid sinus and aortic arch also affect ADH release. A fall in pressure at these sites stimulates ADH release. Relatively large changes in blood volume (about 10%) are required to affect ADH release. A large blood loss (e.g., 15% to 20% of blood volume), however, produces very large increases in plasma ADH. With severe hemorrhage, the high circulating levels of ADH may cause a significant rise in blood pressure due to the vasoconstrictor properties of this peptide.

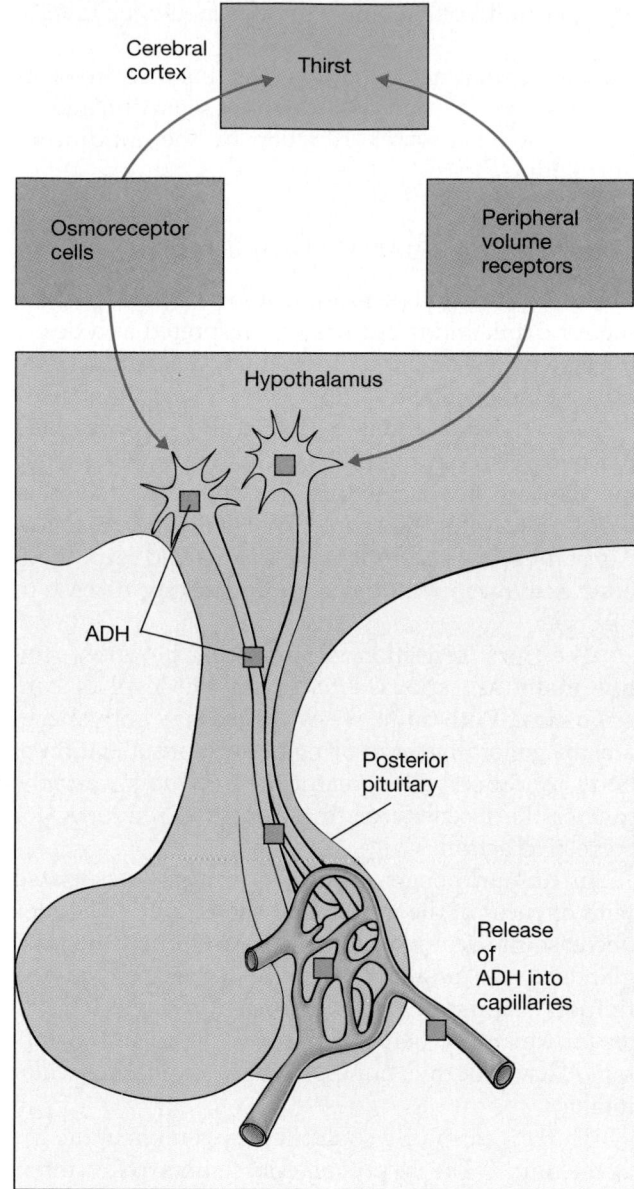

Figure 24–3

Schema showing sites controlling thirst and ADH release. The cell bodies of the ADH-producing cells are located in the supraoptic and paraventricular nuclei of the hypothalamus, and their nerve terminals are located in the posterior pituitary.

The two stimuli, cellular and extracellular dehydration, often work synergistically in promoting ADH release. But (as discussed previously) in certain conditions (e.g., congestive heart failure) they may act in opposite directions. Other stimuli, primarily unrelated to water balance control, may also affect ADH release, and were discussed in Chapter 13. Figure 24–3 is a schema for the control of thirst and ADH release by osmo- and volume receptors.

Figure 24–4

Mechanism of action of ADH on the collecting duct epithelium.

The primary target for ADH action in the kidney is the collecting duct epithelium (Fig. 24–4). In the absence of ADH, the luminal membrane is relatively water impermeable, and fluid in the lumen is hypoosmotic to the cell or blood. ADH combines with a receptor in the basolateral cell membrane. By way of a guanine nucleotide stimulatory protein, the membrane-bound enzyme adenylate cyclase is activated. This enzyme catalyzes the formation of cyclic AMP from ATP. In turn, cyclic AMP activates a protein kinase that phosphorylates other proteins. The subsequent steps are not entirely clear, but the final result is an increase in water permeability of the luminal cell membrane. This appears to be due to fusion of intracellular vesicles, which contain water channel proteins, with the luminal cell membrane. When the collecting duct epithelium becomes water permeable, fluid in the duct lumen tends to become concentrated by osmotic withdrawal of water.

Renal Responses to Extra Water Intake or to Dehydration

Renal excretion of water is continuously adjusted by changes in plasma ADH levels, so that plasma osmolality and extracellular fluid volume are kept at normal levels. To illustrate this adjustment, we will consider the response to two extreme states: (1) ingestion of a liter of tap water and (2) dehydration due to sweating.

Figure 24–5*a* summarizes the response to ingesting a liter of tap water. In this case, the body is threat-ened by dilution and excess volume, and the appropriate renal response is excretion of a large volume of osmotically dilute urine, called a **water diuresis.** The ingested water is absorbed chiefly in the small intestine, which leads to a fall in plasma osmolality. The fall in plasma osmolality causes osmoreceptors in the hypothalamus to swell, resulting in decreased release of ADH from the posterior pituitary. The increased volume of the extracellular fluid, resulting from the water ingestion, also acts (by way of stretch receptors in the cardiovascular system) to inhibit ADH release. (This pathway appears to be less important under normal conditions.) When ADH release is decreased, plasma ADH falls, because this hormone is continuously destroyed in the body. With a low plasma ADH level, the collecting duct expresses its intrinsically low water permeability. Filtered water is then not reabsorbed at this site. The result is the excretion of a large volume (up to 20 ml per minute) of osmotically dilute urine. Urine osmolality may be as low as 30 to 40 mosm per kg H_2O. The increase in urine flow starts a few minutes after water ingestion, and the entire extra load of water is excreted in a few hours. By getting rid of the extra water, the kidneys are able to restore plasma osmolality and volume back to normal.

Consider now what happens in the opposite case, dehydration (see Fig. 24–5*b*). Suppose you exercise in hot weather, and you sweat a lot. The sweat you lose is a hypo-osmotic fluid, and so the body fluids become hyperosmotic. In addition to water lost in sweat, there is also insensible water loss from the lungs and skin. These losses lead to a rise in plasma osmolality, stimulation of hypothalamic osmoreceptors, and increased release of ADH. Also, a fall in extracellular volume causes the cardiovascular stretch receptors to increase ADH release reflexly. The rise in plasma ADH levels causes increased water reabsorption from the collecting ducts, with resulting excretion of the urinary solutes in a hyperosmotic urine. By excreting the urinary solutes in a minimum volume, the body saves water, which can delay or minimize the effects of dehydration. To correct the dehydration, the thirst mechanism, stimulated by the same changes that affect ADH release, is engaged. Thirst stimulates water intake, and so the water deficit is repaired.

The Kidneys Are the Major Site of Control of Sodium Balance

Sodium Input and Output

In a normal person on a normal diet, the amount of sodium ingested is closely matched by the amount

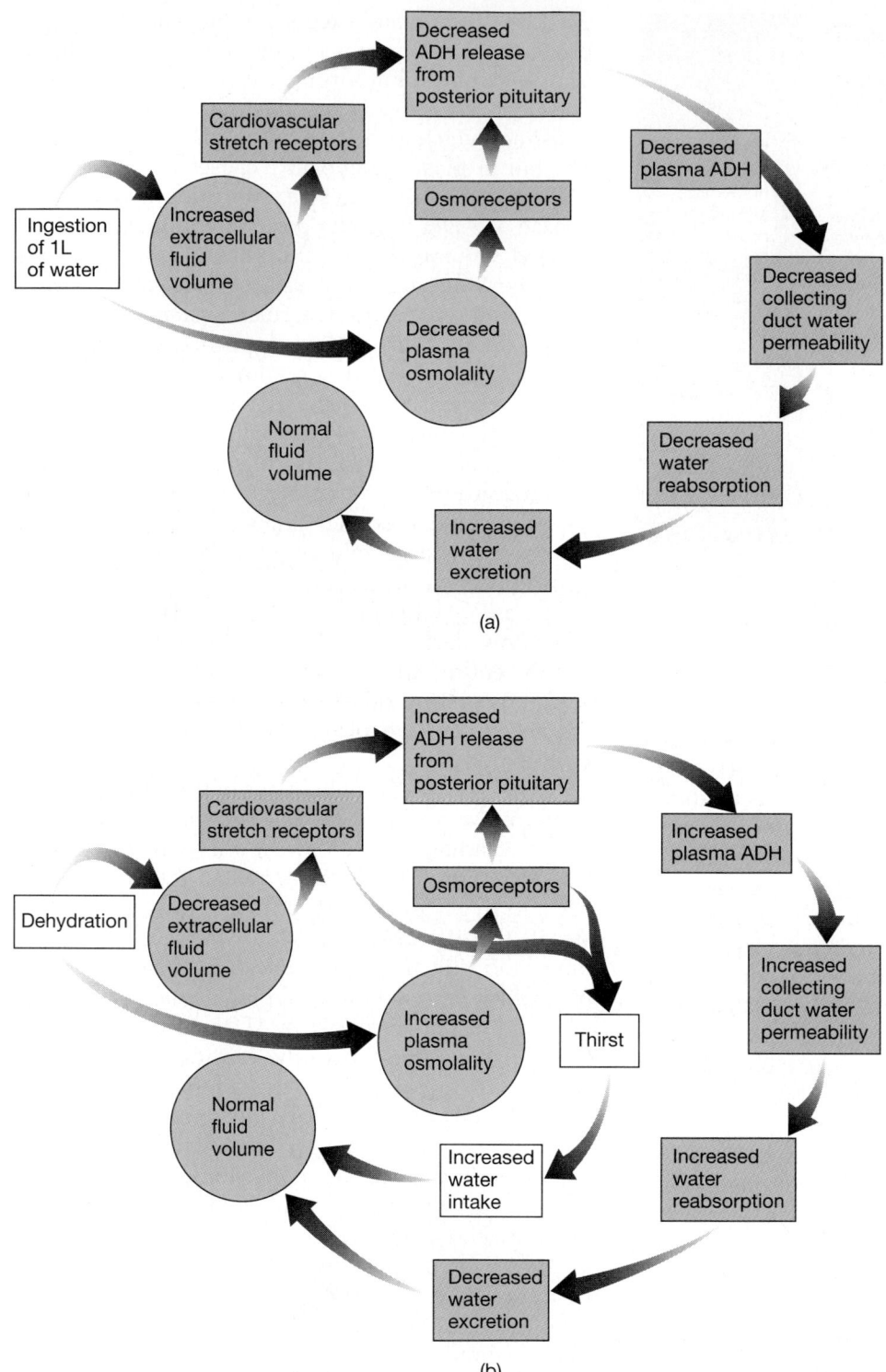

(a)

(b)

Figure 24–5

(*a*) Response to excess hydration. (*b*) Response to dehydration.

of sodium lost from the body. Figure 24–6 summarizes sodium balance relationships. On the input side, sodium intake in the diet may be quite variable, but in an average American diet, sodium amounts to about 100 to 300 mEq per day (6 to 18 g NaCl per day). In a hospital setting, intravenous fluids can also be a source of sodium.

On the output side, sodium may be lost from the skin, gastrointestinal tract, or kidneys. Loss of

sodium from the skin may occur with sweating, burns, or hemorrhage. Sweat contains 5 to 80 mEq of sodium per liter, and so with heavy exercise or in warm environments, an appreciable amount of sodium may be lost in sweat. Gastrointestinal losses of sodium are normally small, but with diarrhea or vomiting, large losses may result. The urine normally accounts for about 95% of the total output of sodium from the body. The kidneys constitute the major site

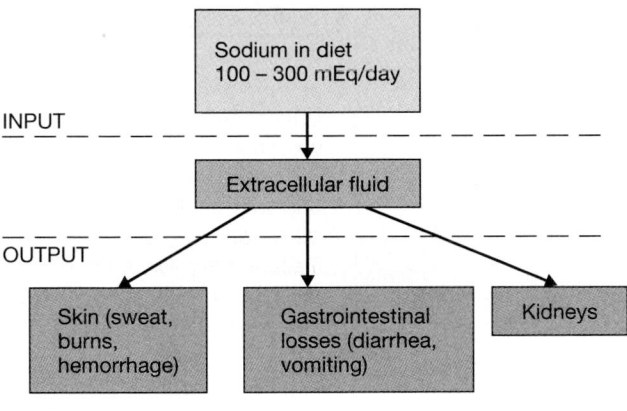

Figure 24–6

Sodium balance. The kidneys are usually responsible for 95% of the sodium output.

of control of sodium output and normally adjust excretion to maintain balance.

A positive sodium balance results if sodium input exceeds output. Since water is retained along with sodium, and since sodium is primarily an extracellular ion, extracellular fluid volume will be increased. If sufficient salt and water are retained, expansion of the interstitial fluid space will lead to generalized edema. This condition is associated with weight gain (due to the extra water) and skin puffiness, especially in the feet and ankles (due to the effects of gravity). A positive sodium balance and generalized edema occur in a number of clinically important conditions, including congestive heart failure, hepatic cirrhosis (a liver disease), and nephrotic syndrome (a kidney disease). In order to reduce the edema, it is common practice to (1) put the patient on a low salt diet and (2) use diuretic drugs to promote sodium excretion by the kidneys. In this way, the physician tries to relieve the edema and bring the patient back into sodium balance.

A negative sodium balance results if sodium output exceeds input. This may occur if there is excessive loss of sodium via the skin, gastrointestinal tract, or kidneys. A negative sodium balance leads to a reduction of extracellular fluid volume. In turn, this produces a decrease in plasma volume and consequently may lead to an inadequate circulation, hypotension, and even death.

It should be clear that sodium balance is closely related to regulation of extracellular fluid volume. This is so because sodium salts are the major solutes of the extracellular fluid, as discussed earlier. Since the osmolality of the extracellular fluid is closely regulated by ADH, the kidneys, and thirst, it follows that retention or loss of sodium from the body will be ac-

companied by parallel changes in the amount of water in (and hence volume of) the extracellular fluid. The regulation of sodium balance falls largely upon the kidneys, although there is some evidence for a **sodium appetite** (which stimulates sodium intake) when a sodium deficit exists. Since the kidneys are intimately concerned with control of sodium balance and regulation of extracellular fluid volume, it should not be surprising that extracellular fluid volume is a major determinant of renal sodium excretion. We will elaborate on this later.

There is compelling evidence that hypertension (high blood pressure) may often be due to a disturbance in sodium (salt) balance. Excessive dietary intake of salt or inadequate renal excretion of salt tends to increase intravascular fluid volume; somehow this is translated into an increase in blood pressure. Avoidance of a high salt intake may help to prevent high blood pressure. A reduced salt intake and diuretic drugs are used to keep blood pressure under control in people with hypertension.

The Renal Responses to Changes in Dietary Intake of Sodium

Let us consider the renal response to changes in dietary intake of sodium. Figure 24–7 illustrates a balance study. A person was put on a fixed sodium intake of 100 mEq per day. During the first three days of the study (the control period), the person was in a stable balance, as indicated by the fact that urinary excretion matched the sodium intake (in this study, we will ignore extrarenal losses of sodium, which are usually small). On day 4, the sodium intake was increased to 300 mEq per day and was maintained at that level until the end of day 10. What happened? Sodium excretion rose, but for the first few days (days 4 to 7) was less than the sodium input. Consequently, there was a phase of positive sodium balance. During this time, the person retained salt, and since the subject was allowed free access to water, water intake increased, and salt and water were retained as an isosmotic solution. If we measured plasma sodium concentration, we would find that it was essentially constant throughout this study. The sodium content of the body, the extracellular fluid volume, and body weight, however, increased during days 4 to 7. During days 8 to 10, the subject was back in a stable balance; that is, input and output of sodium were once again equal. When the person switched to a low sodium diet (10 mEq per day), sodium excretion fell. During days 11 to 14, sodium excretion was higher than sodium intake, and so a negative sodium balance resulted. During this time, extracellular fluid volume and body weight decreased. Finally, a new stable balance on the low salt diet was established (days 15 and 16).

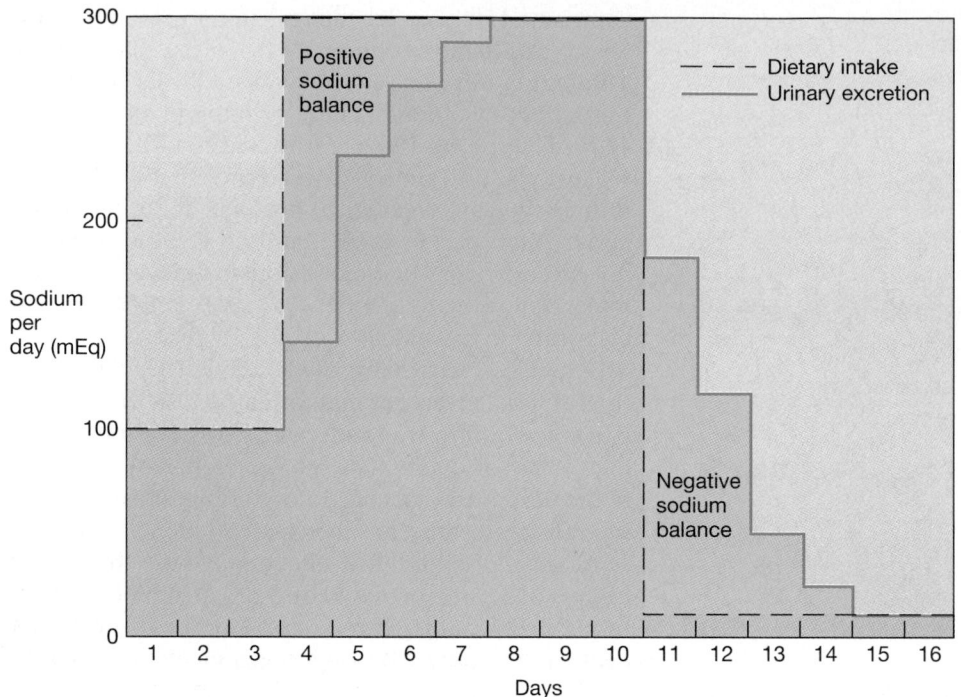

Figure 24–7

A sodium balance study. Dietary intake of sodium was initially 100 mEq per day, was then increased to 300 mEq per day (days 4 to 10), and was finally reduced to 10 mEq per day (days 11 to 16). Daily urinary excretion of sodium (solid blue lines) was measured.

This study demonstrates that changing the dietary sodium intake leads to appropriate changes in sodium excretion by the kidneys. An increase in sodium intake leads to increased urinary sodium excretion, and a decrease in sodium intake results in decreased sodium excretion. The kidneys can adjust sodium excretion over a wide range. Note, however, that the renal response is sluggish, and so it takes a few days to reach the appropriate rate of renal sodium excretion. During the time that renal sodium excretion is less than sodium intake (days 4 to 7), extracellular fluid volume is increasing. Extracellular fluid volume is stable, but higher than the control volume, during days 8 to 10. The increased extracellular fluid volume may be thought of as the stimulus that promotes renal sodium loss. When the sodium intake is decreased, a fall in extracellular fluid volume may be thought of as the cause of the reduced sodium excretion.

Effector Mechanisms Activated by Altered Extracellular Fluid Volume

What mechanisms allow the kidneys to adjust their output of sodium to maintain balance and relative stability of extracellular fluid volume? Figure 24–8 summarizes how sodium excretion may be controlled by a negative feedback system. According to this scheme, the regulated variable is extracellular fluid volume. Changes in extracellular fluid volume are sensed by cardiovascular volume receptors and by the kidneys. The effector mechanisms include changes in (1) glomerular filtration rate, (2) plasma aldosterone levels, (3) peritubular capillary Starling forces, (4) renal sympathetic nerve activity, (5) intrarenal blood flow distribution, and (6) plasma atrial natriuretic peptide. There may be additional effector mechanisms. Changes in these mechanisms lead to changes in sodium excretion and bring extracellular fluid volume back to normal.

We will first discuss the six effector mechanisms just listed and then the nature of the sensors and what they may actually be sensing. Sodium excretion represents the difference between filtered and reabsorbed amounts, and so factors that affect sodium excretion may act on the glomeruli, kidney tubules, or both.

Glomerular Filtration Rate. Glomerular filtration, as discussed in Chapter 23 delivers an enormous amount of sodium to the tubules. Most of this is reabsorbed. Increases or decreases in glomerular filtration rate (GFR) lead to directionally similar changes in urinary sodium excretion. Expansion of the extracellular fluid volume with isotonic saline tends to raise GFR and so may contribute to increased sodium excretion. An increase in GFR actually leads to only a slight increase in sodium excretion, due to a phenomenon called **glomerulotubular balance.** In response to an increase in GFR, the proximal convoluted tubule and loop of Henle increase their rates of sodium reabsorption so that excessive loss of sodium does not result.

Severe hemorrhage is accompanied by a fall in blood pressure and renal vasoconstriction and hence a fall in GFR. The kidney tubules can then reabsorb

Figure 24–8

Regulation of extracellular fluid volume (or effective arterial blood volume) by a negative feedback control system. Changes in effective arterial blood volume are sensed by cardiovascular and renal sensors, and are translated into changes in renal effector pathways, leading to changes in sodium excretion that bring effective arterial blood volume back to normal.

the reduced filtered sodium load more completely, and sodium excretion therefore decreases. This is a valuable adjustment, because, with hemorrhage, the kidneys ideally should retain sodium to aid restoration of the depleted extracellular volume.

Aldosterone and the Renin-Angiotensin System. Aldosterone is the most important mineralocorticoid hormone in human beings. It is a salt-retaining steroid produced by the zona glomerulosa region of the adrenal cortex (Chapter 14). It increases the reabsorption of sodium (and secretion of potassium) by a variety of epithelial tissues, including sweat glands, salivary glands, and intestine. In the kidney, it acts chiefly on the collecting ducts. In the absence of aldosterone (in animals whose adrenal glands have been removed), the collecting ducts fail to lower the sodium concentration in the urine, and an abnormal quantity of sodium is lost. Instead of reabsorbing

99.6% of the filtered sodium load, 98.0% of the filtered sodium load may be reabsorbed in the absence of aldosterone. This may seem like a small difference. But remember that if 24,000 mmoles of sodium are filtered in a day, a difference of 1.6% amounts to 384 mmoles sodium, roughly the quantity of sodium in 2.5 liters of extracellular fluid. Maintained excretion of this amount of sodium over a period of days would result in serious (and even fatal) depletion of the extracellular fluid. To some extent, the effects of such salt loss in animals without their adrenal glands can be reversed by a high sodium chloride intake.

The most important pathway controlling aldosterone secretion in response to volume changes is the renin-angiotensin system (Chapter 14). Briefly, a decrease in extracellular fluid volume (or to be more exact, a decrease in effective arterial blood volume) stimulates renin release from the granular cells of the juxtaglomerular apparatus (see Fig. 23–5). Increased renin release may be due to three mechanisms: (1) a decrease in blood pressure at the level of the afferent arteriole (renal baroreceptor mechanism), (2) an increase in activity of renal sympathetic nerves that innervate the granular cells, and (3) a decrease in sodium chloride delivery to and transport by the macula densa cells. All three mechanisms are activated by volume depletion—for example, due to hemorrhage. **Renin,** a proteolytic enzyme, acts on a plasma protein, **angiotensinogen,** to produce **angiotensin I** (Fig. 24–9). Angiotensin I is converted to the active peptide, **angiotensin II,** by an enzyme called **angiotensin converting enzyme,** mainly as blood passes through the lungs. Angiotensin II has several important actions. First, it is a powerful vasoconstrictor and so raises blood pressure. Second, it increases the synthesis and release of aldosterone by a direct action on the adrenal cortex. Aldosterone causes enhanced sodium reabsorption by distal parts of the nephron. Third, angiotensin II directly stimulates sodium reabsorption by the kidney proximal tubule. Fourth, angiotensin II acts on the thirst center in the hypothalamus, stimulating increased water intake. Fifth, angiotensin II may stimulate ADH release from the posterior pituitary, resulting in renal water retention. In all of these actions, angiotensin II helps restore a normal blood pressure and extracellular fluid volume. In the day-to-day control of sodium balance, changes in aldosterone secretion, mediated by changes in renin release, are probably most important in bringing about changes in sodium excretion. Aldosterone secretion is also stimulated directly by a fall in plasma sodium concentration of the blood perfusing the adrenal cortex.

Peritubular Capillary Starling Forces. In Chapter 23 we learned that changes in hydrostatic and colloid

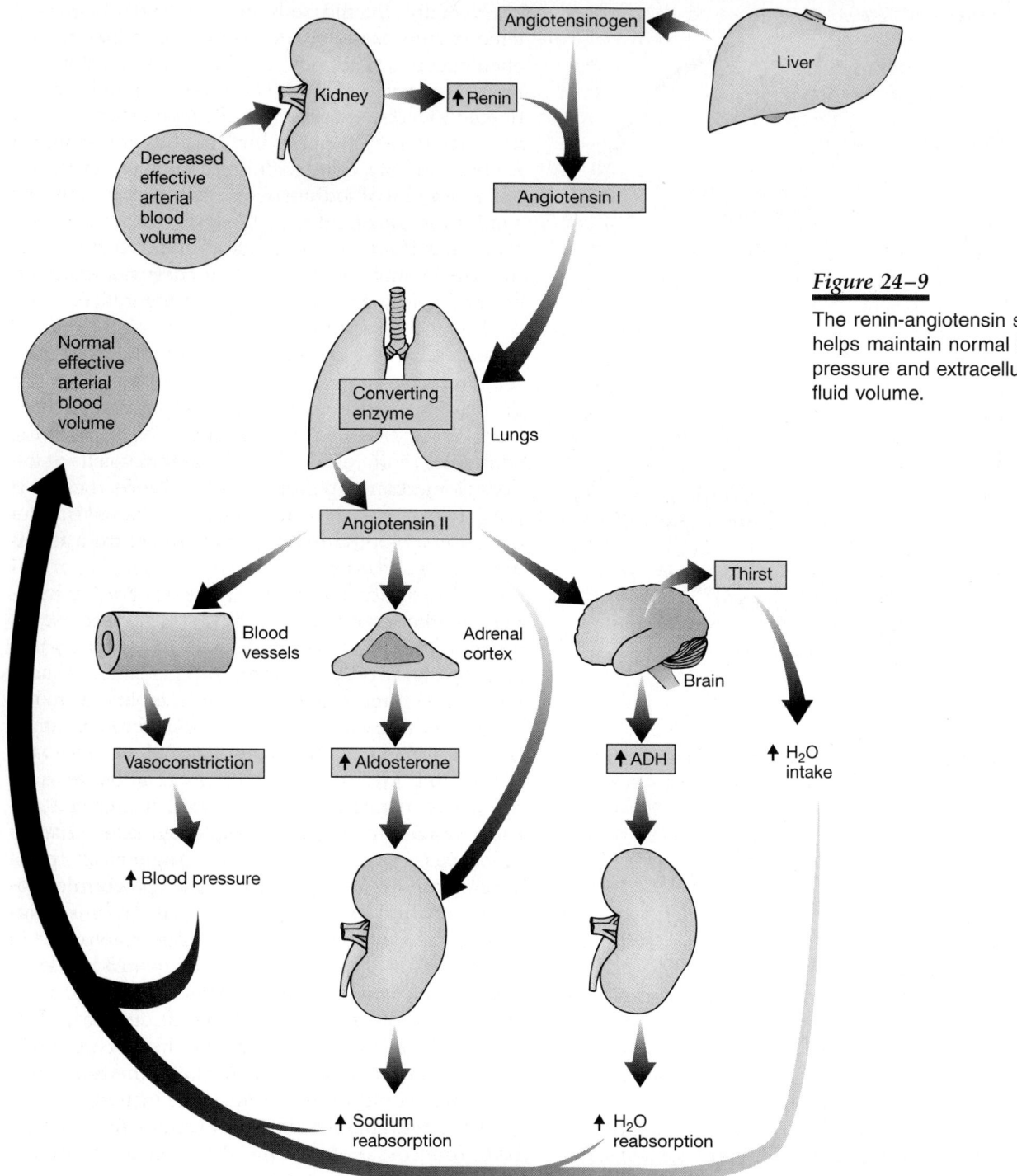

Figure 24–9

The renin-angiotensin system helps maintain normal blood pressure and extracellular fluid volume.

osmotic pressures in the peritubular capillaries affect the net reabsorption of salt and water by the proximal convoluted tubule. Specifically, an increase in hydrostatic pressure or a decrease in colloid osmotic pressure reduces fluid uptake. Such a change occurs, for example, when a large volume of isotonic saline is added to the body. The result is an increase in sodium excretion, an appropriate response.

Renal Sympathetic Nerve Activity.   Stimulation of renal sympathetic nerves results in a decrease in renal sodium excretion. If stimulation is intense, the fall in sodium excretion may be explained largely by hemodynamic changes—that is, by a fall in glomerular filtration rate and renal blood flow. Low levels of renal sympathetic nerve stimulation, at intensities below those needed to produce measurable hemodynamic changes, also result in decreased sodium excretion. This latter response may be due to direct stimulation of tubular cells, so that they reabsorb more sodium. There is anatomical evidence that renal sympathetic nerve fibers lie in close proximity to tubular cells. Inhibition of renal sympathetic nerve activity—for example, by volume expansion—accordingly promotes sodium excretion.

Change in Intrarenal Blood Flow Distribution. Changes in filtration rate and blood flow in different populations of nephrons could conceivably alter sodium excretion. This idea is consistent with the notion that not all nephrons in the kidney are alike, so a shift from cortical to juxtamedullary nephron populations could result in different rates of sodium excretion. At present, it is not certain whether this mechanism is important.

Atrial Natriuretic Peptide. Recent studies indicate that the atria of the heart produce a peptide called **atrial natriuretic peptide (ANP).** ANP is released by the atria in response to mechanical stretch—for example, by an increase in blood volume—and it increases sodium excretion (see the Research Focus Box on p. 752).

Clearly, multiple effector mechanisms influence sodium excretion. The relative importance of these mechanisms under normal conditions and in various disease states is not fully understood.

The Nature of the Stimuli and the Location of Receptors Affecting Sodium Excretion

What exactly is the sensed variable that leads to changes in renal sodium excretion? The idea that it is the extracellular fluid volume seems a satisfactory explanation in a normal person. There are, however, a number of disease states in which generalized edema develops because the kidneys retain salt and water; sodium is conserved despite an expanded extracellular fluid volume. The argument has been presented that it is not extracellular fluid volume per se, but a component thereof, the plasma volume, that is regulated. But this argument is not convincing, because blood volume is commonly increased in patients with congestive heart failure whose kidneys avidly retain sodium.

To explain these difficulties, researchers have postulated that the effective arterial blood volume is the regulated variable that affects sodium excretion. The effective arterial blood volume is an index of the degree of filling of the systemic arterial tree. It determines whether blood flow will be adequate to meet the metabolic demands of the tissues. In congestive heart failure, the heart muscle is weak and does not adequately transfer blood from the venous to the arterial side of the circulation. Blood tends to "dam up" in back of a failing ventricle, which decreases effective arterial blood volume. Effective arterial blood volume is also decreased in other conditions in which generalized edema occurs, such as liver cirrhosis, malnutrition, arteriovenous fistula, and nephrotic syndrome. Effective arterial blood volume or extracellular fluid volume is decreased with hemorrhage, diarrhea, severe sweating, and low salt intake. Therefore, all of these conditions are accompanied by renal salt retention.

Where are the volume receptors? There is convincing evidence that various cardiovascular stretch receptors can affect sodium excretion. These include stretch receptors in the wall of the left atrium of the heart and in intrathoracic veins. An increase in volume of blood in the thorax is a well-known stimulus that increases sodium excretion. The arterial baroreceptors may also function as stretch receptors affecting sodium excretion.

The kidneys may act as volume sensors. When effective arterial blood volume is decreased, renal blood flow is threatened. The kidneys act to retain salt and water in an attempt to improve their blood flow. How might changes in effective arterial blood volume be translated into the appropriate renal response? Three possible pathways may be involved. In the first one, increases or decreases in effective arterial blood volume lead to increases or decreases, respectively, in glomerular filtration rate, due to changes in glomerular blood flow and capillary pressure. Directionally similar changes in sodium excretion ensue. In the second, there is an intrarenal baroreceptor in the afferent arteriole. A decrease in pressure at this site causes increased renin release and ultimately increased sodium reabsorption. Finally, in the third pathway, changes in peritubular capillary Starling forces could result directly from changes in extracellular fluid volume and once again would lead to appropriate changes in sodium excretion.

Re-examination of Figure 24–8 will help to review some of the foregoing points. When effective arterial blood volume is decreased (e.g., after hemorrhage), renal sodium excretion is decreased. This makes sense, since renal conservation of sodium aids in restoring effective arterial blood volume to normal. The reduction in renal sodium excretion may be due to a reduced glomerular filtration rate, increased plasma angiotensin II and aldosterone levels, decreased peritubular capillary hydrostatic pressure, increased renal sympathetic nerve activity, changes in intrarenal blood flow distribution, and decreased plasma levels of the atrial natriuretic peptide. When

RESEARCH FOCUS

Atrial Natriuretic Peptide: A Hormone from the Heart

One of the most exciting recent developments in medicine has been the discovery of a new hormone, atrial natriuretic peptide (ANP). This hormone, which increases sodium excretion and lowers blood pressure, is produced by the cardiac atria. It now appears that the heart, in addition to pumping the blood, is an endocrine organ. The hormone it produces has important effects on the kidneys and vascular system, and thereby influences cardiac function.

The original observations were made by a group of investigators in Canada, led by Adolpho de Bold. In 1981, they reported that intravenous injection of an extract of rat atria into anesthetized rats produced a dramatic increase in renal excretion of sodium and water. Extracts of rat ventricle were without effect. Injection of crude tissue extracts is a classic method that has led to the discovery of a number of hormones.

Within two years of this discovery, the peptide sequence of the natriuretic factor was determined. The circulating hormone contains 28 amino acids. In a short time, the peptide was chemically synthesized, and rat, mouse, and human genes for ANP were cloned and their nucleotide sequences determined. The extreme rapidity of these developments made this hormone available for study all over the world. Because many different groups had simultaneously isolated it, a variety of names were applied to ANP, including atrial natriuretic factor (ANF), atrin, atriopeptin, auriculin, and cardionatrin.

ANP is stored in secretorylike granules within ordinary atrial muscle cells. These granules were discovered more than thirty years ago; however, their function was not appreciated until recently. The number of atrial granules changes in response to changes in body fluid balance; this finding supports the idea that ANP production or secretion is related to fluid balance.

The main stimulus causing ANP release into the circulation is stretching of the cardiac atria. So, for example, if there is too much blood in the thorax, and the atria are stretched, ANP is released into the circulation. It acts on the kidneys to increase excretion of salt and water, and on arterioles, which it dilates. The hormone is part of a negative feedback control system, and helps to restore blood volume and blood pressure to normal.

ANP has several actions, all consistent with its overall effect of promoting sodium excretion and lowering blood pressure. It increases renal blood flow and glomerular filtration rate. It also inhibits sodium reabsorption by inner medullary collecting ducts. It inhibits the release of aldosterone by acting directly on the adrenal cortex and by inhibiting renin release. It relaxes arteriolar smooth muscle, thereby causing vasodilation and a fall in blood pressure.

The role of ANP in the normal control of renal sodium excretion and in a variety of pathophysiological states is being intensively investigated. Simply eating a bag of lightly salted potato chips leads to a rise in plasma ANP levels, which suggests that plasma levels change appropriately in response to small forcings. Intravenous infusion of very low doses of ANP into people produces a sustained increase in sodium excretion and a fall in blood pressure. These observations support a physiological role for this hormone. In congestive heart failure, plasma ANP levels may be markedly increased; this suggests that salt retention in this disorder is not due to deficient ANP secretion. Deficiency of ANP may explain some cases of hypertension. ANP itself, an ANP analogue, or a drug that inhibits the breakdown of ANP might be useful in treating congestive heart failure and other edema-forming states, or hypertension.

A problem in elucidating the role of ANP is that it is only one of many factors that affect renal sodium excretion. There is growing evidence that several other hormones, still incompletely characterized, also affect sodium excretion.

■ **Scientists have suggested that there are other circulating natriuretic factors. What information would you want before deciding that a factor is a physiologically important hormone?**

effective arterial blood volume is increased (e.g., because of increased dietary sodium intake), sodium excretion is enhanced because of an increase in glomerular filtration rate (GFR), decreased plasma angiotensin II and aldosterone levels, increased hydrostatic pressure and decreased colloid osmotic pressure in the peritubular capillaries, reduced sympathetic nerve activity, changes in intrarenal blood flow distribution, and increased plasma levels of the atrial natriuretic peptide. The relative importance of the various effector mechanisms varies with the circumstances and is not always known. Since regulation of extracellular fluid volume or effective arterial blood volume is so important, it is not surprising that multiple mechanisms, which interact and overlap, control renal sodium excretion.

Other Factors That Influence Sodium Excretion

A number of other factors are known to affect sodium excretion. These include (1) glucocorticoids, (2) estrogens, (3) osmotic diuretics, (4) poorly reabsorbed anions, and (5) diuretic drugs.

Glucocorticoids. The glucocorticoids (e.g., cortisol) are steroid hormones produced by the adrenal cortex (Chapter 14). They increase tubular sodium reabsorption but also cause an increase in GFR, which may mask the tubular effect. In general, glucocorticoids cause a decrease in sodium excretion.

Estrogens. Estrogens may increase sodium reabsorption by the renal tubules. High plasma levels of estrogens, together with numerous other hormonal and hemodynamic changes, contribute to sodium retention during pregnancy.

Osmotic Diuretics. Osmotic diuretics are solutes excreted in the urine that obligate the excretion of water. Urine flow rate, sodium excretion, and potassium excretion are all increased.

Clinically important osmotic diuretics include (1) mannitol, (2) glucose, and (3) urea. Mannitol is a six-carbon sugar alcohol filtered by the glomerulus but not reabsorbed. Increased excretion of mannitol leads to increased excretion of salt and water. A concentrated mannitol solution is sometimes given intravenously to promote urine output. In uncontrolled diabetes mellitus, more glucose is filtered than can be reabsorbed by the tubules. Again, increased excretion of solute (glucose) leads to enhanced excretion of salt and water. The loss of salt leads to a fall in extracellular fluid volume and so stimulates thirst.

Urea excretion also promotes renal excretion of salt. This fact may be important in sustaining sodium excretion in chronic kidney disease.

Poorly Reabsorbed Anions. Poorly reabsorbed anions result in increased sodium excretion. Because of the requirement for electroneutrality in solutions, whenever more anion is excreted, more cation must also be excreted. Increased excretion of phosphate, bicarbonate, ketone body acids (as occurs in uncontrolled diabetes mellitus), or sulfate, results in increased sodium excretion. To some extent, sodium can be replaced by other cations such as potassium, ammonium, and hydrogen ions.

Diuretic Drugs. Diuretic drugs inhibit tubular reabsorption of sodium. The increase in urine volume that ensues is basically due to impaired sodium reabsorption. Remember that tubular reabsorption of water depends on sodium reabsorption.

There are additional factors that can affect sodium excretion. Some investigators have found circulating natriuretic factors distinct from the atrial natriuretic peptide. Various substances formed within the kidneys (e.g., prostaglandins, kinins, a polypeptide called urodilation) increase sodium excretion.

The Kidneys Are the Major Site of Control of Potassium Balance

Most of the body's potassium is found within cells. The concentration of potassium in cells is typically about 150 mEq per liter of cell water. In extracellular fluid, the potassium concentration averages about 4.5 mEq per liter.

Potassium plays a number of important roles in the body. First, it influences the electrical excitability of cells. As discussed in Chapter 7, the resting membrane potential of cells is largely determined by the unequal distribution of potassium inside and outside of cells and by a high membrane permeability to potassium ions. Disturbances in potassium balance lead to altered excitability of nerves and muscles.

Second, potassium is the major osmotically active solute in cells, and so the amount of potassium in cells affects cell volume. We are already familiar with the idea that the amount of sodium influences the volume of the extracellular compartment. Loss of potassium from cells leads to a decrease in cell volume.

Third, potassium balance is intimately related to acid-base balance. The relationships are complex, but there is abundant evidence to suggest that disturbances in acid-base status can affect potassium balance, and vice versa.

Finally, intracellular potassium affects cell metabolism. Tissue growth and repair require potassium. Conversely, tissue breakdown or increased protein catabolism leads to release of cell potassium into the extracellular fluid.

The normal range for plasma potassium concentration is 3.5 to 5.5 mEq per liter. A plasma potassium concentration below 3.5 mEq per liter is called **hypokalemia.** This condition may cause skeletal muscle weakness and even paralysis. A plasma potassium concentration above 5.5 mEq per liter is called **hyperkalemia.** Plasma potassium concentrations above 7 mEq per liter demand immediate attention because of possible cardiac irregularities. A plasma potassium concentration greater than 10 to 12 mEq per liter is usually fatal; the cause of death is cardiac arrhythmia or arrest. The changes in skeletal and cardiac muscle function that occur with abnormal plasma potassium levels result from the influence of potassium on cell membrane potentials.

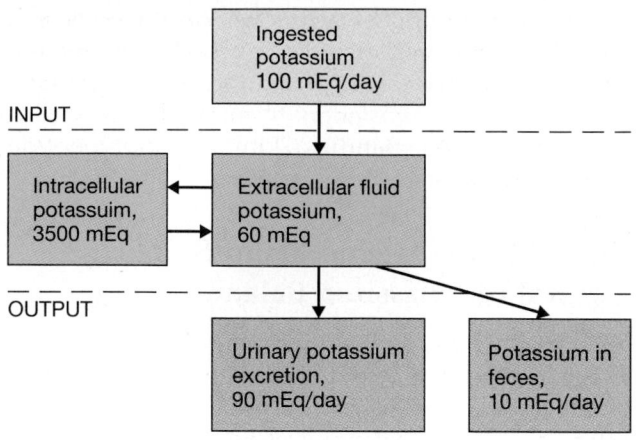

Figure 24–10

Potassium balance (input-output) and the distribution of potassium between the two major body fluid compartments.

Figure 24–10 summarizes potassium balance. Normally, we ingest about 100 mEq per day of potassium. The amount we take in can be quite variable, depending on what we eat and drink. About 10% of the daily potassium intake is excreted in the feces, and about 90% in the urine. Loss of potassium in sweat (which contains 5 to 10 mEq potassium per liter) is usually negligible. The kidneys are the major site of potassium loss from the body, and they are the major site of control of potassium balance.

Figure 24–10 also reemphasizes that most of the body's potassium is inside cells. The amount of potassium in extracellular fluid can be calculated from its concentration (4.5 mEq per liter) and the extracellular fluid volume (about 14 liters), and is equal to about 60 mEq. This amounts to only 2% of the body's potassium.

Several factors affect the distribution of potassium between cells and extracellular fluid. First, activity of the cell membrane Na^+/K^+-ATPase, which pumps potassium into cells, is of key importance. Impaired cell metabolism or cardiac glycosides (e.g, digitalis) inhibit the pump, and tend to produce lower cell and higher extracellular potassium levels. Second, acid-base status affects the distribution of potassium between intracellular and extracellular compartments. A fall in plasma pH causes a shift of hydrogen ions into cells in exchange for cell potassium ions. The consequence is a rise of plasma potassium concentration. Third, the availability of insulin is an important factor affecting potassium distribution. Insulin promotes the movement of potassium into skeletal muscle and liver cells; lack of insulin causes more of the body's potassium to stay outside of cells. Fourth, shifts of potassium from cells to extracellular fluid occur with cell breakdown, due to tissue trauma, in-

fection, ischemia (inadequate blood flow), or heavy exercise. Since so much of the body's potassium is within cells, a small loss from or gain by this compartment could profoundly affect the plasma potassium concentration.

Plasma potassium must be closely regulated. Part of this regulation involves the hormones insulin, epinephrine, and aldosterone, which stimulate the uptake of potassium by cells. The kidneys also play an important part in regulating plasma potassium. Normally, when increased potassium loads are presented to the body, the kidneys can rapidly excrete excess potassium.

In Chapter 23, we discussed some of the factors that affect potassium excretion by the kidneys. Recall that most of the filtered potassium is reabsorbed by early portions of the nephron and that the collecting ducts secrete potassium. Some of the factors that affect potassium excretion include intracellular potassium concentration, aldosterone, excretion of anions, and urine flow rate.

Figure 24–11 shows the mechanisms whereby an increase in potassium intake in the diet leads to increased renal potassium excretion. Increased potassium intake tends to raise the plasma potassium concentration. By a direct effect on the adrenal cortex, an elevated extracellular potassium concentration stimulates the release of aldosterone. Aldosterone then travels in the blood to the kidneys, where it acts on the principal cells of the collecting duct. It increases the sodium permeability of the luminal cell membrane, the activity of the Na^+/K^+-ATPase in the basolateral cell membrane, and potassium permeability of the luminal membrane of these cells, thereby favoring potassium secretion (see Figure 23–17). An increased potassium intake raises the potassium concentration in most body cells, including the principal cells, and so potassium secretion is directly increased in this way.

The opposite changes occur if potassium intake is reduced. Plasma and cell potassium levels tend to fall, aldosterone secretion is inhibited, and secretion of potassium by the collecting duct is diminished. The result is a decreased rate of potassium excretion.

The kidneys normally do a remarkable job of maintaining potassium balance. But abnormal renal excretion of potassium (either too much or too little) is a common cause of disturbed potassium balance.

Too little renal excretion of potassium causes hyperkalemia. Most foods are rich in potassium, so continued food intake with inadequate renal excretion leads to a positive potassium balance. In acute renal failure—that is, with sudden renal shutdown—a life-threatening hyperkalemia may develop. Often compounding inadequate renal excretion (in patients with acute renal failure) are tissue trauma, infection,

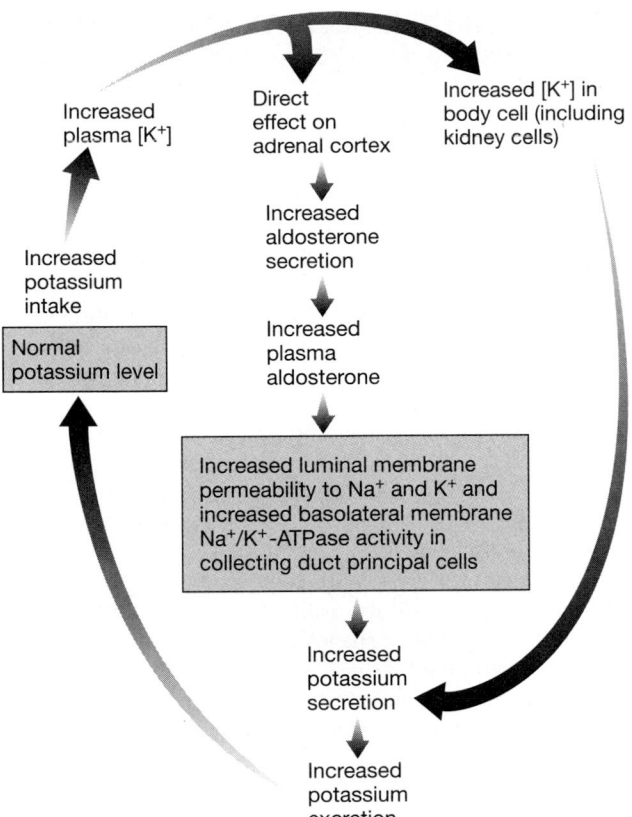

Figure 24–11

Pathways by which an increase in dietary potassium intake leads to increased renal potassium excretion.

and acidosis, all of which tend to raise the plasma potassium level. Chronic renal disease leads to hyperkalemia when the glomerular filtration rate falls below 15 to 20 ml/min. Up to that point, hyperkalemia does not develop because each surviving nephron can excrete a greater-than-normal quota of potassium. In adrenocortical insufficiency (Addison's disease), secretion of aldosterone is inadequate. A lowered plasma aldosterone level results in decreased stimulation of renal potassium secretion and excretion, so that ingested potassium tends to be retained in the body.

Too much excretion of potassium by the kidneys leads to hypokalemia. With adrenocortical hyperfunction, aldosterone levels are abnormally high, and potassium secretion is stimulated. Treatment with glucocorticoids can also result in excessive potassium excretion. In a number of kidney diseases, the ability of the tubules to reabsorb filtered potassium is impaired, and excessive potassium excretion results. Excessive renal potassium loss occurs in uncontrolled diabetes mellitus, for several reasons: (1) an osmotic diuretic effect of glucose (increased urine output produces potassium loss), (2) increased production of ke-

tone body acids (excretion of anions results in increased excretion of cations), and (3) increased plasma aldosterone levels (secondary to volume contraction caused by excessive loss of salt and water in the urine). Diuretic drugs (used in the treatment of hypertension and edema) are the most common cause of abnormally high renal potassium loss. This effect is probably due mostly to the increased rate of fluid flow through the collecting ducts that these agents produce. Patients taking diuretics are often advised to eat bananas, a rich source of potassium.

Excessive amounts of potassium can also be lost from the body through the gastrointestinal tract. Diarrheal fluid may have a potassium concentration of 50 to 60 mEq per liter, so a disturbance of the lower gastrointestinal tract may rapidly lead to potassium depletion. Potassium depletion also develops during vomiting.

Kidney Failure and Its Treatment

Kidney failure can occur at any age. It may develop suddenly (acute renal failure) and lead to death in a matter of days, as was the case with the young men described at the beginning of this chapter. Or it may be due to a loss of functional nephrons that occurs over decades (chronic renal disease). Kidney disease has a variety of causes, among which are bacterial infections, congenital defects, immunological injury, impaired renal blood flow, toxins, tumors, and urinary tract obstruction. The specific type of kidney disease can usually be diagnosed from the patient's history and clinical tests. In all forms of renal failure, however, the signs and symptoms are very similar. This symptom complex has been called **uremia** (urine in the blood). Basically, uremia indicates a disturbed internal environment, due to renal failure, that impairs the function of all body cells.

Renal Failure Results in Disturbances in Many Body Systems

Following are some of the disturbances that occur in uremia. This alphabetical list is not comprehensive but should serve to reemphasize many of the important functions of normal, healthy kidneys.

Anemia, with a fall in blood hematocrit to 20% to 35%, may develop in chronic kidney failure. The anemia is mainly due to deficient production of **erythropoietin,** a hormone produced mainly by the kidneys. Erythropoietin stimulates the bone marrow to produce red blood cells (erythropoiesis). Human erythropoietin has been synthesized in the laboratory by recombinant DNA techniques. When administered to renal failure patients, it corrects the anemia.

C L I N I C A L　　F O C U S

Polyuria

Polyuria is usually defined as a urine output of more than 3 L/day in adults. Such a urine output requires that the urinary bladder be emptied more frequently than the usual 4 to 6 times in a day. Polyuria can be distressing because of frequent bladder emptying and disturbed sleep. It may also be an important sign of some underlying pathology.

Frequency of urination may occur without polyuria; in other words, a person may frequently void small volumes. This is a symptom of decreased urinary bladder filling capacity, and can be caused by urinary tract infection, stones, tumors, prostatic hyperplasia, or other conditions. Polyuria is best substantiated by making a 24-hour urine collection.

Causes of polyuria include neurogenic diabetes insipidus, nephrogenic diabetes insipidus, primary polydipsia, solute diuresis, and salt-losing syndromes. In *neurogenic diabetes insipidus*, polyuria arises because of deficient production or secretion of antidiuretic hormone (ADH), so that filtered water is poorly reabsorbed in the collecting ducts of the kidneys. It is usually caused by damage to the hypothalamus or pituitary, the normal sites of synthesis and secretion, respectively, of ADH. In *nephrogenic diabetes insipidus,* the response of the collecting duct to ADH is impaired. This disorder may be caused by various drugs (such as lithium carbonate, methoxyflurane anesthetic, the antibiotic amphotericin), diseases of the renal medulla, inherited defects, or electrolyte disturbances (hypercalcemia and hypokalemia). In *psychogenic diabetes insipidus* (often called *primary polydipsia*), the polyuria is a consequence of increased water intake, as a result of habit, a psychiatric disorder, a lesion in the brain, or certain medications. A common form of polyuria is due to a solute diuresis, for example, as occurs in uncontrolled diabetes mellitus. Abnormal excretion of glucose in the urine leads to enhanced urinary loss of salt and water, dehydration, and thirst. In various salt-losing conditions—for example, diuretic drug abuse—the polyuria simply reflects decreased sodium reabsorption. Recall that tubular water reabsorption depends on sodium reabsorption, and when the latter is inhibited, increased excretion of water naturally follows.

To distinguish among the various causes of polyuria, a patient history and physical examination, analysis of the urine for glucose, and measurements of plasma creatinine and osmolality are helpful. Polyuria due to uncontrolled diabetes mellitus is easily identified by measuring glucose in the urine. Knowledge of the medications a patient may be taking may supply important information. A person with primary polydipsia tends to have a reduced plasma osmolality, because of excessive water intake. On the other hand, persons with neurogenic or nephrogenic diabetes insipidus tend to have elevated plasma osmolalities, due to excretion of a dilute urine.

A common test used to distinguish the various forms of diabetes insipidus is the *water deprivation test.* This test is easily interpretable if the lesions are severe, but is less conclusive with partial (incomplete) lesions. Under constant observation, a patient is deprived of water for 8 to 14 hours, and urine osmolality is measured. The person is then given 5 units vasopressin (ADH) subcutaneously, and two hours later another urine sample is collected. If the patient has complete neurogenic diabetes insipidus, the urine osmolality will be persistently dilute (less than 200 mosm/kg H_2O) during dehydration, but will increase substantially in response to injected vasopressin. If the person has complete nephrogenic diabetes, the urine osmolality will stay dilute in the presence of both dehydration and injected vasopressin. With primary polydipsia, the urine should become hyperosmotic (more than 400 mosm/kg H_2O) during simple dehydration, and injected vasopressin produces little additional increase in urine osmolality since plasma levels of ADH are already high due to the dehydration protocol. Once the underlying cause of polyuria has been identified, a proper mode of treatment can be instituted.

■ **Compared to a normal person, what is the plasma ADH concentration in people with neurogenic diabetes insipidus, nephrogenic diabetes insipidus, and primary polydipsia, and why?**

Azotemia, the accumulation of nitrogenous waste products in the blood, develops as a consequence of a markedly decreased GFR. Plasma urea, creatinine, and uric acid levels are elevated.

Calcium and phosphate metabolism are altered with renal failure. In uremia, plasma calcium is typically decreased, and plasma phosphate is increased. Secondary hyperparathyroidism, deficient synthesis

of the active form of vitamin D by the kidneys, and acidosis all lead to bone disease.

Hypertension (high blood pressure) is often present in renal failure. Its origin is not always known, but there is evidence for several possible pathways. The diseased kidney may produce excessive amounts of renin. Renin causes increased production of angiotensin II (a potent vasoconstrictor) and aldo-

Figure 24–12

Arrangement for hemodialysis. An arteriovenous fistula is created between the radial artery and a nearby superficial wrist vein. The vein is punctured with needles, which allows blood to be carried to and from the dialyzer. Alternatively, a shunt made of silicone rubber tubing can be permanently implanted and can be opened and connected to the blood lines of the dialyzer. In the dialyzer ("artificial kidney") the blood is separated from dialysis fluid by a thin membrane. Urea, other waste products, and potassium ions diffuse out of the blood into the dialysis fluid. The blood may also be ultrafiltered, to remove excess fluid, by applying a pressure gradient across the membrane.

sterone (which favors salt retention). The inability of the kidneys to excrete a normal amount of sodium could also lead to hypertension, owing to excess volume. Finally, the diseased kidneys may fail to produce vasodilator substances.

Metabolic acidosis typically accompanies severe renal failure. Acid end products of metabolism are not excreted at a normal rate and hence accumulate in the body. Impaired renal synthesis of ammonia contributes to the decreased ability to excrete hydrogen ions.

Potassium balance is usually well maintained until GFR falls below 15 to 20 ml/min. Retention of potassium can cause plasma levels to rise to lethal levels.

Sodium balance is typically impaired in patients with severe renal failure. On a normal sodium diet, excessive salt and water may be retained, leading to generalized edema. On a low-salt diet, patients may become salt-depleted.

Urinary concentrating ability is impaired with renal failure. The osmotic gradient in the medulla is reduced or abolished. Urine with an osmolality close to that of plasma is excreted.

Numerous other cardiovascular, gastrointestinal, muscular, and nervous abnormalities occur in the uremic state. In many instances, it is not understood why renal failure causes these disturbances.

Dialysis and Transplantation Are Used to Treat Patients with Terminal Renal Failure

Most of the signs and symptoms of uremia can be relieved by dialysis. **Dialysis** is the separation of smaller from larger molecules in solution by diffusion of the small molecules through a selectively permeable membrane. In treating uremia, two methods of dialysis may be used.

In **peritoneal dialysis,** about 2 liters of a balanced salt solution are introduced into the abdominal cavity. The peritoneum acts as a dialyzing membrane. Small molecules and ions (e.g., urea, potassium) diffuse into the injected solution, which is then drained and discarded. This procedure is used by some patients with chronic renal failure on an outpatient basis and is called **continuous ambulatory peritoneal dialysis.** The patient can perform self-dialysis and usually does this several times in a day.

Hemodialysis (Fig. 24–12) is a more efficient process involving an **artificial kidney.** The patient's blood is pumped through cellophanelike tubing immersed in a bath of balanced salt solution. Urea, potassium, and other substances in excess in the body diffuse through the tubing membrane out of the blood into the bath. Patients are usually dialyzed three times a week, and they can often undergo this procedure at home for 4 to 6 hours while watching television or reading.

Dialysis can allow patients with otherwise fatal chronic renal failure to live useful and productive lives. It has some drawbacks, however. It does not correct the anemia. It controls hypertension only with difficulty. Uremia develops between the periods of dialysis. There is a constant risk of infection and, with hemodialysis, hemorrhage. Dialysis does not maintain normal growth and development in children. Finally, it is costly. Dialysis is, however, invaluable in keeping patients alive and functioning until suitable kidney transplants become available.

Renal transplantation is the only "cure" for patients with terminal chronic renal disease. It may restore complete health and function. In 1993, about 8000 renal transplant operations were performed in the United States. At present, about 85% of kidneys grafted from a living donor function well for at least two years; grafts derived from an unrelated person who has just died (cadaver donors) have a one-year survival of about 80%.

A major problem in transplantation today is an inadequate number of kidney donors; the organ transplant waiting list has reached all-time highs. The immunological rejection of the transplanted kidney by the host is a major biological problem. Finally, transplantation (or dialysis) remains quite costly.

CHAPTER REVIEW

Summary

- The human body is about 60% water by weight. Two thirds of body water is in the intracellular fluid compartment, and one third is in the extracellular fluid compartment.
- Potassium ions are the major osmotically active solutes in cells; potassium influences cell volume, excitability, and metabolism.
- Sodium and its accompanying anions, chloride and bicarbonate, are the major osmotically active solutes in extracellular fluid. The amount of water in, and hence volume of, the extracellular fluid space is determined primarily by the amount of sodium in this compartment.
- In general, cell membranes are highly permeable to water, and so osmotic equilibrium between intracellular and extracellular fluids results.
- In the vascular system, the colloid osmotic pressure exerted by the plasma proteins keeps fluid within the circulation.
- The kidneys play a major role in keeping us in a stable balance with respect to water, sodium and potassium ions, and numerous other substances. Abnormal renal excretion is a common cause of imbalances (positive or negative balances).
- Plasma osmolality is regulated by renal excretion of water, which is controlled by ADH, and the thirst mechanism. ADH release and thirst are stimulated by cellular dehydration and by extracellular dehydration. Thirst is an emergency mechanism. Under ordinary circumstances, the main stimulus for ADH release is a rise in plasma osmolality detected by osmoreceptor cells in the hypothalamus.
- The kidneys are the major site of sodium output and regulation of extracellular fluid volume. An increase in extracellular fluid volume (or effective arterial blood volume) leads to increased renal sodium excretion and vice versa. Renal excretion of sodium represents the difference between filtered and reabsorbed amounts. The following factors influence sodium excretion: (1) glomerular filtration rate, (2) mineralocorticoids (aldosterone), (3) peritubular capillary Starling forces, (4) renal sympathetic nerve activity, (5) intrarenal blood flow distribution, (6) atrial natriuretic peptide, (7) glucocorticoids, (8) estrogens, (9) osmotic diuretics, (10) poorly reabsorbed anions, and (11) diuretic drugs.
- Most of the body's potassium is within cells. The kidneys normally maintain potassium balance by excreting most of the ingested potassium. Changes in dietary potassium intake cause changes in plasma aldosterone levels and kidney cell potassium concentration, and in this way change secretion and excretion of potassium by the kidneys.
- The symptom complex (uremia) of renal failure is much

the same no matter what the cause, and reflects an abnormal internal environment.
- Dialysis can correct some of the abnormalities that occur in uremia (e.g., acidosis, hyperkalemia, high plasma urea and creatinine concentrations), but it does not effectively correct anemia and hypertension, so it is not a cure for renal failure.
- Successful renal transplantation is the best hope for patients with chronic renal failure.

Review Questions

Identify Each with a Term
1. A below-normal blood volume.
2. Specialized neurons in the anterior hypothalamus that are stimulated by an increase in effective plasma osmotic pressure.
3. Excretion of a large volume of osmotically dilute urine.
4. A craving for salt stimulated by a sodium deficit.
5. Separation of smaller from larger molecules in solution by diffusion of the small molecules through a selectively permeable membrane.

Define Each Term
6. renin
7. converting enzyme
8. extracellular fluid
9. negative balance
10. insensible water loss
11. hypokalemia

Choose the Correct Answer
12. The greatest fraction of the body's water is contained within:
    a. blood plasma
    b. cells
    c. extracellular fluid
    d. transcellular fluid
13. Intravenous infusion of 1 liter of isotonic saline will cause:
    a. a 1-liter increase in intracellular fluid volume
    b. a 1-liter increase in extracellular fluid volume
    c. a 0.5-liter increase in intracellular fluid volume and a 0.5-liter increase in extracellular fluid volume
14. Intravenous infusion of 1 liter of hypertonic saline will cause:
    a. a decrease in intracellular fluid volume
    b. an increase in extracellular fluid volume
    c. an increase in plasma osmolality
    d. all of the above
15. Antidiuretic hormone increases epithelial water permeability of:
    a. collecting ducts
    b. proximal tubules
    c. thick ascending limbs
    d. all of the above
16. Antidiuretic hormone release from the posterior pituitary is stimulated by:
    a. a fall in plasma osmolality
    b. severe hemorrhage
    c. stimulation of arterial baroreceptors
    d. stretch of left atrial receptors

17. Which of the following produces a decrease in renal sodium excretion?
    a. decreased plasma aldosterone level
    b. increased plasma level of atrial natriuretic peptide
    c. increased GFR
    d. increased renal sympathetic nerve activity
18. Which of the following produces an increase in renal sodium excretion?
    a. administration of glucocorticoids
    b. decreased peritubular capillary hydrostatic pressure
    c. increased plasma estrogen levels
    d. uncontrolled diabetes mellitus
19. Renin release is stimulated by:
    a. increased blood pressure in afferent arterioles
    b. increased effective arterial blood volume
    c. increased sodium chloride transport by macula densa cells
    d. stimulation of renal sympathetic nerves
20. Which of the following are commonly seen in patients with severe congestive heart failure?
    a. elevated plasma ADH levels
    b. generalized edema
    c. hyponatremia
    d. thirst
    e. all of the above
21. The most abundant intracellular cation is:
    a. calcium   c. potassium
    b. chloride   d. sodium
22. Which of the following promotes a shift of potassium from cells to extracellular fluid?
    a. a fall in plasma pH
    b. an overdose of digitalis
    c. inadequate blood flow
    d. lack of insulin
    e. all of the above
23. Which of the following produces excessive urinary excretion of potassium?
    a. acute renal failure
    b. inadequate aldosterone secretion
    c. severe chronic renal failure (e.g., GFR = 10 ml/min)
    d. uncontrolled diabetes mellitus

Calculate
24. The following measurements were obtained in a 3.40 kg newborn infant:

Total body water	2600 ml
Extracellular water	1490 ml
Plasma water	155 ml

What are the volumes of (1) the intracellular water and (2) the interstitial fluid-lymph water?

25. If a person with severe congestive heart failure consumes 100 mmoles of sodium chloride every day, but excretes only 10 mmoles of sodium chloride per day, how much weight will be gained in one week? Assume free access to water and that 1 liter of isotonic saline (150 mM NaCl) weighs 1 kg (2.2 lbs).

Answer Each Question in One or Two Sentences

26. How does plasma differ in composition from interstitial fluid-lymph?
27. How does transcellular fluid differ in composition from plasma?
28. Which of the following is present in cells at a higher concentration than in extracellular fluid? Na^+, K^+, Ca^{2+}, Mg^{2+}, Cl^-, HCO_3^-, proteins.
29. Why is thirst (like shivering) considered to be mainly an emergency mechanism?
30. Why does drinking a large volume of water lead to excretion of a large volume of osmotically dilute urine?
31. How is ADH thought to produce its effects on collecting duct water permeability?

32. List six factors that *might* contribute to the enhanced sodium excretion that results from increased dietary intake of sodium.
33. What are four effects of angiotensin II that help the body recover from hemorrhage?
34. What is meant by "effective arterial blood volume"?
35. What two factors contribute to decreased renal potassium excretion on a low potassium diet?

Problem Solving in Physiology

36. The plasma sodium concentration in a person with severe congestive heart failure may be below normal even though the amount of sodium in the extracellular fluid is greatly increased. Explain.
37. In the emergency treatment of hyperkalemia, physicians may administer insulin and a glucose solution. Explain.
38. Adrenal cortical insufficiency may be fatal unless a person increases dietary intake of sodium chloride. Explain.

Suggested Readings

Brenner, B.M., and Rector, F.C., Jr. *The Kidney*, 4th ed. Philadelphia: W.B. Saunders Company, 1991.

Narins, R.G. (ed.) *Maxwell and Kleeman's Clinical Disorders of Fluid and Electrolyte Metabolism*, 5th ed. New York: McGraw-Hill, 1993.

Rose, B.D. *Clinical Physiology of Acid-Base and Electrolyte Disorders*, 4th ed. New York: McGraw-Hill, 1994.

Seldin, D.W., and Giebisch, G.H. (ed.) *The Kidney: Physiology and Pathophysiology*, 2nd ed. New York: Raven Press, 1992.

Windhager, E.E. (ed.) *Handbook of Physiology Section 8: Renal Physiology*. New York: Oxford University Press, 1992.

Key Terms

antidiuretic hormone (p. 744)
atrial natriuretic peptide (p. 751)
dialysis (p. 757)
effective arterial blood volume (p. 743)

extracellular fluid (p. 737)
glomerulotubular balance (p. 748)
hyperkalemia (p. 753)
hypokalemia (p. 753)
hyponatremia (p. 743)
hypovolemia (p. 743)

intracellular fluid (p. 737)
milliequivalent (p. 738)
negative balance (p. 742)
osmoreceptor cells (p. 743)
positive balance (p. 741)
renin-angiotensin system (p. 749)

sodium appetite (p. 747)
thirst (p. 743)
transcellular fluid (p. 738)
uremia (p. 755)
water diuresis (p. 745)

Chemicals and instruments used to measure the pH of a solution and determine the concentrations of acid and base.   *(© Fisher Supply Catalog)*

Chapter 25

Regulation of Acid-Base Physiology

Key Concepts

Stability of pH (or H^+ concentration) is achieved by buffers.

The HCO_3^-/CO_2 buffer pair is especially important in the body. The lungs influence plasma pH by controlling the arterial blood P_{CO_2} and the kidneys influence plasma pH by controlling the plasma bicarbonate concentration.

There are four fundamental types of acid-base disturbance.

Intracellular pH is kept within narrow limits by cell buffers and plasma membrane transporters.

Chapter Outline

Principles of Acid-Base Physiology

Regulation of Extracellular pH: Buffering Mechanisms in the Body

Disturbances of Acid-Base Balance

Intracellular pH Regulation

A young woman with diabetes neglects her insulin shots for two days, and is brought in a coma to the hospital emergency room. Arterial blood pH is 7.20, plasma bicarbonate concentration is 8 mEq per liter, and P_{CO_2} is 21 mm Hg. She appears to be dehydrated and is breathing slowly but deeply. Her urine is positive for glucose.

An elderly man is brought to the emergency room by his wife. He has chronic pulmonary disease and has been sick in bed with pneumonia. He has a bad cough and complains of shortness of breath. His blood pH is 7.22, plasma bicarbonate concentration is 32 mEq per liter, and P_{CO_2} is 80 mm Hg.

Both of these individuals have an abnormally low blood pH, but how this came about is different in the two cases, and will be apparent after you have studied this chapter. In this chapter, we will explain why pH is important, how pH changes are minimized by chemical buffers, how the lungs and kidneys defend a normal blood pH, what types of acid-base disturbances occur in people, and how intracellular pH is regulated.

Most of this chapter will focus on the regulation of extracellular fluid pH or hydrogen ion concentration

([H$^+$]). A normal extracellular fluid pH is an important part of the ideal internal environment that our cells need (Chapter 1). Intracellular pH regulation is probably even more important for normal cell function. pH influences the speed of chemical reactions (catalyzed by protein enzymes), cell permeability properties, and cell structure. Disturbances in pH profoundly affect function and can be lethal.

Principles of Acid-Base Physiology

An Acid Is a Proton (H$^+$) Donor and a Base Is a Proton Acceptor

As first presented in Chapter 2, an **acid** is a substance that can donate hydrogen ions. A hydrogen ion (H$^+$) consists of a bare proton, a hydrogen atom without its orbiting electron. Examples of acids include **hydrochloric acid (HCl), sulfuric acid (H$_2$SO$_4$), nitric acid (HNO$_3$), phosphoric acid (H$_3$PO$_4$), ammonium ion (NH$_4$$^+$), lactic acid, acetic acid,** and **carbonic acid (H$_2$CO$_3$).** An acid will donate its hydrogen ion to a base.

Accordingly, a **base** is a substance that can accept or bind hydrogen ions. Examples of bases include **sodium hydroxide (NaOH), potassium hydroxide (KOH), ammonia (NH$_3$),** and **lactate, acetate,** and **bicarbonate (HCO$_3$$^-$) ions.**

Not all acids have the word *acid* in their names. For example, the ammonium ion (NH$_4$$^+$) is an acid. If we mix aqueous solutions of the salts ammonium chloride (NH$_4$Cl) and sodium bicarbonate (NaHCO$_3$), the following reversible reaction takes place:

$$NH_4^+ \; Cl^- + Na^+ \; HCO_3^- \; \rightleftharpoons$$

$$NH_3 + H_2CO_3 + Na^+Cl^-$$

In this reaction (left to right), the ammonium ion is an acid; it donates a hydrogen ion to the base bicarbonate.

In a **neutralization reaction,** an acid reacts with a base to form a salt and water. For example, in the reaction

$$HCl + NaOH \; \rightarrow \; NaCl + H_2O$$

the acid HCl donates its hydrogen ion to the hydroxide ion of NaOH to form water and sodium chloride (a salt).

The amount of an acid or base is often expressed in terms of **equivalents** or **milliequivalents (mEq)** (1 mEq is 1/1000 of an equivalent). One equivalent of acid will neutralize one equivalent of base. One

mole of HCl contains one equivalent of acid. One mole of H$_2$SO$_4$ contains two equivalents of acid because sulfuric acid can donate two hydrogen ions. Therefore, two moles (two equivalents) of NaOH are required to neutralize one mole (two equivalents) of sulfuric acid, according to the following two steps:

$$H_2SO_4 + NaOH \; \rightarrow \; NaHSO_4 + H_2O$$

$$NaHSO_4 + NaOH \; \rightarrow \; Na_2SO_4 + H_2O$$

Amphoteric Substances Can Function as Acids and Bases

Some chemical substances can function as both an acid and a base and are referred to as **amphoteric substances** (*amphoteros* in Greek means "pertaining to both"). These include amino acids and proteins. For example, the amino acid **glycine,** $^+$H$_3$N—CH$_2$—COO$^-$, acts as an acid in the following reaction:

$$NaOH + \; ^+H_3N\text{—}CH_2\text{—}COO^- \; \rightarrow$$

$$H_2N\text{—}CH_2\text{—}COO^-Na^+ + H_2O$$

The ammonium group of glycine donates a hydrogen ion to the hydroxide ion of NaOH. Glycine acts as a base in the following reaction:

$$HCL + \; ^+H_3N\text{—}CH_2\text{—}COO^- \; \rightarrow$$

$$Cl^- \; ^+H_3N\text{—}CH_2\text{—} COOH$$

The carboxylate group of glycine accepts a hydrogen ion from HCl. Proteins contain many different side groups that can act as acids or bases.

The Acid Dissociation Constant Is Directly Related to the Strength of an Acid

When an acid (generically indicated as **HA**) is added to water, the following reversible reaction takes place:

$$HA + H_2O \rightleftharpoons H_3O^+ + A^-$$

The species H$_3$O$^+$ is a hydrated hydrogen ion and is called the **hydronium ion.** For simplicity, it is common practice to ignore water in this reaction and to write

$$HA \rightleftharpoons H^+ + A^-$$

The reaction going from left to right is called the **dissociation reaction,** and the reaction going from right to left is the **association reaction.** The rate of the dissociation reaction is equal to the product of the concentration of HA and the **dissociation rate constant** k_1, a specific value for this reaction. The rate of the association reaction is equal to the product of the concentrations of H^+ and A^- and the **association rate constant** k_2, a specific value for this reaction. At equilibrium, the rates of dissociation and association reactions are equal. We can therefore write:

$$k_1 \times [HA] = k_2 \times [H^+] \times [A^-]$$

Rearranging, we get $[H^+] \times [A^-]/[HA] = k_1/k_2$. We define a new constant K_a as equal to this ratio k_1/k_2. K_a is the **equilibrium constant** for this reaction and is called the **dissociation constant** or **ionization constant** for the acid.

The higher the acid dissociation constant, the more completely an acid is dissociated or ionized and, consequently, the more acidic the solution, because more free (unbound) hydrogen ions are present. Acids with high dissociation constants are **strong acids.** These include hydrochloric, sulfuric, nitric, and phosphoric acids. Hydrochloric acid, for example, is essentially completely dissociated into hydrogen ions and chloride ions in dilute aqueous solution. There is very little undissociated HCl. In a 0.1 M HCl solution, the free hydrogen ion concentration is nearly 0.1 M.

A low acid dissociation constant is characteristic of **weak acids.** In other words, a weak acid is incompletely dissociated in solution. For example, the acid dissociation constant of acetic acid is equal to 1.8×10^{-5}. If we have a 0.100 M solution of acetic acid in water, the concentration of undissociated acid (CH_3COOH) is 0.0987 M, and the concentrations of free hydrogen and acetate (CH_3COO^-) ions are both equal to 0.0013 M. In other words, nearly 99% of the acetic acid molecules are not dissociated in this solution. Note especially that the concentration of free hydrogen ions (or acidity) is low for a solution of a weak acid.

Acid dissociation constants are usually small and awkward to manipulate mathematically, so they are often presented in a logarithmic form. The pK_a is defined as the negative logarithm to the base 10 of K_a or

$$pK_a = -\log K_a$$

For example, for acetic acid, $pK_a = -\log 1.8 \times 10^{-5} = 4.7$. The negative logarithm of a number is equal to the logarithm of the inverse of the number. So the equation can be written as

$$pK_a = \log (1/K_a)$$

This equation shows more clearly that pK_a and K_a are inversely related. A low pK_a corresponds to a high dissociation constant (strong acid), and a high pK_a corresponds to a low dissociation constant (weak acid).

The pH of Aqueous Solutions Ranges Widely

As discussed in Chapter 2, we can also define **pH** as the negative logarithm to the base 10 of the molar activity (concentration) of hydrogen ions. In equation form:

$$pH = -\log [H^+] = \log (1/[H^+])$$

The pH of a solution is inversely related to the free hydrogen ion concentration. A low pH indicates a high hydrogen ion concentration, and a high pH indicates a low hydrogen ion concentration. The pH scale is useful because it compresses the wide range of hydrogen ion concentrations possible in aqueous solutions into a more manageable scale. The pH scale usually goes from 0 to 14. A 1 M solution of the strong acid HCl has a pH of about 0 ($\log 1 = 0$). A 1 M solution of the strong base NaOH has a pH of about 14 ($[H^+] = 10^{-14}$ M). A pH of 7 corresponds to a $[H^+]$ of 10^{-7} M. If the pH of a solution is below 7, it is considered to be **acidic;** if equal to 7, **neutral;** and if greater than 7, **alkaline** or **basic.**

The Henderson-Hasselbalch Equation

The equilibrium equation for dissociation of an acid is often expressed in logarithmic form. We can rearrange the equation

$$[H^+] \times [A^-]/[HA] = K_a$$

and solve for $[H^+]$:

$$[H^+] = K_a \times [HA]/[A^-]$$

Taking the logarithms of this equation, we get:

$$\log [H^+] = \log K_a + \log ([HA]/[A^-])$$

Multiplying both sides of the equation by -1, we get:

$$-\log [H^+] = -\log K_a + \log ([A^-]/[HA])$$

Since $pH = -\log [H^+]$ and $pK_a = -\log K_a$, we can write the equation

$$pH = pK_a + \log ([A^-]/[HA])$$

This logarithmic form of the acid dissociation equation is known as the **Henderson-Hasselbalch equation** and is very useful, as we will see later. From this equation, we can see that when $[A^-] = [HA]$ (i.e., when half of the acid is present in its dissociated form), the solution pH equals the pK_a (since the logarithm of one is zero).

The Ion Product of Water

Water reversibly dissociates into hydrogen and hydroxide ions according to the reaction:

$$H_2O + H_2O \rightleftharpoons \boxed{H_3O^+} + \boxed{OH^-}$$

In this reaction, water is acting as both an acid (proton donor) and base (proton acceptor). It is customary to simplify this reaction and write:

$$H_2O \rightleftharpoons \boxed{H^+} + \boxed{OH^-}$$

The dissociation constant (K) for this reaction is equal to

$$K = [H^+] \times [OH^-]/[H_2O]$$

The tendency of water to dissociate is very small. The concentration of water ($[H_2O]$) is large and is essentially constant. We can therefore define a new constant, K_w, as equal to $K \times [H_2O]$, and write

$$K_w = [H^+] \times [OH^-]$$

This equation indicates that in any aqueous solution, the product of the hydrogen and hydroxide ion concentrations is constant. This means that $[H^+]$ and $[OH^-]$ are inversely related. K_w, called the **ion product of water,** has a value of 1×10^{-14} M^2 at room temperature (25°C). In *pure* water, $[H^+] = [OH^-]$. Therefore, $[H^+]^2 = 1 \times 10^{-14}$ M^2 or $[H^+] = 10^{-7}$ M (pH 7). Hence, the pH of pure water is 7. If we add a strong base (NaOH) to water, OH^- ions will combine with hydrogen ions supplied by water, thereby shifting the equilibrium toward undissociated water and lowering the free hydrogen ion concentration. For example, in a 1 M solution of NaOH, $[OH^-] = 1$ M, and so $[H^+] = 10^{-14}$ $M^2/1$ M $= 10^{-14}$ M (pH 14). In a 0.1 M solution of HCl, $[H^+]$ is essentially 0.1 M, so $[OH^-] = 1 \times 10^{-14}$ $M^2/0.1$ M or 1×10^{-13} M.

pH Values for Solutions of Acids and Bases, Body Fluids, and Beverages

Table 25-1 presents pH values for some aqueous solutions of chemicals, body fluids, and beverages. Notice that the pH of various solutions of acids depends not only on the concentration of acid, but also on the dissociation constant of the acid. Strong acids produce solutions with a lower pH than do the same concentrations of weak acids, because the stronger acid is dissociated (ionized) more and so yields a greater concentration of free hydrogen ions in solution.

The pH of arterial blood averages 7.40, which corresponds to a $[H^+]$ of 4×10^{-8} M or 40 nanomolar. The pH of venous blood is slightly lower (average 7.35), owing to its higher content of carbon dioxide

Table 25–1

pH Values of Aqueous Solutions of Chemicals, Body Fluids, and Beverages

Chemicals	pH
0.100 M hydrochloric acid	1.1
0.010 M hydrochloric acid	2.0
0.001 M hydrochloric acid	3.0
0.100 M lactic acid (pK_a = 3.9)	2.5
0.100 M acetic acid (pK_a = 4.7)	2.9
Pure water	7.0
0.100 M sodium bicarbonate	8.4
0.100 M ammonia	11.1
0.100 M sodium hydroxide	13.0
Body fluids	
Arterial blood	7.40
Venous blood	7.35
Cerebrospinal fluid	7.35
Skeletal muscle cell cytoplasm	6.9
Saliva	5.8–7.1
Gastric juice	0.7–3.8
Gallbladder bile	5.6–8.0
Pancreatic juice	7.5–8.8
Intestinal juice	7.0–8.0
Feces	5.9–8.5
Urine	4.5–8.0
Beverages	
Lemon juice	2.2–2.4
Sodas	2.8–3.7
Orange juice	3.7
Beer	4.4
Coffee	5.0
Milk	6.7
Tea	6.9

(hence the acid carbonic acid). Most body fluids, with the exception of gastric juice, have pH values in the range 5 to 8. Cell cytoplasm is more acidic than extracellular fluid; the pH of skeletal muscle cell cytoplasm averages 6.9. pH may differ strikingly from one organelle to another within a cell; for example, in lysosomes, which contain enzymes that operate best in an acidic environment, pH is about 4.5. Gastric juice is acidic, owing to its high concentration of hydrochloric acid. Pancreatic juice is slightly more alkaline than blood because of its high bicarbonate concentration. The pH of urine can vary between 4.5 and 8.0.

Many foods are acidic or alkaline. When we consume food, the load of acid or base depends not only on the free hydrogen ion concentration, but also on how much undissociated acid (or base) is present. An undissociated acid may liberate hydrogen ions (or a base may bind hydrogen ions) at the pH of body fluids. The effect of a chemical substance on acid-base balance also depends on how the substance is metabolized in the body, a subject we will discuss later.

Buffers Promote Stability of pH

A buffer, by definition, is something that minimizes a shock and thereby promotes relative stability. A **pH buffer** minimizes the change in pH when either acid or base is added to a solution.

A pH buffer *does not prevent* a pH change. Whenever an acid is added, pH will fall. If a base is added, pH will rise. The extent of change in pH depends on the amount and nature of added acid or base, and on the amount and nature of the pH buffer.

Previously, we wrote the equation for dissociation of an acid: $HA \rightleftharpoons H^+ + A^-$. In this reaction, HA is the acid and A^- is the **conjugate base.** *Conjugate* means "joined in a pair." A **chemical buffer** consists of a weak acid and its conjugate base (or a weak base and its conjugate acid). Following are some examples of conjugate acid-base buffer pairs:

Acid		Conjugate base		
$H_2PO_4^-$	$\rightleftharpoons$	HPO_4^{2-}	$+$	H^+
Dihydrogen phosphate		Monohydrogen phosphate		
H_2CO_3	$\rightleftharpoons$	HCO_3^-	$+$	H^+
Carbonic acid		Bicarbonate		
NH_4^+	$\rightleftharpoons$	NH_3	$+$	H^+
Ammonium		Ammonia		

The equilibrium expression for dissociation of an acid can be written in the Henderson-Hasselbalch equation form:

$$pH = pK_a + \log ([\text{conjugate base}]/[\text{acid}])$$

For example, for a mixture of monohydrogen and dihydrogen phosphate,

$$pH = 6.8 + \log ([HPO_4^{2-}]/[H_2PO_4^-])$$

This equation demonstrates that if the ratio of conjugate base to acid is defined for a buffer of known pK, then the pH is automatically determined. This principle is useful in making up pH buffer solutions.

What, for example, is the pH of an aqueous solution containing a mixture of 0.1 M Na_2HPO_4 (sodium monohydrogen phosphate) and 0.1 M NaH_2PO_4 (sodium dihydrogen phosphate)? Since the ratio $HPO_4^2/H_2PO_4^-$ is 1, and the logarithm of 1 is 0, the pH is 6.8 + 0 = 6.8. If the ratio of $HPO_4^{2-}/H_2PO_4^-$

is 4, then the pH of this solution is 6.8 + log 4 = 6.8 + 0.6 = 7.4.

Chemical buffers stabilize the pH of aqueous solutions. For example, if we add strong acid (HCl) to a phosphate buffer solution, the following reaction takes place:

$$HCl + Na_2HPO_4 \rightarrow Na H_2PO_4 + NaCl$$

In this reaction, the base component of the buffer pair (HPO_4^{2-}) binds hydrogen ions. The strong acid (HCl) is, in effect, converted into a weak acid ($H_2PO_4^-$), thereby minimizing the increase in free hydrogen ion concentration (fall in pH). If we add strong base (NaOH) to the phosphate buffer solution, the following reaction takes place:

$$NaOH + Na H_2PO_4 \rightarrow Na_2 HPO_4 + H_2O$$

The acid component of the buffer pair ($H_2PO_4^-$) liberates hydrogen ions, thereby diminishing the rise in pH caused by addition of strong base. The strong base (NaOH) is, in effect, converted into a weak base (HPO_4^{2-}).

Figure 25–1 shows a titration curve for a phosphate buffer solution. The y axis gives the pH. The x axis gives the amount of strong acid or base added to the solution. Going from left to right along the curve, $H_2PO_4^-$ is converted to HPO_4^{2-} by addition of strong base. Going from right to left along the curve, HPO_4^{2-} is converted to $H_2PO_4^-$ by the addition of strong acid. The slope of the titration curve is an index of the effectiveness of the buffer in resisting a change in pH. The slope of the curve is flattest when the pH is equal to the pK_a of the phosphate buffer. This means that for a given amount of added acid or base, the pH change will be least near the pK_a of the buffer. As we move away from the pK_a, the effectiveness of the buffer declines. In choosing a buffer to stabilize a certain pH, it is always best to select a buffer pair with a pK_a close to the desired pH.

The effectiveness of a buffer in minimizing pH changes depends on two factors. The first factor is the pK_a of the buffer in relation to the desired pH. The pK_a of the imidazole group of histidine (an amino acid found in hemoglobin) is close to 7.4, and hence this is an ideal buffer in blood. The pK_a of phosphate (6.8) is also effective for buffering blood. The second factor that determines the effectiveness of a buffer is the amount (or concentration) of buffer present. The greater the concentration of buffer, the greater the ability to bind (when acid is added) or release (when base is added) hydrogen ions. If no buffers are present, as in pure water or in an aqueous solution of

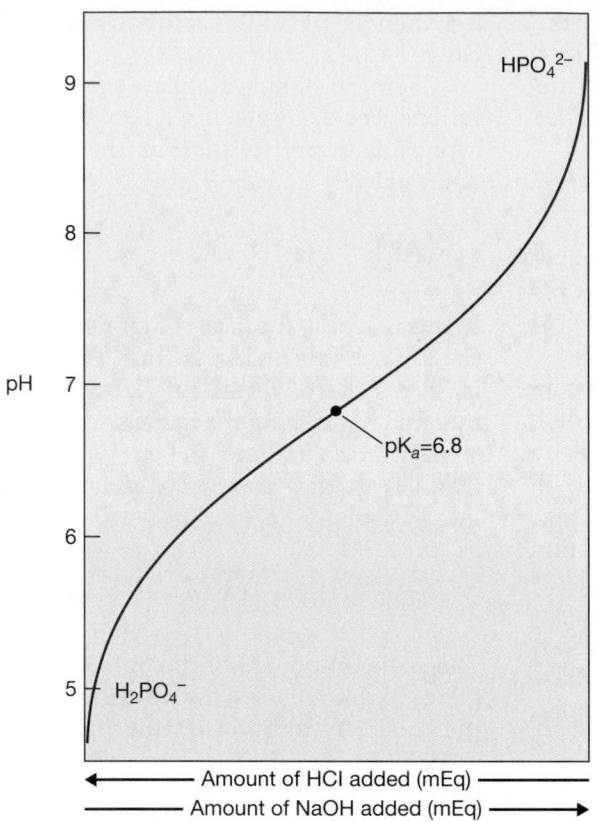

Figure 25–1

Titration curve for inorganic phosphate buffer. HPO_4^{2-} predominates at alkaline pHs. When strong acid (HCl) is added, the reversible reaction $H^+ + HPO_4^{2-} \rightarrow H_2PO_4^-$ takes place. When strong base (NaOH) is added, $H_2PO_4^-$ is converted to HPO_4^{2-}; $OH^- + H_2PO_4^- \rightarrow HPO_4^{2-} + H_2O$. The change in pH produced by a given amount of acid or base is least when the solution pH is at the pK_a, (6.8) of the buffer pair $HPO_4^{2-}/H_2PO_4^-$.

pure NaCl, large pH changes will be produced by addition of very small amounts of acid or base.

Hydrogen Ions Are Produced in Metabolic Reactions

The metabolism of food consumed in our diet is usually the major source of hydrogen ions. For most people eating a mixed diet of meat and vegetables, metabolism results in net addition of acid to the body. Vegetarians are faced with the opposite situation, net addition of base. Since the usual problem is one of getting rid of acid, we will emphasize that situation for most of this chapter.

Metabolic Production of CO_2 as a Source of Hydrogen Ions

In the course of metabolism, a normal adult produces about 300 liters of carbon dioxide per day. Carbon dioxide, added to the blood in the tissues, reacts with water to produce an acid, carbonic acid:

$$CO_2 + H_2O \rightleftharpoons H_2CO_3 \rightleftharpoons H^+ + HCO_3^-$$

In the lungs, these reactions are reversed, since carbon dioxide is expired. No acid burden is imposed

on the body, as long as carbon dioxide is expired by the lungs as rapidly as it is produced. Carbon dioxide production and expiration are usually matched so that arterial blood carbon dioxide tension and pH do not change.

Metabolism of Carbohydrates and Fatty Acids as a Source of Hydrogen Ions

Most carbohydrates (e.g., glucose) and fatty acids are normally completely oxidized to carbon dioxide and water. No net addition of acid to the body results, provided that carbon dioxide does not accumulate.

Incomplete oxidation of carbohydrates and fats produces nonvolatile acids. If glucose is not oxidized completely, then lactic acid is produced. With severe exercise or an inadequate circulation, insufficient delivery of oxygen to the tissues results in lactic acid production and an acidic blood pH. If fatty acids are incompletely oxidized, as occurs in uncontrolled diabetes mellitus (for example, the young woman described at the beginning of this chapter), starvation, and alcoholism, ketone body acids (acetoacetic acid, β-hydroxybutyric acid) are produced. This situation also results in an acidic blood pH.

Metabolism of Proteins as a Source of Hydrogen Ions

The metabolism of dietary proteins results in production of strong acids. Sulfuric acid is produced by the oxidation of sulfur-containing amino acids (cysteine, cystine, methionine). Hydrochloric acid is produced by oxidation of cationic amino acids (e.g., lysine, arginine). Phosphoric acid is produced by the oxidation of phosphorus-containing proteins.

To some extent, these acid-producing reactions are neutralized by acid-consuming reactions. Thus, the oxidation of dietary organic anions (e.g., citrate, lactate, acetate) uses up hydrogen ions. For example, citrate, which is contained in fruits and vegetables, is oxidized in the body according to the following simplified reaction:

$$\text{citrate}^- + O_2 + \boxed{H^+} \rightleftharpoons CO_2 + H_2O$$

In this reaction, a hydrogen ion is consumed. The net result of acid-producing and acid-consuming reactions on the usual American diet is such that an excess of about 50 to 100 mEq of hydrogen ions is produced in a day.

Regulation of Extracellular pH: Buffering Mechanisms in the Body

Despite the metabolic production of acids in the body, a normal pH in blood and interstitial fluid is preserved in a healthy person. The pH is stabilized by buffers in the body. The first line of defense of pH consists of chemical buffers in extracellular and intracellular fluids and in bone. The second is the lungs, which buffer blood pH by disposing of carbon dioxide as rapidly as it is formed. When there is excessive production of acids (or bases) in the body, lung ventilation changes in order to reduce the deviation from a normal blood pH. The third is the kidneys, which buffer blood pH by excreting hydrogen ions in the urine. The kidneys are in charge of eliminating excess hydrogen ions and anions from the body. Figure 25–2 sketches the pathways for maintaining a stable blood pH despite the acid threat posed by cell metabolism of foodstuffs.

Many Chemical Buffers Stabilize Extracellular Fluid pH

We will consider chemical pH buffers in the body. These include (1) phosphates, (2) proteins, and (3) the bicarbonate-CO_2 system. Chemical buffers consist of

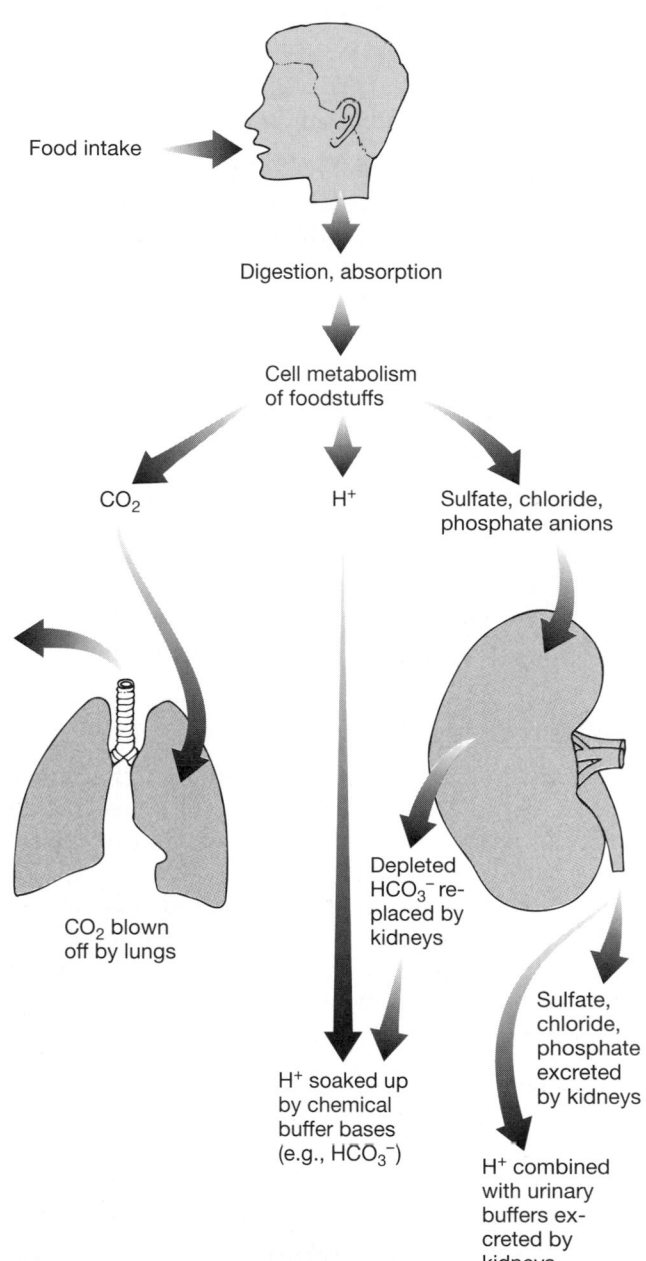

Figure 25–2

Overall schema for maintenance of acid-base balance. On the usual mixed diet, pH is threatened by production of strong acids (sulfuric, hydrochloric, phosphoric), which result mainly from protein metabolism. These strong acids are buffered by chemical buffers in the body. Removal of extra hydrogen ions and the accompanying anions from the body is accomplished by renal excretion. When the kidneys excrete hydrogen ions, they add new bicarbonate to the blood, thereby restoring depleted body buffer bases. The respiratory system eliminates CO_2 produced by metabolism. CO_2 is not a threat to acid-base balance, provided it is expired at the same rate it is produced.

conjugate acid-base pairs that bind hydrogen ions (when acid is added) or release hydrogen ions (when base is added).

Phosphate as a Chemical Buffer

Phosphate, by virtue of its pK of 6.8, is a good buffer for stabilizing the pH of the blood close to its normal value of 7.4. The concentration of inorganic phosphate in plasma or extracellular fluid, however, is rather low (about 1.0 mmole per liter). Because of its low concentration, inorganic phosphate has a limited capacity to buffer added acid or base. There are, however, large quantities of phosphate salts in bone, so bone can act to buffer pH. Also, the cellular compartment contains many organic phosphate compounds, such as adenosine triphosphate (ATP), adenosine diphosphate (ADP), and creatine phosphate. These compounds primarily serve in energy metabolism, but they also bind or release hydrogen ions in response to excess acid or base. Inorganic phosphate is a significant buffer in the urine.

Proteins as Chemical Buffers

Proteins comprise the largest amount of buffer in the body and are excellent pH buffers. Proteins contain many different ionizable groups, which can donate or accept hydrogen ions.

Serum albumin and plasma globulins act as pH buffers in blood plasma. The large amounts of proteins in cells also participate in buffering hydrogen ions. An important example of an intracellular protein buffer is hemoglobin, found in red blood cells. When carbon dioxide is added to the blood in the tissues, hydrogen ions liberated from carbonic acid are mainly bound by hemoglobin. This combination with hydrogen ions, incidentally, also favors the release of oxygen from hemoglobin (Chapter 21). These reactions are summarized in Figure 25–3.

The Bicarbonate-CO_2 System as a Chemical Buffer

The bicarbonate-CO_2 system is of special importance in pH buffering in extracellular fluid for several reasons. First, the components of this buffer system are present in large quantities. A practically limitless supply of CO_2 is available from metabolism. There is an appreciable concentration of bicarbonate in extracellular fluid, about 24 mEq per liter. Second, despite a pK of 6.10, which is somewhat removed from the desired pH of 7.40, this buffer pair is quite effective, as will be explained later. Third, the physiological buffers, lungs and kidneys, affect blood pH by acting on components of this buffer pair.

We will now examine the bicarbonate-CO_2 relationship in the light of acid-base chemistry. We can write the following reversible reactions:

$$CO_2(g) \rightleftharpoons CO_2(d)$$

$$+$$

$$H_2O \underset{\substack{\text{Carbonic} \\ \text{anhydrase}}}{\rightleftharpoons} H_2CO_3 \rightleftharpoons H^+ + HCO_3^-$$

$$\Updownarrow$$

$$H^+ + CO_3^{2-}$$

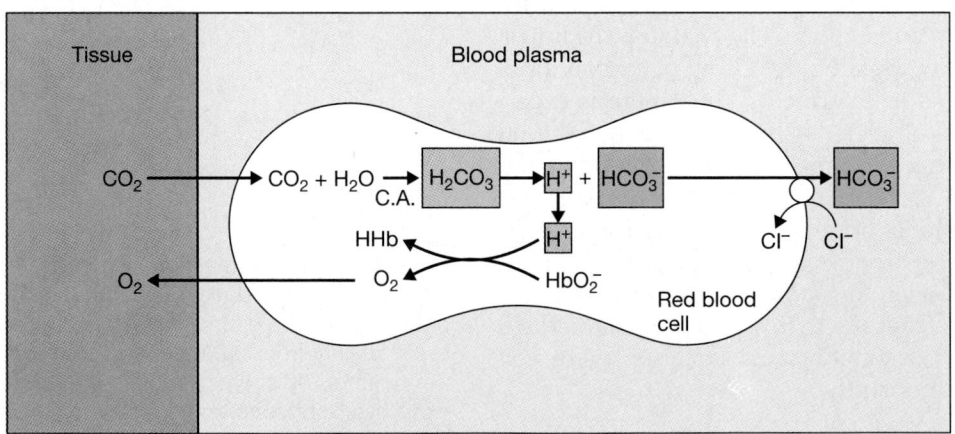

Figure 25–3

Buffering of carbonic acid by hemoglobin. The hydrogen ion formed by dissociation of carbonic acid combines with oxyhemoglobin (abbreviated HbO_2^-), resulting in formation of reduced hemoglobin (HHb) and release of oxygen. C.A. is the enzyme carbonic anhydrase. Bicarbonate exchanges with chloride across the red cell membrane via an anion exchanger. The reverse reactions occur when P_{CO_2} is lowered, as happens in the capillaries of the lungs.

(a) Normal condition

$[HCO_3^-] = 24$ mmoles per liter
$[$dissolved $CO_2] = 1.2$ mmoles per liter
($P_{CO_2} = 40$ mm Hg)
pH = 7.40

(b) Closed system (after adding 10 mmoles strong acid per liter)

$[HCO_3^-] = 14$ mmoles per liter
$[$dissolved $CO_2] = 11.2$ mmoles per liter
($P_{CO_2} = 373$ mm Hg)
pH = 6.20

(c) Open system (after adding 10 mmoles strong acid per liter)

$[HCO_3^-] = 14$ mmoles per liter
$[$dissolved $CO_2] = 1.2$ mmoles per liter
($P_{CO_2} = 40$ mm Hg)
pH = 7.17

(d) Open system + hyperventilation (after adding 10 mmoles strong acid per liter)

$[HCO_3^-] = 14$ mmoles per liter
$[$dissolved $CO_2] = 0.90$ mmoles per liter
($P_{CO_2} = 30$ mm Hg)
pH = 7.29

Figure 25–4

Schema to illustrate the effectiveness of the bicarbonate/CO_2 system in buffering added strong acid. Note particularly how in an open system (*c* and *d*), which allows removal of CO_2, the fall in pH is much less than in the closed system (*b*). (*a*) is the normal condition.

in which $CO_2(g)$ represents CO_2 in the lung alveoli, or gaseous CO_2, and $CO_2(d)$ represents CO_2 dissolved in pulmonary capillary blood. CO_2 in the lung alveoli ($CO_2(g)$) equilibrates with CO_2 dissolved in pulmonary capillary blood. $CO_2(d)$ reacts with water in a **hydration reaction** to form carbonic acid. The reverse reaction, the formation of $CO_2(d)$ and water from H_2CO_3, is called the **dehydration reaction.**

Both of these reactions are slow if uncatalyzed. As mentioned in previous discussions, they are speeded up by the zinc-containing enzyme **carbonic anhydrase.** This enzyme is present in red blood cells and in cells of the kidneys, pancreas, stomach, ciliary body of the eye, and choroid plexuses of the brain. The dissociation of carbonic acid to hydrogen ions and bicarbonate or the association of hydrogen ions and bicarbonate to form carbonic acid occur essentially instantaneously. The dissociation of bicarbonate ions into hydrogen and carbonate (CO_3^{2-}) ions occurs to an appreciable extent only at pH values much more alkaline than those encountered in the body fluids, and so we will not discuss this last reaction. It should be noted, however, that bone contains considerable amounts of calcium carbonate, which is important in buffering acids.

For the bicarbonate-CO_2 system, we can write the Henderson-Hasselbalch equation as

$$pH = 6.10 + \log ([HCO_3^-]/0.03 P_{CO_2})$$

Carbonic acid does not appear explicitly in this expression, but is implicit. The denominator of the log term is the concentration of dissolved CO_2. In blood plasma at a temperature of 37°C, the concentration of dissolved CO_2 is related to the P_{CO_2} by the equation

$$[CO_2(d)] = 0.03 \times P_{CO_2}$$

In this equation, $CO_2(d)$ is in mmoles per liter, P_{CO_2} is in millimeters of Hg, and 0.03 is the solubility coefficient in mmoles per liter per millimeter of Hg. $CO_2(d)$ and H_2CO_3 are usually in equilibrium (remember carbonic anhydrase). In blood plasma at 37°C, the ratio of these two chemical species is about 400 to 1 (1.2 mmoles per liter versus 3 μmoles per liter). Because the H_2CO_3 concentration is so low and hard to measure, $CO_2(d)$ is substituted for H_2CO_3 in the Henderson-Hasselbalch equation just given. The constant (6.10) incorporates both the acid dissociation constant for H_2CO_3 and the constant for the H_2CO_3-$CO_2(d)$ equilibrium. The real source of hydrogen ions is H_2CO_3, not CO_2; the H_2CO_3 concentration is directly proportional to the P_{CO_2} by way of dissolved CO_2.

As stated previously, the pK of the bicarbonate-CO_2 buffer pair, 6.10, is remote from the pH of 7.40 that we want to maintain in blood. From this alone, we might conclude that this buffer pair is not effective. This is not true, however, because it operates in an "open system." In an **open system,** a component can be removed or added. An example of how this works is presented in Figure 25–4. Suppose we had blood containing 24 mmoles of bicarbonate per liter plasma and a P_{CO_2} of 40 mm Hg ($CO_2(d) = 1.2$ mmoles per liter). The pH of the blood can be calculated as follows: pH = 6.10 + log ($[HCO_3^-]/0.03 P_{CO_2}$) = 6.10 + log (24/[0.03 × 40]) = 6.10 + log 20 = 6.10 + 1.30 = 7.40. Suppose we now add 10 mmoles of strong acid to each liter of blood. For simplicity, we will ignore the fact that bicarbonate is not the only buffer base that combines with the added hydrogen ions. According to the following reactions, 10 mmoles of dissolved CO_2 will be formed:

$$H^+ + HCO_3^- \rightarrow H_2CO_3 \rightarrow H_2O + CO_2$$

In a closed container (Fig. 25–4b), from which none of the CO_2 formed can escape, the resulting pH is equal to $6.10 + \log ([24 - 10]/[1.2 + 10]) = 6.20$. In an open system (see Fig. 25–4c), the CO_2 produced can be blown off as it is formed, and P_{CO_2} can be maintained constant at 40 mm Hg ($CO_2(d) = 1.2$ mmoles per liter). In this case, the pH is equal to $6.10 + \log ([14]/[1.2]) = 7.17$, which is much better than a pH of 6.20 (a lethal pH). A fall in blood pH causes hyperventilation so that the P_{CO_2} is not maintained constant, but actually falls—for example, to 30 mm Hg ($CO_2(d) = 0.90$ mmoles per liter) (see Fig. 25–4d). In this case, the pH is equal to $6.10 + \log ([14]/[0.90]) = 7.29$. We can see, therefore, that the ability to blow off CO_2 and to lower its level in the blood by hyperventilation diminishes the fall in pH produced by adding strong acid.

The bicarbonate-CO_2 buffer system is "open" in other respects, too. There is a continuous source of CO_2 from metabolism, which can replace CO_2 consumed when strong base is added to the body. The kidneys can change the amount of bicarbonate in the extracellular fluid by excreting bicarbonate in the urine (when there is excess base in the body) or by synthesizing new bicarbonate and adding it to the blood (when there is excess acid in the body). The ability of the body to change the amounts of components of this particular buffer pair makes the bicarbonate-CO_2 system a remarkably effective buffer.

The Isohydric Principle

We have discussed the various buffer systems separately, but in reality they all work together. In fact, in any fluid compartment, all are in equilibrium with the same hydrogen ion concentration. This idea is called the **isohydric principle**. The term *isohydric* means "same hydrogen ion." We can therefore write:

$$pH = 6.8 + \log([HPO_4^{2-}]/[H_2PO_4^-])$$
$$= pK_{protein} + \log ([proteinate^-]/[H\text{-}proteinate])$$
$$= 6.1 + \log ([HCO_3^-]/0.03\ P_{CO_2})$$

When an acid or base is added to the body, all buffer pairs participate in buffering. The importance of each depends on its pK, amount, and accessibility. The expression just given reemphasizes the point that the ratio of conjugate base to acid for any buffer pair of known pK defines the pH. For the remainder of this chapter, we will stress the bicarbonate-CO_2 system and its effects on blood pH. Recognize, however, that other buffers are also important and interact with the bicarbonate-CO_2 buffer pair. Changes in the ratio HCO_3^-/CO_2 indicate that other body buffers are

changing too. The reason for emphasizing the bicarbonate-CO_2 system is mainly that the lungs and kidneys act on its components. The respiratory system controls arterial blood pH by regulating the P_{CO_2}, and the kidneys control blood pH by regulating plasma bicarbonate concentration.

The Lungs and Kidneys Normally Stabilize Extracellular Fluid pH

The Respiratory System as a Buffer System: Control of the P_{CO_2}

The respiratory system controls the partial pressure of CO_2 in arterial blood. Recall from Chapter 20 that the level of alveolar ventilation is a determinant of the P_{CO_2} in the alveolar spaces of the lung. CO_2 pressures in alveoli and arterial blood are equal because of equilibration of CO_2 between alveolar gas and pulmonary capillary blood. For a constant rate of CO_2 production, alveolar ventilation and alveolar P_{CO_2} are inversely related. The higher the rate of alveolar ventilation, the lower the alveolar P_{CO_2}, and vice versa. Normally, CO_2 is expired at the same rate that it is produced from metabolism, and the P_{CO_2} remains near 40 mm Hg.

In a state of hyperventilation, CO_2 is flushed out of the alveolar spaces at a rate greater than the rate at which it is produced and added to the alveolar spaces. Consequently, the P_{CO_2} of alveolar gas and arterial blood falls to a new lower-than-normal level, and the reactions

$$CO_2 + H_2O \rightleftharpoons H_2CO_3 \rightleftharpoons H^+ + HCO_3^-$$

are pulled to the left. The hydrogen ion concentration of the blood falls; in other words, the blood becomes more alkaline.

During hypoventilation, CO_2 added to the alveolar spaces is not removed at an adequate rate; P_{CO_2} in the alveoli and arterial blood rises. The above reactions are pushed to the right, so that a more acidic blood pH results. This explains the low pH in the elderly man described at the beginning of this chapter.

The respiratory system normally acts to minimize pH changes in the blood. Such changes might be produced by adding a "fixed" acid or a base to the blood or by adding or removing too much CO_2.

A **fixed acid** is an acid other than carbonic acid. In contrast to the concentration of carbonic acid, the concentration of a fixed acid in blood is not affected by lung function. If, for example, hydrochloric acid is added to or produced in the body, it will be neutralized by chemical buffers, and then the hydrogen ions (combined with urinary buffers) and chloride will be excreted by the kidneys. A fixed acid cannot be blown off by the lungs.

If the blood is made more acidic by addition of fixed acid, pulmonary ventilation is increased. The receptors that sense the fall in pH and reflexly increase ventilation are mainly the **peripheral chemoreceptors,** the carotid and aortic bodies (Chapter 10). A large pH drop may also stimulate central chemoreceptors or the medullary respiratory center itself. Increased ventilation serves to lower the arterial blood P_{CO_2} and carbonic acid concentration, and so reduces the acidic shift in blood pH. Figure 25–4*d* illustrates how hyperventilation in response to added acid minimizes a fall in blood pH.

If the blood is made more alkaline by infusing base, the opposite changes occur. In this case, hypoventilation results, which leads to retention of CO_2, a higher carbonic acid concentration in the blood, and a less alkaline shift in blood pH. Hypoventilation is limited because it causes retention of CO_2 and a fall in arterial oxygen, and both of these changes stimulate ventilation.

If excess CO_2 accumulates in the body because of inadequate alveolar ventilation or breathing of CO_2-enriched air, ventilation will be stimulated. CO_2 is a very powerful stimulus to ventilation; an increase in arterial blood P_{CO_2} of 2 to 3 mm Hg causes the rate of ventilation to double. The receptors mediating this increase in ventilation are chiefly **central chemoreceptors** in the medulla of the brain. Carbon dioxide diffuses from the blood into brain interstitial and cerebrospinal fluids, where it forms carbonic acid, which liberates hydrogen ions. It is generally believed that chemoreceptors respond to hydrogen ion concentration (or pH) and not to molecular CO_2 directly. The peripheral chemoreceptors are also stimulated by an increase in arterial blood P_{CO_2}, but are less important than the central chemoreceptors in the ventilatory response to CO_2. The increase in ventilation caused by CO_2 is useful in bringing about removal of CO_2, thereby minimizing carbonic acid accumulation in the blood and an acidic shift in blood pH.

Buffering of acid-base disturbances by the respiratory system is rapid but relatively coarse. Respiratory responses begin within minutes and are maximal in about 12 to 24 hours. The respiratory system brings blood pH back closer to normal, but cannot eliminate a fixed acid or base from the body, and so cannot restore a normal pH. By contrast, renal responses to acid-base disturbances are slower and more complete. If an excess of fixed acid or base is added to the body, normal kidneys will slowly (over many hours or days) eliminate it from the body and return the blood pH to normal. The kidneys exert a fine control over acid-base balance.

The Kidneys as a Buffer System: Acidification of the Urine

The kidneys play a major role in acid-base regulation by getting rid of acid or base in the urine when there is an excess or deficit of hydrogen ions in the body. As mentioned previously, the usual condition is one of excess acid, since metabolism of food produces about 50 to 100 mEq of strong acid in a day. This acid is first neutralized by bicarbonate (and other chemical buffer bases) in the body. Then the kidneys get rid of the acid in the urine. In the process of excreting acid, the kidneys add new bicarbonate to the blood and so bring the blood pH back to normal.

Most of the hydrogen ions excreted in the urine are combined with urinary buffers. The limits of urine pH are from a maximum of 8.0 to a minimum of 4.5. Suppose we excrete 1.5 liters of pH 4.5 urine in a day. How many *free* hydrogen ions are excreted? This can be calculated by multiplying the urine volume (1.5 liters) by the hydrogen ion concentration ($10^{-4.5}$ mEq per liter) and amounts to about 0.05 mEq. This amount is clearly inadequate to take care of the amount of acid that needs to be excreted. Most of the hydrogen ions excreted in the urine are not free but are combined with buffers, such as ammonia and inorganic phosphate.

Titratable acid is measured in the clinical laboratory by determining how many milliequivalents of strong base (NaOH) are needed to bring a urine sample back to the pH of arterial blood (usually 7.4). Most of the titratable acid is normally inorganic phosphate. Depending on the circumstances, other substances in the urine (creatinine, various organic acids) may also be included. In the test tube, the following reaction takes place:

$$H_2PO_4^- + OH^- \longrightarrow HPO_4^{2-} + H_2O$$

This reverses the reaction that took place along the kidney tubules when hydrogen ions were added to the tubular urine:

$$H^+ + HPO_4^{2-} \longrightarrow H_2PO_4^-$$

Ammonia also buffers secreted hydrogen ions in the urine, according to the following reaction:

$$H^+ + NH_3 \rightleftharpoons NH_4^+$$

The pK_a for the ammonium ion (about 9.0) is so high that it is not included when titratable acid is measured. In blood plasma and urine, most ammonia is in the form of ammonium ion and not NH_3. Of the hydrogen ions excreted in urine, normally nearly one third are in the form of titratable acid, close to two

(a) Reabsorption of filtered bicarbonate

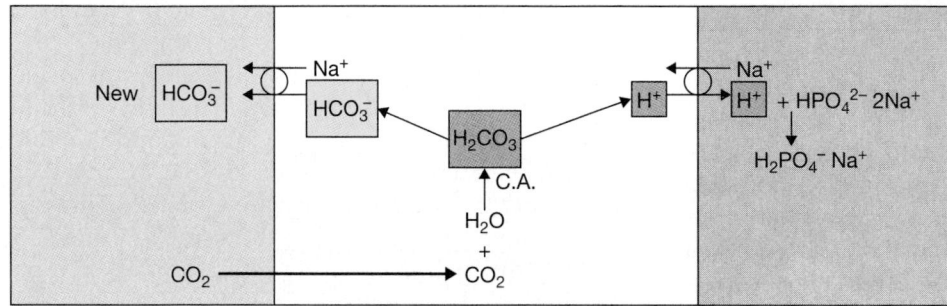

(b) Formation of titratable acid

(c) Excretion of ammonia

Figure 25–5

Cell models of the three processes involved in urinary acidification: (*a*) reabsorption of filtered bicarbonate, (*b*) formation of titratable acid, and (*c*) excretion of ammonia. In the distal nephron, H⁺ secretion occurs via a H⁺-ATPase or H⁺/K⁺-ATPase, and not via the N₂⁺/H⁺ exchanger illustrated here. C.A. = carbonic anhydrase.

thirds are present as ammonia (ammonium ion), and a trivial amount is free.

The urine is acidified by secretion of hydrogen ions by the tubular epithelium (Fig. 25–5). The nature of the secretory processes, the amounts of hydrogen ions secreted, and the pH changes produced differ in different segments of the nephron. For example, most of the hydrogen ions are secreted by the proximal convoluted tubule. This secretion is accomplished by a brush border membrane carrier that exchanges one sodium for one hydrogen ion (Na⁺/H⁺ exchanger), an example of a secondary active transport process. Because the glomerular filtrate contains large amounts of buffer (especially bicarbonate) and because the proximal convoluted tubule cannot develop steep [H⁺] gradients, the fall in tubular fluid

pH along the proximal convoluted tubule is modest (e.g., to pH 6.8). More distal portions of the nephron (distal convoluted tubule, collecting duct) secrete smaller amounts of hydrogen ions. Distal hydrogen ion secretion involves primary active transport by a H⁺-ATPase and H⁺/K⁺-ATPase in the luminal cell membrane. The distal portions of the nephron can establish steep transepithelial pH gradients. A final urine pH of 4.4, compared to a blood pH of 7.4, means that the urine [H⁺] is 1000 times higher than in plasma. For simplicity, we treat the tubular epithelium as a whole in the cell model in Figure 25–5 and do not consider differences in different nephron segments.

Three processes are involved in urinary acidification: (1) reabsorption of filtered bicarbonate, (2) excretion of titratable acid, and (3) excretion of ammonia (see Fig. 25–5). All three involve hydrogen ion secretion; furthermore, all three processes are associated with addition of bicarbonate to the peritubular capillary blood by the kidney tubules. Hydrogen ion secretion into the lumen and addition of bicarbonate ion to the blood by the tubules are best viewed as opposite sides of the same coin. Reabsorption of filtered bicarbonate is important in conserving the normal amount of bicarbonate in the extracellular fluid. In the process of glomerular filtration, large quantities (4500 mEq per day) of bicarbonate are filtered, and only traces (e.g., 2 mEq per day) are excreted in the usual acidic urine. Reabsorption of filtered bicarbonate reclaims bicarbonate for the body. Most of the hydrogen ions secreted are used to bring about reabsorption of filtered bicarbonate and are not excreted in the final urine. Those that are excreted are combined with phosphate (titratable acid) and ammonia. In these latter processes, *new* bicarbonate is generated in the kidney tubule cells. This new bicarbonate is added to the blood and replaces bicarbonate consumed in the buffering of strong acids produced in the body.

Reabsorption of Filtered Bicarbonate in Urinary Acidification.

Figure 25–5a depicts the reabsorption of filtered bicarbonate. This occurs mainly in the proximal convoluted tubule. Hydrogen ions are secreted into the tubular urine in exchange for sodium ions (Na^+/H^+ exchanger) and combine with bicarbonate that has been filtered by the glomeruli. This leads to formation of carbonic acid in the tubular urine. The dehydration of this acid to CO_2 and water is catalyzed in the proximal convoluted tubule lumen by carbonic anhydrase in the brush border membrane. P_{CO_2} values in tubular urine, cells, and blood are essentially identical since CO_2 diffuses rapidly through cell membranes. In the cell, CO_2 combines with water to form carbonic acid. This reaction is catalyzed by carbonic anhydrase inside the cell. Carbonic acid dissociates into hydrogen ions and bicarbonate. The hydrogen ions are secreted into the lumen. Bicarbonate ions exit from the basolateral side of the cell along with sodium ions. In the proximal tubule, this appears to involve a coupled Na/HCO_3 transport mechanism in the basolateral cell membrane. In this scheme, it should be apparent that filtered bicarbonate is not reabsorbed by direct transport out of the tubular urine; rather, filtered bicarbonate is reabsorbed indirectly by hydrogen ion secretion.

Excretion of Titratable Acid.

Normally, we ingest phosphate in our diet. Filtered phosphate is incompletely reabsorbed by the kidney tubules, and the unreabsorbed phosphate acts as a pH buffer in the urine.

The cell scheme for formation of titratable acid is shown in Figure 25–5b. Hydrogen ions are secreted into the tubular urine, and they titrate buffers in the urine in the acidic direction. For example, the basic form of phosphate (HPO_4^{2-}) is converted to the acidic form ($H_2PO_4^-$). One of the sodium ions accompanying HPO_4^{2-} is reabsorbed. The intracellular and membrane events are similar to those described in Figure 25–5a. For each mEq of hydrogen ion excreted as titratable acid, a *new* bicarbonate ion is added to the blood. Excretion of titratable acid, therefore, eliminates hydrogen ions from the body and, at the same time, restores depleted plasma bicarbonate reserves.

Excretion of Ammonia.

Figure 25–5c summarizes the role of ammonia in urinary acidification. Ammonia is synthesized by kidney tubule cells (mainly in the proximal tubule). The metabolism of glutamine results in formation of two ammonium ions and one α-ketoglutarate molecule. Each ammonium ion yields one molecule of free base ammonia (NH_3) and one hydrogen ion. Ammonia is secreted into the urine by two mechanisms. In the proximal tubule, ammonia is actively secreted by the Na^+/H^+ exchanger (operating in a Na^+/NH_4^+ exchange mode). NH_3 also diffuses readily through cell membranes into the urine and combines with a secreted hydrogen ion to form an ammonium ion. Recall from Chapter 4 that, mainly because of its charge, the ammonium ion does not readily diffuse through the lipid phase of plasma membranes, and so is in effect "trapped" in an acidic urine, a process called **diffusion trapping.** Diffusion trapping of ammonia is especially important in the medullary collecting ducts because they have the lowest intraluminal pH. The ammonium ion is accompanied by an anion (e.g., chloride) in the urine and allows the body to conserve a sodium ion, which it replaces. In the cell, hydration of CO_2 produces hydrogen ions and bicarbonate ions, as described previously. These hydrogen ions are consumed when α-ketoglutarate is metabolized to form glucose or CO_2. The bicarbonate ions are reabsorbed together with sodium ions. Note that for each hydrogen ion excreted in the urine as ammonia (ammonium ion), an equivalent quantity of *new* bicarbonate is added to the blood. The kidneys' ability to synthesize ammonia from glutamine is increased under conditions of excess acid in the body. This adaptive increase in ammonia synthesis takes several days to develop fully but allows the body to dispose of large acid loads.

Effects of CO_2 on Urinary Acidification.

The effects of CO_2 on urinary acidification are also indicated in Figure 25–5. A high blood P_{CO_2}, by mass action,

(text continues on page 775)

R E S E A R C H   F O C U S

Collecting Duct Intercalated Cells

Scattered among the majority (principal) cells of the collecting ducts are cells called *intercalated cells*. These cells are distinguished ultrastructurally by their dark cytoplasm, numerous mitochondria, large numbers of small apical vesicles, and characteristic luminal membrane surface projections. The intercalated cells play an important role in renal regulation of acid-base balance. Recent studies, using electrophysiological, immunological, and optical techniques, have helped to define key features of these cells, although many questions remain unanswered.

Intercalated cells have traditionally been divided into two subtypes: acid-secreting (α) and base or bicarbonate-secreting (β). Models for these two cell types are illustrated below.

The α-intercalated cell has a H^+-ATPase in its luminal membrane and a chloride-bicarbonate exchanger in its basolateral membrane. The luminal membrane H^+-ATPase is a vacuolar type of ATPase (the same kind as is found in lysosomes, endosomes, and secretory vesicles). In the presence of an acidosis, the membranes of intracellular vacuoles, which possess these H^+ pumps, fuse with the luminal membrane, thereby inserting more acid-secreting pumps. The basolateral membrane of α-intercalated cells contains a chloride-bicarbonate exchanger which is immunologically and functionally identical to the chloride-bicarbonate exchanger (the so-called "band 3 protein") found in red blood cells.

The β-intercalated cell has the opposite polarity. In this case, the H^+-ATPase is in the basolateral membrane and the chloride-bicarbonate exchanger is in the luminal membrane. This cell secretes bicarbonate ions into collecting duct urine. The notion that tubular cells can secrete this ion is relatively new; for many years it was thought that tubules could only reabsorb bicarbonate. In an alkalotic animal, the rate of bicarbonate secretion by β-intercalated cells is increased. The chloride-bicarbonate exchanger of β-intercalated cells differs immunologically and in its sensitivity to transport inhibitors from the red cell exchanger.

These two subtypes of intercalated cells look almost like mirror images. Some investigators have proposed that they can reverse their polarity and interconvert, thereby changing the direction of H^+ and HCO_3^- transport. Convincing evidence for this idea has, however, been hard to come by. Recent studies indicate that there are significant numbers of intermediate forms of intercalated cells that possess features of both α and β subtypes. Perhaps these "hybrids" can change into α or β subtypes, depending on the acid-base status of an animal.

The cell models shown above also depict another novel feature of intercalated cells, the presence of a H^+/K^+-ATPase. This pump, when in the luminal membrane, may be important not only for collecting duct H^+ secretion, but also for K^+ reabsorption (for example, during potassium depletion). The collecting duct H^+/K^+-ATPase is remarkably similar to the H^+/K^+-ATPase found in the gastric parietal cell (Chapter 22). Future studies will undoubtedly focus on the regulation of this transporter in the kidney and its role in urinary acidification and potassium balance.

■ **Why might a defect in intercalated cell H^+ secretion lead to hypokalemia?**

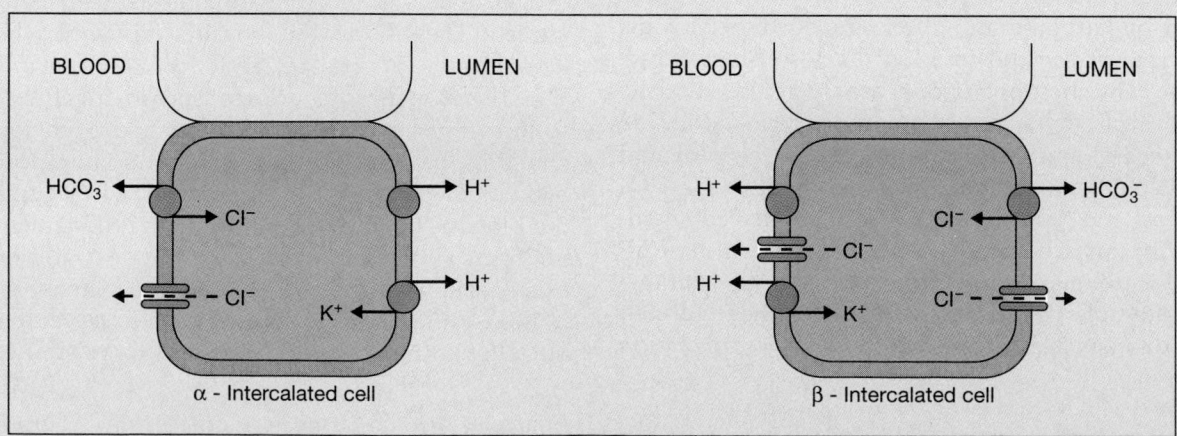

(*a*) A model of acid-secreting and (*b*) bicarbonate-secreting subtypes of the intercalate.

favors generation of hydrogen ions in the cell. This results in more complete reabsorption of filtered bicarbonate and increased excretion of hydrogen ions combined with titratable acids and ammonia. Conversely, a low blood P_{CO_2} results in a lower rate of hydrogen ion secretion; filtered bicarbonate is less completely reabsorbed, and excretion of an alkaline (bicarbonate-rich) urine is favored. Fewer hydrogen ions will be excreted in combination with titratable acids and ammonia. These effects of CO_2 on kidney tubule hydrogen ion secretion are key to understanding the renal compensations for respiratory disturbances of acid-base balance.

If we reconsider the Henderson-Hasselbalch equation $pH = 6.10 + \log ([HCO_3^-]/0.03 \, P_{CO_2})$, we can see that blood pH is affected by changes in plasma bicarbonate concentration. In cases of acid excess, the kidneys increase acid excretion in the urine, and raise the plasma $[HCO_3^-]$. This helps to make the blood less acidic; that is, it raises the pH toward a normal value. In cases of base excess, the kidneys increase the output of bicarbonate in the urine, thereby lowering the plasma bicarbonate concentration toward normal. This helps to make the blood less alkaline; that is, it lowers the pH toward a normal value. From these relations, we can see that it is useful to think of the kidneys as regulating plasma pH by changing the plasma bicarbonate concentration in the appropriate direction. We will discuss later the renal compensation for various acid-base disturbances.

Disturbances of Acid-Base Balance

In numerous conditions, the pH of arterial blood may not be within the normal range, 7.35 to 7.45. If arterial blood is more acidic than pH 7.35, **acidemia** exists. If the blood is more alkaline than pH 7.45, **alkalemia** exists. The range of pH values compatible with life is approximately 6.8 to 7.8.

There are four fundamental types of acid-base disturbance: (1) **respiratory acidosis,** (2) **respiratory alkalosis,** (3) **metabolic acidosis,** and (4) **metabolic alkalosis.** This classification is based on the nature of the process leading to the disturbed pH, and on the direction of pH change. If the problem is too much or too little CO_2, then a **respiratory disturbance** of acid-base balance is present. If the problem is an excess or deficit of a fixed acid or base, then a **metabolic (nonrespiratory) disturbance** of acid-base balance is present. Acidosis leads to acidemia, and alkalosis leads to alkalemia. These four disturbances of acid-base balance are each accompanied by characteristic changes in arterial blood pH, P_{CO_2}, and $[HCO_3^-]$ (Table 25–2). We will describe for each disturbance the nature of the abnormal process, characteristic changes in blood chemistry, some common causes, and how buffering mechanisms in the body minimize the deviation from normal pH.

Accumulation of CO_2 Results in Respiratory Acidosis

Respiratory acidosis is an abnormal process characterized by accumulation of CO_2. Most often it is due to failure to expire (blow off) metabolically produced CO_2 at an adequate rate. In other words, alveolar ventilation is inadequate, and CO_2 accumulates in the blood. The problem may be caused by (1) insufficient neural drive for ventilation, (2) inadequate movements of the respiratory muscles or thoracic cage, (3) airway obstruction, or (4) pulmonary disease. Respiratory acidosis is also produced if a person breathes CO_2-enriched air. If CO_2 is allowed to build up in the

---

Table 25–2
Summary of Directional Changes in Blood Values in the Four Fundamental Types of Acid-Base Disturbance

Disturbance	Arterial Blood			Compensatory Response
	pH	Plasma [HCO$_3^-$]	P$_{CO_2}$	
Respiratory acidosis	↓	↑	⇑	Kidneys increase H$^+$ excretion (increase plasma [HCO$_3^-$])
Respiratory alkalosis	↑	↓	⇓	Kidneys increase HCO$_3^-$ excretion (decrease plasma [HCO$_3^-$])
Metabolic acidosis	↓	⇓	↓	Alveolar hyperventilation
Metabolic alkalosis	↑	⇑	↑	Alveolar hypoventilation

Heavy arrows indicate the main change.

arterial blood, the following reactions are pushed to the right:

$$CO_2 + H_2O \rightleftharpoons \boxed{H_2CO_3} \rightleftharpoons \boxed{H^+} + \boxed{HCO_3^-}$$

Clearly, retention of CO_2 (the primary chemical change) leads to an increase in $[H^+]$ and so to a fall in pH. The plasma $[HCO_3^-]$ will also be elevated, since body buffer bases (e.g., proteins and phosphates) combine with hydrogen ions generated in the above reactions, leaving behind bicarbonate. The increase in $[HCO_3^-]$ is proportionately less than the increase in P_{CO_2}. (Otherwise, the pH would not fall. See the Henderson-Hasselbalch equation.)

Buffering of hydrogen ions produced by excess CO_2 is first accomplished by chemical buffers in the body. These buffers include the large amounts of protein and organic phosphate buffers in cells and phosphate salts in bone. For example, hemoglobin in red blood cells binds hydrogen ions, thereby reducing the increase in free $[H^+]$.

Respiratory acidosis is most often due to inadequate ventilation. The retention of CO_2 and fall in blood pH that ensue, however, both act as important stimuli to enhance ventilation. In addition, hypoxia due to inadequate ventilation may also be present, and a low arterial P_{O_2} acts on the peripheral chemoreceptors to stimulate ventilation. The respiratory system is the problem, however, so it cannot return blood pH to normal.

The kidneys compensate for respiratory acidosis by increasing the output of hydrogen ions in the urine. As was mentioned, a high P_{CO_2} stimulates the renal tubular epithelium to secrete hydrogen ions into the tubular urine. The result is that all of the filtered bicarbonate is reabsorbed, and increased amounts of hydrogen ions, combined with urinary buffers, are excreted in the urine. By getting rid of hydrogen ions in the urine, the kidneys raise the plasma bicarbonate concentration. In other words, they add buffer base (bicarbonate) to the blood, diminishing the severity of the acidemia. The renal response takes a few days, so with chronic respiratory acidosis, higher plasma levels of bicarbonate are achieved than those seen with respiratory acidosis of short duration.

Excessive Loss of CO_2 Results in Respiratory Alkalosis

Respiratory alkalosis is easily understood as the opposite of respiratory acidosis. In this case, we have an abnormal process leading to excessive loss of CO_2. This causes a fall in arterial blood P_{CO_2}, a fall in plasma $[H^+]$ (a rise in pH), and a decline in plasma $[HCO_3^-]$. Respiratory alkalosis is produced by alve-

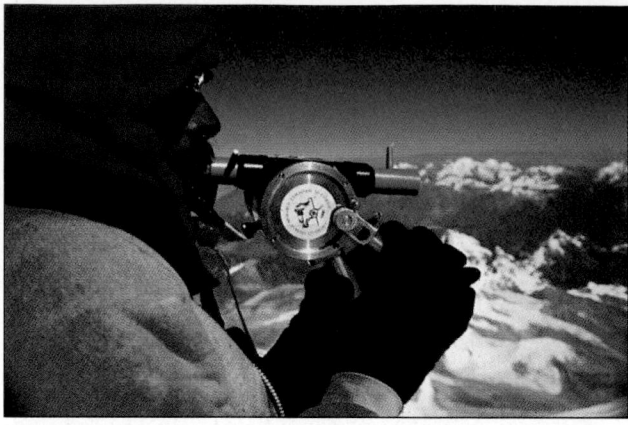

Figure 25-6

High altitude results in respiratory alkalosis, since breathing is stimulated by the low oxygen tension. The photograph shows a scientist at the summit of Mount Everest collecting an alveolar gas sample. *(© Dr. John B. West, M.D./University of California at San Diego.)*

olar hyperventilation, which results in excessive removal of CO_2 from the blood. Hyperventilation may be caused by voluntary effort, anxiety, or stimulation of respiratory centers owing to some abnormal influence acting directly on the brain (e.g., aspirin intoxication, fever, meningitis). Hyperventilation may also be produced reflexly. For example, hyperventilation and respiratory alkalosis are observed at high altitudes because of reflex stimulation of ventilation by low P_{O_2} (Figure 25-6).

The alkaline shift in pH produced during respiratory alkalosis is diminished by the action of buffers in the body. Chemical buffers (e.g., hemoglobin) are titrated in the alkaline direction and liberate hydrogen ions, thereby tending to diminish the fall in free $[H^+]$. As is the case for respiratory acidosis, most ($> 95\%$) of the chemical buffering occurs within cells. Although the origin of respiratory alkalosis is hyperventilation, the resulting loss of CO_2 and rise in pH limit the extent of hyperventilation. CO_2 and hydrogen ions are two important stimuli for ventilation, and a decrease in these stimuli leads to diminished ventilatory drive.

The kidneys compensate for respiratory alkalosis by excreting bicarbonate in the urine. A low P_{CO_2} leads to decreased hydrogen ion secretion by the tubular epithelium. The result is that filtered bicarbonate is incompletely reabsorbed, and so is excreted. The loss of bicarbonate in the urine results in a further decrease in plasma bicarbonate. In other words, the kidneys reduce the amount of base (bicarbonate) in the blood and so diminish the severity of the alkalemia. The kidneys work relatively slowly, so it

CLINICAL FOCUS

Renal Tubular Acidosis

Normal kidneys reabsorb nearly all of the filtered bicarbonate and excrete in the urine the net amount of strong acids produced by metabolism. An adult on the usual mixed diet of meat and vegetables has a net daily production of about 1 mmole H^+/kg body weight, and children usually produce and need to dispose of 2 to 3 mmoles H^+/kg body weight every day. If the kidney tubules fail to reabsorb filtered bicarbonate or have an impaired ability to secrete H^+, this leads to a type of metabolic acidosis called *renal tubular acidosis*, abbreviated *RTA*. Several types of RTA have been described; we will focus on two of these.

Distal (type I) RTA is characterized by inability to lower the final urine pH below 5.5, even in the face of severe acidosis. Normal kidneys can lower urine pH to 4.5 (a 10-fold higher H^+ concentration). Distal RTA may be the result of an inherited defect, autoimmune disease, treatment with lithium or the antibiotic amphotericin B, or the result of diseases of the kidney medulla. The fundamental defect may be impaired secretion by the collecting duct H^+-ATPase or abnormal leakiness of the collecting ducts to H^+ (so that steep transepithelial pH gradients are not maintained). The amount of ammonia trapped in the urine is diminished, so less NH_4Cl is excreted. The consequence is a fall in plasma bicarbonate and rise in plasma chloride. The acidic blood pH is buffered in part by bone; this may cause bone abnormalities, rickets in children and osteomalacia in adults. Excess liberation of calcium from bone, an alkaline urine, and low levels of citrate in the urine favor deposition of calcium phosphate in the kidneys (nephrocal-

cinosis) and renal stone formation. Diagnosis can be confirmed by an *acid load test:* ingestion of ammonium chloride tablets (0.1 g/kg body weight) leads to a urine pH less than 5.2 within 3 to 6 hr in normal people; in distal RTA, urine pH remains greater than 5.5. Treatment of this disorder can often be achieved by daily oral administration of a modest amount of sodium bicarbonate (0.5 to 3 mmoles/kg body weight) or a sodium citrate solution.

In *proximal (type II) RTA* the capacity of the proximal tubules to reabsorb filtered bicarbonate is impaired, leading to loss of this base in the urine and a fall in plasma bicarbonate and low blood pH. Proximal RTA often occurs in the setting of a generalized disorder of proximal tubule function and is seen in several inherited diseases, after renal transplantation or intoxication with heavy metals, or after treatment with carbonic anhydrase inhibitors or certain antibiotics. A *bicarbonate titration test* can be used to confirm the diagnosis. Sodium bicarbonate is slowly infused intravenously or given by mouth; in proximal RTA, bicarbonate will be excreted at a plasma concentration below that observed in normal individuals or patients with distal RTA. If the acid loading test is performed, the urine pH may fall below 5.5. If the acidosis is severe, treatment may require large daily doses of sodium bicarbonate (e.g., 5 to 15 mmoles/kg body weight) to replace the bicarbonate lost in the urine.

■ **Why is renal ammonium ion excretion decreased in distal RTA?**

takes them a few days to compensate fully for a persistent respiratory alkalosis.

Metabolic (Nonrespiratory) Acidosis Results from Gain of Fixed Acid or Loss of Bicarbonate

Metabolic acidosis is an abnormal process characterized by gain of fixed acid (i.e., an acid other than H_2CO_3) or loss of bicarbonate. Many conditions can produce metabolic acidosis. In renal failure, the acid end-products of metabolism are not excreted at a rate that keeps up with production. In uncontrolled diabetes mellitus, large quantities of ketone body acids are produced. If the circulation is inadequate—for example, due to cardiogenic or hemorrhagic shock—

glucose is converted anaerobically to lactic acid. In heavy exercise, too, lactic acid production is increased. Ingestion of certain acidifying agents (e.g., ammonium chloride) produces a metabolic acidosis. Loss of bicarbonate from the body, due to (1) diarrhea (loss of alkaline intestinal fluids) or (2) abnormal excretion of bicarbonate in the urine, also results in metabolic acidosis.

If strong acid is added to the body, the reactions

$$H^+ + HCO_3^- \rightleftharpoons H_2CO_3 \rightleftharpoons H_2O + CO_2$$

are pushed to the right. We can then correctly predict that in this type of metabolic acidosis, the hydrogen ion concentration should be increased and the plasma bicarbonate concentration decreased. Further, we might predict that P_{CO_2} should be elevated (but recall that CO_2 operates in an open system). P_{CO_2} is increased only transiently during acute infusion of large amounts of strong acid; in established metabolic acidosis, P_{CO_2} is actually below normal due to respiratory compensation (see Table 25–2).

Buffering during metabolic acidosis is accomplished by chemical buffers present in extracellular fluid, cells, and bone. Roughly one half of the buffering occurs in extracellular fluid, where bicarbonate is the principal buffer. Due to the acidic blood pH, the respiratory system is stimulated and lowers the P_{CO_2}, and hence $[H_2CO_3]$, in the blood, thereby making the blood less acidic (more alkaline). Respiratory compensation for a metabolic acidosis is prompt (minutes to hours), but usually does not restore the blood pH to normal. Figure 25–4 illustrates how the bicarbonate-CO_2 system participates in buffering strong acid.

Renal compensation is slower (takes days) but more complete than respiratory compensation. The kidneys increase the output of hydrogen ions in the urine. Since the plasma bicarbonate is mainly decreased, the filtered bicarbonate load is reduced and all of the filtered bicarbonate is reabsorbed. Increased amounts of hydrogen ions, together with urinary buffers (phosphate, ammonia), are excreted in the urine, so new bicarbonate is added to the blood by the kidneys to replenish depleted bicarbonate reserves. If the underlying cause of the metabolic acidosis is corrected, then normal kidneys bring the pH of the blood back to normal.

Metabolic (Nonrespiratory) Alkalosis Results from Gain of Strong Base or Bicarbonate or Loss of Acid

Metabolic alkalosis is an abnormal process characterized by gain of strong base or bicarbonate or loss of acid (other than H_2CO_3). Ingestion of excess sodium bicarbonate (used in treating gastric ulcers) results in a metabolic alkalosis. Oxidation of organic anions (e.g., citrate, lactate, acetate) has the same effect. A common cause of metabolic alkalosis is vomiting of acidic gastric juice. Every time a hydrogen ion is secreted into the lumen of the stomach, a bicarbonate ion is added to the blood (Chapter 22). Loss of the stomach contents through vomiting results in net addition of bicarbonate to the body. The characteristic arterial blood changes include a primary increase in plasma $[HCO_3^-]$, a rise in pH, and a compensatory rise in P_{CO_2} (see Table 25–2).

Chemical buffers in the body act to limit the alkaline shift in pH by liberating hydrogen ions as they are titrated in the alkaline direction. Respiratory compensation for a metabolic alkalosis is alveolar hypoventilation, brought about by the alkaline pH of the blood (inhibits ventilation). Respiratory compensation is limited, however, because hypoventilation leads to a rise in P_{CO_2} and a fall in P_{O_2}, both of which stimulate ventilation.

Renal compensation involves increased excretion of bicarbonate in the urine. Since the plasma bicarbonate is primarily elevated, more bicarbonate is filtered than can be reabsorbed by the tubules. Increased excretion of bicarbonate in the urine leads to a fall in plasma bicarbonate concentration and a return of blood pH back toward normal.

Clinical Evaluation of Acid-Base Status Involves a History, Physical Examination, and Arterial Blood Chemistry Values

We have described individually each of the four fundamental disturbances of acid-base balance. Proper identification of an acid-base disturbance in a patient requires clinical evidence (medical history and examination) and supporting laboratory studies. From blood acid-base data (pH, P_{CO_2}, and plasma $[HCO_3^-]$), it is often possible to suggest what type of disturbance might be present. In physiology and clinical medicine, systemic arterial blood is used to evaluate acid-base status. The pH of whole blood, measured with a pH meter, is actually the pH of plasma (not the cells) and so is a measurement of extracellular fluid pH. Blood samples need to be handled anaerobically (without exposure to air), so as to prevent CO_2 loss and an alkaline shift in pH.

Suppose, for example, that the following systemic arterial acid-base data were obtained for a patient who had been vomiting: pH 7.48, P_{CO_2} 45 mm Hg, and plasma $[HCO_3^-]$ 32 mEq per liter. What type of acid-base disturbance is present? We already know that vomiting of acidic gastric juice produces a metabolic alkalosis. To evaluate the blood chemistry values, we have to know normal values for these blood variables. These are:

pH	7.35–7.45
P_{CO_2}	35–45 mm Hg
$[HCO_3^-]$	22–26 mEq per liter

Clearly, a pH of 7.48 is above normal, and so an alkalosis is present. If it were a respiratory alkalosis, P_{CO_2} would be below normal. In this case it is not, so we conclude that a metabolic alkalosis is present. The fact that plasma bicarbonate is above normal confirms our conclusion. The P_{CO_2} on the high side of normal is probably due to respiratory compensation.

Sometimes, two independent disturbances of acid-base balance, usually one respiratory and the other metabolic, are present together in the same person. This is called a **mixed acid-base disturbance.** For example, a patient with chronic pulmonary disease may develop diarrhea. The mixed respiratory acidosis and metabolic acidosis in this patient could result in severe acidemia. Sometimes, two primary acid-base disturbances are simultaneously present and offset each other. For example, a mixed respiratory acidosis and metabolic alkalosis might be accompanied by a normal pH. Treatment of mixed acid-base disturbances is more complicated than when a single disorder is present.

Intracellular pH Regulation

Intracellular (cytoplasmic) pH is closely regulated. It is important because it affects the activity of enzymes that control metabolic reactions. Intracellular pH affects many physiological activities, including (1) muscle contraction, (2) the opening and closing of cell membrane ion channels, (3) secretion by endocrine gland cells, and (4) cell division and proliferation.

Cell cytoplasm is generally more alkaline than we would predict if hydrogen ions were passively distributed across the cell plasma membrane. With a cell membrane potential of 60 mv, inside negative, passive distribution of H^+ across the cell membrane would lead to a cell $[H^+]$ 10 times higher than that in extracellular fluid (Nernst equation; see Chapter 7). In other words, if the extracellular $[H^+]$ is 40 nM (pH = 7.4), intracellular $[H^+]$ should be 400 nM (pH = 6.4) at equilibrium. Intracellular pH in skeletal muscle cells, as determined with pH-sensitive microelectrodes, is about 6.9 ($[H^+]$ = 125 nM). Clearly, hydrogen ions are not distributed passively across the cell membrane, but are maintained in cells *below* the equilibrium value. Cells tend to gain H^+ because of metabolic production of acids and because hydrogen ions diffuse into the cells from the extracellular fluid along a favorable electrical gradient.

Intracellular $[H^+]$ is kept low by various membrane transport mechanisms that remove hydrogen ions (or add bicarbonate). One such mechanism that appears to be present in nearly all body cells is a **Na^+/H^+ exchanger.** This exchanger exchanges one H^+ for one Na^+ and is electrically neutral. Transport of H^+ out of the cell is uphill and is coupled to the downhill movement of sodium into the cell. Thus, the Na^+/H^+ exchanger is a secondary active transport system. Several forms of the Na^+/H^+ exchanger have recently been cloned.

Activity of the Na^+/H^+ exchanger is increased when the cell cytoplasm is made more acidic. In part, this reflects an increase in substrate $[H^+]$ concentration. In addition, however, H^+ also stimulates the exchanger by combining with an activator site on its cytoplasmic side. This makes the exchanger more effective in pumping H^+ out into the extracellular fluid so that the cell can counteract the threat of intracellular acidosis.

Many cells also possess various cell membrane bicarbonate transporting mechanisms—for example, sodium-independent and sodium-dependent Cl^-/HCO_3^- exchangers. These mechanisms remove bicarbonate if the cell becomes too alkaline, or add bicarbonate if the cell becomes too acidic. Different pathways are activated in response to a rise or fall in pH.

Cells have additional means to deal with excesses of acid or base. They contain large stores of protein and organic phosphate buffers that can bind or release H^+. Many biochemical reactions in cells use up or release H^+. For example, conversion of lactic acid to CO_2 and water, or to glucose, effectively disposes of acid. Various cell organelles may sequester hydrogen ions. For example, a H^+-ATPase in endosomes and lysosomes pumps H^+ out of the cytoplasm into these organelles. All of these processes assure a relatively constant cytoplasmic pH.

CHAPTER REVIEW

Summary

- An acid is a substance that can donate hydrogen ions. A base is a substance that can accept or bind hydrogen ions.
- An acid in aqueous solution reversibly dissociates into a hydrogen ion and conjugate base. A high acid dissociation constant (low pK_a) is characteristic of a strong acid; a low acid dissociation constant (high pK_a) is characteristic of a weak acid.

- The pH of a solution is equal to the negative logarithm to the base 10 of the molar hydrogen ion activity (concentration). A pH value of 7 is neutral; below 7, acidic; and above 7, alkaline. Arterial blood pH averages 7.40; various body fluids have different pH values.
- pH buffers minimize changes in pH produced when acid or base is added to a solution.

- The body obtains H^+ ions from the metabolism of foods, and from carbonic acid formed when CO_2 reacts with water.
- Chemical buffers consist of a weak acid and its conjugate base (or a weak base and its conjugate acid). Important chemical buffers in the extracellular fluid include the bicarbonate-CO_2 system, plasma proteins, and inorganic phosphate.
- The Henderson-Hasselbalch equation for the bicarbonate-CO_2 system is $pH = 6.10 + \log$ ($[HCO_3^-]/0.03 \times P_{CO_2}$). The bicarbonate-$CO_2$ buffer system has special importance in acid-base regulation, because the physiological buffers (lungs and kidneys) act on the components of this buffer pair. The respiratory system influences plasma pH by regulating the partial pressure of CO_2 (P_{CO_2}) in arterial blood by varying the level of alveolar ventilation. The kidneys influence plasma pH by getting rid of acid or base in the urine—in other words, by adding new bicarbonate to the blood or getting rid of bicarbonate.
- On the usual mixed diet, people are challenged by excess production of nonvolatile (fixed) acids. Although chemical buffers can minimize a fall in pH, the burden of eliminating hydrogen ions from the body falls upon the kidneys. Urinary acidification involves three processes: reabsorption of filtered bicarbonate, excretion of titratable acid, and excretion of ammonia. All involve hydrogen ion secretion by the kidney tubule cells. Renal reabsorption of filtered bicarbonate returns to the blood bicarbonate that was temporarily lost in the process of filtration.
- Respiratory acidosis is characterized by accumulation of CO_2 and a fall in blood pH. The renal compensation is enhanced excretion of hydrogen ions in the urine, which results in addition of bicarbonate to the blood, thereby reducing the acidic shift in blood pH.
- Respiratory alkalosis is characterized by excessive loss of CO_2 and a consequent rise in blood pH. The renal compensation is increased excretion of bicarbonate in the urine, which lowers the blood pH.
- Metabolic acidosis involves an abnormal gain of fixed acid or loss of bicarbonate, and is characterized by a fall in plasma bicarbonate and pH. Respiratory compensation lowers the blood P_{CO_2}, making the blood less acidic.
- Metabolic alkalosis is associated with an abnormal gain of base or loss of fixed acid, and is characterized by a rise in plasma bicarbonate and pH. Respiratory compensation raises the blood P_{CO_2}, bringing the blood pH closer to normal.
- Intracellular pH is stabilized by cell buffering mechanisms and by ion transport processes (e.g., Na^+/H^+ exchanger) in plasma membranes.

Review Questions

Identify Each with a Term

1. An enzyme that catalyzes the reversible reaction $CO_2 + H_2O \rightleftharpoons H_2CO_3$.
2. An arterial blood pH less than 7.35.
3. A substance that can accept or bind hydrogen ions.
4. An agent that minimizes the change in pH when either acid or base is added to a solution.
5. A chemical substance that can function as both an acid and a base.
6. An acid with a high pK_a.

Define Each Term

7. pH
8. Henderson-Hasselbalch equation
9. open system
10. respiratory acidosis
11. respiratory alkalosis
12. metabolic acidosis
13. metabolic alkalosis

Choose the Correct Answer

14. Which of the following is an acid?
    a. HCO_3^-
    b. NaOH
    c. NH_3
    d. NH_4^+

15. Most of the hydrogen ions secreted by the kidney tubules are:
    a. consumed by filtered bicarbonate
    b. excreted as ammonium ions
    c. excreted as free hydrogen ions
    d. excreted as titratable acid

16. Which process in the kidney leads to generation of *new* bicarbonate to replenish depleted bicarbonate reserves?
    a. excretion of ammonia
    b. excretion of titratable acid
    c. reabsorption of filtered bicarbonate
    d. a and b
    e. all of the above

17. The nephron segment which secretes the greatest number of hydrogen ions is the:
    a. collecting duct
    b. distal convoluted tubule
    c. proximal convoluted tubule
    d. thick ascending limb

18. The nephron segment that can establish the steepest pH gradient (urine-to-blood) is the:
    a. collecting duct
    b. distal convoluted tubule
    c. proximal convoluted tubule
    d. thick ascending limb

19. A person has been vomiting for several days. Arterial

blood values are: pH = 7.53, P_{CO_2} = 50 mm Hg, and plasma $[HCO_3^-]$ = 40 mEq/liter. What type of acid-base disturbance is present?

a. respiratory acidosis
b. respiratory alkalosis
c. metabolic acidosis
d. metabolic alkalosis

20. A person on a mountain-climbing expedition has been living at high altitude for several days. Arterial blood values are: pH = 7.48, P_{CO_2} = 25 mm Hg, and plasma $[HCO_3^-]$ = 18 mEq/liter. What type of acid-base disturbance is present?

a. respiratory acidosis
b. respiratory alkalosis
c. metabolic acidosis
d. metabolic alkalosis

21. The renal compensation for chronic respiratory alkalosis is:

a. increased excretion of bicarbonate
b. increased excretion of titratable acid
c. increased synthesis and excretion of ammonia
d. both b and c
e. all of the above

22. A person with uncontrolled diabetes mellitus also has emphysema (pulmonary disease). Arterial blood values are: pH = 7.21, P_{CO_2} = 60 mm Hg, and plasma $[HCO_3^-]$ = 23 mEq/liter. This subject has a:

a. metabolic acidosis
b. mixed metabolic acidosis and respiratory acidosis
c. mixed metabolic acidosis and respiratory alkalosis
d. respiratory acidosis

23. Which of the following plays a role in the regulation of intracellular (cytosolic) pH?

a. cell membrane Na^+/H^+ exchanger
b. cell membrane bicarbonate transport mechanisms
c. chemical buffering by intracellular proteins and phosphates
d. metabolic reactions that consume or liberate hydrogen ions
e. all of the above

Calculate

24. What is the approximate pH of a 0.01 M HCl solution?
25. What is the approximate pH of a 0.01 M NaOH solution?

26. If the pK_a of NH_4^+ is 9.0 and if urine pH is 5.0, what fraction of the ammonia present in the urine is NH_3?

27. If a person with uncontrolled diabetes mellitus excretes 200 milliequivalents of titratable acid and 500 milliequivalents of NH_4^+, how many milliequivalents of new bicarbonate have the kidneys added to the blood?

28. A person with severe, uncontrolled diabetes mellitus is found to have an arterial plasma pH of 7.15 and a P_{CO_2} of 13 mm Hg. What is the plasma bicarbonate concentration?

Answer Each Question in One or Two Sentences

29. Does a 0.1 M hydrochloric acid solution have the same pH as a 0.1 M acetic acid solution? Why or why not?
30. What two factors determine the effectiveness of a buffer?
31. Why is a person who eats a mixed diet (meat and vegetables) threatened by metabolic acidosis?
32. What are the important chemical buffers in (1) extracellular fluid, (2) intracellular fluid, (3) bone, and (4) urine?
33. Why is the bicarbonate-CO_2 buffer pair, despite a low pK of 6.10, an effective buffer in defending pH?
34. What is the isohydric principle?
35. What is the location of the receptors that stimulate ventilation when CO_2 accumulates in the body?

Problem Solving in Physiology

36. Ten mmoles of HCl are added to a liter of water and to one liter of 0.1 M pH 7.4 phosphate buffer. The resulting pHs are 2.0 and 7.2, respectively. Explain the different degrees to which the pH fell.
37. In a person with severe metabolic acidosis, why doesn't the respiratory system return blood pH to normal?
38. A person has had chronic airway obstruction and respiratory acidosis. The airway obstruction is suddenly alleviated by giving a potent bronchodilator, and an arterial blood sample is taken and has a pH of 7.50 (greater than normal). Explain this finding.

Suggested Readings

Davenport, H.W. *The ABC of Acid-Base Chemistry*, 6th ed. Chicago: University of Chicago Press, 1974.

Frelin, C., Vigne, P., Ladoux, A., and Lazdunski, M. "The regulation of intracellular pH in cells from vertebrates." *European Journal of Biochemistry*, 174:3–14, 1988.

Knepper, M.A., Packer, R., and Good, D.W. "Ammonium transport in the kidney." *Physiological Reviews*, 69:179–249, 1989.

Lowenstein, J. *Acid and Basics, A Guide to Understanding Acid-Base Disorders*. New York: Oxford, 1993.

Madshus, I.H. "Regulation of intracellular pH in eukaryotic cells." *Biochemical Journal*, 250:1–8, 1988.

Rose, B.D. *Clinical Physiology of Acid-Base and Electrolyte Disorders*, 4th ed. New York: McGraw-Hill Book Company, 1994.

Valtin, H., and Gennari, F.J. *Acid-Base Disorders. Basic Concepts and Clinical Management*. Boston: Little, Brown and Company, 1987.

Winters, R.W., and Dell, R.B. *Acid-Base Physiology in Medicine. A Self-Instruction Program*, 3rd ed. Boston: Little, Brown and Company, 1982.

Key Terms

acid (p. 762)
acid dissociation constant
   (p. 763)
acidemia (p. 775)
alkalemia (p. 775)
base (p. 762)
carbonic anhydrase (p. 769)
chemical pH buffer (p. 765)

diffusion trapping (p. 773)
fixed acid (p. 770)
Henderson-Hasselbalch
   equation (p. 763)
intracellular pH (p. 779)
ion product of water
   (p. 764)

isohydric principle (p. 770)
metabolic acidosis (p. 775)
metabolic alkalosis (p. 775)
neutralization (p. 762)
open system (p. 769)
pH (p. 763)

respiratory acidosis (p. 775)
respiratory alkalosis
   (p. 775)
strong acid (p. 763)
titratable acid (p. 771)
weak acid (p. 763)

Histological section of femur bone tissue. *(© Cabisco/Visuals Unlimited)*

Chapter 26

Calcium, Phosphate, and Bone Metabolism

Key Concepts

Calcium and phosphate ions serve important roles in the body. Each ion exists in several forms and is regulated by mechanisms involving several tissues and organs.

The majority of ingested calcium is excreted in the feces, whereas the majority of ingested phosphate is absorbed into the blood.

Calcium and phosphate metabolism in the gastrointestinal tract, kidneys, and bone is coordinated by hormonal signals. Cells involved in bone formation and bone resorption are regulated by hormones. Not only does bone serve an important structural role, but also it serves as an important reservoir of body calcium.

Hormonal mechanisms are less rapid than nonhormonal mechanisms, but can provide for changes of large capacity over long periods of time. Three hormones are involved; parathyroid hormone, calcitonin, and metabolites of vitamin D.

The most common abnormalities of bone mineral metabolism are grouped together under the term metabolic bone disease, of which there are two principal types. Osteoporosis results from an equivalent loss of bone mineral and organic matrix. Osteomalacia and rickets result from inadequate mineralization of new bone.

Chapter Outline

Overview: Functions of Calcium and Phosphate

Pathways of Calcium and Phosphate Homeostasis

Mechanisms Involved in Calcium and Phosphate Homeostasis

Regulation of Calcium and Phosphate in Plasma

Abnormalities of Bone Mineral Metabolism

It has been almost two years since Marge, an older lady in the neighborhood, had fallen and broken her hip. The break was of such a nature that she required an artificial replacement, using a stainless steel ball and socket. She was pretty uncomfortable for a while, but after several months of physical therapy she was back on her feet and as active as ever. The doctor's diagnosis was that the break was the result of something that has drawn much attention in the public media in recent years: postmenopausal osteoporosis.

It was a bit of a surprise, however, when just a few days ago Marge's husband, Henry, also fell and broke his hip. Henry is 73 and had retired from his job as a foreman of a construction firm about ten years ago. Even though he was retired, he still remained as active as ever, enjoying his fishing, vegetable garden, and outings with the grandchildren. In fact, the broken hip occurred when Henry was working in the garden. He stepped in a small hole, lost his balance, and took what initially seemed just a minor tumble—like the kind he had taken many times during his life. However, in this case there was something different and his hip gave way. The break turned out not to be as bad as Marge's, and doctors were able to repair it with a rod placed in Henry's upper thigh bone. However, the diagnosis was similar: the break was the result of osteoporosis.

The surprising aspect of Henry's experience is that osteoporosis is often portrayed as being exclusively a disease of older women. It is true that females are much more prone to development of the disease than are males, but it is not an exclusively female problem.

In this chapter, mechanisms involved in bone mineral metabolism and the primary causes of osteoporosis will be presented. Total bone mass begins to decline around age 40 to 45 in both males and females. Depending on the "peak" or starting point and the rate of bone loss, bones may become so weakened in either sex that they break with relatively minor or inconsequential trauma; thus the diagnosis of osteoporosis.

As described in several earlier chapters, a number of different homeostatic mechanisms are at work in the body. The primary focus of these is maintenance of a constant environment in which cells can develop and function. If we were to monitor a number of different physiological parameters in the body—including, for example, blood sugar concentration, core body temperature, and blood carbon dioxide concentration—results would show that one of the most closely regulated of all parameters is the plasma calcium concentration. Such strict regulation usually suggests that the factor in question plays a vital role in one or more processes in the body. This chapter will describe calcium homeostasis in the body. Although not as closely regulated, phosphate also plays several important roles in the body, and these will be described as well.

Overview: Functions of Calcium and Phosphate

Calcium, or more correctly the Ca^{2+} ion, is involved in a number of different physiological processes, several of which are listed in Table 26–1. The first of these, the effect of calcium on nerve function, is probably the most important with regard to an acute dependency on normal calcium ion concentrations. External calcium ions dramatically influence the permeability of nerves to sodium ions. Therefore, maintenance of extracellular calcium concentration is crucial if the resting membrane potential of nerves is to be maintained and nerve impulse transmission is to occur normally. When extracellular calcium ion concentrations are significantly decreased, spontaneous action potentials can be generated in nerves. If a motor neuron is involved, the result can be **tetany** of the muscles in that motor unit. Figure 26–1 illustrates a classic clinical sign indicating the first stages of **hypocalcemic (low-calcium) tetany.** Muscles in the hand and forearm contract to produce the characteristic appearance shown. With prolonged hypocalcemia, muscles of the larynx become tetanized to the extent that the individual dies from suffocation.

Other functions of calcium ion in the body are also listed in Table 26–1. At neuromuscular junctions, an influx of calcium into nerve endings serves to trigger release of acetylcholine. As described in Chapter 16, calcium fluxes in muscle provide the signal for excitation-contraction coupling. In other cell types, calcium influx provides a signal analogous to a second messenger; thus, calcium ion is important in the

---

Table 26–1

Some of the Physiological Actions of Calcium

1. Required for the maintenance of normal sodium permeability in nerves
2. Involved in triggering the release of acetylcholine from nerve endings at the neuromuscular junction
3. Involved in excitation-contraction coupling in muscle cells
4. Serves as an intracellular signal for some hormones
5. Required by some enzymes for normal activity
6. Required for protein secretion
7. Required for blood clotting to occur normally
8. Constituent of bone

Figure 26–1

Typical position of the hand in hypocalcemic tetany. The spastic contractions of muscles in the forearm and hand are the result of hyperexcitability in nerves, which occurs when the plasma calcium concentration drops below normal.

mechanism of action of some hormones. In addition, a number of cellular enzymes and proteins are known to require calcium to function properly, and the process of protein secretion is also known to depend on the presence of calcium ion.

A specialized case in which calcium ion is needed for enzymatic activity involves enzymes responsible for blood clotting (see Table 26–1). Some of the compounds used clinically to prevent clotting in blood samples taken from patients (such as citrate or EDTA) produce their effect by binding calcium and preventing its participation in the clotting reactions.

Finally, calcium is a constituent of bone, and thus in this regard it serves an important structural role in the body. As later sections will describe, the bones provide an important reservoir of calcium and therefore play an important role in the regulation of extracellular calcium ion homeostasis.

Phosphorus is also present in considerable amounts in bone. As will be described later, the majority of phosphorus in the body exists in the form of **phosphate ion (PO$_4$)**. Extracellular phosphate is not as strictly regulated as is calcium, however. As indicated in Table 26–2, phosphate serves the important role of providing a major portion of the intracellular buffering capacity. In addition, when coupled in organic form, phosphate is a component of a number of important cellular constituents and macromolecules. Nucleic acids, phospholipids, and various metabolic intermediates, to name a few, all contain phosphate. Chapter 5 introduced another important aspect of phosphate metabolism involving the reg-

Table 26–2
Some of the Physiological Actions of Phosphate

1. Functions as part of the intracellular buffer system
2. Important constituent of a variety of macromolecules, such as nucleic acids, phospholipids, metabolic intermediates, and phosphoproteins
3. Constituent of bone

ulation of certain metabolic processes by phosphorylation or dephosphorylation of key regulatory enzymes.

Calcium and Phosphorus Are Distributed Differently in Tissues

Relative distributions of calcium and phosphate among bone and other tissues in the body are presented in Table 26–3. As this table indicates, approximately 99% of the total calcium in the body is located in bone. Even muscle, in which calcium plays such a vital role in excitation-contraction coupling, contains only about 0.3% of the total. The remaining 0.7% is distributed between the extracellular fluid and the rest of the cells in the body. Phosphate is also located predominantly in bone, although not nearly to the extent as calcium; about 85% of the total body phosphate is present in bone.

Although the majority of body calcium and phosphate is present in bone, these are not the only con-

Table 26–3
Typical Body Content and Tissue Distribution of Calcium and Phosphorus

	Calcium	Phosphorus
Total body content*	~1000 g	~500 g
Tissue distribution	Relative distribution (% of total body content)	
Bone	99	85
Muscle	0.3	6
Other tissues	0.7	9

*Based on a body weight of 70 kg (150 lbs) and an average body composition.

Table 26–4
Major Inorganic Constituents of Bone

Constituent	Total Body Content Present in Bone (%)
Calcium	99
Phosphate	85
Carbonate	80
Magnesium	50
Sodium	35
Water	9

stituents found in relatively high abundance there. As Table 26–4 indicates, considerable amounts of total body carbonate, magnesium, and sodium are also located in bone. The large amount of carbonate present is of particular importance in that bone can contribute this carbonate to the extracellular fluid to combat acidosis, and thus bone provides a third line of defense in acid-base regulation along with the lungs and kidneys. Longstanding, uncorrected acidosis can lead to a loss of bone mineral for this very reason. Perhaps surprisingly, bone also contains a considerable amount of water. About 20% of the weight of bone is actually due to water, and thus about 9% of all the water present in the body is located in bone (see Table 26–4).

Table 26–5
Distribution of Plasma Calcium

Form	Amount (mg/100 ml)	Percentage of Total (approximate)
Filterable		
Ionized calcium	4.8	50
Complexed, ultra-filterable, bicarbonate (0.5), citrate (0.3), phosphate (0.4), and others	1.2	10
Protein-bound	4.0	40
Albumin (3.0)		
Globulins (1.0)		
Total	10.0	100

Numbers in parentheses represent the contribution (in mg/100 ml) of each of several forms within a category.

Calcium and Phosphorus Exist in Several Forms in the Blood

Forms of Calcium in the Blood

As indicated previously, calcium content of the extracellular fluid is maintained within fairly narrow limits. Total calcium content of plasma in humans is about 10 mg/100 ml. Of this total, calcium can exist in one of several different forms. Generally, there are two major forms of calcium in plasma. As indicated in Figure 26–2, these are **filterable calcium** and the **protein-bound** calcium. Each of these forms behaves differently in physiological systems.

Filterable Calcium: Ionized or Complexed. **Filterable calcium** represents about 60% of total calcium in the plasma (see Fig. 26–2 and Table 26–5) and includes those forms of calcium that are readily filtered in the glomerulus of the kidney. This calcium must therefore be reabsorbed in later segments of the renal tubules if it is not to be lost in the urine. Filterable calcium can be further subdivided into the **ionized** and the **complexed** forms.

Free Ca^{2+} ion constitutes roughly half of the total calcium present in plasma (see Table 26–5). Ionized calcium is the **physiologically active** form of cal-

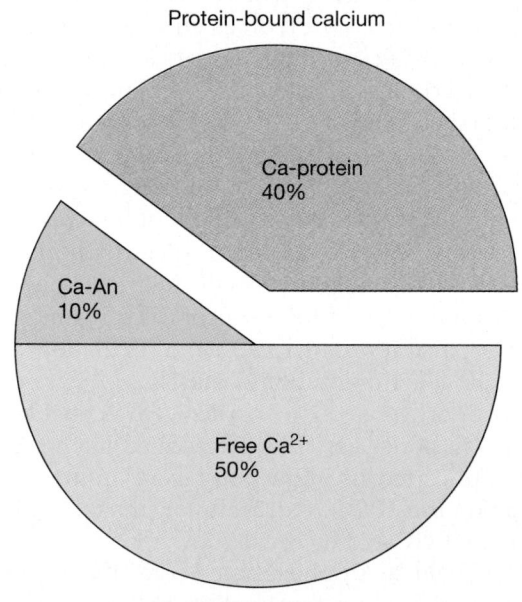

Figure 26–2

Relative distribution of different forms of calcium in the plasma. Calcium bound to plasma proteins is represented by "Ca-protein." Calcium bound to an anion, such as bicarbonate or citrate, is represented by "Ca-An," and the free ionized calcium in the plasma is represented by "Free Ca^{2+}."

cium. As the free ion, it is able to diffuse across plasma membranes and participate in processes such as those listed in Table 26–1. Note that this is the form of calcium that is strictly regulated in plasma.

A smaller portion of filterable calcium is complexed with an anion (see Table 26–5). **Complexed calcium** normally represents about 10% of the total calcium present in plasma. As indicated in Table 26–5, bicarbonate is the major anion to which calcium is bound, although significant amounts are also found in association with citrate. A variety of other anions, including phosphate, make up the remainder of this category.

Protein-Bound Calcium. The second major category is that of **protein-bound calcium** (see Fig. 26–2 and Table 26–5). Calcium bound to relatively large plasma proteins cannot freely pass through plasma membranes, nor is it filtered in the kidney. As indicated in Table 26–5, the majority of protein-bound calcium is bound to **albumin.** Albumin is a relatively large protein produced by the liver and is normally the most abundant protein present in the blood and interstitial fluid. A smaller proportion of protein-bound calcium is associated with various globulins in the blood.

Forms of Phosphorus in the Blood

In contrast to the plasma calcium concentration, which is normally regulated within very narrow limits, the phosphate concentration in blood may fluctuate on a daily basis from as low as one half of normal up to one and a half times normal. A plasma concentration ranging from 3.0 to 4.5 mg/100 ml (expressed as milligrams of **phosphorus**) is considered to be normal in humans.

In plasma, phosphorus exists primarily as **orthophosphate,** or **PO4.** At a normal blood pH, about 80% of the phosphate is present as HPO_4^{2-} and about 20% $H_2PO_4^-$. When present as the phosphate ion, the majority of the phosphorus readily diffuses across membranes and is also filterable in kidney glomeruli. In addition to the phosphate ion, small amounts of phosphorus are also present in blood in organic form, such as in hexose or triose phosphates, but these represent a relatively small proportion of the total.

Pathways of Calcium and Phosphate Homeostasis

Diet is, of course, the ultimate source of both calcium and phosphate in the body. Once these ions are ingested, **intestines, kidneys,** and **bone** are primarily involved in determining their plasma concentrations. As we will see, however, the overall patterns of movement in the body of the two ions differ considerably.

Understanding Pathways of Calcium Homeostasis Furthers Understanding of How Plasma Calcium is Regulated

Figure 26–3 shows both tissue distribution and average daily flow of calcium among different tissues. In this figure, the quantity of calcium present is indicated by blue numerals, and daily flow of calcium among tissues is indicated by red numerals.

Average calcium intake is about 1 gram per day (see Fig. 26–3). Most normal people can consume up to twice that amount without any deleterious effects. However, excessive calcium intake can result in problems with deposition of calcium in tissues (soft tissue **calcification**) or kidney stone formation. Of the 1 gram of calcium that is ingested, normally about one third to one half is absorbed from the intestines (see Fig. 26–3). The percentage can change with differing calcium intake and with differing plasma calcium status, however. For example, in growing children and pregnant or nursing women, this percentage tends to be somewhat higher, while in people of advanced age it tends to be lower.

Note also in Figure 26–3 that about 150 mg of calcium enters the intestines from the body each day. This occurs in part as a result of sloughing of cells that line the digestive tract and also secretion of fluids and various digestive enzymes into the intestines. This component of calcium flux into and out of the intestines tends to change relatively little with varying calcium intake or plasma calcium status. Therefore, as Figure 26–3 indicates, the major route of excretion of *ingested* calcium is via the feces, simply because the majority of ingested calcium is not absorbed. However, as we will see later, regulation of calcium absorption from the intestines is an important aspect of overall calcium homeostasis.

Figure 26–3 also indicates that the kidneys excrete about 150 mg of calcium into the urine each day. This represents only about 1% of the total amount of calcium filtered through kidney glomeruli, as the remaining 99% is reabsorbed and returned to the blood. A small change in calcium reabsorption by the kidneys can therefore dramatically affect calcium homeostasis.

Although we may think of bone as being relatively inert, bone mineral is constantly being turned over and remodeled to meet differing mechanical needs, as well as contributing to calcium homeostasis on both a short-term and long-term basis. Roughly

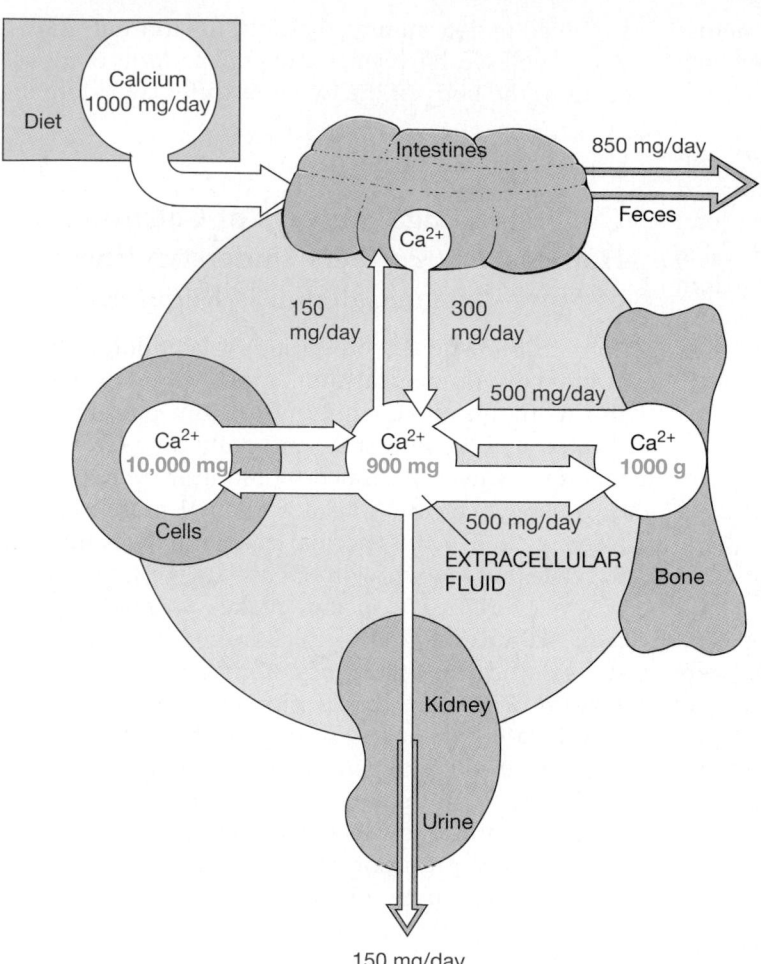

Figure 26–3

Diagram of the typical daily exchanges of calcium that occur between tissue compartments of a normal adult. Numbers shown in blue represent the amount of calcium in each compartment in *grams* or *milligrams*. Numbers in red represent the exchange of calcium between compartments in *milligrams/day*. Note that most calcium ingested each day in the diet is excreted in the feces; also note the high daily exchange of calcium between bone and the extracellular fluid. [1000 milligrams (mg) = 1 gram]

500 mg of calcium both enter and leave bone each day (see Fig. 26–3), indicating that bone mineral turnover is rather extensive. In a nongrowing adult, the fluxes of calcium into and out of bone will of course be equal, as indicated in Figure 26–3. In growing children, the rate of calcium deposition is greater than the rate of calcium loss, and as bone mineral is lost with advancing age, the converse becomes true.

Each of these processes—calcium absorption from the intestines, calcium reabsorption by the kidneys, and bone mineral formation and resorption—is precisely controlled. The mechanisms involved in this control will be discussed in later sections.

Understanding Pathways of Phosphate Homeostasis Also Provides Important Basic Concepts

Figure 26–4 shows the various pathways of phosphate metabolism in the body. As previously indicated, the extracellular phosphate concentration is not regulated as tightly as the calcium concentration. Normally, about 1.4 grams (1400 mg) of phosphorus is ingested each day. Because phosphate is relatively abundant in most foods, a nutritional phosphorus deficiency is generally not possible as long as sufficient food is taken in to meet caloric and protein needs. As Figure 26–4 shows, the majority of ingested phosphate is absorbed from the intestines. This is in direct contrast to calcium metabolism, since, as shown in the previous figure, the majority of ingested calcium is not absorbed. As with calcium, however, a small amount of phosphate is also secreted into the intestines each day. Thus, there is a *net* uptake of approximately 1100 mg of phosphorus each day.

The primary route for excretion of this daily phosphate load is the kidneys (see Fig. 26–4). Remember that most of the phosphorus present in the plasma exists as the phosphate ions, which are readily filtered in the glomeruli of the kidneys. Phosphate would thus be readily lost from the body if not for mechanisms that reabsorb and return it to the plasma. As will be described later, the reabsorption of phosphate by kidney tubule cells is a major mechanism of the hormonal regulation of phosphate homeostasis.

As with calcium, considerable amounts of phosphate flow both into and out of bone each day. Again, this reflects the very active turnover and remodeling of bone mineral that normally occurs.

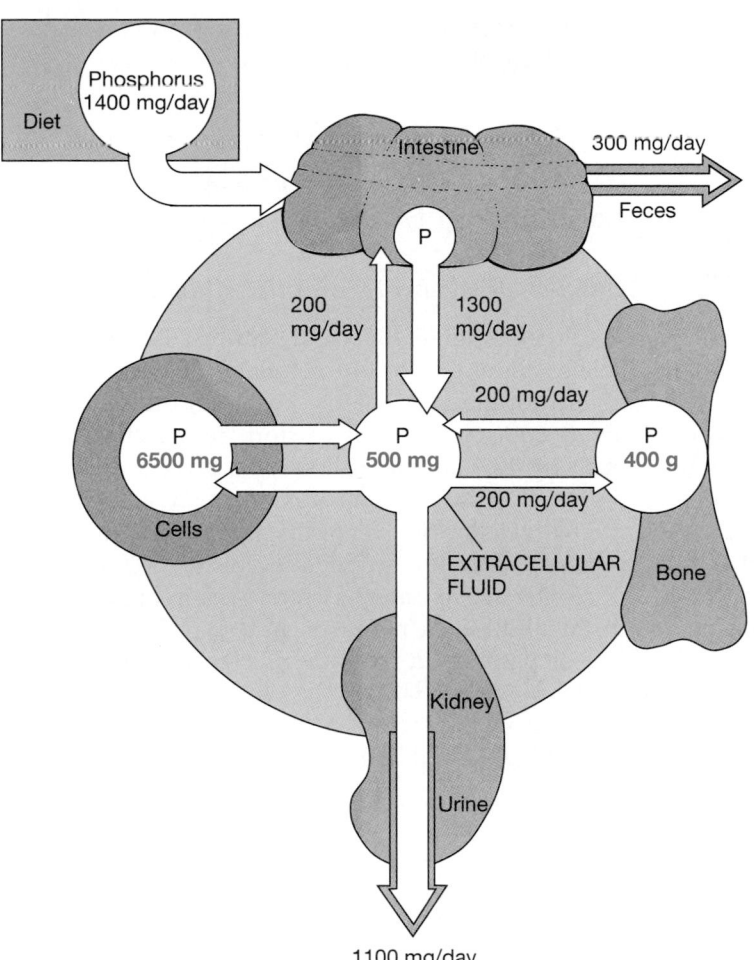

Figure 26-4

Diagram of the typical daily exchanges of phosphate that occur between tissue compartments in a normal adult. As in the previous figure, blue numbers represent amounts of phosphate in each compartment, and red numbers represent the exchange of phosphate between compartments. Note that most of the phosphate ingested each day in the diet is absorbed from the gastrointestinal tract and is excreted in the urine. Values are given in terms of grams or milligrams of phosphorus.

Mechanisms Involved in Calcium and Phosphate Homeostasis

Previous sections referred to processes of absorption, reabsorption, and resorption of calcium and phosphate as they occur in the gastrointestinal tract, kidney, and bone, respectively. We will now describe in more detail specific mechanisms involved in handling calcium and phosphate in each of these tissues.

Calcium and Phosphate Absorption in the Gastrointestinal Tract Differ in Magnitude

Calcium Absorption in the Gastrointestinal Tract

In the gastrointestinal tract, calcium is absorbed predominantly from the small intestine. Calcium is taken up from the small intestine by both active transport and simple passive diffusion. Relative contributions of each of these two processes depend on the region of small intestine. Active transport predominates in the duodenum and jejunum, whereas simple diffusion predominates in the ileum. Because it is longer than the other two regions of the intestine, the ileum is where the most calcium is absorbed.

As with other active transport processes, calcium transport exhibits characteristics of saturability. It requires energy, and moves calcium against a concentration gradient. Calcium uptake by passive diffusion is directly related to the concentration gradient of calcium between the lumen of the small intestine and the plasma.

The relative importance of active transport and passive diffusion depends on a number of factors. For example, with low calcium intake, active transport predominates because uptake by diffusion is low. With ingestion of large amounts of calcium, the active transport system becomes saturated, and uptake by diffusion predominates.

Calcium uptake from intestines is regulated through regulation of active calcium transport mechanisms. As we will see later, metabolites of vitamin D that function as hormones regulate calcium absorption by intestinal mucosal cells. Specific calcium-binding proteins involved in calcium transport are present in intestinal mucosal cells. Under the influence of vitamin D metabolites, these mucosal cells produce more calcium-binding proteins, thereby increasing their capacity for calcium transport.

Phosphate Absorption in the Gastrointestinal Tract

Phosphate absorption also occurs by both active transport processes and passive diffusion, with active transport predominating. The small intestine is the primary site of phosphate absorption from the gastrointestinal tract. As Figure 26–4 indicates, phosphate is very efficiently absorbed from the intestines, as usually 70% to 90% of ingested phosphate is eventually absorbed. Active transport of phosphate in the small intestine is in part coupled to calcium transport. Therefore, in situations in which active transport of calcium is low (e.g., with vitamin D deficiency) phosphate absorption is also low. The coupling of phosphate transport to calcium transport offers certain physiological advantages that we will discuss later in this chapter. However, phosphate absorption from the small intestine differs from calcium absorption in that there is only minor regulation. As

will be described subsequently, the primary site of regulation of phosphate metabolism is the kidney.

Calcium and Phosphate Reabsorption in Kidneys Differ by Tubular Location and Extent

Calcium Reabsorption in the Kidney

Recall from Table 26–5 and Figure 26–2 that filterable calcium consists of free calcium ion and calcium bound to filterable anions. Together, these make up about 60% of the total calcium in plasma; the remaining 40% circulates bound to protein and is not filtered in kidney glomeruli. Ordinarily, about 99% of filtered calcium is eventually reabsorbed by tubule cells and does not appear in the urine. Reabsorption occurs by active transport processes in proximal and distal tubules. The nature of calcium reabsorption (active or passive) in the loop of Henle is not well established. About 60% of the filtered calcium is reabsorbed from the proximal tubule, about 30% from the loop of Henle, and another 9% from the distal tubule;

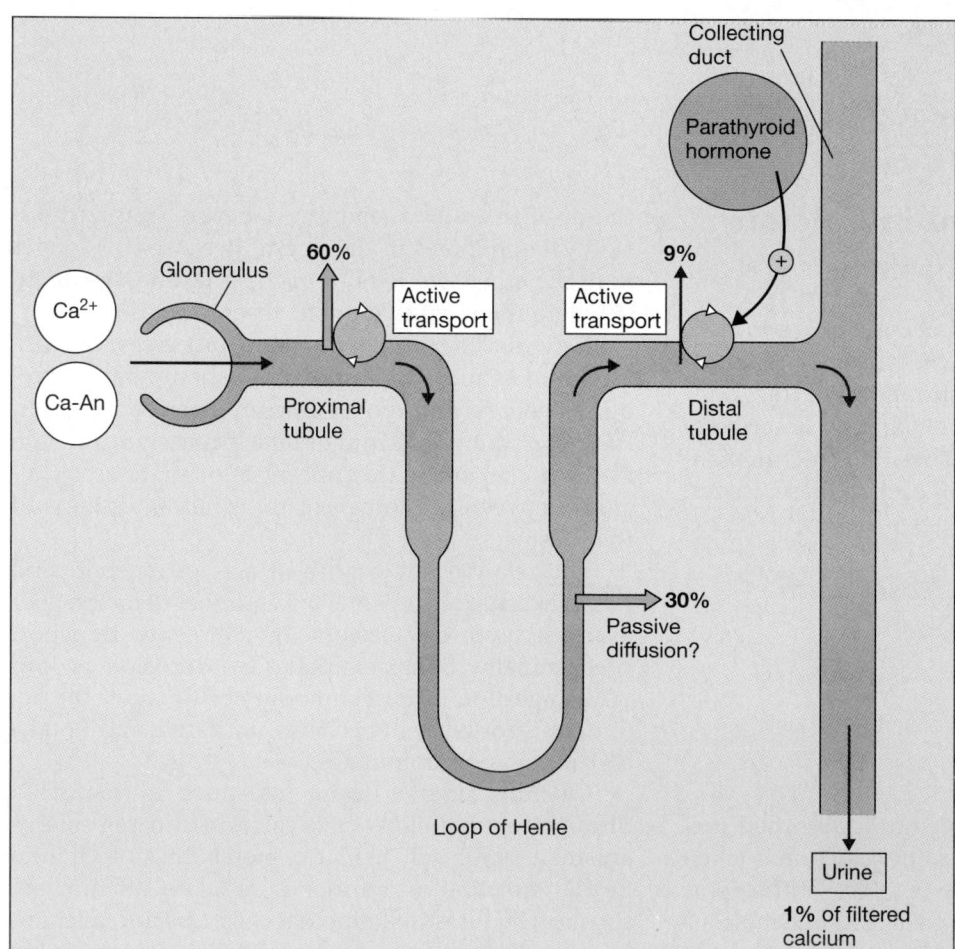

Figure 26–5

Diagram illustrating calcium filtration and reabsorption in a kidney nephron. Note that only free ionized calcium or calcium bound to an anion is filtered in the glomerulus. Calcium is reabsorbed by active transport processes in both the proximal and distal tubules. In the distal tubule, calcium reabsorption is stimulated by parathyroid hormone.

the remaining 1% is excreted in the urine (Fig. 26–5). Although calcium reabsorption in the distal tubule is a small percentage of the total, major adjustments in calcium excretion can occur at this site as a result of an action of parathyroid hormone to stimulate calcium reabsorption. Parathyroid hormone is one of three hormones involved in regulating calcium and phosphate homeostasis, as later sections of this chapter will describe. The effect of parathyroid hormone to increase transfer of calcium from tubular fluid to plasma is mediated by an increase in cAMP in distal tubule cells (refer back to Table 12–3).

Phosphate Reabsorption in the Kidney

As was indicated in Figure 26–4, a major route by which ingested phosphate leaves the body is via the urine. The kidney is therefore an important site of regulation of extracellular phosphate. Ordinarily, about 85% of filtered phosphate is reabsorbed, and the remaining 15% is excreted in the urine. Reabsorption of phosphate is due to active transport processes with the principal site of phosphate reabsorption being the proximal tubules (Fig. 26–6), but some reabsorption occurs in distal tubules as well (see Fig. 26–6). Parathyroid hormone exerts a major regulatory effect on phosphate excretion by inhibiting phosphate reabsorption in proximal tubules. As a result, parathyroid hormone causes an increase in urinary phosphate excretion with an accompanying decrease in plasma phosphate.

Calcium and Phosphate Homeostasis in Bone Is a Highly Regulated Process

The Composition of Bone

Simply described, mature bone consists of inorganic bone mineral deposited on a framework of organic support material. Imagine what would happen if you dipped a sponge into wet plaster and then allowed the plaster to harden. The sponge is analogous to the organic support material, and the hardened plaster is analogous to the bone mineral.

The mineral fraction of bone is composed largely of calcium phosphate, in the form of what are known

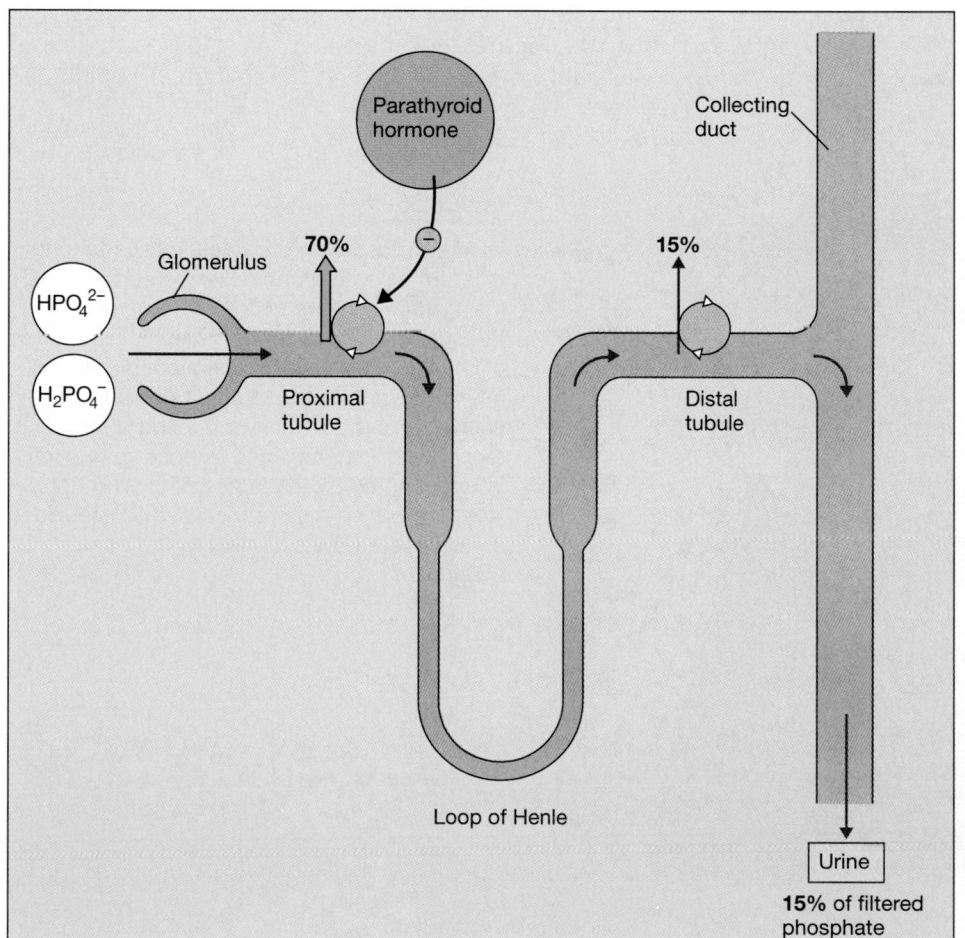

Figure 26–6

Diagram illustrating the filtration and reabsorption of phosphate in a kidney nephron. Most phosphate reabsorption occurs by active transport in the proximal tubule. As indicated, this process is inhibited by parathyroid hormone.

as **hydroxyapatite crystals.** The hydroxyapatite crystals in bone have the general formula:

$$Ca_{10}(PO_4)_6(OH)_2$$

Normally, the mineral portion of bone makes up about one quarter of the bone volume, but because of its high density, the mineral fraction makes up about one half of bone weight. Recall from Table 26–4 that, in addition to calcium phosphate, inorganic matter of bone also contains a considerable amount of carbonate, magnesium, and sodium.

The organic matrix of bone is known as **osteoid,** 95% of which is **collagen.** The type of collagen in bone is similar to that present in tendons and in skin, although collagen in bone does have biochemical properties that give it greater mechanical strength than other types of collagen. The remaining 5%, or noncollagen portion of organic matter, is known as **ground substance,** which consists of a mixture of various **proteoglycans,** which are high-molecular-weight compounds made up of carbohydrate and protein.

If one closely examines bone structure using an electron microscope, one sees needlelike hydroxyapatite crystals lying alongside the collagen fibers and surrounded by the somewhat amorphous ground substance. Association of hydroxyapatite with the collagen fibers is responsible for the strength and hardness characteristic of bones. If a bone is **de-**

mineralized—that is, if all the inorganic material is removed—the shape of the bone will be preserved because of the collagen that remains, but it will be as flexible as a tendon. Conversely, removal of the organic matter will also leave the bone with its original shape, but in this case the mineral that remains will be very brittle and easily broken. Thus, it is the combination of organic and inorganic material that gives bone its high degree of mechanical strength.

Bone Cells

There are three primary cell types involved in formation and resorption of mature bone. These are **osteoblasts, osteocytes,** and **osteoclasts** (Fig. 26–7).

Osteoblasts.   Located side by side on the surfaces of bone, **osteoblasts** are the cells responsible for synthesis of osteoid. Osteoblasts that are actively producing osteoid have a cuboidal shape and exhibit an abundant rough endoplasmic reticulum and Golgi complex, two characteristics typical of cells actively synthesizing proteins for export. Osteoblasts not actively engaged in bone formation have a considerably flattened appearance. Osteoblasts also have numerous cytoplasmic processes that serve to bring them into contact with neighboring osteoblasts.

As Figure 26–7 indicates, osteoid secreted from osteoblasts is deposited on the surface of bone. Bone

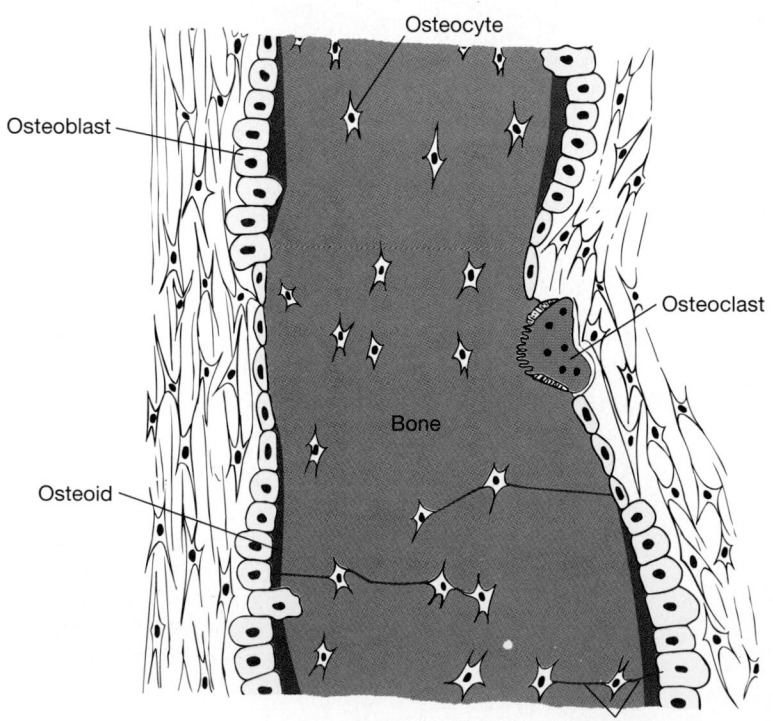

Figure 26–7

Diagram illustrating the relationships among different types of bone cells. Osteoblasts line the surface of bone and secrete osteoid into the space adjacent to the bone. For the sake of simplicity, only a few of the canaliculi that interconnect osteocytes and osteoblasts are shown near the bottom of the figure. An osteoclast actively engaged in bone resorption is represented by the large cell on the right. Note the ruffled appearance of the osteoclast membrane on the side next to the bone surface.

mineral is then deposited on the organic osteoid matrix, and as this process occurs, the osteoblasts eventually become surrounded by mineralized bone.

Osteocytes. At this stage, osteoblasts progressively lose their bone-forming capability and are termed **osteocytes.** Cell-to-cell connections that were present among osteoblasts are maintained through the osteocyte stage of development. These interconnections are channels known as **canaliculi** in the bone and represent direct contacts between cells. This coupling among cells provides a "bucket brigade" mechanism whereby nutrients and ions can travel from blood vessels to distant osteocytes deeply embedded in bone. Hormones that control bone growth and development are also believed to pass from one osteocyte to another in this way.

Osteoclasts. The process of bone resorption is carried out by **osteoclasts.** These are large, multinucleated cells, each containing about 4 to 6 nuclei. The surface of osteoclasts adjacent to bone has a very ruffled appearance (see Fig. 26–7). This ruffling increases surface area and allows cells to perform the task of bone resorption more effectively. When stimulated, osteoclasts are thought to secrete both acids and proteolytic enzymes into the extracellular space adjacent to bone. The acidic environment increases the solubility of bone mineral, while the proteolytic enzymes attack the organic matrix of bone. Together, these two factors promote bone resorption.

Formation of Bone

During early fetal development, the skeleton consists solely of cartilage, which is laid down in the shape and form of the bony skeleton. This cartilage "model" serves as the template for development of the bony skeleton and is eventually replaced by bone.

The Epiphyseal Plate. The process of cartilage replacement begins in the middle or center and progresses out toward each end of what will later form the bone (Fig. 26–8*a–c*). As this progression occurs, the bone increases in both length and thickness. Figure 26–8*d* illustrates the typical structure of a long bone in a growing adolescent. The histological appearance of bone is the result of an extension of a series of events shown in parts *a* to *c* of Figure 26–8. The area of bone designated as the **epiphyseal plate** is of particular interest, since it is here that elongation or growth of bone occurs after birth.

Chondrocytes. As indicated in Figure 26–9, the epiphyseal plate shows considerable differences between its leading and trailing edges. On the leading edge are cells known as **chondrocytes,** which are actively engaged in synthesis of cartilage of the epiphyseal plate. As cartilage is formed, it surrounds the chondrocytes and gradually becomes calcified. The trapped chondrocytes are replaced by new cells on the surface of the cartilage, so that active synthesis of cartilage continues. The calcified cartilage that is formed *does not* itself become bone, but is later re-

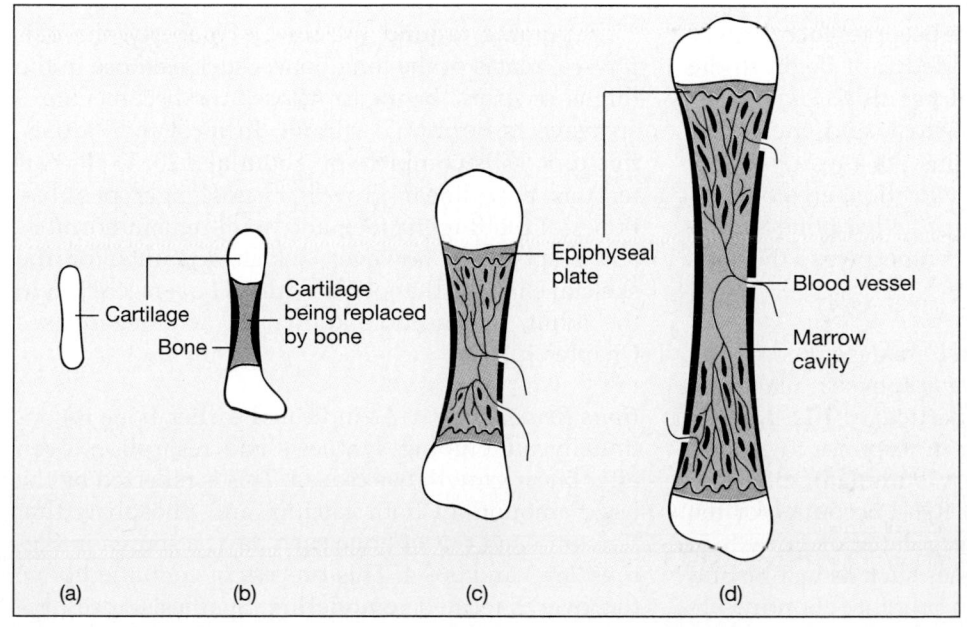

(a)   (b)   (c)   (d)

Figure 26–8

The process of growth and elongation of a typical long bone. (*a*) During fetal development, cartilage "models" of the bone are formed. (*b, c*) The cartilage is then replaced by mineralized bone, beginning in the middle of the bone and progressing toward each end. In the process, the bone becomes thicker and longer. (*d*) The center of the bone shaft becomes hollowed to form the marrow cavity. The epiphyseal plate is the zone when elongation of bone occurs.

Labels in figure: Cartilage; Bone; Cartilage being replaced by bone; Epiphyseal plate; Blood vessel; Marrow cavity

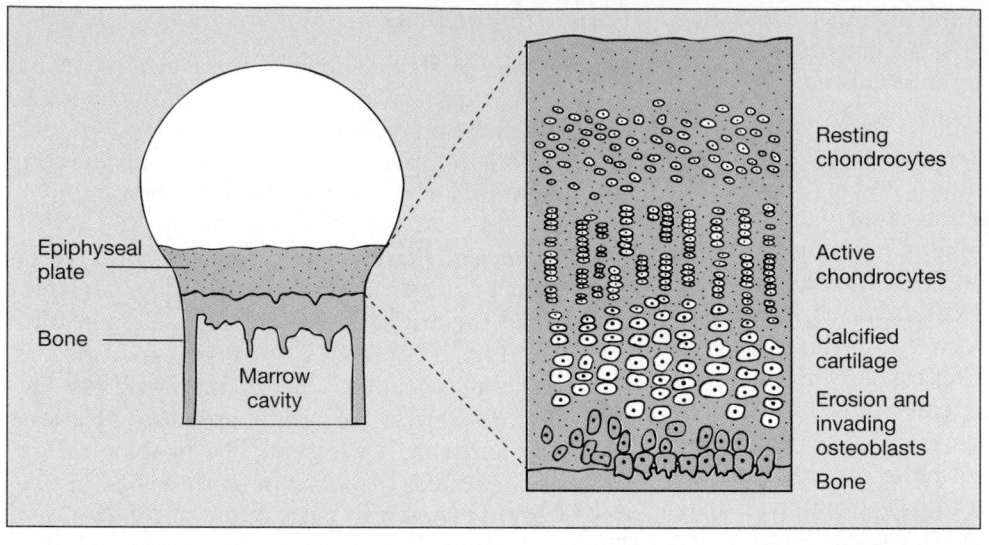

(a)

Figure 26–9

(*a*) The appearance of an epiphyseal plate. On the right, zones of different cellular activities within the epiphyseal plate are illustrated. Active chondrocytes secrete collagen, which eventually becomes calcified. The calcified cartilage erodes and is replaced by osteoblasts, which then form new bone. The entire sequence is repeated many times as the zone of new bone formation gradually moves upward. (*b*) Histological photo of cellular activities of an epiphyseal plate showing chondrocytes, calcified cartilage, and osteoblasts forming new bone.   (© Biophoto Associates/Photo Researchers)

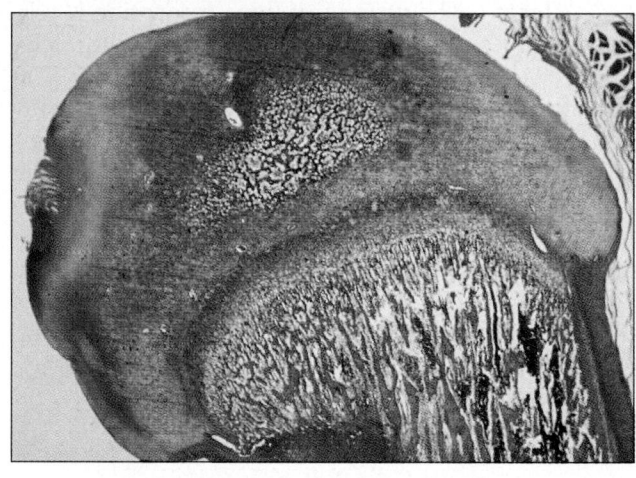

(b)

placed by bone. As the cartilage becomes increasingly calcified, chondrocytes embedded in it begin to die off, and the calcified material begins to erode. Osteoblasts then move into the area, and the actual process of bone formation begins. As a result of continuing cartilage synthesis, calcification, erosion, and osteoblast invasion, the zone of active bone formation moves outward from the center toward the ends of the bones.

Hormones in Bone Formation.   As described earlier in Chapter 13, chondrocytes of epiphyseal plates are under hormonal control. In particular, **IGF-I,** a somatomedin produced by liver in response to growth hormone (GH), is the primary stimulator of chondrocyte activity. This effect of IGF-I accounts for the role of GH as an important regulator of growth in young children. Other hormones such as insulin and thyroid hormones also tend to promote chondrocyte activity and therefore promote bone growth as well.

Beginning around the time of puberty, the epiphyseal plates of the long bones, such as those in the thighs or arms, begin to "close," or become unresponsive to hormonal stimuli. In most individuals, this process is complete by about age 20, so that after this time linear growth is no longer possible. Bones of the fingers, feet, and skull remain hormonally responsive, however, which accounts for the skeletal changes that occur with GH oversecretion in the adult, a condition known as acromegaly (see Chapter 13).

Bone Remodeling.   As indicated earlier, bone is continually undergoing synthesis and resorption, even after linear growth has ceased. This is reflected by the large amounts of both calcium and phosphate that flow into and out of bone each day, as shown in Figures 26–3 and 26–4. This process of continued bone turnover is termed **remodeling.** Synthesis or resorption of bone can occur along most of the outer surface of bones, making them thicker or thinner, respectively. In the long bones, synthesis and resorption can also occur along inner surfaces of the bone shaft, adjacent to the bone marrow cavity. Remodeling reshapes and re-forms bone to meet the changing mechanical needs that occur throughout life. A good illustration of such a case is a cowboy's bowed legs, in which the curvature of the legs results partly from the continued pressure on the legs as he rides his horse. In addition, because of the role of bone as a reservoir for both calcium and phosphate, it must be able to store and rapidly mobilize these two ions as required. The factors involved in regulating these processes will be described shortly.

Regulation of Calcium and Phosphate in Plasma

Because calcium ions are so important in physiological processes, and because disturbances in the plasma concentration of calcium have such grave consequences, most of the regulatory processes to be discussed are directed toward controlling the plasma concentration of this ion. Calcium ions can move between extracellular fluid and bone as a result of either rapid nonhormonal processes or slower hormonally regulated mechanisms. Both processes are important in maintaining a constant extracellular fluid calcium ion concentration.

Nonhormonal Mechanisms for Calcium and Phosphate Regulation Provide a Rapid, but Low-Capacity System of Regulation

Recall from Figure 26–3 and Table 26–3 that about 1000 g, or 99%, of the total body calcium is present in bone. The vast majority of this calcium exists as hardened bone mineral, but about 1% of the calcium present in bone exists in a readily exchangeable form that can rapidly move between bone and extracellular fluid. This rapidly exchangeable source of calcium is primarily located on the surface of bone and in newly formed bone. Calcium can move rapidly by this mechanism, since it involves purely physical events. Free ionized calcium in the extracellular fluid exists in chemical equilibrium with calcium present in hydroxyapatite crystals of bone. Any decrease in free calcium in the extracellular fluid therefore directly results in movement of calcium out of hydroxyapatite and into extracellular fluid until a new chemical equilibrium is attained.

This mechanism, while permitting very rapid movement of calcium both into and out of bone, has a very limited capacity because of the small proportion of calcium that exists in this readily exchangeable pool.

Hormonal Mechanisms for Calcium and Phosphate Regulation Provide for Long-Term Regulation

The second mechanism involves hormones, and although these are somewhat slower than nonhormonal processes, they can cause large shifts in calcium between bone and the extracellular fluid over a prolonged period of time. Three hormones are primarily involved in regulating calcium ion concentrations in the extracellular fluid: (1) parathyroid hormone, (2) calcitonin, and (3) metabolites of vitamin D. As will be described, the actions of these hormones are intimately interrelated by a series of feedback loops, with the concentration of **ionized calcium** (Ca^{2+}) in the plasma as the factor being regulated.

Chemistry and Regulation of Hormone Production and Secretion

Production and Secretion of Parathyroid Hormone. **Parathyroid hormone (PTH, parathormone)** is a polypeptide hormone produced by the parathyroid glands. Humans have four parathyroid glands, found in pairs on the dorsal surface of each of the two lobes of the thyroid gland (Fig. 26–10). As is true of other

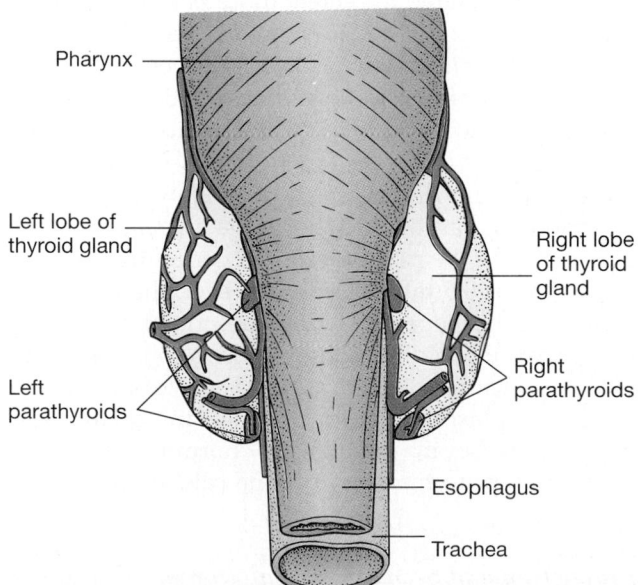

Figure 26–10

The anatomical location of the four parathyroid glands in humans. The view depicts the dorsal surface of the two lobes of the thyroid gland.

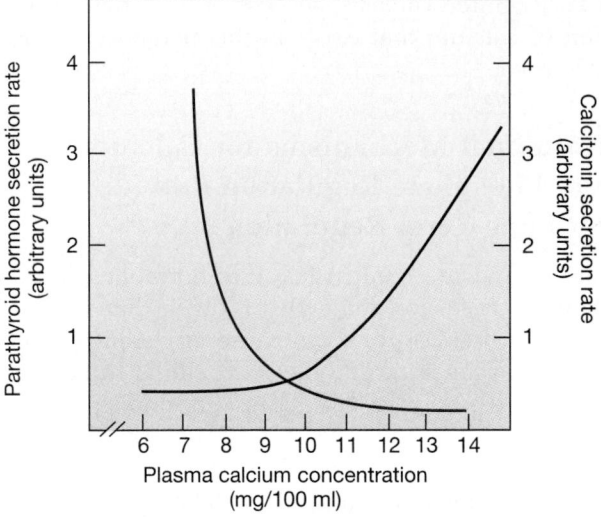

Figure 26–11

Relationship between the total plasma calcium concentration, parathyroid hormone secretion, and calcitonin secretion. The normal plasma calcium concentration is about 10 mg/100 ml.

endocrine glands, the parathyroid glands have a very rich blood supply.

The primary stimulus for parathyroid hormone secretion is a *decrease* in the plasma calcium concentration. As Figure 26–11 illustrates, as the calcium concentration in the plasma decreases below its normal value of about 10 mg/100 ml, the parathyroid gland is stimulated to secrete more parathyroid hormone. Although Figure 26–11 shows the relationship between the *total* plasma calcium concentration and the rate of parathyroid hormone secretion, the parathyroid glands actually respond to changes in free *ionized* calcium. Note also in Figure 26–11 that as the plasma calcium first begins to fall below normal, the secretion of parathyroid hormone increases rather gradually. As the plasma calcium concentration falls below about 8.5 mg/100 ml, however, the curve becomes much steeper. As will be described, the net effect of parathyroid hormone action is to increase the plasma calcium concentration. Thus, dramatic decreases in plasma calcium produce an equally dramatic increase in parathyroid hormone secretion in an attempt to maintain plasma calcium at normal values.

Production and Secretion of Calcitonin. Calcitonin (sometimes abbreviated as CT) is a peptide hormone that, in humans, is produced by cells known as **parafollicular cells** or **C-cells** of the thyroid gland (the "C" stands for *clear*, which is how they appear under the microscope). As described in Chapter 13,

thyroid follicle cells are the site of thyroid hormone production. Parafollicular cells are located in spaces between the thyroid follicles and are distinct from the follicle cells. Thus, the thyroid gland produces both thyroid hormones and calcitonin. Calcitonin is therefore sometimes referred to as **thyrocalcitonin,** to indicate its tissue of origin.

Like parathyroid hormone, calcitonin is secreted in response to changes in the plasma concentration of ionized calcium, except that an *increase* rather than a decrease in plasma calcium stimulates calcitonin secretion (Fig. 26–11). Calcitonin secretion also increases in response to several gastrointestinal hormones, particularly gastrin. The net effect of calcitonin actions, which will be described in detail shortly, is to cause an increased deposition of calcium in bones, with an accompanying decrease in the plasma calcium concentration. Stimulation of calcitonin secretion by gastrointestinal hormones coordinates increased calcium uptake into bone with ingestion of calcium associated with a meal.

Production and Secretion of Vitamin D Metabolites. The third hormone involved in regulating plasma calcium is **vitamin D,** or, more precisely, a metabolite of the vitamin that functions as the active hormone. D vitamins are a group of lipid-soluble vitamins long known to prevent the childhood disease of **rickets.** Only in the last several decades has a hormonal role of vitamin D metabolites in regulating plasma calcium been determined.

D vitamins have steroidlike structures and are derived from cholesterol, as are other steroids. The natural form is **vitamin D_3, or cholecalciferol,** which is shown in Figure 26–12. Vitamin D_3 can either be obtained from the diet or formed in the skin by the action of **ultraviolet light** on a precursor, **7-dehydrocholesterol,** which is derived from cholesterol (Fig. 26–13). Thus, with adequate exposure to sunlight,

Vitamin D_3
(Cholecalciferol)

Vitamin D_2
(Ergocalciferol)

Figure 26–12

Structures of vitamin D_3 and vitamin D_2.

Figure 26–13

Illustration of the pathway of formation of vitamin D_3, beginning with cholesterol.

sufficient vitamin D_3 can be formed in the skin to supply the body's needs without any dietary supplementation.

Another related compound of interest is **vitamin D_2**, or ergocalciferol (see Fig. 26–12). Ergocalciferol is a form of the vitamin found in plants and yeasts, and thus it can be a significant dietary source of D vitamin. Vitamin D_2 is metabolized and has actions essentially identical to those of vitamin D_3. Because it can be obtained relatively inexpensively from plant sources, it is often used to fortify foods, such as milk, with vitamin D. The following sections will discuss the metabolism and actions of vitamin D_3, since this is the natural form of the vitamin. But keep in mind that vitamin D_2 undergoes the same modifications and has the same actions as does D_3.

Vitamins D_2 and D_3 by themselves have essentially no biological activity. The following paragraphs describe steps by which these materials are transformed into potent hormones involved in regulating plasma calcium concentrations.

The hormonally active form of vitamin D is produced when inactive molecules are modified by liver and kidney. The modifications involve addition of a hydroxyl (OH) group first to carbon number 25 and then to carbon number 1 of vitamin D_3. As indicated in Figure 26–14, the first step involved in activation of vitamin D occurs in liver. Vitamin D_3, derived either from the diet or formed in the skin, is carried to the liver via the blood. In liver, a specific enzyme catalyzes the addition of a hydroxyl group to carbon number 25 to form 25-hydroxy-vitamin D_3. 25-hydroxy-vitamin D_3 then leaves the liver and travels via the blood to the kidneys. In proximal tubule cells of the kidney, a specific enzyme known as **1α-hydroxylase** catalyzes the conversion of 25-hydroxy-vitamin D_3 to 1,25-dihydroxy-vitamin D_3 (1,25-OH-vitamin D_3). It is this form of the vitamin that functions as a hormone and is involved in regulating plasma calcium concentrations.

Formation of 1,25-OH-vitamin D_3 is a highly regulated process. The primary site of regulation is the kidney. Activity of 1α-hydroxylase is markedly increased by parathyroid hormone (see Fig. 26–14). Thus, activity of the enzyme changes indirectly as a

result of changes in the plasma calcium concentration. (As mentioned, if the plasma calcium concentration falls below normal, parathyroid hormone secretion is stimulated. This results in an increase in 1α-hydroxylase activity, leading to increased formation of 1,25-OH-vitamin D_3. See Fig. 26–14).

In addition, a decrease in plasma phosphate causes an increase in activity of 1α-hydroxylase, which also leads to increased formation of 1,25-OH-vitamin D_3. In this case, phosphate appears to directly modulate activity of the enzyme. Therefore, either

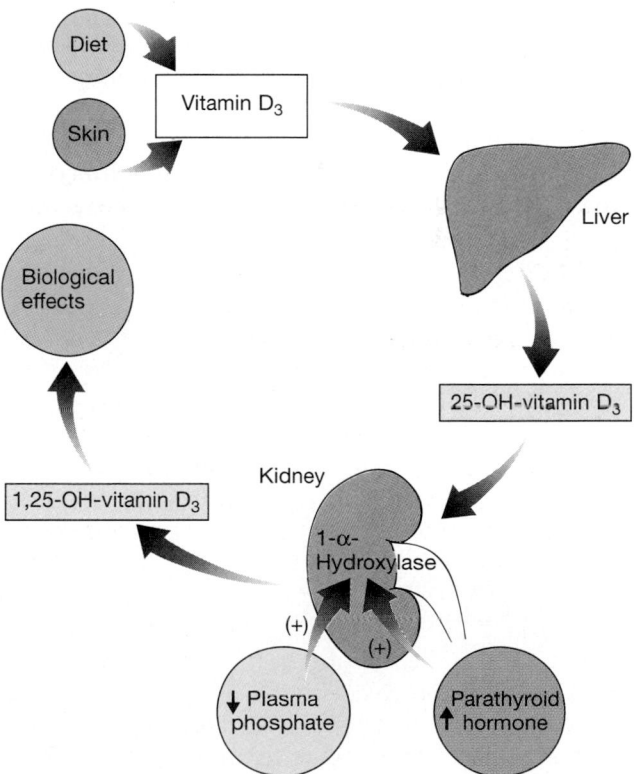

Figure 26–14

Diagram of the pathway for conversion of vitamin D_3 into 1,25-dihydroxy-vitamin D_3. Hydroxylation of vitamin D_3 on the 25 position occurs first in the liver, followed by hydroxylation on the 1 position in the kidney. Parathyroid hormone stimulates the kidney enzyme, which catalyzes the addition of a hydroxyl group to carbon number 1.

a decrease in plasma calcium, with its associated increase in parathyroid hormone, or a decrease in plasma phosphate results in increased production of the hormonally active form of vitamin D (see Fig. 26–14).

Actions of Hormones Involved in Regulating Calcium and Phosphate Homeostasis

Actions of parathyroid hormone, calcitonin, and 1,25-OH-vitamin D_3 are summarized in Table 26–6. The specific effects of these hormones will be described in the following sections.

Actions of Parathyroid Hormone in Calcium and Phosphate Homeostasis.

As a result of its actions in regulating plasma calcium concentrations, parathyroid hormone is *acutely* required for life. Removal of the parathyroid glands without any subsequent intervention usually leads to death from hypocalcemic tetany within 48 to 72 hours. Figure 26–1 shows the initial signs of hypocalcemic tetany, involving spasm of muscles of the hand and forearm.

As indicated in Table 26–6, the overall effect of parathyroid hormone is to increase plasma calcium and decrease plasma phosphate. With these effects in mind, what specific responses does the hormone elicit that lead to these changes?

One important effect of parathyroid hormone in kidneys is to increase activity of 1α-hydroxylase, which catalyzes the formation of 1,25-OH-vitamin D_3, the active metabolite of vitamin D. Parathyroid hormone also stimulates active transport mechanisms for calcium reabsorption in distal tubules, leading to an increase in calcium retention and a decrease in urinary excretion of calcium. Parathyroid hormone also decreases proximal tubular reabsorption of phosphate, resulting in **phosphaturia** (increased phosphate excretion in the urine). This phosphaturic effect of parathyroid hormone is an important one, since other effects of parathyroid hormone tend to increase the flow of *both* calcium and phosphate into the blood. Without increased phosphate excretion in the urine, both calcium and phosphate would accumulate in plasma and then simply recrystallize in bone mineral. This would counteract the desired effect, which is to raise the plasma calcium concentration.

In bone, parathyroid hormone has several effects, all of which lead to increased net resorption of bone. The primary effect is to stimulate dissolution of bone by increasing osteoclast activity. In addition to increasing the activity of existing osteoclasts, parathyroid hormone also stimulates maturation of immature osteoclasts, resulting in a greater number of active osteoclasts. A third effect of parathyroid hormone in bone is to decrease collagen synthesis in osteoblasts, resulting in a decrease in bone matrix formation. Thus, parathyroid hormone impairs new bone formation while simultaneously stimulating bone resorption. Together, these effects of parathyroid hormone result in an increased net transfer of calcium from bone into plasma.

Table 26–6
Summary of the Effects of Hormones Involved in Regulating Calcium and Phosphate Metabolism

Parameter or Target Tissue	Parathyroid Hormone	Calcitonin	1,25-OH-Vitamin D_3
Plasma calcium concentration	↑	↓	↑
Plasma phosphate concentration	↓	↓	↑
Kidney	↑ Reabsorption of calcium	↓ Reabsorption of calcium	↑ Reabsorption of calcium
	↓ Reabsorption of phosphate	↓ Reabsorption of phosphate	↑ Reabsorption of phosphate
	↑ Activity of 1α-hydroxylase		
Bone	↑ Resorption of bone	↓ Resorption of bone	Promotion of PTH actions
Gastrointestinal tract	Indirect effects by increasing 1,25-OH-vitamin D_3 formation	No effect	↑ Absorption of calcium and phosphate

Parathyroid hormone does not appear to have any direct effects on the gastrointestinal tract, but indirectly results in increased absorption of both calcium and phosphate. These effects are due to the fact that it stimulates formation of 1,25-OH-vitamin D_3, which in turn acts on the gastrointestinal tract. A summary of the effects of parathyroid hormone is shown in Figure 26–15. Note that the net effect of PTH is to restore plasma calcium to normal.

Actions of Calcitonin in Calcium and Phosphate Homeostasis. The overall effect of calcitonin is to decrease both plasma calcium and phosphate (see Table 26–6). These changes occur as the result of an effect of the hormone primarily in bone, although it has some weak effects in the kidney as well.

In kidney, calcitonin decreases tubular reabsorption of both calcium and phosphate, which results in an increased urinary excretion of both ions. In bone, calcitonin decreases resorption by inhibiting the activity of osteoclasts. Therefore, this effect on osteoclasts is opposite to that of parathyroid hormone and results in increased bone formation. These effects of calcitonin on bone and kidney each contribute to the overall effect of the hormone to decrease the concentrations of these two ions in plasma.

Despite these distinct and demonstrated biological effects of calcitonin, this hormone probably plays only a minor role in the regulation of plasma calcium and phosphate in humans. This general conclusion results from observations made on persons having marked abnormalities of calcitonin secretion. Neither individuals with a complete deficiency of calcitonin secretion nor those with a marked oversecretion of the hormone display any overt abnormalities of calcium or phosphate metabolism. Thus, it is believed that the role of calcitonin may only be to provide a "fine tuning" for the regulation of calcium and phosphate metabolism.

Actions of 1,25-OH-Vitamin D_3 in Calcium and Phosphate Homeostasis. As indicated in Table 26–6, the net effect of 1,25-OH-vitamin D_3 action is to increase plasma concentrations of both calcium and phosphate. Its primary target tissue is the small intestine, although it also has some actions in kidney and bone.

In kidneys, 1,25-OH-vitamin D_3 increases tubular reabsorption of both calcium and phosphate, leading to a decreased urinary excretion of both ions. However, this appears to be only a minor effect of the hormone. In bone, the primary effect of 1,25-OH-vitamin D_3 is to promote the actions of parathyroid hormone. Parathyroid hormone has a greater effect in the presence of 1,25-OH-vitamin D_3 than it does alone. Therefore, the net effect of 1,25-OH-vitamin D_3 in bone is to promote bone resorption. This in turn contributes to an increase in plasma calcium and phosphate concentrations.

As indicated, 1,25-OH-vitamin D_3 has its primary effect in the gastrointestinal tract, where it stimulates both calcium and phosphate absorption from the small intestine. Again, this contributes to an increased concentration of these two ions in plasma. The effect of this hormone on calcium absorption is seen only after a delay of several hours, which is associated with increased production of calcium transport proteins in intestinal mucosal cells. Increased numbers of calcium transport proteins in turn lead to greater calcium absorption from the small intestine.

Figure 26–15

Diagram showing the changes that occur in response to a decrease in plasma calcium. The end result is that plasma calcium concentration is restored to its normal value.

Abnormalities of Bone Mineral Metabolism

The most common abnormality of bone metabolism is not a single disorder but a group of disorders termed **metabolic bone disease,** in which there is a

Vitamin D in Health and Disease

In the accompanying sections, the classical view of vitamin D physiology with regard to the regulation of calcium and phosphate metabolism was presented. Recent research indicates that vitamin D may have a wider biologic role than simply the maintenance of mineral homeostasis, and suggests that it may have important actions in a variety of tissues.

Considerable evidence indicates that the mechanism of action of 1,25-OH-vitamin D_3 is similar to that of other steroid hormones. 1,25-OH-vitamin D_3 interacts with an intracellular receptor and appears to produce its effects by interaction of the hormone-receptor complex with the genome of target cells. Analysis of the amino acid sequence of the 1,25-OH-vitamin D_3 receptor suggests that it belongs to the same family of proteins as the receptors for steroid and thyroid hormones. Not surprisingly, vitamin D receptors can be found in intestine, kidney, and bone, the primary targets for the calcium-regulating hormones. Of interest, however, is the large number of other tissues in which this receptor has been found. Normal tissues or cell types containing the receptor include skeletal and cardiac muscle, mammary tissue, gonads, pituitary gland, pancreas, skin, various blood cell types, and parathyroid gland. It has also been discovered in significant numbers in various malignant cell types, including cancers of the breast, bladder, pancreas, cervix, and colon.

While the physiological basis for the presence of vitamin D receptors in many of the tissues just mentioned remains obscure, their presence in the parathyroid gland provides the basis for a feedback loop involved in calcium regulation. By virtue of its effect to raise plasma calcium levels, 1,25-OH-vitamin D_3 indirectly provides an inhibitory signal for parathyroid hormone production. However, it was recently demonstrated that 1,25-OH-vitamin D_3 also directly acts on the parathyroid gland by inhibiting transcription of the parathyroid hormone gene. The end result of this effect is the reduction of parathyroid hormone production. Thus, 1,25-OH-vitamin D_3 exerts both direct and indirect effects that limit the production of parathyroid hormone, an antagonist of its own actions.

One area of active investigation into the actions of 1,25-OH-vitamin D_3 involves those effects that occur within the hematopoietic/immune system. The hormone has been shown to promote the differentiation of precursors toward cells with the properties of mature macrophages. Receptors for 1,25-OH-vitamin D_3 are also present in lymphocytes, and have been shown to mediate a large spectrum of effects of the hormone on these cells. Among these is repression of interleukin-2 (IL-2) production via a reduction in IL-2 mRNA levels.

As indicated, receptors for 1,25-OH-vitamin D_3 are present in a number of different cancer cell types. Several studies have shown the hormone to have an antiproliferative action, although the exact effects of the hormone probably vary with the particular type of cancer. Additional studies are needed to establish whether 1,25-OH-vitamin D_3 or suitable analogs have a useful role in the therapy of malignant diseases. The skin also appears to be a target for 1,25-OH-vitamin D_3 action, and some evidence suggests that under certain conditions the hormone may even be produced locally. It appears capable of affecting epidermal differentiation and has been tested with some success in the treatment of psoriasis, a disorder in which the normal maturation and differentiation of epidermal cells is disturbed.

Additional studies will be needed to more clearly define the role of vitamin D_3 in normal physiology and to assess its usefulness in various clinical situations.

■ **If hormonally active vitamin D metabolites do prove useful in the treatment of psoriasis, what do you suspect might be the best means for delivering this to the affected tissues?**

---

generalized abnormality of the ongoing processes of bone formation and bone resorption. The two major categories of metabolic bone disease are (1) **osteoporosis** and (2) **osteomalacia and rickets.**

Osteoporosis Results from an Equivalent Loss of Bone Mineral and Organic Matrix

Osteoporosis is a condition in which there is a reduction in overall bone mass, with equal losses of both mineral and organic matrix. The ratio of mineral relative to the amount of organic matrix therefore does not change. Since there is a net decrease in the amount of bone, it follows that bone resorption must be occurring at a faster rate than bone formation.

Although a number of factors can contribute to osteoporosis, in many cases the exact cause of this imbalance is not known. Long-term calcium deficiency can lead to osteoporosis, owing to mobilization of bone mineral in an attempt to maintain normal plasma calcium concentrations. In addition, pro-

CLINICAL FOCUS

Osteoporosis

As described in the accompanying sections of this chapter, osteoporosis is a major public health concern in the United States and other countries. In the United States alone, the total cost of osteoporosis and its associated bone fractures has been estimated to be over $6 billion each year. In addition to this tremendous economic cost, the bone fractures that result from osteoporosis often lead to considerable disability and in many cases can actually result in premature death.

Although in many cases the exact cause of osteoporosis cannot be pinpointed, in a number of situations it is known to be caused by a particular factor or condition. One such situation involves prolonged treatment with large doses of glucocorticoid, as might be used for immunosuppressive therapy (refer to Chapter 14). High doses of glucocorticoids antagonize the actions of vitamin D, resulting in a vitamin D deficiency, which in turn leads to a relative calcium deficiency.

Another factor known to cause osteoporosis is immobilization. When a limb is placed in a cast, the immobilization can result in a localized loss of bone mineral in that limb. Similarly, immobilization of a limb as a result of a neurologic deficit, such as paralysis, can lead to osteoporosis in the affected limb. Generalized immobilization, such as that experienced by a person confined to bed for a prolonged period of time, can produce a somewhat generalized loss of mineral from all bones in the body. In the weightless conditions of space travel, astronauts also suffer from a generalized osteoporosis. Although they are not immobilized, the lack of weight-bearing activity by their muscles and bones leads to osteoporosis.

In addition to the direct causes just mentioned, a number of other factors, while apparently not causing osteoporosis, appear to increase the risk of developing the disorder. Both the ancestry and sex of the individual are risk factors. Women with ancestors from northern Europe are generally more likely to develop osteoporosis than the average male or female in the general population. For reasons not completely understood but under investigation by researchers, persons with lean builds have a greater tendancy to develop osteoporosis than do obese individuals. Several factors associated with lifestyle can also have some influence. Smoking, excessive consumption of alcohol or caffeine, and a sedentary lifestyle can all contribute to an increased rate of bone loss and the development of osteoporosis. Although no clear cause and effect relationship has been shown, it is interesting to note that 75% of all women who develop osteoporosis are smokers.

At present, much remains to be learned about the development, treatment, and prevention of osteoporosis. In postmenopausal women, treatment with estrogens has been shown to be quite effective in combating bone loss. However, associated with estrogen treatment is an increased risk for the development of endometrial (uterine) cancer. Researchers are currently investigating whether treatment with low doses of estrogen significantly reduces the risk for developing cancer while still being effective in preventing osteoporosis.

For many years, exercise has been known to promote bone mineral accumulation. Presumably, this effect is due in large part to the increased weight bearing that occurs during exercise, but other factors, such as changes in hormone levels, may also contribute. An interesting example showing that weight bearing promotes accumulation of bone mineral is the fact that tennis players have significantly more bone mineral in the arm with which they swing their racquet than in the other arm. Also of interest, recent studies have shown exercise to be effective in slowing the rate of bone loss in postmenopausal women. Thus, exercise may be a relatively safe and inexpensive way of combating osteoporosis.

■ **In recent years, calcium supplements have been heavily promoted by commercial supplier as being useful in the prevention and treatment of osteoporosis. Suppose a middle-aged friend of the family came to you and said she was considering taking calcium supplements and asked your opinion, knowing that you were taking a physiology course and had learned about calcium homeostasis. What might be your response?**

nounced vitamin C deficiency can also result in a net loss of bone. Vitamin C is required for normal synthesis of organic bone matrix, and thus without synthesis of new matrix bone formation cannot occur. Immobilization, such as might occur after injury of a limb, also can lead to osteoporosis.

Most commonly, osteoporosis is associated with advancing age in both men and women. However, women tend to suffer from osteoporosis far more frequently than men. Figure 26–16 shows the amount of bone present in males and females according to age. Note that up until about 10 to 12 years of age, males and females have roughly equivalent amounts of bone. At about this time, which coincides with the onset of puberty, males begin to acquire bone at a more rapid rate than females. At the point between

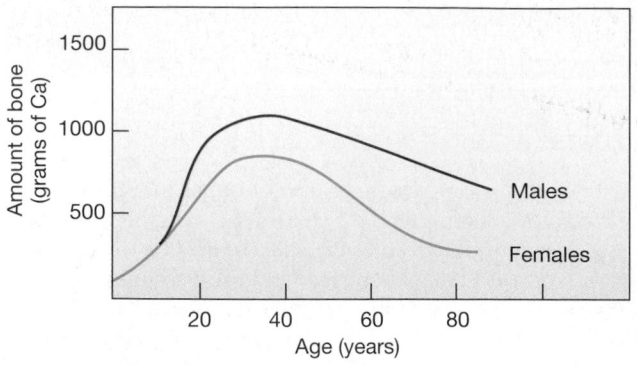

Figure 26–16

Diagram showing the amount of bone present in males and females of different ages. Note that maximal bone mass is higher in males than in females. Also illustrated is the more rapid loss of bone experienced by women compared to men, beginning around age 50.

ages 30 and 40 when maximal bone mass is reached, males have 20% to 30% more bone than females, on average. After about age 40, both males and females begin to lose bone mass and do so at roughly equivalent rates up until about age 50 (see Fig. 26–16). At menopause (around age 50), women begin to lose bone more rapidly, as the result of decreasing estrogen levels (see Chapter 31). This more rapid loss of bone can result in **postmenopausal osteoporosis.**

Osteoporosis is a major public health problem, owing to the increased risk of fractures occurring with only minimal stress or trauma placed on the bone and the resulting disability that occurs.

Osteomalacia and Rickets Result from Inadequate Mineralization of New Bone

The second major category of metabolic bone disease is that of osteomalacia and rickets. Both diseases are characterized by **inadequate mineralization** of new bone matrix, such that the ratio of mineral to organic matrix is lower than normal. When this disorder occurs in children, it is referred to as **rickets** (Fig. 26–17). In adults, the same disorder is termed **osteomalacia.** In children with rickets, bones have decreased mechanical strength and are subject to distortion, which may lead to a bowed appearance, particularly in the long bones of the legs. In adults, osteomalacia can produce severe bone pain.

Rickets and osteomalacia are most commonly caused by a deficiency in vitamin D *activity,* as a result of one of three general causes. The first is a combined dietary deficiency of vitamin D and insufficient exposure to sunlight. Because foods are fortified with

Figure 26–17

Classic appearance of rickets in a child.   (*© Biophoto Associates/Photo Researchers, Inc.*)

vitamin D, severe nutritional deficiencies of vitamin D are relatively rare in the United States, although this can be a health problem in many other parts of the world. A second cause of rickets, or osteomalacia, involves an impairment in converting vitamin D in liver and kidney to its hormonally active form. Numerous factors can contribute to this type of deficiency, including liver disease, kidney disease, or an absence of parathyroid hormone, which normally stimulates activity of 1α-hydroxylase in the kidney. The third cause of impaired vitamin D activity involves an inability of 1,25-OH-vitamin D_3 to act on its target tissues. Certain anticonvulsant drugs are known to interfere with vitamin D action, but the mechanisms involved are poorly understood. However, long-term treatment with these drugs, which are used to treat conditions such as epilepsy, may lead to ostomalacia or rickets.

CHAPTER REVIEW

Summary

- Because calcium is involved in a number of important physiological processes in the body, its concentration in extracellular fluid must be maintained within fairly narrow limits. Phosphate also serves several important functions, but regulation of plasma phosphate concentrations is not as critical as that of calcium. Approximately 99% of the total body calcium is located in bone. The majority (85%) of phosphate is also present in bone.

- Total plasma calcium is normally about 10 mg/100 ml. Two primary forms of calcium in plasma are filterable calcium and protein-bound calcium. Filterable calcium consists of free calcium ion and calcium complexed with an anion, such as bicarbonate. Free calcium is the physiologically active form of calcium. Most phosphorus in plasma exists as phosphate (PO_4). Normal plasma concentrations range from 3.0 to 4.5 mg of phosphorus/100 ml.

- Three tissues are primarily involved in regulating calcium and phosphate metabolism: the gastrointestinal tract, kidneys, and bone. Most calcium taken in via diet is not absorbed and leaves the body in the feces; absorbed calcium is eventually excreted in the urine. The majority of ingested phosphate is absorbed from the gastrointestinal tract; its primary route for excretion is via the urine.

- Calcium absorption from the gastrointestinal tract occurs primarily in the small intestine and involves active transport and passive diffusion; regulation occurs via changes in active transport. Phosphate absorption from the small intestine occurs by active transport and passive diffusion and does not appear to be regulated.

- Approximately 99% of calcium filtered by the kidneys is reabsorbed. Reabsorption involves both active transport and passive diffusion. Active transport in distal tubules is stimulated by parathyroid hormone. In the case of phosphate, normally about 85% of that filtered is absorbed. Reabsorption occurs primarily by active transport in proximal tubules and is inhibited by parathyroid hormone.

- Bone consists of inorganic bone mineral deposited on an organic support matrix. The mineral fraction is composed primarily of hydroxyapatite; the organic matrix consists primarily of collagen. Three primary cell types in mature bone are osteoblasts, osteocytes, and osteoclasts. Osteoblasts and osteocytes are involved in bone formation and bone maintenance; osteoclasts are responsible for bone resorption. In growing bone, chondrocytes of the epiphyseal plate are under hormonal control. Stimulation of chondrocytes leads to bone growth.

- Three hormones are involved in regulating plasma calcium and phosphate concentrations: (1) parathyroid hormone, (2) calcitonin, and (3) vitamin D metabolites. Parathyroid hormone is secreted in response to decreased plasma calcium. Its target tissues are kidney, bone, and, indirectly, the gastrointestinal tract, with a net effect of increasing calcium and decreasing phosphate in plasma.

- Calcitonin plays only a minor role in regulation. It is secreted in response to increased plasma calcium. Acting primarily on bone, its net effect is to decrease both calcium and phosphate in plasma.

- Vitamin D3 is converted into 1,25-OH-vitamin D_3 by sequential conversion in liver and kidney. The final step of vitamin D activation in kidney is stimulated by parathyroid hormone. The net effect of 1,25-OH-vitamin D_3 action is to increase plasma calcium and phosphate, acting primarily on the intestines to increase calcium absorption.

- The most common abnormality of bone metabolism is a group of disorders known as metabolic bone disease. One category of metabolic bone disease is osteoporosis, which is categorized by an overall loss of bone mass with equal loss of bone mineral and organic matrix. Osteomalacia and rickets make up the second major category of metabolic bone diseases. These are characterized by inadequate mineralization of bone, most commonly caused by a deficiency of vitamin D activity.

Review Questions

Identify Each with a Term
1. Form of plasma calcium consisting of ionized and complexed calcium.
2. Crystals of bone mineral, composed mainly of calcium phosphate.
3. Organic matrix of bone.
4. Cells of bone responsible for synthesizing osteoid.
5. Cell type responsible for synthesis and secretion of calcitonin.
6. The condition in which a reduction in overall bone mass occurs, with equal losses of both mineral and organic matrix.
7. Cells of the epiphyseal plate responsible for collagen synthesis.
8. The ongoing process of bone turnover.

Define Each Term
9. hypocalcemia
10. ionized calcium
11. ground substance

12. osteoclasts
13. parathyroid hormone
14. 1,25-OH-vitamin D_3
15. ergocalciferol

Choose the Correct Answer

16. The physiologically active form of calcium in the body is:
    a. calcium bound to albumin
    b. calcium citrate
    c. calcium ion (Ca^{2+})
    d. unable to cross membranes
17. At a pH of 7.4, most of the phosphate in the blood will be present as:
    a. hydroxyapatite   c. hexose phosphate
    b. HPO_4^{2-} d. $H_2PO_4^-$
18. The majority of ingested calcium is excreted in the:
    a. feces   c. sweat
    b. urine   d. plasma
19. Parathyroid hormone acts to _____ the reabsorption of calcium in the _____ convoluted tubules of the kidney.
    a. increase, proximal   c. increase, distal
    b. decrease, proximal   d. decrease, distal
20. The conversion of 7-dehydrocholesterol to vitamin D_3 occurs in the:
    a. epiphyseal plate   c. liver
    b. kidney   d. skin
21. The kidney enzyme 1α-hydroxylase catalyzes the conversion of:
    a. calcium to hydroxyapatite

b. 25-OH-vitamin D_3 to 1,25-OH vitamin D_3
c. vitamin D_2 to vitamin D_3
d. cholecalciferol to dehydrocholesterol

22. A deficiency in vitamin D action can lead to:
    a. epilepsy
    b. osteomalacia
    c. osteoporosis
    d. postmenopausal osteoporosis

Answer Each Question in One or Two Sentences

23. What is the mechanism of hypocalcemic tetany?
24. What are the major forms of calcium in the plasma and what percent of the total does each of these compose?
25. What are the actions of parathyroid hormone that tend to raise plasma calcium and lower plasma phosphate?
26. What conditions lead to increased secretion of parathyroid hormone? of calcitonin?

Problem Solving in Physiology

27. Numerous physiological mechanisms exist to defend against either an increase or a decrease in normal plasma calcium levels. However, the mechanisms that protect against hypocalcemia appear much more potent than those that defend against hypercalcemia. What might be the physiological basis for this difference?
28. By what mechanisms might kidney disease lead to abnormalities in calcium and phosphate homeostasis?

Suggested Readings

Ganong, William F. *Review of Medical Physiology*, 12th ed. Los Altos: Lange Medical Publications, 1985.

Goodman, H. Maurice. *Basic Medical Endocrinology.* New York: Raven Press, 1988.

Griffin, James E., and Ojeda, Sergio R. *Textbook of Endocrine Physiology*, 2nd ed. New York: Oxford University Press, 1992.

Guyton, Arthur C. *Human Physiology and Mechanisms of Disease*, 4th ed. Philadelphia: W.B. Saunders Company, 1987.

Hedge, George A., Colby, Howard D., and Goodman, Robert L. *Clinical Endocrine Physiology.* Philadelphia: W.B. Saunders Company, 1987.

Norman, Anthony W., and Litwack, Gerald. *Hormones.* Orlando: Academic Press, 1987.

Reichel, Helmut, Koeffler, H. Phillip, and Norman, Anthony W. "The Role of the Vitamin D Endocrine System in Health and Disease." *New England Journal of Medicine*, 320:980–991, 1989.

Tepperman, Jay, and Tepperman, Helen M. *Metabolic and Endocrine Physiology*, 5th ed. Chicago: Year Book Medical Publishers, Inc., 1987.

Walters, Marian R. "Newly identified actions of the vitamin D endocrine system." *Endocrine Reviews*, 13:719–764, 1992.

Wilson, Jean D., and Foster, Daniel W., eds. *Williams' Textbook of Endocrinology*, 7th ed. Philadelphia: W.B. Saunders Company, 1985.

Key Terms

C-cells (p. 796)
calcitonin (p. 796)
calcium (p. 784)
canaliculi (p. 793)
chondrocytes (p. 793)
epiphyseal plate (p. 793)
ground substance (p. 792)

hydroxyapatite (p. 792)
hypocalcemia (p. 784)
metabolic bone disease (p. 799)
osteoblast (p. 792)
osteoclast (p. 792)
osteocyte (p. 792)

osteoid (p. 792)
osteomalacia (p. 802)
osteoporosis (p. 800)
parathyroid gland (p. 795)
parathyroid hormone (PTH) (p. 795)

phosphate (p. 785)
phosphaturia (p. 798)
proteoglycan (p. 792)
rickets (p. 800)
tetany (p. 784)
vitamin D_3 (p. 796)

Photomicrograph of a human sweat gland. *(© John D. Cunningham/Visuals Unlimited)*

Chapter 27

Regulation of Body Temperature

Key Concepts

Thermal homeostasis in humans is maintained by physiological and physical mechanisms that continuously balance metabolic heat production and heat loss to the environment. The basal metabolic rate (BMR) is a standard measure under specific conditions of metabolic rate that allows comparison between individuals.

Body temperature is regulated within a relatively narrow range of temperatures when humans are exposed to a wide range of environmental temperatures.

The physiological control system that regulates body temperature is a biological negative feedback system and has at least three basic components. Diseases that affect temperature regulation produce both regulated and unregulated changes in body temperature.

Chapter Outline

The Balance Between Heat Production and Heat Loss
Normal Body Temperature
Regulation of Body Temperature

Edith, attending summer school, leaves the air-conditioned classroom, which is maintained at 27°C and 30% humidity. She walks outside where the temperature is 35°C and the humidity is 65%. She immediately senses a warm skin followed by reddening of the skin. As Edith walks to her next class she also experiences a wetting of the skin, especially in the exposed forehead and hands. The responses activated by walking into a hot, humid environment are physiological effector-produced responses, which increase heat loss from the body and match the increased heat gained from the hot environment to maintain the student's core temperature relatively constant.

In this chapter we will discuss in detail the physiological mechanisms that continuously balance metabolic heat production and body heat loss to maintain normal body temperature. We will describe the elements of the precise negative feedback control system that regulates body temperature of healthy individuals within a relatively narrow range when humans are exposed to a wide measure of environmental temperatures. Finally, we will examine the disease states associated with changes and/or malfunctions in the regulation of body temperature.

Humans are **homeotherms** and normally regulate deep "core" body temperature within a relatively narrow range (36 to 39°C) when exposed to a wide range of climatic conditions. We share this characteristic with all mammals and birds. (The range can be expanded in disease.) Other animals—for example, fish and reptiles—are **poikilotherms.** Their body temperatures may fluctuate with changes in ambient temperature.

A nude human can be exposed to temperatures as low as 12°C or as high as 60°C in dry air and still maintain a core body temperature of 36 to 39°C. Higher core body temperatures of 40 to 41°C, such as those associated with strenuous exercise, fever, and hyperthermia, can be tolerated for only short periods. Body temperatures higher than 42°C are associated with a rapid breakdown of vital cellular proteins and death. Humans, therefore, regulate core body temperature only a few degrees below the point of thermal death.

The regulation of body temperature in humans near the higher limit of thermal tolerance ensures optimal cellular function. All cellular, biochemical, and enzymatic reaction rates are temperature dependent. At normal body temperature, cellular enzymatic reactions work optimally. If the cellular temperature were to drop by 10°C, however, these reaction rates would decrease by a factor of approximately 2.5. Human metabolic activity, therefore, changes about 12% for every degree centigrade of change in body temperature. Thus, the effective regulation of body temperature is necessary in homeotherms to ensure the capacity for rapid metabolism that makes intense physiological activity such as exercise possible. In addition, the regulation of the body temperature by humans ensures thermal homeostasis of the internal environment and provides considerable independence from any possible detrimental effects of changes in ambient temperature.

The Balance Between Heat Production and Heat Loss

Human body temperature is maintained at a relatively constant level because of a balance between heat production and heat loss. Thermal balance is maintained by a combination of physical and physiological mechanisms (Fig. 27–1). If heat production is greater than heat loss, the body temperature increases; conversely, when heat loss is greater than heat production, body temperature decreases. In humans, physiological regulatory mechanisms operate continuously to keep heat production and heat loss

Figure 27–1

Balance between factors affecting heat production and heat loss, illustrating that body temperature remains in the normal range as long as these two are equal.

approximately equal, thereby maintaining **thermal homeostasis** and a normal body temperature.

Regulation of Body Temperature Involves a Balance Between Heat Production and Heat Loss

Heat is being produced continuously in the body as a by-product of the **catabolism** of food. The energy liberated by catabolic processes within the body can be converted into three forms: **chemical storage, external work,** and **metabolic heat.**

Food energy =
   chemical storage + external work + metabolic heat

Metabolic Rate and Heat Production

The amount of food energy converted per unit of time by an individual is the person's **metabolic rate.** The energy systems of the body are about 20% efficient. That is, at maximum efficiency, 20% of the chemical energy of food is available to do external work, and 80% appears as heat (Fig. 27–2). In a resting human doing no external work, essentially all of the chemical energy of food eventually appears as heat.

Energy = Heat

The basic unit of heat energy is the **calorie,** which is the amount of heat required to raise 1 g of water one degree centigrade. A **Calorie** or kilocalorie is equivalent to 1000 calories and is used when discussing

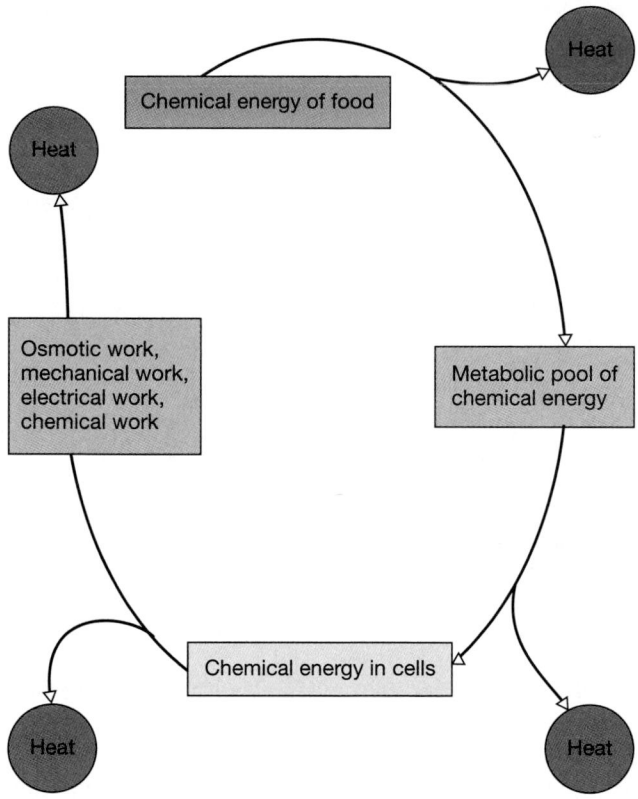

Figure 27–2

Summary of transformation of food energy within the body and its transfer as heat energy to the environment.

heat energy in humans. Calories (calories × 1000) will be used in this chapter.

In a resting adult human, the metabolic rate averages 70 Calories of heat per hour. If there were no heat loss responses, the resting metabolic rate would raise the body temperature approximately 1°C per hour. The rate of heat storage of the body can be estimated by:

Heat storage (Cal/hr) =

$$\frac{\text{mass (kg)} \times \text{specific heat (0.83)} \times \text{temp. change (°C)}}{\text{Time (hr)}}$$

In a resting person, the vital organs in the head, thorax, abdomen, and pelvis are the major areas of metabolic heat production (see Fig. 27–14). The pattern changes during work, when the skeletal muscle mass becomes a major source of heat production.

Factors Affecting Metabolic Rate

Any factor that increases or decreases the energy released from food also affects a person's metabolic rate. Some of the most important are listed in Table 27–1.

Table 27–1

Some Factors Affecting Metabolic Rate

Exercise	Shivering
Ingestion of food	Body size
Age	Gender
Temperature	Climate
Various hormones	Emotional state
Disease	Fever

Exercise and Metabolic Rate. **Exercise,** the most powerful stimulus for increasing body heat production, has a dramatic effect on metabolic rate. Muscle contraction requires chemical energy, which, during strenuous exercise in a trained athlete, can increase heat production from a resting level of 1500 to 1700 Calories per day to 20 times that amount for several minutes. Table 27–2 illustrates the energy expenditure associated with different types of activities. Prolonged human activities that are classified as **light work** (e.g., writing) raise metabolic rates by approximately 500 Calories per day, while those classified as **moderate work** (e.g., walking) raise metabolic rates by 1000 Calories per day. **Heavy work** (e.g., bicycling) raises metabolic heat production by 2000 Calories per day, while **heavy** occupational **physical labor** (e.g., lumbering) on the average requires 4800 Calories per day and corresponds roughly to three times the resting metabolic rate. If no heat were lost from the body during the heaviest work, the heat produced would be sufficient to raise the body temperature by 4°C per hour and result in death. The limit for maximal metabolic rate for women is lower (around 3700 Cal/day) because of their lower aver-

Table 27–2

Rate of Energy Expenditure for 70-Kg Person During Different Types of Activities

Type of Activity	Energy Expenditure (Cal/hr)
Sitting at rest	100
Walking slowly on level	200
Sexual intercourse	280
Swimming	500
Jogging	325
Rowing	828

age body weight and muscle mass. They therefore have similar maximal rates on a per gram basis.

Specific Dynamic Action and Metabolic Rate. The metabolic rate of a resting person increases following the ingestion of food and remains elevated for several hours. The increased heat production that results from processing a meal is called the **specific dynamic action** of foods. A meal containing a large quantity of carbohydrates and fats increases the metabolic rate by only 4%. A meal containing large quantities of proteins, however, can increase metabolic rate by as much as 30%. The major source of the increased heat appears to be the liver and its metabolism. This is supported by the observation that metabolic rate increases in dogs when amino acids are infused intravenously. If the liver is removed prior to infusion, then heat production increases only slightly.

Hormones and Metabolic Rate. Many **hormones** influence energy metabolism. Of particular importance are the **catecholamines** and **thyroxin.** Stimulation of the sympathetic nervous system causes the release of norepinephrine and epinephrine. These catecholamines increase the breakdown of glycogen to glucose. They also increase the rates of enzymatic reactions, which increase cellular metabolism. Maximal stimulation of the sympathetic system can increase metabolic rate by 100%, but it remains elevated for only a few minutes.

Maximal secretion of thyroxin by the thyroid gland also increases most cellular enzymes. Cellular metabolism increases to about 100% above normal. Unlike the effects of catecholamines, the thyroid hormones continue to act for several weeks. Complete lack of thyroid hormones, on the other hand, can reduce metabolic rate by as much as 50%.

Many other hormones have minor effects on metabolic rate. For example, testosterone from the testes, insulin from the pancreas, and growth hormone from the pituitary gland can increase the overall metabolic rate of a person by only 5% to 15%.

Body Size and Metabolic Rate. Because of the greater tissue mass, the total energy exchange of an adult male weighing 160 kilograms is greater than that of an adult male weighing 65 kilograms. Metabolic rate, however, is not directly related to body weight, but is roughly proportional to surface area. Consequently, metabolic rate is commonly expressed as a function of body surface area ($Cal/m^2/hr$). When heat production is expressed in this manner, adults of different body sizes have similar metabolic rates. The rationale for this observation is that metabolic heat is lost at the surface of the body and that an amount of heat equivalent to what is lost must be

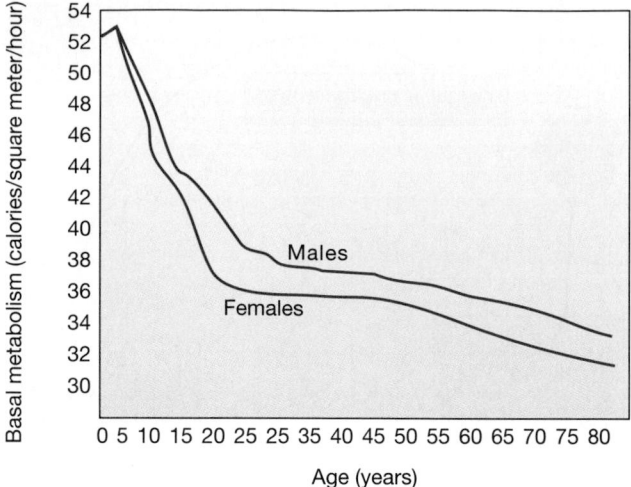

Figure 27–3

Average basal metabolic rates in normal males and females at different ages.

produced in order to maintain thermal homeostasis and a constant body temperature.

Age, Sex, and Metabolic Rate. Metabolic rate ($Cal/m^2/hr$) reaches a peak at birth and gradually declines with age in both males and females (Fig. 27–3). The metabolic rate of growing children is higher than that of adults because large amounts of chemical energy are needed during growth for the synthesis of carbohydrates, fats, and proteins. Simultaneously, more energy is liberated as heat by the catabolic reactions that necessarily accompany synthetic reactions.

The metabolic rate of females is 5% to 10% lower than that of males of the same size and age. The cause for this difference is not certain, but it may be related to a greater proportion of relatively inactive tissue such as subcutaneous fat in females compared to males. The metabolic rate of the female increases during pregnancy as a result, in part, of the increased synthetic reactions associated with growth of the fetus.

Temperature and Metabolic Rate. Fever, regardless of its cause, increases metabolic rate. As previously mentioned, all chemical reactions are temperature-dependent and are increased during fever.

Acute exposure to cold induces shivering, which consists of a series of rapidly oscillating muscle tremors. Because no external work is performed, all of the energy liberated appears as internal heat. Shivering can increase metabolic rates to about five times above resting levels within minutes. Metabolic rates in studies of long-term residents in Arctic regions have been shown to be 10% to 20% higher than the

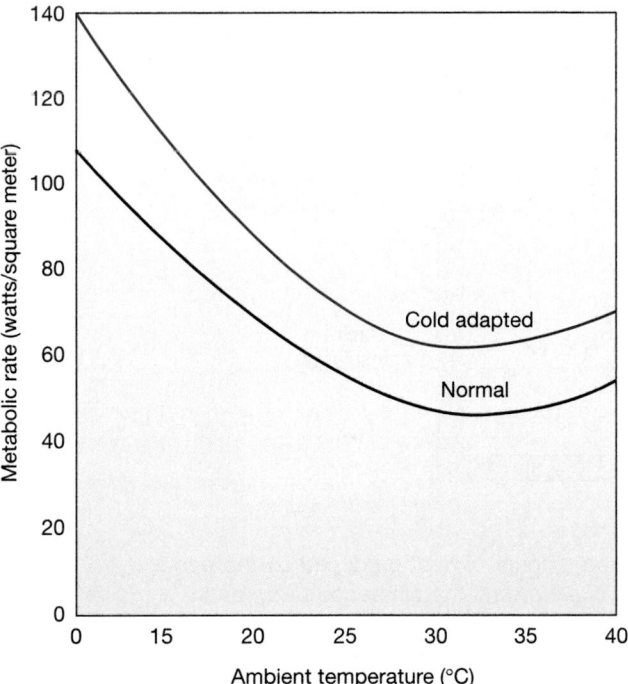

Figure 27–4

Effect of ambient temperature on resting metabolic rate before and after chronic cold exposure.

rates of residents in tropical areas. The cause for the difference is not clear, since in these studies the effects of atmospheric temperature cannot be isolated from the possible effects of other variables such as diets and light-dark cycles. Controlled studies of monkeys, however, also show that long-term exposure to cold produces a 20% to 30% increase in metabolic rate (Fig. 27–4). The increase is caused, in part, by adaptation of the thyroid gland. Thyroxin secretion increases in cold climates and decreases in hot climates.

Basal Metabolic Rate

The concept of **basal metabolic rate (BMR)** has evolved because of our increasing knowledge of the many factors that can influence a person's metabolic rate. BMR is a standard measure of metabolic rate that allows comparison between individuals. To record BMR, a set of standard conditions has been established to measure the so-called "basal conditions." Actually, the BMR is not truly "basal," since metabolic rate during sleep is lower than BMR. BMR is more appropriately described as the "metabolic cost of living." The following four conditions are usually specified for measurements of BMR.

1. The person must have fasted for 12 hours.
2. The measurement is usually made in the morning after a night of restful sleep.

3. The person must remain at complete rest for 30 to 60 minutes before the test.
4. The atmospheric temperature must be comfortable, about 25°C.

Measurements of Metabolic Rate

The metabolic rate is a measure of the total energy used by the body per unit of time. In a person completely at rest and sitting in a comfortable temperature, nearly all of the metabolic energy is transformed to body heat. Furthermore, a normal person under these conditions keeps body temperature constant by releasing an amount of body heat into the environment equal to the heat produced inside the body. A person's metabolic rate, therefore, can be determined by measuring the heat given off by the body in a known period of time or the amount of oxygen consumed by the body per unit of time.

Direct Calorimetry. A person's metabolic rate can be measured by direct or indirect methods. In the direct method, a person is placed in a large insulated chamber called a **human calorimeter** (Fig. 27–5). The temperature of the calorimeter is kept constant by a water cooling radiator system that picks up the heat lost from the person's body. The rate of heat gain by the radiator system can be accurately measured with a thermometer and is equal to the rate of heat liberated from the person's body. Although direct calorimetry is a very accurate technique, since it measures the heat output of the body directly, it is a time-consuming, expensive, and cumbersome procedure. It is therefore used most commonly for research purposes and to validate less costly and more convenient methods for measuring metabolic rate.

Indirect Calorimetry. A commonly used indirect method for measuring metabolic rate is based on the determination of a person's **oxygen consumption** per unit of time. This method is based on the concept that since 95% to 100% of the energy available to the body is derived from cellular reactions with oxygen, metabolic rate can be accurately estimated from oxygen use. Studies comparing metabolic rates of humans using both indirect and direct methods have shown good agreement, with differences of less than 5%. The errors in the indirect method are not large enough to outweigh its advantages, which include low cost, speed, and convenience.

The Energy Equivalent of Oxygen. In practice, the indirect method for measuring metabolic rate is dependent on the concept of the energy equivalent of oxygen and a special instrument for the accurate measurement of oxygen consumption called a **respirom-**

Figure 27–5

Direct method for measuring metabolic rate. The amount of heat produced by the person's body is calculated from the total volume of water and the difference between inflow and outflow temperature.

eter (Fig. 27–6). The energy equivalent of oxygen is the quantity of energy liberated when the body uses one liter of oxygen to produce metabolic energy. For example, the amount of energy released when the body uses one liter of oxygen to oxidize **carbohydrates** is 5.0 Calories. When the body uses one liter of oxygen to burn **fats,** the energy released is 4.7 Calories. The corresponding value for **protein** is 4.6 Calories. Since the energy equivalent for all three food types is about the same, an approximate average of these values is commonly used and is equal to 4.825 Calories of heat produced per liter of oxygen con-

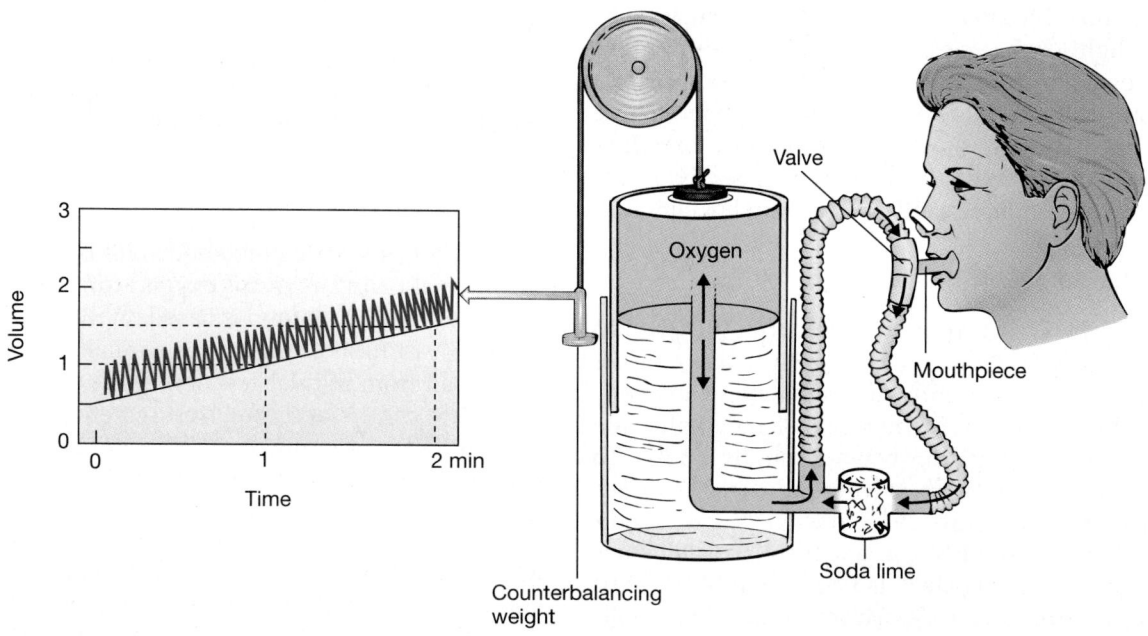

Figure 27–6

A respirometer for measuring the rate of oxygen consumption. The subject inspires oxygen from a bell gasometer through a mouthpiece and valve. CO_2 is removed from the expired air by soda lime before it returns to the bell.

sumed. Metabolic rate, therefore, is commonly determined by the following equation:

Metabolic rate (Cal/m^2/hr) =
   liters of O$_2$ consumed (per hr) × 4.825 Calories
      per liter of O$_2$ ÷ surface area (meters^2)

Heat Loss from the Body Involves Both Physical and Physiological Processes

Humans and other homeotherms must continuously exchange heat with their environment to maintain thermal homeostasis. If the heat produced by metabolism were not continually released, the body temperature would rise indefinitely. The body exchanges heat with its surroundings through four physical processes, **radiation, conduction, convection,** and **evaporation,** as well as the physiological process of sweating.

Heat Loss by Radiation

Radiation is heat transfer by electromagnetic radiation between objects that are not in contact (Fig. 27–7). The greater the temperature difference, the larger the heat transfer from the warmer to the cooler object. Approximately 60% of the heat from a nude person sitting in a comfortable 21°C temperature is lost by radiation. At high atmospheric temperatures (35°C),

the body may actually gain heat by radiation. The position of the body is an important factor in determining radiant heat exchange. When a person assumes the fetal position, radiant heat exchange is reduced because less surface area is exposed to the environment, whereas when a person is standing with arms and legs held away from the body, maximum heat is exchanged. Typically, most people consider air temperatures between 21 and 28°C to be comfortable because these temperatures permit the body to radiate heat at approximately the rate of metabolic heat production, thus keeping body temperature constant. **Thermal comfort** occurs when no thermoregulatory mechanisms are activated and blood flow through the skin is at an intermediate rate. The atmospheric temperature for thermal comfort is not constant and changes with changes in humidity, wind speed, and the level of activity.

Heat Loss by Conduction

Conduction is heat exchange in the form of kinetic energy between molecules of objects in direct contact. As in radiation, the greater the temperature difference, the larger the heat transfer by conduction from warmer to cooler objects. In humans, only relatively small amounts of heat are usually lost from the body by direct contact with other objects (see Fig. 27–7). Heat loss or gain by conduction, however, is very sizable in persons exposed to cold water because water is 20 times more effective than air as a conductor of heat. The rapid loss of body heat in cold water and subsequent hypothermia can lead to life-threatening conditions in boating and shipping accidents.

Heat Loss by Convection

Convection is the transfer of heat between the body and a moving gas or liquid. In fact, conduction is part of the convective process. For example, when air temperature is lower than that of the skin, heat is transferred by conduction to the cool air adjacent to the skin. The heated air expands, becomes less dense, rises, and is replaced by cooler air in a continuous process called **free convection. Forced convection** considerably increases the rate of heat exchange by the body. When the body is exposed to wind, for example, the air adjacent to the body is replaced by new air much more rapidly than normal, increasing convective heat loss. The cooling effect of wind is approximately proportional to the square root of the velocity of the wind and accounts for the **"wind chill factor,"** which increases discomfort in the cold and can lead to the freezing of tissue at higher ambient temperatures than occurs in calm air.

Figure 27–7

Mechanisms of heat exchange between the body and the environment.

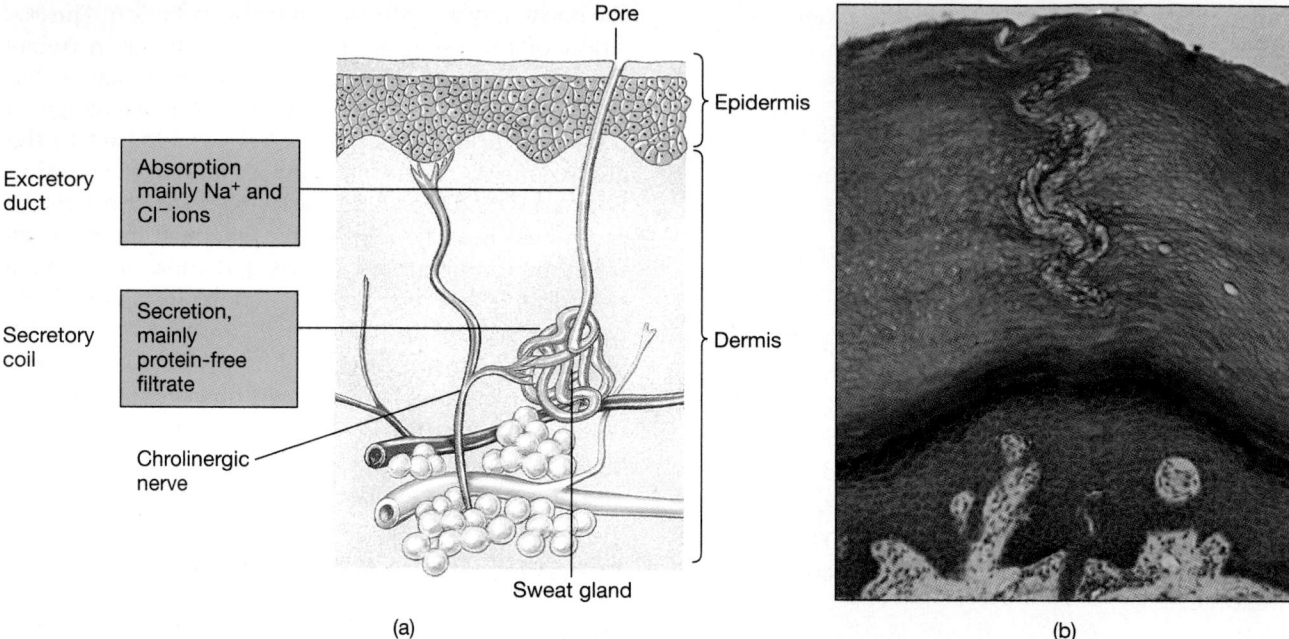

Pore

Epidermis

Excretory
duct

Absorption
mainly Na⁺ and
Cl⁻ ions

Secretory
coil

Secretion,
mainly
protein-free
filtrate

Dermis

Chrolinergic
nerve

Sweat gland

(a)

(b)

Figure 27–8

(*a*) A human sweat gland innervated by a sympathetic nerve. A protein-free filtrate is
formed by the secretory coil, and most of the electrolytes are reabsorbed by the excretory duct, producing a dilute, watery sweat. (*b*) Photomicrograph of human sweat gland.
(×35.)   *(© John D. Cunningham/Visuals Unlimited.)*

Heat Loss by Evaporation

Evaporation is the process of converting water from
a liquid to a gas. Thermal energy is required for the
transformation. When water evaporates from the surface of the body, 0.58 Calories of heat are removed
from the skin for each gram of water that evaporates.
Evaporation is a continuous heat-loss mechanism,
since small amounts of water are continuously diffusing from the skin, the respiratory passages, and
the mucous membranes of the mouth. In a person
resting in a comfortable atmosphere, the total water
loss from evaporation of diffused water is small and
equals approximately 600 ml per day. Evaporative
heat loss under these conditions is relatively small
and averages about 20% of the total heat exchange
(see Fig. 27–7). Since we are usually unaware of this
loss, it is referred to as heat loss due to **insensible
perspiration.**

Heat Loss by Sweating

Sweating, in contrast with passive diffusion of water through the skin, is a physiological process that
involves the active secretion of water by specialized
sweat glands in the skin (Fig. 27–8). Sweat secretion
is activated by sympathetic nerves in response to

overheating, and increases linearly with abnormal increases in body temperature (Fig. 27–9). Evaporation
of sweat is the only way for the body to lose heat
when it is exposed to hot environments, since radiation and conduction at air temperatures greater than
skin temperature actually transfer heat into the body.
The evaporative capacity of the body decreases with
increasing humidity under these conditions.

Under extreme conditions the body can secrete
about 1.5 liters of sweat per hour. For sweating to be
an effective cooling process, sweat must rapidly
evaporate from the skin. Under optimal conditions,
the evaporation of 1.5 liters of sweat can remove as
much as 800 Calories of body heat per hour, or about
12 times the rate of basal heat production.

Heat Transfer Within the Body Is Essential in Order to Maintain a Relative Constant Body Temperature

To keep the body temperature constant, the heat produced by metabolism must be transferred from the
core cells that are producing it to the body shell,
where it can be transferred to the environment.
Within the body, the transfer of heat is accomplished

Figure 27–9

Effect of changes in body temperature on evaporative heat loss due to sweating. This figure illustrates the critical temperature at which sweating begins, and the linear increase that occurs with increasing body temperature.

by a combination of **internal conduction** and **circulatory convection.**

Heat Transfer by Internal Conduction

The rate of heat transfer due to internal conduction is slow because of the small temperature gradient that exists between adjacent cells. In addition, most of the tissues of the body are relatively poor heat conductors. The thermal conductivity of unperfused tissue is equal to that of cork, a good insulator and poor conductor. Thus, internal conduction alone is not adequate to maintain a balance between the production of heat and the loss of heat needed to maintain a constant body temperature.

Heat Transfer by Circulatory Convection

Convective heat transfer occurs within the body with the movement of body fluids. Forced convection by the circulatory system is very important for thermal balance. Cooler blood, returning from the shell, is warmed by the core tissues that are producing metabolic heat; the warmed blood returns to the periphery and is cooled by the tissues of the shell, which are in contact with atmospheric air. Vasomotor responses that control peripheral blood flow determine

the **effective insulation** of the body by determining the body proportions that are maintained at core and shell temperatures. At comfortable temperatures, vasomotor responses regulate peripheral blood flow to achieve thermal balance without sweating or changing metabolic heat production. At the lower limit of the thermal comfort zone, peripheral vessels constrict and decrease blood flow and heat flow to a minimum (Fig. 27–10). At the upper limits of the thermal comfort zone, peripheral vessels dilate and expose large quantities of warm blood to peripheral tissues, increasing heat loss severalfold. The skin therefore can be a heat radiator for the body, and the blood flow to the skin determines the rate of heat transfer by circulatory convection from the body core to the skin.

Normal Body Temperature

The metabolic heat produced within the body is normally lost to the environment at the surface. The physical laws of heat flow therefore dictate that the temperature of parts of the body near the surface must be lower than the temperature of the central parts. Furthermore, the fluctuations in body temperature produced by changes in atmospheric temperature are greater at the body surface. The temperature gradients of the human body can be conceived as a **"poikilothermic" shell** at the surface and a

Figure 27–10

Effect of changes in environmental temperature on the heat conductance from the body core to the skin surface.

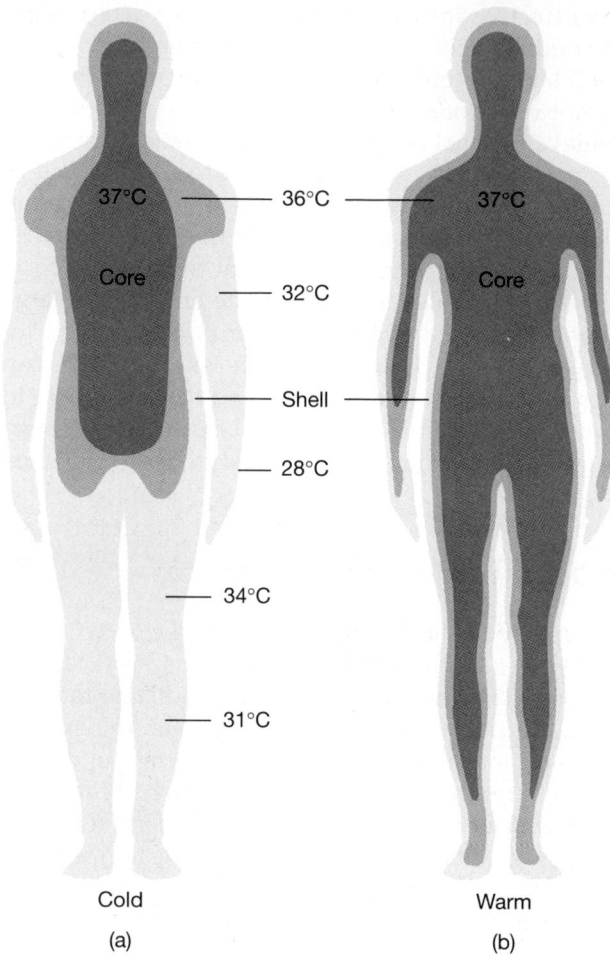

Figure 27–11

Core-shell concept in the human body. The temperature of the surface and underlying areas (shell) varies with atmospheric temperature. The shell is "thicker" in a cold environment (*a*) and "thinner" in a warm one (*b*).

"homeothermic" core in the center (Fig. 27–11). The core temperature is kept constant over a wide range of atmospheric temperatures, whereas the temperature of the shell varies with atmospheric temperature.

Core Temperature Is Variable in Healthy, Normal Individuals

No single body temperature can be specified as *normal* since the human body has both **longitudinal** and **radial temperature gradients.** Even the temperature of the core has temperature gradients from 0.2 to 1.2°C. For practical and comparative purposes, however, the most widely accepted measures of "normal" body temperatures are **rectal, oral,** and, in some cases, **axillary temperatures.**

In a resting human, the rectal temperature obtained at a depth of 5 to 10 cm is slightly higher than that of arterial blood, slightly lower than that of the

Figure 27–12

Range of rectal temperatures measured under well-controlled conditions in a group of healthy young students. The mean rectal temperature is 37.1°C.

brain, and equal to that of the liver. Rectal temperature obtained at a standard depth, therefore, can be regarded as representative of the body core.

The oral or **"sublingual"** (below-the-tongue) temperature is approximately 0.6°C lower than rectal temperature. Oral temperature can be easily affected by the inspired air and the temperature of food and drink.

The axillary (under-arm) temperature can also represent the core temperature and is approximately 0.7°C lower than rectal temperature. When the upper arm is held tightly to the thorax, the core boundary moves to the armpit (see Fig. 27–11). This shift, how-

Figure 27–13

Survival range of body temperature in human subjects.

ever, requires that the shell regions of the body gain considerable amounts of heat before a stable temperature is reached. At least one half hour must be allowed for equilibration to ensure that axillary temperature truly represents the core body temperature.

Another problem with the concept of a single value for normal body temperature is that body temperature varies considerably among resting healthy individuals (Fig. 27–12). At any given time, the rectal temperature during well-controlled conditions can vary by 1.1°C in a group of normal, resting, healthy students. The variation is even greater in less controlled situations.

Normal body temperature should be conceived, therefore, as a *range* of core temperatures that may be expected in a group of normal subjects under various conditions (Fig. 27–13). It is perfectly normal, for example, for an athlete who has just run a race to have a rectal temperature of 40°C. The same rectal temperature in the same person resting the next day in a physician's office would be considered abnormal and could be used as a possible index of infection and fever. The range of body temperature observed in normal humans has a lower limit of 35°C and an upper limit of 41°C. Body temperatures outside of this range are associated with an impaired ability to regulate body temperature effectively.

Skin Temperature

Skin temperature is the accepted representative of the temperature of the body shell. The skin temperature, which is easily measured, is very different at different body sites and at different atmospheric temperatures. Surfaces covering the areas of high resting heat production (Fig. 27–14) have the highest skin tem-

(a)

(b)

Figure 27–14

(*a*) Regional skin temperatures and heat production at rest in a comfortable ambient temperature of 25°C. (*b*) Thermograph of a woman standing in front of a tub of hot water. Color coding runs from white (hottest) through to blue and black (coldest).

(© Dr. Ray Clark & Mervyn Goff/Science Photo Library/Photo Researchers, Inc.)

peratures. In a person in a comfortable room at 24 to 25°C, the skin temperature of the head, chest, and abdomen is about 35°C; the surfaces covering the large muscle masses of the arms and legs have a lower skin temperature of 31°C; whereas areas that cover very little muscle, such as the hands and feet, have the lowest average skin temperature of 29°C.

Mean Skin Temperature

For practical purposes, the **mean skin temperature** is regarded as representative of the temperature of the body shell and is calculated as the mean of the temperature at several representative body sites. In this calculation, the individual temperatures are weighted in accordance with the surface area they represent. One such equation is as follows:

Mean skin temperature =

0.3 (chest + arm temperature)

+ 0.2 (thigh + calf temperature)

Using this equation for a person in a comfortable atmospheric temperature, the mean skin temperature calculated is about 33 to 34°C.

Thermography

Thermography is a technique for measuring mean skin temperature in humans. It consists of taking pictures of the skin with an infrared camera. The apparatus scans the skin for several minutes and records the emitted infrared rays, which are directly related to the temperature of the skin. The integrated thermogram that is generated provides an accurate measure of mean skin temperature. Thermography has also proved to be a useful clinical tool for detecting areas of abnormal metabolism and growth under the skin. For example, tumors near the skin are associated with skin temperatures several degrees higher than those in corresponding normal areas.

Mean Body Temperature Is Estimated by the Fractional Sum of the Temperature of the Body Core and the Shell

The **mean body temperature** of a human resting in a comfortable temperature is a complex function of the relatively high temperature of the internal organ areas, the intermediate temperature of muscle areas, and the low temperature of the skin (see Fig. 27–14). For practical purposes, the mean body temperature is estimated by the fractional sum of the core and the shell. The core temperature is most often represented by rectal temperature, and the temperature of the shell is most often represented by mean skin temperature. For example, the mean body temperature

of a person in a comfortable atmospheric temperature can be estimated from the following equation:

Mean body temperature =

0.3 × mean skin temperature +

0.7 × rectal temperature

The equation is based on the observation that, in a comfortable temperature, approximately two thirds of the body mass is at core temperature, and one third is at the temperature of the shell (see Fig. 27–11). Appropriate adjustments in the weighing factors for rectal and mean skin temperature are made to account for the relative changes in the ratio of the core temperature to shell temperature in persons exposed to hot or cold atmospheric temperatures.

The concept of mean body temperature is important because it allows estimation of the amount of heat gained or lost from the body when exposed to hot or cold environmental conditions. Furthermore, it can be used to estimate the rate of heat transfer through the skin associated with the regulation of body temperature.

Factors Affecting Body Temperature Are Variable and Multiple

Many factors affect the body temperature of an individual, as illustrated in Table 27–3, including exercise, circadian rhythms, menstrual cycles, age, and disease.

Exercise and Body Temperature

Exercise can increase the core temperature by 2°C or more depending on how vigorously the muscles are working and how much metabolic heat is generated. During prolonged, exhausting exercise such as marathon running, rectal temperatures of 41°C have been measured. The mean skin temperature, by contrast, falls because the sweat glands are secreting sweat, which cools the skin and counteracts the increased metabolic heat generated by active muscles. The precise control mechanisms for the adjustments in the regulation of body temperature during exercise have

Table 27–3
Some Factors Affecting Body Temperature

Exercise	Shivering
Diurnal rhythms	Menstrual cycle
Fever	Sweating
Age	Shift in blood distribution

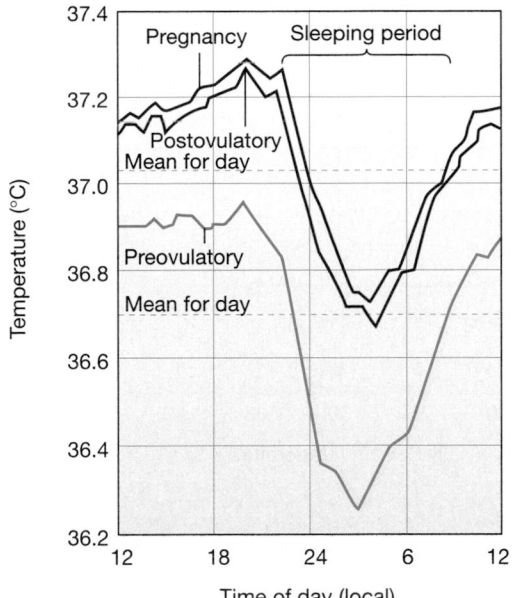

Figure 27–15

Daily variation in body temperature of normal female subjects during the menstrual cycle and pregnancy.

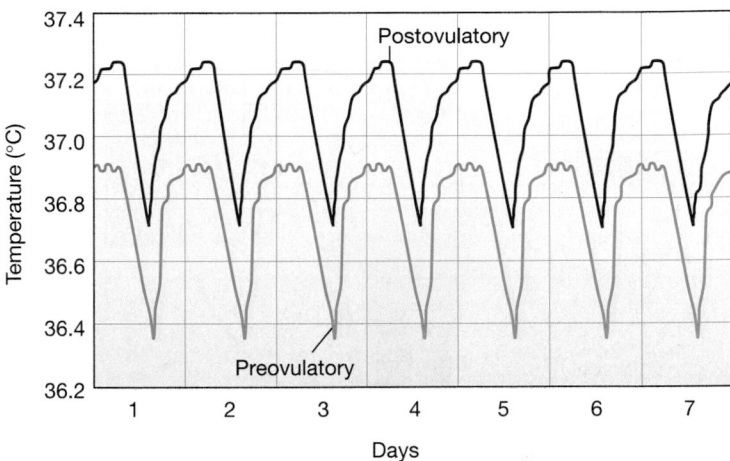

not been fully elucidated and remain an area of intensive investigation and controversy.

Circadian Rhythms and Body Temperature

Recall from Chapter 11 that the body temperature of a human reaches a minimum during sleep and increases slightly in the early morning hours (see Fig. 11–2). Body temperature reaches a maximum, sometimes with two peaks, in the early afternoon or midafternoon when activity is highest. The increase in temperature, however, is not simply the result of increased activity. Even in a sedentary person with all external entraining clues removed, the body temperature continues to cycle in a **circadian** or **diurnal (daily) rhythm,** with fluctuations in temperature that average 1°C every twenty-four hours. Most thermal physiologists believe that inherent body temperature rhythmicity is the result of the central nervous system changing the regulatory set point. Negative feedback control mechanisms actively operate to maintain whatever set point temperature is established for that time of day.

The Menstrual Cycle and Body Temperature

The changes in temperature associated with the **menstrual cycles** of most women are superimposed on diurnal temperature changes (Fig. 27–15). Several days before menstruation, the body temperature usually drops about 0.6°C and is maintained at the lower level until just before ovulation, when it may drop an additional 0.2°C. After ovulation, body temperature quickly rises approximately 1°C and remains at this normal level until the onset of the next cycle.

The magnitude of the oscillations in body temperature associated with the menstrual cycle varies with the individual. Therefore, measurement of core temperature as a method of contraception has been consistently ineffective. If pregnancy occurs, the premenstrual drop in body temperature is inhibited, and body temperature may actually increase to above-normal levels during pregnancy.

Age and Body Temperature

The increased body temperature of children, who tend to have higher body temperatures than adults, is the product of increases in both their physical and metabolic activity. Premature babies do not have well-developed control systems for body temperature. They are, therefore, routinely kept in incubators controlled at optimal temperature and humidity. The elderly also have body temperatures that tend to fluctuate with atmospheric temperature. Their temperatures are more variable because of age-related decreases in the onset and sensitivity of the cardiovascular responses to thermal stress.

Regulation of Body Temperature

Thermoregulation Is an Example of Precise Biological Negative Feedback Control and Homeostasis

The regulation of body temperature in humans is a classic example of a precise biological control system that has evolved to maintain thermal homeostasis. The control of body temperature within narrow limits is dependent upon the coordinated activities of several subsystems of the body, including sweating, vasomotor control, and metabolic rate.

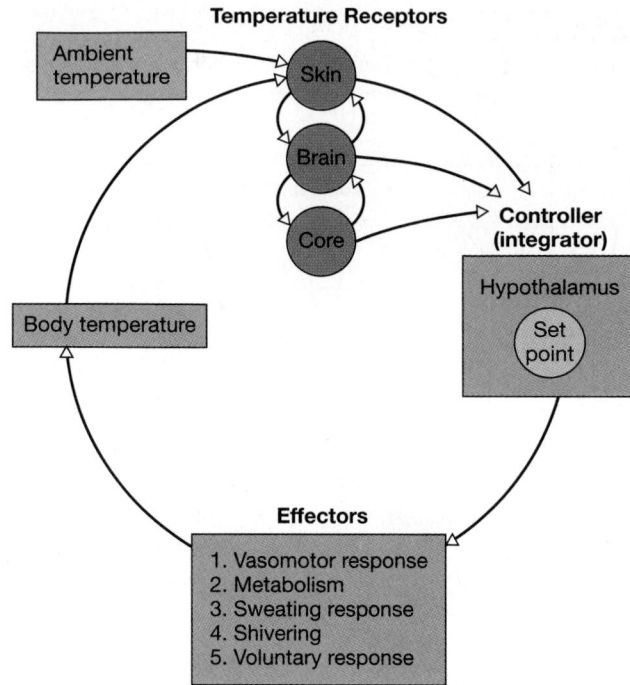

Figure 27–16

Basic components of the negative feedback control system for regulation of body temperature.

Figure 27–17

Static discharge frequency of cold and warm nerve fibers as a function of skin temperature.

The physiological control system that coordinates the responses for effective regulation of body temperature is a **negative feedback system** with at least three major components (Fig. 27–16). It has **temperature receptors** that sense the existing body temperature, **effector organ systems** that determine heat production or heat loss, and an **integrator** or **controller** that compares the sensed temperature to a "normal" or "set point" temperature to determine whether it is too high or too low. The physiological control system then activates the appropriate effector systems, which return the body temperature back to normal.

Peripheral Thermoreceptors in Temperature Regulation

Temperature receptors that measure skin and shell temperature are located just beneath the skin. Peripheral temperature receptors are naked nerve endings that are very sensitive to temperature and are classified as **cold receptors** or **warm receptors** (Fig. 27–17). Cold receptors are characterized by increasing steady-state discharge rates as the skin is cooled. Warm receptors, in contrast, are those whose steady-state discharge rates increase in response to warming of the skin. While both types of temperature receptors are found throughout the body surface, cold receptors are about 10 times more numerous than warm

receptors. Nerve impulses from peripheral receptors enter the spinal cord at all levels and ascend to the brain.

Central Thermoreceptors in Temperature Regulation

Central (core) thermoreceptors are found in deep body areas including the hypothalamus, spinal cord, abdominal viscera, and great veins. Central thermoreceptors include both **cold receptors** and **warm receptors** that are uniquely sensitive to decreases or increases in body core temperature. Nerve impulses from core receptors feed into the brain, where they are integrated with thermal information from peripheral nerves.

Central Integration of Thermal Information

Animal experiments have shown that a part of the hypothalamus is the brain center for body temperature regulation. This area, through its various interactions with other brain areas as well as with central and peripheral temperature receptors, acts as a **thermostat** or controller to prevent both excessive body heating or cooling. Stimulation of the hypothalamus by overheating initiates **antirise responses** such as vasodilatation and sweating. Conversely, stimulation of the hypothalamus by excessive cooling stimulates **antidrop responses** such as vasoconstriction and shivering. In all cases appropriate voluntary responses are also activated.

The neural pathways that deliver the hypothalamic signals to the effector organ systems include the **somatic** and **autonomic** portions of the nervous system. The sympathetic portion of the autonomic

R E S E A R C H   F O C U S

Ethanol and Thermoregulation

Clinical reports of accidental hypothermia in alcohol-intoxicated individuals have commonly occurred since 1875. In 1984, a case study of 51 individuals who died of accidental hypothermia indicated that 33 had blood ethanol concentrations ranging up to 310 mg/dl, and all were legally intoxicated. These case studies do not permit any conclusion about the role of ethanol in the production of hypothermia, even though intoxicated individuals are consistently overrepresented among victims of accidental hypothermia.

Experimental studies on healthy human volunteers and experimental animals have recently clarified the effects of alcohol on body temperature regulation. Studies in both humans and animals have consistently shown that in thermal neutral environmental temperatures ethanol does tend to lower body temperature and that the response is somewhat dose dependent. Even at high doses, however, under these conditions the effects of ethanol on body temperature are relatively minor. More striking results have been observed when the effects of alcohol have been studied at high environmental temperatures. These studies have shown that the combination of ethanol and high environmental temperature not only prevents the drop in body temperature described above but results in a significant increase in body temperature when compared to control subjects.

While the physiological mechanisms have not be elucidated, the findings are consistent with the idea that alcohol prevents effective dissipation of an increased heat load on the body. The most striking and consistent results of the effects of ethanol have been observed at low environmental temperatures. The most common effect of alcohol at low temperature has been a significant inhibition of shivering and, therefore, a reduction in metabolic heat production. At the same time, alcohol has been shown to induce skin vasodilatation and increased heat loss. This combination has been shown to result in a greater feeling of comfort in intoxicated experimental subjects, but it also simultaneously increases the rate of fall in core temperature. The combination of decreased heat production and increased cutaneous heat loss despite falling body temperature is suggestive that ethanol may lower the set-point for temperature regulation in the cold. This combination places intoxicated individuals at a greater risk of suffering from life-threatening hypothermia.

■ **What are the observed physiological changes that support the hypothesis that ethanol may change the set-point temperature for thermoregulation?**

nervous system plays an important role because it controls vasomotor responses and sweating. The somatic portion, on the other hand, controls gross voluntary muscle movements such as those used in adding or removing clothing and the small muscle movements associated with increased muscle tone and shivering.

Diseases That Affect Temperature Regulation Can Result in Either Regulated or Unregulated Changes in Body Temperature

Many disease states cause **fever,** which is a **regulated increase** in body temperature beyond the normal range. Therefore, measurement of a person's body temperature is a common way to assess the severity of an illness. Some pathophysiological conditions produce either **hypothermia** or **hyperthermia,** which are **unregulated** decreases or increases, respectively, in body temperature.

Fever: A Regulated Increase in Body Temperature

Fever is usually caused by chemical agents called **pyrogens** released into the body during the disease processes. The exact way these agents work is not known, but they appear to **reset** the controlled level of the hypothalamic thermostat at a higher temperature level (Fig. 27–18). The hypothalamic integrating center compares existing peripheral and core temperature thermoreceptor information with the new elevated set-point temperature, activating heat production and conservation responses and inhibiting heat loss responses. Simultaneously, the person feels cold and may react behaviorally by turning up the heat or putting on more clothing. After several hours, the body temperature will rise to the new reference temperature, at which point the hypothalamus will equalize heat loss and heat conservation responses. The body temperature continues to be regulated at the elevated level as long as the pyrogen is present. When the pyrogen is no longer effective or present,

CLINICAL FOCUS

Pathogenesis of Fever

The pathophysiology of fever is initiated with the introduction of foreign substances called **exogenous pyrogens** into a host. The most common exogenous pyrogens include endotoxins, gram-positive bacteria, and viruses. The exogenous pyrogens are absorbed and destroyed by phagocytes within the host, including leucocytes and neutrophils. Activated phagocytes produce and release a polypeptide called interleukin 1 (IL1). IL1 is one of a group of **cytokines** which are hormonelike substances with multiple immunological and other functions. IL1 and other cytokines are called **endogenous pyrogens.** Microinjections of endogenous pyrogens into various brain areas have shown that the hypothalamus is the major target area for the thermal effects of endogenous pyrogens. The action of IL1 and other endogenous pyrogens at the level of the hypothalamus is due in part to the synthesis and release of prostaglandin E. Other mechanisms are known to be involved, but their role in the pathogenesis of fever is uncertain. Agents that prevent the development of fever such as aspirin are called **antipyretics.** Most antipyretics have also been shown to inhibit the synthesis of prostaglandin E in the brain. Microinjections of prostaglandin E into the hypothalamus results in an increase in the "set-point" for body temperature regulation. As the set-point is raised, the hypothalamus signals an increase in heat production and heat conservation to raise body temperature to the new level. Peripheral vasoconstriction occurs; epinephrine release increases metabolic rate; and shivering may occur. Body temperature is maintained at the new higher level until the rate of production of endogenous pyrogens is less than their rate of destruction. The fever "breaks" and normal body temperature is restored by accelerated heat-loss mechanisms.

Fever production has several benefits and aids the responses to infections in several ways. Raising the body temperature in and of itself kills some microorganisms and adversely affects the growth and replication of others. Higher body temperature also causes lysosomal breakdown and autodestruction of cells, thus preventing viral replication in infected cells. Heat increases the rate of activity of all elements of the immune response. Finally, phagocytosis is increased and the production of antiviral interferon is enhanced. Because fever is a beneficial response to infection, the suppression of fever by treatment with antipyretic medication should be carefully reviewed on a case-by-case basis. The pathophysiology of fever can be outlined as follows:

Exogenous Pyrogens → Phagocytosis → Endogenous Pyrogen → Prostaglandins → Brain (increased "set-point") → Fever

■ **Many clinicians tend to use the title of fever for any rise in body temperature. What are the fundamental physiological differences between fever and other hyperthermias?**

---

the so-called **crisis** occurs, and the reference temperature returns to the lower, normal range. When this happens, the hypothalamic integrator activates heat loss and inhibits heat gain processes that gradually return the body temperature to normal levels.

Hyperthermia: Unregulated Increases in Body Temperature

Heat Stroke.   **Heat stroke** is a form of unregulated hyperthermia or heat illness associated with a breakdown of the body's temperature-regulating system. Heat stroke usually occurs during heavy work or exposure to hot, humid conditions. In these situations, the rate of heat gain can exceed the maximal rate of heat loss so that the body temperature continues to rise unless the level of work is decreased or the person moves to a cooler, less humid atmosphere.

If, however, the body temperature rises beyond a critical temperature in the range of 41 to 42°C, the person is likely to develop heat stroke. Symptoms in-

Figure 27–18

Time course of a typical febrile episode. Effects of changing the setting of the "hypothalamic thermostat."

clude dizziness, abdominal distress, absence of sweating, sometimes delirium, and eventually loss of consciousness and death if the body temperature is not decreased quickly. Some of the symptoms result from circulatory shock or insufficiency brought about by fluid and electrolyte loss before the onset of the hyperthermia. The high temperature itself may damage body tissues, particularly in the brain, leading to loss of sweating, and a vicious cycle is started in which core temperature is raised continually toward lethal levels.

Heat Exhaustion. **Heat exhaustion,** a less serious heat illness caused by **circulatory insufficiency** and **hypotension,** often takes the form of fainting due to decreased blood pressure. This is brought about by depletion of plasma volume secondary to sweating, and by vasodilatation of skin blood vessels, which is a normal vasomotor response to heat stress. Unlike heat stroke, heat exhaustion can occur under relatively mild heat stress and results from overactivity of heat loss mechanisms. People with deficiencies in blood pressure regulation, such as the elderly, are particularly susceptible to heat exhaustion.

Hypothermia: Unregulated Decreases in Body Temperature

Cold illness or disorders in humans usually involve peripheral circulatory or cardiac defects. Hypersensitivity of the arteries in the skin to cold (Raynaud's disease) can reduce blood flow to such critical levels that pain and local asphyxia may occur if the stress is continued. When the body mechanisms for heat production and conservation are exceeded by exposure to severe cold, **hypothermia** results, and body temperature drops below the normal range. At body temperature of about 26 to 28°C, death can occur due to **myocardial fibrillation.** Older persons are particularly susceptible to hypothermia, since, in some elderly individuals, the hypothalamic reference temperature can be reset to a lower level. Some elderly people can maintain a core temperature of 35°C or even less without the onset of shivering. In the elderly, this appears to be a regulated hypothermia. To some extent this condition can be considered a counterpart of fever, and it makes the elderly more susceptible to hypothermia.

CHAPTER REVIEW

Summary

- Humans are homeotherms and regulate body temperature within a relatively narrow range.
- The regulation of body temperature ensures thermal homeostasis of the internal environment.
- Physiological and physical processes continuously operate to balance metabolic heat production with heat loss to the environment, in order to maintain body temperature constant.
- Metabolic rate is the amount of food energy converted to heat per unit of time.
- Basal metabolic rate (BMR) is a measure of the metabolic cost of living measured under standard conditions.
- A person's metabolic rate can be measured by direct or indirect methods.
- The energy equivalent of oxygen is the quantity of energy liberated when the body uses one liter of oxygen to produce metabolic energy.
- The body can exchange heat with the environment through the four physical processes of radiation, conduction, convection, and evaporation.
- Sweating, the physiological process of heat loss, is the active secretion of mostly water by specialized glands in the skin. Sweat must be rapidly evaporated from the skin for cooling to occur and is the only process for heat loss when exposed to hot environments.
- To keep the body temperature relatively constant, the metabolic heat produced in the core must be transferred to the body shell for exchange to the environment.
- Heat is transferred from the core to the shell by the processes of internal conduction and circulatory convection, which determine the effective insulation layer of the body.
- Normal body temperature should be considered as a range of temperature in healthy individuals.
- The body has a poikilothermic shell and a homeothermic core.
- The most widely accepted measures of body temperature are rectal, oral, and axillary.
- The mean skin temperature is the accepted measure of the temperature of the shell.
- The mean body temperature is determined by weighing mean temperatures of the core and the shell and allows estimation of the amount of heat gained or lost from the body when exposed to hot or cold environments.
- The physiological control system that regulates body temperature is a negative feedback system.
- The negative feedback system consists of temperature receptors that monitor existing shell and core temperatures, and effector organ systems that regulate the rate of heat production or heat loss from the body. Finally, the system has a hypothalamic integrator or controller that compares the existing temperature to a reference set

point temperature and activates appropriate effector responses to maintain thermal homeostasis.
- Fever is a regulated increase in body temperature caused by pyrogens which increase the set-point for body temperature regulation.
- Heat stroke is a form of unregulated heat illness associated with failure of the negative feedback control system and can be life threatening.

- Heat exhaustion is caused by circulatory insufficiency and often results in fainting due to circulatory hypotension.
- Hypothermia or cold illnesses involve peripheral circulatory defects or cardiac malfunctions. During severe hypothermia death can occur due to myocardial fibrillation.

Review Questions

Identify Each with a Term
1. The brain center for integration of body temperature regulation.
2. The disease state characterized by unregulated hyperthermia.
3. Organisms whose body temperatures fluctuate with changes in the ambient temperature.
4. The increased heat production associated with ingestion of food.
5. The exchange of heat using electromagnetic energy.

Define Each Term
6. catabolism
7. thermography
8. mean skin temperature
9. homeotherm
10. fever

Choose the Correct Answer
11. The process by which a human loses heat when the environmental temperature exceeds the body temperature is:
    a. convection
    b. evaporation
    c. conduction
    d. radiation
12. Metabolic rate is *not* increased by:
    a. exercise
    b. eating
    c. age
    d. catecholamines
13. Which of the following is *not* a homeotherm?
    a. snake
    b. cat
    c. dog
    d. horse
14. How many Calories per hour does the average adult's metabolic rate burn?
    a. 300   c. 70
    b. 40   d. 120

Answer Each Question in One or Two Sentences
15. A 70-kilogram person at rest is consuming 0.25 liter of oxygen per minute. What is the metabolic rate in Calories per hour?
16. A person in a comfortable room has a mean skin temperature of 34°C and a rectal temperature of 38°C. What is the mean body temperature?
17. A 70-kilogram adult with a specific heat of 0.83 walks from a comfortable room into a hot room. In one hour, core body temperature increases by 0.5°C. What was the rate of heat storage during that hour of heat exposure?
18. A person in a hot environment evaporates 700 grams of water as sweat per hour. What is the rate of heat exchange per hour?
19. A person in a comfortable room has the following skin temperatures: arm-31°C, chest-35°C, calf-30°C, and thigh-31°C. What is the mean skin temperature?
20. Describe the physical mechanisms that the body uses to exchange thermal energy with the environment.
21. What are the conditions for the measurement of BMR?
22. Define the term *normal body temperature* as it applies to humans.
23. What is the role of the hypothalamus in the control of body temperature?
24. Explain the major elements of a negative feedback system for control of body temperature as one goes from a comfortable room of 25°C to a hot room of 35°C.

Problem Solving in Physiology
25. The core temperature of your body is regulated only a few degrees below the point of thermal death. Explain why this situation has evolved in homeotherms.
26. A soldier standing at attention on a hot sunny summer day suddenly feels dizzy and faints. Describe the thermoregulatory mechanisms that may have contributed to his fainting.
27. Explain why, as a patient with a fever, you would not ask your physician to prescribe antipyretics.

Suggested Readings

Bligh, J. "Regulation of body temperature in man and other mammals." In Shitzer, A., and Eberhart, R., (eds.) *Heat Transfer in Medicine and Biology*, Vol. 1. New York: Plenum Press, 1985, p. 15.

Bringelmann, George. "Specialized brain cooling in humans." *The FASEB Journal*, 7:1148–53, 1993.
Cabanac, Michael. "Selective brain cooling in humans: "Fancy or fact." *The FASEB Journal*, 7:1143–7, 1993.

Crawshaw, L.I. "Temperature regulation in vertebrates." *Annual Review of Physiology*, 42:473, 1980.

Elizondo, R.S. "Temperature regulation in primates." In Robertshaw, D., (ed.) *International Review of Physiology: Environmental Physiology II*, Vol. 15. Baltimore: University Park Press, 1977, p. 71.

Elizondo, R.S., and Johnson, G.S. "Peripheral effector mechanism of temperature regulation." In Szelenyi, Z., and Szekely, M., (eds.) *Advances in Physiological Sciences*, Vol. 32. New York: Pergamon Press, 1980. p. 397.

Gordon, C.J., and Heath, J.E. "Integration and central processing in temperature regulation." *American Review of Physiology*, 48:595, 1986.

Heller, H.C., et al. "The thermostat of vertebrate animals." *Scientific American*, 239 (2):102, 1978.

Hulbert, A.J., and Else, P.L. "Evolution of mammalian endothermic metabolism." *American Journal of Physiology*, 256:R63–R69, 1989.

Kluger, Matthew. "Fever: Role of pyrogens and cryogens." *Physiological Review*, 17:93–127, 1991.

Nielsen, B., Hales, J.R.S., and Strange, S. "Human circulatory and thermoregulation adaptations with heat acclimation and exercise in a hot environment." *Journal of Physiology*, 460:467–85, 1993.

Schonbaum, E., and Lomax, P. "Temperature regulation and drugs: An introduction." *In Thermoregulation, Physiology, and Biochemistry*. New York: Pergamon Press, 1990, P.I.

Veale, W.L., Ruwe, W.D., and Cooper, K.E. "Mechanism of fever and antipyresis." In Khogali, M., and Hales, J.R.S., (eds.) *Heat Stroke and Temperature Regulation*. New York: Academic Press, 1983, p. 79.

Key Terms

antipyretics (p. 820)

basal metabolic rate (p. 809)

carbohydrates (p. 810)

catabolism (p. 806)

circadian rhythms (p. 817)

conduction (p. 811)

convection (p. 811)

evaporation (p. 812)

fever (p. 819)

homeotherm (p. 806)

hyperthermia (p. 819)

hypotension (p. 821)

hypothermia (p. 819)

mean body temperature (p. 816)

mean skin temperature (p. 816)

menstrual cycle (p. 817)

negative feedback (p. 818)

poikilotherms (p. 806)

pyrogens (p. 820)

radiation (p. 811)

respirometer (p. 809)

sublingual (p. 814)

sweating (p. 812)

thermography (p. 816)

thermostat (p. 818)

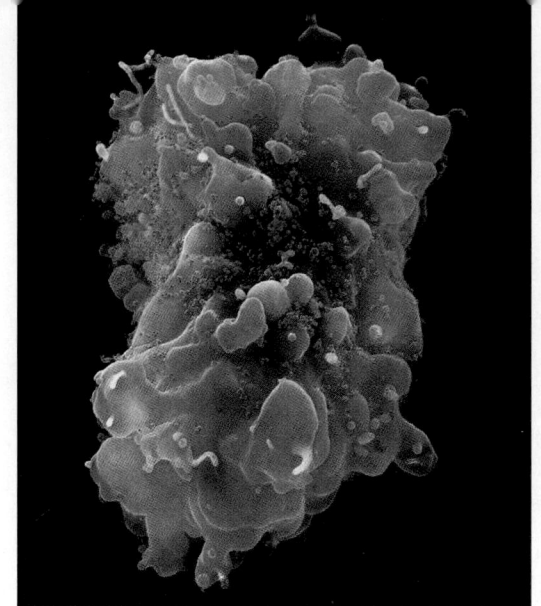

A CD4+ lymphocyte called a helper T cell being attacked by the AIDS virus (blue particles). *(© Boehringer Ingelheim International GmbH. Photo by L. Nilsson)*

Chapter 28

Body Defense and the Immune Response

Key Concepts

Many of the cells that function in the immune response originate in lymphoid tissue, mature and become competent in body defense, and encounter antigens within the lymphatic system.

Internal defense against pathogens includes nonspecific mechanisms, which are general defenses against a wide variety of pathogens and specific mechanisms such as antibody-mediated or cell-mediated immunity that targets specific antigens.

Chapter Outline

The Lymphatic System

Internal Defense

Paul, age 20, was an all-around student-athlete during his high school years. President and valedictorian of his senior class, he also lettered in football, basketball, and track. He kept in shape by working as a brick mason's helper during the summer after graduation and then entered college to begin premedical studies.

By the middle of his sophomore year, Paul began to pay more attention to the fact that, physically, he tired more easily than he used to. At first he attributed his fatigue and greater need for sleep to the pressure and rigor of maintaining his excellent academic record, but when moderate exercise such as walking back to the dormitory after classes left him exhausted and in need of sleep, he realized something was wrong and called his family physician to schedule a physical exam.

Paul complained of physical weakness and periods of exhaustion alleviated only by 10 to 20 hours of sleep, as well as drooping eyelids (ptosis) and double vision (diplopia). An evaluation of Paul's physical history coupled with skeletal muscle endurance testing before and after intravenous injection of a cholinesterase inhibitor suggested a diagnosis of generalized myasthenia gravis. Treatment with anticholinesterase medication restored 90% of Paul's former strength and endurance, thus supporting the initial diagnosis.

Myasthenia gravis is an autoimmune disease. Autoimmune disease is characterized by production of antibodies, called autoantibodies, which react with the body's own proteins rather than foreign protein. In myasthenia gravis, blockage of cholinergic receptors by autoantibodies cause receptor loss, leading to generalized skeletal muscle weakness and chronic fatigue. Weakness in muscles that elevate the upper eyelids results in ptosis, and an inability to control extraocular muscles leads to diplopia—as presented in Paul's case.

Myasthenia gravis is an example of misdirected immune defense. Some defenders within the immune system have become traitors attacking self rather than nonself. The first step in understanding autoimmune disease, and indeed many other forms of human disease, is to examine normal immune system function.

Many cells of the immune system originate, mature, and function defensively in the lymphatic system. Therefore, it is appropriate to begin with the lymphatic system and then proceed to more specific aspects of body defense.

The Lymphatic System

The lymphatic system consists of (1) lymphatic vessels, which transport lymphocytes, (2) lymph, a clear, faint amber fluid derived from serum and interstitial fluid, (3) lymphoid tissue, and (4) lymphoid organs, such as the lymph nodes and spleen, which modify lymph. The lymphatic system has three principal functions: (1) it returns filtered plasma proteins and excess interstitial fluid to the blood (see Chapter 19); (2) it transports lipids absorbed into intestinal lymphatic vessels to the blood (see Chapter 23); and (3) it helps defend the body against pathogens. In this chapter, we will focus on the third function.

Lymph Is Chemically Similar to Plasma and Interstitial Fluid

Usually, **lymph** is a clear, amber-tinted fluid and has a chemical composition similar to plasma and interstitial fluid (Table 28–1). Lymph has a lower protein content than plasma, with an albumin/globulin ratio near 1.0, and generally contains less carbohydrate and more lipid. Lymph in intestinal lymphatic vessels may be opalescent due to high fat content. It is unusual to find erythrocytes in lymph unless blood vessels have been damaged; however, lymph contains all kinds of leukocytes. Some leukocytes emigrate from blood into the lymphatic vessels and others enter from the lymph nodes.

Lymph Is Drained into Systemic Venous Blood

Beginning with the vessel in which lymph forms and in order of increasing diameter, there are three principal kinds of lymphatic vessels: **lymphatic capillaries, collecting vessels,** and **lymphatic ducts.**

The lymphatic system begins as a network of tiny, porous, closed-end channels called lymphatic capil-

Table 28–1
Chemical Composition of Lymph and Plasma

Constituent	Lymph	Plasma
Calcium	4.2 mEq/L	5.0 mEq/L
Chloride	98 mEq/L	98 mEq/L
Sodium	141 mEq/L	141 mEq/L
Potassium	4.7 mEq/L	5.0 mEq/L
Phosphorus	2.5 mEq/L	2.5 mEq/L
Protein	4 gm/100 ml	7 gm/100 ml
Albumin	2 gm/100 ml	4 gm/100 ml
Globulin	2 gm/100 ml	3 gm/100 ml
Lipid	7 gm/100 ml (intestinal)	0.5 gm/100 ml

Figure 28–1

Diagram of terminal lymphatic capillaries and the formation and flow of lymph.

laries (Fig. 28-1). Lymphatic capillaries are microscopic endothelial tissues that resemble blood capillaries, except for the blind end of the terminal lymphatic capillary, and they are more porous, enabling large molecules such as proteins to enter the lymph. The flattened cells that form the wall of the terminal lymphatic capillary loosely overlap. The free end of the overlapping cell acts as a flapper valve allowing interstitial fluid to enter the lymphatic capillary when the pressure is greater on the outside, and preventing fluid from leaving by the same route when pressure is greater inside the capillary. Lymphatic capillaries are interconnected, forming extensive networks within connective tissue before emptying into collecting vessels.

Collecting vessels are of two kinds: **afferent lymphatic vessels** and **efferent lymphatic vessels.** Afferent vessels collect lymph from lymphatic capillaries and carry it to lymph nodes. Efferent vessels carry lymph away from nodes and toward collecting

trunks or lymphatic ducts. Collecting vessels have a connective tissue covering containing scattered smooth muscle cells and elastic fibers that surround the endothelium. These components enable the vessels to contract and relax, helping to move lymph toward the lymphatic ducts. The collecting lymphatic vessels are also equipped with valves that prevent the backward flow of lymph in the same way that venous valves prevent the backward flow of blood (see Chapter 19).

Two main lymphatic ducts, the **thoracic duct** and the **right lymphatic duct,** collect all of the body's lymph from efferent lymphatic vessels and empty it into the subclavian veins near the base of the neck. The right lymphatic duct drains lymph from the right half of the head and neck, the right half of the upper thorax, and the right upper extremity, and empties into the right subclavian vein (Fig. 28-2). The remainder of the body is drained by the thoracic duct, which empties into the left subclavian vein. Each lym-

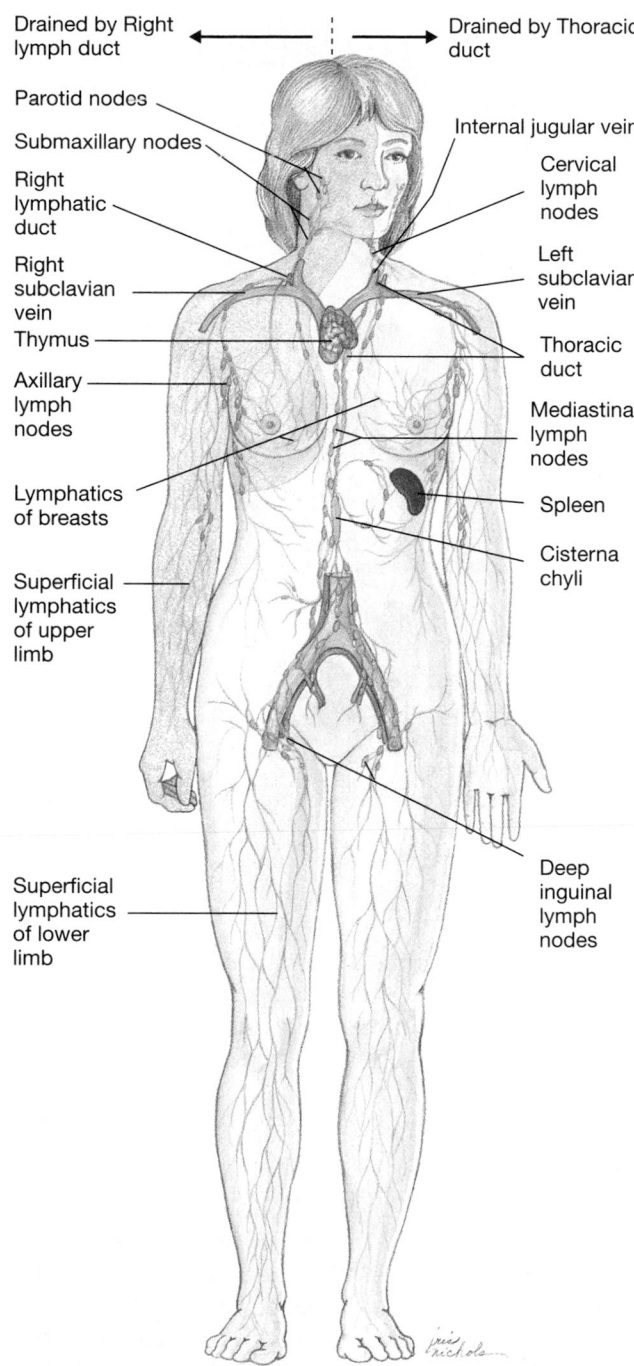

Drained by Right lymph duct

Drained by Thoracic duct

Parotid nodes

Submaxillary nodes

Right lymphatic duct

Right subclavian vein

Thymus

Axillary lymph nodes

Lymphatics of breasts

Superficial lymphatics of upper limb

Superficial lymphatics of lower limb

Internal jugular vein

Cervical lymph nodes

Left subclavian vein

Thoracic duct

Mediastinal lymph nodes

Spleen

Cisterna chyli

Deep inguinal lymph nodes

Figure 28–2

Major lymphatic drainage routes. All lymph formed is eventually emptied into the left and right subclavian veins by the thoracic duct and right lymphatic duct respectively.

phatic duct is equipped with a valve at its termination so that blood cannot enter the duct. Forces causing lymph to be formed and to move through lymphatic vessels toward the subclavian veins were outlined in Chapter 19.

Lymphoid Tissue Contains Free and Fixed Cells

Lymphoid tissue consists of fixed, interconnected reticular cells and reticular fibers that form a loose network (reticulum) called stroma. The interstices of the stroma contain free (unfixed) cells such as T-lymphocytes, B-lymphocytes, other leukocytes (e.g., eosinophil), and macrophages.

Five distinct kinds of lymphoid tissue are present in the body: (1) diffuse unencapsulated lymphoid tissue found just beneath the epithelia of organs; (2) lymphoid nodules, which are dense, discrete, unencapsulated solitary aggregates of lymphocytes and other cells (nodules are found in the loose connective tissue beneath the epithelium of the respiratory passage ways and in the walls of the alimentary canal and genitourinary tract; generally, diffused lymphoid tissue and lymphoid nodules are located in areas of the body most likely to be invaded by pathogens, and thus they form a primary line of defense); (3) small encapsulated collections of nodules which form a lymph node (lymph nodes are always located along the course of a lymphatic vessel); (4) large encapsulated collections of lymphoid cells, which form the lobules of each lobe of the thymus gland; (5) nodules that form the white pulp of the spleen.

Primary Lymphoid Tissues and Organs Are Sites Where Lymphocytes Develop and Mature

During fetal development, lymphocytes are produced in the yolk sac and liver, but after birth, the red marrow of bone becomes the major source of lymphocytes. Other primary lymphoid tissues make up the thymus gland in the neck and Peyer's patches in the wall of the intestine.

Bone Marrow

The **bone marrow** (Fig. 28–3) is the source of both red and white blood cells in the adult. More than a billion white blood cells are produced and released each day, replacing those that die in defense of the body and those that die of old age. Within the marrow some of the developing lymphocytes mature and become competent in body defense before being released into blood. Other lymphocytes are released earlier in their development and travel by blood to other lymphoid tissues where they complete their development and become competent to defend the body against pathogens.

In addition to lymphocyte and other blood cell production, the bone marrow serves as an impor-

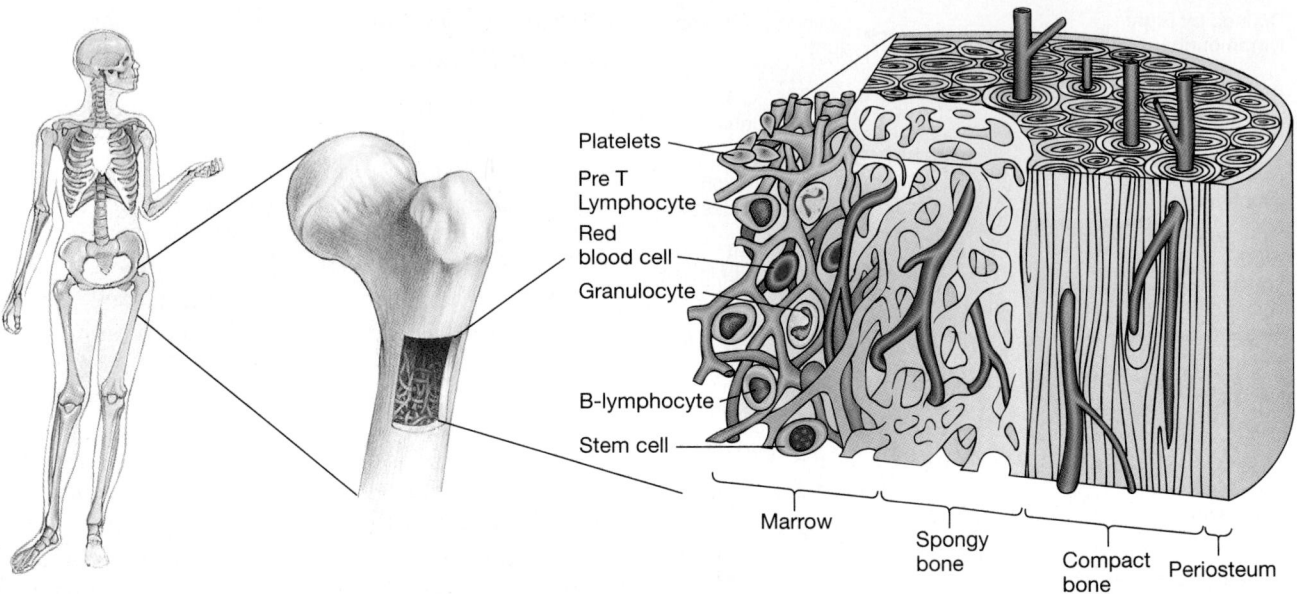

Figure 28–3

The bone marrow is the principal source of red blood cells, white blood cells, and platelets. Within the marrow, precursors develop into B-lymphocytes and pre-T-lymphocytes. Some of the B-lymphocytes remain in the marrow and develop into antibody-secreting plasma cells. Others enter the blood of the sinusoidal capillaries and travel to lymphoid organs such as the spleen and lymph nodes, where they colonize and clone plasma cells in response to antigen. Pre-T-lymphocytes enter the blood and travel to the thymus to complete their development before recirculating in blood and lymphoid tissue.

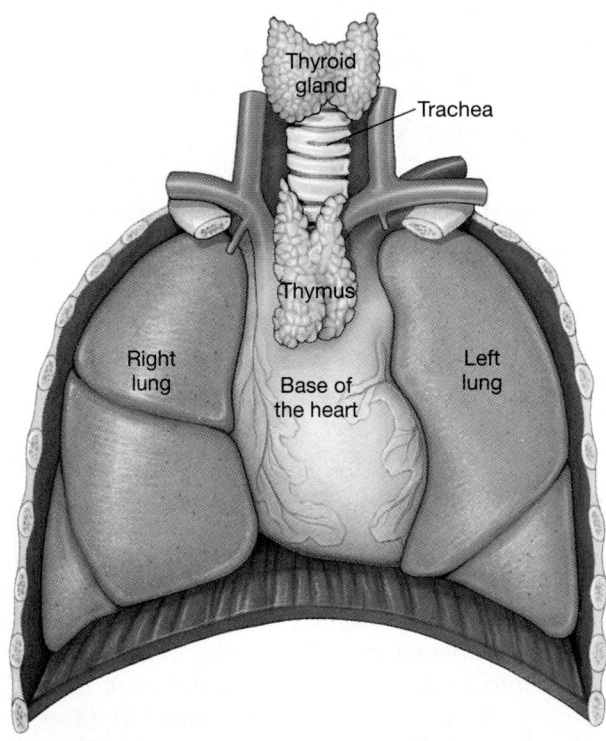

Figure 28–4

The thymus gland has two lobes and is located at the base of the heart anterior to the great blood vessels and immediately behind the sternum.

tant site for B-lymphocyte antibody production. The blood sinuses of the marrow also contain numerous resident phagocytic cells that remove foreign materials from the blood as it circulates through the marrow.

Thymus

The **thymus** is a soft, pale, bilobed gland located in the mediastinum (middle chest cavity) near the base of the heart (Fig. 28–4). It is situated anterior to the aortic arch and behind the upper part of the sternum (breastbone). The size of the thymus varies inversely with age. Relative to surrounding structures, the thymus is large in the infant and young child, but before puberty the gland begins gradually to decrease in size (involute) and is partially replaced with fatty tissue. In the adult the gland is very small, and in the elderly it has been so reduced as to be nearly absent, except for scattered bits of thymic remnants among the adipose replacement.

The internal structure of the thymus (Fig. 28–5) consists of lymphoid tissue loosely arranged into irregularly shaped interconnected lobules by connective tissue partitions that extend inward from the outer connective tissue capsule. Each lobule has a pe-

Figure 28–5

(*a*) The thymus consists of irregularly shaped lobules separated from one another by connective tissue septae. (*b*) Each lobule has an outer cortex and inner medulla. Pre-T-lymphocytes begin their development in the outer cortex and complete it in the medulla.   *(© Vu/Fred E. Hossler)* (*c*) Developing T cells test their immunological compe-tence against dendritic cells, macrophages, and reticular cells. T cells that fail to develop into competent defenders of the body self-destruct (undergo apoptosis) and remnants are removed by macrophages.

ripheral cortex and a central medulla. The medulla consists of densely packed epithelial cells and smaller numbers of lymphocytes and occasionally granulo-cytes. The cortex contains scattered but intercon-nected reticular cells that form a network within which are found large numbers of rapidly dividing immature lymphocytes called thymocytes. The cor-tex also contains macrophages and small numbers of plasma cells.

Thymocytes develop from immature lymphocytes released from bone marrow into circulating blood. The precursors, called **pre-T cells,** are attracted from thymic blood by **thymotaxin,** a chemotactic peptide secreted by reticular cells in the thymic capsule. Within the thymic lobules, the **thymocytes** mature into **T-lymphocytes (T cells)** and reenter the blood prepared to defend the body against foreign mater-ial. Maturation of T cells will be discussed in greater detail later in the chapter.

Peyer's Patches

Great numbers of lymphocytes are found in the mu-cosa of the intestinal tract, where they help protect the body against bacterial invasion through the wall of the intestine. Some lymphocytes are individually

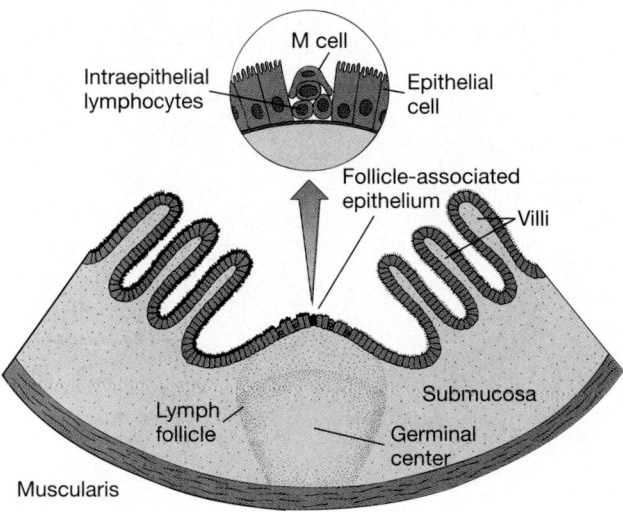

Figure 28-6

Peyer's patches in the cat ileum. *(© Biophoto Associates/ Photo Researchers, Inc.)*

scattered along the entire length of the intestinal tract, but others aggregate to form lymphoid follicles or nodules. Many of the lymphoid nodules are small and individual but some coalesce to form large aggregates of the nodules known as **Peyer's patches** (Fig. 28-6). Peyer's patches are numerous in the mucosa of the ileum but are also located to a lesser extent in other parts of the intestine such as the jejunum. The lymphoid nodules contain a germinal center of proliferating **B-lymphocytes (plasma cells)** specialized to secrete antibodies, macrophages, dendritic cells, and different kinds of **T-lymphocytes** that help coordinate defense against antigens contacting the intestinal epithelium.

The intestinal epithelium covering Peyer's patches contains specialized antigen-binding cells called M cells. These cells express receptors on their luminal surface to which antigens bind. Bound antigens are internalized by endocytosis, transported across the M cell's cytoplasm, and released by exocytosis into the interstitium, where antigen processing is completed by local macrophages, T-lymphocytes, and B-lymphocytes.

Secondary Lymphoid Tissue and Organs Are Where Mature Lymphocytes Reside and Encounter Antigen

Secondary lymphoid tissue and organs develop late in fetal life and continue to develop and function through adulthood. They include lymph nodes and the spleen, as well as previously mentioned bone marrow and Peyer's patches. There are also sites where lymphoid cells reside and take position for body defense.

Lymph Nodes

Lymph nodes, sometimes erroneously called "glands," are small (1 to 2 cm) round or slightly flattened oval encapsulated bodies located in series with lymphatic vessels (Fig. 28-7). Groups of lymph nodes are located in specific regions of the body and are named after the region. Examples are: cervical nodes of the neck, inguinal nodes of the groin, and axillary nodes of the arm. Many others occur irregularly in the course of lymphatic vessels and are not named.

The lymph node is surrounded by a dense white fibrous connective tissue capsule that projects into the node as trabecular partitions. The partitions support reticular cells and reticular fibers which form the stroma of the node, and separate lymphoid nodules. Lymphoid nodules are located in the cortex of the lymph node and contain germinal centers that are active sites of B-lymphocyte proliferation and macrophage phagocytosis. The central medulla is made up of medullary sinuses and cords of lymphocytes, macrophages, and reticular cells. The area between the cortex and the medulla, called the paracortical region or thymic dependent zone, contains densely packed T-lymphocytes, whereas the area immediately beneath the capsule that forms the subcapsular sinus is relatively cell free.

Lymph is carried from drainage sites in the body to the periphery of the lymph node by valve-equipped afferent lymphatic vessels (see Fig. 28-7). Lymph enters the subcapsular sinus, then flows through cortical and medullary sinuses before exiting the node by way of valve-equipped efferent lymphatic vessels. In its one-way passage through the lymph node, lymph is modified. Antibodies pro-

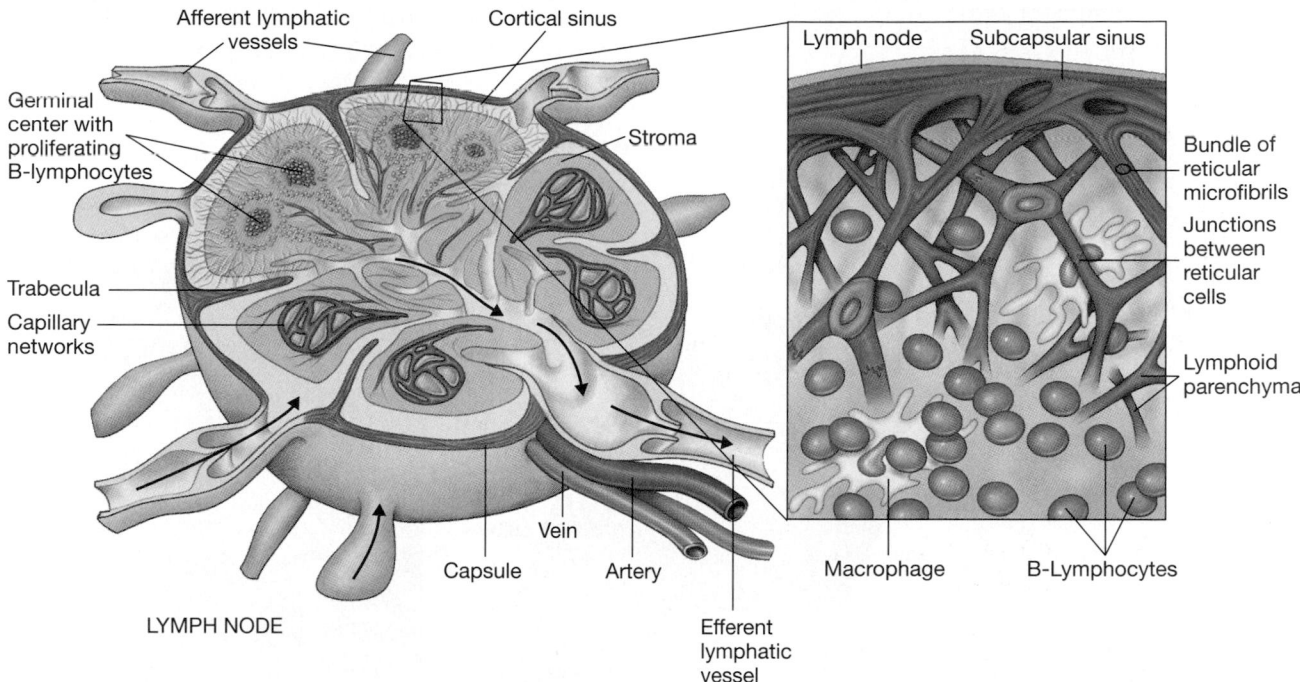

Figure 28–7

Flow through a lymph node. Lymph is brought to the node by afferent lymphatic vessels and enters the subcapsular sinus. Lymph flows through cortical sinuses to medullary sinuses and exits the node by way of efferent lymphatic vessels. As lymph passes through the node, B cells, T cells, and macrophages remove antigens. T cells and antibodies produced by plasma cells also enter blood capillaries within the node and leave by way of veins.

duced by B-lymphocytes (plasma cells) may be added to the lymph. Foreign materials such as bacteria and particulate matter are removed by phagocytosis. B-lymphocytes and T-lymphocytes enter the lymph and are carried to the blood to perform defensively at other sites in the body. Basically, the lymph node performs three distinct yet related functions in body defense: (1) it is a center for lymphocyte proliferation, (2) it acts as a phagocytic organ for purification of lymph, and (3) it is an important site for antibody production and release.

Spleen

The **spleen** is located in the upper left quadrant of the abdominal cavity below the dome of the diaphragm and protected by the lower ribs. The spleen is the largest lymphoid organ in the body and its construction is similar to a large lymph node; however, the spleen is designed to filter blood, not lymph. No afferent lymphatic vessels bring lymph to the spleen.

The splenic capsule (Fig. 28–8) contains elastic connective tissue and some smooth muscle, suggesting the possibility of slight distension and contraction, as when storing and releasing blood into the sys-

temic circulation. Contraction of the splenic capsule is probably unimportant in the human, due to the small amount of smooth muscle present. Internally, trabeculae incompletely partition the spleen into lobules, each containing red pulp and white pulp.

Red pulp is the major component and consists of large thin-walled sinuses filled with blood separated by cords of lymphoid tissue containing numerous macrophages. As blood enters the spleen and passes through the red pulp, aged erythrocytes are destroyed by resident macrophages. During fetal development, the red pulp is a temporary site of hematopoiesis.

The **white pulp** consists of clusters of lymphocytes anchored by reticular cells and fibers around a small central artery. Numerous T-lymphocytes are located immediately adjacent to the central artery. In response to a blood-borne antigen, they can quickly enter the blood and mount a cell-mediated defense. Most of the white pulp consists of germinal centers (splenic nodules) containing densely packed B-lymphocytes. Since B-lymphocytes develop into antibody-secreting plasma cells in response to an antigen, the spleen is important for antibody-mediated defense as well as cell-mediated defense.

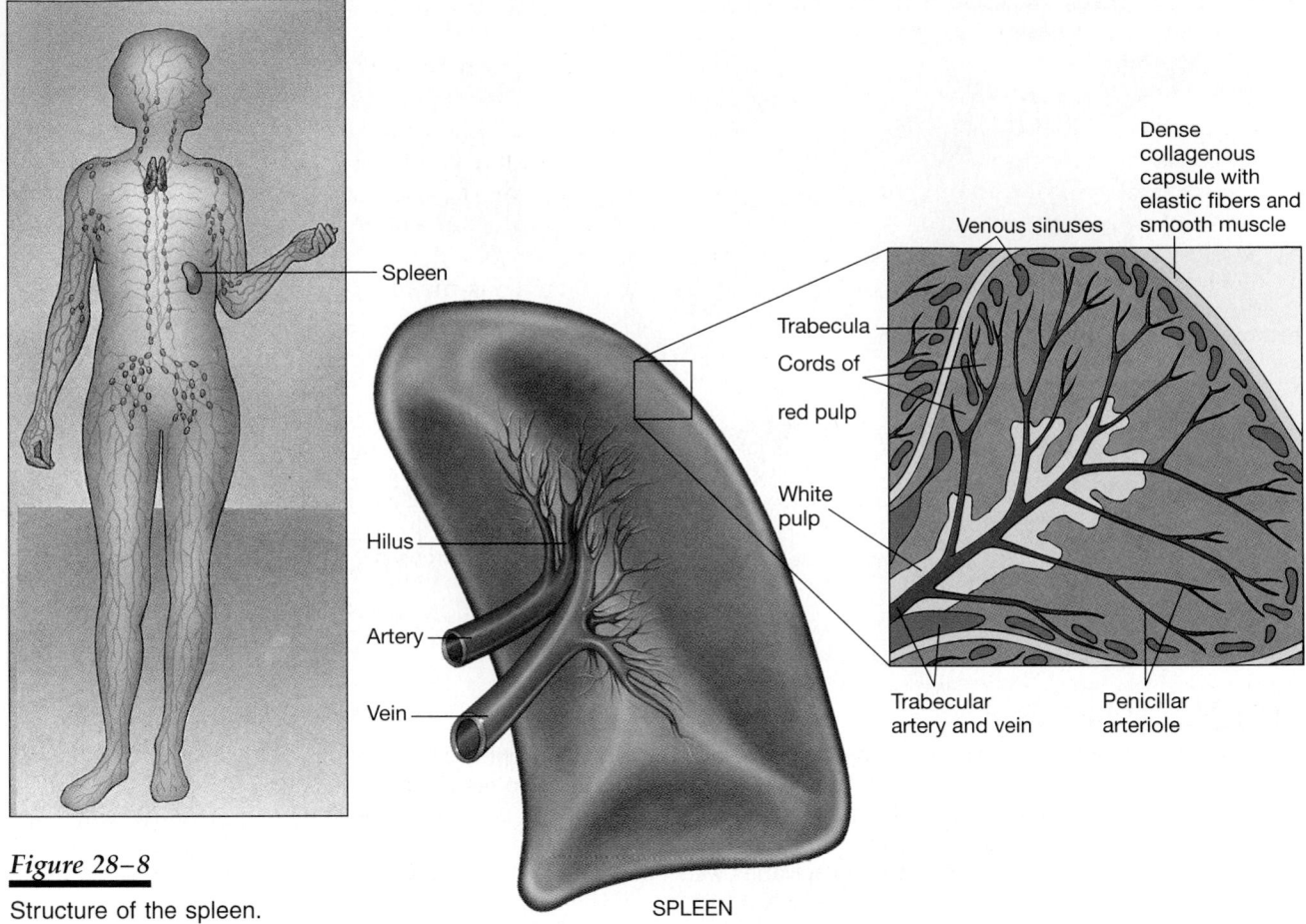

Figure 28–8

Structure of the spleen.

Internal Defense

Internal Defenses Comprise Mechanisms That Deter Pathogen Invasion of the Body and Mechanisms That Destroy Pathogens After Invasion

Disease-producing microorganisms, called **pathogens,** are always present in our environment. We are exposed to these organisms through the food we eat and drink, the air we breathe, and the animate and inanimate objects that we contact in daily life. Bodily defense against these pathogens, their toxins, and other harmful chemicals is comprised of both nonspecific and specific responses.

Nonspecific defense does not depend on specific antigenic determinant recognition and includes mechanisms that deter a variety of pathogens by preventing their entrance into the body as well as mechanisms that quickly destroy an invader if it gets in. A neutrophil encountering, phagocytizing, and destroying a bacterium is an example of general, nonspecific defense.

Specific defense includes mechanisms that are selectively geared toward destruction of specific pathogens or toxins. The sythesis and secretion of highly specific antibodies by a plasma cell is an example. Although other pathogens may be present in the body, the antibody reacts only with a specific pathogen, rather than reacting with any or all pathogens present. When an antibody binds, the pathogen's ability to do harm is reduced. For example, some antibodies react with a pathogen and make the pathogen more likely to be recognized, ingested, and destroyed by a macrophage. Other kinds of antibody-antigen reactions also serve to reduce the pathogen's harmful influence.

Internal Defense Mechanisms Must Allow for Recognition of Self Versus Nonself

All specific defenses and many nonspecific defenses require the body to be able to recognize substances that are foreign—that is, chemicals not usually present in or a part of the body. The ability to recognize "self" and "nonself" is possible because each organism is chemically unique: No two organisms, even though they may be individuals of the same species, are chemically identical. Plasma membranes of human and other animal cells, as well as cell walls of bacteria and other forms of life, contain surface molecules of protein, glycolipid, and glycoprotein that allow the cell to be recognized as "foreign" (nonself) or "self." A molecule, or part of a molecule, that is recognized by the body as foreign is called an **antigen** because it stimulates specific defense mechanisms that include the production and secretion of a specific **antibody.** Many human cells, such as the erythrocyte, have hundreds of different antigens on their surface; some bacteria have more than a thousand.

Nonspecific Defenses Deter or Destroy a Variety of Pathogens Regardless of Pathogen Type

Nonspecific defenses include surface-protective mechanisms, groups of defensive proteins such as interferons and complement, and defensive processes such as inflammation and phagocytosis.

Surface-Protective Mechanisms

The human skin is the first line of defense against pathogens. Most pathogens, as well as many harmful chemicals, cannot penetrate the densely packed keratinized cells of the epidermis. Sweat and sebum (a secretion of the oil glands of the skin) contain chemicals that inhibit the growth of certain pathogenic bacteria. Many nonpathogenic bacteria called commensal bacteria normally inhabit the skin, and their presence, owing to competition for nutrients/environment, inhibits the growth of pathogenic bacteria. Also, the pH of the skin is normally acidic because of the chemicals in sebum and sweat and the chemical products of commensal bacterial metabolism; an acid environment inhibits the growth of many pathogenic bacteria.

Mucous membranes line body cavities that open to the exterior and, like the skin, are potential sites for invasion by pathogens. The linings of the alimentary canal, respiratory passageways, urinary tract, and the reproductive tract are examples of frequently invaded sites. Mucus is a sticky, viscous glycoprotein secretion of the epithelial cells that form part of the mucous membrane. It traps microorganisms as well as prevents the membrane from drying out. The mucus and trapped microbes in the upper respiratory passageways are moved by ciliated epithelial cells toward the nasal and oral cavities (via the aptyl named "ciliary escalatur") from which they

may be discharged or swallowed. Swallowed microbes are destroyed by the highly acidic secretions of the stomach. Microbes trapped in the mucus of the lower respiratory tract and the genitourinary tracts are usually destroyed by macrophages. In addition, the continual secretion and swallowing of saliva and the periodic flow of urine and feces help prevent pathogens from colonizing on associated mucous membranes. Millions of commensal microbes also inhabit mucous membranes and, as on the skin, they effectively reduce the chances that pathogens will multiply and invade the body. Commensal bacteria also help to inhibit pathogen invasion by producing acids and peroxides which form an unfavorable environment for potential pathogens.

Finally, most of the body's secretions contain antimicrobial chemicals. Sweat, saliva, tears, and urine for example, contain **lysozyme,** an enzyme that effectively digests the cell wall of many pathogenic bacteria. Figure 28–9 summarizes primary physical and chemical defense mechanisms.

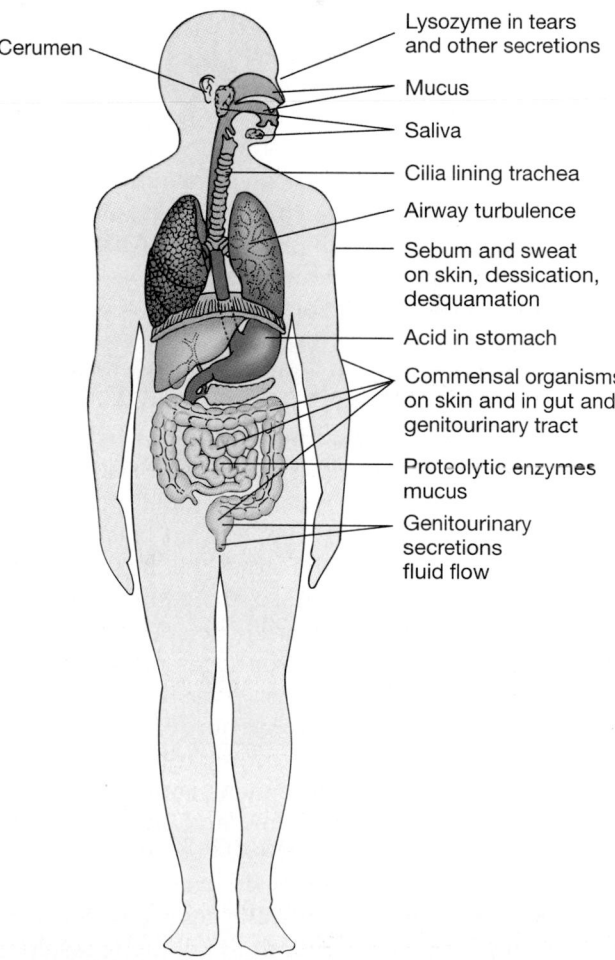

Cerumen

Lysozyme in tears and other secretions

Mucus

Saliva

Cilia lining trachea

Airway turbulence

Sebum and sweat on skin, dessication, desquamation

Acid in stomach

Commensal organisms on skin and in gut and genitourinary tract

Proteolytic enzymes mucus

Genitourinary secretions fluid flow

Figure 28–9

Primary physical and chemical defenses of the body.

Interferons

Interferons (INF) are a group of small proteins that have antiviral activities. Interferons are released by many types of cells as part of a secondary defense system that is activated when a virus invades the body. They were discovered in 1957 and named for their ability to "interfere" with viral replication. There are two main types of interferon: **Type I** interferons (alpha α, beta β, tau τ, and omega ω), each comprised of a single chain of amino acids, and **Type II** (gamma γ) interferon, a dimer of two identical proteins.

Interferon-alpha (INFα) and **interferon-beta** (INFβ) are the two principal Type I interferons. Alpha is made by almost any type of cell infected with a virus; fibroblasts (fiber-forming cells found in connective tissues) are the principal source of INFβ. Type I interferons bind to Type I receptors on the plasma membranes of noninfected cells, triggering the cells to synthesize enzymes that break down viral mRNA thereby making the cells virus-resistant. Since viruses need host cells to replicate, interferons interfere with viral replication, thereby helping to prevent viral spread. The Food and Drug Administration recently licensed interferon-alpha for use in treating genital warts, hairy cell leukemia, AIDS-related Kaposi's sarcoma, hepatitis B, and hepatitis C. INFβ is used to treat multiple sclerosis (MS), an autoimmune disease in which various lymphocytes and macrophages collaborate to destroy myelin in the CNS. Although the exact mechanism of action is not clear, INFβ is believed to suppress the involved immune cells thereby alleviating symptoms of the disease.

Interferon-gamma is produced mainly by activated T-lymphocytes and natural killer cells. These cells need not be virally infected, only alerted to the presence of bacteria, viruses, parasites, or cancer cells in order to secrete interferon-gamma. Interferon-gamma binds to Type II receptors on plasma membranes, inducing within the target cell the production of proteins that inhibit viral replication, inhibit the growth of tumors, and orchestrate specific immune defense. INFγ induces macrophages to kill cells infected by viruses, bacteria, or parasites, and tumor cells. It also stimulates production of Class I or II MHC proteins (major histocompatibility complex surface proteins) and subsequent antigen presentation, key steps in cell-mediated and antibody-mediated immune responses. INFγ is currently used to treat chronic granulomatous disease, a rare hereditary disease of blood granulocytes in which neutrophils ingest bacteria but fail to kill them, resulting in severe chronic infections. INFγ stimulates macrophage destruction of bacteria, thereby boosting defense against bacteria.

Complement

Complement is a set of about 20 proteins, found in normal plasma or serum, which interact in a defined sequence to promote the lysis of certain microbes and the release of chemical signals. The system is called "complement" because its functions complete—that is, are complementary to—other internal defense mechanisms. When an antibody reacts with an antigen, for example, an immune complex is formed that initiates sequential interactions between complement proteins in a manner similar to the way in which chemicals initiate the reaction sequences in Stage I of blood coagulation (see Chapter 17). Activated complement proteins (Fig. 28–10) exhibit four biological actions: (1) some lyse cell walls of certain microbes; (2) some increase capillary permeability by stimulating histamine release from tissue cells and leukocytes; (3) some coat the surface of the microbe, and signal the neutrophils and macrophages to ingest and destroy the microbe (this process is called **opsonization,** and the complement proteins that function in this matter are called **opsonins**); and (4) some attract leukocytes (by chemotaxis) to the site of the reactions.

Inflammation

Inflammation is usually a localized body response to tissue damage or injury such as occurs when pathogens invade tissues. It signifies the body's attempt to defend itself against the pathogen and prevent the pathogen from spreading and damaging other tissues (Fig. 28–11). The clinical signs of inflammation are redness, heat, edema, and pain. They are caused by the following inflammatory responses:

1. Injured cells, basophils, mast cells, and others release histamine and other chemicals that dilate blood vessels and increase the blood supply to the injured area. This process makes Caucasian skin look red and feel warm.

2. **Histamine** and other chemicals increase the permeability of local capillaries, which allows antibody proteins and other chemicals that mediate an immune response to pass out of the capillary and enter the tissues. The increased flow of materials out of the blood and into the tissues causes local swelling or edema (fluid accumulation), which, in turn with chemicals released from damaged cells, causes pain.

3. **Neutrophils,** and to a lesser extent other leukocytes, migrate out of the capillaries **(diapedesis)** and into the surrounding tissues (Fig. 28–12). Neutrophils and tissue macrophages are attracted **(chemotaxis)** to the area of tissue damage by chemicals released from damaged

FUNCTIONS OF COMPLEMENT PROTEINS

Lysis of bacterial membranes

Increased capillary permeability and diapedesis

Opsonization and phagocytosis of bacteria

Chemotaxis

Figure 28–10

Four different ways in which complement proteins function.

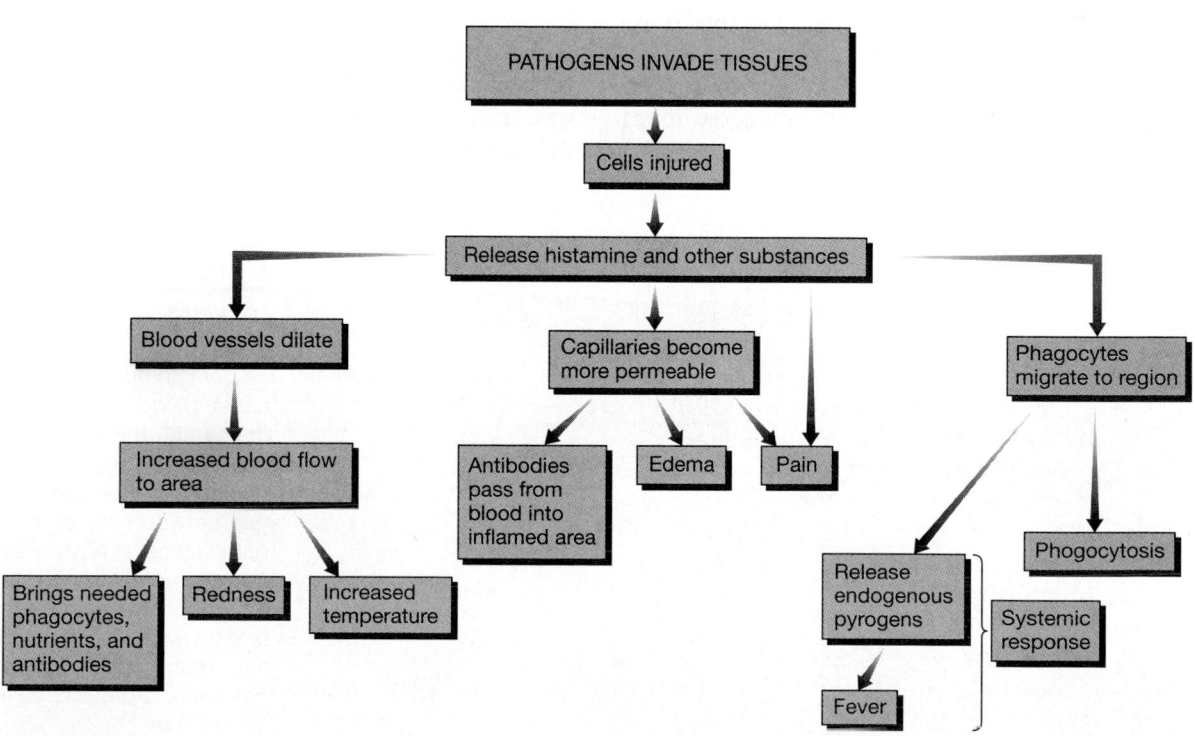

Figure 28–11

The inflammatory response.

cells, activated complement proteins, lymphokines (from lymphocytes), and other substances.

Phagocytosis

After a pathogen penetrates the body's epithelium, it encounters **phagocytes,** cells that ingest and destroy microorganisms by the processes of phagocytosis. Although there are several types of phagocytes, they are all derived from a common stem cell in the bone marrow. Phagocytes are divided into two groups: the neutrophils, or **polymorphonuclear leukocytes,** and the monocyte/macrophages.

Neutrophils, as previously discussed in Chapter 17, are the most numerous of the leukocytes, making up between 50% and 70% of the total in normal peripheral blood. They are highly phagocytic, engulfing all kinds of particles (including pathogens) and destroying them. They are short-lived, existing for just 1 to 5 days. They are continually being replaced and are the first phagocytes to arrive at an invasion site.

Monocytes migrate out of the blood and into the tissues, where they develop into macrophages. Tissue macrophages include the Kupffer cells of the liver, microglia in the brain, alveolar macrophages in the lung, and others. They are strategically located (near the blood, near the alveoli of the lung, near the lumen of the intestine, etc.) where they are most likely to encounter foreign substances, including infectious agents. Some, such as **Kupffer cells,** remain relatively stationary (fixed macrophages) in the organs where they settle; others (wandering macrophages) roam about through the connective tissues.

The monocyte/macrophage differs from the neutrophil in that it is not as phagocytic unless it is "activated" by complement proteins, antigen-antibody complexes, interferon, or other chemicals. Once activated, the macrophage becomes highly phagocytic and is known as an "activated" macrophage. **Macrophages** are usually slower in arriving at an invasion site but live longer than neutrophils (1 to 3 weeks) and can ingest and destroy more material.

Phagocytosis of a bacterium by a macrophage is illustrated in Figure 28–13. The macrophage attaches to the bacterium by way of nonspecific receptors and extends pseudopodia around the attached bacterium, engulfing it and forming a membrane-bounded intracellular vesicle called a **phagosome.** Lysosomes bind to the phagosome and release digestive enzymes into it, a process known as degranulation. The bacterium is digested, and the end-products of the digestion are released or in some cases used by the macrophage. For example, antigenic parts of the bacterium may be displayed with MHC proteins on the surface of the macrophage so as to activate specific immune defense mechanisms.

Specific Defenses Engage and Destroy Specific Kinds of Pathogens

The term immunity is derived from the Latin *immunis* meaning "to make safe." There are two kinds of protective immunity: natural, or innate, immunity, and acquired, or adaptive, immunity. The nonspecific defense mechanisms previously discussed, such as physical and chemical barriers, provide us with natural immunity. Specific defense mechanisms, geared toward the identification and destruction of specific

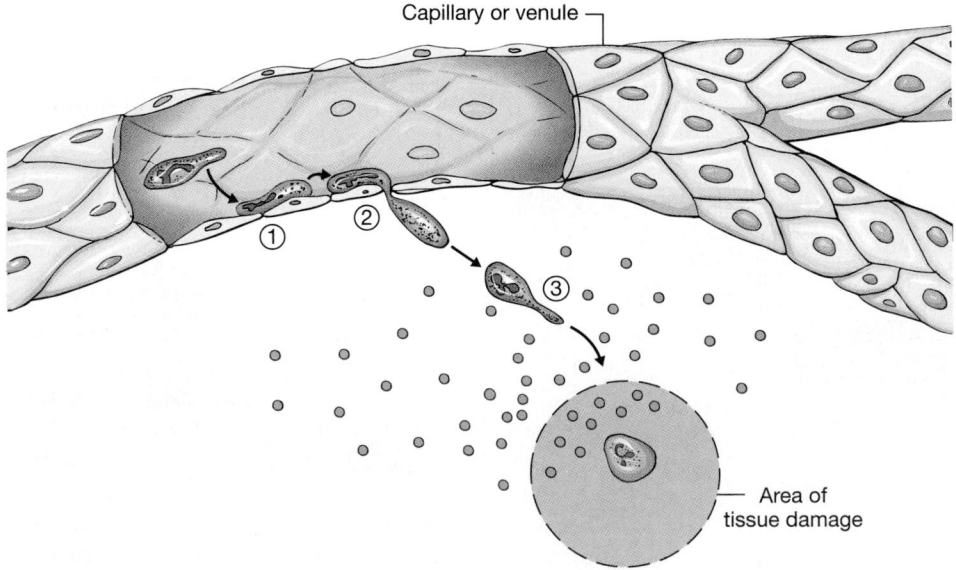

Capillary or venule

① ② ③

Area of tissue damage

Figure 28–12

Diapedesis and chemotaxis *(1)* Chemotactic factors such as complement protein C5a cause blood neutrophil to marginate and adhere to endothelial cell. *(2)* The neutrophil migrates through a gap between adjacent endothelial cells, a process called diapedesis. *(3)* The neutrophil, once outside the blood vessel, moves toward the source of the attractant chemical, a process called chemotaxis.

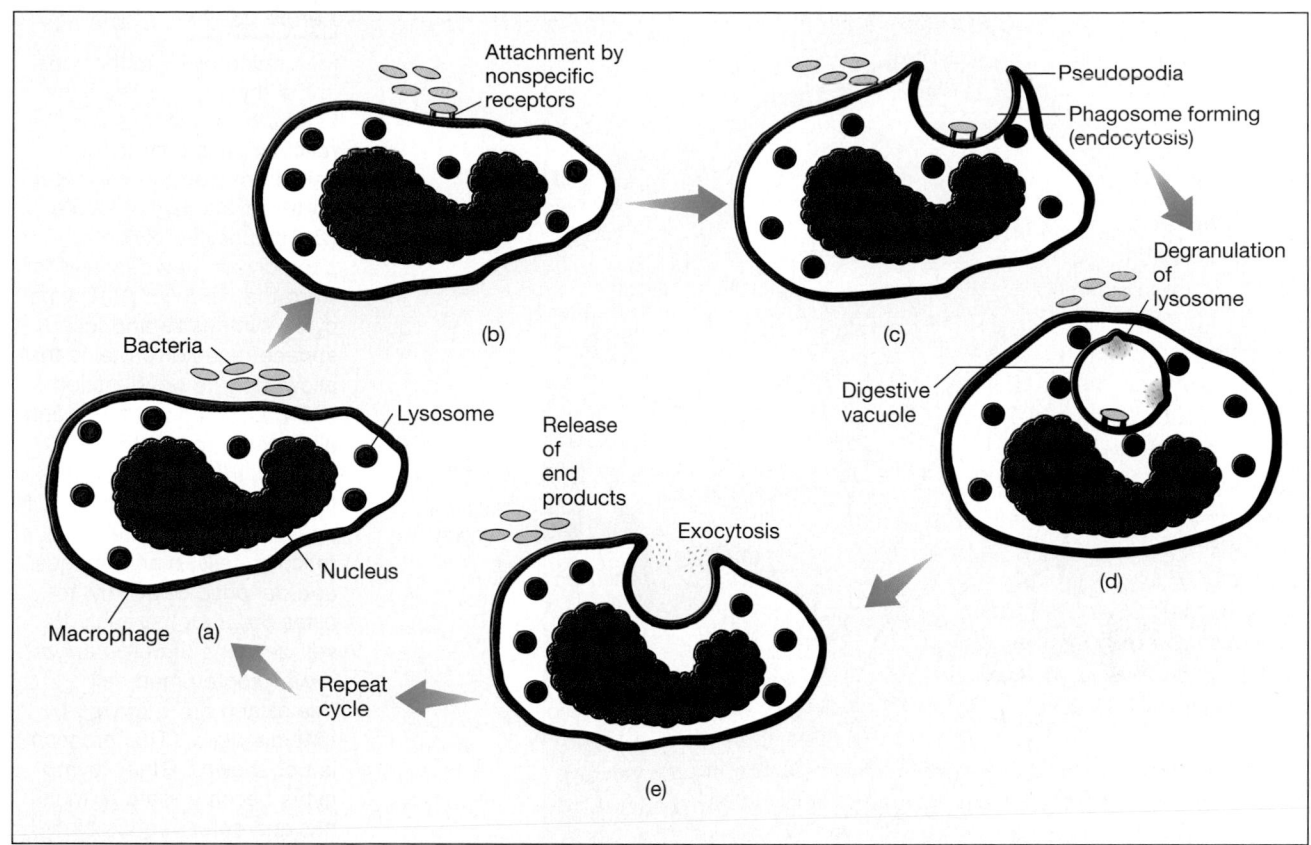

Figure 28–13

Phagocytosis of a bacterium by a macrophage.

pathogens, give us acquired immunity. There are two main types of specific defense responses: cell-mediated immune responses and antibody-mediated immune responses.

Cell-Mediated Immune Response

The **cell-mediated immune response** is the primary responsibility of the macrophages and the T-lymphocytes (T cells): both types of cells collaborate in the direct destruction of viruses and foreign cells that enter the body. Collaboration is required because most T cells cannot by themselves recognize free antigens in tissue fluids; the antigen must be presented to the T cell by another cell called an antigen-presenting cell. Usually, the antigen-presenting cell is a macrophage, but B cells and dendritic cells can also present antigen. Macrophages and lymphocytes coordinate their defensive functions and communicate by sending and receiving molecular signals called cytokines, a family of small proteins that act as intercellular messengers. **Cytokines** include **interferons, monokines,** and **lymphokines.** Macrophages secrete monokines that stimulate T cell development or help

destroy a pathogen, and various T cells secrete lymphokines that regulate T- and B-lymphocyte function, attract macrophages, and in other ways help with specific defense. Before examining cell-to-cell interactions of the immune response, we must understand how a T cell becomes competent to function and learn the various types of T cells.

Maturation of T-lymphocytes takes place in an orderly manner. Pre-T cells migrate from the blood into the cortex of the thymus glands and descend through a web of reticular cells, dendritic cells, and macrophages (Fig. 28–14) toward the medulla. As they descend, they divide repeatedly and acquire distinctive surface proteins that contain receptor sites. As we shall see, some receptors allow the T cell to be identified and other receptors allow the T cell to identify and distinguish between the body's own cells and antigens.

T-lymphocyte precursor cells are carried in the blood from the yolk sac, embryonic liver, and bone marrow to the thymus where they are induced by thymotaxin to leave the blood and enter the outer cortex of the thymic lobule. In the outer cortex, the developing T-lymphocytes (thymocytes) begin to ex-

Thymic capsule
Capillary
Subcapsular reticular cells
Reticular cell
Dendritic cell
Subcapsular reticular cells
Reticular fibers
Macrophage
Cortex
Medulla
CD4+
CD8+

Figure 28–14

Maturation of T-lymphocytes in the thymus. *(1)* Thymosin secreted by subcapsular reticular cells attracts pre-T cells from blood to the outer cortex of the thymic lobule. *(2)* Thymocytes divide repeatedly as they descend toward the medulla. *(3)* Thymocytes synthesize and insert surface receptor proteins that allow them to be identified (CD4, CD8, and others), and allow them to identify MHC antigen. *(4)* Developing thymocytes test themselves against surface proteins of reticular cells, macrophages, and dendritic cells. Thymocytes that react strongly to self antigens undergo apoptosis (programmed cell death) and are removed by macrophages. (This process is not shown.) Other thymocytes become mature, multiply, and enter venous blood.

press surface proteins on their plasma membranes, which act as receptors or cell markers. Developing T-lymphocytes possessing a surface marker designated as CD4 are called **CD4+ lymphocytes,** or simply T$_4$ cells. Developing T-lymphocytes possessing a surface marker designated as CD8 are called **CD8+ lymphocytes** or simply T8 cells. The (+) means the CD4 or CD8 can be detected when it binds an appropriate monoclonal antibody. The binding means the presence of (+) of the CD4 or CD8 marker. T-lymphoctes also acquire surface proteins called MHC **(major histocompatibility complex)**, two classes of which we will consider here. **Class I MHC** molecules are present on the surface of all nucleated cells in the body. **Class II MHC** molecules are present on the surface of some cells of the immune system. CD8+ cells express Class I MHC, whereas CD4+ cells express Class II MHC.

As CD4+ and CD8+ cells move through the inner cortex, they begin to synthesize and insert into their plasma membranes T cell receptors for recogni-

tion of foreign antigens. These receptors (TCR) consist of two peptide chains, usually alpha and beta chains, which determine the antigen and MHC specificity of the receptor. Extensive rearrangement of the genes coding for the alpha and beta peptides during T cell development results in the expression of a great variety of T cell receptors in the lymphocyte population of the thymic cortex. T cells that by chance express receptors that react with self MHC in combination with other self proteins undergo programmed self-destruction **(apoptosis)** and their remains are removed by macrophages and other cells in the thymic cortex. This is a process of **negative selection.** It has been estimated that approximately 96% to 99% of the developing T cells are negatively selected and destroyed or rendered unresponsive (anergic) in the thymus, leaving 1% to 4% to mature into immunocompetent lymphocytes. The latter move from the inner cortex through the medulla, reenter the blood, and leave the thymus, ready to colonize other lymphoid tissue and battle antigens.

Figure 28–15

Cell-mediated immune response to bacterial infection.

An immunocompetent T cell possesses surface receptors that allow the cell to recognize, bind, and react to an antigen presented in combination with either Class I MHC or Class II MHC. There are thousands of different kinds of competent T cells, each capable of responding to a different antigen. Many of these T cell clones are also capable of rapidly expanding their numbers (clonal expansion) when stimulated by an appropriate antigen.

Functionally, there are two types of CD8+ cells: **cytotoxic T cells (T_C)** and **suppressor T cells (T_S)**. Cytotoxic/supressor cells are CD8+ Class I MHC restricted cells. Restriction means the cell can recognize an antigen only if the antigen is presented in combination with an appropriate MHC protein on the surface of another cell. Functionally, there are two kinds of CD4+ cells: **helper (T_H)** and **memory cells (T_M)**. Helper/memory cells are CD4+ Class II MHC restricted cells.

The cell-mediated immune response depends upon cooperation among the various types of T cells and macrophages. We can gain a basic understanding of the interplay between cellular elements of the cell-mediated immune response by looking at the responses to bacterial and viral infection.

When a pathogen such as a bacterium invades the body tissues, it is ingested by a macrophage, an **antigen-presenting cell (APC)**. Within the macrophage, bacterial proteins are broken down into peptides or epitopes (antigenic determinants) which are then combined with newly synthesized Class II MHC molecules and transported to the plasma membrane for display on the surface of the membrane as MHC II associated antigen. The macrophage presents the MHC II antigen to a competent resting CD4+ cell (Fig. 28–15) which binds to the MHC II antigen. To activate the bound CD4+ cell, the macrophage synthesizes a protein called CD80 and inserts it into its plasma membrane. CD80 binds to a CD4+ cell surface protein (designated CD28) and together these interactions activate the bound CD4+ cell. The binding also stimulates the macrophage to secrete **interleukin-1** (IL-1), a cytokine which stimulates the activated T_4 cell to become a T_4 helper cell designated T_HO. Stimulated by another lymphokine produced during the immune response, **interleukin-2** (IL-2), T_HO clonally expands to produce some memory T cells and cells which are stimulated by **interleukin-12** (IL-12) to become TH1 cells. CD4+ TH1 cells secrete INFγ

CLINICAL FOCUS

Acquired Immune Deficiency Syndrome (AIDS)

Acquired Immune Deficiency Syndrome (AIDS) is a viral disease first recognized by the Centers for Disease Control in Atlanta in the summer of 1981. By the end of 1985, approximately 14,000 cases had been reported; by June, 1987, the number had risen to around 37,000. By June 1994 CDC reported more than 62,000 HIV infected United States residents and more than 400,000 cases of AIDS. More than 6 million persons in the rest of the world have been reported to be infected with the AIDS virus as of January 1995. By the end of 1994, the number of people infected since the beginning of the pandemic had exceeded 20 million and may exceed 50 million by the turn of the century. No cure has been discovered, no human vaccine has been developed that will prevent infection, and the disease is fatal.

The cause of AIDS is an RNA retrovirus called the human immunodeficiency virus (HIV). Thus far, two genetically related variants of the virus are known: HIV-1 and HIV-2. When a retrovirus infects a host cell, the viral RNA is used as a template to create a DNA copy— in a reverse ("retro") sequence to that ordinarily used by cells. Using the host cells' enzymes, the viral DNA directs the synthesis of new viral RNA and proteins, which are assembled into new viruses. While doing so, the virus disrupts normal cell functions, ultimately causing the cell to die and release the new viruses, which then infect more cells.

As the number of viruses and infected cells increases, the rate of cell death increases, and overt symptoms of the disease appear. In the case of AIDS, the incubation period or the time between the initial infection and the appearance of symptoms may vary from several months to decades, but typically it is around 7 to 8 years, with death occurring 9 to 10 years after initial infection.

The principal (but not exclusive) target cells of HIV are the CD4+ helper cells which play central roles in orchestrating both cell-mediated and antibody-mediated immune responses (see text). CD4+ helper cells bear a surface receptor, designated CD4, which HIV uses to attach to and infect the cells. A healthy adult has about 1000 CD4+ helper cells per cubic millimeter of blood. HIV infection causes the number of CD4+ helper cells to decline to about 800 after the first year, to about 400 after the seventh year, and to near zero after the ninth year. As the CD4+ helper cell population decreases, immune defenses are gradually compromised, allowing opportunistic infections or malignancies to develop and cause death.

Although HIV infects and kills CD4+ cells, this does not appear to be the sole reason for the decrease in CD4+ cell population in the AIDS patient. It is possible that the viral infection triggers an autoimmune response in which uninfected killer T cells destroy uninfected as well as infected CD4+ cells. Other suggested mechanisms for the decline of the CD4+ cell population include inhibition of CD4+ cell clonal expansion and a triggering of CD4+ cell apoptosis (genetically programmed cell death).

When the CD4+ helper cell count falls to about 300 per cubic millimeter, minor skin and mucous membrane infections develop. These include fungal infections of the mouth (thrush) and feet (athletes foot), whitish patches on the tongue (leukoplakia), and shingles, a viral infection of nerves and skin caused by varicella zoster virus (VZV). Associated symptoms include chronic low-grade fever, fatigue, diarrhea, muscle aches, night sweats, and unexplained weight loss. Collectively, these early infections and associated symptoms are known as ARC (AIDS-related complex). ARC signifies

(INF-γ) which activates macrophages, stimulating them to destroy ingested antigens and to present the digested antigens more efficiently on the macrophage plasma membrane.

When a virus invades the body tissues, it enters cells and moves to the infected cell's nucleus. Using the host cell's DNA, RNA, and endoplasmic reticulum, viral proteins are produced. Some of the viral proteins are broken down into peptides (Fig. 28–16), returned to the endoplasmic reticulum, combined with Class I MHC, and transported to the plasma membrane for incorporation as MHC I associated antigen. A competent resting CD8+ cell binds to the presented MHC I antigen and becomes activated

when the antigen-presenting cell synthesizes CD80 protein, inserts it into its plasma membrane, and binds to CD28 of the CD8+ cell. An activated T_H1 cell secretes interleukin-2, which stimulates the CD8+ cell to differentiate into CD8+ suppressor cells and CD8+ cytotoxic cells. The cytotoxic CD8+ lymphocytes bind to virally infected cells, which present Class I MHC+ antigen and secrete cytotoxic chemicals, which disrupt the plasma membrane and kill the infected cells, thereby limiting the spread of the virus.

Suppressor T cells develop more slowly from activated CD8+ cells and, with the aid of helper T cells, eventually suppress or turn off cytotoxic cells when the pathogen is no longer a threat. Suppressor

T-lymphocyte infected with the human immunodeficiency virus (HIV). The green granularlike structures in this electron photomicrograph are the virus particles in the process of budding. *(© NIBSC/Science Photo Library)*

the interval between initial HIV infection and the development of more serious AIDS-defining infections.

When the CD4+ helper cell count falls to about 200 per cubic millimeter (about 8 years after initial HIV infection), fatal opportunistic infections develop. These include Pneumocystis carinii pneumonia (a protozoan-caused pneumonia), toxoplasmosis (a parasitic infection of the brain), cryptococcal meningitis (a fungal inflammation of the brain) and a relatively rare form of skin cancer called Kaposi's sarcoma. Shortness of breath, dry coughing, sore throat, sharp chest pains, enlarged lymph nodes in the neck, axilla, and groin, fever, and the appearance of painless purple or brownish spots on the skin or mucous membranes of the mouth and rectum may accompany the infections.

The AIDS virus does not appear to penetrate intact skin or membranes. There is no epidemiological evidence to suggest that insects, air, water, food, or casual contact spreads AIDS. The common modes of transmission are sexual contact (homosexual and/or heterosexual) and blood contact such as by infusion of infected blood, use of infected syringes, spattering of infected blood on skin cuts or abrasions, or transmission from maternal to fetal blood.

Current AIDS research is being concentrated in three major endeavors: (1) the development of a vaccine that will prevent HIV infections, (2) the development of drugs that will suppress and/or cure HIV infection, and (3) the development of drugs and therapies that will suppress and/or cure AIDS-related infections, thereby improving survival time and the quality of the AIDS patient's life.

Several antiviral drugs have been approved for use in the U.S. in the treatment of AIDS: dideoxycytidine (ddC), dideoxyinosine (ddI), and azidothymidine (AZT). These drugs block reverse transcription by hindering reverse transcriptase, the enzyme required to produce DNA from the viral RNA. Unfortunately, the effectiveness of these drugs is short-lived, because HIV mutates producing variant forms of reverse transcriptase that are able to bypass the blocking drugs.

■ **What other steps in the life cycle of HIV could be exploited in an attempt to block HIV infection?**

T cells also act to turn off helper T cells, thereby preventing a "runaway" immune response.

Following suppression of the cell-mediated immune response, some of the CD4+ cells and some of the CD8+ cells remain and function as memory T cells, which are capable of initiating an accelerated response to subsequent encounters with the antigen.

We have seen that the entire cell-mediated immune response is orchestrated by the CD4+ cells; in fact, a deficiency in CD4+ cells caused by viral infection is ultimately responsible for death in patients with acquired immune deficiency syndrome (AIDS). CD4+ cells are also required to initiate antibody-mediated immune responses.

Antibodies

Antibodies are glycoproteins called immunoglobulins that are produced and secreted by specialized B-lymphocytes in response to specific antigens. The basic unit of all immunoglobulins is a structure consisting of two identical **light polypeptide chains** and two identical **heavy polypeptide chains,** stabilized and linked by disulfide (—s—s—) bonds between cysteines to form a Y-shaped molecule (Fig. 28–17). The light (smaller) chains exist in two forms called kappa and lambda. The heavy (larger) chains exist in five major forms (alpha, gamma, delta, epsilon, and mu) and are used to designate the class of antibody

Figure 28–16

Cell-mediated immune response to viral infection.

(e.g., IgA contains two alpha heavy chains, IgD contains two delta heavy chains, etc.). The heavy chains also contribute significantly to the functional character of that antibody class. Both heavy chains and light chains have a constant portion (C_H and C_L respectively) and a variable portion of (V_H and V_L respectively).

Various constant regions of the heavy chains bind complement proteins and thus can initiate complement reactions. They also bind to receptors on macrophages, monocytes, neutrophils, and NK cells.

The variable portions of the heavy and light chains also contain the antigen-specific binding sites,

one on each arm of the Y-shaped molecule. A typical antibody molecule can therefore combine with two of the same kind of antigens, forming an antigen-antibody complex (see Fig. 28–17), as well as binding to receptors on cells such as macrophages, which in turn ingest and destroy the immune complex.

Immunoglobulins are grouped into five classes (and several subclasses) according to the structure of the heavy chains. Using Ig as an abbreviation for immunoglobulin, the five major classes are designated **IgG; IgA; IgM; IgD;** and **IgE.**

IgG is the major immunoglobulin in normal blood, and accounts for 70% to 75% of all immuno-

Figure 28–17

General structure of an antibody and agglutination of red cells.

globulins present. It contributes immunity against many kinds of pathogens, including bacteria, viruses, and fungi, as well as chemical defense against toxins. Usually, IgG molecules are distributed evenly between the blood and extravascular fluids.

IgA immunoglobulins represent about 15% to 20% of the total immunoglobulin pool. They are the principal antibodies found in various secretions such as mucus (respiratory and digestive tract), saliva, milk, tears, and fluids secreted into the genitourinary and digestive tracts. IgA antibodies provide defense against pathogens that contact the body surface or are ingested or inhaled.

IgM immunoglobulins account for approximately 10% of the body's immunoglobulins. Almost all IgM molecules are in the blood. Their primary importance is in bacterial defense, working in conjunction with complement and blood phagocytes. IgM molecules are also the antibodies that are used to characterize ABO blood type (e.g., Anti-A, Anti-B, etc.).

IgD immunoglobulins make up less than 1% of the blood immunoglobulins but are present on the plasma membranes of many circulating B-lymphocytes. IgD antibodies are believed to be involved in antigen-triggered lymphocyte differentiation—that is, the development of plasma cells and memory cells from B-lymphocytes.

IgE immunoglobulins are rarely found as free circulating antibodies but are commonly found on the surface of the plasma membrane of basophils and mast cells of connective tissue. When engaged by an antigen, IgE molecules stimulate basophils and mast cells to release histamine and other chemicals that mediate the local, often allergic, response to the presence of an antigen. Thus, IgE molecules are directly involved in diseases characterized by hypersensitivity, an exaggerated immune response to certain antigens; asthma and hayfever are well known examples.

IgE molecules also play an important role in defense against helminthic parasites (worms). For example, IgE binds to mast cells, stimulating the mast cell to secrete eosinophilic chemotactic factor A which attracts eosinophils to the infested area to destroy the parasites.

All antibodies work by first combining with the specific antigen for which the antibody is designed. The formation of an antigen-antibody complex may then result in one or more of the following:

1. The antigen-antibody complex stimulates phagocytosis by neutrophils, monocytes, lymphocytes, macrophages, and other phagocytic cells.

2. In reacting with the antigen, the antibody may neutralize the antigen's ability to harm. Some IgG antibodies called antitoxins, for example, block the toxin's (antigen) binding to cell receptors, thereby preventing the toxin from exerting its effect.

3. The antigen-antibody complex activates complement, which in turn attracts phagocytes, opsonizes the pathogen, perforates membranes, and stimulates other defense cells to release chemicals that mediate general and specific defensive responses to the antigen.

Antibody-Mediated Immune Response

The B-lymphocytes, with assistance from T_4 helper cells, are responsible for the **antibody-mediated immune response.** As with the T-lymphocytes, B-lymphocytes are able to recognize and bind to a specific antigen once they become competent and, as with T-lymphocytes, there are thousands of different kinds of B-lymphocytes. In contrast to the T-lymphocyte, most B-lymphocytes develop compe-

Figure 28–18

Antibody-mediated immune response to bacterial infection.

tency in the bone marrow before entering the blood and lymph. As an example of the antibody-mediated immune response, let us look again at the invasion of the body by a pathogenic bacterium (Fig. 28–18).

A macrophage recognizes the bacterial antigen, binds to it, phagocytizes the bacterium, chemically removes the antigen, and displays the bacterial antigen along with Class II MHC protein on its own plasma membrane. A competent CD4+ cell binds to the antigen + MHC II complex on the antigen-presenting cell, and is activated when its CD28 protein binds to CD80 of the macrophage. Interleukin-1, secreted by the macrophage, stimulates the CD4+ cell to develop into a helper cell (T_HO). **Interleukin-4** causes further differentiation of the helper cell into T_H2, a CD4 + helper cell that mediates the antibody-related immune response.

A competent B-lymphocyte binds to an antigen, processes it internally and displays it with MHC II on its surface. A T_H2 cell binds to the MHC II antigen and synthesizes a protein called CD40 ligand which binds to CD40 protein of the antigen-presenting B-lymphocyte, thereby activating the B cell. Interleukins 4,5 and 10 stimulate the activated B-lymphocyte to differentiate into and clone plasma cells and memory B cells.

Plasma cells are enlarged B-lymphocytes with a highly developed and extensive endoplasmic reticulum. Plasma cells remain in the lymphatic system and secrete antibody molecules that are chemically identical to the antigen receptors and thus specific for the antigen. The antibodies travel via the lymph and blood to the invasion site, where they combine with the antigen, thereby marking it for destruction or reducing its capacity to cause disease.

Memory B cells function as do memory T cells; they permit a rapid mobilization of the immune response if subsequent exposure to the same antigen occurs. Memory B cells secrete small amounts of antibodies for several years after the initial infection is overcome. These antibodies circulate as part of the gamma globulin proteins in plasma and are available for immediate defense should reinfection occur. Reinfection also activates memory T cells, which in turn help memory B cells form a new clone of antibody-secreting plasma cells. Memory B and memory T cells form the basis of lasting immunity and often prevent us from suffering twice from the effects of the same pathogen: we may contract measles (a viral disease) as children, for example, but it is uncommon for us to suffer from it again as adults.

Immunity Against Pathogens Can Be Induced

The immune system can be manipulated to prevent a person from becoming ill with disease or to lessen the severity of the disease and speed recovery. Immunization, a process of conferring immunity against disease, is an example. Passive immunity confers temporary, short-term protection. Active immunity confers long-lasting continuing protection. In either case, the protection conferred is called acquired immunity.

Active Immunity

Active immunization procedures involve giving an antigen to a recipient to stimulate the person's immune system so that an immune response is provoked and resistance to the antigen builds up. Usually, the antigen is administered in the form of a vaccine by a process called **vaccination.** Pathogenic bacteria and viruses that have been treated in such a way as to be no longer virulent (able to cause disease) yet retain their antigenic properties are often used in vaccines. Such pathogens are said to be attenuated. The vaccines for tuberculosis, measles, and mumps are examples of vaccines containing attenuated organisms. Recombinant DNA technology has been used to mass-produce antigenic proteins that can be incorporated into a vaccine. One of the vaccines for hepatitis B (an inflammatory, sometimes fatal liver disease caused by a virus) is a recombinant type of vaccine.

Antibody titer, a measure of the amount of a specific antibody present in a sample of serum, can be periodically determined in order to follow the development of active immunity after antigen exposure. After first exposure to an antigen, the antibody titer rises and then falls within the first six weeks (Fig. 28–19). This is termed the **primary response.** If antigen is given again during the decline of the primary response, a **secondary response** may occur in which the antibody titer rises more rapidly to a level higher

Figure 28–19

Passive versus active immunity.

than before. Generally, the decline of the secondary response is much slower than for the primary response, conferring a longer-lasting immunity. A third "booster" vaccination during decline of the secondary response elevates the antibody titer higher than achieved with the second "booster" and usually confers prolonged immunity against the disease caused by the antigen-bearing organism.

Passive Immunity

Passive immunity occurs when a person is given antibodies to help him or her combat disease. This confers an immediate but temporary state of immunity that is termed passive because the recipient's immune system is not the source of the antibodies. The duration of passive immunity is determined by the amount of antibody given, the frequency of administration, and the rate at which the recipient's body metabolizes the antibody. Generally, for an adult, a single dose of **gamma globulin** (the antibody fraction of serum) confers passive immunity for up to 3 months (see Fig. 28–19).

Usually, the source of antibodies used to confer passive immunity is another human who has active immunity against the specific disease. Occasionally, horses are used to produce immunoglobulins, which can be harvested and given to humans. Examples of equine antisera include those used to treat people exposed to rabies, tetanus, and botulism. Since horse serum contains proteins recognized as antigens by the human body, a recipient of equine immunoglobulin may develop an immune response against the serum and become ill, especially after a repeated dose

(text continues on page 848)

Hypersensitivity

Generally, the word hypersensitive means overreactive, that is, an excessive or exaggerated response to a stimulus which most persons react to in a more subdued manner. The immunological definition of hypersensitivity is similar: an exaggerated adaptive immune response to an antigen, resulting in tissue damage.

Hypersensitivity begins with initial exposure to an antigen and stimulation of the body's normal adaptive immune response to the antigen, including activation of cell-mediated and antibody-mediated defense mechanisms. This initial "gearing-up" of defenses against the particular antigen is called "sensitization." In a susceptible person, hypersensitivity reactions occur when a second or subsequent exposure to the antigen evokes an excessive immune response, which often includes inflammation and tissue damage. Hypersensitivity reactions are classified into four types (Types I, II, III, and IV). These reactions may occur singly or in multiple and are characteristic of the individual.

Type I hypersensitivity (immediate hypersensitivity) occurs after an initial exposure to an antigen (often an innocuous antigen such as pollen) stimulates IgE production. Large numbers of IgE antibodies bind to plasma membrane receptors on basophils and mast cells during the sensitization process. Exposure to the same antigen at a later date cross-links the bound IgE molecules, thereby stimulating these cells to degranulate, releasing histamine (a potent local va-

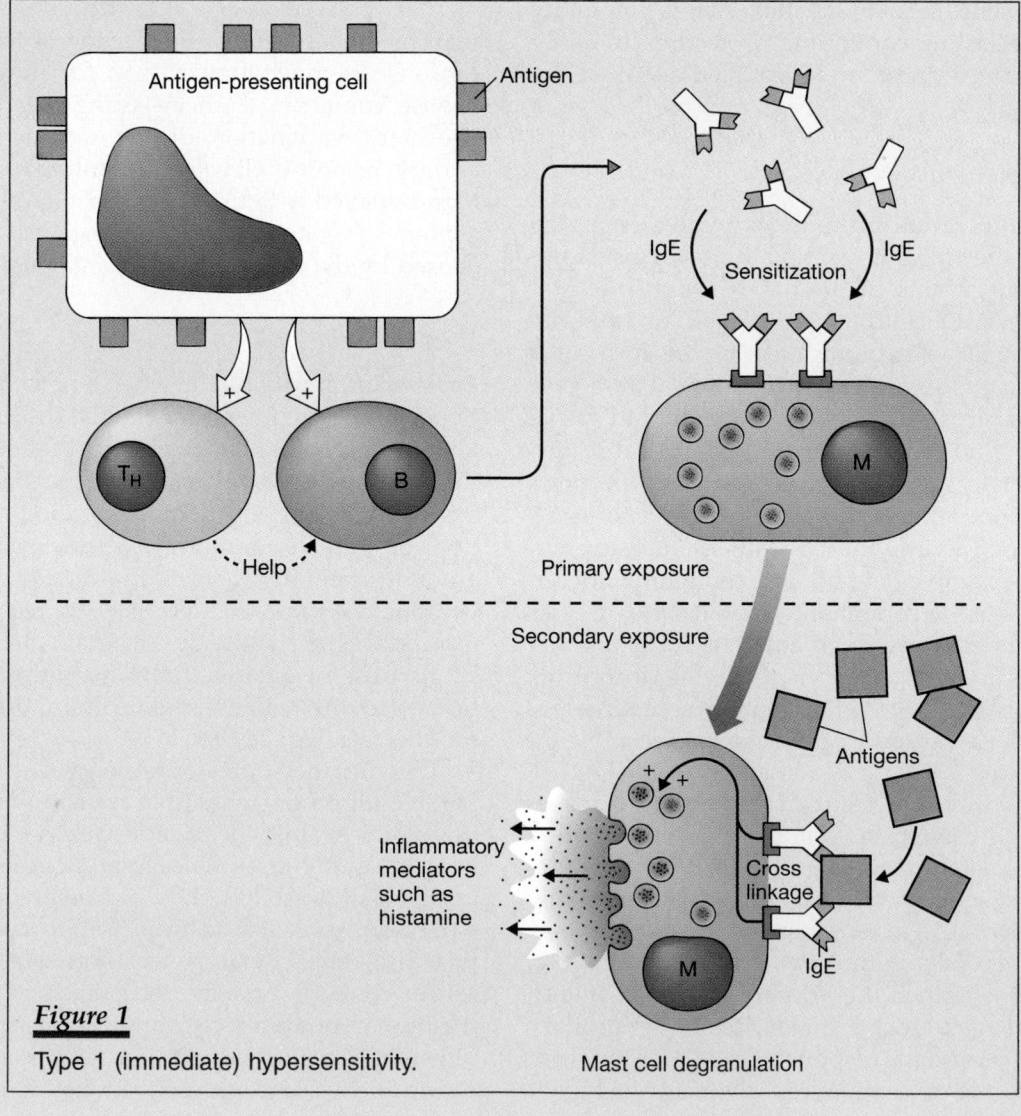

Figure 1

Type 1 (immediate) hypersensitivity.

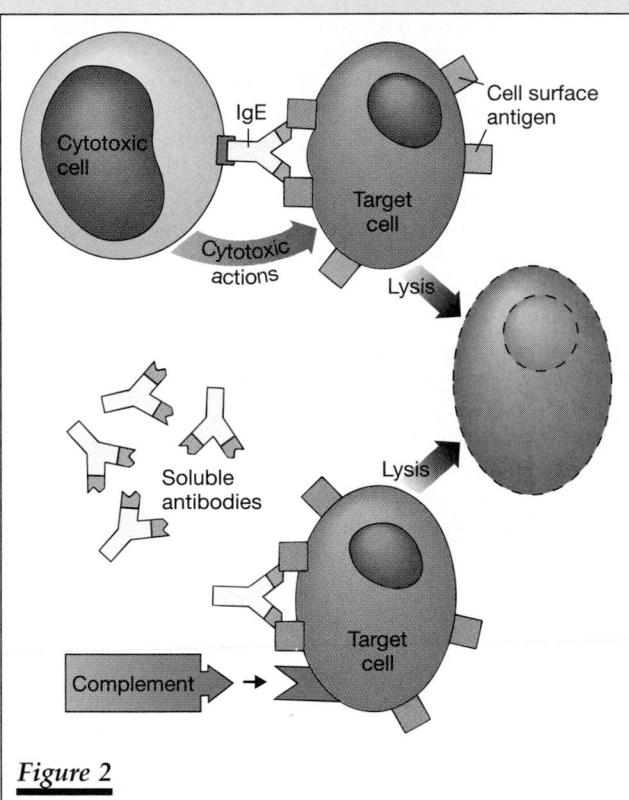

Figure 2

Type II (cytotoxic) hypersensitivity.

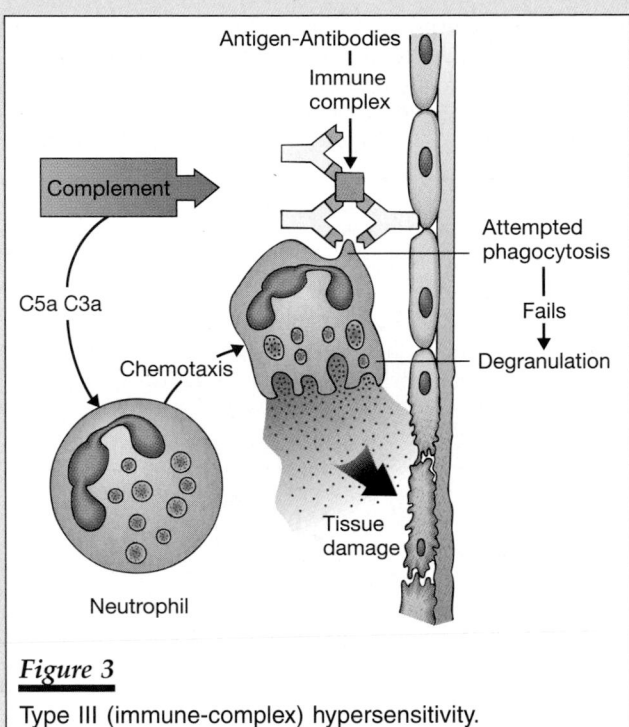

Figure 3

Type III (immune-complex) hypersensitivity.

sodilator), leukocyte chemotactic factors, and other chemicals which promote local acute inflammation (Figure 1). Symptoms of local acute inflammation include swelling in the affected area (raised welts or hives on the skin, congestion in nasal passages and airways, etc.), redness and increased local temperature of the skin, itching, watery eyes, and sneezing (see text Fig. 28–11). Since the inflammatory response to the particular antigen occurs within minutes of the secondary exposure, the response is termed immediate hypersensitivity. Type I hypersensitivities include several kinds of allergies (Type I responses to pollen, dust mite feces, animal dander, etc.), asthma, and rhinitis.

If exposure to the antigen occurs at the body surface (skin) respiratory epithelium, gastrointestinal mucosa, etc., the response may remain localized, but if the antigen enters the circulatory system, the response will be widespread and could be dangerous. In anaphylaxis (*ana* = throughout, *phylax* = guard) circulating antigens of the particular type bind to mast cells and basophils throughout the body, producing systemic effects within a matter of minutes. Hives and intense itching of the skin all over the body, accompanied by swelling and increased skin temperature are cardinal signs of anaphylaxis. In severe cases, increased mucus secretion and

contraction of smooth muscles along the airways restrict airflow and make breathing more difficult. Extensive peripheral vasodilation accelerates heat loss and causes blood pressure to fall, leading toward circulatory collapse and death. These more severe effects of the immune response characterize anaphylactic shock.

Type II hypersensitivity (cytotoxic hypersensitivity) is antibody-dependent and occurs when IgG and IgM antibody bound to a cytotoxic cell also binds to antigen on the surface of a target cell thereby initiating cytotoxic cell destruction of the target cell (Figure 2). The cytotoxic cell is usually a CD8+ lymphocyte but may be a monocyte, eosinophil, neutrophil, B-lymphocyte, or NK cell. Type II reaction also occurs when soluble antibody binds to cell surface antigen and initiates complement-mediated lysis of the target cell. Inflammation occurs when the destroyed target cells release inflammatory chemicals. Examples of Type II hypersensitivity include immune reactions against blood cells (transfusion reactions, hemolytic disease of the newborn, etc.) and graft rejection by a transplant recipient.

Type III hypersensitivity (immune-complex hypersensitivity) develops when large numbers of immune complexes form faster than they can be cleared by phagocytes. The excess immune complexes are deposited in tissues and activate complement (Figure 3). Complement proteins C3a and C5a attract neutrophils that attempt to phagocytize the immune complexes, fail,

(continued on next page)

CLINICAL FOCUS

Hypersensitivity (continued)

and release digestive enzymes by exocytosis, causing tissue damage. Activated complement also stimulates histamine release from mast cells and basophils, increasing blood flow, capillary permeability, and inflammation. Systemic lupus erythematosus (SLE), an autoimmune disease, and farmers lung disease, characterized by repetitive exposure and an allergic response to fungal spores in hay, are typical of Type III hypersensitivity.

Type IV hypersensitivity (delayed hypersensitivity) occurs when antigen-sensitized T cells release vasoactive and chemotactic lymphokines following secondary contact with the same antigen. The lymphokines induce local inflammatory reactions and attract and ac-

tivate macrophages that amplify the inflammatory response by releasing additional inflammatory mediators (Figure 4). The response takes at least 24 hours to reach maximal intensity and is therefore called delayed hypersensitivity. Type IV reactions are seen in response to graft rejection by a transplant recipient, and in several bacterial (e.g., tuberculosis) and viral diseases.

In many instances, the symptoms of hypersensitivity reactions can be prevented or at least minimized. For example, a person allergic to a certain food can avoid the food and thus the allergic response.

■ **Suggest another method of preventing many of the symptoms of Type I hypersensitivity.**

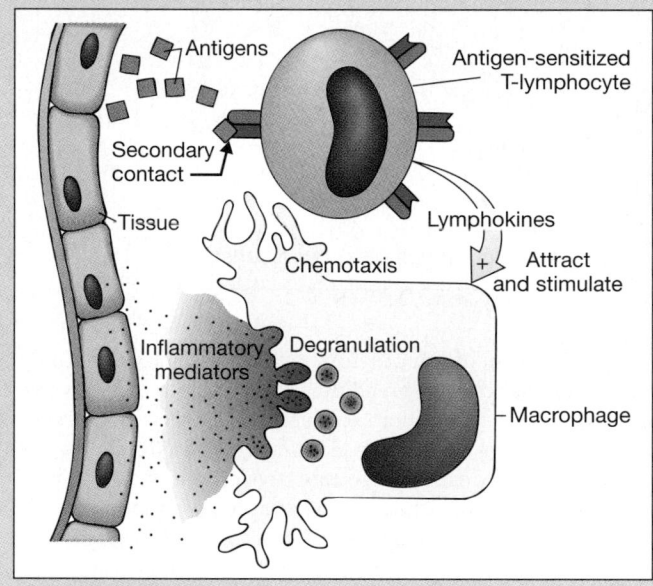

Figure 4

Type IV (delayed) hypersensitivity.

---

of the serum is given a few weeks after the first. The illness is termed serum sickness.

Development of techniques for harvesting human monoclonal antibodies could eliminate the use of equine immunoglobulins for conferring passive immunity in humans. **Monoclonal antibodies** are identical antibodies against one specific antigen, and they are produced by plasma cells (hybridomas) cloned from the same B-lymphocyte. At present, monoclonal antibodies are available and in use for diagnostic purposes such as pregnancy testing, but they are usually derived from mice, and therefore could cause serum sickness if used to confer passive immunity.

An important kind of passive immunity naturally occurs in newborn infants as a result of antibodies that have crossed the placenta from the mother's

blood. Although this passive immunity lasts for just a few months, it is important in helping infants resist infection until their immune systems become more capable. Breast feeding prolongs passive immunity because the infant receives antibodies in the mother's milk.

Autoimmunity Occurs When Lymphocytes and Antibodies React with Normal Components of the Body

Autoimmune literally means to protect the body against itself. An **autoimmune response** arises from and is directed against the individual's own tissues. Most often autoimmunity is thought of solely in

Table 28–2
Antibody Targets in Autoimmune Diseases

Organ-Specific Diseases

Organ	Disease	Antibody Target
Skeletal muscle	Myasthenia gravis	Acetylcholine receptors
Thyroid	Graves' disease	TSH receptor
	Thyroiditis	Thyroglobulin, thyroid peroxidase
Pancreas	Type I diabetes	Islet cells, insulin, GAD, ICA69, GM2-1
	Insulin-resistant diabetes with acanthosis nigricans	Insulin receptor
Adrenal	Addison's disease	Adrenal cell 17α hydroxylase
Parathyroid	Idiopathic hypoparathyroidism	Parathyroid cells
Gastrointestinal tract	Pernicious anemia	Parietal cell K^+, H^+-ATPase, intrinsic factor
Biliary tract	Primary biliary cirrhosis	Mitochondrial dihydrolipoyl acetyltransferase complex
Liver	Chronic active hepatitis	Cytochrome P450
Skin	Pemphigus	Keratinocyte surface antigens
	Vitiligo	Melanocytes
Blood	Autoimmune hemolytic anemia	Erythrocyte antigens
	Idiopathic thrombocytopenic purpura	Platelet membrane glycoproteins
Collagen-vascular system	Ménière's disease	Type II collagen

Systemic Diseases

	Goodpasture's syndrome	Basement membrane, type IV collagen
	Rheumatoid arthritis	Gamma globulin
	Sjögren's syndrome	Ribonuclear proteins, gamma globulin
	Systemic lupus erythematosus	Nucleosome, spliceosomal complex, Ro La particle
	Polymyositis	Aminoacyl-tRNA synthases, tRNA, EF-1, nuclear MI-2
	Rheumatic fever	Myocardium, heart valves, choroid plexus

Source: From George Eisenbarth and Donald Bellgrau, "Autoimmunity," in *Science & Medicine*, 1(2):May/June 1994.

terms of disease, in which the immune system attacks normal tissues as though they were foreign. But normal or physiological autoimmune responses are also part of the body's daily maintenance of homeostasis. A good example of physiological autoimmunity is the mechanism whereby senescent erythrocytes are removed from circulating blood and destroyed.

Red blood cells normally live about 125 days before being removed from circulation and destroyed, usually within the spleen. When a red blood cell nears the end of its life span, changes occur in some of the proteins on its plasma membrane. The changes expose a binding site for an IgG molecule which binds to the red blood cell and induces its phagocytosis by a macrophage in the spleen. The macrophage destroys the aged red blood cell and recycles some of its constituents (see Chapter 17). This kind of autoimmune response to aged red blood cells causes the destruction of approximately 2 1/2 billion red blood cells each day.

Autoimmune disease occurs when there is a breakdown in self tolerance—that is, when self proteins that are normally not recognized as being foreign (one could say they are "tolerated") begin to be treated as though they were foreign. The immune system mounts an attack resulting in inflammation and destruction of affected tissues. The disease may be widespread and involve many organs, such as in systemic lupus erythematosus, or it may be specific for a single organ such as Type I diabetes mellitus (Table 28–2). The specific molecular events that trigger autoimmune disease remain unclear but may include events like exposure to toxic chemicals and infection by certain pathogens.

As discussed earlier, maturation of B-lymphocytes and T-lymphocytes involves a process of negative selection whereby many autoreactive lymphocytes are killed prior to release, and a positive selection process in which lymphocytes that do not react to self antigens are permitted to mature and enter circulation. In fact, some autoreactive cells (a very small number) do escape and enter circulation; however, they do not respond to self antigens unless there is a triggering event. Possible inducers of autoimmune disease include loss of suppressor T cells, which suppress autoreactive lymphocytes, and viral infections that activate autoreactive T cells or directly infect B cells, stimulating B cell poliferation and polyclonal antibody production—the antibodies reacting with self antigens to cause disease.

The development and the severity of autoimmune disease is known to be determined in part by the type of MHC molecules present in the affected individual. Class II MHC molecules exist in many forms; the type present in an individual is determined genetically. The type of Class II MHC molecules present determines which peptides (part of an antigenic protein) can be bound and presented to T cell receptors. Autoimmune disease may occur when an individual inherits genes that code for a Class II MHC protein that allows presentation of self antigens to activated T cells.

CHAPTER REVIEW

Summary

- The lymphatic system functions to return filtered proteins and interstitial fluid to the blood, to transport absorbed lipids from the intestine to the blood, and to defend the body against pathogens.
- Normal lymph is usually a clear, light amber fluid similar in composition to plasma and interstitial fluid.
- Lymphoid tissue contains fixed reticular cells and fibers and free cells such as T cells, B cells, granulocytes, and macrophages.
- Primary lymphoid organs, where lymphocytes initially develop and mature, include red bone marrow, the thymus gland, and Peyer's patches.
- Secondary lymphoid tissue and organs, where lymphocytes mature, reside, and encounter antigens, include lymph nodes and the spleen.
- Internal defense refers to physiological mechanisms that defend the body against toxins and pathogens.
- Internal defense mechanisms allow for recognition of self versus nonself.
- Nonspecific defenses are general in nature and deter a wide variety of pathogens. These defenses include surface-protective mechanisms, interferons, complement, and processes of inflammation and phagocytosis.
- Surface-protective mechanisms include physical and chemical barriers against pathogen invasion such as the skin, sweat, sebum, saliva, and the linings of the gastrointestinal, urinary, respiratory, and reproductive tracts.
- Interferons are antiviral proteins that help prevent spread of viruses and stimulate leukocytes that destroy virally infected cells. There are two major types: Type I (alpha, beta), and Type II (gamma) interferons.
- Complement is a set of proteins in the blood that interact to promote lysis of microbes and signal leukocytes for defense.
- Inflammation is a localized response to an invading microbe that signals active defense; it is characterized by redness, heat, edema, and pain.
- Phagocytosis is a process whereby a phagocyte ingests and destroys toxins, invading microbes, foreign matter, and immune complexes. Neutrophils, monocytes, and macrophages are phagocytes.
- Specific defenses are geared toward identification and destruction of specific pathogens. Two types of specific defense responses are cell-mediated and antibody-mediated responses.
- Cell-mediated immune responses involve coordinated efforts of antigen-presenting cells (macrophages and B-lymphocytes) and two kinds of T-lymphocytes: CD4+ (T4 cells) and CD8+ (T8 cells). Coordination between these cells involves cytokines such as interleukins and interferons.
- Antibodies are proteins produced and released by B-lymphocytes called plasma cells in response to the presence of an antigen. Antibodies are usually specific for a given antigen and help to destroy the antigen's harmful effect.
- Antibody-mediated immune responses involve coordinated efforts to T_4 cells, antigen presenting cells, and B-lymphocytes. Coordination involves cytokines and results in the clonal expansion of plasma cells that secrete antibody specific to the antigen present.
- Immunization is a process of conferring immunity against disease. Active immunity occurs when an individual is given an antigen that stimulates specific antibody production by plasma cells. Repeat doses elevate antibody titer and confer long-lasting immunity against the disease. Passive immunity occurs when a person is given antibodies to help him or her combat disease or minimize infection. This form of protection is short-lived.
- Autoimmunity refers to physiological and pathological processes in which self proteins are recognized by the immune system and attacked. An example of physiological autoimmunity is the recognition and destruction of aged red blood cells. An example of pathological autoimmunity (autoimmune disease) is myasthenia gravis, a disease of the neuromuscular junction in which cholinergic receptors are destroyed.

Review Questions

Identify Each with a Term

1. The term for the smallest vessel in the lymphatic system.
2. The term for organs where lymphocytes develop and mature.
3. The term for immature lymphocytes that mature in the thymus.
4. The term for antibody-secreting B-lymphocyte.
5. The term for a set of plasma proteins that interact in a defined sequence to promote microbial lysis.
6. The term for neutrophil ingestion and destruction of a foreign microbe.
7. The term for the type of lymphocyte that recognizes antigen + MHC II.

Define Each Term

8. IgE
9. autoimmunity
10. IgG
11. induced immunity
12. interleukin
13. interferon
14. Class II MHC protein
15. antigen-presenting cell

Choose the Correct Answer

16. The largest leukocyte cell is part of the body's macrophage defense system. It is the:
    a. eosinophil
    b. basophil
    c. neutrophil
    d. monocyte
    e. lymphocyte
17. CD4+ helper cells:
    a. Act as memory cells for antibody production.
    b. Suppress the immune response.
    c. Are involved in both cell-mediated and antibody-mediated immune responses.
    d. Recognize antigen + MHC I.
    e. Two of the preceding
18. The following cell recognizes an antigen presented in combination with MHC II:
    a. cytotoxic cell
    b. suppressor cell
    c. neutrophil cell
    d. helper cell
    e. basophil
19. The lymphocyte most responsible for orchestrating the body's immune response is the:
    a. CD8+ lymphocyte
    b. plasma cell
    c. CD4+ lymphocyte
    d. memory cell
    e. suppressor cell
20. Which cell secretes interleukins that stimulate B-lymphocyte development?
    a. T_H cell
    b. CD8+ cytotoxic cell
    c. liver macrophage
    d. neutrophil
    e. basophil
21. The principal antibody in circulation, it attacks microorganisms and their toxins. It is class _____.
    a. IgE   d. IgM
    b. IgD   e. IgA
    c. IgG
22. The type of immunity conferred by intravenous infusion of gamma globulins (antibodies) is:
    a. active
    b. passive
    c. cell-mediated
    d. autoimmunity
    e. secondary
23. The term describing the movement of a leukocyte through a thin-walled blood vessel is:
    a. chemotaxis
    b. inflammation
    c. diapedesis
    d. cytokinesis
    e. leukopoiesis
24. Antibodies against the red cell antigens A and B are _____ immunoglobulins.
    a. IgA   d. IgE
    b. IgG   e. IgM
    c. IgD
25. Complement proteins that coat the surface of a microbe, thereby making it for phagocytosis are called:
    a. interleukins
    b. cytokines
    c. monokines
    d. agglutinins
    e. opsonins

Problem Solving in Physiology

26. Rejection of a transplanted organ is an immune response. Discuss two ways to minimize the chance for rejection.
27. Why are CD4+ helper cells critical to mounting cell-mediated and antibody-mediated immune responses?
28. What advantages do multiple defenses (nonspecific and specific) against the same antigen afford?

Suggested Readings

Akbar, A., and Salmon, M. "Selection and survival of activated lymphocytes." *Science & Medicine*, 2(2):48–57, March/April 1995.

Eisenbarth, G.S., and Bellgrau, D. "Autoimmunity." *Science & Medicine*, 1(2):38–47, May/June 1994.

Engelhard, V. "How cells present antigens." *Scientific American*, 271(2):54–61, August 1994.

Heumann, D., and Glauser, M. "Pathogenesis of sepsis." *Science & Medicine*, 1(5):28–37, November/December 1994.

Johnson, H., et. al. "How interferons fight disease." *Scientific American.* 270(5):68–75, May 1994.

Mills, J., and Masur, H. "AIDS-related infections." *Scientific American,* 263(2):50–59, August 1990.

Nossal, G., et. al. "Life, death, and the immune system." *Scientific American,* 269(3):52–144, September 1993.

Romagani, S. "T_{h1} and T_{h2} subsets of CD4$^+$ T lymphocytes." *Science & Medicine,* 1(2):68–77, May/June 1994.

Telford, I.R., and Bridgman, C.F. "Lymphatic system." In New York: *Introduction to Functional Histology,* 2nd ed. pp. 235–255. Harper Collins, 1995.

Tizard, I.R. "Immunology, an introduction," 4th ed. Philadelphia: Saunders College Publishing, 1995.

Key Terms

active immunity (p. 845)

antibody-mediated immunity (p. 843)

antigen (p. 833)

antigen-presenting cell (p. 837)

autoimmunity (p. 848)

B cell (p. 844)

bone marrow (p. 827)

CD4+ cell (p. 838)

CD8+ cell (p. 838)

cell-mediated immunity (p. 837)

class I MHC (p. 838)

class II MHC (p. 838)

complement (p. 834)

cytokines (p. 837)

histamine (p. 834)

immunity (p. 836)

immunoglobulins (p. 841)

inflammation (p. 834)

interferons (p. 834)

interleukins (p. 839)

leukocyte (p. 827)

lymph (p. 825)

lymph node (p. 830)

lymphokines (p. 837)

lysozyme (p. 833)

monoclonal antibody (p. 848)

monokine (p. 837)

nonspecific defense (p. 833)

passive immunity (p. 845)

pathogen (p. 832)

Peyer's patches (p. 829)

phagocytosis (p. 836)

plasma cell (p. 830)

specific defense (p. 832)

T cell (p. 837)

thymus (p. 828)

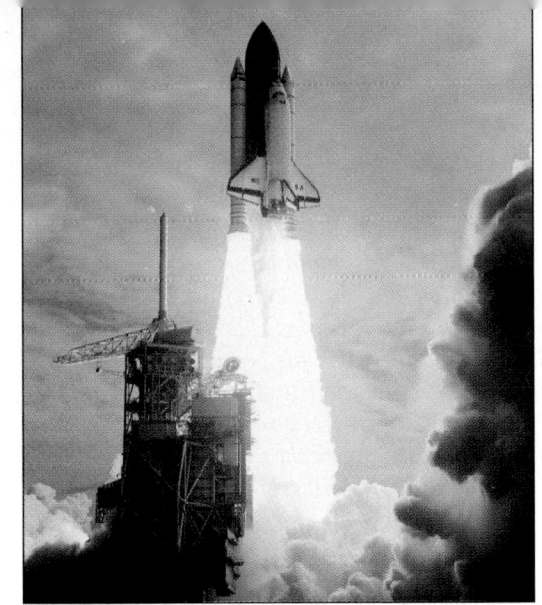

Take-off of the space shuttle Columbia, the first space mission dedicated to life science research. *(Courtesy of NASA)*

Chapter 29
Environmental Physiology

Key Concepts

Environmental stress can lead to genetic changes (adaptation) that are transmitted to the next generation, or to physiological changes (acclimatization) that benefit the individual but are not passed on to the next generation.

Rhythmic physiological activity occurs at both the cellular and whole body level. These rhythms are controlled by intrinsic clocks via the neuroendocrine system. Circadian rhythms are daily rhythms that are typically set to the light-dark cycle.

Acclimatization in response to the physical environment is most apparent at high altitude, during deep-sea diving, and during cold and heat stress.

Pollutants are classified as either respiratory irritants or systemic poisons. Both types of pollutants have different modes of action that can lead to specific diseases.

Chapter Outline

Responses to the Environment
Biological Clocks
Responses to the Physical Environment
Responses to the Chemical Environment

In February, the elite fighting branch of the U.S. Army known as the Rangers hold the swamp phase of their annual military training exercises just outside of Eglin Air Force Base, located in the Panhandle of Florida. The swamp phase is the fourth and final leg of a physically challenging program in which Ranger candidates have to live through the rigors of combat–simulating exercises under the most stressful environmental conditions, bearing arms and equipment with very restricted food rations, sleep deprivation, and little protective gear for a period of two months. The program rotates the soldiers through the forest (Phase I), the desert (Phase II), the mountains (Phase III), and the swamp (Phase IV).

Paul was one of 102 "unphased out" Rangers from the original 334 in his class who made it to the swamp phase, having succeeded at the earlier training phases. He had endured frostbite, a sandstorm, and nausea from altitude, and was feeling optimistic and somewhat invulnerable.

During the earlier part of the week, prior to Paul's arrival, the Panhandle had suffered torrential, cold rains causing the Yellow River, located 15 miles from the training swamp, to rise seven feet above normal and to flood the swamp area. The water temperature was measured at 52°F.

In the middle of a sunless day, Paul and his training school comrades were sent out in zodiac boats to execute an attack on a fictitious enemy. With his fellow soldiers he got out of the boat into the water, which was chest deep and sometimes over his head, carrying rucksacks and weapons and trying to build a rope bridge. Paul was cold and hungry and shivering. Working through the cold swampy waters was grueling. He had been in the water well over an hour when his body began to shake uncontrollably. Every muscle felt stiff, and he could not carry his weapons any more. His vision was hazy and he was not sure of the location of the platoon. The original plans had been so well laid out. He could not understand how he was losing his sense of direction. What were the coordinates? He became drowsy and just wanted to sleep.

Five hours later, Paul awoke in the hospital to learn that he had nearly succumbed to acute hypothermia in that swamp water. If it had not been for the rescue efforts of one of his buddies, he would have died, as was the fate of four of his classmates on this mission. When he was rescued, Paul's body temperature was 80°F, his skin was extremely pale, and his lips were purple. Death from hypothermia generally occurs at 75°F, which Paul was approaching.

In this situation, healthy individuals' bodies reacted to environmental extremes. Ironically, the "fighting Rangers" were in additional danger from hypothermia due to their high level of fitness. Because of their relatively vigorous activities and low insulating fat levels their chances of hypothermia survival were lower at the start of the mission than were those of other people. Why do cold exposure and other environmental stresses such as heat, pressure, high altitude, and deep-sea diving kill some people, whereas others survive? In this chapter on environmental physiology, we will examine questions about fundamental physiological responses to environmental stress. Environmental stresses such as heat, cold, light, atmospheric pressure, and pollutants will be considered.

Responses to the Environment

Environmental physiology studies functional mechanisms of healthy individuals who are exposed to an environmental stress. Environmental **stress** refers to the body's mobilizing response to external stimuli (stressors) in the environment like noxious chemicals or abnormal conditions such as extreme cold or high altitudes, which are potential disruptions to life and health. To maintain life and health, the integrated nervous and endocrine systems must stimulate homeostatic adjustments. Environmental physiologists are particularly interested in the immediate physiological adjustments induced by the environment, as well as the long-term effects that occur with environmental challenges. Environmental Physiology focuses more on individuals, rather than on members of populations or communities.

Environmental Stresses Cause Adaptation

In the study of environmental physiology, four key concepts are important: *homeostasis, adaptation, acclimation* and *acclimatization.* Recall from Chapter One that homeostasis involves processes that maintain a relatively constant internal environment in the face of changes in the external environment. This stability is achieved when regulatory and integrative mechanisms of tissues and organs are automatically initiated to minimize the change in the internal or external environment. An example of a homeostatic mechanism is increased heart rate and vasoconstriction to maintain a constant blood pressure in response to standing. Such homeostatic mechanisms are the focus of physiology.

Adaptation involves a genetic change. An adaptive change arises from mutation or recombination, and individuals who possess it may be more likely to survive. Sweat glands and sweating are examples of adaptations, because the genes that code them are passed on from one generation to the other. The genes were formed as a result of a mutation that occurred by chance, and then were selected.

Environment-Induced Changes Lead to Acclimatization

Two terms often used in environmental physiology are **acclimation** and **acclimatization.** Acclimation refers to a functional change that occurs in response to a single environmental stress that is artificially induced in the laboratory. Acclimation may last months, years, or a lifetime, but does not involve genetic alteration and is not passed on to the next generation. Acclimation is illustrated by a study using identical twins exposed to heat stress in the laboratory. In this experiment, both girls are exposed for 30 minutes to a 95°F temperature and are asked to complete a standardized exercise test. In the hot environment, both will produce sweat in very similar amounts during the heat stress testing. Then, a second experiment is performed on the identical twins. For two weeks twin A enters the heat chamber for two hours/day for three weeks and exercises. Twin B maintains her normal activity and does not participate in the second experiment. At the end of the three-week period, both twins will again be brought into the heat chamber and tested for their sweating rates. Twin B's sweating rate is about the same as before. In contrast, twin A begins to sweat earlier and more profusely than twin B. Twin A has acclimated to the heat stress. That is, she has undergone functional changes induced by the environment that allow her to adjust better, but with no permanent change in her genome. Thus, acclimation can be viewed as an environment-induced improvement in the functioning of an already existing genetically based homeostatic mechanism. In most cases, acclimation is reversible. For example, if the heat stress is discontinued, the sweating rate of twin A will eventually return to pre-acclimated values.

Acclimatization refers to physiological changes that occur in response to natural environmental stresses. Acclimatization usually involves multiple environmental stresses in the natural environment, whereas acclimation involves a single environmental factor artificially induced in the laboratory. Acclimatization occurs in climbing a mountain, a situation in which multiple environmental factors are present (cold stress, altitude, and physical stress), whereas the twin study is an example of acclimation, because there is only a single environmental factor involved (heat stress). Acclimatization, like acclimation, can last months to years but with no permanent change in the genome.

Developing organ systems are most vulnerable to environmental stress. Often acclimatization is most pronounced during a critical period of organ development. For example, early malnutrition will cause permanent stunting of growth. Another example is a newborn exposed to high altitude, an environmental stress leading to permanent changes in some lung functions, whereas an adult exposed to the same altitude will experience changes in lung function that return to pre-acclimatized values upon return to sea level.

Not all types of acclimatization are beneficial, and in some cases they can be harmful. An example is exposure to a low level of air pollutants (dust, pollen, chemicals) that irritate the lining of the airways. This irritation causes hypersecretion in the airways, leading to excessive production of mucus, which protects the airways against the irritant pollutants. However, chronic mucus production can lead to airway plugging, infection, and a persistent cough that can develop into bronchitis and emphysema—a disease that leads to permanent lung damage.

Biological Clocks

Biological Rhythms Have Intrinsic Clocks

There are countless examples of rhythmic activities in humans. These rhythms occur at the cellular as well as the whole-body level. Examples at the cellular level include the "pacemaker cells" of the heart and the neuronal activity controlling breathing rhythms. At the whole-body level, examples include metabolic rate, temperature, and daily activity. These rhythms are referred to as *biological rhythms* and they vary in time from seconds to minutes to days. Biological rhythms do not simply track light-dark, seasonal, or lunar cycles, but often occur in the absence of these environmental cycles. The mechanisms of biological rhythms are poorly understood, but in many instances they are neuronal in origin and often are located in some component of the nervous system.

Circadian Rhythms Track the Light-Dark Cycles

One of the most common rhythms seen in nature is the *circadian rhythm,* that can be detected in physiological responses of humans and other animals. The word circadian means "nearly 24 hours" and was introduced because daily activity does not always encompass exactly 24 hours.

Many common circadian rhythms are tied to the light-dark cycle. The light-dark cycle has a significant effect on mammals, since it is the most error free and is the sole environmental factor common to all species. For example, metabolic rate, temperature, and heart rate are highest during the day or light (L) cycle and lowest during the night or dark (D) cycle (Fig. 29–1). In nocturnal animals, these rhythmic activities are reversed. Many hormones are secreted in a circadian rhythm. In humans, for example, melatonin, a hormone produced by the pineal gland, is secreted only during the dark period. The melatonin in the urine is highest from 11:00 PM to 7:00 AM. The concentration of cortisol, one of the principal hormones produced by the adrenal cortex, also follows a circadian rhythm (see Fig. 29–1).

In some mammals, many of the daily rhythms persist even if the L-D cycles are altered, either in constant light (LL) or constant darkness (DD), as seen in Figure 29–1. Examples in nature are the Arctic and Antarctic regions, where each year there are 82 days of continuous light. A number of studies indicate that radiotelemetry, heart rate, metabolism, and activity followed a circadian rhythm despite the continued light in foxes, wolves, and ground squirrels. Also, continuous darkness has been suggested to have an adverse effect which is most apparent in the Arctic and Antarctic regions as well as in some of the Scandinavian countries. One example is seasonal effective disorder (SAD), which is linked to the lack of adequate daylight and is a problem in the winter months with extended periods of darkness. This disorder has a negative effect on personality and performance.

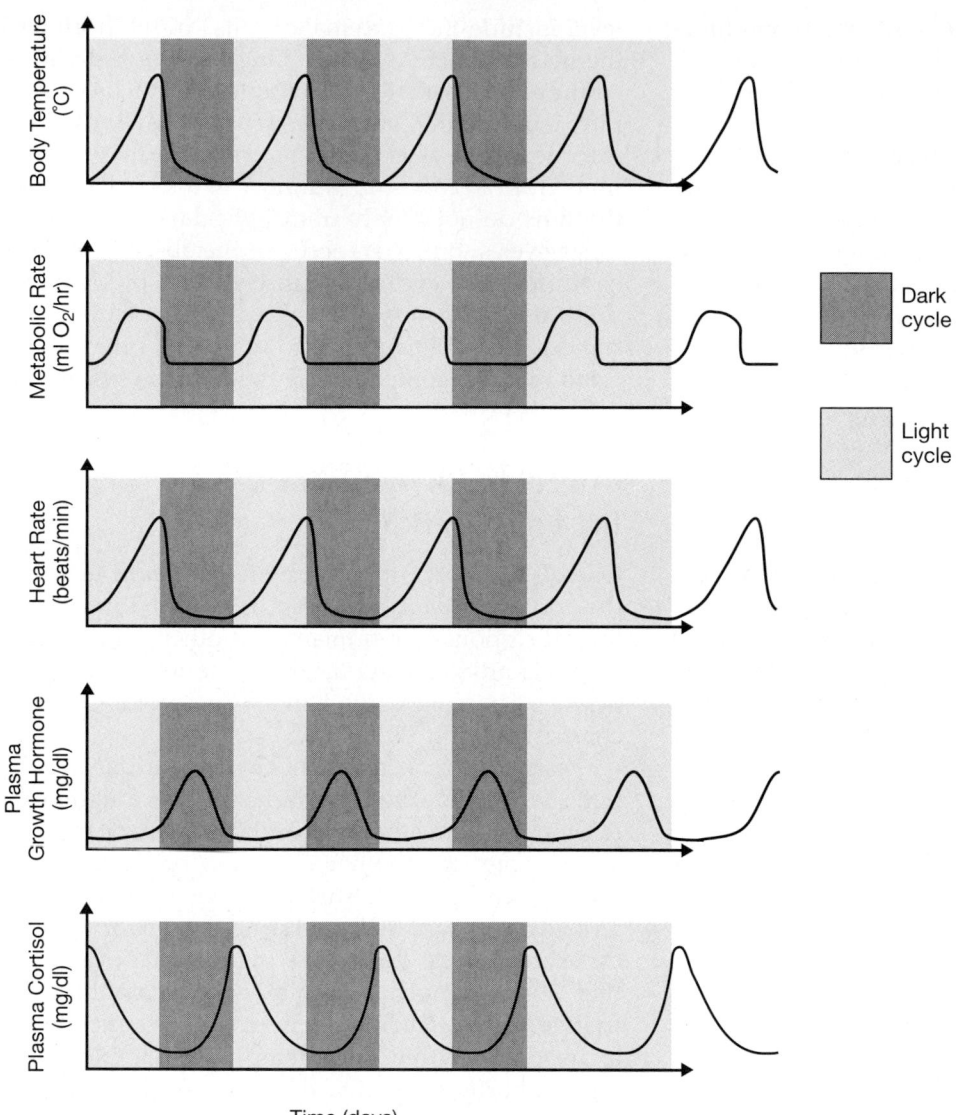

Figure 29–1

Effect of circadian rhythm on body functions. Most circadian rhythms continue in the absence of the light-dark cycle.

Dark cycle

Light cycle

Time (days)

Milk production in dairy cattle is also affected in the Scandinavian countries during prolonged periods of darkness. There is recent evidence that SAD may be opioid mediated in the CNS, and opioid antagonists like naloxone may help people with SAD.

The cellular mechanisms underlying circadian rhythms are not well understood. In mammals, the central biological clock is thought to be located in the suprachiasmatic nucleus of the hypothalamus. Transplantation of the suprachiasmatic nucleus in animals transfers the biological rhythm of the donor to the recipient. The nervous connection of the eye to the suprachiasmatic nucleus provides the light-dark signal for the circadian rhythm. The suprachiasmatic-eye connection is an independent nervous pathway of the normal visual pathway. The pineal gland is also involved with circadian rhythmicity. In mammals,

neural connections can be observed between the pineal gland and the suprachiasmatic nucleus.

The lunar cycle has a periodicity of approximately 29.5 days. The phase of the moon affects the light intensity at night and has some effect on the circadian rhythm of nocturnal animals. In humans, the lunar cycle has little effect on biological rhythms. Interestingly, the 28-day menstrual cycle follows the lunar cycle, as does the length of human gestation—10 lunar months or 40 weeks. Most of the evolved life occurred before the invention of artificial light. Babies conceived during full moon would clearly stand a better chance of survival if they were born on a full moon, when their parents could see 24 hours a day to protect them against predators during the vulnerable time. Mating can now occur at all times of the month without regard to the phase of the moon. However, the 28-day menstrual cycle still reminds us of exactly how and when the length of the cycle was selected in our species.

Responses to the Physical Environment

For much of our history, humans have been exposed to a range of environmental stresses. These were encountered on ascents to high altitude where alveolar oxygen tension dropped as barometric pressure fell as well as on exposure to temperature extremes. There were also some cases, beginning in Grecian times, of underwater activities in primitive diving bells in which divers must have encountered air pressures greater than one atmosphere (1 atm). In the eighteenth century, the chemistry of gases blossomed, and individual gases such as oxygen, nitrogen, and carbon dioxide were isolated. Then, the study of unusual gases began in earnest. Joseph Priestley said of pure oxygen, "Only two mice and myself have had the privilege of breathing it." Today we breathe in high-oxygen environments, low-oxygen environments, polluted environments, and artificial environments such as those in space capsules and underwater devices. We have contrived all sorts of artificial breathing devices to keep up alive in hostile environments and during times when our natural ability to breathe is inadequate.

High Altitude Causes Hypoxia

The condition of low oxygen (**hypoxia**) is most commonly encountered during travel to high altitudes. As altitude increases, barometric pressure decreases. The percentage of oxygen in the air, however, remains constant at near 21%. The decreasing barometric pressure causes the partial pressure of oxygen to fall. This fact was not appreciated until the last century, when the French physiologist Paul Bert was shocked by the deaths of balloonists who ascended to altitudes of more than 15,000 feet. Bert's lengthy scientific work promoted by this tragedy was published in the classic book *La Pression Barometricque*, in which he emphasized, "Oxygen tension is everything; barometric pressure in itself is nothing or almost nothing." Bert has been acclaimed as the founder of aerospace medicine; he also did important work on the effects of high pressure experienced by divers, and discovered the "diving reflex."

Immediate Effects of Hypoxia

The detrimental effects of hypoxia are amplified if the exposure is sudden. For example, a pilot who loses his or her oxygen supply at 30,000 feet has just over one minute to descend to lower altitude before losing consciousness. The effect of altitude on alveolar gas tension and arterial O_2 saturation is shown in Table 29–1. In contrast, after careful acclimatization, several climbers have been able to reach the summit of Mount Everest, nearly 30,000 feet, without the use of supplemental oxygen. This amazing achievement emphasizes that the body is more able to tolerate hypoxia if the ascent is slow so the body can acclimate. The ideal of successful acclimatization through slow ascent is highlighted repeatedly in the literature on high altitude and mountain climbing. Acclimatization to altitude is not only to hypoxia, but also to the cold and physical stress. For example, at the 12,500-foot White Mountain Research Station in California, the mean temperature is below freezing 8 months of the year.

Human respiration, as pointed out in Chapters 20 and 21, are very well adapted for gas exchange at

---

Table 29–1

Effects of Altitude on Atmospheric Pressure (P_B) and Alveolar Gas Tensions

Altitude (feet)	P_B (mm Hg)	Percent O_2	P_{AO_2} (mm Hg)	P_{ACO_2} (mm Hg)	Arterial O_2 Saturation (%)
0	760	21	104	40	97
10,000	523	21	67	36	90
20,000	349	21	40	24	20
30,000	226	21	21	24	20
40,000	141	21	8	24	5

P_{AO_2} = Alveolar partial pressure of oxygen; P_{ACO_2} = Alveolar partial pressure of carbon dioxide.

CLINICAL FOCUS

Altitude Sickness

Altitude sickness, sometimes referred to as acute mountain sickness, is a condition caused by lack of oxygen. The hypoxia-induced sickness is due to a decrease in partial pressure of oxygen (P_{O_2}) and not to a decrease in the percentage of oxygen. With increasing altitude, the drop in barometric pressure makes the air thinner and lowers the P_{O_2}.

Altitude can catch the unwary by surprise. For example, college students who take an extended weekend to go to a ski resort at an elevation of 8,000 to 10,000 feet, and start pounding down the slopes as soon as they arrive, may wake up the next morning with what feels like a terrible hangover and a strong desire to vomit. These students are experiencing the early onset of acute mountain sickness. The tourist industry estimates that a fourth of the 25 million people who visit the Rockies each year experience mild symptoms of altitude sickness, which costs the tourist industry 40 million dollars a year because they feel too ill to ski, eat at a restaurant, or go to the bar and drink.

How does hypoxia affect the body? How do natives who live at high altitude become acclimatized and immune to altitude sickness? Symptoms of altitude sickness date back to 30 BC when Chinese documents describe the "Great Headache Mountain." The early symptoms of altitude sickness, which include headaches, dizziness, and nausea, are due to increased pressure on the brain. Hypoxia causes cerebral vessels to dilate and also causes cerebral vessels to leak fluid that seeps into the brain, both of which increase cerebral pressure. How fluid leaks into the brain is still open for debate. Some physiologists think that hypoxia causes the capillary endothelial cells simply to "run out of gas" and the energy-dependent transport mechanisms for salt and water falter. Others believe that, with hypoxia, cerebral vessels overdilate and in the process endothelial cells are stretched and start leaking. Both mechanisms may occur, and the fluid that leaks out of the cerebral vessels puts added pressure on the brain that triggers headaches, dizziness, nausea, and muscle spasms. Cerebral edema can also cause a victim to lose motor skills, resulting in staggering and slurring of words as with intoxication.

Pulmonary edema is even more serious and can strike at 9,000 feet. Pulmonary endothelial cells become leaky, and recent studies have shown that severe hypoxia constricts the pulmonary vein, which adds further insult by increasing capillary hydrostatic pressure. The problem of pumping blood through water-soaked lungs as well as trying to oxygenate the blood in the edematous lung becomes life threatening. As lungs fill up with fluid, victims become increasingly short of breath and often spit up foamy blood. Severe edema causes death from heart failure due to hypoxemia and the added strain of pumping blood through edematous lungs.

Altitude sickness can be avoided with a bit of care. One precaution is to ascend the mountain slowly. Most climbers go up the mountain in plateaus and work patiently to acclimatize. Acclimatization includes step-up breathing to capture more O_2, a faster heart rate to pump more oxygenated blood to the tissues, renal compensation for respiratory alkalosis, and increased production of red blood cells to carry more O_2. In addition, there are other acclimatization processes that take place at the cellular level. One is the increase in the number of mitochondria, the power plants of cells that transform glucose into usable ATP that energizes all cellular activity. The second is glucose utilization. The body's two biggest energy consumers are the brain and heart. Unlike the brain, which uses only glucose, the heart typically burns fat and some glucose as well. However, at higher altitudes the heart switches substrates and has a higher preference for glucose because fat is a far less efficient fuel to burn when oxygen levels are low. The brain also develops a lower utilization of glucose. Recent studies have shown that, with acclimatization, the brain down-regulates its demand for glucose as well and therefore uses less O_2. Third, there is also mitochondrial acclimatization to altitude. The cell burns glucose two ways. One is anaerobically (without O_2) and the other is aerobically. The anaerobic step produces lactate, and is not very efficient. The aerobic step produces 18 times more energy with O_2. Recent studies have shown that individuals who have acclimatized to altitude produce less lactate than nonacclimatized individuals. This means that the mitochondrial enzymes that control aerobic metabolism have developed a way of keeping this pathway open during prolonged hypoxia.

■ **Why do caffeine and alcohol exacerbate altitude sickness?**

---

a normal partial pressure of oxygen. However, human respiration does not function as well as that of birds that fly at high altitudes or fish that live in low-oxygen environments. The reason is that humans have tidal flow with concurrent exchange in their lungs. In contrast, birds have continuous flow with cross-current exchange and fish have continuous flow with counter-current exchange. Humans approach continuous flow only with high ventilatory rates.

Figure 29–2

Ventilatory response to hypoxia. Note that minute ventilation is unchanged until the arterial P_{O_2} drops below 50 mm Hg.

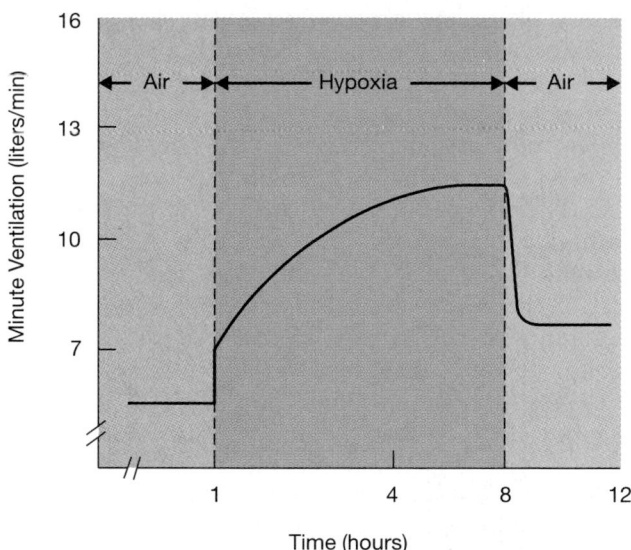

Figure 29–3

Hypoxic drive occurs in two stages. The first phase is a rapid increase in minute ventilation, and the second phase is a slow sustained increase in ventilation.

As altitude increases, however, the body makes noticeable efforts to deliver normal amounts of oxygen to the tissues. Chief among these responses to altitude is **hyperventilation**—that is, deeper breaths taken more rapidly. The sequence of events for hyperventilation is as follows: (1) a fall in inspired oxygen tension (P_{IO_2}), (2) decreased alveolar tension (P_{AO_2}), (3) decreased arterial oxygen tension (P_{AO_2}), (4) increased firing of peripheral chemoreceptors (the carotid bodies), (5) hyperventilation, and (6) increased alveolar and arterial oxygen tension. The loop is regulated in part by a negative feedback limb: (1) hyperventilation, (2) elimination of alveolar carbon dioxide, (3) decreased arterial carbon dioxide tension, (4) increased arterial and cerebrospinal fluid (CSF) pH, (5) decreased firing of both central and peripheral chemoreceptors, and (6) decreased ventilation. The increased ventilatory response to hypoxia is mediated directly through the peripheral chemoreceptors, namely the carotid bodies (see Chapter 21 for control of breathing). The stimulus for ventilation during hypoxia is a decrease in arterial P_{AO_2} rather than O_2 content or % O_2 saturation. Although the ventilatory response to hypoxia is immediate, the arterial P_{O_2} must drop below 50 mm Hg before ventilation is stimulated (Fig. 29–2). Ventilation is essentially unchanged at values above 50 mm Hg. It is important to realize that there is also an immediate rise in cardiac output to match the increase in ventilation.

Hypoxic hyperventilation appears in two stages. In the first stage there is an immediate increase in ventilation. However, the increase in ventilation is small compared to the second stage, in which ventilation continues to rise slowly over 8 to 12 hours with a sustained ventilation after 12 hours (Fig. 29–3). The reason for the two phases is due to the acid-base disturbance seen with hyperventilation (Fig. 29–4). In

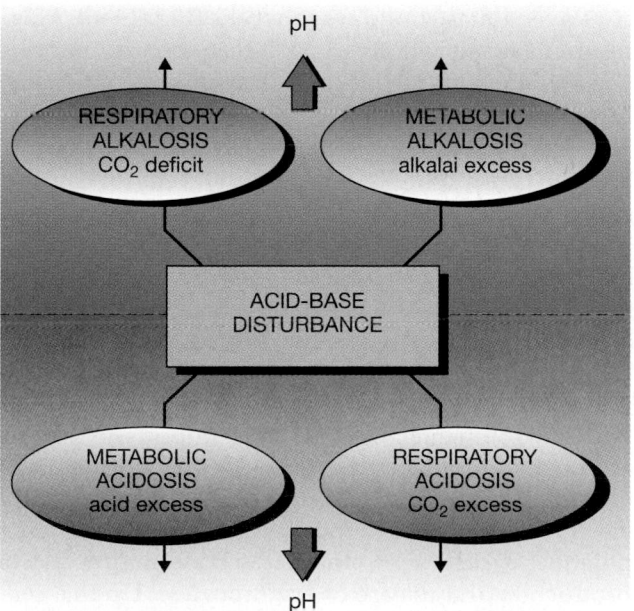

Figure 29–4

Various pathways of acid-base disturbance. High altitude takes the pathway of respiratory alkalosis because hyperventilation creates a carbon dioxide deficit.

the first phase, the increase in ventilation is rapid but small, because of the excess CO_2 which is blown off, causing respiratory alkalosis. The increased pH depresses respiration, and respiratory alkalosis becomes antagonistic to the hypoxic drive. During the second phase, the kidneys begin to return the acid-base disturbance back to normal. Recall from Chapter 26 that the kidneys regulate HCO^-_3 and the lungs regulate arterial P_{CO_2}, and that the ratio of $HCO^-_3/PaCO_2$ in the blood is a key component in regulating pH (Henderson-Hasselbach equation) and not the absolute amounts. This is illustrated in Figure 29–5, where the normal ratio of $HCO^-_3/PaCO_2$ is about 20 : 1. With hypoxia, the ratio is increased during hyperventilation and the blood becomes alkalotic. The kidney compensates by excreting bicarbonate, which brings the ratio back toward normal. The antagonistic effect of increased pH is minimized, and this allows the hypoxic drive to further increase ventilation. Renal compensation occurs within 10 to 12 hours and corresponds to the second phase of hypoxic hyperventilation.

Long-Term Effects of Hypoxia

Since millions of people travel to high altitudes annually for recreation, the acute effects of hypoxia are of both physiological and of clinical interest. Some effects are present at even moderate elevations of 5000 feet, an altitude at which major cities are found throughout the world. There are noticeable changes in ventilation, especially during exercise, even at these relatively low elevations. Acclimatization occurs over a matter of days to weeks (Fig. 29–6). Most residents of elevations of 5000 feet soon do not notice any effect of the altitude. The highest settlement in the world is a mining camp at 17,500 feet in Peru. The residents work a mine that is at 19,000 feet, with a daily climb of 1,500 feet. Interestingly, when the camp was moved to 18,500 feet, the miners complained that they had no appetite, could not sleep, and lost weight. From these observations, it appears that 17,500 feet is the highest altitude at which even acclimatized humans can live permanently.

In addition to ventilatory adjustments, the body undergoes other physiological changes to adjust to low environmental oxygen. One important step in acclimatization is that more red blood cells are manufactured in the red bone marrow, which causes an elevated hematocrit (hypoxia → stimulates erythropoietin by kidney → stimulates bone marrow → ↑ production and release of erythrocytes). Up to a certain point, this change enables the blood to carry more oxygen to the tissues. However, the higher the hematocrit, the more viscous (thickened) the blood becomes, a change that increases the work load on

Figure 29–5

Renal compensation to high altitude respiratory alkalosis.

the heart. Normal hematocrit is near 45%. In some cases, high-altitude residents develop a condition called **polycythemia,** in which hematocrits reach levels of over 70%. These individuals must have blood withdrawn periodically to reduce their hematocrit to a range that permits the heart to pump effectively.

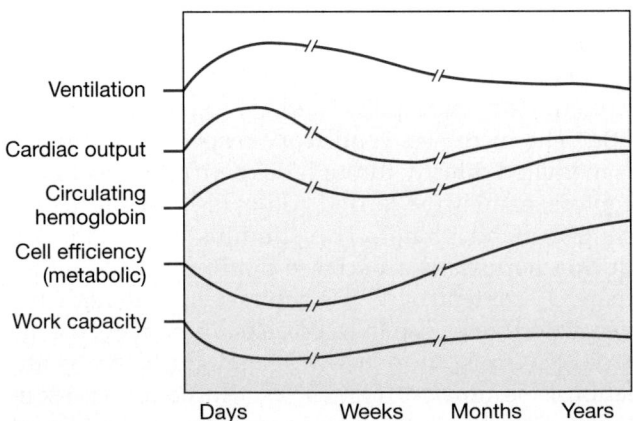

Figure 29–6

Acclimatization that takes place at high altitude (14,000 to 16,000 feet). From Houston, C.S. *Hypoxia: Man at Altitude,* New York: Thieme-Stratton, 1982.

The body increases ventilation, cardiac output, and circulating hemoglobin (Fig. 29–6) during the first few weeks to months at high altitude. Over a long period, cardiac output and ventilation tend to return to low-altitude values, while hematocrit continues to climb. The cells also change to become metabolically more efficient. For example, the mitochondria increase in number, and there are alterations of metabolic pathways. Even with these changes, the maximum work capacity never reaches sea-level values. Nevertheless, the acclimatization of long-term residents can be impressive. In the Andes, the natives return from the high mines in the evenings to their "low" altitude homes for a rousing soccer game at 15,000 feet.

The best acclimatized of all high-altitude residents are the Sherpas in Nepal, whose climbing feats are legendary (Fig. 29–7). Other people—for example, the residents of high altitudes (10,000 feet) in Colorado—are often not as well adapted to low oxygen. The percentage of Coloradans aged 60 years and older who live at high altitude is much lower than the percentage of older people who live at low altitude. High-altitude residents tend to leave for health reasons, usually because of heart and lung disease. The former residents report that their health is improved after they settle at low altitude.

Korean Women Are Champion Divers

Off the shores of Korea and Southern Japan, over 30,000 women make their living by diving to harvest shellfish and seaweed for food (Fig. 29–8). This practice has continued over the last 1,500 years without the use of any mechanical equipment. These divers, called Ama, can descend to depths of 80 feet and hold their breath for up to two minutes. The Ama dive repeatedly to these depths, often 30 times per hour, coming up only for a few breaths of air and a brief rest. In warm weather they work four hours a day with one-hour rest intervals. The Ama work in the winter in a water temperature of 50°F. Diving is a lifelong profession for these women, often starting at 10 to 12 years of age and continuing to age 65. Pregnancy does not interrupt their work. They often dive up to the day of delivery and nurse their babies between diving shifts. As we shall see, the female is better suited to this work than the male.

Other diving groups who make a similar living include the sponge divers in the Mediterranean area

Figure 29–8

The Ama are Korean and Japanese women who make a living descending unassisted to a depth of 60 to 80 feet to collect shellfish. At the end of each dive, a helper in a boat at the surface pulls the Ama up by means of a long rope attached to her waist. (The other rope belongs to another diver.) In warmer weather, the Ama wear only loincloths during their dives. In the winter, they wear only light cotton bathing suits. *(From* Scientific American, May 1967. "The Diving Women of Korea and Japan" by Suk Ki Hong and Hermann Rahn. Photo by Fosco Maraini.)

Figure 29–7

Sherpas are mountain-dwelling people who have highly adapted to the low oxygen environment of high altitudes (Nepal). *(© Steve Covello/Mountain Stock/1990.)*

Table 29–2
Breath-Holding Capacities of Some Mammals

Species	Time, Min
Polar bear	1.5
Pearl divers (human)	2.5
Sea otter	5
Hippopotomus	15
Beaver	20
Seals	30
Sperm whale	90

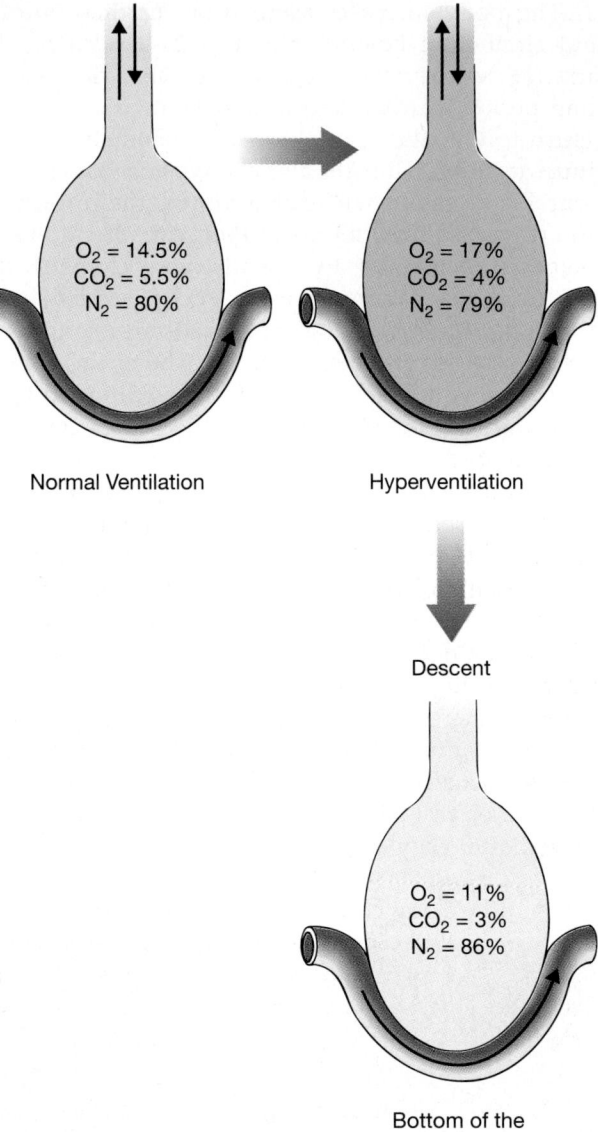

Figure 29–9

Changes in gas concentration of O_2, N_2, and CO_2 during diving. During descent, pressure on the lungs causes all gases to enter the lungs; during ascent, the situation is reversed.

and the pearl divers off the coast of Australia. These divers are mostly men, and the record descent was held by a Greek sponge diver who, in 1913, dived down to 200 feet to put a line on a boat anchor. This record has recently been broken by a U.S. Navy diver. Although human divers do not stay submerged as long as other diving mammals (Table 29–2), they have similar physiological responses.

Preparing for the dive, divers hyperventilate for 5 to 10 seconds before they make the plunge. Hyperventilation does two things: it blows off considerable amounts of CO_2 and increases the concentration of O_2 in the lungs (Fig. 29–9). The final breath, however, is about 85% of their total lung capacity. This is done to avoid the uncomfortable pressure on the lungs during the dive and to restrict the body's buoyancy. If divers go deeper than the level of lung compression (residual lung volume), they are subjected to a painful "lung squeeze." More importantly, if the hydrostatic pressure in the pulmonary vessels exceeds the air pressure in the lungs, small pulmonary blood vessels can rupture. Surprisingly, the seal also exhausts part of its lung volume before a dive.

The Ama have evolved another technique. During hyperventilation they purse their lips and emit a loud whistle with each forced expiration. The whistling is a trademark of the Ama and can be heard for long distances. This practice helps to protect their lungs by preventing excessive hyperventilation which can lead to underwater blackout (unconsciousness) during the dive (see Chapter 21). Not all divers whistle when they hyperventilate. Nevertheless, this is another classic example of how humans, without any knowledge of respiratory physiology, learned that: (1) hyperventilation blows off excess CO_2 which extends their breath holding, (2) whistling during hyperventilation prevents too much CO_2 from being blown off, and (3) not taking a full breath before diving can extend the depth of the dive.

The effect of diving on blood gases is seen in Figure 29–9. When divers reach the bottom at a depth of 40 feet, the O_2 concentration is reduced to approximately 11% because of breath holding and the continual uptake of O_2 by the blood. However, at this depth the pressure has compressed the lungs to about half of their predive volume, but the alveolar partial pressure of O_2 (P_{AO_2}) has increased from 100 mm Hg before the dive to 149 mm Hg at the bottom of the dive. Another paradox occurs in CO_2 uptake during the dive. Normally, when someone holds his breath, the carbon dioxide in the lungs increases. However, during a dive, CO_2 decreases in the lungs. As seen in Figure 29–9, alveolar carbon dioxide drops from 4% after hyperventilation to 3% at the end of the dive. The reason for this is that at a depth of 40 feet or greater the compression of lung raises the partial pressure of CO_2 in the alveoli that is greater than the partial pressure of CO_2 in the pulmonary capillary blood. As a result, CO_2 is actually taken up from the lungs and retained in the tissues.

Upon ascent from the bottom, the diver's lungs expand, drastically reversing the situation of compression. With the reduction in pressure, CO_2 leaves the blood very rapidly. However, the situation with O_2 transport is quite different. With the precipitous drop in pressure, partial pressure of O_2 in the alveoli goes from 149 to 40 mm Hg in a matter of 30 seconds. The P_{AO_2} of 40 mm Hg is no greater than the partial pressure of O_2 in the pulmonary capillary blood. Hence, hemoglobin cannot pick up O_2, and the O_2 content of the arterial blood drops. In some cases the blood actually loses O_2 to the lungs. Thus, the cumulative O_2 deficiency in tissues, especially the brain, is sharply accentuated during ascent and can be the cause of death among divers (especially sport divers) returning to the surface after a deep and lengthy dive.

Another important physiologic event that occurs is the "diving reflex," in which blood vessels to all parts of the body constrict except those to the heart, brain, and exercising muscles that require undiminished O_2 supply to be maintained. The heart rate also decreases, thereby decreasing myocardial O_2 consumption (Fig. 29–10). Most diving mammals have

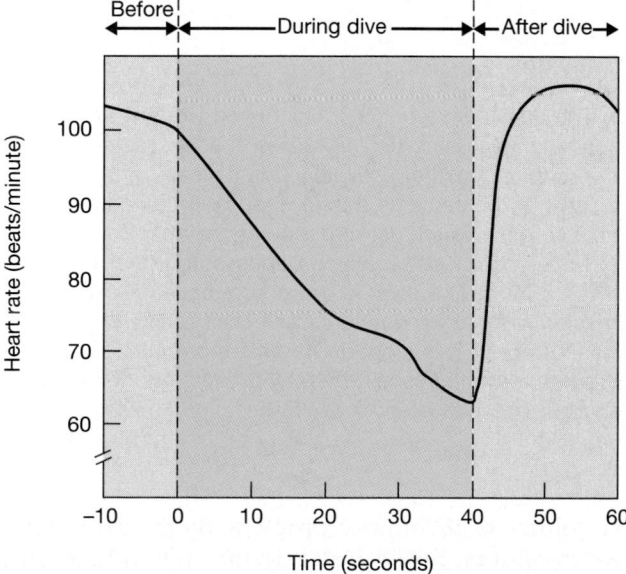

Figure 29–10

Diving reflex causes a slowing of the heart rate during descent.

the diving reflex that assists in the conservation of O_2 and extends their diving time. At a depth of 300 feet, a seal's heart rate is only 4 beats/min, whereas on the surface it is 120 beats/min. Interestingly, breath holding alone in humans will not initiate the diving reflex. The reflex is initiated by submerging of the face in water and is part of the gasp reflex seen in the newborn human and other mammals.

Acclimatization to Diving

Many divers, especially the Ama, have acclimatized to diving. The Ama have a larger vital capacity than nondiving Korean women—due, in part, to development of the diver's inspiratory muscles, which also helps to resist compression of the chest and lungs under water. The acclimatization is acquired through training. Daughters of Korean farmers can become just as competent divers as the daughters of divers. The other physiologic change that occurs in the acclimatization process is the elevated basal metabolic rate to compensate for the heat loss during diving. The Ama on the average lose about 1,000 Calories per day from cold exposure. They compensate by eating more than their nondiving sisters and by having a significantly higher basal metabolic rate. Normally, humans all over the world—in cold climates and in warm climates—have the same basal metabolic rate. In the winter months, the Ama basal metabolic rate is about 25% higher than nondiving women in the same community with the same economic background.

Heat production is one defense against cold. The other is prevention of heat loss. Women have better protection against heat loss because they have more subcutaneous fat, which explains why women dominate the diving profession of Korea and Japan. However, the diving women have the same thickness of subcutaneous fat as nondiving women, but have less heat loss. This means that, as part of the acclimatization to diving, vasoconstriction restricts the heat loss from the blood vessels to the skin.

Increased Gas Pressures Can Be Lethal

As divers submerge under water, the pressure surrounding their bodies rises because of the weight of the water. To visualize how the pressure rise takes place, imagine diving with a balloon that is filled with air at a pressure of 1 atm. The deeper the balloon goes, the greater the water pressure pushing in on it. This increased pressure causes the volume of the balloon to decrease. Every 33 feet below the surface, the pressure increases by 1 atm. As the balloon gets smaller and smaller, the air molecules are compressed closer and closer together, increasing the pressure in the bal-

Figure 29–11

Air in a container is compressed progressively as the container is lowered into water, where the surrounding pressure increases.

(a)

(b)

Figure 29–12

(*a*) The development of Self Contained Underwater Breathing Apparatus (SCUBA) has enabled many people to pursue underwater activities. It is critically important, however, to be cautious during deep or prolonged dives because gases such as nitrogen can be absorbed into the body and cause problems during both the dive and the ascent. (*b*) The diver in the photo makes a decompression safety stop during ascent from a dive of 37 meters (120 feet). This stop facilitates the gradual release of nitrogen from the tissues, thereby preventing the bends. (*Courtesy of C. Petree and J. Holtmeier.*)

loon. Similarly, as a diver submerges, the lungs become compressed and the air inside the body also becomes compressed (Fig. 29–11). This means if a diver goes 33 feet underwater, the alveolar pressure is two atmospheres instead of one. If he goes 330 feet the pressure is 11 atm.

For thousands of years, people have used equipment to help them remain underwater for prolonged periods. The earliest devices were probably snorkels (Fig. 29–12). These devices provide a practical way to breathe to a depth of about 18 inches. Below that depth, two problems occur. First, the water pressure makes it difficult to expand the chest to draw in fresh air. Secondly, the snorkel adds dead space through which the diver must draw air. If the dead space becomes too large, then the diver is simply moving air back and forth in the snorkel and does not get any fresh air into the alveoli.

The next apparatus to be invented was most likely the diving bell. These devices have been used for centuries. The first ones were simple, inverted chambers, open at the bottom, in which the diver stood and breathed the air in the chamber as it submerged (Fig. 29–13). These diving bells had many limitations, including air that become rapidly fouled by carbon dioxide. Also, the air volume was compressed to a smaller and smaller volume as the pressure increased during descent. Alexander the Great is said to have used a diving bell in the fourth century B.C. Modern diving bells are completely enclosed, so compressed

air volume is no longer a problem. By the early nineteenth century, the enclosed diving suit had been invented (Fig. 29–13). Fresh air was fed from the surface into the suit by a pump to sustain the diver for long periods under water. However, pressure plays an important role in delivering compressed air un-

Figure 29–13

(*a*) Diving bells have been used since the time of the ancient Greeks. (*b*) During the eighteenth century, the advent of pumps allowed for the development of diving suits, which gave the diver greater mobility under water. (© *The Bettmann Archive.*)

Figure 29–14

Although a swimmer can remain submerged for long periods with a snorkel, the depth of a dive is limited because the snorkel increases respiratory dead space and because the increasing pressure on the chest makes it difficult to draw in fresh air. (© *Maria Paraskevas*)

derwater. If, for example, 1 cubic foot of air is pumped down to a depth equivalent to 4 atm (approximately 100 feet), then this 1 cubic foot, when it reaches the diver's helmet, will be only 1/4 cubic foot. Therefore, the fixed amount of air that must be pumped to the helmet per minute to wash out the carbon dioxide is directly proportional to the depth to which the diver descends.

During World War II, work began in earnest on the Self Contained Underwater Breathing Apparatus, "SCUBA" (Fig. 29–14). Much of the pioneering work has been credited to Jacque Cousteau. This kind of gear has become very popular for recreation, exploration, and work on underwater oil rigs and pipelines.

Effects of High Pressures on Respiration

A characteristic of all these devices is that air enters the lungs at higher-than-normal pressure. Unfortunately, breathing such high pressures has some potentially serious consequences. The problems come not during the descent, but when the diver ascends. When the pressure becomes high enough, particularly at depths below 100 feet, significant amounts of nitrogen in the compressed air are absorbed into the tissue. The longer the time spent under water, the more nitrogen is dissolved. As the diver rises to the

surface, the nitrogen is released from the tissues. If the ascent is slow, the nitrogen can be picked up by the capillary blood and returned to the lungs to be exhaled. However, if the ascent is rapid, bubbles can form, both in the blood and in the tissue. When bubbles form near the joints, they cause agonizing pain. Since the diver bends over to relieve the pain, this syndrome has been called the "bends." The risk of the bends increases as both the depth of the dive and the length of time at depth increase. The treatment for the bends is to spend time in a decompression chamber in which the barometric pressure is raised until the bubbles are compressed and the nitrogen redissolves into the tissue. Pressure in the chamber is gradually lowered, giving the nitrogen time to be exhaled through the respiratory system instead of forming bubbles in the body.

At depths below about 200 feet, the concentration of nitrogen molecules in the body rises to a point at which they have a narcotic effect, causing "raptures of the deep." Under these conditions, divers can often make unintelligent, life-threatening decisions. One way they can avoid this is to switch from a nitrogen-oxygen mixture to a helium-oxygen mixture. Helium is less soluble in tissues than nitrogen, so fewer helium molecules dissolve in the body. This reduces the narcotizing effects as well as the number of molecules that must be removed from the tissue during ascent. Also, helium has a smaller molecular size than nitrogen, so it diffuses more rapidly, which aids in elimination of the gas.

Respiration in Submarines

The first practical submarines were built in the 1800s, although crude, hand-propelled models were fabricated in the 1700s. By the start of World War I, Britain had 77 submarines, France 45, the United States 35, Germany 29, and Russia 28. The submarines in the early 1900s could stay submerged only for several hours and operated exclusively in shallow coastal waters. The effectiveness of submarines in warfare made them popular with every major power. To date, thousands of these vessels have been built. During World War II, a big problem was the building up of CO_2 either from "CO_2 Scrubbers" or in battle, when submersion for long periods of time was required. It was not uncommon for CO_2 to build up to 4% inside the submarine (sea-level concentration is 0.04%).

One of the unpleasant aspects of submarines is that, when they sink, they usually take all hands aboard and trap them below. That grim reality has prompted much work on methods of escape. All sorts of scuba-type gear have been developed, but usually to no avail. During World War II, only five men were saved by mechanical lungs. Ironically, a simple, excellent escape technique effective to depths of 300 feet had been known for nearly a century. It was used by Wilhelm Bauer, in 1851, when his hand-propelled submarine sank in 60 feet of water and the hull began to collapse from the pressure (Fig. 29–15). Bauer waited until the air pressure in the vessel was equalized with the water pressure and then opened the hatch. The captain and his crew members by then were breathing air at 3 atm, but not for long enough to build up nitrogen pressure in the tissue. Thus, they could ascend very rapidly without experiencing the bends. As they rose to the surface, the compressed air in their lungs expanded. This caused them to exhale steadily during the ascent. They never ran out of air even though they exhaled the whole time. As Captain Bauer explained, "We came to the surface like bubbles in a glass of champagne."

This technique was ignored and forgotten. It was accidentally repeated in several instances during World War II, but was never developed. Instead, with technology gone awry, navies continued to develop impractical mechanical contrivances that did not work, while all the time the body was superbly adapted to escape. In the 1950s, the British and U.S. navies recognized that the best technology for escape was no technology. Water-filled escape training towers have been built, and unaided escapes at depths to 300 feet are now a routine part of submarine training.

One step has been added to the natural escape method: sailors now inflate small life vests at the moment of departure. This extra buoyancy causes them

Figure 29–15

Wilhelm Bauer's submarine sank in 60 feet of water, but the captain and crew escaped by swimming easily to the surface. (Photo courtesy of: Submarine Force Library and Museum, Groton, CT.)

to ascent very rapidly to the surface. The decompression is so rapid under these circumstances that the sailors exhale as fast as they can. The air they breath while at 300 feet is so highly compressed that it expands rapidly in their lungs as they ascend and, even with continual maximal forced expiration, they never run out of air on the way up. In navy jargon, this technique is known as "blow and go."

"Homo Aquaticus"

Two methods currently in the research stage offer promise for underwater use. One is the actual breathing of liquid. If saline (salt solution) or some other fluid such as fluorocarbon is put under several atmospheres of pressure by 100% oxygen, then the oxygen tension of the saline rises to thousands of millimeters of mercury. If an experimental animal such as a rat is held under the surface, it will eventually be forced to breathe the hyperoxygenated saline. Rats can breathe such fluid and swim around for periods of up to 18 hours.

There are two major problems with liquid breathing. The first is that the removal of carbon dioxide is difficult because the respiration depends on the bellows action of the lungs to wash carbon dioxide out. If the lungs are filled with fluid, it is much more difficult for the bellows to work effectively, and carbon dioxide builds up in the blood. The second problem is that the fluid washes the surfactant out of the lungs, leading to alveolar collapse when the animal attempts to return to air. This makes the transition from fluid back to air difficult. It has been done successfully, however, by a number of rats, mice, and even some dogs.

Another possibility for prolonged undersea activities is the development of artificial gills. This is not a small engineering feat, but it seems possible. If a gill could be developed that could trail behind a diver while being held in the mouth, its large surface area could extract oxygen from the water and excrete carbon dioxide just as the gills of fish do. Animals such as the mudpuppy have external gills as well as lungs. They use the gills to stay successfully under water for prolonged periods. Liquid breathing techniques, artificial gills, or some similar method might lead to the development of what Jacques Cousteau describes as *Homo aquaticus,* an underwater man capable of living indefinitely beneath the surface of the sea.

Respiration in High-Oxygen Environments

The usual problem in breathing is getting enough oxygen into the body. However, with modern chemistry and technology it is possible to breathe 100% oxygen and actually get too much. High-oxygen environments are often found in hospitals, where oxygen is administered to patients with oxygen deficiency, and in aviation and space flights, where the pilot or astronaut breathes air with high oxygen concentration because of extremely low environmental oxygen tension.

Atelectasis: Collapse of Alveoli

One of the physical dangers of breathing pure oxygen is **atelectasis** (collapsed alveoli), which can occur in pilots, astronauts, and hospital patients. This respiratory complication occurs when small airways become blocked by mucus; the oxygen in the plugged alveoli continues to be absorbed, and the alveoli rapidly collapse (Fig. 29–16). The total pressure of the trapped gas is close to 760 mm Hg, but O_2 pressure in the venous blood is only 40 to 45 mm Hg. Consequently the oxygen is continuously absorbed, and the alveoli collapse (Fig. 29–16*a, b*). Normally, when gas is trapped in the course of breathing air, nitrogen acts to deter collapse because it is not utilized and therefore is not absorbed into the blood. However, when one is breathing pure oxygen, there is no nitrogen.

Postoperative atelectasis occurs commonly in patients who are ventilated with high-oxygen mixtures. Atelectasis is also prevalent in fighter pilots who breathe high oxygen and experience large gravitational forces (see the section on space travel). The added gravitational effects augment alveolar collapse. Atelectasis is most likely to occur at the bottom of the lung where alveoli are poorly expanded and airways are partially closed. Reopening atelectic areas is difficult because the surface tension effects are more pronounced on small units. Atelectasis can, however, be reopened by deep breaths.

Free-Radical–Induced Toxic Effect of Breathing High Oxygen

The second problem associated with high oxygen, much more serious than atelectasis, is toxicity. This occurs when a person breathes high concentrations of oxygen for more than one day. The primary target organ for oxygen toxicity is the lung, where tissue injury can occur within 24 hours and can cause pulmonary edema.

The first site of injury is endothelial cells of the pulmonary capillary. The mechanism for tissue damage is the formation of **free radicals.** These highly reactive molecules are formed as by-products from molecular oxygen. Before molecular oxygen (O_2) can be used in the oxidation of fuels for energy, it first must be converted to "active forms." These active forms are called **oxidizing free radicals,** and the most

Figure 29–16

Cause of atelectasis as a result of a plugged airway. When (*a*) air is breathed and (*b*) oxygen is breathed, alveolar pressure in both cases is the same, but the sum of the gas pressures in the mixed venous blood is smaller. (*c*) Atelectasis caused by airway obstruction. Trapped air is absorbed into the pulmonary capillaries, causing alveoli to collapse.

important one in the cell is the **superoxide anion radical** ($O_2^-\cdot$). Free radicals are continuously being produced from dissolved molecular oxygen under normal oxygen tensions (tissue $P_{O_2} = 40$ mm Hg). However, under these normoxic conditions, the free radicals are rapidly removed by protective cellular enzymes—namely, **superoxide dismutase, catalase,** and **peroxidase.** As long as P_{O_2} is normal, the oxidizing free radicals are rapidly removed and no cellular damage occurs. When tissue P_{O_2} is greatly increased—such as occurs when an individual breathes a high concentration of oxygen—excess free radicals are formed that overwhelm the protective enzymes. As a result, these free radicals cause destructive and lethal cellular effects by (1) oxidizing membrane lipids, thus altering essential properties of membranous structures of the cells; (2) oxidizing cellular proteins and DNA; and (3) oxidizing enzymes, thus shutting down cellular respiration.

High oxygen is not the only source of oxidant damage from free radicals. Breathing ozone (O_3) or nitrogen dioxide (NO_2) from polluted air will also cause free radical injury to the lung. Some herbicides, such as paraquat, are extremely toxic because of the free-radical-induced injury. If someone smokes tobacco or marijuana that has been treated with paraquat, the paraquat in the smoke penetrates the lungs and forms free radicals. Crop dusters and mi-grant workers are also at risk because of the exposure to paraquat through the lungs or skin.

Retrolental Fibrosis

Another hazard of breathing high concentrations of oxygen is seen in intensive care units in which premature infants develop blindness from **retrolental fibrosis.** The mechanism for the blindness is deterioration of the retina and inadequate development of retinal blood vessels. The poor vascularization leads to the formation of fibrosis tissue behind the lens (retrolental fibrosis). If the P_{O_2} is kept below 140 mm Hg in the incubator, retrolental fibrosis can be avoided.

Toxic Effects of High-Pressure Oxygen

A third type of oxygen toxicity occurs when oxygen is breathed at pressures exceeding 760 mm Hg. Oxygen is breathed at pressures greater than 1 atm in deep-sea diving or in hyperbaric (high-pressure) chambers. When oxygen is breathed at 2 to 4 atm, it stimulates the central nervous system (CNS), leading to convulsions. The exact cause for the toxic effect on brain tissue is not fully known, but the high oxygen pressure is thought to inactivate dehydrogenase enzymes. The convulsions are often preceded by twitching of the face, ringing in the ears, and nausea. Convulsions can occur when an amateur scuba diver inadvertently uses a tank filled with O_2.

Respiration in Unusual Gaseous Environments

Breathing high carbon dioxide levels usually does not occur, because the normal P_{CO_2} in the air we breathe has a partial pressure of 0.3 mm Hg (0.04%), which is negligible. However, in certain types of diving gear and closed environments (e.g., submarines and spacecraft), carbon dioxide can build up and be rebreathed because of defects in equipment. Rebreathing higher-than-normal levels of carbon dioxide will stimulate ventilation (see Chapter 22). An alveolar P_{CO_2} of 80 mm Hg, which is about twice normal, is the maximum that an individual can tolerate. With these high concentrations, hyperventilation is maximum, and the individual becomes severely acidotic. When P_{CO_2} levels are greater than 80 mm Hg, the situation becomes intolerable, and the respiratory centers are no longer stimulated but instead are inhibited. Respiration begins to fail, and acidosis becomes more sever. Eventually, the CO_2 acts like a narcotic, and the individual becomes lethargic and subsequently becomes unconscious.

Carbon Monoxide and Respiration

Carbon monoxide (CO), unlike carbon dioxide, does not stimulate respiration. The problem with carbon monoxide (an odorless, colorless, and poisonous gas coming from incomplete combustion of fuels), is that it cannot be detected by the body. Carbon monoxide becomes lethal because it competitively binds at the same site as oxygen on the hemoglobin molecule to form **carboxyhemoglobin (COHb).** The formation of COHb interferes with the transport of oxygen to the cells. The reaction (Hb + CO $\rightleftharpoons$ HbCO) is reversible and is a function of the partial pressure of CO. This means that breathing a higher concentration of CO will shift the reaction to the right to form more HbCO, and breathing lower concentrations of CO will shift the reaction to the left to form less HbCO. The problem is further complicated because the binding affinity of CO for hemoglobin is about 210 times that of O_2. Because of this greater tendency for binding to Hb, even trace amounts of CO in the respired air can be extremely dangerous. For example, an alveolar P_{CO} of 0.5 mm Hg (1/210 that of alveolar P_{O_2}) will cause half the hemoglobin in the blood to bind with CO and half to bind with O_2. Under these conditions, the amount of oxygen carried by the blood to the cells is halved. Thus, a P_{CO} of 0.7 mm Hg (0.1% of air) can be lethal because the cells become **anoxic** (deprived of oxygen). The first organ to become sensitive to the lack of oxygen is the brain. Alteration of vision, reaction times, and coordination are the first signs of cerebral hypoxia and often explains why there are more accidents on the freeway when CO levels rise. The real danger of CO poisoning is that it cannot be detected in the body. With hypoxemia, there are physical signs; the individual will turn blue around fingertips and lips. However, the formation of HbCO makes the blood crimson red and masks the bluish tint of hypoxemia. There are no chemoreceptors to detect CO, the problem is further compounded because arterial P_{O_2} is still normal (P_{AO_2} 95 mm Hg) and the chemoreceptors that detect changes in P_{O_2} are not stimulated.

An individual severely poisoned with CO can best be treated by administration of 100% O_2. Administering pure oxygen elevates the blood P_{O_2}, which displaces CO from the hemoglobin. Also beneficial is the simultaneous administration of 5% to 7% carbon dioxide. Breathing the high concentration of carbon dioxide stimulates ventilation and reduces alveolar CO, which concomitantly releases CO from the blood. When the combination of high oxygen and carbon dioxide is breathed, CO can be removed 10 to 15 times faster than when just fresh air is breathed.

Space Travel Imposes Unusual Stresses

Because of the development of spacecraft and supersonic planes, the human body can now be subjected to acceleratory forces resulting from sudden changes in velocity or direction. For example, a spacecraft must attain a speed of 25,000 miles/hr in order to sufficiently escape the earth's gravitational field and attain a permanent orbit. High acceleration rates are required to attain these speeds. When an aircraft or spacecraft takes off, the human body inside is subjected to **positive acceleration.** (Inertial forces move in a direction that may be described as head-to-foot with respect to the pilot.) When an aircraft or spacecraft slows down at the end of a flight, **negative acceleration** occurs (inertial forces move in a foot-to-head direction).

High G Forces and the Human Body

Any time an aircraft turns, centrifugal and acceleratory forces come into play as a result of gravitational attraction. The human body's sensation of weight is due to **gravitational attraction** and is referred to as **G force.** The G force is a unit used to express the magnitude of the acceleratory force. One G is equal to a force of magnitude equal to that of the normal gravitational force of the earth. For example, when an individual is sitting in a chair, the force that presses the body against the chair seat is equal to 1 G (one times

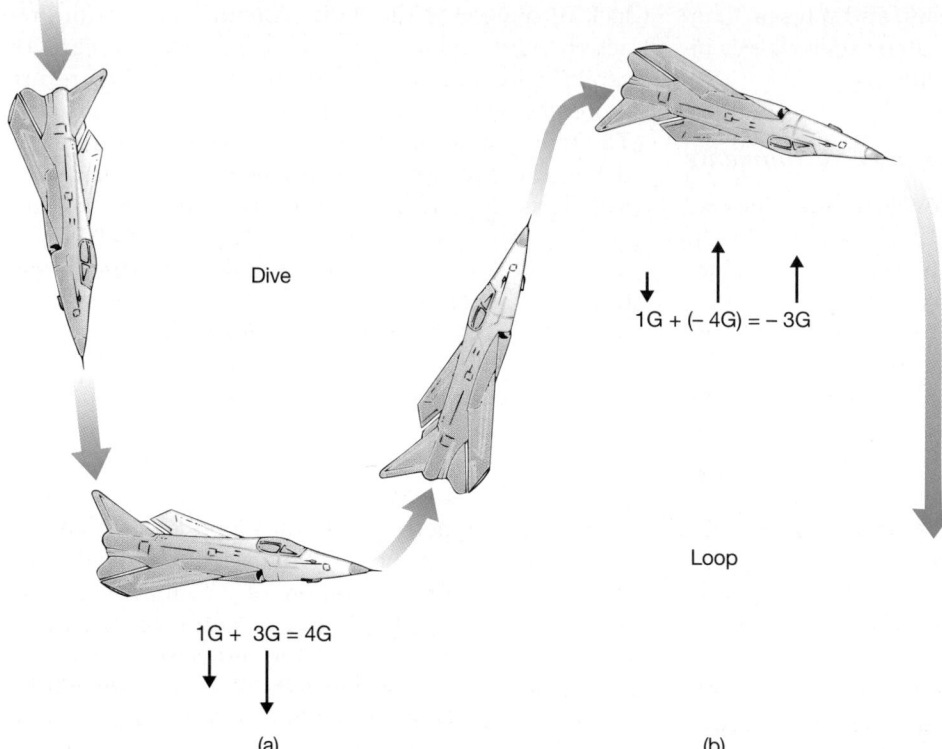

Figure 29–17
Illustration of positive and negative G forces on an airplane.

Dive

1G + (– 4G) = – 3G

Loop

1G + 3G = 4G

(a)

(b)

the pull of gravity). If the force that presses against the seat is four times the normal weight of the body, the force is 4 G. If force is less than the body weight, then the force is a negative G. Positive and negative G forces are illustrated in Figure 29–17, in which an aircraft pulls out of a dive or goes through an outside loop so that the pilot is held down by the seat belt. In the dive, the pilot experiences positive G force, with inertial forces acting in a head-to-foot direction. In the loop, the pilot experiences a negative G force, in which the inertial forces are directed from foot to head.

The physiological effects of these acceleratory forces are due to the fact that some tissues (blood in particular) are "loose" with respect to body. When a positive G force is exerted in a head-to-foot direction, blood is pulled toward the lower parts of the body. In Figure 29–17, when the pilot "pulls 4 Gs," venous pressure in the leg veins is four times the normal pressure. The heart cannot pump unless venous blood is returned, and consequently there is considerable pooling of blood in the lower extremities. At 4 G, a pilot in the seated position experiences a fall in arterial systemic pressure, at the level of the heart, of approximately 40 mm Hg. Under these conditions, blood flow to the brain almost stops. Consequently, loss of vision, referred to as **blackout,** occurs. Other symptoms are also prevalent. At 3 G to 4 G, a pilot or astronaut experiences difficulty in the use of muscles.

Facial tissue sags and eyelids droop. At 5 G, movement of the body is almost impossible, and respiration stops because the respiratory muscles cease to function. Between 5 G and 9 G, legs become congested and calf muscle cramps are prevalent. Pulmonary edema occurs because of venous engorgement in the base of the lung, and vision as well as hearing is lost. Finally, consciousness is lost due to cerebral hypoxia.

Most of these symptoms quickly vanish when the positive G force is removed. To minimize the G force, astronauts lie down during takeoff or re-entry and wear special antigravity suits. On acceleration, these suits automatically pump up, and increased pressure exerted on the body prevents pooling of the blood. With negative G force (foot-to-head), the motion of blood is toward the head. Up to −3 G, there is pressure in the eye sockets. Neck veins are overdistended and the face becomes puffy. Ringing in the ears occurs, and headaches may ensue. Retinal vessels and vision become engorged and vision is impaired at −3 to −4 G. The light filtering through the blood gives a sensation of redness, a phenomenon known as **redout.** Often, at high negative G (greater than −3 G), retinal and cerebral hemorrhages can occur. The physiological effects of negative G are much more lasting than those of positive G, and problems with vision and headache may persist for hours following a negative G exposure.

Figure 29–18
Astronaut James Bagian floats through the Spacelab Life Sciences module aboard the Earth-orbiting Columbia (STS-40 mission). Bagian and six other crewmembers spent over nine days in space conducting life science research. *(Courtesy of NASA, Lyndon B. Johnson Space Center, Houston, TX.)*

Weightlessness and Respiration

Once a spacecraft leaves the earth's gravitational forces, astronauts are free of the problem of acceleratory forces but are faced with the problem of weightlessness (Fig. 29–18). The physiological problems associated with weightlessness do not appear to be severe. One concern is the lack of hydrostatic pressures, which results in decreased blood volume, red cell mass, and cardiac output. The second physiological problem related to weightlessness is diminished physical activity, because very little muscle contraction is required for movement. The decrease in physical activity leads to a decrease in work capacity and loss of calcium from bone.

Artificial Atmospheres in Spacecraft

There is no atmosphere in space, consequently astronauts must be provided with an artificial atmosphere that is pressurized and equipped with life-sustaining oxygen. The types of atmospheric conditions selected depend, in part, on the sophistication of the engineering. For example, the Russian space program used 21% oxygen in nitrogen at a pressure equivalent to sea level, while the early Apollo missions used 100% oxygen at one third of atmospheric pressure, which gives a P_{O_2} equivalent of 1 atm. The reason for these two different gas mixtures was based on weight and space. In the earlier Apollo missions, NASA did not have booster rockets sufficient to launch a heavy spacecraft. Using 100% oxygen at 1/3 atm gives a lighter-weight load than a thick-shelled spacecraft using air at 1 atm. One of the dangers of 100% oxygen is the possibility of an explosion or a fire. Such an accident did occur aboard one of the early Apollo crafts on the launch pad, resulting in a fatal fire and forcing NASA to change gas mixtures. Now all U.S. space flights use air as a gas mixture.

Temperature Extremes Cause Injury

Modern heating and air conditioning allows cars, homes and offices to maintain temperatures at 72°F. The average North American is not aware of the physiological responses involved in temperature extremes. Crossing the desert at temperatures of 115°F or the plains at temperatures of −30°F can now be done without much discomfort.

Even our culture somewhat influences the degree of cold or heat exposure that we experience. For example, back in the 1950s, before the energy crisis, a study of children showed that the average temperature for dressing for school in the winter for British children of university-educated parents was 50°F while the room temperature averaged 75°F for U.S. children with similar background. Also, office temperatures in the United States are maintained at much lower temperatures in the summer and at higher temperatures in the winter compared to many other countries.

Response to Cold

The mechanisms of heat loss have been discussed previously, in Chapter 28. Environmental physiology deals with these mechanisms but focuses primarily on acute and chronic effects of cold stress.

(text continues on page 874)

Space Shuttle Columbia: First Space Laboratory Dedicated to Life Science Research

When the United States made the decision to explore space, many Americans were very skeptical whether humans could live in outer space, and many raised questions. Could cosmic particles affect the brain? Did space harbor foreign organisms that could be lethal to humans back on earth? Could humans live for long periods of time in weightlessness? Since the National Aeronautics and Space Administration (NASA) was formed over 30 years ago, we have learned that people can indeed survive—and more importantly, can live and work productively—in space.

Successful exploration of space depends on the health and well-being of astronauts who travel and work there. For this reason, NASA has dedicated several shuttle missions to examine how space flight affects the human body. Space-Lab Life Sciences-1 (SLS-1) as the first of these dedicated missions. It was launched as the eleventh flight of the Space Shuttle Columbia and completed its mission in June 1991. The SLS-1 was the first space laboratory dedicated to life science research. Its goal was to study how the human body adapted to microgravity.

In previous space flights, body fluid shifts, fluid losses, baro reflex deconditioning, deconditioning of muscles, calcium loss from bones, and reduction of red blood cell production were just some of the effects observed in astronauts. Missing from the previous studies were data on the interrelationships among different body systems in microgravity. Space Shuttle Columbia's crew of seven used themselves as human guinea pigs to enable scientists to understand how the body acclimates to weightlessness. Twenty different studies were carried out on five body systems during the SLS-1 mission. These systems included: (1) cardiopulmonary, (2) renal/endocrine, (3) blood–immune, (4) musculoskeletal, and (5) neurovestibular (brain, nerves, eyes, and inner ear). Specific experiments were designed to determine how much blood was delivered to the various organs, how pulmonary function changes, and what reflex mechanisms regulate blood pressure. The problem of motion sickness in space was investigated by subjecting the crew to conflicting visual, vestibular, and tactile information to determine the effect of sense on orientation. Studies on the other systems included muscle metabolism, inhibition of bone formation, and blood and lymphocyte production.

In addition to the human studies, over 2000 jellyfish and 29 laboratory rats were carried on the mission and were used for complementary studies. In the jellyfish ex-periments, thousands of jellyfish polyps were encased in flasks containing artificial seawater. Early in the mission, thyroxine or iodine was injected into the flasks to induce the polyps to metamorphose into free-swimming ephyrae. The tiny jellyfish were filmed to observe their swimming motions compared to a control on earth to determine if there is any difference between jellyfish gravity receptors that develop in space and those that develop on earth.

Columbia shuttle lift-off, June 5, 1991. The Spacelab Life Sciences-1 Laboratory was the first Spacelab focusing solely on life sciences research. During the trip the crew explored physiological changes that occur during space flight, as well as the consequences of the body's adaptation to microgravity and the readjustment to gravity upon return to Earth. *(Courtesy of NASA, John F. Kennedy Space Center.)*

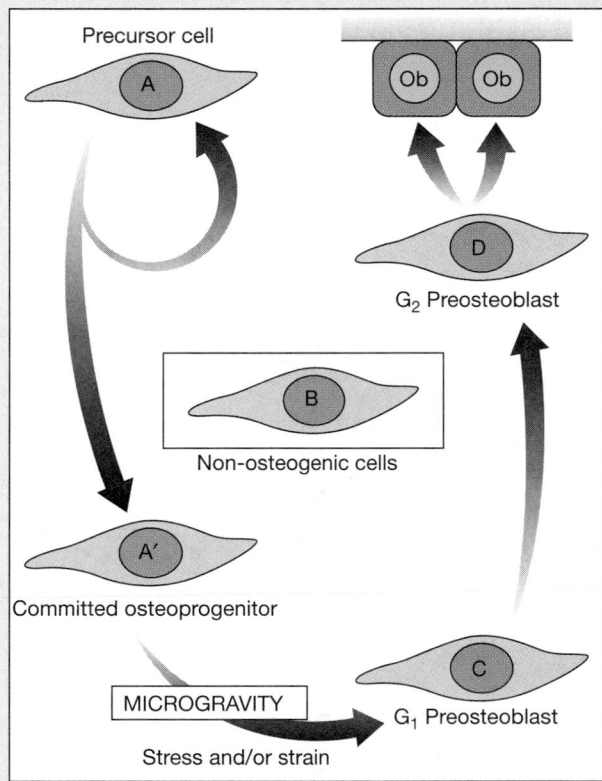

STS-40 Mission Insignia: The STS-40 patch, like the space mission, focused on human beings living and working in space. Against a background of the universe, seven silver stars pictured around the orbital path of Columbia represent the seven crewmembers. The orbiter's flight path forms the double-helix of the DNA molecule that is common to all living creatures. In the words of a crew spokesman, ". . . (the helix) affirms the ceaseless expansion of human life and American involvement in space while simultaneously emphasizing the medical and biological studies to which this flight is dedicated." Leonardo's Vitruvian man, silhouetted against the blue darkness of the heavens, also serves as a powerful embodiment of the extension of human inquiry from the boundaries of Earth to the limitless laboratory of space. Drawn by artist Sean Collins, the STS-40 Space Shuttle patch was designed by the crewmembers for the flight. *(Courtesy of NASA, Lyndon B. Johnson Space Center, Houston, TX.)*

Microgravity inhibits conversion of the committed osteoprogenitor cell to a G_1 preosteoblast. The result is decreased bone formation. The mechanisms for this inhibition in microgravity conditions are under investigation.

The rodents were used for research on muscle, bone, and inner ear function. In one particular study, young rats in rapid growth stage were studied for alterations in bone physiology. Two scientists from Indiana University School of Dentistry, Drs. W.E. Roberts and L. Garetto, in conjunction with Dr. E.M. Holton at NASA Ames Research Center are studying microgravity-induced osteopenia (loss of bone) in the jawbones. Under normal conditions, the conversion of the committed osteoprogenitor cell to a G_1 preosteoblast requires a certain amount of mechanical loading (stress/strain). In space, however, this conversion step is inhibited, and bone formation is decreased (see figure). Drs. Roberts and Garetto are investigating two possible mechanisms that could inhibit this process in microgravity. First, is there a general decrease in mechanical loading, which fails to induce committed osteoprogenitor cells to differentiate into preosteoblast cells? Second, is the change in tissue geometry due to the loss of extracellular water? Differentiation of preosteoblasts is inhibited by excessive cell contact. Data from the SLS-1 mission will provide critical information on how microgravity affects bone physiology.

■ **Would calcium supplementation help space-induced osteopenia?**

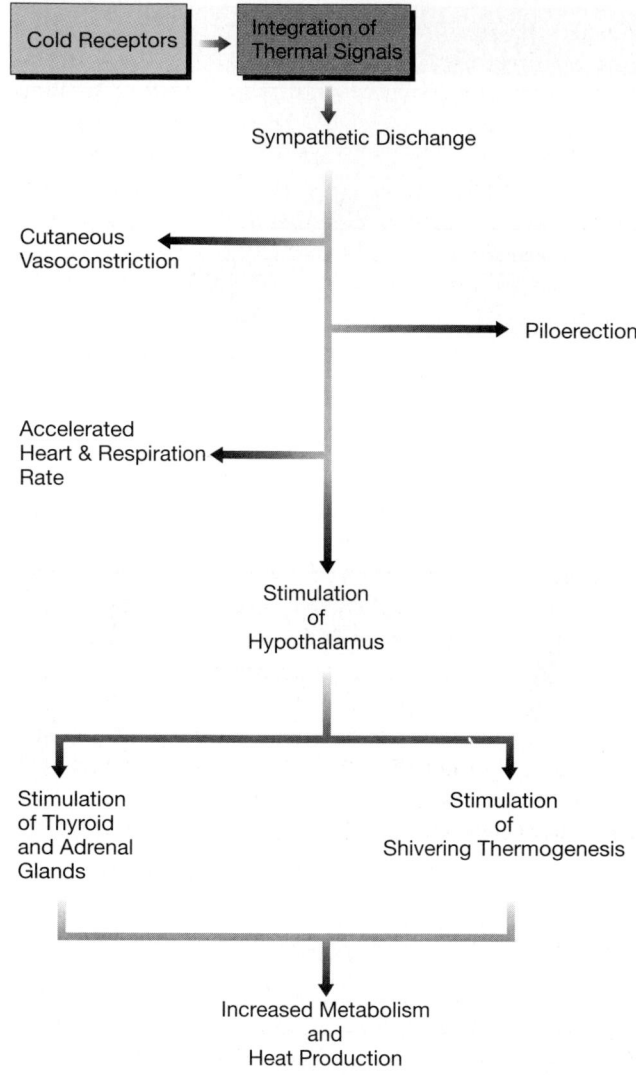

Figure 29–19

Integration of thermal signals in response to cold stress.

When a lightly clad human subject is placed in a cold room at −20°F, a number of physiological events take place very quickly (Fig. 29–19). (1) **Cutaneous vasoconstriction** shifts blood from the body surface to the core and decreases heat conductivity from the interior by lowering heat loss from skin to the environment. (2) **Piloerection** (erection of hair) in the skin, often called "goose bumps" or "goose flesh," increases insulation by trapping air on the skin. In mammals such as dogs and cats, piloerection is very effective in trapping air in fur, but not very effective in humans, who have little body hair. (3) Neurohormonal activation of the hypothalamus triggers the release of hormones from the anterior pituitary, especially those that stimulate the thyroid and adrenal cortex. (4) Increased electrical activity of skeletal muscle results in shivering and often occurs as soon as a

lightly clad subject is exposed to cold. Shivering increases metabolic rate and can lead to three- to fourfold increase in heat production as a result of the energy expended by this type of muscular activity. Because all of this energy goes into producing heat and none into work, shivering thermogenesis is a very economical way of elevating heat production. Interestingly, *hypothermia* cannot be induced in experimental animals unless shivering is blocked, for example, by a light anesthetic. Women have a higher shivering threshold than men; i.e., men shiver before women at the same environmental temperature, because women's extra subcutaneous fat layer insulates against the cold. Similarly, subjects who are heavily built, with extracutaneous fat, do not shiver as quickly as those who are lean. (5) Body position by both humans and animals is changed in order to reduce heat loss. This is accomplished by tucking in arms and legs and curling up in a ball position. (6) Heart rate and ventilation increase. These responses are somewhat paradoxical, because increased ventilation accelerates heat loss from the lungs, and increased heart rate places an added stress on the cardiovascular system because of the massive peripheral vasoconstriction that occurs. This means that there is not only increased blood flow to the internal organs, but also increased systolic and diastolic blood pressures.

Most of these responses to cold exposure are initiated through the sympathetic stimulation, which tends to conserve heat. Cold receptors located on the skin trigger a sympathetic discharge that causes vasoconstriction, piloerection, increased heart/ventilation rate, and increased shivering thermogenesis (Fig. 29–19).

Cold Acclimatization

Acclimatization to cold occurs over a 2 to 3 week period. One of the first criteria of cold acclimatization is the switch from shivering thermogenesis to nonshivering thermogenesis. During continuous exposure to cold, nonshivering thermogenesis is optimal at the end of 2 to 3 weeks. At the end of this period, shivering disappears and the sustained heat production is now maintained solely by the elevated metabolism of visceral organs. Metabolic rate is elevated about 20% above normal during cold exposure due to the level of t_3 and t_4 (Fig. 29–20). If acclimatized individuals are removed from a cold environment and placed into a warm environment, metabolism and O_2 consumption remain elevated for several weeks. The next step in acclimatization is an increase in food intake followed by a rise in metabolic rate. Other features of acclimatization are a rise in skin

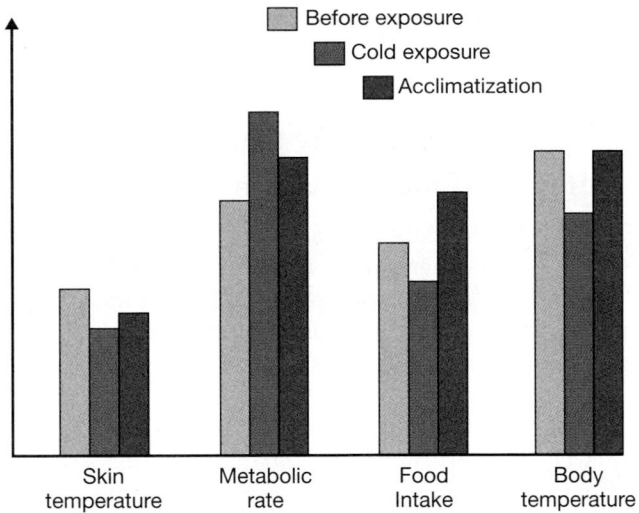

Legend:
☐ Before exposure
■ Cold exposure
■ Acclimatization

Categories: Skin temperature, Metabolic rate, Food Intake, Body temperature

Figure 29–20

Acclimatization to cold.

temperature, warmer hands and feet, and a higher core temperature (Fig. 29–20).

Response to Heat Stress

Temperature extremes have been recorded all over the world. A temperature of 127°F has been reported in Queensland, Australia, and a temperature of 136°F recorded in Libya. Death Valley, California, has similar temperatures. Unfortunately, these high temperatures are also in areas where water is not available for evaporative cooling and for osmotic regulation. In a dramatic account of a man lost in the Mexican desert for 8 days with only 1 day's supply of water, he survived by drinking his own urine, chewing the end of cactus spines for water, and drinking blood from scorpions.

Humans suffer more from heat than from cold exposure, because of our high internal heat production and the limited avenues for unloading heat. Essentially, the only physiological mechanisms for cooling the body are sweating, vasodilation, and panting (see Chapter 28 for mechanisms of heat loss). Heat stress is usually classified as hot-dry or hot-humid. For hot-dry, the temperature is greater than 95°F with a low humidity. For hot-humid, the upper limit for ambient temperature is 90 to 95°F with a relative humidity greater than 75%. Abrupt functional changes occur in the first two days of heat stress. There is an immediate increase in sweating, followed by cutaneous vasodilation. Second, there is a transitory increase in plasma volume as fluid shifts from the interstitial space to the vascular bed. When water lost by sweat is not replaced, the blood volume decreases.

Third, there is an increase in oxygen consumption due to the effect of elevated body temperature on metabolic rate. Fourth, there is an increase in food consumption, even though body weight decreases with heat stress. Fifth, there is dehydration because the thirst mechanism is not sufficient to replenish the body's loss of water. Dehydration progresses with heat stress. Surprisingly, sweating continues at the same rate regardless of the dehydration. At 2% weight loss, dehydration is very severe. At 4% weight loss, the throat and mouth become extremely dry, and at 8% weight loss salivary function stops and speech is slurred. Humans are unable to function if dehydration continues and there is a 10% weight loss. Much of the pioneering work in this area was done in the United States during World War II on conscientious objectors.

Acclimatization to heat takes 7 to 14 days. Specific changes include: (1) increased peripheral thermal (heat) conductance, (2) increased sweating capacity with a fall in salt concentration in the sweat, and (3) a rise in skin temperature, a fall in core temperature, and more even distribution of sweat over the body. Women do not acclimatize to heat as well as men. This is due, in part, to a lower thermal gradient to remove metabolic heat, lower capacity to increase blood flow, and a lower rate of sweating.

Heat Exhaustion

Failure to tolerate heat stress leads to **heat exhaustion,** a condition marked by a subnormal body temperature, dizziness, headache, nausea, and sometimes collapse. Each year several hundred deaths occur in this country from heat exhaustion. Mild heat exhaustion is a common occurrence in hot work environments, and exposure to direct sunlight. Heat exhaustion is prevalent in the military, especially during training or troop mobilization in hot environments, and physical fitness alone does not protect against it. There are many instances of highly trained combat troops who suffered heat exhaustion even though they were in top physical condition. Interestingly, increases in core temperature that result from high external environmental temperatures are not tolerated as well as similar core temperature increases that result from internal causes. For example, heat exhaustion does not occur when core temperatures reach 42°C (normal 37°C) with a fever. During strenuous exercise, rectal temperatures can reach 40°C without harm. Yet, if a rectal temperature reaches 39°C when heated from the external environment, a person can lose consciousness.

The first stage of heat exhaustion, termed *heat shock,* occurs when thermal regulation fails and there is a critical rise in body temperature. Heat shock can also occur through dehydration without a rise in body temperature. One of the early symptoms of heat shock is *heat-induced tetany* (muscle cramps and uncontrolled muscle spasms). This appears to be induced by cellular changes in CO_2 tension and pH rather than a rise in temperature or increased intracellular calcium.

The second stage of heat exhaustion is *heat stroke* (sometimes referred to as sunstroke) and is marked by convulsions, coma, and brain damage. Death occurs from cardiovascular failure. Individuals usually go into circulatory shock with internal massive vasodilation and a concomitant decrease in central blood volume.

Responses to the Chemical Environment

The human body is continually subjected to a wide variety of chemicals in the environment. Some toxic poisons such as lead have been known for 2,000 years. However, with the industrial revolution the release of new pollutants into the environment has been staggering. There are over 10,000 new chemicals produced annually that are "foreign" and not normally found in nature. These and other environmental chemicals are in the air, water, and food that we use.

What are the effects of these environmental chemicals on the body? How does the body handle them? How successfully does the body acclimatize to chemical stress? These questions and other closely related questions will be the focus of this section.

Three important issues with chemical pollutants are mode of entry, removal, and tissue storage. Chemicals gain entry into the body through the lungs, gastrointestinal tract, and skin. Many organic compounds enter through the lining of the lung and gut by either simple diffusion or carrier-mediated transport. Chemicals are removed by excretion in the urine and feces. The liver helps to remove these molecules by transforming them to the less lipid-soluble form so that passive reabsorption by the kidney is prevented, resulting in urinary excretion. The liver is one of the most important organs in the body's defense mechanisms against toxic agents. The liver is capable of detoxifying many chemicals or transforming them so that they can be readily eliminated from the body. However, transformation by the liver is not always beneficial. In some cases, foreign chemicals are nontoxic until liver enzymes transform them. Many of these transformed chemicals are carcinogenic. A classic example is DDT. This pesticide is essentially harmless until the liver enzymes transform DDT into toxic

molecules that cause most of the adverse biological effects of DDT.

Respiratory Pollutants Exhibit Different Modes of Action

Pollutants are classified as respiratory or systemic (Table 29–3). There are a number of pollutants that irritate and damage the lungs. Each exhibits a different mode of action. The gaseous pollutants (e.g., sulfur dioxide, nitrogen dioxide, ozone, chlorine, and ammonia) damage the epithelial and mucous lining of the respiratory tract. Dust particles (e.g., silica, quartz, carbon, asbestos, cobalt, and iron oxides) damage the pulmonary interstitial tissue, and other irritants such as beryllium, nickel, and chromium are carcinogenic and affect many different types of lung cells.

Gaseous Pollutants

Sulphur Oxides. Sulfur oxide is one of the oldest air pollutants. Long before coal was used as a fuel, hydrogen sulfide (H_2S) was added to the atmosphere from volcanoes, swamps (swamp gas), and shallow lakes. However, sulfur oxides became a major pollutant with the advent of the industrial revolution and the burning of coal and oil, both of which contain sulfur as an impurity. Sulfur content of oil and coal is about 3%, with soft coal having a higher content than hard coal. The major sulfur pollutants in the atmosphere are hydrogen sulfide (H_2S), sulfur dioxide (SO_2), sulfuric acid (H_2SO_4) and sulfates (SO_4^{-2}). Hydrogen sulfide readily combines with O_2 to form sulfur dioxide ($H_2S + O_2 \rightarrow SO_2 + H_2O$). Sulfur dioxide can undergo a series of reactions to form sulfuric acid ($SO_2 + O_2 \rightarrow SO_3$ which immediately reacts with water vapor: $SO_3 + H_2O \rightarrow H_2SO_4$). These reactions are extremely fast in the sunlight. Sulfur dioxide is a colorless gas with a pungent irritating ordor that can be detected at 0.5 to 0.8 parts per million (ppm). Sulfur dioxide is very soluble, and more than 95% of the inhaled SO_2 is absorbed in the moist lining of the upper airways and never reaches the small airways or alveoli. SO_2 penetrates the alveoli only when it is absorbed onto the surface of a small dust particle (≤ 5 μm in diameter) or is converted to H_2SO_4. SO_2 at 1 ppm causes irritation of eyes, nose, and throat and causes constriction of the bronchi at higher concentrations. During the 2-day London air pollution crisis in 1952, the average concentration of SO_2 was 1.7 ppm. The most serious of the sulfur oxides is H_2SO_4. In the atmosphere this exists as sulfuric acid mist, and in the presence of small dust particles H_2SO_4 can be absorbed and transported to the alve-

Table 29–3
Respiratory and Systemic Pollutants and Their Principal Effects

Classification	Pollutants	Effects
Respiratory pollutants		
Gaseous irritants	Sulfur dioxide	Respiratory lining
	Nitrogen dioxide	Respiratory lining; bronchitis and emphysema
	Ozone	
	Choline	Pulmonary edema
	Ammonia	
Granuloma producing	Beryllium	Lung cells
	Nickel	
	Chromium	
Respiratory dusts	Silica	Pulmonary interstitial tissue
	Quartz	
	Carbon	
	Asbestos	
	Cobalt	
	Iron oxides	
Systemic pollutants		
	Lead	Nerve tissue
	Mercury	Brain and gut
	Fluoride	Bones and teeth
	Cadmium	Kidneys and blood vessels
	Chlorinated hydrocarbons	Liver and fatty tissues
	Organic phosphates	Neuromuscular junction

oli. H_2SO_4 causes swelling of mucous membranes, increased mucous secretion, narrowing of airway passages, and alveolar thickening.

Nitrogen Oxides. Both ozone and the oxides of nitrogen are oxidants and are capable of attacking biological membranes. The principal industrial source of nitrogen oxides is from the burning of oil or natural gas, and motor vehicle fuel combustion. Over 20 million tons of nitrogen oxides are produced annually in the United States. When nitrogen is exposed to high temperatures (e.g., furnace, combustion engine), nitrogen is converted to nitric oxide (NO; the new nomenclature is nitrogen monoxide). When NO is vented and cooled, a fraction is converted to nitrogen dioxide (NO_2). Further complex reactions in the presence of sunlight result in the formation of ozone (O_3).

Nitrogen dioxide is a reddish-brown gas that is highly toxic and irritating, with a pungent odor. Nitrogen dioxide poisoning was first recognized in farmers who filled silos with silage (chopped green corn). NO_2 builds up shortly after filling, and if a farmer ventured into the top of the silo after filling he could be exposed to lethal concentrations of NO_2.

Similar accidents have been reported by firefighters fighting burning chemical plants, welders, and workers inside of boilers. The symptoms of NO_2 poisoning include headache and a dry hacking cough shortly after exposure, followed by pulmonary edema 1 to 2 days later. With pulmonary edema, these workers experience a shortness of breath, fever, wheezing, and rattling throughout the lungs. The increased burden of oxygenating the blood as well as pushing blood through water-soaked lungs leads to hypoxemia and a life-threatening strain on the right heart. If pulmonary edema is not reversed, these individuals can become comatose and die from severe hypoxemia and right heart failure. One of the most tragic incidents of NO_2 poisoning occurred in a fire at the Cleveland Clinic in 1929. The burning of X-ray films killed 97 persons within 2 hours, and 26 other victims died within 3 weeks.

Nitrogen dioxide can be detected at concentrations of 1 to 2 ppm, which are known to occur in polluted air. Low-level exposure (of 1 to 3 ppm) leads to irritation of the lung lining. A high incidence of bronchitis and emphysema has been reported in cities with high NO_2 levels. Nitrogen dioxide in cities is often part of photochemical **smog** that makes epidemiologic studies difficult. Nitrogen dioxide is also an ingredient of tobacco smoke. Cigarette, pipe, and cigar smoke contain approximately 300, 900, and 1,200 ppm NO_2, respectively. Tobacco smoke also contains other harmful compounds, which again makes it difficult to separate the NO_2 effects from other long-term hazards of smoking. However, smokers working in an environment already polluted by auto exhaust are at high risk for the harmful effects of NO_2.

Ozone.   Ozone is an unstable blue gas with an oxidizing power that is surpassed only by fluorine. The name comes from its pungent odor, which is derived from the Greek "Ozein": the smell that occurs from lightning during a severe thunderstorm. Ozone constitutes over 90% of the oxidants in urban smog.

Ozone, like NO_2, acts on the lung, where it can harm the respiratory lining. Ozone can be detected at 0.05 to 0.1 ppm and at 0.1 ppm can cause irritation of the eyes, nose, and throat. Higher concentrations can cause bronchitis and emphysema. In Los Angeles, a concentration of 0.5 ppm constitutes a "first alert." A second effect of chronic low-level exposure to ozone is that it is involved in premature aging. A third effect is its leading to a higher incidence of lung tumors. Both the premature aging and high incidence of tumors are due to the formation of free radicals from ozone.

Respiratory Dusts

When a small glass mirror covered with Vaseline is placed in the backyard on the picnic table of any urban home, a number of particles of all sizes and shapes accumulate on the glass. These include an abundance of carbon, quartz, and silica particles ranging from 500 μm down to 0.5 μm. The glass can also show rubber particles, and trace amounts of cobalt and iron oxide. These are examples of the kinds of particulate matter in our city parks, shopping centers, and factories that become part of our respired air. Dust particles produce a disease known as *pneumoconiosis*, a condition characterized by fibrosis (scarring of lung tissue). Pneumoconiosis caused by a certain type of particulate is given a specific name. For example, silicosis is a specific type of pneumoconiosis caused by silica.

Inhaled particles in a range from 0.5 to 5.0 μm constitute **respiratory dust.** These are the particles suspended in air and often are seen in the light beam from a projector shining onto a screen. Particles $\geq$15 μm are subject to gravitational settling. With air pollution control, the large particles (>15 μm) have decreased. Removal of these large particles has caused a shift in the particle size distribution in the atmosphere and has actually increased the number of respiratory dust particles (0.5 to 5.0 μm).

Particles from 5 to 10 μm are stopped by the hairs in the nose and moist membranes in the nasopharyngeal passage. Particles ranging in diameter from 2 to 5 μm are caught by a sheet of mucous lining the tracheobronchial tree (Fig. 29–21). These particles are removed by this mucociliary transport system and are either coughed up or swallowed. Particles $\leq$2 μm are capable of reaching the alveolar/capillary bed. Alveoli have neither mucus nor cilia, and can only remove these small particles by *alveolar macrophages.* These are phagocytic cells that creep along the floor of the alveoli and scavenge foreign particulate in-

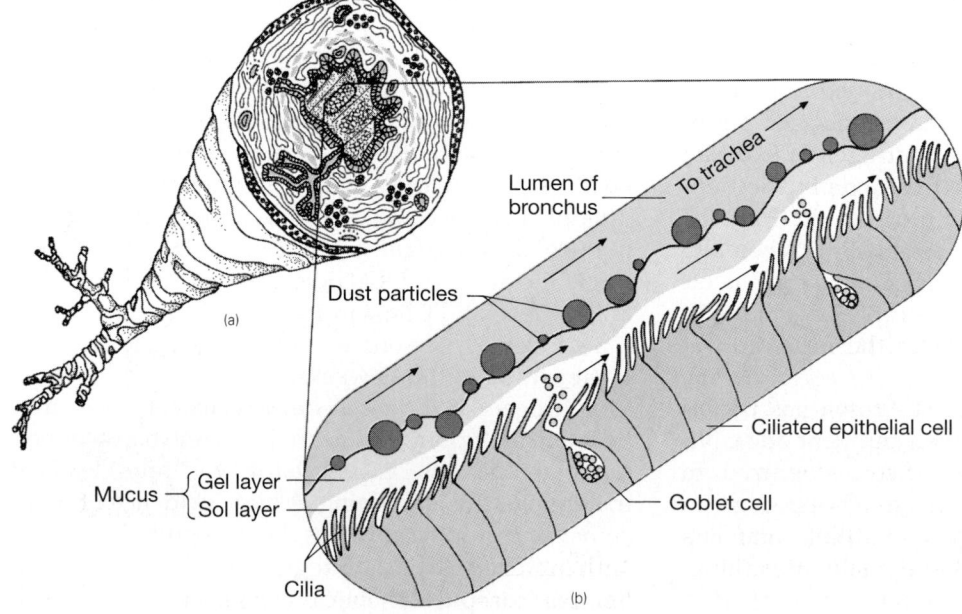

Lumen of bronchus

To trachea

Dust particles

Ciliated epithelial cell

Goblet cell

Mucus { Gel layer / Sol layer

Cilia

(a)

(b)

Figure 29–21

The mucociliary transport system acts as a biological scrubber to trap and remove inhaled dust particles and chemical irritants.

cluding bacteria. The foreign matter is ingested and then destroyed by proteolytic enzymes. Because of their mobility, macrophages move up the respiratory tract and are also removed by the mucociliary transport system.

Many of the particles that penetrate the alveolar/capillary bed are not removed and eventually get into the lung interstitial tissue, causing injury.

Silica and Other Dusts. The disease caused by silicon dioxide (usually called silica) is *silicosis*. At one time silica was thought to be responsible for the specific disease coal miners get from breathing coal dust ("black lung disease"). Coal miners' pneumoconiosis is now recognized as a distinct disease due to coal dust alone and not to silica.

Asbestos is present in nature, and, if handled improperly, becomes a respiratory pollutant. Inhalation of asbestos dust has long been recognized as the cause of *asbestosis*. Asbestos is a general name given to a variety of minerals that can unravel into fibers and are found in nearly every country in the world. The most important deposits in the United States are California, Arizona, and Vermont. The value of asbestos comes from its indestructibility and heat-resistance properties. Before the ban on asbestos, it was manufactured into more than 3,000 different products. One microgram of asbestos suspended in air can unravel into more than 200 million tiny fibers and, because of the geometry, asbestos fibers 200 μm long and 3.5 μm in diameter can penetrate the lung. Asbestos is marked by an excess of collagen synthesis which leads to fibrosis.

Iron oxide is another particulate that will cause pneumoconiosis. At autopsy, lungs show a brick-red discoloration. Usually atmospheric iron binds with other particulate, namely silica, to cause lung injury. Iron oxide has also been shown to be carcinogenic.

Cobalt is another inhaled metal that causes pneumoconiosis. Cobalt exists in nature, and many species of bacteria, protozoa, fungi, and insects require cobalt to sustain life (it is an essential part of vitamin B_{12}). Only during the 20th century has cobalt become an important air pollutant, particularly since World War II.

Pneumoconiosis, regardless of its etiology (dust from silica, coal, asbestos, etc.) has similar pathophysiology. The first stage leads to scarring of lung tissue (fibrosis), increased lung stiffness, shortness of breath, and a decline in pulmonary function (difficulty inflating the lungs). The latter stage leads to hypoxemia. The hypoxemia triggers vasoconstriction of the pulmonary arterial vessels, causing pulmonary hypertension and eventually right heart failure.

Some Pollutants Act as Systemic Poisons

In contrast to pollutants that mainly involve the lung, there is another class of pollutants that affect numerous organs and are known as systemic pollutants (Table 29–3). These include such compounds as lead, mercury, fluoride, and cadmium. They are emitted from both moving and stationary sources and constitute a great threat to our environment because of their toxicity and their ability to accumulate in the body.

Lead. Lead is a heavy soft metal that is used in a number of utensils, paints, and glazes in pottery. The toxic properties have been known for more than 2,000 years. The Greeks knew about lead poisoning, but the Romans apparently did not. Major sources of lead in the environment and its path into the body are shown in Figure 29–22.

The sources of entry of lead into the body are the lungs and digestive tract. Only 5% to 10% of the ingested lead is actually absorbed, with the remaining amount excreted in the feces. However, over 40% of the lead inhaled by the lungs enters the blood and is deposited in tissue. Lead has four sites of action (Fig. 29–23): blood, gastrointestinal tract and, in the advanced stages, the nervous system (brain and spinal cord) and kidneys. Early symptoms are anemia, headaches, and muscle pain. Persons affected also be-

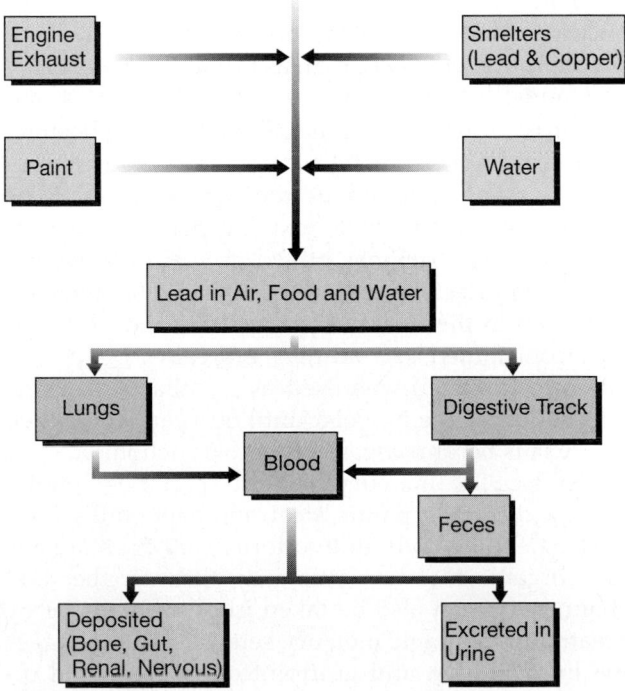

Figure 29–22

Sources of lead and its path into the body.

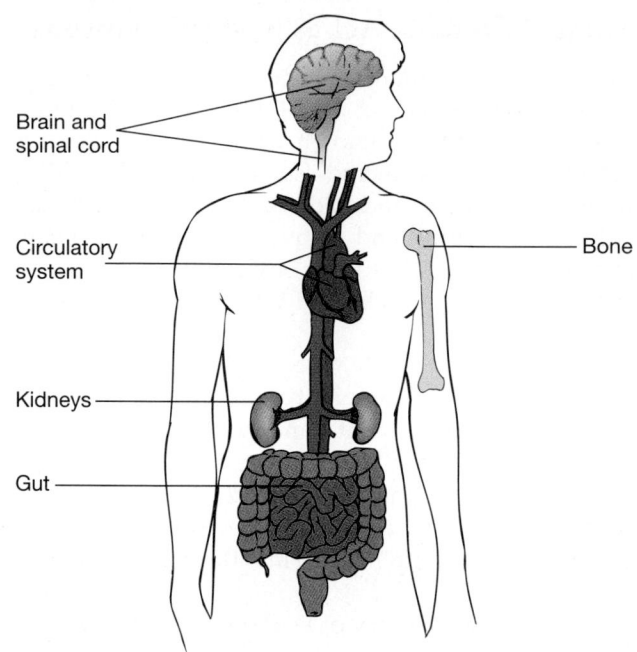

Figure 29–23

Principal sites of action of ingested and inhaled lead.

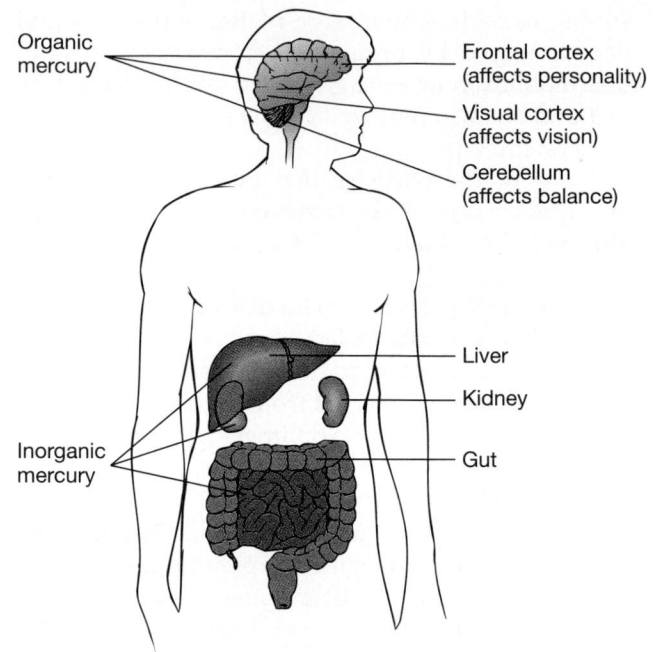

Figure 29–24

Sites of action of mercury.

come clumsy and irritable. In the later stages, symptoms include vomiting, abdominal pain, and black stools from bleeding of the colon. Advanced stages include brain damage, convulsions, and renal failure.

Lead contamination in the environment has been dramatically reduced by removing it from gasoline and paints. However, lead poisoning is still a major problem in metropolitan areas around the world.

Mercury.   Mercury, or quicksilver, has long been regarded as a special metal because of its unique property of being a liquid at room temperature. The Greeks were aware of its toxic properties, and mercury ore (cinnabar) was mined in Spain for twenty-seven centuries. As early as the year 1500, mercury was used in the treatment of syphilis and undoubtedly contributed to a number of deaths. Mercurous chloride (calomel) was used as a cathartic (causing evacuation of the bowels) until 50 years ago, when physicians became aware of its toxic action.

Sources of mercury include batteries, photographic materials, paints, steel, and paper mills. Mercury enters the body in two forms, either as organic or inorganic mercury, primarily through the food chain, but it can also be taken in by drinking or by breathing. Inorganic mercury settles in and damages the liver, kidney, and gastrointestinal tract, and organic mercury settles in the brain and destroys neuronal cells that control coordination, vision, and personality (Fig. 29–24).

Cadmium.   Cadmium is a soft, silver-white metal that resists corrosion and is used in batteries, electrical wires, engine parts, and radio/tv parts, as well as in nuclear reactors to control the rate of nuclear fission. In the early 1900s cadmium was used in the treatment of syphilis and malaria, until its toxic properties became recognized.

Cadmium enters the body primarily through food and water, and is absorbed through the intestinal wall from the gut lumen. Cadmium also enters the lungs from polluted air and from smoking cigarettes. Cadmium is deposited in a variety of tissues, but is concentrated in three organs: kidney, liver, and lungs (Fig. 29–25). Long-term cadmium poisoning leads to pneumonia, pulmonary emphysema, and cirrhosis of the liver. Cadmium is also linked to hypertension (high blood pressure). Humans with high concentration of cadmium in the kidney have a high incidence of arterial hypertension and excrete 40 times more cadmium in their urine than individuals with normal blood pressures.

The cellular action of cadmium has been clearly pinpointed to the membrane of the mitochondria. Recall from Chapter 6 that two major energy-producing pathways exist in the mitochondria: the citric acid cycle and the electron transport (respiratory) chain. The citric acid cycle generates nicotinamide and flavin, which contain electrons and hydrogen atoms. The electron transport chain takes these electrons and hydrogen atoms and converts them to usable ATP. Cad-

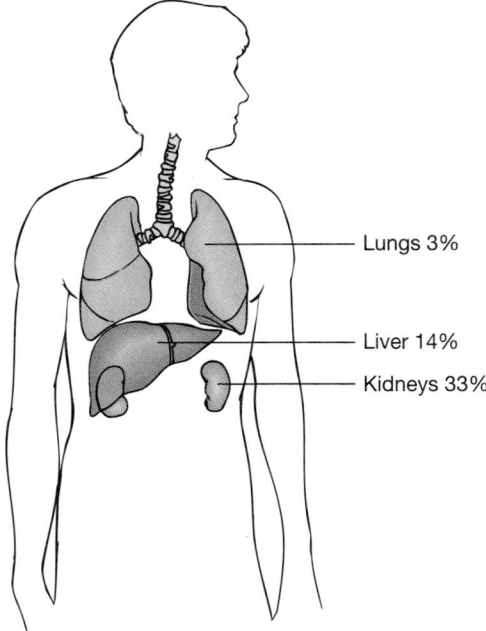

Figure 29–25

Major sites of cadmium deposition in the body. Almost half of the cadmium in the body is stored in kidney and liver.

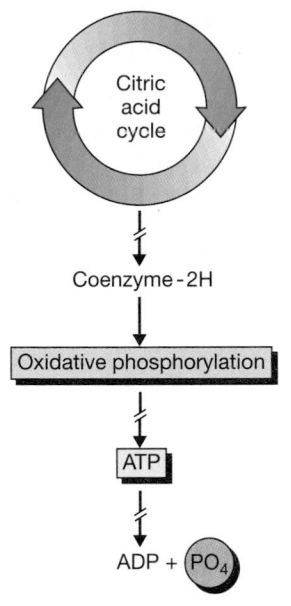

Figure 29–26

Sites of action of cadmium on cellular ATP. Cadmium (1) blocks transfer of electrons from citric acid cycle, (2) blocks ATP synthesis, and (3) inhibits ATPase (i.e., the breakdown of ATP to ADP).

mium acts on mitochondrial enzymes and blocks ATP synthesis (Fig. 29–26). Cadmium also binds to the enzyme adenosine triphosphatase (ATPase), which is required to split a phosphate from ATP. Inhibition of ATPase prevents the body from utilizing ATP as a source of cellular energy.

Is Environmental Toxicology a Modern Issue?

Systemic poisons vividly illustrate how difficult it is to answer the central question of modern environmental toxicology: what are the effects of long-term exposure to small doses of a chemical pollutant? It is relatively easy to identify symptoms of acute poisoning (e.g., lead), but extremely difficult to detect chronic low-level effects. For most environmental pollutants we don't know whether every potential toxic agent has a threshold below which there is no damage.

CHAPTER REVIEW

Summary

- Environmental physiology is the study of physiological mechanisms of healthy individuals that occur in response to their external environment.
- Environmental stress can lead to: (a) adaptation is a genetically induced change that is passed on to the next generation, and (b) acclimatization, which is environmentally induced and leads to improvement of an existing physiological mechanism, but with no change in the genome.
- Biological clocks lead to intrinsic rhythms that vary from minutes to days.
- Circadian rhythms follow a 24-hour cycle and are often linked to a light-dark cycle.
- High altitude causes hypoxia, and rapid ascent to high altitude can lead to altitude sickness.

- Acclimatization to altitude allows the body to adjust to hypoxia and includes improved regulation of pH as well as increased cardiac output, ventilation, and hematocrit.
- Underwater divers have improved diving reflexes, improved respiration, increased metabolism, and increased body insulation.
- Deep-sea diving leads to an increase in gas pressures and causes nitrogen to be absorbed by blood and tissues. Rapid ascent leads to "the bends."
- Breathing pure oxygen leads to pulmonary edema due to damage of pulmonary capillaries. Cellular damage is due to the formation of free radicals which oxidize membrane lipids, proteins, and enzymes. High oxygen also causes blindness in the newborn (retrolental fibrosis); oxygen under high pressure causes convulsion.

- Space travel leads to unusual stresses. Astronauts are subjected to G forces, artificial environments, and weightlessness.
- Temperature extremes lead to a number of physiologic changes. Response to cold stress leads to improved neuroendocrine function, cardiovascular changes, and heat production. Humans have lower options in response to heat stress, which often leads to heat exhaustion.
- The lung and liver are the most important organs in the body's defense mechanism against environmental pollutants.

- Gaseous pollutants (e.g., nitrogen dioxide, sulfur dioxide, ozone) damage epithelial and mucous linings of the respiratory tract, whereas dust particles (e.g., silica, carbon, quartz) damage the interstitium to cause fibrosis and cancer.
- Other pollutants such as lead, mercury, and cadmium are classified as systemic pollutants because they affect numerous organs.

Review Questions

Identify Each with a Term

1. A genetically induced change in response to environmental stress.
2. A biological rhythm with a periodicity of about 24 hours.
3. The change in ventilation that occurs at high altitude.
4. Loss of vision during high G force.
5. The compounds formed from breathing high concentrations of oxygen, ozone, or nitrogen dioxide that cause oxidant damage to biological tissues.

Define Each Term

6. acclimatization
7. biological clocks
8. heat stroke
9. pneumoconioses
10. G force

Choose the Correct Answer

11. Which of the following will lower body temperature with cold stress?
    a. curling up
    b. increased food intake
    c. increased circulating epinephrine
    d. increased peripheral dilation
12. Compared to an unacclimatized person, one who is acclimatized has:
    a. a higher metabolic rate in the cold to produce more heat.
    b. a lower peripheral blood flow in the cold to retain the heat.
    c. higher blood flow in the hands and feet in the cold to preserve their function.
    d. increased food intake to produce more heat.
    e. all of the above.
13. Dust particles that are inhaled have a size range of:
    a. 15 to 20 μm
    b. 10 to 15 μm
    c. 5 to 10 μm
    d. 1 to 5 μm
14. The main detoxifying organ in the body is the:
    a. spleen.
    b. kidney.
    c. lung.
    d. liver.

15. For every 33 feet below sea level, barometric pressure will increase by:
    a. 1/4 atmosphere
    b. 1/3 atmosphere
    c. 1 atmosphere
    d. 2 atmospheres
16. The stimulus for hyperventilation at altitude is:
    a. low blood pH
    b. low blood oxygen tension
    c. low blood bicarbonate
    d. low blood [H$^+$]
17. Bends associated with deep-sea diving are due to:
    a. excess oxygen dissolved in brain tissue.
    b. excess carbon dioxide dissolved in brain tissue.
    c. excess gas in gut tissue.
    d. excess nitrogen dissolved in fat tissue.
18. Which of the following *does not* occur upon acute exposure to cold?
    a. shivering thermogenesis
    b. vasoconstriction to skeletal muscles
    c. sympathetic stimulation to heart
    d. release of epinephrine
19. As one ascends to high altitude, which of the following decrease(s)?
    a. partial pressure of O$_2$
    b. barometric pressure
    c. partial pressure of N$_2$
    d. a and b
    e. all of the above
20. During acclimatization to altitude, which of the following will increase?
    a. vital capacity
    b. minute ventilation
    c. cardiac output
    d. b and c
    e. all of the above

Calculate

21. What is the partial pressure of oxygen on top of Mt. Everest (barometric pressure = 231 mm Hg)?
22. Estimate how much oxygen is delivered to the tissues for a cardiac output of 5 liters in a high mountain dweller (15,000 feet) if arterial oxygen tension is 40 and venous oxygen tension is 20 mm Hg.
23. A diver goes 99 ft underwater. What is the barometric pressure at this depth?

24. Calculate the total G force of a pilot who pulls a negative G force of 3.
25. What is the partial pressure of O_2 in a space cabin filled with 50% O_2 at 1/2 atm?

Answer Each Question in One or Two Sentences

26. What are the effects on respiration of acute exposure to high altitude?
27. Describe what happens to gases in the body during an underwater dive.
28. List some of the effects of breathing high oxygen at sea level.

29. List some of the effects of high G forces on the body.
30. Briefly describe the stages of heat exhaustion.

Problem Solving in Physiology

31. The temperature is $-30°F$ with a wind chill factor of $-60°F$. You come upon a car that is stranded and find that the driver is suffering from hypothermia and is almost unconscious. How would you treat this person?
32. Why is a statistical observation used to determine long-term effects of a low-level pollutant rather than a quantitative effect?

Suggested Readings

Edmunds, L.N. *Cellular and Molecular Bases of Biological Clocks: Models and Mechanisms for Circadian Timekeeping.* Berlin: Springer-Verlag, 1988.

Fenn, W.O., and Rahn, H., eds. *Handbook of Physiology: Respiration (Vol II).* Bethesda, MD: American Physiological Society, 1965, sect 3.

Gantenbein, D. "The high life." *Discover,* October 1993, pp. 115–122.

Gelfond, J.A., Dinarello, C.A., Sheldon, M.W. "Fever." In Iseelbacher, Braunwald, Wilson, Martin, Fauci, Kasper, eds. *Harrison's Principles of Internal Medicine,* 13th ed. New York: McGraw-Hill, 1994, pp. 81–90.

Gordon, C.J., and Heath, J.E. "Integration and central processing in temperature regulation." *Annu. Rev. Physiol.,* 48:595–612, 1986.

Hong, S.K., and Rahn, H. "The diving women of Korea and Japan." In *Human Physiology and the Environment in Health and Disease.* New York: Scientific American/W.H. Freeman & Co., 1976, pp. 92–102.

Hook, G.E., ed. *Environmental Health Perspectives.* Volumes 55 and 56, 1984.

Hubbard, R.W., Gaffin, S.L., and Squire, D. "Heat-related illnesses." In Auerbach, P.S., ed. *Management of Wilderness and Environmental Emergencies,* 3rd ed. St. Louis; Mosby-Year Book, 1995, Chap. 8, 1994.

Lambertson, C.J. *Underwater Physiology.* Baltimore: Williams & Wilkins, 1967.

Mangum, C.P., and Towle, D.W. "Physiological adaptation to unstable environments." *Am. Sci.,* 65:67–75, 1977.

Marrow, P.E. "Animals in toxic environments: Mammals in polluted air." In Dill, D.B., Adolph, E.F., and Wilbur, C.G., eds. *Handbook of Physiology: Adaptation to the Environment.* Bethesda, MD: American Physiological Society, 1964, pp. 773–794.

Pandolf, K.B., Sawka, M.N., and Gonzalez, R.R., eds. *Human Performance Physiology and Environmental Medicine at Terrestrial Extremes.* Indianapolis, Benchmark, 1988.

Rowell, L.B. "Cardiovascular aspects of human thermoregulation." *Circ Res,* 52:367–379, 1983.

Key Terms

acclimatization (p. 854)
adaptation (p. 854)
asbestosis (p. 879)
atelectasis (p. 867)
biological rhythms (p. 855)
blackout (p. 870)

carboxyhemoglobin (p. 869)
circadian rhythms (p. 852)
diving reflex (p. 863)
free radicals (p. 867)
G force (p. 869)

heat stroke (p. 876)
hypoxia (p. 854)
hypothermia (p. 874)
oxygen radicals (p. 867)
oxygen toxicity (p. 868)
pneumoconioses (p. 878)

polycythemia (p. 860)
SCUBA (p. 865)
silicosis (p. 879)
weightlessness (p. 871)

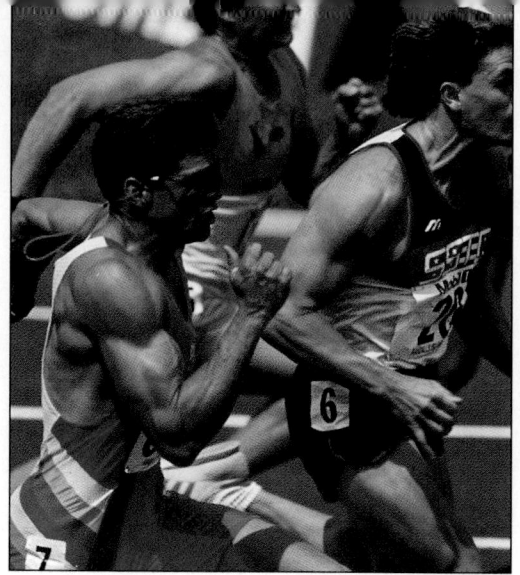

Muscle in action: Dave Johnson at the Olympics. *(© David Madison, 1991)*

Chapter 30

Physiology of Exercise

Key Concepts

The primary energy source for all types of muscle contraction is high-energy phosphate compounds. The increase in pulmonary ventilation during exercise is proportional to the increased oxygen demands. Maximum oxygen uptake is used to quantitate aerobic exercise. Ventilation in exercise matches metabolic demands.

Blood flow is preperentially directed to exercising tissue for O_2 delivery, CO_2 removal, and removal of other waste products. This requires a remarkable coordination among cardiac output, muscle blood flow, oxygen consumption, and work output.

Chapter Outline

Metabolic Aspects of Exercise
Respiration During Exercise
Circulation During Exercise

Ray is visiting New York City as a tourist during spring break. Late one afternoon he decides to visit the Empire State Building. As he is leaving the building at dusk, he sees a male in his late forties snatch the purse of an elderly woman and start to run down Fifth Avenue. Ray assesses the situation and instinctively decides to run after the robber. He suddenly realizes that he is breathing heavily, he is sweating, his heart is beating rapidly, and his legs are becoming fatigued. Ray, however, seeing that he is gaining ground on the perpetrator, decides to continue the chase. Shortly thereafter, he catches the criminal just as two New York City police approach the area. They apprehend the robber and Ray returns the purse to the elderly woman.

Sudden intensive exercise such as running requires activation and coordination of numerous body systems and physiological functions. In this chapter, we will examine in detail the supporting role of the central nervous system, which is necessary for coordinated skeletal muscle movement. The respiratory system supplies increased oxygen and removes CO_2 from the body. The cardiovascular system increases its activity, resulting in increased blood supply to the working muscle to ensure the appropriate supply of oxygen and nutrients and the removal of metabolic waste products. Finally, exercising muscles increase metabolic heat production, which is compensated for by activation of the temperature-regulating system.

In all types of exercise, the final common **physiological event** is the **contraction** of skeletal muscle. But exercise also places demands on other body systems that play a supportive role. The basic biochemical alterations at the muscle cell level that convert chemical energy into mechanical energy are accompanied by increases in the integrated activity of many organ systems, including the **central nervous system** (CNS), which initiates coordinated skeletal muscle activity. The **respiratory system** and **circulatory system** increase their activity during exercise to provide increased oxygen and nutrients to active muscle cells and to remove CO_2, other metabolites, and heat. Contracting muscle cells may increase heat production to 20 times that of resting levels; thus, exercise can also place extra demands on the **temperature-regulating system.** The extent of the physiological adjustments in organ systems in response to exercise is often a direct measure of the **strain** or **stress** of exercise on the body. Exercise induces a disturbance in homeostasis that activates multiple physiological responses designed to restore homeostasis.

Metabolic Aspects of Exercise

The basic concepts of exercise physiology, which this chapter will discuss, have been derived mostly from studies on young adult male subjects, because relatively complete measurements of the physiology of exercise have traditionally been made only on this group of athletes. However, recent studies in which measurements have been made on females indicate that the same basic principles apply. The cellular mechanisms controlling most physiological and biochemical responses to exercise are the same in males and females. However, some **quantitative differences** are found, such as differences in body size, body composition, and sex hormones. In general, most quantitative values for women related to muscle mass—such as ventilation, cardiac output, and muscle strength—will vary from 65% to 75% of the values recorded in males. Differences in athletic performance between the sexes, on the other hand, are smaller because the female body is on average smaller, requiring smaller muscles to perform the same activities. The differences reported to date are differences in magnitude rather than in mechanism.

Exercise is often classified as either **dynamic exercise** or **static exercise** (Fig. 30–1). Dynamic exercise is performed anytime muscle length changes. In a physical sense, resistances are overcome along a certain distance. In this type of exercise, which includes such activities as bicycling, running, stair climbing, or mountain climbing, exercise is quantified in physical units such as foot-pounds or watts. In **positive dynamic exercise,** such as ascending a mountain, the muscles act as a "motor." In **negative dynamic exercise,** such as descending a mountain, muscles act as "brakes." Static exercise is performed during **isometric muscle contraction.** During isometric exercise, shortening of the muscle does not occur and no distance is traversed, so it cannot be defined as work in the physical sense. Nevertheless, static exercise is associated with increased metabolic demands. The **elevated metabolic demand** is the common feature of all types of exercise, and, in most instances, this factor is more important for physiological adjustments than any other factor associated with exercise.

Energy Requirements of Exercise Are Provided by High-Energy Phosphate Compounds

The primary energy source for skeletal muscle contraction, as for other types of cellular work, is high-energy phosphate compounds such as **adenosine triphosphate (ATP).** These are produced by both anaerobic and aerobic metabolic processes (Fig. 30–2). Three important metabolic systems that supply energy for muscle contraction are (1) the phosphagen system, (2) the glycogen–lactic acid system, and (3) the aerobic system **(the citric acid cycle).**

The Phosphagen Energy System

The term *phosphagen* is another name for a high-energy phosphate molecule, one that serves as a reservoir of phosphate bond energy. The **phosphagen energy system** of muscle includes ATP and **phosphocreatine.** When one phosphate group is removed from ATP (producing ADP), approximately 11,000 calories of energy become available for the contractile process. When a second phosphate is removed, leaving AMP, an additional 11,000 calories become available. The energy released during the conversion of ATP to ADP and AMP is the immediate source of energy that can be used by the muscle cell to perform its work. The amount of energy present as ATP in the muscle is enough to maintain maximal exercise, however, for only 5 to 6 seconds.

Phosphocreatine has slightly more energy than the bond energy of ATP and can reconstitute ATP by donating phosphate groups to AMP and ADP. Muscle cells have two to three times more phosphocreatine than ATP, and together the phosphagens ATP and phosphocreatine can maintain maximal muscle contraction for only 10 to 15 seconds. Thus, the energy supply from the phosphagen system is used for *very* short bouts of maximal exercise. The phosphagen system is the most rapidly available source of ATP for

(a)

(b)

(c)

Figure 30–1

(a) Climbing a mountain—positive dynamic exercise (Mt. Everest). *(© John Roskelly/Photo Researchers, Inc.)* (b) Descending a mountain—negative dynamic exercise (Alps, France). *(© Didier Givois/Photo Researchers, Inc.)* (c) Static or isometric exercise. *(© David Weinstein & Associates/Custom Medical Stock Photo.)*

use by the muscles to perform work. The phosphagen system does not depend on transporting oxygen to the working muscle. It also does not depend on a long series of chemical reactions and is thus efficient and readily available. Finally, both ATP and phosphocreatine are stored directly within the contractile elements of the muscles.

The Glycogen–Lactic Acid System

The glycogen–lactic acid system in muscle can also supply energy for contraction. The stored muscle glycogen can be split into glucose (**glycogenolysis**) and then used for energy. Recall from Chapter 6 that glycolysis occurs in the absence of oxygen and is

Figure 30–2

Summary of the three primary metabolic pathways that supply energy for muscle contraction.

RESEARCH FOCUS

Exercise and Diabetes Mellitus

Exercise was a main form of therapy for diabetes mellitus before insulin was discovered. Today, proper management of the diabetic patient should include regulation of diet, medications, and exercise. As described in detail in Chapter 15, the major hormones secreted by the pancreas are insulin and glucagon. Insulin lowers the blood glucose level by increasing the rate of glucose transport through the membranes of most cells of the body, including muscle cells. A lack of insulin results in diabetes mellitus, which is a clinical condition characterized by impaired glucose transport and metabolism. Insulin also inhibits glucose release from the liver and fatty acid release from adipose tissue. Glycogen, on the other hand, causes the opposite effects, for example, glucose and fatty acid mobilization.

Numerous recent studies have shown that during periods of exercise a diabetic patient often decreases his or her insulin requirement. Conversely it is a clinical fact that when a diabetic patient is inactive the insulin requirement increases. A change in insulin requirement with exercise does not necessarily indicate a change in the severity of the diabetes but may only represent a temporary alteration in glucose metabolism.

Moderate exercise in and of itself has been shown to be associated with a decrease in the circulating levels of plasma insulin. Simultaneously, exercise results in increased levels of circulating glucagon in the plasma. Studies in diabetic patients have shown that the decreased insulin levels during exercise do not mean that glucose uptake by the muscle cells is reduced. In fact, glucose uptake in diabetics during exercise is significantly increased. Recent studies have shown that a small amount of insulin is adequate and necessary to promote the exercise-related increase in glucose uptake. This exercise-related increased sensitivity to the effects of insulin have been shown to be present for at least 48 hours after one hour of moderate exercise. Much more research is needed to determine the physiological mechanisms responsible for the increased sensitivity to insulin during exercise. The fact remains, however, that proper management of the diabetic patient must include proper regulation of diet, medication, and exercise.

■ **Based on your knowledge of glucose transport into cells, how might insulin promote the exercise-related increase in glucose uptake by muscle cells?**

therefore **anaerobic.** During glycolysis, each glucose molecule is converted into two molecules of pyruvic acid, and energy is released to form two molecules of ATP. If oxygen is available, pyruvic acid enters the citric acid cycle, and much more ATP is formed. If there is insufficient oxygen to support the oxidative stage of glucose metabolism, however, the pyruvic acid is converted to lactic acid. The glycogen–lactic acid system can produce ATP molecules more rapidly than the oxidative mechanisms in the mitochondria but less rapidly than the phosphagen system. The glycogen–lactic acid system, used as a *relatively* rapid source of energy for muscle contraction, can provide the energy for maximal exercise for only 30 to 40 seconds.

The Aerobic Energy System

Recall from Chapter 6 that the aerobic energy system involves the use of oxygen within the mitochondria for the metabolism of glucose, lipids, and proteins in the citric acid cycle and electron-transport chain to produce 36 ATP molecules for every molecule of glucose. Only 20% of the energy released, however, is used by the muscles as energy of contraction; 80% is released as heat energy. The heat released during exercise places additional demands on the cardiovascular system and the sweat response, which play major roles in temperature regulation and the transfer of heat from the body to the environment (see Chapter 27).

Aerobic metabolism is normally the source of energy for prolonged light work. During light exercise, energy is obtained anaerobically only during the first few minutes when muscle blood flow is gradually increasing (Fig. 30–3). Thereafter, metabolism is entirely aerobic, utilizing glucose, fatty acids, and amino acids. During heavy exercise, by contrast, part of the energy is always obtained anaerobically, and if large amounts of lactic acid are produced, muscle fatigue and cramps result.

Oxygen Consumption During Exercise Corresponds to the Intensity of Exercise

During exercise, oxygen consumption increases by an amount corresponding to the intensity of the exercise. As illustrated in Figure 30–4, oxygen consumption increases during the first few minutes of exercise

Figure 30–3

Relative contribution of various energy sources to the total energy used by the muscle at the beginning of light exercise.

and then reaches a "steady state" in which the oxygen uptake corresponds to the demands of the tissues. The average person in the basal state consumes about 0.25 liter (L) of oxygen per minute. An untrained person can increase O_2 consumption threefold (0.75 liter/min) during light work and eightfold to tenfold (2 to 3 liters/min) during heavy exercise. Well-trained athletes can increase their O_2 consumption to over 20 times (5 to 6 liters/min) above basal level. Oxygen consumption is often standardized in units of body weight (O_2 ml/kg/min), and averages 44 to 51 ml/kg/min for untrained males. Values between 50 and 60 ml/kg/min are often obtained in young, well-trained males. Untrained females normally have O_2 consumption during exercise that is about 20% less than that of untrained males.

Oxygen consumption increases linearly with increasing levels of exercise up to a point at which the maximum for oxygen transport and uptake in the tissues appears to be reached, and a plateau, or the so-called **maximal oxygen consumption (V_{O_2}max),** is attained (Fig. 30–5). The V_{O_2}max is generally assumed to represent a person's **maximal aerobic power** and thus is an important factor in determining one's ability to sustain high-intensity exercise. Three main criteria have been established to certify that a measurement of V_{O_2}max actually represents a person's maximal aerobic power: (1) there is no further increase in oxygen consumption despite further increases in the rate of exercise, (2) the blood lactate concentration exceeds 8 mM, and (3) large muscle groups are exercised for at least 3 minutes. Figure 30–6 illustrates that the V_{O_2}max peak obtained during maximal exercise is maximal only when a rela-

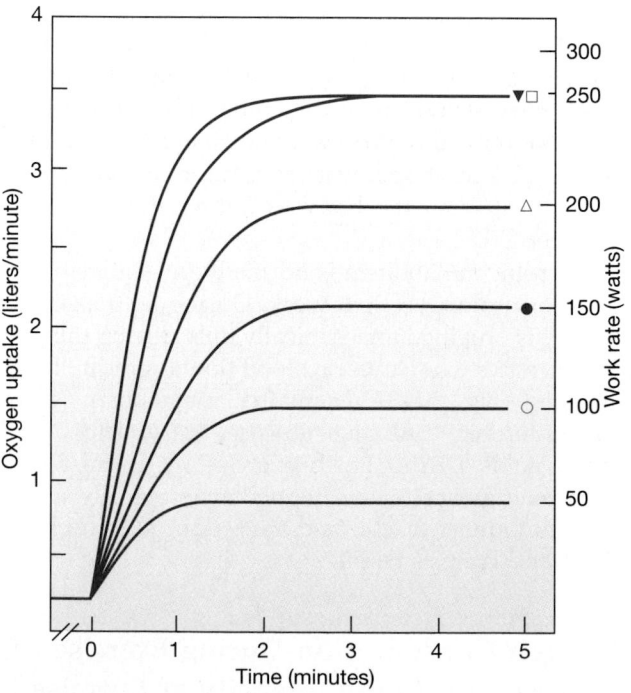

Figure 30–4

Oxygen uptake during increasing levels of steady-state exercise.

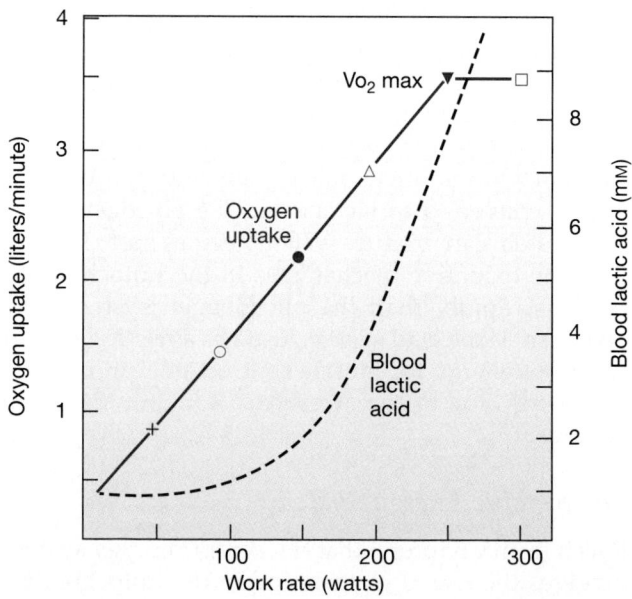

Figure 30–5

Oxygen uptake and blood lactic acid during increasing levels of exercise.

Figure 30-6

Peak oxygen consumption during maximal exercise using different sizes of muscle mass.

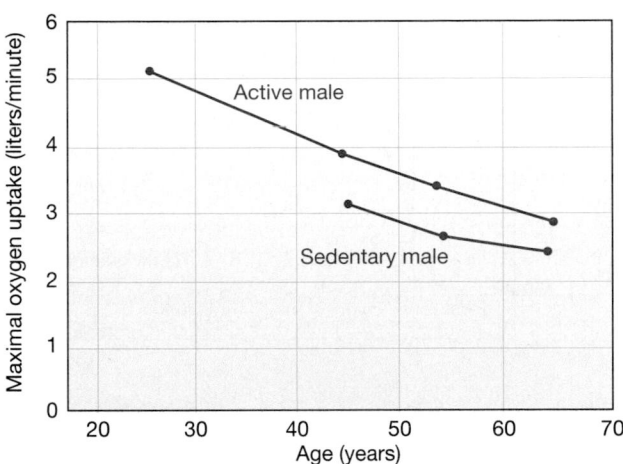

Figure 30-8

Maximal oxygen consumption as a function of age and level of activity.

tively large muscle mass (e.g., two legs) is being exercised.

Numerous factors can affect an individual's V_{O_2}max. For example, with age, V_{O_2}max decreases in both males and females. As shown in Figure 30-7, maximal uptake in liters per minute increases during the growth stages of the early years and reaches a peak between 18 and 25 years of age. This early increase in oxygen consumption is directly related to the increase in body mass that occurs at this age. Thus, the apparent increase in aerobic capacity is essentially eliminated when the V_{O_2}max is expressed relative to body weight as ml/kg/min. After age 25, the V_{O_2}max in liters per minute, as well as relative to

body weight, steadily declines so that at age 55 it is about 72% below values obtained for 20-year-olds. While active adults tend to retain relatively higher V_{O_2}max than inactive adults, a progressive decline in this physiological capacity also occurs in this group with advancing age (Fig. 30-8).

Training increases V_{O_2}max and can partially counteract the effects of aging (Fig. 30-9). For example, the mean V_{O_2}max for nine males on an initial exposure to running 7 mph (9% grade) on a treadmill averaged 53 ml/kg/min. After 2 months of training by running on a treadmill at a rate of 8 mph (zero incline) for 10 to 15 minutes, the V_{O_2}max of the nine males was increased to 56 ml/kg/min. The physiological factors that may limit human O_2 consumption include (1) the rate of cardiovascular transport to ac-

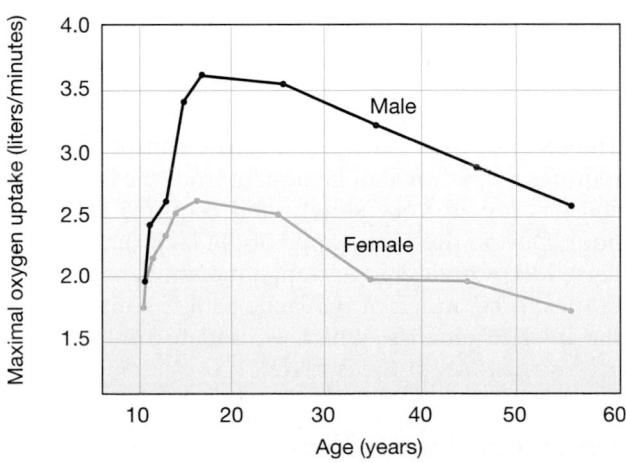

Figure 30-7

Maximal oxygen consumption as a function of age in males and females.

Figure 30-9

Change in maximal oxygen consumption during 7 to 13 weeks of athletic training.

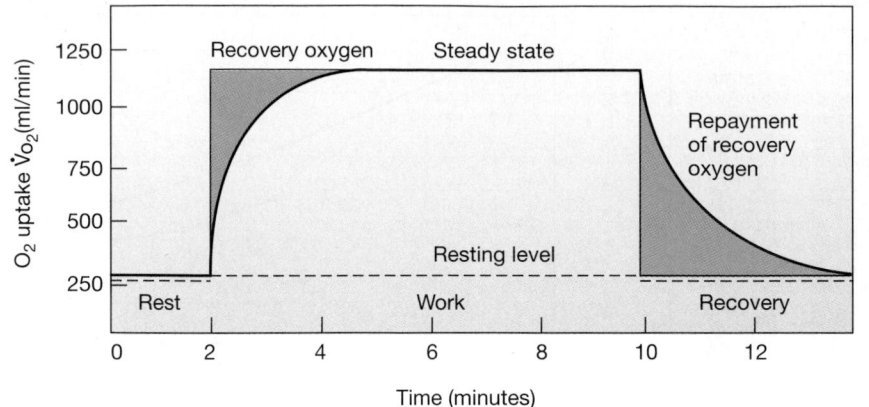

Figure 30–10

Oxygen consumption during light steady-state exercise.

tive tissue, (2) the rate of O_2 use by active tissue, and (3) the O_2 diffusion capacity of the lungs. There appears to be a dual basis for an increased V_{O_2}max with training, including (1) an increase in maximal cardiac output and (2) an increase in blood arterial–venous oxygen difference. Training, on the other hand, has a relatively small effect on pulmonary function, which normally is not a limiting factor to exercise capacity. Most training benefits are lost within weeks or months after training has stopped.

Oxygen Recovery of Exercise Is the Result of a Temporary Disruption in Homeostasis

When exercise begins, the demand for oxygen increases immediately. However, tissue blood flow and aerobic oxidative metabolism require some time to adjust to the new demands, and **recovery oxygen** is

incurred as oxygen is used from existing body stores (Fig. 30–10). During light work, the recovery oxygen remains constant after a steady state is reached in which oxygen uptake and consumption are in balance. During heavy exercise, the recovery oxygen continues to build up until the exercise ends. At the end of exercise, although the demand for oxygen by contracting tissue decreases abruptly, the rate of oxygen uptake remains above the resting level for several minutes as the recovery oxygen is being repaid. The recovery oxygen is often defined as the amount of oxygen above the pre-exercise or resting level that must be taken into the body after exercise. The recovery oxygen repaid after the end of exercise is often greater than that incurred during the work itself and is affected by other factors such as increases in body temperature and hydrogen ion concentration. After light work, the recovery oxygen amounts to as much as 4 liters, and after heavy exercise it can be as high as 20 liters.

The repayment of the recovery oxygen occurs in two stages (Fig. 30–11). The first stage is called the **fast component,** the term referring to the oxygen needed to replenish the oxygen stores of the body as well as the phosphagen system. The fast component is usually repaid within 2 to 3 minutes after exercise. The second stage is called the **slow component** and requires the removal of lactic acid from the body. This stage is repaid very slowly and requires at least 1 hour. During this time, some of the lactic acid is converted to pyruvic acid and then metabolized by body tissues. Also, much of the lactic acid is converted by the liver to glucose, which is used to replenish the glycogen stores in the muscle.

Figure 30–11

The rate of oxygen uptake during and following four minutes of maximal exercise. Following exercise, oxygen uptake remains above resting levels as the fast and slow components of recovery oxygen are repaid.

Several Nutrients Fuel the Work of Exercise

Carbohydrates, fatty acids, and amino acids can be utilized as fuel during exercise in proportions that vary depending on the **intensity** and **duration** of the exercise and the **availability** of stored fuels. At low rates of exercise of short duration, such as short walks, lipids supply a large portion of the required energy. With increasing intensity and duration of exercise, there is a gradual change toward a proportionally greater share of energy yield from carbohydrates. However, if moderately heavy exercise continues for 4 to 5 hours, the glycogen stores of the muscle become depleted; the muscle then depends on the limited supply of glucose that can be absorbed from the gut, or on energy from other sources, mainly fats and to a much lesser extent protein. Clearly, the choice of fuel for exercising muscle is mainly limited to carbohydrates and fat.

The relative usage of carbohydrates and fat for energy during prolonged, exhaustive exercise is illustrated in Figure 30–12. Note that in an individual on a mixed diet, most of the energy at this level of exercise is derived from carbohydrates during the first few minutes; however, as the exercise continues there is a gradual shift toward fat use, and at exhaustion as much as 60% to 70% of the energy is being derived from fats rather than carbohydrates. It is also clear that diet can influence the relative proportion of stored fuels and thus influence the relative usage of fat and carbohydrates. A high-carbohydrate diet is associated with a relatively greater share of the energy yield from carbohydrates and longer endurance, whereas a high-fat diet results in a proportionally greater share of the energy yield from fat and relatively shorter endurance.

Diet can also affect glycogen recovery following exhaustive exercise. Figure 30–13 shows the rate of muscle glycogen replenishment in three different sit-

Figure 30–13

Rate of replenishment with different diets following prolonged exercise.

uations: a high-carbohydrate diet, a high-fat/high-protein diet, and starvation. Note that even in individuals on high-carbohydrate diets, muscle recovery is a relatively slow process, lasting approximately two days. Individuals on high-fat and high-protein diets, as well as persons eating no food at all, showed very little recovery even after 5 days. It seems clear that to achieve optimal performance, approximately 48 hours of recovery after exhausting exercise are essential.

Respiration During Exercise

Pulmonary Ventilation During Exercise Is Proportional to Oxygen Consumption

The increased aerobic metabolism in muscle during exercise places extra demands on the respiratory system. The increased oxygen demands of an exercising individual are nicely paralleled by increased **pulmonary ventilation**. During exercise, the increase in ventilation is proportional to the metabolic demands (O_2 consumption), as illustrated in Figure 30–14. Pulmonary ventilation, which is normally 5 to 6 liters per minute, may approach 150 liters per minute in severe, short-term exercise. The change in total ventilation is associated with a fourfold increase in respiratory rate and a sixfold increase in tidal volume. These changes require greater work by the respiratory muscles, and the work of breathing is increased by one-hundredfold (Fig. 30–15). Note, however, that maximal breathing capacity in a healthy young individual is approximately 170 liters per minute and is therefore 15% greater than the pulmonary ventilation achieved

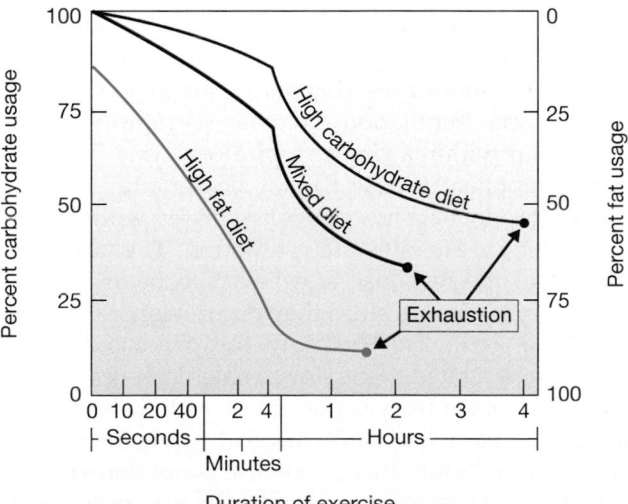

Figure 30–12

The relative use of fats and carbohydrates as a function of diet and duration of exercise.

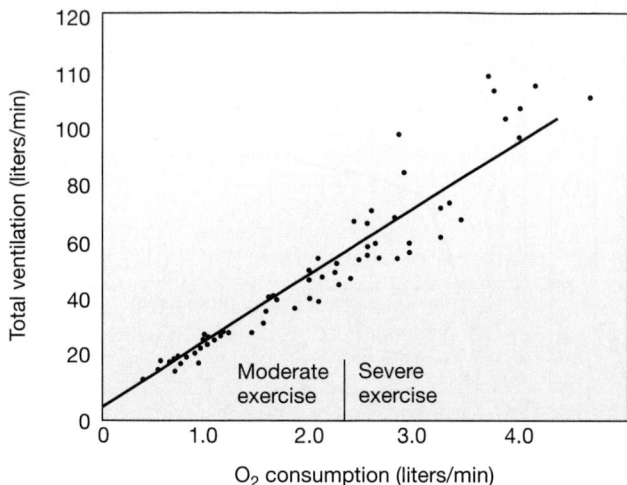

Figure 30–14

Effect of exercise on oxygen uptake and total ventilation.

during maximal exercise. The point is that the respiratory system is not normally the factor limiting the delivery of oxygen to the muscles during maximal exercise.

The Diffusing Capacity of the Lung During Exercise Increases as a Result of Increased Lung Blood Flow

The **diffusing capacity of the lung** is defined as the milliliters of a gas diffusing across the pulmonary membrane per minute per millimeter of mercury pressure difference between the alveolar air and the pulmonary blood. The diffusing capacity of the lung

Figure 30–15

Work of breathing as a function of minute volume (pulmonary ventilation).

for the respiratory gases provides an index of the dimensions of the pulmonary capillary bed, the pulmonary membrane, and the overall efficiency of the system in the exchange of respiratory gases. The diffusion capacity of oxygen and carbon dioxide increases during exercise. Oxygen diffusion capacity in a highly trained athlete increases from an average of 23 ml/mm Hg at rest to about 80 ml/mm Hg at maximum exercise. The increased diffusing capacity in the lung is largely the result of the increased pulmonary blood flow that accompanies exercise and ensures maximal perfusion of pulmonary capillaries, thus providing a greater surface area for gas diffusion.

Control of Pulmonary Ventilation During Exercise Involves Three Primary Homeostatic Systems

Pulmonary ventilation at rest, as described in Chapter 20, exhibits inherent rhythmicity. Its control involves interactions between CNS inspiratory and expiratory centers, pulmonary stretch receptors, and peripheral and central chemoreceptors sensitive to blood gas pressure and H^+ concentrations.

The mechanisms that control ventilation at rest cannot entirely explain the increased ventilation or **hyperpnea** during exercise. For example, artificial changes in P_{O_2}, P_{CO_2}, and acidity in the body at rest do not lead to the high levels of ventilation observed during exercise. How this increase in ventilation is elicited during exercise is unknown. Several theories have been advanced that suggest interactions between numerous stimuli falling into two main categories: **chemical** or **humoral** (fluid-related) **stimuli,** and **neural stimuli.** Figure 30–16 illustrates the time course of ventilation changes before and during moderate steady state exercise. The first component (Phase I) is a rapid rise that occurs just before the exercise starts. Ventilation increases so rapidly that it may occur within a single respiratory cycle. The second component (Phase II) is a slower rise, and the third component (Phase III) is associated with the development of steady state ventilation. The fast component of the response is believed to be associated with neural stimuli, since they occur sooner than the metabolic changes. The slow components are believed to be related to the slower chemical stimuli that alter CNS respiratory center activity. While the exact nature of the neural and chemical mechanisms that control ventilation during exercise is not known, it is clear that they interact to provide the proper increase in pulmonary ventilation to meet the metabolic demands of the exercise and thus keep blood respiratory gases almost normal. Some of the factors that

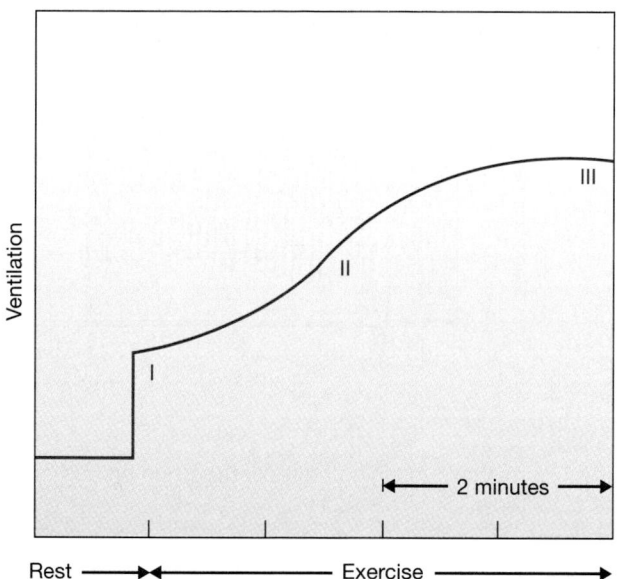

Figure 30–16

The time course of change in ventilation before and during moderate exercise.

have been studied as possible stimuli for the exercise hyperpnea will now be briefly discussed.

Chemical Stimulation of Pulmonary Ventilation

Increased Arterial Carbon Dioxide Concentration During Exercise. The close correlation between ventilation rate and metabolic rate during steady state exercise suggests that the ventilatory drive may be associated with a chemical constituent that is altered at rates proportional to metabolism. Carbon dioxide production is greatly increased in muscle during exercise and could theoretically increase arterial P_{CO_2} enough to stimulate CNS respiratory centers. Actual measurements of arterial P_{CO_2} indicate that the increases are nonexistent during moderate exercise and that a sizable decrease can occur during very heavy exercise. Furthermore, much larger increases in arterial P_{CO_2} than have been measured would have to occur if P_{CO_2} alone were the primary stimulus.

Increased Arterial Hydrogen Ion Concentration During Exercise. Lactic acid production is greatly increased in muscles during exercise and could lead to an increase in arterial hydrogen concentrations sufficient to stimulate CNS respiratory centers. Actual measurements have shown that significant changes in arterial lactic acid concentration occur only during very strenuous exercise. Furthermore, at the end of exercise, lactic acid levels may increase slightly while

ventilation rate rapidly decreases. It is therefore unlikely that the blood hydrogen concentration alone provides the major stimulus for exercise hyperpnea.

Decreased Arterial Oxygen Concentration During Exercise. Arterial **hypoxemia** can stimulate respiration by eliciting chemoreceptor activity from the carotid and aortic bodies. The threshold for these receptors, however, is an arterial blood oxygen pressure of approximately 60 mm Hg. Even during strenuous exercise, arterial oxygen pressure usually remains absolutely constant, and this mechanism alone cannot provide the primary stimulus for the increased ventilation during exercise.

Neural Stimulation of Pulmonary Ventilation

Impulses from Central Nervous System Centers During Exercise. It has been suggested that the increased ventilation just before and at the onset of exercise is at least partly due to neural stimuli arising from the motor cortex. The effect is the result of the direct stimulation of the respiratory center by the same neural signals that radiate from the motor cortex to the muscles to cause the exercise. While neural signals may be involved in the hyperpnea of exercise, it is difficult to envision how purely neural activity can be so finely tuned to the metabolic demands of the exercise.

Impulses from Peripheral Receptors During Exercise. Passive movement of the limbs is often associated with an increase in ventilation. The ventilatory response is closely related to the number of joints and muscles involved in the movement. Researchers have suggested that muscle contraction and movement of the extremities may stimulate peripheral proprioreceptors in the joints and in muscle spindles. These receptors, in turn, are believed to send afferent neural signals to the respiratory center and stimulate ventilation. The role of peripheral receptors in fine tuning ventilation to metabolic demands is unknown.

Impulses from Lung Stretch Receptors During Exercise. Lung stretch receptors are known to be stimulated by filling of the lungs and to send afferent neural signals to the respiratory center, resulting in an inhibition of inspiration and initiation of expiration. Lung stretch receptors may play a role in the hyperpnea of exercise by increasing respiratory rate. This mechanism, however, is not considered the primary stimulus for exercise hyperpnea, since voluntary increases in respiratory rate are not self-perpetuated, as is required during exercise.

Increased Sensitivity of the Respiratory Centers During Exercise. An increased sensitivity of the respiratory centers to arterial Pco_2 could account for the hyperpnea of exercise. The change in sensitivity, however, would have to be very large, since even during strenuous exercise arterial O_2 and Pco_2 change very little. The fact remains, however, that when breath-holding time is measured during moderate exercise, rebreathing occurs at a lower alveolar CO_2 and a higher O_2 than occurs at rest. It is not possible to ascribe this effect only to increased receptor center sensitivity, because the neural factors described previously may be operating to increase the activity of the respiratory center.

Increased Body Temperature During Exercise. An increase in body temperature increases ventilation and could play a role in exercise hyperpnea. The increase in ventilation during exercise, however, is much greater and occurs before any significant rise in body temperature. While an increase in body temperature is not the primary stimulus for exercise hyperpnea, the possibility remains that the hyperthermia of exercise may have a direct effect on the activity of joint and muscle receptors as well as on the respiratory center during exercise.

We must conclude that neither neural nor chemical stimuli alone can explain the hyperventilation of exercise. Furthermore, while neurogenic factors cannot be considered as a regulating stimulus, they may act as an activator of the response. The humoral factors, on the other hand, appear to be critical for the finer adjustments of ventilation to keep the blood respiratory gases almost normal during exercise. The control of the respiratory system during exercise remains an important unanswered physiological question.

Circulation During Exercise Increases to Deliver More Nutrients to the Working Muscles and to Remove Metabolic By-Products

Muscle Blood Flow During Exercise

One of the most important functions of the cardiovascular system during exercise is to deliver oxygen and other nutrients to the working muscles and remove CO_2 and other metabolic by-products. As illustrated in Figure 30–17, this function is facilitated by a marked increase in muscle blood flow during dynamic exercise. Muscle blood flow can increase approximately twenty-fivefold during maximal exercise. Approximately 50% of the increase in flow is the result of local vasodilatation caused by the direct ef-

Figure 30–17

Effect of rhythmic muscle contraction on human calf muscle blood flow.

fects of the increased muscle metabolism. The rest of the increase in muscle blood flow during exercise is the result of several factors, an important one being a moderate (about 30%) increase in arterial blood pressure. Figure 30–17 also shows that, during contraction, muscle blood flow temporarily decreases as the intramuscular vessels are compressed. Strong tonic contractions are known to cause rapid muscle fatigue, owing to the inadequate delivery of oxygen and nutrients during the continuous contraction.

Cardiac Output During Exercise

Cardiovascular regulation during exercise is designed to provide an adequate cardiac output to ensure an adequate oxygen and nutrient supply to the working muscles. As illustrated in Figure 30–18, there is a remarkable consistency among cardiac out-

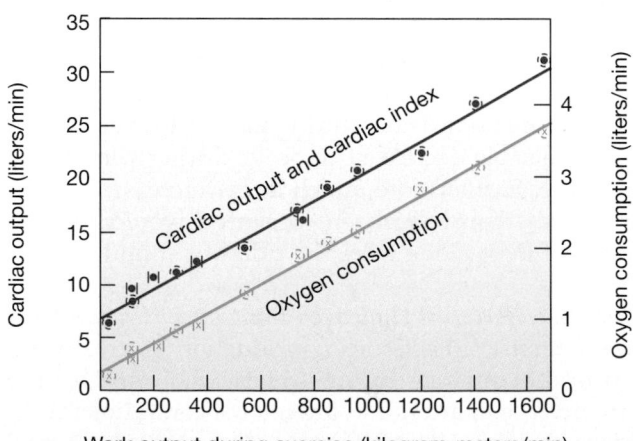

Figure 30–18

Relationship among cardiac output, oxygen consumption, and work output during different levels of exercise.

put, oxygen consumption, and work output. As work output increases during light or moderate exercise, there is a proportional increase in oxygen consumption, which dilates the muscle blood vessels. This dilation increases venous return, resulting in a linear increase in cardiac output. There is a deviation from linearity as oxygen consumption approaches V_{O_2}max, related to thermoregulation at high work rates. As body temperature rises during strenuous exercise, skin blood flow is increased so that the high internal heat loads can be transferred to the skin and dissipated to the environment.

The increased blood flow to the skin under these conditions places an additional demand on the available cardiac output. Typical cardiac outputs for nontrained and highly trained athletes are illustrated in Table 30–1. The table shows that, during maximal exercise, an untrained individual can increase cardiac output to approximately four times the resting level, while the well-trained person can increase cardiac output to about six times the resting level. The greater cardiac output in trained athletes is mainly due to **hypertrophy** (enlargement) of the heart. The hypertrophy in endurance athletes is characterized by a large ventricular cavity with normal wall thickness. Nonendurance athletes exhibit hypertrophy characterized by thick ventricular walls with a normal ventricular cavity. Exercise cardiac hypertrophy has not been shown to have any clinical consequences or significance.

Stroke Volume and Heart Rate During Exercise

Table 30–1 also compares stroke volume, heart rate, and cardiac output in nonathletic and endurance-trained individuals at several levels of exercise. It shows that both at rest and during maximal exercise, the trained individual has a significantly greater stroke volume and a lower heart rate than the un-

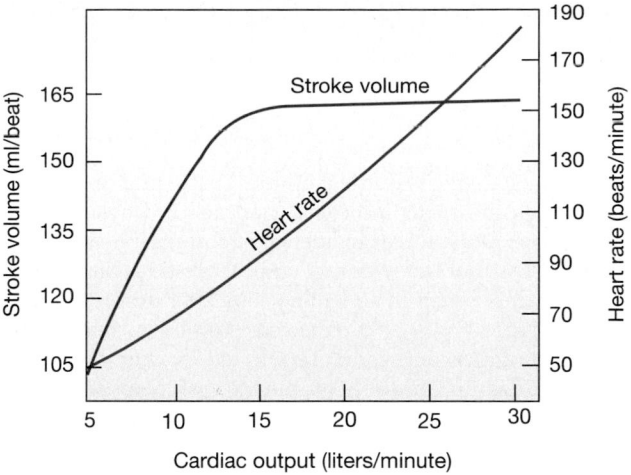

Figure 30–19

Relationship among stroke volume, heart rate, and different levels of cardiac output during exercise in an endurance-trained athlete.

trained person. Because of the greater pumping effectiveness of the heart, the trained individual can achieve the same cardiac output both at rest and during exercise with a lower heart rate. Figure 30–19 illustrates the interrelationships among stroke volume, heart rate, and cardiac output in a trained person with a resting cardiac output of 5.5 liters per minute and a maximum exercise output of 30 liters per minute. The heart rate increases more than 250%, while stroke volume increases about 50% and reaches a maximum when the cardiac output approaches 15 liters per minute. Further increases in cardiac output are the result of increases in heart rate. Therefore, in strenuous exercise, the increase in heart rate accounts for the major increase in cardiac output.

During maximal exercise, cardiac output increases to about 90% of the maximum that the per-

Table 30–1

Comparison of Stroke Volume, Heart Rate, and Cardiac Output in a Trained Athlete and a Nonathlete

State	Stroke Volume (ml/beat)	Heart Rate (beats/min)	Cardiac Output (liters/min)
Resting			
Nonathlete	75	75	5.63
Athlete	105	50	5.25
Maximum exercise			
Nonathlete	110	195	21.5
Athlete	162	185	29.9

Exercise and Coronary Heart Disease

Initial statistical studies in the 1940s and extending through the 1950s indicated that men in physically active jobs had a lower incidence of coronary heart disease—that is, fewer first clinical episodes than comparable men in sedentary jobs. The problem with these types of prospective studies is clear. Different men select and are selected for different jobs. What proportion of the risk of coronary heart disease is due to selection factors and what proportion is due to the activity level of the job?

Epidemiological studies in the 1960s and 1970s that controlled for selection factors as well as longitudinal studies on selected matched populations support the hypothesis that habitual exercise is a significant factor in reducing the incidence and severity of coronary heart disease.

By the 1980s it was evident that if exercise was to contribute to the prevention of the epidemic of coronary heart disease it would be through the use of exercise programs utilized in leisure time. It was clear that strenuous physical activity was being progressively eliminated from the workplace and people were becoming increasingly sedentary. Automation, robots, and computers in conjunction with the growth of the service industry were largely responsible for our sedentary lifestyle.

More than 50% of all the deaths in the United States are due to cardiovascular disease, and heart attacks account for 34% of the total. In the last 30 years the incidence of heart disease has declined by approximately 25%. A combination of improved early treatment and prevention programs that include exercise are largely responsible for the decrease in the death rates from coronary heart disease.

The critical pathophysiological mechanisms by which exercise reduces the occurrence or severity of coronary heart disease have not been elucidated. For example, it has been known for a long time that the size of the coronary vasculature increased in rats following a regular exercise program and facilitated coronary blood flow. And autopsy studies on marathon runners have shown small increases in the size of coronary vessels. The fact that genetic factors may have also been involved cannot be ruled out. Some recent human studies have shown that exercise lowers total blood cholesterol and also increases high-density lipoproteins (HDL), which are thought to protect against coronary heart disease. Subsequent exercise studies on humans have been less consistent and it is difficult to indicate a clear relationship between exercise and blood cholesterol or HDL.

Numerous clinical studies indicate that exercise training causes a decrease in blood pressure, particularly in humans who are hypertensive. Exercise training, on the other hand, has relatively little effect on blood pressure in normotensive subjects. The physiological benefits of exercise and the lowering of blood pressure in hypertensive populations are sufficient to recommend the inclusion of exercise in most therapeutic programs designed to manage hypertensive and coronary disease patients. Finally, although the exact physiological mechanisms are not clear, it can be concluded that exercise training significantly reduces the risk of coronary heart disease.

■ **What are some of the cardiovascular changes associated with exercise training that may account for the reduced risk of coronary heart disease in well-conditioned athletes?**

son can achieve. Therefore, the capacity of the cardiovascular system is normally a limiting factor in terms of maximal oxygen consumption and exercise performance. Maximal cardiac output decreases with age and significantly contributes to the decrease in the maximal exercise performance of the elderly.

Local Blood Flow to Organs During Exercise

The active muscles, including the respiratory muscles, receive the bulk of the increased cardiac output during exercise. The increase in muscle blood flow is associated with drastic alterations in the blood flow distribution to other organs. Blood flow and the percentage of the cardiac output perfusing various tissues during rest and strenuous exercise are illustrated in Table 30–2. As the table shows, brain blood flow is maintained in heavy exercise. Splanchnic and renal blood flow, on the other hand, are decreased, while skeletal muscle, heart, and skin blood flow are increased. Exercise has a dual effect on skin blood flow. At the start of exercise, there is a decrease in cutaneous blood flow that facilitates increased muscle blood flow. However, as exercise continues, skin blood flow increases above resting levels to permit effective transfer and dissipation of body heat. Clearly, during heavy exercise some of the large vascular beds in the viscera receive reduced blood flow

Table 30–2

Blood Flow Distribution During Rest and Exercise in an Athlete

Area	Rest ml/min	Rest %	Heavy Exercise ml/min	Heavy Exercise %
Splanchnic	1400	24.0	300	1.0
Renal	1100	19.0	900	4.0
Brain	750	13.0	750	3.0
Coronary	250	4.0	1000	4.0
Skeletal muscle	1200	21.0	22000	85.5
Skin	500	9.0	600	2.0
Others	600	10.0	100	0.5
Total cardiac output	5800	100.0	25650	100.0

so that the flow through more vital and active areas can be maintained or increased.

Body Temperature Increases as a Well-Regulated Response to Exercise

Prolonged, strenuous exercise markedly increases body temperature. Since the metabolic efficiency of the human body is less than 25%, more than 75% of the metabolic energy of exercise is converted to heat. The greater the exercise intensity, the greater the total amount of heat produced, and at maximal exercise it can be 10 to 20 times greater than resting levels. The excessive heat must be removed and dissipated from the body to prevent overheating and tissue damage from excessive hyperthermia.

Figure 30–20 illustrates the relationship between muscle and rectal temperature as a function of the percentage of a person's maximal oxygen uptake. As expected, the highest temperatures occur in the exercising muscles, where most of the increased heat is being produced. It should be noted that the temperatures depend not on the absolute magnitude of the energy output, but rather on the level of metabolism relative to the individual's maximal oxygen consumption. Rectal temperatures may reach 40 to 41°C in highly trained athletes during prolonged exertion. Body temperature increases during exercise are independent of environmental temperature except at very extreme levels. Normally, internal body temperature levels off at a few degrees above normal regardless of the atmospheric temperature. The increase in body temperature during exercise is not due to a failure in the mechanisms for effective temperature regulation. On the contrary, it is a well-regulated response that occurs even during exercise in a cold

environment. The primary stimulus for the increase in body temperature during exercise is not known, but it may involve a change in the set-point for body temperature regulation.

A key advantage of allowing body temperature to rise during exercise is that less strain is placed on the thermoregulatory mechanisms. Maintenance of a resting body temperature during exercise would necessitate a greater increase in skin blood flow and

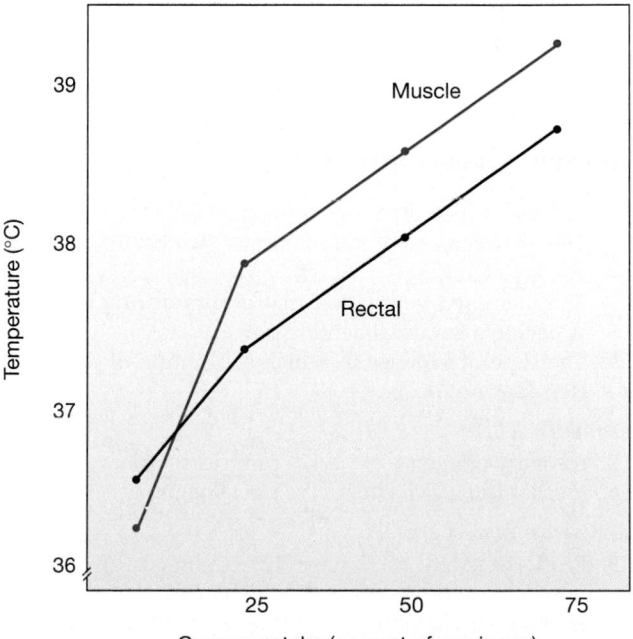

Figure 30–20

Change in exercising muscle and rectal temperature in relation to the oxygen uptake as a percentage of a person's maximum.

could reduce essential blood flow to the working muscles. The increased blood flow to the skin for temperature regulation and the increased blood flow to the muscles during exercise place competing demands on the cardiovascular system. The rate of sweating would also have to be increased, resulting in possible dehydration earlier in prolonged exercise. These considerations do not mean that heat is always beneficial to the athlete. In very hot and humid environments, or great excesses in athletic clothing, nonregulated body temperature increases exceeding 42°C can easily occur. When this happens, multiple symptoms begin to appear, including extreme weakness, dizziness, decreased sweating, headache, nausea, confusion, collapse, and unconsciousness. This whole complex is called **heat stroke** (see Chapter 27), and failure to treat by appropriate body cooling can lead to death.

CHAPTER REVIEW

Summary

- Exercise induces a disturbance in homeostasis that activates multiple physiological responses designed to restore homeostasis.
- An elevated metabolic demand is characteristic of all types of exercise and is a most important factor for proper physiological adjustments.
- The primary sources of energy for exercise are phosphagen supplied by the phosphagen system, the glycogen–lactic acid system, and the aerobic system.
- Oxygen consumption increases linearly with increasing levels of exercise until VO₂max is reached. VO₂max represents a person's maximal aerobic power.
- A recovery oxygen is incurred at the beginning of exercise as oxygen is used from existing body stores. The oxygen recovery is restored at the end of exercise.
- Carbohydrates, fatty acids, and amino acids can be used as energy sources during exercise. They are used in proportions that vary depending on the intensity and duration of exercise and available stores.
- Pulmonary ventilation increases during exercise in proportion to the increased oxygen demand. Ventilation is not normally a limiting factor during maximal exercise.
- Cardiovascular function is regulated during exercise to ensure adequate supply of oxygen and nutrients and removal of increased rates of production of CO₂ and metabolic by-products. The capacity of the cardiovascular system is normally a limiting factor during maximal exercise.
- Increased metabolic heat production is a characteristic of all forms of exercise. Activation of the temperature-regulating system increases heat loss through a combination of physical and physiological mechanisms that maintain thermal homeostasis.

Review Questions

Identify Each with a Term
1. The enlargement of muscle tissue that occurs with repeated exercise.
2. The increased ventilation that occurs during exercise.
3. A person's maximal aerobic power.
4. The type of exercise in which shortening of the muscles does not occur.

Define Each Term
5. recovery oxygen
6. mechanical efficiency
7. proprioreceptors
8. phosphagen

Choose the Correct Answer
9. Proportionately, the largest drop in blood flow during maximal exercise occurs to which area?
 a. renal   c. skin
 b. brain   d. splanchnic
10. Approximately what percent of the energy released during muscle contraction is available to do work?
 a. 70   c. 20
 b. 33   d. 80
11. During maximal exercise, pulmonary ventilation may exceed how many times the resting rate?
 a. 50   c. 10
 b. 30   d. 100
12. Muscle blood flow during exercise can increase to how many times the resting level?
 a. 50   c. 10
 b. 25   d. 100
13. The major factor associated with the increase in muscle blood flow during exercise is:
 a. increased vasodilation
 b. increased heart rate
 c. increased body temperature
 d. increased blood pressure
14. Maximal oxygen uptake would be the same during maximal exercise involving:
 a. two arms versus two legs
 b. one arm versus one leg
 c. two legs versus two arms plus two legs
 d. one leg versus two legs

15. During exercise, which area shows an initial decrease in blood flow, followed by an increase?
    a. skin   c. brain
    b. renal   d. muscle
16. The phosphagen system does not include:
    a. ATP   c. glucagon
    b. phosphocreatine   d. ADP
17. Glycolysis of a single molecule of glucose produces how many molecules of ATP?
    a. 36   c. 6
    b. 2   d. 20
18. When one phosphate group is removed from ATP, approximately how many Calories of energy are released?
    a. 25,000   c. 11,000
    b. 5,000   d. 100,000

Answer Each Question in One or Two Sentences

19. Since energy cannot be created or destroyed, what is the ultimate fate of all of the energy released from the breakdown of ATP?
20. Why is the phosphagen system the most rapidly available source of ATP for muscle contraction?
21. What are the criteria for the determination of maximal oxygen consumption?

22. What is the relationship between rectal temperature and the level of exercise in humans?
23. Define the diffusing capacity of the lung and how it changes during exercise.
24. What is the significance of the size of the fast component of the recovery oxygen?
25. Explain how blood flow is redistributed between body systems during exercise.

Problem Solving in Physiology

26. A well-conditioned male student can have maximal cardiac output values of 30 liters per minute while sedentary students have maximal values of approximately 20 liters per minutes. What are the differences and similarities in cardiac function in the two students?
27. The Boston Marathon has often been run on a hot, humid day. What, in your opinion, was the primary physiological system that failed to meet the demands of many of the runners and caused them to drop out of the race because of muscle fatigue?
28. If you were a coach, what diet would you recommend, how long would you have your long distance runners recover between events, and why?

Suggested Readings

Astrand, P.O., and Rodahl, K. *Textbook of Work Physiology.* New York: McGraw-Hill, 1986.

Bell, A.W., and Hales, J.R.S. "Cardiac output and its distribution during exercise and heat stress." In *Thermoregulation, Pathology, Pharmacology, and Therapy.* New York: Pergamon Press, 1991, p. 105.

Brengelman, G.L. "Circulatory adjustments to exercise and heat stress." *Annual Review of Physiology,* 45:191, 1983.

Fox, E., Bowers, R., and Foss, M. *The Physiological Basis for Exercise and Sport.* 5th ed. Madison, WI: Brown and Benchmark, 1993.

Lamb, D.R. *Physiology of Exercise Responses and Adaptations.* New York: Macmillan Publishing Co., 1984.

Martin, I.K., and Wahren, J. "Glucose metabolism during physical exercise in patients with noninsulin-dependent (type II) diabetes." *Advanced Experimental Medical Biology,* 334:221–33, 1993.

McArdle, W., Katch, W.F., and Katch V. *Exercise Physiology: Nutrition and Human Performance.* 3rd ed. Philadelphia: Lea and Febiger, 1991.

McManus, . *Cardiovascular Complications of Exercise.* Chicago: Year Book Medical Publishers, 1985.

Noble, B.J. *Physiology of Exercise and Sport.* 2nd ed. St. Louis: Times Mirror/Mosby College Publishing, 1991.

Stainby, W.N., and Brooks, G.A. "Control of lactic acid metabolism in contracting muscles and during exercise." *Exercise and Sports Sciences Reviews,* 18:29–63, 1990.

Teitz, C.C. (ed.) *Scientific Foundations of Sports Medicine.* Philadelphia: Decker, Inc. 1989.

Wells, C.L. *Women, Sport, and Performance: A Physiological Perspective.* Champaign, IL: Human Kinetics Publishers, 1991.

Key Terms

adenosine triphosphate (ATP)(p. 885)
coronary disease (p. 896)
diabetes mellitus (p. 887)
diffusing capacity (p. 892)
dynamic exercise (p. 885)
heat stroke (p. 898)
humoral stimuli (p. 892)

hyperpnea (p. 892)
hypertrophy (p. 895)
hypoxemia (p. 893)
isometric muscle contraction (p. 885)
lactic acid (p. 893)
maximal aerobic power (p. 888)

phosphagen system (p. 885)
pulmonary stretch receptors (p. 892)
pulmonary ventilation (p. 891)

pyruvic acid (p. 890)
recovery oxygen (p. 890)
respiratory center (p. 892)
static exercise (p. 885)
stretch receptors (p. 893)
watts (p. 888)

Mature sperm which have been
pushed into the central duct of
the seminiferous tubule.
(© *Lennart Nilsson from* A Child Is
Born, *Dell Publishing Company)*

Chapter 31
Reproductive Physiology

Key Concepts

Two major integrative organ systems of the body, the nervous
and the endocrine systems, act together to control reproductive functions in the male and female. Reproduction in both males and females is controlled by hormonal interactions among the hypothalamus, pituitary,
and gonads.

The ovaries and the testes have two major functions in
common: production of the germ cells and synthesis and
secretion of sex hormones.

Chapter Outline

Reproductive Physiology in the Male
Reproductive Physiology in the Female

*Jim will turn 65 next month and plans to retire as
soon as he can sell the company that he founded 25
years ago. He has always been in reasonably good health
and he was looking forward to some extended traveling
during retirement. Now he's not so sure. For the last six
months he has had frequent problems with urination. On
the one hand he has trouble initiating urination and on
the other hand he has trouble terminating urination*

*which results in dribbling. He also has the sensation of
incomplete emptying and feels he has to urinate all the
time, especially at night. Last Friday evening
when Jim and his wife had dinner with another
couple, Jim had several beers. That night he had
trouble sleeping and was up many times feeling
the need to urinate but unable to do so. His
problem became worse the next day, which
prompted him to call his physician.*

*Jim's diagnosis, by means of a digital rectal
examination and a serum prostate specific antigen
(PSA) test, indicated that he had enlargement of
the prostate (called benign prostate hyperplasia or BPH),
which causes obstruction of the urethra and restricts
urine flow in men.*

*Jim took pride in taking care of himself. He asked his
physician what caused this enlargement and if prostate
hyperplasia leads to cancer. Jim's physician reassured
him that prostatic hyperplasia does not predispose to the
development of prostatic cancer, and that there was
nothing that he did to cause his prostatic hyperplasia.
He learned that prostatic hyperplasia is a common
occurrence in men in their 60's. The prostate weighs
only a few grams at birth but at puberty it undergoes a
testosterone-mediated growth, reaching adult size
between the ages 20–25 and weighing about 20g.
During the next 25 years of life the prostate does not*

900

grow. Then, around age 50, a second growth spurt begins in a majority of men. Consequently, prostate hyperplasia increases in frequency with age and at age 70 over 80% of men have some form of prostatic hyperplasia. Jim asked how it would be corrected. He was informed that it can be corrected with drug therapy or other noninvasive approaches in 90% of men such as balloon dilation of the prostate, microwave therapy, laser prostatectomy, or by a new procedure using focused ultrasound to ablate the prostate.

In this chapter, we will not only examine how glands like the prostate function, but also explore the physiology of reproduction in both males and females. We will learn how complex hormonal interactions between brain, pituitary gland, ovaries, and testes control reproduction. In addition, we will explore the events controlling the formation of sperm and ova, testicular development, menstruation, and menopause.

In preceding chapters we have examined the structural adaptations that allow cells to carry out particular physiological functions. We have seen that cell specialization produces a complex interdependence among the different cell types that control various processes within the human body. We have often observed that a common purpose of many of these processes is homeostasis, or the maintenance of a stable internal environment.

In contrast, the physiological processes we are about to discuss center on cells from two separate individuals that interact to form a third individual. The sperm cell is structurally and chemically designed to leave the male and join with the female ovum. The ovum, while remaining inside the female, is designed to react with the sperm that originates in the male. These are the only two human cell types that are so adapted.

In Chapter 31 we will first examine the male and female anatomy involved in reproduction. Then we will discuss how the two major integrative organ systems of the body, the nervous and the endocrine systems, act together to control reproductive functions. In Chapter 32, we will examine how fertilization (the union of sperm and ovum) occurs in the female. We will follow the alterations that this union produces in the female nervous and endocrine systems, which then allow for the development of the fetus. The special physiological features of the developing fetus will also be examined. In Chapter 33, we will examine sexual physiology as separate from reproductive physiology, and look at the factors that determine an individual's sex.

Reproductive Physiology in the Male

The major male reproductive functions are (1) secretion of sex hormones, (2) production of sperm, and (3) transport of sperm from the male to the female. The first and second functions are carried out by the testes, while the third function is performed by a network of ejaculatory ducts together with the penis.

The Reproductive Anatomical Structures in the Male Include Both the Internal and External Genitalia

Anatomy of the Male External Genitalia

The principal reproductive organs in the male are external. They include the **penis** and the **scrotum** which contains the testes (Fig. 31–1). The penis is covered by thin, loose skin and contains the **urethra, erectile tissue,** nerves, and many blood vessels. The scrotum is a pouch of skin made of connective tissue and muscle. This pouch contains the **testes,** the **epididymis,** and the **spermatic cord,** which contains the vas deferens. The location of the testes in the scrotal compartment outside the body is essential for normal sperm production, which occurs only at temperatures below normal body temperature. In response to cold, the cremaster muscle of the scrotum contracts involuntarily and lifts the testes closer to the body. In warm weather, the muscle of the scrotum relaxes and allows the testes to move away from the body, which reduces the temperature of the testes. These muscular reflexes ensure that the testes are maintained at a temperature conducive to spermatogenesis, or the formation of spermatozoa.

As we will learn in Chapter 33, both the penis and the scrotum contain a dense supply of sensory nerves that make these external organs very sensitive to touch and pressure.

Anatomy of the Testes

The testes of the adult male have two major functions: production of sperm and production of male sex hormones. Each testis contains a large number of tightly coiled tubules called **seminiferous tubules** (Fig. 31–2). The seminiferous tubules contain sperm-producing or **spermatogenic cells,** also referred to as **germinal epithelium.** The cells that synthesize male sex hormones, such as **testosterone,** are called **Leydig cells,** or the **interstitial cells of Leydig,** because they are located in the interstitial areas between the seminiferous tubules in the testes. In addition to the

Figure 31–1

Side view of the male reproductive tract.   *(Redrawn from Masters and Johnson,* Human Sexuality, *Second Edition.)*

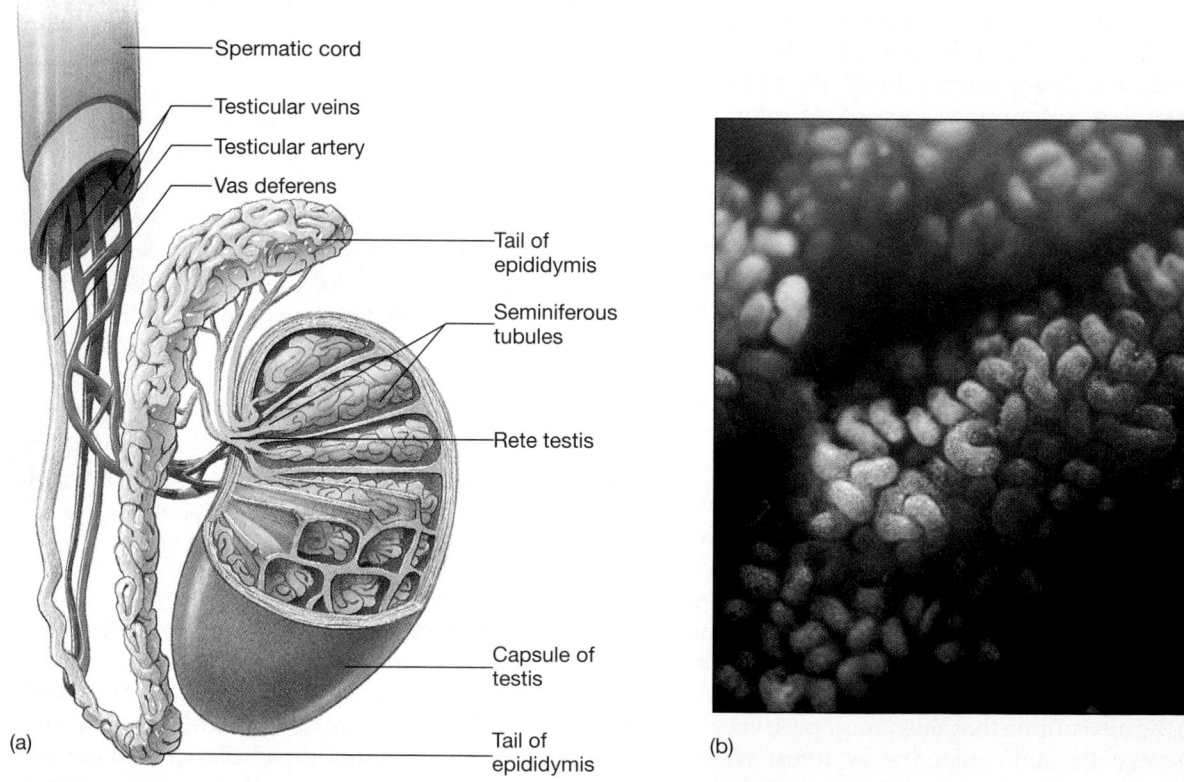

(a)   (b)

Figure 31–2

(*a*) A cross-section of the internal structure of the testis, illustrating compartmentalization of the seminiferous tubules. (*b*) The convolutions of a single seminiferous tubule.
(© Lennart Nilsson, A Child Is Born, *Dell Publishing Company, 1977.)*

Leydig cells, the interstitial space also contains lymphatics, blood vessels, nerve fibers, connective tissue, and extracellular fluid. The nerve fibers and blood vessels enter and leave the testes via the spermatic cord.

Spermatogenesis, or the Formation of Sperm, Takes Place in the Seminiferous Tubules of the Testes

The process of sperm production takes place within the coiled **seminiferous tubules** of the testes. Primitive germ cells, called **spermatogonia,** are present in the testes during embryonic development and childhood. Prior to puberty the spermatogonia divide by mitosis to produce more spermatogonia. However, spermatogenesis, or the formation of mature sperm, does not begin until puberty. Once begun, the process of sperm production never ceases. Although sperm production slows down with age, the number of sperm produced in old age is usually sufficient to maintain fertility.

During spermatogenesis, germ cells undergo a series of divisions and differentiations to become mature sperm. Some of the spermatogonia enlarge and differentiate to become **primary spermatocytes.** Each primary spermatocyte then undergoes meiotic cell division to produce two **secondary spermatocytes.** The secondary spermatocytes undergo a second meiotic division to produce **spermatids.** During the final stage of spermatogenesis, referred to as **spermiogenesis,** the spermatids differentiate to become mature sperm, or **spermatozoa.** This process of spermatogenesis takes about 70 to 80 days. Spermatogenesis is constantly being repeated, and all stages of sperm cell development are present in the seminiferous tubules at any given time.

The magnitude of sperm production in the testes is impressive. As many as several hundred million sperm may be produced daily. Two decades ago the average college age male produced about 90 to 100 million sperm per cc of semen. Today the average has dropped to about 60 million per cc. The cause of this decline is not known, but it has been speculated that some environmental pollutants may inhibit spermatogenesis. If sperm counts drop below 35 million per cc, fertility is markedly impaired. Of the enormous number of sperm produced daily, only a few hundred may reach the distal portion of the female fallopian tube, where fertilization occurs. As we shall see in Chapter 32, sperm loss is high for a number of reasons, and a very high rate of sperm production is necessary in order to overcome the odds against fertilization.

Differentiation of Germ Cells

Two types of cells are involved in spermatogenesis: the **spermatogenic cells,** from which mature sperm arise, and the **Sertoli cells,** which provide mechanical and nutritional support for the spermatogenic cells. Both spermatogenic and Sertoli cells are located inside the seminiferous tubules. Each seminiferous tubule is surrounded by a basement membrane that separates the tubule from the surrounding Leydig cells and connective tissue. The anatomical relationship among the spermatogenic, Sertoli, and Leydig cells is illustrated in Figure 31–3.

The immature or primitive germ cells are found along the outer border of each seminiferous tubule in contact with the basement membrane. The primitive germ cells undergo a mitotic division and differentiate into **spermatogonia** (singular, *spermatogonium*). In the adult, spermatogonia represent the first cell stage of spermatogenesis. Speramtogenesis begins when some of the spermatogonia differentiate into **primary spermatocytes.** As the spermatogonia differentiate, they increase in size, undergo a second mitotic division, and move away from the basement membrane toward the center of the seminiferous tubule (Fig. 31–4). The remaining spermatogonia continue to divide mitotically and provide a constant source of new germ cells that differentiate into generation after generation of primary spermatocytes.

Mitotic Divisions During Spermatogenesis. Spermatogonia, in common with all other cells of the body, are diploid and contain 46 chromosomes, 23 contributed by the mother and 23 by the father. During mitotic division, the chromosomes are duplicated before cell division. When the cell divides, each of the two new cells receives one copy of each of the 46 chromosomes present in the parental cell. Therefore, primary spermatocytes, formed during mitotic division of spermatogonia, have chromosomes identical to those of the spermatogonia that produced them (23 maternal and 23 paternal chromosomes).

Meiotic Divisions During Spermatogenesis. Primary spermatocytes undergo two more divisions through a special form of cell division known as **meiosis.** The first meiotic division (meiosis I) reduces the number of chromosomes in the cell by half. During meiotic division there is a random assortment and recombination of maternal and paternal chromosomes. In addition, homologous chromosomes (similar chromosomes, one of maternal origin and the other of paternal origin) exchange chromosomal segments in a process known as **crossing over** (Fig.

(a)

(b)

Figure 31–3

(*a*) A cross-section of a seminiferous tubule in the testes, illustrating the anatomical relationships among different cell types. (*b*) Scanning electron micrograph of a transverse section through a seminiferous tubule.   *(© Professor P. Motts/Department of Anatomy/University "La Sapienza," Rome/Science Photo Library/Custom Medical Stock Photo.)*

31–5*a*). Crossing over, or exchange of chromosomal segments, can occur at random along the length of the paired homologous chromosomes, and several exchanges can occur at different physical points during a single meiotic division. The term "recombination" refers to any process that leads to new gene combinations. Crossing over produces a recombina-

tion of genetic information or the formation of new gene combinations not present in the parental generation. Crossing over ensures that genetic material is recombined in an infinite number of ways, leading to infinite individual variation.

Before the first meiotic division, DNA synthesis increases within the primary spermatocyte, and each

(a)

(b)

Figure 31–4

(*a*) Cross-section of the seminiferous tubule, illustrating the ring formed by Sertoli cells and the location of the spermatogenic cells. (*b*) Mature sperm that have been pushed into the central duct of a seminiferous tubule.   *(© Lennart Nilsson, A Child Is Born, Dell Publishing Company, 1977.)*

Chromosome of   Chromosome of
maternal origin   paternal origin

(a)

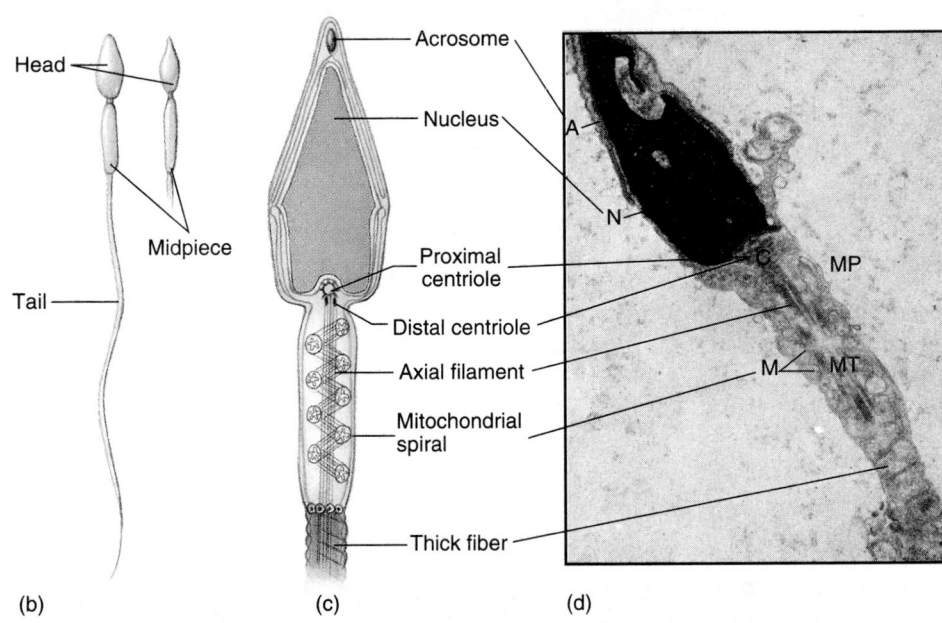

Head

Midpiece

Tail

(b)

Acrosome

Nucleus

Proximal
centriole

Distal centriole

Axial filament

Mitochondrial
spiral

Thick fiber

(c)

A

N

C   MP

M   MT

(d)

Figure 31–5

(*a*) Diagramatic representation of "crossing over," or exchange of chromosomal segments between homologous chromosomes, during meiotic division of the spermatocyte. (*b*) Top and side view of a sperm. (*c*) The head and midpiece of a sperm. (*d*) Electron micrograph of a human sperm. A, acrosome; N, nucleus; MP, midpiece; M, mitochondria; MT, microtubules; C, centrioles.   (© *Dr. Lyle C. Dearden.*)

of the 23 maternal and 23 paternal chromosomes in the cell is duplicated. The homologous maternal and paternal chromosomes pair with each other form 23 pairs of chromosomes, with each chromosomal pair known as a bivalent. Each chromosome has two chromatids and therefore each pair of chromosomes has four chromatids. You may recall that homologous chromosomes are the two members of a given chromosomal pair that correspond in shape, size, and type of genetic information. The primary spermatocyte, which contains 23 pairs of chromosomes, then undergoes two very rapid meiotic divisions that do not involve chromosomal duplication. During the first meiotic division each primary spermatocyte gives rise to two **secondary spermatocytes** which receive a random mixture of maternal and paternal chromosomes. Therefore, individual secondary spermatocytes differ from each other and from primary spermatocytes in terms of their genetic composition.

Each secondary spermatocyte then undergoes a second meiotic division and gives rise to two **spermatids,** each of which contains 23 chromosomes. Through this process four spermatids are formed from the original primary spermatocyte (Fig. 31–6). The spermatids then differentiate into mature sperm.

This process of meiotic cell division ensures great diversity in the genetic composition of sperm.

Physiology of the Mature Sperm

As primary spermatocytes differentiate into spermatids, they migrate away from the basement membrane of the seminiferous tubule toward the fluid-filled lumen of the tubule (see Fig. 31–4). Near the lumen, the spermatids undergo the final phase of spermatogenesis and differentiate into mature **spermatozoa** or **sperm,** which enter the lumen of the seminiferous tubule. From the lumen, the sperm leave the seminiferous tubule and enter the epididymis.

During the final phase of spermatogenesis, the spermatids do not undergo cell division but rather go through a complex series of cytological changes. The **chromatin,** or genetic material of the spermatid, is contained within the nucleus in the head of the sperm and is surrounded by a membrane. As the spermatid differentiates into a mature sperm, it increases in size and becomes elongated (Fig. 31–5*b*). The anterior half of the head of the sperm is covered by a small, caplike structure called the **acrosome,** which differentiates from the Golgi apparatus and

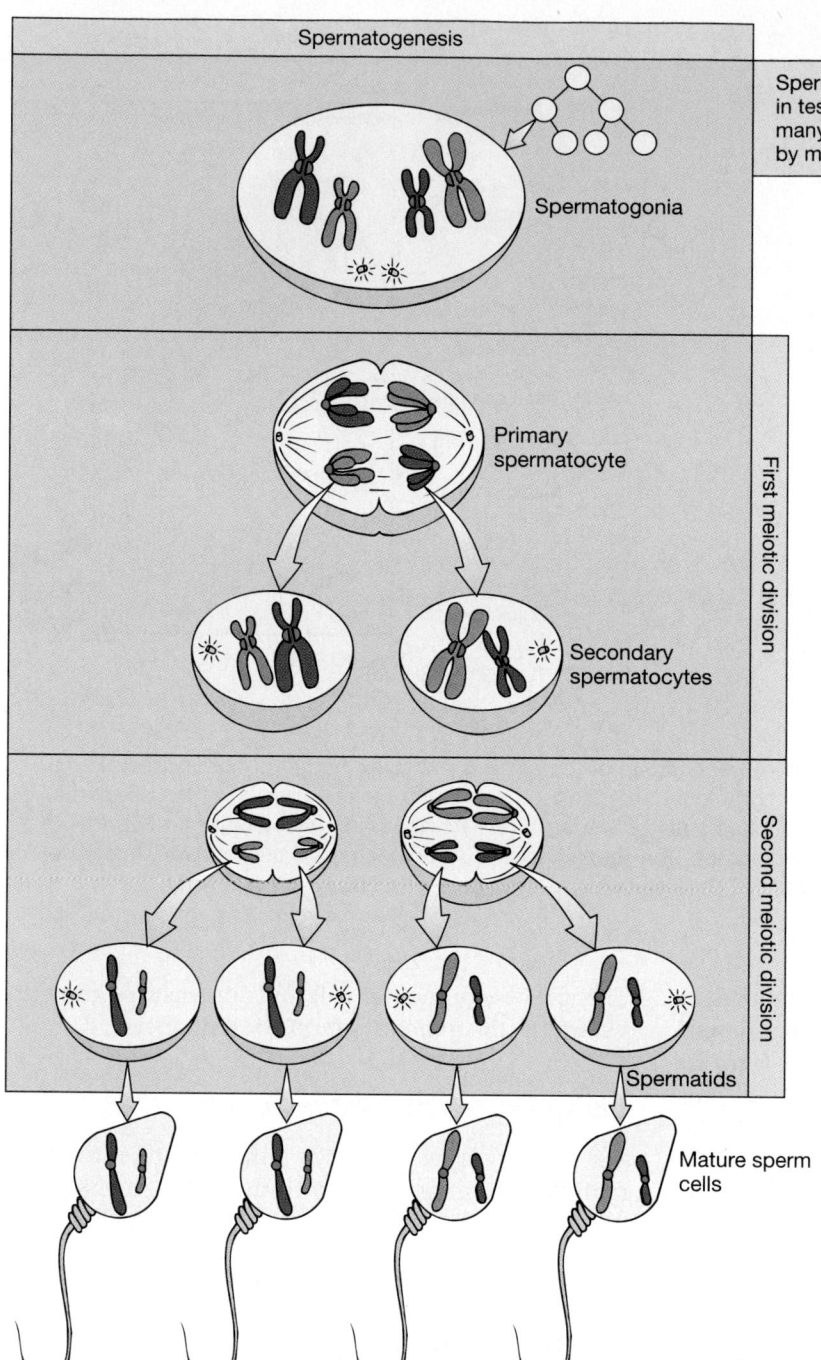

Spermatogenesis

Spermatogonia in testis divide many times by mitosis

Spermatogonia

First meiotic division

Primary spermatocyte

Secondary spermatocytes

Second meiotic division

Spermatids

Mature sperm cells

Figure 31–6

Cell division during spermatogenesis. The primary spermatocyte undergoes meiosis and gives rise to four spermatids.

contains enzymes that aid the sperm during fertilization of the ovum (Fig. 31–5c). The midpiece of the sperm contains a spiral arrangement of mitochondria that generate the energy necessary for sperm movement. The tail of the sperm, or flagellum, contains filaments that slide across one another to produce a back-and-forth movement of the flagellum, which propels the sperm forward (Fig. 31–5c). Movement of sperm through the reproductive tract is discussed in an upcoming section.

Sertoli Cells Serve Several Functions That Are Important for Spermatogenesis

The **Sertoli cells** are located inside the seminiferous tubules. Each Sertoli cell extends from the basement membrane all the way to the fluid-filled lumen of the seminiferous tubule (see Fig. 31–4). The Sertoli cells are joined to one another by specialized attachments called **tight junctions.** As a result, the Sertoli cells form a ring around the inner diameter of the semi-

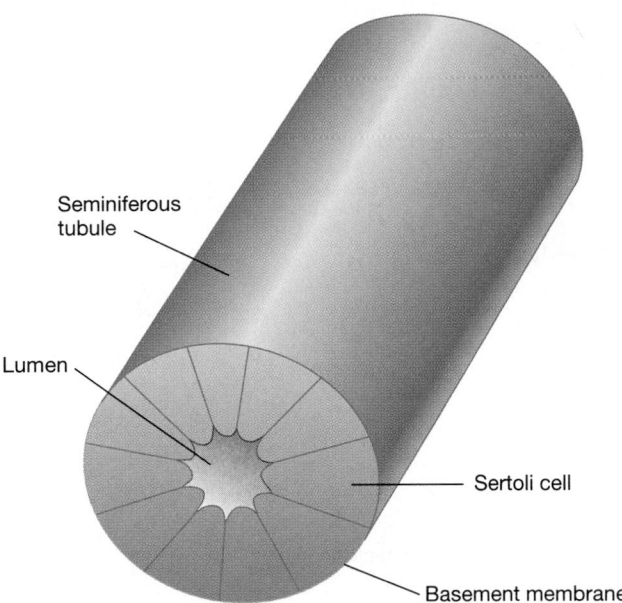

Seminiferous
tubule

Lumen

Sertoli cell

Basement membrane

Figure 31–7

Cross-section of a seminiferous tubule, illustrating the ring formed by Sertoli cells.

niferous tubule (Figure 31–7). All the spermatogenic cells, in different stages of differentiation, are interspersed between and surrounded by the Sertoli cells (see Figure 31–4). This anatomical relationship between the spermatogenic and Sertoli cells is important for a number of reasons (Table 31–1).

The Sertoli cells form a **blood-testes barrier,** which prevents many potentially toxic substances from entering or leaving the seminiferous tubules. For instances, the Sertoli cell barrier excludes from the tubule a variety of proteins, hormones, ions, and drugs, some of which could adversely affect the developing spermatogenic cells. Conversely, this barrier keeps the sperm in the tubules from diffusing into the blood. Mature sperm are very immunogenic. Thus, if they entered the blood, they would be rec-

Table 31–1

Functions of Sertoli Cells

Maintenance of the blood-testes barrier
Nourishment of developing germ cells
Production of seminiferous tubular fluid
Removal of damaged germ cells
Synthesis of androgen-binding protein
Synthesis of inhibin

ognized as foreign antigens, which would result in the production of antibodies specific to them. The blood-testes barrier develops at the time of puberty, just before spermatogenesis begins, and provides a protected environment in which sperm maturation and development can occur.

Sertoli cells secrete most of the fluid in the lumen of the seminiferous tubules. This fluid is rich in proteins, enzymes, nutrients, androgens (such as testosterone), and certain ions such as potassium and bicarbonate. These chemicals stimulate and nourish the developing sperm. The fluid produced by the Sertoli cells also provides a drive for flushing sperm out of the tubules into the epididymis. Another role of the Sertoli cells is the engulfment and removal of any damaged germ cells in the seminiferous tubules.

Sertoli cells synthesize two proteins of particular importance. The first is called **androgen-binding-protein,** or **ABP,** which is secreted into the seminiferous tubular fluid. ABP binds testosterone and increases the concentration of testosterone within the seminiferous tubules of the testes. The second protein is called **inhibin,** which is secreted into the blood and transported to the pituitary gland, where it inhibits the secretion of **follicle-stimulating hormone (FSH).** As we will see, both of these proteins are important for spermatogenesis.

Sperm Are Transported Through the Internal Reproductive Structures

Maturation of Sperm in the Epididymis

The seminiferous tubules meet at the top of the testes in a network of ducts called the **rete testis** (see Fig. 31–2). Sperm move from the tubules through the rete testis into the epididymis as a result of pressure created by the continuous formation of fluid in the tubules. The **epididymis** lies along the back portion of each testis and is lined with smooth muscle. The muscle undergoes rhythmic peristalsis, which transports the sperm through the epididymis toward the vas deferens (see Fig. 31–2). The duct system involved in sperm transport is illustrated in Figure 31–8.

When sperm enter the epididymis, they are immature, nonmotile, and incapable of fertilization. The epididymis secretes a fluid containing nutrients, enzymes, and hormones that aid in sperm maturation. Sperm take about two weeks to pass through the epididymis, during which time they become mature and mobile. From the epididymis, they enter the vas deferens. As the sperm pass through the epididymis, about 90% of the fluid surrounding them is reabsorbed, hence the sperm entering the vas deferens are highly concentrated.

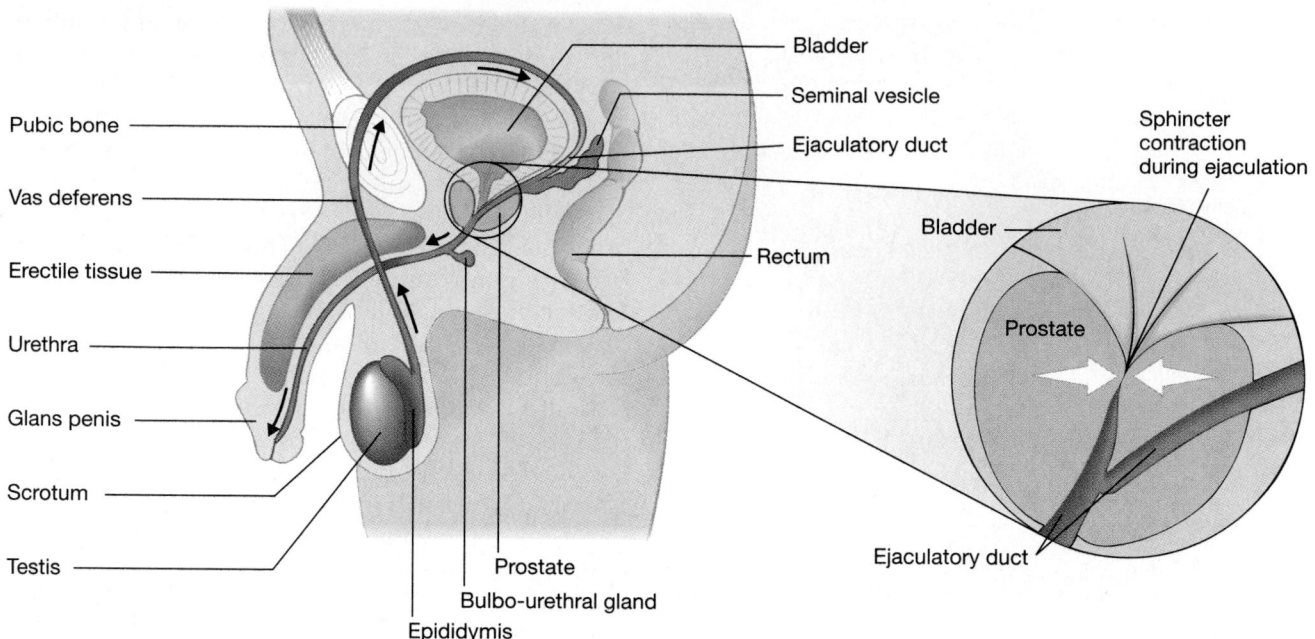

Figure 31–8

Reproductive duct system of the male, illustrating sperm pathway and sphincter contraction in the prostate during ejaculation.

Storage of Mature Sperm

Sperm can be stored in the caudal portion of the epididymis for several months in the absence of ejaculation. Sperm travel from the epididymis to the **vas deferens (ductus deferens),** which are components of the spermatic cords that arise from the tail of the epididymis. The vas deferens are composed of long slender tubes, or ducts, which run from the scrotum up into the abdominal cavity and behind the bladder, where they empty into the **ejaculatory duct** (see Fig. 31–8). The ejaculatory duct runs through the center of the **prostate gland,** located at the base of the bladder surrounding the urethra (see Fig. 31–1). The thick wall of the vas deferens is composed of smooth muscle and innervated by sympathetic nerves. The importance of this innervation for ejaculation will be discussed in Chapter 33. As the vas deferens enter the ejaculatory duct, they are joined by the secretory ducts of the seminal vesicles.

Addition of Nutrient Fluid by the Seminal Vesicles

The **seminal vesicles** are composed of two sacs, one on either side of the bladder. The seminal vesicles secrete a viscous material rich in **prostaglandins,** fructose, and other nutrients. Nutrients in seminal vesical fluid serve as an energy source for the sperm following ejaculation. Seminal vesicle fluid accounts for approximately 60% of the total volume of ejaculated semen. During ejaculation, the contents of the vas deferens are emptied into the ejaculatory duct immediately before the contents of the seminal vesicles. Seminal vesicle fluid washes sperm out of the ejaculatory duct.

Addition of Alkaline Fluid by the Prostate Gland

The **prostate gland** secretes a liquid that contains citric acid, calcium, various enzymes, and **prostaglandins** (see Chapter 5). Prostaglandins derive their name from the fact that they were first isolated from the prostate gland. As we shall see in Chapter 32, the prostaglandins, contained in seminal fluid, stimulate contractions of the uterus in the female; these contractions play a role in moving the sperm through the female reproductive tract following ejaculation. Fluid from the prostate gland adds to the volume of the semen. Prostatic fluid, with a pH of approximately 6.5, is more alkaline than the secretions of the vagina, which may be very important in protecting sperm fertility. Secretions of the vagina are acidic, and sperm motility is inhibited in acidic environments. **Prostatic fluid** may neutralize vaginal acidity and increase sperm motility.

The prostate gland contains an elaborate system of valves. During ejaculation, a valve in the prostate opens, and the semen is ejected into the urethra. A sphincter in the prostate contracts during ejaculation and seals off the exit from the bladder, which prevents the passage of urine from the bladder into the urethra (see Fig. 31–8). Thus, although urine and semen share a common anatomical pathway to the outside of the body, the valves in the prostate prevent these two fluids from mixing. During ejaculation, semen passes from the ejaculatory duct into the urethra, where it is joined by secretions from the bulbo-urethral glands, also called Cowper's glands (see Fig. 31–8).

Addition of Lubricant Fluid by the Bulbo-Urethral Glands

The **bulbo-urethral glands** are located on each side of the urethra. They secrete a clear viscous liquid that adds to the volume of the semen. During sexual arousal, a small amount of fluid from the bulbo-urethral glands may appear at the tip of the penis before ejaculation. This **pre-ejaculatory fluid** occasionally contains a small number of live sperm cells, which makes the practice of withdrawal prior to ejaculation ("coitus interuptus") an unreliable method of contraception.

The **urethra** transports the semen through the shaft of the penis to the outside of the body during ejaculation (see Fig. 31–8). Stimulation of the sympathetic nerves of the epididymis, vas deferens, prostate, and seminal vesicles during sexual arousal results in smooth muscle contractions in these organs and expulsion of the fluids that compose the semen.

The average volume of semen ejaculated is 3 to 5 milliliters (5 milliliters = about one teaspoonful). The normal concentration of sperm in the ejaculate is highly variable (40 to 120 million per milliliter) and depends in part on frequency of ejaculation. Once sperm are ejaculated their maximum life span is only 24 to 72 hours at body temperature.

The major organs involved in sperm production, maturation, and transport are described in Table 31–2.

Neuroendocrine Control of Reproduction in the Male Involves an Interaction Between the Hypothalamus, Pituitary, and Testes

The **pituitary gland,** which lies at the base of the brain, was described in Chapter 13. Two decades ago, it was thought to be the source of the hormonal signals that control reproduction. We now know that the pituitary gland acts as a relay station, receiving neural and hormonal input from the brain and relaying this information, in the form of hormonal messages, to other organs of the body.

Neuroendocrine Control of Spermatogenesis

As described in Chapter 13, the hypothalamus secretes a releasing factor called **gonadotropin-releasing hormone,** or GnRH (Fig. 31–9). GnRH is transported to the anterior pituitary gland via a com-

---

Table 31–2

Major Organs Involved in Sperm Production, Maturation, and Transport

Testes: Sperm production: mitosis, meiosis, and differentiation
Epididymis: Sperm transport and maturation— motility and fertility
Vas deferens: Sperm storage
Seminal vesicles: Production of seminal fluid containing nutrients, fructose, and prostaglandins
Prostate: Production of prostatic fluid that is alkaline pH $\simeq$ 6.5 and contains calcium and citric acid
Bulbo-urethral gland: Production of "pre-ejaculatory" fluid
Penis: Erection and ejaculation

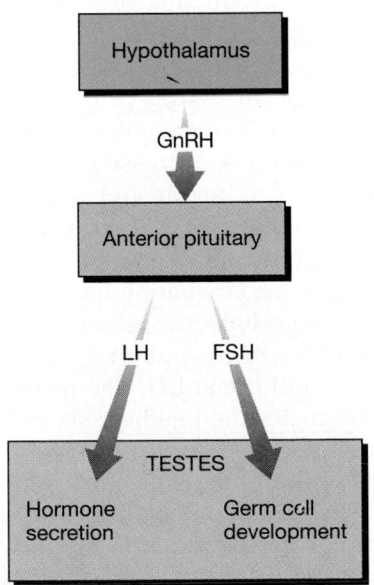

Figure 31–9

Relationship among the hypothalamus, pituitary gland, and testes.

plex set of blood vessels known as the **hypothalamic-pituitary portal vessels** (see Chapter 13). When GnRH reaches the anterior pituitary in the male, it stimulates the release of two major reproductive hormones, termed the **gonadotropins.** One of the gonadotropins, **luteinizing hormone (LH),** is secreted by the anterior pituitary into the blood and is transported to the testes. There it stimulates the Leydig cells to synthesize and secrete testosterone. The other gonadotropin, **follicle-stimulating hormone (FSH),** is secreted by the anterior pituitary into the blood and is also transported to the testes, where it binds to specific receptors on the plasma membranes of Sertoli cells. FSH stimulates the conversion of spermatogonia into spermatocytes in the seminiferous tubules of the testes. Both LH and FSH are required for spermatogenesis (see Fig. 31–9). As we shall see, GnRH, LH, and FSH control reproductive function in the female as well as in the male.

Feedback Regulation of Neuroendocrine Function

The secretion of GnRH, LH, and FSH is fairly constant from day to day in the male, in marked contrast to the cyclic secretion of these hormones in the female. The constancy of hormone secretion in the male is due to the presence of a continuous inhibitory **feedback loop** between the hypothalamus, pituitary gland, and testes (Fig. 31–10).

The hypothalamus is sensitive to changes in the circulating levels of sex hormones. It monitors the level of these hormones and responds by increasing or decreasing the production of its releasing factor, GnRH. For example, if the level of circulating testosterone in the blood is too high, testosterone inhibits further production of GnRH in the hypothalamus, leading to a drop in the secretion of LH and FSH from the anterior pituitary. A decrease in LH in the blood results in reduced production and secretion of testosterone by the Leydig cells of the testes, hence blood levels of testosterone fall.

Conversely, if the circulating level of testosterone is too low, the hypothalamus increases synthesis and release of GnRH, which stimulates the anterior pituitary to secrete additional LH. The increase in circulating LH stimulates the Leydig cells of the testes to increase the synthesis and release of testosterone into the blood. In addition to this feedback loop between testosterone and the hypothalamus, testosterone can also act directly on the anterior pituitary to inhibit the release of LH in response to GnRH secreted by the hypothalamus.

Another hormone secreted by the testes exerts an inhibitory feedback effect on the release of FSH from the pituitary. This peptide hormone, called **inhibin,**

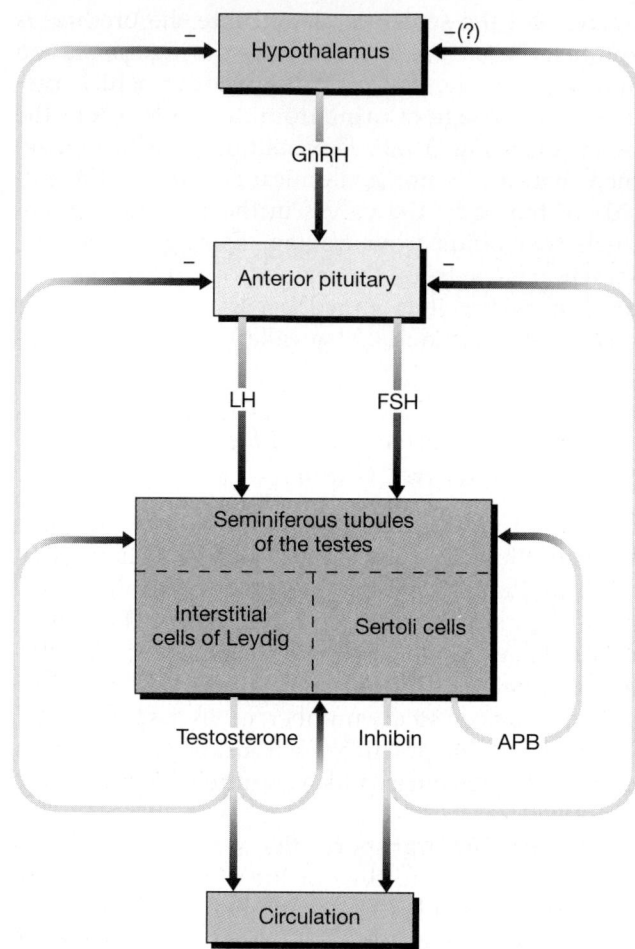

Figure 31–10

Hormonal feedback loops within the hypothalamic-pituitary-gonadal system. The minus sign indicates inhibition.

is secreted by the Sertoli cells in the seminiferous tubules of the testes. Although inhibin is known to directly inhibit the release of FSH from the anterior pituitary, it is not known whether this hormone can inhibit the release of GnRH from the hypothalamus as well (see Fig. 31–10).

Control of Neuroendocrine Function by Sensory Stimuli

The hypothalamus receives neural input from a variety of brain areas involved in processing sensory stimuli. The amount of GnRH released by the hypothalamus is controlled by the excitatory and inhibitory neural signals that the hypothalamus receives. Visual, olfactory, or tactile stimuli are capable of altering GnRH release and hence the release of LH, FSH, and testosterone. In addition, thoughts and moods can alter the secretion of these sex hormones. Chapter 32 discusses the role of the central nervous system in mediating the sexual response to direct or imagined stimulation. Stress, such as protracted aerobic exercise, can inhibit gonadotropin secretion in both males and females, which may result in transient inhibition of sperm production in the male and disruption of menstruation in the female.

Testosterone, the Major Sex Hormone in the Male, Has a Biological Effect on Many Tissues of the Body

Testosterone is the primary **androgen,** or male sex hormone, produced in the testes. Testosterone is a steroid formed from the steroid precursor cholesterol. The general features of testosterone biosynthesis were discussed in Chapter 12. The Leydig cells do not store testosterone, but they do store large quantities of the cholesterol precursor in the form of lipid droplets. Following synthesis, testosterone is secreted into the extracellular fluid surrounding the seminiferous tubules, where it diffuses into the tubules to produce local effects on sperm production. Testosterone is also transported from the extracellular fluid into the blood vessels of the testes and from there to the general circulation, where it produces a variety of systemic effects.

When testosterone enters the blood, it is bound to albumin and to another plasma protein, **sex steroid-binding globulin (SSBG)** or **testosterone-estradiol-binding globulin (TeBG).** As discussed in Chapter 12, hormones bound to plasma proteins are not metabolized by the body as quickly as unbound hormones. A certain proportion of all steroid hormones are "stored" in the circulation in their bound form, while the free or unbound hormone acts on target tissues.

Testosterone has some biological effect on almost every tissue of the body. Unaltered testosterone has direct effects on target tissues such as muscle, kidney, and bone. In other tissues, testosterone is transformed into another steroid by the target tissue before it exerts its effect. For instance, in the prostate and seminal vesicles, testosterone is transformed into **dihydrotestosterone** before acting on these tissues. Note that the action of testosterone on the brain is due to the conversion of testosterone to estradiol by the enzyme aromatase. As we will see in Chapter 33, this conversion of a "male" hormone to a "female" hormone in the central nervous system (CNS) is very important in sexual differentiation.

Effects of Testosterone on Puberty

Before puberty, the hypothalamic-gonadal system is relatively dormant. Puberty begins with the secretion of increasing amounts of GnRH from the hypothala-

---

Table 31–3

Testosterone-Dependent Changes at Puberty

1. Penis and testes enlarge and become pigmented; the scrotal skin becomes wrinkled.
2. Growth of facial hair results in the development of mustache and beard, and scalp line recedes.
3. Pubic hair grows in an upward pattern on the abdomen; hair appears on the chest, armpits, and extremities.
4. Growth of the long bones increases to a rate of about 3 inches per year. Bones increase in size and strength due to an increase in the quantity of bone matrix and increased calcium retention.
5. Vocal cords and larynx enlarge, thereby lowering the pitch of the voice.
6. Muscle mass and skin thickness increase, and the skin deepens in hue due to melanin deposition. Secretions from the sebaceous glands of the face increase, resulting in the development of acne.

---

mus, which stimulates the release of LH and FSH from the pituitary. FSH produces enlargement of the testes, and LH stimulates the secretion of testosterone from the Leydig cells. This increase in testosterone secretion at puberty results in the enlargement of the penis, scrotum, and testes, and stimulates the development of the male secondary sex characteristics. The **testosterone-dependent changes** that occur at puberty are detailed in Table 31–3. Testosterone stimulates muscular development, bone growth, thickening of the skin, and growth of facial and body hair. Testosterone also produces growth and thickening of the vocal cords and enlargement of the larynx, thereby lowering the pitch of the voice. These anatomical changes, stimulated by testosterone, mark the onset of puberty. If testosterone secretion is interrupted prior to puberty, as a result of destruction or removal of the testes, a male will not undergo the anatomical changes characteristic of puberty.

Effects of Testosterone on Spermatogenesis

Spermatogenesis requires the presence of high levels of testosterone in the seminiferous tubules. A decrease in testosterone production can result in sterility. Testosterone is thought to stimulate both the formation of spermatogonia and the second meiotic division that results in the formation of spermatids in the seminiferous tubules. An interesting interdependence exists between testosterone and the Sertoli cells (see Fig. 31–10). Testosterone stimulates protein synthesis and fluid secretion from the Sertoli cells, and the Sertoli cells in turn secrete androgen-binding-protein (ABP), which binds testosterone and keeps testosterone levels high in the seminiferous tubules (see Fig. 31–10).

Reproductive Physiology in the Female

The major reproductive functions in the female are (1) secretion of sex hormones, (2) production of ova, and (3) gestation and delivery of the fetus. The first two functions are carried out by the ovary, while the third is performed by the uterus and cervical canal.

The Principal Reproductive Anatomical Structures in the Female Are Internal

Anatomy of the Female External Genitalia

The external genitalia of the female, which are collectively called the **vulva,** include the **labia majora,** the **labia minora,** and **clitoris,** and the **vaginal opening** (Fig. 31–11). The labia majora consist of folds of skin that cover fat and smooth muscle. The labia majora meet on the midline and protect the urinary opening and the entrance to the vagina. The labia minora are composed of skin folds, rich in small blood vessels, that meet above the clitoris and provide mechanical protection for the clitoris. The only visible part of the clitoris is termed the **glans clitoris.** This structure contains a dense supply of nerve endings, which makes it highly sensitive to touch, pressure, and temperature. Note that while the penis and the clitoris are derived embryologically from the same tissue, the clitoris is the only tissue in either sex that has no known reproduction function. The uniquely sexual function of the clitoris will be discussed in Chapter 33.

Anatomy of the Female Internal Genitalia

The internal reproductive organs of the female include the **vagina, uterus, uterine tubes,** and **ovaries** (Fig. 31–12). In contrast to those in the male, the urinary and reproductive tracts in the female are anatomically separate.

The Vagina.   The vagina is a muscular organ that extends from the external opening in the vulva to the uterus. The muscular vaginal walls have the ability to contract and expand, allowing for the passage of a baby during childbirth. The vaginal lining, called the **vaginal mucosa,** secretes a fluid during sexual arousal that acts as a lubricant. The vagina contains few sensory nerves and is relatively insensitive to touch and pressure.

The Uterus.   The uterus, which is located in the lower abdomen, is a hollow, fist-sized organ with

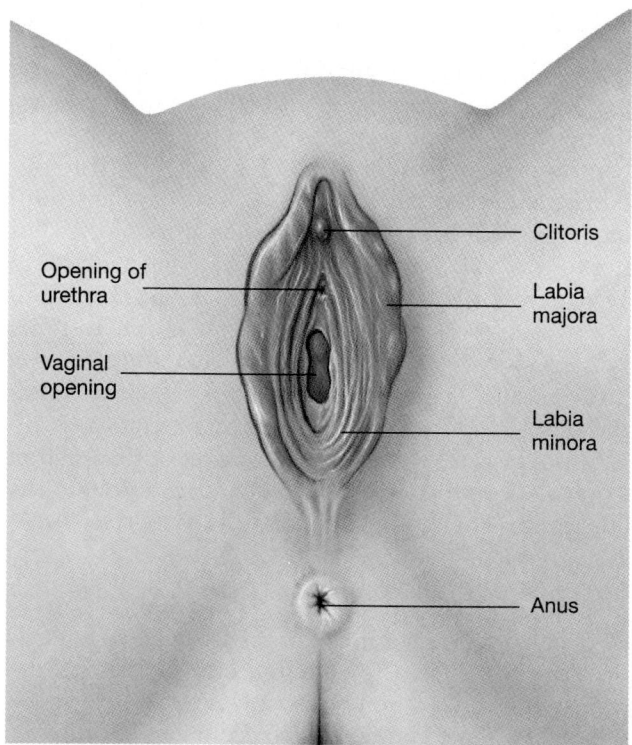

Figure 31–11

External genitalia of the female.   *(Redrawn from Masters and Johnson,* Human Sexuality, *Second Edition.)*

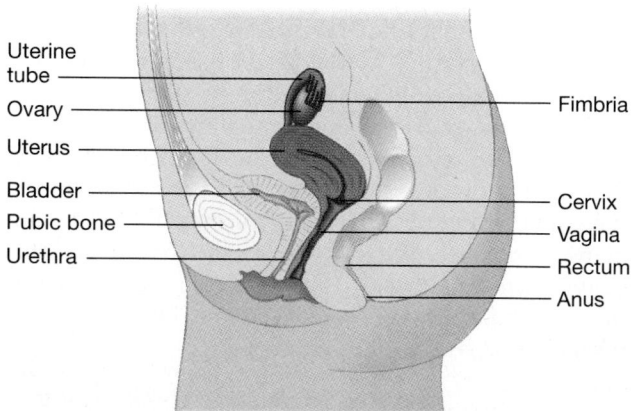

Figure 31–12

Side view of the female reproductive tract.   *(Redrawn from Geer, J., Heiman, J., and Leitenberg, H.,* Human Sexuality.*)*

walls composed of smooth muscle. The term **cervix** applies to the portion of the uterus that extends into the vagina. The lining of the uterus, termed the **endometrium,** changes in thickness during the menstrual cycle. At the beginning of pregnancy, the fertilized ovum implants in the endometrium of the uterus. The muscular walls of the uterus, called the **myometrium,** expand and contract during labor and delivery (see Chapter 32) and during sexual stimulation (see Chapter 33).

The Uterine Tubes.   The uterine tubes, also called **fallopian tubes,** or **oviducts,** arise from the uterus and extend into the abdominal cavity (Fig. 31–13). These muscular, tubelike structures terminate near the ovaries. The end of each uterine tube is surrounded by long, fingerlike extensions called the **fimbria.** The uterine tubes are lined with cilia, which propel the ovum through the uterine tube to the uterus (Fig. 31–13).

The Ovaries.   The ovaries are oval-shaped organs located in the pelvic cavity below the open ends of the uterine tubes (see Fig. 31–13). The ovaries and the testes have similar embryonic origins and similar functions: (1) the production of germ cells, and (2) the production of hormones.

Oogenesis, or the Formation of Ova, Takes Place in the Ovaries

The primary germ cells in the female, termed **oogonia,** are analogous to spermatogonia in the male. During the second and third months of embryonic development, meiosis begins in some oogonia, converting them to primary oocytes. The **primary oocytes** increase in size and are surrounded by a layer of flattened follicular cells called granulosa cells. The entire structure is then called a **primary follicle** or primordial follicle. Primary oocytes that are not converted to primary follicles degenerate.

During the fourth and fifth months of fetal development, the number of primary follicles in the ovary increases, and the number reaches a peak during the sixth month of fetal development (Fig. 31–14). The primary follicles increase in size and are surrounded by a layer of cuboidal granulosa cells called the stratum granulosum. The granulosa cells secrete mucopolysaccharides that form a coating, or membrane, called the **zona pellucida,** which lies between the follicular cells and the oocyte (Fig. 31–15a, b). The granulosa cells provide nutrients and chemical signals that stimulate further maturation of the follicle (Fig. 31–15a, b).

The **granulosa cells** form a barrier around the developing follicle that prevents the entrance of sub-

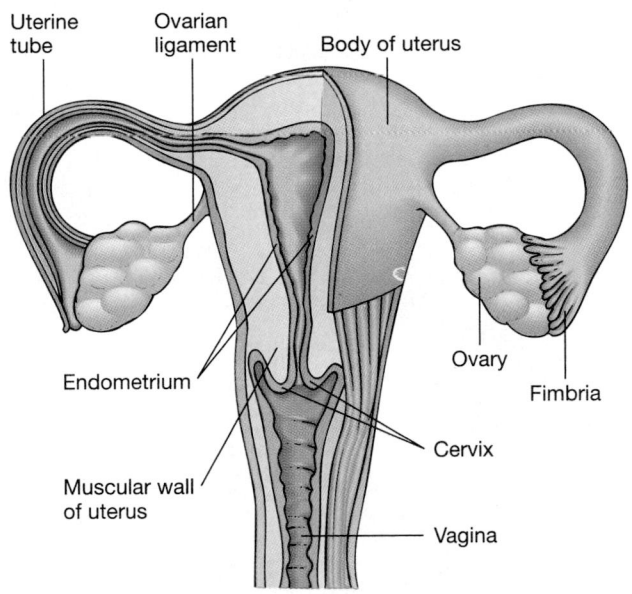

Figure 31–13

Front view of the female reproductive tract.   *(Redrawn from Masters and Johnson,* Human Sexuality, Second Edition.*)*

stances such as proteins, ions, and drugs, which could destroy the developing germ cell. Therefore, the granulosa cells perform a function similar to the function performed by Sertoli cells in the testes. During the fifth and sixth months of gestation, the connective tissue around the follicle differentiates to a form a highly vascular cell layer, the **theca interna,** which surrounds the follicle. The primary follicles remain in this first stage of meiotic division until puberty, when ovulation begins. The first meiotic division of the

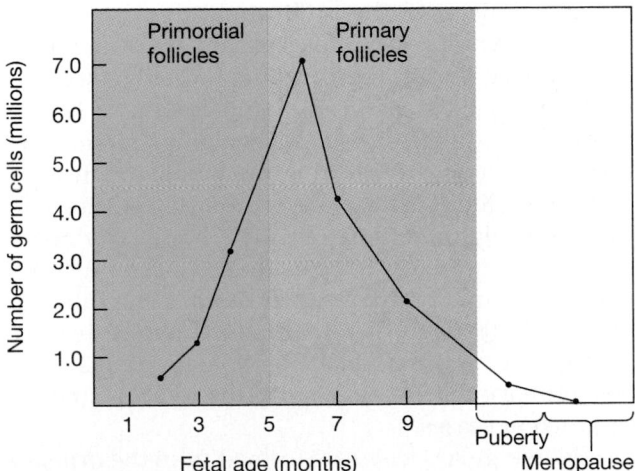

Figure 31–14

Number and type of follicles in the ovary during development.

Figure 31–15

(*a*) Cross-section of the ovary, containing follicles in different stages of development. (*b*) Cross-section of a Graafian follicle. (*c*) Cross-section of a developing follicle in the ovary.   (© *Biophoto Associates.*)

oocyte is completed just before ovulation; the second meiotic division does not occur unless and until the germ cell is fertilized by a spermatozoon.

There are major differences between the growth patterns of germ cells in the male and female. In the male, as we have seen, the spermatogenic cells are constantly undergoing a complicated structural and biochemical transformation into mature sperm in the testes. In contrast, in the female, meiotic division of the germ cells begins during embryonic development but then stops until puberty, when ovulation begins.

Production of primary follicles in the female contrasts sharply with the production of sperm in the male. While millions of sperm may be produced daily in the testes during the entire reproductive lifespan of the male, all of the primary follicles that a female will ever produce, a maximum of about 7 million, are formed in the ovaries during the first six months of embryonic life. Most of these primary follicles will disappear. By the time a female is born, only about two million primary follicles remain in the ovaries, and at the onset of puberty this number has fallen to about 400,000 (Fig. 31–14). When the female reaches menopause, at approximately 50 years of age, very few primary follicles are left in the ovaries, and reproduction ceases.

Ova (singular, *ovum*) are formed from the primary follicles in the ovary. During the middle of each ovulatory cycle, a single ovum is expelled by one of the two ovaries into the abdominal cavity. As we shall see in Chapter 32, the ovum enters one of the uterine tubes and passes through the uterine tube into the uterus. If the ovum has been fertilized by a sperm, it implants in the uterus. If it has not been fertilized, it disintegrates and leaves the uterus along with the menstrual flow. Usually only one ovum is released from one of the two ovaries each month during the reproductive lifespan of the female, which begins at the onset of puberty and ends with menopause. So of the 7 million follicles formed in the ovaries during embryonic development, only about 400 to 500 will ever develop into mature ova and be released from the ovaries during a woman's reproductive lifespan.

Ovulation, or the Release of an Ovum from One of the Two Ovaries, Occurs Once Each Month from Puberty Until Menopause

Once each month, from puberty until menopause, 6 to 20 primary follicles enter the next stage of follicular development which usually culminates in the release of a single ovum from one of the two ovaries.

Growth of the Follicle

Each month the spindle cells, surrounding 6 to 20 of the primary follicles, proliferate to form the **theca externa**. These mature follicles are called **Graafian follicles**. Figure 31–15*a* illustrates the structure of the ovary, which contains follicles in different stages of development. Although 6 to 20 follicles may begin to develop in the ovary each month, only one reaches maturation. As one follicle increases in size faster than the others, the remaining follicles in the ovary degenerate and undergo **atresia.** Degeneration of the remaining follicles can occur at any stage of follicular development, and atretic follicles can always be found in the ovary. As we will see, hormone levels in the ovary may determine whether a particular follicle will ovulate or undergo atresia.

Several days before ovulation, a single Graafian follicle increases dramatically in size to reach 10 to 20 mm in diameter within 48 hours. The granulosa cells surrounding this follicle begin to secrete increasing amounts of fluid that contains proteins, electrolytes, polysaccharides, and hormones. The follicular fluid in the Graafian follicle collects in a single central area called the **antrum** (see Fig. 31–15*b, c*). The fluid-filled antrum displaces the ovum to the side of the follicle, and the follicle itself forms a bulge on the surface of the ovary. The follicle secretes proteolytic enzymes that break down the surface of the ovary, and the antral fluid and the ovum are released from the follicle into the peritoneal cavity in a process known as **ovulation** (see Fig. 31–15*a*). Just before ovulation, the first meiotic division of the ovum is resumed and, following ovulation, the ovum enters the uterine tube.

Growth of the Corpus Luteum

During the first few hours after ovulation, the remaining granulosa and theca cells of the follicle undergo complex physical changes, and the follicle becomes the **corpus luteum** (see Fig. 31–15*a*). The cells of the corpus luteum increase in size and develop a rich vascular supply for 7 to 8 days following ovulation. If fertilization and implantation of the ovum do not occur, the corpus luteum degenerates between days 8 and 21 following ovulation. The corpus luteum secretes large amounts of **progesterone** and smaller amounts of **estrogen,** which interact with pituitary hormones to retard the growth of new follicles in the ovary. When the corpus luteum degenerates, new follicles begin to grow again in the ovary, marking the beginning of a new monthly ovarian cycle.

Neuroendocrine Control of Reproduction in the Female Involves an Interaction Between the Hypothalamus, Pituitary, Ovaries, and Uterus

Neuroendocrine control of reproduction in the female is dependent on a continuous interaction between the hypothalamus, the pituitary gland, and the ovaries. The major difference between hormonal control of reproduction in the male and female is that hormone

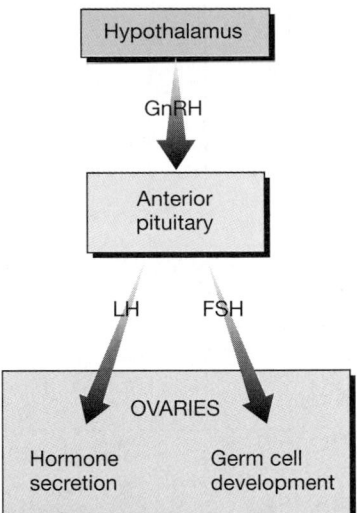

Figure 31–16

Relationship among the hypothalamus, pituitary gland, and ovaries.

secretion and reproductive processes in the female are cyclic rather than continuous.

In the female, as in the male, the hypothalamus secretes gonadotropin-releasing hormone, or GnRH, which is transported to the anterior pituitary, where it stimulates the release of LH and FSH. LH and FSH stimulate germ cell development and hormone secretion in the ovaries (Fig. 31–16).

In the female, the effects of LH and FSH on the ovary vary during different phases of the ovarian cycle. The **ovarian cycle** is approximately 28 days long and can be divided into three phases: (1) the follicular phase, which lasts for approximately 12 days; (2) the luteal phase, which lasts for approximately 11 days; and (3) the menstrual phase, which lasts for approximately 5 days.

During the **follicular phase,** a single follicle in one of the two ovaries matures and is released during ovulation. During the **luteal phase,** which immediately follows ovulation, the post-ovulatory follicular cells in the ovary differentiate into the corpus luteum, which functions for a few days and then degenerates. During the **menstrual phase,** there is little activity in the ovary but much activity in the uterus, where the lining of the uterus, called the endometrium, is shed in a process called **menstruation.** Figure 31–17 illustrates the cyclic hormonal, ovarian, and uterine changes that occur in the female during the ovarian cycle.

Neuroendocrine Control of the Follicular (Pre-ovulatory) Phase

Once each month, just after menstruation ceases, growth and development of follicles begin in the ovary. Follicular growth is stimulated by the pituitary

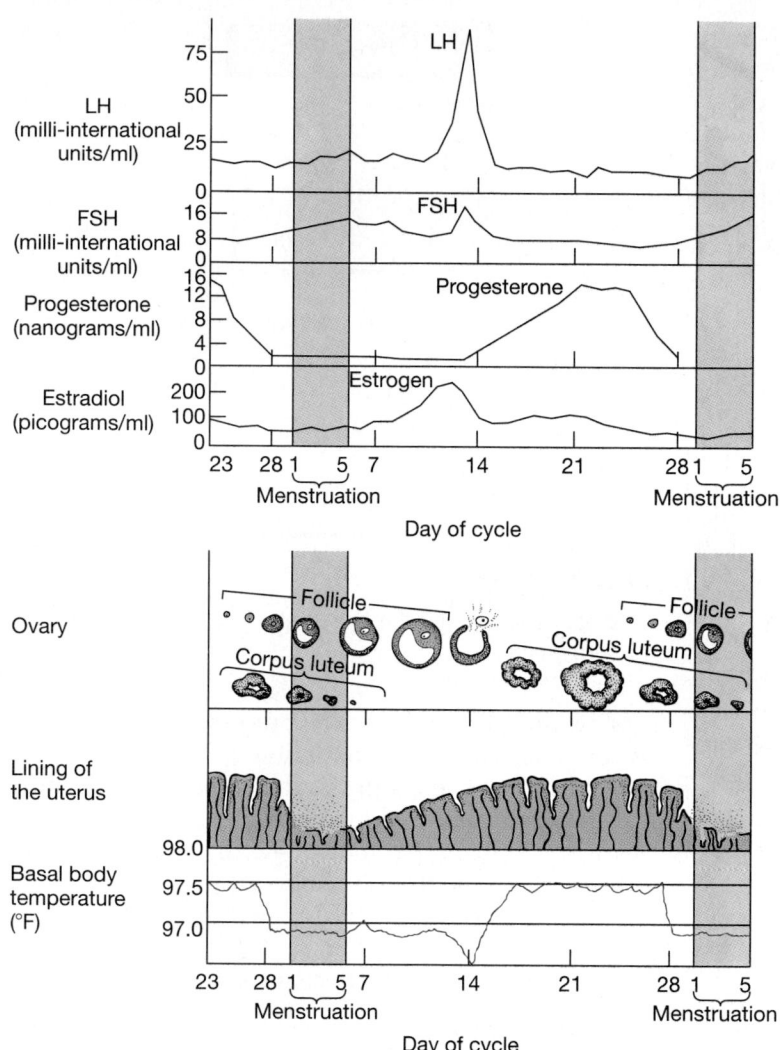

Figure 31–17

Hormonal, ovarian, and uterine changes that occur during the ovarian cycle. *(Courtesy of Dr. Resco, Oregon Health Sciences University.)*

hormone FSH. FSH stimulates both hypertrophy (increase in cell size) and **hyperplasia** (increase in cell-number) of the granulosa cells that surround the primary follicle. In response to FSH, the granulosa cells synthesize estrogen from androgen precursors. Estrogen secretion from the granulosa cells results in rising levels of circulating estrogen (see Fig. 31–17). Estrogen stimulates further GnRH secretion from the hypothalamus and additional LH and FSH secretion from the pituitary (Fig. 31–17).

The hormonal interactions between the pituitary and the ovary during the follicular phase of the ovarian cycle are represented in Figure 31–18. **Estrogen,** which is secreted by the granulosa cells of the follicle, acts locally on the granulosa cells to increase the number of receptors in the follicle for both estrogen and FSH. As the number of estrogen receptors in the follicle rises, more estrogen is retained in the follicle, and follicular estrogen content rises. The estrogen-induced increase in the number of FSH receptors on the granulosa cells in the follicle, serves to augment

the growth-promoting effect of FSH on the follicular granulosa cells. As the follicle grows, it secretes more estrogen, which continues to augment the action of FSH in the follicle.

Follicular growth and development in the ovary can be viewed as a self-perpetuating mechanism that involves a complex interaction between the pituitary gland and the ovary, resulting in a steady rise in follicular growth and estrogen secretion. As discussed in Chapter 12, when a hormone acts on the cell from which it was secreted, it is said to have an **autocrine action.** Estrogen, secreted by the granulosa cells of the follicle, acts on the granulosa cells to augment the further secretion of estrogen and to augment the action of FSH. Therefore, estrogen exerts an autocrine action on the granulosa cells of the developing follicle. FSH secreted by the pituitary and estrogen secreted by the follicle together increase the number of LH receptors on the granulosa cells of the follicle. This increase, as we will see, plays an important role in the formation of the corpus luteum.

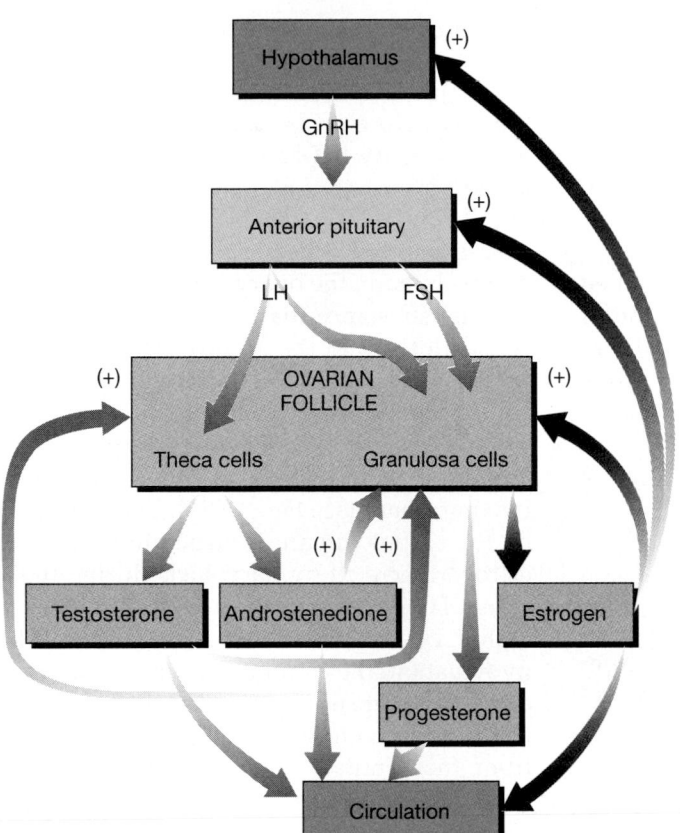

Figure 31–18

Hormonal interactions between the pituitary gland and ovary during the follicular phase of the ovarian cycle. The plus sign indicates stimulation.

During the follicular phase, another interaction between the pituitary gland and the ovaries occurs (see Fig. 31–18). The increasing estrogen, secreted by the granulosa cells of the follicle, stimulates the pituitary to synthesize and secrete additional LH. This rise of LH in the blood stimulates the theca cells of the follicle to synthesize and secrete two androgens: testosterone and **androstenedione.** These androgens exert a **paracrine** action on neighboring cells in the follicle. The androgens diffuse into the follicular granulosa cells and serve as precursors for estrogen production in the granulosa cells.

The rising blood levels of LH stimulate the granulosa cells of the follicle to secrete progesterone. Progesterone enters the circulation, and blood levels of progesterone begin to rise (see Fig. 31–17). In addition, some of the progesterone from the granulosa cells diffuses back into the theca cells and acts as a substrate for further androgen synthesis by the theca cells (see Fig. 31–18).

As we have seen, as many as 20 follicles in the ovary may increase in size each month during the follicular phase, but usually only one follicle reaches maturity. The others undergo atresia. The mechanisms that control the selection of a single follicle for full maturation are not understood. It appears that complex endocrine events in the ovary control the

growth and atresia of the various follicles. For example, atresia may result from an insufficient number of FSH receptors in some follicles, or from an inability of the granulosa cells of certain follicles to produce adequate amounts of estrogen from androgen precursors. Fluid from atretic follicles has been found to contain lower levels of both FSH and estrogen than fluid from mature follicles.

Neuroendocrine Control of Ovulation

During the follicular phase of the ovarian cycle, increased release of estrogen from the ovary stimulates LH and FSH release from the anterior pituitary, which results in continued follicular growth for approximately 10 days (see Fig. 31–17). On day 14, just prior to ovulation, blood levels of LH and FSH are higher than at any other time in the ovarian cycle. These peaks in blood levels of LH and FSH are referred to as the **preovulatory surges** of LH and FSH. The surge of LH induces changes in the structure of the follicle that result in ovulation (see Fig. 31–17). As we have discussed, these LH-induced changes include an increase in the production of antral fluid, compression of the ovum toward the wall of the follicle, protrusion of the follicle against the wall of the ovary, and, finally, enzymatic degeneration of the ovarian wall and release of the antral fluid and ovum into the abdominal cavity during ovulation. One or two days before ovulation, the follicle stops secreting estrogen, and blood levels of estrogen begin to fall (see Fig. 31–17). The mechanism that controls the cessation of estrogen secretion by the follicle is not known. Ovulation marks the end of the follicular phase of the ovarian cycle in the female.

In many women, ovulation may be accompanied by a slight decrease, followed by an upward shift of approximately 0.5 to 1.0 degree, in basal body temperature which persists throughout the luteal phase (see Fig. 31–17). However, since ovulation can occur in the absence of a detectable change in body temperature, measurements of body temperature are not reliable indicators of ovulation.

Neuroendocrine Control of the Luteal (Postovulatory) Phase

Immediately following the release of the ovum from the follicle on day 14 of the ovarian cycle, the remaining granulosa cells of the follicle increase in size

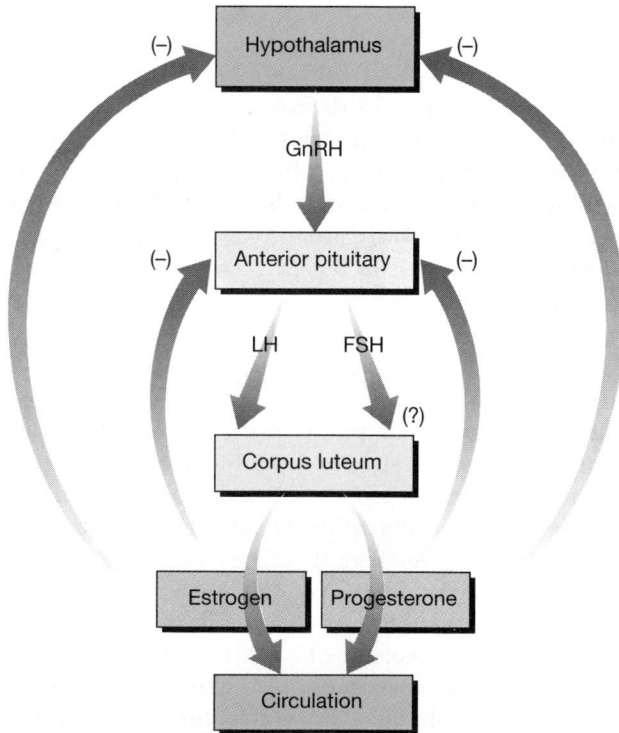

Figure 31–19

Hormonal interactions between the pituitary gland and ovary during the luteal phase of the ovarian cycle. The minus sign indicates inhibition.

and number and undergo chemical changes to form the corpus luteum (see Fig. 31–17). This process is called **luteinization.** Differentiation and growth of the corpus luteum are stimulated by LH from the anterior pituitary. LH derives its name, "luteinizing hormone," from its action on the corpus luteum. The role of FSH in the luteal phase is unknown. Although the presence of LH is necessary for the formation of the corpus luteum, the ovary is thought to secrete an additional substance that inhibits the formation of the corpus luteum in the mature follicle prior to ovulation, as well as in follicles that do not ovulate.

During the luteal phase, the corpus luteum secretes a large amount of **progesterone** and a smaller amount of **estrogen,** increasing blood levels of these two hormones (see Fig. 31–17). The estrogen secreted by the corpus luteum is structurally identical to the estrogen secreted by the follicle during the follicular phase. However, while estrogen stimulates release of GnRH, LH, and FSH during the follicular phase prior to ovulation, the estrogen and progesterone secreted during the luteal phase *inhibit* both GnRH secretion from the hypothalamus and LH and FSH secretion from the pituitary (Fig. 31–19). During the luteal phase, blood levels of LH and FSH fall (see Fig. 31–17). As we will see in Chapter 33, the efficacy of the "birth control pill" is based on this inhibitory action. The fact that estrogen and progesterone have different effects on GnRH, LH, and FSH release during the follicular versus the luteal phase of the ovarian cycle is due to a change in hypothalamic sensitivity to estrogen and progesterone during the follicular and luteal phases.

As LH and FSH blood levels fall during the luteal phase, the signal for follicular growth diminishes, and no new follicles begin to grow in the ovary. The corpus luteum increases in size for about 7 to 8 days following ovulation and then begins to degenerate. The mechanism responsible for degeneration of the corpus luteum is not completely known. Falling blood levels of LH may not provide an adequate signal for luteal growth (see Fig. 31–17). In addition, the corpus luteum itself is thought to secrete hormones, known as **prostaglandins,** which induce further degeneration of the corpus luteum.

As the corpus luteum degenerates, it secretes less progesterone and estrogen, so that blood levels of these hormones begin to fall (see Fig. 31–17). The inhibition of GnRH, LH, and FSH release, which was maintained by progesterone and estrogen from the corpus luteum, is now removed, so that blood levels of LH and FSH begin to rise again. Rising levels of LH and FSH in the blood stimulate growth of another set of follicles in the ovary, which marks the beginning of a new follicular cycle (see Fig. 31–17).

Chemical Termination of Early Pregnancy

Termination of pregnancy normally requires surgical intervention, which is accompanied by the risks of infection and postsurgical trauma. However, medical researchers in Europe have recently developed a new drug, mefipristone or RU 486, that represents a noninvasive alternative to surgical termination of early pregnancy. Mefipristone, known abroad as RU 486, after Roussel Uclaf, the French company that developed it, is a synthetic steroid derivative with a high binding affinity for progesterone receptors. It blocks the action of progesterone and therefore has been termed an *antiprogesterone.*

Progesterone, secreted by the corpus luteum during the ovarian cycle and by the placenta after the sixth week of pregnancy, is essential for the maintenance of pregnancy. Progesterone stimulates growth and proliferation of the endometrial lining of the uterus and strongly inhibits the uterine contractions. In the absence of stimulation from progesterone, the endometrium undergoes hemorrhagic changes; the outer layers of the endometrium separate from the uterus and are discharged, which represents the onset of menstruation. This same process results in natural miscarriage, or spontaneous abortion, which occurs 30% to 40% of the time during the first trimester of pregnancy.

Oral administration of RU 486, together with a prostaglandin to induce uterine contractions, results in the onset of menstruation and evacuation of the contents of the uterus whether or not a fertilized ovum is present. RU 486, which has been used throughout Europe since 1981, has been found to be both safe and effective in terminating early pregnancy and eliminates the risks of anesthesia and uterine perforation that are associated with surgical intervention. RU 486 is currently being tested in clinical trials in the United States as a contragestive agent for fertility control. If the results of the clinical trials indicate that RU 486 is both safe and effective, approval will be sought from the Food and Drug Administration in order to make RU 486 widely available to women in the United States.

■ **Would you expect progesterone levels to be low, moderate, or high during the last month of pregnancy?**

Neuroendocrine Control of the Menstrual Phase

Dramatic changes in the structure and secretory capacity of the endometrium of the uterus occur during the ovarian cycle in the female. These changes are induced by both estrogen and progesterone. As we have seen, during the follicular phase of the ovarian cycle, the growing follicle secretes large amounts of estrogen. Estrogen, among its many effects, stimulates growth and proliferation of the endometrium of the uterus (see Fig. 31–17). The endometrium thickens as cells enlarge and new cells are produced. Blood vessels invade the thickened tissue and tubular endometrial glands are formed. In addition, estrogen increases the number of progesterone receptors in the uterus. At the end of the follicular phase in the ovary, just before ovulation, the follicle stops secreting estrogen, resulting in a drop in circulating levels of estrogen (see Fig. 31–17).

Following ovulation, during the luteal phase of the ovarian cycle, estrogen and progesterone (secreted by the corpus luteum) stimulate continued growth and proliferation of the coiled blood vessels of the endometrium (see Fig. 31–17). Progesterone, from the corpus luteum, stimulates the endometrial glands to secrete a nutrient-rich mucus. The tubular endometrial glands fill with glycogen, and enzymes accumulate in the connective tissue of the endometrium. During the luteal phase, the endometrium swells and becomes approximately 4 to 6 mm thick, owing to the progressive growth of glands, blood vessels, and connective tissue.

The function of the proliferation of endometrial cells is to provide a continuing source of nutrients for the ovum if it is fertilized and implants in the uterus. If the ovum is not fertilized, the corpus luteum begins to degenerate on day 8 following ovulation and stops secreting progesterone and estrogen.

A drop in circulating levels of estrogen and progesterone, resulting from degeneration of the corpus luteum, causes **hemorrhagic changes** in the uterine endometrium that result in the onset of menstruation (see Fig. 31–17). In the absence of stimulation from estrogen and progesterone, the endometrium becomes **necrotic.** Arterioles of the endometrium constrict, which results in a slowing of the circulation. Blood enters the vascular layer of the endometrium, resulting in blood pooling. Gradually, the outer layers of the endometrium separate from the uterus at the site of hemorrhage.

Prostaglandins, produced by the uterus, stimulate the uterine smooth muscles to contract rhythmically. These uterine contractions result in the discharge of blood and endometrial cells that constitutes the **menstrual flow.** The menstrual flow lasts for approximately 5 to 7 days (see Fig. 31–17), during which time the uterus expels an average of 20 to 200 milliliters of blood. An overproduction of prostaglandins in the uterus during the menstrual phase of the ovarian cycle can lead to very strong and protracted uterine contractions in many women causing menstrual cramps, a condition called **dysmenorrhea.** Prostaglandin action on smooth muscles in other areas of the body can also result in the nausea, vomiting, and headache that sometimes accompany the menstrual phase.

A summary of the hormonal, ovarian, and uterine changes that occur during the ovarian cycle is contained in Table 31–4.

Estrogen, the Primary Sex Hormone in the Female, Has a Biological Effect on Many Tissues of the Body

The major sex hormones in the female are **estrogen** and **progesterone.** As described in Chapter 12, both of these hormones are steroids. The term **estrogen** refers to a group of steroids that includes estradiol, estrone, and estriol. In addition to their roles in regulating the ovarian cycle, estrogen and progesterone have widespread effects on many other physiological and metabolic processes. The effects of estrogen are summarized in Table 31–5.

Effects of Estrogen at Puberty

Estrogen stimulates the development of the sex organs and the appearance of the secondary sex characteristics in the female. Therefore, the effects of estrogen in the female are analogous to the effects of testosterone in the male. At puberty, estrogen stimulates growth of the vagina, uterus, and fallopian tubes as well as the external genitalia, breasts, and body hair. Estrogen stimulates linear growth but also accelerates closure of the epiphyses of the long bones, which results in cessation of bone growth—which is why many women are shorter than men. Stimulation by estrogen results in the particular pattern of distribution of body fat around the hips, buttocks, and abdomen that characterizes the female form. When the ovary is absent, or when it secretes inadequate amounts of estrogen, puberty either fails to occur or progresses very slowly.

Table 31–4
Summary of the Hormonal, Ovarian, and Uterine Changes That Occur During the Ovarian Cycle

1. During the early part of the follicular phase, blood levels of LH and FSH begin to rise.
2. This rise in LH and FSH stimulates growth of 6 to 12 primary follicles and the full maturation of a single Graafian follicle.
3. The growing follicles secrete large amounts of estrogen, lesser amounts of progesterone, and small amounts of testosterone and androstenedione.
4. Rising blood estrogen stimulates growth of the endometrium of the uterus.
5. Rising blood estrogen also stimulates further secretion of GnRH, which results in the preovulatory surges of LH and FSH.
6. Just prior to ovulation the follicle stops secreting estrogen and blood estrogen levels fall.
7. The surges of LH and FSH at midcycle induce ovulation.
8. Following ovulation, the remaining follicular cells in the ovary become the corpus luteum.
9. The corpus luteum secretes large amounts of progesterone and lesser amounts of estrogen. Blood levels of these hormones begin to rise.
10. Progesterone and estrogen from the corpus luteum stimulate further growth of the endometrium of the uterus.
11. Progesterone and estrogen also inhibit secretion of LH and FSH, and blood levels of these pituitary hormones begin to fall.
12. As blood levels of LH and FSH fall, the signal for follicular growth diminishes and no new follicles being to grow.
13. On day 8 following ovulation, the corpus luteum begins to degenerate and secretes less progesterone and estrogen.
14. Falling blood levels of progesterone and estrogen result in hemorrhagic changes in the uterine endometrium, which culminate with the onset of menstruation.
15. As blood levels of progesterone and estrogen fall at the end of the luteal phase, the inhibition of GnRH, LH, and FSH, which was exerted by progesterone and estrogen, is removed. Blood levels of LH and FSH begin to rise, thus marking the beginning of another follicular phase (see point 1).

Metabolic Effects of Estrogen

Estrogen interacts with a variety of other hormones that control metabolic processes. For instance, estrogen decreases the sensitivity of peripheral tissues to

Table 31–5
Effects of Estrogen

Puberty
  Stimulates growth of internal reproductive
    organs, external genitalia, and breasts
  Stimulates growth of body hair
  Stimulates growth of long bones and early
    closure of the epiphyses
  Stimulates "female" pattern of fat distribution
Hormonal
  Stimulates secretion of GnRH, LH, and FSH
    during the follicular phase and inhibits
    secretion of these same hormones during the
    luteal phase
  Decreases sensitivity of peripheral tissues to
    insulin
  Increases circulating levels of renin and
    angiotensin II
Metabolic
  Protects against bone loss

(a)

(b)

Figure 31–20

Photograph of (*a*) normal bone matrix and (*b*) erosion of bone matrix which characterizes osteoporosis. *(With permission from Dempster et al., J. Bone Min. Res. 1, 15–21, 1986.)*

insulin, which can result in decreased glucose tolerance. Estrogen increases the levels of plasma proteins that bind thyroxine, cortisol, and progesterone. In addition, estrogen increases circulating levels of renin and thus angiotensin II, which can result in hypertension.

Evidence is accumulating that estrogen protects against bone loss in women. For instance, removal of the ovaries, or inadequate ovarian secretion of estrogen, results in the acceleration of bone loss known as **osteoporosis** (Fig. 31–20). In such cases, replacement of estrogen can dramatically slow the rate of bone loss. Postmenopausal women, in whom ovarian secretion of estrogen is very low, exhibit osteoporosis most often. Although men have very low circulating levels of estrogen, secreted by the adrenal corex, they are not particularly prone to osteoporosis. However, impairment of gonadal function in the male, a condition called **hypogonadism,** is often accompanied by low bone mass. It has been suggested that androgens are converted to estrogen in bone and hence may protect against bone loss in the male.

Effects of Progesterone

Progesterone is the most active naturally occurring antagonist of androgens. Progesterone antagonizes the effects of both testosterone and dihydrotestosterone by competing for the androgen-binding sites present in androgen-dependent tissues. Progesterone prevents target tissues from responding to the nor-

mally low levels of circulating androgens in the female. Progesterone and its derivatives are very useful in treating abnormally heavy hair growth in the female, a condition termed **hirsutism,** which is thought to be due to an overproduction of androgens from the adrenal cortex.

Progesterone, secreted by the corpus luteum, may also promote the edema (salt and fluid retention) that occurs in many women at the end of the luteal phase of the ovarian cycle prior to the onset of menstruation. Progesterone is thought to increase circulating levels of renin and angiotensin, which stimulates aldosterone secretion and results in salt and fluid retention. However, the severity of premenstrual fluid retention does not appear to correlate with circulating levels of renin or aldosterone. The role of ovarian hormones in controlling fluid and electrolyte levels in the female is not completely understood.

CLINICAL FOCUS

Menopausal Osteoporosis

Menopause is accompanied by decreased ovarian production of estrogen. One of the long-term metabolic effects of decreased estrogen is a reduction of total bone mass or bone density. Following menopause, women lose approximately 1% to 2% of their bone mass per year, because during this time more bone matrix erodes than is being replaced. Consequently, bones become thin and brittle, and fractures occur easily. Bone pain is often the first symptom indicating the presence of osteoporosis. Intense pain may develop suddenly because of compression fractures of the vertebrae or fractures of the long bones. In X-ray images, the bones appear abnormally lucent (transparent), owing to thinness, and the spine is curved because of compression. Such curvature of the spine may result in decreased height over a period of years.

Treatment of menopausal osteoporosis is difficult, because no pharmacological agent has been discovered that is both safe and effective in inducing the formation of new bone. At present, the best treatment of menopausal osteoporosis involves maintaining estrogen levels through hormone replacement therapy (HRT). When begun at the onset of menopause, estrogen replacement decreases the rate of bone loss that normally occurs following menopause. Estrogen is usually administered in the lowest dose found to relieve symptoms, since high doses are associated with increased risk for certain types of cancer.

In treatment of menopausal osteoporosis, replacement of estrogen is accompanied by maintenance of an adequate dietary intake of calcium. However, excessive calcium intake is of limited value since absorption of calcium from the gastrointestinal tract is never complete. Dietary calcium is derived from dairy products such as milk, cheese, and eggs, as well as from white bread. Adults on normal diets consume approximately 1000 mg of calcium daily. Of this amount, less than half is absorbed from the gastrointestinal tract, and the rest is lost in the feces and urine. Most women with menopausal osteoporosis excrete normal amounts of calcium into the urine.

Although a cure for menopausal osteoporosis remains elusive, prevention involves ensuring that excessive bone loss does not occur before the onset of menopause. Prevention also involves slowing the rate of bone loss following menopause. Maintenance of physical activity and an adequate diet throughout life, together with estrogen replacement at the onset of menopause, will decrease bone loss in later years.

■ **Are women more susceptible to developing osteoporosis following a total hysterectomy?**

Menopause Refers to the Cessation of Menstruation in the Female

Menopause, the cessation of menstruation in the female, occurs on the average at age 50. However, there is considerable variation in the age of onset of menopause. For several years before the onset of menopause, menstruation occurs less frequently and at variable intervals. As previously discussed, the number of follicles in the ovary declines with age, hence estrogen secretion declines as well. When most of the follicles have disappeared from the ovary, the ovary stops secreting estrogen. The low circulating levels of estrogen found in postmenopausal women are due to secretion of estrogen precursors by the adrenal cortex, which are then converted to estrogen in the periphery.

During menopause, the remaining follicles in the ovary become less sensitive to LH and FSH, and to compensate, blood levels of these pituitary hormones begin to rise. The decline in circulating estrogen during menopause results in a loss of vaginal epithelium, a decrease in breast mass, and an increase in bone loss. Other symptoms that accompany menopause include vascular flushing ("hot flashes"), rapid shifts in mood and emotion, and an increased risk of cardiovascular diseases.

CHAPTER REVIEW

Summary

- The reproductive organs in the male are the penis, testes, epididymis, vas deferens, seminal vesicles, and prostate gland.
- Spermatogenesis takes place in seminiferous tubules of the testes.
- During spermatogenesis, germ cells undergo a series of divisions and differentiate from spermatogonia into mature spermatozoa.
- The Sertoli cells, which form a protective blood-testes barrier around the inside of the seminiferous tubules, ensure that the sperm develop in a protected environment.
- Sperm mature in the epididymis and are stored in the vas deferens.
- The seminal vesicles, prostate gland, and bulbo-urethral gland contribute important fluids that nourish sperm and add volume to the ejaculate.
- Neuroendocrine control of reproduction in the male involves the secretion of gonadotropin-releasing hormone from the hypothalamus, which in turn stimulates the release of luteinizing hormone and follicle-stimulating hormone from the pituitary.
- LH stimulates the Leydig cells of the testes to secrete testosterone. FSH stimulates spermatogenesis in the testes. Release of GnRH, LH, and FSH is controlled through feedback loops involving both testosterone and inhibin.
- Testosterone stimulates the onset of puberty and promotes spermatogenesis in the testes.
- The external reproductive organs in the female are the labia majora, the labia minora, and the clitoris. The internal reproductive organs include the vagina, uterus, ovaries, and uterine tubes. In contrast to the male, the urinary and reproductive tracts in the female are anatomically separate.

- Oogenesis takes place in the ovaries. In contrast to the male, germ cell division in the female begins during embryonic development but is then arrested until puberty.
- Oogenesis begins with the growth of primary follicles in the ovary. Once each month, beginning at puberty, one of the primary follicles in the ovaries grows at a rapid rate and differentiates into a mature Graafian follicle containing an ovum.
- At midcycle, this mature follicle releases its ovum in a process termed ovulation. Following ovulation, the ovarian follicle becomes the corpus luteum.
- Neuroendocrine control of reproduction in the female involves the secretion of both GnRH from the hypothalamus and LH and FSH from the pituitary.
- During the follicular phase of the ovarian cycle, FSH stimulates the growth of ovarian follicles, which in turn secrete estrogen.
- The luteal phase of the ovarian cycle, which begins following ovulation, is marked by growth of the corpus luteum in response to LH stimulation. The corpus luteum secretes both estrogen and progesterone, which in turn stimulate growth and proliferation of the endometrium of the uterus.
- As the corpus luteum begins to degenerate at the end of the luteal phase, secretion of estrogen and progesterone diminishes, and this results in necrotic changes in the endometrium that culminate in the onset of menstruation.
- Estrogen, secreted first by the developing follicles and later by the corpus luteum, has a wide variety of effects on other endocrine systems and on many tissues of the body.
- Menopause, the cessation of menstruation, marks the end of the reproductive years in the female.

Review Questions

Identify Each with a Term

1. The process resulting in a reduction of total bone mass or bone density.
2. The tightly coiled tubes in the testes that contain sperm-producing cells.
3. The ovarian follicle after ovulation.
4. The lining of the uterus.

Define Each Term

5. crossing over
6. menopause
7. mitotic division
8. spermatogonia

Choose the Correct Answer

9. The acrosome contains:

a. enzymes   c. chromatin
b. mitochondria   d. secondary spermatocytes

10. Sperm become motile in the:

a. seminiferous tubules   c. urethra
b. epididymis   d. vas deferens

11. At the onset of puberty, the ovary contains the following number of follicles, or germ cells:

a. 400   c. 400,000
b. 4,000   d. 400,000,000

12. Mature sperm are produced in:

a. one day   c. one month
b. one week   d. three months

Answer Each Question in One or Two Sentences

13. Why is the location of the testes in the scrotal compartment important for sperm production?

14. Why does the endometrium thicken during the luteal phase of the ovarian cycle?
15. Describe the changes that occur in the follicle following ovulation and explain why the resulting structure is important.
16. In what ways are Sertoli cells important for spermatogenesis?

Problem Solving in Physiology

17. Why is it important that events in the uterus and ovary be precisely synchronized, and how is this synchrony regulated?
18. What factors can interfere with proper functioning of the reproductive system in males and females?
19. Why is formation of the corpus luteum important for reproduction?

Suggested Readings

Bulletti, C., Gurpide, E., and Flamigni, C., eds. "The Human Endometrium." *Annals of the New York Academy of Sciences,* Vol. 734, 1994.

Flint, M., Kronenberg, F., and Utian, W.H., eds. "Multidisciplinary perspectives on menopause." *Annals of the New York Academy of Sciences,* Vol. 592, 1990.

Robaire, B., ed. "The male germ cell: Spermatogonium to fertilization." *Annals of the New York Academy of Sciences,* Vol. 637, 1991.

Seppala, M., and Hamberger, L., eds. "Frontiers in Human Reproduction." *Annals of the New York Academy of Sciences,* Vol. 626, 1991.

Wassarman, P.M., "Fertilization in mammals." *Scientific American,* Vol. 259, 1988.

Yen, S.S.C., and Jaffe, R.B., eds. *Reproductive Endocrinology.* Philadelphia: W.B. Saunders Co., 1986.

Key Terms

androgen-binding protein (ABP) (p. 907)
corpus luteum (p. 915)
crossing over (p. 903)
dysmenorrhea (p. 920)
endometrium (p. 913)
epididymis (p. 907)

estrogen (p. 915)
uterine (fallopian) tube (p. 913)
gonadotropin releasing hormone (GnRH) (p. 909)
inhibin (p. 907)

luteinization (p. 918)
meiotic division (p. 903)
menopause (p. 922)
oogenesis (p. 913)
ovary (p. 912)
ovum (p. 914)
progesterone (p. 915)

puberty (p. 911)
sperm (p. 905)
spermatogenesis (p. 903)
testis (p. 907)
testosterone (p. 911)
vas deferens (p. 908)

The embryo at 4–5 weeks of development. The chorion, which is shaggy in appearance, and the amniotic sac have been opened to visualize the embryo.
(© Lennart Nilsson from A Child Is Born, *Dell Publishing Company)*

Chapter 32

Pregnancy, Fetal Development, and Lactation

Key Concepts

Because both sperm and ovum are viable for a limited period of time, fertilization will only occur when sperm are deposited in the vagina close to the time of ovulation.

Cell division, which begins prior to implantation, marks the beginning of embryonic development. Embryonic cells are the only cells, other than cancer cells, that divide on their own initiative. All other cells divide only in response to a need for tissue replacement.

The placenta serves many functions during pregnancy. It delivers oxygen and nourishment to the developing fetus, absorbs and removes waste products, delivers immunity to various diseases, and secretes hormones necessary for both the maintenance of pregnancy and delivery of the fetus.

The plasma hormonal profile of the pregnant woman is very different from the hormonal profile seen in nonpregnant women during the ovarian cycle.

Parturition and lactation are under hormonal control.

Chapter Outline

The Process of Fertilization

    Cell Division and Implantation Following Fertilization

    The Placenta

    Hormonal and Physical Changes in the Female During Pregnancy

    Parturition

The sun is just rising. Barbara is tired after only three hours of sleep, but her one-month old daughter is restless and hungry again. Just hearing her daughter's cry stimulates Barbara's breasts to produce milk. As she stands to get the baby, her breasts feel full. It is time to breast-feed.

    Barbara remembers when her daughter was first placed in her arms and how anxious she felt about breast-feeding. She worried that she would not be able to produce enough milk; or that the baby would reject her milk; and about the best way to feed her baby. After all, Barbara's mother had bottle-fed her and her younger

925

siblings. Sensing Barbara's growing anxiety, the maternity nurse gave her some comforting information, "Breastfeeding is a wonderful way of bonding. It also gives your infant the benefits of your immune system. Don't worry. The baby will know what to do. Ninety-five percent of the mothers who try breastfeeding can do it."

As Barbara now feels the comfortable sensation of the baby nuzzling her breast and "rooting around" for her nipple, she recalls their initial struggle. For the first two weeks her nipples always felt sore and cracked, and her milk did not seem to be available in sufficient quantity. Barbara felt humiliated and inadequate. Years ago, when she was in the Peace Corps in Paraguay, she had witnessed births in the most humble of circumstances and had watched mothers of all ages carrying babies at their breasts, nudging them to feed. It all looked so effortless. Barbara had even watched wet nurses breast-feeding children whose mothers were too far away to feed them. If it was so easy for all those women, what was wrong with her?

One week after bringing her daughter home, Barbara called her obstetrician in frustration to discuss the problem. The obstetrician examined her to rule out the possibility of a tumor, and in the process found that Barbara had mastitis, or breast inflammation, which resulted in blocked milk ducts. The baby was simply placing too much pressure on the nipple itself rather than around the alveolar ring, the dark area around the nipple. Barbara's breasts were so engorged with milk that the baby could not fit her tiny mouth around the alveolar ring. The baby's sucking was not stimulating the nerves that lie under the alveolar ring and control milk expulsion. In addition, Barbara's own inability to relax and enjoy breast-feeding, due to her pain, was interfering with the "let-down reflex," which signals the breasts to expel milk.

The obstetrician suggested that Barbara limit the amount of time at each feeding, increase the frequency of the feedings, and alternate breasts more often. Barbara was also told to manually express her milk before feedings in order to decrease the size of the breast's alveolar area so it would more easily fit into the baby's mouth. In between feedings, she was to keep her breasts clean and dry, and to rub the nipples with lanolin. Also, massages and hot baths would help the blockage in the milk ducts to disappear. The obstetrician told Barbara that "Air and sunlight are the best medicine for sore nipples. Enjoy."

Recently, Barbara started attending lactation courses, which are available to all women through the La Leche League, a national organization that helps educate new mothers and also offers advice to working mothers who have to be away from their babies during the workday. In the lactation course Barbara also learned about the importance of maternal nutrition and the effects on infants of pollutants, caffeine, and alcohol, which pass from mother to infant by way of breast milk.

Barbara is still exhausted from nighttime feedings, but she has learned how to relax and has come to appreciate this special bonding opportunity. She would like to be this close to her child forever.

In this chapter, we will explore the physiological changes associated with pregnancy, which include breast development and lactation. We will also discuss the mechanisms underlying fetal development.

In Chapter 31, we investigated the cyclic hormonal and physiological changes that occur in the female and culminate in the release of an ovum once each month during the years between puberty and menopause. We also examined the processes that control production and expulsion of sperm in the male. We will now investigate the processes of fertilization of the ovum and maturation and development of the embryo, as well as the physiological changes that occur in the female during pregnancy, parturition (childbirth), and lactation (milk production and release).

The Process of Fertilization

Because both sperm and ova are viable only for short periods of time, fertilization of an ovum will occur only when sperm are deposited in the vagina close to the time of ovulation. Following ejaculation, sperm normally retain their capacity to fertilize an ovum in the female reproductive tract for approximately 24 to 72 hours, although longer periods of sperm viability have been reported. Following ovulation, the ovum is "receptive" to fertilization for approximately 10 to 15 hours. Therefore, in order for the ovum to be fertilized, sperm must be present in the female reproductive tract during the 72 hours before or the 15 hours after ovulation. This restricted time interval makes the odds against fertilization rather high, and only one out of every four women becomes pregnant after one month of repeated intercourse without contraception.

Following Ovulation, the Ovum Is Transported Through the Uterine Tube to the Uterus

As described in Chapter 31, ovulation occurs in response to stimulation from high blood levels of LH and FSH approximately 14 days after the beginning of the previous menstrual cycle. Except in rare in-

(a)

(b)

(c)

Figure 32–1

(*a*) Path of the ovum through the fallopian tube and into the uterus following ovulation. (*b*) An unfertilized ovum in the folds of the fallopian tube. Photograph taken with an electron microscope (×100). (© *Lennart Nilsson*, A Child Is Born, *Dell Publishing Company, 1977.*) (*c*) A sperm cell migrates through the fallopian tube toward the ovum. (© *Lennart Nilsson from* Behold Man, *published by Little, Brown and Company.*)

stances, ovulation occurs from only one of the two ovaries each month.

During ovulation, a single ovum is usually released from an ovary into the abdominal cavity at a point directly below the open end of the uterine tube (Fig. 32–1*a*). As we shall see, rarely (less than 1% of the time) more than one ovum is released during ovulation. The open end of the uterine tube is composed of long, fingerlike projections called **fimbriae.** The fimbriae and uterine tubes are lined with **cilia,** which move or sway in a rhythmic pattern toward the uterus and induce wavelike motions on the inner sur-

face of the uterine tube. The uterus and cervix also contain ciliated cells, but not as many as are found in the uterine tubes. Following ovulation, the ovum is drawn from the abdominal cavity into the uterine tube by the current induced by the sweeping motion of the cilia (see Fig. 32–1*a* and *b*).

After entering the uterine tube, the ovum is transported to the uterus by the wavelike motion of the cilia (Fig. 32–1*a*). The ovum takes approximately 3 to 5 days to travel from the ovary through the uterine tube to the uterus. Because the ovum is receptive to fertilization only for approximately 10 to 15 hours following ovulation, fertilization normally takes place in the uterine tube before the ovum has reached the uterus.

Following Ejaculation into the Vagina, Sperm Are Transported by Way of the Vagina, Cervix, and Uterus into the Uterine Tubes

Following ejaculation into the vagina, sperm are transported by way of the vagina, cervix, and uterus into the uterine tubes (Fig. 32–1c). The time required for sperm to reach the uterine tubes is not precisely known, but it is thought to be a few hours, with some sperm reaching the uterine tubes within minutes following ejaculation.

Sperm are propelled through the female reproductive tract toward the fimbrial ends of the uterine tubes by wavelike movements in the tail portion of the sperm. However, the transport time of the sperm through the uterus and the uterine tubes is much too short to be attributed solely to the movement in the tail of the sperm. Sperm are also moved through the uterus and uterine tubes by contractions of the muscles of the uterus and uterine tubes. In the female, vaginal and cervical stimulation during intercourse results in the release of **oxytocin** from the posterior pituitary, which in turn may stimulate uterine contractions. In addition, seminal fluid contains a high concentration of **prostaglandins,** which also act to stimulate contractions of the uterus.

Cervical Mucus and Sperm Transport

The ease with which sperm migrate through the cervix into the uterus is partially determined by secretions of the cervix. The cervix contains glands that secrete a mucus composed of glycoproteins, salts, and water. The volume and consistency of the cervical mucus change during the ovarian cycle in response to changes in the circulating levels of estrogen and progesterone.

Cervical Mucous Secretion at Midcycle.   As was discussed in Chapter 31, estrogen secretion by the ovary increases at midcycle just prior to ovulation. Estrogen stimulates the cervix to secrete large amounts of mucus, which is clear, watery, and nonviscous. Upon estrogen stimulation, the glycoproteins in the cervical mucus assemble to form elongated fibers arranged in channels that allow sperm to penetrate the mucus and pass upward through the cervix. These changes in the fluidity and glycoprotein structure of cervical mucus at midcycle facilitate sperm migration through the cervix into the uterus at a time that coincides with ovulation. Note that many struc-

turally abnormal sperm do not penetrate the cervical mucus at midcycle. Whether this cervical exclusion of abnormal sperm is due to decreased motility of the sperm or to immunological factors present in the cervical mucus is not known.

Cervical Mucous Secretion before and after Midcycle.   Following ovulation, circulating levels of progesterone are higher than at midcycle. **Progesterone** stimulates the cervix to secrete a thick, viscous, sticky mucus that lacks glycoprotein channels. This mucus plugs the cervical opening into the uterus and acts as a barrier that impedes sperm migration into the uterus following ovulation. As we will see in Chapter 33, the efficacy of many oral contraceptives containing steroids is due in part to steroid-induced alterations of the physical and chemical properties of cervical mucus.

Sperm Loss in the Female

Although the number of sperm found in ejaculated semen depends on many factors—including the quantity of the ejaculate and the time elapsed since the previous ejaculation—an average of 100 to 500 million sperm are deposited in the vagina during intercourse. Since only one sperm can fertilize the ovum, this number may seem much larger than is necessary to ensure fertilization. However, of this enormous number, only a few thousand sperm (less than 0.1% of sperm contained in the total volume of ejaculated semen) reach the uterine tubes, and only a few hundred reach the vicinity of the ovum.

Sperm loss is high for many reasons. Up to 50% of the sperm in a normal ejaculate may be incapable of fertilization because of abnormal shape or reduced motility. When deposited into the vagina, some sperm are destroyed by the acidic secretions of the vagina. Upon reaching the cervix, sperm entry into the uterus may be blocked by the cervical mucus at certain times in the ovarian cycle, as we have previously seen. Of the sperm that manage to reach the uterus, about half will enter the uterine tube that does not contain an ovum.

Although these factors result in a large reduction in the number of sperm that reach the ovum, only one sperm is required for fertilization of the ovum.

Sperm Must Undergo the Acrosomal Reaction Before They Are Capable of Fertilizing an Ovum

As stated in Chapter 31, sperm are not motile when they leave the seminiferous tubules of the testes and enter the epididymis. It is in the epididymis that sperm mature and acquire the ability to move. However, mature sperm from the epididymis are not ca-

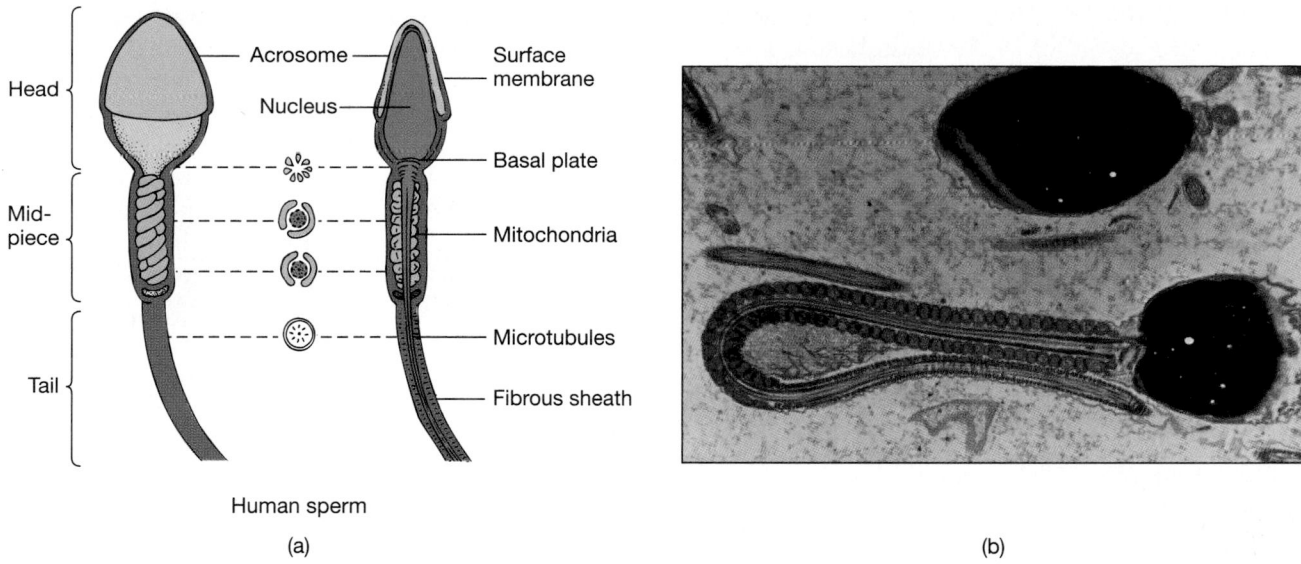

Head

Acrosome

Surface membrane

Nucleus

Basal plate

Mid-piece

Mitochondria

Tail

Microtubules

Fibrous sheath

Human sperm

(a)

(b)

Figure 32–2

(*a*) Structure of a human sperm. (Redrawn from Warsaw, J.B., ed., *The Biological Basis of Reproductive and Developmental Medicine*, 1983.) (*b*) Transmission electron micrograph of a human sperm showing the nucleus, acrosome, midpiece, mitochondria, and flagella. (© *David M. Phillips/Visuals Unlimited.*)

pable of fertilizing an ovum. Sperm must first attain the capacitated state, which occurs in the female reproductive tract prior to fusion of the sperm and ovum. During **capacitation,** the sperm's plasma membrane is altered, loses its acrosomal cap, and increases its motility. Capacitation enables the sperm to undergo a specialized exocytotic reaction, called the **acrosome reaction,** which is necessary for fertilization. The acrosome reaction enables the sperm to penetrate the ovum.

Fertilization Refers to the Fusion of Sperm and Ovum

Before we examine the process of sperm-ovum fusion, let us review the anatomy of the mature sperm. Figure 32–2*a* and *b* illustrates the general structure of a mature sperm. The head of the sperm is composed of a dense, compact nucleus, an acrosome that caps the nucleus, and a surface membrane. The **acrosomal portion** of the sperm head contains enzymes that play an important role in the process of fertilization. The midpiece of the sperm contains spirals of mitochondria, while the tail of the sperm is composed of microtubules surrounded by a fibrous sheath (Fig. 32–4*b*).

When the sperm reaches the ovum, glycoproteins that make up the ovum's extracellular matrix coat, called the **zona pellucida,** induce the sperm to undergo the acrosome reaction. The acrosome reaction (Fig. 32–3) involves a series of fusions between the

outer membranes of the acrosome, resulting in the formation of pores or channels in the outer acrosomal membrane. During the acrosome reaction these pores expand to allow macromolecules to pass in and out of the pores. Extracellular calcium has access to the acrosomal matrix, which then swells and begins to disperse (Fig. 32–3*a* and *b*). The sperm is now ready to penetrate the zona pellucida of the ovum. Proteolytic acrosomal enzymes, which disperse the cells surrounding the ovum and aid the sperm in penetrating the zona pellucida, are released from the acrosomal matrix. Following penetration of the zona pellucida, the sperm binds to the plasma membrane of the ovum. The acrosome-reacted sperm is now capable of fusing with the ovum plasma membrane.

Once fusion is complete, the sperm enters the ovum and fertilization is underway. The term *fertilization* refers to the fusion of the sperm and ovum. Figure 32–4 illustrates the initial meeting of sperm and ovum. The fertilized ovum, or zygote, is the first cell of a new organism which will divide and eventually differentiate to become an embryo.

Polyspermy, or the Fertilization of the Ovum by More than One Sperm, Is Lethal for the Ovum

Fertilization of the ovum by more than one sperm, a condition called **polyspermy,** is lethal for the fertilized ovum. The prevention of polyspermy is therefore very important for reproduction. Although many

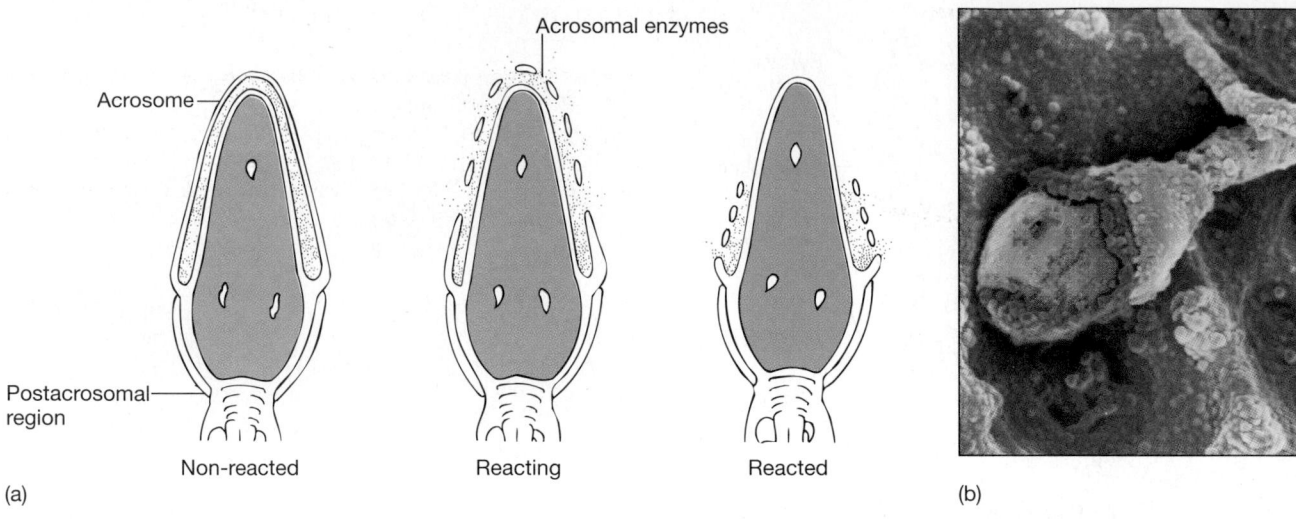

(a)   (b)

Figure 32–3

(*a*) Successive stages of the acrosome reaction of the human sperm. *(Redrawn from* Atlas of Fine Structure of Human Sperm Penetration, Eggs, and Embryos Cultured in Vitro, *Sathananthan et al., 1986.)* (*b*) The cap of the sperm gradually dissolves during the acrosome reaction, allowing enzymes to be released, which help the sperm to penetrate the ovum. *(© Lennart Nilsson from* A Child Is Born, *Dell Publishing Company.)*

sperm may attach themselves to the surface of the ovum, normally only one sperm penetrates to the inner portion of the zona pellucida surrounding the ovum.

Figure 32–4

Photograph of a sperm beginning to penetrate an ovum. *(© David M. Phillips/Visuals Unlimited.)*

The ovum has efficient ways of blocking the entry of more than one sperm. First, attachment of a sperm to the surface of the ovum induces membrane depolarization and increased permeability to calcium which triggers an electrical block on the surface of the ovum and prevents other sperm from fusing with the ovum. In addition, a large structural change in the ovum occurs shortly after sperm-ovum fusion. Fusion of the first sperm with the ovum results in the exocytosis of cortical granules from the periphery of the ovum into the space between the plasma membrane and the zona pellucida. These granules alter the surface membrane of the ovum, making penetration by additional sperm more difficult. In addition, the cortical granules released from the ovum during fertilization alter the binding sites for sperm on the zona pellucida and the plasma membrane of the ovum. These processes prevent more than one sperm from fusing with the ovum. As Figure 32–5 illustrates, after one sperm enters the inner portion of the zona pellucida, all other sperm are virtually stopped before they reach the inner zona pellucida.

Multiple Births May Be Monozygotic, Dizygotic, or a Combination of Both

Normally only one ovum is released at midcycle each month. However, occasionally two or more ova may be released during ovulation. Fertilization of each of these ova by individual sperm results in the formation of **dizygotic** (or **fraternal**) **twins,** triplets, or

Figure 32–5

Fertilization of the human ovum. Note that as one sperm enters the ovum, the entrance of all other sperm is blocked, which prevents polyspermy. Phase contrast microscopy (original magnification: ×400).

(© Dr. Landrum B. Shettles.)

quadruplets. Dizygotic siblings are as genetically different from each other as any other siblings. **Monozygotic** (or **identical**) **twins**, triplets, or quadruplets, on the other hand, result from the fertilization of a single ovum by a single sperm followed by cell division that results in a single duplication or multiple duplications of the fertilized ovum prior to implantation. Monozygotic siblings are genetically identical. Multiple births may be monozygotic, dizygotic, or a combination of both.

Cell Division and Implantation Following Fertilization

As discussed in Chapter 31, when the ovum is released from the follicle at midcycle, it has completed the first meiotic division. Fertilization stimulates the ovum to complete the second meiotic division, which results in the formation of the **haploid ovum** containing 23 chromosomes. During fertilization, a number of events occur within the ovum in a short time. The ovum releases cortical granules that act to prevent polyspermy, as previously described. In addition, metabolic changes occur that initiate protein synthesis and the transformation of the haploid set of chromosomes to form the **female pronucleus.** This process occurs within hours after fertilization. At the same time, the sperm head swells to form a **male pronucleus.**

Once the sperm has entered the ovum, the nucleus of the sperm migrates toward the nucleus of the ovum. The two sets of unpaired chromosomes, 23 from the ovum and 23 from the sperm, align themselves to form a complete set of 46 chromosomes in the **diploid** ovum. This alignment is such that each parent contributes half of the genetic material that will determine the characteristics of the embryo.

The Fertilized Ovum Undergoes Cell Division During Transport Through the Uterine Tube

The Zygote

Initially, the fertilized ovum is a single cell containing 46 chromosomes and is called the **zygote.** The zygote begins to divide mitotically about 30 hours after fertilization. Cell division, which begins prior to implantation, marks the beginning of embryonic development. Note that embryonic cells are the only cells, other than cancer cells, that divide on their own initiative. All other cells divide only in response to a need for tissue replacement. The signal for division of embryonic cells is contained in the genetic material of the zygote.

The zygote undergoes cleavage and first divides into two cells, then these two cells divide into four cells, and so on as illustrated in Figure 32–6. As cell division progresses, the size of each individual cell decreases. During mitotic division, each of the 46 chromosomes in the dividing cell duplicates itself before division, which is why all the cells of the body contain the same 46 chromosomes.

(a)

(b)

(c)

(d)

Figure 32–6

Stages in the early development of the fertilized ovum. (*a*) The single cell contains all the genetic information necessary to produce a complete human being. (*b*) The two-cell stage of development. (*c*) The eight-cell stage of development. (*d*) Cell division continues and gives rise to a cluster of cells called the morula.   (*© Lennart Nilsson from* Being Born, *pp. 14, 15, 17.*)

When the zygote is composed of 8 or more cells, it is called a **morula** (see Fig. 32–6). As cells of the morula continue to divide, a hollow area begins to form in the center of the cell mass. When the hollow inner portion becomes filled with fluid, the morula is called a **blastocyst** (see Fig. 32–6). Note that the cell divisions resulting in the formation of the zygote, morula, and blastocyst occur while the fertilized ovum is being transported through the uterine tube.

The Blastocyst

The blastocyst is the precursor of the embryo. Blastocysts are composed of two major cell types. The outer layer of cells form the **trophoblast,** which maintains the fluid-filled cavity of the blastocyst. Cells of the trophoblast are tightly connected to each other by membranes and form a protective shell around the blastocyst. This layer of trophoblastic cells plays an important role in the initiation of implantation because these cells are sticky and adhere to the endometrium of the uterus. The trophoblastic cells will eventually differentiate into the placenta. The inner cell mass of the blastocyst is composed of cells called **embryoblasts** because they give rise to the embryo itself. Hence, both the placenta and the embryo arise from cells of the fertilized ovum (Figure 32–7).

The Blastocyst Implants in the Endometrium of the Uterus

As stated earlier, the fertilized ovum, while undergoing cell division, is transported through the uterine tube to the uterus over the course of approximately 3 to 5 days. By the time the fertilized ovum has reached the uterus, it has undergone many cell divisions and has reached the blastocyst stage of development. The blastocyst is moved by contractions of the uterine tube and wavelike motions of the cilia through the uterine tube and into the uterus, where it remains and undergoes continuous cell divisions for an additional 2 to 4 days before becoming implanted in the endometrium. Once the blastocyst has fully implanted in the endometrium, at approximately 2 weeks after fertilization, it is called an **embryo.** After approximately 10 weeks of embryonic development the embryo is called a **fetus** (see Fig. 32–10).

Secretion of Enzymes and Hormones by the Blastocyst

Prior to implantation, while the blastocyst is floating free in the uterus, it begins to secrete hormones and proteolytic enzymes. These enzymes cause some of the neighboring endometrial cells to degenerate, forming small cavities in the endometrium, a process referred to as the "decidual reaction." The cavities in the endometrium become filled with blood from the arteries of the uterus. In addition, within 9 days following fertilization, the trophoblast cells of the blastocyst begin to secrete **human chorionic gonadotropin (hCG)** which can be detected in maternal plasma and urine. As we shall see, maternal plasma levels of hCG increase and reach a peak at 7 to 12 weeks, and then decline to a stable plateau for the remainder of pregnancy. Detection of hCG is the most commonly used and most specific test for pregnancy. Human chorionic gonadotropin, which binds to LH receptors on luteal cells, prevents the degeneration of the corpus luteum that normally occurs during the ovarian cycle, so the corpus luteum continues to secrete estrogen and progesterone. Estrogen, secreted by the corpus luteum, stimulates the secretion of prostaglandins from the endometrium of the uterus. The prostaglandins increase vascular permeability and swelling of the endometrium. As we will see later, prostaglandins also stimulate the maternal pituitary gland and corpus luteum to secrete hormones necessary for the continuation of pregnancy.

Formation of Embryonic Membranes

During implantation, the trophoblast cells form microvilli on the outer surface of the blastocyst. These microvilli begin to invade the endometrium of the uterus. The trophoblasts penetrate to the level of the first muscle layer, and the blastocyst becomes buried in the endometrium. The trophoblasts differentiate into two cell layers that will form the **amnion** and the **chorion.** The amnion is a tough extraembryonic membrane that surrounds and protects the developing embryo. The chorion, which is the outermost exctraembryonic membrane, gives rise to the placenta (see Fig. 32–7).

Secretion of Progesterone by the Corpus Luteum

Implantation is completed around day 9 following ovulation, which corresponds to day 23 of the ovarian cycle. As we learned in Chapter 31, on day 23 the corpus luteum is secreting large amounts of progesterone. In response to progesterone stimulation, the endometrial cells of the uterus have proliferated, and the endometrium has become thick and vascular in preparation for receiving and nourishing the blastocyst (Fig. 32–8). Therefore, progesterone helps prepare the uterus for implantation. The heightened secretion of progesterone following ovulation is an example of the importance of timing in the reproductive cycle.

Abnormal Implantation

Implantation is not accompanied by any physical sensations, so it is impossible for the female to know when it has occurred. Occasionally, transport of the fertilized ovum from the uterine tube to the uterus is blocked or slowed down, and the fertilized ovum does not reach the uterus but instead becomes implanted in the uterine tube. Abnormal ovum trans-

(text continues on page 935)

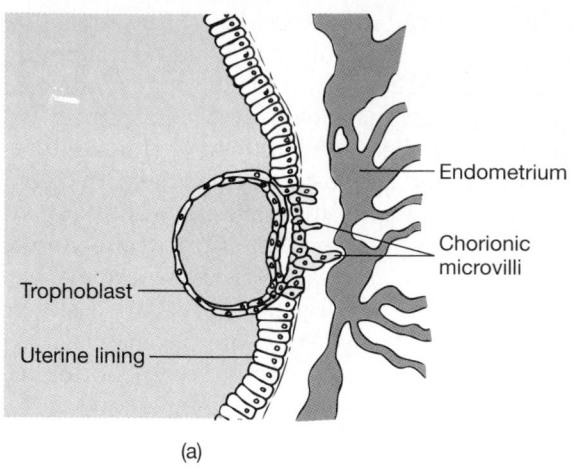

Endometrium

Chorionic microvilli

Trophoblast

Uterine lining

(a)

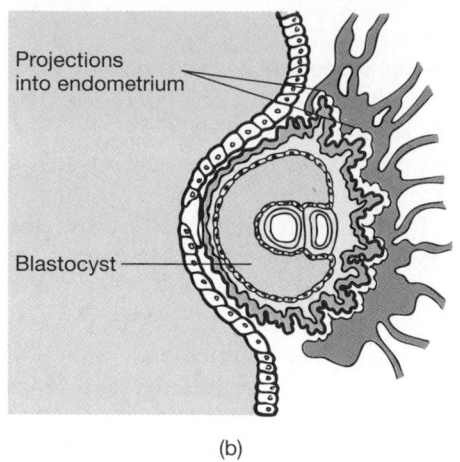

Projections into endometrium

Blastocyst

(b)

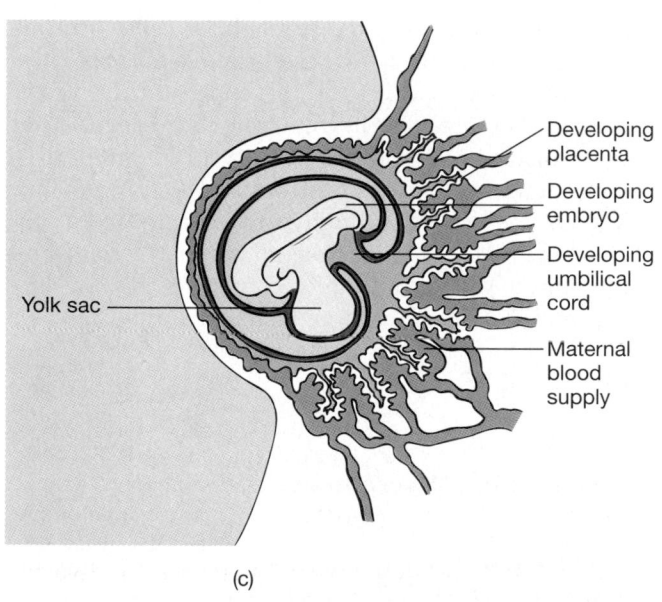

Developing placenta

Developing embryo

Developing umbilical cord

Maternal blood supply

Yolk sac

(c)

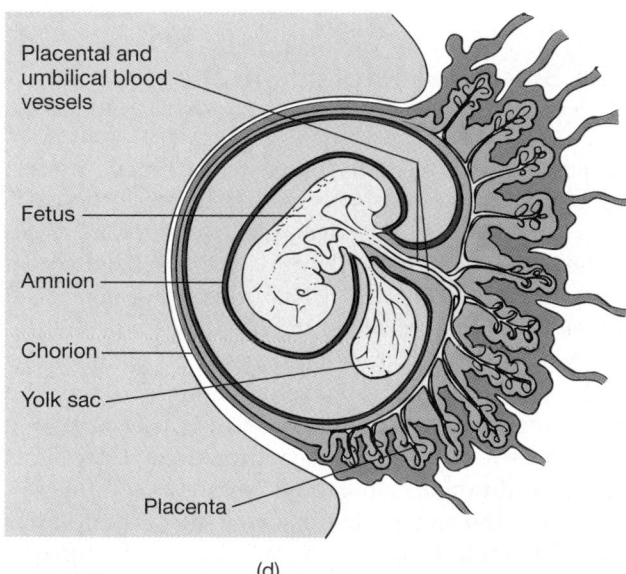

Placental and umbilical blood vessels

Fetus

Amnion

Chorion

Yolk sac

Placenta

(d)

(e)

(f)

◀ *Figure 32–7*

Sequential development of the human placenta. (*a*) Initial implantation of the blastocyst in the wall of the uterus. (*b*) Blastocystic invasion of the endometrium. (*c*) Development of the placenta and the umbilical cord, which connects the fetus and the endometrium. (*d*) Formation of the blood vessels of the placenta and the umbilical cord. *(Adapted from "The placenta," Beaconsfield, Birdwood, and Beaconstield, Scientific American, 243:94, 1980.)* (*e*) On the twelfth day following fertilization, the blastocyst is securely anchored in the endometrium. *(© Lennart Nilsson from A Child Is Born, Dell Publishing Company.)* (*f*) The embryo at 4 to 5 weeks of development. The chorion, which is shaggy in appearance, and the amniotic sac have been opened to visualize the embryo. *(© Lennart Nilsson from A Child Is Born, Dell Publishing Company.)*

Figure 32–8

The menstrual cycle is interrupted when pregnancy occurs. The corpus luteum does not degenerate, and menstruation does not take place. Instead, the wall of the uterus remains thickened so that the embryo can develop within it.

port can result from anatomical abnormalities, from the presence of scar tissue in the uterine tube following infections or surgery, or from tumors located in the uterine tube. In rare cases, the fertilized ovum may be expelled from the uterine tube back into the abdominal cavity and may become implanted there. Both of these cases are called **ectopic pregnancies;** the term *ectopic* refers to implantation outside of the uterus.

Ectopic pregnancy occurs in approximately one out of every one hundred pregnancies. About 97% of ectopic pregnancies occur in the uterine tubes and are termed **tubal pregnancies.** Tubal pregnancies do not continue to term because there is not enough space in the uterine tube for fetal growth. Most tubal pregnancies abort at a relatively early stage. When they do not, surgical removal may be necessary in order to reduce the risk of tubal rupture and bleeding.

The Placenta

The needs of the embryo, and later of the fetus, are met by the **placenta,** which delivers oxygen and nourishment, absorbs and removes waste products, provides for transfer of immunity to various diseases, and secretes hormones necessary for both the maintenance of pregnancy and the expulsion of the fetus at term (Table 32–1).

The Embryo Is Dependent on the Maternal Circulation for Delivery of Nutrients and Oxygen and for Removal of Waste Products

Following implantation, the embryo is no longer independent. From this point on, it must take everything it needs—including oxygen, nutrients, salts,

Table 32–1
Functions of the Placenta

Exchange of gases between fetus and mother
Delivery of nutrients from mother to fetus
Delivery of antibodies from mother to fetus
Removal of fetal waste
Secretion of hormones including human chorionic
   gonadotropin (hCG), progesterone, estrogen,
   and human chorionic somatomammotropin
   (hCS)

and fluid—from the maternal circulation. The embryo also depends on the maternal circulation for removal of its waste products.

Note that while the fetus is totally dependent on the placenta for survival, the placenta is not dependent on the presence of the fetus for survival. The placenta can continue to function in the uterus even after a fetus dies or is removed. However, the placenta is normally expelled along with the fetus during delivery and therefore is the only organ of the body with a life span of less than one year.

The Placenta Serves as an Interface Between the Fetal and Maternal Circulatory Systems

The **placenta** is a flat organ that is connected to the uterus on one side and to the embryo, via the umbilical cord, on the other side. The **umbilical cord** is formed from trophoblastic cells that surround the blastocyst. The amniotic cavity is also formed from trophoblastic cells. The cells that line the inner surface of the amniotic cavity form the amnion, which expands and forms a fluid-filled cavity surrounding the entire embryo. The clear amniotic fluid keeps the embryo moist and serves as a shock absorber to protect it from physical injury. The placenta is derived from the embryonic chorion and from the maternal endometrium. When the blastocyst implants in the endometrium, the tissue and blood vessels of the endometrium break down at the site of implantation to form small spaces in the endometrium that fill with maternal blood. The developing embryo sends out rootlike villi into the pools of maternal blood. In the first few weeks following implantation, the villi are composed of thick columns of cells that contain capillaries. These capillaries serve as the route for exchange of nutrients from the mother to the fetus and of waste products from the fetus to the mother. These small vessels become the arteries and veins of the um-

bilical cord and placenta. Within 5 weeks after implantation, the vasculature of the umbilical cord and placenta is well established. Figure 32–7 illustrates the different stages in the development of the placenta.

The placenta is heavily supplied with fetal blood vessels on one side and with maternal blood vessels on the other side, but the two blood supplies normally never mix. The placenta anatomically separates the circulatory systems of the fetus and mother and acts as an interface between these two circulatory systems. As shown in Figure 32–9, blood from the mother enters the placenta through the uterine artery, perfuses the placenta, and exits through the uterine vein. Fetal blood enters the placenta through the umbilical artery, perfuses the placenta, and exits through the umbilical vein. As illustrated in Figure 32–9, the side of the placenta that faces the fetus is smooth and highly vascularized, with membranes that surround the fetal arteries and veins. The maternal side of the placenta, on the other hand, is divided into thick, highly vascularized lobes that provide a very large surface area for exchange of materials between the fetal and maternal blood supplies. The exchange of all materials between the fetus and the mother takes place through the placenta.

Exchange of Materials Through the Placenta

Placental Exchange of Gases.   Oxygen diffuses from the maternal blood into the placenta, where it is picked up by the fetal blood and carried through the umbilical cord to the fetus (Fig. 32–9). Fetal blood is capable of carrying more oxygen than maternal blood, for at least two reasons. First, the concentration of hemoglobin is 50% higher in fetal blood than in maternal blood. Second, fetal hemoglobin has a higher affinity for oxygen than does maternal hemoglobin and, therefore, is more saturated with oxygen.

Carbon dioxide, which is continually formed in fetal tissues, enters the fetal blood and is transported to the placenta, where it diffuses into the maternal circulation.

Placental Delivery of Nutrients.   Some substances that the fetus needs pass from maternal blood to fetal blood by simple diffusion in the placenta. These substances include glucose, sodium, potassium, and chloride.

The fetus grows at a more rapid rate than any other tissue in the adult body. To sustain this rapid growth, the fetus needs a variety of structural proteins that will be incorporated into fetal tissues. The placenta extracts amino acids from the maternal blood and synthesizes proteins which are then trans-

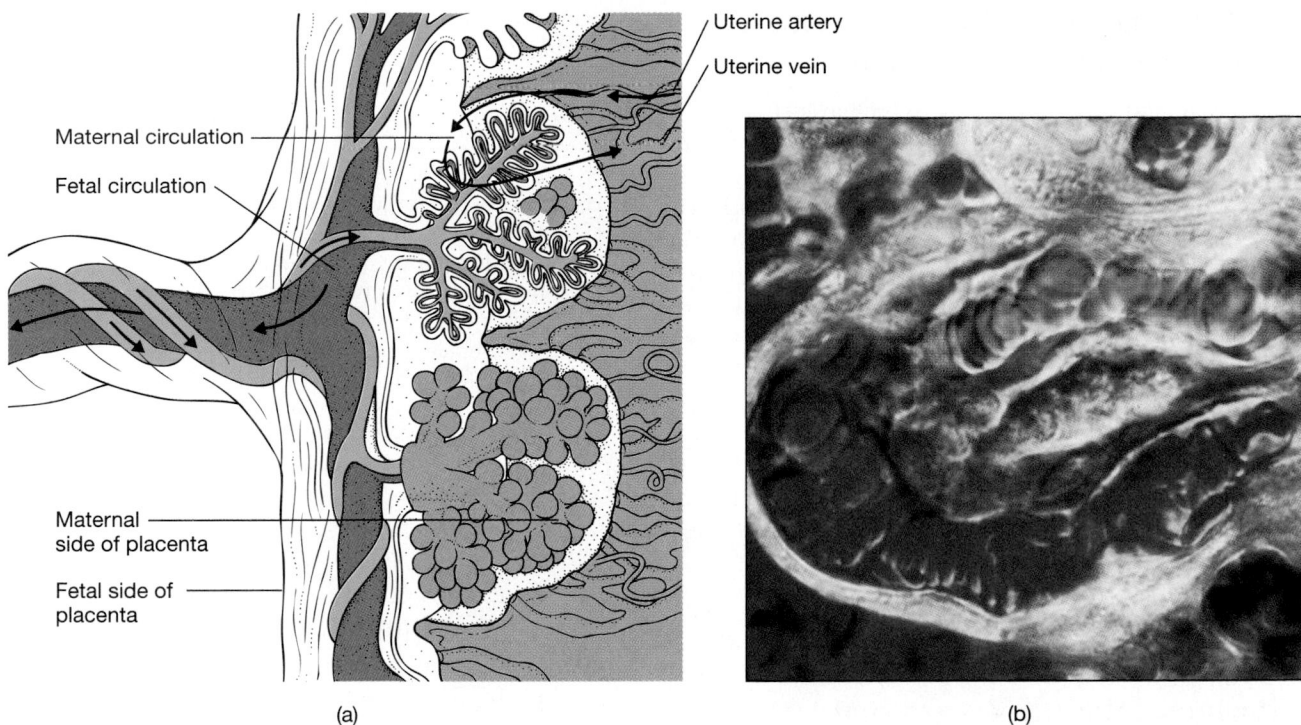

Maternal circulation

Fetal circulation

Uterine artery

Uterine vein

Maternal side of placenta

Fetal side of placenta

(a)

(b)

Figure 32-9

(*a*) Cross-section of placental tissue illustrating the directions of fetal and maternal blood flow. *(Adapted from "The placenta," Beaconsfield, Birdwood, and Beaconsfield,* Scientific American, *243:100, 1980)* (*b*) Fetal blood cells in a U-shaped capillary protruding from the placenta. The fetal blood cells (blue) are separated from the maternal blood cells (red) by a thin membrane that transfers oxygen and nutrients from the maternal blood to the fetal blood. As fetal blood cells are oxygenated they turn from blue to red. *(© Lennart Nilsson from* A Child Is Born, *Dell Publishing Company.)*

ported into the fetal circulation and delivered to the developing fetus. In fact, the placenta synthesizes proteins at a higher rate than any other organ of the body. Although the demands of the developing fetus may vary throughout gestation, on average the placenta uses approximately one third of all the oxygen and glucose that the maternal blood supplies in order to meet the metabolic needs of the fetus and the placenta itself.

Placental Removal of Excretory Products

Fetal waste products, including urea, uric acid, and nonprotein nitrogens, pass by simple diffusion from fetal blood into maternal blood in the placenta. These fetal waste products are then excreted by the mother along with her own wastes.

Placental Transport of Drugs

While the efficiency of the transport system from mother to fetus is essential for nutrient delivery to the fetus, this same transport system exposes the fetus to many of the drugs and pollutants that the pregnant woman is exposed to. Such exposure is particularly dangerous during the first three months of pregnancy, when fetal tissue is still differentiating and is especially susceptible to damage. Alcohol, aspirin, and the chemicals contained in tobacco can reach the fetus via the placenta and influence its growth and development. Perhaps the most striking example of this transport of drugs to the fetus is seen in the offspring of women who use heroin during pregnancy. These babies experience severe heroin withdrawal symptoms after birth.

Placental Transport of Antibodies

Another function of the placenta is to provide the fetus with maternal antibodies. The mother has formed these antibodies during her contact with a variety of different antigens. Antibodies are too large to diffuse across the placenta, but the placenta has highly selective active transport systems that convey these large antibodies to the fetus. In this way, the newborn infant is protected against a variety of infections.

The Placenta and Embryo Are Protected from Immunological Rejection

Embryonic and placental tissue (which arise from the embryo) are immunologically foreign to the mother and hence might be expected to stimulate the synthesis of antibodies by the maternal system and induce tissue rejection of the embryo. This does not occur for a variety of reasons. First, during the early stages of development the blastocyst does not express antigens that would stimulate the production of antibodies by the mother. Second, the placenta secretes many hormones, such as **human chorionic gonadotropin (hCG),** that suppress lymphocytes, the cells that induce tissue rejection. Third, the embryo is physically protected from the maternal system by membranes that surround it, and, as noted, the fetal and maternal blood supplies do not mix.

Hormonal and Physical Changes During Pregnancy

The Placenta Secretes Four Hormones That Are Important for the Continuation of Pregnancy

During pregnancy, the placenta secretes large amounts of four important hormones that regulate the physiological changes associated with early pregnancy: human chorionic gonadotropin (hCG), progesterone, estrogen, and human chorionic somatomammotropin (hCS), also called human placental lactogen (hPL).

Placental Secretion of Human Chorionic Gonadotropin

Effects of hCG on Endometrial Growth. As discussed in Chapter 31, in the absence of fertilization of the ovum, the corpus luteum secretes progesterone for about 7 days following ovulation, and then progesterone levels fall as the corpus luteum degenerates. The rise in progesterone, secreted by the corpus luteum, stimulates growth of the endometrium of the uterus, and the fall in progesterone following degeneration of the corpus luteum results in necrosis of the endometrium and the onset of menstruation. If these events were to occur following fertilization of the ovum, the fertilized ovum would be expelled from the uterus along with the menstrual flow. These events are prevented by the secretion of **human chorionic gonadotropin** (hCG) (Table 32–2). hCG is a

Table 32–2

Physiological Functions of Human Chorionic Gonadotropin

Prevents degeneration of the corpus luteum
Stimulates the corpus luteum to secrete estrogen and progesterone, which in turn stimulate continued growth of endometrium
Stimulates steroid synthesis in the developing fetal adrenals
Stimulates fetal gonads, especially testosterone production by the fetal testes
Suppresses maternal lymphocytes and reduces the possibility of immunorejection of the fetus

glycoprotein, which has a molecular structure very similar to that of luteinizing hormone (secreted by the pituitary). hCG is secreted first by trophoblast cells of the blastocyst before implantation in the endometrium and later by the placenta. hCG diffuses from fetal blood into maternal circulation and elevates maternal plasma hCG content. hCG stimulates the corpus luteum of the pregnant woman and prevents the degeneration and involution of the corpus luteum that normally occur during the ovarian cycle, thus ensuring the continued secretion of progesterone. In response to hCG, the corpus luteum doubles in size during the first 2 months of pregnancy. As pregnancy continues, hCG stimulates the corpus luteum to secrete large amounts of both estrogen and progesterone. These two hormones stimulate continued growth of the endometrium, which is necessary for adequate development of placental and fetal tissue. If the corpus luteum is removed during the first 7 to 10 weeks of pregnancy, spontaneous abortion will occur because progesterone and estrogen are no longer stimulating the endometrium. During the second month of pregnancy, the placenta becomes capable of secreting estrogen and progesterone on its own, and the corpus luteum is no longer needed. At this time, the placenta stops secreting hCG, and as maternal plasma levels of hCG fall, the corpus luteum becomes less functional. During the remaining months of pregnancy, the high levels of estrogen and progesterone in the maternal circulation are mainly the result of secretion by the *placenta.*

Effects of hCG on Fetal Adrenal Glands and Testes. It is thought that hCG also stimulates steroid synthesis in the developing fetal adrenal gland (see Table 32–2). In the male fetus, hCG stimulates the synthesis of testosterone by the fetal testes. As we will see in Chapter 33, fetal testosterone stimulates development of the internal genitalia of the male fetus.

Placental Secretion of Progesterone

The placenta begins to secrete progesterone at about the sixth week of pregnancy, and by weeks 12 to 14 the placenta produces enough progesterone to replace that previously secreted by the corpus luteum during the ovarian cycle. In the later stages of pregnancy, the placenta secretes 10 times as much progesterone as the corpus luteum is capable of producing during the ovarian cycle. The placenta extracts cholesterol from the maternal circulation and uses this cholesterol to form progesterone, which then diffuses back into maternal circulation.

Progesterone, secreted by the placenta, is essential for the maintenance of pregnancy for many reasons. First, progesterone stimulates growth of the endometrium and secretion of nutrients from glands in the endometrium, which nourish the developing embryo. Second, high circulating levels of progesterone serve to inhibit contractions of the uterus, which prevents premature expulsion of the fetus. Third, as we will see, progesterone, together with oxytocin from the posterior pituitary, stimulates the breast in preparation for milk production and release following delivery (Figure 32–14). Fourth, placental progesterone serves as a precursor for the synthesis of estrogen by the fetoplacental unit.

Placental Secretion of Estrogen

During the first month of pregnancy, the corpus luteum secretes estrogen in response to hCG stimulation. After the first month, hCG begins to stimulate estrogen secretion from the placenta (see Table 32–2). Circulating levels of estrogen in maternal plasma continue to rise up to the time of delivery. In fact, the concentration of estrogen in maternal plasma at term is several hundred times higher than the highest concentration seen during the ovarian cycle.

Estrogen, secreted by the placenta, stimulates enlargement of the uterus in the pregnant woman, which is necessary to accommodate the growing fetus. Estrogen also stimulates breast growth in the pregnant woman and, as we will see, development of the duct system within the breasts. In preparation for delivery, estrogen stimulates enlargement of the external genitalia and relaxation of the pelvic ligaments, which facilitates the passage of the fetus through the birth canal.

Estrogen production in the placenta is unique. The synthesis of estrogen requires a complex interaction between the placenta and the adrenal gland of the fetus. As we learned in Chapter 14, the formation of estrogen involves the synthesis of androgens from cholesterol, followed by the conversion of these androgens to estrogens. The placenta lacks the enzymes necessary to form androgens from cholesterol but contains those necessary to convert certain androgens into estrogens. On the other hand, the fetal adrenal cortex contains the enzymes needed to form androgens from cholesterol (specifically, **dehydroepiandrosterone-sulfate,** or **DHEA-sulfate**). The placenta and the fetal adrenal cortex work together to produce estrogens in the following way: the fetal adrenal synthesizes the androgen DHEA-sulfate from cholesterol. This androgen is then transported to the placenta, where it is converted to **estriol,** a form of estrogen. Estriol is then released from the placenta into maternal circulation. Since estrogen synthesis is dependent on the participation of both the fetal adrenal and the placenta, it is sometimes said to be secreted by the **fetoplacental unit.**

Placental Secretion of Human Chorionic Somatomamotropin

Another hormone that begins to be secreted by the placenta at about four weeks of gestation is **human chorionic somatomammotropin (hCS)**. Placental secretion of hCS rises steadily during pregnancy. hCS is a protein with a molecular structure similar to that of human growth hormone. Human chorionic somatomammotropin is believed to be involved in metabolism and nutrition of both the pregnant woman and her fetus. It has weak growth-promoting effects that may play a role in fetal development and maternal breast development. It also has effects on both glucose and fat metabolism in the pregnant woman. For instance, it decreases maternal utilization of glucose, making more glucose available to the developing fetus. In addition, it promotes the release of free fatty acids from maternal fat stores, which the pregnant woman may then use for energy, sparing other nutrients, such as glucose, for the fetus.

In Addition to the Hormones Secreted by the Fetoplacental Unit, a Number of Other Hormonal Changes Occur in the Pregnant Woman During Gestation of the Fetus

In addition to the hormones being secreted by the fetoplacental unit, a number of other hormonal changes are occurring in the pregnant woman during gestation of the fetus.

Inhibition of Pituitary Gonadotropins

High circulating levels of estrogen and progesterone in maternal plasma during pregnancy inhibit the secretion of **gonadotropin-releasing hormone (GnRH)** from the maternal hypothalamus and hence inhibit release of **luteinizing hormone** and **follicle-stimulating hormone** from the maternal pituitary. Low circulating levels of LH and FSH during pregnancy are not adequate to stimulate growth of ovarian follicles. Therefore, follicular development, ovulation, and menstruation are all eliminated for the duration of pregnancy.

Stimulation of Prolactin

The maternal pituitary secretes an abundance of prolactin during pregnancy. As we will see, prolactin is essential for estrogen and progesterone stimulation of maternal breast development and for the synthesis of breast milk. Circulating prolactin levels rise progressively throughout pregnancy and reach very high concentrations at birth.

Stimulation of Relaxin

During the first two months of pregnancy, **relaxin** is secreted by the corpus luteum in response to hCG stimulation. Later in pregnancy, relaxin is secreted by the endometrium of the uterus. Relaxin, as its name suggests, relaxes pelvic ligaments in preparation for parturition. Relaxin, together with progesterone, inhibits uterine contractions and prevents premature expulsion of the fetus.

Stimulation of Insulin

Secretion of **insulin** from the maternal pancreas increases after the third month of pregnancy in response to decreased maternal sensitivity to insulin. Insulin insensitivity during pregnancy may become so severe that it produces a temporary form of diabetes, called **gestational diabetes.** After delivery, maternal sensitivity to insulin usually returns to normal.

Stimulation of Aldosterone

Increased activity within the renin-angiotensin system during pregnancy results in increased secretion of **aldosterone** from the cortex of the maternal adrenal gland. Hypersecretion of aldosterone is needed during pregnancy to maintain adequate sodium in the maternal circulation, much of which is being provided to the developing fetus. Sodium retention and accompanying water retention during pregnancy help to maintain high maternal and fetal plasma volumes.

Stimulation of Cortisol

Levels of both free and protein-bound **cortisol** increase in the maternal plasma during pregnancy. These changes are due in part to increased secretion of cortisol by the maternal adrenal gland, and also to a threefold increase in **cortisol binding globulin (CBG)** in maternal plasma. The liver increases synthesis of CBG during pregnancy in response to estrogen stimulation. Increased plasma cortisol contributes to an increase in maternal adipose tissue volume and development of the mammary glands.

Stimulation of Thyroxine (T4) and Triiodothyronine (T3)

During pregnancy, high circulating levels of estrogen stimulate TRH release from the hypothalamus which, in turn, stimulates TSH release from the anterior pituitary. TSH stimulation results in an increase in maternal thyroid function during pregnancy as evidenced by increases in thyroid gland size, thyroid secretion of T_3 and T_4, basal metabolic rate, and cardiopulmonary function. T_3 and T_4 increase resting pulse rate in the mother, which results in an increase in the rate of delivery of oxygen and nutrients to the developing fetus.

Pregnancy Is Divided into Three-Month Periods Called Trimesters, Each Marked by Dramatic Physical Changes in Many Tissues of the Body

Pregnancy begins when a fertilized ovum becomes implanted in the endometrium of the uterus. Pregnancy, which usually lasts for approximately 266 days, is divided into three-month periods called **trimesters.** The first trimester is composed of the first three months after fertilization of the ovum; the second trimester includes months four to six of pregnancy; and the third trimester refers to the time between month seven of pregnancy and parturition or delivery.

Physical Changes During the First Trimester

Tiredness, nausea, and vomiting often accompany the first trimester of pregnancy. These symptoms, referred to as **morning sickness,** usually disappear within the first few months. Other physical changes that occur during this time are frequent urination due to compression of the bladder by the enlarging uterus, breast swelling and tenderness, and increased vaginal secretions. A pregnant woman cannot feel the changes occurring in her uterus during the first trimester.

Physical Changes During the Second Trimester

During the second trimester of pregnancy, a woman becomes very aware of the changes in her uterus, because the fetus begins to move. Toward the end of the fourth month, the fetus begins to twist and kick, and these movements intensify during the following weeks. In response to aldosterone and estrogens, most women begin to retain fluid during the second trimester. Increased fluid retention results in swelling, or edema, of the hands and feet. Increased production of red blood cells by maternal bone marrow, together with fluid retention, results in a large increase in maternal blood volume. The increased nutritional demands of the fetus during this time increase the mother's appetite. An increased demand for oxygen by the fetus is accompanied by an increase in maternal blood flow to the placenta and an increase of 30% to 40% in maternal cardiac output in response to increased maternal thyroid function, as previously discussed on p. 949.

Growth of the Uterus.

A pregnant woman's uterus grows at a rapid rate during the second trimester. Expansion of the uterus results in many physical changes, including protrusion of the abdomen and widening of the waist. Rapid growth of the abdomen can cause stretch marks to form in the skin covering the abdomen. As the uterus grows, it compresses other internal organs such as the stomach and intestines, which can lead to indigestion, constipation, and hemorrhoids.

Development of the Breasts.

In response to estrogen and progesterone, secreted by the placenta, and prolactin, secreted by the maternal pituitary, the breasts increase dramatically in size during the second trimester. This increase in size occurs so quickly that the skin covering the breast, like that covering the abdomen, is stretched, which can result in the formation of stretch marks on the surface of the breasts. Breast development during pregnancy will be discussed in more detail later. During the second trimester of pregnancy, the breasts begin to secrete a thin, yellowish fluid called **colostrum,** which is the precursor of breast milk and contains antibodies which, when ingested and absorbed, give the newborn a measure of immune protection.

Physical Changes During the Third Trimester

The largest weight gain of pregnancy usually occurs during the third trimester. Although individual weight gains vary widely, the average woman gains approximately 20 to 25 pounds. The fetus, placenta, and amniotic fluid, all of which are expelled during delivery, account for approximately 10 to 15 pounds of this added weight.

During the third trimester, the uterus reaches its largest size. In a standing position, a woman's uterus protrudes forward, which alters her center of gravity. To compensate for this, a pregnant woman often walks with head and shoulders held back, which can result in backache.

The increased size of the uterus puts constant pressure on the bladder, which results in the need to urinate four to five times during the night, disrupting sleep patterns and increasing fatigue. Increased pressure on the blood vessels of the lower part of the body results in muscle cramps in the legs, and compression of the diaphragm can cause shortness of breath.

During the third trimester, the muscles of the uterus undergo short periods of spasmodic contractions, called **Braxton-Hicks contractions.** The fetus changes position in the uterus during the last few weeks of pregnancy, and the head of the fetus commonly becomes oriented toward the pelvis in preparation for parturition.

Development of the Fetus Is Divided into Three Stages: The Germinal, Embryonic, and Fetal Stages

Prenatal development, which occurs throughout the nine months of pregnancy, can be divided into three stages: (1) the **germinal stage,** the first two weeks of development following fertilization of the ovum, (2) the **embryonic stage,** weeks 2 to 8 following fertilization, and (3) the **fetal stage,** the time between week eight following fertilization and birth.

The Germinal Stage of Development (Weeks 1 and 2)

During the first two weeks of pregnancy, the fertilized ovum, though microscopic in size, has reached the blastocyst stage of development. As previously stated, two major fetal membranes arise from the blastocyst, the inner **amnion** and the outer **chorion.** The amnion forms a thin sac of tissue that surrounds the fetus and is filled with a liquid called **amniotic fluid.** Amniotic fluid is constantly absorbed by the placenta and replaced with new fluid every 2 to 3 hours throughout pregnancy. This fluid-filled sac provides a constant temperature for fetal development and serves as a shock absorber to protect the fetus against physical injury. The chorion gives rise to the placenta.

The inner cells of the blastocyst differentiate into three cell layers. The outer layer of cells will become

RESEARCH FOCUS

Alcohol Drinking During Pregnancy

It has long been recognized that drinking alcohol during pregnancy is harmful to the developing fetus. The most severe hazard associated with heavy drinking by pregnant women is the development of fetal alcohol syndrome (FAS). FAS is characterized by growth retardation and abnormalities of facial development, as well as central nervous system dysfunction including hyperactivity, tremors, and impairment of intellectual development.

The exact relationship between the amount of alcohol consumed by the mother and the degree of impairment of the fetus is hard to establish because of the difficulty in separating the effects of alcohol from the contribution of other factors such as smoking, drugs, and malnutrition, which can also increase the risk of fetal impairment. However, the effects of alcohol on the fetus are commonly regarded as a continuum from severe impairment associated with heavy drinking to minor impairment associated with light drinking.

Heavy alcohol drinking as early as around the time of conception can produce craniofacial deformity and impairment of the central nervous system. Therefore, although discontinuation of heavy drinking after the first trimester of pregnancy will be beneficial to fetal growth, it cannot fully overcome the increased risk of congenital abnormalities.

Maternal alcohol drinking is also associated with increased risk of spontaneous abortion, premature labor, and neonatal death. The precise way in which alcohol damages the fetus is still not known. However, it has been determined that alcohol can impair transport of amino acids and glucose to the fetus, and can induce fetal acidosis and hypoxia through constriction of blood flow to the fetus.

■ **In addition to the effects of alcohol on the developing fetus, is there any evidence that parental alcoholism influences alcohol drinking in the offspring?**

---

the skin, sense organs, and nervous system of the developing fetus. The middle layer will form the skeleton, muscles, heart, and blood vessels. The inner cell layer will develop into the liver, pancreas, and digestive system of the fetus.

The Embryonic Stage of Development (Weeks 2 Through 8)

This is a period of remarkable growth for the developing embryo even though it is less than an inch long by the end of the eighth week of development (Table 32–3). During this time, all of the embryonic organs become partially formed, although cellular development is far from complete. By the end of the first month the primitive heart has begun to beat; the digestive system, brain, spinal cord, and nervous system are developing; and buds of tissue that will become the arms and legs begin to form (Fig. 32–10). By the sixth week, the head of the embryo is quite large, and the outlines of eyes can be seen. During the sixth and seventh weeks, the eyes and ears develop more fully, and the teeth, bones, and facial muscles begin to form.

The Fetal Stage of Development (Week 8 Through Birth)

At about 9 weeks of development, the embryo is now called a **fetus.** At the beginning of the ninth week of development, the head makes up almost half of the total length of the fetus. All primitive organs and structures become more differentiated.

By the second month, distinct hands and feet have formed, and major blood vessels are forming (see Fig. 32–10). By the third month, the fetal stomach is producing digestive fluids. The heart is pumping blood through the fetal arteries and veins, the kidneys have begun to function, and the bone marrow is manufacturing red blood cells. Also by the third month,

Table 32–3

Average Fetal Growth Patterns During Gestation

Age of Fetus (in weeks)	Crown to Rump Length (in inches)	Weight (in pounds)
5–8	0.2–.10	0.01–0.03
13–16	3.5–5.5	0.12–0.43
21–24	7.9–9.1	1.06–1.81
29–32	11.0–11.8	3.10–4.60
37–40	13.8–14.2	6.60–7.50

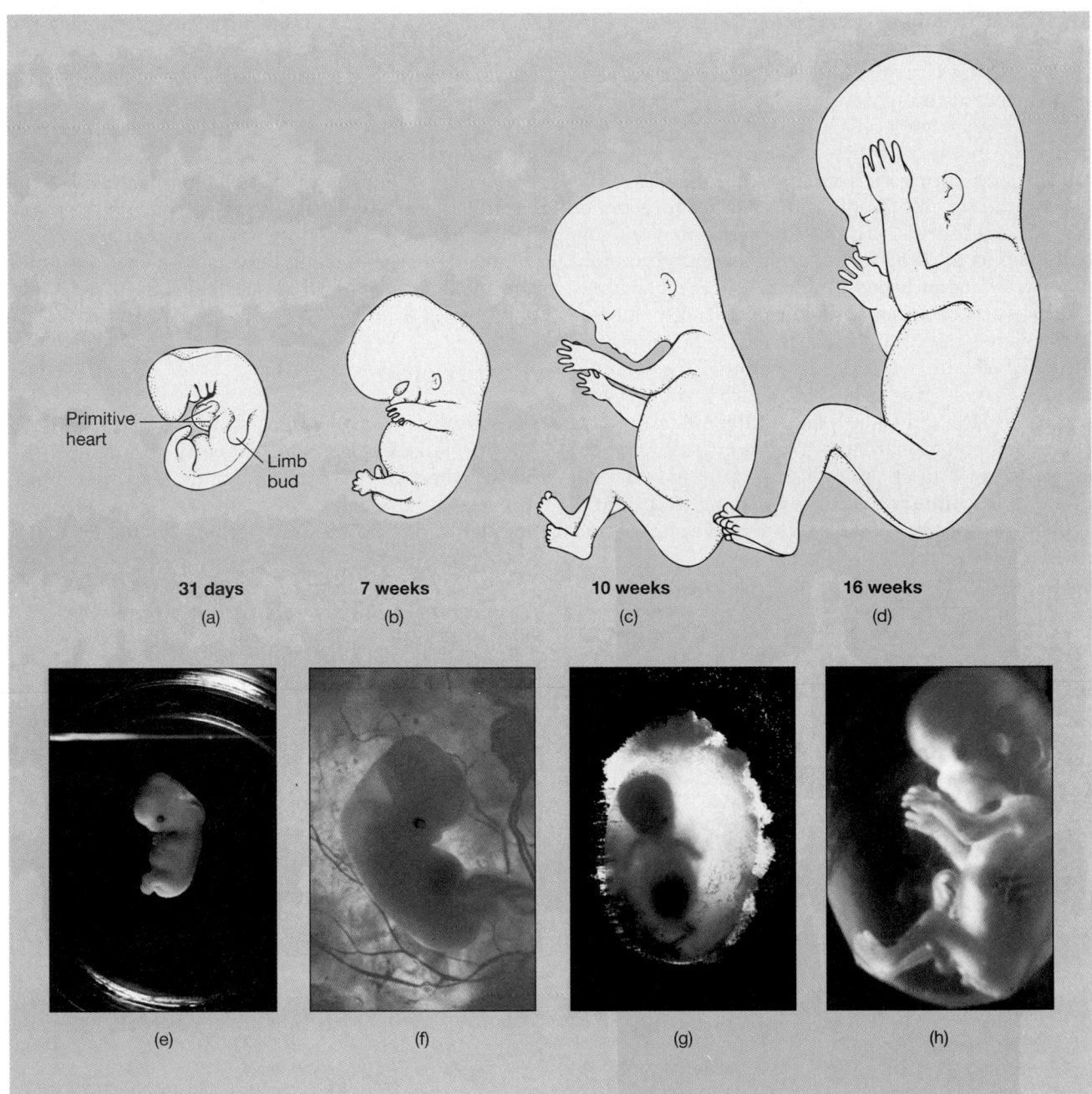

Figure 32–10

Appearance of the embryo (*a* and *b*) and the fetus (*c* and *d*) at various times during gestation. (*e*) The embryo at 31 days. *(© Dr. Landrum B. Shettles.)* (*f*) The embryo at 7 weeks. *(© Petit Format and Guigoz.)* (*g*) The fetus at 10 weeks. *(© CNRI/Phototake, N.Y.)* (*h*) The fetus at 16 weeks. *(© Petit Format/Science Source.)*

the external genitalia of the fetus appear, and the sex of the fetus becomes anatomically distinguishable.

By the fourth month, the fetus has developed vocal cords, lips, fingernails, toenails, and hair on its head (see Fig. 32–10). Sucking motions and respiratory movements begin during the fourth month, and the fetus swallows and absorbs amniotic fluid. Many peripheral reflexes are well developed by this time, though the central nervous system is still very underdeveloped and incapable of higher functions.

During the fifth month, the fetal heartbeat can be heard through the maternal abdomen. The body of the fetus becomes covered with fine, downy hair called **lanugo.** At the end of the sixth month, the fetus is only about 12 inches long and weighs about 1.5 pounds (see Table 32–3). If the fetus is born prematurely at 6 months of development, its chances for survival are very poor.

During the seventh month, the brain and nervous system become more developed, although many important higher functions of the central nervous system will undergo further development throughout fetal life and during the first 6 months following birth. During the seventh month, fatty tissue forms under the skin of the fetus, and the downy hair covering its body begins to disappear. During the eighth month, many fetal organs become mature, and the fetal lungs produce a substance called **surfactant** (see Chapter 21), which is important for survival of the fetus after birth. Surfactant is produced by epithelial cells in fetal lung alveoli. This substance, which has a very low surface tension, coats the inside of the lungs and prevents their collapse at low lung volumes. When fetal lungs do not contain enough surfactant, the infant is prone to **respiratory distress syndrome (RDS).** RDS is characterized by decreased lung compliance and instability of lung alveoli, which cause prolonged periods of respiratory depression that can result in death. Infants born prematurely during the eighth month of pregnancy have a good chance of survival because the fetal lungs have already begun to produce surfactant.

During the ninth month of pregnancy, the fetus migrates to a lower part of the uterus and normally positions itself head downward in preparation for parturition. The average fetus is 20 inches long and weighs 7 pounds at birth (see Table 32–3).

Parturition

As stated earlier, weak, spasmodic contractions of the uterine muscles occur infrequently throughout the third trimester of pregnancy. As parturition approaches, these contractions become stronger, more regular, and more frequent. Delivery of the fetus and

(a)

(b)

(c)

(d)

Figure 32–11

Birth of a baby. In about 95% of all human births the baby descends through the vagina head downward. (*a*) The abdominal muscles of the mother are used to push the baby out through the vagina. (*b* and *c*) When the baby emerges, the mouth and pharynx are gently aspirated to remove any amniotic fluid, mucus, or blood that could impede respiration. (*d*) The placenta emerges following delivery of the baby.

expulsion of the placenta result from strong contractions of the uterus that push the fetus and the placenta through the cervix and out of the vagina (Fig. 32–11). **Labor** is said to begin when uterine contractions start to occur once every 10 to 15 minutes. In some women, the amniotic membrane surrounding the fetus ruptures, and amniotic fluid flows out of the vagina before delivery begins. But in most women the amniotic membrane ruptures during delivery.

Parturition, or Delivery, Involves Both Cervical Dilation and Uterine Contractions Which Are Under Hormonal Control

Uterine Contractions During Parturition

Although the smooth muscles of the uterus are capable of autonomous contractions, uterine contractions at delivery are facilitated by a number of hormones, which will be discussed shortly. Muscular contractions begin in the upper portion of the uterus and move down toward the cervix. These contractions force the fetus against the cervix. Under hormonal stimulation, and with pressure from the fetus, the cervix widens or dilates to a maximum diameter of about 10 cm. **Cervical dilation,** combined with pressure generated by the mother, forces the fetus and placenta through the cervix and out of the vaginal canal (see Fig. 32–11). At birth, the infant is still connected by the umbilical cord to the placenta. Within minutes following delivery, the maternal and infant blood vessels in the placenta and umbilical cord constrict, and the infant is separated from the maternal blood supply. The placenta becomes separated from the wall of the uterus, and continued uterine contractions expel the placenta through the birth canal (see Fig. 32–11). Because the placenta leaves the vagina after the infant is delivered, it is sometimes referred to as the **afterbirth.**

When the fetus and placenta are expelled from the uterus, all of the hormones secreted by the fetoplacental unit begin to decrease markedly in maternal plasma. Within 3 days after delivery, the concentration of steroid and protein hormones in maternal plasma returns to nonpregnant levels.

Many hormones, including estrogen, progesterone, relaxin, oxytocin, and the prostaglandins, have been found to play important roles in the initiation of parturition (Fig. 32–12).

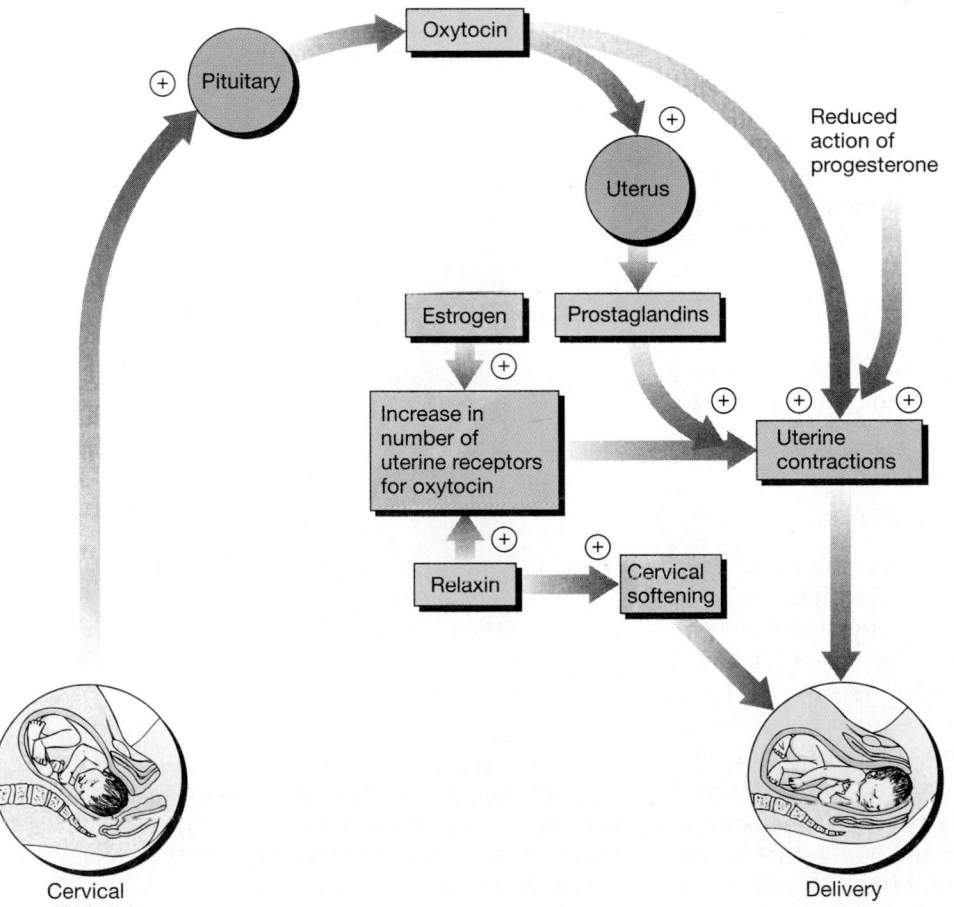

Figure 32–12

Hormonal factors that contribute to the onset of parturition. The plus sign indicates stimulation.

Effects of Oxytocin on Parturition

Oxytocin stimulates contraction of uterine muscles. Dilation of the cervix ("cervical stretch") before delivery provides afferent neural input to the hypothalamus, which results in the release of oxytocin from the posterior pituitary gland. In addition—and perhaps more importantly—in response to estrogen and relaxin stimulation, the number of receptors for oxytocin in the uterus increases progressively during the later part of pregnancy. This increases uterine sensitivity to oxytocin at the end of pregnancy. In response to oxytocin, uterine contractions become strong and rhythmic.

Effects of Prostaglandins on Parturition

Prostaglandins also stimulate contractions of uterine smooth muscle. In response to oxytocin, the myometrial or smooth muscle cells of the uterus secrete prostaglandins, which, through a paracrine action, stimulate further uterine contractions. The prostaglandins are so powerful in stimulating uterine contractions that exogenous administration of prostaglandins can induce labor and terminate pregnancy at almost any stage of gestation.

Effects of Progesterone on Parturition

As stated earlier, progesterone strongly inhibits uterine contractions. In some women, circulating levels of progesterone decline just prior to delivery, which removes progesterone inhibition of uterine contractions and facilitates the onset of labor. However, a decrease in total circulating progesterone prior to delivery has not been found in all women. Recent evidence indicates that the placenta may secrete a progesterone-binding protein before delivery, which reduces the amount of free progesterone in maternal circulation. This could serve functionally to reduce the action of progesterone on uterine muscle and facilitate delivery.

Effects of Relaxin on Parturition

Relaxin, secreted first by the corpus luteum and later by the endometrium, may facilitate delivery in at least three ways. First, as previously stated, relaxin increases the number of oxytocin receptors in the uterus and makes the uterus more sensitive to the effects of oxytocin at the end of pregnancy. Second, relaxin softens the cervix, making it more pliable and facilitating passage of the fetus through the cervix during delivery. Third, as previously mentioned, relaxin causes relaxation of the pelvic ligaments, easing passage of the fetus through the birth canal.

(a)

(b)

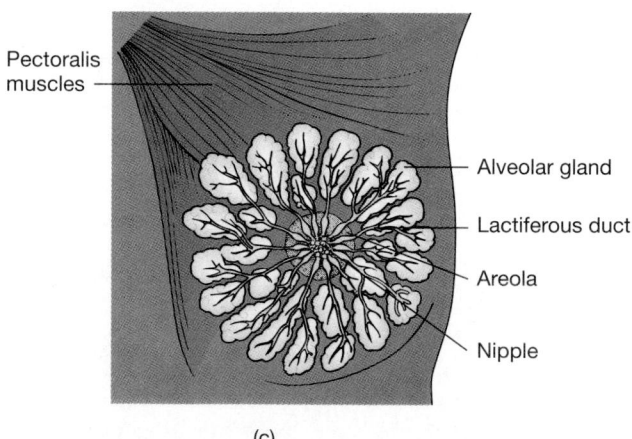

(c)

Figure 32–13

Anatomical features of the human breast. (*a*) External view. (*b*) Internal side view. (*c*) Internal front view. *(a is redrawn from Masters, Johnson, and Kolodny,* Human Sexuality. *b and c are redrawn from Worthington-Roberts,* Nutrition in Pregnancy and Lactation.*)*

Contaminants of Breast Milk

Many substances to which a lactating woman is exposed are transferred to her breast milk. Drugs, caffeine, viruses, environmental pollutants, and alcohol can pass into breast milk by simple diffusion, by carrier-mediated diffusion, or by active transport. The entry of these substances into breast milk depends on many factors, including the molecular weight and solubility properties of the particular substance. In general, the amount of a drug that enters breast milk is not more than 2% of the maternal dose. Although moderate amounts of many substances pose no risk to the suckling infant, some drugs can be harm-ful. For instance, many sedatives, including lithium, reserpine, and Valium (diazepam), produce drowsiness and lethargy in breast-fed infants. Anticoagulants can induce bleeding problems in the infant. More powerful narcotics, such as heroin, or the painkiller Darvon (dextropropoxyphene), can lead to addiction in breast-fed infants.

■ **Can heavy alcohol consumption during pregnancy lead to alcohol addiction in the infant?**

Lactation, the Synthesis and Release of Breast Milk, Is Under Hormonal Control

Anatomy of the Breast

The outer surface of the female breast is illustrated in Figure 32–13*a*. In the center of the breast is a pigmented area, called the **areola,** which surrounds the nipple. The internal structure of the breast is illustrated in Figure 32–13*b* and *c*. The mammary or breast tissue lies directly over the pectoralis major muscle and is separated from this muscle by a layer of fat (see Fig. 32–13*c*). The breast is composed of **glandular epithelium** and a **ductal system** embedded in in-terstitial tissue and fat. The ducts of the breast, which terminate in the nipple, branch through the breast tissue and are connected to glands called the alveoli (see Fig. 32–13*b*). The **alveoli,** which secrete breast milk, have been described as resembling clusters of grapes on stems. The alveoli are surrounded by contractile cells called **myoepithelial cells,** which, as we will see, are involved in the expulsion of milk from the breasts.

Development of the Breast

A number of hormones, including estrogen, progesterone, human chorionic somatomammotropin, and prolactin stimulate breast development during

Table 32–4
Hormones Involved in Breast Development During Pregnancy

Hormone	Origin	Function During Pregnancy
Prolactin	Anterior pituitary	Stimulates mammary growth
Oxytocin	Posterior pituitary	Generally no effect on mammary growth, but sensitivity of myoepithlial cell to oxytocin increases during pregnancy
Estrogen	Ovary and placenta	Stimulates proliferation of glandular tissue and ducts of the breast; stimulates prolactin release but blocks action of prolactin on the breast
Progesterone	Ovary and placenta	Stimulates proliferation of glandular tissue and ducts of the breast; blocks action of prolactin on the breast
Human chorionic somatomammotropin (hCS)	Placenta	Stimulates mammary growth
Human chorionic gonadtropin (hCG)	Placenta	Stimulates mammary growth

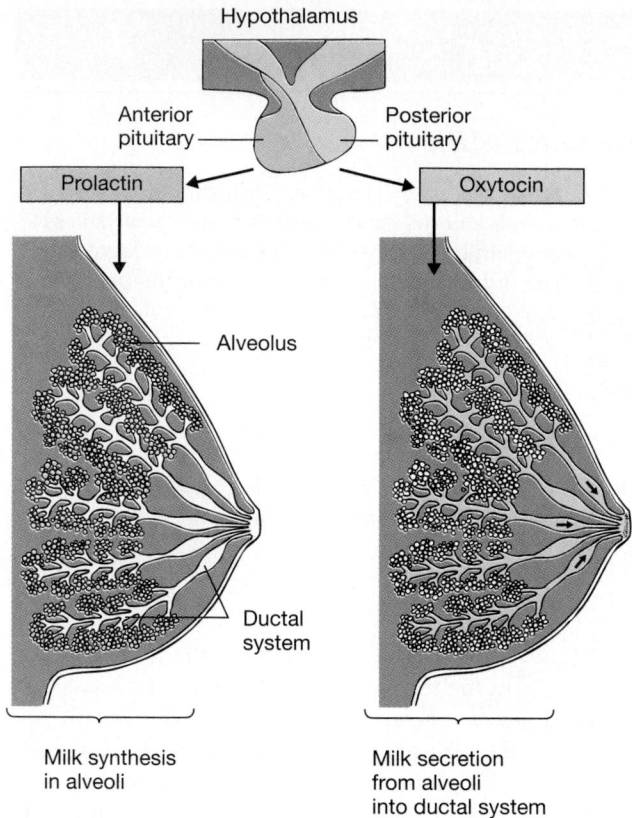

Figure 32–14

Hormonal control of milk production and expulsion in the breast.

pregnancy. Under the influence of these hormones, the breasts increase in size, and the ductal system and alveolar structures become more complex. A summary of various hormones and their effects on breast development prior to parturition is shown in Table 32–4. (p. 947)

Under hormonal stimulation, the alveolar cells of the breast extract a variety of substances, including glucose, amino acids, fatty acids, and glycerol, from the maternal circulation and convert these materials into breast milk. During lactation, the breasts may produce as much as 1.5 liters of milk each day.

Effects of Estrogen and Progesterone on Lactation

High circulating levels of estrogen and progesterone, secreted by the placenta during pregnancy, stimulate breast development. Estrogen also stimulates prolactin release; prolactin in turn stimulates additional breast growth. However, both estrogen and progesterone inhibit milk production in the breast prior to parturition by blocking the milk-inducing effects of prolactin on the breasts. After the placenta is expelled from the uterus at parturition, the concentrations of estrogen and progesterone in maternal plasma fall. The removal of estrogen and progesterone leaves the breasts under the influence of prolactin and oxytocin. As we will see, prolactin, together with oxytocin, controls the production and release of milk from the breasts following parturition (Fig. 32–14).

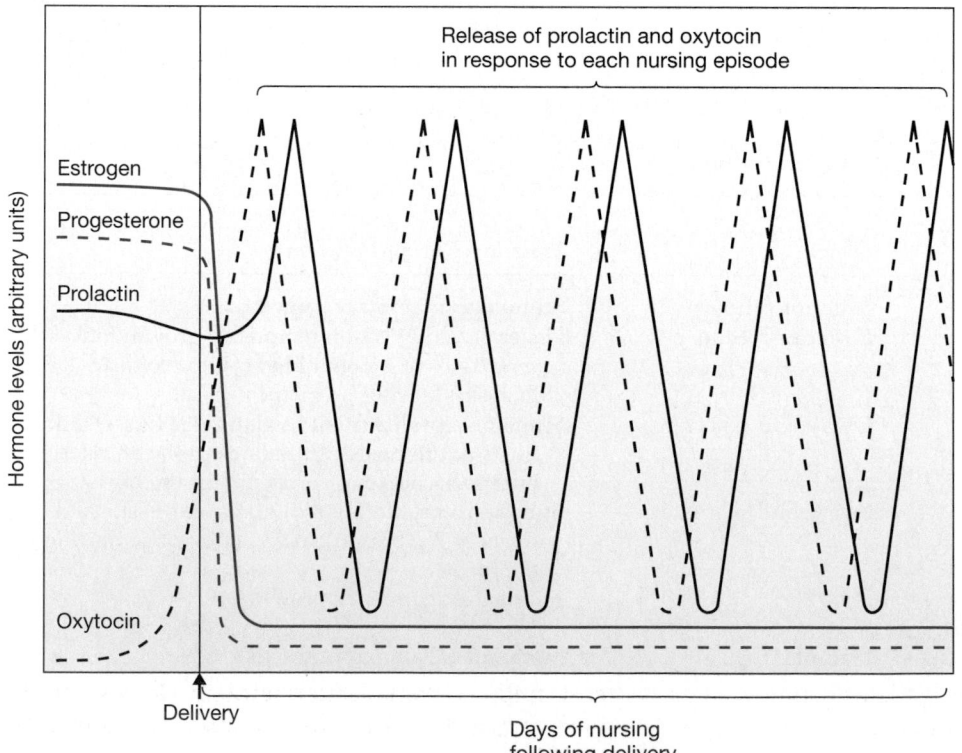

Figure 32–15

Changes in maternal hormones following delivery and during episodes of nursing.

Effects of Prolactin on Lactation

Prolactin stimulates synthesis of milk in the breast and secretion of milk into the alveoli of the breasts (see Fig. 32–14). High circulating levels of estrogen during pregnancy stimulate the release of prolactin from the maternal pituitary gland. Prolactin levels in maternal plasma rise during pregnancy and reach very high concentrations at birth. After birth, plasma prolactin levels begin to decline, but during this time the hypothalamus is very sensitive to neural signals it receives from the breasts. Breast stimulation during **suckling** stimulates afferent neural impulses that travel from the breast, through the spinal cord, to the hypothalamus and stimulate release of a large amount of prolactin from the maternal pituitary. Suckling results in a tenfold rise in plasma prolactin that lasts for about 1 hour, after which plasma prolactin levels return to normal. Therefore, prolactin concentrations in maternal plasma rise and fall with every episode of nursing (Fig. 32–15). Prolactin, released in response to a single nursing episode, stimulates **milk production** in the breast, which makes additional milk available for the next nursing episode. If an infant is not breast-fed or when nursing is terminated, there is no stimulus for prolactin release; hence, milk synthesis in the breast is not stimulated, and the breast stops secreting milk within a few days. With prolonged nursing, milk production in the breast begins to decline within 7 to 9 months after birth, but the breast can still produce significant quantities of milk for several years.

As we discussed in Chapter 13, prolactin inhibits the release of GnRH from the hypothalamus and hence inhibits LH and FSH release from the pituitary. Low levels of LH and FSH result in the absence of follicular growth and ovulation. Therefore, breast-feeding, which stimulates the episodic release of prolactin, tends to inhibit ovulation and reduce fertility. However, there is a large amount of individual variation in the time during which prolactin inhibits GnRH release with prolonged nursing; breast-feeding is thus not a reliable method of birth control.

Effects of Oxytocin on Lactation

Oxytocin stimulates the release of milk from the breast, a process termed **milk let-down.** Oxytocin induces milk let-down by stimulating contraction of the myoepithelial cells that surround the outer walls of the alveoli. This pressure forces milk from the alveoli into the ductal system of the breast (see Fig. 32–14). Once milk reaches the ductal system, it can be pulled from the breast by the negative pressure that results from suckling.

As with prolactin, oxytocin is released in response to breast stimulation during nursing. Suckling stimulates sensory nerves in the breast that transmit af-

Table 32–5
Nutrient Content of Human Milk and Cow's Milk

Constituent (per liter)	Human Milk	Cow's Milk
Energy (kcal)	690	660
Protein (g)	9	35
Fat (g)	40	38
Lactose (g)	68	49
Vitamins		
Vitamin A (IU)	1898	1025
Vitamin D (activity)	40	14
Vitamin E (IU)	3.2	0.4
Vitamin K (μg)	34	170
Thiamin (μg)	150	370
Riboflavin (μg)	380	1700
Niacin (mg)	1.7	0.9
Pyridoxine (μg)	130	460
Folic acid (μg)	41–84.6	2.9–68
Cobalamine (μg)	0.5	4
Ascorbic acid (μg)	44	17
Minerals		
Calcium (mg)	241–340	1200
Phosphorus (mg)	150	920
Sodium (mg)	160	506
Potassium (mg)	530	1570
Chlorine (mg)	400	1028
Magnesium (mg)	38–41	120
Sulfur (mg)	140	300
Iron (mg)	0.56–0.3	0.5
Iodine (mg)	200	80
Manganese (μg)	5.9–4.0	20–40
Copper (μg)	60	110
Zinc (mg)	4–0.5	3–5
Selenium (μg)	20	5–50
Fluoride (mg)	0.05	0.03–0.1
Chromium (μg)	4	2

Adapted from Worthington-Roberts, B.S., Vermeersch, J., and Williams, S.R. *Nutrition in Pregnancy and Lactation.* St. Louis: Times Mirror/Mosby, 1985, p. 257.

ferent neural impulses to the spinal cord and from there to the hypothalamus, resulting in the release of oxytocin from the pituitary gland. As with prolactin, oxytocin is released episodically in response to nursing episodes, as illustrated in Figure 32–15. In addition to breast stimulation, certain auditory stimuli, such as the sound of a baby crying, can also stimulate the release of oxytocin and hence milk ejection from the breast.

The Composition of Breast Milk

Breast milk contains water, protein, **lactose** (milk sugar), fat, and a variety of vitamins and minerals. Table 32–5 compares the nutrient content of human milk and cow's milk. One of the major differences between human milk and cow's milk is the amount and type of protein that each contains. The protein in human milk is largely in the form of **α-lactalbumin,** whereas that of cow's milk is in the form of **casein.**

The caseins are milk proteins that form coagula, or curds, when exposed to digestive enzymes, pepsin, or changes in pH. The lower casein content of human milk makes it easier for the infant to digest than cow's milk.

CHAPTER REVIEW

Summary

- Ovulation occurs approximately 14 days after the beginning of the last menstrual cycle in response to stimulation from high circulating levels of LH and FSH. Following ovulation, the ovum enters one of the two uterine tubes and is transported to the uterus over the course of 3 to 5 days.

- Following ejaculation into the vagina, sperm are transported through the vagina, cervix, and uterus into the uterine tubes. Sperm loss following ejaculation is very high owing to many factors.

- Capacitation and the acrosome reaction prepare the sperm to fertilize the ovum. Fertilization of the ovum normally occurs in the uterine tube before the ovum has reached the uterus.

- Sperm-ovum fusion results in the release of cortical granules from the ovum, which alter the surface membrane of the ovum and block entry of more than one sperm into the ovum (polyspermy).

- Initially, the fertilized ovum is a single cell, called a zygote, which contains 46 chromosomes.

- The zygote begins to divide mitotically about 30 hours after fertilization, while it is still in the uterine tube. Cell division marks the beginning of embryonic development.

- By the time the fertilized ovum reaches the uterus, it has divided many times and is called a blastocyst. The blastocyst sends out tissue projections that invade the endometrium, resulting in the implantation of the blastocyst in the endometrium.

- When the blastocyst is fully implanted in the endometrium it is called an embryo. After the second month of development, the embryo is called a fetus.

- The placenta, which is formed from cells of the early embryo, acts as an interface between the fetal and maternal blood supplies. The embryo is dependent on maternal circulation for its supply of oxygen, fluid, nutrients, salts, and antibodies, as well as for the removal of carbon dioxide, urea, and nitrogen. These substances pass through the placenta.

- hCG is secreted by the blastocyst prior to implantation and then by the placenta during the first two months of pregnancy. hCG stimulates growth of the corpus luteum during early pregnancy and ensures continued secretion of estrogen and progesterone from the corpus luteum.

- Estrogen and progesterone stimulate continued growth of the endometrium of the uterus. During the second month of pregnancy, the placenta secretes estrogen and progesterone on its own, and the corpus luteum is no longer needed.

- As the placenta secretes more estrogen and progesterone, it stops secreting hCG, leading to the degeneration of the corpus luteum. Throughout the remaining months of pregnancy, the placenta continues to secrete estrogen and progesterone.

- Progesterone secreted by the placenta is essential for the maintenance of pregnancy. It stimulates endometrial growth and secretion of nutrients from the endometrium, which nourish the developing embryo. Progesterone inhibits uterine contractions, helping to prevent premature birth. Progesterone also stimulates breast development in preparation for nursing.

- Estrogen secreted by the placenta stimulates enlargement of the uterus to accommodate the growing fetus, and development of the ducts in the breasts in preparation for nursing. At birth, estrogen induces relaxation of pelvic ligaments, facilitating delivery.

- High circulating levels of estrogen and progesterone in maternal plasma during pregnancy inhibit the secretion of hypothalamic gonadotropin-releasing hormone (GnRH), and hence pituitary secretion of luteinizing hormone (LH) and follicle-stimulating hormone (FSH). In the absence of stimulation from LH and FSH, no new follicles develop in the ovaries, resulting in the inhibition of ovulation and menstruation during pregnancy.

- Maternal plasma prolactin levels rise progressively throughout pregnancy, resulting in growth of the breasts. Maternal secretion of insulin, aldosterone, cortisol, and thyroxine also increases during pregnancy.

- Pregnancy, which usually lasts nine months, is divided into three-month periods called trimesters. Major physical changes occur in the pregnant woman during the second and third trimesters of pregnancy.

- The germinal stage of fetal development is marked by the formation of the blastocyst and implantation of the blastocyst in the endometrium.

- During the embryonic stage of development, internal organs including the heart, digestive system, and nervous system, as well as external organs including the eyes and ears, begin to form in the developing embryo.

- During the fetal stage of development, the fetus develops from a partially formed organism, completely dependent on the maternal blood supply, into an organism capable of sustaining life outside the uterus.

- During delivery, uterine contractions force the fetus through the dilated cervix and through the birth canal.

- As delivery approaches, oxytocin released by the pituitary and prostaglandins secreted by the uterus stimulate uterine contractions and cervical dilation. In addi-

tion, relaxin, secreted by the endometrium of the uterus, softens the cervix and relaxes pelvic ligaments.
- The breast is composed of secretory structures called alveoli, which synthesize and secrete milk, and a ductal system for the transport of milk.
- A variety of hormones including prolactin, estrogen, progesterone, hCG, and hCS stimulate breast development during pregnancy in preparation for nursing.
- After delivery, production and release of milk from the breasts are largely controlled by prolactin, which is secreted by the anterior pituitary, and oxytocin, which is secreted by the posterior pituitary.

Review Questions

Identify Each with a Term
1. Fertilization of the ovum by more than one sperm.
2. Implantation of a fertilized ovum outside of the uterus.
3. The contractile cells in the breast that are responsible for milk expulsion.

Define Each Term
4. dizygotic twins
5. zygote
6. chorion

Choose the Correct Answer
7. Fertilization of the ovum normally occurs in the:
   a. vagina   c. uterus
   b. abdominal cavity   d. uterine tube
8. Polyspermy is prevented by:
   a. mitotic division of the ovum prior to fertilization
   b. secretion of estrogen by the ovum
   c. a brief electrical block on the surface of the ovum
   d. formation of the zygote
9. The time between fertilization of the ovum and implantation of the fertilized ovum in the uterus is approximately:
   a. 5 to 10 days   c. 3 weeks
   b. fewer than 24 hours   d. 2 days
10. Which of the following is *not* an action of the placenta?
   a. delivery of waste from fetus to mother
   b. exchange of gases between fetus and mother
   c. delivery of antibodies from mother to fetus
   d. delivery of nutrients from fetus to mother
11. An increase in which of the following hormones would serve to inhibit the onset of parturition?
   a. relaxin   c. oxytocin
   b. progesterone   d. prostaglandins

Answer Each Question in One or Two Sentences
12. Which hormones are secreted by the corpus luteum during pregnancy?
13. Why is the acrosomal reaction important for fertilization?
14. How does sucking increase milk availability in the breast?

Problem Solving in Physiology
15. Why are estrogen and progesterone important for the maintenance of pregnancy?
16. Why do ovulation and menstruation cease during pregnancy?
17. Why do sperm enter the uterus more easily at midcycle?

Suggested Readings

Cowie, A.T., Forsyth, I.A., and Hart, I.C. *Hormonal Control of Lactation,* Chapters 3 and 4. New York: Springer-Verlag, 1980.

Diamond, M.P., and Decherney, A.H., eds. *Ectopic Pregnancy.* Philadelphia: W.B. Saunders Co., 1991.

Keen, C.L., Bendich, A., and Willhite, C.C., eds. "Maternal nutrition and pregnancy outcome." *Annals of the New York Academy of Sciences,* Vol. 678, 1992.

Lagercrantz, H., and Slotkin, T.A. "The 'stress' of being born." *Scientific American,* 254:100–107, 1986.

Neville, M., and Daniel, C.W., eds. *The Mammary Gland.* New York: Plenum Press, 1987.

Nilsson, L. *A Child Is Born.* New York: Delacorte Press, 1990.

Simpson, J.L., and Elias, S., eds. "Fetal cells in maternal blood: Prospects for noninvasive prenatal diagnosis." *Annals of the New York Academy of Sciences,* Vol. 731, 1994.

Key Terms

acrosome reaction (p. 929)
alveoli, alveolus (p. 947)
amniotic fluid (p. 941)
areola (p. 947)
blastocyst (p. 932)
Braxton-Hicks contractions (p. 941)
capacitation (p. 929)
colostrum (p. 941)
dizygotic twins (p. 930)
embryo (p. 933)
fertilization (p. 926)
fetus (p. 933)
human chorionic gonadotropin (hCG) (p. 933)
lactation (p. 948)
monozygotic twins (p. 931)
myoepithelial cells (p. 947)
oxytocin (p. 946)
parturition (p. 944)
placenta (p. 935)
polyspermy (p. 929)
pregnancy (p. 938)
prolactin (p. 940)
relaxin (p. 940)
umbilical cord (p. 936)
zygote (p. 931)

Human zygote at the
4-cell stage of cell division.
(© CNRI/Phototake, NYC)

Chapter 33

Sexual Physiology

Key Concepts

Male and female embryos are identical during the first six weeks of embryonic life. The characteristics that define a male and female are established through sexual determination and differentiation.

Sexual and reproductive physiology are intertwined, and most of the reproductive organs in both sexes serve both sexual and reproductive functions.

Four stages of sexual arousal occur in both males and females.

Fertility in males and females depends on the integrity of the hypothalamus, pituitary, and gonads. A disruption of normal function at any of these levels can result in infertility.

Chapter Outline

Sexual Determination and Differentiation
Human Sexual Response
Sexual Dysfunction
Contraceptive Methods

Jeff, a father of two boys, elected to have a vasectomy on his thirtieth birthday, fifteen years ago. At the time, most of his friends advised him against it saying that he was too young to make such an irreversible decision. It was, after all, permanent sterilization. And, no one really knew what the long-term health effects were. Even his wife, Barbara, had appeared hesitant. Jeff and Barbara had decided that two children were their limit — physically, emotionally, and financially. Two careers, long commutes, and significant credit card debt were taking their toll. Permanent sterilization or permanent abstinence were the only ways to ensure they would stick to their two-child plan. Also, fewer risks were associated with vasectomy than with tubal ligation, which he and Barbara had discussed. The vasectomy seemed to be the best and most responsible decision for Jeff to make at the time. Unfortunately, Jeff's and Barbara's marriage deteriorated.

The vasectomy seems like a lifetime ago to Jeff as he pulls into the parking lot of the hospital for a three-day stay. Remarried for two years now, Jeff and his new wife, Marianne, have decided to try to have a child of their own. While a vasectomy reversal was not a condition of their marriage, Marianne would like to have children, and she is ten years younger than Jeff. Marianne has been very pleased to be the step-mother to Jeff's two high-school sons, but she welcomes the idea of having a baby of her own. When Jeff suggested the vasectomy reversal, she was thrilled.

The cost of a vasovasotomy, or vasectomy reversal, is high. Jeff's vasectomy had been a relatively simple operation that was performed in a doctor's office, using a local anesthetic, and took only 20 minutes. Reattaching the severed vas deferens, termed vasovasotomy, on the other hand, involves bilayer microsurgery of both vas deferens, as well as a two- to three-day hospital stay. Jeff recalls how swollen and painful his scrotum was around the area of incision after the vasectomy. Will there be more severe complications after the vasovasotomy? Will the surgical operation be successful?

The urologist informed Jeff during the initial consultation that the rate of success in reattaching the vas deferens with the subsequent return of sperm in the semen is fairly high, about 70% to 90%. However, fertility rates are low, about 30% in men who undergo vasectomy reversals ten or more years after their original vasectomies. The reason for the low rate of pregnancies is the subject of much speculation. One theory is that, after a vasectomy, some men develop antibodies to their own sperm since the sperm are continually reabsorbed by the testes instead of being ejaculated, and the antibodies destroy the sperm.

During the month following the surgery, Jeff will visit the urologist for a semen analysis to determine sperm concentration and sperm motility. Whatever the outcome of the reversal, Jeff feels that the operation will have been worth it. If he and Marianne cannot have children, they still have two great sons. If the vasectomy reversal does allow them to have children, it will be cause for a great celebration.

In this chapter we will discuss various forms of birth control, including vasectomy and tubal ligation. We will also examine the physiology of sexual determination and differentiation, the phases of the human sexual response, and sexual dysfunction and infertility.

In Chapter 31, we discussed the physiological processes that control reproduction in the male and female. In Chapter 32, we examined the events that occur during pregnancy, fetal development, parturition, and lactation. We will now discuss the physiological processes that control human sexuality, beginning with an examination of how the biological sex of an individual is established. We will also investigate the physiological basis of the human sexual response, sexual dysfunction, and how contraceptives work.

Sexual Determination and Differentiation

When we consider the number of differences in anatomical structure, function, and hormonal regulation of the male and female, it is surprising that male and female embryos are identical during the first six weeks of life. The characteristics that define an individual's sex are established through the processes of **genetic sexual determination** and **sexual differentiation,** which is controlled by specific hormones. Determination of sex begins with the fusion of sperm and ovum, which establishes the **chromosomal** (**genetic,** or **genotypic**) sex of the zygote. The genetic information contributed to the zygote determines whether the **primitive gonads** will differentiate into **testes** or **ovaries,** which constitute the **gonadal sex** of the individual. The differentiated gonads secrete hormones that, in turn, control development of the genital ducts, external genitalia, secondary sex characteristics, and endocrine profiles that constitute the **phenotypic sex** of the individual. Therefore "maleness" or "femaleness" begins with genetic or chromosomal determination of sex, followed by gonadal differentiation and the development of phenotypic sex characteristics.

The Chromosomal Sex of the Individual Is Established When the Sperm and Ovum Fuse

Sexual Determination: The Sex Chromosomes

Every cell in the body, except the sperm and ovum, contains 23 pairs of chromosomes. Twenty-two of these chromosomal pairs are termed **autosomes;** the twenty-third pair are the **sex chromosomes.** Autosomes provide the genetic information for the development and differentiation of most of the cells in the body. The sex chromosomes provide the genetic information that directs the differentiation of gonadal cells into ovaries or testes; hence, the sex chromosomes determine the gonadal sex of the individual.

Meiosis and the Sex Chromosomes

The two identical sex chromosomes in the female are designated **XX;** the two nonidentical sex chromosomes in the male are designated **XY.** As discussed in Chapter 30, prior to fertilization the male germ cells (sperm) and the female germ cells (ova) undergo a special form of cell division known as **meiosis.** During meiotic division, each daughter cell receives only 23 chromosomes, one from each of the 23 chromosomal pairs. Given that the twenty-third pair (the sex chromosomes) in the male are XY, half of the sperm formed during meiosis will receive 22 autosomes and an X sex chromosome, while the other half will receive 22 autosomes and a Y sex chromosome. In contrast, the twenty-third pair of chromosomes in the female are XX, so all the ova formed during meiosis will each receive 22 autosomes and an X sex chromosome.

The Sex Chromosomes Provide the Genetic Information That Directs Differentiation of the Gonads into Ovaries or Testes

During fertilization, when the sperm and ovum fuse, the 23 chromosomes from the sperm (22 autosomes and an X or Y sex chromosome) and the 23 chromosomes from the ovum (22 autosomes and an X sex chromosome) combine to produce a single-celled zygote that contains a total of 46 chromosomes. When a sperm containing an X chromosome fertilizes the X-bearing ovum, the chromosomal sex of the zygote will be XX, or **female.** When a sperm containing a Y chromosome fertilizes the X-bearing ovum, the chromosomal sex of the zygote will be XY, or **male.** Do males have a male phenotype because they have a Y chromosome or because they have only one X chromosome? The evidence is that the Y chromosome carries genes, most notably the SRY gene, that direct the gonads to differentiate as testes. If an ovum is not fertilized by a Y-bearing sperm, the testes, male genital ducts, and male external genitalia will not develop. Therefore, the Y chromosome, contributed by the father, determines the male phenotype or "maleness" (Fig. 33–1). This is why a person with Klinefelter's syndrome, a chromosomal abnormality in which a genetic male receives an extra X chromosome (XXY), will appear as a nearly normal male. Evidence is accumulating that the X chromosome carries genes for femaleness just as the Y chromosome carries genes for maleness. These genes, one of which has preliminarily been termed the DSS gene, may direct the gonads to differentiate as ovaries. In the female, both of the X chromosomes are essential for normal differentiation of the ovaries. A person with Turner's syndrome, or a single X chromosome, will have the appearance of a female but the development of the ovaries is impaired and, in the absence of hormonal replacement, the development of secondary sex characteristics is impaired and menstruation does not occur (see Table 33–2).

The X chromosome contains many genes that are required by both sexes. In addition, some traits are determined by genes that are located only on the X

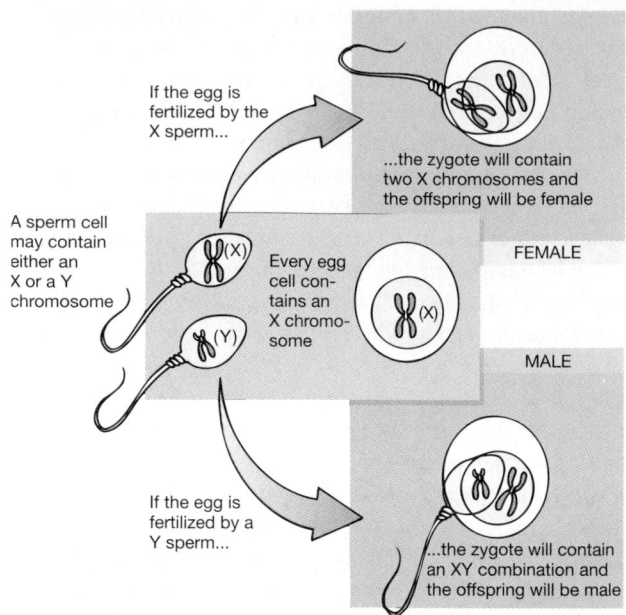

If the egg is fertilized by the X sperm...

...the zygote will contain two X chromosomes and the offspring will be female

FEMALE

A sperm cell may contain either an X or a Y chromosome

Every egg cell contains an X chromosome

MALE

If the egg is fertilized by a Y sperm...

...the zygote will contain an XY combination and the offspring will be male

Figure 33–1

Sex of the zygote is determined at the time of fertilization by the genetic material contributed by the sperm. An X-bearing sperm produces a female; a Y-bearing sperm produces a male.

chromosome. These are termed sex-linked traits and they follow the pattern of transmission of the X chromosome. Color blindness and hemophilia are two such sex-linked traits. For most X-linked chromosomal loci, the uncommon allele is recessive and the common allele is dominant in the female. Therefore, the uncommon phenotype is not usually expressed in the female. In order for a female to express an uncommon phenotype, such as color-blindness, she must inherit two recessive X-linked alleles, one on the X chromosome from her mother and one on the X chromosome from her father. This is a rare combination. In contrast, the male receives only one X chromosome from his mother and a Y chromosome from his father. In the male, every X chromosome allele present is expressed whether the allele was dominant or recessive in the mother. Therefore, an uncommon recessive allele that may have been carried by the mother, but not expressed by the mother, will be expressed in her male offspring. In consequence, X-linked recessive traits are much more common in males than in females.

The Role of the Y Chromosome and the H-Y Antigen

A question of current interest to molecular geneticists is how the Y chromosome directs the embryonic gonad to differentiate as a testis. Although this process is not completely understood, it appears that the Y chromosome contains genetic loci, such as the SRY gene, which either code for, or regulate the expression of, a specific protein found on the membrane of male cells. This protein is antigenic and induces immunological rejection in females. Hence, it is termed the **histocompatibility-Y antigen** or **H-Y antigen.** The H-Y antigen is thought to be responsible for inducing testicular differentiation in the primitive gonad. The exact chromosomal location and the mechanism of expression of the gene (or genes) that code for the H-Y antigen are not known. Current evidence suggests that the H-Y antigen, disseminated by cells in the developing gonads, binds to H-Y receptors in the gonads and induces gonadal cells to differentiate as testes. Whether the Y chromosome in the female contains genetic loci that either code for, or regulate the expression of, substances that direct the gonads to differentiate as ovaries is not yet known, but the existence of a female counterpart to the H-Y antigen is very likely.

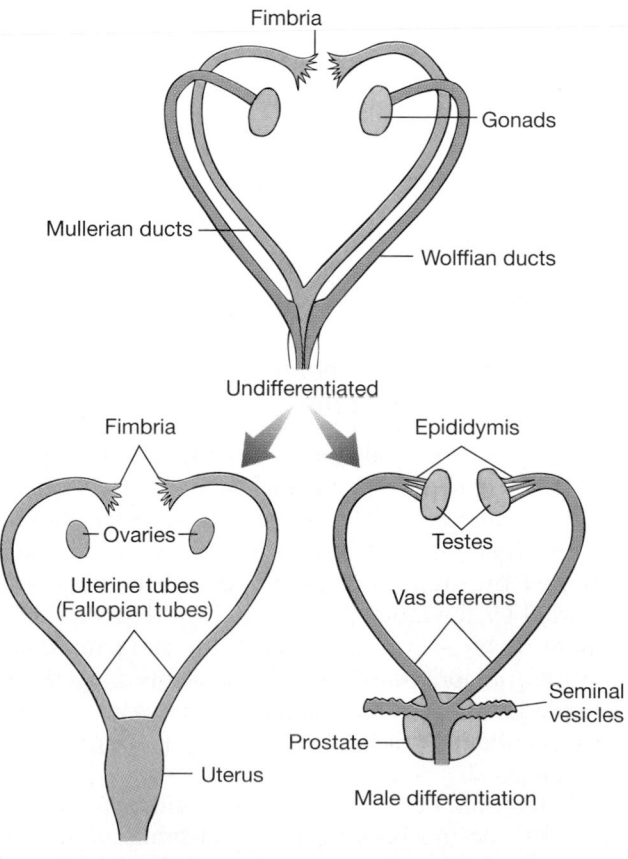

Figure 33–2

Embryonic differentiation of the male and female internal genitalia.

Differentiation of the Internal Genitalia

During the first six weeks of embryonic development, the gonads of a genetically male or female embryo are anatomically and morphologically identical. During the third to the seventh week of embryonic development, two sexually undifferentiated gonads (one on each side) develop, which will differentiate into ovaries in the female or testes in the male (Fig. 33–2).

The Embryonic Ductal Systems

In addition, two distinct ductal systems develop in the embryo regardless of genetic sex (XX or XY). One ductal system, termed the **Mullerian duct system,** is the precursor of the internal genitalia of the female. The other duct system, called the **Wolffian duct system,** is the precursor of the internal genitalia of the male (see Fig. 33–2). In the genetically female embryo, the Mullerian duct system will differentiate into uterine tubes, uterus, cervix, and vagina, and the Wolffian duct system will regress (see Fig. 33–2). In the genetically male embryo, the Wolffian duct system will differentiate into the epididymis, vas deferens, seminal vesicles, and ejaculatory duct, and the Mullerian duct system will regress (see Fig. 33–2). As we will see, differentiation and development of one or the other of the duct systems is due to the presence or absence of hormones secreted by the developing gonads.

Differentiation of the Internal and External Genitalia Is Under Hormonal Control

Differentiation of Internal Genitalia in the Male

The **timing** of gonadal differentiation in the male embryo is different from that in the female. Presence of the H-Y antigen in the male stimulates differentiation of the testes at 6 to 7 weeks of gestation, while in the female, the ovaries do not begin to differentiate until week 12 of gestation.

Anti-Mullerian Hormone and Gonadal Differentiation. As the testes develop in the male embryo, under the direction of the H-Y antigen, the Sertoli cells of the testes begin to secrete a substance called **Anti-Mullerian Hormone (AMH).** AMH is a glycoprotein that has local effects, through paracrine action, on cells surrounding the gonads. AMH induces regression, or involution, of the Mullerian duct system. The paracrine action of AMH is evidenced by the fact that if one of the two testes is removed at an early stage of development, Mullerian regression will occur only on the side of the body that has an intact testis. Although the Sertoli cells secrete AMH for about the first two months of embryonic life, secretory rates are much higher during the first month, when regression of the Mullerian ducts is required in the male.

Testosterone and Gonadal Differentiation.   Although AMH has been identified as the factor responsible for regression of the Mullerian ducts, it does not stimulate the Wolffian ducts to differentiate into the internal genitalia of the male. Differentiation of the male internal genitalia is under the control of **testosterone.** Leydig cells of the developing testes begin to secrete testosterone at about 7 to 9 weeks of gestation. Testicular content and secretion of testosterone rise from week 6 to week 14 of gestation and then begin to decline during weeks 15 to 20 (Fig. 33–3). However, even at birth, when testosterone levels are lower than during early development, the male fetus is exposed to considerably higher concentrations of testosterone than the female. Testosterone, which diffuses from the testes, acts locally to stimulate differentiation of the male or Wolffian duct system during weeks 9 to 16

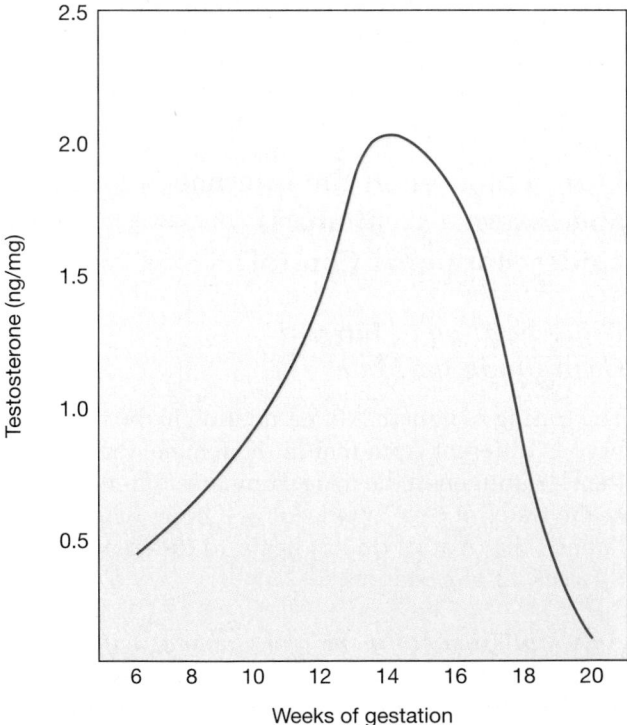

Figure 33–3

Testicular content of testosterone in the male fetus during gestation. (Adapted from Shearman, *Clinical Reproductive Endocrinology*, 1985, p. 349, originally reproduced from Tapanainen, et al., 1981.)

of embryonic development. Therefore, along with AMH, testosterone has paracrine actions on the developing gonads. While AMH induces Mullerian duct regression, testosterone stimulates Wolffian duct development (see Fig. 33–2). As we will see, when testosterone biosynthesis is deficient in the developing testes, or when the Wolffian ducts do not respond to testosterone, the internal genitalia of the male embryo do not develop, but remain rudimentary, resulting in a condition called **testicular feminization.**

Differentiation of Internal Genitalia in the Female

Ovarian differentiation in the female embryo begins long after the testes have differentiated in the male embryo. At about 12 weeks of gestation, the germ cells (oogonia) in the ovary begin to divide meiotically to form oocytes, marking the beginning of ovarian differentiation. By week 20 to 25 of gestation, the female gonads can be structurally characterized as ovaries. During the third month of gestation, the Mullerian duct system in the female embryo undergoes further development, while the Wolffian duct system degenerates, owing to the absence of testosterone stimulation (see Fig. 33–2). The presence of the ovaries is not necessary for Mullerian duct development in the female embryo, since differentiation of the uterus and uterine tubes occurs even when no ovaries are present. Therefore, it appears that the ovaries (unlike the testes) do not contribute hormonal factors that govern differentiation of the female genital tract.

Differentiation of External Genitalia in the Female

As we have seen with the internal genitalia, the external genitalia of both the male and female embryo are identical during the first 6 weeks of gestation. The external genitalia consist of a urogenital slit surrounded by urethral folds, which are in turn surrounded by the labioscrotal swellings (Fig. 33–4). A genital tubercle is located above the urogenital slit, and the anal indentation is located below. Sexual differentiation of these embryonic structures into the external genitalia of the male and female is summarized in Table 33–1.

Differentiation of the external genitalia in the female also occurs because of the **absence of testosterone.** In the female embryo, the urethral folds and the labioscrotal swellings remain separate and differentiate into the labia minora and labia majora, respectively (Fig. 33–4). The genital tubercle differen-

Undifferentiated

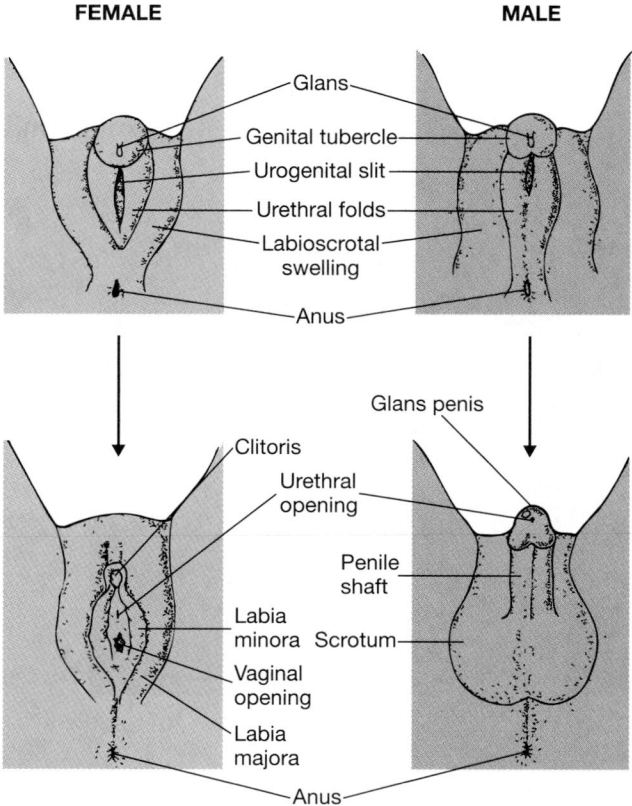

FEMALE   **MALE**

Figure 33–4

Embryonic differentiation of the male and female external genitalia.

tiates first into the glans and then into the clitoris. The urogenital slit differentiates into the vaginal opening and the urethra (see Fig. 33–4).

Differentiation of External Genitalia in the Male

Secretion of testosterone by the developing testes stimulates differentiation of the external genitalia in the male embryo. Testosterone reaches the external genitalia via the systemic circulation and hence exerts an endocrine action on differentiation of the external genitalia. Note that the external genitalia are androgen-sensitive tissues that contain the enzyme **5-alpha-reductase,** which converts testosterone to **dihydrotestosterone,** or **DHT.** Therefore, while testosterone directly stimulates differentiation of the Wolffian ducts, DHT, rather than testosterone itself, stimulates sexual differentiation of the external genitalia in the male. Deficiencies of testosterone secretion, 5-alpha-reductase activity, or receptors for DHT in external genital tissues can, as we will see, result in impaired masculinazation of the male external genitalia.

In response to stimulation from dihydrotestosterone, the urethral folds fuse and surround the urethra. Fusion of the urethral folds forms the shaft of the penis (see Fig. 33–4). Dihydrotestosterone also stimulates fusion of the labioscrotal swellings on the midline, forming the scrotum and the skin that covers the ventral portion of the penis (see Fig. 33–4). Although the clitoris and the glans penis originate from similar embryonic tissue, the penis grows 3 to 4 times faster than the clitoris. Therefore, the sex of the embryo can be determined by the size of the glans as well as by the presence or absence of urethral fold fusion. In the male embryo, dihydrotestosterone also induces differentiation of the prostate and inhibits the development of a vagina.

Table 33–1
Structural Origin of Male and Female External Genitalia

Undifferentiated Embryonic Structure	Differentiation in the Female	Differentiation in the Male
Genital tubercle	Clitoris	Glans of penis
Urethral folds	Labia minora	Shaft of penis
Labioscrotal swellings	Labia majora	Scrotum
Urogenital slit	Urethra and vaginal opening	Urethra
Anal indentation	Anus	Anus

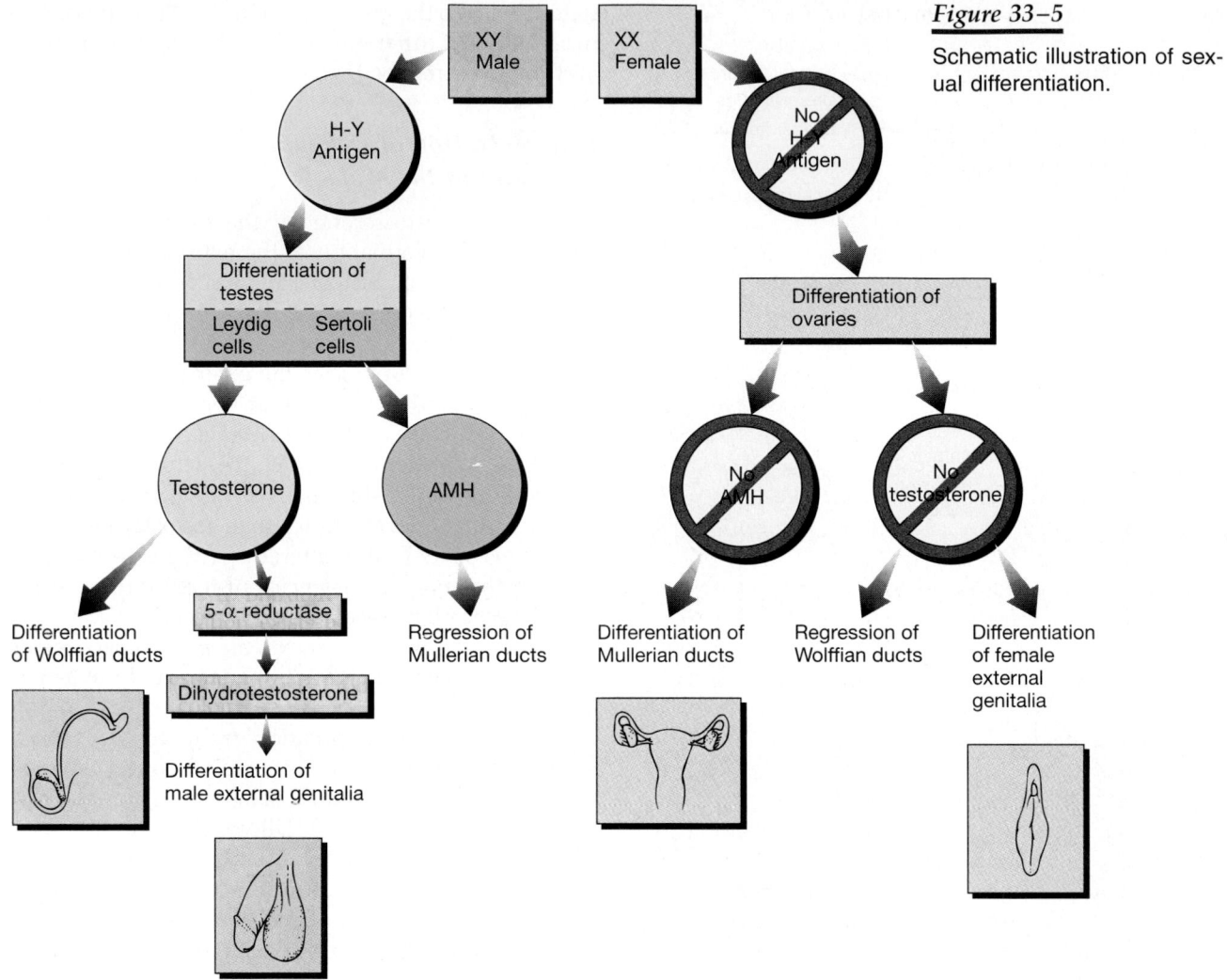

Figure 33–5

Schematic illustration of sexual differentiation.

The factors controlling sexual determination and differentiation are schematically illustrated in Figure 33–5.

Abnormalities of Sexual Differentiation Can Occur at Several Levels

Normal patterns of sexual differentiation can be altered because of abnormalities at the level of chromosomal sex (genotype), gonadal sex, internal genital sex, or external genital sex (phenotype). Although many variations and degrees of variation in patterns of sexual differentiation are possible, very few are common. We will discuss only those that are best understood.

Chromosomal Abnormalities in the Female

As we have previously discussed, the normal female embryo has 23 pairs of chromosomes (22 autosomal pairs and one pair of sex chromosomes, XX) (Fig. 33–6). As previously discussed, both of the X chromosomes are essential for normal differentiation of the ovaries, but only one X is necessary for differentiation of the Mullerian ducts and the female external genitalia. One chromosomal abnormality that has been identified in the female is the presence of only one X sex chromosome, designated as XO, which is termed **Turner's syndrome** (Table 33–2). This chromosomal deficiency is characterized by impaired development of the female gonads, which results in the appearance of an incompletely formed gonad, termed a **vestigal gonadal streak.** The uterus, uterine tubes, and external genitalia develop normally.

Chromosomal Abnormalities in the Male

The normal male embryo has 23 pairs of chromosomes (22 autosomal pairs and one pair of sex chromosomes, XY) (see Fig. 33–6). Some genetically male individuals receive an extra X chromosome (XXY), which results in **Klinefelter's syndrome** (see Table

33–2). In these cases the presence of the Y chromosome directs differentiation of the testes, but they are reduced in size. As in normal males, testosterone secretion from the testes in XXY males stimulates development of the Wolffian duct system and development of the male external genitalia, while AMH secretion from the testes induces Mullerian regression. However, the presence of the extra X chromosome leads to testicular dysfunction, owing to impaired development of the seminiferous tubules and a deficiency in spermatogenesis.

Enzyme Deficiencies in the Female

As we have previously learned, the presence of androgens (testosterone and dihydrotestosterone) during early embryonic development results in the differentiation of male external genitalia. Therefore, it is not surprising that exposure of a genetically female embryo (XX) to androgens during embryonic development can result in **masculinization**, or **virilization**, of the female embryo's external genitalia.

Some genetic females (XX) lack the enzymes **11-hydroxylase** or **21-hydroxylase** in their adrenal glands. These enzymes are necessary for the synthesis of gluco- and mineralocorticoids. In the absence of these enzymes, the adrenal glands use the steroid precursor cholesterol to synthesize androgens. Hypersecretion of adrenal androgens results in the development of the **adrenogenital syndrome,** also called **congenital adrenal hyperplasia** (Table 33–2). These individuals have normal ovaries, owing to a full chromosomal XX complement. The absence of a Y chromosome precludes the development of testes and hence testicular secretion of AMH, and the absence of AMH ensures normal development of the Mullerian duct system (including the uterus and uterine tubes). However, hypersecretion of adrenal androgens can result in masculinization of the external genitalia. The degree of masculinization depends on the severity of the adrenal enzymatic deficiency, it can range from mild hypertrophy of the clitoris to fusion of the urethral folds and the labioscrotal swellings, which then resemble a penis and scrotum. Severe masculinization of the female external genitalia, a condition termed "female pseudohermaphroditism," can lead to mistakes in identifying the gender of the infant.

Enzyme and Receptor Deficiencies in the Male

Genetic males (XY) who lack androgen receptors cannot respond to testosterone or dihydrotestosterone, which results in a condition known as **testicular feminization syndrome** (see Table 33–2). In these individuals, the presence of the Y chromosome results in

Figure 33–6

Photograph of normal human karyotype. *(© Leonard Lessin/Photo Researchers, Inc.)*

development of the testes, and testicular secretion of AMH induces regression of the Mullerian ducts. However, androgen insensitivity results in a lack of development of the Wolffian duct system and the absence of masculinization of the external genitalia. Therefore, these individuals have no internal genitalia at all, because AMH induces regression of the Mullerian ducts while the lack of androgen receptors precludes development of the Wolffian ducts.

As with androgen receptor deficiencies, a deficiency of the enzymes necessary for testosterone biosynthesis also results in varying degrees of Wolffian duct underdevelopment and reduced masculinization of external genitalia (see Table 33–2). In such cases, the presence of the Y chromosome and of AMH ensures the development of the testes and regression

Table 33–2
Abnormalities of Sexual Differentiation

Chromosomal Complement	Enzyme or Receptor Defects	Gonads	Internal Genitalia	External Genitalia
XX Normal female	—	Ovaries	Female	Female
XO Turner's syndrome	—	Ovaries with impaired development— "streak gonads"	Female	Female
XXY Klinefelter's syndrome	—	Testes with impaired development of seminiferous tubules and deficient spermatogenesis	Male	Male
XX Adrenogenital syndrome	Adrenal 11- or 21-hydroxylase deficiency	Ovaries	Female	Male/Female
XY Testicular feminization	Lacks androgen receptors	Testes	None	Female
XY	Lacks ability to synthesize testosterone	Testes	Male with impaired development	Female/Male
XY	5-Alpha-reductase deficiency	Testes	Male	Female/Male

of the Mullerian ducts. A deficiency of the enzyme 5-alpha-reductase, necessary for converting testosterone to dihydrotestosterone, results in reduced masculinization of the external genitalia (which respond to dihydrotestosterone). However, this enzyme deficiency does not interfere with the formation of normal testes due to the presence of the Y chromosome. Mullerian regression occurs because of the presence of AMH, and the Wolffian duct system develops in response to testosterone stimulation (see Table 33–2).

A summary of the important events that control sexual determination and differentiation is given in Figure 33–5 and Table 33–3.

Human Sexual Response

With the increased availability and popularity of contraceptives, it is now useful to draw a distinction between reproductive and nonreproductive, or sexual, physiology. Note, however, that sexual and reproductive physiology are inextricably intertwined. Most of the reproductive organs in both sexes serve

both sexual and reproductive functions. In addition, sexual arousal in the male is necessary for reproduction since ejaculation, which results in sperm transmission, does not occur in the absence of sexual arousal.

The Physiology of the Sexual Response in both Males and Females Can Be Divided into Four Stages of Sexual Arousal: Excitement, Plateau, Orgasm, and Resolution

Sexual arousal, like other physiological responses, can result from either **physical stimulation** or **mental stimulation.** For instance, salivation can occur in response to the thought or sight of food as well as to eating. In much the same way, sexual arousal can result from mental stimulation provided by movies or books, as well as from physical stimulation such as touching or kissing. Therefore, the human sexual response can be significantly altered by thoughts, feelings, and learned values. However, the physiological responses to sexual arousal are the same whether

Table 33–3
Summary of Events Controlling Sexual Determination and Differentiation

Female

1. Two X chromosomes are needed for normal differentiation of the ovaries.
2. Only one X chromosome is required for differentiation of the Mullerian ducts and the female external genitalia.
3. The absence of ovaries will not preclude development of the Mullerian ducts or the female external genitalia.
4. Exposure of the genetically female embryo to androgens during early development will result in varying degrees of masculinization of the external genitalia, depending on when during development the exposure occurs. Exposure to androgens will not alter development of the ovaries or the Mullerian ducts.

Male

1. The Y chromosome, which codes for or regulates the expression of the H-Y antigen, is required for differentiation of the testes.
2. Mullerian Inhibiting Substance, secreted by the Sertoli cells of the testes, induces regression of the Mullerian ducts.
3. Testosterone, secreted by Leydig cells of the testes, stimulates Wolffian duct development but has no effect on the Mullerian ducts.
4. Testosterone is converted to dihydrotestosterone (DHT) in tissues of the external genitalia, and DHT stimulates differentiation of the male external genitalia. An absence of DHT formation results in reduced masculinization of the external genitalia but does not affect normal development of the Wolffian ducts, which respond directly to testosterone.

arousal is due to physical or mental stimulation. Sexual arousal, which occurs throughout life from infancy into old age, involves activation of the nervous system, the endocrine system, and the sex organs.

The four stages of sexual arousal, as defined by Masters and Johnson are, **excitement, plateau, orgasm,** and **resolution.** While distinctions among the four stages are not clearcut and individual variations are many, all four stages of sexual arousal occur in both males and females and in both heterosexuals and homosexuals.

Excitement Phase in the Male

Sexual arousal during the excitement phase increases parasympathetic nerve activity, which causes the arteries in the penis to dilate, resulting in increased blood flow to the penis. As blood becomes concentrated in the tissues of the penis, a process called **vasocongestion,** the penis increases in diameter, length, and firmness, which defines **penile erection** (Fig. 33–7). Although the penis of some lower animals contains a bone, human males do not have a penile bone. The human penis does contain a thin skeletal muscle, but this muscle does not contribute significantly to penile erection, as evidenced by the fact that normal erection occurs in the absence of penile muscle activity following certain spinal cord injuries. Instead, penile erection depends on increased blood flow to the penis and increased intrapenile pressure. Penile vascoconstriction during sexual arousal in the male is not constant. Decreased blood flow to the penis can result in a reduction in penile size and firmness even though sexual stimulation and neuromuscular tension have not diminished. Such fluctuations in blood flow to the penis are usually short-lived.

Sexual arousal also results in increased blood flow to the testes, causing an increase in testicular size (see Fig. 33–7). During the excitement phase, the muscles surrounding the vas deferens contract, lifting the testes closer to the body and smoothing the scrotal skin (see Fig. 33–7). An increase in general muscle tone or tension, termed **myotonia,** also accompanies sexual arousal. Erection of the nipples of the breast occurs in some (but not all) men.

Excitement Phase in the Female

One of the physiological responses to sexual arousal during the excitement phase in the female is **vaginal lubrication** (Fig. 33–8). Both penile erection in the male, and vaginal lubrication in the female, occur in response to vasocongestion. Sexual arousal results in increased blood flow to the vagina, and vasocongestion of the vagina leads to release of moisture from the vaginal lining, a process termed **transudation.** As vaginal secretion of moisture increases, some moisture may reach the vaginal opening and the surrounding labia. Vaginal secretion of moisture lubricates the vagina and facilitates insertion of the penis without discomfort. Note that there are great individual variations in the amount of moisture secreted by the vagina during sexual arousal. The amount of moisture present cannot be used as an index of sexual arousal since heightened sexual arousal may be accompanied by sparse vaginal lubrication.

During the excitement phase, the vagina expands, and the cervix and uterus are lifted upward away

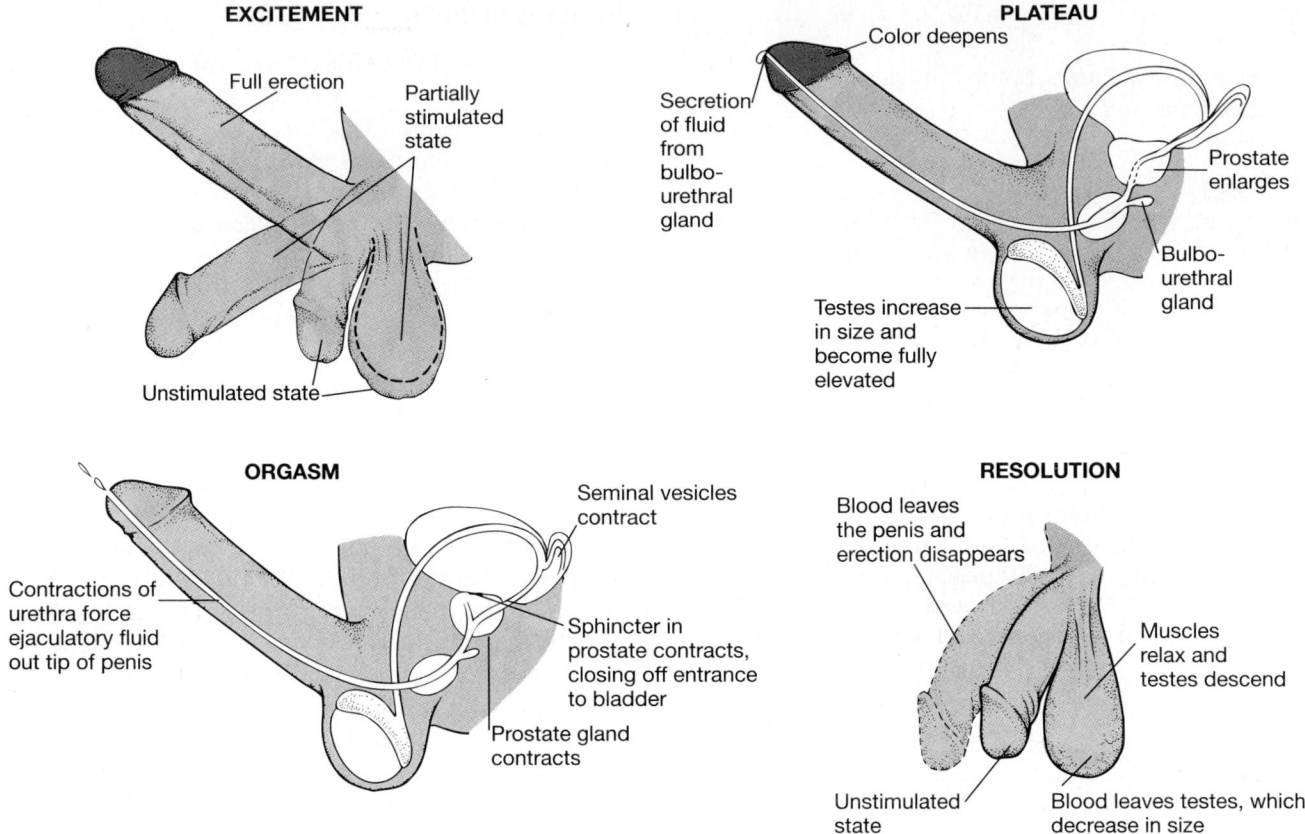

EXCITEMENT

Full erection

Partially stimulated state

Unstimulated state

PLATEAU

Color deepens

Secretion of fluid from bulbo-urethral gland

Prostate enlarges

Bulbo-urethral gland

Testes increase in size and become fully elevated

ORGASM

Seminal vesicles contract

Contractions of urethra force ejaculatory fluid out tip of penis

Sphincter in prostate contracts, closing off entrance to bladder

Prostate gland contracts

RESOLUTION

Blood leaves the penis and erection disappears

Muscles relax and testes descend

Unstimulated state

Blood leaves testes, which decrease in size

Figure 33–7

Changes in the external and internal genitalia of the male during the sexual response cycle.   *(Redrawn from Masters, Johnson, and Kolodny,* Human Sexuality, *Little, Brown, 1985.)*

from the vagina (see Fig. 33–8). Increased blood flow to the external genitalia results in increased size of the clitoris, labia minora, and labia majora (Fig. 33–9a). Muscle contractions in the breast tissue often result in nipple erection, and increased blood flow to the breast causes breast enlargement (see Fig. 33–9b). As in the male, there is an increase in general muscle tension, or myotonia, in the female during the excitement phase.

Plateau Phase in the Male

The plateau phase of sexual arousal precedes orgasm. Time spent in the plateau phase varies widely within and among individuals. Men who ejaculate quickly spend little time in this phase. During the plateau phase, blood pooling in the head of the penis causes this area to enlarge and become darker in color (see Fig. 33–7). The testes continue to increase in size because of increased blood flow, and muscle contractions in the scrotum lift the testes up and back toward the anus (see Fig. 33–7). As was discussed in Chap-

ter 31, pre-ejaculatory fluid from the bulbo-urethral glands, which may contain viable sperm, is often secreted at the tip of the penis during the plateau phase of sexual arousal (see Fig. 33–7). Heart rate, blood pressure, respiration rate, and muscle tension all increase significantly during this phase.

Plateau Phase in the Female

In the female, blood flow to the vagina increases, resulting in swelling of the walls of the vagina, which reduces the size of the vaginal opening (see Fig. 33–8). This increases the vaginal surface area that contacts the penis during intercourse. The uterus is elevated away from the vagina in a process termed *tenting* (see Fig. 33–8). The clitoris, which is highly sensitive to touch during this phase, is retracted against the pubic bone and becomes partially covered by the skin of the clitoral hood, which reduces direct physical contact of the clitoris (see Fig. 33–9a). Increased blood flow to the labia produces a deepening in the color of the labia minora and labia majora.

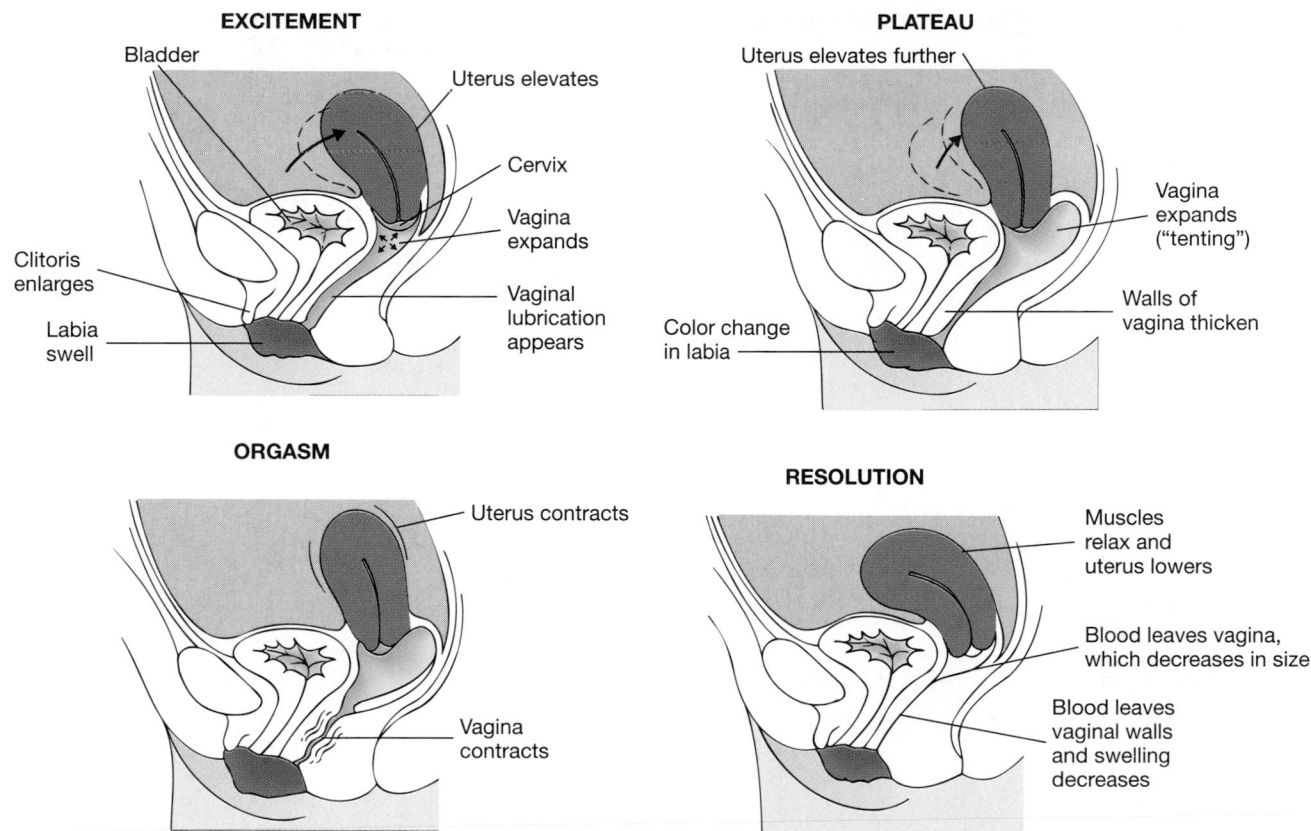

EXCITEMENT

Bladder

Uterus elevates

Cervix

Vagina expands

Clitoris enlarges

Labia swell

Vaginal lubrication appears

PLATEAU

Uterus elevates further

Vagina expands ("tenting")

Walls of vagina thicken

Color change in labia

ORGASM

Uterus contracts

Vagina contracts

RESOLUTION

Muscles relax and uterus lowers

Blood leaves vagina, which decreases in size

Blood leaves vaginal walls and swelling decreases

Figure 33–8

Changes in the external and internal genitalia of the female during the sexual response cycle. *(Redrawn from Masters, Johnson, and Kolodny, Human Sexuality, Little, Brown, 1985.)*

Blood flow to the breasts results in increased size of the areoli and reddening or darkening of the skin color in areas on the breasts, referred to as the "sexual flush" (see Fig. 33–9*b*). Reddening or darkening of the skin on other areas of the body may occur as well. As is seen in the male, heart rate, blood pressure, respiration rate, and muscle tension all increase during the plateau phase in the female.

Orgasmic Phase in the Male

Ejaculation in the male involves two processes: the **emission** and the **expulsion** of ejaculatory fluid (see Fig. 33–7). During emission, increased sympathetic nerve activity stimulates rhythmic muscular contractions of the vas deferens, seminal vesicles, and prostate, forcing ejaculatory fluid into the urethra. Once these contractions have begun, they cannot be stopped until ejaculation is complete. During the expulsion phase, rhythmic muscular contractions of the urethra force ejaculatory fluid out the tip of the penis. These two phases of ejaculation usually last only a few seconds. As was discussed in Chapter 30, a sphincter in the prostate contracts during ejaculation, sealing off the entrance to the bladder and preventing urine from mixing with ejaculatory fluid (see Fig. 33–7).

Orgasmic Phase in the Female

Orgasm in the female is characterized by rhythmic muscular contractions of the uterus and vagina that may last for as long as 15 seconds (see Fig. 33–8). As in the male, muscles in various parts of the body involuntarily contract during female orgasm, resulting in muscular rigidity and spasms.

Some controversy surrounds the importance of clitoral stimulation in inducing orgasm in the female, and whether females ejaculate. It appears that most women, while capable of achieving orgasm in response to vaginal stimulation, experience more intense orgasms when clitoral and vaginal stimulation are combined. Although most females do not ejaculate fluid from the urethra during orgasm, ejacula-

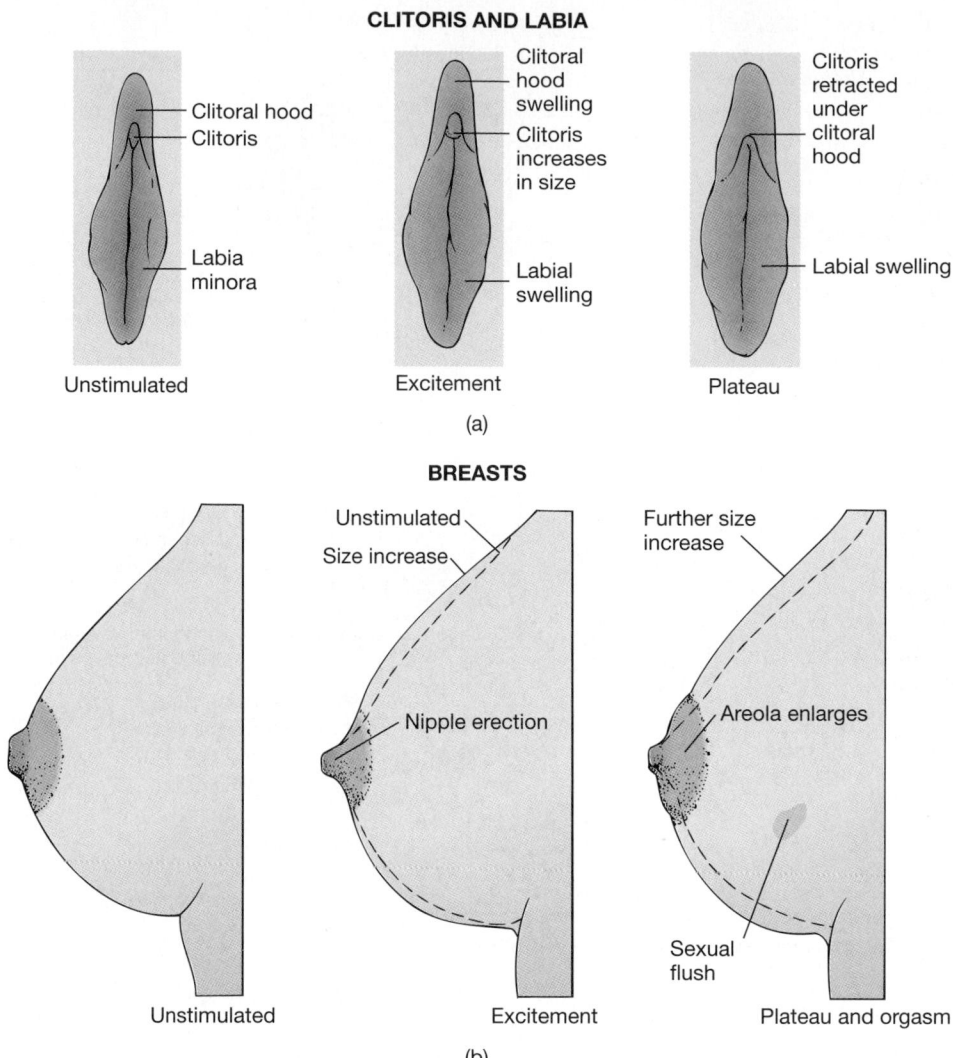

CLITORIS AND LABIA

Clitoral hood
Clitoris

Labia
minora

Unstimulated

Clitoral
hood
swelling

Clitoris
increases
in size

Labial
swelling

Excitement

Clitoris
retracted
under
clitoral
hood

Labial swelling

Plateau

(a)

BREASTS

Unstimulated

Unstimulated
Size increase

Nipple erection

Excitement

Further size
increase

Areola enlarges

Sexual
flush

Plateau and orgasm

(b)

Figure 33–9

Changes in the (*a*) clitoris, labia, and (*b*) breasts during the female sexual response cycle. *(Redrawn from Masters, Johnson, and Kolodny,* Human Sexuality, *Little, Brown, 1985.)*

tion-like responses have been reported by some females.

Note that female orgasm is not essential for reproduction, since insemination of the female can occur in the absence of female sexual arousal.

Resolution Phase in the Male

Following ejaculation, most males enter a refractory period during which no further orgasm or ejaculation is possible. The length of the refractory period is highly variable (from a few minutes to many hours) and is prolonged by fatigue, time elapsed since the last orgasm, and age. During the refractory period, increased sympathetic nerve activity constricts the arteries in the penis, and blood leaves the penis through the veins, resulting in a loss of penile erection (see Fig. 33–7). Decreased blood flow to the testes also results in a decrease in testicular size. As muscles relax, the testes are lowered away from the body (see

Fig. 33–7). Heart rate, blood pressure, respiration, and muscle tension all return to normal during the resolution phase. When intense sexual arousal is not followed by orgasm, the length of the resolution phase is prolonged and can be accompanied by aching sensations in the testes due to continued vasoconstriction.

Resolution Phase in the Female

Many females, unlike most males, are **multiorgasmic** (capable of experiencing more than one orgasm before entering the resolution phase) if intense sexual arousal is maintained. During the resolution phase, muscular relaxation results in a lowering of the uterus and cervix toward the vagina, which increases the probability of cervical contact with any ejaculatory fluid present in the vagina (see Fig. 33–8). Blood flow away from the breasts and genitals results in a reduction in the size of the breasts, clitoris, labia, and vagina. As in the male, the female's heart rate, blood pressure, respiration, muscle tension, and skin flushing all return to normal during the resolution phase. Breast and genital tissues are extremely sensitive to touch following orgasm in both sexes, and continued physical stimulation of these tissues during the resolution phase can be irritating or painful.

Sexual Response During Infancy

Penile erection, in response to physical stimulation, occurs even before birth. Penile erections have been detected by ultrasound in the male fetus during the last few months of gestation. At birth, many male infants exhibit penile erections, while female infants manifest vaginal lubrication and clitoral enlargement. Following birth, certain forms of physical contact, including breastfeeding, can result in penile erection in male infants and vaginal lubrication and enlargement of the clitoris in female infants. Although both males and females are capable of achieving orgasm from birth, the ability of the male to ejaculate does not occur until the internal and external genitalia have matured at puberty.

Changes Occur in the Male and Female Sexual Response During Aging

Although physiological changes in the sexual response occur in both males and females during aging, the ability to become sexually aroused does not normally disappear.

Aging and the Female Sexual Response

In the female, the breasts, clitoris, and vagina remain sensitive to stimulation in old age. However, after menopause, the ovaries stop secreting estrogen, which leads to changes in the vagina. A low level of circulating estrogen results in diminished blood flow to the vagina during sexual arousal, which in turn leads to decreased vaginal lubrication. Use of an artificial lubricant can circumvent the pain that may result from intercourse in the absence of sufficient vagina lubrication, a condition termed **dyspareunia.** Some evidence suggests that regular sexual activity following menopause slows the decline in vaginal lubrication that occurs during aging. A decrease in the elasticity of the vaginal walls in aging women leads to reduced vaginal expansion during sexual arousal. In addition, enlargement of the breast, development of the sexual flush, and increased muscle tension during arousal are significantly reduced in older women. Changes in the female response during aging are summarized in Table 33–4.

Table 33–4
Changes in the Sexual Response During Aging

Female

1. A decrease in circulating estrogen following menopause results in decreased blood flow to the vagina, which in turn results in diminished production of vaginal lubrication. Decreased vaginal lubrication can result in painful intercourse, or dyspareunia.
2. The vaginal walls lose elasticity, and sexual arousal produces less vaginal expansion.
3. Many responses to sexual arousal—including enlargement of the breast, reddening of skin color (sexual flush), and muscle tension—are significantly reduced.

Male

1. Penile erection occurs more slowly and may require more prolonged stimulation.
2. The penis is less firm when fully erect.
3. Time spent in the refractory period following orgasm, during which no erection or ejaculation is possible, is prolonged.
4. The volume of ejaculatory fluid expelled at orgasm is decreased.
5. Testicular size and elevation during sexual arousal are reduced.
6. Many responses to sexual arousal—including reddening of skin color (sexual flush) and muscle tension—are reduced.

Aging and the Male Sexual Response

Although circulating testosterone levels begin to decline gradually after age 55 to 60 (see Chapter 30), there is no dramatic drop in testosterone levels during aging in the male that parallels the rapid decline in circulating estrogen levels in the female following menopause. Healthy males can maintain the capacity for penile erection throughout their entire life spans, and sperm production continues till age 80 or 90, although the production rate slows down after age 40.

However, physiological changes in the male sexual response do occur with aging. The sensitivity of penile sensory nerves to tactile stimulation decreases with age, and hence more prolonged and direct penile stimulation may be needed to induce erection. Blood flow to the penis during sexual arousal may also be reduced with age, and so the penis may be less firm when fully erect. Testicular elevation during sexual arousal is slower and less complete in the older male. The amount of fluid ejaculated at orgasm is often reduced, and the refractory period following orgasm is prolonged. As in the female, muscle tension during sexual arousal in the older male is also reduced. A summary of the changes in the male sexual response during aging is contained in Table 33–4.

Sexual Dysfunction

Sexual Dysfunction Refers to Any Condition in which the Normal Physical Responses to Sexual Stimulation Are Impaired, and Can Result from Hormonal or Neural Disorders, Certain Diseases, or Drugs

The term *sexual dysfunction* refers to any condition in which the normal physical responses to sexual stimulation are impaired. Sexual dysfunction can be a source of anxiety, frustration, and distress disrupting personal relationships and family life.

Estrogen Deficiencies in the Female

Estrogen plays an important role in maintaining vaginal lubrication. Estrogen stimulates proliferation of the vascular bed beneath the vaginal epithelium and increases blood flow to the vagina, which results in secretion of moisture from the vaginal walls. Inadequate estrogen stimulation can result in **atrophic vaginitis,** which is characterized by dryness and thinning of the vaginal mucosal membrane and decreased elasticity and muscle tone of the vagina. Following menopause, when circulating estrogen levels are decreased, vaginal elasticity and vascularization are reduced, and blood flow to the vagina can be decreased by as much as 80%. Estrogen replacement therapy following menopause can increase blood flow to the vagina by fourfold.

Premenopausal women with very low levels of circulating estrogen (for instance, following ovarian disease or surgical removal of the ovaries) can also experience atrophic vaginitis with accompanying loss of vaginal lubrication and elasticity. Inadequate vaginal lubrication can produce discomfort or pain during sexual intercourse (dyspareunia). Women with very low levels of circulating estrogen can also experience symptoms of peripheral neural dysfunctions, which can include slowed response to tactile stimulation or loss of clitoral sensation. Estrogen replacement therapy can reduce the severity of these symptoms.

Testosterone Deficiencies in the Male

Some evidence suggests that testosterone is necessary for normal sexual activity in the male. Although males with low circulating levels of testosterone can attain penile erection, the capacity to ejaculate during orgasm may be delayed, diminished, or absent. In addition, low levels of testosterone are often associated with reduced sexual desire or sexual libido. Androgen replacement therapy often increases both **ejaculatory capacity** and **sexual desire.** The exact mechanism by which testosterone facilitates ejaculation or sexual desire is not known. It has been proposed that androgens are involved in sensory perception and the capacity for development of muscle tension within the pelvis.

Neural Disorders and Sexual Dysfunction

Afferent sensory impulses from the internal and external genitalia enter the spinal cord and travel to the brain. Efferent nerve impulses from the brain leave the spinal cord and activate muscles and blood vessels of the internal and external genitalia. The excitement phase of sexual arousal in both males and females involves increased afferent neural input from sensory nerves in the genitals and efferent neural induction of vascular changes in the internal and external genitalia. Orgasm involves increased efferent motor activity, which induces muscular contractions in the internal and external genitalia. Therefore, both mechanical and disease-related injuries to the spinal cord or brain can result in impairment of sexual arousal and orgasm in both males and females. Interruption of afferent sensory pathways can interfere with sexual arousal. Interruption of efferent pathways may disrupt increased blood flow to the genitals during sexual arousal and hence may reduce penile erection in the male and vaginal swelling and lubrication in the female, as well as muscle contractions during orgasm in both sexes.

Multiple sclerosis, a disease characterized by patches of demyelinization in the spinal cord and brain, can be accompanied by a reduction or elimination of sexual arousal and orgasmic capacity in both males and females because of lesions in the spinal cord that disrupt afferent and efferent nerve transmission.

Severe **diabetes mellitus** can be accompanied by dysfunctions of the sensory fibers that innervate the penis and clitoris, resulting in a reduction or absence of the genital sensations necessary for sexual arousal and orgasm. Diabetic neural dysfunction can also result in decreased blood flow to the genitals, which can result in insufficient penile rigidity to achieve vaginal penetration.

CLINICAL FOCUS

Amenorrhea

Amenorrhea, or the absence of menstruation, is classified as either primary or secondary. Primary amenorrhea is the absence of menstruation in females who have never menstruated. It is rare and usually results from developmental abnormalities of the ovaries or the reproductive tract, or from the formation of scar tissue in these structures in response to physical injury or infections occurring before the first menstrual cycle would normally begin.

Secondary amenorrhea is more common and refers to the cessation of menstruation in females who have previously menstruated. Diagnosis can be problematic, because the length of the menstrual cycle varies greatly among women. A general rule used to diagnose this disorder is the absence of menstruation for an interval greater than three times the individual's normal cycle length.

Secondary amenorrhea can result from malfunctions at the level of the ovary, uterus, pituitary, or hypothalamus. During adulthood, autoimmune diseases, radiotherapy, chemotherapy, and surgery or infections that cause scar tissue can result in ovarian failure and a reduction in the number of viable follicles. At the level of the pituitary, the most common cause of secondary amenorrhea is the presence of a prolactin-secreting tumor, which results in hyperprolactinemia and consequent suppression of GnRH release (see text). At the level of the hypothalamus, many factors can contribute to a suppression of GnRH release and the onset of secondary amenorrhea. Stress, which often accompanies intense programs of physical exercise, can suppress GnRH

release and result in secondary amenorrhea. Extreme weight loss, which can occur during dieting; a response to drugs such as alcohol, narcotics, or amphetamines; and the condition termed "anorexial nervosa" can also suppress GnRH release and result in secondary amenorrhea.

Extreme weight gain can result in secondary amenorrhea through a different mechanism. An increase in adipose tissue results in increased conversion of adrenal androgens to estrogen, and this increase in circulating estrogen disrupts the normal feedback signals to the hypothalamus (see text). Psychological stresses of many types also result in amenorrhea.

Secondary amenorrhea can be seen following discontinuation of birth control pills in approximately 2% of the women who have taken the pills for two or more months, a condition termed "post-pill amenorrhea." It is not clear whether this type of amenorrhea is caused by a hypothalamic disturbance in response to the steroid content of the pills, or whether the women displaying post-pill amenorrhea have a prior history of menstrual irregularities.

Secondary amenorrhea is often reversible and tends to disappear with prolonged discontinuation of oral contraceptives, when normal dietary patterns are resumed, or when the source of stress is removed. It should be noted that the most common nonpathological condition that results in secondary amenorrhea is pregnancy.

■ **Is amenorrhea seen in postmenopausal women?**

Drugs and Sexual Dysfunction

Although small amounts of **alcohol** are often used to reduce sexual inhibitions, large amounts have a depressant effect on the central nervous system, which can interfere with sexual arousal. In addition, chronic and excessive consumption of alcohol, as seen in alcoholics, can lead to damage of the CNS, which results in alcoholic neuropathy in males and females. Alcohol-induced injury to the somatic and autonomic nerves in the periphery, spinal cord, or brain can result in a reduction or absence of sexual arousal and orgasmic capacity.

Sedatives (e.g., barbiturates, chloral hydrate), **narcotics** (e.g., codeine, morphine), and **tranquilizers** (e.g., diazepam, meprobamate), when taken in high doses, can retard or inhibit sexual arousal and orgasm in both males and females because of a de-

pression of nerve conductance in the central nervous system.

Some drugs used to treat hypertension can alter neural control of blood flow to the internal and external genitalia in both males and females, resulting in decreased vaginal swelling and lubrication in the female and difficulty in achieving erection in the male.

Some antihistamines that contain ephedrine decrease blood flow to the vagina and vaginal lubrication, resulting in dyspareunia.

Anorgasmia, Infection, and Vaginismus

Difficulty in reaching orgasm, which is referred to as **anorgasmia,** represents the largest category of sexual dysfunction in the female. However, organic disorders that can interfere with orgasm, such as hormone

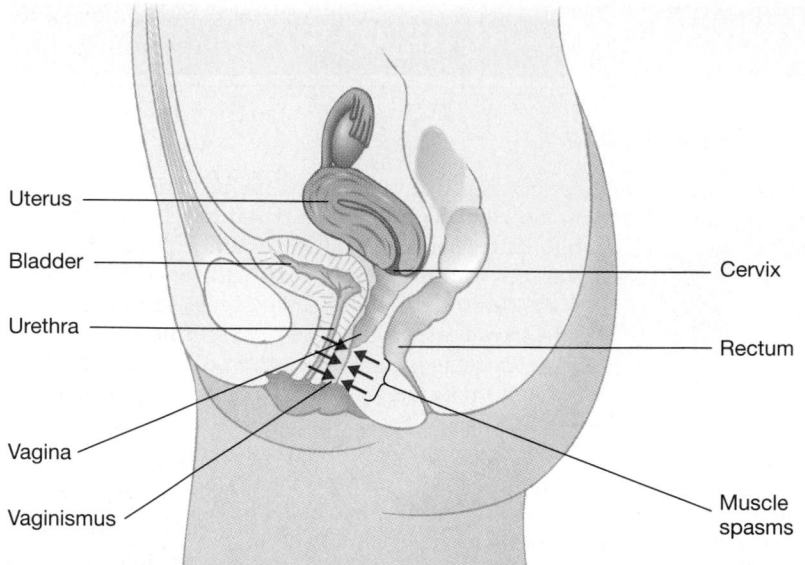

Figure 33–10

The involuntary muscle spasms of vaginismus. *(Redrawn from Masters, Johnson, and Kolodny, Human Sexuality, Little, Brown, 1985.)*

Uterus

Bladder

Urethra

Cervix

Rectum

Vagina

Vaginismus

Muscle spasms

deficiencies, alcoholism, diabetes, or neurological disorders, account for fewer than 5% of the cases of anorgasmia. Anorgasmia appears to be largely due to psychological factors, which may include sexual inhibitions acquired during early training, unrealistic performance expectations for oneself or one's partner, difficulty with sexual communication, and negative self-image. For further information on these issues, which are beyond the scope of this chapter, consult the volumes on human sexuality listed at the end of this chapter.

Vaginal infections, allergic reactions, or chemical irritants can produce a condition termed **nonatrophic vaginitis.** This condition, which is quite common and usually of short duration, can result in vaginal dryness or tenderness.

Endometriosis (see "Infertility") and **pelvic inflammatory disease** can both cause pain during sexual arousal and intercourse. This is due to the fact that vasocongestion and pelvic muscle contractions during sexual arousal place increased pressure on diseased tissues, in the case of endometriosis, or on chronically inflamed tissues, in the case of pelvic inflammatory disease. Further pressure due to intercourse can result in deep pelvic pain. Pain associated with sexual arousal and intercourse can result in decreased vaginal lubrication, loss of orgasmic capacity, and a loss of sexual desire—or sexual aversion.

Vaginismus is defined as involuntary contractions and spasms of the vaginal (or pubococcygeal) muscles, which make vaginal penetration very difficult or impossible (Fig. 33–10). Vaginismus may stem from emotional problems or traumatic or painful sexual experiences; or it may arise as a protective reflex to avoid pain in the case of pelvic disease. Prolonged dyspareunia, or painful intercourse, can sometimes also lead to vaginismus. Note that dyspareunia and/or vaginismus in the female can cause sexual problems in male partners when they feel responsible for unsuccessful attempts at intercourse. Feelings of in-

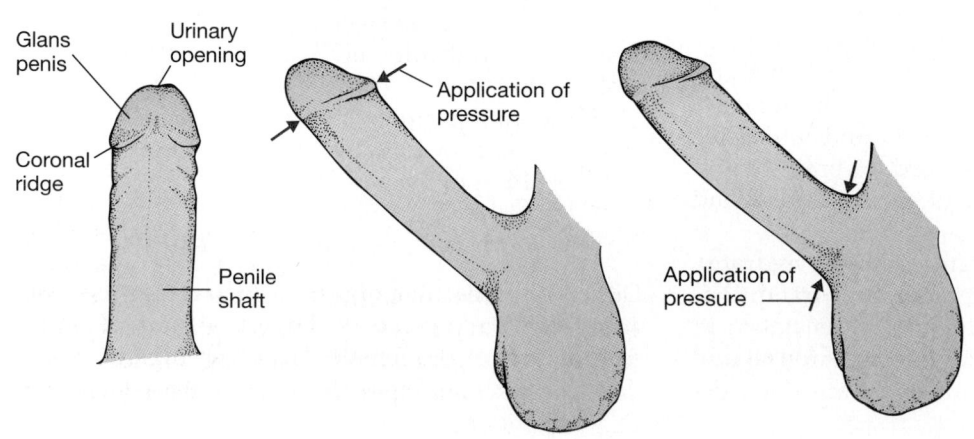

Glans penis

Urinary opening

Application of pressure

Coronal ridge

Penile shaft

Application of pressure

Figure 33–11

Application of manual pressure to the penis, used to prevent premature ejaculation. *(Redrawn from Masters, Johnson, and Kolodny, Human Sexuality, Little, Brown, 1985.)*

adequacy and guilt can result in erectile difficulty and loss of sexual desire.

Erectile Dysfunction, Premature Ejaculation, and Dyspareunia

Erectile dysfunction, or **impotence,** is defined as the inability to achieve penile erection sufficient for vaginal penetration. Erectile dysfunction is classified as either primary or secondary. **Primary erectile dys-**

Table 33–5
Origins of Sexual Dysfunction

Male

1. Neurological disorders resulting from physical or disease-related injury (multiple sclerosis, diabetes mellitus) to the brain, spinal cord, or peripheral nerves.
2. Testosterone deficiency characterized by decreased penile erection, capacity to ejaculate, or sexual desire.
3. Excessive use of drugs including alcohol, sedatives, narcotics, tranquilizers, antihistamines, and antihypertensive agents.
4. Primary or secondary erectile dysfunction resulting from physiological or psychological causes.
5. Premature ejaculation.
6. Dyspareunia due to physical, chemical, or infectious damage to the penis or urethra.

Female

1. Neurological disorders resulting from physical or disease-related injury (multiple sclerosis, diabetes mellitus) to the brain, spinal cord, or peripheral nerves.
2. Chronic estrogen deficiency that results in atrophic vaginitis characterized by vaginal dryness, thinning, and decreased elasticity.
3. Acute estrogen deficiency that results in nonatrophic vaginitis characterized by vaginal dryness.
4. Excessive use of drugs including alcohol, sedatives, narcotics, tranquilizers, antihistamines, and antihypertensive agents.
5. Dyspareunia due to physical, chemical, or infectious damage to the vagina, clitoris, or labia.
6. Endometriosis and pelvic inflammatory disease.
7. Involuntary contractions of the vaginal muscles (vaginismus).

---

function, which is very rare, defines the male who has never been able to achieve erection sufficient for vaginal penetration. **Secondary erectile dysfunction** is defined as a period of inability to achieve or maintain erection. Secondary erectile dysfunction can occur in response to anxiety, stress, illness, fatigue, lack of privacy, alcohol consumption, or change of sexual partner. Episodes of secondary erectile dysfunction are usually short-lived, and isolated episodes occur in most men.

Premature ejaculation can be defined as consistent, unintentional ejaculation prior to vaginal penetration. When defined in this way, very few men experience premature ejaculation. More typically, loss of ejaculatory control results in ejaculation soon after vaginal penetration. The probability of premature ejaculation is higher in young men than in older men and increases with time elapsed since last orgasm, novelty of the sexual partner, and prolonged stimulation prior to vaginal intromission. Manual application of pressure to the penis for a few seconds, from front to back just below the coronal ridge or at the base of the penis, can reduce the urge to ejaculate and help to control premature ejaculation (Fig. 33–11). Application of pressure to the penis may also cause a partial loss of erection that is only temporary.

Dyspareunia in the male is classified as painful erection, intromission, or ejaculation. Painful erection and/or pain on intromission can result from many factors, including inflammation of the foreskin of the penis (termed **balanitis**), reduced elasticity of the skin covering the penis, and penile scarring. These conditions can result from physical trauma to, or infection of, penile tissue. Painful ejaculation can result from physical, infectious, or chemical irritation of the urethra, termed **urethritis.**

Sources of sexual dysfunction in the male and female are summarized in Table 33–5.

Infertility, or the Inability to Reproduce, Can Be Due to Anatomical, Hormonal, Chromosomal, Immunological or Psychological Factors

Fertility in the male depends on several factors: adequate secretion of the gonadotropins that act to stimulate spermatogenesis; testes capable of responding to gonadotropins by producing sperm; glandular production of seminal fluid; an intact ductal system for sperm delivery; and an intact nervous system for the control of penile erection and ejaculation. A disruption of normal function at any of these levels can result in infertility.

CLINICAL FOCUS

Sexually Transmitted Diseases (STDs)

Sexually transmitted diseases (STDs) include a variety of diseases that range from mild illnesses that are easily treated to human immunodeficiency virus (HIV) which is incurable and ultimately fatal (see Chapter 28). STDs are so defined because they share a common mode of transmission, namely sexual activity. STDs include gonorrhea, chlamydia, syphilis, genital herpes, human papillomaviruses, and chancroid. Coinfection, or infection with two or more STDs, is not uncommon. For instance, 15% to 20% of men and 35% to 40% of women with gonorrhea are also infected with chlamydia. Three of the most common STDs are described below:

Chlamydia: Chlamydia is caused by the chlamydia trachomatis bacteria. Initial infection is often asymptomatic (without symptoms). When symptoms do occur, they most often include vaginal discharge and dysuria (painful urination). In women, both chlamydia and gonorrhea can progress asymptomatically from a lower genital tract infection to an ascending tubal infection. Chlamydial infections are a major cause of pelvic inflammatory disease (PID), which is characterized by inflammation of the uterine tubes (termed salpingitis), and is associated with an increased risk of both ectopic pregnancy and infertility. The risk of infertility increases with each episode of tubal inflammation. The cervix is the most common female organ infected by chlamydia trachomatis although the urethra and rectum may also be infected. Use of a barrier contraceptive is effective in preventing cervical infection. Secondary prevention in-

volves screening for, and treating, the cervical infection before it ascends to the uterine tubes. Tertiary prevention consists of treating the inflammation of the uterine tubes. Asymptomatic chlamydial infections are detectable only by screening pelvic examinations. Chlamydial infections in men can be asymptomatic as well, while producing inflammation of the urethra (termed urethritis) and increased risk of pain and swelling of the epididymis (termed epididymitis). Because chlamydial infections are accompanied by few, if any, symptoms, rate of sexual transmission is high. Antibiotics are effective in treating chlamydial infections.

Genital Herpes: Genital herpes, caused by herpes simplex virus type 2, is the most common cause of genital ulcers in the United States. Lesions can occur on the skin or epithelium of the vulva, vagina cervix, urethra or rectum and can be spread, by touch, to other skin areas or to the eyes. Initial infection is characterized by clusters of vesicles that rupture to form painful shallow ulcers. These ulcers dry up, form a crust, and disappear in about 3 weeks. The virus then establishes a permanent latent infection in dorsal root ganglia, which results in recurrent episodes which are usually milder, and of shorter duration, than the first. Recurrent episodes are often preceded by pain, burning, itching or tingling of the skin in the area of subsequent outbreak. Risk of infection is highest when active lesions are present, but asymptomatic shedding of the virus occurs between episodes and serves as a frequent mode of transmission.

In addition, three aspects of the sperm themselves are important for sperm penetration and fertilization of the ovum: sperm number, sperm morphology (shape and structure), and sperm motility. Although it is always difficult to establish "normal" ranges for physiological processes, it is generally accepted that 15 to 250 million sperm per milliliter of semen represents a normal sperm count. A further definition of sperm normality requires that at least 60% of the sperm have a normal appearance and that at least 60% are active or motile.

A low sperm count and/or decreased sperm motility can result from gonadotropin deficiencies or from damage to the testes. Testicular damage can result from infections, inflammation, impaired circulation, and certain diseases such as mumps. In addition, a variety of environmental factors such as fever, prolonged heat exposure, chemical agents, and irra-

diation can result in testicular damage and impaired spermatogenesis. The extent of testicular damage produced by these agents is determined by the intensity and duration of testicular exposure. Abnormal sperm shape or structure, which includes a non-oval head section, a broken midpiece, or a bent or coiled tail, can result in infertility by preventing sperm migration through the cervical mucus or penetration of the ovum. Infection and inflammation of the epididymis, vas deferens, or ejaculatory ducts can reduce or prevent sperm transport, which can also result in infertility.

As in the male, infertility in the female can be due to a number of factors. One of the most common causes is failure to ovulate. Ovulatory failure can result from inadequate secretion of gonadotropins, failure of the ovary to respond to gonadotropins, variations in the anatomical structure of the ovaries, and

Many adult men and women have serologic evidence of past infection, are capable of periodically shedding the virus and thereby transmitting genital herpes to a sex partner, and yet have no clinical history of outbreaks. While there is no known cure for genital herpes, drugs are often effective in suppressing outbreaks, and some patients inexplicably will cease having recurrences.

Human Papillomaviruses: There are approximately 70 types of papillomaviruses (HPVs). The majority of HPVs cause genital warts that are usually found on the skin in the genital and rectal areas or on the mucosa of the urethra, vagina, or rectum and occasionally in the oropharyngeal region. These warts are self-limiting, hyperproliferative lesions. Because there is no effective antiviral treatment for HPV disease, warts are usually removed by cryotherapy using liquid nitrogen or a cryoprobe. Intravaginal or perianal warts are surgically removed. Some HPVs are associated with malignant disease. For instance, certain strains of HPVs are a major cause of cervical cancer in women although not all women infected with an oncogenic HPV develop malignant cervical cancer. In men, some strains of HPVs cause malignant disease of the penis, or penile cancer, which is rare in the United States but is much more common in Central and South America and parts of Africa. Some strains of HPVs are associated with anal cancer in homosexual men.

Many STDs are accompanied by symptoms that alert the patient to the presence of the disease, which allows for medical diagnosis and treatment. The presence of symptoms also provides the patient with the opportunity to identify potentially infected sexual partners. All states require reporting of HIV, syphilis, and gonorrhea, and many states require reporting of chlamydia. Public health authorities can usually assist in notification, diagnosis, and treatment of infected individuals, which serves to limit further disease transmission. However, some STDs can be initially asymptomatic (see chlamydia above) and hence may go undetected. Therefore, screening for asymptomatic infections is suggested for all sexually active populations and is particularly important for those individuals engaging in high-risk behaviors.

Given that STDs are transmitted via sexual activity, one of the most effective approaches for the prevention and control of STDs is to change behaviors that increase the risk of STD transmission. Factors that influence the risk of acquiring STDs include number of sexual partners, rate of acquiring new partners, incidence and diversity of sexual activity of the partners, sexual preference, and type of sexual practice. Behaviors that can contribute to increased transmissions of STDs include failure to use condoms, delay in seeking medical care following onset of infection, noncompliance with suggested medical treatment, and failure to refer infected partners.

■ **How might coinfection increase the rate of HIV transmission?**

ovarian damage resulting from infection or inflammation. Other factors that can cause female infertility include overproduction of thick cervical mucus, which can block sperm entry into the uterus, and infection or disease of the uterus or uterine tubes, with accompanying scarring that can block ovum transport to, or implantation in, the uterus. For instance, **endometriosis** is a disease characterized by the presence of endometrial-like tissue in locations other than the uterus, such as the uterine tubes or ovaries. Tissue formation, which may appear as cysts or nodules, can reduce the diameter of the uterine tubes and block ovum transport to the uterus (Fig. 33–12).

Immunological factors may also contribute to infertility. Antisperm antibodies have been found in both males and females. Antibodies, produced by the male, coat the surface of the sperm and reduce its ability to penetrate the cervical mucus. Antibodies produced by the female can induce **agglutination,** or aggregation of sperm into clumps, which also reduces their ability to penetrate the cervical mucus. Whether immunological factors play an important role in infertility remains to be determined.

Contraceptive Methods

Being able to block conception ("contra-ception") has resulted in increased independence of human sexuality from reproduction. Various methods are used to interfere with conception, but all of them fall within four basic categories: **barrier methods, chemical methods, surgical methods,** and **abstinence.** In addition, all methods of contraception work by disrupting one of the three major steps in the reproduc-

(a)

(b)

Figure 33–12

Radiographic image of the uterus and (*a*) normal bilateral uterine tubes and (*b*) bilateral uterine tube blockage. (© *Hysterosalpingography: A Text and Atlas by Ott and Fayez. Published by Urban & Schwarzenberg, 1991.*)

tive process: ovulation, sperm and ovum transport, or implantation of the fertilized ovum in the uterus. The sites of actions of various methods of contraception are illustrated in Figure 33–13.

Barrier Methods of Contraception Prevent Sperm Transport

All barrier methods of contraception prevent sperm transport. Barrier methods include condoms, diaphragms, cervical caps, and contraceptive sponges. Condoms are thin sheaths of rubber that fit over the penis and prevent sperm from entering the vagina. Use of condoms has increased in recent years because they not only prevent sperm transport, but also provide some degree of protection against sexually transmitted diseases. Diaphragms, cervical caps, and sponges are rubber or polyurethane devices that cover the entrance to the cervix and prevent sperm from entering the uterus. Condoms, diaphragms, and sponges are used once and then must be replaced (or, in the case of the diaphragm, re-inserted) prior to each sexual encounter. Cervical caps, which work similarly to the diaphragm, can be worn for longer periods prior to removal; they are held in place by suction.

Barrier methods of contraception vary in effectiveness and are often combined with chemical methods to increase their effectiveness. For instance, the

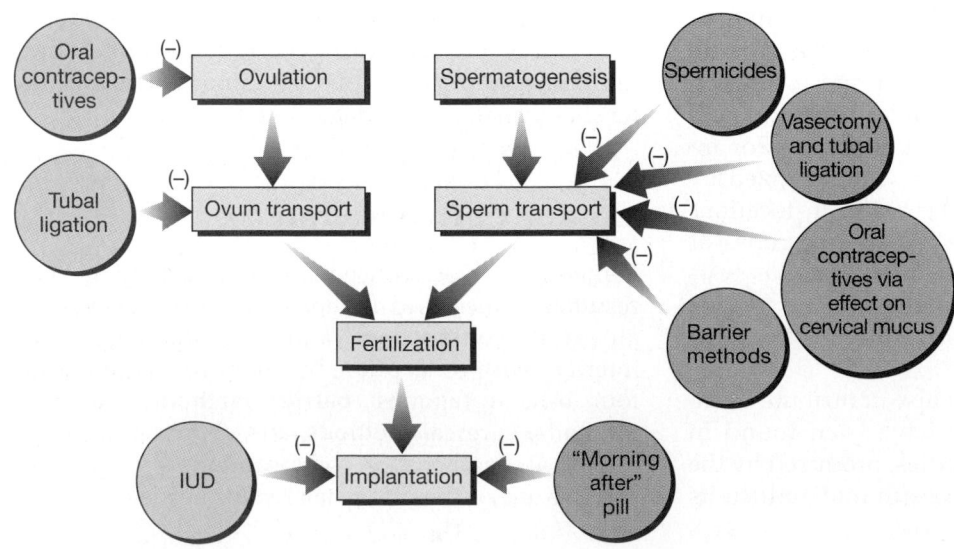

Figure 33–13

Sites of action of various methods of contraception. Minus sign indicates inhibition.

R E S E A R C H   F O C U S

Infertility

A variety of methods are currently available for the diagnosis and treatment of infertility. When male infertility is suspected, a microscopic examination of a sperm sample can be used to determine if infertility is due to a low sperm count, abnormal sperm morphology, or decreased sperm motility. When sperm counts are low, sperm can be collected and introduced directly into the uterus or uterine tubes via a catheter to avoid the sperm loss that normally occurs during passage of the sperm through the cervix and uterus. This process, termed artificial insemination, can also be used to circumvent immunological infertility, which can occur when cervical mucus induces an allergic reaction in sperm resulting in sperm death.

Many cases of female infertility are due to damage or obstruction of the uterine tubes resulting from previous infections. Reconstructive surgery has been successful in repairing damage to the uterine tubes and restoring the capacity to conceive in approximately a third of these cases. Dilation of the cervix, a relatively simple operation, can also be used to treat infertility resulting from cervical constriction. Another cause of female infertility is inadequate secretion of gonadotropins from the pituitary gland, which results in a failure to ovulate. In such cases, oral or intramuscular administration of drugs can be used to induce gonadotropin secretion by the pituitary, which in turn stimulates estrogen secretion from the ovary and facilitates follicular maturation and ovulation. Hormonal induction of ovulation is often so successful that multiple births result.

In vitro or "test-tube" fertilization is an approach used to treat both male and female infertility. This method involves removing one or more ova from a woman's ovary by means of aspiration through a needle inserted into the ovarian follicle. Collection of ova is carefully timed to coincide with ovulation. The conditions used for storage and treatment of ova and sperm are designed to approximate the conditions normally found in the woman's body following ovulation. For example, following collection, the ova are stored for a few hours in a nutrient solution at 37°C (body temperature) prior to the introduction of sperm. The ova and sperm are then incubated together at body temperature for two days prior to implantation of the fertilized ovum in the uterus. *In vitro* fertilization is useful in treating male infertility resulting from a low sperm count because it requires far fewer sperm than does fertilization resulting from sexual intercourse. *In vitro* fertilization can also be used to circumvent uterine tube obstruction in the female. Research efforts are currently being directed toward developing a method of injecting a single sperm directly into an ovum. This process, if feasible, would eliminate many cases of infertility that arise from low sperm number, poor sperm motility, uterine tube damage, or immunological blockade.

■ **Can a bacterial infection result in infertility?**

contraceptive sponge contains **spermicides,** chemicals that kill sperm. Spermicides, contained in cream or jelly, are also applied to the inside of the diaphragm to kill sperm that may pass the rim of the diaphragm.

Barrier methods of contraception have minimal side effects, which include the possibility of an allergic reaction to the spermicide or to the material used in construction, and a chance of introducing infection into the vagina.

Chemical Methods of Contraception Include Birth Control Pills, Which Contain Synthetic Variations of Gonadal Hormones

The most popular method of contraception, the **birth control pill,** is taken orally and acts to prevent preg-

nancy by blocking ovulation (see Fig. 33–13). There are many types of birth control pills, but they all contain synthetic variations of the gonadal hormones estrogen and/or progesterone, and they act to inhibit the release of follicle-stimulating hormone (FSH) and luteinizing hormone (LH) from the pituitary. As we learned in Chapter 30 (see Fig. 30–17), a rise in circulating levels of LH and FSH at the midpoint of the ovarian cycle stimulates maturation of the ovarian follicle, resulting in ovulation. When LH and FSH release are suppressed by oral contraceptives, as illustrated in Figure 33–14, follicular maturation does not occur, and ovulation is inhibited. Continuous use of oral contraceptives will inhibit follicular maturation, ovulation, and the onset of menstruation. Discontinuation of oral contraceptives for one week each month results in a drop in circulating levels of estrogen and progesterone, which stimulates the onset of menstruation. The contraceptive efficacy of birth con-

(a)

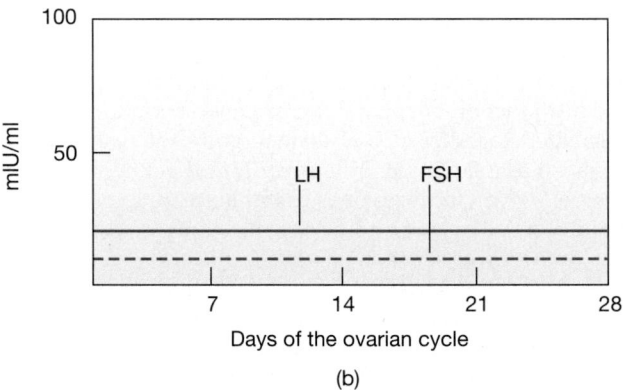

(b)

Figure 33–14

Plasma levels of gonadotropins during the ovarian cycle in women (*a*) not using oral contraceptives and (*b*) using oral contraceptives.

trol pills does not require monthly discontinuation of use.

Progesterone, contained in oral contraceptives, also interferes with conception in two additional ways. First, progesterone increases the viscosity of cervical mucus, which, as we have seen in Chapter 31 (see Fig. 31–3), blocks the entry of sperm into the uterus. Second, progesterone interferes with implantation of the ovum in the uterus by inhibiting estrogen-induced proliferation of the endometrial lining of the uterus, which is required for implantation.

Birth control pills can produce a variety of side effects, including nausea, constipation, elevated blood pressure, skin rashes, and salt retention with accompanying weight gain. However, the most serious side effect is increased risk of cardiovascular disease, which becomes significant in women over 35 who use both oral contraceptives and tobacco.

Another type of oral contraceptive, termed the **morning-after pill,** contains a very high level of **estrogen,** which alters the endometrial lining of the uterus and prevents implantation of the fertilized

ovum. However, the high levels of estrogen contained in the morning-after pill often induce severe nausea and vomiting, which have limited the use of this type of oral contraceptive.

Another type of contraceptive method is the **intrauterine device (IUD),** which can be worn for extended periods. The IUD is made of plastic or metal and is inserted by a physician into the uterus through the vagina and cervix. Although the mechanism of action of the IUD is not completely understood, it is thought to induce a local inflammatory reaction in the endometrium of the uterus, which prevents implantation of the ovum. Side effects of the IUD include uterine bleeding, spasms and cramping of uterine muscles, pelvic infection, and perforation of the uterine wall.

Surgical Methods of Contraception Include Vasectomy in the Male and Tubal Ligation in the Female

The most effective form of birth control, with the exception of abstinence, is surgical sterilization. Sterilization procedures, which include vasectomy in the male and tubal ligation in the female, have enjoyed a recent increase in popularity because they are safe, effective, and permanent.

Vasectomy

Vasectomy is a simple surgical procedure performed within 15 to 20 minutes under local anesthesia. As illustrated in Figure 33–15, vasectomy consists of a small surgical incision on each side of the scrotum to expose a small section of the vas deferens. The vas deferens are severed, and the ends are tied, clamped, or cauterized to prevent sperm transport.

Vasectomy does not interfere with sperm production. Following vasectomy, sperm accumulate in the epididymis of the testes, where they are engulfed and destroyed by phagocytosis. Vasectomy does not interfere with the synthesis or release of testosterone from the testes and does not affect erection, ejaculation, or orgasm in any way. Vasectomy is a very effective method of contraception, with failure rates below 0.15%. Postsurgical complications, which can include bleeding, infection, and swelling of scrotal tissue due to sperm leakage, are rare. Few long-range health risks appear to be associated with vasectomy.

Tubal Ligation

Tubal ligation consists of an abdominal incision to permit insertion of an instrument called a **laparoscope,** which is used to locate and sever the uterine

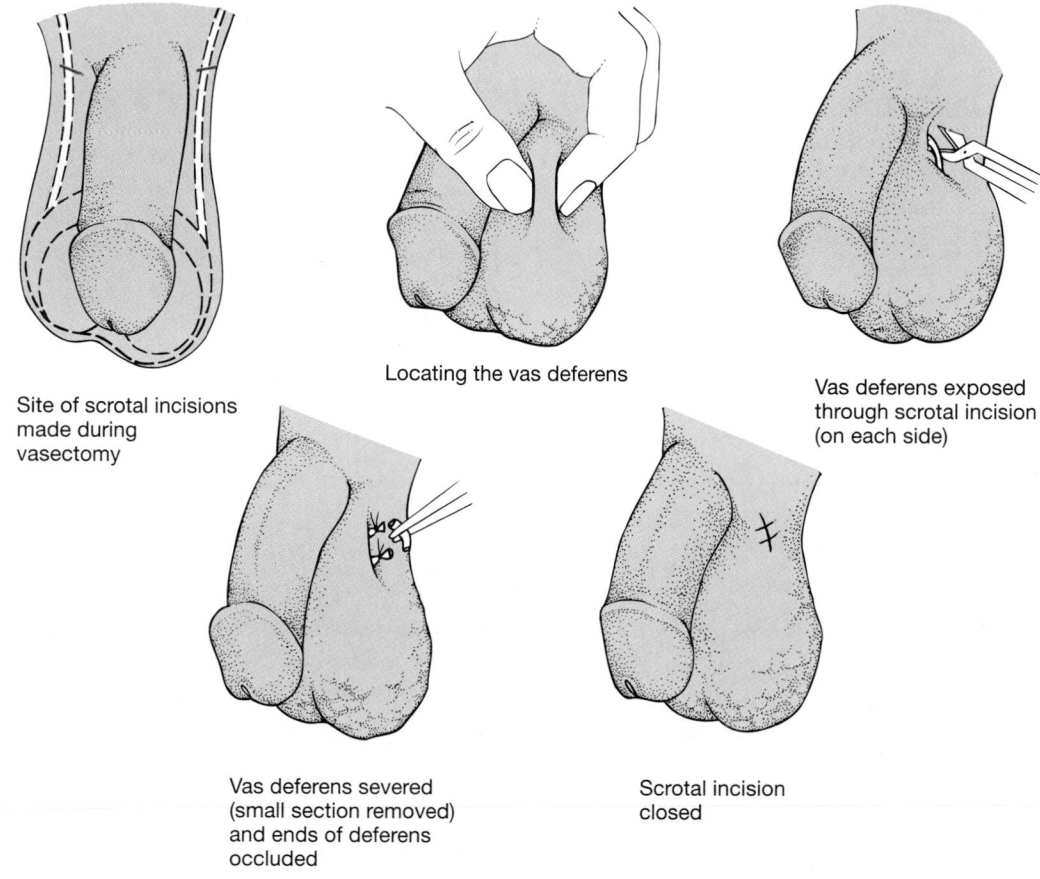

Site of scrotal incisions
made during
vasectomy

Locating the vas deferens

Vas deferens exposed
through scrotal incision
(on each side)

Vas deferens severed
(small section removed)
and ends of deferens
occluded

Scrotal incision
closed

Figure 33–15

Steps involved in a vasectomy.

tubes on each side of the body. As illustrated in Figure 33–16, the cut ends of the uterine tubes are occluded with rings, clips, or cautery, blocking both ovum transport to the uterus and sperm transport through the uterine tubes. Tubal ligation is a very effective method of preventing pregnancy, and postsurgical complications, which can include bleeding

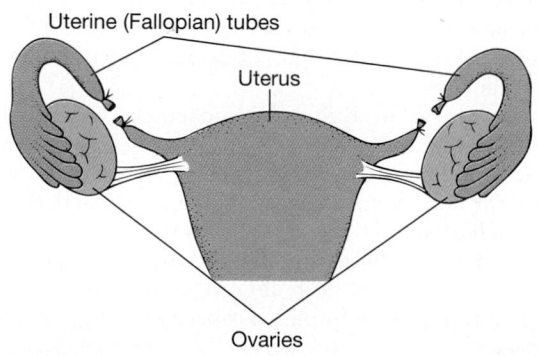

Uterine (Fallopian) tubes

Uterus

Ovaries

Figure 33–16

Illustration of a tubal ligation.

and infection, are rare. As with vasectomy, few long-range health risks appear to be associated with tubal ligation.

Abstinence Is the Only Completely Effective Form of Birth Control

Complete abstinence, which is the only completely effective form of birth control, is unpopular for obvious reasons. However, many people practice **periodic abstinence** (termed the **rhythm method**) during the time of ovulation when fertility is the greatest. The efficacy of this method depends on the accuracy of predicting the time of ovulation.

Three methods are used to predict ovulation. The first involves monitoring the ovarian cycle and length of menstruation for at least 6 months and avoiding intercourse on the days preceding and following ovulation (at midcycle). This method has a high failure rate because many women exhibit a large variation in length of monthly cycles. The second method involves daily monitoring of basal body temperature and avoiding intercourse on the days preceding and following ovulation. As discussed in Chapter 31 (see Fig. 31–17), body temperature decreases prior to ovulation and rises again following ovulation. This method also has a high failure rate because the temperature shifts that occur during the ovarian cycle are small (usually less than 1 degree) and are difficult to identify accurately. The third method of identifying the time of ovulation requires monitoring the consistency of cervical mucus. As described in Chapter 31, cervical mucus becomes clear, thin, and stringy around the time of ovulation. Predicting ovulation on the basis of consistency of cervical mucus has a high failure rate because of the difficulty of accurately identifying these changes.

Several New Approaches to Contraception Are Currently Being Developed and Tested

Several new approaches to contraception are currently being developed and clinically tested. The important variables that should be identified for all new methods of contraception are efficacy, duration of action, reversibility, ease of use, and incidence of both short- and long-term side effects. In the female, research efforts are being directed toward blocking ovulation and implantation. In the male, the focus is on blocking production, maturation, motility, and/or transport of sperm.

Female Contraception

Long-acting forms of estrogen and progesterone are being administered by subcutaneous implantation and by injection, which can prolong the effectiveness of these substances for months. Other hormonal analogs that suppress LH and FSH release are also being investigated, as are substances that antagonize the actions of progesterone and human chorionic gonadotropin (hCG) on the endometrium and hence interfere with implantation of the ovum. Synthetic analogs of the prostaglandins, which induce uterine contractions, are being investigated for their efficacy in inducing the onset of menstruation, which would evacuate the contents of the uterus whether or not a fertilized ovum is present.

Male Contraception

Combinations of estrogen, progesterone, and testosterone are currently being examined for their ability to transiently block sperm production. Other naturally occurring substances, such as an extract from cottonseed oil (termed *gossypol*), which are capable of immobilizing or killing sperm, are also being investigated. Efforts are focused as well on identifying and testing substances, such as inhibin, that inhibit FSH release from the pituitary and hence remove the hormonal signal for sperm production in the testes.

CHAPTER REVIEW

Summary

- The presence or absence of a Y chromosome determines the sex of the zygote. In the absence of a Y chromosome (XX), the gonads develop into ovaries.
- During the first six weeks of embryonic development, the gonads in all embryos, whether genetically male or female, are anatomically and morphologically identical.
- Hormones secreted by the developing gonads direct gonadal differentiation and the development of the internal genitalia. In the male embryo, the developing testes secrete Anti-Mullerian Hormone (AMH), which induces regression of the female internal genitalia. Testosterone, secreted by the developing testes, stimulates development of the male internal genitalia (the Wolffian duct system).
- The external genitalia in the male and female differentiate in response to the presence and absence of testosterone, respectively.
- Abnormal patterns of sexual differentiation can occur at

four levels: chromosomal sex (genotype), gonadal sex, internal genital sex, and external genital sex.
- The ability to become sexually aroused is present before birth and remains throughout the life span of both males and females.
- Sexual arousal in adult males and females is divided into four phases: excitement, plateau, orgasm, and resolution.
- Endocrine deficiencies can result in reduced sexual arousal and decreased capacity for orgasm in both males and females.
- Mechanical or disease-related damage to the peripheral nerves, spinal cord, or brain can result in impaired sexual arousal and/or impaired capacity for orgasm in both sexes.
- Drugs such as alcohol, antihistamines, sedatives, narcotics, and tranquilizers can retard or inhibit sexual arousal.
- Infertility can result from hormonal deficiencies, de-

struction of gonadal tissue resulting from inflammation or disease, and the production of antisperm antibodies.
- Barrier methods of contraception, which include diaphragms, cervical caps, contraceptive sponges, and condoms, interfere with sperm transport.
- Birth control pills contain synthetic hormones that inhibit the secretion of pituitary hormones that control ovulation.
- Surgical methods of contraception include vasectomy in the male, which interferes with sperm transport, and

tubal ligation in the female, which interferes with both sperm and ovum transport.
- The rhythm method of contraception refers to periodic abstinence from intercourse at midcycle, around the time of ovulation.
- Future trends in contraception include the development of long-acting natural and synthetic chemicals that inhibit ovulation, spermatogenesis, sperm transport, ovum transport, and/or implantation of the ovum in the uterus.

Review Questions

Identify Each with a Term

1. The release of moisture from the lining of the vagina during sexual arousal.
2. The elevation of the uterus away from the vagina during sexual arousal.
3. Any condition in which the normal physical responses to sexual stimulation are impaired.
4. The protein thought to be responsible for inducing testicular differentiation in the primitive gonad.

Define Each Term

5. vaginismus
6. impotence
7. Anti-Mullerian Hormone (AMH)
8. myotonia
9. Turner's syndrome
10. amenorrhea

Choose the Correct Answer

11. Which of the following does *not* result from testosterone stimulation?
    a. masculinization of the female embryo's external genitalia
    b. stimulation of Mullerian duct regression
    c. masculinization of the male embryo's external genitalia
    d. stimulation of Wolffian duct development
12. The chromosomal make-up of an individual with Klinefelter's syndrome is:
    a. XO
    b. XYY
    c. XXY
    d. XX

13. Which of the following can help control premature ejaculation?
    a. testosterone therapy
    b. increasing the time since last orgasm
    c. vasectomy
    d. application of pressure at the base of the penis
14. Which of the following is *not* a barrier method of contraception?
    a. birth control pill
    b. condom
    c. diaphragm
    d. contraceptive sponge

Answer Each Question in One or Two Sentences

15. What is the importance of the sex chromosomes?
16. How can intense programs of physical exercise result in amenorrhea?

Problem Solving in Physiology

17. Do males have a male phenotype because they have a Y chromosome or because they have only one X chromosome? Provide evidence to support your answer based on the phenotypes that accompany abnormal sex chromosome constitutions.
18. Explain why a female who is color-blind must have a color-blind father while a color-blind male would not. Compare and contrast the factors that can result in sexual dysfunction in males and females. Compare and contrast the physiological changes that occur in the male and female during the sexual response cycle.

Suggested Readings

Acosta, A.A., ed. *Human Spermatozoa in Assisted Reproduction.* Baltimore: Williams and Wilkins, 1990.

Centers for Disease Control and Prevention. "Sexually transmitted diseases treatment guidelines." MMWR 42 (Number RR-14):1–102, 1993.

Cohen, M.S., Hook, E.W., and Hitchcock, P.J., eds. "Sexually transmitted diseases in the AIDS era: Part I." In *Infectious Diseases Clinics of North America,* 7(4):739–873, 1993.

Cohen, M.S., Hook, E.W., and Hitchcock, P.J., eds. "Sexually transmitted diseases in the AIDS era: Part II." In *Infectious Diseases Clinics of North America,* 8(4):751–821, 1994.

Geer, J., Heiman, J., and Leitenberg, H., eds. *Human Sexuality.* Englewood Cliffs, N.J.: Prentice Hall, Inc., 1984.

Glover, T.D., ed. *Human Male Fertility and Semen Analysis.* Academic Press, 1990.

Mandell, G.L., Bennett, J.E., and Dolin, R., eds. *Principles and Practice of Infectious Diseases*, 4th ed. New York: Churchill Livingstone, 1995.

Martin, D.H., ed. "Sexually transmitted diseases." In *Medical Clinics of North America*, 74(6):1339–1697, 1990.

Masters, W.H., Johnson, V.E., and Kolodny, R.C., eds. *Human Sexuality*, 4E. New York: Harper Collins, 1991.

Nimmons, D. "Sex and the brain." *Discover*, March 1994.

Reinisch, J.M. *The Kinsey Institute New Report on Sex*. New York: St. Martin's Press, 1990.

Shearman, R.P., ed. *Clinical Reproductive Endocrinology*. New York: Churchill Livingstone, 1985.

Key Terms

amenorrhea (p. 967)
autosomes (p. 953)
chromosomal sex (p. 953)
clitoris (p. 963)
contraception (p. 971)
dyspareunia (p. 965)

ejaculation (p. 963)
endometriosis (p. 971)
gonadal sex (p. 953)
histocompatibility-Y antigen (H-Y antigen) (p. 955)

impotence (p. 969)
infertility (p. 969)
Mullerian duct system (p. 955)
orgasm (p. 961)
spermicides (p. 973)

transudation (p. 961)
vaginismus (p. 968)
vasectomy (p. 974)
Wolffian duct system (p. 955)

Glossary

A-band A transverse band of thick filaments of cardiac and skeletal muscle located in the center portion of the sarcomere. (Ch. 16)

absolute refractory period The period of time in which an axon is unable to generate an action potential regardless of the applied stimulus strength. (Ch. 7)

absorption (ab-**sorp**-shun) The movement of digestion products and other nutrients across the wall of the small intestine to the blood or lymph. (Ch. 22)

acclimatization The process of becoming accustomed to a new environment. (Ch. 29)

accumbens (a-**kum**-binz) **nucleus** A region of the basal ganglia that is considered as a reward center of the brain. (Ch. 11)

A-cell A glucagon-secreting cell of islets of Langerhans; also called alphacell. (Ch. 14)

acetone (**ass**-ih-tone) A volatile, sweet-smelling ketone body produced from acetyl CoA during prolonged fasting or untreated severe diabetes mellitus. (Ch. 25)

acetyl coenzyme A (*as*-**ee**-til koh-**ehn**-zym) An important intermediate in the Krebs cycle and the main precursor to lipids; it is formed in the mitochondrial matrix when all of the acetyl groups produced from food molecules are coupled to the terminal —SH group of coenzymes. (Ch. 6)

acetylcholine (*as*-ee-til-**koe**-leen) (Often abbreviated as Ach.) A low-molecular weight neurotransmitter; found in the central and peripheral nervous systems. (Ch. 10)

acetylcholinesterase (*as*-ee-til-**koe**-leen-s-tur-aze) Enzyme essential for breaking down Ach to an inactive form once it has engaged a receptor. (Ch. 16)

acid (**as**-id) Any substance whose dissociation in water releases hydrogen ions. (Ch. 2)

acid dissociation constant The ionization constant for an acid. (Ch. 25)

acidemia (*as*-i-**dehm**-ee-uh) An arterial blood pH less than 7.35. (Ch. 25)

acidosis (*as*-i-**doe**-sis) An abnormal process that tends to produce acidemia. (Ch. 25)

acinus (**as**-ih-nuhs) Saclike clusters of exocrine secretory cells in the pancreas that secrete digestive enzymes. (Ch. 22)

acromegaly (*ak*-roe-**meg**-ah-lee) A disfiguring condition resulting from the overproduction of GH in adults whose epiphyses have closed; stimulates excessive growth of soft tissues and transverse diameter of bones leading to bony thickening and deformity. (Ch. 13)

acrosome reaction (**ak**-roe-sowm) A controlled disruption of the head of a sperm, which allows it to penetrate the cumulus oophorus and the zona pellucida of the ovum. (Ch. 32)

ACTH Adrenocorticotropic hormone, secreted by the anterior lobe of the pituitary. Also called corticotropin, the hormone regulates the synthesis and secretion of cortisol by the adrenal cortex. (Ch. 14)

actin (**ak**-tin) A globular protein to which myosin cross bridges bind; located in muscle thin filaments and microfilaments of cytoskeleton. (Ch. 16)

actin-linked regulation Muscle regulation that is accomplished through some action on the thin filaments. (Ch. 16)

action potential The mechanism by which impulses are transmitted, thereby carrying information along the nerve fiber membranes. The action potential is a rapid change in the voltage of the membrane potential that occurs when adequate stimulus is applied. The all-or-none response causes the voltage to change from its negative resting state to a positive value. (Ch. 7)

activation energy (ac-ti-**va**-tion **en**-er-gy) The initial energy required to break the existing bonds in a molecule and form new ones. (Ch. 2)

activation state One of three states of the sodium channel; if the nerve cell is stimulated and the membrane potential is depolarized to the threshold value, the sodium channel switches from its resting state to its activating state; sodium ions will then flow until the inactivating gate closes the channel. (Ch. 7)

active immunity Long-lasting continuing protection against disease conferred through immunization with an antigen. (Ch. 28)

active site Areas on the surfaces of enzymes that permit reactant molecules to bind. (Ch. 2)

active tension The force (tension) in a muscle produced in response to stimulation. (Ch. 16)

active transport A situation in which a solute must move from a less concentrated region outside the cell to a more concentrated region inside, against the prevailing gradient and thus requiring energy. When the movement of solute is coupled directly to an energy-yielding reaction, the process is termed **primary active transport;** when the active transport of a solute is not coupled directly to the energy-yielding reaction, it is described as **secondary active transport.** (Ch. 4)

active zone Restricted areas of the presynaptic terminal. (Ch. 7)

acute mountain sickness An illness typically seen in high-altitude mountain climbers, in which too much carbon dioxide is blown off during hyperventilation. (Ch. 29)

adaptation (ad-ap-**tay**-shun) A genetic change that enhances the ability of an organism to adjust to the environment. (Ch. 29)

Addison's (**add**-i-sonz) **disease** A condition caused by an overall deficiency in steroid production by the adrenal cortex. It is a progressive disease with symptoms such as diarrhea, hypotension, increased skin pigmentation, fatigue, and digestive difficulties. (Ch. 14)

adenohypophysis (*ad*-e-no-**high-pof**-i-sis) The anterior lobe of the pituitary gland. (Ch. 13)

adenosine (a-**den**-oh-seen) A nucleoside that has been shown to promote vasodilation of arteriolar smooth muscle. (Ch. 19)

adenosine diphosphate (ADP) (a-**den**-oh-seen die-**fos**-fate) An important molecule that is formed from ATP when the terminal phosphate is removed. This reaction produces the energy that drives other energy-requiring reactions. (Ch. 6)

adenosine monophosphate (AMP) (a-**den**-oh-seen *mah*-noe-**fos**-fate) A molecule that is formed from ATP when pyrophosphate is removed (also known as adenosine phosphate. (Ch. 6)

adenosine triphosphate (ATP) (a-**den**-oh-seen try-**fos**-fate) The universal energy compound that is produced in all cells and is the source of energy in most of the energy-requiring reactions of metabolism. ATP is produced by adenosine — plus three phosphate groups attached. (Ch. 6)

adenylate cyclase (a-**den**-ih-late **sye**-klase) An enzyme located in the inner surfaces of cell membranes that catalyzes the conversion of ATP to cAMP following the binding of a signal molecule to a membrane-bound receptor. (Ch. 12)

adequate stimulus The type of stimulus to which a sensory receptor is most sensitive. (Ch. 8)

adipose tissue (**ad**-i-poz) Tissue composed mainly of fat-storing cells. (Ch. 11)

adrenal cortex (ad-**ree**-nal **kore**-teks) The outer two-thirds shell of the adrenal gland that produces the steroid hormones—cortisol, aldosterone, and androgens. (Ch. 12)

adrenal glands A pair of endocrine glands, one located above each kidney: Each consists of outer adrenal cortex and inner adrenal medulla. (Ch. 12)

adrenal medulla (ad-**ree**-nal meh-**dul**-ah) The inner core of the adrenal gland that secretes the catecholamines, epinephrine, and norepinephrine. (Ch. 12)

adrenergic (*ad*-ren-**er**-jik) Pertaining to epinephrine or norepinephrine. (Ch. 10)

adrenocorticotropic (ad-*ree*-no-kore-ti-kow-**trope**-ik) **hormone (ACTH)** A polypeptide hormone that is synthesized, stored, and released from the anterior pituitary gland. It stimulates the production and secretion of cortisol by the zona fasciculata of the adrenal cortex. (Ch. 11)

adrenogenital (ad-ree-no-*gehn*-ih-tahl) **syndrome** A genetic disorder in females in which hypersecretion of adrenal androgens results in the development of female internal genitalia but male/female external genitalia. (Ch. 33)

adult respiratory distress syndrome A form of pulmonary edema caused by increased permeability of the capillaries; damage is caused by leakage of oxygen radicals. (Ch. 21)

aerobic (air-**oh**-bik) Any oxygen-requiring process. (Ch. 6)

afferent (**af**-er-ent) Carrying toward. (Ch. 9)

afferent arteriole (af-er-ent ar-**tee**-ree-ole) The arteriole that carries blood toward the glomerulus. (Ch. 23)

afferent nerve fiber Nerve fiber that transmits stimuli to the CNS. (Ch. 9)

affinity chromatography (ah-**fih**-ih-tee kro-muh-**tog**-raph-ee) A technique that uses the specificity of an antigen-antibody reaction to study antibodies. (Ch. 3)

afterload The force a weight on a lever system will provide during an isotonic contraction. (Ch. 16)

agglutination (a-*glue*-ti-**nay**-shun) The process that is described by the aggregation or clumping of red blood cells which is caused by the attachment of substances to cell surfaces that, in turn, bind the cells together. (Ch. 17)

agglutinin (a-**glue**-ti-nin) Specific antibodies in the blood that cause clumping or agglutination. (Ch. 17)

agglutinogen (a-**glue**-ti-nin-oh-gin) Specific antigens on red blood cell membrane surfaces that causes the pro-

duction of an agglutinin-type of immunoglobulin. (Ch. 17)

agonist (ah-**goh**-nist) Any ligand capable of both binding to a plasma membrane receptor and evoking a physiologic response. (Ch. 5)

agranulocyte (a-**grahn**-you-low-site) Leukocyte that does not have a granular cytoplasm; monocyte and lymphocyte. (Ch. 17)

airway A series of highly branched hollow tubes within the lung that carry gas to and from the alveoli, trachea, bronchii, bronchioles. (Ch. 20)

alactacid oxygen debt The first stage of repayment of the oxygen debt during exercise; refers to the oxygen needed to replenish the oxygen stores of the body as well as the phosphagen system. (Ch. 30)

albumin (al-**byou**-min) The most common plasma protein; produced in the liver. It regulates plasma osmotic pressure. (Ch. 12)

aldosterone (al-**dos**-ter-own) A mineralocorticoid (steroid) hormone produced by the zona glomerulosa of the adrenal cortex that helps to regulate body fluid volume by controlling sodium chloride and potassium excretion from the kidney tubular cells. (Ch. 12)

alkalemia (*al*-kuh-**lee**-me-uh) An arterial blood pH greater than 7.45. (Ch. 25)

alkaline (**al**-kuh-line) Basic; a pH greater than 7. (Ch. 25)

alkalosis (*al*-kuh-**low**-sis) An abnormal process that tends to produce alkalemia. (Ch. 25)

allosteric enzyme (*al*-o-**steer**-ik **en**-zime) A catalytic protein that has two reactive sites. One site activates the enzyme. The other one, the allosteric site, can enable a substance other than the normal substrate to bind to the enzyme molecule, thereby changing the shape of the molecule and activity of the enzyme. (Ch. 6)

alpha adrenergic (**al**-fah *ad*-ree-**ner**-jik) **receptor** A type of receptor site for norepinephrine and epinephrine that, when activated, opens a calcium channel in the plasma membrane where it is located. (Ch. 7)

alpha cells of the pancreas One cell type found in the islets of Langerhans; produces glucagon and is responsible for regulation of blood glucose concentrations and fuel homeostasis. (Ch. 15)

alpha motor neuron An efferent neuron originating in the CNS that synapses with extrafusal muscle fibers. (Ch. 9)

alpha-helix A type of conformation that results when a single polypeptide chain coils about itself to form a uniform cylindrical structure. (Ch. 2)

alpha receptor A receptor to which norepinephrine binds. (Ch. 10)

alveolar (al-**vee**-oh-lar) **capillary membrane** The membrane through which gas exchange takes place between the alveolus and the blood in the pulmonary capillaries. (Ch. 21)

alveolar pressure The pressure of air inside the alveoli. (Ch. 20)

alveolar surface area The site of gas exchange in the lung; roughly 80 times greater than the external body surface. (Ch. 20)

alveolar ventilation The volume of atmospheric air entering alveoli each minute. (Ch. 20)

alveolus (al-**vee**-oh-lus) A microscopic air sac in the lung; effects gas exchange (plural: *alveoli*). (Ch. 20)

ameboid (a-**mee**-boyd) **movement** Movement similar to that of an amoeba, in which the cell projects protoplasmic extensions ahead and then follows them. (Ch. 17)

amenorrhea (a-men-oh-**ree**-ah) The absence of menstruation. (Ch. 33)

amino acid The basic unit of protein molecules whose chemical substrate can be written as R—CHNH$_2$COOH. There are 20 naturally occurring amino acids that appear in biological proteins. (Ch. 2)

amino group —NH$_2$. (Ch. 2)

aminoacyl attachment site of tRNA A feature of transfer RNA in which an amino acid destined to join a growing polypeptide chain attaches for transport to a ribosome. (Ch. 5)

ammonia NH$_3$. (Ch. 25)

amniotic (am-**nee**-ot-ik) **fluid** The liquid that fills the amnion. (Ch. 32)

amphetamine (am-**fet**-uh-meen) A sympathomimetic drug that stimulates the CNS, increases blood pressure, and reduces the appetite. (Ch. 11)

amphipathic (*am*-fee-**path**-ik) **molecule** Molecule that has a polar region at one location and a nonpolar region at another location, making it both hydrophobic and hydrophilic. Examples of amphipathic molecules are the phospholipids in cell membranes. (Ch. 4)

amphoteric (am-**fo**-tear-ik) Capable of acting as either acid or base. (Ch. 25)

amylase (**am**-uh-laze) A digestive enzyme that helps break down starch. (Ch. 22)

amylin (**am**-uh-lyn) Amylopectin; the insoluble constituent of starch. (Ch. 15)

anabolic (an-a-**ball**-ik) **hormone** A hormone that promotes tissue growth. (Ch. 13)

anabolic steroid A steroid hormone having anabolic effects.

anabolism (biosynthesis) (a-**nab**-o-lizm) A series of steps in which the energy extracted in catabolism is used to assemble building block molecules into proteins, nucleic acids, lipids, and other macromolecules for the cell. (Ch. 6)

anaerobic (an-air-**oh**-bik) Refers to processes that can occur in the absence of molecular oxygen. (Ch. 6)

analgesia (an-al-**jee**-zee-ah) The first stage of anesthesia; a reduced response to pain without loss of consciousness. (Ch. 29)

anatomical planes Imaginary lines used to divide the body into parts. (Ch. 1)

anatomy Structure and organization of the body. (Ch. 1)

androgen (**an**-drow-jen) Any chemical with testosteronelike action. (Ch. 11)

androgen-binding-protein (ABP) A protein synthesized by Sertoli cells that binds testosterone and increases the concentration of testosterone in seminiferous tubules of the testes. (Ch. 31)

anemia (ah-**nee**-mee-uh) A condition characterized by an abnormally low oxygen-carrying capacity of the blood, resulting from marked deficiencies in the number of circulating red blood cells, the amount of hemoglobin, or both. (Ch. 17)

anesthesia (an-es-**thee**-zee-ah) Loss of feeling or sensation. (Ch. 29)

angina pectoris (an-**jie**-nah **pek**-toe-ris) Chest pain due to hypoxia of the myocardium that may occur because of overexertion, stress, or excitement. (Ch. 19)

angiogram (**an**-jee-oh-gram) A roentgenogram of blood vessels filled with a contrast medium. (Ch. 21)

angiotensin I (*an*-jee-o-**ten**-sin) A compound in the blood formed by the action of renin, an enzyme secreted by the kidney. Angiotensin I becomes angiotensin II by means of a converting enzyme of the pulmonary capillaries. (Ch. 10)

angiotensin II A hormone that causes the hypothalamus to release ADH into the circulatory system; stimulates aldosterone secretion from adrenal cortex, vascular smooth muscle contraction, and thirst. (Ch. 10)

angiotensin-converting enzyme An enzyme that converts angiotensin I into angiotensin II through the removal of two amino acids. (Ch. 14)

angiotensinogen A plasma protein synthesized by the liver and secreted into the blood; precursor for angiotensin I. (Ch. 14)

angular gyrus The association cortex thought to be involved with the association of visual, auditory, and tactile sensations. (Ch. 11)

anion (**an**-eye-on) An atom that gains one or more electrons, thus acquiring a negative charge. (Ch. 2)

anorgasmia (*an*-ore-**gaz**-me-uh) Difficulty in reaching orgasm. (Ch. 33)

anoxia (ah-**nok**-see-uh) Reduced oxygen supply. (Ch. 29)

antagonist (an-**tag**-o-nist) A ligand that binds with high affinity to a receptor but evokes no response. (Ch. 5)

antagonist muscle A muscle that opposes an agonist muscle at a joint where two bones meet; contraction of the agonist muscle results in the extension of the antagonist and vice versa. (Ch. 9)

anterior hypothalamus The anterior portion of the hypothalamus; appears to be the center for sexual drive in both males and females. (Ch. 11)

anterior pituitary gland Anterior portion of the pituitary gland. (Also called *adenohypophysis*.) (Ch. 13)

anterograde fast transport The movement of organelles and materials from the soma to the axon terminal of a nerve cell. (Ch. 7)

anterolateral pathway One of two major pathways for the transmission of somatic sensory information from the surface of the body to the neocortex. (Ch. 8)

antibiotic (*an*-ti-bye-**ot**-ik) Molecules produced by microorganisms that kill or inhibit the growth of other microorganisms. (Ch. 5)

antibody (**an**-ti-*bod*-ee) A highly specific protein produced by plasma cells in response to the introduction of foreign matter (antigens). The antibody reacts against the antigen to remove or destroy it. The new term for antibody is *immunoglobulin*. (Ch. 3)

antibody-mediated immunity Humoral immunity; major protection against bacteria and viruses in extracellular fluid. (Ch. 28)

anticoagulant (*an*-ti-koh-**ag**-you-lant) A chemical that blocks or inhibits the coagulation process. (Ch. 17)

anticodon A set of three nucleotides in tRNA complementary to a codon on an mRNA molecule. (Ch. 5)

antidiuretic (an-ti-die-your-**et**-ik) **hormone** (ADH) A hormone that directs kidneys to increase the absorption of water; *see also* vasopressin. (Ch. 10)

antidrop response Vasoconstriction and shivering due to stimulation of the hypothalamus by excessive cooling. (Ch. 27)

antigen (**an**-tih-jen) Any molecule that stimulates the production of antibodies. (Ch. 3)

antigen-presenting cell A cell, usually a macrophage or B-lymphocyte, that displays an antigen in combination with MHC proteins so as to activate specific immune defense mechanisms. (Ch. 28)

anti-Mullerian hormone (AMH) A glycoprotein, secreted by the Sertoli cells of the testes, which induces regression of the Mullerian duct system; earlier known as müllerian inhibiting substance or MIS. (Ch. 33)

antiparallel DNA strands When the 3′ ends of the two complementary nucleic acid strands of the DNA molecule are at opposite ends of the double helix. (Ch. 5)

antiport The movement of solutes in the opposite direction of Na^+. (Ch. 4)

antipyretics Agents that prevent or reduce the fever response. (Ch. 27)

antirise response Vasodilation and sweating due to stimulation of the hypothalamus by overheating. (Ch. 27)

antithrombin (ahn-tee-**throm**-bin) A natural coagulation inhibitor. Prevents thrombin from converting fibrinogen to fibrin. (Ch. 17)

antithromboplastin (ahn-tee-**throm**-bo-plas-tin) A natural coagulation inhibitor. Inactivates tissue thromboplastin. (Ch. 17)

antrum (ahn-**trum**) The lower portion of the stomach closest to the pyloric sphincter. (Ch. 22)

aorta (ay-**or**-tah) The artery that receives all the blood the left heart pumps out; branches into a number of smaller arteries that direct blood flow to organs. (Ch. 18)

aortic arch The part of the aorta in which baroreceptors that sense blood pressure are located. (Ch. 10)

aortic bodies Chemoreceptors of the peripheral nervous system within the wall of the aorta; sensitive to levels of O_2 and CO_2 in blood. (Ch. 10)

aortic valve The valve between the left ventricle and the aorta. (Ch. 18)

aphasia (ah-**fay**-zee-ah) Disorders of language resulting from damage to the brain. (Ch. 11)

apoptosis An organized series of steps which, once triggered, leads to cell death. (Ch. 3)

appendicitis An inflammation of the appendix; usually preceded by obstruction of the lumen of the appendix. (Ch. 22)

appendix A small, fingerlike blind tube that extends from the cecum. (Ch. 22)

aprosodia (a-**pros**-o-dea) Damage to the areas of the nondominant hemisphere of the brain; results in an inability to understand or express intonation. (Ch. 11)

aqueous (**ak**-wee-us) Prepared with water. (Ch. 2)

areola (ah-**ree**-o-lah) The pigmented area of skin that surrounds the nipple of the breast. (Ch. 32)

arm projection neuron A neuron of the posterior parietal cortex that generates action potentials when the arm reaches for an object. (Ch. 9)

arrhythmia (ah-**ryth**-mee-ah) Conduction and rhythm disturbances of the heart. (Ch. 18)

arterial baroreceptors (ar-**teer**-ee-al) Nerve endings sensitive to stretch or distortion produced by arterial blood pressure changes; located in carotid sinus or aortic arch arteries. (Ch. 19)

arterial pressure The pressure within arteries. (Ch. 19)

arterial system The system of blood vessels that carry blood from the heart to the capillaries. (Ch. 18)

arteriole (ar-**tee**-ree-ole) The smaller blood vessels that branch off an artery once the artery has entered an organ. (Ch. 18)

arteriosclerosis (ar-**teer**-ee-o-skleh-roe-sis) A group of diseases characterized by thickening and loss of elasticity of arterial walls. (Ch. 17)

artery (**ar**-ter-ee) A blood vessel that transports blood away from the heart to the various organs of the body. (Ch. 1)

asbestosis A lung disease caused by breathing asbestos fibers. (Ch. 29)

associative learning A type of learning in which a subject acquires an understanding of the relationships between stimuli or between a behavior and subsequent reward or punishment. (Ch. 11)

asthma (**az**-muh) A breathing disorder brought on by narrowing of the bronchi. (Ch. 20)

astrocytes Star-shaped glial cells of the CNS that form the blood-brain barrier and provide nutrients, support, and insulation for CNS neurons (also called *astroglia*). (Ch. 7)

atelectasis (*at*-ee-**lek**-tah-sis) Alveolar collapse due to high surface tension in the lungs. (Ch. 20)

atherosclerosis (*ath*-er-or-skleh-**roe**-sis) A disease process of arteries that results in the progressive occlusion of the lumen of the vessel. (Ch. 17)

atherosclerotic plaque Accumulation of lipid and fibrous tissue inside blood vessels resulting in narrowing of the vessel lumen. (Ch. 19)

atmospheric pressure Air pressure surrounding the body; equal to 760 mm Hg at sea level. (Ch. 20)

atom (**a**-tom) The smallest unit of an element that retains the properties of the element. (Ch. 2)

atomic number (a-**tom**-ic **num**-ber) A specific number of protons in the nucleus of an atom, by which it can be identified. (Ch. 2)

ATP synthetase An enzyme that transforms the energy of the proton electrochemical gradient into chemical energy in the form of ATP. (Ch. 6)

atrial (**ay**-tree-al) **natriuretic factor** A peptide secreted from the cells in the walls of the atria; causes both vasodilation and increased secretion of salt and water by the kidneys. (Ch. 19)

atrial septal defect A congenital birth defect in which the septum between the atria is not completely formed. (Ch. 21)

atrioventricular (AV) (*ay*-tree-o-ven-**trik**-you-lar) **node** A region of specialized cells located in the right atrial wall near the interventricular septum. It relays the contractile stimulus to the bundle of His, the Purkinje fibers, and the ventricular myocardium. (Ch. 18)

atrioventricular valve Valve located between an atrium and a ventricle. The left AV valve is called the bicuspid or mitral valve. The right AV valve is called the tricuspid valve. (Ch. 18)

atrium (**ay**-tree-um) A thin-walled chamber of the heart that receives blood returning to the heart from either the vena cava or the pulmonary vein. (Ch. 18)

atrophic (a-**tro**-fik) **vaginitis** Drying and thinning of vaginal mucosal membrane and decreased elasticity and muscle tone of the vagina due to inadequate estrogen stimulation. (Ch. 33)

atrophy (**at**-roe-fee) Wasting away (Ch. 16)

auditory cortex (**aw**-di-*toe*-ree **kor**-teks) That part of the neocortex that detects sounds via the ear. (Ch. 7)

auditory sensory aphasia A condition produced by lesions in the auditory association cortex or Wemicke's area, in which voices can be heard but the words appear to have no meaning. (Ch. 8)

auditory system The system of the body that pertains to the sense of hearing. (Ch. 8)

autocrine (*aw*-**toe**-krin) Mechanism by which a hormonelike substance is released into the interstitial fluid surrounding the cells in which it is produced, and engages receptors on the same cells. (Ch. 12)

autoimmunity (*aw*-toe-ih-**mew**-ni-tee) An immune response arising from and directed against a person's own tissues. (Ch. 28)

autonomic (aw-toe-**nom**-ik) **nervous system** The portion of the nervous system that controls the visceral functions of the body such as regulation of the cardiac and smooth muscles and glands. The ANS comprises the sympathetic and parasympathetic divisions. (Ch. 10)

autophagy (aw-toe-*fay*-jee) A process in which lysosomes break down the cell's own organelles and other matter; can be a response to cell injury or can occur during remodelling of the cell or during adverse nutritional conditions. (Ch. 3)

autoradiography (*aw*-toe-ray-**dee**-og-ra-fee) A technique that locates the radioactive molecules within cells or tissues; has been useful in discovering the secretory pathways in cells and in determining the role of the Golgi complex in this process. (Ch. 3)

autoreceptor Alpha-2 and beta-2 receptors located on the membrane of the presynaptic terminal that appear to regulate the amount of NE released. (Ch. 7)

autoregulation of blood flow The ability of individual vascular beds to maintain constant blood flow despite changes in arterial blood pressure; the intrinsic ability of the heart to regulate the strength of ventricular contraction. (Ch. 19)

autosome (**aw**-toe-sowm) Chromosomes that provide genetic information for the development and differentiation of most of the cells of the body. (Ch. 33)

auxotonic (**ox**-o-tohn-ik) **contraction** Muscle movement in which the force continually increases as the muscle shortens. (Ch. 16)

axial (**ak**-si-al) **muscles** Muscles that produce movement of the skull, backbone, and ribs. (Ch. 9)

axoaxonic (**ak**-so-ak-son-ik) **synapse** Synaptic contact made between two axons. (Ch. 7)

axodendritic synapse Synaptic contact made between axons and dendrites. (Ch. 7)

axon (**ak**-son) A long fiberlike structure that extends from the soma to make contact with other nerve cells. (Ch. 7)

axon terminal The ends of axons; the point of contact with other nerve cells at synapses to transmit information. (Ch. 7)

axonal transport The movement of cellular components from the cell body to the terminals. (Ch. 7)

axosomatic synapse Synaptic contacts made between axons and somata. (Ch. 7)

Babinski (bah-**bin**-skee) **sign** Extension of the big toe and fanning of the other toes in response to stroking the bottom of the foot; due to upper-motor-neuron lesions. (Ch. 9)

bacteria (bak-**teer**-ee-ah) Unicellular organisms that have an outer cell wall and a plasma membrane. (Ch. 17)

barometric pressure Atmospheric pressure. The pressure on the surface of the earth due to the weight of the atmosphere. At sea level, atmospheric pressure is approximately 760 mm Hg. (Ch. 20)

baroreceptor (*bar*-oh-re-**sept**-tor) A stretch receptor sensitive to pressure and to rate of change in pressure. (Ch. 10)

basal (**bay**-sal) Refers to resting level. (Ch. 3)

basal ganglia (**bay**-sal **gang**-lee-ah) A

group of nuclei in the brain involved in the programming of motor patterns. (Ch. 9)

basal metabolic rate (BMR) Metabolic rate measured under basal conditions. (Ch. 27)

base Any substance that can accept or bind hydrogen ions. (Ch. 2)

basement membrane The thin proteinaceous layer of extracellular material upon which epithelial and endothelial cells fit. (Ch. 3)

basic electrical rhythm Cyclic spontaneous depolarization and repolarization of pacemaker cells in the stomach causing gastric peristalsis at a frequency of three contractions per minute. (Ch. 22)

basilar (**bas**-ih-lar) **membrane** Thin membrane that divides the scala tympani from the cochlear duct in the inner ear. It supports the organ of Corti and helps to record the frequency of sound waves. (Ch. 8)

basolateral membrane The side of the epithelial cell facing away from the lumen. (Ch. 22)

basophil (**bay**-so-fil) The rarest type of leukocyte; found in greater numbers outside the blood in loose connective tissue. (Ch. 17)

B cell A lymphocyte that can become an antibody-secreting cell. (Ch. 17)

belly The wide or thick portion of a muscle. (Ch. 16)

beta cells of the pancreas One cell type found in the islets of Langerhans; produces insulin and is responsible for regulation of blood glucose concentrations and fuel homeostasis. (Ch. 15)

beta (**bay**-tah) **receptor** One of two types of plasma membrane receptor sites for norepinephrine and epinephrine. When stimulated, the effects include accelerated heart rate, dilation of bronchioles, and glycogenolysis. (Ch. 10)

beta sheet A secondary structural conformation of a protein molecule that results when the polypeptide chain runs back and forth upon itself. (Ch. 2)

beta-lipotropin (**bay**-tah **lipo**-tro-pin) A large polyprotein that breaks down into a number of opiates that can inhibit the sensation of pain. (Ch. 7)

bicuspid (bye-**kus**-pid) **valve** The atrioventricular valve located between the atrium and the ventricle of the left heart; also called the mitral valve. (Ch. 18)

bile (**byl**) The fluid secreted by the liver; contains bicarbonate, bile salts, cholesterol, lecithin, and bile pigments. (Ch. 22)

bile canaliculi (kan-ah-**lik**-u-li) Small ducts next to liver cells into which bile is secreted. (Ch. 22)

bile salt An organic component secreted by the liver; promotes solubilization and digestion of fat in the small intestine. (Ch. 22)

bilirubin (*bil*-ee-**roo**-bin) A red-colored product of the breakdown of hemoglobin by the liver cells. It is then excreted as waste in the bile. (Ch. 17)

biliverdin (*bil*-ee-**vur**-din) A green-colored product of the breakdown of hemoglobin by the liver cells. It is the first product of catabolism and is then converted to bilirubin. It is excreted as waste. (Ch. 17)

biological rhythms Biological processes that occur in a rhythmic or regular periodic manner; for example, the menstrual cycle (once a month) or the circadian (24-hour) cycle of core body temperature. (Ch. 29)

bipolar cell A type of nerve cell found in the retina. (Ch. 8)

bipolar neuron A type of neuron that has one axon and one dendrite, each extending from opposite sides of the soma. (Ch. 7)

birth control pill An oral contraceptive containing synthetic estrogen and/or progesterone that prevents pregnancy by blocking ovulation. (Ch. 33)

blackout Loss of vision; often due to reduced blood flow to the brain. (Ch. 29)

blastocyst (**blas**-toe-sist) The precursor of the embryo; composed of a ball of developing cells surrounding a central cavity. (Ch. 32)

blood group A system of related blood types, such as the ABO blood group or the Rh blood group. (Ch. 17)

blood pressure The force that causes the blood to flow through the vessels; usually expressed in units of mm Hg. (Ch. 18)

blood type Method of classifying blood by determining the presence or absence of the specific antigens on the red blood cells. (Ch. 17)

blood vessels A series of tubes that convey blood from the heart throughout the body and back to the heart. (Ch. 18)

blood-brain barrier A group of anatomical barriers and transport systems that controls kinds of substances entering the brain extracellular space from blood and their rates of entry. (Ch. 7)

B-lymphocyte A blood cell that develops in the bone marrow and that is the basis of humoral or antibody-mediated immunity. (Ch. 28)

bolus A rounded mass of food. (Ch. 22)

Bowman's capsule A double-walled cup that surrounds the glomerulus and collects the glomerular filtrate. (Ch. 23)

bradycardia An abnormally low heart rate. (Ch. 18)

bradykinin (**brad**-i-kin-in) A hormone produced by the kidneys, whose primary biological effect is vasodilation. (Ch. 10)

brain The organ of the nervous system encased in the skull. (Ch. 7)

brain stem The section of the brain consisting of the medulla oblongata, pons, and midbrain; located between the spinal cord and the forebrain. (Ch. 7)

Braxton-Hicks contractions Spasmodic muscle contractions of the uterus that occur during the third trimester of pregnancy. (Ch. 32)

breast alveolus A gland located in the breast that stores and secretes breast milk. (Ch. 32)

Broca's (**broe**-kaz) **area** Region of the neocortex responsible for the motor control of speech. Usually located in the left frontal lobe. (Ch. 7)

Brodmann area 3 The location of cells in the somatosensory cortex that exhibit the annular receptive field. (Ch. 8)

bronchiole (**brong**-kee-ol) A small division of a bronchus. (Ch. 20)

bronchitis (**brong**-kye-tis) A condition in which the lining of the bronchioles becomes inflamed and swollen, thereby reducing airway diameter. (Ch. 20)

bronchus A large-diameter airway branching off the trachea and entering the lung. (Ch. 20)

brush border membrane The microvilli of all the epithelial cells lining the small intestine. (Ch. 22)

buffer An agent that minimizes the change in pH when either acid or base is added to a solution. (Ch. 2)

bulk Material that cannot be digested. (Ch. 22)

bulk flow The movement of fluids or gases from a region of higher pressure to one of lower pressure. (Ch. 19)

bundle branches Right and left branches of the AV bundle going to each ventricle. (Ch. 18)

bundle of His Group of modified muscle cells between the right atrium and the interventricular septum specialized for very rapid conduction of impulses between the AV node and the Purkinje network; also called the *atrioventricular bundle (AV bundle)*. (Ch. 18)

calcitonin (*kal*-sih-**toe**-nin) A peptide hormone produced by c-cells of the thyroid gland; secreted in response to increases in plasma calcium concentration causing increased deposition of calcium in bones. (Ch. 26)

calcium (**kal**-see-um) A chemical element essential for nerve function, muscle contraction, blood coagulation, bone formation, and many other physiological processes. (Ch. 26)

calcium antagonist A class of drugs that block the entry of calcium into the smooth muscle cells and dilate coronary arteries. (Ch. 19)

calcium ion A chemical that acts as a second messenger to initiate the intracellular reactions that modify the behavior of the target cell. (Ch. 5)

calmodulin (kal-**mod**-you-lin) An intracellular calcium-binding protein that mediates many of calcium's second-messenger functions. (Ch. 11)

caloric (kal-**or**-ik) **density** The amount of calories per gram of basic foodstuff. (Ch. 15)

calyx (**kayl**-ikz) An extension of the renal pelvis; collects urine from the renal papillae. (Ch. 23)

cAMP *See* cyclic AMP.

canaliculus (*kan*-ah-**lik**-you-lus) A narrow tubular passage or channel between cells. (Ch. 26)

canaliculi (*kan*-ah-**lik**-you-lye) Plural of canaliculus. (Ch. 26)

capacitation A process involving physiological change in the plasma membrane of the sperm that increases motility and permits penetration of the surface of an ovum. (Ch. 32)

capillary (**kap**-ih-lar-ee) The smallest type of blood vessel between arteriole and venule. (Ch. 1)

capillary recruitment An increase in the number of capillaries that have blood flowing through them. (Ch. 21)

capillary transit time The amount of time necessary for blood to move through the capillaries; normal transit time is 1 second. (Ch. 21)

carbaminoglobin (*kar*-bam-**in**-oh-globe-in) The binding of carbon dioxide and hemoglobin to facilitate the transport of carbon dioxide in venous blood. (Ch. 21)

carbohydrate (*kar*-boe-**hye**-drayt) A class of organic compounds containing carbon, hydrogen, and oxygen in a ratio of 1:2:1. Sugars, starches, and cellulose are carbohydrates. This class is divided into monosaccharides, disaccharides, and polysaccharides. (Ch. 2)

carbon dioxide dissociation curve A graph of the relationship between blood P_{CO_2} and CO_2 content. (Ch. 21)

carbon monoxide CO, a colorless, odorless gas that readily binds to hemoglobin, thereby decreasing blood-oxygen capacity. (Ch. 22)

carbonic anhydrase (kar-**bon**-ik an-**hi**-draz) An enzyme that catalyzes the reversible reaction $CO_2 + H_2O \longleftrightarrow H_2CO_3$. (Ch. 7)

carboxyhemoglobin (COhb) (kar-*bok*-see-hee-moh-**glo**-bin) The binding of carbon monoxide, rather than oxygen, to hemoglobin, interfering with transport of oxygen to the cells. (Ch. 29)

carboxyl group COOH. (Ch. 2)

carboxypeptidase (car-*bok*-see-**pep**-tih-daze) An enzyme, the inactive precursor of which is procarboxypepti-

dase, secreted by the exocrine pancreas into the small intestine; breaks the peptide bond at the carboxyl end of the peptide chain. (Ch. 22)

carcinogen (car-**sin**-oh-jin) Any agent that induces cancerous transformation of cells. (Ch. 3)

cardiac cycle One heartbeat sequence of the heart. (Ch. 18)

cardiac muscles Striated, involuntary muscles that form the myocardium of the heart. (Ch. 9)

cardiac output The amount of blood pumped out of one ventricle in 1 minute. (Ch. 18)

cardiovascular control center A region of the medulla that integrates sensory input from many sources and subsequently modulates the autonomic nerve activity to the heart and blood vessels to maintain a relatively constant arterial blood pressure. (Ch. 19)

carotid (kar-**rot**-id) **body** A chemoreceptor located at the bifurcation of the carotid artery. It is sensitive to changes in levels of oxygen, carbon dioxide, and pH in the blood. (Ch. 10)

carotid sinus The dilated portion of the internal carotid artery located just above the branching of the common carotid artery. Pressure receptors that help regulate blood pressure are located here. (Ch. 10)

carrier proteins Membrane proteins that allow ions and large water-soluble molecules to pass through the lipid bilayer via facilitated diffusion. (Ch. 4)

catabolic hormone A hormone that promotes tissue breakdown. (Ch. 13)

catabolism (kah-**tab**-o-lizm) The breakdown of organic molecules into simpler compounds accompanied by the release of energy. (Ch. 6)

catalase (**kah**-tuh-laz) A cellular enzyme that catalyzes the decomposition of hydrogen peroxide. (Ch. 23)

catecholamine (*kat*-e-**kole**-ah-mean) A hormone that acts on the heart to increase both heart rate and the volume of blood pumped, thereby increasing blood pressure; dopamine, epinephrine, and norepinephrine. (Ch. 10)

cathartic A purgative; facilitates defecation by increasing bulk. (Ch. 22)

catheter (**kath**-eh-ter) A tube that is inserted into a body cavity or along a blood vessel for the introduction or removal of body fluids. (Ch. 23)

cation (**cat**-ye-on) An atom that loses one or more electrons, thus acquiring a positive charge. (Ch. 2)

caudal Anatomical term of position that means nearer to the tail, or further from the head; also called *inferior*. (Ch. 1)

caudate (**kaw**-dayt) One type of basal ganglia; receives inputs from the neocortex, thalamus, and substantia nigra

of the brain stem. The **caudate nucleus** is one pair of basal ganglia associated with the reward system in the brain. (Ch. 9, 11)

c-cells Parafollicular cells in the thyroid gland, cells that produce the peptide hormone calcitonin. (Ch. 26)

CD4+ cell A T-lymphocyte possessing a surface marker (receptor) designated as DC4; called also T4 cell. (Ch. 28)

CD8+ cell A T-lymphocyte possessing a surface marker (receptor) designated as CD8; called also T_8 cell. (Ch. 28)

cecum (**see**-come) The blind sac that marks the first part of the large intestine. (Ch. 22)

cell adhesion molecule (CAM) A glycoprotein important in nervous system development. (Ch. 7)

cell cycle The period of time from the beginning of one cell division to the beginning of the next division. (Ch. 3)

cell differentiation *See* differentiation.

cell fractionation The separation of different organelles of the cell in order to study each on its own in a strictly controlled system. (Ch. 3)

cell-mediated immunity A type of specific defense response wherein several kinds of T-cells and macrophages work together to destroy an antigen. (Ch. 28)

cell theory The broad basis for all modern biology and physiology; key concepts are (1) all organisms are made up of cells; (2) new cells arise only from preexisting cells; (3) all cells have the same fundamental chemical makeup and metabolic processes; and (4) activities and processes of an organism result from interdependent and cooperative workings of groups of cells. (Ch. 1)

cellular respiration The process by which organic molecules are catabolized by reactions that require atmospheric O_2, producing CO_2 and H_2O. The reverse of photosynthesis. (Ch. 6)

cellulose A polysaccharide composed of glucose subunits; found in plant cells. (Ch. 22)

central nervous system (CNS) The brain and spinal cord. (Ch. 7)

centrifuge A machine that allows materials to be spun at high speeds, thus separating them by density. (Ch. 3)

cerebellar (*ser*-eh-**bel**-ar) **anterior lobe** The portion of the cerebellum involved with planning, initiation, and control of limb movement. (Ch. 9)

cerebellar flocculonodular (flock-u-low-nod-you-lar) **lobe** The portion of the cerebellum involved with the maintenance of equilibrium and posture. (Ch. 9)

cerebellar granule cell Fibers in the cerebellum that synapse with the dendrites of the Purkinje cells, which in turn generate simple patterns of action potentials. (Ch. 9)

cerebellar posterior lobe The portion of the cerebellum involved with planning, initiation, and control of limb movement. (Ch. 9)

cerebellum (*ser*-eh-**bel**-um) The area of the brain responsible for locomotor coordination. (Ch. 7)

cerebrospinal (se-*ree*-broe-**spy**-nal) **fluid (CSF)** The fluid that fills the cerebral ventricles and the subarachnoid space surrounding the brain and the spinal cord, provides an extracellular environment for brain cells. (Ch. 10)

cervical mucosa (**ser**-vih-kul mew-**ko**-sah) Secretions of the cervix composed of glycoproteins, salt, and water. (Ch. 32)

cervix (**ser**-viks) The lower portion of the uterus; connects uterus and vagina. (Ch. 32)

channel gating Mechanism by which the opening and closing of ion channels is regulated. Channel gating is regulated by channel proteins. (Ch. 4)

channel protein Membrane proteins that allow ions and large water-soluble molecules to pass through the lipid bilayer via facilitated diffusion. (Ch. 4)

characteristic frequency, f_0 The frequency that most effectively activates a particular auditory nerve fiber. (Ch. 8)

chemical energy The energy of the cell used to do work; released by reactions that take place within the cell. (Ch. 2)

chemical equation A representation using chemical symbols to show a chemical reaction. (Ch. 2)

chemical pH buffer A weak acid and its conjugate base. (Ch. 25)

chemical reaction Process in which chemical bonds are broken down and recombine into new compounds. (Ch. 2)

chemical stimuli Humoral or fluid-related stimuli. (Ch. 29)

chemical symbol An abbreviation for an element. (Ch. 2)

chemiosmotic (*kem*-ee-oz-*moh*-tik) **mechanism** Mitochondrial process by which movement of protons down an electrochemical gradient is used to convert ADP into ATP. (Ch. 6)

chemiosmotic synthesis of DNA The overall process by which oxidative phosphorylation generates ATP. (Ch. 6)

chemoreceptor An afferent nerve ending (or cells) that is sensitive to concentrations of certain chemicals. (Ch. 10)

chemotaxis (kee-moe-*tak*-sis) The attraction by and subsequent movement toward a chemical by a leukocyte or macrophage. (Ch. 28)

chloride shift The exchange of chloride ions for bicarbonate ions by the erythrocyte as carbon dioxide enters and

leaves the erythrocyte. (Ch. 21)

chloroform An inhalation anesthetic, first used for women in childbirth. (Ch. 29)

cholecalciferol (**koe**-lee-**kal**-sif-er-al) Natural vitamin D. (Ch. 26)

cholecystectomy (**koe**-lee-sis-**tek**-toe-mee) The surgical removal of the gall bladder. (Ch. 22)

cholecystokinin (CCK) A polypeptide hormone produced by the gastrointestinal tract; stimulates enzyme secretion and growth in the pancreas, stimulates gallbladder contraction, and relaxes the sphincter of Oddi. (Ch. 22)

cholesterol (koe-**les**-ter-ol) The most abundantly found steroid located in cell membranes; it is actually classified as a lipid. Cholesterol is a component of gallstones and is a precursor in the production of many steroid hormones. (Ch. 4)

cholinergic (koe-lin-**er**-jik) **receptor** A receptor of acetylcholine. (Ch. 7)

chondrocyte (**kon**-droe-site) A cell involved in cartilage formation. (Ch. 13)

chromaffin (kroe-**maf**-in) **cell** The functional unit of the adrenal medulla; functions as a neuroendocrine cell. A cell that is readily stained with solutions containing chromium salts. (Ch. 14)

chromaffin granules Dense, membrane-bound vesicles in chromaffin cells that contain epinephrine. (Ch. 14)

chromatid (**kroe**-mah-tyd) One of two identical strands of chromatin that result from DNA replication during meiosis and mitosis. (Ch. 3)

chromatin (**kroe**-mah-tin) A combination of DNA and nuclear proteins that is the principal component of chromosomes. (Ch. 3)

chromosomal sex The genotypic sex of an individual. (Ch. 33)

chromosome (**kroe**-moe-sowm) A highly coiled, condensed form of chromatin formed in the cell nucleus during mitosis and meiosis. (Ch. 3)

chylomicron (**kie**-loe-**my**-kron) An aggregate of triglycerides, cholesterol, and phospholipids surrounded by protein that is absorbed from the intestinal cells into the lacteal of a villus. (Ch. 17, 22)

chymotrypsin (kie-moe-**trip**-sin) An enzyme secreted by the exocrine pancreas; breaks down certain peptide bonds in proteins and polypeptides. (Ch. 22)

circadian (*ser*-**kay**-dee-an) **rhythms** Rhythms of the body that vary on a daily basis. (Ch. 11)

circulatory convection A method of transfer of heat within the body via the movement of body fluids. (Ch. 27)

circulatory shock The progressive dete-

rioration of cardiovascular function; generally caused by great losses of blood, severe tissue damage, and irreversible circulatory collapse. (Ch. 19)

citric acid cycle A series of reactions occurring in the mitochondrion that represent the final common pathway for the breakdown of all the fuel molecules in the cell. Also called the Krebs cycle or the tricarboxylic acid cycle. (Ch. 6)

classical conditioning One type of associative learning in which an animal acquires an understanding about the relationship between one stimulus and another. (Ch. 11)

clearance The volume of plasma per unit time needed to account for the rate of removal of a particular substance. (Ch. 23)

clitoris (**kli**-tor-iss) A small body of erectile tissue in the female external genitalia; homologous to the penis in males. (Ch. 33)

cloning vector Plasmids used in DNA cloning studies. (Ch. 3)

clot A semisolid mass similar in consistency to gelatin. (Ch. 17)

clot retraction The decrease in the size of a clot; caused by thrombosethenin. (Ch. 17)

CNS Central nervous system. (Ch. 7)

CO_2 fixation The series of steps in which CO_2 from the atmosphere is used in the creation of carbohydrates and other high-energy organic compounds. (Ch. 6)

coagulation (koh-*ag*-you-**lay**-shun) The process of clot formation. (Ch. 17)

cocaine (coh-**cane**) A drug that enhances release of dopamine and augments the rate of self-stimulation. (Ch. 11)

cochlea (**koke**-lee-ah) A spirally coiled, fluid-filled inner tube of the inner ear; contains the organ of Corti, where nerve impulses are generated in response to sound waves. (Ch. 8)

cochlear hair cells Cells in the cochlea that transduce the vibrations of the basilar membranes into action potentials. (Ch. 8)

codon (**koh**-don) Three-base sequence in a polynucleotide of DNA or mRNA that specifies the location of a single amino acid in a polypeptide chain. (Ch. 5)

coenzyme (koe-**en**-zime) Complex organic cofactors that serve to transfer atoms and small groups from one enzymatic substrate to another acceptor. Vitamins function as coenzymes. (Ch. 2)

coenzyme A A coenzyme also known as pantothenic acid; involved in many reactions, such as the oxidation of pyruvate or fatty acids to acetyl CoA in the citric acid cycle. (Ch. 6)

cofactor (**koe**-fak-tor) The nonprotein component that some enzymes need to function as catalysts. (Ch. 2)

cold receptors Peripheral temperature receptors that are naked nerve endings sensitive to temperature; characterized by increasing steady-state discharge rates as the skin is cooled. (Ch. 27)

collagen (kol-a-jen) A fibrous protein found in many connective tissues. (Ch. 26)

colloid (**kah**-loyd) A mixture in which there is suspended material of large molecular size. The protein-containing material found in the interior of the follicles of the thyroid gland. (Ch. 13)

colloidal osmotic pressure (also called **oncotic pressure**) Refers to the portion of the *total osmotic pressure* that is due to the presence of plasma proteins. (Ch. 19)

colon (**koe**-lon) Part of the large intestine between the cecum and the rectum; subdivided into the ascending colon, transverse colon, descending colon, and sigmoid colon. (Ch. 22)

colony-forming unit A group of progenitor cells committed to forming a colony of one kind of blood cell. (Ch. 17)

color-blind Refers to visual defect in which an individual is unable to distinguish between certain colors. (Ch. 8)

colorigenesis (ko-lo-ri-**gen**-i-sis) The production of heat as a result of chemical reactions of cell metabolism. (Ch. 13)

colostrum (koe-**las**-strum) Thin yellowish fluid excreted by the breast; precursor to breast milk. (Ch. 32)

combined white blood cell count The number of leukocytes per cubic millimeter of whole blood. (Ch. 28)

common bundle *See* bundle of His.

complement (**kom**-plih-ment) A system of proteins that are normally present in plasma and that are necessary for the lysis of foreign cells in immunological defenses. (Ch. 28)

compliance The change in volume of a hollow organ that results from a change in pressure within the organ. (CL = delta volume/delta pressure.) (Ch. 20)

compound A combination of two or more elements. (Ch. 2)

concentration gradient The gradation in concentration that occurs between two regions having different concentrations of the same chemical. Expressed as a difference divided by the distance that separates the two regions. (Ch. 7)

conditioned response The behavior that results from repeated use of conditioned stimulus. (Ch. 11)

conditioned stimulus The stimulus used to create an association with another stimulus to acquire the desired response. (Ch. 11)

conductance The ease with which ions can flow through an ion channel; measured in Siemens. (Ch. 7)

conduction Heat exchange in the form of kinetic energy between molecules of objects that are in direct contact. (Ch. 27)

conductive hearing loss Hearing loss resulting from damage to the middle ear. (Ch. 8)

cone photoreceptor A light-sensitive cell located in the retina that is cone-shaped; responsible for color vision. (Ch. 8)

congestive heart failure Heart failure that results from the inability of the heart to maintain adequate cardiac output due to progressive congestion of the veins and heart. (Ch. 19)

conjugation (*kon*-ja-**gay**-shun) To join together. For example, a process by which a steroid is coupled to a polar sulfate or glucuronide group to make it more water-soluble to facilitate elimination. (Ch. 12)

connective tissue A loosely arranged group of cells, separated by an intercellular matrix, that support, bind, protect, and partition all bodily components. (Ch. 1)

connexin A protein channel that spans the gap between presynaptic and postsynaptic cells. (Ch. 7)

connexon A protein subunit of connexin; there are six, arranged in hexagonal assembly. (Ch. 7)

constipation (*kon*-stih-**pay**-shun) The accumulation of feces beyond what is normal in the large intestine. (Ch. 22)

continence (**kon**-tih-nence) The ability to keep urine in the bladder and to retain fecal material within the colon and rectum. (Ch. 10)

contraception (*kon*-trah-**sep**-shun) The ability to block or prevent conception. (Ch. 33)

contractility (*kon*-trak-**til**-ih-tee) The ability of muscle to shorten and contract. (Ch. 16)

contraction (kon-**trak**-shun) Any form of muscle activity in response to stimulation. (Ch. 16)

control systems Mechanisms by which the body maintains homeostasis. (Ch. 1)

convection (kon-**vek**-shun) The transfer of heat between the body and a gas or liquid moving over the body surface. (Ch. 27)

coronary angioplasty A procedure in which a catheter is inserted into the occluded portion of a coronary artery and inflated, compressing the plaque and expanding the vessel's diameter. (Ch. 19)

coronary bypass surgery Surgery in which a vein from elsewhere in the body is grafted around the occluded segment of the coronary artery so that blood flow can bypass the diseased area. (Ch. 19)

coronary circulation The arrangement of blood vessels and their distribution through the myocardium. (Ch. 18)

coronary disease Disease that compromises blood flow through arteries which supply heart muscle. (Ch. 30)

corpus luteum (**kor**-pus **loo**-tee-um) An ovarian structure formed from the follicle after ovulation; secretes estrogen and progesterone. (Ch. 31)

cortex (**kor**-teks) The outer part of an organ. (Ch. 23)

cortical efferent zone (CEZ) Functional columns of cells in the motor cortex; all the cells in a given zone are involved in the contraction of a given muscle. (Ch. 9)

cortical nephron A nephron whose renal corpuscle is in the outer cortex. (Ch. 23)

corticobulbar (kor-ti-koe-**bul**-bar) **pathway** The avenue by which information from the motor cortex is transmitted to the brain stem. (Ch. 9)

corticospinal (kor-ti-koe-**spy**-nal) **pathway** The major motor pathway within the brain and spinal cord, used for voluntarily activating skeletal muscle. (Ch. 9)

corticospinal tract lesion A damaged area of the spinal cord that disrupts inputs to motor neurons; characterized by loss of strength and movement in muscle groups, loss of strength in voluntary muscles, and the Babinski sign. (Ch. 9)

corticosteroid-binding (kor-tih-koe-**stear**-oid) **globulin (CBG)** A protein that binds with cortisol to facilitate transport in blood. (Ch. 14)

corticotropin-releasing hormone (CRH) A hormone produced by the hypothalamus that regulates the secretion of ACTH. (Ch. 10)

cortisol (**kor**-ti-sol) A glucocorticoid secreted by the adrenal glands. (Ch. 12)

cortisol-binding globulin (CBG) A carrier protein that binds cortisol for transport in the blood. (Ch. 12)

cortisone (**kor**-tih-zone) A hormone regulated by the hypothalamus. (Ch. 11)

countercurrent exchange The passive exchange of solutes, water, or heat between two adjacent streams flowing in opposite directions, as in the vasa recta of the kidney medulla. (Ch. 23)

countercurrent multiplication An energy-demanding process that sets up a solute (osmotic) gradient along the length of two streams flowing in opposite directions, as in Henle's loop in the kidney medulla. (Ch. 23)

coupled reactions Endergonic reactions that are paired with exergonic reactions to obtain the free energy necessary to proceed. (Ch. 6)

covalent (koe-**vay**-lint) **bond**   The chemical bond formed when two or more atoms share electrons. (Ch. 2)

c-peptide   A peptide involved in insulin synthesis. (Ch. 15)

cranial   An anatomical term of position that means nearer to the head; also called superior or caudal. (Ch. 1)

creatine phosphate (CP)   A high-energy compound capable of transferring its phosphate and energy to ADP to create ATP. (Ch. 6)

creatinine (kree-**at**-ih-nine)   An end product of muscle metabolism excreted in the urine; its renal clearance can be used to estimate glomerular filtration rate. (Ch. 23)

crenation   The shriveling of an erythrocyte as it loses an excessive amount of water. (Ch. 17)

cretinism (**kree**-tin-izm)   A mental handicap that results from an untreated thyroid hormone deficiency; linear growth is also impaired, resulting in dwarfism. (Ch. 13)

CRM   Corticotropin-releasing hormone, produced by the hypothalamus, that controls the release of ACTH from the anterior pituitary. (Ch. 14)

cross bridge   Myosin molecule head that projects from the ends of the thick myofilaments within a muscle cell; can bind to an active site on a thin myofilament in the presence of calcium, causing the filaments to slide past one another. (Ch. 16)

crossing over   A process in mitosis in which homologous chromosomes exchange chromosomal segments, ensuring the infinite recombination of genetic material. (Ch. 31)

crossmatching   A process by which agglutinin in the serum of blood is tested to confirm blood type; the presence or absence of red cell agglutination indicates the type of antibody present, thus confirming the type of antigen present on unknown red cells. (Ch. 17)

current flow   The movement of electric charge; in biologic systems the current flow is the movement of ions. (Ch. 7)

Cushing's syndrome   A condition resulting from excessive secretion of the adrenal cortex. General appearance of the patient is think arms and legs, increased fat in the face, trunk, shoulder blades, and base of the neck, as well as masculinization of the female. (Ch. 14)

cutaneous (cue-**tay**-nee-us)   Pertaining to the skin. (Ch. 8)

cyclic AMP (cAMP)   Cyclic 3′, 5′-adenosine monophosphate; the compound derived from nucleotide that mediates the action of many hormones in cells. (Ch. 2)

cyclic GMP   Cyclic 3′, 5′ quanosine monophosphate, an intracellular phosphate that opens sodium channels in photoreceptor plasma membranes. (Ch. 12)

cyclins   Specialized activating proteins that undergo a cycle of synthesis and degradation during the cell cycle. They help control progression from one stage of the cell cycle to the next. (Ch. 3)

cyclosporine (*sye*-cloe-**spore**-in)   An immunosuppressive agent that reduces the possibilities of rejection of foreign substances by the immune system. (Ch. 18)

cystic fibrosis (*sis*-tik fi-**broe**-sis)   An inherited genetic disorder in which secretions of body glands are abnormally thick and sticky, clogging glands and preventing them from working properly. (Ch. 4)

cytochemistry (*sye*-to-**khem**-is-tree)   A research technique used with microscopy to produce color or enhanced contrast at the sites of specific molecules. (Ch. 3)

cytochrome (**sye**-toe-krome)   A protein that serves as an electron carrier in oxidative phosphorylation. (Ch. 6)

cytokine (**sye**-toe-keen)   A general term for a protein messenger secreted by macrophages and monocytes or by lymphocytes. (Ch. 30)

cytokinesis (*sye*-toe-kih-**nee**-sis)   The process by which the cytoplasm divides, each half taking with it one of the two new nuclei to form a new cell. (Ch. 3)

cytoplasm (**sye**-toe-plazm)   One of two major compartments of the cell; composed of the cytosol and the organelles. (Ch. 3)

cytoskeleton (**sye**-toe-skel-eh-ton)   The internal structure of the cell responsible for maintaining cell shape. (Ch. 3)

cytosol (**sye**-toe-sol)   The fluid surrounding the nucleus. (Ch. 3)

data interpretation   The comparison of amounts, rates, times, and other measurable evidence of the body processes that reveals relationships between function and structure. (Ch. 1)

dead space air   The amount of air in conducting airways, approximately 150 ml for each 500 ml inhaled, that does not participate in gas exchange. (Ch. 20)

deamination (*dee*-am-in-ay-shun)   **of amino acids**   Removal of an amino acid group (NH$_2$) from a molecule. A process by which amino acids are broken down by acetyl compounds so they can be used to generate ATP. (Ch. 6)

decremental conduction   The slowed travel of a pacemaker signal as it passes through the atrioventricular node. (Ch. 18)

defecation reflex-defecation (deaf-ih-**kay**-shun)   A sacral spinal reflex elicited by distension of the rectum and, unless voluntarily inhibited, resulting in propulsion of feces through the anus. (Ch. 22)

dehydration (dee-hi-**dray**-shun)   A deficiency of water in the body. (Ch. 15)

7-dehydrocholesterol   A precursor to cholecalciferol, which itself is derived from cholesterol; changes to cholecalciferol through the action of ultraviolet light. (Ch. 26)

dehydroepiandrosterone (DHEA) (dee-hy-dro-ep-i-an-dro-stir-**own**)   A steroid hormone secreted by the adrenal cortex. (Ch. 12)

dehydrogenation reaction   The removal of an electron during oxidation through the loss of an entire atom, almost always hydrogen. (Ch. 6)

deiodinase   A compound involved in the breakdown of thyroglobulin through the removal of iodide from MIT and DIT to produce iodine and tyrosine. (Ch. 13)

delta cells of the pancreas   Cells found in the islet of Langerhans responsible for the production of somatostatin. (Ch. 15)

dendrite (**den**-drite)   A fiberlike structure that branches out from the cell body. (Ch. 7)

dentate nucleus   A deep cerebellar nucleus that forms synaptic connections with the axons of Purkinje cells. (Ch. 9)

deoxyribonucleic (dee-ok-see-*ri*-boe-nyoo-**klee**-ik) **acid (DNA)**   A double-helical molecule consisting of two strands of four different nucleotides. Contains the genetic information that is duplicated in cell division. (Ch. 3)

deoxyribonucleic (DNA) ligase   An enzyme that joins together broken strands of different DNA molecules, allowing any DNA fragment to be grafted into the DNA of a bacterium. (Ch. 3)

dependent variable   The variable that changes as a result of changes in the independent variable; usually represented on the vertical axis. (Ch. 1)

depolarization (dee-*poe*-lar-i-**zay**-shun)   The gradual movement of the membrane potential toward a more positive value. (Ch. 8)

dermatome   A particular area of the skin whose receptors send sensory information to specific spinal cord segments. (Ch. 8)

dermis (**der**-mis)   The thick layer of skin located beneath the epidermis; composed of irregular, dense connective tissue. (Ch. 27)

desmosome (**dez**-moe-zome)   Type of cell-to-cell junction where membrane differentiation takes place. (Ch. 16)

DHEA   Dehydroepiandrosterone, a steroid hormone secreted by the adrenal cortex and converted to testosterone in peripheral tissues. (Ch. 14)

diabetes insipidus (*dye*-ah-**bee**-teez in-**sip**-ih-des) A disease due to a lack of ADH; marked by great thirst and excretion of a large volume of dilute urine. (Ch. 13)

diabetes mellitus (**mel**-ih-tus) A disease in which plasma glucose control is defective because of insulin deficiency or decreased target cell response to insulin. (Ch. 15)

diabetic neuropathy A secondary complication of diabetes involving the impairment of nerve function, resulting in abnormalities in the bladder or gastrointestinal tract function, impotence, or loss of feeling in the lower limbs. (Ch. 15)

diabetic retinopathy (*reh*-tin-**op**-uh-thee) A secondary complication of diabetes that results in the deterioration of blood vessels that nourish the retina, eventually resulting in blindness. (Ch. 15)

diacylglycerol (DG) (dye-*ace*-ill-**glis**-er-all) The second messenger that activates the protein kinase C, which then phosphorylates a large number of other proteins. (Ch. 5)

dialysis (dye-**al**-ih-sis) The separation of smaller from larger molecules in solution by diffusion of the small molecules through a selectively permeable membrane. (Ch. 24)

diapedesis (*dye*-ah-peh-**dee**-sis) The process by which leukocytes pass through the capillary wall to enter spaces around blood vessels. (Ch. 17)

diaphragm (**dye**-ah-fram) A large, dome-shaped skeletal muscle that separates the chest cavity from the abdomen. (Ch. 20)

diarrhea (*dye*-ah-**ree**-ah) Frequent defecation of highly fluid material. (Ch. 22)

diastole (dye-**as**-toe-lee) The period of cardiac cycle during which the ventricles are filling with blood. (Ch. 18)

dichromat (dye-**kroh**-mat) Males who are color-blind due to possessing only two types of cones; a sex-linked trait. (Ch. 8)

dichromatic color vision Color vision that relies on two types of cones. (Ch. 8)

diencephalon (die-en-**sef**-ah-lon) The core of the anterior part of the brain; lies beneath the cerebral hemispheres and contains the thalamus and the hypothalamus. (Ch. 9)

dietary fiber Cellulose and other indigestible polysaccharides found in plant food sources. (Ch. 22)

differential white blood cell count The number of each kind of white blood cell, usually expressed as a percentage of the total or combined white blood cell count. (Ch. 17)

differentiation (dif-er-en-she-**aye**-shun) The process by which a single cell type develops into many different cell types. (Ch. 1)

diffuse synapse A chemical synapse in which the neurotransmitter release is not limited to specific active zones. (Ch. 7)

diffusing capacity of the lung The milliliters of a gas diffusing across the pulmonary membrane per minute per millimeter of mercury pressure difference between the alveolar air and the pulmonary blood. (Ch. 30)

diffusion (dif-**you**-zhun) A process that involves the random movement of particles, such as ions and molecules in solution, such that the substances become evenly mixed. (Ch. 2)

diffusion coefficient $D = 1.0 \times 10^{-5}$ cm^2/sec. (Ch. 4)

diffusion trapping A phenomenon in which the neutral form of a weak base will diffuse across a cell membrane, whereas the charged form will not. (Ch. 4)

digestion (dye-**jes**-chun) The conversion of large organic molecules in the diet to smaller molecules. (Ch. 23)

digestive enzymes Organic compounds that break down large organic molecules into simpler, smaller components. (Ch. 22)

digestive tract A tube about thirty feet long that traverses the body from lips to anus and is composed of the mouth, pharynx, esophagus, stomach, and small and large intestines; also called *alimentary canal*. (Ch. 22)

digitalis A drug that tends to increase the contractility of cardiac muscle by increasing intracellular calcium concentrations. (Ch. 19)

dihydrotestosterone (DHT) (dye-*hi*-droe-tes-**tos**-ter-own) A steroid that stimulates sexual differentiation of the external genitalia in the male. (Ch. 33)

1,25 dihydroxyvitamin Vitamin D$_3$. (Ch. 12)

diiodotyrosine (DIT) An iodinated compound found primarily within the thyroid follicle cell; used to make T$_3$ and T$_4$. (Ch. 13)

direct calorimetry A method of measuring metabolic rate that uses a large insulated chamber to determine the heat loss from a person's body. (Ch. 27)

disaccharidase (dye-**sak**-er-ide-aze) An enzyme that can hydrolyze disaccharides. (Ch. 22)

disaccharide (dye-**sak**-er-ide) A carbohydrate molecule composed of two monosaccharides. (For example, maltose = glucose-glucose.) (Ch. 2)

discrete synapse A chemical synapse in which the chemical neurotransmitter is released from restricted areas of the presynaptic terminal into a small synaptic cleft. (Ch. 7)

distal (**dis**-tal) Refers to the insertion, or less stationary point of attachment, of a muscle. (Ch. 16)

distal convoluted tubule A short nephron segment between the thick ascending limb of Henle's loop and the connecting segment. (Ch. 23)

distal muscles Muscles involved with the movement of the skeleton, primarily the limbs. (Ch. 9)

distal tubule The distal convoluted tubule, connecting segment, and first part of the cortical collecting duct. (Ch. 23)

diuresis (dye-you-**ree**-sis) An increased urine flow rate. (Ch. 23)

diuretic (dye-you-**ret**-ik) An agent that increases urine output. (Ch. 24)

diurnal (**die**-urn-al) Occurring on a 24-hour cycle. (Ch. 11)

diving bell A chamber, either open at the bottom or enclosed, which holds air for a diver to breathe. (Ch. 29)

diving reflex A protective reflex, elicited when the face is submerged in water, that shunts blood away from peripheral tissues and allows for heart-brain perfusion. (Ch. 20)

dizygotic twins Siblings resulting from fertilization of two ova by individual sperm. (Ch. 32)

DNA polymerase An enzyme that during DNA replication forms a new DNA strand by joining together nucleotides already base-paired with an existing DNA strand. (Ch. 5)

DNA replication The copying of the chromatin in the nucleus. (Ch. 5)

DNA template An existing strand of DNA that guides synthesis of a new strand. During DNA synthesis the template strand is incorporated into the new DNA molecule. During RNA synthesis the DNA template is not incorporated. (Ch. 5)

DNA transcription The formation of mRNA containing, in linear sequence of its nucleotides, genetic information in a specific gene; the first stage of protein synthesis. (Ch. 5)

dominant hemisphere The side of the brain in which the language centers are located. (Ch. 11)

dopamine (**dope**-ah-mean) A catecholamine synthesized from tyrosine; a precursor of epinephrine and norepinephrine. (Ch. 7)

dopamine receptor A receptor that causes the hyperpolarization of the postsynaptic potential through the activation of chemically sensitive potassium ion channels. (Ch. 7)

dorsal column nuclei A collection of neurons located on the dorsal column. (Ch. 8)

dorsal column pathway A major pathway for the transmission of somatic sensory information from the surface of the body to the neocortex. (Ch. 8)

dorsal hypothalamus A central component of the autonomic nervous system. (Ch. 10)

dorsal root ganglion A cluster of sensory cell bodies located along the posterior root of a spinal nerve. (Ch. 8)

double helix A configuration in which two nucleotide strands of DNA are coiled around each other. (Ch. 2)

down regulation of receptor number A sustained elevation of the concentration of an agonist circulating in the blood, leading to a decrease in the number of receptors present in target cells. (Ch. 5)

ductus arteriosus (**duk**-tus ar-*tee*-ree-**o**-sus) A fetal bypass vessel that connects the pulmonary artery to the aorta. (Ch. 21)

duodenum (**doo**-o-dee-num) The initial 30 cm segment of the small intestine. (Ch. 22)

dwarfism The state of being an abnormally undersized person. (Ch. 13)

dynamic exercise Activities such as bicycling or running in which the length of muscles changes and which can be quantified in physical units such as watts. (Ch. 30)

dynamic neurons Neurons that change their activity when there is an increase in the level of force applied. (Ch. 9)

dynein A protein from the microtubules of cilia and flagella; functions as an ATP-splitting enzyme. (Ch. 7)

dyslexia (dis-**lek**-see-ah) A congenital disorder that affects an individual's ability to read, write, or spell. (Ch. 11)

dysmenorrhea (*dis*-men-o-ree-ah) Excessive uterine contractions or menstrual cramps. (Ch. 31)

dyspareunia (*dis*-**par**-ee-oo-nee-ah) The absence of sufficient vaginal lubrication. (Ch. 33)

dysphagia Difficulty in swallowing. (Ch. 22)

E_m The difference in electrical potential across a membrane. (Ch. 7)

ear drum The membrane located between outer and middle ear that vibrates in response to sound waves; also called *tympanic membrane*. (Ch. 8)

ECG, EKG Electrocardiogram (electro − + G, *kardia*, heart + *gamma*, a drawing). See electrocardiogram. (Ch. 18)

ectopic pregnancy (ek-**top**-ik) The implantation and development of a fetus at a site other than the uterus. (Ch. 32)

edema (e-**dee**-mah) An abnormal accumulation of fluid in the interstitial fluid spaces. (Ch. 23)

EEG *See* electroencephalogram.

effective arterial blood volume The degree of filling of the arterial system. (Ch. 24)

efferent (**ef**-er-ent) Carrying away from. (Ch. 9)

efferent arteriole The arteriole that carries blood away from the glomerulus. (Ch. 23)

efferent neuron Nerve cell that carries signals *away* from the CNS. (Ch. 7)

efficiency The amount of work done for a certain amount of energy. (Ch. 16)

ejaculation (e-**jak**-yoo-*lay*-shun) The emission and expulsion of ejaculatory fluid. (Ch. 33)

elastic recoil The property of stretched tissue that allows it to return to a pre-stretched state. (Ch. 20)

elasticity The ability of stretched material to return to its unstretched position. (Ch. 20)

electrical potential (E) The difference in net charge between the inside and the outside of a cell. (Ch. 7)

electrocardiogram (ECG or EKG) (e-*lek*-troe-**kar**-dee-o-gram) A record of the electrical activity of the heart. (Ch. 18)

electrochemical gradient The combination of chemical and electrical gradients of a solution that determines the flux of positive and negative ions. (Ch. 4)

electroencephalogram (EEG) (e-lek-troe-en-**sef**-ah-loe-gram) A record of the frequency and amplitude of the brain's electrical activity. (Ch. 11)

electrolyte (e-**lek**-troe-lite) An ionized substance in solution; capable of conducting electricity. (Ch. 23)

electromagnetic radiation A continuum of electromagnetic energy of varying wavelengths, ranging from radio waves to X-rays. (Ch. 8)

electron Negatively charged particle of an atom. (Ch. 2)

electron carrier A molecule that donates electrons to proteins. (Ch. 6)

electron transport chain (respiratory chain) The transfer of electrons from reduced coenzymes to O_2, a process that is coupled with phosphorylation of ADP to ATP. (Ch. 6)

element A substance that cannot be broken down into simpler substances by chemical reactions. (Ch. 2)

embolus (**em**-boe-lus) A foreign object floating in the blood; for example, a clot (thromboembolus) or an air bubble (airembolus). (Ch. 17)

embryo (**em**-bree-o) A blastocyst after it has fully implanted in the endometrium of the uterus. (Ch. 32)

embryonic stage Weeks 2 to 8 following fertilization. (Ch. 32)

emetic (**em**-et-ik) Chemical agent that causes vomiting by stimulating chemoreceptors in the stomach and small intestine or brain. (Ch. 22)

emphysema (**em**-fih-**see**-mah) A lung disease that is characterized by destruction of the alveolar wall and consequent impairment of gas exchange. (Ch. 20)

emulsion (ee-**mul**-shun) A preparation of one liquid distributed in globules throughout a second liquid. (Ch. 22)

en passant An overlapping system of synapses in passage. (Ch. 7)

end-diastolic (dye-ah-**stol**-ik) **volume (EDV)** The volume of blood in the ventricle at the end of diastole. (Ch. 18)

end-plate current A small local current that flows across the postsynaptic membrane at the neuromuscular junction. (Ch. 16)

end-plate potential The depolarization of motor end-plate of skeletal muscle fiber in response to acetylcholine; initiates action potential in muscle plasma membrane. (Ch. 16)

end-systolic volume The volume of blood in the ventricle at the end of systole. (Ch. 18)

endergonic reaction A reaction that cannot proceed without the input of free energy. (Ch. 6)

endocardium (en-doe-**kar**-dee-um) A thin layer of endothelial cells that lines the inner surface of the myocardium, which forms the inside walls of the atria and ventricles. (Ch. 18)

endocrine (en-doe-krin) **gland** Any of the several ductless-type glands that secrete hormones into the blood stream, thereby influencing the activity of cells located away from the gland. (Ch. 12)

endocrine pancreas Small portion of the pancreas that secretes insulin and glucagon and other hormones (also called *islets of Langerhans*). (Ch. 15)

endocrine system All of the body's hormone-secreting glands. (Ch. 12)

endocytosis (*en*-doe-sye-**toe**-sis) The process that allows the entrance into a cell of substances that are otherwise unable to penetrate the cell membrane; includes pinocytosis and phagocytosis. (Ch. 3) **Receptor-mediated endocytosis** involves larger peptide hormones that are internalized by their receptor tissue cells. (Ch. 12)

endogenous (en-**doj**-e-nus) Produced within the body or due to internal causes. (Ch. 23)

endogenous creatinine clearance The renal clearance of endogenous (i.e., not administered) creatinine; it is used clinically to estimate glomerular filtration rate. (Ch. 23)

endogenous protein Digestive enzymes, epithelial cells, mucoproteins, and plasma proteins that leak into the digestive tract and are absorbed. (Ch. 23)

endometriosis (*en*-doe-**mee**-tree-oh-sis) A disease characterized by the presence of endometrial-like tissue in locations other than the uterus, such as the fallopian tubes or ovaries. (Ch. 33)

endometrium (*en*-doe-**mee**-tree-um) The lining of the uterus. (Ch. 30)

endomysium (*en*-doe-**meez**-ee-um) The connective tissue sheath surrounding individual muscle fibers. (Ch. 16)

endoplastic reticulum (ER) (*en*-doe-**plaz**-mik re-**tik**-you-lum) A network of membranous tubules, flattened sacs, and vesicles located in the cytoplasm that function in intracellular transport, synthesis, storage, packaging, and secretion. (Ch. 3)

endorphin (en-**dor**-fin) Any of the "endogenous-morphine" family of peptides; they function as a neurotransmitter at synapses activated by opiate drugs and as a paracrine and hormone. (Ch. 12)

endothelial (*en*-doe-**thee**-lee-al) **clefts (or pores)** Small openings between endothelial cells that allow for the passage of molecules. (Ch. 19)

endothelium-derived relaxing factor (EDRF) Vasodilator released from the endothelial cells that line the blood vessels; currently thought to be nitric oxide (NO). (Ch. 19)

energy transformation Conversion of energy from one form to another. Plants convert radiant energy into chemical energy stored in carbohydrates. Animals convert this chemical energy into a form (ATP) that can be used for biological work. (Ch. 6)

enkephalin (en-**kef**-ah-lin) A peptide of the nervous system that affects the perception of pain. (Ch. 8)

enterohepatic (*en*-teh-roe-he-**pat**-ik) **circulation of bile salts** Recycling of bile salts between liver and small intestine. (Ch. 22)

enterokinase (*en*-teh-roe-**kin**-aze) An enzyme present in the luminal membrane of the epithelial lining of the small intestine; converts inactive trypsinogen to active trypsin. (Ch. 22)

enzyme (**en**-zime) A natural catalyst that increases the rate of chemical reactions in the body. (Ch. 2)

eosinophil (*ee*-o-**sin**-o-fil) A leukocyte whose primary function is detoxification of foreign proteins and other substances. (Ch. 17)

epicardium (*ep*-i-**kard**-ee-um) The outer surface of the myocardium, which consists primarily of connective tissue and fat. (Ch. 18)

epididymis (*ep*-i-**did**-i-mis) The portion of the male reproductive duct system in which sperm mature between the seminiferous tubules and the vas deferens. (Ch. 31)

epiglottis (*ep*-ih-**glot**-is) Cartilage guarding the superior opening into the

larynx. (Ch. 22)

epimysium (*ep*-ih-**meez**-ee-um) A connective tissue sheath that covers the entire surface of a muscle. (Ch. 16)

epinephrine (*ep*-ih-**nef**-rin) One of the two catecholamines that is produced by the adrenal medulla; stimulates the sympathetic nervous systems, increasing blood pressure, stimulating the cardiac muscle, and causing glycogenolysis. Also known as *adrenaline*, it acts together with norepinephrine to prepare the body for "fight or flight" response. (Ch. 14)

epiphyseal (*ep*-ih-**fizz**-e-al) **plate** The region near the end of a bone where linear growth occurs. (Ch. 13)

epithelial tissue Closely packed cells arranged in flat sheets ranging from one to several layers in thickness that cover exposed body surfaces and line body cavities, tubes, and organs. (Ch. 1)

equilibrium A state at which the rate of formation of products of a chemical reaction is exactly balanced by the rate of reformation of the original reactants. (Ch. 2)

equilibrium potential The voltage of electrical force required to draw ions across a cell membrane at the same rate as they pass because of the concentration gradient. (Ch. 7)

ergocalciferol Vitamin D_2, the form of vitamin D found in plants and yeasts. (Ch. 26)

erythrocytosis (eh-*rith*-roe-sigh-**toe**-sis) An increase in the number of red blood cells in response to low oxygen environments and other states in which oxygenation of blood is reduced. (Ch. 17)

erythropoiesis (eh-*rith*-roe-**poy**-ee-sis) Erythrocyte formation. (Ch. 17)

erythropoietin (eh-*rith*-roe-**poy**-ih-tin) A glycoprotein hormone, produced chiefly by the kidneys, that stimulates the rate of production, maturation, and release of red blood cells from bone marrow. (Ch. 23)

esophagus (eh-**sof**-ah-gus) The muscular tube located mostly in the thorax that conducts material from the pharynx to the stomach. (Ch. 22)

essential amino acids Amino acids that cannot be synthesized and must be ingested as food. (Ch. 6)

essential hypertension Chronically elevated arterial blood pressure due to primary disturbance of the cardiovascular system. (Ch. 19)

estradiol (es-trah-die-ol) The principal sex steroid in females. (Ch. 13)

estrogen (**es**-trow-jen) A hormone secreted by the ovaries and the placenta that stimulates the growth and development of tissues in the reproductive

tract. (Ch. 12, 32)

ether An anesthetic gas. (Ch. 22)

eukaryotic cell A cell that has a well-defined nucleus contained within a nuclear membrane. (Ch. 3)

evaporation The process of converting a liquid to a gas. (Ch. 27)

excision repair of DNA A process under enzyme control by which a section of damaged DNA is selectively removed and replaced. (Ch. 5)

excitatory postsynaptic potentials (EPSP) Potentials that depolarize the postsynaptic membrane and tend to excite nerve cells to discharge action potentials. (Ch. 7)

excretion (eks-**kree**-shun) The elimination of a substance from the body in the urine or feces. (Ch. 23)

exergonic reaction A chemical reaction that releases free energy and is likely to proceed spontaneously. (Ch. 6)

exocrine (ek-so-krin) **gland** A type of gland that secretes its products by way of ducts that open onto lining or covering epithelium. (Ch. 22)

exocrine pancreas A portion of the pancreas responsible for the secretion of fluid and various enzymes involved in the process of digestion. (Ch. 15, 22)

exocytosis (**eks**-oh-sih-**toe**-sis) A process in which a secretory vesicle fuses with the plasma membrane and materials are expelled into the extracellular fluid. (Ch. 3)

exogenous (eks-**oj**-en-us) Foreign to the body. (Ch. 23)

exon (**eks**-on) Segment of the eukaryotic gene that consists of DNA coding for a sequence of nucleotides in a specific molecule of messenger RNA. (Ch. 5)

experimental design Construction of experiments so that specific conditions are created deliberately to produce the observations in question, ideally so that only one explanation exists for the observation. (Ch. 1)

expiration (eks-pih-**ray**-shun) Movement of air out of the lungs. (Ch. 10)

expiratory neurons Neurons in the medulla that stimulate expiratory muscles during exercise. (Ch. 20)

extension The movement of one body part away from another; for example, movement of the forearm away from the arm when extension occurs at the elbow. (Ch. 16)

extensor A muscle that, by contracting, causes extension; for example, the triceps brachius is an extensor of the forearm. (Ch. 9)

external Of two sites, the one farther from the center of the body. (Ch. 1)

external anal sphincter Skeletal muscles at the end of the large intestine that are relaxed during defecation. (Ch. 22)

external intercostal muscles A set of inspiratory muscles that pull the rib cage upward and outward. (Ch. 20)

external sphincter A group of striated muscles that close off the exit from the bladder. (Ch. 10)

extracellular fluid The fluid outside of cells. (Ch. 1, 23)

extrafusal fiber Skeletal muscle fibers that form the bulk of skeletal muscles. Contraction is controlled by alpha motor neurons. (Ch. 9)

extrapolation The extension of a relationship beyond the range of available experimental data. (Ch. 1)

extrinsic (eks-**trin**-sik) Having an external origin, arising from outside the body. (Ch. 1)

extrinsic (peripheral) proteins Membrane proteins that do not penetrate the lipid bilayer. (Ch. 4)

extrinsic clotting pathway The process by which clotting begins after blood comes in contact with injured tissue. (Ch. 17)

F cells of the pancreas Cells that produce pancreatic polypeptide. (Ch. 15)

facilitated (**fah**-sil-ih-**tay**-ted) **diffusion** Downhill movement of substances, in or out of the cell, that is enabled by the presence of a carrier protein. (Ch. 4)

fallopian (fal-**low**-pee-an) **tubes** Muscular, tubelike structures that arise from the uterus and extend into the abdominal cavity to terminate near the ovaries. Also known as the uterine tubes. (Ch. 31)

fasciculi Bundles of individual muscle fibers. (Ch. 16)

fast axonal transport The movement of newly produced membranous organelles down the axon. (Ch. 7)

fast skeletal muscles Muscles that contract rapidly but fatigue quickly because they are poorly vascularized. (Ch. 9)

fastigial nucleus One of three deep cerebellar nuclei involved in the maintenance of balance and posture. (Ch. 9)

fat cells Adipose tissue that stores triglycerides, a major source of energy for the body. (Ch. 11)

fats Lipids that are poorly soluble in water and highly soluble in organic liquids; *see also* triacylglycerol. (Ch. 2)

fatty acid A long chain of covalently linked carbon atoms to which only hydrogen atoms are attached, except for a carboxylic acid (—COOH) group at one end. (Ch. 2)

fatty acid oxidation The breakdown of fatty acids in the mitochondrial matrix. (Ch. 5)

feedback excitation A process by which the activation of beta-2 receptors increases the release of norepinephrine from the presynaptic terminal. (Ch. 7)

feedback inhibition A process by which

the activation of alpha-2 autoreceptors inhibits the release of norepinephrine from the presynaptic terminal. (Ch. 7)

female pseudohermaphroditism Severe mascularization of the female external genitalia. (Ch. 33)

ferritin (**fer**-ih-tin) An iron-protein complex that stores iron in the liver. (Ch. 17)

fertilization The union of sperm and ovum. (Ch. 32)

fetal stage The period between week 8 after fertilization and birth. (Ch. 32)

fetoplacental (fee-toe-plah-*sen*-tahl) **unit** The fetal adrenal gland and placenta, which together stimulate estrogen synthesis. (Ch. 31)

fetus (**fee**-tus) An embryo after second month of development. (Ch. 32)

fever (**fee**-ver) A regulated increase in body temperature beyond the normal range. (Ch. 27)

fibrin (**fye**-brin) An insoluble protein derived from fibrinogen, which forms the foundation of a blood clot. (Ch. 17)

fibrinogen (fye-**brin**-o-jin) A soluble plasma protein manufactured in the liver. (Ch. 17)

fibrinolysis The enzymatic dissolution of a blood clot. (Ch. 17)

"fight or flight" response The physiological changes that prepare the body for the stress of threatening situations. (Ch. 10)

filterable calcium One of two major forms of calcium in extracellular fluid. (Ch. 26)

filtered load The quantity of a substance, per unit of time, that is filtered by the glomeruli and is therefore presented to the tubules; for a freely filterable substance, this is equal to the product of the plasma concentration and the glomerular filtration rate. (Ch. 23)

filtration (fil-**tray**-shun) The loss of capillary water to the extravascular space; *see* filtration pressure. (Ch. 19)

filtration pressure A net pressure leading to loss of capillary water; occurs when the hydrostatic pressure in the capillaries exceeds the oncotic pressure of the blood. (Ch. 19)

first heart sound The closure of the atrioventricular vales that marks the beginning of systole. (Ch. 18)

fixed acid Physiologically speaking, a nonvolatile acid (i.e., an acid other than carbonic acid). (Ch. 25)

flatus (**flay**-tus) Gas present in the large intestine. (Ch. 22)

flavin adenine dinucleotide (FAD) A coenzyme derived from riboflavin. (Ch. 5)

flexion (**flek**-shun) A muscle action that causes a joint to close and a limb to assume a more withdrawn position. (Ch. 16)

flexor (**flek**-sor) A muscle that by contracting causes a joint to close up. (Ch. 9)

fluid mosaic model A generalized model of the structure of cell plasma membranes. (Ch. 4)

follicle stimulating hormone (FSH) A glycoprotein hormone, created by the anterior pituitary, that stimulates growth and development of the gonads. (Ch. 12)

foramen ovale (foe-**ray**-men o-**va**-lee) A passage that allows blood to flow between the right and left atria in the fetal heart (Ch. 21)

force-velocity curve A plot of the velocity of muscle shortening over a range of forces ranging from zero to full isometric. (Ch. 16)

forced vital capacity The maximum amount of air that can be forcefully and rapidly exhaled after a deep breath. (Ch. 20)

fovea (**foe**-vee-ah) The most sensitive part of the retina. (Ch. 8)

Frank-Starling law of the heart The ability of the heart to adjust its output (stroke volume) in response to the input (venous return). (Ch. 18)

free energy The energy available to be put to work. (Ch. 2)

free radicals Highly reactive molecules formed as byproducts from molecular oxygen. (Ch. 29)

frequency code of contractile force The control of muscle tension as determined by the frequency of action potentials generated by upper motor neurons. (Ch. 9)

frequency code of stimulus intensity The number of action potentials generated per unit of time as a function of intensity of a stimulus. (Ch. 8)

frontal plane The plane that divides the body into front and back portions (also called the coronal plane). (Ch. 1)

functional residual capacity The volume of air remaining in the lungs after a normal expiration; the sum of expiratory reserve volume and residual volume. (Ch. 20)

gallbladder A small sac that stores bile between meals; propels bile to the small intestine where it participates in the emulsification and absorption of fat. (Ch. 22)

gallstones Small crystals of cholesterol that form in the gallbladder, this formation occurs when the capacity of bile salts and lecithin to dissolve cholesterol is exceeded. (Ch. 22)

gamma amino butyric acid (GAB) A neurotransmitter found throughout the CNS; a potent inhibitory neurotransmitter synthesized from glutamate by glutamic acid decarboxylase. (Ch. 7)

gamma motor neuron Motor neurons that synapse with intrafusal muscle fibers. (Ch. 9)

ganglion cells (gang-lee-on)   Nerve cells in the retina that generate action potentials even in the absence of input from the bipolar cells. (Ch. 8)

gap junction   The intercellular junction that allows ions and small molecules to flow between the cytoplasm of adjacent cells; the two adjacent plasma membranes are joined by small tubes. (Ch. 5)

gas diffusion   The net movement of gas molecules from an area of higher to lower concentration. (Ch. 21)

gas exchange   Process involved with delivering oxygen to and removing carbon dioxide from the body's cells. (Ch. 20)

gas uptake   The process whereby the blood absorbs oxygen from alveolar air. (Ch. 21)

gastric (gas-trik) **chief cells**   Cells, located in the basal region of exocrine secretory glands, that secrete pepsinogen, the precursor to pepsin, the enzyme that initiates the digestion of protein in the digestive tract. (Ch. 22)

gastric inhibitory peptide (GIP)   A hormone or gastrointestinal origin that is important in the control of motility and secretion; released in response to intraluminal stimuli acting directly on the endocrine cells in the luminal lining. (Ch. 22)

gastric mucosal barrier   The ability of the cell membranes of the gastric mucosa to restrict entrance of hydrogen ions to the interior of these cells because of the low permeability of the cell membrane to hydrogen ions. (Ch. 22)

gastric parietal cells   Cells that secrete an isosmotic solution of HCl. (Ch. 22)

gastrin   A hormone released by the gastric antrum in response to either impulses over vagal nerves or the stretching of the gastric wall and the contacting of chemical substances with the gastric antral mucosa; excites secretion of gastric juice. (Ch. 22)

gastrointestinal hormone   Gastrin, cholecystokinin, secretin, and gastric inhibitory peptide. (Ch. 22)

gastrointestinal (gas-tro-in-tes-tin-al) **system**   A collection of organs specialized to handle food through the processes of motility, secretion, digestion, and absorption. (Ch. 22)

G-cells   Cells in the epithelial lining of the antral mucosa, which secrete the hormone gastrin into the portal blood. (Ch. 22)

gene   Each section of chromosomal DNA that carries the information necessary for synthesis of a single polypeptide. (Ch. 3)

generator potential   A change in the membrane potential of a sensory receptor cell that generates action potentials. (Ch. 8)

gene amplification   An increase in the number of copies of a specific gene on a chromosome. (Ch. 5)

genetic code   A reference to the way in which information about protein synthesis is stored in DNA, translated into RNA, and then used to make proteins. (Ch. 5)

genetic expression   The phenomenon of the fact that only certain genes are tapped for information about how to make specific proteins they describe. (Ch. 5)

genome   The sum of all the chromosomal genes of a cell. (Ch. 3)

germinal stage   In prenatal development, the first two weeks of development following fertilization of the ovum. (Ch. 32)

gestation   The period of development of the young from the time of fertilization of the ovum until birth. (Ch. 11)

gestational diabetes   A transient form of diabetes that occurs in about 3% of pregnant women. (Ch. 15)

G-force   The human body's sensation of weight due to gravitational attraction. (Ch. 29)

gigantism   *See* acromegaly.

glia (glee-al)   Nonneuronal cellular elements of nervous tissue; provide physical and functional support to adjacent neurons. (Ch. 7)

glial scar   Formed by reactive astrocytes in conjunction with fibroblasts after a brain injury; the scar blocks the regeneration of new axons. (Ch. 7)

globulin (glob-you-lin)   A type of plasma; classified into alpha, beta, and gamma groups. (Ch. 17)

glomerular (glo-mer-yoo-lar) **filtration**   The process in the kidney glomeruli in which an ultrafiltration of plasma is pushed out of the capillaries into the urinary space of Bowman's capsule. (Ch. 23)

glomerular filtration coefficient   A coefficient which depends on the surface area and fluid permeability of the glomerular filtration barrier; it relates glomerular filtration rate to the net filtration pressure. (Ch. 23)

glomerular filtration rate (GFR)   The rate at which the plasma is filtered by the kidney glomeruli; usually @ 125 ml/min. (Ch. 23)

glomerulotubular balance   The proportionate change in sodium reabsorption by the proximal convoluted tubule and Henle's loop when glomerular filtration rate is changed. (Ch. 23)

glomerulus (glo-mer-you-lus)   In the kidney, the tuft of capillaries, surrounded by Bowman's capsule, where the blood is filtered. (Ch. 23)

glottis (glot-iss)   The true vocal cords and the opening between them. (Ch. 22)

glucagon (gloo-kuh-gon)   Hormone released by the islets of Langerhans of the pancreas; elevates the level of blood sugar. (Ch. 15)

glucagonlike peptide   Peptides that are synthesized in the small intestine and hypothalamus. (Ch. 15)

glucocorticoid (*gloo*-koe-**kor**-tih-koyd)   A steroid hormone produced by the adrenal cortex and having major effects on glucose metabolism. (Ch. 10)

gluconeogenesis (*gloo*-koe-*nee*-oh-**jen**-ih-sis)   The formation of glucose by the liver and kidney from noncarbohydrate precursors. (Ch. 6)

glucose threshold   The plasma level at which glucose first appears in the urine. (Ch. 23)

glucosidase   An enzyme that splits a glucoside. α-Glucosidase (maltase) occurs in intestinal juice and β-glucosidase (cellobiase) in the kidney, liver, and intestinal mucosa. (Ch. 2)

glucosuria (*gloo*-koe-**soo**-ree-uh)   A condition in which glucose is excreted in the urine. (Ch. 15)

glutamate (**gloo**-tah-mate)   One of the most potent excitatory neurotransmitters in the nervous system. (Ch. 7)

glycerol-phosphate   A three-carbon molecule that is combined with free fatty acids in triaclyglycerol synthesis. (Ch. 15)

glycogen (**gly**-ko-jen)   A polysaccharide carbohydrate resembling starch that occurs in many tissues, particularly the liver and skeletal muscles; a polymer of glucose. (Ch. 6)

glycogenolysis (*gly*-koe-jeh-**nol**-ih-sis)   The process by which glycogen is broken down into units of glucose-1-phosphate, which enter gluconeogenesis after conversion of glucose-6-phosphate. (Ch. 6)

glycolipid (*gly*-koe-**lip**-id)   Lipids, occurring in cell membranes, that are derived from sphinogosine and contain one or more monosaccharide units. (Ch. 2)

glycolysis (gly-**kol**-ih-sis)   A process by which glucose obtained from food is converted to pyruvic acid. (Ch. 6)

glycophorin   An intrinsic glycoprotein in the red blood cell membrane. (Ch. 4)

glycoprotein (*gly*-koe-**pro**-teen)   A member of a class of conjugated proteins composed of protein with an attached carbohydrate group. (Ch. 2)

glycoprotein hormones   Hormones that contain peptide and carbohydrate parts. Examples: thyroid stimulating hormone (TSH), follicle stimulating hormone (FSH). (Ch. 12)

goiter   Any abnormal enlargement of the thyroid gland. (Ch. 13)

goitrogen An agent that inhibits iodide uptake or organification reactions, resulting in a blockage of thyroid hormone formation and the formation of a thyroid goiter. (Ch. 13)

Goldman equation The relationship between the intra- and extracellular concentrations of Na^+, K^+, and Cl^- ions, as well as their membrane permeabilities, P_{Na}, P_K, and P_{Cl}, and membrane potential. (Ch. 7)

Golgi (*goal*-jee) **complex** Part of the cell responsible for modifying macromolecules synthesized in the ER; processes and packages the macromolecules inside vesicles to be transported to the cell surface or cytosol. (Ch. 3)

Golgi tendon organ A type of muscle receptor located in series with the extrafusal muscle fiber and that transmits information about the force or tension produced by the contraction of a muscle. (Ch. 9)

gonadal sex The presence of testes or ovaries that determines the gender of the embryo. (Ch. 33)

gonadotropin (*gon*-ad-oh-**trow**-pin) Luteinizing hormone and follicle-stimulating hormone; hormones whose target tissue is the gonads. (Ch. 13)

gonadotropin-releasing hormone (GnRH) A hormone secreted by the hypothalamus that stimulates the release of luteinizing hormone and follicle-stimulating hormone. (Ch. 31)

gout A condition in which plasma uric acid is elevated and uric acid crystals precipitate in the joints, causing inflammation and painful swelling of the affected joints (often in the big toe). (Ch. 23)

G-proteins A group of integral membrane proteins that bind and require guanosine triphosphate to function. (Ch. 12)

gradient (**gray**-dee-ent) The continuous decrease or increase of a variable over a distance. (Ch. 2)

granulocyte (**gran**-yoo-low-site) A leukocyte that contains large granules and has a complex lobulated nucleus; also called polymorphonuclear granulocyte. (Ch. 17)

graph A diagram that expresses a relationship between two or more quantities. (Ch. 1)

graphic interpretation Learning and understanding the relationship between the variable quantities depicted in a graph. (Ch. 1)

Graves' disease Hyperthyroidism that occurs due to abnormality in the immune system that leads to the production of antibodies that recognize and bind to sites on the surface of thyroid follicle cells. (Ch. 13)

ground substance The noncollagenic portion of organic matter in bone; consists of a mixture of various proteoglycans. (Ch. 26)

growth cone A specialized enlargement at the top of the growing neuron. (Ch. 7)

growth hormone A large peptide hormone that stimulates body growth; also called somatotropin. (Ch. 12)

gustatory receptors Taste receptors; a type of epithelial cell clustered with other cells to compose a taste bud. (Ch. 8)

gustatory system The peripheral and central parts of the nervous system involved in the processing of taste information. (Ch. 8)

habituation (hah-bit-you-**a**-shun) A type of behavior in which the individual learns to decrease a response to a repeated stimulus. (Ch. 11)

hair receptor A mechanoreceptor that is a hair follicle on the surface of the skin and that has a nerve fiber wrapped around its base. (Ch. 8)

Haldane effect Describes the amount of carbon dioxide the blood can load, which is greater the more the hemoglobin is deoxygenated. (Ch. 21)

hand manipulation neurons Neurons in the posterior parietal cortex that increase their firing rate when the hand explores an object. (Ch. 9)

hand-eye coordination neurons Neurons in the posterior parietal cortex that are most active when the eye fixates on an object that is being touched. (Ch. 9)

haptoglobin A protein that binds with hemoglobin; prevents hemoglobin from being excreted in the urine. (Ch. 17)

heart block A situation in which the ventricles are not stimulated by the atrial impulse; due to failure of the AV node or AV bundle to conduct the impulse. (Ch. 18)

heart failure The inability of the heart to maintain adequate cardiac output (Ch. 19)

heart sounds A series of distinct sounds that correlate with specific events of the cardiac cycle; four heart sounds are distinguishable in most healthy individuals. (Ch. 18)

heartburn An unpleasant sensation due to irritation of the esophagus caused by reflux of gastric acid. (Ch. 22)

heat exhaustion An illness caused by circulatory insufficiency and hypotension; takes the form of fainting due to decreased blood pressure. (Ch. 27)

heat stroke A form of unregulated hyperthermia or heat illness associated with a breakdown of the body's temperature-regulating system. (Ch. 27)

hematocrit (hee-**mat**-o-krit) The volume occupied by packed red blood cells in a centrifuged 100-ml sample of whole blood. (Ch. 17)

hematopoeisis (*hem*-ah-toe-poy-**ee**-sis) The formation of blood cells in the red bone marrow. (Ch. 17)

hemocytoblast (*hee*-moh-**sigh**-tow-blast) A colony-forming unit or stem cell of the bone marrow that gives rise to several types of committed stem cells. (Ch. 17)

hemoglobin (*hee*-moh-**glo**-bin) A pigment molecule that gives the red blood cells its color; can take up oxygen or release it. (Ch. 17)

hemolysis (he-**mol**-ih-sis) The destruction of erythrocytes and the resultant escape of hemoglobin. (Ch. 17)

hemolytic disease of the newborn Hypoprothrombinemia causing possible life-threatening hemorrhage in newborns. (Ch. 17)

hemophilia (*hee*-moh-**fil**-ee-ah) A disease in which one of the clotting factors is absent due to a genetic mutation. (Ch. 17)

hemorrhage (**hem**-or-ij) Loss of blood. (Ch. 19)

hemostasis (*hee*-moh-**stay**-sis) Mechanisms that minimize or prevent the loss of blood when a blood vessel is opened. (Ch. 17)

Henderson-Hasselbalch equation The logarithmic form of the acid dissociation equation: $pH = pK_a + \log([A^-]/[HA])$. (Ch. 25)

heparin (**hep**-ah-rin) A mucopolysaccharide acid, abundant in the lungs and liver; inhibits the clotting process. (Ch. 17)

hierarchial motor organization Organization of the components of the somatic motor system into serial levels of complexity beginning with the simplest. (Ch. 9)

hierarchial organization The range of complexity from lowest level to highest; also called serial organization. (Ch. 9)

high altitude cerebral edema Cerebral edema that occurs at altitudes of 10,000 to 12,000 feet; characterized by severe headache, hallucinations, ataxia, weakness, impaired mental ability, stupor, and death. (Ch. 29)

high altitude pulmonary edema Edema that occurs at altitudes of 9000 to 10,000 feet; characterized by dyspnea, cough, weakness, headache, stupor, and rarely death. (Ch. 21)

high energy phosphate compound A compound that has a strong tendency to transfer its phosphate group to an acceptor molecule; the transfer is accompanied by the release of a large amount of free energy. (Ch. 6)

histamine A vasoactive substance released from mast cells during an inflammatory response that stimulates vasodilation. (Ch. 19)

histocompatibility-Y antigen (H-Y antigen) An antigenic protein that induces an immunologic rejection in females; thought to be responsible for inducing testicular differentiation in the primitive gonad. (Ch. 33)

histone A specialized protein that binds to DNA and organizes its structure in the nucleus. (Ch. 3)

homeostasis (*ho*-me-o-**sta**-sis) The stable state of the internal environment, which is maintained by regulatory physiologic processes, despite changes that may occur in the external environment. (Ch. 1)

homeothermy (*ho*-mee-oh-**thur**-me) The maintenance of a constant body temperature despite changes in environmental temperature. (Ch. 20)

"homo aquaticus" Jacques Cousteau's description of an underwater human capable of living indefinitely beneath the surface of the sea due to liquid breathing techniques or artificial gills. (Ch. 29)

hormone (**hoar**-moan) A chemical messenger that helps regulate the activity of other tissues and organs; secreted by endocrine or ductless glands. (Ch. 3)

hormone receptor A hormone that allows the signaling molecule to recognize and bind to the target cell. (Ch. 12)

hormone sensitive lipase An enzyme that catalyzes the breakdown of triglycerides into their component fatty acids and glycerol. (Ch. 15)

human calorimeter A large insulated chamber used for measuring metabolic rate. (Ch. 27)

human chorionic gonadotropin (hCG) (*ko*-ree-**on**-ik *gon*-ah-do-**trow**-pin) A glycoprotein that is secreted by cells of the trophoblast, that signals the corpus luteum that pregnancy has begun. (Ch. 12)

human chorionic somatomammotropin (hCS) A peptide hormone secreted by the placenta during pregnancy. (Ch. 13)

humoral stimuli Chemical or fluid-related stimuli. (Ch. 30)

hydration The addition of water molecules to an ion. (Ch. 7)

hydrochloric acid (HCl) A highly dissociated acid (dissociates to H and chloride ions) produced by parietal cells in the gastric gland. HCl provides an optimal pH for the activity of the pepsin enzyme and is, therefore, an aid in digestion. HCl kills bacteria that enter the digestive system through the mouth. It contributes to the breakdown of muscle fiber and connective tissue and is a major factor in the formation of ulcers in the upper digestive tract. (Ch. 22)

hydrogen bond Similar to van der Waals bond, but slightly stronger; forms when the negatively charged electron cloud formed by an asymmetrical orbit of electrons in one atom is attracted to the positively charged nucleus of a neighboring atom. (Ch. 2)

hydrogen ion A hydrogen atom without its orbiting electron; a proton. (Ch. 2)

hydrogenation reaction The reduction process in which a hydrogen atom is gained. (Ch. 6)

hydrophilic (*high*-droe-**fil**-ik) **compound** (literally, "water loving") A compound that readily absorbs water or readily dissolves in water. (Ch. 2)

hydrophobic (*high*-droe-**foe**-bik) **compound** (literally, "water fearing") A compound that is insoluble in water; repels water molecules. (Ch. 2)

hydrostatic (*high*-droe-**stat**-ik) **pressure** Pressure created by a fluid that is either moving or stationary. (Ch. 19)

hydroxyapatite Calcium phosphate deposits usually found in crystalline form in the bone. (Ch. 26)

α-hydroxylase A specific enzyme in the proximal tubule cells of the kidneys that catalyzes the conversation of 25-hydroxy-vitamin D_3 to 1,25-dihydroxy-vitamin D_3. (Ch. 26)

5-hydroxytryptamine (5-HT) *See* serotonin.

hyperalgesic (*high*-pur-al-**jee**-zik) **zones** Painful areas on the surface of the body that are associated with inflammations of the internal organs. (Ch. 10)

hyperbilirubinemia An elevation of blood bilirubin above normally low levels. (Ch. 17)

hypercalcemia (*high*-pur-kal-**see**-mee-ah) An excessive amount of calcium in the blood. (Ch. 26)

hyperchromic A morphologic description of red cells; indicates an erythrocyte that is deeper in color than normal. (Ch. 17)

hyperglycemia (*high*-pur-gly-**see**-me-ah) An above-normal increase in the blood glucose concentration due to the absence of either insulin secretion or insulin action. (Ch. 15)

hyperkalemia (*high*-pur-kah-**lee**-me-ah) A plasma potassium concentration greater than 5.5 mEq/liter. (Ch. 24)

hypernatremia (*high*-pur-nah-**tree**-me-ah) A plasma sodium ion concentration greater than 145 mEq/liter. (Ch. 24)

hyperpnea Increased ventilation. (Ch. 30)

hyperpolarization (high-per-*pohl*-er-ih-zay-shun) When a membrane becomes more negative on its inner surface than it was in the resting state. (Ch. 7)

hypertension (**high**-pur-*ten*-shun) A chronically elevated blood pressure

above 145 systolic or 95 diastolic. (Ch. 24)

hyperthermia (*high*-pur-**ther**-mee-ah) An elevation above normal body temperature. (Ch. 27)

hyperthyroidism The excess secretion of thyroid hormones that results in markedly increased heart rate, considerable weight loss, and high responsiveness to external stimuli. (Ch. 13)

hypertonic (*high*-per-**ton**-ik) Having an osmotic pressure or solute concentration greater than that of some other solution, which is taken as standard. (Ch. 17)

hypertrophy (high-**per**-trow-fee) An increase in the mass of a muscle without an increase in the actual number of cells. (Ch. 16)

hyperventilation (*high*-pur-ven-tih-**lay**-shun) Increased alveolar ventilation. (Ch. 20)

hypervolemia (high-pur-vo-**lee**-mee-ah) An abnormal increase in the volume of circulating blood in the body. (Ch. 19)

hypocalcemia (*high*-poe-kal-**see**-me-ah) Abnormally low calcium concentration in blood. (Ch. 26)

hypochromic A morphologic description of red cells; indicates an erythrocyte that is paler in color than normal. (Ch. 17)

hypoglycemia (*high*-poe-gly-**see**-me-ah) A decrease in the blood glucose concentration to below normal. (Ch. 15)

hypokalemia (*high*-poe-kahl-**ee**-me-ah) A plasma potassium concentration level less than 3.5 mEq/liter. (Ch. 24)

hyponatremia (*high*-poe-nay-**tree**-me-ah) A plasma sodium concentration less than 137 mEq/liter. (Ch. 24)

hypoosmotic Having an osmolality less than a reference solution (usually plasma). (Ch. 24)

hypophysis (high-**pof**-ih-sis) The pituitary gland; located at the base of the brain. (Ch. 13)

hypotension Low blood pressure; less than 100 systolic, 60 diastolic. (Ch. 19)

hypothalamic-pituitary portal system A complex of blood vessels that consists of the primary capillary network, the secondary capillary network, and the vessels that connect them; carries factors that regulate the secretion of anterior pituitary hormones from the hypothalamus to the anterior pituitary. (Ch. 13)

hypothalamic-pituitary-adrenal axis The relationship between CRH, ACTH, and the adrenal cortex. (Ch. 13)

hypothalamus (*high*-poe-**thal**-ah-mus) A portion of the brain that regulates the pituitary gland, the autonomic nervous system, emotional response, body temperature, water balance, and appetite. (Ch. 10)

hypothermia (*high*-poe-**ther**-mee-ah) The lowering of body temperature below normal. (Ch. 27)

hypothesis (high-**poth**-e-sis) A statement that tentatively explains an observation. (Ch. 2)

hypothyroidism A condition that exists when there is a deficiency in thyroid hormone production; affected individuals cannot tolerate cold temperatures and exhibit lethargy, sometimes accompanied by myxedema. (Ch. 13)

hypotonic (high-poe-**ton**-ik) Refers to a solution whose osmotic pressure or solute content is less than that of some standard of comparison. (Ch. 17)

hypoventilation (*high*-poe-ven-tih-**lay**-shun) Decreased alveolar ventilation below normal. (Ch. 20)

hypovolemia (**high**-poe-vo-**lee**-mee-ah) An abnormally decreased volume of circulating blood in the body. (Ch. 19)

hypoxemia (high-pock-**see**-me-ah) The condition of low blood oxygen. (Ch. 21)

hypoxia (high-**pock**-see-ah) Oxygen deficiency. (Ch. 21)

hypoxic pulmonary vasoconstriction A mechanism in the lung that balances the flow or perfusion of blood with the availability of regional alveolar oxygen. (Ch. 21)

H-zone A somewhat lighter region in the sarcomere; found in the center region of the A-band where the thin filaments of the I-band have not penetrated. (Ch. 16)

I-band A region of a muscle where only thin filaments are present. (Ch. 16)

IGF-I, IGF-II Insulin-like growth factor one and two; peptides produced by the liver that exhibit many of the growth-promoting activities of insulin; part of a group of growth-promoting factors called somatomedins. (Ch. 13)

ileocecal (ill-ee-oh-**see**-kal) **sphincter** The last 2–3 cm of the musculature of the ileum; controls passage of material to the large intestine. (Ch. 22)

ileum (**il**-lee-um) The distal portion of the small intestine; extends from the jejunum to the cecum. (Ch. 22)

immune (ih-**mewn**) **response** Any reaction that is designed to defend the body against pathogens or any other foreign substances. (Ch. 28)

immunity (ih-**mew**-nih-tee) Either non-specific (natural) or specific (adaptive) physical and chemical barriers or defense mechanisms geared toward the identification and destruction of specific pathogens. (Ch. 28)

immunoglobulin (Ig) (*im*-yoo-no-**glob**-yoo-lin) any glycoprotein secreted by B-lymphocytes in response to specific antigens. (Ch. 28)

immunosuppression The inhibition of the body's ability to mount an effective immune response; may be due to certain drugs or exposure to ionizing radiation. (Ch. 28)

implantation (*im*-plan-**tay**-shun) The penetration of the epithelial lining of the uterus by a blastocyst to embed in the endometrium and form an embryo. (Ch. 32)

impotence (*im*-poe-tence) The inability to achieve penile erection sufficient for vaginal penetration. (Ch. 33)

in utero (**yoo**-ter-oh) Within the uterus. (Ch. 32)

in vitro (**vee**-tro) In an artificial environment, such as of cells grown in culture dishes. (Ch. 2)

in vivo (**vee**-vo) Within the body. (Ch. 2)

inactivating state One of three states of operation of the sodium channel of membranes; the time-dependent gate is closed. (Ch. 7)

incontinence (in-**kon**-tih-nence) The inability to retain urine or feces due to loss of sphincter control. (Ch. 22)

incus The middle of the three ossicles of the ear; serves to conduct vibrations from the tympanic membrane to the inner ear; also known as the anvil. (Ch. 8)

independent variable A datum plotted on the horizontal axis of a line graph. (Ch. 1)

indirect calorimetry An estimate of metabolic rate based on oxygen consumption. (Ch. 27)

infarct (**in**-farkt) An area of tissue death resulting from circulatory obstruction, usually a thrombus or embolus. (Ch. 19)

inferior (in-**fe**-ree-or) Nearer to the lower end of the body. (Ch. 1)

inferior vena cava The large vein that returns blood from the lower part of the body to the heart. (Ch. 18)

infertility A diminished or absent ability to reproduce. (Ch. 33)

inflammation (*in*-flah-**may**-shun) A localized body response to tissue damage or injury; the clinical signs are redness, heat, edema, and pain. (Ch. 28)

inhibin (in-**hib**-in) A protein secreted by Sertoli cells of the seminiferous tubules that regulates the secretion of follicle stimulating hormone (FSH). (Ch. 31)

inhibitory postsynaptic potential (IPSP) A potential that hyperpolarizes the membrane potential and inhibits the nerve cell from generating an action potential. (Ch. 7)

initial segment The first portion of an axon plus part of the cell body where the axon joins. (Ch. 7)

inner ear The part of the ear containing receptors for hearing and equilibrium. (Ch. 8)

inner mitochondrial membrane A substrate for proteins in the electron transport chain of oxidative phosphorylation. (Ch. 6)

inositol triphosphate (IP_3) One of a pair (with diacylglycerol) of intracellular second messenger molecules. (Ch. 5)

inotropic agent A compound that affects the contractility of heart muscle. (Ch. 16)

insensible perspiration Water loss by evaporation from the lungs and skin. (Ch. 27)

insertion (in-**sir**-shun) The distal, or less stationary, point of attachment of a skeletal muscle. (Ch. 16)

insomnia (in-**som**-nee-uh) The inability to sleep. (Ch. 11)

inspiration (*in*-spih-**ray**-shun) The act of drawing air into the lungs. (Ch. 10)

inspiratory neurons Neurons within the medulla oblongata responsible for rhythmic breathing when the body is at rest. (Ch. 20)

insulin (**in**-suh-lin) A double-chain protein hormone produced by the beta cells of the islets of Langerhans in the pancreas that allows cells to take in glucose in response to an increase in blood sugar levels. (Ch. 15)

insulin/glucagon ration (I/G) The ratio of the blood concentrations of insulin and glucagon, pancreatic hormones that have opposing actions on cell metabolism. The ratio is used to assess metabolic status in the body. A high ratio suggests an anabolic state; a low ratio suggests a catabolic state. (Ch. 15)

insulin resistance A lack of response by tissues to insulin in the blood. (Ch. 15)

insulin-dependent diabetes mellitus (IDDM) A disease caused by inadequate insulin secretion; type 1 diabetes. (Ch. 15)

insulin-like growth factor Somatomedins produced by the liver; IGF-I stimulates skeletal growth; IGF-II stimulates tissue growth and repair. (Ch. 13)

integument (in-**teg**-yoo-ment) Any covering, especially the skin. (Ch. 28)

intercalated (in-**ter**-kah-lay-ted) **disc** A specialized structure that provides mechanical and electrical connection between cells of cardiac muscle. (Ch. 16)

intercalated duct System of small ducts that delivers the exocrine secretion of the pancreas to larger ducts that conduct pancreatic juice into the small intestine; also contributes a water-bicarbonate secretion to pancreatic juice. (Ch. 22)

intercellular Between cells. (Ch. 23)

intercostal muscles Skeletal muscles between the ribs. Internal intercostals depress the rib cage and contract during forced expiration. External intercostals elevate the rib cage and contract during inspiration. (Ch. 20)

interferon (*in*-ter-**feer**-on) A group of proteins secreted by cells that have been infected by a virus, inducing noninfected cells to form antiviral proteins that inhibit viral multiplication. (Ch. 28)

interleukins (*in*-ter-**loo**-kins) Hormone-like messengers that stimulate T lymphocytes; **IL-1** stimulates activated T_4 lymphocytes to differentiate and reproduce, forming a clone of inducer T cells and helper T cells; **IL-2** induces T_8 lymphocytes to differentiate and reproduce, forming cytotoxic T cells and suppressor T cells. (Ch. 28)

internal Nearer the center of an organ, cavity, or part of the body. (Ch. 1)

internal anal sphincter The smooth muscle that is relaxed during defecation. (Ch. 22)

internal environment The extracellular fluid that surrounds the cell, outside the cell but inside the organism. (Ch. 1)

internal intercostal muscle The expiratory muscle that contracts to pull the rib cage down. (Ch. 20)

internal sphincter A group of smooth muscles that close off the exit from the bladder. (Ch. 10)

interneuron Any neuron in a chain of neurons situated between the sensory and the final motor neuron. (Ch. 7)

interoceptor (*in*-ter-o-**sep**-ter) A sensory receptor that transmits information from viscera to CNS. (Ch. 8)

interomedial lateral nuclei (ILN) The location in the spinal cord of the preganglionic nerve cells of the sympathetic nervous system. (Ch. 10)

interphase (**in**-ter-faze) The period between the end of one cell division and the beginning of the next. (Ch. 3)

interpolation Prediction of data values that fall between existing data points. (Ch. 1)

interstitial (*in*-ter-**stish**-al) **fluid** The fluid outside blood vessels that directly bathes most body cells. (Ch. 24)

interstitium (*in*-ter-**stish**-ee-um) The fluid-filled space between tissue cells. (Ch. 23)

intracellular (*in*-tra-**sell**-yoo-lar) Within a cell. (Ch. 23)

intracellular fluid The volume of fluid within cells. (Ch. 24)

intracellular pH The pH level within the cell; generally lower than extracellular pH. (Ch. 25)

intrafusal muscle fibers Fibers in the inside of skeletal muscles; arranged parallel to extrafusal fibers. The fibers are part of the neuromuscular spindle and supplied by gamma motor neurons.

(Ch. 9)

intrauterine (*in*-trah-**yoo**-ter-in) **device (IUD)** A contraceptive device inserted into the uterus to prevent implantation of the ovum. (Ch. 33)

intrinsic clotting pathway The initiation of clotting in the absence of tissue damage; can occur inside or outside the body. (Ch. 17)

intrinsic factor A glycoprotein, secreted by the lining of the stomach, that is essential for absorption of vitamin B_{12} in the small intestine. (Ch. 17)

intrinsic (in-**trin**-sik) **protein** A protein that partially or completely penetrates the lipid bilayer of a plasma membrane. (Ch. 4)

intron (**in**-trahn) Noncoding region of a eucaryotic gene that is transcribed into RNA but removed by splicing during production of messenger RNA. (Ch. 5)

inulin A fructose polymer whose renal clearance is used to measure glomerular filtration rate. (Ch. 23)

inverse myotatic reflex A reflex that inhibits skeletal muscle contraction when the Golgi tendon organ detects excessive force that might damage the muscle. (Ch. 9)

inverse relationship The relationship between data such that y-values get smaller as x-values get larger. (Ch. 1)

involuntary muscle The smooth or cardiac muscle whose actions are controlled by the autonomic nervous system in response to some internal reflex. (Ch. 16)

iodide trap The transport process that pumps iodide into cells of the thyroid. (Ch. 13)

ion (**eye**-on) **channel** The pores in a nerve cell membrane that allow passage of certain ions. A **passive ion channel** allows a specific ion to pass through; a **chemically activated ion channel** can be opened or closed by a chemical transmitter; a **voltage-activated ion channel** is opened when it detects a certain voltage. (Ch. 7)

ion product of water The product of $[H^+]$ and $[OH^-]$ in aqueous solutions; it has a value of 1×10^{-14} M^2 at 25°C. (Ch. 25)

ionic bond The equal and opposite electrical charge that holds two atoms together. (Ch. 2)

ionic current The flow of ions. (Ch. 7)

ionization (*eye*-on-eye-**zay**-shun) The dissociation of a substance in solution into ions. (Ch. 26)

ionized calcium A free Ca^{2+} ion, the physiologically active form of calcium. (Ch. 26)

ischemia (is-**kee**-me-ah) Reduced blood flow to an organ due to functional constrictions or obstruction of a blood vessel. (Ch. 17)

islets of Langerhans (**eye**-lits of **Lahng**-er-hanz) Small groups of cells interspersed within the cells of the exocrine pancreas; secrete insulin, glucagon, and somatostatin. (Ch. 15)

isoelectric line A line of an electrocardiograph with no deflection. (Ch. 18)

isohydric principle The idea that all buffer systems are in equilibrium with the same hydrogen ion concentration. (Ch. 25)

isometric contraction A contraction in which a muscle does not shorten but develops force or tension while pulling against immovable attachments. (Ch. 16)

iso-osmotic (*eye*-so-oz-**mat**-ik) Having the same osmotic pressure. (Ch. 4)

isotonic (*eye*-so-**tahn**-ik) Having the same osmolality, equal to that of a reference solution. (Ch. 16)

isotonic (*eye*-so-**tahn**-ik) **contraction** A contraction in which the force in a muscle remains constant while the muscle shortens. (Ch. 16)

isotope (**eye**-so-tope) An atom of an element that has a differing number of neutrons. (Ch. 2)

isovolumetric ventricular contraction Early phase of the cardiac cycle in which both the atrioventricular valves and the semilunar valves are closed and ventricular volume does change. (Ch. 18)

isovolumetric ventricular relaxation Early phase ventricular relaxation when atrioventricular and aortic valves are closed. (Ch. 18)

jaundice (**jawn**-dis) A condition in which the skin is yellowish due to hyperbilirubinemia and deposition of bile pigment in the skin. (Ch. 17)

jejunum (jeh-**joo**-num) The middle portion of the small intestine; extends from the duodenum to the ileum. (Ch. 22)

joint The site of junction or union between two or more bones of the skeleton. (Ch. 16)

junctional complex The cellular structure that prevents the lateral diffusion of proteins in the plasma membrane. (Ch. 4)

junctional folds A series of deep groves of the muscle plasma membrane. (Ch. 4)

juxtaglomerular (*juks*-tah-glo-**mer**-you-lar) **apparatus** A structure in the kidney consisting of macula densa, extraglomerular mesangial cells, and granular cells, located in the region where the distal straight tubule touches afferent and efferent arterioles of its parent glomerulus. (Ch. 23)

juxtamedullary nephron Nephrons with renal corpuscles in the inner cortex; have large glomeruli, long thin limbs, and long loops; 15% of all nephrons are of this type. (Ch. 23)

kainate One of the three subtypes of glutamate receptors. Its activation results in excitatory postsynaptic potentials by opening ion channels that increase in Na^+ and K^+ conductance. (Ch. 7)

kallikrein (kal-li-**kre**-in) A proteolytic enzyme that leads to the formation of kinins. (Ch. 23)

keloid (**key**-loyd) An elevated, enlarging scar due to excessive amounts of collagen in the dermis during connective tissue repair. (Ch. 8)

keto (**key**-toe) **acid** Acid formed by the removal of an amino group from amino acid. (Ch. 15)

ketoacidosis An acid/base imbalance caused by an increase in the concentration of ketones in the blood. (Ch. 15)

ketogenesis The formation of ketones in the liver; stimulated by glucagon. (Ch. 15)

ketone (**key**-tone) **bodies** Four-carbon compounds that are released by the accumulate in the blood during starvation and untreated diabetes mellitus; acetoacetate, acetone, or β-hydroxybutyrate acid. (Ch. 15)

ketosis A condition characterized by an abnormally elevated concentration of ketone bodies in body tissue and fluids; a complication of diabetes mellitus and starvation. (Ch. 15)

kinesthesia (kin-es-**the**-ze-ah) The sense of movement. (Ch. 8)

kinetic energy Energy in use that is producing motion. (Ch. 2)

kinin (**kye**-nin) An endogenous peptide derived from kallikrein that causes vasodilation. (Ch. 23)

knee jerk reflex A stretch reflex activated by tapping the patellar tendon below the knee. (Ch. 9)

Krebs cycle Aerobic cycle of respiratory chemical reactions that occur in the mitochondria and produce cellular ATP (also called *the citric acid cycle*). (Ch. 3)

labor Strong contractions of the uterus that occur with increasing frequency and intensity during the latter part of pregnancy; results in expulsion of the fetus. (Ch. 32)

lactase (**lak**-tase) β-Galactosidase. An enzyme that digests lactose into glucose and galactose. (Ch. 22)

lactation (lak-**tay**-shun) The production and secretion of milk by the breast. (Ch. 32)

lacteal (**lak**-tee-al) Pertaining to milk; any of the intestinal lymphatics that transport chyle. (Ch. 22)

lactic (**lak**-tik) **acid** A three-carbon molecule formed as a byproduct of anaerobic metabolism of the energy pathways of cells. (Ch. 16)

lactic acid oxygen debt The second stage of repayment of the oxygen debt that involves removal of lactic acid from the body. (Ch. 30)

lactose (**lak**-tose) A disaccharide composed of one molecule of glucose and one molecule of galactose. (Ch. 22)

laminar air flow The streamlined flow of air in the lungs that parallels the sides of the airways. (Ch. 20)

larynx (**lar**-inks) The upper part of the airway between the pharynx and the trachea that contains the vocal cords. (Ch. 22)

latent period Period of time (several milliseconds) between stimulus delivery and muscle contraction. (Ch. 16)

lateral (**lat**-er-al) Position farther from the midline or midsagittal plane. (Ch. 1)

lateral reticulospinal tract A descending pathway from the brain stem to the spinal cord; used to control flexor muscles involved in the maintenance of posture. (Ch. 9)

laxative A cathartic such as mineral oil that hydrates or lubricates material in the large intestine, promoting evacuation of the bowel. (Ch. 22)

L-dopa Dopamine precursor that is often administered to patients with Parkinson's disease. (Ch. 9)

lead I An electrocardiograph recording made with electrical connections between the right and left arms; **lead II** with electrical connections between the left leg and the right arm; **lead III** with electrical connections between the left leg and left arm. (Ch. 18)

length-tension curve The relationship between the force capability of a muscle and its length. (Ch. 16)

leukemia (loo-**key**-mee-ah) A malignant disease characterized by proliferation of hematopoietic cells that are immature and therefore functionally impaired. (Ch. 17)

leukocyte (**loo**-ko-site) A white blood cell (WBC). (Ch. 17)

leukocyte inducing factor A specific factor in plasma that can be demonstrated when leukocytes are depleted; stimulates leukopoiesis. (Ch. 17)

leukocytosis (*loo*-ko-sigh-**toe**-sis) An abnormal increase in the total number of circulating leukocytes. (Ch. 17)

leukopenia (*loo*-ko-**pee**-nee-ah) An abnormal reduction in the total number of circulating leukocytes: ≥5000 mm³. (Ch. 17)

leukopoiesis (*loo*-ko-**poe**-ee-sis) The development of white blood cells. (Ch. 17)

ligand (**lihg**-and) Any compound that binds with a high degree of specificity to a receptor. (Ch. 5)

limb leads An electrocardiograph recording made between any two of the standard connections: the right and left arms and the left leg. (Ch. 18)

limbic system Parts of the prefrontal cortex, hippocampus, amygdala, and hypothalamus that collectively function in the expression of emotional behavior. (Ch. 8)

linear relationship A relationship that can be described by a straight line. (Ch. 1)

lipase (**lye**-pase) An enzyme produced by the pancreas that digests triacylglycerols to monoglycerides and free fatty acids. (Ch. 22)

lipids A group of fat and fatlike substances that are water-insoluble and highly soluble in organic liquids. (Ch. 2)

lipogenesis (*lipe*-oh-**jen**-eh-sis) The production of lipids from precursors such as glucose and amino acids. (Ch. 15)

lipolysis The breakdown of triglycerides to fatty acids and glycerol. (Ch. 14)

lipoprotein lipase An enzyme that promotes the uptake of lipoproteins from the blood. (Ch. 15)

lipoproteins A complex of lipids and proteins, synthesized in the endoplasmic reticulum lumen, that has a critical role in the transport and distribution of lipids. (Ch. 3)

local chemical mediator A chemical signal released by a cell that affects only nearby cells and that is rapidly destroyed if not used. (Ch. 5)

long-term memory Memory that is stable, less volatile, and long-lasting. (Ch. 11)

loop of Henle The hairpin portion of the nephron situated between proximal convoluted and distal convoluted tubules. (Ch. 23)

lordosis A receptive posture adopted by female animals during mating. (Ch. 11)

low density lipoprotein A lipoprotein aggregate that transports most of the cholesterol of the cell. (Ch. 17)

lower esophageal sphincter A ring of smooth muscle that acts to restrict entrance of gastric contents into the esophagus; also called the gastroesophageal sphincter. (Ch. 22)

lower-motor-neuron lesion Occurs with spinal cord or brain stem injuries that affect motor neurons directly; characterized by ipsilateral hypoactive reflexes, paralysis, and limp, atrophic muscles. (Ch. 9)

lumen (**loo**-men) The cavity within a tubelike structure such as a blood vessel, lymphatic vessel, or intestine. (Ch. 23)

luminal membrane The cell surface that contacts fluid in the tubule lumen of the kidney. (Ch. 23)

lung volumes Four primary subdivisions of air in the lungs: inspiratory reserve volume, tidal volume, expiratory reserve volume, and residual volume. (Ch. 20)

luteinization The formation of the corpus luteum from the remnants of the ovarian follicle after ovulation. (Ch. 31)

luteinizing (**loot**-eh-*nigh*-zing) **hormone (LH)** Gonadotropin hormone released by the anterior pituitary that initiates ovulation. (Ch. 11)

lymph (limph) A transparent, slightly yellow fluid of alkaline reaction that is found in lymphatic vessels and derived from tissue fluids. (Ch. 19)

lymph node Small organelle located along lymph vessels that contain lymphocytes; filters microorganisms and other foreign material. (Ch. 11)

lymphatic system (lim-**fat**-ik) A system of vessels that directs fluid and plasma protein from interstitial spaces to the blood. (Ch. 19)

lymphocyte (**lim**-foe-site) An agranular leukocyte, the second most common type of white blood cell. (Ch. 17)

lymphokines (**lim**-foe-kines) Chemicals secreted by lymphocytes that regulate T- and B-leukocyte function, attract macrophages, activate complement proteins, and help with specific defense. (Ch. 28)

lysis (**lye**-sis) The destruction of cells by a specific antibody. (Ch. 4)

lysosome (**lye**-so-sowm) A cytoplasmic, membrane-bound organelle that contains digestive enzymes used for intracellular digestion. (Ch. 3)

lysozyme (**lye**-so-zime) An enzyme that digests the cell wall of many pathogenic bacteria. Present in sweat, saliva, tears, and urine. (Ch. 28)

macrocytosis (mak-roe-sye-**toe**-sis) The production of red blood cells that are larger than normal; also called macrocythemia. (Ch. 17)

macromolecule A very large molecule having a polymeric chain structure; all contain carbon and assume huge proportions and a great variety of shapes. (Ch. 2)

macrophage (**mak**-roe-faje) A large phagocytic cell that forms part of the body's defense against microorganisms and harmful chemicals. (Ch. 17)

macula densa (**mak**-you-la **den**-sa) Portion of the straight distal tubular epithelium that helps make up the juxtaglomerular apparatus of the kidney. (Ch. 23)

major histocompatibility complex II (MCHII) A protein present on the plasma membrane of antigen-presenting cells; involved in cell-mediated immune response. (Ch. 28)

malignant hyperthermia An inherited defect in skeletal muscle metabolism. (Ch. 27)

malleus (**mal**-lee-us) The largest of the auditory ossicles, the one attached to the tympanic membrane; also known as the hammer. (Ch. 8)

maltase (**mawl**-tase) α-Glucosidase. A digestive enzyme that breaks down maltose into glucose. *See also* glucosidase. (Ch. 22)

mass movement A propulsive contraction of the colon, moving material into the distal large intestine. (Ch. 22)

mass number The total number of protons and neutrons in the nucleus of an element. (Ch. 2)

mast cell A specialized type of cell in connective tissue; contains histamine and is important in allergic reactions. (Ch. 5)

mastocytosis (*mas*-toe-sign-**toe**-sis) A condition in which greatly elevated numbers of mast cells accumulate in body tissues. (Ch. 5)

matrix (**ma**-triks) The intercellular substance of a tissue or the tissue from which a structure develops. (Ch. 1)

maximal aerobic power A means of determining an individual's ability to sustain high-intensity exercise; the maximal oxygen consumption is an indicator of this state. (Ch. 30)

maximal oxygen consumption (V_{O_2max}) The state at which the maximum oxygen transport and uptake in the tissues are reached and a plateau is attained. (Ch. 30)

mean arterial pressure A calculation of a single value for blood pressure that factors both the pulse and diastolic pressures. (Ch. 19)

mean body temperature An estimate of the fractional sum of core and mean skin body temperatures. (Ch. 27)

mean corpuscular hemoglobin (MCH) An index used to determine the concentration of hemoglobin in red blood cells. (Ch. 17)

mean corpuscular volume (MCV) A quantitative assessment of the volume of red blood cells, reported in mm^3. (Ch. 17)

mean skin temperature The temperature of the body shell, calculated as the mean of the temperatures at several representative parts of the body. (Ch. 27)

mechanoreceptor A sensory receptor that is excited by mechanical pressures or distortions, such as bending, twisting, or compressing. (Ch. 8)

medial (**mee**-dee-al) A term of position; literally means nearer to the midsagittal plane. (Ch. 1)

medial lemniscal pathway A pathway in the brain stem through which sensory fibers of the dorsal column ascend toward the thalamus. (Ch. 8)

medial reticulospinal tract A component of the ventromedial pathway; originates in the reticular formation and is involved in maintaining posture by activation of extensor muscles. (Ch. 9)

median eminence A specialized region in the hypothalamus where arterial blood vessels branch into a network of capillaries. (Ch. 13)

medulla (mah-**dull**-ah) The innermost portion of an organ or structure. (Ch. 10)

megakaryocyte (*meg*-ah-**kare**-ee-oh-site) A type of blood cell from which thrombocytes, or platelets, are derived. (Ch. 17)

meiotic (my-ah-tik) **division** A division of germ cells that reduces the number of chromosomes in the cell by half; random assortment and recombination of maternal and paternal chromosomes occur as well as crossing over. (Ch. 31)

meiotonic (my-oh-**tahn**-ik) **contraction** A muscle contraction in which the force becomes progressively less as the contraction goes on. (Ch. 16)

melatonin (**mel**-ah-toe-nin) A hormone derived from tryptophan; suspected to have a role in puberty onset and control of body rhythms. (Ch. 12)

membrane conductance The ease with which ions can flow through channels in the plasma membrane. (Ch. 7)

membrane potential Voltage difference between inside and outside of cell due to an ionic concentration gradient. (Ch. 7)

memory cell A B-lymphocyte cell that permits rapid mobilization of immune response upon subsequent exposure to an antigen. (Ch. 28)

menopause (**men**-oh-pawz) The cessation of menstruation in the human female; occurs on the average at age 50. (Ch. 31)

menstrual (**men**-stroo-al) **cycle** The approximately 28-day ovarian cycle; consists of the follicular phase, the luteal phase, and the menstrual phase. (Ch. 31)

menstruation (*men*-streh-**way**-shun or *men*-**stray**-shun) Normal flow of menstrual fluid from the uterus that occurs approximately once a month in most nonpregnant females. (Ch. 31)

mesentery (**mez**-en-*ter*-ee) A membrane attaching various organs to the body wall, especially the peritoneum connecting the intestine to the posterior abdominal wall. (Ch. 10)

messenger RNA The transcribed segment of DNA containing the information specifying the amino acid sequence of a polypeptide chain. (Ch. 5)

metabolic acidosis An abnormal process characterized by a gain of acid which leads to a fall in blood pH; due to any cause other than elevated carbon dioxide. (Ch. 22, 24, 25)

metabolic alkalosis An abnormal process characterized by a gain of strong base, which leads to a rise in blood pH; due to any cause other than a decrease in blood carbon dioxide levels. (Ch. 22, 24, 25)

metabolic bone disease A group of disorders in which there is generalized abnormality of the ongoing process of bone formation and bone resorption. (Ch. 26)

metabolic pathway The sequence of enzyme-mediated chemical reactions by which molecules are synthesized and broken down in cells. (Ch. 6)

metabolic rate Total energy production of an individual per unit time. (Ch. 3)

metabolism (meh-**tab**-oh-*lizm*) The sum of all physical and chemical processes by which living organized substance is produced and maintained; also the transformation by which energy is made available for the organism's use. (Ch. 3)

metarteriole Blood vessels that are slightly larger than capillaries. (Ch. 19)

MHC Major histocompatibility complex; a group of proteins on the surface of nucleated cells. Class I MHC molecules are present on the surface of all nucleated cells in the body. Class II MHC molecules are present on the surface of some cells (e.g., CD4 + cells) of the immune system. MHC proteins play important roles in cell recognition and antigen recognition. (Ch. 28)

micelle (my-**sel**) Small aggregates of the bile salts, monoglycerides, and free fatty acids that diffuse in aqueous solution of intestinal contents. (Ch. 22)

microcirculation The circulation of blood through the arterioles, metarterioles, capillaries, and venules. (Ch. 19)

microcytosis (*my*-kroe-si-**toe**-sis) The production of red blood cells that are smaller than normal; also called microcythemia. (Ch. 17)

microfilament A component of cytoskeletal structure; a rodlike cytoplasmic protein believed to have a supportive and cytoskeletal function and/or mediate movement of the cell and the organelles within it. (Ch. 3)

microglia (my-**krog**-lee-ah) Small, nonneural, interstitial cells of mesodermal origin that form part of the supporting structure of the CNS. (Ch. 7)

microtubule (*my*-krow-**too**-bewl) A tubular cytoplasmic filament that provides internal support for cells; allows change in cell shape and organelle movement in cell. (Ch. 7)

microtubule-associated protein (MAPS) Accessory proteins that are thought to be responsible for the distribution of material to dendrites and axons. Dendrites have high-molecular-weight MAPS, while axons have low-molecular-weight MAPS. (Ch. 7)

microvilli (*mi*-krow-**vill**-ee) Small projections that cover epithelial cells of the digestive tract. (Ch. 22)

micturition (*mik*-tu-**rish**-un) Emptying of the bladder. (Ch. 10)

middle ear The portion of the ear between the tympanic membrane and the cochlea; contains the ossicles and internal auditory meatus (eustachian tube). (Ch. 8)

midsagittal plane An imaginary plane that passes directly along the midline of the body, dividing it into right and left halves. Also called median sagittal plane. (Ch. 1)

milliequivalent (mil-ee-**kwiv**-a-lent) For an ion, an amount equal to the product of millimoles and valence (e.g., 2.5 mmoles Ca^{2+} is the same as 5 milliequivalents of Ca^{2+}). (Ch. 24)

mineral A nonorganic homogeneous solid substance, usually a constituent of the earth's crust. (Ch. 2)

mineralocorticoid (*min*-er-al-oh-**kor**-ti-koyd) A steroid hormone produced by the adrenal cortex, involved in regulation of electrolyte and water balance. (Ch. 12)

minute ventilation The total volume of inspired air that enters the lungs per minute; equals tidal volume times frequency. (Ch. 20)

mitochondrion (*my*-tow-**kon**-dree-un) An organelle of the cell enclosed in a double membrane; the site of ATP production through the process of aerobic respiration. (Ch. 3)

mitosis (my-**tow**-sis) A process of cell division whereby each of the two daughter cells receives a complete set of chromosomes. (Ch. 3)

mitral cell (**my**-tral) Cells located in the olfactory bulbs that form the olfactory tract. (Ch. 8)

mitral valve The atrioventricular valve on the left side of the heart; also called the bicuspid. (Ch. 18)

M-line A structure in muscle cells that holds the thick filaments in side-by-side alignment. (Ch. 16)

molality A unit of concentration; the number of moles of solute per kilogram of solvent in a solution. (Ch. 2)

molarity A unit of concentration; the number of moles of solute per liter of solution. (Ch. 2)

molecular formula A symbol that shows the number of each type of atom in a molecule or group of atoms. (Ch. 2)

molecular mass The sum of the atomic masses of the elements in a compound. (Ch. 2)

monochromatic color vision Color vision that relies on one type of cone. (Ch. 8)

monoclonal antibody A specific antibody produced in vitro by hybridoma cell lines; the hybridoma cell lines are produced through fusion of activated β cells with cancer cells. (Ch. 3)

monocyte (**mon**-oh-site) One of two types of agranulocyte white blood cells; they migrate out of blood and into tissues to form macrophages. (Ch. 17)

monoiodotyrosine (MIT) A compound found within the thyroid follicle cell; used in the synthesis of T_3 and T_4. (Ch. 13)

monokine A chemical secreted by macrophages that stimulate T-lymphocyte development. (Ch. 28)

monosaccharide (*mon*-oh-**sak**-ah-ride) A simple sugar; cannot be degraded by hydrolysis to a simpler sugar. (Ch. 2)

monozygotic twins Twins that result from the fertilization of a single ovum by a single sperm, followed by cell division that results in a single duplication or multiple duplications of the fertilized ovum prior to implantation; also called identical twins. (Ch. 32)

mossy fiber Neurons that originate in the neocortex and spinal cord; sensory stimuli and voluntary movements enhance activity. (Ch. 9)

motility (mow-**til**-i-tee) Mixing and propulsive movements of the digestive tract. (Ch. 22)

motor cortex The area of the brain that controls movement. (Ch. 7)

motor end-plate The connection between a muscle fiber and the myelinated axon of a motor neuron; also called myoneural junction. (Ch. 9)

motor neuron A neuron that carries instructions about muscle movements from the CNS in response to sensory stimuli; also called an efferent neuron. (Ch. 7)

motor neuron pool A functional grouping in the spinal cord of motor neurons that innervate a single muscle. (Ch. 9)

motor system Collectively, the areas of the brain and spinal cord that control and coordinate skeletal muscle contractions. (Ch. 7)

motor unit A motor neuron plus the muscle fibers it innervates. (Ch. 9)

mRNA Messenger ribonucleic acid.

mucin (mew-sin) A nonenzymatic protein found in mucus; lubricates and protects the lining of the digestive tract. (Ch. 22)

mucosa (mew-ko-sah) A mucous membrane. (Ch. 22)

mucus (mew-kus) A viscous mixture of mucin and watery secretions from mucus and serous cells of salivary glands. (Ch. 22)

müllerian duct system An embryonic

ductal system that is the precursor of the internal genitalia of the female. (Ch. 33)

multipolar neuron A type of neuron with several short dendrites and one long axon. (Ch. 7)

multiunit smooth muscle A class of muscles that are closely controlled by the nervous system. (Ch. 16)

muscarinic acetylcholine receptor A category of cholinergic receptor, a receptor on target cells such as muscle. (Ch. 10)

muscle (**mus**-ul) A number of muscle fibers bound together by connective tissue. (Ch. 9)

muscle contraction The shortening of muscle tissue. (Ch. 9)

muscle fiber A type of contractile cell that is elongated in shape; also called muscle cell. (Ch. 9)

muscle relaxation The lengthening of muscle tissue. (Ch. 9)

muscle spindle A receptor complex attached to somatic sensory neurons that detect the length of the muscle and its velocity of contraction. (Ch. 9)

muscle spindle receptor A capsule-enclosed arrangement of afferent nerve fiber endings in skeletal muscle; sensitive to stretch. (Ch. 8)

muscle tissues Tissues whose cells are elongated and contain many myofibrils; specialized for contraction; functions to accomplish the organism's movement. (Ch. 1)

muscle tone The amount of resistance of muscle to passive stretch. (Ch. 9)

myasthenia (my-ass-**then**-ee-ya **gra**-vis) **gravis** A disorder of neuromuscular function characterized by muscle weakness; may affect any muscle of the body, but especially those of the eye, face, lips, tongue, throat, and neck. (Ch. 16)

myelin (**my**-eh-lin) An insulating lipid elaborated by oligodendrocytes (centrally) or Schwann cells (peripherally) that surrounds the fiber extension of neurons in order to speed the transmission of information. (Ch. 7)

myenteric plexus A network of nerve cells that lie between the two layers of the muscularis externa of the digestive tract. (Ch. 22)

myocardial infarction (my-oh-**kar**-dee-al in-**fark**-shun) A heart attack; necrosis of the myocardium when a coronary blood vessel is occluded. (Ch. 18)

myocardium (my-oh-**kar**-dee-um) The middle and thickest layer of the heart; composed of cardiac muscle. (Ch. 18)

myoepithelial (my-ah-ep-ah-**thee**-lee-yel) **cells** Contractile cells in the breast that help the flow of milk. (Ch. 32)

myofibril (my-oh-**fye**-bril) A functional unit of muscle cells formed by hun-

dreds of sarcomeres connected in series. (Ch. 16)

myofilament (my-oh-**fil**-ah-ment) A protein filament that provides for contraction of skeletal, cardiac, or smooth muscle. (Ch. 16)

myogenic activity The increase in the contractile activity of smooth muscle cells in response to stretch. (Ch. 19)

myoglobin (my-oh-**glow**-bin) A respiratory pigment that transports oxygen and imparts a red color to muscle tissue. (Ch. 16)

myoneural (my-oh-**new**-ral) **junction** The connection between a muscle fiber and the myelinated axon of a motor neuron; also called motor end-plate. (Ch. 9)

myosin (**my**-oh-sin) A very large protein that composes the thick filaments of muscle cells; 68% of protein in muscle is myosin. (Ch. 16)

myosin light-chain kinase (MLCK) An enzyme involved in phosphorylation and the regulation of muscle contraction. (Ch. 16)

myosin-linked regulation The regulation of muscle contraction that takes place via thick filaments. (Ch. 16)

myotonia (my-oh-**toe**-nee-ah) A condition that is characterized by tonic spasms or temporary rigidity of muscle. (Ch. 33)

myxedema (mik-seh-**dee**-mah) The accumulation of water in the skin, leading to a thickened and puffy appearance. (Ch. 13)

narcolepsy A severe disorder of sleep in which an individual suddenly lapses into sleep during the middle of the day; thought to be the result from activation of the REM-sleep generator. (Ch. 11)

natriuretic An agent that promotes the excretion of sodium in urine. (Ch. 23)

negative balance The condition in which there is net loss of a substance from the body. (Ch. 23)

negative feedback A class of control systems active in the body; works to restore normal values of a variable and thus to exert a stabilizing influence. (Ch. 1)

neocortex The outer gray matter of the cerebrum. (Ch. 7)

nephron (**nef**-ron) The basic unit of kidney structure and function; consists of a renal corpuscle and its attached tubule. (Ch. 23)

Nernst equation A chemical equation that allows the equilibrium potential for any ion to be calculated. (Ch. 7)

nerve A group of many nerve fibers traveling together in the peripheral nervous system. (Ch. 7)

nerve cell A cell specialized to initiate, integrate, and conduct electric signals; also called neuron. (Ch. 7)

nervous tissue One of four primary types of tissue; made up of neurons and glial cells. (Ch. 1)

neural fat *See* triacylglycerol.

neuroendocrinology An area of study that covers the overlap between the nervous and endocrine systems. (Ch. 12)

neurofilament A rodlike cytoplasmic protein that forms a major component of the cytoskeleton of neurons. (Ch. 7)

neurohormone Chemical messenger released by a neuron and carried by the blood to its target cell. (Ch. 12)

neurohypophyseal (new-row-high-**pawf**-is-ee-al) **peptide** A neuropeptide released from the posterior pituitary that functions to regulate plasma osmolarity and lactation. (Ch. 7)

neuromuscular junction The location of synaptic connections between alpha motor neurons and skeletal muscles where acetylcholine is released from the nerve fiber terminal. (Ch. 9)

neuron (**new**-ron) *See* nerve cell.

neuropeptide transmitter A family of at least 50 neurotransmitters composed of two or more amino acids; often function as chemical messengers in nonneural tissues. (Ch. 7)

neurotransmitter A chemical messenger used by neurons to communicate with each other or with effectors. (Ch. 5)

neutralization The reaction of an acid and base to produce a salt and water. (Ch. 25)

neutron (**new**-tron) The electrically neutral particle found in the nucleus of the atom. (Ch. 2)

neutrophil (**new**-tro-fil) A polymorphonuclear granulocytic leukocyte whose granules show preference for neither eosin nor basic dyes; functions as a phagocyte and releases chemicals involved in inflammation. (Ch. 17)

nexus (**nek**-sus) A structure that connects numerous portions of smooth muscle cells; involves a fusion of the outer layers of the membrane of each cell so that there is no extracellular space at the region between cells; also called gap junction. (Ch. 16)

NH$_3$ Ammonia; a base. (Ch. 23)

NH$_4$ Ammonium ion; an acid. (Ch. 23)

nicotinamide adenine dinucleotide (NAD) A coenzyme that is most frequently employed as a hydrogen acceptor in metabolic reactions. (Ch. 6)

nicotinic A cholinergic receptor that also responds to nicotine; located on postganglionic autonomic neurons. (Ch. 10)

nitroglycerin (nigh-troh-**glis**-eh-rin) A therapeutic drug that provides relief from angina by dilating the coronary arteries to increase blood flow. (Ch. 19)

nitrous oxide A colorless, sweetish gas

used as an anesthetic; commonly called laughing gas. (Ch. 29)

NMDA (*N*-methyl-*D*-aspartate) One of three subtypes of glutamate receptors; its activation results in an increase in Ca^{2+} conductance. It is both a ligand and a voltage-gated channel. (Ch. 7)

nociceptor (**no**-sih-sep-tor) A receptor that detects painful stimuli; a **mechanical nociceptor** is activated by intense mechanical stimulation such as a slap to the head; a **heat nociceptor** responds to temperatures above 45°C. (Ch. 8)

nodes of Ranvier A space between adjacent myelin-forming cells along a myelinated axon where axonal plasma membrane is exposed to extracellular fluid. (Ch. 7)

nonassociative learning A type of learning in which we adapt to a single type of stimulus according to its relevance to our survival. (Ch. 11)

nonatrophic vaginitis Vaginal dryness or tenderness due to allergic reactions or chemical irritations. (Ch. 33)

nonessential amino acids Amino acids that can be synthesized within the body by relatively simple pathways. (Ch. 6)

non-insulin-dependent diabetes mellitus (NIDDM) A form of diabetes in which the target tissues of insulin lose their responsiveness to the hormone; also known as type II diabetes. (Ch. 15)

nonspecific defense Defense mechanisms that deter or destroy a variety of pathogens regardless of pathogen type (e.g., phagocytosis). (Ch. 28)

norepinephrine (*nor*-ep-ih-**nef**-rin) A catecholamine neurotransmitter released from the adrenal medulla and at most sympathetic postganglionic endings and in many CNS regions. (Ch. 7)

normovolemia The normal blood volume. (Ch. 17)

nuclear bag fiber Intrafusal fibers that help make up muscle spindle; innervated by type Ia nerve fibers that transmit information about muscle length and velocity of contraction to the CNS. (Ch. 9)

nuclear chain fibers Intrafusal fibers that help make up muscle spindle; innervated by type II nerve fibers that transmit information about muscle length to the CNS. (Ch. 9)

nuclear envelope The double membrane that surrounds the cell nucleus. (Ch. 3)

nucleic (new-**klee**-ik) **acid** A nucleotide polymer in which phosphate of one nucleotide is linked to the sugar of the adjacent one; stores and transmits genetic information; includes DNA and RNA. (Ch. 2)

nucleolus (new-**klee**-oh-lus) Densely stained structure within the nucleus that contains the DNA that codes for rRNA; site of ribosome assembly.

(Ch. 3)

nucleotide A compound consisting of a pentose sugar, one of a set of special groups called bases, and one to three phosphate groups. (Ch. 2)

nucleus (**new**-klee-us) A large membrane-bound organelle that contains the cell's DNA; (neural) cluster of neuron cell bodies in the CNS. (Ch. 3)

nucleus accumbens The lower forebrain structure that is part of the reward system. (Ch. 11)

nucleus of the vagus A region of the brain stem that, when activated, directly reduces the heart rate. (Ch. 10)

ocular (ah-**kew**-lar) **dominance column** The organization within the visual cortex in which columns of cells respond to light impinging on either the right or left eye. (Ch. 8)

"off" center and "on" surround cells Ganglion cells for which the annulus is inhibitory and periphery is excitatory. (Ch. 8)

olfactory (ohl-**fak**-toe-ree) **bulb** The site of the mitral cells of the olfactory system. (Ch. 8)

olfactory receptor cells Cells within the nasal cavity that are activated by odors; enervate mitral cells in the olfactory bulb. (Ch. 8)

olfactory system The physiological system that provides the ability to smell odors. (Ch. 8)

olfactory tract A bundle of fibers that lead from mitral cells in the olfactory bulb and innervate areas of the cortex. (Ch. 8)

oligodendroglia (ohl-ih-go-den-**dro**-glee-ah) One of two types of glial cells that wrap around the fiber extensions of neurons to form myelin. (Ch. 7)

oligosaccharide A carbohydrate that on hydrolysis yields a small number of monosaccharides. (Ch. 22)

oliguria (ohl-ih-**goo**-ree-ah) The excretion of a small amount of urine in relation to fluid intake. (Ch. 24)

"on" center and "off" surround cells Ganglion cells for which the center of the annulus is excitatory and the periphery is inhibitory. (Ch. 8)

oncotic (on-**kot**-ik) **pressure** The osmotic pressure due to the presence of colloids in a solution; in the case of plasma interstitial fluid interaction, it is the force that tends to counterbalance the capillary blood pressure. (Ch. 17)

oogenesis (oh-ah-**gen**-eh-sis) The formation of ova. (Ch. 31)

operant conditioning A type of associative learning that involves the relationship between a behavior and a subsequent reward or punishment. (Ch. 11)

opioid (**oh**-pea-oid) One of a family of neuropeptides that are structurally similar to opiates and that modulate

the perception of pain. (Ch. 7)

opsonin (op-**so**-nin) A complement protein that is part of the internal defense mechanism; coats bacteria and other cells so that they are susceptible to phagocytosis. (Ch. 28)

ordinate The vertical axis of a graph; used to plot a dependent variable. (Ch. 1)

organ (**or**-gan) A group of tissues that have been combined for the performance of a specific function or a series of related functions. (Ch. 1)

organelle An organized, reproducible, permanent subcellular structure, membranous or nonmembranous, that provides for specific functions of the cell. (Ch. 3)

orgasm (**or**-gazm) The apex and culmination of sexual excitement; characterized in the male by ejaculation and in the female by rhythmic muscular contractions in the uterus and vagina. (Ch. 31)

orientation column Nerve cells of the primary visual cortex that are sensitive to bars of light oriented at different angles. (Ch. 8)

origin The more stationary and proximal point of the attachment of a muscle. (Ch. 16)

orthophosphate PO_4; the primary form of phosphorus in plasma. (Ch. 26)

osmolality Total solute concentration, expressed as osmoles per kg of H_2O. (Ch. 23)

osmolarity The concentration of osmotically active particles in solution, expressed as osmoles of a solute per liter of solution. (Ch. 2)

osmoreceptor Specialized neurons in the anterior hypothalamus that are stimulated by increased osmolality of the extracellular fluid, their excitation promotes the output of antidiuretic hormone and stimulates the thirst sensation. (Ch. 9)

osmosis (oz-**mow**-sis) The net movement of solvent (usually water) from a solution of lesser to one of greater solute concentration when the two solutions are separated by a membrane that selectively prevents passage of solute molecules, but is permeable to the solvent. (Ch. 2)

osmotic (oz-**mot**-ik) **concentration** Total concentration of all osmotically active solutes present in the aqueous solution. (Ch. 4)

osmotic fragility The ability of a cell to withstand an osmotic stress. (Ch. 4)

osmotic pressure A property of a solution that is proportional to the solute concentration; a cause of diffusion across membranes. (Ch. 2)

ossicle (**os**-ih-kul) A small bone within the inner ear. (Ch. 8)

osteoblast (**os**-tee-oh-*blast*) A cell that arises from a fibroblast and which at maturity is associated with the production of bone. (Ch. 26)

osteoclast (**os**-tee-oh-*klast*) A cell that carries out the process of bone resorption. (Ch. 26)

osteocyte (**os**-tee-oh-*site*) An osteoblast that has progressively lost its bone-forming capability. (Ch. 26)

osteoid (**os**-tee-oyd) The organic matrix of bone. (Ch. 26)

osteomalacia (**os**-tee-oh-mal-*ay*-see-ah) A metabolic bone disease in which there is softening of the bone due to inadequate mineralization, with excess accumulation of osteoid. (Ch. 26)

osteoporosis (*os*-tee-oh-poe-**row**-sis) A metabolic bone disease in which there is reduction of overall bone mass. (Ch. 26)

otolithic membrane (oh-toe-*lith*-ik) A membrane in which the stereocilia of hair cells in the utricle of the inner ear are embedded; contains calcium carbonate crystals called statoconia. (Ch. 9)

otolithic organ The utricle and saccule of the inner ear; detects linear acceleration. (Ch. 8, 9)

otosclerosis (**oh**-toe-skler-**oh**-sis) Abnormal growth of bones in the middle ear, causing conductive hearing loss. (Ch. 8)

outer ear The portion of the ear that channels sound waves to the ear drum. (Ch. 8)

oval window A membranous structure of the inner ear that vibrates with movement of the ossicles. (Ch. 8)

ovarian (oh-**var**-ee-an) **cycle** The approximately 28-day cycle of the female reproductive system; also called the menstrual cycle. (Ch. 31)

ovary (**oh**-var-ee) The female gonad; a oval-shaped organ located in the pelvic cavity below the open ends of the fallopian tubes; produces germ cells and hormones. (Ch. 31)

Overton's rules Relationships that characterize the permeability of cell membranes to polar and nonpolar solutes. (Ch. 4)

ovulation (*ov*-vu-**lay**-shun) The release of a single ovum from one of the two ovaries. (Ch. 31)

ovum (**oh**-vum) The female reproductive cell, which after fertilization develops into an embryo. (Ch. 31)

oxidation (oks-ih-**day**-shun) **reaction** A chemical reaction in which electrons are lost by a reactant. (Ch. 6)

oxidation-reduction reaction A class of chemical reactions in which one reactant accepts electrons that are removed from another. (Ch. 2)

oxidative phosphorylation The chemical pathway that provides ATP to the cell by phosphorylation of ADP in the mitochondrion. (Ch. 6)

oxygen An element with a high affinity for electrons; the primary receptor of electrons in the mitochondrial respiratory chain. (Ch. 6)

oxygen debt The amount of oxygen required to metabolize excess lactic acid formed in skeletal muscle during exercise. (Ch. 16)

oxygen radicals Highly reactive byproducts of molecular oxygen that are toxic to cells. (Ch. 29)

oxygen tension The partial pressure of oxygen in a liquid or a gas. (Ch. 21)

oxygen toxicity Damage to the lung and other body parts that occurs after breathing high concentrations of oxygen for more than 1 day. (Ch. 29)

oxygenated blood Blood that has passed through the pulmonary capillaries and carries about 20 ml of O_2 per deciliter of whole blood. (Ch. 21)

oxygen-carrying capacity The amount of oxygen that the hemoglobin in 100 ml of whole body is capable of transporting. (Ch. 17)

oxyhemoglobin (*ok*-see-**he**-mow-glow-bin) Hemoglobin that is saturated with oxygen after passing through the lungs. (Ch. 17)

oxyhemoglobin dissociation curve An S-shaped curve that illustrates the way that hemoglobin binds to oxygen in the lung and unloads oxygen in the tissues. (Ch. 21)

oxytocin (*ok*-see-**toe**-sin) A hormone secreted by the posterior pituitary; stimulates contraction of uterine muscles during labor and milk ejection actions. (Ch. 13)

pacemaker Neurons that set the rhythm of biological clocks independent of external cues; any nerve or muscle cell that has an inherent auto-rhythmicity and determines activity patterns of other cells. (Ch. 18)

pacemaker cell A cardiac cell that regulates heart rate. (Ch. 10)

Pacinian corpuscle A somatic sensory mechanoreceptor, located beneath the skin and in walls of viscera, that responds to pressure. (Ch. 9)

pain receptor A mechanical or heat nociceptor. (Ch. 9)

palmitic acid A fatty acid; one of the most common types found in phospholipids. (Ch. 4)

pancreas (**pan**-kree-as) An organ of the digestive system; located in the abdomen near the stomach; connected by ducts to the small intestine; contains endocrine gland cells, which secrete insulin, glucagon, and somatostatin into the bloodstream, and exocrine gland cells, which secrete digestive enzymes and bicarbonate into the intestine. (Ch. 15, 22)

pancreatic polypeptide A polypeptide produced by the F-cells of the pancreas; its physiological function is not fully characterized. (Ch. 15)

pancreatitis (*pan*-kree-ah-**tie**-tis) An acute or chronic inflammation of the pancreas; due to autodigestion of pancreatic tissue by its own enzymes. (Ch. 22)

paracellular transport pathway Junctions that are relatively permeable and allow water and ions to move between the epithelial cells. (Ch. 4)

paracrine (**pear**-ah-krin) The mechanism whereby a hormone secreted by one cell affects an adjacent or nearby target cell. (Ch. 12)

parafollicular cells Thyroid cells that produce calcitonin; also called C-cells. (Ch. 26)

parallel organization The arrangement of the motor system whereby information from multiple sources can simultaneously and independently affect the same target. (Ch. 9)

parasympathetic nervous system The portion of the autonomic nervous system that coordinates the body's vegetative activities. (Ch. 10)

parathormone *See* parathyroid hormone.

parathyroid (*par*-ah-**thy**-roid) **gland** Two pairs of glands located on the dorsal surface of each of the two lobes of the thyroid glands. (Ch. 26)

parathyroid hormone (PTH) A polypeptide protein produced by the parathyroid glands. (Ch. 26)

paraventricular nucleus The location of osmoreceptors in the hypothalamus. (Ch. 11)

partial pressure That part of the total gas pressure that is due to molecules of one gas species; a measure of concentration of a gas in a gas mixture. (Ch. 2)

parturition (*par*-too-**rish**-in) Birth; the delivery of an infant and placenta. (Ch. 32)

passive immunity Short-term protection against disease as a result of receiving antibodies. (Ch. 28)

passive tension The force in a muscle at rest that is due to the elastic nature of its anatomical structure. (Ch. 16)

passive transport Nonmediated movement of solute molecules down an electrochemical gradient. (Ch. 4)

patellar (pah-**tell**-ar) **tendon** A tendon of the anterior thigh muscles that passes through the knee, contains the patella (knee cap), and is inserted into the anterior tibia of the leg. (Ch. 9)

pathogen (**path**-oh-jen) Any disease-producing organism or agent. (Ch. 28)

P_{CO_2} Carbon dioxide partial pressure or tension. (Ch. 30)

penile (pee-nile) erection The increase in diameter, length, and firmness in the penis; due to vasocongestion. (Ch. 33)

pentose phosphate pathway The pathway for oxidation of glucose in the cytosol; also called the hexose monophosphate shunt. (Ch. 6)

pepsin (pep-sin) An enzyme, secreted by the stomach, that initiates digestion of protein. (Ch. 22)

peptidase Any enzyme that hydrolyzes specific peptides to their constituent amino acids. (Ch. 22)

peptide (pep-tide) A compound that is composed of two or more amino acids; forms the constituent parts of proteins. (Ch. 7)

peptide bond A polar covalent chemical bond joining two amino acids; forms the protein backbone. (Ch. 2)

perfusion Capillary blood flow. (Ch. 19)

perimysium (per-ih-mis-ee-um) A type of connective tissue that groups bundles of muscle fibers. (Ch. 16)

peripheral nervous system (PNS) Nerve fibers that emerge from the brain stem and spinal cord; composed of somatic and autonomic portions. (Ch. 7)

peripheral proprioreceptors Neuroreceptors in joints and muscle spindles that detect movement of limbs. (Ch. 30)

peripheral thermoreceptors Temperature receptors that detect changes in skin and shell temperature; located just beneath the skin. (Ch. 27)

peristalsis (per-ih-stal-sis) The propulsive movement of semisolid food mass through the digestive system by a wave of muscle contractions. (Ch. 22)

peritoneum (per-ih-tow-nee-um) A membrane that lines the abdominal cavity and covers some of the abdominal viscera. (Ch. 22)

peritubular capillary The capillary network surrounding the tubules of the kidney. (Ch. 23)

peritubular capillary starling forces The hydrostatic and colloid osmotic pressures that influence fluid movement across the capillaries that surround the kidney tubules. (Ch. 23)

permeability coefficient A number expressing the ease with which a solute can penetrate a plasma membrane; governed by the thickness of the membrane and the lipid solubility of the solute; can be expressed as P. (Ch. 4)

permissive action The amplifying effect of one hormone on another. (Ch. 14)

peroxidase (per-ox-ih-dase) A cellular enzyme that readily reacts with oxidizing free radicals. (Ch. 29)

peroxide An oxide of any element that contains more oxygen than any other. (Ch. 29)

peroxisome (per-ox-ih-zome) A small, membrane-bounded bag, found in a cell, that is involved in digestion and food waste; also important in detoxifying a number of molecules. (Ch. 3)

peyer's patches Aggregates of lymphoid nodules in the mucosa of the intestine. (Ch. 28)

pH A measure of the acidity or alkalinity of a solution; it is equal to the negative logarithm (to base 10) of the hydrogen ion concentration (the latter measured in moles/liter). (Ch. 2)

phagocytosis (fag-oh-sigh-toe-sis) A special type of enocytosis involving internalization of particulate matter. (Ch. 3)

pharynx (fare-inks) The portion of the digestive tract between the mouth and the esophagus. (Ch. 22)

phase contrast microscope A microscope that enhances the contrast between the internal organelles so that structures can be seen clearly in living, unstained, and nonfixed cells. (Ch. 3)

phasic (faze-ik) contraction A contraction that occurs intermittently. (Ch. 16)

phasic discharge A very brief burst of action potentials by a proprioceptor. (Ch. 9)

phenotype (fee-no-type) The visible expression of the genotype; physical or other characteristics, which are produced by genes and influenced by the interaction between genes and environmental factors. (Ch. 11)

pheromone A substance secreted to the outside of the body by an individual and perceived (as by smell) by a second individual. (Ch. 10)

phonocardiogram The electronic recording of heart sounds. (Ch. 18)

phosphagen system The energy system of muscles based on ATP and phosphacreatine; used for short bouts of maximal exercise. (Ch. 30)

phosphate PO_4. (Ch. 26)

phosphatidylinositol (fass-fah-tid-el-inos-i-tahl) A minor phospholipid constituent of the plasma membrane. (Ch. 12)

phosphatidylinositol 4,5-bisphosphate (PIP2) The immediate precursor for the generation of two intracellular second messengers, derived from phosphatidylinositol. (Ch. 12)

phosphaturia (fos-fa-tur-ee-ah) Increased phosphate excretion in the urine. (Ch. 26)

phosphocreatine A high-energy phosphate molecule used as an energy resource in the metabolism of muscle cells. (Ch. 30) *See also* phosphate.

phosphodiesterase (fass-foe-die-es-teraze) An enzyme present in the disc membrane of photoreceptors; catalyzes the conversion of cyclic AMP to AMP. (Ch. 9)

phospholipase C (fos-fo-lie-pace) A receptor-controlled plasma-membrane enzyme that catalyzes phosphatidyl bisphosphate breakdown to inositol trisphosphate and diacylglycerol. (Ch. 5)

phospholipid (fos-fo-li-pid) A lipid subclass similar to tracylglycerol except that a phosphate group and a small nitrogen-containing molecule are attached to the third hydroxyl group of glycerol; a major component of the cell membrane. (Ch. 2)

phosphoprotein phosphatase An enzyme that removes phosphate from protein, restoring the protein's original shape. (Ch. 5)

phosphorylation The addition of a phosphate group to an organic molecule. (Ch. 6)

photon A particle of light; photons have wavelengths and energies associated with different colors. (Ch. 9)

photoreceptor (fo-toe-re-sep-tor) cones Conically shaped photoreceptors that are the basis of color perception and day vision. (Ch. 9)

photoreceptor rods Cylindrically shaped photoreceptors that are very sensitive to low levels of illumination. (Ch. 9)

photoreceptors Receptor cells of the retina that absorb photons. (Ch. 9)

photosynthesis The use of radiant energy to manufacture ATP, which drives biosynthesis in plant cells. (Ch. 6)

phrenic (free-nik) nerve The nerve that sends respiratory messages to the diaphragm. (Ch. 20)

physiology (fiz-ee-ol-oh-jee) A branch of biology dealing with the mechanisms by which the body functions. (Ch. 1)

Pickwickian syndrome A type of altered breathing pattern that affects ventilation; occurs in severely obese individuals who often suffer hypoventilation. (Ch. 20)

pituitary (pit-too-ih-terr-ee) dwarfism Dwarfism that is due to a deficiency in GH secretion, which occurs as a result of a defect in pituitary function. (Ch. 13)

pituitary gland The endocrine gland that lies in a bony pocket below the hypothalamus; includes the anterior, middle, and posterior pituitary. (Ch. 13)

pituitary stalk A thin segment of tissue that connects the pituitary gland to the hypothalamus. (Ch. 13)

pituitary stalk A thin segment of tissue that connects the pituitary gland to the hypothalamus. (Ch. 13)

placenta (plah-sen-tah) An interlocking fetal and maternal organ of molecular exchange between fetal and maternal circulation. (Ch. 32)

plasma (plas-muh) The liquid portion of blood; a component of extracellular fluid. (Ch. 17)

plasma cell A cell that differentiates from activated β-lymphocytes and secretes antibodies. (Ch. 28)

plasma membrane The membrane that forms the outer surface of the cell and separates the cell's contents from extracellular fluid. (Ch. 1)

plasmid (**plas**-mid) A small accessory DNA molecule used for DNA cloning. (Ch. 3)

plateau The phase of sexual excitement that precedes orgasm. (Ch. 33)

platelet (**plate**-let) Thrombocyte; a cell fragment, suspended in plasma, that functions in blood clotting. (Ch. 17)

platelet plug An aggregation of platelets at the site of vascular damage; effectively plugs the defect end. (Ch. 17)

pleural (**ploor**-ahl) **pressure** The pressure in the pleural fluid between the lung and the chest wall. (Ch. 20)

pluripotential hematopoietic stem cell An undifferentiated cell that is capable of developing into progenitor cells of the erythrocyte, leukocyte, or megakaryocyte. (Ch. 17)

pneumatology An early theory of physiology put forward by Galen; involved the belief that the blood took on natural spirits and carried them through the veins to the bodily organs. (Ch. 1)

pneumoconioses (*new*-mo-koe-ni-**oh**-seis) A disease that causes scarring of the lung from breathing dust particles. (Ch. 29)

pneumothorax The condition when air enters the pleural space and the lungs collapse. (Ch. 20)

P$_{O_2}$ Partial pressure of oxygen. (Ch. 30)

poikilocytosis (poy-**kee**-lo-sigh-**to**-sis) A variation in the shape of an erythrocyte. (Ch. 17)

poikilotherm (poy-**kee**-lo-therm) An animal whose body temperature may fluctuate with ambient temperature. (Ch. 27)

polycythemia (*pol*-ee-sie-**thee**-me-ah) An increase in the number of red blood cells in the blood. (Ch. 17)

polypeptide A polymer consisting of amino acid subunits joined by peptide bonds; also called peptide and protein. (Ch. 2)

polysome A structure composed of ribosomes attached to a single strand of mRNA, playing a role in peptide synthesis; also called a polyribosome. (Ch. 5)

polyspermy Fertilization of the ovum by more than one sperm; lethal for the fertilized ovum. (Ch. 32)

polyuria (*poly*-ee-**yoo**-ree-ah) The excessive excretion of urine; characteristic of diabetes. (Ch. 15)

population code A method of informing the brain about the intensity of a stimulus; the strength of a stimulus is re-lated to the number of receptor cells that are activated. (Ch. 9)

population code of contractile force A means of control of the amount of force that a muscle generates during a contraction; an increase in the number of motor neurons and motor units can increase a muscle's force of contraction. (Ch. 9)

portal circulation Blood flow from the stomach, intestines, and pancreas through the portal vein and to the liver. (Ch. 22)

positive balance The condition in which there is net gain of a substance in the body. (Ch. 24)

positive feedback A class of control system that destabilizes normal values of a variable and thus has limited value in maintaining homeostasis. (Ch. 1)

posterior (pos-**teer**-ee-or) A descriptive term of position that means nearer to the back of the body. (Ch. 1)

posterior parietal cortex An area of the brain that processes sensory stimuli, leading to purposeful movement. (Ch. 9)

posterior pituitary gland The posterior lobe of the pituitary gland from which ADH and oxytocin are released into blood. (Ch. 13)

postganglionic (post-gang-glee-**on**-ik) **cell** The autonomic neuron that directly innervates the effector cell. (Ch. 10)

postganglionic fibers Axons of postganglionic autonomic neurons. (Ch. 14)

postmenopausal osteoporosis The rapid loss of bone due to decreasing estrogen levels in postmenopausal women. (Ch. 26)

post-translational modification of proteins A series of reactions in which newly synthesized RNA molecules are covalently modified; also known as RNA processing. (Ch. 5)

potential energy Stored energy. (Ch. 2)

power output The rate of doing work. (Ch. 16)

precapillary sphincter (**sfink**-ter) A smooth muscle ring around the arteriole that regulates the flow of blood into the capillaries supplied by that arteriole. (Ch. 19)

preganglionic cells Efferent sympathetic and parasympathetic neurons whose cell bodies are located within the CNS but whose fibers leave the CNS and synapse within peripheral autonomic ganglia. (Ch. 10)

pregnancy The condition of having a developing embryo or fetus within the body. (Ch. 32)

pregnenolone A steroid hormone that is an early constituent of the synthesis of all steroids; its precursor is cholesterol. (Ch. 12)

premature ejaculation Consistent, unin-tentional ejaculation prior to vaginal penetration; more commonly defined as loss of ejaculatory control resulting in ejaculation soon after vaginal penetration. (Ch. 33)

premise Given information stated in the form of an absolute rule using a word such as "all" or "always." (Ch. 1)

premotor cortex A component of the frontal, cerebral cortex where movements are planned and programmed. (Ch. 9)

preoptic area An area of the hypothalamus in which electrical stimulation induces non-REM sleep; destruction of this area of the brain causes insomnia in laboratory rats. (Ch. 11)

pressure gradient $P_1 - P_2$ or ΔP; the difference in pressure between two points. (Ch. 18)

presynaptic facilitation An underlying mechanism that enhances an animal's response to a particular stimulus; causes sensitization to the stimuli. (Ch. 11)

presynaptic (*pre*-sin-**ap**-tik) **inhibition** The action of an interneuron to inhibit the release of substance P by the release of the peptide enkephalin. (Ch. 9)

primary somatosensory area A narrow strip of cortical tissue, the location of cells in the somatosensory cortex that exhibit the annular receptive field; also called Brodmann area 3. (Ch. 9)

primitive gonads Gonads found in the zygote before they have differentiated into testes or ovaries. (Ch. 33)

procarboxypeptidase (**pro**-car-boxy-**pep**-tid-aze) An inactive precursor of carboxypeptidase, which is a proteolytic enzyme secreted by the pancreas. (Ch. 22)

progenitor cell A committed stem cell that is capable of developing into precursor cells of a specific blood cell. (Ch. 17)

progesterone (pro-**jes**-ter-own) A steroid hormone secreted by the corpus luteum and placenta; stimulates uterine-gland secretions, inhibits uterine smooth-muscle contractions, and stimulates breast growth. (Ch. 12)

progestin (pro-**jes**-tin) One of the major categories of steroid hormones; involved in the maintenance of pregnancy as well as being produced by the ovaries and the placenta. (Ch. 12)

proglucagon (pro-**glue**-ka-gon) The immediate precursor of glucagon; synthesized in the pancreas, small intestine, and hypothalamus, but only converted to glucagon in the pancreas. (Ch. 15)

proinsulin A large precursor molecule from which insulin is derived; synthesized in the rough endoplasmic reticulum of the beta cells in the pancreas. (Ch. 15)

prokaryotic cells Cells in which the DNA is not in a separate compartment from the rest of the cell. (Ch. 3)

prolactin (pro-**lak**-tin) A peptide hormone secreted by the anterior pituitary; stimulates milk secretion in the mammary glands. (Ch. 13)

promoter site The place where transcription of a given gene begins when the RNA polymerase binds on a DNA template. (Ch. 5)

pro-opiomelanocortin (pro-**opeo**-mellano-cort-in) **(POMC)** A large protein precursor for ACTH, endorphins, and several other hormones. (Ch. 13)

proprioception (*pro*-pre-oh-**sep**-shun) The sensation of a change in the position of the limbs. (Ch. 9)

proprioceptor (*pro*-pre-oh-**sep**-tor) A somatic sensory receptor activated by movement of the limbs. (Ch. 9)

prosomatostatin (pro-so-**mato**-stat-in) The immediate precursor to somatostatin. (Ch. 15)

prostaglandins (*pros*-tah-**glan**-dins) A group of naturally occurring hydroxy fatty acids derived from arachidonic acid; affect a wide variety of physiological processes. (Ch. 23)

protease (pro-tee-ase) A general term for a proteolytic enzyme; *see* peptidase. (Ch. 22)

protein (pro-teen) A large polymer consisting of one or more sequences of amino acid subunits joined by peptide bonds. (Ch. 2)

protein asymmetry in membranes The condition when extrinsic and intrinsic membrane proteins are distributed unequally between the two halves of the bilayer. (Ch. 4)

protein conformation The distinctive three-dimensional shape of a protein molecule formed when the linear polypeptide chain folds up in a precise way. (Ch. 1)

protein kinase One of a family of enzymes that phosphorylate certain other proteins by transferring them to a phosphate group from ATP. (Ch. 5)

protein-bound calcium One of two major forms of calcium in plasma; calcium bound to relatively large plasma proteins that cannot pass through plasma membranes. (Ch. 26)

proteinuria (pro-tee-**nyoo**-re-ah) The presence of an excess of proteins in the urine. (Ch. 23)

proteoglycan High-molecular-weight compounds that form the noncollagen portion of bone. (Ch. 26)

prothrombin (pro-**throm**-bin) An inactive precursor of thrombin; produced by the liver and normally present in plasma. (Ch. 17)

prothrombin converting factor The activated Stuart factor, in the presence of

calcium ions; forms complexes with accelerin on phospholipid micelles provided by tissue thromboplastin; converts prothrombin to thrombin. (Ch. 17)

proton A positive charged subatomic particle. (Ch. 2)

protoplasm A complex mixture of chemicals that displays the attributes of life; that is, is organized into specific kinds of structural units, has the ability to enter into chemical activities that include the transformation of energy and the maintenance or synthesis of protoplasm, the ability to respond to changes in the environment, and the ability to grow and reproduce. (Ch. 1)

proximal Nearer, closer to a reference point; the opposite of distal. (Ch. 16)

proximal convoluted tubule The first part of the renal tubule. (Ch. 23)

pseudopodia Footlike extensions of the cytoplasm. (Ch. 17)

puberty The attainment of sexual maturity, when conception becomes possible; as commonly used, refers to 3 to 5 years of sexual development that culminates in sexual maturity. (Ch. 31)

pulmonary (**pul**-mow-*ner*-ee) **artery** The large vessel leading from the right ventricle that carries systemic venous blood to the lungs. (Ch. 18)

pulmonary circulation The circulation through the lungs; the portion of the cardiovascular system between the pulmonary trunk, as it leaves the right ventricle, and the pulmonary veins, as they enter the left atrium. (Ch. 18)

pulmonary edema An abnormal accumulation of fluid in the lung interstitium and alveoli. (Ch. 21)

pulmonary embolism Blockage in the pulmonary circulation due to emboli which are often blood clots (thrombi) that have formed in the systemic venous system. (Ch. 21)

pulmonary hypertension Abnormally elevated pulmonary arterial pressure. (Ch. 21)

pulmonary hypotension High pulmonary artery pressure; can cause vessel damage and eventual failure of the right ventricle. (Ch. 21)

pulmonary stretch receptors Afferent nerve endings lying between the smooth muscle cells of the airway and activated by lung inflation. (Ch. 20)

pulmonary trunk A large artery that carries blood from the right ventricle of the heart. (Ch. 18)

pulmonary valve The valve between the right ventricle of the heart and the pulmonary trunk. (Ch. 18)

pulmonary veins Veins that take oxygenated blood into the left heart. (Ch. 18)

pulmonary ventilation Tidal volume × respiratory rate; normally 5 to 6 liters

per minute at rest. (Ch. 30)

pulmonic stenosis Narrowing of the right atrioventricular orifice or pulmonary trunk resulting in hypertrophy of the right ventricle. (Ch. 18)

pulse pressure The difference between systolic and diastolic arterial blood pressures. (Ch. 19)

Purkinje (per-**kin**-jee) **cell** The most prominent nerve cell in the cerebellum. (Ch. 9)

Purkinje fiber A specialized myocardial cell that constitutes part of the conducting system of the heart; conveys excitation from bundle branches to ventricular muscle. (Ch. 18)

putamen One of the basal ganglia involved with involuntary control of skeletal muscle. (Ch. 9)

P-wave A component of an electrocardiogram that reflects atrial depolarization. (Ch. 18)

pyloric (pie-**lor**-ik) **sphincter** A thick bundle of circular smooth muscle and connective tissue that controls the passage of chyme from the stomach to the small intestine. (Ch. 22)

pylorus (pie-**lor**-us) The distal funnel-shaped end of the stomach; stomach contents are emptied through it to the duodenum. (Ch. 22)

pyramidal tract Nerve fibers that descend from the motor cortex to the brain stem and down the lateral corticospinal tract. (Ch. 9)

pyrogens Chemical agents that cause fever. (Ch. 27)

pyruvic acid A three-carbon intermediate in the glycolytic pathway that, in the absence of oxygen, forms lactic acid or, in the presence of oxygen, enters the Krebs cycle. (Ch. 30)

QRS complex The component of an electrocardiogram corresponding to ventricular depolarization. (Ch. 18)

qualitative Pertaining to quality. (Ch. 1)

quantal release Refers to the release of a fixed number of neurotransmitter molecules from synaptic vesicles. (Ch. 7)

quantitative Denoting or expressible as a quantity; relating to the proportionate quantities or the amount of the constituents of a compound. (Ch. 1)

quisqualate One of three subtypes of glutamate receptors that, when activated, results in excitatory postsynaptic potentials by opening ion channels that increase in Na^+ and K^4 conductance. (Ch. 7)

radiation Heat transfer by electromagnetic radiation between objects that are not in contact. (Ch. 27)

radioimmunoassay (**ray**-dee-o-im-mew-no-**ass**-ay) A procedure used to measure very small amounts of molecules with great accuracy; uses radioisotopes to label cyclic AMP in a sample. (Ch. 3)

radioisotope An unstable isotope of an element that undergoes radioactive decay. (Ch. 3)

raphe nucleus A portion of the brain stem that is responsible for the generation of REM sleep. (Ch. 11)

rapid eye movement (REM) sleep A phase of sleep during which eye movement becomes rapid, the heart rate and blood pressure increase, and gastrointestinal motility decreases. (Ch. 11)

rate-limiting enzyme An enzyme in the metabolic pathway that is most easily saturated with substrate; determines the rate of the entire metabolic pathway. (Ch. 6)

rate-limiting reaction The slowest reaction in a metabolic pathway. (Ch. 6)

Rathke's pouch A pocket of cells from the primitive oral cavity that are involved in the embryological development of the pituitary. (Ch. 13)

ratio An expression that compares two numbers or quantities by division. (Ch. 1)

reactive hyperemia (hi-per-ee-me-ah) Extreme reactive vasodilation. (Ch. 19)

receptive field The small area of the skin that can activate the receptor of a nerve cell in the dorsal root ganglion. (Ch. 8)

receptor A macromolecule, usually a protein, that binds a signal molecule with high specificity and triggers intracellular events leading to a change in cell function. (Ch. 1)

receptor-mediated endocytosis A process of endocytosis whereby a hormone and its plasma membrane receptor are internalized. The hormone is degraded and the receptor is recycled to the plasma membrane. (Ch. 12)

recombinant (re-kom-bih-nent) **DNA rDNA** A DNA molecule created by combining pieces of DNA from markedly different organisms. (Ch. 3)

recovery oxygen Net amount of oxygen consumed during recovery from exercise. (Ch. 30)

rectum (rek-tum) The portion of the large intestine between the sigmoid colon and the anus. (Ch. 22)

red cell indices Quantitative assessments of erythrocyte size, shape, and color using the hematocrit, hemoglobin content, and RBC count of blood. (Ch. 17)

redout A loss of vision due to high negative G-forces. (Ch. 29)

redox potential The oxidation–reduction potential; a measure of the affinity of molecules to give up or accept electrons. (Ch. 6)

reduction reaction A chemical reaction in which a molecule is reduced by gaining electrons. (Ch. 6)

referred pain Pain that is attributed to the surface of the body rather than its internal source. (Ch. 10)

reflex (ree-flex) An automatic, involuntary action. (Ch. 9)

refractory (re-frak-tor-ee) **period** The period between the peak of an action potential and the return of a membrane to the resting state. (Ch. 16)

regenerative process The basis for the generation of an action potential; the depolarization of a membrane after a threshold is reached. (Ch. 7)

relative pressure Pressure relative to the atmospheric pressure. (Ch. 20)

relative refractory period The period near the end of an action potential during which the threshold voltage needed to initiate a second action potential is large. (Ch. 7)

relaxin (re-lax-in) A hormone that relaxes pelvic ligaments in preparation for parturition. (Ch. 32)

REM generator *See* raphe nucleus.

REM sleep *See* rapid eye movement sleep.

renal (ree-nul) Pertaining to the kidneys. (Ch. 23)

renal hilum The indentation of the medial aspect of the kidneys. (Ch. 23)

renal pelvis The expanded upper portion of the ureter that collects the urine from the renal calyces. (Ch. 23)

renal plasma clearance The volume of plasma per unit time that must be completely freed (cleared) of a substance to supply the quantity of the substance excreted per unit time in the urine. (Ch. 23)

renin (reh-nin) A proteolytic enzyme, produced by granular cells in the kidney, that is involved in the regulation of thirst, osmolarity, and water volume. (Ch. 10)

renin-angiotensin system Sometimes called "renin-angiotensin-aldosterone system"; a hormonal system that plays an important role in blood pressure and extracellular fluid volume regulation. (Ch. 14, 24)

replication fork A structure that is produced by the separation of the two original DNA strands at the site of active synthesis. Many replication forks can exist on a DNA molecule at once. (Ch. 5)

repolarization The period during an action potential in which the membrane potential is restored to its original value. (Ch. 18)

residual volume The amount of air remaining in the lungs after a person expires as forcefully as possible. (Ch. 20)

respiration (res-per-ay-shun) The act or function of breathing; the act by which air is drawn in and expelled from the lungs; also called external respiration. (Ch. 20)

respiratory acidosis An abnormal process characterized by CO_2 accumulation, which leads to a fall in blood pH. (Ch. 25)

respiratory alkalosis An abnormal process characterized by the loss of too much CO_2, which leads to a rise in blood pH. (Ch. 25)

respiratory center The portion of the medulla that is responsible for the control of breathing. (Ch. 20)

respiratory distress syndrome Labored breathing that afflicts premature neonates born with insufficient amounts of surfactants in the lungs. (Ch. 20)

respirometer An instrument for the accurate measurement of oxygen consumption. (Ch. 27)

restriction endonuclease An enzyme that can break a strand of DNA at a specific site. (Ch. 3)

resting tension *See* passive tension. (Ch. 16)

retching A process by which the stomach contents are forced into the esophagus but not expelled. (Ch. 22)

reticular (reh-tik-yoo-lur) **formation** A diffuse network of neurons in the brain stem; involved in maintaining posture by the activation of extensor muscles. (Ch. 9)

reticulocytes Stem cells that have lost their nuclei but are not yet mature erythrocytes. (Ch. 17)

retina (ret-ih-na) The interior surface of the eye; contains photoreceptors and other visual system neurons. (Ch. 8)

retinotopic map The organization of the visual cortex in the form of a map of the external visual field. (Ch. 8)

retrograde (ret-row-grade) **transport** Transport against the main direction of flow. (Ch. 7)

retrolental fibrosis Blindness due to deterioration of the retina and inadequate development of the retinal blood vessels; the formation of fibrosis tissue behind the lens. (Ch. 29)

retroperitoneal (reh-trow-per-ih-tow-nee-al) External to the peritoneal lining of the abdominal cavity. (Ch. 22)

retropulsion The closure of the pyloric sphincter and forcible squirting of stomach contents back into the body of the stomach. (Ch. 22)

re-uptake A mechanism by which transmitter molecules are returned into the presynaptic terminal. (Ch. 7)

reward system Brain structures that provide pleasurable inner emotions in response to certain behaviors and drugs. (Ch. 11)

Rh factor Antigen D, which is genetically determined to be present or absent on red blood cells. (Ch. 17)

rhodopsin (row-**dop**-sin) A visual pigment; a photosensitive purple-red chromoprotein in the retinal rods that is bleached to visual yellow by light. (Ch. 8)

ribonucleic (rye-bo-new-**clay**-ik) **acid (RNA)** A single strand of nucleotides that represent a copy of a limited region of the nucleotide sequence of a strand of DNA. (Ch. 3)

ribosomal (rye-bo-**zome**-al) **RNA (rRNA)** One of three kinds of RNA synthesized; the nucleotide sequence on rRNA molecules provides information for protein synthesis. (Ch. 5)

ribosome (rye-bo-**sown**) Cellular organelles that make proteins according to the instructions carried by RNA. (Ch. 3)

rickets (**rik**-ets) A metabolic bone disease of children, characterized by inadequate mineralization of new bone matrix; caused by deficiency of vitamin D. (Ch. 26)

right heart The chambers of the heart that receive systemic venous blood and pump it to the lungs. (Ch. 18)

rigor mortis After death, a stiffness of skeletal muscles resulting from loss of ATP. (Ch. 7)

RNA polymerase Enzyme that synthesizes RNA during the transcription of DNA. (Ch. 5)

RNA processing (post-transcriptional modification) A series of reactions that modify newly synthesized RNA before it leaves the nucleus. (Ch. 5)

RNA splicing A type of RNA processing in which axon segments are joined together to form functional mRNA. (Ch. 5)

RNA translation The translation of codons on mRNA in the synthesis of proteins. (Ch. 5)

rod photoreceptors Cylindrically shaped photoreceptors that contain one type of pigment capable of absorbing photons representing a broad range of wavelengths. (Ch. 8)

Ruffini capsule A slowly adapting mechanoreceptor sensitive to the stretching of skin. (Ch. 8)

saccule (**sak**-yool) One of two sacs in the membranous labyrinth of the inner ear. (Ch. 9)

saltatory (**sal**-tah-*tow*-ree) **conduction** Conduction of a nerve impulse (action potential) by a myelinated fiber in which the impulse jumps from one node of Ranvier to the next. (Ch. 7)

sanguine Abounding in blood. (Ch. 1)

sarcolemma (*sar*-koe-**lem**-ma) The plasma membrane of a muscle cell, especially a skeletal muscle cell. (Ch. 16)

sarcomere (*sar*-koe-mere) A repeating structural unit of myofibril; composed of thick and thin filaments and extending between two adjacent Z lines. (Ch. 16)

satiety (sah-**tie**-ih-tee) A feeling of fullness or of not being hungry. (Ch. 11)

saturation The degree to which protein-binding sites are occupied by ligands. (Ch. 2)

scanning electron microscope A microscope that images the deflection of electrons by a specimen. (Ch. 3)

Schwann (shvon) **cell** A type of cell that forms the myelin sheath around a motor axon. (Ch. 7)

scientific measurements Quantitation of functions and processes, using the metric system of units. (Ch. 1)

scientific method A form of inquiry that involves posing questions and then searching for answers to those questions through experimentation and testing to validate answers. (Ch. 1)

scientific notation Quantities expressed as products of numbers plus a power of ten. (Ch. 1)

SCUBA An acronym for Self-Contained-Underwater-Breathing-Apparatus. (Ch. 29)

second heart sound A heart sound that occurs at the time of closure of the semilunar valves and defines the end of systole; the sound is caused by the vibrations in the heart and chest that are initiated by the closure of the heart valves and the sudden cessation of blood flow. (Ch. 18)

second messenger An extracellular signal that travels from a cell to a target cell, where it reacts with receptor molecules. (Ch. 5)

secondary somatosensory area The area of the somatosensory cortex that does not exhibit an annular receptive field; also called Brodmann areas 1 and 2. (Ch. 8)

secretin (**seek**-reh-tin) A strongly basic polypeptide hormone secreted by the upper small intestine; stimulates the pancreas to secrete bicarbonate into the small intestine. (Ch. 12, 22)

secretion (se-**kree**-shun) The elaboration and release of organic molecules, ions, and water by cells in response to specific stimuli. (Ch. 22)

sedimentation rate The rate at which red cells settle; the normal rate is between 2 and 10 mm/hr. (Ch. 17)

segmenting contraction The most frequently occurring type of muscular movement in the small intestine; moves the contents of different regions of the small intestine back and forth over short distances. (Ch. 22)

semicircular canal The three sensory organs of the vestibular apparatus of the ear that detect position and motion of the head in space; the **superior, inferior,** and **horizontal semicircular canals** are each situated perpendicular to the other two. (Ch. 9)

semiconservative replication A process in DNA replication in which the structure of the parent DNA molecule is disrupted but the structure of the nucleotide sequence of each strand is conserved. (Ch. 5)

semilunar (*sem*-ee-**loo**-nar) **valve** A valve located at the junction of the ventricle with the pulmonary and systemic arteries; consists of flaps shaped like half-moons. (Ch. 18)

sense organ Any organ of the sensory system, such as the eyes or ears. (Ch. 8)

sensitization An increase in the response to a stimulus. (Ch. 11)

sensor A mechanism that detects changes to bodily or environmental conditions. (Ch. 1)

sensory adaptation The gradual loss of perception of a stimulus that is continuously applied. (Ch. 8)

sensory neural hearing loss Hearing loss due to the damage to hair cells located between the basilar and tectorial membranes of the ear. (Ch. 8)

sensory neuron A neuron that carries information about external or internal stimuli from sense receptors to the central nervous system; also called an afferent neuron. (Ch. 7)

sensory system Parts of the peripheral and central nervous systems involved with the reception and processing of sensory information; sometimes subdivided into special sensory systems—vision, audition, equilibrium, gustation, olfaction, and somatic sensory systems—pain, touch, temperature, pressure, proprioception. (Ch. 8)

sensory transduction The conversion of energy (light, heat, sound, etc.) into nerve impulses by a sensory receptor. For example, the process by which the sensory receptors of the eye convert various wavelengths of light into electrical impulses that nerve cells transmit to the thalamus. (Ch. 8)

septal defect An opening in the wall (septum) that separates blood in the right heart from blood in the left heart. (Ch. 21)

serotonin (sair-oh-**toe**-nin) A vasoconstrictor synthesized in the brain stem; inhibits gastric secretion, stimulates smooth muscle, and serves as a central neurotransmitter. (Ch. 7)

sex chromosomes The X and Y chromosomes; females have two X chromosomes and males have one X and one Y; provide the genetic information for the development and differentiation of most of the cells in the body. (Ch. 33)

sex steroid binding globulin (SSBG) A plasma protein that binds with testosterone in the blood; aids in storing hormones in the blood. (Ch. 31)

sexual dysfunction Any condition in which the normal physical responses to sexual stimulation are impaired. (Ch. 33)

sexual flush A darkening of the body's skin color during sexual intercourse. (Ch. 33)

shivering Involuntary trembling or quivering of the body caused by contraction or twitching of the muscles; a physiologic method of heat production. (Ch. 27)

short portal system A vascular system that connects the posterior and anterior lobes of the pituitary; may carry substances secreted by the posterior lobe to the anterior lobe where they regulate hormone secretion from the anterior pituitary. (Ch. 13)

short-reflex A regulatory reflex that is mediated over sensory neurons, interneurons, and motor and secretory neurons that lie entirely within the wall of the digestive tract. (Ch. 22)

short-term memory A section of the brain's memory bank that is easily disrupted and short-lived. (Ch. 11)

Siemens Dimensional units of conductance. (Ch. 7)

silicosis (sil-ih-**koe**-sis) A lung disease caused by breathing silicon (sand) dust. (Ch. 29)

single channel current The number of ions that flow through a channel per unit of time. (Ch. 7)

sino-atrial (*sigh*-no-**ay**-tree-al) **node (SA node)** The region, in the right atrium of the heart, containing specialized cardiac cells that depolarize spontaneously faster than other heart cells; determines heart rate; also called pacemaker of the heart. (Ch. 18)

skeletal muscle The muscle that produces voluntary movement of body parts. (Ch. 9)

sleep apnea A condition in which respiration can stop for long periods of time (often 30 to 60 seconds) during sleep. (Ch. 20)

sliding-filament hypothesis The hypothesis that explains muscle contraction on the basis of overlapping, interdigitating, thick (myosin) and thin (actin) filaments that are drawn past each other by the action of myosin cross bridges, with ATP providing the energy. (Ch. 16)

slow axonal transport The movement of proteins along the axon from the soma to the terminal region. (Ch. 7)

slow skeletal muscle The muscle that contracts slowly and is more resistant to fatigue. (Ch. 9)

slowly adapting tactile mechanoreceptor A type of receptor that continues to generate action potentials as long as a stimulus is applied. (Ch. 8)

small intestine A portion of the gastrointestinal system consisting of the duodenum, jejunum, and ileum; digestion is completed here by enzymes in the epithelial cells that line the inner surface, and the products of digestion are absorbed into the blood and lymph. (Ch. 22)

smooth muscle A muscle type associated with blood vessels and other organ systems. (Ch. 9)

sodium appetite A craving for salt. (Ch. 24)

sodium pump The Na/K-ATPase that uses energy from ATP hydrolysis to drive active transport of Na^+ ions out of cells and K^+ ions in. (Ch. 4)

soft tissue calcification The deposition of calcium in tissues. (Ch. 26)

solute (**sol**-yoot) A substance that is uniformly distributed in a solvent to form a solution. (Ch. 2)

solution A type of mixture in which one substance is uniformly distributed throughout another substance, which is usually a liquid. (Ch. 2)

solvent (**sol**-vent) A substance in which a solute is uniformly distributed to form a solution. (Ch. 2)

soma The body and metabolic center of a neuron. (Ch. 7)

somatic Pertaining to the skin and skeletal muscles of the body. (Ch. 7)

somatic (so-**mat**-ik) **nervous system** Sensory nerves that transmit information from sense receptors to the central nervous system and motor nerves that transmit instructions about appropriate responses of skeletal muscles. (Ch. 7)

somatic sensory cortex The portion of the neocortex that responds to sensations felt on the surface of the skin and the position of limbs relative to the body. (Ch. 7)

somatomammotropic hormones A group of related peptide hormones that stimulate body growth and breast development; human growth hormone (HGH) and human chorionic somatomammotropin (HCS) are examples. (Ch. 13)

somatomedin (*so*-mah-tow-**me**-din) A growth factor found in many tissues, including the liver, that mediate the effect of growth hormone on cartilage. (Ch. 13)

somatostatin (*so*-mat-oh-**stat**-in) A hypothalamic hormone that inhibits growth hormone and TSH secretion by the anterior pituitary; a possible neurotransmitter; also in the stomach and pancreatic islets. (Ch. 13)

somatotropin *See* growth hormone.

spatial summation The additive effects of two simultaneously active synaptic inputs. (Ch. 7)

specific defense Defense mechanisms that engage and destroy specific kinds of pathogens (e.g., antibody production). (Ch. 28)

specific dynamic action The increased heat production that results from processing food. (Ch. 26)

specific gravity The ratio between the weight of a given volume of a liquid and the weight of an equal volume of pure water. (Ch. 17)

specificity The ability of a binding site to react with only one (or a limited number of) type of molecule. (Ch. 5)

spectrin An extrinsic protein found only on the cytoplasmic side of the red blood cell membrane; thought to be important in determination of red blood cell shape. (Ch. 4)

sperm The male germ cell; also called spermatozoa. (Ch. 31)

sperm loss Nonviable sperm. (Ch. 32)

sperm viability The ability of a sperm to fertilize an ovum. (Ch. 33)

spermatogenesis (*spur*-mah-tow-**jen**-eh-sis) The formation of sperm. (Ch. 31)

spermicide (**spur**-muh-side) A chemical that kills sperm; one form of contraception. (Ch. 33)

spherical micelle A small sphere formed by phospholipids in solution. (Ch. 4)

sphincter (**sfink**-ter) **of Oddi** A smooth muscle ring surrounding the bile duct at its entrance into the duodenum. (Ch. 22)

spinal (**spy**-nal) **autonomic reflex arc** A loop of nerve cell activity; begins with the detection of visceral information by autonomic sensory receptors in an organ that in turn activates interneurons within the spinal cord. These neurons synapse with the cells of the IML; the IML cells integrate incoming information and discharge to activate postganglionic neurons. These neurons convey information to the target organ, causing it to make the appropriate response. (Ch. 10)

spinal cord The part of the central nervous system that transmits information to and receives information from most of the body's internal organs, muscles, and skin; located caudal to the brain stem. Consists of 31 segments, each with a pair of spinal nerves. (Ch. 7)

spirogram The record of lung volumes measured with a spirometer. (Ch. 20)

spirometry The measurement of pulmonary volumes and capacities. (Ch. 20)

splanchnic circulation The circulation in the stomach, small and large intestines, pancreas, and liver. (Ch. 22)

stable balance The condition in which there is no net gain or loss of a substance. (Ch. 3)

stance phase The phase of the walk cycle in which the foot is on the ground and supporting the weight of the body. (Ch. 9)

standard free energy change $\Delta G°$, or the difference in total free energy between reactants and products of a chemical reaction under a set of standard conditions: temperature of 25°C, pressure of 1 atmosphere, and a concentration of 1.0 M for all reactants and products. (Ch. 6)

stapes (**stay**-peas) The innermost of the auditory ossicles, shaped somewhat like a stirrup; also called stirrup. (Ch. 8)

Starling's hypothesis Description of the principle that net filtration pressure is proportional to the capillary hydrostatic pressure minus the oncotic pressure of the blood; not to be confused with the Frank-Starling law of the heart. (Ch. 19)

start codon A codon that initiates translation of nucleotide triplets. (Ch. 5)

stasis (**stay**-sis) A slowing down or stoppage of flow, such as of blood or other body fluid. (Ch. 17)

static exercise Exercise that involves isometric muscle contractions. (Ch. 30)

statistics The science that deals with the collection, tabulation, and analysis of numerical facts. (Ch. 1)

statoconia Calcium carbonate crystals contained within the otolithic membrane. (Ch. 9)

steady state A condition unvarying with respect to time. (Ch. 5)

steatorrhea The appearance of excessive amounts of undigested fat in the feces; usually due to the absence of lipase in the gastrointestinal system. (Ch. 22)

stereocilia Fiberlike elements that protrude from hair cells located between the basilar and tectorial membranes of the ear. (Ch. 8)

steroid (**steer**-oid) A lipid subclass; the molecule consists of four interconnected carbon rings to which polar groups may be attached. (Ch. 11)

steroid hormones Hormones derived from cholesterol. They are the glucocorticoids, mineralocorticoids, androgens, estrogens, and progestins. (Ch. 12)

stomach (**stum**-ak) The portion of the gastrointestinal system located between the esophagus and the small intestine; functions as a storage organ for food. (Ch. 22)

stop codon A codon that terminates translation of nucleotide triplets. (Ch. 5)

stretch receptors Afferent nerve endings that are depolarized by stretching; *see also* muscle spindle receptor. (Ch. 30)

stretch reflex Reflex contraction of skeletal muscle in response to its stretch. (Ch. 9)

stria A narrow bandlike structure; a general term for such longitudinal collections of nerve fibers in the brain. (Ch. 14)

striate (**stri**-ate) **muscle** *See* skeletal muscle.

striatum An area beneath the neocortex that is involved in the initiation and coordination of movement. (Ch. 7)

stroke volume The blood volume ejected by a ventricle during one heartbeat. (Ch. 18)

strong acid An acid with a high dissociation constant, which consequently is highly ionized in solution. (Ch. 25)

structural formula A formulaic illustration of the arrangement in space of the atoms of a molecule. (Ch. 2)

subcutaneous (*sub*-kyoo-**tay**-nee-us) Beneath the skin. (Ch. 27)

sublingual (sub-**ling**-gwal) Below the tongue. (Ch. 27)

submucosa (*sub*-myoo-**koe**-sah) A connective tissue layer under the mucosa in the gastrointestinal tract. (Ch. 22)

submucosal plexus A nerve-cell network in the submucosa of the esophageal, stomach, and intestinal walls. (Ch. 22)

substantia gelatinosa A group of nerve cells located in the dorsal horn of the spinal cord. (Ch. 8)

substantia nigra An area of the brain associated with the reward system. (Ch. 11)

substrate-level (**sub**-straight) **phosphorylation** The direct formation of high-energy phosphate compounds such as ATP by a chemical reaction that incorporates a phosphate group. (Ch. 6)

suckle To derive or to provide nourishment by feeding. (Ch. 33)

sucrase An enzyme that digests sucrose to glucose and fructose. (Ch. 22)

sudden infant death syndrome (SIDS) A disease in which a seemingly normal, healthy infant is found dead in its crib; sleep apnea is being studied extensively as a causative link. (Ch. 20)

sulfonyl ureas-sulfonylureas Oral hypoglycemic agents that act by stimulating insulin secretion from pancreatic beta cells. Used to treat type II diabetes mellitus. (Ch. 15)

summation The cumulative effects of a number of stimuli applied to a muscle nerve or reflex arc. (Ch. 16)

superior Literally, nearer to the head. (Ch. 1)

superior vena cava (**vee**-nah **kay**-vah) A large vein that carries blood from the upper half of the body to the right atrium of the heart. (Ch. 18)

superoxide anion radical An active form of free radical. (Ch. 29)

superoxide dismutase A protective cellular enzyme that removes free radicals. (Ch. 29)

supplemental motor cortex An area of the brain that designs the necessary movements required to accomplish a task; also called Brodmann area 6. (Ch. 9)

suprachiasmatic nucleus The nucleus of the hypothalamus, thought to be responsible for coordinating various biological rhythms. (Ch. 11)

supraoptic nucleus Cells in the hypothalamus that secrete vasopressin into the blood in response to increases in osmolarity. (Ch. 10)

surface tension A molecular force created at the gas–liquid interface; specifically, the attractive forces between the liquid and air molecules. (Ch. 20)

surfactant (ser-**fak**-tant) A lipoprotein produced by pulmonary alveolar cells; a detergentlike material that reduces surface tension of fluid film lining alveoli. (Ch. 20)

swallowing The propulsion of a food bolus through the pharynx and esophagus into the stomach; the bolus moves in response to skeletal and smooth muscle contractions that occur sequentially from above to below. (Ch. 22)

sweating The active secretion of water and electrolytes by specialized sweat glands in the skin for the purpose of increasing evaporative heat loss; activated by sympathetic nerves in response to overheating; increases linearly with abnormal increases in body temperature. (Ch. 27)

swing phase The phase of the walk cycle in which the foot is off the ground and swinging forward. (Ch. 9)

sympathetic (*sim*-pah-**thet**-ik) **nervous system** One portion of the autonomic nervous system; preganglionic fibers leave the CNS at the thoracic and lumbar portions of the spinal cord. (Ch. 10)

sympathoadrenal (*sim*-pah-thow-ah-**dree**-nal) **system** The sympathetic nervous system and the adrenal medulla together. (Ch. 14)

sympathomimetic (*sim*-pah-thow-my-**met**-ik) Adrenergic; producing effects resembling those produced by the sympathetic nervous system. (Ch. 10)

symport The movement of solutes in the same direction as Na^+. (Ch. 4)

synapse (**sin**-aps) The junction between nerve cells; **electrical synapses** consist of gap junctions that permit direct passage of ions, and **chemical synapses** prevent the direct passage of ions. (Ch. 7)

synaptic (sin-**ap**-tik) **cleft** In a chemical synapse, the space, between one cell and another, that prevents the direct passage of ions. (Ch. 7)

synaptic transmission The process by which there is communication between nerve cells. (Ch. 7)

synaptic vesicles Small structures filled with neurotransmitting chemicals that fuse with the presynaptic membrane. (Ch. 7)

syncytium (sin-she-um) A multinucleate mass of protoplasm produced by the merging of cells. (Ch. 16)

synergistic   (sin-er-**jis**-tik)   **muscle** A muscle that works together with another to produce a given movement. (Ch. 9)

systemic (sis-tem-ik) **circulation** The circulation from the left ventricle through all the organs except the lungs and back to the heart. (Ch. 18)

systemic hypertension A complex disease state resulting in chronically elevated arterial blood pressure. (Ch. 19)

systole (sis-toe-lee) A period of ventricular contraction. (Ch. 18)

systolic pressure The peak arterial pressure, which occurs during contraction of the ventricle. (Ch. 19)

T cell A lymphocyte derived from a precursor that differentiated in thymus; *see also* cytotoxic T cell, helper T cell, suppressor T cell. (Ch. 28)

tachycardia (*tak*-ih-**kar**-dee-ah) An abnormally rapid heart rate, usually defined as more than 100 beats per minute. (Ch. 18)

tactile (tak-tile) **receptor** A mechanoreceptor that detects touch, pressure, and vibrations applied to the skin. (Ch. 8)

target cell A cell whose activity is affected by a specific hormone. (Ch. 5)

target tissue Tissue containing cells that have receptors for a particular neurotransmitter or hormone. (Ch. 12)

taste bud A cluster of gustatory receptors, basal cells, and supporting cells. (Ch. 8)

tectorial (tek-toe-ree-al) **membrane** A membranous structure in which the hair cells of the cochlea are embedded. (Ch. 8)

tectospinal tract A component of the ventromedial pathway that originates in the tectum. (Ch. 9)

tectum (tek-tum) A structure involved in the coordinated control of heat and eye movements. (Ch. 9)

temperature The metabolic heat produced within the body. (Ch. 27)

temperature regulating system A system that works to maintain thermal homeostasis in the body. (Ch. 30)

temporal summation The addition of postsynaptic potentials that arise from activation at one synaptic input. (Ch. 7)

tendon A collagen fiber bundle that connects muscle to bone and transmits muscle contractile force to the bone. (Ch. 16)

tenting The elevation of the uterus away from the vagina as the blood flow to the vagina increases during the plateau phase of sexual arousal. (Ch. 33)

terminal cisterna Enlarged end portions of the sarcoplasmic reticulum; also called lateral sacs. (Ch. 16)

terminator site on DNA A sequence of nucleotides immediately following the 5' end of the gene; acts as a recognition site for the ending of the RNA strand. (Ch. 5)

testicular feminization syndrome A condition in which the internal genitalia of the male do not develop; usually due to a deficiency of testosterone biosynthesis or when Wolffian ducts do not respond to testosterone. (Ch. 33)

testis The male gonad. (Ch. 11)

testosterone (tes-tos-teh-rone) A steroid hormone produced in interstitial cells of the testes; essential for spermatogenesis and maintaining growth and development of reproductive organs and secondary sexual characteristics of males. (Ch. 12)

tetanic fusion frequency The lowest frequency that will produce a fused tetanus; around 20 to 60 times per second for most skeletal muscles. (Ch. 16)

tetanus (tet-ah-nus) A maintained mechanical response of muscle to high-frequency stimulation; the disease lockjaw. (Ch. 16)

tetany (tet-ah-nee) A syndrome manifested by sharp flexion of the wrist and ankle joints; occurs in response to low concentrations of extracellular calcium ion. (Ch. 26)

thalamus (thal-uh-mus) An area of the brain that receives all types of sensory information. (Ch. 8)

thalassemia (*thal*-ah-**see**-me-ah) A group of inherited hemolytic anemias characterized by decreased synthesis of one or more hemoglobin polypeptide chains. (Ch. 17)

theory A formulated hypothesis. (Ch. 1)

thermal homeostasis Physiological regulatory mechanisms continuously operating to keep heat production and heat loss approximately equal. (Ch. 27)

thermal receptor A sensory receptor that responds to temperature. (Ch. 8)

thermography A technique for measuring mean skin temperature in humans. (Ch. 27)

thermoreceptor (*ther*-mow-ree-**sep**-tor) A receptor that detects changes in temperature. (Ch. 1)

thermostat A controller to prevent excessive body heating or cooling. (Ch. 27)

thick myofilament A 12- to 18-nm myosin filament in a muscle cell. (Ch. 16)

thin myofilament A 5- to 8-nm myosin filament in a muscle cell; consists of actin, troponin, and tropomyosin. (Ch. 16)

thirst A conscious sensation associated with a craving for drink; ordinarily interpreted as a desire for water. (Ch. 24)

thoracic (thow-ras-ik) **cavity** The chest cavity. (Ch. 20)

threshold The point when a membrane potential is depolarized sufficiently to produce an action potential. (Ch. 18)

thrombin An enzyme that catalyzes conversion of fibrinogen to fibrin. (Ch. 17)

thrombocyte (throm-bow-site) A fragment of cytoplasm, enclosed in plasma membrane, that plays a role in blood clotting; found in the circulation. Also called platelet. (Ch. 17)

thrombocytopenia (throm-boe-sigh-toe-**pee**-nee-ah) A deficiency in circulating platelets; results in excessive bleeding. (Ch. 17)

thromboplastin (throm-bow-plas-tin) A complex of several phospholipids and a proteolytic enzyme released from damaged tissue; important in the clotting process. (Ch. 17)

thrombosis (throm-**bow**-sis) The formation or presence of a clot in an unbroken blood vessel or heart chamber. (Ch. 17)

thrombus A blood clot. (Ch. 17)

thymus A primary lymphoid gland located in the lower anterior neck. T-lymphocytes mature in the thymus. (Ch. 28)

thyrocalcitonin A peptide hormone, secreted by the parafollicular cells of the thyroid, that causes increased deposition of calcium in bones; also called calcitonin. (Ch. 26)

thyroglobulin (*thy*-row-**glob**-you-lin) A large protein to which thyroid hormones bind in the thyroid gland; a storage form of thyroid hormones. (Ch. 13)

thyroid (thy-roid) **follicle** The internal structure of the thyroid gland as a collection of closely packed follicles; normal thyroid contains about 3 million follicles. (Ch. 13)

thyroid hormone (TH) A collective term for amine hormones released from the thyroid gland; that is, thyroxine (T4) and triiodothyronine (T3). (Ch. 13)

thyroid-stimulating hormone (TSH) A glycoprotein hormone secreted by the anterior pituitary; induces secretion of thyroid hormone. Also called thyrotropin. (Ch. 13)

thyrotropin-releasing hormone (TRH) A hypothalamic hormone that stimulates thyrotropin and prolactin secretion by the anterior pituitary. (Ch. 13)

thyroxin (thy-rok-sin) A hormone, secreted by the thyroid gland, that accelerates the rate at which cells consume oxygen, metabolize glucose, and produce heat. (Ch. 11)

thyroxine-binding globulin (CBG) A carrier protein that binds and carries mainly thyroxine in the blood. (Ch. 12)

tidal volume The volume of air inhaled

or exhaled during a single breath. (Ch. 20)

"tight" epithelia Epithelia that do not permit leakage between the cells. (Ch. 4)

tight junction A specialized attachment that connects Sertoli cells to one another inside the seminiferous tubules; the area where plasma membranes are joined. (Ch. 31)

time period The time between the peaks of a sound wave of a tone; the inverse of the frequency of a sound wave. (Ch. 8)

tinnitus (tie-**nie**-tis) A constant ringing in the ear due to the inappropriate discharge of action potentials by particular nerve fibers associated with specific frequencies. (Ch. 8)

tissue An aggregate of differentiated cells of similar type united in the performance of a particular function; also denotes the general cellular fabric of a given organ. (Ch. 1)

titratable acid A solution whose acid concentration can be determined by measuring the milliequivalents of strong base needed to bring the pH to a certain level; for example, the amount of base needed to bring a urine sample back to the pH of arterial blood. (Ch. 25)

tone *See* muscle tone.

tonic contraction A long, sustained contraction in response to single or continued stimulation by nerves, drugs, or hormones. (Ch. 16)

tonic discharge An action potential generated at one specific frequency in response to the position of a joint. (Ch. 8)

tonotopic map The organization of auditory nerve fibers such that fibers with a high characteristic frequency innervate hair cells at the base of the cochlea, while those with a low characteristic frequency innervate hair cells near the apex of the cochlea. (Ch. 8)

tonus The constant steady force of contraction of most smooth muscles; also present in skeletal muscles. (Ch. 16)

total peripheral resistance (TPR) The total resistance to flow in systemic blood vessels from the beginning of the aorta to the end of the venae cavae. (Ch. 19)

trachea (**tray**-kee-ah) The main airway of the respiratory system, which branches into two bronchi. (Ch. 20)

transcellular fluid Specialized extracellular fluid that is separated from the blood plasma not only by a capillary endothelium but by a continuous layer of epithelial cells; it includes aqueous humor, cerebrospinal fluid, digestive secretions, sweat, and renal tubular and bladder urine. (Ch. 24)

transcellular transport pathway The movement of water and solutes through epithelial cells; entry and exit pathways are on opposite sides of the cell. (Ch. 4)

transcription (tran-**skrip**-shun) The synthesis of mRNA using a DNA template; the first step in the transfer of genetic information from DNA to proteins. (Ch. 2)

transcytosis (tran-si-**toe**-sis) An active process by which relatively large molecules can cross the capillary wall by moving through the epithelial cells either in discrete shuttling vesicles or through transient vesicle-derived channels. (Ch. 19)

transducin A protein present in the disc membrane of photoreceptors. (Ch. 8)

transfer RNA (tRNA) A type of RNA; different tRNAs combine with different amino acids and with codon on mRNA specific for that amino acid, thus arranging amino acids in sequence to form specified proteins. (Ch. 5)

transferrin An iron-binding protein carrier for iron in plasma. (Ch. 17)

translation The synthesis of a polypeptide using mRNA as a template. (Ch. 2)

transmission electron microscope A microscope that uses magnetic lenses to focus a beam of electrons in much the same way that glass lenses in ordinary microscopes focus a beam of light; specimens must be fixed and stained to be seen under a transmission electron microscope. (Ch. 3)

transmitter *See* neurotransmitter.

transpulmonic pressure The pressure across the lung; measured by subtracting the pleural pressure from the alveolar pressure. (Ch. 20)

transthyretin (trans-**thigh**-re-tin) A plasma protein that acts as a transporter for thyroxine. (Ch. 12)

transudation The vasoconstriction of the vagina, leading to the release of moisture from the vaginal lining; lubricates the vagina to facilitate insertion of the penis without discomfort. (Ch. 33)

transverse plane An anatomical position that describes a plane dividing the body into upper and lower portions. (Ch. 1)

transverse tubule (t-tubule) A tubule extending from striated muscle plasma membrane into the fiber, passing between opposed sarcoplasmic-reticulum segments; conducts muscle action potential into muscle fiber. (Ch. 16)

triacylglycerol (tri-ace-il-**gliss**-er-al) A subclass of lipids composed of glycerol and three fatty acids; also called fat, neural fat, or triglyceride. (Ch. 6)

triad The region of contact in skeletal muscles of one t-tubule and the terminal cisternae from two adjacent sarcomeres. (Ch. 16)

trichromatic color vision The ability to discriminate between different colors on the basis of three types of photoreceptors: blue, red, and green cones.

tricuspid valve The valve between the right atrium and the right ventricle of the heart. (Ch. 18)

triglyceride *See* triaclyglycerol.

triiodothyronine (*try*-eye-oh-dow-**thy**-row-neen) **(T3)** An iodine-containing amine hormone secreted by the thyroid gland. (Ch. 13)

trimesters The three 3-month periods of pregnancy. (Ch. 32)

trophic Of or pertaining to nutrition. (Ch. 7)

tropomyosin A regulatory protein of the I-band that inhibits contraction unless its position is modified by troponin so that myosin molecules can make contact with actin. (Ch. 16)

troponin A regulatory protein bound to actin and tropomyosin of striate-muscle thin filaments; a site of calcium binding that initiates contractile activity. (Ch. 16)

trypsin (**trip**-sin) A protein-digesting enzyme secreted by the pancreas. (Ch. 2, 22)

tryptophan (**trip**-toe-fan) An essential amino acid; a precursor to serotonin. (Ch. 7)

tubal ligation (lie-**gay**-shun) The cutting and tying-off of both fallopian tubes, resulting in sterilization. (Ch. 33)

tubular reabsorption The transport of materials by the kidney tubule epithelium out of the tubular urine. (Ch. 23)

tubular secretion The transport of materials by the kidney tubule epithelium into the tubular urine. (Ch. 23)

tubular transport maximum (T_m) The maximum rate at which a particular substance is reabsorbed (e.g., glucose) or secreted (e.g., PAH) by the kidney tubules. (Ch. 23)

turbulent flow Completely disorganized patterns of air flow, which occur at high flow rates. (Ch. 20)

Turner's syndrome A chromosomal abnormality in females in which there is only one X sex chromosome; characterized by impaired development of female gonads, resulting in the appearance of an incompletely formed gonad. (Ch. 33)

T-wave A component of an electrocardiogram, corresponding to ventricular repolarization. (Ch. 18)

twitch The mechanical response of muscle to a single maximal stimulus. (Ch. 16)

tympanic (tim-**pan**-ik) **membrane** A thin, semitransparent membrane that stretches across the ear canal; separates outer and middle ear. (Ch. 8)

type I diabetes Diabetes in which insulin

is completely or almost completely absent; also called insulin-dependent diabetes. (Ch. 15)

type II diabetes Diabetes in which insulin is present at near-normal or even above-normal levels, but the tissues do not respond optimally to it; also called insulin-independent diabetes. (Ch. 15)

tyrosine An amino acid; the precursor of catecholamines and thyroid hormones. (Ch. 12)

tyrosine hydroxylase An enzyme that converts tyrosine to dihydroxyphenylalanine. (Ch. 14)

ulcer (**ul**-ser) An erosion of the mucosa of certain regions of the digestive tract. (Ch. 22)

ultrafiltrate (ul-tra-**fill**-trate) An essentially protein-free fluid formed from plasma as it is forced through capillary walls by a pressure gradient. (Ch. 23)

ultrafiltration Filtration through a membrane with exceedingly fine pores. (Ch. 23)

umbilical (um-**bil**-ih-kal) **cord** An artery or vein that transports blood between fetus and placenta. (Ch. 32)

unconditioned stimulus A stimulus that is paired with a conditioned stimulus to evoke a conditioned response. (Ch. 11)

unitary postsynaptic potential A change in the membrane potential due to the flow of synaptic current resulting from the release of transmitters. (Ch. 7)

unitary smooth muscle The smooth muscle that makes up the bulk of the visceral organs; less strictly controlled by the nervous system; a large number of its cells function together as a single unit. (Ch. 16)

upper esophageal sphincter A band of skeletal muscle between the pharynx and the body of the esophagus. (Ch. 22)

urea (you-**ree**-ah) The major nitrogen-containing end product of metabolism; formed chiefly by the liver. (Ch. 23)

urea cycle A cyclic series of reactions that produce urine; a major route for removal of ammonia. (Ch. 14)

uremia (you-**ree**-mee-ah) The symptom that results from renal failure. (Ch. 23)

ureter (you-**ree**-ter) The tube that carries urine from the kidney to the urinary bladder. (Ch. 23)

urethra (yoo-**ree**-thra) The tube that carries urine from the urinary bladder to the outside. (Ch. 23)

uric acid An end product of purine metabolism, secreted by the kidneys. (Ch. 23)

uricosuric An agent that promotes excretion of uric acid in the urine. (Ch. 23)

urinary bladder A hollow, muscular organ that collects the urine and is periodically emptied. (Ch. 23)

utricle (**yoo**-trih-kul) The largest of the two divisions of the membranous labyrinth of the inner ear. (Ch. 9)

vacuole (**vac**-you-ohl) A cavity enclosed by a membrane and located in cytoplasm. (Ch. 3)

vaginismus Involuntary contractions and spasms of the vaginal or pubococcygeal muscles; makes vaginal penetration difficult or impossible. (Ch. 33)

vagus (**vay**-gus) The tenth cranial nerve. Motor functions include control of laryngeal and pharyngeal muscles and muscles of the esophagus, stomach, heart, lungs, and intestines (swallowing, control of heart rate, gastric secretion, etc.). Sensory functions include input from viscera or thorax and abdomen (blood pressure from aorta, pain from stomach, etc.). (Ch. 10)

Van Allen belt Particles of cosmic radiation trapped in the earth's magnetic field. (Ch. 29)

van der Waals bond Weak attractive forces between nonpolar regions of molecules. (Ch. 2)

variables The related quantities displayed on a graph. (Ch. 1)

vas deferens (**def**-ur-enz) Components of the spermatic cords that arise from the tail of the epididymis; also called ductus deferens. (Ch. 31)

vasa recta The blood vessels of the medulla that supply the nutritional needs of the kidney medulla as well as act as countercurrent exchangers. (Ch. 23)

vascular (**vas**-kyoo-lar) **bed** Many separate circulatory pathways that supply blood to various organ systems within the body. (Ch. 18)

vascular reactivity The degree of responsiveness or sensitivity of vascular smooth muscle to agents which stimulate contraction or elevate blood pressure. Glucocorticoids, for example, enhance vascular reactivity. (Ch. 14)

vascular resistance The total frictional force opposing blood flow through a particular vascular system. See total peripheral resistance (TPR). (Ch. 18)

vasectomy (vah-**sec**-tow-mee) The cutting and tying-off of both ductus deferens, which results in sterilization of the male without loss of testosterone. (Ch. 33)

vasocongestion The concentration of blood in the tissues of the penis, which causes an increase in diameter, length, and firmness. (Ch. 33)

vasoconstrictor (vas-oh-kon-**strik**-tor) A substance that causes contraction of the smooth muscle associated with blood vessels leading to a decrease in vessel diameter. (Ch. 19)

vasodilator (*vas*-oh-die-**lay**-tor) A substance that produces relaxation of the smooth muscle associated with blood vessels leading to an increase in vessel diameter. (Ch. 19)

vasomotion (vas-oh-**moe**-shun) Contraction and relaxation of the precapillary sphincter smooth muscle which typically results in intermittent blood flow through capillaries. (Ch. 19)

vasopressin (*vas*-oh-**press**-in) A hormone, secreted by the paraventricular nucleus of the hypothalamus, that causes the kidneys to reabsorb more water. (Ch. 10)

vein (vane) Any vessel that returns blood to the heart. (Ch. 1)

venomotor tone The contraction of venous smooth muscle. (Ch. 19)

venous capacitance The total blood volume in the veins. (Ch. 19)

venous return The rate (ml/min) of blood return to the right atrium; usually equal to cardiac output. (Ch. 19)

venous system The vessels that carry the blood back to the heart; begins at the point where the capillaries reunite to form venules. (Ch. 18)

ventilation The process of moving air in and out of the lungs. (Ch. 20)

ventilation-perfusion balance A mechanism for balancing regional blood flow with regional alveolar ventilation. (Ch. 21)

ventral An anatomical position that describes the area of the body; specifically, toward, or at the front of, the body. (Ch. 1)

ventral corticospinal tract Fibers from the motor cortex that descend to the spinal cord without crossing the midline of the body; primarily innervate motor neurons in the medial region of the ventral horn associated with axial muscles of the body. (Ch. 9)

ventral posterior lateral (VPL) thalamus The region of the thalamus that receives somatic sensory information. (Ch. 8)

ventral posterior medial (VPM) thalamus The region of the thalamus that receives information from taste receptors. (Ch. 8)

ventricle (**ven**-trih-kul) Any cavity. (Ch. 18)

ventricular ejection Part of the pumping process of the heart when the semilunar valve is forced open and blood flows out of the ventricle; this occurs when ventricular pressure exceeds the pressure in the outflow tract. (Ch. 18)

ventricular fibrillation The uncoordinated stimulation of the ventricular muscle; useless for pumping blood. (Ch. 18)

ventricular filling Part of the pumping process of the heart when the ventricle is being filled with blood from the atrium. (Ch. 18)

ventricular relaxation Part of the pumping process of the heart when the ventricle relaxes and pressure inside the ventricle falls; the semilunar valve is forced closed as blood attempts to flow from the outflow track back into the ventricle. (Ch. 18)

ventromedial pathway A pathway that descends along the ventral and medial region of the spinal cord. (Ch. 9)

venule (ven-yule) A small vessel that carries blood from the capillary network to a vein. (Ch. 18)

vestibular (ves-tib-you-lar) **nerve fiber** A nerve fiber that innervates the hair cells of the vestibular apparatus of the ear. (Ch. 9)

vestibular system Parts of the peripheral and central nervous systems involved with the reception and processing of sensory information regarding body balance, orientation in space, linear acceleration and rotational (angular) acceleration. (Ch. 9)

vestibulospinal tract A component of the ventromedial tract that originates in the vestibular nucleus and carries information for the reflex control of equilibrium. (Ch. 9)

villus A small, fingerlike projection that creates the convoluted surface of the small intestine. (Ch. 22)

viscera (vis-ur-uh) Organs of the thoracic and abdominal cavities. (Ch. 10)

visceral (vis-er-al) **afferent fiber** Nerve fibers that carry information from an organ to the spinal cord. (Ch. 10)

visceral efferent fiber Nerve fibers that carry information from the spinal cord to an organ. (Ch. 10)

visceral muscles Muscles that line the walls of the internal organs. (Ch. 16)

viscosity (vis-kos-ih-tee) A property of fluid that makes it resist flow. (Ch. 17)

visual agnosia Difficulty recognizing familiar objects or colors due to lesions in the secondary visual cortices. (Ch. 8)

visual cortex The area of the neocortex that responds to visual images seen by the eye. (Ch. 7)

visual field The area that can be seen through either the right or the left eye. (Ch. 8)

visual orientation columns Columns of nerve cells in the visual cortex that are sensitive to bars of light oriented at different angles. (Ch. 8)

visual pigment The photoreceptor pigment that absorbs photons. (Ch. 8)

visual system The physiological system that perceives light to determine the shape, color, and grouping of objects in the environment. (Ch. 8)

visuotopic map A precise mapping of the visual field onto the visual cortex. (Ch. 8)

vitamin A general term for a number of unrelated organic molecules that occur in many foods in trace amounts and that are necessary for normal growth and development. They may be water-soluble or fat soluble. (Ch. 26)

vitamin D$_3$ Cholecalciferol, the natural form of vitamin D, obtained from the diet or formed in the skin by the action of ultraviolet light on 7-dehydrocholesterol. (Ch. 26)

volt A unit of electric potential. (Ch. 7)

volume regulation Mechanism by which most cells are able to maintain normal volume during periods of osmotic stress. (Ch. 4)

voluntary muscles Muscles that move in response to conscious effort. (Ch. 16)

vomiting The rapid expulsion of gastric contents through the mouth. (Ch. 22)

warm receptors Peripheral temperature receptors sensitive to temperature; their steady-state discharge rates increase in response to warming of the skin. (Ch. 27)

warm-blooded *See* homeothermy.

water diuresis Excretion of a large volume of osmotically dilute urine. (Ch. 24)

watt A unit measurement of power. (Ch. 30)

wavelength A characteristic of electromagnetic radiation that is unique for each part of the continuum of radio waves to X-rays. (Ch. 8)

weak acid An acid that does not dissociate completely when dissolved in water. (Ch. 4)

weak base A base that does not dissociate completely when dissolved in water. (Ch. 4)

"wedge" pressure A pressure measurement made during cardiac catheterization; near-left atrial pressure. (Ch. 21)

weightlessness Lacking gravitational pull. (Ch. 29)

Wernicke's area An area of the neocortex responsible for the interpretation of language. (Ch. 7)

Wolffian duct system A duct system in the embryo that is the precursor to the internal genitalia of the male. (Ch. 33)

x-**axis** The abscissa or horizontal axis of a graph. (Ch. 1)

y-**axis** The ordinate or vertical axis of a graph. (Ch. 1)

Z-disc Another term for Z-line to emphasize its two-dimensional nature. (Ch. 16)

Z-line A structure running across a myofibril at each end of a striated muscle sarcomere; anchors one end of the thin filaments. (Ch. 16)

zona fasciculata (zow-nah fah-*sik*-you-**lay-**tah) The thick middle layer of the adrenal cortex that secretes glucocorticoid hormones. (Ch. 14)

zona glomerulosa (glow-*mer*-you-**low-**sah) The outer layer of the adrenal cortex that secretes mineralocorticoid hormones. (Ch. 14)

zona pellucida (peh-**loo-**sih-da) The thick, clear layer separating the ovum from surrounding granulosa cells. (Ch. 31)

zona reticularis (reh-**tik-**you-lar-is) The inner layer of the adrenal cortex that secretes sex hormones, mainly androgens. (Ch. 14)

zygote (zye-goat) A fertilized ovum; the cell resulting from the fusion of male and female gametes. (Ch. 31)

zymogen (zye-mow-jen) **granules** Vesicles of the enzyme salivary amylase that are stored in the serous cells. (Ch. 22)

Appendix A
Answers to Review Questions

Chapter 1

1. mm Hg 2. exponents 3. internal environment
4. extracellular fluid 5. chemoreceptors 6. tissue
7. abscissa or x-axis
8. homeostasis: a term used to describe a constant internal environment despite changes in the external environment
9. hypothesis: an idea proposed to explain a scientific observation
10. a condition that does not change with time
11. receptor 12. hormone 13. sensors
14. sagittal 15. Harvey
16. a. 1×10^6 cells b. 5×10^{-4} mm
17. a. 0.005 moles/liter b. 0.0001 μm
18. 12×10^{-6} kg 19. one to ten or 1/10
20. The concept that cells originate from other cells and not from nonliving matter laid the foundation for the idea that all cells have the same fundamental chemical makeup and that all organs are made up of cells.
21. The scientific method is a quantitative approach that allows scientists to devise experiments to validate their answers. The scientific method also allows other scientists to reproduce their results.
22. All enzymatic reactions as well as protein conformation are dependent on a constant environment (i.e., temperature, pH, ionic concentration).
23. Hypothalamic cells, adrenal cells.
24. Fever, blood clotting.
25. Ancient ideas persisted because they were based on dogma and not on quantitative information or a scientific approach.
26. New ideas in science are still not readily accepted. Many scientists find themselves without research funding if their proposed ideas stray too far from "main stream science." New ideas should be approached with objectivity, sound data, and willingness to accept a different way of thinking.
27. The answer requires knowledge only about homeostatic control and not about calcium regulation. Plasma calcium is homeostatically regulated because a high or low calcium level is life threatening. The answer is 2.5 mM/L. Doubling the individual's dietary calcium intake will activate a negative feedback to prevent plasma calcium from rising.

Chapter 2

1. exergonic 2. DNA (deoxyribonucleic acid)
3. Vmax (maximum velocity) 4. glycoprotein
5. molecular formula: shows the number of each type of atom present in a molecule
6. atomic number: the number of protons in the nucleus of an atom
7. covalent bond: a very strong bond formed when two or more atoms share electrons
8. solution: usually a mixture of a liquid (solvent) containing another substance (solute), which is distributed uniformly
9. hexose: a monosaccharide containing six carbon atoms
10. a 11. c 12. d 13. d 14. c 15. 18 daltons
16. 1×10^{-5} moles/liter
17. A peptide bond is formed between the alpha carboxyl group (—COOH) of one amino acid and the alpha amino group (—NH$_2$) of the other amino acid. The peptide bond has the structure —CONH—.
18. (a) Hydrogen bond formation between two adjacent peptide bonds or between polar side chains of amino acids.
(b) Hydrophobic interactions between nonpolar side chains of amino acids. (c) Disulfide bonds between —SH groups of cysteine amino acids.
19. The two DNA strands in the double helix are not identical. The base adenine in one strand always lies opposite thymine in the other strand. Likewise, cytosine in one strand is always opposite guanine in the other strand.
20. The crystal structure of KCl, a hydrophilic compound, is stabilized by ionic bonds. Small polar water molecules enter the crystal structure and form shells around the K$^+$ and Cl$^-$ ions. This breaks the ionic bonds, and the K$^+$ and Cl$^-$ ions separate or dissociate from one another.
21. An enzyme is a protein that lowers the activation energy level for a chemical reaction. It provides binding sites (active sites) on its surface so that reactant molecules can be brought close together and in the proper orientation for the reaction to occur. These encounters between reactant molecules do not occur at a high rate in the absence of the enzyme.

Chapter 3

1. cytosol 2. chromatin 3. exocytosis
4. cytoskeleton 5. cytokinesis
6. eukaryotic cell: type of cell characterized by the presence of a well-defined nucleus confined within a nuclear membrane
7. genome: the sum of all the different chromosomal genes
8. autoradiography: a technique for determining the location of radioactive molecules in cells and tissues
9. density gradient centrifugation: a procedure that allows separation of cell organelles based on differences in their densities
10. recombinant DNA: a DNA molecule formed experimentally by combining pieces of DNA from different organisms
11. d 12. b 13. a 14. b 15. c 16. c
17. (a) Degradative enzymes can be stored safely within lysosomes. (b) Physical separation of synthetic and degradative pathways allows independent control. (c) Some newly synthesized molecules can be processed further and sorted according to final destinations.
18. DNA binds to histone proteins, and parts of the double helix wrap around the histones to form a nucleosome, the basic packing unit. Close packing of nucleosomes forms a chromatin fiber. This, together with further levels of folding, allows the double helix to be packed in a highly organized and compact way.

19. (a) Ribosomal RNA joins with other proteins to form ribosomes. (b) Messenger RNA contains information about the sequence of amino acids which must be joined together at the ribosomes. (c) Transfer RNA brings the required amino acids to the ribosomes.

20. Aerobic respiration is the main metabolic pathway for extracting the energy stored in fuel molecules. When the energy is released, it is trapped in the form of ATP. Aerobic respiration is combined in the mitochondrion, which is the major site of ATP production within the cell.

21. Primary cultures are formed by transferring cells directly from a tissue sample to culture medium. Most of the cells will divide and grow only a limited number of times in culture before dying. In contrast, cells from a cell line will divide and grow indefinitely in culture.

22. All physiological phenomena have a molecular basis. Use of molecular biological techniques can lead to determination of molecular structures of various proteins or interest to physiologists, e.g., hormones, membrane proteins, contractile proteins. These structures can be correlated with the known function of the molecule. Once the DNA sequence of the gene coding of these proteins is known, specific mutations can be introduced *in vitro* and the mutant gene can be reintroduced into the organism. The effect of the mutation on protein function then can be analyzed. Hereditary disorders in human patients provide naturally occurring mutations. In this way it is possible to determine which amino acids are critical to the physiological function of the protein.

23. Lysosomes can degrade the cell's organelles, in addition to material imported from outside the cell. This can occur during remodelling of the cell or when the cell is short of fuel. During cell death, the lysosomes rupture and their digestive enzymes destroy the cell itself.

24. Protein synthesis occurs on ribosomes in the cytosol. The new polypeptide contains an ER signal that targets it to a receptor on the ER membrane. The polypeptide is then transferred to the ER lumen as synthesis continues. Monosaccharide units are attached to the protein within the ER lumen and the resultant glycoprotein is passed to the Golgi complex via transition vesicles. The polysaccharide component is modified during passage through the Golgi and the glycoprotein is packaged into a secretory vesicle for transport to the plasma membrane. The vesicle fuses with the plasma membrane (exocytosis) and the glycoprotein is expelled into the extracellular fluid.

Chapter 4

1. integral (or intrinsic)   2. osmosis
3. facilitated diffusion   4. antiport
5. regulatory volume decrease
6. amphipathic molecule: a molecule, such as a phospholipid or intrinsic membrane protein, that has both hydrophobic and hydrophilic regions in its structure
7. Overton's rules: (a) Membrane permeability to small nonpolar solutes is directly proportional to their lipid solubility. (b) Membrane permeability to polar solutes is inversely proportional to their molecular size.
8. diffusion trapping: the neutral form of a weak base (or acid) diffuses across a cell membrane, but the charged form will not. The neutral form can be "trapped" on one side of a membrane if it reverts to its charged form after crossing the membrane.
9. secondary active transport: carrier-dependent transport of solute against the electrochemical gradient. Solute transport is not directly coupled to the energy-yielding reaction.

10. tight epithelium: an epithelium that does not permit leakage of water and solutes between the cells. This is because the cell junctions are highly impermeable.
11. c   12. a   13. d   14. d   15. b   16. a
17. (a) The core of the plasma membrane is a double layer of phospholipid molecules. (b) Proteins are closely associated with, and often embedded in, the lipid bilayer. (c) The interior of the lipid bilayer is like a viscous fluid. (d) Both phospholipid and protein molecules move laterally within the bilayer. (e) Both phospholipids and proteins are distributed asymmetrically between the two halves of the bilayer.
18. Amino acids form only linear chains, but monosaccharides can be assembled in branched chain structures. This provides almost unlimited possibilities for making a membrane receptor that will bind only one specific type of extracellular molecule.
19. The normal erythrocyte has a biconcave shape. When water enters from a hypo-osmotic solution, the cell can swell to become more like a sphere. In this way the cell can accommodate a large increase in its normal volume.
20. This describes the response of mammalian cells to an increase in intracellular volume. The increase in volume activates mechanisms which, in most cells, increase the efflux of K^+ and Cl^- from the cell. This decreases intracellular osmolarity and stimulates efflux of water, returning cell volume to normal.
21. The tight junctions between the cells of a leaky epithelium allow passage of small ions and water. This "leak" pathway allows paracellular transport across the epithelium in addition to transcellular transport. The tight junctions of a tight epithelium do not permit leakage between the cells. Transcellular transport is the only pathway across a tight epithelium.
22. Osmosis is the diffusion of water across a cell membrane in response to a difference in solute concentration in the intracellular and extracellular fluid. Dialysis is the separation of small molecules from large molecules in solution by diffusion of the small molecules across an artificial membrane. The small molecules diffuse in response to a difference in their concentrations on either side of the membrane. In kidney dialysis, for example, waste products are small molecules such as urea and are removed from the bloodstream by diffusion across the dialysis membrane. Blood cells, proteins and other large molecules remain in the blood because they are too large to diffuse across the membrane.
23. As the ATP already present in the cell is used up, the activity of the Na^+/K^+-ATPase will begin to decrease and its ability to pump Na^+ out of the cell will be lost. Since Na^+ continues to enter the cell the net result will be an increase in the intracellular level of Na^+. This will diminish the transmembrane Na^+ gradient because the extracellular Na^+ concentration remains unchanged. The Na^+ gradient is the main driving force for Na^+-coupled transport of solutes such as glucose, amino acids, and phosphate. Hence, the transport of these solutes into the cell will decrease.

Chapter 5

1. degenerate   2. semiconservative   3. intron
4. gap junction   5. ligand
6. start codon: the mRNA codon that is involved in initiating translation of the nucleotide triplets in the mRNA molecule
7. RNA polymerases: the group of enzymes that use a DNA template to synthesize molecules of RNA
8. post-translational modifications: alteration of newly

synthesized polypeptides, e.g., to produce a biologically active polypeptide from a larger precursor

9. antagonist: a ligand that binds with high affinity to a receptor but produces no response
10. cyclic AMP phosphodiesterase: the enzyme that rapidly breaks down intracellular cyclic AMP
11. c   12. d   13. b   14. d   15. a   16. a   17. c
18. This is possible because the genetic code is a triplet code. The four different nucleotides, when combined in groups of three, allow 64 different combinations (4^3). A doublet code would provide only 16 different combinations (4^2), which is not sufficient for 20 amino acids.
19. (a) RNA synthesis requires ribose instead of deoxyribose, (b) uracil instead of thymine, (c) and RNA polymerase instead of DNA polymerase. (d) RNA synthesis produces a single nucleotide strand instead of two, and (e) the DNA template is not incorporated into the RNA product.
20. The reason is that expression of the genetic information in a cell is controlled so that different genes are expressed in different cell types. Most of the genetic information in any cell remains suppressed and is not used. The genes which are being expressed, in the form of specific protein products, may change during the life cycle or in response to a specific stimulus such as a hormone.
21. An agonist is any ligand which is capable of both binding to a receptor and producing a physiological response. An antagonist will bind to the same receptor but cannot evoke a response. By occupying the receptors, the antagonist will block the action of the agonist.
22. (a) Local mediators (e.g., prostaglandins), (b) neurotransmitters, and (c) hormones.
23. A common mechanism is to activate a protein kinase, an enzyme which will catalyze addition of phosphate groups to specific protein targets. Phosphorylation often changes the function of the protein. For example, if the protein target is another enzyme, it may be activated (or inactivated) by the phosphorylation step.
24. (a) If the product of the gene is required in large amounts, the chromosome may contain multiple copies of the gene. Thus the cell can make huge amounts of this one gene product very quickly. (b) Genes can be suppressed by changes in chromosome structure. Many inactive genes are located in regions of chromosomes where the chromatin is highly coiled and compacted. Active genes are associated with areas where the chromatin is more loosely packed because this increases the accessibility of the DNA to RNA polymerase.
25. Polypeptide hormones are water soluble and are broken down rapidly by enzymes present in the blood. This provides a way to control the concentrations of these hormones. They can be removed quickly once the stimuli that released them have subsided. Cessation of release coupled with degradation of hormone already in the blood produces a rapid fall in the concentration circulating in the blood. This means that the response of the target cell can be turned off quickly. It is important that rapid responses, such as control of blood glucose by insulin, can be switched off as soon as the required effect has been achieved. For example, insulin is released after a meal to return the elevated blood glucose to a normal level. Prolonged action of insulin, due to slow removal, would decrease blood glucose to a dangerously low level. Steroid hormones, in contrast, are lipid-soluble and must be carried in the blood by specialized carrier molecules. They are not degraded quickly and their effects may last for days once they are released. They primarily regulate processes that occur over days or weeks, such as growth and development.

Chapter 6

1. anabolism or biosynthesis
2. endergonic reaction
3. oxidized nicotinamide adenine dinucleotide (NAD^+)
4. gluconeogenesis   5. urea
6. chemiosmotic mechanism: synthesis of ATP by a process that is driven by the flow of protons across the inner mitochondrial membrane
7. pentose phosphate pathway: a cytosolic pathway for glucose oxidation, which is important in cells heavily involved in biosynthesis. The pathway produces no ATP but generates NADPH to drive biosynthetic reactions.
8. oxidative phosphorylation: synthesis of ATP by coupling oxidation to phosphorylation of ADP. The oxidation steps occur in the mitochondrial electron transport chain as electrons are transferred from reduced coenzymes to O_2.
9. allosteric inhibitor: a compound that inhibits an enzyme activity by binding to a site on the enzyme which is separate from the active site. Binding of the inhibitor produces a conformational change in the active site which impairs its function.
10. high-energy phosphate compound: a compound with a strong tendency to transfer a phosphate group to an acceptor. This transfer is accompanied by release of a large amount of free energy.
11. a   12. d   13. c   14. c
15. (a) 0 (b) 12   16. (a) 31 (b) 15
17. NADH is oxidized by the mitochondrial electron transport pathway to drive synthesis of ATP. Oxidation of NADPH is used to drive reactions in various biosynthetic pathways in the cytosol, e.g., synthesis of fatty acids and steroids.
18. Catabolism of proteins, carbohydrates, and fats. Biosynthesis of fatty acids, cholesterol, and steroids.
19. When the supply of ATP is high, ATP acts as an allosteric inhibitor of citrate synthetase to reduce the rate of formation of citric acid. This reaction is the entry point in the cycle for acetyl groups from acetyl coenzyme A.
20. (a) Plants use light from the sun to make ATP by photosynthesis. (b) ATP is used by plant cells during CO_2 fixation. (c) This process produces high-energy organic compounds and O_2. (d) These organic "fuel" molecules are a source of chemical energy and building-block molecules for animals that eat the plants. (e) In animals the molecules are catabolized by cellular respiration, a process which requires O_2, and the chemical energy is trapped as ATP. (f) Water and CO_2 are produced as by-products.
21. A mobile animal, unlike a plant, must be able to carry its energy store with it. The amount of ATP produced by a given weight of fat far exceeds that produced by the same weight of carbohydrate. In addition, more fat can be stored than carbohydrate. Thus, it is much more efficient for an animal to use fat as an energy store.
22. There would be a flow of mass if the utilizing cell converted the glucose molecule into a different molecule. There would be a flow of energy if the glucose molecule was oxidized to provide metabolic energy for the utilizing cell. There would be a flow of information if the utilizing cell sensed an increase in glucose concentration and responded by changing a specific function.
23. (a) Reactant molecules do not have sufficient activation energy under mild conditions at room temperature. (b) Use of extreme conditions, such as high temperature and pressure, and inorganic catalysts. (c) Extreme conditions would denature proteins, nucleic acids, and other macromolecules in the cells. (d) Cells use specific proteins as enzymes to catalyze specific intracellular reactions.

Chapter 7

1. cation   2. anion
3. depolarization   4. hyperpolarization
5. The membrane potential is the voltage across the membrane that results from an imbalance in electrical charge between the inside and outside of the cell.
6. The equilibrium potential for an ion is the potential at which the influx of the ion across the membrane is equal to its efflux so that there is no net current flow.
7. The action potential is an electrical impulse generated by nerve cells and some muscle cells. It is characterized by a depolarization that quickly reaches a peak positive value which is rapidly followed by a repolarization of the membrane.
8. Saltatory conduction is the apparent jumping of the action potential from one node of Ranvier to the next as it propagates along the axon.
9. d   10. d   11. c   12. b
13. During the absolute refractory period, sodium channel inactivation gates close, reducing inward sodium ion conductance, and voltage-gated potassium channels open, increasing outward potassium ion conductance. These conductance changes account for the first 1/3 of the repolarization period. The membrane remains in a state of absolute refractoriness until the sodium channel inactivation gates begin to reopen, at which time the membrane becomes relatively refractory.
14. Increases in membrane conductance to Na^+ ions and decreases to K^+ can cause EPSPs.
15. Increases in membrane conductance to K^+ and Cl^- ions as well as decreases to Na^+ can cause IPSPs.
16. Ions will move from a region of high concentration to a region of lower concentration. Anions will move to regions that are positively charged while cations will move to negatively charged regions.
17. −85.7 mV   18. 75 meters per second

Chapter 8

1. ocular dominance columns   2. proprioceptor
3. slowly adapting receptor   4. somatic sensory map
5. generator potential: a change in the membrane potential of a sensory cell in response to a sensory stimulus
6. conductive hearing loss: results from damage to the bones of the middle ear
7. labeled-line code of sensory stimuli: different types of sensory information are transmitted along different pathways
8. a   9. d   10. c
11. The transduction of sensory information requires an adequate or effective stimulus, which causes an alteration in membrane conductance. The change in membrane conductance, in turn, produces a generator potential, leading to the generation or inhibition of action potentials.
12. Stimulus intensity is encoded in the nervous system by a frequency code and by a population code.
13. Receptive fields are regions on a transducing surface that activate and/or inhibit a neuron with the presence of the sensory stimulus upon the transducing surface.
14. The trichromatic theory of color vision states that the total spectrum of colors that we perceive can be explained by the differential absorption of photons by three types of photoreceptors, which optimally absorb photons with wavelengths of 443 nm, 535 nm, and 570 nm, respectively.
15. Within the inner ear the basilar membrane is the first structure involved in decomposing sound information into its constituent components. The basilar membrane is designed such that travelling waves associated with high frequencies cause maximal deformations of the membrane in the region near the base of the cochlea. This region of the basilar membrane is wider than that found in the apical region. To increase the frequency response of the basilar membrane, one could therefore conceptually increase its overall length and proportionally increase its width near the base of the cochlea.
16. Complete lesions of the spinal cord and lower medulla result in bilateral sensory loss below the level of the lesion. Hemisections produce loss of pain/temperature, and touch/vibration in an alternating pattern below the level of the lesion. Ipsilaterally touch/vibration sense is lost, while the side of the body contralateral to the site of the lesion would display deficits in pain/temperature sensation.

Chapter 9

1. flexor   2. cortical efferent zone
3. muscle spindle   4. semicircular canal
5. An extensor is a muscle that increases joint angle
6. An alpha motor neuron is a nerve cell in the spinal cord that directly innervates and activates skeletal muscles.
7. The Babinski sign is a fanning of the toes when the sole of the foot is stroked and is an indication of a lesion of the corticospinal tract.
8. Utricles are otolithic organs that detect linear acceleration and are also capable of detecting the position of the head with respect to gravity.
9. a   10. b   11. c   12. c
13. Alpha motor neurons can increase the force of muscle contraction by increasing their frequency of discharge and/or by increasing the number of alpha motor neurons that are activated.
14. Alpha motor neurons that innervate fast skeletal muscles have rapid conduction velocities and high frequencies of action potential discharge. Those that innervate slow skeletal muscles have slower conduction velocities and lower discharge rates.
15. Gamma motor neurons maintain muscle spindle sensitivity as the extrafusal muscle fiber is shortening.
16. Motor units are composed of a single alpha motor neuron and the number of individual muscle fibers that it innervates. A motor neuron pool is a cluster of alpha motor neurons that innervate motor units within the same skeletal muscle.
17. Output fibers from the deep cerebellar nuclei innervate the thalamus on the contralateral side of the brain which in turn, gives rise to a thalamocortical projection to the motor cortex. The descending fibers which form the cortico-spinal tract crossover to the spinal cord on the opposite side of the body. Consequently, information from the cerebellum to the spinal cord decussate the midline twice producing ipsilateral effects while information from motor cortex decussate once producing contralateral effects.
18. Lesions of the spinal cord can damage descending corticospinal nerve fibers that are needed for movement. All motor neurons below the level of the lesion would lose this input thus resulting in paralysis. In contrast, autonomic processes such as bladder control depend in part on motor reflexes. These reflexes are maintained as long as the motor neurons which innervate the bladder are not damaged.

Chapter 10

1. "fight-or-flight" response
2. epinephrine and norepinephrine

3. acetylcholine **4.** norepinephrine
5. The sympathetic nervous system is a subcomponent of the autonomic nervous system that coordinates the body's response to stress.
6. The parasympathetic nervous system coordinates the body's response for vegetative and digestive functions.
7. Referred pain is the sensation of pain that appears to come from the surface of the body, but the source of the pain originates from an internal organ.
8. Viscera are internal organs of the body.
9. b **10.** d **11.** c **12.** a
13. The optimal conditions for cellular function occur at a temperature of 37°C, pH 7.4, and an osmolarity of 300 mOsm. These conditions are maintained by central autonomic regulatory systems in the hypothalamus and brain stem. Thermo- and osmoreceptors in the hypothalamus detect changes in temperature and osmolarity while chemoreceptors in the brain stem monitor pH.
14. The urge to urinate is detected by stretch receptors in the bladder. Information conveyed by these receptors enter the spinal cord and activate parasympathetic pelvic nerve fibers resulting in the contraction of the smooth muscles of the bladder wall. In addition there is a coordinated inhibition of sympathetic and somatic outflow resulting in the relaxation of the internal and external sphincters respectively.
15. The urge to consume water is initiated by osmoreceptors in the hypothalamus which monitor plasma osmolarity. Increases in osmolarity above 300 mOsm activate somatic motor responses leading to drinking behavior and autonomic responses resulting in the retention of water.
16. The hypothalamus coordinates the physiological changes seen in the "fight-or-flight" response. These bodily alterations act in concert to provide more oxygen to the skeletal muscles and brain by increasing heart rate and contractility, and the depth of respiration. Changes also occur to provide more fuel and energy for critical organ systems.
17. Drugs that block nicotinic receptors inhibit both sympathetic and parasympathetic activity, while drugs that block muscarinic receptors inhibit the outflow of parasympathetic regulation.
18. Blood pressure can be reduced by decreasing the constriction of blood vessels. Drug therapies focused on blocking the actions of norepinephrine would therefore result in blood pressure reduction.

Chapter 11

1. dominant hemisphere **2.** planum temporale
3. narcolepsy **4.** raphe nucleus
5. synaptic facilitation—the process which increases neurotransmitter release
6. synaptic habituation—the process which decreases neurotransmitter release
7. associative learning—the process by which an understanding is acquired between two stimuli as in classical conditioning, or a behavioral response and a reward as in operant conditioning.
8. REM sleep—a period of sleep associated with the state of dreaming during which the eyes are rapidly moving
9. b **10.** c **11.** a
12. There are five stages of sleep: Stages I–IV and REM. REM sleep is associated with dreaming.
13. Several areas of brain are involved in the sleep process. The preoptic area of the hypothalamus controls non-REM sleep while the raphe nucleus controls REM sleep. The suprachiasmatic nucleus, the body's biological clock, has inputs to the preoptic area to control the wake-sleep cycle.

14. Prenatal exposure to androgens during critical periods of development can alter adult sexual behavior. Fetal male brains that have not been exposed to androgens result in "feminized" sexual responses. Fetal female brains that have been exposed to androgens result in "masculinized" sexual responses.
15. Short-term memory can be easily disrupted and limited in time while long-term memory is more stable and long-lasting. In addition, long-term memory and its consolidation from short-term memory requires protein synthesis.
16. Information about images in the temporal aspects of the left visual field are transmitted to the contralateral non-dominant hemisphere on the right side of the brain. Because of the lesion of the corpus callosum, information cannot be transmitted between the two hemispheres. Consequently, the language centers responsible for speech do not have access to information received by the non-dominant hemisphere. Likewise, since the right hand is controlled by the left hemisphere, it too is unable to select the object placed in the temporal portion of the left visual field. In contrast, objects placed in the nasal aspect of the left visual field would be perceived by the left dominant hemisphere resulting in the patient's ability to verbally identify the object as well as selecting it with his right hand.
17. Question for discussion—no answer given.

Chapter 12

1. autocrine, or paracrine **2.** neuroendocrine cell
3. cholesterol **4.** radioimmunoassay, or RIA
5. adenylate cyclase
6. inositol trisphosphate (IP$_3$) and diacylglycerol (DG)
7. hormone receptor: a chemical entity, usually a protein, that is present either on or in a cell that specifically recognizes a hormone. Binding of the hormone to the receptor triggers a series of events that result in the specific biological effects characteristic of that hormone.
8. neuroendocrinology: that branch of science that deals with the study of the interaction of the nervous and endocrine systems.
9. mineralocorticoids: a class of steroid hormones produced by cells of the adrenal cortex that act to regulate sodium balance by the kidneys.
10. conjugation of steroids: the process by which steroid metabolites are coupled to a sulfate or glucuronide group, resulting in improved solubility of the conjugated steroid and facilitation of its excretion.
11. G-proteins: a group of GTP-binding proteins that act to couple cell surface receptors with their respective intracellular signal transduction systems. For example, stimulation of G$_S$ proteins by the appropriate receptor results in activation of adenylate cyclase.
12. receptor-mediated endocytosis: the process by which peptide hormones are taken up by their target cells after binding to specific cell surface receptors. Typically, receptor-mediated endocytosis results in hormone degradation.
13. phosphodiesterase: a cellular enzyme that is responsible for the conversion of cAMP to AMP and thus the termination of intracellular signals mediated by an increase in cAMP
14. b **15.** c **16.** d **17.** d **18.** d **19.** e **20.** a
21. b **22.** d **23.** e **24.** b
25. Receptors act as signal transducers in the endocrine system by converting extracellular hormone signals into intracellular signals that ultimately result in altered cell function. Specificity results from the fact that receptors typically only recognize and bind one hormone, thus preserving the faithfulness of the signalling process.
26. Steroid and thyroid hormones readily enter their target cells

and bind to specific receptors in or on the nucleus. The hormone-receptor complex interacts with DNA and alters the transcription of a few specific genes coding for proteins responsible for the actions of that particular hormone.

27. Hormones that modulate intracellular cAMP interact with cell surface receptors that are functionally coupled to adenylate cyclase through either stimulatory (G_S) or inhibitory (G_i) G-proteins. Adenylate cyclase forms cAMP from ATP, and cAMP alters cell function by activating specific cAMP-dependent protein kinases. Several forms of cAMP-dependent protein kinase are known, each with its own particular protein substrates. Changes in intracellular cAMP in different cells may lead to different effects, depending on the particular form of cAMP-dependent protein kinase that is present and on the particular protein substrates that are present.

28. Activation of certain cell surface receptors results in the activation of a form of phospholipase C which cleaves phosphatidylinositol 4, 5-bisphosphate (PIP_2) into diacylglycerol (DG) and inositol trisphosphate (IP_3). IP_3 acts as a second messenger inside the cell by promoting the release of calcium from intracellular stores, and DG activates a membrane-bound protein kinase, C-kinase, which catalyzes the phosphorylation of specific cellular proteins, leading to alterations in their activity and specific changes in cell function.

29. Hormones act via mechanisms that involve amplification of the original hormone "signal." Binding of a single hormone molecule to its receptor initiates a series of events (a cascade) whereby multiple enzymes or steps are activated in a domino or multiplying effect. As a result of the amplification, hormones play a potent role in regulating tissue and cell function.

30. Multiple steps are involved related to hormone production and action. Possible defects might include:
   a. Lack of appropriate synthesis and secretion of the hormone due potentially to a defect in the hormone gene, or due to a defect in an enzyme responsible for metabolism and/or activation of the hormone (i.e., not enough active hormone is produced).
   b. A defect at the level of the receptor. There may be insufficient receptors, or the receptors that are present may have reduced function (i.e., the hormone signal is not producing an appropriate effect in target cells).
   c. The signal transduction pathway for a particular hormone may be defective due to a deficiency in one of the enzymes or proteins in the intracellular hormone action pathway (i.e., adequate signal is being produced and reaches the cells, but they lack the mechanisms to respond appropriately).

31. Peptide hormones typically act through second messenger systems. Many of their biological effects are evident within seconds or minutes after they reach the target cell. This is because they only need to activate or inactivate enzymes or proteins already present in the cell. Steroids typically act by altering gene expression in the nucleus. Responses of target cells to steroid hormones are usually not apparent for hours or even days after the hormone reaches the cell.

Chapter 13

1. Rathke's pouch   2. median eminence   3. somatostatin
4. diabetes insipidus   5. acromegaly   6. thyroglobulin
7. the hypothalamic-pituitary portal system: a vascular system consisting of a primary capillary network in the median eminence of the hypothalamus and a secondary capillary network in the anterior pituitary, which are connected by long portal veins. The pituitary portal system carries factors from the hypothalamus to the anterior pituitary that serve to regulate the secretion of anterior pituitary hormones.
8. corticotropes: ACTH-secreting cells of the anterior pituitary; normally they comprise about 20% of the cells of the gland.
9. Gonadotropins: a collective term that refers to both LH and FSH, two hormones secreted by the anterior pituitary, which are of primary importance in regulating gonadal function.
10. Somatomedins: a collective term referring to growth factors that are responsible for many of the growth-promoting actions of growth hormone. The principal somatomedins in the body are IGF-I and IGF-II.
11. "iodide trap": the active transport mechanism present on the basal surface of thyroid follicle cells, which concentrates iodide within the thyroid gland
12. a   13. b   14. a   15. d   16. c   17. d   18. c
19. d   20. b
21. The two posterior pituitary hormones, oxytocin and ADH, are synthesized in cell bodies of neuroendocrine cells in the hypothalamus and are transported via their axonal projections to the posterior pituitary. Upon appropriate stimulation of these neuroendocrine cells, the hormones are released from nerve termini in the posterior pituitary, where they enter capillaries and are carried to the general systemic circulation.
22. The two primary stimuli for ADH secretion are (1) an increase in osmolality of the blood or extracellular fluid and (2) a large decrease in blood volume. The primary effect of ADH is to increase water retention by the kidneys.
23. The hypothalamic-pituitary-thyroid axis consists of three hormones: TRH, TSH, and T_4. Hypothalamic TRH stimulates TSH secretion by the anterior pituitary; TSH in turn stimulates T_4 secretion by the thyroid gland. In addition to its biological effects, T_4 has a negative feedback effect on the anterior pituitary to inhibit TSH secretion.
24. The shortened stature associated with dwarfism results from a deficiency in GH secretion or action prior to puberty. Gigantism results when increased GH secretion occurs during childhood; acromegaly results when overproduction of GH begins after puberty. Persons exhibiting gigantism have disproportionately long arms and legs, owing to the overstimulation of the bones in the limbs. Persons with acromegaly typically exhibit an enlarged jawbone, hands, and feet, owing to the continued stimulation of these bones, even after the closure of the epiphyseal plates of the long bones.
25. Thyroid hormones generally promote glycogen formation in liver and glucose uptake into adipose tissue and muscle. Thyroid hormones promote all aspects of lipid metabolism but in excess tend to promote the loss of most fat stores in the body. The effects of thyroid hormones on protein metabolism are biphasic: At normal physiological concentrations, thyroid hormones increase protein synthesis and promote an overall accumulation of protein in the body; when present in excess, thyroid hormones promote an overall loss of protein.

Chapter 14

1. zona fasciculata   2. aldosterone   3. renin
4. Cushing's syndrome   5. tyrosine hydroxylase
6. chromaffin cells
7. zona glomerulosa: outermost zone of the adrenal cortex. It is the predominant site of aldosterone synthesis in the body.
8. corticotropin-releasing hormone: a hypothalamic peptide responsible for the regulation of ACTH secretion by the anterior pituitary
9. pharmacological effects of glucocorticoids: those effects produced by high concentrations of the hormone; these

effects can be classified primarily as being immunosuppressive or anti-inflammatory in nature.

10. angiotensin II: an eight amino acid peptide hormone formed from angiotensin I, which acts to regulate aldosterone secretion from the zona glomerulosa of the adrenal cortex

11. Addison's disease: a disorder resulting from insufficient steroid hormone production by the adrenal cortex

12. lipolysis: the process whereby stored triglycerides are broken down into their component glycerol and free fatty acids

13. renin-angiotensin system: a multi-step process that serves to couple changes in blood pressure or volume to appropriate changes in aldosterone secretion. Several tissues, enzymes, and peptide or protein factors are involved.

14. c   **15.** b   **16.** d   **17.** b   **18.** b   **19.** d   **20.** c

21. "Tropic" effects refer to the fact that ACTH promotes the growth of all three zones of the adrenal cortex. The steroidogenic effect of ACTH involves the regulation of cortisol secretion by cells of the zona fasciculata.

22. Corticotropin-releasing hormone is produced by the hypothalamus and regulates ACTH secretion by the anterior pituitary; ACTH in turn regulates cortisol secretion by the adrenal cortex. Various inputs from higher brain centers serve to regulate CRH production by the hypothalamus. Negative feedback regulation by cortisol occurs at the level of the hypothalamus and anterior pituitary to inhibit CRH and ACTH secretion, respectively, and ACTH itself has negative feedback effects in the hypothalamus to inhibit CRH release.

23. The adrenal medulla and the sympathetic nervous system, sometimes referred to collectively as the sympathoadrenal system, have influence over a number of organ systems in the body and are responsible for activation of the "fight or flight" responses. Functionally, cells of the adrenal medulla are analogous to postganglionic fibers of the sympathetic nervous system, but rather than secreting catecholamines into a synapse, these cells secrete epinephrine directly into the bloodstream.

24. The protein-catabolic and lipid-mobilizing effects of cortisol lead to a loss of muscle mass and fat in the extremities with a redistribution of fat to the trunk, shoulder blades, and face, which result in the physical appearance of thinned arms and legs, a large abdomen, and a round face. The catabolic effects of excess cortisol also leads to a loss of connective tissue in skin and vascular beds, resulting in the formation of striae, reddening of the cheeks, and easy bruising.

25. The excess skin pigmentation that results in these conditions is the result of overproduction of ACTH. The peptide sequence of ACTH bears considerable similarity to that for melanocyte-stimulating hormone (MSH). Adrenal cortical hypersecretion could result from a tumor of anterior pituitary cells that are continuously producing ACTH at high levels; thus both ACTH and adrenal steroids would be elevated. A defect at the level of the adrenal gland would eliminate adrenal steroid production and the negative feedback effects they have on ACTH production, allowing ACTH levels to rise and pigmentation to occur.

26. Glucocorticoids primarily decrease glucose uptake into peripheral adipose tissue cells. Thus there is loss of fat in the arms and legs. For reasons that are not entirely understood, fat stores in the trunk of the body and in particular the abdomen increase in size in response to excess glucocorticoids. This leads to the rather distinct physical appearance of patients with Cushing's disease.

Chapter 15

1. islets of Langerhans   2. beta cells   3. proinsulin
4. lipoprotein lipase   5. gluconeogenesis

6. Type II diabetes, or NIDDM   7. exocrine pancreas

8. C-peptide: the peptide segment of proinsulin that connects what will eventually form the A-chain and B-chain of insulin. It is removed from proinsulin by proteolytic cleavage and is secreted from beta cells in equal amounts as insulin.

9. hormone-sensitive lipase: an enzyme in adipose tissue and liver that breaks triglycerides into fatty acids and glycerol. It is inhibited in adipose tissue by insulin and stimulated in liver by glucagon.

10. glycogenolysis: the process of breaking down stored glycogen into glucose-6-phosphate, which can be used for energy or converted into glucose and released into the blood.

11. ketogenesis: the name for the process of ketone body formation from fatty acids. It occurs primarily in liver and is stimulated by glucagon.

12. caloric density: the number of calories in a substance per unit weight of the substance

13. Type I diabetes mellitus (IDDM): the form of diabetes generally caused by an autoimmune disorder that leads to destruction of beta cells and a lack of circulating insulin. Patients with Type I diabetes are typically treated with insulin injections.

14. sulfonylureas: a group of oral drugs often used in the treatment of Type II diabetes mellitus (NIDDM). Sulfonylureas promote insulin action in target tissues and insulin secretion from the pancreas.

15. hyperglycemia: the term used to describe a higher than normal blood glucose concentration

16. b   **17.** b   **18.** a   **19.** d   **20.** a   **21.** d   **22.** d

23. Insulin-secreting beta cells are most numerous and are located toward the center of the islet; glucagon-secreting alpha cells are located around the outside of the islet; and somatostatin-secreting delta cells tend to be located between alpha and beta cells. This arrangement of cells may serve to allow paracrine interactions among pancreatic hormones, as in the inhibition of insulin and glucagon secretion by somatostatin.

24. Pure triglycerides have about twice the caloric density as does glycogen. In addition, glycogen is hydrophilic, and thus considerable water is found associated with stored glycogen, greatly reducing the number of calories for a given weight of stored fuel as compared to triglycerides.

25. Type I diabetes is characterized by little or no circulating insulin as a result of autoimmune destruction of beta cells. Type II diabetics have apparently adequate circulating insulin levels, but target cells do not respond to insulin. Both forms of diabetes are characterized by hyperglycemia.

26. Acute complications of diabetes include hyperglycemia with glucosuria, increased urine volume and frequency of urination. Type I diabetics often also suffer from episodes of ketoacidosis and electrolyte imbalance.

27. In insulin target tissues such as adipose and muscle, glucose transporters exist both in an active form in the plasma membrane and in an inactive pool inside of cells. Insulin stimulates the movement of glucose transporters from their intracellular site to the membrane, thus making a greater number of active transporters available and increasing glucose uptake.

Chapter 16

1. sarcomere   2. myofilaments
3. smooth muscle (visceral muscle, involuntary muscle)
4. isometric contraction
5. myoneural junction (motor end-plate)
6. ATP (adenosine triphosphate)   7. force-velocity curve

8. twitch: a single brief contraction of a skeletal muscle in response to a single stimulus
9. tetanus: a series of twitch contractions that occur so close together in time that relaxation between them does not take place, and the force produced is a summation of the forces of the individual twitches
10. sarcoplasmic reticulum: the membranous cellular organelle in muscle that is responsible for the release and uptake of the calcium ions that regulate the activity of the contractile proteins
11. troponin: a complex of three proteins that can bind calcium ions and regulate the interaction of the contractile proteins in skeletal and cardiac muscle
12. crossbridges: the projections of the myosin molecules from the thick myofilaments that make contact with the thin (actin) myofilaments. The relative motion of the crossbridges and the actin myofilaments produces the muscle contraction, and the association between the actin and myosin molecules is the site where the energy of ATP is released.
13. end-plate potential: the electrical response in the postsynaptic (end-plate) membrane of skeletal muscle. It occurs when the membrane channels become open to sodium and potassium ions through the action of the acetylcholine that has been released from the presynaptic (nerve) membrane.
14. acetylcholine esterase: an enzyme located in the postsynaptic region of the myoneural junction that hydrolyzes acetylcholine and terminates its action on the receptor sites of the ion channels
15. rigor mortis: a rigidity of muscle that occurs when ATP is not available to bind with myosin crossbridges to allow their release from the actin filaments. It occurs progressively after death as energy stores become depleted.
16. myoglobin: a protein found predominantly in red skeletal muscle fibers that is capable of binding oxygen in a manner similar to that found in hemoglobin
17. disuse atrophy: a progressive loss of muscle tissue that occurs when a muscle is prevented from contraction by immobilization, denervation, and so on.
18. b   19. b   20. a   21. a   22. a   23. b   24. c
25. c   26. b   27. b   28. a   29. c
30. The three types of muscle are (a) skeletal (striated, voluntary), (b) cardiac (heart), and (c) smooth (nonstriated, involuntary, visceral).
31. Skeletal muscle undergoes hypertrophy in response to a long-term increase in its work load or contractile activity, and atrophy occurs when the muscle is used very little or is immobilized.
32. The speed of contraction of cardiac muscle lies between that of the other two types; it is slower than most skeletal muscles and faster than smooth muscle.
33. Curare mimics acetylcholine in that it can bind to the postsynaptic receptors and prevent the binding of any other molecule; however, it cannot cause the channels to open, and it is not broken down by acetylcholinesterase.
34. During a tetanus a skeletal muscle produces more force because relaxation from the individual twitches is prevented, and the force due to subsequent stimulations adds to the force already present.
35. In all types of muscle, relaxation is associated with a reduction in the number of calcium ions that are free to interact with the regulatory proteins. In some cases the calcium is taken back up into the sarcoplasmic reticulum, while in other cases it is actively transported out of the cells.
36. Since its metabolism is primarily aerobic and its activation mechanisms do not permit sustained tetanic contraction, its use would result in an unsteady posture with a high expenditure of metabolic energy.

37. Because calcium must be withdrawn from the region of the myofilaments, an impaired sarcoplasmic reticulum mechanism would result in twitch contractions (in response to single stimuli) greater in duration and force. The maximal force during a fused tetanus would be little affected.
38. In both cases the relative myofilament sliding velocity in each sarcomere would be the same, but with more sarcomeres in series adding up their shortening, the muscle with shorter sarcomeres should shorten faster.
39. Shivering, which is essentially a series of incomplete isometric tetanic contractions of large skeletal muscles, does not result in external work; however, because of the inefficiency of the force-production mechanism, virtually all of the metabolic energy appears as heat.
40. Since white muscle has a considerable anaerobic capacity, it could continue its activity longer than red muscle, which depends on an adequate oxygen supply (which implies an adequate blood supply).

Chapter 17

1. Hematocrit   2. Colloidal osmotic pressure
3. Hypotonic solution   4. Hemolysis   5. Hematopoiesis
6. Polycythemia   7. Diapedesis
8. Mean Corpuscular Hemoglobin Concentration; an index of the proportion of hemoglobin (weight/volume) measured per average red cell in a sample of blood. Normal adult value is 34% plus or minus 2%.
9. Mean Corpuscular Volume: an index of average red cell volume. Normal adult value is 87 μm^3 plus or minus 5 μm^3.
10. Mean Corpuscular Hemoglobin: an index of the average concentration of hemoglobin per red cell. Normal adult value is 29 $\mu \mu g$ plus or minus 5 $\mu \mu g$.
11. Anemia: a condition in which the number of circulating red cells and/or the amount of hemoglobin in the blood are below normal, resulting in reduced oxygen-carrying capacity of the blood.
12. Low Density Lipoproteins: transport proteins in the plasma that favor the deposition of cholesterol in arterial walls and thereby contribute to the development of vascular disease.
13. Hemostasis: refers to mechanisms that minimize or prevent the loss of blood. Four interrelated events constitute hemostasis: (1) local vasoconstriction, (2) platelet clumping, (3) formation of a clot, and (4) clot retraction and dissolution.
14. Phagocytosis: a process whereby a cell, called a phagocyte, recognizes, ingests, and enzymatically breaks down (digests) foreign substances.
15. Thrombin: an enzyme formed from prothrombin during the clotting process. Thrombin converts fibrinogen, a soluble plasma protein, into fibrin, an insoluble plasma protein which forms a clot.
16. Heparin: a naturally occurring anticoagulant produced by basophils and mast cells, especially in the connective tissues of the liver and lungs. Heparin prevents intravascular clotting by acting as an antithrombin agent.
17. c   18. b   19. e   20. d   21. c   22. d   23. e
24. $4/9 \times 100 = 44\%$
25. $16g/dl \times 1.34$ ml $O_2/gm = 21.4$ ml O_2/dl
26. $(45 \times 10) \div 5 = 90$ μm^3
27. $(5$ million$/\mu l)(1000$ $\mu l/cc)$ $(1000$ cc/liter$)(5$ liters$) = 25 \times 10^{12}$ cells present. $(25 \times 10^{12}$ cells$) \div (23 \times 10^{10}$ cells per day$) =$ 109 days.
28. Swell. Reduction of a plasma protein concentration reduces colloidal osmotic pressure of the plasma leading to increased capillary filtration and reduced capillary absorption. Fluid

leaves the blood vessels and begins to accumulate in the interstitium, mostly in tissues of the abdomen, producing a bloated appearance.

29. Major agglutination occurs when a recipient is infused with a foreign red cell antigen for which the recipient has developed antibodies. An O recipient's plasma contains anti-A and anti-B antibodies; therefore, an infusion of whole blood type AB in which the red cells contain both A and B antigens would result in major agglutination of the donated red cells. The agglutination would be major because of the large number of antigens infused (more than 5 million per microliter of donated blood) and because the antibodies are present to immediately react. Minor agglutination occurs when a recipient is infused with whole blood, plasma, or serum containing antibodies which react with the recipient's red cell antigens. An AB recipient would therefore experience minor agglutination if transfused with type A, type B, or type O blood since the plasma of each contains anti-A, or anti-B, or both antibodies. The agglutination would be minor because the amount of donated antibody is relatively small compared to the recipient's amount of antigen and the antibodies are diluted upon entering the recipient's blood.

Chapter 18

1. contractility or change in contractility
2. stroke volume   3. isovolumetric contraction
4. end-diastolic volume   5. pacemaker
6. blood flow: the volume of blood that moves past a particular point in the cardiovascular system during a given interval of time

$$flow = volume/time$$

7. systole: that portion of the cardiac cycle that occurs after the first heart sound and before the second heart sound
8. Frank-Starling law of the heart: An increase in end-diastolic volume will result in an increase in stroke volume, whereas a decrease in end-diastolic volume will decrease stroke volume.
9. electrocardiogram (also ECG or EKG): an electrical recording, made at the body surface, that provides a record of the electrical activity of the heart. It is produced by very small electrical currents caused by the action potentials of the atrial and ventricular muscles.
10. arrhythmia: a disturbance of the normal rate or pathway of the heartbeat
11. c   12. a   13. b   14. c   15. a   16. a   17. d
18. b   19. a   20. d
21. stroke volume = end-diastolic volume − end-systolic volume
    70 ml = 100 ml − 30 ml
22. cardiac output = stroke volume × heart rate
    5600 ml/min = 70 ml/beat × 80 beats/min
23. If the interval between beats is 0.8 second, the rate is 1 beat/0.8 second, or 1.25 beats/second. Since there are 60 seconds in a minute, the rate is:

    1.25 beats/second × 60 seconds/minute,

    or 75 beats/minute.
24. Vasoconstriction increases the resistance for blood flow to a particular tissue bed and therefore decreases blood flow to that tissue. If blood pressure is below normal, as with hemorrhage, vasoconstriction would increase total peripheral resistance and consequently increase blood pressure; this would then increase flow to tissues such as the heart and brain.

25. At normal heart rates, there is sufficient time for complete filling of the ventricle before the atria are activated. At higher heart rates, ventricular filling time is required and atrial contraction occurs before the ventricles have filled completely. The atrial contraction accelerates the flow of blood into the incompletely filled ventricle.
26. Contractility changes are primarily the result of changes in the intracellular calcium concentration.
27. SA node to atrial muscle to AV node to common bundle (or Bundle of His) to right and left bundle branches to Purkinje fibers to ventricular muscle.
28. No, because the EKG measures electrical activity only and cannot directly determine anything about the mechanical function of the heart; however, it can provide information by which a physician can make inferences about cardiac function.
29. The fact that resting heart rates in well trained athletes are as low as 40–50 beats/min indicates that, to achieve normal cardiac outputs, the stroke volume must be greater than 100 ml. With aerobic training the heart not only gets larger (by as much as 50% larger), which would increase EDV, but it also contracts more vigorously at rest so that ESV is reduced. These changes both contribute to the increased resting stroke volume. A champion marathon runner might have a resting stroke volume of 120 ml that would increase to 180 ml (1.5-fold increase) during maximum exercise, and a resting heart rate of 45 beats/min that would increase to 200 beats/min (5-fold increase). This would result in a 7-fold increase in cardiac output from a resting level of 5.4 L/min to 36 L/min during maximum exercise.
30. a. A pacemaker in a location somewhere lower than the atrial muscle would normally not be functional, since the faster pacemaker in the atria would force it to "fire off" before it did it on its own. Only when the rate of the atria pacemaker drops below the inherent rate of the lower pacemaker would it show its spontaneous activity.
    b. The QRS complex, since it is associated with the spread of ventricular depolarization, would be the portion affected. Slowing of the spread, as would occur during a conduction defect, would cause a widening (and shape change) in the QRS complex, since this represents a graph of voltage versus time.
31. Heart failure with occlusion of the coronary arteries is the result of inadequate perfusion for a given level of cardiac activity. Let's first consider a patient with less severe coronary occlusion. This individual would typically experience chest pain or angina during periods of intense exercise. The angina could be relieved by ceasing the exercise. In more severe cases, coronary vasodilators such as nitroglycerine might also be taken sublingually. An important aspect of the overall response to angina was the cessation of physical activity, which lessens the metabolic demands of the tissues, and, consequently, lessens the demands on the heart, which lessens the demand for increased coronary perfusion. In contrast, a patient with severe coronary occlusion is probably already bedridden, so decreasing the level of physical activity is not an option. However, it is possible to decrease the metabolic demands of the heart in some cases by reducing arterial pressure with the use of either calcium channel blockers or angiotensin converting enzyme (ACE) inhibitors as vasodilators. These drugs will also dilate the coronary arteries if they are not already maximally dilated. Because of the relatively high energy cost of pumping blood against an elevated arterial pressure, unloading the heart (by reducing mean arterial pressure) will bring the metabolic demands of the heart more in line with the available coronary blood flow.

Chapter 19

1. reactive hyperemia   2. tone or vascular tone
3. myogenic activity   4. precapillary sphincter
5. carotid baroreceptor
6. colloidal osmotic pressure: that portion of the total osmotic pressure of blood due to plasma proteins that cannot readily diffuse across the capillary wall. This is also referred to as the "oncotic pressure."
7. mean arterial pressure: a calculated pressure that provides the same effective driving force as a particular pulsatile pressure. For example, the blood flow through a particular artery in response to a pulsatile pressure of 120/90 mm Hg would be about the same as the calculated mean pressure to 100 mm Hg.

   mean pressure = 1/3 pulse pressure + diastolic pressure

8. Starling's hypothesis: a mathematical description of the balance of hydrostatic and osmotic pressures across the capillary endothelium
9. vasomotion: the contraction and relaxation of arteriolar and precapillary sphincter smooth muscle. Vasomotion is strongly influenced by local tissue factors.
10. cardiovascular control center: region in the medulla that coordinates the actions of the heart and blood vessels to maintain a constant blood pressure. Afferent inputs from the baroreceptors initiate appropriate changes in sympathetic and parasympathetic efferent nerve activity to the heart and blood vessels to maintain normal arterial blood pressure.
11. d   12. d   13. b   14. d   15. c   16. b   17. a
18. d   19. c   20. d
21. mean pressure = 1/3 pulse pressure + diastolic pressure

    100 mm Hg = 1/3 30 mm Hg + 90 mm Hg

22. total blood flow = MAP/TPR − 100 mm Hg/20 mm Hg · min · liter^{-1} = 5 liters/min
23. net pressure = capillary hydrostatic + interstitial fluid colloidal osmotic pressure − capillary colloidal osmotic pressure
    = 35 mm Hg + 3 mm Hg − 28 mm Hg
    = +10 (positive = net filtration)
24. pulse pressure = systolic pressure − diastolic pressure

    diastolic pressure = systolic pressure − pulse pressure

    diastolic pressure = 120 mm Hg − 50 mm Hg = 70 mm Hg

25. During systole the left ventricle rapidly ejects blood into the elastic aorta, which expands to accommodate the increased volume. During diastole, when the ventricle is no longer pumping blood, the walls of the arteries recoil and pump blood, momentarily stored in the arterial system to the capillaries.
26. The distribution of blood is determined by the relative resistances of the blood vessels leading to any particular bed. Tissue beds with higher resistances will receive less blood than those with less resistance.
27. An increase in arterial pressure will stretch the arterioles and associated smooth muscle. The stretch will activate smooth muscle contraction (myogenic activity), which will cause the vessel diameter to decrease until flow is returned to its normal value.
28. During hemorrhage, the circulating blood volume is reduced. This leads to reduced end-diastolic volume and low cardiac output in spite of increased heart rate and contractility. During exercise, end-diastolic volume is maintained, and therefore the increased heart rate and increased contractility lead to a proportional increase in cardiac output.

29. The weakened pumping action of the heart leads to accumulation of blood in the venous circulation and in the heart. The increased end-diastolic volume of the heart results in increased vigor of contraction as described by the Frank-Starling Law of the heart.
30. Recall what the major factors are that govern cardiac output: venous return, heart rate, and mean arterial pressure. Recall also that Frank-Starling law of the heart says that the heart will pump whatever blood returns to it as a result of the automatic adjustments of the end-diastolic volume of the heart. The major change with exercise is that the vascular beds open up in skeletal muscle as part of the auto-regulatory response of muscle to exercise. This will tend to cause a fall in arterial pressure and an increase in venous return as blood volume is shifted to the venous side of the systemic circulation. The reduction in arterial pressure will increase the stroke volume due to the decrease in afterload. The increase in venous volume will increase venous return and end-diastolic volume, leading to a further increase in stroke volume. As you will learn more about in Chapter 29, one of the major hormonal responses to exercise is the release of catecholamines from the adrenal medulla. Recall from Chapter 10 that roughly 80% of the released catecholamines are epinephrine and 20% norepinephrine. This will increase both the contractility of the heart and the heart rate. Taken together, these intrinsic and extrinsic regulatory mechanisms will largely compensate for the increased skeletal muscle blood flow. Keep in mind that only the innervation to the heart has been disrupted, the baroreceptor reflex, as well as the sympathetic innervation of the peripheral vasculature, is intact. Consequently, the normal redistribution of blood flow from the gut and other organs to skeletal muscle will take place normally during exercise. Regulation of coronary blood should be relatively normal. Remember, the coronary blood flow is largely autoregulated. The absence of autonomic innervation to the heart has removed any sympathetic vasoconstrictor innervation to the coronary arteries. So all-in-all, the coronary blood flow and cardiac output of heart transplant patients should, and do, respond well to exercise by virtue of the built-in autoregulatory mechanisms.
31. The pressure measured in the feet is due to two factors: 1) distension of the arteries by blood that has been pumped out of the heart, and 2) the weight of the column of blood between the arch of the aorta and the arteries in the feet. The pressure component due to the weight of blood increases in proportion to the distance below the heart, and is about 85 to 90 mm Hg at the level of the feet. The only two pressure measurements that are relative to the flow of blood to and from the heart, however, are the pressures measured in the aorta and in the right atrium. Demonstrate this for yourself by holding a 200 cm length of clear plastic tubing in front of you with the two open ends at the same level so that it forms a U. Now fill the tubing with water. The water level will be the same at both ends of the tubing and there will be no flow because there is no pressure difference between the two ends, even though the pressure at the bottom of the U will be 100 cm H_2O. Have a friend move the loop of tubing up and down while you continue to hold the two ends at the same level. Even if your friend elevates the loop above the two open ends, there should be no flow. Now have your friend release the tubing so it hangs down, and then, quickly raise one of the open ends a few centimeters. For a moment, the fluid level will be higher in the raised end of the tubing and water will flow out the other end because of the pressure difference that you have created between the two ends (i.e., a few cm H_2O). As soon as the water in the

elevated end falls to the same level as the stationary end, flow will cease, demonstrating that flow between the two ends is determined only by the pressure difference between the two ends. To more realistically approximate the vascular system, cut the tubing at the bottom of the U and tie a section of very compliant tubing between the two ends (e.g., a balloon with an extra hole cut in the top or a condom with the tip cut off). Now fill the tubing again while your friend holds the bottom of the U a couple of cm below the open ends. What happens when your friend lowers the tubing? The fall in the water level in the tube as the loop is lowered is due to the dilation of the balloon in response to the weight of the column of water. This illustrates that the high pressure measured in your feet, while it has no direct effect on flow, will reduce cardiac output by redistributing blood volume away from the heart. The compliant balloon in this example is analogous to the small veins and veinules. To stimulate the effects of venous smooth muscle contraction, have your friend squeeze the balloon and watch what happens to the water level in the tubing.

Chapter 20

1. forced vital capacity (FVC)
2. alveolar ventilation   3. pneumothorax
4. pulmonary surfactant: a lipoprotein that coats the inner surface of the alveoli and acts as a surface reducing agent
5. hypoventilation: a condition in which alveolar ventilation is below normal, causing high CO_2 in the blood (hypercapnia)
6. b   7. a   8. d   9. a   10. c   11. a   12. b
13. d   14. a   15. 6.2 liters   16. 6.2 liters/min
17. 200 ml   18. a. 4.0 liters/min b. 0.333 liter
19. FVC is one of the most useful measurements in lung spirometry because it is a dynamic measurement that provides important information about airflow and elastic properties of the lung.
20. The lung inflates when the chest cavity expands, causing pressure inside the alveoli to drop below atmospheric pressure and thereby creating a pressure gradient between mouth and alveoli that causes air to flow into the lung. Deflation occurs when the chest cavity decreases in size, causing alveolar pressure to rise above atmospheric pressure and creating a pressure gradient that causes air to flow out of the lung.
21. Lung elastic recoil is essential for the exhalation of air from the lung during expiration. During inflation the lung is stretched, and during deflation the elastic recoil from the stretched lung causes alveolar pressure to be greater than atmospheric pressure, thus causing airflow out of the lungs.
22. Emphysema causes destruction of alveoli and connective tissue of the lung. As a result, the lung is not as stiff. Thus lung compliance is abnormally high in an emphysema patient.
23. Surfactant lowers surface tension in the alveoli, making the lung more compliant and thereby reducing the work required to inflate it. Also, by reducing surface tension, surfactant prevents edema by reducing the tendency for the alveoli to absorb fluid from the pulmonary capillaries. At low lung volumes, surfactant promotes alveolar stability.
24. The normal ratio of inspiratory/expiratory (I/E) time is approximately 1. Individuals with increased airway resistance (e.g., asthma) typically take longer to exhale because of the increased airway resistance. Consequently, these individuals have a prolonged expiratory time which results in a reduced I/E ratio.
25. If TLC = 6.0 L and FVC = 4.0 L, residual volume would equal 2.0 L. Thus, the initial RV/TLC ratio is 2.0/6.0 = 0.33,

which is a normal value. If FVC is reduced by 30%, FVC would equal 2.8 L. If TLC is unchanged, the new RV would be 6.0 − 2.8 or 3.2 L. Thus, the new RV/TLC ratio is 3.2/6.0 = 0.53. Such a marked increase in the RV/TLC ratio, with a reduced FVC, is a characteristic pattern of chronic obstructive pulmonary disorder (COPD).
26. The metabolic rate does not appreciably change with hyperventilation because the amount of O_2 used by the respiratory muscles is very small. Consequently, oxygen uptake will increase very little. For the same reasons, CO_2 production will also increase only slightly. However, since alveolar ventilation has doubled, carbon dioxide output from the lungs will almost double. Therefore, metabolic V_{O_2} will equal lung V_{O_2}, but metabolic V_{CO_2} will be approximately one half of the lung V_{CO_2}.

Chapter 21

1. cardiac catheterization   2. pulmonary edema
3. hemoglobin   4. embolus   5. placenta
6. ductus arteriosus: a fetal blood vessel connecting the pulmonary artery directly to the descending aorta
7. hypoxic vasoconstriction: vasoconstriction of the pulmonary artery caused by alveolar hypoxia
8. Sleep apnea: a condition in which breathing stops for a long period of time (30–60 seconds) during sleep
9. carbaminohemoglobin: the chemical term for carbon dioxide bound to hemoglobin
10. hypoxemia: abnormally low P_{O_2} in the blood
11. b   12. a   13. a   14. b   15. b   16. d   17. b
18. d   19. c   20. d   21. b   22. b   23. a   24. d
25. 159.6 mm Hg (760 mm Hg × 0.21 = 159.6 mm Hg)
26. 200 ml/min. In a normal resting individual, 4 vol% (4 m/ 100 ml blood) are transported to the lungs each minute (40 ml/liter × 5 liters/min = 200 ml/min).
27. 11 mm Hg $\left(15 \text{ cm} = 150 \text{ mm} \times \dfrac{1 \text{ mm Hg}}{13.6 \text{ mm}}\right)$
28. 134.4 mm Hg (640 mm Hg × 0.21 = 134.4 mm Hg)
29. 420 ml/min (70 ml/liter × 6 liters/min = 420 ml/min)
30. Pulmonary lymphatics are auxiliary drainage vessels to remove excess interstitial fluid.
31. The metabolic functions of the lung include the metabolism (synthesis, release, and conversion) of vasoactive mediators such as angiotension, histamine, prostaglandin, and serotonin.
32. Mean pulmonary arterial pressure at rest is 15 mm Hg, systolic is 25 mm Hg, and diastolic is 10 mm Hg.
33. Ventilation-perfusion balance is maintained by directing blood away from poorly ventilated regions in the lungs by hypoxic vasoconstriction.
34. Major changes in pulmonary circulation include an increase in blood flow, increased pulmonary arterial pressure, a decrease in capillary transmit time, and capillary recruitment.
35. In an otherwise normal individual who is anemic, the transit time does not change (i.e., cardiac output normal). However, the time it takes to completely oxygenate the blood is decreased (<0.25 seconds). The reason is that each 100 ml of blood passing through capillaries has less hemoglobin (fewer red cells) and it takes less time to oxygenate the blood.
36. With hemorrhage, blood volume is lost but also the hematocrit is decreased. However, the arterial P_{O_2} and the O_2 saturation are normal. Remember arterial P_{O_2} is partial pressure of dissolved oxygen in the plasma which has not changed. O_2 saturation does not change because both the O_2 carrying capacity and O_2 content are decreased (O_2 content/O_2 capacity × 100 = % saturation). What keeps us

alive is O_2 content, not Pa_{O_2} or O_2 saturation. Thus the real danger with hemorrhaging is the decrease in O_2 content of the blood.

37. During lung inflation alveolar diameter increases, which compresses alveolar vessels.

Chapter 22

1. absorption   2. basic electrical rhythm
3. enterohepatic circulation   4. microvilli
5. micelles   6. diarrhea
7. digestive tract: the hollow tube extending from the lips to the anus that is composed of the mouth, pharynx, esophagus, stomach, and small and large intestines
8. portal circulation: the segment of gastrointestinal circulation (portal vein) that carries blood from the stomach, intestines, and pancreas to the liver
9. short reflex: a reflex that is mediated over sensory neurons, interneurons, and motor and secretory neurons that lie entirely within the wall of the digestive tract
10. secretin: a peptide hormone, secreted by endocrine cells in the upper small intestine, that inhibits gastric HCl secretion and motility and stimulates pancreatic and biliary secretion of bicarbonate
11. dysphagia: any condition in which there is difficulty swallowing
12. G cell: endocrine cells in the mucosa of the gastric pyloric antrum that secrete the hormone gastrin
13. cephalic phase: the regulation of gastrointestinal function through neural and hormonal mechanisms that are activated by stimulation of receptors in the head
14. emetic: a chemical agent that causes vomiting by stimulation of chemoreceptors in the stomach and small intestine or in the brain
15. trypsin: a digestive enzyme, secreted by the pancreas, that breaks certain amino acid linkages in proteins and peptides
16. pancreatic lipase: a digestive enzyme, secreted by the pancreas, that splits triacylglycerols into monoglycerides and free fatty acids
17. peristalsis: a wave of smooth muscle contraction that moves along the wall of the digestive tract, propelling the contents in the lumen of the tract ahead of it
18. lower esophegal sphinctor: a smooth muscle portion of the esophagus lying below the body; the last portion lies below the diaphragm in the abdominal cavity.
19. steatorrhea: the condition in which there are excessive amounts of fat in the feces
20. flatus: gas that is present in, or expelled from, the large intestine
21. b   22. a   23. c   24. c   25. a   26. d   27. b
28. d   29. b   30. b
31. When stretched as a meal enters the stomach, the smooth muscle of the body and fundus of the stomach adjusts its resting length without large changes in resting tension (plasticity). In addition, with each swallow as a meal is eaten, a long reflex over the vagus nerves causes a slight relaxation of the body and fundus.
32. HCl in the gastric pyloric antrum inhibits release of gastrin, and HCl in the duodenum causes release of secretin, a hormone that inhibits secretion by the gastric parietal cells.
33. Bicarbonate of pancreatic juice neutralizes HCl entering the duodenum from the stomach and in so doing (1) protects against damage by HCl to the lining of the upper small intestine and (2) increases the pH of the duodenal contents to an optimal level for the activity of pancreatic digestive enzymes.
34. Bile salts (1) promote digestion of triacylglycerols by participating in the formation of an emulsion, consisting of very small fat particles, that provides a large surface area on which lipase can act to split triacylglycerols and (2) facilitate absorption by participating in the micellar solubilization of monoglycerides and free fatty acids, the digestion products of triacylglycerols.

35. Dietary fiber (bulk) helps maintain regular bowel movements by contributing to the volume of contents of the large intestine, which increases the degree of distension and thus the contractile activity of this organ.
36. Hydrochloric acid is the major culprit in the production of ulcers. In order to minimize the destructive actions of acid, antacids and periodic snacks are ingested to neutralize acid as it is secreted. In addition, the amount of acid secreted is reduced by avoiding acid-stimulating agents and by a drug resulting in a reduction in the activity of the vagus nerves, impulses over which excite secretion of acid. Other drugs that could be used are cimetidine (blocks action of histamine) or omiprazole (blocks H^+ pump).
37. The patient has a malabsorption problem, because fat (oil droplets) and protein (meat fibers), constituents of the diet that are not normally lost in the feces, are being excreted in large quantities. It might be suspected that the malabsorption is due to maldigestion, digestion being a prerequisite to absorption. That a deficiency in digestion exists is supported by the observation that the pancreas secreted only a small amount of juice and not much bicarbonate in response to secretin, a powerful hormonal stimulus of water—bicarbonate secretion of this organ. Since the pancreas shows a marked inability to secrete its water-electrolyte juice, it might logically be reasoned that this organ is also deficient in secreting its digestive enzymes in quantities sufficient for the adequate digestion of food. The deficiency in pancreatic enzymes, a condition that sometimes results from pancreatitis, can be corrected by the oral administration of pancreatic enzyme.
38. The terminal ileum is the site of absorption of bile salts. Because this segment of intestine was removed, bile salts are not absorbed. Instead, they move into the large intestine and are excreted in the feces. In this circumstance, the liver cannot synthesize bile salts rapidly enough to maintain normal levels of these molecules in intestinal contents. Bile salts are required for normal fat digestion through their participation in the formation of emulsion particles and normal fat absorption by solubilizing fat digestion products through the formation of micelles. These processes are impaired in a bile salt deficiency, and excessive quantities of fat are excreted in the feces. The therapy is to substitute medium chain triacylglycerols, the fatty acids of which contain 8–10 carbon atoms, for long chain triacylglycerols in the diet. Since medium chain triacylglycerols and their digestion products are water-soluble, there is no requirement for bile salts in the digestion and absorption of these molecules.

Chapter 23

1. nephron   2. ultrafiltrate   3. glomerular filtration barrier
4. renal plasma clearance   5. glucose threshold
6. tubular transport maximum for PAH (Tm_{PAH})
7. countercurrent multiplication
8. neurogenic diabetes insipidus
9. afferent arteriole: the arteriole that carries blood toward the glomerulus
10. juxtaglomerular apparatus: the region in which the distal straight tubule touches afferent and efferent arterioles of its parent glomerulus; consists of macula densa, extraglomerular mesangial cells, and granular cells

11. proteinuria: the presence of an excess of proteins in the urine
12. urethra: the tube that carries urine from the urinary bladder to the outside of the body
13. glomerular filtration rate (GFR): the rate at which plasma is filtered in the kidney glomeruli
14. tubular reabsorption: the transport of substances out of the tubular urine
15. tubular secretion: the transport of substances into tubular urine
16. filtered load: the quantity of a substance, per unit of time, which is filtered by the glomeruli
17. renal autoregulation: the maintenance of a relatively constant renal blood flow and glomerular filtration rate despite changes in arterial blood pressure
18. juxtamedullary nephron: a nephron whose glomerulus lies deep in the cortex, adjacent to the medulla; such nephrons typically have long loops of Henle.
19. a   20. c   21. d   22. c   23. c   24. c   25. d
26. d   27. 80 ml plasma/min
28. 12.5 ml/min per mm Hg
29. a. 125 ml plasma/min b. 31 ml plasma/min c. 375 mg/min d. 660 ml plasma/min e. 733 ml plasma/min f. 1.33 liters/min
30. See Figures 23–2 and 23–4.
31. Serum albumin and many other plasma proteins bear a net negative charge at physiological pH. Passage of these macromolecules through the glomerular filtration barrier is normally impeded by fixed negative charges in the glomerular capillary membrane. Loss of this electrostatic barrier leads to filtration and hence abnormal excretion of plasma proteins (proteinuria).
32. The high rate of filtration of plasma in the human kidney is explained by (a) a high filtration coefficient (high fluid permeability, large surface area), (b) a high hydrostatic pressure in the glomerular capillaries, and (c) a high renal blood flow.
33. All of the changes will result in a decrease in GFR.
34. Creatinine is continuously produced in the body as an end-product of muscle metabolism. In a steady state, its production and urinary excretion rates are equal. The excretion of creatinine depends on filtration of plasma in the kidneys. If GFR falls, creatinine production will exceed excretion, and so the plasma creatinine concentration will rise.
35. In uncontrolled diabetes mellitus, due to a deficiency in the insulin system, plasma glucose levels are abnormally increased. This results in an increase in the amount of glucose filtered in the kidney glomeruli. When the filtered glucose load exceeds the reabsorptive capacity of the tubules for glucose, some of the filtered glucose will be excreted in the urine (glucosuria).
36. PAH, in addition to being filtered by the kidney glomeruli, is so vigorously secreted by the kidney (proximal) tubules that it is nearly completely cleared from all of the plasma in one passage of blood through the kidneys. Its renal plasma clearance, therefore, approximates the renal plasma flow.
37. Sodium enters proximal tubule cells by symport with glucose, amino acids, other organic solutes, phosphate, and chloride, and by exchange with cellular hydrogen ions. Sodium is pumped out of the basolateral cell membrane by its Na^+/K^+-ATPase. Sodium ions also traverse the epithelium by passive movement through leaky "tight junctions."
38. The proximal tubule reabsorbs large quantities of sodium and water as an isosmotic solution. The collecting ducts reabsorb smaller amounts of sodium and water. Unlike the proximal tubule, they can establish steep concentration gradients for sodium. Collecting duct fluid can be rendered hypo-osmotic (due to the low intrinsic water permeability of its epithelium, expressed in the absence of ADH) or hyperosmotic (due to equilibration of the collecting duct urine with the medullary interstitium, in the presence of ADH). ADH and aldosterone increase, respectively, water and sodium reabsorption by the collecting ducts but do not affect reabsorption by the proximal tubule.
39. All of these factors increase renal potassium excretion.
40. The vasa recta (a) supply the nutritional needs of the medulla, (b) act as countercurrent exchangers, and (c) remove water from the medulla.
41. Urea, added to the inner medulla by its collecting ducts and from the calyceal urine, osmotically promotes withdrawal of water along the descending thin limb of Henle's loop. This results in a rise in NaCl concentration in this fluid. When the tubule fluid moves up the ascending thin limb, sodium and chloride move passively into the medullary interstitium. This comprises a "single effect" and permits countercurrent multiplication in the inner medulla. The concentrated urea solution which enters the medulla furnishes the potential energy for establishing an osmotic gradient in the inner medulla.
42. The urinary bladder (a) stores the urine and (b) is emptied when this is appropriate.
43. A substance that is used to measure GFR should be only filtered, i.e., not reabsorbed or secreted by the kidney tubules. Urea is reabsorbed by the kidney tubules and so its clearance is less than the GFR.
44. If the architecture of the kidney medulla, its blood flow, or countercurrent multiplication by its loops of Henle are disturbed by disease, then the osmotic gradient in the medulla cannot be established and maintained. The result is excretion of a urine which is isosmotic to plasma. If the urinary solutes are not concentrated osmotically, more water is excreted in the urine, and polyuria results.
45. The major solute that is filtered and reabsorbed in the kidneys is Na^+. Reabsorption of sodium involves active transport and is powered by hydrolysis of ATP, via the Na/K-ATPase in the basolateral cell membrane of renal epithelial cells. Reabsorption of water and other solutes is coupled directly or indirectly to sodium reabsorption.

Chapter 24

1. hypovolemia   2. osmoreceptors   3. water diuresis
4. sodium appetite   5. dialysis
6. renin: a proteolytic enzyme, produced by granular cells in the kidney, that splits off angiotensin I from angiotensinogen
7. converting enzyme: an enzyme that catalyzes the conversion of angiotensin I to angiotensin II
8. extracellular fluid: the fluid outside of cells
9. negative balance: the condition in which there is net loss of a substance from the body
10. insensible water loss: water loss by evaporation from the lungs and skin
11. hypokalemia: a plasma potassium concentration below the normal range, i.e., below 3.5 mEq/liter
12. b   13. b   14. d   15. a   16. b   17. d   18. d
19. d   20. e   21. c   22. e   23. d
24. (a) Intracellular water = 1110 ml. (b) Interstitial fluid-lymph water = 1335 ml.
25. 4.2 kg (or 9 lbs)
26. Interstitial fluid-lymph is an ultrafiltrate of plasma and has practically the same composition as plasma except for the higher protein concentration in plasma and the effects this has on the distribution of small ions across capillary walls. The concentrations of small cations are slightly higher in

plasma than in interstitial fluid-lymph, and the opposite is true for small anions.

27. Transcellular fluids (e.g., cerebrospinal fluid, digestive secretions, tubular urine) can differ strikingly in composition from plasma, because these fluids are formed by or acted upon by epithelial cells, which transport various ions.

28. Potassium, magnesium, and proteins are present at higher concentrations in cells than in extracellular fluid.

29. We drink mostly by habit, and this intake covers our water needs under ordinary circumstances. We usually operate below the threshold for thirst. In the same way, when it is cold we regulate our body temperature by behavioral adjustments and by cutaneous vasoconstriction; rarely do we get cold enough to shiver.

30. Drinking a large volume of water results in a fall in plasma osmolality, which is detected by osmoreceptors in the anterior hypothalamus. They cause a decrease in release of ADH from the posterior pituitary, which allows the collecting ducts to express their intrinsically low water permeability. Tubular urine becomes osmotically dilute in the thick ascending limb; continued reabsorption of solute (mostly NaCl) out of the collecting ducts, without water following, leads to excretion of a large volume of dilute urine.

31. ADH combines with a receptor in the basolateral cell membrane of collecting duct cells. Via a guanine nucleotide stimulatory protein, adenylate cyclase is activated. This enzyme catalyzes the formation of cyclic AMP, which in turn activates a protein kinase. The subsequent steps are not entirely clear, but the final result is an increase in water permeability of the luminal cell membrane.

32. (a) an increase in GFR, (b) a decrease in plasma aldosterone levels, (c) an increase in peritubular capillary hydrostatic pressure or decrease in peritubular capillary colloid osmotic pressure, (d) a decrease in renal sympathetic nerve activity, (e) a change in intrarenal blood flow distribution, and (f) an increase in plasma atrial natriuretic peptide level.

33. Angiotensin II causes (a) vasoconstriction, (b) increased aldosterone secretion, (c) thirst, and (d) increased ADH secretion.

34. Effective arterial blood volume is an index of the degree of filling of the systemic arterial system, which determines whether blood flow needs of the tissues will be met. It is sometimes called the "effective circulating blood volume."

35. (a) Reduced plasma potassium concentration and (b) reduced plasma aldosterone level.

36. The amount of sodium in the extracellular fluid is increased with heart failure because of renal retention of sodium in response to a decrease in effective arterial blood volume. Sodium output in the urine is substantially below the dietary intake of sodium. The decrease in effective arterial blood volume also stimulates thirst and ADH release; this combination leads to accumulation of water in the body. The retained sodium is diluted by excess water, resulting in hyponatremia.

37. The hormone insulin promotes the uptake of potassium by skeletal muscle and liver cells, thereby lowering the extracellular (plasma) $[K^+]$. The glucose is administered along with the insulin to prevent hypoglycemia, since insulin also lowers the plasma glucose concentration.

38. Adrenal cortical insufficiency results in inadequate secretion of aldosterone, a mineralocorticoid hormone that stimulates sodium reabsorption by the distal nephron. With less aldosterone, less filtered sodium is reabsorbed. This can result in depletion of extracellular sodium and water, a decrease in plasma volume, circulatory collapse, and even death. The excessive loss of sodium in the urine can be compensated for by increased dietary intake of salt.

Chapter 25

1. carbonic anhydrase   2. acidemia   3. base
4. pH buffer   5. amphoteric substance   6. weak acid
7. pH: $-\log [H^+]$
8. Henderson-Hasselbalch equation: $pH = pK_a + \log ([A^-]/[HA])$
9. open system: a system in which a component can be added or removed
10. respiratory acidosis: an abnormal process characterized by CO_2 accumulation, which leads to a fall in blood pH
11. respiratory alkalosis: an abnormal process characterized by the loss of too much CO_2, which leads to a rise in blood pH
12. metabolic acidosis: an abnormal process characterized by a gain of acid (other than H_2CO_3) or loss of bicarbonate, which leads to a fall in blood pH
13. metabolic alkalosis: an abnormal process characterized by a gain of strong base or bicarbonate or a loss of acid (other than H_2CO_3), which leads to a rise in blood pH
14. d   15. a   16. d   17. c   18. a   19. d   20. b
21. a   22. b   23. e   24. 2   25. 12
26. 1/10,000   27. 700 milliequivalents   28. 4.4 mEq/liter
29. No. Hydrochloric acid, a strong acid, will be essentially completely dissociated (ionized); a 0.1 M solution has a pH of 1. Acetic acid is a weak acid and so will be incompletely dissociated; a 0.1 M solution has a pH of 2.9.
30. The effectiveness of a buffer depends on (a) how close the pK_a of the buffer is to the desired pH and (b) on the amount of buffer present.
31. Metabolism of proteins produces strong acids (sulfuric, hydrochloric, and phosphoric) that are not matched by sources of base in the diet.
32. (a) bicarbonate-CO_2, plasma proteins, inorganic phosphate, (b) proteins, organic phosphates, bicarbonate-CO_2; (c) carbonate and phosphate salts; (d) bicarbonate-CO_2, ammonia, inorganic phosphate.
33. The bicarbonate-CO_2 buffer pair operates in an open system. The respiratory system can change arterial blood P_{CO_2} and the kidneys can change the plasma bicarbonate concentration.
34. The isohydric principle states that in any fluid compartment all buffer pairs are in equilibrium with the same hydrogen ion concentration.
35. The main receptors which cause an increase in ventilation in response to elevated P_{CO_2} are central chemoreceptors located in the medulla of the brain. Peripheral chemoreceptors (carotid and aortic bodies) play a lesser role.
36. HCl is a strong acid and is essentially completely ionized in aqueous solutions. Pure water is unbuffered, so the $[H^+]$ after adding the acid is 10^{-2} M (pH 2.0). Addition of the same amount of acid to the phosphate buffer results in a much smaller fall in pH, because the basic form of phosphate (HPO_4^{2-}) soaks up most of the added H^+, thereby preventing a large increase in free $[H^+]$.
37. The driving force that stimulates hyperventilation (the compensatory response for the low blood pH) would be abolished if the pH were returned to normal. In other words, there has to be some offset to stimulate ventilation.
38. As a compensation for chronic respiratory acidosis, the kidneys had added new bicarbonate to the blood, thereby raising the plasma bicarbonate. When the CO_2 accumulation is suddenly relieved by giving a bronchodilator, then the persistently elevated plasma bicarbonate makes the blood pH abnormally alkaline.

Chapter 26

1. filterable calcium   2. hydroxyapatite   3. osteoid
4. osteoblasts   5. parafollicular, or C-cells
6. osteoporosis   7. chondrocytes   8. remodelling

9. hypocalcemia: the condition in which there is a lower than normal plasma calcium concentration
10. ionized calcium: this form of calcium represents about 50% of the total calcium in blood and is the physiologically active form of calcium
11. ground substance: this comprises about 5% of the organic material of bone and is composed primarily of proteoglycans
12. osteoclasts: large, multinucleated cells that secrete acids and proteolytic enzymes and are responsible for bone resorption. The activity and maturation of osteoclasts are stimulated by parathyroid hormone.
13. parathyroid hormone: a peptide hormone secreted by the parathyroid glands in response to decreased plasma calcium. It acts on bone, kidney, and indirectly on the gastrointestinal tract; its overall effects are to increase plasma calcium and decrease phosphate.
14. 1,25-OH-vitamin D_3: the hormonally active form of the vitamin. It is formed in the kidney and acts primarily on the gastrointestinal tract, where it promotes calcium absorption.
15. ergocalciferol (or vitamin D_2): found in many plants and yeasts. It is metabolized, has actions similar to those of vitamin D_3, and is often used for the supplementation of foods with vitamin D.
16. c   17. b   18. a   19. c   20. d   21. b   22. b
23. Hypocalcemia results in altered sodium permeability in nerves and the generation of spontaneous action potentials. This can lead to continuous stimulation and tetanization of the muscles of the motor unit.
24. In plasma, calcium exists primarily as free ionized calcium ion (Ca^{2+}), calcium complexed to an anion, and calcium bound to plasma proteins. The first two forms comprise the filterable calcium and represent 50% and 10% of the total, respectively; calcium bound to plasma proteins is not filterable and constitutes approximately 40% of the total.
25. In bone, parathyroid hormone stimulates osteoclasts, leading to bone resorption. In the kidney, parathyroid hormone stimulates calcium reabsorption in distal tubules and inhibits phosphate reabsortion in proximal tubules. PTH also stimulates the enzyme responsible for the final step of vitamin D activation in kidney, leading to an increase in calcium and phosphate absorption by the gastrointestinal tract.
26. Parathyroid hormone and calcitonin secretion are reciprocally regulated in response to changes in plasma Ca^{2+} concentrations; a decrease in calcium stimulates parathyroid hormone secretion, and an increase stimulates calcitonin secretion. Several gastrointestinal hormones, particularly gastrin, also promote calcitonin secretion.
27. Modest increases in plasma calcium are reasonably well tolerated. If calcium levels remain elevated for many weeks or months, calcium phosphate may crystalize out of solution with stone formation in sites such as kidney, bladder, joints, or muscle. Typically, this does not present as a life threatening situation. Owing the role of calcium in the maintenance of resting membrane potential, even a brief period of hypocalcemia can lead to tetanization of the affected muscle and death. Thus the need to vigorously protect against hypocalcemia.
28. Calcium and phosphate reabsorption by kidney tubules is a key site in determining overall balance of these two ions. Any impairment in the transport process per se or in the ability to respond to appropriate hormonal signals could lead to abnormalities in homeostasis. Furthermore, kidneys are the site of 1-α-hydroxylase, which forms the active metabolite of vitamin D. Impaired activation of vitamin D could result in osteomalacia or rickets.

Chapter 27

1. hypothalamus   2. heat stroke   3. poikilotherm
4. specific dynamic action   5. radiation
6. catabolism: biochemical reactions that break down complex substances into simpler compounds
7. thermography: a technique using infrared ray detection to measure skin temperature
8. mean skin temperature: a representative temperature of the body shell
9. homeotherm: an organism that regulates its body temperature within a relatively narrow range
10. fever: a regulated increase in body temperature associated with disease
11. b   12. c   13. a   14. c   15. 72.4 Cal/hr
16. 36.8°C   17. 29.1 Cal/hr   18. 406 Cal/hr   19. 32.0°C
20. The physical mechanisms that the body uses to exchange thermal energy include:
    a. radiation—heat transfer by electromagnetic radiation
    b. conduction—transfer of heat by direct contact
    c. convection—transfer of heat by conversion of water to a moving liquid or gas
    d. evaporation—transfer of heat by conversion of water to a gas
21. The conditions for measuring BMR are:
    a. measurement made after a person has fasted for 12 hours
    b. measurement usually made in morning
    c. measurement made after 30 minutes of complete rest
    d. measurement made in a comfortable temperature
22. Normal body temperature in humans is a range of body temperature between 35°C and 41°C.
23. The hypothalamus is the central controller and integrator of all the body responses for effective regulation of body temperature.
24. As an individual goes from an ambient temperature of 25°C to 35°C, the skin temperature increases. This increase is detected by thermoreceptors in the skin, which send neural information to the hypothalamus. Additionally, core temperature may increase slightly. This increase is detected by the central thermoreceptors, which also send neural information to the hypothalamus. The hypothalamus, acting as a central controller, integrates the neural information from the thermoreceptors and activates antirise responses, which correct the initial rises in skin and core temperature.
25. All chemical reactions and processes in the body are temperature dependent and accelerate with increasing temperature. The regulation of body temperature near the higher level of thermal tolerance ensures maximum optimal cellular function and provides homeotherms with the capacity for rapid changes in metabolism and other physiological activities such as exercise.
26. The increased heat load imposed on the body by exposure to a hot ambient temperature results in a shift in blood flow from the core to the periphery. The blood flow shift results in a decrease in central blood volume, a decrease in venous return, a decrease in stroke volume, and systemic hypotension. The hypotension results in decreased blood flow to the brain and can be associated with fainting.
27. In adult humans fever is a regulated increase in body temperature that does not produce any damage to the body. In addition, fever triggers numerous responses that are beneficial to the host and promote recovery from infections and other disease states.

Chapter 28

1. terminal lymphatic capillary   2. primary lymphoid organs
3. pre-T cells   4. plasma cell   5. complement

6. phagocytosis   7. CD4+ or T_4 cell
8. A class of immunoglobulins commonly found on the surface of basophils and mast cells. When engaged with antigen, IgE stimulates the cells to release histamine.
9. Autoimmunity means to make safe against one's self, and includes physiological as well as pathological mechanisms.
10. IgG is the major class of blood immunoglobulins but is also present in interstitial fluids. It provides protection against bacteria, viruses, toxins, and other antigens.
11. A process of conferring immune protection against disease. The protection acquired may be passive and thus short-lived, or active and long-lasting.
12. The name applied to a class of cytokines used to carry signals between leukocytes and other defensive cells for the purpose of coordinating their efforts in immune defense.
13. Interferons are antiviral proteins that help prevent spread of viruses and stimulate leukocytes that destroy virally infected cells. There are two major types: Type I (alpha, beta, tau, omega) and Type II (gamma).
14. Major histocompatibility complex, Class II. A plasma membrane protein used by antigen-presenting cells to present antigen to T_4 lymphocytes.
15. APC. A macrophage or a B-lymphocyte which presents antigen-MHC protein to a T-lymphocyte.
16. d   17. c   18. d   19. c   20. a   21. c   22. b
23. c   24. e   25. e
26. Graft rejection can be minimized by tissue typing, a process that assesses the degree of similarity between MHC proteins of two different individuals. The greater the similarity between donor and host MHC proteins, the lesser the change for graft rejection. Lymphocytes are collected and used for tissue typing because they're easily obtained and they bear Class I MHC and Class II MHC proteins. Another way to minimize graft rejection is to suppress the sensitivity of the immune system with administration of immunosuppressive drugs, such as cyclosporin A (CsA), that block specific mechanisms of immune defense. A disadvantage to this alternative is the host becoming more susceptible to bacterial and viral infection.
27. CD4+ helper cells are required for activation and differentiation of B lymphocytes into antibody-secreting plasma cells. CD4+ helper cells are also essential for development and activation of CD8+ cytotoxic cells and macrophages. CD4+ helper cells orchestrate cell-mediated and antibody-mediated immune responses to antigen.
28. Nonspecific defenses prevent an infectious agent from infecting by blocking its entry into tissues, by chemically neutralizing or destroying it, or by phagocytizing and digesting it. If these first-line defenses are breached, adaptive immune defenses—cell-mediated immunity and antibody-mediated immunity—are activated, destroying the infector and helping the body to become more resistant to repeated infection.

Chapter 29

1. adaptation   2. circadian rhythm   3. hyperventilation
4. blackout   5. free radicals
6. acclimatization: An environmentally-induced physiological change that leads to long lasting improvement to the change in the environment.
7. biologic clock: are intrinsic rhythms initiated by nervous tissue which causes physiologic activity to occur at regular intervals.
8. heat stroke: a severe stage of heat exhaustion that can cause unconsciousness and permanent brain damage.
9. pneumoconioses: a general classification of diseases that

causes scarring of lung tissue from dust particles (e.g., silica causes silicosis).
10. G-force: the unit used to express the amount of gravitational force on the body.
11. d   12. e   13. c   14. liver
15. c   16. b   17. d   18. b   19. e   20. e
21. 48.6 mm Hg   22. 1,000 ml   23. 4 atm   24. −2G
25. 190 mm Hg
26. Acute effects of altitude on respiration include increased ventilation, respiratory alkalosis, and increased pulmonary arterial vasoconstriction.
27. Nitrogen gases during diving dissolve into the tissue. If the ascent is too rapid, bubbles form in both blood and tissue, causing "chokes" and "bends."
28. One problem associated with breathing high concentrations of oxygen is atelectasis, and the other is the toxic effects of free radical formation.
29. High G-forces alter blood hemodynamics in the eye, causing either blackout or redout.
30. There are two stages of heat exhaustion. The first stage is characterized by heat-induced tetany, headache, nausea. The second stage is characterized by heat stroke which results in loss of consciousness, brain damage, and death from cardiovascular failure.
31. Hypothermia is a condition that occurs when core temperature cannot be maintained. A person who is unconscious from cold exposure is in severe hypothermia with core temperature below 28°F. Treatment consists of preventing further cooling and rushing them to a hospital. The most serious complication associated with treating hypothermia is the development of ventricular fibrillation. Vigorous handling of a patient may trigger fibrillation. The reason hypothermia is so dangerous to the heart is that cold depresses conduction velocity in Purkinji fibers more than in ventricular muscles, favoring movement propagation of action potentials.
32. Obtaining quantitative information on low level pollutants is very difficult because of individual differences in response to age, sex, etc. A statistical observation shows a correlation rather than cause-and-effect. For example, smokers show a 30% higher incidence in lung cancers. However, if one randomly took 10 individuals from 100 smokers to look for lung cancer, it is conceivable that none would develop cancer since 70% of the population does not develop lung cancer.

Chapter 30

1. hypertrophy   2. hyperpnea
3. maximal oxygen consumption   4. isometric exercise
5. recovery oxygen: the use of oxygen from body stores, associated with the beginning of exercise
6. mechanical efficiency: the fraction of the available chemical energy converted to mechanical energy during exercise
7. proprioreceptors: receptors in joints and muscle spindles that detect motion of the extremities
8. phosphagen: high-energy phosphate molecule
9. d   10. c   11. b   12. b   13. a   14. c   15. a
16. c   17. b   18. c
19. Since energy cannot be created or destroyed ultimately all of the energy released by the body is converted to heat energy.
20. Phosphagen is the most rapidly available source of ATP for muscle contraction because it does not depend on transporting oxygen from the lungs to the muscles and secondly the phosphagen is stored directly within the contractile elements of the muscle.
21. The criteria for determination of VO_2 may include (a) no further increase in oxygen consumption with increases in the

rate of exercise; (b) blood lactate in excess of 8 mM; (c) large muscle mass is exercised for at least 3 minutes.

22. The body temperature during steady-state exercise is not related to the absolute level of exercise, but instead is related to the level of exercise relative to the person's maximal exercise capacity.

23. The diffusing capacity of the lung is the milliliters of a gas diffusing across the pulmonary membrane per millimeter pressure difference between alveolus and the blood. The diffusing capacity of the lung increases during exercise and is related to increased perfusion of pulmonary capillaries.

24. The fast component of the recovery oxygen response is significant because it is a measure of the oxygen needed to replenish the oxygen stores and the phosphagen system of the body to pre-exercise levels.

25. Strenuous exercise is associated with increased blood flow to vital organs including the brain, muscle, heart, and decreases in visceral organs such as the intestine and the kidney.

26. In both students the stroke volume increases progressively until levels of moderate exercise are reached. In both students, furthermore, stroke volume reaches a maximal value when oxygen consumption is only 40% of VO_2 max. Once stroke volume reaches a maximum further increases in cardiac output can be achieved only by increase in heart rate. A major difference, therefore, between the athlete and the sedentary student in terms of cardiac output is the ability of the athlete to achieve and maintain significantly higher heart rates during maximal exercise conditions.

27. The circulating levels of plasma oxygen and carbon dioxide are nearly normal during maximal exercise. In addition maximal ventilation in a resting subject is greater than that associated with maximal exercise. The respiratory system, therefore, does not limit exercise during maximal conditions. The fact that exercise fatigue is associated with increases in metabolic by-products such as lactic acid would suggest that the limiting faction in maximal exercise is related to a failure of the cardiovascular system to provide adequate oxygen and nutrients to the working muscles and remove metabolic by-products.

28. Glycogen represents a major fuel for the aerobic system during various stages of endurance running. Recent studies have shown only a small amount of muscle glycogen is replenished during the first two hours following long distance exercise. Furthermore, that complete resynthesis of muscle glycogen is optimized by a high dietary intake of carbohydrates over a two day recovery period.

Chapter 31

1. osteoporosis   2. seminiferous tubules
3. corpus luteum   4. endometrium
5. crossing over: the exchange of maternal and paternal homologous chromosomal segments during meiotic division.
6. menopause: the phase in a woman's life when menstruation ceases.
7. mitotic division: duplication of chromosomes prior to cell division, which results in the formation of two cells, each of which have one copy of the chromosomes present in the original cell.
8. spermatogonia: the first cell stage of spermatogenesis in the adult male. At puberty, the primitive germ cells undergo a mitotic division and differentiate into spermatogonia.
9. a   10. b   11. c   12. d
13. Sperm production occurs only at temperatures below normal body temperature. The musculature of the scrotum moves the testes away from the body in warm weather and lifts the

testes closer to the body in cold weather to ensure that the temperature in the testes is conducive to spermatogenesis.

14. During the luteal phase of the ovarian cycle the corpus luteum secretes estrogen and progesterone, which stimulate growth and proliferation of blood vessels and accumulation of glycogen and enzymes in the endometrium, which results in an increase in endometrial thickness.

15. Following ovulation, the cells of the follicle undergo complex physical changes, increase in size, and develop a rich vascular supply, and the follicle becomes the corpus luteum.

16. The Sertoli cells form the blood-testes barrier, nourish the developing germ cells, produce seminiferous tubular fluid, remove damaged germ cells, and synthesize and secrete androgen-binding protein (ABP) and inhibin. (a) The blood-testes barrier prevents substances which could destroy the developing spermatogenic cells from entering the seminiferous tubules. (b) The blood-testes barrier keeps sperm from diffusing from the seminiferous tubules into the blood and hence prevents the production of antibodies specific to sperm.

17. In response to progesterone, secreted by the corpus luteum in the ovary, the endometrial cells of the uterus proliferate and become vascular in preparation for implantation of the fertilized ovum. Progesterone, secreted by the corpus luteum, prepares the uterus for implantation, nourishes the fertilized ovum, and stimulates cell division of the fertilized ovum. Events in the uterus and ovary are synchronized because they are controlled by precisely timed hormonal fluctuations.

18. Anatomical abnormalities of the reproductive organs, neural disorders, hormonal imbalances and certain diseases and drugs can interfere with reproductive function in both males and females.

19. The corpus luteum is important for reproduction because it secretes large amounts of progesterone and smaller amounts of estrogen. This stimulates growth and development of the endometrium which feedback to inhibit hormonal secretion from the hypothalamus and pituitary resulting in the absence of growth of new follicles in the ovary during the luteal phase of the ovarian cycle.

Chapter 32

1. polyspermy   2. ectopic pregnancy   3. myoepithelial cells
4. dizygotic twins: dizygotic or fraternal twins result from the fertilization of two ova by individual sperm.
5. zygote: the single-cell fertilized ovum, which contains 46 chromosomes.
6. chorion: trophoblast cells which form a membrane that completely surrounds the developing embryo.
7. d   8. c   9. a   10. d   11. b
12. During pregnancy the corpus luteum secretes progesterone.
13. The acrosomal reaction results in the release of enzymes from the head of the sperm, which allows sperm to penetrate the ovum during fertilization.
14. Suckling stimulates afferent neural impulses that travel from the breast, through the spinal cord, to the hypothalamus and stimulate prolactin release from the pituitary. Prolactin stimulates milk production in the breast.
15. Progesterone is important for the maintenance of pregnancy for many reasons. Progesterone stimulates growth of the endometrium and secretion of nutrients from glands in the endometrium, which nourish the developing embryo. Progesterone inhibits contractions of the uterus, which prevents premature expulsion of the fetus. Progesterone stimulates the breast in preparation for milk production and milk release following delivery. Progesterone serves as a

precursor for the synthesis of estrogen by the fetoplacental unit.

16. Ovulation and menstruation cease during pregnancy because high levels of estrogen and progesterone, secreted by the placenta, inhibit the secretion of gonadotropin-releasing hormone (GnRH) from the maternal hypothalamus which results in no signal for release of luteinizing hormone (LH) or follicle-stimulating hormone (FSH) from the maternal pituitary. Low circulating levels of LH and FSH during pregnancy are not adequate to stimulate growth of ovarian follicles. Therefore, follicular growth, ovulation, and menstruation are all eliminated during pregnancy.

17. Secretion of estrogen by the ovary increases at midcycle, and estrogen stimulates the formation of glycoprotein channels in cervical mucus, which facilitates sperm migration through the cervix into the uterus.

Chapter 33

1. transduction   2. tenting   3. sexual dysfunction
4. histocompatibility-Y (H-Y) antigen
5. vaginismus: involuntary contractions of the vaginal muscles, which make vaginal penetration very difficult or impossible
6. impotence: also termed erectile dysfunction; the inability to achieve penile erection sufficient for vaginal penetration
7. Anti Mullerian Hormone (AMH): a glycoprotein, secreted by the Sertoli cells of the testes, which induces regression of the Mullerian duct system of the embryo
8. myotonia: an increase in general muscle tone or tension
9. Turner's syndrome: a chromosomal abnormality in the female characterized by only one X sex chromosome (XO), which results in the development of an incompletely formed gonad, termed a vestigal gonadal streak
10. amenorrhea: the absence of menstruation
11. b   12. c   13. d   14. a
15. The sex chromosomes (XX in the female and XY in the male) provide the genetic information that directs the differentiation of gonadal cells into ovaries or testes and hence determine the gonadal sex of the individual.
16. Stress, which often accompanies intense programs of physical exercise, can suppress GnRH release and hence LH and FSH release, which can result in secondary amenorrhea.
17. The Y chromosome determines maleness and codes for differentiation of the testes. Much evidence for this comes from studies of persons with abnormal sex chromosome constitution. A person with an XXY constitution is a nearly normal male in external appearance but has underdeveloped gonads, referred to as Klinefelter's syndrome. A person with one X but no Y chromosome has the appearance of an immature female, referred to as Turner's syndrome.
18. In order for a recessive sex-linked trait to be expressed in the female, a recessive X-linked allele must be present on both X chromosomes she inherits. That is, the alleles must be inherited from both parents. Therefore, a color-blind female must have a color-blind father and a mother who is at least heterozygous for color blindness. In contrast, because every X chromosome allele present is expressed in the male, whether it was dominant or recessive in the mother, a color-blind male need only have a mother who is heterozygous for color blindness in order to be color-blind.

Appendix B
Normal Physiological Values for the Adult

Total Body Water (% body weight)

Female	50
Male	60

Core Temperature

°C	°F
36–38	96.8–100.4

Blood

Hematocrit (volume % RBC in blood)	37–52
Hemoglobin content (g/dl)	12–18
Erythrocytes (million/mm^3)	4.6–6.2
Leukocytes (thousand/mm^3)	7–10
Plasma glucose (mg/dl)	80–120
Plasma calcium (ionized—mEq/liter)	2.0–2.5
Plasma iron (μg/dl)	50–150
Plasma sodium (mEq/liter)	136–145
Plasma osmolarity (mOsm/kg H_2O)	290–300

Cardiovascular

Resting heart rate (beats/min)	65–75
Systolic pressure (mm Hg)	100–140
Diastolic pressure (mm Hg)	60–95
Mean systemic arterial pressure (mm Hg)	80–100
Mean pulmonary arterial pressure (mm Hg)	10–15
Cardiac output (liter/min)	5–8

Renal

Glomerular filtration rate (ml/min)	115–130
Urine output (ml/day)	600–6000
Urine pH	4.5–8.0
Urine osmolarity (mOsm/kg H_2O)	500–800

Respiratory

Alveolar ventilation (liter/min)	4.2
Respiratory rate (breaths/min)	12–15
Tidal volume (ml)	500
Total lung capacity (liter)	6
Vital capacity (liter)	4.8
Oxygen uptake (ml/mm)	240
Carbon dioxide production (ml/min)	192
Respiratory exchange ratio (CO_2 output/O_2 uptake)	0.8
Alveolar gas tension	
O_2 (mm Hg)	104
CO_2 (mm Hg)	40
Arterial blood gases	
pH	7.35–7.45
O_2 saturation (% saturation of HB with O_2)	95–98
O_2 tension (mm Hg)	85–100
CO_2 tension (mm Hg)	35–45
Oxygen content (ml/dl)	20
Carbon dioxide content (ml/dl)	50
Bicarbonate (mEq/liter)	23–28

Appendix C
Medical Terminology

Prefixes

Prefix	Meaning (and Example)	Prefix	Meaning (and Example)
a-, ab-	away from (*abnormal*)	dacry-	tears, lacrimal (*dacryostenosis*)
a-, an-	negative, not, without (*anaerobic*)	dactyl-	pertaining to fingers (*dactylospasm*)
abdomin-	pertaining to abdomen (*abdominal*)	de-	from, not, down (*decongestant*)
ad-	toward, adhere to (*adduction*)	deci-	tenth (*decibel*)
aer-	pertaining to air (*aerobic*)	dent-	pertaining to teeth (*dentition*)
af-	movement toward a point (*afferent nerve*)	derm-	skin (*dermatitus*)
algi-	pain (*algesic*)	di-	two, double (*digastric*)
all-	other, different (*allele*)	dia-	through, between, separate (*diastema*)
amb-	both, on both sides (*ambidextrous*)	dipl-	double (*diplococcus*)
amph-	on both sides (*amphiarthrosis*)	dis-	apart, away from, negative (*dislocation*)
ana-	up (*anaphylaxis*)	duct-	to lead or conduct (*ductus arteriosus*)
angio-	pertaining to blood or lymph vessels (*angiogram*)	dys-	difficult, bad, painful (*dysmenorrhea*)
ante-	before (*antebrachium*)		
anti-	against (*antibiotic*)	ecto-	external, outside (*ectoderm*)
apo-	from, opposed to (*apocrine*)	ef-	out (*effusion*)
aqua-	water (*aqueous*)	endo-	internal, inside (*endoderm*)
arche(i)-	first (*archetype*)	entero-	intestine (*enteroscope*)
arthr-	joint (*arthritis*)	ento-	within (*entocyte*)
aud-	pertaining to hearing (*audiology*)	epi-	upon, over, in addition to (*epicardium*)
auto-	pertaining to self (*autoimmunity*)	erythro-	red (*erythrocyte*)
		eu-	well, healthy, normal (*euphoria*)
bi-	double (*binocular*)	ex-	from, out of (*excrement*)
bio-	life, living (*biofeedback*)	exo-	outside (*exoskeleton*)
blast-	germinal, bud (*blastocyst*)	extra-	beyond, outside of (*extraembryonic*)
brachi-	arm (*brachium*)		
brachy-	short (*brachycephalic*)	fasci-	band (*fasciculus*)
brady-	slow (*bradycardia*)	febr-	pertaining to fever (*febricide*)
bucc-	cheek (*buccinator muscle*)	feto-	pertaining to a fetus (*fetoscope*)
		fibr-	fibrous (*fibroma*)
cac-	bad, evil (*cachexia*)	for-	opening (*foramen rotundum*)
calc-	pertaining to calcium, lime, or stone (*calcification*)	fore-	in front of (*forebrain*)
carcin-	cancer (*carcinogen*)		
cardi-	heart (*cardiac*)	galact-	pertaining to milk (*galactosemia*)
cata-	down, lower (*catatonia*)	gastr-	pertaining to the stomach (*gastroenteritis*)
caud-	tail (*caudal*)	gloss-	tongue (*glossectomy*)
centri(o)-	center, middle (*centrilobular*)	gluco-	pertaining to sugar (*glucogenesis*)
cephal-	pertaining to head (*cephalomeningitis*)	glyco-	pertaining to sugar (*glycosemia*)
cerebro-	brain (*cerebrovascular accident*)	gonado-	pertaining to the gonads (*gonadotropin*)
chol-	bile (*cholangiography*)	gran-	particle, grain (*granulocyte*)
chondr-	cartilage (*chondrocranium*)	gravi-	heavy (*gravimetric*)
chromo-	color (*chromosome*)	gyn-	female sex (*gynecologist*)
circum-	around (*circumcision*)		
co-	together (*coarctation*)	haplo-	simple, single (*haploid*)
coel-	hollow cavity (*coelom*)	hema(o)-	blood (*hemapoiesis*)
com-	together (*communicable*)	hemato-	blood (*hematologist*)
con-	together, with (*congested*)	hemi-	half (*hemisphere*)
contra-	against (*contraceptive*)	hepat-	liver (*hepatitis*)
corn-	hardened, horny (*cornea*)	hetero-	different, other (*heterosexual*)
corp-	body (*corpus cavernosum*)	histo-	tissue (*histocompatibility*)
cryo-	pertaining to cold (*cryopreservation*)	holo-	whole (*holocrine*)
crypt-	hidden (*cryptocephalus*)	homo-	same (*homosexual*)
cyan-	blue in color (*cyanotic*)	hydro-	water (*hydrocephalic*)
cyst-	bladder, baglike (*cystocarcinoma*)	hyper-	excessive, above (*hypertension*)
cyto-	cell (*cytoplasm*)	hypo-	under, below (*hypoglycemia*)

Prefix	Meaning (and Example)
idio-	individual, distinct (*idiopathic*)
ilei(o)-	pertaining to the ileum (*ileitis*)
ilio-	pertaining to the ilium (*iliosacral*)
in-	in, within, negative (*incontinence*)
infra-	beneath (*infraorbital*)
inter-	between, among (*intercoastal*)
intra-	within, inside (*intramuscular*)
irid-	pertaining to the iris (*iridectomy*)
iso-	like, equal (*isometric*)
karyo-	nucleus (*karyotype*)
kerato-	pertaining to tough substances, cornea (*keratosis*)
keto-	pertaining to ketone bodies (*ketosis*)
kine-	movement (*kinesiology*)
kypho-	humped (*kyphosis*)
labi-	lip (*labia majora*)
lacri-	tears (*lacrimal duct*)
lacto-	pertaining to milk (*lactogenic*)
laparo-	pertaining to loin or abdomen (*laparoscopy*)
laryngi(o)-	pertaining to larynx (*laryngoscope*)
later-	side (*lateral*)
leuk-	white (*leukocyte*)
lipo-	pertaining to fat (*lipocyte*)
lith-	stone (*lithodialysis*)
lymph-	pertaining to lymphatic system (*lymphoma*)
macro-	large (*macrophage*)
mal-	abnormal, bad, ill (*malnutrition*)
mamm-	pertaining to breast (*mammography*)
mast-	pertaining to breast (*mastectomy*)
medi-	middle (*mediastinum*)
mega-	large (*megakaryocyte*)
mela-	black (*melanoma*)
meningo-	meninges (*meningopathy*)
meso-	middle (*mesoderm*)
meta-	beyond, after (*metacarpal*)
metra(o)-	pertaining to uterus (*metrocystosis*)
micro-	small (*microcephaly*)
mio-	less, smaller (*mioplasmia*)
mito-	thread (*mitosome*)
mono-	one, alone (*mononucleosis*)
morph-	shape (*morphogenesis*)
muco-	pertaining to mucous (*mucoenteritis*)
multi-	many (*multicellular*)
myelo-	pertaining to bone marrow (*myeloma*)
myo-	muscle (*myocardium*)
narc-	numbness, stupor (*narcolepsy*)
naso-	nose (*nasoscope*)
necro-	dead, corpse (*necrosis*)
neo-	young, new (*neonatal*)
nephro-	kidney (*nephropathy*)
neuro-	nerve (*neurosurgery*)
nucleo-	nucleus (*nucleotide*)
ob-	against, in front of (*obstruction*)
oc-	against (*occlusion*)
oculo-	eye (*oculomotor*)
odont-	tooth (*odontoblast*)
oligo-	small, few (*oligospermia*)
omo-	shoulder (*omoclavicular*)
oo-	egg (*oocyte*)
oophoro-	ovary (*oophorotomy*)
opisth-	backward, behind (*opisthotic*)
orchi-	testis (*orchidectomy*)

Prefix	Meaning (and Example)
ortho-	straight, normal (*orthopedics*)
osteo-	bone (*osteocyte*)
oto-	ear (*otopathy*)
ovi(o)-	egg (*oviduct*)
pachy-	thick, heavy (*pachyderma*)
pan-	all (*panarthritis*)
para-	beside, near (*paranasal*)
path-	disease (*pathology*)
pedia-	children (*pediatrics*)
per-	through (*perfusion*)
peri-	surrounding, near (*periosteum*)
phag-	pertaining to eating (*phagocyte*)
pharmaco-	drugs, medicine (*pharmacotherapy*)
phlebi(o)-	vein (*phlebitis*)
photo-	pertaining to light (*photoreceptor*)
platy-	flat, broad (*platycephaly*)
pleur-	pertaining to pleura (*pleurisy*)
pneumo-	air, lung (*pneumonia*)
pod-	foot (*podiatry*)
poly-	much, many (*polydactyly*)
post-	after, behind (*postmortem*)
pre-	before (*prenatal*)
pro-	before (*prochondral*)
proct-	anus (*proctology*)
proto-	first (*protoplasm*)
pseudo-	false (*pseudogout*)
psycho-	mental (*psychogenesis*)
pyo-	pus (*pyodermatitis*)
quad-	pertaining to four (*quadriplegia*)
re-	again, back (*reabsorb*)
reno-	pertaining to the kidney (*renography*)
rete(i)-	network (*reticulocyte*)
retro-	behind, backward (*retroperitoneal*)
rhino-	nose (*rhinopathy*)
saccharo-	sugar (*saccharuria*)
sarco-	flesh (*sarcolemma*)
sclero-	hard (*scleroderma*)
semi-	half (*semilunar*)
steno-	narrow (*stenosis*)
sub-	under, below, beneath (*subcutaneous*)
super-	beyond, above, upper (*supervirulent*)
supra-	over, above (*supramandibular*)
sym-	joined together (*symphysis*)
syn-	joined together (*synarthrosis*)
tachy-	fast (*tachycardia*)
tele-	far (*telencephalon*)
tetra-	four (*tetraploid*)
therm-	pertaining to heat (*thermogenesis*)
thorac-	chest (*thoracic cavity*)
thrombo-	clot, lump (*thrombosis*)
tox-	poison (*toxicology*)
trans-	over, across (*transfuse*)
tri-	three (*triceps brachii*)
tropho-	pertaining to nourishment (*trophoblast*)
ultra-	excess, beyond (*ultrasonogram*)
uni-	one (*unicellular*)
uro-	urinary organs, urine (*urogenital*)
vas-	vessel (*vasodilation*)
viscer-	internal organs (*visceroreceptor*)

Prefix	Meaning (and Example)
vit-	life (*vitality*)
xanth-	pertaining to yellow (*xanthoderma*)
zoo-	animal (*zoology*)
zygo-	union, join (*zygote*)

Suffixes

Suffix	Meaning (and Example)
-able	capable of (*digestable*)
-a(e)sthesia	sensation (*anesthesia*)
-algia	pain (*gastralgia*)
-ase	enzyme (*lactase*)
-asis	condition (*homeostasis*)
-cele	tumor, cyst (*hydrocele*)
-cide	destroy (*fungicide*)
-coele	enlarge cavity, swelling (*blastocoele*)
-cyte	cell (*erythrocyte*)
-dynia	pain (*ophthalmodynia*)
-ectomy	surgical removal (*appendectomy*)
-emesis	vomiting (*hyperemesis*)
-emia	pertaining to the blood (*glycemia*)
-ferent	moving, carrying (*afferent*)
-form	shape (*penniform*)
-fuge	drive away (*centrifuge*)
-gen	causative agent (*pathogen*)
-genic	causing (*carcinogenic*)
-gram	recording, record (*electrocardiogram*)
-ia	condition (*uremia*)
-iatrics	medical specialties (*geriatrics*)

Suffix	Meaning (and Example)
-ism	condition (*rheumatism*)
-itis	inflammation (*arthritis*)
-kinesis	movement (*orthokinesis*)
-lith	stone (*otolith*)
-logy	study of (*osteology*)
-lysis	dissolve, break apart (*hemolysis*)
-megaly	enlarged (*acromegaly*)
-oid	resembling (*nephroid*)
-oma	tumor (*melanoma*)
-ory	pertaining to (*sensory*)
-ose	full of (*adipose*)
-osis	condition (*scoliosis*)
-pathy	disease, abnormality (*colonopathy*)
-penia	deficiency (*leukopenia*)
-phil	attraction to (*neutrophil*)
-phobia	abnormal fear (*hematophobia*)
-phylaxis	protection (*prophylaxis*)
-plasty	to reconstruct (*rhinoplasty*)
-plegia	stroke, paralysis (*paraplegia*)
-pnea	pertaining to breathing (*apnea*)
-poiesis	formation of (*hematopoiesis*)
-rrhage	overflow (*hemorrhage*)
-rrhea	discharge, flow (*menorrhea*)
-sclerosis	hardness, dryness (*arteriosclerosis*)
-sis	action, process (*dialysis*)
-tomy	cut, incision (*tracheostomy*)
-trophy	pertaining to nutrition (*hypertrophy*)
-tropic	changing, influencing (*gonadotropic*)
-uria	pertaining to urine (*glycosuria*)

Latin and Greek Terms for Body Parts

ankle: (L): *tars-, tarsi-, -tarsus*

anus: (L): *ano-, -anus*; (G): *procto-, -proctus*

arm: (G): *brachi-, -brachium*

back: (L): *dors-, dorsi-, -dorsum*; (G): *noto-, -notus*; (L): *terg-, -tergum*

bag: See *bladder.*

beak: (G): *rhyncho-, -rhychus*; (L): *rostr-, -rostrum*

belly: (L): *-venter, ventr-, ventro-*

bladder: (G): *asco-, -ascus*; (G): *cysto-, -cystis*; (G): *-physa, physo-*; (L): *-vesica, vesico-*

blood: (G): *-haema, haemato-, haemo-*; (L): *sanguini-, -sanguis*

body: (L): *corporo-, -corpus*; (G): *-soma, somato-*

bone: (L): *-os, ossi-*; (G): *osteo-, -osteum*

border: (G): *chilo-, -chilus*; (G): *craspedo-, -craspedum*

brain: (L): *cereb-, cerebr-, -cerebrum*; (G): *encephalo-, -encephalus*

breast: (L): *pector-, -pectus*; (L): *stern-, -sternum*; see also *chest.*

bristle: (G): *-chaeta, chaeto-*; (L): *-seta, seti-*

cartilage: (G): *chondro-, -chondrus*

cell: (L): *-cella, celli-*; (G): *cyto-, -cytus*

cheek: (L): *bucc-, -bucca*; (L): *gen-, -gena, geno-*

chest: (G): *stetho-, -stethus*; see also *breast* and *thorax.*

claw: (G): *chel-, chela, chelo-*; (G): *onycho-, -onyx*; (L): *ungui-, -unguis*

crest: (L): *crist-, -crista*; (G): *lopho-, -lophus*

crown: (L): *coron-, -corona*; (G): *stephano-, -stephanus*

digit: (G): *dactylo-, -dactylus*; (L): *digiti-, -digitus*

ear: (L): *auri-, -auris*; (G): *otido-, oto-, -ous*

egg: (G): *oo-, -oum*; (L): *ovi-, -ovum*

eye: (L): *oculi-, -oculus*; (G): *-omma, ommato-*; (G): *ophthalmo-, -ophthalmus*; (G): *opo-, -ops, opto-*

eyelash: (G): *-blepharis, blepharo-*

eyelid: (L): *cili-, -cilium*

face: (L): *faci-, -facies*; (G): *-ops*

feather: (L): *-pinna, pinni-*; (L): *plum-, -pluma, plumi-*; (G): *-pteryla, pterylo-*; (G): *ptilo-, -ptilum*

finger: See *digit.*

flesh: (L): *carni-, -caro*; (G): *sacro-, -sarx*

foot: (L): *pedi-, -pes*; (G): *podo-, -pus*

forehead: *-frons, front-*; (G): *metopo-, -metopus, -metopius*

gill: (G): *branch-, -branchium, brancho-*

gland: (G): *-aden, adeno-*; (L): *glandi-, -glans*

groin: *-inguen, inguini-*

hair: (L): *capill-, -capillus*; (L): *crini-, -crinis*; (L): *pil-, pili-, -pilus*; (G): *-thrix, tricho-*

hand: (G): *-chir, chiro-*; (L): *mani, manu-, -manus*

head: (L): *capit-, capiti-, -caput*; (G): *-cephala, cephalo-*

heart: (G): *cardi-, -cardia*; (L): *-cor, cordi-*

heel: (L): *calcan-, calcane-, -calcaneum*; (L): *talari-, tali-, -talus*

horn: (G): *-cera, cerato-*; (L): *corn-, -cornus*

jaw: (G): *genyo-, -genys*; (G): *gnatho-, -gnathus*; (L): *maxill-, -maxilla*

joint: (G): *arthro-, -arthrum*; (L): *articuli-, articulus, -artus*

kidney: (G): *nephro-, -nephrus*; (L): *ren-, -ren, reni-*

knee: (L): *genu-, -genu*; (G): *gony-, gonyo-, -gonys*; (G): *-gonatium, gonato-*

knuckle: (G): *condylo-, -condylus*

leg: (G): *cnemi-, -cnemis*; (L): *crur-, -crus*; (G): *-scelis, scelo-, scelido-*

lip: (G): *chilo-, -chilus*; (L): *labi-, labio-, -labium, labr-, -labrum*

liver: (G): *-hepar, hepato-*; (L): *jecori-, -jecur*

lung: (G): *-pneuma, pneumo-*; (L): *pulmo-, -pulmo, pulmono-*

membrane: (G): *chorio, -chorium*; (G): *-hymen, hymeno-*; (L): *membran-, -membrana*; (G): *meningo-, -meninx*

mouth: (L): *ora-, ori-, -os*; (G): *-stoma, stomato-*

mucus: (G): *blenno-, -blennus*

muscle: (G): *myo-, -mys*; see also *flesh.*

neck: (G): *-auchen, aucheno-*; (L): *cervic-, -cervix*; (L): *coll-, -collum*; (G): *-dera, dero-*; (G): *trachelo-, -trachelus*

nose: (L): *nasi-, -nasus*; (G): *rhino-, -rhis*

rib: (L): *cost-, -costa, costi-*; (G): *scelido-, -scelis*

rump: (G): *gluteo-, -gluteus*; (G): *-pyga, pygo-*

scale: (G): *lepido-, -lepis*; (L): *squam-, -squama, squami-*

shell: (G): *-concha, concho-*; (G): *ostraco-, ostracum*

shoulder: (G): *omo-, -omus*

skin: (L): *-byrsa, byrso-*; (G): *chorio-, -chorium*; (I.): *cutan-, cuti-, -cutis*; (G): *derm-, -derma, dermo-, dermato-*; (G): *scyto-, -scytus*

skull: (G): *cranio-, -cranium*

snout: See *beak.*

sperm: (L): *-semen, semin-*; (G): *-sperma, spermato-*

spine: (G): *-acantha, acantho-*; (G): *rhachi-, -rhachis*; (L): *-spina, spini-*

stomach: (G): *-gaster, gastro-*; (L): *-venter, ventr-*

suture: (G): *-rhapha, rhapho-*

tail: (L): *caud-, -cauda*; (G): *cerco-, -cercus*; (G): *-ura, uro-*

thigh: (L): *femor-, -femur*; (G): *mero-, -merus*

thorax: (G): *thoraco-, -thorax*

throat: (L): *gula-, -gula*; (L): *guttur-, -guttur*; (G): *laemo-, -laemus*; (G): *pharyngo-, -pharynx*; (G): *trachelo-, -trachelus*

tissue: (G): *histo-, -histus*; (L): *tel-, -tela, teli-*

toe: See *digit.*

tongue: (G): *-glossa, glosso-, -glotta, glotto-*; (L): *-lingu, -lingua*

tooth: (L): *-dens, dent-, denti-*; (G): *odonto-, -odous, -odus*

vein: (G): *phlebo-, -phleps*; (L): *ven-, -vena, veni-*

windpipe: (G): *broncho-, -bronchus*; (G): *trache-, -trachea*

wing: (L): *ala-, -ala, ali-*; (G): *ptero-, -pterum*; (G): *pterygo-, -pteryxs*

wrist: (L): *carpo-, -carpus*

INDEX

Clinical Abbreviations

a.c.	before eating	CoA	coenzyme A	G6PD	glucose 6-phosphate dehydrogenase
abs. feb.	without fever	COAD	chronic obstructive airways disease		
ACTH	adrenocorticotrophic hormone			GABA	gamma-aminobutyric acid
		comp	compound	GFR	glomerular filtration rate
ad effect.	until effective	contra	against	GI	gastrointestinal
ad lib.	as desired	COPD	chronic obstructive pulmonary disease	GnRH	gonadotrophin-releasing hormone
ad us.	as customary				
ad us. ext.	for external use	CPAP	continuous positive airway pressure	gr	grain
ADH	antidiuretic hormone			GTP	guanosine triphosphate
ADP	adenosine diphosphate	CPR	cardiopulmonary resuscitation	gtt	drops
adst. feb.	when fever is present			GU	genitourinary
ag. feb.	when fever increases	Cr	creatinine	Hb	hemoglobin
AIDS	acquired immuno-deficiency syndrome	CRF	chronic renal failure	Hcy	hematocrit
		CRH	corticotrophin-releasing hormone	HD	Huntington's disease
alt. dieb.	every other day			HDL	high density lipoprotein
alt. hor.	every other hour	CSF	cerebrospinal fluid	hgb	hemoglobin
ANS	autonomic nervous system	CT	computed tomography	HGH	human growth hormone
AQ	aqueous	CV	cardiovascular	HIV	human immunodeficiency virus
ARC	AIDS-related complex	CVA	cerebrovascular accident		
ARDS	adult respiratory distress syndrome	CVP	central venous pressure	HLA	human leukocyte group A
		CVS	chorionic villus sampling	HTLV-III	human T-lymphotrophic virus Type III
ARF	acute respiratory failure; acute renal failure	d	dexto (right); day		
		D & C	dilation and curettage	hypo	hypodermic
AS	aortic stenosis	da	day	HZ	hertz (cycles per second)
atm	atmosphere	DBP	diastolic blood pressure	IBS	irritable bowel syndrome
ATP	adenosine triphosphate	def	defecation	ICF	intracellular fluid
AV	atrioventricular	DH	delayed hypersensitivity	IDL	intermediate density lipoprotein
BBB	bundle branch block; blood-brain barrier	DHAP	dihydroxyacetone		
		dl	deciliter	Ig	immunoglobulin
bid	twice daily	DNA	deoxyribonucleic acid	IM	intramuscular
bis in 7d	twice a week	dr	dram	inf	infusion
BMR	basal metabolic rate	DTP	diphtheria-tetanus-pertussis	inhal	inhalation
BP	blood pressure			inj	injection
BSA	body surface area	ECF	extracellular fluid	IPPB	inspiratory positive pressure breathing
BUN	blood urea nitrogen	ECG	electrocardiogram		
C	celsius; complement; gallon	ECT	electroconvulsive therapy	IPSP	inhibitory postsynaptic potential
c̄	with	EDV	end-diastolic volume		
C$_{cr}$	renal creatinine clearance	EEG	electroencephalogram	IRV	inspiratory reserve volume
C1, etc.	first cervical vertebra, etc.	EM	electron micrograph		
ca	about	EMG	electromyogram	IU	international unit
CAD	coronary artery disease	ENS	enteric nervous system	IUD	intrauterine device
cal	calorie	ENT	ear, nose, and throat	IV	intravenous
Cal	kilocalorie	EPSP	excitatory postsynaptic potential	IVP	intravenous pyelogram
cAMP	cyclic AMP			s̄	without
CAPD	continuous ambulatory peritoneal dialysis	ER	endoplasmic reticulum	kcal	kilocalorie
		ERV	expiratory reserve volume	kg	kilogram
CAT	computerized axial tomography	ESR	erythrocyte sedimentation rate	l	laevo (left)
				L	liter
CBC	complete blood count	ESRD	end-stage renal disease	L1, etc.	first lumbar vertebra, etc.
cc	cubic centimeter (cm^3)	ESV	end-systolic volume	LD$_{50}$	median lethal dose
CCK,	cholecystokinin-pancreozymin	F	Fahrenheit	LDH	lactic dehydrogenase
CCK-PZ		F1, F2	first, second filial generation	LDL	low density lipoprotein
CDC	Centers for Disease Control			LE	lupus erythematosus
CF	complement fixation; cystic fibrosis	FDA	Food and Drug Administration	LH	luteinizing hormone
		FEV	forced expiratory volume	LHRH	luteinizing hormone–releasing hormone
CHD	congenital heart disease				
CHF	congestive heart failure	fl oz	fluid ounce	liq.	liquid
cm	centimeter	FSH	follicle-stimulating hormone	m	meter; minimum
CNS	central nervous system			MCH	mean corpuscular hemoglobin
CO	cardiac output	FUO	fever, unknown origin		